FLORA OF WEST TROPICAL AFRICA

FLORA

OF

WEST TROPICAL AFRICA

THE BRITISH WEST AFRICAN TERRITORIES,
LIBERIA, THE FRENCH AND PORTUGUESE
TERRITORIES SOUTH OF LATITUDE 18° N. TO
LAKE CHAD, AND FERNANDO PO.

BY

J. HUTCHINSON, LL.D., F.R.S., V.M.H., F.L.S.

FORMERLY ASSISTANT IN THE HERBARIUM, ROYAL BOTANIC GARDENS, KEW

AND

J. M. DALZIEL, M.D., B.Sc., F.L.S.

FORMERLY OF THE WEST AFRICAN MEDICAL SERVICE, AND
ASSISTANT FOR WEST AFRICA, ROYAL BOTANIC GARDENS, KEW.

SECOND EDITION

REVISED BY

R. W. J. KEAY, M.A., B.Sc., F.L.S.

DEPARTMENT OF FOREST RESEARCH, NIGERIA.

PREPARED AND REVISED AT THE
HERBARIUM, ROYAL BOTANIC GARDENS,
KEW, UNDER THE SUPERVISION OF THE
DIRECTOR.

VOL. I. PART 2

27TH MARCH, 1958

PUBLISHED ON BEHALF OF THE GOVERN-
MENTS OF NIGERIA, GHANA, SIERRA
LEONE AND THE GAMBIA

BY THE

CROWN AGENTS FOR OVERSEA GOVERNMENTS
AND ADMINISTRATIONS

MILLBANK, LONDON, S.W.1

Price £2.75

ISBN 0 85592 027 0

Reprinted 1973

Printed in Great Britain by
The Whitefriars Press, Ltd., London and Tonbridge

CONTENTS

PHYLOGENETIC SEQUENCE OF ORDERS (COHORTS) AND FAMILIES CONTAINED IN VOL. I, PART 2

(The cross-lines show breaks in affinity)

ARCHICHLAMYDEAE (contd.

TILIALES

73. Scytopetalaceae, p. 299.
74. Tiliaceae, p. 300.
75. Sterculiaceae, p. 310.
76. Bombacaceae, p. 332.

MALVALES

77. Malvaceae, p. 335.

MALPIGHIALES

78. Malpighiaceae, p. 350.
79. Humiriaceae, p. 354.
80. Ixonanthaceae, p. 355.
81. Erythroxylaceae, p. 355.
82. Lepidobotryaceae, p. 356.
83. Ctenolophonaceae, p. 357.
84. Linaceae, p. 358.
85. Zygophyllaceae, p. 361.

EUPHORBIALES

86. Euphorbiaceae, p. 364.

ROSALES

87. Rosaceae, p. 423.
88. Chailletiaceae, p. 433.

LEGUMINOSAE

89. Caesalpiniaceae, p. 439.
90. Mimosaceae, p. 484.
91. Papilionaceae, p. 505.
(by F. N. Hepper)

HAMAMELIDALES

92. Buxaceae, p. 587.

SALICALES

93. Salicaceae, p. 588.
(by R. D. Meikle)

MYRICALES

94. Myricaceae, p. 589.

URTICALES

95. Ulmaceae, p. 591.
96. Moraceae, p. 593.
97. Urticaceae, p. 616.
98. Cannabinaceae, p. 623.

CELASTRALES

99. Aquifoliaceae, p. 623.
100. Celastraceae, p. 623.
(with R. A. Blakelock)
101. Pandaceae, p. 634.
102. Icacinaceae, p. 636.
103. Salvadoraceae, p. 644.

OLACALES

104. Olacaceae, p. 644.
105. Pentadiplandraceae, p. 649.
106. Opiliaceae, p. 651.

MEDUSANDRALES

107. Medusandraceae, p. 652.

SANTALALES

108. Octoknemataceae, p. 656.
109. Loranthaceae, p. 658.
(by S. Balle)
110. Santalaceae, p. 665.
(with F. N. Hepper)
111. Balanophoraceae, p. 666.

RHAMNALES

112. Rhamnaceae, p. 667.
113. Vitaceae, p. 672.

RUTALES

114. Rutaceae, p. 683.
115. Simaroubaceae, p. 689.
116. Irvingiaceae, p. 692.
117. Burseraceae, p. 694.

MELIALES

118. Meliaceae, p. 697.

SAPINDALES

119. Sapindaceae, p. 709.
120. Melianthaceae, p. 725.
121. Anacardiaceae, p. 726.
122. Connaraceae, p. 739.
(by F. N. Hepper)

UMBELLIFLORAE

123. Alangiaceae, p. 749.
124. Araliaceae, p. 750.
125. Umbelliferae, p. 751.
(by J. F. M. Cannon)

73. SCYTOPETALACEAE

Trees ; branchlets sometimes winged. Leaves alternate, simple, exstipulate. Flowers hermaphrodite, actinomorphic, in terminal panicles or axillary racemes or fasciculate on the old wood. Calyx cupular, entire or toothed. Petals 3–10, valvate, free or shortly connate. Stamens numerous, in several series on the margin of or on the disk, free or united towards the base ; anthers 2-celled, opening by a pore or slit at the side or towards the top. Ovary superior, 3–6-celled ; ovules 2 to several in each cell, axile. Fruit woody. Seeds with ruminate or uniform copious endosperm and linear embryo.

A small family confined to tropical West Africa and the Congo region ; closely allied to *Tiliaceae*.

Corolla furrowed in bud, subsequently splitting into 5–14 petals ; anthers opening by longitudinal slits :
  Flowers in long, lax, axillary panicles ; fruit a 1-seeded, oblong or globose capsule ; endosperm uniform .. .. **1. Oubanguia**
  Flowers in short axillary racemes ; fruit a 1-seeded, ovoid-conical drupe ; endosperm ruminate .. .. **2. Scytopetalum**
Corolla not furrowed in bud, subsequently splitting into 3–5 petals ; anthers opening by apical pores :
  Filaments much shorter than the anthers ; fruit a 1-seeded, ellipsoid drupe ; endosperm ruminate ; flowers in very short axillary racemes .. .. **3. Rhaptopetalum**
  Filaments longer than the anthers ; fruit a several-seeded capsule ; endosperm not ruminate ; flowers in fascicles on the main stem and older branches **4. Brazzeia**

1. OUBANGUIA Baill. in Bull. Soc. Linn. Paris 2 : 869 (1890).

Branchlets with foliaceous wings, the wings expanded into an obtusely triangular blade beneath each node ; leaves oblong-elliptic, subacute at base, obtusely long-caudate-acuminate, 18–25 cm. long, 7–8 cm. broad, slightly undulate on the margin, glabrous ; lateral nerves about 8 pairs, very prominent beneath and looped ; flowers small, in loose axillary panicles ; calyx cupular, 3 mm. long ; petals about 6, oblanceolate, about 8–10 mm. long ; stamens numerous, connate at base ; fruits oblong, glaucous, about 1·5 cm. long .. .. 1. *alata*
Branchlets not winged, angular ; leaves elliptic, cuneate at base, abruptly long-caudate-acuminate, 8–13 cm. long, 3·5–4·5 cm. broad, glabrous ; lateral nerves 5–8 on each side of midrib, only slightly prominent beneath ; inflorescence and flowers similar to above ; fruits globose, about 1·5 cm. long .. .. 2. *laurifolia*

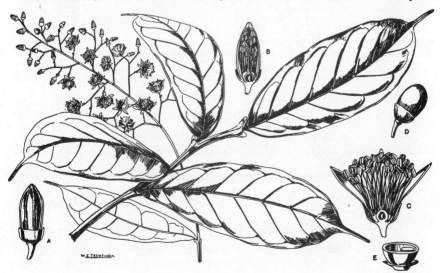

FIG. 113.—OUBANGUIA ALATA *Bak. f.* (SCYTOPETALACEAE).
A, flower-bud. B, longitudinal section of bud. C, section of flower. D, fruit. E, cross-section of fruit.

1. **O. alata** *Bak. f.* in Cat. Talb. 15 (1913). A forest tree, to 60 ft. high ; petals pink inside, white outside ; anthers " bright aureolin."
   S.Nig.: Obutong Road, Oban *Talbot* 200 ! 1513 ! Orem, Calabar (fr. May) *Ejiofo* FHI 21892 ! **Br.Cam.:** Ndian, Kumba (Mar.) *Smith* Cam. 86/36 ! Bibundi *Mildbr* 10649 !
2. **O. laurifolia** (*Pierre*) *Van Tiegh.* in Ann. Sci. Nat. sér. 9, 1 : 327 (Apr. 1905) ; Pierre in De Wild. Miss. Laurent 150 (Oct. 1905). *Ega.sea laurifolia* Pierre in De Wild. Ann. Mus. Congo sér. 5, 1 : 31, t. 17 (1903). A forest tree.
   S.Nig.: Oban *Talbot* 1693 ! Also in Gabon.

2. SCYTOPETALUM Pierre ex Engl. in E. & P. Pflanzenfam., Nachtr. 244 (1897).

Leaves narrowly oblong-elliptic, acutely cuneate at base, long-tailed-acuminate, 4·5–9 cm. long, 2–3·5 cm. broad, papery, with about 5 pairs of rather obscure lateral nerves and less prominent secondary ones ; flowers in short axillary racemes ; pedicels flattened and twisted, 3 mm. long ; petals several, connate below, 5–6 mm. long ; fruits ovoid, acute, 2 cm. long, ribbed   ..   .. *tieghemii*

S. tieghemii (*A. Chev.*) *Hutch. & Dalz.* F.W.T.A. ed. 1, 1 : 238 (1927) ; Kew Bull. 1928 : 228 ; Aubrév. Fl. For. C. Iv. 2 : 266, t. 232. *Rhaptopetalum tieghemii* A. Chev. Vég. Util. 5 : 220 (1909) ; Chev. Bot. 122. A forest tree, to 60 ft. high, branchlets " weeping " ; sometimes abundant in relatively moist places ; flowers white, fragrant when first open in the morning, petals and stamens shrivelling and falling the same evening ; fruit red when ripe.
   S.L.: Njala (Mar., Apr.) *Deighton* 1157 ! 4283 ! Matotaka (fr. July) *Thomas* 1269 ! 1341 ! Gola Forest (fl. & young fr. Apr.) *Small* 627 ! **Lib.:** Dukwia R. (fl. Feb., young fr. June) *Cooper* 247 ! 356 ! Barclayville (young fr. Mar.) *Baldwin* 11134 ! Zeahtown, Tchien (fr. Aug.) *Baldwin* 6982 ! **Iv.C.:** Malamalasso (fr. Mar.) *Chev.* 16252 ! Abidjan *Aubrév.* 94 ! **G.C.:** Bonsaso, Tarkwa (May) *Chipp* 209 ! Tarkwa (Feb.) *Thompson* 44 ! Bawdia (fl. & young fr. Apr.) *Vigne* FH 953 ! Axim (Feb.) *Irvine* 2236 ! Prestea Road, Enchi Dist. (fl. & young fr. Mar.) *Andoh* FH 5631 ! (See Appendix, p. 95.)

3. RHAPTOPETALUM Oliv.—F.T.A. 1 : 351, under *Olacaceae*.

Leaves broadly elliptic, very broadly and obtusely acuminate, 8–12 cm. long, 4–6·5 cm. broad, leathery, with about 5 pairs of opposite or subopposite lateral nerves ; flowers in very short axillary racemes ; pedicels articulating at the top and persisting ; calyx saucer-shaped, 2 mm. long ; petals 3, leathery, about 7 mm. long ; fruit ellipsoid, 2 cm. long   ..   .. *coriaceum*

R. coriaceum *Oliv.*—F.T.A. 1 : 351. A tree, 30 ft. high ; calyx light green ; petals pink ; anthers yellow.
   S.Nig.: Calabar *Thomson* 40 ! Ibum-Netim Road, Oban (Apr.) *Talbot* 562 ! Babiri, Obudu (May) *Latilo* FHI 30920 ! **F.Po:** *Mann* 1443 !

Imperfectly known species.

R. sp.—Tree ; flowers not known.
   G.C.: Ankobra Junction (fr. Jan.) *Kitson* 1020 ! Abra, W. Prov. (fr. Jan.) *Akpabla* 814 !

4. BRAZZEIA Baill. in Bull. Soc. Linn. Paris 1 : 609 (1886).

A glabrous shrub ; branchlets narrowly winged ; leaves obovate-elliptic or elliptic, narrowed to the unequal base, usually with one side more or less cuneate and the other rounded, rather long-acuminate, margin crenate to subentire, 6–17 cm. long, 2·7–6·2 cm. broad, with 4–7 main lateral looping nerves on each side of midrib prominent beneath ; fascicles few-flowered ; calyx shallowly cupular, subtruncate ; corolla ovoid in bud, about 1 cm. long ; capsules subglobose, up to 4 cm. long and broad, 4–8-valved ; seeds adhering by sticky hairs to form a solid subglobose mass which becomes loose in the capsule ..   .. *soyauxii*

B. soyauxii (*Oliv.*) *Van Tiegh.* in Ann. Sci. Nat. sér. 9, 1 : 356 (1905). *Rhaptopetalum soyauxii* Oliv. in Hook. Ic. Pl. t. 1405 (1883). An erect forest shrub, to 6 ft. high ; leaves deep green, glossy ; flowers on older wood below leaves and down stem to the base ; corolla pinkish-purple ; stamens yellow ; fruits red.
   Br.Cam.: Banga, S. Bakundu F.R. (Mar.) *Brenan & Onochie* 9479 ! Also in French Cameroons and Gabon.

74. TILIACEAE

Trees or shrubs, sometimes herbs, often clothed with stellate hairs. Leaves usually alternate, simple ; stipules paired or absent. Flowers mostly cymose, actinomorphic, mostly hermaphrodite. Sepals valvate. Petals free or absent, contorted, imbricate or valvate. Stamens usually numerous, free or rarely connate into 5–10 bundles ; anthers 2-celled. Ovary syncarpous or very rarely apocarpous, superior, 2–10-celled ; style usually simple ; ovules on axile placentas. Fruit baccate or drupaceous or variously dehiscent. Seeds sometimes hairy, mostly with copious endosperm and usually straight embryo.

A fairly large family well represented in tropical Africa, readily recognised by the stipulate leaves mostly with stellate hairs, valvate calyx, the numerous free stamens, and 2-celled anthers.

Carpels free, usually 2–5, rarely 1 ; sepals united into a campanulate 3–5-fid calyx ; leaves large, cordate and 7-nerved at the base   ..   ..   ..   **1. Christiana**

Carpels united, sepals free or nearly so :
Fruits not prickly :
Trees or shrubs ; leaves not tailed at the base :
Ovary 1–4-celled ; fruits entire, or 1–4-lobed, not exceeding 3 cm. ong ; andro-
gynophore usually present ; stamens free   ..   .. **2. Grewia**
Ovary 5–10-celled ; fruits longitudinally grooved or ridged, exceeding 3 cm. long :
Anthers short, much shorter than the filaments, without a sterile apical appendage ;
flowers sometimes enclosed in an involucre of 3 or 4 valvate bracts ; fruit
ellipsoid, often exceeding 2 cm. diam. :
Androgynophore present ; stamens free ; flowers always enclosed in an involucre
of bracts ; fruit woody, 6–8-ribbed, about 4 cm. long   .. **3. Duboscia**
Androgynophore absent ; stamens united into a short tube ; flowers not always
enclosed in an involucre of bracts (*D. subericarpa* and *D. dewevrei*) ; fruit
fibrous, with 5–10 longitudinal grooves, 6–15 cm. long, 3–10 cm. diam.
4. Desplatsia
Anthers long, as long as or longer than the filaments, with a sterile apical appendage;
flowers not in an involucre of bracts ; fruit spindle-shaped, 5–7 cm. long, 1–1·5
cm. diam., longitudinally ribbed..   ..   .. **5. Glyphaea**
Herbs ; leaves often tailed-setose at the base ; ovary 2–5-celled ; fruit spindle-
shaped   ..   .. **6. Corchorus**
Fruits prickly :
Anthers short, very much shorter than the filaments ; stamens free, not in bundles ;
herbs or undershrubs :
Fruits globose or ovoid, up to 1·5 cm. long ; petals mostly yellowish, glandular at
base ; androgynophore present   ..   .. **7. Triumfetta**
Fruits cylindrical, 2–6 cm. long ; petals purple to pink (or white), not glandular at
base ; androgynophore absent ; outer stamens barren   .. **8. Clappertonia**
Anthers long ; stamens united into 4 bundles ; petals not glandular at base ; sepals
apiculate ; androgynophore absent ; fruits globose, 3–4 cm. diam. ; a small tree
9. Ancistrocarpus

Besides the above indigenous genera, *Berria cordifolia* (Willd.) Burret, a native of tropical Asia, has been
introduced to several localities ; species of *Schoutenia* Korth. and of *Elaeocarpus* Linn. have also been introduced.

1. CHRISTIANA DC.—F.T.A. 1 : 241.

A tree ; leaves broadly ovate, cordate and 7-nerved at the base, acuminate, about
20 cm. long and 15 cm. broad, entire, stellate-tomentellous and prominently nerved
beneath ; cymes lateral, stalked ; bracts linear, deciduous ; carpels 1–5, free, obovoid,
about 1 cm. long, in fruit splitting into 2 boat-shaped valves, tomentose ; seeds
marbled ..   .. *africana*

C. africana *DC.*—F.T.A. 1 : 241 ; Chev. Bot. 86 ; Burret in Notizbl. Bot. Gart. Berl. 9 : 611 (1926). Aubrév.
Fl. For. C. Iv. 2 : 212, t. 209. Tree, to 40 ft. high, with spreading crown ; often by streams, mainly in
the drier parts of the forest regions.
Sen.: Manpalago, Casamance (fr. Feb.) *Chev.* 3071! *Chev.* 3071! **S.L.:** Kambia (fr. Jan.) *Sc. Elliot* 4707! Njala (July)
Deighton 758! Makene (fr. Mar.) *Deighton* 3899! Musaia (fr. Sept.) *Deighton* 4883! Mayoso (Aug.)
Thomas 1453! **Iv.C.:** Anoumaba (fr. Nov.) *Chev.* 22392! Aniasué, Comoé *Aubrév.* 657! **G.C.:** Atabadi,
Elmina (July) *Box* 2735! Nteinseri (June) *McAinsh* 911! Jabo, Upper Wassaw F.R. (June) *Vigne*
964! Bana Hill, Krobo (fr. Jan.) *Irvine* 1922! **N.Nig.:** Lokoja *T. Vogel* 200! **S.Nig.:** Awba Hills F.R.
(fr. Oct.) *Jones* FHI 7061! Oban *Talbot*! **Br.Cam.:** Likomba *Mildbr.* 10748! Also in Sharī, A.-E.
Sudan, Belgian Congo, Angola and in eastern tropical S. America. (See Appendix, p. 95.)

2. GREWIA Linn.—F.T.A. 1 : 242.

*Flowers in terminal (and sometimes also axillary), several- to many-flowered panicles ;
ovary not lobed, bilocular ; stigma not or scarcely lobed ; fruit not lobed :
Lower surface of leaves clearly visible between the more or less laxly arranged stellate
hairs, or glabrous :
Leaves glabrous beneath, obovate-elliptic, rounded or slightly cuneate and 3-nerved
at base, long-caudate-acuminate, 15–30 cm. long, 6–14 cm. broad ; fruits obovoid,
shortly beaked, about 3 cm. long, glabrous   ..   .. 1. *coriacea*
Leaves stellate-pubescent or pilose beneath, up to 18 cm. long and 7 cm. broad ; fruits
pubescent :
Primary bracts entire or digitately divided to the base :
Primary bracts digitately divided to the base, with linear-filiform segments,
persistent ; leaves ovate or ovate-elliptic, rounded at base, rather gradually
acuminate, 10–15 cm. long, 5–7 cm. broad, sparingly stellate-pubescent beneath ;
fruits spindle-shaped, acute at both ends, 3–3·5 cm. long, with scattered stellate
hairs..   ..   .. 2. *barombiensis*
Primary bracts entire, or 2-lobed, often early caducous ; fruits obovoid, about
2 cm. long, scabrid-pubescent ;

Branchlets, petioles and lower surface of leaves shortly tomentellous with brown
scaly hairs, sometimes with a few long white hairs ; leaves elliptic, rounded at
base, acuminate, 6–12 cm. long, 2·5–5 cm. broad ..   ..   3. *hookerana*
Branchlets, petioles and lower surface of leaves rather densely clothed with long
white hairs ; leaves oblong-elliptic, rounded at base, rather long-acuminate,
8–14 cm. long, 4–7 cm. broad ; ..   .. 4. *brunnea*
Primary bracts pinnately divided ; leaves narrowly oblong-elliptic, gradually
acuminate, unequal-sided at base, 12–18 cm. long, 4–5·5 cm. broad, very minutely
pubescent beneath and pilose on the midrib and petiole   ..   5. *oligoneura*
Lower surface of leaves completely hidden by the soft felt of hoary indumentum ;
stipules linear-subulate, early caducous ; leaves oblong-elliptic, rounded at base,
acutely acuminate, 6–14 cm. long, 2–6 cm. broad ; glabrous above ; prominently
3-nerved at base, sepals about 5 mm. long ; fruits ellipsoid, flattened, about 1·5 cm.
long, softly tomentellous   ..   .. 6. *malacocarpa*
*Flowers in axillary or lateral (and rarely terminal), usually few- to several-flowered
cymes, or solitary ; stigma always lobed :
†Flowers 2 or more together :
Leaves digitately 5–7-nerved and more or less cordate at base, very broadly ovate or
reniform-orbicular, mostly as broad or nearly as broad as long :
Cymes leaf-opposed ; common peduncle solitary and shorter than the petiole, more
or less branched above:
Pedicels slender, about 1·5 cm. long ; stipules linear or filiform, 1 cm. long ; leaves
broadly ovate-elliptic, serrulate, up to 15 cm. long and 11 cm. broad, almost
scabrid above, softly tomentose beneath ; cymes with numerous flowers ; buds
bristly-tomentose ; fruits with 3–4 almost free carpels, tuberculate.. 7. *cissoides*
Pedicels very short and stout ; stipules broadly oblanceolate ; leaves reniform or
suborbicular, up to 12 cm. broad, denticulate and often lobulate ; tertiary nerves
parallel and conspicuous ; flowers few and crowded ; buds densely villous ;
fruits depressed-globose, scarcely lobed, tuberculate and pilose   .. 8. *villosa*
Cymes axillary ; common peduncles usually 2–3 in each leaf axil, longer than the
petiole ; leaves very broadly ovate to suborbicular, serrate, up to 13 cm. long and
broad, softly tomentose on both surfaces ; stipules narrowly lanceolate, about
1 cm. long ; cymes 3–6-flowered ; buds softly tomentose ; fruits 1–3-lobed,
shortly tomentose and pilose   ..   .. 9. *barteri*
Leaves 3-nerved and at the most rounded (or rarely subcordate) at base, ovate to
elliptic or lanceolate, usually much longer than broad :
Lower surface of leaves completely hidden by the soft felt of hoary indumentum :
Leaves distinctly dentate, with prominent lateral nerves on the lower surface ;
tertiary nerves numerous, parallel, distinct ; blade oblong-lanceolate or oblong-
elliptic, sometimes oblique at base, 5–10 cm. long, 2–5 cm. broad, with very
minute stellate hairs on the upper surface ; sepals nearly 1 cm. long.. 10. *mollis*
Leaves minutely crenate, with very obscure lateral nerves beneath ; tertiary nerves
scarcely visible ; blade broadly lanceolate, very obtuse or slightly acuminate,
2–5 cm. long, 1–2 cm. broad, glabrous or nearly so above ; sepals up to 8 mm.
long ..   .. 11. *bicolor*
Lower surface of the leaves clearly visible between the more or less laxly arranged
stellate hairs :
Sepals at length 2·5 cm. long ; petals pale pinkish or white ; peduncle 2–3 cm.
long, pedicels 1–2 cm. long ; fruits deeply 4-lobed, pilose in the grooves ; stem,
petioles, undersurface of leaves, inflorescence and outside of sepals rather long-
pilose ; leaves lanceolate-elliptic, rounded to cuneate at base, gradually acuminate, 8–17 cm. long, 2·5–8 cm. broad, denticulate, especially towards the base
12. *pubescens*
Sepals up to 1·8 cm. long ; petals yellow to reddish-brown ; peduncle up to 1 cm.
long, pedicels up to 1 cm. long :
Ovary stipitate above the glandular torus ; fruit entire, with both long and short
pubescence ; leaves oblong-elliptic, rounded at both ends, 2·5–10 cm. long,
1·5–5·5 cm. broad, slightly stellate-pubescent above, shortly so beneath, closely
serrate ; peduncle 3–4 mm. long, pedicels 4–7 mm. long ; sepals 8–12 mm. long
13. *lasiodiscus*
Ovary sessile above the glandular torus :
Fruits not lobed (rarely obscurely 2-lobed), 8–12 mm. diam., scabrid-pubescent ;
flower-buds elongate, slightly constricted above the broad base, shortly
stellate-pubescent ; leaves oblong-ovate to oblong-obovate, 5–14 cm. long,
2·5–6·5 cm. broad, very laxly and minutely pubescent beneath
14. *carpinifolia*
Fruits 2–4-lobed (rarely some not lobed) ; flower-buds densely bristly tomentose :
Fruits scabrid-pubescent, not deeply lobed ; flower-buds elongate
15. *flavescens*
Fruits glabrous, deeply lobed ; flower-buds globose   .. 16. *megalocarpa*

†Flowers solitary ; leaves broadly ovate, acute at apex, rounded at base, dentate, 2–4 cm. long, up to 4 cm. broad, strongly 5-nerved at base, laxly stellate-pubescent beneath ; pedicels stellate-tomentose ; sepals linear-lanceolate, 1·5 cm. long ; fruits 3–4-lobed, glabrous .. .. 17. *tenax*

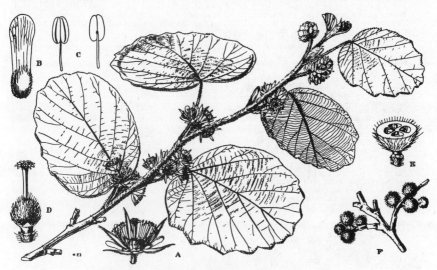

FIG. 114.—GREWIA VILLOSA *Willd.* (TILIACEAE).

A, flower. B, petal, inner face. C, stamens, front and back. D, ovary with style. E, cross-section of ovary. F, fruits.

1. **G. coriacea** *Mast.* in F.T.A. 1 : 252 (1868) ; Burret in Engl. Bot. Jahrb. 45 : 162. *Microcos coriacea* (Mast.) Burret in Notizbl. Bot. Gart. Berl. 9 : 759 (1926). Shrub or small tree in understorey of rain forest, to 50 ft. high ; larger trees with stilt roots ; sepals pink, petals yellow ; fruits purplish when ripe ; sometimes in swamp forest.
 S.Nig.: Shasha F.R. (Feb.) *Richards* 3155 ! Okomu F.R. (Jan.) *A. F. Ross* 239 ! Sapoba (June-Nov.) *Kennedy* 363 ! *Meikle* 608 ! Oban *Talbot* 1660 ! **Br.Cam.:** Likomba *Mildbr.* 10608 ! S. Bakundu F.R. *Ejiofor* FHI 14065 ! Also in French Cameroons, Gabon, Spanish Guinea, Belgian Congo and Angola. (See Appendix, p. 98.)
2. **G. barombiensis** *K. Schum.* in Engl. Bot. Jahrb. 15 : 124 (1892) ; Chev. Bot. 86. *G. africana* of Burret in Engl. Bot. Jahrb. 45 : 165, partly, not of (Hook. f.) Mast. A shrub, up to 25 ft., sometimes scrambling ; flowers white.
 Iv.C.: *fide* Chev. *l.c.* **S.Nig.:** Ajilite *Millen* 171 ! Cross River North F.R. (June) *Latilo* FHI 31848 ! Degema *Talbot* 3613 ! 3793 ! Oban *Talbot* 1270 ! **Br.Cam.:** Barombi *Preuss* 385 ! Also in French Cameroons, Gabon and Angola.
3. **G. hookerana** *Exell & Mendonça* in Consp. Fl. Angol. 1 : 215 (1951). *Omphacarpus africanus* Hook. f. (1849). *Grewia africana* (Hook. f.) Mast. in F.T.A. 1 : 253 (1868), partly, not of Mill. (1768) ; Burret l.c. 165, partly ; F.W.T.A., ed. 1, 1 : 245, partly (excl. syn. *G. brunnea* K. Schum.). *Microcos africana* (Hook. f.) Burret in Notizbl. Bot. Gart. Berl. 9 : 760 (1926). A shrub, to 20 ft. high, often scandent ; flowers white inside, brown outside ; fruits red or russet.
 S.L.: *Don* ! Njama, Kowa (May) *Deighton* 5764 ! Mabonto (fr. Oct.) *Thomas* 3549 ! **Lib.:** Du R. (fr. Aug.) *Linder* 249 ! Gbanga (fr. Sept.) *Linder* 735 ! **G.C.:** Kumasi (fl. May, fr. Dec.) *Vigne* FH 1717 ! *Andoh* FH 5158 ! Agogo (Apr.) *Bally* FH 4979 ! **S.Nig.:** Akilla (fr. Jan.) *Onochie* FHI 20675 !
4. **G. brunnea** *K. Schum.* in Engl. Bot. Jahrb. 33 : 301 (1903). *G. africana* of Burret in Engl. Bot. Jahrb. 45 : 165, partly ; F.W.T.A., ed. 1, 1 : 245, partly. *Microcos iodocarpa* Burret in Notizbl. Bot. Gart. Berl. 9 : 761 (1926). Scandent shrub, to 30 ft. high ; flowers pinkish-brown, petals with white edges.
 Lib.: Dukwia R. (Apr.) *Cooper* 407 ! **Togo:** Misahöhe (fr. Nov.) *Mildbr.* 7290 ! **S.Nig.:** Lagos *Batten-Poole* 22 ! R. Ogbesse, Ondo (May) *Jones* FHI 3591 ! Also in French Cameroons.
5. **G. oligoneura** *Sprague* in Kew Bull. 1909 : 22 ; Burret in Engl. Bot. Jahrb. 45 : 165. *Microcos oligoneura* (Sprague) Burret in Notizbl. Bot. Gart. Berl. 9 : 763 (1926). A tree, up to 30 ft.
 S.Nig.: Oban *Talbot* 1460 ! **F.Po:** *Mann* 210 ! Also in French Cameroons and A.-E. Sudan.
6. **G. malacocarpa** *Mast.* in F.T.A. 1 : 253 (1868) ; Burret in Engl. Bot. Jahrb. 45 : 163. *G. schlechteri* K. Schum. (1901). *G. dependens* K. Schum. (1903). *Microcos malacocarpa* (Mast.) Burret in Notizbl. Bot. Gart. Berl. 9 : 760 (1926). Scandent shrub or liane, in forest regions ; sepals pinkish-grey, petals orange ; fruits grey-green ; leaves almost white beneath.
 S.L.: Bagroo R. (Apr.) *Mann* 820 ! Kahreni, Limba (Apr.) *Sc. Elliot* 5582 ! Kongohun (Apr.) *Deighton* 3677 ! Roruks (fr. July) *Deighton* 2509 ! Gola Forest (fr. May) *Small* 685 ! **Lib.:** Kakatown *Whyte* ! Sinoe *Whyte* ! Suen (Nov.) *Linder* 1400 ! **G.C.:** S. Fomang Su F.R. (Oct.) *Andoh* FH 4247 ! Bou, Axim *Vigne* FH 1466 ! Abra (fr. Jan.) *Akpabla* 809 ! **Togo:** Day R., Cwegbe *Baumann* 441 ! **N.Nig.:** Lokoja *Barter* 447 ! **S.Nig.:** Ishazama to Ibadan *Schlechter* 12316 ! Ibadan South F.R. (Apr.) *Keay* FHI 22831 ! Oban *Talbot* 1743 ! **Br.Cam.:** Victoria *Winkler* 22a. Mbo *Ledermann* 1504. **F.Po:** San Carlos *Mildbr.* 6841. Bokoko *Mildbr.* 6931. Also in French Cameroons.
7. **G. cissoides** *Hutch. & Dalz.* F.W.T.A., ed. 1, 1 : 244 (1927) ; Kew Bull. 1928 : 229 ; Aubrév. Fl. For. Soud.-Guin. 154, t. 25, 5. *G. herbacea* of Burret in Engl. Bot. Jahrb. 45 : 176, partly ; Chev. Bot. 87. *G. venusta* of F.T.A. 1 : 249, partly. A tomentose, savannah shrub (or subherbaceous after burning) with wrinkled leaves and yellow flowers.
 Sen.: Dienoundiella *Berhaut* 1816 ; 4166. **Fr. Sud.:** near Bama (June) *Chev.* 941. **G.C.:** Bambisi (May) *Kitson* ! Kayoro (June) *Vigne* 4713 ! Wa (Dec.) *Adams* GC 4475 ! **Dah.:** Djougou (May) *Chev.*

23851! **N.Nig.:** Nupe *Barter* 1662! 3438! Alagbede F.R., Ilorin (Feb.) *Ejiofor* FHI 19818! Kontagora (Jan.) *Dalz.* 108! Yola (Apr.) *Dalz.* 54!

8. **G. villosa** *Willd.*—F.T.A. 1 : 249 ; Burret in Engl. Bot. Jahrb. 45 : 198 ; Burret in Notizbl. Bot. Gart. Berl. 9 : 721 ; Aubrév. l.c. 154, t. 25, 5 ; Ic. Pl. Afr. t. 5. A coarse-leaved shrub of the drier savannah regions, often on rocky hills ; 10–15 ft. high ; flowers reddish-brown.
Sen.: *Roger*! Cape Verde (Mar.) *Perrottet* 118! Dakar (Jan.) *Hagerup* 777! **Fr.Sud.:** Gao *De Wailly* 4762. Sumpi (Aug.) *Chev.* 3085. **Iv.C.:** Kaya *Aubrév.* 2176. Yako *Aubrév.* 2751. **G.C.:** Cape Coast *Brass*! Accra *Brown* 415! Prampram (Dec.) *Robertson* 41! **N.Nig.:** Kwarre, Sokoto (June) *Keay* FHI 7561! Yola (fl. & fr. June) *Dalz.* 53! Throughout the drier parts of tropical Africa and in Arabia and India. (See Appendix, p. 99.)

9. **G. barteri** *Burret* in Engl. Bot. Jahrb. 45 : 186 (1910) ; Aubrév. l.c. 154, t. 25, 7. *G. asiatica* of F.T.A. 1 : 248, partly. A savannah shrub, erect, up to 10–12 ft., or more or less scrambling, with stout densely yellow-tomentose branches ; petals bright yellow turning orange.
Fr.Sud.: Birgo, Arbala *Dubois* 69. **G.C.:** Mojonoiri, Bawku (May) *Vigne* FH 4503! Nangodi (Feb.) *Vigne* FH 4728! Zuarungu (July) *C. J. Taylor* FH 5356! **Togo:** Sokode-Basari *Kersting* A 318. **Dah.:** Kandi *De Gironcourt* 125. **N.Nig.:** Nupe *Barter* 1721! Zungeru (Aug.) *Elliott* 8! Jebba (fr. Apr.) *Meikle* 1386! Kontagora (Feb.) *Dalz.* 110! Gwari Hill, Minna (Dec.) *Meikle* 692! Auchang (Feb.) *Barter* 344! Bauchi (July) *Kennedy* FHI 7277! Sherifuri (Apr.) *Thornewill* 155! Also in French Cameroons. (See Appendix, p. 98.)

10. **G. mollis** *Juss.*—F.T.A. 1 : 248 ; Burret in Engl. Bot. Jahrb. 45 : 174 ; Aubrév. l.c. 154, t. 25, 4. A shrub or small tree, up to 20 ft. high, in savannah ; leaves pale greenish-white beneath ; flowers yellow ; fruits black when ripe ; very variable.
Sen.: Tambacounda *Berhaut* 1294. **Fr.Sud.:** Faragaran (Mar.) *Chev.* 668! **Fr.G.:** Kouroussa (July) *Pobéguin* 224! **S.L.:** Falaba (Mar.) *Sc. Elliot* 5163! **Iv.C.:** various localities *fide* Aubrév. *l.c.* **G.C.:** Krepi Plains (Jan.) *Johnson* 536! Kintampo (Mar.) *Dalz.* 30! Kwahu Tafo (May) *Kitson* 1147! Tamale (Apr.) *Williams* 123! **Togo:** Kusuntu to Palime *Baumann* 466! Sokode-Basari *Kersting* A. 581! Yendi (Apr.) *Vigne* FH 1705! **Dah.:** Savalou *Aubrév.* 44-D. Birni *Aubrév.* 93-D. **N.Nig.:** Sokoto (May) *Lely* 831! Zamfara F.R. (fl. & fr. Apr.) *Keay* FHI 18014! Zaria (June) *Keay* FHI 25894! 25895! Kontagora (Jan.-Feb.) *Dalz.* 109! *Meikle* 1082! Nupe *Barter* 1097! Bauchi Plateau (Mar.) *Lely* P. 175! Mbakon F.R., Tiv (Feb.) *Jones* FHI 6292! **S.Nig.:** Old Oyo F.R. (Feb.) *Keay* FHI 16011! Widespread in tropical Africa. (See Appendix, p. 99.)

FIG. 115.—GREWIA PUBESCENS *P. Beauv.* (TILIACEAE).

A, under-surface of leaf. B, vertical section of flower. C, stamens, front and back. D, cross-section of ovary. E, 4-lobed fruit. F, fruit-lobe in vertical section showing 3 chambers.

11. **G. bicolor** *Juss.* in Ann. Mus. Paris 4 : 90, t. 50, fig. 2 (1804) ; Burret l.c. 45 : 176 ; Aubrév. l.c. 154, t. 25, 3. *G. salvifolia* Heyne ex Roth (1821)—F.T.A. 1 : 247 ; not of Linn. f. (1781). A shrub or tree, up to 40 ft. high, in the drier savannahs ; leaves almost white beneath ; flowers yellow.
Sen.: *Roger*! *Sieber*! *Heudelot*! Kouma (Oct.) *Perrottet* 116! **Fr.Sud.:** Gassa to Tacadji (Aug.) *Chev.* 3088! **Iv.C.:** Kaya *Aubrév.* 2180. **Dah.:** Abomey *Chev.* 23143. **N.Nig.:** Nupe *Barter*! Mamu (May) *Lely* 161! Daddara F.R., Katsina (May) *Onwudinjoh* FHI 22371! Widespread in the drier parts of tropical Africa, and in Arabia and India. (See Appendix, p. 98.)

12. **G. pubescens** *P. Beauv.* Fl. Oware 2 : 76, t. 108 (1819) ; F.T.A. 1 : 250 ; Burret in Engl. Bot. Jahrb. 45 : 187 ; Aubrév. l.c. 154, t. 26, 1–2. *G. tetragastis* R. Br. ex Mast. in F.T.A. 1 : 252 ; Chev. Bot. 88. *G. gigantiflora* K. Schum. (1903). Shrub or small tree, more or less scandent, to 30 ft. high ; flowers fairly large, pale pinkish or white ; fruits 4-lobed, pilose.
Fr.G.: Kindia *Pobéguin* 1300. Timbo *Maclaud* 268. **S.L.:** Binkolo (Aug.) *Thomas* 1783! Musaia (fr. Dec.) *Deighton* 4546! Yetaya (Sept.) *Thomas* 2455! Bumban (Aug.) *Deighton* 1228! **Lib.:** Bonuta (Oct.) *Linder* 871! Sanokwele (Sept.) *Baldwin* 9530! Soplima, Vonjama (fr. Nov.) *Baldwin* 10068! **Iv.C.:** Toumodi (Aug.) *Chev.* 22397! **G.C.:** Cape Coast *Brass*! Sekondi (Oct.) *Howes* 976! Nsawam (Feb.) *Dalz.* 102! Kumasi (Sept., Oct.) *Vigne* 1335! *Darko* 348! **Togo:** Lome *Warnecke* 469! Misahöhe (Nov.) *Mildbr.* 7257! **Dah.:** Adja Oumé *Le Testu* 58! **S.Nig.:** Agege, Lagos *Millen* 1! Abeokuta *Irving*! Ibadan (fl. Sept., fr. Jan.) *Newberry* 77! *Meikle* 957! Degema *Talbot* 3754! Oban *Talbot* 142! 1345! Also in French Cameroons, Ubangi-Shari, Gabon and Angola. (See Appendix, p. 99.)

13. **G. lasiodiscus** *K. Schum.* in Notizbl. Bot. Gart. Berl. 3 : 100 (1901) ; Aubrév. l.c. 154, t. 26, 4–5. *Vinticena lasiodiscus* (K. Schum.) Burret in Notizbl.'Bot. Gart. Berl. 12 : 715 (1935). *Grewia kerstingii* Burret in Engl. Bot. Jahrb. 45 : 172 (1910) ; Chev. Bot. 87 ; F.W.T.A., ed. 1, 1 : 245. *Vinticena kerstingii* (Burret) Burret in Notizbl. Bot. Gart. Berl. 12 : 715 (1935). Savannah shrub ; flowers yellow.
   Sen.: *Leprieur.* Bakel *Perrottet* 119 ! **Gam.:** *Park* ! *D. H. Saunders* 47 ! Genieri (July) *Fox* 116 ! 171 ! **Fr.Sud.:** Darabougou (Apr.) *Chev.* 726 ! Bamako (Apr.) *Hagerup* 19 ! **Port.G.:** Bafatá (June) *Esp. Santo* 2394 ! **Fr.G.:** Sulemania, Farana (Mar.) *Sc. Elliot* 5277 ! Kouroussa (Mar.) *Pobéguin* 184 ! Timbo *Pobéguin* 1587 ! **Iv.C.:** Ouangolo *Aubrév.* 1406. Boromo *Aubrév.* 2178. **G.C.:** Navere to Delmizen (Apr.) *Kitson* 536 ! Cherepong, Wa (fl. & fr. Mar.) *Kitson* 802 ! Laura-Tumu Dist. (Mar.) *Kitson* 813 ! **Togo:** Jakuga Tamberma (fl. & fr. Apr.) *Kersting* A. 582 ! **N.Nig.:** Bida (Apr.) *Keay* FHI 25719 ! Mokwa (Apr.) *Keay* FHI 25714 ! Anara F.R. (May) *Keay* FHI 22993 ! (See Appendix, p. 98.)

14. **G. carpinifolia** *Juss.*—F.T.A. 1 : 247 ; Burret in Engl. Bot. Jahrb. 45 : 167, incl. var. *rowlandii* Burret ; Aubrév. l.c. 154, t. 26, 3. *Vinticena carpinifolia* (Juss.) Burret in Notizbl. Bot. Gart. Berl. 12 : 715 (1935). Scrambling shrub to 15 ft. high ; sepals pale yellow-green petals and stamens, golden-yellow ; flowers sweet-scented ; fruits yellow.
   S.L.: *Afzelius.* **Iv.C.:** Toumodi *Pobéguin* 265. **G.C.:** Cape Coast (July) *Brass* ! *Don* ! *T. Vogel* 67 ! Princetown (fl. & fr. Apr.) *Chipp* 172 ! Accra (fl. & fr. Feb.-Aug.) *Moloney* ! *Dalz.* 61 ! *Irvine* 20 ! Aburi (fl. & fr. Mar., June) *Brown* 395 ! *Johnson* 900 ! Kumasi (Apr.) *Vigne* 1097 ! Kwahu Hills (Apr.) *Johnson* 909 ! **Togo:** Lome *Warnecke* 34 ! 374 ! *Mildbr.* 7480 ! **Dah.:** Adja Oumé *Le Testu* 285 ! Cotonou (Mar.) *Debeaux* 351 ! **S.Nig.:** Lagos (Mar.) *Millen* 44 ! *Foster* 36 ! Ibadan (Apr.) *Meikle* 1471 ! Ibadan North F.R. (May) *Keay* FHI 16207 ! Oshogbo (Apr.) *Millson* ! Also in Belgian Congo, Angola and S. Tomé. (See Appendix, p. 98.)

15. **G. flavescens** *Juss.* in Ann. Mus. Paris 4 : 91 (1804) ; Burret in Engl. Bot. Jahrb. 45 : 168 ; Aubrév. l.c. 154, t. 26, 6–7. *Vinticena flavescens* (Juss.) Burret in Notizbl. Bot. Gart. Berl. 12 : 715 (1935). *Grewia pilosa* of F.T.A. 1 : 250, partly. *G. guazumifolia* of Chev. Bot. 87. A savannah shrub ; flowers yellow ; ripe fruits brown.
   Sen.: Cape Verde *Perrottet* 120 ! Bakel *Heudelot* 283 ! **Fr.Sud.:** Gourma (July) *Chev.* 24376 ! **Iv.C.:** Kaya *Aubrév.* 2177. **Fr.Nig.:** Niamey *De Wailly* 4920. **N.Nig.:** Sokoto (fr. Oct.) *Dalz.* 419 ! Katagum (fl. & fr. Sept.) *Dalz.* 84 ! Damaturu (fr. Oct.) *Daggash* FHI 22013 ! Mandara (Aug.) *E. Vogel* 23 ! Widespread in the drier parts of tropical Africa and in India.

16. **G. megalocarpa** *Juss.* l.c. 4 : 91 (1804) ; P. Beauv. Fl. Oware 2 : 69, t. 102 (1818) ; F.T.A. 1 : 245, partly (*Beauvois* spec.). *G. ferruginea* of F.T.A. 1 : 251, partly (*Brass* spec.). Scandent shrub ; sepals green outside, white within ; petals greenish-yellow.
   G.C.: Cape Coast (Apr.) *Brass* ! *Scholes* 286 ! Achimota (fl. & fr. May, fr. July) *Milne-Redhead* 5093 ! *Irvine* 678 ! [I suggest that Beauvois' specimen, a photograph of which I have seen, came from what is now the Gold Coast, and not from Oware. The species appears to be peculiar to the coastal scrub of the Cape Coast to Accra region.—R.W.J.K.]

17. **G. tenax** (*Forsk.*) *Fiori* in Agric. Colon. 5, Suppl. 23 (1912) ; Burret in Notizbl. Bot. Gart. Berl. 9 : 689 ; Aubrév. l.c. 151, t. 25, 1–2. *Chadara tenax* Forsk. (1775). *Grewia chadara* Lam. (1789). *G. populifolia* Vahl (1790)—Burret in Engl. Bot. Jahrb. 45 : 192 ; Chev. Bot. 88 ; Ic. Pl. Afr. 27. *G. betulaefolia* Juss. (1804)—F.T.A. 1 : 246 ; F.W.T.A., ed 1, 1 : 245. Shrub, to 10 ft. high, in the driest savannahs ; flowers white ; fruits reddish.
   Sen.: *Sieber* ! Richard Tol (fl. Jan., fr. Sept.) *Roger* ! *Perrottet* 117 ! Walo *Heudelot* ! **Fr.Sud.:** Gao (fl. & fr. Sept.) *Hagerup* 366 ! **Fr.Nig.:** Niamey (Oct.) *Hagerup* 552 ! From Mauritania throughout the drier parts of northern Africa to Arabia, Persia and India. (See Appendix, p. 98.)

3. DUBOSCIA Bocq.—F.T.A. 1 : 265.

Involucre composed of 3 valvate bracts enclosing 3 flowers ; leaves obliquely ovate or ovate-oblong, rounded or cordate at base, acutely acuminate, 8–16 cm. long, 4–8 cm. broad, denticulate, prominently 3-nerved at base, softly stellate-pubescent with a closer felt beneath ; stipules early deciduous, subulate-lanceolate ; inflorescence leaf-opposed ; peduncle stout, 3·5 cm. long ; fruits woody, 8-ribbed, ellipsoid-globose, 4 cm. long, tomentose .. .. 1. *macrocarpa*

Involucre composed of 4 valvate bracts enclosing 2 flowers ; leaves ovate or ovate-lanceolate, cordate at base, acutely acuminate, 8–17 cm. long, 3·5–7 cm. broad, denticulate, 3-nerved at base, hairy as above ; stipules at length deciduous, linear-lanceolate, 1 cm. long ; inflorescence as above but peduncle more slender and longer ; fruits woody, 6–7-ribbed, 4 cm. long, tomentellous. .. .. 2. *viridiflora*

1. **D. macrocarpa** *Bocq.*—F.T.A. 1 : 266 ; Burret in Notizbl. Bot. Gart. Berl. 9 : 817. A forest tree, to 30 ft. high ; bracts pale green, petals pink with brown hairs.
   S.Nig.: Oban (Jan.) *Talbot* 1293 ! Afi River F.R., Ikom (Dec.) *Keay* FHI 28256 ! Also in French Cameroons, Gabon, Spanish Guinea and Cabinda. (See Appendix, p. 97.)
   [*Chev.* 22618 cited as this species in Chev. Bot. 89 is *Desplatsia dewevrei* (De Wild. & Th. Dur.) Burret (q.v.). The sterile specimen *Chev.* 22583 cited as this species in F.W.T.A., ed. 1, 1 : 243 is probably also a *Desplatsia.*]

2. **D. viridiflora** (*K. Schum.*) *Mildbr.* in Wiss. Ergebn. 1910–11, 2 : 59 (1922) ; Burret l.c. 818 ; Hutch. & Dalz. in Kew Bull. 1928 : 229 ; Aubrév. Fl. For. C. Iv. 2 : 218, t. 212, 3–8. *Diplanthemum viridiflorum* K. Schum. (1897). A forest tree, to 85 ft. high, with fluted bole and narrow buttresses ; flowers greenish-white.
   Iv.C.: *fide* Aubrév. l.c. **G.C.:** Ofin Headwaters F.R. (Nov.) *Vigne* FH 3419 ! Damang *Akpabla* 928 ! **S.Nig.:** Sapoba *Kennedy* 2326 ! Oban *Talbot* 1729 ! Cross River North F.R. (June) *Latilo* FHI 31824 ! **Br.Cam.:** Buea (Mar.) *Maitland* 584 ! Victoria (Feb.) *Winkler* 1089. Mombo, S. Bakossi (fr. May) *Olorunfemi* FHI 30569 ! Also in French Cameroons and Belgian Congo. (See Appendix, p. 97.)

4. DESPLATSIA Bocq.—F.T.A. 1 : 266 ; Burret in Notizbl. Bot. Gart. Berl. 9 : 818 (1926).

Bracts suborbicular, about 1 cm. long, forming an involucre which encloses the flowers in bud ; inflorescence, bracts and outside of flowers densely villous ; stem and petioles densely rusty villous ; stipules deeply divided into linear, ciliate lobes, 5–15 mm. long ; leaves oblong-oblanceolate to oblong-elliptic, rather oblique, truncate to subcordate or sometimes obtuse at base, acuminate at apex, 15–35 cm. long, 5–13 cm. broad, irregularly and rather sharply toothed on the margin, long-stellate-pubescent beneath and on midrib above ; fruit suborbicular, to about 10 cm. long and 8–9 cm. broad, 5–6-celled and rather obscurely longitudinally grooved
   1. *chrysochlamys*

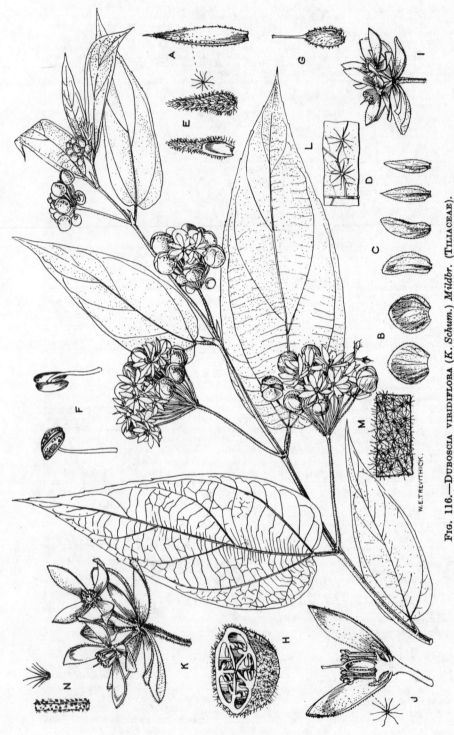

W.E.TREVITHICK.

Fig. 116.—Duboscia viridiflora (K. Schum.) Mildbr. (Tiliaceae).

A, stipule. B, bracts. C, D, bracteoles. E, petals. F, stamens. G, pistil. H, transverse section of ovary. I, K, partial inflorescences. J, longitudinal section of flower. L, upper surface of leaf. M, lower surface of leaf. N, portion of branchlet.

Bracts linear or ovate-lanceolate, up to 5 mm. long, not forming an involucre, flowers not enclosed in bud ; stem and petioles shortly pilose or pubescent :

Bracts, bracteoles and lobes of stipules linear, ciliate ; outside of flowers densely velutinous ; leaves cordate on each side at base, oblong-ovate to oblong-obovate, acutely long-acuminate, 10–16 cm. long, 3–6 cm. broad, irregularly and rather sharply toothed on the margin, sparingly stellate-pubescent on nerves and veins beneath and on midrib above ; fruit oblong-ellipsoid, 6–10 cm. long, 5–6-celled, longitudinally grooved   ..   ..   2. *subericarpa*

Bracts, bracteoles and lobes of stipules ovate-lanceolate, puberulous ; outside of flowers densely tomentose ; leaves unequal-sided at base, rounded to subcordate on one side, obtuse to rounded on the other, oblong-elliptic, long-acuminate, 12–25 cm. long, 4–9 cm. broad, coarsely and irregularly toothed in the upper part, sparingly stellate-pubescent on nerves and veins beneath, glabrous on midrib above ; fruit oblong-ellipsoid, 10–15 cm. long, 8–10 cm. broad, 6–10-celled, longitudinally grooved   ..   ..   3. *dewevrei*

1. **D. chrysochlamys** (*Mildbr. & Burret*) *Mildbr. & Burret* in Notizbl. Bot. Gart. Berl. 9 : 819 (1926). *Ledermannia chrysochlamys* Mildbr. & Burret in Mildbr. Wiss. Ergebn. 1907–8, 2 : 497 (1912) ; Aubrév. Fl. For. C. Iv. 2 : 216, t. 211. *Pleianthemum macrophyllum* K. Schum. ex A. Chev. Bot. 92, name only. Forest tree or shrub, to 20 ft. high ; flowers pinkish.
S.L.: Levuma, Koya (fr. Dec.) *Deighton* 3850 ! **Lib.:** (fl. Aug., fr. Nov.) *Harley* 995 ! Gbanga (Sept.) *Linder* 781 ! Diebla, Webo Dist. (July) *Baldwin* 6367 ! **Iv.C.:** San Pedro *Thoiré* ! Abidjan *Chev.* 15441 ! Banco *Martineau* 307 ! Rasso *Aubrév.* 155 ! **G.C.:** Achimkrom, Prestea (July) *Vigne* FH 1260 ! Anibil *Andoh* FH 3252 ! Subiri F.R., Benso (Sept.) *Andoh* FH 5569 ! Sefwi Bekwai (young fr. Oct.) *Akpabla* 890 ! **Br.Cam.:** Mopanya *Kalbreyer* 107 ! Bulifambo, Buea (Mar.) *Maitland* 562 ! Also in French Cameroons, Belgian Congo and Uganda.

2. **D. subericarpa** *Bocq.*—F.T.A. 1 : 266 ; Chev. Bot. 88 ; Burret l.c. 9 : 821 ; Aubrév. l.c. 220, t. 213, 1–3. Understorey tree, to 30 ft. high, or shrub ; flowers pink.
S.L.: Mabonto (Oct.) *Thomas* 3525 ! **Iv.C.:** Bouroukrou (fl. & fr. Jan., Dec.) *Chev.* 16697 ! 16837. Tai *Aubrév.* 1217. **G.C.:** Kwahu Prasu (fl. & fr. Feb.) *Vigne* FH 1599 ! Tafo (fl. Sept.-Nov., fr. Nov.) *Darko* 149 ! *Lovi* WH 3844 ! Bobiri F.R. (Jan.) *Andoh* FH 5457 ! **S.Nig.:** Shasha F.R. (Feb.) *Jones & Onochie* FHI 15373 ! Sapoba (Nov.-Jan.) FHI 1259 ! *Kennedy* 2349 ! Afi River F.R. (Dec.) *Keay* FHI 28254 ! Oban *Talbot* 1430 ! Also in French Cameroons, Spanish Guinea, Gabon, Belgian Congo and Cabinda. (See Appendix, p. 97.)

3. **D. dewevrei** (*De Wild. & Th. Dur.*) *Burret* in Mildbr. Wiss. Ergebn. 1907–8, 2 : 496 (1912) ; Engl. Pflanzenw. Afr. 3, 2 : 354, t. 171 ; Burret in Notizbl. Bot. Gart. Berl. 9 : 821 ; Aubrév. l.c. 220, t. 213, 4–5. *Grewiopsis dewevrei* De Wild. & Th. Dur. (1899). *Duboscia acuminata* A. Chev. in Bull. Soc. Bot. Fr. 58, Mém. 8 : 139 (1912) ; Chev. Bot. 89, partly. *Duboscia macrocarpa* of Chev. Bot. 89, not of Bocq. *Desplatsia lutea* A. Chev. ex Hutch. & Dalz. F.W.T.A., ed. 1, 1 : 240 (1927) ; Kew Bull. 1928 : 229 ; Chev. Bot 88, name only. Understorey tree, to 45 ft. high, with wide spreading crown ; flowers white or yellow ; fruits greenish-yellow.
Iv.C.: Dyola country (Apr.) *Chev.* 21323 ! Zaranou (young fr. Mar.) *Chev.* 16272 ! Ebrinahoué to Diangobo (Dec.) *Chev.* 22618 ! Man *Aubrév.* 959 ! **G.C.:** Tafo (Nov.) *Brewu* WH 6831 ! Atroni (Jan.) *Vigne* FH 2469 ! Owabi (Dec.) *Lyon* FH 2622 ! **S.Nig.:** Ibadan F.R. (Mar., Nov.) *Punch* 50 ! Akure F.R. (fl. & fr. Oct., Nov.) *Keay* FHI 25494 ! 25532 ! Sapoba (fr. Nov.) *A. F. Ross* 233 ! *Kennedy* 1649 ! 1861 ! Ikom *Catterall* 71 ! Degema *Talbot* 3637 ! **Br.Cam.:** Victoria (Nov.) *Maitland* 797 ! Johann-Albrechtshöhe (Jan.) *Staudt* 570 ! Also in French Cameroons, Belgian Congo and Uganda. (See Appendix, p. 97.)

5. **GLYPHAEA** Hook. f. ex Planch.—F.T.A. 1 : 267.

A tree ; branchlets stellate-puberulous, at length glabrous ; leaves oblong or oblong-obovate, rounded and 3-nerved at base, caudate-acuminate, denticulate or subentire, nearly glabrous, 7–16 cm. long, 2·5–9 cm. broad, thin ; stipules subulate, small, deciduous ; flowers yellow, few in terminal or leaf-opposed inflorescences ; pedicels up to 1·5 cm. long ; sepals oblong, 1·5–

Fig. 117.—GLYPHAEA BREVIS (*Spreng.*) *Monachino* (TILIACEAE).

A, flower.   B, fruit.   C, cross-section of fruit.   D, longitudinal section of fruit.

1·75 cm. long, stellate-tomentellous ; petals slightly shorter than the sepals ; fruits spindle-shaped, beaked, longitudinally ribbed, 5–6 cm. long   ..   ..   *brevis*

G. brevis (*Spreng.*) *Monachino* in Phytolog. 2 : 484 (1948). *Capparis brevis* Spreng. (1807). *Grewia lateriflora* G. Don (1831). *Glyphaea grewioides* Hook. f. ex Planch. (1848) ; Chev. Bot. 92. *G. lateriflora* (G. Don) Hutch. & Dalz. F.W.T.A., ed. 1, 1 : 239 (1927) ; Kew Bull. 1928 : 229 ; Aubrév. Fl. For. C. Iv. 2 : 216. Mainly in regrowth forest.
Port. G.: Bissau *Esp. Santo* 1892 ! **Fr.G.:** *Heudelot* 862 ! Kaba to Haut-Mamou (Apr.) *Chev.* 12755. **S.L.:** *Don* ! *T. Vogel* 101 ! Sherbro *Sc. Elliot* 5766 ! Kahreni, Limba (Apr.) *Sc. Elliot* 5579 ! Njala (May, Aug.) *Deighton* 636 ! 1835 ! Binkolo (Aug.) *Thomas* 1858 ! **Lib.:** Kakatown *Whyte* ! Ganta (fl. Apr., fr. Sept.) *Harley* ! *Baldwin* 9286 ! Sodu (Jan.) *Bequaert* 50 ! **Iv.C.:** San Pedro *Thoiré* ! Soumié *Aubrév.* 447 ! **G.C.:** Abetifi, Kwahu (fl. & young fr. June) *Thompson* 82 ! Aburi Hills (Nov.) *Johnson* 281 ! Kumasi (Mar.) *Chipp* 142 ! Nkoranza (fl. & fr. Apr.) *Dalz.* 3 ! Togo: *Baumann* 574 ! **N.Nig.:** Gurara R. (July) *Elliott* 178 ! Sanga River F.R. (fl. & young fr. Nov.) *Keay* FHI 21046 ! **S.Nig.:** Lagos (Apr.) *Moloney* ! *Dalz.* 926 ! Abeokuta *Irving* 97 ! Benin *Unwin* 46 ! *Dennett* 81 ! Oban *Talbot* 433 ! **Br.Cam.:** Buea (fl. & fr. Mar.) *Maitland* 578 ! Tiko (Jan.) *Dunlap* 183 ! 301 ! Johann-Albrechtshöhe *Staudt* 500 ! 540 ! Fonfuka, Bamenda (fl. & fr. May) *Maitland* 1374 ! **F.Po:** *T. Vogel* 119 ! *Mann* 18 ! Widespread in tropical Africa. (See Appendix, p. 97.)

6. CORCHORUS Linn.—F.T.A. 1 : 261.

Leaf-blade 0·5–2 cm. long, elliptic or obovate-elliptic, about as long as or shorter than the petiole, without basal hair-like auricles ; prostrate woody perennial in dry regions ; branches tortuous ; flowers solitary or paired ; fruits beaked, breaking into 4 valves, about 2 cm. long, subsessile .. ..   1. *depressus*
Leaf-blade much longer than the petiole ; annuals or rarely perennials :
  Leaves without basal hair-like auricles, up to 10 cm. long, 0·5–1·5 cm. broad ; fruits shortly beaked, about 1–1·5 cm. long, fasciculate ; valves not septate within
                                               2. *fascicularis*
  Leaves usually with basal hair-like auricles :
    Capsule with 3 divergent beaks at the apex, the valves not or indistinctly transversely ridged inside :
      Leaves ovate or ovate-lanceolate, 3–6 cm. long, 2–3 cm. broad ; pods acutely winged, 1·5–2·5 cm. long, fairly stout .. ..   3. *aestuans*
      Leaves lanceolate or linear-lanceolate, 3–9 cm. long, 1–1·5 cm. broad ; pods faintly ridged, slender, 2·5–3·5 cm. long .. ..   4. *tridens*
    Capsule with an entire beak, the valves with distinct transverse ridges inside :
      Capsule usually 5-valved, stout, 3–6 cm. long, rather abruptly narrowed to the apex ; leaves lanceolate or ovate-lanceolate, up to 20 cm. long and 7 cm. broad
                                           5. *olitorius*
      Capsule usually 3-valved, slender, 3–6 cm. long, gradually narrowed to the apex ; leaves elliptic to oblong-lanceolate, 4–8 cm. long, 1–1·5 cm. broad
                                           6. *trilocularis*

1. **C. depressus** (*Linn.*) *C. Christensen* in Dansk. Bot. Arkiv. 4, 3 : 34 (1922) ; Chatterjee in Kew Bull. 1948 : 372. *Antichorus depressus* Linn. (1761). *C. antichorus* Raeuschel (1797)—F.T.A. 1 : 263 ; Chev. Bot. 91 ; F.W.T.A., ed. 1, 1 : 241. Branches prostrate, a few inches long, from a stout rootstock ; flowers yellow ; a plant of desert soils.
Fr.Sud.: Timbuktu (Aug.) *Chev.* 3529. Gao (Nov.) *Hagerup* 348 ! From Cape Verde Islands and Teneriffe to N. Africa, Arabia, Central and N.W. India.

2. **C. fascicularis** *Lam.*—F.T.A. 1 : 263. Annual or sometimes perennial ; branches lax, up to 4 ft. high ; flowers yellow.
Sen.: *Roger* 6 ! Walo *Perrottet* ! **Fr.G.:** Kouroussa (July) *Pobéguin* 332 ! **S.L.:** Port Loko (fr. Dec.) *Thomas* 6659 ! Njala (Nov.) *Deighton* 1786 ! Musaia (fr. Dec.) *Deighton* 4439 ! 5702 ! Mange, Bure (fl. & fr. Dec.) *Jordan* 727 ! **G.C.:** Red Volta R., Nangodi (fr. Dec.) *Adams & Akpabla* GC 4288 ! **N.Nig.:** Katagum *Dalz.* 85 ! Widespread in tropical Africa, and in Arabia, India and Australia.

3. **C. aestuans** *Linn.* Syst. Nat., ed. 10, 2 : 1079 (1759). *C. acutangulus* Lam. (1786)—F.T.A. 1 : 264 ; Chev. Bot. 90 ; F.W.T.A., ed. 1, 1 : 241. Annual or perennial, branched, erect or prostrate ; flowers golden-yellow.
Sen.: *Heudelot* ! **Fr.Sud.:** Macina (fl. & fr. Sept.) *Chev.* 24865 ! **Fr.G.:** Conakry (fr. May) *Debeaux* 313 ! **S.L.:** Njala (Sept.) *Deighton* 2120 ! Makene (fr. Mar.) *Deighton* 3900 ! Musaia (fl. & fr. Dec.) *Deighton* 4509 ! Jigaya (fl. & fr. Nov.) *Thomas* 2726 ! **Lib.:** Peahtah (fl. & fr. Oct.) *Linder* 1014 ! Bumbuma (Oct.) *Linder* 1321 ! **Iv.C.:** Bouaké (July) *Chev.* 22108 ! **G.C.:** Cape Coast (July) *T. Vogel* 59 ! Accra *Brown* 420 ! Achimota (fl. & fr. June) *Irvine* 742 ! Sandema (Sept.) *Hughes* FH 5349 ! **Togo:** Lome *Warnecke* 32 ! **Dah.:** *Le Testu* 171 ! **N. Nig.:** Lokoja (Sept.) *T. Vogel* 162 ! Nupe *Barter* 802 ! **S.Nig.:** Olokemeji (Aug.) *Foster* 322 ! Ibadan (fl. & fr. Oct.) *Jones* FHI 13726 ! Widely distributed in the tropics. (See Appendix, p. 96.)

4. **C. tridens** *Linn.*—F.T.A. 1 : 264 ; Chev. Bot. 91. A weed of cultivation ; flowers yellow.
Sen.: *Perrottet* ! **Fr.Sud.:** *Hagerup* 1635 ! Timbuktu (July) *Chev.* 1240 ! **Fr.G.:** Conakry (Feb.) *Chev.* 12015. **Iv.C.:** Bouaké (July) *Chev.* 22109. **G.C.:** *Saunders* 1 ! Accra (fl. & fr. Oct.) *Brown* 341 ! **N.Nig.:** Nupe *Barter* 935 ! Sokoto *Moiser* 84 ! Zungeru (Aug.) *Dalz.* 111 ! Keana, Nassarawa (June) *Hepburn* 25 ! 26 ! Biu (Sept.) *Noble* 34 ! Kalkala (fl. & fr. Sept.) *Gwynn* 133 ! 134 ! **S.Nig.:** Lagos (Apr.) *Dalz.* 1391 ! Widespread in tropical Africa ; also in India and Australia. (See Appendix, p. 96.)

5. **C. olitorius** *Linn.*—F.T.A. 1 : 262 ; Chev. Bot. 91. A more or less glabrous herb, often woody at the base ; flowers yellow ; cultivated or subspontaneous ; one of the plants which yield jute. A common form has laciniate leaves.
Sen.: Kaolak *Kaichinger.* **Fr.Sud.:** Koro, Macina (fl. & fr. Sept.) *Chev.* 24823 ! Gao (Aug.) *De Wailly* 5160 ! **Fr.G.:** Kouroussa *Pobéguin* 336 ! **S.L.:** *Barter* ! Falaba (Mar.) *Sc. Elliot* 5228 ! Mabang *Dudgeon* 2 ! Kenema (fr. Nov.) *Deighton* 388 ! Makump (fr. Dec.) *Glanville* 409 ! **Lib.:** Kakatown *Whyte* ! **Iv.C.:** Pehiri *Chev.* 19190. **G.C.:** Assuantsi (fl. & fr. Mar.) *Irvine* 1541 ! Achimota (fl. & fr. Aug.) *Irvine* 790 ! **Togo:** Lome *Warnecke* 294 ! **Dah.:** *Le Testu* 173 ! **N.Nig.:** R. Niger (fl. & fr. Sept.) *T. Vogel* 119 ! Mongu (July) *Lely* 401 ! Katagum *Dalz.* 86 ! **S. Nig.:** Lagos *Dodd* 436 ! Cross R. (Mar.) *Johnston* ! Widespread throughout the tropics ; also in Australia. (See Appendix, p. 96.)

6. **C. trilocularis** *Linn.*—F.T.A. 1 : 262 ; Chev. Bot. 92. Herbaceous or half-woody annual, 1–2 ft. high ; flowers yellow.
Sen.: Niayes (Dec.) *Chev.* 3094. **Fr.Sud.:** Kabara (July) *Chev.* 1241 ! **G.C.:** Krobo Plains (fr. Dec.)

Johnson 495 ! Achimota (fl. & fr. Oct.) *Irvine* 837 ! Prampram (fl. & fr. Dec.) *Robertson* 44 ! **N.Nig.:** Takwara (May) *Lely* 100 ! Kano *Adelodun* FHI 5417 ! Widespread in the tropics of the Old World ; also in Australia. (See Appendix, p. 97.)

Doubtfully recorded species.

C. urticifolius *Wight & Arn.*—F.T.A. 1   263.   Recorded from Gambia (see Appendix, p. 97), but no specimen seen.

7. **TRIUMFETTA** Linn.—F.T.A. 1 : 254 ; Sprague & Hutch. in J. Linn. Soc. 39 : 231–276 (1909).

Sepals lepidote outside ; upper leaves oblanceolate, lower suborbicular and subtrilobed, up to 14 cm. in diam. ; petiolar glands 2–4 on each side ; stamens about 60 ; tubercles of the ovary pilose, terminated by several spinules. .   . .   . 1. *lepidota*
Sepals not lepidote :
   Prickles of the ovary and fruit terminating in several small spines stellately arranged ; fruits globose ; leaves lanceolate or ovate-lanceolate, base subcuneate, acute at the apex, 3–9 cm. long, up to 3·5 cm. broad, serrate, stellate-pilose or tomentose beneath 2. *dubia*
   Prickles of the ovary and fruit terminating in a single straight or curved but not hooked spine :
      Prickles slender, dilated only at base ; leaves mostly undivided, mostly ovate, subcordate at base, acute, 5–12 cm. long, 2·5–7 cm. broad, tomentose beneath ; sepals 5–9 mm. long, horn 0·5 mm. long ; stamens 8–10 ; fruit 1–1·5 cm. in diam., prickles pilose or nearly glabrous . .   . .   . .   3. *tomentosa*
      Prickles stout, much widened at base ; leaves very like *T. tomentosa*   4. *heudelotii*
   Prickles terminating in a single hooked or falcate spine :
      Fruit ovoid or globose, indehiscent, 4–10 mm. in diam. including the prickles :
         Stamens 15 ; fruit globose or subglobose :
            Fruit 4–5 mm. in diam. including the prickles, tomentose, the prickles glabrous ; leaves very variable   . .   . .   5. *rhomboidea*
            Fruit 8–10 mm. in diam. including the prickles, tomentose, the prickles densely pilose ; leaves ovate to lanceolate, undivided or 3-lobed, sharply serrate, setulose with appressed simple hairs on both surfaces ; stamens usually 15 6. *eriophlebia*
         Stamens 5–13 ; fruit ovoid, prickles ciliate ; leaves rhomboid-orbicular to hexagonal, broadly cuneate or truncate at base, acute or subacute, undivided or 3-lobed, coarsely serrate, stellate-pubescent beneath   . .   . .   7. *pentandra*
      Fruit globose, dehiscent, 1–1·5 cm. in diam. including the prickles ; stamens 10–12 ; leaves undivided or trilobed or sub-5-lobed   . .   . .   . .   8. *cordifolia*

1. **T. lepidota** *K. Schum.* in Engl. Bot. Jahrb. 15 : 127 (1892) ; Sprague & Hutch. in J. Linn. Soc. 39 : 245. *Ancistrocarpus tomentosus* A. Chev. Bot. 92.   Erect shrub, 3–5 ft. ; flowers yellow ; in savannah.
   G.C.: Yeji *Anderson* 21 ! **Togo:** Yendi (May) *Akpabla* 509 ! **Dah.:** Djougou (June) *Chev.* 23915 ! **N. Nig.:** Sokoto Prov. (June) *Dalz.* 535 ! Funtua to Mahuta (Aug.) *Keay* FHI 25995 ! Nupe *Barter* ! Gurara R., Gornapara (Dec.) *Elliott* 205 ! Nabardo (May) *Lely* 217 ! Extends to A.-E. Sudan.
2. **T. dubia** *De Wild.* in Ann. Mus. Congo, sér. 5, 1 : 54 (1903) ; Sprague & Hutch. l.c. 259, incl. var. *tomentosa.* An undershrub, 3–4 ft. ; stem stellate-hirsute ; flowers small, yellow.
   Fr.G.: Kouroussa (Sept.) *Pobéguin* 450 ! **S.L.:** Kabala (Sept.) *Thomas* 2189 ! 2215 ! Jigaya (Sept.) *Thomas* 2746 ! Musaia (fr. Dec.) *Deighton* 5708 ! Warantamba (Oct.) *Small* 357 ! **G.C.:** Techiman (fr. Dec.) *Adams* GC 4489 ! **Togo:** Sokode *Schröder* 88 ; 112. Bismarckburg *Büttner* 194. **N.Nig.:** Nupe *Barter* 1550 ! Patti Lokoja (Oct.) *Dalz.* 55 ! Naraguta (June–Aug.) *Lely* 334 ! 535 ! P. 626 ! Vom *Dent Young* 21 !   Also in Belgian Congo and E. Africa.
3. **T. tomentosa** *Boj.*—F.T.A. 1 : 258 ; Sprague & Hutch. l.c. 260.   An undershrub, 2–5 ft., more or less tomentose all over ; flowers yellow.
   S.L.: Port Loko (fl. & fr. Dec.) *Thomas* 6646 ! Njala (fl. & fr. Dec.) *Deighton* 1516 ! 5291 ! Musaia (fl. & fr. Dec.) *Deighton* 4554 ! **Iv.C.:** Anoumaba (Nov.) *Chev.* 22403 ! **N.Nig.:** Naraguta (Aug.) *Lely* 511 ! P. 665 ! Vom *Dent Young* 22 ! **Br.Cam.:** Buea (fl. & fr. Dec.) *Maitland* 135 ! Njinikom, Bamenda (June) *Maitland* 1767 !   Widespread in tropical Africa.
4. **T. heudelotii** *Planch. ex Mast.* in F.T.A. 1 : 259 (1868) ; Sprague & Hutch. l.c. 261.   Like the last, with smaller·fruits and stouter prickles.
   Fr.G.: *Heudelot* 681 !
5. **T. rhomboidea** *Jacq.*—F.T.A. 1 : 257, partly ; Sprague & Hutch. l.c. 266 ; Chev. Bot. 89.   A weedy undershrub, variable in habit and foliage ; flowers orange or yellow.
   Sen.: Dakar (Dec.) *Chev.* 3074. **Gam.:** Koto (fl. & fr. May) *Fox* 102 ! **Fr.Sud.:** *fide* Chev. l.c. **S.L.:** Freetown *Smythe* 51 ! Kumrabai (fr. Dec.) *Thomas* 7059 ! Mano Salija (Nov.) *Deighton* 289 ! Gorahun (Nov.) *Deighton* 359 ! Magbile (Dec.) *Thomas* 6195 ! **Lib.:** Grand Bassa *T. Vogel* 88 ! Monrovia (Nov.) *Whyte* ! *Linder* 1435 ! 1524 ! *Barker* 1457 ! Peahtah (fr. Dec.) *Baldwin* 10598 ! **Iv.C.:** *fide* Chev. l.c. **G.C.:** Cape Coast *Don* ! Amuni (Dec.) *Chipp* 44 ! Mansu (Aug.) *Howes* 960 ! Shai Plains (fr. Jan.) *Johnson* 571 ! Atwabo *Fishlock* 24 ! Kumasi (Dec.) *Darko* 649 ! **Togo:** Lome *Warnecke* 268 ! **Dah.:** *Burton* ! *Le Testu* 62 ! **N.Nig.:** Nupe *Barter* 1058 ! Lokoja (Oct.) *Parsons* 50 ! Dogon Kurmi, Jemaa (Nov.) *Keay* FHI 21045 ! Bauchi Plateau (Oct.) *Lely* P. 827 ! **S.Nig.:** Lagos *Dalz.* 1084 ! Ibadan (fl. & fr. Dec., Jan.) *Meikle* 860 ! 936 ! Enugu (fl. & fr. Feb.) *Jones* FHI 4572 ! Ikom (Jan.) *Holland* 207 ! **Br. Cam.:** Buea (fr. Dec.) *Maitland* 134 !   Distributed throughout the tropics. (See Appendix, p. 100.)
6. **T. eriophlebia** *Hook. f.* in Fl. Nigrit. 235 (1849) ; Sprague & Hutch. l.c. 267.   *T. rhomboidea* of F.T.A. 1 : 257, partly.   An erect, branched undershrub.
   N.Nig.: Bauchi Plateau (July) *Lely* P. 353 ! **S.Nig.:** Benin *Imp. Inst.* 7 ! Agulu *Thomas* 195 ! Calabar (Sept.) *Holland* 145 ! *Farquhar* 45 ! Owhy-Ikure (fr. Jan.) *Holland* 217 ! **Br.Cam.:** Buea (Jan., Apr.) *Maitland* 279 ! *Hutch. & Metcalfe* 114 ! Bamenda (June) *Maitland* 1739 ! **F.Po.** (Oct.) *T. Vogel* 9 ! Mioka (fl. & fr. Dec.) *Boughey* 88 !
7. **T. pentandra** *A. Rich.* in Guill. & Perr. Fl. Seneg. 1 : 93, t. 19 (1831) ; Sprague & Hutch. l.c. 267.   *T. neglecta* Wight & Arn.—F.T.A. 1 : 255.   An erect, branched annual, slightly pubescent.
   Sen.: *Roger* ! **Fr.Sud.:** Djenne (Sept.) *Chev.* 3078 ! **S.L.:** Waterloo (fl. & fr. Dec.) *Deighton* 3359 ! Magbile (Dec.) *Thomas* 5910 ! Manankon, Musaia (fl. & fr. Apr.) *Deighton* 5421 ! **G.C.:** Sandema (fl. & fr. Sept.)

Hughes FH 5346! **Fr.Nig.**: Niamey (fl. & fr. Oct.) *Hagerup* 470! **N.Nig.**: Sokoto (Nov.) *Moiser* 22! Katagum (fl. & fr. Sept.) *Dalz.* 83! **S. Nig.**: Aguku *Thomas* 596! Extends to Abyssinia, Ngamiland and S.W. Africa.

8. **T. cordifolia** *A. Rich.* l.c. 1 : 91, t. 18 (1831) ; Sprague & Hutch. l.c. 270, incl. vars. ; Chev. Bot. 89. A shrub, sometimes up to 12 or 15 ft. high, pubescent, hairy or nearly glabrous, with long-stalked ovate cordate leaves ; fruits ¼ inch across.

Sen.: *Farmar! Perrottet.* Komboland *Heudelot.* **Gam.**: *Boteler! Ingram!* **Port.G.**: *Esp. Santo!* **Fr.G.**: Kouria (Oct.) *Chev.* 14968! Macenta (Oct.) *Baldwin* 9829! **S.L.**: Bagroo R. *Mann* 869! Kumrabai (fr. Dec.) *Thomas* 7104! Kenema (fl. Nov., fr. Jan.) *Deighton* 389! *Thomas* 7560! **Lib.**: Monrovia *Whyte!* Peahtah (Oct.) *Linder* 1033! Ganta (Nov.) *Harley!* Vonjama (Oct.) *Baldwin* 9941! **Iv.C.**: *fide* Chev.*l.c.* **G.C.**: Arreboo *Dudgeon* 109! Axim *Cort Dev. Synd.!* Aburi (Oct.) *Johnson* 476! Assuantsi (fl. & fr. Mar.) *Irvine* 1556! **Togo:** Agu Mt., Nyambo *Busse* 3356. Misahöhe *Baumann* 306. **N.Nig.** : Anara F.R. (Oct.) *Keay* FHI 20125! **S.Nig.**: Lagos *Barter* 2186! *Millen* 5! *Dalz.* 1269! Abeokuta *Irving!* Ibadan (fl. & fr. Dec.) *Meikle* 861! Oban *Talbot!* Okimi (fr. Jan.) *Holland* 170! Mt. Koloishe, Obudu (fl. & fr. Dec.) *Keay & Savory* FHI 25070! **Br.Cam.**: Victoria *Preuss* 1356! Buea (fr. Jan.) *Maitland* 322! Mamfe *Rudatis* 93. Njinikom (June) *Maitland* 1766! **F.Po** : *Barter!* Widespread in the moister parts of tropical Africa. (See Appendix, p. 99.)

Imperfectly known species.

T.sp.—A low herb with pink hairs ; possibly a new species. **S.Nig.**: Idanre Hills (Oct.) *Keay* FHI 22652!

8. CLAPPERTONIA Meisn. Pl. Vasc. Gen. tabl. diagn. 36, comm. 28 (1837). *Honckenya* Willd. (1793)—F.T.A. 1 : 260, not of Ehrh. (1788).

A small erect shrub ; branches stellate-tomentose ; leaves more or less broadly ovate in outline, rounded or cordate at base, variously 3–5-lobed or sometimes not lobed, serrate, up to about 13 cm. long and 7 cm. broad, stellate-pubescent or tomentose beneath ; stipules persistent, lanceolate, about 1 cm. long ; flowers few in terminal leafy racemes ; fruits narrowly oblong, 4–6 cm. long, very bristly, bristles pilose .. 1. *ficifolia*

A prostrate woody herb ; branches laxly pilose ; leaves mostly 3–5-lobed, 2–5 cm. long, thin, serrate, pilose only on the nerves ; stipules linear-subulate, about 0·5 cm. long ; flowers 2–3 together ; fruits ellipsoid, 2 cm. long, densely bristly, bristles laxly pilose   ..   ..   2. *minor*

1. **C. ficifolia** (*Willd.*) *Decne.* in Deless. Ic. Select. Pl. 5 : 1, t. 1 (1846) ; Becherer in Fedde Rep. 28 : 58 (1930). *Honckenya ficifolia* Willd.—F.T.A. 1 : 260 : Chev. Bot. 90 ; Burret in Notizbl. Bot. Gart. Berl. 9 : 862 (1926) ; F.W.T.A., ed. 1, 1 : 241. To 7 ft. high ; usually in marshy places ; flowers conspicuous red-purple, rarely white, 2–3 in. diam. when opened ; fruits reddish-brown.
Fr.G.: *Heudelot* 791! Farana (Mar.) *Sc. Elliot* 5372! Macenta (fr. Oct.) *Baldwin* 9776! **S.L.**: Kitchom *Sc. Elliot* 4309! Hangha (Sept.) *Aylmer* 604! Njala (Sept.) *Deighton* 112! **Lib.**: Grand Bassa (July) *T. Vogel* 57! 91! Kakatown *Whyte!* Congotown (Aug.) *Barker* 1397! **Iv.C.**: Mankono (July) *Chev.* 21973! **G.C.**: Accra *Don!* Essiama (Apr.) *Williams* 435! Aburi (May) *Johnson* 957! Kumasi (Nov.) *Irvine* 1848! **Dah.**: Porto Novo (Jan.) *Chev.* 22712. **N.Nig.**: Nupe *Barter* 1226! **S.Nig.**: Lagos (fl. & fr. Dec.) *Moloney!* *Hagerup* 845! Agulu *Thomas* 153! Eket *Talbot* 3009! Oban *Talbot* 1235! **Br.Cam.**: Cameroons R. (Jan.) *Mann* 724! Kentu, Bamenda (June) *Maitland* 1553! Widespread in tropical Africa. (See Appendix, p. 95.)

2. **C. minor** (*Baill.*) *Becherer* l.c. 58 (1930). *Honckenya minor* Baill. in Adansonia 10 : 183 (1872) ; Burret l.c. 862 ; F.W.T.A., ed. 1, 1 : 241. *H. parva* K. Schum. in Engl. Bot. Jahrb. 15 : 115 (1893) ; Chev. Bot. 90. Flowers white or rose, scarcely 1 in. diam. ; in dry places.
S.L.: *Afzelius!* Regent (Oct.) *Deighton* 209! Heddle's Farm (Dec.) *Sc. Elliot* 3915! Koribundu (fl. & fr. Nov.) *Deighton* 4669! Yonibana (Nov.) *Thomas* 4801! **Lib.**: Monrovia (fl. & fr. Nov.–Jan.) *Krause!* *Dalz.* 8114! *Barker* 1465! Gbanga (Sept.) *Linder* 697! **Iv.C.**: Alépé (Feb.) *Chev.* 17378. Mt. Niénokoué (July) *Chev.* 19490. **G.C.**: Axim *Joly.* (See Appendix, p. 96.)

9. ANCISTROCARPUS Oliv.[1]—F.T.A. 1 : 265.

Leaves oblong-elliptic, rounded at base, acutely acuminate, up to 20 cm. long and 7 cm. broad, serrulate ; with about 8 pairs of lateral nerves, 3-nerved at base ; stamens united in bundles opposite the petals, fruit globose, 3–4 cm. diam. including the prickles which are up to half as long as the fruit-body   ..   ..   *densispinosus*

A. densispinosus *Oliv.*—F.T.A. 1 : 265. *A. brevispinosus* Oliv.—F.T.A. 1 : 265 ; F.W.T.A., ed. 1, 1 : 240. A small tree with long half-scrambling branches ; flowers white, turning yellow.
S.Nig.: Lagos *Lamborn* 264! Epe *Barter* 3276! Abeku, Shasha F.R. (Mar.) *Jones & Onochie* FHI 17196! Oji R. (Mar.) *Kitson!* Agbani, Udi (July) *Cousens* FHI 4851! Aking, Calabar (Apr.) *Ejiofor* FHI 21881! Eket *Talbot!* Oban *Talbot* 7. **Br.Cam.**: *Kalbreyer* 207! Cameroons R. (Jan.) *Mann* 2204! Victoria (fl. May, fr. Oct.) *Maitland* 672! 751! Johann-Albrechtshöhe *Staudt* 671! Abonando, Mamfe (Mar.) *Rudatis* 29! Also in French Cameroons, Gabon, Belgian Congo and Cabinda. (See Appendix, p. 95.)

75. STERCULIACEAE

Trees or shrubs, often with soft wood, or very rarely herbs, mostly with stellate hairs. Leaves alternate, simple or digitately compound ; stipules usually present. Flowers variously arranged, hermaphrodite or unisexual, actinomorphic. Sepals valvate, mostly partly connate or rarely spathaceous. Petals 5, or absent, contorted imbricate, often hooded. Stamens free or

---

[1] The name *Ancistrocarpus* Oliv. has been conserved against *Ancistrocarpus* H. B. & K. (1817) ; see Rehder in Kew Bull. 1935 : 355. *A. brevispinosus* Oliv. was adopted as the standard species of the genus, following Rehder who stated that it was " the better known and more widely distributed of the two original species ". This was, however, clearly an error, and in uniting the two species I have chosen the name *A. densispinosus* Oliv. which is in fact the better known and more widely distributed species.—R.W.J.K.

connate into a column, sometimes with staminodes ; anthers 2-celled. Ovary superior, of 2–12 united carpels or of one carpel ; ovules on axile placentas ; style simple or rarely the styles free to the base. Fruit various. Seeds with or without endosperm and straight or curved embryo.

A large tropical family, with the stamens often united into a column ; indumentum mostly stellate or lepidote ; petals sometimes absent.

Flowers with petals, hermaphrodite (rarely polygamous) :
  Staminodes present ; mostly trees and shrubs :
    Androgynophore well-developed ; fruiting carpels free, with a terminal wing, samara-like ; petaloid staminodes present between the stamens and carpels :
      Calyx spathaceous ; stamens 10 ; leaves not lobed   ..   ..   1. **Mansonia**
      Calyx 5-lobed ; stamens 30 or more, filaments united at base in pairs ; leaves digitately 5–7-lobed   ..   ..   2. **Triplochiton**
    Androgynophore absent or very short ; fruit capsular :
      Petals hooded over the androecium ; anthers sessile on the staminal tube :
        Petals with a long apical appendage ; fruits spiny   ..   ..   4. **Byttneria**
        Petals without a long apical appendage ; fruits not spiny   5. **Scaphopetalum**
      Petals not hooded :
        Petals much shorter than the sepals, pilose towards margins ; staminodes in groups or phalanges with the fertile stamens ; staminal tube short with small staminodes within it and alternating with the phalanges, or absent ; seeds conspicuously arillate   ..   ..   6. **Leptonychia**
        Petals as long as or longer than the sepals :
          Bracteoles neither persistent nor enlarging in fruit :
            Petaloid staminodes 5, inserted between the stamens and carpels ; filaments much shorter than the anthers ; staminal tube absent ; petals falling off ; seeds with a unilateral wing   ..   ..   3. **Nesogordonia**
            Petaloid staminodes between the stamens and carpels absent ; staminal tube present, bearing the staminodes and stamens usually with distinct filaments ; petals persistent, often becoming dry or scarious   ..   ..   7. **Dombeya**
          Bracteoles persistent and enlarging in fruit, suborbicular   ..   8. **Melhania**
  Staminodes absent ; mostly herbs or undershrubs ; androgynophore absent (or rarely very short) :
    Ovules numerous ; capsule loculicidally 5-valved, apex often with horny appendages ; carpels opposite the sepals   ..   ..   9. **Hermannia**
    Ovules 2 ; carpels opposite the petals :
      Ovary 5-celled ; capsule loculicidally 5-valved ; cells 1-seeded ..   10. **Melochia**
      Ovary 1-celled ; capsule 2-valved, 1-seeded   ..   ..   ..   11. **Waltheria**
Flowers without petals, unisexual or sometimes polygamous ; trees or shrubs :
  Carpels numerous, multiseriate ; calyx-lobes normally 8 ; anther-cells arranged in 1 whorl ..   ..   12. **Octolobus**
  Carpels up to 12, in a single whorl ; calyx-lobes (3–) 4–5 (–6) :
    Fruiting carpels neither winged nor inflated and submembranous, usually 2–many-seeded, rarely 1-seeded :
      Seeds with a large unilateral wing ; anther-cells (in our species) arranged in 1 whorl ; leaves simple   ..   ..   13. **Pterygota**
      Seeds not winged :
        Anther-cells arranged in an irregular mass ; seeds usually with endosperm   14. **Sterculia**
        Anther-cells arranged in 1 or 2 whorls, rarely at more than 2 levels owing to lobing of the cupshaped upper part of the synandrium (*C. digitata* and *C. megalophylla*) ; seeds usually without endosperm   ..   ..   15. **Cola**
    Fruiting carpels winged, or inflated and submembranous, normally 1-seeded :
      Fruiting carpels inflated, submembranous ; calyx-tube long ; leaves simple, broadly ovate-cordate, not densely lepidote beneath ; anther-cells arranged in an irregular mass   ..   ..   16. **Hildegardia**
      Fruiting carpels samara-like ; calyx-tube short ; leaves digitately 5–7-foliolate, or 1-foliolate, densely lepidote beneath ; anther cells arranged in 1 whorl
      17. **Tarrietia**

Besides the above, other genera are represented by the following introduced plants cultivated in our area : *Guazuma ulmifolia* Lam., *Helicteres isora* Linn., *Kleinhovia hospita* Linn., *Pterospermum* spp. and *Theobroma cacao* Linn. (the Cocoa tree).

1. **MANSONIA** J. R. Drumm. ex Prain in J. Linn. Soc. 37 : 260 (1905) ; Brenan in Hook. Ic. Pl., sub. t. 3451 (1947) ; Chatterjee & Brenan in Hook. Ic. Pl., sub t. 3484 (1951).

Branchlets and lower surface of leaves densely and softly pubescent with stellate hairs ; leaves obovate-orbicular, cordate at the base, slightly pointed, 15–30 cm. long, 8–15

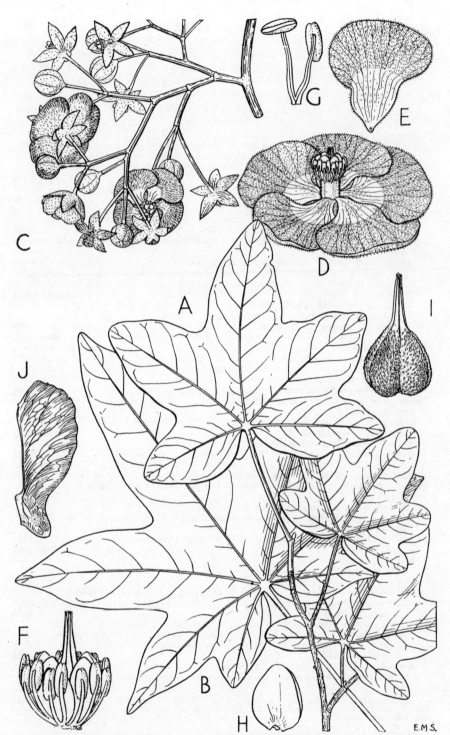

FIG. 118.—TRIPLOCHITON SCLEROXYLON *K. Schum.* (STERCULIACEAE).

A, leafy shoot, × ⅔.   B, young leaf, × ⅔.   C, inflorescence, × 1.   D, flower, × 2.   E, petal, × 2.
F, stamens, staminodes and ovary, × 6.   G, pair of stamens, × 6.   H, staminode, × 6.
I, carpels, × 6.   J, fruit, × ⅔.
A is drawn from *Chev.* 16105 ; B from *Kennedy* 739 ; C–E from *Mildbr.* 7945 ; F–I from *Foster*
370 ; J from *Rosevear* 8.

cm. broad, thinly papery, crenulate-denticulate, 6–7-nerved at the base, with about 5 pairs of lateral nerves ; stipules deciduous ; cymes pedunculate, many-flowered ; calyx spathaceous, densely stellate-tomentose, 1 cm. long, apiculate ; petals obovate, nearly 1·5 cm. long ; androgynophore nearly. 1 cm. long ; fruits winged, body 1·5 cm. long, obovoid, reticulate, wing unilaterally obovate-oblanceolate, about 6 cm. long

1. altissima var. altissima

Branchlets and lower surface of leaves sparingly pubescent or nearly glabrous ; otherwise similar to above ..   ..   1a. <i>altissima</i> var. <i>kamerunica</i>

1. **M. altissima** (*A. Chev.*) *A. Chev.* var. **altissima**—in Bull. Soc. Bot. Fr. 58, Mém. 8 : 138 (1912) ; Aubrév. Fl. For. C. Iv. 2 : 261, t. 230 ; Kennedy For. Fl. S. Nig. 62 ; Brenan in Hook. Ic. Pl., sub t. 3451 ; Chatterjee & Brenan in Hook. Ic. Pl., sub t. 3484. *Achantia altissima* A. Chev. in Bull. Mus. Hist. Nat. 15 : 548 (1910). A forest tree, to 100 ft. high, deciduous, with white fragrant flowers and winged maple-like fruits.
    Iv.C.: Mbayakro, Nzi (Aug.) *Chev.* 22281 ! Anoumaba (fr. Nov.) *Chev.* B. 22345 ! **G.C.:** Nkwanta (fr. Jan.) *Chipp* 65 ! Sunyani (fr. Jan.) *Chipp* 72 ! Kumasi (fr. Oct.) *Brent* 936 ! Jimira F.R. (July) *Vigne* FH 3013 ! **S.Nig.:** Olokemeji (fr. Jan.) *A. F. Ross* 50 ! Mamu F.R., Ibadan (July) *Ahmed & Chizea* FHI 20004 ! Akure F.R. (July) *Onwudinjoh* FHI 23419 ! Agbabu, Ondo Prov. *Kennedy* 2373 ! (See Appendix, p. 108.)
1a. **M. altissima** var. **kamerunica** *Jac.-Fél.* in Rev. Bot. Appliq. 25 : 236 (1945) ; Chatterjee & Brenan in Hook. Ic. Pl., sub t. 3484. Forest tree, to 100 ft. high.
    Br.Cam.: Barombi Kotto, Kumba Div. (fr. Dec.) *Smith* 129/36 ! Boviongo, Kumba Div. (July) *Smith* 129/41 ! Also in French Cameroons.
    [Jacques-Félix in Rev. Bot. Appliq. 27 : 516 (1947) suggests that this variety is the same as *M. nymphaeifolia* Mildbr. in Notizbl. Bot. Gart. Berl. 7 : 489 (1921) from the French Cameroons. Brenan (ll.cc.), however, is doubtful of this. The type of *M. nymphaeifolia* has, presumably, been destroyed, but even if it does prove to be the same as Jacques-Félix suggests, it seems wiser to treat our plant as a variety.]

2. **TRIPLOCHITON** K. Schum.[1] in Engl. Bot. Jahrb. 28 : 330 (1900). *Samba* Roberty (1953).

A large tree, with glabrous branchlets and leaves ; mature leaves 5–7-lobed, 5–7-nerved at base, lobes broadly ovate or triangular to oblong, obtusely acuminate, lamina averaging 10–15 cm. diam., leaves of seedlings larger and more deeply lobed ; flowers hermaphrodite or polygamous in axillary cymose-panicles up to 7·5 cm. long ; calyx 7–9 mm. long, deeply lobed, tomentellous outside ; petals obovate or obcordate, about 1 cm. long and broad, densely pilose on both surfaces ; androgynophore 3–3·5 mm. long ; stamens 30–46, with filaments 2–3 mm. long, connate at the base in pairs ; staminodes 5, broadly oblong-ovate, antisepalous, surrounding the 5 antipetalous carpels ; fruiting carpels about 6 cm. long, including the obliquely oblong-obovate, membranous wing ..   ..   *scleroxylon*

T. scleroxylon *K. Schum.* in Engl. Bot. Jahrb. 28 : 331 (1900) ; Chev. Bot. 75 ; Aubrév. Fl. For. C. Iv. 2 : 260, t. 229 ; Kennedy For. Fl. S. Nig. 59. *T. johnsoni* C. H. Wright in Hook. Ic. Pl. t. 2758 (1903) ; Chev. Bot. 75. *T. nigericum* Sprague in Kew Bull. 1909 : 212 ; F.W.T.A., ed. 1, 1 : 248. *Samba scleroxylon* (K. Schum.) Roberty (1953). A large forest tree, to 200 ft. high, with large buttresses, deciduous ; leaves and fruits maple-like ; petals white, red-purple at base.
    Fr.G.: Koulikoro *Chev.* 20649. **S.L.:** Koinadugu, Loma Mts. *Frith* 26 ! **Iv.C.:** Bouroukrou (Dec.–Jan.) *Chev.* 16105 ! Agboville *Chev.* 22341 ! Rasso *Aubrév.* 548 ! **G.C.:** Anum (Dec.) *Johnson* 813 ! Aburi (Jan.) *Irvine* 1898 ! Owabi (Nov.) *Vigne* FH 2597 ! Agogo *Chipp* 450 ! **Dah.:** Sakété (Jan.) *Chev.* 22819 ! **S.Nig.:** Olokemeji (Dec.) *Foster* 370 ! Ibadan F.R. (Jan.) *Punch* 125 ! Ijebu-Ode (fl. & young fr. Jan.) *Punch* ! Ojojwo, Benin *Foster* 188 ! Sapoba (fl. Dec., fr. Feb.–Mar.) *Kennedy* 739 ! 1679 ! Also in French Cameroons, Spanish Guinea and Belgian Congo. (See Appendix, p. 111.)

3. **NESOGORDONIA** Baill. in Bull. Soc. Linn. Paris 1 : 555 (1886) ; R. Capuron in Notulae Syst. 14 : 258 (1953). *Cistanthera* K. Schum. (1897)—F.W.T.A., ed. 1, 1 : 240 (under *Tiliaceae*) ; Burret in Notizbl. Bot. Gart. Berl. 9 : 865 (1926).

A tree ; leaves ovate-elliptic, acuminate, mucronate, cuneate at base, up to 11 cm. long and 5 cm. broad, entire, very minutely stellate-puberulous beneath, at length glabrous ; stipules linear, deciduous ; petiole slender, 1·5–4 cm. long ; flowers few in slender axillary cymes ; peduncle a little longer than the petiole ; sepals ovate-lanceolate, 8–10 mm. long, tomentellous ; petals glabrous, about 1·3 cm. long ; fruits broadly turbinate, truncate, woody, about 3 cm. long, each valve with 2 blunt horns, stellate-tomentellous ; seeds winged, wings oblique, 1 cm. long *papaverifera*

N. papaverifera (*A. Chev.*) *R. Capuron* in Notulae Syst. 14 : 259 (1953). *Cistanthera papaverifera* A. Chev. in Bull. Soc. Bot. Fr. 58, Mém. 8 : 141 (1912) ; F.W.T.A., ed. 1, 1 : 240 ; Aubrév. Fl. For. C. Iv. 2 : 214, t. 210 ; Kennedy For. Fl. S. Nig. 58. A forest tree, to about 100 ft. high, with clean straight bole ; flowers yellowish-white, fragrant.
    S.L.: Kasewe F.R. (fl. Mar., fr. Dec.) *Deighton* 4594 ! *King* 126 ! **Iv.C.:** Bangouanou, Morénou (Nov.) *Chev.* 22459 ! Sahoua to Bangouanou (fr. Nov.) *Chev.* 22442 ! Bouroukrou (Dec.) *Chev.* 16108 ! 16123 ! Abidjan *Aubrév.* 109 ! Man *Aubrév.* 404 ! **G.C.:** Dunkwa (fl. June, fr. Feb.) *Vigne* 267 ! FH 893 ! Abuasi (Jan.) *Soward* 666 ! Koforidua (June) *Moor* 73 ! Awaso, Sefwi Bekwai (fr. Oct.) *Akpabla* 882 ! **Dah.:** Le Testu 159 ! **S.Nig.:** Olokemeji (Dec.) *Foster* 373 ! *Chev.* 14001. Akure F.R. (young fr. Apr.) *Onochie* FHI 3354 ! (See Appendix, p. 95.)

---

[1] This name has been conserved against the earlier *Triplochiton* Alef. (1863).

FIG. 119.—NESOGORDONIA PAPAVERIFERA (*A. Chev.*) *R. Capuron* (TILIACEAE).

A, flowering shoot. B, stipule. C, flower-bud. D, open flower. E, F, stamens. G, petal.
H, gynaecium. I, section of ovary. J, fruiting branch. K, segment of fruit. L, seed.

4. BYTTNERIA Loefl.—F.T.A. 1 : 239.

Branchlets not prickly, stellate-puberulous ; leaves with petiole up to about 23 cm.
long, lamina broadly ovate to suborbicular-ovate or ovate, widely truncate-cordate at
base, acutely acuminate, up to about 18 cm. long and broad, thin, entire, 5-nerved
at base, nearly glabrous ; flowers small in slender leaf-opposed cymes, sometimes with
an additional subaxillary smaller cyme ; cymes about as long as the petiole ; sepals
lanceolate, about 6 mm. long, finely tomentellous ; petals a little longer, obovate with
a long linear-lanceolate appendage ; body of fruit about 3 cm. diam., with numerous
spines ..   ..   1. *catalpifolia* subsp. *africana*
Branchlets prickly, puberulous with simple or fascicled hairs ; leaves with petiole up
to 9 mm. long, lamina oblanceolate-oblong, rounded at base, acuminate at apex, up to
6 cm. long and 2·3 cm. broad, margin markedly crenate in the upper part ; flowers in
small umbels or groups of umbels ; calyx glabrous ; petals with an incurved claw,
2 small lateral processes and a long ligulate appendage :
Umbels 2 together, each with (1–) 2 (–3–4) flowers ; petioles up to 5 mm. long, body of
fruit about 1 cm. diam. with unequal-sized spines ..   ..   2. *guineensis*
Umbels up to 6 together, each with 6–9 flowers ; petioles up to 9 mm. long ; fruit not
known ..   ..   3. *sp. A*

1. **B. catalpifolia** *Jacq.* subsp. **africana** (*Mast.*) *Exell & Mendonça* in Consp. Fl. Ang. 1 : 197 (1951). *Buettneria
africana* Mast. in F.T.A. 1 : 239 (1868) ; K. Schum. in Engl. Monogr. Afr. Sterculiac. 90, t. 5 B ; F.W.T.A.,
ed. 1, 1 : 249. Scandent shrub or liane, mainly in secondary forest ; flowers white, sweet-scented ; cut
stem yielding water copiously.
    G.C.: Agogo (fr. Dec.) *Chipp* 443 ! *Irvine* 874 ! Kumasi (July) *Akwa* FH 1766 ! Akroso (Aug.) *Howes*
947 ! Jimira F.R. (Aug.) *Vigne* FH 3997 ! Agona (Nov.) *Darko* 718 ! **S.Nig.**: Oban *Talbot* 438 ! 1388 !
Br.Cam.: Nkom-Wum, Bamenda Prov. (July) *Ujor* FHI 30465 ! Also in French Cameroons, Belgian
Congo, Angola, Uganda and A.-E. Sudan. The subsp. *catalpifolia* is in tropical S. America.
2. **B. guineensis** *Keay & Milne-Redhead* in Kew Bull. 1954 : 263. Prickly scrambling shrub ; flowers red ;
fruits green with reddish tinge.
    S.L.: Njala (fl. & fr. Apr.) *Deighton* 4740 ! 5030 !
3. **B. sp. A.** Prickly climber, similar to *B. guineensis* but differing by the following characters as well as those
shown in the key : (i) primary branches obscurely angled, hollow, with larger prickles, (ii) leaves with less
hairy petioles and somewhat longer acumens, (iii) sepals apparently narrower. Better material of this
plant is needed before making a decision about its taxonomical status.
    S.Nig.: R. Addo, Igbessa, Lagos Colony (Dec.) *Millen* 194 !

5. SCAPHOPETALUM Mast. in J. Linn. Soc. 10 : 28 (1869).

Sepals up to 8 mm. long, at anthesis usually adherent in two pairs with the fifth sepal
free, calyx thus appearing 3-lobed, or rarely adhering 3 and 2 together, calyx thus
appearing 2-lobed :
Sepals entirely glabrous, more or less flat, elliptic, acuminate at apex about 7 mm. long
and 2·5 mm. broad, rather thin, with a strong midrib ; sepals or pairs of sepals
separating to the base ; petals glabrous, with broad lateral lobes ; 3-cuspidate at
apex, strongly veined ; anthers at top of staminal tube ; ovary glabrous ; flowers

generally solitary, with pedicels about 1·5 mm. long ; branchlets minutely puberulous when very young, soon glabrous ; leaves glabrous, oblong to elliptic, rounded
to cuneate at base, caudate-acuminate, 9–17·5 cm. long, 3–6·7 cm. broad

1. *parvifolium*

Sepals distinctly pubescent or puberulous outside, concave, ovate to oblong, 5–8 mm.
long, 3–3·5 mm. broad, united at base ; petals not 3-cuspidate ; ovary densely
pubescent ; flowers several together ; branchlets distinctly hairy :

Branchlets puberulous ; leaves only sparsely puberulous beneath ; flowers axillary
and on bole, bracts not usually conspicuous :

Petals glabrous outside except at the apex, attached to the staminal tube at the base
only and thus easily detached ; sepals puberulous outside, densely so on the
margins thus making them distinct in bud ; staminal tube with a second lower
whorl of anthers (? always) ; leaves elliptic-lanceolate, to lanceolate, rounded or
obtuse at base, gradually acuminate, 14–33 cm. long, 4–10·5 cm. broad

2. *cf. zenkeri*

Petals puberulous outside, firmly attached to the staminal tube for 1–1·5 mm. ;
sepals densely puberulous outside, margins not clearly marked in bud by lines of
hairs, pairs of sepals firmly connate ; anthers at top of staminal tube ; leaves
elongate-oblong-elliptic to oblong-oblanceolate, obtuse to rounded (rarely cuneate)
at base, shortly acuminate, 10–34 cm. long, 3·5–12·5 cm. broad   3. *amoenum*

Branchlets densely tomentellous ; leaves oblong or oblong-lanceolate, rounded at
base, shortly acuminate at apex, 16–35 cm. long, 5–12·5 cm. broad, finely puberulous with stellate and fascicled hairs beneath ; flowers usually in axils of leaves,
with numerous often conspicuous bracts ; sepals densely puberulous outside,
appearing distinct in bud ; petals puberulous outside ; anthers at top of staminal
tube   ..   ..   4. *letestui*

Sepals about 15 mm. long, quite separate at anthesis, subglabrous outside ; petals about
10 mm. long, 3-toothed at apex ; pedicels 1–2 cm. long ; flowers on the old wood ;
branchlets puberulous ; leaves oblong, abruptly long-caudate-acuminate at apex,
margin more or less undulate, 10–22 cm. long, 4–8 cm. broad ..   .. 5. *talbotii*

1. **S. parvifolium** *Bak. f.* in Cat. Talb. 14 (1913). Small tree, to 30 ft. high ; "flowers bright yellow, sepals
bright palish green " ; in forest.
  S.Nig.: Obutong Road, Oban *Talbot* 1264 ! Oban *Talbot* ! Uwet—Budeng Road (Nov.) *Rosevear* C. 5 !
Between Ndiumo and Orora Road (Nov.) *Rosevear* C. 15 !

2. **S. cf. zenkeri** *K. Schum.* in E. & P. Pflanzenfam. Nachtr. 1 : 241 (1897) ; Engl. Monogr. Afr. Sterculiac. 93,
t. 7 B. Small forest tree, to 45 ft. high, or shrub ; flowers on leafy shoots and on branches down to base
of bole ; sepals greenish-purple outside, very pale green inside ; petals hyaline yellow at apex, with lines
which are white outside and purple inside ; staminal tube rich deep purple ; fruits ellipsoid, knobbly and
puberulous outside, about 2·8 cm. long and 1·5 cm. diam.
  S.Nig.: Oban *Talbot* ! **Br.Cam.:** S. Bakundu F.R., Kumba (fl. Mar., Apr., fr. Aug.) *Brenan* 9291 ! 9427 !
Ejiofor FHI 29314 ! *Olorunfemi* FHI 30704 ! Also in French Cameroons. [The duplicate of the type at
Kew is not in a fit state to dissect so I cannot confirm that the androecium of *S. zenkeri* is as figured by
Schumann. This figure does not show the second lower whorl of anthers which is present in our specimens
from the British Cameroons which in other respects closely resemble the type material.]

3. **S. amoenum** *A. Chev.* in Bull Soc. Bot. Fr. 61, Mém. 8 : 255 (1917) ; Hutch. & Dalz. in Kew Bull. 1928 :
296 ; Aubrév. Fl. For. C. Iv. 2 : 262, t. 231, 1–7. Forest tree, to 45 ft. high, bole usually leaning ; sepals
brownish-puberulous ; petals yellowish-green with deep purple lines and markings ; staminal tube deep
purple or red ; fruits ellipsoid, about 1·5 cm. long.
  Lib.: Dukwia R. (Feb.–Mar.) *Cooper* 57 ! 237 ! 264 ! Moala (Nov.) *Linder* 1379 ! Truo (Mar.) *Baldwin*
12570 ! **Iv.C.:** *Aubrév.* 1690 ! Songan, Indénié (Mar.) *Chev.* 17788 ! Massa-Mé *Aubrév.* 138 ! Bouroukrou
(Dec.–Jan.) *Chev.* 16578 ! 16586 ; 16896. **G.C.:** Abiabo, Axim (Nov.) *Vigne* FH 1421 ! Abura F.R.
(Nov.) *Tinsley* WH 3854 ! (See Appendix, p. 108.)

4. **S. letestui** *Pellegr.* in Bull. Mus. Hist. Nat. 27 : 196 (1921) ; Fl. Mayombe 1 : 31, fig. 1 (1924). *S. blackii* of
F.W.T.A., ed. 1, 1 : 250, not of Mast. Forest shrub.
  S.Nig.: Oban *Talbot* s.n. ! 437 ! 476 !

5. **S. talbotii** *Bak. f.* l.c. 14 (1913). *S. macranthum* of F.W.T.A., ed. 1, 1 : 249, not of K. Schum. Small forest
tree, to 30 ft. high ; sepals dull pink outside, dull purplish-red on inside ; petals ribbed, yellow ; anthers
cream ; androecium deep wine-red at base ; stigma pale dull pink.
  S.Nig.: Oban *Talbot* 1562 ! **Br.Cam.:** Rio del Rey *Johnston* !

    Several species have been described from the French Cameroons by Engler and K. Krause, but I have
not been able to see material of them. The types have presumably been destroyed. With further material
from the type localities it may be possible to prove that some of these species are the same as some of those
described above.

6. LEPTONYCHIA Turcz.—F.T.A. 1 : 238.

*Flowers axillary ; leaves up to 25 cm. long ; ovary (3–) 5-celled :

Staminal tube short but more or less well-developed, with 5 lanceolate or ovate-
lanceolate staminodes inserted inside it so that one of them appears between each
bundle or group of fertile stamens and filament-like staminodes ; sepals 8 mm. or
more long ; ovary densely pilose ; fruits densely hairy :

Fertile stamens 15–20 :

Petals 2-lobed, pilose towards the margins ; sepals 10–11 mm. long ; filament-like
staminodes 2 in each group of 4 fertile stamens ; branchlets soon nearly glabrous ;
petioles sparingly puberulous ; leaves elliptic to obovate-elliptic, gradually long-
acuminate, 12–22 cm. long, 3·3–7 cm. broad   ..   ..   .. 1. *occidentalis*

Petals not 2-lobed ; sepals 13–18 mm. long :
  Petioles rusty-pubescent ; branchlets densely rusty-pubescent when young, glabrescent ; sepals 13–15 mm. long ; petals deltoid, long-pilose towards the margins ; staminodes 4 in each group of 4 fertile stamens, only shortly united at base ; style more or less glabrous ; leaves broadly elliptic to obovate-elliptic, gradually long-acuminate, 9–19 cm. long, 3·5–7 cm. broad . .   2. *pubescens*
  Petioles and branchlets sparingly whitish puberulous, soon nearly glabrous ; sepals 15–18 mm. long ; petals suborbicular-elliptic, pilose on both surfaces as well as towards the margins ; staminodes 2–3 in each bundle of 3–4 fertile stamens, each bundle united at base into a common stipe above the staminal tube ; style pilose in lower half ; leaves obovate-elliptic, gradually long-acuminate, 14–23 cm. long, 4·5–8 cm. broad . .   . .   3. *batangensis*
Fertile stamens 10 ; style pilose ; petals not 2-lobed :
  Sepals about 15 mm. long, covered with greyish-brown hairs ; petals suborbicular, densely pilose towards the margins, pilose inside and sparingly so outside ; staminodes 3–5 in each bundle of 2 fertile stamens ; leaves elliptic, acuminate, 9–18 cm. long, 4–7 cm. broad . .   . .   4. *macrantha*
  Sepals about 8–9 mm. long, covered with yellow-brown hairs ; petals elliptic, densely pilose towards the margins only ; staminodes 3 in each bundle of 2 fertile stamens ; leaves more or less oblong, abruptly long-acuminate, 11–19 cm. long, 4·5–7·7 cm. broad   . .   . .   5. *multiflora*
Staminal tube and inner whorl of lanceolate staminodes absent ; fertile stamens 15, with 2 filament-like staminodes between every 3 fertile stamens and in the same whorl as them ; sepals 7–8 mm. long, only sparingly puberulous outside ; petals pilose towards the margins and also on both surfaces ; ovary puberulous ; style glabrous ; fruits muricate, glabrous except for a few hairs on the tips of the warts ; leaves narrowly oblong-lanceolate, long-caudate-acuminate, 13–23 cm. long, 3·5–6·5 cm. broad, very thin . .   . .   6. *sp. nr. echinocarpa*
*Flowers in fascicles on the older wood well below the leaves which are usually over 25 cm. long ; branchlets and leaves soon glabrous ; sepals 6–8 mm. long, sparingly puberulous outside ; petals suborbicular, pilose inside and towards the margins, nearly glabrous outside ; staminal tube short but well-developed with a lanceolate staminode between each group of 2 fertile stamens and 1–2 filament-like staminodes ; ovary 5-celled, densely pilose ; style more or less pilose ; fruits finely puberulous (? always):
Leaves broadly oblong-elliptic, rounded at base, long-acuminate, 25–35 cm. long, 11–14 cm. broad, with 7–10 main lateral nerves on each side of midrib . .   7. *pallida*
Leaves elongate-oblong-oblanceolate, slightly rounded at base, gradually acuminate, 27–32·5 cm. long, 6–7 cm. broad, with 6–7 main lateral nerves on each side of midrib . .   . .   8. *lanceolata*

1. **L. occidentalis** *Keay* in Kew Bull. 1954 : 264. *L. urophylla* of F.W.T.A., ed. 1, 1 : 249, partly, not of Welw. ex Mast. Shrub or small tree to 20 ft. high, rarely more ; flowers greenish-white ; in forest.
   S.L.: *Smythe* 215 ! *Thomas* 9778 ! Hangha (Jan.) *Thomas* 7782 ! Njala (Feb.) *Deighton* 4702 ! Kambui F.R. (Mar.) *Lane-Poole* 187 ! Gola Forest (fr. Mar.) *Small* 540 ! **Lib.:** Dukwia R. (fr. Apr.) *Cooper* 406 ! Vahon, Kolahun (fr. Nov.) *Baldwin* 10245 ! Tappita (fr. Aug.) *Baldwin* 9114 ! Jabroke, Webo (fr. July) *Baldwin* 6463 ! (See Appendix, p. 107.)
2. **L. pubescens** *Keay* l.c. 265 (1954). *L. urophylla* of F.T.A. 1 : 238, partly (*Barter* spec.) ; F.W.T.A., ed. 1, 1 : 249, partly ; Aubrév. Fl. For. C. Iv. 2 : 264, t. 231, 8–12 ; not of Welw. ex Mast. Small tree to 30 ft. high, or shrub ; flowers green, white or yellow ; fruits greyish-green with black seeds and scarlet arils.
   Iv.C.: Anoumaba (Nov.) *Chev.* 22415 ! **G.C.:** Juaben (Mar.) *Johnson* 631 ! Mampong (fr. Apr.) *Vigne* FH 1931 ! Konongo (Apr.) *Vigne* FH 1932 ! Ampan (fr. Nov.) *Vigne* FH 4275 ! **Dah.:** Allada (young fr. Mar.) *Chev.* 23404 ! **S.Nig.:** Agege *Foster* 222 ! Ilugun, Abeokuta Prov. (fr. Feb.) *Keay* FHI 22484 ! Ibadan South F.R. (Jan.) *Keay* FHI 19803 ! Attah *Barter* 3419 ! Appiapam, Obubra (fl. & fr. Apr.) *Jones* FHI 12159 !
3. **L. batangensis** (*C. H. Wright*) *Burret* in Notizbl. Bot. Gart. Berl. 9 : 728 (1926). *Grewia batangensis* C. H. Wright in Kew Bull. 1896 : 158. *L. urophylla* of F.T.A. 1 : 238 (1868), partly (*Mann* spec.) ; K. Schum. In Engl. Monogr. Afr. Sterculiac. 96, partly ; F.W.T.A., ed. 1, 1 : 249, partly ; not of Welw. ex Mast., see Exell & Mendonça in Consp. Fl. Ang. 1 : 198. A small forest tree, 15–20 ft. high, with greenish flowers.
   Br.Cam.: Ambas Bay (Jan.) *Mann* 13 ! **F.Po:** (Jan.) *Mann* 166 ! Also in French Cameroons.
4. **L. macrantha** *K. Schum.* in E. & P. Pflanzenfam. Nachtr. 1 : 241 (1897) ; Monogr. 96. Shrub or small tree, in forest ; flowers green.
   S.Nig.: Ologbo, Ishan Dist. (Mar.) *Ebuade* FHI 22759 ! Ikom (Dec.) *Rosevear* 73/30a ! Also in French Cameroons.
5. **L. multiflora** *K. Schum.* in E. & P. Pflanzenfam. Nachtr. 1 : 241 (1897) ; Monogr. 95, t. 8 C ; Pellegr. in Mém. Soc. Bot. Fr. 1950–51 : 49. Small tree, in forest.
   Br.Cam.: Abonando, Mamfe (Mar.) *Rudatis* 20 ! Also in French Cameroons and Belgian Congo.
6. **L. sp. nr. echinocarpa** *K. Schum.* in E. & P. Pflanzenfam. Nachtr. 1 : 241 (1897) ; Monogr. 95. Erect shrub, to 15 ft. high ; sepals green, stamens and petals cream ; in forest.
   S.Nig.: Oban *Talbot* 604 ! Afi River F.R. (Dec.) *Keay* FHI 28221 ! **Br.Cam.:** Korup F.R., Kumba Dist. (fr. June) *Olorunfemi* FHI 30641 ! [This differs slightly from *L. echinocarpa* K. Schum., from French Cameroons and Gabon, as known to me from the description and the syntype *Soyaux* 216 in Kew Hb., but may well prove conspecific.]
7. **L. pallida** *K. Schum.* in E. & P. Pflanzenfam. Nachtr. 1 : 241 (1897) ; Monogr. 98, t. 8 E. Small tree, to 30 ft. high ; cauliflorous ; sepals green, filaments and petals cream ; foliage drying pale green ; fruits green with grey indumentum and seeds covered with bright orange aril ; in forest.
   S.Nig.: Afi River F.R. (fl. & young fr. Dec., fr. May) *Keay* FHI 28223 ! 28235 ! *Jones & Onochie* FHI 5836 ! **Br.Cam.:** Barombi (June, Sept.) *Preuss* 313 ! 483 ! Bambuko F.R., Kumba Div. (Sept.) *Olorunfemi* FHI 30759 ! **F.Po:** *Mildbr.* 6354. Also in French Cameroons.
8. **L. lanceolata** *Mast.* in F.T.A. 1 : 239 (1868). A shrub 12–15 ft. high, with pale leaves and glabrous branchlets.
   F.Po : (Jan.) *Mann* 190 !

Imperfectly known species.

1. **L. adolfi-friederici** *Engl. & K. Krause* in Engl. Bot. Jahrb. 48 : 556 (1912). Erect shrub. The androecium is not sufficiently accurately described for this species to be included in the key ; the type specimen has presumably been destroyed.
   F.Po: S.W. coast, near Bokoko (Oct.) *Mildbr.* 6958.
2. **L. densivenia** *Engl. & K. Krause* l.c. 557 (1912). Small erect shrub. Included in the first edition as a synonym of *L. urophylla* Welw. ex Mast., but a specimen collected by Barter in Fernando Po agrees reasonably well with the description and appears distinct from *L. urophylla* and all the above species. The androecium is not sufficiently well described for the species to be included in the key ; the type has presumably been destroyed. The *Barter* specimen (cited by K. Schum. in Engl. Monogr. Afr. Sterculiac. 97 as *L. urophylla*) is not good enough for dissection.
   F.Po: *Barter!* S.W. coast, near Bokoko (Oct.) *Mildbr.* 6898.
3. **L. fernandopoana** *Engl. & K. Krause ex Mildbr.* Wiss Ergebn. 1910–11, 2 : 187 (1922), name only.
   F.Po: Bokoko *Mildbr.* 6823 ; 6862.

7. DOMBEYA Cav.—F.T.A. 1 : 226.

Flowers arranged in a subglobose rather dense subumbellate head, with a peduncle 6–22 cm. long ; sepals lanceolate, about 1 cm. long, villous-pilose ; petals about 1·6 cm. long and 1 cm. broad, not dry and hardened in fruit ; ovary 5-celled, with about 5 ovules in each cell ; branchlets, peduncles, petioles and nerves beneath coarsely pilose with simple hairs ; leaves digitately lobulate, cordate at base, 10–25 cm. long and broad ; stipules long-subulate-acuminate, loosely villous   1. *buettneri*

Flowers in lax, well-branched cymes ; sepals up to 6 (–8) mm. long ; petals considerably longer than broad, becoming dry and papery in fruit ; ovary 3-celled, with about 2 ovules in each cell ; branchlets, peduncles and petioles glabrous, or puberulous to tomentose with stellate hairs :

  Flowers produced on woody branchlets before the leaves ; young branchlets, petioles and inflorescence axes glabrous or whitish-puberulous to tomentellous ; leaves rounded-ovate, cordate at base, irregularly dentate at margin, up to 17 cm. long and broad :

    Inflorescence axes, young branchlets and petioles densely stellate-puberulous ; sepals woolly-tomentose outside ..   .. 2. *quinqueseta* var. *quinqueseta*

    Inflorescence axes, young branchlets and petioles glabrous or sparingly pubescent ; sepals puberulous-tomentellous outside ..   .. 2a. *quinqueseta* var. *senegalensis*

  Flowers terminal or axillary, produced after or with the leaves ; young branchlets and petioles densely reddish-brown stellate-tomentose ; inflorescence axes and outside of sepals woolly-tomentose ; leaves ovate to suborbicular, cordate at base, 8–23 cm. long, 6·5–19 cm. broad, undersurface densely covered with a close felt of grey tomentellous indumentum beneath, also with larger brownish stellate hairs which are scattered over the lamina but become abundant on the nerves, upper-surface green, puberulous   ..   .. 3. cf. *ledermannii*

1. **D. buettneri** *K. Schum.* in Engl. Bot. Jahrb. 15 : 133 (1892) ; Engl. Monogr. Afr. Sterculiac. 27, t. 2 B ; Aubrév. Fl. For. Soud.-Guin. 163, t. 29, 3. *D. mastersii* of F.T.A. 1 : 228, partly (*Irving* spec.). A weedy shrub, or small tree, to 15 ft. high, with pithy branches ; leaves with very long channelled petioles ; flowers white, usually with red centre, about 1 in. across, turning yellow in drying.
   Fr.G.: Farana (Feb.) *Chev.* 20660. Niger sources *Chev.* 20576. **S.L.:** Sefadu (Dec.) *Deighton* 3475 ! Jagbwema, Faiama (Dec.) *Deighton* 5286 ! Bintumane, 3,000 ft. (Jan.) *Glanville* 462 ! **Iv.C.:** Beoumi (Dec.) *Lowe* ! Bouroukrou *Aubrév.* 702 ! **G.C.:** Essiama *Cummins* ! Kumasi (Jan.) *Dalz.* 66 ! Anum (Nov., Jan.) *Johnson* 553 ! 818 ! Sekodumase (Jan.) *Kitson* 11 ! **Togo:** Ketschenkobache (Nov.) *Büttner* 347 ! Ketschenki (Nov.) *Büttner* 345 ! Misahöhe (Nov.) *Mildbr.* 7341 ! **N.Nig.:** Kurmin Ninte, Jemaa (Nov.) *Keay* FHI 22258 ! **S.Nig.:** Lagos *Millen* 39 ! Abeokuta *Irving* ! Olokemeji *Foster* 146 ! Ibadan (Nov.) *Jones & Keay* FHI 14161 ! Ikom (Jan.) *Holland* ! (See Appendix, p. 107.)
2. **D. quinqueseta** (*Del.*) *Exell* var. *quinqueseta*— in J. Bot. 73 : 263 (1935). *Xeropetalum quinquesetum* Del. (1826). *D. reticulata* Mast. (1868)—F.W.T.A., ed. 1, 1 : 248. *D. multiflora* (Endl.) Planch. var. *vestita* K. Schum. in Engl. Monogr. Afr. Sterculiac. 34 (1900) ; Aubrév. l.c. 163, partly.
   Dah.: Kandi (Jan.) *Aubrév* 4d ! **N.Nig.:** *Talbot* ! Widespread in the savannah regions of tropical Africa.
2a. **D. quinqueseta var. senegalensis** (*Planch.*) *Keay* in Kew Bull. 1954 : 263. *D. senegalensis* Planch. (1851)— F.W.T.A., ed. 1, 1 : 248. *D. multiflora* of F.T.A. 1 : 227, partly (*Heudelot* 160). *D. multiflora* var. *senegalensis* (Planch.) Aubrév. l.c. 163, t. 29, 1–2. *D. atacorensis* A. Chev. Bot. 83, name only. Shrub or tree, to 15 ft. high ; flowers white, appearing before the leaves ; in savannah.
   Sen.: *Heudelot* 160 ! **Gam.:** Georgetown (Jan.) *Dalz.* 8051 ! Jarume Koto, Niani (Jan.) *Frith* 161 ! **Fr.Sud.:** Kati (Jan.) *Chev.* 176 ! Birgo (Feb.) *Dubois* 58 ! **Port.G.:** Buruntuma to Pitche, Gabú (Jan.) *Esp. Santo* 2372 ! **Fr.G.:** *Maclaud* 20 ! Dinguiraye (Dec.) *Jac.-Fél.* 1443 ! **Togo:** Kete Kratchi (Jan.) *Vigne* FH 1579 ! **Dah.:** Konkobiri (fr. July) *Chev.* 24314 (partly) ! **N.Nig.:** Bauchi Plateau (Nov.– Jan.) *Lely* P. 2 ! *W. D. MacGregor* 385 ! *Costello* P. 17/28 ! **S.Nig.:** N. of Shepeteri (Feb.) *Keay* FHI 22493 ! 22524 !
3. **D. cf. ledermannii** *Engl.* Bot. Jahrb. 45 : 319 (1910). Tree, to 25 ft. high in tall grass savannah; flowers white.
   N.Nig.: above Masenge, W. escarpment of Bauchi Plateau, 3,000 ft. (Feb.) *Hepburn* 134 ! **Br.Cam.:** Bamenda Prov. : Bamenda *Johnstone* 276/31 ! Bambui road (Jan.) *Keay* FHI 28333 ! Babanki, 4,000 ft. (Jan.) *Johnstone* 8/31 ! Kumbo (Jan.) *Johnstone* 31/31 ! [I have been unable to trace authentic material of this species ; the type, *Ledermann* 1557, was collected between Fossong in Mamfe Division of the British Cameroons and Dchang in the French Cameroons. *D. discolor* Engl. l.c., from Babudju (probably in the French Cameroons), is also known now only from the description ; it appears to be close to our plant.]

Imperfectly known species.

D. squarrosa *Engl.* Bot. Jahrb. 45 : 318 (1910). No authentic material available ; possibly a form of *D. buettneri* K. Schum.
Br.Cam.: Bamenda Prov. : Babungo (Dec.) *Ledermann* 1967. Ngom (Jan.) *Ledermann* 2089.

8. MELHANIA Forsk.—F.T.A. 1 : 230.

A low shrublet ; branches and leaves softly stellate-tomentose ; leaves ovate-elliptic, subcordate at base, rounded or truncate at apex, about 3 cm. long and 1·5 cm. broad, crenate-serrate ; stipules subulate-filiform, 3 mm. long, becoming glabrous ; flowers axillary, solitary ; bracteoles 3, much enlarged, membranous, and wing-like in fruit, about 3 cm. diam. ; stamens 5, with alternating staminodes ; fruit a small loculicidal capsule, 1–2-seeded ..   ..   1. *denhamii*
A low shrublet as above, but bracteoles linear-subulate and cells of fruit 3-seeded
2. *ovata*

1. **M. denhamii** *R.Br.*—F.T.A. 1 : 230 ; K. Schum. in Engl. Monogr. Afr. Sterculiac. 15, t. 1 A ; Chev. Bot., 84. A small shrublet, woody at the base, the flowers enclosed in large bracteoles.
    Sen.: Walo and Cayor (Dec.–Mar.) *Perrottet* 90 ! **Fr.Sud.:** Timbuktu (fl. & fr. June–July) *Hagerup* 119 ! 213 ! Extends from Mauritania to India.
2. **M. ovata** (*Cav.*) *Spreng.* Syst. Veg. 3 : 32 (1826) ; K. Schum. l.c. 6, t. 1 D. *Brotera ovata* Cav. (1799). *B. leprieurii* Guill. & Perr. Fl. Seneg. 85 (1831). *M. abyssinica* A. Rich. (1847)—F.T.A. 1 : 231. *M. leprieuri* Webb in Hook. Nig. Fl. 110, t. 4–5 (1849). Our plant is var. *oblongata* (Hochst.) K. Schum. l.c. 7.
    Sen.: St. Louis, Gandiole and Four-à-Chaux (Sept.–Oct.) *Perrottet.* Sandy deserts in N. and S. tropical Africa generally ; also in India.

9. HERMANNIA Linn.—F.T.A. 1 : 232.

Herb, woody only at the base ; stems ascending ; leaves lanceolate or lanceolate-elliptic, subacute at each end, 2–6 cm. long, 1–2·5 cm. broad, loosely stellate-pubescent on both surfaces ; petiole up to 1 cm. long ; stipules linear-lanceolate, acute, 4–6 mm. long, often leafy ; flowers axillary ; pedicels very slender, 3–4 cm. long ; calyx-lobes triangular-lanceolate, 1·5 mm. long, strongly nervose ; fruits broadly turbinate, truncate and with several horns at the apex, 5–6 mm. long, loosely stellate-pubescent   ..   ..   *tigreensis*

H. tigreensis *Hochst. ex A. Rich.*—F.T.A. 1 : 233 ; K. Schum. in Engl. Monogr. Afr. Sterculiac. 85. A herb of farms, 1 ft. high, not common ; flowers small, deep red.
    Sen.: Cayor *Berhaut* 1314. **N.Nig.:** Sokoto (Aug.) *Dalz.* 414 ! Bichikki, 13 miles W. of Bauchi (Sept.) *Lely* 626 ! Also in N. Cameroons, extending to Abyssinia, Eritrea and Angola.

10. MELOCHIA Linn.—F.T.A. 1 : 235.

Flowers in terminal clusters ; stems with lines of stellate hairs on the young parts ; leaves ovate, rounded or very slightly cuneate at base, acute, 3–6 cm. long, up to 4 cm. broad, 5-nerved at the base, nearly glabrous ; petiole ⅓–½ as long as the blade ; bracts linear, ciliate   ..   ..   1. *corchorifolia*
Flowers axillary ; petioles very short :
Stems and young parts densely covered with long silky hairs ; leaves ovate-lanceolate, acute, rounded to truncate at the base, 4–8 cm. long, 2–3·5 cm. broad, pilose ; petiole about ¼ as long as the blades ; bracts very long-linear, densely plumose-pilose   ..   ..   2a. *melissifolia* var. *mollis*
Stems at most pubescent with rather short hairs :
   Bracts about 1 cm. long ; leaves ovate-lanceolate, pointed, 4–7 cm. long, 2–3 cm. broad ..   ..   2b. *melissifolia* var. *bracteosa*
   Bracts about 0·5 cm. long ; leaves rather broadly ovate, scarcely pointed, 2–4 cm. long, about 1·5 cm. broad ..   ..   ..   2c. *melissifolia* var. *microphylla*

1. **M. corchorifolia** *Linn.*—F.T.A. 1 : 236 ; K. Schum. in Engl. Monogr. Afr. Sterculiac. 41, t. 39 ; Chev. Bot. 84. Herb or undershrub, 1½–2 ft. or more, with hollow stems, erect or sometimes prostrate ; flowers small, white or sometimes yellowish or pinkish ; more common in wet places.
    Sen.: *Heudelot* 494 ! *Roger* 119 ! M'bidjem (Oct.) *Thierry* 84 ! **Gam.:** Wallikunda (Sept.) *Macluskie* 14 ! **Fr.Sud.:** Tacadji (Aug.) *Chev.* 2747 ! Sansanding (Sept.) *Chev.* 24965 ! **Fr.G.:** Farana (Mar.) *Sc. Elliot* 5320 ! **S.L.:** Mahela (fr. Dec.) *Sc. Elliot* 3911 ! Pendembu (July) *Thomas* 831 ! Konta (Sept.) *Jordan* 103 ! **Lib.:** Cape Palmas (July) *T. Vogel* 30 ! 57 ! **G.C.:** Navrongo (fr. Dec.) *Adams & Akpabla* GC 4357 ! **Togo:** Yendi (Dec.) *Adams & Akpabla* GC 4095 ! **Fr.Nig.:** Niamey (Oct.) *Hagerup* 528 ! **N.Nig.:** Sokoto (Oct.) *Moiser* 118 ! Kontagora (Nov.) *Dalz.* 125 ! Nupe *Barter* 1529 ! Damaturu Kabau (fl. & fr. Oct.) *Daggash* FHI 22018 ! **S.Nig.:** Gambari, Ibadan Prov. (Sept.) *Onochie* FHI 7900 ! Widespread in the tropics of the Old World. (See Appendix, p. 108.)
2. **M. melissifolia** *Benth.* in Hook. Journ. Bot. 4 : 124 (1841) ; K. Schum. l.c. 42. Several varieties are widespread in tropical Africa ; var. *melissifolia* is in tropical America. For this edition I have followed K. Schumann, but I am not altogether satisfied that these varieties are really distinct. There seems no justification for raising them to specific rank as in the first edition. A thorough study of this variable species throughout its range is needed.
2a. **M. melissifolia** var. **mollis** *K. Schum.* l.c. 43 (1900). *M. mollis* (K. Schum.) Hutch. & Dalz. F.W.T.A., ed. 1, 1 : 250 (1928) ; Kew Bull. 1928 : 296. Herb with axillary clusters of small white or yellowish flowers.
    Lib.: Monrovia *Whyte* ! Vonjama (Oct.) *Baldwin* 9865 ! **S.Nig.:** Eket *Talbot* ! **Br.Cam.:** Rio del Rey *Johnston* ! Extends to Shari, Upper Nile.
2b. **M. melissifolia** var. **bracteosa** (*F. Hoffm.*) *K. Schum.* l.c. 43 (1900). *M. bracteosa* F. Hoffm. (1889)— F.W.T.A., ed. 1, 1 : 250. *M. melissifolia* var. *brachyphylla* K. Schum. l.c., partly (*Sc. Elliot* 5095). Habit of the preceding, up to 5 ft. high.
    Sen.: *Berhaut* 1335 ; 2324 ; 4354. **S.L.:** Kambia (Dec.) *Sc. Elliot* 4351 ! *Deighton* 863 ! Pujehun (Dec.) *Deighton* 268 ! Yonibana (Nov.) *Thomas* 4915 ! 5061 ! Musaia (Dec.) *Deighton* 4550 ! **Lib.:** Gbanga (Oct.) *Linder* 1174 ! Tawata (Nov.) *Baldwin* 10310 ! **N.Nig.:** Bauchi Plateau (Sept.) *Lely* P. 716 ! **S.Nig.:** Lagos *Millen* 108 ! Ibadan (Mar.) *Schlechter* 12337 ! Ibuzo (Nov.) *Thomas* 2025 ! Onitsha *Barter* 1798 ! Okumi (Jan.) *Holland* 176 ! **Br.Cam.:** Rio del Rey *Johnston* ! Widespread in tropical Africa. (See Appendix, p. 108.)

2c. **M. melissifolia** var. **microphylla** K. Schum. l.c. 43 (1900). M. melissifolia of F.W.T.A., ed. 1, 1 : 250. A branched, prostrate, woody weed, with white or mauve flowers in axillary clusters.
Sen.: Heudelot 564 ! **S.L.:** Sc. Elliot 4163 ! Picket Hill (Nov.) T. S. Jones 206 ! Port Loko (Dec.) Thomas 6554 ! Rokupr (Oct.) Jordan 131 ! **Lib.:** Monrovia (Nov.) Linder 1551 ! Barker 1469 ! Zigida (Oct.) Baldwin 9987 ! Medina, Bumbuma (Oct.) Linder 1319 ! **N.Nig.:** Vom Dent Young 19 ! Plains between Hepham and Ropp (July) Lely 370 ! **S.Nig.:** Ebute Metta (Mar.) Dalz. 1191 ! Idanre Hills (Oct.) Keay FHI 22600 ! Ikom (Jan.) Rosevear 7/31 !

<h3 align="center">11. WALTHERIA Linn.—F.T.A. 1 : 235.</h3>

Bracts and calyx puberulous and with larger stellate hairs having conspicuous reddish stalks ; calyx 6–6·5 mm. long, the lobes triangular, about 2 mm. broad at base ; cymes often stalked ; stem and undersurface of leaves shortly stellate-pubescent or stellate puberulous, becoming less so with age ; leaves lanceolate to ovate-lanceolate, obtuse at base, acute at apex, up to 8 cm. long and 3·5 cm. broad, crenate   1. lanceolata
Bracts and calyx villous, with long simple hairs ; calyx 4–4·5 mm. long, the lobes long-attenuate, about 1 mm. broad at base ; cymes congested, usually many-flowered, usually sessile ; stem and undersurface of leaves densely stellate-tomentose ; leaves ovate to lanceolate, rounded or subcordate at base, obtuse or subacute at apex, up to 14 cm. long, and 6 cm. broad, rather closely serrulate-crenate .. .. 2. indica

1. **W. lanceolata** R.Br. ex Mast. in F.T.A. 1 : 235 (1868) ; K. Schum. in Engl. Monogr. Afr. Sterculiac. 44, t. 3 H ; Chev. Bot. 85. W. incana A. Chev. Bot. 85, name only. Straggling undershrub, to 10 ft. high ; flowers yellow.
Sen.: Heudelot 68 ! Sedhiou, Casamance (Feb.) Chev. ! **Gam.:** Ingram ! **Fr.G.:** Heudelot 692 ! Pobéguin 377 bis ! Farana (Jan.) Chev. 20403 ! Kouria (Oct.) Caille in Hb. Chev. 14986. **S.L.:** Heddles Farm (Dec.) Sc. Elliot 3906 ! Freetown (Oct.) Deighton 249 ! Njala (Feb.) Deighton 1067 ! Port Loko (Dec.) Thomas 6709 ! **Lib.:** Bonuta (Oct.) Linder 882 ! Zigida, Vonjama (Oct.) Baldwin 9991 ! Loffa R., Boporo (Dec.) Baldwin 10679 !
2. **W. indica** Linn. Sp. Pl. 2 : 673 (1753) ; Exell & Mendonça Consp. Fl. Ang. 1 : 192 ; Int. Code Bot. Nomenclat. 38 (1952). W. americana Linn. (1753)—F.T.A. 1 : 235 ; Chev. Bot. 85 ; K. Schum. l.c. 45 ; F.W.T.A., ed. 1, 1 : 250. Erect herb or shrub to 7 ft. high ; petals yellow, turning orange or brown with age ; common in open places.
Sen.: Kaolak Chev. **Gam.:** Brooks 25 ! Genieri (Feb.) Fox 25 ! **Fr.Sud.:** Oualia (Dec.) Chev. 33 ! Labbézenga (Nov.) Hagerup 450 ! Timbuktu (July) Hagerup 226 ! **Fr.G.:** Pobéguin 377 ! **S.L.:** Freetown Deighton 1056 ! Rokupr (Apr.) Jordan 7 ! Magbile (Dec.) Thomas 5893 ! Karina (Feb.) Glanville 169 ! **Lib.:** Cinco, Monrovia (Nov.) Linder 1547 ! **Iv.C.:** fide Chev. l.c. **G.C.:** Abra, Cape Coast (Oct.) Howes 993 ! Achimota Irvine 460 ! Krobo Plains (Jan.) Johnson 538 ! Damongo, N.T. (July) Andoh FH 5195 ! **Togo:** Lome Warnecke 30 ! **Da°.:** Burton ! Poisson ! **N.Nig.:** Lokoja Ansell ! Katsina (fl. & fr. Apr.) Meikle 1359 ! Naraguta (June) Lely 339 ! **S.Nig.:** Lagos Millen 178 ! Dawodu 8 ! Abeokuta Irving ! Old Oyo F.R. (Sept.) Latilo FHI 23541 ! Uguoka, Awka (fl. & fr. Mar.) Jones FHI 7353 ! Widespread in the tropics. (See Appendix, p. 112.)

12. OCTOLOBUS Welw. ex Benth. & Hook. f. Gen. Pl. 1 : 982 (1867) ; Hutch. in Kew Bull. 1937 : 394.

Shrub or small tree ; branchlets shortly tomentose when very young, soon glabrous ; stipules subulate, caducous ; leaves unifoliolate, with petiole up to 5 cm. long, swollen at apex ; lamina oblanceolate to elliptic-obovate, narrowed to the cuneate or rounded base, gradually long-acuminate, 6–24 cm. long, 2·5–7·5 cm. broad, glabrous, lateral nerves arcuate and prominent beneath, 7–9 pairs, venation closely reticulate ; flowers unisexual, solitary and surrounded at base by an involucre of bracts, in axils of lower leaves, or on stem just below the leaves ; calyx up to 2·8 cm. long, more or less densely tomentose outside, with 8 lanceolate lobes, 1·2–1·5 cm. long, each with a crisped marginal fringe, tube about 1–1·3 cm. long, narrow in male flower but broader in female ; anthers about 4 mm. long, in 1 whorl with stipe about 5 mm. long ; carpels numerous ; fruiting carpels oblique ellipsoid, 2–2·5 cm. long, about 1·5 cm. broad, with stipe 1–1·5 cm. long, densely crowded, each with 5–6 seeds   angustatus

O. **angustatus** Hutch. in Kew. Bull. 1937 : 395. O. spectabilis of Aubrév. Fl. For. C. Iv. 2 : 252, t. 225 A (1936), not of Welw. Cola rolandi-principis A. Chev. Bot. 83, name only. Cola angustifolia of F.W.T.A., ed. 1, 1 : 255, partly (as to syn.). C. flavo-velutina of F.W.T.A., ed. 1, 1 : 255, partly (Unwin & Smythe 53). Small tree, to 25 ft. (rarely to 50 ft.) high, or shrub, in forest ; flowers yellow or cream ; fruiting carpels red, with shining black seeds.
S.L.: Gola Forest (fr. Jan.) Unwin & Smythe 53 ! **Iv.C.:** Adzopé to Boudépé (fr. Dec.) Chev. 22672 ! **G.C.:** Kwahu Prasu (Feb.) Vigne FH 1602 ! S. Fomang Su F.R. (Jan.) Vigne FH 2674 ! Akotui (Oct.) Vigne FH 4028 ! Aketewia (fl. & fr. Nov.) Vigne FH 4268 ! Akrum Ravine, Begoro (Oct.) Box 3497 ! **S.Nig.:** Etemi, Omo F.R. (Mar.) Jones & Onochie FHI 16613 ! Shasha F.R. (Feb.) Richards 3115 ! Owo F.R. (May) Jones FHI 3561a ! Akure F.R. (Nov.) Keay FHI 25538 ! Okomu F.R. (Mar.) Akpabla 1118 ! Ekem, Nsukka (June) Cousens FHI 3561 ! Aboabam (Dec.) Keay ! [O. spectabilis Welw. from Gabon and Angola is closely related.]

Imperfectly known species.

O. **sp.** Forest tree, to 35 ft. high ; flowers not known.
Br.Cam.: Mombo, S. Bakosi Forest, Kumba Dist. (fr. May) Olorunfemi FHI 30565 !

13. PTERYGOTA Schott & Endl. Meletem. Bot. 32 (1832).

Undersurface of young leaves and shoots and inflorescence axes with very short appressed stellate hairs, often whitish or olivaceous in colour ; leaves usually 7-nerved at base, more or less pentagonal-ovate, sometimes slightly lobed, widely to deeply cordate at base, 10–38 cm. long and broad ; flowers 17–20 mm. long, distinctly pedicellate, sepals tomentose outside but subglabrous within ; fruiting carpels 12–18 cm. long, 10–13 cm. broad, ellipsoid or subglobose, rounded on the side opposite the suture,

with a stout stipe 1·5–4·5 cm. long and 1·5–2 cm. diam. ; seeds (including wing) 8–12
cm. long, 3·5–5·5 cm. broad .. .. 1. *macrocarpa*
Undersurface of young leaves and shoots and inflorescence axes densely brown stellate-
tomentose ; leaves usually 5-nerved at base, oblong-ovate to suborbicular, truncate
to widely cordate at base, 5–17 cm. long and broad ; flowers 10–12 mm. long, sessile
or subsessile, sepals tomentose outside and inside ; fruiting carpels tapered on the
side opposite the centre of the suture, 7·5–10 cm. broad from the centre of the suture
to the tip of the tapered back, 5–7 cm. long, attenuate at base into the 3–5 cm. long
stipe which is about 4–5 mm. diam. in the middle ; seeds (including wing) 7–7·5 cm.
long, 2·5–3 cm. broad .. .. 2. *bequaertii*

1. **P. macrocarpa** *K. Schum.* in Engl. Monogr. Afr. Sterculiac. 135, fig. 1, A–B (1900) ; Aubrév. Fl. For. C. Iv. 2 :
254, t. 226 ; Milne-Redhead in Kew Bull. 1950 : 348, partly (excl. ref. frts., *Maitland* 445 and carpel walls
associated with *Maitland* 90). *P. cordifolia* A. Chev. Vég. Util. 5 : 252 (1909). Forest tree, to 120 ft. high,
with buttresses.
   S.L.: *Smythe* 258 ! (sterile spec., ident. provis.). **Iv.C.:** *Aubrév.* 957 ! Morénou (Nov.) *Chev.* 22462 ! Man
Aubrév. 405 ! Zaranou (fr. Mar.) *Chev.* 16271. **G.C.:** Nkawe *Chipp* 143 ! Dedeswa (fr. Mar.) *Chipp* 144 !
Ntenseri (fr. Mar.) *Mc.Ainsh* 872 ! Kotokron (June) *Vigne* FH 1198 ! Owabi (fr. Nov.) *Vigne* FH 2589 !
S.Nig.: Olokemeji F.R. (fr. Feb.) *Corbin* 522 ! *Ejiofor* FHI 33451 ! Ibadan North F.R. *Ubogu* FHI 7152 !
Shasha F.R. (fr. Feb.) *Jones* FHI 16974 ! Akure F.R. (fl. Nov., fr. Oct.) *Latilo* FHI 15284 ! *Keay* FHI
19660 ! Kataban, Ikom (Nov.) *Catterall* ! Oban *Talbot* ! **Br.Cam.:** Victoria (fl. & fr. Nov.) *Maitland* 90 !
769 ! Likomba (fl. Nov., fr. Dec.) *Mildbr.* 10701 ! 10778 ! (See Appendix, p. 108.)
   [Small fragments of *Zenker* 2652 (the type-number of *P. kamerunensis* K. Schum. (1907)), from Bipinde
in the French Cameroons in the herbaria at Kew and the British Museum are probably equal to *P.
macrocarpa* K. Schum.]

2. **P. bequaertii** *De Wild.* in Comp. Rend. Soc. Biol. (Réunion, Dec. (1919)) 82 : 1397 (1919). *P. aubrevillei*
Pellegr. in Bull. Soc. Bot. Fr. 78 : 441 (1931) ; Aubrév. Fl. For. C. Iv. 2 : 256, t. 227. *Sterculia thompsonii*
Hutch. & Dalz. F.W.T.A., ed. 1, 1 : 251 (1928), partly (lvs. only). Forest tree, to 100 ft. high ; less
common than *P. macrocarpa*, and usually in rather moister regions ; flowers cream when young, with
ginger-brown hairs.
   Iv.C.: (May) *Aubrév.* 1926 ! Mé (Oct.) *Aubrév.* 166. Yapo (fr. Dec.) *Aubrév.* 607 ! **S.Nig.:** W. Prov.
Thompson 63 ! Abeku, Omo F.R. (fl. & fr. Feb.) *Jones & Onochie* FHI 16738 ! Sapoba *Kennedy* 1633 !
1857 ! 1905 ! Usonigbe F.R. (fr. Aug.) *Adekunle* FHI 22637 ! 22644 ! Mamu River F.R. (fl. & fr. Nov.)
Keay FHI 21529 ! Degema (fl. & fr. Aug.) *Rosevear* PH. 1/39 ! *Talbot* 3641 ! Oban *Talbot* 217 ! Also in
French Cameroons, Gabon and Belgian Congo.

Besides the above indigenous species, *P. alata* (Roxb.) R. Br. from tropical Asia has been collected (*Maitland*
445) at Victoria, British Cameroons ; the references to the fruit in Milne-Redhead's paper (Kew Bull. 1950 :
348) are based on this specimen and not on *P. macrocarpa* K. Schum.

14. STERCULIA Linn.—F.T.A. 1 : 215.

Leaves digitately nerved and lobed, cordate at base, lobes triangular-acuminate, up to
20 cm. long and broad, softly stellate-tomentellous beneath, becoming less so with
age ; flowers in short crowded cymes on the previous year's shoots ; calyx widely
campanulate, lobed to about the middle, about 1 cm. long, tomentose outside ;
fruiting carpels boat-shaped, 6 or more cm. long, very hairy inside ; seeds ellipsoid,
slate-coloured, 10–14 mm. long, with a yellow fleshy aril at base.. .. 1. *setigera*
Leaves pinnately nerved and entire :
   Calyx divided to the base, lobes about 5 mm. long, woolly inside; fruiting carpels with
   very woody walls about 1 cm. thick, ellipsoid, with a short stout stipe, acuminate at
   apex, 10–16 cm. long, with about 16 seeds each completely enclosed in yellow aril ;
   panicles with 1–5-flowered lateral branches up to about 2 cm. long ; leaves elliptic
   or oblong, rounded to obtuse at base, shortly acuminate at apex, 7–14 cm. long,
   4–8 cm. broad, with 10–14 pairs of parallel lateral nerves, thinly stellate-pubescent
   to subglabrous beneath, petioles 2–7 cm. long .. .. .. 2. *oblonga*
   Calyx campanulate, lobed in the upper half ; fruiting carpels with walls up to 5 mm.
   thick, boat-shaped, distinctly stipitate, 5–7 cm. long ; seeds not covered with yellow
   aril ; lateral branches of inflorescence several-flowered, up to 5 cm. long :
      Calyx-lobes not adhering at apex, with apical incurved hairy appendages ; inflore-
      scence axes sparingly stellate puberulous ; leaves glabrous or nearly so beneath,
      more or less oblong, rounded at base, scarcely acuminate, 10–30 cm. long, 4–10 cm.
      broad, petioles 3–11 cm. long ; fruiting carpels glabrous or puberulous inside, with
      about 8 seeds each completely enclosed in bright red aril .. 3. *rhinopetala*
      Calyx-lobes adhering at apex, without appendages ; inflorescence axes densely
      stellate-tomentose ; leaves stellate-tomentellous beneath, ovate-elliptic or slightly
      obovate, rounded to slightly cordate at base, obtusely and shortly acuminate,
      10–18 cm. long, 5–11 cm. broad, petioles 2–7·5 cm. long ; fruiting carpels bristly-
      tomentose inside, with slate-coloured seeds, not arillate .. 4. *tragacantha*

1. **S. setigera** *Del.* Cent. Pl. Afr. 61 (1826) ; Aubrév. Fl. For. Soud.-Guin. 159, t. 28. *S. tomentosa* Guill. & Perr.
Fl. Seneg. 1 : 81, t. 16 (1831) ; F.T.A. 1 : 217 ; K. Schum. in Engl. Monogr. Afr. Sterculiac. 106 ;
F.W.T.A., ed. 1, 1 : 251 ; Kennedy For. Fl. S. Nig. 63 ; not of Thunb. (1794). *S. cinerea* A. Rich.—F.T.A.
1 : 218. A deciduous, savannah tree, to 40 ft. high ; bark rough, yielding a white gum ; young branches
softly tomentose ; flowers dull red or yellowish-green with red streaks, appearing before the new foliage ;
fruiting carpels usually 4–5, grey to brownish outside, beaked, velvety inside with pungent bristles along
the placenta line.
   Sen.: *Perrottet* 88 ! Bakel *Leprieur*. Bondou *Heudelot* 156. **Gam.:** *Dawe* 47 ! **Fr.Sud.:** Bougouni *Chev.*
678. And various localities *fide* Aubrév. *l.c.* **Fr.G.:** Kadé *Pobéguin* 1983. **Iv.C.:** Korhogo *Aubrév.* 1865.
Ferkéssédougou *Aubrév.* 1619 ! **G.C.:** Ejura (fr. Aug.) *Chipp* 761 ! Wa (Mar.) *McLeod* 841 ! Salaga (fr.
June) *Saunders* 804 ! **Togo:** Sokode (Apr.) *Kersting* ! **Dah.:** Ouessé *Poisson* 79 ! **N.Nig.:** Nupe *Barter*

1115! 1307! Zungeru (Jan.) *Elliott* 28! Ilorin (Feb., Mar.) *Lamb* 41! Yola (fr. Oct.) *Shaw* 93! Musgu (Mar.) *E. Vogel* 98! **S.Nig.**: Okure Okwe *Unwin* 152! Widespread in tropical Africa. (See Appendix, p. 109.)

2. **S. oblonga** *Mast.* in F.T.A. 1 : 216 (1868) ; K. Schum. l.c. 101, fig. 1 C ; Chev. Bot. 76 ; Kennedy l.c. *S. elegantiflora* Hutch. & Dalz. F.W.T.A., ed. 1, 1 : 251 (1928) ; Kew Bull. 1928 : 297 ; Aubrév. Fl. For. C. Iv. 2 : 236, t. 218, 1–5. *S. thompsonii* Hutch. & Dalz. ll.cc., partly (flowers only ; the leaves are *Pterygota bequaertii* De Wild.). Forest tree, to 120 ft. high, with buttresses and clean bole ; bark-slash cream ; flowers cream or greenish-yellow ; fruiting carpels greyish-brown.
Iv.C.: Soubiré, Sanvi (Mar.) *Chev.* 17693! Yapo *Chev.* 22316! Abidjan *Aubrév.* 182! Bouroukrou *Chev.* 16137. Mid. Comoé *Chev.* 22633. Attié *Chev.* 22668. **G.C.**: Amentia (Sept.) *Vigne* FH 1359! N. Agogo, Ejan (Jan.) *Vigne* FH 1789 (partly)! Jimira F.R. (Oct.) *Vigne* FH 3108! **S.Nig.**: (fr. Dec.) *Foster* 369! W. Prov. *Thompson* 6 (partly)! Sapoba (Nov.) *Kennedy* 410! 1632! Usonigbe F.R. (fr. Aug.) *Adekunle* FHI 22639! **Br.Cam.**: Likomba *Mildbr.* 10616! **F. Po**: *Mann* 1162! Also in French Cameroons and Gabon. (See Appendix, p. 108.)

3. **S. rhinopetala** *K. Schum.* l.c. 102 (1900) ; Aubrév. l.c. 236, t. 217 ; Kennedy l.c. 64. *S. oblongifolia* A. Chev. Bot. 76, name only. Forest tree, to 120 ft. high, with narrow buttresses and straight bole ; bark-slash pinkish-brown ; flowers cream or greenish-yellow ; fruiting carpels brown.
Iv.C.: Morénou (fr. Nov.) *Chev.* 22455! And various localities *fide* Aubrév. *l.c.* **G.C.**: Dunkwa (Aug.– Sept.) *Chipp* 717! *Vigne* FH 157! Sefwi-Bibiani (fl. & fr. Aug.) *Vigne* FH 1326! Abuasi (Sept.) *Vigne* FH 925! **Togo**: Worawora (Oct.) *St. C. Thompson* FH 3595! **S.Nig.**: Ibadan F.R. (Nov.) *Punch* 106! Akure F.R. (Oct.) *Keay* FHI 16250! Okomu F.R. (fr. Feb.) *A. F. Ross* 189! Ikom (Sept.) *Catterall* 34! Also in French Cameroons. (See Appendix, p. 109.)

4. **S. tragacantha** *Lindl.*—F.T.A. 1 : 216 ; K. Schum. l.c. 102 ; Chev. Bot. 77 ; Aubrév. l.c. 234, t. 218, 6–9 ; Kennedy l.c. 63. A tree of the open parts or edges of forests ; up to 80 ft. high ; bole 30–40 ft., sometimes buttressed ; grey corky bark yielding a coloured gum ; conspicuous with red-purple flowers about Jan.– Feb. ; fruit bright red turning brown.
Fr.Sud.: Banfara (Mar.) *Chev.* 536! **Port.G.**: Bissalanca, Bissau (Feb.) *Esp. Santo* 1728! **Fr.G.**: Fouta Djalon *Heudelot* 657! Conakry *Maclaud*! Mamou to Dalaba (fr. Mar.) *Dalz.* 8387! **S.L.**: *Don*! Buya-buya (Feb.) *Sc. Elliot* 4465! Panguma (Mar.) *Lane-Poole* 74! Kosso Town *Smythe* 249! Rosino (Feb.) *Deighton* 4975! **Lib.**: Peahtah (fr. Oct.) *Linder* 1106! Genna Tanyehun (Dec.) *Baldwin* 10760! Kle, Boporo (fl. & fr. Dec.) *Baldwin* 10625! **Iv.C.**: Azaguié (young fr. Sept.) *Chev.* 22305! **G.C.**: Cape Coast *Brass*! Fiapere (Feb.) *Chipp* 84! Peki Plains (Jan.) *Johnson* 544! Asofa, Achimota (fl. & fr. Jan.) *Irvine* 1967! Odumase (fr. Mar.) *Irvine* 896! Tafo (fl. & fr. Nov.) *N. F. Robertson* 4! 5! *Lovi* WH 3848! **Togo**: Lome *Warnecke* 456! **Dah.**: *Le Testu* 169! **N.Nig.**: Nupe *Barter* 1228! Agaie (Feb.) *Yates* 32! Selfwe R., Plateau Prov. *Thornewell* 59! **S.Nig.**: Lagos *Rowland*! Benin City *Unwin* 36! Oban (May) *Talbot*! *Ujor* FHI 30820! **Br.Cam.**: Ambas Bay (Feb.) *Mann* 759! Victoria (Jan., Feb.) *Maitland* 366! 372! Johann-Albrechtshöhe *Staudt* 887! Widespread in tropical Africa. (See Appendix, p. 109.)

Besides the above, *S. foetida* Linn., a native of tropical Asia, and *S. apetala* (Jacq.) Karst. (syn. *S. cartha-ginensis* Cav.) a native of tropical America have been introduced into cultivation in our area.

15. COLA Schott & Endl.—F.T.A. 1 : 220. *Chlamydocola* (K. Schum.) Bodard in J. Agr. Trop. Bot. Appliq. 1 : 313 (1954).

With J. P. M. Brenan

*Leaves compound, digitate; anthers in 1 whorl (N.B.—rudimentary anthers can always be observed below the carpels in the female flowers) :
Leaflets with undersurface completely hidden (or almost so) by scales clearly visible under × 10 lens, and giving the surface a silvery appearance ; leaflets 3, up to 50 cm. long and 20 cm. broad, shortly petiolulate ; flowers in densely clustered cymes on the older wood ; pedicels jointed, distinct, scaly ; calyx-tube campanu-late, 7–10 mm. long, densely scaly outside, lobes with crisped marginal frill
1. *lepidota*
Leaflets not scaly, or only minutely, sparsely and inconspicuously so ; leaflets 5–7 (–9) :
Calyx 6–9 cm. long, elongate-urceolate, ribbed, brown-woolly in bud, later only minutely stellate-puberulous ; plant cauliflorous ; leaflets usually deeply lobed, subsessile, central leaflets up to 45 cm. long, basal leaflets much smaller   2. *gigas*
Calyx up to 4 cm. long :
Flowers in axillary clusters among the leaves, sometimes also a few clusters just below the leaves :
Leaflets, at least the central ones, usually deeply lobed, never abruptly caudate-acuminate ; stipules caducous ; perianth without a crisped marginal fringe, lobes more or less erect ; stamens grouped in bundles within the whorl ; fruiting carpels long-stipitate, smooth :
Leaves papery to subcoriaceous, gradually acuminate at apices of leaflets and often also of upper lobes, up to 55 cm. long, almost glabrous beneath ; petioles 2·5–7 mm. diam. ; male calyx about 15 mm. long, lobed to below the middle or to base ; fruiting carpels opening out almost flat with the seeds borne on the margins ..   .. 3. *digitata*
Leaves thickly coriaceous and rigid, obtuse or hardly acuminate at apices of leaflets and lobes, up to 85 cm. long, densely stellate-puberulous beneath ; petioles 8–15 mm. diam. ; male calyx 16–22 mm. long, lobed to the middle ; fruiting carpels apparently not opening out flat, boat-shaped 4. *megalophylla*
Leaflets oblanceolate to obovate, not lobed, very abruptly long-caudate-acuminate at apex, up to 40 cm. long and 11 cm. broad ; branchlets, pedicels and outside of flowers brown-tomentellous, petiole and midrib beneath puberulous ; stipules ovate-lanceolate, erect and rigid, 1·5–4·5 cm. long ; calyx-tube about 8–12 mm. long, lobes reflexed, about 14–16 mm. long, with a crisped almost laciniate

marginal fringe towards the apex ; stamens in a continuous whorl, not grouped
or twisted ; fruiting carpels apparently sessile, with irregular conical warts,
ellipsoid to subglobose, 4–5 cm. diam., 3–8 cm. long   ..   .. 5. *rostrata*
Flowers borne on the older wood below the leaves :
Pedicels 1·8–5 cm. long ; calyx subglabrous or minutely puberulous outside,
campanulate to shortly funnel-shaped, 1·3–3·5 cm. long ; leaflets broadly
oblanceolate to obovate, always petiolulate, not lobed, but sometimes sinuate
especially towards the acuminate apex, up to 35 cm. long and 14·5 cm. broad,
subglabrous beneath ; fruiting carpels shortly stipitate, ellipsoid or oblong,
stout, glabrous, up to 20 cm. long and 10 cm. thick ..   ..   6. *pachycarpa*
Pedicels up to 1 cm. long ; calyx densely pubescent or puberulous outside :
Leaflets all sessile, basal ones very asymmetrical towards base (proximal side more
or less rounded, distal side cuneate) more or less glaucous-grey beneath ;
petiole, midrib and lateral nerves at first with coarse plumose or stellate hairs
which are easily rubbed off ; central leaflets to 35 cm. long and 15 cm. broad ;
pedicels up to 3 mm. long with small ovate bracts at base ; calyx campanulate
9–15 mm. long ..   .. 7. *argentea*
Leaflets petiolulate, rarely some sessile, entire or rarely sporadically sinuately
lobed, not glaucous beneath, basal ones cuneate and not or only very slightly
asymmetrical at base :
Carpels 3–5 ; calyx-tube up to 9 mm. long, lobes up to 13 mm. long ; pedicels
up to 1 cm. long in flower ; leaflets papery to thinly coriaceous, always long-
petiolulate :
Calyx-tube 2–3 mm. long, lobes 6–7 mm. long ; anthers 8–10 ; carpels 3–4 ;
flowers fasciculate ; fruiting carpels when dry thin-walled, shortly boat-
shaped to cylindrical, 4–9 cm. long, longitudinally ridged, otherwise smooth
and subglabrous, with stipe up to 8 mm. long and beak up to 15 mm. long ;
leaflets up to 30 cm. long, gradually acuminate, ultimate venation obscure
                                                           8. *umbratilis*
Calyx-tube 7–9 mm. long, lobes 8–13 mm. long ; anthers about 15 ; carpels
4–5 ; flowers in short panicles, with common axis up to 3 cm. long ; leaflets
up to 42 cm. long, gradually acuminate, ultimate venation a prominulous
reticulum   ..   .. 9. *buntingii*
Carpels 9–12 ; calyx-tube 1·1–1·9 cm. long, lobes 1·3–2·5 cm. long ; stamens
17–18 ; flowers sessile or subsessile ; fruiting carpels thick-walled, obliquely
oblong, 9–12 (–16) cm. long, densely tomentellous, subsessile, with a short,
curved beak ; leaflets up to 60 cm. long and 20 cm. broad, petiolulate or
subsessile, acuminate ..   .. 10. *chlamydantha*
*Leaves simple, lobed or entire :
†Leaves mostly deeply digitately lobed and nerved ; anthers in 1 whorl (N.B.—
rudimentary anthers can always be observed below the carpels in the female flowers):
Stipules persistent, linear, 1·4–2·8 cm. long, rigid ; leaves closely crowded at the top
of the unbranched main stem, very variable in size, petioles of lower leaves up to
53 cm. long ; lamina deeply 3-lobed, sometimes with 1–2 additional basal lobes,
acumen abrupt, very narrow, up to 4 cm. long, central lobe with 8–14 main lateral
nerves on each side of midrib ; petioles and main nerves beneath scurfy-puberulous,
surface of lamina otherwise glabrous except for closely set minute reddish sessile
glands ; flowers sessile, bracteate, in clusters in leaf-axils and down to base of trunk;
calyx about 6 mm. long, densely rusty-pubescent outside, deep plum-purple inside
with numerous minute papillae ; carpels 3 ; fruiting carpels ellipsoid, acuminate
at apex, 4–7 cm. long, 1–3-seeded   ..   .. 11. *ficifolia*
Stipules soon caducous, up to 1 (–1·5) cm. long ; main stem usually branched, leaves
not closely crowded at top ; leaves (2–) 3–5 (–7)-lobed or rarely entire, gradually
acuminate, central lobe with up to 7 main lateral nerves on each side of midrib ;
flowers mostly in cymes ; calyx 9–20 mm. long :
Calyx with numerous minute papillae inside ; carpels 4–6 ; fruiting carpels ellipsoid
to subcylindrical, with a long apical beak ; leaves 2–3-lobed, or rarely entire ;
young branchlets and petioles with short stellate pubescence ; flowers in slender
few-flowered cymes, pedicels well-developed   ..   .. 12. *heterophylla*
Calyx with elongate more or less flexuous simple hairs inside ; carpels (5–) 8–11 :
Carpels 5–7, in fruit elongate, subcylindrical, up to 8 cm. long, including the 2–3 cm.
long beak ; leaves not more than 3-lobed, sometimes simple, often with long
hairs on petiole-apex and main nerves ; flowers yellowish ; calyx with usually
rather sparse stellate hairs outside ; small tree to 3·6 m. high ; see below
                                                             13. *brevipes*
Carpels (6–) 8–11, in fruit (where known) without a long apical beak ; leaves 3–7-
lobed, never simple :
Petioles and main basal nerves of leaves beneath pubescent, puberulous or almost
glabrous ; hairs short, stellate or simple, not hispid ; lobes of leaves usually
entire :

Calyx-lobes 1–2 mm. thick, 4 or 5, sometimes 3 ; flowers yellowish inside, with dense usually brownish stellate indumentum outside ; inflorescence-axes and pedicels very short ; inflorescences few-flowered ; fruits not known :
    Leaves 3-lobed, sometimes with 2 additional basal lobes, the lobes separated by open sinuses, their margins not overlapping ; leaves green ; tree to 5 m. high
                                         14. *flaviflora*
    Leaves 5-, very rarely 3-lobed, the lobes close to one another and usually with their margins overlapping ; leaves reddish or purplish beneath when fresh
                                       15. *ricinifolia*
Calyx-lobes 1 mm. or less thick, usually 5, rarely 4 ; flowers normally red or purplish-red inside (very rarely uniformly pale yellow), with short stellate indumentum outside, usually rather sparse on the lobes ; inflorescence-axes and pedicels usually distinct ; inflorescences several- to many-flowered ; leaves usually 5-lobed, sometimes 3- or 7-lobed ; petioles and main basal nerves of leaves beneath more or less stellate-pubescent ; fruiting carpels sessile, obovoid to ellipsoid, rounded at apex, up to 8·5 cm. long and 3–5 cm. broad, glabrous ; tree 4–12 (–18) m. high   ..   ..   ..   16. *millenii*
Petioles and main basal nerves of leaves beneath hispid with long spreading simple or fascicled hairs :
    Calyx outside with sparse to moderately dense indumentum of small stellate hairs, lobes 4–7·5 mm. long ; leaves usually 5-lobed, sometimes with secondary lobes ; fruiting carpels with stipe about 1·5 cm. long, subcylindrical, narrowed to apex, with prominent suture, up to 10 cm. long and 3 cm. broad, glabrous
                                       17. *caricaefolia*
    Calyx outside densely and shortly brown-tomentose, lobes 3·5–5 mm. long ; leaves 3- or 5-lobed, usually without secondary lobes ; fruiting carpels sub-sessile, obovoid to ellipsoid, rounded at apex, up to 5 cm. long and 3 cm. broad, densely pubescent to shortly tomentose   ..   .. 18. *hispida*
†Leaves entire, sometimes obscurely lobed but then anthers in 2 whorls :
‡Anthers in 1 whorl (N.B.—rudimentary anthers can always be observed the carpels in the female flowers) :
    Leaves cordate at base, often with a pair of small pouches near base of lamina, obovate-elliptic, abruptly acutely acuminate, up to 35 cm. long and 15 cm. broad ; stems, stipules, petioles and midrib of leaves beneath densely clothed with long spreading hairs ; flowers crowded in the upper axils, shortly pedicellate ; calyx 5–12·5 mm. long, more or less spreading-pilose outside, with minute simple elongate flexuose hairs inside ; carpels 5–6 ; fruiting carpels subcylindrical, beaked, up to about 7 cm. long and 1·5 cm. diam. (when dry)   19. *marsupium*
    Leaves not cordate, lamina without small pouches at base :
        Calyx 3–5 cm. long, campanulate, stellate-puberulous outside, tube glabrous within except for microscopic papillae near base, lobes densely stellate-puberulous within ; flowers solitary or fasciculate in leaf-axils, pedicels about 2–3·5 cm. long ; carpels 4–5, densely and minutely tomentellous, with an androgynophore up to 3 mm. long ; leaves obovate, more or less rounded at base, shortly acuminate, up to 40 cm. long and 16 cm. broad, coriaceous, glabrous or nearly so when mature   ..   ..   20. *altissima*
        Calyx up to 2·5 cm. long :
        Calyx with minute simple elongate more or less flexuose hairs inside ; petioles usually with long spreading simple or fascicled hairs ; flowers shortly pedicellate in the upper axils ; calyx 7–13 mm. long, sparsely stellate-pubescent outside ; carpels 5–7, shortly tomentose ; fruiting carpels subcylindrical, up to 8 cm. long including the 2–3 cm. long beak ; leaves obovate or obovate-elliptic, acute or rounded at base, more or less acuminate at apex, up to 35 cm. long and 13 cm. broad, sometimes lobed ; see above   ..   ..   ..   13. *brevipes*
        Calyx with minute papillae inside ; young branchlets and petioles without long spreading hairs :
            Flowers in cymes or panicles, with main axis well-developed :
                Inflorescence and flowers densely and shortly rusty-tomentose, paniculate, many-flowered ; calyx 4–6 mm. long with very short tube ; leaves and shoots when young with dense easily detached ginger-brown stellate indumentum ; leaves elliptic to lanceolate-elliptic, usually obtuse at each end, to 25 cm. long and 13 cm. broad, coriaceous ; fruiting carpels 4–6, subglobose, rounded at apex, up to 5 cm. long   ..   ..   ..   21. *laurifolia*
                Inflorescence and flowers rather sparsely and very shortly grey-puberulous, cymose, rather few-flowered ; calyx 7–20 mm. long ; leaves and shoots when young with indumentum as on inflorescence ; leaves more or less acuminate at apex, membranous ; fruiting carpels ellipsoid to ellipsoid-oblong, beaked
                                      12. *heterophylla*
        Flowers fasciculate, main axis of inflorescence absent or nearly so ; mature leaves glabrous or nearly so :

Stipules large and persistent, 2·5–3·5 cm. long ; flowers sessile ; calyx campanulate, about 5–7 mm. long, tomentellous outside ; leaves elongate-obovate or elliptic-lanceolate, abruptly acuminate at apex, up to 58 cm. long and 20 cm. broad, woolly-tomentose when young, glabrescent, petiole up to 36·5 cm. long   ..   ..   22. *semecarpophylla*

Stipules usually up to 1·5 cm. long, caducous (but in No. 23 up to 3 cm. long and more or less persistent) ; flowers distinctly pedicellate :

  Calyx 6–25 mm. long :

    Calyx glabrous outside, 16–25 mm. long, with very short tube ; anthers 9–10 ; carpels 8, glabrous except for microscopic papillae ; ovules 6–7 per carpel ; styles bifid at apex ; flowers borne on cushions along whole of trunk ; pedicels up to 14 mm. long, jointed 2·5–5 mm. above base, glabrous ; leaves crowded at ends of shoots, with stipules up to 3 cm. long and more or less persistent ; lamina oblanceolate-elliptic, up to 35 cm. long and 14 cm. broad, often smaller ; fruiting carpels subglobose, stipitate 23. *glabra*

    Calyx more or less puberulous to tomentose outside, 6–17 mm. long ; carpels (3–) 4–5, tomentellous to tomentose ; pedicels hairy :

      Petioles 2·5–8 cm. long, but sometimes shorter near the apices of shoots :

        Stipules lanceolate, caducous ; upper pulvinus of petiole glabrous, or with a few minute simple hairs on upper side ; calyx 9–10 mm. long, stellate-puberulous outside ; leaves elliptic or elongate-elliptic, up to 27 cm. long and 12·5 cm. broad   ..   ..   24. *nigerica*

        Stipules subulate, up to 7 (–14) mm. long, persistent ; upper pulvinus of petiole densely and shortly puberulous all round ; calyx 6–17 mm. long, coarsely brown-pubescent outside ; leaves elliptic or oblanceolate-elliptic, up to 30 cm. long and 10 cm. broad ..   25. *flavo-velutina*

      Petioles up to 1 cm. long, rarely a few up to 2·8 cm. :

        Outside of calyx with sparse stellate hairs ; calyx about 10 mm. long, with tube about 5 mm. long ; androphore and base of calyx inside glabrous except for minute papillae on latter ; carpels whitish-pubescent ; leaves oblong-elliptic to oblong-oblanceolate or lanceolate, up to 14 (–20) cm. long and 5 (–8) cm. broad ; fruiting carpels boat-shaped, 3·5–4 cm. long, subsessile, with short deflexed beak about 5 mm. long   26. *reticulata*

        Outside of calyx densely tomentellous ; calyx about 6–8 mm. long, with tube about 1–2 mm. long ; androphore and base of calyx inside pubescent ; carpels rusty-pubescent ; leaves oblanceolate to elliptic-oblanceolate or obovate, up to 20 cm. long and 5 cm. broad, coriaceous

                                         27. *angustifolia*

Calyx 2·5–5 (–6) mm. long ; carpels 3–5 :

  Acumen of leaves long, enlarged towards the blunt apex ; lamina ovate-elliptic or elliptic, up to 11·5 cm. long and 4·5 cm. broad ; petiole up to 8 mm. long, puberulous ; flowers solitary or paired in leaf-axils on slender puberulous pedicels about 8–10 mm. long ; calyx densely puberulous outside   ..   ..   28. *philipi-jonesii*

  Acumen of leaves not enlarged towards the apex :

    Petioles up to 1·2 cm. long, rarely longer ; lamina oblanceolate or obovate, long-attenuate to a usually narrow obtuse base ; leaves aggregated near ends of shoots, mostly 10–22 (–30) cm. long and 3–9 (–10) cm. broad ; flowers fascicled on leafy twigs and all down stems, on slender sparsely puberulous pedicels up to 16 mm. long ; calyx sparsely puberulous or subglabrous outside ; fruiting carpels 5, up to 6 cm. long and 4·5 cm. broad, narrowed to base, shortly beaked   ..   ..   29. *cauliflora*

    Petioles (1·2–) 2–9 cm. long ; lamina not long-attenuate at base ; leaves not conspicuously aggregated near ends of shoots :

      Calyx glabrous or almost so except on lobe-margins and for microscopic papillae inside ; carpels glabrous ; fruiting carpels obliquely subglobose to obovoid, 1·5–3 cm. long, 1·4–2·2 cm. broad, usually with 2 seeds in each ; flowers numerous in dense fascicles among and just below leaves ; pedicels 4–13 mm. long, glabrous ; leaves oblanceolate to elliptic, up to about 15 cm. long and 6·5 cm. broad ..   ..   ..   30. *boxiana*

      Calyx densely puberulous to pubescent outside ; carpels pubescent :

        Leaves narrowed to a rounded or obtuse base, lanceolate to slightly oblanceolate, gradually very long-acuminate, 15–40 cm. long, 4–12 cm. broad ; flower-fascicles among leaves and below ; pedicels slender, up to about 13 mm. long ; ovules 4–6 per carpel   ..   31. *praeacuta*

        Leaves narrowed to an acute base, more shortly and abruptly acuminate, acumen to 1·5 cm. long ; flowers on old wood :

          Acumen of leaves tapering to a very narrow (0·75–1 mm.) apex ; pedicels short, about 3–6 mm. long ; ovules 2 per carpel ; leaves oblanceolate to obovate, 8–22 cm. long, 3·2–10 cm. broad   ..   32. *bodardii*

Acumen of leaves 2–3 mm. broad near apex ; pedicels about 2 cm. long ;
   ovules 1 per carpel ; leaves broadly oblanceolate, 12–26 cm. long,
   5–10 cm. broad ; fruiting carpels subglobose, apiculate, stipitate, about
   1·5 cm. long, glabrous..   ..   ..   **33.** *attiensis*

‡Anthers in 2 superposed whorls (N.B.—rudimentary anthers can always be observed
below the carpels in the female flowers) :
Leaves elliptic to oblong-lanceolate or obovate, rarely ovate-elliptic, cuneate or
rounded at base, never lobed, more or less clearly 3-nerved at base :
Leaves not verticillate at nodes, though sometimes appearing in a false verticil
terminating the shoot but with alternate leaves below ; flowers and bracts
(except in No. 36) non-verticillate ; cotyledons 2 or more :
Flowers and bracts non-verticillate ; inflorescence puberulous, sometimes densely
so, not conspicuously elongated, up to 9 cm. long ; flowers up to 2 (–3) cm. long:
Fruiting carpels green and shiny, surface smooth to touch, but knobbly with
large tubercles, up to about 13 cm. long and 7 cm. broad, upper suture forming
a conspicuous ridge produced into a short deflexed beak ; seeds 4–8 (–10) per
carpel, each seed with 2 cotyledons ; leaves broadly oblong to broadly elliptic,
sometimes elliptic-oblanceolate, abruptly and shortly acuminate at apex, up
to 33 cm. long and 13 cm. broad, usually smaller, often discolorous on lower
surface when dry ; indumentum on inflorescence often dense   .. **34.** *nitida*
Fruiting carpels russet-brown or olivaceous, surface rough to touch owing to a
minute indumentum, but not knobbly, up to 20 cm. long by 6 cm. broad,
narrowed to apex, upper suture not conspicuously ridged, apex not deflexed ;
seeds up to 14 per carpel, each seed with 3–4 (rarely 2 or 5–6) cotyledons ;
leaves oblanceolate to narrowly oblong or elliptic, sometimes narrowly obovate, gradually long-acuminate at apex, the acumen often twisted downwards,
up to 22 cm. long and 8 cm. broad ; indumentum on inflorescence often
comparatively sparse and fine   ..   ..   **35.** *acuminata*
Flowers and bracts irregularly verticillate ; inflorescence very densely shortly
tomentose all over (brown when dry), often conspicuously elongated (sometimes
growing into leafy shoots) up to 17 cm. long, with long bare axes bearing 2–3
pairs of long-pedunculate lateral cymules, the latter small and dense with short
pedicels ; calyx not more than 1 cm. long, apparently not opening widely ;
leaves oblong-elliptic, up to 26 cm. long and about 9 cm. broad, with acumen
which is often abrupt and usually rather long ; fruiting carpels with surface as
in No. 35 and without tubercles ; seeds (in only carpel opened) 4, each with 2
cotyledons..   ..   ..   **36.** *simiarum*
Leaves verticillate but sometimes a few not clearly whorled ; flowers and bracts
tending to be verticillate ; inflorescence very densely tomentellous, up to 7 cm.
long, with short pedicels, densely flowered ; calyx-lobes densely tomentellous
within as well as without ; cotyledons 3–6 (rarely 2) :
Leaves usually in whorls of 4 ; fruiting carpels similar to No. 35, up to 20 cm. long
and 9 cm. broad ; seeds about 6–8 per carpel, each with thick testa and (2–)
3–4 cotyledons ; leaves elliptic-lanceolate or obovate, shortly and rather
abruptly narrowly acuminate, 11–25 cm. long, 3·5–9 cm. broad

                                                 37. *verticillata*
Leaves usually in whorls of 3 ; fruiting carpels green, surface with close rounded
knobs and ridges and with a more or less straight beak at apex, keeled on upper
suture, up to 12 cm. long ; seeds (in only carpel opened) 2 per carpel, each with
(3–) 4–5 (–6) cotyledons ; leaves elliptic to obovate-elliptic, rarely ovate-
elliptic, up to 17 cm. long and 7·5 cm. broad, with a usually short acumen,
lamina usually less elongate than in No. 37, with close lateral nerves and
prominent venation   ..   ..   **38.** *anomala*
Leaves ovate to suborbicular, mostly truncate to cordate at base :
Leaves golden beneath with dense scales, rigidly coriaceous, obtuse to subcordate
at base, obtuse to rounded at apex, sometimes apiculate, 7·5–40 cm. long, 5–30
cm. broad, lateral nerves (including basal) 4–5 ; panicles axillary, golden-scaly,
sparingly branched near the base with spiciform branches much shorter than the
leaves, or appearing unbranched ; flowers sessile or subsessile, up to 5 mm. long ;
fruiting carpels 4–5, sessile, 3–7 cm. long, 2–4 cm. broad, very shortly and
abruptly beaked and when dry with interrupted longitudinal ridges

                                                 39. *hypochrysea*
Leaves glabrous or puberulous to shortly tomentose with stellate hairs beneath, not
scaly, lamina truncate to cordate at base ; panicles not scaly, well branched
throughout :
Inflorescence-axes and outside of calyx puberulous, the surface usually visible
between the small stellate hairs ; calyx thin ; leaves sparingly puberulous
beneath when young, nearly glabrous when mature ; fruiting carpels glabrous,
ellipsoid to suborbicular, knobbly or smooth outside, 5–10 cm. long :
Pedicels well-developed, about 4·5–15 mm. long, each breaking at the joint,

leaving the 1·5–11 mm. long slender basal portion, the latter dilated at apex ;
leaves acute to shortly acuminate at apex, 8–35 cm. long, 5·5–30 cm. broad

40. lateritia var. lateritia

Pedicels mostly 0·5–2 mm. long, rarely some (especially towards base of inflorescence) up to 6·5 mm., each breaking at the joint leaving the basal portion up
to about 1·5 mm. long (rarely up to 5 mm.) ; leaves shortly acuminate at apex,
9–45 cm. long, 8–40 cm. broad .. .. 40a. lateritia var. maclaudi

Inflorescence-axes and outside of calyx densely puberulous to shortly tomentose
all over ; calyx thick ; leaves sparingly puberulous to shortly tomentose
beneath ; fruiting carpels glabrous or tomentose :

Leaves sparingly and minutely puberulous beneath when young, nearly glabrous
when mature, the surface not grey ; leaves 7–45 (–68) cm. long,
7–44 cm. broad ; inflorescence shortly brown-tomentose ; fruiting carpels
tomentose outside, up to 15 cm. long .. 41a. gigantea var. glabrescens

Leaves densely tomentellous or shortly tomentose beneath when young, the
surface completely covered, grey to brownish-grey :

Young branchlets, petioles, inflorescence-axes and outside of calyx densely and
shortly brown-tomentose ; leaves densely and shortly tomentose beneath,
even when mature, lamina 13–37 cm. long, 10·5–30 cm. broad ; fruiting
carpels densely brown-tomentose outside, scarcely beaked, up to 9 cm. long

41. gigantea var. gigantea

Young branchlets, petioles, inflorescence-axes and outside of calyx densely and
very shortly grey-tomentellous ; leaves densely and very shortly tomentellous
beneath when young, surface becoming clearly visible between the hairs when
mature, lamina often shallowly 3-lobed, rounded to subacute at apex, 8–35
cm. long, 8–30 cm. broad ; fruiting carpels glabrous, crescent-shaped,
11·5–15 cm. long, tapered into a long, recurved beak at apex

42. cordifolia

1. **C. lepidota** *K. Schum.* in Engl. Monogr. Afr. Sterculiac. 121, t. 13 (1900) ; Pellegr. in Mém. Soc. Bot. Fr. 1950–51 : 41. Forest tree, to 60 ft. high, cauliflorous ; flowers brownish-red and scaly outside, red-purple within ; fruiting carpels about 20 cm. long and 5 cm. broad, subcylindric, shortly acuminate, roughly scaly ; seeds smooth, brownish.
    S.Nig.: Oban *Talbot* 124 ! 1382 ! 1570 ! 1643 ! Arochuku *Box* 3549 ! Cross River North F.R., Ikom (fr. June) *Latilo* FHI 31849 ! **Br.Cam.:** Rio del Rey *Johnston* ! Barombi (Oct.) *Preuss* 43 ! Johann-Albrechtshöhe (Apr.) *Staudt* 925 ! 936 ! S. Bakundu F.R. (Mar.) *Brenan* 9444 ! Also in French Cameroons and Gabon.

2. **C. gigas** *Bak. f.* in Cat. Talb. 10, t. 3, 1–4 (1913). *C. lanata* Bak. f. ex Hutch. & Dalz. F.W.T.A., ed. 1, 1 : 256 (1928) ; Kew Bull. 1928 : 298. *Chlamydocola gigas* (Bak. f.) Bodard in J. Agr. Trop. Bot. Appliq. 1 : 314 (1954), excl. descr. & fig. of fruit (*Cola gigas* has 4 carpels in flower and is scarcely likely to have a fruit with 12 carpels as shown by Bodard). Forest tree, to 40 ft. high, cauliflorous ; flowers deep red ; young shoots densely woolly, becoming nearly glabrous.
    S.Nig.: Oban (Oct.) *Talbot* 160 ! 411 ! 1244 ! 1299 ! [The type of *C. lanata* at Kew is numbered *Talbot* 1299, but it is clearly a duplicate of *Talbot* 1244 at the British Museum, the number 1299 being an error of copying.]

3. **C. digitata** *Mast.* in F.T.A. 1 : 224 (1868) ; K. Schum. l.c. 123, t. 15 D ; Chev. Bot. 79 ; Aubrév. Fl. For. C. Iv. 2 : 250, t. 225, 7 ; Pellegr. l.c. 41. *Sterculia digitata* (Mast.) Roberty (1953). *C. schizandra* Bak. f. (1913). Small forest tree to 30 ft. high, leaves in a bunch at the top ; flowers dark purple ; fruiting carpels greyish-green outside, crimson-red within ; seeds black.
    Lib.: Monrovia *Whyte* ! Dukwia R. *Cooper* 455 ! *fide* Chev. l.c. **G.C.:** Simpa *Vigne* FH 2871 ! Ankasa F.R. (fr. Dec.) *Vigne* FH 3167 ! **S.Nig.:** Aboabam, Ikom (fl. Dec., fr. May) *Keay* FHI 28216 ! *Latilo* FHI 30977 ! Oban *Talbot* 1598 ! **Br.Cam.:** Kumba Dist. (fr. Apr., June) *Ejiofor* FHI 14018 ! *Olorunfemi* FHI 30620 ! Missellele, Tiko (fl. & fr. Jan.) *Boz* 3567 ! Also in French Cameroons, Gabon, Cabinda, Belgian Congo and the islands of Principe and S. Tomé.

4. **C. megalophylla** *Brenan & Keay* in Hook. Ic. Pl. t. 3532 (1955). Forest tree, to 125 ft. high, with buttresses ; bole straight, cylindrical, usually quite unbranched ; fruiting carpels scarlet within, seeds black when ripe.
    S.Nig.: Okomu F.R. (fl. Jan., fr. Jan., Feb.) *A. F. Ross* 251 ! *Brenan* 8571 ! 9176 ! Ndekke turning, Oban *Talbot* 192 ! **Br.Cam.:** Mulonga, Kumba (Mar.) *Jeme* 15/38 !

5. **C. rostrata** *K. Schum.* in Engl. Bot. Jahrb. 33 : 314 (1903) ; Pellegr. l.c. 43. Forest tree, with wide crown, to 50 ft. high ; flowers creamy yellow within.
    S.Nig.: Degema *Talbot* 3244 ! 3642 ! Oban *Talbot* 1626 ! Omon-Ndeliche F.R. (fr. June) *Ujor* FHI 30172 ! **Br.Cam.:** S. Bakundu F.R. (Mar.) *Brenan* 9433 ! Also in French Cameroons and Gabon. (See Appendix, p. 106.)

6. **C. pachycarpa** *K. Schum.* in Engl. Monogr. Afr. Sterculiac. 122, t. 12, E–H (1900) ; Pellegr. l.c. 42. Forest tree, to 30 ft. high, little-branched ; flowers pale red, white within ; fruiting carpels brown and orange-yellow ; seeds with a thick yellowish waxy coat.
    S.Nig.: Rosevear 17/29 ! 23/32 ! Oban *Talbot* 1738 ! **Br.Cam.:** Barombi (fl. Mar., fr. Aug.) *Preuss* 22 ! 370 ! Abonando (Apr.) *Rudatis* 40 ! S. Bakundu F.R. (fl. Mar., Apr., fr. Nov.) *Dundas* FHI 8356 ! 8486 ! *Brenan* 9475 ! Also in French Cameroons and Gabon.

7. **C. argentea** *Mast.* in F.T.A. 1 : 224 (1868). *C. talbotii* Bak. f. in Cat. Talb. 11 (1913). Forest shrub or small tree, to 18 ft. high ; flowers red ; fruiting carpels about 4 in. long (*fide* Bak. f.).
    S.Nig.: Kwa R., Oban–Nsan Road (Apr., May) *Talbot* 4 ! Oban (fr. May) *Talbot* 439 ! 1267 ! 1608 ! *Onochie* FHI 7724 ! *Ujor* FHI 30842 ! **Br.Cam.:** S. Bakundu F.R. *Clayton* 226 ! Also in Gabon.

8. **C. umbratilis** *Brenan & Keay* l.c. t. 3537 (1955). Small forest tree, to 20 ft. high, cauliflorous ; flowers red ; fruits bright pink with large raised veins on the surface.
    G.C.: Prestea (Sept., Oct.) *Vigne* FH 167 ! 3076 ! Simpa (fr. Feb.) *Vigne* FH 2797 ! Axim (fr. Feb.) *Irvine* 2205 !

9. **C. buntingii** *Bak. f.* in Cat. Talb. 12 (1913) ; Aubrév. l.c. 252. *C. chlamydantha* of F.W.T.A., ed. 1, 1 : 256, partly (*Linder* 290). Forest tree, to 40 ft. high ; cauliflorous ; fruiting carpels about 14 cm. long, pinkish-red on outside and glossy (*fide* Bak. f.) ; seed-coat white, glistening waxy, exterior of cotyledons purplish-red, interior darkish crimson (*fide* Bak. f.).
    Lib.: Begwai (Aug.) *Bunting* 1 ! Tappi (fr. Dec.) *Bunting* 36 ! Dukwia R. (Aug., Oct.) *Linder* 290 ! *Bequaert* ! *Cooper* 63 ! Gbanga (Sept.) *Linder* 579 ! **Iv.C.:** Tabou (fr. Dec.) *Aubrév.* 1700 ! Guiglo (Oct.) *Aubrév.* 2040 !

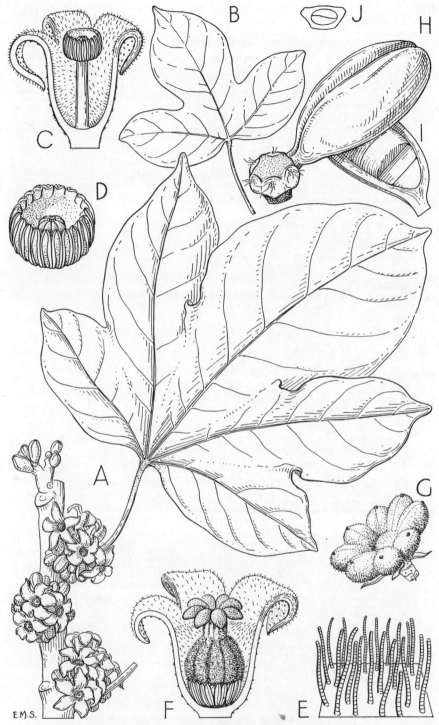

FIG. 120.—COLA MILLENII *K. Schum.* (STERCULIACEAE).

A, flowering shoot and leaf, × 1. B, leaf, × 1. C, male flower, calyx longitudinally divided, × 4. D, anthers and rudimentary carpels, × 8. E, hairs on inside of calyx, × 60. F, female flower, calyx longitudinally divided, × 4. G, very young fruit, × 1½. H, fruit with only one carpel remaining, × ⅔. I, carpel cut longitudinally to show seeds, × ⅔. J, transverse section of seed, × ⅔.

A is drawn from *Mildbr.* 7263; B from *Milne-Redhead* 5102a; C–G from *Meikle* 972; H–J from *Box* 3518.

10. **C. chlamydantha** *K. Schum.* in Engl. Monogr. Afr. Sterculiac. 112 (1900) ; Pellegr. l.c. 42. *C. mirabilis* A. Chev. in Vég. Util. 5 : 249 (1909) ; Aubrév. l.c. 2 : 250, t. 224. *Sterculia mirabilis* (A. Chev.) Roberty (1953). *Chlamydocola chlamydantha* (K. Schum.) Bodard l.c. 313 (1954). Forest tree, to 60 ft. high, cauliflorous ; flowers deep red ; carpels green to reddish outside, brown within, each with 20–24 scarlet seeds.
    S.L.: Tikonko, Dama (Jan.) *Deighton* 3887 ! Panguma (fr. Apr.) *Deighton* 3931 ! **Lib.:** Vahon, Kolahun (Nov.) *Baldwin* 10241 ! **Iv.C.:** *Aubrév.* 199 ! Alépé (fr. Dec., Jan.) *Chev.* 16241 ! **G.C.:** Tuapem (fl. & fr. Nov.) *Chipp* 8 ! Prestea (Oct.) *Vigne* FH 165 ! Awiabo, Axim (Nov.) *Vigne* 1434 ! Simpa (fr. Feb.) *Vigne* FH 2782 ! Sefwi Bekwai (fl. & fr. Oct.) *Darko* 62 ! Opon Valley (fr. Jan.) *Irvine* 1092 ! Nanankwa (Nov.) *Tinsley* WH 3855 ! 3856 ! Axim (fr. Feb.) *Irvine* 2180 ! **S.Nig.:** Oban *Talbot* 6 ! 1645 ! Also in French Cameroons and Ubangi-Shari. (See Appendix, p. 105.)
11. **C. ficifolia** *Mast.* in F.T.A. 1 : 223 (1868) ; K. Schum. l.c. 118. *C. preussii* K. Schum. l.c. 120 (1900). *Schubea heterophylla* Pax (1899), partly (leaves only, flowers are *Trichoscypha acuminata* Engl.). Unbranched tree, to 20 ft. high ; lamina of leaves up to 36 cm. long and 32 cm. broad ; in forest ; fruits red.
    Br.Cam.: Victoria (Apr.) *Maitland* 601 ! *Preuss* 1330 (partly). S. Bakundi F.R. (Apr.) *Brenan* 9295 ! Kumba to Kiliwindi (fr. Aug.) *Preuss* 394. **F.Po:** *Mann* 1157 !
12. **C. heterophylla** (*P. Beauv.*) *Schott & Endl.* Meletem. Bot. 33 (1832) ; K. Schum. l.c. 118, partly ; F.W.T.A., ed. 1, 1 : 255, partly. *Sterculia heterophylla* P. Beauv. Fl. Oware 1 : 67, t. 40 (1806). *C. gabonensis* of Aubrév. l.c. 2 : 244, t. 221, 7–8, not of Mast. Shrub or small tree, usually 2–10 ft., rarely to 16 ft., high ; leaves commonly simple, but there is a tendency for (2-) 3-lobed leaves to occur in Gold Coast and Nigeria ; simple leaves 7–32 cm. long, 2·6–12·5 cm. broad, to 35 cm. long and 32 cm. broad when lobed ; flowers pale yellow or yellowish green to dull reddish, or purplish-red inside and yellowish outside ; fruiting carpels bright red, ellipsoid to ellipsoid-oblong, 1·5–7·5 cm. long, including the rather slender 0·5–2 cm. long hooked apical beak ; in forest.
    S.L.: Peri, Gaura (Oct.) *Deighton* 5215 ! Kenema–Bo road (Dec.) *King* 79b ! Gola Forest (fr. Jan.) *Unwin & Smythe* 54. **Lib.:** Dukwia (fl. Oct., Nov., fr. Feb.) *Cooper* 31 ! 154 ! Zeahtown, Tchien (July, Aug.) *Baldwin* 6992 ! 7004 ! Tawata, Boporo (Nov.) *Baldwin* 10292 ! Ganta (fl. Sept., fr. Dec.–Mar.) *Baldwin* 9247 ! *Barker* 1138 ! **Iv.C.:** *fide* Aubrév. l.c. **G.C.:** Prestea (Sept.) *Vigne* FH 3077 ! Asuansi (Aug.–Sept.) *Box* 2797 ! 3300 ! Asankragwa, W. Prov. (Aug.) *Vigne* FH 1293 ! Ntrentreso, W. of Sefwi Bekwai (Oct., Nov.) *Darko* 61 ! 81 ! **S.Nig.:** Lagos (Apr.) *Rowland* ! Abeokuta *Irving* 106 ! Okeigbo, Ondo Prov. (fr. Feb.) *Jones* FHI 14716 ! 14731 ! Owena F.R. *Kennedy* 2464 ! 2465 ! Okomu F.R. (fr. Feb.) *Brenan* 9114 ! Sapoba (Nov.) *Meikle & Keay* 574 ! *Kennedy* 1858 ! 1945 ! Oban *Talbot* 1571 ! Also in Gabon. (See Appendix, p. 106.)
13. **C. brevipes** *K. Schum.* in Engl. Monogr. Afr. Sterculiac. 119 (1900). *C. flavescens* Engl. (1907). *C. heterophylla* of F.W.T.A., ed. 1, 1 : 255, partly (*Kalbreyer* 206, *Mann* 2286, *Talbot* 1566 ; 1587). *C. togoensis* of F.W.T.A., ed. 1, 1 : 255, partly (*Talbot* 3671). Small tree, to 12 ft. high.
    S.Nig.: Degema *Talbot* 3671 ! Calabar (fr. Feb., May) *Kalbreyer* 206 ! *Mann* 2286 ! Oban *Talbot* 1566 ! 1587 ! **Br.Cam.:** Victoria *Braun* 25. Also in French Cameroons.
14. **C. flaviflora** *Engl. & K. Krause* in Engl. Bot. Jahrb. 45 : 327, fig. 1 (1911). Small tree, to 16 ft. high, in rain forest ; flowers among or below the leaves, olive-green outside, yellow inside.
    S.Nig.: Aboabam, Ikom (May, Dec.) *Keay* FHI 28184 ! 28326 ! *Latilo* FHI 30974 ! Bendiga Ayuk to Ikom (May) *Jones & Onochie* FHI 18895 ! Cross River North F.R. (Dec.) *Keay* FHI 28158 ! Oban *Talbot* 1340 ! **Br.Cam.:** Abonando, Mamfe (May) *Rudatis* 60 ! Also in French Cameroons.
15. **C. ricinifolia** *Engl. & K. Krause* l.c. 334 (1911). *C. viridiflora* Engl. & K. Krause l.c. (1911). Small tree about 6 ft. high, in forest ; flowers among or below the leaves, brownish outside, dull yellow inside.
    S.Nig.: Aboabam, Ikom (May) *Jones & Onochie* FHI 16547 ! Oban *Talbot* 15 ! **Br.Cam.:** Mundame, nr. Kumba (Jan.) *Darko* in Hb. *Box* 3591 ! Also in Spanish Guinea.
16. **C. millenii** *K. Schum.* in Engl. Bot. Jahrb. 33 : 313 (1903). *C. togoensis* Engl. & K. Krause in op. cit. 45: 33, fig. 2 (1911) ; F.W.T.A., ed. 1, 1 : 255. ? *C. heterophylla* of F.T.A. 1 : 223, partly (*Barter* spec.). Tree, 12–40 ft., sometimes to 60 ft. high, in drier parts of the forest region, especially in secondary forest ; leaves up to 30 cm. long and 35 cm. broad, but usually smaller ; flowers greenish outside, red-brown to purplish inside, outside varying to dull purple, inside to purple and greenish-yellow, rarely flowers wholly pale yellow ; fruiting carpels rich pink to orange-red.
    Iv.C.: Morénou (Nov.) *Chev.* 22450 ! **G.C.:** Bususo (Sept.) *Darko* 145 ! R. Kakum, Asuansi (Sept.) *Box* 3476 ! 3477 ! Tobiasi (Sept.) *Box* 3479 ! Tafo (Oct., Nov.) *Box* 3644 ! *Lovi* WH 3850 ! Bunso (Oct., Nov.) *Scholes* 324 ! *Box* 3487 ! 3488 ! 3489 ! 3490 ! 3491 ! 3494 ! Achimota (May, Sept.) *Milne-Redhead* 5102 ! 5102a ! *Box* 3486 ! Togo: Lome *Warnecke* 414 ! Fume (Dec.) *Box* 3513 ! Kafeiro, Hohoe (Nov.) *St. C. Thompson* FH 3679 ! Misahöhe (Nov.) *Mildbr.* 7263 ! **Dah.:** *Le Testu* 61 ! **S.Nig.:** Ebute Metta *Millen* 31 ! Olokemeji (Nov.) *A. F. Ross* 27 ! Ibadan (Nov.) *Newberry & Etim* 179 ! *Jones & Keay* FHI 14163 ! Ibadan F.R. (Nov.) *Punch* 136 !
17. **C. caricaefolia** (*G. Don*) *K. Schum.* in Engl. Monogr. Afr. Sterculiac. 111 (1900), partly (*Afzelius* spec. only), excl. t. 11 B (?) ; Aubrév. Fl. For. C. Iv. 2 : 248, t. 223, 1–4. *Sterculia caricaefolia* G. Don Gen. Syst. 1 : 517 (1831). *Courtenia afzelii* R.Br. (1844). *Cola afzelii* (R.Br.) Mast. in F.T.A. 1 : 223 (1868). Shrub or small tree, to 20 ft. high, in forest ; leaves up to 32 cm. long and 46 cm. broad ; flowers brown outside, purplish inside, outside varying to yellow-green, inside to red ; fruiting carpels orange-pink.
    S.L.: *Afzelius* ! Limba *Oldfield* 8 ! Njala (Jan., Feb.) *Deighton* 2604 ! 5904 ! Rokupr (Apr.) *Jordan* 204 ! Roruks (Apr.) *Deighton* 3967 ! Mayoso (Aug.) *Thomas* 1449 ! **Lib.:** Ganta (fl. Sept., Oct., fr. Apr.) *Harley* 445 ! 683 ! *Baldwin* 12516 ! Dukwia R. (Sept.) *Cooper* 364 ! Vonjama Dist. (Oct.) *Baldwin* 10032 ! 10078 ! **Iv.C.:** Man *Aubrév.* 1082 ! **G.C.:** Amentia (Sept.) *Box* 3484 ! Asuansi (Sept.) *Box* 3478 ! Upper Wassaw F.R. (Oct.) *Darko* 69 ! 70 ! Wassaw Akropong (fr. Oct.) *Darko* 71 ! Aguna (Aug.) *Adjei* FH 1306 !
18. **C. hispida** *Brenan & Keay* l.c. t. 3531 (1955). *C. caricaefolia* of K. Schum. l.c., partly ; F.W.T.A., ed. 1, 1 : 255, partly. Shrub or tree, to 40 ft. high, in forest ; leaves up to 30 cm. long and 24 cm. broad ; flowers greenish-yellow inside with brown pubescence outside, inside varying to reddish-pink.
    N.Nig.: (fr. Aug.) *Shaw* 42 ! Jemaa Dist. (Nov.) *Keay* FHI 21723 ! 22266 ! **S.Nig.:** Onitsha (Nov.) *Barter* 1306 ! 1774 ! *Kennedy* 2511 ! *Onochie* FHI 21623 ! Degema *Talbot* 3629 ! 3666 ! Arochuku (Feb.) *Box* 3547 ! Oban *Talbot* 1341 ! Iso Bendiga to Bendiga Afi (Dec.) *Keay* FHI 28263 ! **Br.Cam.:** Kentu, Bamenda (fr. June) *Maitland* 1782 ! Also in Gabon. (See Appendix, p. 104.)
19. **C. marsupium** *K. Schum.* in Ber. Deutsch. Bot. Ges. 9 : 68 (1891) ; Monogr. 113, t. 12, A–D. A shrub or small tree up to about 30 ft. high ; young shoots with pale brown hairs ; flowers cream, pale reddish at base ; fruiting carpels red ; in forest.
    S.Nig.: Oban *Talbot* 24 ! 1338 ! **Br.Cam.:** Victoria (Apr.) *Maitland* 603 ! Misellele, Tiko (Jan.) *Box* 3566 ! S. Bakundu F.R. (Mar.) *Brenan* 9451 ! Johann-Albrechtshöhe (Jan.) *Staudt* 519 ! 795 ! Also in French Cameroons, Gabon, Spanish Guinea and Belgian Congo.
20. **C. altissima** *Engl.* Bot. Jahrb. 39 : 594 (1907). A forest tree, to 100 ft. high ; flowers green outside carmine inside.
    S.Nig.: Ondo (Mar.) *A. F. A. Lamb* 149 ! **Br.Cam.:** Nyanga, Kumba Div. (Feb.) *Jeme* 9/38 ! Also in French Cameroons and Gabon.
21. **C. laurifolia** *Mast.* in F.T.A. 1 : 222 (1868) ; K. Schum. in Engl. Monogr. Afr. Sterculiac. 115, t. 14 B ; Chev. Bot. 80. *Sterculia laurina* Roberty (1953). *S. laurifolia* of Chev. Bot. 76, not of F. Muell. Tree, 20–90 ft. high, on river banks ; flowers yellow or brownish.
    Sen.: *Berhaut* 782 ; 3234 ; 4303 ; 4335. **Fr.Sud.:** Macina (fr. Sept.) *Chev.* 24868 ! And other localities *fide* Chev. l.c. **Port.G.:** R. Corubal, Saltinho (May) *Esp. Santo* 2482 ! **Fr.G.:** Kouroussa *Pobéguin*

670 ! **S.L.:** R. Dantilia, Falaba (Mar.) *Sc. Elliot* 5292 ! **Iv.C.:** Soubré *Chev.* 19124 ! R. Sassandra, Dolou *Chev.* 21779. **G.C.:** R. Volta, Bamyoi Ferry (fl. & fr. Apr.) *Kinloch* FH 3273 ! Navrongo (fr. Sept.) *Vigne* FH 4571 ! R. Kru, Prang (Jan.) *Vigne* 1543 ! **Togo:** Katscha stream *Kersting* A. 563 ! **Dah.:** R. Ouémé, between Savé and Agouagon *Chev.* 23590. Atacora Mts. *Chev.* 24279. **N.Nig.:** Nupe *Barter* 1304 ! Alagbede F.R., Ilorin (Feb.) *Latilo* FHI 32957 ! Abinsi (Aug.) *Dalz.* 739 ! **S.Nig.:** Lagos *Rowland* ! *Foster* 47 ! R. Ogun *Millen* 138 ! Cross R., Itu (fr. June) *Ujor* FHI 27993 ! Calabar *Robb* ! (See Appendix, p. 106.)

22. **C. semecarpophylla** *K. Schum.* in Notizbl. Bot. Gart. Berl. 2 : 306 (1899) ; Monogr. 117, t. 11 A. Forest tree, to 40 ft. high ; flowers red, small, abundant, on the old wood ; fruits red.
S.Nig.: Oban *Talbot* 1585 ! **Br.Cam.:** N. Korup F.R., Kumba (young fr. July) *Olorunfemi* FHI 30669 ! Also in French Cameroons.

23. **C. glabra** *Brenan & Keay* in Hook. Ic. Pl. t. 3530 (1955). Forest tree, 10–40 ft. high ; flowers bright orange-pink inside and outside.
S.Nig.: Akure F.R. (fr. Apr.) *Symington* FHI 4147 ! *Onyeagocha* FHI 7145 ! Owena, Akure F.R. (Dec.) *Box* 3520 ! Owena F.R. *Kennedy* 2459 !

24. **C. nigerica** *Brenan & Keay* l.c. t. 3533 (1955). *C. flavo-velutina* of F.W.T.A., ed. 1, 1 : 255, partly (*Talbot* 1344). Forest tree, to 25 ft. high, or shrub ; flowers white, in cushions on the stem below the leaves, or with some male flowers among the leaves ; fruits red.
S.Nig.: Shasha F.R. (Dec.) *A. F. A. Lamb* 164 ! 3 miles from Abeku on path to Owosomole, Shasha F.R. (fl. & fr. Mar.) *Jones & Onochie* FHI 16672 ! Etemi, Omo F.R. (Nov.) *Tamajong* FHI 20998 ! Oban *Talbot* 1332 ! 1344 !

25. **C. flavo-velutina** *K. Schum.* in Notizbl. Bot. Gart. Berl. 2 : 306 (1899) ; Monogr. 116, t. 15 C ; F.W.T.A., ed. 1, 1 : 255, partly (only as to *Talbot* 29 ; 31 ; 1266). *C. arcuata* Bak. f. (1913). Forest tree, 10–30 ft. high ; flowers yellow.
G.C.: Tano-Offin F.R. (Jan.) *Vigne* FH 1005 ! S. Scarp F.R., Kwahu (Jan.) *Moor* FH 1028 ! **S.Nig.:** Ohosu F.R. *Hide* 13/38 ! Oban *Talbot* 29 ! 31 ! 1266 ! 1736 ! 1737 ! **Br.Cam.:** S. Bakundu F.R. (Mar., Apr.) *Brenan* 9416 ! *Ejiofor* FHI 29328 ! *Olorunfemi* FHI 30503 ! Also in French Cameroons and Gabon. (See Appendix, p. 106.)

26. **C. reticulata** *A. Chev.* in Bull. Soc. Bot. Fr. 61, Mém. 8 : 254 (1917) ; Aubrév. Fl. For. C. Iv. 2 : 246, t. 220, 5–7. *Sterculia reticulata* (A. Chev.) Roberty (1953). *C. johnsonii* Hutch. & Dalz. F.W.T.A., ed. 1, 1 : 255 (1928) ; Kew Bull. 1928 : 297. *C. basaltica* A. Chev. Bot. 78, name only. *C. flavo-velutina* of F.W.T.A., ed. 1, 1 : 255, partly. Shrub or small tree, to 15 ft. high ; flowers yellow, solitary, or few together, among or just below the leaves.
Fr.G.: Mt. Nzo, Guerzés country (Mar.) *Chev.* 21018. **Iv.C.:** foot of Mt. Nouba, by sources of R. Nuon (Apr.) *Fleury* in *Hb. Chev.* 21105 ! Mt. Dou *Aubrév.* 1053 ! Mt. Copé, Grabo (July) *Chev.* 19723 ! Mbasso (fr. Dec.) *Chev.* 22653 ! **G.C.:** Kwahu (Mar.) *Johnson* 619 ! Pompo Headwaters F.R. *S. T. Jackson* FH 2250 !

27. **C. angustifolia** *K. Schum.* in Engl. Monogr. Afr. Sterculiac. 115, t. 16 D (1900) ; F.W.T.A., ed. 1, 1: 255 (excl. syn. *C. rolandiprincipis* A. Chev.). Small or medium-sized tree ; fruits orange-red.
S.L.: *Afzelius* ! Kessewe Hills (Apr.) *Lane-Poole* 119 ! Dambaye, Kambui F.R. *King* 180 ! Tosien, Tonkia (young fr. Mar.) *King* 334 !

28. **C. philipi-jonesii** *Brenan & Keay* l.c. t. 3534 (1955). Forest shrub, 6–10 ft. high ; flowers dull cream.
S.Nig.: Boje, Afi River F.R., Ikom Dist. (May) *Jones & Onochie* FHI 5838 !

29. **C. cauliflora** *Mast.* in F.T.A. 1 : 221 (1868), partly ; K. Schum. Monogr. 117. *Sterculia cauliflora* (Mast.) Roberty (1953). *C. micrantha* K. Schum. (1899)—l.c. 116, t. 14 D. Straggling shrub or small tree, 4–15 ft. high ; flowers pinkish, with orange tinge ; in forest.
S.Nig.: Oban *Talbot* 188 ! **Br.Cam.:** Ambas Bay (Feb.) *Mann* 772 ! S. Bakundu F.R. (Mar.) *Onochie* in *Hb. Brenan* 9304 ! Johann-Albrechtshöhe (Feb.) *Staudt* 602 ! Also in French Cameroons and Gabon.

30. **C. boxiana** *Brenan & Keay* l.c. t. 3529 (1955). Understorey forest tree, up to 50 ft. high and 5 ft. girth ; flowers creamy yellow, profuse ; fruiting carpels bright red.
G.C.: Atewa Range, E. Prov. (Feb.) *Vigne* FH 4331 ! 4825 ! Akrum Ravine, near Begoro, E. Prov. (fl. & fr. Oct.) *Box* 3495 !

31. **C. praeacuta** *Brenan & Keay* l.c. t. 3535 (1955). Shrub in forest.
Br.Cam.: Musongiseli, Kumba Dist. (Feb.) *Smith* 78/36 !

32. **C. bodardii** *Pellegr.* in Bull. Soc. Bot. Fr. 97 : 189 (1950). Small tree, about 15 ft. high and 8 in. girth ; " flowers pale translucent cream " ; in forest.
S.Nig.: near Ibeku–Ayetoro crossing of Owena R., Shasha F.R. (Feb.) *Jones & Onochie* FHI 16718 ! Also in Gabon. [The Nigerian specimen has thinner, less glossy leaves than the type.]

33. **C. attiensis** *Aubrév. & Pellegr.* in Bull. Soc. Bot. Fr. 92 : 255 (1946) ; Aubrév. Fl. For. C. Iv. 2 : 244, t. 221, 4–6 (1936), French descr. only. Shrub or small tree.
Iv.C.: R. Agnéby *Aubrév.* 1785 ! Aouabo *Aubrév.* 645 !

34. **C. nitida** (*Vent.*) *Schott & Endl.* Meletem. Bot. 33 (1832) ; Chev. Vég. Util. 6 : 120, figs. 2, 10, 27, 31, 32, etc., tt. 1–5, 10–13, 15 (1911) ; Aubrév. Fl. For. C. Iv. 2 : 242, t. 220, 8–10. *Sterculia nitida* Vent. Jard. Malm., in note under t. 91 (1805). *C. vera* K. Schum. in Notizbl. Bot. Gart. Berl. 3 : 15 (1900) ; K. Schum. in Engl. Monogr. Afr. Sterculiac. 124, fig. 2, A–F. *C. acuminata* var. *latifolia* K. Schum. l.c. 127 (1900). *C. acuminata* of Engl. Pflanzenw. Afr. 3, 2 : 465, not of (P. Beauv.) Schott & Endl. Forest tree, to 80 ft. high ; flowers cream, usually with dark reddish markings within ; often cultivated, but native only as far east as the Gold Coast, introduced elsewhere.
Sen.: cult. *fide* Berhaut. **Fr.Sud.:** Babela *Chev.* 371 ! **Fr.G.:** Massadou, Macenta (May) *Collenette* 21 ! **S.L.:** Regent (Dec.) *Sc. Elliot* 4011 ! Samu country (Dec.) *Sc. Elliot* 4213 ! Njala (Jan.) *Dalz.* 8133 ! Kanya (Oct.) *Thomas* 2928 ! Mayoso (Aug.) *Thomas* 1394 ! **Lib.:** Monrovia *Whyte* ! Gbanga (Sept.) *Linder* 811 ! Webo Dist. (June, July) *Baldwin* 6139 ! 6332a ! Vonjama (Oct.) *Baldwin* 9892 ! **Iv.C.:** *fide* Chev. *l.c.* **G.C.:** Aburi Bot. Gdns. (fl. Oct., fr. Aug.) *Johnson* 484 ! 1050 ! Kumasi *Cummins* 227 ! Dunkwa (fl. Feb., Aug.) *Vigne* 15/26 ! *Chipp* 718 ! Asamankese, E.P. (Aug.) *Howes* K. 1 ! Opon Valley, W.P. (July) *Foggie* FH 211 ! Kakum F.R. *Hughes* 127 ! **Togo:** Tove, Palime (Oct., cult.) *Darko* 152 ! **N.Nig.:** Labozhi (Jan., cult.) *Yates* 53 ! *Elliott* ! Fahji (fr. Dec.) *Elliott* ! **S.Nig.:** Odo Ona, Ibadan (July, cult.) *Russell* FHI 14971 ! Agodi, Ibadan (July, cult.) *Russell* FHI 14970 ! Epe (Jan., cult.) *Onochie & Ladipo* FHI 20654 ! **F.Po:** *Mann* 8 ! *T. Vogel* 174 ! (See Appendix, p. 100.)

35. **C. acuminata** (*P. Beauv.*) *Schott & Endl.* l.c. (1832) ; K. Schum. l.c. 125, fig. 2, G–J, incl. vars. *kamerunensis, ballayi* (partly), *grandiflora* & *trichandra* ; Chev. l.c. 6 : 129, figs. 13, 25, 29, 30, tt. 6, 9, 14 (1911), q.v. for full synonymy. *Sterculia acuminata* P. Beauv. Fl. Oware 1 : 41, t. 24 (1805). *C. pseudo-acuminata* Engl. Pflanzenw. Afr. 3, 2 : 465 (1921). Forest tree, to 60 ft. high, very similar to *C. nitida* ; sometimes cultivated, but native only from Togo eastwards and southwards.
S.L.: Newton (Nov., cult.) *Deighton* 1490 ! **Lib.:** Gbanga (Sept.) *Okeke* 17 ! Zigida, Vonjama (young fr. Oct., cult.) *Baldwin* 10019a ! (?) **Iv.C.:** cult. *fide* Chev. *l.c.* **Togo:** *fide* Chev. *l.c.* **Dah.:** Sakété *Chev.* 22872 ! **N.Nig.:** *Yates* ! Agaie (Feb.) *Yates* 42 ! **S.Nig.:** Agodi, Ibadan (Jan., cult.) *Russell* FHI 14977 ! 14978 ! Okomu F.R. (Dec.) *Brenan* 8573 ! Akure (fr. June, cult.) *Russell* FHI 14983 ! Shasha F.R. *Lancaster* 7 ! Sapoba *Kennedy* 1648 ! 1746 ! Warri Prov. *Beauvois* ! Benin City *Unwin* 55 ! Oban *Talbot* ! Ikom (Jan.) *Holland* 174 ! **F.Po:** *Barter* 2075 ! Also in French Cameroons and southwards to Angola. (See Appendix, p. 100.)

[*Maitland* 758 from Victoria Bot. Gdns. has shortly acuminate leaves but appears to be *C. acuminata* and may be var. *kamerunensis* K. Schum. Hybrids between *C. nitida* and *C. acuminata* are known and may well occur wherever the two species occur together. For convenience we keep *C. nitida* and *C.*

acuminata as species but the characters are not altogether constant and when better known may show that the plants are no more than subspecies or geographical variants.]

36. **C. simiarum** *Sprague ex Brenan & Keay* l.c. t. 3536 (1955). Tree, 20–60 ft. high, with reddish-brown rough bark ; in forest by streams.
   S.L.: Kenema (Mar.) *Lane-Poole* 195 ! Baiima (fr. Sept.) *Deighton* 3079 ! Mahwe R., Zimi (Apr.) *King* 179 ! **Lib.:** Dukwia R. (Mar.) *Cooper* 336 ! **Iv.C.:** Toulépleu Road, Cavally *Aubrév.* 1191 !

37. **C. verticillata** (*Thonn.*) *Stapf ex A. Chev.* Vég. Util. 6 : 136, fig. 3, t. 7 (1911), excl. syn. *C. anomala.* *Sterculia verticillata* Thonn. in Schum. Beskr. Guin. Pl. 240 (1827).. Forest tree, to 80 ft. high ; flowers cream with dark reddish markings within.
   G.C.: *Farmar* 559 ! Aquapim *Thonning.* Aburi (fl. & fr. Oct., Nov.) *Johnson* 485 ! 1051 ! 1052 ! Konomngo, Akim (May) *Thompson* 75 ! Tafo (fl. Aug.–Oct., fr. Oct.) *Benstead* in *Hb. Darko* 335 ! *Lovi* WH 3843 ! **Dah.:** Sakété *Chev.* 22837 ! **N.Nig.:** Agaie (fl. irregularly Aug.–May, principally Aug.–Sept.) *Yates* 41 ! **S.Nig.:** Ibadan (Dec., May, cult.) *Russell* FHI 14952 ! 14982 ! Akure F.R. (fr. Nov.) *Keay* FHI 25535 ! 25537 ! Sapoba *Kennedy* 418 ! Also in French Cameroons (*Mildbr.* 8122—?).
   [*Maitland* 592, from Bulifambo, Buea, British Cameroons, is near *verticillata* but with usually coriaceous elliptic leaves ; fruits are required for certain identification.]

38. **C. anomala** *K. Schum.* in Engl. Monogr. Afr. Sterculiac. 134, t. 16 B (1900). Tree to 60 ft. high, with dense crown ; flowers yellow, with no red markings.
   Br.Cam.: Bamenda Prov. : Bangwe (June) *Conrau* 192. Nchan (May) *Maitland* 1450 ! Bamenda (Jan.) *Johnstone* 275/32 ! Bambui (Jan., June, cult.) *Russell* FHI 14964 ! 14984 ! Banso (Nov., cult.) *Russell* FHI 14974 !

39. **C. hypochrysea** *K. Schum.* in Notizbl. Bot. Gart. Berl. 2 : 306 (1899) ; Monogr. 131, t. 15 A. Forest tree of medium size ; " the red fruits are as if lacquered ".
   S.Nig.: Eket *Talbot* 3046 ! Also in French Cameroons.

40. **C. lateritia** *K. Schum.* var. **lateritia**—in Notizbl. Bot. Gart. Berl. 2 : 305 (1899) ; Monogr. 131, t. 15 B. *C. rhodoxantha* K. Schum. (1899). *C. sereti* De Wild. (1908). *C. lateritia* var. *nigerica* Bak. f. in Cat. Talb. 13 (1913). A forest tree of medium size ; fruiting carpels red.
   S.Nig.: Oban *Talbot* 1313 ! Upper Cross River *Catterall* 75 ! **Br.Cam.:** Victoria (fr. Feb., Mar.) *Maitland* 530 ! *Dalz.* 8257 ! Johann-Albrechtshöhe (Jan.) *Staudt* 510 ! 822. Also in French Cameroons, Gabon and Belgian Congo. (See Appendix, p. 106.)

40a. **C. lateritia** var. **maclaudi** (*A. Chev.*) *Brenan & Keay* in Kew Bull. 1954 : 266. *C. cordifolia* var. *maclaudi* A. Chev. in Bull. Soc. Bot. Fr. 55, Mém. 8 : 32 (1908), excl. *Chev.* 16149. *C. maclaudi* (A. Chev.) Aubrév. Fl. For. C. Iv. 2 : 246–8, t. 222, 1–4. *C. leonensis* Hutch. ex Lane-Poole (1916), name only. *C. lateritia* of F.W.T.A., ed. 1, 1 : 255, partly (as to S.L., Lib., G.C. and Togo specimens). A forest tree, to 125 ft. high ; flowers yellow changing to red, rarely white turning greenish-yellow (*Darko* 153) ; fruiting carpels red.
   Fr.G.: Yombo to Orobé, Fouta Djalon (Nov.) *Caille* in *Hb. Chev.* 14766 ! **S.L.:** Sinjo (Feb.) *Lane-Poole* 144 ! Wallia, Scarcies *Sc. Elliot* 4750 ! Njala (Feb., May) *Deighton* 698 ! 2457 ! Bumbuna (Oct.) *Thomas* 3121 ! 3181 ! Bogboabu (Oct.) *D. G. Thomas* 114 ! **Lib.:** Gbanga (Sept.) *Linder* 537 ! Ganta (fr. Apr.) *Harley* 658 ! Zolopla, Sanokwele (Sept.) *Baldwin* 9377 ! **Iv.C.:** Beriby (fr. Aug.) *Chev.* 20014 ! Aboisso (Apr.) *Chev.* 16317 ! Abidjan *Aubrév.* 173 ! **G.C.:** Dunkwa (Aug.–Sept.) *Vigne* FH 147 ! Subiri F.R. (fl. Sept., young fr. Dec.) *Andoh* FH 5441 ! 5568 ! Axim (Nov.) *Brent* 29 ! Akumadan (Oct.) *Vigne* FH 2567 ! Tobiasi (Sept.) *Box* 3480 ! Awaso, Sefwi-Anwiawso (Oct.) *Darko* 54 ! **Togo:** Misahöhe (Nov.) *Mildbr.* 7209 ! Kpeve (Oct.) *Chev.* 153 ! Wiawia, Hohoe (Oct.) *St. C. Thompson* FH 3617 ! **S.Nig.:** Ohosu F.R. *Hide* 12/38 ! Okomu F.R. (Dec.) *Brenan* 8442 ! Sapoba *Kennedy* 2543 ! Also in Cabinda.

41. **C. gigantea** *A. Chev.* var. **gigantea**—Bull. Soc. Bot. Fr. 55, Mém. 8 : 32 (1908) ; Engl. Pflanzenw. Afr. 3, 2 : 467. *C. cordifolia* of K. Schum. in Engl. Monogr. Afr. Sterculiac. 132, partly ; of F.W.T.A., ed. 1, 1 : 255, partly. Tree of forest and fringing forest, to 100 ft. high ; flowers white to pinkish or purple within ; fruiting carpels scarlet-tomentose, drying brown ; seeds scarlet.
   S.Nig.: OC 105 ! OC 106 ! Oju (fr. Apr.) *Dundas* 1933/5 ! Emege, Brass *Akpata* FHI 3944 ! **Br.Cam.:** Fongom, Bamenda, 3,500 ft. (Apr.) *Maitland* 1590 ! Also in French Cameroons, Ubangi-Shari, Belgian Congo, Uganda and A.-E. Sudan.

41a. **C. gigantea** var. **glabrescens** *Brenan & Keay* l.c. (1954). *C. cordifolia* of K. Schum. in Engl. Monogr. Afr. Sterculiac. 132, partly ; F.W.T.A., ed. 1, 1 : 255, partly ; Aubrév. Fl. For. C. Iv. 2 : 246–8, t. 222, 5–8 ; not of (Cav.) R.Br. *C. cordifolia* var. *maclaudi* A. Chev. l.c. 8 : 32 (1908), partly (*Chev.* 16149). Forest tree, to 100 ft. high ; " flowers uniformly dull creamy-white inside, darker yellowish, tinged reddish outside, changing to orange, rich pink and red " (*Box* 3485) ; fruiting carpels light tawny-brown outside, pinkish inside.
   Iv.C.: Agboville (Dec.) *Chev.* 22334 ! **G.C.:** Asebu, Cape Coast (Apr.) *Box* 3175 ! *Scholes* 285 ! S. Fomang Su F.R. (Oct.) *Brown* FH 2073 ! Asuboi, Nsawam (Sept.) *Box* 3485 ! Amentia (fl. & fr. Sept.) *Vigne* FH 1377 ! Tafo (Sept.) *Darko* 201 ! **Togo:** Misahöhe (Feb., Nov.) *Baumann* 431 ! *Mildbr.* 7316 ! **Dah.:** *Poisson* ! **N.Nig.:** Agaie (Dec.) *Yates* ! **S.Nig.:** Lagos (Dec.) *Moloney* ! *Millen* 36 ! Olokemeji (Nov.) *Jones* FHI 14191 ! Onda to Apoge, Shasha F.R. (fr. Feb.) *Jones & Onochie* FHI 17576 !

42. **C. cordifolia** (*Cav.*) *R.Br.* in Benn. Pl. Jav. Rar. 237 (1844) ; K. Schum. l.c. 132, partly ; F.W.T.A., ed. 1, 1 : 255, partly ; Aubrév. Fl. For. Soud.-Guin. 160, t. 27. *Sterculia cordifolia* Cav. Diss. 5 : 286, t. 143, 2 (1788), excl. fruit ; Guill. & Perr. Fl. Seneg. 79 (1831). *C. cordifolia* var. *puberula* A. Chev. l.c. 31 (1908). A tree, to 50 ft. high, with buttresses, in savannah country ; fruiting carpels gamboge-yellow, irregularly suffused and blotched with cherry-red, especially towards back ; seeds 8–9, rose-pink.
   Sen.: *Adanson* 214. *Perrottet* 87 ! *Talmy* ! Casamance (May) *Leprieur* ! Tambanaba, Casamance (Feb.) *Chev.* 3070 ! **Gam.:** *Hayes* 516 ! Albreda *Heudelot* 82 ! Fajara, Bathurst (fr. July) *Box* ! Kudang (Jan.) *Dalz.* 8053 ! Gunjur *Dawe* 7 ! 62 ! **Fr.Sud.:** Bamako (Jan.) *Chev.* 200 ! Sikoro (Jan.) *Chev.* 248 ! Koulikoro *Chev.* 3069 ! **Fr.G.:** Labé (Mar.) *Chev.* 12408 !
   [Pellegrin in a letter dated 16 July 1945 describes in detail the indumentum of the leaves of *Adanson* 214 which evidently agrees closely with more recent material from the same area.]

Imperfectly known species.

1. **C. winkleri** *Engl. & K. Krause* in Engl. Bot. Jahrb. 45 : 333 (1911). Flowers not described ; type presumably destroyed ; possibly near *C. flaviflora* Engl. & K. Krause.
   Br.Cam.: Johann-Albrechtshöhe (fr. Dec.) *Winkler* 1045.

2. **C. triloba** (*R.Br.*) *K. Schum.* in Engl. Monogr. Afr. Sterculiac. 119 (1900). *Courtenia triloba* R.Br. in Benn. Pl. Jav. Rar. 236 (1844). *Cola heterophylla* of F.T.A. 1 : 223, partly (*Heudelot* spec.) ; ? Aubrév. Fl. For. C. Iv. 2 : 250, t. 221, 1–3. Tree, to 30 ft. high ; previously regarded as conspecific with *C. heterophylla* (P. Beauv.) Schott & Endl., but differs in the dense rusty pubescence on young shoots and apparently in the calyx being densely rusty-pubescent outside ; the leaves are mostly shallowly to deeply 3-lobed. Trilobed forms of *heterophylla* are not known from French Guinea and Sierra Leone.
   Fr.G.: *Heudelot* 720 ! **S.L.:** Gberia Fotombu (old fr. Oct.) *Small* 405 ! **Iv.C.:** Abidjan *Aubrév.* 163 !

3. **C. sp. A.** Near *C. anomala* K. Schum., but leaves not verticillate.
   S.Nig.: *Rosevear* 14/30 ! **Br.Cam.:** Ossidinge (Oct.) *Rudatis* 78 !

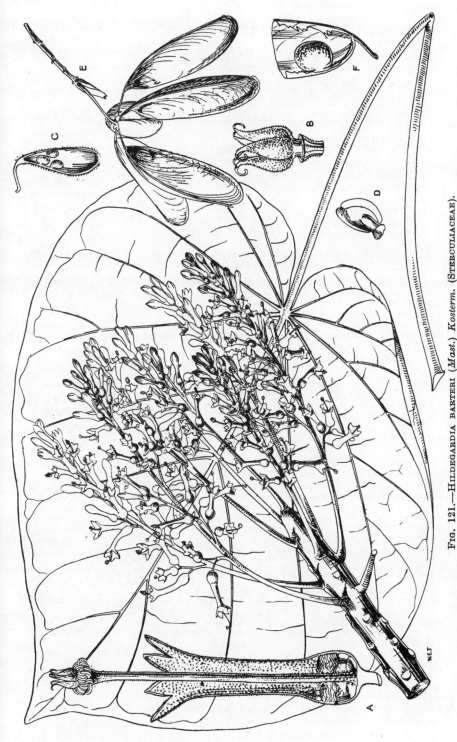

FIG. 121.—HILDEGARDIA BARTERI (*Mast.*) *Kosterm.* (STERCULIACEAE).

A, longitudinal section of flower. B, pistil. C, longitudinal section of a carpel. D, anther from back. E, fruit. F, seed.

331

4. **C. sp. B.**   Shrub ; leaves obovate, about 21–24 cm. long and 10·5–11·5 cm. broad, shortly lobed near apex or entire, glabrous, coriaceous ; petioles 8·5–10 cm. long ; fruiting carpels 12, red, shortly stipitate, 4·3 cm. long, 2·5 cm. broad, shortly more or less stellate-pubescent, 6-seeded.
   Lib.: Ganta (fr. Feb.) *Harley* 852 !

5. **C. sp. C.**   Shrub or tree, to 15 ft., with red fruit ; on flooded ground ; leaves narrowly obovate up to 15 cm. long and 5·5 cm. broad ; fruiting carpels probably about 6, with stipe 1·25–1·5 cm. long, shortly and broadly ellipsoid, shortly beaked, about 3·5 cm. long and 2·5 cm. broad, more or less rusty stellate-tomentellous.
   Lib.: Piatah (fr. Dec.) *Baldwin* 10597a !

6. **C. sp. D.**   Forest tree ; " flowers growing from cushions on the old wood, purple " ; leaves digitate, leaflets oblanceolate, sessile, long-acuminate, entire, up to 30 cm. long and 7 cm. broad.
   Br.Cam.: Mopanya, Cam. Mt., 2,400 ft. (Apr.) *Maitland* 1069 !

7. **C. sp. E.**   Forest tree, to 60 ft. high and 3 ft. girth ; fruits borne on stem ; leaves digitate, leaflets oblanceolate, more or less petiolulate, acuminate, up to 27 cm. long and 8 cm. broad, coriaceous, with scurfy stellate-puberulence beneath, especially on midrib, stipules subulate, up to 7 cm. long ; fruiting carpels, tomentellous.
   Br.Cam.: Korup F.R., Kumba Dist. (fr. July) *Olorunfemi* FHI 30672 !

8. **C. sp. F.**   Small forest tree, 12–15 ft. high ; fruits borne on stem near base ; leaves digitate ; leaflets oblanceolate, acuminate, shortly petiolulate, up to 30 cm. long and 9·5 cm. broad ; fruiting carpels 7–9, with stipe about 1·5 cm. long, up to about 15 cm. long and 4·5 cm. broad, bright orange-red.
   S.L.: *Smythe* 53 ! Gola Forest (fr. May) *Small* 602 ! **Lib.:** Jaurazon, Sinoe (young fr. **Mar.**) *Baldwin* 11476 !

9. **C. sp. G.**   *C. heterophylla* of F.W.T.A., ed. 1, 1 : 255, partly. Stems, petioles and midrib of leaves above hirsute with rather long spreading hairs ; stipules filiform, up to 1 cm. long ; petioles slender, to 6 cm. long ; leaves thin, ovate-elliptic, very acutely acuminate, sometimes with 1 or 2 acute small marginal lobes ; flowers in short cymes among the leaves, about 10–12 mm. long, 4-lobed, more or less densely stellate-pubescent outside, papillose inside.
   S.Nig.: Degema *Talbot* 3741 !

16. HILDEGARDIA Schott & Endl. Meletem. Bot. 33 (1832) ; Kostermans in Bull. Jard. Bot. Brux. 24 : 335 (1954).

Branchlets rather thick, glabrous ; leaves very broadly ovate-cordate, shortly acuminate, 15–25 cm. long and broad, very thin, about 7-nerved at base, thinly pilose when quite young ; petiole up to 25 cm. long ; flowers in numerous slender panicles towards the ends of the branchlets ; calyx long-tubular, narrower in the middle, 2 cm. long, minutely puberulous ; lobes 5, lanceolate, acute, pubescent inside ; stamens and carpels exserted ; fruit broadly and obliquely inflated, submembranous, red, 7 cm. long, 3 cm. broad ..   ..   *barteri*

H. barteri (*Mast.*) *Kosterm.* l.c. 337 (1954).   *Sterculia barteri* Mast. in F.T.A. 1 : 218 (1868) ; Hook. Ic. Pl. t. 2277.   *Firmiana barteri* (Mast.) K. Schum. in E. & P. Pflanzenfam. 3, 6 : 97 (1893) ; Engl. Monogr. Afr. Sterculiac. 109, t. 10 B ; Chev. Bot. 77 ; Aubrév. Fl. For. C. Iv. 2 : 238, t. 219 ; Kennedy For. Fl. S. Nig. 65.   *Erythropsis barteri* (Mast.) Ridley (1934)—F.W.T.A., ed. 1, 2 : 606. Tree, 80–100 ft. or more high, buttressed, with 30–40 ft. bole and girth up to 12 ft. ; bark green, wood very soft ; deciduous ; flowers scarlet, appearing before the leaves ; in drier parts of the forest regions, especially on rocky hills. **Iv.C.:** Man *Aubrév.* 402 ! **G.C.:** N. of Agogo (Dec.) *Chipp* 603 !   Ejan (Jan.) *Vigne* FH 1795 !   Achimota (Dec.) *Irvine* 1861 !   Dunkwa (fl. & fr. Dec.) *Vigne* FH 945 (partly) !   New Tafo (Dec.) *Lovi* WH 3868 ! **Togo:** Chito (Dec.) *Howes* 1060 !   **Dah.:** Hollis country (fr. Feb.) *Chev.* 22942 !   **N.Nig.:** Nupe *Barter* 1085 !   Vom *Dent Young* !   **S.Nig.:** Abeokuta (Jan.) *Rowland* !   Olokemeji (fl. Dec., fr. Feb.) *A. F. Ross* 40 !   Owo F.R. *Jones* FHI 3517 !   Lem, Okigwe (fl. & fr. Jan.) *Jones* FHI 517 !   (See Appendix, p. 107.)

17. TARRIETIA Blume Bijdr. 227 (1825).

Leaves digitately 5–7-foliolate, or by reduction the upper leaves 1-foliolate, long-petiolate ; leaflets elongate-oblanceolate or oblong-elliptic, caudate-acuminate, cuneate at base, 10–20 cm. long, 3–7 cm. broad, glabrous and with a raised midrib above, densely lepidote beneath ; lateral nerves 10–15 pairs ; flowers numerous in axillary panicles ; calyx-lobes narrowly lanceolate, subacute, 3 mm. long, softly stellate-tomentellous ; fruits unilaterally winged ; body ellipsoid, 2·5 cm. long, wing about 7 cm. long and 3 cm. broad ..   ..   *utilis*

T. utilis (*Sprague*) *Sprague* in Kew Bull. 1916 : 851 ; Aubrév. Fl. For. C. Iv. 2 : 256, t. 228 ; Pellegr. in Bull. Soc. Bot. Fr. 88 : 380 (as var. *laxiflora* Pellegr. which is var. *utilis*).   *Triplochiton utile* Sprague in Kew Bull. 1908 : 257.   *Cola proteiformis* A. Chev. Vég. Util. 5 : 250 (1909).   *Heritiera utilis* (Sprague) Sprague (1909).   A large forest tree 75–90 ft. high or more, with bole 40–60 ft. and flying buttresses which separate from the trunk shortly below the point at which they are given off ; leaves coriaceous, simple on seedlings in the first year and on the flowering twigs, otherwise digitate ; winged fruits 2–5 together.
   S.L.: Babadoori Valley (Sept.) *Lane-Poole* 399 !   Kenema (Oct.) *Aylmer* 254 !   Kambui Hills, Dambaye Valley (Sept.) *Small* 388 !   Gola Forest *Small* 646 !   **Lib.:** Gbanga (Sept.) *Linder* 664 !   Dukwia R. (Oct.) *Cooper* 61 !   103 !   **Iv.C.:** Azaguié (Sept.) *Chev.* 22293 !   San Pedro *Serv. For.* 1663 !   **G.C.:** Hunisu (fr. Jan.) *Thompson* 1 !   Imbraim (Mar.) *Thompson* 25 !   Ajakwa, N. of Chama *Chipp* 204 !   Wassidimo, E. of Bensu *Chipp* 210 !   Prestea (Oct.) *Vigne* 231 !   The species also occurs in Gabon, as var. *densiflora* Pellegr. l.c.   (See Appendix, p. 110.)

76. BOMBACACEAE

Trees sometimes very large.   Leaves simple or digitate, alternate, often lepidote ; stipules deciduous.   Flowers hermaphrodite, large and showy.   Calyx closed and valvate in bud, often with an epicalyx.   Petals often elongated, sometimes absent.   Stamens numerous free or united into a tube ; anthers reniform to linear, 1-celled ; pollen smooth.   Ovary superior, 2–5-celled ; style simple,

FIG. 122.—TARRIETIA UTILIS (*Sprague*) *Sprague* (STERCULIACEAE).

A, flowering shoot. B, leaf. C, flower laid open. D, pistil. E, carpel. F, seed. G, undersurface of leaf. G₁, peltate scale, surface view. G₂, peltate scale from below.

333

capitate or lobed ; ovules 2 or more on the inner angle of each cell. Capsule loculicidally dehiscent or not dehiscent, the valves rarely falling away. Seeds often embedded in hairs from the wall of the fruit, with little or no endosperm and flat or contorted or plicate cotyledons.

Arboreal relatives of the *Malvaceae*, with which they have much in common ; confined to the tropics.

Calyx deeply 5-lobed, 5·5–8·5 cm. long ; fruit indehiscent, not woolly inside ; flowers large, solitary, pendulous on pedicels up to 25 cm. long ; stamens very numerous
1. **Adansonia**
Calyx truncate to slightly 5-lobed, up to about 3 cm. long ; fruit filled with softly woolly floss inside ; pedicels up to about 3 cm. long :
Stamens numerous ; flowers 4–9 cm. long, solitary ..   .. 2. **Bombax**
Stamens 5–15, in 5 bundles of 1–3 ; flowers generally 3 cm. long or less, in fascicles
3. **Ceiba**

Besides the above genera, *Ochroma lagopus* Sw. (syn. *O. pyramidale* (Cav.) Urb.), the Balsa wood tree, a native of central America has been introduced into our area ; for note on nomenclature see Little in Caribbean Forester 5 : 108–114 (1944).

1. ADANSONIA Linn.—F.T.A. 1 : 212.

A large tree (see frontispiece in Vol. I, Part 1) ; leaves digitately 5-foliolate, long-petiolate ; leaflets obovate, acutely acuminate, subsessile, up to about 12 cm. long and 5 cm. broad, entire or denticulate, stellate-pubescent or nearly glabrous beneath ; flowers large, pendulous on long stalks ; calyx deeply 5-lobed, very hirsute-tomentose on both sides ; petals setose outside, adnate to the base of the staminal column ; stamens very numerous ; fruits oblong-ellipsoid to globose   ..   .. *digitata*

A. digitata *Linn.*—F.T.A. 1 : 212 ; Chev. Bot. 73 ; Bakhuizen in Bull. Jard. Bot. Buitenz., sér. 3, 167 ; Aubrév. Fl. For. Soud.-Guin. 165, t. 30, 1–4. *A. sphaerocarpa* A. Chev. (1901). Stem comparatively short and stout, 40–60 ft. high, 30–40 or more feet in girth, with short thick branches ; bark unarmed, often purplish ; fruit 6–9 in. long, pendulous on long stalks, yellowish-felted outside ; seeds embedded in a dry acid pulp ; often planted near villages, mainly in the drier regions.
Sen.: *Roger*! Fr.Sud.: San (June) *Chev.* 1104 ! Fr.G.: Farana (Jan.) *Chev.* 20453 ! Conakry *Chev.* 25691 ! S.L.: Makasi, Russell (May) *Deighton* 2759 ! Musaia (Oct.) *Thomas* 2626 ! Jigaya (Sept.) *Thomas* 2799 ! G.C.: Amansare (July) *Chipp* 513 ! Nangodi (July) *Vigne* FH 3301 ! Togo & Dah.: *fide* Aubrév. l.c. N.Nig.: Nupe *Baikie*! Maiduguri (July) *Ujor* FHI 21948 ! S.Nig.: Attah *T. Vogel* 34 ! Widespread in the drier regions of Africa. (See Appendix, p. 112.)

2. BOMBAX Linn.—F.T.A. 1 : 213. *Salmalia* Schott & Endl. (1832). *Rhodognaphalon* (Ulbr.) Roberty (1953).

Petals red, orange or pink (rarely yellow), oblong ; calyx silky-pilose within, separating from the receptacle and falling with the petals and stamens, broadly cup-shaped ; stamens grouped in 5 bundles ; floss white or greyish :
Young leaves and shoots glabrous ; leaflets with about 15–25 main lateral nerves on each side of the midrib ; flowers 6·5–9 cm. long ; fruits cylindrical, 10–18 (–25) cm. long, 4·3–6 (–8) cm. broad, with valves more or less flat when mature, fruit thus angular in section ; tree of forest regions ..   .. 1. *buonopozense*
Young leaves and shoots densely stellate-pubescent, but soon glabrescent ; leaflets with about 8–10 main lateral nerves on each side of the midrib ; flowers 4–7 cm. long ; fruits cylindrical to subglobose, 6–16 cm. long, 4–6 cm. broad, with convex valves ; tree of savannah regions   ..   .. 2. *costatum*
Petals white, lanceolate, 4·8–5·2 cm. long, about 1 cm. broad ; calyx glabrous within, remaining attached to the receptacle after the fall of the petals and stamens, campanulate ; stamens not grouped in bundles ; fruits obovoid, 5–8 cm. long, 3–4 cm. broad, with brownish floss ; leaflets broadly obovate, rounded or shortly acuminate at apex, with up to 10 pairs of main lateral nerves ; tree of forest regions..   .. 3. *brevicuspe*

1. B. buonopozense *P. Beauv.* Fl. Oware 2 : 42, t. 83, 1 (1816) ; Aubrév. Fl. For. Soud.-Guin. 170 ; F.W.T.A., ed. 1, 1 : 258, in part only. *Gossampinus buonopozensis* (P. Beauv.) Bakh. (1924). *B. flammeum* Ulbr. in Engl. Bot. Jahrb. 49 : 530, fig. 1 (1913) ; F.W.T.A., ed. 1, 1 : 258 ; Aubrév. Fl. For. C. Iv. 2 : 224, t. 214, 1–6. *G. flammea* (Ulbr.) Bakh. (1924). *B. buesgenii* Ulbr. l.c. 537 (1913)—F.W.T.A., ed. 1, 1 : 258. *G. buesgenii* (Ulbr.) Bakh. (1924). *B. angulicarpum* Ulbr. l.c. 534, fig. 2 (1913)—F.W.T.A., ed. 1, 1 : 259. *G. angulicarpa* (Ulbr.) Bakh. (1924). Large tree of forest regions, to 120 ft. high ; bark with stout conical spines mainly on the younger branches ; calyx yellowish-green ; petals deep pink or red ; flowering when leafless.
S.L.: *Hart*! *Sc. Elliot*! Njala (Feb.) *Deighton* 1584 ! 1584a ! Iv.C.: various localities *fide* Aubrév. l.c. G.C.: Tanosu (Feb.) *Chipp* 99 ! Kumasi *Vigne* FH 1881 ! Dunkwa (Feb.) *Vigne* 259 ! Nchira (Dec.) *Sam* 782 ! Axim *Irvine*! Togo: Esime, Atakpame (Jan.) *Doering* 351 ! N.Nig.: Agaie (Jan.) *Yates* 62 ! S.Nig.: *Foster* 371 ! Ikirun, Lagos (Feb.) *Millson*! Olokemeji (fr. Feb.) *A. F. Ross* 84 ! Ibadan (Feb., Mar.) *Keay* FHI 25632 ! *Latilo* FHI 22749 ! Buonopozo, Benin *Beauvois*. Br.Cam.: Victoria (Jan.–Mar.) *Maitland* 175a ! *Preuss* 1372. Also in French Cameroons, Gabon and Spanish Guinea. (See Appendix, p. 116.)
2. B. costatum *Pellegr. & Vuillet* in Notulae Syst. 3 : 88 (1914) ; Aubrév. Fl. For. Soud.-Guin. 170, t. 31 (incl. vars.). *B. andrieui* Pellegr. & Vuillet (1914). *B. houardii* Pellegr. & Vuillet (1914). *B. vuilleti* Pellegr. (1919). *B. buonopozense* of Ulbr. in Engl. Bot. Jahrb. 49 : 527 ; F.W.T.A., ed. 1, 1 : 258, for the most part ; not of P. Beauv. Tree of savannah regions, 10–50 ft. high ; bark usually rough and spiny ; petals red to orange, rarely yellow ; flowering when leafless.

Sen.: *fide* Aubrév. *l.c.* **Gam.**[1]: Kudang (Jan.) *Dalz.* 8111 ! Sankuli Kunda (Jan.) *Dalz.* 8112 ! **Fr.Sud.:** Badumbé (Dec.) *Chev.* 46 ! Koulikoro *Vuillet* 692 ; 731. **Port.G.:** Pussubé, Bissau (fl. & young fr. Feb.) *Esp. Santo* 1468 ! **Fr.G.:** Farana *Chev.* 20437. Ditinn *Chev.* 18023. **S.L.**[1]: Materboi (Feb.) *Deighton* 3622 ! Kasikeri (fr. May) *Thomas* 162 ! **Iv.C.:** Banfora, Niangho and Bobo-Dioulasso *fide* Aubrév. *l.c.* **G.C.:** Samba to Pabia (fl. & young fr. Feb.) *McLeod* ! Savelugu to Walwale *Kitson* 542 ! Navrongo (Dec.) *Vigne* FH 4683 ! **Togo:** Sokodé-Basari (Dec.) *Kersting* 464 ! **Dah.:** Savalou (May) *Chev.* 23712 ! **N.Nig.:** *Farquhar* ! Wawa, Borgu (Dec.) *Barter* 731 ! Ejidongari F.R., Ilorin (Feb.) *Latilo* FHI 32961 ! Zungeru (Dec.) *Elliott* 18 ! Bauchi Plateau *Batten-Poole* 234 ! Also in Ubangi-Shari.

3. B. brevicuspe *Sprague* in Kew Bull. 1909 : 306 ; Ulbr. l.c. 538 ; Aubrév. Fl. For. C. Iv. 2 : 226, t. 215. *Rhodognaphalon brevicuspe* (Sprague) Roberty in Bull. I.F.A.N. 15 : 1404 (1953). A tall tree of rain forest, to 120 ft. high, with straight bole and small buttresses ; deciduous ; petals white, flowers fragrant. **S.L.:** Jagbwema—Njala path, Gaura *King* 134 ! Kambui Hills (Nov.) *Small* 831 ! **Lib.:** Dukwia R. *Cooper* 421 ! **Iv.C.:** Abidjan *Aubrév.* 60 ! And other localities *fide* Aubrév. *l.c.* **G.C.:** Subirinsu (fr. Feb.) *Thompson* 10 ! 12 ! Imbraim (fr. Mar.) *Thompson* 31 ! Dunkwa (fr. June) *Vigne* FH 887 ! Begoro (Dec.) *Brown* FH 2137 ! **S.Nig.:** Sapoba *Kennedy* 1125 ! **Br.Cam.:** Johann-Albrechtshöhe (fr. Mar.) *Schultze* 105 ! Ngombe, Kumba (fr. Nov.) *Dundas* FHI 13902 ! Also in French Cameroons. (See Appendix, p. 116.)

In addition to the above, *Bombax sessile* (Benth.) Bakh. (syn. *Pachira sessilis* Benth., syn. *Pachira oleaginea* Decne.), a native of central America is cultivated in our area.

3. CEIBA Mill. Gard. Dict. Abridg. ed. 4 (1754).
Eriodendron DC. (1824)—F.T.A. 1 : 213.

Leaves digitately 8–15-foliolate, long-petiolate ; leaflets sessile, lanceolate to oblanceolate, acutely acuminate, acute at base, 10–20 cm. long, 3–4 cm. broad, entire or slightly toothed in the upper part, glabrous ; flowers fasciculate, small ; pedicels about 3 cm. long, articulated at the top ; calyx about 1 cm. long, shortly 5-lobed, lobes broad, glabrous ; petals united at base, 2–2·5 cm. long, densely silky-villous outside ; stamens 15, united into 5 separate slender bundles ; anthers sinuous ; fruits ellipsoid or elongate-fusiform, narrowed at both ends, 4–12 in. long, pale brownish .. `.`. .. .. *pentandra*

C. pentandra (*Linn.*) *Gaertn.* Fruct. 2 : 244, t. 133, fig. 1 (1791); Bakhuizen in Bull. Jard. Bot. Buitenz., sér. 3, 6 : 194 (q.v. for synonymy); Aubrév. Fl. For. C. Iv. 2 : 228, t. 216 ; Fl. For. Soud.-Guin. 169. *Bombax pentandrum* Linn. (1753). *Eriodendron anfractuosum* DC.—F.T.A. 1 : 214 ; Chev. Bot. 74. *Ceiba thonningii* A. Chev. in Rev. Bot. Appliq. 17 : 249 (1937). *C. guineensis* (Thonn.) A. Chev. l.c. 261 (1937), incl. vars. *C. caribaea* (DC.) A. Chev. l.c. 266 (1937). A tree, up to 200 ft. high, in forest and planted by villages ; trunk with large buttresses ; branches in different forms without prickles or with obtuse or acutely pointed stout prickles, generally prickly in the young state ; flowers white, appearing either when the whole tree is bare or on leafless branches ; fruits in some forms bursting on the tree, in others often falling ; floss white or grey. **Sen.:** Sedhiou, Casamance *Chev.* 3068. **Fr.Sud.:** Koubita (Feb.) *Davey* 45 ! **Port.G.:** Bissalanca (Feb.) *Esp. Santo* 1730 ! **S.L.:** *Afzelius* ! Batkanu (Jan.) *Deighton* 3615 ! **Iv.C.:** *fide* Aubrév. *l.c.* **G.C.:** Accra Plains (Jan.–Feb.) *Irvine* 1832 ! Krepe Plains (Jan.) *Johnson* 550 ! Ejura *Vigne* FH 3469 ! **Togo:** Sokodé-Basari (Jan.) *Kersting* A. 709 ! 710 ! Kete Kratchi (Jan.) *Vigne* FH 1539 ! **Dah.:** *fide* Chev. *l.c.* **N.Nig.:** various localities *fide* Keay. **S.Nig.:** *Unwin* 21 ! *Foster* 372 ! Probably of ancient introduction in Africa ; distributed throughout the tropics ; the genus is native in tropical America. (See Appendix, p. 118.)

77. MALVACEAE

Herbs often with fibrous stems or rarely shrubs ; hairs usually stellate or lepidote. Leaves alternate, often palmately nerved or divided ; stipules present. Flowers actinomorphic, hermaphrodite or rarely unisexual. Sepals valvate, with or without an epicalyx of bracteoles. Petals 5, free from each other but often adnate at the base to the staminal column, contorted or imbricate. Stamens numerous, monadelphous ; anthers 1-celled. Ovary syncarpous, rarely of 1 carpel, rarely the carpels in vertical rows ; style one. Ovules on axile placentas. Fruit a capsule or breaking into separate compartments. Seeds usually with some endosperm and straight or curved embryo, the cotyledons often folded.

A large cosmopolitan family, generally recognised by the usually stellate indumentum, showy flowers with valvate calyx, monadelphous stamens and 1-celled anthers.

Epicalyx absent :
Carpels transversely divided, contracted in the middle, the upper half spreading stellately in fruit, 2-seeded, the lower seed tomentose ; leaves entire, cordate ; flowers small in panicles .. .. **1. Wissadula**
Carpels not transversely divided :
Staminal column provided with anthers up to the top or nearly so :
Carpels 10 or more, with more than 1 ovule in each cell ; leaves always cordate
2. Abutilon
Carpels up to 10, with 1 ovule in each cell ; leaves usually not cordate **3. Sida**
Staminal column destitute of anthers at the top ; some of the leaves trifoliolate
10. Hibiscus

---

[1] The specimens from Gambia and Sierra Leone lack leaves and are identified only provisionally.

Epicalyx present :
  Styles twice as many as the carpels ; staminal column destitute of anthers at the apex :
    Flowers not in heads or at any rate not surrounded by an involucre :
      Bracteoles of epicalyx connate at the base and adnate to the calyx ; carpels with
        hooked spines   ..   ..   **4. Urena**
      Bracteoles of epicalyx free to the base or nearly so ; carpels without hooked spines,
        but sometimes with barbate bristles   ..   ..   **5. Pavonia**
    Flowers in heads surrounded by an involucre of bracts   ..   ..   **6. Malachra**
  Styles the same number as the carpels or style more or less undivided :
    Style undivided or nearly so :
      Calyx truncate :
        Leaves with minute scales, otherwise glabrous ; epicalyx of 3–5 bracteoles ; fruit
          indehiscent ; seeds pubescent   ..   ..   **7. Thespesia**
        Leaves with abundant stellate hairs ; epicalyx of (5–) 9–15 bracteoles ; fruit
          dehiscent ; seeds glabrescent   ..   ..   **8. Azanza**
      Calyx 5-lobed ; seeds cottony-hairy ; dwarf shrublet with glandular leaves
                                                                9. Cienfuegosia
    Style divided into separate stigmas :
      Epicalyx of 5 or more free or partially connate usually narrow bracteoles ; ovules 3
        or more in each cell ; carpels 5, dehiscent, not separating from the axis ; seeds
        various   ..   ..   **10. Hibiscus**
      Epicalyx of 3 broad foliaceous more or less cordate usually deeply toothed or
        laciniate bracteoles ; calyx truncate or nearly so ; carpels usually 5, with
        numerous ovules, dehiscent ; seeds covered with long cottony hairs
                                                                11. Gossypium
      Epicalyx of 7–10 small or obsolete linear bracteoles ; ovary 5-celled with 1 ovule in
        each cell ; fruits depressed, prominently 5-angled ; carpels dehiscent
                                                                12. Kosteletzkya
      Epicalyx of 1–3 narrow free bracteoles ; carpels 8–12, 1-ovulate, subindehiscent,
        separating from the central axis ..   ..   **13. Malvastrum**

1. WISSADULA Medic.—F.T.A. 1 : 182.

An undershrub with slender branches, the latter puberulous with very short hairs and
often also pubescent with stellate hairs ; leaves ovate, cordate, acutely acuminate,
5–10 cm. long, 4–8 cm. broad, entire, minutely puberulous above, densely stellate-
tomentellous beneath ; flowers small in terminal panicles ; epicalyx none ; calyx
5-lobed to about the middle ; fruiting carpels 5, about 1 cm. long, abruptly spreading
in the upper half, with rather long sharp beaks, minutely puberulous, constricted in
the middle, the upper seed puberulous, the lower tomentose

                                                      amplissima var. *rostrata*

W. amplissima (*Linn.*) *R. E. Fries* in Kungl. Sv. Vet. Akad. Handl. 43, 4 : 48 (1908). *Sida amplissima*
    Linn. (1753).
W. amplissima var. **rostrata** (*Schum. & Thonn.*) *R. E. Fries* l.c. 51, t. 6, 13–14 (1908) ; Exell Cat. S. Tomé 116
    (q.v. for synonymy). *Sida rostrata* Schum. & Thonn. (1827). *W. rostrata* (Schum.) Hook.f.—F.T.A. 1 :
    182. *W. hernandioides* and *W. periplocifolia* of Chev. Bot. 59. Erect, 3–7 ft. ; leaves hoary beneath ;
    flowers yellow or orange to reddish ; in savannah and cleared ground.
      Sen.: *Roger* 7 ! *Heudelot* 222 ! **Fr.Sud.:** Mopti to Djenné (Sept.) *Chev.* 2744 ! **Port.G.:** Pussubé, Bissau
      (fl. & fr. Dec.) *Esp. Santo* 1437 ! **Fr.G.:** Kouria (Nov.) *Chev.* 14813 ! Nzérékoré (Oct.) *Baldwin* 9738 !
      S.L.: Regent (Dec.) *Sc. Elliot* 3986 ! Njala (fr. Jan.) *Deighton* 2944 ! Port Loko (Dec.) *Thomas* 5852 !
      6669 ! Kenema (fr. Jan.) *Thomas* 7526 ! Kle, Boporo (fl. & fr. Dec.) *Baldwin* 10638 ! **Iv.C.:** Bouroukrou
      Chev. 16856 ! **G.C.:** Cape Coast (July) *T. Vogel* 81 ! Aburi (fr. Jan.) *Brown* 879 ! Wenchi (Nov.) *Thorold*
      46 ! **Togo:** Lome *Warnecke* 176 ! Kpedsu (fr. Dec.) *Howes* 1068 ! **Dah.:** Zagnanado *Chev.* 22975 !
      Fr.Nig.: Niamey (fl. & fr. Oct.) *Hagerup* 532 ! **N.Nig.:** Jebba *Barter* ! Nupe *Barter* 1043 ! 1717 ! Bauchi
      (Sept.) *Lely* 631 ! **S.Nig.:** Ogun R., Lagos *Millen* 74 ! Ibadan (fl. & fr. Nov.) *Obi* FHI 14462 ! Aboh
      Barter 304 ! Cross R. (Jan.) *Holland* 263 ! Throughout tropical Africa including Cape Verde Islands and
      S. Tomé ; var. *amplissima* is found in tropical America. (See Appendix, p. 133.)

2. ABUTILON Mill.—F.T.A. 1 : 183.

Carpels numerous, 20 or more :
  Pedicels and branchlets eglandular, usually densely tomentellous or tomentose, with or
    without long weak hairs :
    Carpels rounded at apex :
      Branches terete or subterete, softly tomentellous ; leaves ovate-rounded, cordate at
        base, triangular-acute at apex, 3–7 cm. diam., doubly dentate, softly tomentellous
        on both sides ; flowers usually paired, shortly pedicellate ; calyx-lobes acuminate,
        tomentose   ..   ..   1. *pannosum*
      Branches angular, tomentellous, leaves broadly ovate, acuminate, widely cordate,
        5–12 cm. long, up to 10 cm. broad, slightly crenulate, minutely tomentellous ;
        flowers in leafy panicles ; calyx-lobes acute, tomentellous   ..   2. *angulatum*
    Carpels acute or acuminate ; flowers solitary ; pedicels long ; especially in fruit,
      jointed near the top ; calyx-lobes acuminate ; leaves broadly ovate, stellate-
      tomentellous beneath, less so above :

Leaves not acuminate, crenate, 3–7 cm. long ; carpels abruptly acute, remaining
   more or less erect and villous   ..   ..   3. *guineense*
Leaves conspicuously acuminate, slightly dentate, 4–12 cm. long ; carpels gradually
   acuminate, spreading stellately in fruit, at first densely pilose, at length nearly
   glabrous and blackish   ..   ..   4. *mauritianum*
Pedicels and branchlets with gland-tipped hairs as well as longer spreading eglandular
   hairs ; carpels acute or shortly acuminate at apex ; flowers solitary ; calyx-lobes
   acuminate ; leaves ovate-rounded, cordate at base, acute to acuminate at apex,
   5–15 cm. long, 4–11 cm. broad, doubly dentate, shortly stellate-tomentellous
   beneath, puberulous above   ..   ..   5. *hirtum*
Carpels about 10 :
  Stems densely pilose with long rather weak, more or less reflexed hairs ; calyx-lobes
   broadly ovate, as long as the carpels ; carpels 10, acute, villous, 3–5-seeded ; petioles
   becoming reflexed ; leaves broadly ovate, cordate, gradually acute, up to 10 cm. long
   and 8 cm. broad, crenate, slightly stellate-pubescent above, more densely so beneath ;
   flowers solitary or paired on each peduncle   ..   6. *grandifolium*
  Stems at most shortly stellate-tomentose, rarely with a few longer, very weak ascending
   hairs :
   Leaves acuminate ; indumentum lax and soft :
    Stipules ovate to broadly lanceolate, subcordate at base, 1–2 cm. long, up to 1·3 cm.
     broad ; leaves ovate-rounded, denticulate, rather widely cordate, 5–12 cm. long,
     with long spreading hairs on the nerves beneath ; flowers in terminal panicles ;
     pedicels jointed about the middle ; carpels acute   ..   7. *auritum*
    Stipules filiform, 1 cm. long, ciliate ; leaves rounded-ovate, slightly trilobate, up to
     12 cm. broad, coarsely crenate, pilose on the nerves ; flowers in axillary and
     terminal cymes ; pedicels jointed near the apex ; carpels with long subulate
     points   ..   ..   8. *ramosum*
   Leaves not acuminate, cordate at base ; indumentum very short and tomentellous ;
    stipules linear, caducous ; pedicels jointed near the apex :
    Leaves ovate, subentire or denticulate, 2·5–7 cm. long, 2–5 cm. broad ; flowers
     solitary or paired ; carpels densely tomentose, acute ..   9. *fruticosum*
    Leaves subreniform, shortly 3–5-lobulate, about 2·5 cm. broad ; flowers axillary and
     terminal, 1–3 together ; carpels with subulate points ..   10. *macropodum*

1. **A. pannosum** (*Forst. f.*) *Schlechtend.* in Bot. Zeit. 9 : 828 (1851). *Sida pannosa* Forst. f. (1787). *S. mutica*
Del. ex DC. (1824). *A. muticum* (Del. ex DC.) Sweet (1830)—F.W.T.A., ed. 1, 1 : 260 ; Chev. Bot. 60.
A. glaucum of F.T.A. 1 : 185. An undershrub, 3–6 ft. high ; flowers yellow or pink with a darker
centre ; carpels 2–3-seeded.
  Sen.: *Perrottet* 76 ! Dagana, Walo *Heudelot* ! **Fr.Sud.:** Dianté (July) *Chev.* 1162 ! Timbuktu (July)
Hagerup 221 ! **N.Nig.:** Gashua, Bornu (June) *Onochie* FHI 23367 ! Kukawa (fl. & fr. Feb.) *E. Vogel* 34 !
Also in Cape Verde Islands, and in the drier regions of tropical Africa and Asia.
2. **A. angulatum** (*Guill. & Perr.*) *Mast.* in F.T.A. 1 : 183 (1868). *Bastardia angulata* Guill. & Perr. (1831).
A. intermedium Hochst. ex Garcke (1867)—Chev. Bot. 60. A tall perennial, 6–12 ft. high, whitish-downy
all over ; flowers yellow or lilac with dark purple centre ; carpels 1-seeded.
  Sen.: *Perrottet* 73 ! Richard Tol (Feb.) *Roger* 138 ! Thiès *Chev.* **Gam.:** Albreda *Perrottet*. **N.Nig.:**
Katagum (Sept.) *Dalz.* 73 ! Widespread in the drier parts of Africa, and in Madagascar.
3. **A. guineense** (*Schum. & Thonn.*) *Bak. f. & Exell* in J. Bot. 74, Suppl. Polypet., Addend. 22 (1936). *Sida
guineensis* Schum. & Thonn. (1827). *A. asiaticum* of F.T.A. 1 : 184, partly, and of F.W.T.A., ed. 1,
1 : 261. An erect, more or less villous perennial with wrinkled leaves and yellow-orange flowers.
  G.C.: Accra (fl. Mar., fl. & fr. Sept., Oct.) *T. Vogel* 71 ! *Dalz.* 125 ! *Deighton* 590 ! *Morton* GC 6018 !
Also in Angola and Rhodesia. (See Appendix, p. 122.)
4. **A. mauritianum** (*Jacq.*) *Medic.* Malv. 28 (1787). *Sida mauritiana* Jacq. (1781). *A. zanzibaricum* Boj. ex
Mast. in F.T.A. 1 : 186 (1868). *A. indicum* of F.T.A. 1 : 186, partly ; Chev. Bot. 60 (as var. *australiense*).
A whitish-downy perennial, 3–5 ft. high ; flowers yellow or orange.
  Sen.: *Talmy* 32. **G.C.:** Achiobuana (Apr.) *Chipp* 183 ! Kumasi *Cummins* 15 ! Aburi (Jan.) *Johnson*
846 ! **Togo:** Kpeve (Mar.) *Williams* 93 ! **Dah.:** Sakété (fl. & fr. Jan.) *Chev.* 22818 ! **S.Nig.:** Lagos
(Apr.) *Dalz.* 1196 ! Abeokuta *Barter* 3347 ! Ibadan (fl. & fr. Feb., Oct.) *Marks* 1 ! *Jones* FHI 13749 !
Idanre (fl. & fr. Jan.) *Brenan* 8710 ! Oban *Talbot* 1389 ! **Br.Cam.:** Tiko (fl. & fr. Feb.) *Maitland* 975a !
Widespread in tropical Africa, and in Mauritius. (See Appendix, p. 122.)
5. **A. hirtum** (*Lam.*) *Sweet* Hort. Brit., ed. 1 : 53 (1826) ; Brenan in Kew Bull. 1953 : 91. *Sida hirta* Lam.
(1783). *A. graveolens* (Roxb. ex Hornem.) Wight & Arn. (1833).
  N.Nig.: *Gosling* ! Widespread in the tropics.
6. **A. grandifolium** (*Willd.*) *Sweet* l.c. (1826) ; Brenan l.c. 90. *Sida grandifolia* Willd. (1809). *A. tortuosum*
Guill. & Perr. (1831) ; Chev. Bot. 60. *A. graveolens* of F.T.A. 1 : 184. A tall perennial, 3–6 ft. high,
with more or less zigzag hairy branches ; flowers yellow to orange.
  Sen.: *Roger* ! St. Louis *Chev.* Walo, Richard Tol, etc. (Sept., Oct.) *Perrottet* 75 ! **Fr.Sud.:** Macina
(Sept.) *Chev.* 24828. A native of S. America, introduced into parts of Africa and India, and Madeira.
7. **A. auritum** (*Wall. ex Link*) *Sweet* l.c. (1826) ; Bot. Mag. t. 2495 ; F.T.A. 1 : 187. *Sida aurita* Wall. ex Link
(1822). A tall perennial covered with down and spreading hairs ; leaves on long hispid petioles, with
conspicuous ear-like stipules ; seeds with scattered groups of short stellate hairs ; flowers orange-red.
  S.L.: Wellington (fl. & fr. Mar.) *Kirk* 4 ! Leicester (fl. & fr. Dec.) *Sc. Elliot* 3892 ! *Deighton* 3354 !
Murray Town (fl. & fr. Apr.) *Deighton* 5761 ! Probably native in the East Indies, introduced into various
parts of the world.
8. **A. ramosum** (*Cav.*) *Guill. & Perr.*—F.T.A. 1 : 186. *Sida ramosa* Cav. (1785). An erect branching perennial,
somewhat glandular-tomentose and with spreading hairs ; leaves long-stalked ; flowers yellow or
whitish ; carpels 2–3-seeded.
  Sen.: *Heudelot* 303 ! Walo (Nov.) *Perrottet* 77 ! 78 ! Richard Tol (Dec.) *Roger* 137 ! **N.Nig.:** Katagum
(Sept.) *Dalz.* 71 ! Widespread in the drier parts of tropical Africa.
9. **A. fruticosum** *Guill. & Perr.* Fl. Seneg. 1 : 70 (1831) ; F.T.A. 1 : 187 ; Chev. Bot. 59. A hoary, much-
branched, woody perennial, 3–6 ft. high, with comparatively small leaves ; flowers yellow ; carpels
2–3-seeded.
  Sen.: *Roger* ! Saffal Isl., R. Senegal, near St. Louis (Sept., Oct.) *Perrottet* 80 ! Widespread in the drier
parts of Africa, extending to Arabia and N.W. India.

10. **A. macropodum** *Guill. & Perr.* l.c. 69, t. 14 (1831) ; F.T.A. 1 : 185. Low, much-branched, suffrutescent, downy on the younger parts ; flowers yellow ; seeds large, 3 in each carpel.
Sen.: Walo, Richard Tol, etc. (Sept.–Oct.) *Perrottet* !

3. SIDA Linn.—F.T.A. 1 : 178.

Prostrate, with long slender branches, often rooting at the lower nodes ; stems weakly
pilose ; stipules linear-filiform ; leaves broadly ovate, widely cordate, acuminate,
2–5 cm. long, up to 4 cm. broad, doubly crenate, sparingly pubescent ; petiole nearly as
long as the lamina ; flowers solitary or paired ; pedicels filiform ..   1. *veronicifolia*
More or less erect herbs or small shrubs, rarely procumbent, but then stems woody ;
not rooting at the nodes :
Leaves linear or linear-lanceolate, entire ; flowers in lax terminal cymes ; leaf-lamina
up to 12 cm. long and 1·5 cm. broad, loosely pilose ; petiole short ; stipules linear ;
pedicels jointed above the middle ; carpels 5 or 6, scarcely beaked   2. *linifolia*
Leaves lanceolate to broadly ovate or rounded, toothed or crenate ; flowers mostly
axillary :
Leaves cordate at base, broadly ovate ; flowers clustered or rarely subsolitary :
Indumentum of long pilose loose hairs ; leaves crenate-serrate, 3–7 cm. long, 2–4 cm.
broad, pilose ; petiole fairly long ; stipules subulate ; flowers clustered or rarely
solitary, shortly pedicellate ; carpels 5, scarcely beaked   ..   ..   3. *urens*
Indumentum of short dense stellate hairs ; leaves dentate, not acuminate, 3–7 cm.
long, 3–5 cm. broad, softly tomentose ; petiole stout, nearly as long as the lamina ;
stipules filiform, short ; flowers solitary in the lower axils, more crowded above ;
carpels 10, with long sharp awns covered with reflexed hairs ..   4. *cordifolia*
Leaves rounded to acute at base ; flowers mostly solitary or paired :
Carpels 5, each breaking at the base to release the seed, awned ; flowers solitary to
clustered ; leaves oblong to rounded, 1·5–3 cm. long, 0·5–2 cm. broad, crenate,
softly tomentellous beneath   ..   ..   5. *alba*
Carpels 7–10, dehiscent along the dorsal side, or indehiscent :
Carpels indehiscent, usually 7–8, rugose, more or less connivent ; leaves ovate-
elliptic, rounded at apex, 2·5–4·5 cm. long, 1·5–3 cm. broad, crenate, shortly
tomentellous beneath   ..   ..   6. *ovata*
Carpels dehiscent (often only gaping) along the dorsal side, usually 8–10, not
connivent :
Leaves finely stellate-tomentellous beneath, at least when young, the hairs cover-
ing the surface of the lamina :
Sides of carpels rugose ; leaves more or less obovate-elliptic or obovate ; flowers
often rather long-pedicellate and solitary   ..   7a. *rhombifolia* var. α
Sides of carpels smooth ; leaves more or less elliptic ; flowers usually shortly
pedicellate, often clustered :
Leaves with minute stellate hairs only on upper surface

         7b. *rhombifolia* var. β
Leaves with long simple hairs on upper surface   7c. *rhombifolia* var. γ
Leaves pilose, pubescent, or more or less glabrescent beneath, with the surface of
the lamina clearly visible beneath the hairs ; carpels strongly rugose on sides as
well as back :
Flowers about 2·5–2·8 cm. diam. when open ; stems often procumbent ; leaves
elliptic, up to about 3 cm. long and 1 cm. broad, rarely more ; carpels awned,
pubescent towards the apex ..   ..   8. *scabrida*
Flowers up to about 1·5 cm. diam. when open ; stems erect :
Carpels awned (sometimes only shortly so) ; stem and leaves with minute
stellate hairs and rather sparse simple or fascicled hairs ; leaves up to 7·5 cm.
long and 2·5 cm. broad :
Carpels longer than broad, gradually narrowed into the rather short awns,
usually ochre-coloured ; leaves more or less lanceolate, up to 6·5 (–7·5) cm.
long and 1·2 (–2) cm. broad..   ..   ..   9. *acuta*
Carpels as broad as long, D-shaped, rather abruptly awned, awns often long ;
carpels usually brown ; leaves elliptic-ovate to elliptic-lanceolate, up to
7·5 cm. long and 2·5 cm. broad   ..   ..   ..   10. *stipulata*
Carpels more or less beaked, but not awned ; stem and leaves clothed, often
densely, with long simple hairs ; leaves rhombic-elliptic to ovate-lanceolate
up to 14 cm. long and 5·8 cm. broad, rather coarsely serrate in the upper
half   ..   ..   11. *corymbosa*

1. **S. veronicifolia** *Lam.* Encycl. 1 : 5 (1783). *S. humilis* Cav. (1788)—F.T.A. 1 : 179 ; Chev. Bot. 59. A
straggling wayside plant, sometimes matting the sandy soil, with pale yellow flowers.
S.L.: Talla Hills (Feb.) *Sc. Elliot* 4857 ! Njala (Nov.) *Deighton* 1509 ! Musaia (Dec.) *Deighton* 5694 !
Lib.: Peahtah (Oct.) *Bequaert* 1058 ! Vonjama (Oct.) *Baldwin* 9863 ! **Iv.C.:** Bingerville (Dec.) *Chev.*
16054. Bouroukrou (Dec.–Jan.) *Chev.* 16674. **G.C.:** Fumso (Mar.) *Darko* 531 ! Afram Plains (Dec.)
Morton GC 6090 ! **S.Nig.:** Lagos *Barter* 2139 ! *Millen* 73 ! Ibadan (Dec.) *Umana* FHI 26750 ! Idanre
(Jan.) *Brenan* 8719 ! **Br.Cam.:** Buea (Dec.) *Maitland* 905 ! Distributed in the tropics generally.

2. **S. linifolia** *Juss. ex Cav.*—F.T.A. 1 : 179 ; Chev. Bot. 58. A stiff, erect, more or less hairy weed, 1½–2 ft. ; flowers white, sometimes pink, ½–¾ in. across ; petals red-purple at the base.
Sen.: *Heudelot* 492 ! **Gam.:** *Hayes* 581 ! **Fr.Sud.:** Tabacco (Jan.) *Chev.* 126. **Fr.G.:** *Maclaud* ! **S.L.:** *Barter* ! Freetown (Oct.) *Deighton* 247 ! Port Loko (Nov.) *Thomas* 6532 ! Makump (Jan.) *Glanville* 145 ! **Lib.:** Tawata, Boporo (Nov.) *Baldwin* 10312 ! Soplima, Kolahun (Nov.) *Baldwin* 10048 ! **Iv.C.:** Dabou (Feb.) *Chev.* 17248. **G.C.:** Cape Coast (July) *T. Vogel* 19 ! Nchenenchena to Kwahu Tafo (May) *Kitson* 1166 ! Accra *Brown* 380 ! **Togo:** Lome *Warnecke* 267 ! **Dah.:** Sakété to Pedjilé (Feb.) *Chev.* 22911. **N.Nig.:** Zungeru (Oct.) *Dalz.* 124 ! Bauchi Plateau (June, July) *Lely* 279 ! P. 344 ! Lokoja (Oct.) *Parsons* 43 ! **S.Nig.:** Lagos (June) *Dalz.* 1193 ! Ibadan (Dec.) *Meikle* 898 ! Oban *Talbot* 1408 ! Widespread in tropical Africa and America. (See Appendix, p. 132.)

3. **S. urens** *Linn.*—F.T.A. 1 : 179 ; Chev. Bot. 59. A hirsute perennial ; branches sometimes lax and trailing ; flowers whitish or pale yellow.
Sen.: *Roger* 142 ! *Heudelot* 553 ! **S.L.:** Heddle's Farm (Dec.) *Sc. Elliot* 3907 ! Kumrabai (Dec.–Jan.) *Thomas* 6901 ! Magbile (Dec.) *Thomas* 6222 ! Kenema (Nov.) *Deighton* 411 ! **G.C.:** Aburi (May) *Hughes* 812 ! Kumasi (Oct.) *Darko* 350 ! **N.Nig.:** Ilorin Prov. *Clarke* 13 ! **S.Nig.:** Ibadan (Jan.) *Meikle* 939 ! Aguku Dist. *Thomas* 737 ! Widespread in the tropics and subtropics. (See Appendix, p. 133.)

4. **S. cordifolia** *Linn.*—F.T.A. 1 : 181 ; Chev. Bot. 57. An erect, softly downy perennial, up to 5 ft. high, with bright yellow flowers ; common in waste and sandy places.
Sen.: *Heudelot* 490 ! **Fr.Sud.:** Bara (Sept.) *Hagerup* 370a ! Ténétou (Mar.) *Chev.* 657 ! **Port.G.:** Barro to Sedengal (Nov.) *Esp. Santo* 3126 ! **Fr.G.:** Timbo (Feb.) *Pobéguin* 135 ! Kouroussa (July) *Pobéguin* 320 ! Dalaba-Diaguissa (Oct.) *Chev.* 18702 ! **S.L.:** Regent (Dec.) *Sc. Elliot* 4147 ! Magburaka (Oct.) *Glanville* 52 ! Mano Salija (Dec.) *Deighton* 431 ! **Lib.:** Monrovia (Nov.) *Whyte* ! *Linder* 1560 ! Peahtah (Oct.) *Linder* 951 ! Suacoco, Gbanga (Nov.) *Daniel* 35 ! **G.C.:** Abra (Oct.) *Howes* 994 ! Paliba (Dec.) *Adams & Akpabla* GC 4123 ! Tamale (Nov.) *Williams* 396 ! **Togo:** Lome *Warnecke* 237 ! **Dah.:** Cotonou (Mar.) *Debeaux* ! **N.Nig.:** Nupe *Barter* ! Kontagora (Nov.) *Dalz.* 129 ! Sokoto (Aug.) *Dalz.* 428 ! Panyam (July) *Lely* 437 ! **S.Nig.:** Lagos *Millen* 123 ! Ibadan (Dec.) *Meikle* 900 ! Awka (Oct.) *Jones* FHI 6788 ! Widespread in the tropics and subtropics. (See Appendix, p. 132.)

5. **S. alba** *Linn.* Sp. Pl., ed. 2, 2 : 960 (1763). *S. spinosa* of Linn. (1753), partly ; F.T.A. 1 : 180 ; Chev. Bot. 58. A woody herb, with branches sometimes lax and trailing ; flowers small, white.
Sen.: *Roger* ! Dakar (May) *Baldwin* 5709 ! **Gam.:** Genieri (Feb.) *Fox* 42 ! **Fr.Sud.:** Bara, Amongo (Sept.) *Hagerup* 370 ! Sompi, and Mopti to Djenné, *fide* Chev. *l.c.* **S.L.:** Bonthe Isl. (Mar.) *Deighton* 2478 ! **G.C.:** Accra (Mar.) *Deighton* 565 ! **Togo:** Lome *Warnecke* 150 ! **N.Nig.:** Sokoto *Lely* 150 ! Fodama (Dec.) *Moiser* 251 ! Lemme (May) *Lely* 140 ! Biu (Aug.) *Noble* 13 ! Widespread in the tropics.

6. **S. ovata** *Forsk.* Fl. Aegypt.-Arab. 124 (1775). *S. grewioides* Guill. & Perr. (1831)—F.T.A. 1 : 182 ; Chev. Bot. 57 ; F.W.T.A., ed. 1, 1 : 263. A rather hoary woody herb, 2 ft. or more high ; flowers yellow.
Sen.: *Roger* 8 ! *Heudelot* 413 ! Walo (Sept., Oct.) *Perrottet.* **Fr.Sud.:** L. Télé, Goundam (Aug.) *Chev.* 2748 ! Timbuktu (July) *Hagerup* 219 ! **Iv.C.:** *fide* Chev. *l.c.* **N.Nig.:** Bindawa valley, Katsina (July) *Grove* 4 ! Katagum (Oct.) *Dalz.* 69 ! Karaswa, Bornu (June) *Onochie* FHI 23341 ! In the drier parts of Africa, and extending to Arabia and India.

7. **S. rhombifolia** *Linn.* Sp. Pl. 2 : 684 (1753) ; F.T.A. 1 : 181 ; Chev. Bot. 58. A widespread perennial weed ; requiring further taxonomical study throughout its range. There appear to be three varieties in our area:—

7a. **S. rhombifolia var. α.** Flowers pale yellow.
Fr.G.: *Heudelot* 697 (partly) ! Kouria (Oct.) *Caille* in *Hb. Chev.* 15029 ! **S.L.:** Rokupr (Jan.) *Deighton* 2959 ! *Jordan* 704 ! Bureh town (Dec.) *Deighton* 3289 ! Bumbuna (Oct.) *Thomas* 3712 ! **G.C.:** Akwadum (July) *Deighton* 5281 ! **Fr.Nig.:** Niamey (Oct.) *Hagerup* 504 ! **N.Nig.:** Sokoto (Nov.) *Moiser* 35 ! Katagum (Oct.) *Dalz.* 70 ! Nupe *Barter* 1718 ! **S.Nig.:** Ibadan (Apr., July, Dec.) *Meikle* 909 ! 1448 ! Widespread in the tropics.

7b. **S. rhombifolia var. β.** Flowers pale yellow.
Fr.G.: *Heudelot* 697 (partly) ! **S.L.:** Njala (Dec.) *Deighton* 1795 ! Baoma, Koya (Dec.) *Deighton* 3799 ! Musaia (Dec.) *Deighton* 5695 ! **Lib.:** Kakatown *Whyte* ! **G.C.:** Benso (June) *Andoh* FH 5527 ! Kaasi, Kumasi (Feb.) *Darko* 524 ! **Togo:** Lome *Warnecke* 429 ! **N.Nig.:** Kontagora (Nov.) *Dalz.* 126 ! Lokoja *Ansell* ! **S.Nig.:** Ibadan (Jan.) *Meikle* 938 ! Okomu F.R. (Feb.) *Brenan* 9164 ! Nun R. *T. Vogel* 26 ! Ikom (Dec.) *Rosevear* 44/30a ! **Br.Cam.:** Buea (Nov., Jan.) *Maitland* 118 ! 254 ! Bamenda (Feb., Apr.) *Maitland* 1729 ! *Migeod* 462 ! **F.Po:** (Dec.) *Guinea* 289. Widespread in tropical Africa and (?) in the tropics generally.

7c. **S. rhombifolia var. γ.** Possibly a montane form ; flowers yellow or pinkish-yellow.
Br.Cam.: Cam. Mt., 4,400–6,000 ft. (Dec., Jan., Apr.) *Maitland* 809 ! 1119 ! *Dunlap* 30 ! Bamenda 6,600 ft. (Jan.) *Keay* FHI 28411 ! Also in N.E. tropical Africa.

8. **S. scabrida** *Wight & Arn.* Prodr. Fl. Pen. Ind. Or. 1 : 57 (1834). Woody herb, to 18 in. high, or procumbent ; flowers pale orange, larger than the other related species.
S.L.: Makunde, Limba (Apr.) *Sc. Elliot* 5714 ! Rokupr (Mar.) *Jordan* 426 ! **Lib.:** Monrovia (Nov.) *Linder* 1526 ! **S.Nig.:** Ibadan (Apr.) *Meikle* 1410 ! Also in India.

9. **S. acuta** *Burm. f.* Fl. Ind. 147 (1768). *S. carpinifolia* of F.W.T.A., ed. 1, 1 : 263, partly, not of Linn. f. *S. vogelii* Hook. f. in Fl. Nigrit. 231. A shrubby perennial, much-branched, to 3 ft. high ; flowers pale yellow.
S.L.: Mano (Nov., Dec.) *Deighton* 5661 ! 5662 ! Bumpe (Nov.) *Deighton* 5874 ! **Lib.:** Sinoe Basin *Whyte* ! **G.C.:** Accra *Moloney* ! Kaasi, Kumasi (Feb.) *Darko* 519 ! **Togo:** Lome *Warnecke* 188 ! **N.Nig.:** Fodama (Dec.) *Moiser* 161 ! Aguji, Ilorin *Thornton* ! Bauchi Plateau (Aug.) *Lely* P. 610 ! **S.Nig.:** Ibadan (Aug., Dec.) *Golding* 4 ! *Meikle* 882 ! 887 ! 892 ! Olokemeji (May) *Foster* 287 ! **F.Po** : (Nov.) *T. Vogel* 214 ! Widespread in the tropics. (See Appendix, p. 131.)

10. **S. stipulata** *Cav.* Diss. 1 : 22, t. 3, 10 (1785). *S. carpinifolia* of F.W.T.A., ed. 1, 1 : 263, partly, not of Linn. f. Shrubby perennial, to 4 ft. high ; flowers yellow.
S.L.: (June) *T. Vogel* 26 ! 83 ! Leicester Peak (Dec.) *Sc. Elliot* 3897 ! Kenema (Nov.) *Deighton* 418 ! Musaia (Apr.) *Deighton* 5463 ! **Lib.:** Cinco, Monrovia (Nov.) *Linder* 1546 ! Sinkor (June) *Barker* 1360 ! **G.C.:** Cape Coast (July) *T. Vogel* 13 ! Woase (June) *Chipp* 487 ! Asuansi (Feb.) *Williams* 282 ! Kumasi (Feb.) *Andoh* FH 4134 ! **Togo:** Amedzofe (Feb.) *Irvine* 180 ! **S.Nig.:** Idu, Sapoba (Nov.) *A. F. Ross* 220 ! Calabar (Apr.) *Holland* 114 ! **Br.Cam.:** Victoria (June) *Maitland* 49 ! **F.Po:** (Dec.) *Guinea* 283. Also in tropical America.

11. **S. corymbosa** *R. E. Fries* in Bull. Herb. Boiss., sér. 2, 7 : 998 (1907). A shrubby perennial, to 7 ft. high ; flowers yellow, often apricot-coloured.
S.L.: Sumbuya (Nov.) *Deighton* 2272 ! Mattru (Nov.) *Deighton* 2350 ! Gegbwema, Tunkia (Oct.) *Deighton* 5202 ! Njala (Oct., Dec.) *Deighton* 5189 ! 5293 ! **Lib.:** Gbanga (Sept.) *Linder* 595 ! **S.Nig.:** Ibadan (Nov., Dec.) *Meikle* 880 ! 888 ! 897 ! *Keay* FHI 28297 ! Idanre (Jan.) *Brenan* 8708 ! Also in central America.

Hybrids probably occur between several of the above species. Meikle has collected specimens from Ibadan, Nigeria, which appear to be hybrids between *S. acuta* and *S. corymbosa*. A specimen from Lagos (*Foster* 326) may be a hybrid between *S. stipulata* and *S. corymbosa*.

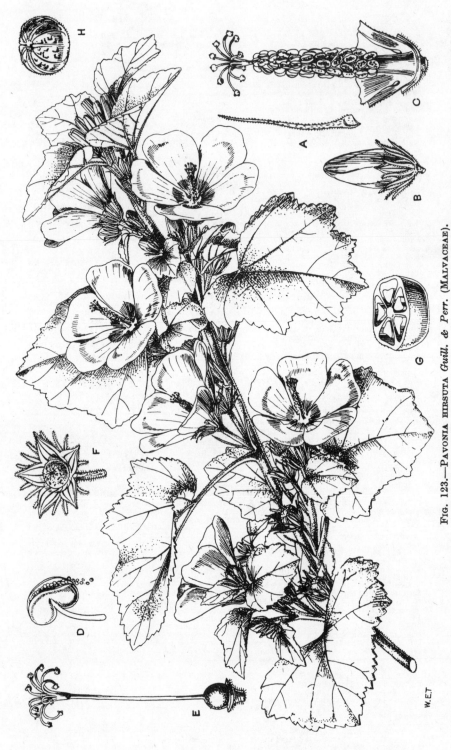

W.E.T

Fig. 123.—Pavonia hirsuta *Guill. & Perr.* (Malvaceae).

A, lobe of epicalyx.   B, flower-bud.   C, portion of flower.   D, anther.   E, pistil.   F, ovary, calyx and epicalyx.   G, cross-section of fruit.   H, mature fruit.

4. URENA Linn.—F.T.A. 1 : 189.

A fibrous undershrub covered with stellate hairs ; leaves very variable, usually more
or less 3–5-lobed, sometimes deeply so, whitish beneath ; middle nerve often with a
large pitted gland near the base ; flowers axillary, usually solitary; epicalyx of 5
linear-lanceolate bracteoles united at the base and persistent in fruit ; fruit depressed-
globose, covered with rigid hooked bristles .. 　　.. 　　.. 　　.. 　　.. 　　*lobata*

U. lobata *Linn.*—F.T.A. 1 : 189 ; Chev. Bot. 61 (incl. vars.). Varying from a low woody weed to a shrub
8–10 ft. high ; flowers rose-pink or yellow, 1 in. or less long ; cultivated in some of its forms.
Sen.: (Aug.) *Roger* 118! Gam.: *Brooks* 13! Fr.Sud.: Ianfola (Mar.) *Chev.* 450! Fr.G.: Conakry *Lecerf*!
Farana (Mar.) *Sc. Elliot* 5321! S.L.: Freetown (Nov.) *Deighton* 233! Yetaya (Sept.) *Thomas* 2289!
Gberia Timbako (Sept.) *Small* 312! Lib.: Kakatown *Whyte*! Vonjama (Oct.) *Baldwin* 9927! Suacoco,
Gbanga (Nov.) *Daniel* 32! G.C.: Cape Coast (July) *T. Vogel* 32! Wanki (June) *Chipp* 479! Accra (Oct.)
Morton GC 6021! Tamale (Oct.) *Williams* 381! Togo: Lome *Warnecke* 232! Fr.Nig.: Niamey (Oct.)
Hagerup 556! N.Nig.: Kebbi, Sokoto Prov. (Oct.) *Dalz.* 422! Nupe *Barter* 849! Bauchi Plateau (Aug.)
Lely P. 574! 581! Yola (Sept.) *Shaw* 68! S.Nig.: Ibadan (Dec.) *Meikle* 916! Idu, Sapoba (Nov.) *A. F.
Ross* 224! Nun R. (Aug.) *T. Vogel* 5! Oban *Talbot* 1265! Br.Cam.: Buea (Nov.) *Migeod* 224! Bambili,
Bamenda (Aug.) *Ujor* FHI 29984! F.Po: *Barter*! Widely distributed in the tropics and subtropics.
See Appendix, p. 133.)

5. PAVONIA Cav.—F.T.A. 1 : 189 ; Ulbrich in Engl. Bot. Jahrb. 57 : 54–184 (1920).

Fruiting carpels with a narrow membranous wing on the margin, not awned, not
enclosed by the calyx ; epicalyx of 10 linear-filiform bracteoles, much longer than the
calyx, forming a little open basket around the carpels in fruit ; leaves up to 2·5 cm.
long, with long and short stellate hairs beneath ; flowers solitary on long pedicels
jointed towards the top :
　　Leaves digitately divided almost to the base ; lobes 3–5, oblanceolate, slightly toothed,
　　　　up to 2·5 cm. long 　.. 　　.. 　　.. 　　.. 　　.. 　　.. 　　1. *zeylanica*
　　Leaves elliptic-suborbicular, not lobed, rounded at base, 1–2 cm. long 　　2. *kotschyi*
Fruiting carpels not winged, enclosed by the persistent calyx ; epicalyx of 10–12 linear
or linear-lanceolate bracteoles, shorter than or about as long as the calyx ; leaves
usually much exceeding 2·5 cm. long, usually cordate at base ; flowers solitary or
clustered :
　　Fruiting carpels not awned, with reflexed hairs on the margins ; leaves slightly lobed,
　　lobes triangular, doubly dentate ; stem and leaves densely tomentose beneath ;
　　flowers solitary, or crowded above .. 　　.. 　　.. 　　.. 　　.. 　　3. *hirsuta*
　　Fruiting carpels with long awns armed with short reflexed bristles ; leaves more or less
　　distinctly digitately lobed, usually coarsely toothed ; flowers clustered in leaf-axils
　　on short pedicels, or solitary towards ends of shoots :
　　　　Stem and leaves more or less densely long-hirsute .. 　　.. 　　4. *urens* var. *urens*
　　　　Stem and leaves pubescent or glabrescent .. 　　.. 　　4a. *urens* var. *glabrescens*

1. P. zeylanica *Cav.*—F.T.A. 1 : 192 ; Ulbr. in Engl. Bot. Jahrb. 57 : 153 (as var. *microphylla* Ulbr.). *P.
arabica* of Chev. Bot. 62. Perennial with hispid branches 1–1½ ft. high from a thick woody stock ;
flowers yellow or reddish, exceeding the epicalyx.
　　Sen.: *Perrottet* 69! Richard Tol *Lelièvre*! Walo *Heudelot*! Fr.Sud.: Tacadji (Aug.) *Chev.*! Extends to
Abyssinia and Somaliland, S.E. Arabia and India.
2. P. kotschyi *Hochst. ex Webb*—F.T.A. 1 : 192 ; Ulbr. l.c. 171 ; Chev. Bot. 62. A low woody perennial of
semi-desert regions, with short densely pubescent branchlets ; flowers yellow, less than 1 in. across,
shorter than the epicalyx.
　　Fr.Sud.: Timbuktu (fl. & fr. July) *Chev.* 1246! *Hagerup* 188! N.Nig.: Kukawa (fl. & fr. Jan.) *E. Vogel*
27! Extends to Abyssinia and Somaliland ; also in Arabia.
3. P. hirsuta *Guill. & Perr.*—F.T.A. 1 : 191 ; Ulbr. l.c. 116 (incl. var. *microphylla* Ulbr.) ; Chev. Bot. 62
(excl. *Chev.* 18504 which is *Kosteletzkya grantii*). An erect or sometimes prostrate undershrub, softly
yellowish tomentose all over ; flowers large, yellow with a red-purple centre 2 in. or more across ; epicalyx
of 12 downy free bracts, calyx with 3–5 green veins.
　　Sen.: *Heudelot* 410! Walo (Jan.) *Perrottet* 71! Fr.Sud.: Timbuktu (July, Sept.) *Chev.* 1244! *Hagerup*
201! N.Nig.: Sokoto (June, Nov.) *Dalz.* 420! *Moiser* 59! Katagum (Nov.) *Dalz.* 72! L. Chad (Feb.,
Mar., Oct., Dec.) *E. Vogel* 42! *Elliott* 148! Widespread in the drier parts of tropical Africa. (See
Appendix, p. 131.)
4. P. urens *Cav.* var. urens—Diss. 3 : 137, t. 49, 1 (1787) ; Ulbr. l.c. 104 ; Brenan in Mem. N.Y. Bot. Gard. 8,
3 : 223. *P. schimperiana* var. *hirsuta* of F.W.T.A., ed. 1, 1 : 263 ; Chev. Bot. 62. Coarse woody herb,
to 10 ft. high ; flowers white, with pink or purplish markings.
　　Fr.G.: Dalaba to Diaguissa (Sept.) *Chev.* 18588! S.L.: Bintumane Peak, 5,400 ft. (Jan.) *T. S. Jones*
140! N.Nig.: Bauchi Plateau (Sept.) *Lely* P. 709! S.Nig.: Mt. Koloishe, Obudu, 4,800 ft. (Dec.) *Keay &
Savory* FHI 25124! Br.Cam.: Cam. Mt., 4,000 ft. (Dec.) *Maitland* 840! Bambili, Bamenda (Aug.)
Ujor FHI 29976! Nchan, Bamenda (June) *Maitland* 1681! F.Po: Mioka, near Moka (Feb.) *Exell* 849!
Widespread in tropical Africa, also in Madagascar and the Mascarenes.
4a. P. urens var. glabrescens (*Ulbr.*) Brenan in F. W. Andr. Fl. Pl. A.-E. Sudan 2 : 36 (1952). *P. schimperiana*
Hochst. ex A. Rich. var. *glabrescens* Ulbr. l.c. 108 (1920). *P. schimperiana* of F.W.T.A., ed. 1, 1 : 263.
Woody herb ; flowers white with pink or reddish markings.
　　Fr.G.: Longuery (Dec.) *Chev.* 14828! Togo: Bismarckburg (fl. Oct., Nov., fr. Nov.) *Büttner* 261!
Br.Cam.: Buea (Jan.) *Dunlap* 126! *Maitland* 318! F.Po: Finca Puente (Jan.) *Guinea* 1710! Wide-
spread in tropical Africa.

6. MALACHRA Linn.—F.T.A. 1 : 188.

Leaves deeply 3–7-lobed, lobes oblanceolate, toothed, coarsely strigose beneath like the
stems, about 6 cm. long ; flowers in dense subglobose heads subtended by leafy
suborbicular bracts ; epicalyx of several linear-filiform plumose bracteoles ; calyx
long-pilose ; fruiting carpels reticulate, about ½ as long as the calyx 　　1. *radiata*
Leaves slightly 3–7-lobed, lobes broad at the base, slightly pubescent beneath and on
the stems, 5–9 cm. long ; flowers few in heads ; bracts ovate, cordate, pale with

4

distinct veins ; calyx markedly veined, pilose ; carpels shorter than the calyx, finely
reticulate, glabrous .. .. 2. *capitata*

1. **M. radiata** (*Linn.*) *Linn.* Syst., ed. 12, 2 : 459 (1767) ; F.T.A. 1 : 188 ; Chev. Bot. 61. *Sida radiata* Linn.
(1763). An erect, coarsely hairy perennial, about 4 ft. high ; in wet places ; flowers light red.
Sen.: Néma *Berhaut* 823. **Iv.C.:** between Toumodi and Nzaakro (Aug.) *Chev.* 22422 ! **G.C.:** Hatogodo,
near Quittah (Sept.) *Akpabla* 330 ! **N.Nig.:** Nupe *Barter* 1547 ! Extends to A.-E. Sudan, also in tropical
America.
2. **M. capitata** (*Linn.*) *Linn.* l.c. 458 (1767) ; F.T.A. 1 : 188. *Sida capitata* Linn. (1753). Like the last, with
roundish leaves and yellow or white flowers.
Sen.: Richard Tol *Perrottet* 64 ! A plant of the West Indies and central America.

7. THESPESIA Soland. ex Corr.—F.T.A. 1 : 209.

Leaves ovate, cordate at base, acuminate at apex, entire, with minute scales on both
surfaces, otherwise glabrous, 6–15 cm. long, 5–10 cm. broad ; flowers axillary,
solitary ; epicalyx of 3–5 oblong-lanceolate segments, early caducous ; calyx about
1 cm. long, scaly outside ; fruit depressed-globose, woody, finely scaly, about 2·5 cm.
diam. .. .. *populnea*

T. populnea (*Linn.*) *Soland. ex Corr.*—F.T.A. 1 : 209 ; Chev. Bot. 70 ; J. B. Hutch. in New Phytol. 46 : 135.
Hibiscus populneus Linn. (1753). A coastal tree or shrub ; petals yellow, large and showy ; generally
planted in the streets of the coastal towns.
Sen.: St. Louis *Brunner* ! *Farmar* 110 ! **G.C.:** Cape Coast *T. Vogel* ! Dixcove (Apr.) *Chipp* 179 !
Achimota (Jan.) *Irvine* 1946 ! **Togo:** Lome *Warnecke* 357 ! **Dah.:** Zagnanado *Chev.* 23087. **S.Nig.:**
Calabar (fr. Jan.) *Goldie* ! Widely distributed in the tropics. (See Appendix, p. 133.)

8. AZANZA Alef. in Bot. Zeit. 29 : 298 (1861) ; Exell & Hillcoat in Contrib. Conhec. Fl. Moçamb. 2 : 59 (1954).

Leaves palmately 3–5-lobed, cordate at base, lobes usually obtuse, more or less stellate-
pubescent above, densely stellate-pubescent to tomentose and reticulate beneath,
4–13 cm. long, 5–14 cm. broad ; flowers axillary, solitary, sometimes crowded towards
ends of shoots ; epicalyx of (5–) 9–15 linear segments, more or less caducous ; calyx
about 1 cm. long, densely stellate-tomentose or tomentellous outside ; fruits
tomentose .. .. *garckeana*

A. garckeana (*F. Hoffm.*) *Exell & Hillcoat* l.c. (1954). *Thespesia garckeana* F. Hoffm. Beitr. Fl. Centr. Ost-Afr.
12 (1889). Tree to 30 ft. high, or more, in savannah regions ; possibly introduced in our area ; flowers
large, yellow.
N.Nig.: Tula Dist., Bauchi Prov. *Kennedy* FHI 7276 ! Widespread in east and south tropical Africa.

Besides the above *A. lampas* (Cav.) Alef., a native of tropical Asia, has been introduced into gardens.

9. CIENFUEGOSIA Cav. Diss. 3 : 174 (1787) ; J. B. Hutchinson in New Phytol. 46 : 125 (1947). *Fugosia* Juss. (1789)—F.T.A. 1 : 208.

Leaves undivided or slightly lobulate, obovate, very shortly acuminate, 8–13 cm. long,
4–6·5 cm. broad, 5-nerved at the base, glabrous but gland-dotted beneath, finely
reticulate on both surfaces ; petiole very short ; stipules linear-subulate, about

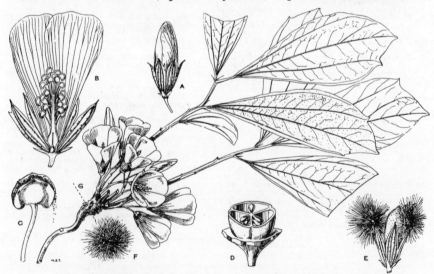

Fig. 124.—Cienfuegosia heteroclada *Sprague* (Malvaceae).

A, flower-bud. B, longitudinal section of flower. C, stamen. D, cross-section of ovary.
E, dehiscence of fruit. F, seed. G, level of soil.

5 mm. long ; flowers from near the base of the previous season's stem ; pedicels up
to 3 cm. long ; epicalyx of about 3 short unequal bracteoles ; calyx turbinate, 1·5 cm.
long, 5-lobed, lobes acuminate, nerved and gland-dotted ..   ..   1. *heteroclada*
Leaves digitately divided to the base, lobes 5–7, linear or oblanceolate, up to 4 cm.
long, glabrous ; petiole about ⅓ as long as the leaf ; stipules linear, slightly pubescent ; flowers on the young shoots, axillary, solitary ; pedicels longer than the leaves ;
epicalyx of 10–12 short linear bracteoles ; calyx 1·5 cm. long, lobes lanceolate,
acuminate, 3-nerved and gland-dotted   ..   .. 2. *digitata*

1. **C. heteroclada** *Sprague* in Kew Bull. 1907 : 48 ; J. B. Hutch. in New Phytol. 46 : 127. A suffrutex, in
savannah, sending up branches from a woody stock after burning ; stems angular, glabrous, dark gland-
dotted ; flowers pink or purplish, 1 in. or more long ; capsule obovoid, ¾ in. long, gland-punctate ; seeds
densely clothed with long rusty hairs.
  G.C.: Zuarungu Dist. (Jan.) *Vigne* FH 4665 ! **N.Nig.:** Kontagora (fl. & fr. Dec.–Jan.) *Dalz.* 122 ! *Meikle*
1034 ! Naraguta (Mar.) *Hill* 25 ! (See Appendix, p. 122.)
2. **C. digitata** *Cav.* Diss. 3 : 174, t. 72, 2 (1787) ; Chev. Bot. 69 ; J. B. Hutch. l.c. 128. *Fugosia digitata* (Cav.)
Pers.—F.T.A. 1 : 209. A low shrubby plant a foot or more high, with angular branches ; flowers yellow
with red-purple centre, 1–2 in. in diam.
  Sen.: *Sieber* 58 ! Richard Tol (Sept.) *Perrottet* 66 ! *Roger* ! Walo *Heudelot* ! **Fr.Sud.:** Amango (Sept.)
Hagerup 385a ! **G.C.:** Bawku (Nov.) *Vigne* FH 4664 ! **Fr.Nig.:** Fada to Koupéla, Gourma (July) *Chev.*
24535 ! **N.Nig.:** Kano (Mar., May) *Buxton* ! *Ejiofor* FHI 26150a ! Bauchi Plateau (July) *Lely* P. 563 !
Widespread in the drier parts of tropical Africa. (See Appendix, p. 122.)

10. HIBISCUS Linn.—F.T.A. 1 : 194.

*Calyx persistent, not splitting down one side, usually more or less deeply lobed :
 †Calyx more or less deeply lobed :
  Bracteoles of epicalyx united for more than half their length ; trees with broadly
  ovate or pentagonal, rather large, more or less cordate leaves :
   Leaves broadly ovate-rounded, deeply cordate, acuminate, up to 15 cm. long and
   broad, entire or crenulate, at length glabrous, canous-tomentellous beneath,
   digitately nerved from the base ; flowers solitary, large and showy   1. *tiliaceus*
   Leaves pentagonal, with triangular denticulate lobes, rather widely cordate, up to
   15 cm. broad, minutely stellate-puberulous beneath, 5–7-nerved at the base ;
   flowers comparatively small in cymes   ..   .. 2. *sterculiifolius*
  Bracteoles of epicalyx free, or united only at base, or absent ; shrubs, straggling
  climbers or herbs :
   Epicalyx absent ; annual herbs ; leaves very variable, simple to trifoliolate sometimes on the same plant, long-petiolate ; flowers solitary, axillary ; seeds strongly
   tuberculate :
    Calyx longer than the fruit ; leaf-margin entire or slightly dentate ; stems more or
    less pubescent   ..   .. 3. *sidiformis*
    Calyx shorter than the fruit ; leaf-margin distinctly dentate ; stems with long
    weak hairs ..   .. 4. *lobatus*
   Epicalyx present :
    Bracteoles of epicalyx forked or provided with a foliaceous appendage :
     Stipules foliaceous, reniform-orbicular, 1–1·5 cm. long ; stems prickly ; leaves
     digitately 3–5-lobed, lobes coarsely toothed, up to 7 cm. long, prickly on the
     nerves beneath and on the petioles ; flowers solitary ; bracteoles of the epicalyx
     nearly as long as the calyx, linear with a foliaceous appendage on the back
                                                              5. *surattensis*
     Stipules not foliaceous :
      Stems prickly ; flowers axillary ; bracteoles of epicalyx 8–12, shorter than the
      calyx :
       Pedicels up to 2 cm. long ; sepals ovate-lanceolate, long-acuminate ; leaves
       mostly deeply digitately 3–5-lobed, lobes coarsely toothed, up to 6·5 cm.
       long ; bracteoles of epicalyx with a foliaceous appendage but often without
       lateral lobes ; stems prickly and more or less shortly hirsute   .. 6. *noldeae*
       Pedicels 2–13 cm. long ; sepals ovate, acute ; leaves usually palmately 3–5-
       lobed, or triangular or ovate, up to about 17 cm. long and broad ; bracteoles
       of epicalyx forked ; stems prickly and more or less long-hirsute
                                                              7. *rostellatus*
      Stems not prickly ; flowers umbellate ; pedicels up to 2·5 cm. long ; bracteoles
      of epicalyx 5–6, entire, spathulate-oblanceolate, much longer than the calyx ;
      leaves 3–7-lobed, suborbicular, up to 18 cm. broad, crenate, softly pubescent
      beneath ..   .. 8. *comoensis*
    Bracteoles of epicalyx neither forked nor provided with a foliaceous appendage :
     Calyx inflated ; bracteoles of epicalyx 7–12, free, linear, ciliate ; upper leaves
     mostly deeply 3–5-lobed with the lobes variable in shape ; much-branched
     hispid annual   ..   .. 9. *trionum*
     Calyx not inflated :
      ‡Epicalyx of more than 5 bracteoles :
       Bracteoles of epicalyx spathulate, somewhat widened above the middle, nearly
       as long as the calyx ; leaves ovate to orbicular, often somewhat pentagonal,
       5–8 cm. long and broad, doubly crenate, densely and softly tomentose beneath ;

flowers usually solitary, shortly stalked ; fruit a little longer than the calyx,
very villous   ..   ..   10. *panduriformis*
Bracteoles of epicalyx narrowed to the apex :
  Epicalyx of subulate or filiform, more or less terete bracteoles :
    Leaves broad, 5-lobed or pentagonal, with long petioles ; flowers on slender
      pedicels :
      Capsule depressed, distinctly winged, setose on the wings ; leaves without
        incrustations at the base of the nerves beneath :
        Stems densely pubescent, with prickles ; leaves ovate to palmately 3- or
          slightly 5-lobed   ..   ..   11. *vitifolius* var. *vitifolius*
        Stems with small prickles ; leaves deeply lobed
                                            11a. *vitifolius* var. *ricinifolius*
      Capsule pointed, not winged, densely villous ; leaves long-petiolate, usually
        rather sharply 5-lobed, dentate, softly pubescent, with chalk-like incrusta-
        tions at the base of the nerves beneath ; bracteoles of epicalyx very
        slender :
        Epicalyx about ½ as long as the calyx :
          Leaves more or less 5-lobed ; indumentum of rather long and short hairs
                                                          12. *physaloides*
          Leaves obscurely 3-lobed or not lobed ; indumentum short and harsh
                                                          13. *pseudohirtus*
        Epicalyx about ¼ as long as the calyx   ..   ..   ..   14. *longisepalus*
    Leaves lanceolate to ovate, distinctly but rather shortly petiolate, serrate-
      dentate, 3–5 cm. long, 1–3·5 cm. broad, scabrid ; flowers solitary on slender
      pedicels ; bracteoles of the epicalyx small and inconspicuous
                                                          15. *micranthus*
  Epicalyx of linear to lanceolate, more or less flat bracteoles :
    Leaves not lobed, linear or lanceolate, subsessile or shortly petiolate, obtusely
      serrate :
      Leaves linear, 5–13 cm. long, softly pubescent beneath ; seeds covered with
        small scales ; bracteoles of epicalyx about 10, linear, shorter than the
        calyx ; calyx-lobes lanceolate, setose ; capsule about 1·5 cm. long,
        setose-pilose   ..   ..   16. *squamosus*
      Leaves lanceolate or linear-lanceolate, thinly pubescent beneath or glabrous ;
        seeds not scaly :
        Petals up to 1 cm. long ; calyx-lobes triangular, about 6 mm. long ;
          bracteoles of epicalyx about 7, lanceolate, about ⅓ as long as calyx ;
          capsule scarcely 1 cm. long ; leaves narrowly lanceolate, 5–6 cm. long,
          thinly stellate-pubescent beneath ; suffrutex with erect shoots up to
          15 cm. long   ..   ..   17. *gourmania*
        Petals at least 2 cm. long ; calyx-lobes lanceolate, 1 cm. long or more :
          Stems trailing or suberect, up to 60 cm. long, hispid with yellowish
            spreading hairs ; leaves linear-lanceolate towards the ends of shoots,
            but usually more or less lobed below   18. *articulatus* var. *articulatus*
          Stems erect, up to 2 m. high, not hispid, or hispid only below ;   leaves
            lanceolate, apparently never lobed   18a. *articulatus* var. *glabrescens*
    Leaves lobed or broadly ovate, usually with a fairly long petiole :
      Stems very hispid with yellowish spreading hairs ; leaves deeply 3-lobed, or
        scarcely lobed, 5–7 cm. long, lobes oblong-lanceolate, crenate, slightly
        setulose on the nerves beneath ; pedicels longer than the calyces ;
        bracteoles of epicalyx 8–10, pilose ; capsule about 1·5 cm. long, sparingly
        setose ; seeds nearly smooth   ..   ..   18. *articulatus* var. *articulatus*
      Stems not hispid, sometimes prickly ; flowers shortly pedicellate :
        Calyx fleshy at maturity ; stems not prickly ; leaves ovate or 3-lobed,
          glabrous or nearly so   ..   ..   19. *sabdariffa*
        Calyx not fleshy at maturity :
          Bracteoles of the epicalyx much shorter than the calyx in fruit ; petiole
            usually long, often as long as the lamina :
            Calyx woolly as well as setose ; leaves deeply lobed or broadly ovate,
              rather sharply serrate, glabrescent ; stems smooth or sometimes
              slightly prickly   ..   ..   20. *cannabinus*
            Calyx not woolly, sometimes setose, or puberulous with small stellate
              hairs ; leaves often deeply lobed, puberulous to tomentellous beneath ;
              stems often prickly or sometimes scabrid :
              Calyx densely setose, but only sparingly puberulous and later glab-
                rescent between the broad-based simple setae ; stems more or less
                prickly ; leaves puberulous beneath   ..   ..   ..   21. *asper*
              Calyx stellate-puberulous, with fascicles of longer hairs, but scarcely
                setose ; stems slightly scabrid and densely pubescent ; leaves
                tomentellous beneath   ..   ..   22. *scotellii*

Bracteoles of the epicalyx nearly as long as the calyx in fruit ; petioles short ; leaves ovate-triangular, irregularly dentate, slightly setulose
23. *whytei*

‡Epicalyx of 5 bracteoles :
Leaves linear-lanceolate, often hastate at the base, up to 18 cm. long and 4 cm. broad, serrulate, slightly pubescent beneath ; flowers crowded at the apex of the shoots, very shortly stalked ; bracteoles of epicalyx linear, about as long as the calyx, ciliate   ..   .. 24. *congestiflorus*
Leaves broadly ovate to more or less pentagonal, rounded or cordate at base ; flowers axillary, shortly pedicellate :
Bracteoles of epicalyx linear-subulate about 1 cm. long and scarcely 1·5 mm. broad, bristly pilose, shorter than the calyx ; calyx-lobes lanceolate, contracted towards the base ; corolla about 6 cm. long ; leaves ovate, 4–6·5 cm. long, 3·5–5 cm. broad, rather coarsely doubly dentate, stellate-hirsute beneath   ..   .. 25. *lonchosepalus*
Bracteoles of epicalyx linear to lanceolate, 1·8–2·8 cm. long, as long as or longer than the calyx :
Leaves permanently and softly stellate-tomentose beneath, ovate or obscurely 3-lobed ; fruits stellate-tomentose but not setose ; seeds long-pilose
26. *owariensis*
Leaves beneath stellate-hirsute, or glabrescent and pubescent mainly on the nerves ; fruits setose at first ; seeds shortly pubescent or papillose :
Bracteoles of epicalyx linear, about 1·5 mm. broad, not touching each other towards the base ; flowers crowded towards the apex of the stem ; leaves more or less glabrescent and pubescent mainly on the nerves beneath
27. *lunarifolius*
Bracteoles of epicalyx rather broadly lanceolate, 5–7 mm. broad and touching each other towards the base ; flowers not crowded ; leaves rather coarsely stellate-hirsute beneath   ..   ..   .. 28. *macranthus*
†Calyx cupuliform, about 7 mm. long, with very short lobes ; epicalyx cupuliform, 5-dentate ; corolla about 4·5 cm. long ; branchlets, petioles and midrib and nerves on underside of leaves hispid-pilose ; stipules filiform ; leaves obovate-oblong, rounded or obtuse at base, acuminate at apex, margin remotely dentate in upper part   ..   .. 29. *grewioides*
*Calyx spathaceous, splitting down one side and falling off at anthesis or in fruit :
Bracteoles of the epicalyx linear, 6 or more, not early caducous ; flowers large :
Pedicel as long as the fruit or longer ; fruits narrowly ovoid, narrowed to the base, acuminate at apex, more or less membranous, up to 8 cm. long ; stems covered with reflexed bristly hairs   ..   .. 30. *abelmoschus*
Pedicel much shorter than the fruit ; fruits elongate-conical, truncate at base, ribbed, becoming fibrous, up to 30 cm. long ; stems glabrous or shortly pubescent
31. *esculentus*
Bracteoles of the epicalyx ovate to lanceolate, 4–8, persistent, or caducous before anthesis :
Lobes of leaves obtuse, narrowed to their base ; bracteoles 5–6, lanceolate, early caducous ; upper flowers in terminal leafless racemes ; flowers small, about 3 cm. long ; fruits ovoid, up to about 4 cm. long ; stems pubescent or glabrescent, with minute thorns   ..   .. 32. *ficulneus*
Lobes of leaves acute (or leaves sometimes not lobed) ; bracteoles 4–8, ovate to ovate-lanceolate, persistent ; flowers large ; fruits ovoid, up to about 6 cm. long :
Stems, pedicels and fruits covered with long bristly hairs ; pedicels about 3 cm. long ; bracteoles usually 4   ..   ..   .. 33. *manihot* var. *manihot*
Stems and pedicels glabrous or glabrescent ; pedicels 1–1·5 cm. long ; bracteoles 6–8, long-ciliate on the margins, otherwise glabrous   .. 33a. *manihot* var. *caillei*

1. **H. tiliaceus** *Linn.*—F.T.A. 1 : 207 ; Chev. Bot. 69 (incl. var. *elatus* Hochr.). Epicalyx not enclosing the calyx in bud, with very short lobes much shorter than the tube. A small tree with purplish branches ; stipules leafy, leaving annular scars ; flowers 2–3 in. long, yellow with a purple centre. Common near the sea.
Sen.: *Perrottet* 48 ! *Heudelot* ! *Sieber* 34 ! Carabane (fr. Feb.) Chev. 2741 ! **Gam.:** (fl. & fr. July) *Ruxton* 3 ! **Port.G.:** Bissau (Feb.) *Esp. Santo* 1784 ! **S.L.:** *T. Vogel* 58 ! Rotomba Isl. (Mar.) *Kirk* 30 ! Lumley (fl. & fr. Feb.) *Dudgeon* 5a ! Bureh town (Nov.) *Lane-Poole* 82 ! **Lib.:** Kru, Fishtown (fr. Mar.) *Baldwin* 11612 ! **Iv.C.:** Bliéron *Chev.* 19915 ! **G.C.:** mouth of Ancobra R. (fl. & fr. July) *Chipp* 312 ! **S.Nig.:** Lagos (Feb.) *Dalz.* 1060 ! 1194 ! Nun R. (fl. & fr. Aug.) *T. Vogel* 9 ! **F.Po:** (fr. Dec.) *Barter* 1549 ! *Mann* 15 ! Throughout the tropics. (See Appendix, p. 130.)
2. **H. sterculiifolius** (*Guill. & Perr.*) *Steud.* Nom., ed. 2, 1 : 760 (1840) ; Chev. Bot. 67. *Paritium sterculii-folius* Guill. & Perr. Fl. Seneg. 60, t. 13 (Apr. 1831). *H. quinquelobus* G. Don (1831, after Aug.)—F.T.A. 1 : 208. *H. perrottetii* Steud.—Chev. Bot. 66. A small tree ; branches stellate-tomentose ; epicalyx cup-shaped with 10 teeth ½ as long as the calyx.
Sen.: *Farmar* ! Casamance (Apr.) *Perrottet* 49 ! **Gam.:** *Brown-Lester* 19 ! Albreda (Mar.) *Perrottet* 50 ! *Heudelot* 640 ! **Fr.Sud.:** Balani (Jan.) *Chev.* 146 ! Sarakouroba (fr. Feb.) *Chev.* 478 ! **Fr.G.:** *fide* Chev. l.c. **S.L.:** *Don* ! Heddle's Farm (Dec.) *Sc. Elliot* 3995 ! Kumrabai (Dec.–Jan.) *Thomas* 6875 ! Mt. Horton (Jan.) *T. S. Jones* 303 ! **Lib.:** Jabrocca (Dec.) *Baldwin* 10856 ! **S.Nig.:** Benin City *Dennett* 42 ! (See Appendix, p. 130.)

3. **H. sidiformis** *Baill.* in Bull. Soc. Linn. Paris 1 : 518 (1885). *Solandra ternata* Cav. (1788). *H. ternatus* (Cav.) Mast. in F.T.A. 1 : 206 (1868) ; Chev. Bot. 68 ; F.W.T.A., ed. 1, 1 : 267 ; not of Cav. (1787). *H. ambongoensis* Baill. (1885). *H. pamanzianus* Baill. (1885). *H. ternifoliolus* F. W. Andr. (1952). An erect, branching, pubescent herb, woody below ; flowers pale yellow less than 1 in. long.
Sen.: *Heudelot* 466 ! Richard Tol (Oct.) *Roger* 109 ! Walo (Oct.) *Perrottet* 72 ! Fr.Sud.: Sompi *Chev.* Amongo (fl. & fr. Sept.) *Hagerup* 378 ! Iv.C.: Yako to Ouahigouya (Aug.) *Chev.* 24734 ! N.Nig.: Sokoto (fl. & fr. June) *Dalz.* 421 ! Jebba *Barter* ! Yola (fl. & fr. June) *Dalz.* 81 ! Widespread in the drier parts of tropical Africa and in Madagascar.

4. **H. lobatus** (*Murr.*) *O. Ktze.* Rev. Gen. Pl. 3, 2 : 19 (1898). *Solandra lobata* Murr. (1785). *Hibiscus solandra* L'Hérit. (1788)—F.T.A. 1 : 206 ; Hochr. l.c. 128 (as var. *genuinus*). Erect, slightly villous, rather laxly branched herb, to 5 ft. high ; flowers cream.
Sen.: *Perrottet.* N.Nig.: Dogon Kurmi, Jemaa, Plateau Prov. (fl. & fr. Nov.) *Keay* FHI 21044 ! Widespread in tropical Africa, extending to Madagascar and India.

5. **H. surattensis** *Linn.*—F.T.A. 1 : 201 ; Chev. Bot. 68. Herbaceous, trailing or climbing ; stems with spreading hairs and recurved prickles ; flowers yellow with red centre.
Sen.: Khann and Kounoun (Sept.–Oct.) *Perrottet* 57 ! Fr.G.: Conakry and Kouria, *fide* Chev. *l.c.* S.L.: Kumrabai (fr. Dec.) *Thomas* 6803 ! York Pass Lane-Poole 437 ! Njala (Dec.) *Deighton* 1576 ! Mambolo (Jan.) *Deighton* 963 ! Lib.: Monrovia (fl. Nov., fr. Jan.) *Baldwin* 10990 ! *Linder* 1434 ! Kakatown *Whyte* ! Iv.C.: Bouroukrou *Chev.* 16613. G.C.: Atwabo *Fishlock* 21 ! Kumasi (Sept.) *Vigne* FH 1381 ! Togo: Lome *Warnecke* 243 ! Dah.: *Le Testu* 188 ! S.Nig.: Ikoyi, Lagos (Dec.) *Dalz.* 1192 ! Olokemeji *Foster* 138 ! Sapoba (Nov.) *Kennedy* 1639 ! *Meikle* 588 ! Oban *Talbot* 1410 ! Br.Cam.: Buea (Apr.) *Hutch. & Metcalfe* 103 ! Widespread in the tropics of the Old World. (See Appendix, p. 130.)

6. **H. noldeae** *Bak. f.* in J. Bot. 77 : 20 (1939). Shrubby perennial herb, sometimes scandent ; flowers yellow with deep red centre.
S.Nig.: Mt. Koloishe, 6,200 ft. (fl. & fr. Dec.) *Keay & Savory* FHI 25094 ! Br.Cam.: Bamenda, 5,000 ft. (fl. & fr. Feb.–Apr.) *Migeod* 463 ! *Ujor* FHI 30075 ! Also in Uganda, Tanganyika, Belgian Congo and Angola.

7. **H. rostellatus** *Guill. & Perr.*—F.T.A. 1 : 201 ; Chev. Bot. 67. *H. furcellatoides* Hochr. (1917)—Chev. Bot. 66. *H. furcatus* of F.W.T.A., ed. 1, 1 : 267, not of Willd. Undershrub, to 10 ft. high, or half-climbing, more or less hispid and with recurved prickles ; flowers large, yellow, with a dark purple centre.
Sen.: Kounoun (Sept.–Mar.) *Perrottet* 53 ! Gam.: *Ingram* ! *Dudgeon* ! Port.G.: Bissalanca, Bissau (Jan.) *Esp. Santo* 1663 ! Fr.G.: Koukouré to Timbo (Mar.) *Chev.* 12504 ! Nzérékoré (Oct.) *Baldwin* 9700 ! S.L.: Kambia (fl. & fr. Dec.) *Sc. Elliot* 4349 ! *Jordan* 755 ! Mano *Smythe* 60 ! Njala (Nov.–Feb.) *Deighton* 2436 ! 4947 ! Makump (Dec.) *Glanville* 109 ! Lib.: Kakatown *Whyte* ! Peahtah (Oct.) *Linder* 1098 ! Sanokwele (Sept.) *Baldwin* 9640 ! G.C.: Kintampo (Nov.) *Andoh* FH 5438 ! Dah.: Abbo to Massé (Feb.) *Chev.* 22969. N.Nig.: Zaria (Feb.) *Dalz.* 360 ! Vom *Dent Young* 18 ! S.Nig.: Lagos (Nov.) *Dalz.* 1059 ! 1195 ! Ibadan (Jan.) *Meikle* 959 ! Akilla (Jan.) *Onochie* FHI 20674 ! Onitsha (Feb.) *Jones* FHI 7475 ! Br.Cam.: Bulifambo, Buea (fr. Mar.) *Maitland* 577 ! Extends to central, east and north-east tropical Africa. (See Appendix, p. 129.)

8. **H. comoensis** *A. Chev. ex Hutch. & Dalz.* F.W.T.A., ed. 1, 1 : 267 (1928) ; Kew Bull. 1928 : 298 ; Chev. Bot. 65.
Iv.C.: Akabossué to Ebrinahoué, Mid. Comoé (Dec.) *Chev.* 22613 !

9. **H. trionum** *Linn.*—F.T.A. 1 : 196 ; Hochr. l.c. 144 ; Chev. Bot. 69. Much-branched hispid annual ; flowers yellow with a purple centre.
Sen.: Richard Tol *Berhaut* 1412. Fr.Sud.: Sompi (Aug.) *Chev.* ! Widespread in the tropics of the Old World.

10. **H. panduriformis** *Burm. f.*—F.T.A. 1 : 203 ; Hochr. in Ann. Conserv. Jard. Bot. Genèv. 4 : 95 ; Chev. Bot. 66. Erect branched undershrub, 4–8 ft., covered with fine down mixed with bristles ; flowers large, yellow turning orange, or purple-streaked, with a red or purple centre ; seeds striped.
Sen.: *Roger* 12 ! Walo (Sept.–Oct.) *Perrottet* 58 ! 60 ! G.C.: Shai Plains (Feb.) *Irvine* 1979 ! Burufo, Lawra (fl. & fr. Dec.) *Adams* GC 4426 ! Dah.: Abbo to Massé *Chev.* 22974 ! N.Nig.: Kontagora (fl. & fr. Nov.) *Dalz.* 131 ! Maigana, Zaria (Dec.) *Lamb* 68 ! Nabardo (fl. & fr. Sept.) *Lely* 615 ! Bauchi Plateau (fl. & fr. Aug.) *Lely* P. 658 ! Widespread in tropical Africa, also in Madagascar, tropical Asia and Australia. (See Appendix, p. 129.)

11. **H. vitifolius** *Linn.* var. **vitifolius**—F.T.A. 1 : 197, partly ; Hochr. l.c. 169 (as var. *genuinus*) ; Chev. Bot. 69. Low shrub, or scrambling herb, to 6 ft. high ; flowers bright yellow with a purple centre, about 2 in. long ; recognisable by the 5-winged fruit.
G.C.: Cape Coast (fl. & fr. July) *T. Vogel* 66 ! Aburi (Nov.) *Johnson* 847 ! Accra Plains (fl. & fr. Aug.) *Irvine* 70 ! Togo: Lome *Warnecke* 204 ! Dah.: Adjaonéré to Abbo (fl. & fr. Feb.) *Chev.* 22966 ! N.Nig.: Jebba *Barter* ! Katagum (fl. & fr. Sept.) *Dalz.* 62 ! S.Nig.: Abeokuta *Millen* 95 ! Olokemeji (fl. & fr. Sept.) *Onochie* FHI 20227 ! Ibadan (fl. & fr. Aug.) *Golding* 5 ! Widespread in tropical Africa, Madagascar, Asia and Australia, also in West Indies (? introduced).

11a. **H. vitifolius** var. **ricinifolius** (*E. Mey. ex Harv.*) *Hochr.* in Ann. Conserv. Jard. Bot. Genèv. 4 : 170 (1900) ; Brenan in Mem. N.Y. Bot. Gard. 8, 3 : 225. *H. ricinifolius* E. Mey. ex Harv. (1860).
Br.Cam.: Mopanya, Cam. Mt., 2,400 ft. (fr. Apr.) *Maitland* FHI 12332 ! Also in Uganda, Abyssinia, Eritrea, Nyasaland, Angola and Natal.

12. **H. physaloides** *Guill. & Perr.*—F.T.A. 1 : 199, partly ; Hochr. l.c. 161 (as var. *genuinus*) ; Chev. Bot. 67. A woody herb, to 6 ft. high, with spreading pungent hairs ; flowers up to 4 in. diam., yellow, with a red base turning violet.
Sen.: Dagana, Walo (Sept.–Oct.) *Perrottet* 61 ! 62 ! Gam.: Albreda *Perrottet.* Fr.Sud.: Kati to Kita (Oct.) *Chev.* 2745 ! Port.G.: Picle, Bissau *Esp. Santo* 1737 ! S.L.: Gene (Nov.) *Deighton* 446 ! Makump (Dec.) *Glanville* 110 ! Njala (Dec.) *Deighton* 1577 ! Lib.: Kakatown *Whyte* ! Karmadhun (Nov.) *Baldwin* 10187 ! Iv.C.: Ano (Dec.) *Chev.* 22534 ! Beoumi (Dec.) *Lowe* ! N.Nig.: Jebba (fl. & fr. Sept.) *Elliott* 211 ! S.Nig.: Lagos *Millen* 124 ! Akure F.R. (fl. & fr. Oct., Nov.) *Keay* FHI 25466 ! Usonigbe F.R. (Nov.) *A. F. Ross* 259 ! Sapele *Darker* ! Widespread in tropical Africa, and in Canary, Cape Verde and Comoro Islands, Madagascar and Seychelles.

13. **H. pseudohirtus** *Hochr.* in Bull. Soc. Bot. Fr. 61, Mém. 8 : 247 (1916) ; Chev. Bot. 67 ; Cufodontis in Ann. Naturhist. Mus. Wein 56 : 55. An undershrub with shortly pilose and scabrid branchlets ; flowers white, rather small, solitary, on long peduncles.
Fr.Sud.: Macina (fl. & fr. Sept.) *Chev.* 24866 !

14. **H. longisepalus** *Hochr.* l.c. 250 (1916) ; Chev. Bot. 66. An undershrub, tomentose and with spreading hairs ; flowers yellow, 1–1½ in. long ; epicalyx very small, setaceous.
Fr.Sud.: Macina (Sept.) *Chev.* 24842 !

15. **H. micranthus** *Linn. f.* Suppl. Pl. 308 (1781) ; F.T.A. 1 : 205 ; Hochr. in Ann. Conserv. Jard. Bot. Genèv. 5 : 60 ; Cufodontis l.c. 46 (q.v. for vars.). Undershrub, 5–6 ft. high, with stiff, straight, scabrid branches ; stipules almost spiny ; flowers small, about ½ in. long, white turning pink ; seeds cottony.
Sen.: *Heudelot* 539 ! Walo (Sept.–Nov.) *Perrottet.* Fr.Sud.: Gao (Sept.) *Hagerup* 365 ! G.C.: Cape Coast *Brass.* Accra (Mar., May) *Deighton* 587 ! *Irvine* 721 ! Togo: Lome *Warnecke* 218 ! 339 ! Widespread in the drier parts of tropical Africa, S. Africa, Madagascar, Arabia and India. (See Appendix, p. 129.)

16. **H. squamosus** *Hochr.* in Ann. Conserv. Jard. Bot. Genèv. 4 : 165 (1900). *H. epidospermus* Mast. in F.T.A. 1 : 197 (1868), not of Miq. (1859). Stem erect, little-branched, 3 ft. or more, scabrid-pubescent or downy ; flowers conspicuous, pale yellow with dark purple centre ; common in marshy places.
G.C.: *Imp. Inst.*! Tamale (Oct.) *Williams* 401! **N.Nig.:** Nupe *Barter* 1165! Kontagora (fl. & fr. Nov.) *Dalz.* 127! **S.Nig.:** Awba R., Ibadan (fl. & fr. Oct.) *Jones* FHI 6365!

17. **H. gourmania** *Hutch. & Dalz.* F.W.T.A., ed. 1, 1 : 267 (1928) ; Kew Bull. 1928 : 298. *Gourmania grewioides* A. Chev. Bot. 86, name only. Suffrutex, with stems up to 6 in. high, from short erect woody rhizomes ; flowers white.
Iv.C.: Ouagadougou to Ouahigouya (Aug.) *Chev.* 24717. **Fr.Nig.:** Fada to Koupéla, Gourma (July) *Chev.* 24514! **N.Nig.:** Ejidogari F.R., Ilorin (Feb.) *Ejiofor* FHI 19831!

18. **H. articulatus** *Hochst. ex A. Rich.* var. **articulatus**—F.T.A. 1 : 200 ; Hochr. l.c. 159 (as var. *genuinus*) ; Chev. Bot. 63. Stems trailing or suberect, to 2 ft. long, from woody rhizome ; leaves very variable in shape ; flowers white or yellow (? sometimes turning pink) ; in marshy ground.
Iv.C.: Toura country (May) *Chev.* 21746! **G.C.:** Nangodi (June) *Vigne* FH 4514! **Fr.Nig.:** Gourma (July) *Chev.* 24428! **N.Nig.:** Aguji, Ilorin (Dec.) *Thornton*! Nabardo (May) *Lely* 206! Kwoya Dist., Bornu (Aug.) *Daggash* FHI 24859! Widespread in tropical Africa.

18a. **H. articulatus** var. **glabrescens** *Hochr.* l.c. 159 (1900). Erect herb, to 6 ft. high ; leaves apparently not lobed ; flowers white or yellow or pink.
Togo: Misahöhe *Baumann* 48! **N.Nig.:** *Kentish Rankin*! Lema, Laflagi Dist., Ilorin (Aug.) *Onyeagocha* FHI 27266! Also in A.-E. Sudan.

19. **H. sabdariffa** *Linn.*—F.T.A. 1 : 204 ; Hochr. l.c. 116 ; Chev. Bot. 67 ; Berhaut Fl. Sen. 160. Erect, slightly branched with smooth or slightly hispid stem, often coloured ; epicalyx and calyx sometimes with scattered hairs and tubercles, becoming succulent ; flowers an inch or more long, yellow with a purple centre ; widely cultivated and variable in form and size.
Sen.: *fide* Berhaut *l.c.* **Fr.Sud.:** Timbuktu *Chev.* 1250! **Fr.G.:** *fide* Chev. *l.c.* **S.L.:** *Sc. Elliot* 4173! Njala (Nov.) *Deighton* 1849! Yonguru (Jan.) *Thomas* 7149! **G.C.:** Tamale *Dudgeon*! **Dah.:** *Burton*! **N.Nig.:** Katagum (Sept.) *Dalz.* 68! Kukawa (Feb.) *E. Vogel* 42! 52! **S.Nig.:** Olokemeji (Nov.) *Onochie* FHI 8156! Sapoba *Kennedy* 1654! Widely cultivated in the tropics. (See Appendix, p. 129.)

20. **H. cannabinus** *Linn.*—F.T.A. 1 : 204, partly ; Hochr. l.c. 114, partly ; Chev. Bot. 64–65 (excl. var. *chevalieri*). An erect annual, to 15 ft. high ; flowers large, yellow with a red-purple centre ; cultivated as a fibre plant, and naturalized.
Sen.: Richard Tol (Feb.) *Döllinger*! **Gam.:** MacCarthy Isl. (fl. & fr. Jan.) *Dalz.* 8129! **Fr.Sud.:** Sansanding (Sept.) *Chev.* 24967! Koulikoro (Sept.) *Chev.* 25026! Labbézenga (Sept.) *Hagerup* 458! And many other localities *fide* Chev. *l.c.* **S.L.:** *Deighton* 2935! **G.C.:** Kpong (fl. & fr. Oct.) *Johnson* 827! **Togo:** Lome *Warnecke* 282! **N.Nig.:** Sokoto *Dalz.* 424! 425! Katagum (fl. & fr. Sept.–Oct.) *Dalz.* 63! 64! Kalkala, L. Chad (fl. & fr. Mar.) *Gwynn* 93! **S.Nig.:** Lagos *Dodd* 442! Cultivated in most tropical countries. (See Appendix, p. 127.)

21. **H. asper** *Hook f.* in Fl. Nigrit. 228 (1849) ; Berhaut Fl. Sen. 160 (? incl. var.). *H. cannabinus* var. *chevalieri* Hochr.—Chev. Bot. 65. *H. cannabinus* of F.T.A. 1 : 204, partly ; Hochr. l.c., partly. Erect fibrous herb to 8 ft. high, often red-tinged in parts and armed with small spine-tipped tubercles ; flowers yellow with purple centre ; probably native in savannah areas.
Sen.: *fide* Berhaut *l.c.* **Gam.:** (July) *Brooks* 14! **Fr.Sud.:** Sarédina (May) *Davey* 130! Bure, Amongo (fl. & fr. Sept.) *Hagerup* 409! **Fr.G.:** Kouria (Oct.) *Caille* in *Hb. Chev.* 14909! **S.L.:** *Turner*! Waterloo (fl. & fr. Dec.) *Deighton* 3346! Kora, Scarcies R. (fr. Feb.) *Sc. Elliot* 4592! Mahela, Samu (fl. & fr. Dec.) *Sc. Elliot* 4041! Musaia (fl. & fr. Dec.) *Deighton* 4499! **G.C.:** Dukwesein (fl. & fr. Dec.) *Chipp* 599! Lumbunga (fl. & fr. Dec.) *Morton* GC 6225! **Dah.:** *Burton*! **N.Nig.:** Sokoto (fl. & fr. Oct.) *Dalz.* 426! Nupe *Barter* 1026! Zungeru (Oct.) *Dalz.* 128! Katagum (Sept.) *Dalz.* 65 (partly)! **S.Nig.:** *Foster* 362! *Higginson* 11! Probably widespread in tropical Africa ; perhaps no more than varietally distinct from *H. cannabinus* Linn. (See Appendix, p. 127.)

22. **H. scotellii** *Bak. f.* in J. Linn. Soc. 30 : 74 (1894) ; Hochr. l.c. 117. Erect fibrous herb, to 6 ft. high; flowers yellow with a dark red or purplish centre ; often on rocky hills ; probably native.
S.L.: Sasseni, Scarcies R. *Sc. Elliot* 4535! Lungi (fl. & fr. Nov.) *Glanville* 111! Waterloo (fl. & fr. Dec.) *Deighton* 3347! Tisana, Bonthe Isl. (fl. & fr. Nov.) *Deighton* 2318! **Lib.:** Vonjama (Oct.) *Baldwin* 9887! Kolahun (fl. & fr. Nov.) *Baldwin* 10148! **G.C.:** Kwahu-Pepease (fl. & fr. Nov.) *Irvine* 1713! **N.Nig.:** Anara F.R. (fl. & fr. Sept., Oct.) *Keay* FHI 20117! 24354! Bauchi Plateau (Sept.) *Lely* P. 686! Panshanu (fl. & fr. Sept.) *Lely* 609! **S.Nig.:** Idanre Hills (fl. & fr. Oct.) *Keay* FHI 22587! [The specimens from Liberia, Gold Coast, and Nigeria differ somewhat from the Sierra Leone material and are rather intermediate between typical *H. scotellii* and typical *H. asper*.] (See Appendix, p. 130.)

23. **H. whytei** *Stapf* in J. Linn. Soc. 37 : 79 (1905) ; Chev. Bot. 69. A tall woody herb, sparingly stellate-pubescent ; leaves 1½–3½ in. long ; bracteoles of epicalyx 10, narrow linear-lanceolate ; corolla yellow with purple centre, about 1½ in. long.
Fr.G.: Kindia *Chev.* 13031; 13224. Kouria to Ymbo *Chev.* 14750. **Lib.:** Kakatown *Whyte*! **Iv.C.:** Bouroukrou *Chev.* 16614 ; 16787. Bingerville *Chev.* 17355.

24. **H. congestiflorus** *Hochr.* in Ann. Conserv. Jard. Bot. Genèv. 10 : 21 (1906) ; Chev. Bot. 65. *H. baumannii* Ulbr. (1922). Erect herb, to 10 ft. high, often unbranched ; stems pilose ; flowers yellow with purple centre, 1½–2½ in. long ; in savannah.
G.C.: *Irvine*! Kintampo (Nov.) *Andoh* FH 5440! **Togo:** Kpedsu (fl. & fr. Dec.) *Howes* 1052! **Dah.:** Massé to Kétou (Feb.) *Chev.* 23006! **N.Nig.:** Lokoja (Oct.) *Dalz.* 80! Bauchi Plateau (Oct.) *Lely* P. 831! Yola (Oct.) *Shaw* 70! **S.Nig.:** Eruwa Road, Olokemeji (Nov.) *Onochie* FHI 8163! Oyo (fl. & fr. Jan.) *Keay* FHI 25646! Owo *G. Burton* 3! Also in Ubangi-Shari.

25. **H. lonchosepalus** *Hochr.* in Bull. Soc. Bot. Fr. 61, Mém. 8 : 251 (1917) ; Chev. Bot. 66. A tall undershrub ; flowers probably yellow with purple centre.
Dah.: Agouagon (May) *Chev.* 23540!

26. **H. owariensis** *P. Beauv.* Fl. Oware 2 : 88, t. 117 (1820). *H. calycinus* of F.T.A. 1 : 202, partly. *H. calyphyllus* of Chev. Bot. 64. An undershrub, to 6 ft. high ; flowers large, yellow with red-purple centre.
Iv.C.: Kouroukoro to Touna (May) *Chev.* 21790! **G.C.:** Afaben (fl. & fr. Aug.) *Johnson* 110! Akwapim Hills (fl. & fr. Aug.) *Johnson* 772! Aburi (July) *Deighton* 3384! **Togo:** Lome *Warnecke* 222! **S.Nig.:** *Foster* 365! Lagos Colony (fl. & fr. Aug.) *Rowland*! Ijaiye F.R. (fr. Apr.) *Onochie* FHI 21956! Warri *Beauvois*. Also in A.-E. Sudan.

27. **H. lunarifolius** *Willd.*—F.T.A. 1 : 202, partly. A tall undershrub, stellate-hirsute becoming glabrous; flowers large, yellow with purple centre.
S.L.: *Deighton* 5712! Kofiu Mt. (fr. Jan.) *Sc. Elliot* 4612! **Lib.:** St. Paul R. *Reynolds*! **S.Nig.:** Lagos Colony *Foster* 358! Calabar *Maitland* 13! An Indian plant found also in parts of tropical Africa. (See Appendix, p. 129.)

28. **H. macranthus** *Hochst. ex A. Rich.* Fl. Abyss. 1 : 55 (1847). *H. lunarifolius* of F.T.A. 1 : 202, partly. Shrub, or woody herb, to 6 ft. high or more ; flowers large, yellow with deep purple centre.
Br.Cam.: Bamenda, 5,000 ft. (fl. & fr. Jan.) *Migeod* 402! Also in Abyssinia and Eritrea.

29. **H. grewioides** *Bak. f.* in Cat. Talb. 9 (1913). A small tree, up to 16 ft. high ; flowers yellow.
S.Nig.: near Owum, Oban *Talbot* 1343!

30. **H. abelmoschus** *Linn.*—F.T.A. 1 : 207 ; Hochr. in Ann. Conserv. Jard. Bot. Genèv. 4 : 150 ; Chev. Bot. 63 ; Rev. Bot. Appliq. 20 : 325. To 6 ft. high ; leaves 3–5-lobed ; flowers about 4 in. diam., yellow, with crimson centre ; seeds musk-scented ; an introduced plant, found near habitation.

Sen.: *Roger* ! R. Senegal (Sept.) *Perrottet* 51 ! **Fr.G.:** Farana circle *Chev.* 20400. **S.L.:** Luseniya, Samu (fr. Dec.) *Sc. Elliot* 4087 ! Futa, Masakoi (fl. & fr. Sept.) *Deighton* 2240 ! Ronietta (Nov.) *Thomas* 5380 ! **Lib.:** Grand Bassa (July) *T. Vogel* 26 ! Sinoe Basin *Whyte* ! Jabroke, Webo (July) *Baldwin* 6685 ! **Iv.C.:** Baoulé-Nord circle *Chev.* 22307. **G.C.:** Aburi *Brown* 784 ! Ancobra Junction (Jan.) *Irvine* 1072 ! **Dah.:** Savalou *Chev.* 23690. **S.Nig.:** Lagos (Aug.) *Dodd* 423 ! Onitsha Prov. *Thomas* 607 ! Widely distributed in the tropics. (See Appendix, p. 126.)

31. **H. esculentus** *Linn.*—F.T.A. 1 : 207 ; Hochr. l.c. 150 ; Chev. Bot. 65 ; Rev. Bot. Appliq. 20 : 323 (incl. vars.). ? *H. hispidissimus* A. Chev. l.c. 326, t. 10 (1940). A cultivated, or subspontaneous herb, with long-stalked, usually 5-lobed leaves ; flowers large, yellow, with purple or red centre.
    Sen., Fr.Sud. & Dah.: *fide* Chev. *l.c.* **S.L.:** Mayoso (Aug.) *Thomas* 1421 ! **N.Nig.:** *Yates* 42 ! **S.Nig.:** Lagos *Millen* 5 ! Widespread in the tropics. (See Appendix, p. 128.)

32. **H. ficulneus** *Linn.* Sp. Pl. 2 : 695 (1753) ; Hochr. l.c. 153 ; Chev. in Rev. Bot. Appliq. 20 : 320. A fibrous herb, becoming woody below ; flowers about 1 in. diam., pink or white, with rosy centre.
    Fr.Nig.: *fide* Chev. *l.c.* **N.Nig.:** *Gosling* ! Mud flats near Dikoa, L. Chad *Lamb* 111 ! Also in A.-E. Sudan, Madagascar, Asia and Australia ; possibly introduced and naturalized in Africa, but stated by Lamb to be wild.

33. **H. manihot** *Linn.* var. **manihot**—Sp. Pl. 2 : 696 (1753) ; Hochr. l.c. 153 ; Chev. in Rev. Bot. Appliq. 20 : 322. *H. abelmoschus* of F.W.T.A., ed. 1, 1 : 268, partly, not of Linn. To 6 ft. high ; leaves 3–5-lobed ; flowers 4–9 in. diam., pale yellow (or white), with purple centre ; an introduced plant, found near habitation.
    S.L.: Likuru, Tamisso (fr. Feb.) *Sc. Elliot* 4954 ! Mabonto (Oct.) *Thomas* 3578 ! **Br.Cam.:** Sofo, Buea (Dec.) *Maitland* 142 ! Widespread and native in Asia.

33a. **H. manihot** var. **caillei** *A. Chev.* in Rev. Bot. Appliq. 20 : 322, t. 8 (1940). Similar in appearance to *H. esculentus*, but with large bracteoles.
    Fr.G.: Kouria (Oct.) *Caille* in *Hb. Chev.* 14910. **S.L.:** Bumbuna (Oct.) *Thomas* 3341 ! **Lib.:** Kakatown *Whyte* !

Imperfectly known species.

H. clandestinus *Cav.* Icon. 1 : 1, t. 2 (1791) ; Cufodontis l.c. 35 (1948). Based on a plant from Senegal cultivated by Cavanilles ; flowers scarcely 2 mm. long. Usually regarded as a synonym of *H. micranthus* Linn. f., but maintained by Cufodontis.

H. (sect. *Abelmoschus*) sp. **A.** To 8 ft. high, in cultivated ground ; flowers in a long leafless raceme.
    G.C.: Kumasi *Cummins* 200 !

Besides the above, several species are cultivated, including *H. mutabilis* Linn., *H. rosa-sinensis* Linn. and *H. schizopetalus* (Mast.) Hook. f.

11. GOSSYPIUM Linn.[1]—F.T.A. 1 : 210 ; J. B. Hutchinson, Silow & Stephens, The Evolution of *Gossypium* (1947) ; see also Watt, Wild & Cultivated Cotton Pl. (1907) ; Watt in Kew Bull. 1926 : 193–210, and op. cit. 1927 : 321–356 ; Wouters in Publ. I.N.É.A.C., Sér. Sci. 34 (1948) ; Roberty in Candollea 9 : 19–103 (1942), and op. cit. 13 : 9–165 (1950). (See Appendix, p. 122.)

Bracteoles linear-oblong or narrowly triangular, not more than 4 mm. broad, with up to 4 teeth, or entire ; capsules with hairs on the sutures ; seeds with a single coat of fine brown hairs, but without long lint-hairs (" cotton-wool ") .. 1. *anomalum*
Bracteoles suborbicular or broadly triangular, more than 1 cm. broad ; capsules without hairs on the sutures ; seeds with long lint-hairs (" cotton-wool "), with or without a layer of fuzz-hairs below :
  Bracteoles entire or dentate, the teeth usually less than 3 times longer than broad :
    Bracteoles closely investing the flower, longer than broad, entire or with 3–4 teeth near the apex ; capsules tapering .. .. 2. *arboreum*
    Bracteoles flaring widely from the flower, usually broader than long, the upper margin usually with 6–8 teeth ; capsules rounded, or with prominent shoulders ; plants covered with a dense even coat of long hairs
                                           3. *herbaceum* var. *acerifolium*
  Bracteoles deeply laciniate, the laciniae usually more than 3 times as long as broad :
    Staminal column short ; anthers loosely arranged, the filaments longer in the upper part of the column than below ; capsule surface smooth, sometimes with oil-glands conspicuous beneath the surface .. .. 4. *hirsutum*
    Staminal column long ; anthers compactly arranged, on short subequal filaments ; capsule surface usually coarsely pitted with black oil-glands in the pits :
      Capsules usually less than 6 cm. long, broadest near the base ; seeds free
                                           5. *barbadense* var. *barbadense*
      Capsules usually more than 6 cm. long, broadest near the middle and tapering to the base ; seeds connate .. .. .. 5a. *barbadense* var. *brasiliense*

1. **G. anomalum** *Wawra* in Sitzungsber. Math.-Nat. Akad. Wiss. Wien 38 : 561 (1860) ; F.T.A. 1 : 211 ; Chev. in Rev. Bot. Appliq. 13 : 190–194 ; J. B. Hutch. in J. B. Hutch. et al. Evol. Gossyp. 28, t. 4 (q.v. for syn.).
    Fr.Sud. & Fr.Nig.: *fide* Chev. *l.c.* Extends to A.-E. Sudan, Eritrea, Somaliland, S.W. Africa and Angola.
2. **G. arboreum** *Linn.*—F.T.A. 1 : 211 ; J. B. Hutch. l.c. 32, t. 4 (q.v. for syn. ; specimens from our area are referred to the race *soudanense*). *G. simpsonii* and *G. africanum* of F.W.T.A., ed. 1, 1 : 269. Perennial shrubs, or annual subshrubs ; flowers red or purple, or yellow.
    G.C.: Gambaga *Imp. Inst.* 48487 ! **Togo:** Kpedsu (Jan.) *Howes* 1099 ! **Dah.:** Zagnanado (Feb.) *Chev.* 23069 ! **N.Nig.:** Nupe *Barter* 1182 ! Kazaure (fl. & fr. Feb.) *Imp. Inst.* H 1258/1913. Ilorin (Nov.) *Lamb* 3 ! *Thornton* ! Bassa (June) *Elliott* 249 ! Kano (Jan.) *Andrew* ! *Lamb* 8 ! Geidam (Dec.) *Elliott*

---

[1] It is impossible to give a full account of the taxonomy of *Gossypium* in a Flora of this type. The treatment here is based on that given by J. B. Hutchinson in The Evolution of *Gossypium* (1947). This work and the other references given above should therefore be carefully consulted when identifying a cotton plant. Numerous cultivated forms and hybrids occur.

135! **S.Nig.:** Lagos *Rowland*! Abeokuta *Irving*! *Millen*! Oba *Foster* 188! Widespread in tropical and subtropical Africa, Madagascar and Asia.

3. **G. herbaceum** *Linn.* var. **acerifolium** (*Guill. & Perr.*) *A. Chev.* in Compt. Rend. Acad. Sci. 208 : 25 (1939) ; Rev. Bot. Appliq. 19 : 540 ; J. B. Hutch. l.c. 36, t. 5 (q.v. for syn.). *G. punctatum* var. *acerifolium* Guill. & Perr. (1831). *G. obtusifolium* var. *wightianum* (Tod.) Watt—F.W.T.A., ed. 1, 1 : 269. A hairy shrub, to 6 ft. high ; flowers yellow usually purple-blotched in centre.
Sen.: *fide* Guill. & Perr. *l.c.* **Dah.:** Dassa-Zoumé (fl. & fr. May) *Chev.* 23614! **N.Nig.:** Nupe *Barter*! Kano Prov. *Imp. Inst.* 46323! Ilorin *Thornton*! **S.Nig.:** Abeokuta (Feb.) *Irving*! Extends to India and Nyasaland.

4. **G. hirsutum** *Linn.* Sp. Pl., ed. 2, 2 : 975 (1763) ; J. B. Hutch. l.c. 40, 6–7 (q.v. for syn.), including var. *punctatum* (Schum. & Thonn.) J. B. Hutch. *G. mexicanum* Tod. and *G. purpurascens* Poir.—F.W.T.A., ed. 1, 1 : 269. Cultivated and subspontaneous.
Fr.Sud.: Goundam (Aug.) *Chev.* 2732! **Gam.:** *Dudgeon*! Georgetown (Jan.) *Dalz.* 8130! **S.L.:** *Barter*! Moyamba (fl. & fr. Feb.) *Dudgeon* 7! **G.C.:** Winnebah (Feb.) *Dalz.* 8283! Anum (Feb.) *Johnson*! Akuse (fl. & fr. Apr.) *Johnson* 1073! Labolabo *Dudgeon* 136! **Togo:** Kpedsu *Howes* 1100! **Dah.:** Dassa-Zoumè (May) *Chev.* 23617! **N.Nig.:** Nupe *Barter* 1184! Kukawa (Feb.) *E. Vogel* 35! **S.Nig.:** Badagry *Higginson*! Wild and cultivated in America, introduced in many parts of the world.

5. **G. barbadense** *Linn.* var. **barbadense**—Sp. Pl. 2 : 693 (1753) ; J. B. Hutch. l.c. 48, t. 8 (q.v. for syn.). *G. peruvianum* Cav. and *G. vitifolium* Lam.—F.W.T.A., ed. 1, 1 : 269.
Sen.: *Roger*! **S.L.:** Moyamba (Feb.) *Dudgeon* 8! **Lib.:** Kakatown *Whyte*! **G.C.:** Kumasi *Cummins* 137! Anum (Dec.) *Johnson* 1! Sittam, E. Akim (fr. Mar.) *Dudgeon* 1! **Togo:** Lome *Warnecke* 18! Kpedsu (Jan.) *Howes* 1095! **N.Nig.:** Ankpa (Feb.) *Lamb* 17! Mawpa, Kabba (Nov.) *Lamb* 6! **S.Nig.:** Abeokuta (Feb.) *Irving*! Ikure *Holland* 250! Native in tropical S. America and widely cultivated in the tropics.

5a. **G. barbadense** var. **brasiliense** (*Macfadyen*) *J. B. Hutch.* l.c. 50 (1947). *G. brasiliense* Macfadyen (1837)—F.W.T.A., ed. 1, 1 : 270.
S.Nig.: Lagos Colony *Millen* 28! Abeokuta *Irving* 102! **F.Po:** *T. Vogel* 20! Native and cultivated in tropical S. America, now distributed throughout central America and the Antilles, sporadic in Africa and India.

12. KOSTELETZKYA C. Presl—F.T.A. 1 : 193.

Flowers densely clustered at the ends of the shoots, forming a panicle of spike-like racemes ; fruits enclosed by the calyx, 5-angled, tomentose all over and bristly pilose mainly on the angles ; epicalyx of 8–10 filiform pilose bracteoles as long as the calyx ; leaves suborbicular to ovate, widely cordate at base, slightly 3-lobed, up to about 12 cm. long and broad, stellate pilose, doubly dentate   ..   .. 1. *grantii*
Flowers solitary on well-developed pedicels, or in lax axillary fascicles or cymes ; fruits not enclosed by the calyx :
Leaves linear to linear-lanceolate, up to 12 cm. long and 1·3 cm. broad, the lower ones sometimes hastate at the base, serrulated, scabrid on both surfaces ; stems scabrid ; flowers solitary ; pedicels up to 4 cm. long ; epicalyx of 7–10 linear bracteoles a little shorter than the calyx ; fruits depressed, pentagonal, bristly pilose all over
2. *buettneri*
Leaves ovate, pentagonal or 3 (–5)-lobed :
Fruits with recurved bristles on the very prominent, almost winged angles, the body tomentose and about 7 mm. diam. ; flowers in lax fascicles or cymes ; epicalyx of about 7 linear pubescent bracteoles, shorter than the calyx ; leaves pentagonal, crenate about 7 cm. diam., shortly stellate-pubescent above, stellate-pubescent and with longer stellate hairs beneath, somewhat glabrescent ; stems stellate-pubescent
3. *stellata*
Fruits with spreading, not recurved bristles, mainly on the angles, the body stellate-pubescent and about 5 mm. diam. ; flowers in lax axillary fascicles ; epicalyx of about 7 linear-filiform long-pilose bracteoles, about as long as the calyx, or longer ; leaves ovate to 3 (–5)-lobed, dentate, up to about 10 cm. long and 8 cm. broad, more or less densely hirsute on both surfaces ; stems more or less densely hirsute or somewhat glabrescent   ..   .. 4. *adoensis*

1. **K. grantii** (*Mast.*) *Garcke* in Linnaea 43 : 53 (1881). *Hibiscus grantii* Mast. in F.T.A. 1 : 203 (1868). *K. chevalieri* Hochr. (1906)—Chev. Bot. 63 ; F.W.T.A., ed. 1, 1 : 270. Erect coarse herb, to 6 ft. high, densely strigoso-hispid ; flowers pale pink.
Port.G.: Piche to Canquelifa (Oct.) *Esp. Santo* 3107! **Fr.G.:** Timbo to Ditinn (Sept.) *Chev.* 18504! Soumbalako to Boulivel (Sept.) *Chev.* 18661! **S.L.:** Musaia (Oct.–Dec.) *Thomas* 2609! *Deighton* 4452! Warantamba (Oct.) *Small* 338! **N.Nig.:** Kontagora (Oct.) *Dalz.* 130! Extends to A.-E. Sudan, Uganda, Belgian Congo and Angola. (See Appendix, p. 131.)

2. **K. buettneri** *Gürke* in Verh. Bot. Verein. Brand. 31 : 92 (1889). *K. flava* Bak. f. (1894)—F.W.T.A., ed. 1, 1 : 270. *K. augusti* Hochr. (1917)—Chev. Bot. 63 (incl. vars.). *Hibiscus rhodanthus* of F.W.T.A., ed. 1, 1 : 267, not of Gürke. An erect, scabrid fibrous perennial, 2–3 ft. high ; flowers white with pink veins, turning yellow, over 1 in. diam.
Gam.: Kuntaur *Ruxton* 94! **Fr.G.:** *Pobéguin*! Farana *Chev.* 20465. **S.L.:** Erimakuna (Mar.) *Sc. Elliot* 5256! Rotenbana (fl. & fr. Jan.) *Jordan* 188! **Iv.C.:** Dialakora to Kénégoué *Chev.* 21970. **N.Nig.:** Anara F.R., Zaria (May) *Keay* FHI 19196! Panyam (fl. & fr. July) *Lely* P. 416! Extends to A.-E. Sudan, Angola, S.W. Africa and Rhodesia.

3. **K. stellata** *Hutch. & Dalz.* F.W.T.A., ed. 1, 1 : 270 (1928) ; Kew Bull. 1928 : 209. *Hibiscus solandra* of Chev. Bot. 67, not of L'Hérit. Erect, or scrambling, fibrous perennial, up to 14 ft. high ; flowers pink, about ¾ in. diam.
G.C.: Anum (Nov.) *Johnson* 816! **Dah.:** Abbo (Feb.) *Chev.* 22965! **S.Nig.:** Iddo Isl. (Dec.) *Millen* 44! Ebute Metta *Millen* 53! Akilla, Ijebu Ode Prov. (fl. & fr. Jan.) *Onochie & Ladipo* FHI 20676!

4. **K. adoensis** (*Hochst. ex A. Rich*) *Mast.* in F.T.A. 1 : 194 (1868). *Hibiscus adoensis* Hochst. ex A. Rich. (1847). Prostrate or scrambling ; flowers pink or white, with reddish or purplish centre.
Br.Cam.: Cam. Mt., 3,000–6,000 ft. (fl. & fr. Nov.–Apr.) *Dunlap* 35! 55! *Maitland* 224! 821! 1238! Bamenda (fl. & fr. Sept.) *Ujor* FHI 30208! Also in A.-E. Sudan, Abyssinia, Kenya, Uganda, Tanganyika, Nyasaland, N. Rhodesia, Belgian Congo, Angola and Madagascar.

13. MALVASTRUM A. Gray—F.T.A. 1 : 177.

Woody herb ; branches with loose long appressed branched hairs ; leaves with slender petioles, ovate, coarsely serrate, more or less rounded at the base, 3–5 cm. long, 2–2·5 cm. broad, laxly pilose on the upper surface with simple hairs, hairs branched beneath ; flowers single or crowded, shortly stalked ; epicalyx of 3 narrow bracteoles ; calyx-lobes broadly lanceolate, acute ; carpels 8–12, with 3 sharp points at the apex and on the sides, laxly pilose. ..   .. *coromandelianum*

M. coromandelianum (*Linn.*) *Garcke* in Bonplandia 5 : 297 (1857) ; Chev. Bot. 56. *Malva coromandeliana* Linn. (1753). *Malvastrum tricuspidatum* A. Gray (1852)—F.W.T.A., ed. 1, 1 : 270. Erect, 2–3 ft. ; flowers pale yellow.
G.C.: W. of Axim (Apr.) *Morton* GC 6649 ! Obrachera (Nov.) *Morton* GC 7890 ! Kumasi (Sept.) *Darko* 594 ! Accra (Mar.) *Deighton* 564 ! **Dah.**: Adjara, Porto Novo (Jan.) *Chev.* 22751. **N.Nig.**: Aguji, Ilorin *Thornton* ! **S.Nig.**: Lagos *Rowland* ! Ibadan (Jan.–Apr., Aug.) *Golding* 3 ! 18 ! *Meikle* 935 ! 1145 ! 1455 ! Native of America, widely distributed in the tropics.

78. MALPIGHIACEAE

Trees, shrubs or climbers, usually clothed with appressed medifixed hairs. Leaves opposite or rarely alternate, simple ; a pair of glands often present at the base of the blade ; stipules present or absent, sometimes connate. Flowers mostly hermaphrodite, actinomorphic. Sepals 5, imbricate or rarely valvate, often biglandular outside. Petals 5, clawed. Disk small. Stamens 10, hypogynous ; filaments free or connate at the base ; anthers 2-celled, opening lengthwise. Carpels usually 3, free or connate into a 3-celled ovary ; cells 1-ovuled ; styles usually separate and spreading. Fruiting carpels mostly winged or carpels connate into a fleshy or woody drupe. Seeds without endosperm and with straight, curved or uncinate, rarely circinate embryo.

A distinct and mainly tropical family, most highly represented in tropical America ; the medifixed hairs, glandular calyx-lobes and samaroid fruit are almost constant features of the few represented in our area.

Leaves alternate :
Samara more or less oblanceolate or obovate, the wing arising from the upper part of the fruit body ..   .. 1. **Acridocarpus**
Samara elliptic-suborbicular, the wing surrounding the outer side of the fruit body
2. **Rhinopterys**
Leaves opposite :
Sepals closed over the petals in bud ..   .. 3. **Flabellaria**
Sepals open in bud, the petals conspicuous :
Wing of samara very small, the latter nut-like   ..   ..   .. 4. **Stigmaphyllon**
Wing of samara large :
Samara broadly elliptic ; calycine glands large and fleshy   .. 5. **Heteropteris**
Samara shield-like, orbicular ; calycine glands absent   ..   .. 6. **Triaspis**

Besides the above indigenous genera, *Thryallis glauca* (Cav.) O. Ktze. (syn. *Galphimia glauca* Cav.), a native of Central America, is commonly cultivated as a hedge plant. Species of *Malpighia* Linn. are also cultivated in our area.

1. ACRIDOCARPUS Guill. & Perr.—F.T.A. 1 : 277.

*Bracts and bracteoles erect, usually small and subulate (rarely short and ovate), up to 3 mm. long ; inflorescence not or only a little branched ; mature leaves glabrous beneath, or the midrib sometimes shortly pubescent :
Bracts and bracteoles subulate, without glands :
Leaves more or less oblong, rounded or very often obtuse at base, distinctly acuminate ; samara at least 2·5 cm. long :
Flowers racemose, numerous, floriferous part of axis at least 5 cm. long ; calyx with a single large gland below the lobes and opposite the outer petal ; leaves more or less oblong, 12–27 cm. long, 7–12 cm. broad, lateral nerves looped and very prominent beneath, coarsely reticulate ; samara 3–5 cm. long, 1·2–2 cm. broad
1. *longifolius*
Flowers subcorymbose, few, floriferous part of axis up to 5 cm. long ; calyx with 2 glands ; leaves narrowly oblong, 8–12 cm. long, 3·5–5 cm. broad, rather finely reticulate beneath ; samara up to 3·5 cm. long and 1·5 cm. broad
2. *alternifolius*
Leaves oblanceolate, or oblong-oblanceolate, cuneate at base, rounded and sometimes very shortly cuspidate at apex, 8–13 cm. long, 2–5 cm. broad ; flowers in long racemes ; calyx with 2–3 glands ; samara up to 2·8 cm. long ..   3. *chevalieri*

Fig. 125.—Acridocarpus smeathmannii (*DC*.) *Guill. & Perr.* (Malpighiaceae).

A, base of leaf showing glands.　B, flower-bud.　C, flower.　D, petal.　E, stamen from front.　F, stamen from back.　G, pistil.
H, cross-section of ovary.　I, fruit.

W. E. T.

Bracts and bracteoles glandular ; leaves more or less cuneate at base ; flowers in long
racemes or subpaniculate ; samara oblanceolate about 5 cm. long and 1·5 cm. broad :
Leaves oblong or narrowly obovate-oblanceolate, 6–12 cm. long, 3–5 cm. broad,
lateral nerves about 6 pairs, much branched ; pedicels up to about 1·5 cm. long
                                                                          4. *smeathmannii*
Leaves broadly obovate, about 18 cm. long and 8 cm. broad, with about 9 pairs of
strong lateral nerves and few tertiary nerves ; pedicels up to about 8 mm. long
                                                                          5. *staudtii*
*Bracts and bracteoles soon becoming reflexed, spathulate-oblanceolate to broadly
lanceolate or ovate, 4–6 mm. long, persistent ; inflorescence much-branched, with
curved spreading branches ; leaves more or less obovate to oblanceolate-obovate,
cuneate, rounded or subcordate at base, rounded and often apiculate at apex :
Sepals 3–3·5 mm. long ; anthers acute at apex, not 2-horned ; samara obliquely
oblanceolate, up to 7 cm. long and 2 cm. broad ; leaves 8–30 cm. long, 4–14 cm.
broad, glabrous or pubescent beneath   ..   ..   6. *plagiopterus*
Sepals 6–6·5 mm. long ; anthers with two divergent sterile horns at the apex ; samara
up to 6 cm. long and 2·2 cm. broad ; leaves 10–30 cm. long, 5–17 cm. broad, more
or less pilose beneath ..   ..   7. *macrocalyx*

1. **A. longifolius** (*G. Don*) *Hook. f.* in Fl. Nigrit. 244 (1849) ; Niedenzu in Engl. Pflanzenr. Malpighiac. 271
(incl. vars.) *Anomalopteris longifolia* G. Don (1831). *Acridocarpus smeathmannii* var. β, Oliv. F.T.A. 1 :
278. A glabrous scandent shrub ; flowers yellow ; fruits reddish ; often growing near water.
    Lib.: Fishtown *Dinklage* 2040. Monrovia (June) *Baldwin* 5873 ! *Dinklage* 2218. Dimei (fl. & fr. Apr.)
Barker 1260 ! **S.Nig.:** Lagos (Feb., July) *Barter* 2196 ! *Millen* 145 ! *Dalz.* 1262 ! Nun R. (fl. & fr.
Aug.) *Mann* 467 ! Brass *Barter* 3701 ! Eket *Talbot* 3096 ! **Br.Cam.:** Victoria *Winkler* 1228. **F.Po:**
(Dec.) *T. Vogel* 125 ! 195 ! Also in French Cameroons, Gabon, Angola, Principe and S. Tomé.
2. **A. alternifolius** (*Schum. & Thonn.*) *Niedenzu* in Arb. Bot. Inst. Ak. Braunsberg 6 : 53 (1915) ; Engl.
Pflanzenr. Malpighiac. 269. *Malpighia alternifolia* Schum. & Thonn. (1827). *Acridocarpus corymbosus*
Hook. f.—F.T.A. 1 : 278 ; Chev. Bot. 95. A glabrous scrambling shrub, like the preceding ; sometimes
on the sea coast.
    S.L.: *Afzelius.* Freetown (June) *T. Vogel* 177 ! Kumrabai (fr. Dec.) *Thomas* 7084 ! Njala (fl. Mar., fr.
Apr.) *Deighton* 1709 ! 1832 ! **Iv.C.:** Dabou (Feb.) *Chev.* 17219. **G.C.:** Kratté *Thonning.* Cape Coast
(Sept.) *Brass* ! *T. Vogel* 12 ! Dixcove (Apr.) *Chipp* 177 ! Abokobi (Oct.) *Vigne* FH4263 ! Aburi Scarp (Oct.)
Scholes 300 ! **N.Nig.:** Jebba (Dec.) *Barter* ! *Hagerup* 746 *Meikle* 845 ! Nupe *Barter* 507 ! Lokoja (Oct.)
Dalz. 12 !
3. **A. chevalieri** *Sprague* in Journ. de Bot. 22 : 24 (1909). *A. smeathmannii* of Niedenzu in Engl. Pflanzenr.
Malpighiac. 274, partly. A climbing shrub with yellowish, or rusty, pubescent branchlets and inflorescence;
flowers and fruits yellow.
    Fr.Sud.: Koulikoro (Oct.) *Chev.* 3144 ! **G.C.:** Aboma F.R., Ashanti (fl. & fr. Nov.) *Vigne* FH 3437 !
4. **A. smeathmannii** (*DC.*) *Guill. & Perr.*—F.T.A. 1 : 277, partly (var. α) ; Niedenzu l.c. 274, partly (excl. syn.
A. chevalieri Sprague ; excl. var. *staudtii* Engl.). *Heteropteris smeathmannii* DC. (1824). Large scandent
shrub with abundant yellow flowers and reddish fruits.
    Port.G.: Suzana to S. Domingos (fl. & fr. Mar.) *Esp. Santo* 2246 ! **S.L.:** *Smeathmann* ! Freetown (Jan.)
Johnston 48 ! *Dalz.* 8086 ! Kent *Smythe* 247 ! Layah, Scarcies (Jan.) *Sc. Elliot* 4661 ! Kenema *Thomas*
7551 ! Musaia (fl. & fr. Feb.) *Deighton* 4211 ! **Lib.:** Dobli Isl., St. Paul R. *Bequaert* 26 ! Gbanga (Dec.)
Baldwin 10523 ! Péahtah (Dec.) *Baldwin* 10614 ! **G.C.:** Koforidua (Dec.) *Johnson* 497 ! Achimota (fl. &
fr. Feb.) *Irvine* 143 ! Sampa *Vigne* FH 3479 ! **Togo:** Lome *Warnecke* 401 ! **N.Nig.:** Abinsi *Dalz.* ! **S.Nig.:**
Lagos (Oct.) *Barter* 2137 ! 2217 ! *Rowland* ! *Dalz.* 1264 ! Ibadan (fl. & fr. Jan., Feb.) *Meikle* 991 ! 1158 !
Idanre (fl. & fr. Jan.) *Brenan* 8678 ! Newi (Jan.) *Kitson* ! Also in French Cameroons. (See Appendix,
p. 134.)
5. **A. staudtii** (*Engl.*) *Engl. ex Hutch. & Dalz.* F.W.T.A., ed. 1, 1 : 271 (1928). *A. smeathmannii* var. *staudtii*
Engl. Bot. Jahrb. 36 : 251 (1905) ; Niedenzu l.c. 275.
    Br.Cam.: Johann-Albrechtshöhe, by the lake (fl. & fr. Dec.–Jan.) *Staudt* 498 !
6. **A. plagiopterus** *Guill. & Perr.*—F.T.A. 1 : 278 ; Chev. Bot. 95 ; Niedenzu l.c. 276. *A. hirundo* S. Moore in
J. Bot. 1880 : 1 ; F.W.T.A., ed. 1, 1 : 273 ; Niedenzu l.c. 277. Scandent shrub, up to 20 ft. high, with
rusty puberulous branchlets, yellow or orange flowers, and reddish fruits.
    Sen.: *Perrottet* 94 ! **Port.G.:** Canchungo (Apr.) *Esp. Santo* 1236 ! Bissau (fl. & fr. Feb.) *Esp. Santo* 1758 !
Fr.G.: *Heudelot* 761 ! Labé (fr. Apr.) *Chev.* 12375 ! Timbo (Nov.) *Chev.* 14767 ! **S.L.:** *Don* ! Leicester
Peak (fl. & fr. Dec.) *Sc. Elliot* 3870 ! Regent (fr. Jan.) *Dalz.* 8085 ! York Pass (Dec.) *Dawe* 414 ! Njala
(Oct.) *Deighton* 4912 ! **Lib.:** *Carder* ! Taniné (Sept.) *Linder* 824 ! Vonjama (fl. & fr. Feb.) *Bequaert* 84 !
Mt. Wolagwisa, 4,500 ft. (fr. Mar.) *Bequaert* 101 ! (See Appendix, p. 134.)
7. **A. macrocalyx** *Engl.* Bot. Jahrb. 36 : 250 (1905) ; Niedenzu l.c. 277. A large forest liane.
    Togo: Kabo River F.R. (fl. Nov., fr. Dec.) *St. C. Thompson* FH 3672. **S.Nig.:** R. Jamieson, Sapoba (Sept.)
Kennedy 1764 ! 1963 ! Oban *Talbot* ! **Br.Cam.:** Mamfe (Mar.) *Onochie* FHI 30895 ! Also in French
Cameroons.

Imperfectly known species.

A. sp. Similar to *A. longifolius* but calyx with 3 glands instead of 1. More material required.
    S.Nig.: Utanga, Obudu Div. (fl. & fr. Dec.) *Keay & Savory* FHI 25147 !

2. RHINOPTERYS Niedenzu in E. & P. Pflanzenfam. 3, 4 : 352 (1897) ; Engl. Pflanzenr. Malpighiac. 279.

Sepals obovate to oblanceolate-obovate :
Branchlets pubescent ; leaves alternate, sessile, oblong-elliptic, or elliptic, rounded at
base, rounded or very slightly acuminate at apex, 8–15 cm. long, 4–8 cm. broad, very
coarsely reticulate and thinly pubescent beneath, with about 8–9 pairs of lateral
nerves ; racemes stout, about 30 cm. long, densely flowered ; pedicels 2 cm. long,
pubescent ; petals nearly 1·5 cm. long ; samara elliptic-orbicular, the wing all
round the side of the fruit-body, slightly pubescent, 4 cm. long ..   1. *spectabilis*
Branchlets nearly glabrous ; leaves alternate, subsessile, narrowly oblong or oblong-
oblanceolate, subacute at base, triangular-acuminate, 10–17 cm. long, 2·5–4·5 cm.
broad, coarsely reticulate and slightly pubescent beneath, with about 10 pairs of

much-branched lateral nerves ; racemes about 20 cm. long, thinly pubescent ; pedicels 2 cm. long, pubescent ; samara orbicular-elliptic, the wing all round one side of the fruit body, slightly pubescent, about 6 cm. long   ..   2. *angustifolia*

Sepals ovate ; branchlets shortly pilose ; leaves oblong, acute, 13 cm. long, 3–5·5 cm. broad, densely reticulate ; racemes lax-flowered, densely pilose ; petals 1·2 cm. long ; samara semicordate, 6 cm. long and 3·5 cm. broad..   ..   3. *kerstingii*

1. **R. spectabilis** *Niedenzu* in. E. & P. Pflanzenfam. 3, 4 : 352 (1897) ; Engl. Pflanzenr. Malpighiac. 280. *Acridocarpus hemicyclopterus* Sprague (1906). Suffrutex, or shrub.
    Gam.: *Brown-Lester* 47 ! 50 ! *Ozanne* 5 ! Genieri (Feb.) *Fox* 41 !
2. **R. angustifolia** *Sprague* in Kew Bull. 1922 : 194. *R. spectabilis* subsp. *angustifolia* (Sprague) Niedenzu l.c. 281. A small tree in savannah, with red-brown branchlets and white flowers.
    G.C.: Lawra, N.T. (fr. Feb.) *Macleod* 815 !
3. **R. kerstingii** (*Engl.*) *Niedenzu* in Arb. Bot. Inst. Ak. Braunsberg 7 : 20 (1921) ; Engl. Pflanzenr. Malpighiac. 280. *Acridocarpus kerstingii* Engl. Bot. Jahrb. 43 : 383 (1909). Suffrutex, in savannah.
    Fr.G.: *Pobéguin* 2143 ! **Togo:** Bangeli (fr. Mar.) *Kersting* 507.

Imperfectly known species.

R. sp. A straggling plant in savannah.
    G.C.: Gambaga (fr. June) *Akpabla* 680.

More thorough field work is required to determine just how far the above species are really distinct.

3. FLABELLARIA Cav.—F.T.A. 1 : 282.

Climbing shrub ; branchlets appressed-pubescent ; leaves opposite, ovate or ovate-orbicular, obtuse or subacute, 10–14 cm. long, up to 10 cm. broad, glabrous above, appressed-silky-tomentose beneath ; lateral nerves about 4–5 pairs ; petiole 1·5–2 cm. long, tomentose ; panicles terminal, lax, many-flowered ; bracts oblanceolate ; pedicels 8–10 mm. long ; petals closed in bud, 4 mm. long ; petals entire, glabrous ; wings of fruit together suborbicular, about 3–6 cm. diam.   ..   *paniculata*

F. paniculata *Cav.*—F.T.A. 1 : 282 ; Niedenzu in Engl. Pflanzenr. Malpighiac. 38, t. 10. A climbing shrub with grey-silky branchlets and leaves, and white to light pink flowers. **Sen.:** Messira *Berhaut* 1792. **Port.G.:** Formosa (Apr.) *Esp. Santo* 1964 ! **Fr.G.:** Rio Nunez *Heudelot* 848 ! Kouroussa (fl. & fr. Mar.) *Pobéguin* 672 ! Diaguissa to Timbo (Apr.) *Chev.* 13441 ! **S.L.:** *Afzelius* ! Erimakuna (Mar.) *Sc. Elliot* 5728 ! Limba (fr. Apr.) *Sc. Elliot* 5640 ! Kent (May) *Deighton* 5519 ! Musaia (fr. Apr.) *Deighton* 5431 ! **Lib.:** Sinoe *Whyte* ! *Baldwin* 11558 ! **Iv.C.:** Abégo *Jolly* 285 ! Mt. Goula, Danané (fr. Apr.) *Chev.* 21209 ! **G.C.:** Akropong (fl. & fr. Mar.) *Johnson* 904 ! Achimota (fr. Dec.) *Irvine* 1868 ! Dodowah (Feb.) *Irvine* 1974 ! **Togo:** Kekedjandi *Kersting* 621. **N.Nig.:** Nupe *Barter* 1337 ! **S.Nig.:** Lagos *Foster* 62 ! 63 ! Agbemia *Barter* 3422 ! Ibadan (Apr.) *Meikle* 1414 ! Degema *Talbot* 3722 ! Calabar *Thomson* 99 ! Ikwette, Obudu (Dec.) *Keay & Savory* FHI 25170 ! **Br.Cam.:** Mbo *Ledermann* 1521. **F.Po:** *T. Vogel* 99 ! Also in French Cameroons, Gabon, Belgian Congo, A.-E. Sudan, Uganda, Tanganyika and Angola. (See Appendix, p. 134.)

4. STIGMAPHYLLON A. Juss. in St.-Hil. Fl. Bras. Mer. 3 : 48 (1832). *Brachypterys* A. Juss. (1837)—F.W.T.A., ed. 1, 1 : 273.

Branchlets finely appressed-pubescent ; leaves opposite, lanceolate, obtusely pointed, shortly cuneate at base, about 7 cm. long and 2·5 cm. broad, minutely appressed-pubescent beneath ; lateral nerves about 6 pairs, slender and inconspicuous ; flowers few, at the ends of short lateral branchlets ; pedicels about 2 cm. long, glabrous ; sepals with 2 large contiguous glands in the lower part ; petals with a long claw, suborbicular, about 1·5 cm. long, crenulate ; stamens 10 ; styles 3, distinct ; samaras 2–3 together, nut-like, with a very small wing   ..   ..   ..   *ovatum*

S. ovatum (*Cav.*) *Niedenzu* in Ind. lect. Lyc. Braunsberg. p. aest. 1900 : 31 (1900) ; Engl. Pflanzenr. Malpighiac. 515. *Banisteria ovata* Cav. (1790). *Brachypterys ovata* (Cav.) Small (1910)—F.W.T.A., ed. 1, 1 : 273. A soft-wooded straggling climber in mangrove swamps and among rocks on sandy sea shores ; flowers yellow. **Fr.G.:** Rio Nunez *Whitfield* ! **S.L.:** Mahela (Nov., Dec.) *Sc. Elliot* 4089 ! *T. S. Jones* 49 ! Rokupr (fl. & fr. July) *Jordan* 54 ! Tombo (May) *Deighton* 2768 ! Kent (fr. Nov.) *Tindall* 21 ! Bonthe Isl. (Nov.) *Deighton* 2273 ! Also in Central and East Tropical America.

5. HETEROPTERIS Kunth Nov. Gen. Sp. Pl. 5 : 163 (1822).

Branchlets lenticellate ; leaves opposite, oblong-elliptic, shortly acuminate, rounded at base, 12–20 cm. long, 5–8 cm. broad, thinly pubescent beneath, at length glabrous but densely papillose ; lateral nerves about 10 pairs, much branched ; petiole 1 cm. long, thicker at the base ; panicles terminal, about 10 cm. long, rusty-tomentose ; bracts ovate, 4 mm. long ; calycine glands very large and fleshy ; stamens 10 ; samaras 1–3, broadly obliquely elliptic, about 2·5 cm. long, tomentellous   *leona*

H. leona (*Cav.*) *Exell* Cat. S. Tomé 123 (1944). *Banisteria leona* Cav. (1790), partly—F.W.T.A., ed. 1, 1 : 274. *Heteropteris africana* A. Juss. (1843)—F.T.A. 1 : 276 ; Chev. Bot. 95 ; Niedenzu in Engl. Pflanzenr. Malpighiac. 375, incl. vars. A climbing shrub with coriaceous leaves, yellow flowers and red-winged fruits ; chiefly in coastal districts. **Port.G.:** Catió (fl. & fr. July) *Esp. Santo* 2138 ! **Fr.G.:** Rio Nunez *Heudelot* 892 ! **S.L.:** *Afzelius* ! *Don* ! *T. Vogel* 62 ! Kent *Smythe* 245 ! Goderich (Apr.) *Deighton* 4752 ! Njala (fl. May–July, fr. Aug.–Sept.) *Dawe* 564 ! *Deighton* 1753 ! **Lib.:** Grand Bassa (fr. July) *T. Vogel* 117 ! Monrovia (fl. & fr. Apr.) *Barker* 1282 ! Sinoe *Whyte* ! *Baldwin* 11556 ! **Iv.C.:** Assinié *Chev.* 17863. Bliéron (fl. & fr. Mar.) *Chev.* 19893 ; 19918. **G.C.:** Princes, W. Prov. (Jan.) *Akpabla* 798 ! **S.Nig.:** Lagos (Aug., Sept.) *Dalz.* 927 ! 1263 ! Akilla (Apr.) *Jones & Onochie* FHI 17413 ! Nun R. (fl. & fr. Aug.,-Sept.) *Mann* 462 ! Brass *Barter* 29 ! Oban *Talbot* 1271 ! **Br.Cam.:** Victoria *Preuss* 1332 ! Tiko (fl. & fr. Apr.) *Maitland* 1148 ! Barombi *Preuss* 282 ! Mamfe *Mansfeld* 19. Also in French Cameroons, Gabon, Belgian Congo, Angola, Principe and S. Tomé. (See Appendix, p. 134.)

6. TRIASPIS Burchell—F.T.A. 1 : 280.

Stipules persistent, foliaceous, reniform, 1–1·5 cm. long, silky below ; leaves opposite,
ovate to ovate-lanceolate, acuminate, rounded at base, 8–15 cm. long, 5–9 cm.
broad, thinly long-appressed-pilose beneath ; petiole about 1 cm. long ; panicles
rather lax-flowered, with large leafy bracts ; pedicels about 1·5 cm. long ; sepals
elliptic, pubescent, 3 mm. long ; petals glabrous ; wings of fruit together orbicular,
about 5–6 cm. diam., glabrous .. .. 1. *stipulata*
Stipules small and deciduous ; leaves ovate-elliptic, very shortly acuminate, cordate
at base, 7–8 cm. long, 4·5–5·5 cm. broad, densely appressed-villous beneath ; petiole
1·5 cm. long, tomentose ; panicles rather small and densely flowered ; pedicels
about 1 cm. long, tomentose ; sepals oblong, tomentose ; petals glabrous ; wings of
fruit together suborbicular, cordate at base, about 2·5 cm. diam., glabrous
$\qquad$ 2. *odorata*

1. **T. stipulata** *Oliv.* F.T.A. 1 : 281 (1868) ; Niedenzu in Engl. Pflanzenr. Malpighiac. 49. A woody climber
with brownish-pilose branchlets and inflorescence ; flowers in abundant panicles ; calyx pinkish brown,
petals white to yellow, and pink.
Togo: Misahöhe (Nov.) *Mildbr.* 7323 ! **S.Nig.:** Lagos *Rowland* ! *Millen* 71 ! Abeokuta *Irving* 99 !
Usonigbe F.R. (Nov.) *Keay* FHI 25583 ! Awgu (Nov.) *Keay* FHI 21532 ! **Br.Cam.:** Victoria (fl. & fr.
Dec.–Feb.) *Maitland* 911 ! *Dalz.* 8173 ! Also in French Cameroons.
2. **T. odorata** (*Willd.*) *A. Juss.*—F.T.A. 1 : 280 ; Niedenzu l.c. 45. *Hiraea odorata* Willd. (1799). *T. aurea*
Niedenzu (1915)—l.c. 45. A climbing shrub with rusty-tomentose branchlets and pink or reddish flowers.
S.L.: *Afzelius.* **G.C.:** *Isert.* Ada *Thonning.* Cape Coast *Brass* ! Shai Plains (fl. & fr. Jan.) *Johnson* 575 !
Akropong (Oct.) *Johnson* 800 ! Mansu *Cummins* 9 ! Kumasi (fl. & fr. Dec.) *Vigne* FH 1778 ! Aboma
F.R. (Nov.) *Vigne* FH 3436 ! **Togo:** Lome (fl. & fr. Nov.) *Warnecke* 251 ! *Mildbr.* 7511 ! Kpandu
Baumann 328.

79. HUMIRIACEAE

Trees or shrubs. Leaves alternate, simple ; stipules absent. Flowers
hermaphrodite, actinomorphic. Sepals 5, imbricate, shortly or wholly connate.
Petals 5, free, soon falling away, slightly imbricate-contorted. Stamens 10 or
more, hypogynous, more or less connate in the lower part ; anthers versatile,
2–4-celled, opening lengthwise. Disk annular and often toothed, or of separate
glands, surrounding the base of the ovary. Ovary free, sessile, 5–7-celled ;
style simple ; ovules 1–3, pendulous from the apex, anatropous. Fruit a drupe
with rather thin fleshy pericarp and hard endocarp, sometimes with numerous
resin-filled cavities. Seeds 1–2 in each cell ; embryo straight in the middle of
copious endosperm ; cotyledons short.

A small family confined to tropical America and Africa. Recognized by the exstipulate
leaves, stamens 10 or more, partially connate, style undivided.

Fig. 126.—Sacoglottis gabonensis (*Baill.*) *Urb.* (Humiriaceae).

A, flower-bud. B, longitudinal section of flower. C, D, stamen front and back. E, cross-section
of ovary. F, fruit.

SACOGLOTTIS Mart. Nov. Gen. & Sp. 2 : 146 (1827). *Aubrya* Baill.—F.T.A. 1 : 275.

Trees ; leaves leathery, glabrous, toothed. Flowers in axillary cymes. Ovary 5-celled ; style simple ; ovules solitary. Fruit with a thick woody pericarp with resin-filled cavities.

A large tree, 25 m. or more ; branchlets with the bark splitting lengthwise ; leaves oblong-elliptic, shortly cuneate at base, obtusely acuminate, 10–15 cm. long, 3–6 cm. broad, coriaceous, crenulate, glabrous, reticulate on both surfaces, with rather numerous much-branched lateral nerves ; petiole about 0·5 cm. long, swollen at the base ; cymes axillary, many-flowered, about ½ as long as the leaves ; pedicels very short, stout ; calyx pubescent, 2 mm. long, with broad, almost truncate lobes ; petals slightly imbricate, shortly pubescent except on the margin ; fruit hard, ellipsoid, 3–4 cm. long, 2·5 cm. diam., with numerous unequal resin-filled cavities

gabonensis

S. gabonensis (*Baill.*) *Urb.* in Mart. Fl. Bras. 12, 2 : 449 (1877) ; Aubrév. Fl. For. C. Iv. 1 : 306, t. 123. *Aubrya gabonensis* Baill.—F.T.A. 1 : 275. *A. occidentalis* Chev. Bot. 94, name only. A forest tree, with red-brown shaggy bark ; flowers yellow.
S.L.: Gola Forest (Jan.) *Unwin & Smythe* 55 ! **Lib.:** Dukwia R. (fr. Aug.) *Linder* 219 ! *Cooper* 274 ! Duport (Nov.) *Linder* 1469 ! **Iv.C.:** Dabou (Feb.) *Chev.* 16195 ! Abidjan *Aubrév.* 92 ! **G.C.:** Essiama *Irvine* 2321. Simpa *Vigne* FH 2800 ! **S.Nig.:** Lagos *Moloney* 30 ! Oni *Sankey* 8 ! Sapoba (Nov.) *Keay* FHI 28077 ! Brass *Barter* 68 ! Oban *Talbot* 1744 ! **F.Po:** *Mann* 1417 ! Also in French Cameroons, Gabon and Angola. (See Appendix, p, 134.)

80. IXONANTHACEAE

Trees and shrubs. Leaves alternate, simple, entire or toothed, pinnately nerved ; stipules very small, caducous. Flowers hermaphrodite, in axillary fascicles or cymes. Sepals 5, shortly connate, imbricate. Petals 5, free, contorted, persistent and indurated in fruit. Stamens 5–20, inserted on an annular disk or shortly united at base, hypogynous or subperigynous ; filaments otherwise free ; anthers 2-celled, short. Ovary 5–3-celled ; ovules 2 in each cell, pendulous from the central axis ; style simple or shortly 5–3-fid. Fruit a capsule, woody or coriaceous, septicidal, the carpels often spuriously septate. Seeds arillate, with fleshy endosperm.

OCHTHOCOSMUS Benth. in Hook. Lond. Journ. Bot. 2 : 366 (1843). *Phyllocosmus* Klotzsch—F.T.A. 1 : 272.

Leaves acuminate, with black glandular teeth on the margin of the acumen, oblong lanceolate or obovate, acute at base, 8–11 cm. long, 2–4 cm. broad, shining above ; lateral nerves few, ascending, with numerous obscure fine parallel tertiary nerves ; racemes of fascicles clustered, axillary, slender, up to 7 cm. long ; rhachis puberulous ; pedicels slender, fasciculate, 2–3 mm. long ..   ..   1. *africanus*
Leaves rounded at the apex, shortly crenate-dentate nearly to the base, obovate-oblanceolate, acute or obtuse at base, 8–15 cm. long, 2·5–6 cm. broad, rigidly coriaceous, dull or slightly shining, with about 10 pairs of lateral nerves rather obscure beneath and the tertiary nerves not evident ; petiole swollen, transversely split ; panicles clustered near the ends of the shoots, up to 15 cm. long ; rhachis and pedicels glabrous, the latter 5 mm. long ; flowers white, turning pink   ..   2. *chippii*

1. **O. africanus** *Hook. f.* in Fl. Nigrit. 240, t. 23 (1849) ; Aubrév. Fl. For. C. Iv. 1 : 302. *Phyllocosmus africanus* (Hook. f.) Klotzsch—F.T.A. 1 : 273 ; Stapf 583 ; Chev. Bot. 94. A small tree or shrub, with sarmentose branches, to 50 ft. high ; flowers very abundant, small, white or yellowish, fragrant ; capsules 1–2 seeded, 5–6 mm. long.
   Port.G. : Mansoa (Oct.) *Esp. Santo* 2319 ! **Fr.G.:** Conrakry (Dec.) *Chev.* 25687. **S.L.:** *Don* ! Heddle's Farm (Mar.) *Lane-Poole* 360 ! Kukuna (Jan.) *Sc. Elliot* 4665 ! Roruks (Sept.) *Deighton* 4844 ! And many other localities. **Lib.:** Grand Bassa (Aug.) *Dinklage* 2034 ! Gbanga (Sept.–Oct.) *Linder* 541 ! 1166 ! **Iv.C.:** Abidjan (Aug.) *Aubrév.* 58 ! **G.C.:** Axim (Sept.) *Vigne* FH 4790 ! Otrokpe (Sept.) *Vigne* FH 4005 ! **S.Nig.:** Sapoba *Kennedy* 1843 ! 2384 ! Obunike, Onitsha (Apr.) *Chesters* 168 ! **Br.Cam.:** R. Mente, Wum (fr. June) *Ujor* FHI 29259 ! Also in French Cameroons and Congo. (See Appendix, p. 38.)
2. **O. chippii** *Sprague & Hutch. ex Hutch. & Dalz.* F.W.T.A., ed. 1, 1 : 134 (1927) ; Kew Bull. 1928 : 29. *Phyllocosmus sessiliflorus* of Chev. Bot. 94, not of Oliv. A shrub about 10 ft. high, flowers white turning pink.
   G.C.: Axim (Apr.) *Chipp* 424 ! Princestown (Apr.) *Chipp* 171 ! Kickham to Essiama *Fishlock* 15 ! 41 ! Beyin (Nov.) *Vigne* 1451 !

81. ERYTHROXYLACEAE

Trees, shrubs, or undershrubs. Leaves alternate, rarely opposite, simple, entire ; stipules intrapetiolar, rarely extrapetiolar, often caducous. Flowers fasciculate, hermaphrodite, rarely subdioecious, hypogynous, actinomorphic.

Calyx persistent, campanulate, lobes 5, imbricate. Petals 5, free, deciduous, imbricate, mostly ligulate on the inside. Stamens 10, in 2 series, more or less connate at the base ; anthers ellipsoid, 2-celled, opening lengthwise. Ovary of 3 carpels, 3-celled, mostly two of the cells sterile, fertile cells 1–2-ovuled ; ovules pendulous ; styles 3, free or more or less connate ; stigmas oblique, depressed-capitate or clavate. Fruit drupaceous ; seeds with or without endosperm ; embryo straight.

A small family sparsely represented in Africa. Stipules mostly intrapetiolar ; petals with a scale on the inside.

ERYTHROXYLUM Browne—F.T.A. 1 : 273.

Shrubs or small trees. Leaves alternate, entire ; stipules intrapetiolar. Flowers small, pedicellate, in axillary fascicles. Sepals usually 5. Petals as many, imbricate, with a scale on the inner side. Stamens 10. Ovary 3–4-celled ; styles 3 or 4. Fruit a drupe.

Leaves oblanceolate to narrowly obovate, slightly narrowed to the very obtuse emarginate apex, 3–8 cm. long, 2–3 cm. broad, shining above and reticulate on both surfaces ; stipules triangular-ovate, keeled, 4 mm. long ; flowers small, 1–4 in clusters on stout pedicels about 5 mm. long ; sepals triangular, obtuse   1. *emarginatum*
Leaves obovate, rounded at the apex, 3–6 cm. long, 1·5–2·5 cm. broad, slightly shining above and laxly reticulate on both surfaces ; stipules curved inside the petiole, becoming hardened ; flowers several in clusters, on rather slender pedicels about 1 cm. long ; sepals triangular, acute ; young fruits arcuate, about 1 cm. long
          2. *mannii*

1. **E. emarginatum** *Thonn.*—F.T.A. 1 : 274 ; O. E. Schulz in Engl. Pflanzenr. Erythroxylac. 135 ; Aubrév. Fl. For. C. Iv. 1 : 304. A glabrous shrub or small tree ; flowers white, fragrant ; fruit red.
   Fr.G.: Kouroussa *Pobéguin* 671 ! **G.C.:** *Thonning*. Accra Plains *Irvine* 424. Krepi (Jan.) *Johnson* 560 ! Afram Plains (fr. May) *Kitson* 1097 ! Abokobi (Oct.) *Vigne* FH 4260 ! **Dah.:** Abomey (Feb.) *Chev.* 23128 ! **N.Nig.:** Nupe *Barter* 1650 ! Kachiya (Feb.) *Meikle* 1205 ! **S.Nig.:** Ijaye *Barter* 3421 ! Olokemeji *Foster* 277 ! Idanre Hills (Oct.–Jan.) *Keay* FHI 22681 ! *Brenan* 8663 ! Widespread in tropical Africa.
2. **E. mannii** *Oliv.* F.T.A. 1 : 274 (1868) ; O. E. Schulz l.c. 124, t. 22 ; Aubrév. l.c. 1 : 304, t. 122. A forest tree, to 80 ft. high and 6 ft. girth, with straight bole and open spreading crown, deciduous ; profuse white flowers.
   S.L.: Bagroo R. (Apr.) *Mann* 850 ! Gorahun (fl. & young fr. June) *Small* 719 ! **Iv.C.:** *Aubrév.* 662 ! 1379 ! **G.C.:** Ofin Headwaters (Apr.) *Vigne* FH 1907. **Br.Cam.:** Johann-Albrechtshöhe (Apr.) *Staudt* 941 ! Also in French Cameroons.

Besides the above indigenous species, *E. coca* Lam., the Cocaine plant, a native of tropical S. America, is cultivated in our area.

82. LEPIDOBOTRYACEAE

Trees, shrubs or climbers. Leaves alternate, pinnate trifoliolate or unifoliolate with a jointed petiole, leaflets entire, pinnately nerved ; stipules and stipels caducous. Flowers hermaphrodite or unisexual, actinomorphic ; in axillary or terminal panicles or slender or short catkin-like racemes. Sepals 5, imbricate, united in the lower part. Petals 5, imbricate, free. Stamens 10, inserted on the margin of a fleshy disk ; filaments more or less connate at the base into a short tube, sterile in female flowers ; anthers 2-celled, dorsifixed, opening by slits lengthwise. Ovary superior, 5- or 3-celled ; styles 5 or 3, free or shortly connate at the base ; ovules 2 in each cell, on axile placentas, collateral or superposed. Fruit berry-like, indehiscent or septicidally dehiscent. Seeds covered by a fleshy, sometimes laciniate aril, with endosperm ; embryo straight or oblique, with 2 fleshy foliaceous cotyledons.

A family of three genera in tropical Asia and Africa.

LEPIDOBOTRYS Engl. Bot. Jahrb. 32 : 108 (1902).

Small tree ; leaves alternate, unifoliolate, oblong-elliptic, acuminate, 6–10 cm. long, 3–4·5 cm. broad, pinnately nerved ; petiolule jointed at the base ; stipules and stipels caducous, the former small and suborbicular, the latter oblong ; flowers unisexual, dioecious, the males arranged in short dense sessile catkin-like axillary inflorescences ; bracts and bracteoles small, hairy, persistent ; female flowers fewer in short fascicle-like racemes ; pedicels jointed at the apex, subpersistent ; male flowers : sepals 5, imbricate, shortly connate at the base ; petals 5, free, imbricate ; stamens 10, inserted on the margin of a fleshy disk ; rudimentary ovary small, with 3 stigmas ; female flowers like the male but with sterile stamens ; ovary 3-celled ; ovules 2, collateral on axile placentas, pendulous ; styles 3, connate at the base ; capsule 1-seeded, dehiscing septicidally ; seeds nearly covered by a fleshy aril ; embryo straight .. .. *staudtii*

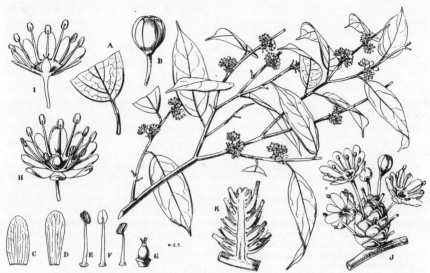

FIG. 127.—LEPIDOBOTRYS STAUDTII *Engl.* (LEPIDOBOTRYACEAE).

A, base of leaf and petiole showing joint. B, flower-bud. C, sepal. D, petal. E, F, stamens. G, rudimentary ovary. H, open flower. I, longitudinal section of same. J, inflorescence. K, longitudinal section of same.

L. staudtii *Engl.* Bot. Jahrb. 32 : 108 (1902). Small tree of forest.
　　S.Nig.: Sapoba (Nov.) *Ross* 221 ! *Kennedy* 2827 ! Kwale, Warri (Jan.) *Ainslie* FHI 2827 ! **Br.Cam.:** Johann-Albrechtshöhe *Staudt* 911 ! Also in French Cameroons, Ubangi-Shari, Gabon and Belgian Congo.

83. CTENOLOPHONACEAE

Trees. Leaves opposite, undivided, oblong-elliptic, acuminate, pinnately nerved, coriaceous ; stipules interpetiolar, united in pairs, deciduous. Flowers hermaphrodite, actinomorphic ; in short terminal panicles ; bracts very small and caducous. Sepals 5, rounded, broadly elliptic, imbricate. Petals 5, free, oblong, imbricate. Stamens 10, hypogynous ; filaments free, alternately shorter; anthers 2-celled, introrse, opening lengthwise. Ovary superior, 2-celled ; style 2-fid at the apex ; ovules 2 in each cell, collateral, pendulous from a rather long funicle attached to the central axis. Fruit woody, 1-seeded, splitting unilaterally. Seed pendulous from the slender free central axis, nearly covered by a fibrous comb-like aril.

A family with a single genus of two species, one in West Africa, the other in South-East Asia.

CTENOLOPHON Oliv. in Trans. Linn. Soc. 28 : 516 (1873).

Tree ; with shortly stellate-pubescent branchlets ; leaves coriaceous, broadly ovate to elliptic-ovate, obtuse at base, broadly acuminate, 4–9 cm. long, 2·5–5 cm. broad, glabrous ; flowers in short terminal panicles, pedicels short ; sepals suborbicular to broadly transverse elliptic, about 4 mm. broad ; petals about 1 cm. long and 2 mm. broad, densely stellate-lepidote outside ; fruit a bivalved capsule, about 2·5 cm. long and 8 mm. broad, with a single seed 1·5 cm. long, bearing a dorsal fibrillate aril
englerianus

C. englerianus *Mildbr.* in Notizbl. Bot. Gart. Berl. 8 : 706 (1924) ; Kennedy For. Fl. S. Nig. 27 ; Exell & Mendonça in Consp. Fl. Angol. 1 : 248. Riparian forest tree, to 80 ft. high ; flowers pink.
　　S.Nig.: Sapoba (Mar., Sept.) *Kennedy* 1574 ! 1689 ! *Jones* FHI 7442 ! Calabar (Apr.) *Amachi* FHI 24306 ! Also in Angola ; the other species of the genus occur in S.E. Asia. (See Appendix, p. 38.)

5

84. LINACEAE

Trees, shrubs, rarely herbs ; branches sometimes climbing by hooks. Leaves simple, alternate or opposite ; stipules present or absent, sometimes gland-like. Flowers actinomorphic, hermaphrodite. Sepals 4–5, free or partly united, imbricate. Petals free, contorted, fugacious. Stamens as many as and alternate with the petals, sometimes alternating with small staminodes ; filaments connate at base ; anthers introrse, 2-celled, opening lengthwise. Ovary superior, 3–5-celled, cells often again partially subdivided ; ovules 2 in each cell, pendulous ; styles 3–5, free or partly united, with simple capitate stigmas. Fruit a septicidal capsule. Seeds compressed, shining, with or without endosperm ; embryo straight, cotyledons flat.

Woody shrubs, mostly scandent ; stems often with recurved hooks ; stipules deeply
    divided ; leaves alternate ; flowers 5-merous ; sepals not lobed ; petals longer than
    the sepals   ..   ..   1. **Hugonia**
Delicate annual herb ; leaves opposite, 3 mm. long or less ; flowers 4-merous, sepals
    lobed or toothed at the apex ; petals about equalling the sepals ..   ..2. **Radiola**

1. HUGONIA Linn.—F.T.A. 1 : 270.

Flowers in leafless racemes or panicles, terminal above the leaves, or terminating
    leafless lateral shoots ; outer sepals ovate, about 5 mm. long, tomentose outside ;
    petals about 1 cm. long, glabrous ; fruit globose, about 15 mm. diam., glabrous ;
    stipules early caducous ; leaves elliptic or ovate-elliptic, broadly rounded at base,
    subacute at apex, 12–22 cm. long, 5–11 cm. broad, distinctly crenate, with 12–18 pairs
    of lateral nerves :
Mature leaves densely tomentose beneath, venation obscured   1. *spicata* var. *spicata*
Mature leaves puberulous to glabrescent beneath, venation not obscured
                                      1a. *spicata* var. *glabrescens*
Flowers axillary among the leaves, in cymes or solitary, along the shoots or clustered
    towards the ends of the branches among numerous young leaves :
Young branchlets, petioles and midribs glabrous ; flowers in few-flowered axillary
    cymes towards the ends of the branches ; outer sepals suborbicular, about 3 mm.
    long, pubescent outside ; petals about 12 mm. long ; stipules early caducous ;
    leaves obovate-oblanceolate, cuneate at base, obtuse at apex, 6–14 cm. long, 4–6 cm.
    broad, glabrous   ..   ..   2. *obtusifolia*
Young branchlets, petioles and midribs, and often leaf-surface as well, hairy, sometimes
    very densely so ; outer sepals 5 mm. or more long :
Outer sepals suborbicular, 5–6 mm. long, striate, with a broad membranous recurved
    margin, spreading in bud ; petals 15–20 mm. long ; flowers 2–5 together in axillary
    cymes ; leaves oblong to obovate or oblanceolate, somewhat narrowed to the base,
    obtuse or subacute at apex, 7–14 cm. long, 3–7 cm. broad   ..   3. *platysepala*
Outer sepals without a membranous margin, acute :
Outer sepals spreading in bud ; flowers in long-pedunculate pseudo-corymbs,
    axillary towards the ends of the branches ; whole plant pilose with golden hairs ;
    leaves obovate to obovate-oblanceolate, long-cuneate at base, acute at apex,
    11–23 cm. long, 5–9 cm. broad, coarsely dentate ; stipules with triangular centre
    and long filiform lobes   ..   ..   4. *talbotii*
Outer sepals not spreading in bud :
Petals densely pilose outside ; flowers crowded at the ends of the shoots, amongst
    long-pectinate stipules and the young leaves, not axillary along the shoots ;
    stems densely and thickly tomentose ; leaves elongate-obovate, narrowed to the
    very short hirsute petiole, densely rusty-hirsute on midrib above and beneath
    and on the nerves beneath :
Petals triangular, about 26 mm. long, 12 mm. broad at apex, narrowed gradually
    at the base into a small claw ; stipules palmate, with 5–9 subulate, long-plumose
    segments, joined only at the base ; leaves 12–30 cm. long, 5–10 cm. broad,
    gradually acuminate, margin entire, with 15–22 pairs of lateral nerves
                                          5. *macrophylla*
Petals elongate-oblong, about 16 mm. long, 5 mm. broad at apex, abruptly long-
    clawed at base ; stipules pinnately divided with a flattened, subglabrous central
    portion and 5–9 subulate, pubescent segments, not long-plumose ; leaves
    10–20 cm. long, 3·5–7·5 cm. broad, rounded at apex, with an abrupt short
    acumen, or mucronate ; margin obscurely denticulate to dentate, with tufts of
    hairs at intervals all round the margin ; 12–17 pairs of lateral nerves
                                          6. *rufipilis*

Petals glabrous ; flowers mostly axillary along the shoots, rarely crowded at the
apex as well : leaves not elongate-obovate, densely tomentose or shortly
pubescent, not long-hirsute :
Branchlets and undersurface of leaves on young shoots pubescent with short
brown hairs ; stipules more or less persistent, pinnately divided into subulate
lobes ; leaves elliptic to oblanceolate, cuneate at base, acutely acuminate,
7–15 cm. long, 3–6 cm. broad, mature leaves glabrescent except for the midrib ;
flowers in short axillary 2–4-flowered cymes ; flower buds very long-acuminate,
sepals ovate-lanceolate about 10 mm. long, densely brown pubescent outside,
becoming reflexed ; petals triangular, 2–2·5 cm. long   ..   7. *planchonii*
Branchlets and undersurface of leaves on young shoots, densely white-woolly-
tomentose ; flower-buds acute or shortly and bluntly beaked, sepals narrowly
ovate, about 10 mm. long, white woolly-tomentose outside ; flowers 2–3,
shortly pedicellate in leaf-axils along the shoots :
   Mature leaves glabrous beneath except for the puberulous midrib and nerves,
   venation obvious ; leaves elliptic, or oblong-elliptic, rounded or broadly
   cuneate at base, abruptly acuminate, 15–17 cm. long, 5–9 cm. broad, irregularly
   serrate   ..   .. 8. *foliosa*
   Mature leaves densely white-tomentose or -tomentellous beneath, elliptic to
   oblong-elliptic, rounded or narrowed at base, acute to acuminate, 8–18 cm.
   long, 3–8 cm. broad, crenate-dentate ; 7–16 pairs of lateral nerves   9. *afzelii*

1. **H. spicata** *Oliv.* var. **spicata**—F.T.A. 1 : 270 (1868). A woody climber, stems to 30 ft. long.
 F.Po: (Jan.) *Mann* 224 !
1a. **H. spicata** var. **glabrescens** *Keay* in Bull. Jard. Bot. Brux. 26 : 183 (1956). Woody climber as above ;
 sepals olive-brown, petals white.
 S.Nig.: Okomu F.R. (Feb.) *Brenan* 9084 ! Onicha Olona (Oct.) *Thomas* 1875 ! Also in Belgian Congo.
 (See Appendix, p. 38.)
2. **H. obtusifolia** *C. H. Wright* in Kew Bull. 1901 : 119. A scrambling shrub ; flowers bright yellow, " waxy-
 looking ".
 S.Nig.: Imo R. *Kennedy* 2563 ! Eket *Talbot* 3023 ! Cross R. (Jan.–Mar.) *Johnston* ! Also in French
 Cameroons and Belgian Congo.
3. **H. platysepala** *Welw. ex Oliv.* F.T.A. 1 : 272 (1868) ; Chev. Bot. 93. *H. baumannii* Engl. (1902) ; F.W.T.A.,
 ed. 1, 1 : 132. Scandent shrub, flowers bright yellow ; mainly in regrowth forest.
 Fr.G.: Macenta (May) *Collenette* 11 ! **S.L.:** Limba (Apr.) *Sc. Elliot* 5504 ! 5523 ! Sherbro *Sc. Elliot*
 5757 ! Njala (May) *Deighton* 667 ! Daru (Apr.) *Deighton* 3146 ! **Lib.:** Gola (Apr.) *Bunting* ! St. Paul
 R., Dobli Isl. (Apr.) *Bequaert* 168 ! **Iv.C.:** Abiosso (Apr.) *Chev.* 17813 ! Dyolas (Apr.) *Chev.* 21118 !
 G.C.: Kumasi (Mar., Apr.) *Andoh* FH 4170 ! *Scholes* 271 ! Pra R., Krobo (June) *Chipp* 238 ! **Togo:**
 Misahöhe (May) *Baumann* 530 ! **N.Nig.:** Koton Karifi (Apr.) *Ainslie* 394 ! **S.Nig.:** Sapoba *Kennedy*
 2314 ! Kukuruku (Apr.) *Dundas* FHI 21459 ! Obubra *Rosevear* 31/29 ! Degema *Talbot* 3753 ! **Br.Cam.:**
 Barombi *Preuss* 283 ! Bum, Bamenda (May) *Maitland* 1533 ! **F.Po:** *Mann* ! Extends to A.-E .Sudan,
 Uganda and Angola. (See Appendix, p. 38.)
4. **H. talbotii** *De Wild.* in Pl. Bequaert. 4 : 287 (1927).
 S.Nig.: Sapoba *Kennedy* 2250 ! Oron to Eket *Talbot* !
5. **H. macrophylla** *Oliv.* F.T.A. 1 : 271 (1868). *H. macrophylla* forma *acuminata* De Wild. l.c. 4 : 273 (1927).
 An extensive woody climber ; leaves rather crowded at the ends of the branches.
 S.Nig.: Degema *Talbot* ! Calabar (Mar.) *Thomson* 78 ! Holland 88 ! Oban *Talbot* 1545 ! [Pellegrin and
 De Wildeman record this species from Gabon, the latter regarding the Gabon specimen as a distinct forma
 obtusecuneato-apiculata De Wild. l.c.]
6. **H. rufipilis** *A. Chev. ex Hutch. & Dalz.* F.W.T.A., ed. 1, 1 : 132 (1927) ; Chev. Bot. 93, name only. Scan-
 dent shrub with yellow flowers.
 S.L.: Kenema *Thomas* 7928 ! Mamanasu *Deighton* 4066 ! Gola Forest *Deighton* 4099 ! **Lib.:** Grand
 Kola R. (Mar.) *Baldwin* 11210 ! **Iv.C.:** Bingerville *Chev.* 15443 ! 17266 ! **G.C.:** Axim (Feb.) *Irvine*
 2224 ! Also in Belgian Congo.
7. **H. planchonii** *Hook. f.*—F.T.A. 1 : 272 ; Chev. Bot. 93. *H. acuminata* Engl. (1902) ; Chev. Bot. 93.
 Scandent shrub or straggling tree ; flowers bright yellow.
 Port.G.: Bissau (Mar.) *Esp. Santo* 1893 ! Cacine *Esp. Santo* 2119 ! **Fr.G.:** Pobéguin 818 ! Dalaba (Mar.)
 Chev. 18109 ! **S.L.:** T. *Vogel* 45 ! Bagroo R. (Apr.) *Mann* 873 ! Berria (Mar.) *Sc. Elliot* 5426 ! Kenema
 (May) *Lane-Poole* 245 ! Vevehun (Apr.) *Deighton* 1623 ! **Lib.:** Dukwia R. (May) *Cooper* 456 ! Ganta
 (fr. Sept.) *Baldwin* 9258 ! **G.C.:** Accra T. *Vogel* 27 ! Kwahu (Apr.) *Johnson* 670 ! Assuantsi *Fishlock*
 32 ! Tafo (May) *Kitson* 1159 ! **Togo:** Kpandu (Mar.) *Robertson* 134 ! **S.Nig. :** Lagos (Apr.) *Dalz.* 1065 !
 1396 ! Ijaiye F.R. (May) *Onochie* FHI 21992 ! Oban *Talbot* 1640 ! Also in French Cameroons, Gabon
 and Belgian Congo. (See Appendix, p. 38.)
8. **H. foliosa** *Oliv.* F.T.A. 1 : 271 (1868) ; Chev. Bot. 93. A woody climber.
 S.L.: Bagroo R. (Apr.) *Mann* 816 ! **Iv.C.:** Byanouan to Soubiré (Mar.) *Chev.* 17739 ! **S.Nig.:** Akure F.R.
 Keay FHI 25584 !
9. **H. afzelii** *R.Br. ex Planch.*—F.T.A. 1 : 270 ; Chev. Bot. 93 ; Mildbr. in Fedde Rep. 41 : 254 (1937).
 H. chevalieri Hutch. & Dalz. F.W.T.A., ed. 1, 1 : 132 (1927) ; Kew Bull. 1928 : 29. A shrub, 15–18 ft.
 high, with long, loose branches ; flowers yellow.
 S.L.: *Afzelius* ! Kasewe Hills F.R. (June) *Deighton* 5832 ! **Lib.:** Monrovia (Mar.) *Dinklage* 3069 ! **Iv.C.:**
 Guidéko (May) *Chev.* 16371 ! **G.C.:** Axim (Feb.) *Akpabla* 104 ! Also in Belgian Congo.

Imperfectly known species.

1. **H. dinklagei** *Engl. ex Mildbr.* in Fedde Rep. 41 : 254 (1937), name only.
 Lib.: *Dinklage* 2535.
2. **H. sp. A.** Leaves oblanceolate, 20–25 cm. long, 6–6·5 cm. broad, glabrous beneath ; flowers and fruits not
 known.
 S.L.: *Thomas* 4539 !
3. **H. sp. B.** Small scandent shrub ; leaves oblanceolate, 25–40 cm. long, 6–11 cm. broad, pubescent beneath,
 very densely so when young ; flowers and fruits not known.
 S.Nig.: Shasha F.R. *Tamajong* FHI 20287 ! Akure F.R. *Keay* FHI 25612 !

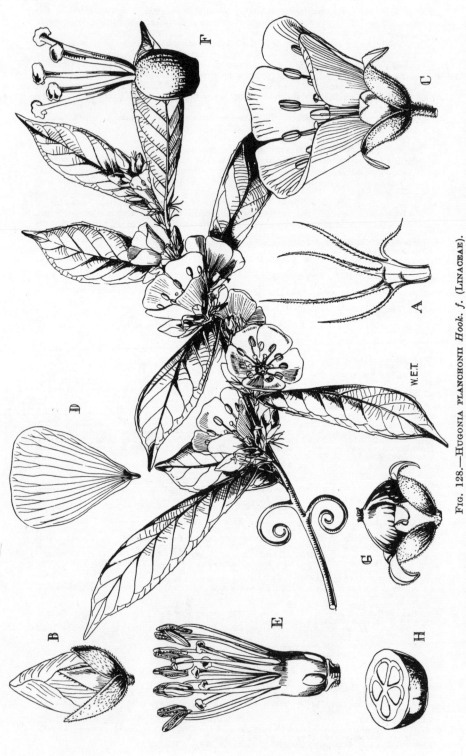

FIG. 128.—HUGONIA PLANCHONII *Hook. f.* (LINACEAE).

A, stipule.   B, flower-bud.   C, flower.   D, petal.   E, flower after removal of sepals and petals.   F, ovary with styles.   G, fruit.   H, cross-section of fruit.

W.E.T.

2. RADIOLA Hill—F.T.A. 1 : 268.

A very small glabrous annual herb ; leaves opposite, ovate or elliptic, about 4 mm. long and 2–3 mm. broad ; flowers very small, numerous, in leafy dichotomous cymes ; sepals ovate, acute ; valves of the capsule tipped by the persistent styles   *linoides*

R. linoides *Roth* Tent. Fl. Germ. 1 : 71 (1788) ; Clapham, Tutin & Warburg Fl. Brit. Isl. 377. *R. millegrana* Sm. (1800)—F.T.A. 1 : 268. A very small delicate annual, up to 3 in. high ; petals white.
   Br.Cam.: Cam. Mt., 7–12,000 ft. (Nov., Jan.) *Mann* 1334 ! 2021 ! *Maitland* 816 ! 1293 ! Also in Europe, Madeira, Teneriffe, N. Africa and temperate Africa.

85. ZYGOPHYLLACEAE

Shrubs or herbs woody at the base, rarely trees ; branches often jointed at the nodes. Leaves opposite or alternate, simple, 2-foliolate or pinnate, rarely 3-foliolate, not gland-dotted ; stipules present, persistent, often spinescent. Flowers hermaphrodite, actinomorphic or zygomorphic. Sepals 5, rarely 4, free or rarely connate at the base, imbricate, rarely valvate. Petals 4–5, rarely absent, hypogynous, free, imbricate or contorted, rarely valvate. Disk mostly present. Stamens the same number to three times the number of the petals, often unequal in length ; filaments free, often with a scale inside ; anthers 2-celled, opening lengthwise. Ovary superior, sessile or rarely stipitate, usually 4–5-celled ; cells rarely transversely locellate ; style simple, short, or stigmas sessile ; ovules 2 or more in each cell, axile. Fruits various. Seeds mostly with some endosperm ; embryo as long as the seed, straight or slightly curved.

Mainly in the tropics and subtropics, often in the drier parts.

Fruits not drupaceous :
   Leaves pinnate, with 3–7 pairs of leaflets, anisophyllous ; spreading herbs ; filaments
      without scales at the base, but the alternate ones each subtended by a gland :
   Fruit dividing into usually 10 (twice as many as original carpels) 1-seeded nutlets
                                                                          1. **Kallstroemia**
   Fruit dividing into usually 5 carpels, each with 3–5 compartments, the compartments
      1-seeded   ..   ..   2. **Tribulus**
   Leaves 1–3-foliolate :
      Unarmed shrublets with fleshy leaves ; filaments each with a scale at the base ; fruit
         sharply 5-angled, ellipsoid, opening lengthwise   ..   ..   ..   3. **Zygophyllum**
      Spiny diffuse subwoody annual ; filaments without scales at the base ; fruit pyramidal,
         opening from below into five 1-seeded cocci tipped by the persistent style
                                                                          4. **Fagonia**

Fruits drupaceous :
   Leaves simple ; branchlets spiny ; stamens (10–) 15 ; ovary and fruit 3-celled
                                                                          5. **Nitraria**
   Leaves 2-foliolate ; branchlets often armed with supra-axillary spines ; flowers in
      supra-axillary fascicles or racemes ; stamens 10 ; ovary 5-celled ; fruit a 1-seeded
      drupe   ..   ..   6. **Balanites**

Besides the above, *Guaiacum officinale* Linn., a native of tropical America, known as " Lignum Vitae," is cultivated in our area.

1. KALLSTROEMIA Scop. Introd. 212 (1777).

Stems diffuse or ascending, striate ; stipules lanceolate, 3 mm. long ; leaves 3–5 cm. long ; petioles shorter than the leaflets ; leaflets usually 3 pairs, obliquely elliptic or oval, rounded and mucronate at apex, 1–2 cm. long ; pedicels in fruit about 2 cm. long, strongly ribbed and thickened upwards ; sepals narrowly linear-lanceolate, hispid ; petals yellow, obovate, 6–7 mm. long ; fruit pyramidal, ribbed and muricate, beaked by the persistent style, strigose   ..   ..   *pubescens*

K. pubescens (*G. Don*) *Dandy* in Kew Bull. 1955 : 138. *Tribulus pubescens* G. Don Gen. Syst. 1 : 769 (1831). *K. caribaea* Rydb. in N. Am. Fl. 25, 2 : 111 (1910) ; Engl. in E. & P. Pflanzenfam. 19A : 177 (1931). *Tribulus maximus* of F.T.A. 1 : 284 and of F.W.T.A., ed. 1, 1 : 136, not of Linn. A spreading annual herb of waste places ; introduced.
   G.C.: Cape Coast (July) *Don* ! *T. Vogel* 58 ! Achimota *Irvine* 657 ! S.Nig.: Lagos (June) *Dalz.* 1039 ! A native of tropical America. (See Appendix, p. 38.)

2. TRIBULUS Linn.—F.T.A. 1 : 283.

Stems diffuse and trailing, hirsute, slightly swollen at nodes ; stipules subulate, caducous ; petioles about as long as leaflets ; leaves 2–5 cm. long ; leaflets 5–7 pairs, oblong or elliptic, somewhat oblique, rather obtuse, 3–13 mm. long ; pedicels in fruit

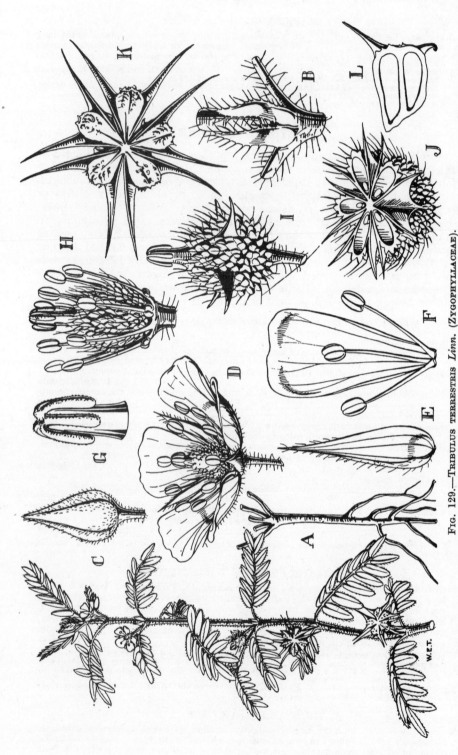

Fig. 129.—Tribulus terrestris *Linn.* (Zygophyllaceae).

A, root. B, leaf-bases with stipules. C, flower-bud. D, flower open. E, sepal. F, petal with 3 stamens. G, style with adnate stigmas. H, flower with sepals and petals removed. I, ovary with tubercles and spines. J, the same in cross-section. K, fruit from above. L, vertical section of a fruiting carpel.

about 1 cm. long, stout ; sepals caducous ; petals obovate, pale yellow, about 4 mm. long ; fruit depressed, spiny, each carpel with two large rigid divergent spines and a few tubercles, thinly setose-pilose .. .. *terrestris*

T. terrestris *Linn.*—F.T.A. 1 : 283 ; Chev. Bot. 96 ; Engl. in E. & P. Pflanzenfam. 19A : 176 (1931). *T. saharae* Chev. Bot. 96, name only. A trailing herb, woody below, from pubescent to tomentose ; a pest of footpaths and fields.
Sen.: *Sieber* 14 ! *Heudelot* ! **Fr.Sud.:** Timbuktu (July) *Chev.* 1287 ! **G.C.:** Accra (Mar.) *Deighton* 562 ! **Togo:** Lome *Warnecke* 12 ! **Dah.:** *Debeaux* 161 ! **Fr.Nig.:** Zinder (Oct.) *Hagerup* 589a ! **N.Nig.:** Sokoto (Sept.) *Moiser* 98 ! Maiduguri (May) *Ujor* FHI 32966 ! Also common throughout the tropical and warm temperate regions of the world. (See Appendix, p. 39.)

3. ZYGOPHYLLUM Linn.—F.T.A. 1 : 285.

Leaves 2-foliolate, opposite, fleshy, internodes short ; flowers solitary or paired, axillary :
Fruits villous, linear, rounded at apex, 5-angled, about 10 mm. long 1. *waterlotii*
Fruits glabrous, oblong-ellipsoid, rounded at apex, 5-angled, about 5 mm. long
2. *fontanesii*
Leaves 1-foliolate, opposite, slightly fleshy ; flowers solitary, axillary ; fruits glabrous, globular, depressed at apex with the style persistent, deeply 5-angled, about 2 mm. long .. .. 3. *simplex*

1. **Z. waterlotii** *Maire* in Bull. Soc. Hist. Nat. Afr. Nord 28 : 348 (1937) ; Trochain Contrib. Étude Vég. Sénégal 386 (1940). A fleshy shrublet.
  Sen.: *fide* Trochain *l.c.* Also in Mauritania and S.W. Sahara.
2. **Z. fontanesii** *Webb & Berth.* Phytog. Canar. 3, 1 : 17, t. 1 (1836) ; Chev. Bot. 97 ; Berhaut Fl. Sen. 9. An undershrub about 1 ft. high with a woody rootstock ; flowers white ; a maritime plant.
  Sen.: Oualo *Berhaut* 1355. Also in Mauritania, N. Africa, Canary and Cape Verde Islands.
3. **Z. simplex** *Linn.*—F.T.A. 1 : 285 ; Berhaut l.c. 76. A diffuse wiry undershrub a few inches high ; flowers yellow.
  Sen.: Gandiole *Berhaut* 1747. Also in Cape Verde Islands, through the arid regions of northern tropical Africa to Arabia and Baluchistan and in S.W. Africa.

4. FAGONIA Linn.—F.T.A. 1 : 286.

A diffuse woody spiny annual, puberulous-glandular or nearly glabrous ; leaves very variable, 1–3-foliolate, opposite, leaflets linear to lanceolate, acute, with a callous tip, bright green, up to 2 cm. long ; stipules spinose, often longer than the leaves, very acute ; flowers solitary, axillary ; pedicels nearly 0·5 cm. long ; sepals ovate, very acute, 3 mm. long ; petals more than twice the length of the sepals, rose or lilac, narrowly clawed ; fruit pyramidal, shortly pubescent, about 0·5 cm. long, tipped by the hard persistent style, opening from below into five 1-seeded cocci .. *cretica*

F. cretica *Linn.*—F.T.A. 1 : 287. *F. bruguieri* DC.—Berhaut Fl. Sen. 53, 75. A variable desert plant sometimes forming meadows in hollow places.
Sen.: Gandiole *Berhaut* 1745. Also throughout the warmer dry parts of the world.

5. NITRARIA Linn.—F.T.A. 1 : 288 ; Engl. in E. & P. Pflanzenfam. 19A : 178 (1931).

A shrub 2–3 m. high ; branchlets becoming spiny ; leaves mostly borne on short arrested branchlets, obovate, entire, 1·5–3 cm. long, up to 1 cm. broad, rather fleshy, glabrous ; petiole slender, about 4 mm. long ; flowers in short cymes, nearly sessile, the terminal ones pedicellate ; sepals ovate-lanceolate, fleshy, slightly puberulous, about 1·5 mm. long ; petals about twice as long ; fruits narrowly conical, triangular, 1 cm. long, capped by the sessile stigma .. .. *retusa*

N. retusa (*Forsk.*) *Aschers.* in Verh. Bot. Ver. Brandenb. 18 : 94 (1876) ; Engl. l.c., fig. 85, H–P. *Peganum retusum* Forsk. (1775). *Nitraria senegalensis* Lam. (1798)—F.W.T.A., ed. 1, 1 : 484 ; Berhaut Fl. Sen. 135. *N. tridentata* Desf. (1798)—Chev. Bot. 97. *N. schoberi* of F.T.A. 1 : 288. A plant of dry and saline areas ; commonly known as " Nitre Bush."
Sen.: On banks of the lower R. Senegal (Jan.) *Döllinger* ! *Heudelot* 432 ! Extends from Mauritania and desert regions of northern Africa to Palestine and Arabia.

6. BALANITES Del.—F.T.A. 1 : 314 ; Engl. in E. & P. Pflanzenfam. 19A : 179 (1931).

A forest tree ; flowering branchlets mostly unarmed, sterile branchlets with simple or forked spines 1–4 cm. long ; branchlets and petioles more or less densely, or sparingly, minutely puberulous ; petioles 1–2·5 cm. long ; petiolules 0·4–1·4 cm. long ; leaflets broadly ovate, rounded or obtuse and often unequal at base, shortly acuminate, 5–8·5 (–12) cm. long, 4–6 (–7) cm. broad, glabrous or nearly so ; flowers in supra-axillary puberulous or tomentellous pseudoumbels ; peduncles 0·5–1·8 cm. long, pedicels 1–1·7 cm. long ; buds nearly globose, tomentellous ; petals about 8 mm. long, more or less long-pilose inside ; fruits ellipsoid or ovoid-ellipsoid, about 4·5–10 cm. long, 3·5–5 cm. broad .. .. .. 1. *wilsoniana*
A savannah tree ; all branchlets with stout, rigid, simple spines up to 8 cm. long ; branchlets and petioles pubescent and often glabrescent ; petioles up to 1 cm. long ; leaflets subsessile, obovate to orbicular-rhomboid, 0·7–5 cm. long and 0·4–4·5 cm. broad, pubescent beneath, then glabrescent ; flowers in tomentose supra-axillary fascicles or rarely subracemose ; pedicels up to 1 cm. long ; buds ovoid tomentose ; petals about 5 mm. long, glabrous inside ; ovary tomentose ; fruits broadly oblong-

ellipsoid, rounded at each end, 3–4 cm. long, smooth or wrinkled, tomentose when young ..   ..   2. *aegyptiaca*

1. **B. wilsoniana** *Dawe & Sprague* in J. Linn. Soc. 37 : 506 (1906) ; Aubrév. Fl. For. C. Iv. 2 : 99, t. 166 ; Kennedy For. Fl. S. Nig. 152. *B. tieghemi* A. Chev. (1912)—Chev. Bot. 105. *B. mayumbensis* Exell (1927). A forest tree to 120 ft. high, with high buttresses, sometimes spiny, continuing upwards as twisted fluting ; young trees with forked spines up to 4 in. long ; flowers greenish-yellow, ripe fruits yellow with strong smelling pulp and a single woody seed.
    Iv.C.: Endé to Yaboisso, Morénou *Chev.* 22482 ! And other localities *fide* Chev. *l.c.* **G.C.:** Adeambra, W. Ashanti (young fr. May) *Vigne* FH 859 ! Mampong Scarp (Mar.) *Vigne* FH 1059 ! **Dah.:** Pedjilé to Pobé, Zagnanado *Chev.* 22920. **S.Nig.:** Sapoba (Dec.) *Kennedy* 1658 ! 1949 ! 1975 ! 2080 ! Also in French Cameroons, Ubangi-Shari, Gabon, Cabinda, Belgian Congo and E. Africa. (See Appendix, p. 311.)
    [A sterile specimen (*Keay* FHI 28198) from Afi River F.R., Ikom, S. Nig., may also be this species.]
2. **B. aegyptiaca** (*Linn.*) *Del.* in Descr. Égypt. Hist. Nat. 2 : 221, t. 28, 1 (1813) ; F.T.A. 1 : 315 ; Aubrév. Fl. For. Soud.-Guin. 366, t. 76, 1–5. *Ximenia aegyptiaca* Linn. (1753). *Agialida senegalensis* Van Tiegh. (1906). *A. barteri* Van Tiegh. (1906). *A. tombouctensis* Van Tiegh. (1906). *Balanites zizyphoides* Mildbr. & Schlechter (1913). A tree 20–30 ft. high, with greenish stems and branches and yellow edible fruits each with a single hard seed.
    Sen.: *Perrottet* 1261. Dara *Dubois* 276 ! **Fr.Sud.:** Timbuktu (fl. & fr. July) *Chev.* 1197 ! **Port.G.:** Bafata (fl. & fr. Aug.) *Esp. Santo* 2722 ! **G.C.:** Ayikuma, Shai (fl. & fr. Feb.) *Irvine* 1978 ! Tamale (Apr.) *Williams* 145 ! Prang (fl. & fr. Jan.) *Vigne* FH 1553 ! **Togo:** N. Basari (fl. Jan., Feb., fr. July) *Kersting* 533 ! **Dah.:** Atacora Mts. *Chev.* 24073 ; 24230. **N.Nig.:** Zurmi (Apr.) *Keay* FHI 18013 ! Nupe *Barter* 739 ! Samae (Mar.) *Foster* 24 ! Yola (fr. Oct.) *Shaw* 105 ! Widespread in the drier parts of tropical Africa, extending from Mauritania and the Central Sahara to Palestine, Arabia and India. (See Appendix, p. 309.)

✛ 86. EUPHORBIACEAE

Trees, shrubs or herbs, occasionally with milky juice. Leaves alternate or rarely opposite, simple or digitately compound, sometimes reduced, mostly stipulate. Flowers unisexual, mostly monoecious. Sepals imbricate or valvate, or in very specialized inflorescences much reduced or absent. Petals absent or rarely present and sometimes united. Stamens 1–1,000, free or variously connate ; anthers 2–4-celled, erect or inflexed in bud, opening lengthwise, rarely by pores. Rudimentary ovary often present in the male flowers. Ovary mostly 3-celled ; styles free or united at the base ; ovules solitary or paired, pendulous from the inner angle of the cells. Fruit a capsule or drupe. Seeds often with a conspicuous caruncle ; endosperm copious, fleshy ; embryo straight.

A large and heterogeneous family, mainly in the tropics and subtropics ; flowers much reduced and mostly apetalous ; allied to *Sterculiaceae, Flacourtiaceae, Celastraceae,* and perhaps *Sapindaceae.*

KEY TO THE TRIBES.

Male and female flowers not enclosed in a common involucre ; stamens usually more than 1 :
  Ovary-cells 2-ovuled ; plants without milky juice ..   ..   I. PHYLLANTHEAE
  Ovary-cells 1-ovuled :
    Petals present ; stamens 1-seriate, opposite the sepals ; flowers very small in axillary fascicles ; plants without milky juice   ..   .. II. GALEARIEAE
    Petals present or absent ; stamens 1–2-seriate, the outer alternate with the sepals or all more or less central, sometimes very numerous ; plants with or without milky juice ..   .. III. CROTONEAE
Male and female flowers much reduced and enclosed in a common involucre ; stamens 1 ; perianth usually absent or rim-like, rarely cupular ; plants with milky juice
                                    IV. EUPHORBIEAE

Tribe I.—PHYLLANTHEAE.

Leaves digitately compound, opposite ; leaflets stalked ; flowers dioecious ; petals absent ; stamens 4–10 ..   ..   **1. Oldfieldia**
Leaves simple :
  Sepals of the male flowers valvate in bud ; petals present (in our species), 5 ; stamens 5, connate in their lower part :
    Flowers in axillary fascicles or glomerules ; ovary 2-celled ; fruit drupaceous, 1–2-celled ; tertiary nerves of the leaves usually parallel   ..   **2. Bridelia**
    Flowers in short axillary racemes ; ovary 3-celled ; fruit capsular, 3-celled ; tertiary nerves not parallel ..   ..   **3. Cleistanthus**
  Sepals of the male flowers imbricate in bud :
    *Petals present in the male flowers ; stamens 5 (–6) :
      Inflorescence spicate, racemose or fasciculate :
        Styles absent ; stigmas sessile, thick, entire ; flowers monoecious, with conspicuous coriaceous bracts ; filaments free ; fruit capsular ..   ..   .. **4. Amanoa**

Styles well-developed, usually divided ; fruit a 3-lobed capsule ; bracts not large and coriaceous :
 Flowers in axillary fascicles, monoecious ; filaments shortly connate at the base
                   5. **Pentabrachion**
 Flowers in spikes or racemes, dioecious ; filaments free ; fruit with 1 seed per cell ; see also below   ..   ..   6. **Thecacoris**
 Inflorescence paniculate, flowers dioecious ; capsule ellipsoid, entire, loculicidal ; stigmas reflexed, entire ; leaves very unequal in size, borne on petioles of very unequal length   ..   ..   7. **Spondianthus**
*Petals absent from the male flowers ; stamens 2 to numerous :
Male and female flowers not surrounded by an involucre of calycine bracts :
 Flowers in spikes or racemes (see also No. 16) :
  Fruits not winged ; disk present in both sexes, or rarely absent from the female ; leaves not glandular beneath :
   Stems perennial ; much-branched trees or shrubs :
    Ovary 2-celled, each cell divided by a false septum ; styles 2, connate at the base, undivided   ..   ..   8. **Martretia**
    Ovary cells not divided ; styles bifid or bilobed :
     Ovary 4-celled ; male spikes fasciculate, axillary   ..   ..   9. **Apodiscus**
     Ovary 2–3-celled :
      Bracts subtending the male flowers in 3 series, the middle bract cupular and enclosing the flower in bud ; disk of the female flower adnate to the sepals ; flowers dioecious ; capsule large, septicidal   ..   10. **Protomegabaria**
      Bract subtending the male flowers solitary ; disk in the female not adnate to the sepals :
       Fruits depressed-globose, 3-lobed, 3-celled, dehiscing into 2-valved cocci ; see also above   ..   ..   6. **Thecacoris**
       Fruits entire, 1-celled by abortion ..   ..   ..   11. **Maesobotrya**
     Ovary 1-celled ; fruit drupaceous ; flowers small, in slender catkin-like inflorescences ..   ..   12. **Antidesma**
   Stems annual, mostly simple, arising from a woody rhizome ; flowers dioecious ; capsule 3-lobed ..   ..   13. **Cyathogyne**
  Fruits 2-celled, broadly winged, flat ; leaves glandular beneath ; male inflorescence a catkin-like spike, female shortly racemose ; styles 2, free from the base
                 14. **Hymenocardia**
 Flowers in fascicles or glomerules, or solitary in the leaf-axils, or (*Phyllanthus muellerianus* and *P. profusus*) in racemes of fascicles :
  Disk in the male flowers central ; fruit indehiscent ; flowers dioecious ; ovary and fruit 1–7-celled ; stamens 4–30, filaments free, variously inserted around and in the folds of the disk ; trees and shrubs   ..   ..   15. **Drypetes**
  Disk in the male flowers not central ; fruit usually dehiscent ; ovary and fruit 3-celled ; shrubs and herbs, rarely trees :
   Rudimentary ovary absent from the male flower ; flowers usually monoecious ; stamens 2–5, free or variously united   ..   ..   16. **Phyllanthus**
   Rudimentary ovary present in the male flower, deeply 2–3-partite ; flowers dioecious ; stamens 5, free   ..   ..   17. **Securinega**
Male flowers in a globose head surrounded by calycine bracts ; female solitary in an involucre of bracts ; fruit indehiscent, containing 3–4 pyrenes   .. 18. **Uapaca**

Tribe II.—GALEARIEAE.

Flowers small in axillary fascicles ; fruit a small indehiscent drupe ; rudimentary ovary present in the male flowers ; shrub with pellucid-dotted leaves ..   19. **Microdesmis**

Tribe III.—CROTONEAE.

Leaves digitately compound, with separate leaflets ; flowers paniculate :
 Petals absent ; fruit capsular ; the introduced Para rubber tree ..   ..   20. **Hevea**
 Petals present and connate ; fruit indehiscent ; indigenous tree, not rubber-producing ; stipules broad, reniform, toothed   ..   ..   ..   21. **Ricinodendron**
Leaves simple, sometimes deeply lobed but not into separate leaflets :
 *Petals present in the male flowers :
  Anthers inflexed in bud ; calyx-lobes valvate ; male petals free ; leaves often lepidote or stellate-pubescent beneath ; flowers spicate or racemose .. 22. **Croton**
  Anthers erect in bud :
   Male calyx-lobes imbricate :
    Flowers in axillary fascicles or the females solitary ; shrub with narrow leaves ; filaments partially connate ; fruit a small capsule   ..   ..   23. **Clutia**
    Flowers cymose or paniculate :
     Inflorescence bisexual ; male petals free or united into a tube ; stamens 8,

filaments partially connate ; shrubs or herbs with entire or digitately lobed
leaves   ..   ..   24. **Jatropha**
Inflorescence unisexual ; male petals free, imbricate ; stamens 10–20, on a hairy
receptacle ; fruit a 3-lobed capsule ; a shrub   ..   ..   25. **Mildbraedia**
Male calyx-lobes valvate :
  Herbs, monoecious ; calyx splitting regularly into 5–7-lobes ; filaments connate
  into a column :
    Rudimentary ovary present in the male flowers ; styles lacerate ; flowers in lax
    axillary racemes ; capsule covered with processes   ..   ..   26. **Caperonia**
    Rudimentary ovary absent ; style 2-fid ; flowers in rather dense racemes from
    the axils of the upper leaves ; capsule covered with stellate hairs or scales
                                                27. **Chrozophora**
  Trees, shrubs and lianes, usually dioecious ; calyx splitting irregularly into 2–3-lobes :
    Plants without stellate hairs or scales :
      Disk of male flowers with 5 glands outside the stamens and alternate with the
      petals ; stamens 13–40, filaments (in our species) irregularly connate at the
      base ; flowers in racemes or panicles..   ..   ..   ..   28. **Grossera**
      Disk of male flowers without extrastaminal glands ; stamens 20–30, filaments
      free ; flowers in racemes   ..   ..   ..   29. **Pseudagrostistachys**
    Plants with stellate hairs and/or scales at least on the inflorescence ; leaves
    2-glandular or multiglandular at the base ; filaments free :
      Lower surface of leaves concealed by a dense covering of silvery scales ; stamens
      about 30, accompanied by an equal number of interstaminal glands ; male
      petals free ; inflorescence paniculate   ..   ..   ..   30. **Cyrtogonone**
      Lower surface of leaves not concealed by scales ; stamens 7–25, not accompanied
      by interstaminal glands :
        Leaves pinnately nerved ; plants with scales at least on the inflorescence ;
        stamens 7–19 ; male petals mostly united (free in *C. preussii*) ; inflorescence
        racemose   ..   ..   31. **Crotonogyne**
        Leaves palmately nerved ; plants with stellate hairs, but without scales ;
        stamens 10–25 ; male petals united ; inflorescence paniculate
                                                32. **Manniophyton**
* Petals absent from the male flower :
  Male calyx closed in bud and enveloping the stamens ; plants mostly without milky
  juice (but see Nos. 53 and 56) :
    †Calyx-lobes of male flowers valvate :
      ‡Styles free, or if connate into a column then continuous with the central axis :
        Filaments unbranched, usually free :
          Anther-cells not long-cylindrical and worm-like :
            The anther-cells erect and separate, 2-celled :
              Buds perulate ; flowers in racemes or glomerules ; pedicels capillary, jointed
              near the base ; stamens 2–60, free ; capsule usually 2-lobed
                                                33. **Erythrococca**
              Buds not perulate :
                Flowers regularly distributed along the rhachis of the spikes or racemes ;
                fruits 2-lobed ; stamens 6–12, free ; trees or shrubs   ..   34. **Claoxylon**
                Flowers in distant glomerules or spikelets on a filiform rhachis ; stamens
                3–20, free ; fruits 3- or rarely 2-lobed ; annual herbs ..   35. **Micrococca**
            The anther-cells pendulous, either separate or adnate to the connective :
              Anthers 2-celled :
                Leaves opposite ; stamens without receptacular glands, numerous, free ;
                flowers generally dioecious ; plants with stellate hairs ; leaves gland-dotted
                or not beneath   ..   ..   36. **Mallotus**
                Leaves alternate :
                  Stamens 7–8 ; flowers dioecious ; styles simple :
                    Interstaminal glands absent ; male flowers paniculate or racemose ;
                    ovary 2- or 3-celled ; styles 5–15 mm. long ..   ..   37. **Alchornea**
                    Interstaminal glands present ; ovary 2- or 3-celled ; styles not exceeding
                    1 mm. long, recurved :
                      Flower spikes in large terminal panicles ; interstaminal glands numerous,
                      hairy ; fruits dehiscent   ..   ..   ..   38. **Discoglypremna**
                      Flower spikes fasciculate or solitary, from nodes on the old wood or some-
                      times axillary ; glands among the outer stamens only ; fruits inde-
                      hiscent   ..   ..   39. **Mareyopsis**
                  Stamens 9–120 :
                    Leaves palmately nerved, orbicular-cordate ; flowers in terminal panicles,
                    usually dioecious ; stamens 15–40 ; disk-glands extrastaminal ; leaves
                    glandular and stellate-pubescent beneath   ..   ..   40. **Neoboutonia**
                    Leaves pinnately nerved, not orbicular-cordate ; flowers racemose or
                    spicate, mostly monoecious :

Stamens 9–25 :
    Leaves with distinct petiole 1–5 cm. long, cuneate or obtuse at base ; connective of stamens broad ; styles plumose   ..   41. **Mareya**
    Leaves subsessile, auriculate-cordate at base ; connective of stamens narrow ; styles lacerate at the tips ..   ..   42. **Crotonogynopsis**
Stamens 30–120 :
    Leaves oblanceolate, with numerous lateral nerves ; styles simple, often more or less connate :
        Ovary and fruit smooth   ..   ..   ..   43. **Argomuellera**
        Ovary and fruit with 6 wing- or horn-like appendages ; see also below   44. **Pycnocoma**
    Leaves obovate-elliptic, with 7–10 lateral nerves on each side of the midrib ; styles bifid ..   ..   45. **Necepsia**
Anthers 4-celled :
    Stamens numerous ; anther-cells superposed, connective produced ; styles bifid ; leaves not gland-dotted beneath ..   ..   ..   46. **Cleidion**
    Stamens 1–6 (in our species) ; anther-cells collateral, connective not produced ; styles simple ; leaves gland-dotted beneath ; fruits glandular-granular   ..   ..   47. **Macaranga**
Anther-cells long-cylindrical or worm-like ; stamens 8 or fewer ; flowers monoecious or rarely dioecious ; female flowers within a variously toothed or lobed, sometimes foliaceous bract ; styles often laciniate ..   48. **Acalypha**
Filaments repeatedly branched, columnar below ; anthers 2-celled ; leaves peltate, palmately lobed ; capsule smooth or echinate   ..   49. **Ricinus**
‡Styles connate into a stout column and continuous with the body of the carpels :
Flowers racemose or spicate, not in involucrate heads :
    Ovary 4-celled ; stamens 8–40, free, mixed with interstaminal glands ; racemes axillary or leaf-opposed   ..   ..   50. **Tetracarpidium**
    Ovary 3-celled :
        Stamens numerous ; female flowers at the top of the inflorescence ; stems not twining ; see also above   ..   ..   44. **Pycnocoma**
        Stamens 3, free ; female flowers at the base ; stems often twining ; plants often with stinging hairs   ..   ..   51. **Tragia**
Flowers in dense involucrate heads ; stamens 20–30, connate below ; stems usually twining   ..   ..   52. **Dalechampia**
†Calyx-lobes of male flowers imbricate :
Leaves digitately lobed, rarely some simple ; flowers monoecious, in terminal or subterminal racemes or panicles ; calyx large, often coloured ; stamens 10, in two series, free ; cultivated plants   ..   ..   ..   53. **Manihot**
Leaves not lobed, or rarely some slightly 3-lobed (see No. 56) ; calyx small ; flowers dioecious :
    Calyx 4–6-lobed ; stamens 8–30 ; anthers 2-celled :
        Flowers in axillary and terminal panicles ; leaves distinctly petiolate ; stamens 8 or 10 in two series   ..   ..   ..   54. **Klaineanthus**
        Flowers in fascicles or racemes up to 4 cm. long ; stamens 18–30, filaments very short :
            Flowers in small leaf-opposed fascicles ; leaves sessile or very shortly petiolate ; plant without milky juice   ..   ..   ..   55. **Gelonium**
            Flowers in axillary racemes ; leaves distinctly petiolate, sometimes slightly 3-lobed ; plant with milky juice   ..   ..   56. **Hamilcoa**
    Calyx 3-lobed ; stamens 3 ; anthers 4-celled ; male flowers in catkin-like spikes ; female flowers in very short racemes or solitary ; leaves opposite or alternate   57. **Tetrorchidium**
Male calyx open in bud, not covering the stamens ; plants mostly with milky juice :
Flowers racemose, dioecious ; stamens 15–32 ; calyx 5(–6)-lobed ; ovary 1-celled ; fruit indehiscent   ..   ..   58. **Plagiostyles**
Flowers spicate, mostly monoecious, with a few female flowers below the male flowers at the base of the spike ; stamens 2–3 (rarely 1 or 4) ; calyx 2–3(–5)-lobed ; ovary 2–3-celled ; fruits mostly dehiscent :
    Stamens free ; spikes slender :
        Seeds strophiolate ; ovary 3-celled ; fruits dehiscent ..   ..   59. **Sebastiania**
        Seeds not strophiolate ; ovary 2–3-celled ; fruits dehiscent or indehiscent   60. **Sapium**
    Stamens monadelphous ; male flowers in densely ovoid or subglobose spikes   61. **Maprounea**

Tribe IV.—EUPHORBIEAE.

Ovary (female flower) with a cup-like calyx at its base, half surrounded by 4 involucres formed of free bracts, containing the stamens ; male flowers usually about 8 in each

involucre, each consisting of a single stamen with a cup-like perianth at its base,
jointed to the pedicel ; fruit 3-lobed ; trees   ..   ..   ..   62. **Anthostema**
Ovary (female flower) enclosed in the same involucre as the stamens or in a separate
involucre, the latter cupular or 4-angled, more or less lobed and often bearing variously
shaped glands on the margin :
Involucre 4-angled, divided into 4 compartments by 4 large partition-like glands,
without glands on the margin ; small tree with milky juice   ..   63. **Dichostemma**
Involucre not 4-angled, with 1–8 distinct and often very conspicuous glands on the
margin :
Fruit a capsule ; trees, shrubs, fleshy plants or small herbs   ..   64. **Euphorbia**
Fruit indehiscent, drupaceous, thick and fleshy ; trees with succulent angular
branches   ..   ..   65. **Elaeophorbia**

Besides the above, other genera are represented by the following introduced species : *Hura crepitans* Linn.,
the Sandbox Tree, a native of tropical America ; *Aleurites cordata* (Thunb.) R.Br. ex Steud., the Tung-oil tree,
a native of S.E. Asia, and *A. moluccana* (Linn.) Willd., the Candle Nut, probably native of the Malaya region
but now widely cultivated in the tropics ; *Codiaeum variegatum* (Linn.) Blume, shrubs widely grown, in
numerous varieties, for their coloured leaves, commonly called " Croton " ; *Hippomane₁ mancinella* Linn., a
native of central America ; *Putranjiva roxburghii* Wall., a native of India ; *Breynia nivosa* (W. G. Sm.) Small
(syn. *Phyllanthus nivosus* W. G. Sm.), apparently a native of Polynesia, widely cultivated as a hedge-plant on
account of its decorative foliage, the " Snow bush " ; *Pedilanthus tithymaloides* (Linn.) Poit., a native of
tropical America.

1. **OLDFIELDIA** Benth. & Hook. f.—F.T.A. 6, 1 : 625 ; Pax & K. Hoffm. in Engl.
Pflanzenr. Euph. 15 : 297 (1922).

Leaves opposite, digitately 5–8-foliolate ; leaflets oblanceolate or oblong-elliptic, more
or less acuminate, very acute at base, up to 17 cm. long and 6 cm. broad, glabrous ;
male flowers in axillary cymes ; calyx pubescent ; capsule depressed-globose, 2·5 cm.
diam., woody   ..   ..   *africana*

O. africana *Benth. & Hook. f.*—F.T.A. 6, 1 : 625 ; Chev. Bot. 561 ; Aubrév. Fl. For. C. Iv. 2 : 22, t. 128. A
tall forest tree, up to 100 ft. or more in height, with straight bole of 50 ft. and girth 16 ft. ; bark scaly,
flaking ; young branchlets rusty-puberulous ; fruit hard, brown, nut-like, with 3 seeds.
S.L.: *Oldfield* ! *Welw.* 478 ! Rotumba Isl. *Kirk* ! Gola Forest (fl. & fr. Apr.–May) *Small* 621 ! 707 ! **Lib.:**
Brewers Landing (fr. Nov.) *Linder* 1395 ! Dukwia R. (fr. May, Oct.) *Cooper* 88 ! 344 ! 439 ! **Iv.C.:**
Malamalasso (Mar.) *Chev.* 16250 !   (See Appendix, p. 156.)

2. **BRIDELIA** Willd.—F.T.A. 6, 1 : 611 ; Jablonszky in Engl. Pflanzenr. Euph. 8 :
54–88 (1915).

Fruits 2-celled, subglobose, black, about 8 mm. diam. ; branchlets pubescent ; leaves
oblong or oblanceolate, rounded at apex, 3–11 cm. long, 1–5 cm. broad, stiffly
coriaceous, more or less pubescent on the nerves beneath, very closely and strongly
reticulate beneath, the lateral nerves continued to the margin   ..   1. *scleroneura*
Fruits 1-celled :
Lateral nerves not continued to the margin to form a marginal nerve ; branchlets
flexuose, slightly pubescent ; leaves oblong-elliptic, acuminate, 8–16 cm. long,
4–7 cm. broad, glabrescent on both surfaces, not prominently reticulate, drying
black ; fruits oblong-ellipsoid, about 6 mm. long, black ..   ..   2. *atroviridis*
Lateral nerves continued to the margin and forming a marginal nerve :
Acumen of leaves conspicuously mucronate ; leaves glabrous or glabrescent on both
surfaces, veins inconspicuous beneath, lamina oblong-elliptic, long-acuminate,
7–14 cm. long, 4–6·5 cm. broad ; branchlets glabrous or puberulous ; tree in
montane forest   ..   ..   3. *speciosa*
Acumen of leaves not mucronate ; branchlets and undersurface of leaves puberulous
to pubescent or tomentose :
Undersurface of leaves minutely puberulous with appressed hairs on the surface of
the lamina as well as on the midrib, nerves and veins, sometimes also with longer
hairs on the midrib and nerves, at length glabrescent ; veins not prominent
beneath ; branchlets pubescent to glabrescent ; leaves elliptic to obovate,
acuminate, 4–17 (–30) cm. long, 1·5–8 (–12) cm. broad ..   ..   4. *micrantha*
Undersurface of leaves pubescent to pilose, the hairs not appressed ; branchlets
pubescent to tomentose :
Hairs spread over the undersurface of the leaf and not confined to the nerves and
veins ; leaves elongate-elliptic to oblong-oblanceolate or oblong-lanceolate,
gradually long-acuminate, 7–14 cm. long, 2–5·5 cm. broad, lateral nerves (8–)
10–13 on each side of the midrib ; forest tree with black bark on stem and
branches   ..   ..   5. *grandis*
Hairs confined to the midrib, nerves and veins beneath, often very dense when
young ; venation very prominent beneath ; leaves oblong, to oblong-obovate or
suborbicular, shortly acuminate ; bark more or less greyish-brown :
Leaves (3–) 3·5–10·5 cm. long, (1·8–) 2·5–5·5 (–7) cm. broad ; lateral nerves 5–9
on each side of the midrib ; densely to rather sparsely pubescent beneath ;
savannah tree or shrub ..   ..   6. *ferruginea*

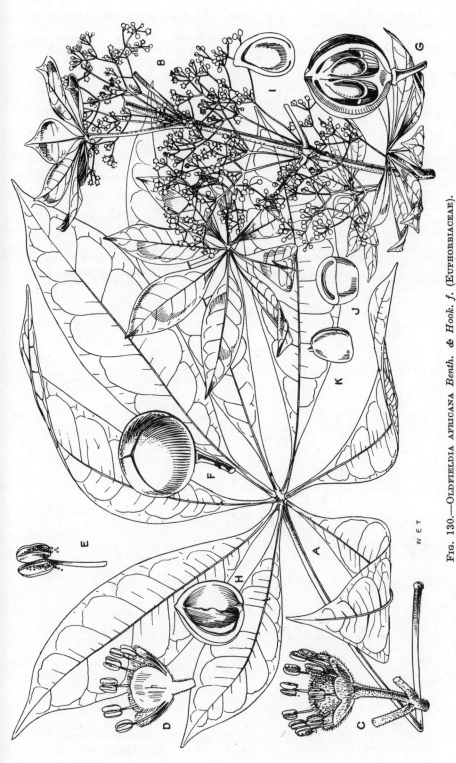

Fig. 130.—Oldfieldia africana *Benth. & Hook. f.* (Euphorbiaceae).

A, leaf. B, male inflorescence. C, male flower with rudimentary ovary. D, cross-section of same. E, stamen. F, fruit. G, fruit with portion of wall removed. H, same with seeds removed. I, longitudinal section of seed. J, cross-section of seed. K, embryo.

369

Leaves 6–20 cm. long, 3·8–10 cm. broad ; lateral nerves 9–13 on each side of the midrib ; indumentum on midrib and nerves redder (in dried specimens) and denser than on the veins ; tree of fringing forest   ..   ..   7. *ndellensis*

1. **B. scleroneura** *Müll. Arg.* in Flora 47 : 515 (1864) ; F.T.A. 6, 1 : 614 ; Chev. Bot. 554. *B. scleroneuroides* Pax (1893). *B. tenuifolia* of Aubrév. Fl. For. Soud.-Guin. 179–182, t. 4. A savannah shrub or tree, to 15 ft. high, or more ; flowers in small axillary clusters, red in bud, yellow when open.
Fr.G.: Irébéléya to Timbo (Sept.) *Chev.* 18295. **Iv.C.:** Boromo *Aubrév.* 2198. **G.C.:** Sambisi *Kitson* 544 ! Prang (June) *Vigne* FH 3903 ! **Togo:** Sansanné Mango *Aubrév.* 131d. **Fr.Nig. :** Gourma *Chev.* 24515. **N.Nig.:** Nupe *Barter* 908 ! Sokoto (May) *Lely* 827 ! Zaria (Aug.) *Keay* FHI 28009 ! Abinsi (Oct.) *Dalz.* 784 ! Yola (fl. & fr. July) *Dalz.* 162 ! **S.Nig.:** Onitsha *Barter* 577 ! Extends to A.-E. Sudan, E. Africa and Belgian Congo. (See Appendix, p. 138.)
2. **B. atroviridis** *Müll. Arg.*—F.T.A. 6, 1 : 614 ; Chev. Bot. 553 ; Aubrév. Fl. For. C. Iv. 2 : 34, t. 133, 1–3 ; Fl. For. Soud.-Guin. 179–182, t. 32, 3. A forest shrub or tree, to 20 ft. high, with spiny stem.
S.L.: Buyabuya (fl. & fr. Feb.) *Sc. Elliot* 4778 ! Pendembu (July) *Thomas* 706 ! 739 ! **Iv.C.:** Dimbokro (Apr.) *Aubrév.* 425 ! Agboville *Aubrév.* 551. **G.C.:** Tinte Hills, Wioso (fl. & fr. Dec.) *Chipp* 784 ! Aberem, Birrim (fr. Nov.) *Fishlock* 72 ! Asin-Edubiasi (Apr.) *Andoh* FH 3289 ! **Togo:** Atakpame *Doering* 236 ! **Dah.:** Sakété (fr. Jan.) *Chev.* 22850 ! **S.Nig.:** Ibadan to Abeokuta *Schlechter* 12356 ! Benin City (fl. Sept., fr. Jan., Sept.) *Dennett* 14 ! 43 ! Sapoba *Kennedy* 228 ! 1124 ! 2283 ! Ogwashi (fr. Nov.) *Thomas* 2031 ! Idumuje (fr. Jan.) *Thomas* 2163 ! **Br.Cam.:** Johann-Albrechtshöhe *Staudt* 948 ! Mamfe (fr. Aug.) *Johnstone* 145/31 ! Also in French Cameroons, Spanish Guinea, Belgian Congo, Uganda, Kenya, Tanganyika, S. Rhodesia and Angola. (See Appendix, p. 137.)
3. **B. speciosa** *Müll. Arg.*—F.T.A. 6, 1 : 618. Tree, to 30 ft. high, in montane forest ; flowers greenish-yellow.
S.Nig.: Mt. Koloishe, 4,800 ft., Obudu Div. (Dec.) *Keay & Savory* FHI 25127 ! **Br.Cam.:** Cam. Mt., 4,500–5,000 ft. (Dec.–Feb.) *Mann* 1215 ! *Maitland* 1201 ! Bamenda-Nkwe (fr. Apr.) *Ujor* FHI 30059 ! Bamenda, 6,000 ft. (Jan.) *Keay & Lightbody* FHI 28362 ! Also in Ubangi (*fide* Hutch.).
4. **B. micrantha** (*Hochst.*) *Baill.*—F.T.A. 6, 1 : 620 ; Chev. Bot. 554 ; Aubrév. Fl. For. C. Iv. 2 : 36, t. 132 ; Fl. For. Soud.-Guin. 179–182. *Candelabria micrantha* Hochst. (1843). *B. speciosa, B. mollis, B. tenuifolia* of Chev. Bot. 554–5 (*fide* Hutch.). Shrub or tree, to 50 ft. high, with a dense widely spreading crown ; branches often spiny ; in secondary forest regrowth and in savannahs of the moister regions.
Sen.: *Sieber* ! Tambana, Sedhiou *Chev.* 2625. **Gam.:** *Heudelot* 38 ! **Fr.Sud.:** Birgo *Dubois* 20. **Port.G.:** Sedengal to Barro (fr. Aug.) *Esp. Santo* 3091 ! **Fr.G.:** Rio Nunez *Heudelot* 605 ! Kouroussa (Apr.) *Pobéguin* 683 ! **S.L.:** Bagroo R. (Apr.) *Mann* 883 ! Erimakuna (Mar.) *Sc. Elliot* 5397 ! Manjoro, Biriwa (Mar.) *Deighton* 5396 ! Rokupr (May) *Jordan* 259 ! **Lib.:** Robertsport (fr. Dec.) *Baldwin* 10940 ! **Iv.C.:** Danipleu *Aubrév.* 1153 ! Aboisso (Apr.) *Chev.* 16302 ! **G.C.:** Aburi Hills (Feb.) *Brown* 931 ! Kwahu (Apr.) *Johnson* 657 ! Kumasi (Mar.) *Vigne* FH 1650 ! **Togo:** Amedzofe (Feb.) *Irvine* 159 ! **Dah.:** Natitingou *Aubrév.* 73d. **S.Nig.:** Lagos *Rowland* ! Idi-Ishin, Ibadan (Mar.) *Abubuka* FHI 3036 ! Calabar *Thomson* 85 ! Oban *Talbot* 2331 ! **Br.Cam.:** Buea (Dec.–Jan.) *Deistel* 615 ! *Maitland* 248 ! 903 ! Abonando (Mar.–Apr.) *Rudatis* 26 ! 45 ! Widespread in tropical Africa and in S. Africa. (See Appendix, p. 137.)

FIG. 131.—BRIDELIA MICRANTHA (*Hochst.*) *Baill.* (EUPHORBIACEAE).

A, male shoot. B, female shoot. C, flower-bud. D, male flower. E, anther. F, female flower. G, longitudinal section of female flower. H, cross-section of ovary. I, fruit.

5. **B. grandis** *Pierre ex Hutch.* in F.T.A. 6, 1 : 1042 (1913). *B. aubrevillei* Pellegr. in Bull. Soc. Bot. Fr. 78 : 683 (1931) ; Aubrév. Fl. For. C. Iv. 2 : 36, t. 133, 4–6. A forest tree, to 100 ft. high, with black longitudinally fissured bark ; aerial roots at the base of stem ; spines on branches and on young trees.
S.L.: Colonial Reserve (fr. Jan.) *King-Church* 9 ! Bumpe (fr. Dec.) *Deighton* 4693 ! Kasewe (fr. Dec.) *King* 127 ! **Lib.:** Brewersville (fr. Nov.) *Barker* 1081 ! Peahtah (Oct.) *Linder* 1101 ! **Iv.C.:** Fleury 12 ! Yapo (fr. Oct.) *Chev.* 22359 ! Agboville (fr. June) *Serv. For.* 423 ! *Aubrév.* 1391 ! **G.C.:** Axim (Mar.) *Chipp* 394 ! Sunyani *Vigne* FH 2472 ! **S.Nig.:** Onda, Omo F.R. *Jones & Onochie* FHI 16966 ! Sapoba *Kennedy* 2292 ! Also in Gabon and Belgian Congo.
6. **B. ferruginea** *Benth.*—F.T.A. 6, 1 : 619, partly ; Aubrév. l.c. 2 : 34 ; Fl. For. Soud.-Guin. 181, t. 32, 1. A savannah shrub or tree, to 20 ft. high, often with spines ; flowers greenish-yellow, with reddish disk ; the more sparsely pubescent forms come mainly from the moister regions.
Fr.Sud.: Bambana Tounba *Chev.* 559 ! Kayes to Nioro *Aubrév.* 3020. **Fr.G.:** Kouroussa (Apr.) *Pobéguin*

684 ! **S.L.**: Gberia Timbako (Oct.) *Small* 458 ! Loma Mts. (Apr.) *Frith* 22 ! **Iv.C.**: Mt. Boko, Haut Sassandra (May) *Chev.* 21460 ! Dimbokro (Apr.) *Aubrév.* 426 ! Touba (Apr.) *Aubrév.* 1239 ! **G.C.**: Nebiowalli to Samboro (May) *Kitson* 769 ! Chama, Sekondi (Apr.) *Chipp* 189 ! Kumawo (June) *Chipp* 468 ! Funsi to Bofiama (Apr.) *Kitson* 548 ! **Togo**: Lome *Warnecke* 151 ! **Dah.**: Agouagon (fr. May) *Cher.* 23521 ! Cotonou (fr. Sept.) *Chev.* 4442 ! **N.Nig.**: Nupe *Barter* 1685 ! 1686 ! 1711 ! Zungeru (Aug.) *Elliott* 6 ! Bauchi Plateau (Feb., May) *Lely* P. 170 ! P. 302 ! Idah (Sept.) *T. Vogel* 62 ! **S.Nig.**: Lagos *Rowland* ! *Millen* 104 ! Ibadan (fr. Nov.) *Jones & Keay* FHI 14171 ! Onitsha Dist. *Rosevear* 15/29 ! **Br.Cam.**: Bamenda (May) *Lightbody* FHI 26304 ! Also in French Cameroons, Ubangi-Shari, Belgian Congo and Angola. (See Appendix, p. 137.)

7. **B. ndellensis** *Beille* in Bull. Soc. Bot. Fr. 55, Mém. 8 : 69 (1908). *B. ferruginea* var. *orientalis* Hutch. in F.T.A. 6, 1 : 620 (1912) ; Aubrév. Fl. For. Soud.-Guin. 179–182, t. 32, 2. *B. aubrevillei* of F. W. Andr. in Fl. Pl. A.-E. Sud. 2 : 56, not of Pellegr. Tree of fringing forest in savannah regions, to 45 ft. high, with small stilt-roots and spiny stem and branches ; flowers greenish-yellow.

Dah.: Birni (Feb.) *Aubrév.* 99d ! **S.Nig.**: Old Oyo F.R. Oyo Prov. (Feb.) *Keay* FHI 16289 ! Also in Ubangi-Shari, A.-E. Sudan, Belgian Congo and Uganda.

3. CLEISTANTHUS Hook. f.—F.T.A. 6, 1 : 621 ; Jablonszky in Engl. Pflanzenr. Euph. 8 : 8 (1915).

Branchlets rusty-pubescent, then glabrescent, grey ; leaves coriaceous oblong-lanceolate or oblong-elliptic, caudate-acuminate, 5–9 cm. long, 1·3–4 cm. broad, acumen 0·5–2 cm. long ; flowers in short axillary rusty-tomentose racemes ; bracts 2–5 mm. long ; ovary densely rusty-tomentose ; fruits subglobose, muricate   ..   *polystachyus*

C. polystachyus *Hook. f. ex Planch.*—F.T.A. 6, 1 : 624 ; Jablonszky l.c. 8 : 47 ; Aubrév. Fl. For. C. Iv. 2 : 54, t. 133, 7–9. *C. libericus* N.E. Br.—F.T.A. 6, 1 : 624 ; Jablonszky l.c. 8 : 46 ; Chev. Bot. 555 ; F.W.T.A., ed. 1, 1 : 282. A shrub or small tree, to 30 ft. high, in forest.
S.L.: *Whitfield* ! Madina, Limba (Apr.) *Sc. Elliot* 5597 ! Laoma (Apr.) *Aylmer* 589 ! Falaba (Apr.) *Aylmer* 54 ! Njala (Apr.) *Deighton* 1138 ! **Lib.**: Sinoe Basin *Whyte* ! Harbel Plantation, Kingsville (Mar.) *Bequaert* 159 ! **Iv.C.**: near Guidéko (June) *Chev.* 19023 ! Danané (Mar.) *Aubrév.* 1100 ! Nzo (Apr.) *Aubrév.* 1166 ! **G.C.**: Bonsasu (May) *Vigne* FH 1971 ! Amentia (Apr.) *Vigne* FH 2902 ! Kangkang (fr. Aug.) *Moor* 269 ! **S.Nig.**: Lower Enyong F.R., Calabar (fr. May) *Onochie* FHI 33213 ! Also on S. Tomé and widespread in tropical Africa.

4. AMANOA Aubl.—F.T.A. 6, 1 : 630 ; Pax & K. Hoffm. in Engl. Pflanzenr. Euph. 15 : 195 (1922).

Flower clusters spicate ; leaves oblong-lanceolate, obtusely acuminate, deltoid at the base, 9–18 cm. long, 3–6 cm. broad, rigidly coriaceous ; lateral nerves 8–10 pairs, slightly raised beneath ; capsule subglobose, about 2·5 cm. long ; seeds shining
     1. *bracteosa*

Flower clusters solitary ; leaves oblanceolate or oblong-elliptic, 8–23 cm. long, 2·5–9·5 cm. broad, coriaceous ; lateral nerves 8–12 pairs, conspicuous below ; capsule up to 3·8 cm. long   ..   ..   2. *strobilacea*

1. **A. bracteosa** *Planch.*—F.T.A. 6, 1 : 631 ; Pax & K. Hoffm. in Engl. Pflanzenr. Euph. 15 : 200 ; Aubrév. Fl. For. C. Iv. 2 : 54, t. 139. A glabrous tree, up to 60 ft. high and 5 ft. diam., with stiff, bracteate, bright yellow, flower-spikes and hard-shelled fruit.
S.L.: *Afzelius* ! *Unwin & Smythe* 12 ! Near Freetown *Don* ! York Pass (fr. Apr.) *Lane-Poole* 78 ! Kambui F.R. (Apr., May) *Lane-Poole* 248 ! *King* 241 ! Magibisi to Baiima (fr. Sept.) *Deighton* 3064 ! **Lib.**: Monrovia (fr. June) *Whyte* ! *Baldwin* 5909 ! Sinoe Basin *Whyte* ! Loffa R., Grand Cape Mount (fr. Dec.) *Baldwin* 10939 ! **Iv.C.**: Sangouiné (fr. Jan.) *Aubrév.* 2107 ! Mt. Nimba *Aubrév.* 1144. Mt. Momy *Aubrév.* 1188. (See Appendix, p. 136.)

2. **A. strobilacea** *Müll. Arg.*—F.T.A. 6, 1 : 631 ; Pax & K. Hoffm. l.c. 198, fig. 16. A glabrous tree, 30 ft. high, with flowers in small bracteate sessile subglobose heads.
Lib.: Sinoe Basin *Whyte* ! **G.C.**: Mumuni Camp, Prestea road, Enchi Dist. (Mar.) *Andoh* FH 5621 ! **Br.Cam.**: Cameroons R. (Jan.) *Mann* 745 ! 2220 ! Muyaku (fr. May) *Maitland* 1144 ! Also in French Cameroons and Cabinda.

5. PENTABRACHION Müll. Arg. in Flora 47 : 532 (1864) ; Pax & K. Hoffm. in Engl. Pflanzenr. Euph. 15 : 188 (1922).

Small tree or shrub ; branchlets and leaves glabrous or very nearly so ; leaves oblong, rarely obovate, acute or obtuse at base, long-caudate-acuminate, 9–24 cm. long, 3·5–11 cm. broad, strongly reticulate on both surfaces ; petiole 2–7 mm. long ; flowers monoecious in axillary fascicles, the females very few ; sepals imbricate, about 2 mm. long ; petals present, ⅓ length of sepals ; male flowers with 5–6 stamens, filaments shortly connate at base, inserted within the disk, rudimentary ovary 3-partite ; ovary 3-celled ; capsule 3-lobed, about 1–1·2 cm. long and broad, on pedicel about 2·5 cm. long, glabrous outside ; seeds 2 in each cell   *reticulatum*

P. reticulatum *Müll. Arg.* l.c. 47 : 533 (1864) ; Pax & K. Hoffm. l.c. fig. 15. *Actephila reticulata* (Müll. Arg.) Pax (1899)—F.T.A. 6, 1 : 632. A forest tree, to 25 ft. high, or shrub.
Br.Cam.: Mombo-Bakosi, Kumba (fr. May) *Olorunfemi* FHI 30556 ! Also in French Cameroons and Gabon.

6. THECACORIS A. Juss.—F.T.A. 6, 1 : 658 ; Pax & K. Hoffm. in Engl. Pflanzenr. Euph. 15 : 8 (1922).

Ovary glabrous or nearly so :
Leaves minutely puberulous on the midrib beneath or glabrescent ; petals absent from male flowers ; leaves elliptic-lanceolate or elliptic-oblanceolate, rarely oblong, unequally cuneate or sometimes rounded at base, long-acuminate, 9·5–27 cm. long, 3·2–12 (–15) cm. broad ; male inflorescences very slender, 4–16 cm. long, sparingly pubescent ; female inflorescences up to 18 cm. long in fruit ; capsules 3-lobed, 6–7 mm. diam.   ..   ..   1. *leptobotrya*

Leaves with setose hairs on the midrib beneath, sometimes very sparse ; petals present
  in male flowers, linear-oblanceolate, ciliate at apex ; leaves elliptic or obovate-
  elliptic, cuneate to rounded at base, acuminate or obtuse at apex, 5–18 cm. long,
  2–9 cm. broad ; male inflorescences 2–8 cm. long, pubescent ; female inflorescences
  up to 10 cm. long in fruit ; capsules 3-lobed, about 9 mm. diam.   2. *stenopetala*
Ovary densely setose-pilose ; leaves oblong-elliptic, unequally rounded or obtuse at
  base, acuminate, 18–34 cm. long, 8–15 cm. broad, with very sparse setose-hairs on
  midrib beneath ; inflorescences up to 25 cm. long, densely pubescent on the rhachis ;
  petals absent from male flowers ; capsules 3-lobed, about 9 mm. diam.
                                                                  3. *cf. annobonae*

1. **T. leptobotrya** (*Müll. Arg.*) *Brenan* in Kew Bull. 1952 : 446 (1953). *Antidesma leptobotryum* Müll. Arg.
    (1864)—F.T.A. 6, 1 : 645 ; Pax & K. Hoffm. in Engl. Pflanzenr. Euph. 15 : 149. *Thecacoris gymnogyne*
    Pax (1899)—F.T.A. 6, 1 : 662 ; Pax & K. Hoffm. l.c. 15 : 12. *T. obanensis* Hutch. in F.W.T.A., ed. 1,
    1 : 283 (1928). *T. talbotae* Hutch. l.c. (1928), excl. *Chev.* 19502. A small forest tree, to 35 ft. high, or
    shrub ; inflorescences drooping, with green axes and yellowish flowers.
    S.Nig.: Oban *Talbot* 1503 ! 2330 ! s.n. ! Kwa Falls, Calabar (Mar.) *Brenan* 9232 ! New Ndebiji (fr. **May**)
    Ujor FHI 30153 ! 30159 ! **Br.Cam.:** Korup F.R., Kumba (fr. June) *Olorunfemi* FHI 30638 ! Also in
    French Cameroons, Spanish Guinea and Gabon.
2. **T. stenopetala** (*Müll. Arg.*) *Müll. Arg.*—F.T.A. 6, 1 : 659 ; Pax & K. Hoffm. l.c. 15 : 11. *Antidesma*
    stenopetalum Müll. Arg. (1864). *Thecacoris chevalieri* Beille (1910)—F.T.A. 6, 1 : 660 ; Chev. Bot. 567 ;
    Pax & K. Hoffm. l.c. ; F.W.T.A., ed. 1, 1 : 283. A forest shrub, 5–12 ft. high ; flowers greenish-yellow.
    S.L.: Bagroo R. (Apr.) *Mann* 833 ! Rowala (fr. July) *Thomas* 1034 ! Dumbuna (fl. & fr. Oct., fr. Jan.)
    Thomas 3132 ! 3322 ! 3324 ! *Deighton* 3593 ! Roruks (July, Nov.) *Thomas* 5783 ! *Deighton* 3262 ! **Lib.:**
    Monrovia (fl. & fr. Nov.) *Dinklage* 3291 ! 3292 ! *Barker* 1461 ! Kolahun (Nov.) *Baldwin* 10228 ! Loffa
    R., Boporo (fr. Dec.) *Baldwin* 10672 ! **Iv.C.:** Fort Binger to Mt. Niénokué (July) *Chev.* 19502 *bis* ! Man
    R., Mid. Cavally (Aug.) *Chev.* Malamalasso (Mar.) *Chev.* 17489. **G.C.:** Prestea (fl. & fr. Sept.) *Vigne*
    FH 3084 ! Simpa (May) *Vigne* FH 1954 ! Suhuma F.R. (fl. & fr. Nov.) *Cansdale* FH 3970 ! 3971 ! 3972 !
    Br.Cam.: Mokonyong, Mamfe (fr. Nov.) *Tamajong* FHI 22117 ! Kembong F.R., Mamfe (fr. Mar.)
    Onochie FHI 31184 ! **F.Po:** *T. Vogel* 228 ! Also in Principe and S. Tomé.
    [*T. batesii* Hutch. from the French Cameroons is doubtfully distinct from this rather variable species.]
3. **T. cf. annobonae** Pax & K. Hoffm. l.c. 15 : 9 (1922). An understorey forest tree, to 30 ft. high ; flowers
    greenish-yellow.
    Br.Cam.: S. Bakundu F.R. (Jan.) *Keay & Russell* FHI 28679 ! Mungo R., Kumba (Apr.) *Olorunfemi*
    FHI 30531 ! Kiliwindi—Mamfe road (fr. July) *Akuo* FHI 15153 ! [The type specimen from Annobon
    Island is not available but the description suggests that it may be conspecific with our material.]

7. **SPONDIANTHUS** Engl. Bot. Jahrb. 36 : 215 (1905) ; F.T.A. 6, 1 : 1044 ; Pax & K.
    Hoffm. in Engl. Pflanzenr. Euph. 15 : 13–15 (1922). *Megabaria* Pierre ex Hutch.
    (1910)—F.T.A. 6, 1 : 627.

Leaves obovate or elliptic, obtuse at apex, glabrous, up to 36 cm. long and 18 cm. broad,
  very unequal in size, borne on petioles of very unequal length ; inflorescence terminal,
  paniculate ; capsule ellipsoid, horny, 1·5–2 cm. long and 1–1·5 cm. broad ; seeds
  ellipsoid, about 1 cm. long, red, shining :
Inflorescence pubescent, paniculate, terminal ; leaves obovate or elliptic, obtuse,
  9–18 cm. long, 4–11 cm. broad, glabrous, on petioles of very unequal length ; lateral
  nerves 6–8 pairs ; petiole up to 8 cm. long, wrinkled at the apex ; capsule subsessile
  or shortly pedicellate, ellipsoid, horny, 2 cm. long, 1·5 cm. broad ; seeds ellipsoid,
  1 cm. long, red, shining   ..   ..   ..   1. *preussii* var. *preussii*
Inflorescence glabrous ; leaves more or less as above, but with 6–14 lateral nerves ;
  fruits ellipsoid, angular, 2·5 cm. long, red-brown, slightly pointed at the ends
                                                          1a. *preussii* var. *glaber*

1. **S. preussii** *Engl.* var. **preussii**—Bot. Jahrb. 36 : 216 (1905) ; Engl. in Notizbl. Bot. Gart. Berl. 5 : 241
    (1911) ; F.T.A. 6, 1 : 1044 ; Pax & K. Hoffm. l.c. 15 : 13, (as var. *genuinus*) ; Aubrév. Fl. For. C.
    Iv. 2 : 52, t. 138. *Megabaria trillesii* Pierre ex Hutch. (1910), partly—F.T.A. 6, 1 : 627, partly. *M.*
    ugandensis of Chev. Bot. 586 and of F.W.T.A., ed. 1, 1 : 282, partly. Tree, to 90 ft. high, in swampy areas
    within the rain forest ; flowers cream.
    Lib.: Fisherman's Lake, Grand Cape Mount (fr. Dec.) *Baldwin* 10894 ! Brewersville (fr. Dec.) *Baldwin*
    10973 ! **Iv.C.:** Agboville (fr. Nov.) *Chev.* 22381 ! Banco *Martineau* 309 ! Abidjan *Aubrév.* 123 ! Yapo
    (fr. Dec.) *Aubrév.* 590 ! **G.C.:** Atwabo (fr. Feb.) *Irvine* 2289 ! Aguna (Aug.) *Vigne* FH 1309 ! Benso (fr.
    Aug.) *Andoh* FH 5396 ! **Dah.:** Niaouli (fr. Feb.) *Aubrév.* 45d ! **N.Nig.:** Ibaji-Ojoku F.R. (fr. Feb.)
    Taylor L.10 ! **S.Nig.:** Akilla (fr. Jan.) *Thornewill* 204 ! Ama Ekpelima, Awka (Sept.) *Jones* FHI 6719 !
    Br.Cam.: Victoria (young fr. Apr., Oct.) *Maitland* 600 ! 739 ! Likomba (Oct.) *Mildbr.* 10587 ! Johann-
    Albrechtshöhe (fl. Aug., fr. Dec.) *Preuss* 426 ! *Staudt* 778. **F.Po:** Bokoko de San Carlos (fr. Jan.) *Guinea*
    834 ! Also in French Cameroons, Spanish Guinea, Gabon and (?) Belgian Congo. (See Appendix, p. 164.)
1a. **S. preussii** var. **glaber** (*Engl.*). *Engl.* Notizbl. Bot. Gart. Berl. 5 : 242 (1911) ; Pax & K. Hoffm. l.c. *S.*
    glaber Engl. Bot. Jahrb. 36 : 216 (1905). *Megabaria ugandensis* Hutch. (1910)—F.T.A. 6, 1 : 628.
    Spondianthus ugandensis (Hutch.) Hutch. in F.T.A. 6, 1 : 1044 (1913), partly ; F.W.T.A., ed. 1, 1 : 282,
    partly ; Aubrév. l.c. Tree, to 80 ft. high, in fringing forest.
    Fr.G.: Socoroula, Farana (Jan.) *Chev.* 20495 ! Dantilia (Mar.) *Sc. Elliot* 5390 ! **Iv.C.:** Kani (fl. & fr.
    Feb.) *Serv. For.* 2826 ! 2827 ! **S.Nig.:** R. Ata, Bebi, Obudu Div. (Dec.) *Keay & Savory* FHI 25008 !
    Br.Cam.: Ololo bush, Assumbo, Mamfe Div. (Feb.) *Lobe Babute* 5/37 ! Babungo to Babesse, Bamenda
    Prov. (Dec.) *Ledermann* 1981. Also in French Cameroons, Ubangi-Shari, A.-E. Sudan, Uganda, Tangan-
    yika, Belgian Congo and Angola. (See Appendix, p. 164.)

8. **MARTRETIA** Beille—F.T.A. 6, 1 : 655 ; Pax & K. Hoffm. in Engl. Pflanzenr.
    Euph. 15 : 79 (1922).

Leaves lanceolate to ovate-elliptic, gradually acuminate, shortly cuneate at base,
  10–22 cm. long, 3–9 cm. broad, glabrous, with 10–15 pairs of nerves slightly prominent
  on both surfaces ; flowers dioecious ; male racemes axillary, 5 cm. long ; sepals 4 ;

stamens 5–7 ; female sepals 5 or 6 ; ovary 2-lobed, each lobe slightly bifid ; styles linear, undivided   ..   ..   *quadricornis*

M. quadricornis *Beille*—F.T.A. 6, 1 : 655 ; Chev. Bot. 585 ; Pax & K. Hoffm. l.c. ; Aubrév. Fl. For. C. Iv. 2 : 60, t. 142. A small forest tree of swampy places, with glabrous branchlets.
S.L.: Mano *Thomas* 9913 ! Senihun, Kamagai (Sept.) *Deighton* 3051 ! Dimbaia, Samu (Sept.) *Adames* 251 ! **Iv.C.:** Bingerville (fr. Feb.) *Chev.* 17360 ! Dabou *Chev.* 16215. **S.Nig.:** Onitsha Dist. *Cons. of For.* OC 31 ! Also in Upper Ubangi region and Belgian Congo.

9. APODISCUS Hutch.—F.T.A. 6, 1 : 1045 ; Pax & K. Hoffm. in Engl. Pflanzenr. Euph. 15 : 45 (1922).

A tree ; branchlets glabrous ; leaves oblong-elliptic, acuminate, subcuneate at base, 8–17 cm. long, 3–8 cm. broad, entire, leathery, glabrous ; lateral nerves 8–14 pairs with intermediate ones less conspicuous ; spikes bisexual, with 1–2 shortly pedicellate females towards the base ; axis tomentose ; male sepals tomentellous ; stamens 5 ; ovary 4-celled, pubescent   ..   ..   *chevalieri*

A. chevalieri *Hutch.*—F.T.A. 6, 1 : 1046 ; Hook. Ic. Pl. 3032 ; Chev. Bot. 586 ; Pax & K. Hoffm. l.c. A tree, 25–30 ft. high.
Fr.G.: between Santa R. and Timbo (Mar.) *Chev.* 12588. By tributary of Bowali R. (Nov.) *Chev.* 14997 ! 15057 !

10. PROTOMEGABARIA Hutch.—F.T.A. 6, 1 : 656 ; Pax & K. Hoffm. in Engl. Pflanzenr. Euph. 15 : 43 (1922).

Leaves pubescent on the midrib and lateral nerves beneath ; rudimentary ovary bipartite ; leaf-blade elongate-oblong or obovate, rounded or shortly acuminate, 15–37 cm. long, 8–15 cm. broad, thinly papery ; lateral nerves 10–17 pairs ; male racemes about 7 cm. long, slender ; capsule subglobose or oblong-ellipsoid, 3–4 cm. diam.   ..   ..   1. *stapfiana*
Leaves glabrous on the midrib and lateral nerves ; rudimentary ovary undivided ; otherwise more or less as above, but male racemes 10–13 cm. long   2. *macrophylla*

1. **P. stapfiana** (*Beille*) *Hutch.* in Hook. Ic. Pl. t. 2929 (1911) ; F.T.A. 6, 1 : 656 ; Chev. Bot. 586 ; Pax & K. Hoffm. in Engl. Pflanzenr. Euph. 15 : 44 ; Aubrév. Fl. For. C. Iv. 2 : 58, t. 141. *Maesobotrya stapfiana* Beille (1910). *Spondianthus obovatus* Engl. (1911). A forest tree, to 80 ft. high, with stout glabrous branchlets, low branching and buttressed stem ; locally abundant in swampy parts of the rain forest.
S.L.: Falaba-Pujehun (Mar.) *Aylmer* 18 ! Gorahun (Apr.) *Deighton* 3650 ! 4106 ! **Lib.:** Dukwia R. (fr. Mar.) *Cooper* 109 ! 294 ! Webo Dist. (fr. June–July) *Baldwin* 6215 ! 6672 ! **Iv.C.:** Malamalasso (Mar.) *Chev.* 16249 ! Soubiré to Yaou, Sanvi (Mar.) *Chev.* 17782. **G.C.:** Kankan, E. Akim (Mar.) *Johnson* 596 ! Kwahu Dist. *Thompson* 92 ! Bawdia (fl. & fr. Apr.) *Vigne* 156 ! **S.Nig.:** Oban *Talbot* 2326 ! Calabar (Apr.) *Ejiofor* FHI 21889 ! **Br.Cam.:** Mamfe (Apr.) *Ndep Enoh* 42/38 ! Also in Gabon. (See Appendix, p. 158.)
2. **P. macrophylla** *Hutch.*—F.T.A. 6, 1 : 657 ; Pax & K. Hoffm. l.c. *Baccaurea macrophylla* Pax (1899), not of Müll. Arg. (1866). A forest tree, to 50 ft. high ; like the preceding and perhaps only varietally distinct from it.
G.C.: Benso, Tarkwa (Apr.) *Andoh* FH 5474 ! **S.Nig.:** Oban *Talbot* ! **Br.Cam.:** N. Korup F.R. (fr. June) *Olorunfemi* FHI 30654 ! Also in French Cameroons, Gabon, Principe and S. Tomé.

11. MAESOBOTRYA Benth.—F.T.A. 6, 1 : 663 ; Pax & K. Hoffm. in Engl. Pflanzenr. Euph. 15 : 17–26 (1922).

Stipules foliaceous, falcate, up to 2·5 cm. long, persistent or subpersistent ; leaves elliptic or oblong-elliptic, rounded at base, caudate-acuminate, 9–25 cm. long, 4–10 cm. broad, obscurely denticulate, setulose on the midrib and nerves beneath ; male and female racemes, up to 17 cm. long in dense clusters on the older wood, the former sometimes extending on to the leafy shoots ; ovary glabrous ; fruits obovoid, 8–9 mm. long, about 7 mm. broad, glabrous   ..   ..   1. *dusenii*
Stipules not foliaceous, very small, symmetrical, soon caducous :
Leaves distinctly pilose or setose on the upper surface ; ovary glabrous ; male inflorescences crowded on young woody shoots below the leaves ; female inflorescences in clusters, arising from the stem and larger branches, up to 9 cm. long ; fruits obovoid-ellipsoid, up to about 6 mm. long and 5 mm. broad ; leaves elliptic to obovate-elliptic, mostly rounded at base, 11–25 cm. long, 7–10 cm. broad ; branchlets, petioles and both surfaces of leaves hispid   ..   ..   ..   2. *staudtii*
Leaves glabrous above except on the midrib ; ovary glabrous or pilose :
Ovary and fruit glabrous ; axes of male racemes puberulous, usually sparingly so ; male inflorescences mostly on young shoots among or just below the leaves or on bole ; female inflorescences mostly on the bole and large branches ; young branchlets, petioles and underside of midribs sparsely to distinctly pilose with weak, yellowish (in dried specimens) hairs ; petioles up to 8 cm. long ; lamina membranous elliptic to elliptic-obovate, gradually acutely acuminate, 7–22·5 cm. long, 2·7–10 cm. broad ; fruits fleshy, about 12–14 mm. long and broad
3a. *barteri* var. *sparsiflora*
Ovary densely setose, fruit sparingly setulose ; axes of male racemes densely pubescent :
Midrib and main nerves beneath pale yellowish brown when dry, like the petioles and young shoots, sparsely to distinctly pilose ; male and female inflorescences

extending on to the bole, the female not normally found among the leaves ; flesh surrounding the seed red ; otherwise as above .. .. 3. *barteri* var. *barteri*
Midrib and main nerves beneath reddish when dry, densely pilose with long, patent hairs ; petiole and young shoots patent-pilose ; male and female inflorescences on the young shoots amongst the leaves ; flesh surrounding the seed bright blue ; leaves oblong-obovate or oblong-elliptic, abruptly acuminate, 7·5–17 cm. long, 4–7·5 cm. broad .. .. 4. *floribunda* var. *vermeuleni*

1. **M. dusenii** (*Pax*) *Hutch.* in F.T.A. 6, 1 : 664 (1912) ; Pax & K. Hoffm. in Engl. Pflanzenr. Euph. 15 : 19, fig. 4, A. *Staphysora dusenii* Pax (1897). A forest tree or shrub, to 45 ft. high ; branchlets pilose ; flowers and ripe fruits red.
   S.Nig.: Eket *Talbot*! Calabar (Mar.) *Brenan* 9230! Oban *Talbot* 11! 16! 609! 1581! R. Njujua, Shakwa, Obudu (May) *Latilo* FHI 30931! **Br.Cam.:** S. Bakundu F.R. (fl. Apr., fr. June) *Olorunfemi* FHI 30544! *Dundas* FHI 8358! Mombo-Bakosi, Kumba Div. (fr. May) *Olorunfemi* FHI 30557! **F.Po:** Bokoko *Mildbr.* 6832. Also in French Cameroons, Spanish Guinea.
2. **M. staudtii** (*Pax*) *Hutch.* in F.T.A. 6, 1 : 668 (1912) ; Pax & K. Hoffm. l.c. 15 : 23, fig. 4, B–E. *Baccaurea staudtii* Pax (1897). A forest tree, to 25 ft. high ; flowers white.
   S.Nig.: Ikom Div. (fl. July, fr. May) *Catterall* 28! *Latilo* FHI 30981! Baleghete, Obudu Div. (May) *Latilo* FHI 30953! **Br.Cam.:** Victoria *Preuss* 1336! Likomba (fr. Oct.) *Mildbr.* 10546! Abonando, Mamfe (May) *Rudatis* 63! Nkom-Wum F.R., Bamenda (fr. June) *Ujor* FHI 29251! Also in French Cameroons, Gabon and Belgian Congo.
3. **M. barteri** (*Baill.*) *Hutch.* in F.T.A. 6, 1 : 669 (1912) ; Pax & K. Hoffm. l.c. 15 : 24. *Pierardia barteri* Baill. (1863–64). A small forest tree with crooked bole, to 30 ft. high.
   S.L.: Bumbuna (fr. Oct.) *Thomas* 3395! **S.Nig.:** Lagos (Apr., June, Dec.) *Millen* 32! 41! 153! Epe *Barter* 3289! 3296! Idanre Hills (Jan.) *Brenan & Keay* 8667! Sapoba (young fr. Nov.) *Meikle & Keay* 575! Okomu F.R. (Mar.) *Richards* 3284! Calabar (Feb.) *Mann* 2299! **Br.Cam.:** Kumba Div. (fr. May) *Olorunfemi* FHI 30581! Also in French Cameroons and Spanish Guinea. (See Appendix, p. 149.)
3a. **M. barteri** var. **sparsiflora** (*Sc. Elliot*) *Keay* in Kew Bull. 1955 : 139. *M. sparsiflora* (Sc. Elliot) Hutch. in F.T.A. 6, 1 : 665 (1912) ; Chev. Bot. 565 ; Aubrév. Fl. For. C. Iv. 2 : 62, t. 144. *Baccaurea sparsiflora* Sc. Elliot (1894). *M. floribunda* var. *sparsiflora* (Sc. Elliot) Pax & K. Hoffm. l.c. 15 : 20 (1922). *M. edulis* Hutch. & Dalz. F.W.T.A., ed. 1, 1 : 284 (1928) ; Kew Bull. 1928 : 299. *Baccaurea edulis* A. Chev. Bot. 563, name only. A straggling forest tree, to 30 ft. high ; flowers reddish in bud ; fruits red with a glaucous bloom.
   Fr.G.: Farana (Mar.) *Sc. Elliot* 5311! Kouria, etc., *fide* Chev. *l.c.* **S.L.:** Luseniya (fr. Dec.) *Sc. Elliot* 4035! Sasseni, Scarcies (Jan.) *Sc. Elliot* 4543! Wallia (Jan.) *Sc. Elliot* 4746! Njala (fr. Oct.) *Deighton* 2532! **Lib.:** Dukwia R. (fr. Jan.) *Cooper* 53! Gbanga (fr. Sept.) *Linder* 736! Tappita (fl. & fr. Aug.) *Baldwin* 9066! 9107! **Iv.C.:** Agboville (fr. Nov.) *Chev.* 22376! Soubiré to Yaou (fr. Mar.) *Chev.* 17746! And several other localities *fide* Chev. *l.c.* **G.C.:** Tanosu (Feb.) *Brent* 7 *dd*! Aburi (fl. Aug., fr. Sept.) *Johnson* 1067! *Deighton* 3422 (partly)! (See Appendix, p. 150.)
4. **M. floribunda** Benth. var. **vermeuleni** (*De Wild.*) *J. Léonard* in Bull. Jard. Bot. Brux. 17 : 258 (1945). *Baccaurea vermeuleni* De Wild. (1910). Tree, about 12 ft. high ; flowers greenish-yellow ; fruit dehiscent, exposing the bright blue flesh surrounding the seed.
   S.Nig.: Oban (Mar.) *Onochie* FHI 34824! **Br.Cam.:** Bambui, Bamenda (fl. & fr. Jan.) *Keay* FHI 28420! Also in French Cameroons, Gabon, Cabinda and Belgian Congo.

12. ANTIDESMA Linn.—F.T.A. 6, 1 : 642 ; Pax & K. Hoffm. in Engl. Pflanzenr. Euph. 15 : 107 (1922).

Stipules laciniate, deeply 3–8-partite ; leaves oblong to oblanceolate-elliptic, rounded at base, acuminate at apex, 10–25 cm. long, 4–11 cm. broad ; racemes up to 25 cm. long ; ovary glabrous ; fruits ellipsoid, 5–7 mm. long :
Midribs and nerves beneath, petioles and branchlets shortly pubescent ; lobes of stipules 3–6, lanceolate .. .. 1. *laciniatum* var. *laciniatum*
Midribs and nerves beneath, petioles and branchlets densely clothed with long spreading hairs ; lobes of stipules 6–9, filiform, sometimes branched
          1a. *laciniatum* var. *membranaceum*
Stipules entire :
Leaves mostly rounded, obtuse or very shortly and obtusely acuminate at apex, obovate, or obovate-elliptic, or obovate-oblong, 3–10·5 cm. long, 1·5–5 cm. broad, rarely larger, tomentose to nearly glabrous beneath ; male inflorescences up to 8·5 cm. long, female up to 7·5 cm. long ; fruits ellipsoid, 6–8 mm. long, 5–7 mm. broad ; mostly in savannah .. .. 2. *venosum*
Leaves mostly distinctly and often long and acutely acuminate, up to 25 cm. long and 12 cm. broad ; male inflorescences up to 24 cm. long, female up to 42 cm. long ; in rain forest and fringing forest outliers :
Fruits 3–5 mm. long, 2–4 mm. broad ; leaves usually rather densely pubescent beneath, more or less oblong-elliptic, 4–15 cm. long, 2–6 cm. broad ; inflorescences up to 24 cm. long ; mostly in fringing forest outliers .. 3. *membranaceum*
Fruits 7–8 mm. long, 4–5 mm. broad ; leaves sparingly pilose mainly on midrib and main nerves beneath ; in rain forest :
Leaves obovate to elliptic-oblanceolate, usually caudate-acuminate, 7–25 cm. long, 3–12 cm. broad ; female inflorescences up to 42 cm. long, male up to 13 cm. long ; female calyx cup-shaped, very shortly lobed .. .. .. 4. *vogelianum*
Leaves elongate-oblong, usually rather broadly triangular-acuminate, 7·5–25 cm. long, 2·7–5·6 (–7) cm. broad ; female inflorescences up to 11·5 cm. long, male up to 22 cm. long ; female calyx deeply lobed .. .. 5. *oblonga*

1. **A. laciniatum** *Müll. Arg.* var. **laciniatum** in Flora 47 : 520 (1864) ; F.T.A. 6, 1 : 643 ; Pax & K. Hoffm. in Engl. Pflanzenr. Euph. 15 : 145, fig. 13 (as var. *genuinum*) ; Chev. Bot. 564. Small forest tree, to 30 ft. high ; flowers in pale red spikes ; fruits red when ripe.
   Iv.C.: *fide* Chev. *l.c.* **S.Nig.:** Lagos *Rowland*! Osun F.R. (Apr.) *Ejiofor* FHI 26120! Okomu F.R. (fl. & fr. Jan., Feb.) *Brenan* 8906! 8927! Oban *Talbot* 612! 2336! **Br.Cam.:** Victoria (Feb.–Apr.)

Preuss 1104! *Schlechter* 12363! Buea (fr. Mar.) *Maitland* 561! **F.Po:** (Jan., Feb., June) *Mann* 201! 256! *Barter*! Ureka (Feb.) *Guinea* 2339! Also in French Cameroons, Spanish Guinea and Belgian Congo. (See Appendix, p. 136.)

1a. **A. laciniatum var. membranaceum** *Müll. Arg.* l.c. (1864) ; F.T.A. 6, 1 : 644 ; Chev. Bot. 564 ; Pax & K. Hoffm. l.c. *A. laciniatum* of Aubrév. Fl. For. C. Iv. 2 : 60, t. 143, 2–6.
    Fr.G.: Lola to Nzo (Mar.) *Chev.* 20983! Macenta (May) *Collenette* 8! **S.L.:** Bagroo R. *Mann*! Kafogo, Limba (Apr.) *Sc. Elliot* 5607! Bandajuma (May) *Aylmer* 75! Gola Forest (fr. May) *Small* 695! **Lib.:** Peahtah (Oct.) *Linder* 964! Karmadun, Kolahun (Nov.) *Baldwin* 10180! Zuie, Boporo (fr. Dec.) *Baldwin* 10686! **Iv.C.:** Abidjan *Aubrév.* 200! **G.C.:** Kwahu (Mar.) *Johnson* 637! Axim Dist. (Apr.) *Chipp* 153! Owabi (fr. Sept.) *Andoh* FH 4424! Kumasi (Oct.) *Vigne* 1395! **Togo:** Misahöhe *Bauman* 529. Also in French Cameroons, Spanish Guinea and Belgian Congo.

2. **A. venosum** *Tul.* in Ann. Sci. Nat., sér. 3, 15 : 232 (1851) ; F.T.A. 6, 1 : 646, partly ; Chev. Bot. 565 ; Pax & K. Hoffm. l.c. 139, partly ; Aubrév. l.c. 2 : 60 ; Fl. For. Soud.-Guin. 182, t. 33, 4–5, partly. A shrub or small tree, mostly in savannah ; flowers yellow ; ripe fruits black, edible ; inflorescence often attacked by a gall insect and then more or less paniculate.
    Gam.: *Hayes*! Albreda *Leprieur*! *Perrottet* 726! **Fr.Sud.:** Koulikoro & Bandiagara *fide* Aubrév. *l.c.* **Fr.G.:** Kouroussa (May) *Pobéguin* 246! Mamou, Timbo, Labé & Ditinn *fide* Aubrév. *l.c.* **S.L.:** Njala (fl. May, young fr. June) *Deighton* 2998! 3243! 4019! Musaia *Sc. Elliot* 5105! Erimakuna (Mar.) *Sc. Elliot* 5259! Batkanu (young fr. Apr.) *Thomas* 48! **Iv.C.:** several localities *fide* Aubrév. *l.c.* **G.C.:** Afram Plains (May) *Kitson* 1119! Near Kumbungu (May) *Kitson* 754! 757! **Togo:** Kpandu (Apr.) *Robertson* 137! Lome *Warnecke* 468! **Dah.:** Savalou, Bassila & Djougou *fide* Aubrév. *l.c.* **Fr.Nig.:** Diapaga *Chev.* 24402. **N.Nig.:** Nupe *Barter* 1696! Kontagora (May) *Dalz.* 389! Riruwe F.R., Kano (Mar.) *Onwudinjoh* FHI 22356! Randa (fl. & fr. June) *Hepburn* 64! Yola (May) *Dalz.* 200! **S.Nig.:** Lagos (Apr.) *Foster* 97! Olokemeji *Foster* 92! Nchan, Bamenda (Apr.) *Maitland* 1475! Widespread in tropical and S. Africa. (See Appendix, p. 136.)

3. **A. membranaceum** *Müll. Arg.* in Linnaea 34 : 68 (1865), excl. var. *molle* ; in DC. Prod. 15, 2 : 261 (1866), var. *glabrescens* only ; F.T.A. 1 : 645, partly ; Chev. Bot. 564, partly ; Pax & K. Hoffm. l.c. 141, partly. *A. meiocarpum* J. Léonard in Bull. Jard. Bot. Brux. 27 : 260, fig. 23 (1945). *A. venosum* of F.T.A. 6, 1 : 646, partly ; of Pax & K. Hoffm. l.c. 139, partly ; of F.W.T.A., ed. 1, 1 : 284, partly ; not of Tul. A shrub or tree, up to 40 ft. high ; flowers yellowish-green ; ripe fruits blackish ; mostly in fringing forest in the savannah regions.
    [This is best distinguished from *A. venosum* and the following species by the very small fruits ; male plants are often difficult to name.]
    Port.G.: Catió (July) *Esp. Santo* 2126! 2127! **Fr.G.:** Boulivel to Dalaba (fr. Sept.) *Chev.* 18683! **S.L.:** Binkolo (fr. Aug.) *Thomas* 1691! Erimakuna (Mar.) *Sc. Elliot* 5393! **Iv.C.:** Mt. Boho, Zoanlé (May) *Chev.* 21490! Ouodé to Gouréni, Toura country *Chev.* 21633! **G.C.:** Kintampo (fr. July) *Chipp* 527! Abetifi-Kwahu (Mar., Apr.) *Johnson* 634! 636! Mampong (fr. May) *Kitson* 557! Ejura (fr. Aug.) *Chipp* 748! **Togo:** *Baumann* 242! Bo Mts. *Kersting* A. 635! **N.Nig.:** Lokoja (fr. Aug.) *Elliott* 189! Nupe *Barter* 1557! **S.Nig.:** Olokemeji *Foster* 92! Ilaro Dist. (fr. Oct.) *Onochie* FHI 13528! Ibadan (fr. Aug.) *Adekunle* FHI 24215! Also in French Cameroons, Belgian Congo, Angola, Uganda, Tanganyika and Mozambique. (See Appendix, p. 136.)
    [This is best distinguished from *A. venosum* and the following species by the very small fruits ; male plants are often difficult to name.]

4. **A. vogelianum** *Müll. Arg.* in Flora 47 : 529 (1864) ; F.T.A. 6, 1 : 645 ; Pax & K. Hoffm. l.c. 119. *A. membranaceum* of F.T.A. 6, 1 : 645, partly ; of Pax & K. Hoffm. l.c. 141, partly ; not of Müll. Arg. A shrub or tree, to 30 ft. high ; inflorescences pendulous, very conspicuous ; filaments bright red ; ovary red ; fruits red ripening to purple ; in rain forest.
    S.Nig.: Ibo country (fr. Aug.) *T. Vogel* 23! Eket Dist. (fr. May) *Talbot* 3072! 3114! *Onochie* FHI 32098! 32100! Aboabam, Ikom (Dec.) *Keay* FHI 28207! Oban *Talbot* 611! 670! **Br.Cam.:** Victoria *Preuss* 1321! Mamfe (young fr. Jan.) *Maitland* 1158! Abonando, Mamfe (Apr.) *Rudatis* 43! Also in French Cameroons, Gabon, Belgian Congo and E. Africa.

5. **A. oblonga** (*Hutch.*) *Keay* in Bull. Jard. Bot. Brux. 26 : 184 (1956). *Maesobotrya oblonga* Hutch. in F.T.A. 6, 1 : 668 (1912) ; Pax & K. Hoffm. l.c. 22 ; F.W.T.A., ed. 1, 1 : 283. *Antidesma membranaceum* of Chev. Bot. 564, partly ; of F.W.T.A., ed. 1, 1 : 284, partly (*Linder* 802) ; not of Müll. Arg. A shrub or small tree, 4–10 ft. high ; inflorescences pendulous ; flowers yellowish ; fruits reddish ; in rain forest.
    Lib.: Kakatown *Whyte*! Gbanga (Sept.) *Linder* 568! 783! 802! Dukwia R. (fl. & fr. Feb.) *Cooper* 149! 158! 159! 212! Sangwin, Sinoe Co. (Mar.) *Baldwin* 11319! Gletown, Tchien Dist. (fr. July) *Baldwin* 6753! **Iv.C.:** Guidéko (fr. May) *Chev.* 16415! Mbago, Agnéby valley (fr. Feb.) *Chev.* 17129! Mid. Sassandra to Mid. Cavally (fr. June) *Chev.* 19204!

Imperfectly known species.

A. sp. A small tree, up to 25 ft. high ; female inflorescences shorter than the leaves. Fruits and male flowers not known.
    Br.Cam.: Buea, 3,000 ft. (Jan., Mar.) *Maitland* 191! 451! Bali-Ngemba F.R., Bamenda (June) *Ujor* FHI 30410!

13. CYATHOGYNE Müll. Arg.—F.T.A. 6, 1 : 653 ; Pax & K. Hoffm. in Engl. Pflanzenr. Euph. 15 : 40 (1922).

A herb ; leaves obovate-elliptic, shortly acuminate, acute or subacute at base, 7–11 cm. long, 4–5 cm. broad, coarsely repand-dentate except in the lower third, glabrous or nearly so ; lateral nerves 5–6 pairs ; female racemes 4 cm. long..   *viridis*

C. viridis *Müll. Arg.*—F.T.A. 6, 1 : 654 ; Pax & K. Hoffm. l.c. 15 : 41 (incl. var. *preussii*). *C. preussii* Pax (1897)—F.T.A. 6, 1 : 655 (1912) ; F.W.T.A., ed. 1, 1 : 284. Forest herb, about 1 ft. high, from a horizontal rhizome.
    S.Nig.: Umon Ndealichi F.R., Calabar (Apr.) *Akpata* FHI 3936! **Br.Cam.:** Victoria *Preuss* 1210! Also in French Cameroons, Spanish Guinea, Gabon and Belgian Congo.

14. HYMENOCARDIA Wall. ex Lindl.—F.T.A. 6, 1 : 648 ; Pax & K. Hoffm. in Engl. Pflanzenr. Euph. 15 : 72–78 (1922).

Axis of the fruit projecting from the middle of an obscure sinus, the wings spreading horizontally ; leaves elliptic to broadly ovate or rounded, cordate at base, obtuse at apex, 3·5–9 cm. long, 2·2–7 cm. broad, papery, midrib and lateral nerves pubescent beneath ; male spikes 2·5–5 cm. long ; fruits 2·5–3 cm. broad, 1·4–1·6 cm. long, reticulate, glabrous or pubescent   ..   ..   1. *heudelotii*

Axis of the fruit not projecting from the middle of the sinus, the wings well-developed and ascending :

W.E.T.

FIG. 132.—HYMENOCARDIA ACIDA *Tul.* (EUPHORBIACEAE).

A, male shoot.   B, male flower with rudimentary ovary.   C, D, stamen back and front.   E, female shoot.   F, female flower.   G, fruiting branch.

Leaves ovate-elliptic, rounded at base, acuminate apex, papery, 3·5–8·5 cm. long, 1–4 cm. broad, laxly glandular beneath, hairy only in axils of nerves ; wings of the cocci almost touching at the top, forming a very narrow sinus, fruits lyrate, 2–2·5 cm. long ; forest species   ..   ..   2. *lyrata*

Leaves elliptic oblong, obtuse or rounded at both ends, coriaceous, 4–9 cm. long, 1·5–3 (–4) cm. broad, densely glandular beneath, pubescent when young but usually becoming glabrous ; wings of the cocci divergent, forming a broad sinus at the top ; fruits obcordate, about 2·5 cm. long ; savannah species ..   ..   ..   3. *acida*

1. **H. heudelotii** *Müll. Arg.*—F.T.A. 6, 1 : 650 ; Chev. Bot. 567 ; Pax & K. Hoffm. in Engl. Pflanzenr. Euph. 15 : 74, fig. 8, C ; Aubrév. Fl. For. C. Iv. 2 : 48, t. 137, A, B, E ; Fl. For. Soud.-Guin. 184, t. 34, 8–9. *H. chevalieri* Beille (1908)—F.W.T.A., ed. 1, 1 : 286 ; Pax & K. Hoffm. l.c. 75, fig. 8, D ; Aubrév. ll.c. ; Berhaut Fl. Sen. 131. Shrub or small tree, to 30 ft. high, locally abundant in riverain forest.
Sen.: Niokolo-Koba *Berhaut* 1471 ; 4502. **Fr.Sud.**: Kita *Dubois* 197 ! **Port.G.**: Saltinho (fr. May) *Esp. Santo* 2488 ! Pirada, Gabu (fr. June) *Esp. Santo* 3038 ! **Fr.G.**: *Heudelot* 838 ! Konkouré to Timbo (Mar.) *Chev.* 12458 ! **S.L.**: Pujehun (fr. Apr.) *Aylmer* 66 ! Njala (fl. Mar., fr. Sept.) *Deighton* 1935 ! 1996 ! 5011 ! Musaia (fl. Mar., fr. Oct., Dec.) *Sc. Elliot* 5104 ! *Thomas* 2620 ! *Deighton* 4438 ! Berria to Falaba (Feb.) *Sc. Elliot* 4976 ! Madina, Limba (fr. Apr.) *Sc. Elliot* 5899 ! **Iv.C.**: various localities *fide* Aubrév. *l.c.* **G.C.**: *Easmon* ! R. Volta below Kpong (fr. Apr.) *Irvine* 2886 ! **Togo**: Kete Krachi *Vigne* FH 3881 ! **N.Nig.**: Nupe *Barter* 1150 ! Lokoja *Barter* 383 ! Idah *Barter* 327 ! **S.Nig.**: Onitsha *Jones* FHI 4838 ! Also in Belgian Congo.
[The specimens with pubescent fruits, previously named *H. chevalieri*, seem, at the most, no more than varietally distinct from those with glabrous fruits. The forms are not otherwise distinguishable.]

2. **H. lyrata** *Tul.*—F.T.A. 6, 1 : 651 ; Pax & K. Hoffm. l.c. 15 : 75, fig. 8, F ; Aubrév. Fl. For. C. Iv. 2 : 48, t. 137, C ; Fl. For. Soud.-Guin. 184, t. 34, 6–7. *H. beillei* A. Chev. Bot. 567, name only. *Thouiniu ? dicarpa* Turcz. (1863). Shrub or small tree, to 50 ft. high, often in riverain forest.
Sen.: Casamance (fr. Dec.) *Heudelot* 581 ! **Port.G.**: Bubaque, Bijagos (fr. Nov.) *Esp. Santo* 2329 ! **Fr.G.**: Dyeke (fr. Oct.) *Baldwin* 9671 ! **S.L.**: *Smeathmann* ! Waterloo (fr. May) *Deighton* 2654 ! Mano Bonjema (fr. Apr.) *Deighton* 3711 ! 3712 ! Njala (fl. & fr. Apr.) *Deighton* 648 ! 1123 ! Rowala (fr. July) *Thomas* 1024 ! **Lib.**: Peahtah (fr. Oct.) *Linder* 902 ! Ganta (fr. May) *Harley* 1159 ! Browntown, Tchien (fr. Aug.) *Baldwin* 7052 ! Robertsport (fr. Dec.) *Baldwin* 10891 ! **Iv.C.**: Mt. Niénoku, Mid. Cavally *Chev.* 16454. And other localities *fide* Aubrév. *ll.c.* **G.C.**: Aframso (fr. Oct.) *Vigne* FH 4037 ! Otrokpe (fr. Oct.) *Vigne* FH 4012 !
[Milne-Redhead in Kew Bull. 1947 : 46, described *Deighton* 3710 from Mano Bonjema and *Deighton* 3185 from Mano (Luawa), Sierra Leone as in all probability natural hybrids between *H. lyrata* and *H. heudelotii*. In addition *Deighton* 4284 from Njala may be a hybrid between these parents, and *Linder* 930 from Peahtah, Liberia has fruits which appear intermediate between *H. lyrata* and pubescent-fruited forms of *H. heudelotii*.]

3. **H. acida** *Tul.*—F.T.A. 6, 1 : 651 ; Chev. Bot. 566 ; Pax & K. Hoffm. l.c. 15 : 76, fig. 9, A–D ; Aubrév. Fl. For. C. Iv. 2 : 48, t. 137, D ; Fl. For. Soud.-Guin. 186, t. 34, 1–3. A savannah shrub or small tree about 20 ft. high ; branchlets becoming rusty-powdery when the bark peels off.
Sen.: Sedhiou (fr. Feb.) *Chev.* 2626 ! **Gam.**: Albreda *Leprieur* ! **Fr.Sud.**: Guélia *Chev.* 311. **Port.G.**: Uno-Am-Onho (fr. May) *Esp. Santo* 2011 ! **Fr.G.**: *Heudelot* 779 ! Kouroussa (Feb.) *Pobéguin* 658 ! **S.L.**: Musaia (fl. Mar., fr. Mar.–Feb.) *Deighton* 5356 ! 5465 ! Kabala (Feb.) *Deighton* 4163 ! 4164 ! Talla Hills (Feb.) *Sc. Elliot* 4886 ! **Iv.C.**: Zoanlé *Chev.* 21489. **G.C.**: Abetifi-Kwahu *Thompson* 86 ! Kumawo (fr. June) *Chipp* 459 ! Grube (fr. Mar.) *Rea* FH 1667 ! **Togo**: Kpedsu (fr. Dec.) *Howes* 1050 ! Amedzofe (fr. May) *Scholes* 56 ! **Dah.**: various localities *fide* Chev. *l.c.* **N.Nig.**: Nupe *Barter* 1098 ! 1207 ! 1558 ! Kontagora (Jan., Feb.) *Dalz.* 78 ! Sokoto (Mar.) *Lely* 852 ! Bauchi (Jan.) *Lely* P. 61 ! **S.Nig.**: Lagos *Moloney* ! Olokemeji (fr. Nov.) *A. F. Ross* 19 ! Ubiaja (fr. Aug.) *Oyebade* FHI 20402 ! **Br.Cam.**: Bamenda (fr. May) *Maitland* 1534 ! Widespread in tropical Africa. (See Appendix, p. 146.)

Imperfectly known species.

H. intermedia *Dinkl. ex Mildbr.* in Fedde Rep. 41 : 255 (1937), name only. *H. chevalieri* of Pax & K. Hoffm. l.c. 15 : 75, partly (*Dinklage* 2483). Possibly *H. lyrata*, or a hybrid.
Lib.: *Dinklage* 2483 ; 2686.

15. **DRYPETES** Vahl—F.T.A. 6, 1 : 674 ; Pax & K. Hoffm. in Engl. Pflanzenr. Euph. 15 : 229–279. *Cyclostemon* Blume (1825).

KEY TO SPECIMENS WITH MALE FLOWERS.

*Stamens inserted around the margin of the disk, free, or enclosed by marginal folds but not in the central part of the disk ; flowers in several- to many-flowered fascicles :
 †Stamens inserted outside the thin cup-shaped disk, not enclosed by folds ; flowers in axillary fascicles on leafy shoots ; stamens 4–8 ; sepals 4, rarely 5, up to 3 mm. long ; leaves cuneate or obtuse at base :
   Sepals puberulous or tomentellous outside ; stamens 4–6 :
    Disk glabrous within :
     Sepals normally 5, tomentellous outside, puberulous within ; stamens 4–6 ; leaves elliptic to lanceolate, obtusely acuminate, 10–25 cm. long, 5–10·5 cm. broad, crenate-serrulate to subentire, lateral nerves 5–7 on each side of midrib, long-ascending ; branchlets and leaves glabrous or nearly so   ..   1. *pellegrini*
     Sepals normally 4 ; stamens 4 ; leaves up to 12 cm. long and 6 cm. broad, branch-lets and petioles pubescent, rarely glabrous ; leaves entire or subentire :
      Disk with a small conical projection in the middle ; leaves rhomboid-lanceolate, shortly obtusely acuminate, 2·5–6 cm. long, 0·8–2·5 cm. broad, main lateral nerves 3–4 on each side of midrib   ..   ..   2. *klainei*
      Disk without a conical projection in the middle ; leaves up to 12 cm. long and 6 cm. broad :
       Leaves with numerous close, parallel, lateral nerves ; lamina ovate-elliptic to elliptic-lanceolate, caudate-acuminate, 6–11 cm. long, 1·5–4·5 cm. broad
                                                                              3. *aubrevillei*

Leaves with 4–7 main lateral nerves on each side of midrib, the secondary nerves
and venation not parallel ; lamina elliptic to elliptic-lanceolate, 3·5–12 cm.
long, 1·5–6 cm. broad ..   ..   **4.** *leonensis*
Disk and sepals pubescent within ; leaves oblong to oblong-elliptic, caudate-
acuminate, 7·5–12·5 cm. long, 3·5–4·5 cm. broad, entire, lateral nerves 10–11 on
each side of midrib, with parallel secondary nerves ; branchlets and petioles
puberulous ..   ..   **5.** *paxii*
Sepals, disk, branchlets and leaves glabrous ; sepals 4 ; stamens 7–8 ; leaves elliptic
to elliptic-lanceolate, unequal at base, shortly acuminate, 5–19 cm. long, 2·5–9 cm.
broad, entire ..   ..   **6.** *aframensis*
†Stamens inserted between the marginal lobes of the disk ; sepals usually 5 :
Disk more or less plano-convex and rugose ; sepals not exceeding 9 mm. long and
6 mm. broad ; trees and shrubs to 25 m. high, not, so far as is known, smelling
strongly of horse-radish :
Stamens 4–22 :
Leaves more or less spiny-dentate, coriaceous, ovate-elliptic to oblong or oblong-
lanceolate, rounded or obtuse, and often unequally sided at base, shortly acuminate, 3·5–11·5 cm. long, 1–5·5 cm. broad ; branchlets and petioles shortly
pubescent ; flowers on main stem and branches, rarely among the leaves ; sepals
glabrous outside except for ciliate margins ; stamens 10–15 ; disk glabrous
7. *floribunda*
Leaves not spiny-dentate :
Stamens 4–10 :
Sepals oblong-lanceolate or oblong, about 3 mm. long and 1·5 mm. broad,
glabrous or sparsely puberulous outside ; stamens 4–6 ; flowers in fascicles on
the branches among and below the leaves ; stipules woody, subpersistent,
6–10 mm. long ; leaves oblong-elliptic, cuneate to obtuse at base, obtuse at
apex, 30–43 cm. long, 9–15 cm. broad, glabrous, coriaceous, margin undulate,
main lateral nerves 8–11 on each side of midrib   ..   ..   **8.** *similis*
Sepals orbiculate to orbicular-ovate, at least 2·5 mm. broad ; stipules not woody ;
leaves up to 28 cm. long and 11 cm. broad :
Sepals puberulous outside ; stipules early caducous ; flower fascicles axillary on
leafy twigs or just below the leaves ; trees to 10 m. high :
Disk glabrous ; leaves glabrous and coriaceous :
Sepals 2·5–3 mm. long, 2·5 mm. broad ; stamens 5–8 ; leaves oblong to
oblong-elliptic, cuneate to obtuse at base, acuminate, 12–23 cm. long,
3·5–8·5 cm. broad, margin obscurely denticulate, main lateral nerves 5–6
on each side of midrib ; flowers mostly below the leaves   .. **9.** *aylmeri*
Sepals about 3·5 mm. long and 3 mm. broad ; stamens 9 ; leaves oblong-
elliptic, obtuse to cuneate and unequal at base, shortly acuminate, 15–25
cm. long, 6·5–11 cm. broad, margin undulate, main lateral nerves 7–9 on
each side of midrib ; flowers among the leaves   **10.** *fernandopoana*
Disk pilose in centre ; stamens 8–10 ; branchlets, petioles and midribs beneath
more or less densely pubescent at first, then glabrescent ; leaves obovate-
oblong to oblong-lanceolate, cuneate to obtuse at base, shortly acuminate,
9–23 cm. long, 2–10 cm. broad, crenate-dentate to subentire

11. *principum*
Sepals glabrous outside, except for the ciliate margins ; stamens 8 ; flowers on
the main stem ; branchlets, petioles and midribs beneath densely pubescent,
the latter glabrescent ; stipules subulate, subpersistent, about 5 mm. long ;
leaves lanceolate-elliptic to elliptic or oblong-obovate, unequal at base, one
side cuneate, the other rounded or obtuse, rather long-acuminate, 10–28 cm.
long, 4–8 cm. broad, denticulate to subentire ; small tree, to 4 m. high
12. *ivorensis*
Stamens 12–22 ; flowers in fascicles on the main stem or on leafless branches :
Stipules persistent, 5–10 mm. long ; sepals ovate, about 6–9 mm. long and
4–6 mm. broad, sparingly pubescent outside ; stamens 12–18 ; flowers on the
main stem ; leaves elongate-oblong to obovate-oblong, cuneate to obtuse
at base, abruptly long-acuminate, 11–27 cm. long, 4–9 cm. broad, lateral
nerves rather strong beneath and looping, impressed above, margin crenate
13. *molunduana*
Stipules early caducous ; sepals glabrous outside except for the ciliate margins ;
leaves with 4–7 main lateral nerves on each side of midrib :
Branchlets, petioles and leaves glabrous ; sepals 5–7 mm. long, 5–5·5 mm.
broad ; disk and anthers pubescent ; flowers on main stem and lateral
branches ; pedicels about 1·5 cm. long ; leaves oblong to oblong-elliptic,
cuneate at base, acutely acuminate, 10–20 cm. long, 4–8 cm. broad, margin
subentire ..   ..   **14.** *afzelii*
Branchlets and petioles pubescent ; sepals up to 4 mm. long and broad ; anthers
glabrous ; leaves subobtuse to rounded at base, margins denticulate :

Pedicels scarcely 3 mm. long ; sepals 3·5 mm. long and 3 mm. broad ; disk
pubescent ; leaves ovate-oblong to lanceolate-ovate, cuspidate-acuminate,
5–11 cm. long, 1·5–5·5 cm. broad, midrib pubescent above, glabrous beneath ;
flowers on the older branches   ..   ..   15. *obanensis*
Pedicels (5-) 10–12 mm. long ; sepals 4 mm. diam. ; disk glabrous ; leaves
oblong to obovate-lanceolate, acuminate, 9–20 cm. long, 2·5–7 cm. broad,
midrib glabrous above, pubescent beneath; flowers on the main stem
16. *preussii*

Stamens about 30 ; sepals ovate, 7–8 mm. long, about 5 mm. broad, glabrous
outside ; flowers in fascicles on woody branches ; branchlets angular, glabrous ;
petiole and leaves glabrous ; leaves oblong-elliptic, unequal at base, one side
cuneate or obtuse, the other rounded to cordate, gradually acuminate at apex,
26–40 cm. long, 8·5–13 cm. broad   ..   .. 17. *staudtii*
Disk cup-shaped, the folded margins higher than the centre which has rudimentary
linear styles ; stamens about 30 ; sepals ovate, 12–16 mm. long, 8–10 mm. broad,
densely brownish tomentellous ; pedicels up to about 2 cm. long ; flowers in axils
of lower leaves and just below the leafy shoots ; leaves more or less oblong,
unequally obtuse or cuneate at base, abruptly acuminate, 11–24 cm. long, 3·6–7·5
cm. broad, crenate-dentate, glabrous ; tree to 42 m. high, cut parts emitting a
strong smell of horse-radish   ..   .. 18. *gossweileri*
*Stamens deeply enclosed by folds near the centre as well as at the margin of the much-
folded disk ; flowers axillary on leafy shoots ; leaves distichous, with very short
petioles, usually oblique and markedly unequal at the base (but see No. 24), margins
usually serrate :
Pedicels 3·5–10 cm. long, slender ; sepals 5, glabrous outside, stamens about 25 ;
branchlets and leaves glabrous ; leaves oblong-elliptic or elliptic-lanceolate, rounded
at base with proximal side of the lamina inserted on the petiole slightly below the
distal, acutely caudate-acuminate, 7–18 cm. long, 3·5–6·8 cm. broad   19. *capillipes*
Pedicels up to 8 mm. long ; stamens 8–15 (–20) :
Flowers distinctly pedicellate, solitary or 2–3 together :
Sepals tomentellous or villous outside ; young branchlets densely pubescent to
pilose :
Proximal side of leaf-lamina narrowed to an obtuse or subcordate base inserted on
the petiole below the much broader distal side which is rounded at base ; lamina
obliquely oblong or ovate, acutely caudate-acuminate, 5–11 cm. long, 2·5–4·5
cm. broad ; branchlets spreading-pilose ; sepals 5, villous ..   20. *inaequalis*
Proximal side of leaf-lamina broadly auriculate at base, but inserted on the petiole
at the same level as the distal side which is rounded or subcordate at base ;
lamina ovate to oblong-lanceolate, gradually long-acuminate, 5–18 cm. long,
2–7·5 cm. broad ; young lateral shoots with suborbicular amplexicaul basal
leaves ; branchlets spreading-pubescent ; sepals 4, tomentellous
21. *chevalieri*
Sepals glabrous outside except for the margins, 4 ; young branchlets sparingly
pubescent to glabrescent :
Proximal side of leaf-lamina inserted on the petiole below the distal, both sides
obtuse to subcordate ; lamina ovate- to oblong-elliptic, caudate-acuminate,
4–12 cm. long, 1·5–6 cm. broad ; pedicels up to 8 mm. long   22. *gilgiana*
Proximal and distal sides of the leaf-lamina inserted on the petiole at the same
level, rounded to subcordate ; lamina ovate to ovate-lanceolate, subacutely
acuminate, 4–12 cm. long, 1·3–4·5 cm. broad ; pedicels up to 2 mm. long
23. *parvifolia*
Flowers sessile or subsessile, in dense glomerules ; sepals 4, pubescent outside ;
stamens 8–10 ; branchlets and leaves glabrous ; leaves obliquely oblong-lanceolate,
proximal side of lamina inserted on the petiole below the level of the distal side,
6–11 cm. long, 3–5 cm. broad   ..   .. 24. *tessmanniana*

See also No. 25 D. *occidentalis*, a species represented only by the type specimen which has no male flowers.

KEY TO SPECIMENS WITH FEMALE FLOWERS AND FRUITS.

Ovary and fruit 5–7-celled ; fruits 3·3–10 cm. diam., depressed-subglobose :
Leaves 11–24 cm. long, 3·6–7·5 cm. broad, unequally obtuse or cuneate at base,
abruptly acuminate ; fruits 8–10 cm. diam. 5–6 cm. long, tomentellous ; flowers in
axils of lower leaves and just below the leafy shoots ; tree to 42 m. high, cut parts
emitting a strong smell of horse-radish   ..   .. 18. *gossweileri*
Leaves 26–40 cm. long, 8·5–13 cm. broad, unequal at base, one side cuneate or obtuse,
the other rounded to cordate, gradually acuminate ; fruits about 3·3 cm. diam.,
scurfy-puberulous ; flowers on main stem ; tree to 11 m. high, not, so far as is
known, smelling strongly of horseradish   ..   .. 17. *staudtii*
Ovary and fruit 1–3 (–4)-celled ; fruits (where known) up to 2·7 cm. diam. ; trees and
shrubs up to 25 m. high, not, so far as is known, smelling strongly of horse-radish :
*Ovary and fruit 3 (–4)-celled ; flowers on main stem and older branches :

Fruits 3-lobed, ellipsoid, about 1·8 cm. long and 1·1 cm. diam., glabrous

                                    16. *preussii*

Fruits subglobose, not lobed :

  Fruits glabrous, about 1·5 cm. diam. ; leaves more or less spiny-dentate

                                    7. *floribunda*

  Fruits puberulous, 2–2·7 cm. diam. ; leaves not spiny-dentate   12. *ivorensis*

*Ovary and fruit 1–2-celled :

Ovary and fruit 2-celled :

  Pedicels 3·5–10 cm. long ; flowers 1–2 in leaf-axils ; fruits ellipsoid up to 2·2 cm.

    long and 1·7 cm. broad, minutely tomentellous, later glabrescent 19. *capillipes*

  Pedicels up to 2·5 cm. long :

    Sepals normally 5 :

      Flowers solitary in leaf-axils ; branchlets, petioles and midribs beneath densely

        hairy ; leaves serrate :

        Leaves subsessile, obliquely oblong or ovate, proximal side narrowed to an

          obtuse or subcordate base, inserted on the petiole below the much broader

          distal side which is rounded at base, acutely caudate-acuminate, 5–11 cm. long,

          2·5–4·5 cm. broad ; branchlets spreading-pilose ; fruits subglobose, up to

          3 cm. diam. densely villous, stigmas close together..   20. *inaequalis*

        Leaves with petiole up to 1 cm. long, oblong-ovate, unequal and rounded or

          obtuse at base, acuminate, 10–16 cm. long, 4–8 cm. broad ; branchlets

          tomentose ; ovary villous ; styles linear, spreading ; fruits not known

                            25. *occidentalis*

      Flowers in fascicles in leaf-axils or on leafless branches or on main stem :

        Stipules persistent, 5–10 mm. long ; sepals ovate, about 6–9 mm. long and 4–6

          mm. broad, sparingly pubescent outside ; flowers on the main stem ; fruits

          ovoid, or ellipsoid, 2·8–3 cm. long, up to 2·4 cm. diam. 13. *molunduana*

        Stipules caducous :

        Sepals glabrous outside except for the ciliate margins, 5–7 mm. long, 5–5·5 mm.

          broad ; flowers on main stem and leafless branches ; fruits ellipsoid, about

          2 cm. long and 1·7 cm. diam., densely puberulous   14. *afzelii*

        Sepals puberulous or tomentellous outside ; sepals up to 4 mm. long and broad :

          Sepals 2·5–3 mm. long ; flower fascicles in axils of leaves ; leaves elliptic to

            lanceolate, obtusely acuminate ; branchlets and leaves glabrous or nearly

            so ..   ..   1. *pellegrini*

          Sepals 4 mm. long ; flower fascicles in axils of leaves and on leafless branches ;

            leaves obovate-oblong to oblong-lanceolate, shortly acuminate ; branchlets,

            petioles and midribs beneath more or less densely pubescent at first, then

            glabrescent ..   ..   11. *principum*

    Sepals normally 4 ; flowers mostly axillary on leafy shoots (but see No. 6) :

      Outside of sepals and mature fruits glabrous ; young branchlets sparingly

      pubescent to glabrescent or glabrous ; fruits globose :

      Flowers in fascicles on woody branchlets ; leaves with distinct petioles 4–6 mm.

        long, unequally cuneate at base ; fruits with fleshy pericarp, about 1·2 cm.

        diam. ..   ..   6. *aframensis*

      Flowers solitary, or 2–3 together, axillary on leafy shoots ; leaves very shortly

        petiolate :

        Proximal side of leaf-lamina inserted on the petiole below the distal, both sides

          obtuse to subcordate ; fruits hard, crustaceous, up to 1·8 cm. diam.

                            22. *gilgiana*

        Proximal and distal sides of leaf-lamina inserted on the petiole at the same level,

          rounded to subcordate ; fruits with fleshy pericarp, up to 1·2 cm. diam.

                            23. *parvifolia*

      Outside of sepals and mature fruits hairy ; branchlets pubescent :

      Leaves very unequal at base, proximal side broadly auriculate inserted on the

        petiole at same level as the distal side which is rounded or subcordate at base,

        subsessile, serrate, distichous ; lamina ovate to oblong-lanceolate, gradually

        long-acuminate ; branchlets spreading-pubescent ; flowers solitary ; fruits

        ovoid, 2-lobed stigmas separated, up to 2·5 cm. long, tomentellous

                            21. *chevalieri*

      Leaves cuneate to obtuse at base, distinctly petiolate, subentire or crenate ;

        flowers a few together :

        Leaves with numerous close, parallel, lateral nerves, caudate-acuminate at

          apex :

          Lamina ovate-elliptic to elliptic-lanceolate ; fruits globose, about 1·3 diam.,

            tomentellous ..   ..   3. *aubrevillei*

          Lamina oblong to oblong-elliptic ..   ..   ..   5. *paxii*

        Leaves with 3–7 main lateral nerves on each side of midrib ; secondary nerves

          not parallel :

          Fruits 2-lobed, rugose and tomentose on the surface ; leaves rhomboid-

lanceolate, 2·5–6 cm. long, 0·8–2·5 cm. broad ; main lateral nerves 3–4 on
each side of midrib .. .. *2. klainei*
Fruits globose, tomentellous, about 1 cm. diam. ; leaves elliptic to elliptic-
lanceolate, 3·5–12 cm. long, 1·5–6 cm. broad ; main lateral nerves 4–7 on
each side of midrib .. .. *4. leonensis*
Ovary and fruit 1-celled ; fruits ellipsoid, glabrous ; leaves glabrous, coriaceous :
Leaves 30–43 cm. long, 9–15 cm. broad, main lateral nerves 8–11 on each side of
midrib ; sepals oblong-lanceolate or oblong, glabrous ; fruits 1·2–1·5 cm. long
8. similis
Leaves 12–23 cm. long, 3·5–8·5 cm. broad, main lateral nerves 5–6 on each side of
midrib ; sepals orbicular, pubescent ; fruits about 1·5 cm. long .. *9. aylmeri*

See also the following species whose female flowers and fruits are not yet known : No. 10 *D. fernandopoana*,
No. 15 *D. obanensis* and No. 24 *D. tessmanniana*.

1. **D. pellegrini** *Léandri* in Bull. Soc. Bot. Fr. 81 : 459 (1934). *D. vignei* Hoyle in Kew Bull. 1935 : 257. A
forest tree, 25–40 ft. high, with spreading branches ; locally abundant ; flowers white.
 Iv.C.: Soubré (Jan.) *Aubrév.* 864 ! **G.C.:** Mampong Scarp (Mar.) *Vigne* FH 1062 ! S. Scarp F.R. (fr.
May) *Moor* FH 2344 ! Atewa Range (Feb.) *Vigne* FH 4344 ! Konkong, E. Akim (young fr. Mar.)
Johnson 595 ! (See Appendix, p. 141.)
2. **D. klainei** *Pierre ex Pax* in Engl. Bot. Jahrb. 43 : 218–219 (1909) ; F.T.A. 6, 1 : 681 ; Pax & K. Hoffm. in
Engl. Pflanzenr. Euph. 15 : 252 ; Aubrév. Fl. For. C. Iv. 2 : 42, t. 134, 4–5. A forest tree, to 80 ft. high.
 Iv.C.: Tai, in flood-plain of R. Cavally (young fr. Apr.) *Aubrév.* 1209 ! Also in Gabon.
3. **D. aubrevillei** *Léandri* l.c. 81 : 458 (1934) ; Aubrév. l.c. 2 : 42, t. 134, 1–3. A forest tree, to 45 ft., rarely
to 80 ft. high.
 S.L.: Kambui F.R. *Sawyerr* FHI 13556 ! **Lib.:** Dukwia R. (fr. May) *Cooper* 227 ! 458 ! R. Cess, Grand
Bassa (Mar.) 12530 ! **Iv.C.:** Agnéby *Aubrév.* 2117 ! Tabou (fr. Dec.) *Aubrév.* 1667 ! Nimba Mts. (Mar.)
Aubrév. 1146 ! Banco (Feb.) *Aubrév.* 885 ! Mt. Momy *Aubrév.* 1182 !
4. **D. leonensis** *Pax* in Engl. Bot. Jahrb. 43 : 219 (1909), not of (Pax) Pax & K. Hoffm. in Engl. Pflanzenr.
Euph. 15 : 262 (1922) ; F.T.A. 6, 1 : 682. *D. hutchinsonii* Pax & K. Hoffm. l.c. 15 : 254 (1912). *D.
rowlandii* Pax l.c. 43 : 219 (1909). *D. kamerunica* Pax & K. Hoffm. l.c. 15 : 254 (1922). *D. urophylla*
Pax & K. Hoffm. l.c. (1922). A forest tree, to 70 ft. high ; flowers heavily scented.
 Fr.G.: massif de Fon, Beyla (Aug.) *Schnell* 3313 ! **S.L.:** Duniya, Talla Hills (Feb.) *Sc. Elliot* 4981 ! **Lib.:**
Truo, Sinoe (young fr. Mar.) *Baldwin* 11385 ! (?). **G.C.:** Mansiso (Mar.) *Vigne* 92 ! **Togo:** *Mildbr.* 7211.
S.Nig.: Lagos *Rowland* ! Olokemeji F.R. (fr. Mar.) *A. F. Ross* 99 ! Shasha F.R. (young fr. Jan.) *A. F. A.
Lamb* 181 ! Okomu F.R. (young fr. Feb.) *Onochie* in *Hb. Brenan* 9141 ! Oluwa F.R. *Kennedy* 2474 !
Also in French Cameroons.
 [*Brenan & Keay* 8653 from the Idanre Hills, Nigeria, is close to this species but lacks mature flowers
and fruits ; better material is needed.]
5. **D. paxii** *Hutch.* in F.T.A. 6, 1 : 681 (1912) ; Pax & K. Hoffm. l.c. 15 : 276. A forest shrub or small tree.
 Br.Cam.: Bimbia road, Victoria (Feb.) *Maitland* 423 ! Also in French Cameroons and Gabon.
6. **D. aframensis** *Hutch.* in F.T.A. 6, 1 : 682 (1912) ; Pax & K. Hoffm. l.c. 15 : 237. A forest tree, to 40 ft.
high ; flowers white ; fruits orange.
 G.C.: Afram Plains (Mar.) *Johnson* 714 ! Pra-Anum F.R., Oda (fr. Apr.) *C. J. Taylor* FH 5270 !
Abrimasu F.R., Mampong (Feb.) *Andoh* FH 4734 ! 5633 ! **S.Nig.:** Gambari F.R., Ibadan *W. D.
MacGregor* 584 ! Omo F.R., Ijebu Ode Prov. (Mar.) *Jones & Onochie* FHI 17010 ! 17013 ! 17544 !
Sunmoge (Mar.) *Keay* FHI 16062 !
7. **D. floribunda** (*Müll. Arg.*) *Hutch.* in F.T.A. 6, 1 : 687 (1912) ; Pax & K. Hoffm. l.c. 15 : 237 ; Aubrév. l.c.
2 : 46, t. 136, 10. *Cyclostemon floribundus* Müll. Arg. (1864)—Chev. Bot. 561. *D. ovata* Hutch. 6,
1 : 688 (1912) ; Pax & K. Hoffm. l.c. ; F.W.T.A., ed. 1, 1 : 287. A cauliflorous tree, to 30 ft. high, in
the drier parts of the forest regions, extending also into the savannah regions ; flowers cream ; fruits
yellow when ripe.
 Sen.: Ravin des Voleurs *Berhaut* 1313. **Port.G.:** Jangada, Xitoli (June) *Esp. Santo* 2267 ! **Iv.C.:** Danané
(Apr.) *Chev.* 21289. Groumania *Aubrév.* 776 ; 787. **G.C.:** Achimota (fl. & fr. Apr.) *Irvine* 651 ! *Milne-
Redhead* 5061 ! Agogo—Ejian road (Apr.) *Vigne* FH 1103 ! 1104 ! **Togo:** Lome *Warnecke* 154 ! **N.Nig.:**
Idah *Barter* 1673 ! **S.Nig.:** Lagos *Foster* 80 ! Olokemeji (fl. Jan., fr. Mar.) *Keay* FHI 22469 ! 22533 !
Owo F.R. (Apr.) *Jones* FHI 3519 ! Also in Belgian Congo. (See Appendix, p. 140.)
8. **D. similis** *Hutch.* in F.T.A. 6, 1 : 679 (1912) ; Pax & K. Hoffm. in Engl. Pflanzenr. Euph. 15 : 267. A
forest shrub or tree, to 20 ft. high.
 S.Nig.: Oban *Talbot* 619 ! 1495 ! 1710 ! **Br.Cam.:** Kumba (Feb.) *Smith* 74/36 ! Also in French
Cameroons.
9. **D. aylmeri** *Hutch. & Dalz.* F.W.T.A., ed. 1, 1 : 288 (1928) ; Kew Bull. 1928 : 300. *D. mottikoro* Léandri
in Bull. Soc. Bot. Fr. 81 : 459 (1934) ; Aubrév. Fl. For. C. Iv. 2 : 44, t. 136, 1–4. A forest tree, to 40 ft.
high, with white easily split timber ; flowers yellow, profuse on the branches ; fruits white, 1-seeded.
 S.L.: Gerihun (Sept.) *Aylmer* 246 ! **Lib.:** Dukwia R. *Cooper* 315 ! (?—sterile). **Iv.C.:** *Aubrév.* 1659 !
2081 ! Djibi (fl. Sept., fr. Dec.) *Aubrév.* 533 ! 1642 ! Azaguié (Sept.) *Chev.* 22274 ! **G.C.:** Ankasa F.R.
(fr. Dec.) *Vigne* FH 3181 ! Subiri F.R., Benso (Sept.) *Andoh* FH 5567 ! (See Appendix, p. 140.)
10. **D. fernandopoana** *Brenan* in Kew Bull. 1953 : 92. *Cyclostemon glomeratus* Müll. Arg. (1864). *D. glomerata*
(Müll. Arg.) Hutch. in F.T.A. 6, 1 : 679 (1912) ; Pax & K. Hoffm. l.c. 15 : 242 ; not of Griseb. (1857).
A forest shrub, about 20 ft. high, with glabrous terete branchlets.
 F.Po: *Mann* 278 !
11. **D. principum** (*Müll. Arg.*) *Hutch.* in F.T.A. 6, 1 : 684 (1912) ; Pax & K. Hoffm. l.c. 15 : 241 ; Aubrév. l.c.
2 : 43, t. 135, 9–10. *Cyclostemon principum* Müll. Arg. (1864). A forest tree, to 30 ft. high ; flowers
cream, borne on leafy shoots and the woody branches ; fruits dark brown, velvety.
 Fr.G.: Longuéry, Kouria (Dec.) *Caille* in *Hb. Chev.* 14826 ! **Lib.:** Ganta (May) *Harley* 1191 ! **Iv.C.:** Mt.
Tonkoui *Aubrév.* 1007. Mt. Dou *Aubrév.* 1076. **G.C.:** Ofin Headwaters F.R. (Feb.) *Vigne* FH 3423 !
Akumadan (Oct.) *Vigne* FH 2554 ! 2555 ! **S.Nig.:** Omo F.R. (fl. Oct., fr. Feb.) *Tamajong* FHI 20982 !
Jones & Onochie FHI 17586 ! Akure F.R. (Oct., Nov.) *Keay* FHI 16246 ! Latilo FHI 15285 ! Oban
Talbot 579 ! **Br.Cam.:** Victoria *Winkler* 551 ; 552. Johann-Albrechtshöhe *Staudt* 849 ! **F.Po:** San
Carlos *Mildbr.* 6791. Also in French Cameroons and Principe.
12. **D. ivorensis** *Hutch. & Dalz.* F.W.T.A., ed. 1, 1 : 287 (1928) ; Kew Bull. 1928 : 299 ; Aubrév. l.c. 2 : 46,
t. 135, 6–8. *D. pierreana* of Chev. Bot. 561, not of Hutch. A small forest tree, to 15 ft. high ; flowers
greenish-yellow or white, on the main stem ; fruits red, or yellow or orange.
 Lib.: Dukwia R. (fr. May) *Cooper* 418 ! Truo, Sinoe (fr. Mar.) *Baldwin* 11387 ! 11396 ! Diebla, Webo
(fr. July) *Baldwin* 6280 ! **Iv.C.:** between Mid. Sassandra and Mid. Cavally (July) *Chev.* 19227 ! Tébo,
Yabas country (fr. July) *Chev.* 19416 ! **G.C.:** Subiri F.R., Benso (Sept.) *Andoh* FH 5562 ! (See Appendix,
p. 140.)
13. **D. molunduana** *Pax & K. Hoffm.* in Engl. Pflanzenr. Euph. 15 : 258 (1922). A small forest tree, to 12 ft.
high with deeply grooved pubescent branchlets ; flowers greenish-yellow-white, in dense fascicles all
down the main stem, not on the twigs.

S.Nig.: Onda, Omo F.R. (Feb.) *Jones & Onochie* FHI 16985 ! Okomu F.R. (Dec.) *Brenan* 3574 ! Oban *Talbot* 1645 ! **Br.Cam.:** S. Bakundu F.R. (Mar.) *Brenan* 9296 ! 9297 ! 9310 ! 9311 ! Mayuku (Feb.) *Dalz.* 8239 ! Johann-Albrechtshöhe *Staudt* 739 ! 741. Also in French Cameroons.

14. **D. afzelii** (*Pax*) *Hutch.* in F.T.A. 6, 1 : 685 (1912) ; Pax & K. Hoffm. l.c. 15 : 249 ; Aubrév. l.c. 2 : 46, t. 136, 5–9. *Cyclostemon afzelii* Pax (1897). A forest tree, to 80 ft. high ; flowers on bole and branches, fragrant, male white or yellow, female green ; fruits bright orange, abundant.
    S.L.: *Afzelius* ! Gerihun (Sept.) *Aylmer* 245 ! Jepihun (fr. Jan.) *Smythe* 236 ! Njala (fl. Oct., fr. Jan., Mar.) *Deighton* 1414 ! 2904 ! 4718 ! *Dalz.* 8102 ! Yonibana (Oct.) *Thomas* 4111 ! Kambui Hills (fl. Oct., fr. Dec.) *Small* 825 ! 854 ! **Lib.:** Yeh R., 2 miles above St. Paul R. (Oct.) *Linder* 990 ! **Iv.C.:** Mé *Aubrév.* 167. Guiglo *Aubrév.* 2075. **G.C.:** Enchie (Oct.) *Tolmie* FH 2381 ! (See Appendix, p. 140.)

15. **D. obanensis** *S. Moore* in Cat. Talb. 97 (1913) ; Pax & K. Hoffm. l.c. 15 : 240. A small forest tree with pubescent branchlets and crenate leaves ; flowers small, colourless.
    S.Nig.: Oban *Talbot* 677 ! 2310 !

16. **D. preussii** (*Pax*) *Hutch.* in F.T.A. 6, 1 : 686 (1912) ; Pax & K. Hoffm. l.c. 15 : 247. *Cyclostemon preussii* Pax (1899). A forest tree, to 65 ft. high ; flowers yellow, borne on the stem.
    S.Nig.: Oban *Coombe* 182 ! **Br.Cam.:** Barombi *Preuss* 23. S. Bakundu F.R. (fr. Aug.) *Olorunfemi* FHI 30739 !

17. **D. staudtii** (*Pax*) *Hutch.* in F.T.A. 6, 1 : 688 (1912) ; Pax & K. Hoffm. l.c. 15 : 263. *Cyclostemon staudtii* Pax (1899). A forest tree, to 40 ft. high : flowers heavily scented.
    S.Nig.: Etemi, Omo F.R. (Mar.) *Jones & Onochie* FHI 16609 ! 16624 ! 16625 ! 17007 ! Ikom (fr. June) *Latilo* FHI 31851 ! Umon Ndealichi F.R., Calabar (young fr. May) *Onochie* FHI 33230 ! **Br.Cam.:** Meghom, Wum, Bamenda Prov. (fr. July) *Ujor* FHI 30488 ! Also in French Cameroons.

18. **D. gossweileri** *S. Moore* in J. Bot. 58 : 271 (1920) ; Brenan in Kew Bull. 1952 : 445. *D. armoracia* Pax & K. Hoffm. l.c. 15 : 275 (1922). *D. "arinozacia"* of Kennedy For. Fl. S. Nig. 75. Forest tree, to 125 ft. high, with straight clear bole ; cut parts emitting a strong smell of horse-radish ; flowers brownish-green.
    S.Nig.: Okomu F.R., Benin (fl. & fr. Dec.) *Brenan* 8474 ! 8475 ! Sapoba (fl. May, Dec., fr. June) *Kennedy* 1986 ! 2075 ! Also in French Cameroons, Cabinda and Belgian Congo. (See Appendix, p. 140.)

19. **D. capillipes** (*Pax*) *Pax & K. Hoffm.* l.c. 15 : 260 (1922) ; Brenan in Kew Bull. 1952 : 444. *Lingelsheimia capillipes* Pax (1909)—F.T.A. 6, 1 : 691. A forest shrub or small tree, to 12 ft. high, with grey branches ; pedicels drooping ; flowers greenish-white ; ripe fruits rather hard and orange.
    S.Nig.: Fowa, Omo F.R. (young fr. Mar.) *Jones & Onochie* FHI 17205 ! Okomu F.R. (fl. & fr. Dec.–Jan.) *Brenan* 8403 ! 8411 ! 8466 ! 8751 ! Also in French Cameroons and Belgian Congo.

20. **D. inaequalis** *Hutch.* in F.T.A. 6, 1 : 684 (1912). *Cyclostemon leonensis* Pax (1903). *D. leonensis* (Pax) Pax & K. Hoffm. l.c. 15 : 262 (1922), not of Pax (1909). A forest shrub, to 16 ft. high ; branchlets angular, with spreading yellowish pubescence ; fruit orange.
    S.L.: *Afzelius* ! Ninia, Talla Hills (fr. Feb.) *Sc. Elliot* 4895 ! **Lib.:** Ganta (fr. Feb., May) *Harley* 1184 ! *Baldwin* 11003 !

21. **D. chevalieri** *Beille* in Bull. Soc. Bot. Fr. 61, Mém. 8 : 293 (1917) ; Aubrév. Fl. For. C. Iv. 2 : 44, t. 135, 1–3. A forest shrub or small tree, to 15 ft. high ; flowers greenish-cream ; ripe fruits bright orange, firmly fleshy.
    Lib.: Ganta (fr. Apr., Sept.) *Harley* 1140 ! *Baldwin* 9249 ! **Iv.C.:** Danané (Apr.) *Chev.* 21290 ! Dyolas country (Apr.) *Chev.* 21128 ! And several other localities *fide* Beille *l.c.* **G.C.:** Awiabo (fr. Nov.) *Vigne* FH 1463 ! Banka (Sept.) *Vigne* FH 1369 ! Yakasi, Boin River F.R. (fr. Dec.) *Vigne* FH 3192 ! **S.Nig.:** Akinjare, Owo (fr. Mar.) *Jones* FHI 3068 ! Okomu F.R. (fl. Dec., fr. Jan.) *Brenan* 8409 ! 8436 ! 8757 ! 8892 ! Oban *Talbot* 678 !

22. **D. gilgiana** (*Pax*) *Pax & K. Hoffm.* l.c. 15 : 261 (1922) ; Aubrév. l.c. 2 : 43, t. 135, 4–5. *Cyclostemon gilgianus* Pax (1903). *Lingelsheimia gilgiana* (Pax) Hutch. (1912)—F.W.T.A., ed. 1, 1 : 288 ; Chev. Bot. 587. A small forest tree, to 30 ft. high, with pale grey-brown bark ; flowers yellowish ; ripe fruits bright scarlet or orange.
    Port.G.: Ponte de Dada, Fulacunda (Oct.) *Esp. Santo* 2205 ! Pobresa, Cubisseque (fr. Feb.) *Esp. Santo* 2376 ! **S.L.:** *Afzelius* ! Yonibana (Oct.) *Thomas* 4292 ! **Lib.:** Yeh R. (Oct.) *Linder* 992 ! Peahtah (Oct.) *Linder* 1075 ! Ganta (fr. Feb.) *Baldwin* 11019 ! **Iv.C.:** Akoupé (fr. Jan.) *Serv. For.* 1782 ! Man (fr. Mar.) *Aubrév.* 1045 ! **G.C.:** Jomura, Tano R. (fr. Jan.) *Brent* 6 DD ! Amentia (Sept.) *Vigne* FH 1360 ! **Dah.:** Kétou to L. Azri *Chev.* 23033. **N.Nig.:** Kurmin Ninte, Jemaa Div. (Nov.) *Keay* FHI 21050 ! 22252 ! **S.Nig.:** Lagos (Oct.) *Foster* 353 ! Oshogbo (fr. Apr.) *Millson* 139 ! Mamu F.R., Ibadan (Oct.) *Onochie* FHI 8101 ! Omo F.R. (Oct.) *Tamajong* FHI 20977 ! **Br.Cam.:** Likomba (fr. Dec.) *Mildbr.* 10758 ! Also in French Cameroons. (See Appendix, p. 140.)

23. **D. parvifolia** (*Müll. Arg.*) *Pax & K. Hoffm.* l.c. 15 : 262 (1922). *Cyclostemon parvifolius* Müll. Arg. (1864). *Lingelsheimia parvifolia* (Müll. Arg.) Hutch. (1912)—F.W.T.A., ed. 1, 1 : 288 ; Aubrév. l.c. 2 : 44 (in note). A shrub or small tree, with pale grey branches ; ripe fruits red.
    S.L.: *Afzelius.* **Iv.C.:** Guiglo *Aubrév.* 2025 ! Mt. Nienokoué *Chev.* 19453 ! **G.C.:** Agboho to Dedesu, Afram Plains (fr. May) *Kitson* 1124 ! Alabadi (fr. Apr.) *Scholes* 295 ! Akatin (Oct.) *Vigne* FH 4040 ! Kwabia (Oct.) *Vigne* FH 4022 ! **N.Nig.:** Nupe *Barter* 1032 ! 1700 !
    [*Aubrév.* 2025 has branchlets, petioles and pedicels more pubescent than the type, but it does not seem to me specifically distinct.]

24. **D. tessmanniana** (*Pax*) *Pax & K. Hoffm.* l.c. 15 : 262 (1922). *Cyclostemon tessmannianus* Pax (1910). *Lingelsheimia tessmanniana* (Pax) Hutch. in F.T.A. 6, 1 : 692 (1912). A forest shrub or small tree.
    Br.Cam.: Likomba (Nov.) *Mildbr.* 10600 ! Also in Spanish Guinea.

25. **D. occidentalis** (*Müll. Arg.*) *Hutch.* in F.T.A. 6, 1 : 683 (1912) ; Pax & K. Hoffm. l.c. 15 : 276. *Cyclostemon occidentalis* Müll. Arg. (1864). A forest tree, 20 ft. high.
    F. Po: *Mann* 1158 !

Imperfectly known species.

D. sp. Probably a new species near to *D. chevalieri* Beille ; male flowers are required.
    G.C.: Adeiso, E.P. (Mar.) *Irvine* 2413 !

I have seen a number of sterile specimens which have been provisionally referred to this genus, but cannot be named with any certainty. Field workers are reminded that both sexes of flowers and mature fruits should be collected whenever possible.—R.W.J.K.

Excluded species.

D. sassandraensis *Aubrév.* Fl. For. C. Iv. 2 : 48 (1936), French descr. only. (**Iv.C.:** *Aubrév.* 863 ! 1110 !). This is **Casearia bridelioides** Mildbr. ex Hutch & Dalz. (*Samydaceae*), a species not recorded from the Ivory Coast when Vol. I, part 1 of the revised Flora was published.

D. talbotii *S. Moore* in Cat. Talb. 97 (1913) ; Pax & K. Hoffm. l.c. 241. The type specimen (*Talbot* 8 from Oban, Nigeria) consists of a leafy shoot of **Uvariopsis bakeriana** (Hutch. & Dalz.) Robyns & Ghesq. (*Annonaceae*) and detached cauliflorous inflorescences of a *Drypetes* which cannot be determined with any certainty. The name *D. talbotii* S. Moore must therefore, be rejected under Art. 76 of the International Code of Botanical Nomenclature (1952). A second specimen (*Talbot* 2337) cited in F.W.T.A., ed. 1 as *D. talbotii* consists of a leafy shoot of a *Napoleona* and very poor female flowers of a *Drypetes*.

Cyclostemon gabonensis of Chev. Bot. 561, not of Pierre. The specimen cited (**Iv.C.:** *Chev.* 19993 !) is **Diospyros canaliculata** De Wild. (*Ebenaceae*), and not a species of *Enantia* as suggested in F.W.T.A., ed. 1, 1 : 286.

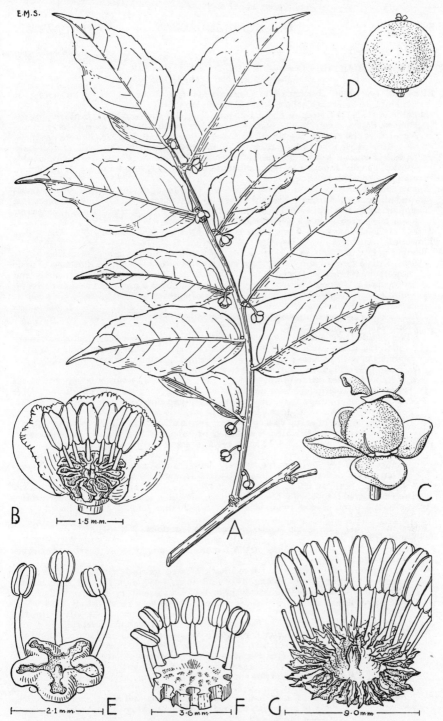

FIG. 133.—DRYPETES SPP. (EUPHORBIACEAE).

D. gilgiana (Pax) Pax & K. Hoffm.—A, flowering shoot, × 1.   B, male flower with 1 sepal and 5 stamens removed, × 12.   C, female flower, showing rudimentary stamen, × 4.   D, fruit, × 1½.
  A & B are drawn from *Foster* 353 ;   C from *Keay* FHI 21050 ;   D from *Millson* 139.
D. pellegrini Léandri—E, disk and stamens.
  Drawn from *Vigne* FH 1062.
D. principum (Müll. Arg.) Hutch.—F, disk and stamens.
  Drawn from *Latilo* FHI 15285.
D. gossweileri S. Moore—G, disk and stamens.
  Drawn from *Brenan* 8475.

383

16. PHYLLANTHUS Linn.—F.T.A. 6, 1 : 692 ; Pax & K. Hoffm. in E. & P. Pflanzenfam. 19C : 60 (1931).

Flowering branchlets fasciculate ; filaments all free ; shrubs, often straggling or scandent, or sometimes arborescent ; leaves glabrous :
Sepals and stamens 5 ; sepals about 1·25 mm. long, not drying white ; fruits depressed-globose, fleshy, about 3 mm. diam. ; branchlets reddish-brown ; barren leafy shoots crowded at the tops of the shoots and resembling pinnate leaves ; flowers produced on clustered slender leafless racemes of fascicles arising in the axils of the leafy flowerless shoots and/or on the older leafless branchlets ; leaves ovate or ovate-elliptic, usually rounded at base, subacute at apex, 2–7 cm. long, 1·5–4 cm. broad, glaucous beneath   ..   ..   1. *muellerianus*
Sepals and stamens 4 (–5) ; sepals about 1·5 mm. long, drying white ; young fruits conical, beaked, fibrous ; branchlets grey ; barren leafy shoots and inflorescences similar to above, upper inflorescences all male and sometimes with a few leaves, lower ones often bisexual ; leaves ovate, rounded, obtuse or slightly cuneate at base, acutely acuminate, 2·5–7 cm. long, 2–4 cm. broad, paler beneath   .. 2. *profusus*
Flowering branchlets not in fascicles :
Filaments free to the base or rarely (No. 5) some free and the others connate :
Disk of male flowers annular ; large shrub to tree to 30 m. high, dioecious ; male flowers numerous in fascicles in the leaf-axils, the females paired ; sepals and stamens 4 ; filaments all free ; fruits 3–4-lobed, about 8 mm. diam. ; leaves ovate-elliptic to obovate-oblanceolate, rounded to cuneate at base, mostly rounded at apex, 2·5–10 cm. long, 2–4 cm. broad, glabrous or pubescent ..   3. *discoideus*
Disk of male flowers composed of separate glands ; herbs, or shrubs up to 3 m. high :
Stamens and sepals 5, rarely 4 :
Two or three of the filaments connate to the apex, the others free ; fruits depressed-globose, about 7 mm. diam., 8–16-seeded, fleshy ; 1 female flower and 2–3 or more males in each fascicle ; leaves oblong or elliptic, rounded at both ends, 1·5–4 cm. long, 0·7–1·5 cm. broad ; shrubs to 3 m. high :
Stem and leaves crisped-pubescent   ..   ..   4. *reticulatus* var. *reticulatus*
Stem and leaves glabrous or nearly so ..   ..   4a. *reticulatus* var. *glaber*
All the filaments quite free ; leaves glabrous, glaucous beneath ; fruits depressed-globose, about 2 mm. diam., capsular :
Pedicels very slender, up to 2 cm. long ; leaves obovate or elliptic-obovate, obtuse or slightly cuneate at base, rounded or subacute and slightly mucronate at apex, 8–15 mm. long, 6–12 mm. broad ; small shrub, to 1·5 m. high
5. *capillaris*
Pedicels shorter and stouter, usually less than 3 mm. long ; leaves oblong-lanceolate or linear-lanceolate, subacute at apex, up to 2 cm. long and 8 mm. broad ; herb, to 0·5 m. high ..   ..   6. *pentandrus*
Stamens 3, free ; sepals 6 ; plants dioecious, shrubby, to 2 m. high ; leaves variable in shape, often oblong, ovate or orbicular, 8–30 mm. long, 4–21 mm. broad
7. *alpestris*
Filaments connate the whole length or nearly so ; stamens 2–3 (–4) :
Stamens (2–) 3 (–4) :
*Sepals of both sexes 5, those of the females in a single series ; herbs, sometimes subwoody :
Ovary prominently warted ; curved back of seed marked with 12–14 fine longitudinal ridges ; leaves oblong, 4–10 mm. long, 1·5–3 mm. broad ; male flowers 2–3 together in lower leaf-axils of branches, female solitary in upper axils ; disk of female flower stellate with 5 deep lobes ; styles very short, deeply bifid
8. *niruroides*
Ovary smooth ; curved back of seed marked with 5–10 longitudinal ridges :
Sepals in fruit about 2·5 mm. long, more or less embracing and subequalling the ripe fruit ; back of seed with 9–10 faint longitudinal ridges ; plant with few (up to 10) lateral branches ; leaves oblong-elliptic to oblong-obovate, 5–9 mm. long, 3–5 mm. broad ; flowers arranged as in No. 8 ; anther-thecae horizontally dehiscing ; disk of female flowers shallowly 9-crenate ; styles very short, deeply bifid at apex ..   ..   9. *nigericus*
Sepals in fruit 0·75–1·5 mm. long, scarcely embracing and much shorter than the ripe fruit ; back of seed with 5–7 more prominent longitudinal ridges ; plants normally profusely branched :
Disk of female flowers annular, not lobed ; stigmas quite sessile, closely packed into a minute circle at top of ovary, each stigma when viewed from above appearing reniform with an emarginate apex ; anther-thecae dehiscing vertically ; male flowers 1–2 together in lower axils of flowering branches, female solitary in upper axils ; leaves elliptic-oblong, about 5 mm. long, 1·5–2 mm. broad   ..   ..   10. *sublanatus*

Disk of female flowers stellate with normally 5 deep radiating lobes ; styles short but distinct, deeply bifid at apex, not forming a closely packed circle at top of ovary ; anther-thecae with markedly oblique dehiscence ; 1 female flower together with 1 male flower in each axil ; leaves elliptic-oblong, 5–10 mm. long, 2–4·5 mm. broad   ..   .. 11. *amarus*

*Sepals of both sexes 6, those of the female often in two series :

Ovary warted ; seeds transversely ridged ; branchlets narrowly winged, asperulate; leaves oblong, rounded at base, shortly pointed and minutely mucronate at apex, 6–15 mm. long, margins asperulate ; monoecious ; flowers solitary and subsessile
12. *urinaria*

Ovary smooth ; seeds longitudinally ridged :

Disk of female flowers consisting of separate glands ; seeds marked with 9–10 fine longitudinal lines of dots on back ; branchlets sharply angular ; leaves linear to oblanceolate, up to 4 cm. long and 1·3 cm. broad ; stipules broad and auriculate at base ; male and female flowers sometimes together in the leaf-axils or the female solitary and larger   ..   ..   .. 13. *maderaspatensis*

Disk of female flowers annular, saucer-shaped or cupular :

Seeds 1–1·25 mm. long, marked with 6–7 longitudinal lines on the back ; herbs and shrublets up to 1·5 m. high :

Disk of female flowers deeply toothed ; styles very short ; stem and branchlets glabrous ; leaves oblong-elliptic, rounded at both ends, 6–14 mm. long, 2·5–5·5 mm. broad ; flowers monoecious, solitary, the males in the lower, the females in the upper axils ; annual herb   ..   ..   .. 14. *niruri*

Disk of female flowers only slightly crenate or undulate :

Leaves suborbicular or rounded-obovate, mostly about 5 mm. diam. ; stem, branchlets and often the leaves asperulate with numerous pointed papillae; 2–3 male flowers and 1 female together in each leaf-axil ; styles distinct, spreading   ..   .. 15. *rotundifolius*

Leaves not as above ; stem and branchlets glabrous or pubescent :

Stem densely puberulous, woody ; branchlets winged, puberulous ; leaves oblong-linear-lanceolate, 12–21 mm. long, 4–5·5 mm. broad ; flowers monoecious, the males occupying most of the branchlet, usually a solitary female in the axil of the uppermost leaf ..   ..   .. 16. *dusenii*

Stem and branchlets glabrous, or nearly so ; leaves more or less oblong to obovate, 8–25 mm. long, 4–11 mm. broad :

Stipules 2·5–4 mm. long, brownish, conspicuous ; branchlets not or scarcely winged ; leaves obovate or obovate-elliptic, subacute at base, rounded and mucronate at apex ; flowers dioecious (? always) ; male flowers 1–3 in each leaf-axil, females solitary, sepals broad, enclosing the fruit ; ovary sessile ; shrublet, to 1·5 m. high, in montane areas
17. *mannianus*

Stipules up to 2 mm. long, not conspicuous ; branchlets flattened and winged ; leaves oblong, rounded at base, obtuse at apex ; flowers monoecious ; male flowers few together in axils of lower leaves, females solitary in upper axils and larger, sepals oblong-linear, not completely enclosing the fruit ; ovary very shortly stipitate ; ruderal herb
18. *odontadenius*

Seeds 2·5–3 mm., dotted all over with minute papillae which can easily be scraped off ; flowers monoecious, the males quite small, the females much larger with thick reticulate sepals ; branches with membranous peeling bark ; leaves elongate-oblong or oblong-elliptic, rounded at base, rounded or truncate at apex, 1–4·5 cm. long, 0·5–2·5 cm. broad ; shrub to 3 m. high   19. *beillei*

Stamens 2 ; sepals 4 in male flower, (5–) 6 in female ; flowers dioecious, the males several in each fascicle with numerous membranous laciniate bracts at the base, the females few or solitary ; branches with membranous peeling bark ; leaves elliptic to oblanceolate, obtuse at subcuneate at base, obtusely pointed at apex, 2·5–10 cm. long, 1–4 cm. broad ; shrub to 2 m. high ..   .. 20. *petraeus*

1. **P. muellerianus** (*O. Ktze.*) *Exell* in Cat. S. Tomé 290 (1944). *Diasperus muellerianus* O. Ktze. (1891). *Kirganelia floribunda* Baill. (1860), not of Spreng. (1826). *Phyllanthus floribundus* Müll. Arg. (1863), not of Kunth (1817) ; F.T.A. 6, 1 : 701 ; F.W.T.A., ed. 1, 1 : 290 ; Chev. Bot. 556 ; Aubrév. Fl. For. Soud.-Guin. 189. A glabrous shrub or woody climber, sometimes arborescent, often armed with recurved stipular spines, with copious inflorescences of minute greenish flowers, and small berry-like red fruit.
Fr.Sud.: Bamako *Waterlot* 1368. Birgo *Dubois* 210. **Port.G.:** Granja, Catió (June) *Esp. Santo* 2088 ! **Fr.G.:** Rio Nunez *Heudelot* 659 ! Kouroussa (June) *Pobéguin* 264 ! 812 ! **S.L.:** Wilberforce (fr. Mar.) *Johnston* 96 ! Mofari, Scarcies R. (Jan.) *Sc. Elliot* 4402 ! Njala (fl. Mar., fr. Apr.) *Deighton* 1097 ! 1824 ! 4742 ! Batkanu (Jan.) *Deighton* 2863 ! **Lib.:** Tappita (fr. Aug.) *Baldwin* 9100 ! Flumpa, Sanokwele (fr. Sept.) *Baldwin* 9354 ! Kulo, Sinoe (fr. Mar.) *Baldwin* 11437 ! Soplima, Vonjama (Nov.) *Baldwin* 10111a ! **Iv.C.:** Baoulé, Ferkéssédougou, Bobo Dioulasso and Kampti *fide* Aubrév. *l.c.* **G.C.:** Axim (Mar.) *Chipp* 395 ! Kumasi (fl. & fr. Feb.) *Irvine* 119 ! Assuantsi (Jan., Mar.) *Fishlock* 6 ! *Irvine* 1561 ! **Dah.:** Djougou *Chev.* 23875. **N.Nig.:** Nupe *Barter* 1666 ! Zungeru (July) *Dalz.* 64 ! **S.Nig.:** Lagos *Rowland* ! Ibadan (Feb.) *Meikle* 1149 ! Okomu F.R. (Feb.) *Brenan* 9039 ! Old Calabar (Feb.) *Mann* 2262 ! **Br.Cam.:** Buea (Mar.) *Maitland* 499 ! **F.Po:** *Mann* 12 ! Widespread in tropical Africa. (See Appendix, p. 157.)

Fig. 134.—Phyllanthus muellerianus (*O. Ktze.*) *Exell* (Euphorbiaceae).

A, male shoot. B, male flower. C, stamen. D, female flower. E, pistil with disk-glands. F, fascicle of female flowers. G, fruit. H, transverse section of fruit.

W.E.T.

2. **P. profusus** *N. E. Br.* in J. Linn. Soc. 37 : 113 (1905) ; F.T.A. 6, 1 : 704. *P. wildemannii* Beille in Bull. Soc. Bot. Fr. 61, Mém. 8 ; 293 (1917) ; Chev. Bot. 558 ; F.W.T.A., ed. 1, 1 : 290 (excl. *Irvine* s.n. from Gold Coast). Shrub with profuse white flowers.
   Fr.G.: Conakry (Feb.) *Chev.* 12721 ! Los Isl. (Feb.) *Chev.* 13305 ! **Lib.:** Sinoe Basin *Whyte* ! Kulo, Sinoe (Mar.) *Baldwin* 1430 ! Kakatown *Whyte* ! Dukwia R. (Feb.) *Cooper* 246 ! **G.C.:** Birrim F.R. (Apr.) *Vigne* FH 2887 ! (See Appendix, p. 158.)

3. **P. discoideus** (*Baill.*) *Müll. Arg.*—F.T.A. 6, 1 : 707 ; Aubrév. Fl. For. C. Iv. 2 : 56, t. 140 ; Fl. For. Soud.-Guin. 187, t. 35, 1–3 (incl. var. *pubescens*). *Cicca discoidea* Baill. (1860) Chev. Bot. 559. *Fluggea klaineana* Pierre ex A. Chev. Bot. 559, name only. *F. obovata* var. *luxurians* Beille. A large deciduous shrub or tree, to 100 ft. high, in moister parts of the savannah regions and drier parts of the forest ; especially on old farmlands ; flowers yellowish-green.
   Sen.: *Leprieur* ! M'bidjem *Thierry* 209 ! **Gam.:** *Heudelot* 102. **Fr.Sud.:** Médinani *Chev.* 514. **Port.G.:** Botelhe, Caravela (May) *Esp. Santo* 2002 ! S. Domingos (fr. Aug.) *Esp. Santo* 3078 ! **Fr.G.:** Rio Nunez *Heudelot* 789 ! Moribadou to Niomorodougou (Mar.) *Chev.* 20944 ! Nzo (June) *Collenette* 40 ! **S.L.:** Bagroo R. *Mann* 882 ! Newton (Feb.) *Deighton* 5358 ! 5359 ! Kenema (Mar.) *Lane-Poole* 192 ! Musala (fl. Mar., fr. Sept., Dec.) *Deighton* 5393 ! 5707 ! **Lib.:** Dukwia R. (fr. May) *Cooper* 436 ! Yratoke, Webo (fr. July) *Baldwin* 6239 ! **Iv.C.:** Anoumaba (fr. Nov.) *Chev.* 22348 ! Groumania *Aubrév.* 793 ! **G.C.:** Axim (Mar.) *Chipp* 418 ! Kintampo (Mar.) *Dalz.* 35 ! Akwapim Hills (Mar.) *Irvine* 1600 ! **Togo:** Sokode (Mar.) *Kersting* 29 ! **Dah.:** *Poisson* ! **N.Nig.:** Nupe *Barter* 1498 ! Bauchi Plateau (Apr.) *Lely* P. 215 ! **S.Nig.:** Lagos *Rowland* ! Olokemeji (Mar.) *A. F. Ross* 102 ! 117 ! Oban *Talbot* ! **Br.Cam.:** Buea (Mar.) *Maitland* 443 ! Mopanya (fr. Apr.) *Maitland* 1138 ! Widespread in tropical Africa. (See Appendix, p. 156.)

4. **P. reticulatus** *Poir.* var. **reticulatus**—F.T.A. 6, 1 : 700 ; Aubrév. Fl. For. Soud.-Guin. 189. *P. prieurianus* Müll. Arg.—Chev. Bot. 558. Erect or scrambling shrub, to 10 ft. high ; flowers cream or reddish ; fruits black when ripe ; often growing by rivers.
   Sen.: *Leprieur* ! *Sieber* ! **Fr.Sud.:** Sumpi (Aug.) *Chev.* 3461 ! Beragungu (fl. & fr. Feb.) *Hagerup* 312 ! Ansongo (fl. & fr. Mar.) *De Wailly* 5365 ! **Fr.G.:** Farana *Chev.* 20467. **S.L.:** Njala (fl. July, fl. & fr. Oct.) *Deighton* 739 ! 2539 ! Yetaya (Sept.) *Thomas* 2318 ! Mabonto (fr. Oct.) *Thomas* 3571 ! **G.C.:** R. Oti, Paliba (fl. & fr. Dec.) *Adams & Akpabla* GC 4114 ! **Dah.:** Djougou *Chev.* 23872. **N.Nig.:** Koton Karifi (fl. & fr. Nov.) *Keay* FHI 28087 ! Kano (fl. & fr. Aug.) *Keay* FHI 19180 ! Duchi, Bornu (Dec.) *Elliott* 143 ! Yola (fl. & fr. Dec.) *Dalz.* 161 ! **S.Nig.:** Ogun R., Olokemeji (fl. & fr. Nov.) *Jones* FHI 14232 ! Widespread in tropical Africa. (See Appendix, p. 158.)

4a. **P. reticulatus** var. **glaber** *Müll. Arg.*—F.T.A. 6, 1 : 701. Shrub, by rivers ; similar to the preceding.
   S.L.: Njala (Sept.) *Deighton* 1997 ! Magbile (Dec.) *Thomas* 6127 ! Sasa, Tonko Limba (Sept.) *Jordan* 337 ! **Iv.C.:** Tano R. (fr. Sept.) *Chipp* 347 ! **G.C.:** Kwitta *Thonning*. Juaso (fl. & fr. Aug.) *Irvine* 76 ! **S.Nig.:** Lagos *Rowland* ! *Millen* 53 ! Lower Niger R. *T. Vogel* ! Widespread in tropical Africa.

5. **P. capillaris** *Schum. & Thonn.*—F.T.A. 6, 1 : 709 ; Chev. Bot. 556. A small shrub, to 5 ft. high, with slender stems ; flowers white, in small axillary fascicles, the female often solitary ; in open grassy places.
   Fr.G.: Ditinn *Chev.* 13533. **S.L.:** Erimakuna (Mar.) *Sc. Elliot* 5245 ! Zimmi (Nov.) *Deighton* 375 ! Rokupr (fl. & fr. May) *Jordan* 253 ! Bafodea (fl. & fr. Apr.) *Deighton* 5506 ! **Lib.:** Sinoe Basin *Whyte* ! Gbanga (fl. & fr. Sept.) *Linder* 587 ! Nyaake, Webo (fl. & fr. June) *Baldwin* 6165 ! Flumpa, Sanokwele (fl. & fr. Sept.) *Baldwin* 9356 ! **Iv.C.:** Té, Man (fl. & fr. July) *Collenette* 52 ! **G.C.:** Accra (fl. & fr. June) *Dalz.* 138 ! Kpokoase (Mar.) *Irvine* 199 ! **Togo:** Klingfall *Kersting* 649 ! Bismarckburg *Büttner* 333 ! **N.Nig.:** Bauchi Plateau (fl. & fr. Apr.) *Lely* P. 238 ! Hepham to Ropp (fl. & fr. July) *Lely* 371 ! **S.Nig.:** Lagos *Rowland* ! Bonny (fl. & fr. Feb.) *Kalbreyer* 44 ! Aguku *Thomas* 835 ! Calabar (fl. & fr. Apr.) *Holland* 85 ! **Br.Cam.:** Cameroon R. (fl. & fr. Jan.) *Mann* 758 ! 2226 ! **F.Po:** Moka (fl. & fr. Jan.) *Exell* 838 ! Widespread in tropical Africa.

6. **P. pentandrus** *Schum. & Thonn.*—F.T.A. 6, 1 : 710 ; Chev. Bot. 558. A glabrous, much-branched woody herb, up to 18 in. high ; flowers white, small, males 2–3 together in the lower axils on the branchlets ; females solitary in the upper axils.
   Sen.: *Heudelot* 466 ! *Richard* ! **Gam.:** Genieri (fl. & fr. Feb.) *Fox* 81 ! 126 ! **Fr.Sud.:** Sindou (May) *Chev.* 870. San *Chev.* 2616. **Port.G.:** Praia Varela, Suzana (fl. & fr. Oct.) *Esp. Santo* 2283 ! Buruntuma, Gabu (fl. & fr. July) *Esp. Santo* 2705 ! **G.C.:** Atwabo *Fishlock* 75 ! Accra (fl. & fr. June) *Irvine* 1793 ! Aburi *Anderson* 2 ! **Togo:** Lome *Warnecke* 2 ! **Dah.:** *Burton* ! **Fr.Nig.:** Niamey (fl. & fr. Oct.) *Hagerup* 462 ! **N.Nig.:** Nupe *Barter* ! Sokoto (fl. & fr. Oct.) *Dalz.* 390 ! Bida (fl. & fr. Jan.) *Meikle* 1099 ! Katagum *Dalz.* 206 ! **S.Nig.:** Achalla, Onitsha (fl. & fr. Sept.) *Onochie* FHI 34058 ! Widespread in tropical Africa. (See Appendix, p. 158.)

7. **P. alpestris** *Beille*—F.T.A. 6, 1 : 712 ; Chev. Bot. 556. *P. leonensis* Hutch. (1917)—F.W.T.A., ed. 1, 1 : 291. *P. monticola* Hutch. & Dalz. F.W.T.A., ed. 1, 1 : 291 (1928). *P. virgatus* of Chev. Bot. 558. Shrub to 6 ft. high, or suffrutex.
   Fr.G.: Ditinn to Diaguissa (Apr.) *Chev.* 12907 ! Diaguissa (Apr.) *Chev.* 12943 ! Mt. Nzo (Mar.) *Chev.* 21006 ! 21030 ! Grandes Chutes (Dec.) *Chev.* 20241 ! 20317 ! Labé (Apr.) *Chev.* 12305 ! **S.L.:** Soukourala, foot of Mt. Tafaro (Jan.) *Chev.* 20586 ! Sugar Loaf Mt. (Nov.) *Sc. Elliot* 3962 ! *Deighton* 5637 ! 5638 ! York Pass *Lane-Poole* 424 ! Regent (Apr.) *Sc. Elliot* 5819 ! Sendugu (June) *Thomas* 580 ! Sankan-Biriwa (July) *Edwardson* 75 ! **Iv.C.:** Nimba Mts. (fl. & fr. Aug.) *Boughey* GC 18095 !

8. **P. niruroides** *Müll. Arg.*—F.T.A. 6, 1 : 715 ; Chev. Bot. 558. A glabrous subwoody herb, up to 2½ ft. high ; flowers white, small, on short pedicels ; a weed of cultivated land and waste ground.
   Fr.G.: Kouria *Chev.* 14839. **S.L.:** Freetown *Welw.* 316 ! Kambia (fl. & fr. Dec.) *Sc. Elliot* 4346 ! Newton (fl. & fr. Sept.) *Deighton* 4878 ! Rokupr (fl. & fr. Apr., Aug.) *Deighton* 3019 ! *Jordan* 235 ! Bo (fl. & fr. June) *Deighton* 5102 ! **Lib.:** Gbanga (fl. & fr. Sept.) *Linder* 522 ! Monrovia (fl. & fr. June, Nov.) *Baldwin* 5871 ! *Barker* 1462 ! Mt. Barclay (fl. & fr. Apr.) *Bunting* ! **Togo:** Misahöhe *Baumann* 144 (partly). **S.Nig.:** Calabar (fl. & fr. July) *Holland* 44 ! Also in French Cameroons, Gabon, Belgian Congo, Tanganyika and S. Rhodesia.

9. **P. nigericus** *Brenan* in Kew Bull. 1950 : 215. Herb with slender stems decumbent at base ; sepals pale green in centre, with white margins.
   S.Nig.: Carter's Peak, Idanre, Ondo Prov. (fl. & fr. Jan.) *Brenan & Keay* 8693 ! **F.Po:** Moka, 4,000 ft. (fl. & fr. Dec.) *Boughey* 15 ! Mioka, 5,000 ft. (fl. & fr. Dec.) *Boughey* 159 !
   [The Fernando Po specimens have the filaments rather less completely connate than in the type.]

10. **P. sublanatus** *Schum. & Thonn.*—F.T.A. 6, 1 : 715 (excl. spec. *Ansell*) ; Brenan l.c. 217. A glabrous herb with slender subwoody stems ; a weed.
   Fr.Sud.: Sikasso (fl. & fr. May) *Chev.* 804 ! **S.L.:** Buyabuya, Scarcies R. (fl. & fr. Feb.) *Sc. Elliot* 4267 ! Rokupr (fl. & fr. Apr.) *Jordan* 228 ! 239 ! Makene (fl. & fr. Apr.) *Deighton* 5509 ! **G.C.:** *Thonning*. Kumbungu to Mabilo, N.T. (fl. & fr. May) *Kitson* ! **Togo:** Togba Toto (fl. & fr. June) *Howes* 1005 ! **N.Nig.:** Jebba (fl. & fr. Dec.) *Meikle* 679 ! Abinsi (fl. & fr. Nov.) *Dalz.* 763 ! Onitsha Prov. *Thomas* 622 ! Also in Lower Shari.

11. **P. amarus** *Schum. & Thonn.*—F.T.A. 6, 1 : 717 ; Chev. Bot. 556. A glabrous herb with slender subwoody stems ; a weed.
   S.L.: Bonthe (fl. & fr. Nov.) *Deighton* 2265 ! Kent (fl. & fr. May) *Deighton* 2657 ! Rokupr (fl. & fr. Apr.) *Jordan* 237 ! Njala (fl & fr. May) *Deighton* 1945 ! **Lib.:** Monrovia *Whyte* ! Sinoe Basin *Whyte* ! **Iv.C.:** *fide* Chev. l.c. **G.C.:** *Thonning*. Axim (Jan.) *Chipp* 56 ! Accra *Ansell* ! **N.Nig.:** Idah *T. Vogel* ! **S.Nig.:** Lagos *Dawodu* 24 ! 365 ! *Dalz.* 1360 ! Sapoba *Kennedy* 2698 ! Awka *Thomas* 50 ! Calabar (fl. & fr. Mar.) *Brenan* 9210 ! **F.Po:** *T. Vogel* ! Widespread in the tropics. (See Appendix, p. 156.)

12. **P. urinaria** *Linn.*—F.T.A. 6, 1 : 721. A herb with slender subwoody stems.
    S.L.: Regent, on rocks (Dec.) *Sc. Elliot* 4102 ! Freetown (Dec.) *Deighton* 499 ! Moyamba (fl. & fr. Aug.) *Deighton* 2215 ! Sembehun (fl. & fr. Aug.) *Deighton* 3794 ! Rokupr (fl. & fr. Apr.) *Jordan* 238 ! **G.C.:** Benso, Tarkwa (fl. & fr. June) *Andoh* FH 5525 ! **S.Nig.:** Sapoba (fr. Sept.) *Onochie* FHI 34311 ! A common tropical weed but apparently rare in Africa.

13. **P. maderaspatensis** *Linn.*—F.T.A. 6, 1 : 722 ; Chev. Bot. 557. A woody undershrub or herb, of variable habit ; glabrous or slightly asperulate ; a weed in cultivated ground.
    Sen.: Dakar (fl. & fr. May) *Baldwin* 5707 ! **Iv.C.:** Koupéla, Mossi *Chev.* 24544. **G.C.:** Accra (fl. & fr. Mar., June, July) *Dalz.* 139 ! *Deighton* 561 ! *Irvine* 726 ! **Togo:** Yendi (fl. & fr. Dec.) *Adams & Akpabla* GC 4065 ! **N.Nig.:** Sokoto (Aug.) *Dalz.* 391 ! Throughout the tropical and subtropical regions of the Old World.

14. **P. niruri** *Linn.*—F.T.A. 6, 1 : 731. A glabrous annual weed, up to about 18 in. high, with smooth sulcate stem and slightly winged branchlets.
    Sen.: *Perrottet* ! **N.Nig.:** Nupe *Barter* 825 ! Common in India, but apparently rare elsewhere. (See Appendix, p. 157.)

15. **P. rotundifolius** *Klein ex Willd.*—F.T.A. 6, 1 : 731. *P. niruri* var. *genuinus* Beille—Chev. Bot. 557. An annual herb, to 18 in. high ; seeds dark brown.
    Sen.: *Trochain.* **Fr.Sud.:** Sumpi (fl. & fr. Aug.) *Chev.* 3462 ! Timbuktu (fl. & fr. July) *Hagerup* 147 ! 158a ! Extends from Cape Verde Islands to India.

16. **P. dusenii** *Hutch.*—F.T.A. 6, 1 : 723. Stems flexuous, about 1 ft. high, arising from a stout woody root, the flowering branchlets crowded towards the top.
    S.Nig.: Oban *Talbot* 2323 ! **Br.Cam.:** *Dusen* 296 ! On rocks in Ndian R., Kumba (Mar.) *Smith* 85/36 !

17. **P. mannianus** *Müll. Arg.*—F.T.A. 6, 1 : 730 ; Chev. Bot. 557. Erect shrublet, to 4 ft. high, in montane forest, woodland and grassland ; sepals white with broad green central line.
    Fr.G.: Fouta Djalon *fide* Chev. *l.c.* **Iv.C.:** Mt. Gbon, Haut Cavally, 3,500 ft. (fl. & fr. May) *Fleury* in *Hb. Chev.* 21415 ! **Br.Cam.:** Cam. Mt., 5,000–9,000 ft. (fl. & fr. Nov.–Jan.) *Mann* 1231 ! 1998 ! *Migeod* 208 ! Mt. Mba Kokeka, Bamenda, 7,500 ft. (Jan.) *Keay* FHI 28407 ! L. Oku, Bamenda, 7,200 ft. (fr. Jan.) *Keay* FHI 28493 !

18. **P. odontadenius** *Müll. Arg.*—F.T.A. 6, 1 : 727. A subwoody herb, to 3 ft. high ; a common weed.
    Port.G.: Bissau (Nov.) *Esp. Santo* 897 ! **S.L.:** Njala (fl. & fr. May) *Deighton* 639 ! Rokupr (fl. & fr. Apr.) *Jordan* 236 ! Bafodea (fl. & fr. Apr.) *Deighton* 5507 ! Musaia (fl. & fr. Dec.) *Deighton* 4480 ! **Lib.:** Kakatown *Whyte* ! **G.C.:** Aburi Hills (fl. & fr. Oct.) *Johnson* 471 ! Kumasi *Cummins* 77 ! **S.Nig.:** Lagos (fl. & fr. July) *Dalz.* 1361 ! Ibadan (fl. & fr. Nov.–Apr.) *Meikle* 652 ! 1306 ! 1454 ! Okomu F.R. (fl. & fr. Feb.) *Brenan* 9135 ! 9148 ! Calabar (fl. & fr. Mar.) *Brenan* 9209 ! **Br.Cam.:** Buea (fl. & fr. July) *Dundas* FHI 15237 ! **F.Po:** *Barter* ! Extends to A.-E. Sudan and Angola, also on S. Tomé.

19. **P. beillei** *Hutch.* in F.T.A. 6, 1 : 733 (1912). Shrub, to 10 ft. high, with bright green leaves and white flowers ; on rocks, and by streams in savannah.
    S.L.: Sugar Loaf Mt., 1,800 ft. (Nov.–Dec.) *Sc. Elliot* 3961 ! *Deighton* 5635 ! York Pass *Lane-Poole* 438 ! Picket Hill, 2,100 ft. (Nov.) *T. S. Jones* 191 ! **Togo:** *Kersting* 44. **N.Nig.:** Kaduna (fr. Dec.) *Meikle* 781 ! Bauchi Plateau (Oct.) *Lely* P. 819 ! **S.Nig.:** Little Osse R. (Aug.) *Symington* FHI 5636 ! Also in French Cameroons and Upper Shari region.

20. **P. petraeus** *A. Chev. ex Beille*—F.T.A. 6, 1 : 733. A shrub, to 7 ft. high ; flowers yellowish-green ; in moist places by streams.
    Fr.G.: between Santa R. and Timbo *Chev.* 12610. Kouria to Longuéry *Caille* in *Hb. Chev.* 14628 ! 14646. Kouria *Caille* in *Hb. Chev.* 14991 ! **S.L.:** Sasseni, Scarcies R. (Jan.) *Sc. Elliot* 4419 ! Wallia, Scarcies R. (Jan.) *Sc. Elliot* 4637 ! Mabum (Aug.) *Thomas* 1627 ! Njala (Feb., Oct.) *Deighton* 1087 ! 2535 ! **Lib.:** between Nua and St. John rivers, Tappita Dist. (Aug.) *Baldwin* 9140 !

Imperfectly known species.

P. sp. Dioecious shrub, to 15 ft. high ; flowers solitary in leaf-axils, pedicels slender, up to about 1 cm. long in fruit ; sepals 6 ; stamens 3, filaments united for about half their length.
    S.L.: *Lane-Poole* 443 ! Edge of grassfield, Russell (May) *Deighton* 2647 ! **Lib.:** Sanokwele (Sept.) *Baldwin* 9413 ! Monrovia (Dec.) *Baldwin* 10510 !

Besides the above *P. acidus* (Linn.) Skeel, a native of India and Madagascar has been introduced and is cultivated for its fruit : "Indian Gooseberry".

FIG. 135.—SECURINEGA VIROSA (*Roxb. ex Willd.*) *Baill.* (EUPHORBIACEAE).

A, male flower with rudimentary ovary. B, C, perianth segments. D, stamens from side and back. E, female flower. F, fruit. G, cross-section of fruit. H, longitudinal section of fruit.

17. SECURINEGA Commers. in Juss. Gen. 388 (1789) ; Pax & K. Hoffm. in E. & P. Pflanzenfam. 19C : 60 (1931). *Fluggea* Willd.—F.T.A. 6, 1 : 736.

Shrub or small tree ; leaves elliptic or obovate, cuneate at base, up to 6 cm. long and 3 cm. broad, rigidly membranous, glabrous, often glaucous below ; flowers dioecious, the males in axillary fascicles on slender pedicels up to 5 mm. long ; stamens 5, free ; rudimentary ovary tripartite ; female disk annular ; fruits depressed-globose, about 5 mm. diam. ; seeds shining with several lines of pits on the back   ..   *virosa*

S. virosa (*Roxb. ex Willd.*) *Baill.* in Adansonia 6 : 334 (1866). *Phyllanthus virosus* Roxb. ex Willd. (1805). *Fluggea virosa* (Roxb. ex Willd.) Baill. (1858)—F.W.T.A., ed. 1, 1 : 291. *F. microcarpa* Blume (1825)— F.T.A. 6, 1 : 736 ; Chev. Bot. 559. *Securinega microcarpa* (Blume) Pax & K. Hoffm. ex Aubrév. Fl. For. Soud.-Guin. 190 (1950). Up to 12 ft. high, with somewhat angular glabrous branchlets ; fruit a small white berry ; locally common in open regrowth vegetation.   Liable to be mistaken for a species of *Phyllanthus*, but distinguished by the large tripartite rudimentary ovary in the male flower.
Sen.: *Leprieur* ! *Sieber* 49 ! **Gam.:** Genieri (July) *Fox* 132 ! **Fr.Sud.:** various localities *fide* Aubrév. *l.c.* **Fr.G.:** Kouroussa (Apr.) *Pobéguin* 233 ! **S.L.:** *Sc.* Elliot ! *Sc.* Elliot ! **Iv.C.:** *fide* Aubrév. *l.c.* **G.C.:** Cape Coast Castle (July) *T. Vogel* 86 ! Accra Plains (Feb., Mar.) *Brown* 934 ! *Irvine* 215 ! Bjury (July) *Chipp* 498 ! Yegi (July) *Pomeroy* 1247 ! **Togo:** Lome *Warnecke* 297 ! Kpedsu (Dec.) *Howes* 1066 ! **Fr.Nig.:** Maïné Soroa *Aubrév.* **N.Nig.:** Nupe *Barter* 1496 ! Katagum *Dalz.* 207 ! Aguji, Ilorin *Thornton* ! Bauchi Plateau (Aug.) *Lely* P. 656 ! **S.Nig.:** Lagos *Rowland* ! Abeokuta *Irving* ! Widespread in tropical Africa, S. Africa, the Mascarene Islands, tropical and subtropical Asia, and in Australia. (See Appendix, p. 146.)

18. UAPACA Baill.—F.T.A. 6, 1 : 634 ; Pax & K. Hoffm. in Engl. Pflanzenr. Euph. 15 : 298–311 (1922).

Stipules foliaceous, lanceolate to ovate, persistent or subpersistent, up to 3 cm. long and 2 cm. broad ; young shoots pubescent ; trees of swamps in rain forest :
Leaves glabrous or with the midrib minutely puberulous beneath, oblanceolate or obovate-oblanceolate, cuneate to rounded at base, (8–) 19–48 cm. long, (3·2–) 7–17 cm. broad, main lateral nerves 8–14 pairs ; petiole 1–7 cm. long ; bracts glabrous ; heads of male flowers about 12 mm. diam. ; male calyx glabrous ; ovary puberulous ; fruits ovoid or subglobose, 2–2·5 cm. diam., warted, puberulous, 3-seeded
1. *staudtii*
Leaves shortly pilose beneath on midrib, nerves and veins, obovate to oblong-obovate, rounded to cordate at base, 16–50 cm. long, 10–25 cm. broad, main lateral nerves 10–20 pairs ; petiole 3–10 cm. long, stout ; bracts pubescent outside ; heads of male flowers 8–10 mm. diam. ; male calyx pubescent ; ovary velutinous ; fruits ovoid, 2·5–3 cm. long, (3-) 4-seeded   ..   .. 2. *paludosa*
Stipules linear or linear-subulate, mostly early caducous (but see No. 3) :
Leaves pilose beneath, especially on midrib and nerves ; fruits globose :
Stipules persistent, linear-subulate, up to 1 cm. long ; branchlets pilose ; leaves obovate-oblanceolate, rounded to cuneate at base, 6–16 (–27) cm. long, 2·7–7·5 (–12) cm. broad, main lateral nerves 7–9 (–12) pairs ; petiole 1·5–3·8 (–7·5) cm. long ; fruits about 2 cm. diam., glabrous   ..   ..   ..   3. *vanhouttei*
Stipules early caducous ; branchlets sparsely pubescent ; leaves obovate to obovate-suborbicular, rounded at base, 7·5–17 cm. long, 4–11·5 cm. broad, main lateral nerves 6–10 pairs ; petiole 2–4 cm. long ; bracts obovate-oblanceolate, pilose outside ; heads of male flowers about 5 mm. diam. ; male calyx pilose ; ovary pubescent ; fruits 2–2·5 cm. diam., warted, pubescent, 3-seeded   4. *chevalieri*
Leaves glabrous or rather sparsely puberulous beneath :
Calyx of male flowers pilose ; heads of male flowers 5–6·5 mm. diam. ; fruits subglobose up to about 2 cm. diam. :
Ovary densely pubescent ; fruits pubescent, always with a ring of long hairs at the base ; stipules and young branchlets pubescent, usually rather densely so ; leaves usually oblong-obovate, entire or irregularly undulate, cuneate, obtuse or rounded at base, 11–25 cm. long, 5–15·5 cm. broad, sparingly puberulous or glabrescent beneath, coriaceous, rough to the touch ; petiole 1–8 cm. long ; tree of savannah normally without stilt-roots, sometimes in fringing forest with stilt-roots
5. *togoensis*
Ovary glandular, glabrous or nearly so ; fruits glabrous, warted ; stipules and young branchlets glabrous, or pubescent and soon glabrescent ; leaves obovate, markedly undulate, cuneate at base, 7–26 cm. long, 2·8–13 cm. broad, glabrous beneath, thinly coriaceous ; petiole 1·5–7 cm. long ; tree with stilt-roots, in rain forest
6. *guineensis*
Calyx of male flowers glabrous or ciliate on the margin ; heads of male flowers 7–13 mm. diam. ; fruits ovoid or ellipsoid, 2·5–5 cm. long, glabrous ; leaves cuneate at base, glabrous :
Heads of male flowers 7–8 mm. diam., bracts glabrous ; fruits ellipsoid, 2·5–3·5 cm. long, 1·2–2·5 cm. diam., 3-seeded ; leaves oblanceolate to obovate-oblong, 8–20 cm. long, 3–6·5 cm. broad, papery ; petiole 1–3·8 cm. long :
Leaves rounded or obtuse at apex, lateral nerves 10–15 pairs ; tufts of reddish curly hairs in leaf-axils   ..   ..   7. *heudelotii*
Leaves acuminate at apex, lateral nerves 7–8 pairs ; tufts of curly hairs absent
8. *acuminata*

7

Heads of male flowers about 13 mm. diam., bracts puberulous ; calyx of male
flowers ciliate on the margin ; fruits ovoid about 5 cm. long and 4 cm. broad,
5-seeded ; leaves oblanceolate to oblong-obovate, 9–30 cm. long, 3–15 cm. broad,
thickly coriaceous ; lateral nerves 5–8 pairs ; petiole 2–4·5 cm. long ; young
shoots glabrous, not waxy   ..   ..   9. *esculenta*

1. **U. staudtii** *Pax*—F.T.A. 6, 1 : 638 ; Pax & K. Hoffm. in Engl. Pflanzenr. Euph. 15 : 300.   Tree to 90 ft.
   high, with stilt-roots, in swamp forest ; flowers white or pale yellowish.
   S.Nig.: Oni *Sankey* 14 ! Shasha F.R. (Apr.) *Richards* 3416 ! Boje, Ikom (May) *Jones & Onochie* FHI
   17335 ! Okwango F.R., Obudu (May) *Latilo* FHI 30964 ! **Br.Cam.:** Victoria (fl. Apr., fr. Feb.) *Preuss*
   1171 ! *Maitland* 397 ! 613 ! 630 ! Bambuko F.R., Kumba (fr. Sept.) *Olorunfemi* FHI 30773 ! **F.Po:**
   Bokoko *Mildbr.* 6854.   Also in French Cameroons. (See Appendix, p. 166.)
2. **U. paludosa** *Aubrév. & Léandri* in Bull. Soc. Bot. Fr. 82 : 50, fig. 1 (1935) ; Aubrév. Fl. For. C. Iv. 2 : 30,
   t. 131.   *U. guineensis* of F.W.T.A., ed. 1, 1 : 292, partly (*Barter* 1833, *Talbot* 3050). Tree to 60 ft. high,
   with stilt-roots, in swamp forest.
   Iv.C.: Banco (fr. Jan., Aug.) *Aubrév.* 361 ! 502 ! 875 ! 1510 ! Tabou (Dec.) *Aubrev.* 1671 ! **G.C.:** Huniso
   Irvine 1062 ! **S.Nig.:** Omodofen, Ifon (Aug.) *Farquhar* 57 ! Okomu F.R. *Brenan* 9010 ! Sapoba *Kennedy*
   1967 ! Brass *Barter* 1833 ! Eket *Talbot* 3050 ! Also in French Cameroons. (See Appendix, p. 166.)
3. **U. vanhouttei** *De Wild.*   Et. Fl. Bas-et Moy. Congo 2 : 275 (1908) ; Pax & K. Hoffm. l.c. 15 : 303. Tree,
   to 30 ft. high, in rain forest.
   S.Nig.: Aking, Oban (fr. Feb.) *Aninze* FHI 15429 ! Aboabam, Ikom (fr. May) *Jones & Onochie* FHI 18705 !
   Oban *Talbot* 671 ! Also in Belgian Congo.
4. **U. chevalieri** *Beille* in Bull. Soc. Bot. Fr. 55, Mém. 8 : 68 (1908) ; Aubrév. Fl. For. Soud.-Guin. 190–193.
   U. togoensis Pax [1] in Engl. Bot. Jahrb. 34 : 371 (1904), partly (*Sc. Elliot* 4828 only) ; F.T.A. 6, 1 : 638 ;
   F.W.T.A., ed. 1, 1 : 292 (excl. distrib. Togo & Dah.) ; Aubrév. Fl. For. C. Iv. 2 : 32, t. 130B.   Tree to
   60 ft. high, usually in wet places near streams.
   Fr.G.: Fouta Djalon : Ditinn to Diaguissa (Apr.) *Chev.* 12670 ! Kala to Dalana (fl. & fr. Apr.) *Chev.*
   13477 ! Pita (Apr.) *Pobéguin* 2305 ! **S.L.:** Mt. Gonkwi, Talla (fr. Feb.) *Sc. Elliot* 4828 ! Worawaia (Apr.)
   Lane-Poole 3 ! Loma Mts. (Apr.) *Jaeger* 1309 ! *Frith* 21 ! L. Sonfon (Apr.) *Deighton* 5069 ! **Iv.C.:** *Aubrév.*
   1131 ! Nimba Mts. *Schnell* !
5. **U. togoensis** *Pax* [1] in Engl. Bot. Jahrb. 34 : 371 (1904), excl. *Sc. Elliot* 4828 ; in Engl. Pflanzenr. Euph. 15 :
   304 (excl. *Sc. Elliot* 4828, excl. syn. *U. chevalieri* Beille) ; F.W.T.A., ed. 1, 1 : 292, in part only (distrib.
   Togo & Dah.).   *U. guignardi* A. Chev. ex Beille (1908).   *U. somon* Aubrév. & Léandri (1935) ; Aubrév.
   Fl. For. C. Iv. 2 : 34 ; Fl. For. Soud.-Guin. 190–193, t. 45, 4–5.   *U. guineensis* of F.T.A. 6, 1 : 640 and of
   F.W.T.A., ed. 1, 1 : 292, partly.   A tree to 40 ft. high, common in savannah ; growing also as a taller tree
   with stilt-roots in fringing forest in the savannah regions ; flowers yellow ; ripe fruits orange-yellow ;
   leaves shining dark green, clustered at the ends of the branches.
   Fr.Sud.: Segouna *Dubois* 192c ! **Port.G.:** Gabu (young fr. Oct.) *Esp. Santo* 3114 ! **Fr.G.:** *Farmar* 325 !
   Farana (fr. Feb.) *Chev.* 20633 ! Kouroussa *Pobéguin* 432.   **S.L.:** Erimakuna (fr. Mar.) *Sc. Elliot* 5395 !
   Musaia (fl. Sept., Oct., fr. Feb.) *Thomas* 2660 ! *Deighton* 4169 ! *Small* 205 ! Port Loko (fr. Feb.) *Deighton*
   3609 ! Foot of Loma Mts. (Sept.) *Jaeger* 1991 ! **Iv.C.:** Bondoukou (fr. Jan.) *Aubrév.* 728 ! Tafiré (Dec.)
   Aubrév. 614 ! **G.C.:** Ejura (Aug.) *Chipp* 725 ! Kintampo (Nov.) *Andoh* FH 5436 ! **Togo:** Sokodé (fr.
   Mar.) *Kersting* 22 ! **Dah.:** Poisson.   **N.Nig.:** Agaie (Nov.) *Yates* 24 ! 24a ! Jos (Oct.) *Costello* 15/28 !
   Katagum *Dalz.* 356 ! **S.Nig.:** Awba Hills, Ibadan *Jones & Keay* FHI 4893 ! Enugu (Sept.) *Onochie*
   FHI 34083 ! Bansara to Ikom (fr. Dec.) *Keay* FHI 28146 ! Obudu (young fr. Dec.) *Keay & Savory*
   FHI 25004 ! 25005 ! 25006 ! **Br.Cam.:** Babati *Rudatis* 36.   Also in French Cameroons and Ubangi-
   Shari. (See Appendix, p. 166.)
   [I have seen trees of this species in savannah woodland, in Ogoja Province, Nigeria, producing aerial
   roots.—R.W.J.K.]
6. **U. guineensis** *Müll. Arg.* in Flora 47 : 517 (1864) ; F.T.A. 6, 1 : 640, partly ; F.W.T.A., ed. 1, 1 : 292,
   partly (excl. *Sc. Elliot* 5395, *Yates* 24, 24a, *Barter* 1833 and *Talbot* 3050) ; Pax & K. Hoffm. in Engl.
   Pflanzenr. Euph. 15 : 306 ; Aubrév. Fl. For. C. Iv. 2 : 30, t. 129, 7–10.   *U. bingervillensis* Beille (1908).
   A tree to 90 ft. high, with stilt-roots ; in rain forest ; flowers yellow ; fruit edible.
   S.L.: *Unwin & Smythe* 38 ! Kenema (Jan.) *Thomas* 7728 ! Mayoso (fr. Aug.) *Thomas* 1484 ! Mofari
   (Jan.) *Sc. Elliot* 4407 ! **Lib.:** Kakatown *Whyte* ! Dukwia R. (fr. Feb.) *Cooper* 298 ! Gbanga (fr. Sept.)
   Linder 696 ! Cape Palmas (May) *Hunting* 1 ! **Iv.C.:** Tano R. (fr. July) *Chipp* 15415 ! Rasso *Aubrév.* 562 ! Banco *Aubrév.*
   616 ! **G.C.:** Simpa (May) *Vigne* FH 2010 ! Awiabo (Nov.) *Vigne* FH 1433.   **S.Nig.:** Lagos *Moloney* !
   Nun R. *Barter* 2116 ! Brass *Barter* 1834 ! Itu (Oct.) *King-Church* 49 ! **Br.Cam.:** Victoria (fl. Apr., fr.
   Mar.) *Maitland* 596 ! 1141 ! **F.Po:** *T. Vogel* 194 ! *Mann* 74 ! Also in French Cameroons, Gabon and
   Belgian Congo, but specimens from Uganda recorded as this species appear to be distinct. (See Appendix,
   p. 165.)
   [*Barter* 1693 from Nupe, N. Nigeria which is one of the syntypes of *U. guineensis* Müll. Arg. is probably
   the fringing forest form of *U. togoensis* (see above), but the specimen lacks female flowers and fruits so its
   determination is somewhat doubtful.]
7. **U. heudelotii** *Baill.*—F.T.A. 6, 1 : 639 ; Aubrév. l.c. 2 : 29, t. 130 ; Pax & K. Hoffm. l.c. 15 : 308, fig. 25.
   U. benguelensis of Chev. Bot. 562.   A forest tree, to 90 ft. high and 9 ft. girth, with stilt-roots ; always
   near water, in forest regions and in fringing forest in the savannah regions ; flowers white or greenish.
   Fr.Sud.: Diendema (Feb.) *Chev.* 423 ! **Fr.G.:** Fouta Djalon *Heudelot* 836 ! Ditinn (Apr.) *Chev.* 12702 !
   S.L.: Moa R., Kambui F.R. (Mar.) *Lane-Poole* 327 ! Talla Hills (Feb.) *Sc. Elliot* 4946 ! Musaia (Feb.)
   Sc. Elliot 5150 ! *Deighton* 4239 ! 4240 ! **Lib.:** Dukwia R. (Feb.) *Cooper* 179 ! Zolopla, Sanokwele (fr.
   Sept.) *Baldwin* 9379 ! **Iv.C.:** Tano R. (fr. July) *Chipp* 300 ! Various localities *fide* Aubrév. *l.c.* **G.C.:**
   Zantana *Kitson* 560 ! Anibil, Ancobra R. (Mar.) *Irvine* 2390 ! **Togo:** *fide* Pax & K. Hoffm. *l.c.* **Dah.:**
   Djougou *Chev.* 23860.   **N.Nig.:** Nupe *Barter* 1151 ! Agaie (Jan.) *Yates* 25 ! **S.Nig.:** Lagos *Rowland* !
   Oni R. *Sankey* 3 ! Calabar R. (Feb.) *Mann* 2268 ! **Br.Cam.:** *fide* Pax & K. Hoffm. *l.c.* Also in French
   Cameroons, Spanish Guinea, Gabon and Belgian Congo. (See Appendix, p. 166.)
8. **U. acuminata** (*Hutch.*) *Pax & K. Hoffm.* l.c. 15 : 308 (1922).   *U. heudelotii* var. *acuminata* Hutch. in F.T.A.
   6, 1 : 639 (1912), excl. *Preuss* 8 (see Pax & K. Hoffm. l.c.). A tree, to 75 ft. high, in rain forest.
   S.Nig.: Ikom *Catterall* 74 ! Bendiga Ayuk, Ikom (fr. May) *Jones & Onochie* FHI 14127 ! **Br.Cam.:**
   Johann-Albrechtshöhe *Staudt* 577 ! Also in French Cameroons and Cabinda.

----

[1] In his original description Pax cited seven specimens none of which was designated as the holotype.
Hutchinson in F.T.A. 6, 1 : 638 (1912), realizing that these syntypes of *U. togoensis* represented at least two
species, chose *Sc. Elliot* 4828 as the lectotype, thus excluding the species from Togoland. This choice must
however be rejected because the leaves of *Sc. Elliot* 4828, at any rate in the Kew sheet, are pilose beneath and
therein quite at variance with Pax's original description and diagnosis. I am therefore basing my interpretation
of *Uapaca togoensis* Pax on the Kew duplicate of the syntype *Sc. Elliot* 5395, which agrees with the original
description, and on duplicates of *Kersting* 22 and *Chevalier* 6154 cited as this species by Pax in the Pflanzenreich
(1922).   Aubréville (1950) rejects *U. togoensis* Pax on the grounds of its being a *nomen confusum*. However,
apart from *Sc. Elliot* 4828, Pax's original description and his treatment in the Pflanzenreich clearly applies to
the plant Aubréville & Léandri later (1935) described as *U. somon*.   Even if *U. togoensis* is rejected, the name
U. somon Aubrév. & Léandri, which has become well known in recent years, canno tbe maintained, as an earlier
legitimate name *U. guignardi* A. Chev. ex Beille (1908) is available for the same plant.

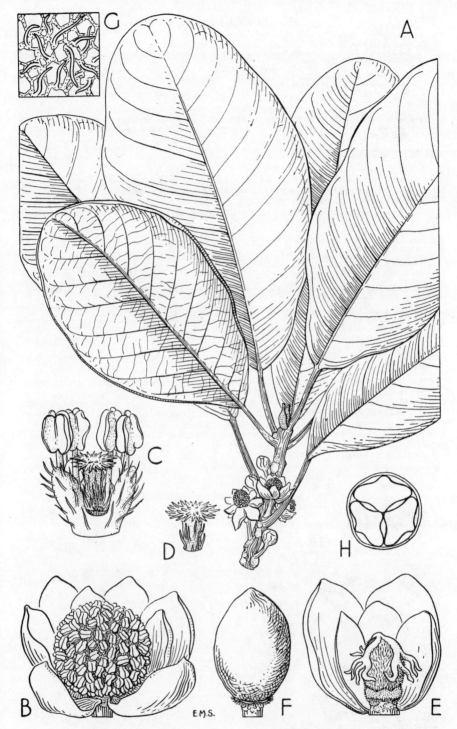

FIG. 136.—UAPACA TOGOENSIS *Pax* (EUPHORBIACEAE).

A, leafy shoot with male inflorescences, × ⅔. B, male inflorescence, × 4. C, male flower with one sepal and one stamen removed, × 12. D, rudimentary ovary from male flower, × 12. E, female inflorescence, × 4. F, fruit, × 1½. G, part of surface of fruit, with rather dense pubescence, × 24. H, diagram of fruit in transverse section, × 1½.
A–D are drawn from *Chipp* 725   E from *Yates* 24   F–H from *Dalz.* 356.

9. **U. esculenta** *A. Chev. ex Aubrév. & Léandri* in Bull. Soc. Bot. Fr. 82 : 52, fig. 2 (1935) ; Aubrév. Fl. For. C. Iv. 2 : 30, t. 129, 1–6. A large tree with stilt-roots in rain forest ; flowers yellow ; fruits edible.
S.L.: Commendi—Gengaru road (Nov.) *Aylmer* 255 ! Kambui Hills (fr. Sept.) *Edwardson* 36 ! **Lib.:** Dukwia R. *Cooper* 147 ! **Iv.C.:** Yapo (Oct.) *Chev.* 22320 ! Abidjan *Aubrév.* 212 ! Banco *Aubrév.* 355 ! 498 ! And other localities *fide* Aubrév. *l.c.* **G.C.:** Simpa (fr. Feb.) *Vigne* FH 2803 ! **S.Nig.:** Jamieson R., Sapoba *Kennedy* 1876 ! 1895 ! (See Appendix, p. 165.)
[*U. corbisieri* De Wild. from the Belgian Congo may well prove conspecific ; it is the older name.]

Doubtfully recorded species.

A sterile species (*Ogua* FHI 7850) from Banki, Anchau Dist., Zaria Prov., N. Nigeria, appears to be different from all the above species. It may possibly be the same as *U. pilosa* Hutch. from central Africa ; see also *Mildbr.* 9327 from Buar, French Cameroons.

19. MICRODESMIS Hook. f. ex Planch.—F.T.A. 6, 1 : 741 ; Pax & K. Hoffm. in Engl. Pflanzenr. Euph. 3 : 105 (1911).

Branchlets densely to sparsely pubescent ; leaves more or less oblong or oblong-ovate, cuneate and unequal-sided at base, more or less caudate-acuminate, up to 15 cm. long and 5 cm. broad, dentate or subentire ; flowers in axillary fascicles ; fruit globose, muricate, about 7 mm. diam. ..   ..   *puberula*

M. puberula *Hook. f. ex Planch.*—F.T.A. 6, 1 : 741 ; Pax & K. Hoffm. l.c. 3 : 106, fig. 34 ; Chev. Bot. 568 ; Aubrév. Fl. For. C. Iv. 2 : 63, t. 145. *M. zenkeri* Pax (1897)—F.T.A. 6, 1 : 742 ; F.W.T.A., ed. 1, 1 : 292. A forest shrub, or small tree to 50 ft. high ; flowers greenish yellow or orange ; fruits red ; variable in indumentum and leaf-shape.
Fr.G.: Fouta Djalon *Heudelot* 832 ! Massadou, Macenta (May) *Collenette* 25 ! **S.L.:** Heddle's Farm (Dec.–Jan.) *Lane-Poole* 178 ! Kenema (Nov.) *Deighton* 436 ! Njala (fl. May–July, fr. Dec.) *Deighton* 692 ! 1857 ! Kakuna, Scarcies (Jan.) *Sc. Elliot* 4729 ! **Lib.:** Dukwia R. (Feb., Nov.) *Cooper* 124 ! 207 ! Ganta (Sept.) *Baldwin* 9225 ! Zeahtown (Aug.) *Baldwin* 6965 ! Tappita (Aug.) *Baldwin* 9058 ! **Iv.C.:** Bingerville—Abidjan—Dabou region *Chev.* 15220 ! 15223 ! Azaguié (Sept.) *Chev.* 22285 ! **G.C.:** Axim *Chipp* ! Aburi (Apr., May, Oct.) *Johnson* 479 ! 955 ! Kumasi *Cummins* 161 ! **Togo:** Misahöhe (Nov.) *Mildbr.* 7259 ! **Dah.:** Allada *Chev.* 23244 ! **N.Nig.:** Idah *Barter* 1640 ! **S.Nig.:** Lagos *Foster* 114 ! Ibadan (Oct.) *Jones & Keay* FHI 13785 ! Okomu F.R. (Dec.) *Brenan* 8394 ! Sapoba *Kennedy* 1414 ! Oban *Talbot* 692 ! Onitsha *Barter* 1805 ! Eket *Talbot* 3376 ! **Br.Cam.:** Victoria (Nov., Jan.) *Kalbreyer* 28 ! *Maitland* 779 ! Likomba (Oct.) *Mildbr.* 10588 ! Johann-Albrechtshöhe *Staudt* 545 ! **F.Po:** *T. Vogel* 169 ! 175 ! Also in French Cameroons, Ubangi-Shari, Gabon, Belgian Congo, Uganda and Angola. (See Appendix, p. 155.)

20. HEVEA Aubl.—F.T.A. 6, 1 : 743 ; Pax & K. Hoffm. in E. & P. Pflanzenfam. 19C : 102 (1931).

Leaves digitately trifoliolate ; leaflets stalked, elliptic or elliptic-oblanceolate, acuminate, narrowed to the base, 8–20 cm. long, 2·5–8 cm. broad, entire, glabrous, often glaucous beneath ; lateral nerves about 20 pairs, looped near the margin, prominent beneath ; stalks of leaflets each with a large flat orbicular gland at the base, the glands almost merged into one ; petiole up to 25 cm. long ; panicles below the leaves on each young shoot ; male flowers numerous, females fewer ; capsule 3-lobed, about 4·5 cm. diam. ; seeds broadly oblong or oblong-ellipsoid, about 2·5 cm. long, 2 cm. broad ; testa mottled and speckled, slightly shining   ..   ..   *brasiliensis*

H. brasiliensis (*Kunth*) *Müll. Arg.*—F.T.A. 6, 1 : 743 ; Holl. 4 : 585. *Siphonia brasiliensis* Kunth (1825). The Para rubber tree ; cultivated in plantations in many parts of the forest regions ; native of tropical S. America.

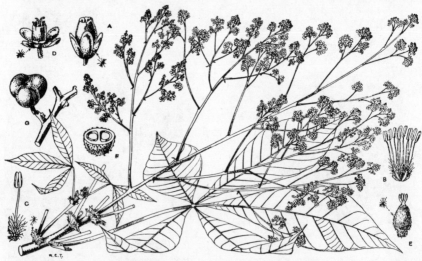

FIG. 137.—RICINODENDRON HEUDELOTII (*Baill.*) *Pierre ex Pax* (EUPHORBIACEAE). A, male flower. B, same with disk-glands removed. C, stamen. D, female flower. E, pistil. F cross-section of ovary. G, fruit.

21. RICINODENDRON Müll. Arg.—F.T.A. 6, 1 : 744 ; Pax & K. Hoffm. in Engl. Pflanzenr. Euph. 3 : 45 (1911).

Leaves digitately 3–5-foliolate ; leaflets sessile, obovate to obovate-elliptic, acuminate, narrowed to the base, the lateral ones often smaller, 6–20 cm. long, 2·5–12 cm. broad, glandular-denticulate, stellate-puberulous when young ; lateral nerves 10–16 pairs; petiole up to 20 cm. long ; stipules large, foliaceous, persistent, suborbicular, up to 2·5 cm. long, deeply toothed ; male panicles slender, about 30 cm. long ; stamens 10 ; disk-glands erect ; fruit 2-lobed, 2-celled, 2 cm. long, 3 cm. broad   ..   ..   *heudelotii*

R. heudelotii (*Baill.*) *Pierre ex Pax* in Engl. Pflanzenr. Euph. 3: 46, figs. 13, C–D, & 16 (1911). *Jatropha heudelotii* Baill. (1860).   *Ricinodendron africanum* Müll. Arg. (1864)—F.T.A. 6, 1 : 745 ; F.W.T.A., ed. 1, 1 : 294 ; Aubrév. Fl. For. C. Iv. 2 : 63, t. 146.   A large deciduous tree, to 150 ft. high ; in drier types of forest, especially in secondary regrowth ; inflorescence yellow-tomentose, petals white ; fruits indehiscent, yellow.
Port.G.: Fulacunda (May) *Esp. Santo* 2032 ! **Fr.G.**: Fouta Djalon *Heudelot* 857 ! **S.L.**: Bagroo R. (Apr.) *Mann* 825 ! Njala (Apr.) *Deighton* 1121 ! Ky Yema (Apr.) *Lane-Poole* 24 ! **Lib.**: Dukwia R. (fr. May) *Cooper* 457 ! **Iv.C.**: Attéou to Mbago (Feb.) *Chev.* 16185 ! Mt. Kouan, Danané (fr. Apr.) *Chev.* 21250 ! Azaguié *Chev.* 22203 ! **G.C.**: Aburi (Dec.) *Johnson* 448 ! Bunsu, Akim (Mar.) *Irvine* 1755 ! Kumasi (Feb., Mar.) *Vigne* FH 1801 ! *Andoh* FH 5489 ! **S.Nig.**: Olokemeji F.R. (Mar.) *A. F. Ross* 128 ! Sapoba *Kennedy* 170 ! Oban *Talbot* 2333 ! **F.Po**: (Jan.) *Mann* 229 ! Also in French Cameroons, Gabon, Spanish Guinea, Cabinda, Belgian Congo, A.-E. Sudan, Uganda, Tanganyika and Angola. (See Appendix, p. 159.)

22. CROTON Linn.—F.T.A. 6, 1 : 746 ; Pax & K. Hoffm. in E. & P. Pflanzenfam. 19C : 83 (1931).

Leaves digitately 3–5-lobed to near the base, up to 10 cm. long and broad ; segments oblanceolate or obovate, acuminate, narrowed to the base, crenate or crenate-serrate ; flowers monoecious, the lower half of the inflorescence consisting of subsessile female flowers, the upper part of much smaller males ; fruit setose, 8 mm. long ; herb, sometimes woody at base   ..   ..   1. *lobatus*
Leaves not lobed or rarely slightly so ; mostly trees and shrubs (but see Nos. 2 & 3) :
  Stems hirsute, densely covered with stellate hairs of which one or more rays are longer and more or less erect or deflexed, the others much shorter and appressed ; flowers monoecious ; racemes up to 5 cm. long, lower part with the female flowers, upper part with the much smaller males ; herbs ; leaves ovate, stellate-pubescent beneath, upper surface pilose with stellate hairs with only one long ray :
    Leaves serrate, the larger ones sometimes doubly crenate-serrate, 2·5–7 cm. long, 1–4·5 cm. broad, with a pair of long-stipitate glands at base ; racemes up to 3 cm. long   ..   ..   2. *hirtus*
    Leaves entire, up to 7·5 cm. long and 5 cm. broad, without glands at base
        3. *membranaceus*
  Stems not hirsute ; trees and shrubs :
    *Leaves densely covered beneath, at least when young, with silvery-grey, often fimbriate, scales or scale-like many-rayed stellate hairs :
      Leaves entire, permanently densely scaly beneath :
        Upper surface of leaves entirely glabrous ; petals well-developed in the female flowers ; stamens 13–20 ; branchlets, undersurface of leaves and inflorescences densely covered with silvery and rust-coloured scales :
          Racemes much abbreviated, giving the flowers a clustered appearance, up to 1 cm. long ; leaves subverticillate, lanceolate, elliptic-lanceolate or ovate, obtuse at apex, 1·5–9 cm. long, 0·8–2·5 cm. broad, basal glands absent or very obscure ; petiole up to 2·5 cm. long   ..   ..   4. *pseudopulchellus*
          Racemes elongated, 3–15 cm. long ; leaves not or scarcely verticillate, lanceolate, oblong-lanceolate or oblong-elliptic, acuminate or obtuse at apex, 4–15 cm. long, 1·5–6·5 cm. broad, with a pair of long-stipitate glands at base ; petiole up to 4 cm. long   ..   ..   5. *zambesicus*
      Upper surface of leaves with scales or stellate hairs :
        Upper surface of leaves with orbicular unfringed scales ; petals absent from the female flowers ; stamens about 7 ; leaves oblong-lanceolate or elliptic-oblong, very acute at apex, 6–12 cm. long, 2–4 cm. broad ; petiole up to 2·5 cm. long, with a pair of flat sessile glands at apex ; branchlets, leaves and inflorescence densely covered with scales, those on undersurface of leaves fimbriate ; racemes unisexual, terminal, up to 13 cm. long ; capsule densely scaly   6. *leonensis*
        Upper surface of leaves minutely stellate-puberulous, dark when dry, undersurface densely covered with silvery and rusty fimbriate scales ; petals present in the female flowers ; stamens about 30 ; leaves ovate-elliptic, cordate and with 2 large glands at base, acuminate at apex, 4–16 cm. long, 2–8 cm. broad ; petiole 2·5–7·5 cm. long ; branchlets and inflorescence densely scaly ; flowers monoecious ; racemes 7–20 cm. long ; styles 3, each one 3 times bifid ; fruits subglobose, about 9 mm. diam., scaly   ..   ..   ..   7. *sp. w. mubango*
    Leaves with obscurely serrate margins, undersurface densely covered with silvery fimbriate scale-like many-rayed stellate hairs, upper surface also with scale-like hairs ; leaves ovate, cordate or rounded and with 2–4 stipitate glands at base,

shortly and obtusely acuminate at apex, 6·5–18·5 cm. long, 4·5–13 cm. broad ;
petiole 3·5–12 cm. long ; racemes normally unisexual, up to 25 cm. long, many-
flowered .. .. 8. *macrostachyus*
*Leaves glabrous or sparingly puberulous beneath, not scaly ; petals absent from the
female flowers, or very rudimentary ; inflorescences, outside of flowers and fruits
stellate-pubescent or glabrescent, not scaly :
Leaves clustered on much-shortened branchlets, glabrous, entire, oblanceolate to
obovate, rounded at apex, 1·3–4 cm. long, up to 1·8 cm. broad, membranous ;
male racemes very short, about 5-flowered, 6 mm. long ; female flowers solitary
on a branchlet separate from the males ; stamens 5–6 .. 9. *scarciesii*
Leaves laxly arranged on more or less elongated branchlets, obscurely or distinctly
toothed, glabrous or pubescent ; stamens 10–12 ; racemes elongated :
Petiole up to 1 (–3·2) cm. long ; leaves lanceolate to elliptic, obtuse at both ends,
crenate or serrate, 3–10·5 cm. long, 1·3–5·3 cm. broad, glabrous or pubescent, with
a pair of shortly stipitate glands at the base ; racemes up to 15 cm. long, slender,
with female flowers at the base and males towards the top ; calyx of female
flowers up to 1·5 mm. long ; fruits subglobose, 3-lobed, about 5 mm. diam.
10. *nigritanus*
Petioles 1·2–8·5 cm. long ; racemes 15–42 cm. long ; calyx of female flowers 2–5
mm. long :
Basal glands on leaves long-stipitate, 2–3 mm. long ; leaves elliptic to ovate-
elliptic, rounded or cuneate at base, shortly and abruptly acuminate, crenate,
6–14 cm. long, 3·5–8 cm. broad ; petiole 2–5·5 cm. long ; racemes solitary,
terminal, 28–42 cm. long, bisexual ; calyx of female flowers 2–3 mm. long
11. *penduliflorus*
Basal glands on leaves sessile :
Leaf-margin distinctly crenate or crenate-serrate ; petiole 2–8·5 cm. long ;
lamina ovate to lanceolate-elliptic, obtuse or cuneate at base, rather long-
acuminate, 8–19 cm. long, 4–9 cm. broad, lateral nerves 5–9 on each side of
midrib ; male racemes up to 40 cm. long ; bisexual racemes up to 16 cm. long ;
calyx of female flowers about 5 mm. long ; bracts 4–7 mm. long, subpersistent ;
fruits 3-lobed, about 2·7 cm. long and broad .. 12. *longiracemosus*
Leaf-margin subentire, with sessile glands at intervals, often nearly hyaline ;
petiole 1·2–3 (–4) cm. long ; lamina ovate, round to cuneate at base, abruptly
acuminate, 5–9·5 cm. long, 3–6·5 cm. broad, lateral nerves 4–5 on each side of
midrib ; racemes bisexual, 15–27 cm. long ; bracts up to 3 mm. long, mostly
caducous ; calyx of female flowers about 2 mm. long ; fruits ovoid-ellipsoid,
about 1 cm. long .. .. 13. *dispar*

1. **C. lobatus** *Linn.*—F.T.A. 6, 1 : 750 ; Chev. Bot. 570. An erect branched pubescent annual, 2–3 ft. high,
sometimes woody at base.
    Sen.: *Heudelot* 465 ! Cape Verde *Brunner* 163 ! **Gam.:** *Hayes* 588 ! Genieri (July) *Fox* 134 ! **Fr.Sud.:**
Dendéla (Mar.) *Chev.* 631. Sikasso (Apr.) *Chev.* 750. **Port.G.:** Antula, Bissau (fl. & fr. Oct.) *Esp. Santo*
2550 ! **S.L.:** Kenema (fl. & fr. Nov.) *Deighton* 440 ! Bo (Jan.) *Thomas* 7424 ! **Iv.C.:** Bingerville *Chev.*
17317. **G.C.:** Achimota (Nov.) *Andoh* FH 4468 ! Bomase (fl. & fr. Apr.) *A. S. Thomas* 181 ! Tamale
(fl. & fr. Mar., Apr.) *Williams* 129 ! 815 ! **Dah.:** *Burton* ! **Fr.Nig.:** Niamey (fl. & fr. Oct.) *Hagerup* 542 !
N.Nig.: Nupe *Barter* 1158 ! Sokoto (fl. & fr. Sept.–Nov.) *Moiser* 55 ! 129 ! Katagum *Dalz.* 211 ! 212 !
Yola Prov. (fl. & fr. May) *Dalz.* 158 ! **S.Nig.:** Lagos *Moloney* ! Iguoriakhi, Benin (fl. & fr. Jan.) *Brenan*
8929 ! Awka *Thomas* 6 ! **F.Po :** !(fl. & fr. June) *Mann* 405 ! Extends to A.-E. Sudan, Abyssinia, Eritrea
and Arabia. (See Appendix, p. 139.)
2. **C. hirtus** *L'Hérit.* Stirp. Nov. 17, t. 9 (1784) ; Fawcett & Rendle Fl. Jamaica 4, 2 : 285. *C. glandulosus*
Linn. var. *hirtus* (L'Hérit.) Müll. Arg. in DC. Prod. 15, 2 : 684. Herb, to 2 ft. high ; an introduced weed.
    S.L.: Masanki (fl. & fr. Jan.) *Deighton* 4584 ! Newton (fl. & fr. June) *Deighton* 5551 ! Njala (fl. & fr.
June) *Deighton* 6069 ! Widespread in the West Indies and tropical America.
3. **C. membranaceus** *Müll. Arg.*—F.T.A. 6, 1 : 755. A herb, or undershrub, up to 3 ft. high.
    N.Nig.: near Lokoja *Barter* 615 ! Wuru, Nupe *Barter* 814 ! **S.Nig.:** Onitsha *Barter* 574 !
4. **C. pseudopulchellus** *Pax*—F.T.A. 6, 1 : 755 ; Aubrév. Fl. For. Soud.-Guin. 195–196, t. 37, 8. Shrub, to
12 ft. high.
    Fr.Sud.: Fo (fr. June) *Chev.* 961 ! **N.Nig.:** near Gombe (Feb.) *Thornewill* 15 ! Also in Kenya, Tangan-
yika, Zanzibar, Portuguese East Africa and S. Rhodesia.
5. **C. zambesicus** *Müll. Arg.* in Flora 47 : 483 (Oct. 1864) ; F.T.A. 6, 1 : 758 ; Aubrév. l.c. 193–196, t. 37,
5–7. *C. amabilis* Müll. Arg. l.c. 47 : 537 (Nov. 1864) ; F.W.T.A., ed. 1, 1 : 296. A small tree, 20–50 ft.
high ; branchlets sulcate ; capsule scaly, ⅓ in. diam. ; often pollarded in villages and towns.
    Gam.: Georgetown (Jan.) *Dalz.* 8055 ! **S.L.:** Freetown (June) *Daniell* ! Waterloo (Mar.) *Kirk* 27 !
Allen Town (Dec.) *Dawe* 410 ! **Iv.C.:** Groumania (Jan.) *Aubrév.* 788 ! **G.C.:** Cape Coast *Brass* ! Wenchi
(Oct.) *Vigne* FH 2534 ! Mampong (July) *Vigne* FH 1222 ! **Togo:** Fulandi-Kedia *Kersting* A. 516 !
Dah.: Adjara, Porto Novo (Jan.) *Chev.* 22756 ! Djougou (fr. June) *Chev.* 23876 ! **Fr.Nig.:** Niamey to
Zinder (Oct.) *Hagerup* 568 ! Gaya (Oct.) *Aubrév.* N. 20 ! **N.Nig.:** Ilorin *Barter* 3323 ! Kontagora (Feb.)
Dalz. 281 ! Sokoto (Feb.) *Lely* 802 ! Bauchi Plateau (Jan.) *Lely* P. 105 ! **S.Nig.:** Lagos *Punch* 20 !
Foster 33 ! Abeokuta (Feb.) *Irving* 62 ! Okuni (Jan.) *Holland* 173 ! Oban *Talbot* 2302 ! Widespread in
tropical Africa. (See Appendix, p. 139.)
6. **C. leonensis** *Hutch.* in F.T.A. 6, 1 : 1050 (1913). A shrub, 5–6 ft. high ; branchlets sulcate when young.
    S.L.: Makunde, Limba (Apr.) *Sc. Elliot* 5716 ! Njala, in somewhat moist place (Feb.) *Deighton* 1086 !
7. **C. sp. nr. mubango** *Müll. Arg.*—F.T.A. 6, 1 : 659 ; Aubrév. Fl. For. C. Iv. 2 : 74, t. 152, 3–6. A small
tere, in forest.
    Iv.C.: Anoumaba (Nov.) *Chev.* 22399 ! Nzi, Comoé (Jan.) *Aubrév.* 2121 ! True *C. mubango* occurs in
Belgian Congo and Angola
8. **C. macrostachyus** *Hochst. ex Del.* in Ferret & Galinier Voy. Abyss. 3 : 158 (1847) ; F.T.A. 6, 1 : 772 ;
Aubrév. Fl. For. Soud.-Guin. 193–196, t. 36, 4–5. *C. guerzesiensis* Beille in Bull. Soc. Bot. Fr. 61, Mém. 8 :
294 (1917) ; Chev. Bot. 570. A tree, in savannah, up to 30–50 ft high ; flowers white ; fruit slightly
3-lobed, about ½ in. diam.

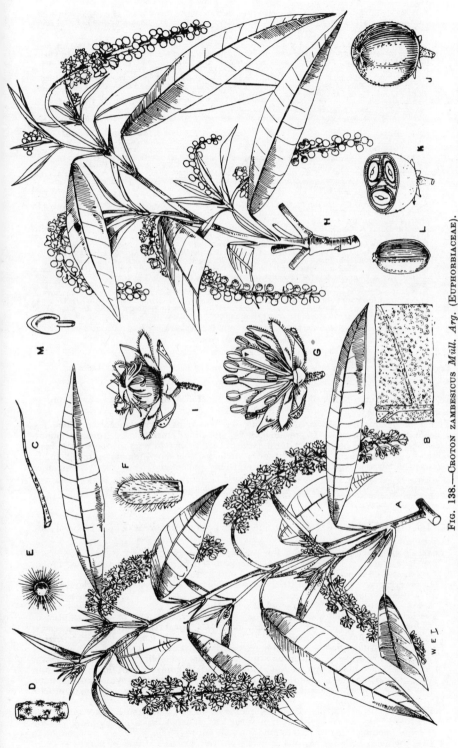

Fig. 138.—Croton zambesicus Müll. Arg. (Euphorbiaceae).

A, male shoot. B, undersurface of leaf showing peltate scales. C, stipule. D, portion of stem. E, peltate scale. F, bract. G, male flower. H, female shoot. I, female flower. J, fruit. K, cross-section of fruit. L, seed. M, embryo.

Fr.G.: Lola to Nzo, Guerze country (Mar.) *Chev.* 20979! Beyla *Adam* 124! **Iv.C.:** Danipleu (Mar.) *Aubrév.* 1150! 1151! **N.Nig.:** Bauchi Plateau (May) *Lely* P. 314! Pankshin *Kennedy* FHI 8064! **Br.Cam.:** Bamenda Prov.: Njinikom (May) *Maitland* 1506! Bambui (fr. Jan.) *Keay* FHI 28334! Kumbo (May) *Johnstone* 138/31! Widespread in tropical Africa.

9. **C. scarciesii** *Sc. Elliot*—F.T.A. 6, 1 : 752 ; Chev. Bot. 571 ; Berhaut Fl. Sen. 129. A shrub 2–10 ft. high, growing on rocks by streams, under water in the rainy season ; capsule bristly with tubercle-based hairs. **Sen.:** *fide* Berhaut *l.c.* **Port.G.:** Cusselinta, Chitole (fr. Dec.) *Esp. Santo* 3172! **S.L.:** Mofari, Scarcies (Jan.) *Sc. Elliot* 4432! Sasseni, Scarcies (Jan.) *Sc. Elliot* 4518! Njala, R. Taiya (Jan., May) *Dalz.* 8059! *Deighton* 505! 702! Makump (Jan.) *Glanville* 135! Mange, Bure (fl. & fr. Jan.) *Jordan* 851! 861! **Iv.C.:** Alépé to Malamalasso, R. Comoé (Mar.) *Chev.* 17513. Abiati (Mar.) *Chev.* 17539. **G.C.:** Pong Tamale (June) *Vigne* FH 3864!

10. **C. nigritanus** *Sc. Elliot* in J. Linn. Soc. 30 : 97 (1894) ; F.T.A. 6, 1 : 770 ; Aubrév. Fl. For. Soud.-Guin. 193–196, t. 37, 3–4. *C. chevalieri* Beille (1910)—F.W.T.A., ed. 1, 1 : 297. *C. nudifolius* Bak. & Hutch. (1912)—F.W.T.A., ed. 1, 1 : 297. Shrub, to 10 ft. high, with white flowers ; usually in moist places by rivers.
    Fr.G.: Farana (Mar.) *Sc. Elliot* 5379! 5380! Labé (May) *Pobéguin* 2115! Macenta *Adam* 84! **S.L.:** Kahreni, Limba (Apr.) *Sc. Elliot* 5583! Sasseni, Scarcies (Jan.) *Sc. Elliot* 4429! Njala (fl. & fr. Mar., May, July) *Deighton* 1088! 1098! 1952! 4020! 5013! Pujehun (Apr.) *Aylmer* 60! **Lib.:** Peahtah (Oct.) *Linder* 953! Dobli Isl., St. Paul R. *Bequaert* 21! **Iv.C.:** Prolo, Lower Cavally (fl. & fr. Aug.) *Chev.* 19861! **G.C.:** Zantana to Lungbungu (May) *Kitson* 728! **N.Nig.:** Lokoja *T. Vogel* 165!

11. **C. penduliflorus** *Hutch.* in F.W.T.A., ed. 1, 1 : 297 (1928) ; Kew Bull. 1928 : 300. *C. rubinoensis* Aubrév. l.c. 195, t. 36, 1–3 (1950), French descr. only. A tree to 70 ft. high ; racemes pendulous.
    S.L.: Kenema (May) *Aylmer* 138! *Lane-Poole* 269! Sendugu (June) *Thomas* 669! Danbara *King* 261! Bandajuma (May) *Deighton* 3729! Gola Forest (May) *Small* 651! **Iv.C.:** Rubino (young fr. Dec.) *Serv. For.* 2855! **G.C.:** Kwahu Plateau (May) *Kitson* 1170! Sra (June) *Moor* 305! Pepease (young fr. July) *Akpabla* 173! **S.Nig.:** Idanre F.R. (fr. July) *Onochie* FHI 33394! Onitsha Circle *Cons. of For.* 169! (See Appendix, p. 140.)

12. **C. longiracemosus** *Hutch.* in F.T.A. 6, 1 : 1052 (1913). *C. lehmbachii* Hutch. l.c. (1913). Tree, to 40 ft. high ; flowers greenish-white ; racemes apparently bisexual or entirely male.
    Br.Cam.: Buea (fl. Mar., Apr., fr. June) *Deistel* ! *Lehmbach* 19! *Reder* 24! *Maitland* 59! 465! 1107! Mimbia (Apr.) *Maitland* 1106! Moliko (Apr.) *Hutch. & Metcalfe* 86! (See Appendix, p. 139.)
    [*Sankey* 10 from Ondo Prov., Nigeria was cited as this species in the first edition, but it appears to be distinct. It is described as a tree 60–80 ft. high. The fruits which may or may not belong to the specimen are only about 7 mm. long. Further specimens are needed.]

13. **C. dispar** *N. E. Br.*—F.T.A. 6, 1 : 763. *C. collenettei* Hutch. & Dalz. F.W.T.A., ed. 1, 1 : 297 (1928) ; Kew Bull. 1928 : 300. A straggling bush, to 10 ft. high (tree 25 ft. high *fide* Collenette) ; flowers cream. **Fr.G.:** Macenta (May) *Collenette* 15! **S.L.:** Njala (fl. May–July, fr. Sept.) *Deighton* 670! 753! 1734! 2638–2641! *Cole* 61! Mayoso (fr. Aug.) *Thomas* 1416! Mattru to Sumbuya (June) *Deighton* 4775! **Lib.:** Monrovia *Whyte* ! Tappita (fr. Aug.) *Baldwin* 9090! Vonjama (fr. Oct.) *Baldwin* 9895!

Besides the above, *C. eluteria* (Linn.) Sw. a native of the Bahamas, has been introduced into **Nigeria.**

Imperfectly known species.

C. sp. A. Forest tree, about 20 ft. high ; fruits bright orange.
    S.Nig.: Ubiaja (fr. Aug.) *Onochie* FHI 33254!
C. sp. B. A poor specimen recorded as *C. oxypetalus* Müll. Arg. by Aubrév. Fl. For. Soud.-Guin. 196.
    Fr.G.: Mt. Ziama, Nialé (young fr. May) *Adam* 47!

---

23. CLUTIA Linn.—F.T.A. 6, 1 : 805 ; Pax & K. Hoffm. in Engl. Pflanzenr. Euph. 3 : 50 (1911).

Young branchlets densely woolly-tomentose ; leaves narrowly lanceolate, subacute, narrowed to the base, 6–12 cm. long, 1–2·5 cm. broad, membranous, densely pubescent on the nerves beneath ; lateral nerves 11–15 pairs ; flowers monoecious, several males and 1–2 females in each leaf-axil ; male pedicels short, female up to 1·3 cm. in fruit, the latter subglobose, 6 mm. diam., thinly pubescent   ..   *kamerunica*

C. kamerunica *Pax*—F.T.A. 6, 1 : 805 ; Pax & K. Hoffm. l.c. 3 : 58. An erect shrub, to 15 ft. high, with densely tomentose branchlets and small yellowish axillary flowers ; in streamside forest. **Br.Cam.:** Bamenda Prov.: Mt. Mba Kokeka, 6,000 ft. (fl. & fr. Jan.) *Keay* FHI 28385! Bafut-Ngemba F.R. (Aug.) *Tamajong* FHI 26895! Also in French Cameroons.

---

24. JATROPHA Linn.—F.T.A. 6, 1 : 775 ; Pax in Engl. Pflanzenr. Euph. 1 : 21 (1910).

Leaves not peltate :
Leaves penninerved and undivided or 3-lobed :
  The leaves sessile, linear or linear-lanceolate, 9–15 cm. long, up to 3 cm. broad, serrulate, or sometimes broader (6 cm.) and more coarsely toothed :
    Leaves shortly pilose beneath ; cymes short, few-flowered, with gland-tipped laciniate bracts   ..   ..   1. *neriifolia*
    Leaves glabrous beneath ; bracts linear, toothed ..   ..   ..   2. *atacorensis*
  The leaves with petioles 0·5–2 cm. long ; leaves ovate-oblong or ovate-elliptic, or 3-lobed in the upper half, 12–15 cm. long, 6–10 cm. broad, serrulate ; flowers in terminal few-flowered panicles   ..   ..   3. *kamerunica* var. *trochainii*
Leaves variously divided or digitately nerved, long-petiolate :
  Petioles furnished with numerous stipitate glands resembling the stipules ; leaves digitately 3–5-lobed beyond the middle, up to 13 cm. in diam., lobes obovate or obovate-oblanceolate, gland-toothed and shortly pubescent ; cymes pedunculate, lax ; bracts linear-lanceolate, with stipitate glands on the margin   4. *gossypiifolia*
  Petioles not glandular :
    Leaf-segments 11–12, pinnatisect or pinnately lobed, lobes acute, apex tailed, 10–15 cm. long, 2·5–5 cm. broad, more or less glaucous beneath ; cymes 2·5–5 cm. long ; peduncle up to 16 cm. long ; sepals entire   ..   ..   .. 5. *multifida*
    Leaf-segments or lobes toothed or rarely entire ; leaves glabrous or nearly so :

Leaf-segments toothed, ovate, sharply 5–8-dentate ; stipules divided into filiform glabrous gland-tipped segments ; sepals entire or with a few glandular teeth
6. *chevalieri*

Leaves undulately 5-lobed or entire ; stipules very small ; sepals entire 7. *curcas*

Leaves peltate ; orbicular-ovate, palmately 3–5-lobed, 10–20 cm. long and broad, lobes more or less ovate, entire, glabrous, pale and glaucous beneath ; petiole about 10 cm. long, glabrous ; cymes congested, long-pedunculate ; stems much swollen at the base   ..   ..   8. *podagrica*

1. **J. neriifolia** *Müll. Arg.*—F.T.A. 6, 1 : 781 ; Pax in Engl. Pflanzenr. Euph. 1 : 65. Stems pubescent, about 1 ft. high from a woody base, with viscid greenish sap ; flowers small, brick-red.
   N.Nig.: Nupe *Barter* 1679 ! Abinsi, on laterite (May) *Dalz.* 752 !
2. **J. atacorensis** *A. Chev.* in Bull. Soc. Bot. Fr. 58, Mém. 8 : 206 (1912). Habit of the last.
   Dah.: Somba country *Chev.* 24015. Farfa to Toukountouna (fr. June) *Chev.* 24070 ! Toukountouna to Kototengou *Chev.* 24129.
3. **J. kamerunica** *Pax & K. Hoffm.* var. **trochainii** *Léandri* in Bull. Soc. Bot. Fr. 83 : 525 (1936). *J. kamerunica* of Berhaut Fl. Sen. 164. An erect herb, to 2½ ft. high, with a large carrot-like rhizome.
   Sen.: Tambacounda (fl. June, Sept., fr. June) *Berhaut* 1655 ! *Trochain* 3636.
4. **J. gossypiifolia** *Linn.*—F.T.A. 6, 1 : 783 ; Pax l.c. 1 : 26 (as var. *elegans* (Klotzsch) Müll. Arg.) ; Chev. Bot. 569 ; Berhaut Fl. Sen. 144. A shrub, up to 6 ft. high with stout glabrous branches and glandular, often purple-tinged foliage ; flowers deep red-purple ; planted in villages.
   Sen.: *fide* Berhaut *l.c.* **Fr.Sud.:** Ouassaia *Chev.* 515. **Fr.G.:** Conakry ! **S.L.:** Kayima Kono (fl. & fr. July–Aug.) *Dawe* 532 ! Mano (May) *Deighton* 1761 ! Bonthe (fl. & fr. Feb.) *Dalz.* 942 ! **G.C.:** Cape Coast Castle (July) *T. Vogel* ! Nkwanta (Jan.) *Chipp* 67 ! Accra (fl. & fr. Apr.) *Bally* 19 ! *Don* ! **Togo:** Lome *Warnecke* 286 ! **Dah.:** Agouagon *Chev.* 23509 ! Dassa Zoumé *Chev.* 23645 ! **N.Nig.:** Abinsi *Dalz.* 11 ! **S.Nig.:** Lagos *Dawodu* 256 ! Yoruba country *Barter* 3353 ! Olokemeji (Apr.) *Foster* 273 ! Widely distributed in the tropics. (See Appendix, p. 148.)
5. **J. multifida** *Linn.*—F.T.A. 6, 1 : 784 ; Pax l.c. 1 : 40 ; Chev. Bot. 569. A shrub or tree, up to 20 ft. high with stout glabrous branches and coral-red flowers ; planted in villages.
   Sen.: *Roger* ! **S.L.:** Freetown *Barter* ! Kambia, R. Scarcies (Jan.) *Sc. Elliot* 4708 ! Njala (May) *Deighton* 1987 ! **Iv.C.:** Makougnié *Chev.* 16955 ! **G.C.:** *Irvine* 1414 ! **Togo:** Lome *Warnecke* 373 ! **S.Nig.:** Lagos *Phillips* 43 ! Annye, Benin (Sept.) *Unwin* 136 ! An American plant widely cultivated. (See Appendix, p. 148.)
6. **J. chevalieri** *Beille*—F.T.A. 6, 1 : 788 ; Pax l.c. 1 : 36 ; Chev. Bot. 569. A shrub about 3 ft. high, with glabrous striate branches ; fruit glabrous, ½ in. long, tipped by the base of the style.
   Sen.: L. Panie-Foul (Jan.) *Roger* 143 ! Cape Verde *Brunner* 138 ! Niayes, on sand dunes near the sea (Dec.) *Chev.* 2629. Also in Mauritania.
7. **J. curcas** *Linn.*—F.T.A. 6, 1 : 791 ; Pax l.c. 1 : 77, fig. 30 ; Chev. Bot. 569 ; Berhaut l.c. 128, 143. A shrub or small tree, up to 20 ft. high, with thick glabrous branchlets ; flowers yellowish-green ; fruit ellipsoid, scarcely lobed, about 1 in. long, black when ripe ; commonly cultivated in our area.
   Sen.: *fide* Berhaut *l.c.* **Fr. Sud.:** *fide* Chev. *l.c.* **S.L.:** Waterloo (June) *Lane-Poole* 259 ! Laminaiya (Apr.) *Thomas* 136 ! **Lib.:** Peahtah (Oct.) *Linder* 936 ! **Iv.C.:** Bobo-Dioulasso *Chev.* 912. **G.C.:** near Cape Coast Castle *Cummins* ! Axim (Apr.) *Chipp* 427 ! Afram Plains (May) *Kitson* 1143 ! **Togo:** Lome *Warnecke* 358 ! **N.Nig.:** Anara F.R., Zaria (May) *Keay* FHI 22873 ! Vom *Dent Young* 226 ! **S.Nig.:** Obu *Thomas* 121 ! Agulu *Thomas* 324 ! Oban *Talbot* 1372 ! **F.Po:** (Jan.) *Mann* 169 ! An American plant, common in the tropics. (See Appendix, p. 147.)
8. **J. podagrica** *Hook.* in Bot. Mag. t. 4376 (1848) ; Pax l.c. 1 : 44. Stem up to 2 ft. high, much swollen at the base ; flowers red ; cultivated in our area.
   S.L.: Njala (May) *Deighton* 1986 ! **G.C.:** *fide* Deighton. **S.Nig.:** Ibadan *fide* Keay. A native of Central America.

25. **MILDBRAEDIA** Pax—F.T.A. 6, 1 : 798 ; Pax & K. Hoffm. in Engl. Pflanzenr. Euph. 3 : 11 (1911).

A tree ; young branches stellate-tomentellous, at length glabrous ; leaves lanceolate-obovate or oblanceolate-elliptic, very obtuse at base, shortly and acutely acuminate, up to 15 cm. long, 1·5–6 cm. broad, acutely repand-dentate ; lateral nerves 7–10 pairs ; stipules subulate, setulose ; male cymes few-flowered, the female cymules laxly racemose, only the terminal flower of each developed ; stamens 20 ; ovary pubescent   ..   ..   *paniculata*

M. paniculata *Pax*—F.T.A. 6, 1 : 800 ; Pax & K. Hoffm. l.c. 3 : 12, fig. 2 (1911) ; Chev. Bot. 587. A forest shrub, to 10 ft. high ; flowers greenish-white.
   Lib.: Peahtah (Oct.) *Linder* 1124 ! Yangi (Sept.) *Linder* 847 ! Gletown, Tchien (fr. July) *Baldwin* 6917 ! **Iv.C.:** Kéeta, Oubi country *Chev.* 19332. Soubiré to Yaou *Chev.* 17802. Oubi to Fort Binger *Chev.* 19283. Also in Belgian Congo.

26. **CAPERONIA** St.-Hil.—F.T.A. 6, 1 : 829 ; Pax & K. Hoffm. in Engl. Pflanzenr. Euph. 6 : 27 (1912).

Leaves palmately 5-nerved at the base, ovate, crenate, slightly cordate, acute, 8–11 cm. long, 4·5–6 cm. broad, sparingly hispid on the nerves ; stipules caducous ; racemes 4–6 cm. long ; male petals subequal ; female sepals 6 ; ovary bristly ; fruit muricate
1. *latifolia*

Leaves pinnately nerved, narrowly lanceolate, serrate :
Leaves linear, remotely serrate or sometimes subentire, up to 15 cm. long ; young stems not or very sparsely clothed with gland-tipped hairs ; racemes up to 15 cm. long ; male petals unequal ; female sepals 5–6   ..   ..   ..   2. *senegalensis*
Leaves lanceolate, closely serrate, up to 15 cm. long and 3·5 cm. broad ; young stems very densely setose with stiff gland-tipped hairs ; racemes up to 7·5 cm. long ; male petals unequal ; female sepals 6   ..   ..   3. *palustris*

1. **C. latifolia** *Pax*—F.T.A. 6, 1 : 830 ; Pax & K. Hoffm. in Engl. Pflanzenr. Euph. 6 : 35. A weak herb softly hispid in the upper parts, up to 3 ft. high ; in and by ponds.
   S.Nig.: Okomu F.R., Benin (Dec.) *Brenan* 8587 ! **Br.Cam.:** Victoria *Dusen* 281. Also in S. Tomé and (*fide* Prain) in S. America and the West Indies.

2. **C. senegalensis** *Müll. Arg.*—F.T.A. 6, 1 : 832 ; Pax & K. Hoffm. l.c. 40 ; Chev. Bot. 572. *C. chevalieri*
Beille (1908)—Pax & K. Hoffm. l.c. 38. A slender herb, erect or half procumbent, up to 3 ft. high, with
small white flowers and roughly setose capsules ; usually in moist ground, sometimes a weed in rice fields.
Sen.: Bakel *Leprieur.* Casamance R. *Perrottet* 737. **Gam.:** Wallikunda (fl. Aug., Sept., fr. Sept.) *Macluskie*
11 ! 17 ! Kuntuar *Ruxton* 96 ! **Fr.Sud.:** Nafadié (fl. & fr. Jan.) *Chev.* 157 ! Sansanding (Feb.) *Chev.* 2623.
Sarédina (May) *Davey* 134 ! **Fr.G.:** Tristao Isl. *Paroisse* 46. Ditinn *Chev.* 12999. **S.L.:** Scarcies R.
(fl. & fr. Feb.) *Sc. Elliot* 4271 ! Rokupr (fl. & fr. Jan.) *Deighton* 2949 ! Kichom (July) *Deighton* 3756 !
G.C.: Walewale (fl. & fr. Dec.) *Adams & Akpabla* GC 4204 ! Burufo (fl. & fr. Dec.) *Adams & Akpabla*
GC 4387 ! **Togo:** Lome *Warnecke* 430. Kpedsu (fl. & fr. Dec.) *Howes* 1044 ! **Dah.:** Savalou (May)
Chev. 23709 ! **N.Nig.:** Nupe *Barter* ! Zungeru (Aug.) *Dalz.* 65 ! Bida Dist. (fl. & fr. Dec.) *Savory* UCI
101 ! Kumu (fl. & fr. Oct.) *Lely* 669 ! Abinsi (fl. & fr. Nov.) *Dalz.* 783 ! **S.Nig.:** Lagos *Rowland* ! Also
in French Cameroons and Ubangi-Shari.

3. **C. palustris** (*Linn.*) *St.-Hil.*—F.T.A. 6, 1 : 832 ; Pax & K. Hoffm. l.c. 33, figs. 1, A–B & 6, D–E ; Chev.
Bot. 572, partly. *Croton palustris* Linn. (1753). *Caperonia macrocarpa* Pax & K. Hoffm. l.c. 39 (1912).
C. fistulosa Beille (1908)—Pax & K. Hoffm. l.c. 37. *C. hirtella* Beille (1908)—Pax & K. Hoffm. l.c. 38.
A fairly stout herb, erect or half procumbent, up to 3 ft. high, with small white flowers and roughly setose
capsules ; in moist ground.
Fr.Sud.: Gao (Sept.) *Hagerup* 367a ! Cotaga (July) *Chev.* 1158. Sébi to Mopti (fl. & fr. Aug., Sept.) *Chev.*
15763. **G.C.:** R. Pawmpawm, E. of Anyaboni (fl. & fr. Dec.) *Morton* GC 6092 ! **N.Nig.:** Minna (fl. & fr.
Dec.) *Meikle* 700 ! **S.Nig.:** Olokemeji (fl. & fr. Nov.) *Keay* FHI 14160 ! Awba Hills F.R., Ibadan (fl. & fr.
Oct.) *Jones* FHI 5992 ! Lagos *Rowland.* Widespread in tropical Africa, and in S. Africa, Madagascar
and S. America. (See Appendix, p. 138.)

A thorough revision of this, mainly American, genus is much needed. The treatment given by Prain (in
F.T.A.) and by Pax & K. Hoffmann is by no means satisfactory ; I suspect that far too many species have
been described.—R.W.J.K.

27. CHROZOPHORA Neck.—F.T.A. 6, 1 : 834 ; Pax & K. Hoffm. in Engl. Pflanzenr.
Euph. 6 : 17 (1912).

Prostrate undershrub, covered with soft white stellate hairs ; leaves ovate, rounded
or very obtuse at apex, long-petiolate, biglandular at base, 1·3–3 cm. long, 1–2 cm.
broad ; stipules subulate ; racemes short, leafy at base ; stamens 12–15 ; ovary
tomentose ; fruits stellate-tomentose, deeply 3-lobed ; seeds rough   1. *plicata*
Usually ascending or suberect undershrubs ; fruits lepidote ; stamens up to 10 :
Indumentum of the shoots and lower surface of the leaves very woolly ; petioles fairly
long ; leaves ovate-rhomboid, 2·5–4 cm. long, up to 3 cm. broad, biglandular at
base, sometimes slightly 3-lobed ; sepals in fruit sometimes elongating to 6 mm.
long   ..   ..   2. *brocchiana*
Indumentum of the shoots and lower surface of the leaves short and close ; petioles
comparatively short ; leaves similar to preceding ; sepals not elongating in fruit
3. *senegalensis*

1. **C. plicata** (*Vahl*) *A. Juss. ex Spreng.*—F.T.A. 6, 1 : 834 (incl. var. *obliquifolia* (Vis.) Prain) ; Pax & K. Hoffm.
in Engl. Pflanzenr. Euph. 6 : 19. *Croton plicatus* Vahl (1790).
Sen.: Dagana *Leprieur.* Podar *Mathieu.* **N.Nig.:** Kukawa (fl. & fr. Jan.–Feb.) *E. Vogel* 3 ! 32 ! Extends
to A.-E. Sudan and Abyssinia, Mozambique and Rhodesia ; also in Egypt and Palestine. (See Appendix,
p. 138.)

2. **C. brocchiana** *Vis.*—F.T.A. 6, 1 : 838 (incl. *C. senegalensis* var. *lanigera* Prain) ; Pax & K. Hoffm. l.c. 6 : 20 ;
Chev. Bot. 573. A low undershrub with stout, more or less erect stems ; scales covering the capsules
whitish or violet-tinged ; in sandy soil.
Sen.: *Perrottet* 42 ! *Roger* ! Dagana *Leprieur* ! Cape Verde *Brunner* 108 ! **Fr.Sud.:** Timbuktu *Chev.* 1302.
N.Nig.: by the Niger *Barter* ! Borgu *Barter* 812 ! Bornu *Oudney* ! Also in Mauritania, and extends to
the Red Sea.

3. **C. senegalensis** (*Lam.*) *A. Juss. ex Spreng.*—F.T.A. 6, 1 : 837 ; Pax & K. Hoffm. l.c. 6 : 21 ; Chev. Bot. 573.
Croton senegalensis Lam. (1786). A low undershrub, sometimes prostrate ; flowers scarlet ; capsules as
in the preceding.
Sen.: Dagana (Apr.) *Leprieur* 1194 ! Walo and Cayor *Heudelot* 419 ! Dakar (fl. & fr. May) *Baldwin* 5751 !
Gam.: Kudang (fl. & fr. Jan.) *Dalz.* 8054 ! Yundum (fl. & fr. Dec.) *Austin* 26 ! **Fr.Sud.:** Bougouni (Apr.)
Chev. 687. Sienso, San (Apr.) *Chev.* 1051. **G.C.:** Gambaga (fl. & fr. May) *Vigne* FH 3913 ! Navrongo
(fl. & fr. May) *Andoh* FH 5175 ! Burufo (fl. & fr. Dec.) *Adams* GC 4414 ! **Dah.:** Atacora Mts. *Chev.* 24078.
N.Nig.: Sokoto (fl. & fr. Oct., Dec.) *Moiser* 60 ! 114 ! Kano (fl. & fr. Mar.) *Bally* 5 ! Katagum (fl. & fr.
July) *Dalz.* 205 ! Nguru (fl. & fr. June) *Onochie* FHI 23327 ! Also in Shari region. (See Appendix, p. 139.)

28. GROSSERA Pax—F.T.A. 6, 1 : 816 ; Pax & K. Hoffm. in Engl. Pflanzenr. Euph.
6 : 105 (1912) ; Cavaco in Bull. Mus. Hist. Nat. 21 : 272 (1949).

Inflorescences paniculate, terminal and axillary, up to 22 cm. long, with pubescent axes,
flowers in fascicles, with pedicels 3–7 mm. long ; male calyx ovoid in bud, splitting
into 3–4 valvate segments about 2 mm. long ; petals 5, about 1 mm. diam. ; disk-
glands 5 ; stamens 13–16 ; leaves oblanceolate, obovate or elliptic, obtuse or cuneate
at base, obtuse to obtusely acuminate at apex, margin with small glandular teeth,
8–30 cm. long, 3–12 cm. broad, sparingly pubescent on the midrib and nerves beneath,
otherwise glabrous ; petiole 2–7 cm. long, densely pubescent on the lower side
1. *vignei*
Inflorescences racemose, terminal, up to 2 cm. long, sparingly pubescent, flowers 1–3
together, with pedicels 1–3 mm. long ; male calyx ovoid in bud, about 2 mm. long ;
leaves narrowly elliptic, long-attenuate at base, gradually acuminate at apex, margin
entire, up to 13 cm. long and 5·1 cm. broad, glabrous ; petiole up to 1 cm. long,
sparingly pubescent   ..   ..   2. *baldwinii*

1. **G. vignei** *Hoyle* in Kew Bull. 1935 : 259 ; Cavaco in Bull. Mus. Hist. Nat. 21 : 276 (excl. distrib. Nigeria).
" Espèce indéterminée ", Aubrév. Fl. For. C. Iv. 2 : 82. A dioecious understorey forest tree, 5–50 ft. high ;
flowers pale yellow or white.
Iv.C.: Man *Aubrév.* 1091. Danané *Aubrév.* 1034. Mudjika *Serv. For.* 2781. Lyola country, Haut

Sassandra *Chev.* 21573. **G.C.:** Ofin Headwaters F.R. (Apr.) *Vigne* FH 1915 ! Mampong Scarp (Feb., Dec.) *Vigne* FH 2752 ! 4094 ! 4095 ! Abofaw (Nov.) *Vigne* FH 3135 ! Atewa Range (Feb.) *Vigne* FH 4343 ! Tano-Offin F.R. (Jan.) *Lyon* FH 2862 ! **Br.Cam.:** Missellele, Tiko (Jan.) *Box* 3565 !
2. **G. baldwinii** *Keay & Cavaco* in Port. Acta Biol. (B), 6 : 1, t. 1 (1955). A dioecious shrub, to 8 ft. high ; female flowers green ; male flowers yellow.
S.L.: between Gola F.R. boundary and Lalahun (Jan.) *King* 122b ! **Lib.:** Grand Cape Mount Co. : Genna Tanyehun (Dec.) *Baldwin* 10747 ! Jabrocca (Dec.) *Baldwin* 12567 !

29. PSEUDAGROSTISTACHYS Pax & K. Hoffm. in Engl. Pflanzenr. Euph. 6 : 96 (1912) ; Lebrun in Bull. Soc. Roy. Bot. Belg. 67 : 97 (1934).

A small tree, 8 m. high, glabrous ; stipules connate into a sheath, soon falling off and leaving a scar ; leaves oblong-elliptic, acute, shortly cuneate at base, 17–40 cm. long, up to 15 cm. broad, glabrous, with 2 basal glands on the upper surface at base ; lateral nerves 18–20 pairs, raised on both surfaces ; racemes axillary, up to 7·5 cm. long ; flowers white ; bracts striate ; stamens about 20, free ; ovary densely pilose   ..   ..   *africana*

P. africana (*Müll. Arg.*) *Pax & K. Hoffm.* l.c. 97, fig. 18 (1912) ; Exell Cat. S. Tomé 296. *Agrostistachys africana* Müll. Arg. (1864)—F.T.A. 6, 1 : 829.
S.Nig.: Ikwette Plateau, Obudu Div., 4,200 ft. (fr. Dec.) *Keay & Savory* FHI 25207 ! **F.Po:** 4,000 ft. (Nov.) *Mann* 582 ! Also on S. Tomé.

30. CYRTOGONONE Prain—F.T.A. 6, 1 : 815 ; Pax & K. Hoffm. in Engl. Pflanzenr. Euph. 6 : 111 (1912).

A tree ; branchlets and undersurface of the leaves densely silvery-lepidote ; leaves long-petioled, elliptic to obovate, widely cuneate at base, abruptly acuminate, entire or lobulate-toothed, 12–25 cm. long, 8–13 cm. broad, with a pair of glands at base ; flowers dioecious, in large terminal narrow panicles up to 30 cm. long ; calyx globose in bud, at length splitting into 3 sepals, lepidote ; stamens numerous, erect in bud ; fruits 3-lobed, about 3·5 cm. long and 4 cm. broad, scurfy brown-tomentellous ; seeds about 2·5 cm. long and broad   ..   ..   *argentea*

C. argentea (*Pax*) *Prain*—F.T.A. 6, 1 : 815 ; Pax l.c. 6 : 111, fig. 23 ; Prain in Hook. Ic. Pl. t. 3008. *Crotonogyne ? argentea* Pax (1903). A forest tree, 30–100 ft. high ; flowers white.
S.Nig.: Oban *Talbot* 1590 ! Afi River F.R., Ikom (May) *Latilo* FHI 30988 ! **Br.Cam.:** Likomba *Mildbr.* 10523 ! Bambuko F.R., Kumba (fr. Sept.) *Olorunfemi* FHI 30770 ! R. Metschum, Wum (young fr. July) *Ujor* FHI 29284 ! Also in French Cameroons and Spanish Guinea.
[In the first edition this species was recorded from the Ivory Coast (*fide* Chev. Bot. 588) by error. The specimen upon which the record was based is *Crotonogyne caterviflora* N. E. Br., q.v.]

31. CROTONOGYNE Müll. Arg.—F.T.A. 6, 1 : 819 ; Pax & K. Hoffm. in E. & P. Pflanzenfam. 19C : 97 (1931) ; Pflanzenr. Euph. 6 : 111 (1912). *Neomanniophyton* Pax & K. Hoffm. l.c. 6 : 115 (1912), see F.T.A. 6, 1 : 1054.

Leaves elongate-oblanceolate, narrowed to a cordate-auriculate base, acuminate at apex, up to 55 cm. long, 5–13 cm. broad ; leaves crowded ; petioles up to 1·8 cm. long ; racemes up to 60 cm. long, the flowers in both sexes occupying the greater part of the rhachis ; male pedicel longer than the calyx ; male petals free ; indumentum of plant lepidote   ..   ..   1. *preussii*
Leaves not cordate-auriculate at base, not crowded ; female flowers few and solitary towards the apex of the raceme ; male pedicel not longer than the calyx ; male petals united :
  Indumentum lepidote only or none ; ovary lepidote :
    Bracts usually eglandular ; leaves oblong, rounded at base, gradually acuminate, 10–17 cm. long, 4–7 cm. broad, very loosely scaly beneath ; racemes 5–15 cm. long, the males sometimes in clusters on the old wood ; female calyx-lobes eglandular   ..   ..   2. *caterviflora*
    Bracts always 2-glandular ; leaves obovate, long-narrowed to an acute base, abruptly and shortly acuminate, 16–28 cm. long, 5–10 cm. broad, with very few scales beneath ; racemes up to 30 cm. long ; female calyx-lobes with glands between
        3. *manniana*
  Indumentum of setae or stellate hairs or both, associated with or replacing the flattened scales ; ovary setose :
    Branchlets and stipules stellate-pubescent and sparingly lepidote ; leaves oblong-obovate, abruptly narrowed into a distinct " pseudopetiole " longer than the petiole proper, shortly acuminate, 26–30 cm. long, 6–9 cm. broad ; stipules ovate, acuminate, 5 mm. long ; racemes very long ; bracts 2-glandular ; male calyx glabrous, female lepidote and densely stellate-pubescent   ..   ..   ..   4. *impedita*
    Branchlets, stipules, petioles, undersurface of leaves, racemes and calyces at least when young with stiff spreading hairs ; stipules lanceolate, 0·6–1·5 cm. long ; leaves acutely long-acuminate :
      Indumentum composed of scales and long stiff hairs ; petioles 1·8–6 cm. long ; lamina elliptic to oblanceolate, 15–23 cm. long, 3·5–9 cm. broad   5. *chevalieri*
      Indumentum strigose, scales absent except on the inflorescence ; petioles up to 1·2 cm. long ; lamina elongate-obovate, 17–30 cm. long, 6–10 cm. broad   6. *strigosa*

1. **C. preussii** *Pax*—F.T.A. 6, 1 : 821 ; Pax & K. Hoffm. in Engl. Pflanzenr. Euph. 6 : 113, fig. 24. *Pycnocoma macrophylla* of F.W.T.A., ed. 1, 1 : 307, partly (*Talbot* 691). A shrub or small tree, 10–25 ft. high, in rain forest ; flowers greenish-yellow or white ; inflorescence erect.
   S.Nig.: Owena, Ondo Prov. (Mar.) *Darko* 190 ! Idanre F.R. (fl. & young fr. July) *Onochie* FHI 33380 ! Oban *Talbot* 691 ! 694. **Br.Cam.:** Victoria (Apr.) *Preuss* 1220. Likomba (Oct.) *Mildbr.* 10544 ! S. Bakundu F.R. (Jan., Mar.) *Brenan* 9455 ! *Keay* FHI 28564 ! Also in French Cameroons.
2. **C. caterviflora** *N. E. Br.*—F.T.A. 6, 1 : 821 ; Chev. Bot. 573. *Neomanniophyton caterviflorum* (N. E. Br.) Pax in Engl. Pflanzenr. Euph. 6 : 118 (1912). *Cyrtogonone argentea* of Chev. Bot. 588, not of Prain. A forest shrub, to 18 ft. high, scaly.
   S.L.: Yonibana (Oct., Nov.) *Thomas* 4136 ! 4889 ! 4890 ! Nyandehun (Apr.) *Deighton* 3667 ! Falama (Jan.) *Deighton* 3874 ! **Lib.:** *Dinklage* 3362 ! Sinoe Basin *Whyte* ! Moylakwelli (Oct.) *Linder* 1274 ! Ganta (Sept.) *Baldwin* 9227 ! 9306 ! Boporo Dist. (fl. Dec., fr. Nov.) *Baldwin* 10278 ! 10667 ! Bobel Mt. (fr. Sept.) *Baldwin* 9564 ! 9586 ! **Iv.C.:** Malamalasso (Mar.) *Chev.* 17520 !
3. **C. manniana** *Müll. Arg.*—F.T.A. 6, 1 : 822 ; Pax & K. Hoffm. l.c. 113. A forest shrub, to 12 ft. high, scaly ; flowers whitish.
   G.C.: Ankasa F.R. (Dec.) *Vigne* FH 3183 ! **S.Nig.:** Eket *Talbot* 3258 ! Stubbs Creek F.R., Eket (May) *Onochie* FHI 32096 ! 32950 ! **F.Po:** (Jan.) *Mann* 219 ! 220 !
   [*C. zenkeri* Pax from French Cameroons and Gabon is probably conspecific with this species.]
4. **C. impedita** *Prain*—F.T.A. 6, 1 : 823. *Neomanniophyton impeditum* (Prain) Pax l.c. 6 : 116 (1912). A forest shrub, 6–8 ft. high.
   Br.Cam.: Johann-Albrechtshöhe *Büsgen* 163. Also in French Cameroons.
5. **C. chevalieri** (*Beille*) *Keay* in Kew Bull. 1955 : 139. *Neomanniophyton chevalieri* Beille in Bull. Soc. Bot. Fr. 61, Mém. 8 : 295 (1917) ; Chev. Bot. 588. *C. caterviflora* of F.W.T.A., ed. 1, 1 : 299, partly (*Chev.* 17783). *C. strigosa* of Aubrév. Fl. For. C. Iv. 2 : 70, t. 151, 6–11. A forest shrub, to 12 ft. high.
   Iv.C.: Soubiré to Yaou, Sanvi (Mar.) *Chev.* 17783 ! Mt. Tonko (July) *Chev.* 19707 ! Soumié (Apr.) *Aubrév.* 448 ! **G.C.:** Konongo (Feb.) *Akpabla* 267 ! Opon Mansi F.R. (Oct.) *Akpabla* 921 ! Simpa (Feb.) *Vigne* FH 2780 ! Prestea (Sept.) *Vigne* FH 3093 !
6. **C. strigosa** *Prain* in Kew Bull. 1912 : 191 (May 1912) ; F.T.A. 6, 1 : 826. *Neomanniophyton ledermannianum* Pax & K. Hoffm. l.c. 6 : 116 (Oct. 1912). *C. ledermanniana* (Pax & K. Hoffm.) Pax & K. Hoffm. in op. cit. 7 : 427 (1914). A forest shrub, to 6 ft. high, all parts with stiff spreading hairs.
   S.Nig.: Oban *Talbot* 658 ! 659 ! Kwa Falls, Calabar (Mar.) *Brenan* 9243 ! Also in French Cameroons.

32. MANNIOPHYTON Müll. Arg.—F.T.A. 6, 1 : 818 ; Pax & K. Hoffm. in Engl. Pflanzenr. Euph. 6 : 120 (1912).

A shrub or climber ; branches scabrous with short stellate hairs ; leaves polymorphous, entire and ovate or more or less 2–3-lobed, cordate at base, tips acutely acuminate, up to 25 cm. long and broad, 5-nerved at base, scabrous-setose with stellate hairs on both surfaces ; male panicles slender, up to 25 cm. long ; flowers small, clustered ; stamens 10–20 ; female panicles much smaller ; ovary densely setose ; capsule deeply 3-lobed, about 2·5 cm. long, rusty-tomentose ..   ..   *fulvum*

M. fulvum *Mull. Arg.* in J. Bot. 2 : 332 (1 Nov. 1864) ; Léonard in Bull. Jard. Bot. Brux. 25 : 290. *M. africanum* Müll. Arg. in Flora 47 : 531 (9 Nov. 1864) ; F.T.A. 6, 1 : 818 (incl. var.) ; Pax & K. Hoffm. l.c. 121, fig. 25 A–E (incl. vars.) ; Chev. Bot. 572. A forest shrub or climber ; flowers pale yellow.
S.L.: Freetown (Dec.) *Smythe* 204 ! Pujehun (Feb.) *Aylmer* 91 ! Kenema (fl. & fr. Jan.) *Thomas* 7612 ! 7673 ! Mattru (Nov.) *Deighton* 2336 ! Yonibana (Oct., Nov.) *Thomas* 4899 ! **Lib.:** Monrovia *Whyte* ! Grand Bassa (Oct.) *Dinklage* 1758 ! Peahtah (Oct.) *Linder* 981 ! Ganta (Dec.) *Barker* 1136 ! Belleyella, Boporo (fl. & fr. Dec.) *Baldwin* 10646 ! **Iv.C.:** Bouroukrou (Dec., Jan.) *Chev.* 16879. Grabo (July, Aug.) *Chev.* 19634. **G.C.:** Akwapim Hills (Oct.) *Johnson* 795 ! Foso Juaso F.R. (Apr.) *Scholes* 240 ! Ateiku (May) *Vigne* FH 1939 ! **S.Nig.:** Oluwa F.R. (Apr.) *Symington* FHI 3391 ! Ugbojiobo, Benin (May) *Farquhar* 3 ! Awka (July) *Onyeagocha* FHI 16556 ! Eket *Talbot* 3074 ! 3248 ! Calabar (fr. Feb.) *Mann* 2308 ! **Br.Cam.:** S. Bakundu F.R. (Apr.) *Olorunfemi* FHI 30540 ! Also in French Cameroons, Principe, S. Tomé, Gabon, Spanish Guinea, Belgian Congo, A.-E. Sudan and Angola. (See Appendix, p. 154.)

33. ERYTHROCOCCA Benth.—F.T.A. 6, 1 : 847 ; Pax & K. Hoffm. in Engl. Pflanzenr. Euph. 7 : 86 (1914). *Chloropatane* Engl. (1899). *Athroandra* (Hook. f.) Pax & K. Hoffm. l.c. 7 : 76 (1914).

Branchlets armed with a pair of sharp stipular spines at the base of each leaf ; leaves ovate or oblong, acuminate, 4–5 cm. long, 2–3 cm. broad, faintly crenate, glabrous ; lateral nerves about 2 pairs ; male racemes very short and slender ; calyx globose in bud, glabrous ; stamens 6–11 ; stigmas laciniate ; fruits 1–3-lobed, sparingly setose
                                                       1. *anomala*
Branchlets not spiny ; stigmas entire, smooth or papillose ; stamens 24–30 :
Filaments longer than the anthers ; branchlets long-pilose with spreading hairs ; leaves very shortly petiolate, oblong-obovate, coarsely jagged-toothed, 5–15 cm. long, 1·5–5 cm. broad, membranous, pilose with stiff slender hairs on both surfaces ; stipules small and hyaline ; male pedicels up to 1 cm. long ; sepals glabrous ; stamens about 27 ; stigmas recurved and closely appressed on the ovary
                                                       2. *membranacea*
Filaments shorter than the anthers ; branchlets pubescent, puberulous, often glabrescent, or glabrous :
Undersurface of leaves remaining densely pilose or softly pubescent :
Leaves ovate-oblong, obtuse or cuneate at base, long-acuminate, 7–20 cm. long, 2·2–7 cm. broad, distantly crenate-serrate, densely pilose beneath ; petiole 5–16 mm. long ; stigmas lanceolate, spreading, style distinct ; fruits 12–15 mm. broad ..   ..   3. *hispida*
Leaves ovate, or ovate-lanceolate, obtuse at base, cuspidate-acuminate, 3–5 cm. long and nearly 2 cm. broad, denticulate, softly pubescent beneath ; petiole about 6 mm. long ; stigmas suborbicular, connate at base, smooth ..   4. *chevalieri*
Undersurface of leaves puberulous and glabrescent, or glabrous :

Male buds rounded ; stigmas usually suberect or spreading ; branchlets, petioles and leaves puberulous, at least when young :
Stigmas subsessile, suborbicular, spreading, papillose ; leaves ovate, 4–6·5 cm. long, 2·5–4 cm. broad, crenate ; male pedicels up to 1 cm. long .. .. 5. *africana*
Stigmas on a distinct style, smooth ; male pedicels slender, 1–2 cm. long :
Stigmas suborbicular, small, much shorter than the style ; leaves oblong to elliptic-obovate, 10–15 cm. long, 5–6 cm. broad, remotely toothed
6. *welwitschiana*
Stigmas ovate-lanceolate, longer than the style ; leaves oblong-lanceolate, 8–15 cm. long, 3–5 cm. broad, glandular-dentate .. .. .. 7. *pallidifolia*
Male buds ovoid, pointed ; stigmas sessile, closely appressed on the ovary ; branchlets, petioles and leaves glabrous ; leaves oblong-lanceolate, acuminate, 7–20 cm. long, 2·5–6 cm. broad, distantly serrulate ; male pedicels subumbellate, up to 1·5 cm. long .. .. 8. *mannii*

1. **E. anomala** (*Juss. ex Poir.*) *Prain* in Ann. Bot. 25 : 614 (1911) ; Pax & K. Hoffm. in Engl. Pflanzenr. Euph. 7 : 90. *Adelia anomala* Juss. ex Poir. (1810). *Erythrococca aculeata* Benth. (1849)—F.T.A. 6, 1 : 858 ; Chev. Bot. 575. A spiny shrub, 3–10 ft. high ; fruit (as in the other species) usually 2-lobed and splitting to expose the pitted spherical seeds enclosed in a thin scarlet aril.
   Fr.G.: Boquet *Heudelot* 856 ! **S.L.:** *Smeathmann* ! Bagroo R. (Apr.) *Mann* 879 ! Mt. Aureole (Mar.) *Dalz.* 936 ! Ndilajula, Njala (Apr.) *Deighton* 3733 ! **Lib.:** Monrovia *Whyte* ! Peahtah (Oct.) *Linder* 1052 ! 1097 ! Jabroke, Palipo (July) *Baldwin* 6605 ! **Iv.C.:** Bouroukrou, Capiékrou, Sanvi and Soubré *fide* Chev. *l.c.* **G.C.:** Achimota (July) *Milne-Redhead* 5128 ! Dunkwa (Sept.) *Brown* FH 2357 ! **Togo:** Kete Krachi (June) FH 3876 ! **N.Nig.:** Okene (young fr. Sept.) *Symington* FHI 5685 ! **S.Nig.:** Lagos *Rowland* ! Olokemeji (Apr.) *Foster* 285 ! Ibadan (Apr.) *Meikle* 1432 ! *Schlechter* 13012 ! Oban *Talbot* ! **Br.Cam.:** Victoria (Apr.) *Maitland* 624 ! Johann-Albrechtshöhe *Staudt* 466 ! **F.Po:** *Mildbr.* 6881 ; 6998. (See Appendix, p. 142.)
2. **E. membranacea** (*Müll. Arg.*) *Prain*—F.T.A. 6, 1 : 864. *Claoxylon membranaceum* Müll. Arg. (1864). *Athroandra membranacea* (Müll. Arg.) Pax & K. Hoffm. l.c. 7 : 78 (1914). A forest shrub, to 10 ft. high.
   S.Nig.: Oloji area, Omo F.R., Ijebu Ode (fr. Apr.) *Onochie* FHI 15515 ! **Br.Cam.:** Cam. Mt., 4,000 ft. (Feb.) *Mann* 1197 ! Buea *Lehmbach* 212.
3. **E. hispida** (*Pax*) *Prain*—F.T.A. 6, 1 : 873. *Claoxylon hispidum* Pax (1894). *Athroandra hispida* (Pax) Pax & K. Hoffm. l.c. 7 : 84, fig. 13 (1914). A shrub or tree, to 15 ft. high ; flowers yellowish-green ; in lower montane forest.
   Br.Cam.: Cam. Mt., 4,000–5,000 ft. (fl. Nov.–May, fr. Jan.) *Preuss* 908 ! *Dunlap* 19 ! 100 ! *Maitland* 214 ! 677 ! *Keay* FHI 28649 ! Also in French Cameroons.
4. **E. chevalieri** (*Beille*) *Prain*—F.T.A. 6, 1 : 866 ; Chev. Bot. 575. *Claoxylon chevalieri* Beille (1908). *Athroandra chevalieri* (Beille) Pax & K. Hoffm. l.c. 7 : 79 (1914). A shrub.
   Fr.G.: Labé Plateau (Apr.) *Chev.* 12296 ! Diaguissa (Mar.) *Chev.* 12643 ! Fassakoidou to Kesséridou (Feb.) *Chev.* 20811 ! **Dah.:** Cabolé to Bassila, Savalou (May) *Chev.* 23786 !
5. **E. africana** (*Baill.*) *Prain*—F.T.A. 6, 1 : 866. *Trewia ? africana* Baill. (1860). *Claoxylon barteri* Hook. f. (1862)—Chev. Bot. 576. *Athroandra africana* (Baill.) Pax & K. Hoffm. l.c. 7 : 79 (1914). A shrub or small tree, with pale leaves, and slender, more or less pendulous, flower-racemes.
   Sen.: Casamance R. *Perrottet* 748 ! **Port.G.:** Catió (June) *Esp. Santo* 2098 ! **S.L.:** Crawford's Isl. *Afzelius.* Bafodeya (Apr.) *Sc. Elliot* 5505 ! Mano (Apr.) *Deighton* 3957 ! **Lib.:** Sinoe Basin *Whyte* ! Ganta *Harley* ! **Iv.C.:** Zaon to Ayame *Chev.* 17805 *ter.* Zaramou *Chev.* 17631. Cavally Basin *Chev.* 19987. **G.C.:** Sekondi (Nov.) *Vigne* FH 1404 ! Amentia (Mar.) *Vigne* FH 1857 ! Bunsu, Akim (Mar.) *Irvine* 1811 ! Mangvase (Feb.) *Darko* 244 ! **Dah.:** Koussi, Allada (Feb.) *Chev.* 23235 ! **S.Nig.:** Lagos *Barter* 2223 ! *Rowland* ! Epe *Barter* 3285 ! Lekki (Feb.) *Millen* 163 ! Ipetu *Foster* 184 ! Yoruba forests *Barter* 3344 ! (See Appendix, p. 142.)
6. **E. welwitschiana** (*Müll. Arg.*) *Prain*—F.T.A. 6, 1 : 868. *Claoxylon welwitschianum* Müll. Arg. (1864). *Athroandra welwitschiana* (Müll. Arg.) Pax & K. Hoffm. l.c. 7 : 81, fig. 12 (1914). A forest shrub, 4–15 ft. high.
   S.Nig.: Oban *Talbot* 663 ! Also in French Cameroons, extending to Belgian Congo and Angola.
7. **E. pallidifolia** (*Pax & K. Hoffm.*) *Keay* in Kew Bull. 1955 : 139. *Athroandra pallidifolia* Pax & K. Hoffm. l.c. 7 : 84 (1914). A shrub, to 6 ft. high, on forest-margin.
   F.Po: Musola *Mildbr.* 7060.
8. **E. mannii** (*Hook. f.*) *Prain*—F.T.A. 6, 1 : 865. *Claoxylon mannii* Hook. f. (1862). *Athroandra mannii* (Hook. f.) Pax & K. Hoffm. l.c. 7 : 78 (1914). A shrub, 15 ft. high.
   F.Po: Clarence Peak, 5,000 ft. (Dec.) *Mann* 260 ; 633 !

34. CLAOXYLON A. Juss.—F.T.A. 6, 1 :, 874. *Discoclaoxylon* (Müll. Arg.) Pax & K. Hoffm. in Engl. Pflanzenr. Euph. 7 : 137 (1914).

Leaves long-cuneate at base, elongate obovate-oblanceolate, 15–25 cm. long, 5–6·5 cm. broad, purplish when young, remotely denticulate, nearly glabrous beneath ; male flowers in lax racemes ; stamens 11–12 .. .. 1. *pedicellare*
Leaves rounded or very widely cuneate at base, elliptic-obovate, 17–35 cm. long, 8–15 cm. broad, closely serrate, the petioles purplish when young ; male flowers in rather close racemes resembling catkins when young ; stamens (3–) 6–8
2. *hexandrum*

1. **C. pedicellare** *Müll. Arg.*—F.T.A. 6, 1 : 874. *Discoclaoxylon pedicellare* (Müll. Arg.) Pax & K. Hoffm. in Engl. Pflanzenr. Euph. 7 : 137 (1914). A small tree with rather long-petiolate leaves and green glabrescent branchlets.
   F.Po: *Mann* !
2. **C. hexandrum** *Müll. Arg.*—F.T.A. 6, 1 : 875 ; Aubrév. Fl. For. C. Iv. 2 : 77, t. 151, 1–5. *Discoclaoxylon hexandrum* (Müll. Arg.) Pax & K. Hoffm. l.c. 7 : 139, fig. 19 (1914). A soft-wooded forest tree, to 60 ft. high, mostly in secondary regrowth ; flowers greenish-yellow, ripe seeds red.
   Lib.: Gbanga (Sept.) *Linder* 734 ! Peahtah (Oct.) *Linder* 1104 ! Ganta (Sept.) *Baldwin* 9240 ! **Iv.C.:** Yapo (Oct.) *Chev.* B. 22363 ! Agboville *Aubrév.* 1927 ! **G.C.:** Manso, Akim (Dec.) *Irvine* 2688 ! Bobiri F.R. (fr. Feb.) *Foggie* FH 4948 ! Benso, Tarkwa (Dec.) *Andoh* FH 5393 ! **S.Nig.:** Omo F.R. (Mar.) *Jones & Onochie* FHI 17034 ! Okomu F.R. (Jan.–Mar.) *Brenan* 9119 ! Akpabla 1115 ! Oban *Talbot* 1352 ! **Br.Cam.:** Buea (Jan.–Mar.) *Maitland* 234 ! 403 ! Bali-Ngemba F.R. (fr. May) *Ujor* FHI 30350 ! **F.Po:** (Jan.) *Mann* 186 ! Also in French Cameroons, Belgian Congo and (?) Uganda. (See Appendix, p. 139.)

35. MICROCOCCA Benth.—F.T.A. 6, 1 : 876 ; Pax & K. Hoffm. in Engl. Pflanzenr. Euph. 7 : 131 (1914).

Annual herb ; leaves ovate, acuminate, 2·5–5 cm. long, 1·5–2·5 cm. broad, obtusely crenate, glabrous or nearly so, sometimes purple-tinged ; flowers monoecious, in slender axillary racemes, 5–8 cm. long ; male sepals 3 ; stamens usually 6, intermixed with staminal glands ; female sepals 3–5 ; fruit deeply 3-lobed, strigose *mercurialis*

M. mercurialis (*Linn.*) *Benth.*—F.T.A. 6, 1 : 878 ; Pax & K. Hoffm. l.c. 7 : 133, fig. 18, D–F ; Chev. Bot. 576. *Tragia mercurialis* Linn. (1753). A weed of cultivation, ½–2 ft. high, stems simple or branched, pubescent, woody below.
Sen.: *Perrottet* 715 ! **Fr.G.:** Conakry (June) *Chev.* 4425. **S.L.:** Regent (Dec.) *Sc. Elliot* 4117 ! Kent (fl. & fr. May) *Deighton* 2658 ! Binkolo (fl. & fr. Aug.) *Thomas* 1695 ! 1874 ! Rokupr (fl. & fr. July) *Jordan* 49 ! **Lib.:** Monrovia Cinco (fl. & fr. Nov.) *Linder* 1552 ! **G.C.:** Owabi (fl. & fr. Apr.) *Andoh* FH 4189 ! **Togo:** Lome *Warnecke* 187 ! **N.Nig.:** Nupe *Barter* 1494 ! Lokoja (Sept., Nov.) *T. Vogel* 194 ! *Dalz.* 164 ! Abinsi (fl. & fr. Oct.) *Dalz.* 785 ! **S.Nig.:** Ibadan (fl. & fr. July) *Ahmed & Chizea* FHI 20002 ! **Br.Cam.:** Mamfe (fl. & fr. Apr.) *Ejiofor* FHI 29375 ! Throughout tropical Africa, also in Madagascar and Asia.

36. MALLOTUS Lour.—F.T.A. 6, 1 : 927 ; Pax & K. Hoffm. in Engl. Pflanzenr. Euph. 7 : 145 (1914).

Leaves not golden-glandular beneath ; stipules persistent, rigid-subulate ; fruit bristly ; leaf-blades ovate or ovate-elliptic, rounded or subcordate at base, acuminate, 10– 18 cm. long, 8–10 cm. broad, often entire ..   ..   1. *subulatus*
Leaves golden-glandular beneath ; stipules small, caducous ; fruits smooth ; leaf-blades broadly ovate, rounded to quite cordate at base, up to 18 cm. long and 10 cm. broad, often toothed, glabrescent or pubescent beneath ..   ..   2. *oppositifolius*

1. **M. subulatus** *Müll. Arg.*—F.T.A. 6, 1 : 927 ; Pax & K. Hoffm. in Engl. Pflanzenr. Euph. 7 : 153 ; Chev. Bot. 580. A dioecious forest shrub or small tree, 3–15 ft. high, with stellate-pubescent branchlets and racemes ; flowers cream.
S.L.: Nganyama, Bo (fl. Dec., fr. Mar.) *Deighton* 3514 ! 3580 ! Zimi to Gorahun, Gola Forest (Mar.) *Deighton* 4107 ! **Lib.:** Jabrocca (fr. Dec.) *Baldwin* 10859 ! **Iv.C.:** Abiosso *Chev.* 17825. **G.C.:** Axim (Dec.) *Chipp* 50 ! Princes (fr. Jan.) *Akpabla* 760 ! **S.Nig.:** Sapoba *Kennedy* 2090 ! Ubuluku (Dec.) *Thomas* 2074 ! 2075 ! Awka (Nov.) *Keay* FHI 22281 ! Umuahia (young fr. Feb.) *Carpenter* 407 ! Oban *Talbot* 1311 ! **Br.Cam.:** Victoria (young fr. Feb.) *Maitland* 368 ! Tiko (Jan.) *Dunlap* 252 ! Kembong F.R., Mamfe (fr. Mar.) *Onochie* FHI 31175 ! **F.Po:** *Mann* 260 ! Also in French Cameroons, Spanish Guinea, Gabon and Belgian Congo. (See Appendix, p. 150.)
2. **M. oppositifolius** (*Geisel.*) *Müll. Arg.*—F.T.A. 6, 1 : 928 (incl. vars.) ; Pax & K. Hoffm. l.c. 7 : 158, fig. 23 A; Chev. Bot. 579. *Croton oppositifolius* Geisel. (1807). *Mallotus beillei* A. Chev. Bot. 579, name only. A dioecious shrub or tree, 6–40 ft. high, common in forest regrowth and drier types of forest ; flowers creamy-white. Several forms or varieties exist, but the treatments given in the Pflanzenreich and in the F.T.A. are not satisfactory and more thorough research is required.
Sen.: *Berhaut* 1256 ; 4337. **Gam.:** *Hayes* 550 ! **S.L.:** Farana to Falaba (Mar.) *Sc. Elliot* 4461 ! Makump (July) *Thomas* 888 ! Juring (Dec.) *Deighton* 443 ! Njala (fl. & fr. July) *Deighton* 759 ! 760 ! 761 ! **Lib.:** Dinklage. **Iv.C.:** Bouroukrou, Nzi & Mt. Dou *fide* Chev. l.c. **G.C.:** Cape Coast *Brass* ! Bompata (May) *Chipp* 441 ! Accra (Apr., May) *Dalz.* 127 ! *Bally* 14 ! Dawa (May) *Howes* 905 ! **Togo:** Lome (Nov.) *Warnecke* 51 ! 319 ! 7481 ! Aguna *Kersting* A. 290 ! **Dah.:** Zagnanado and Abomey *fide* Chev. l.c. **N.Nig.:** Nupe *Barter* 1702 ! 1742 ! Dogon Kurmi, Jemaa (Nov.) *Keay* FHI 22269 ! Abinsi *Dalz.* 725 ! **S.Nig.:** Lagos *Rowland* ! *Dawodu* 75 ! 299 ! Abeokuta *Barter* 3381 ! Ibadan (fl. & fr. Apr.) *Meikle* 1465 ! 1466 ! Abo (Aug.) *T. Vogel* 31 ! 36 ! 48 ! Aguku *Thomas* 637 ! Calabar (July) *Holland* 51 ! **Br.Cam.:** Tiko (Jan.) *Dunlap* 251 ! Fonfuka, Bamenda (Apr.) *Maitland* 1763 ! Widespread in tropical Africa and in Madagascar. (See Appendix, p. 150.)

37. ALCHORNEA Sw.—F.T.A. 6, 1 : 914 ; Pax & K. Hoffm. in Engl. Pflanzenr. Euph. 7 : 220 (1914). *Lepidoturus* Boj. ex Baill.—F.T.A. 6, 1 : 913 ; F.W.T.A., ed. 1, 1 : 303.

Leaves without stipel-like appendages, but with sessile glands at the base ; bracts up to 1 mm. long, inconspicuous ; male flowers in panicles of spikes :
Petioles 5–14 cm. long ; ovary 2-celled ; styles 2 ; leaves broadly ovate, cordate at base, shortly acuminate, repand-dentate or subentire, 10–28 cm. long, 6·5–16·5 cm. broad, finely stellate-puberulous or glabrescent beneath ; male panicles axillary, 8–36 cm. long ; female inflorescences axillary, branched or simple ; fruits 2-celled, about 1 cm. broad, stellate-pubescent ..   ..   1. *cordifolia*
Petioles 0·5–3 cm. long ; ovary 3-celled ; styles 3 ; leaves not cordate at base ; indumentum not stellate :
Leaves elongate-obovate-oblanceolate, long-attenuated to the base, shortly acuminate, repand-denticulate, 14–31 cm. long, 6–12 cm. broad, lateral nerves 12–19 pairs ; branchlets, petioles and undersurface of leaves minutely puberulous ; male panicles 10–25 cm. long, terminal, axillary and on the old wood ; female inflorescences terminal, simple or branched, up to 11–40 cm. long ; fruits 3-celled, about 8–11 mm. broad, smooth, pubescent ..   ..   2. *floribunda*
Leaves oblong-elliptic, rounded at base, obtusely acuminate, crenate or entire, 7–13 cm. long, 2·5–6 cm. broad, lateral nerves 7–10 pairs ; branchlets, petioles and undersurface of leaves with spreading pilose hairs ; male panicles up to 17 cm. long, terminal and axillary, female inflorescences terminal, simple or branched, up to 8 cm. long ; fruits 3-celled, 8–9 mm. broad, tuberculate ..   ..   3. *hirtella*
Leaves with a pair of stipel-like appendages and sessile glands at the base ; bracts 4–6 mm. long, conspicuous ; male flowers in simple spikes, 4–10 cm. long, from perulate

buds on the 1-year-old shoots, appearing before the leaves ; female spikes axillary on the young shoots, flowers very few ; ovary 3-celled ; styles 3 ; fruit 3-celled, about 8 mm. broad, smooth, glabrescent ; leaves ovate-elliptic, rounded at base, acutely acuminate, crenate or subentire, 8–13 (–16) cm. long, 3–5 (–10) cm. broad, sparsely pilose beneath ; petioles 1–4 (–7·5) cm. long ; branchlets and young leaves purplish

4. *laxiflora*

1. **A. cordifolia** (*Schum. & Thonn.*) *Müll. Arg.* in Linnaea 34 : 170 (1865) ; Pax & K. Hoffm. in Engl. Pflanzenr. Euph. 7 : 230, fig. 34, A–C ; Chev. Bot. 577. *Schousboea cordifolia* Schum. & Thonn. (1827). *Alchornea cordata* Benth. (1849)—F.T.A. 6, 1 : 915. A shrub or small tree, erect or half climbing ; flowers greenish-white ; styles long and persistent on the ripening fruits in lax pendulous spikes or racemes.
Sen.: *Roger* ! **Gam.:** *Heudelot* 345 ! Genieri (Feb.) *Fox* 28 ! **Fr.Sud.:** Banamiane (Jan.) *Chev.* 151 ! Moussaia (Feb.) *Chev.* 456 ! **Port.G.:** *Esp. Santo* 428 ! **Fr.G.:** Timbo (Jan.) *Pobéguin* 124 ! Kindia (Mar.) *Chev.* 12748. **S.L.:** *Don* ! Hastings (young fr. Mar.) *Kirk* ! Kenema (Jan.) *Thomas* 7528 ! Waterloo to York Pass (Jan.) *Tindall* 25 ! **Lib.:** Gbanga (fl. & young fr. Sept.) *Linder* 483 ! Nyaake, Webo (June) *Baldwin* 6158 ! Brewersville (Dec.) *Barker* 1107 ! **Iv.C.:** various localities *fide* Chev. *l.c.* **G.C.:** Kumasi *Cummins* 179 ! Aburi (Jan.) *Brown* 937 ! Essuasu (Nov.) *Chipp* 3 ! 4 ! **Togo:** Akpafu-Todji *St. C. Thompson* FH 3682 ! **Dah.:** *Le Testu* 204 ; 205. **N.Nig.:** Kontagora (Dec.) *Dalz.* 88 ! Katagum *Dalz.* 405 ! Abinsi (Apr.) *Dalz.* 703 ! **S.Nig.:** Lagos *Moloney* ! Benin City *Dennett* 69 ! Oban *Talbot* 682 ! 1319 ! **Br.Cam.:** Victoria (fl. & fr. Jan.) *Maitland* 181 ! Johann-Albrechtshöhe *Staudt* 551 ! **F.Po:** (Nov., Dec.) *T. Vogel* 73 ! *Mann* 81 ! Widespread in tropical Africa. (See Appendix, p. 135.)

2. **A. floribunda** *Müll. Arg.*—F.T.A. 6, 1 : 916 ; Pax & K. Hoffm. l.c. 7 : 239, fig. 36 ; Chev. Bot. 578, partly. A leaning shrub or small tree, sometimes subscandent, to 30 ft. high ; in forest undergrowth ; flowers pale green.
Fr.Sud.: Moussaia *fide* Chev. *l.c.* (?). **Lib.:** Ganta (May, Sept.) *Harley* ! *Baldwin* 9246 ! Bangee, Boporo (fr. Nov.) *Baldwin* 10358 ! Zeahtown (Aug.) *Baldwin* 6928 ! **Iv.C.:** various localities *fide* Chev. *l.c.* **S.Nig.:** Okomu F.R. (Dec., Mar.) *Brenan* 8559 ! Eket *Talbot* 2304 ! Oban *Talbot* 2304 ! Boje, Ikom (fr. May) *Latilo* FHI 31803 ! **Br.Cam.:** Likomba (Apr.) *Maitland* 1103 ! Bulifambo (Mar.) *Maitland* 564 ! S. Bakundu F.R. (Mar.) *Latilo* in *Hb. Brenan* 9400 ! **F.Po:** *Mann* 306 ! Also in French Cameroons, Spanish Guinea, Gabon, Cabinda, Belgian Congo, Uganda and A.-E. Sudan. (See Appendix, p. 136.)

3. **A. hirtella** *Benth.*—F.T.A. 6, 1 : 917 ; Pax & K. Hoffm. l.c. 7 : 241 ; Chev. Bot. 579. A forest shrub or tree, 2–30 ft. high ; flowers reddish.
Sen.: Djikoy *Adam* 1042. **Port.G.:** Brene, Bissau (Jan.) *Esp. Santo* 1683 ! **Fr.G.:** Rio Nunez *Heudelot* 746 ! Banacoro (Feb.) *Chev.* 499. And other localities *fide* Chev. *l.c.* **S.L.:** *Smeathmann* ! Kambia, Scarcies R. (Dec.) *Sc. Elliot* 4230 ! Njala (fl. Feb., Sept., fr. Mar.) *Deighton* 521 ! 2131 ! 2869 ! Musaia (fl. & fr. Dec.) *Deighton* 4415 ! Kambui Hills (Mar.) *Small* 516 ! **Lib.:** Grand Bassa (July) *T. Vogel* 35 ! Peahtah (Oct.) *Linder* 1085 ! Bushrod Isl. (Apr.) *Barker* 1299 ! 1300 ! Soplima Kolahun (Nov.) *Baldwin* 10021 ! **Iv.C.:** various localities *fide* Chev. *l.c.* Also in French Cameroons, Spanish Guinea, Gabon, Belgian Congo, Uganda, Tanganyika and Angola. (See Appendix, p. 136.)
[*Winkler* 664 from Victoria, cited as this species in the first edition is *Mareya micrantha* (Benth.) Müll. Arg.]

4. **A. laxiflora** (*Benth.*) *Pax & K. Hoffm.* l.c. 7 : 245, fig. 37 (1914). *Lepidoturus laxiflorus* Benth. (1879)—F.T.A. 6, 1 : 913 ; F.W.T.A., ed. 1, 1 : 303. A shrub or small tree, 15–20 ft. high, with leaves 3-nerved at the base.
N.Nig.: Naraguta *Hill* 33 ! Bauchi Plateau (Apr.) *Lely* P. 221 ! **S.Nig.:** Lagos *Rowland* ! Ikirun (Feb.) *Millson* 135 ! Ibadan (Jan., Feb.) *Ejiofor* FHI 29399 ! Idanre Hills (Jan.) *Brenan & Keay* 8632 ! Benin City *Unwin* 39 ! Sapoba *Kennedy* 2403 ! **Br.Cam.:** Buea (Mar.) *Maitland* 454 ! Bambui, Bamenda (Apr.) *Maitland* 1463 ! Also on S. Tomé and widespread in central, eastern and southern tropical Africa. (See Appendix, p. 148.)

38. DISCOGLYPREMNA Prain—F.T.A. 6, 1 : 931 ; Pax & K. Hoffm. in Engl. Pflanzenr. Euph. 7 : 18 (1914).

Leaves long-petiolate, broadly elliptic or suborbicular, rounded or broadly cuneate at base, shortly and abruptly acuminate, 7–15 cm. long, 5–10 cm. broad, 3-nerved at the base, prominently nervose ; flower-spikes paniculate, terminal, up to 20 cm. long, shortly pubescent ; stamens 7–8, mixed with hirsute glands ; fruits deeply 3-lobed, rather fleshy, slightly hairy .. .. *caloneura*

D. caloneura (*Pax*) *Prain*—F.T.A. 6, 1 : 932 ; Pax & K. Hoffm. l.c. 7 : 19, fig. 1 (incl. vars.) ; Aubrév. Fl. For. C. Iv. 2 : 78, t. 156. *Alchornea caloneura* Pax (1909). A forest tree, to 90 ft. high or more, with cylindrical stem ; dioecious ; flowers white ; seeds bright red.
Fr.G.: Moussadougou to Lola, Guerzès (fr. Mar.) *Chev.* 20972 ! **S.L.:** Bo (Apr.) *Deighton* 3137 ! Tonboli F.R. (Nov.) *Edwardson* 255 ! **Lib.:** Dukwia R. (Nov.) *Cooper* 126 ! **Iv.C.:** *Aubrév.* 999 ! Abidjan *Aubrév.* 208 ! Morénou (fr. Dec.) *Chev.* 22531 ! **G.C.:** Ancobra R. (Dec.) *Johnson* 919 ! Esubompang (fr. Feb.) *Chipp* 100 ! Dunkwa (fr. Feb.) *Vigne* FH 213 ! Opon Valley (fr. Jan.) *Irvine* 1086 ! **Togo:** Okrabi (Dec.) *St. C. Thompson* FH 3701 ! **S.Nig.:** 70 miles east of Lagos *Lamborn* 123 ! Sapoba *Kennedy* 621 ! 1660 ! 2232 ! Ekpoma, Ishan (fr. June) *Dundas* FHI 21484 ! **Br.Cam.:** Victoria (Mar.) *Maitland* 590 ! Bisong Abang, Kumba (Apr.) *Rosevear* 1/36 ! Also in French Cameroons, Spanish Guinea, Gabon, Cabinda, Belgian Congo and S. Tomé. (See Appendix, p. 140.)

39. MAREYOPSIS Pax & K. Hoffm. in Engl. Pflanzenr. Euph. 14 : 13 (1919).

Shrub or tree, dioecious ; branchlets and petioles shortly pubescent ; leaves obovate, obovate-lanceolate or oblanceolate, cuneate or obtuse at base, long-acuminate, serrate-dentate, 14–32 cm. long, 6–13 cm. broad, puberulous beneath ; spikes from nodes on the old wood, or females sometimes axillary, males several, females solitary, up to 20 cm. long ; fruits indehiscent, depressed-globose, 3-lobed, up to 2·5 cm. broad and 1·5 cm. long .. .. *longifolia*

M. longifolia (*Pax*) *Pax & K. Hoffm.* l.c. (1919). *Mareya longifolia* Pax (1903)—F.T.A. 6, 1 : 912. A forest shrub or tree, to 30 ft. high.
S.Nig.: Uquo, Eket (young fr. May) *Onochie* FHI 33187 ! **Br.Cam.:** N. Korup F.R., Kumba (young fr. July) *Olorunfemi* FHI 30672 ! Also in French Cameroons and Belgian Congo.

40. NEOBOUTONIA Müll. Arg.—F.T.A. 6, 1 : 918 ; Pax & K. Hoffm. in Engl. Pflanzenr. Euph. 7 : 71–75 (1914).

Pubescence or scurf not continuous over the lower surface of the leaves :
  Leaves with long soft spreading hairs as well as scurf on the nerves beneath ; male calyx pubescent :
    Twigs and male panicles softly hairy with long simple hairs ; leaves orbicular-cordate, crenate, 13–15 cm. long and broad. .   . .   . 1. *diaguissensis*
    Twigs and male panicles only scurfy-stellate-puberulous ; leaves orbicular-cordate, entire, 10–20 cm. long and broad ; styles narrow, with 2 linear lobes   2. *mannii*
  Leaves without long hairs beneath ; male calyx glabrous ; leaf-blade suborbicular, shallowly cordate, 10–20 cm. long and broad ; styles narrow, with 2 linear lobes
                                                                                      3. *glabrescens*
Pubescence or scurf continuous and felted over the whole undersurface of the leaves ; male calyx hairy ; leaves orbicular-cordate, 10–20 cm. long and broad, entire, softly hairy on the nerves above ; styles narrow, with 2 linear lobes ; ovary and capsule densely velvety . .   . . 4. *velutina*

1. **N. diaguissensis** *Beille*—F.T.A. 6, 1 : 920 ; Chev. Bot. 579. *N. africana* (Müll. Arg.) Pax var. *diaguissensis* (Beille) Pax & K. Hoffm. in Engl. Pflanzenr. Euph. 7 : 75 (1914). A small tree.
   Fr.G.: high plateau of Diaguissa (Apr.) *Chev.* 12691. Koumi (June) *Pobéguin* 1603 ! Labé, Ditinn, Dalaba and Toukan to Bouria *fide* Chev. *l.c.*

2. **N. mannii** *Benth.*—F.T.A. 6, 1 : 920 ; Exell Cat. S. Tomé 300 (q.v. for synonymy). *N. africana* (Müll. Arg.) Pax var. *mannii* (Benth.) Pax & K. Hoffm. l.c. (1914). A tree, to 70 ft. high, in secondary forest regrowth ; flowers yellowish-green.
   S.Nig.: Ukpon F.R., Obubra (fr. July) *Latilo* FHI 31871 ! **Br.Cam.:** Buea (May) *Deistel* 643 ! *Maitland* 689 ! Big Kotto, Kumba *Rosevear* 90/37 ! **F.Po:** *Mildbr.* 6407. Also in French Cameroons, Principe and S. Tomé.

3. **N. glabrescens** *Prain*—F.T.A. 6, 1 : 921. *N. africana* (Müll. Arg.) Pax var. *glabrescens* (Prain) Pax & K. Hoffm. l.c. (1914). A soft-wooded tree, to 55 ft. high, growing in open spaces in forest.
   Lib.: Dukwia R. (May) *Cooper* 438 ! **S.Nig.:** Shasha F.R. (Mar.) *Richards* 3203 ! Ikom *Catterall* 7 ! **Br.Cam.:** Bimbia *Preuss* 1288 ! Big Kotto, Kumba (June) *Rosevear* 89/37 ! Etawo, Mamfe (May) *Ndep Enoh* 48/38 ! Also in French Cameroons and Spanish Guinea. (See Appendix, p. 156.)

4. **N. velutina** *Prain*—F.T.A. 6, 1 : 921. *N. melleri* (Müll. Arg.) Prain var. *velutina* (Prain) Pax & K. Hoffm. l.c. (1914). A shrub or small tree, 15–20 ft. high with stellate-scurfy branchlets and panicles ; flowers yellow ; in streamside forest.
   Br.Cam.: Bamenda, 5,500 ft. *Ledermann* 1924 ! Bum Su, Bamenda (young fr. May) *Maitland* 1737 ! Nkom-Wum forest, Wum (fl. & fr. July) *Ujor* FHI 29279 ! 30474 ! Balibagam, Bamenda (fr. Jan.) *Johnstone* 277/32 ! Also in French Cameroons. (See Appendix, p. 156.)

   A thorough revision of this genus, which has a wide distribution in tropical Africa, is needed to determine how far the taxa recognized by Prain in the F.T.A. can still be maintained as distinct species.

41. MAREYA Baill.—F.T.A. 6, 1 : 910 ; Pax & K. Hoffm. in Engl. Pflanzenr. Euph. 14 : 11 (1919).

Shrub or tree ; branchlets and petioles puberulous or pubescent ; leaves oblong-elliptic, obovate or oblanceolate, cuneate or obtuse at base, acuminate, slightly to coarsely toothed, 8–24 cm. long, 3–9 cm. broad, glabrous or pubescent beneath ; flowers monoecious, in rather numerous interrupted spikes up to 39 cm. long ; fruits 3-lobed, 3–4 mm. diam.   . .   . . *micrantha*

M. micrantha (*Benth.*) *Müll. Arg.* in DC. Prod. 15, 2 : 792 (1866), incl. vars. ; Benth. in Hook. Ic. Pl. t. 1281 ; F.T.A. 6, 1 : 911 ; Chev. Bot. 577. *Acalypha micrantha* Benth. (1849). *A. leonensis* Benth. (1849). *Mareya leonensis* (Benth.) Baill. (1874). *M. spicata* Baill. (1860)—F.T.A. 6, 1 : 911 ; Pax & K. Hoffm. l.c. 14 : 11 (1919), incl. vars. ; Chev. Bot. 577 ; F.W.T.A., ed. 1, 1 : 303 ; Aubrév. Fl. For. C. Iv. 2 : 76, t. 154. A forest shrub or tree, up to 40 ft. high, rarely more ; flowers white, or greenish ; ripe fruits red.
   Fr.G.: Landouma (Jan.) *Heudelot* 752 ! Kouria (Nov.) *Caille* in Hb. *Chev.* 14936 ! **S.L.:** *Don* ! (May, June) *T. Vogel* 137 ! *Barter* ! Wossin Highlands (fr. Mar.) *Sc. Elliot* 5055 ! Njala (Aug.) *Dawe* 567 ! Ronietta (Nov.) *Thomas* 5475 ! **Lib.:** Kakatown *Whyte* ! Ganta (May) *Harley* ! Nyandamolahou (Feb.) *Bequaert* 78 ! **Iv.C.:** Assikasso (Dec.) *Chev.* 22578 ! Rasso *Aubrév.* 146 ! **G.C.:** Koforidua (Dec.) *Johnson* 447 ! Aburi (Apr.) *Howes* 1187 ! Kwahu Plateau (fl. & fr. May) *Kitson* 1185 ! Begoro (Nov.) *Beveridge* 98 ! **Togo:** Misahöhe (Nov.) *Mildbr.* 7295 ! **S.Nig.:** Usonigbe F.R. (Oct.) *A. F. Ross* 257 ! Oban *Talbot* 45 ! 667 ! 2303 ! **Br.Cam.:** Victoria (Jan., Feb.) *Winkler* 664 ! *Maitland* 387 ! 605 ! *Preuss* 1319 ! Bulfambo, Buea (Mar.) *Maitland* 582 ! **F.Po:** (Jan., June) *Mann* 209 ! 235 ! 435 ! Goderich Bay (Nov., Dec.) *T. Vogel* 210 ! *Daniell* ! Also in French Cameroons, Spanish Guinea, Gabon, Cabinda and Belgian Congo. (See Appendix, p. 154.)

42. CROTONOGYNOPSIS Pax—F.T.A. 6, 1 : 924 ; Pax & K. Hoffm. in Engl. Pflanzenr. Euph. 7 : 14 (1914).

Leaves crowded, sessile, elongate-oblanceolate, much narrowed to and auriculate-cordate at the base, about 30 cm. long and up to 8 cm. broad, rather coarsely dentate, glabrous ; lateral nerves numerous ; flowers dioecious, apetalous ; male racemes fascicled on the old wood ; pedicels pubescent, jointed above the base ; calyx glabrous ; stamens 12–15 ; female calyx ciliate ; styles with laciniate tips
                                                                                      usambarica

C. usambarica *Pax*—F.T.A. 6, 1 : 924 ; Pax & K. Hoffm. l.c. 7 : 15. A shrub or small tree, 12–30 ft. high, with yellowish bark.
   Br.Cam.: Buea *Deistel* 218. Also in Tanganyika.

43. ARGOMUELLERA Pax—F.T.A. 6, 1 : 925 ; Pax & K. Hoffm. in E. & P. Pflanzenfam. 19C : 107 (1931). *Wetriaria* (Müll. Arg.) O. Ktze. (1903).

Branchlets softly tomentose ; leaves obovate-oblanceolate, acute, 20–40 cm. long, 5–13 cm. broad, sharply serrate, villous on the midrib and nerves beneath ; lateral nerves numerous ; racemes axillary, up to 20 cm. long ; flowers monoecious, apetalous ; male calyx splitting into 3–4 valvate segments ; stamens numerous, inserted in pits of the fleshy receptacle ; female sepals 5–6, imbricate ; ovary densely pubescent ; styles recurved, nearly free or shortly connate .. ..   *macrophylla*

A. **macrophylla** *Pax*—F.T.A. 6, 1 : 925. *Wetriaria macrophylla* (Pax) Pax in Engl. Pflanzenr. Euph. 7 : 50, fig. 30, E–G (1914). *Pycnocoma hutchinsonii* Beille in Bull. Soc. Bot. Fr. 61, Mém. 8 : 295 (1917) ; Chev. Bot. 582. *P. sassandrae* Beille l.c. 296 (1917), *fide* Hutch. A forest shrub, 2–12 ft. high.
  S.L.: Tiama (Jan.) *Dalz.* 8103 ! Njala (Nov.–Feb.) *Deighton* 504 ! 2439 ! 3584 ! **Iv.C.:** Morénou (Dec.) *Chev.* 22491 ; 22505 ! Zaranou (Mar.) *Chev.* 17634 ! Bâle to Diguoalé *Chev.* 21502. **G.C.:** Atuma (Dec.) *Vigne* FH 3518 ! Bia F.R. (Feb.) *Foggie* FH 4444 ! **Br.Cam.:** Likomba (Nov.) *Mildbr.* 10618 ! 10776 ! Also in French Cameroons, Spanish Guinea, Gabon, Belgian Congo, A.-E. Sudan, Uganda, Kenya, Tanganyika, Angola, Nyasaland, N. and S. Rhodesia and Mozambique.

44. PYCNOCOMA Benth.—F.T.A. 6, 1 : 955 ; Pax & K. Hoffm. in Engl. Pflanzenr. Euph. 7 : 52 (1914).

Styles connate nearly to the top ; leaves oblanceolate, narrowed into a distinct petiole, acute, 20–40 cm. long, 3–8 cm. broad, remotely denticulate in the upper part, glabrous, strongly reticulate ; bracts lanceolate, acuminate, spreading or recurved, about 6 mm. long, shortly pubescent ; male pedicels up to 2 cm. long, 1–3 to each bract   1. *angustifolia*
Styles connate only in the lower third ; bracts concave, not recurved :
  Rhachis of inflorescence slightly pubescent or glabrescent ; leaves elongate-oblanceolate, narrowed into a winged petiole at the base, acutely acuminate, up to 50 cm. long and 11 cm. broad, glabrous, laxly reticulate ; bracts broadly ovate-rounded, shortly appressed-pubescent ; male pedicels up to 2 cm. long ; fruits shortly horned   2. *macrophylla*
  Rhachis of inflorescence rather densely pubescent ; leaves obovate-oblanceolate, gradually narrowed to the base, rather long-acuminate, up to 45 cm. long and 10 cm. broad, rather closely reticulate ; bracts broadly ovate, villous ; male pedicels up to 2 cm. long, 1–3 to each bract ; fruits with large horn-like wings..   .. 3. *cornuta*

1. **P. angustifolia** *Prain*—F.T.A. 6, 1 : 960 ; Pax & K. Hoffm. in Engl. Pflanzenr. Euph. 7 : 58. *P. beillei* A. Chev. Bot. 582, name only. A shrub, glabrous except on the inflorescence, the leaves clustered towards the ends of the branchlets ; racemes peduncled in the upper axils ; capsule (young) velvety, at length glabrous, 4 cm. diam., each lobe with a pair of wing-like horns ; often in forest undergrowth.
  S.L.: Zimi (Apr.) *Deighton* 3631 ! Nyandehun (Apr.) *Deighton* 3656 ! **Lib.:** Kakatown *Whyte* ! Begwai (Nov.) *Bunting* ! Gletown, Tchien (July) *Baldwin* 6918 ! 6958 ! **Iv.C.:** various localities *fide* Chev. l.c. Mid. Sassandra to Mid. Cavally (young fr. July) *Chev.* 19269 !

2. **P. macrophylla** *Benth.*—F.T.A. 6, 1 : 959 ; Pax & K. Hoffm. l.c. 7 : 54, fig. 7, A–G (incl. vars.) ; Chev. Bot. 582. A shrub, about 8 ft. high, with long leaves clustered at the ends of the soft branchlets ; capsule 3-lobed, velvety, with 6 horn-like wings.
  Iv.C.: Agboville *Aubrév.* 1922 ! Yacassé to Adzopé, Attie (Dec.) *Chev.* 22665 ! **G.C.:** Banka F.R. (Dec.) *Chipp* 604 ! Pra River Station (Feb.) *Vigne* FH 1033 ! Benso (Apr.) *Andoh* FH 5476 ! **S.Nig.:** Oban *Talbot* ! **Br.Cam.:** Victoria (fr. Dec.) *Maitland* 789 ! Likomba (Oct.) *Mildbr.* 10543 ! Kumba *Büsgen* 49. Ossidinge *Rudatis* 96. **F.Po:** (Oct.) *T. Vogel* 18 ! Bokoko *Mildbr.* 6911. Also in French Cameroons and Belgian Congo. (See Appendix, p. 159.)

3. **P. cornuta** *Müll. Arg.*—F.T.A. 6, 1 : 958 ; Pax & K. Hoffm. l.c. 7 : 56. *P. brachystachya* of F.T.A. 6, 1 : 959, partly (*Millen* 149). A stout shrub with leaves clustered at the ends of the branchlets.
  G.C.: *Farmar* 424 ! *Irvine* 872 ! Aburi (Dec.) *Eady* ! **N.Nig.:** Sanga River, Jemaa (fl. & young fr. Nov.) *Keay* FHI 22261 ! 22268 ! Kurmin Damisa, Jemaa (Nov.) *Keay & Onochie* FHI 21712 ! **S.Nig.:** Abeokuta, rocky places *Barter* 3385 ! Obumo Rock, Abeokuta *Millen* 149 ! Idanre Hills (Dec.) *Brenan* 8665 ! Idumuje (young fr. Jan.) *Thomas* 2173 !

Imperfectly known species.

P. **brachystachya** *Pax* in Engl. Bot. Jahrb. 43 : 82 (1909) ; F.T.A. 6, 1 : 959, partly ; Pax & K. Hoffm. in Engl. Pflanzenr. Euph. 7 : 55.
  Br.Cam.: Victoria (Sept.) *Winkler* 367.

45. NECEPSIA Prain—F.T.A. 6, 1 : 923 ; Pax & K. Hoffm. in Engl. Pflanzenr. Euph. 7 : 16 (1914).

Leaves obovate-elliptic, acutely acuminate, widely cuneate at base, 10–25 cm. long, 6–9 cm. broad, glabrous, remotely denticulate ; lateral nerves about 10 pairs, with distinct parallel tertiary nerves ; flowers monoecious, apetalous ; spikes solitary, slender, about as long as the leaves ; bracts clustered, persistent ; stamens numerous, with pubescent glands between ; styles fimbriate-papillose ; fruits setose, 3-lobed, about 1 cm. diam. ..   ..   *afzelii*

N. **afzelii** *Prain*—F.T.A. 6, 1 : 923 ; Pax & K. Hoffm. l.c. 7 : 17. A forest shrub or tree, to 40 ft. high ; branchlets yellowish-pubescent, becoming glabrous.
  S.L.: *Afzelius*. *Thomas* 7938 ! Colony F.R. (fl. & fr. Dec.) *Vigne* SL 27 ! Gorahun (fl. & fr. Apr.) *Deighton* 3651 ! Gola Forest (Mar.) *Small* 561 ! **Lib.:** Dukwia R. (Mar., Apr.) *Cooper* 76 ! 263 ! Gbanga (Sept.) *Linder* 450 ! Monrovia (Aug.) *Baldwin* 8021 ! **Iv.C.:** Grabo *Chev.* 19595 ! **Br.Cam.:** Bakundu to Meanja *Winkler* 1078. Also in French Cameroons. (See Appendix, p. 155.)

8

46. CLEIDION Blume—F.T.A. 6, 1 : 930 ; Pax & K. Hoffm. in Engl. Pflanzenr. Euph.
7 : 288 (1914).

Leaves shortly petiolate, oblong-obovate, narrowed to the base, 10–17 cm. long, 4–
6·5 cm. broad, crenate, glabrous ; male spikes simple ; flowers clustered, sessile
gabonicum

C. gabonicum *Baill.*—F.T.A. 6, 1 : 930 ; Pax & K. Hoffm. l.c. 7 : 297. A dioecious shrub or a tree reaching
about 50 ft., with slightly pubescent branchlets and male inflorescences several inches long.
G.C.: Dansu R. (Dec.) *Johnson* 437 ! Tappa (Feb.) *Chipp* 101 ! Kumasi (Mar.) *Chipp* 141 ! Amentia
(Dec.) *Vigne* FH 1493 ! Juaso (Feb.) *Akpabla* 285 ! Also in French Cameroons and Gabon. (See Appendix, p. 139.)

47. MACARANGA Thouars—F.T.A. 6, 1 : 932 ; Pax & K. Hoffm. in Engl. Pflanzenr.
Euph. 7 : 298 (1914).

Leaves palmately 5–9-nerved from the more or less cordate and auricled base, lamina
usually more or less distinctly digitately or palmately lobed ; petioles 7–46 cm. long ;
stipules 2·5–5·5 cm. long (but see No. 3), 1–3 cm. broad ; ovary 2-celled ; fruits
2-lobed :
Bracts subtending the flower-clusters 3–4 mm. long, glandular-denticulate to subentire,
densely glandular-puberulous ; leaves rather shallowly palmately 3-lobed, lobes
acute or shortly acuminate, 13–60 cm. long and nearly as broad ; male panicles
5–23 cm. long ; female panicles about 8 cm. long, little branched
1. *schweinfurthii*
Bracts subtending the flower-clusters 5–10 mm. long, sinuate-dentate or deeply
laciniate :
Bracts loosely pubescent or pilose, deeply laciniate ; leaves cordate to subtruncate at
base, caudate-acuminate at apices of leaf and lobes :
Leaves usually rather deeply digitately 3–7-lobed, rarely not lobed, coriaceous,
midrib and main nerves beneath sparingly pubescent, 13–32 cm. long and broad ;
stipules more than 2·5 cm. long ; male panicles 15–30 cm. long ; female flowers
in spike-like racemes or narrow sparsely branched panicles, up to 14 cm. long ;
fruits 12–20 mm. broad   ..   ..   2. *heterophylla*
Leaves not lobed or only shallowly 3–lobed, membranous, midrib and main nerves
beneath with long spreading hairs, 10–17 cm. long, 11–17 cm. broad ; stipules
lanceolate, about 1·5 cm. long ; male panicles about 10–15 cm. long ; female
flowers in spike-like racemes, up to 6 cm. long ; fruits 6–10 mm. broad 3. *beillei*
*Bracts densely tomentellous, sinuate-dentate ; leaves very shallowly 3-lobed, rarely
not lobed, lobes cuspidate, the lateral ones sometimes no bigger than large teeth,
margin repand-dentate, deeply overlapping-cordate at base, 14–30 cm. long and
broad ; male panicles 7–14 cm. long ; female panicles 8–13 cm. long
4. *occidentalis*
Leaves 3-nerved at base, or pinnately nerved, not lobed ; stipules up to 2·5 cm. long ;
ovary, where known, 1-celled :
*Bracts well-developed, 2–10 mm. long, more or less persistent in the male panicles :
Bracts orbicular, entire, sometimes apiculate, 3–4 mm. long ; leaves thickly coriaceous,
orbicular-obovate to broadly elliptic, obtuse at the glandular-auricled base, obtuse
to shortly acuminate at apex, margin shallowly repand-dentate especially towards
the apex, 12–16 cm. long, 6–11 cm. broad ; petiole stout 4–11 cm. long ; male
panicles 4–7 cm. long, rusty-tomentellous ; female not known   .. 5. *staudtii*
Bracts ovate, mostly dentate ; leaves not as above, mostly rather thin :
Leaves distinctly toothed, mostly ovate :
Branchlets, petioles and midrib beneath pubescent or pilose with whitish hairs or
nearly glabrous ; bracts 2–5 mm. long, dentate :
Stipules about 2·5 cm. long ; branchlets, petioles and midrib beneath very
sparsely pubescent or glabrescent ; bracts 2–2·5 mm. long ; leaves ovate to
orbicular-ovate, rounded at base, long-acuminate, margin sinuate-dentate,
9–25 cm. long, 5·5–13 cm. broad ; petiole 6–10 cm. long ; male panicles 6–7 cm.
long ; female racemes 2–3 cm. long   ..   ..   6. *paxii*
Stipules up to 1 cm. long, early caducous ; branchlets, petioles and midrib beneath
more or less densely pubescent or pilose with whitish hairs ; bracts 3–5 mm.
long ; leaves very broadly ovate, rounded or truncate at base, shortly acuminate,
margin coarsely repand-dentate, 9–18 cm. long, 6–15 cm. broad ; very densely
glandular beneath ; petiole 3–18 cm. long ; panicles up to 10 cm. long, usually
borne on leafless parts of the stem ; fruits globose, about 3 mm. diam.
7. *hurifolia*
Branchlets, petioles and midrib beneath densely rusty-tomentellous at least when
young ; bracts 4–10 mm. long, dentate ; stipules 5–6 mm. long, early caducous ;
leaves ovate to oblong, rounded to cuneate at base, acute to acuminate, margin
coarsely repand-dentate, 8–20 cm. long, 3·5–12 cm. broad, densely covered with
rather small glands beneath ; petiole 1·5–10·5 cm. long ; panicles up to 10 cm.

long, axillary, among the leaves ; fruits obliquely ovoid, 6–8 mm. long and broad ; stamens 2, rarely 1   ..   ..   8. *monandra*

Leaves entire or at most undulate, elliptic-obovate to oblong-oblanceolate, broadly cuneate or rounded and usually auriculate-glandular at base, shortly acuminate, 7–16 cm. long, 4–9 cm. broad, minutely and closely glandular beneath ; petiole 2–10 cm. long ; branchlets, petioles and midrib beneath densely rusty-tomentellous at least when young ; bracts 3–8 mm. long, entire or somewhat dentate ; male panicles 5–8 cm. long, axillary, among the leaves ; female panicles 2–6 cm. long ; fruits obliquely ovoid or subglobose, about 5 mm. diam., with pedicels up to 2·5 cm. long   ..   ..   9. *barteri*

*Bracts less than 2 mm. long, caducous :

Branchlets, petioles and midrib beneath densely rusty-tomentellous at least when young, without spreading whitish hairs :

Leaves ovate or ovate-elliptic, rather long-acuminate at apex, margin distinctly repand-dentate, rounded at base, 7–15 cm. long, 3·5–8·3 cm. broad ; petiole 1·5–8 cm. long ; male panicles up to 6 cm. long ; female flowers and fruits not known   ..   ..   10. sp. *A*

Leaves oblong-oblanceolate, abruptly and shortly acuminate at apex, margin entire or undulate, or denticulate in the upper half, obtuse at base, 4·5–12 cm. long, 2–5·5 cm. broad ; petiole 0·5–2 (–3) cm. long ; panicles up to 5 cm. long ; fruits subglobose, about 3 mm. diam., with pedicels up to 1 cm. long   11. *heudelotii*

Branchlets, petioles and midrib beneath with long spreading whitish hairs ; leaves elliptic-oblong, rounded at base, rather long-acuminate, margin undulate, 6–14 cm. long, 3·5–6 cm. broad ; petiole 0·7–6 cm. long ; panicles 3–6 cm. long ; fruits subglobose, about 3 mm. diam., with pedicels up to 6 mm. long   12. *spinosa*

1. **M. schweinfurthii** *Pax* in Engl. Bot. Jahrb. 19 : 92 (1894) ; F.T.A. 6, 1 : 935 ; Pax & K. Hoffm. in Engl. Pflanzenr. Euph. 7 : 313. *M. rosea* Pax (1899)—F.T.A. 6, 1 : 935 ; F.W.T.A., ed. 1, 1 : 306. A straggling tree or shrub, in forest ; fruits with a yellow granular covering. **S.Nig.:** Oban *Talbot* 112 ! 1377 ! Kanyang, Ikom (fr. May) *Latilo* FHI 30991 ! **Br.Cam.:** Johann-Albrechtshöhe *Büsgen* 39. Also in French Cameroons, Ubangi-Shari, A.-E. Sudan, Uganda, Belgian Congo, Gabon, Angola, N. Rhodesia and Tanganyika.

2. **M. heterophylla** (*Müll. Arg.*) *Müll. Arg.*—F.T.A. 6, 1 : 936 ; Pax & K. Hoffm. l.c. 7 : 314 ; Chev. Bot. 580 ; Aubrév. Fl. For. C. Iv. 2 : 70. *Mappa heterophylla* Müll. Arg. (1864). A forest shrub or tree, 10–30 ft. high ; branchlets armed with spines ; flowers tinged pink ; fruits bilobed, half-succulent, pink or red, dusted with sticky golden glands. **Sen.:** Sindialone, Casamance *Chev.* 15770. **Port.G.:** Formosa (fr. Apr.) *Esp. Santo* 1970 ! **Fr.G.:** Santa to Timbo *Chev.* 12596 ; 12813. R. Koukouré, Kouria *Chev.* 15003. Mamou to Soya *Chev.* 20365. Batagya (Oct.) *Small* 495 ! **S.L.:** *Afzelius* ! *Don* ! Sugar Loaf Mt. *Welw.* 464 ! *Barter* ! Heddle's Farm (Sept.) *Lane-Poole* 359 ! Kaballa (Sept.) *Thomas* 2588 ! **Lib.:** Reppue's Town (Aug.) *Linder* 363 ! Monrovia (Aug.) *Baldwin* 8011 ! Ganta (Sept.) *Baldwin* 12511 ! **Iv.C.:** Abidjan *Chev.* 15250 ! Voguié (Jan.) *Chev.* 17136. **G.C.:** Ashanti *Cummins* 132 ! Foso, Cape Coast (Sept.) *Chipp* 568 ! Ankassa (Aug.) *Chipp* 326 ! Anyimam (fr. Jan.) *Brent* FH 401 ! **Togo:** Alvanyo (fl. Sept., fr. Nov.) *St. C. Thompson* FH 3565 ! (See Appendix, p. 149.)

3. **M. beillei** *Prain*—F.T.A. 6, 1 : 939 ; Pax & K. Hoffm. l.c. 7 : 326 ; Chev. Bot. 580 ; Aubrév. l.c. 2 : 68, t. 149, 1–5. A shrub or small tree, 10–15 ft. high ; flowers yellow-green ; fruits red ; in forest. **Iv.C.:** Abidjan to Dabou *Chev.* 15551 ! Bingerville (fr. Dec.) *Chev.* 16050 ! Dabou (Feb.) *Chev.* 17213 ! Accrédiou (Feb.) *Chev.* 17067 ! Banco *Martineau* 306 ! *Serv. For.* 370 !
[The specimen (*Lecomte* A. 89) from French Congo cited as this species in the F.T.A. appears to be *M. monandra* Müll. Arg.]

4. **M. occidentalis** (*Müll. Arg.*) *Müll. Arg.*—F.T.A. 6, 1 : 936 ; Pax & K. Hoffm. l.c. 7 : 315. *Mappa occidentalis* Müll. Arg. (1864). A tree, 30–60 ft. high, at margin of montane forest ; stem and branches often with spines ; flowers green ; fruits with a yellow granular covering. **S.Nig.:** Oban *Talbot* 1431 ! **Br.Cam.:** Cam. Mt., 2,500–6,000 ft. (Dec.–Feb.) *Mann* 771 ! 1189 ! 2155 ! *Maitland* 117 ! 296 ! *Rosevear* 29/37 ! Bali-Ngemba F.R., Bamenda (fr. May) *Ujor* FHI 30341 ! **F.Po:** *Mann* 303 ! Moka (Feb.) *Exell* 850 ! (See Appendix, p. 149.)

5. **M. staudtii** *Pax*—F.T.A. 6, 1 : 939 ; Pax & K. Hoffm. l.c. 7 : 316. A tree, to 50 ft. high, in swamp forest. **S.Nig.:** Osho, Omo F.R. *Jones* FHI 17281 ! Okomu F.R. *Ebuade* in *Hb. Brenan* 9056 ! Also in French Cameroons, Gabon and Cabinda.

6. **M. paxii** *Prain*—F.T.A. 6, 1 : 939 ; Pax & K. Hoffm. l.c. 7 : 326. A forest shrub or small tree, to 30 ft. high ; branches armed with patent white spines. **S.Nig.:** Oban *Talbot* 610 ! Etomi, Ikom (young fr. May) *Jones* FHI 14120 ! Also in French Cameroons.

7. **M. hurifolia** *Beille* in Bull. Soc. Bot. Fr. 55, Mém. 8 : 80 (1908) ; Pax & K. Hoffm. l.c. 7 : 327 ; Aubrév. l.c. 2 : 68, t. 150. *M. togoensis* Pax (1909)—F.T.A. 6, 1 : 939 ; Chev. Bot. 581. *M. monandra* of F.W.T.A., ed. 1, 1 : 306, partly (*Chipp* 317). A forest shrub or tree, to 40 ft. high ; inflorescences greenish, mostly borne on leafless branches ; branchlets armed with spines ; fruits waxy-glandular. **S.L.:** Scarcies R., Sasseni (Jan.) *Sc. Elliot* 4530 ! Njala (July) *Deighton* 725 ! Bumbuna (young fr. Oct.) *Thomas* 3904 ! Gola Forest (June) *Small* 732 ! **Lib.:** Du R. (July) *Linder* 73 ! Gbanga (fr. Sept.) *Linder* 498 ! Woeme, Vonjama (fr. Oct.) *Baldwin* 10079 ! **Iv.C.:** *Aubrév.* 396 ! Bingerville-Abidjan-Dabou region (May, June) *Chev.* 15319 ; 15449. **G.C.:** Arnokokrom, W. Apollonia (Aug.) *Chipp* 317 ! Bataba, N. Beyin (July) *Chipp* 266 ! Boankro, Ashanti (May) *Chipp* 440 ! **Togo:** *Döring* 291. Misahöhe (Nov.) *Mildbr.* 7258 ! **S.Nig.:** Ajilite *Millen* 164 ! Okomu F.R. (Jan., Feb.) *A. F. Ross* 248 ! *Brenan* 9037 ! Idumuje (Jan.) *Thomas* 2186 ! Oban *Talbot* 610 ! **Br.Cam.:** Abonse, Bamenda (May) *Johnstone* 130/31 ! (See Appendix, p. 149.)

8. **M. monandra** *Müll. Arg.*—F.T.A. 6, 1 : 940 ; Pax & K. Hoffm. l.c. 7 : 327, fig. 53, A–D. *M. zenkeri* Pax (1897)—F.T.A. 6, 1 : 940 ; F.W.T.A., ed. 1, 1 : 306. A forest shrub or tree, to 50 ft. high ; inflorescences pinkish. **S.Nig.:** Oban *Talbot* 2311 ! Old Calabar (young fr. Mar.) *Holland* 99 ! Ikom (young fr. June) *Catterall* 22 ! Shakwa, Obudu (fr. May) *Latilo* FHI 30927 ! **Br.Cam.:** Bulifambo, Buea (Mar.) *Maitland* 522 ! S. Bakundu (Mar.) *Onochie* FHI 32055 ! Abonando (Mar.) *Rudatis* 12 ! Also in French Cameroons, Gabon, Spanish Guinea, Belgian Congo, Uganda, Tanganyika and Angola. (See Appendix, p. 149.)

9. **M. barteri** *Müll. Arg.*—F.T.A. 6, 1 : 942 ; Pax & K. Hoffm. l.c. 7 : 329 ; Chev. Bot. 580 ; Aubrév. Fl. For. C. Iv. 2 : 66, t. 147. *M. rowlandii* Prain (1910)—F.T.A. 6, 1 : 942 ; Pax & K. Hoffm. l c. ; F.W.T.A., ed. 1, 1 : 306. *M. heudelotii* of Chev. Bot. 581, partly (*Chev.* 16197 ; 16279). A forest shrub or tree, to 50 ft. high ; branchlets armed with spines.

Fr.G.: Dyeke (fr. Oct.) *Baldwin* 9677! **S.L.:** Bumban, Limba (Apr.) *Sc. Elliot* 5668! Njala (May) *Deighton* 691! 5947! Yonibana (fr. Nov.) *Thomas* 5237! **Lib.:** Bonuta (fr. Oct.) *Linder* 889! Gletown (July) *Baldwin* 6779! Ganta (May) *Harley* 1196! **Iv.C.:** Dabou (Feb.) *Chev.* 16197! Zaranou (Mar.) *Chev.* 16279! **G.C.:** Akwapim (fl. Aug., fr. June) *Irvine* 764! *Johnson* 748! Agogo (Apr.) *Irvine* 953! *Vigne* 1119! Essiama (Apr.) *Williams* 432! **Togo:** Misahöhe *Baumann* 162. **S.Nig.:** Lagos *Millen* 125! *Rowland*! Ijuwoye (Dec.) *Millen* 28! Ibadan South F.R. (Feb.) *Keay & Meikle* FHI 25666! Sapoba *Kennedy* 2321! Onitsha *Barter* 1654! Cross River *Johnston*! Also in Spanish Guinea. (See Appendix, p. 149.)

10. **M. sp. A.** Possibly new, but more material required. *M. monandra* of F.W.T.A., ed. 1, 1 : 306, partly (*Aylmer* 71), not of Müll. Arg. A tree.
   S.L.: Bandajuma (May) *Aylmer* 71! Giema, Koya (Dec.) *Deighton* 3844! **Lib.:** Fortsville (Mar.) *Baldwin* 11150!

11. **M. heudelotii** *Baill.*—F.T.A. 6, 1 : 945 ; Pax & K. Hoffm. l.c. 7 : 331, fig. 53, E ; Chev. Bot. 581, partly ; Aubrév. l.c. 2 : 68, t 149, 6–9. A shrub or tree, to 25 ft. high ,in swamp forest ; branchlets with patent spines ; fruits covered with yellow scales.
   Sen.: Cape Verde *Leprieur*. **Gam.:** Albreda *Heudelot* 106. **Fr.Sud.:** Medinani (Feb.) *Chev.* 513! **Port.G.:** Bissalanca, Bissau *Esp. Santo* 1662! **Fr.G.:** Rio Nunez *Heudelot* 784! Kindia (Mar.) *Chev.* 13028! **S.L.:** Scarcies R. (Dec.–Jan.) *Sc. Elliot* 4390! 4517! Erimakuna (Mar.) *Sc. Elliot* 5258! Sherbro (Feb.) *Dalz.* 938! Njala (Feb.) *Deighton* 1083! 1084! **Lib.:** Péahtah (Oct.) *Linder* 1077! Mao R., Maa (fl. & fr. Feb.) *Bequaert* 53! Jabrocca (Dec.) *Baldwin* 10862! Ganta (fr. Feb.) *Baldwin* 11026! **Iv.C.:** *fide* Aubrév. *l.c.* **G.C.:** Essiama *Fishlock* 8! Tano R (Sept.) *McAinsh* FH 378! **Togo:** Lome *Warnecke* 455! **S.Nig.:** Lagos *Moloney*! Epe (fr Mar.) *Barter* 3271! *Keay* FHI 16058! (See Appendix, p. 149.)

12. **M. spinosa** *Müll. Arg.*—F.T.A. 6, 1 : 944 ; Pax & K. Hoffm. l.c. 7 : 331 ; Aubrév. l.c. 2 : 66, t. 148. A forest shrub or tree, to 40 ft. or more high ; branchlets armed with spines.
   Lib.: Nekabozu, Vonjama (fr. Oct.) *Baldwin* 9969! **Iv.C.:** Abidjan *Aubrév.* 46! Man *Aubrév.* 1087. **S.Nig.:** Mamu River F.R., Awka (Feb.) *Jones* FHI 5014! Abaragba (Apr.) *Jones* FHI 1546! **Br.Cam.:** Johann-Albrechtshöhe *Schultze* 95 ; 97. **F.Po:** *Mann* 1160! Also in French Cameroons, Spanish Guinea, Gabon and Angola.

48. ACALYPHA Linn.—F.T.A. 6, 1 : 880 ; Pax & K. Hoffm. in Engl. Pflanzenr. Euph. 16 : 12 (1924).

Male and female flowers on separate inflorescences ; male racemes axillary, slender, female inflorescences terminal :
   Female inflorescence paniculate, 10–36 cm. long, flowers pedicellate ; stems pubescent ; leaves ovate, subcordate at base, acuminate, 4–15 cm. long, 2·5–10 cm. broad, coarsely crenate-serrate, 5-nerved at base ; petioles 5–16 cm. long ; male racemes 6–14 cm. long .. .. 1. *racemosa*
   Female inflorescence spicate, up to 15 cm. long, flowers sessile :
      Female spikes very short, ovoid ; leaves more or less lanceolate, villous on both sides ; stems very villous with reflexed hairs ; styles laciniate :
         Styles 2–3 mm. long ; leaves lanceolate to triangular-lanceolate, 4–8 cm. long, 1·5–2·5 cm. broad ; petiole 1–1·5 cm. long .. .. .. 2. *senegalensis*
         Styles 10–20 mm. long ; ovary with gland-tipped hairs ; leaves lanceolate, linear-lanceolate or oblong-lanceolate, 3·5–11 cm. long, 1–2 (–3·5) cm. broad ; petiole 1·5–5·5 cm. long .. .. 3. *senensis*
      Female spikes cylindrical, 5–15 cm. long ; ovary without gland-tipped hairs ; leaves more or less ovate :
         Bracts of the female flowers with numerous gland-tipped hairs on the teeth or within ; bracts with short triangular teeth on either side of the acumen :
            Styles bilobed and not laciniate ; female bracts about 5-toothed on each side of the acumen ; stems shortly pubescent ; leaves ovate, rounded at base, caudate-acuminate, crenate-serrate, 3–6 cm. long, 2·5–4 cm. broad, softly pubescent beneath ; petiole up to 4 cm. long .. .. 4. *nigritiana*
            Styles laciniate ; female bracts about 10-toothed on each side of the acumen, 8–12 mm. broad in fruit ; stems shortly pubescent and sometimes villous ; leaves ovate, rounded or subcordate at base, rather coarsely serrate, 5–15 cm. long, 3–10 cm. broad, spreadingly pilose beneath ; petiole 1–13 cm. long .. 5. *ornata*
         Bracts of the female flowers without gland-tipped hairs ; bracts deeply divided into lanceolate lobes or laciniae :
            Bracts of the female flowers about 7 mm. broad in fruit, with about 17 unequal lanceolate lobes up to about 1 mm. long, pubescent outside ; female inflorescences terminal on short lateral branches ; stems and petioles softly pubescent ; leaves ovate rounded at base, acute at apex, crenate-serrate, up to 10 cm. long and 5 cm. broad, pubescent on both surfaces ; petiole up to 3 cm. long   6. *sp. A*
            Bracts of the female flowers 12–16 mm. broad in fruit, with 12–20 filiform laciniae 2–3 mm. long, densely hirsute ; leaves ovate to ovate-lanceolate, rounded to cordate at base, gradually acuminate, crenate-serrate, 3–10 cm. long, 2·5–6 cm. broad, pubescent on both surfaces ; petiole 3–6 cm. long .. 7. *manniana*
Male and female flowers on the same inflorescences or in the same leaf-axil ; the males above the females :
   Leaves with numerous raised resinous glands beneath, ovate-rhomboid, cuneate at base, acuminate, obtusely serrate, 6–9 cm. long, 1·5–5 cm. broad ; petiole 2–4 cm. long ; inflorescence bisexual axillary, the upper part male, with a few 2-flowered denticulate female bracts at the base enlarging in fruit to about 1·5 cm. broad
                                                                      8. *ceraceopunctata*

Leaves without resinous glands beneath :

Female bracts digitately 3-partite, with gland-tipped hairs on the margins ; leaves ovate, subcordate at base, acuminate, crenate, 2–5 cm. long, 1·5–4 cm. broad, sparingly pubescent ; petiole 1·5–7 cm. long ; inflorescence very short, with 6–9 female flowers below ; annual herb　　..　　..　　..　　.. 9. *brachystachya*

Female bracts dentate, fimbriate or laciniate with numerous teeth :

Annual herbs ; leaves membranous ; inflorescences axillary :

Female bracts fimbriate, with about 20 shortly ciliate teeth, ribbed ; leaves ovate to rhomboid-ovate, obtuse to cordate at base, cuspidate-acuminate, crenate-serrate, 4–11 cm. long, 2–6 cm. broad, slightly scabrous ; petiole 2–9 cm. long ; inflorescences with about 10 female flowers in the lower part　　.. 10. *ciliata*

Female bracts crenate or dentate :

Bracts of the female flowers pilose inside and outside, more or less triangular, narrowed to the base, in fruit about 6–7 mm. long and broad, with 8–9 teeth about 1·5 mm. long ; stems and leaves pubescent and long-spreading pilose ; leaves ovate, rounded at base, acute at apex, rather coarsely crenate-serrate, 2–7·5 cm. long, 1–4 cm. broad ; petiole 0·8–5·5 cm. long ; inflorescences with few to many female flowers in the lower part ; bracts crowded in fruit

11. sp. *B*

Bracts of the female flowers glabrous or nearly so, except for the ciliate crenate or triangular dentate margins, broad at the base ; stems and leaves shortly crisped-pubescent and sometimes with long hairs as well :

Female bracts in fruit 4–4·5 mm. long, 7–8 mm. broad, dentate with 10–18 triangular teeth crowded and usually overlapping ; leaves ovate, rounded at base, obtuse or subacute, lightly crenate, 2–5 cm. long, 1·3–3 cm. broad ; petiole 2–6 cm. long　　..　　..　　..　　..　　.. 12. *crenata*

Female bracts in fruit up to 13 mm. long and broad, ovate when spread out, crenate, very laxly arranged ; inflorescence about 2 cm. long, with 2–4 female flowers remote from the male ; leaves ovate, obtuse at base, acute or subacute at apex, crenate, 1·5–6 cm. long, 1–3 cm. broad ; petiole 1–7 cm. long

13. *segetalis*

Shrub or small tree, 1·5–7 m. high ; inflorescence bisexual, axillary, 2·5–12 cm. long, with 1–3 female flowers at the base ; female bracts 4–6 mm. diam., repand-dentate ; leaves oblanceolate-elliptic, obovate or subrhomboid, more or less narrowed to the subcordate base, shortly caudate-acuminate, serrate, 8–20 cm. long, 3·5–10 cm. broad ; petiole 2–8 cm. long　..　　..　　.. 14. *neptunica*

1. **A. racemosa** *Wall. ex Baill.* Etud. Gén. Euphorb. 443 (1858). *A. paniculata* Miq. (1859)—F.T.A. 6, 1 : 886 ; Chev. Bot. 577 ; Pax & K. Hoffm. in Engl. Pflanzenr. Euph. 16 : 14, fig. 3 ; F.W.T.A., ed. 1, 1 : 302. A perennial herb or shrub, to 10 ft. high, with a copious female inflorescence of small flowers, often deep red in colour.
Iv.C.: Bingerville *Chev.* 15348. **G.C.:** Kumasi (Jan.) *Dalz.* 141 ! Abofuo (July) *Chipp* 535 ! Begoro, Akim (fr. Apr.) *Irvine* 1192 ! **Togo:** Kratchi (May) *Krause* ! **S.Nig.:** Lagos *Dawodu* 212 ! Odiani, Asaba (Aug.) *Onochie* FHI 33438 ! Agulu *Thomas* 1018 ! Oban *Talbot* 693 ! **Br.Cam.:** Ambas Bay (Feb.) *Mann* 767 ! Cam. Mt., 1,000 ft. (Dec.) *Mann* 1254 ! R. Malu, Wum (fl. & fr. July) *Ujor* FHI 30455 ! **F.Po:** *Barter* ! Widespread in tropical Africa and Asia. (See Appendix, p. 135.)
2. **A. senegalensis** *Pax & K. Hoffm.* l.c. 16 : 79 (1924). Perennial herb ; stems to 16 in. high.
Sen.: *Lecard* 301. R. Bucoy *Lecard* 45. [Hutchinson suggests this may be the same as *A. senensis*.]
3. **A. senensis** *Klotzsch*—F.T.A. 6, 1 : 888 (incl. var. *chariensis* (Beille) Hutch.) ; Chev. Bot. 577 ; Pax & K. Hoffm. l.c. 16 : 79 ; Berhaut Fl. Sen. 167. A hairy herb, 2–3 ft. or more high, woody at the base, with red-purple female flowers.
Sen.: Niokolo-Koba *Berhaut* 1544 ; 4595. **Fr.Sud.:** Dioubéba to Oualia (Dec.) *Chev.* 23. **Port.G.:** Buruntuma, Gabu (July) *Esp. Santo* 2700 ! **N.Nig.:** Farin Dutse (Oct.) *Onwudinjoh* FHI 24035 ! Bauchi Plateau (Aug.) *Lely* P. 660 ! Panyam (July) *Lely* 412 ! Matyoro *Thornewill* 135 ! Yola (May) *Dalz.* 159 ! Widespread in tropical and southern Africa.
4. **A. nigritiana** *Müll. Arg.*—F.T.A. 6, 1 : 890 ; Pax & K. Hoffm. l.c. 16 : 57. A small shrub.
S.Nig.: Yoruba forests, Lagos *Barter* 3425 ! [Hutchinson suggests this may be only a variety of *A. ornata*.]
5. **A. ornata** *Hochst. ex A. Rich.*—F.T.A. 6, 1 : 890 ; Pax & K. Hoffm. l.c. 16 : 58. A shrub, to 10 ft. high ; by streams and in open places.
N.Nig.: Tilde Filani (fr. May) *Lely* 230 ! Bauchi Plateau (May) *Lely* P. 279 ! **S.Nig.:** Lagos *Millen* 20 ! 113 ! 135 ! Ajilite *Millen* 176 ! Olokemeji *Foster* 288 ! Idanre (fr. Apr.) *Symington* FHI 3357 ! **Br.Cam.:** *Preuss* 570 ! Widespread in tropical Africa.
6. **A. sp. A.** Possibly new, but better material required. *A. manniana* of Chev. Bot. 576, not of Müll. Arg.
Fr.G.: Soumbalako to Boulivel, Fouta Djalon (fr. Sept.) *Chev.* 18652 ! **S.L.:** Yetaya (fl. & fr. Sept.) *Thomas* 2432 !
7. **A. manniana** *Müll. Arg.*—F.T.A. 6, 1 : 893 ; Pax & K. Hoffm. l.c. 16 : 70. A slender, climbing shrub, to 6 ft. high, at edge of forest ; flowers pinkish.
Br.Cam.: Cam. Mt., 3,000–5,700 ft. (fl. Nov.–Apr., fr. Feb.) *Mann* 1270 ! *Maitland* 210 ! 1305 ! Bamenda, 6,000 ft. (fl. & fr. Jan.) *Migeod* 425 ! Lakom (fl. & fr. Apr.) *Maitland* 1732 ! **F.Po:** Musola (Jan.) *Guinea* 1104 ! 1614 ! Moka, 3,500–4,000 ft. (Dec.) *Boughey* 65 ! Also in French Cameroons.
8. **A. ceraceopunctata** *Pax*—F.T.A. 6, 1 : 896 ; Pax & K. Hoffm. l.c. 16 : 136. A shrub to 5 ft. high, in savannah or fringing forest ; fruits hidden by the enlarged pale bracts, 3-lobed and dotted with yellow scale-like glands.
G.C.: Sirigu, N.T. (fr. Oct.) *Vigne* FH 4587 ! **N.Nig.:** Rimingata, Kano (fl. & fr. Aug.) *Onwudinjoh* FHI 24019 ! Katagum (fr. Sept.) *Dalz.* 209 ! Yola (June) *Dalz.* 151 ! Also in the French Cameroons.
9. **A. brachystachya** *Hornem.*—F.T.A. 6, 1 : 899 ; Pax & K. Hoffm. l.c. 16 : 101. A small slender-branched annual 6–12 in. high.
Br.Cam.: Buea *Lehmbach* 190 ; *Preuss* 1002 ! Also from French Cameroons to Red Sea, Abyssinia, Uganda, Belgian Congo, Angola and E. Africa ; also in tropical Asia.

10. **A. ciliata** *Forsk.*—F.T.A. 6, 1 : 901 ; Pax & K. Hoffm. l.c. 16 : 98. *A. fimbriata* Schum. & Thonn. (1827)—
Chev. Bot. 576. *A. vahliana* Müll. Arg. (1865)—Chev. Bot. 577. An erect, slightly pubescent branched
annual, to 2½ ft. high.
    Sen.: *Roger* 54 ! **Fr.Sud.:** Ouahigouya to Koro, Yatenga *Chev.* 24819. Diafarabé to Sansanding *Chev.*
2619. **Port.G.:** Ingora to Barro, S. Domingos (fl. & fr. Aug.) *Esp. Santo* 3087 ! **Fr.G.:** Mamou *Chev.*
15046. **S.L.:** Yetaya (fl. & fr. Sept.) *Thomas* 2398 ! Kukuna, Bramaia (fr. Oct.) *Jordan* 939 ! **Lib.:**
Gbanga (fl. & fr. Sept.) *Linder* 555 ! **Iv.C.:** Sokomanta to Sampleu, Dyola *Chev.* 21087. **G.C.:** Aburi
Brown 414 ! Amentia *Irvine* 525 ! **Togo:** Lome *Warnecke* 368 ! Kpandu (fr. June) *Andoh* FH 5282 !
N.Nig.: Nupe *Barter* 1047 ! Zungeru (fr. Aug.) *Dalz.* 100 ! Bauchi Plateau (fr. July) *Lely* P. 400 ! Yola
(fl. & fr. June) *Dalz.* 153 ! **S.Nig.:** Lagos *Millen* 58 ! Okomu F.R. (Feb.) *Brenan* 9171 ! Agulu *Thomas*
223 ! **Br.Cam.:** Rio del Rey *Johnston* ! Widespread in tropical Africa, in Arabia and in India ; also in
Cape Verde Islands. (See Appendix, p. 134.)
11. **A. sp. B.** Possibly new, or else perhaps introduced. A weed up to 2½ ft. high, well-branched and straggly.
    S.L.: Tower Hill, Freetown (fl. & fr. June) *Deighton* 2988 ! Mano railway station (fl. & fr. Nov.) *Deighton*
3280 !
12. **A. crenata** *Hochst. ex A. Rich.*—F.T.A. 6, 1 : 903 ; Pax & K. Hoffm. l.c. 16 : 97. An annual herb, 1–1½ ft.
high.
    N.Nig.: Lokoja (fl. & fr. Nov.) *Ansell* ! *Barter* ! *Dalz.* 163 ! Aguji, Ilorin *Thornton* ! Also in French
Cameroons, A.-E. Sudan, Somaliland, Abyssinia and Uganda.
13. **A. segetalis** *Müll. Arg.*—F.T.A. 6, 1 : 904 ; Pax & K. Hoffm. l.c. 16 : 36 ; Berhaut l.c. 168. *A. crenata* of
F.W.T.A., ed. 1, 1 : 303, partly. An annual herb, to 1½ ft. high ; flowers reddish ; a weed.
    Sen.: *fide* Berhaut *l.c.* **S.L.:** Musaia (fl. & fr. July) *Deighton* 4807 ! **Togo:** Misahöhe *Baumann* 227.
Dah.: *Le Testu* 95. **N.Nig.:** Nupe *Barter* 1490 ! 1495 ! Zaria (fl. & fr. Aug.) *Keay* FHI 28014 ! Kyana,
Nassarawa (fl. & fr. July) *Hepburn* 28 ! Nabardo (fl. & fr. May) *Lely* 214 ! Bauchi Plateau (fl. & fr.
June) *Lely* P. 335. Vom *Dent Young* 225 ! **S.Nig.:** Lagos *Millen* 37 ! **Br.Cam.:** Bamenda Prov. (May)
Maitland 1761 ! Mamfe (fl. & fr. Apr.) *Ejiofor* FHI 29374 ! Widespread in tropical Africa.
14. **A. neptunica** *Müll. Arg.*—F.T.A. 6, 1 : 907 ; Pax & K. Hoffm. l.c. 16 : 109. A forest shrub. The only
specimen from our area has one twig with long spreading hairs on the stem and leaves, while the other
twigs have a short curly indumentum ; two varieties may be represented here.
    G.C.: Akatui (Oct.) *Vigne* FH 4029 ! Also widespread in central and eastern Africa.

Besides the above, various ornamental species are cultivated, including *Acalypha hispida* Burm. and *A. wilkesiana* Müll. Arg., natives of tropical Asia.

Imperfectly known species.

A. sp. Herb, about 1 ft. high, in montane forest ; leaves reddish below. Female flowers and fruits apparently
solitary and sessile in the leaf-axils. More material required.
    S.Nig.: Ikwette Plateau, Obudu, 5,200 ft. (fr. Dec.) *Keay & Savory* FHI 25264 !

49. RICINUS Linn.—F.T.A. 6, 1 : 945 ; Pax & K. Hoffm. in Engl. Pflanzenr. Euph. 11 : 119 (1919).

Leaves alternate, long-petiolate, orbicular in outline, peltate, up to 60 cm. broad, deeply
palmately lobed ; lobes 7 or more, glandular-toothed, glaucous, glabrous, green or
reddish ; stipules large, ovate, connate into a sheath and enclosing the bud ; flowers
in large panicles, the males below, the females above ; capsule smooth or prickly,
about 2·5 cm. diam. .. .. *communis*

R. communis *Linn.*—F.T.A. 6, 1 : 945 ; Pax & K. Hoffm. l.c. 11 : 119, fig. 29, q.v. for varieties and forms.
The common castor-oil plant, widely cultivated and very variable ; several varieties are distinguished
with variously prickly fruits. (See Appendix, p. 160.)

50. TETRACARPIDIUM Pax in Engl. Bot. Jahrb. 26 : 329 (1899). *Angostylidium* (Müll. Arg.) Pax & K. Hoffm. in Engl. Pflanzenr. Euph. 9 : 17 (1919).

Leaves ovate, abruptly acuminate, rounded at base, about 10 cm. long, 4·5–5 cm.
broad, crenulate, 3-nerved at base, glabrous ; petiole up to 5 cm. long ; male flowers
in narrow raceme-like panicles about as long as the leaves, with 1–2 females near
the base ; pedicels jointed above the base ; sepals glabrous ; stamens about 40 ;
style very stout, at length quadrangular, with 4 spreading stigmas ; ovary 4-lobed ;
capsule about 7 cm. across, 4-winged, ridged between the wings .. *conophorum*

T. conophorum (*Müll. Arg.*) *Hutch. & Dalz.* F.W.T.A., ed. 1, 1 : 307 (1928) ; Kew Bull. 1928 : 300 (q.v. for
full synonymy). *Plukenetia conophora* Müll. Arg. (1864)—F.T.A. 6, 1 : 949 ; Chev. Bot. 582. *Ango-
stylidium conophorum* (Müll. Arg.) Pax & K. Hoffm. l.c. 9 : 17, fig. 5 (1919). *Cleidion preussii* (Pax) Bak.
(1910)—F.T.A. 6, 1 : 931. A climbing shrub 10–20 ft. long, glabrous ; male flowers deciduous, leaving
the females at the base of the raceme. Cultivated for its oil-rich fruits. Apparently not indigenous in
Sierra Leone.
    S.L.: *Sc. Elliot* 4118 ! Allen Town (Dec.) *Deighton* 3304 ! Port Loko (Jan.) *Deighton* 4041 ! Njala (Jan.)
Deighton 5018 ! **Dah.:** Saketé (Jan.) *Chev.* 22870 ! **S.Nig.:** Igbajo Dist. (Jan.) *Latilo* FHI 31764 !
Modakeke *Foster* 205 ! Sapoba (Aug.) *Kennedy* 1731 ! Ibuzo (Nov.) *Thomas* 1997 ! Aguku *Thomas* 888 !
Oban *Talbot* 615 ! 1371 ! 1384 ! 2307 ! **Br.Cam.:** R. Cameroon (Jan.) *Mann* 2202 ! Victoria (Feb.)
Maitland 184 ! 375 ! Barombi *Preuss* 420 ! **F.Po:** *Mildbr.* 6968. Also in French Cameroons, Spanish
Guinea, Gabon and Belgian Congo. (See Appendix, p. 164.)

51. TRAGIA Linn.—F.T.A. 6, 1 : 964 ; Pax & K. Hoffm. in Engl. Pflanzenr. Euph. 9 : 32 (1919).

Female flower on a long slender pedicel as long as the male raceme ; female calyx hardly
accrescent in fruit ; leaves ovate-oblong, rounded or subcordate at the base, acute
at the apex, 5–7·5 cm. long, 2–4 cm. broad, serrate, slightly bristly ; fruits deeply
3-lobed, 7 mm. broad, bristly-pilose .. .. 1. *volubilis*
Female flowers on very short pedicels (sometimes extending in fruit) ; female calyx
usually accrescent in fruit :
Leaves linear or linear-lanceolate, sessile or subsessile ; inflorescence long and slender,
15 cm. or more :

Racemes unisexual ; leaves 4–5 cm. long, 1 cm. broad, nearly glabrous, slightly
   serrulate near the apex ; stems glabrescent   ..   ..   ..   2. *akwapimensis*
Racemes bisexual ; leaves acute, 4–5 cm. long, 1 cm. broad, softly pubescent on both
   sides and with stinging hairs, sharply serrate ; stems pilose ; female calyx-segments
   3, pectinately toothed   ..   ..   3. *wildemanii*
Leaves lanceolate to ovate-elliptic, distinctly petiolate ; inflorescence usually rather
   short :
Female calyx-lobes 3 (rarely 4–5) in a single series, orbicular-ovate, pectinate all
   round, densely setose :
Leaves ovate, deeply cordate at base, long-caudate-acuminate, up to 12 cm. long
   and 6·5 cm. broad, margins rather coarsely serrate, densely setose-pilose on both
   surfaces ; petioles up to 5 cm. long   ..   ..   4. *senegalensis*
Leaves triangular-lanceolate, widely cordate at base, acute but scarcely acuminate
   at apex, 6–14·5 cm. long, 1–4·2 cm. broad, margins serrate, often only denticulate
   in the upper part of the leaves, setose-pubescent, very shortly so above, more or
   less glabrescent ; petioles up to 2·5 cm. long ..   ..   5. *vogelii*
Female calyx-lobes 6, usually more or less distinctly 2-seriate :
Calyx-lobes of female flowers without a foliaceous apex, deeply pectinate :
Leaves sharply serrate ; stem, petioles and leaves with pungent setose hairs ;
   leaves ovate, cordate at base, acutely acuminate, 5–14 cm. long, 2·5–9 cm. broad ;
   petioles 2–10 cm. long   ..   ..   6. *benthami*
Leaves subentire ; stem, petioles and leaves without pungent setose hairs, minutely
   puberulous or glabrescent ; leaves oblong-lanceolate or ovate-elliptic, cordate
   or subhastate at base, acutely acuminate, 8–20 cm. long, 2·5–8 cm. broad ;
   petioles 2·5–6 cm. long   ..   ..   7. *preussii*
Calyx-lobes of female flowers with a foliaceous undivided apex, entire or variously
   lobed or pectinate below :
Female calyx-lobes pectinately divided on each side into 4–6 teeth below the
   foliaceous apical portion :
Lateral teeth of female calyx-lobes bristly with numerous white stinging hairs ;
   leaves ovate-oblong, widely cordate, acutely acuminate, slightly toothed,
   6–13 cm. long, 2·5–6 cm. broad, bristly pubescent mainly on the nerves
                         8. *tenuifolia*
Lateral teeth of female calyx-lobes setose but not bristly ; leaves ovate-lanceolate,
   cordate at base, acute at apex, entire or undulate, 6–10 cm. long, 2–4·5 cm.
   broad, bristly pilose on both surfaces   ..   ..   9. *chevalieri*
Female calyx-lobes with 0–1 (–2) teeth on each side below the foliaceous apical
   portion :
Claw of female calyx-lobes at least twice as long as the terminal 1 mm. diam.
   foliaceous apical portion which is entire or with 1 tooth on each side at base,
   the calyx-lobes 2·5–3·5 mm. long in fruit ; leaves oblong-oblanceolate to ovate,
   cordate or truncate at base, rather long-acuminate, margin coarsely and
   irregularly toothed or subentire, 4–12 cm. long, 1·5–5·3 cm. broad, bristly
   pilose on both surfaces   ..   ..   10. *laminularis*
Claw of female calyx-lobes less than twice as long as the 3–4 (–6) mm. broad
   terminal foliaceous apical portion, the calyx-lobes 6–10 (–13) mm. long in fruit :
Leaves glabrous except for a few appressed hairs on the nerves above and very
   sparingly pubescent beneath ; ovate-elliptic, deeply cordate, obtusely
   acuminate, entire, 6–10 cm. long, 2–3·5 cm. broad ; female calyx-lobes obovate,
   entire, sparingly hirsute   ..   ..   11. *polygonoides*
Leaves pilose or pubescent on both surfaces :
Female calyx-lobes densely bristly but without teeth below the broad foliaceous
   apical portion ; leaves ovate-elliptic, to obovate-elliptic, rather shallowly
   cordate at base, acutely acuminate, margin shortly crenate or serrate,
   4·5–8 cm. long, 1·8–4·8 cm. broad, shortly bristly-pilose above, rather
   densely pubescent beneath   ..   ..   ..   12. *spathulata*
Female calyx-lobes setose but not bristly, with a tooth (or rarely 2) each side
   below the broad foliaceous apical portion ; leaves elliptic to obovate-elliptic
   or oblong-elliptic, more or less narrowed to the obtuse or narrowly cordate
   base, acuminate, margin usually serrate in the upper half, 6·5–16 cm. long,
   2·5–9·5 cm. broad, rather densely bristly-pilose on both surfaces 13. sp. *A*

1. **T. volubilis** *Linn.*—F.T.A. 6, 1 : 969 ; Pax & K. Hoffm. in Engl. Pflanzenr. Euph. 9 : 47–50 (as v ar
genuina), fig. 14 ; Chev. Bot. 583. A slender twiner armed with stinging hairs.
S.L.: Kurusu (Apr.) *Sc. Elliot* 5514 ! Rowala (July) *Thomas* 1156 ! Sewa River ferry, Nganyama
(fl. & fr. Nov.) *Deighton* 4660 ! **Iv.C.:** Soubré *Chev.* 19134. Zagoué to Soucourala *Chev.* 21577. **S.Nig.:**
Chama [1] *Beauvois.* Awka *Thomas* 1325 ; 1332. **F.Po:** (Dec.) *Mann* 75 ! Extends to Belgian Congo
and Angola ; also widespread in tropical America from whence it has probably been introduced to
Africa. (See Appendix, p. 165.)

[1] Beauvois says Chama is in the kingdoms of Warri and Benin ; see footnote on p. 638.

2. **T. akwapimensis** *Prain*—F.T.A. 6, 1 : 992 ; Pax & K. Hoffm. l.c. 9 : 87. Stems suberect from a woody base, 1½–2 ft. high, branched, sparingly armed with stinging hairs ; dioecious.
G.C.: Aburi *Anderson* 54 !

3. **T. wildemanii** *Beille*—F.T.A. 6, 1 : 993 ; Pax & K. Hoffm. l.c. 9 : 92 ; Chev. Bot. 583. Erect from a woody base, 1–1½ ft. high, sparingly branched, softly pubescent and armed with stinging hairs.
Fr.Sud.: Katou (Mar.) *Chev.* 539. Fr.G.: Timbo t˄ Farana *Chev.* 13307.

4. **T. senegalensis** *Müll. Arg.*—F.T.A. 6, 1 : 999, partly ; Pax & K. Hoffm. l.c. 9 : 92, partly ; Chev. Bot. 583 ; F.W.T.A., ed. 1, 1 : 308, partly. *T. angustifolia* of Chev. Bot. 583. A twiner, in savannah regions ; flowers yellow-green with stinging hairs on inflorescence.
Gam.: Albreda, in ricefields *Leprieur* ! Fr.G.: Timbo *Maclaud* 30 ! S.L.: Musaia (Mar . July, Aug., Dec.) *Deighton* 4482 ! 4798 ! 5395 ! *Jordan* 497 ! G.C.: Salaga *Krause* ! Togo: Kekejandé *Kersting* A. 615. Kete Krachi *Zech* 301. Dah.: Massé to Kétou, Zagnanado (fl. & fr. Feb.) *Chev.* 23009 ! Atacora Mts. *Chev.* 23943.

5. **T. vogelii** *Keay* in Kew Bull. 1955 : 139. *T. angustifolia* Benth. (1849)—F.T.A. 6, 1 : 999 ; Pax & K. Hoffm. l.c. 9 : 93 ; F.W.T.A., ed. 1, 1 : 308 ; not of Nutt. (1837). *T. senegalensis* of F.T.A. 6, 1 : 999, partly ; of Pax & K. Hoffm. l.c. 9 : 92, partly ; of Chev. Bot. 583 ; of F.W.T.A., ed. 1, 1 : 308, partly ; not of Müll. Arg. Erect, then scrambling or twining stems from a perennial stock ; in savannah, especially on rocky hillsides.
N.Nig.: Nupe *Barter* 1732 ! Kontagora (Jan.) *Meikle* 1080 ! Addaenda, R. Niger (Sept.) *T. Vogel* 108 ! Lokoja (fl. & fr. Sept.) *Parsons* 45 ! *Jones* FHI 5000 ! Yola (fl. & fr. Apr.) *Dalz.* 157 ! S.Nig.: Lagos *Dawodu* 205 ! Also in French Cameroons.

6. **T. benthami** *Bak.*—F.T.A. 6, 1 : 984. *T. cordifolia* of Benth. (1849)—Pax & K. Hoffm. l.c. 9 : 76 ; not of Vahl (1790). *T. kassiliensis* Beille (1910)—Chev. Bot. 583. Herbaceous climber or trailer, with stinging hairs.
Iv.C.: Gottoro to Diahbo (July) *Chev.* 22037 ! Kassilé (Jan.) *Chev.* 17050 ! G.C.: Cape Coast (July) *T. Vogel* 53 ! Aburi (July–Aug.) *Deighton* 3383 ! *Irvine* 845 ! Owabi (fl. & fr. May) *Andoh* FH 4202 ! Togo: Misahöhe *Baumann* 283. Dah.: Adja Onéré *Le Testu* 11 ; 12. S.Nig.: Abeokuta *Millen* 80 ! Isijere, Ibadan (fl. & fr. Nov.) *Keay* FHI 28140 ! Oban *Talbot* 613 ! Br.Cam.: Cam. Mt., 1,000 ft. (fl. & fr. Dec.) *Mann* 1255 ! Buea (fr. Nov.–Jan.) *Maitland* 238 ! *Migeod* 104 ! F.Po: Clarence (fr. Oct.) *T. Vogel* 26 ! Also in French Cameroons and extends to A.-E. Sudan, Uganda, Belgian Congo, Angola and E. Africa.

7. **T. preussii** *Pax*—F.T.A. 6, 1 : 989 ; Pax & K. Hoffm. l.c. 9 : 75. Stems twining ; apparently without stinging hairs.
S.Nig.: Oban *Talbot* 624 ! Br.Cam.: Victoria *Winkler* 343. Barombi *Preuss* 467 ! Also in Belgian Congo.

8. **T. tenuifolia** *Benth.*—F.T.A. 6, 1 : 973 ; Pax & K. Hoffm. l.c. 9 : 96. Stems twining ; flowers greenish, the female calyx-lobes with white stinging hairs.
Fr.G.: Dyeke (fl. & fr. Oct.) *Baldwin* 9652 ! S.L.: Bagroo R. (fl. & fr. Apr.) *Mann* 834 ! Kambia (Jan.) *Sc. Elliot* 4371 ! Yonibana (Nov.) *Thomas* 4731 ! 4838 ! Gola Forest, Gaura (fl. & fr. Dec.) *King* 935 ! Gegbwema—Gorahun road, Tunkia (fl. & fr. Dec.) *Deighton* 3826 ! Lib.: Nékabozu, Vonjama (Oct.) *Baldwin* 9970 ! Pallilah, Gbanga (fl. & fr. Sept.) *Baldwin* 13227 ! G.C.: Abetifi (fl. & fr. Jan.) *Irvine* 1696 ! Togo: Bismarckburg *Kling* 140. Misahöhe *Baumann* 534. N.Nig.: Neill's Valley, Jos (fl. & fr. June) *Lely* 265 ! S.Nig.: 70 miles east of Lagos *Lamborn* 260 ! Ibadan (fl. & fr. Feb.) *Meikle* 1130 ! Okomu F.R. (Dec.) *Brenan* 8435 ! Onitsha (fl. & fr. Dec.) *Thomas* 1855 ! Br.Cam.: Johann-Albrecht-shöhe *Büsgen* 52. Bonge *Dusen* 318. Also in French Cameroons, Gabon, Belgian Congo, Uganda and S. Tomé.

9. **T. chevalieri** *Beille*—F.T.A. 6, 1 : 972 ; Pax & K. Hoffm. l.c. 9 : 95 ; Chev. Bot. 583. A slender twiner armed with stinging hairs.
Iv.C.: Guidéko (June) *Chev.* 16471 ! S.Nig.: Ajilite *Millen* 172 ! Oban *Talbot* 1494 !
[The type specimen is extremely poor, and the other two specimens cited above which were originally identified by Prain for the F.T.A. are little better. Better material is needed to decide whether this is really a good species.]

10. **T. laminularis** *Müll. Arg.*—F.T.A. 6, 1 : 971 ; Pax & K. Hoffm. l.c. 99. Stem slender, woody and erect below, twining above, armed with stinging hairs.
S.L.: *Afzelius* ; *Smeathmann* ! Hill Station, Freetown (fl. & fr. May) *Deighton* 1126 ! Sugar Loaf Mt. (fl. & fr. Nov.) *Deighton* 5622 !

11. **T. polygonoides** *Prain*—F.T.A. 6, 1 : 970 ; Chev. Bot. 583 ; Pax & K. Hoffm. l.c. 9 : 98. A slender twiner without stinging hairs.
Iv.C.: Bouroukrou (Dec., Jan.) *Chev.* 16860.

12. **T. spathulata** *Benth.*—F.T.A. 6, 2 : 971 ; Pax & K. Hoffm. l.c. 9 : 98, fig. 22, A–B. A slender pubescent twiner armed with stinging hairs ; capsule densely bristly.
S.L.: Njala (Nov.) *Deighton* 2548 ! Lib.: Gbanga (fl. & fr. Sept.) *Linder* 702 ! G.C.: Cape Coast (July) *T. Vogel* 8 ! Odumase, Krobo Plains (fl. & fr. Aug.) *Johnson* 1089 ! Aburi (fl. & fr. Aug.) *Deighton* 3396 ! Ananekurom (fl. & fr. Jan.) *Irvine* 634 ! Gida Abdulker *Krause* ! Togo: Misahöhe *Baumann* 450 (partly). S.Nig.: Abeokuta *Irving* 72 ! Olokemeji (fl. & fr. Apr., Aug.) *Foster* 278 ! *Symington* FHI 5065 !

13. **T. sp. A.** Possibly this is **T. mildbraediana** Pax & K. Hoffm. l.c. 96, fig. 22, C–E (1919), but the type has presumably been destroyed and the female calyx-lobes in our material do not match the figure in the Pflanzenreich very well. A twiner with bristly-pilose stems and leaves, stinging : in forest vegetation.
S.L.: Bumbuna (fl. & fr. Oct.) *Thomas* 3174 ! Mabonto (Oct.) *Thomas* 3570 ! Bobobu, Giewahun (fl. & fr. Dec.) *Deighton* 3808 ! Peri, Gaura (fl. & fr. Oct.) *Deighton* 5199 ! 5200 ! Lib.: Soplima, Vonjama (fl. & fr. Nov.) *Baldwin* 10019 ! Vahun, Kolahun (fl. & fr. Nov.) *Baldwin* 10213 ! S.Nig.: Akure F.R. (fl. & fr. Oct.) *Keay* FHI 25478 ! Owam F.R., Benin (fl. & fr. Nov.) *Meikle* 515 ! Okomu F.R., Benin (fl. & fr. Jan.) *Brenan* 8853 !

Imperfectly known species.

T. monadelpha *Schum. & Thonn.* Beskr. Guin. Pl. 404 (1827) ; Pax & K. Hoffm. l.c. 9 : 100. The type specimen could not be traced at Copenhagen in 1954.
G.C.: Aquapim *Thonning.*

52. DALECHAMPIA Linn.—F.T.A. 6, 1 : 952 ; Pax & K. Hoffm. in Engl. Pflanzenr. Euph. 12 : 3 (1919).

Involucral bracts ovate, entire ; leaves ovate and not lobed or deeply 3-lobed ; stipules subulate-lanceolate ; female sepals accrescent, pinnatifid .. .. 1. *ipomoeifolia*
Involucral bracts 3-lobed and serrulate ; stipules ovate-lanceolate ; female sepals as above .. .. 2. *scandens*

1. **D. ipomoeifolia** *Benth.*—F.T.A. 6, 1 : 953 ; Pax & K. Hoffm. in Engl. Pflanzenr. Euph. 12 : 39 ; Chev. Bot. 584. A slender twiner with pubescent stems and flowers in close clusters within two leafy bracts.
S.L.: Bumbuna (Oct.) *Thomas* 3752 ! Njala (fr. Jan.) *Deighton* 2965 ! Segbwema (Dec.) *Deighton* 3463 ! Lib.: Monrovia (fl. & fr. July) *Linder* 23 ! Yratoke, Webo (July) *Baldwin* 6252 ! Iv.C.: Bouroukrou *Chev.* 16783. Zaranou *Chev.* 17637. G.C.: Princes (fr. Jan.) *Akpabla* 764 ! Sutah, Ashanti *Cummins*

48! Achimota (fr. Aug.) *Irvine* 861! **Togo:** Kpandu *Robertson* 69! **S.Nig.:** Lagos (Dec.) *Dalz.* 1270!
Hagerup 836! Lower Niger *T. Vogel* 9! Degema *Talbot* 3802! Eket *Talbot* 3365! Oban *Talbot* 683!
Also in French Cameroons, Spanish Guinea, Gabon and Belgian Congo.
2. **D. scandens** *Linn.* var. **cordofana** (*Hochst. ex Webb*) *Müll. Arg.* in DC. Prod. 15, 2 : 1245 (1866) ; F.T.A.
6, 1 : 954 (incl. var. *parvifolia*) ; Pax & K. Hoffm. l.c. 10 : 34 ; F. W. Andr. Fl. Pl. A.-E. Sud. 2 : 61,
fig. 22. *D. cordofana* Hochst. ex Webb (1849). Like the last but with leaves always deeply 3–5-partite.
Sen.: Dakar *Adanson.* Dagana *Perrottet* 740. Richard Tol *Lelièvre.* Also in Cape Verde Islands, Chad
area and widespread in north-eastern, central and southern tropical Africa and in Arabia.

53. MANIHOT Mill.—F.T.A. 6, 1 : 839 ; Pax & K. Hoffm. in E. & P. Pflanzenfam. 19C : 174 (1931).

Leaves peltate ; fruits without wings or ridges ; bracts small .. 1. *glaziovii*
Leaves not peltate ; fruits with 6 distinct wings or ridges :
 Bracts large and leafy, at first concealing the flowers :
  Fruits dehiscent.. .. .. 2. *piauhyensis*
  Fruits indehiscent .. .. 3. *teissonnieri*
 Bracts small, not longer than the pedicels :
  Lobes of the leaves sometimes more or less lobulate, the middle lobe the longest with
   the midrib continued to the top, abruptly acuminate ; rubber plant 4. *dichotoma*
  Lobes of the leaves entire ; food plants .. .. 5. *esculenta*

1. **M. glaziovii** *Müll. Arg.*—F.T.A. 6, 1 : 839 ; Holl. 4 : 598 ; Chev. Bot. 574. The Ceara rubber tree, native of
Brazil, cultivated for rubber and planted as an ornamental tree. (See Appendix, p. 150.)
2. **M. piauhyensis** *Ule*—F.T.A. 6, 1 : 840 ; Holl. 4 : 600. A shrub or small tree, Piauhy rubber, native of
Brazil ; cultivated.
3. **M. teissonnieri** *A. Chev.*—F.T.A. 6, 1 : 841 ; Chev. Bot. 574. Similar to the last ; cultivated in French West
Africa.
4. **M. dichotoma** *Ule*—F.T.A. 6, 1 : 841 ; Holl. 4 : 597. The Jequié rubber tree, native of Brazil, cultivated
as a rubber plant.
5. **M. esculenta** *Crantz* Inst. Rei Herb. 1 : 167 (1766). *M. utilissima* Pohl (1828)—F.T.A. 6, 1 : 842 (incl. vars.) ;
Holl. 4 : 600 ; Chev. Bot. 574 ; F.W.T.A., ed. 1, 1 : 300 (incl. var.). A shrub 6–10 ft. high, glabrous,
with digitate leaves glaucous beneath, and tuberous roots ; very variable in colour of bark, tint of foliage,
etc. Cassava, a cultivated food-plant ; native of Brazil. (See Appendix, p. 150.)

54. KLAINEANTHUS Pierre ex Prain—F.T.A. 6, 1 : 963 ; Pax & K. Hoffm. in Engl. Pflanzenr. Euph. 7 : 408 (1914).

Tree ; branchlets pubescent when young, at length glabrous ; leaves oblong or obovate,
obtuse to cuneate at base, shortly acuminate, entire, 10–32 cm. long, 4–12 cm. broad,
glabrous ; petiole 2·5–5 cm. long, distinctly pulvinate and geniculate at apex ;
panicles axillary and terminal, puberulous and often with long spreading hairs as well ;
flowers about 2 mm. long ; sepals strongly imbricate ; petals absent ; male flower
with 8 or 10 stamens in two series, surrounded by a ring of 4 or 5 extrastaminal glands
alternate with the outer stamens and opposite the sepals, rudimentary ovary usually
3-lobed ; female flowers with cupular, denticulate and 4–5-lobed disk, staminodes
often present, ovary 3-celled, style 3, deeply 2-partite ; capsule trigonous-sub-
orbicular about 1·5 cm. long and broad, glabrous and shining outside, dehiscent ;
seeds solitary in each cell, clothed with a yellow fleshy aril .. .. *gaboniae*

K. gaboniae *Pierre ex Prain*—F.T.A. 6, 1 : 963 ; Hook. Ic. Pl. t. 2985 ; Pax & K. Hoffm. l.c. A forest tree,
30–90 ft. high, with rough, fluted bole ; flowers white ; fruits bright red.
 S.Nig.: Sapoba *Kennedy* 2268! 2280! 2312! Oban (fr. May) *Ujor* FHI 31800! Uwet Odot F.R., Itu (fr.
May) *Onochie* FHI 33223! Obudu Dist. (May) *Latilo* FHI 30922! Ikom Dist. (fr. May, June) *Latilo*
FHI 30998! 31827! **Br.Cam.:** S. Bakundu F.R. (fr. Aug.) *Olorunfemi* FHI 30711! Near Mungo R.,
Kumba Dist. (Apr.) *Olorunfemi* FHI 30532! Also in French Cameroons, Gabon and S. Tomé.

55. GELONIUM Roxb.—F.T.A. 6, 1 : 947 ; Pax & K. Hoffm. in Engl. Pflanzenr. Euph. 4 : 14 (1912).

A glabrous shrub or treelet with flexuous branchlets, angular when young ; leaves
sessile or very shortly petiolate, oblanceolate, obovate or oblong-elliptic, rounded to
subcordate at base, shortly acuminate to subcaudate-acuminate, acumen itself obtuse,
10–18 cm. long, 3·5–8 cm. broad ; flowers dioecious, subsessile, in small, leaf-opposed
fascicles ; male flowers 3–6 per fascicle, sepals 5–6, imbricate, the outer 3 suborbicular
and about 3 mm. diam. ; stamens 20–30, with short filaments ; female flowers
apparently 1–2 per fascicle ; ovary glabrous, 3-celled, ovules solitary in each cell ;
capsule depressed 3-lobed, about 1·4 cm. diam., seeds suborbicular *occidentale*

G. occidentale *Hoyle* in Kew Bull. 1935 : 258. A forest shrub or treelet, to 10 ft. high ; outer sepals green, inner
sepals and stamens creamy-white.
 G.C.: Mampong Scarp (fl. & fr. Feb., Dec.) *Vigne* FH 2754! 2755! 4092! 4093! Ofin Headwaters F.R.
(Oct.) *Vigne* FH 3124! **N.Nig.:** Kurmin Damisa, Jemaa (Nov.) *Keay & Onochie* FHI 21725! **S.Nig.:**
Etemi Odo, Ijebu Ode Prov. (Nov.) *Tamajong* FHI 20992! Akure F.R. (Nov.) *Keay & Trochain* FHI
25605! Owo (May) *Jones* FHI 3567!

56. HAMILCOA Prain—F.T.A. 6, 1 : 1000 ; Pax & K. Hoffm. in Engl. Pflanzenr. Euph. 7 : 419 (1914).

A glabrous shrub ; leaves nearly sessile or long-petiolate on the same branch, entire or
slightly 3-lobed ; lamina elliptic- or ovate-oblong to lanceolate, rounded to subacute

at base, caudate-acuminate, 10–26 cm. long, 3·5–10 cm. broad ; petiole up to 6 cm. long ; racemes axillary, 6–8-flowered, up to 4 cm. long ; male sepals imbricate, about 3 mm. long, petals and rudimentary ovary absent ; stamens 18–30, subsessile ; female sepals 1 mm. long, disk urceolate, ovary 3-celled, glabrous, style thick, with 3 free stout stigmas ; capsule deeply 3-lobed, about 1·2 cm. long and 2·2 cm. broad ; seeds solitary in each cell, large, with smooth mottled testa .. .. .. *zenkeri*

H. zenkeri (*Pax*) *Prain*—F.T.A. 6, 1 : 1000 ; Hook. Ic. Pl. t. 3009 ; Pax & K. Hoffm. l.c. *Plukenetia zenkeri* Pax (1909). A scandent forest shrub, stems 20–30 ft. high, or tree to 45 ft. high (*fide* Brenan), with white latex.
Br.Cam.: S. Bakundu F.R., Kumba (fl. Mar., fr. Mar., Aug.) *Brenan* 9430 ! *Olorunfemi* FHI 30705 ! Also in French Cameroons.

57. TETRORCHIDIUM Poepp. & Endl. Nov. Gen. & Sp. 3 : 23 (1842) ; Pax in Engl. Pflanzenr. Euph. 4 : 29 (1912) ; Pax & K. Hoffm. l.c. 14 : 53 (1919). *Hasskarlia* Baill. (1860)—F.T.A. 6, 1 : 846 ; Pax l.c. 7 : 416 (1914).

Plants entirely glabrous ; leaves of the upper branchlets alternate, lower leaves sometimes opposite, leaf-base cuneate ; inflorescences leaf-opposed :
Male inflorescences 3·5–9·5 cm. long, flowers crowded ; leaves oblong-elliptic to obovate, cuspidate-acuminate, 8–22 cm. long, 4–11 cm. broad, margin entire or denticulate ; female flowers solitary or few ; shrub, or a tree to 25 m. high .. 1. *didymostemon*
Male inflorescences up to 6 mm. long, flowers crowded ; leaves obovate to oblanceolate, long-acuminate, 3–8 cm. long, 1–3 cm. broad, markedly serrate in the upper half ; female flowers not known ; shrub, or a tree to 6 m. high .. .. 2. *minus*
Plants pubescent on branchlets, petioles and inflorescences ; leaves mostly opposite, elongate-oblong-elliptic to oblong-elliptic, cuneate to obtuse at base, long-acuminate, 8–19 cm. long, 2·5–8 cm. broad, margin irregularly serrate or undulate ; male inflorescences 4·5–11 cm. long, interpetiolar, the spike interrupted ; female flowers not known ; shrub, to 2·5 m. high .. .. 3. *oppositifolium*

1. **T. didymostemon** (*Baill.*) *Pax & K. Hoffm.* in Engl. Pflanzenr. Euph. 14 : 53 (1919) ; Aubrév. Fl. For. C. Iv. 2 : 77, t. 155. *Hasskarlia didymostemon* Baill. (1860)—F.T.A. 6, 1 : 846 ; Chev. Bot. 575. A shrub or a tree up to 80 ft. high with zigzag branchlets, pale yellowish-green foliage and slender yellow male flower-spikes ; in secondary forest regrowth.
Fr.G.: Fouta Djalon (Apr., May) *Heudelot* 835 ! **S.L.:** Berria, Falaba (Mar.) *Sc. Elliot* 5430 ! Freetown (Mar.) *Dalz.* 960 ! Zimmi (Nov.) *Deighton* 383 ! Rokupr (fl. & young fr. May) *Jordan* 242 ! **Dukwia** R. (fr. May) *Cooper* 437 ! Gbanga (Sept.) *Linder* 806 ! Kingsville (Mar.) *Bequaert* 163 ! **Iv.C.:** Bouroukrou (fr. Dec., Jan.) *Chev.* 16862 ! Bingerville (fr. Feb.) *Chev.* 15321 ! 16219 ! **G.C.:** Kumasi (Mar.) *Chipp* 138 ! Obuasi (Nov.) *Chipp* 589 ! Oda (Sept.) *Fishlock* 52 ! **Togo:** Misahöhe (Nov.) *Mildbr.* 7227 ! Amedzofe (Feb.) *Irvine* 176 ! **S.Nig.:** Nikrowa (Dec., Jan.) *A. F. Ross* 240 ! *Brenan* 8400 ! Brass *Barter* 1848 ! Eket *Talbot* ! Obanliko, Obudu (fl. & fr. Apr.) *Latilo* FHI 30904 ! **Br.Cam.:** Bulifambo, Buea (Mar.) *Maitland* 594 ! S. Bakundu F.R. (Mar.) *Onochie* FHI 31191 ! 31200 ! **F.Po:** *Mildbr.* 7051. Also in French Cameroons, Gabon, Belgian Congo, Uganda, Tanganyika, Angola and S. Tomé. (See Appendix, p. 164.)
2. **T. minus** (*Prain*) *Pax & K. Hoffm.* l.c. (1919). *Hasskarlia minor* Prain (1912)—F.T.A. 6, 1 : 1055. A glabrous shrub or small tree, to 20 ft. high, with very slender branchlets. There are no collectors' notes to support Prain's statement that this plant is a herb.
S.L.: near Pendembu (Apr.) *Sc. Elliot* 5680 ! **S.Nig.:** Okeigbo, Ondo Prov. (Feb.) *Jones & Onochie* FHI 14714 ! Benin (Feb.) *Onochie* FHI 31172 ! Degema *Talbot* 3795 !
3. **T. oppositifolium** (*Pax*) *Pax & K. Hoffm.* l.c. (1919). *Hasskarlia oppositifolia* Pax (1909)—F.T.A. 6, 1 : 847 ; Chev. Bot. 576. A forest shrub, to 9 ft. high ; flowers cream.
Lib.: Sinoe Basin *Whyte* ! Monrovia *Dinklage* 2213. Jaurazon (Mar.) *Baldwin* 11450 ! Nyaake, Webo (June) *Baldwin* 6108 ! **Iv.C.:** Mt. Copé, Grabo (July) *Chev.* 19695. **S.Nig.:** Oban *Talbot* 617 ! 640 ! 664 ! Kwa Falls, Calabar (Mar.) *Brenan* 9233 ! **Br.Cam.:** Kembong F.R., Mamfe (Mar.) *Onochie* FHI 32051 ! Also in Gabon.

58. PLAGIOSTYLES Pierre—F.T.A. 6, 1 : 170 & 1001 ; Pax & K. Hoffm. in Engl. Pflanzenr. Euph. 7 : 420 (1914).

Branchlets glabrous ; leaves oblong-elliptic, abruptly acuminate, slightly cuneate or rounded at base, 15–20 cm. long, 5–7 cm. broad, obscurely toothed, thinly coriaceous, markedly reticulate ; lateral nerves 5 pairs ; petiole 1–1·5 cm. long, pulvinate at base and apex ; flowers dioecious ; male racemes about 5 cm. long ; pedicels up to 8 mm. long ; stamens 15–21, anthers papillous ; fruit transversely oblong, 2 cm. across, with a lateral stigma .. .. *africana*

P. africana (*Müll. Arg.*) *Prain* in Kew Bull. 1912 : 107 ; Pax & K. Hoffm. l.c. 7 : 421. *Daphniphyllum africanum* Müll. Arg. (1864). *Plagiostyles klaineana* Pierre (1897)—F.T.A. 6, 1 : 171 & 1001. A forest tree, 30–50 ft. high, with pale yellow-green viscid sap.
S.Nig.: Oban *Talbot* ! **Br.Cam.:** S. Bakundu F.R. (fr. Mar.) *Brenan* 9330 ! Barombi *Preuss* 421 ! Johann-Albrechtshöhe *Staudt* 496. Also in French Cameroons, Spanish Guinea, Gabon, French and Belgian Congo.

59. SEBASTIANIA Spreng.—F.T.A. 6, 1 : 1007 ; Pax & K. Hoffm. in Engl. Pflanzenr. Euph. 5 : 88 (1912).

Shrub ; leaves oblong-lanceolate, caudate-acuminate, rounded at base, 6–13 cm. long, 3–5 cm. broad, minutely denticulate with glandular teeth ; male spikes branched, axillary, about ¼ as long as the leaves, female inflorescences mixed with the male ; stamens 2–4 ; ovary smooth .. .. 1. *inopinata*

Herb ; leaves linear, not acuminate, narrowed at base, 3–6 cm. long, up to 0·8 cm.
broad, closely serrulate and cartilaginous on the margin ; spikes simple, terminal or
leaf-opposed, short ; stamens 3 ; ovary and fruit muricate with 2 rows of bristles
on each cell   ..   ..   2. *chamaelea*

1. **S. inopinata** *Prain*—F.T.A. 6, 1 : 1007 ; Pax & K. Hoffm. in Engl. Pflanzenr. Euph. 5 : 120, fig. 23. A glabrous forest shrub, 8–15 ft. high.
   Br.Cam.: R. Cameroon (Jan.) *Mann* 755 ! 2225 ! Also in French Cameroons.
2. **S. chamaelea** (*Linn.*) *Müll. Arg.*—F.T.A. 6, 1 : 1008 ; Pax & K. Hoffm. l.c. 5 : 116 (as var. *africana*) ; Chev. Bot. 584. *Tragia chamaelea* Linn. (1753). A more or less woody herb, glabrous, branched, 1–2 ft. high.
   G.C.: Achimota (June) *Irvine* 569 ! **Togo:** Lome *Warnecke* 186 ! Agome *Schlechter* 14965 ! **Dah.:** Ouidah to Adjounja *Chev.* 23461. Agouagon *Chev.* 23519. **N.Nig.:** Nupe *Barter* 1492 ! Zungeru (July) *Dalz.* 72 ! Lokoja (Sept.) *Ansell* ! *Parsons* 7 ! Abinsi (fr. Nov.) *Dalz.* 748 ! Yola (fr. Sept.) *Dalz.* 160 ! **S.Nig.:** Lagos *Dawodu* 287 ! Ibadan (fl. & fr. July) *Onochie* FHI 7566 ! Awka Dist. *Thomas* 711 ! 1064 ! Also in French Cameroons and Ubangi-Shari ; widely spread throughout S.E. Asia to N. Australia.

60. SAPIUM Browne—F.T.A. 6, 1 : 1009 ; Pax & K. Hoffm. in Engl. Pflanzenr. Euph. 5 : 199 (1912).

Ovary and fruit normally 2-celled, rarely 3- or 1-celled ; fruits tardily dehiscent or
indehiscent ; trees 7–35 m. high :
Fruits 2-lobed, about 8–10 mm. broad, on pedicels up to 3·3 cm. long ; female flowers
  on pedicels 3–5 mm. long at the base of the more or less puberulous 5–12 cm. long
  spikes ; leaves elliptic or oblong-elliptic, obtuse to rounded at base, acute or shortly
  acuminate at apex, 4–15 cm. long, 1·6–7·2 cm. broad, margin crenate, with 1 or 2
  pairs of sessile glands near the base ; petioles up to 1 cm. long ..   1. *ellipticum*
Fruits not 2-lobed, ovoid to subglobose, 15–18 mm. broad, on short pedicels ; female
  flowers subsessile or very shortly pedicellate at the base of puberulous 6–9 cm. long
  spikes ; leaves oblong-elliptic, obtuse to rounded at base, subacuminate, 7·5–15 cm.
  long, 4–6·5 cm. broad, margin crenate, with a pair of sessile glands near the base ;
  petioles 1–3 cm. long ..   ..   2. *aubrevillei*
Ovary and fruit normally 3-celled, dehiscent ; glabrous shrubs and suffrutices :
Leaves long-acuminate ; forest shrubs :
Valves of the fruits each with a projecting dorsal wing or horn ; male flowers several
  to each bract ; leaves elongate-elliptic to oblanceolate-elliptic, rounded at base,
  6·5–21 cm. long, 2·5–5·5 cm. broad, margin entire or subentire ; petioles 0·5–12 cm.
  long ; fruits about 1·2 cm. diam. ..   ..   3. *cornutum*
Valves of the fruits without dorsal wings or horns ; male flowers solitary in each
  bract ; leaves oblanceolate rounded to cuneate at base, caudate-acuminate, 5–22
  cm. long, 4–7 cm. broad, margin crenulate or subentire ; petioles 1–1·6 cm. long ;
  fruits about 1 cm. diam. ..   ..   4. *guineense*
Leaves not or only slightly acuminate ; savannah suffrutices ; male flowers solitary in
each bract ; fruits not horned :
Leaves elliptic to obovate-elliptic :
Lateral nerves about 12 on each side of midrib ; leaves 4–13 cm. long, 3–5 cm. broad,
  margin denticulate, biglandular at the base ; spikes 1–5 cm. long ; fruits about
  2 cm. diam. ..   ..   5. *grahamii*
Lateral nerves 3–5 on each side of midrib, obscure ; leaves 2·5–3 cm. long, 1·5–2 cm.
  broad, margin minutely denticulate, not glandular at the base ; spikes 4–6 cm.
  long ..   ..   6. *faradianense*
Leaves linear or oblong-linear, acute, 2–4·5 cm. long, up to 1 cm. broad, nerves
scarcely visible, margin cartilaginous and undulate with a pair of glands on each
side at the base ; spikes slender, up to 10 cm. long ; fruits 7–8 mm. diam.
7. *dalzielii*

1. **S. ellipticum** (*Hochst.*) *Pax* in Engl. Pflanzenr. Euph. 5 : 253, fig. 49 (1912) ; Chev. Bot. 584, partly ; Aubrév. Fl. For. C. Iv. 2 : 80, t. 157 ; Aubrév. Fl. For. Soud.-Guin. 196, t. 38, 5–8. *Sclerocroton ellipticus* Hochst. (1845). *Excoecaria manniana* Müll. Arg. (1864). *Sapium mannianum* (Müll. Arg.) Benth. ex Pax (1893)— F.T.A. 6, 1 : 1016. A tree, 25–120 ft. high with drooping branches ; flowers yellow-green ; fruits orange-green, tardily dehiscent ; in savannah and in streamside forest in both savannah and forest regions.
   Fr.G.: Farana (fl. Jan., fr. Mar.) *Chev.* 20441 ! *Sc. Elliot* 5342 ! Pita *Pobéguin* 864. **S.L.:** Kabala (Jan.) *King* 187b ! Kurusu (fr. Apr.) *Sc. Elliot* 5508 ! Musaia (fr. Apr.) *Deighton* 5476 ! **Iv.C.:** Ferkéssédougou, Bondoukou, Bobo Dioulasso & Fétékoro *fide* Aubrév. l.c. **G.C.:** S. Fomang Su F.R. (Jan.) *Vigne* FH 2693. **Togo:** Sokode *Kersting* 391 ! **Dah.:** Atacora Mts. *Chev.* 24151. **N.Nig.:** Kaduna (Dec.) *Meikle* 748 ! S. Zaria *Shaw* 104 ! Jos (Oct., Nov.) *Keay* FHI 21030 ! 21426 ! |ₛ **S.Nig.:** Shasha F.R. *Lancaster* 10 ! Ondo (Mar.) *Lamb* 115 ! **Br.Cam.:** Buea (fl. Dec., fr. Mar.) *Maitland* 116 ! 459 ! **F.Po:** *Mildbr.* 6783. Widespread in tropical Africa and in Natal. (See Appendix, p. 163.)
2. **S. aubrevillei** *Léandri* in Bull. Soc. Bot. Fr. 81 : 449 (1934) ; Aubrév. Fl. For. C. Iv. 2 : 80, t. 158. *S. ellipticum* of Chev. Bot. 584, partly (*Chev.* 16190). A forest tree of medium to large size, bole up to 4 ft. diam. ; fruit a drupe.
   Iv.C.: Mbago, Agnéby valley (fr. Feb.) *Chev.* 16190 ! Mt. Tonkoui *Aubrév.* 1004 ! Mt. Momy *Aubrév.* 1184. Taï *Aubrév.* 1210.
3. **S. cornutum** *Pax*—F.T.A. 6, 1 : 1013 ; Pax l.c. 5 : 246. A forest shrub, to 10 ft. high ; spikes pendulous, red ; fruits reddish.
   S.L.: Njala (fl. Feb., May, fr. Feb.) *Deighton* 635 ! 4700 ! Nganyama Tikonko (fr. Nov.) *Deighton* 5262 ! **Lib.:** Gbanga (fr. Sept.) *Linder* 563 ! Mecca, Boporo (fr. Nov.) *Baldwin* 10433 ! 10439 ! Also in French Cameroons, Spanish Guinea, French and Belgian Congo and Angola.
4. **S. guineense** (*Benth.*) *O. Ktze.*—F.T.A. 6, 1 : 1011. *Stillingia guineensis* Benth. (1849). *Excoecaria guineensis* (Benth.) Müll. Arg. (1863)—Pax l.c. 5 : 164 ; Chev. Bot. 585. *E. comoensis* (Beille) A. Chev. Bot. 584. A forest shrub, usually 2–5 ft. high.

S.L. : *Smeathmann*! Freetown (Dec.) *Sc. Elliot*! *Dalz.* 975 (partly)! Sugar Loaf Mt. (fl. & fr. Nov.) *Deighton* 5643! **Lib.**: Monrovia *Dinklage* 2418. **Iv.C.**: Bettié (Mar.) *Chev.* 17577! And other localities *fide* Chev. *l.c.* **S.Nig.**: Oban *Talbot* 666! **Br.Cam.**: Likomba (fl. & fr. Oct.) *Mildbr.* 10548! Johann-Albrechtshöhe *Staudt* 609. Also in French Cameroons, French and Belgian Congo and Principe.

5. **S. grahamii** *(Stapf) Prain* in F.T.A. 6, 1 : 1012 (1913) ; Chev. Bot. 584. *Excoecaria grahamii* Stapf (1906)—Pax l.c. 5 : 164 ; Chev. Bot. 585. A savannah suffrutex sending up half-woody stems to 2 ft. high from a creeping rootstock, with milky juice ; capsule pale, crustaceous, breaking up into three 2-valved cocci. **Fr.Sud.**: Pénia (May) *Chev.* 823. **Fr.G.**: Kollangui *Chev.* 12200. **Iv.C.**: Gampela to Ouagadougou *Chev.* 24661. **G.C.**: Gambaga (fl. & fr. Apr.) *Graham*! Kintampo (Mar.) *Dalz.* 37! Tumu (fl. & fr. Apr.) *Vigne* FH 3783! **Togo:** Gabotawe *Kersting* 276. Basari *Kersting* 605. **Dah.:** Djougou *Chev.* 23895! (See Appendix, p. 163.)

6. **S. faradianense** *(Beille) Pax* in Engl. Pflanzenr. Euph. 5 : 247 (1912) ; F.T.A. 6, 1 : 1012. *Excoecaria faradianense* Beille (1910)—Chev. Bot. 585. A small savannah suffrutex with slender woody stems. **Sen.:** Karakoro to Niana, Casamance (Feb.) *Chev.* 2611. **Fr.G.:** Faradiana (Mar.) *Chev.* 640!

7. **S. dalzielii** *Hutch.* in Kew Bull. 1917 : 234 ; Pax l.c. 17 : 204 (1924). A savannah suffrutex with slender, slightly branched 1–2 ft. high, erect stems, leafy from base to apex. **G.C.:** Tamale (fl. & fr. Dec.) *Adams & Akpabla* GC 4157! **N.Nig.:** Katsina Ala (June) *Dalz.* 749!

61. MAPROUNEA Aubl.—F.T.A. 6, 1 : 1002 ; Pax & K. Hoffm. in Engl. Pflanzenr. Euph. 5 : 174 (1912).

Leaves membranous, laxly reticulate, ovate or ovate-elliptic to lanceolate, 3·5–9·5 cm. long, 1·7–4 cm. broad, glaucous beneath ; petiole 7–10 mm. long, slender ; male inflorescences ovoid, 5 mm. long, shortly pedicellate, usually with 2 or more pedicellate female flowers below ; stamens 2 ; capsule globose, 3-lobed ; seeds rugulose, scarcely 4 mm. long, with a caruncle as large as the body of the seed ; tree of forest or savannah   1. *membranacea*

Leaves thinly coriaceous, closely and conspicuously reticulate, ovate, oblong-ovate, lanceolate or orbicular-ovate, 1·5–5·5 cm. long, 1·5–3 cm. broad, glaucous beneath ; petiole 7–15 mm. long, slender ; male inflorescences oblong, 3–12 mm. long, with 1–3 pedicellate female flowers below ; stamens 2–3 ; capsule globose, 3-lobed ; seeds smooth, 6–8 mm. long, with a caruncle less than ¼ the size of the body of the seed ; savannah tree ..   .. 2. *africana*

1. **M. membranacea** *Pax & K. Hoffm.* in Engl. Pflanzenr. Euph. 5 : 178, fig. 33 (1912) ; F.T.A. 6, 1 : 1002 ; Kennedy For. Fl. S. Nig. 82. A tree, up to 50 ft. high, usually in forest, but sometimes in savannah ; inflorescence and fruit red.
S.Nig.: Ajebandele, Omo F.R. (Apr.) *Onochie* FHI 15536! Sapoba *Kennedy* 2301! Udi Plateau (Feb., Mar.) *Ainslie*! *Cons. of For.* 102! Eket *Talbot* 3253! Akparabong to Adijinkpon, Ikom (young fr. June) *Latilo* FHI 31821! Also in French Cameroons, Spanish Guinea, Gabon and Belgian Congo.

2. **M. africana** *Müll. Arg.*—F.T.A. 6, 1 : 1004 ; Pax & K. Hoffm. l.c. 5 : 178 (incl. vars) ; Aubrév. Fl. For. Soud.-Guin. 198, t. 38, 2–4. A savannah tree, up to 20 ft. high, rarely more.
N.Nig.: Tambari Hill, Birnin Gwari *Keay* FHI 28053! Anara F.R., Zaria *fide* Keay. **Br.Cam.:** Mai Idoanu, Gashaka Dist. (Feb.) *Latilo & Daramola* FHI 34474! Also in French Cameroons, Ubangi-Shari and widespread in central and east Africa.

62. ANTHOSTEMA A. Juss.—F.T.A. 6, 1 : 607 ; Pax & K. Hoffm. in E. & P. Pflanzenfam. 19C : 207 (1931).

Leaves cuneate at base, elliptic with rounded margins, obtusely acuminate, 7·5–13 cm. long, 3–4·5 cm. broad, with numerous lateral nerves spreading at right angles ; flowers cymose ; cymes subsessile, axillary, rather densely flowered ; fruits deeply 3-lobed, nearly 2·5 cm. broad ..   .. 1. *senegalense*

Leaves rounded or very broadly cuneate at base, oblong with more or less straight margins, 8–15 cm. long, 2·5–4·5 cm. broad ; cymes rather densely flowered ; fruits up to 3 cm. broad ..   .. 2. *aubryanum*

1. **A. senegalense** *A. Juss.*—F.T.A. 6, 1 : 607 ; Chev. Bot. 553 ; Aubrév. Fl. For. C. Iv. 2 : 20–22. A forest tree with glabrous branchlets and rather leathery leaves ; juice milky.
Sen.: *Leprieur*! Sangalkans (young fr. Oct.) *Berhaut* 996! **Gam.:** Abuko (Jan.) *Dalz.* 8057! **Fr.Sud.:** Fangala Kouta (fl. Nov., fr. Dec.–Jan.) *Dubois* 218! Moussaia *Chev.* 406. Nono *Chev.* 429. **Fr.G.:** *Heudelot* 592! Tristao Isl. *Paroisse* 1! **S.L.:** Kessewe F.R. (fr. Feb.) *Macdonald* 6! Njala (Oct.) *Deighton* 1342! Ronietta (Nov.) *Thomas* 5416! Newton (Nov.) *T. S. Jones* 290! **Lib.:** Monrovia (Nov.) *Linder* 1409! Grand Bassa *Dinklage* 1769! Duport (Oct.) *Barker* 1442! **Iv.C.:** *fide* Chev. *l.c.* (See Appendix, p. 136.)

2. **A. aubryanum** *Baill.*—F.T.A. 6, 1 : 608 ; Aubrév.'l.c. 2 : 20–22, t. 127 (?). A forest tree, 30–40 ft. high or more, with glabrous branchlets ; leaves shining above ; juice milky ; in swamp forest.
Iv.C.: *fide* Aubrév. *l.c.* **G.C.:** W. Apollonia (Aug.) *Chipp* 324! Abosso (Aug.) *Vigne* FH 3027! **S.Nig.:** Ajilite to Addo *Millen* 178! Sapoba (Aug., Sept.) *Kennedy* 1737! *Ajayi* FHI 26923! Also in Gabon, Cabinda and Principe. (See Appendix, p. 136.)

63. DICHOSTEMMA Pierre—F.T.A. 6, 1 : 605 ; Pax & K. Hoffm. in E. & P. Pflanzenfam. 19C : 207 (1931).

A small tree or climber ; branchlets glabrous ; leaves obliquely oblong-lanceolate, shortly cuneate at base, acuminate, 7–16 cm. long, 2·5–6 cm. broad, entire, glabrous, glaucous beneath ; lateral nerves 6–8 pairs, looped ; panicles terminal, up to 40 cm. long, puberulous ; involucres cup-like, obtusely 4-angled, up to 5 mm. broad ; male flowers reduced to a single stamen ; fruits depressed-quadrangular, 4-lobed, 3 cm. broad, rusty-tomentellous ..   ..   *glaucescens*

D. glaucescens *Pierre*—F.T.A. 6, 1 : 606. A forest tree, to 40 ft. high, or said sometimes to be climbing ; with abundant milky juice ; flowers yellow.
S.Nig.: Oban (young fr. May) *Talbot* 2328! *Ujor* FHI 31792! New Ndebiji, Calabar (May) *Ejiofor* FHI 21890! **Br.Cam.:** Victoria to Kumba (fr. Aug.) *Dundas* FHI 8403! Also in French Cameroons, Spanish Guinea, Gabon, Cabinda and Belgian Congo. (See Appendix, p. 140.)

64. EUPHORBIA Linn.—F.T.A. 6, 1 : 470.[1]

*Herbs, annual or perennial, or suffrutices ; rarely fleshy, but never prickly :
Inflorescences appearing directly from the apices of fleshy underground or partially
    exposed stems 1–1·5 cm. thick ; peduncles 2–4 cm. long, involucres glabrous, 2–3 mm.
    diam. ; leaves appearing later, few, oblanceolate, up to 14 cm. long and 4 cm. broad
                                                                            21. *baga*
Inflorescences not as above :
†Leaves all strictly opposite, usually markedly unequal at base, up to 5 cm. long and
    2 cm. broad ; stipules present, mostly conspicuous ; mostly erect or prostrate
    annual herbs, but some perennial (see Nos. 1 and 13) :
Ovary and capsule hairy :
Involucres borne in dense, rather long-pedunculate, axillary and terminal leafless
    glomerules ; stems rather coarsely spreading-pilose with yellowish hairs over a
    shorter indumentum ; leaves obliquely ovate to lanceolate, rounded on one side,
    cuneate on the other at base, acute at apex, up to 5 cm. long and 2 cm. broad,
    serrate   ..   ..   2. *hirta*
Involucres solitary or on short leafy lateral branches :
Capsules pubescent only on the angles ; stems subglabrous except for a line of
    short curly hairs on the upper side ; seeds with minute sharp transverse ridges ;
    leaves broadly elliptic to elliptic-oblong, obtuse or rounded at apex, mostly
    4–8 mm. long   ..   ..   12. *prostrata*
Capsules with the hairs not confined to the margins ; stems pubescent or
    tomentose :
Stems erect, sparingly branched ; leaves lanceolate, very unequal at base,
    20–40 mm. long and 6–13 mm. broad, serrulate or subentire ; stems and
    inflorescences, at least when young, densely woolly-tomentose ; capsules,
    2–2·5 mm. broad, densely tomentose ..   ..   ..   9. *convolvuloides*
Stems more or less prostrate, or ascending, well-branched ; leaves more or less
    oblong-elliptic, 2·5–14 mm. long, 3–8 mm. broad :
Capsules all exserted from the involucre on well-developed pedicels ; capsules
    1·5–2 mm. long and broad ; seeds usually with a whitish waxy coat ; stems
    pubescent or pilose all round :
Glands on the involucre with broad petal-like appendages ; stems, leaves and
    inflorescences very densely long spreading-pilose ; stipules minute ; leaves
    conspicuously serrulate, rounded at apex, 7–14 mm. long, 3–8 mm. broad
                                                                            8. *scordifolia*
Glands on the involucre without broad appendages ; stems, leaves and
    inflorescences with curly pubescence ; stipules well-developed ; leaves
    subentire or minutely serrulate, 3–12 mm. long, 1·5–8 mm. broad
                                                                            7. *aegyptiaca*
Capsules mostly included at the base in the involucre which is thus distended
    at maturity, pedicels very short ; capsules about 1 mm. long and broad ;
    stems pilose mainly on the upper side ; stipules lanceolate, conspicuous ;
    leaves conspicuously serrulate, rounded to obtuse at apex, 2·5–10 mm. long,
    1–6 mm. broad ..   ..   11. *thymifolia*
Ovary and capsule glabrous :
Leaves suborbicular, 3–5 mm. long and broad, subcordate at base, emarginate at
    apex ; stipules united into a white, glabrous, membranous scale ; glabrous
    annual with prostrate branches ; involucres solitary at the ends of the shoots
                                                                            6. *serpens*
Leaves not suborbicular :
Stems arising annually after fires from a woody rootstock, flowering when only a
    few centimetres high but eventually attaining up to 16 cm. long, sparingly
    puberulous ; lower leaves more or less oblong, 3–9 mm. long, 2–5 mm. broad,
    upper leaves linear-lanceolate, 12–24 (–30) mm. long, 2·5–4 mm. broad, entire
    except at the very tip ; involucres solitary, axillary, with peduncles elongating
    to 13 mm. long   ..   ..   13. *kerstingii*
Stems not as above :
Leaves not or very obscurely nerved at the base, linear-lanceolate, cordate on
    one side at base, toothed towards apex, 3–21 mm. long, 1–4 mm. broad ;
    involucres solitary, peduncles short ; erect glabrous annual
                                                                            10. *polycnemoides*
Leaves with distinct lateral nerves at the base ; involucres in cymes :
Perennial glabrous herb of seashore ; stems dark purplish, prostrate, with short
    internodes (usually 1–2·5 cm. long) ; leaves glaucous, oblong-elliptic, cordate
    on one side at base, rounded and mucronate at apex, 5–20 mm. long, 3·5–
    10 mm. broad, entire ; stipules deeply dentate ; involucres several together
                                                                            1. *glaucophylla*

---

[1] Species Nos. 22–30 revised in collaboration with E. Milne-Redhead.

Annual herbs ; main stems erect, with internodes 2–5 cm. long, or if procumbent
with shorter internodes then puberulous :

Involucres few in lax slender dichotomously branched cymes accompanied by
leafy bracts ; stipules short, ciliate, broad in outline ; stem and leaves
often pubescent when young, glabrescent, leaves obliquely obovate-oblong
or oblong, rounded on one side, acute on the other at base, rounded to
subacute at apex, up to 2·8 (–4) cm. long, and 1·3 (–1·8) cm. broad, distinctly
serrulate ; capsules about 2 mm. broad   ..   ..   ..   4. *hyssopifolia*

Involucres in several- to many-flowered rather close cymes :

Stems erect, like the leaves quite glabrous ; stipules ovate to lanceolate,
about 1 mm. long ; leaves obliquely lanceolate-elliptic, unequally rounded
and obtuse at base, obtuse or subacute at apex, up to 3·5 cm. long and
1·2 cm. broad, serrulate ; capsules about 1·3–1·4 mm. broad   3. *glomerifera*

Stems decumbent, like the undersurface of leaves rather densely pubescent ;
stipules ovate, scarcely 1 mm. long ; leaves obliquely oblong-elliptic,
unequally rounded and cuneate at base, obtuse or subacute at apex, up
to 12 mm. long and 5 mm. broad, serrulate ; capsules about 1·4 mm.
broad ..   ..   5. *sp. A*

†Leaves alternate below, but opposite or whorled above where they subtend inflorescences, usually exceeding 5 cm. long and 1 cm. broad ; stipules absent or
glandular ; mostly erect perennial herbs :

Involucres solitary, 6–9 mm. diam., with 5 broad flap-like glands transversely
elliptic with crenulate margins ; perennial herbs in savannah, sending up annual
shoots to 30 cm. long from a woody stock ; stem and leaves glabrous, subsucculent;
leaves rounded or subacute at apex, usually mucronulate, lateral nerves inconspicuous beneath :

Peduncles 2–6 mm. long ; leaves broadly linear to elongate-elliptic, narrowed to
an obtuse subsessile base, 5–14 cm. long, 1–2 cm. broad   .. 14. *ledermanniana*

Peduncles 8–26 mm. long ; leaves obovate to obovate-oblanceolate, long-cuneate
at base, 4–8 cm. long, 2·5–3·5 cm. broad   ..   ..   .. 15. *kouandenensis*

Involucres in terminal cymes or umbels ; involucral glands mostly not as above,
but see No. 16 ; habit of plant not as above :

Involucres in lax several- to many-rayed umbels or cymes ; each involucre normally
with 4–5 glands :

Ovary not tuberculate :

Leaves obovate or broadly oblanceolate ; few ; glands of the involucre not
crescent-shaped or 2-horned :

Bracts subtending the involucres about 3 cm. long, rounded-ovate, subcoriaceous ; glands of the involucre flap-like, denticulate on the margin ; leaves
subcoriaceous, apiculate, 7–15 cm. long, 3–5·5 cm. broad, with rather obscure
lateral nerves ; capsule 1·5 cm. broad   ..   ..   .. 16. *macrophylla*

Bracts subtending the involucres about 1 cm. long, broadly obovate, mucronate
at apex ; glands of the involucre with 2–5 filiform forked processes ; leaves
slightly acuminate, 10–15 cm. long, 4–6 cm. broad, with several pairs of
distinct lateral nerves ..   .. 17. *cervicornu*

Leaves lanceolate or narrowly oblanceolate, numerous ; 4–10 cm. long, 1–2·5 cm.
broad ; leafy bracts ovate, up to 4 cm. long ; glands of the involucre crescent-
shaped or 2-horned ; capsule 3–4 mm. broad   ..   .. 19. *schimperiana*

Ovary densely and conspicuously tuberculate ; leaves numerous, varying from
linear-lanceolate to elliptic or subsessile, those of the whorl below the inflorescence usually shorter and broader than the stem leaves ; bracts broadly
ovate, about 1 cm. long and broad ; capsule far-exserted, about 6 mm. broad
20. *depauperata*

Involucres in close terminal cymes ; each involucre with only a single large fleshy
substipitate gland ; introduced annual weeds ; leaves variable in shape and
size, upper leaves sometimes panduriform ; often with whitish or red blotches
at the base of the upper leaves and bracts ; stipules glandular 18. *heterophylla*

*Shrubs or trees, with fleshy glabrous stems, often prickly (except Nos. 22–24) :

Inflorescences and leaves and spines when present, arranged spirally, or inflorescences
terminal ; stems more or less terete ; leaves present, often well-developed (except
No. 23), crowded towards the ends of the branches, normally deciduous :

Spines absent, or sometimes (No. 24) with solitary spines rudimentary or present only
on young plants :

Flowering branchlets 0·25–0·6 cm. diam. ; leaves linear up to 0·7 cm. broad, with a
single main nerve :

Involucres unisexual, with stout peduncles up to 2 mm. long, solitary in upper
leaf-axils ; ovary and fruit pubescent ; leaves numerous, usually deciduous,
up to 10 cm. long and 7 mm. broad   ..   .. 22. *balsamifera*

Involucres bisexual, with slender peduncles 6–20 mm. long, arranged in terminal umbels ; ovary and fruit glabrous ; leaves rather sparse and early deciduous, up to 3 cm. long and 2·5 mm. broad .. .. 23. *lateriflora*

Flowering branchlets 2·5–4 cm. diam. with leaves in 8–10 spiral ranks ; leaves obovate-spathulate, long-cuneate at base, usually widely emarginate at apex, 5–14 cm. long, 3–7 cm. broad, pinnately nerved ; involucres shortly pedunculate, usually paired, axillary, numerous at the ends of the shoots ; fruits deeply 3-lobed, about 5–6 mm. diam., glabrous, each coccus subglobose ; fruiting pedicels about 5–12 mm. long .. .. 24. *poissoni*

Spines present, well-developed :

Spines solitary, subulate from a very broad shield-like base, about 1 cm. long, glaucous grey ; stems 1–2 cm. diam., with leaves (as far as can be judged from available material) in 4–5 spiral ranks ; leaves oblong-spathulate, sometimes 2-lobed at apex, or mucronate, subentire or fringed, 5–11·5 cm. long, 1·5–5 cm. broad ; involucres and fruits apparently as in No. 24 but more scattered

25. *unispina*

Spines paired :

Flowering branchlets 2–5 cm. diam. ; fruits subglobose or ovoid about 15 mm. long and 12 mm. broad, on rather short pedicels ; seeds oblong, about 8 mm. long and 3 mm. broad ; leaves triangular-obovate to spathulate, rounded or truncate at the apex and mucronate or emarginate, up to 7 cm. long and 6 cm. broad

26. *paganorum*

Flowering branchlets 1–1·5 cm. diam. ; fruits deeply 3-lobed, about 3 mm. long and 5 mm. broad, on pedicels 5–7 mm. long ; seeds subglobose about 1·5 mm. diam. ; leaves spathulate to triangular-obovate, dentate to lacerate at apex, 2–5 cm. long, 1·5–2 cm. broad .. .. 27. *sudanica*

Inflorescences and spines, and leaves when present, arranged along 3–6 angles or wings of the branchlets which are not terete (note however that old branches may become more or less terete) ; spine-shields with a pair of well-developed or vestigial spines :

Leaves present, deciduous, fleshy-coriaceous, spathulate-obcordate, up to 17 cm. long and 12 cm. broad ; branchlets 3–5-angled, not or scarcely winged, not segmented ; involucres solitary or paired, protogynous, on a common peduncle 2–4·5 cm. long ; fruits trigonous, about 1 cm. long, with sides about 1·4 cm. wide ; fruiting pedicels up to 1·3 cm. long, recurved .. .. 28. *desmondi*

Leaves absent, or very rudimentary and caducous ; branchlets 3–4 (–6)-winged, segmented ; involucres protandrous, on common peduncles up to only 1 cm. long, even in fruit :

Calyx of female flower truncate ; fruits sharply 3-angled, about 5 mm. long, with sides about 1 cm. wide, slightly concave, without subsidiary lobes ; fruiting pedicels about 4 mm. long .. .. 29. *deightonii*

Calyx of female flower laciniate ; fruits 3-angled, about 7 mm. long, with sides about 1·2 cm. wide each with fleshy subsidiary lobes while ripening ; fruiting pedicels about 3 mm. long .. .. 30. *kamerunica*

1. **E. glaucophylla** *Poir.*—F.T.A. 6, 2 : 499. Perennial herb with prostrate branches ; woody stems dark purplish, leaves pale green ; just above high-water-mark on sandy seashores.
Sen.: *Perrottet* ! *Sieber* 18 ! **Gam.**: Bathurst *Don* ! **Port.G.**: Paia Varela, Suzana (fl. & fr. Oct.) *Esp. Santo* 2288 ! **S.L.**: Lumley (fl. & fr. Aug.) *Jordan* 778 ! Yele, Turtle Isl. (fl. & fr. Nov.) *Deighton* 2323 ! Mahera (fl. & fr. Nov.) *Deighton* 5672 ! **Lib.**: Grand Bassa *Dinklage* 2094 ! *T. Vogel* 54 ! Cess R. (fl. & fr. Mar.) *Baldwin* 11298 ! **G.C.**: Half Assinie (fl. & fr. July) *Chipp* 271 ! Teshi (fl. & fr. Nov.) *Irvine* 808 ! **S.Nig.**: Victoria Beach, Lagos (fl. & fr. Mar.) *Dalz.* 1357 ! Nun R. (fl. & fr. Aug.) *T. Vogel* 10 ! Also along the coast from French Cameroons to Angola.

2. **E. hirta** *Linn.*—F.T.A. 6, 1 : 496 ; Schnell in Ic. Pl. Afr. t. 6. *E. pilulifera* of Chev. Bot. 552. An erect or decumbent herb, to 18 in. high, sometimes purple-tinged ; a common weed.
Sen.: Dakar (fl. & fr. May) *Baldwin* 5706 ! **Fr.Sud.**: Oualia *Chev.* 30. **Fr.G.**: Kouroussa *Chev.* 389. **S.L.**: Freetown *Welw.* 288 ! Kenema (fl. & fr. Nov.) *Deighton* 404 ! Rokupr (fl. & fr. June) *Jordan* 271 ! Musaia (fl. & fr. Apr.) *Deighton* 5455 ! **Lib.**: Grand Bassa (fl. & fr. July) *T. Vogel* 38 ! Ganta *Harley* ! Kakatown *Whyte* ! **Iv.C.**: Bingerville *Chev.* 16072. **G.C.**: Axim (Apr.) *Chipp* 421 ! Aburi (fl. & fr. Nov.) *Brown* 386 ! **Togo**: Lome *Warnecke* 39 ! **Fr.Nig.**: Niamey (fl. & fr. Oct.) *Hagerup* 535 ! **N.Nig.**: Sokoto (fl. & fr. Nov.) *Moiser* 29 ! Zaria (fl. & fr. Mar.) *Hill* 45 ! Katagum *Dalz.* 414 ! **S.Nig.**: Lagos *MacGregor* 360 ! Okomu F.R. (fl. & fr. Feb.) *Brenan* 9147 ! **Br.Cam.**: Ambas Bay (fl. & fr. Apr.) *Hutch. & Metcalfe* 135 ! **F.Po**: *T. Vogel* 2 ! *Mann* 237 ! Widespread in tropical and subtropical countries. (See Appendix, p. 143.)

3. **E. glomerifera** (*Millsp.*) *Wheeler* in Contrib. Gray Hb. 127 : 78 (1939). *Chamaesyce glomerifera* Millsp. (1913). *Euphorbia hypericifolia* Linn. (1753), partly, and of various authors, including F.W.T.A., ed. 1, 1 : 312, partly. An erect, branched annual herb, to 18 in. high ; an introduced weed.
Sen.: St. Louis (fl. & fr. May–June) *Chev.* 3495 ! **S.L.**: Hill Station (fl. & fr. May) *Deighton* 2688 ! **Lib.**: Monrovia (fl. & fr. June) *Baldwin* 5881 ! **G.C.**: Aburi Gardens (fl. & fr. June, Sept.) *Deighton* 3418 ! *Irvine* 2830 ! A native of tropical America.

4. **E. hyssopifolia** *Linn.* Syst., ed. 10 : 1048 (1759). *E. brasiliensis* Lam. (1786)—Boissier in DC. Prod. 15, 2 : 24. *E. hypericifolia* of F.T.A. 6, 1 : 498, partly, and of F.W.T.A., ed. 1, 1 : 312, partly, not of Linn. An erect, branched, rather graceful, annual herb, to 18 in. or more high ; an introduced weed.
S.L.: Hill Station (fl. and fr. May) *Deighton* 5350 ! Musaia (fl. & fr. July) *Deighton* 4809 ! Mateboi, Sanda Tenraran (fl. & fr. Nov.) *Jordan* 823 ! **Lib.**: Cape Palmas (fl. & fr. July) *Ansell* ! **G.C.**: Black Volta R. (fl. & fr. Dec.) *Adams* GC 4455 ! **S.Nig.**: Ogurude (fl. & fr. Jan.) *Holland* 267 ! **Br.Cam.**: *Preuss* 1236 ! Also in French Cameroons and on the islands of Principe and S. Tomé ; a native of tropical America.

5. **E. sp. A.** Better material required ; perhaps an introduced weed. *E. hypericifolia* of F.W.T.A., ed. 1, 1 : 312, partly. A decumbent annual herb.
G.C.: Accra (fl. & fr. May) *Dalz.* 140 !

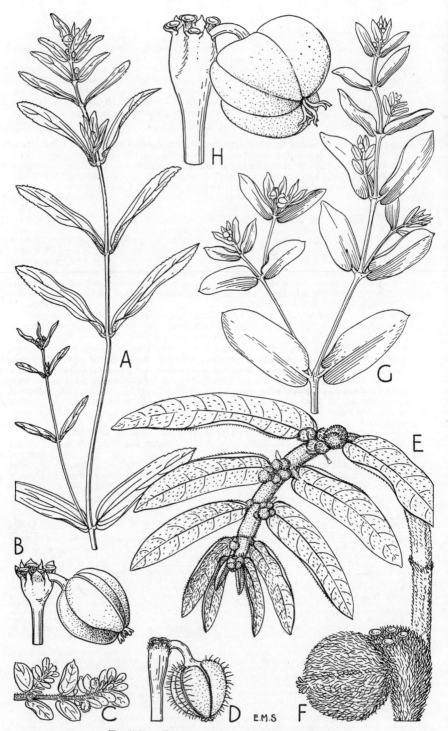

Fig. 139.—Euphorbia spp. (Euphorbiaceae).

E. *polycnemoides* Hochst. ex Boiss.—A, shoot, × 2. B, involucre and capsule, × 10.
 Drawn from *Keay* FHI 25971.
E. *prostrata* Ait.—C, shoot, × 2. D, involucre and capsule, × 10.
 Drawn from *Brenan* 8741.
E. *convolvuloides* Hochst. ex Benth.—E, shoot, × 2. F, involucre and capsule, × 10.
 Drawn from *Lely* P. 655 and *Moiser* s.n., respectively.
E. *glaucophylla* Poir.—G, shoot, × 2. H, involucre and capsule × 5.
 Drawn from *Chipp* 271.

6. **E. serpens** *H.B. & K.*—F.T.A. 6, 1 : 511. A prostrate spreading annual, 2–7 in. long, glabrous.
S.L.: Freetown (June) *T. Vogel* 29 ! A native of tropical America.

7. **E. aegyptiaca** *Boiss.*—F.T.A. 6, 1 : 507, partly ; Chev. Bot. 550. *E. granulata* of Chev. Bot. 551. A prostrate herb, stems up to 1 ft. long ; on river banks and other sandy places.
Sen.: *Perrottet* 744 ! R. Senegal, Galam *Heudelot* 226 ! Dagana *Leprieur* ! **Fr.Sud.:** Timbuktu (fl. & fr. July, Aug.) *Chev.* 1283 ! *Hagerup* 170 ! **G.C.:** Red Volta R., Nangodi (fl. & fr. Dec.) *Adams & Akpabla* GC 4281 ! **N.Nig.:** Nupe *Barter* 316 ! 834 ! Katagum *Dalz.* 306 ! Kukawa (fl. & fr. Jan.) *E. Vogel* 2 ! Yola *Dalz.* 154 ! Also in Cape Verde Islands, the drier parts of Africa, and in Arabia, Syria and N. India. (See Appendix, p. 142.)

8. **E. scordifolia** *Jacq.*—F.T.A. 6, 1 : 501 ; Chev. Bot. 552. *E. thymifolia* of Chev. Bot. 552. An erect or prostrate herb with rather woody stems and root ; densely tomentose on all parts ; a weed.
Sen.: *Heudelot* 474 ! *Perrottet* 177 ! *Roger* ! *Sieber* 17 ! Sor Isl. *Brunner* 11a ! **Fr.Sud.:** Timbuktu *Hagerup* 125 ! *Chev.* 1281 ; 1282. **Iv.C.:** Lower Comoé *Chev.* 17584. **N.Nig.:** Lokoja (fl. & fr. Aug.) *Elliott* 195 ! Kukawa (fl. & fr. Jan.) *E. Vogel* 6 ! Bure, L. Chad (fl. & fr. Dec.) *Elliott* 109 ! Also in Cape Verde Islands and Mauritania ; extending to A.-E. Sudan and Red Sea.

9. **E. convolvuloides** *Hochst. ex Benth.* in Hook. Fl. Nigrit. 499 (1849) ; F.T.A. 6, 1 : 495 ; Chev. Bot. 551. A usually erect herb with branches drooping at the tips, to 2 ft. high, usually tinged pink ; in savannah and waste places.
Sen.: *Berhaut* 1225. Upper Senegal R. *Lecard* 136. **Iv.C.:** Bobo Dioulasso *Chev.* 906. Sassandra and Bafing R. confluence *Chev.* 21780. **G.C.:** Damongo, N.T. (fl. & fr. July) *Andoh* FH 5194 ! Tamale (fl. & fr. Dec.) *Adams* GC 4150 ! **Togo:** Yendi (fl. & fr. Dec.) *Adams* GC 4062 ! **Dah.:** Atacora Mts. *Chev.* 24017. **N.Nig.:** Sokoto (fl. & fr. Aug.) *Dalz.* 393 ! 394 ! Nupe *Barter* 1489 ! Stirling Hill *Ansell* ! Bauchi Plateau (fl. & fr. Aug.) *Lely* P. 655 ! **S.Nig.:** Lagos *Dawodu* 272 ! Olokemeji (fl. & fr. Sept.) *Keay* FHI 25388 ! Extends to A.-E. Sudan and E. Africa. (See Appendix, p. 143.)

10. **E. polycnemoides** *Hochst. ex Boiss.*—F.T.A. 6, 1 : 506 ; Chev. Bot. 552. An erect, slender, usually much-branched annual herb, to 18 in. high ; stems often tinged crimson, leaves glaucous ; in savannah and as a farm weed.
Sen.: Niokolo-Koba *Berhaut* 1393. **Fr.Sud.:** Koulikoro *Chev.* 3466. **G.C.:** Deloro to Tankara, N.T. (fl. & fr. Dec.) *Adams* GC 4326 ! **Fr.Nig.:** Gourma *Chev.* 24512. Zinder *Hagerup* 606 ! **N.Nig.:** Sokoto (fl. & fr. Aug.) *Dalz.* 395 ! *Moiser* 99 ! Liruwen-Kano Hills *Carpenter* ! Anara F.R., Zaria (fl. & fr. July) *Keay* FHI 25971 ! Jos (fl. & fr. Aug.) *Keay* FHI 12710 ! Extends to the Red Sea and into E. Africa. (See Appendix, p. 145.)

11. **E. thymifolia** *Linn.* Sp. Pl. 1 : 454 (1753) ; Chev. Bot. 552. *E. burmanniana* Gay (1847), partly ; Berhaut Fl. Sen. 66. (?) *E. afzelii* N.E. Br. in F.T.A. 6, 1 : 506 (1911) ; F.W.T.A., ed. 1, 1 : 312. *E. aegyptiaca* and *E. scordifolia* of F.T.A. 6, 1 : 501, 507, partly, and of F.W.T.A., ed. 1, 1 : 312, partly. A prostrate herb ; leaves often purplish ; involucral glands red ; an introduced weed.
Sen.: *Leprieur* ! **S.L.:** *Afzelius* ! *Sc. Elliot* 4636 ! Freetown (fl. & fr. May) *Deighton* 1190 ! Ninia, Talla Hills *Sc. Elliot* 4993 ! Rokupr (fl. & fr. June) *Jordan* 272 ! Kenema (fl. & fr. Nov., Jan.) *Deighton* 419 ! *Thomas* 7593 ! 7727 ! **Iv.C.:** Bettié to Mbasso, Lower Comoé (fl. & fr. Mar.) *Chev.* 17584 ! **G.C.:** Accra (fl. & fr. Mar.) *Deighton* 571 ! **N.Nig.:** Kano (fl. & fr. Dec.) *Hagerup* 649 ! **S.Nig.:** Olokemeji, Ogun R. (fl. & fr. Dec.) *Meikle* 925 ! Ijaiye F.R., Ogun R. (fl. & fr. Mar.) *Keay* FHI 21185 ! Sapoba *Kennedy* 2333 ! 2684 ! **Br.Cam.:** Rio del Rey *Johnston* ! Widespread in the warmer parts of the world.

12. **E. prostrata** *Ait.*—F.T.A. 6, 1 : 510, 1036 ; Chev. Bot. 552. A prostrate spreading herb ; stems slender 2–8 in. long radiating from the tough root ; a weed of footpaths, etc.
Sen.: *Farmar* 8 ! **Fr.G.:** Conakry *Maclaud* 15. **S.L.:** *Don* ! *T. Vogel* ! Freetown (fl. & fr. Feb.) *Deighton* 5312 ! Murray Town (fl. & fr. Aug.) *Deighton* 5756 ! **Lib.:** Cape Palmas *Dinklage* 67. **Iv.C.:** Malamalasso *Chev.* 17530. **G.C.:** Aburi (fl. & fr. Nov.) *Johnson* 238 ! Adafo *Krause* 96. **S.Nig.:** Lagos (fl. & fr. May, Nov.) *Dalz.* 1358 ! *Meikle* 500 ! *Millen* 23 ! Idanre (fl. & fr. Jan.) *Brenan* 8741 ! Ogwashi (fl. & fr. Dec.) *Thomas* 2062 ! Calabar (fl. & fr. May) *Holland* 123 ! **F.Po:** *Mann* 213 ! A native of tropical America ; widespread as a weed in the tropics. (See Appendix, p. 145.)

13. **E. kerstingii** *Pax*—F.T.A. 6, 1 : 500. *E. polycnemoides* of F.W.T.A., ed. 1, 1 : 312, partly (*Dalz.* 33). A perennial, with short annual stems arising from the woody rootstock after grass fires in savannah.
G.C.: Kintampo (fl. & fr. Nov.) *Dalz.* 33 ! **Togo:** Trogode, Königsdorf (Jan.) *Kersting* 35. Kete Krachi *Zech* 45. **S.Nig.:** between Igbetti and Kishi, Oyo Prov. (Feb.) *Collier* FHI 16272 !

14. **E. ledermanniana** *Pax & K. Hoffm.* in Engl. Bot. Jahrb. 45 : 241 (1910) ; F.T.A. 6, 1 : 524. *E. calva* N.E. Br. in F.T.A. 6, 1 : 522 (1911) ; F.W.T.A., ed. 1, 1 : 312. Perennial herb in savannah, sending up rather fleshy shoots from a woody stock after grass fires.
N.Nig.: Mando, Birnin Gwari (fr. Aug.) *Keay* FHI 28001 ! Kyana, Nassarawa (fl. & fr. Mar.) *Hepburn* ! Bauchi Plateau (Mar.) *Lely* P. 190 ! Katsina Ala (June) *Dalz.* 754 ! Also in French Cameroons.

15. **E. kouandenensis** *Beille* in Bull. Soc. Bot. Fr. 61, Mém. 8 : 296 (1917) ; Chev. Bot. 552. Perennial savannah herb, like the preceding.
G.C.: Bahare, Gambaga (June) *Akpabla* 668 ! **Togo:** Yendi (fr. May) *Akpabla* 518 ! **Dah.:** Kouandé (June) *Chev.* 24248 !

16. **E. macrophylla** *Pax*—F.T.A. 6, 1 : 529. Erect perennial, 1½ ft. or more high, with a whorl of 3–5 leaves below the umbel of bracteate involucres ; lower leaves alternate.
Sen.: Tambacounda *Berhaut* 1656 ; 2954 ; 3267. **G.C.:** Gambaga, N.T. (fr. Aug.) *Norman* ! Also in A.-E. Sudan and Tanganyika.

17. **E. cervicornu** *Baill.*—F.T.A. 6, 2 : 528. Erect, perennial, 2–3 ft. high, with rather soft half-woody branches, glabrous.
S.Nig.: Calabar R. (Feb.) *Mann* 2315 ! Also in Belgian Congo and S. Tomé.

18. **E. heterophylla** *Linn.* Sp. Pl. 1 : 453 (1753) ; Berhaut Fl. Sen. 196. *E. geniculata* Orteg. (1797)—Milne-Redhead in Kew Bull. 1948 : 457. A rather lax erect annual weed, to 3 ft. high ; locally abundant. In most specimens from our area the upper leaves and bracts are either entirely green, or with whitish blotches at the base, or tinged purplish on the petiole, nerves and margins ; *Deighton* 4505 and *Meikle* 877 are specimens of the form with red or orange blotches which is sometimes cultivated.
Sen.: cult. *fide* Berhaut *l.c.* **S.L.:** Njala, cult. (fl. & fr. Oct.) *Deighton* 4505 ! **Lib.:** Monrovia (fl. & fr. May) *Barker* 1320 ! **G.C.:** Amanase, Suhum (fl. & fr. Apr.) *Hughes* 80 ! **N.Nig.:** Jebba (fl. & fr. Dec.) *Hagerup* 715 ! Zaria (fl. & fr. Aug.) *Keay* FHI 28024 ! Kafanchan (fl. & fr. Apr.) *Jones* FHI 6425 ! **S.Nig.:** Ibadan (fl. & fr. Nov., Dec.) *Meikle* 655 ! 877 ! 878 ! A native of tropical and subtropical America, now widespread in the tropics.

19. **E. schimperiana** *Scheele* in Linnaea 17 : 344 (1843) ; F.T.A. 6, 1 : 533. *E. ampla* Hook. f. (1862)—F.T.A. 6, 1 : 532 ; F.W.T.A., ed. 1, 1 : 312. Stems 2–4 ft. high, probably perennial ; leaves alternate with a whorl of 4–10 below the umbel ; in montane vegetation.
Br.Cam.: Cam. Mt., 5,000–9,000 ft. (fl. & fr. Nov.–Mar.) *Mann* 1265 ! 2006 ! *Maitland* 197 ! *Mildbr.* 10829 ! Manenguba Mt. *Thornbecke.* Bafut-Ngemba F.R., Bamenda, 5,500–7,000 ft. (fl. & fr. Jan.) *Keay* FHI 28348 ! *Tiku* FHI 22243 ! Banso *Ledermann* 2017. Widespread in tropical Africa, from Abyssinia to S. Rhodesia, also in Arabia.

20. **E. depauperata** *Hochst. ex A. Rich.*—F.T.A. 6, 1 : 537. Rootstock perennial, woody ; stems 2 to several, herbaceous, or, if not burnt by fires, enduring for 2 or more years and becoming woody and leafless below simple or branched, 4 to 36 in. high ; in montane grassland.
S.L.: Bintumane Peak, 5,500 ft. (Jan.) *Glanville* 469 ! **Iv.C.:** Nimba Mt., 4,500 ft. *Schnell* 3535 ! **Br.Cam.:** Bamenda Prov. : Banso, 6,000 ft. *Ledermann* 2006. Lakom, 6,000 ft. (June) *Maitland* 1502 ! Bambili (fr. Aug.) *Ujor* FHI 29979 ! Widespread in tropical Africa.

9

21. **E. baga** *A. Chev.* in Rev. Bot. Appliq. 13 : 569, fig. 18 (1933). A perennial herb with stout rootstock and short erect underground or partially exposed stems ; inflorescences appearing when leafless ; flowers red. The association of *Lely* P. 63 (a specimen with flowers but no leaves) with *Chev.* 935 (the type specimen which has leaves but no flowers) is only tentative ; specimens showing both leaves and flowers from the same plant are needed to confirm this.
 Iv.C.: Banankolidoro to Bama, Bobo Dioulasso *Chev.* 935 ! **N.Nig.:** on flat-topped lateritic hills, Bauchi Plateau (Jan.) *Lely* P. 63 ! (See Appendix, p. 145.)

22. **E. balsamifera** *Ait.* Hort. Kew., ed. 1, 2 : 137 (1789) ; Chev. Bot. 550 ; Chev. in Rev. Bot. Appliq. 13 : 545 ; Aubrév. Fl. For. Soud.-Guin. 177 ; Schnell in Ic. Pl. Afr. t. 28. *E. sepium* N.E. Br. in F.T.A. 6, 1 : 551, 1040 (1911). *E. rogeri* N.E. Br. l.c. 551 (1911) ; F.W.T.A., ed. 1, 1 : 313. A low shrub with glabrous succulent branches ; commonly grown in hedges in dry regions ; leaves deciduous, or evergreen on plants growing in especially favourable situations.
 Maur.: *fide* Chev. *l.c.* **Sen.:** *Farmar* 54 ! Sor Isl. *Brunner* 21. Lampsor and Maka (Dec.–Jan.) *Roger* ! Cayor & Dakar *Paroisse* 43. **Fr.Sud.:** Danga *Chev.* 1315. Gao (Feb.) *De Wailly* 4673 ! **G.C.:** Yendi (fr. May) *Akpabla* 541 ! **Togo:** *Kersting* 739. **Dah.:** Dassa-Zoumé *Chev.* 23620 ! **Fr.Nig.:** *fide* Aubrév. *l.c.* **N.Nig.:** Sokoto *Dalz.* 528 ! Katsina to Daura *Sampson* 8 ! Katagum *Dalz.* 320 ! Also in the Canary Islands and Rio de Oro ; apparently not known east of L. Chad. (See Appendix, p. 142.)
 [The record from Lagos in the first edition is based on a sterile leafless specimen which may well be *E. lateriflora*.]

23. **E. lateriflora** *Schum. & Thonn.*—F.T.A. 6, 1 : 552 ; Chev. in Rev. Bot. Appliq. 13 : 546 ; Aubrév. l.c. ; Monod in Mém. I.F.A.N. 18 : 19. A low shrub with smooth-glaucous more or less erect branches ; flowers yellow ; commonly grown in hedges ; leaves usually deciduous.
 S.L.: Newton *Deighton* 1958 ! **G.C.:** Chama (Apr.) *Chipp* 188 ! Achimota *Irvine* 1010 ! **N.Nig.:** Sokoto *Dalz.* 392 ! Katsina (Feb.) *Meikle* ! Kano (Mar., Oct.) *Bally* ! *Keay* FHI 21139 ! Katagum *Dalz.* 327 ! Jos (Mar.) *Hill* 21 ! Pedong, Jos (fl. & fr. Dec.) *Monod* 9705 ! **S.Nig.:** Epe *Barter* 3309 ! Ibadan *fide* Keay. **Br.Cam.:** Gwoza, Dikwa Div. (fr. Dec.) *McClintock* 58 !
 [The record from Kratschi, Togoland, given in the first edition seems to be based on a mistranslation of the Danish phrase " I Krati " which means " In thickets " ; Thonning does not cite an exact locality. This species is close to the common and variable S. African species known as *E. mauritanica* Linn. ; comparison of living specimens would be desirable in determining the taxonomic status of the W. African plant ; *E. nubica* N.E. Br. of north-eastern Africa is also close.]

24. **E. poissoni** *Pax*—F.T.A. 6, 1 : 560 ; Chev. Bot. 552 ; Chev. in Rev. Bot. Appliq. 13 : 549 ; Aubrév. l.c. 177. An erect shrub to 6 ft. high, with candelabriform branching ; branches silvery-grey, stout ; leaves pale green, deciduous ; flowers pale greenish with red stamens ; on rocky hills in savannah and planted ; sometimes with rudimentary spines, or with spines present only on young plants.
 Fr.G.: *Pobéguin.* **Iv.C.:** Mossi *fide* Chev. *l.c.* **G.C.:** *Anderson* ! Tamale *Bally* 5192 ! **Togo:** various localities *Kersting* 88 ; 91 ; 414 ; 574. **Dah.:** *Poisson.* Bassila, Savalou *Chev.* 23792. **Fr.Nig.:** Gourma *fide* Chev. *l.c.* **N.Nig.:** 44 miles from Katsina on Daura road (Apr.) *Meikle* 1370 ! Gwari Hill, Minna *Meikle* ! Mile 36 on Ilorin–Jebba road (fl. & fr. Dec.) *Meikle* 682 ! Lokoja (fl. & fr. Dec.) *Dalz.* 917 ! Near Jos *Batten-Poole* FHI 12894 ! Kasaje to Lom *Barter* 1491 ! (See Appendix, p. 145.)
 [The forms with rudimentary spines may be distinguished from *E. unispina* by the stouter flowering branchlets which have the leaves in 8–10 spiral ranks instead of about 5, by the smaller spines, and by the more numerous inflorescences densely clustered at the ends of the shoots.]

25. **E. unispina** *N.E. Br.* in F.T.A. 6, 1 : 561 (1911) ; Chev. in Rev. Bot. Appliq. 13 : 550 ; Aubrév. l.c. 178. *E. "venenifica"* of Chev. l.c. 549, partly ; of Aubrév. l.c. 178, partly. An erect shrub to 10 ft. high with candelabriform branching ; on rocky hills in savannah.
 Fr.Sud.: Sikasso to Bobo Dioulasso (June) *Chev.* **Iv.C.:** Soucourala to Sanrou *Chev.* 21583 ! **G.C.:** Navrongo (fr. Oct.) *Vigne* FH 4616 ! **Togo:** Loso *Kersting* 569. Difale (fr. Dec.–Feb.) *Kersting* A. 573 ! **N.Nig.:** Katagum *Dalz.* 329 ! **Br.Cam.:** Gwoza, Dikwa Div. (Jan.) *McClintock* 142 !
 [Certain forms of *E. poissoni* may be confused with this species ; see note under *E. poissoni*.]

26. **E. paganorum** *A. Chev.* in Rev. Bot. Appliq. 13 : 556, fig. 16 (1933) ; op. cit. 28 : 351, t. 8 ; Aubrév. l.c. 178. A low fleshy shrub, with pale green stems and brown spines.
 Fr.Sud.: Souloula, Bougouni (Mar.) *Chev.* 672 ! Sindou rocks, Sikasso (fr. Jan.) *Chev.* **N.Nig.:** Zaria (fl. buds Feb.) *Milne-Redhead* 5039 ! Ibeto, Kontagora (fl. buds Jan.) *Meikle* 1049 ! (See Appendix, p. 146.)

27. **E. sudanica** *A. Chev.* in Bull. Mus. Hist. Nat. sér. 2, 4 : 589, fig. 2 (1932) ; Rev. Bot. Appliq. 13 : 554, t. 9 & fig. 15 ; Aubrév. l.c. 178 ; Berhaut Fl. Sen. 3, 122. *E. trapaeifolia* A. Chev. l.c. 556, t. 10 (1933) ; Aubrév. l.c. *E.tellieri* A. Chev. l.c. 555 (1933) ; Aubrév. l.c. 178. A fleshy shrub, to 6 ft. high ; on rocky hills in dry savannah.
 Sen.: *Berhaut* 258 ; 4270. Bakel *Trochain* 1097. **Fr.Sud.:** Koulikoro (fl. & fr. Mar.) *Chev.* 26023 ! 44039 ! Koulouba *Chev.* 45093 ! Gao to Ansongo *Chev.* 43180 *bis* ! **Fr.Nig.:** Kodjar rocks, Gourma *Chev.* 24417 ! Niamey (cult. in Paris, fls. June) *Tellier* ! (See Appendix, p. 145.)

28. **E. desmondi** *Keay & Milne-Redhead* in Kew Bull. 1955 : 139. A shrub or tree, with erect rather stiff branching and large deciduous leaves ; inflorescences reddish ; involucral-glands dull crimson ; ovary pale green with bright red stigmas.
 N.Nig.: about 33 miles from Katsina on Kano road (fl. & fr. Dec.) *Meikle* 813 ! Jos *Batten-Poole* FHI 12893 ! Bargesh, Plateau Prov. (fl. & fr. Dec.) *Monod* ! Pedong, Plateau Prov. (Dec.) *Monod* 9709 ! **Br.Cam.:** Gwoza, Dikwa Div. (fr. Dec.) *McClintock* 68 !

29. **E. deightonii** *Croizat* in Kew Bull. 1938 : 58. A shrub forming large clumps, apparently without a distinct trunk ; bracts tinged reddish, involucral-glands green ; fruits dull red ; often planted.
 S.L.: Njala (Nov., Jan.) *Deighton* 2609 ! 3109 ! 3985 ! **G.C.:** Achimota (fl. & fr. July) *Milne-Redhead* 5118 ! **S.Nig.:** Ikare (fl. & fr. Oct.) *G. Burton* !

30. **E. kamerunica** *Pax*—F.T.A. 6, 1 : 586 ; Chev. in Rev. Bot. Appliq. 13 : 565 ; Croizat in Kew Bull. 1938 : 53–56 ; Aubrév. l.c. 177. *E. barteri* N.E. Br. in F.T.A. 6, 1 : 597 (1912) ; Croizat l.c. *E. kamerunica* var. *barteri* (N.E. Br.) A. Chev. in op. cit. 28 : 348 (1948), as to name. *E. garudna* N.E. Br. (1912). A shrub or tree with candelabriform branching, to 25 ft. high or more, normally with a distinct trunk ; often planted in towns and villages.
 N.Nig.: Ketsa Rock, Jebba (shrivelled fls. & frts. Mar.) *Barter* 1012 ! *Meikle* 1314 ! **Br.Cam.:** Barombi (Sept.) *Preuss* 511 ! Mamfe (fr. Jan.) *Keay* FHI 28323 ! Also in French Cameroons.

Besides the above, a few introduced species are cultivated, including *E. pulcherrima* Willd. ex Klotzsch, *E. milii* Des Moul. (syn. *E. splendens* Boj. ex Hook.) and *E. tirucalli* Linn.

Imperfectly known species.

1. **E. purpurascens** *Schum. & Thonn.* Beskr. Guin. Pl. 252 (1827) ; Boissier in DC. Prod. 15, 2 : 22, cites this as *E. indica* Lam. (1786) ; this is probably correct, but the Thonning specimen, which I have seen, is too poor for certainty. No comparable material has been collected in our area since Thonning's specimen (1799–1802).
 G.C.: *Thonning* !

2. **E. sp.** Near to *E. ledermanniana* and *E. kouandenensis* but probably new ; further material desired.
 Fr.G.: Kandika (Apr.) *Roberty* 17578 !

65. ELAEOPHORBIA Stapf—F.T.A. 6, 1 : 604 ; Croizat in Bull. Jard. Bot. Brux. 15 : 109 (1938).

Peduncles forked, with a sessile involucre in the fork, sometimes with the lateral branches forked again ; common peduncles 2·5–4·5 cm. long, axillary, in threes, lateral branches of inflorescence 1–2 cm. long ; fruits ellipsoid, with a very short stipe, usually about 2–3 cm. long and 1·5–1·8 cm. broad ; involucre with 5 transversely oblong, denticulate lobes and 5 fleshy similar-shaped glands ; male flowers numerous ; female solitary, subsessile ; branchlets fleshy, armed with pairs of broad-based prickles about 5 mm. long ; leaves oblanceolate to obovate, sometimes widely emarginate at apex, up to 20 cm. long and 9 cm. broad, fleshy, entire   1. *drupifera*
Peduncles not forked, each with a solitary involucre, axillary, in threes, up to 5·8 cm. long in fruit ; fruits ellipsoid or obovoid-ellipsoid, 3·5–4 cm. long, 2·8–3·5 cm. diam., with a stipe 4–7 mm. (or rarely 12–20 mm.) long ; leaves up to 30 cm. long and 7·5 cm. broad ; otherwise similar to above ..   ..   2. *grandifolia*

1. **E. drupifera** (*Thonn.*) *Stapf* in Hook. Ic. Pl. t. 2823 (1906) ; F.T.A. 6, 1 : 604, partly ; Croizat in Bull. Jard. Bot. Brux. 15 : 113. *Euphorbia drupifera* Thonn. (1827)—Chev. in Rev. Bot. Appliq. 13 : 560, partly. *Elaeophorbia neriifolia* of Chev. in op. cit. 28 : 348, partly, not of (Linn.) Chev. A tree with woody stem and fleshy branches containing abundant, very caustic, white latex ; branched low down in open situations but with a long clear bole in forest ; 10–50 ft. or more high ; ripe fruits yellow.
   G.C.: *Eady* ! Accra Plains (fl. & fr. Feb.) *Johnson* 605 ! 1053 ! Achimota (fr. Mar.) *Milne-Redhead* 5051 ! Newtown (fr. July) *Chipp* 279 ! Aburi Gardens (July) *Deighton* 3446 ! **Togo:** N. of Yendi *fide* Morton. **Dah.:** *Poisson.* **S.Nig.:** Gambari, Ibadan (Oct.) *W. D. MacGregor* 548 ! Ezi (Feb.) *Thomas* 2345 ! Ikare (fl. & young fr. Oct.) *G. Burton* ! (See Appendix, p. 141.)
2. **E. grandifolia** (*Haw.*) *Croizat* l.c. 109 (1938). *Euphorbia grandifolia* Haw. (1812). *Eu. leonensis* N.E. Br. in F.T.A. 6, 1 : 563 (1911) ; F.W.T.A., ed. 1, 1 : 313. *Elaeophorbia drupifera* of F.T.A. 6, 1 : 604, partly ; of Aubrév. Fl. For. C. Iv. 2 : 20, t. 126 ; and of various other authors in part, not of (Thonn.) Stapf. *Euphorbia drupifera* of Chev. in Rev. Bot. Appliq. 13 : 560, partly. *Elaeophorbia neriifolia* of Chev. in op. cit. 28 : 348, partly. A tree, to 40 ft. high, with woody stem and fleshy branches containing abundant very caustic latex ; mainly in forest, more branched when in open conditions ; ripe fruits yellow.
   Fr.G.: Fouta Djalon (fl. & young fr. Apr.) *Chev.* 12882 ! 12905 ! 13443 ! **S.L.:** Regent (Dec.) *Sc. Elliot* 4112 ! MacDonald Town (cult. Njala, fl. June, July, fr. July) *Deighton* 5110a ! 5110b ! Bambuibu *Deighton* 5047 ! Pandobu, Luawa (cult. Njala, fl. July, Sept., Dec., fr. Dec.) *Deighton* 5111 ! 5136 ! 5721 ! Silema, Dodo (fl. & fr. Apr.) *Deighton* 3954 ! Kangahun, near Jama, Kowa (fr. Feb.) *Deighton* 5048 ! **Iv.C.:** Guiglo *Aubrév.* 2048 ! Agnéby *Aubrév.* 2118 ! **G.C.:** Akropong, W.P. (May) *Vigne* 191 ! Kibbi Morton. Also in Ubangi-Shari.

87. ROSACEAE

Trees, shrubs or herbs. Leaves various, simple or compound, alternate or rarely opposite ; stipules mostly present and paired, sometimes adnate to the petiole. Flowers actinomorphic or subzygomorphic, usually hermaphrodite. Calyx free or adnate to the ovary ; lobes imbricate. Petals present, rarely absent, imbricate. Stamens usually numerous ; filaments free or connate ; anthers small, 2-celled. Carpels 1 or more, free or connate, superior or inferior ; styles free or rarely connate. Ovules 2 or more in each carpel, superposed. Fruit superior or inferior, drupaceous, pomaceous, follicular or achenial. Seeds without endosperm.

A cosmopolitan family rather poorly represented in the tropics ; distinguished from *Ranunculaceae* by the perigynous type of flower and absence of endosperm. The tribe *Chrysobalaneae* closely connects with the *Caesalpiniaceae.*

Herbs with prostrate, not prickly, stems ; carpels more than 1 :
  Leaves digitately lobed ; calyx without setiform bracts ; petals 0 ; stamens usually 4 ;
    carpels 5, free, not all fertile ..   ..   1. **Alchemilla**
  Leaves pinnately lobulate ; calyx-tube surrounded with setiform bracteoles ; petals 5,
    small ; stamens about 10 ; carpels 5–10, more or less consolidated with the calyx-
    tube   ..   ..   2. **Neurada**
Scrambling shrubs with prickly stems and compound leaves ; ovary apocarpous, carpels
    numerous on a subglobose torus   ..   ..   3. **Rubus**
Trees or shrubs ; carpels usually solitary :
  Style or stigma terminal ; flowers in simple racemes   ..   ..   4. **Pygeum**
  Style basal or gynobasic :
    Flowers actinomorphic ; ovary at the base and in the centre of the receptacle (calyx-
      tube) ; stamens inserted all round the receptacle :
        Petals present ; flowers hermaphrodite, in short cymes ..   ..   5. **Chrysobalanus**
        Petals absent ; flowers polygamous, in panicles   ..   ..   ..   6. **Afrolicania**
    Flowers zygomorphic ; ovary inserted laterally near the mouth of the receptacle
      (calyx-tube) :
        Carpels 2-celled   ..   ..   7. **Parinari**

Carpels 1-celled :
　Stamens free or united only at the base　..　..　..　..　.. 8. **Hirtella**
　Stamens united into a long unilateral strap-shaped bundle　..　.. 9. **Acioa**

Besides the above indigenous genera, *Eriobotrya japonica* (Thunb.) Lindl. and *Prunus persica* (Linn.) Batsch, natives of China and Japan, are sometimes cultivated in our area.

1. ALCHEMILLA Linn.—F.T.A. 2 : 377.

Leaves usually with 5 short lobes rounded and dentate at apex, the sinuses not reaching half the radius of the leaf ; petioles of radical leaves 2·5–7·5 cm. long ; petioles of cauline leaves 1–3 cm. long ; inflorescences up to 15 cm. long, but normal inflorescences often absent ; subsessile flowers hidden under the stipules of the stolons very frequent ..　..　..　..　..　..　..　..　..　.. 1. *cryptantha*
Leaves usually with 7 elliptic-lanceolate lobes, the sinuses reaching more than half the radius of the leaf, margin dentate almost to the base of the lobes ; petioles of radical leaves 4–18 cm. long ; petioles of cauline leaves 3–5 cm. long ; inflorescences in axillary many-flowered cymes 10–15 cm. long, or in a panicle up to 35 cm. long ; subsessile flowers under the stipules absent or very rare　..　..　.. 2. *kiwuensis*

1. **A. cryptantha** *Steud. ex A Rich.*—F.T.A. 2 : 377 ; Hauman in Fl. Congo Belge 3 : 7 (1952). *A. tenuicaulis* Hook. f.—F.T.A. 2 : 378 ; F.W.T.A., ed. 1, 1 : 314. Low creeping herb ; flowers greenish-yellow. **Br.Cam.**: Cam. Mt., 5,000–12,000 ft. (Dec.–Mar.) *Mann* 1984 ! *Maitland* 485 ! 842 ! 1294 ! *Brenan* 9391*a* ! Bamenda, 6,000 ft. (Jan.) *Migeod* 448 ! Belo, Bamenda (Apr.) *Maitland* 1760 ! L. Oku, Bamenda, 7,700 ft. (Jan.) *Keay* FHI 28467 ! **F.Po:** 7,500 ft. *Mann* 1447 ! Moka (Jan.) *Guinea* 1967 ! Widespread on the mountains of eastern and central Africa from Abyssinia to Transvaal, of S. Tomé and of Madagascar.
2. **A. kiwuensis** *Engl.* Bot. Jahrb. 46 : 129 (1911) ; Hauman l.c. 6, t. 1. Herb with trailing stems ; flowers pale green. **Br.Cam.**: below Liwonge, Cam. Mt., 6,600 ft. (Mar.) *Brenan* 9549 ! Also on the mountains of Belgian Congo, Uganda, Kenya, Tanganyika and S. Rhodesia.

2. NEURADA Linn.—F.T.A. 2 : 381.

A woody tomentose annual with short spreading prostrate branches, the fruit persisting around the base of the stem ; leaves pinnately lobulate or deeply toothed, rather long-petiolate, about 1·5 cm. long, densely white-woolly-tomentose ; flowers axillary, solitary, roundish, surrounded with setiform bracteoles ; fruit woody, orbicular, about 1·5 cm. diam., spinose ..　..　..　..　..　..　.. *procumbens*

N. procumbens *Linn.*—F.T.A. 2 : 382 ; Chev. Bot. 253. **Fr.Sud.:** Timbuktu (June–Aug.) *Hagerup* 126 ! 240 ! *Chev.* 1343 ! Common in desert regions from Mauritania, across N. Africa to India. (See Appendix, p. 168.)

3. RUBUS Linn.—F.T.A. 2 : 374 ; Gustafsson in Arkiv. för Bot. 26A, 7 : 1–68 (1934) ; Bull. Jard. Bot. Brux. 13 : 267–276 (1935) ; Kew Bull. 1938 : 177–187.

Carpels at least 100, forming an elongate-ovoid fruit 1·5–2 cm. long ; flowers solitary or few together ; petals suborbicular, about 1 cm. long, showy ; branchlets laxly villous, prickles few, recurved ; leaves pinnate, those of main stems usually with 5–7 (rarely 9–11) leaflets, those of branches usually with 5 leaflets (uppermost with 3, or simple) ; leaflets ovate- or oblong-lanceolate, rounded at base, coarsely biserrate, hairy and green beneath　..　..　..　..　..　.. 1. *rosifolius*
Carpels much fewer ; panicles many-flowered, prickly ; petals small, often caducous, or absent :
　Lower surface of leaves obscured by a dense pilose-tomentum or by a dense tomentellous felt, the leaves consequently white beneath :
　　Lower surface of leaves densely pilose-tomentose ; leaves mostly trifoliolate ; leaflets ovate-rotundate, cordate at base, coarsely biserrate, pilose above ; branchlets pilose, prickles recurved ; panicles stout with prickles extending on to base of sepals ; carpels glabrous　..　..　..　..　..　..　.. 2. *apetalus*
　　Lower surface of leaves densely and closely white-tomentellous with longer pilose hairs as well ; leaves mostly pinnate, 5-foliolate, upper leaves sometimes trifoliolate :
　　　Branchlets, leaf-rhachides and inflorescence densely villous ; upper surface of leaflets pilose ; leaflets ovate-elliptic, rounded or subcordate at base, coarsely biserrate ; upper pair of lateral leaflets very shortly petiolulate ; branchlets with numerous large recurved prickles ; prickles extending on to base of sepals ; carpels glabrous 3. *exsuccus*
　　　Branchlets, leaf-rhachides and inflorescence tomentellous and shortly pilose ; upper surface of leaflets almost glabrous except for the nerves ; leaflets ovate-elliptic, obtuse or rounded at base, coarsely biserrate ; upper pair of leaflets distinctly petiolulate ; branchlets with recurved prickles ; carpels pubescent .. 4. *fellatae*
　Lower surface of leaves clearly visible between the hairs, the leaves consequently green beneath ; branchlets shortly tomentose, with recurved prickles ; leaves pinnate, mostly with 5–7 (rarely 9) leaflets ; leaflets narrowly ovate, rounded at base, biserrate, pilose mainly on the nerves beneath, usually glabrous (rarely pilose) above ; carpels pubescent　..　..　..　..　..　.. 5. *pinnatus* var. *afrotropicus*

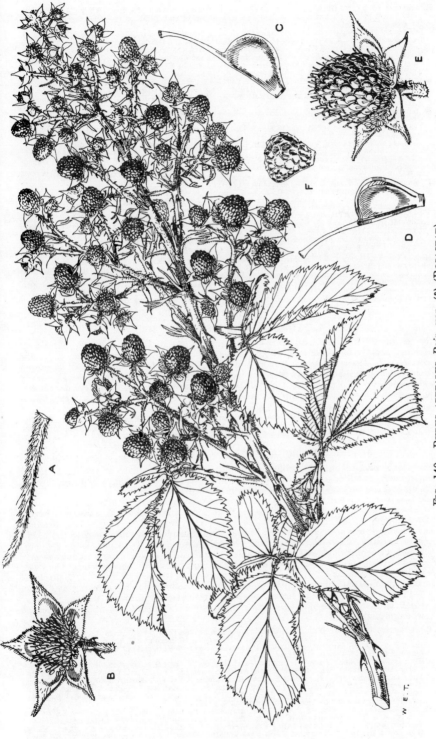

FIG. 140.—RUBUS APETALUS *Poir.* var. (?) (ROSACEAE).

A, stipule.   B, flower.   C, carpel.   D, longitudinal section of carpel.   E, young fruit.   F, seed.

425

1. **R. rosifolius** *Sm.* Pl. Ic. Hact. Ined. 3, t. 60 (1791) ; Gustafsson in Arkiv. för Bot. 26A, 7 : 11. Flowers white ; fruit red, edible but rather insipid. Cultivated and often naturalized in certain localities within our area.
  Br.Cam.: Buea *Dundas* FHI 15232 ! Bamenda (fl. & fr. Dec.) *Egbuta* FHI 3764 ! **F.Po:** Moka (fl. & fr. Dec.) *Boughey* 112 ! A native of Asia (? also of Mascarene Isls. and the Cape).

2. **R. apetalus** *Poir.* var. (?)—F.T.A. 2 : 374. Petals usually absent ; fruit edible.
  N.Nig.: Bauchi Plateau *Lely* 731 ! [This specimen cited and illustrated in F.W.T.A., ed. 1 as *R. apetalus* Poir. is not identical with the other specimens of the species at Kew, but may well be a variety of this widespread species.—R.W.J.K.] (See Appendix, p. 170.)

3. **R. exsuccus** *Steud. ex A. Rich.* Tent. Fl. Abyss. 1 : 256 (1848) ; Gustafsson l.c. 36. Low scrambling bramble ; sepals green ; petals caducous ; stamens crimson ; fruit black when ripe.
  Br.Cam.: Bambui, Bamenda (Jan.) *Keay* FHI 28338 ! Also in Uganda, Kenya, Abyssinia, Nyasaland and S. Rhodesia.

4. **R. fellatae** *A. Chev.* in Bull. Soc. Bot. Fr. 54, Mém. 8 : 18 (1907) ; Chev. Bot. 253. *R. interjungens* of Gustafsson in Bull. Jard. Bot. Brux. 13 : 269, partly. Fruits red when ripe, edible.
  Fr.G.: Ditinn to Diaguissa (fr. Apr.) *Chev.* 12905 ! Macenta (fl. & fr. Oct.) *Baldwin* 9850 ! **S.L.:** Bintumane, 5,400 ft. (fl. & fr. Jan.) *Nichols* 1 ! **Lib.:** Mt. Wolagwisi, Pandamai, 4,500 ft. (fl. & fr. Mar.) *Bequaert* 115 ! Nekabozu, Vonjama (fl. & fr. Oct.) *Baldwin* 9965 ! **Iv.C.:** Mt. Dou, 4,000 ft. (May) *Chev.* 21481 ! (See Appendix, p. 171.)
  [This species is not mentioned in the papers by Gustafsson cited above, but a specimen (*Chev.* 12908) from French Guinea is cited as *R. interjungens* Gust. (1934). The two species may prove to be identical.—R.W.J.K.]

5. **R. pinnatus** *Willd.*—F.T.A. 2 : 374 ; Gustafsson in Arkiv. för Bot. 26A, 7 : 45.
  R. pinnatus var. **afrotropicus** (*Engl.*) *C. E. Gust.* l.c. 46 (1934). *R. pinnatus* subsp. *afrotropicus* Engl. Pflanzenw. Afr. 3, 1 : 292 (1915). Scandent shrub, to 15 ft. high ; sepals grey-green ; petals small, mauve-pink ; stamens deep red-mauve ; fruits red when ripe, edible.
  Fr.G.: Cabara, near Liberian border *Baldwin* 9855 ! **Lib.:** Vonjama (fr. Oct.) *Baldwin* 9952 ! **Br.Cam.:** Cam. Mt., 4,000–9,000 ft. (fl. & fr. Nov.–Apr.) *Mann* 1221 ! 2026 ! *Brenan* 9578 ! Bali, Bamenda (May) *Ujor* FHI 30393 ! **F.Po:** 7,000 ft. *Mann* 295 ! Mioka (Dec.) *Boughey* 155 ! Widespread on the mountains of tropical Africa, also on S. Tomé and Annobon ; the species also occurs in S. Africa, Ascension and St. Helena. (See Appendix, p. 171.)

Most of the West African *Rubi* bear their inflorescences on leafy branches arising from axils of the leaves of a primary or main stem (*turion*). This main stem appears to be usually suberect or scandent. In order that this difficult genus may be better understood and to facilitate accurate determination, collectors should be careful not only to preserve an inflorescence but also a piece of the main stem bearing at least one leaf. Care should be taken to ensure that the main stem and inflorescence belong to the same plant. Notes on the fruits are much desired (e.g. colour and whether separating from receptacle as in the raspberry or adhering to it as in the blackberry).

4. PYGEUM Gaertn.—F.T.A. 2 : 373.

Tree to 25 m. or more, with dark longitudinally fissured bark ; leaves elliptic or oblong, shortly acuminate, 8–13 cm. long, 3–5 cm. broad, crenate, glabrous, thinly coriaceous ; lateral nerves 7–8 pairs ; petiole 1–2 cm. long ; flowers small in simple racemes shorter than the leaves arising from the lower part of the branchlets ; pedicels spreading, up to 5 mm. long ; petals woolly on the margin ; fruit dry, depressed-globose, 1–1·5 cm. diam., glabrous, with a short style ..   ..   .. *africanum*

P. africanum *Hook. f.*—F.T.A. 2 : 373. Flowers white ; petioles red.
  Br.Cam.: Cam. Mt., 7,000–7,500 ft. (Dec., Jan.) *Mann* 1207 ! 2165 ! L. Bambulue, Bamenda, 7,000 ft. (Jan.) *Keay* FHI 28377 ! **F.Po:** *Mildbr.* 7151. Widely distributed in the mountainous parts of tropical Africa, extending south to Angola and Transkei (Cape Prov.). (See Appendix, p. 170.)

5. CHRYSOBALANUS Linn.—F.T.A. 2 : 365.

Leaves orbicular to obovate-orbicular, usually widely emarginate at the apex, never acuminate, rounded or slightly cuneate at base, 5–9 cm. long, 3·5–6 cm. broad, thick and leathery, glabrous, reticulate on both surfaces ; petiole about 3 mm. long ; flowers very few in axillary cymes, appressed-tomentose ; bracts ovate ; petals obovate, glabrous ; filaments pilose ; fruit indehiscent, obovoid, glabrous, about 2·5 cm. long ..   .. 1. *orbicularis*
Leaves more or less acuminate, obovate to oblong, conspicuously narrowed to the base :
  Leaves broadly obovate, shortly and obtusely acuminate, 5–10 cm. long, 3–5 cm. broad, closely reticulate and often shining, glabrous ; flowers more numerous and smaller than above, often becoming galled ; fruit obovoid, dehiscent from the base, ribbed, crustaceous ..   .. 2. *ellipticus*
  Leaves oblong or oblong-elliptic, conspicuously acuminate, cuneate at the base, 7–12 cm. long, 3·5–5 cm. broad, obscurely reticulate, glabrous ; flowers few and small, similar to those of *C. ellipticus* ; fruit obovoid, more or less 5-ribbed, pubescent when young ..   .. 3. *atacorensis*

1. **C. orbicularis** *Schum.* Beskr. Guin. Pl. 232 (1827) ; Aubrév. Fl. For. C. Iv. 1 : 137, t. 48 B ; Hauman in Bull. Jard. Bot. Brux. 21 : 172 (1951) ; Fl. Congo Belge 3 : 34. *C. icaco* of F.T.A. 2 : 365 ; not of Linn.; Chev. Bot. 250. A shrub or small tree of the coastal areas ; flowers white or pinkish ; fruit plum-like, deep red-purple ; always near the sea.
  Sen.: *Sieber* 31 ! **Gam.:** *Dawe* 77 ! Sukuko (Jan.) *Dalz.* 8042 ! **Port.G.:** *Esp. Santo* 707 ! **Fr.G.:** Los Isl. (May) *Pobéguin* 1211 ! **S.L.:** Lumley Beach (Aug.) *Deighton* 18 ! Yele, Turtle Isl. (Nov.) *Deighton* 2330 ! **Lib.:** Cape Palmas (July) *T. Vogel* 34 ! Grand Bassa (July) *T. Vogel* 8 ! Monrovia *Whyte* ! Sinkor (Sept.) *Barker* 1066 ! **Iv.C.:** *fide* Aubrév. l.c. **G.C.:** Axim (Feb.) *Irvine* 2308 ! **Togo:** Lome *Warnecke* 382 ! **Dah.:** Cotonou (Mar.) *Debeaux* 350 ! **S.Nig.:** Badagri (Nov.) *A. F. Ross* 168 ! Lagos (Oct.) *Dalz.* 981 ! Extends along the coast to Angola. (See Appendix, p. 167.)

2. **C. ellipticus** *Soland. ex Sabine*—F.T.A. 2 : 366 ; Hauman in Fl. Congo Belge 3 : 35 ; Aubrév. l.c. 1 : 138, t. 48 A. A shrub or tree up to 30 ft. high, in habit like the preceding ; usually in coastal areas.
  Port.G.: Empacaca, Canchungo (Apr.) *Esp. Santo* 1947 ! **Fr.G.:** *Heudelot* 897 ! **S.L.:** *Don* ! Mahela (Dec.) *Sc. Elliot* 4181 ! Kumrabai (Dec.) *Thomas* 6919 ! Goderich (Apr.) *Deighton* 4750 ! Yele, Turtle Isl. (Nov.) *Deighton* 2331 ! **Lib.:** Grand Bassa *Ansell* ! Dukwia R. (Feb.) *Cooper* 191 ! Duport (Nov.)

Linder 1478! Monrovia (Dec.) *Baldwin* 10500! **Iv.C.:** Dabou *Aubrév.* 899! **G.C.:** Ancobra R. (July) *Chipp* 314! Beyin (July) *Chipp* 309! **S.Nig.:** Epe (May) *Thompson* 2! Olokemeji *Foster* 296! Nun R. (Sept.) *Barter* 2100! *Mann* 498! **Br.Cam.:** *Preuss* 1267! 1284! Victoria (Jan.) *Kalbreyer* 30! **F.Po:** San Carlos (Jan.) *Guinea* 800! Extends along the coast to Belgian Congo and (?) Angola. (See Appendix, p. 167.)

3. **C. atacorensis** *A. Chev.* in Bull. Soc. Bot. Fr. 58, Mém. 8 : 169 (1912) ; Aubrév. Fl. For. Soud.-Guin. 199, t. 38, 1 ; Hauman l.c. 36. A shrub or tree, up to 40 ft. high, branching low down ; in galley forests and fresh-water swamp forest.

Dah.: Atacora Mts. (June) *Chev.* 24175! **S.Nig.:** Jamieson R., Sapoba (Nov.–Jan.) *Kennedy* 1729! *Brenan* 8935! Also in Ubangi-Shari and Belgian Congo.

For the purposes of this edition I have followed Hauman's treatment in Fl. Congo Belge 3 : 34 (1952). Further investigation, especially in the field, is, however, needed to determine just how far the above species are really distinct and whether or not they should be regarded as subspecies or varieties of the American *C. icaco* Linn.

6. AFROLICANIA Mildbr. in Notizbl. Bot. Gart. Berl. 7 : 483 (1921).

A medium-sized tree ; branchlets soon glabrous and purplish ; leaves elliptic, cuneate at base, acutely acuminate, 7–14 cm. long, 3–6 cm. broad, glabrous, laxly reticulate ; petiole 5 mm. long ; stipules subulate, persistent, 5–7 mm. long ; inflorescence a slender terminal panicle ; flowers very small, in small cymules on the main branches, softly tomentose ; stamens about 20 ; fruit ovoid, 3–4 cm. long, warty   *elaeosperma*

A. elaeosperma *Mildbr.* in Notizbl. Bot. Gart. Berl. 7 : 483 (1921) ; Aubrév. Fl. For. C. Iv. 1 : 138, t. 49, 1–5. *Homalium zenkeri* Gilg ex Mildbr. l.c. 484, name only. *Parinarium glabrum* of F.W.T.A., ed. 1, 1 : 317, partly. A tree with drooping branches and foliage, and greenish-white flowers ; young leaves densely white-tomentellous, older leaves glabrous.

S.L.: Yoni and Walai, Bonthe Isl. (fr. Mar., June) *Deighton* 1971! 2465 *a–f*! Njala (Nov.) *Deighton* 1508! **Lib.:** Grand Bassa (fl. Aug., fr. Feb.) *Dinklage* 1969! 2761! **Iv.C.:** (fl. Nov., fr. Aug.) *Aubrév.* 1494 ; 1651. **G.C.:** Half Assinie (fr. July) *Chipp* 299! Ankasa F.R. (Dec.) *Vigne* FH 3176! **S.Nig.:** Eba Isl. (Jan.) *Thornewill* 195! 201! Eket *Talbot* 3098! Also in French Cameroons and Gabon. (See Appendix, p. 167.)

7. PARINARI Aubl.—F.T.A. 2 : 366 ; Hauman in Bull. Jard. Bot. Brux. 21 : 184–198 (1951).

*Sepals rounded ; flower buds not enclosed by a pair of bracts as long as themselves ; stamens 15–20, or more :

†Flowers in much-branched cymes ; fertile stamens at least 20 ; leaves glabrous or more or less densely hairy beneath, and not strongly and closely reticulate ; lateral nerves few (less than 13) and rather distant :

Receptacle (calyx-tube) 3–15 mm. long :

Undersurface of leaf green or grey, glabrous or hairy :

Leaves obtuse at apex, or very shortly and obtusely acuminate, rounded or obtuse at base, oblong or elliptic, 8–16 cm. long, 5–9 cm. broad ; basal glands conspicuous ; inflorescence and outside of flowers densely tomentose ; receptacle about 8 mm. long, sepals about 4 mm. long ; fruits broadly ovoid-ellipsoid, 2–2·5 cm. long, puberulous ; savannah trees :

Mature leaves glabrous or sparingly pilose beneath   1. *polyandra* var. *polyandra*

Mature leaves with a cottony white felt beneath   1a. *polyandra* var. *cinerea*

Leaves distinctly acuminate :

Leaves not denticulate towards the apex, cuneate to rounded at base ; without a bracteole at the base of the receptacle :

Receptacle elongate-turbinate, at least 5 mm. long, glabrous ; exposed outer surfaces of sepals glabrous :

Receptacle 6–7 mm. long ; branches of inflorescence puberulous ; carpels solitary ; leaves elliptic, cuneate at base, 6–11 cm. long, 3–4·5 cm. broad ; basal glands obscure   ..   ..   2. *glabra*

Receptacle about 1 cm. long ; branches of inflorescence glabrous ; carpels 2–3 ; leaves ovate-elliptic, cuneate to rounded at base, 8–13 cm. long, 3·5–8·5 cm. broad ; basal glands more or less conspicuous   ..   ..   3. *kerstingii*

Receptacle very shortly turbinate, about 3 mm. long ; inflorescence branches and outside of flowers densely pubescent or tomentose ; leaves more or less mealy-puberulous all over the lower surface, elliptic, 7–12 cm. long, 3·5–6 cm. broad ; basal glands conspicuous   ..   ..   4. *robusta*

Leaves denticulate towards the apex, cordate at base ; receptacle densely tomentose, 3–5 mm. long, accompanied by a more or less persistent bracteole ; leaves elliptic, 7–16 cm. long, 4–7 cm. broad, lower surface with a thin cottony felt when young   ..   ..   5. *aubrevillei*

Undersurface of leaves, even when mature, covered with a dense reddish-brown tomentellous felt ; leaves rather broadly oblong-elliptic, rounded or subcordate at base, shortly acuminate, 10–23 cm. long, 4·5–12 cm. broad ; inflorescence and outside of flowers densely reddish-brown tomentose ; receptacle conical, 12–15 mm. long   ..   ..   6. *chrysophylla*

Receptacle (calyx-tube) 3–4 cm. long, strongly curved ; inflorescence and exposed parts of flower glabrous ; leaves elliptic or obovate-elliptic, obtuse and slightly decurrent at base, abruptly acuminate, 12–34 cm. long, 4·5–11 cm. broad, glabrous   7. *gabunensis*

†Flowers in terminal, subspiciform few-branched panicles ; inflorescence and flowers
   densely tomentose ; fertile stamens 15 ; leaves ovate or elliptic, cordate at base,
   rounded or subacute at apex, 10–25 cm. long, 5–15 cm. broad, densely white-
   tomentellous and conspicuously reticulate beneath with 15–20 pairs of prominent
   lateral nerves ; fruit ellipsoid, about 5 cm. long, finely warted ..   8. *macrophylla*
*Sepals acute ; flower-buds 2–3 together, enclosed by a pair of bracts as long as them-
   selves ; inflorescence and flowers densely tomentose ; fertile stamens 7–8 ; leaves
   with 12 or more pairs of lateral nerves prominent beneath ; undersurface white
   tomentose with prominent close reticulation :
 Receptacle (calyx-tube) 2–5 mm. long ; lateral nerves 15–25 pairs ; inflorescence and
    flowers silky-tomentose :
  Leaves rounded at apex, rounded or slightly cuneate at base (sometimes cordate on
     coppice shoots), oblong-elliptic, 5–17 cm. long, 3–8 cm. broad ; inflorescence a lax
     open many-flowered panicle ; fruit ovoid, up to 3·5 cm. long ; a savannah tree or
     shrub ..   ..   9. *curatellifolia*
  Leaves conspicuously acuminate, narrowed at base, elliptic or ovate-elliptic to
     lanceolate, 5–12 cm. long, 2–5 cm. broad ; flowers in a leafy terminal panicle ; fruit
     obliquely ellipsoid, 3–4 cm. long ; a forest tree ..   ..   ..   10. *excelsa*
  Leaves shortly acuminate, cordate at base, ovate-oblong, 5–10 cm. long, 2·5–5 cm.
     broad ; flowers in a leafy terminal panicle ; fruit ellipsoid, about 3 cm. long ; a
     forest tree   ..   ..   11. *congensis*
 Receptacle (calyx-tube) about 8 mm. long ; lateral nerves about 12 pairs ; inflor-
    escence a terminal corymbose, much-branched cyme, shorter than the leaves, like the
    flowers shortly tomentose ; leaves broadly elliptic, cuneate to cordate at base,
    rounded at apex, 6–10 cm. long, 5–7 cm. broad ; young fruits ovoid, about 2 cm.
    long, slightly pubescent   ..   ..   12. *benna*

1. **P. polyandra** *Benth.* var. polyandra—F.T.A. 2 : 370 ; Aubrév.[1] Fl. For. Soud.-Guin. 207, t. 40, 6–7, incl.
   var. *villosa* Aubrév. ; Hauman in Bull. Jard. Bot. Brux. 21 : 187. Savannah tree, to 25 ft. high ;
   leaves coriaceous, dark green and glossy above ; flowers white or pink in flattened corymbs ; fruits
   deep red to blackish purple.
       G.C. : Ejura (fr. Aug.) *Chipp* 724 ! Masera (Apr.) *Thompson* 69 ! Yegi (July) *Pomeroy* FH 1243 !
   Bujan to Sekaii (Apr.) *Kitson* 561 ! Yendi (fl. & fr. Feb.) *McLeod* 814 ! **Togo**: Sokode (fl. Apr., fr. Oct.)
   Kersting 84 ! A. 690 ! **Dah.**: Zagnanado (Feb.) *Chev.* 23103 ! **N.Nig.**: Nupe *Barter* 1105 ! Kontagora
   (Feb.) *Dalz.* 278 ! Bida (Jan.) *Meikle* 1018 ! Lokoja (fl. & fr. Mar.) *Elliott* 39 ! **S.Nig.**: Idah (Sept.)
   T. Vogel 3 ! Olokemeji (Mar.) *A. F. Ross* 141 ! Ogoja Dist. *Rosevear* 53/29 ! Also in French Cameroons,
   Ubangi-Shari and A.-E. Sudan. (See Appendix, p. 170.)
1a. **P. polyandra** var. **cinerea** *Engl.* Bot. Jahrb. 17 : 87 (1893) ; Hauman l.c. 188. *P. polyandra* var. *argentea*
    Aubrév.[1] l.c. 207 (1950).
       Togo: Glei *Busse* 3565. **S.Nig.**: Lagos *Foster* 23 ! Ibode Bere Road (fr. May) *Denton* 8 ! Also in A.-E.
    Sudan.
2. **P. glabra** *Oliv.* F.T.A. 2 : 370 (1871) ; Hauman l.c. 186 ; Fl. Congo Belge 3 : 54. *P. vassonii* A. Chev.
   Bot. 250, name only. *P. kerstingii* of Aubrév. Fl. For. C. Iv. 1 : 148, t. 51 and of F.W.T.A., ed. 1, 1 :
   317, partly. A forest tree, up to 100 ft. high and 12 ft. girth ; calyx green ; petals white ; fruit ovoid,
   somewhat flattened, about 1½ in. long.
       S.L.: Bobobu (June) *Aylmer* 577 ! Yoni, Bonthe Isl. (Mar.) *Deighton* 2482 ! **Lib.**: Gbanga (Sept.)
   Linder 538 ! Dukwia R. (May) *Cooper* 454 ! Harbel, Montserrado (July) *Baldwin* 6661 ! **Iv.C.**: Yapo
   Chev. 22328 ! Banco *Aubrév.* 627 ! Abidjan *Chev.* 127 ! **G.C.**: Abosso (Aug.) *Vigne* FH 981 ! Dunkwa
   (Oct.) *Vigne* FH 162 ! **S.Nig.**: Okomu F.R. (Jan.–Mar.) *A. F. Ross* 193 ! *Brenan* 9126 ! Sapoba
   Kennedy 2129 ! Awka *Thomas* 108 ! Also in French Cameroons, Ubangi-Shari, Spanish Guinea and
   Belgian Congo. (See Appendix, p. 169.)
       [The specimens cited above, other than those from Nigeria, differ from the type by having the young
   branchlets covered with shortly puberulous and long-pilose hairs ; they may represent a distinct
   variety.]
3. **P. kerstingii** *Engl.* Bot. Jahrb. 46 : 140, t. 3 (1911) ; Hauman in Bull. Jard. Bot. Brux. 21 : 186 ; Fl.
   Congo Belge 3 : 56 ; F.W.T.A., ed. 1, 1 : 317, partly (excl. syn.). Tree to 60 ft. high, in fringing forests
   in the savannah regions ; flowers white.
       Togo: Sokode-Basari (fl. & fr. Jan., Feb.) *Kersting* A. 320 ! A. 547 ! Atakpame (Feb.) *Doering* 297.
   N.Nig.: Bauchi Plateau (Jan.) *Lely* 737 ! P. 119 ! *Batten-Poole* 106 ! **S.Nig.**: Nsukka (fl. & fr. Feb.)
   Cousens FHI 7456 ! Udi (Feb.) *Chesters* 193 ! Also in French Cameroons, Ubangi-Shari and Belgian
   Congo. (See Appendix, p. 169.)
4. **P. robusta** *Oliv.* F.T.A. 2 : 370 (1871) ; Aubrév. Fl. For. C. Iv. 1 : 150, t. 52, 1–5 ; Hauman in Bull. Jard.
   Bot. Brux. 21 : 186 (N.B.—var. *glabrifolia* Hauman is var. *robusta*). Tree, to 40 ft. high, or more ; in
   coastal areas and swamp forest.
       Iv.C.: Sassandra, Abidjan, Bingerville, etc., *fide* Aubrév, *l.c.* **G.C.**: Abbontiakoon (July) *Brown* 1 !
   S.Nig.: Badagri (Nov.) *A. F. Ross* 171 ! Lagos (Feb.) *Dalz.* 1338 ! Nun R. (fl. & fr. Sept.) *Mann* 481 !
   Okeluseh (Nov.) *Farquhar* 42 ! Onitsha *Unwin* 72 ! Abokam (Jan.) *Holland* 200 ! (See Appendix,
   p. 170.)
5. **P. aubrevillei** *Pellegr.* in Bull. Soc. Bot. Fr. 78 : 140 (1931) ; Aubrév. l.c. 1 : 150, t. 53 ; Hauman l.c. 187.
   P. polyandra and *P. robusta* of F.W.T.A., ed. 1, 1 : 317, partly. A large forest tree.
       S.L.: Kambui (Oct.) *King* 283 ! *Aylmer* 252 ! **Iv.C.**: Azaguié (Sept.) *Chev.* 22355 ! Dakpadou (Oct.)
   Aubrév. 847 ! (See Appendix, p. 168.)
6. **P. chrysophylla** *Oliv.* F.T.A. 2 : 369 (1871) ; Aubrév. l.c. 152, t. 54 B ; Hauman l.c. 187. Forest tree, to
   75 ft. high, with very characteristic rusty-brown indumentum on lower surface of leaves.
       Lib.: Dukwia R. (Mar.–May) *Cooper* 297 ! 446 ! Sangwin (Mar.) *Baldwin* 11331 ! **Iv.C.**: Tabou
   Aubrév. 1678 ! **G.C.**: Prestea *Vigne* FH 3081 ! **S.Nig.**: Oban *Talbot* ! Afi River F.R. *Jones* FHI 14147 !
   Keay ! Also in French Cameroons and Gabon. (See Appendix, p. 170.)
7. **P. gabunensis** *Engl.* Bot. Jahrb. 27 : 87 (1893) ; Hauman l.c. 187 ; Fl. Congo Belge 3 : 58. A glabrous
   forest tree, to 60 ft. high.
       Br.Cam.: Kumba to Banga (Mar.) *Brenan* 9470 ! Also in French Cameroons, Gabon and Mayumbe.

---

[1] Aubréville does not indicate to which variety the specimens he cites belong. He records the species as a
whole from French Sudan, Ivory Coast and Ubangi-Shari.

8. **P. macrophylla** *Sabine*—F.T.A. 2 : 369 ; Aubrév. Fl. For. Soud.-Guin. 205, t. 39, 1–3 ; Hauman in Bull. Jard. Bot. Brux. 21 : 189. Shrub or tree to 30 ft. high, with stout densely tomentose branchlets, and gnarled bole ; flowers crowded, white or pinkish ; fruit rough-skinned edible ; on sandy beaches and also inland in the driest savannah regions.
   Sen.: *Sieber* 24 ! *Döllinger* 30 ! Panie-Foul *Perrottet* ! St. Louis *Roger* ! **Gam.:** *Brown-Lester* 219 ! Genieri *Fox* 82 ! **Port.G.:** Pussubé, Bissau (Mar.) *Esp. Santo* 1154 ! **Fr.G.:** *Farmar* 303 ! **S.L.:** *Don* ! Bawbaw (Dec., Jan.) *Smythe* 219 ! *Lane-Poole* 44 ! Gbap (Feb.) *Dawe* 470 ! Lumley (Feb., Aug.) *Deighton* 15 ! *Jones* FHI 12581 ! Fotani, Simiria (Apr.) *Glanville* in *Hb. Deighton* 2492 ! **Lib.:** Monrovia (Apr., June) *Whyte* ! *Bequaert* 181 ! *Baldwin* 5844 ! *Barker* 1242 ! Congotown (July) *Linder* 45 ! **N.Nig.:** Sokoto (fl. all year) *Lely* 804 ! *Foster* 1 ! Katagum *Dalz.* 324 ! Also recorded very doubtfully from S. Tomé. (See Appendix, p. 169.)

9. **P. curatellifolia** *Planch. ex Benth.*—F.T.A. 2 : 368 ; Aubrév. l.c. 253, t. 39, 4–6 ; Hauman l.c. 193 ; Fl. Congo Belge 3 : 66. A savannah tree, to 25 ft. high, with black fissured bark ; leaves pale green beneath ; petals white.
   Sen.: *Heudelot* 362 ! Fouladou, Casamance *Etesse* 63. **Fr.Sud.:** Arbala *Dubois* 63. **Port.G.:** Farim (Apr.) *Esp. Santo* 2451 ! Coiada, Gabu (June) *Esp. Santo* 2507 ! **Fr.G.:** Kouroussa (Feb.) *Pobéguin* 653 ! **S.L.:** Falaba (Mar.) *Sc. Elliot* 5108 ! Manankon, Musaia (Apr.) *Deighton* 5427 ! **Iv.C.:** Baoulé, Ferkéssédougou, Bondoukou and Nzi Comoé, *fide* Aubrev. *l.c.* **G.C.:** Jema (Feb.) *Brown* FH 2159 ! Durimo to Yijia (Mar.) *Kitson* 939 ! Bujan to Sekaii (Apr.) *Kitson* 919 ! **Togo:** Ho (Feb.) *Howes* 1106 ! Kete Krachi (Jan.) *Vigne* FH 1535 ! **Dah.:** *Poisson* ! Zaganado (Feb.) *Chev.* 23016 ! **N.Nig.:** Nupe *Barter* 1106 ! Patti Lokoja (fl. Jan., fr. Sept.) *T. Vogel* 177 ! *Elliott* 224 ! Kontagora (Jan.) *Dalz.* 86 ! Bauchi Plateau (Jan.) *Lely* P. 13 ! Otobi, Idoma (Feb.) *Jones* FHI 702 ! **S.Nig.:** Lagos (Feb.) *Foster* 24 ! Olokemeji (Feb.) *Ejiofor* FHI 31768 ! Also in French Cameroons, Ubangi-Shari, A.-E. Sudan, Uganda and Belgian Congo. (See Appendix, p. 168.)

10. **P. excelsa** *Sabine*—F.T.A. 2 : 367 ; Aubrév. Fl. For. C. iv. 1 : 146, t. 50 B ; Hauman in Bull. Jard. Bot. Brux. 21 : 191. *P. holstii* Engl. (1895)—Hauman l.c. 191 ; Fl. Congo Belge 3 : 59. *P. elliotii* Engl. (1899). *P. tenuifolia* A. Chev. (1909)—F.W.T.A., ed. 1, 1 : 318 ; Aubrév. l.c. A large evergreen forest tree with pale tomentose branchlets and inflorescence ; fruit rough-skinned, sometimes eaten.
   Sen.: *Berhaut* 1787. **Gam.:** *Dawe* 84 ! **Port.G.:** Pussubé (Feb.) *Esp. Santo* 1134 ! **Fr.G.:** Rio Nunez *Heudelot* 633 ! Conakry (Jan.) *Dalz.* 8384 ! Konkouré to Timbo (Mar.) *Chev.* 12486 ! Kouria (Jan.) *Chev.* 15109 ! **S.L.:** *Don* ! Bagroo R. (Apr.) *Mann* 885 ! Njala (Jan., Feb.) *Deighton* 514 ! 1568 ! Berria, Falaba *Sc. Elliot* 4998 ! **Lib.:** Dukwia R. (Feb.) *Cooper* 270 ! **Iv.C.:** *Fleury* 6 ! Dabou *Chev.* 16204 ! Azaguié *Chev.* 22295 ! **G.C.:** Prestea *Green* FH 899 ! Axim (Jan.) *Vigne* FH 4845 ! Subirisu (Feb.) *Thompson* 14 ! Obuasi (fr. Sept.) *Vigne* FH 926 ! Oni *Sankey* 19 ! Sapoba *Kennedy* 1643 ! Okigwi *Cousens* FHI 4856 ! Widespread in tropical Africa. (See Appendix, p. 168.)

11. **P. congensis** *F. Didr.* Kjoeb. Vidensk. Medd. 197 (1854) ; Hauman in Bull. Jard. Bot. Brux. 21 : 192 ; Fl. Congo Belge 3 : 62, t. 3. *P. subcordata* Oliv. F.T.A. 2 : 367 (1871) ; F.W.T.A., ed. 1, 1 : 318 ; Aubrév. l.c. 148. An evergreen tree, to 100 ft. high ; on river banks and in fringing forest in savannah country ; flowers white, with a pink tinge, fragrant.
   Fr.G.: R. Dentilia (fr. Mar.) *Sc. Elliot* 4274 ! **S.L.:** Njala (Nov.) *Deighton* 4953 ! Bonjema (fr. Sept.) *Deighton* 4634 ! **Iv.C.:** Touba *Aubrév.* 1241 ! **G.C.:** Grube (Jan.) *Vigne* FH 1548 ! Senchi, R. Volta (fr. Apr.) *Irvine* 2876 ! **Togo:** R. Tamboma *Kersting* A. 683 ! **Dah.:** *Le Testu* 416 ! **N.Nig.:** Nupe *Barter* 1001 ! 1694 ! Jebba (Dec.) *Hagerup* 722 ! R. Benue (Dec.) *Dalz.* 718 ! **S.Nig.:** Olokemeji (Nov.) *Jones* FHI 14231 ! Old Oyo F.R. *Keay* FHI 16269 ! Enyon Creek, Arochuku (Sept.) *Corbin* 50 ! Also in Congo Basin and Ubangi. (See Appendix, p. 170.)

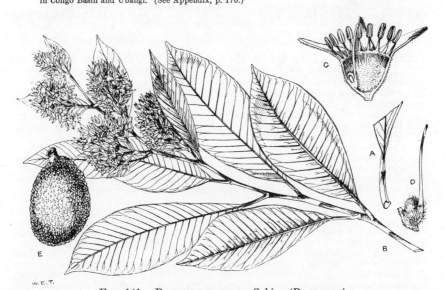

FIG. 141.—PARINARI EXCELSA *Sabine* (ROSACEAE).

A, base of leaf. B, flowering branch. C, longitudinal section of flower. D, pistil. E, fruit.

12. **P. benna** *Sc. Elliot* in J. Linn. Soc. 30 : 78 (1894) ; Aubrév. Fl. For. Soud.-Guin. 207, t. 40 ; Hauman in Bull. Jard. Bot. Brux. 21 : 194. *P. mobola* of Chev. Bot. 251. A savannah shrub or small tree to 20 ft. high ; flowers white with pink centre.
   Sen.: Kaolak *Kaichinger.* **Fr.G.:** *Farmar* 181 ! 228 ! Kindia (Mar.) *Chev.* 13068 ! Follo, Kouria (Dec.) *Chev.* 14824 ! Mamou (Mar.) *Dalz.* 8386 ! **S.L.:** Talla Hills (Mar.) *Sc. Elliot* 5052 ! Bafodeya (Apr.) *Sc. Elliot* 5480 ! Jigaya (fr. Sept.) *Thomas* 2723 ! Musaia (fl. & fr. Mar., fr. Sept.) *Deighton* 5445 ! *Small* 277 ! 330 ! (See Appendix, p. 168.)

Imperfectly known species.

P. sp. Rain forest tree to 95 ft. high ; near *P. congensis* F. Didr., but leaves larger and stipules foliaceous, up to 3 in. long ; flowers and fruits not known.
Br.Cam.: S. Bakundu F.R. *Ejiofor* FHI 15253 !   *Olorunfemi* FHI 30502 !

8. HIRTELLA Linn. Sp. Pl. 1 : 34 (1753) ; A. Chev. in Bull. Mus. Hist. Nat. Sér. 2, 3 : 192–195 (1931).

Stipules large, ovate, 2–4 cm. long, 1·5–2 cm. broad ; leaves oblong, obtuse at base, acuminate at apex, 15–20 cm. long, about 10 cm. broad, petiole 1–1·2 cm. long ; panicles glabrous ; receptacle 5–6 mm. long ; staminodes united into a narrow tongue, denticulate at apex   ..   ..   1. *conrauana*
Stipules small, linear-oblong, up to 1 cm. long and 2 mm. broad ; inflorescences rusty-pubescent :
   Leaves large, oblong-ovate, cordate at base, rounded and abruptly apiculate at apex, 18–40 cm. long, 8–18 cm. broad, subsessile or petiole up to 5 mm. long ; inflorescence a panicle of long racemes, up to 40 cm. long ; receptacle about 12 mm. long ; staminodes united into a narrow tongue, denticulate at apex   ..   2. *fleuryana*
   Leaves smaller, up to 20 cm. long and 8 cm. broad ; receptacle 3–5 mm. long ; staminodes in the form of a comb :
     Leaves rounded or subcordate at base, broadly oblong-lanceolate, acute or rounded at apex, 10–20 cm. long, 4–8 cm. broad ; inflorescence a terminal many-flowered ample panicle, up to 20 cm. long ; receptacle about 3 mm. long   ..   3. *butayei*
     Leaves cuneate at base, narrowly obovate-elliptic, gradually acuminate, 8–18 cm. long, 3–5 cm. broad ; flowers in axillary, lax, few-flowered racemes, up to 8 cm. long ; receptacle about 5 mm. long   ..   ..   4. *cupheiflora*

1. **H. conrauana** (*Engl.*) *A. Chev.* in Bull. Mus. Hist. Nat. Sér. 2, 3 : 194 (1931). *Magnistipula conrauana* Engl. Bot. Jahrb. 36 : 226 (1905) ; Hauman in Bull. Jard. Bot. Brux. 21 : 174. A glabrous shrub.
   Br.Cam.: Bangwe (Mar.) *Conrau* 65.
2. **H. fleuryana** *A. Chev.* l.c. 192 (1931). *Parinari fleuryana* (A. Chev.) Aubrév. Fl. For. C. Iv. 1 : 153 (1936). *Magnistipula fleuryana* (A. Chev.) Hauman l.c. 175 (1951). Forest tree, to 45 ft. high ; in swampy places ; flowers white, fragrant ; inhabited by ants.
   Lib.: Zwedru, Tchien (Aug.) *Baldwin* 7017 !   Sarbo, Webo (July) *Baldwin* 12522 !   **Iv.C.**: Gagnoa to Issia, Sassandra (fl. & fr. Oct.) *Chev.*
3. **H. butayei** (*De Wild.*) *Brenan* in Trop. Woods 86 : 4 (1946) ; Hauman in Fl. Congo Belge 3 : 41, t. 2. *Magnistipula butayei* De Wild. (1908). *Parinari sargosii* Pellegr. (1920)—Aubrév. Fl. For. C. Iv. 1 : 152, t. 54 A. *P. tisserantii* Aubrév. & Pellegr. (1950)—Aubrév. Fl. For. Soud.-Guin. 208, t. 39, 7–14. Forest tree, to 100 ft. high.
   Iv.C.: Guiglo to Toulépleu *Aubrév.* 1231 !   Also in Ubangi-Shari, Gabon and Belgian Congo.
4. **H. cupheiflora** (*Mildbr.*) *Mildbr. ex A. Chev.* l.c. 195 (1931) ; Hauman in Bull. Jard. Bot. Brux. 21 : 184. *Magnistipula cupheiflora* Mildbr. in Notizbl. Bot. Gart. Berl. 7 : 57 (1921). A medium-sized forest tree, with irregular bark and rough bole.
   S.L.: Gola East F.R. (Oct.) *D. G. Thomas* 118 !   *King* 172 !   Also in French Cameroons.

9. ACIOA Aubl. Pl. Gui. 2 : 698 (1775) ; De Wild. in Bull. Jard. Bot. Brux. 7 : 188 (1920). *Griffonia* Hook. f. (1865)—F.T.A. 2 : 371.

*Calyx-tube glabrous or very shortly and thinly pubescent outside :
   Bracteoles on the pedicels more or less ovate, entire, not or rarely slightly gland-toothed :
     Inflorescence a simple raceme, rarely the racemes more than 1 in the leaf-axils ; the bracts usually much larger than the bracteoles :
       Calyx-tube (receptacle) distinctly puberulous outside, 1·5 cm. long ; branchlets glabrescent ; leaves elliptic, cuneate at base, 6–10 cm. long, 3·5–5 cm. broad, glabrous ; lateral nerves 4–5 pairs ; stipules subulate, entire ; flowers numerous, small ; staminal column 1·5–2 cm. long ..   ..   1. *barteri*
       Calyx-tube (receptacle) quite glabrous outside ; leaves more or less scabrid above :
       Leaves cuneate at base ; characters otherwise much as in *A. barteri*
                                         2. *scabrifolia*
     Leaves rounded at base, ovate-lanceolate or oblong-lanceolate, gradually acuminate, 6–15 cm. long, 2·5–6 cm. broad, more or less hispid on the nerves ; lateral nerves 5–6 pairs ; stipules ovate-lanceolate, toothed at base ; axis of inflorescence up to 3 cm. long ; flowers crowded ; calyx-tube (receptacle) 1·5–2 cm. long ; fruit broadly ovoid, 3 cm. long, tomentose   ..   ..   ..   3. *whytei*
     Inflorescence a panicle ; the bracts scarcely larger than the bracteoles ; leaves broadly elliptic, slightly cuneate at base, 12–16 cm. long, 5–8 cm. broad, glabrous ; lateral nerves 7 pairs ; stipules obliquely ovate, nervose ; calyx-tube (receptacle) up to 1·5–2 cm. long, puberulous ; staminal column 2·5 cm. long   ..   4. *mannii*
   Bracteoles on the pedicels deeply gland-toothed or palmatisect :
     Lateral nerves up to 8 pairs ; leaves ovate or shortly elliptic :
       Bracteoles divided to the base, gland-tipped ; leaves ovate, rounded at the base, acutely acuminate, 6–11 cm. long, 4–5·5 cm. broad, long-pilose on the nerves beneath ; lateral nerves 6 pairs ; branchlets bristly hairy ; stipules very small ; axis of inflorescence 4 cm. long ; bracteoles laciniate nearly to the base, gland-tipped ; calyx-tube (receptacle) glabrous, 1·5 cm. long, falling away and leaving long peg-like persistent pedicels ; staminal column 1·5 cm. long   ..   5. *unwinii*

Bracteoles divided up to the middle :
Calyx-tube (receptacle) glabrous outside ; bracts not glandular ; stipules entire :
Leaves rounded or subcuneate at base, more or less elliptic, shortly acuminate,
7–13 cm. long, 4–6 cm. broad, pubescent on the nerves ; lateral nerves about 7
pairs ; bracteoles ovate, with short glandular teeth ; stipules linear-subulate ;
branchlets with short hairs ; calyx-tube glabrous, 2·5 cm. long ; staminal
column 2 cm. long   ..   ..   6. *rudatisii*
Leaves subcordate at base, oblong, abruptly acuminate, 6–10 cm. long, 3–4 cm.
broad, the midrib with long spreading hairs ; bracteoles with long gland-tipped
teeth ; branchlets spreadingly pilose ; calyx-tube glabrous, 2 cm. long ;
staminal column 3 cm. long   ..   ..   7. *hirsuta*
Calyx-tube (receptacle) setose-pilose outside ; bracts as well as the bracteoles with
glandular teeth ; stipules with glandular teeth ; leaves ovate-elliptic to lanceolate, rounded to broadly cuneate at base, gradually acutely acuminate, 5·5–11·5
cm. long, 1·7–4 cm. broad ; branchlets and undersurface of leaves setose-pilose ;
fruit ovoid, 2–3·5 cm. long, 2–3 cm. broad, densely tomentellous and with longer
hairs..   ..   ..·..   ..   ..   ..   8. *johnstonei*
Lateral nerves 10–16 pairs ; leaves elongate-oblong, cordate at base, up to 30 cm.
long and 10 cm. broad, glabrous or hirsute on nerves beneath ; stipules ovate or
ovate-lanceolate, foliaceous ; branchlets with long spreading hairs ; bracteoles
deeply palmatisect with gland-tipped segments ; calyx-tube (receptacle) 2·5–3·5
cm. long, glabrous ; staminal column 4 cm. long ; fruit ovoid, 6–7 cm. long,
tomentose and bristly   ..   ..   9. *klaineana*
*Calyx-tube densely tomentose outside :
Leaves cordate at base :
Midrib rather densely pubescent beneath ; hairs on the calyx-tube very short ;
bracts and bracteoles conspicuous ; leaf-blades oblong-elliptic, abruptly acuminate,
10–15 cm. long, 5–7 cm. broad ; lateral nerves about 7 pairs ; stipules subulate ;
calyx-tube 2–2·5 cm. long ; staminal column about 3·5 cm. long ; fruits ellipsoid,
about 4 cm. long, tomentose with short hairs and also sparsely pilose with long
hairs   ..   ..   10. *pallescens*
Midrib glabrous beneath ; leaves broadly oblong-elliptic, shortly acuminate, about
15 cm. long, 8–9 cm. broad, with about 7 pairs of lateral nerves ; inflorescence
paniculate ; calyx-tube 1 cm. long ; staminal column 2·5 cm. long, shortly
pubescent   ..   ..   11. *eketensis*
Leaves rounded or slightly cuneate at base ; hairs on the calyx-tube fairly long :
Inflorescence racemose :
Inflorescence lax-flowered, the axis about 10 cm. long ; bracts ovate-lanceolate, 5–6
mm. long, glabrous inside ; bracteoles ovate-lanceolate, conspicuous ; leaf-blades
oblong-elliptic, 8–15 cm. long, 5–6·5 cm. broad ; midrib glabrescent ; lateral
nerves 8 pairs ; stigmas ovate, long-acuminate ; calyx-tube 3 cm. long ; staminal
column 3 cm. long ..   ..   12. *dinklagei*
Inflorescence rather close-flowered, the axis 3–4 cm. long ; bracts subulate ; about
2 mm. long, hairy all over ; bracteoles minute ; leaf-blades elliptic, acutely
acuminate, 9–14 cm. long, 4–7 cm. broad ; midrib tomentose beneath ; lateral
nerves about 10 pairs ; stipules lanceolate, short ; calyx-tube 2 cm. long ;
staminal column 3 cm. long   ..   ..   13. *talbotii*
Inflorescence dichotomously cymose ; leaves broadly elliptic, rounded at base,
shortly acuminate, up to 15 cm. long and 9 cm. broad, glabrous, with about 8 pairs
of lateral nerves ; branchlets lenticellate ; calyx 1·5 cm. long, softly pubescent ;
staminal column 2 cm. long   ..   ..   14. *dichotoma*

1. **A. barteri** (*Hook. f. ex Oliv.*) Engl. Bot. Jahrb. 26 : 382 (1899) ; De Wild. in Bull. Jard. Bot. Brux. 7 : 210.
Griffonia barteri Hook. f. ex Oliv. F.T.A. 2 : 373 (1871). *Acioa tenuiflora* Dinkl. & Engl. (1899). *A.
eketensis* of F.W.T.A., ed. 1, 1 : 320, partly (*Kitson* s.n.). A shrub or small tree, with slender branches
more or less climbing ; leaves dark glossy green, reddish-brown when dry ; flowers greenish-white,
fragrant ; fruit 1½ in. long, green, turgid, tubercled, more or less pear-shaped and pointed ; sometimes
planted.
S.L.: Makeni (introduced from Nigeria in 1938 ; fl. Mar.) *Deighton* 4259 ! **Lib.:** Grand Bassa *Dinklage*
2101 ! Gbanga (Sept.) *Linder* 613 ! Gbau, Sanokwele (Sept.) *Baldwin* 9398 ! Tappita (Aug.) *Baldwin*
9108 ! **Iv.C.:** Banco *Aubrév.* 375 ! **G.C.:** Benso, Tarkwa (fl. Aug., fr. Nov.) *Andoh* FH 5388 ! **S.Nig.:**
Lagos (fl. Oct.–Feb., fr. Mar.) *Barter* 2183 ! *Dalz.* 1013 ! Agege *Kitson* ! Agulu *Thomas* 177 ! Calabar
(Jan.) *Thomson* 66 ! Eket *Talbot* 3338 ! Oban *Talbot* 1372 ! 1625 ! Also in Belgian Congo (?).
2. **A. scabrifolia** Hua in Bull. Mus. Hist. Nat. 3 : 328 (1897) ; De Wild. l.c. 206 ; Aubrév. Fl. For. C. Iv. 1 :
154, t. 55. *A. lehmbachii* Engl. (1899), *fide* Hutch. & Dalz. A tree up to 30 ft. high ; flowers white,
sometimes pink.
Fr.G.: Timbo (Mar.) *Miquel* 24. R. Konkouré, Kouria (Sept., Dec.) *Chev.* 14889 ! *Pobéguin* K. 14 !
S.L.: Ninia, Talla Hills (Feb.) *Sc. Elliot* 4894 ! Mamboma (Nov.) *Deighton* 3779 ! Makali (Feb.) *Deighton*
4056 ! Musaia (fl. Dec., fr. Apr.) *Deighton* 4477 ! 5405 ! **Lib.:** Peahtah (Oct.) *Linder* 1114 ! 1117 !
Zigida, Vonjama (Oct.) *Baldwin* 10010 ! Kondessu, Boporo (fr. Dec.) *Baldwin* 10665 ! Gbanga (Dec.)
Baldwin 10515 ! **Iv.C.:** *fide* Aubrév. l.c. **Br.Cam.:** Buea *Lehmbach* 115 (possibly distinct). (See
Appendix, p. 167.)
3. **A. whytei** Stapf in J. Linn. Soc. 37 : 97 (1905) ; De Wild. l.c. 204. *A. stapfiana* De Wild. l.c. 204 (1920) ;
F.W.T.A., ed. 1, 1 : 320. Shrub or small tree, to 30 ft. high, branchlets slender, long and pendulous ;
flowers white.
S.L.: Sherbro (Oct.) *Garret* ! Kafogo, Limba (Apr.) *Sc. Elliot* 5521 ! Bumbuna (fr. Oct.) *Thomas* 3358 !

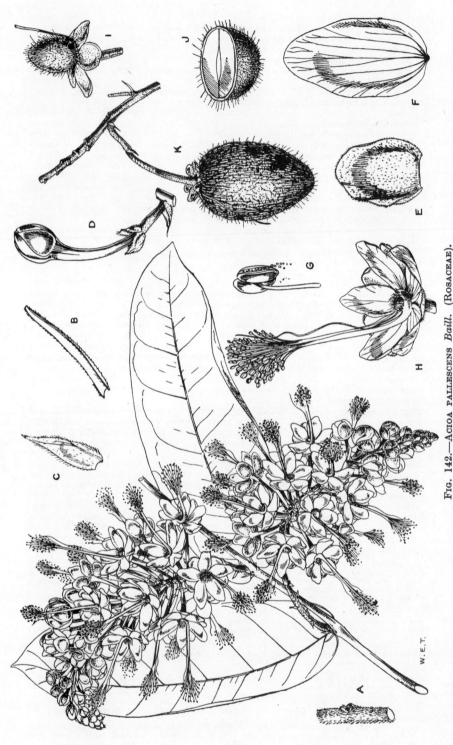

Fig. 142.—Acioa pallescens *Baill.* (Rosaceae).

A, portion of stem showing leaf-scar and axillary bud. B, stipule. C, bract. D, flower-bud. E, sepal. F, petal. G, stamen. H, flower. I, young fruit. J, cross-section of same. K, mature fruit.

W. E. T.

Njala (Sept.) *Deighton* 1319 ! Bo F.R. (Aug.) *Small* 191 ! Pujehun, Valunia (July) *Deighton* 580 ! **Lib.:** Monrovia *Whyte*! Kakatown *Whyte*! Moala (fr. Nov.) *Linder* 1381 ! Grand Cape Mount (fr. Dec.) *Baldwin* 10780 ! 10823 !

4. **A. mannii** (*Oliv.*) *Engl.* l.c. 381 (1899) ; De Wild. l.c. 210. *Griffonia mannii* Oliv. F.T.A. 2 : 372 (1871). A climbing shrub or small tree, 15–20 ft., with glabrous branchlets.
   Br.Cam.: Buea (Mar.) *Maitland* 467 ! Johann-Albrechtshöhe (Jan., Mar.) *Staudt* 513 ! 888 ! Sonje-Pflamung (Feb.) *Winkler* 1104 ! **F.Po:** *Mann* 1427 !

5. **A. unwinii** *De Wild.* l.c. 198 (1920). *A. parvifolia* of F.W.T.A., ed. 1, 1 : 320, partly. A shrub, hairy at first, but becoming more or less glabrous ; flowers white.
   S.L.: Buri town, on sea beach (Dec.) *Unwin & Smythe* 36 ! Ronietta (Nov.) *Thomas* 5621 ! **S.Nig.:** Sapoba (Oct.) *Kennedy* 2058 ! FHI 892 ! *A. F. Ross* 210 !

6. **A. rudatisii** *De Wild.* l.c. 215 (1920). A forest tree, to 60 ft. high.
   S.Nig.: Bateriko *Jones* FHI 18992 ! Calabar *Robb* ! **Br.Cam.:** Buea (Mar., May) *Deistel* 654 ! *Maitland* 440 ! 545 ! 659 ! Abonando (Mar.) *Rudatis* 30 !

7. **A. hirsuta** *A. Chev. ex De Wild.* l.c. 200 (1920). *A tenuiflora* of Chev. Bot. 253, not of Dinkl. & Engl.
   Iv.C.: Grabo (Aug.) *Chev.* 19738 ! Taté to Tabou (Aug.) *Chev.* 19818 !

8. **A. johnstonei** *Hoyle* in Kew Bull. 1932 : 258 ; l.c. 1947 : 71. A small tree or shrub at edge of forest, or in savannah.
   Br.Cam.: Bum, Bamenda (Jan.) *Johnstone* 74/31 ! Fonfuka, Bamenda (June) *Maitland* 1401 ! (See Appendix, p. 167.)

9. **A. klaineana** *Pierre ex De Wild.* l.c. 196 (1920). A shrub or tree ; inflorescence copious, up to 6 or more inches long.
   S.Nig.: Eket *Talbot* 3123 ! 3153 ! Oban (fr. May) *Talbot* 1529 ! Ujor FHI 31795 ! **Br.Cam.:** Barombi *Preuss* 263 ! Also in French Cameroons and Gabon.

10. **A. pallescens** *Baill.* in Adansonia 7 : 224 (1867) ; De Wild. l.c. 207. *Griffonia pallescens* (Baill.) Oliv. F.T.A. 2 : 372, incl. vars. A small tree or shrub, more or less climbing, with pale silky racemes of white flowers and brown fruits.
    S.Nig.: Calabar (Feb.) *Thomson* 2 ! *Mann* 2288 ! Itu (Feb., Apr.) *Holland* 35 ! *Smith* 75 ! Oban *Talbot* 1599 ! Degema *Talbot* ! **Br.Cam.:** Johann-Albrechtshöhe *Staudt* 618 ! 944 ! Also in French Cameroons and Spanish Guinea.

11. **A. eketensis** *De Wild.* l.c. 214 (1920). Branchlets glabrous.
    S.Nig.: Eket *Talbot* 3239 !

12. **A. dinklagei** *Engl.* Bot. Jahrb. 26 : 381 (1899) ; De Wild. l.c. 211. A tall shrub or tree, with drooping tomentose branchlets and hoary racemes of white flowers.
    Lib.: Grand Bassa (Sept.) *Dinklage* 2023 ! Tchien Dist. (Aug.) *Baldwin* 6980 ! 7078 ! **G.C.:** Sa, Bonsa R., Tarkwa (Aug.–Sept.) *Vigne* FH 156 ! 973 ! (See Appendix, p. 167.)

13. **A. talbotii** *Bak. f.* in Cat. Talb. 29 (1913) ; De Wild. l.c. 206. A small tree, about 20 ft. high, with pale tomentose branchlets and brownish-velvety inflorescence.
    S.Nig.: Oban *Talbot* 1533 !

14. **A. dichotoma** *De Wild.* l.c. 216 (1920). Branchlets tomentose.
    S.Nig.: Eket, along the rivers *Talbot* 3048 !

Imperfectly known species.

A. parvifolia *Engl.* Bot. Jahrb. 26 : 380 (1899) ; De Wild. l.c. 198.
　S.L.: *Afzelius.* [I have been unable to study the type of this species, but from the description it does not seem to be the same as *A. unwinii* De Wild. as stated in F.W.T.A., ed. 1.—R.W.J.K.]

88. CHAILLETIACEAE

Small trees or shrubs, sometimes climbing. Leaves alternate, simple ; stipules present. Flowers small, mostly hermaphrodite, actinomorphic or slightly zygomorphic. Sepals 5, free or partially connate, imbricate. Petals mostly 2-lobed or 2-partite, free or united with the stamens into a tube. Stamens 5, alternate with the petals, free or united ; anthers 2-celled, opening lengthwise, the connective often dorsally thickened. Hypogynous glands opposite to the petals, free or connate. Ovary superior to quite inferior, 2–3-celled ; style mostly simple, 2–3-fid at the apex ; ovules 2 in each cell, pendulous. Fruit a drupe, dry or rarely fleshy, sometimes the epicarp splitting. Seeds without endosperm ; embryo large, straight.

A small tropical family with the aspect of some *Rosaceae* ; recognized by the usually bilobed petals ; peduncles sometimes adnate to the petiole.

Petals equal, free, or rarely united ; fertile stamens 5 ; peduncle free, or sometimes adnate to the petiole　..　　..　　..　　..　　..　　..　　1. **Dichapetalum**
Petals unequal, 1 or 2 of them larger than the others, united into a tube ; fertile stamens 2 or 3 ; peduncle always adnate to the petiole　..　　..　　..　　..　2. **Tapura**

1. **DICHAPETALUM** Thouars Nov. Gen. Madag. 23 (1806). *Chailletia* DC. (1811)— F.T.A. 1 : 339. *Icacinopsis* Roberty (1953).

*Petals free :
　†Stipules pinnatisect, about 1 cm. long ; branchlets densely tomentose ; sepals reflexed ; petals deeply lobed ; inflorescences lax, well-branched :
　Peduncle never adnate to the petiole ; inflorescence a large terminal paniculate cyme about 15 cm. long ; leaves with a close-felted indumentum beneath, elliptic-obovate, cordate at base, shortly acuminate, 10–15 cm. long, 5–8 cm. broad
　　　　　　　　　　　　　　　　　　　　　　　　　　　　1. *petersianum*
　Peduncle adnate to the petiole ; inflorescence a lax, divaricate paniculate cyme ;

leaves densely pubescent beneath, but with the surface visible between the hairs ;
oblong-elliptic, rounded or subcordate at base, gradually acuminate, 9–17 cm. long,
4–9 cm. broad   ..   ..   2. *angolense*
†Stipules entire :
　Leaves with a close-felted whitish indumentum beneath, the hairs entirely covering
　　the lower surface :
　　Midrib and lateral nerves beneath covered with long spreading hairs quite different
　　　from the whitish wool of the surface ; leaves elliptic or oblong, unequally rounded
　　　at base, shortly acuminate, 5–12 cm. long, 3–5 cm. broad ; cymes short, crowded,
　　　pedunculate ..   ..   3. *argenteum*
　　Midrib and lateral nerves beneath covered with the same close-felted whitish
　　　indumentum as the lower surface ; leaves oblong-elliptic to oblong-oblanceolate
　　　or obovate, obtuse at base, acutely acuminate, 6–17 cm. long, 3–7 cm. broad,
　　　lateral nerves 6–9 pairs, prominent beneath ; cymes crowded, with a peduncle
　　　0·5–3 cm. long   ..   ..   4. *pallidum*
　Leaves glabrous, or laxly pilose or pubescent beneath :
　　‡Peduncle never adnate to the petiole :
　　　Petals entire (or shortly bifid), about as long as the 6–7 mm. long erect sepals ;
　　　　leaves elliptic or slightly obovate-oblong, rounded to cuneate at base, bluntly
　　　　acuminate, 7–12 cm. long, 4–7 cm. broad, glabrous, coriaceous ; branchlets
　　　　rusty-tomentellous ; cymes lax, pedunculate, sometimes borne on leafless shoots
　　　　　　　　　　　　　　　　　　　　　　　　　　　　　　5. *barteri*
　　Petals more or less deeply bilobed, usually longer than the sepals (but see No. 12) ;
　　　sepals erect or reflexed, usually less than 5 mm. long :
　　　Inflorescences pedunculate and more or less laxly cymose : ·
　　　　Leaves cordate to subcordate at base ; branchlets glabrous or nearly so ; leaves
　　　　　very broadly elliptic, obtuse to acute at apex, 7–15 cm. long, 5–10 cm. broad,
　　　　　slightly pubescent on the nerves beneath ; cymes few-flowered, about 1·5 cm.
　　　　　long, bracteate ..   ..   6. *subcordatum*
　　　　Leaves not cordate at base :
　　　　　Sepals reflexed ; cymes very slender ; branchlets and leaves glabrous or nearly
　　　　　　so ; leaves shining, with about 6 pairs of lateral nerves prominently looped
　　　　　　beneath :
　　　　　　Petals about 2 mm. long ; pedicels very short ; leaves oblong-elliptic, subacute
　　　　　　　at base, acutely acuminate, 7–12 cm. long, 3–5 cm. broad ; cymes many-
　　　　　　　flowered   ..   ..   7. *zenkeri*
　　　　　　Petals 3–4 mm. long ; pedicels about 5 mm. long ; leaves oblong-oblanceolate
　　　　　　　or obovate-oblong, acutely acuminate, 6–12 cm. long, 2–5 cm. broad ; cymes
　　　　　　　rather few-flowered ..   ..   8. *cymulosum*
　　　　　Sepals not reflexed :
　　　　　　Pedicels scarcely 1 mm. long ; sepals broadly oblong ; petals about 3 mm.
　　　　　　　long, shortly bilobed ; cymes relatively short and densely flowered ; some-
　　　　　　　times subcapitate, more usually branched and often borne on leafless shoots
　　　　　　　thus appearing paniculate ; leaves obovate-elliptic to oblanceolate, obtuse
　　　　　　　to rounded or sometimes cuneate at base, shortly obtusely acuminate, 8–16
　　　　　　　cm. long, 3–7 cm. broad, pubescent beneath (rarely nearly glabrous) ; stigma
　　　　　　　deeply 3-lobed ; fruit ellipsoid, 1–1·5 cm. long ..   ..   9. *guineense*
　　　　　　Pedicels 2–3 mm. long ; sepals narrowly oblong ; petals about 4 mm. long,
　　　　　　　deeply bilobed and much narrowed below ; cymes lax, well-branched and
　　　　　　　usually at least 3 cm. long ; stigma shortly or deeply 3-lobed :
　　　　　　　Cymes few-flowered, slender ; leaves oblong-elliptic, 6–14 cm. long, 2·5–6 cm.
　　　　　　　　broad, with 5–6 lateral nerves prominent beneath and conspicuously looped
　　　　　　　　well within the margin, usually drying green ..   ..   10. *oblongum*
　　　　　　　Cymes many-flowered, large and lax ; leaves broadly elliptic or obovate,
　　　　　　　　10–22 cm. long, 5–11 cm. broad, with 6–7 lateral nerves more or less
　　　　　　　　prominent beneath and looped near the margin, usually drying brownish
　　　　　　　　　　　　　　　　　　　　　　　　　　　　　　11. *floribundum*
　　Inflorescences sessile, or densely subcapitate and pedunculate :
　　　Inflorescences subcapitate and pedunculate :
　　　　Petals equal to or shorter than the sepals ; pedicels longer than the flowers ;
　　　　　peduncle stout, up to 2 cm. long and 2 mm. broad ; branchlets, inflorescences,
　　　　　flowers and petioles densely tomentose ; leaves broadly oblong-elliptic to
　　　　　oblong-obovate, acute to cordate at base, acuminate, 8–18 cm. long, 5–10 cm.
　　　　　broad, strongly and rather closely reticulate beneath ; stigma very shortly
　　　　　3-fid   ..   ..   12. *reticulatum*
　　　　Petals longer than the sepals ; pedicels shorter than the flowers ; stigma
　　　　　3-lobed :
　　　　　Leaves cordate at base :
　　　　　　Leaves spreadingly pilose on the nerves of both surfaces even when mature,
　　　　　　　obovate-elliptic, 6–10 cm. long, 4–5 cm. broad ; peduncle 0·5–2 cm. long ;

cymes rather few-flowered ; stamens and style much longer than the petals. .   ..   .   13. *umbellatum*
Leaves almost glabrous on both surfaces when mature, rather strongly reticulate beneath ; cymes many-flowered, usually sessile, but sometimes with a peduncle up to 1 cm. long ; stamens and style about as long as the petals ; for other characters see below ..   ..   ..   17. *johnstonii*
Leaves not cordate at base :
  Stamens and style much longer than the petals ; peduncles always branched, but sometimes appearing subcapitate ; for other characters see above
                                                    9. *guineense*
  Stamens and style about as long as the petals ; peduncles 2–5 cm. long, usually unbranched ; leaves obovate-elliptic, rounded to obtuse at base ; acutely acuminate, 8–13 cm. long, 3·5–5·5 cm. broad ..   14. *tomentosum*
Inflorescences sessile, or very nearly so :
Leaves cordate and often glandular at the base :
  Flowers sessile or subsessile ; petals only a little longer than the sepals ; for other characters see below ..   ..   22. *heudelotii*
  Flowers distinctly pedicellate :
    Young stems, midrib and main lateral nerves beneath very long spreading-pilose ; petals only a little longer than the sepals ; leaves oblong, unequally cordate at base, obtuse or shortly acuminate at apex, 8–20 cm. long, 3·5–7·5 cm. broad, with 6–10 pairs of looped main lateral nerves ; fruits densely velutinous, up to 3 cm. diam. and 2 cm. long   15. *martineaui*
    Young stems, midrib and main lateral nerves beneath not long spreading-pilose ; petals considerably longer than the sepals :
      Leaves oblong-obovate, long-acuminate, 8–15 cm. long, 3·5–6 cm. broad, main lateral nerves 4–5 pairs, setulose on the nerves beneath 16. *linderi*
      Leaves oblong-elliptic, shortly acuminate or obtuse, 5–15 cm. long, 2·5–6 cm. broad, main lateral nerves about 7 pairs, almost glabrous beneath when mature ; for other characters see above ..   ..   17. *johnstonii*
      Leaves obovate-elliptic, long-acuminate, 10–25 cm. long, 5–10 cm. broad, main lateral nerves 6–10 pairs, almost glabrous beneath
                                        18. *subauriculatum*

Leaves not cordate at base :
  Sepals reflexed ; petals only shortly bilobed ; branchlets and leaves sparsely puberulous ; leaves oblong-elliptic, cuneate at base, shortly acuminate, 6–10 cm. long, 2·5–4 cm. broad, main lateral nerves 4–5 pairs ; flowers pedicellate in sessile clusters ..   ..   19. *staudtii*
  Sepals not reflexed ; petals deeply bilobed ; branchlets mostly hispid or tomentose, or if sparsely pubescent or glabrous then leaves much larger than above :
    Branchlets very hispid with spreading, rather long hairs, or densely tomentose:
      Flowers pedicellate :
        Branchlets softly tomentose ; leaves more or less elliptic, shortly acuminate, 5–14 cm. long, 2·5–5·5 cm. broad, sparingly puberulous on the 5–6 pairs of lateral nerves beneath ..   ..   20. *kumasiense*
        Branchlets and undersurface of leaves hispid with long spreading hairs ; leaves oblong-elliptic, acutely acuminate, 6–8 cm. long, 2–3 cm. broad ; lateral nerves about 6 pairs conspicuously looped and prominent beneath
                                        21. *acutisepalum*

    Flowers sessile :
      Leaves rounded at base :
        Lateral nerves 5–6 pairs ; leaves oblong, rather narrow and subauriculate at base, acuminate and mucronate at apex, 7–14 cm. long, 2·5–4·5 cm. broad, glabrous beneath except on the nerves ; branchlets loosely tomentose ..   ..   22. *heudelotii*
        Lateral nerves about 10 pairs ; leaves as above ; branchlets very densely tomentose ..   ..   23. *scabrum*
      Leaves cuneate and unequal-sided at base, oblong, gradually acuminate, 11–15 cm. long, 4–5 cm. broad, loosely pubescent beneath ; lateral nerves about 8 pairs ..   ..   24. *ferrugineum*
    Branchlets glabrous or shortly appressed-pubescent ; leaves glabrous or very nearly so :
      Leaves elongate-oblong, very shortly cuneate, acuminate at apex, 18 cm. long, 6–7 cm. broad ; lateral nerves 6–7 pairs ; flowers few, crowded, distinctly pedicellate ; petals bilobed to the middle ..   25. *thomsonii*
      Leaves elliptic, rounded at base, obtusely acuminate, 12–15 cm. long, 6–7·5 cm. broad, leathery ; lateral nerves 5–6 pairs ; flowers numerous, clustered, with rather slender pedicels ; petals bilobed ⅓ their length
                                        26. *obanense*

‡Peduncle (or at least the upper ones) adnate to the petiole :
  Sepals sparingly puberulous ; petals about as long as the sepals ; cymes short ;
    branchlets and leaves glabrous or nearly so ; leaves more or less oblong-elliptic,
    acute to rounded at base, gradually long-acuminate, 9–15 cm. long, 3·5–6 cm.
    broad, lateral nerves 5–6 pairs, prominently looped well within the margin
                                                                    27. *nitidulum*
  Sepals densely white-tomentellous :
  Cymes lax, with the free portion of the peduncle at least 4 mm. long, lower
    peduncles free from the petioles ; flowers apparently very small (but not known
    at maturity) ; branchlets nearly glabrous ; leaves oblong-elliptic or obovate-
    oblong, shortly cuneate, abruptly and shortly acuminate, 4–8 cm. long, 2–4 cm.
    broad, glabrous and shining   ..   ..   28. *chrysobalanoides*
  Cymes short and dense, peduncle less than 4 mm. long ; flowers at least 4 mm.
    long :
    Branchlets puberulous ; lower inflorescences often borne on short axillary lateral
      shoots with small leaves, hence appearing free from the petioles ; leaves
      oblong or elliptic, cuneate at base, obtusely acuminate, 10–15 cm. long, 4–6 cm.
      broad, glabrous ; axils of main lateral nerves beneath often with small glands
      and tufts of hair ; fruits ellipsoid, 3–4 cm. long, hoary, not beaked and not
      ridged   ..   ..   29. *toxicarium*
    Branchlets densely rusty-tomentose ; flowers all subfasciculate at the apex of
      the petiole of well-developed leaves ; leaves oblong-elliptic, obtuse to cuneate
      at base, obliquely acuminate, 9–15 cm. long, 4–8 cm. broad, glabrous ; fruits
      ellipsoid, about 3 cm. long, tomentellous, long-beaked, longitudinally ridged
                                                                    30. *rudatisii*
*Petals united into a tube with the stamens ; cymes subsessile :
  Branchlets rusty-pilose ; leaves rounded to subcordate at base, oblong-lanceolate,
    gradually long-acuminate, 7–14 cm. long, 3–7 cm. broad ; cymes rather few-flowered ;
    petals bilobed ; fruits ellipsoid rusty pilose   ..   ..   31. *longitubulosum*
  Branchlets finely puberulous ; leaves cuneate at base, narrowly elliptic, long-caudate-
    acuminate, 5·5–12 cm. long, 1·7–4 cm. broad ; cymes many-flowered ; petals entire
                                                                    32. *sp. A.*

1. **D. petersianum** *Dinkl. & Engl.* in Engl. Bot. Jahrb. 46 : 572 (1912).   An erect shrub with more or less
    pendulous branchlets, softly tomentose.
    Lib.: Grand Bassa (Aug.) *Dinklage* 1970 ! Monrovia (July) *Dinklage* 2231. Sarbo, Webo (young fr.
    July) *Baldwin* 6392 !
2. **D. angolense** *Chodat* in Bull. Herb. Boiss. 3 : 672 (1895). *D. ferrugineo-tomentosum* Engl. (1896) ; Chev.
    Bot. 120. Shrub, up to 12 ft.
    Iv.C.: Dabou *Chev.* 15560 *bis* ! **S.Nig.:** Oban (May) *Ujor* FHI 31795 ! **Br.Cam.:** Buea (fr. Apr.) *Dunlap*
    157 ! *Maitland* 1173 ! Also in French Cameroons, Gabon and Angola.
3. **D. argenteum** *Engl.* Bot. Jahrb. 33 : 82 (1902). *Chailletia rufipilis* of F.T.A. 1 : 342, partly, not of Turcz.
    A climbing shrub, up to 15–20 ft. long ; calyx green ; petals white.
    Br.Cam.: Cameroons R. (Jan.) *Mann* 747 ! Also in French Cameroons, Gabon and Cabinda.
4. **D. pallidum** *(Oliv.) Engl.* in E. & P. Pflanzenfam. 3, 4 : 249 (1896) ; Chev. Bot. 121. *Chailletia pallida*
    Oliv. F.T.A. 1 : 343 (1868). *D. liberiae* Engl. & Dinkl. in Engl. Bot. Jahrb. 33 : 84 (1902) ; Chev. Bot.
    120 ; F.W.T.A., ed. 1, 1 : 324. *D. warneckei* Engl. l.c. 83. *D. cinereum* Engl. l.c. 85 ; Chev. Bot. 119.
    D. whytei Stapf in Johnston Lib. 2 : 586 (1906). *D. bussei* Engl. Bot. Jahrb. 46 : 574, fig. 1 (1912). *D.
    albidum* A. Chev. Bot. 119, name only. A shrub, erect or scandent ; branchlets pale yellowish tomentose.
    Fr.G.: R. Konkouré, Kouria (July) *Chev.* 15086 ! Konanké country (Feb.) *Chev.* 20826. Kindia (Mar.)
    Chev. 13368. **S.L.:** Mabum (fr. Aug.) *Thomas* 1604 ! Mayoso (Aug.) *Thomas* 1431 ! Mabonto (young
    fr. July) *Deighton* 3269 ! Pujehun (young fr. July) *Deighton* 5794 ! **Lib.:** Kakatown *Whyte* ! Grand
    Bassa (May) *Dinklage* 1832 ! Nekabozu, Vonjama (fr. Oct.) *Baldwin* 9961 ! Gbanga (Sept.) *Linder* 674 !
    Iv.C.: Grabia (June) *Chev.* 19159 ! **G.C.:** Axim (Feb.) *Irvine* 2229 ! Assuantsi (Apr.) *Scholes* 282 ! **Togo:**
    Lome (Nov., Dec.) *Warnecke* 13 ! *Mildbr.* 7520 ! Kpeme (fl. & fr. Jan.) *Busse* 3639 ! **Dah.:** Sakété
    (Jan.) *Chev.* 22821. Abomey *Chev.* 23161. **S.Nig.:** Lagos *Moloney* ! *Millen* 5 ! Apapa (Dec.) *Dalz.*
    1336 ! Epe (Apr.) *Barter* 3299 ! *Jones & Onochie* FHI 17428 ! Also in French Cameroons and Gabon.
    (See Appendix, p. 171.)
      The following specimens seem to represent a distinct variety with tuberculate fruits :—
    S.L.: Gbesebu, Njala (fr. Sept.) *Deighton* 3792 ! Gola Forest (fr. Apr.) *Deighton* 3639 ! Taiama (fr. Oct.)
    Deighton 4504 ! **Lib.:** Peahtah (fr. Oct.) *Linder* 897 ! Mecca, Boporo (fr. Nov.) *Baldwin* 10420 ! Kolahun
    Dist. (fr. Nov.) *Baldwin* 10163 ! 10206 ! **G.C.:** Kumasi (July) *Vigne* 3290 !
5. **D. barteri** *Engl.* Bot. Jahrb. 23 : 134 (1896). *Icacinopsis annonoides* Roberty in Bull. I.F.A.N. 15 : 1420,
    fig. 6 (1953). A tree, to 35 ft. high, with dense bushy crown ; flowers cream ; fruit apparently tuberculate.
    G.C.: Cape Coast Castle (Nov.) *Roberty* 12829 ! Komenda (fr. Jan.) *Andoh* FH 5597 ! **S.Nig.:** Onitsha
    Barter 1781 ! Orlu (Dec.) *Jones* FHI 6202 ! Agulu (Nov.) *Keay* FHI 25593 ! Udi (Nov.–Feb.) *Jones*
    FHI 7431 ! *Keay* FHI 22280 ! Also in French Cameroons. (See Appendix, p. 171.)
6. **D. subcordatum** *(Hook. f.) Engl.* in E. & P. Pflanzenfam. 3, 4 : 349 (1896). *Chailletia subcordata* Hook. f.—
    F.T.A. 1 : 341. A shrub about 15 ft. high, with puberulous branchlets and fairly large leaves.
    F.Po: (June, Nov.) *T. Vogel* 207 ! *Mann* 432 !
7. **D. zenkeri** *Engl.* Bot. Jahrb. 23 : 138 (1896). A climbing shrub ; branchlets yellowish-tomentellous
    becoming glabrous ; flowers small, white, fragrant.
    S.Nig.: Oban *Talbot* 1360 ! 1631 ! Also in French Cameroons, Gabon and Belgian Congo.
8. **D. cymulosum** *(Oliv.) Engl.* in E. & P. Pflanzenfam. 3, 4 : 349 (1896) ; Chev. Bot. 120. *Chailletia cymulosa*
    Oliv. F.T.A. 1 : 340 (1868). A shrub, sometimes climbing ; branchlets pubescent, becoming glabrous ;
    fruit plum-like, 1–1½ in. long, glabrous.
    Iv.C.: Dabou (Feb.) *Chev.* 17165 ! **Br.Cam.:** Cameroons R. (Jan.) *Mann* 2200 ! Also in French Cameroons
    and Gabon.
9. **D. guineense** *(DC.) Keay* in Kew Bull. 1955 : 137 (q.v. for full synonymy). *Ceanothus ? guineensis* DC.
    (1825). *Rhamnus paniculatus* Thonn. (1827). *Chailletia flexuosa* Oliv. F.T.A. 1 : 340 (1869). *C. floribunda*
    var. β, Oliv. F.T.A. 1 : 341. *Dichapetalum flexuosum* (Oliv.) Engl. (1896)—F.W.T.A., ed. 1, 1 : 324 ;
    Aubrév. Fl. For. C. Iv. 2 : 4, t. 125, A. *D. paniculatum* (Thonn.) De Wild. (1919). *D. rowlandii* Hutch.

& Dalz. F.W.T.A., ed. 1, 1 : 324 (1928). *D. bocageanum* of Chev. Bot. 119. A shrub or small tree ;
flowers cream, fragrant ; fruits bright orange, tomentose.
S.L.: Kafogo (Apr.) *Sc. Elliot* 5601*b* ! Rowala (July) *Thomas* 1017 ! **Iv.C.:** *Aubrév.* 1129 ! Dyolas (fr.
May) *Chev.* 21549 ! **G.C.:** *Thonning* ! Accra Plains (Jan.–Apr.) *Brown* 935 ! *Dalz.* 124 ! *Irvine* 1964 !
Kpong-Bomm (Apr.) *A. S. Thomas* M. 5 ! Jimira F.R. (Aug.) *Andoh* FH 4418 ! **Togo:** Lome (Nov.)
Warnecke 88 ! *Mildbr.* 7482 ! **Dah.:** Porto Novo (Mar.) *Chev.* 23332 ! Zagnanado (Mar.) *Chev.* 23298 !
S.Nig.: Lagos *Barter* 2142 ! *Rowland* ! Abeokuta *Irving* ! *Barter* 3355 ! 3372 ! Ibadan (fl. Dec.–Apr.,
fr. Mar., Apr.) *Meikle* 1259 ! 1463 ! *Keay* FHI 25681 ! 25696 ! (See Appendix, p. 171.)

FIG. 143.—DICHAPETALUM TOXICARIUM (*G. Don*) *Baill.* (CHAILLETIACEAE).

A, flower. B, longitudinal section of flower. C, petal. D, stamen.   E, pistil showing glands.
F, fruit.   G, cross-section of fruit.

10. **D. oblongum** (*Hook. f.*) Engl. l.c. 3, 4 : 349 (1896) ; Chev. Bot. 121. *Chailletia oblonga* Hook. f.—F.T.A.
1 : 342. A shrub, more or less scandent ; flowers white, fragrant ; fruits yellow, tomentellous, wrinkled,
oblanceolate to ovate, up to 1½ in. long.
S.L.: Njala (fl. Mar., fr. Jan.–Mar.) *Deighton* 4701 ! 4714 ! 5349 ! Bumbuna *Thomas* 3457 ! **Lib.:** Kan-
nadhun, Kolahun (Nov.) *Baldwin* 10162 ! Tawata, Boporo (Nov.) *Baldwin* 10331 ! Barclayville (fr.
Mar.) *Baldwin* 11117 ! **Iv.C.:** Aboisso (Apr.) *Chev.* 17908 ! **G.C.:** Axim (fr. Feb.) *Irvine* 2227 ! Senchi
(Oct.) *Vigne* FH 4051 ! **N.Nig.:** Igala Div., Kabba (fl. & fr. Feb.) *Jones* FH 5019 ! **S.Nig.:** Akilla
(fr. Jan.) *Onochie* FHI 20671 ! **Br.Cam.:** Likomba (Nov.) *Mildbr.* 10686 ! **F.Po:** (Nov., Dec.) *Ansell* !
T. Vogel 36 ! 113 ! *Barter* 2056 ! *Mann* 46 ! Also in French Cameroons.
  [The specimens from Sierra Leone and Liberia tend to have the branchlets clothed with long-spreading
hairs, broader based leaves, and ovate fruits ; they may prove to represent a distinct species, but there
appear to be intermediates.]
11. **D. floribundum** (*Planch.*) Engl. l.c. 3, 4 : 348 (1896) ; Bot. Jahrb. 23 : 137, incl. var. *preussii* Engl. *D.*
floribundum Planch.—F.T.A. 1 : 340, partly (excl. vars.). Small tree, or shrub, often scandent, with grey
or ochrey-pubescerulous branchlets ; flowers white.
S.Nig.: R. Njujua, Obudu (May) *Latilo* FHI 30932 ! **Br.Cam.:** Buea (Mar.) *Preuss* 904 ! *Maitland* 466 !
Moliko (Apr.) *Hutch. & Metcalfe* 85 ! Mamfe (Apr.) *Babute* Cam. 34/38 ! **F.Po:** (June, Nov., Dec.) *T.*
Vogel 105 ! 137 ! 175 ! *Barter* ! *Mann* 16 !

10

12. **D. reticulatum** *Engl.* Bot. Jahrb. 33 : 82 (1902). *D. cordifolium* Hutch. & Dalz. F.W.T.A., ed. 1, 1 : 324 (1928) ; Kew Bull. 1928 : 380. Scandent shrub with densely tomentose branchlets and inflorescences and rather large leaves.
    S.Nig.: Ibadan South F.R. (Mar.) *Chizea* FHI 24485 ! Akure *Foster* 192 ! Udo, Ubiaja (young fr. May) *Umana* FHI 29104 ! Also in French Cameroons.
13. **D. umbellatum** *Chodat* in Bull. Herb. Boiss. 3 : 671 (1895). A woody climber with short branchlets spreading at right angles ; flowers creamy-white.
    S.Nig.: Ajilite *Millen* 167 ! Okomu F.R. (Feb.) *Brenan* 9140 ! *Akpabla* 1087 ! Also in Angola.
14. **D. tomentosum** *Engl.* Bot. Jahrb. 23 : 138 (1896). *D. acutifolium* Engl. l.c. 136 (1896) ; F.W.T.A., ed. 1, 1 : 325. Scandent shrub or woody climber, rusty-pilose ; calyx pale green ; petals cream, turning yellow ; fruit obliquely ovoid, about 1 in. long, rusty-tomentose.
    S.Nig.: Abontakon, Ikom (Dec.) *Keay* FHI 28172 ! **Br.Cam.:** Victoria (Apr., May) *Preuss* 1103 ! *Maitland* 663 ! Barombi (Apr.) *Preuss* 55 ! Also in Gabon. (See Appendix, p. 171.)
15. **D. martineaui** *Aubrév. & Pellegr.* in Aubrév. Fl. For. C. Iv. 2 : 4, t. 124, 7–10 (1936), French descr. only. A small tree in the understorey of rain forest ; fruits orange.
    Iv.C.: Banco (fl. June, fr. Aug.) *Martineau* 290 ! *Serv. For.* 483 !
16. **D. linderi** *Hutch. & Dalz.* F.W.T.A., ed. 1, 1 : 324 (1928) ; Kew Bull. 1928 : 380. Low bush ; flowers yellowish.
    Lib.: Suen (Nov.) *Linder* 1407 !
17. **D. johnstonii** *Engl.* l.c. 23 : 141 (1896). A shrub, sometimes climbing ; young branchlets densely rusty-tomentose, becoming glabrous ; inflorescence close-clustered.
    G.C.: Axim (Feb.) *Irvine* 2260 ! **S.Nig.:** Ebute Metta (Nov.) *Millen* 22 ! Cross R. (Apr.) *Johnston* ! Oban *Talbot* 1763 ! Eket *Talbot* !
18. **D. subauriculatum** (*Oliv.*) *Engl.* in E. & P. Pflanzenfam. 3, 4 : 349 (1896). *Chailletia subauriculatum* Oliv. F.T.A. 1 : 344 (1868). A shrub 6–8 ft. high ; branchlets slightly puberulous, becoming soon glabrous.
    S.Nig.: Calabar (Feb.) *Mann* 2260 ! **Br.Cam.:** Bimbia (Feb.) *Maitland* 394 ! Buea (Mar.) *Maitland* 573 !
19. **D. staudtii** *Engl.* Bot. Jahrb. 23 : 139 (1896). Scrambling shrub ; petals greenish-cream.
    S.Nig.: Omo F.R. (Mar.) *Jones & Onochie* FHI 17037 ! Also in French Cameroons.
20. **D. kumasiense** *Hoyle* in Kew Bull. 1932 : 260. Straggling shrub.
    G.C.: Kumasi (Mar.) *Vigne* 1635 ! **Togo:** Okrabi, Hohoe Dist. (Nov.) *St. C. Thompson* FH 3700 ! **S.Nig.:** Shasha F.R. (young fr. Apr.) *Tamajong* FHI 16848 !
21. **D. acutisepalum** *Engl.* l.c. 23 : 140 (1896). A rusty hairy shrub.
    S.L.: near Kurusu (Apr.) *Sc. Elliot* 5526 !
22. **D. heudelotii** (*Planch. ex Oliv.*) *Baill.* Hist. des Pl. 5 : 140 (1874) ; Chev. Bot. 120. *Chailletia heudelotii* Planch. ex Oliv. F.T.A. 1 : 344 (1868). A shrub, branchlets and young leaves rusty-hairy.
    Fr.G.: *Heudelot* 770 ! Fouta Djalon (Mar.) *Chev.* 12599. **S.L.:** Bagroo R. (Apr.) *Mann* 800 ! Njala (Mar.–May) *Deighton* 1580 ! 2999 ! Dodo (Apr.) *Deighton* 3918 ! Gima, Valunia (Apr.) *Deighton* 5859 !
23. **D. scabrum** *Engl.* Bot. Jahrb. 33 : 86 (1902). A climbing shrub ; branchlets yellow-tomentose.
    Br.Cam.: Johann-Albrechtshöhe (Jan.) *Staudt* 542.
24. **D. ferrugineum** *Engl.* Bot. Jahrb. 23 : 139 (1896). A shrub with densely rusty-tomentose branchlets.
    Br.Cam.: Victoria (May) *Preuss* 1275 !
25. **D. thomsonii** (*Oliv.*) *Engl.* in E. & P. Pflanzenfam. 3, 4 : 349 (1896). *Chailletia thomsonii* Oliv. F.T.A. 1 : 342 (1868). A small tree.
    S.Nig.: Calabar *Thomson* 79 !
26. **D. obanense** (*Bak. f.*) *Hutch. & Dalz.* F.W.T.A., ed. 1, 1 : 324 (1928). *D. thomsonii* var. *obanense* Bak. f. in Cat. Talb. 19 (1913). A shrub or small tree ; petals orange.
    S.Nig.: Oban *Talbot* 1627 !
27. **D. nitidulum** *Engl. & Ruhl.* in Engl. Bot. Jahrb. 33 : 77 (1902). Small tree.
    S.Nig.: Oban *Talbot* ! **Br.Cam.:** Missellele, Tiko (Jan.) *Boz* 3552 ! Also in French Cameroons.
28. **D. chrysobalanoides** *Hutch. & Dalz.* F.W.T.A., ed 1, 1 : 325 (1928) ; Kew Bull. 1928 : 380.
    S.L.: *Sc. Elliot* ! Magbile (Dec.) *Thomas* 6019 ! 6110 ! 6119 !
29. **D. toxicarium** (*G. Don*) *Baill.* Hist. des Pl. 5 : 139 (1874) ; Chev. Bot. 121 ; Aubrév. Fl. For. C. Iv. 2 : 6, t. 125 B. *Chailletia toxicaria* G. Don (1824)—F.T.A. 1 : 341, incl. var. *elliptica* Oliv. *D. suboblongum* Engl. (1912), partly (*Sc. Elliot* 5601a). A shrub or small tree, glabrous except for the hoary-inflorescence and fruit.
    Fr.G.: *Heudelot* 651 ! 814 ! *Pobéguin* 2017 ! Boké *Chittou* ! **S.L.:** *Don* ! *T. Vogel* 100 ! R. Bagroo (Apr.) *Mann* 826 ! Regent (Apr.) *Sc. Elliot* 5820 ! Kafogo (Apr.) *Sc. Elliot* 5601a ! Mofari (fl. & fr. Jan.) *Sc. Elliot* 4401 ! 4408 ! Rokupr (Apr.) *Jordan* 231 ! Gola Forest (Apr.) *Small* 573 ! **Lib.:** Vonjama (fr. Oct.) *Baldwin* 9885 ! Gletown, Tchien (fr. July) *Baldwin* 6947a ! Tappita (fr. Aug.) *Baldwin* 9113 ! **Iv.C.:** Guidéko (May) *Chev.* 16408 ; and various other localities *fide* Chev. l.c. **G.C.:** Axim (fl. Feb., fr. Aug.) *Beavan* FH 3129 ! *Irvine* 2162 ! Pamu-Berekum F.R. (Sept.) *Vigne* FH 2485 ! Dompim, Bonsa Su (May) *Jackson* FH 1979 ! (See Appendix, p. 171.)
30. **D. rudatisii** *Engl.* Bot. Jahrb. 46 : 582 (1912). A shrub, erect or climbing.
    S.Nig.: Degema *Talbot* ! Oban *Talbot* 19 ! 482 ! **Br.Cam.:** Abonando, Mamfe (Mar.) *Rudatis* 16 ! Also in French Cameroons.
31. **D. longitubulosum** *Engl.* Bot. Jahrb. 33 : 90 (1902). Scandent shrub, to 12 ft. high ; branchlets rusty-hirsute ; calyx pale green ; petals cream, becoming yellow at tips ; fruit bright orange with white hairs.
    S.Nig.: Bendiga Ayuk, Ikom (fl. & fr. Dec.) *Keay* FHI 28165 ! Also in French Cameroons.
32. **D. sp. A.** Near to *D. integripetalum* Engl., but probably new. Understorey tree, to 15 ft. high. Better specimens required.
    S.Nig.: Okwango F.R., Obudu (May) *Latilo* FHI 30960 !

Imperfectly known species.

1. **D. affinis** *Planch. ex Benth.* in Fl. Nigrit. 276 (1849). Included in *D. toxicarium* (G. Don) Baill. by Oliver (F.T.A. 1 : 341), but probably distinct. Perfect flowers and mature fruits not known.
    F.Po: *T. Vogel* 149 !
2. **D. sp. B.** Liane ; flowers white. Near *D. umbellatum* Chodat.
    S.Nig.: Shasha F.R. (May) *Richards* 3461 ! *Ross* 91 !
3. **D. sp. C.** " Straggling climbing shrub in forest."
    G.C.: Abosso (Aug.) *Vigne* FH 3033 !
4. **D. sp. D.** Shrub 10 ft. high ; flowers white. Probably near *D. heudelotii*.
    Port.G.: Catió (June, July) *Esp. Santo* 2092 ! 2101 !

Note. A number of specimens of this difficult genus have come to hand since the above key was sent to the printer and it is evident that there is still a lot more work to be done before the genus is satisfactorily understood in our area.

2. TAPURA Aubl.—F.T.A. 1 : 344.

Leaves 10–22 cm. long, 4–11 cm. broad, main lateral nerves 3–4 on each side of the midrib, lamina obovate-elliptic, shortly cuneate at base, abruptly acuminate, puberulous only on the nerves beneath, laxly reticulate ; flowers shortly pedicellate, clustered ; peduncle completely adnate to the petiole ; corolla with 2 large bilobed petals and 3 small lanceolate petals ; fertile stamens 3 ; fruit ellipsoid, about 3 cm. long   ..   .. 1. *africana*

Leaves 2·5–10 cm. long, 1·2–4 cm. broad, main lateral nerves 4–6 on each side of the midrib, lamina elliptic to elliptic-lanceolate, shortly cuneate at base, gradually sub-acuminate, pubescent to glabrescent beneath ; flowers on pedicels about 2 mm. long ; peduncle more or less free from the petiole in its upper part ; corolla with 1 or 2 larger bilobed petals, and 3 or 2 smaller bilobed petals, and a single lanceolate or bifid petal ; fertile stamens 2 ; fruit 3-lobed, about 4 mm. long.. .. .. .. 2. *fischeri*

1. **T. africana** *Oliv.* F.T.A. 1 : 344 (1868). A forest tree, to 80 ft. high ; flowers yellow.
   S.Nig.: Aboabam, Ikom (fl. & fr. May) *Latilo* FHI 30978 ! Cross River North F.R., Ikom (fl. & fr. June) *Latilo* FHI 31829 ! **Br.Cam.:** Bimbia Road, Victoria (Apr.) *Maitland* 611 ! 614 ! **F.Po:** *Mann* 1161 !
2. **T. fischeri** *Engl.* Pflanzenw. Ost-Afr. C. 423 (1895). *T. africana* Engl. l.c. 235 (1895), not of Oliv. Small much-branched tree, to 40 ft. high ; flowers white.
   Iv.C.: Akakoumoekron (Dec.) *Chev.* 22562 ! **G.C.:** Afram Headwaters F.R. (Apr.) *Andoh* FH 4733 ! Ijang F.R., Ashanti (Apr.) *Williams* 255 ! **S.Nig.:** Olokemeji (Apr.) *A. F. Ross* 145 ! Also in A.-E. Sudan, Belgian Congo, Kenya, Uganda and Tanganyika.

89. CAESALPINIACEAE

Trees, shrubs or very rarely herbs or scramblers. Leaves pinnate or bipinnate, rarely simple or 1-foliolate ; stipels mostly absent. Flowers mostly showy, racemose, spicate or rarely cymose, zygomorphic, rarely subactinomorphic ; bracteoles sometimes large and enclosing the flower in bud. Sepals 5, or the two upper connate, mostly free, imbricate or valvate, sometimes much reduced and then replaced by the large bracteoles. Petals 5 or fewer or absent, the adaxial (upper) one inside in bud, the others variously imbricate. Stamens mostly 10, rarely numerous, free or variously connate ; anthers various, some-times opening by terminal pores. Ovary superior, 1-celled, of 1 carpel. Seeds with or without endosperm and large embryo.

Mainly tropical and subtropical ; recognized amongst its allies by the adaxial petal being inside the others in bud. A large number of the African genera have a much-reduced calyx, a condition then compensated for by the large pair of opposite valvate bracteoles which completely envelop the flower in bud. These bracteoles may easily be mistaken for a true calyx.

Leaves simple or unifoliolate :
  Leaves more or less deeply bilobed :
    Plants monoecious, flowers hermaphrodite ; ovules and seeds in 1 series
                                              1. **Bauhinia**
    Plants dioecious, flowers unisexual ; ovules in several series ; seeds scattered in the pulp of the subwoody indehiscent fruits .. .. 2. **Piliostigma**
  Leaves not bilobed :
    Petals present :
      Calyx with 4–5 distinct lobes in bud ; petals 5 ; stamens 10 :
        Fruits inflated, bladder-like ; flowers numerous in pyramidal racemes, at length reflexed ; calyx with 5 lobes, united at the base into a long tube   3. **Griffonia**
        Fruits not bladder-like ; flowers in very short few-flowered racemes ; sepals 4, free, imbricate .. .. 22. **Zenkerella**
      Calyx in one piece and closed in bud, splitting irregularly ; petals 6 ; stamens 14–41 ; flowers in short racemes or umbels .. .. 4. **Baphiopsis**
    Petals absent or if present then sepals absent or much reduced :
      Bracteoles not valvate, early caducous ; sepals 4, well-developed ; see also below
                                        30. **Guibourtia**
      Bracteoles valvate completely enclosing the bud, persistent ; sepals absent or much-reduced :
        Flowers in racemes ; petal 1 (or 2–3), well-developed, or absent ; fertile stamens 3–6 (–7–8) ; ovary stipitate ; fruits without longitudinal ridges
                                        49. **Cryptosepalum**
        Flowers in narrow panicles ; petals (4–) 5, filiform ; fertile stamens (4–) 5 ; ovary subsessile ; fruits with 1–2 longitudinal ridges .. .. 50. **Didelotia**
Leaves compound ; 2 or more foliolate :
  *Leaves once pinnate :
    †Bracteoles not valvate, usually early caducous ; leaflets alternate or opposite :
      Calyx in one piece and closed in bud, splitting irregularly ; stamens 16 to very numerous ; petals 0 or 1, rarely 3 ; ovary stipitate ; fruits indehiscent ; leaves with an odd terminal leaflet (imparipinnate) :

Flowers with a single large posterior petal, rarely with a pair of minute lateral petals ; fruits cylindrical with numerous seeds ; ovary shortly stipitate ; stamens numerous, anthers basifixed ; leaflets 5–19, without translucent glands   .. 5. **Swartzia**
Flowers without petals ; fruits subglobose or ovoid ; ovary long-stipitate ; anthers dorsifixed ; leaflets 18–38, with translucent glands :
Stamens very numerous, inserted halfway up the campanulate calyx-tube ; racemes (in our species) arising on the woody shoots before the leaves ; savannah trees   ..   .. 6. **Cordyla**
Stamens 12–18, inserted round the edge of the disk at the base of the open calyx ; racemes arising on young shoots with the leaves ; rain forest tree

7. **Mildbraediodendron**

Calyx with the sepals, lobes or teeth distinct in bud, splitting regularly ; stamens mostly 10 or less, rarely more :
Petals all deeply bilobed to below the middle, the lobes narrow or ligulate ; fruit flat, elliptic, indehiscent, winged all round, 1–2-seeded ; leaves with an odd terminal leaflet (imparipinnate), stipels well-developed   ..   8. **Amphimas**
Petals present or absent, not as above ; stipels absent :
Anthers attached at or near the base, often opening by terminal pores, sometimes dorsifixed in which case 2–3 of the filaments are bent into the form of a hook and the fruits are cylindrical with transverse partitions (certain species of *Cassia*) :
Leaves with an odd terminal leaflet (imparipinnate) :
Fruit an elongate, woody, dehiscent pod with a ridge near each margin of the valves ; inflorescence a stout simple raceme, or panicle of racemes ; sepals 4, the largest one encircling the others in bud ; stamens 4 (–5), with the adaxial pair more or less connate   ..   ..   ..   ;   ..   9. **Duparquetia**
Fruit drupaceous, indehiscent, orbicular or ovoid ; inflorescence a panicle of cymes ; sepals 5, subequal ; stamens 2, or 4–5, free   ..   10. **Dialium**
Leaves without an odd terminal leaflet (paripinnate) :
Petals 3 ; fertile stamens 2 ; fruit thin, flat, elliptic, indehiscent, several-seeded

11. **Distemonanthus**

Petals 5 ; stamens 10, fertile or partly abortive ; fruits various .. 12. **Cassia**
Anthers attached at the middle, never opening by pores ; filaments never bent into the form of a hook :
Sepals subvalvate in bud (i.e. valvate, but with very narrow membranous, slightly imbricate margins), or sepals reduced to short teeth ; flowers often arranged distichously :
Petals 5, well-developed, subequal in length :
Sepals 4, well-developed :
Flowers relatively small ; receptacle very short ; sepals up to 1·5 cm. long ; leaflets with translucent dots, alternate :
Stamens 10, free ; lateral nerves mostly joined to a marginal nerve ; leaflets not, or very obscurely, emarginate at apex

13. **Gilletiodendron**

Stamens 10, of which 9 are connate into a tube at the base ; lateral nerves looping before the margin ; leaflets usually emarginate at apex

14. **Tessmannia**

Flowers large ; receptacle elongate ; sepals at least 7 cm. long ; stamens 10, of which 9 are connate into a tube at the base ; leaflets without translucent dots, opposite or alternate   ..   ..   15. **Baikiaea**
Sepals 5, teeth-like or up to 1 mm. long, calyx-tube campanulate ; petals up to 7 mm. long ; leaflets opposite :
Flowers in long narrow panicles ; disk fleshy, campanulate, bearing the petals and 10 equal stamens ; seeds up to 15 ..   ..   ..   16. **Chidlowia**
Flowers in spike-like racemes ; disk absent ; 5 stamens long, 5 stamens short ; seeds 1–2   ..   .. 17. **Kaoue**
Petals absent ; sepals 4 ; stamens 10, free :
Young shoots with long, rolled, stipule-like scales opposite the leaves, leaving an almost annular scar on the stem after falling ; leaflets alternate ; disk absent ; ovules 10–12 ; fruits indehiscent, with a proximal wing

18. **Hylodendron**

Young shoots without long, stipule-like leaf-opposed scales ; ovules 2 (rarely 4–7) ; fruits not winged :
Disk absent ; fruits indehiscent, drupaceous, with fibrous mesocarp ; leaflets alternate   ..   ..   19. **Detarium**
Disk present ; fruits dehiscent ; seeds arillate ; leaflets opposite to alternate

20. **Copaifera**

Sepals imbricate in bud ; flowers usually arranged in more than 2 ranks :
†Petals present ; leaflets opposite :
Corolla of 5 petals subequal in length :

Stipe of the ovary free, inserted at the base of the shallow receptacle ; **sepals 4–5** .. .. **21. Cynometra**
Stipe of the ovary completely adnate to one side of the tubular receptacle ; sepals 4 .. .. **23. Schotia**
Corolla of 1–5 petals very unequal in length :
Fertile stamens 7 (–8) ; 1 petal well-developed with a long claw, the other 4 minute or absent ; fruit thick, woody ; seeds black, with red, orange or yellow arils ; petiolules twisted .. .. **24. Afzelia**
Fertile stamens 10 (rarely 16–26) :
Bracteoles not petaloid ; leaflets not emarginate at apex :
Leaflets 3 or more pairs ; flowers in well-developed racemes or panicles ; bracteoles sepal-like, imbricate, surrounding the buds :
Petiolules twisted ; 3 petals well-developed, clawed, the other 2 (or rarely 1) small or minute ; fruit woody .. .. **25. Loesenera**
Petiolules not twisted ; petals not clawed ; 1 petal well-developed and the other 4 minute, or 3 petals well-developed and the other 2 minute ; fruit with coriaceous valves, 1-seeded .. .. **26. Daniellia**
Leaflets 1 pair ; 1 petal well-developed, not clawed, the other 4 minute ; flowers in very short racemes ; buds obovoid, not surrounded by the minute bracteoles .. .. **27. Eurypetalum**
Bracteoles petaloid, coloured, ovate or lanceolate :
Leaflets emarginate ; flowers distichous ; disk present as a swelling prominent at the base of the receptacle ; stipe of ovary completely adnate to one side of the receptacle ; petals 5, 3 large and 2 smaller (in our species) **28. Plagiosiphon**
Leaflets not emarginate (except in *H. aubrevillei*) ; flowers not arranged distichously ; disk absent or very obscure ; stipe of ovary adnate to one side of the receptacle only in the lower part ; petals 3–5, unequal, or absent ; see also below .. .. .. **29. Hymenostegia**
‡Petals absent ; leaflets alternate or opposite :
Leaflets with translucent dots ; ovary sessile, or with stipe not fused to the side of the receptacle ; stamens 4, or (8–) 10 (–12) ; fruits mostly indehiscent :
Stamens (8–) 10 (–12), free :
Disk present, fleshy :
Sepals 4 ; ovary sessile or stipitate ; fruits thin and indehiscent, or thick and dehiscent with arillate seeds ; leaflets always 1 pair, opposite ; see also above .. .. **30. Guibourtia**
Sepals 5 (–6) ; ovary sessile ; fruits indehiscent, sometimes with a proximal wing ; leaflets 1 pair, or leaflets up to 9, subopposite or alternate **31. Oxystigma**
Disk absent ; sepals 4 (–5) ; ovary stipitate ; fruits indehiscent, with a proximal wing ; leaflets 6–10, alternate .. .. **32. Gossweilerodendron**
Stamens 4, united below ; sepals 4 ; disk present ; ovary shortly stipitate ; fruits indehiscent, flat, papery, 1–2-seeded ; leaflets 4–10, alternate **33. Stemonocoleus**
Leaflets without translucent dots ; ovary stipitate, with the stipe adnate in its lower part to the side of the receptacle ; sepals 4 (rarely 5) ; fruits dehiscent :
Leaflets alternate, petiolules twisted ; disk present ; stamens 8–10 ; bracteoles not petaloid .. .. **34. Crudia**
Leaflets opposite ; disk absent or very obscure ; stamens 8–10, or 16–26 ; bracteoles linear to linear-oblong, petaloid :
Stamens 8–10 ; leaflets 5–20 pairs .. .. **35. Talbotiella**
Stamens 16–26 ; leaflets 3–4 pairs ; petals apparently absent in *H. talbotii*, for other species see above .. .. **29. Hymenostegia**
†Bracteoles valvate, well-developed, completely enclosing the bud, persistent (except in *Tamarindus*) ; leaflets opposite, mostly without translucent dots ; flowers arranged in more than 2 ranks :
Stamens very numerous ; anthers deeply sagittate ; bracteoles with a gland-like knob at the apex ; sepals 4 ; petals 5, subequal .. **36. Polystemonanthus**
Stamens up to 15 :
**Sepals well-developed :
††Androecium of 10 (–13) stamens, all fertile, exserted and subequal in length :
Receptacle elongated into a tube 2·5–18 mm. long ; sepals not ciliate, petals 5 :
Sepals 5, free ; leaflets up to 5 pairs, at least 6 cm. long, rhachis not winged :
Stamens 10 (–13), free to the base ; petals scarcely exceeding the bracteoles, subequal in length (in our species) ; flowers not exceeding 2 cm. in length **37. Isoberlinia**
Stamens 10, of which 9 are connate at the base ; petals much exceeding the bracteoles, subequal in length or the lateral ones smaller than the posterior ; flowers 4 cm. or more long .. .. **38. Berlinia**

Sepals 3, free and 2 united ; stamens 10, of which 9 are connate at the base ; petals 5, subequal in length ; flowers less than 1 cm. long ; leaflets 12–18 pairs, up to 3·5 cm. long, rhachis narrowly winged between the leaflets
                                                                        39. **Microberlinia**

Receptacle flat or only slightly concave ; sepals 5, free, densely ciliate ; stamens 10, of which 9 are connate at the base ; flowers up to 2 cm. long :
Leaflets with conspicuous translucent dots, 2–5 pairs ; posterior petal well-developed, clawed, lateral petals minute   ..   ..   .. 40. **Julbernardia**
Leaflets without conspicuous translucent dots, 1 pair ; all petals minute
                                                                        41. **Paraberlinia**

††Androecium of 3 large, fertile, exserted stamens, the others reduced, included, staminodal or absent, rarely of 9 large fertile stamens (*Gilbertiodendron splendidum*) :
Leaflets more than 3 cm. long ; fruits dehiscent ; large stamens not unilateral, free or shortly united :
Intrastaminal tube absent ; leaflets without marginal glands ; fruits without longitudinal ridges :
Petals not blue ; stipe of ovary up to 1 mm. long ; bracteoles up to 1·5 cm. long, tomentellous to puberulous outside ; fruits with more or less prominent transverse ridges ; petiolules not twisted ..   ..   .. 42. **Anthonotha**
Petals blue ; stipe of ovary 10–12 mm. long ; bracteoles 2·5 cm. or more long, glabrous ; fruits without transverse ridges ; petiolules twisted
                                                                        43. **Paramacrolobium**
Intrastaminal tube present (see fig. 152), bearing staminodes some of which may be reduced to mere protuberances or absent ; leaflets with glands on or near the margin :
Fruits without longitudinal ridges ; leaflets 1 pair ; stipels present ; glands near the margin of the leaflets   ..   ..   .. 44. **Pellegriniodendron**
Fruits with longitudinal ridges ; leaflets 2 or more pairs ; stipels absent ; glands actually on the margins of the leaflets   .. 45. **Gilbertiodendron**
Leaflets up to 3 cm. long ; fruits indehiscent ; the 3 large stamens unilateral and united up to the middle   ..   .. 46. **Tamarindus**
**Sepals absent or much reduced :
Fertile stamens (8–) 10 (–18), almost free, or connate at base, or 9 connate and 1 free :
Sepals 0–5 ; petal 1, well-developed (in our species), the other 4 minute
                                                                        47. **Monopetalanthus**
Sepals and petals not differentiated ; tepals 0–10   ..   .. 48. **Brachystegia**
Fertile stamens 3–6 (–7–8), free :
Flowers in racemes ; petal 1 (–2–3), well-developed, or absent ; fertile stamens 3–6 (–7–8) ; ovary stipitate ; fruits without longitudinal ridges
                                                                        49. **Cryptosepalum**
Flowers in narrow panicles ; petals (4–) 5, filiform ; fertile stamens (4–) 5 ; ovary subsessile ; fruits with 1–2 longitudinal ridges   .. 50. **Didelotia**
*Leaves bipinnate :
Sepals free to the base or nearly so ; leaflets opposite or alternate :
Common petiole normally developed, not ending in a spine ; rhachis not leaf-like :
Branchlets and often the leaf-rhachis prickly ; sepals imbricate :
Fruits indehiscent, flat and thin, with an intramarginal nerve, not prickly ; stipules small and inconspicuous   ..   ..   .. 51. **Mezoneuron**
Fruits (in our species) dehiscent, not flat, prickly ; stipules leafy, bilobed, usually persistent   ..   .. 52. **Caesalpinia**
Branchlets and leaf-rhachis not prickly :
Sepals valvate ; petals red, with long claws ; fruits many-seeded ; leaflets small, opposite, very numerous   ..   .. 53. **Delonix**
Sepals imbricate ; petals yellow, shortly clawed ; fruits 1–2-seeded ; leaflets relatively large, alternate   ..   .. 54. **Bussea**
Common petiole very short, ending in a sharp spine ; rhachis of pinnae flat and leaf-like, leaflets very small   ..   .. 55. **Parkinsonia**
Sepals connate into a campanulate tube ; leaflets alternate :
Fruits 1-seeded, indehiscent, flat and thin ; ovules 1–2 ..   ..   .. 56. **Burkea**
Fruits 2–18-seeded ; ovules 6–20 :
Fruits dehiscent, valves subcoriaceous to woody-leathery, with projecting sutures ; seeds 2–10 ..   .. 57. **Erythrophleum**
Fruits indehiscent, valves thickly woody with thick projecting sutures ; seeds 10–18   ..   .. 58. **Pachyelasma**

Besides the above, other genera are represented by the following introduced species :—*Brownea* spp. from tropical America ; *Haematoxylum campechianum* Linn., the " Logwood," from tropical America ; *Peltophorum pterocarpum* (DC.) Backer (syn. *P. ferrugineum* (Decne.) Benth.), a native of tropical Asia and Australia, and *P. dasyrrhachis* (Miq.) Kurz ex Bak., a native of the East Indies ; *Trachylobium verrucosum* (Gaertn.) Oliv., a " Copal tree " from East Africa.

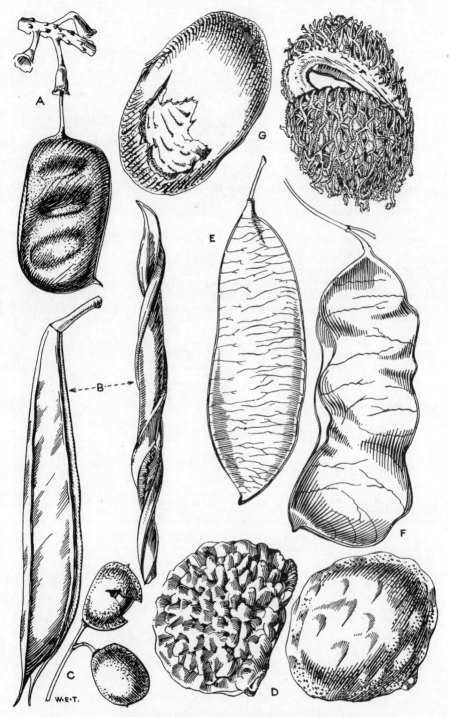

FIG. 144.—FRUITS OF CAESALPINIACEAE.

A, *Griffonia simplicifolia* (Vahl ex DC.) Baill. B, *Duparquetia orchidacea* Baill. C, *Dialium guineense* Willd. D, *Cynometra megalophylla* Harms. E, *Distemonanthus benthamianus* Baill. F, *Crudia senegalensis* Planch. ex Benth. G, *Detarium senegalense* J. F. Gmel.

1. BAUHINIA Linn.—F.T.A. 2 : 285.

Leaves very small, bilobed almost to the base, rarely over 2 cm. long, lobes semicircular, glaucous, glabrous ; petiole about ⅓ as long as the leaf ; racemes few-flowered ; calyx spathaceous, the limb twice as long as the tube, finely puberulous ; petals spathulate ; stamens 10, all perfect, filaments hairy at the base ; anthers thinly pilose ; ovary long-stipitate, glabrous ; fruit linear, curved, obliquely constricted, 6–9 cm. long, glabrous .. .. *rufescens*

B. rufescens *Lam.*—F.T.A. 2 : 289 ; Aubrév. Fl. For. Soud.-Guin. 214, t. 41, 6–9. *B. adansoniana* of Guill. & Perr. 1 : 265, partly (notes only), see Pellegrin in Bull. Soc. Bot. Fr. 74 : 622 (1927). *Piliostigma rufescens* (Lam.) Benth. (1852). A much-branched shrub or small tree, up to 25 ft. high, with white flowers and blackish, often twisted, several-seeded pods which persist on the plants a long time ; branchlets often armed with pointed short lateral shoots ; in the drier savannahs especially on river banks, and sometimes planted in villages.
Sen.: *Roger* ! *Perrottet* 259 ! *Walo Heudelot* ! **Fr.Sud.:** Fô (fl. & fr. June) *Chev.* 959 ! Ansongo (fl. & fr. Mar.) *De Wailly* 5364 ! Timbuktu (fl. & fr. July) *Chev.* 1182 ! *Hagerup* 278 ! **Port.G.:** Canchungo (fl. & fr. Feb.) *Esp. Santo* 1321 ! **Fr.G.:** Timbo *Pobéguin* 101. **S.L.:** Manna *Dawe* 514 ! Kukuna, Bramaia (Sept.) *Jordan* 343 ! **Iv.C.:** Kaya, Ouagadougou & Bobo Dioulasso *fide* Aubrév. *l.c.* **G.C.:** Navrongo (fl. & fr. Apr.) *Vigne FH* 3755 ! **Dah.:** Kouandé to Konkobiri *Chev.* 24244. **Fr.Nig.:** Zinder (fl. & fr. Sept.) *Hagerup* 617 ! **N.Nig.:** Sokoto (Dec.) *Ryan* 36 ! 65 ! Wawa, Borgu *Barter* 734 ! Katagum (fl. & fr. July) *Dalz.* 32 ! Kukawa (fl. & fr. Jan.) *E. Vogel* 26 ! **S.Nig.:** Abeokuta (fl. & fr. Jan.) *Rowland* ! Extends to A.-E. Sudan. (See Appendix, p. 174.)

Besides the above, various introduced species are cultivated in W. Africa, including the following : *B. acuminata* Linn., *B. monandra* Kurz, *B. purpurea* Linn., *B. racemosa* Linn., *B. tomentosa* Linn. and *B. variegata* Linn.

2. PILIOSTIGMA Hochst.[1] in Flora 29 : 598 (1846) ; Milne-Redhead in Hook. Ic. Pl. t. 3460 (1947). *Elayuna* Raf. Sylva Tellur. 145 (1838).

Leaves glabrous beneath, 5–8 cm. long, 5–11 cm. broad ; inflorescence with racemose branches, or racemose ; mature fruit glabrous, pruinose .. .. 1. *reticulatum*
Leaves finely pubescent beneath, 7·5–16 cm. long, 10–18 cm. broad ; inflorescence narrowly paniculate ; mature fruit shortly ferrugineous pubescent 2. *thonningii*

1. **P. reticulatum** (*DC.*) *Hochst.* (1846)—Milne-Redhead in Hook. Ic. Pl. sub. t. 3460. *Bauhinia reticulata* DC. (1825)—F.W.T.A., ed. 1, 1 : 330 ; Aubrév. Fl. For. Soud.-Guin. 214, t. 41, 1. *B. benzoin* Kotschy (1868). *B. glabra* and *B. glauca* A. Chev. Bot. 226, names only. *Elayuna biloba* Raf. (1838). Tree, to 30 ft. high, with spreading crown ; petals white ; in the drier savannah regions.
Sen.: *Perrottet* 66 ! *Roger* ! Cape Verde *Brunner* 114 ! **Fr.Sud. & Iv.C.:** various localities *fide* Aubrév. *l.c.* **G.C.:** Akpabla 708 ! **Fr.Nig.:** Niamey (Oct.) *Hagerup* 546 ! **N.Nig.:** Zamfara F.R. (fr. Apr.) *Keay FHI* 16161 ! 18036 ! Katagum (Oct.) *Dalz.* 33 ! Duchi, Bornu (fl. & fr. Dec.) *Elliott* 114 ! Kukawa (fr. Feb.) *E. Vogel* 50 ! Also in French Cameroons, Ubangi-Shari and A.-E. Sudan. (See Appendix, p. 174.)
2. **P. thonningii** (*Schum.*) *Milne-Redhead* in Hook. Ic. Pl. t. 3460 (1947). *Bauhinia thonningii* Schum. (1827)—F.W.T.A., ed. 1, 1 : 330 ; Aubrév. l.c. 215, t. 41, 2-5 ; Wilczek in Fl. Congo Belge 3 : 275, t. 22. Shrub or small tree, to 25 ft. high ; often of crooked growth and occasionally scrambling ; petals white ; in savannah regions moister than those of the preceding species.
Sen.: *Derrien* 10 ! *Heudelot* 159 ! Bignona, Casamance *Aubrév.* 3030. **Gam.:** *Dawe* 61 ! **Fr.Sud.:** Kaarta *Dubois* 51 ! **Port.G.:** Safim, Bissau (fr. Jan.) *Esp. Santo* 1672 ! **Fr.G.:** Sabodougou, Touba (July) *Collenette* 62 ! **S.L.:** Senneya, Scarcies (fr. Jan.) *Sc. Elliot* 4658 ! Musaia (Sept.) *Deighton* 4887 ! Kotahun (Oct.) *Glanville* 16 ! **Iv.C.:** Koudougou & Kampti *fide* Aubrév. *l.c.* **G.C.:** Aquapim *Thonning* ! Shai Plains (fl. & fr. Jan.) *Johnson* 573 ! Ejura (Aug.) *Chipp* 734 ! Wanki (June) *Chipp* 481 ! Tamale *Anderson* 26 ! **Togo:** Lome *Warnecke* 435 ! Kpedsu (fr. Dec.) *Howes* 1071 ! **Dah.:** *Poisson* ! **N.Nig.:** Nupe *Barter* 981 ! Zaria (Oct.) *Keay FHI* 21424 ! Zungeru (Aug.) *Elliott* 4 ! Bauchi Plateau (May) *Lely* P. 303 ! **S.Nig.:** Lagos *MacGregor* 231 ! Abeokuta *Irving* ! Olokemeji (Aug.) *Punch* 5 ! **Br.Cam.:** Lakka, Bamenda (June) *Maitland* 1570 ! Widespread in tropical Africa from Abyssinia to Transvaal. (See Appendix, p. 174.)

The areas occupied by these two species are well-marked, *P. reticulatum* occurring in the drier and *P. thonningii* in the moister savannah regions. At Zaria in N. Nigeria where the areas overlap Milne-Redhead (l.c.) noted a mixed population with strong indications of hybridity.

3. GRIFFONIA Baill.[2] in Adansonia 6 : 188 (7 Oct., 1865) ; Taub. in E. & P. Pflanzenfam. 3, 3 : 147 (1892). *Bandeiraea* Welw. ex Benth. in Benth. & Hook. f. Gen. Pl. 1 : 577 (1865, about 19 Oct.) ; F.T.A. 2 : 284 ; F.W.T.A., ed. 1, 1 : 328.

Calyx-tube softly tomentose outside, about 1 cm. long ; leaves more or less ovate, rounded or widely cordate at base, rounded or shortly obtusely acuminate at apex, 6–12 cm. long, 3–6 cm. broad, glabrous and shining, prominently 3-nerved at base ; inflorescence softly tomentose, with a curled hook-like branch at base ; fruits obliquely oblong, inflated, up to 4–5 cm. long, blackish and reticulate ; stipe slender, 1–1·5 cm. long, pubescent .. .. 1. *simplicifolia*
Calyx-tube glabrous outside, about 3 cm. long ; leaves oblong-elliptic, a little cuneate at base, slightly acuminate at apex, 6–14 cm. long, 3–6 cm. broad, glabrous, less prominently 3-nerved at base ; inflorescence glabrous, without a modified branch at base ; fruits probably smaller than above, but otherwise similar 2. *physocarpa*

---

[1] This name has been proposed for conservation.
[2] Bentham's diary clearly indicates that Vol. I part 2 of the Genera Plantarum was published after 7 Oct., 1865, the date given for Baillon's publication in Adansonia. I am therefore following Taubert (l.c.), Dalla Torre & Harms (1907) and Macbride (Contrib. Gray Herb. 59 : 21 (1919)) in using the name *Griffonia* Baill. I can see no valid grounds for conserving *Bandeiraea* Welw. ex Benth. and rejecting *Griffonia* Baill., as suggested by Pellegrin (l.c.), seeing that the later homonym *Griffonia* Hook. f. (*Rosaceae*) is universally regarded as a synonym of *Acioa* Aubl.

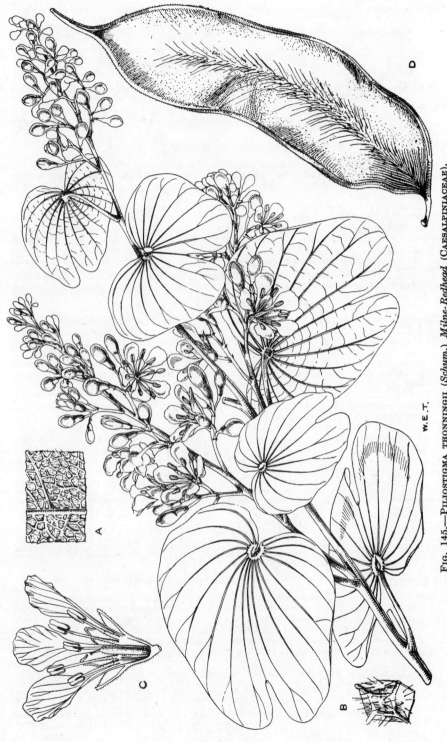

w.e.t.

Fig. 145.—Piliostigma thonningii (*Schum.*) *Milne-Redhead* (Caesalpiniaceae).

A, undersurface of leaf. B, same, more highly magnified. C, longitudinal section of flower. D, fruit.

1. **G. simplicifolia** (*Vahl ex DC.*) *Baill.* in Adansonia 6 : 197 (1866). *Schotia simplicifolia* Vahl ex DC. (1825). *Bandeiraea simplicifolia* (Vahl ex DC.) Benth. in Trans. Linn. Soc. 25 : 306 (1865) ; F.T.A. 2 : 285 ; F.W.T.A., ed. 1, 1 : 328 ; Chev. Bot. 228 ; Pellegr. Lég. Gab. 135.   A hard-wooded shrub, usually climbing, with greenish flowers and inflated black pods.
Lib.: Ganta (Oct.) *Harley* ! Tchien Dist. (Aug.) *Baldwin* 9023 ! 9024 ! **Iv.C.:** San Pedro *Thoiré* 318 ! Bouroukrou (fr. Dec.) *Chev.* 16543. Sassandra Port *Chev.* 17939. **G.C.:** Christianborg & Aquapim *Thonning.* Cape Coast (fl. & fr. Oct.) *Brass* ! *Howes* 992 ! Winnebah (Sept.) *Dalz.* 95 ! Accra *Moloney* ! Kumasi *Cummins* 134 ! Aburi Hills (Oct.) *Johnson* 475 ! Anwhiaso F.R. (Sept.) *Vigne* 226 ! **Togo:** Lome (Nov.) *Warnecke* 231 ! *Mildbr.* 7510 ! Kpeve (Oct.) *Williams* 712 ! **S.Nig.:** Epe (Feb.) *Millen* 158 ! Ijebu Ode (Aug.) *Punch* ! Also in Gabon. (See Appendix, p. 174.)
2. **G. physocarpa** *Baill.* l.c. 6 : 188 (7 Oct., 1865). *Bandeiraea tenuiflora* Benth. l.c. 25 : 307 (1865, after 2 Nov.) ; F.T.A. 2 : 285 ; F.W.T.A., ed. 1, 1 : 328 ; Pellegr. l.c. 136. A climbing or erect shrub ; calyx yellow or red, petals green with dull red markings.
S.Nig.: Umuahia (fl. & fr. Sept.) *Carpenter* 69 ! Ikom (fl. & fr. Dec.) *Keay* FHI 28160 ! Oban (Apr.) *Ejiofor* FHI 21877 ! **Br.Cam.:** Johann-Albrechtshöhe *Staudt* 951 ! **F.Po:** *Mann* 1163 ! Also in French Cameroons, Gabon and Belgian Congo.

4. BAPHIOPSIS Benth. ex Bak.—F.T.A. 2 : 256.

A tree ; leaves unifoliolate, broadly elliptic or ovate-elliptic, obtusely acuminate, 9–20 cm. long, 4–10 cm. broad, closely reticulate on both surfaces, glabrous ; lateral nerves 5–6 pairs ; petiolule 1–4 cm. long, pulvinate and transversely rugose at each end ; flowers in very short axillary racemes ; axis of inflorescence up to 2·5 cm. long, slender, shortly pubescent ; pedicels up to 1 cm. long, slightly pubescent ; buds broadly ellipsoid, hairy at the tip, about 3·5 mm. long ; calyx closed in bud, at length splitting irregularly ; petals 6, about as long as the calyx, subequal ; stamens 16–18, short, free ; ovary sessile, 2-ovuled ; fruits not known   ..   ..   .. *parviflora*

B. parviflora *Benth. ex Bak.* in F.T.A. 2 : 256 (1871) ; Bak. f. Leg. Trop. Afr. 2 : 590 ; Gilbert & Boutique in Fl. Congo Belge 3 : 554. An understorey forest tree, 20–50 ft. high ; flowers white, with an unpleasant smell.
Br.Cam.: Ambas Bay (Jan.) *Mann* 715 ! Likomba (Oct.) *Mildbr.* 10561 ! Also in Gabon, Belgian Congo, Uganda and Tanganyika.

5. SWARTZIA Schreb.—F.T.A. 2 : 256.

Leaflets obtuse or emarginate, not apiculate, alternate or opposite, lanceolate to elliptic, 3–7 cm. long, 1·5–4·5 cm. broad, closely reticulate above, thinly pubescent beneath ; flowers few in short racemes ; pedicels up to 4 cm. long, tomentose ; calyx globose in bud, finally splitting to the base and spreading or reflexed ; petal normally 1, orbicular, about 2·5 cm. broad, shortly clawed, densely villous outside ; stamens numerous, free ; anthers unequal, oblong ; ovary shortly stipitate ; fruit up to 30 cm. long, cylindrical, glabrous, black ; seeds numerous ; savannah species
1. *madagascariensis*
Leaflets apiculate, alternate, lanceolate, 5–9 cm. long, 3–4·5 cm. broad, glabrous ; flowers in lax axillary racemes, peduncles about 20 cm. long, pedicels 1–2 cm. long, shortly appressed-pubescent ; flowers and fruits similar to the last species ; rain forest species   ..   .. 2. *fistuloides*

1. **S. madagascariensis** *Desv.*—F.T.A. 2 : 257 ; Bak. f. Leg. Trop. Afr. 2 : 605 ; Aubrév. in Fl. For. Soud.-Guin. 302, t. 60, 5–8 ; Gilbert & Boutique in Fl. Congo Belge 3 : 551 ; Berhaut Fl. Sen. 42. A tree, 20–30 ft. high with conspicuous white sweetly-scented flowers, and pods a foot long rather like those of *Cassia sieberiana*.
Sen.: *fide* Berhaut *l.c.* **Gam.:** *Saunders* 45 ! **Fr.Sud.:** Bougouni (Apr.) *Chev.* 675 ! **Port.G.:** Gabu (June) *Esp. Santo* 2506 ! **Fr.G.:** Kouroussa (Apr.) *Pobéguin* 185 ! **Iv.C.:** Bobo Dioulasso, Banfara and Tiengara *fide* Aubrév. *l.c.* **G.C.:** Bujan (Apr.) *Kitson* 506 ! 922 ! Parria (May) *Kitson* 574 ! 743 ! Gambaga *Akpabla* 732 ! **Dah.:** Atacora Mts. *Chev.* 23948. **N.Nig.:** Nupe *Barter* 1649 ! Sokoto (June) *Dalz.* 531 ! Katsina (May) *MacElderry* FHI 16468 ! L. Chad (May) *E. Vogel* 95 ! Widespread in tropical Africa and the Mascarene Islands. (See Appendix, p. 263.)
2. **S. fistuloides** *Harms* in Engl. Bot. Jahrb. 45 : 305 (1910) ; Bak. f. l.c. 605 ; Aubrév. Fl. For. C. Iv. 1 : 280, t. 110 ; Gilbert & Boutique l.c. A forest tree to 80 ft. high ; the one petal of each flower white or pale pink.
Iv.C.: Abengourou to Bondoukou *Aubrév.* 709 ! Agboville *Aubrév.* 1497 ! 1913 ! **S.Nig.:** Sapoba *Kennedy* 1696 ! **Br.Cam.:** Abonando, Mamfe (May) *Rudatis* 65 ! Also in French Cameroons, Spanish Guinea, Gabon, Cabinda and Belgian Congo. (See Appendix, p. 262.)

6. CORDYLA Lour.—F.T.A. 2 : 257 ; Milne-Redhead in Fedde Rep. 41 : 227 (1937).

A tree ; leaflets opposite or subopposite, about 9–10 pairs, ovate, obtuse, 5–6 cm. long, 2–2·5 cm. broad, glabrous or slightly pubescent, often glaucous ; rhachis and petiolules pubescent ; flowers in rather dense softly woolly-tomentose racemes up to about 8 cm. long ; pedicels about 1 cm. long ; calyx closed in bud, at length splitting into 4–5 lobes ; petals none ; stamens numerous, nearly free ; ovary long-stipitate, glabrous, multiovulate ; fruit stipitate, ellipsoid, about 5 cm. long, 2–3-seeded
pinnata

C. pinnata (*Lepr. ex A. Rich.*) *Milne-Redhead* [1] l.c. 232 (1937) ; Aubrév. Fl. For. Soud.-Guin. 304, t. 61, 1–3. *Calycandra pinnata* Lepr. ex A. Rich. (1831). *Cordyla africana* of F.T.A. 2 : 257, partly ; of Chev. Bot. 219 ; of F.W.T.A., ed. 1, 1 : 370, partly ; of Bak. f. Leg. Trop. Afr. 2 : 606, partly ; not of Lour. A tree 40–50 ft. high, with corky bark, grey pubescent branchlets, pale foliage and white or yellow flowers (stamens, not petals) appearing on one-year-old shoots before the leaves ; in savannah.
Sen.: *Heudelot* 367 ! Sedhiou (Feb.) *Chev.* 3537 ! *Trochain* 1403. Kaolack *Kaichinger* 63. **Gam.:** *Brown-Lester* 70 ! *Dawe* 41 ! *Pirie* 37 ! Gounat *Perrottet* 32. Georgetown (Jan.) *Dalz.* 8029 ! **Fr.Sud.:** Ségou to Barouéli *Chev.* 44017. Diondiou *Chev.* 1006. (See Appendix, p. 235.)

---

[1] See fig. 161, E-F, on p. 509.

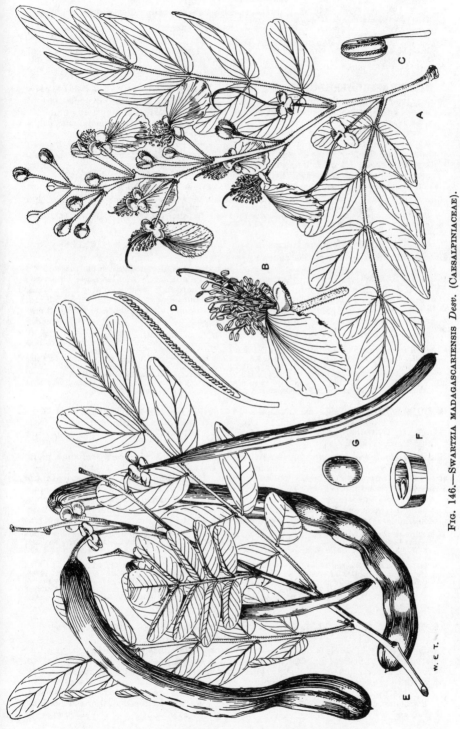

Fig. 146.—Swartzia madagascariensis *Desv.* (Caesalpiniaceae).

A, flowering shoot. B, flower. C, stamen. D, longitudinal section of ovary. E, fruiting shoot. F, transverse section of ovary. G, seed.

W. E. T.

447

Imperfectly known species.

1. **C. sp. A.**   See Milne-Redhead l.c. 228.   Leaves minutely pubescent beneath ; fruit poisonous.
  Togo: Basari *Kersting* A. 517 !
2. **C. sp. B.**   See Milne-Redhead l.c.   Leaves minutely pubescent beneath.
  N.Nig.: Abinsi *Dalz.* 705 !
3. **C. sp. C.**   Leaves with a few long curled hairs beneath.
  Iv.C.: Batié (fl. & fr. Feb.) *Serv. For.* 2837 ! [Aubrév. in Fl. For. Soud.-Guin. 306, cites this as *C. richardii* Planch. ex Milne-Redhead, but it appears to me to be an undescribed species. Further material of this and of the other imperfectly known species is much needed.—F.N.H.]

7. **MILDBRAEDIODENDRON** Harms in Mildbr. Wiss. Ergebn. 1907–08, 2 : 241 (1911).

A large tree ; leaflets alternate or opposite, 12–19 on each side of rhachis, lanceolate to oblong, 2·5–5 cm. long, nearly 1 cm. broad, pellucid-punctate and minutely pubescent on both surfaces ; flowers few in short racemes about 3 cm. long ; rhachis and calyx on both sides covered with bright yellow appressed pubescence ; calyx subglobose in bud, splitting into usually 3 broad, deltoid lobes, subequal, more or less reflexed ; about 4 mm. long ; petals absent ; stamens 12–18 in a ring on the margin of the disk, filaments about 12 mm. long, anthers small, ovate, dorsifixed ; ovary stipitate, oblong, 1–8-ovuled, glabrous ; fruit stipitate, obliquely ellipsoid, up to 10 cm. diam., coriaceous outside, fleshy inside, with 4–5 transversely arranged seeds   *excelsum*

M. excelsum *Harms* l.c., t. 27 (1911) ; Bak. f. Leg. Trop. Afr. 2 : 607. A tall forest tree with buttresses, a straight trunk and spreading crown ; leaves deciduous, reappearing with the flowers.
  G.C.: Bosumkese, Sunyani Dist. (Dec.) *Lawton* FH 5769 ! **Br.Cam.:** Issongo (Nov.) *Mildbr.* 10643 ! Also in French Cameroons, A.-E. Sudan and Belgian Congo.

8. **AMPHIMAS** Pierre ex Harms in Nat. Pflanzenfam. Nachtr. 3 : 157 (1906).

Branchlets stout, rusty-tomentellous ; leaflets 5–6 pairs, opposite or subopposite, lanceolate to narrowly obovate, rounded or broadly acuminate and mucronate, 6–12 cm. long, 2–3·5 cm. broad, glabrous, with rather numerous slender lateral nerves ; stipels filiform, about 7 mm. long ; panicles bracteate ; bracts ovate, acuminate, about 7 mm. long, tomentellous ; fruits flat, oblong-elliptic, 15–18 cm. long, 4–6 cm. broad, the body lanceolate and reticulate, surrounded by a smooth wing about 2 cm. broad ; seed solitary, reniform, about 2 cm. broad..   ..   *pterocarpoides*

A. pterocarpoides *Harms* in Fedde Rep. 12 : 12 (1913) ; Bak. f. Leg. Trop. Afr. 3 : 609 ; Aubrév. Fl. For. C. Iv. 1 : 280, t. 111 ; Wilczek in Fl. Congo Belge 3 : 548, t. 40. A forest tree, to 150 ft. high, with cylindrical stem and spreading crown ; slash with a red exudate ; flowers white, fragrant ; fruits papery.
  Fr.G.: Kissidougou (fr. Feb.) *Chev.* 20716 ! **S.L.:** Colony F.R. *Formah* 27 ! Njala (fl. Oct., Nov., fr. Feb.) *Deighton* 4914 ! 4952 ! **Lib.:** Banga (Oct.) *Linder* 1227 ! Monrovia *Cooper* 123 ! 354 ! **Iv.C.:** Guiglo *Aubrév.* 1896 ! And other localities *fide* Aubrév. *l.c.* **G.C.:** Anwhiaso F.R. *Vigne* FH 225 ! Kumasi *Vigne* FH 1390 ! Insuta (fr. Jan.) *Chipp* 62 ! Aburi (fr. Jan.) *Irvine* 1900 ! **S.Nig.:** Lagos *Foster* 160 ! Also in French Cameroons, Belgian Congo and A.-E. Sudan. (See Appendix, p. 227.)

9. **DUPARQUETIA** Baill.[1] in Adansonia 6 : 189 (7 Oct., 1865) ; Taubert in E. & P. Pflanzenfam. 3, 3 : 166 (1892). *Oligostemon* Benth. in Benth. & Hook. f. Gen. Pl. 1 : 570 (1865, about 19 Oct.) ; F.T.A. 2 : 267 ; F.W.T.A., ed. 1, 1 : 330 ; not of Turcz. (1858).

Shrub or small tree ; young parts rusty-pubescent ; leaves pinnate ; leaflets 3 pairs, opposite, ovate or elliptic to oblanceolate, abruptly acuminate, rounded at base, 6–12 cm. long, 4–6 cm. broad, pubescent beneath, the terminal leaflet larger and more obovate ; racemes terminal, erect, many-flowered ; pedicels up to 2 cm. long ; sepals 4, the outer enclosing the others and about 2 cm. long, tomentellous ; petals 5, veiny, unequal ; stamens 4, the adaxial pair more or less connate ; anthers 1·3 cm. long, acuminate ; fruit 10 cm. long, dehiscent, with 4 ridges, glabrous   ..   *orchidacea*

D. orchidacea *Baill.* in Adansonia 6 : 189, t. 4 (7 Oct., 1865) ; Steyaert in Fl. Congo Belge 3 : 545. *Oligostemon pictus* Benth. in Trans. Linn. Soc. 25 : 305, t. 39 (1865, after 2 Nov.) ; F.T.A. 2 : 267 ; Chev. Bot. 221 ; F.W.T.A., ed. 1, 1 : 330. A shrub, usually scandent, or small tree, to 20 ft. high, with showy pink and white flowers.
  Lib.: Kakatown *Whyte* ! Sinoe Basin *Whyte* ! Dukwia R. (fl. Oct., Nov., fr. Feb.) *Cooper* 64 ! 135 ! 245 ! Harbel (fl. & fr. Dec.) *Bequaert* 13 ! **Iv.C.:** Cavally Basin *Chev.* 19563 ; 19642. Grabo *Chev.* 19766. **G.C.:** Essuasu (fl. & fr. Nov.) *Chipp* ! Tarkwa (Sept.) *Dalz.* 97 ! Anaje, Sekondi (Oct.) *Howes* 977 ! **S.Nig.:** Calabar *Williams* 22 ! Cross R. (fl. & fr. Jan.) *Holland* 249 ! Oban *Talbot* 1709 ! **Br.Cam.:** Victoria to Kumba (fl. & fr. Apr.) *Olorunfemi* FHI 30533 ! Cameroons R. (Jan.) *Mann* 751 ! 2210 ! Also in French Cameroons, Gabon and Belgian Congo. (See Appendix, p. 198.)

10. **DIALIUM** Linn.—F.T.A. 2 : 282 ; Steyaert in Bull. Soc. Roy. Bot. Belg. 84 : 29–45 (1951).

Leaflets usually 5, rarely 3 or 7, usually alternate, sometimes subopposite :
  Reticulation of leaflets lax, lateral nerves about 4 pairs ; leaflets 2–7 cm. long, 2–4 cm. broad, rhachis 3·5–5 cm. long ; lateral leaflets ovate, obtuse to rounded or sub-cordate at base, obtusely acuminate at apex, terminal leaflet elliptic, acute at base ; rhachis, petioles and undersurface of leaflets pubescent ; petals absent ; stamens 4–5, filaments straight ; fruits subglobose, 1·2–1·5 cm. diam., densely tomentellous, brown..   ..   ..   1. *pobeguinii*

---

[1] See footnote on p. 444 for remarks on date of publication.

Reticulation of leaflets close, lateral nerves 8–15 pairs ; petals 1 (rarely 2), caducous ; stamens 2, rarely 3 :

Stamens with straight filaments ; fruits suborbicular, flattened, 2–2·5 cm. diam., densely velvety-tomentellous, black ; leaflets elliptic to lanceolate, 3–12 cm. long, 1·5–5 cm. broad, rounded or obtuse at base, obtuse or shortly and obtusely acuminate at apex, often puberulous beneath ; rhachis 4–12 cm. long   *2. guineense*

Stamens with bent filaments ; leaflets glabrous and shining :

Fruits glabrous, suborbicular, flattened, 1·7–2 cm. diam. ; leaflets elliptic to oblong- or ovate-elliptic, cuneate at base, rather long-acuminate, 7–13 cm. long, 2–5·5 cm. broad ; rhachis 4–7 cm. long ..   ..   *3. aubrevillei*

Fruits densely tomentellous, obovoid to subglobose, slightly flattened, about 2·5 cm. long and 1·5 cm. broad ; leaflets oblong-elliptic or sometimes obovate, cuneate at base, acuminate at apex, 8–23 cm. long, 3·5–8 cm. broad ; rhachis 8–17 cm. long   *4. pachyphyllum*

Leaflets 11–17 (rarely up to 21), opposite or subopposite ; lateral leaflets oblong rounded at base, acute or obtuse at apex, 1·5–5 cm. long, 0·6–2·5 cm. broad, terminal leaflet elliptic more or less acute at each end, 3–6 cm. long, 1·3–2 cm. broad ; rhachis 7–13 cm. long, together with petiolules and undersurface of leaflets rusty pubescent ; petals absent ; stamens 2, filaments straight ; fruits subglobose, 1·5–2 cm. diam., densely tomentellous, brown ..   ..   *5. dinklagei*

1. **D. pobeguinii** *Pellegr.* in Bull. Soc. Bot. Fr. 88 : 449 (1941) ; Aubrév. Fl. For. Soud.-Guin. 217, t. 46, 7–8 ; Steyaert in Bull. Soc. Roy. Bot. Belg. 84 : 37. *D. ovatum* Hutch. & Dalz. in Dalz. Us. Pl. W.T.A. 190 (1937), name only. Tree, to 45 ft. high, on river banks close to water ; flowers white.

   Fr.G.: R. Koumi (June) *Pobéguin* 1629. Labé *Chev.* 12383. **S.L.:** Njala (fl. June, fr. Mar.) *Deighton* 1752 ! 1823 ! Batkanu (fr. Jan.) *Deighton* 2849 ! (See Appendix, p. 190.)

2. **D. guineense** *Willd.*—F.T.A. 2 : 283 ; Aubrév. Fl. For. C. Iv. 1 : 206 ; Steyaert l.c. 39. Forest tree, to 60 ft. high, but often shrubby, sometimes persisting in savannah vegetation ; flowers small in copious brownish flat panicles ; petals white or pinkish ; fruits black, seeds embedded in reddish pulp.

   Sen.: *Heudelot* 584 ! Casamance (fl. & fr. Apr.) *Perrottet* 299 ! Cape Verde to St. Louis (Mar.) *Döllinger* 39 ! R. Salum *Brunner* 135 ! N'guer (Oct.) *Chev.* 2944 ! **Gam.:** *Hayes* 503 ! *Saunders* 18 ! **Port.G.:** *Esp. Santo* 1624 ! Formosa (fr. Apr.) *Esp. Santo* 1977 ! **Fr.G.:** Kouroussa (fl. & fr. Oct.) *Pobéguin* 584 ! Fouta Djalon (Sept., Oct.) *Maclaud* 447 ! *Chev.* 18252 ! 20165 ! **S.L.:** *Smeathmann* ! *Don* ! Regent *Sc. Elliot* 4178 ! Njala (fr. Feb.) *Deighton* 1933 ! Rokupr (fl. & fr. Nov., Dec.) *Jordan* 703 ! 713 ! **Lib.:** Gbanga (Sept.) *Linder* 768 ! **Iv.C.:** Morénou *Chev.* 22427. Abidjan *Aubrév.* 108. Rasso *Aubrév.* 150 ! 559 ! **G.C.:** Cape Coast *Hove* ! Sekondi (Nov.) *Vigne* FH 1400 ! Accra Plains (fl. Oct., fr. Mar.) *Brown* 371 ! *Howes* 1138 ! Larte Hills (Nov.) *Akobi* Abokobi (Oct.) *Vigne* FH 4257 ! Togo: Lome (Nov.) *Warnecke* 211 ! *Mildbr.* 7494 ! **Dah.:** Cotonou *Chev.* 4440. **N.Nig.:** Agaie (Dec.) *Yates* 17 ! Lokoja (Sept.) *Ansell* ! *T. Vogel* 168 ! Naraguta (Dec.–Jan.) *Lely* 735 ! P. 12 ! **S.Nig.:** Abeokuta *Irving* 77 ! Ogba (young fr. Dec.) *Unwin* 174 ! Onitsha *Barter* 1795 ! Also in Principe and S. Tomé. (See Appendix, p. 190.)

3. **D. aubrevillei** *Pellegr.* in Bull. Soc. Bot. Fr. 80 : 463 (1933) ; Aubrév. l.c. 1 : 206, t. 76, A ; Steyaert l.c. 41. Forest tree, to 200 ft. high, with buttresses ; flowers white.

   S.L.: Gola North F.R. (young fr. Nov.) *King* 294 ! Colony F.R. (fr. Nov.) *Vigne* 10 ! **Lib.:** Du R. (Aug.) *Linder* 214 ! **Iv.C.:** Abidjan *Aubrév.* 191 ! Djibi *Aubrév.* 534. Basse Me *Martineau* 219. **G.C.:** Tanosu (Sept.) *Chipp* 360 ! Asiakwa (fr. Jan.) *Brent* 398 ! Benso, Tarkwa (fr. Aug.) *Andoh* FH 5386 ! (See Appendix, p. 190.)

   [Aubréville (l.c.) records probable hybrids between this species and *D. guineense* Willd.]

4. **D. pachyphyllum** *Harms* in Engl. Bot. Jahrb. 53 : 468 (1915) ; Steyaert l.c. 42 ; Fl. Congo Belge 3 : 538. Forest tree, to 65 ft. high ; petals yellow ; fruits blackish-brown.

   S.Nig.: Ikom (fr. July) *Catterall* 15 ! Also in French Cameroons, Gabon, Belgian Congo and Angola.

5. **D. dinklagei** *Harms* in Engl. Bot. Jahrb. 26 : 275 (1899) ; Aubrév. l.c. 1 : 204, t. 75 ; Steyaert in Bull. Soc. Roy. Bot. Belg. 84 : 37 ; Fl. Congo Belge 3 : 544. *D. staudtii* Harms l.c. ; F.W.T.A. ed. 1, 1 : 337. Forest tree, to 65 ft. high with buttressed stem and spreading crown with pendulous branches ; flowers greenish-yellow or greenish-brown ; fruits brown.

   Fr.G.: Kissidougou (fr. Feb.) *Chev.* 20709 ! **S.L.:** Potoru, Barri (fr. Apr.) *Deighton* 1666 ! Kailahun (fr. Apr.) *Deighton* 3138 ! Njala (Sept.) *Deighton* 5171 ! **Lib.:** Grand Bassa (May) *Dinklage* 1813 ! Du R. (July) *Linder* 68 ! Brewersville (June) *Barker* 1322 ! Harbel, Farmington R. (fr. Mar.) *Baldwin* 11097. **Iv.C.:** Abidjan (Sept.) *Aubrév.* 112 ! Dabou *Chev.* 16206 ! **G.C.:** Aburi (Aug.) *Johnson* 1087 ! Dunkwa (Aug.) *Vigne* 972 ! Kumasi (Mar.) *Andoh* FH 5636 ! **S.Nig.:** Sapoba (May) *Kennedy* 766 ! Afikpo (Oct.) *Espley* 11 ! **Br.Cam.:** Johann-Albrechtshöhe *Staudt* 492 ! Also in French Cameroons, Gabon and Belgian Congo. (See Appendix, p. 190.)

11. DISTEMONANTHUS Benth.—F.T.A. 2 : 282.

A large tree ; leaves pinnate ; leaflets 9–11, alternate, ovate-lanceolate to narrowly lanceolate, acutely acuminate, rounded at base, 5–10 cm. long, 2–4 cm. broad, slightly pubescent beneath, with numerous conspicuous lateral nerves ; flowers in lax paniculate cymes ; bracts small and deciduous ; pedicels pubescent ; sepals 5, the outer 3 valvate, about 1 cm. long ; petals 3, narrow, a little longer than the calyx ; fertile stamens 2, anthers opening by pores ; ovary tomentose ; fruit narrowly elliptic, acute, flat, about 9 cm. long and 3·5 cm. broad, at length glabrous

benthamianus

D. benthamianus *Baill.* Hist. des Pl. 2 : 135 (1870) ; Engl. Pflanzenw. Afr. 3, 1 : 492, fig. 263 ; Aubrév. Fl. For. C. Iv. 1 : 202, t. 74 ; Kennedy For. Fl. S. Nig. 90. *D. laxus* Oliv. F.T.A. 2 : 282 (1871). A forest tree, to 120 ft. high, with conspicuous orange to red bark, particularly on the upper bole and branches. **S.L.:** *King* 248 ! Daru (fr. Apr.) *Deighton* 3215 ! Roruks *Deighton* 4139 ! **Iv.C.:** *fide* Aubrév. l.c. **G.C.:** Dunkwa (fl. May, fr. July, Aug.) *Chipp* 714 ! *Vigne* FH 907 ! *Brent* 885 ! Kumasi (fr. Mar.) *Vigne* FH 1653 ! Bunsu, Akim (fr. Mar.) *Irvine* 1804 ! Duabayie (fr. Mar.) *Thompson* 26 ! Koforidua (Oct.) *Vigne* FH 4265 ! **Togo:** Atakpame *Doering.* **S.Nig.:** Okomu F.R. (fr. Jan.) *Richards & Onochie* 8888 ! Benin City (Jan.) *Dennett* 50 ! Sapoba (Sept.–Nov.) *Kennedy* 406 ! 1615 ! Oban *Talbot* 1497 ! **Br.Cam.:** Johann-Albrechtshöhe *Staudt* 638 ! 926 ! Likomba *Mildbr.* 10738 ! Also in French Cameroons, Spanish Guinea and Gabon. (See Appendix, p. 191.)

12. CASSIA Linn.—F.T.A. 2 : 268 ; Benth. in Trans. Linn. Soc. 27 : 503–591 (1871).

Leaflets 1 pair, sessile, obovate, very unequal at base, 7–22 mm. long, 5–14 mm. broad ; petiole up to 5 mm. long, eglandular ; stems prostrate, like the leaves long-spreading-pilose ; stipules foliaceous, obliquely ovate, persistent, 5–12 mm. long ; flowers axillary, solitary or paired ; pedicels longer than the leaves, up to 3·5 cm. long in fruit ; fruits flat, linear-oblong, slightly curved, 1·8–3·5 cm. long, 2·5–4 mm. broad, valves thin, pubescent, the septae clearly visible from the outside, curling on dehiscence ; seeds 8–16   ..   ..   1. *rotundifolia*

Leaflets 2 or more pairs :
  Leaflets sessile ; petioles with at least one gland towards the top but below the lowest pair of leaflets ; leaflets up to 2·6 cm. long and 6 mm. broad ; flowers solitary, or in few-flowered racemes, usually supra-axillary (the peduncle adnate to the stem) ; fruits flat, linear-oblong, slightly curved, valves thin, curling on dehiscence ; seeds in 1 series :
    Leaflets (3–) 4–6 (–12), narrowly and obliquely oblong, midrib nearly central, unequal at base, rounded and mucronate at apex, 7–18 (–25) mm. long, 2·5–6 mm. broad, petiole and rhachis 1·8–3·5 (–7·8) cm. long, petiole with a stipitate gland about 1 mm. long just below the lowest pair of leaflets ; flowers solitary, supra-axillary ; pedicels up to 3·8 cm. long in fruit ; fruits about 3·5 cm. long and 4 mm. broad, pubescent ; seeds about 12   ..   ..   2. *jaegeri*
  Leaflets 10–70 pairs :
    Rhachis of leaf channelled above between each pair of leaflets :
      Midrib central, leaflets symmetrical or nearly so, 10–18 pairs, oblong-elliptic, rounded and apiculate at apex, 15–26 mm. long, 5–6 mm. broad ; rhachis and petiole about 7 cm. long, petiole with an ellipsoidal gland about 2 mm. long ; racemes very short, 5–6-flowered, solitary, or paired with one raceme axillary and the other supra-axillary ; pedicels 1–3 mm. long ; fruits 2·5–4 cm. long ; seeds about 10   ..   ..   3. *nigricans*
      Midrib lateral, leaflets asymmetrical, 25–30 pairs, oblong-linear, obliquely acute and mucronate at apex, 9–17 mm. long, 1·5–4 mm. broad ; rhachis and petiole 7–14 cm. long, petiole with a sessile gland just below the lowest pair of leaflets ; flowers solitary or 2–3 together, supra-axillary, pedicels about 1 cm. long ; fruits 6–7·5 cm. long ; seeds 14–20   ..   ..   4. *kirkii*
    Rhachis of leaf with a serrated ridge along the top, having a tooth between each pair of leaflets ; leaflets asymmetrical, 30–70 pairs, linear, obliquely acute and mucronate at apex, 2–8 mm. long, 0·7–2 mm. broad ; rhachis and petiole 4–9 cm. long, petiole with a sessile gland just below the lowest pair of leaflets ; flowers solitary, or 2–3 together, supra-axillary ; pedicels 1–2 cm. long ; fruits 3–5 cm. long ; seeds 13–24 ..   ..   5. *mimosoides*

Leaflets petiolulate ; petiole and rhachis with or without glands ; leaflets less than 15 pairs, more than 1·5 cm. long and 1 cm. broad, never linear, at most lanceolate ; flowers axillary or terminal ; fruits indehiscent or dehiscent, valves not curling on dehiscence (except in *C. absus*) :
  *Petiole and rhachis without glands :
    Stems not viscid-glandular ; leaflets 3 pairs, or more ; valves of fruit not curling on dehiscence, or fruit indehiscent :
      †Filaments of 2–3 lower stamens much longer than the petals, bent into the shape of a hook ; anthers dorsifixed ; fruits more or less terete, with numerous transverse septae :
        Petals pink or white ; bracts caducous ; inflorescences terminating short axillary branches, racemose or subcorymbose, 5–12 cm. long, with about 20 flowers ; pedicels about 4·5 cm. long ; fruits indehiscent, straight, subterete or slightly compressed dorsally, up to 90 cm. long and 2–3 cm. diam., with a longitudinal septum, the seeds thus in 2 series ; leaflets 7–10 pairs, elliptic to elliptic-ovate, acute or shortly acuminate at apex, 6–8 cm. long, 2·5–3·5 cm. broad ; petiole and rhachis up to about 30 cm. long ..   ..   6. *mannii*
        Petals yellow ; bracts more or less persistent :
          Flowers in racemes up to 40 cm. long, pedicels up to 5 cm. long ; fruits indehiscent :
            Fruits with a longitudinal septum, the seeds thus in 2 series ; fruit straight, subterete or slightly compressed dorsally, 60–70 cm. long, 2·5–3 cm. broad, 2–2·5 cm. thick ; bracts obliquely lanceolate, about 1 cm. long, densely velutinous ; pedicels velutinous ; leaflets 6–12 pairs, oblong-lanceolate, obtuse at apex, 3–10 cm. long, 1–4 cm. broad ; petiole and rhachis up to 35 cm. long   ..   ..   7. *aubrevillei*
            Fruits without a longitudinal septum, the seeds thus in 1 series ; fruits seldom straight, terete, 50–80 cm. long, about 1·5 cm. diam. ; bracts linear-lanceolate, 1–2 cm. long, pubescent ; pedicels pubescent ; leaflets 5–8 pairs, elliptic or oblong, shortly subacute or emarginate at apex, 5–10 cm. long, 2·5–5 cm. broad ; petiole and rhachis up to 20 cm. long   ..   ..   8. *sieberiana*

Flowers in terminal bracteate corymbs, the lower pedicels the longest and up to 10 cm. long ; fruits dehiscent along one suture, seeds in 1 series ; leaflets 5–9 pairs, ovate-elliptic, cuneate at base and long petiolulate, acute or slightly acuminate at apex, 3–4·5 cm. long, 1–2·5 cm. broad ; petiole and rhachis up to 26 cm. long   ..   ..   9. *arereh*

†Filaments of all stamens straight and shorter than the petals ; fruits flat or winged :
Rhachis narrowly winged, with a transverse ridge connecting the leaflets ; leaflets 8–14 pairs, oblong to obovate, rounded at each end, 5–15 cm. long, 3–8 cm. broad ; petiole and rhachis up to 60 cm. long ; flowers in erect dense, terminal racemes, with large enveloping bracts ; fruits straight, oblong, 15–25 cm. long, about 1·8 cm. broad, each valve with a prominent broad crenate wing along the middle   ..   ..   10. *alata*
Rhachis not winged ; fruits flat not winged :
Fruits 10 cm. long or more, up to 1·5 cm. broad, straight, attached centrally ; leaflets at least 5 cm. long ; petiole and rhachis up to about 30 cm. long :
Flowers in dense, erect, spike-like terminal racemes ; leaflets 4–5 pairs, elliptic, more or less narrowed at each end, 6–12 cm. long, 3·5–6 cm. broad ; fruits about 10–12 cm. long and 1·5 cm. broad, with transverse ridges, indehiscent   11. *podocarpa*
Flowers in panicles of axillary or terminal corymbs, numerous at the ends of the shoots ; leaflets 8–12 pairs, oblong-elliptic, more or less rounded at each end, usually emarginate and mucronate at apex, 5–6 cm. long, 2–2·5 cm. broad ; fruits 15–30 cm. long, about 1 cm. broad, dehiscent   ..   12. *siamea*
Fruits up to 6 cm. long, 1·8–2·5 cm. broad, broadly oblong, attached towards the ventral suture ; leaflets up to 4 cm. long ; petiole and rhachis up to about 16 cm. long ; flowers in simple racemes :
Fruits with an undulate crest up the middle of each valve, curved, 3·5–5·5 cm. long, about 2 cm. broad ; leaflets 5–6 pairs, obliquely oblong-obovate, rounded and mucronate at apex, 1·5–4 cm. long, 0·8–2·5 cm. broad
            13. *italica*
Fruits without an undulate crest, more or less straight, 4–6 cm. long, 1·8–2·5 cm. broad ; leaflets 3–9 pairs, elliptic-lanceolate or lanceolate, 1·5–3·5 cm. long, 0·5–1·2 cm. broad   ..   ..   14. *senna*
Stems and leaves viscid-glandular-pubescent ; leaflets 2 pairs, close together at the end of a 3–5 cm. long petiole ; lamina obliquely obovate or elliptic, 1·5–3 cm. long, 1–1·5 cm. broad ; flowers in terminal or leaf-opposed racemes ; fruits oblong, 2·5–4 cm. long, 7–8 mm. broad, strigose, valves curling on dehiscence.. 15. *absus*
*Petiole and rhachis with conspicuous glands :
Glands on the rhachis at the insertion of 1 or several pairs of leaflets :
Leaflets 5–8 pairs, elliptic, rounded at each end, emarginate and mucronate at apex, 2–5 cm. long, 1–3 cm. broad ; petiole and rhachis to 20 cm. with a slender club-shaped gland between each pair of leaflets ; flowers in corymbose racemes, clustered towards the end of the branches ; fruits subturgid and subtorulose at maturity, beaked at apex, 10–15 cm. long, 5–8 mm. diam. ; seeds in 1 series
            16. *singueana*
Leaflets 3–5 pairs, or if more then seeds in 2 series and fruits not subtorulose :
Fruits subcylindrical, 1–1·5 cm. diam., 7–10 cm. long ; seeds in 2 series, transversely arranged ; inflorescences 4–10-flowered axillary or terminal racemes ; leaflets 3–6 pairs :
Leaflets acutely acuminate at apex, ovate-elliptic to lanceolate, 4–9 cm. long, 2–5 cm. broad ; petiole and rhachis 10–20 cm. long, with a gland between each pair of leaflets ..   ..   17. *laevigata*
Leaflets rounded and often mucronate at apex, obovate, 1·5–3 cm. long, 1–2 cm. broad ; petiole and rhachis 3–7 cm. long, with a gland at the insertion of the lowest pair of leaflets only   ..   ..   ..   18. *bicapsularis*
Fruits more or less compressed, 3–6 mm. broad, 15–25 cm. long, curved, valves longitudinally ridged ; seeds in 1 series, arranged longitudinally ; inflorescences 2-flowered, axillary, or flowers solitary ; leaflets 3 pairs, obovate, mucronate at apex, up to 6 cm. long and 3 cm. broad ; petiole and rhachis up to 15 cm. long, with a gland at the insertion of the lowest pair, or 2 lowest pairs, of leaflets
            19. *tora*
Glands towards the base of the petiole only ; leaflets acuminate ; seeds in 1 series ; racemes few-flowered :
Leaflets and fruits glabrous, or shortly pubescent :
Leaflets 5–9 pairs, the top pair not the largest, narrowly lanceolate, very acute, 4–7 cm. long, 1·5–2 cm. broad ; fruits subturgid with prominent sutures, 7–8 cm. long, about 1 cm. broad, shortly beaked   ..   ..   20. *sophera*
Leaflets 4–5 pairs, the top pair the largest, broadly lanceolate or ovate, 3·5–10 cm. long, 2–4 cm. broad ; fruits flattish, up to 14 cm. long, abruptly beaked
            21. *occidentalis*

Leaves, stem and fruits long-pilose or hirsute ; leaflets 4 pairs, lanceolate or ovate-lanceolate, 4–7 cm. long, 1·5–2·5 cm. broad ; fruits flattish, about 12 cm. long and 5 mm. broad, very densely long-hirsute   ..   .. 22. *hirsuta*

1. **C. rotundifolia** *Pers.* Syn. 1 : 456 (1805) ; Benth. in Trans. Linn. Soc. 27 : 570. Subwoody herb, with prostrate branches up to 18 in. long radiating from the rootstock ; flowers yellow ; fruits blackish-brown when ripe ; a weed locally common, introduced into our area.
   G.C.: Afringba, Keta (fl. & fr. Sept.) *Akpabla* 335 ! **N.Nig.:** Zaria (fl. & fr. Dec.) *Daggash* FHI 31406 ! 51 miles SE of Ilorin *Clarke* 24 ! **S.Nig.:** Lagos (fl. & fr. Aug.) *Rowland* ! Abeokuta (fl. & fr. Feb.) *Burtt* B. 32 ! Ibadan (fl. & fr. Aug.) *Chizea* FHI 20021 ! A native of tropical America.

2. **C. jaegeri** *Keay* in Bull. I.F.A.N. 18 : 375 (1956). *C. grantii* var. *pilosula* of Berhaut Fl. Sen. 33, not of Oliv. Slender subwoody herb, erect, to 3 ft. high.
   Sen.: without precise locality *Berhaut* 1232. **Fr.Sud.:** Kita (fl. & fr. Oct.) *Jaeger* ! **S.L.:** Musaia (fr. Dec.) *Deighton* 4424 !

3. **C. nigricans** *Vahl*—F.T.A. 2 : 280 ; Benth. l.c. 577 ; Chev. Bot. 223 ; Steyaert in Fl. Congo Belge 3 : 518. A woody herb or undershrub, up to 4 or 5 ft. high, with small yellow flowers.
   Sen.: *Heudelot* 256 ! **Gam.:** *Park* ! *Pirie* 326 ! **Fr.Sud.:** Bilma (fl. & fr. Dec.) *E. Vogel* ! Toukoto (fl. & fr. Dec.) *Chev.* 71 ! Gao (Sept.) *Hagerup* 361 ! **Port.G.:** Sama, Bafata (fl. & fr. Jan.) *Esp. Santo* 437 ! **G.C.:** Ahinsa, Kumasi (fl. & fr. Oct.) *Darko* 415 ! Gambaga (fl. & fr. Sept.–Dec.) *Vigne* FH 4586 ! *Adams & Akpabla* GC 4219 ! 4242 ! **Dah.:** Atacora Mts. *Chev.* 24119. **N.Nig.:** Sokoto *Moiser* ! Kano (fl. & fr. Feb.) *Dalz.* 339 ! Kontagora (fl. & fr. Oct.) *Dalz.* 21 ! Anara F.R. (Sept.) *Olorunfemi* FHI 24358 ! Kukawa (Jan.) *E. Vogel* 21 ! Widespread in tropical Africa, also in Arabia and India. [*Barter* 534 cited as this species in F.W.T.A., ed. 1, 1 : 335, is a species of *Aeschynomene.*] (See Appendix, p. 180.)

4. **C. kirkii** *Oliv.* F.T.A. 2 : 281 (1871) ; Benth. l.c. 580 ; Steyaert l.c. 525, t. 38. Erect shrub, to 5 ft. high ; flowers yellow. Steyaert (l.c.) distinguishes 3 varieties : var. *kirkii*—stems with long, spreading, woolly hairs mixed with the shorter appressed hairs ; var. *guineensis* Steyaert—stems with appressed hairs only ; var. *glabra* Steyaert—plant quite glabrous. The first two varieties occur in our area, but several specimens are intermediate.
   S.L.: Jigaya (Nov.) *Thomas* 2785 ! Jokibu (fl. & fr. Nov.) *Deighton* 3781 ! **Lib.:** Bakratown (Oct.) *Linder* 876 ! Jabroke, Webo (fr. July) *Baldwin* 6678 ! Ganta (fl. & fr. Sept.) *Harley* ! **G.C.:** *Miles* ! Ejura (Nov.) *Vigne* FH 3472 ! *Darko* 750 ! Accra (fl. & fr. May) *Irvine* 2838 ! **N.Nig.:** Naraguta, Bauchi Plateau (Aug.) *Vigne* FH 747 ! Vom *Dent Young* 86 ! **S.Nig.:** Ibadan (Sept., Oct.) *Adekunle* FHI 24221 ! Ondo (Sept.) *Keay* FHI 22563 ! Newi, Onitsha (Oct.) *Jones* FHI 6804 ! Eket *Talbot* ! **Br.Cam.:** Buea (fr. Dec.) *Maitland* 810 ! Bamenda, 5,000 ft. (fl. & fr. Feb.) *Migeod* 459 ! L. Bambulue, Bamenda (Sept.) *Ujor* FHI 30207 ! **F.Po:** Moka (Dec.) *Boughey* 83 ! Widespread in tropical Africa.

5. **C. mimosoides** *Linn.* Sp. Pl. 1 : 379 (1753) ; F.T.A. 2 : 280, partly ; Benth. l.c. 579 ; Chev. Bot. 222 ; Steyaert l.c. 514, t. 37. Herb or low shrub to 5 ft. high, or diffuse ; flowers yellow ; a variable plant, locally common on sandy soil.
   Sen.: *Perrottet* 283 ! *Heudelot* 251 ! Casamance *fide* Chev. *l.c.* **Gam.:** Wallikunda (Aug.) *Macluskie* 7 ! Kuntaur *Ruxton* 81 ! 133 ! **Fr.Sud.:** *fide* Chev. *l.c.* **Fr.G.:** Tristao Isl. *Paroisse* 49 ! **S.L.:** Bonthe Isl. (fl. & fr. Nov.) *Deighton* 2278 ! Nyaflanda, Masakoi (Sept.) *Deighton* 2234 ! Roruks (fl. & fr. Sept.) *Deighton* 4842 ! Juring (fl. & fr. Dec.) *Deighton* 330 ! **Lib.:** Monrovia (fl. & fr. May–Dec.) *Linder* 1522 ! *Baldwin* 5810 ! 10502 ! Sinkor (fl. & fr. July–Sept.) *Barker* 1064 ! 1385 ! **Iv.C.:** R. Férédougouba, Touba (July) *Collenette* 63 ! And *fide* Chev. *l.c.* **G.C.:** Cape Coast (July) *T. Vogel* 78 ! Amansare (fl. & fr. July) *Chipp* 521 ! Tamale *Anderson* 33 ! **Togo:** Lome *Warnecke* 37 ! Sokode-Basari *Schröder* 65 ! **Dah.:** *fide* Chev. *l.c.* **Fr.Nig.:** Niamey (Oct.) *Hagerup* 465 ! **N.Nig.:** Pandiaki (Sept.) *Ansell* ! Nupe *Barter* 1619 ! 1627 ! Sokoto (Sept., Nov.) *Moiser* 13 ! 113 ! Biu (Aug., Sept.) *Noble* 8 ! 17 ! **S.Nig.:** Lagos *MacGregor* 155 ! Ondo (Sept.) *Keay* FHI 22564 ! Sobo Plains, Jesse (Sept.) *Butler-Cole* 20 ! **Br.Cam.:** Bambuko F.R. (Sept.) *Olorunfemi* FHI 30775 ! Widespread in tropical Africa and Asia. (See Appendix, p. 180.)

6. **C. mannii** *Oliv.* F.T.A. 2 : 272 (1871) ; Benth. l.c. 514 ; Milne-Redhead in Hook. Ic. Pl. t. 3368 (1938), partly (excl. *Aubrév.* 655) ; Pellegr. Lég. Gab. 94 ; Steyaert l.c. 499, t. 35. A forest tree, to 80 ft. high, deciduous ; petals pink or white ; flowers sweet-scented.
   S.Nig.: Umu-Akpo, Owerri (Nov.) *Rosevear* 20/28 ! **Br.Cam.:** Victoria Bot. Gdns. (Dec.) *Winkler* 602 ! Also in French Cameroons, Gabon, Belgian Congo, Uganda, A.-E. Sudan, S. Tomé and Principe.

7. **C. aubrevillei** *Pellegr.* in Bull. Soc. Bot. Fr. 94 : 5 (1947) ; Lég. Gab. 94. *C. sp. aff. mannii* of Aubrév. Fl. For. C. Iv. 1 : 210, partly (excl. descr. fls.), t. 77, 1–3 only. A forest tree, to 2 ft. diam., bole without buttresses, deciduous ; petals yellow.
   Iv.C.: Adikokoi, Mid. Comoé (fr. Jan.) *Aubrév.* 655 ! Mudjika (May) *Serv. For.* 2785 ! Also in Gabon.

8. **C. sieberiana** *DC.*—F.T.A. 2 : 270 ; Benth. l.c. 516 ; Chev. Bot. 224 (incl. var. *saheliensis*) ; Aubrév. Fl. For. Soud.-Guin. 217, t. 45, 1–3 ; Steyaert l.c. 500. *C. kotschyana* Oliv. F.T.A. 2 : 271 (1871). A tree, to 50 ft. high, with pendulous laburnum-like yellow flower-racemes, and black fruits ; flowering in dry season, often when leafless ; common in savannah.
   Sen.: *Sieber* ! **Gam.:** *Heudelot* 348 ! Albreda (Mar.) *Perrottet* 281 ! Genieri (Mar.) *Fox* 80 ! **Fr.Sud.:** Koulouba, Bamako (Mar.) *Waterlot* 1067 ! Kouroussa (Feb.) *Pobéguin* 123 ! **S.L.:** Bagroo R. (Apr.) *Mann* 799 ! Kukuna (Jan.) *Sc. Elliot* 4645a ! Giema (Mar.) *Dawe* 459 ! Njala (Mar.) *Deighton* 1096 ! **Lib.:** Zigida (Mar.) *Bequaert* 137 ! **Iv.C.:** Bouroukrou *Aubrév.* 913. Man *Aubrév.* 947. **G.C.:** Fiapere (Feb.) *Chipp* 79 ! Kintampo *Dalz.* 42 ! Lawra (Mar.) *MacLeod* 829 ! **Dah.:** Sakété (Jan.) *Chev.* 22815 ! **N.Nig.:** Sokoto (May) *Lely* 813 ! Nupe *Barter* 1171 ! Agaie (Feb.) *Yates* 40 ! Katagum *Dalz.* 332 ! Abinsi (June) *Dalz.* 586 ! Musgu (May) *E. Vogel* 72 ! **S.Nig.:** Abeokuta (Jan.) *Rowland* ! Agbemia, Egbaland *Barter* 3361 ! Olokemeji (Feb.) *Obaseki* FHI 27911 ! Also in Ubangi-Shari, A.-E. Sudan, Uganda, Belgian Congo and Tanganyika. (See Appendix, p. 182.)

9. **C. arereh** *Del.*—F.T.A. 2 : 270 ; Benth. l.c. 517 ; Aubrév. l.c. 221, t. 45, 4–5. Savannah tree, to 30 ft. high, with corymbose yellow flowers and brown cylindrical pods 1–2 ft. long, splitting lengthwise.
   N.Nig.: Sokoto (Feb.) *Lely* 847 ! Garin Gwenchi, 30 miles E. of Zaria (Mar.–Apr.) *Ryan* 39 ! Anara F.R. (fr. May) *Keay* FHI 22987 ! Bauchi Plateau (Mar.) *Lely* P. 199 ! Biu (Mar.) *Kennedy* FHI 8052 ! Yola (fl. & fr. Mar.) *Dalz.* 8 ! Extends to Abyssinia and the Red Sea. (See Appendix, p. 179.)

10. **C. alata** *Linn.*—F.T.A. 2 : 275 ; Benth. l.c. 550 ; Chev. Bot. 222 ; Aubrév. l.c. 221, t. 44, 4–5 ; Steyaert l.c. 507. A soft-wooded shrub, up to 15 ft. or more high, with yellow flowers in stout dense erect racemes ; fruits black when ripe ; in villages and clearings, chiefly in forest regions.
   Sen.: *Berhaut* 1783. **Fr.Sud.:** *fide* Chev. *l.c.* **Fr.G.:** Conakry (Jan.) *Maclaud* ! **S.L.:** *Barter* ! Ronietta (Nov.) *Thomas* 5575 ! Sembihun (fr. Dec.) *Deighton* 350 ! Kagbantama, Kasse (fl. & fr. Dec.) *T. S. Jones* 50 ! **Lib.:** Gbanga (Sept.) *Linder* 809 ! Robertsport (Dec.) *Baldwin* 10960 ! **Iv.C.:** *fide* Chev. l.c. **G.C.:** Kumasi *Cummins* 6 ! Anwhiaso F.R. (July) *Foggie* FH 4439 ! **N.Nig.:** Yola (Oct.) *Shaw* 106 ! **S.Nig.:** Ibadan (Oct.) *Newberry* 138 ! Okomu F.R. (Jan.) *Napoleon* in Hb. *Brenan* 8793 ! Oban *Talbot* 184 ! A native of America, widely spread in the tropics. (See Appendix, p. 179.)

11. **C. podocarpa** *Guill. & Perr.*—F.T.A. 2 : 276 ; Benth. l.c. 552 ; Chev. Bot. 224 ; Aubrév. l.c. 219, t. 42, 3. A glabrous shrub, up to 15 ft. high, with rather dense racemes of yellow flowers and thin flat pods ; locally common on old farmland, mainly in forest regions.
   Sen.: Kounoun (fr. Mar.) *Perrottet*. **Port.G.:** *Esp. Santo* 409 ! **Fr.G.:** Karkandy *Heudelot* 794. Kouroussa *Pobéguin* 966. **S.L.:** *Afzelius* ! Sherbro *Sc. Elliot* 5854 ! Regent (fl. & fr. Dec.) *Sc. Elliot* 4098 !

Commendi (Nov.) *Aylmer* 259 ! Njala (fl. & fr. Nov.) *Deighton* 1850 ! **Lib.:** Kakatown *Whyte* ! Peahtah (Oct.) *Linder* 941 ! Harbel (fl. & fr. Dec.) *Bequaert* 1 ! Ganta (Oct.) *Harley* ! **Iv.C.:** Niangbo, Ferkéssé-dougou & Baoulé *fide* Aubrév. *l.c.* **G.C.:** Aburi (fl. & fr. Oct., Nov.) *Johnson* 466 ! Afram Plains (fr. Jan.) *Chipp* 596 ! **S.Nig.:** Lagos *Barter* 2203 ! Olokemeji *Foster* 152 ! Ibadan (fl. Oct., fr. Dec.) *Jones* FHI 13763 ! 14701 ! Ibuzo (Nov.) *Thomas* 1991 ! **F.Po:** (fl. & fr. Dec.) *Mann* 638 ! (See Appendix, p. 182.)

12. **C. siamea** *Lam.* Encycl. Méth. Bot. 1 : 648 (1783) ; Benth. l.c. 549 ; Steyaert l.c. 506. A tree, to 70 ft. high ; flowers yellow ; frequently planted for fuel, sometimes subspontaneous.
Sen.: *Berhaut* 128. **S.L.:** *Deighton* 402 ! **Iv.C.:** Abengourou *Aubrév.* 677 ! **G.C.:** Achimota *Irvine* 1416 ! **Dah.:** Porto Novo (Jan.) *Chev.* 22710 ! **N.Nig., S.Nig. & Br.Cam.:** many localities *fide* Keay. A native of tropical Asia, now widely spread in the tropics.

13. **C. italica** (*Mill.*) *Lam. ex F. W. Andr.* Fl. Pl. A.-E. Sud. 2 : 117, fig. 49 (1952). *Senna italica* Mill. (1768). *C. aschrek* Forsk. (1775)—F.W.T.A., ed. 1, 2 : 606. *C. obovata* Collad. (1816)—F.T.A. 2 : 277 ; Benth. l.c. 553 ; Chev. Bot. 223 ; F.W.T.A., ed. 1, 1 : 335 ; Aubrév. l.c. 217, t. 42, 1. Undershrub, often with herbaceous branches from a woody stock, more or less glaucous. Italian Senna.
Sen. (Jan.) *Roger* ! Walo *Perrottet* 277 ! 279 ! **Fr.Sud.** Timbuktu (fl. & fr. July) *Chev.* 1263 ! *Hagerup* 267a ! Ségou *Chev.* 3342. **Fr.Nig.** Gourma *Chev.* 24480. Agades *De Wailly* 4938. **N.Nig.** Zurmi (fl. & fr. Apr.) *Keay* FHI 15667 ! Bichikki (fl. & fr. May) *Lely* 174 ! Katagum (fl. & fr. July) *Dalz.* 4 ! Kalkala, L. Chad *Golding* 67 ! Kukawa *E. Vogel* 13 ! Widespread in the dry regions of Africa ; extends to N.W. India. (See Appendix, p. 180.)

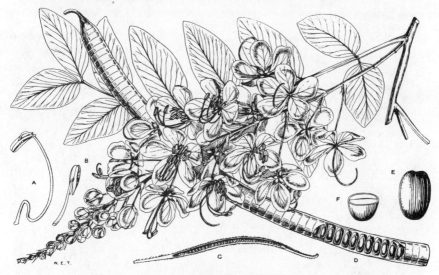

FIG. 147.—CASSIA SIEBERIANA *DC.* (CAESALPINIACEAE).

A, B, two types of stamen. C, longitudinal section of pistil. D, mature fruit. E, seed. F, cross-section of seed.

14. **C. senna** *Linn.* Sp. Pl. 1 : 377 (1753) ; F. W. Andr. l.c. 118, fig. 50. *C. acutifolia* Del. (1802)—F.T.A. 2 : 278 ; Benth. l.c. 553 ; F.W.T.A., ed. 1, 1 : 335. *Senna acutifolia* (Del.) Batka Monogr. Senn. 41 (1866). An undershrub with pale more or less zigzag branchlets and erect racemes of yellow flowers. Alexandrian Senna.
Fr.Sud.: Aïr (Nov.) *Francis Rodd.* Timbuktu *fide* Batka *l.c.* **N.Nig.:** Sokoto *fide* Batka *l.c.* Also in Sahara, Upper Egypt, A.-E. Sudan and Somaliland. (See Appendix, p. 178.)

15. **C. absus** *Linn.*—F.T.A. 2 : 279 ; Benth. l.c. 558 ; Chev. Bot. 221 ; Steyaert l.c. 507. A viscid herb or weak shrub, 1–2 ft. or more high, erect or prostrate ; flowers small, red or yellow, petals dark veined ; in waste places.
Sen.: *Heudelot* 414 ! Walo *Perrottet* 280 ! Ziguinchor, Casamance *Chev.* 3343. **Gam.:** Kuntaur *Ruxton* 132 ! **Fr.Sud.:** Koulikoro *Chev.* 3344. **Fr.G.:** Kouria *Chev.* 14776. **S.L.:** Kitchom (fl. & fr. Dec., Jan.) *Sc. Elliot* 4315 ! *Deighton* 960 ! Musaia (fl. & fr. Sept.–Dec.) *Deighton* 4426 ! 4851 ! Fintonia (Sept.) *Deighton* 5178 ! Jigaya (fl. & fr. Nov.) *Thomas* 2742 ! **Iv.C.:** Mossi (Aug.) *Chev.* 24723. Bingerville *Chev.* 20072. **G.C.:** Accra (fl. & fr. May) *Brown* 57 ! *Irvine* 1761 ! Tamale (fl. & fr. Aug.) *Williams* 842 ! **Dah.:** Cotonou (fl. & fr. Apr.) *Debeaux* 147 ! **Fr.Nig.:** Niamey *Hagerup* 545 ! **N.Nig.:** Sokoto *Lely* 107 ! Nupe *Barter* 1620 ! Idah (Sept.) *T. Vogel* 13 ! Anara F.R. (May) *Keay* FHI 22997 ! Katagum (fl. & fr. Sept.) *Dalz.* 1 ! Biu (fl. & fr. Aug.) *Noble* 12 ! Dikwa (Aug.) *E. Vogel* 68 ! Widespread in tropical Africa, Asia and Australia. (See Appendix, p. 178.)

16. **C. singueana** *Del.* Cent. Pl. Afr. 28 (1826) ; Aubrév. l.c. 217, t. 42, 2 (incl. var. *kethulleana* (De Wild.) Ghesq.) ; Steyaert l.c. 509. *C. goratensis* Fres. (1839)—F.T.A. 2 : 273 ; Chev. Bot. 222. A shrub, or small tree, up to 30 ft. high, with copious yellow flowers.
Fr.Sud.: Negala (Jan.) *Chev.* 163 ! Kati *Chev.* 170. Bamako *Waterlot* 1465. **Iv.C.:** Koudougou *Aubrév.* 2150. **G.C.:** Lawra (Nov.) *Vigne* FH 4661 ! Burufo (Dec.) *Adams* GC 4434 ! **Fr.Nig.:** Gourma (July) *Chev.* 24175 ! Niamey (Oct.) *Hagerup* 561 ! **N.Nig.:** Katsina (fl. & fr. Mar.) *Meikle* 1347 ! Kontagora (Jan.) *Dalz.* 29 ! Bauchi Plateau (fl. & fr. Jan.) *Lely* P. 142 ! Katagum (fl. & fr. Dec.) *Dalz.* 3 ! Musgu (fr. May) *E. Vogel* 96 ! Yola (Oct.) *Shaw* 79 ! Widespread in tropical Africa. (See Appendix, p. 180.)

17. **C. laevigata** *Willd.*—F.T.A. 2 : 275 ; Benth. l.c. 527 ; Steyaert l.c. 511. A glabrous shrub, 6–10 ft. high, with smooth green branchlets and fairly large yellow flowers ; introduced and subspontaneous in some places.
S.L.: (fr. June) *T. Vogel* 126 ! **G.C.:** Begoro, Akim (Mar.) *Irvine* 1181 ! Abetifi Kwahu (June) *Irvine* 308 ! **Br.Cam.:** Bambui (fl. & fr. Jan.) *Keay* FHI 28339 ! **F.Po:** (fr. Nov.) *T. Vogel* 200 ! *Mann* 1168 ! A native of tropical America, widespread in the tropics.

11

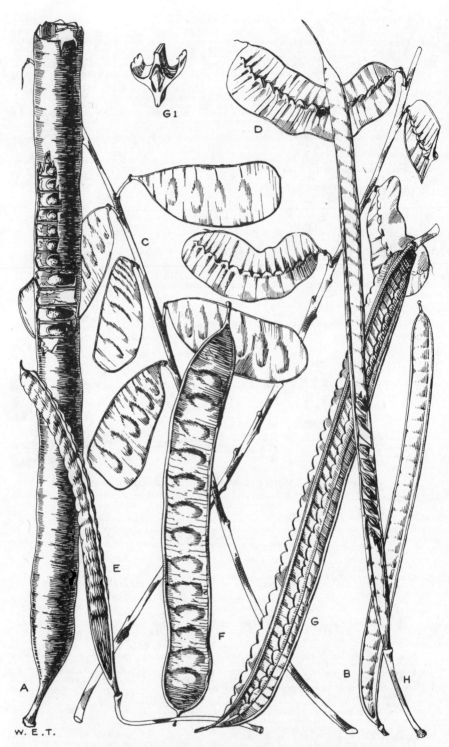

FIG. 148.—FRUITS OF CASSIA (CAESALPINIACEAE).

A, *Cassia sieberiana* DC. B, *C. occidentalis* Linn. C, *C. senna* Linn. D, *C. italica* (Mill.) Lam.
ex F. W. Andr. E, *C. singueana* Del. F, *C. podocarpa* Guill. & Perr. G, *C. alata* Linn.
G1, section of *C. alata* Linn. H, *C. tora* Linn.

18. **C. bicapsularis** *Linn.* Sp. Pl. 1 : 376 (1753) ; Benth. l.c. 525 ; Steyaert l.c. 511. A shrub, to about 5 ft. high, with yellow flowers ; introduced, sometimes used in hedges, subspontaneous in some places.
Sen.: *Berhaut* 374. **S.L.:** Dublin, Banana Isl. (fr. May) *Deighton* 2760 ! **G.C.:** Kumasi (fl. & fr. Feb.) *Chipp* 116 ! Akroful *Cummins* 35 ! Inundated flats 5 miles N. of Saltpond (Mar.) *Irvine* 1542 ! A native of tropical America, widespread in the tropics.

19. **C. tora** *Linn.*—F.T.A. 2 : 275 ; Benth. l.c. 535 ; Chev. Bot. 225 ; Steyaert l.c. 512. ? *C. rogeonii* Ghesq. in Rev. Bot. Appliq. 14 : 238 (1934). An undershrub, 3–10 ft. high, with yellow flowers in pairs or singly.
Sen.: *Talmy* ! **Gam.:** *Pirie* 51 ! **Fr.Sud.:** Saredina (May) *Davey* 94 ! **Port.G.:** Bissau (fl. Oct., fr. Dec.) *Esp. Santo* 1080 ! 2553 ! **Fr.G.:** Kouria (Oct.) *Chev.* 14985 ! **S.L.:** Port Loko (Dec.) *Thomas* 6692 ! Luti (fr. Feb.) *Deighton* 5008 ! Mano (Nov.) *Deighton* 1434 ! Musaia (Sept.) *Small* 270 ! **Lib.:** Cape Palmas *T. Vogel* 12 ! **G.C.:** Accra *Brown* 432 ! Kumasi (Nov.) *Vigne* FH 4081 ! Offinso (Oct.) *Darko* 722 ! **Togo:** Lome *Warnecke* 283 ! **Fr.Nig.:** Niamey (Oct.) *Hagerup* 485a ! **N.Nig.:** Nupe *Barter* 1630 ! Katagum (fl. & fr. Nov.) *Dalz.* 2 ! Panyam (July) *Lely* 417 ! Dikwa (fl. & fr. Nov.) *H. B. Johnston* 80 ! 91 ! **S.Nig.:** Sapoba *Kennedy* 1652 ! Newi, Onitsha (Oct.) *Jones* FHI 608 ! **F.Po:** *T. Vogel* 38 ! Widespread in the tropics. (See Appendix, p. 183.)

20. **C. sophera** *Linn.*—F.T.A. 2 : 274 ; Benth. l.c. 532 ; Chev. Bot. 225. A weak shrub, 3–5 ft. high, with yellow flowers in short racemes or fascicles ; introduced and found mainly in villages and farmlands ; very close to the next species.
S.L.: York (May) *Deighton* 2687 ! Waterloo (June) *Lane-Poole* 304 ! **Lib.:** Vahun, Kolahun (fl. & fr. Nov.) *Baldwin* 10230 ! **Iv.C.:** Ebrinahoué to Diangobo (fl. & fr. Dec.) *Chev.* 22623 ! **G.C.:** Cape Coast (July) *T. Vogel* 24 ! Tarkwa (fl. & fr. Jan.) *Chipp* 61 ! Accra Plains (fl. & fr. Apr.) *Bally* 28 ! **N.Nig.:** Share, Ilorin (Mar.) *Ejiofor* FHI 19844 ! Bassa (Oct.) *Lamb* 86 ! **S.Nig.:** Lagos *Punch* 74 ! Ibadan (Oct.) *Newberry* 89 ! **F.Po:** *T. Vogel* 211 ! *Mann* 69 ! A native of India, widespread in the tropics. (See Appendix, p. 183.)

21. **C. occidentalis** *Linn.*—F.T.A. 2 : 274 ; Benth. l.c. 532 ; Chev. Bot. 224 ; Aubrév. l.c. 221, t. 44, 1–2 ; Steyaert l.c. 513 ; Ic. Pl. Afr. t. 7. A glabrous herb or undershrub, woody below, annual or of 2–3 years' duration ; found mainly near villages ; much like the preceding.
Sen.: *Perrottet* 285 ! *Heudelot* 479 ! Tamboukané *Chev.* 3341. **Fr.Sud.:** Timbuktu (fl. & fr. July) *Hagerup* 153 ! Ansongo (Apr.) *De Wailly* 5038 ! And various localities *fide* Chev. *l.c.* **Fr.G.:** Kankan to Kouroussa *Chev.* 15742. Dalaba (Apr.) *Chev.* 18156. **S.L.:** Sherbro *Sc. Elliot* 5786 ! Samaia (fl. & fr. May) *Thomas* 232 ! Rowala (fl. & fr. Sept.) *Thomas* 1100 ! **Lib.:** Gbanga (Sept.) *Linder* 656 ! Ganta (fl. & fr. Oct.) *Harley* ! Bonuta (fl. & fr. Oct.) *Linder* 872 ! **Iv.C.:** Bobo Dioulasso (May) *Chev.* 94 ! **G.C.:** Amorkrom (fl. & fr. Dec.) *Chipp* 38 ! **Togo:** Lome *Warnecke* 208 ! **Fr.Nig.:** Niamey to Gao *Ryff* ! **Dah.:** Abomey *Chev.* 23284. **N.Nig.:** *Barter* 1602 ! Sokoto (Oct.) *Moiser* 122 ! *Lely* 142 ! Yo, N. Bornu (Dec.) *Elliott* 150 ! Kukawa (fl. & fr. Feb.) *E. Vogel* 55 ! **S.Nig.:** Lagos *Dawodu* 44 ! Okomu F.R. (fl. & fr. Feb.) *Brenan* 9170 ! Awka *Thomas* 4 ! **F.Po:** *T. Vogel* 21 ! Widespread in the tropics. (See Appendix, p. 180.)

22. **C. hirsuta** *Linn.* Sp. Pl. 1 : 378 (1753) ; Benth. l.c. 534 ; Steyaert l.c. 513. Undershrub, to 7 ft. high ; flowers yellow ; introduced and found mainly near habitation.
Fr.G.: Macenta (fl. & fr. Oct.) *Baldwin* 9848 ! **Lib.:** Diebla, Webo (fl. & fr. July) *Baldwin* 6346 ! **G.C.:** Offinso (fl. & fr. Nov.) *Darko* 640 ! **S.Nig.:** Ibadan (July) *Tamajong* FHI 16761 ! A native of tropical America, introduced into Africa.

Besides the above indigenous and subspontaneous species, several other species are cultivated in gardens in our area. They include the following : (*a*) with yellow flowers : *C. auriculata* Linn., *C. fistula* Linn., *C. fruticosa* Mill. (syn. *C. bacillaris* Linn. f.), *C. ligustrina* Linn., *C. multijuga* Linn., *C. spectabilis* DC. and *C. surattensis* Burm. f. (syn. *C. glauca* Lam.) ; (*b*) with pink flowers : *C. grandis* Linn. f., *C. javanica* Linn., *C. roxburghii* DC. and *C. nodosa* Buch.-Ham. ex Roxb.

13. GILLETIODENDRON Vermoesen Man. Ess. For. Congo Belge 85 (1923) ; Léonard in Bull. Jard. Bot. Brux. 21 : 402 (1951). *Cymonetra* Roberty (1954).

Leaflets 9–20 on each side of the rhachis, linear, 0·7–1·8 cm. long, 0·2–0·5 cm. broad ; fruits 3·5–5 cm. long, 2–2·5 cm. broad, valves woody .. .. 1. *glandulosum*
Leaflets 4–6 (rarely 3 or 7) on each side of the rhachis, rhombic-elliptic, often obliquely so, 1·5–7·2 cm. long, 0·6–3·5 cm. broad ; fruits 2·5–4 cm. long, 1·5–2·5 cm. broad, valves coriaceous .. .. 2. *kisantuense*

1. **G. glandulosum** (*Portères*) *J. Léonard* in Bull. Jard. Bot. Brux. 21 : 404, t. 4 (1951). *Cymonetra glandulosa* Portères in Rev. Bot. Appliq. 19 : 785, t. 5 (1939) ; Aubrév. Fl. For. Soud.-Guin. 227, t. 46. *Cymonetra glandulosa* (Portères) Roberty (1954). Evergreen tree, to 65 ft. high ; fruits with prominent glands.
Fr.Sud.: Kita Massif (fl. July, fr. Oct.) *Dubois* 105 ; *Chev.* 28934 ; *Jaeger* 14 !

2. **G. kisantuense** (*Vermoesen ex De Wild.*) *J. Léonard* l.c. 405, t. 109 (1951). *Cynometra kisantuense* Vermoesen ex De Wild. Pl. Bequaert 3 : 99 (1925). *C. pierreana* of Aubrév. Fl. For. C. Iv. 1 : 244 & 246, t. 93 C, not of Harms. Forest tree, to 50 ft. high.
Iv.C.: Morénou, Daoukrou *Chev.* 22492. Also in Gabon and Belgian Congo.

14. TESSMANNIA Harms in Engl. Bot. Jahrb. 45 : 295 (1910) ; Léonard in Publ. I.N.É.A.C., Sér. Sci. 45 : 35–55 (1950).

Outside of sepals, and usually the pedicels and rhachis of inflorescence, covered with glandular excrescences as well as with hairs ; pedicels 0·5–1·2 cm. long ; the 4 larger petals 1·7–2·5 cm. long ; leaflets 7–12, alternate, elliptic or oblong-elliptic, obtuse or acuminate and slightly emarginate at apex, 3·5–9·5 cm. long, 1·5–4·5 cm. broad ; fruits obliquely suborbicular or oblong, 4·5–6 cm. long, 3·5–5 cm. broad, flattened, glabrescent, studded with large spinescent warts .. .. 1. *africana*
Outside of sepals, pedicels and rhachis of inflorescence densely pubescent, but without glandular excrescences ; pedicels 1·5–2 cm. long ; the 4 larger petals 2·5–3 cm. long ; leaflets 4–6, alternate, obliquely obovate-oblanceolate to narrowly elliptic, 4–10 cm. long, 1·8–4 cm. broad ; fruits obliquely oblong-elliptic, 5–7 cm. long, 3–3·5 cm. broad, flattened, densely tomentose ; studded with small glandular warts 2. *baikiaeoides*

1. **T. africana** *Harms* in Engl. Bot. Jahrb. 45 : 295, t. 2 (1910) ; Léonard in Publ. I.N.É.A.C., Sér. Sci. 45 : 41 ; Fl. Congo Belge 3 : 288. Large forest tree, to 160 ft. high with tall straight cylindrical bole ; calyx green ; petals and filaments white.
Br.Cam.: S. Bakundu F.R. (fl. Apr., fr. Jan.) *Ejiofor* FHI 29316 ! *Keay* ! Also in Spanish Guinea, Gabon, French and Belgian Congo.

2. **T. baikiaeoides** *Hutch. & Dalz.* F.W.T.A., ed. 1, 1 : 336 (1928) ; Kew Bull. 1928 : 382 ; Aubrév. Fl. For. C. Iv. 1 : 252, t. 96, 1–2. Forest tree, to 7 ft. in girth ; with slight buttresses and smooth, round straight bole.
S.L.: York Pass (Dec.) *Aylmer* 137 ! *For. Dept.* ! Dodo Hills F.R. (fr. Jan.) *Sawyerr* FHK 13576 !
Iv.C.: Mt. Dou *Aubrév.* 1046. (See Appendix, p. 202.)

15. BAIKIAEA Benth.—F.T.A. 2 : 309.

Leaflets 2–5, alternate or subopposite, obliquely oblong-elliptic, gradually acuminate and rather blunt, base rounded and umbonately thickened on one side, 12–30 cm. long, 4·5–8 cm. broad, entire, leathery, glabrous, with indistinct nerves ; petiolules 1 cm. long, transversely wrinkled ; stipules small, scale-like ; flowers few, nodding, clustered at the tips of the twigs, densely velvety in bud ; pedicels stout, 2 cm. long, velvety ; calyx-tube turbinate ; lobes broadly linear, up to 11 cm. long, velvety outside, at length recurved ; petals up to 20 cm. long, obovate, hairy down the middle, the upper petal yellow inside, the others white, veiny ; stamens 10, upper one free, the others unequally connate and pubescent below ; anthers 1·3 cm. long ; ovary long-stipitate, densely villous ; fruit flat, curved, large, wrinkled and finely velvety . .   . .   *insignis*

B. **insignis** *Benth.*—F.T.A. 2 : 309 ; Bot. Mag. t. 8819 ; Léonard in Fl. Congo Belge 3 : 298, t. 23. A tree, up to 80 ft. high, in rain forest, with very large flowers.
S.Nig.: Lagos Colony (Feb.) *Moloney* 6 ! *Punch* ! Oban *Talbot* 1749 ! R. Imo *Kennedy* 2556 ! Afi River F.R. *Keay* FHI 28684 ! **Br.Cam.:** Victoria (Jan.) *Maitland* 1307 ! **F.Po:** (Mar.) *Mann* 2342 ! Also in French Cameroons, Gabon, French and Belgian Congo, Uganda and Angola. (See Appendix, p. 174.)

16. CHIDLOWIA Hoyle in Kew Bull. 1932 : 101 (1932).

A tree, wholly glabrous ; leaves paripinnate, leaflets 4–6 pairs, opposite or subopposite, the lowest ovate, the middle pairs oblong-elliptic, the distal pair obovate-elliptic, 4–12 cm. long, 2–5 cm. broad, cuneate to rounded and oblique at base, rather abruptly long-acuminate, main lateral nerves 5–7 on each side of midrib, lower surface dull ; panicles very slender, pendulous, 1–2 together on the old wood or occasionally terminal on young branchlets, to 40 cm. long, lateral branchlets numerous, up to 4 mm. long, 5–7-flowered ; pedicels 3–3·5 mm. long ; calyx campanulate, 2 mm. long, with 5 short rounded teeth ; disk fleshy, campanulate, bearing 5 subequal petals and 10 stamens ; fruit oblong-linear, acute at both ends, up to 60 cm. long and 6 cm. broad, valves coriaceous-woody and very elastic, dehiscing along both sutures ; seeds up to 15, suborbicular, flat . .   . .   *sanguinea*

C. **sanguinea** *Hoyle* l.c. 101, t. 1 (1932) ; Aubrév. Fl. For. C. Iv. 1 : 98. Forest tree, to 80 ft. high, locally abundant ; flowers wine-red in very slender, long, pendulous panicles.
S.L.: York Pass (Feb.) *Aylmer* 102 ! *Lane-Poole* 447 ! Kenema (Apr.) *Aylmer* 225 ! Kambui (May) *Lane-Poole* 198 ! Gola Forest (young fr. Mar.) *Small* 558 ! **Iv.C.:** *Aubrév.* 929 ! And various localities *fide* Aubrév. *l.c.* **G.C.:** Ofin Headwaters F.R. (Mar.) *Vigne* FH 1056 ! Juaso (fl. Jan., fr. June) *Irvine* 386 ! *Vigne* FH 1796 ! Puso Puso (Jan.) *Scholes* 447 ! Also in Gabon. (See Appendix, p. 183.)

17. KAOUE Pellegr. in Bull. Soc. Bot. Fr. 80 : 464 (1933).

A large tree ; leaflets 3–4 pairs, the lowest pair elliptic, smaller, the uppermost pair broadly oblanceolate, 8–25 cm. long, 3–9 cm. broad, glabrous, reticulate, with 5–8 pairs of lateral nerves, rhachis 6·5–20 cm. long ; flowers small, in spikes 6–10 cm. long, crowded at the ends of the branches ; sepals 5, valvate, united into a tube ; petals 5, conspicuous in bud ; stamens 10, 5 long and 5 short ; fruit dehiscent, woody, flat, more or less falcate, with 1–2 large seeds near the apex, gradually tapered below, up to 18 cm. long and 4·5 cm. broad . .   . .   *stapfiana*

K. **stapfiana** (*A. Chev.*) *Pellegr.* l.c. 465 (1933) ; Aubrév. Fl. For. C. Iv. 1 : 252, t. 97. *Oxystigma stapfiana* A. Chev. in Bull. Soc. Bot. Fr. 58, Mém. 8 : 166 (1912) ; Chev. Bot. 234 ; F.W.T.A., ed. 1, 1 : 336. A forest tree, to 100 ft. high, with buttresses and sometimes with aerial roots ; flowers white.
S.L.: Mahwe R., Zimi (Apr.) *King* 178 ! Gola Forest (Apr., May) *Small* 590 ! 710 ! **Lib.:** Dukwia R. (Mar.) *Cooper* 256 ! Logantown, New Cess R. (Mar.) *Baldwin* 1184 ! Paynesville (Apr.) *Barker* 1244 ! Mecca, Cape Grand Mount Co. (fr. Dec.) *Baldwin* 10843 ! **Iv.C.:** Cavally Basin (July) *Chev.* 19528 ! *Aubrév.* 1305. Tabou *Aubrév.* 1672 !

18. HYLODENDRON Taub. in E. & P. Pflanzenfam. 3, 3 : 386 (1894).

Tree up to 32 m. high ; branchlets glabrous ; leaves pinnate ; leaflets alternate, about 7 pairs, oblong, acuminate, rounded or obtuse at base, 10–12 cm. long, 3–5 cm. broad, with numerous lateral nerves on each side of the midrib prominent on both surfaces ; stipules very long, leathery, linear, 5–6 cm. long, closely longitudinally striate, glabrous ; flowers small, in short lateral racemes, covered in bud by leathery, closely imbricate, striate, broadly ovate bracts, minutely ciliate on the margin ; fruits narrowly oblong, slightly curved, flat, 10–12 cm. long, 2–3 cm. broad, laxly reticulate, 1–2-seeded, glabrous . .   . .   *gabunense*

H. **gabunense** *Taub.* l.c. ; Kennedy For. Fl. S. Nig. 92 ; Léonard in Fl. Congo Belge 3 : 302. Tall thin boled forest tree with sharp buttresses armed with woody spines.
S.Nig.: Sapoba (fr. Aug.) *Kennedy* 1127 ! **Br.Cam.:** Victoria (Apr.) *Winkler* 1164 ! *Maitland* 609 ! N. Korup F.R. (fr. July) *Olorunfemi* FHI 30693 ! Johann-Albrechtshöhe *Staudt* 939. Also in French Cameroons, Gabon and Belgian Congo. (See Appendix, p. 194.)

19. DETARIUM Juss.—F.T.A. 2 : 312.

Leaflets rounded and usually emarginate at apex, oblong-elliptic, 6–12, alternate :
Calyx in bud densely pubescent outside ; inflorescences congested ; leaflets thickly
coriaceous, usually about 7–11 cm. long and 3·5–5 cm. broad, but sometimes smaller,
with numerous translucid gland-dots ; fruits suborbicular and flattened, about
4 cm. diam. and 2·5 cm. thick, not very fleshy ; often forming small abortive or
galled fruits   ..   ..   1. *microcarpum*
Calyx in bud glabrous or sparsely pubescent outside ; inflorescences lax ; leaflets
thinly coriaceous or papery, usually 4–6 cm. long and 2·5–3 cm. broad, with rather
few translucid gland-dots ; fruits subglobose, slightly flattened, about 5–6 cm. diam.
and 3·5–4 cm. thick, fleshy   ..   ..   2. *senegalense*
Leaflets acuminate at apex, ovate to oblong-elliptic, 8–20, alternate ; lamina 4–8 cm.
long, 2–4·5 cm. broad, with numerous translucid gland-dots ; inflorescence lax, calyx
in bud glabrous outside ; fruits subglobose, slightly flattened, 7–8 cm. diam.
      3. *macrocarpum*

1. **D. microcarpum** *Guill. & Perr.* Fl. Seneg. 271, t. 59, p (1832) ; Aubrév. & Trochain in Bull. Soc. Bot. Fr.
84 : 487–494 (1938) ; Aubrév. Fl. For. Soud.-Guin. 229, t. 47, 1–2. *D. senegalense* of F.T.A. 2 : 313,
partly, and of F.W.T.A., ed. 1, 1 : 338, partly, not of Gmel. Savannah tree, to 30 ft. high, with reddish-
brown scaly bark ; leaves glaucous beneath ; flowers cream in dense inflorescences ; fruits edible,
branchlets deciduous ; a plant of the drier savannahs.
Sen.: Tielimane, Cayor *fide* Guill. & Perr. *l.c.* **Gam.:** Albreda *fide* Guill. & Perr. *l.c.* **Fr.Sud.:** Bamako
(Sept.) *Waterlot* 1398 ! Sébékoro, Kita (Apr.) *Dubois* 146 ! **Port.G.:** Cajambarim to Fajanquito (Oct.)
Esp. Santo 2300 ! **Fr.G.:** Kouroussa (Aug.) *Pobéguin* 427 ! **Iv.C.:** Samankono, Ferkéssédougou & Léo
fide Aubrév. *l.c.* **G.C.:** Ejura (Aug.) *Chipp* 769 ! *Tolmie* FH 4423 ! Yendi (July) *Pomeroy* 1342 !
Amansare (July) *Chipp* 515 ! Yeji (July) *Vigne* FH 1221 ! **Togo:** *Kersting* 13 ! **Fr.Nig.:** Gaya *Aubrév.*
22. **N.Nig.:** Nupe *Barter* 1073 ! Zungeru (July) *Dalz.* 94 ! Kaduna (May) *Keay* FHI 25801 ! Bauchi
Plateau (Aug.) *Lely* P. 627 ! Yola (Apr.) *Dalz.* 5 ! **S.Nig.:** *Rosevear* 35/29 ! Ihapo (May) *Denton* 26 !
Also in French Cameroons, Ubangi-Shari and A.-E. Sudan. (See Appendix, p. 188.)
  [*Ogua* FHI 7822 from Anchau, N. Nigeria is possibly a distinct, small-leaved variety.]
2. **D. senegalense** *J. F. Gmel.* in Linn. Syst. Nat., ed. 13, 2 : 700 (1791) ; Guill. & Perr. l.c. 269, t. 59, a–o ;
Aubrév. & Trochain l.c. ; Aubrév. Fl. For. Soud.-Guin. 229 ; F.T.A. 2 : 313, partly ; F.W.T.A., ed. 1,
1 : 338, partly. *D. heudelotianum* Baill. (1866)—Aubrév. Fl. For. C. Iv. 1 : 266, t. 104. Forest tree,
30–120 ft. high, with large crown ; leaves bright green ; flowers creamy white ; fruits edible or poisonous
(in the form known as *D. heudelotianum* Baill.) ; in the forest regions and in forest outliers in the moister
savannah regions.
Sen.: Cape Verde (Mar.–May) *Perrottet* ! Portudal (Feb.) *Döllinger* 47 ! **Gam.:** *Brown-Lester* 1 ! *Hayes*
512 ! Albreda *fide* Guill. & Perr. *l.c.* **Port.G.:** Safrin, Bissau (Mar.) *Esp. Santo* 1904 ! **Fr.G.:** Rio Nunez
Heudelot 575 ! 822 ! Tristao Isl. *Paroisse* 68 ! Kouroussa (Apr.) *Pobéguin* 689 ! Ditinn (Apr.) *Chev.*
12992 ! Fourboia, Farana (Feb.) *Chev.* 20644 ! **S.L.:** Bandajuma (Feb.) *Lane-Poole* 160 ! Gene, Njala
(Feb.) *Deighton* 3110a ! 3110b ! Kambia (fr. Oct.) *Small* 478 ! **Lib.:** Dukwia R. (fr. Apr.) *Cooper* 401 !
Iv.C.: Dinderesso, Abidjan, Gaoua to Banfara, & Morénou *fide* Aubrév. *l.c.* **Togo:** Blita *Kersting* A. 538 !
N.Nig.: Agaie (Jan.–Feb.) *Yates* 31 ! **S.Nig.:** Sapoba (Nov.) *Kennedy* 1883 ! 1934 ! Degema *Talbot* !
Also in Ubangi-Shari, Belgian Congo and A.-E. Sudan. (See Appendix, p. 188.)
3. **D. macrocarpum** *Harms* in Engl. Bot. Jahrb. 30 : 78 & fig. (1901) ; Pellegr. Lég. Gab. 133, t. 4, 9. Forest
tree, to 180 ft. high, with large buttresses.
S.Nig.: Afi River F.R., Ikom (fr. Dec.) *Keay* FHI 28234 ! Also in French Cameroons and Gabon.

20. COPAIFERA Linn. Sp. Pl., ed. 2, 1 : 557 (1762) ; Léonard in Bull. Jard. Bot.
Brux. 19 : 391–398 (1949).

Leaflets 10–14 (rarely to 18), opposite, or subopposite, oblong or elliptic, rounded and
emarginate and very slightly apiculate at apex, 1·8–4·3 cm. long, 1·1–2·6 cm. broad,
rhachis 6–13 cm. long ; mature leaflets without translucent dots ; inflorescence 3–12
cm. long ; fruit obliquely oblong, 3·2–3·8 (–4·5) cm. long, 2·4–3 cm. broad, with a
single large seed covered by a waxy red aril   ..   ..   ..   1. *salikounda*
Leaflets 20–40, alternate or subopposite, oblong, rounded and emarginate and very
slightly apiculate at apex, 1·5–4 (–6) cm. long, 1–1·8 (2·3) cm. broad, rhachis 13–25
cm. long ; mature leaflets with translucent dots ; inflorescence 8–30 cm. long ; fruit
and seed similar to above   ..   ..   2. *mildbraedii*

1. **C. salikounda** *Heckel* in Ann. Fac. Sci. Marseille 3, 4 : 4, t. 16 (1893) ; Aubrév. Fl. For. C. Iv. 1 : 262, t. 102 ;
Léonard in Publ. I.N.É.A.C., Sér. Sci. 45 : 58. *Detarium chevalieri* Harms (1909). A large forest tree,
often more or less gregarious, with rugose bark ; pods brown, woody.
Fr.G.: Lower Koukouré *Paroisse.* Sombouya *Maclaud.* **S.L.:** Sherbro *Poisson* ! Baiima, Gbo (fr. May)
Deighton 5768 ! **Iv.C.:** Bingerville (Feb.) *Chev.* ! Abidjan *Aubrév.* 499 ! **G.C.:** *Evans* 15 ! Owabi (fr.
Nov.) *Vigne* FH 2592 ! Amentia (fr. Dec.) *Vigne* FH 1491 ! (See Appendix, p. 184.)
2. **C. mildbraedii** *Harms* in Notizbl. Bot. Gart. Berl. 8 : 147 (1922) ; Léonard l.c. 65, with photo ; Fl. Congo
Belge 3 : 307. *C. salikounda* of Kennedy For. Fl. S. Nig. 95, not of Heckel. Large forest tree, to 200 ft.
high, with clean cylindrical bole.
S.Nig.: Sapoba (fr. Feb.) *Kennedy* 2093 ! 2306 ! Urhuehue *Keay* FHI 21591 ! **Br.Cam.:** Likomba
Mildbr. 10572 ! S. Bakundu F.R. (fr. Feb.) *Ngalame* FHI 24807 ! Also in Ubangi-Shari, French Cameroons, Gabon and Belgian Congo.

21. CYNOMETRA Linn.—F.T.A. 2 : 316.

Leaflets 6–10 pairs, more or less truncate or emarginate, oblong, up to 3 cm. long and
1·5 cm. broad :
Leaflets about 10 pairs, shortly pubescent on the midrib beneath ; flowers very small,
in panicles shorter than the leaves ; pedicels 5–8 mm. long, shortly pubescent ;
bracts ovate, long-acuminate, about 6 mm. long ..   ..   1. *hankei*
Leaflets about 6 pairs, glabrous beneath, otherwise more or less as in the preceding
      2. *leonensis*

Leaflets 4–1 pairs :
  Leaflets 3–4 pairs, acuminate :
    Acumen very short and abrupt ; bracts shortly pubescent ; leaflets very obliquely
      oblanceolate, the uppermost pair the longest, up to 12 cm. long and 3·5 cm. broad,
      leathery, stiff, reticulate, nerves equally prominent on both sides ; inflorescence
      short and dense ; fruits about 4–6 cm. diam. ..   ..   .. 3. *megalophylla*
    Acumen rather long and gradual, deeply emarginate ; bracts glabrous and closely
      striate ; leaflets obliquely obovate-oblanceolate, the uppermost pair the largest,
      up to 6 cm. long and 2·5 cm. broad, glabrous ; inflorescence very short and densely
      bracteate ; pedicels rather densely pubescent ; fruits elliptic, about 4 cm. long,
      very rugose ..   .. 4. *mannii*
  Leaflets 1 pair (very rarely 2 pairs) :
    Leaflets deeply emarginate, not acuminate, obovate-elliptic, up to 9 cm. long and
      4 cm. broad, glabrous and shining ; flowers subfasciculate ; pedicels up to 1·5 cm.
      long, softly pubescent ; bracts small, strongly striate ; fruits shortly falcate,
      3·5 cm. long, 2 cm. broad, wrinkled ..   .. 5. *vogelii*
    Leaflets acuminate, broadly sickle-shaped, 6–10 cm. long, 2–4 cm. broad, glabrous ;
      flowers small and rather crowded in short cymes ; pedicels very short, rusty-
      pubescent ; bracts small ; fruits broadly oblong, about 9 cm. long and 5 cm.
      broad, slightly nervose ..   .. 6. *ananta*

1. **C. hankei** *Harms* in Notizbl. Bot. Gart. Berl. App. 21 : 39 (1911) ; Pellegr. Lég. Gab. 106 ; Léonard in Bull.
    Jard. Bot. Brux. 21 : 392, t. 106 ; Fl. Congo Belge 3 : 318, t. 20 & photo. Forest tree, to 150 ft. high ;
    rhachis of leaves slightly winged between the leaflets ; flowers white.
    S.Nig.: Eket *Talbot* 3279 ! Oban *Talbot* 1639 ! **Br.Cam.:** Banga, Kumba *Keay* ! *Ejiofor* FHI 29361 !
    Also in French Cameroons and Belgian Congo.
2. **C. leonensis** *Hutch. & Dalz.* F.W.T.A., ed. 1, 1 : 331 (1928) ; Kew Bull. 1928 : 381 ; Léonard in Bull. Jard.
    Bot. Brux. 21 : 392. A large forest tree, near water ; flowers small, white.
    S.L.: Yandahun (Jan.) *Unwin & Smythe* 52 ! Bandama (Mar.) *Aylmer* 576 ! Kenema (fl. Mar., fr. Apr.)
    Aylmer 136 ! *King* 10 ! Njala (Dec., Mar.) *Deighton* 1106 ! 1861 ! **Lib.:** Dobli Isl., St. Paul. R. *Bequaert*
    25 ! (See Appendix, p. 185.)
3. **C. megalophylla** *Harms* in Engl. Bot. Jahrb. 26 : 262 (1899) ; Aubrév. Fl. For. C. Iv. 1 : 244, 246, t. 93 A ;
    Pellegr. l.c. 110 ; Léonard l.c. 385, t. 102, 103. A large riverside and forest tree, the trunk sometimes
    divided near the base ; young foliage drooping, often wine-red ; flowers white ; fruits warted.
    S.L.: *Afzelius.* **Iv.C.:** Bettié, Lower Comoé (fr. Mar.) *Chev.* 17259 ! Groumania *Aubrév.* 784 ! **G.C.:**
    Anamabou *Hove* ! Cape Coast (Dec.) *Taylor* FH 5448 ! Pra R. (fl. Dec., fr. Apr.) *Johnson* 926 ! *Vigne*
    FH 937 ! Odumasi *Johnson* 122 ! Dixcove (fr. Apr.) *Chipp* 181 ! Achimota (fr. Feb.) *Irvine* 2002 ! **Dah.:**
    Zagnanado (Feb.) *Chev.* 23105 ! **S.Nig.:** Lagos *Foster* 48 ! Ebute Metta (May) *Millen* 69 ! Olokemeji
    Kennedy 148 ! (See Appendix, p. 185.)
    [The specimen recorded by Harms from Sierra Leone may well have been a duplicate from Hove's
    gathering which Afzelius apparently studied ; the species is not otherwise known from Sierra Leone.]
4. **C. mannii** *Oliv.* F.T.A. 2 : 317 (1871) ; Pellegr. l.c. 109 ; Léonard l.c. 386, t. 102 ; Fl. Congo Belge 3 : 313,
    t. 18 A. A tree of river banks and fringing forest, up to 60 ft. high, the trunk sometimes short and divided
    near the base, occasionally prop-rooted ; young foliage red, leaflets notched at the tip like a claw-hammer ;
    flowers small, white.
    S.Nig.: Oban *Talbot* 1708 ! Calabar *Thomson* 1 ! **Br.Cam.:** Ambas Bay (Jan.) *Mann* 707 ! Victoria
    (Mar.) *Kalbreyer* 83 ! Kumba *Lobe* FHI 2803 ! Also in French Cameroons, Gabon, S. Tomé and Belgian
    Congo. (See Appendix, p. 185.)
5. **C. vogelii** *Hook. f.*—F.T.A. 2 : 317'; Chev. Bot. 235 ; Aubrév. Fl. For. C. Iv. 1: 246, t. 93 B ; Fl. For.
    Soud.-Guin. 227 ; Léonard in Bull. Jard. Bot. Brux. 21 : 384, t. 102. A tree of spreading habit sometimes
    attaining 60 ft. in height, or branching almost to the base.
    Sen.: Niéri-ko *Berhaut* 1292. **Gam.:** *Heudelot* 340 ! **Fr.Sud.:** various localities *fide* Chev. *l.c.* **Port.G.:**
    R. Corubal, Jangada (fl. Jan., fr. June) *Esp. Santo* 2268 ! 2363 ! **Fr.G.:** Rio Nunez *Heudelot* 699 ! Kadé
    Pobéguin 1998. Kouroussa *Pobéguin* 644. **S.L.:** Scarcies R., Mofari (Jan.) *Sc. Elliot* 4434 ! Kasanko,
    Mafore (Dec.) *T. S. Jones* 53 ! Mano *Smythe* 67 ! Njala (Dec.) *Deighton* 1862 ! Bumpe (Nov.) *Deighton*
    4944 ! **Iv.C.:** *fide* Aubrév. *l.c.* **G.C.:** Yeji, R. Volta (young fr. July) *Pomeroy* FH 1244 ! Prang, R. Pra
    (fr. Jan.) *Vigne* FH 1544 ! **N.Nig.:** Lokoja (fl. & fr. Sept.) *Ansell* ! T. Vogel 105 ! Nupe *Barter* 901 !
    Abinsi (Dec.) *Dalz.* 714 ! **S.Nig.:** Illushi *Farquhar* 54 ! Akilla *Kennedy* 2005 ! Aboh *Barter* 296 ! Aguleri
    (Sept.) *Chesters* OC 27 ! (See Appendix, p. 186.)
6. **C. ananta** *Hutch. & Dalz.* F.W.T.A., ed. 1, 1 : 331 (1928) ; Kew Bull. 1928 : 381 ; Aubrév. Fl. For. C. Iv.
    1 : 244, t. 92 ; Léonard l.c. 397. A tall forest tree, the crown covered with light grey flowers ; often
    gregarious.
    Lib.: Dukwia R. (fr. July) *Cooper* 411 ! **Iv.C.:** Abidjan *Aubrév.* 189 ! **G.C.:** Ancobra River Basin (Nov.)
    Chipp 11 ! Dunkwa (fr. June) *Vigne* 875 ! (See Appendix, p. 185.)

Besides the above, *C. trinitensis* Oliv., a native of the W. Indies has been cultivated in our area.

22. ZENKERELLA Taub. in E. & P. Pflanzenfam. 3, 3 : 386 (1894) ; Léonard in Bull. Jard. Bot. Brux. 21 : 408 (1951).

Leaves 1-foliolate ; leaflet very shortly petiolulate, elliptic, shortly acuminate, rounded
  at base, 7–14 cm. long, 4–7 cm. broad, glabrous ; racemes very short and glomerule-
  like, few-flowered ; pedicels 1·5–3 mm. long ; sepals 4–5 mm. long, glabrous ; petals 5
                                                                          citrina

Z. citrina *Taub.* l.c. 386 (1894) ; Léonard l.c. 410, t. 110. *Cynometra citrina* (Taub.) Harms (1915)—Bak. f.
    Leg. Trop. Afr. 3 : 757 ; Pellegr. Lég. Gab. 112. *C. leptoclada* Harms (1915)—F.W.T.A., ed. 1, 1 : 331,
    fide Bak. f. l.c. and *fide* Pellegr. l.c. ; type specimen destroyed, see Léonard l.c. 412. A shrub or small tree
    with white flowers.
    S.Nig.: Oban *Talbot* 1303 ! **F.Po:** Clarence Peak (Aug.) *Mildbr.* 6346. Also in French Cameroons and
    Gabon.

23. SCHOTIA Jacq.—F.T.A. 2 : 309.

A tree 8–10 m. high ; branchlets often swollen and hollow under each node (occupied
  by ants) ; leaves paripinnate, 2–4 pairs, opposite, the lowest pair at the base of the

petiole, obliquely oblong-elliptic, obtusely caudate-acuminate, unequal-sided at base, 10–18 cm. long, 4·5–7 cm. broad, with about 8 pairs of prominently looped lateral nerves, glabrous ; racemes lateral, solitary or 2–3 together, very short, densely flowered ; pedicels 2–3 mm. long, bracteolate ; calyx-tube 7–8 mm. long, puberulous ; lobes 4, nearly as long as the tube, ovate-elliptic ; petals 5, subequal ; stamens 10, shortly coherent at the base ; filaments glabrous ; ovary and gynophore pilose ; ovules 4–5   ..   ..   *africana*

8. africana (*Baill.*) *Keay* in Kew Bull. 1953 : 490 (1954). *Humboldtia africana* Baill. (1870). *Schotia humboldtioides* Oliv. F.T.A. 2 : 310 (1871) ; Harms in Notizbl. Bot. Gart. Berl. App. 20 : 50 & fig. ; F.W.T.A., ed. 1, 1 : 331 ; Bak. f. Leg. Trop. Afr. 3 : 710. *Theodora africana* (Baill.) Taub. (1892). Calyx green, petals purple.
S.Nig.: Oban *Talbot* 1440 ! **Br.Cam.:** Cameroons R. (Jan.) *Mann* 726 ! R. Bombe, Kumba *Obiorah* FHI 29511 ! Iwasa, Kumba (Mar.) *Sangama* 12/38 ! Also in French Cameroons, Spanish Guinea and Gabon.

24. AFZELIA Sm.—F.T.A. 2 : 301 ; Léonard in Reinwardtia 1 : 61–66 (1950).

Receptacle 5–7 mm. long ; petal 13–15 mm. long ; flowers numerous in spreading panicles ; fruits oblong, straight or only slightly curved, 12–17 cm. long, 5–8 cm. broad, valves not curling after dehiscence, seeds arranged in a more or less straight row down the middle of the valves so that points of attachment are some way from the upper suture ; seeds 2–2·3 cm. long, with a cupular aril covering ¼ to ⅓ of the seed ; leaflets 3–5 pairs, broadly elliptic, acuminate or obtuse to rounded at apex, 5–15 cm. long, 3·5–8·5 cm. broad, glabrous, conspicuously reticulate on both surfaces ; rhachis 10–30 cm. long, petiolules 4–10 mm. long   ..   .. 1. *africana*
Receptacle 1–3·5 cm. long ; petal 2·5–6·5 cm. long ; valves of fruit curling after dehiscence :
  Leaflets pubescent beneath, 5–10 pairs, oblong or oblong-lanceolate, rounded or emarginate or rarely subacute at apex, 2–6·2 cm. long, 0·9–2·5 cm. broad, reticulate on both surfaces, shining above, matt beneath ; rhachis 6–20 cm. long ; petiolules 2–3 mm. long ; receptacle 1·5–2 cm. long ; petal 2·5–4 cm. long, with the claw rather densely pubescent ; fruits pale brown, matt, obliquely elliptic, reniform, 13–20 cm. long, 9–13 cm. broad, valves 1·5–2 cm. thick, curling after dehiscence ; seeds 3–4·5 (–5·5) cm. long, with a cupular, slightly 4-lobed aril not covering more than ¼ of the seed   ..   ..   2. *pachyloba*
  Leaflets glabrous beneath ; petiolules 3–7 mm. long ; fruits black and shining, oblong, curved ; seeds attached close to the upper suture ; arils 2-lobed, covering about ½ one side of seed and ¾ to ⅘ of the other side :
    Bracts large, up to 1 cm. long, persistent up to the time of flowering or longer, soon reflexed ; receptacle 1–1·5 cm. ; petal 3–4 cm. long, red ; leaflets 3–5 pairs, oblong, slightly curved, obtuse or rounded at apex, 6–11 cm. long, 3–4 cm. broad, shining above, matt beneath, rather faintly reticulate ; rhachis 5·5–15 cm. long ; fruits 7–10 cm. long, 4–4·5 cm. broad ; seeds 1·5–2 cm. long ..   3. *bracteata*
    Bracts caducous before flowering, usually very small and inconspicuous ; receptacle 1·5–4 cm. long ; leaflets conspicuously reticulate on both surfaces :
      Tall tree ; leaflets 5–8 pairs (rarely 4), oblong, acuminate to obtuse or rounded at apex, 5–13 (–20) cm. long, 2·5–5·7 (–7) cm. broad ; rhachis 14–27 cm. long ; receptacle 1·5–4 cm. long ; petal 3·4–5 cm. long, 2–3·5 cm. broad, with a pubescent claw ; fruits 8–19·5 cm. long, 5·5–8 cm. broad ; seeds 2·5–4·5 cm. long
                                                      4. *bipindensis*
      Small tree or shrub ; leaflets 3–5 pairs (rarely 6), elliptic or elliptic-lanceolate or elliptic-obovate, generally rather oblique, acuminate at apex, 4·5–16 cm. long, 2·2–7·3 cm. broad ; rhachis 6–18 cm. long ; receptacle 1–2 cm. long ; petal (3·8–) 4–6·5 cm. long, (2·2–) 2·5–4 cm. broad, with a glabrous or slightly pubescent claw ; fruits 6–15 cm. long, 3–6 cm. broad ; seeds 2–3·1 cm. long
                                                 5. *bella* var. *bella*
      Tall tree ; leaflets 4–6 pairs, obliquely oblong-elliptic, obtusely acuminate, 4–9 cm. long, 1·5–3·8 cm. broad ; rhachis 10–15 cm. long ; petal (2·3–) 3·4–3·8 (–4·3) cm. long, (1·2–) 1·9–2·1 cm. broad, with a pubescent claw ; fruits up to 12 cm. long and 5 cm. broad, often smaller and only 1–3-seeded   ..   5a. *bella* var. *gracilior*

1. **A. africana** *Sm.*—F.T.A. 2 : 302 ; Kennedy For. Fl. S. Nig. 100 ; Aubrév. Fl. For. C. Iv. 1 : 214 ; For. Trees & Timbers Brit. Emp. 2 : 15, fig. 2 ; Léonard in Fl. Congo Belge 3 : 355, fig. 27 C. A tree, to 100 ft. high, in savannah, in fringing forest and in the drier parts of the forest regions ; flowers very fragrant, green, except for the petal which is white with reddish markings ; fruits and seeds black, shining ; aril orange.
Sen.: Samatite, Casamance (Mar., Apr.) *Perrottet* 293 ! **Fr.G.:** *Heudelot* 767 ! **Port.G.:** Empandja, Bissau (Mar.) *Esp. Santo* 1508 ! **S.L.:** Musaia (fl. Mar., fr. Apr.) *Sc. Elliot* 5106 ! *Deighton* 5475 ! Rokupr (Mar.) *Jordan* 774 ! Koyeima (fl. & fr. Mar.) *Deighton* 3141 ! **Iv.C.:** Anoumaba (fr. Nov.) *Chev.* 22346 ! And various localities *fide* Aubrév. l.c. **G.C.:** Afram Plains (fl. & fr. Apr.) *Johnson* 698 ! Kintampo *W. T. S. Brown* FH 2160 ! Bjury (fr. July) *Chipp* 510 ! Tamale (Feb.) *McLeod* 809 ! **Togo:** Atakpame (fr. Nov.) *Mildbr.* 7457 ! Kete Kratchi (fl. & fr. Apr.) *Kitson* ! **Dah.:** *Poisson* 98 ! Abomey (Feb.) *Chev.* 23154 ! **N.Nig.:** Nupe *Barter* 1218 ! Zaria (Apr.) *Collier* ! Lokoja (May) *Elliott* 234 ! Katagum *Dalz.* 425 ! **S.Nig.:** Lagos *Rowland* ! *Foster* 10 ! Olokemeji (Feb.) *Ejiofor* FHI 31767 ! **Br.Cam.:** Mufung, Bamenda (Feb.) *Johnstone* 79/31 ! Also in Ubangi-Shari, French Cameroons, A.-E. Sudan, Uganda, Belgian Congo and Tanganyika. (See Appendix, p. 172.)

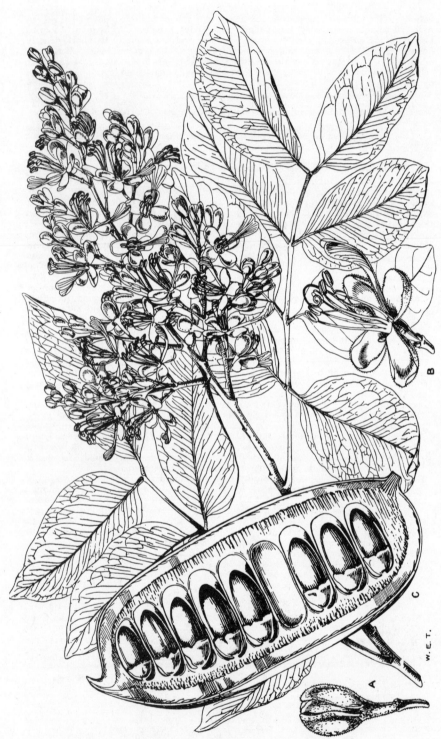

Fig. 149.—AFZELIA AFRICANA *Sm.* (CAESALPINIACEAE).
A, flower-bud. B, flower. C, longitudinal section of fruit.

W. E. T.

460

2. **A. pachyloba** *Harms* in Engl. Bot. Jahrb. 49 : 426 (1913) ; Léonard l.c. 3 : 351, fig. 27 A. *A. zenkeri* Harms l.c. 427 (1913) ; F.W.T.A., ed. 1, 1 : 344. *A. caudata* Hoyle in Kew Bull. 1933 : 170, partly (fruits & seeds only). Forest tree, to 140 ft. high ; petal white with reddish markings ; fruits pale brown, dull ; seeds black and shining with yellow aril.
    S.Nig.: Olomitutu, Shasha F.R. (Feb.) *Ross* 25 ! Akilla *Kennedy* 2375 (partly) ! Alagum, Omo F.R. (fr. Feb.) *Jones & Onochie* FHI 16716 ! Ondo Dist. (seed Mar.) *Unwin* 91 ! Okomu F.R. (fr. Mar.) *A. F. Ross* 197 ! Oban *Talbot* 1468 ! Also in French Cameroons, Gabon, Cabinda and Belgian Congo. (See Appendix, pp. 173-174.)
3. **A. bracteata** *T. Vogel ex Benth.* in Hook. Ic. Pl. t. 790 (1848) ; F.T.A. 2 : 301 ; Aubrév. l.c. 216, t. 80 B. Forest tree, to 50 ft. high, with spreading crown ; petal red ; seeds black and shining with red aril; usually found by streams.
    Fr.G.: Fouta Djalon *Heudelot* 882 ! **S.L.:** *Smeathmann* ! Freetown (June–July) *T. Vogel* ! Bagroo R. (Apr.) *Mann* 890 ! Talla Hills (Apr.) *Sc. Elliot* 5533 ! Kumrabai Mamila (fr. Nov.) *Deighton* 4950 ! Rokupr (Apr.) *Jordan* 241 ! Daru (Apr.) *Deighton* 3214 ! Makump (July) *Thomas* 911 ! Gola Forest (June) *Small* 733 ! **Lib.:** Cess R., Grand Bassa (Mar.) *Baldwin* 11266 ! Suen, Montserrado (fr. Nov.) *Baldwin* 10461 ! **Iv.C.:** Tabou valley *Aubrév.* 1325. (See Appendix, p. 173.)
    [The record of this species from Morénou (Ivory Coast) in F.W.T.A., ed. 1, 1 : 344 refers to *Chev.* 22478 which is *A. bella* var. *gracilior* (q.v.).]
4. **A. bipindensis** *Harms* l.c. 426 (1913) ; *Kennedy* l.c. 101 ; For. Trees & Timbers Brit. Emp. 2 : 21, fig. 3 ; Léonard l.c. 3 : 355, t. 26, fig. 27 D. *A. caudata* Hoyle l.c. (1933), partly (flowers and leaves). Tall forest tree, to 130 ft. high ; petal with red-purple claw and white crumpled lamina turning pink ; seeds black and shining with orange aril.
    S.Nig.: Okomu F.R. (Jan.) *Brenan* 8889 ! Sapoba (Feb.–May) *Kennedy* 1128 ! *McIntosh* FHI 12413 ! Oban *Talbot* 1650. Also in French Cameroons, Ubangi-Shari, Gabon, Belgian Congo and Angola. (See Appendix, p. 173.)
    [The leaves of *A. caudata* Hoyle (*W. D. MacGregor* 540 and *Kennedy* 2375) both from S. Nigeria are from seedlings and they are conspicuously caudate-acuminate. They appear, however, to be only the juvenile stage of *A. bipindensis*. The flowers of *MacGregor* 540 agree well with those of *A. bipindensis* and the fruits of *Kennedy* 2375 are evidently *A. pachyloba*.]
5. **A. bella** *Harms* var. **bella**—l.c. 425 (1913) ; *Kennedy* l.c. 101 ; Léonard l.c. 3 : 358, fig. 27 F. *A. africana* of F.T.A. 2 : 302, partly. Shrub or small tree, to 35 ft. high, deciduous ; petal white with red markings ; seeds black and shining with orange aril.
    S.Nig.: Ogba *Kennedy* 399 ! Sapoba *Kennedy* 1618 ! 2722 ! Owam R. (Sept.) *Unwin* 127 ! Aguku *Thomas* 917 ! Eket *Talbot* ! Calabar (July) *Holland* 56 ! **Br.Cam.:** Victoria (Oct.) *Maitland* 742 ! Banga (Aug.) *Olorunfemi* FHI 30713 ! Mamfe (Apr.) *Lobe Babute* 47/38 ! Also in French Cameroons, Gabon and Belgian Congo. (See Appendix, p. 173.)
5a. **A. bella** var. **gracilior** *Keay* in Kew Bull. 1954 : 266. *A. microcarpa* A. Chev. in Vég. Util. 5 : 172 (1909), name only ; Aubrév. in Rev. Bois et For. Trop. 30 : 54 (1954). *A. bella* of F.W.T.A., ed. 1, 1 : 344, partly, and of Aubrév. Fl. For. C. Iv. 1 : 214, t. 79, not of Harms. *A. bracteata* of F.W.T.A., ed. 1, 1 : 344, partly (*Chev.* 22478). Tall forest tree, to 120 ft. high ; seeds black and shining with red aril.
    Iv.C.: Dabou (fr. Feb.) *Chev.* 16207 ! Morénou (Dec.) *Chev.* 22478 ! Abidjan *Aubrév.* 88 ! **G.C.:** Ancobra R. (fr. Jan.) *Thompson* 7 ! Jamang (fr. Apr.) *Vigne* 96 ! Axim (fr. Feb.) *Irvine* 2151 ! Kumasi (fl. Oct., fr. Nov.) *Andoh* FH 5811 ! 5815 !

25. LOESENERA Harms in E. & P. Pflanzenfam. Nachtr. 1 : 197 (1897) ; Engl. Bot. Jahrb. 26 : 268 (1899) ; Léonard in Bull. Jard. Bot. Brux. 21 : 429 (1951).

Inflorescence racemose-paniculate ; receptacle about 4 mm. long ; leaves mostly with 3 pairs of slightly falcate, elliptic to elliptic-oblong leaflets, rounded to broadly cuneate at base, acuminate at apex, 8–10 cm. long, 3–3·5 cm. broad ; fruit not known
                                   1. *talbotii*

Inflorescence racemose, terminal, receptacle 6–9 mm. long ; leaves with 3–4 pairs of leaflets, the end pair the largest, the lowermost ovate-lanceolate, about 4–5 cm. long, the uppermost obliquely oblong-oblanceolate, 6–8 cm. long, acute to acuminate at apex ; fruits woody, broadly oblong, oblique at base, acuminate, about 17 cm. long and 7 cm. broad, light-brown-tomentellous   ..   ..   ..   2. *kalantha*

1. **L. talbotii** *Bak. f.* in J. Bot. 67 : 195 (1929) ; Léonard in Bull. Jard. Bot. Brux. 21 : 432. A forest tree (?).
    S.Nig.: Oban *Talbot* 1459 !
2. **L. kalantha** *Harms* in E. & P. Pflanzenfam. Nachtr. 1 : 197 (1897) ; Engl. Bot. Jahrb. 26 : 26 (1899) ; Aubrév. Fl. For. C. Iv. 1 : 201, t. 91, 3–5 ; Léonard l.c. A tall shrub or tree up to 65 ft. high ; leaves pale-rusty beneath ; flowers pinkish.
    Lib.: Grand Bassa (Apr.–May) *Dinklage* 1634 ! 1805 ! Du R. (fr. July) *Linder* 129 ! (See Appendix, p. 196.)

26. DANIELLIA Benn.—F.T.A. 2 : 299.

Flowers white or greenish-white ; filaments glabrous, free ; 1 petal well-developed, 8–16 mm. long, 2–6 mm. broad, the other 4 petals minute, up to 2 mm. long ; flowers in rather flattened panicles, the rhachides, pedicels and receptacles glabrous ; sepals 1·5–2 cm. long, 0·8–1·2 cm. broad ; leaflets 4–11 pairs, ovate, unequal-sided and acute to subcordate at base, obtusely acuminate, 6·5–17 cm. long, 3·5–10 cm. broad, glabrous, or pubescent mainly on nerves beneath and becoming glabrescent ; fruit obliquely elliptic, flat, 5–9·5 cm. long, 2·5–5 cm. broad, glabrous..   ..   1. *oliveri*
Flowers blue to reddish-purple ; filaments pilose below and more or less united at base ; 3 petals well-developed (1 rather smaller than the other 2) the other 2 petals minute :
    Leaves with spreading hairs on the rhachides, petiolules and midribs beneath ; leaflets 5–9 pairs, oblong to elliptic or lanceolate, unequal-sided and acute to subcordate at base, acuminate, 5–13 cm. long, 2–5 cm. broad ; inflorescence glabrous or nearly so ; receptacle and pedicel together 1·5–2 cm. long, glabrous ; sepals 1·2–2 cm. long, glabrous outside except at the apex..   ..   ..   2. *pynaertii*
    Leaves entirely glabrous :
        Inflorescence, pedicels, outside of receptacle and exposed parts of sepals glabrous or nearly so ; ovary glabrous :

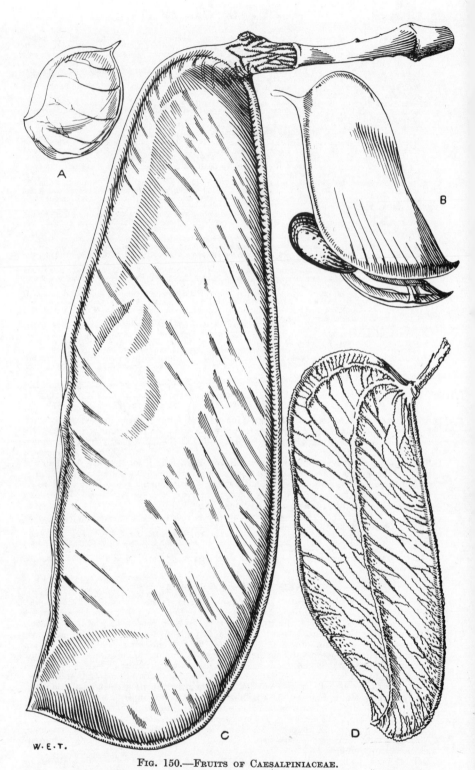

W.E.T.

Fig. 150.—Fruits of Caesalpiniaceae.

A, *Guibourtia copallifera* Benn. B, *Daniellia oliveri* (Rolfe) Hutch. & Dalz. C, *Berlinia grandi-flora* (Vahl) Hutch. & Dalz. D, *Gilbertiodendron limba* (Sc. Elliot) J. Léonard.

Sepals 1·6–2·2 cm. long, 11–17 mm. broad ; pedicel and receptacle together 1·5–2
cm. long, receptacle conical ; filaments 3–4·3 cm. long ; leaflets 7–9 pairs, oblong
or oblong-elliptic, unequal-sided at base, rather abruptly acuminate, in crown
foliage 6–9 cm. long, 2·5–3·5 cm. broad, in saplings, etc., up to 19 cm. long and
6 cm. broad ..   ..   3. *thurifera*
Sepals 1–1·2 cm. long, about 8 mm. broad ; pedicel and receptacle together about
1·2 cm. long, receptacle distinctly turbinate ; filaments 2–2·7 cm. long ; leaflets
about 8 pairs, oblong, unequal-sided at base, rather abruptly acuminate, 6–12 cm.
long, 2·5–4 cm. broad   ..   ..   4. *oblonga*
Inflorescence, pedicels, outside of receptacle and exposed parts of sepals densely
pubescent ; ovary densely pubescent, at least on the margins, sooner or later
glabrescent ; sepals 1·2–1·7 cm. long, 6–11 mm. broad ; pedicel and receptacle
together 1–1·5 cm. long, receptacle distinctly turbinate ; filaments 2–3 cm. long ;
leaflets 6–9 pairs, oblong, unequal-sided at base, acuminate, in crown foliage 6–11
cm. long, 2·5–3·5 cm. broad, considerably larger in saplings, etc.   ..   5. *ogea*

1. **D. oliveri** (*Rolfe*) Hutch. & Dalz. F.W.T.A., ed. 1, 1 : 341 (1928) ; Kew Bull. 1928 : 382 ; Léonard in Publ.
   I.N.É.A.C., Sér. Sci. 45 : 118 ; Aubrév. Fl. For. Soud.-Guin. 235, t. 46, 3–6 ; Léonard in Fl. Congo Belge
   3 : 348, t. 26. *Paradaniellia oliveri* Rolfe in Kew Bull. 1912 : 96 (q.v. for synonymy). *D. thurifera* of
   F.T.A. 2 : 300, partly, not of Benn. ; Chev. Bot. 231. *D. thurifera* Benn. var. *chevalieri* J. Léonard in
   Bull. Jard. Bot. Brux. 19 : 407 (1949), partly (leaves only). A large savannah tree, to 100 ft. high, with
   copious rather flat panicles of white, scented flowers and smooth pale flat 1-seeded fruits of horny con-
   sistence, tardily dehiscent.
   Sen.: *Heudelot* 364 ! Koulaye-Kouraye (Feb.) *Chev.* 2968 ! 2969 (partly) ! Bignona, Casamance *Serv. For.*
   44. **Gam.:** Genieri (Feb.) *Fox* 27 ! **Fr.Sud.:** Kita region (Jan.–Feb.) *Chev.* 96 ; *Dubois* 7. **Port.G.:**
   Bissau (fl. Jan., fr. Apr.) *Esp. Santo* 1514 ! 1675 ! **Fr.G.:** Kouroussa *Pobéguin* 861. Fouta Djalon
   Maclaud 40. Mabina, Boké *Chillou* ! **S.L.:** Musaia (fr. Feb.) *Deighton* 4217 ! Magbuke *Deighton* 4083.
   Iv.C.: *fide* Aubrév. *l.c.* **G.C.:** Kwahu (fr. Jan.) *Chipp* 623 ! Krobo Plains (Dec.) *Johnson* 501 ! **Togo:**
   Yendi *Williams* 421 ! Kete Kratchi (Jan.) *Vigne* FH 1530 ! **Dah.:** Adja-Ouere *Poisson* 21. **N.Nig.:**
   Nupe *Barter* 978 ! Kontagora (fl. & fr. Dec.) *Dalz.* 16 ! Lokoja (fl. & fr. Dec.) *Elliott* 14 ! Bauchi Plateau
   Lely P. 62 ! **S.Nig.:** Lagos *Foster* 151 ! Ishan (fr. Jan.) *Dennett* 102 ! Onitsha (fl. & fr. Feb.) *Johnson* !
   Also in French Cameroons, Ubangi-Shari, Belgian Congo, A.-E. Sudan and Angola. (See Appendix,
   p. 186.)
   [The specimen " *Mann* 978 " (in Hb. Paris), cited by Léonard l.c. 120–121 and said to come from
   Fernando Po, is almost certainly *Barter* 978 from Nigeria ; *Mann* 978 at Kew is in fact *Parinari chryso-
   phylla* Oliv., from Gabon. I have seen in the Paris herbarium a number of Mann's and Barter's specimens
   on which the labels have been interchanged in this way.—R.W.J.K.]
2. **D. pynaertii** De Wild. in Ann. Mus. Congo Belge, sér. 5, 3 : 193 (1910) ; Léonard in Publ. I.N.É.A.C., Sér.
   Sci. 45 : 109 ; Fl. Congo Belge 3 : 343. *D. pubescens* Hutch. & Dalz. F.W.T.A., ed. 1, 1 : 341 (1928) ;
   Kew Bull. 1928 : 382. Forest tree, to 110 ft. high ; flowers mauve, very abundant.
   S.Nig.: Lagos *Moloney* ! Also in French Cameroons (?), Gabon and Belgian Congo. (See Appendix,
   p. 187.)
3. **D. thurifera** Benn. in Pharm. Journ. 14 : 252 (1854) ; F.T.A. 2 : 300, partly ; Aubrév. Fl. For. C. Iv. 1 :
   210, t. 78 A ; Léonard in Publ. I.N.É.A.C., Sér. Sci. 45 : 98, partly (excl. syn. *D. oblonga* Oliv.). *D.*
   oblonga of Chev. Bot. 231, partly, not of Oliv. *D. caillei* A. Chev. Bot. 230, name only. Forest tree, to
   150 ft. high.
   Port.G.: Catió (fr. June) *Esp. Santo* 2099 ! **Fr.G.:** Longuery (Dec.) *Caille* in *Hb. Chev.* 14827 ! **S.L.:**
   Afzelius ! *Daniell* ! *Melville* ! Waterloo (Oct.) *Lane-Poole* ! Commendi to Gengaru (Nov.) *Aylmer* 256 !
   Lib.: Dukwia R. (Oct.) *Cooper* 77 ! 95 ! **Iv.C.:** Yapo *Chev.* 22360 ! Abidjan *Aubrév.* 197 ! **G.C.:** Awiabo,
   Axim (Nov.) *Vigne* FH 1422 ! (See Appendix, p. 187.)
4. **D. oblonga** Oliv. F.T.A. 2 : 301 (1871). *D. thurifera* of F.W.T.A., ed. 1, 1 : 341 and of Léonard l.c. and partly,
   not of Benn. Forest tree to 200 ft. high, flowers blue.
   S.Nig.: Iso Bendiga to Bendiga Afi, Ikom (Dec.) *Keay* ! **Br.Cam.:** Likomba (Dec.) *Mildbr.* 10759 ! **F.Po:**
   Barter 2074 ! (See Appendix, p. 187.)
5. **D. ogea** (*Harms*) Rolfe ex Holl. in Kew Bull. Add. Ser. 9 : 268 (1911) ; Rolfe in Kew Bull. 1912 : 97 ;
   Léonard in Publ. I.N.É.A.C., Sér. Sci. 45 : 102, t. 11 C ; Keay in Kew Bull. 1953 : 491–492. *Cyanothyrsus*
   ogea Harms in Engl. Bot. Jahrb. 26 : 270 (1899). *D. caudata* Craib ex Holl. l.c. 268 (1911), partly (flowers
   only, the leaves are *Clitandra barteri* Stapf) ; Craib in Kew Bull. 1912 : 94 ; F.W.T.A., ed. 1, 1 : 341 ;
   Léonard l.c. 97, t. 11 A. *D. punchii* Craib ex Holl. l.c. 269 (1911), partly (flowers only, the leaves are
   Crudia sp.) ; Craib l.c. 95 ; F.W.T.A., ed. 1, 1 : 341 ; Léonard l.c. 102, t. 11 B. *D. fosteri* Craib ex Holl.
   l.c. 268 (1911) ; Craib l.c. 95 ; F.W.T.A., ed. 1, 1 : 341 ; Léonard l.c. 106. *D. similis* Craib ex Holl. l.c.
   269 (1911) ; Craib l.c. 95 ; F.W.T.A., ed. 1, 1 : 341 ; Aubrév. Fl. For. C. Iv. 1 : 212, t. 78 B ; Léonard
   l.c. 108. *D. thurifera* var. *chevalieri* J. Léonard in Bull. Jard. Bot. Brux. 19 : 407 (1949), partly (flowers
   only, the leaves are *D. oliveri*). Forest tree, to 180 ft. high, with clean cylindrical bole and very short
   rounded buttresses ; flowers blue to purple.
   Sen.: Koulaye-Kouraye (fl. & young fr. Feb.) *Chev.* 2969 (partly) ! **Port.G.:** Teixeira Pinto (Jan.) *D'Orey*
   175 ! **S.L.:** Gola Forest (fl. & fr. Jan.) *Unwin & Smythe* 45 ! **Lib.:** Gbanga (Oct.) *Linder* 1233 ! **Iv.C.:**
   Agboville *Aubrév.* 2775 ! **G.C.:** *Imp. Inst.* ! Ashanti (Feb.) *Armitage* 107 ! Birrim (fl. & fr. Nov.) *Chipp*
   670 ! Owabi (Dec.) *Lyon* FH 2644 ! Mpraeso *Beveridge* FH 3210 ! **S.Nig.:** Ijebu Ode *Millen* 191 !
   Millson 37 ! Ibadan F.R. (Nov.) *Punch* 115 ! Mamu F.R. (fl. & fr.
   Dec.) *Unwin* 179 ! Sapoba (Dec.–Jan.) *Kennedy* 328 ! 1981 ! Okomu F.R. (Jan.) *Onochie* in *Hb. Brenan*
   8817 ! Degema *Talbot* 3677 ! **F.Po:** (Jan.) *Mann* 166 ! Also in Gabon. (See Appendix, p. 186.)

27. **EURYPETALUM** Harms in Engl. Bot. Jahrb. 45 : 293 (1910).

A glabrous tree ; petioles 3–9 mm. long with a subulate caducous tip between the
2–4 mm. long turgid petiolules ; leaflets 1 pair, lanceolate or oblong-lanceolate,
somewhat falcate, unequally cuneate or obtuse at base, gradually acuminate at apex,
12–16 cm. long, 4·5–6·5 cm. broad ; flowers in solitary or paired axillary, glabrous
racemes up to 2·5 cm. long ; pedicels slender, 1·5–2 cm. long ; bracts and bracteoles
much reduced, caducous ; sepals 4, unequal, about 5 mm. long, one ovate and rather
larger than the others which are oblong ; large petal about 6 mm. long and 10 mm.
broad, the other 4 petals minute ; stamens 10, with subulate filaments, shortly
connate at base ; fruit stipitate, oblong, up to about 10 cm. long and 4 cm. broad,
valves thin, glabrous   ..   ..   *unijugum*

E. unijugum *Harms* in Engl. Bot. Jahrb. 49 : 222 (1913) ; Bak. f. Leg. Trop. Afr. 3 : 770 ; Pellegr. Lég. Gab. 114. Forest tree, to 70 ft. high.
Br.Cam.: S. Bakundu F.R. (fr. Dec.) *Ejiofor* FHI 29308 ! *Ezeilo* FHI 29551 ! Also in French Cameroons and (?) Gabon.

28. PLAGIOSIPHON Harms in E. & P. Pflanzenfam. Nachtr. 1 : 194 (1897) ; Engl. Bot. Jahrb. 26 : 263 (1899) ; Léonard in Bull. Jard. Bot. Brux. 21 : 425 (1951).

Branchlets rusty-tomentose ; leaflets 8–12 pairs, narrowly oblong, emarginate at apex, 6–18 mm. long, 2–5 mm. broad ; flowers in short racemes ; pedicels about 6 mm. long ; bracteoles ovate about 8 mm. long and 6 mm. broad ; petals 5, 3 large (or rarely 4) and 2 smaller (or rarely 1) ; stamens 10 ; ovary stipitate the stipe adnate up to the apex of the receptacle, ovules 2 ; fruits oblong-elliptic, about 6 cm. long and 3 cm. broad ; rusty-tomentose   ..   ..   *emarginatus*

P. emarginatus (*Hutch. & Dalz.*) *J. Léonard* l.c. 426 (1951). *Monopetalanthus emarginatus* Hutch. & Dalz. F.W.T.A., ed. 1, 1 : 342 (1928) ; Kew Bull. 1928 : 397. *Hymenostegia emarginata* (Hutch. & Dalz.) Milne-Redhead ex Hutch. & Dalz. F.W.T.A., ed. 1, 2 : 606 (Feb. 1936) ; Aubrév. & Pellegr. in Bull. Soc. Bot. Fr. 82 : 603 (Apr. 1936) ; Aubrév. Fl. For. C. Iv. 1 : 250, t. 95 B. *Tripetalanthus emarginatus* (Hutch. & Dalz.) A. Chev. in Rev. Bot. Appliq. 26 : 597, t. 19 A (1946). Forest tree, to 100 ft. high ; flowers white ; by river banks.
S.L.: Morro R. (fr. Jan.) *Unwin & Smythe* 42 ! Njala (fl. July–Sept., fr. Jan.) *Deighton* 1950 ! 2836 ! Bumpe (fl. July, fr. Nov.) *Deighton* 4835 ! 4927 ! Lib.: St. John R., Tappita (Aug.) *Baldwin* 9139 ! Iv.C.: Sassandra & Cavally Basins *Aubrév.* 860 ! 2798 !
A sterile specimen (*Keay* s.n.) of a tall tree from Banga in the British Cameroons probably represents another species of this genus ; flowering and fruiting material is required.

29. HYMENOSTEGIA (Benth.) Harms in E. & P. Pflanzenfam. Nachtr. 1 : 193 (1897) ; Léonard in Bull. Jard. Bot. Brux. 21 : 433 (1951).

Leaflets 1–2 pairs, rhachis winged ; the lower pair of leaflets (if present) smaller and more or less ovate-rhombic, about 3 cm. long, the upper pair obliquely obovate, narrowed at both ends, obtusely acuminate, 7–10 cm. long, 2–5 cm. broad, glabrous ; racemes about as long as the leaves ; pedicels up to 1 cm. long ; bracteoles broadly ovate, 6–12 mm. long ; petals 5, 3 larger, 2 much smaller ; stamens 10 ; fruits obliquely obovate, 7–9 cm. long, 3–3·5 cm. broad   ..   ..   1. *afzelii*
Leaflets 3–10 pairs (rarely 2), rhachis not winged :
　Petals 5, 3 larger, 2 very much smaller ; stamens 10 ; bracteoles ovate, about 1·7 cm. long and 1·1 cm. broad ; pedicels up to 1·5 cm. long ; leaflets 3–4 pairs (rarely 2), oblong-elliptic, obtuse and slightly emarginate at apex, 2·5–9 cm. long, 1–4 cm. broad, finely pubescent beneath ; fruits subtrapezoid, 8–10 cm. long, 4–5 cm. broad   ..   ..   2. *aubrevillei*
　Petals 3, 1 or 2 of which are large and 2 or 1 very small, or absent :
　　Stamens 10 ; flowers with only 1 larger petal ; bracteoles oblong-elliptic about 1 cm. long ; pedicels 2–2·5 cm. long ; leaflets 3–4 pairs, obliquely oblong-oblanceolate, triangular-acuminate, 4–8 cm. long, 2–3·5 cm. broad, glabrous ; fruits not known   3. *gracilipes*
　　Stamens 16–26 ; flowers with 2 larger petals and 1 smaller petal or absent ; bracteoles lanceolate or linear-oblong :
　　　Leaflets 3–4 pairs, obliquely oblong, obtusely acuminate, 8–15 cm. long, 3·5–5 cm. broad ; pedicels 1–2 cm. long, glabrous ; bracteoles linear-oblong, acute, about 2 cm. long ; petals either absent or very minute   ..   ..   ..   4. *talbotii*
　　　Leaflets 8–10 pairs, oblong, obtusely acuminate, 4–8 cm. long, 1·5–2·5 cm. broad ; racemes very densely flowered ; pedicels about 2 cm. long, velutinous ; bracteoles linear-oblanceolate, about 1 cm. long, pilose outside ; petals 3, 2 larger and 1 minute   ..   ..   5. *bakeriana*

1. **H. afzelii** (*Oliv.*) *Harms* in E. & P. Pflanzenfam. Nachtr. 1 : 193 (1897) ; Chev. Bot. 235 ; Aubrév. Fl. For. C. Iv. 1 : 248, t. 94 A ; For. Trees & Timbers Brit. Emp. 2 : 24, fig 4 ; Léonard in Bull. Jard. Bot. Brux. 21 : 438. *Cynometra afzelii* Oliv. F.T.A. 2 : 318 (1871). A forest tree, to 60 ft. high ; bracteoles white, lilac or pinkish ; petals yellow.
Fr.G.: Farana *Chev.* 13432 ! S.L.: *Afzelius.* Lib.: Monrovia *Whyte* ! Iv.C.: Indenié (Dec.) *Chev.* 22646 ! Tabou *Aubrév.* 1682 ! And other localities *fide* Aubrév. *l.c.* G.C.: Cape Coast (Nov.) *Brass* ! *Vigne* 946 ! Akyease (Oct.) *Fishlock* 62 ! Dunkwa (Jan.) *Dalz.* 68 ! Aburi Hills (Nov.–Mar.) *Johnson* 615 ! 621 ! Togo: Misahöhe (Nov.) *Mildbr.* 7340 ! S.Nig.: Lagos *Moloney* ! Sapoba *Kennedy* 413 ! Oban *Talbot* 1309 ! 1720 ! Ikom *Catterall* 70 ! Br.Cam.: Likomba *Mildbr.* 10537 ! Also in French Cameroons. (See Appendix, p. 194.)
2. **H. aubrevillei** *Pellegr.* in Bull. Soc. Bot. Fr. 80 : 464 (1933) ; Aubrév. l.c. 250, t. 94 B ; Léonard l.c. 440. *Brachystegia eurycoma* of F.W.T.A., ed. 1, 1 : 348, partly (*Chev.* 16283). Forest tree of medium size with short bole and dense rounded crown ; by river banks ; bracteoles mauve, sepals white, petals reddish.
Iv.C.: Abidjan (Nov.–May) *Aubrév.* 83 ! Zaranou *Chev.* 16283 ! G.C.: Krokosana (Feb.) *Cansdale* FH 3985 !
3. **H. gracilipes** *Hutch. & Dalz.* F.W.T.A., ed. 1, 1 : 332 (1928) ; Kew Bull. 1928 : 381 ; Léonard l.c. 442. Riverside forest tree, to 45 ft. high ; flowers white.
G.C.: Sa, Bonsa R. (July, Aug.) *Vigne* FH 154 ! 978 ! 4765 ! Obin *Vigne* FH 1280 ! (See Appendix, p. 195.)
[A sterile specimen (*Fobes* LF 1) from Bomi Hills, Montserrado, Liberia, also appears to be this species.]
4. **H. talbotii** *Bak. f.* in J. Bot. 67 : 196 (1929) ; Léonard l.c. 442. A forest tree (?).
S.Nig.: Eket Dist. *Talbot* 3141 !
5. **H. bakeriana** *Hutch. & Dalz.* F.W.T.A., ed. 1, 1 : 332 (1928) ; Kew Bull. 1928 : 381 ; Léonard l.c. 442. *Cynometra longituba* of Bak. f. in Cat. Talb. 127, not of Harms. A tree with rather densely pubescent branchlets ; flowers red (?) or drying reddish.
S.Nig.: Oban *Talbot* 1567 !

30. GUIBOURTIA Benn. in J. Linn. Soc. 1 : 149 (1857) ; Léonard in Bull. Jard. Bot.
Brux. 19 : 391–406 (1949) ; op. cit. 20 : 269–284 (1950).

Bracteoles persistent forming a calyx-like cup at the base of the ellipsoidal flower bud ;
fruits indehiscent, flat, thin ; seeds not arillate ; leaflets glabrous :
Leaflets with 3–5 prominent longitudinal nerves from the base, the strong nerves not
    arranged pinnately ; lamina falcate, cuneate or unequal at base, obtusely acuminate,
    3–12 cm. long, 1·5–5 cm. broad ; petiole 3–5 mm. long ; stipules foliaceous, 5–7 mm.
    long, subpersistent ; fruits obliquely elliptic, 3·8–4·2 cm. long, 2·7–3 cm. broad,
    with a nerve parallel to the margin on one side .. .. ..   1. *copallifera*
Leaflets with pinnately arranged strong lateral nerves ; lamina ovate-falcate, strongly
    asymmetrical, one side cuneate, the other rounded at base, acute or acuminate at
    apex, 6·5–20 cm. long, 3–8 cm. broad ; petiole 1·5–3 cm. long ; stipules early
    caducous ; fruits suborbicular or elliptic, 2·2–3·8 cm. long, 1·8–2·9 cm. broad, with a
    rather indistinct nerve running all round parallel to the margin   .. 2. *demeusei*
Bracteoles caducous :
Buds globose ; fruits, where known, indehiscent, flat, thin ; seeds, where known, not
    arillate ; young leaflets pubescent on nerves beneath :
Leaflets with 2 strong longitudinal ascending nerves on one or both sides of the mid-
    rib ; upper leaves usually 2-foliolate with falcate leaflets up to 11·5 cm. long and
    3–9 cm. broad, long- or caudate-acuminate at apex ; lower leaves 1-foliolate with
    an ovate to ovate-rhombic leaflet, rounded at base, long-acuminate at apex, 7–13
    cm. long, 4–7 cm. broad ; petioles 3–4 mm. long ; fruits not known   3. *dinklagei*
Leaflets with 5–9 pinnately arranged strong lateral nerves ; leaves all 2-foliolate with
    falcate leaflets, one side cuneate, the other obtuse at base, acuminate at apex, 5–10
    cm. long, 1·5–5 cm. broad ; petiole 4–10 mm. long ; fruits obliquely rhombic-
    elliptic, 4–6 cm. long, 2·5–3·5 cm. broad, with a prominent curved nerve towards
    the upper margin dividing the narrow winged portion from the reticulate body of
    the fruit   ..   ..   4. *ehie*
Buds, where known, ellipsoidal ; fruits dehiscent, thick, without marginal or longi-
    tudinal nerves ; seeds arillate ; leaflets glabrous :
Bark of branchlets exfoliating in thin reddish-brown pieces ; rhachis of inflorescence
    up to 1·5 mm. diam. ; fruits black when dry, 2–2·5 cm. long, 1·7–1·8 cm. broad,
    with a stipe 2–7 mm. long ; leaflets elliptic-falcate, abruptly acuminate at apex
    6–11·5 cm. long, 2·5–5·5 cm. broad ; petiole 1·5–2·5 cm. long   5. *pellegriniana*
Bark of branchlets not exfoliating in thin pieces ; rhachis of inflorescence more than
    2 mm. diam. ; fruits reddish-brown when dry :
Leaflets coriaceous, acumen about 5 mm. long ; fruits with very short pedicel and
    3–5 mm. long stipe ; body of fruit 3–3·5 cm. long, 2–2·5 cm. broad ; leaflets
    falcate, 7–15 cm. long, 3–6 cm. broad ; petiole 1·5–2·5 cm. long   6. *tessmannii*
Leaflets membranous, abruptly acuminate, acumen 5–20 mm. long ; fruits with
    pedicel about 2 mm. long and stipe 9–11 mm. long ; body of fruit 2·6–2·8 cm. long,
    1·7–1·8 cm. broad ; leaflets falcate-elliptic, 5–13 cm. long, 2–5 cm. broad ; petiole
    1·2–2·7 cm. long   ..   .. 7. *leonensis*

1. **G. copallifera** *Benn.* in J. Linn. Soc. 1 : 150 (1857) ; Léonard in Publ. I.N.É.A.C., Sér. Sci. 45 : 72. *Copaifera
guibourtiana* Benth. (1865)—F.T.A. 2 : 314 ; F.W.T.A., ed. 1, 1 : 338 (excl. syn. *C. ehie* A. Chev.) ;
Aubrév. Fl. For. C. Iv. 1 : 262, t. 103 B ; Kennedy For. Fl. S. Nig. 96. *C. copallifera* (Benn.) Milne-
Redhead in Kew Bull. 1934 : 400 ; Aubrév. in Agron. Trop. 3 : 18 (1948). *C. vuilletiana* A. Chev. in Bull.
Soc. Bot. Fr. 61, Mém. 8 : 258 (1917). *Guibourtia vuilletiana* (A. Chev.) A. Chev. (1950). *Copaifera
vuilletii* A. Chev. Bot. 234, name only ; Aubrév. Fl. For. C. Iv. 1 : 260. A tree to 80 ft. high, or shrub ;
flowers white ; sometimes planted.
    Fr.Sud.: Koulikoro *Vuillet* in *Hb. Chev.* 28936 ! Kita *Dubois* 223 ! **Port.G.:** Empacaca, Canchungo (fr.
Apr.) *Esp. Santo* 1939 ! **Fr.G.:** Conakry *Maclaud* ! Boké *Chillou* ! **S.L.:** Goderich & Lumley (Aug.)
Daniell 2 ! R. Scarcies *Sc. Elliot* 4537 ! Kenema (fr. Dec.—planted) *Burns* ! *Deighton* 3570 ! **Mano**
(Oct.—planted) *Deighton* 3793 ! **Iv.C.:** Ferkéssédougou *Aubrév.* 2271. **S.Nig.:** *Farquhar* 53 ! Olokemeji
(planted) *Onochie* FHI 8147 ! Benin City (planted) *Kennedy* 2253 ! [*Farquhar* 53 from S. Nigeria has no
precise locality ; it was probably collected from an introduced tree ; the species is not, apparently,
indigenous in Nigeria.] (See Appendix, p. 184.)
2. **G. demeusei** (*Harms*) *J. Léonard* in Bull. Jard. Bot. Brux. 19 : 403 (1949) ; Publ. I.N.É.A.C., Sér. Sci. 45 :
82 ; Fl. Congo Belge 3 : 361. *Copaifera demeusei* Harms in Engl. Bot. Jahrb. 26 : 264 (1899) ; not of
F.W.T.A., ed. 1, 1 : 338. A forest tree, to 130 ft. high ; flowers white.
    Br.Cam.: Victoria (fr. Oct.) *Maitland* 755 ! Likomba *Mildbr.* 10753 ! 10768 ! Also in French Cameroons,
Ubangi-Shari, Gabon, French and Belgian Congo. (See Appendix, p. 184.)
3. **G. dinklagei** (*Harms*) *J. Léonard* in Bull. Jard. Bot. Brux. 19 : 403 (1949) ; Publ. I.N.É.A.C., Sér. Sci. 45 :
72. *Copaifera dinklagei* Harms in Engl. Bot. Jahrb. 26 : 265 (1899) ; F.W.T.A., ed. 1, 1 : 338. *G.
liberiensis* J. Léonard in Bull. Jard. Bot. Brux. 20 : 276, fig. 27, E & 28 (1950). A glabrous shrub or small
tree, to 15 ft. ; flowers white.
    Lib.: *Sim* ! Banks of Biso R., Grand Bassa (Aug.) *Dinklage* 1695 ! Cess R., Grand Bassa (Mar.) *Baldwin*
11235 !
4. **G. ehie** (*A. Chev.*) *J. Léonard* in Bull. Jard. Bot. Brux. 19 : 404 (1949) ; Publ. I.N.É.A.C., Sér. Sci. 45 : 78.
Copaifera ehie A. Chev. in Bull. Soc. Bot. Fr. 61, Mém. 8 : 258 (1917) ; Bot. 234 ; Aubrév. Fl. For. C. Iv.
1 : 264, t. 103 A ; Pellegr. Lég. Gab. 117. *C. guibourtiana* of F.W.T.A., ed. 1, 1 : 338, partly (*Chev.*
22447). Forest tree, to 150 ft. high, with buttresses and straight bole ; flowers white, scented.
    Iv.C.: Morénou (fr. Nov.) *Chev.* 22447 ! Adikokoi *Aubrév.* 658. Guiglo *Aubrév.* 2038. **G.C.:** Adiembra,
Ashanti (fl. & fr. Oct.) *Vigne* FH 4916 ! Jimira F.R. (fl. Oct., fr. Jan.) *Vigne* FH 3115 ! **S.Nig.:** Okomu
F.R. *Hide* 30/37 ! Ikom *Catterall* 73 ! Afi River F.R. (fr. May) *Jones & Onochie* FHI 17313 ! Latilo
FHI 30996 ! Also in Gabon.
5. **G. pellegriniana** *J. Léonard* in Bull. Jard. Bot. Brux. 19 : 405, fig. 38 (1949) ; op. cit. 20 : 273 ; Publ.

I.N.É.A.C., Sér. Sci. 45 : 75. *Copaifera demeusei* of F.W.T.A., ed. 1, 1 : 338, not of Harms. Large forest tree.
S.Nig.: Old Calabar *Milne* ! **Br.Cam.:** S. Bakundu F.R. *Ejiofor* FHI 15252 ! Also in Gabon, French Congo, and Cabinda. (See Appendix, p. 184.)

6. **G. tessmannii** (*Harms*) J. *Léonard* in Bull. Jard. Bot. Brux. 19 : 404 (1949) ; op. cit. 22 : 271 ; Publ. I.N.É.A.C., Sér. Sci. 45 : 74. *Copaifera tessmannii* Harms in Notizbl. Bot. Gart. Berl. 5 : 181 (1910) ; Pellegr. Lég. Gab. 116, t. 6, fig. 9. A large forest tree.
Br.Cam.: Likomba (Dec.) *Mildbr.* 10771 ! Mbalange, Kumba (fr. Jan.) *Rosevear* 1/33 ! Also in French Cameroons, Gabon and Spanish Guinea.

7. **G. leonensis** J. *Léonard* in Bull. Jard. Bot. Brux. 20 : 271, fig. 27, C-D (1950). Forest tree, to 80 ft. high.
Port.G.: S. Domingos (fr. Jan.) *D'Orey* 174 ! Suzana (fr. Mar.) *D'Orey* 354 ! **S.L.:** Kambui Hills *Wallace* 89 ! 114 ! 147 ! Dambaye (young fr. Sept.) *Edwardson* 133 !

31. OXYSTIGMA Harms in E. & P. Pflanzenfam. Nachtr. 1 : 195 (1897) ; Léonard in Bull. Inst. Roy. Col. Belg. 21 : 744–753 (1950).

Fruits not winged, bulky, bilobed at apex, up to 7·5 cm. long, 6·5 cm. broad and 3·5 cm. thick, irregularly knobbly, pericarp woody, containing numerous small resinous cavities ; leaflets 1 pair, or occasionally 3–4 leaflets, oblong, rounded at both ends, 7–15 cm. long, 3–5 cm. broad, rhachis up to 3 cm. long   ..   .. 1. *mannii*
Fruits samaroid, ovate-lanceolate to lanceolate, 8–13 cm. long, 2–4 cm. broad, the proximal part consisting of the well-developed wing 6–10 cm. long, pericarp with prominent longitudinal nerves ; leaflets 5–9 (rarely 2), alternate or rarely subopposite, elliptic or elliptic-lanceolate, slightly falcate, acute at base, acuminate at apex, 4–14 cm. long, 1·5–4 cm. broad, rhachis up to 16 cm. long   .. 2. *oxyphyllum*

1. **O. mannii** (*Baill.*) *Harms* in E. & P. Pflanzenfam. Nachtr. 1 : 195 (1897) ; Engl. Bot. Jahrb. 26 : 264 ; Engl. Pflanzenw. Afr. 3, 1 : 437, t. 241 ; Léonard in Bull. Inst. Roy. Col. Belg. 21 : 748. *Copaifera ? mannii* Baill. (1866). *Hardwickia ? mannii* (Baill.) Oliv. F.T.A. 2 : 316 (1871). *Eriander engleri* H. Winkl. (1908). *Macrolobium diphyllum* of F.W.T.A., ed. 1, 1 : 347, not of Harms. A forest tree, to 80 ft. high with glabrous branchlets and spike-like racemes of white or pinkish flowers"; in freshwater swamps.
S.Nig.: Eket *Talbot* s.n. ! 3014 ! Calabar R. (fl. Jan.–Mar., fr. Oct.) *Holland* 92 ! *Espley* 1 ! *Kennedy* 1812 ! **Br.Cam.:** Cameroons R. (Jan.) *Mann* 754 ! 2194 ! Victoria (Nov.–Jan.) *Winkler* 669 ! *Maitland* 109 ! 786 ! 1150 ! Also in French Cameroons and Spanish Guinea. (See Appendix, p. 198.)

2. **O. oxyphyllum** (*Harms*) J. *Léonard* l.c. 753, fig. 1, B, fig. 2, G–H (1950) ; Fl. Congo Belge 3 : 373. *Pterygopodium oxyphyllum* Harms in Engl. Bot. Jahrb. 49 : 439 (1913) ; Kennedy For. Fl. S. Nig. 94 ; Pellegr. Lég. Gab. 128. A forest tree, to 150 ft. high ; similar to *Gossweilerodendron*.
S.Nig.: S. Ondo Prov. *Kennedy* 2422 ! *Ainslie* 10 ! Also in French Cameroons, Gabon, Cabinda and Belgian Congo.

32. GOSSWEILERODENDRON Harms in Notizbl. Bot. Gart. Berl. 9 : 457 (1925).

A large tree ; branchlets glabrous ; leaves pinnate ; leaflets alternate, with an end one, oblong-elliptic, obtuse or obtusely pointed, 4–9 cm. long, 2–4 cm. broad, densely pellucid-glandular, otherwise glabrous ; racemes spiciform, simple or branched, up to 10 cm. long ; axis slightly pubescent ; calyx slightly pubescent ; petals absent ; stamens 8–10, free, filaments hairy at the base ; ovary stipitate, villous ; ovule solitary, pendulous ; fruit 1-seeded, with a broad basal oblique wing about 8 cm. long and 3·5 cm. broad, the fruit body ellipsoid, about 2·5 cm. long   ..   *balsamiferum*

G. balsamiferum (*Verm.*) *Harms* l.c. fig. 9, F–L (1925) ; Kennedy For. Fl. S. Nig. 93 ; Léonard in Fl. Congo Belge 3 : 375, t. 28. *Pterygopodium balsamiferum* Verm. (1923). Forest tree, to 200 ft. high, with cylindrical bole without buttresses or basal thickening ; crown spreading with light foliage ; flowers in whitish spikes, the sepals pellucid-dotted ; fruit samara-like with the seed at the distal end.
S.Nig.: Benin (fr. June) *St. Barbe Baker* ! Sapoba (fl. Dec.–Jan., fr. Mar.) *Kennedy* 327 ! 554 ! 1677 ! Udi (Jan.–Feb.) *Chesters* 201 ! Degema Dist. *Chesters* OC 25 ! Also in Cabinda and Belgian Congo. (See Appendix, p. 194.)

33. STEMONOCOLEUS Harms in Engl. Bot. Jahrb. 38 : 77 (1907).

A tree, wholly glabrous ; leaflets 4–10, alternate, oblong or elliptic, rounded or obtuse at base, rounded and emarginate at apex, 5–11 cm. long, 3–5 cm. broad, with 6–10 lateral nerves, prominulous above and beneath ; rhachis 6–15 cm. long ; inflorescence a panicle of short racemes ; flowers 3–4 mm. long, crowded towards the ends of the racemes, pedicels up to 2 mm. long ; flower buds covered by an ovate bract and two small bracteoles ; calyx tubular with 4 lobes ; petals absent ; stamens 4, united below ; fruit indehiscent, a large papery samara, oblong or oblong-elliptic, cuneate and twisted at base, rounded at apex, 12–16 cm. long, 4–8 cm. broad ; seeds 1–2, flat, suboblong, 1·5–2 cm. long   ..   .. *micranthus*

S. micranthus *Harms* l.c. fig. 7 (1907) ; Aubrév. Fl. For. C. Iv. 1 : 258, t. 101 ; Kennedy For. Fl. S. Nig. 87 ; Pellegr. Lég. Gab. 127, t. 3, 4. Forest tree, to 150 ft. high, with large straight bole and buttresses.
Iv.C.: Abidjan *Aubrév.* 29 ! Banco *Aubrév.* 372 ! **S.Nig.:** Benin (fl. & fr. Jan.) *Lancaster* FHI 2887 ! Sapoba (fl. Oct., fr. Dec.) *Kennedy* 2490 ! 2544 ! Akpaka F.R., Onitsha (Dec.) *Chesters* 182 ! Ukpor, Onitsha (fl. & fr. Aug.) *Cousens* FHI 6932 ! Awka (Aug.) *Cousens* FHI 6933 ! Calabar *Mackay* 227/38 ! Also in French Cameroons and Gabon. (See Appendix, p. 199.)

34. CRUDIA Schreb.—F.T.A. 2 : 312.

Leaflets quite glabrous beneath ; stipules persistent, lanceolate, 2–2·5 cm. long ; leaflets 5–11, oblong-elliptic to obovate-elliptic, falcate, cuneate and unequal at base, obtusely acuminate, 5–10 cm. long, 2–4·5 cm. broad ; pedicels slender, 1·5–2·5 cm. long, glabrous, with 2 small bracteoles towards the base ; sepals obovate, about 5 mm.

long ; ovary densely woolly ; fruits shortly oblong to suborbicular, bulgy, densely
tomentose outside, 5–10 cm. long, 5–6 cm. broad, containing 1–2 large seeds ; seeds
ellipsoid to suborbicular, 4·5–6·5 cm. long, 4–5 cm. broad, 2–2·8 cm. thick, brown,
glabrous, the contents loose within the testa when dry   ..   ..   1. *senegalensis*
Leaflets minutely puberulous beneath (good lens needed) ; fruits relatively flat :
Pedicels glabrous, 5–18 mm. long, with minute bracteoles at the base ; stipules
    persistent, ovate-lanceolate, up to 3 cm. long and 2·5 cm. broad ; leaflets 5–7,
    broadly elliptic, obtuse at base, obtusely acuminate, 8–13 cm. long, 2·5–7 cm. broad ;
    fruits oblong, transversely nerved, minutely puberulous or glabrescent, 15–18 cm.
    long, 4–6 cm. broad, containing 3–4 seeds ; seeds suborbicular or reniform, flat,
    3·5–4·5 cm. diam., about 0·5 cm. thick   ..   ..   ..   2. *klainei*
Pedicels puberulous, with 2 bracteoles towards the middle ; stipules caducous :
Flowering pedicels 1–2·2 cm. long ; leaflets 5–9, ovate to oblong-elliptic, obtuse or
    rounded at base, abruptly long caudate-acuminate, 6–15 cm. long, 2·5–5·5 cm.
    broad ; fruits oblong, tomentose, to 30 cm. long and 6–7 cm. broad, containing
    2–5 seeds   ..   ..   3. *gabonensis*
Flowering pedicels 4–6 mm. long ; leaflets 6–8, oblong to oblong obovate, obtuse to
    rounded at base, shortly and obtusely acuminate, up to 10 cm. long and 4·5 cm.
    broad, the lower leaflets smaller than the upper ones and sometimes less than 1 cm.
    long and broad ; fruits broadly and obliquely oblanceolate, minutely puberulous
    or glabrescent, about 16 cm. long and up to 6 cm. broad   ..   ..   4. *sp. A*

1. **C. senegalensis** *Planch. ex Benth.*—F.T.A. 2 : 312, partly ; Chev. Bot. 233 ; Pellegr. Lég. Gab. 129 ; Berhaut
Fl. Sen. 35. *C.* sp. aff. *senegalensis* of Aubrév. Fl. For. C. Iv. 1 : 256, t. 100 A. Forest tree, to 60 ft. high,
with spreading crown ; often by rivers.
Sen.: banks of R. Gambia *fide* Berhaut *l.c.* **Port.G.:** Bor, Bissau (Feb.) *Esp. Santo* 1787 ! Fulacunda
(young fr. May) *Esp. Santo* 2028 ! **Fr.G.:** Rio Nunez *Heudelot* 708 ! **S.L.:** *Afzelius* ! Masao R., Potoro
(Oct.) *Fisher* 41 ! Gene, Kamagai (fr. Feb., Apr.) *Deighton* 3363 ! 4764 ! Bandajuma ferry (Oct.) *King*
223 ! **Lib.:** Fisherman's Lake (Dec.) *Baldwin* 10895 ! Buchanan, Grand Bassa (young fr. Mar.) *Baldwin*
11174 ! **Iv.C.:** *Aubrév.* 1692 ! 1695 ! And various localities *fide* Aubrév. *l.c.* [**G.C.:** Axim *Irvine* 2222 !
S.Nig.: Lagos *Punch* !—these, and other specimens from S. Nigeria lack flowers and fruits and cannot,
therefore, be determined certainly. The type specimen of *Daniellia punchii* Craib ex Holl. consists of the
leaves of this *Crudia* mixed with the flowers of *Daniellia ogea* (Harms) Rolfe ex Holl. (q.v.).] Also in Gabon.
2. **C. klainei** *Pierre ex De Wild.* in Bull. Jard. Bot. Brux. 7 : 250 (1920) ; Aubrév. *l.c.* 1 : 256, t. 99, 100 B ;
Kennedy For. Fl. S. Nig. 92 ; Pellegr. *l.c.* *C. senegalensis* of Oliv. in Hook. Ic. Pl. t. 2378, and of F.T.A.
2 : 312, partly, not of Planch. ex Benth. Tree to 25 ft. high, with spreading crown, in freshwater swamp
forest by creeks and lagoons ; stipules purple ; calyx mauve, filaments and ovary pale green, anthers
yellow.
Iv.C.: Île de Petit Bassam (fr. June) *Aubrév.* 538 ! **S.Nig.:** Lagos *Millen* 130 ! Epe *Barter* 3270 ! Eba
Isl. (fr. Jan.) *Thornewill* 242 ! Sapoba (Nov.) *Kennedy* 412 ! *A. F. Ross* 226 ! Nun R. (Sept.) *Mann* 500 !
Eket *Talbot* 3240 ! **F.Po:** *Mann* 14 ! Also in French Cameroons and Gabon. (See Appendix, p. 185.)
3. **C. gabonensis** *Pierre ex Harms* in Notizbl. Bot. Gart. Berl. App. 21 : 49 (1911) ; De Wild. *l.c.* 7 : 248 (1920) ;
Pellegr. *l.c.* *C.* sp. aff. *gabonensis* of Aubrév. *l.c.* 1 : 256, t. 100 C. Forest tree to 80 ft. high, with straight
bole and buttresses.
Iv.C.: Djibi *Aubrév.* 856 ! **G.C.:** Dunkwa (Sept.) *Vigne* FH 4740 ! Also in Gabon. (See Appendix,
p. 185.)
4. **C. sp. A.** *C. klainei* of F.W.T.A., ed. 1, 1 : 337, partly (*Dalz.* 8246), not of Pierre ex De Wild. Tall forest
tree. Noted in MS. by Harms as a new species ; further material and a thorough revision of the genus
are needed before deciding whether it is really new.
S.Nig.: Oban *Talbot* 580 ! **Br.Cam.:** Bibundi, very common (Nov.) *Mildbr.* 10645 ! Victoria to Buea (fr.
Feb.) *Dalz.* 8246 ! Also in French Cameroons.

35. **TALBOTIELLA** Bak. f. in J. Bot. 52 : 2 (1914) ; Léonard in Bull. Jard. Bot. Brux. 21 : 445 (1951).

A shrub up to 6 m. high ; leaves paripinnate ; leaflets about 15–20 pairs, opposite,
1·3–1·5 cm. long, auriculate on the lower side, obtuse at apex, glabrous ; racemes
axillary and terminal, slender, softly pilose ; pedicels 1·5 cm. long, very slender,
pilose ; flowers about 5 mm. long ; bracteoles linear ; sepals glabrous, elliptic ;
stamens 8–10, free ; ovary stipitate, villous   ..   ..   ..   1. *eketensis*
A tree ; leaves paripinnate ; leaflets about 6 pairs, opposite, about 2 cm. long and
1 cm. broad, auriculate on the lower side, rounded at apex, glabrous ; racemes
axillary, about 3 cm. long, pilose, with thick leathery perulae at the base ; pedicels
1 cm. long, pubescent ; sepals glabrous ; stamens 8–9 ; ovary stipitate, villous
                                                                                    2. *gentii*

1. **T. eketensis** *Bak. f.* in J. Bot. 52 : 2, t. 529 (1914) ; Leg. Trop. Afr. 3 : 768. A bushy shrub with dark
glossy leaves and white mature ; white flowers with orange stamens and pink bracteoles.
S.Nig.: Ibeno, at the estuary of the Kwa Ibo River *Talbot* 3188 ! Degema *Talbot* 3625 !
2. **T. gentii** *Hutch. & Greenway* in F.W.T.A., ed. 1, 1 : 337 (1928) ; Kew Bull. 1928 : 382 ; Bak. f. Leg. Trop.
Afr. 3 : 769. A heavy-foliaged tree, to 50 ft. ; flowers pink and white.
G.C.: *Irvine* 957 ! N. Scarp, Kwahu (Dec.) *Gent* FH 184 ! Abetifi, Kwahu (fr. June) *Thompson* 87 !
Huhunia *A. S. Thomas* D 57 ! (See Appendix, p. 199.)

36. **POLYSTEMONANTHUS** Harms in E. & P. Pflanzenfam. Nachtr. 1 : 197 (1897).

Tree with rusty-tomentellous branchlets and inflorescence ; leaves simply paripinnate ;
leaflets about 6 pairs, opposite, oblong, abruptly acuminate, rounded at base,
5–12 cm. long, 2·5–5 cm. broad, glabrous above, finely rusty-puberulous beneath ;
lateral nerves 10–12 pairs, prominent beneath ; panicles terminal ; bracteoles large,
persistent for a long time, enclosing the flower-bud, valvate, 2–2·5 cm. long, with a

gland-like horn at the apex ; sepals 4, imbricate, silky-villous outside ; petals 5, subequal ; stamens very numerous ; ovary shortly stipitate, villous ; fruit about 13 cm. long, 3 cm. broad, scythe-shaped, acute at one side   ..   ..   *dinklagei*

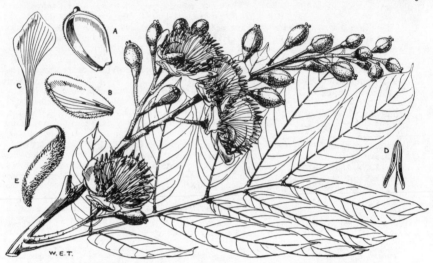

FIG. 151.—POLYSTEMONANTHUS DINKLAGEI *Harms* (CAESALPINIACEAE).
A, bracteole. B, sepal. C, petal. D, anther. E, ovary.

P. dinklagei *Harms* l.c. 1 : 197 (1897) ; Engl. Bot. Jahrb. 26 : 273, t. 7, D–F ; Bak. f. Leg. Trop. Afr. 3 : 693. Flowers fairly large in a stout panicled raceme, sweet-scented ; petals white ; stamens very numerous. **Lib.:** Grand Bassa (Sept.) *Dinklage* 1687 ! 1755 !   **Iv.C.:** *Aubrév.* 2812 ! 2813 !

37. ISOBERLINIA Craib & Stapf in Kew Bull. Add. Ser. 9 : 266 (1911) ; Kew Bull. 1912 : 93 ; Troupin in Bull. Jard. Bot. Brux. 20 : 302–309 (1950).

Branchlets, leaflets and inflorescence softly tomentellous ; leaflets 4 pairs, ovate or elliptic, rounded at apex and base, 8–15 cm. long, 5–9 cm. broad, with 7–8 pairs of lateral nerves ; stipules very large and cordate-auriculate, unequally 2-lobed at the apex, about 3·5 cm. long and broad, tomentose, soon falling off ; bracts ovate, tomentose, 7–8 mm. long ; flowers subsessile, obovoid in bud, the two enclosing bracteoles about 1·3 cm. long, densely tomentose ; petals subequal, scarcely exceeding the bracteoles ; fruit woody, obliquely oblong, flat, about 25 cm. long and 8 cm. broad, brown-velvety-pubescent   ..   ..   **1. *dalzielii***
Branchlets and leaflets glabrous ; leaflets usually 3 pairs and subopposite or occasionally more numerous and alternate, ovate or elliptic, unequally rounded or slightly cuneate at base, up to 20 cm. long and 10 cm. broad, shining and closely reticulate on both sides, with 6–7 pairs of lateral nerves ; bracts ovate, keeled, tomentellous, about 4 mm. long ; flowers subsessile ; bracteoles about 1 cm. long, thick, tomentellous outside ; petals and fruit as above, but the latter glabrous or nearly so   **2. *doka***

1. **I. dalzielii** *Craib & Stapf* in Kew Bull. Add. Ser. 9 : 267 (1911) ; Kew Bull. 1912 : 93 ; Aubrév. Fl. For. Soud.-Guin. 232 ; Troupin in Bull. Jard. Bot. Brux. 20 : 307. *Berlinia dalzielii* (Craib & Stapf) Bak. f. Leg. Trop. Afr. 3 : 690 (1930) ; Aubrév. Fl. For. C. Iv. 1 : 216. Savannah tree, to 60 ft. high, with grey or ochraceous tomentellous foliage and branchlets, bark exuding a red resin, comparatively small fragrant flowers and large 5–6-seeded pods which burst with curling valves scattering the seeds ; gregarious, usually on stony ground.
    Fr.Sud.: Bamako to Bougouni *Aubrév.*   **Fr.G.:** Siguiri (Nov.) *Jac.-Fel.* 1413 !   **Iv.C.:** *Aubrév.* 630 ! Boromo *Aubrév.* 2151.   **G.C.:** Gambaga Scarp, Sakogu (Dec.) *Adams* GC 4240 ! Yijia to Pong *Kitson* 570 !   **Togo:** Sokodé to Basari *Kersting* A. 702 !   **Dah.:** Parakou *Poisson* 50 !   **N.Nig.:** Zamfara F.R. (fr. Apr.) *Keay* FHI 16141 ! Kontagora (Nov.) *Dalz.* 26 ! Zaria Prov. (fr. Apr.) *Collier* ! Anchau (Nov.) *Keay* FHI 21672 ! Bauchi *Shaw* 96 !   **S.Nig.:** Old Oyo *Keay* FHI 16280 ! Also in French Cameroons, Ubangi-Shari. (See Appendix, p. 195.)
2. **I. doka** *Craib & Stapf* in Kew Bull. Add. Ser. 9 : 267 (1911) ; Kew Bull. 1912 : 94 ; Hook. Ic. Pl. t. 3003 ; Aubrév. Fl. For. Soud.-Guin. 232, t. 48 ; Troupin l.c. 20 : 303. *Berlinia doka* (Craib & Stapf) Bak. f. l.c. 3 : 688 (1930) ; Aubrév. Fl. For. C. Iv. 1 : 216. *B. kerstingii* Harms (1915). *B. chevalieri* De Wild. (1925). *B. angolensis* of Chev. Bot. 230, not of Welw. ex Benth. Savannah tree, to 60 ft. high, with glabrous shining foliage ; flowers white ; gregarious.
    Fr.Sud.: Koumana (Mar.) *Chev.* 612 ! Birgo *Dubois* 48 !   **Fr.G.:** Siguiri *Pobéguin* 880 ! *Chev.* 302 ! Kankan *Chev.* 583 ! Kouroussa *Pobéguin* 189 !   **Iv.C.:** *fide* Aubrév. *ll.c.*   **G.C.:** Kintampo (Mar.) *Dalz.* 45 ! Aframsu, Kwahu (young fr. Jan.) *Chipp* 638 ! Kulmasa to Jagalo (Feb.) *Kitson* 568 ! Samboro to Basisan (May) *Kitson* 569 !   **Togo:** *Kersting* A. 389.   **Dah.:** Badagba *Poisson* 105 !   **N.Nig.:** Sokoto Prov. *Dalz.* 334 ! Katagum *Dalz.* 364 ! Zaria (Apr.) *Lamb* 51 ! Kontagora (Feb.) *Dalz.* 12 ! Bauchi Plateau (May) *Lely* P. 287 !   **S.Nig.:** Oyo-Iseyin Road (Feb.) *Keay* FHI 22478 ! Also in French Cameroons, Ubangi-Shari, A.-E. Sudan, Uganda and Belgian Congo. (See Appendix, p. 195.)

38. BERLINIA Soland. ex Hook. f.—F.T.A. 2 : 292 ; Keay in Kew Bull. 1954 : 267.

Petals more or less subequal in length, all about 6–8 cm. long, limb of posterior petal broader than the others ; inflorescence a stout terminal simple raceme ; bracts at least 1·6 cm. long, conspicuous in the unopened part of the inflorescence ; calyx-tube glabrous or sparingly puberulous outside ; fruits glabrous :

Bracts 5–7 cm. long, 1·8–2·5 cm. broad, enclosing the flower-buds until opening, the unopened part of the inflorescence thus head-like and wrapped in the imbricating bracts ; bracteoles elliptic-oblanceolate, 5–8 cm. long, 1–2 cm. broad, sparingly puberulous to glabrous within ; pedicels 1–3 cm. long ; leaflets 4–5 pairs, sub-opposite, obovate-elliptic rounded or obtuse at base, acuminate, 10–36 cm. long, 4–13·5 cm. broad, with 10–15 pairs of lateral nerves ; rhachis up to 30 cm. long, petiolules 6–10 mm. long ; fruits with stipe 2–2·5 cm. long, oblong, 27–36 cm. long, 9–10 cm. broad, upper suture 1 cm. broad, marked by a double groove   1. *bracteosa*

Bracts up to 2·7 cm. long and 1·4 cm. broad, enclosing only the very young buds, the older buds thus conspicuous ; bracteoles obovate-oblanceolate, 4–5 cm. long, 1·5–2 cm. broad, densely puberulous to tomentellous within ; pedicels 2–4·8 cm. long ; leaflets 4–5 pairs (rarely 1–3), subopposite, obovate-elliptic to oblong-elliptic, rounded or obtuse at base, acuminate, 9–25 cm. long, 3·2–9 cm. broad, with 10–13 pairs of lateral nerves ; rhachis up to 30 cm. long, petiolules up to 1 cm. long ; fruits with stipe 2–3 cm. long, oblong, 28–36 cm. long, 9–11 cm. broad, upper suture grooved   ..   ..   2. *occidentalis*

Petals very unequal in length, the posterior one much longer and broader than the others ; inflorescence a cluster of racemes or sparingly branched panicles ; bracts up to 0·5 cm. long ; fruits tomentellous or minutely puberulous, rarely glabrous :

Bracteoles densely silky-villous inside, tomentellous outside, about 3 cm. long ; calyx-tube glabrous or pubescent ; posterior petal about 4–6 cm. long ; leaflets 3–6 pairs, elliptic-obovate or oblong-elliptic, subfalcate, acuminate at apex, 8–16 cm. long, 3–6 cm. broad, puberulous or nearly glabrous beneath ; fruits 20–30 cm. long, 5–7 cm. broad, tomentellous   ..   ..   3. *grandiflora*

Bracteoles shortly pubescent, puberulous, or nearly glabrous inside, usually more densely hairy outside :

Calyx-tube and lobes glabrous or sparingly puberulous outside, or rather densely puberulous down one side only :

Bracteoles 1·8–2·2 (–2·7) cm. long, 0·8–1·1 cm. broad ; calyx-tube 7–8 mm. long ; posterior petal 4·5–5 cm. long, 2·5–3 cm. broad ; inflorescence a cluster of racemes, up to 10 cm. long, pedicels 1–1·5 cm. long ; inflorescence and outside of bracteoles finely puberulous, the surface usually being clearly visible between the hairs ; leaflets 2–3 pairs, thinly coriaceous, obliquely oblong or obovate-oblong or oblanceolate-oblong, (6–) 12–25 cm. long, (3·5–) 4·5–9·5 cm. broad, the upper pair larger than the lower pair or pairs ; fruits not known   ..   4. *auriculata*

Bracteoles 3–4·2 cm. long, 1–2·1 cm. broad ; calyx-tube 1–1·7 cm. long ; posterior petal 3·5–6 cm. broad ; pedicels 1·2–3·5 cm. long :

Outside of bracteoles densely minutely tomentellous ; leaflets papery or very thinly coriaceous :

Leaflets (2–) 3–5 pairs, the surfaces discolorous when dry, ultimate venation usually obscure, obliquely oblong to oblong-ovate, acuminate and usually emarginate at apex, 5·5–12 (–17) cm. long, 3–6·5 (–9) cm. broad ; bracteoles 3–3·7 cm. long, 1–1·5 cm. broad ; calyx-tube usually more or less puberulous ; posterior petal 4·5–5 cm. long, 3·5–4·5 cm. broad ; fruits up to 45 cm. long and 9 cm. broad, minutely scurfy-puberulous   ..   ..   5. *confusa*

Leaflets 2 (–3) pairs, the surfaces concolorous when dry, ultimate venation prominulous on both surfaces, oblong-elliptic or oblong-obovate, shortly and obtusely acuminate, 10–16 cm. long, 6–8 cm. broad ; bracteoles 3·7–4·2 cm. long, 8–11 mm. broad ; calyx-tube usually with a densely puberulous line down the back ; posterior petal 4–4·5 cm. long, about 5 cm. broad ; fruits about 32 cm. long, and 8–9 cm. broad, densely yellow-brown tomentellous

6. *congolensis*

Outside of bracteoles very finely puberulous ; leaflets thickly coriaceous, (1–) 2–3 pairs, oblong, 10–25 cm. long, 5–13 cm. broad ; bracteoles 3·5–3·8 cm. long and 1·7–2·1 cm. broad ; posterior petal 6·5–7 cm. long and 4·5–6 cm. broad ; fruits 33–47 cm. long, 11–12 cm. broad, minutely pustulate, nearly glabrous

7. *coriacea*

Calyx-tube and lobes (at least in part) densely tomentose outside ; bracteoles puberulous or tomentellous within ; inflorescences densely tomentellous :

Bracteoles 2·3–2·8 cm. long, 1–1·5 cm. broad ; pedicels 1·4–2·8 cm. long ; posterior petal (3·4–) 4–4·5 (–5) cm. long, 2·5 cm. broad, deeply 2-lobed ; inflorescence to about 16 cm. long ; leaflets (2–) 3–5 pairs, oblong-elliptic, to obovate-elliptic or ovate-elliptic, often long-acuminate at apex, 7–25 cm. long, 3–10 cm. broad

8. *tomentella*

Bracteoles 4·5–6 cm. long, 1·4–2·8 cm. broad ; pedicels 4–6 cm. long ; posterior petal 6–7 cm. long, 5·5–8 cm. broad ; inflorescence up to 30 cm. long ; midrib and lateral nerves puberulous beneath :

Bracteoles obovate, 4·5–4·8 cm. long, 2–2·8 cm. broad ; leaflets 4–6 pairs, narrowly oblong-elliptic to obovate-elliptic, long-acuminate at apex, 5·5–18 cm. long, 2·8–8 cm. broad ; young fruits brown-puberulous, not known at maturity

                                                    9. *craibiana*

Bracteoles oblanceolate, about 6 cm. long and 1·4 cm. broad ; leaflets 3–4 pairs, elliptic-lanceolate, very long-tapered-acuminate at apex, up to 16 cm. long and 6 cm. broad ; fruits not known ..   ..   10. *hollandii*

1. **B. bracteosa** *Benth.*—F.T.A. 2 : 294 ; Bak. f. Leg. Trop. Afr. 3 : 686 ; Pellegr. Lég. Gab. 62. *Westia bracteosa* (Benth.) J. F. Macbr. in Contrib. Gray Herb. n.s. 3, 59 : 20 (1919). *Macroberlinia bracteosa* (Benth.) Hauman (1952)—Fl. Congo Belge 3 : 386. Forest tree, to 80 ft. high ; glabrous except on the hoary puberulous inflorescence ; flowers white, crowded in a stout terminal raceme.
S.Nig.: Oban *Talbot* 1312 ! Afi River F.R. (June) *Jones & Onochie* FHI 17340 ! Cross River North F.R. (fr. Dec.) *Keay* FHI 28153 ! **Br.Cam.:** Victoria (Jan., Feb.) *Winkler* 1087 ! *Deistel* 460 ! *Maitland* 276 ! Likomba *Mildbr.* 10729 ! Barombi *Preuss* 330 ! Bai Mauya, Kumba (fr. June) *Rosevear* 92/37 ! **F.Po:** *Mann* 1434 ! Also in French Cameroons, Gabon, Cabinda and Belgian Congo. (See Appendix, p. 176.)
2. **B. occidentalis** *Keay* in Kew Bull. 1954 : 269. *B. acuminata* Soland. ex Hook. f. in Fl. Nigrit. 326 (1849), partly (*Ansell* spec.) ; F.T.A. 2 : 293, partly (*Ansell* spec.). *B. bracteosa* of F.W.T.A., ed. 1, 1 : 343, partly (spec. ex Axim) ; of Aubrév. Fl. For. C. Iv. 1 : 222, t. 82, B ; not of Benth. *B. grandiflora* of F.W.T.A., ed. 1, 1 : 343, partly (spec. ex S.L., Lib. & G.C.) ; of Bak. f. l.c. 686, partly ; not of (Vahl) Hutch. & Dalz. Forest tree, to 80 ft. high ; flowers white, in a stout terminal raceme ; fruits dark brown, glabrous.
S.L.: Kambui (Apr.) *Lane-Poole* 244 ! Southern Prov. (Apr.) *Deighton* 1151 ! Gorahun (June) *Small* 725 ! **Lib.:** Bassa Cove *Ansell* ! Dukwia R. (fl. Feb., fr. July) *Linder* 173 ! *Cooper* 177 ! 217 ! Sasstown (Mar.) *Baldwin* 11604 ! White Plains (fl. Apr., fr. Nov.) *Barker* 1271 ! *Baldwin* 10458 ! **Iv.C.:** Aboisso (Apr.) *Chev.* 16300 ! Abidjan *Aubrév.* 100 ! **G.C.:** Axim *Burton & Cameron* ! Princestown (fl. & fr. Apr.) *Chipp* 173 ! (See Appendix, p. 176.)
3. **B. grandiflora** (*Vahl*) *Hutch. & Dalz.* F.W.T.A., ed. 1, 1 : 343 (1928), partly (as to syn. only) ; Kew Bull. 1928 : 398 ; Bak. f. l.c. 686, partly ; Keay l.c. 269. *Westia grandiflora* Vahl in Skrivt. Nat. Selsk. 6 : 118 (1810). *B. acuminata* Soland. ex Hook. f. in Fl. Nigrit. 326 (1849), partly (*Heudelot* spec., and most of description). *B. heudelotiana* Baill. in Adansonia 6 : 185, t. 3, 8–9 (1865) ; F.W.T.A., ed. 1, 1 : 343 ; Bak. f. l.c. 3 : 684 (incl. var. *foliosa* Bak. f.) ; Aubrév. l.c. 220, t. 81 ; Pellegr. l.c. 63, partly ; Hauman l.c. 3 : 396. *B. acuminata* var. *heudelotiana* (Baill.) Oliv. F.T.A. 2 : 294 (1871). Tree to 70 ft. high, or shrub, usually in fringing forest ; bracts and calyx pale green, petals white ; fruits cinnamon-brown, tomentellous.
Fr.Sud.: Bamako (Apr.) *Waterlot* 1108 ! *Hagerup* 40a ! **Fr.G.:** Bangalan, Upper R. Pongo *Heudelot* 886 ! Kouroussa *Pobéguin* 861 ! **S.L.:** Mano (Apr.) *Deighton* 4021 ! Mange Bure (Feb.) *Deighton* 3491 ! **Iv.C.:** Bouaké to Tiébissou *Aubrév.* 826 ! **G.C.:** Ajura (May) *Vigne* FH 1167 ! Lissa to Lambussie (Mar.) *Kitson* 572 ! Gambaga (Mar.) *Williams* 489 ! Salaga to Tamale (Apr.) *Dalz.* 1 ! Bole (May) *Vigne* FH 3834 ! Wa to Black Volta (Apr.) *Kitson* ! **Togo:** *Kersting* 33 ! Misahöhe *Baumann* 122 ! Hohoe (fl. & fr. Feb.) *Beveridge* FH 2925 ! *Kpandu Robertson* 106 ! **Dah.:** Whydah *Isert* ! **N.Nig.:** S. Sokoto Prov. (June) *Lely* 819 ! Zungeru (Apr.) *Dalz.* 34 ! Zaria Prov. (Apr.) *Collier* ! Wamba (Mar.) *Thornewill* 53 ! **S.Nig.:** Lagos (fl. & young fr. Feb.) *Foster* 6 ! Ijaye *Barter* 3327 ! Benin City *Dennet* 62 ! Ububulu (Jan.) *Thomas* 2263 ! Akpaka F.R., Onitsha (fr. Apr.) *Chesters* 20 ! Ekwolobia, Awka (Nov.) *Keay* FHI 22707 ! Ikodu, Ahoada (Feb.) *Chesters* 9 ! 10 ! Ikom *Catterall* 52 ! **Br.Cam.:** Munkep, Bamenda Prov., 3,000 ft. (Feb.) *Johnstone* 35/31 ! Munken, Bamenda Prov., 3,000 ft. (Feb.) *Johnstone* 77/31 ! Also in Ubangi-Shari, French Cameroons and Belgian Congo. (See Appendix, p. 176.)
4. **B. auriculata** *Benth.*—F.T.A. 2 : 294 ; F.W.T.A., ed. 1, 1 : 343, in part only ; Bak. f. l.c. 682, partly ; Pellegr. l.c. 64 (excl. ref. Aubrév. Fl. For. C. Iv.). *B. heudelotiana* of Pellegr. l.c. 63, partly. *Westia auriculata* (Benth.) J. F. Macbr. (1919). Forest tree, to 30 ft. high ("large tree", *fide* Chesters), with low spreading crown ; flowers white.
S.Nig.: Degema (Feb.–Mar.) *Talbot* 3611 ! *Chesters* 11 ! 12 ! 13 ! 14 ! Eket *Talbot* 3341 ! Port Harcourt (Jan.) *Jones* FHI 6196 ! Old Calabar *Robb* ! Oban *Talbot* 79 ! 1556 ! **Br.Cam.:** Rio del Rey *Johnston* ! R. Cameroons (Jan.) *Mann* 2195 ! Also in French Cameroons and Gabon. (See Appendix, p. 176.)
5. **B. confusa** *Hoyle* in Kew Bull. 1934 : 184 ; Keay l.c. 271. *B. acuminata* of Fl. Nigrit. 326 (1849), partly (ref. Hb. Banks. ms.) ; Bak. f. l.c. 683 (excl. var. *velutina*) ; Aubrév. Fl. For. C. Iv. 1 : 220, t. 82, A ; not of Soland. ex Hook. f. *B. auriculata* of F.W.T.A., ed. 1, 1 : 343, partly, not of Benth. Forest tree, 40–120 ft. high ; petals white ; fruits pale brown.
S.L.: *Smeathmann* ! Hangha (May) *Lane-Poole* 229 ! Laoma (Apr.) *Aylmer* 223 ! Njala (Apr.) *Deighton* 647 ! Peri, Guara (Mar.) *King* 330 ! Mongheri, Lunia (Apr.) *King* 234 ! Gola Forest (fr. May) *Small* 711 ! **Lib.:** Monrovia *Whyte* ! Dukwia R. (Feb.) *Cooper* 218 ! 283 ! Fisherman's Lake (Dec.) *Baldwin* 10892 ! **Iv.C.:** *Aubrév.* 3 ! Bingerville (Feb.) *Chev.* 17269 ! **G.C.:** *Burton & Cameron* ! Kumasi (Mar., Apr.) *Vigne* FH 1055 ! 1676 ! Tano Anwia F.R. (Jan.) *Andoh* FH 5634 ! **S.Nig.:** Sapoba *Kennedy* 800 ! Usonigbe F.R. (fl. Aug., fr. Nov.) *Adekunle* FHI 22647 ! Meikle *& Keay* 560 ! Cross R. *McLeod* ! Eket *Talbot* 3340 ! Calabar (Mar.) *Smith* 54 ! Also in French Cameroons and Gabon. (See Appendix, p. 176.)
6. **B. congolensis** (*Bak. f.*) *Keay* in Kew Bull. 1954 : 271. *B. heudelotiana* var. *congolensis* Bak. f. Leg. Trop. Afr. 3 : 684 (1930). *B. auriculata* of F.W.T.A., ed. 1, 1 : 343, partly (*Talbot* 3110). *B. acuminata* of Pellegr. Lég. Gab. 64, partly (*Le Testu* 1713). Forest tree, to 120 ft. high ; fruits mustard-brown ; on swampy ground by rivers.
S.Nig.: Shasha F.R. (fl. Apr.–May, fr. May) *Richards* 3408 ! 3487 ! *Ross* 179 ! Eket *Talbot* 3110 ! Also in Gabon, Cabinda and Belgian Congo.
7. **B. coriacea** *Keay* l.c. 272 (1954). *B. grandiflora* of Kennedy For. Fl. S. Nig. 98. Shrub or small forest tree, to 40 ft. high ; in marshy places.
S.Nig.: Fowa, Omo F.R. (Mar.) *Jones & Onochie* FHI 17031 ! Jamieson R., Sapoba *Kennedy* 1656 ! 1694 ! 1992 ! 2180 !
8. **B. tomentella** *Keay* l.c. 273 (1954). *B. acuminata* var. *velutina* A. Chev. ex Bak. f. l.c. 684 (1930) ; Chev. Bot. 229, name only. *B. auriculata* of Aubrév. l.c. 222, t. 83, not of Benth. Forest tree, to 50 ft. high ; flowers white ; in wet places.
S.L.: Pujehun (Apr.) *Aylmer* 61 ! **Lib.:** Mnanulu, Webo (June) *Baldwin* 6050 ! Sarbo, Webo (fl. & fr. July) *Baldwin* 6376 ! 6402 ! **Iv.C.:** Soubré (June) *Chev.* 19149 ! Djibi (June) *Aubrév.* 1360 ! **G.C.:** Essuasu (May) *Vigne* 116 ! Kumasi (July) *Vigne* FH 3009 ! Ayawura, N.W. Axim (July) *Chipp* 258 ! Nkawkaw (Feb.) *King-Church* 7 !
9. **B. craibiana** *Bak. f.* in Cat. Talb. 27 (1913) ; Leg. Trop. Afr. 3 : 686 ; Pellegr. l.c. 65. *B. preussii* De Wild. (1925). *B. acuminata* of F.T.A. 2 : 293, partly (*Mann* and *Thomson* spec.). *B. grandiflora* of F.W.T.A., ed. 1, 1 : 343, partly (spec. ex Abukpani). Forest tree, to 100 ft. or more high ; bracteoles pale green, petals white.
S.Nig.: Sapoba (Jan.) *Kennedy* 1970 ! 2098 ! 2577 ! Ahoada (Jan.) *Chesters* 23 ! Abukpani, N. of

Calabar (Feb.) *McLeod*! Old Calabar (fl. & young fr. Feb.) *Mann* 2302! *Thomson* 13! Oban *Talbot* 1520! 1524! 1584! 1732! **Br.Cam.**: *Preuss* 1317! R. Cameroons (Jan.) *Mann* 733! Kumba (Jan.) *Jeme* 4/38! Also in French Cameroons and Gabon. [Specimens from N. Rhodesia have been named *B. craibiana*, but they are not identical with our material.]

10. **B. hollandii** *Hutch. & Dalz.* F.W.T.A., ed. 1, 1 : 343 (1928) ; Kew Bull. 1928 : 398 ; Bak. f. l.c. 682. Forest tree, 20–30 ft. high.
    S.Nig.: Old Calabar (May) *Holland* 10! Itu *Holland* 30!

39. MICROBERLINIA A. Chev. in Rev. Bot. Appliq. 26 : 588 (1946).

A large tree ; branchlets pubescent ; leaflets opposite, 12–18 pairs, oblong, with the midrib oblique and the very short petiolule close to the rounded proximal side at the base, distal side of base of leaflets held near the rhachis, apex rounded and emarginate, 1·5–3·5 cm. long, 8–10 mm. broad, both surfaces glabrous, lateral nerves numerous and rather close ; rhachis 15–18 cm. long, narrowly winged between the petioles ; stipules deciduous, foliaceous, ovate-lanceolate, 1·5–2 cm. long, 4–8 mm. broad ; panicles axillary, with a more or less straight 8–16 cm. long central rhachis, bearing secundly and alternately 8–12 lateral racemes of up to 3 cm. in length ; flowers crowded, pedicels up to 3 mm. long ; bracteoles obovate, 4–7 mm. long, tomentellous outside, sparsely puberulous within ; receptacle tube 3·5 mm. long, glabrous ; sepals 4, oblong-lanceolate, the posterior sepal formed by the fusion of 2 and so a little broader than the others ; petals 5, subequal, oblong-spathulate ; pods oblong woody, flat, 16 cm. long, 4 cm. broad, the upper suture thickened and marked by a ridge running parallel to, and 5–10 mm. from, it ; seeds 3–4   ..   ..   *bisulcata*

M. bisulcata *A. Chev.* l.c. 26 : 589, t. 21 B (1946) ; Pellegr. Lég. Gab. 67. *Berlinia bisulcata* (A. Chev.) Troupin (as *bifurcata* by error) in Bull. Jard. Bot. Brux. 20 : 302 (1950). Forest tree, to 130 ft. high ; with extensive buttresses ; bracteoles pale green, petals and stamens white.
    Br.Cam.: *Kennedy* 3127! Malende (Mar.) *Maitland* 1067! *Bikom* FHI 12228! Mbalange (Jan.) *Lobe Babute* 57/36! S. Bakundu F.R. (Mar.) *Brenan* 9319! Also in French Cameroons.

40. JULBERNARDIA Pellegr. in Boissiera 7 : 298 (1943).

Leaflets usually 3 pairs (rarely 2 or 4), with the lowest pair inserted close to the base of the 4–11 cm. long rhachis, petiolules twisted ; leaflets elliptic to obovate-elliptic, acute at each end, 4–20 cm. long, 2–8·5 cm. broad, the lowest pair always the smallest, glabrous, with numerous translucid gland-dots ; inflorescence and outside of flowers densely golden-brown tomentose ; flowers with pedicels 3–5 mm. long ; sepals 5, free, subequal ; petals 5, of which 4 are minute and the other is obovate, about 17 mm. long and 6 mm. broad ; fruit a flat woody pod, obtuse at apex, 12–20 cm. long, 4·5–5·5 cm. broad, glabrous ; seeds 1–5   ..   ..   *seretii*

J. seretii (*De Wild.*) Troupin in Bull. Jard. Bot. Brux. 20 : 312, fig. 32 (1950) ; Hauman in Fl. Congo Belge 3 : 400, fig. 32. *Berlinia seretii* De Wild. (1907) ; Bak. f. Leg. Trop. Afr. 3 : 690. *Julbernardia ogoouensis* Pellegr. (1943)—Lég. Gab. 69. Forest tree, to 130 ft. high, with buttresses ; flowers strongly scented ; petals white or light pink.
    S.Nig.: Oban *Talbot* 1586! **Br.Cam.**: Victoria to Kumba road (Mar.) *Onochie* FHI 32084! S. Bakundu F.R. (Apr.) *Ejiofor* FHI 29310! Also in French Cameroons, Gabon, Cabinda and Belgian Congo.

41. PARABERLINIA Pellegr. in Bull. Soc. Bot. Fr. 90 : 79 (1943).

Leaflets 1 pair, narrowly oblong-falcate, unequal at base, gradually acute at apex, 10–17 cm. long, 3–4 cm. broad, glabrous, petiole about 8 mm. long, petiolules very short ; inflorescence and outside of flowers densely golden-brown tomentose ; flowers mostly sessile or with pedicels up to 5 mm. long ; sepals 5, free, subequal ; petals 5, minute ; fruit a flat woody pod, 13–15 cm. long, 4–4·7 cm. broad, glabrous, seeds 2–3
                                                       bifoliolata

P. bifoliolata *Pellegr.* in Bull. Soc. Bot. Fr. 90 : 79 (1943) ; Lég. Gab. 68. *Julbernardia pellegriniana* Troupin in Bull. Jard. Bot. Brux. 20 : 311 (1950). Forest tree, to about 150 ft. high, with buttresses and straight bole.
    Br.Cam.: Banga, Kumba Div. (fl. buds Mar., fr. Aug.) *Brenan* 9312! *Ejiofor* FHI 15255! *Olorunfemi* FHI 30706! Also in Gabon.

42. ANTHONOTHA P. Beauv. Fl. Oware 1 : 70 (1806).

*Leaflets glabrous beneath (or if puberulous then inflorescence very long) ; inflorescences mostly very long, narrow and pendulous :
Leaf-rhachis winged, 4–28 cm. long ; leaflets 2–4 pairs, the lowest pair very close to the stem, elliptic-oblanceolate, gradually narrowed to the base, acuminate at apex, 8–32 cm. long, 3·5–11 cm. broad ; inflorescence borne on the trunk and older branches, pendulous, axis 30–90 cm. long, lateral branches very short ; bracteoles 9–10 mm. long ; petals 5, subequal, 11–12 mm. long ; fruit oblong, about 23 cm. long and 7·5 cm. broad, brown-tomentellous, with 6–8 seeds ..   1. *leptorrhachis*
Leaf-rhachis not winged :
†Petals 4–6, subequal :
    Inflorescence generally under 20 cm. long ; bracteoles 7–8 mm. long ; leaflets 2–3 pairs, elliptic to lanceolate-elliptic, generally cuneate at base, gradually caudate-acuminate at apex, 7–18·5 cm. long, 3–7 cm. broad, with 5–6 pairs of arcuate,

ascending, lateral nerves, rather prominent beneath ; rhachis 4–10 cm. long ;
fruits not known   ..   ..   2. *isopetala*
Inflorescence 20–120 cm. long :
Stipules persistent, oblanceolate, obtuse, leathery, 5–7 mm. long ; leaflets 2 pairs,
   ovate to obovate, rounded to cuneate at base, 7–12·5 cm. long, 4–6 cm. broad,
   with about 5 pairs of arcuate, ascending, lateral nerves, prominent beneath ;
   rhachis 3·5–5·5 cm. long ; axis of inflorescence about 30 cm. long, lateral branches
   up to 4·5 cm. long ; bracteoles about 1 cm. long ; petals 5 ; fruits not known
                                                                  3. *elongata*
Stipules very early caducous or small and inconspicuous :
Lateral nerves or leaflets 9–16, prominent beneath and conspicuously looped near
   the margin ; leaflets 2–3 pairs, elliptic to obovate-lanceolate, obtuse at base,
   shortly acuminate at apex, 9–27 cm. long, 4–9 cm. broad, puberulous to
   glabrescent ; rhachis 6·5–16 cm. long ; axis of inflorescence 25–90 cm. long ;
   lateral branches up to 1·5 cm. long ; bracteoles 7–11 mm. long ; petals 5,
   8–9 mm. long ; fruits obliquely oblong, about 10 cm. long and 3 cm. broad,
   densely tomentellous, with numerous transverse ridges ; seeds 3–4
                                                                  4. *nigerica*
Lateral nerves of leaflets 7–9, not conspicuously looped ; leaflets 3 (rarely 4)
   pairs, mostly obovate or elliptic rounded or obtuse at base, shortly acuminate
   or rounded at apex, upper leaflets 15–20 cm. long, 6–8 cm. broad, lower leaflets
   6·5–10 cm. long, 3·5–5 cm. broad ; rhachis 9–20 cm. long ; inflorescences on
   shoots with reduced leaves arising from older twigs, axis 25–60 cm. long, lateral
   branches up to 7 cm. long, rather distant ; bracteoles 1–1·4 cm. long ; petals
   4–5, 1–1·5 cm. long ; fruits not known   ..   ..   ..   5. *obanensis*
†Petals 5, of which 3 are 0·7–2 cm. long and subequal, and the other 2 minute :
Bracteoles about 1 cm. long ; developed petals up to 2 cm. long ; inflorescence axis
   to 120 cm. long, with lateral branches up to 4 cm. long ; leaflets 2–3 pairs, obovate-
   oblanceolate to oblong-oblanceolate, cuneate at base, acuminate at apex, upper
   pair up to 22 cm. long and 9·5 cm. broad, lower pair up to 7 cm. long and 3·5 cm.
   broad ; fruit not known   ..   ..   6. *ernae*
Bracteoles 5–8 mm. long ; developed petals 7–8 mm. long ; inflorescence axis
   22–45 cm. long, with very short lateral branches ; leaflets 2–3 pairs, ovate or
   oblong-elliptic, rounded to cuneate at base, acuminate at apex, 4·5–12 cm. long,
   2·5–5 cm. broad, lateral nerves 4–6 pairs ; mature fruit not known   7. *explicans*
*Leaflets very shortly tomentellous or tomentose, or pubescent beneath ; inflorescence
   usually up to 35 cm. long :
Petals 4, all developed and subequal, up to 8 mm. long, or only 1–3 developed and the
   others minute ; leaflets reddish-brown pubescent beneath especially near midrib and
   nerves, sometimes eventually almost glabrous ; leaflets 2–4 pairs, ovate to elliptic,
   rounded to obtuse at base, gradually caudate-acuminate, 7–18 cm. long, 3–5 cm.
   broad, with 5–8 pairs of lateral nerves, ascending, arcuate, looped and rather
   prominent beneath ; rhachis 3–12 cm. long ; flowers numerous in terminal panicles
   5–15 cm. long ; bracteoles 7–10 mm. long ; fruits oblong, thick, 10–12 cm. long,
   4–5·5 cm. broad ; seeds 1–2, large ..   ..   8. *vignei*
Petals 5, only the posterior one developed, the other 4 minute ; leaflets very shortly
   tomentellous, or tomentose beneath :
Undersurface of leaflets ferrugineous-tomentose, with lateral nerves prominent ;
   branchlets densely ferrugineous-tomentose :
Bracteoles 5–8 mm. long ; posterior petal 7–8 mm. long ; inflorescence axillary,
   terminal or on old wood, axis 8–35 cm. long, with numerous lateral racemes or
   panicles 2–12 cm. long ; leaflets 2–5 pairs, elliptic, obovate-elliptic or oblong,
   rounded or obtuse at base, shortly acuminate at apex or more often appearing
   rounded by the destruction of the tip, 6–38 cm. long, 4–13·5 cm. broad, densely
   ferrugineous-tomentose beneath, with 12–24 pairs of lateral nerves prominent
   beneath ; rhachis 5–35 cm. long ; fruits oblong, thick, 3·5–12 cm. long, 4·5–6 cm.
   broad, 1·5–3 cm. thick, tomentellous, with numerous prominent anastomosing
   ridges, tardily dehiscent, valves not curling ; seeds 1–3, large   .. 9. *fragrans*
Bracteoles 11–12 mm. long ; posterior petal up to 1·5 mm. long ; inflorescences on
   the wood of several seasons' growth, axis 9–30 cm. long, with rather isolated
   lateral racemes up to 6 cm. long ; leaflets 4–7 pairs, more or less elongate-oblong,
   rounded or obtuse at base, caudate-acuminate, 15–28 cm. long, 5–7 cm. broad,
   densely silky-tomentose beneath, with 9–16 pairs of lateral nerves, rhachis 25–30
   cm. long ; fruits not known   ..   ..   10. *lamprophylla*
Undersurface of leaflets very shortly tomentellous and silvery ; lateral nerves only
   slightly prominent beneath ; leaflets usually 3 pairs, elliptic, oblong or obovate,
   up to 30 cm. long and 12 cm. broad ; branchlets grey, puberulous or glabrescent ;
   inflorescences on young or 1–2 year old shoots :
Upper suture of fruit about 5 mm. broad, not broader than the lower ; fruits more or
   less flattened, 15–23 cm. long, 4–6·5 cm. broad, usually rather dark brown-

tomentellous, 2–7-seeded, tardily dehiscent ; inflorescences 2–28 cm. long, rather laxly flowered ; bracteoles 6–11 mm. long ; leaflets acuminate to long-acuminate
                                     11. *macrophylla*

Upper suture of fruit at least 1·2 cm. broad, distinctly broader than the lower ; fruits stout, 7–15 cm. long, 3·5–8 cm. broad, pale brown-tomentellous, 1–4-seeded, apparently indehiscent, on short stout stipes ; inflorescences up to 4 cm. long, very densely flowered ; bracteoles 5–6 mm. long ; leaflets very short acuminate or more often appearing rounded by the destruction of the tip .. 12. *crassifolia*

1. **A. leptorrhachis** (*Harms*) *J. Léonard* in Bull. Jard. Bot. Brux. 25 : 202 (1955). *Macrolobium leptorrhachis* Harms in Engl. Bot. Jahrb. 33 : 157 (1902) ; Pellegr. Lég. Gab. 46. Forest tree, to 50 ft. high, cauliflorous.
    Br.Cam.: S. Bakundu F.R. (young fr. Aug.) *Olorunfemi* FHI 30737 ! Also in French Cameroons.
2. **A. isopetala** (*Harms*) *J. Léonard* l.c. 202 (1955). *Macrolobium isopetalum* Harms in Engl. Bot. Jahrb. 40 : 25 (1907) ; Pellegr. l.c. 44. Small forest tree ; flowers white.
    Br.Cam.: Mamfe (Feb.) *Lobe Babute* 6/37 ! Also in French Cameroons and Gabon.
3. **A. elongata** (*Hutch*) *J. Léonard* l.c. 202 (1955). *Macrolobium elongatum* Hutch. in Kew Bull. 1916 : 229. A small branching tree with long narrow finely brownish-puberulous pendulous inflorescences of white flowers.
    S.L.: Pujehun (Feb.) *Lane-Poole* 161 !
4. **A. nigerica** (*Bak. f.*) *J. Léonard* l.c. 202 (1955). *Macrolobium leptorrhachis* Harms var. *nigericum* Bak. f. in Cat. Talb. 29 (1913) ; Leg. Trop. Afr. 3 : 671. *M. nigericum* (Bak. f.) J. Léonard in Fl. Congo Belge 3 : 416 (1952). Small forest tree.
    S.Nig.: Oban *Talbot* 582 ! Also in Belgian Congo.
5. **A. obanensis** (*Bak. f.*) *J. Léonard* in Bull. Jard. Bot. Brux. 25 : 203 (1955). *Macrolobium obanense* Bak. f. in Cat. Talb. 28 (1913) ; Leg. Trop. Afr. 3 : 670. Forest tree, to 32 ft. high, with long, pendulous dark brown inflorescences ; petals white.
    S.Nig.: Idanre Hills (Oct.) *Brenan* 8669 ! Ubulubu (Jan.) *Thomas* 2260 ! Utanga, Obudu Div. (Dec.) *Keay & Savory* FHI 25143 ! Oban *Talbot* 1428 ! 1468 ! Degema *Talbot* 3714 !
6. **A. ernae** (*Dinkl.*) *J. Léonard* l.c. 202 (1955). *Macrolobium ernae* Dinkl. in Fedde Rep. 42 : 157 (1937). Shrub or small tree, near the coast.
    Lib.: Zinkor, Monrovia (Feb.) *Dinklage* 2805.
7. **A. explicans** (*Baill.*) *J. Léonard* l.c. 202 (1955). *Vouapa explicans* Baill. in Adansonia 6 : 181 (Oct. 1865). *Macrolobium heudelotii* Planch. ex Benth. in Trans. Linn. Soc. 25 : 308 (Nov. 1865) ; F.T.A. 2 : 298 ; Chev. Bot. 228 ; F.W.T.A., ed. 1, 1 : 347, partly (excl. *Sc. Elliot* 4800 and frt. descr.) ; Aubrév. Fl. For. Soud.-Guin. 239, t. 49, 1–2. *M. explicans* (Baill.) Keay in Kew Bull. 1953 : 490 (1954). Scandent shrub or small tree to 25 ft. high.
    Fr.G.: Fouta Djalon *Heudelot* 738 ! Kindia *fide* Chev. l.c. **S.L.:** *Afzelius* ! Blama, Juring (Dec.) *Deighton* 309 ! **Lib.:** Dukwia R. (Oct.) *Cooper* 100 ! (See Appendix, p. 196.)
8. **A. vignei** (*Hoyle*) *J. Léonard* l.c. 203 (1955). *Macrolobium vignei* Hoyle in Kew Bull. 1933 : 171. *M.* sp. aff. *obanense* of Aubrév. Fl. For. C. Iv. 1 : 236, t. 89. Forest tree, with spreading crown, to 80 ft. high ; petals white ; fruits velvety, dark brown.
    S.L.: Gola Forest (Apr.) *Small* 624 ! 628 ! 635 ! **Lib.:** Grand Bassa *Dinklage* 1972 ! Lakrata (Apr.) *Bequaert* 187 ! Sarbo (fr. July) *Baldwin* 6384 ! Brewersville (fr. Dec.) *Baldwin* 10948 ! Suen (fr. Nov.) *Baldwin* 10450 ! **Iv.C.:** Banco, Toulépleu & Tabou *fide* Aubrév. l.c. **G.C.:** Dompim, Axim (Nov.) *Vigne* FH 1968 ! Bonsa R. (fr. Nov.) *Vigne* FH 1397 ! Bonsa Su (May) *Vigne* FH 1997 !
9. **A. fragrans** (*Bak. f.*) *Exell & Hillcoat* in Bol. Soc. Brot. 29 : 39 (1955). *Macrolobium fragrans* Bak. f. in J. Bot. 66, Suppl. Polypet. 140 (1928) ; Léonard in Fl. Congo Belge 3 : 419, photo 15. *M. chrysophylloides* Hutch. & Dalz. F.W.T.A., ed. 1, 1 : 347 (1928) ; Kew Bull. 1928 : 400 ; Aubrév. Fl. For. C. Iv. 1 : 228, t. 84. Large tree of rain forest, to 120 ft. high with straight bole, smooth bark, only slight buttresses ; foliage dense, rusty beneath.
    S.L.: Commendi-Gengaru road (Nov.) *Aylmer* 261 ! Gengelu, Kambui F.R. (fl. & young fr. Feb.) *King* 112 ! 147 ! Gola Forest *Small* 667 ! **Lib.:** Dukwia R. (fr. Feb.) *Cooper* 283 (partly) ! **Iv.C.:** Bébou to Mbasso, Mid. Comoé *Chev.* 22648 ! Abidjan *Aubrév.* 47 ! 80 ! **G.C.:** Siahua (fr. Mar.) *Chipp* 135 ! Axim (fr. Feb.) *Vigne* FH 2824 ! **S.Nig.:** Cross River North F.R., Ikom (fr. Dec.) *Keay* FHI 28155 ! Also in French Cameroons, Cabinda and Belgian Congo. (See Appendix, p. 196.)
10. **A. lamprophylla** (*Harms*) *J. Léonard* in Bull. Jard. Bot. Brux. 25 : 202 (1955). *Macrolobium lamprophyllum* Harms in Engl. Bot. Jahrb. 30 : 85 (1901). Forest shrub or tree, to 40 ft. high, with brown-silky usually elongated inflorescences of white flowers borne on woody branches.
    S.Nig.: Oban *Talbot* 1272 ! 1656 ! *Onochie* FHI 7726 ! Also in French Cameroons.
11. **A. macrophylla** P. Beauv. Fl. Oware 1 : 70, t. 42 (1806). *Vouapa macrophylla* (P. Beauv.) Baill. (1865), incl. var. *heudelotiana* Baill. (1865). *Macrolobium palisotii* Benth. (1865)—F.T.A. 2 : 297. *M. macrophyllum* (P. Beauv.) J. F. Macbr. in Contrib. Gray Herb., n.s. 3, 59 : 21 (1919) ; Bak. f. Leg. Trop. Afr. 3 : 672 ; Aubrév. Fl. For. C. Iv. 1 : 230, t. 85 B ; Léonard in Fl. Congo Belge 3 : 422, t. 31. *M. heudelotii* of F.W.T.A., ed. 1, 1 : 347, partly (*Sc. Elliot* 4800 and frt. descr.). Forest shrub or tree ; petals white, turning yellow.
    Fr.G.: Pobéguin 840 ! Rio Nunez *Heudelot* 753 ! Kouria (Nov.) *Chev.* 14937 ! **S.L.:** Freetown (Dec.) *Sc. Elliot* 3826 ! Kora R. (young fr. Feb.) *Sc. Elliot* 4800 ! Rokupr (fl. Dec., fr. Mar.) *Jordan* 406 ! 742 ! Njala (Dec.) *Deighton* 4691 ! Musaia (fr. Apr.) *Deighton* 5477 ! Bintumane (Jan.) *Glanville* 450 ! **Lib.:** Kakatown *Whyte* ! Fortsville (fr. Mar.) *Baldwin* 11147 ! Dimei (fr. Feb.) *Barker* 1217 ! Gbanga (Dec.) *Baldwin* 10518 ! Beiden (Nov.) *Baldwin* 10278a ! **G.C.:** Akyease (Oct.) *Fishlock* 73 ! Kwahu (fl. & young fr. Jan.) *Irvine* 1727 ! **N.Nig.:** Lokoja *Barter* 1624 (partly) ! **S.Nig.:** Lagos (Dec.) *Millen* 30 ! Okomu F.R. (Jan.–Feb.) *Brenan* 8755 ! 9121 ! Sapoba *Kennedy* 2145 ! Ubiaja (Nov.) *Keay* FHI 28125 ! Mt. Koloishe, Obudu (Dec.) *Keay & Savory* FHI 25130 ! Degema *Talbot* 3655 ! **Br.Cam.:** Victoria (Apr.) *Maitland* 623 ! Abonando (Mar.) *Rudatis* 28 ! **F.Po:** *Barter* ! *Mann* 259 ! Extends to Belgian Congo and Angola. (See Appendix, p. 197.)
12. **A. crassifolia** (*Baill.*) *J. Léonard* in Bull. Jard. Bot. Brux. 25 : 202 (1955). *Vouapa crassifolia* Baill. in Adansonia 6 : 179 (1865). *Macrolobium crassifolium* (Baill.) J. Léonard in op. cit. 24 : 61 (1954) ; not of A. Chev. Bot. 228 (1920), name only. *M. heudelotianum* of Aubrév. l.c. 1 : 230, t. 85 C ; Fl. For. Soud.-Guin. 241 ; not *M. heudelotianum* (Baill.) Aubrév. l.c. (1936), based on *Vouapa macrophylla* var. *heudelotiana* Baill. (1865) which is *A. macrophylla* P. Beauv. Tree to 80 ft. high ; in forest and savannah woodland.
    Fr.Sud.: Kita *Dubois* 185 ! **Port.G.:** Buba (Nov.) *Esp. Santo* 1280 ! Candamá, Chitole (fr. Feb.) *Esp. Santo* 2884 ! **Fr.G.:** Rio Nunez *Heudelot* 753 bis. Kouroussa (Nov.) *Pobéguin* 588 ! **S.L.:** Port Loko (Nov.) *King* 224 ! Rokupr (fl. Dec., fr. Mar.) *Jordan* 405 ! 714 ! 717 ! Musaia (fr. Apr.) *Deighton* 5414 ! Makene (fr. Mar.) *Deighton* 3898 ! **Lib.:** Linder 62 ! **Iv.C.:** Banco *Martineau* 330 ! **S.Nig.:** Sapoba *Kennedy* 2600 ! Onypende to Odesige, Obubra (fr. Apr.) *Jones* FHI 12571 !

Imperfectly known species.

A. sp. Tall tree, locally frequent. Better material required.
    S.Nig.: Degema *Talbot* 3656 ! **Br.Cam.:** *Cam.* 46/36 ! Litoka, Cam. Mt. *Maitland* 1076 !

Macrolobium crassifolium *A. Chev.* Bot. 228, name only, not of (Baill.) J. Léonard. The specimen lacks flowers and is probably not an *Anthonotha*.
Fr.G.: Labé *Chev.* 12373 !

43. PARAMACROLOBIUM J. Léonard in Bull. Jard. Bot. Brux. 24 : 348 (1954).

Bracteoles 2·5–3·7 cm. long, glabrous outside ; pedicels 1·5–3 cm. long ; stipe of ovary 1–1·2 cm. long ; petals blue, 5, of which the posterior is 3–4·7 cm. long and 1–2·3 cm. broad, the 2 lateral ones 2–3 cm. long and 1–2·5 mm. broad, the 2 anterior ones filiform 3–7 mm. long ; petiolules twisted ; leaflets 3–5 pairs (rarely 2), elliptic to oblong, unequal at base, acuminate at apex, 2–15 cm. long, 1–6 cm. broad, sub-coriaceous, glabrous ; rhachis 5–22 cm. long ; stipules intrapetiolar, oblong ; fruits obliquely oblong, 10–20 cm. long, 3·5–6 cm. broad, glabrous, finely warted but without transverse ridges ; seeds 3–8 ..   .. *coeruleum*

P. **coeruleum** (*Taub.*) *J. Léonard* l.c. (1954).   *Vouapa coerulea* Taub. (1894).   *Macrolobium coeruleum* (Taub.) Harms in Engl. Pflanzenw. Afr. 3, 1 : 475 (1915) ; Léonard in Fl. Congo Belge 3 : 412. *M. coeruleoides* De Wild. (1907) ; Bak. f. Leg. Trop. Afr. 3 : 679 ; Aubrév. Fl. For. Soud.-Guin. 239, t. 49.   *M. dawei* Hutch. & Burtt Davy in F.W.T.A., ed. 1, 1 : 347 (1928) ; Kew Bull. 1928 : 398.   Forest tree, to 120 ft. high ; petals blue.
Fr.G.: Ditinn *Cochet* 51.   Kindia *Pobéguin* 1265 ; 1289.   S.L.: *Dawe* 28 ! Bunkababe (May) *Lane-Poole* 121 ! Kenema (June) *Aylmer* 88 !   Tabe (young fr. Sept.) *Deighton* 3059 !   Njala (fl. & fr. May) *Small* 27 ! 28 !   Also in French Cameroons, Ubangi-Shari, French and Belgian Congo, Kenya and Tanganyika. (See Appendix, p. 196.)

44. PELLEGRINIODENDRON J. Léonard in Bull. Jard. Bot. Brux. 25 : 203 (1955).

Tree ; branchlets glabrous ; stipules triangular, about 3 mm. long ; leaves glabrous with a single pair of leaflets ; petiole about 5 mm. long, with a pair of stipellae at the top, subtending the very short petiolules ; leaflets obliquely oblong-oblanceolate, unequal at the base (proximal side rounded, the distal side cuneate), rounded towards the apex, apex itself twisted, usually shortly acuminate and with a horny mucro, 6–23·5 cm. long, 1·5–7·5 cm. broad ; both surfaces of leaves strongly reticulate, with two glands within the margin on each side ; panicles terminal, composed of densely flowered racemes 2–3 cm. long ; axis of racemes marked by the scars of the fallen pedicels and the short thick bracts ; pedicels 6–8 mm. long ; bracteoles 7–8 mm. long ; sepals 5, lanceolate, subequal, about 3 mm. long ; petals 5, the posterior one with a 3–4 mm. long claw and a 2-lobed limb, each lobe suborbicular and about 4 mm. broad, the other 4 petals similar to the sepals ; fertile stamens 3, with an intra-staminal ring bearing staminodes ; fruits obliquely oblong, flat, 9–10 cm. long, 3·5–4·5 cm. broad, valves glabrous, with no ridges, curling on dehiscence ; seeds 1–3 *diphyllum*

P. **diphyllum** (*Harms*) *J. Léonard* l.c. (1955).   *Macrolobium diphyllum* Harms in Engl. Bot. Jahrb. 30 : 84 (1901) ; Aubrév. Fl. For. C. Iv. 1 : 232, t. 86 ; Pellegr. Lég. Gab. 48.   *M. reticulatum* Hutch. in Chipp, Gold Coast Trees 20 (1913), name only.   Understorey forest tree, to 80 ft. high ; pedicels and bracteoles pink, large petal white, stamens red.
Iv.C.: Banco & Agboville (fl. Nov., fr. Dec.) *Aubrév.* 368 ; 508 ; 509.   G.C.: Tanosu (Dec.) *Chipp* 35 ! Ankassa, W. Apollonia *Chipp* 327 !   Tano Rapids (fr. Sept.) *Chipp* 376 !   Axim (fr. Feb.) *Irvine* 2182 ! Also in French Cameroons and Gabon.

45. GILBERTIODENDRON J. Léonard in Bull. Jard. Bot. Brux. 22 : 188 (1952).

Bracteoles 3–5 cm. long, or more ; sepals 4 or 5 ; stamens 3 or 9 ; posterior petal 9–15 cm. long, 5·5–15 cm. broad :
Stamens 9 ; bracteoles 4·5–5 cm. long, or more, 2·5–3 cm. broad ; sepals 4 ; posterior petal 9–15 cm. long and broad, pink ; pedicels 4–5·5 cm. long ; stipules ovate-lanceolate, connate for ½ their length, 5–9 cm. long, with reniform appendages 4–10 cm. diam. ; leaflets 2–4 pairs, the upper leaflets up to 55 cm. long and 16 cm. broad, the lowest leaflets as little as 8 cm. long and 3 cm. broad ; fruits oblong, up to 60 cm. long and 13 cm. broad, dark brown-tomentellous, with 3 strong longitudinal ridges 1. *splendidum*
Stamens 3 ; bracteoles 3–4·5 cm. long ; posterior petal white or yellow ; stipules without appendages :
  Leaflets 2 pairs ; sepals 4 ; bracteoles 3·5–4·5 cm. long, 2–2·7 cm. broad, puberulous outside ; posterior petal about 10 cm. long (including the 2·5 cm. long claw) and 12·5 cm. broad, yellow ; pedicels 3–3·5 cm. long ; stipules ovate-lanceolate, connate, 2–4 cm. long, 1·3–2 cm. broad ; upper leaflets 22–45 cm. long, 8–16 cm. broad, lower leaflets 9–21 cm. long, 5–9 cm. broad ; fruits with 2 longitudinal ridges (not known at maturity) ..   .. 2. *grandiflorum*
  Leaflets 4–7 pairs ; sepals 5 ; bracteoles 3–3·5 cm. long, 2–2·3 cm. broad, densely tomentellous outside ; posterior petal 10–12 cm. long (including the 3–6 cm. long claw) and 5·5–7 cm. broad, 2-lobed, white ; pedicels 1·5–4 cm. long ; stipules ovate-lanceolate, 3–3·5 cm. long, 0·8–1·2 cm. broad ; leaflets 8–25 cm. long, 4–9 cm. broad ; fruits oblong-obovate, up to 12 cm. long and 5 cm. broad (? at maturity) glabrescent, with 2 longitudinal ridges ..   .. 3. *ivorense*
Bracteoles up to 2 cm. long ; sepals 5 ; stamens 3 ; posterior petal up to 3 cm. long and broad :

Bracteoles 1–2 cm. long :
    Pedicels 2–4 cm. long ; posterior petal wine-red, 1·7–2·8 cm. long (including the
      0·7–1 cm. long claw), 2·5–2·8 cm. broad ; stipules ovate-lanceolate, 2–8 cm. long,
      1·6–4 cm. broad, with 2 reniform appendages, one 0·6–2·5 cm. long, 1–4·3 cm. broad,
      the other smaller ; leaflets 3 pairs (more rarely 2, or 4–5), densely papillose beneath
      and usually puberulous by the midrib, 9–30 (–50) cm. long, 3·5–18 cm. broad ;
      rhachis 2–15 cm. long ; fruit oblong to oblong-obovate, 15–30 cm. long, 6–9 cm.
      broad, densely brown-tomentellous, with 1 longitudinal ridge ..   .. 4. *dewevrei*
    Pedicels up to 2 cm. long ; colour (where known) of standard petals white or cream :
    Stipules and/or stipular appendages more or less persistent :
      Leaflets 2–4 pairs ; fruits with 1–2 longitudinal ridges :
        Stipules up to 1 cm. long, with reniform appendages up to 3 cm. diam. ; leaflets
          2–3 (rarely 4) pairs, puberulous by midrib beneath ; posterior petal about 2·8
          cm. long (including the 1 cm. long claw), deeply 2-lobed, with the lobes 1·3–1·4
          cm. broad ; fruit glabrescent :
          Fruit with 1 longitudinal submedian ridge, up to 20 cm. long and 5 cm. broad ;
            rhachis of leaves hirsute, 1·5–8 cm. long ; leaflets 9–22 cm. long, 3·5–9 cm.
            broad ; pedicels 1·2–2 cm. long, hirsute ; bracteoles 1·2–1·6 cm. long
                                                                    5. *limba*
          Fruit with 2 longitudinal ridges (one submedian, the other submarginal), 10–12
            cm. long, about 5 cm. broad ; rhachis of leaves glabrous, 1·5–18 cm. long ;
            leaflets 6·5–30 cm. long, 2–10·5 cm. broad ; pedicels 1–1·5 cm. long, shortly
            tomentose ; bracteoles 1–1·4 cm. long   ..   ..   .. 6. *mayombense*
      Stipules 1–5 cm. long, without reniform appendages ; leaflets 3–4 pairs, up to
        22 cm. long and 7·5 cm. broad with numerous looped lateral nerves very
        prominent beneath, midrib glabrous beneath, rhachis stout, 8–12 cm. long,
        glabrous ; pedicels 1–1·5 cm. long, hirsute ; bracteoles about 1 cm. long ;
        posterior petal 1·6 cm. long (including the 0·6 cm. long claw), deeply 2-lobed,
        with the lobes about 1 cm. broad ; fruit glabrescent   .. 7. *demonstrans*
      Leaflets 4–7 pairs, up to 25 cm. long and 7 cm. broad, glabrous beneath, rhachis
        10–25 cm. long, soon glabrous ; stipules up to 5 cm. long and 1 cm. broad, with
        reniform appendages, 1·7–2·2 cm. diam., more or less persistent ; pedicels
        1·5–2 cm. long ; tomentellous ; bracts subpersistent ; bracteoles 1–1·3 cm. long ;
        posterior petal about 2 cm. long (including the 1 cm. long claw), deeply 2-lobed,
        with the lobes about 1 cm. broad ; fruit oblong, up to 30 cm. long and 9 cm.
        broad, tomentellous to glabrescent, with 2–3 strong longitudinal ridges
                                                                    8. *bilineatum*
    Stipules and appendages very early caducous, or absent :
      Leaflets 3–5 pairs ; midrib glabrous beneath ; leaflets up to 18 cm. long and 5·5 cm.
        broad, glossy, rhachis 7–20 cm. long, glabrous ; bracts very early caducous ;
        pedicels about 1·5 cm. long ; bracteoles about 1 cm. long ; fruit oblong, up to
        25 cm. long and 6 cm. broad, glabrescent, with 1 longitudinal ridge and occasion-
        ally a second less distinct ridge near the upper margin   ..   .. 9. *preussii*
      Leaflets 7 pairs ; midrib hirsute beneath ; leaflets up to 15 cm. long and 4·5 cm.
        broad with numerous looped lateral nerves prominent beneath ; rhachis 26 cm.
        long, hirsute ; pedicels and bracteoles each about 1·3 cm. long ; fruit not known
                                                                    10. *obliquum*
Bracteoles 0·5–0·6 cm. long :
    Leaflets 2 pairs ; stipules caducous (not seen) ; leaflets obovate to oblanceolate, up
      to 18 cm. long and 7 cm. broad, sometimes much smaller, but upper pair always
      larger than the lower, glabrous, rhachis 1–4 cm. long ; pedicels 1·2–1·4 cm. long ;
      posterior petal about 1·6 cm. long (including the 0·5 cm. long claw), 2-lobed, the
      lobes about 0·6 cm. broad ; ovary glabrous except for the villous margins ; fruit
      glabrescent, with 1 longitudinal ridge (not known at maturity)..   .. 11. *aylmeri*
    Leaflets 5–9 pairs, usually 7 ; stipules broadly ovate, about 1 cm. long, with reniform
      appendages ; leaflets oblong to ovate-oblong, 2–7·5 cm. long, 1·3–3·5 cm. broad,
      with a single gland on the distal margin just above the petiolule ; pedicels 1·2–1·8
      cm. long ; posterior petal about 0·6 cm. long (including the 0·3 cm. long claw),
      2-lobed, the lobes about 0·5 cm. diam. ; ovary hirsute ; fruit not known
                                                                    12. *brachystegioides*

1. **G. splendidum** (*A. Chev. ex Hutch. & Dalz.*) *J. Léonard* in Bull. Jard. Bot. Brux. 24 : 59 (1954). *Berlinia
    splendida* A. Chev. ex Hutch. & Dalz. F.W.T.A., ed. 1, 1 : 343 (1928) ; Kew Bull. 1928 : 398 ; Chev.
    Bot. 230, name only. *Macrolobium splendidum* (A. Chev. ex Hutch. & Dalz.) Pellegr. in Bull. Soc. Bot.
    Fr. 77 : 667 (1931) ; Aubrév. Fl. For. C. Iv. 1 : 236, t. 88. Forest tree, to 80 ft. high ; petals pink ; in
    swamp forest.
      S.L.: Kenema (fr. Oct.) *King* 284 ! Bagbe, Gola Forest (fl. June, fr. May) *Small* 706 ! 721 ! **Iv.C.:** Tébo
    (July) *Chev.* 19387 ! Massa-Mé *Aubrév.* 136 ! **G.C.:** Namachere, E. of Tarkwa (May) *Chipp* 234 ! Ajakwa
    Chipp 206 ! Cape Coast Dist. (May–June) *Vigne* FH 935 ! Bonsasu, Dompim *Vigne* FH 1970 ! (See
    Appendix, p. 197.)
2. **G. grandiflorum** (*De Wild.*) *J. Léonard* l.c. 22 : 190 (1952) ; Fl. Congo Belge 3 : 433. *Macrolobium grandi-
    florum* De Wild. (1914), partly. Forest tree, to 50 ft. high ; petals lemon-yellow ; in swamp forest by
    rivers.
      Br.Cam.: Tedji, Bamenda (May) *Johnstone* 321 ! Also in French Cameroons, Gabon and Belgian Congo.

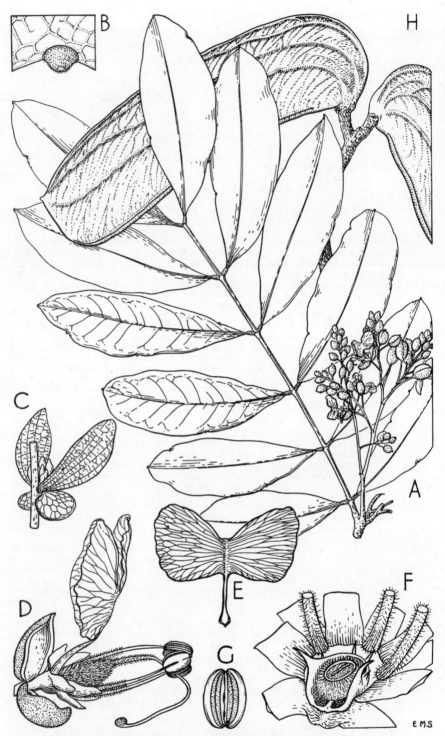

Fig. 152.—GILBERTIODENDRON BILINEATUM (*Hutch. & Dalz.*) *J. Léonard*
(CAESALPINIACEAE).

A, leaf and inflorescence, × ⅓. B, gland on leaf-margin, × 10. C, stipules, × 1. D, flower,
× 2. E, posterior petal, × 1½. F, intrastaminal tube with staminodes and base of stamens,
exposed by removal of ovary and posterior petals, × 6. G, anther, × 4. H, fruits, × ⅓.
A, B, & D–G are drawn from *King* 295 ; C from *Chev.* 22662 ; H from *Small* 709.

3. **G. ivorense** (*A. Chev.*) *J. Léonard* l.c. 24 : 58 (1954). *Berlinia ivorensis* A. Chev. in Bull. Soc. Bot. Fr. 58, Mém. 8 : 165 (1912). *Macrolobium ivorense* (A. Chev.) Pellegr. in Bull. Soc. Bot. Fr. 69 : 745 (1923) ; Aubrév.l.c. 1 : 234. *M. chevalieri* Harms (1922)—F.W.T.A., ed. 1, 1 : 347. Forest tree, to 65 ft. high, with clear bole and no buttresses ; petals white.
  Lib.: Mt. Barclay (June) *Bunting* 158 ! Dukwia R. (May) *Cooper* 427 ! R. Cess, Grand Bassa (fr. **Mar.**) *Baldwin* 11271 ! Jabroke, Webo (July) *Baldwin* 6426 ! Gletown, Tchien (July) *Baldwin* 6752 ! **Iv.C.:** Grabo (July) *Chev.* 19645 ! Tabou *Aubrév.* 1298.
  [The specimens from Sierra Leone (*Sc. Elliot* 4221 and *Thomas* 498) cited as this species in F.W.T.A., ed. 1, 1 : 347 are very doubtful. They are sterile except for detached fruits on *Thomas* 498 which probably do not belong to this species.]
4. **G. dewevrei** (*De Wild.*) *J. Léonard* l.c. 22 : 190 (1952) ; Fl. Congo Belge 3 : 429, photo 15. *Macrolobium dewevrei* De Wild. (1904)—Kennedy For. Fl. S. Nig. 102. Forest tree, to 125 ft. high, with clear bole and no buttresses, bark yellowish-brown, scaly ; petals wine-red ; gregarious, often in swamp forest.
  S.Nig.: Sapoba (Jan.) *Kennedy* 2121 ! 2548 ! *Richards* 3917 ! Degema *Talbot* 3737 ! Also in French Cameroons, Ubangi-Shari, Gabon, French and Belgian Congo and Angola. (See Appendix, p. 196.)
5. **G. limba** (*Sc. Elliot*) *J. Léonard* l.c. 24 : 59 (1954). *Macrolobium limba* Sc. Elliot in J. Linn. Soc. 30 : 77 (1894) ; F.W.T.A., ed. 1, 1 : 347 ; Aubrév. l.c. 1 : 234. Forest tree, to 40 ft. high, with spreading crown and drooping branches ; flowers white ; in wet places.
  Fr.G.: *fide* Aubrév. *l.c.* **S.L.:** Kurusu, Limba (Apr.) *Sc. Elliot* 5539 ! Kafogo, Limba (Apr.) *Sc. Elliot* 5606 ! Giehun (Apr.) *Aylmer* 222 ! Kafoko (fr. Sept.) *Thomas* 2111 ! **Iv.C.:** Abidjan *Aubrév.* 26 ! **G.C.:** Saduso, Huni Valley (fl. & fr. May) *Chipp* 220 ! 221 ! Kwae, Obuasi—Kibbi road (fr. Feb.) *Brent* 405 ! Kumasi (Mar.) *Vigne* FH 1093 ! Pra Anum F.R. (Mar.) *Andoh* FH 4339 ! Bobiri F.R. (Mar.) *Andoh* FH 4960 ! (See Appendix, p. 197.)
6. **G. mayombense** (*Pellegr.*) *J. Léonard* in Bull. Jard. Bot. Brux. 22 : 191' (1952) ; Fl. Congo Belge 3 : 431. *Macrolobium mayombense* Pellegr. in Bull. Mus. Hist. Nat. 26 : 552 (1920) ; Lég. Gab. 54. Forest tree, to 65 ft. high ; posterior petal white or cream with chocolate markings (colour not however recorded from our area).
  S.Nig.: Sapoba *Kennedy* 2238 ! 2252 ! 2549 ! Also in Gabon, Cabinda, French and Belgian Congo.
7. **G. demonstrans** (*Baill.*) *J. Léonard* l.c. 22 : 190 (1952). *Vouapa demonstrans* Baill. (1865). *Macrolobium demonstrans* (Baill.) Oliv. F.T.A. 2 : 299 (1871). *M. demonstrans* (Baill.) Oliv. var. *talbotii* Bak. f. in Cat. Talb. 28 (1913). *M. talbotii* (Bak. f.) Hutch. & Dalz. F.W.T.A., ed. 1, 1 : 347 (1928) ; Kew Bull. 1928 : 398. Forest tree, 20–25 ft. high ; or scandent shrub.
  S.Nig.: Oban *Talbot* 1284 ! 1504 ! Also in Gabon.
8. **G. bilineatum** (*Hutch. & Dalz.*) *J. Léonard* l.c. 24 : 58 (1954). *Macrolobium bilineatum* Hutch. & Dalz. F.W.T.A., ed. 1, 1 : 347 (1928) ; Kew Bull. 1928 : 399 ; Aubrév. l.c. 1 : 234, t. 87. Forest tree, to 100 ft. high, with yellowish-grey, flaking bark ; petals white ; gregarious, often in moist places.
  S.L.: Kahreni, Limba (Apr.) *Sc. Elliot* 5588 ! Falaba *Burbridge* 475 ! Yama Creek, Scarcies R. (fr. Mar.) *Jones* SLFD 652 ! Joru, Gaura (Nov.) *King* 295 ! 308 ! Bumbuna (Jan.) *King* 310 ! Bagbe, Gola Forest (fr. May) *Small* 703 ! 709 ! **Iv.C.:** Yacassé, Attié (fl. & young fr. Dec.) *Chev.* 22662 ! **G.C.:** Enchi (fl. & young fr. Dec.) *Vigne* FH 3156 ! Dompim, Axim (Nov.) *Vigne* FH 1396 !
9. **G. preussii** (*Harms*) *J. Léonard* l.c. 24 : 59 (1954). *Macrolobium preussii* Harms in Engl. Bot. Jahrb. 26 : 272 (1899) ; F.W.T.A., ed. 1, 1 : 347. Forest tree, to 100 ft. high, with straight bole and yellow flaking bark.
  S.L.: Gola Forest (fr. May) *Small* 85 ! **Lib.:** Dukwia R. (Oct.) *Cooper* 94 ! Tawata, Boporo (Nov.) *Baldwin* 10283 ! **G.C.:** Mansi R., Bawdia (Dec.) *Vigne* FH 3144 ! Jema (Dec.) *Vigne* FH 3159 ! **Br.Cam.:** L. Barombi (Aug.) *Preuss* 449 ! Bopo, Kumba (fr. Mar.) *Lobe* FHI 24808 !
10. **G. obliquum** (*Stapf*) *J. Léonard* l.c. 24 : 59 (1954). *Macrolobium obliquum* Stapf in J. Linn. Soc. 37 : 96 (1905) ; F.W.T.A., ed. 1, 1 : 347. A shrub with rusty panicles of white flowers.
  Lib.: Sinoe Basin *Whyte* !
11. **G. aylmeri** (*Hutch. & Dalz.*) *J. Léonard* l.c. 24 : 57 (1954). *Macrolobium aylmeri* Hutch. & Dalz. F.W.T.A., ed. 1, 1 : 347 (1928) ; Kew Bull. 1928 : 399. Small forest tree, often near the sea.
  S.L.: Colonial Reserve, Freetown (Jan., Dec.) *King-Church* 8 ! *Vigne* 13 ! Commendi (Nov.) *Aylmer* 263 ! York (Nov.) *Lane-Poole* 80 ! John Obey *Lane-Poole* 423 ! Buri (Dec.) *Unwin & Smythe* 9 !
12. **G. brachystegioides** (*Harms*) *J. Léonard* l.c. 24 : 58 (1954). *Macrolobium brachystegioides* Harms in Engl. Bot. Jahrb. 40 : 24 (1907). Forest tree, to 120 ft. high.
  Br.Cam.: Mbalange to Bombe (Jan.) *Lobe Babute* 51/36 ! Also in French Cameroons.

Imperfectly known species.

G. sp. A.
  S.Nig.: Ekang stream, Oban *Gray* 8/11 !

46. TAMARINDUS Linn.—F.T.A. 2 : 307.

A large tree ; leaves paripinnate ; leaflets 12–15 pairs, opposite, narrowly oblong, unequally rounded at base, rounded or emarginate at apex, 2–3 cm. long, 0·8–1 cm. broad, glabrous and rather strongly reticulate, often glaucous ; rhachis glabrous or pubescent ; flowers in small terminal glabrous racemes ; pedicels about 5 mm. long, jointed at the apex ; bracts nearly as long as the flower-bud, concave, deciduous ; sepals 4, 7 mm. long ; petals 3, 1 posterior, 2 lateral, slightly exceeding the calyx ; fertile stamens 3, alternating with small staminodes ; ovary stipitate ; fruit curved, oblong, about 10 cm. long, minutely scaly, inner part pulpy, intersected by fibres

indica

T. indica *Linn.*—F.T.A. 2 : 308 ; Chev. Bot. 232 ; Aubrév. Fl. For. Soud.-Guin. 226, t. 57, 6–7. Commonly cultivated ; also in savannah, especially on termite mounds.
  Sen.: Thiès *Chev.* 2949. **Fr.Sud.:** Oualia (fr. Dec.) *Chev.* 40 ! **Port.G.:** Pefiné, Bissau *Esp. Santo* 1701 ! **Fr.G.:** Kouroussa *Chev.* 15644. **S.L.:** Kahreni to Port Loko (Apr.) *Sc. Elliot* 5801 ! Kamalu (May) *Thomas* 319 ! **Lib.:** Fortsville (fr. Mar.) *Baldwin* 11140 ! **Iv.C.:** Dotou *Chev.* 21781. **G.C.:** Bjury (fr. July) *Chipp* 496 ! Lawra (fl. & fr. Mar.) *MacLeod* 836 ! Tamale (Apr.) *Vigne* 1710 ! **Dah.:** Dassa Zoumi (May) *Chev.* 23619 ! Savalou (May) *Chev.* 23744 ! **Fr.Nig.:** *fide* Aubrév. *l.c.* **N.Nig.:** Zamfara F.R. (fl. & fr. May) *Keay* FHI 16177 ! Bauchi Plateau (Mar.) *Lely* P. 186 ! Widespread in the tropics. (See Appendix, p. 200.)

47. MONOPETALANTHUS Harms in E. & P. Pflanzenfam. Nachtr. 1 : 195 (1897).

Leaflets about 20 pairs, opposite, oblong, broadly auriculate on the lower side, curved to a small black sublateral tip at apex, 3 cm. long, 0·8 cm. broad, rather closely reticulate beneath, glabrous ; rhachis densely fulvous-tomentose, the leaflets con-

tiguous ; stipules persistent, 3 cm. long, produced into a broad base, closely venose-striate, glabrous ; racemes about 5 cm. long, densely velvety all over ; pedicels 5-7 mm. long ; bracteoles 8 mm. long ; petal ½ as long again ; young fruit densely villous ..   ..   1. *pteridophyllus*
Leaflets about 20 pairs, as above, but much smaller, up to 1 cm. long and 2·5 mm. broad, rounded at apex, glossy above, venose beneath ; rhachis slightly pubescent ; racemes cone-like when young, covered with densely overlapping striate fringed bracts, the uppermost silky-tomentose ; bracteoles 3–4 mm. long, pubescent ; petal a little longer ; stamens 8 ..   ..   2. *compactus*

FIG. 153.—MONOPETALANTHUS COMPACTUS *Hutch. & Dalz.* (CAESALPINIACEAE).
A, portion of leaf showing attachment of leaflets. B, flower. C, petal. D, E, calyx-lobes.
F, stamens. G, pistil.

1. **M. pteridophyllus** *Harms* in E. & P. Pflanzenfam. Nachtr. 1 : 195 (1897) ; Engl. Bot. Jahrb. 26 : 266 (1899) ; Pellegr. in Bull. Soc. Bot. Fr. 89 : 119 ; Léonard in Fl. Congo Belge 3 : 443, fig. 37 A, B. Forest tree, to 120 ft. high ; flowers white in axillary rusty-tomentose racemes ; by rivers.
S.L.: Taia R., Njala *Dawe* 562 ! Njala (Aug.) *Deighton* 1209 ! Pewa, Gola Forest *Small* 735 ! Lib.: St. John's R., Grand Bassa (Aug.) *Dinklage* 1720 ! Farmington R., Harbel (fr. Mar.) *Baldwin* 11099 ! Also in Belgian Congo. (See Appendix, p. 198.)
2. **M. compactus** *Hutch. & Dalz.* F.W.T.A., ed. 1, 1 : 342 (1928) ; Kew Bull. 1928 : 397 ; Pellegr. l.c. 120. Large forest tree.
S.L.: Falaba (Apr.) *Aylmer* 30 ! Giehun (Apr.) *Aylmer* 226 !

Imperfectly known species.

1. **M. sp. A.** " Toubaouate " of Aubrév. Fl. For. C. Iv. 1 : 242, t. 95 A. Large forest tree ; stipules caducous ; leaflets 15–30 pairs, emarginate at apex, midrib central ; flowers not known.
Iv.C.: Olodio to Grabo *Aubrév.* 1321 !
2. **M. sp. B.** Forest tree to 70 ft. high ; stipules persistent, leafy ; leaflets 28–32 pairs, rounded or slightly emarginate at apex, midrib central ; flowers and fruits unknown.
S.Nig.: Obutong, Calabar *Onochie* FHI 7798 ! Orem, Calabar *Ejiofor* FHI 21893 !
3. **M. sp. C.** Forest tree or shrub ; stipules caducous ; leaflets 19–30 pairs, rounded or obtuse at apex, midrib diagonal from the corner of the base ; flowers not known.
Br.Cam.: FHI 12029 !

48. BRACHYSTEGIA Benth.—F.T.A. 2 : 305.

By A. C. Hoyle

Leaflets progressively larger from base to apex of the whole leaf, the terminal pair usually distinctly the largest, obliquely elliptic to obovate, glabrous or rarely puberulous ; if leaflets obliquely lanceolate, then rarely more than 6 pairs, often less than 5 pairs :

Basal pair of leaflets relatively very minute and appearing vestigial, 0·7–1 cm. long, shortly petiolulate, inserted at or very shortly above the apex of the pulvinus, usually early caducous but leaving a pair of minute scars ; remaining leaflets 3–4 (–6) pairs, the terminal pair (6–) 7–10 (–15) cm. long, (2–) 3–5 (–6) cm. broad, obliquely lanceolate to obovate, more or less falcate and often acuminate, the middle pairs usually smaller or at least narrower ; fruit (15–) 20–25 (–30) cm. long, (5–) 6–8 (–11) cm. broad, smooth, dark purplish or reddish brown, woody, upper suture broadly winged   ..   ..   1. *laurentii*

Basal pair of leaflets not relatively minute and not appearing vestigial, sessile like the others, normally persistent ; if petiole very short and basal pair of leaflets very small (*B. nigerica*), then terminal pair rarely more than 5 cm. long, leaflets usually 7 or more pairs, and fruit rarely exceeding 14 cm. long and 5 cm. broad, coriaceous :

Leaflets (4–) 5 (–6) pairs, those of the terminal pair 5–10 (–15) cm. long, (2·5–) 3–5 (–7) cm. broad, elliptic to oblanceolate ; common petiole (10–) 13–20 (–30) mm. long ; fruit woody, transversely nervose, rather pale brown, normally 16–22 cm. long, 5–6 cm. broad, the upper suture with a conspicuous spreading wing 5–6 mm. broad on each side   ..   ..   2. *eurycoma*

Leaflets (5–) 6–8 (–9) pairs, those of the terminal pair (2–) 3–5 (–7) cm. long, (1–) 1·5–2 (–3·5) cm. broad, broadly elliptic to obovate ; common petiole 4–7 mm. long ; fruit thickly coriaceous, smooth, dark purplish brown, normally 12–14 cm. long, 4–5 cm. broad, the wing of the upper suture not conspicuous, 2–3 mm. broad on each side   ..   ..   3. *nigerica*

Leaflets progressively larger from the base to the middle of the whole leaf, thence usually slowly decreasing to the smaller terminal pair ; leaflets mostly obliquely lanceolate, narrowed to the obtuse or acuminate apex, very rarely oblanceolate or obovate, usually with stiff, spreading hairs on the midrib beneath ; fruit as in *B. laurentii* but smaller :

Leaflets (6–) 8–10 (–12) pairs, mostly obliquely cuneate at base ; usually 2, at most 3, nerves departing fanwise from the midrib on the broader side of the base of the leaflet ; style leaving the ovary at a right angle ; staminal tube scarcely 0·5 mm. long, outside the disk-like rim of the receptacle-cup   ..   ..   4. *leonensis*

Leaflets (5–) 6–7 (–8) pairs, mostly obliquely rounded or subcordate at base ; usually 3, more rarely 4, nerves departing fanwise beside the midrib on the broader side of the base of the leaflet ; style leaving the ovary at an angle of about 45° ; staminal tube more or less irregular, 2–3·5 mm. long, gradually narrowed within to the obscure rim of the receptacle-cup   ..   ..   5. *kennedyi*

1. **B. laurentii** (*De Wild.*) *Louis ex Hoyle* in Fl. Congo Belge 3 : 461 (1952). *Macrolobium laurentii* De Wild. Miss. Laur. 99, t. 37 (1905). *B. zenkeri* Harms in Engl. Bot. Jahrb. 45 : 298 (1910) ; Burtt Davy & Hutch. in Kew Bull. 1923 : 157 ; Bak. f. Leg. Trop. Afr. 3 : 732. A large, heavy-crowned, forest tree, to 150 ft. high, locally common near water, with cylindrical bole attaining 80 ft. high and 9 ft. girth ; bark hard, tough, fibrous, at first smooth, silvery-grey, later brownish-grey, rough and flaking.
   Br.Cam.: S. Bakundu F.R., Kumba (fr. Feb.) *Ngalame* FHI 24805 ! Also in French Cameroons, Gabon and Belgian Congo.
   [The Cameroons (*B. zenkeri*) and Gabon forms of the species differ from some Belgian Congo forms in having fewer, more pointed leaflets and more slender, glabrous rhachis, while the little-known flowers are smaller and the perianth more reduced, but the differences hardly warrant specific rank.]

2. **B. eurycoma** *Harms* in Engl. Bot. Jahrb. 49 : 424 (1913) ; Burtt Davy & Hutch. l.c. ; F.W.T.A., ed. 1, 1 : 348, partly ; Bak. f. l.c. 731, partly ; Kennedy For. Fl. S. Nig. 103. A large-crowned forest tree, to 120 ft. high, common on stream-banks, with buttressed bole to 25 ft. in girth and grey-green bark rough with woody flakes, red when cut, exuding red and yellow gum.
   S.Nig.: Ikirun *Millson* ! Oshogbo *Millson* ! Ijaiye F.R., Ibadan (Apr.) *Onochie* FHI 21983 ! Makori, Ibadan (May) *Tamajong* FHI 23263 ! Owo F.R., Ondo (Apr.) *Jones* FHI 3477 ! Uteru, Onitsha *Chesters* in FHO A.124/30 e–h ! Asaba *Leslie* 40 ! Arochuku (fr. Feb.) *Cons. of For.* 28 ! 31 ! Enugu (Mar.) *Irvine* 3608 ! Br.Cam.: Asu R., Bamenda *Johnstone* 289 ! Also in French Cameroons. (See Appendix, p. 176.)

3. **B. nigerica** *Hoyle & A. P. D. Jones* in Kew Bull. 1947 : 68. *B. eurycoma* of F.W.T.A., ed. 1, 1 : 348, partly, not of Harms. A large forest tree, to 120 ft. high and 12 ft. girth with spreading crown, gregarious near water ; bark smooth and dark at first, later flaking in large, irregular patches, red and fibrous when cut.
   S.Nig.: Gambari, Ibadan *Ladipo* FHI 19061 ! Oluwa F.R., Ondo (fr. Feb.) *Jones* FHI 14732 ! Usonigbe F.R. (fr. Nov.) *Meikle & Keay* 569 ! Sapoba *Kennedy* 1842 ! Ubiaroko, Kwale (fr. Dec.) *Ainslie* 15 ! Onitsha *Unwin* 66 ! Awka (Jan.) *Prior* ! *Thomas* 78 ! Afi River F.R. (fr. May) *Latilo* FHI 31802 ! Port Harcourt (fr. Dec.) *Chesters* 18 ! 19 ! (See Appendix, p. 176.)

4. **B. leonensis** *Burtt Davy & Hutch.* in Kew Bull. 1923 : 156 ; Bak. f. Leg. Trop. Afr. 3 : 724 ; Aubrév. Fl. For. C. Iv. 1 : 242. Not of Topham in Kew Bull. 1930 : 358 ; not of Kennedy l.c. 103. A large forest tree, to 150 ft. high and 6 ft. girth, with wide-buttressed or non-buttressed, cylindrical bole and smooth or scaling, grey-green bark exuding brown resin ; occurs on ridges and in valleys near water.
   S.L.: King 277 ! *Wallace* 158 ! 179 ! Bandama *Aylmer* 200 ! Kambui (Mar., Apr.) *Lane-Poole* 188 ! *Edwardson* 71 ! 72 ! *Wallace* 141 ! *Small* 900 ! Lib.: Du R. (fr. Aug.) *Linder* 229 ! Iv.C.: Grabo (fr. May) *Aubrév.* 1288 ! 1301 ! Nékaounie, Tabou *Aubrév.* 1310 ! (See Appendix, p. 177.)
   [" Two distinct types " of this species are suspected by Small.]

5. **B. kennedyi** *Hoyle* in Bull. Jard. Bot. Brux. 25 : 183, fig. 44 (1955). *B. leonensis* of Kennedy l.c. 102, not of Burtt Davy & Hutch. *B. eurycoma* of Appendix, p. 176, partly. A huge forest tree, to 174 ft. high and 22½ ft. girth above buttresses at 16 ft. from the ground, with cylindrical bole and grey-brown bark smooth at first, later fissured into large, flaking pieces, bright red when cut.
S.Nig.: Sapoba (fr. Nov.) *Keay* FHI 28067 ! *Kennedy* 1747 ! 1840 ! 2233 ! 2451 ! *Meikle & Keay* 581 ! Umuahia *Cousens* FHI 4853 ! Afi River F.R. *Jones & Onochie* FHI 5845 ! 17315 ! Etomi, Ikom (fr. May) *Jones* FHI 18874 ! Arochuku (Apr.) *Cons. of For.* 146 ! 148 ! (See Appendix, pp. 176, 177.)

49. CRYPTOSEPALUM Benth.—F.T.A. 2 : 303. *Pynaertiodendron* De Wild. (1915).

By P. Duvigneaud

Leaves pinnate ; leaflets asymmetrical :
Leaflets 10–15 pairs, narrowly oblong, auriculate on one side, truncate or emarginate at apex, small, 10–14 mm. long and 3–4 mm. broad, glabrous, midrib median ; rhachis pubescent ; pedicels very slender, up to 13 mm. long, glabrous   ..   1. *staudtii*
Leaflets 1–3 pairs, larger, the midrib near the inner side :
Leaflets 2, rarely 3, pairs, very obliquely ovate, shortly acuminate, the lower pair small, the upper ones up to 5 cm. long and 2 cm. broad, midrib pubescent or puberulous on both surfaces, principally in the lower part ; inner side of leaflet ciliate ; petiole and rhachis pubescent ; pedicels up to 10 mm. long, glabrous   .. 2. *tetraphyllum*
Leaflets 1 pair :
Leaflets small, obliquely elliptic, obtusely pointed, 1·8–2·2 cm. long, 8–9 mm. broad, slightly ciliate on the margin and midrib, at length glabrous ; petiole pubescent ; pedicels slender, 6–10 mm. long, glabrous ..   ..   ..   3. *minutifolium*
Leaflets large, obliquely ovate, long-acuminate, 5–6 cm. long, 20–27 mm. broad, glabrous, except for some hairs at the base of the midrib on the upper surface ; petiole glabrous ; pedicels slender, 6–10 mm. long, hairy   ..   4. *diphyllum*
Leaves 1-foliolate ; leaflet symmetrical, ovate, caudate-acuminate, up to 7 cm. long and 2·5 cm. broad, completely glabrous ; petiole glabrous ; pedicels slender, 5–10 mm. long, glabrous   ..   ..   5.ᐟ *pellegrinianum*

1. **C. staudtii** *Harms* in Engl. Bot. Jahrb. 26 : 267 (1899). A shrub with glabrous branchlets, perulate buds, multifoliolate leaves, and numerous racemes at the end of short foliate lateral branchlets.
Br.Cam.: Johann-Albrechtshöhe (Mar.) *Staudt* 907 ! Also in French Cameroons, Gabon and Spanish Guinea.
2. **C. tetraphyllum** (*Hook. f.*) *Benth.*—F.T.A. 2 : 303 ; Aubrév. Fl. For. C. Iv. 1 : 238, t. 90, A. *Cynometra* ? *tetraphylla* Hook. f. (1849). A big timber tree with hard wood, with puberulous branchlets, perulate buds, quadrifoliolate leaves, and short axillary racemes with white or pinkish flowers ; pods suborbicular, glabrous, flat, up to 4 cm. high and 6 cm. broad, with very lateral peduncle.
Fr.G.: Boké (Apr.) *Chillou* ! Ymbo (Apr.) *Pobéguin* 1555 ! Mamou *Aubrév.* 3027 ! **S.L.**: *Barter* ! *Don* ! Sugar Loaf (Apr., May) *Sc. Elliot* 5781 ! Talla Hills, Ninia (Feb.) *Sc. Elliot* 4893 ! Kurusu (Apr.) *Sc. Elliot* 5525 ! Bumbu *Lane-Poole* 169 ! Makene—Kabala road (Mar., Apr.) *Deighton* 5022 ! Gola Forest (Apr.) *Small* 562 ! **Lib.**: mouth of Sinoe R., Greenville (Mar.) *Baldwin* 11561. **Iv.C.**: Abidjan *Aubrév.* 12 ! Guiglo *Aubrév.* 2026 !
3. **C. minutifolium** (*A. Chev.*) *Hutch. & Dalz.* F.W.T.A., ed. 1, 1 : 348 (1928) ; Kew Bull. 1928 : 400 ; Aubrév. l.c. t. 90, B. *Hymenostegia minutifolia* A. Chev. in Bull. Soc. Bot. Fr. 58, Mém. 8 : 165 (1912). A montane forest tree, 80–100 ft. high, with bole of 50–60 ft. and small bifoliolate leaves.
Iv.C.: Toula to Nékaougnié, Middle Cavally (July) *Chev.* 19579 !
4. **C. diphyllum** *Duvign.* sp. nov.[1] A forest tree to 50 ft. high, with reddish-grey, smooth bark, glabrous branchlets, bifoliolate leaves, and short, slightly pubescent, axillary racemes.
S.Nig.: Ekang R., Oderige, Obubra (Apr.) *A. P. D. Jones* 1467 !
5. **C. pellegrinianum** (*J. Léonard*) *J. Léonard* in Bull. Jard. Bot. Brux. 22 : 210 (1952). *Pynaertiodendron pellegrinianum* J. Léonard (1951). A small tree with glabrous branchlets, unifoliolate leaves, and glabrous axillary racemes.
S.Nig.: Ukpon F.R., Ogoga Prov. *Catterall* 1 ! Also in Gabon and Belgian Congo.

50. DIDELOTIA Baill.—F.T.A. 2 : 307.

Leaves 2 or more foliolate :
Leaflets 3–7 pairs :
Leaflets 5–7 pairs, small, up to 3 cm. long and 1·3 cm. broad, oblong, very oblique at the base on one side, emarginate, glabrescent except on the midrib ; racemes short, in the axils of the upper, often much smaller leaves, forming an elongated leafy panicle ; stipules intrapetiolar, united into one ; pedicels up to 1·3 cm. long, tomentose ; bracteoles 2, broadly ovate, tomentose outside, 4–5 mm. long
1. *engleri*
Leaflets 3–4 pairs, larger, 5–8 cm. long, obliquely oblong, emarginate, shining, glabrous ; racemes paniculate, elongated ; pedicels 5–8 mm. long ; bracteoles 2, suborbicular, shortly pilose outside ..   ..   2. *afzelii*
Leaflets 1 pair ; flowers pedicellate, pedicels nearly 1 cm. long, shortly pubescent ; leaflets together forming almost a circle, very oblique, acuminate, 7–13 cm. long, 3–6 cm. broad, glabrous ; petiole 0·5–1 cm. long ; racemes forming a long slender leafless panicle up to 20 or more cm. long ; bracteoles obovate-elliptic   3. *africana*
Leaves 1-foliolate ; petiole 5–7 mm. long, with a pair of small stipellae at the apex, usually early caducous ; lamina ovate or elliptic-ovate, equal and rounded at base, gradually acuminate or acute at apex, 7–15·5 cm. long, 4–8 cm. broad ; panicles narrow, 13–50 cm. long ; pedicels 7–9 mm. long ; bracteoles about 5 mm. long ; fruits oblong with a single median longitudinal ridge, and sometimes a short basal one
4. *sp. nr. unifoliolata*

[1] A *C. minutifolio* foliolis majoribus, pedicellis pubescentibus differt.

1. **D. engleri** *Dinkl. & Harms* in Engl. Bot. Jahrb. 30 : 80 (1901) ; Bak.f. Leg. Trop. Afr. 3 : 736. A shrub or small tree ; inflorescence narrow, drooping, up to 10 in. long ; flowers scented, small, crimson.
   Lib.: Grand Bassa, in moist woods near the sea (Aug.) *Dinklage* 2033 !
2. **D. afzelii** *Taub.* in E. & P. Pflanzenfam. 3, 3 : 387 (1894) ; Harms in Engl. Bot. Jahrb. 26 : 266 ; Bak. f. l.c. Forest tree, to 20 ft. high ; inflorescence pendulous ; stamens and inside of bracts red.
   S.L.: *Afzelius* ! *Dawe* 29 ! *Cons. of For.* 38 ! Njala (fl. Oct., fr. Feb.) *Deighton* 2247 ! 2462 ! Mafindo Falls, Moa R. (fl. & fr. Feb.) *Deighton* 4003 ! (See Appendix, p. 191.)
3. **D. africana** *Baill.* in Adansonia 5 : 367, t. 8 (1865) ; F.T.A. 2 : 307 ; Bak. f. l.c. ; Pellegr. Lég. Gab. 77. A forest tree, with foliage like that of *Hymenaea courbaril* Linn.
   S.Nig.: Eket *Talbot* 3344 ! Oban *Talbot* 1461 ! (See Appendix, p. 191.)
4. **D. sp. nr. unifoliolata** *J. Léonard* in Bull. Jard. Bot. Brux. 22 : 205 (1952) ; Fl. Congo Belge 3 : 495, fig. 43. " Broutou " of Aubrév. Fl. For. C. Iv. 1 : 240, t. 96, 3–5. Large forest tree.
   S.L.: Gola North F.R. *King* 290 ! **Lib.:** Gbanga (fr. Dec.) *Baldwin* 10517 ! **Iv.C.:** Tabou region *Aubrév.* 1309 ! 1561 ! [Not quite the same as the type from Belgian Congo ; flowers not yet known.]

Imperfectly known species.

D. ? sp. Large forest tree ; leaflets 4–6 pairs, obliquely rhombic-elliptic, acuminate and emarginate at apex ; flowers not known.
   S.L.: Gola North F.R. *King* 138 ! **Lib.:** Mecca, Boporo *Baldwin* 10449 ! Mecca, Grand Cape Mount (fr. Dec.) *Baldwin* 10841 !

51. MEZONEURON Desf.—F.T.A. 2 : 260.

Woody climber, with short recurved prickles ; leaves bipinnate ; pinnae about 5 pairs, the axes prickly ; leaflets alternate, about 5 pairs, elliptic, rounded at each end, 3–4 cm. long, 1·5–2·5 cm. broad, glabrous ; racemes terminal, simple or branched, rather dense-flowered, tomentellous, up to about 20 cm. long ; pedicels about 1 cm. long, articulated near the top ; bracts small, very early caducous ; lower calyx-lobe partially enclosing the others in bud, hood-like ; petals 5 ; stamens 10, free ; filaments pilose ; fruits indehiscent, reflexed, flat and thin, linear-oblong, about 10 cm. long and 2–2·5 cm. broad, minutely puberulous, winged along one side   *benthamianum*

M. benthamianum *Baill.*—F.T.A. 2 : 261 ; Chev. Bot. 219. Flowers yellow ; pods bright red.
   Sen.: Bignona *fide* Chev. l.c. **Gam.:** *Brown-Lester* 32 ! **Port.G.:** Bissalanca, Bissau (Jan.) *Esp. Santo* 1646 ! **Fr.G.:** Rio Nunez *Heudelot* 762 ! Mamou (fr. Mar.) *Dalz.* 3406 ! **S.L.:** Sasseni, Tambakka (fl. & fr. Jan.) *Sc. Elliot* 4519 ! Jama-Kangama (Jan.) *Aylmer* 2 ! Njala (Dec.) *Deighton* 2580 ! Kamakui (fr. Feb.) *Deighton* 3501 ! **Lib.:** Ganta *Harley* 34 ! **Iv.C.:** Bouroukrou (Jan.) *Chev.* 16861. **G.C.:** Cape Coast (fl. & fr. Dec.) *C. J. Taylor* FH 5447 ! Afram Plains (fr. Jan.) *Chipp* 594 ! Dodowa (fr. Dec.) *Irvine* 1024 ! **Togo:** Chito (fr. Dec.) *Howes* 1061 ! **Dah.:** Zagnanado (Feb.) *Chev.* 23028 ! **S.Nig.:** Olokemeji (Feb.) *Foster* 145 ! Ibadan (Jan.) *Newberry* 185 ! (See Appendix, p. 198.)

52. CAESALPINIA Linn.—F.T.A. 2 : 262.

Scrambling shrub with very closely prickly branches and slender branchlets ; leaves bipinnate, pinnae 4–7 pairs ; rhachis prickly ; leaflets oblong-elliptic, mucronate, 2·5–4·5 cm. long, 1·5–2·5 cm. broad, slightly pubescent ; flowers paniculate ; bracts linear, soon falling off ; axis and calyx tomentose ; petals 5, oblong, yellow, subequal, about 1 cm. long ; stamens 10, free, filaments villous towards the base ; fruit oblong-elliptic, flattish, 6–7 cm. long, 4 cm. broad, very prickly with slender erect prickles, 1–2-seeded ; seeds spherical, lead-coloured, smooth, about 1·5 cm. diam.   *bonduc*

C. bonduc (*Linn.*) *Roxb.* Fl. Ind. 2 : 362 (1832) ; Dandy & Exell in J. Bot. 76 : 179 (1938), q.v. for full synonymy. *Guilandina bonduc* Linn. (1753). *C. bonducella* (Linn.) Fleming—F.T.A. 2 : 262 ; Chev. Bot. 220. *C. crista* of F.W.T.A., ed. 1, 1 : 348. Originally a native of the coastal regions of both hemispheres, now naturalized near villages in the interior.
   Sen.: *Heudelot* 464 ! Dakar *Chev.* 15821. **Port.G.:** Ilha das Cobras (fr. Apr.) *Esp. Santo* 1928 ! **S.L.:** Nyangai, Turtle Isl. *Deighton* 2381 ! Mano (May) *Deighton* 3974 ! **Lib.:** Cape Palmas (July) *T. Vogel* 9 ! Nana Kru, Sinoe (Mar.) *Baldwin* 11579 ! **Iv.C.:** various localities *fide* Chev. l.c. **G.C.:** Banka (Dec.) *Vigne* FH 1508 ! Peshi, Elmina (Aug.) *Andoh* FH 5556 ! **Togo:** Alibi *Kersting* A. 559 ! **Dah.:** *Le Testu* 157 ! **N.Nig.:** Katsina Ala (June) *Dalz.* 603 ! **S.Nig.:** Olokemeji *Foster* 313 ! Ibadan (Aug.) *Chizea* FHI 20019 ! Aguku *Thomas* 908 ! Degema *Talbot* 3683 ! **Br.Cam.:** Victoria (Nov.) *Maitland* 777 ! **F.Po:** (Mar., June) *Barter* 1612 ! *Mann* 398 ! Widespread in the tropics. (See Appendix, p. 178.)

Besides the above, several other species are cultivated in our area, including : *C. pulcherrima* (Linn.) Sw. (syn. *Poinciana pulcherrima* Linn.), a commonly cultivated shrub, a native of Asia but commonly known as " Pride of Barbados " ; *C. decapetala* (Roth) Alston (syn. *C. sepiaria* Roxb.).

53. DELONIX Raf. Fl. Tellur. 2 : 92 (1837).

A tree ; leaves bipinnate, up to 45 cm. long ; pinnae 11–18 pairs ; leaflets opposite, numerous, oblong, rounded and asymmetrical at base, rounded to mucronate at apex, about 1 cm. long and 2·5–4 mm. broad ; flowers in lax terminal or subterminal racemes, glabrous ; receptacle short, sepals oblong, 2–2·5 cm. long ; petals 5, subequal in length, clawed, 3–5 cm. long, up to 5 cm. broad ; stamens 10, free, up to 3 cm. long ; fruit elongate-oblong, 30–60 cm. long, about 5 cm. broad, valves woody, glabrous, tardily dehiscent ; seeds numerous, up to 2·5 cm. long and 6 mm. diam.   *regia*

D. regia (*Boj. ex Hook.*) *Raf.* l.c. (1837). *Poinciana regia* Boj. ex Hook. in Bot. Mag. t. 2884 (1829) ; F.T.A. 2 : 266. Commonly planted ornamental, known in West Africa as " Flame of the Forest " or " Flamboyant ", now subspontaneous in many localities ; posterior petal variegated red and white, other petals red with orange claw. A native of Madagascar, widely planted in the tropics.

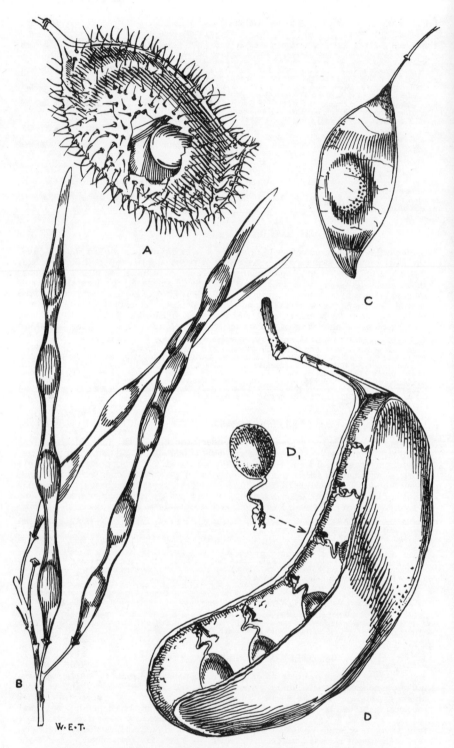

W·E·T·

FIG. 154.—FRUITS OF CAESALPINIACEAE.

A, *Caesalpinia bonduc* (Linn.) Roxb. B, *Parkinsonia aculeata* Linn. C, *Burkea africana* Hook.
D, *Erythrophleum guineense* G. Don. D1, seed of same with funicle.

54. BUSSEA Harms in Engl. Bot. Jahrb. 33 : 159 (1902).

Large tree ; leaves bipinnate ; leaflets alternate, oblong-elliptic, very long-acuminate, 6–10 cm. long, 3–4 cm. broad, laxly reticulate, glossy above, glabrous ; flowers racemose ; racemes about 20 cm. long ; pedicels up to 1 cm. long, tomentose ; sepals triangular-ovate, subacute, about 8 mm. long, rusty-brown-tomentellous when dry ; petals crinkly, obovate, about 2 cm. long ; filaments and style densely villous ; fruit narrowly oblanceolate, woody with strongly recurved valves, about 15 cm. long and 3·5 cm. broad, densely brown-ferruginous when young   ..   ..   *occidentalis*

B. **occidentalis** *Hutch.* in F.W.T.A., ed. 1, 1 : 350 (1928) ; Kew Bull. 1928 : 400 ; Aubrév. Fl. For. C. Iv. 1 : 268, t. 105. *Peltophorum africanum* [1] of Chev. Bot. 219, partly (*Chev.* 19212), and of F.W.T.A., ed. 1, 1 : 350, partly (*Chev.* 19212), not of Sond. ; see Aubrév. Fl. For. Soud.-Guin. 209. Forest tree, to 100 ft. high, with dark brown-tomentose branchlets and inflorescence ; flowers conspicuous, yellow ; pods erect, elastic, curling up after opening.
    S.L.: Hangha (Aug.) *Lane-Poole* 394 ! *Aylmer* 241 ! Gola Forest (fr. Jan.) *Unwin & Smythe* 64 ! **Lib.:** St. Johns R., Begwai (young fr. Sept.) *Bunting* 10 ! Gbanga (Sept.) *Linder* 479 ! Dukwia R. (Apr.) *Cooper* 400 ! **Iv.C.:** Tano R. (July) *Chipp* 305 ! And various localities *fide* Aubrév. *l.c.* **G.C.:** Dunkwa (Aug.) *Chipp* 715 ! Obuasi *Vigne* FH 920 ! Essuasu (May) *Vigne* FH 115 ! Akyempim, NE. of Tarkwa (fr. May) *Chipp* 219 ! Kankan (June) *Thompson* 90 ! (See Appendix, p. 177.)

55. PARKINSONIA Linn.—F.T.A. 2 : 266.

Small tree ; branchlets armed with rigid prickles (modified common petiole) ; leaves bipinnate, pinnae, 2–4, the rhachis elongated, phyllode-like, flattened, spine-pointed, with numerous very small obovate leaflets along the margin ; flowers in small axillary racemes ; pedicels up to 1·3 cm. long, slender, glabrous ; petals 5, spreading ; stamens 10, free, filaments pilose in the lower part ; fruit linear, torulose, up to 13 cm. long, beaked, striate, glabrous   ..   ..   *aculeata*

P. **aculeata** *Linn.*—F.T.A. 2 : 267 ; Chev. Bot. 221. An exotic shrub or small tree, 10–15 ft. high, with fairly conspicuous yellow flowers ; planted in native towns throughout the Sudan and elsewhere in dry localities ; native of tropical America.
    Sen.: *Roger* ! *Perrottet* 251 ! Dakar (May) *Debeaux* 180 ! **Gam.:** *Whitfield* ! **Fr.Sud.:** Diafarabé (Sept. & Lean 102 ! Timbuktu *Hagerup* 224. Sompi *Chev.* 3192. Koulikoro *Chev.* **S.L.:** *Don* ! Yengema (fl. & fr. July) *Deighton* 4131 ! **N.Nig.:** Mashi, Katsina (fl. & fr. Feb.) *Meikle* 1215 ! Kano (Oct.) *Keay* FHI 21149 ! Katagum (fl. & fr. Oct.) *Dalz.* 5 ! Bauchi Plateau (Mar.) *Lely* P. 181 ! Native of tropical America, widely naturalized in Africa. (See Appendix, p. 199.)

56. BURKEA Hook.—F.T.A. 2 : 319.

A tree ; young parts reddish-tomentose ; leaves bipinnate, 2–5-jugate, jugae opposite ; leaflets alternate, up to about 12, ovate to ovate-lanceolate, obtuse or emarginate at the apex, unequal-sided at base, up to 5 cm. long and 3·5 cm. broad, at first silky, at length glabrous and more or less glaucous beneath ; flowers in panicles of slender spikes up to 30 cm. long ; axis pubescent ; calyx 1·5 mm. long, lobes ciliate ; petals about twice as long as the calyx, glabrous ; stamens 10 ; ovary densely villous ; fruits flat and thin, elliptic-lanceolate, long-stipitate, about 6 cm. long, 2·3 cm. broad, shortly appressed-pubescent .. .. *africana*

B. **africana** *Hook.*—F.T.A. 2 : 320 ; Aubrév. Fl. For. Soud.-Guin. 245, t. 50, 4–6. A deciduous tree attaining 50–70 ft. in height, with stout knotted branchlets, blackish corrugated bark, light silky foliage, and pendulous spikes of small creamy fragrant flowers crowded with the leaves at the ends of the branchlets.
    Sen.: *Berhaut* 770. **Fr.Sud.:** Bamako (May) *Hagerup* 50 ! And various localities *fide* Aubrév. *l.c.* **Fr.G.:** Kouroussa *Pobéguin* 871 ! **Iv.C.:** Ferkéssédougou, Bobo-Dioulasso and Black Volta *fide* Aubrév. *l.c.*   **G.C.:** Ejura (fr. Aug.) *Chipp* 726 ! Kintampo (Mar.) *Dalz.* 15 ! Grube (Mar.) *Rea* FH 1668 ! Pan to Bujan (Apr.) *Kitson* 917 ! **Togo:** *Kersting* A. 308 ! **Dah.:** Savé *Chev.* 23577 ! **N.Nig.:** Nupe *Barter* 1134 ! Sokoto (fl. & fr. Mar.) *Lely* 844 ! Kontagora (Dec.) *Dalz.* 73 ! Yola (Feb.) *Dalz.* 188 ! **S.Nig.:** Olokemeji *Foster* 310 ! Awka *Thomas* 11 ! Agbani Road, Enugu (Feb.) *Chesters* 188 ! Widespread in tropical Africa. (See Appendix, p. 177.)

57. ERYTHROPHLEUM Afzel. ex R.Br.—F.T.A. 2 : 320.

Leaflets acuminate ; fruits usually rounded at each end, oblong, stipe more or less lateral, valves woody-leathery ; seeds 5–10 :
  Pedicels 1–2 mm. long ; racemes 1–1·5 cm. diam., yellowish-brown tomentellous, in lax panicles ; pinnae 2–3 (rarely 1 or 4) pairs ; leaflets alternate, 10–13 on each pinna, ovate-elliptic, 5–9 cm. long, 3–5 cm. broad, usually pubescent on petiolule, midrib beneath and rhachis, usually drying pale green or greenish-brown ; fruits up to 15 cm. long and 5 cm. broad .. .. 1. *guineense*
  Pedicels up to 1 mm. long ; racemes less than 1 cm. diam., reddish-brown tomentellous, in rather stiff panicles ; leaves similar to above, but usually quite glabrous and drying blackish ; fruits up to 10 cm. long and 3·5 cm. broad .. .. 2. *ivorense*
Leaflets rounded or emarginate at apex, more or less obliquely oblong, unequal-sided at base, 3–4 cm. long, 1·3–2 cm. broad, softly pubescent when young ; pinnae 2–4 ; leaflets alternate, 8–16 on each pinna ; flowers in densely pubescent panicles of spike-like racemes ; fruits oblong to oblong-elliptic, attenuate-cuneate at base, alternate to rounded at apex, stipe more or less central, valves subcoriaceous, 7–19 cm. long, 2·2–4·5 cm. broad ; seeds 2–5 .. .. 3. *africanum*

---

[1] The other specimens cited by Chevalier and by Hutchinson & Dalziel are of the introduced cultivated *Peltophorum pterocarpum*. The genus *Peltophorum* Walp. is not known to be indigenous in our area.

1. **E. guineense** *G. Don*—F.T.A. 2 : 320 ; Aubrév. Fl. For. C. Iv. 1 : 270 ; Fl. For. Soud.-Guin. 241 ; Wilczek in Fl. Congo Belge 3 : 243, t. 18. Large tree, with spreading crown, to 100 ft. high ; flowers small, very fragrant, crowded in spike-like racemes ; pods opening without scattering the seeds ; in the drier parts of the forest regions and in fringing forest.
Sen.: *Heudelot* 155 ! Tambana (Feb.) *Chev.* 2961. **Gam.:** *Whitfield*! *Daniell*! Albreda (Mar.–Apr.) *Perrottet* 291 ! MacCarthy Isl. *Dalz.* 8067 ! **Fr.Sud.:** Kita *Dubois* 106 ! **Port.G.:** Empacaca, Canchungo (Apr.) *Esp. Santo* 1942 ! **Fr.G.:** Kouroussa (Apr.) *Pobéguin* 692 ! Konkouré Valley *Pobéguin*! Falaba to Farana *Sc. Elliot* 5063 ! **S.L.:** *Afzelius*! *Don*! Musaia (Apr.) *Deighton* 5473 ! Batkanu (fr. Feb., July) *Small* 153 ! Gberia Timbako (fr. Sept.) *Small* 322 ! **Iv.C.:** Baoulé circle *fide* Aubrév. *l.c.* **G.C.:** Ejura (Apr.) *Packham*! *Chipp* 754 ! Bomkrom Kwahu (fr. Jan.) *Chipp* 640 ! Bana Hill, Krobo (fr. Jan.) *Irvine* 1912 ! **Togo:** *Baumann*! *Büttner*! **N.Nig.:** Bassa (fr. Aug.) *Elliott* 92 ! Lapai (Feb.) *Yates* 30 ! Kaduna *Lamb* 96 ! Magamma (fr. Feb.) *Lely* 727 ! **S.Nig.:** *Unwin* 22 ! Ugboha F.R., Ishan (fl. & fr. Feb.) *Umana* FHI 29120 ! Udi (Jan.) *Chesters* 217 ! **Br.Cam.:** Fang, Bamenda (fr. Apr.) *Maitland* 1473 ! Mufung, Bamenda (fl. & fr. Feb.) *Johnstone* 42/31 ! Widespread in tropical Africa. (See Appendix, p. 192.)
2. **E. ivorense** *A. Chev.* Vég. Util. 5 : 178 (1909) ; Aubrév. Fl. For. C. Iv. 1 : 270, t. 106. *E. micranthum* Harms ex Holl. (1911)—F.W.T.A., ed. 1, 1 : 351. Large forest tree, to 130 ft. high ; flowers reddish similar to the above, but found in only the moister parts of the forest regions, and flowering later in the year.
S.L.: *Afzelius*! Kambui F.R. (Oct.) *Edwardson* 145 ! Njala (Sept.) *Deighton* 2905 ! **Lib.:** Dukwia R. *Cooper* 386 ! R. Cess (fr. Mar.) *Baldwin* 11276 ! **Iv.C.:** Bingerville (fr. Feb.) *Chev.* 16220 ! Azaguié *Chev.* 22290 ! And various localities *fide* Aubrév. *l.c.* **G.C.:** Kokum, Axim (July) *Chipp* 259 ! Asientiem (July) *Chipp* 293 ! Dunkwa (fl. & fr. July) *Vigne* FH 913 ! 923 ! Tarkwa (June) *Evans*! Mampong (Aug.) *Andoh* FH 4232 ! **S.Nig.:** Akilla *Kennedy* 2016 ! Brass (June) *Barter* 3 ! *King-Church* 16 ! Nun R. (fl. & fr. Sept.) *Mann* 482 ! Ikom (Sept.) *Catterall* 33 ! **Br.Cam.:** Likomba (fr. Nov.) *Mildbr.* 10696 ! Also in French Cameroons, Gabon and Spanish Guinea. (See Appendix, p. 193.)
3. **E. africanum** (*Welw. ex Benth.*) *Harms* in Fedde Rep. 12 : 298 (1913) ; Aubrév. Fl. For. Soud.-Guin. 243, t. 50, 1–3 ; Wilczek l.c. 3 : 244. *Gleditschia africana* Welw. ex Benth. (1865)—F.T.A. 2 : 265 ; Chev. Bot. 221. A savannah tree, to 40 ft. high, with spreading crown, the young parts softly tomentose and foliage very like that of *Burkea africana*.
Sen.: Casamance *Etesse* 10 ! **Gam.:** *Rosevear* 100 ! 101 ! **Fr.Sud.:** Bougouni to Sikasso *fide* Aubrév. *l.c.* **Port.G.:** Bafata *Esp. Santo* 2342 ! **Fr.G.:** Kadé *Pobéguin* 2005. Dabola *Pobéguin* 1510 ! **Iv.C.:** Bobo Dioulasso *Aubrév.* 1955 ! **G.C.:** Gambaga (fl. Feb., fr. June) *Vigne* FH 4476 ! *Akpabla* 679 ! Pang *Kitson* 587 ! 588 ! Jema (fr. June) *Brown* FH 2309 ! **Togo:** *Kersting* 4 ; 384. **Dah.:** Atacora Mts. (fr. June) *Chev.* 24163 ! **N.Nig.:** Kontagora *Dalz.*! Bakura-Tureta F.R. (fr. Sept.) *Dundas*! Widespread in tropical Africa. (See Appendix, p. 192.)

58. PACHYELASMA Harms in Engl. Bot. Jahrb. 49 : 428 (1913).

Tree, with glabrous branches ; leaves bipinnate, up to 35 cm. long, glabrous ; pinnae 2–5 pairs, opposite or rarely alternate ; leaflets 9–15, alternate, asymmetric, oblong to oblong-oblanceolate, cuneate to obtuse at base, obtuse to rounded or emarginate at apex, 4–10·5 cm. long, 1·5–3·5 cm. broad ; flowers in many-flowered robust spiciform racemes up to 18 cm. long ; bracts small, caducous ; pedicels 2–3 mm. long ; receptacle broadly cupulate, about 2 mm. long ; sepals 5, rounded, about 2 mm. long ; petals 5, obovate-oblong, 4·5–6 mm. long, 2–3 mm. broad ; stamens 10, free ; fruits linear-oblong or oblong-lanceolate, 15–37 cm. long, 3·5–4 cm. broad, 2–2·5 cm. thick, very hard, with thick projecting sutures ; seeds 10–18 .. .. *tessmannii*

P. tessmannii (*Harms*) *Harms* in Engl. Bot. Jahrb. 49 : 430, with fig. (1913) ; Bak. f. Leg. Trop. Afr. 3 : 623 ; Wilczek in Fl. Congo Belge 3 : 240. *Stachyothyrus tessmannii* Harms (1910). Forest tree, to 200 ft. high, with cylindrical bole and well-developed plank-buttresses ; flowers cherry-red ; fruits black.
S.Nig.: Eba (fr. July) *Kennedy* 1720 ! Oban *Gray* 2/6 ! Also Ijebu Ode, Sapoba and Cameroons *fide* Kennedy *l.c.* Also in French Cameroons, Gabon, Spanish Guinea and Belgian Congo. (See Appendix, p. 199.)

90. MIMOSACEAE [1]

Trees or shrubs, very rarely herbs. Leaves mostly bipinnate, rarely simply pinnate, often with large glands on the rhachis. Flowers hermaphrodite, small, spicate, racemose or capitate, actinomorphic, usually 5-merous. Calyx tubular, valvate or very rarely imbricate, 5-lobed or toothed. Petals valvate, free or connate into a short tube. Stamens equal in number to the sepals or more numerous, free or monadelphous ; anthers small, 2-celled, opening lengthwise, often with a deciduous gland at the apex. Ovary superior, of 1 carpel. Fruit mostly dehiscent. Seeds with scanty or no endosperm.

Mainly tropics and subtropics, often in dry regions and then frequently spiny ; easily recognized by the usually bipinnate leaves and small flowers with valvate petals.

Calyx-lobes imbricate :
  Flowers bisexual in elongated paniculate spikes ; stamens 5 ; staminodes several ; fruit elongated, with the valves recurved when mature .. .. 1. **Pentaclethra**
  Flowers in dense short more or less club-shaped heads, the upper ones bisexual, the lower staminate or neuter ; stamens 10 ; no staminodes ; fruits 2-valved
            2. **Parkia**

Calyx-lobes valvate or open from a very early stage :
  Filaments free or united only with the disk at the base ; flowers spicate or capitate :
    Stamens definite, as many as or up to twice as many as the petals :
      Flowers in spikes or racemes :
        All the flowers bisexual or at any rate sexual ; branches without spines or prickles (but see No. 9) :
          Fruits dehiscent :

---

[1] I am much indebted to J. P. M. Brenan for making available to me information which will later be published in his account of this family in the Flora of Tropical East Africa.

Valves of fruit remaining intact, not breaking into 1-seeded segments :
 Leaflets alternate ; seeds flattened, winged :
  Indumentum stellate ; fruits very long, narrow, dehiscent along one margin
   only ; flowers pedicellate, calyx and corolla puberulous outside ; stamens
   with smooth filaments inserted on the side of a thin but conspicuous disk
     5. **Cylicodiscus**
  Indumentum not stellate ; fruits very broadly oblong, dehiscent along both
   margins ; flowers sessile, glabrous outside ; stamens with densely papillose
   filaments inserted on the outside of a low thick disk   .. 8. **Fillaeopsis**
 Leaflets opposite :
  Fruits dehiscent along both margins ; valves woody, spirally twisted or
   recurving on dehiscence, usually conspicuously attenuate below ;   seeds
   not winged :
   Flowers sessile ; petals united below into a tube at least ⅓ as long as corolla;
    leaves with 1 pair of pinnae, rachis glandular at their insertion ; trees
      3. **Calpocalyx**
   Flowers pedicellate ; petals free or nearly so ; most leaves with 2 pairs of
    pinnae, rachis not glandular ; lianes or shrubs   4. **Pseudoprosopis**
  Fruits dehiscent along one margin only ; valves thin ; seeds flattened, winged :
   Leaves without glands on rhachides ; funicle of seeds attached in the middle ;
    ovary and petals glabrous outside   ..   ..   6. **Piptadeniastrum**
   Leaves with glands on rhachides ; funicle of seeds attached at one end ;
    ovary and petals pubescent or puberulous outside   ..   7. **Newtonia**
  Valves of fruit breaking into 1-seeded segments and leaving the continuous
   persistent sutures ; branchlets prickly in *E. scelerata* only   .. 9. **Entada**
 Fruits indehiscent :
  Fruits thin oblong, samaroid ; leaves without glands on rhachides ; leaflets
   opposite ..   ..   10. **Aubrevillea**
  Fruits more or less turgid :
   Leaves with glands on rhachides ; leaflets opposite ; fruits subterete
     11. **Prosopis**
   Leaves without glands on rhachides ; leaflets alternate :
    Fruits tetragonal, not winged ..   ..   12. **Amblygonocarpus**
    Fruits with a fleshy wing-like ridge along the back of each valve
      13. **Tetrapleura**
Upper flowers of the spikes bisexual, the lower neuter ; fruits indehiscent twisted
  and crowded ; leaves with glands ; branches armed with spines terminating
  short lateral branchlets ..   ..   14. **Dichrostachys**
Flowers capitate :
 Trees, shrubs or rarely herbs, not aquatic :
  Anthers tipped by a deciduous apical gland ; valves of fruit woody and strongly
   elastic ; large tree ; pinnae 1 pair   ..   .. 15. **Xylia**
  Anthers without an apical gland ; fruit not woody :
   Tree with smooth branchlets ; pinnae about 5 pairs ; fruits linear, flat, not
    prickly ..   ..   16. **Leucaena**
   Slender trailer with quadrangular stems covered with numerous recurved
    prickles ; pinnae 2–3 pairs ; fruits linear, 4-ridged, prickly   17. **Schrankia**
   Shrubs or herbs ; stems terete ; pinnae several pairs ; fruits flat, prickly or
    setose ..   .. 18. **Mimosa**
 Aquatic herbs with some of the leaves submerged ; leaves sensitive ; pinnae
  2–3 pairs ; flowers capitate, on elongated peduncles, the upper ones bisexual, the
  lower neuter ; anthers with a deciduous apical gland ; fruit flat, glabrous
    19. **Neptunia**
Stamens more than twice as many as the petals ; flowers in spikes or heads ; branch-
  lets usually with spines or prickles..   ..   .. 20. **Acacia**
Filaments united into a short or long tube :
 Flowers in spikes ; branches armed with stipular spines ..   ..   .. 20. **Acacia**
 Flowers capitate ; branches without spines or prickles :
  Valves of fruit not separating elastically ; stipules mostly inconspicuous, or
   caducous :
   Fruit not jointed, oblong, with thin flat valves ..   ..   ..   21. **Albizia**
   Fruit with thick valves, not breaking up into 1-seeded segments   22. **Samanea**
   Fruit jointed and breaking up into 1-seeded segments, often curved or circinate
     23. **Cathormion**
  Valves of fruit separating elastically from the top downwards ; stipules persistent,
   nervose   ..   ..   24. **Calliandra**

In addition to the above, other genera of this family are represented by the following species which have
been introduced in W. Africa : *Pithecellobium dulce* (Roxb.) Benth., a native of tropical America ; *Adenanthera pavonina* Linn., a native of tropical Asia ; *Enterolobium cyclocarpum* (Jacq.) Griseb., a native of tropical
America.

13

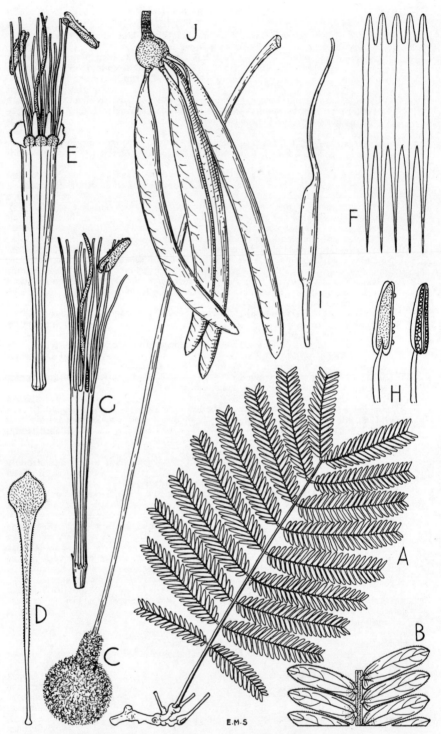

FIG. 155.—PARKIA CLAPPERTONIANA *Keay* (MIMOSACEAE).

A, leaf, × ⅓.   B, upper surface of leaflets, × 1½.   C, inflorescence, × ⅓.   D, bract, × 6.   E. flower, × 6.   F, corolla tube opened out, × 6.   G, staminal tube and style, × 6.   H. anthers, × 10.   I, ovary, × 6.   J, fruits, × ⅓.
A & B are drawn from *A. F. Ross* 115 ; C–G & I from *Barter* 1136 ;   H from *A. F. Ross* 60 : J from *Elliott* 54.

1. PENTACLETHRA Benth.—F.T.A. 2 : 322 ; Benth. in Trans. Linn. Soc. 30 : 360 (1875).

A tree up to 25 m. high ; young parts rusty-tomentellous ; leaves bipinnate ; pinnae opposite, in about 10 pairs ; rhachis angular, scurfy-puberulous ; leaflets about 12 pairs, opposite, obliquely oblong, truncate and unequal-sided at base, emarginate and shortly mucronate at apex, about 2·5 cm. long and 1 cm. broad, glabrous or nearly so ; spikes paniculate, elongated ; calyx very short ; petals about 2 mm. long ; stamens 5, with alternating staminodes ; fruit elongated, 8–10 cm. broad, valves thick and woody, at length recurved ; seeds flattish, varying from oblanceolate to subrhomboid, striate, up to 6–7 cm. long.. .. .. .. *macrophylla*

P. macrophylla *Benth.*—F.T.A. 2 : 322 ; Benth. l.c. 360 ; Bak. f. Leg. Trop. Afr. 3 : 780 ; Aubrév. Fl. For. C. Iv. 1 : 192, t. 71. Forest tree, with spreading crown ; flowers yellowish-creamy ; pods strongly elastic ; often cultivated.
Sen.: Bignona, Casamance *Chev.* 3117 ! **Port.G.:** Bissau *Esp. Santo* 1810 ! **Fr.G.:** Rio Nunez *Heudelot* 825 ! Conakry *Dalz.* 8064 ! **S.L.:** Makong (Apr.) *Macdonald* 27 ! ¡Kenema (Mar.) *Lane-Poole* 211 ! *Smythe* 234 ! **Lib.:** Dukwia R. (Feb.) *Cooper* 193 ! **Iv.C.:** Abidjan *Aubrév.* 19 ! **G.C.:** Imbraim (Mar.) *Thompson* 27 ! Kumasi *Cummins* 219 ! **Dah.:** Porto Novo (Feb.) *Chev.* 22895 ! **S.Nig.:** Sapoba *Kennedy* 1930 ! Eket *Talbot* 3113 ! Oban *Talbot* 1348 ! **Br.Cam.:** Cameroons R. (Jan.) *Mann* 2203 ! **F.Po:** (Nov.) *T. Vogel* 232 ! Extends southwards to Gabon, Angola and Belgian Congo, also in Principe and S. Tomé. (See Appendix, p. 220.)

2. PARKIA R.Br.—F.T.A. 2 : 323.

Corolla lobes very short (about ⅙ total length of corolla) ; leaf-rhachis greyish-puberulous ; leaflets alternate or subopposite, or opposite towards the apices of the pinnae ; fruits 19–30 cm. long, 1·5–2·3 cm. broad, not or only slightly indented between the seeds ; trees of the savannah regions :
  Leaflets 35–65 on each side of the rhachis of each pinna, the median ones about 10 (–12) mm. long and 2 (–3) mm. broad ; pinnae 8–16 on each side of the leaf-rhachis 1. *biglobosa*
  Leaflets 14–30 on each side of the rhachis of each pinna, the median ones (11–) 12–18 (–20) mm. long, (2·5–) 3–5 (–6·5) mm. broad ; pinnae 6–11 on each side of the leaf-rhachis .. .. 2. *clappertoniana*
Corolla lobes long (½–⅘ total length of corolla) ; leaf-rhachis rusty-puberulous or tomentellous ; forest trees :
  Leaflets 20–55 on each side of the rhachis of each pinna, opposite, the median ones 4·5–10·5 (–12·5) mm. long, 1–2 (–3) mm. broad ; pinnae 10–26 pairs ; fruits 25–40 cm. long, 1·8–3 cm. broad, markedly indented between the seeds .. 3. *bicolor*
  Leaflets 12–28 on each side of the rhachis of each pinna, opposite to alternate, the median ones 12–38 mm. long, 4–14 mm. broad ; pinnae 5–11 pairs ; fruits variable 4. *filicoidea*

1. **P. biglobosa** (*Jacq.*) *Benth.*—F.T.A. 2 : 324, partly ; Bak. f. Leg. Trop. Afr. 3 : 782 ; Aubrév. Fl. For. C. Iv. 1 : 192, and Fl. For. Soud.-Guin. 249. *Mimosa biglobosa* Jacq. (1763). Tree 30–50 ft. high, with wide-spreading crown ; flowers red or orange ; seeds and pulp edible ; in savannah country.
Sen.: Tivaouane (May) *De Wailly* 4588 ! Fouladou *Etesse* ! **Gam.:** *Perrottet* 261 ! *Rosevear* 75 ! Kombo *Dawe* 14 ! Genieri *Fox* 50 ! **Fr.Sud.:** Siguiri (Jan.) *Chev.* 292 ! **Fr.G.:** Kouroussa (Jan.) *Pobéguin* 130 ! **S.L.:** *Barter* ! *Sc. Elliot* 3857 ! 4771 ! Heddle's Farm, Freetown (Jan.) *Deighton* 1061 ! Musaia (young fr. Apr.) *Deighton* 5480 ! **Iv.C.:** Bondoukou *Aubrév.* 717 ! **G.C.:** Pong, Wa *Kitson* 698 ! Also cultivated in West Indies, whence it was originally described. (See Appendix, p. 218.)
2. **P. clappertoniana** *Keay* in Bull. Jard. Bot. Brux. 25 : 209 (1955). *P. africana* R.Br. (illegitimate name), partly (*Clapperton* spec. only). *P. filicoidea* Welw. ex Oliv. F.T.A. 2 : 324, partly (*Barter* spec.) ; F.W.T.A., ed. 1, 1 : 352. *P. filicoidea* var. *glauca* Bak. f. l.c. 781 (1930). *P. oliveri* of Aubrév. Fl. For. C. Iv. 1 : 194 and Fl. For. Soud.-Guin. 249, not of J. F. Macbr. Tree, to 70 ft. high, with short bole and wide spreading crown ; flowers deep red ; seeds and pulp edible ; in savannah country.
G.C.: Amansare (fr. July) *Chipp* 514 ! Abene, Kwahu (Jan.) *Chipp* 631 ! **Togo:** Kete Krachi (Jan.) *Vigne* FH 1540 ! *Kitson* ! **Dah.:** *fide* Aubrév. l.c. **N.Nig.:** *Barter* 1136 ! *Clapperton* ! Kaura Namoda (fl. & fr. May) *Keay* FHI 16205 ! Samaru (fr. May) *Baldwin* 12005 ! Zungeru (fl. & fr. Jan.) *Elliott* 54 ! **S.Nig.:** Olokemeji (Feb.) *Foster* 162 ! Also in Ubangi-Shari and A.-E. Sudan. (See Appendix, p. 218.)
3. **P. bicolor** *A. Chev.* in Bull. Soc. Bot. Fr. 55, Mém. 8 : 34 (1908) ; Bak. f. l.c. 782 ; Aubrév. Fl. For. C. Iv. 1 : 196, t. 73 ; Gilbert & Boutique in Fl. Congo Belge 3 : 144, t. 10. *P. agboensis* A. Chev. l.c. 35 (1908). *P. zenkeri* Harms (1911). Tall forest tree, to 100 ft. high, with wide crown and well-developed buttresses ; flowers bluish-red or yellowish-red ; often by river banks.
Fr.G.: Kindia (fr. May) *Chev.* 13547 ! **S.L.:** *Sc. Elliot* ! Njala (Feb.) *Deighton* 1071 ! Simiria *Deighton* 1396 ! Moyamba (fr. Mar.) *Deighton* 1922 ! **Lib.:** Dukwia R. (fr. Feb.) *Cooper* 185 ! 338 ! **Iv.C.:** Azagulé *Chev.* 22292 ! Banco *Aubrév.* 387 ! **G.C.:** Ancobra River mouth (Nov.-Jan.) *Chipp* 23 ! Axim (fr. Nov.) *Chipp* 27 ! Abetifi (Jan.) *Irvine* 1659 ! **Dah.:** Niaouli, Allada *Chev.* 23423 ! **S.Nig.:** Yaba (Dec.) *Millen* 72 ! Sapoba *Kennedy* 1969 ! 2073 ! Degema (fl. & fr. Dec.) *Unwin* 27 ! Eket *Talbot* 3047 ! Oban *Talbot* 1350 ! 1467 ! **Br.Cam.:** Bambuko F.R., Kumba (fr. Sept.) *Olorunfemi* FHI 30774 ! Also in French Cameroons, Gabon, Cabinda and Belgian Congo. (See Appendix, p. 217.)
4. **P. filicoidea** *Welw. ex Oliv.* F.T.A. 2 : 324 (1871), partly (excl. *Barter* s.n.) ; Bak. f. l.c. 781, partly ; Aubrév. Fl. For. C. Iv. 1 : 196, t. 72, and Fl. For. Soud.-Guin. 249 ; Gilbert & Boutique in Fl. Congo Belge 3 : 141. Tree to 100 ft. high, in the drier parts of the forest zone and in gallery forest.
Iv.C.: Bamoro, N. of Bouaké, from Koun to Bondoukou *Aubrév.* 716 ; 2251. Widespread in eastern and central tropical Africa.

3. CALPOCALYX Harms in E. & P. Pflanzenfam. Nachtr. 1 : 191 (1897) ; Engl. Bot. Jahrb. 26 : 257 (1899).

Bracts nearly as long as the flowers, conspicuous in bud, with a broad blade tapered above into a point :
  Spikes about 2·5 cm. long, pedunculate, forming a panicle ; calyx 1·5–2 mm. long ;

calyx and petals tomentellous outside ; bracts persistent ; leaflets 5–6 pairs, oblong-elliptic, unequal-sided and rounded at base, abruptly acuminate, 6–12 cm. long, 3–7 cm. broad ; petiole 1·3–3 cm. long ; petiolules about 5 mm. long
1. *dinklagei*

Spikes 9–11 cm. long, peduncle 3–4 cm. long, solitary or paired in the upper axils, rhachis and peduncle stout (the latter up to 3 mm. diam.) ; calyx 3 mm. long ; calyx and petals densely pubescent outside ; bracts early caducous ; leaflets 5–6 pairs, oblong or obovate-oblong, cuneate or obtuse at base ; caudate-acuminate, 10–19 cm. long, 3·5–6·5 cm. broad ; petiole 1·5–2 cm. long ; petiolules 6–8 mm. long ; fruit obliquely oblanceolate, 23–28 cm. long, 7–9 cm. broad, glabrous, with 3–5 suboblong seeds   ..   ..   2. *aubrevillei*
Bracts much shorter than the flowers, not conspicuous in bud, spathulate, not tapered into a point ; inflorescence of pedunculate spikes forming a panicle :
Inflorescences terminal on leafy shoots only ; calyx densely to sparingly puberulous ; leaflets 4–7 pairs, oblong or oblong-elliptic, rounded at base, long-acuminate, 6–15 cm. long, 2–6 cm. broad ; petiole 1·5–4·5 (rarely to 7) cm. long ; petiolules 3–7 mm. long ; fruit sickle-shaped, tapered to the base, 15–17 cm. long, 3–4·5 cm. broad, glabrous, with 5–10 seeds   ..   ..   3. *brevibracteatus*
Inflorescences terminal on leafy shoots and also on the bole, congested ; calyx sparingly puberulous ; leaflets 4–5 pairs, oblong, unequal at base, shortly acuminate, 5–12 cm. long, 2·5–5 cm. broad ; petiole 5–7 cm. long ; petiolules 4–5 mm. long ; fruit not known   ..   ..   4. *winkleri*
Inflorescences all on the bole, lax ; calyx glabrous ; leaflets 7–8 pairs, oblong-elliptic, obtuse to rounded at base, caudate-acuminate, 10–12 cm. long, 4–6 cm. broad ; petiole 12·5–14 cm. long ; petiolules 3–5 mm. long ; fruit not known 5. *cauliflorus*

1. **C. dinklagei** *Harms* in E. & P. Pflanzenfam. Nachtr. 1 : 191 (1897) ; Engl. Bot. Jahrb. 26 : 257, t. 5, F–G (1899) ; Bak. f. Leg. Trop. Afr. 3 : 799. A forest tree 30–50 ft. high ; flowers white or yellowish.
    S.Nig.: Oban *Talbot* 1641 ! **Br.Cam.:** S. Bakundu F.R. (Mar., Apr.) *Brenan* 9298 ! *Olorunfemi* FHI 30526 ! Also in French Cameroons, Spanish Guinea and Gabon. (See Appendix, p. 213.)
2. **C. aubrevillei** *Pellegr.* in Bull. Soc. Bot. Fr. 80 : 467 (1933) ; Aubrév. Fl. For. C. Iv. 1 : 190, t. 70. A forest tree, to 90 ft. high ; flowers yellowish, fragrant.
    S.L.: Gola Forest (fl. Apr., fr. Oct.) *Sawyerr* in Hb. *King* 222 ! *Small* 570 ! **Lib.:** Dukwia R. (Apr.) *Cooper* 402 ! **Iv.C.:** Zagni *Aubrév.* 1222. Toulépleu to Guiglo *Aubrév.* 1196. Dakpadou *Aubrév.* 858 ! (See Appendix, p. 213.)
3. **C. brevibracteatus** *Harms* in Bull. Soc. Bot. Fr. 58, Mém. 8 : 155 (1912) ; Bak. f. l.c. 799 ; Aubrév. l.c. 1 : 190, t. 69 ; Kennedy For. Fl. S. Nig. 108. *C. macrostachys* Harms l.c. 156 (1912). A forest tree, to 80 ft. high ; flowers orange.
    S.L.: Kambui *Lane-Poole* 459 ! Kenema (Oct.) *Aylmer* 253 ! Gola Forest (fr. Jan.) *Unwin & Smythe* 68 ! **Lib.:** Gbanga (Sept.) *Linder* 471 ! 667 ! Peahtah (Oct.) *Linder* 1050 ! Dukwia R. (Oct.) *Cooper* 56 ! Javajai, Boporo (Nov.) *Baldwin* 10255 ! **Iv.C.:** Yapo (Oct.) *Chev.* 22368 ! Abidjan *Aubrév.* 184 ! **G.C.:** Tano Rapids (Sept.) *Chipp* 371 ! Oda (Sept.) *Vigne* FH 1361 ! Wiawso (Nov.) *Mc.Ainsh* 386 ! Subiri F.R. (Sept.) *Andoh* FH 5571 ! **S.Nig.:** Sapoba *Kennedy* 600 ! 1635 ! 2621 ! (See Appendix, p. 213.)
4. **C. winkleri** (*Harms*) *Harms* in Notizbl. Bot. Gart. Berl. 10 : 971 (1930). *Piptadenia winkleri* Harms in Engl. Bot. Jahrb. 40 : 17 (1907). A forest tree, to 65 ft. high ; flowers among the foliage and also on the stem right down to the ground ; branchlets hollow, inhabited by ants.
    Br.Cam.: Likomba (Nov.) *Mildbr.* 10721 ! Tiko (fl. & fr. Jan.) *Dunlap* 174 ! Also in French Cameroons.
5. **C. cauliflorus** *Hoyle* in Kew Bull. 1933 : 172. Understorey forest tree, to 65 ft. high ; cauliflorous, flowers apparently never terminal among the foliage ; calyx and petals red, stamens yellow ; branchlets hollow, inhabited by ants.
    S.Nig.: Ikom *Rosevear* 39/30a ! 40/30a ! Cross River North F.R., Ikom (Dec.) *Keay* FHI 28151 !

Further collecting to show range in leaf-shape, together with observations on the cauliflorous habit, are needed to determine how far species Nos. 3, 4 and 5 are in fact distinct.

4. PSEUDOPROSOPIS Harms in Engl. Bot. Jahrb. 33 : 152 (1902).

Leaflets closely tomentellous beneath ; pinnae (1–) 2 pairs ; leaflets 3–5 pairs, oblong-lanceolate to obovate-oblanceolate, acute or acuminate, rounded at base, mucronate, 3–7 cm. long, 1·5–3 cm. broad ; panicles elongated, tomentellous ; bracts subulate, persistent ; pedicels 1–1·5 mm. long ; calyx and petals tomentose outside ; fruit sickle-shaped, tapered to the base, glabrous, at least 10 cm. long (mature fruit not known)   ..   ..   *sericeus*

P. sericeus (*Hutch. & Dalz.*) *Brenan* in Kew Bull. 1955 : 184. *Calpocalyx sericeus* Hutch. & Dalz., F.W.T.A., ed. 1, 1 : 353 (1928) ; Kew Bull. 1928 : 400 ; Bak. f. l.c. 800. A strong woody climber with hexagonal branches and yellowish-grey foliage ; flowers yellow, fragrant.
    Fr.G.: Nzérékoré to Macenta (Oct.) *Baldwin* 9740 ! **S.L.:** Giewahun (fl. & young fr. Dec.) *Deighton* 3841 ! **Lib.:** Gbanga (Sept.) *Linder* 478 ! Peahtah (Oct.) *Linder* 478a ! Zigida, Vonjama (Oct.) *Baldwin* 9995 !

5. CYLICODISCUS Harms in E. & P. Pflanzenfam. Nachtr. 1 : 192 (1897) ; Engl. Bot. Jahrb. 26 : 256 (1899).

A large tree ; leaves bipinnate ; pinnae opposite, 1 pair ; leaflets alternate, stalked, ovate, rather long-acuminate, 5–10 cm. long, 3–5 cm. broad, shining above, glabrous and finely reticulate ; primary rhachis 2–2·5 cm. long, secondary up to 12 cm. long ; racemes spike-like, usually paired or several, slender, up to 15 cm. long ; axis shortly pubescent ; pedicels about 1 mm. long, persistent ; calyx nearly 1 mm. long ; petals glabrous, 2 mm. long ; stamens 10, free ; anthers with a small deciduous apical gland   ..   ..   *gabunensis*

C. gabunensis *Harms* in E. & P. Pflanzenfam. Nachtr. 1 : 192 (1897) ; Engl. Bot. Jahrb. 26 : 256 (1899) ; Bak. f. Leg. Trop. Afr. 3 : 796 ; Kennedy For. Fl. S. Nig. 110 ; Aubrév. Fl. For. C. Iv. 1 : 178, **t. 62.** *Cyrtoxiphus staudtii* Harms (1897). *Piptadenia* sp. of Thompson Rep. For. Gold Coast (1910). **A very large tree of rain forest, reaching 200 ft. in height and 37 ft. girth, with straight bole up to 80 ft. and short buttresses ; armed with brown thorns in the sapling stage ; flower-spikes yellowish or greenish-white ; pods up to 3 ft. long, 1¾ in. broad ; seeds flat, thinly winged, sometimes over 3 in. long.**
Iv.C.: Adzopé region *fide* Aubrév. *l.c.* **G.C.:** Imbraim (Mar.) *Thompson* 27 ! Dunkwa (young fr. June) *Vigne* 891 ! Kumasi (Nov., Dec.) *Darko* 743 ! Andoh FH 5600 ! **S.Nig.:** Oni *Sankey* 22 ! Erin *Foster* 181 ! Sapoba *Kennedy* 1673 ! **Br.Cam.:** Johann-Albrechtshöhe *Staudt* 531 ; 766 ! Also in French Cameroons and Gabon. (See Appendix, p. 214.)

6. PIPTADENIASTRUM Brenan in Kew Bull. 1955 : 179.

Leaflets very small and numerous, contiguous, about 3–8·5 mm. long and 0·8–1·25 mm. broad, linear-subfalcate, auriculate at base ; pinnae 10–19 pairs ; panicles of spikes brown-tomentose ; fruits broadly linear, elongated, 17–36 cm. long and 2–3·2 cm. broad, obliquely nervose, glabrous ; seeds 6–8, flat, winged, attached in the middle, about (3–) 5·3–9·5 cm. long and 1·8–2·5 cm. broad .. .. .. *africanum*

P. africanum (*Hook. f.*) *Brenan* l.c. (1955). *Piptadenia africana* Hook. f. (1849)—F.T.A. 2 : 328 ; Bak. f. Leg. Trop. Afr. 3 : 794 ; Aubrév. Fl. For. C. Iv. 1 : 182, t. 65 ; For. Trees & Timbers Brit. Emp. 2 : 71 ; Gilbert & Boutique in Fl. Congo Belge 3 : 226, t. 16. Large tree of rain forest, reaching 150 ft. high, with straight bole, large plank-like buttresses, spreading parasol-like branches, and fine fern-like foliage ; flowers in panicled yellowish-white spikes.
Sen.: Sinédone, Casamance *Chev.* 3360. **Fr.G.:** Kouria *Chev.* 18174. **S.L.:** Kenema (Aug.) *Aylmer* 249 ! Njala (July) *Deighton* 768 ! **Lib.:** Gbanga (Sept.) *Linder* 750 ! Dukwia R. (May) *Cooper* 467 ! Nyaake, Webo (June) *Baldwin* 6125 ! **Iv.C.:** Toura country (May) *Chev.* 21610 ! Aboisso *Chev.* 16304 ! Azaguié (fr. Sept.) *Chev.* 22297 ! **G.C.:** Aburi (Apr.) *Johnson* 690 ! Adibrem (June) *Chipp* 255 ! Kumasi (May) *Vigne* 1647 ! **N.Nig.:** Agaie (fr. Nov.) *Yates* 11 ! **S.Nig.:** Lagos *Millen* 190 ! Sapoba *Kennedy* 282 ! Niger Delta *Ansell* ! *T. Vogel* ! **Br.Cam.:** Victoria (fr. Feb.) *Maitland* 427 ! Also in French Cameroons, Gabon, Belgian Congo, Angola, Ubangi-Shari, A.-E. Sudan and Uganda. (See Appendix, p. 221.)

7. NEWTONIA Baill. in Bull. Soc. Linn. Paris 1 : 721 (1888).

Pinnae with 1 pair of leaflets ; pinnae 1 pair or rarely 2 pairs ; leaflets obovate, cuneate and unequal-sided at base, very slightly and obtusely acuminate, 7–12 cm. long, 4–7 cm. broad, glabrous ; lateral nerves prominent, about 8 pairs ; calyx and corolla densely pubescent outside ; fruit linear-oblong, 15–20 cm. long, about 3·5 cm. broad ; seeds 7–10 cm. long, 2–2·5 cm. broad .. .. .. 1. *duparquetiana*
Pinnae with at least 3 pairs of leaflets ; pinnae 1–7 pairs ; leaflets not exceeding 4·5 cm. long and 2·5 cm. broad :
Pinnae 1 pair ; leaflets 3–4 pairs, obovate, rather narrowed to base, 3–4·5 cm. long, 2–2·5 cm. broad, with numerous looped lateral nerves ; calyx with setose hairs ; young fruits somewhat curved, smooth, 7–8 cm. long, about 2 cm. broad ; mature seeds not known .. .. 2. *elliotii*
Pinnae 2 or more pairs :
Leaflets 3–4 pairs, obovate-elliptic, cuneate on one side and rounded on the other side at base, 2–6·5 cm. long, 1–3·5 cm. broad ; pinnae 3–4 pairs ; calyx nearly glabrous ; fruits linear-oblong, straight, 13–22 cm. long, 2–2·5 cm. broad ; seeds 6–8 cm. long, 1·5–2 cm. broad .. .. 3. *aubrevillei*
Leaflets 5 pairs or more :
Leaflets 5–10 pairs, subrhombic-oblong, more or less truncate at base, with the proximal side slightly auriculate, 1·5–4·5 cm. long, 7–20 mm. broad ; pinnae 3–5 pairs ; calyx and corolla densely pubescent outside ; fruits somewhat curved, 12–17 cm. long, 2–2·5 cm. broad ; seeds 7–8 cm. long, 1·8–2 cm. broad
4. *zenkeri*
Leaflets more than 10 pairs, up to 2 cm. long and 7 mm. broad :
Leaflets 10–13 pairs, 1·5–2 cm. long, 5–7 mm. broad ; pinnae 4 pairs 5. *sp. A*
Leaflets 14–26 pairs, 1·2–1·8 cm. long, 3–5 mm. broad ; pinnae 2 pairs 6. *sp. B*
Leaflets 16–21 pairs, 1–1·4 cm. long, 2–4 mm. broad ; pinnae 5–7 pairs 7. *sp. C*

1. **N. duparquetiana** (*Baill.*) *Keay* in Kew Bull. 1953 : 488 (1954). *Entada ? duparquetiana* Baill. (1866). *Piptadenia duparquetiana* (Baill.) Pellegr. in Bull. Soc. Bot. Fr. 94 : 101 (1947). *Newtonia insignis* Baill. (1888)—F.W.T.A., ed. 1, 1 : 353. *Piptadenia insignis* (Baill.) Bak. f. Leg. Trop. Afr. 3 : 792 (1930) ; Aubrév. Fl. For. C. Iv. 1 : 184, t. 66. A forest tree, up to 80 ft. high, with smooth bark and well-developed root flanges.
S.L.: Kambui F.R. (Apr.) *Lane-Poole* 215 ! **Iv.C.:** *fide* Aubrév. *l.c.* **G.C.:** Ajakwa, Chama (Apr.) *Chipp* 205 ! Maliomo, Esuasu (May) *Chipp* 214 ! **S.Nig.:** Eket Dist. *Gray* SC/2 ! Akah FHI 5769 ! Oban *Talbot* ! Also in French Cameroons and Gabon. (See Appendix, p. 217.)
2. **N. elliotii** (*Harms*) *Keay* l.c. (1954). *Piptadenia elliotii* Harms in Engl. Bot. Jahrb. 26 : 260 (1899) ; F.W.T.A., ed. 1, 1 : 354. Tree, on river banks, to 40 ft. high ; flowers white, sweet-scented.
S.L.: Maria, Scarcies (Dec.) *Sc. Elliot* 4792 ! Madina Limba (young fr. Apr.) *Sc. Elliot* 5660 ! Njala (Dec.) *Deighton* 1854 ! Batkanu (Jan.) *Deighton* 2850 ! Mano (fl. & young fr. Feb.) *Dawe* 481 !
3. **N. aubrevillei** (*Pellegr.*) *Keay* l.c. (1954). *Piptadenia aubrevillei* Pellegr. in Bull. Soc. Bot. Fr. 80 : 466 (1933) ; Aubrév. Fl. For. C. Iv. 1 : 184, t. 67. Forest tree, to 90 ft. high ; flowers creamy.
S.L.: Unwin & Smythe 34 ! Kambui, Hangha (Sept.) *Aylmer* 244 ! Njala (fl. Feb. fr. Oct.) *Deighton* 1409 ! 1813 ! Dambaye, Kenema (fr. July) *King* 259 ! **Lib.:** Dukwia R. *Cooper* 119 ! Suacoco, Gbanga (fr. Jan.) *Daniel* 114 ! **Iv.C.:** Djibi *Aubrév.* 861 ! Yapo *Aubrév.* 600 ! (See Appendix, p. 222.)
4. **N. zenkeri** *Harms* in Engl. Bot. Jahrb. 40 : 17 (1907) ; Bak. f. l.c. 796. *Piptadenia zenkeri* (Harms) Pellegr. l.c. 80 : 466 (1933) ; Lég. Gab. 21 (1948). Tall tree of forest, to 120 ft. high.
Br.Cam.: Victoria (fr. Mar.) *Maitland* 575 ! Also in French Cameroons and Gabon.

5. **N. sp. A.**[1] "Tree common in Makun and more common in Eba Island ; has high buttresses, with compact crown almost dome-shaped."
　　S.Nig.: (fr. Jan.) *Thornewill* 262 !
6. **N. sp. B.**[1] A tall tree.
　　S.Nig.: ? Eket Dist. *Gray* 1/21 !
7. **N. sp. C.**[1] "A tall tree resembling *Piptadenia*, up to 100 ft. high, with large spreading branches."
　　Br.Cam.: Njinikom, Bamenda, 5,000 ft. (young fr. June) *Maitland* 1559 !

8. FILLAEOPSIS Harms in Engl. Bot. Jahrb. 26 : 258 (1899).

A tree ; leaves bipinnate ; pinnae opposite, 1–2 pairs ; leaflets mostly alternate, 4–8 to each pinna, ovate-elliptic, rounded at base, shortly obtusely acuminate, 4–7 cm. long, 2–3·5 cm. broad, glabrous ; spikes rather densely paniculate at the end of the shoots ; flowers very numerous, crowded ; calyx small, 5-toothed, teeth ovate, obtuse, glabrous ; petals 5, free, ovate, acute, valvate, glabrous ; stamens 10 ; anthers glandular at the apex ; disk large, surrounding the ovary ; fruit very large, oblong, up to 40 cm. long and 15 cm. broad, shining, rather faintly nervose ; seeds transverse, oblong, about 7 cm. long and 3 cm. broad, winged, attached by the middle to a slender funicle　..　　..　　..　　..　　..　　..　　..　　.. *discophora*

F. discophora *Harms* l.c. 259, t. 6 (1899) ; Bak. f. Leg. Trop. Afr. 3 : 797 ; Kennedy For. Fl. S. Nig. 108 ; Gilbert & Boutique in Fl. Congo Belge 3 : 216. A forest tree, to 130 ft. high, "strongly branched and with offensive odour".
　　S.Nig.: Sapoba *Kennedy* 1566 ! Boji Hill, Ikom (fr. June) *Unwin* 24 ! **Br.Cam.**: Kumba *Ejiofor* FHI 14058 ! Also in French Cameroons, Gabon, Belgian Congo and Angola. (See Appendix, p. 216.)

9. ENTADA Adans.—F.T.A. 2 : 325 ; Benth. in Trans. Linn. Soc. 30 : 363 (1875).

Branchlets not prickly :
　Leaflets 2–4 pairs on each pinna, exceeding 2·5 cm. long and 1 cm. broad ; pinnae 2–3 pairs, the end pair usually tendriliform ; inflorescence up to 30 cm. long, solitary, axillary or supraxillary ; fruits very large, up to 120 cm. long, 8–10 cm. broad ; seeds 4–6 cm. diam. :
　　Flowers subsessile, or with a robust pedicel not more than 0·5 mm. long ; peduncle borne close to the leaf axil ; fruits straight, very woody ; leaflets 3·5–9 cm. long, 2–4 cm. broad　..　　..　　..　　..　　..　　..　　.. 1. *pursaetha*
　　Flowers on distinct slender pedicels 1–1·5 mm. long ; peduncle borne on the stem about 5 mm. above the leaf axil ; fruits bent into a single or double spiral, often with the sides also twisted, coriaceous or subwoody ; leaflets 2·5–5 cm. long, 1–2·2 cm. broad　..　　..　　..　　..　　..　　..　　.. 2. *gigas*
　Leaflets 8–50 pairs on each pinna, up to 4·2 cm. long and 1·5 cm. broad ; end pair of pinnae not tendriliform, or only rarely so ; fruits chartaceous, up to 45 cm. long ; seeds up to 18 mm. long :
　　Midrib in the middle or near the middle of the leaflets :
　　　Nervation of leaflets obscure on both surfaces ; leaflets 8–13 pairs, oblong, rounded to emarginate at apex, 8–20 mm. long, 3–7 mm. broad, evenly and minutely puberulous beneath ; pinnae 4–6 pairs, end pair sometimes tendriliform ; spikes borne towards the apex of the shoots, sometimes paniculate ; fruits 15–45 cm. long, 6–11 cm. broad ; seeds about 18 mm. long and 9 mm. broad　.. 3. *mannii*
　　　Nervation of leaflets distinct on both surfaces ; leaflets 8–24 pairs, linear or oblong-linear, 9–45 mm. long, 3–15 mm. broad, glabrous or sometimes puberulous ; pinnae (1–) 3–9 pairs ; spikes axillary among the leaves ; fruits up to 38 cm. long and 5–7·3 cm. broad ; seeds about 12 mm. long and 9–10 mm. broad　4. *africana*
　　Midrib near one margin of the leaflets :
　　　A tree of savannah ; flowers cream ; fruit straight, 15–39 cm. long, (3·8–) 5–7·5 (–9) cm. broad ; pinnae (1–) 2–20 (–22) pairs ; leaflets (15–) 22–55 pairs, linear, mucronate, (3–) 4–12 (–14) mm. long, 1–3 (–3·5) mm. broad　.. 5. *abyssinica*
　　　A slender liane ; flowers purple to red ; fruit sickle-shaped, 11–23 cm. long, 2·9–3·8 cm. broad ; pinnae (1–) 2 (–3) pairs ; leaflets 9–18 pairs, linear-oblong, with midrib towards one side at base, 8–16 mm. long, 1·5–4·2 mm. broad 6. *wahlbergii*
Branchlets and leaf-rhachis rather densely armed with short recurved prickles ; pinnae 6–10 pairs ; leaflets 10–12 pairs, oblong, rounded at apex, 10–15 mm. long, 4–7 mm. broad ; spikes 4–6 cm. long, solitary or up to 3 together ; fruits about 25–30 cm. long and 7 cm. broad　..　　..　　..　　..　　..　　..　　..　　.. 7. *scelerata*

1. **E. pursaetha** *DC.* Prod. 2 : 425 (1825) ; Brenan in Kew Bull. 1955 : 164 (q.v. for synonymy). *E. gigas* of F.W.T.A., ed. 1, 1 : 355, partly ; of Gilbert & Boutique in Fl. Congo Belge 3 : 220 ; not of (Linn.) Fawcett & Rendle. A lofty woody climber with very stout stem and enormous straight woody segmented pods.
　　Port.G.: Formosa (Apr.) *Esp. Santo* 1962 ! **Fr.G.**: Landouma country (Apr.) *Heudelot* 850 ! **S.L.**: Kafogo (Apr.) *Sc. Elliot* 5520 ! Makali (fr. Feb.) *Deighton* 4054 ! **G.C.**: Mampong (fl. & fr. Mar.) *Vigne* FH 1069 ! **Togo**: *Baumann* 502 ! **S.Nig.**: Apomu *Foster* 211 ! Owo F.R. (Apr.) *Jones* FHI 3476 ! Awka (fr. Aug.) *Jones* FHI 6145 ! Widespread in tropical Africa and Asia, and in N. Australia. (See Appendix, p. 216.)

---

[1] These specimens appear to represent distinct species, but the material is inadequate for certain identification or for the description of new species. Each specimen has fruits which indicate the genus *Newtonia*.

2. **E. gigas** (*Linn.*) *Fawcett & Rendle* Fl. Jamaica 4, 2 : 124, fig. 38, excl. fruit (1920) ; Brenan l.c. 164 (q.v. for synonymy). *Mimosa gigas* Linn. (1759). *Entada planoseminata* (De Wild.) Gilbert & Boutique l.c. 221 (1953). *E. umbonata* (De Wild.) Gilbert & Boutique l.c. 222 (1953). *E. scandens* of F.T.A. 2 : 325, partly (*Mann* spec.). A liane, similar to the preceding, but with coriaceous or subwoody pods curved into a single or double spiral.
Lib.: Logantown, New Cess R. (Mar.) *Baldwin* 11182 ! **F.Po:** on the beach (Jan.) *Mann* 230 ! 1438 ! Widespread in central tropical Africa and in the West Indies, central and south tropical America. (See Appendix, p. 216.)

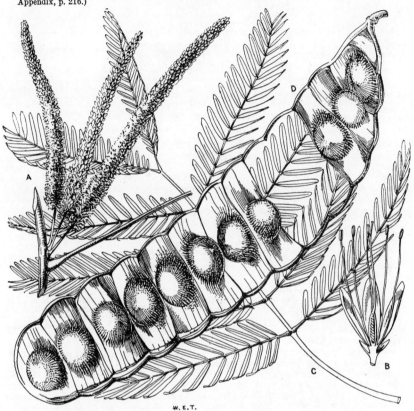

FIG. 156.—ENTADA AFRICANA *Guill. & Perr.* (MIMOSACEAE).

A, inflorescence. B, longitudinal section of flower. C, leaf. D, fruit.

3. **E. mannii** (*Oliv.*) *Tisserant* in Bull. Soc. Bot. Fr. 99 : 257 (1953) ; Berhaut in Bull. Soc. Bot. Fr. 99 : 275. *Piptadenia ? mannii* Oliv. F.T.A. 2 : 329 (1871) ; F.W.T.A., ed 1, 1 : 354. *Entadopsis mannii* (Oliv.) Gilbert & Boutique l.c. 205 (1953). *Entada africana* of F.T.A. 2 : 326, partly ; of F.W.T.A., ed. 1, 1 : 355, partly ; not of Guill. & Perr. A shrub, usually scandent but sometimes arborescent ; in streamside forest and on rocky hills in forest regions.
Sen.: *Berhaut* 763. **Port.G.:** Bijagós (fr. Nov.) *Esp. Santo* 2328 ! **Fr.G.:** Paroisse. **S.L.:** Bumbuna (young fr. Oct.) *Thomas* 3723 ! Giema Ferry, Dama (fr. Nov.) *Deighton* 3986 ! **Lib.:** Browntown, Tchien (Aug.) *Baldwin* 7059 ! **Iv.C.:** Pobéguin 209. Mid. Cavally *Chev.* 19450. **G.C.:** Foso (July) *Hughes* 200 ! Mampong (fr. Nov.) *Vigne* FH 3427 ! Tafo (July) *Moor* 351 ! **S.Nig.:** Lagos *MacGregor* 199 ! Idanre Hills (young fr. Oct.) *Keay* FHI 20243 ! Onitsha *Barter* 1801 ! Aguku Dist. *Thomas* 842 ! **Br.Cam.:** Preuss 364 ! **F.Po:** (fl. June, fr. Dec.) *Mann* 20 ! 414 ! Also in French Cameroons, Ubangi-Shari, Gabon, Belgian Congo and Angola.
4. **E. africana** *Guill. & Perr.*—F.T.A. 326, partly ; Brenan l.c. 165 ; Aubrév. Fl. For. Soud.-Guin. 294, t. 58, 1–5. *E. ubanguiensis* De Wild. (1925)—Aubrév. l.c. 292. *E. sudanica* Schweinf. (1868)—F.T.A. 2 : 327 ; F.W.T.A., ed. 1, 1 : 355 ; Aubrév. l.c. 294. *Entadopsis sudanica* (Schweinf.) Gilbert & Boutique 204 (1953). A savannah tree, to 20 ft. high ; flowers creamy-white.
Sen.: Tielimane, Cayor (June) *Leprieur* ! **Gam.:** Kombo (May) *Heudelot* 53 ! Albreda (Mar.) *Perrottet* ! **Fr.Sud.:** Bamako (May) *Hagerup* 44 ! Ouassana (Mar.) *Chev.* 615 ! **Port.G.:** Buba (May) *Esp. Santo* 2477 ! **Iv.C.:** Bondoukou *Aubrév.* 726 ! Ouangolo *Aubrév.* 1430. **G.C.:** Tili (Apr.) *Vigne* FH 3765 ! Daboya (Mar.) *McLeod* 820 ! Tamale (Apr.) *Williams* 122 ! **Togo:** Yendi *Vigne* 1704 ! **Dah.:** Savi (May) *Chev.* 23549 ! **N.Nig.:** Nupe *Barter* 1056 ! 1122 ! Anara F.R. (fl. Mar., fr. Oct.) *Keay* FHI 21103 ! 24413 ! Zamfara F.R. (Apr.) *Keay* FHI 16123 ! 16200 ! Vom *Dent Young* 87 ! Abinsi (June) *Dalz.* 607 ! **S.Nig.:** Iseyin *MacGregor* 244 ! Also in French Cameroons, Ubangi-Shari, Uganda, Belgian Congo and A.-E. Sudan. (See Appendix, p. 216.)
5. **E. abyssinica** *Steud. ex A. Rich.*—F.T.A. 2 : 327 ; Aubrév. l.c. 292. *Entadopsis abyssinica* (Steud. ex A. Rich.) Gilbert & Boutique l.c. 208, t. 14 (1953). A savannah tree, to 40 ft. high ; flowers cream, fragrant ; in rather moister regions than the two preceding species.
Fr.Sud.: Toukoro *Chev.* 897. **Fr.G.:** Labé *Chev.* 12413. Kouroussa *Pobéguin* 234. Dentilia (Mar.) *Sc. Elliot* 5307 ! **S.L.:** Kabala (fr. Sept.) *Deighton* 3979 ! Musaia (Apr.) *Deighton* 5404 ! **Iv.C.:** Tafiré

Aubrév. 914 ; 1433. Touba *Aubrev.* 1255. **G.C.:** Wanki (fr. June) *Chipp* 478 ! Epala, Bole (Apr.) *McLeod* 844 ! Secodumase (fr. Nov.) *Chipp* 777 ! Bangwon, Lawra (fl. & fr. May) *Vigne* FH 3817 ! Nwereme (May) *Irvine* 2460 ! **Togo:** *Kersting* 49 ! Kpedsu (fr. Dec.) *Howes* 1058 ! **Dah.:** Abomey *Poisson* 86 ! **N.Nig.:** Abinsi to Katsina Ala (June) *Dalz.* 605 ! **S.Nig.:** Olokemeji (fl. Mar., fr. Oct.-Nov.) *A. F. Ross* 10 ! 164 ! Ishoka, Benin (fr. Oct.) *Unwin* 156 ! Ogoja Dist. *Rosevear* 46/29 ! **Br.Cam.:** Nchan, Bamenda (Mar., May) *Johnstone* 93 ! *Maitland* 1742 ! Bambui, Bamenda (fl. & fr. Jan.) *Keay* FHI 28340 ! Widespread in eastern and central tropical Africa, from Abyssinia to Mozambique and Angola. (See Appendix, p. 215.)

[There is some evidence of hybridity between *E. africana* and *E. abyssinica*.]

6. **E. wahlbergii** *Harv.*—F.T.A. 2 : 326 ; Chev. Bot. 239. *E. flexuosa* Hutch. & Dalz. F.W.T.A., ed. 1, 1 : 356 (1928) ; Kew Bull. 1928 : 401. *Entadopsis flexuosa* (Hutch. & Dalz.) Gilbert & Boutique l.c. 206 (1953). Suffrutex, with slender woody stems climbing to a considerable height ; branchlets very zigzag ; flowers brown-purple ; in savannah.

Fr.Sud.: Bamba *Chev.* 943. **Port.G.:** Contubo (fr. Oct.) *Esp. Santo* 2310 ! **Iv.C.:** Mid. Comoé *Chev.* 22612. **G.C.:** Gbongoda, Gambaga (June) *Akpabla* 696 ! **N.Nig.:** Nupe *Barter* 991 ! Yola (July) *Dalz.* 7 ! Also in A.-E. Sudan, Uganda, Belgian Congo, N. Rhodesia (?) and Natal.

7. **E. scelerata** *A. Chev.* in Bull. Soc. Bot. Fr. 55, Mém. 8 : 160 (1912) ; Chev. Bot. 239. *Entadopsis scelerata* (A. Chev.) Gilbert & Boutique l.c. 204 (1953). A prickly scandent shrub or large liane ; in forest regrowth. **Iv.C.:** Morénou *Chev.* 22428 ! **G.C.:** Kumasi (May, July) *Vigne* FH 1169 ! *Andoh* FH 4223 ! Ofuasi to Nsuaem (May) *Kitson* 1209 ! **S.Nig.:** Ibadan Div. *Onochie* FHI 8185 ! Afi River F.R. (May) *Jones & Onochie* FHI 18917 ! Idumuje (fr. Jan.) *Thomas* 2155 ! Also in French Cameroons, Belgian Congo and Cabinda.

10. AUBREVILLEA Pellegr. in Bull. Soc. Bot. Fr. 80 : 466 (1933).

Leaflets 8–12 pairs, oblong to subobovate, 15–35 mm. long, 5–20 mm. broad ; pinnae 4–7 pairs ; ovary puberulous to glabrous ; fruit oblong, 15–22 cm. long, about 5 cm. broad   ..   ..   1. *platycarpa*
Leaflets 16–30 pairs, narrowly oblong, 10–20 mm. long, 2–4 mm. broad ; pinnae 5–8 pairs ; ovary densely pubescent ; fruit oblong, 11–16 cm. long, 2·5–4 cm. broad   2. *kerstingii*

1. **A. platycarpa** *Pellegr.* in Bull. Soc. Bot. Fr. 80 : 467 (1933) ; Aubrév. Fl. For. C. Iv. 1 : 188, t. 68 B ; Gilbert & Boutique in Fl. Congo Belge 3 : 227. Large forest tree, with large buttresses and dense rounded crown.
Fr.G.: Kissidougou (Dec.) *King* in Hb. *Deighton* 4708 ! **S.L.:** Kenema (Dec.) *Edwardson* 267 ! Makeni-Kabala road (Jan.) *King* 317 ! **Iv.C.:** Man region *Aubrév.* 926 ! 990 ! **G.C.:** Yoyo F.R. *C. J. Taylor* 284 ! **S.Nig.:** Sapoba *Kennedy* 2421 ! Also in French Cameroons and Belgian Congo.
2. **A. kerstingii** *(Harms) Pellegr.* l.c. 466 (1933) ; Aubrév. l.c. 1 : 186, t. 68 A ; Gilbert & Boutique in 3 : 228. *Piptadenia kerstingii* Harms in Engl. Bot. Jahrb. 40 : 16 (1907) ; F.W.T.A., ed 1, 1 : 354 ; Bak. f. Leg. Trop. Afr. 3 : 793. Large forest tree, to 120 ft. high, with large buttresses and spreading crown ; especially characteristic of forest outliers in savannah country.
Iv.C.: Bouroukrou *Aubrév.* 704. **Togo:** Sokode-Basari *Kersting* A. 5 ! A. 294 ! A. 511 ! **N.Nig.:** Agaie (Feb.) *Yates* 12 ! **S.Nig.:** Onitsha (Jan.) *Chesters* 202 ! Udi (fl. Jan., fr. Mar.) *Chesters* 203 ! Awka (Feb.) *Jones* FHI 7481 ! Also in Ubangi-Shari, French Cameroons and Belgian Congo. (See Appendix, p. 222.)

11. PROSOPIS Linn.—F.T.A. 2 : 331 ; Benth. in Trans. Linn. Soc. 30 : 376 (1875).

A tree with very hard wood ; leaves bipinnate ; pinnae 2–4 pairs, opposite, with a fleshy pore-like gland between the base of each pair ; leaflets 6–12 pairs, opposite, lanceolate or oblong-lanceolate, mucronate, 1·5–2 cm. long, 5–8 mm. broad, minutely pubescent on both surfaces ; spikes axillary, solitary, shortly pedunculate, 4–6 cm. long ; calyx slightly pubescent ; petals glabrous ; stamens 10, free ; anthers with a small apical gland ; ovary villous ; fruit about 15 cm. long, 2·5 cm. thick, subturgid, black and shining, with a thick pericarp, thinly septate between the seeds ; seeds ellipsoid, 1 cm. long, shining, with a thin intramarginal line all round   *africana*

P. africana *(Guill. & Perr.) Taub.* in E. & P. Pflanzenfam. 3, 3 : 119 (1892) ; Aubrév. Fl. For. Soud.-Guin. 285, t. 57, 1–5 ; Ic. Pl. Afr. t. 29. *Coulteria ? africana* Guill. & Perr. (1832). *Prosopis oblonga* Benth. (1841)—F.T.A. 2 : 331 ; Chev. Bot. 241 ; Bak. f. Leg. Trop. Afr. 3 : 805. A savannah tree, to 40–60 ft. high, with pale drooping foliage and very hard wood ; flower-spikes yellowish, fragrant ; pods stout, roughly cylindrical, with loose rattling seeds.
Sen.: Kounoun (fr. Mar.) *Perrottet.* Cayor *Heudelot* 14 ! **Gam.:** Kombo *Dawe* 34 ! **Fr.Sud.:** Kita *Dubois* 18. Koulouba *Vuillet* 520 ! **Port.G.:** Bor (Aug.) *Esp. Santo* 11 ! **Fr.G.:** Kouroussa (Apr.) *Pobéguin* 591 ! **S.L.:** Falaba (young fr. Oct.) *Small* 446 ! **Iv.C.:** Bobo Dioulasso, Ferkéssédougou & Touba *fide* Aubrév. l.c. **G.C.:** Ejura (fr. Aug.) *Chipp* 741 ! Pong to Kako *Kitson* 701 ! Gambaga (Mar.) *Hughes* FH 4983 ! **N.Nig.:** Nupe *Barter* 1193 ! Zamfara F.R. (fl. & fr. Apr.) *Keay* FHI 16179 ! Kaduna (May) *Lamb* 48 ! Katagum *Dalz.* 362 ! Yola (fr. Sept.) *Shaw* 69 ! **S.Nig.:** Enugu *Rosevear* 5/28 ! Also in French Cameroons, Ubangi-Shari, A.-E. Sudan, Uganda and Belgian Congo. (See Appendix, p. 222.)

Besides the above, *P. chilensis* (Molina) Stuntz (syn. *P. juliflora* (Sw.) DC.), a native of tropical and subtropical America, is cultivated in our area (see Holl. 2 : 285).

12. AMBLYGONOCARPUS Harms in E. & P. Pflanzenfam. Nachtr. 1 : 191 (1897) ; Engl. Bot. Jahrb. 26 : 255 (1899).

A tree ; leaves bipinnate ; pinnae 2–4 pairs, opposite or subopposite, without glands between ; leaflets alternate, broadly obovate-elliptic, truncate or slightly emarginate at apex, 1·5–2·5 cm. long, 1–1·5 cm. broad, glabrous and glaucous ; racemes usually paired, shortly pedunculate, about 7 cm. long ; pedicels 3 mm. long, glabrous ; calyx very short, 5-toothed ; petals glabrous, 3·5 mm. long ; stamens 10 ; anthers with a small deciduous gland at apex ; ovary glabrous ; fruit about 15 cm. long, tetragonal, about 2 cm. broad, with a woody shining pericarp   ..   *andongensis*

A. andongensis *(Welw. ex Oliv.) Exell & Torre* in Bol. Soc. Brot. 29 : 42 (1955). *Tetrapleura andongensis* Welw. ex Oliv. F.T.A. 2 : 331 (1871). *T. obtusangula* Welw. ex Oliv. l.c. (1871). *Amblygonocarpus obtusangulus* (Welw. ex Oliv.) Harms (1899)—Bak. f. Leg. Trop. Afr. 3 : 804 ; Gilbert & Boutique in Fl.

Congo Belge 3 : 217. *A. schweinfurthii* Harms (1897)—F.W.T.A., ed. 1, 1 : 357. *Tetrapleura andongensis* var. *schweinfurthii* (Harms) Aubrév. Fl. For. Soud.-Guin. 287, t. 55, 1–3 (1950). A tree, with graceful pale foliage, 30–40 ft. high in savannah, up to 60 ft. in moister places, with dense spike-like racemes of yellowish-flowers.
G.C.: Parria *Kitson* 582 ! 583 ! **N.Nig.:** Sokoto Prov. (Apr.–June) *Dalz.* 333 ! 537 ! *Lely* 833 ! Bornu *Talbot* ! **S.Nig.:** Old Oyo F.R. (fr. Feb.) *Keay* ! Widespread in central, east and south tropical Africa. (See Appendix, p. 213.)

13. TETRAPLEURA Benth.—F.T.A. 2 : 330 ; Benth. in Trans. Linn. Soc. 30 : 375 (1875).

Pinnae 5–9 pairs ; leaflets 6–12 on each side of rhachis, oblong-elliptic, rounded at each end, sometimes slightly emarginate, 1–2 cm. long, 0·5–1 cm. broad, minutely puberulous beneath ; racemes solitary or paired, axillary, pedunculate, 5–7 cm. long ; fruits 15 cm. long or more, each valve with a longitudinal wing-like rather fleshy ridge about 2 cm. broad .. .. 1. *tetraptera*
Pinnae 2 pairs (rarely 1) ; leaflets 2–4 on each side of rhachis, ovate-elliptic, obtuse at base, obtusely acuminate, 4–8 cm. long, 2–3·5 cm. broad, glabrous ; racemes paniculate at the ends of the branchlets, pedunculate, 5–7 cm. long ; fruits up to 12 cm. long and 2·5 cm. broad, similar to the above species .. .. .. 2. *chevalieri*

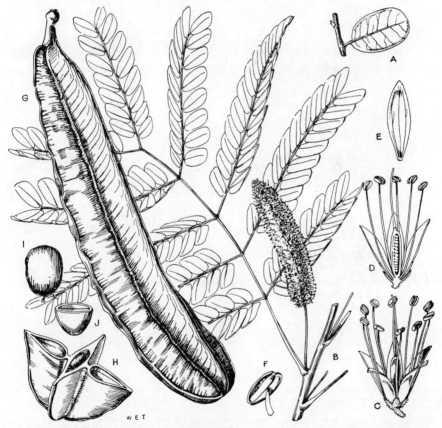

Fig. 157.—Tetrapleura tetraptera (*Schum. & Thonn.*) *Taub.* (Mimosaceae).
A, leaflet. B, portion of stem with leaf and inflorescence. C, flower. D, longitudinal section of flower. E, petal. F, anther. G, fruit. H, cross-section of fruit. I, seed. J, cross-section of seed.

1. **T. tetraptera** (*Schum. & Thonn.*) *Taub.* in Bot. Centralbl. 47 : 395 (1891) ; Bak. f. Leg. Trop. Afr. 3 : 803 ; Aubrév. Fl. For. C. Iv. 1 : 180, t. 64. *Adenanthera tetraptera* Schum. & Thonn. (1827). *Tetrapleura thonningii* Benth.—F.T.A. 2 : 330 ; Benth. Trans. Linn. Soc. 30 : 375 ; Chev. Bot. 241. A forest tree, to 80 ft. high, with dark green fern-like foliage ; flowers creamy or pink, turning orange ; fruits dark purple-brown.
Sen.: Mampalago, Casamance *Chev.* 3114. **Port.G.:** Catió (May) *Esp. Santo* 2060 ! **S.L.:** Buyabuya (Feb.) *Sc. Elliot* 4779 ! Surinuia, Talla Hills (Apr.) *Sc. Elliot* 5534a ! Njala (fr. June) *Deighton* 2527 !

Makene (Mar.) *Deighton* 3901 ! **Lib.:** Palilah, Gbanga (fr. Aug.) *Baldwin* 9159 ! **Iv.C.:** *fide* Aubrév. *l.c.*
G.C.: Tarquah (June) *Chipp* 244 ! Dodowah (June) *Moor* 308 ! Kumasi (Mar.) *Andoh* FH 4167 !
Assuantsi (Mar.) *Irvine* 1570 ! **N.Nig.:** Agaie (Feb.) *Yates* 10 ! **S.Nig.:** Lagos (Jan.) *Dalz.* 1223 ! Abeo-
kuta (Mar.) *Punch* 143 ! Ibadan (Mar.) *Meikle* 1270 ! Onitsha *Unwin* 71 ! Ikom *Rosevear* 83/29 !
Br.Cam.: Tiko (Feb.) *Dunlap* 260 ! Likomba (Nov.) *Mildbr.* 10697 ! Also in French Cameroons, Ubangi-
Shari, Gabon, S. Tomé, Belgian Congo, Uganda and A.-E. Sudan. (See Appendix, p. 223.)

2. **T. chevalieri** (*Harms*) *Bak. f.* Leg. Trop. Afr. 3 : 803 (1930) ; Aubrév. l.c. 180, t. 63. *Piptadenia chevalieri*
Harms in Journ. de Bot. 22 : 112 (1909) ; Chev. Bot. 240 ; F.W.T.A., ed. 1, 1 : 354. *Erythrophleum*
purpurascens A. Chev. (1909). A forest tree.
Lib.: Ganta (fl. May, fr. Sept.) *Harley* 1188 ! *Baldwin* 9305 ! **Iv.C.:** Aboisso, Sanvi (Apr.) *Chev.* 16303
bis ! Also at Yapo, Abengourou and Banco *fide* Aubrév. *l.c.* (See Appendix, p. 223.)
[A sterile specimen, *Small* 666 from the Gola Forest, Sierra Leone, is probably this.]

14. DICHROSTACHYS (DC.) Wight & Arn.—F.T.A. 2 : 332 ; Benth. in Trans. Linn. Soc. 30 : 381 (1875).

A shrub or small tree with axillary spines ; leaves bipinnate ; pinnae about 10 pairs,
opposite, with a long rod-like gland between each ; leaflets numerous, rather variable
in size, linear-oblong, up to 8 mm. long and 2·5 mm. broad, slightly pubescent ;
rhachis pubescent ; spikes usually about 6–8 cm. long, pedunculate, the upper
bisexual flowers yellow, the lower neuter ones pink or mauve ; fruits crowded in a
head, undulate or much curved, glabrous, shining ..   .. *glomerata*

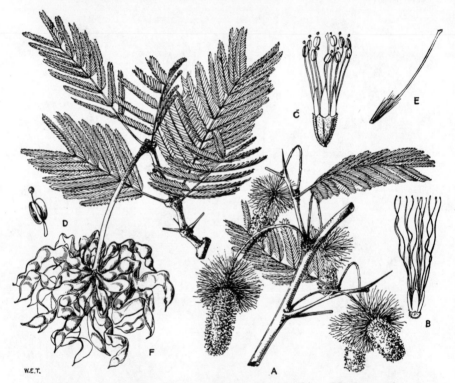

FIG. 158.—DICHROSTACHYS GLOMERATA (*Forsk.*) *Chiov.* (MIMOSACEAE).
A, branch, showing inflorescence.   B, neuter-flower from base of inflorescence.   C, bisexual
flower from upper part of inflorescence.   D, anther..   E, pistil.   F, fruiting branch.

D. glomerata (*Forsk.*) *Chiov.* Ann. Bot. Rom. 13 : 409 (1915) ; Hutch. & Dalz. in Kew Bull. 1928 : 401 ;
Bak. f. Leg. Trop. Afr. 3 : 807 ; Aubrév. Fl. For. Soud.-Guin. 283, t. 56, 1–5 ; Gilbert & Boutique in Fl.
Congo Belge 3 : 202. *Mimosa glomerata* Forsk. (1775). *M. nutans* Pers. (1806). *Dichrostachys nutans* (Pers.)
Benth. (1841)—F.T.A. 2 : 333 ; Chev. Bot. 242.   *D. platycarpa* Welw. ex Oliv. F.T.A. 2 : 333 (1871) ;
Bak. f. l.c. 808 ; Gilbert & Boutique l.c. 199, t. 13.   Shrub or small tree in savannah, often very common
and forming thickets on disturbed scrub lands.
Sen. : Sieber 47 ! *Heudelot* 338 ! Richard Tol (Sept.) *Roger* 149 ! **Gam.:** Koto (May) *Fox* 103 ! Genieri
(July) *Fox* 141 ! **Fr.Sud.:** *fide* Chev. *l.c.* **Port.G.:** Bissau (Apr.) *Esp. Santo* 1165 ! **Fr.G.:** Kouroussa
(Apr.) *Pobéguin* 688 ! **S.L.:** Freetown (Mar.) *Kirk* 41 ! Bagroo R. (fl. & fr. Mar.) *Mann* 795 ! Mt. Gonkwi
(Feb.) *Sc. Elliot* 4829 ! Panguma (Mar.) *Lane-Poole* 39 ! **Lib.:** Kakatown *Whyte* ! Farmington R. (Mar.)
Bequaert 161 ! **Iv.C.:** *fide* Aubrév. *l.c.* **G.C.:** Sekondi (Apr.) *Chipp* 191 ! Accra Plains (Apr.) *Bally* 26 !
Sambisi (May) *Kitson* 590 ! 591 ! **Togo:** Kersting 100 ! Kpedsu (fr. Jan.) *Howes* 1110 ! **Dah.:** Allada
(Mar.) *Chev.* 23383 ! **N.Nig.:** Nupe *Barter* 533 ! Sokoto (Apr.) *Ryan* 15 ! Katagum *Dalz.* 52 ! **S.Nig.:**
Lagos *MacGregor* 148 ! Onitsha *Unwin* 65 ! Widespread in tropical Africa, and in parts of S. Africa. (See
Appendix, p. 215.)

15. XYLIA Benth. in Benth. & Hook. f. Gen. Pl. 1 : 594 (1865) ; Trans. Linn. Soc. 30 : 373 (1875).

A tree ; leaves bipinnate ; pinnae 1 pair, opposite, with a large gland at the base ; leaflets 13–16 pairs, opposite, oblong-lanceolate, acuminate, 3–7 cm. long, 1–2 cm. broad, glabrous above, tomentellous beneath, with distinct lateral nerves ; flowers in pedunculate axillary solitary or paired heads ; heads 2 cm. diam. ; peduncles 3–6 cm. long, tomentellous ; stamens 10 ; anthers with a deciduous apical gland ; ovary pilose ; fruit compressed, oblong, subfalcate, woody, very hard, up to 20 cm. long, 5 cm. broad ; seeds 4, obovate-elliptic, 2 cm. long, shining   ..   ..   *evansii*

X. evansii *Hutch.* in Kew Bull. 1908 : 258 ; Robyns in Rev. Zool. Bot. Afr. 16 : 85 ; Bak. f. Leg. Trop. Afr. 3 : 810 ; Aubrév. Fl. For. C. Iv. 1 : 166, t. 57. A tall tree of rain forest, with yellow flower-heads and hard, slightly bent pods.
S.L.: Kambui *Lane-Poole* 460 ! Bandajuma *Deighton* 3728 ! Iv.C.: Agboville *Chev.* 22338 ! And various localities *fide* Aubrév. *l.c.* G.C.: *Evans* 13 ! Tano R. (Feb.) *Thompson* 15 ! Tenewonan (Apr.) *Thompson* 55 ! Tappa *Chipp* 390 ! Kumasi (Feb.) *Vigne* FH 1593 ! (See Appendix, p. 224.)

Besides the above, *X. xylocarpa* (Roxb.) Taub. (syn. *X. dolabriformis* Benth.), a native of tropical Asia, is occasionally cultivated in our area.

16. LEUCAENA Benth.—F.T.A. 2 : 337 ; Benth. in Trans. Linn. Soc. 30 : 442 (1875).

A small tree ; leaves bipinnate ; pinnae about 5 pairs, sometimes with a gland between the lowest pair ; leaflets about 12 pairs, opposite, lanceolate, slightly falcate, acute, about 1 cm. long and 3 mm. broad, slightly pubescent, the midrib not quite in the middle ; flowers white, in pedunculate globose heads about 2·5 cm. in diam. ; calyx funnel-shaped, teeth pubescent outside ; petals minutely pubescent outside ; stamens 10 ; anthers without an apical gland ; fruits numerous in each head, stellately spreading, linear, flat, 15 cm. long or more, 1·5–2 cm. broad, glabrous, thin, with the numerous seeds slightly oblique to the margins   ..   ..   *glauca*

L. glauca (*Linn.*) *Benth.*—F.T.A. 2 : 337 ; Benth. l.c. 443 ; Chev. Bot. 243 ; Bak. f. Leg. Trop. Afr. 3 : 814 ; Ic. Pl. Afr., t. 8. *Mimosa glauca* Linn. (1753). Introduced ; mostly cultivated or half naturalized by roadsides, etc. ; to 50 ft. high.
Sen.: *Roger* ! Fr.G.: Macenta (fr. Oct.) *Baldwin* 9755 ! S.L.: Freetown *Lane-Poole* 375 ! Magbile (Dec.) *Thomas* 5859 ! N.Nig.: Kaduna (Mar.) *Lamb* 89 ! Jebba (Dec.) *Hagerup* 681 ! S.Nig.: Olokemeji (fl. & fr. Oct.) *A. F. Ross* 5 ! Ibadan (fl. & fr. Apr.) *Meikle* 1429 ! F.Po: (fl. & fr. Nov.) *T. Vogel* 79 ! A native of America ; now common in the warmer parts of both hemispheres. (See Appendix, p. 216.)

Besides the above, *L. esculenta* (DC.) Benth., also a native of tropical America, has been introduced into Senegal (*fide* Berhaut Fl. Sen. 28).

17. SCHRANKIA Willd.—F.T.A. 2 : 336 ; Benth. in Trans. Linn. Soc. 30 : 441 (1875).

A slender trailer with numerous recurved setiform prickles on the quadrangular stems ; leaves bipinnate, sensitive ; pinnae 2–3 pairs, opposite ; leaflets 10–20 pairs, broadly linear, mucronate, about 1 cm. long, glabrous ; flowers in shortly pedunculate axillary heads ; stamens 10, free ; anthers not glandular ; fruits several to each head, linear, up to 13 cm. long, 4-ridged, pubescent and with numerous straight short prickles spreading at right angles   ..   ..   *leptocarpa*

S. leptocarpa *DC.*—F.T.A. 2 : 336 ; Benth. l.c. ; Chev. Bot. 243 ; Bak. f. Leg. Trop. Afr. 3 : 814. A straggling perennial herb resembling the " Sensitive Plant " (*Mimosa pudica*) but with continuous not jointed pods ; introduced.
G.C.: Cape Coast (July) *Ogilvie* ! *T. Vogel* 70 ! Accra *Don* ! *Moloney* ! Dah.: *Le Testu* 58 ! Cotonou (fl. & fr. Jan.) *Chev.* 22692 ! *Debeaux* 149 ! S.Nig.: Lagos *Rowland* ! Abeokuta *Millen* 88 ! A native of tropical America ; also in East Indies.

18. MIMOSA Linn.—F.T.A. 2 : 335 ; Benth. in Trans. Linn. Soc. 30 : 338 (1875).

Pinnae 1–2 pairs, petiole much longer than leaf-rhachis ; leaflets sensitive, linear-oblong, about 1·5 cm. long, more or less bristly-hairy ; branchlets and petioles bristly-pilose or almost glabrous, the former prickly here and there ; peduncles up to 3 cm. long ; stamens 4 ; fruits densely clustered, about 2 cm. long, margined with stiff bristles, segments few   ..   ..   1. *pudica*
Pinnae 6–16 pairs, petiole shorter than leaf-rhachis ; branchlets and leaf-rhachis prickly and roughly hirsute ; leaflets small, linear, margined with bristly hairs and thinly appressed-pubescent on both surfaces ; stamens 8 ; fruits densely bristly all over, 4–6 cm. long, segments narrow and numerous   ..   ..   2. *pigra*

1. M. pudica *Linn.*—F.T.A. 2 : 336 ; Benth. l.c. 397 ; Bak. f. Leg. Trop. Afr. 3 : 812 ; Berhaut Fl. Sen. 26. A straggling prickly plant, the " Sensitive Plant " ; introduced and rather local.
Sen.: *fide* Berhaut *l.c.* Gam.: Bathurst (Dec.) *Lane-Poole* 86 ! S.L.: George Water (Dec.) *Sc. Elliot* 4165 ! Bo (fl. & fr. Apr.) *Deighton* 1588 ! Moyamba (May) *Deighton* 1721 ! Sumbuya (Apr.) *Deighton* 1692 ! Lib.: Monrovia (Nov.) *Whyte* ! *Linder* 1532 ! Belleyella, Boporo (fr. Dec.) *Baldwin* 10639 ! S.Nig.: Oban *Talbot* 1207 ! Degema Dist. *Talbot* ! A native of tropical America ; a common weed of the tropics. (See Appendix, p. 217.)
[Brenan, in Kew Bull. 1955 : 184–189, distinguishes 3 varieties, of which var. *hispida* Brenan, with stipules 8–14 mm. long and bracteoles exceeding the grey-puberulous corolla buds, and var. *unijuga* (Duchass. & Walp.) Griseb., with stipules 4–7 (–8) mm. long and bracteoles shorter than the glabrous corolla buds, are known in our area.]
2. M. pigra *Linn.* Cent. Pl. 1 : 13 (1755). *M. asperata* Linn. (1759)—F.T.A. 2 : 335 ; Benth. l.c. 437 ; Chev. Bot. 243 ; F.W.T.A., ed. I, 1 : 359 ; Bak. f. l.c. A prickly shrub with sensitive leaves and pale mauve flower-balls, forming dense thickets on river banks.

Sen.: *Sieber* 46! *Perrottet* 263! Richard Tol (fr. June) *Döllinger*! **Gam.:** *Hayes* 544! **Fr.Sud.:** Macina (fl. & fr. Apr.) *Lean* 4! L. Debo (May) *Hagerup* 156! **Fr.G.:** Farana *Chev.* 20464! **S.L.:** Musaia (fl. Feb., Mar., fr. Mar.) *Sc. Elliot* 5078! *Deighton* 4166! **Iv.C.:** Tano R. (Sept.) *Chipp* 346! **G.C.:** Black Volta R. (fl. & fr. Mar., Aug.) *Chipp* 500! *Kitson* 952! Kpong (fl. & fr. June) *Irvine* 1779! Dunkwa (fr. Apr.) *Vigne* 1095! **Dah.:** *Le Testu* 221! **N.Nig.:** Nupe *Barter* 1175! Katagum *Dalz.* 49! Lemme (May) *Lely* 141! Bornu (fl. & fr. Apr.) *Oudney* 6! *E. Vogel* 81! **S.Nig.:** Nun R. (fl. & fr. Aug.) *T. Vogel* 15! Onitsha (fl. & fr. May) *Onochie* FHI 7525! Common all over the tropics.

Besides the above, *M. invisa* Mart., a native of tropical America, and *M. rubicaulis* Lam., a native of India, have been introduced and are occasionally met with in our area.

19. NEPTUNIA Lour.—F.T.A. 2 : 333 ; Benth. in Trans. Linn. Soc. 30 : 383 (1875).

Aquatic herb with thickish stems rooting at the nodes ; leaves sensitive, bipinnate ; pinnae 2–3 pairs, opposite ; leaflets in 8 or more pairs, linear-oblong, rounded at apex, about 1·5 cm. long, glabrous ; submerged leaves finely dissected ; stipules broadly ovate ; flowers capitate on elongated stout peduncles up to 20 cm. long, the upper bisexual, the lower neuter ; sepals and petals glabrous ; stamens 10 ; anthers with a deciduous apical gland ; fruit shortly stipitate, oblong, 2–2·5 cm. long, sharply beaked, flat, glabrous ..   .. *oleracea*

N. oleracea *Lour.* Fl. Cochinchin. 2 : 654 (1790) ; F.T.A. 2 : 334 ; Benth. l.c.; Chev. Bot. 242. *N. prostrata* Baill. (1883)—F.W.T.A., ed. 1, 1 : 359 ; Bak. f. Leg. Trop. Afr. 3 : 809. *Mimosa prostrata* Lam. (illegitimate name), partly (1783).
Sen.: *Perrottet* 264! **Gam.:** Bansang, MacCarthy Isl. *Duke* 1! **Fr.Sud.:** Mopti (Sept.) *Chev.* 24964! **Dah.:** Porto Novo Lagoon (fr. Jan.) *Chev.* 22784! **N.Nig.:** R. Yo, Bornu (fl. & fr. Sept.) *Daggash* FHI 24867! **S.Nig.:** Lagos *Moloney*! Widespread in tropical Africa, also in Madagascar, tropical Asia and America.

20. ACACIA Mill.—F.T.A. 2 : 337 ; Benth. in Trans. Linn. Soc. 30 : 444 (1875).

*Flowers in spikes or racemes :
    Leaflets 1 pair on each pinna (rarely 2 pairs on occasional pinnae), obliquely obovate, very unequal-sided at base, 7–18 mm. long, 3–9 mm. broad ; pinnae 3–4, rarely 2 or 5 pairs ; spines in pairs at the nodes, infra-stipular, curved, about 5 mm. long ; flowers distinctly pedicellate ; fruit flat and thin, oblong, cuneate at base, acute or rounded at apex, 3·8–6 cm. long, 1·8–20 cm. broad, reticulate, glabrous ; seeds few, orbicular, about 8 mm. diam.   ..   .. 1. *gourmaensis*
    Leaflets 3 or more pairs on each pinna :
      Prickles or spines only at the nodes of the branchlets (rarely altogether absent) :
        Spines recurved, infra-stipular ; fruits flat and relatively thin, dehiscent :
          Leaflets 3–5 pairs on each pinna, obliquely obovate-oblong, 6–12 mm. long, 2–5 mm. broad, glabrous ; pinnae 2–5 pairs ; spines in pairs, rarely with a third spine ; fruit oblong, often once or twice constricted owing to non-development of seeds, 5–7·5 cm. long, 2–2·5 cm. broad ; seeds few   ..   .. 2. *laeta*
        Leaflets at least 10 pairs on each pinna, up to 4 mm. long and 1·25 mm. broad, pubescent :
          Spines in threes, with the lateral spines curved upwards and the central spine downwards ; leaflets up to 25 pairs on each pinna ; fruit oblong, 4–14 cm. long, 2–2·5 cm. broad :
            Pinnae 3–6 pairs, usually 4 or 5 ; leaflets up to 15 pairs, oblong, narrowed above to an obtuse or slightly pointed apex, not broader at apex except in the upper-most leaflets, about 4 mm. long, 1–1·25 mm. broad ; fruit more or less densely appressed-pubescent when mature, fawn to pale olive-brown ; seeds olive-brown   ..   .. 3. *senegal*
            Pinnae 7–16 pairs ; leaflets up to 25 pairs ; margins slightly revolute, more or less parallel, but leaflets slightly broader at the rounded apex, about 3 mm. long and 0·8 mm. broad ; petiole and rhachis of whole leaf and of pinnae more densely pubescent when young than in the preceding species ; fruit almost glabrous when mature, chestnut to pale chestnut-brown in colour, mottled ; seeds purplish-brown   ..   .. 4. *dudgeoni*
          Spines in pairs, curved downwards with a very broad decurrent base ; pinnae 10–40 pairs ; leaflets 35–60 pairs, midrib close to one side towards the base of the leaflets ; leaf-rhachis with a large disk-like gland towards the base and with a smaller gland between each of the upper pairs ; fruits narrowly oblong, 7·5–15 cm. long, 1·5–2 cm. broad, glabrous, rather coriaceous, with a distinct margin   ..   .. 5. *polyacantha* subsp. *campylacantha*
          Spines straight, thick, up to 1·5 cm. long, in pairs, stipular in origin ; fruits at length sickle-shaped, thick, indehiscent, 10–15 cm. long, about 2·5 cm. broad ; pinnae 3–7 pairs ; leaflets 10–15 pairs, branchlets pale grey   ..   ..   .. 6. *albida*
      Prickles between the nodes of the branchlets, not stipular, the stipules soon deciduous ; calyx glabrous ; fruits thin, oblong, acute at both ends, 7–12 cm. long, 1·5–2 cm. broad :
        Leaf-rhachis with a large flat elliptic gland towards the base scarcely raised above the surface ; pinnae 20–30 pairs ; leaflets 35–50 pairs ; axis of inflorescence rather villous ; fruits slightly pubescent, few-seeded, flat   .. 7. *macrostachya*

Leaf-rhachis with an ovoid gland towards the base standing well up from the surface ;
 pinnae 8–15 pairs ; leaflets 20–30 pairs ; axis of inflorescence tomentellous ;
 fruits glabrous, several-seeded   ..   ..   8. *ataxacantha*
*Flowers in globose heads :
Flower-heads on simple or subsimple axillary peduncles :
 Fruits subterete, more or less turgid when ripe, indehiscent, curved, up to about
  10 cm. long and 1·5 cm. broad, longitudinally striate, becoming glabrous, black ;
  bracteoles near the apex of the peduncle ; flowers yellow ; pinnae 4–10 pairs ;
  leaflets 12–20 pairs, linear-oblong, up to 7 mm. long, glabrous ; spines up to 3 cm.
  long, acicular ; an introduced species   ..   ..   9. *farnesiana*
 Fruits flattened ; bracteoles mostly about or below the middle of the peduncle in open
  flower (but see No. 10) ; indigenous species :
  Valves of fruit thick and woody, straight or somewhat curved, smooth and glabrous,
   margins more or less parallel, 10–20 cm. long, 1·2–3 cm. broad, tardily dehiscent ;
   bracteoles in the upper half of the peduncle ; flowers white or cream ; pinnae
   10–25 pairs ; leaflets 20–40 pairs :
   Branchlets glabrous or sparingly pubescent ; median leaflets 2·5–4·5 mm. long
                                        10. *sieberiana* var. *sieberiana*
   Branchlets densely yellowish-tomentellous ;   median leaflets 2–2·5 (very rarely
    to 3) mm. long   ..   ..   10a. *sieberiana* var. *villosa*
  Valves of fruit thin and membranous or coriaceous, or if thicker then margins
   sinuate or fruit moniliform ; bracteoles about or below the middle of the peduncle :
   Fruits indehiscent, valves coriaceous, 1 cm. or more broad, margins sinuate or
    fruit moniliform ; flowers bright yellow, or pink :
    Segments of fruit very prominently umbonate in the middle ; fruits broadly
     constricted between the seeds ; flowers pink ; pinnae 6–8 pairs ; leaflets 8–15
     pairs, 3–4 mm. long   ..   ..   11. *kirkii*
    Segments of fruit not umbonate ; flowers bright yellow ; pinnae 3–6 pairs ;
     leaflets 10–30 pairs ; spines mostly long and straight, but some short and
     slightly curved :
     Fruits glabrous, narrowly constricted between the seeds
                                        12. *nilotica* var. *nilotica*
     Fruits densely tomentose :
      Fruits straight, narrowly constricted between the seeds
                                        12a. *nilotica* var. *tomentosa*
      Fruits usually curved, with sinuate margins, only slightly constricted between
       the seeds   ..   ..   12b. *nilotica* var. *adansonii*
   Fruits dehiscent, valves membranous or coriaceous, not more than 1 cm. broad :
    Fruit annular or spirally contorted, valves thinly coriaceous, glabrous, 4–6 mm.
     broad, somewhat constricted between the seeds ; flowers white ; some spines
     long and straight, others short and slightly recurved ; pinnae 2–5 pairs ;
     leaflets 6–15 pairs ; branchlets glabrous or puberulous   ..   13. *raddiana*
    Fruit falcate ; flowers yellow or white :
     Valves of fruit coriaceous, more than 6 mm. broad, densely pubescent at first ;
      flowers white ; some spines long and straight, others short and slightly
      recurved ; young branchlets densely pubescent ; pinnae 4–7 pairs ; leaflets
      10–20 pairs   ..   ..   14. *hebecladoides*
     Valves of fruit membranous, up to 6 mm. broad ; flowers bright yellow :
      Stems, except the very youngest branchlets, covered with powder yellow to red
       in colour, bark flaking off in subrectangular pieces ; pinnae 2–7 (rarely 1 or
       8–9) pairs ; leaflets 8–20 pairs, leaflets up to 5 mm. long and 1 mm. broad ;
       fruits up to 14 cm. long, 6 mm. broad, glabrous, somewhat constricted
       between the seeds   ..   ..   15. *seyal*
      Stems not covered with powder :
       Leaflets broadest above the middle, obovate-oblong to oblong-oblanceolate,
        rounded at apex ; leaflets 6–12 pairs ; pinnae 1–4 pairs ; with long straight
        spines up to 6 cm. long, also with much shorter slightly recurved spines ;
        buds and bracts forming a distinct cushion at the nodes, peduncles glabrous,
        not glandular ; fruit glabrous, not glandular, narrowed between the seeds
                                        16. *ehrenbergiana*
       Leaflets more or less parallel-sided but narrowed, towards the obtuse or acute
        apex ; leaflets 12–28 pairs ; pinnae 2–12 pairs ; with straight spines only,
        up to 3 cm. long ; peduncles and fruits usually glandular   ..   17. *hockii*
Flower-heads in terminal much-branched panicles :
 Branchlets with numerous prickles between the nodes and often on the leaf-rhachis
  and inflorescence ; climber ; pinnae in numerous pairs ; leaflets numerous and
  very small, contiguous or nearly so, linear, with the nerve near one margin at the
  base ; fruits linear-oblong, about 12 cm. long and 3 cm. broad, glabrous
                                        18. *pennata*

Branchlets without prickles between the nodes ; tree ; pinnae fairly numerous, leaflets fairly numerous, not contiguous, with the nerve nearly in the middle ; fruits narrowly linear-oblong, about 12 cm. long and 1·5 cm. broad   19. *macrothyrsa*

1. **A. gourmaensis** *A. Chev.* in Bull. Soc. Bot. Fr. 58, Mém. 8 : 167 (1912) ; Aubrév. Fl. For. Soud.-Guin. 264, t. 52, 6 (labelled *A. mellifera* in error). *A. mellifera* of F.T.A. 2 : 340, partly (*Barter* s.n.), not of (Vahl) Benth. ; also of various authors in part, including : F.W.T.A., ed. 1, 1 : 361 ; Bak. f. Leg. Trop. Afr. 3 : 828.   Small tree, to 25 ft. high, often with more or less straight stem and numerous short branches ; bole brown with small scales ; bark of twigs yellowish, peeling ; flowers cream.
    Iv.C.: Mossi *fide* Chev. *l.c.* Ouagadougou *Aubrév.* 2137. Gaoua *Aubrév.* 2138. **G.C.:** Aframsu Kwahu (fr. Jan.) *Chipp* 639 ! Nchenanchene to Kwahu Tafo *Kitson* 1150 ! Sekaii to Boguboi *Kitson* 580 ! Pong-Tamale (fr. Nov.) *Akpabla* 456 ! Navrongo (fl. & fr. Aug.) *Vigne* FH 4563 ! **Dah.:** Konkobiri (July) *Chev.* 24364 ! **Fr.Nig.:** Diapaga to Fada (July) *Chev.* 24482 ! **N.Nig.:** R. Awan, Ilorin Dist. *Barter* 1144 ! Jebba (Mar.) *Dalz.* 1217 ! Morindo, Kontagora (fr. May) *Rosevear* FHI 26612 ! Zungeru (Sept.) *Keay* FHI 28033 ! **S.Nig.:** Old Oyo (fr. Feb.) *Keay* FHI 14643 ! (See Appendix, p. 206.)

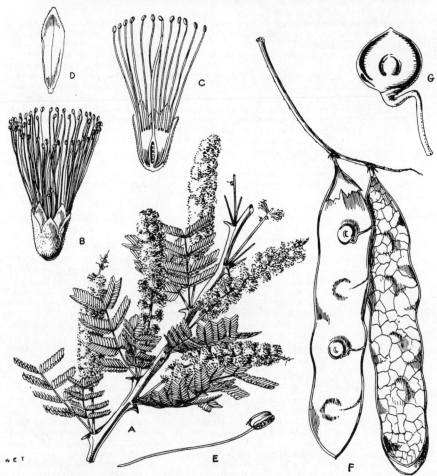

FIG. 159.—ACACIA SENEGAL (*Linn.*) *Willd.* (MIMOSACEAE).

A, flowering branch.   B, flower.   C, longitudinal section of flower.   D, calyx-lobe. E, stamen.   F, mature pods.   G, seed.

2. **A. laeta** *R.Br. ex Benth.*—F.T.A. 2 : 341 ; Chev. Bot. 245 (" *A. lactea* " by error) ; Bak. f. *l.c.* 830 ; Aubrév. *l.c.* 271, t. 52, 4. *A. trentiniani* A. Chev. (1901). A small glabrous tree with black curved spines and spikes of white flowers.
    Fr.Sud.: Timbuktu (July) *Chev.* 1185 ; s.n. ! Bourem *Monod* 565. **Iv.C.:** Ouagadougou *Serv. For.* 2728. **Fr.Nig.:** Gao *De Wailly* 4838. **N.Nig.:** Gaswa, valley of Komadugu *fide* Aubrév. *l.c.* Extends to Egypt, Dead Sea region, Arabia, Abyssinia and E. Africa ; perhaps also in Mauritania (*vide* Aubrév. *l.c.* 272).

3. **A. senegal** (*Linn.*) *Willd.* Sp. Pl. 4 : 1077 (1806) ; Benth. in Trans. Linn. Soc. 30 : 516 ; Aubrév. *l.c.* 266, t. 52, 5, t. 54, 1 ; Ic. Pl. Afr., t. 30. *Mimosa senegal* Linn. (1753). *Acacia verek* Guill. & Perr. (1832)— F.T.A. 2 : 342 ; Chev. Bot. 247. Small tree, to 25 ft. high, with short bole grey and fissured ; bark of twigs yellowish, soon peeling ; flowers cream, in spikes usually longer than the leaves.

Sen.: *Perrottet* 276! *Heudelot*! Sor Isl. *Brunner* 8! M'Bidjem (May) *Thierry* 219! **Gam.:** Yoroberikunda (fl. & fr. Nov.) *Frith* 156! **Fr.Sud.:** Toguere de Banguita (fl. & fr. Jan.) *Davey* 13! Timbuktu (fl. & fr. July, Aug.) *Hagerup* 234! 261! And various localities *fide* Aubrév. *l.c.* **Iv.C.:** Ouagadougou, Bouroum and Kaya *fide* Aubrév. *l.c.* **Fr.Nig.:** Tera *Aubrév.* N. 12. **N.Nig.:** Zamfara F.R. (fl. & fr. Apr.) *Keay* FHI 16145! 16169! Katagum (fl. & fr. July) *Dalz.* 55! Geidam (fr. Oct.) *Dudgeon* 1! Gujba (Dec.) *Foster* 80! Also in Mauritania and extends to the Red Sea and eastern India. (See Appendix, p. 207.)

4. **A. dudgeoni** *Craib ex Holl.* in Kew Bull., Add. Ser. 9 : 291 (1911) ; Kew Bull. 1912 : 98 ; Keay & Brenan in Kew Bull. 1949 : 129. *A. samoryana* A. Chev. in Bull. Soc. Bot. Fr. 58, Mém. 8 : 167 (1912) ; Chev. Bot. 246. *A. senegal* var. *samoryana* (A. Chev.) Roberty in Candollea 11 : 157 (1948) ; Aubrév. l.c. 270 ; Berhaut Fl. Sen. 26. *A. senegal* of F.W.T.A., ed. 1, 1 : 361, partly. Small tree to 25 ft. high with short bole, brown and fissured ; bark of twigs reddish-brown, peeling ; flowers cream, in spikes usually shorter than the leaves ; occurring in much moister regions than *A. senegal*.
Sen.: *fide* Berhaut *l.c.* **Fr.Sud.:** Satadougou (May) *Irvine* 3214! And various localities *fide* Aubrév. *l.c.* **Iv.C.:** Bouroum and Ouangolo *fide* Aubrév. *l.c.* **G.C.:** Pan to Bujan (Apr.) *Kitson* 594! 595! Tamale (Apr.) *Williams* 115! N. Gambaga Dist. (Mar.) *Williams* 490! Yendi (Apr.) *Vigne* FH 1686! Bawku (fl. & fr. Apr.) *Vigne* FH 3754! **Dah.:** Firou to Konkobiri (fl. & fr. June) *Chev.* 24326! Birni *Aubrév.* 94 D! **Fr.Nig.:** Fada N'Gourma *Aubrév.* **N.Nig.:** Borgu *Dudgeon* 58! Kontagora (fl. & fr. Jan.) *Dalz.* 41! Jebba (Jan.) *Meikle* 1107! Kojofa, Biu (fr. May) *Rosevear* FHI 26604! **S.Nig.:** Old Oyo F.R. (fl. & fr. Feb.) *Keay* FHI 16001!

5. **A. polyacantha** *Willd.* subsp. **campylacantha** *(Hochst. ex A. Rich.)* Brenan in Kew Bull. 1956 : 195. *A. campylacantha* Hochst. ex A. Rich. (1847)—F.W.T.A., ed. 1, 1 : 361 ; Bak. f. l.c. 831. *A. catechu* Willd. var. *campylacantha* (Hochst. ex A. Rich.) Roberty l.c. 157 (1948). *A. caffra* Willd. var. *campylacantha* (Hochst. ex A. Rich.) Aubrév. l.c. 272, t. 51, 5, t. 53, 4–5 (1950) ; Gilbert & Boutique in Fl. Congo Belge 3 : 150. *A. catechu* of F.T.A. 2 : 344, not of Willd. *A. suma* of Benth. l.c. 519, partly ; Chev. Bot. 247. A tree, up to 80 ft. high, with a relatively tall bole often with large broad-based thorns ; bark pale yellowish at first, becoming greyish-brown and fissured ; flowers cream, in spikes a little shorter than the leaves ; fruits brown ; usually on moist ground, often gregarious.
Sen.: Sine-Saloum *Serv. For.* 20. Casamance *Etesse* 6. **Gam.:** Genieri (July) *Fox* 122! **Fr.Sud.:** Kaarta *Dubois* 138. **Iv.C.:** Ouangolo *Aubrév.* 1424. Boromo *Aubrév.* 2139. **G.C.:** Bjury (July) *Chipp* 502! Nangodi (June) *Vigne* FH 4513! Sekaii to Boguboi (Apr.) *Kitson* 924! Togo: Kpedsu (fr. Dec.) *Howes* 1069! **Dah.:** *Poisson*! Savalou (May) *Chev.* 23717! **Fr.Nig.:** Gourma (July) *Chev.* 24422! **N.Nig.:** Sokoto (May) *Lely* 837! Katagum *Dalz.* 51! Yola (fr. Dec.) *Shaw* 72! **S.Nig.:** Olokemeji (fl. May, fr. Nov.) *A. F. Ross* 20! 163! Extends eastwards to the Red Sea and southwards to the Transvaal. (See Appendix, p. 205.)

6. **A. albida** *Del.*—F.T.A. 2 : 339 ; Bak. f. l.c. 825 ; Chev. Bot. 243. *Faidherbia albida* (Del.) A. Chev. in Rev. Bot. Appliq. 14 : 876 (1934) ; Aubrév. l.c. 280, t. 51, 3, t. 53, 6 ; Gilbert & Boutique l.c. 3 : 169. A tree, up to 80 ft., with large straight bole, up to 10 ft. in girth, and rounded crown ; bark of twigs pale grey, of bole thick, brown, fissured ; spines white ; leaves glaucous green ; leafless during rainy season ; flowers cream, becoming yellow ; fruit orange when mature.
Sen.: *Roger*! *Perrottet* 261! Cayor *Heudelot* 391! And various localities *fide* Chev. *l.c.* and Aubrév. *l.c.* **Gam.:** Fondi country *Dawe* 80! **Fr.Sud.:** Kita *Dubois* 213! And various localities *fide* Chev. *l.c.* **Port.G.:** Bissau (fl. & fr. Jan.) *Esp. Santo* 1634! *Esp. Santo* 1981! **Iv.C., Togo & Dah.:** *fide* Aubrév. *l.c.* **G.C.:** Navrongo (fl. & fr. Nov.) *Vigne* FH 4658! Tokali *Kitson* 713! Zuarungu (Nov.) *Akpabla* 391! **Fr.Nig.:** Dosso (Oct.) *Hagerup* 571! **N.Nig.:** Sokoto *Dalz.* 314! Kano (fl. Jan., fr. May) *Dudgeon* 4! Katagum *Dalz.* 53! Komadugu Yobe, Duchi (Dec.) *Elliott* 142! Extends throughout the drier parts of N. Africa into Egypt ; also in eastern Africa southwards to the Transvaal. (See Appendix, p. 202.)

7. **A. macrostachya** *Reichenb. ex Benth.*—F.T.A. 2 : 343 ; Bak. f. l.c. 834 ; Aubrév. l.c. 273, t. 53, 3, 9–10. A scrambling shrub, or tree to 20 ft. high (in Sokoto), with rusty-pubescent branchlets, armed with recurved prickles ; flowers cream ; fruits reddish-brown.
Sen.: *Sieber* 44! Bignona, Casamance *Serv. For.* 31. **Fr.Sud.:** Taranoro (May) *Chev.* 835! Macina (Sept.) *Chev.* 24927! Bamako *Hagerup* 56! **Port.G.:** Bissau (fr. Jan.) *Esp. Santo* 1629! Bafata (June) *Esp. Santo* 2517! **Fr.G.:** Kade *Pobéguin* 2075! **S.L.:** Hutton! **Iv.C.:** Bobo Dioulasso, Kaya and Koumi *fide* Aubrév. *l.c.* **Dah.:** Kouande to Konkobiri *Chev.* 24240. **Fr.Nig.:** Niamey *De Wailly.* **N.Nig.:** Sokoto (fl. June, fr. Oct.) *Dalz.* 312! 536! Gwiwa-Korel F.R., Kano (June) *Onwudinjoh* FHI 26416! Also in A.-E. Sudan and Angola. (See Appendix, p. 206.)

8. **A. ataxacantha** *DC.*—F.T.A. 2 : 343 ; Bak. f. l.c. 834 ; Aubrév. l.c. 273, t. 53, 7–8 ; Gilbert & Boutique l.c. 3 : 153. A scrambling shrub or climber forming thickets, in regrowth and on rocky hills within the forest regions as well as in dry savannah vegetation ; branchlets with yellowish bark armed with recurved prickles ; flowers cream ; fruits reddish-brown, shining, surfaces undulate.
Sen.: *Roger*! *Perrottet* 274! Cayor *Heudelot* 536! **Fr.Sud.:** Baranigué (Sept.) *Chev.* 3366! Bamako (Sept.) *Waterlot* 1391! **Fr.G.:** Kankan *Chev.* 586. **S.L.:** Wallia (Feb.) *Sc. Elliot* 4293! Yetaya (Sept.) *Thomas* 2310! Kabala–Musaia road (Sept.) *Deighton* 4871! **Lib.:** Moylakwelli (Oct.) *Linder* 1314! **Iv.C.:** Dedougou *Aubrév.* 2756. **Dah.:** Abomey *Chev.* 23175. **N.Nig.:** Kano *Lamb* 10! Katagum *Dalz.* 50! Geidam (fl. & fr. Dec.) *Elliott* 134! Lokoja (Sept.) *T. Vogel* 187! *Barter*! Bauchi Plateau (Aug.) *Lely* P. 582! **S.Nig.:** Lagos *MacGregor* 261! Olokemeji *Foster* 111! Widespread in tropical Africa. (See Appendix, p. 204.)

9. **A. farnesiana** *(Linn.) Willd.*—F.T.A. 2 : 346 ; Chev. Bot. 245 ; Bak. f. l.c. 835 ; Aubrév. l. c. t. 52, 8. *Mimosa farnesiana* Linn. (1753). A spiny shrub with very fragrant yellow flowers ; introduced.
Sen.: *Perrottet* 273! **Fr.G.:** Los Isl. (Feb.) *Chev.* 12098. **S.L.:** *T. Vogel* 1! Regent (Dec.) *Sc. Elliot* 4121! York (fl. Jan., fr. May) *Lane-Poole* 179! *Deighton* 2672! **Lib.:** Buchanan (Mar.) *Baldwin* 11204! **G.C.:** Accra (fr. Nov.) *Brown* 363! Togo: Lome *Warnecke* 193! **N.Nig.:** Gurara R. (July) *Elliott* 184! **S.Nig.:** Iwaraja (fl. & fr. Mar.) *Foster* 179! Sapoba *Kennedy* 2163! Cultivated throughout the tropics and in S. Europe, probably native in S. America. (See Appendix, p. 205.)

10. **A. sieberiana** *DC.* var. **sieberiana**—F.T.A. 2 : 347 ; Bak. f. 836 ; Aubrév. l.c. 279, t. 51, 1–2. A tree, up to 70 ft. high, usually with a rather short thick bole and wide, spreading, more or less rounded crown ; bark pale greenish-yellow with small grey-brown exfoliating scales ; spines long, straight, white ; flowers cream or white ; fruit brown.
Sen.: *Sieber* 43! Galam *Leprieur*! Richard Tol (May) *Döllinger* 57! **Fr.Sud.:** Kita *Dubois* 100. Bamako *Waterlot* 1073! **Iv.C.:** Ouagadougou *Aubrév.* **Dah.:** Boukombe *Aubrév.* 85–D. **Fr.Nig.:** Maradi *Aubrév.* **N.Nig.:** Nupe *Barter* 1208! Katagum *Dalz.* 300! Kukawa (Apr.) *E. Vogel* 85! Abinsi (May) *Dalz.* 583! **S.Nig.:** Old Oyo F.R. (fr. Feb.) *Keay* FHI 16251! Lagos *Foster* 15! Widespread in tropical Africa. (See Appendix, p. 209.)

10a. **A. sieberiana** var. **villosa** *A. Chev.* in Bull. Soc. Bot. Fr. 74 : 959 (1928) ; Keay & Brenan in Kew Bull. 1950 : 364 ; Aubrév. l.c. *A. rehmanniana* of F.W.T.A., ed. 1, 1 : 361, not of Schinz. *A. verugera* Schweinf. var. of Chev. Bot. 247. A tree, similar to the preceding.
Sen.: Bakel (Mar.) *Serv. For.* 2. **Fr.Sud.:** Ouré (Apr.) *Chev.* 700! Koulikoro (Mar.) *Vuillet* 589! Bandiagara (fl. Apr., fr. Sept.) *Ganay* 4! **Iv.C.:** Niangbo (fl. & fr. Jan.) *Aubrév.* 1702! Po (Jan.) *Aubrév.* 2688! **G.C.:** Lorha (Mar.) *McLeod* 831! Barbitu (fl. & fr. Feb.) *Dudgeon* 100! Pong (fl. & fr. Mar.) *Kitson* 585! 705! Navrongo (fl. & fr. Jan., Feb.) *Vigne* FH 4475! 4657! **Togo:** Sokodé (Feb.) *Kersting* A. 313! Kpedsu (fr. Jan.) *Howes* 1083! **N.Nig.:** Samae (Mar.) *Foster* 25! Zamfara F.R. (Apr.) *Keay* FHI 16136! Jebba (Mar.) *Meikle* 1295! (See Appendix, p. 209.)

11. **A. kirkii** *Oliv.* F.T.A. 2 : 350 (1871) ; Chev. Bot. 245 ; Chev. in Rev. Bot. Appliq. 8 : 206 (1928) ; Aubrév. l.c. 278, t. 51, 6. Small tree, flowers pink.
   Fr.Sud.: Macina (Aug.) *Chev.* R. Bani, Mopti to Djenne *Vuillet* 480. **Fr.G.:** *Maclaud.* Also in Tanganyika, N. Rhodesia, Angola and Bechuanaland. (See Appendix, p. 206.)
12. **A. nilotica** (*Linn.*) *Willd. ex Del.* var. **nilotica**—Fl. Aegypt. Illustr. 79 (1813) ; A. F. Hill (q.v. for full synonymy) in Bot. Mus. Lfts. Harvard Univ. 8 : 97 (1940). *Mimosa nilotica* Linn. (1753). *Acacia arabica* (Lam.) Willd. var. *nilotica* (Linn.) Benth. in Hook. Lond. J. Bot. 1 : 500 (1842). *A. scorpioides* (Linn.) W. F. Wight var. *nilotica* (Linn.) A. Chev. in Bull. Soc. Bot. Fr. 74 : 954 (1927) ; Aubrév. l.c. 277, t. 52, 2, t. 53, 1, t. 54, 2. A tree, usually in moist places.
   Sen.: Kaédi *Serv. For.* 4. **Fr.Sud.:** Gao (Feb.) *De Wailly* 5339 ! **N.Nig.:** Sokoto (fl. & fr. Feb.) *Lely* 810 ! Yo, N. Bornu (Dec.) *Elliott* 151 ! L. Chad *Talbot* 1230 ! Maiduguri *Kennedy* 2922 ! Extends to Eastern Sudan and Nileland. (See Appendix, p. 206.)
12a. **A. nilotica** var. **tomentosa** (*Benth.*) *A. F. Hill* l.c. 98 (1940). *A. arabica* (Lam.) Willd. var. *tomentosa* Benth. l. c. (1842). *A. scorpioides* (Linn.) W. F. Wight var. *pubescens* A. Chev. l.c. 954 (1927) ; Aubrév. l.c. 278, t. 52, 3. *A. arabica* (Lam.) Willd. (1806)—Guill. & Perr. Fl. Seneg. 1 : 250. Rare in our area.
   Sen.: *Roger* ! *Carrey.* **G.C.:** Yeji, N.T. (fl. & fr. Aug.—introduced) *Vigne* FH 3322 ! **N.Nig.:** Katagum *Dalz.* 54 ! Extends to Eastern Sudan, Arabia and India.
12b. **A. nilotica** var. **adansonii** (*Guill. & Perr.*) *O. Ktze.*[1] Rev. Gen. Pl. 1 : 156 (1891). *A. adansonii* Guill. & Perr. (1832). *A. arabica* var. *adansonii* (Guill. & Perr.) A. Chev. Bot. 244 (1920). *Mimosa adstringens* Schum. & Thonn. (1827). *Acacia adstringens* (Schum. & Thonn.) Berhaut (1954), not of Mart. (1823), nor of A. Cunn. ex G. Don (1832). *A. scorpioides* (Linn.) W. F. Wight var. *adstringens* (Schum. & Thonn.) A. Chev. (1927) ; Aubrév. l.c. 277, t. 52, 1. *Acacia arabica* var. *adstringens* (Schum. & Thonn.) Bak. f. l.c. 3 : 849 (1930). *A. nilotica* var. *adstringens* (Schum. & Thonn.) Chiov. (1932). *A. arabica* var. *adansoniana* Dubard (1913). *A. nilotica* var. *adansonian* (Dubard) A. F. Hill l.c. 99 (1940). *A. arabica* of F.W.T.A., ed. 1, 1 : 362, partly. Tree, to 50 ft. high ; bole dark reddish-brown, deeply fissured ; twigs dark reddish-brown, tomentellous ; flowers yellow ; fruits grey, straight or curved. This is the commonest variety in our area.
   Sen.: (spec. without frts., exact identification doubtful) *Heudelot* 488 ! **Gam.:** *Dawe* 16 ! 68 ! **Fr.Sud.:** Koubita (fl. & fr. Feb.) *Davey* 44 ! **Port.G.:** Bolama (fl. & fr. Apr.) *Esp. Santo* 1920 ! **G.C.:** Navaro (fr. Apr.) *Kitson* 592 ! Gambaga *Williams* 473 ! **Togo:** *Kersting* 101 ! Lome *Warnecke* 323 ! **N.Nig.:** Sokoto (fl. & fr. Feb.) *Lely* 809 ! Katsina (fl. & fr. Mar.) *Meikle* 1352 ! Nupe *Barter* 1626 ! Dunjere, N. Bornu *Elliott* 136 ! **S.Nig.:** Igbeti (fr. Feb.) *Keay* FHI 14634 ! Widespread in the northern part of tropical Africa. (See Appendix, p. 206.)
13. **A. raddiana** *Savi* Acac. Egiz. Mem. 1, fig. A–G (1830) ; A. F. Hill l.c. 103 (incl. var. *pubescens*) ; Aubrév. l.c. 274, t. 51, 4, t. 54, 2. *A. tortilis* Hayne (1827), partly—F.T.A 2 : 352 ; Chev. Bot. 247 ; F.W.T.A., ed. 1, 1 : 362 ; Bak. f. l.c. 3 : 841 ; *A. tortilis* Hayne var. *pubescens* A. Chev. in Bull. Soc. Bot. Fr. 74 : 960 (1927). *A. fasciculata* Guill. & Perr. (1832), not of Kunth nor of R.Br. Tree to 30 ft. high, with rounded or irregular crown ; bole and branchlets reddish-brown ; flowers white.
   Sen.: *Perrottet* 267 ! *Heudelot* 495 ! Djolon *Dubois* 272–S ! **Fr.Sud.:** Timbuktu (July) *Chev.* 1186 ! *Hagerup* 89 ! **N.Nig.:** near L. Chad (fl. & fr. Dec., Jan.) *Elliott* 111 ! *Foster* 92 ! Yobe R., Bornu *Weir* 2314 ! 2315 ! Widespread in the dry regions of northern Africa, Palestine, etc. (See Appendix, p. 209.)
14. **A. hebecladoides** *Harms* in Engl. Bot. Jahrb. 36 : 208 (1905) ; Bak. f. l.c. 3 : 846 ; Mildbr. in Engl. Bot. Jahrb. 65 : 20 ; Aubrév. l.c. 276, t. 51, 11. Tree, to 40 ft. high, with flat-topped crown, in low-lying savannah sites ; flowers white, sweet-scented, with numerous flower-heads in each axil.
   N.Nig.: Chad region (fr. Dec.) *Foster* 86 ! Yola (fr. Sept.) *Daggash* FHI 13364 ! Namtari, Adamawa (May) *Rosevear* FHI 26602 ! Also in French Cameroons, Belgian Congo, Uganda, A.-E. Sudan, Kenya, Tanganyika, N. and S. Rhodesia.
15. **A. seyal** *Del.*—F.T.A. 2 : 351 ; Keay in Journ. Ecol. 37 : 357 ; Aubrév. l.c. 275, t. 51, 7–8. *A. stenocarpa* Hochst. ex A. Rich.—Chev. in Rev. Bot. Appliq. 27 : 508 (1947), and in many other papers, not of F.T.A. 2 : 351. Loosely branched, more or less flat-topped tree, to 40 ft. high ; bark of young branchlets reddish-brown, soon flaking off to expose a powdery surface ; bark of older stems covered with pale yellow to red powder, exfoliating in subrectangular pieces green at first but becoming brown ; trees of similar size in the same area often show the full range of colour ; flowers bright yellow ; fruits split and remain on the tree.
   Sen.: *Perrottet* 272 ! *Lelievre* ! *Sieber* 45 ! Dakar *Irvine* 3295 ! Yang Yang *Serv. For.* 6. Thies (fl. & fr. Mar.) *Serv. For.* 14 ! 15 ! **Fr.Sud.:** Ouario to Koubita (Feb.) *Davey* 42 ! 43 ! Kayes and Nioro region *fide* Aubrév. l.c. **Iv.C.:** Ouagadougou (fr. May) *Serv. For.* 2393 ! Bissandero (Nov.) *Serv. For.* 2136 ! **Fr.Nig.:** Dori (Oct.) *Aubrév.* 11–N ! Niamey to Zinder *Hagerup* 565 ! **N.Nig.:** Sokoto Prov. (fl. & fr. Dec.–Feb.) *Dalz.* 347 ! *Lely* 803 ! Zamfara F.R. (fr. Apr.) *Keay* FHI 15633 ! 16180 ! 18063 ! Katagum *Dalz.* 47 ! Mutwe (Dec.) *Foster* 78 ! Kukawa (Dec.) *E. Vogel* 20 ! Widespread in the drier parts of tropical Africa. (See Appendix, p. 208.)
16. **A. ehrenbergiana** *Hayne*—F.T.A. 2 : 352 ; A. F. Hill in Bot. Mus. Lfts. Harvard Univ. 8 : 93 (1940). *Acacia flava* (Forsk.) Schweinf. (1896), not of Spreng. ex DC. (1825) ; Bak. f. l.c. 844 ; Chev. in Rev. Bot. Appliq. 27 : 507 (1947) ; Aubrév. Fl. For. Soud.-Guin. 274. *A. seyal* of Chev. in Rev. Bot. Appliq. 14 : 878 (1934) and in many other papers, not of Del. Shrub to 15 ft. high, in semi-desert regions ; flowers deep yellow.
   Maurit.: M'bout *Serv. For.* 1 ! Mederdra *Serv. For.* 34 ! **Sen.:** Damet, Podor *Trochain* 2920. **Fr.Sud.:** Timbuktu *fide* Chev. l.c. Ansongo to Tillabery *Chev.* 43125. Hombori *Rogeon* 434. Gao to Korogoussou (fl. & fr. Aug.) *De Wailly* 4747 ! **Fr.Nig.:** Agadès *Chev.* 43488. Also in Central Sahara and the Nile region.
17. **A. hockii** *De Wild.* in Fedde Rep. 11: 502 (1913). *A. chariensis* A Chev. in Bull. Soc. Bot. Fr. 74: 958 (1928); Rev. Bot. Appliq. 27 : 508 (1947). *A. stenocarpa* Hochst. ex A. Rich. var. *chariensis* (A. Chev.) Aubrév. Fl. For. Soud.-Guin. 275 (1950). *A. atacorensis* Aubrév. & Pellegr. in Bull. Soc. Bot. Fr. 84 : 465 (1937). *A. flava* (Forsk.) Schweinf. var. *atacorensis* (Aubrév. & Pellegr.) Aubrév. Fl. For. Soud.-Guin. 275 (1950). *A.boboensis* Aubrév. in Rev. Bot. Appliq. 19 : 483 (1939), French descr. only. *A. stenocarpa* var. *boboensis* Aubrév. Fl. For. Soud.-Guin. 275 (1950), French descr. only. *A. stenocarpa* of F.T.A. 2 : 351 and of Bak. f. l.c. 845, not of Hochst. ex A. Rich. *A. seyal* Del. var. *multijuga* Schweinf. ex Bak. f. l.c. 844 (1930). Shrub or small tree, bark reddish brown, flowers deep yellow ; often on rocky hills in the moister savannah regions.
   Fr.G.: Messirah, Fouta Djalon *Maclaud* 67 ! **Iv.C.:** Bobo Dioulasso *Aubrév.* 3–S. **G.C.:** Tamale (Feb., Nov.) *McLeod* 810 ! *Williams* 417 ! Adumasia to Ampoti, Afram Plains (May) *Kitson* 1085 ! **Dah.:** Boukombe, Atacora Mts. *Aubrév.* 83–D ! 84–D ! **N.Nig.:** Isanlu to Egbe, Kabba (fl. & fr. Nov.) *Keay* FHI 28099 ! 28100 ! Lokoja *Barter* ! Bauchi Plateau (Aug.) *Lely* P. 657 ! Biu (Sept.) *Noble* 35 ! Yola (Oct.) *Shaw* 73 ! Widespread in tropical Africa.
   [This appears to be a variable widespread species within which several variants can be recognized, but further field observation and more and better specimens are needed before a more satisfactory classification can be made.]
18. **A. pennata** (*Linn.*) *Willd.* [2]—F.T.A. 2 : 345 ; Benth. in Trans. Linn. Soc. 3 : 530 ; Bak. f. l.c. 3 : 853 ;

---

[1] If this is regarded as a distinct species the correct name is *A. adansonii* Guill. & Perr. ; Berhaut's new combination is illegitimate.

[2] A. W. Exell and J. P. M. Brenan suggest that several distinct species have been included here and that true *A. pennata* occurs only in tropical Asia.

Aubrév. l.c. 274, t. 53, 11–12. *Mimosa pennata* Linn. (1753). A prickly climbing shrub ; flowers in numerous small white balls turning yellowish ; fruits brown ; common in the forest regions, but much less common in savannah.
Fr.Sud.: Debo (Aug.) *Chev.* 3361 ! Bandiagara *Ganay* 2. **Fr.G.:** Macenta (May) *Collenette* 10 ! Pita *Pobéguin* 2188. **S.L.:** Bagroo R. (Apr.) *Mann.* 852 ! Mano *Deighton* 1435 ! Hangha (Sept.) *Aylmer* 236 ! Baiwalla (Oct.) *Deighton* 3767 ! **Lib.:** Gbanga (Sept.) *Linder* 766 ! Flumpa, Sanokwele (Sept.) *Baldwin* 9342 ! **Iv.C.:** Abidjan, Bobo Dioulasso and Ouagadougou *fide* Aubrév. *l.c.* **G.C.:** Adeambra (fr. Mar.) *Chipp* 133 ! Banka (Sept.) *Vigne* 1358 ! Kwadaso (Nov.) *Darko* 721 ! **Togo:** Amedzofe (fr. Feb.) *Irvine* 164 ! **N.Nig.:** Kano (Aug.) *Onwudinjoh* FHI 24020 ! Gurara R. (fr. July) *Elliott* 182 ! **S.Nig.:** Lagos *Dalz.* 1412 ! Benin *Unwin* 137 ! Oban *Talbot* 1252 ! Degema *Talbot* 3620 ! **Br.Cam.:** Buea (May) *Maitland* 658 ! Widespread in tropical Africa. (See Appendix, p. 207.)

19. **A. macrothyrsa** *Harms* in Engl. Bot. Jahrb. 28 : 396 (1900) ; Bak. f. l.c. 3 : 851 ; Aubrév. l.c. 280, t. 54, 4–5. *A. buchananii* Harms (1901)—Bak. f. l.c. 3 : 852. *A. dalzielii* Craib (1912). A tree 20–40 ft. high, sometimes scarcely armed ; pinnae sometimes 20 or more pairs, with 50 or more pairs of leaflets; flower-heads abundant golden yellow ; fruits flat, reddish-brown, remaining on the tree.
Fr.Sud.: Kourala, Sikasso *Vuillet* 700. **Iv.C.:** Botomo, Oradara & Yabo *fide* Aubrév. *l.c.* **G.C.:** Nangruma to Santejan (fr. May) *Kitson* 776 ! **Fr.Nig.:** Fada N'Gourma to Niamey *fide* Aubrév. *l.c.* **N.Nig.:** Sokoto Prov. (fl. & fr. Dec.) *Dalz.* 320 ! *Gilman* 1 ! Dutsen Makurdi F.R., Funtua (Oct.) *Keay* FHI 21143 ! Amo, Plateau Prov. *MacElderry* FHI 16384 ! Yola (Oct.) *Shaw* 85 ! S. Bornu (fr. Dec.) *Foster* 87 ! Widespread in tropical Africa. (See Appendix, p. 206.)

21. ALBIZIA Durazz.—F.T.A. 2 : 355 ; Benth. in Trans. Linn. Soc. 30 : 557 (1875).

Staminal tube included in the corolla-tube or only slightly exserted :
  Leaflets 12–20 pairs, falcate-lanceolate, acute, about 1 cm. long, 2–3·5 mm. broad ; pinnae 6–12 pairs ; all parts greyish pubescent ; fruits 10–15 cm. long, 2–2·5 cm. broad, minutely pubescent or nearly glabrous .. .. ..   1. *chevalieri*
  Leaflets up to 13 pairs, exceeding 5 mm. broad ; pinnae up to 7 pairs :
    Filaments 15–50 mm. long ; calyx (2·5–) 3–6·5 mm. long ; corolla 5–13·5 mm. long ; fruits glabrous or nearly so :
      Filaments red above and white below ; pedicels 0–2 mm. long ; pinnae (2–) 3–6 (–8) pairs ; leaflets (4–) 6–11 (–12) pairs, narrowly oblong, 13–33 mm. long, 5–14 (–17) mm. broad, sparingly pubescent ; fruits (10–) 14–21 cm. long, (2·3–) 3·2–3·7 (–4·8) cm. broad .. ..   2. *coriaria*
      Filaments white to green or greenish-yellow, not red :
        Indumentum on outside of calyx and corolla conspicuously rusty (when dry) ; pedicels 0–3 mm. long ; pinnae 3–6 (–9) pairs ; leaflets 10–14 (–17) pairs on the two distal pairs of pinnae, fewer on the basal pairs ; leaflets oblong, 11–23 mm. long, 4·5–10 mm. broad, terminal pair obovate ; fruits 15–24 cm. long, 3–5 cm. broad .. ..   3. *ferruginea*
        Indumentum on outside of calyx grey to whitish, not rusty :
          Pedicels 1·5–4·5 (–7·5) mm. long ; leaflets not glaucous beneath ; pinnae (1–) 2–4 (–5) pairs ; leaflets 3–11 pairs, elliptic-oblong, 15–48 mm. long, (6–) 8–24 (–33) mm. broad, the terminal pair more or less obovate ; fruits (12–) 15–33 cm. long, (2·4–) 2·9–5·5 (–6) cm. broad .. ..   4. *lebbeck*
          Pedicels 0–0·75 mm. long ; leaflets glaucous beneath ; pinnae 2–6 pairs ; leaflets 3–9 (–10) pairs, obliquely obovate-rhombic, 15–40 mm. long, 10–22 (–30) mm. broad ; fruits 15–20 cm. long, 3·5–4 cm. broad
                                5. *malacophylla* var. *ugandensis*
    Filaments 6–10 mm. long ; calyx 1·5–2·5 (–3) mm. long ; corolla 3–5·5 mm. long ; pedicels 1·5–7 mm. long ; pinnae 1–4 pairs ; leaflets 3–5 pairs, obliquely oblong-elliptic or elliptic, pointed at apex, 10–48 mm. long, 6–20 mm. broad, minutely puberulous on the lamina beneath ; fruits 15–26 cm. long, 2·5–4 cm. broad, puberulous .. ..   6. *glaberrima*
Staminal tube much exserted from the corolla tube :
  Pinnae 1–3 (rarely 4 pairs) ; leaflets 3–5 (rarely 6 pairs), the terminal pair larger than the others, the pinnae thus broadening upwards ; leaflets not auriculate at base on the proximal side, obliquely obovate-elliptic, strongly nerved, glabrous beneath except on midrib and margins, 1·5–8 cm. long, 1–4·5 cm. broad ; young parts minutely puberulous ; bracts at base of peduncles, also stipules, very small and inconspicuous, about 1 mm. broad ; fruits 10–20 cm. long, 2–3·5 cm. broad, glabrous ..   7. *zygia*
  Pinnae 4–9 pairs (rarely 3 on occasional reduced leaves) ; leaflets 9–17 pairs, the terminal pair rather smaller than the others, the pinnae thus more or less narrowing upwards :
    Leaflets not auriculate at base on the proximal side, though the proximal margin may be more or less rounded into the insertion of the petiolule :
      Young parts, rhachides of leaves and pinnae densely fulvous-pubescent ; leaflets more or less pubescent all over lower surface, obliquely rhombic-quadrate or -oblong, 7–17 mm. long, 4–9 mm. broad ; stipules and bracts at base of peduncles ovate, 5–9 mm. long, 3–6 mm. broad ; fruits 9–19 cm. long, 1·9–3·2 cm. broad, more or less densely and persistently pubescent .. ..   8. *adianthifolia*
      Young parts, rhachides of leaves and pinnae finely and shortly pubescent ; leaflets pubescent beneath only on midrib and margins, glabrous between or rarely (especially when young) some occasional hairs on primary lateral nerves ; stipules

14

and bracts at base of peduncles lanceolate, up to 6–7 mm. long and 2–3 mm.
broad ; fruits glabrescent .. .. 9a. *gummifera* var. *ealaënsis*
Leaflets markedly auriculate at base on the proximal side :
Leaflets pubescent beneath only on midrib and margins, glabrous between or rarely
(especially when young) some occasional hairs on primary lateral nerves, obliquely
rhombic-quadrate to rhombic-subfalcate, 10–17 mm. long, 4–8 mm. broad ;
stipules and bracts at base of peduncles lanceolate, up to 6–7 mm. long and 2–3
mm. broad ; fruits 10–18 cm. long, 2–3 cm. broad, glabrescent
9. *gummifera* var. *gummifera*
Leaflets pubescent beneath on midrib and margins, and pubescent or puberulous on
the surface between, often very inconspicuously so, obliquely rhombic-oblong,
9–11 mm. long, 4–6 mm. broad ; stipules and bracts at base of peduncles oblong-
elliptic about 4·5 mm. long and 2 mm. broad ; fruits 11–17 cm. long, 3·2–3·8 cm.
broad, minutely and persistently pubescent .. .. .. 10. *intermedia*

1. **A. chevalieri** *Harms* in Engl. Bot. Jahrb. 40 : 15 (1907) ; Bull. Soc. Bot. Fr. 54, Mém. 8 : 17 (1907) ;
l.c. 8 : 169 (1912) ; Aubrév. Fl. For. Soud.-Guin. 297, t. 59, 1–3. A tree, to 35 ft. high, with grey corky
bark ; petals pale crimson with white margins ; filaments white ; fruits pale brown ; in the drier
savannah regions.
Sen.: *Berhaut* 89. **Fr.Sud.:** Djenné (June) *Chev.* 1132 ! Birgo *Dubois* 56. Koulikoro *Vuillet* 657.
Iv.C.: Sakoinsi *Aubrév.* 2157. **G.C.:** Navrongo (fl. & fr. Mar.) *Vigne* FH 4693 ! **Fr.Nig.:** Gourma *Chev.*
24445. **N.Nig.:** Sokoto Prov. (fl. & fr. Apr.) *Dalz.* 310 ! Zamfara F.R. (fl. & fr. Apr.) *Keay* FHI 16127 !
Katagum *Dalz.* 48 ! L. Chad *Talbot* 1214 ! S. Bornu (fr. Jan.) *Foster* 91 ! Also in French Cameroons.
(See Appendix, p. 210.)

2. **A. coriaria** *Welw. ex Oliv.* F.T.A. 2 : 360 (1871) ; Bak. f. Leg. Trop. Afr. 3 : 861 ; Aubrév. Fl. For. C.
Iv. 1 : 170, t. 60 B ; Fl. For. Soud.-Guin. 299. *A. poissoni* A. Chev. (1912)—Bot. 249. Spreading tree,
to 90 ft. high ; in the drier types of forest and in savannah ; flowers reddish.
Iv.C.: Bondoukou *Aubrév.* 753 ! **G.C.:** Nsuatra (fr. Dec.) *Chipp* 783 ! Jema, Kintampo (Jan.) *J. M.*
Harley FH 5256 ! **Togo:** *Kersting* A. 513 (partly) ! **Dah.:** Gouka, Savalou (May) *Chev.* 23722 ! **N.Nig.:**
Maigana, Zaria (fl. Mar., fr. Dec.) *Lamb* 63 ! 93 ! **S.Nig.:** Nsukka *Rosevear* 12/28 ! **Br.Cam.:** Gayama,
Bamenda (Feb.) *Johnstone* 47 ! Also extends to A.-E. Sudan, Kenya, Uganda, Tanganyika, Belgian
Congo and Angola. (See Appendix, p. 210.)

3. **A. ferruginea** (*Guill. & Perr.*) *Benth.*—F.T.A. 2 : 361, partly ; Hutch. in Kew Bull. 1916 : 238 ; Aubrév.
l.c. 1 : 172, t. 58 A ; Fl. For. Soud.-Guin. 299. *Inga ferruginea* Guill. & Perr. (1832). Forest tree, to
150 ft. high, with rough flaking bark ; flowers greenish-white ; fruits reddish-brown.
Sen.: *Perrottet* 260 ! Tamboukane (fr. Dec.) *Chev.* 2164 ! **Gam.:** Albreda *Leprieur* ! **Port.G.:** Bissau
(fr. Feb.) *Esp. Santo* 1789 ! Cumebu (May) *Esp. Santo* 2473 ! **Fr.G.:** Rio Nunez *Heudelot* 881 ! **S.L.:**
Bonthe (Mar.) *Deighton* 2483 ! Kenema (fr. Jan.) *King* 148 ! Musaia (Feb.) *Deighton* 4252 ! **Iv.C.:** various
localities *fide* Aubrév. *l.c.* **G.C.:** Mampong Scarp (fr. Mar.) *Vigne* 1064 (partly) ! **Togo:** *Baumann* 448 !
Dah.: *fide* Aubrév. *l.c.* **N.Nig.:** Agaie (fl. & fr. Feb.) *Yates* ! **S.Nig.:** Lagos *Moloney* ! Olokemeji (fr.
Jan.) *A. F. Ross* 57 ! Idanre Hills (Jan.) *Brenan* 8675 ! Sapoba *Kennedy* 1638 ! Ahoada *Cons. of For.*
90 ! Also in French Cameroons, Ubangi-Shari, Gabon, Belgian Congo, Uganda and Angola. (See
Appendix, p. 210.)

4. **A. lebbeck** (*Linn.*) *Benth.*—F.T.A. 2 : 358 ; Benth. in Trans. Linn. Soc. 30 : 562 ; Chev. Bot. 249 ;
Aubrév. Fl. For. C. Iv. 1 : 166, t. 58 B ; Fl. For. Soud.-Guin. 295. *Mimosa lebbeck* Linn. (1753). A
tree, to 50 ft. high ; introduced, cultivated and often half naturalized on old farms, etc. ; flowers white ;
fruits pale brown.
Sen.: *Lelievre* ! *Perrottet* 275. **Gam.:** *Brunner* 34 ! **Fr.Sud.:** Ségou *Chev.* 3356 ! **S.L.:** Njala (fl. & fr.
Feb.) *Deighton* 4000 ! **G.C.:** Bole (fl. & fr. Feb.) *Kitson* 878 ! **Togo:** Kete Krachi (fr. Jan.) *Vigne*
FH 1578 ! **Dah.:** Cotonou *Chev.* 23425 ! **S.Nig.:** Olokemeji (Nov.) *A. F. Ross* 17 ! **F.Po:** (June) *Mann*
416 ! A native of tropical Asia, now widespread in the tropics. (See Appendix, p. 211.)

5. **A. malacophylla** (*A. Rich.*) *Walp.* var. *ugandensis* Bak. *f.* l.c. 3 : 860 (1930). *A. boromoensis* Aubrév. &
Pellegr. in Notulae Syst. 14 : 56 (1950) ; Aubrév. Fl. For. Soud.-Guin. 300, t. 54, 6–7. Savannah tree,
to 30 ft. high.
Sen.: Tambacounda *Berhaut* 771. **Fr.Sud.:** Bamako (Mar.) *Waterlot* 1075 ! Sébékoro, Arbala (fl. & fr.
Apr.) *Waterlot* 159 ! **Iv.C.:** Banfora *Aubrév.* 1830 ! Boromo (fr. Nov.) *Aubrév.* 2142 ! 2864 ! Ferkéssé-
dougou *Aubrév.* 2614 ! Léo *Aubrév.* 2772 ! **S.Nig.:** Old Oyo F.R. *Keay* FHI 23435 ! Also in Ubangi-
Shari, A.-E. Sudan and Uganda.

6. **A. glaberrima** (*Schum. & Thonn.*) *Benth.* in Hook. Lond. Journ. Bot. 3 : 88 (1844) ; F.T.A. 2 : 358 ;
Keay in Kew Bull. 1953 : 490. *Mimosa glaberrima* Schum. & Thonn. (1827). *Albizia glabrescens* Oliv.
F.T.A. 2 : 357 (1871) ; Aubrév. Fl. For. Soud.-Guin. 300. *A. warneckei* Harms in Engl. Bot. Jahrb.
30 : 75 (1901) ; Bak. f. Leg. Trop. Afr. 3 : 862 ; Aubrév. Fl. For. C. Iv. 1 : 172, t. 60 C. A shrub or
tree, up to 60 ft. high ; flowers white.
Fr.G.: Kadi *Pobéguin* 2110. **Iv.C.:** Agnéby, Agboville & N'Zo *fide* Aubrév. *l.c.* **G.C.:** Asiama to
Jadafa (Apr.) *Thonning* ! Achimota (Jan.) *Irvine* 1949 ! Kwahu-Nkwanta (Mar.) *Beveridge* FH 3263 !
Bana (Mar.) *Vigne* FH 4355 ! **Togo:** Lome (fl. & fr. Feb.) *Warnecke* 57 ! **S.Nig.:** Olokemeji F.R. (fl. &
fr. Mar.) *A. F. Ross* 106 ! Gambari F.R. *W. D. MacGregor* 574 ! Ala (young fr. Nov.) *Thomas* 1945 !
Widespread in tropical Africa. (See Appendix, p. 211.)

7. **A. zygia** (*DC.*) *J. F. Macbr.* in Contrib. Gray Herb. 59 : 3 (1919) ; Bak. f. l.c. 868 ; Aubrév. Fl. For. C.
Iv. 1 : 172, t. 59 ; Fl. For. Soud.-Guin. 299. *Inga zygia* DC. (1825). *Zygia brownei* Walp. (1842).
Albizia brownei (Walp.) Oliv. F.T.A. 2 : 362 (1871) ; Benth. in Trans. Linn. Soc. 30 : 569. A tree, to
80 ft. high ; flowers white with red staminal tube ; in regrowth forest and fringing forests.
Sen.: Casamance (fr. Feb.) *Chev.* 3121 ! **Gam.:** Albreda *Perrottet* ! **Fr.Sud.:** Kita *Dubois* 150 ! **Port.G.:**
Bissau (fl. Feb., fr. Mar.) *Esp. Santo* 1774 ! 1868 ! **Fr.G.:** Karkandy *Heudelot* 800 ! Kouroussa (Apr.)
Pobéguin 679 ! **S.L.:** *Smeathmann* ! Freetown *Burbridge* 532 ! Rokupr (Mar.) *Jordan* 212 ! Njala (fr.
Jan.) *Deighton* 1808 ! Falaba (Mar.) *Sc. Elliot* 4151 ! **Lib.:** Ganta *Harley* 1139 ! Fortsville, Grand
Bassa (fr. Mar.) *Baldwin* 11161 ! **Iv.C.:** various localities *fide* Aubrév. *l.c.* **G.C.:** Kibbi (fl. & fr. Mar.)
Brent 410 ! Offin R., Nfaense (Feb.) *Chipp* 113 ! Kumasi (Feb., Mar.) *Cansdale* FH 4148 ! **Togo:**
Wora-Wora (Feb.) *Hughes* FH 2986 ! **Dah.:** Abomey (Feb.) *Chev.* 23142 ! **N.Nig.:** Lokoja *Barter* 584 !
Lapai (Mar.) *Yates* 29 ! Sokoto Prov. (fl. & fr. Mar.) *Lely* 855 ! **S.Nig.:** Lagos *Rowland* ! Abeokuta
(Feb.) *Punch* 105 ! Calabar (Feb.) *Mann* 2240 ! **Br.Cam.:** (fl. & fr. Feb.) *Maitland* 409 ! Johann-
Albrechtshöhe *Staudt* 622 ! Widespread in tropical Africa. (See Appendix, p. 212.)

8. **A. adianthifolia** (*Schum.*) *W. F. Wight* in U.S. Dept. Afric. Bur. Pl. Industry, Bull. 137 : 12 (1909) ;
Brenan in Kew Bull. 1952 : 520 (q.v. for synonymy) ; Gilbert & Boutique in Fl. Congo Belge 3 : 178.
Mimosa adianthifolia Schum. (1827). *Albizia fastigiata* (E. Mey.) Oliv. F.T.A. 2 : 361 ; Chev. Bot. 248.
A. sassa of F.W.T.A., ed. 1, 1 : 363, and of Aubrév. Fl. For. C. Iv. 1 : 174, not of (Willd.) Chiov. *A.*
gummifera of F.W.T.A., ed. 1, 2 : 606, of Kennedy For. Fl. S. Nig. 115, and of Aubrév. Fl. For. Soud.-
Guin. 300, partly, not of (Gmel.) C. A. Sm. A tree, to 120 ft. high, mainly in regrowth forest ; flowers
greenish-white with reddish staminal tube.

Sen.: *Heudelot* 16 ! **Gam.:** *Rosevear* 16. **Port.G.:** Acuno, Formosa (Apr.) *Esp. Santo* 1985 ! Bissalanca, Bissau *Esp. Santo* 1626 ! **Fr.G.:** Daragbe, Samu (fr. Dec.) *Sc. Elliot* 4303 ! **S.L.:** Bagroo R. (Apr.) *Mann* 830 ! Njala (Apr.) *Deighton* 1152 ! Kambia (Apr.) *Lane-Poole* 9 ! Falaba (Mar.) *Aylmer* 14 ! R. Sulimania (Mar.) *Sc. Elliot* 5314 ! **Lib.:** Monrovia *Whyte* ! Sangwin, Sinoe (fl. & fr. Mar.) *Baldwin* 11332 ! Ganta (Apr.) *Harley* 1138 ! **Iv.C.:** Zaranou (Mar.) *Chev.* 16266 ! Yapo *Chev.* 22324 ! Bingerville (fr. Feb.) *Chev.* 16218 ! **G.C.:** Bligusso *Thonning* ! Kumasi *Cummins* ! *Chipp* 789 ! Kwahu, Birim (fl. & fr. Mar.) *Johnson* 638 ! S. Scarp F.R. *Beveridge* FH 3279 ! **Dah.:** *Burton* ! Tchoui, Porto Novo (fr. Jan.) *Chev.* 22787 ! **N.Nig.:** Lokoja (Jan.) *Barter* 495 ! *Elliott* 223 ! Agaie (fr. Dec.) *Yates* 13 ! **S.Nig.:** Lagos *Rowland* ! *Dalz.* 1031 ! Ibadan (Mar.) *Ejiofor* FHI 27674 ! Sapoba *Kennedy* 1655 ! Onitsha (fl. & fr. Feb.) *Unwin* 77 ! *Rosevear* 12/29 ! Degema *Talbot* ! Widespread in tropical Africa and in S. Africa. (See Appendix, p. 211.)

[Hybrids between species of this group are known to occur. Within our area, *Vigne* 77 from the Gold Coast has been suggested by Brenan (in Kew Bull. 1952 : 529) as *A. adianthifolia* X *zygia*.]

FIG. 160.—ALBIZIA ADIANTHIFOLIA (*Schum.*) W. F. Wight (MIMOSACEAE).

A, flowering branch. B, C, flowers. D, fruits.

9. **A. gummifera** (*J. F. Gmel.*) *C. A. Sm.* var. **gummifera**—Kew Bull. 1930 : 218, partly (excl. syn. *Mimosa adianthifolia, Zygia fastigiata* and *Albizia fastigiata*) ; Brenan in Kew Bull. 1952 : 511 (q.v. for full synonymy) ; Gilbert & Boutique l.c. 3 : 181. *Sassa gummifera* J. F. Gmel. (1791). Tree to 100 ft. high, mainly at high altitudes ; corolla white, staminal tube white below, crimson above.
  S.Nig.: Mt. Koloishe, Obudu, 4,800 ft. (Dec.) *Keay & Savory* FHI 25125 ! **Br.Cam.:** Bamenda, 5,000–6,500 ft. (Jan.) *Migeod* 416 ! *Johnstone* 272/32 ! *Lightbody* FHI 26280 ! Belo to L. Oku, 5,500 ft. (fl. & fr. Jan.) *Keay* FHI 28463 ! Widespread in eastern and central tropical Africa and in Madagascar.
9a. **A. gummifera** var. **ealaënsis** (*De Wild.*) *Brenan* in Kew Bull. 1952 : 518. *A. ealaënsis* De Wild. (1907)— Gilbert & Boutique l.c. 3 : 177. A forest tree, to 110 ft. high.
  S.Nig.: Oban *Talbot* 1314 (partly) ! Also in French Cameroons, Belgian Congo, Uganda, Tanganyika and Angola.
10. **A. intermedia** *De Wild. & Th. Dur.* in Bull. Herb. Boiss., sér. 2, 1 : 751–2 (1901) ; Brenan l.c. 519 ; Gilbert & Boutique l.c. 3 : 180. A forest tree, to 80 ft. high.
  S.Nig.: Oban *Talbot* 1314 (partly) ! Also in French Cameroons, Spanish Guinea, Gabon, Belgian Congo and Angola.

Besides the above species, *A. falcata* (Linn.) Backer (syn. *A. moluccana* Miq.), a native of tropical Asia, and *A. caribaea* (Urb.) Britton & Rose (syn. *Pithecellobium caribaeum* Urb.), a native of the West Indies, have been introduced into W. Africa.

Imperfectly known species.

A. gigantea *A. Chev.* Vég. Util. 5 : 171 (1909) ; Bot. 249. *Pentaclethra gigantea* A. Chev. Bot. 237, name only.
The specimen in Chevalier's herbarium is a barren leafy shoot ; pinnae 4–5 pairs ; rhachis pubescent with
a fleshy gland near the base ; leaflets 5–10 pairs, subequal, oblong, apiculate, 2·5–4·5 cm. long, 1–2 cm.
broad, pubescent mainly on the midrib below.
Iv.C.: Yapo *Chev.* B. 22326 ! Bouroukrou *Chev.* 16151. (See Appendix, p. 211.)

22. SAMANEA (Benth.) Merrill in Journ. Wash. Acad. Sci. 6 : 46 (1916).

A tall tree ; branchlets and leaf-rhachis pubescent ; leaves bipinnate ; pinnae 20–30
pairs, opposite ; rhachis with a gland towards the base and glands between some of
the upper pinnae ; leaflets up to 40 pairs, very close, linear, auriculate on one side
at the base, about 3–4 mm. long, slightly ciliate when young ; flowers capitate ;
peduncles often paired, axillary, about 2–4 cm. long, rusty-tomentellous ; calyx
tubular, dentate, 3 mm. long, like the corolla tomentellous ; stamens 14–16 ; fruits
more or less straight, woody or subwoody with distinct margins, 13–17 cm. long,
2–2·2 (–2·5) cm. broad .. 　.. 　.. 　.. 　.. 　.. 　.. 　.. 　*dinklagei*

S. dinklagei (*Harms*) *Keay* in Kew Bull. 1953 : 488 (1954). *Mimosa dinklagei* Harms in Engl. Bot. Jahrb. 26 :
253 (1899). *Albizia dinklagei* (Harms) Harms (1915). *Cathormion dinklagei* (Harms) Hutch. & Dandy in
F.W.T.A., ed. 1, 1 : 364 (1928) ; Kew Bull. 1928 : 401 ; partly (excl. descr. frts., excl. distrib. Fr. Cam.,
excl. syn. *Mimosa leptophylla* Harms). *Pithecellobium dinklagei* (Harms) Harms (1918)—Aubrév. Fl. For.
C. Iv. 1 : 176, t. 61 A. A tree, to 50 ft. high, especially in forest near rivers ; flowers white.
Port.G.: Bissau (fl. & fr. Feb., Mar.) *Esp. Santo* 1747 ! Farim (fl. & fr. May) *Esp. Santo* 2385 ! Bafata (fl.
& fr. July) *Esp. Santo* 2697 ! **Fr.G.:** Kouroussa *Pobéguin* 859 ! Konkoure (Mar.) *Pobéguin* 1949 ! *Chev.*
12448 ! 12558 ! **S.L.:** Njala (fl. & fr. Apr.) *Deighton* 2883 ! 2892 ! Bagroo R. (Apr.) *Mann* 836 ! Freetown
(Mar.) *Dalz.* 961 ! R. Kara, Scarcies (Jan.) *Sc. Elliot* 4561 ! **Lib.:** Grand Bassa (May) *Dinklage* 1827 !
Nekabozu, Vonjama *Baldwin* 9974 ! **Iv.C.:** *Aubrév.* 1169 ! And various localities *fide* Aubrév. *l.c.* **G.C.:**
Kumasi (Feb.–Apr.) *Vigne* FH 1901 ! 3726 ! Akrofonso, Ashanti (June) *Andoh* FH 4711 !

[I am not altogether sure that this plant should be included in *Samanea*, but a thorough revision of the
whole group on a pantropical scale is needed before a more satisfactory classification can be produced—
R.W.J.K.]

Besides the above, *Samanea saman* (Jacq.) Merrill (syn. *Pithecellobium saman* (Jacq.) Benth., syn. *Albizia
flavovirens* Hoyle), the "Rain-tree", a native of tropical S. America, has been introduced into the forest regions
of our area, and is widely cultivated in the moister parts of the tropics.

23. CATHORMION Hassk. in Retzia 1 : 231 (1855) ; Merrill in Journ. Wash. Acad. Sci. 6 : 43 (1916).

Pinnae 2 (rarely 3) pairs ; leaflets 3–6 pairs, oblong to obovate, unequal at base, rounded
or obtuse at apex the uppermost pairs the largest (up to 4·5 cm. long and 2·5 cm.
broad) ; pedicels 1·5–2 mm. long ; fruit oblong, slightly curved, margins sinuate,
reticulate on the surface, 8–9 cm. long, about 1·5 cm. broad, breaking into 1-seeded
segments 　.. 　.. 　.. 　.. 　.. 　.. 　.. 　.. 　1. *rhombifolium*
Pinnae 5–7 (rarely 4 or 8) pairs ; leaflets 12–25 pairs, oblong-lanceolate, truncate at base
obtuse at apex, 1–1·5 cm. long, about 4 mm. broad ; flowers subsessile ; fruit falcate
to circinate, jointed, 15–20-seeded, 10–20 cm. long, 1–2 cm. broad, breaking into
1-seeded segments 　.. 　.. 　.. 　.. 　.. 　.. 　.. 　2. *altissimum*

1. **C. rhombifolium** (*Benth.*) *Keay* in Hook Bull. 1953 : 489 (1954). *Albizia rhombifolia* Benth. in Hook. Lond.
Journ. Bot. 3 : 87 (1844) ; F.T.A. 2 : 358 ; Benth. in Trans. Linn. Soc. 30 : 563 (excl. syn. *Mimosa
glaberrima* Schum. & Thonn.). *Albizia glaberrima* of F.W.T.A., ed. 1, 1 : 362, not of (Schum. & Thonn.)
Benth. ; Chev. Bot. 249. *Pithecellobium glaberrimum* of Aubrév. in Notulae Syst. 14 : 57 (1950) and in
Fl. For. Soud.-Guin. 290, t. 59, 4–5, partly, not of (Schum. & Thonn.) Aubrév. Tree, to 30 ft. high ; in
swamp forest ; flowers white.
Sen.: Casamance *Etesse* 71 ! **Port.G.:** Gabu (fr. June) *Esp. Santo* 2500 ! **Fr.G.:** Rio Nunez *Heudelot* 735 !
S.L.: near Tassin and Kukum (Jan.) *Sc. Elliot* 4418 ! Wallia (Jan.) *Sc. Elliot* 4745 ! Mange (Feb.)
Deighton 3618 ! Kasanko (Dec.) *T. S. Jones* 52 ! Rokupr (fr. May) *Deighton* 5925 !
2. **C. altissimum** (*Hook. f.*) *Hutch. & Dandy* in F.W.T.A., ed. 1, 1 : 364 (1928) ; Kew Bull. 1928 : 401. *Albizia
altissima* Hook. f. (1849). *Pithecellobium altissimum* (Hook. f.) Oliv. F.T.A. 2 : 364 (1871) ; Benth. l.c.
586 ; Aubrév. Fl. For. C. Iv. 1 : 178, t. 61 B. *Arthrosamanea altissima* (Hook. f.) Gilbert & Boutique
in Fl. Congo Belge 3 : 193, t. 12. Tree, to 120 ft. high ; flowers white ; in forests by river banks.
S.L.: Njala (fl. Oct., fr. Jan., July, Oct.) *Dalz.* 8070 ! *Deighton* 1331 ! 1957 ! Yonibana (Nov.) *Thomas*
5160 ! **Lib.:** Gbanga (fr. Sept.) *Linder* 689 ! Peahtah (Oct.) *Linder* 933 ! Ganta (fr. Feb.) *Baldwin*
11028 ! **Iv.C.:** Bingerville (Feb.) *Chev.* 17265 ! **G.C.:** Cape Coast (July) *T. Vogel* 18 ! Tano R. (fl. & fr.
Sept.) *Chipp* 336 ! **Togo:** Santrokofi (fl. Oct., fr. Dec.) *St. C. Thompson* FH 3598 ! **S.Nig.:** Okitipupa (fr.
July) *Bassey* FHI 1441 ! Aboh *Barter* ! Bonny (fr. Dec.) *Unwin* 25 ! 30 miles NE. of Benin *Unwin* !
Nun R. (Sept.) *Mann* 469 ! **Br.Cam.:** Cameroon R. (fl. & fr. Jan.) *Mann* 2215 ! Also in French Cameroons,
Ubangi-Shari, A.-E. Sudan, Uganda, Gabon, Belgian Congo and Angola. (See Appendix, p. 213.)

24. CALLIANDRA Benth.—F.T.A. 2 : 354 ; Benth. in Trans. Linn. Soc. 30 : 536 (1875).

Shrub with slender unarmed branches ; stipules persistent, lanceolate, nervose ; leaves
bipinnate ; pinnae 2–5 pairs, opposite ; leaflets 8 or more pairs, linear-oblong,
subacute, 1–1·5 cm. long, about 4–5 mm. broad, slightly pubescent ; flowers in small
axillary heads on very slender peduncles up to 4 cm. long ; stamens numerous, with
very slender filaments ; fruit flat, linear, narrowed to the base, dehiscent, 10–15 cm.
long, with very prominent sutures and bulging seeds 　.. 　.. 　*portoricensis*

C. portoricensis (*Jacq.*) *Benth.*—F.T.A. 2 : 354 ; Benth. l.c. 543 ; Bak. f. Leg. Trop. Afr. 3 : 855. *Mimosa
portoricensis* Jacq. (1790). A glabrous shrub ; flowers cream.
G.C.: *Dalz.* 92 ! Odumase (Aug.) *Howes* 1204 ! Aburi (Aug.) *Irvine* 137 ! **Togo:** Misahöhe (Nov.) *Mildbr.*
7333 ! **S.Nig.:** Lagos *MacGregor* 87 ! Bonny (Oct.) *Mann* 510 ! Aguku *Thomas* 622 ! 730 ! Oban *Talbot*
589 ! 1313 ! Common in the West Indies and on the Atlantic coast of America.

Besides the above, *C. haematocephala* Hassk. and *C. surinamensis* Benth., natives of tropical America, have
been introduced into W. Africa.

91. PAPILIONACEAE

By F. N. Hepper

Trees, shrubs, climbers or herbs. Leaves compound, often imparipinnate or 3-foliolate, or simple, usually stipulate ; leaflets sometimes stipellate. Flowers zygomorphic, mostly hermaphrodite. Sepals usually 5, more or less connate. Petals 5, imbricate, free or rarely partially coherent, the upper (adaxial) exterior and forming the standard (vexillum), the two lateral the wings (alae) more or less parallel with each other, the lower two interior and connate by their lower margins into a keel (carina). Stamens often 10, monadelphous or diadelphous, rarely free, mostly all perfect ; anthers usually opening lengthwise by slits. Fruit dehiscent or not. Seeds without endosperm or very rarely with scanty endosperm.

A large, homogeneous and easily recognized family, generally distributed throughout the world; most numerous in warmer regions.

Key to Tribes

Stamens 10, rarely 11, free (or nearly free) from each other, not united into a tube or sheath ; petals 5 ; trees, tall shrubs or tall climbers (rarely dwarf shrubs) ; leaves imparipinnately multifoliolate or 1-foliolate ; leaflets entire ; fruit sometimes drupaceous .. .. **I. Sophoreae**
Stamens all (or all but one) united into a tube or sheath :
Fruit not jointed :
Trees, woody shrubs or tall woody climbers (see also *Erythrina*) :
Fruit indehiscent ; leaves pinnately 3- or more-foliolate or rarely 1-foliolate, mostly with stipels .. .. **II. Dalbergieae**
Fruit dehiscent :
Leaves pinnately compound ; anthers usually uniform **III. Galegeae**
Leaves digitately 3-foliolate ; anthers usually of 2 kinds ; stamens completely monadelphous .. .. **IV. Genisteae**
Herbs erect, procumbent or climbing ; or subshrubs :
Leaves with a terminal leaflet :
Stipels absent ; mostly erect herbs or subshrubs :
Stamens completely monadelphous ; anthers usually of 2 kinds ; leaves 3-foliolate ; leaflets entire .. .. **IV. Genisteae**
Stamens diadelphous (mostly 9 and 1) :
Leaflets entire, or if toothed (in *Cyamopsis*) pinnate and hairs mainly medifixed, (in *Psoralea* undulate and gland-dotted beneath) :
Flowers in racemes or panicles ; stamens filamentous, not broadened at the apex .. .. **III. Galegeae**
Flowers umbellate ; leaves digitately 3-foliolate, the stipules leaf-like ; stamens (some or all) broadened at the apex.. .. .. .. **V. Loteae**
Leaflets toothed, digitate, hairs basifixed, not gland-dotted **VI. Trifolieae**
Stipels present ; stamens not broadened at the apex :
Ovary not surrounded by a disk ; erect (rarely climbing) herbs ; leaves pinnately 3- or multi-foliolate or simple ; hairs sometimes medifixed **III. Galegeae**
Ovary surrounded by a disk ; procumbent, twining or erect herbs ; leaves pinnately 3-foliolate, rarely 5-foliolate or simple ; hairs never medifixed **VII. Phaseoleae**
Leaves without a terminal leaflet, the rhachis produced into a bristle ; inflorescence axillary ; slender woody twiners .. .. **VIII. Vicieae**
Fruit jointed, constricted between the seeds and breaking transversely into 1-seeded portions ; stipels sometimes present .. .. **IX. Hedysareae**

Tribe I—Sophoreae

* Leaves pinnate or 3-foliolate ; fruit indehiscent or dehiscent :
Leaves pinnate ; flowers small ; fruit indehiscent :
Flowers racemose ; fruit fleshy, more or less torulose :
Sea-shore shrub ; leaflets 11-17, rounded at apex ; racemes terminating the young shoots .. .. **1. Sophora**
Forest shrub or tree ; leaflets 3-9, acute at apex ; racemes on the older wood or terminal .. .. **2. Angylocalyx**
Flowers paniculate ; fruit dry, flat, with a faint intramarginal nerve ; calyx lobed to about the middle .. .. **3. Afrormosia**

Leaves 3-foliolate ; flowers large ; fruit oblong, flattened, dehiscent   4. **Camoënsia**
* Leaves 1-foliolate ; fruit dehiscent :
Calyx cupular, 5-toothed or -lobed :
  Calyx 5-lobed to below the middle, thick ; fruit flat, obovate-elliptic, large, 1-seeded ;
    flowers few in short stout racemes ..   ..   5. **Haplormosia**
  Calyx 5-dentate, the teeth small, membranous ; fruit ovoid, rather small, dehiscing
    to expose 1 (–2) scarlet seeds ; flowers in short slender racemes   6. **Bowringia**
Calyx 2-fid or split down one side, (rarely irregularly 5-fid in *Baphia*) :
  Anthers linear, longer than the filaments ; petals nearly equal ; fruit 1-seeded,
    stipitate ; seed with a large spongy aril ..   ..   7. **Leucomphalos**
  Anthers shorter than the filaments :
    Fruit straight, flat, oblong, oblanceolate or lanceolate, several-seeded   8. **Baphia**
    Fruit twisted, convex, 1–2-seeded, constricted between the seeds   9. **Baphiastrum**

Tribe II—DALBERGIEAE

Leaflets without stipels ; fruit 1- to few-seeded :
  Leaflets alternate ; anthers basifixed, opening by a terminal pore or a longitudinal slit
    10. **Dalbergia**
  Leaflets opposite ; anthers medifixed, opening by a longitudinal slit :
    Fruit oblong, thin   ..   ..   11. **Dalbergiella**
    Fruit orbicular, thick, woody (see also below)   ..   ..   ..   15. **Ostryocarpus**
Leaflets with stipels, alternate or subopposite ; anthers medifixed, versatile, opening
  by longitudinal slits :
  Leaflets alternate ; fruit 1-seeded :
    Stipules not spinescent :
      Fruit suborbicular, winged, smooth, bristly or verrucose ;   inflorescence racemose
        or paniculate ; sap red   ..   ..   12. **Pterocarpus**
      Fruit ovoid, turgid; flowers in lax spreading panicles   ..   ..   13. **Andira**
    Stipules spinescent ;   fruit broadly sickle-shaped, laxly nervose, glabrous ;   flowers
      in small axillary and terminal panicles   ..   ..   ..   14. **Drepanocarpus**
  Leaflets opposite ; fruit 1- to few-seeded :
    Wing-petals free from the keel ; fruit not winged, flat   ..   ..   15. **Ostryocarpus**
    Wing-petals adnate up to the middle of the keel :
      Flowers arranged in clusters on the raceme ; fruit narrowly winged on one side only
        16. **Leptoderris**
      Flowers arranged more or less singly on the axis of the raceme :
        Fruit winged on each side ..   ..   17. **Ostryoderris**
        Fruit not winged   ..   ..   18. **Lonchocarpus**

Tribe III—GALEGEAE

Trees, or tall woody climbers or shrubs ; leaflets with spreading nervation ; the hairs
  (when present) basifixed :
  Fruit at length dehiscent ;   calyx truncate or broadly toothed ;   mostly trees or
    woody climbers :
    Leaflets opposite ; leaves never 3-foliolate :
      Calyx bilabiate, upper lip very large, covering the standard   19. **Platysepalum**
      Calyx equally toothed, or truncate ..   ..   20. **Millettia**
    Leaflets alternate or leaves 3-foliolate   ..   ..   21. **Craibia**
  Fruit indehiscent ;   calyx-teeth linear ;   small shrub ;   flowers in terminal and leaf-
    opposed racemes   ..   ..   22. **Mundulea**
Herbs or undershrubs :
  Hairs on leaves etc. (if present), not medifixed :
    Racemes or panicles terminal or leaf-opposed :
      Leaflets without stipels ; vexillary stamens connate at least to the middle :
        Venation of leaves closely parallel ;   leaflets (1–) 3–many ;   vexillary stamen
          connate about the middle with the others ;   flowers racemose or fasciculate
          23. **Tephrosia**
        Venation of leaves not parallel ; leaves 1-foliolate ; vexillary stamen connate its
          full length ; prostrate herb ; stipules scarious ; flowers glomerate
          79. **Melliniella** [1]
      Leaflets with stipels, with spreading not parallel nerves ; leaves 3-foliolate ; vexil-
        lary stamen free from the base upwards   ..   ..   76. **Pseudarthria** [1]
    Racemes axillary :
      Leaves pinnate, with numerous leaflets ;   fruit many-seeded, elongated, often
        torulose   ..   ..   24. **Sesbania**
      Leaves pinnately 3-foliolate, gland-dotted beneath ; fruit 1-seeded   25. **Psoralea**

---

[1] The genera *Melliniella* and *Pseudarthria* are placed here for convenience in the key owing to their fruits dehiscing, but they are closely allied to other genera in the *Hedysareae*.

Hairs on the leaves medifixed (the arms sometimes very unequal), sometimes with basifixed hairs present as well ; anthers nearly always apiculate ; flowers usually red :

Vexillary stamens free from the others ; standard usually pubescent outside, if glabrous then fruit not 4-ribbed .. .. **26. Indigofera**

Vexillary stamens united with the others ; standard glabrous; fruit stout, erect, 4-ribbed in a short axillary congested raceme .. .. .. **27. Cyamopsis**

Tribe IV—GENISTEAE

Staminal tube slit in the upper part :

Style straight ; fruit inflated ; small branched annual with leaf-opposed subsessile flowers .. .. **28. Rothia**

Style abruptly bent near base ; fruit often turgid but hardly inflated ; leaves 3-foliolate or sometimes 1- or 5-foliolate ; flowers usually yellow .. **29. Crotalaria**

Staminal tube not slit in the upper part :

Calyx-lobes longer than the tube ; leaves digitately 5–7-foliolate ; flowers laxly racemose .. .. **30. Lupinus**

Calyx-teeth shorter than the tube ; leaves 3-foliolate ; flowers in dense sessile terminal heads ; fruit viscid .. .. **31. Adenocarpus**

Tribe V—LOTEAE

Leaves digitately 3-foliolate with leaf-like stipules ; flowers umbellate ; fruit much longer than the calyx .. .. **32. Lotus**

Tribe VI—TRIFOLIEAE

Leaves digitately 3-foliolate, stipules not leaf-like ; inflorescence a condensed raceme or subcapitate ; fruit 1–5-seeded, shorter or but little longer than the calyx
33. Trifolium

Tribe VII—PHASEOLEAE

Leaves beneath (and frequently calyx) gland-dotted :

Ovules 2 :

Fruit compressed :

Stems twining or trailing, with usually spike-like racemes of flowers ; funicle of seed attached in centre of hilum .. .. **34. Rhynchosia**

Stems erect :

Style glabrous ; stigma terminal ; funicle of seed attached at one end of the long hilum .. .. **35. Eriosema**

Style bearded above ; stigma hanging over the inner side at the apex ; funicle of seed attached in the centre of the small hilum.. .. **56. Adenodolichos**

Fruit turgid ; erect shrubs ; hilum of seed small .. .. .. **36. Moghania**

Ovules 4 or more :

Erect shrubs ; flowers in subcapitate axillary racemes ; fruit obliquely subtorulose
37. Cajanus

Slender twiner ; flowers paired in the leaf-axils ; fruit septate between the seeds
38. Atylosia

Leaves not gland-dotted beneath :

*Vexillary stamen free from near the base upwards :

Bracts or bracteoles large, conspicuous, more or less persistent :

Style bearded along the inner face ; leaves pinnate or 3-foliolate ; bracts small; bracteoles broad at base of calyx .. .. **39. Clitoria**

Style glabrous ; leaves 3-foliolate :

Inflorescence a lax, slender raceme, many-flowered with large persistent bracts ; bracteoles absent .. .. **40. Amphicarpa**

Inflorescence a short, rather dense, pedunculate raceme, few-flowered; bracts small and caducous ; bracteoles large, enveloping the calyx .. **41. Centrosema**

Bracts and bracteoles small and inconspicuous, caducous :

Keel longer than the standard petal ; fruit hispid with usually stinging hairs ; flowers in zigzag racemes, short racemes or more or less umbellate **42. Mucuna**

Keel shorter than the standard petal :

Fruit ripening above ground-level :

†Style glabrous ; herbaceous or woody :

Calyx entire or unilateral and irregularly jagged-lobate ; trees or shrubs with large showy flowers ; petals unequal, the standard much exceeding the wings and keel .. .. **43. Erythrina**

Calyx more or less regularly lobed ; petals subequal in length :

Calyx 5-lobed ; standard glabrous :

Stems erect ; for other characters see above .. **76. Pseudarthria**[1]

---

[1] See footnote on p. 506.

Stems twining :
    Stems long-pilose ; fruit septate, densely pilose ..   **44. Calopogonium**
    Stems tomentose ; fruit not septate, pubescent ..   **45. Neorautanenia**
Calyx 4-lobed, upper lobe entire or shortly 2-toothed :
  Stems twining ; fruit several-seeded :
    Nodes of raceme swollen ; standard glabrous ..   ..   **46. Galactia**
    Nodes of raceme not swollen ; standard mainly pubescent   **47. Glycine**
  Stems erect ; fruit 2-seeded ..   ..   **48. Pseudoeriosema**
†Style bearded on one side or around the stigma ; herbaceous twiners :
Stigma laterally oblique :
  Style hooded or flattened :
    Stigma with a large hood-like appendage attached on the outside ; keel
      spirally twisted ..   ..   **49. Physostigma**
    Stigma flat and broad, more or less spathulate but not appendaged ; keel not
      twisted ..   ..   **50. Sphenostylis**
  Style neither hooded nor flattened :
    Glands conspicuous beneath anthers of 5 of the stamens ; keel not spirally
      twisted..   ..   ..   **51. Haydonia**
    Glands absent from all stamens :
      Stipules truncate at base ; keel spirally twisted ; fruit not septate
          52. Phaseolus
      Stipules cordate or appendaged below the base ; keel straight or spirally
        twisted ; fruit septate ..   ..   **53. Vigna**
Stigma terminal :
  Style glabrous ; stigma surrounded by a ring of hairs ..   **54. Dolichos**
  Style bearded down one side ; stigma without a ring of hairs   **55. Lablab**
Fruit ripening below ground-level :
  Style bearded ; calyx with short broad teeth ..   ..   **57. Voandzeia**
  Style glabrous ; calyx deeply divided into narrow lobes ..   **58. Kerstingiella**
*Vexillary stamen connate in the upper part with the others, free below :
Fruit 4-winged, 5–6-seeded ; leaves 1–3-foliolate ..   **59. Psophocarpus**
Fruit not winged ; leaves 3-foliolate :
Fruit many-seeded ; herbaceous twiners :
  Nodes of raceme not swollen ; alternate anthers aborted ; fruit narrow, abruptly
    hooked at apex ..   ..   **60. Teramnus**
  Nodes of raceme swollen ; apex of fruit not hooked ; stamens all fertile :
    Calyx-lobes almost equal in size, all acute ; fruit narrow ..   **61. Pueraria**
    Calyx-lobes unequal in size, upper 2 rounded and larger than the lower 3 ; fruit
    broad, furrowed along the upper suture ..   ..   **62. Canavalia**
Fruit 1–3-seeded ; nodes of raceme swollen ; woody climber ..   **63. Dioclea**

Tribe VIII—Vicieae

Slender woody twiners with pinnate leaves, the rhachis ending in a bristle ; stamens 9
  connate in a sheath ; flowers racemose ..   .. **64. Abrus**

Tribe IX—Hedysareae

Leaflets without stipels :
Stamens monadelphous :
  Fruit not ripening below ground-level :
    Leaflets 3 ; calyx-tube resembling a stalk ; flowers enclosed by 1 or more scarious
    bracts, inflorescence congested ; stipules forming a sheath, not appendaged below
    the base ..   ..   **65. Stylosanthes**
    Leaflets 2 ; calyx-tube not like a stalk ; flowers more or less enclosed by 2 opposite
    green bracts, inflorescence usually lax ; stipules not sheathing, appendaged below
    the base ..   ..   **66. Zornia**
  Fruit ripening below ground-level ; leaflets 2 pairs, without a terminal one ; flowers
    axillary, solitary ..   ..   **67. Arachis**
Stamens diadelphous :
  Calyx subregularly 5-toothed :
    Stamens in 2 bundles of 5 each ; shrubs or small trees ..   **68. Ormocarpum**
    Stamens with 1 free and the rest connate ; a slender herb ..   **69. Antopetitia**
  Calyx bilabiate :
    Fruit well exserted from calyx, or if included then stamens alternately long and short:
      Fruit straight or bent, if coiled then a prickly shrub ..   **70. Aeschynomene**
      Fruit coiled in a ring ; slender shrublet ..   ..   ..   **71. Cyclocarpa**
    Fruit enclosed by the calyx ; stamens all similar :
      Flowers not hidden by the bracts, the latter not reniform :
        Stipules not appendaged below point of insertion ; inflorescence strobilate, a
        congested bracteate raceme, or few flowered ; flowers reflexed   **72. Kotschya**

Stipules appendaged ; inflorescence umbellate, bracts caducous ; flowers erect
73. **Smithia**
Flowers hidden by the large imbricate reniform bracts ..   ..   74. **Bryaspis**
Leaflets with stipels :
Fruit not bent backwards, more or less straight and exserted from the calyx :
Calyx herbaceous ; fruit compressed :
Leaf-rhachis not winged ; alternate filaments not flattened   ..   75. **Desmodium**
Leaf-rhachis winged, wing foliaceous ; alternate filaments flattened
77. **Droogmansia**
Calyx dry and glumaceous ; fruit not compressed   ..   ..   78. **Alysicarpus**
Fruit bent backwards, included in the calyx   ..   ..   ..   80. **Uraria**

Besides the above, *Gliricidia sepium* (Jacq.) Walp. has been widely introduced into our area from tropical America; *Derris microphylla* (Miq.) Jacks., from Burma, and *Myroxylon balsamum* Harms, from tropical America, are both cultivated in Sierra Leone and Gold Coast ; *Trigonella foenum-graecum* Linn. is cultivated at Timbuktu ; *Melilotus indica* (Linn.) All. and *Medicago sativa* Linn. have been recorded from Bilma, French Niger.

1. SOPHORA Linn.—F.T.A. 2 : 253.

Small shrub, covered with short white indumentum ; leaves imparipinnate ; leaflets about 7 pairs, opposite or alternate, elliptic, rounded at each end, about 4 cm. long and 2 cm. broad, shortly appressed-tomentellous beneath ; racemes up to 25 cm. long ; bracts linear-subulate, 6 mm. long ; pedicels up to 1 cm. long ; calyx cupular, nearly 1 cm. long, minutely tomentellous ; corolla 1·5–2 cm. long ; fruits long-stipitate, torulose, about 15 cm. long, minutely tomentellous   ..   ..   *occidentalis*

S. occidentalis *Linn.* Syst., ed. 10, 2 : 1015 (1759) ; Bak. f. Leg. Trop. Afr. 2 : 595.   *S. nitens* Schum. & Thonn. (1827).   *S. tomentosa* of F.T.A. 2 : 254, partly ; Berhaut Fl. Sen. 41.   A maritime plant with hoary foliage, yellow flowers and torulose fruits.
Sen.: *fide* Berhaut *l.c.*   **S.L.:** Kent (fl. & fr. May) *Deighton* 2686 !   York (fl. & fr. May) *Deighton* 5528 !
Lib.: Cape Palmas (fl. & fr. July) *T. Vogel* 1 !   Garaway (fl. & fr. Mar.) *Baldwin* 11631 !   **G.C.:** Sekondi (fr. Dec.) *Johnson* 969 !   Axim (fr. Apr.) *Chipp* 166 !   **Togo:** Lome *Warnecke* 213 !   **S.Nig.:** Lagos (fl. & fr. Apr.) *Dalz.* 1367 !   Also in S. Tomé, West Indies and east tropical S. America.   (See Appendix, p. 261.)

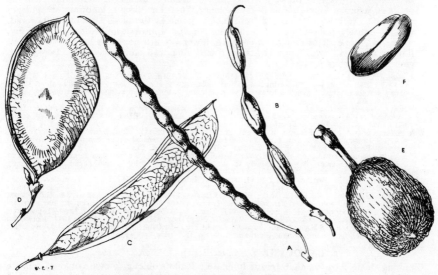

FIG. 161.—FRUITS OF LEGUMINOSAE.

A, *Sophora occidentalis* Linn.   B, *Angylocalyx oligophyllus* (Bak.) Bak. f.   C, *Afrormosia laxiflora* (Benth. ex Bak.) Harms.   D, *Haplormosia monophylla* (Harms) Harms.   E, *Cordyla pinnata* (Lepr. ex A. Rich.) Milne-Redhead.   F, seed of same.

2. ANGYLOCALYX Taub. in Engl. Bot. Jahrb. 23 : 172 (1896).

Flowers amongst the leaves in racemes 10–15 cm. long ; ovary glabrous ; pedicels 10–12 mm. long ; flowers about 1·5 cm. long ; leaves 5–7-foliolate ; leaflets alternate, ovate-lanceolate, rounded at base, acute but hardly acuminate at apex, 8–15 cm. long, 4–6 cm. broad, finely reticulate on both surfaces, lateral nerves obscure   1. *zenkeri*
Flowers cauliflorous ; ovary pubescent ; pedicels about 4 mm. long :
Racemes 4–5 cm. long, becoming twisted ; standard not reflexed ; leaves 3–7-foliolate ; leaflets alternate or subopposite, ovate, cuneate or rounded at base, caudate at apex,

9–23 cm. long, 5–9 cm. broad, laxly reticulate ; fruits torulose, up to 11 cm. long, long-beaked, glabrous. .   . .   .   2. *oligophyllus*

Racemes 1–2 cm. long ; standard reflexed ; leaves 7–9-foliolate ; leaflets alternate, oblong-elliptic, rounded at base, shortly acuminate at apex, 12–23 cm. long, 4·5–9 cm. broad, finely reticulate ; fruits unknown   . .   . .   . .   3. *talbotii*

1. **A. zenkeri** *Harms* in Engl. Bot. Jahrb. 45 : 306 (1910) ; Bak. f. Leg. Trop. Afr. 2 : 598 ; Kennedy For. Fl. S. Nig. 118. A forest tree 50 ft. or more high, with thin brown bark.
   S.Nig.: Sapoba (fl. & fr. Feb., Mar.) *Kennedy* 675 ! 2311 ! Okomu F.R., Benin *Brenan* 8545 ! Also in French Cameroons. (See Appendix, p. 228.)
2. **A. oligophyllus** (*Bak.*) *Bak. f.* in Cat. Talb. 26 (1913) ; Leg. Trop. Afr. 2 : 597. *Sophora oligophylla* Bak. in F.T.A. 2 : 254 (1871). *Angylocalyx ramiflorus* Taub. (1896)—Toussaint in Fl. Congo Belge 4 : 37. *A. trifoliolatus* Bak. f. (1913). A forest shrub 3–6 ft. high, with a single main stem, bark green or grey-brown; calyx and petals greenish-white with reddish spots ; fruits yellow when ripe.
   Lib.: Dukwia R. (Feb.) *Cooper* 203 ! 291 ! Bushrod Isl. (fl. & fr. Jan., Aug.) *Baldwin* 13077 ! 14071 ! **Dah.**: Tori-Kada *Poisson* ! **S.Nig.**: Lagos (fl. & fr. Apr.) *Moloney* ! *Dalz.* 1093 ! Ibadan (Sept.) *Latilo* FHI 23532 ! Okomu F.R., Benin (Feb.) *Brenan* 9062 ! 9076 ! Oban *Talbot* 74 ! Old Calabar R. (fl. & fr. Feb.) *Mann* 2232 ! **Br.Cam.**: Victoria (Mar.) *Preuss* 1374 ; *Maitland* 912 ! Tiko (Jan.) *Dunlap* 246 ! Kumba (fl. Mar., fr. May) *Olorunfemi* FHI 30553 ! *Brenan* 9420 ! Also in French Cameroons, Gabon, Cabinda and Belgian Congo. (See Appendix, p. 228.)
3. **A. talbotii** *Bak. f. ex Hutch. & Dalz.* F.W.T.A., ed. 1, 1 : 371 (1928) ; Kew Bull. 1928 : 402 ; Bak. f. l.c. 598.
   S.Nig.: Oban *Talbot* 239 ! 1205 !

3. AFRORMOSIA Harms in E. & P. Pflanzenfam. Nachtr. 3 : 158 (1906).

Rain forest tree ; leaf-rhachis glabrous or with a few rather long hairs, 8–15 cm. long ; leaflets 7–9 (–11), alternate, ovate-elliptic, rounded at base, shortly acuminate, 3–7 cm. long, 1·5–3 cm. broad, upper leaflets longer than the lower, glabrous ; stipels subulate, 3 mm. long ; panicles terminal, short, many-flowered, with slender pubescent rhachis 2–6 cm. long ; calyx appressed silky-pubescent, 1 cm. long, 5-toothed, upper 2 teeth almost connate, lower 3 teeth twice as long as the tube ; flowers 13–14 mm. long ; fruits 9–14 cm. long and about 2·5 cm. broad, flat, 1–3-seeded, pubescent when young becoming almost glabrous   . .   . .   1. *elata*

Savannah tree ; leaf-rhachis densely tomentose, at least in the groove of older leaves, 12–20 cm. long ; leaflets 9–13, alternate or subopposite, variable in shape from narrowly lanceolate to broadly ovate-elliptic, usually cuneate or sometimes rounded at base, acumen emarginate, 3–7 cm. long, 2·5–5 cm. broad, glabrous or sparsely pubescent ; stipels subulate, 2 mm. long ; panicles terminal, lax, many-flowered, rhachis densely pubescent, 5–10 cm. long ; calyx appressed-pubescent, 1 cm. long, 5-toothed, upper 2 teeth almost connate, lower 3 teeth about as long as the tube ; flowers about 13 mm. long; fruits 5–15 cm. long, 2–3 cm. broad, flat, 1–3-seeded, glabrous, glaucous   . .   . .   2. *laxiflora*

1. **A. elata** *Harms* in Engl. Bot. Jahrb. 49 : 430 (1913) ; Bak. f. Leg. Trop. Afr. 2 : 600 ; Aubrév. Fl. For. C. Iv. 1 : 282, t. 113, 4–5 ; Toussaint in Fl. Congo Belge 4 : 42, t. 3. A large forest tree, to about 160 ft. high ; surface of trunk somewhat fluted and with thin grey flaking bark leaving reddish patches beneath ; in rain forest.
   Iv.C.: *Aubrév.* 690 ! 1592 ; 1804 ! **G.C.**: Gambia, W. Ashanti *Charmant* ! Bia Tano F.R. (fr. Dec.) *Antah* FH 5124 ! Goaso (fl. Apr., May, fr. May, June) *Antah* FH 5156 ! 5183 ! *Charmant* ! *Andoh* FH 5505 ! Also in French Cameroons and Belgian Congo.
2. **A. laxiflora** (*Benth. ex Bak.*) *Harms* in Notizbl. 21, 2 : 64 & fig. (1911) ; Engl. Pflanzenw. Afr. 3, 1 : 527, t. 278 ; Chev. Bot. 218 ; Bak. f. l.c. 2 : 600 ; Aubrév. Fl. For. Soud.-Guin. 306, t. 62, 1–3. *Ormosia laxiflora* Benth. ex Bak. in F.T.A. 2 : 255 (1871) ; Chev. Bot. 218. A tree 30–40 ft. high, with usually tortuous trunk and red- or yellow-blotched bark, grey-pubescent pendulous branchlets and greenish-white or cream flowers ; chiefly in savannah, often on rocky ground, sometimes in fringing forest.
   Sen.: *Berhaut* 913. Bignona *Serv. For.* 51. **Fr.Sud.**: Yatenga *Chev.* 24786 ! **Fr.G.**: Kouroussa *Pobéguin* 663 ! Dentilia, 3,600 ft. (fr. May) *Sc. Elliot* 5302 ! **S.L.**: Erimakuna (Mar.) *Sc. Elliot* 5246 ! Musaia (fr. Mar.) *Deighton* 5484 ! 5751 ! **Iv.C.**: Fétékro, Korhogo, Ferkéssédougou *fide* Aubrév. *l.c.* **G.C.**: Ejura (fr. Aug.) *Chipp* 740 ! Vigne FH 2040 ! Pan (Apr.) *Kitson* 598 ! 605 ! Gambaga (fl. Mar.) *Hughes* FH 4989 ! **Togo**: Sokode (Feb.) *Kersting* 324 ! **Dah.**: Savé (May) *Chev.* 23565 ! **N.Nig.**: Nupe *Barter* 1622 ! Jebba *Dalton* ! Sokoto *Ryan* 10 ! Zaria (fr. May) *Ogwa* FHI 7956 ! *MacGregor* 464 ! Kontagora (fr. Oct.) *Dalz.* 48 ! **S.Nig.**: Aroko (fr. Oct.) *Unwin* 149 ! **Br.Cam.**: Cameroon R. (Jan.) *Mann* 2203 ! Also in French Cameroons, Ubangi-Shari and A.-E. Sudan. (See Appendix, p. 226.)

4. CAMOËNSIA Welw. ex Benth.—F.T.A. 2 : 251.

Climbing shrub ; leaves digitately trifoliolate ; leaflets oblong to ovate-oblong, cuneate at base, subacute at apex, 8–11 cm. long, 4–5 cm. broad, coriaceous, glabrous ; petiole 5–8 cm. long ; flowers in erect, stoutly-peduncolate axillary racemes ; pedicels about 2 cm. long with a pair of small ovate bracteoles at apex ; calyx campanulate nearly 2 cm. long, with 5 broad deltoid teeth 6 mm. long ; corolla about 5 cm. long ; anthers medifixed, 5 mm. long ; ovary pubescent ; fruits unknown   . .   . .   *brevicalyx*

C. brevicalyx *Benth.*—F.T.A. 2 : 251 ; Bak. f. Leg. Trop. Afr. 2 : 602 ; Toussaint in Fl. Congo Belge 4 : 34. A woody forest climber, at least 60 ft. high ; petals white or mauve with a yellow mark in the centre of the standard.
S.Nig.: Aboabam, Ikom Div. (Dec.) *Keay* ! Also in Gabon, Spanish Guinea, Belgian Congo and Angola.

5. HAPLORMOSIA Harms in Engl. Pflanzenw. Afr. 3, 1 : 532 (1915) ; Fedde Rep. 15 : 23 (1919).

A tree ; branchlets glabrous, purplish ; leaves 1-foliolate, broadly elliptic, rounded or shortly cuneate at base, shortly and obtusely acuminate, 7–18 cm. long, 3·5–9 cm.

broad, glabrous ; lateral nerves about 6–8 pairs ; petiole 2–3·5 cm. long ; racemes lateral, few-flowered, up to 10 cm. long ; pedicels 1 cm. long, with a pair of small bracteoles at the top ; calyx glabrous outside except the margins of the 5 lobes ; petals 5 ; stamens 10, free ; fruit flat, obovate-elliptic, beaked, about 8 cm. long and 4–5 cm. broad, glabrous, 1-seeded   ..   ..   *monophylla*

H. monophylla (*Harms*) Harms in Engl. Pflanzenw. Afr. 3, 1 : 532 (1915) ; Bak. f. Leg. Trop. Afr. 2 : 592; Aubrév. Fl. For. C. Iv. 1 : 274, t. 107. *Crudia monophylla* Harms (1901)—Chev. Bot. 233. A tree 20–30 ft. or up to 100 ft. high, sometimes developing buttresses ; flowers showy, blue ; in swamps. **S.L.:** Sherbro *Dawe* 465 ! Falaba (?) (Apr.) *Aylmer* 33 ! Baiama L. (Oct.) *Deighton* 5035 ! **Lib.:** Grand Bassa *Dinklage* 1912 ! 1913 ! Farmington R. (Feb.) *Cooper* 189 ! Brewersville (fr. Dec.) *Baldwin* 10975 ! **Iv.C.:** Abidjan *Chev.* 15420. Bingerville *Chev.* 17349. **S.Nig.:** Lagos *Punch* ! Atijerri (Nov.) *Kennedy* 2004 ! Epe *Barter* 3269 ! Oni *Sankey* 15 ! Eket (fr. Apr.) *Amachi* FHI 24304 ! Also in French Cameroons. (See Appendix, p. 243.)

6. BOWRINGIA Champ. ex Benth. in Hook. Kew Journ. 4 : 75 (1852).

Slender-stemmed climbing shrub ; leaves 1-foliolate, alternate, rounded at base, long-acuminate, 5–10 cm. long, 3·5–5 cm. broad, prominently nerved on both surfaces, glabrous ; petiole about 2 cm. long ; stipules minute ; flowers few, in short axillary racemes ; peduncle 2–5 cm. long ; standard about 11 mm. long, orbicular, emarginate, shortly clawed ; wing falcate-oblong ; keel as long as the wings ; stamens free or slightly joined at the base, anthers 1 mm. long, oblong ; ovary stipitate ; style filiform ; fruits shortly stipitate, woody, 2–3 cm. long, dehiscent, 1–2-seeded

mildbraedii

B. mildbraedii *Harms* in Engl. Bot. Jahrb. 49 : 432, t. 2 (1913) ; Bak. f. Leg. Trop. Afr. 2 : 592 ; in Fl. Congo Belge 4 : 7. A climber ; fruits boat-shaped, gaping to expose 1 or 2 vivid scarlet glossy seeds attached to the suture. **S.Nig.:** Shasha F.R. (fr. Mar.) *Jones & Onochie* FHI 17182 ! Sapoba (fl. & fr. Mar., Oct.) *Jones* FHI 807 ! *Ujor* FHI 23925 ! Idanre (fr. Jan.) *Brenan* 8677 ! 8716 ! Onitsha *Unwin* 76 ! Also in French Cameroons, Ubangi-Shari, Belgian Congo and Angola.

7. LEUCOMPHALOS Benth. ex Planch.—F.T.A. 2 : 251.

A shrub ; leaves 1-foliolate, ovate-elliptic, rounded at base, obtusely acuminate, 5–10 cm. long, 3·5–5 cm. broad, papery, glabrous, closely reticulate on both surfaces, with about 6 pairs of lateral nerves ; petiole up to 6 cm. long, subterete ; flowers in short axillary racemes towards the ends of the branches ; axis of inflorescence pubescent, about 2 cm. long ; pedicels 3–4 mm. long, puberulous ; buds ovoid, 6 mm. long, glabrous ; petals subequal ; anthers longer than the filaments ; fruits dehiscent, obliquely elliptic, beaked, 2–3 cm. long, stipitate, 1-seeded ; seeds with a spongy aril   ..   ..   *capparideus*

L. capparideus *Benth. ex Planch.*—F.T.A. 2 : 251 ; Bak. f. Leg. Trop. Afr. 2 : 591. A shrub or climber slightly branched from the base ; petals and fruit white ; seeds with a large white aril. **S.Nig.:** Oban *Talbot* 1208 ! 1326 ! Akarara, Calabar (May) *Ujor* FHI 30840 ! **F.Po:** in woods (fl. & fr. Dec.) *T. Vogel* 264 ! Also in French Cameroons and Gabon.

8. BAPHIA Lodd.—F.T.A. 2 : 247 ; Lester-Garland in J. Linn. Soc. 45 : 221 (1921).

Calyx divided into two subequal parts :
> Branchlets and midrib shortly puberulous ; inflorescence often branched towards the base ; leaves oblong-elliptic, rounded or subcordate at base, rather long-acuminate, up to 14 cm. long and 7 cm. broad, slightly pubescent beneath ; fruits linear, about 12 cm. long and 1·5 cm. broad, glabrous   ..   ..   ..   1. *polygalacea*
> Branchlets and midrib densely rusty-tomentose ; inflorescence not branched, 5–7 cm. long ; leaves oblong-lanceolate, rounded at base, shortly acuminate, 7–11 cm. long, 2·5–4 cm. broad, hirsute on both surfaces on and towards the midrib ; pedicels 5–7 mm. long ; calyx 1 cm. long   ..   ..   2. *heudelotiana*

Calyx split down one side only (spathaceous) :
> Flowers in racemes, arranged singly on the axis :
>> Bracteoles oblong, nervose, about 2 mm. long ; leaves ovate or oblong-elliptic, subcordate at base, long-acuminate, 9–20 cm. long, 4–6 cm. broad, glabrous, closely reticulate beneath ; racemes axillary ; pedicels nearly 1 cm. long ; calyx about 7 mm. long ; ovary tomentose   ..   ..   ..   3. *leptobotrys*
>> Bracteoles cupular, up to 1 mm. long ; leaves elliptic, rounded at base, abruptly and obtusely acuminate, 8–13 cm. long, 4·5–6·5 cm. broad, glabrous and finely reticulate beneath ; pedicels 6–10 mm. long ; racemes clustered on the old wood, 2–5 (–20 cm.) long ; calyx 8 mm. long ; ovary glabrous   ..   ..   4. *obanensis*

Flowers in axillary fascicles or solitary, or if apparently racemose (by falling of the leaves) then racemes terminal and pedicels fasciculate :
> Ovary glabrous :
>> Petioles 13–25 mm. long (strictly, leaves are 1-foliolate) ; pedicels about 1 cm. long ; calyx slightly pubescent in the upper part, about 1 cm. long ; leaves oblong-elliptic to ovate, rounded at base, acuminate, 8–12 cm. long, 4–5 cm. broad, more

or less pubescent on the nerves beneath ; fruits oblanceolate, up to 15 cm. long,
  glabrous   ..   ..   5. *nitida*
  Petioles 5 mm. long (strictly, leaves are simple) ; pedicels about 1·5 cm. long ;
  calyx glabrous, about 7 mm. long ; leaves obovate, 6–9 cm. long, 2·5–4·5 cm.
  broad, abruptly acuminate, glabrous   ..   ..   6. *gracilipes*
Ovary tomentose:
  Petioles about 1 cm. long ; lamina oblong, ovate-oblong or obovate, acuminate or
  caudate at apex :
    Pedicels 15–22 mm. long, stiff, pubescent ; anthers 3–4 mm. long ; calyx about
    1 cm. long, straight, in bud, appressed-pubescent ; bracteoles small and cup-
    shaped ; leaves oblong or ovate-oblong, 8–15 cm. long, 4–7 cm. broad, finely
    reticulate, leathery, glabrous except on the midrib   ..   ..   7. *laurifolia*
    Pedicels 5–9 mm. long ; anthers 1–2 mm. long :
      Calyx abruptly curved towards apex in bud, about 1·5 cm. long, appressed-
      pubescent ; leaves strongly reticulate and sparsely pubescent beneath, oblong-
      elliptic, rounded at base, acutely acuminate at apex, 8–10 cm. long, 3–4 cm.
      broad, lateral nerves about 6 pairs ; fruits oblong, about 8 cm. long, rusty-pilose,
      becoming glabrous, very acute at apex   ..   ..   ..   8. *spathacea*
      Calyx straight in bud, up to 1 cm. long ; leaves not reticulate, densely to sparsely
      pubescent beneath, often early becoming glabrescent, oblong to obovate-
      oblong, 4–12 cm. long, 2–5 cm. broad, lateral nerves 8–9 pairs ; fruits oblanceo-
      late, about 8 cm. long, glabrous, 1–3-seeded :
        Stipular bracts early caducous, linear-lanceolate, 3–5 mm. long, pubescent ;
        calyx minutely and densely pubescent   ..   ..   ..   9. *pubescens*
        Stipular bracts subpersistent, linear to slightly oblanceolate, 7–10 mm. long,
        strongly nerved, ciliate ; calyx pubescent mainly towards the apex
                                                                    10. *bancoensis*
  Petioles 2–7·5 cm. long ; leaves suborbicular or very broadly elliptic, rounded or
  shortly cordate at base, abruptly and shortly triangular-acuminate at apex,
  10–18 cm. long, 6–8 cm. broad, shortly pubescent on nerves beneath ; lateral
  nerves 6–8 pairs ; flowers axillary, bracteoles broadly ovate, 8 mm. long,
  tomentellous ; calyx 1·4 cm. long ; pedicels stout, 1 cm. long ; ovary villous ;
  fruits oblanceolate, 11–12 cm. long, glabrescent with a thickened margin
                                                                    11. *maxima*

1. **B. polygalacea** (*Hook. f.*) *Bak.* in F.T.A. 2 : 248 (1871), excl. syn. *Delaria pyrifolia* Desv. ; Lester-Garland
    in J. Linn. Soc. 45 : 228 ; Chev. Bot. 217 ; Bak. f. Leg. Trop. Afr. 2 : 573, incl. var. *nigerica* Bak. f.
    (1914) ; Toussaint in Fl. Congo Belge 4 : 12, fig. 2, F–G. *Bracteolaria polygalacea* Hook. f. (1849).
    Baphia pyrifolia Baill. (1885)—Milne-Redhead in Kew Bull. 1947 : 26. A shrub or small tree, sometimes
    scandent ; flowers rather small, fragrant, with a silky calyx and white standard with a yellow spot, in
    copious short racemes or panicles.
    S.L.: Rotamba Isl. (Mar.) *Kirk* ! Bagroo R. *Mann* ! Makump (May) *Deighton* 1705 ! Makene (Apr.)
    Deighton 5501 ! **Lib.**: Grand Bassa (July) *T. Vogel* 46 ! Monrovia (Nov.) *Linder* 1516 ! *Whyte* ! Zorzor
    (Mar.) *Bequaert* 141 ! Belleyella, Boporo (Dec.) *Baldwin* 10657 ! **Iv.C.**: Bingerville *Chev.* 17341. Guidéko
    Chev. 19086. Prolo *Chev.* 19886 . **S.Nig.**: Onitsha (May) *Jones* FHI 492 ! Old Calabar R. (Feb.) *Mann*
    2250 ! *Holland* 24 ! Itu Dist. (May) *Akpata* FHI 3940 ! Eket Dist. (May) *Onochie* FHI 33198 ! Kwa R.
    Talbot 1554 ! **Br.Cam.**: *Preuss* 1257 ! **F.Po**: *Barter* 2071 ! *Mann* 184 ! 396 ! Also in French Cameroons
    and Belgian Congo. (See Appendix, p. 233.)
2. **B. heudelotiana** *Baill.*—F.T.A. 2 : 249 ; Lester-Garland l.c. 229 ; Bak. f. l.c. 574. An erect shrub 12–20 ft.
    high, with white flowers in rusty-tomentose racemes.
    Fr.G.: Rio Pongo *Heudelot* 898 !
3. **B. leptobotrys** *Harms* in Engl. Bot. Jahrb. 26 : 282 (1899) ; Lester-Garland l.c. 232 ; Bak. f. l.c. 583.
    B. silvatica Harms (1913). A glabrous shrub, sometimes more or less scrambling.
    S.Nig.: Oban *Talbot* 591 ! 1761 ! **Br.Cam.**: Tiko (Jan.) *Dunlap* 166 ! Victoria (Dec.) *Mildbr.* 10351 !
    also in French Cameroons and Gabon.
4. **B. obanensis** *Bak. f.* in Cat. Talb. 25 (1913) ; Lester-Garland l.c. 233 ; Bak. f. l.c. 582. A tree about 30 ft.
    high ; flowers white with a yellow base.
    S.Nig.: Oban *Talbot* 1682 !
5. **B. nitida** *Lodd.*—F.T.A. 2 : 249 ; Lester-Garland l.c. 230 ; Chev. Bot 216 ; Bak. f. l.c. 575.
    Delaria pyrifolia Desv. (1826). *B. barombiensis* Taub. (1896). *B .angolensis* of F.W.T.A., ed. 1, 1 : 373,
    not of Welw. ex Bak. A small tree or shrub up to 30 ft. high, with usually glabrous branchlets ; flowers
    usually 1–4 together (occasionally crowded, sometimes fasciated), fragrant, white with a yellow centre.
    Often planted as a dye-wood ; " camwood."
    S.L.: Freetown *Kirk* ! Mesima (fr. Apr.) *Deighton* 3717 ! Newton (Feb.) *Deighton* 4997 ! Bunduel,
    Mende (fl. & fr. June) *Small* 717 ! 717a ! **Lib.**: Mt. Barclay *Bunting* 28 ! Bushrod Isl. (Mar.) *Barker*
    1229 ! Suen (Nov.) *Baldwin* 10459 ! Cess R. (Mar.) *Baldwin* 11270 ! Webo Dist. (July) *Baldwin* 12598 !
    Iv.C.: *Farmar* 359 ! Anoumaba (Nov.) *Chev.* 22410 ! And other localities *fide* Chev. l.c. **G.C.**: Cape
    Coast (July) *T. Vogel* 71 ! Axim (June) *Chipp* 252 ! Kumasi (Feb.) *Kinloch* FH 3245 ! Aburi Hills
    (Nov.) *Howes* 1169 ! **Togo**: Abohire (fl. & fr. Oct.) *St. C. Thompson* FH 3629 ! **Dah.**: Koussi *Chev.*
    23239 ! **S.Nig.**: Lagos Isl. *Barter* 2206 ! Omo F.R. (Mar.) *Jones & Onochie* FHI 17181 ! Cross R.
    (Sept.) *McLeod* ! Old Calabar R. (Feb.) *Mann* 2252 ! **Br.Cam.**: Victoria (Oct.) *Mildbr.* 10533 ! Cameroon
    R. (Jan.) *Mann* 757 ! Barombi *Preuss* 512 ! Kumba (fr. Apr.) *Olorunfemi* FHI 30522 ! **F.Po**: (fl. & fr.
    July) *Mann* 409 ! *Barter* ! Also in French Cameroons. (See Appendix, p. 232.)
6. **B. gracilipes** *Harms* in E. & P. Pflanzenfam. Nachtr. 200 (1897) ; Engl. Bot. Jahrb. 26 : 280 (1899) ;
    Lester-Garland l.c. 231 ; Bak. f. l.c. 575. A small tree with slender glabrous branchlets.
    S.Nig.: Obubra Dist. *Rosevear* 33/29 ! Also in French Cameroons.
7. **B. laurifolia** *Baill.* in Adansonia 6 : 213 (1866) ; Bak. f. l.c. 579 ; Toussaint l.c. 25. *B. crassifolia* Harms
    (1899)—Lester-Garland l.c. 234. *B. myrtifolia* Lester-Garland (1921) ; Bak. f. l.c. 580. *B. densiflora*
    Lester-Garland, partly, not of Harms. A shrub or small tree, 20–50 ft. high ; inflorescence rusty-
    tomentose, flowers white, fairly large, 3–7 together in leafless axils, or apparently paniculate by fall of
    the leaves.
    S.Nig.: Oban *Talbot* 1718 ! 1719 ! Aningeje, Calabar (fl. & fr. May) *Ejiofor* FHI 21900 ! Itu, Calabar

(June) *Ujor* FHI 27978! **Br.Cam.:** Fang, Bamenda, 3,000 ft. (May) *Maitland* 1532! S. Bakundu F.R. *Ejiofor* FHI 15065! Also in French Cameroons, Spanish Guinea, Gabon and Belgian Congo.

8. **B. spathacea** *Hook. f.*—F.T.A. 2 : 250 ; Lester-Garland l.c. 238 ; Bak. f. l.c. 587. *B. dinklagei* Harms (1899)—F.W.T.A., ed. 1, 1 : 373 ; Bak. f. l.c. 581. A straggling shrub, 6–15 ft. high with nearly glabrous branchlets ; flowers fairly large, yellowish-white with rusty-velvet calyx, 2–6 together in leaf axils appearing paniculate by fall of leaves.
S.L.: Nyahlanda (Sept.) *Deighton* 2236! Mano (Oct.) *Deighton* 3787! Njala (fl. Oct., fr. Nov.) *Deighton* 4913! 4925! **Lib.:** Grand Bassa *Dinklage* 1664! Bassa Cove *Ansell*! Du R. (July, Aug.) *Linder* 190! 302! Monrovia (Aug., Sept.) *Baldwin* 9202! 13062!

9. **B. pubescens** *Hook. f.*—F.T.A. 2 : 250 ; Lester-Garland l.c. 238 ; Chev. Bot. 217 ; Bak. f. l.c. 580 ; Toussaint l.c. 24. *B. glauca* A. Chev. Bot. 216, name only. A shrub or tree about 20 ft. high, with brown-tomentose branchlets and inflorescence ; flowers white with a yellow streak, sweet-scented.
Lib.: Jabroke, Webo (July) *Baldwin* 6469! 6495! Gbanga (Mar., June) *Konneh* 183! *Blichenstaff* 34! Farmington R. (Mar.) *Bequaert* 162! **Iv.C.:** Mid. Comoé (Dec.) *Chev.* 22649! Danané (June) *Collenette* 45! Dimbokro *Serv. For.* 444! **G.C.:** Accra Plains (Apr.) *Bally* 37! E. Akim (Mar.) *Johnson* 718! Tarquah (Apr.) *Chipp* 199! Aburi Hills (June) *Williams* 290! **Togo:** *Baumann* 15! **Dah.:** Porto Novo *Chev.* 22792. **N.Nig.:** Gurara R. (July) *Elliott* 177! Dekina (June) *Elliott* 246! Kabba Prov. *Kerr* 2! **S.Nig.:** T. *Vogel*! Ikoyi Plains, Lagos (May) *Dalz.* 1209! 1369! Owo F.R. (Apr.) *Jones* FHI 3524! Okomu F.R. (Jan.) *Brenan* 8923! Sapoba (Mar.) *Jones* FHI 540! Also in French Cameroons, Gabon and Belgian Congo. (See Appendix, p. 233.)

10. **B. bancoensis** *Aubrév.* in Bull. Soc. Bot. Fr. 82 : 602 (1936) ; Fl. For. C. Iv. 1 : 276, t. 109. A small tree or much-branched shrub.
Iv.C.: Abidjan *Aubrév.* 33! 352! Banco *Aubrév.* 1897. Djibi *Aubrév.* 1905. Dabou *Chev.* 17242.

11. **B. maxima** *Bak.* in F.T.A. 2 : 250 (1871) ; Lester-Garland l.c. 238 ; Bak. f. l.c. 588. *B. longipetiolata* Taub. (1896). *B. orbiculata* Bak. f. (1913)—F.W.T.A., ed. 1, 1 : 373 ; Bak. f. l.c. 585. Shrub or tree 10–40 ft. high with leathery leaves ; flowers in axillary yellow-tomentose fascicles ; corolla white with yellow streaks.
S.Nig.: Oron to Eket *Talbot*! Oban *Talbot* 23! 1557! Calabar (Jan.) *Espley* 10! **Br.Cam.:** Cameroon R. (Jan.) *Mann* 2224! Also in Gabon.

Imperfectly known species.

1. **B. sp. A.** Pedicels slender, about 18 mm. long ; ovary pubescent ; stipular bracts linear, persistent, striate, 6–11 mm. long. Allied to *B. bancoensis* Aubrév. but material poor.
S.Nig.: Oban *Talbot* 18!

2. **B. sp. B.** Pedicels about 1 cm. long ; ovary glabrous ; leaves rather large and coriaceous. Material poor.
S.Nig.: Oban *Talbot*!

9. BAPHIASTRUM Harms in Engl. Bot. Jahrb. 49 : 435 (1913).

A shrub ; branchlets and petioles rusty-pubescent ; leaves 1-foliolate, broadly elliptic to elliptic-lanceolate, acutely and shortly acuminate, 7–18 cm. long, 3–7 cm. broad, reticulate and more or less rusty-pubescent beneath, lateral nerves 6–8 pairs ; petiole 1·5–3 cm. long, with 2 small stipels at the top ; stipules lanceolate, about 7 mm. long, caducous ; flowers in axillary fascicles or in apparent racemes ; pedicels about 1 cm. long ; bracteoles about 5 mm. long, broadly ovate, usually reflexed ; calyx 1·3 cm. long, straight in bud, 4-lobed and split down one side in flower ; corolla about 2 cm. long ; anthers 3 mm. long ; ovary densely yellow-pubescent ; fruit 1–3-seeded, twisted   ..   ..   *confusum*

B. confusum (*Hutch. & Dalz.*) *Pellegr.* in Bull. Soc. Bot. Fr. 90 : 162 (1943). *Baphia confusa* Hutch. & Dalz. F.W.T.A., ed. 1, 1 : 373 (1928) ; Kew Bull. 1928 : 402 ; Bak. f. Leg. Trop. Afr. 2 : 587. *Baphia spathacea* of F.T.A. 2 : 250, partly ; Chev. Bot. 217 ; not of Hook. f. *Baphiastrum spathacea* of Staner (1936) ; Fl. Congo Belge 4 : 31 ; not *Baphia spathacea* Hook. f. A shrub or sometimes a climber 6 ft. up to 25 ft. high ; flowers white with a yellow centre ; seeds scarlet.
Iv.C.: Bingerville *Chev.* 15297. **G.C.:** Princes (fr. Jan.) *Akpabla* 771! Wiresu (fr. Dec.) *Brown* 2390! **S.Nig.:** Idanre F.R. (July) *Ejiofor* FHI 26737! Onitsha (Dec.) *Onyeagocha* FHI 7793! Calabar (May) *Onochie* FHI 33162! *Holland* 72! Oban *Talbot* 1209! 1331! 1555! **Br.Cam.:** Cameroon R. (Jan.) *Mann* 746! 2219! **F.Po:** *Mann* 216! *Barter* 2063! Also in Ubangi-Shari and Belgian Congo.

10. DALBERGIA Linn. f.—F.T.A. 2 : 231. *Ecastaphyllum* Browne (1756).

*Fruits suborbicular or obovate-semilunar with one side straight, woody, smooth or rugose ; leaflets 1–3 (–7) :
Leaves 1-foliolate (or very rarely 3-foliolate in *D. ecastaphyllum* f. *trifoliolata*) :
    Leaflet minutely pubescent beneath, lateral nerves about 7, slightly prominent, hardly reticulate between, 8–17 cm. long, 4–7 cm. broad ; petioles 3–8 mm. long ; fruits obovate-orbicular, flat, smooth, about 3 cm. diam.   .. 1. *ecastaphyllum*
    Leaflet glabrous or laxly pubescent beneath, lateral nerves about 12, very prominent, reticulate between, 8–23 cm. long, 4–11 cm. broad ; petioles (5–) 7–15 mm. long ; fruits obliquely ellipsoid, strongly thickened, corrugated, 1·5–3·5 cm. long, 1–2 cm. broad, 5–8 mm. thick   ..   .. 2. *louisii*
Leaves 3–7-foliolate :
    Indumentum on undersurface of leaflets consisting of minute hairs, appressed and lying in one direction ; terminal leaflet usually much larger than the others, 10–15 cm. long ; fruits ovate-semilunar with one side almost straight, young fruit smooth, ripe fruit bullate and wrinkled, about 3 cm. long and 2 cm. broad
                                                            3. *heudelotii*
    Indumentum on undersurface of leaflets tomentose or of crisped hairs not obviously lying in one direction, dense or sparse ; fruits rugose or with a raised centre surrounded by a narrow undulate wing :
        Leaflets beneath and branchlets softly tomentose ; fruits ovate or semilunar, obviously rugose, pubescent   ..   .. 4. *rugosa*

Leaflets beneath with a very few crisped hairs, more on the nerves (young leaflets
pubescent becoming almost glabrous), branchlets slightly pubescent ; fruits
obovate with a raised centre and surrounded by an undulate wing, tomentose
                                                                        5. *crispa*
*Fruits oblong, thin, neither rugose nor woody, but sometimes spiny ; leaflets 5–many :
Branchlets and leaflets densely brown-villous ; flowers in axillary racemes, 1·5–5 cm.
long ; calyx-teeth short ; bracteoles narrow about 1 mm. long ; ovary glabrous :
Leaflets ending in a setose point, lower leaflets ovate, uppermost oblong-elliptic to
obovate, rounded at base, 1–6 cm. long, 1–3·5 cm. broad ; stipules lanceolate,
acuminate, about 1 cm. long ; inflorescences very short, up to 2 cm. long
                                                                        6. *setifera*
Leaflets obtuse or slightly acute, broadly ovate-elliptic, terminal leaflet 9–11 cm. long
and about 6 cm. broad, lower leaflets considerably smaller ; stipules broadly
lanceolate, 1 cm. or more long ; inflorescences 2–5 cm. long ; fruits narrowly
oblong, rounded at each end, shortly stipitate, 6–9 cm. long, 1·5–2 cm. broad,
glabrous and laxly reticulate, thin, 1–3-seeded ..   ..   ..   7. *dalzielii*
Branchlets and leaflets shortly pubescent or tomentose or glabrous :
Leaflets orbicular, or broader than long, broadly cuneate at base :
Leaflets rounded or slightly emarginate at apex, 5–6 cm. long, 4–5 cm. broad,
glabrous ; flowers in axillary panicles 8–12 cm. long ; petals all about 7 mm. long ;
stamens connate into 2 bundles of 5 each ; calyx and peduncles glabrous ; fruits
5–10 cm. long, about 2 cm. broad, glabrous, 1–4-seeded   ..   8. *latifolia*
Leaflets abruptly acuminate, 4–5·5 cm. long, 3–4 cm. broad, sparsely appressed-
pubescent on both sides with hyaline hairs ; flowers in rather lax axillary panicles
(6–) 10–15 cm. long ; standard 9 mm. long, longer than the other petals ; stamens
connate into 1 bundle ; calyx and peduncles sparsely pubescent ; fruits 6–7 cm.
long, about 8 mm. broad, glabrescent, 1–2-seeded   ..   ..   ..   9. *sissoo*
Leaflets oblong or ovate, longer than broad :
Branchlets ending in a spine ; bark light grey ; leaflets 4–6 on each side, obovate-
elliptic, truncate and emarginate at apex, 1–3·5 cm. long, 1–2 cm. broad, becoming
strongly nerved, laxly pubescent beneath ; flowers in lax panicles about 10 cm.
long ; petals 4–5 mm. long ; calyx-tube very slightly pubescent ; fruits acute at
both ends, about 3 cm. long and 1 cm. broad, glabrous   ..   10. *melanoxylon*
Branchlets not ending in a spine (but note axillary spines in No. 12) ; bark brown :
Calyx glabrous, glabrescent or unevenly and shortly pubescent :
Leaflets more or less acute at apex, elliptic or ovate, subopposite, reticulate but
nerves not prominent, 2–3 cm. long, 1·5–2 cm. broad, glabrous ; inflorescences
terminal or axillary, 10–12 cm. long, lax ; flowers about 6 mm. long ; calyx
glabrous ; fruits oblong, rounded at apex, cuneate and stipitate at base, 5–6 cm.
long, 1 cm. broad, glabrous, 1–2-seeded   ..   ..   ..   11. *boehmii*
Leaflets obtuse or emarginate :
Standard petal deltoid, cuneate to a short claw, bifid at apex ; flowers numerous
in pedunculate cymes ; fruits oblong, cuneate at base, 3–3·5 cm. long, about
8 mm. broad, glabrous, 1–2-seeded ; plant with axillary spines (except on the
youngest branchlets) ; leaflets 11–19, oblong to obovate, cuneate at base,
truncate or rounded and rather broadly emarginate at apex, 1·5–4 cm. long,
1–2 cm. broad, young leaflets tomentose becoming glabrescent beneath
                                                                        12. *hostilis*
Standard petal square, truncate to subcordate at base with a distinct claw, bifid
at apex ; flowers in lax terminal or axillary panicles 5–12 cm. long ; fruit
oblong, stipitate, 8–13 cm. long, 3–4 cm. broad, becoming glabrous and laxly
reticulate, 1–2-seeded ; plant without spines ; leaflets 11–13, oblong-elliptic,
rounded at both ends and often slightly emarginate at apex, 2·5–6 cm. long,
1·5–3·5 cm. broad, sparsely pubescent beneath   ..   ..   13. *saxatilis*
Calyx evenly pubescent :
Bracteoles 3 mm. long, ovate, equalling the calyx-tube, persistent ; calyx-lobes
1–2 mm. long ; flowers about 9 mm. long in elongate panicles ; pedicels 3 mm.
long ; leaflets 19–23, oblong or slightly ovate-oblong, rounded at both ends,
3·5–7 cm. long, 1·5–2·5 cm. broad, appressed-pubescent beneath ..   14. *sp. A*
Bracteoles 1 mm. long, inconspicuous at base of bud, shorter than calyx-tube,
caducous :
Calyx-tube strongly ribbed and curved, 4 mm. long, minutely pubescent ; fruits
broadly oblong, about 10 cm. long and 3·5 cm. broad, glabrous, slightly reticulate outside the solitary seed ; leaflets about 19, oblong, truncate or widely
emarginate at apex, 1–1·5 cm. long, about 6 mm. broad, sparsely pubescent
beneath ; flowers few in simple racemes   ..   ..   ..   15. *afzeliana*
Calyx-tube not strongly ribbed or curved, 2–3 mm. long ; pedicels 1–2 mm. long :
Fruits large, 10–15 cm. long, 4–5 cm. broad, glabrous, laxly reticulate, 1 (–2)-
seeded ; inflorescence a many-flowered panicle up to 30 cm. long ; calyx

densely pubescent ; pedicels 1 mm. long ; leaflets 3–7 cm. long, 1·5–2·5 cm. broad, rounded at both ends, slightly emarginate, glabrous to appressed-pubescent .. .. 16. *lactea*

Fruits 5–8 cm. long, 1–2 cm. broad :

Ovary glabrous :

Leaflets 5–7, acute, narrowly elliptic or ovate-elliptic, 4–7 cm. long, 1·5–3 cm. broad, appressed-pubescent beneath ; flowers in dense axillary and terminal fascicles .. .. 17. *oligophylla*

Leaflets 13–21, obtuse and sometimes slightly emarginate at apex, lateral ones oblong, the terminal obovate, 1–4 cm. long, 8–20 mm. broad, rather densely pubescent or tomentose beneath, especially the midrib ; flowers in axillary and terminal racemes 2–6 cm. long .. .. .. 18. *rufa*

Ovary pubescent ; fruits glabrescent, hispid or spiny :

Stamens connate into 2 bundles of 5 each ; inflorescences much shorter than the leaves, 2–5 cm. long ; fruits softly appressed-pubescent to glabrescent, oblong, flat, 6–9 cm. long, 1·5–2·5 cm. broad, 1–2-seeded, seeds not prominent ; leaflets variable in shape and size, elliptic-oblong to broadly ovate, mucronate or acute at apex, 2–6 cm. long, 1–3 cm. broad 19. *oblongifolia*

Stamens connate into 1 bundle ; inflorescences as long as, or much longer than the leaves, 10–30 cm. long ; leaflets broadly obovate-elliptic, 2–4 pairs, 3–5 (–7) cm. long, 2–3 cm. broad, rounded to subcordate at base, ending in a setose point at apex, pubescent beneath, with 6–8 prominent nerves :

Fruits evenly hispid, 1–2-seeded, prominent and reticulate on the seeds, oblong, 5–9 cm. long, about 2 cm. broad ; stipules erect, about 6 mm. long, caducous .. .. 20. *albiflora* subsp. *albiflora*

Fruits densely pubescent and with thick-based, weak-pointed spines concentrated upon the prominent seed, oblong-elliptic, 6–7 cm. long, about 2 cm. broad, 1-seeded ; stipules reflexed, about 10 mm. long, caducous 20a. *albiflora* subsp. *echinocarpa*

1. **D. ecastaphyllum** (*Linn.*) *Taub.* in E. & P. Pflanzenfam. 3, 3 : 335 (1894) ; Bak. f. Leg. Trop. Afr. 2 : 532 (incl. forma *trifoliolata* Stapf). *Hedysarum ecastaphyllum* Linn. (1759). *Ecastaphyllum brownei* Pers. (1807)—F.T.A. 2 : 236 ; Chev. Bot. 211 ; F.W.T.A., ed. 1, 1 : 375. A shrub, up to 20 ft. or more high, with dark purple branchlets and small crowded panicles of fragrant white flowers ; in swamps by the sea.
Sen.: *Heudelot* 510 ! Tambana, Casamance (Feb.) *Chev.* 3427 ! **Gam.:** Bathurst *Don* 5 ! **Port.G.:** Cacine (Sept.) *Esp .Santo* 626 ! Bissau (Mar.) *Esp. Santo* 1885 ! Fulacunda (fr. May) *Esp. Santo* 2026 ! **Fr.G.** *Heudelot* 905 ! **S.L.:** Mahela (fr. Dec.) *Sc. Elliot* 4134 ! Bonthe and Turtle Islands (Nov.) *Deighton* 2332 ! Wellington (fl. & fr. Oct.) *T. S. Jones* 243 ! Luti (July) *Jordan* 281 ! **Lib.:** Grand Bassa (fl. & fr. July) *T. Vogel* 52 ! 81 ! *Baldwin* 11300 ! Harper, Maryland (June) *Baldwin* 5981 ! **G.C.:** Botchiano, Accra *Irvine* 1029 ! Axim (fr. Feb.) *Irvine* 2266 ! **S.Nig.:** Nun R. (Aug.) *T. Vogel* 5 ! Eket *Talbot* 3093 ! Degema *Talbot* ! Brass (fl. & fr. Apr.) *Rhind* FHI 30490 ! *Barter* 1828 ! **Br.Cam.:** Victoria *Preuss* 1303 ! *Maitland* 48 ! 606 ! **F.Po:** *Mann* 262 ! Extends to Angola, also in West Indies and tropical S. America. (See Appendix, p. 241.)

2. **D. louisii** *Cronquist* in Bull. Jard. Bot. Brux. 22 : 217 (1952). A climber or shrub up to 15 ft. high ; flowers white ; beside inland streams.
S.Nig.: Eket Dist. *Talbot* ! Extends southwards to Belgian Congo.

3. **D. heudelotii** *Stapf* in J. Linn. Soc. 37 : 95 (1905) ; Bak. f. l.c. 533. *Ecastaphyllum heudelotii* (Stapf) Hutch. & Dalz. F.W.T.A., ed. 1, 1 : 375 (1928) ; Kew Bull. 1928 : 403, partly. *E. monetaria* of F.T.A. 2 : 236 ; of Chev. Bot. 211 ; not of Pers. A climbing shrub, but sometimes a tree or semi-prostrate shrub, branches purple ; flowers white.
Fr.G.: Rio Nunez *Heudelot* 623 ! **S.L.:** Kowama (fr. Nov.) *Deighton* 5258 ! **Lib.:** Grand Bassa (Mar., Aug.) *Dinklage* 1986 ! *Baldwin* 11220 ! Sanokwele (Sept.) *Baldwin* 12501 ! Duport (Aug.) *Linder* 310 ! **G.C.:** Princes (fl. & fr. Jan.) *Akpabla* 774 ! **S.Nig.:** Jamieson R., Sapoba (fl. & fr. June) *Onochie* FHI 23317 !

4. **D. rugosa** *Hepper* in Kew Bull. 1956 : 133, fig. 9B. *Ecastaphyllum heudelotii* of F.W.T.A., ed. 1, 1 : 375, partly (*MacDonald* 479, *Aylmer* 64). A climber with white flowers turning yellow.
S.L.: Kenema *MacDonald* 479 ! Pujehun (Apr.) *Aylmer* 64 ! Matotoka (fr. July) *Thomas* 1285 ! **Lib.:** Mecca, Grand Cape Mount (Dec.) *Baldwin* 10838 !

5. **D. crispa** *Hepper* in Kew Bull. 1956 : 132, fig. 9A. A climber, similar to the last.
S.L.: *Thomas* 10101 ! 10604 ! Batkanu (Apr.) *Thomas* 18 ! Njala (fl. May, fr. Mar.) *Deighton* 1733 ! 1822 ! Gbap (Mar.) *Adames* 20 !

6. **D. setifera** *Hutch. & Dalz.* F.W.T.A., ed. 1, 1 : 374 (1928) ; Kew Bull. 1928 : 403 ; Bak. f. l.c. 530. A straggling shrub, rusty-villous on the young parts, with short clustered racemes of white flowers.
G.C.: *Evans* 24 ! Sekondi (Feb.) *Vigne* FH 4752 !

7. **D. dalzielii** *Bak. f. ex Hutch. & Dalz.* F.W.T.A., ed. 1, 1 : 374 (1928) ; Kew Bull. 1928 : 402 ; Bak. f. l.c. 529. A shrub, erect or half climbing, with densely clustered white flowers.
S.Nig.: Lagos *Barter* 2141 ! Ebute Metta, Lagos (Jan.) *Millen* 95 ! 211 ! Ikoyi Plains, Lagos (fl. & fr. Feb.) *Dalz.* 1220 ! 1368 ! Bonny *Kalbreyer* 55 ! Also in French Cameroons and Gabon.

8. **D. latifolia** *Roxb.* Pl. Corom. 2 : 7, t. 113 (1798). A planted tree ; flowers white ; sometimes subspontaneous in western Nigeria.
S.Nig.: Ibadan (Feb., Sept.) *Latilo* FHI 27329 ! *Ujor* FHI 30499 ! Olokemeji (fr. Oct.) *A. F. Ross* 14 ! A native of India.

9. **D. sissoo** *Roxb.* Fl. Ind., ed. Carey, 3 : 223 (1832). A planted tree ; flowers pale yellow ; sometimes subspontaneous.
S.L.: Batkanu (Nov.) *Glanville* 447 ! **G.C.:** Gambaga (Apr.) *Morton* GC 9027 ! **N.Nig.:** Samaru, Zaria (fr. May) *Baldwin* 12002 ! Jebba (fl. & fr. Feb.) *Meikle* 1184 ! **Br.Cam.:** Bama, Dikwa Div. (fl. & fr. Jan.) *McClintock* 170 ! A native of India.

10. **D. melanoxylon** *Guill. & Perr.*—F.T.A. 2 233 ; Chev. Bot. 210 ; Bak. f. l.c. 520 ; Aubrév. Fl. For. Soud.-Guin. 311, t. 63, 1–3. A small tree up to 20 ft. high, much-branched, glabrous, armed with woody spines which are the hardened tips of short branches often bearing leaves and flowers ; in savannah country.

Sen.: *Lelièvre*! Kou**m**a *Heudelot*! **Iv.C.:** Yoko to Ouahigouya, Mossi (fr. Aug.) *Chev.* 24715! Kaya *Aubrév.* 2144. **N.Nig.:** Yola (fl. & fr. Apr.) *Dalz.* 6! Katagum *Dalz.* 34! Damaturu Kabau, Bornu (fr. Oct.) *Daggash* FHI 22014! Kiri, Adamawa (May) *Rosevear* FHI 26601! Also in Ubangi-Shari, A.-E. Sudan, Abyssinia and southwards into Transvaal.

11. **D. boehmii** *Taub.* in Pflanzenw. Ost-Afr. C, 218 (1895) ; *Aubrév.* l.c. 311, t. 63, 4–6 ; Bak. f. l.c. 523. *D. elata* Harms (1899)—Bak. f. l.c. 522. *Byrsocarpus caillei* A. Chev., name only. A tree or shrub 12–30 ft. high ; flowers white.
   Sen. : Casamance *Aubrév.* **Port.G.:** Baba (fl. & fr. May) *Esp. Santo* 2479! Also in Ubangi-Shari, Belgian Congo, E. Africa, Mozambique, Nyasaland and N. & S. Rhodesia.

12. **D. hostilis** *Benth.*—F.T.A. 2 : 232 ; Bak. f. l.c. 521. *D. rufa* of F.W.T.A., ed. 1, 1 : 375, partly (*Chev.* 14681). *D. saxatilis* of F.W.T.A., ed. 1, 1 : 375, partly. *D. djalonensis* A. Chev. Bot. 210, name only. A shrub, climbing or erect, usually with axillary spines sometimes clustered and copious rusty-pubescent inflorescences of whitish flowers of variable size ; in dry woodland.
   Port.G.: Bissau (fr. Jan.) *Esp. Santo* 1641! **Fr.G.:** Rio Nunez *Heudelot* 596! Mamou (Nov.) *Caille* in *Hb. Chev.* 14681! Kouria *Caille* in *Hb. Chev.* 14886! Nzérékore (fr. Oct.) *Baldwin* 9685! **S.L.:** Musala (Nov., Dec.) *Miszewski* 19! *Deighton* 4549! Bumban (fr. Jan.) *King* 169b! **Lib.:** Vonjama (fr. Oct.) *Baldwin* 10114! **Iv.C.:** Mt. Kouan, Danané (Apr.) *Chev.* 21271! **G.C.:** Sekodumasi (fr. Oct.) *Vigne* FH 3392! Abetifi, Kwaha (June) *Irvine* 299! Krobo (fl. June, fr. Jan.) *Irvine* 1785! 1911! Abuantem, Mampong (fr. May, Aug.) *Andoh* FH 5023! 5327! **Togo:** *Kersting* A. 513! R. Kabo F.R. (Dec.) *St. C. Thompson* FH 3606! **Dah.:** *Poisson* 102! **N.Nig.:** Naraguta (Dec.) *Lely* 736! **S.Nig.:** Abeokuta *Irving* 84! Oke-Opin, Ilorin (Dec.) *Ajayi* FHI 19273! Enugu (Dec.) *Keay* FHI 28143! Udi Plateau (July) *W. D. MacGregor* 307! Also in Spanish Guinea, Ubangi-Shari, Gabon and Belgian Congo.

13. **D. saxatilis** *Hook. f.*—F.T.A. 2 : 233 ; Bak. f. l.c. 524. *D. macrothyrsa* Harms (1899)—Bak. f. l.c. 526. A more or less glabrous climbing or straggling shrub ; flowers white or pink, in loose axillary or terminal panicles.
   Port.G.: Canchungo (fr. Apr.) *Esp. Santo* 1237! Bissau (fl. Feb., Nov., fr. Mar.) *Esp. Santo* 1828! 1839! 1882! **Fr.G.:** Rio Nunez *Heudelot* 717! Conakry (Apr.) *Debeaux* 317! *Farmar* 148! Fouta Djalon *Maclaud* 379! Nzérékore (fl. & fr. Oct.) *Baldwin* 9705! 9708! **S.L.:** Sugar Loaf Mt. *Barter*! Sherbro *Sc. Elliot* 5794! 5852! Njala (fr. Feb.) *Deighton* 1932! Rokupr (Feb.) *Deighton* 4992! **Lib.:** Cape Palmas *Ansell*! Zigida (Mar.) *Bequaert* 127! Kulo, Sinoe (Mar.) *Baldwin* 11436! **Iv.C.:** Gouékouma, Toura (fr. May) *Chev.* 21657! **G.C.:** Aburi (Jan., Nov.) *Johnson* 298! 547! Nsuta (fl. & fr. May) *Vigne* FH 1734! Cadbury Hall, Kumasi (Oct.) *Darko* 604! **Togo:** Amedzofe (Feb.) *Irvine* 167! **S.Nig.:** Lagos Isl. *Barter* 2233! *Dalz.* 1032! Ubiaja (fr. Aug.) *Oyebade* FHI 20404! Eket *Talbot* 3106! Onitsha (Oct., Nov.) *Parsons* 145! *Keay* FHI 22284! Old Calabar R. (Feb.) *Mann* 2251! **Br.Cam.:** Victoria (Jan.) *Maitland* 285! Extends south into Angola.

14. **D. sp. A.** A woody climber, in derived savannah ; flowers yellow or pale violet ; more mature material with fruits is required.
   S.Nig.: Milliken Hill, Enugu (Nov.) *Keay* FHI 22719! Udi (Nov.) *Ejiofor* FHI 24684!

15. **D. afzeliana** *G. Don* Gen. Syst. 2 : 375 (1832) ; Bak. f. l.c. 528. *D. afzelii* Bak. in F.T.A. 2 : 234 (1871). *D. macrocarpa* Burtt Davy (1937), name only. A hard-wooded climber, rusty-pubescent on the young branchlets and inflorescence ; flowers whitish.
   Port.G.: Canchungo (May) *Esp. Santo* 2047! Bafata (fr. Feb.) *Esp. Santo* 2672! **Fr.G.:** Boullvel, Fouta Djalon (Sept.) *Chev.* 18632! **S.L.:** Lumbaraya, Talla Hills (fr. Feb.) *Sc. Elliot* 4944! Yetaya (Sept.) *Thomas* 2405! Bumbuna (fr. Oct.) *Thomas* 3336! Nganyama, Tikonko (fl. Sept., fr. Nov.) *Deighton* 4026! 5261! **G.C.:** Kumasi (fr. Dec.) *Vigne* FH 1408! **S. Nig.:** Udi, Onitsha (Aug.) *Jones* FHI 6438! Also in Ubangi-Shari, Gabon, Belgian Congo and Angola.

16. **D. lactea** *Vatke* in Oest. Bot. Zeit. 29 : 251 (1879) ; Bak. f. l.c. 525. *D. preussi* Harms (1899)—F.W.T.A., ed. 1, 1 : 375 ; Bak. f. l.c. A scandent shrub ; flowers white ; in upper margins of montane forest.
   S.Nig.: Ikwette Plateau, Obudu, 5,300 ft. (Dec.) *Savory & Keay* FHI 25190! **Br.Cam.:** Buea (fl. Jan., July, Dec.) *Preuss* 897! *Dunlap* 148! *Maitland* 46! 882! Lakom, 6,000 ft., Bamenda (fr. May) *Maitland* 1369! Lakka, 4,000 ft., Bamenda (fr. June) *Maitland* 1566! Also in French Cameroons, Gabon, Belgian Congo and in eastern Africa from A.-E. Sudan and Abyssinia to Mozambique and S. Rhodesia.

17. **D. oligophylla** *Bak. ex Hutch. & Dalz.* F.W.T.A., ed. 1, 1 : 374 (1928) ; Kew Bull. 1928 : 403 ; Bak. f. l.c. 532. *D. bakeri* of F.T.A. 2 : 235, partly. A shrub 25 ft. high ; in gullies at upper edge of montane forest.
   Br.Cam.: Cam. Mt., 5,000–6,000 ft. (fl. Dec., fr. Mar.) *Mann* 2172! *Maitland* 496! 910!

18. **D. rufa** *G. Don* l.c. (1832) ; Bak. f. l.c. 534. *D. pubescens* Hook. f. (1849)—F.T.A. 2 : 234 ; Chev. Bot. 210; Bak. f. l.c. 527. *D. lagosana* Harms (1899)—Bak. f. l.c. 526. A lax-branched scrambling shrub ; flowers white, fragrant ; in secondary regrowth.
   Port.G.: Catió (fr. July) *Esp. Santo* 2102! **Fr.G.:** *Heudelot* 895! Fouta Djalon (Mar.) *Chev.* 12454! **S.L.:** Bonthe (Feb.) *Dalz.* 954! Rokupr (May) *Jordan* 267! Messima (Apr.) *Adames* 29! Njala (Mar. *Deighton* 1579! **S. Nig.:** Lagos (Oct.) *Moloney*! Old Calabar *Thomson* 27! Eket Dist. *Talbot*!

19. **D. oblongifolia** *G. Don* l.c. (1832) ; Bak. f. l.c. 534. *D. dinklagei* Harms (1899)—Chev. Bot. 210 ; Bak. f. l.c. 529. A densely branched shrub or climber with short clusters of fragrant white flowers and rusty flat pods.
   S.L.: Mapaki to Mabonto (Aug.) *Deighton* 1397! Njala (Oct., Dec.) *Deighton* 1538! 5187! Ronletta (fr. Nov.) *Thomas* 5329! **Lib.:** Grand Bassa (fl. Aug., fr. Oct.) *Dinklage* 1724! 1987! 2098! Gbanga (Sept.) *Linder* 526! Moala (fr. Nov.) *Linder* 1371! **Iv.C.:** Zaranou (Mar.) *Chev.* 17618.

20. **D. albiflora** *A. Chev. ex Hutch. & Dalz.* subsp. **albiflora**—F.W.T.A., ed. 1, 1 : 374 (1928) ; Kew Bull. 1928 : 403 ; Bak. f. l.c. 523. A climber up to 40 ft. high ; flowers white, fragrant.
   S.L.: Rowala (July) *Thomas* 1142! 1203! Magburaka (fl. Aug., fr. Jan.) *Deighton* 1367! 3601! Njala (Aug.) *Deighton* 1844! Tikonko (fl. & fr. Nov.) *Deighton* 5263! **Iv.C.:** Soubré to Péhiri (June) *Chev.* 19175!

20a. **D. albiflora** subsp. **echinocarpa** *Hepper* in Kew Bull. 1956 : 131, fig. 8B. A straggling shrub or small tree 10–15 ft. high ; on a sandy river bank and cliff top.
   S.Nig.: Enugu Extension F.R., Ekulu R. (fr. Sept.) *Onochie* FHI 34100! *Cons. of For.* 335! 336! Milliken Hill, Enugu (fl. & fr. May) *Jones* FHI 18863!

Imperfectly known species.

1. **D. sp. nr. macrosperma** *Welw. ex Bak.* *D. pachycarpa* of Kennedy For. Fl. S. Nig. 121. This has the minute indumentum of *D. heudelotii* ; in other characters, such as the cuneate leaflet bases, it resembles *D. macrosperma*, although the fruits are thicker, semi-lunar and the indumentum is short and straight instead of tomentose. Further material is required.
   S.Nig.: Sapoba *Kennedy* 183!

2. **D. sp. nr. pachycarpa** (*De Wild. & Th. Dur.*) *Uhlr. ex De Wild.* (1925), not of Ducke (1922). Terminal leaflet 16 cm. long, 7·5 cm. broad, lateral leaflets smaller, cuneate at base, shortly acuminate at apex, appressed-pubescent beneath ; pedicels 3–4 mm. long ; fruits similar to *D. crispa.* Further material is required.
   G.C.: Kumasi F.R. (fl. & fr. Aug.) *Akwa* FH 1770! Mepe, Volta R. (Aug.) *Adogla* FH 4057!

3. **D. ajudana** *Harms* in Engl. Bot. Jahrb. 26 : 295 (1899). A shrub with rusty-pubescent branchlets and inflorescence ; type presumably destroyed.
   Dah.: Ajuda, on the banks of lagoons *Newton*.

11. DALBERGIELLA Bak. f. in J. Bot. 66, suppl. 1 : 128, t. A.–J. (1928).

A climbing shrub ; leaves imparipinnate ; leaflets 8–13 pairs, opposite or subopposite, oblong, rounded and slightly unequal-sided at base, subacute at apex, 3–7 cm. long, 1·5–3 cm. broad, sparsely pubescent beneath ; inflorescence unbranched, 5–15 cm. long ; flowers fasciculate on the rhachis ; pedicels about 1 cm. long, tomentellous ; calyx deeply lobed, lobes acute, tomentellous ; fruits oblong-lanceolate, 8–11 cm. long, flat, 1-seeded, slightly pubescent, laxly reticulate..   ..   .. *welwitschii*

D. **welwitschii** (*Bak.*) *Bak. f.* l.c. (1928) ; Bak. f. Leg. Trop. Afr. 2 : 535. *Ostryocarpus* ? *welwitschii* Bak. in F.T.A. 2 : 240 (1871) ; F.W.T.A., ed. 1, 1 : 378. *O. racemosus* A. Chev. (1912)—Bot. 213. A wide-climbing shrub, often in riverain forest ; flowers pinkish-white or yellow with a large red streak on the standard, fragrant.
Fr.G.: Mamou to Soya (Jan.) *Chev.* 20368 ! Kouroussa (Nov.) *Pobéguin* 593 ! **S.L.:** Falaba (fr. Apr.) *Sc. Elliot* 5443 ! Musaia (Dec.) *Deighton* 4507 ! **G.C.:** *Lyon* 2643 ! Kumasi *Vigne* FH 1775 ! Nyenahin, Ashanti (Jan.) *Vigne* 1010 ! **Togo:** Misahöhe (Nov.) *Mildbr.* 7388 ! **N.Nig.:** Kagara, Zungeru (fl. & fr. Dec.) *Meikle* 841 ! **S.Nig.:** Abeokuta (fr. Jan.) *Rowland* ! Olokemeji (fr. Jan.) *Dodd* 377 ! Ibadan (Dec.) *Keay* FHI 21151 ! Owo (fl. & fr. Jan.) *G. Burton* FHI 30797 ! **Br.Cam.:** Gashaka Dist. *Latilo & Daramola* FHI 34405 ! Also in Gabon, Ubangi-Shari, Belgian Congo and Angola. (See Appendix, p. 238.)

12. PTEROCARPUS Jacq.—F.T.A. 2 : 237.

Fruit-body smooth or verrucose, glabrous or pubescent but not bristly :
   Leaflets 1–5, opposite or subopposite, oblong-elliptic to suborbicular, emarginate at apex, 4–7·5 cm. long, 2–5 cm. broad, closely reticulate and shining above, glabrous ; calyx about 6 mm. long, shortly lobed, glabrous outside ; fruits obovate to obovate-elliptic, stipitate, 2·5–4·5 cm. long, about 3 cm. broad, reticulate   ..   1. *lucens*
   Leaflets (5–) 7–15, alternate ; fruits orbicular :
     Fruits very broad, 8–9 cm. diam., with broad circular wing and slightly raised fruit-body, stipitate :
       Calyx almost glabrous outside, margins of lobes and inside densely tomentose ; bracts linear-lanceolate, about 10 mm. long, bracteoles similar ; fruits with old style about ¼ round edge from stipe, margin smooth, glabrous ; leaflets 7–9, upper leaflets oblong-elliptic, obtuse or shortly acuminate, 8–13 cm. long, 3·5–6 cm. broad, lower leaflets smaller in size, glabrous, lateral nerves prominent on both surfaces   ..   ..   2. *mildbraedii*
       Calyx densely pubescent outside, almost glabrous inside ; bracts linear-lanceolate, about 10 mm. long, bracteoles linear ; fruits with old style about ⅓ round edge from stipe, margin undulate between, tomentose ; leaflets 11–17, upper leaflets oblong, acuminate, 6–8 cm. long, 2–4 cm. broad, lower leaflets smaller and elliptic, midrib sometimes pubescent, lateral nerves less marked   ..   3. *soyauxii*
     Fruits 3–4 cm. diam., narrowly winged, fruit-body thick and verrucose, tomentellous ; bracts subulate ; calyx deeply lobed, densely pubescent ; leaflets (5–) 7–9, ovate or elliptic-ovate or rarely oblong, rather abruptly acuminate, 5–12 cm. long. 3–3·5 cm. broad, glabrous and closely reticulate on both sides ; stipules falcate   4. *santalinoides*
Fruit-body bristly, fruits orbicular, broadly winged ; leaves 10–15-foliolate :
   Savannah tree ; branchlets not prickly ; leaflets mostly oblong-elliptic, gradually and very shortly acuminate, 6–11 cm. long, 3–6 cm. broad, rather closely and shortly pubescent beneath ; flowers precocious, in lax panicles ; calyx softly tomentellous ; fruits suborbicular, 4–7 cm. diam., the body with long slender bristles   5. *erinaceus*
   Forest tree ; branchlets with numerous short prickles ; leaflets mostly oblong, rounded at each end, shortly and abruptly acuminate, 7–14 cm. long, 3–5 cm. broad, slightly pubescent beneath ; flowers in lax panicles about as long as the leaves ; calyx tomentellous ; fruits orbicular, 9–14 cm. diam., the body shortly bristly   6. *osun*

1. **P. lucens** *Lepr. ex Guill. & Perr.*—F.T.A. 2 : 238 ; Chev. Bot. 212; Bak. f. Leg. Trop. Afr. 2 : 539; Aubrév. Fl. For. Soud.-Guin. 312, t. 63, 1–2. *P. simplicifolius* Bak. in F.T.A. 2 : 238 (1871). *P. lucens* var. *simplicifolius* (Bak.) A. Chev. Bot. 213 (1920) ; Bak. f. l.c. 540. *P. abyssinicus* Hochst. ex A. Rich. (1847)—F.W.T.A., ed. 1, 1 : 376. A small tree, 20–30 ft. high, with light yellow flowers ; in savannah forest.
Sen.: *Perrottet* 257 ! Tambacounda *Berhaut* 708. **Fr.Sud.:** Quiébélé (June) *Chev.* 1003 ! Bamako (June) *Waterlot* 1120 ! **Fr.G.:** Kadé *Pobéguin* 2098. **Iv.C.:** Kaya, Koudougou & Dédougou *fide* Aubrév. *l.c.* **G.C.:** Bantale to Kalpawn R. (Apr.) *Kitson* 836 ! **N.Nig.:** Nupe *Barter* ! Kontagora (Apr.) *Dalz.* 311 ! Gobme Div. (fl. & fr. June) *Kennedy* FHI 7238 ! Yola (fl. & fr. Jan.) *Dalz.* 11 ! Also in French Cameroons, Ubangi-Shari, Belgian Congo, A.-E. Sudan, Abyssinia and Uganda. (See Appendix, p. 258.)
2. **P. mildbraedii** *Harms* in Notizbl. Bot. Gart. Berl. 8 : 152 (1922) ; Bak. f. l.c. 541 ; Aubrév. Fl. For. C. Iv. 1: 286, t. 113, 1–3. A tree 40–50 ft. high, with glabrous branchlets and yellow flowers with conspicuous bracts ; fruits membranous, large, flat and smooth.
Iv.C.: *fide* Aubrév. *l.c.* **G.C.:** Tafo (fr. Sept.) *Andoh* FH 4524 ! **Dah.:** Zagnanado (fr. Feb.) *Chev.* 23091 ! **N.Nig.:** Kurmin Damisa, Jagindi *Keay & Onochie* FHI 21550 ! **S.Nig.:** *Chesters* 196 ! Benin *Kennedy* 1968 ! **Br.Cam.:** Victoria (fr. Feb.) *Maitland* 376 ! 396 ! Dumbo, Bamenda (fr. June) *Maitland* 1556 ! Kung, 4,000 ft. (fr. Mar.) *Johnstone* 299 ! **F.Po:** *Mann* 188 ! Also in French Cameroons and Gabon. (See Appendix, p. 258.)
3. **P. soyauxii** *Taub.* in Hook. Ic. Pl. 24, t. 2369 (1895) ; Bak. f. l.c. 542. A forest tree up to 100 ft. high ; flowers yellow.
S.Nig.: Nikrowa, Benin (fr. Sept.) *Iriah* FHI 23091 ! Ogoja Dist. *Rosevear* 49/29 ! 55/29 ! Ikom (fl. & fr. June) *Rosevear* 60/31 ! **Br.Cam.:** Victoria (fr. Dec.) *Maitland* 100 ! Bimbia (June) *Unwin* 206 ! Also in French Cameroons, Ubangi-Shari, Spanish Guinea, Gabon, Belgian Congo and Cabinda.
4. **P. santalinoides** *L'Hér. ex DC.* Prod. 2 : 419 (1825) ; Bak. f. l.c. 539 ; Aubrév. Fl. For. C. Iv. 1 : 284, t. 112. *P. esculentus* Schum. & Thonn. (1827)—F.T.A. 2 : 238 ; Chev. Bot. 212. A tree 30–40 ft. high, often

15

with short or divided stem, shining foliage and brilliant yellow flowers ; usually evergreen but in some situations briefly deciduous ; on river banks ; fruits float in water.
Sen.: Upper Gambia R. *Heudelot* 363 ! **Gam.:** *Saunders* 2 ! **Fr.Sud.:** Bamako (Jan.) *Chev.* 257 ! R. Bani *Chev.* 1145. L. Débo *Chev.* 3433. **Port.G.:** Cabochanqué, Catió (young fr. June) *Esp. Santo* 2066 ! **Fr.G.:** *Farmar* 224 ! 306 ! Conakry *Lecerf* ! Boké *Chillou* ! Kindia *Chev.* 13026. **S.L.:** *Oldfield* ! Scarcies R. (Jan.) *Sc. Elliot* 4789 ! Bumpe Ferry (fr. Sept.) *Small* 185 ! Bagbe, Gola F.R. (May) *Small* 672 ! **Lib.:** Dukwia R. (fr. May) *Cooper* 466 ! Du R. (fr. July) *Linder* 94 ! Kakata, Salala Dist. (fl. & fr. May) *Baldwin* 5840 ! Bushrod Isl. (Dec.) *Barker* 1098 ! **Iv.C.:** *fide* Chev. *l.c.* **G.C.:** Ankobra R. (Feb.) *Brent* 8 *D.D.* ! Oblogo (fl. & fr. Apr.) *Irvine* 203 ! Axim (Nov.) *Chipp* 28 ! Opro F.R. (Apr.) *Andoh* FH 4710 ! **Togo:** *Kersting* 39 ! Ho (Mar.) *Williams* 54 ! **Dah.:** *Poisson* ! Atacora Mts. *Chev.* 24117. **N.Nig.:** Nupe *Barter* 886 ! Minna (Dec.) *Meikle* 732 ! Wamba Div. (Mar.) *Thornewill* 56 ! Share F.R., Ilorin (Jan.) *Ujor* FHI 31602 ! **S.Nig.:** Ibadan (Jan.) *Meikle* 958 ! Usonigbe F.R., Benin (Nov.) *Adekunle* FHI 24246 ! Eket Dist. *Talbot* 3127 ! Old Calabar R. (Feb.) *Mann* 2236 ! **Br.Cam.:** *Preuss* 1148 ! Ambas Bay (Jan.) *Mann* 713 ! Also in French Cameroons and tropical S. America. (See Appendix, p. 258.)

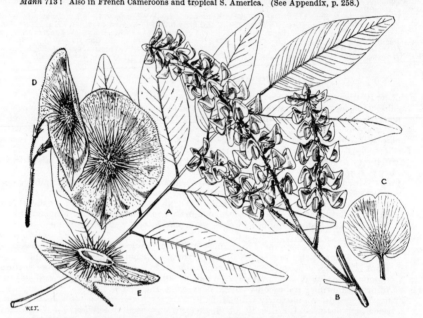

FIG. 162.—Pterocarpus erinaceus *Poir.* (Papilionaceae).

A, leaf. B, inflorescence. C, standard petal. D, fruits. E, fruit opened out.

5. **P. erinaceus** *Poir.*—F.T.A. 2 : 239, excl. syn. *P. angolensis* DC. and *P. echinatus* DC. ; Chev. Bot. 211 ; Bak. f. l.c. 543 ; Aubrév. Fl. For. Soud.-Guin. 314, t. 63, 3–4. A tree up to 40–50 ft. high, often covered with golden-yellow flowers when quite leafless.
Sen.: *Heudelot* 47 ! 373 ! *Perrottet* 256 ! Casamance *Chev.* 3425. **Gam.:** *Mungo Park* ! Sankuli Kunda (Jan.) *Dalz.* 8026 ! **Fr.Sud.:** Kati (Jan.) *Chev.* 173 ! Katou *Chev.* 543. Sansanding *Chev.* 3432. **Port.G.:** Bissau (fl. Jan., fr. Feb.) *Esp. Santo* 1676 ! 1763 ! **Fr.G.:** *Heudelot* 628 ! Aiguiri (Feb.) *Pobéguin* ! Kouroussa (fl. & fr. Jan.) *Pobéguin* 645 ! Kindia (Mar.) *Chev.* 13366 ! 13386. **S.L.:** Wallia to Buyabuya, 3,600 ft. *Sc. Elliot* 4776 ! Musaia (fl. & fr. Feb.) *Deighton* 5357 ! Kambia (Feb.) *Jordan* 395 ! **G.C.:** Afram Plains (fr. Jan.) *Brent* 397 ! Mampong *Brown* 2063 ! Baflawa to Kulkum (fr. Apr.) *Kitson* 606 ! **Togo:** Atakpame (Nov.) *Mildbr.* 7405 ! Sokode *Kersting* A. 223 ! **N.Nig.:** Nupe *Barter* 1190 ! 1628 ! Zungeru (fl. & fr. July) *Dalz.* 47 ! Zamfara F.R. (fr. Apr.) *Keay* FHI 16164 ! Bauchi Plateau (fl. & fr. Jan.) *Lely* P. 1 ! **S.Nig.:** Olokemeji F.R. (fl. & fr. Oct.) *A. F. Ross* 11 ! 82/13/5 ! *Foster* 154 ! *Punch* 148 ! Oyo (fr. Mar.) FHI 20533 ! Also in French Cameroons, Chad and Gabon. (See Appendix, p. 257.)
6. **P. osun** *Craib* in Kew Bull. 1910 : 329 ; Bak. f. l.c. 544. A tree with a comparatively short bole ; branchlets brown-tomentellous and often bristly or prickly ; flowers small, yellow. The Camwood.
S.Nig.: Ibadan F.R. (fr. Dec.) *Punch* 114 ! Ublaja to Okwesan, Benin (Oct.) *Oyebade* FHI 20410 ! Ikom (fr. Jan.) *Holland* 198 ! Obubra (fr. Oct.) *Unwin* 35 ! 36 ! 37 ! Also in French Cameroons, Gabon, Spanish Guinea and French and Belgian Congo. (See Appendix, p. 258.)

The Indian species, *P. indicus* Willd., has been introduced to Njala, Sierra Leone.

13. ANDIRA Lam.—F.T.A. 2 : 246.

A tree ; branchlets shortly tomentellous ; leaves imparipinnate, alternate ; leaflets 5–6 pairs, stipellate, narrowly oblong, slightly acuminate and emarginate, up to 8 cm. long and 4 cm. broad, at first pubescent beneath, soon glabrous, the terminal leaflet rather long-stalked and more obovate ; stipels rigidly subulate ; flowers in large spreading panicles ; pedicels very short ; calyx cupular, 4 mm. long, with very short teeth, pubescent ; corolla about 1·3 cm. long ; stamens diadelphous ; fruits ovoid, shortly pointed, about 4 cm. long and 2·5 cm. thick, glabrous, 1-seeded   *inermis*

A. inermis (*Wright*) *DC.*—F.T.A. 2 : 246 ; Chev. Bot. 216 ; Bak. f. Leg. Trop. Afr. 2 : 563 ; Aubrév. Fl. For. Soud.-Guin. 318, t. 47, 3–7. *Geoffraea inermis* Wright (1787). *Lonchocarpus staudtii* Harms (1899).

Andira jamaicensis Urb. (1905). A savannah tree, 20–30 ft. high, with handsome pink flowers and pendulous fruits, resembling a small mango.
Sen.: *Heudelot* 380! **Gam.:** Galam *Heudelot* 54! **Fr.Sud.:** Kaarta *Dubois* 127. Macina *Chev.* 24917!
Iv.C.: Tiengara, Bamaro & Ferkessédougou *fide* Aubrév. *l.c.* **G.C.:** Nangodi (Mar.) *Vigne* FH 4490!
Togo: Tamberma *Kersting* A 554! **Fr.Nig.:** Gourma Prov. *Chev.* 24356. **N.Nig.:** Sokoto (May) *Lely* 832!
Bauchi Plateau (Feb.) *Lely* P. 192! Selfwe R., Plateau Prov. (Mar.) *Thornewill* 54! Yola (fr. May) *Dalz.*
168! Also in French Cameroons, Ubangi-Shari and A.-E. Sudan. (See Appendix, p. 228.)

14. DREPANOCARPUS G. F. W. Mey.—F.T.A. 2 : 237.

Branchlets armed with pairs of recurved stipular spines ; leaves 5–7-foliolate ; leaflets oblong-elliptic, rounded at each end, up to 4·5 cm. long and 2 cm. broad, minutely pubescent beneath, pitted above, with numerous lateral nerves and a thick margin ; flowers in small axillary and terminal panicles ; calyx glabrous ; fruits broadly sickle-shaped, about 3 cm. in diam., flattened, laxly nervose, glabrous   .. *lunatus*

D. lunatus (*Linn. f.*) *G. F. W. Mey.*—F.T.A. 2 : 237 ; Bak. f. Leg. Trop. Afr. 2 : 536 ; Chev. Bot. 211.
Pterocarpus lunatus Linn. f. (1781). *Drepanocarpus africanus* G. Don (1832). A glabrous shrub, erect
or sometimes straggling, thorny and with purple flowers ; in salt marshes and tidal portions of rivers.
Sen.: *Heudelot*! Cape Nase (Feb.) *Döllinger*! Casamance (Feb.)*Chev.* 3430! **Gam.:** *Hayes* 553! Kudang
(Jan.) *Dalz.* 8068! **Port. G.:** Bissau (fl. Jan., fr. Feb.) *Esp. Santo* 1217! 1691! **Fr.G.:** Conakry *Chev.*
12083 ; 12093 ; 12096 ; 12244. **S.L.:** Kitchom (Dec.) *Sc. Elliot* 4335! Sulima (fr. Mar.) *Small* 525!
Gbabai (fr. Mar.) *Glanville* 247! Gbinti (fr. Jan.) *Deighton* 991! **Lib.:** Grand Bassa (July) *T. Vogel* 49!
Monrovia (Dec.) *Baldwin* 10505! *Whyte*! Fisherman's Lake (Dec.) *Baldwin* 10885! **Iv.C.:** Bingerville
Chev. 17268. Bliéron *Chev.* 19914. **G.C.:** Axim (Mar.) *Morton* GC 6597! **Dah.:** Porto Novo *Chev.* 23311.
S.Nig.: Lagos *Barter* 2147! *Dalz.* 930! Big Town, Eket Dist. (fl. & fr. May) *Onochie & Ujor* FHI 32936!
Br.Cam.: Victoria (Feb.) *Maitland* 399! Along the west coast as far as northern Angola ; also on the
east coast of tropical America and neighbouring islands. (See Appendix, p. 241.)

15. OSTRYOCARPUS Hook. f.—F.T.A. 2 : 240 ; Dunn in Kew Bull. 1911 : 362.

Leaflets 5, elliptic to obovate, rounded at each end, with a twisted apex, 8–20 cm. long, 4–10 cm. broad, glabrous and reticulate on both surfaces ; panicles elongated, many-flowered, up to 30 cm. long, lateral branches rather short, up to 4 cm. long ; corolla about 8 mm. long ; fruits broadly elliptic, flat, about 7 cm. long and 5 cm. broad, glabrous   ..   .. 1. *riparius*
Leaflets 7, elliptic, shortly and obtusely acuminate, acumen twisted, scarcely reticulate ; lateral branches of inflorescence 5–6 cm. long ; corolla 10–12 mm. long   2. *major*

1. **O. riparius** *Hook.f.*—F.T.A. 2 : 240 ; Chev. Bot. 213 ; Bak. f. Leg. Trop. Afr. 2 : 546. *Millettia micrantha*
Harms (1899). A glabrous climbing or straggling shrub, with yellow-white flowers ; habitat semi-aquatic.
Port.G.: Bolama (fr. Mar.) *Esp. Santo* 2051! **Fr.G.:** Rio Nunez *Heudelot* 817! **S.L.:** *T. Vogel* 59! 146!
Bagroo R. (Apr.) *Mann* 832! Pujehun (Mar.) *Lane-Poole* 20! Campbell Town (fl. & fr. July) *Lane-Poole*
308! York (fr. May) *Deighton* 5531! **Lib.:** Dinklage 1725! Sinoe Basin *Whyte*! Dukwia R. (May)
Cooper 460! **Iv.C.:** Assinie *Chev.* 17876. Tabou to Bériby *Chev.* 19983. **N.Nig.:** Ogun R., Lagos (Dec.)
Onochie FHI 26676! R. Niger *T. Vogel*! Jamieson R., Sapoba (fl. Nov., fr. June) *Onochie* FHI 23319!
Keay FHI 28073! **Br.Cam.:** Victoria (Apr.) *Maitland* 604! **F.Po:** Gutnidge Bay (Nov.) *T. Vogel* 107!
Also in Spanish Guinea, Gabon and Belgian Congo. (See Appendix, p. 253.)
2. **O. major** *Stapf* in J. Linn. Soc. 37 : 96 (1905) ; Bak. f. l.c. 546. Similar to the preceding species ; flowers
whitish, fragrant.
Lib.: Monrovia *Whyte*! Dukwia R. (Feb.) *Cooper* 194! Fortsville, Grand Bassa (Mar.) *Baldwin* 11165!

16. LEPTODERRIS Dunn in Kew Bull. 1910 : 386.

Leaves 3-foliolate ; leaflets with 5–7 pairs of lateral nerves, elliptic, rounded at base, subacute at apex, young leaflets densely tomentose on both sides becoming glabrous above and coriaceous, 8–12 cm. long, 5–7 cm. broad ; peduncle about 5 cm. long, densely pubescent at first ; racemes axillary or subterminal, 10–30 cm. long with few-flowered fascicles on the rhachis ; calyx densely grey-tomentose ; keel about 8 mm. long, wing slightly shorter, standard hooded ; fruits flat, 4–7 cm. long, about 1·5 cm. broad, densely yellow-brown, pubescent, very slightly winged on one edge
1. *trifoliolata*

Leaves 5–9-foliolate :
    Lateral nerves 5–7 pairs, 13–25 mm. apart ; stipels 1–3 mm. long :
        Leaflets tomentose beneath, hairs more or less erect ; branchlets and leaf-rhachis softly tomentellous :
            Keel yellow, 11–14 mm. long ; calyx 3–4 mm. long ; racemes of fascicles elongate and rather thick, 20–45 cm. long ; leaflets obovate, broadly emarginate and usually mucronate or more or less rounded at apex, 10–15 cm. long, 6–10 cm. broad
2. *brachyptera*
            Keel red, 6–7 mm. long ; calyx 2 mm. long ; racemes of fascicles slender, paniculate, about 15 cm. long ; leaflets elliptic, rounded or mucronate at apex, 6–8 (–13) cm. long, 4–5·5 (–6·5) cm. broad   ..   .. 3. *micrantha*
        Leaflets sparsely appressed-pubescent, with the hairs lying in one direction, or glabrous ; branchlets and leaf-rhachis shortly pubescent, becoming glabrous :
            Fruits broadly elliptic, 4 cm. long, 3 cm. broad, glabrous, narrowly winged on one side ; flowers 14–15 mm. long ; leaves broadly obovate, shortly cuneate at base, rounded at apex with midrib extended into a mucronate tip, 9–12 cm. long, 4·5–6 cm. broad, sparsely and minutely pubescent beneath   ..   4. *cyclocarpa*
            Fruits oblong, 2–4 times as long as broad :

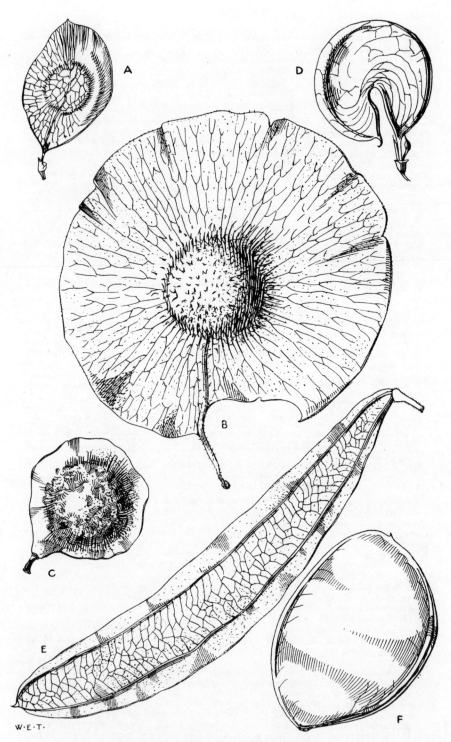

FIG. 163.—FRUITS OF PAPILIONACEAE.

A, *Pterocarpus lucens* Guill. & Perr.   B, *P. osun* Craib.   C, *P. santalinoides* L'Hér. ex DC.
D, *Drepanocarpus lunatus* (Linn. f.) G. F. W. Mey.   E, *Ostryoderris stuhlmannii* (Taub.)
Dunn ex Harms.   F, *Ostryocarpus riparius* Hook. f.

Style glabrous or with a few hairs near base ; keel petals triangular, 8–11 mm. long,
  with a folded pocket at the lower end, claw linear, appendage on upper edge
  conspicuous ; upper edge of wing sometimes ciliate ; standard flat, thin, oblong-
  obovate, more or less tapering to the claw ; stipels 1 mm. long ; leaflets elliptic
  to obovate-elliptic, rounded at base, shortly acuminate, 7–15 cm. long, 5–8 cm.
  broad, shortly appressed-pubescent beneath ..   ..   ..   5. *fasciculata*
Style pubescent with stiff dark hairs ; keel petals obovate-oblong, 7–8 mm. long,
  without a pocket, claw short, adnate to staminal tube, hardly appendiculate ;
  wing glabrous ; standard hooded at apex, rather thick and striate, oblong,
  abruptly narrowed into the claw with 2 small appendages ; stipels subulate,
  2–3 mm. long ; leaflets obovate-elliptic, rounded to subcordate at base, broadly
  retuse at apex, 10–20 cm. long, 8–10 cm. broad ; very minutely and sparsely
  pubescent beneath or glabrous ..   ..   6. *congolensis*
Lateral nerves 11–16 pairs, about 9 mm. apart ; stipels 7–10 mm. long :
Flowers with pedicels 2 mm. long ; bracts lanceolate, 10–12 mm. long, 3–4 mm.
  broad at base, appressed-pilose ; calyx 3 mm. long, shortly 4-lobed, upper lobe
  entire ; bracteoles subulate ; fruits flat, 6–9 cm. long and about 2·5 cm. broad,
  appressed-pubescent, upper edge broadly winged ; leaves 7-foliolate, upper leaflets
  with 14–16 pairs of lateral nerves, oblong to obovate-oblong, rounded at base,
  shortly and abruptly acuminate ..   ..   7. *aurantiaca*
Flowers sessile ; bracts ovate, 10–13 mm. long, about 8 mm. broad at base, densely
  pilose ; calyx 4 mm. long, shortly 4-lobed, upper lobe toothed at apex ; bracteoles
  lanceolate ; young fruits densely yellow-pilose, upper edge narrowly winged ;
  leaves 5–7-foliolate, upper leaflets with 11–12 pairs of lateral nerves, oblong to
  obovate-oblong, rounded to subcordate at base, very shortly acuminate
                                                              8. *ledermannii*

1. **L. trifoliolata** *Hepper* in Kew Bull. 1956 : 133. A woody climber in flooded lowland ; flowers pale yellow
    with wing petals pale mauve.
    S.L.: Njala (fl. Nov., fr. Jan.) *Deighton* 4957 ! 5296 !
2. **L. brachyptera** (*Benth.*) *Dunn* in Kew Bull. 1910 : 388 ; Chev. Bot. 216 ; Bak. f. Leg. Trop. Afr. 2 : 557.
    Lonchocarpus brachypterus Benth. (1860).   *Derris brachyptera* Bak. in F.T.A. 2 : 246 (1871).   *Spatholobus* ?
    africanus Bak. in F.T.A. 2 : 188 (1871).   *Leptoderris reticulata* Dunn (1910)—Berhaut Fl. Sen. 42. A
    widely climbing shrub, sometimes erect, with abundant inflorescences of pink or whitish flowers ; flowering
    infrequently ; in riverain forest.
    Sen.: *Berhaut* 1641. **Port.G.:** Bissau (Mar.) *Esp. Santo* 1863 ! Uno-Am Onhô (fr. May) *Esp. Santo* 2013 !
    Fr.G.: Fouta Djalon *Heudelot* 828 ! Kindia *Chev.* 13078 ; 13090 ! Kissidougou *Chev.* 20722. **S.L.:**
    Musaia, very common (Feb.) *Deighton* 4253 ! **G.C.:** Axim (Feb., Mar.) *Vigne* FH 2834 ! *Irvine* 2168 !
    Morton GC 8466 ! Kumasi (Dec.) *Vigne* FH 2634 ! **Togo:** Amedzofe (Feb.) *Morton* GC 8403 ! **Dah.:**
    Zagnanado (Feb.) *Chev.* 22950 ! **S.Nig.:** Lagos (Aug.) *Rowland* ! Ibadan (fl. & fr. Feb.) *Meikle* 1160 !
    Oban *Talbot* 1511 ! **Br.Cam.:** Victoria (fl. & fr. Feb.) *Maitland* 385 ! **F.Po:** *Mann* 157 ! Extends to
    Angola.
3. **L. micrantha** *Dunn* l.c. 389 (1910) ; Bak. f. l.c. 556. A climbing shrub, with purple flowers.
    G.C.: Aburi Scarp (Nov.) *Morton* A. 132 ! **S.Nig.:** Ijero (Nov.) *Millen* 43 ! Enugu (fl. & fr. Dec.) *Ainslie*
    179 ! 180 !
4. **L. cyclocarpa** *Dunn* in Kew Bull. 1914 : 245 ; Chev. Bot. 216 ; Bak. f. l.c. 558. A climbing shrub, glabrous
    except the inflorescence ; flowers yellow, in short axillary panicles.
    Fr.G.: Ditinn (Apr.) *Chev.* 13578 ! **G.C.:** Ankase F.R. (Dec.) *Vigne* FH 3184 !
5. **L. fasciculata** (*Benth.*) *Dunn* in Kew Bull. 1910 : 390 ; Chev. Bot. 216 ; Bak. f. l.c. 559. *Lonchocarpus* ?
    fasciculatus Benth. (1860)—F.T.A. 2 : 243 ; Chev. Bot. 214. *Leptoderris dinklagei* Harms (1922). A
    lofty woody climber, with stem sometimes 1 ft. in diameter, and rusty-velvety branchlets ; flowers
    rather small but variable in size, fragrant, white or purplish in panicles up to 18 in. long.
    Sen.: Casamance *Chev.* 3423 ; 3424 ! **Port.G.:** Bissau (Mar.) *Esp. Santo* 756 ! 1726 ! Chitole (Feb.)
    Esp. Santo 2881 ! **Fr.G.:** Fouta Djalon *Heudelot* 693 ! Kindia *Chev.* 13089. Farana (Jan.) *Chev.* 20439 !
    S.L.: *Afzelius* ! Regent (Jan.) *Dalz.* 8118 ! Magbile (fr. Dec.) *Thomas* 6307 ! 6469 ! Njala (Jan.)
    Deighton 5016 ! **Lib.:** Peahtah (Oct.) *Linder* 922 ! Vonjama (Dec.) *Baldwin* 9891 ! Gbanga (Oct.)
    Baldwin 10535 ! **S.Nig.:** Shasha F.R. (Mar.) *Jones & Onochie* FHI 16693 ! Also in Spanish Guinea,
    Gabon and Ubangi-Shari.
6. **L. congolensis** (*De Wild.*) *Dunn* in Kew Bull. 1910 : 390 ; Bak. f. l.c. 560. *Derris congolensis* De Wild.
    (1904). *L. fasciculata* of F.W.T.A., ed. 1, 1 : 379, partly (*Talbot* 3342). A climber with leathery leaves.
    S.Nig.: Eket Dist. *Talbot* 3342 ! Also in French Cameroons, Gabon, Belgian Congo and Angola.
7. **L. aurantiaca** *Dunn* in Kew Bull. 1914 : 245 ; Bak. f. l.c. 558. A woody climber with stems covered with
    flaking scales ; flowers white, in large pilose panicles.
    S.Nig.: Omo F.R., Ijebu Ode Prov. (fl. Nov., fr. Mar.) *Jones & Onochie* FHI 16743 ! *Tamajong* FHI 20749 !
    Also in French Cameroons and Gabon.
8. **L. ledermannii** *Harms* in Notizbl. Bot. Gart. Berl. 9 : 295 (1925) ; Bak. f. l.c. 557. Scandent shrub ; flowers
    white in yellow-pilose panicles.
    Br.Cam.: Sabga Range, Bamenda (Jan.) *Keay* FHI 28461 ! Muti Mt., Bamenda (Oct.) *Ledermann* 5852.

17. OSTRYODERRIS Dunn in Kew Bull. 1911 : 363.

Flowers precocious ; branchlets very thick, marked by the large scars of the fallen
  leaves ; panicles softly tomentose ; calyx about 5 mm. long, tomentose ; leaflets
  glaucous-grey, about 6 pairs, ovate-elliptic, rounded and emarginate at apex, 8–11 cm.
  long, 3–5·5 cm. broad, with about 5–7 pairs of lateral nerves ; stipels inconspicuous ;
  fruits flat and thin, with a line parallel within each margin, broadly oblong-linear,
  up to 15 cm. long and 3 cm. broad, reticulate within the parallel lines, slightly
  pubescent   ..   ..   1. *stuhlmannii*
Flowers not precocious ; leaflets acuminate :
  Ovary densely pubescent all over ; fruits glabrescent ; leaflets rounded or subcordate
    at base :

Bracts large, ovate, covering the buds, semi-persistent :
  Leaf-rhachis glabrous ; leaflets about 5 pairs, oblong-elliptic or oblong-lanceolate,
    obliquely abruptly acuminate, 6–10 cm. long, 2–4 cm. broad, glabrous ; stipels
    subulate, persistent ; panicles elongate, terminal ; bracts subacute ; fruit flat,
    oblong-elliptic, 7–12 cm. long, 3–4 cm. broad .. .. .. 2. *leucobotrya*
  Leaf-rhachis softly pubescent ; leaflets 4–5 pairs, oblong-elliptic, abruptly acumi-
    nate, subcordate at base, 6–10 cm. long, 2·5–4·5 cm. broad, pubescent on the nerves
    beneath ; stipels subulate, persistent ; panicles elongated ; bracts ovate,
    acuminate .. .. 3. *impressa*
Bracts small and narrow, not covering the buds, early caducous ; leaflets about
    4 pairs, ovate-elliptic, obtusely acuminate, rounded or subcordate at base, 9–15 cm.
    long, 4–8 cm. broad, glabrous ; stipels subulate, persistent ; panicles terminal,
    elongated ; fruits similar to No. 1 but a little broader .. .. 4. *gabonica*
Ovary pubescent only along upper suture ; leaflets cuneate at base, acute to slightly
    acuminate at apex, oblong or elliptic-oblong, 7–11 cm. long, 2–3 cm. broad ; bracts
    early caducous, probably lanceolate ; fruit unknown .. .. 5. *brownii*

1. **O. stuhlmannii** (*Taub.*) *Dunn ex Harms* in Engl. Pflanzenw. Afr. 3, 1 : 644 (1915) ; Bak. f. Leg. Trop. Afr.
  2 : 563. *Deguelia stuhlmannii* Taub. (1895). *Ostryoderris chevalieri* Dunn (1914)—F.W.T.A., ed. 1, 1 :
  379 ; Bak. f. l.c. 562 ; Aubrév. Fl. For. Soud.-Guin. 317, t. 62, 4–7. *Xeroderris chevalieri* (Dunn) G.
  Roberty (1954). *Lonchocarpus argentea* A. Chev. Bot. 213, name only. A deciduous tree of savannah
  woodland, up to 40 or 50 ft. high, with bark separating in thick scales ; trunk exuding blood-red sap when
  cut ; flowers white in crowded grey- or rusty-tomentose panicles appearing before the leaves.
    Sen. : *Heudelot* 347 ! Bakel (June) *Chev.* 26047 ! **Fr.Sud.** : Kita *Dubois* 50. **Port.G.** : Catió to Guebu (fr.
    June) *Esp. Santo* 2079 ! **Fr.G.** : Siguiri (Feb.) *Chev.* 298 ! Mamou (Mar.) *Dalz.* 8396 ! **S.L.** : Falaba (fr.
    Mar.) *Sc. Elliot* 5225 ! **Iv.C.** : Tafiré *Aubrév.* 1446 ! Bobo Dioulasso *Aubrév.* 1258. **G.C.** : Mangkurugu to
    Yagbung (fr. May) *Kitson* 740 ! 741 ! **Togo** : Sokode-Basari *Kersting* 19. **Dah.** : Atacora Mts. *Chev.* 24023.
    N.Nig. : Sokoto (fl. Feb., fr. May) *Lely* 835 ! *Dalz.* 309 ! Bakuru-Tureta F.R. (fr. May) *Jaiyesimi*
    FHI 15825 ! Zamfara F.R. (fr. Apr.) *Keay* FHI 16172 ! Haskia, Zaria Prov. (fr. June) *Ogua* FHI 7817 !
    Also in Belgian Congo, Kenya, Tanganyika, Mozambique and N. and S. Rhodesia. (See Appendix, p. 254.)
2. **O. leucobotrya** *Dunn* in Kew Bull. 1911 : 364 ; Chev. Bot. 216 ; Bak. f. l.c. 562. *Derris leucobotrya* (Dunn)
  G. Roberty (1954). A forest tree or shrub, sometimes climbing ; flowers white, rarely pink in red-brown
  tomentellous panicles.
    S.L. : near Leicester Peak (Dec.) *Sc. Elliot* 3842 ! Njala (fl. Dec., fr. Jan.) *Deighton* 1506 ! 2444 ! Bawama
    (Jan.) *King-Church* 5 ! Bandajuma *Burbidge* 476 ! **Lib.** : Monrovia (fl. & fr. Nov.) *Linder* 1542 !
    Vonjama (Feb.) *Bequaert* 86 ! Gbanga (Dec.) *Baldwin* 10516 ! Sinkor (Nov.) *Barker* 1096 ! **Iv.C.** :
    Bingerville (fr. Dec.) *Chev.* 17304 ! **G.C.** : Axim (Nov.) *Vigne* FH 1426 ! Tarkwa (fr. Dec.) *Andoh* FH 5431 !
    S.Nig. : Ekwulobia, Onitsha (Oct.) *Onochie* FHI 34124 ! (See Appendix, p. 254.)
3. **O. impressa** *Dunn* in Kew Bull. 1911 : 363 ; Bak. f. l.c. A climbing shrub or small tree, up to 10 ft. high ;
  flowers white.
    S.Nig. : Ugbojiobo, Benin (May) *Farquhar* 5 ! Ibo country *Cons. of For.* 171 ! Calabar *Williams* 4 !
    Aboabam, Ikom (fl. & fr. May) *Latilo* FHI 30984 ! Degema *Talbot* 3738 ! Also in French Cameroons,
    Gabon and Belgian Congo. (See Appendix, p. 254.)
4. **O. gabonica** (*Baill.*) *Dunn* in Kew Bull. 1911 : 364 ; Bak. f. l.c. *Andira ? gabonica* Baill. (1866). *Lonchocarpus macrostachyus* of F.T.A. 2 : 243, partly (spec. Mann), not of Hook. f. A straggling or climbing
  shrub ; flowers white or pinkish in narrow brown-red panicles.
    S.Nig. : Ila (Oct.) *Thomas* 1919 ! Eket Dist. *Talbot* 3070 ! Old Calabar R. (Feb.) *Mann* 2244 ! **Br.Cam.** :
    Cameroon R. (Jan.) *Mann* 2192 ! Also in Gabon and Belgian Congo.
5. **O. brownii** *Hoyle* in Kew Bull. 1932 : 262. A small tree about 25 ft. high ; flowers pinkish ; in savannah.
    G.C. : Jema (Feb.) *Brown* FH 2163 ! Drabonso (Feb.) *Cansdale* FH 4142 !

18. LONCHOCARPUS H. B. & K.—F.T.A. 2 : 241.

Standard petal densely silky outside ; flowers densely arranged in pendulous axillary
    racemes ; pedicels usually paired ; calyx truncate, cupular, 5 mm. long, tomentose;
    fruits 1–several-seeded, up to 12 cm. long, woody, tomentellous ; branchlets flexuous ;
    leaflets 7–11, ovate or elliptic, rounded or subcordate at base, obtusely acuminate,
    6–14 cm. long and 3–9 cm. broad, shortly pubescent beneath, with prominent lateral
    nerves .. .. 1. *sericeus*
Standard petal glabrous outside :
  Inflorescence a simple slender axillary raceme, either solitary or fasciculate ; leaflets
    about 9 with the upper ones larger than the lower, more or less elliptic, markedly
    acuminate, 5–10 cm. long, 2–5 cm. broad, glabrescent ; stipels subulate, persistent ;
    calyx shallowly lobed, shortly pubescent ; fruits flat, linear-lanceolate, 6–9 cm. long,
    glabrous .. .. 2. *griffonianus*
  Inflorescence a much-branched panicle ; leaflets at most very slightly acuminate ;
    stipels inconspicuous :
    Tree or erect shrub ; flowers in panicles on older wood ; calyx toothed, ciliate and
      shortly pubescent outside ; fruits flat and thin, 7–12 cm. long, acute or rounded
      at apex, 1–4 seeded ; leaflets (3–) 5, elliptic or ovate-elliptic, rounded or shortly
      cuneate at base, subacute to slightly emarginate at apex, 3–16 cm. long, 1·5–7 cm.
      broad, glaucous, minutely pubescent on both surfaces .. .. 3. *laxiflorus*
    Climber ; flowers in panicles accompanying the leaves ; calyx shortly lobed, pube-
      scent outside ; fruits 1–5-seeded, flat, glabrous, reticulate, the seeds very pro-
      minent ; lateral branchlets of panicles very short ; leaflets 9–11, ovate or elliptic,
      rounded at base, shortly acuminate, up to 16 cm. long and 8 cm. broad, glabrous
                                4. *cyanescens*

1. **L. sericeus** (*Poir.*) *H. B. & K.*—F.T.A. 2 : 241 ; Chev. Bot. 215 ; Bak. f. Leg. Trop. Afr. 2 : 548 ; Aubrév.
  Fl. For. C. Iv. 1 : 288, t. 114 · Fl. For. Soud.-Guin. 315. *Robinia sericea* Poir. (1804). A tree up to

40 ft. or more high, with brownish downy branchlets and pale purple or lilac fragrant flowers ; **commonly** flowering in the deciduous state ; in fringing and semi-deciduous forest, usually near water.
Sen.: *Perrottet* 294 ! Casamance *Chev.* 3421. **Gam.:** *Skues* ! **Port.G.:** Bissau (fr. Jan.) *Esp. Santo* 1682 ! **Fr.G.:** Rio Nunez *Heudelot* 623a ! Los Isl. *Pobéguin* 1172 ! **S.L.:** *T. Vogel* 169 ! Mahela (fr. Dec.) *Sc. Elliot* 4092 ! Bonthe (fr. Nov.) *Deighton* 2334 ! **Lib.:** Cape Palmas (July) *T. Vogel* 21 ! Grand Bassa (July) *T. Vogel* 67 ! **Iv.C.:** Agboville (fr. Nov.) *Chev.* 22343 ! Abidjan (Feb.) *Aubrév.* 51 ! **G.C.:** Axim (Jan.) *Chipp* 52 ! Saltpond (Sept.) *Dalz.* 96 ! Ejura (fr. Aug.) *Chipp* 753 ! Odumase *Howes* 1206 ! **Dah.:** *fide Chev. l.c.* **N.Nig.:** Attah (Sept.) *T. Vogel* 35 ! 60 ! 61 ! Lokoja *Barter* 1598 ! Stirling, Lokoja (Sept.) *T. Vogel* 201 ! Nupe *Barter* 1608 ! Zungeru (July) *Dalz.* 33 ! Abinsi (Aug.) *Dalz.* 600 ! **S.Nig.:** Abeokuta *Irving* ! Kukuruku Dist. (June) *Dundas* FHI 23555 ! **Br.Cam.:** Victoria (fl. Mar., fr. Jan.) *Maitland* 288 ! 531 ! **F.Po:** *Barter* 1611 ! 1614 ! *Mann* 205 ! *Mildbr.* 6979. Extends south to Angola and in Principe and S. Tomé ; also in tropical Africa. (See Appendix, p. 250.)

2. **L. griffonianus** (*Baill.*) *Dunn* in J. Bot. 49 : 15 (1911) ; Hook. Ic. Pl. t. 3040 (1915) ; Chev. Bot. 214. *Millettia griffoniana* Baill. (1866)—Bak. f. l.c. 229 ; Aubrév. Fl. For. C. Iv. 1 : 292, t. 117 A. *M. thonningii* of F.T.A. 2 : 128, partly. *Derris leptorhachis* Harms (1899). A tree up to 40 ft. or more high, with glabrous branchlets and erect racemes ; flowers reddish-purple ; easily confused with *Millettia thonningii* from which it differs in having indehiscent fruits and lacking the brush of hairs on the lower part of the midrib beneath the leaflets.
Iv.C.: Prolon to Bliéron *Chev.* 19882. **Dah.:** *Le Testu* 211 ! Tohoué (Jan.) *Chev.* 22773 ! **N.Nig.:** Lokoja (Mar.) *Shaw* 14 ! **S.Nig.:** Onitsha *Barter* 1623 ! Epe *Barter* 3265 ! Shasha F.R. (fl. & fr. Apr.) *Keay* FHI 16099 ! *Jones & Onochie* FHI 17412 ! R. Owam, Benin (Jan.) *Brenan* 8946 ! Eket *Talbot* ! **F.Po:** *Mann* 1436 ! Extends southwards into Angola. (See Appendix, p. 250.)

3. **L. laxiflorus** *Guill. & Perr.*—F.T.A. 2 : 242 ; Chev. Bot. 214 ; Bak. f. l.c. 549 ; Aubrév. Fl. For. Soud.-Guin. 315, t. 60, 1–4. *Philenoptera laxiflora* (Guill. & Perr.) G. Roberty (1954). *Lonchocarpus philenoptera* Benth. (1860)—F.W.T.A., ed. 1, 1 : 380. A tree 15–20 ft. high, with lilac or purplish flowers in copious lax panicles ; in savannah woodland.
Sen.: Bondou *Heudelot* 152 ! **Gam.:** *Dawe* 63 ! *Brooks* 6 ! *Pirie* 40 ! **Fr.Sud.:** Macina *Chev.* 24827. Bamako (Jan.) *Waterlot* 1497 ! **Port.G.:** Bafata (Feb.) *Esp. Santo* 471 ! **Fr.G.:** Tominé *Pobéguin* 2150. **Iv.C.:** Bobo Dioulasso *Serv. For.* 1948. Pâ *Serv. For.* 2145. **G.C.:** Doninga (Jan.) *Vigne* FH 4660 ! Pong to Kako (fl. & fr. Mar.) *Kitson* 870 ! **Fr.Nig.:** Niamey *Aubrév.* **N.Nig.:** Zamfara F.R. (fl. & fr. Apr.) *Keay* FHI 16162 ! 18065 ! Dangora F.R. (Apr.) *Latilo* FHI 27431 ! Bauchi Plateau (fl. & fr. Feb.) *Lely* P. 148 ! L. Chad *Talbot* 1237 ! Also in French Chad, Belgian Congo, A.-E. Sudan, Abyssinia, Eritrea, Kenya and Uganda. (See Appendix, p. 250.)

Fig. 164.—LONCHOCARPUS CYANESCENS (*Schum. & Thonn.*) *Benth.* (PAPILIONACEAE).
A, standard petal. B, keel petal. C, wing petal. D, calyx and stamens. E, pistil in longitudinal section. F ,fruit. G, seed.

4. **L. cyanescens** (*Schum. & Thonn.*) *Benth.*—F.T.A. 2 : 243 ; Chev. Bot. 214 ; Bak. f. l.c. 551 ; Aubrév. Fl. For. C. Iv. 1 : 286. *Robinia cyanescens* Schum. & Thonn. (1827). *Philenoptera cyanescens* (Schum. & Thonn.) G. Roberty (1954). A woody climbing or straggling shrub, in cultivation forming a shrub 7–8 ft. high ; branchlets silky when young, the young leaves turning bluish-black in drying ; flowers reddish, turning blue, in panicles up to a foot long ; in coastal districts and in forests. The Yoruba Indigo.
Port.G.: Fulacunda (fl. & fr. May) *Esp. Santo* 2025 ! **Fr.G.:** Rio Nunez *Heudelot* 825 ! Sandenia *Chev.* 13140. Kaba R. *Chev.* 13187. Farana *Chev.* 13197 ; 13407 ; 13500. Sareya *Chev.* 444 ; 455 ! **S.L.:** *Morson* ! Njala (Apr.) *Deighton* 1817 ! Musaia (fr. Feb.) *Deighton* 4242 ! Yetaya (fr. Sept.) *Thomas* 2464 ! **Lib.:** Ganta (May) *Harley* ! Monrovia *Cooper* 132 ! **Iv.C.:** Soubré *Chev.* 19119. Morénou *Chev.* 22458. **G.C.:** Accra Plains (Feb.) *Brown* 927 ! Achimota (fr. Oct.) *Milne-Redhead* 5172 ! Wenchi (fr. Oct.) *Vigne* FH 2529 ! **Togo:** Lome *Warnecke* 195 ! **Dah.:** Abomey *Chev.* 23135. **N.Nig.:** Nupe *Barter* 1593 ! **S.Nig.:** Ibadan (Apr.) *Meikle* 1428 ! Onitsha (May) *Jones* FHI 484 ! Eket Dist. *Talbot* 3339 ! Old Calabar R. *Mann* 2280 ! **Br.Cam.:** Abonando, Mamfe (Mar.) *Rudatis* 31 ! **F.Po:** *Mann* 401 ! Also in French Cameroons. (See Appendix, p. 248.)

Imperfectly known species.

L. sp. A. A climbing shrub up to 10 ft. high ; leaflets 7, about 14 cm. long and 5 cm. broad ; fruits oblong 5–8 cm. long, 2 cm. broad, thin, flat, 1–2-seeded, shortly brown-pubescent. Flowering material is required.
Port.G.: Safim, Bissau(fr. Mar.) *Esp. Santo* 1910 !

19. PLATYSEPALUM Welw. ex Bak. in F.T.A. 2 : 131 (1871).

Leaves rusty-pilose beneath, at least on the nerves, midrib puberulous above ; branchlets and peduncles densely pilose, at length glabrescent ; leaves 7–9-foliolate, leaflets obovate-oblong, rounded or cuneate at base, acutely acuminate, 7–11 cm. long, 2·5–3·5 cm. broad ; inflorescence a terminal panicle, 20–35 cm. long, rhachis densely rusty-pilose ; bracteoles conspicuous, each enclosing a bud, ovate, about 1 cm. long ; flowers about 2 cm. long ; fruits about 7 cm. long and 2·5 cm. broad, with thick woody margins, densely rusty-pilose with stiff hairs ..   ..   ..   1. hirsutum
Leaves minutely pubescent beneath, midrib glabrous above ; branchlets and peduncles shortly pubescent, at length glabrescent ; leaves 7-foliolate, leaflets obovate, cuneate at base, caudate, (6–) 8–13 cm. long, 2·5–5 cm. broad ; flowers in axillary and terminal panicles, 13–20 cm. long ; rhachis shortly pubescent ; bracteoles about 4 mm. long; upper lip of calyx prominently veined, silky ; flowers 1–1·5 cm. long ; fruits about 5 cm. long and 1·5 cm. broad, with rather thick margins, densely tomentose
2. violaceum var. vanhouttei

1. **P. hirsutum** (*Dunn*) *Hepper* in Kew Bull. 1956 : 122. *Millettia hirsuta* Dunn in J. Linn. Soc. 41 : 208 (1912) ; F.W.T.A., ed. 1, 1 : 382 ; Bak. f. Leg. Trop. Afr. 2 : 234. A large climber ; flowers purple and white. **Fr.G.:** Dyeke (Oct.) *Baldwin* 9679 ! **S.L.:** *For. Dept.* 166 ! *Thomas* ! Biawalla (Oct.) *Deighton* 3771 ! Kambui F.R. (Sept.) *Edwardson* 120 ! **Lib.:** Monrovia (fl. & fr. Nov.) *Cooper* 131 ! Gbanga (Aug., Sept.) *Linder* 472 ! *Baldwin* 9129 ! Peahtah (Oct.) *Linder* 1109 ! Wohmen, Vonjama Dist. (Oct.) *Baldwin* 10069 ! **Iv.C.:** Agbo Bridge (fr. Jan.) *Chev.* 16829 ! Indénié, Mbasso *Chev.* 22640. Bébou to Mbasso *Chev.* 22644 ! **G.C.:** Jimika F.R. (fl. & fr. Oct.) *Vigne* FH 3387 ! Agogo (Sept.) *Beavan* 3389 !
2. **P. violaceum** *Welw. ex Bak.* var. **vanhouttei** (*De Wild.*) *Hauman* in Fl. Congo Belge 5 : 65 (1954). *P. vanhouttei* De Wild. (1906). *P. ledermannii* Harms (1913)—Bak. f. l.c. 2 : 254. *P. tessmannii* Harms (1913). *P. polyanthum* Harms (1913). A shrub or small spreading tree up to 40 ft. high. **S.Nig.:** Udi F.R. (Nov.) *Smith* 14 ! *Kennedy* 3114 ! Ibuzo (Nov.) *Thomas* 1994 ! Idumuje (fr. Jan.) *Thomas* 2154 ! Ikom *Catterall* 31 ! **Br.Cam.:** Buea (fl. Dec., fr. Mar.) *Maitland* 897 ! 1171 ! Oji, Mamfe (Oct.) *Johnstone* 245/31 ! Also in French Cameroons, Spanish Guinea, Gabon, Belgian Congo and Angola.

20. MILLETTIA Wight & Arn.—F.T.A. 2 : 126 ; Dunn in J. Linn. Soc. 41 : 123 (1912).

Lower surface of leaflets completely covered or nearly so by a close felt of appressed silky hairs :
  Calyx lobed nearly to the middle :
    Flowers sessile or pedicel up to 1 mm. long :
      Lateral nerves 7–9 pairs ; calyx-lobes not keeled ; leaflets about 5 pairs, oblong, obtusely long-acuminate, 5–8 cm. long, 1·5–2·5 cm. broad ; panicle longer than the leaves, narrow, the lateral branches short, tomentellous ; fruits unknown
1. lucens
      Lateral nerves 13–15 pairs ; calyx-lobes markedly keeled especially in bud ; leaflets 3–6 pairs, oblong-lanceolate, shortly and acutely acuminate, 4–8 cm. long, 1·5–3 cm. broad ; panicle terminal, broad, branches tomentellous ; fruits oblanceolate about 9 cm. long and 2 cm. broad, densely tomentose ..   ..   2. dinklagei
    Flowers on pedicels about 3 mm. long ; calyx-lobes keeled in bud ; leaflets 3–4 pairs, oblong-oblanceolate, long-acuminate, 9–15 cm. long, 3–6 cm. broad ; panicle terminal, with tomentose branches ..   ..   3. hypolampra
  Calyx very shortly toothed, dark-brown pubescent ; corolla 1·4 cm. long ; pedicel 1 mm. long ; panicle terminal or subterminal ; leaflets 2–4 pairs, oblong to oblong-obovate, shortly and abruptly acuminate at apex, 8–15 cm. long, 3–5 cm. broad, lateral nerves about 15 ; fruits about 10 cm. long and 2 cm. broad, beaked
4. chrysophylla
Lower surface of leaflets visible between the hairs, or glabrous :
  *Standard glabrous outside (or pubescent towards the apex, but not wholly sericeous, in No. 9) :
    Calyx-tube glabrous outside ; leaflets oblong-oblanceolate, long- and more or less triangular-acuminate, 4–5 cm. long, glabrous or nearly so ; stipels very prominent, linear-subulate, persistent even after fall of leaflets ; fruits oblanceolate, about 14 cm. long, glabrous ..   ..   5. lane-poolei
    Calyx-tube pubescent outside :
      Stipels present ; flowers 1–1·5 cm. long :
        Bracts caducous :
          Inflorescence simple, unbranched, longer than the leaves, 10–15 cm. long ; calyx shortly brown-pubescent ; fruits oblanceolate, 7–10 mm. long, about 2 cm. broad, shortly pubescent or glabrescent, 1–3-seeded ; leaflets more or less oblong-oblanceolate, cuneate at base, abruptly and obtusely acuminate, 4·5–5·5 cm. long, 1–2 cm. broad, pubescent beneath, especially on the midrib ; stipels subulate, persistent..   ..   ..   6. rhodantha
          Inflorescence branched, more or less pyramidal :
            Indumentum of calyx brown or yellowish ; leaflets oblong-oblanceolate, acuminate and mucronate, 4–6 (–12) cm. long, about 2 cm. broad ; stipels 3 mm. long, subulate ; fruits 8–10 cm. long, tomentose, 1–3-seeded
7. warneckei var. warneckei

Indumentum of calyx black-purple ; otherwise as above

                                             7a. *warneckei* var. *porphyrocalyx*

Bracts persistent, ovate-lanceolate, pubescent ; leaflets about 5 pairs, oblong or oblong-lanceolate, acuminate, 4–6 cm. long, 1·5–2·5 cm. broad, silky pubescent beneath ; calyx 5 mm. long ; fruits curved, linear, 10–12 cm. long, densely velvety-tomentose .. ..   8. *irvinei*

Stipels absent ; flowers 2–2·5 cm. long :

Calyx 5 mm. long ; standard 2 cm. long and 1·4 cm. broad ; fruits sparsely appressed-pubescent ; leaflets oblong, abruptly and shortly acuminate, 6–15 cm. long, about 4 cm. broad, rather loosely to finely pubescent beneath 9. *mannii*

Calyx 8 mm. long ; standard 2·5 cm. long and 1·8 cm. broad :

Leaflets minutely appressed-pubescent beneath, 5–6 pairs, broadly elliptic, shortly acuminate, rounded at base, 10–17 (–22) cm. long, 3–10 cm. broad ; fruits oblanceolate, 15 cm. long, thinly tomentellous ..   10. *macrophylla*

Leaflets rusty-pubescent beneath, about 7–8 pairs, narrowly oblong-elliptic, shortly acuminate, 8–15 cm. long, 3–4·5 cm. broad ; fruits linear-oblanceolate, beaked, about 15 cm. long, densely rusty-tomentose .. .. 11. *aboensis*

*Standard pubescent outside :

Stipels absent or minute :

Flowers 2·5–3·5 cm. long ; leaflets 7–8 pairs, oblong, caudate-acuminate, 5–12 cm. long, 1·5–6 cm. broad, with about 6 pairs of lateral nerves minutely pubescent beneath ; leaf-rhachis pubescent to almost glabrous ; fruits oblanceolate, about 10 cm. long and 2 cm. broad, shortly pubescent .. .. .. 12. *zechiana*

Flowers about 1·5 cm. long ; leaflets 2–4 pairs :

Pedicels with a pair of bracteoles in the middle ; fruits up to 16 cm. long and 2 cm. broad, glabrous ; leaflets ovate or ovate-elliptic, 4–7 cm. long, 2–4 cm. broad, sparsely appressed-pubescent beneath and usually with a brush of stiff white hairs on either side of the midrib near the base .. .. 20. *thonningii*

Pedicels not jointed, bracteoles at the base of the calyx ; fruits pubescent :

Petioles and nerves beneath pilose with dark-brown spreading hairs ; leaflets obovate-elliptic, very abruptly acuminate, 8–11 cm. long, 3·5–5 cm. broad ; inflorescence simple, slender, 20–25 cm. long .. .. .. 13. *pilosa*

Petioles and nerves beneath slightly pubescent with pale hairs ; leaflets elliptic or obovate-elliptic, abruptly and obtusely acuminate, 9–15 cm. long, 4–8 cm. broad, inflorescence simple or branched, slender, 15–35 cm. long ; fruits about 7 cm. long and 2 cm. broad, with thick margins, brown-pubescent

                                               14. *barteri*

Stipels present, 1–3 mm. long :

Leaflets 2–3 pairs :

Calyx truncate ; standard pubescent with whitish hairs ; fruits linear, about 13 cm. long and 3 cm. broad, minutely pubescent to glabrescent ; leaflets oblong-elliptic, rounded at base, abruptly acuminate, 10–20 cm. long, 5–8 cm. broad, glabrous .. .. 15. *sanagana*

Calyx shortly lobed ; standard densely pubescent with purplish hairs ; young fruits densely pilose ; leaflets ovate-elliptic, acutely acuminate, 5–9 cm. long, 2·5–3·5 cm. broad, appressed-pubescent beneath.. .. .. 16. *leonensis*

Leaflets 5–12 pairs, up to 10 cm. long and 4 cm. broad ; stipels 2 mm. long, subulate :

Bracteoles at base of calyx, broadly ovate, 2–3 mm. long, persistent ; standard markedly furrowed ; young fruits curved, tomentose ; rhachis of inflorescence about 3 mm. thick, angular ; leaflets about 6 pairs, oblong-elliptic, caudate, 5–10 cm. long, 3–4 cm. broad, glabrous .. .. 17. *conraui*

Bracteoles narrow, 1 mm. long, persistent or caducous ; standard slightly or not furrowed :

Rhachis of inflorescence angular, shortly brown-tomentose, 2–3 mm. thick, 14–20 cm. long ; leaflets 7–9 pairs, oblong, cuneate at base, acuminate, 4–7 cm. long, 1·5–2·5 cm. broad, glabrous or sparsely appressed-pubescent beneath .. .. 18. *drastica*

Rhachis of inflorescence almost terete, glabrous, slender, about 1 mm. thick (thickening slightly when in fruit), 7–15 cm. long ; leaflets about 6 pairs, obovate-oblong, obtuse at base, caudate-acuminate, 5–11 cm. long, 3·5–5 cm. broad, glabrous .. .. 19. *pallens*

1. **M. lucens** (*Sc. Elliot*) *Dunn* in J. Bot. 49 : 220 (1911) ; J. Linn. Soc. 41 : 193 ; Bak. f. Leg. Trop. Afr. 2 : 225. *Lonchocarpus lucens* Sc. Elliot (1894). A straggling climber 30–40 ft. high, with white flowers. **S.L.:** Samu country (Dec.) *Sc. Elliot* 4241 ! **Lib.:** Peahtah (Oct.) *Linder* 904 ! 1064 ! Vonjama (Oct.) *Baldwin* 9935 ! 9992 !

2. **M. dinklagei** *Harms* in Engl. Bot. Jahrb. 26 : 287 (1899) ; Dunn l.c. 41 : 195 ; Bak. f. l.c. 223. *M. kennedyi* Hoyle (1934). A shrub or small tree 15 ft. high, with white flowers in ample panicles. **S.L.:** Gola Forest (May) *Small* 668 ! **Lib.:** Grand Bassa, in littoral bush (Aug.-Oct.) *Dinklage* 1672 ! 1964 ! **S.Nig.:** Jamieson R. *Kennedy* 1784 ! 1841 ! 2794 ! Eket Dist. *Talbot* 3343 ! Degema Dist. *Talbot* !

3. **M. hypolampra** *Harms* in Engl. Bot. Jahrb. 33 : 168 ; Dunn l.c. 41 : 194 ; Bak. f. l.c. 228. A climber with silvery leaves beneath and green above ; flowers white.
S.Nig.: Oban *Talbot* 578 ! 1709 ! Akim, Calabar (Apr.) *Ejiofor* FHI 21880 ! **Br.Cam.**: *Rudatis* 13 !

4. **M. chrysophylla** *Dunn* in J. Bot. 49 : 220 (1911) ; J. Linn. Soc. 41 : 199 ; Chev. Bot. 182 ; Bak. f. l.c. 226 ; Aubrév. Fl. For. C. Iv. 1 : 292, t. 116 C. *M. melanocalyx* Dunn (1912)—F.W.T.A., ed. 1, 1 : 383 ; Bak. f. l.c. 230. *Lonchocarpus macrostachyus* Hook. f. (1849). *Millettia macrostachyus* (Hook. f.) Dunn (1911), not of Coll. & Hemsl. (1890). A climber or tree up to 50 ft. high, with white flowers ; beside streams and rivers.
Fr.G.: Kouria (Oct.) *Chev.* 15002 ! S.L.: Njagbema (Sept.) *Deighton* 2229 ! Njala (fl. Sept., fr. Dec.) *Deighton* 2250 ! 3993 ! Jigaya (Sept.) *Thomas* 2537 ! Bumbuna (Oct.) *Thomas* 3134 ! Lib.: Gbanga (Sept.) *Linder* 605 ! Iv.C.: Lower Cavally (Aug.) *Chev.* 19895. G.C.: Tafo (July) *Vigne* FH 4407 ! S.Nig.: on a branch of the Niger near Ibaddi *T. Vogel* ! Owenna, Akure Dist. (July) *Onochie & Latilo* FHI 33454 ! Abedemagu (Sept.) *Farquhar* 70 ! Br.Cam.: Barombi *Preuss* 411.

5. **M. lane-poolei** *Dunn* in Kew Bull. 1914 : 79 ; Bak. f. l.c. 228 ; Aubrév. l.c. 290, t. 116 A. A small tree, 15–20 ft. high ; flowers white or pale purplish.
S.L.: Falaba (Apr.) *Lane-Poole* 42 ! 43 ! Kenema (fl. Apr., fr. May) *Lane-Poole* 140 ! 202 ! 203 ! *Small* 79 ! Mano (Apr.) *Deighton* 3148 ! Baoma (fr. Apr.) *Deighton* 3213 ! Lib.: Barclayville (Mar.) *Baldwin* 11124 ! Iv.C.: Yapo *Aubrév.* 594 ! (See Appendix, p. 251.)

6. **M. rhodantha** *Baill.*—F.T.A. 2 : 131, partly ; Dunn in J. Linn. Soc. 41 : 198 ; Chev. Bot. 183 ; Aubrév. l.c. 292, t. 116 B. *Lonchocarpus multifolius* Dunn (1911). A tree 30–40 ft. high, with fragrant bright bluish-pink flowers.
Fr.G.: Rio Nunez *Heudelot* 815 ! S.L.: Njala (fl. Apr., fr. May) *Deighton* 1160 ! 1946 ! Kenema (Apr.) *King* 182 ! Pujehun (Feb.) *Dawe* 457 ! Iv.C.: *Aubrév.* 942 ! G.C.: Owabi, Ashanti (May) *Andoh* FH 4204 ! Dunkwa (May) *Vigne* FH 880 ! Obomeng (Apr.) *Beveridge* 113 ! (See Appendix, p. 251.)

7. **M. warneckei** *Harms* var. **warneckei**—in Engl. Bot. Jahrb. 30 : 87 (1901) ; Dunn l.c. 41 : 200 ; Bak. f. l.c. 225. *M. aureocalyx* Dunn (1911) ; Chev. Bot. 182. A climber or tree about 20 ft. high, with white or pink flowers and golden or rusty-brown pubescence on the leaflets and inflorescence.
Fr.G.: Fouta Djalon (Mar., May) *Chev.* 12399 ! 25823 ! S.L.: Jigaya (fr. Sept.) *Thomas* 2529 ! Musaia (fr. Oct.) *Thomas* 2629 ! R. Seli, Kamadugu (fr. Dec.) *T. S. Jones* 124 ! Lib.: Vonjama (fr. Oct.) *Baldwin* 9890 ! G.C.: Accra (Mar.) *Dalz.* 84 ! *Irvine* 684 ! Togo: Lome *Warnecke* 108 !

7a. **M. warneckei** var. **porphyrocalyx** (*Dunn*) *Hepper* in Kew Bull. 1956 : 122. *M. porphyrocalyx* Dunn in J. Bot. 49 : 220 (1911) ; Chev. Bot. 183 ; F.W.T.A., ed. 1, 1 : 383 ; Bak. f. l.c. 229. *M. scott-ellioti* Dunn (1911). A tree with branchlets and inflorescence clothed with dark coloured hairs.
Fr.G.: R. Konkouré to Timbo (Mar.) *Chev.* 12507 ! S.L.: Musaia (Mar.) *Sc. Elliot* 5126a !

8. **M. irvinei** *Hutch. & Dalz.* F.W.T.A., ed. 1, 1 : 383 (1928) ; Kew Bull. 1928 : 404 ; Bak. f. l.c. 224. *Robinia multiflora* Schum. & Thonn. (1827), not *Millettia multiflora* Coll. & Hemsl. (1890). A shrub or small tree with branchlets, etc., brown-pubescent, and white flowers.
G.C.: *Thonning.* Accra Plains (fl. Apr., fr. June) *Irvine* 47 ! 99 ! 211 ! Ada road (June) *Morton* GC 9270 ! (See Appendix, p. 251.)

9. **M. mannii** *Bak.* in F.T.A. 2 : 127 (1871) ; Bak. f. l.c. 236. A tree.
Br.Cam.: Barombi *Preuss* 258 ! Also in French Cameroons and Gabon.

10. **M. macrophylla** *Benth.* in Hook. Ic. Pl. t. 788–9 (1848) ; F.T.A. 2 : 127 ; Dunn in J. Linn. Soc. 41 : 212 ; Bak. f. l.c. 236. *M. hookeriana* Taub. (1894). A tree 15–30 ft. high of secondary forest, with rather large leaflets and pink or purple flowers fasciculate on erect racemes 1–2 ft. long.
Br.Cam.: Victoria (Nov.) *Maitland* 782 ! Buea (Dec.) *Maitland* 883 ! Etam Bakossi, Kumba (fr. Dec.) *Bamenda* FHI 8434 ! F.Po: *T. Vogel* 117 ! Also in French Cameroons and French Congo.

11. **M. aboensis** (*Hook. f.*) *Bak.* in F.T.A. 2 : 130 (1871) ; Dunn l.c. 41 : 214 ; Bak. f. l.c. 236. *M. macrophylla* var. *aboensis* Hook. f. (1849) ; *M. macrophylla* of F.W.T.A., ed. 1, 1 : 382, partly (*Talbot*). A tree 30–40 ft. high, with reddish-brown pubescence on the petioles, branchlets, inflorescence and fruits ; flowers purple in erect woody racemes up to 18 in. long.
S.Nig.: Benin (fl. & fr. Nov.) *Chizea* FHI 24477 ! Angiama *Barter* 148 ! Aboh *Ansell* ! Onitsha (Sept.) *Barter* 1806 ! *Onochie* FHI 34053 ! 34054 ! 34109 ! Degema *Talbot* ! Also in French Cameroons and Spanish Guinea.

12. **M. zechiana** *Harms* in Engl. Bot. Jahrb. 40 : 36 (1907) ; Dunn in l.c. 41 : 217 ; Chev. Bot. 184 ; Bak. f. l.c. 241. *M. stapfiana* Dunn (1912)—F.W.T.A., ed. 1, 1 : 382 ; Bak. f. l.c. 242 ; Aubrév. l.c. 1 : 290, t. 115. *M. ivorensis* A. Chev. Bot. 183, name only. A shrub or tree 15–30 ft. high ; flowers purple, standard brown outside ; in grassland and secondary forest.
Fr.G.: Dyeke (Oct.) *Baldwin* 9680 ! S.L.: Panguma *Smythe* 127 ! Daru (fl. & fr. Apr.) *Deighton* 3144 ! Sefadu (Dec.) *Deighton* 3575 ! Lib.: Ganta (fr. Dec.) *Barker* 1112 ! Kondessu, Bopote (fr. Dec.) *Baldwin* 10681 ! Belefani, Gbanga (fr. Dec.) *Baldwin* 10541 ! Zeahtown, Tchien (fl. & fr. Aug.) *Baldwin* 6964 ! Iv.C.: Abidjan *Aubrév.* 56 ! Cavally (Aug.) *Chev.* 19740 ! Morenou (fr. Feb.) *Chev.* 22473 ! G.C.: Akwapim (Aug.) *Irvine* 770 ! Pokwase, Accra (fl. & fr. Jan.) *Irvine* 1957 ! Achimota (Oct.) *Milne-Redhead* 5162 ! Amentia *Brown* 2136 ! S.Nig.: Oban *Talbot* ! Ikom (Nov.) *Catterall* 56 ! Onitsha (June) *Jones* FHI 489 ! Sonkwala, Obudu Div. (Dec.) *Savory & Keay* FHI 25169 ! Br.Cam.: Cameroon R. (Jan.) *Mann* 2216 ! Ntai (Oct.) *Bamenda* 91 ! Kumba (Nov.) *Smith* 127/36 ! (See Appendix, p. 252.)

13. **M. pilosa** *Hutch. & Dalz.* F.W.T.A., ed. 1, 1 : 382 (1928) ; Kew Bull. 1928 : 403 ; Bak. f. l.c. 235. A shrub about 6 ft. high with pale purple flowers.
S.Nig.: Oban *Talbot* 583 ! 1308 ! Umuahia (Mar.) *Carpenter* 193 !

14. **M. barteri** (*Benth.*) *Dunn* in J. Bot. 49 : 221 (1911) ; J. Linn. Soc. 41 : 210, partly ; Chev. Bot. 182 ; Bak. f. l.c. 235 ; Hauman in Fl. Congo Belge 5 : 38. *Lonchocarpus barteri* Benth. (1860)—F.T.A. 2 : 243. *L. heudelotianus* Baill. (1866). *Millettia calabarica* Dunn (1911). A lofty woody climber with pith in the stems ; flowers pink or red turning purple.
Sen.: Rio Nunez *Heudelot* 803 ! Port.G.: Bissau (fr. June) *Esp. Santo* 1189 ! Catió (June) *Esp. Santo* 2085 ! Farim (May) *Esp. Santo* 2087 ! 3184 ! Fr.G.: Kaba *Chev.* 13244. Kouria *Chev.* 18206. Farana (May) *Chev.* 13174 ! S.L.: Njala (Dec.) *Deighton* 2583 ! 2886 ! Samu country (Dec.) *Sc. Elliot* 4202 ! Magbile (Apr.) *Deighton* 5921 ! Lib.: Kakatown *Whyte* ! Iv.C.: Bouroukrou *Chev.* 16984. Bingerville *Chev.* 17263. G.C.: Kumasi (Feb.) *Darko* 381 ! Nsawam (June) *Irvine* 2835 ! Ancobra R. (Feb.) *Vigne* FH 2819 ! N.Nig.: Nupe *Barter* 1609 ! Lokoja (Mar.) *Shaw* 23 ! S.Nig.: Brass *Barter* 67 ! R. Obba *Barter* 3401 ! Old Calabar (Feb.) *Mann* 2259 ! Omo F.R. (Apr.) *Onochie* FHI 15542 ! Ibadan (Apr.) *Meikle* 1421 ! Br.Cam.: *Preuss* 1187 ! Victoria (Mar., Apr.) *Maitland* 521 ! 610 ! Lakka, Bamenda, 4,000 ft. (fr. June) *Maitland* 1567 ! Also in French Cameroons, Gabon, Spanish Guinea, French and Belgian Congo, A.-E. Sudan and S. Tomé.

15. **M. sanagana** *Harms* in Engl. Bot. Jahrb. 26 : 288 (1899) ; Dunn in J. Linn. Soc. 41 : 216 ; Chev. Bot. 183 ; Bak. f. l.c. 242. A shrub or small tree up to 15 ft. high with large leaflets, and reddish-purple flowers more or less fasciculate on the rhachis.
Fr.G.: Conakry (Jan.) *Dalz.* 8066 ! *Chev.* 12053 ; 12165 ; 12737. Santa *Chev.* 13091. S.L.: Koßu, R. Scarcies (fr. Jan.) *Sc. Elliot* 4598 ! Mattru (Jan.) *Deighton* 2254 ! Mange Bure (Feb.) *Deighton* 3492 ! Mahera (Dec.) *Deighton* 5686 ! Lib.: Boporo (fr. Sept., Dec.) *Baldwin* 10627 ! 10868a ! Monroviatown, Tchien Dist. (Aug.) *Baldwin* 8004 ! F.Po: *Barter* 1597. Also in French Cameroons. (See Appendix, p. 251.)

16. **M. leonensis** *Hepper* in Kew Bull. 1956 : 120, fig. 2. A climber or shrub in the forest ; flowers purple.
S.L.: Njala (Jan.) *Dalz.* 8110 ! Bumpe (Nov.) *Deighton* 4926 !

17. **M. conraui** *Harms* in Engl. Bot. Jahrb. 33 : 168 (1902) ; Dunn l.c. 41 : 219 ; Bak. f. l.c. 238. A small tree with pale purple flowers and yellow-pubescent calyx and standard.
 S.Nig.: Waliwo, Onitsha (Apr.) *Onochie* FHI 7186 ! **Br.Cam.:** Bamenda, 4,500 ft. (Apr.) *Maitland* 1420 ! Kumbo, 5,000 ft. (Jan.) *Johnstone* 23/31 ! Also in French Cameroons. (See Appendix, p. 251.)

18. **M. drastica** *Welw. ex Bak.* in F.T.A. 2 : 128 (1871) ; Dunn l.c. 41 : 220 ; Bak. f. l.c. 240 ; Hauman l.c. 25, t. 2, fig. 2B. A small tree or shrub with pale blue flowers.
 S.Nig.: Oji R., Onitsha (fl. & fr. Sept.) *Chesters* 221 ! *Cons. of For.* FHI 245 ! **Br.Cam.:** Esu, Bamenda, 5,000 ft. (fl. Feb., fr. Mar.) *Johnstone* 60/31 ! 293 ! Also in French Cameroons, Spanish Guinea, Gabon, Belgian Congo, A.-E. Sudan and Angola.

19. **M. pallens** *Stapf* in Johnston Lib. 2 : 593 (1906) ; Dunn l.c. 41 : 218 ; Chev. Bot. 183 ; Bak. f. l.c. 239. A glabrous shrub or small tree, 15 ft. high with a spreading crown ; flowers purple ; on sandy soil in savannah country.
 Fr.G.: *Farmar* 259 ! Karkandy *Heudelot* 793 ! Timbo (Mar.) *Chev.* 12637 ! **S.L.:** Kofiu Mt., Scarcies (Jan.) *Sc. Elliot* 4609 ! Sefadu (fr. Apr.) *Deighton* 3212 ! Sulima (Mar.) *Deighton* 3646 ! Mando (Apr.) *Jordan* 216 ! **Lib.:** Monrovia *Whyte* ! Suen, Montserrado (Nov.) *Baldwin* 10460 ! Jabrocca (Dec.) *Baldwin* 10860 ! Barclayville (fr. Mar.) *Baldwin* 11123 ! (See Appendix, p. 251.)

20. **M. thonningii** (*Schum. & Thonn.*) *Bak.* in F.T.A. 2 : 128 (1827), partly ; Dunn l.c. 41 : 215 ; Chev. Bot. 183 ; Bak. f. l.c. 238 ; Aubrév. l.c. *Robinia thonningii* Schum. & Thonn. (1827). *Millettia atite* Harms (1902). A deciduous tree, 30–40 ft. high, the purple flowers appearing with the young leaves ; easily confused with *Lonchocarpus griffonianus* ; by streams in savannah country and often planted.
 G.C.: Dodowa Plains (Feb.) *Irvine* 1501 ! Nchenenchene (fr. May) *Kitson* 1162 ! Akim (Nov.) *Gould* ! **Togo:** Lome *Warnecke* 66 ! Ho (fl. & fr. Feb.) *Howes* 1111 ! **N.Nig.:** Lokoja (Mar.) *Barter* 481 ! 1624 ! *Elliott* 36 ! 37 ! Sokoto (Mar.) *Lely* 843 ! Bauchi Plateau (Feb.) *Lely* P. 161 ! Selfwe R., Plateau Prov. (Mar.) *Thornewill* 49 ! **S.Nig.:** Lagos (fl. & fr. Mar.) *Foster* 21 ! Ibadan (fl. Jan., fr. Mar.) *Newberry & Etim* 186 ! *Onochie* FHI 3001 ! *Akpata* FHI 3002 ! **F.Po:** *fide* Exell. Also in S. Tomé (probably introduced), Belgian Congo (?) and Angola (introduced). (See Appendix, p. 251.)

21. CRAIBIA Harms & Dunn in J. Bot. 49 : 107 (1911).

Leaves pinnate, with 5 alternate leaflets (or if 3 leaflets then a pair opposite), ovate-elliptic, shortly cuneate at base, subcaudate-acuminate, 8–10 cm. long, 4–6 cm. broad, glabrous ; youngest branchlets sparsely pilose ; racemes terminal, short, brown-tomentose ; pedicels 5 mm. long ; calyx cupular, 3 mm. long with short triangular teeth ; corolla 1·5 cm. long ; fruits obliquely obovate-lanceolate, about 5 cm. long, 2·5 cm. broad, 1-seeded, glabrous .. .. 1. *atlantica*

Leaves 1-foliolate, ovate, sharply acuminate, 6–8 cm. long, 3–5 cm. broad, closely reticulate on both surfaces ; branchlets glabrous ; racemes glabrous ; pedicels about 1 cm. long, slender ; calyx 5 mm. long ; corolla 2 cm. long .. 2. *simplex*

1. **C. atlantica** *Dunn* in J. Bot. 49 : 108 (1911) ; Bak. f. Leg. Trop. Afr. 2 : 246 ; Aubrév. Fl. For. C. Iv. 1 : 294, t. 117 C. A small tree about 30 ft. high. with crowded racemes of white flowers.
 Iv.C.: Gankapleu, Man region *Aubrév.* 1052 ! **N.Nig.:** Lokoja *Barter* 501 ! **S.Nig.:** Lagos *Foster* 19 ! Ogbesse R., Owo F.R. (May) *Jones* FHI 3583 ! Olokemeji (fl. Mar., fr. Apr.) *Macdonald* 45 ! Ibadan (Apr.) *Meikle* 1437 ! Also in French Cameroons. (See Appendix, p. 235.)

2. **C. simplex** *Dunn* l.c. 107 (1911) ; Bak. f. l.c. 246. A glabrous shrub or perhaps small tree with rather conspicuous flowers in a small panicle.
 S.Nig.: Old Calabar *Milne* 1866 ! Also in Gabon.

22. MUNDULEA (DC.) Benth.—F.T.A. 2 : 126.

Small shrub with corky bark ; branchlets flexuose, softly tomentellous ; leaflets 6–10 pairs, lanceolate, 2·5–5 cm. long, 1–2 cm. broad, glabrous above, appressed-pubescent beneath ; flowers in terminal and leaf-opposed racemes up to 15 cm. long ; pedicels 1–1·5 cm. long, pubescent ; calyx with linear teeth, 3 mm. long ; corolla about 2 cm. long, standard pubescent ; fruit 7–12 cm. long, nearly 1 cm. broad, tomentose, about 6–8-seeded .. .. *sericea*

M. sericea (*Willd.*) *A. Chev.* in Compt. Rend. 180 : 1521 (1925). *Cytisus sericeus* Willd. (1803). *Tephrosia suberosa* DC. (1825). *Mundulea suberosa* (DC.) Benth. (1852)—F.T.A. 2 : 126 ; Chev. Bot. 182 ; F.W.T.A., ed. 1, 1 : 283 ; Bak. f. Leg. Trop. Afr. 2 : 216. A shrub or small bushy tree with lilac or purplish flowers.
 Fr.Sud.: Banan *Chev.* 527. **Fr.G.:** Beyla *Chev.* 20880. **Iv.C.:** Upper Sassandra *Chev.* 21500. **Dah.:** Atacora Mts. (fr. June) *Chev.* 24202 ! **N.Nig.:** Kaduna *Lamb* 97 ! Ilorin (Feb.) *Ejiofor* FHI 19834 ! **S.Nig.:** Abeokuta *Irving* 138 ! Lagos (Sept.) *Dodd* 425 ! *Rowland* ! **Br.Cam.:** Wum (fr. June) *Ujor* FHI 29273 ! Widespread in tropical and parts of S. Africa, the Mascarene Islands, Madagascar and East Indies ; often planted. (See Appendix, p. 253.)

23. TEPHROSIA Pers.—F.T.A. 2 : 104. *Requienia* DC. (1825).

Leaves 1–3-foliolate, or occasionally a few 5-foliolate:

Leaflets solitary, broadly obovate, 1–3 cm. long, 8–20 mm. broad, truncate or rounded and sharply mucronate at apex, cuneate at base, with 5–6 pairs of prominent looped lateral nerves, both surfaces covered with whitish appressed hairs ; flowers solitary or paired, subsessile, 5–6 mm. long ; fruits ovoid, about 5 mm. long, villous, 1-seeded .. .. 1. *obcordata*

Leaflets 1–3 (–5), linear-oblanceolate, 3–9 cm. long, 6–12 mm. broad, deeply emarginate or mucronate at apex, cuneate at base, with numerous ascending lateral nerves nearly hidden by dense silky white hairs beneath, glabrous above ; flowers solitary or paired, sessile, about 1 cm. long ; fruits about 6 cm. long and 8 mm. broad, densely villous .. .. 2. *elegans*

Leaves all pinnately 5- or more-foliolate :

*Flowers solitary or subsolitary, axillary :

 †Fruits about 12 times as long as broad, or if shorter than leaflets about 1 cm. long :

 **Leaflets obovate or obovate-lanceolate :

Leaflets obovate-lanceolate, rounded or slightly emarginate at apex, 1·5–4 cm. long, lower leaflets much smaller than the upper, villous beneath, glabrous or pubescent above ; flowers about 8 mm. long ; calyx lobes subulate, 4 mm. long ; fruits about 6 cm. long, densely pilose, 10–12-seeded ..   3. *uniflora*

Leaflets obovate, truncate and widely and sometimes deeply emarginate at apex, 5–11 mm. long, all leaflets more or less the same size, equally villous on both surfaces ; flowers about 6 mm. long ; calyx lobes subulate, 2 mm. long ; fruits 3·5 cm. long, densely pilose, 8–9-seeded ..   ..   4. *vicioides*

**Leaflets linear, 1–3 cm. long, 1–3 mm. broad ; for other characters see below 19. *linearis*

†Fruits about 6 times as long as broad, thin and flat :

Branchlets villous with weak spreading hairs ; leaflets rounded or acute at apex, mostly oblanceolate, 2·5–3·5 cm. long, thinly appressed-pilose beneath, midrib villous ; corolla about 6 cm. long, slightly longer than calyx-teeth ; fruits 2–3 cm. long, 4 mm. broad, sparsely appressed-pubescent with hairs of equal length 5. *pedicellata*

Branchlets pubescent with appressed hairs ; leaflets truncate or emarginate at apex, broadly linear, 3–4 cm. long, thinly appressed-pilose beneath, midrib pilose ; corolla about 11 mm. long, much longer than calyx teeth ; fruits 3–4 cm. long, 6 mm. broad, sparsely appressed-pubescent with 2 kinds of hairs   6. *platycarpa*

*Flowers racemose, usually fairly numerous :

Stems procumbent :

Leaves pinnate :

Stipules broadly ovate ; leaflets mostly 3 pairs, obovate-oblanceolate, emarginate and mucronate, about 2 cm. long, appressed-pilose beneath ; flowers nearly 1 cm. long, few, in short racemes a little longer than the leaves ; fruits 2 cm. long, 7–8 mm. broad, 2–3-seeded, very slightly pubescent ..   7. *radicans*

Stipules subulate ; leaflets 2 pairs, obovate, emarginate and mucronate, 2–2·5 cm. long, nearly glabrous beneath ; flowers 8 mm. long, few, in very short racemes 8. *djalonica*

Leaves digitate ; leaflets linear-lanceolate, acute, 4–5 cm. long, with numerous ascending lateral nerves, more or less densely hairy ; stems covered with soft spreading hairs (or shortly pubescent in var. *digitata*) ; flowers few, in lax slender racemes ; fruits 2·5–3 cm. long ..   ..   9. *lupinifolia*

Stems more or less erect :

‡Leaflets 5 or more pairs :

Flowers 2 cm. or more long, densely crowded ; fruits large, 8–12 cm. long, very densely villous or tomentose :

Calyx-lobes 5 mm. long, oblong, the lowest lobe lanceolate, keeled, 7 mm. long ; bracts broadly ovate, 10 mm. long, shortly acuminate, early caducous ; leaflets oblanceolate, rounded or truncate and mucronate at apex, 4–7 cm. long, with numerous lateral ascending nerves appressed-silky-villous beneath   10. *vogelii*

Calyx-lobes 2 mm. long, triangular, the lowest lobe subulate, 5 mm. long ; bracts linear-lanceolate, 7 mm. long, reflexed in flower, finally caducous ; leaflets oblong-obovate, broadly emarginate and shortly mucronate at apex, 2·5–5 cm. long, with numerous lateral ascending nerves appressed-silky-pubescent beneath 11. *densiflora*

Flowers at most about 1 cm. long or rarely 1·8 cm. long, usually loosely arranged ; fruits smaller, not more than 6 cm. long :

Hairs on the stems rather long and spreading, about 1 mm. long ; leaflets oblanceolate to almost linear, mostly emarginate and mucronate, up to 7 cm. long, densely silky beneath ; fruits erect and downward curving :

Fruits about 4·5 cm. long, nearly 1 cm. broad, shortly pubescent with hairs of 2 lengths, usually about 8-seeded ; leaflets sparsely pubescent above with fine white hairs ..   ..   12. *flexuosa*

Fruits about 5–6 cm. long, 5 mm. broad, shortly pubescent with hairs of the same length, about 15-seeded ; leaflets often glabrous above ..   13. *barbigera*

Hairs on stems appressed or tomentose, not exceeding 0·5 mm. long (hairs on leaf-rhachis sometimes longer) :

Inflorescence shorter than the leaves :

Flowers numerous, crowded, with many persistent lanceolate bracts about 1·5 cm. long ; fruits linear-oblong, 3 cm. long, 4–5 mm. broad, shortly pubescent, sharply deflexed ; leaflets oblanceolate, widely emarginate and mucronate at apex, 3–6 cm. long, 7–15 mm. broad, thinly pubescent beneath, glabrous above ..   ..   14. *deflexa*

Flowers few not densely crowded ; bracts subulate about 7 mm. long ; fruits erect ; leaflets pubescent on both sides ; for other characters see above 6. *platycarpa*

Inflorescence longer than the leaves :

Leaflets 4·5–8 cm. long, 3–5 (–7) mm. broad, linear, silky-pubescent beneath, glabrous above ; bracteoles broadly ovate, 5 mm. long, enclosing the buds ; flowers in rather lax long-pedunculate racemes about 20 cm. long ; fruits narrowly linear, 5–9 cm. long, 4 mm. broad, erect and usually curved upwards, minutely pubescent ..   ..   15. *bracteolata*

Leaflets 1–3·5 cm. long ; bracteoles linear or subulate, up to 3 mm. long :

Fruits densely pilose with long brown hairs, about 5 cm. long and 6 mm. broad, about 10-seeded ; leaflets thinly appressed-pubescent beneath, surface visible between hairs, glabrous above, obovate-lanceolate, slightly emarginate and mucronate at apex, 2·5–3·5 cm. long, about 8 mm. broad ; inflorescence lax, 20–30 cm. long   ..   ..   ..   16. *noctiflora*

Fruits with short grey indumentum ; leaflets densely pubescent beneath :

Flowers numerous, 12 mm. long, rather crowded in the racemes ; leaflets obovate, 2 cm. long, 7–15 mm. broad, tomentose and strongly reticulate beneath ; fruits 6–7 cm. long, 8 mm. broad, densely tomentose

17. *mossiensis*

Flowers few, 6–8 mm. long, in lax racemes ; fruits 4–5 cm. long, 3–4 mm. broad, pubescent :

Leaflets narrowly obovate, about 1·5 cm. long, 3–7 mm. broad, rounded or emarginate at apex, lateral nerves beneath usually reddish, visible beneath hairs ; stems angular, densely pubescent to subglabrous

18. *purpurea*

Leaflets linear, 1·5–3·5 cm. long, 2–3 mm. broad, rounded at apex, lateral nerves beneath completely obscured by the dense silky pubescence ; stems ribbed, appressed-pubescent..   ..   ..   ..   19. *linearis*

‡Leaflets 2–4 pairs :

Inflorescence overtopping the leaves :

Calyx up to 10 mm. long, with long linear-subulate teeth, densely blackish pilose ; flowers 15–20 mm. long in stout mostly terminal inflorescences ; fruits very densely blackish-pilose ; leaflets oblanceolate, mucronate, 2·5–5 cm. long, 1–2 cm. broad, with about 6–8 prominent lateral nerves   ..   20. *preussii*

Calyx about 2 mm. long, with subulate teeth, sparsely pubescent ; flowers about 5 mm. long in very slender axillary racemes ; fruits sparsely pubescent ; leaflets narrowly oblanceolate, 2–2·5 cm. long, 3 mm. broad ..   21. *gracilipes*

Inflorescence shorter than the leaves ; for other characters see above

6. *platycarpa*

1. **T. obcordata** (*Lam. ex Poir.*) *Bak.* in F.T.A. 2 : 125 (1871) ; Chev. Bot. 181 ; Bak. f. Leg. Trop. Afr. 1 : 215. *Podalyria obcordata* Lam. ex Poir. (1804). *Requienia obcordata* (Lam. ex Poir.) DC. (1825)—F.W.T.A., ed. 1, 1 : 387. *Crotalaria arenaria* of Chev. Bot. 167, partly (*Chev.* 1264). A subwoody herb with rather long subsimple branches, 6 in. to 3 ft. long, covered all over with whitish hairs. **Sen.:** Walo *Perrottet* 201 ! *Heudelot* ! M'bidjem *Thierry* 65 ! **Fr.Sud.:** Timbuktu (fl. & fr. July) *Chev.* 1264 ! 212 ! **Fr.Nig.:** Niamey (fl. & fr. Oct.) *Hagerup* 516 ! **N.Nig.:** Sokoto (fl. & fr. July) *Dalz.* 336 ! Also in Ubangi-Shari and A.-E. Sudan.

2. **T. elegans** *Schum.*—F.T.A. 2 : 118 ; Chev. Bot. 179 ; Bak. f. l.c. 183 ; Cronquist in Fl. Congo Belge 5 : 96, t. 6. An erect shrubby herb 2–5 ft. high, leaves silvery beneath and green above ; flowers yellow ; usually in moist grassland. **Sen.:** *Berhaut* 1818. **Fr.Sud.:** Ségou (fr. Oct.) *Chev.* 3258 ! **Port.G.:** Cubisseque (fr. Dec.) *Esp. Santo* 2333 ! Bafata (fr. Nov.) *Esp. Santo* 2805 ! **Fr.G.:** Pobéguin 461 ! **S.L.:** Wallia (fr. Feb.) *Sc. Elliot* 4258 ! Yetaya (Sept.) *Thomas* 2323 ! Mamunta (Nov.) *Deighton* 4934 ! Musaia (fr. Dec.) *Deighton* 5698 ! **Iv.C.:** Anoumaba (Nov.) *Chev.* 22395 ! Mt. Orumbo-Boka, Toumodi (fr. Dec.) *Boughey* C.3 ! **G.C.:** Kumbunga Farm (bud May) *Kitson* 597 ! Burufo (fr. Dec.) *Adams* GC 4428 ! Togo: *Büttner* 153 ! Kpandu *Robertson* 71 ! **N.Nig.:** R. Niger (Sept.) *T. Vogel* 111 ! Lokoja (Oct.) *Barter* 541 ! *Parsons* 28 ! *Dalz.* 23 ! **S.Nig.:** Lagos (Oct.) *Punch* 34 ! Olokemeji (fr. Nov., Dec.) *Jones & Keay* FHI 4874 ! Onochie FHI 8162 ! Ogun R., Oyo (Sept.) *Latilo* FHI 23534 ! Onitsha *Barter* 1783 ! Also in A.-E. Sudan, Uganda, Belgian Congo and Angola.

3. **T. uniflora** *Pers.* Syn. 2 : 329 (1807) ; Bak. f. l.c. 1 : 184. *T. anthylloides* Hochst. ex Bak. in F.T.A. 2 : 118 (1871) ; Chev. Bot. 179. *T. hirsuta* Schum. & Thonn. (1827). An undershrub, much-branched from the base, often decumbent, grey-silky, with white flowers fading to pink ; in dry waste places. **Sen.:** *Heudelot* 214 ! *Brunner* 178 ! *Perrottet* 207 ! **Fr.Sud.:** Yatenga (fl. & fr. Aug.) *Chev.* 24814 ! Bamoa (Sept.) *Hagerup* 326a ! Gao (fl. & fr. Sept.) *Hagerup* 360a ! **G.C.:** Sakumo Lagoon, Accra *Morton* GC 6115 ! Accra (fl. & fr. Mar., Oct.) *Brown* 349 ! *Deighton* 601 ! **N.Nig.:** Sokoto (fl. & fr. July) *Dalz.* 331 ! Bornu *Talbot* 1221 ! Also in Cape Verde Islands, A.-E. Sudan, Abyssinia, Eritrea, Somaliland, Kenya, Tanganyika and Socotra.

4. **T. vicioides** *A. Rich.*—F.T.A. 2 : 117 ; Bak. f. l.c. 185. A diffuse prostrate herb 1–2 ft. long ; flowers pale yellow, green-veined. **Fr.Sud.:** Gao (Sept.) *Hagerup* 332 ! 343 ! Also in A.-E. Sudan, Eritrea and Abyssinia.

5. **T. pedicellata** *Bak.* in F.T.A. 2 : 117 (1871) ; Chev. Bot. 181 ; Bak. f. l.c. 184. *T. ilorinensis* J. R. Drum. ex Bak. f. l.c. (1926). *T. hirsuta* of F.W.T.A., ed. 1, 1 : 385, not of Schum. & Thonn. A half-woody prostrate plant of rough places and roadsides ; flowers pink on short pedicels. **Sen.:** *Berhaut* 1221. **Port.G.:** Antula, Bissau (fr. Oct.) *Esp. Santo* 2589 ! **S.L.:** Luseniya (fr. Dec.) *Sc. Elliot* 4313 ! Magbema, Rokupr (fl. Oct., fr. Nov.) *Jordan* 657 ! 695 ! **Iv.C.:** Ouagadougou to Ouhigouya (Aug.) *Chev.* 24726 ! **G.C.:** *Williams* 600 ! Zuarungu (Sept.) *Hughes* FH 5331 ! Senchi Ferry (Oct.) *Morton* GC 7734 ! **Togo:** *Büttner* 282 ! Ho (fl. & fr. Oct.) *A. S. Thomas* K. 6 ! **N.Nig.:** Lokoja *Barter* ! Kadaura (fl. & fr. Nov.) *Lely* 593 ! Benue R. *Talbot* ! Ilorin (May) *Bailey* 8 ! **S.Nig.:** Olokemeji F.R. (Oct.) *Keay* FHI 28047 ! **Br.Cam.:** N. Cameroons *Talbot* 1221 ! Also in Cape Verde Islands.

6. **T. platycarpa** *Guill. & Perr.*—F.T.A. 2 : 109 ; Bak. f. l.c. 187. *T. humilis* Guill. & Perr.—F.T.A. 2 : 109 ; Bak. f. l.c. Stems annual 6–12 in. high, woody at base, thinly grey-silky with pink or purple flowers ; in grassy roadsides, etc. **Sen.:** M'bidjem *Thierry* 75 ! Cayor *Heudelot* 363 ! 463 ! **Gam.:** *Ingram* ! **N.Nig.:** Bauchi Plateau *Batten Poole* 152 ! Naraguta & Jos (fl. & fr. Nov.) *Lely* 584 ! Anara F.R., Zaria (Sept.) *Olcrunfemi* FHI 24362 ! Also in A.-E. Sudan. (See Appendix, p. 263.)

7. **T. radicans** *Welw. ex Bak.* in F.T.A. 2 : 121 (1871) ; Bak. f. l.c. 214 ; Cronq. l.c. 106. An annual herb with prostrate branches 1–3 ft. long ; flowers pink ; on rocks and bare soil.
N.Nig.: Vodni, 4,600 ft. (fl. & fr. July) *Lely* 422 ! Bauchi Plateau (July) *Lely* P. 375 ! Also in Belgian Congo, Angola, Transvaal and N. and S. Rhodesia.

8. **T. djalonica** *A. Chev. ex Hutch. & Dalz.* F.W.T.A., ed. 1, 1 : 385 (1928) ; Kew Bull. 1928 : 404 ; Chev. Bot. 179, name only. A prostrate herb with slender striate nearly glabrous stems and reddish-purple flowers.
Fr.G.: Timbo to Ditinn, Fouta Djalon (Sept.) *Chev.* 18447 !

9. **T. lupinifolia** *DC.*—F.T.A. 2 : 107 ; Bak. f. l.c. 183 ; Cronq. l.c. 90. *T. digitata* DC. (1825)—F.W.T.A., ed. 1, 1 : 385. *T. lupinifolia* DC. var. *digitata* (DC.) Bak. f. l.c. 183 (1926). Wide-climbing, slender, perennial herb, with tuberous roots and pink flowers ; in sandy places.
Sen.: M'bidjem *Thierry* 88 ! Cape Verde (fl. & fr. Feb.) *Döllinger* ! Walo *Perrottet* ! Senegal R. *Perrottet* 209 ! **N.Nig.**: Nupe *Barter* 976 ! Jebba *Barter* ! Also in A.-E. Sudan, French and Belgian Congo, Angola, N. and S. Rhodesia and Nyasaland.

10. **T. vogelii** *Hook. f.*—F.T.A. 2 : 110 ; Chev. Bot. 181 ; Bak. f. l.c. 209 ; Cronq. l.c. 108, t. 7. An erect shrub 6–10 ft. high, clothed with dense yellowish or rusty tomentum ; flowers conspicuous, red or red-purple, in dense racemes ; frequently cultivated as a fish-poison.
S.L.: *Afzelius* ! *Deighton* 2000 ! Regent *Sc. Elliot* 3989 ! Samu country (Dec.) *Sc. Elliot* 4207 ! **Lib.**: Gbawia, Webo (July) *Baldwin* 6705 ! Yopla *Chev.* 19845. **Iv.C.**: Zoanlé (fr. May) *Chev.* 21473 ! Bouaké *Chev.* 22145. **G.C.**: Peapease (fl. & fr. Jan.) *Brent* FH 396 ! Yeji (Aug.) *Pomeroy* FH 1344 ! 1349 ! Abetifi (June) *Irvine* 304 ! **N.Nig.**: *T. Vogel* ! Nupe *Barter* 1083 ! Katsina Ala (Aug.) *Dalz.* 579 ! Vom, Bauchi Plateau *Dent Young* 56 ! *Lely* P. 617 ! **S.Nig.**: *T. Vogel* 28 ! Abeokuta *Irving* 96 ! Oban *Talbot* 143 ! Lagos *Foster* 20 ! *MacGregor* 181 ! **Br.Cam.**: Buea (fl. & fr. Apr.) *Hutch. & Metcalfe* 109 ! Bamenda, 5,000 ft. (fr. Dec.) *Baldwin* 13855 ! **F.Po**: *Ansell* ! *T. Vogel* 27 ! Widespread in tropical Africa and often cultivated. (See Appendix, p. 264.)

11. **T. densiflora** *Hook. f.*—F.T.A. 2 : 114 ; Bak. f. l.c. 203. *T. interrupta* of Chev. Bot. 180. *T. vogelii* of F.W.T.A., ed. 1, 1 : 385, partly (*Millen* 140). An undershrub like the last, with grey-pubescent striate branches ; fruits densely brownish-grey tomentose ; amongst rocks by streams, also cultivated.
Fr.Sud.: Damese, Baféné *Chev.* 325 ! **Fr.G.**: Konkouré to Timbo *Chev.* 12495. **S.L.**: Waterloo (Oct.) *Lane-Poole* 388 ! Bure (Oct.) *Glanville* 39 ! **G.C.**: Gowani *Morris* C. 5 ! **N.Nig.**: Jebba *Barter* ! Naraguta (Sept.) *Lely* 488 ! Bauchi Plateau (Aug.) *Lely* P. 618 ! Patti Lokoja (fl. & fr. Sept.) *T. Vogel* 169 ! **S.Nig.**: Ogun R. *Millen* 140 ! (See Appendix, p. 264.)

12. **T. flexuosa** *G. Don* Gen. Syst. 2 : 232 (1832) ; F.T.A. 2 : 111 ; Bak. f. l.c. 203. *T. ansellii* Hook. f. (1849)—F.T.A. 2 : 115 ; Bak. f. l.c. 201 ; F.W.T.A., ed. 1, 1 : 385 ; Chev. Bot. 179. *T. platycarpa* of Bak. f. l.c. 187, partly (*Schröder* 109). An undershrub, low or sometimes reaching 5 ft. in height, silky grey or rusty hairy, with red-purple flowers in lax terminal or sometimes axillary racemes ; in savannah country.
Sen.: *Berhaut* 523. **Gam.**: *Hayes* 579 ! **Fr.Sud.**: Macina (Sept.) *Chev.* 24856 ! **Fr.G.**: Kouroussa (Oct.) *Pobéguin* 561 ! **S.L.**: Kunrabai to Mamila (Nov.) *Deighton* 4940 ! **Iv.C.**: S. Baoulé (Aug.) *Chev.* 22265 ! **G.C.**: Afram Plains (Nov.) *Irvine* 1716 ! Mampong (fl. & fr. Nov.) *Irvine* 1859 ! Ejura (fl. & fr. Nov.) *Vigne* FH 3452 ! **Togo**: Bando (fr. Dec.) *Adams & Akpabla* GC 4045 ! Sokode *Schröder* 109 ! Lome *Warnecke* 246 ! Amedzofe (Jan.) *Irvine* 3369 ! **N.Nig.**: Stirling Hill, Lokoja (Sept.) *Ansell* ! Nupe *Baikie* ! Jebba *Barter* ! Zungeru (Oct.) *Dalz.* 28 ! **S.Nig.**: Lagos *Dodd* 429 ! *MacGregor* 49 ! *Moloney* ! Abeokuta *Irving* ! Olokemeji F.R. (fl. Sept., fr. Nov.) *Onochie* FHI 8158 ! *Keay* FHI 25386 ! Onitsha (fr. Nov.) *Baldwin* 13735 ! Also in S. Tomé, French and Belgian Congo. (See Appendix, p. 263.)

13. **T. barbigera** *Welw. ex Bak.* in F.T.A. 2 : 113 (1871) ; Bak. f. l.c. 196 ; Cronq. l.c. 111. *T. polysperma* Bak. in F.T.A. l.c. (1871). Habit similar to the last, branches with spreading tawny pubescence ; in savannah.
Gam.: *Hayes* 565 ! **Port.G.**: Cacine (Oct.) *Esp. Santo* 662 ! **Fr.G.**: *Pobéguin* 561a ! Mt. Koiré, Nzérékoré (fr. Sept.) *Baldwin* 13288 ! **S.L.**: *Ansell* ! Jigaya (Sept.) *Thomas* 2675 ! Kanya (Oct.) *Thomas* 3032 ! Gberia Timbako (Sept.) *Small* 321 ! Kambaia (fl. & fr. Oct.) *Small* 475 ! **Lib.**: Vonjama (fr. Oct.) *Baldwin* 9934 ! **N.Nig.**: Vom *Dent Young* 53 ! 57 ! Bauchi Plateau (Sept.) *Lely* P. 714 ! **S.Nig.**: Lagos *Rowland* ! Ondo (fl. & fr. Sept.) *Keay* FHI 22566 ! Ala (fr. Nov.) *Thomas* 1978 ! Awka, Onitsha (Sept.) *Jones* FHI 6714 ! Also in French and Belgian Congo, Angola, A.-E. Sudan, Abyssinia, Kenya, Uganda and Tanganyika.

14. **T. deflexa** *Bak.* in F.T.A. 2 : 111 (1871) ; Bak. f. l.c. 194. A woody erect undershrub 1–2 ft. high, more or less thinly silky-pubescent, with pink flowers.
Sen.: Bondou *Heudelot* 194 ! **Gam.**: *Saunders* 67 ! Kuntaur *Ruxton* 3 ! **Port.G.**: Bissau (Oct.) *Esp. Santo* 950 ! *Dinklage* 3205 !

15. **T. bracteolata** *Guill. & Perr.*—F.T.A. 2 : 116 ; Chev. Bot. 179 ; Bak. f. l.c. 201 ; Cronq. l.c. 112, t. 8. *T. elongata* Hook. f. (1849)—F.T.A. 2 : 111. *T. fasciculata* Hook. f. (1849). *T. concinna* Bak. in F.T.A. 2 : 112 (1871) ; Bak. f. l.c. 190. *T. nigerica* Bak. f. l.c. 198 (1926). An erect branched undershrub, usually 2–3 ft. high, but sometimes up to 8 ft. high, with long straight thinly silky branches and bright pink to purple flowers ; in open country.
Sen.: Cayor *Heudelot* 402 ! **Fr.Sud.**: Labbézenga (fr. Sept.) *Hagerup* 441 ! **Port.G.**: Antula, Bissau (Oct.) *Esp. Santo* 2561 ! **Fr.G.**: *Pobéguin* 437 ! *Scaëtta* 3155 ! **S.L.**: Wallia, Scarcies R. (fr. Feb.) *Sc. Elliot* 4281 ! Berria (fr. Feb.) *Sc. Elliot* 4959 ! Musaia (fl. Sept., fr. Dec.) *Deighton* 4413 ! 4855 ! **Iv.C.**: Mt. Kankaniboka (fl. & fr. July) *Chev.* 22163 ! Beoumi (fl. & fr. Dec.) *Lowe* ! **G.C.**: Tamale (Nov.) *Williams* 386 ! Tumu (fr. Dec.) *Adams & Akpabla* GC 4368 ! Achimota (fl. & fr. May) *Irvine* 1768 ! **Togo**: Lome *Warnecke* 59 ! Amedzofe (fl. & fr. May) *Morton* GC 7242 ! Kpandu *Robertson* 63 ! **Dah.**: Agouagon (fr. May) *Chev.* 23545 ! **N.Nig.**: Lokoja *Barter* 147 ! Kontagora (fl. Sept., fr. Oct.) *Dalz.* 35 ! Anara F.R., Zaria (Sept.) *Olorunfemi* FHI 24359 ! Biu, Bornu, 2,500 ft. (fr. Sept.) *Noble* 27 ! **S.Nig.**: Abeokuta (fr. Jan.) *Burtt* B. 22 ! Awba Hills F.R. (Oct.) *Jones* FHI 6350 ! Olokemeji (fr. Aug., Sept.) *Okigbo* FHI 24106 ! *Oladoyinbo* FHI 24263 ! Widespread in tropical Africa. (See Appendix, p. 263.)

16. **T. noctiflora** *Bojer ex Bak.* in F.T.A. 2 : 112 (1871) ; Bak. f. l.c. 193. A rather spreading undershrub 2–3 ft. high ; flowers yellow-green to white with a purple centre.
S.L.: Newton (fl. & fr. Aug.) *Deighton* 2207 ! **Lib.**: Gletown, Tchien Dist. (fl. & fr. July) *Baldwin* 6705a ! **G.C.**: Achimota (fl. & fr. June, Nov.) *Irvine* 1797 ! *Morton* GC 6099 ! **Br.Cam.**: Victoria (fl. & fr. Mar.) *Maitland* FHI 8768 ! Also in Uganda, Kenya, Tanganyika, Zanzibar and Nyasaland where it is probably native ; introduced into our area, Belgian Congo and the West Indies.

17. **T. mossiensis** *A. Chev.* in Bull. Soc. Bot. Fr. 57, Mém. 8 : 159 (1912) ; Chev. Bot. 180 ; Bak. f. l.c. 193. A branched woody undershrub, 2–6 ft. high, shortly tomentose ; flowers deep red in racemes 2–7 in.long ; amongst rocks.
Sen.: Niokolo-Koba *Berhaut* 1571. **Fr.Sud.**: Macina *Chev.* 24844 ! Kita Massif (fl. July, fr. Sept.) *Jaeger* ! **Iv.C.**: Mt. Zongapignié, Mossi (July) *Chev.* 24646 ! **G.C.**: Jowani, Gambaga Dist. (fl. & fr. June) *Akpabla* 671 ! **Fr.Nig.**: Kodjar rocks, Diapaga *Chev.* 24414. **N.Nig.**: Bauchi Plateau (July) *Lely* P. 503 ! *Batten Poole* 115 !

18. **T. purpurea** *(Linn.) Pers.*—F.T.A. 2 : 124 (incl. var. *pubescens* Bak.) ; Bak. f. l.c. 190 ; Cronq. l.c. 98. *Cracca purpurea* Linn. (1753). *Tephrosia leptostachya* DC. (1825)—F.W.T.A., ed. 1, 1 : 385. A variable perennial, 6–18 in. high, more or less erect, woody at the base ; flowers purple or pink in very lax racemes with the lowest flowers appearing axillary and solitary ; in dry waste places.

Sen.: *Roger! Perrottet* 202! *Heudelot* 408! Walo *Heudelot*! M'bidjem *Thierry* 47! **Gam.:** *Boteler*!
Fr.Sud.: Timbuktu (fl. & fr. June, July) *Hagerup* 135! 217! 246b! **G.C.:** Prampram (fr. May) *Irvine*
1437! *Morton* GC 6087! Labadi (fl. & fr. Mar.) *Deighton* 580! **Dah.:** Cotonou (fr. Feb.) *Debeaux* 158!
N.Nig.: Mokoye, Zaria (fl. & fr. Aug.) *Keay* FHI 28026! Sokoto (fl. & fr. Sept., Oct.) *Moiser* 116! 128!
Bauchi Plateau (fl. & fr. Aug.) *Lely* P. 569! Katagum (fl. & fr. Aug.) *Dalz.* 29! L. Chad (fl. & fr. Feb.)
E. Vogel 38! **Br.Cam.:** Bama, Dikwa Div. (Nov.) *McClintock* 29! Widely spread throughout tropical
Africa and Asia. (See Appendix, p. 264.)

19. **T. linearis** (*Willd.*) *Pers.*—F.T.A. 2 : 120 ; Chev. Bot. 180 ; Bak. f. l.c. 189 ; Cronq. l.c. 94. *Galega
linearis* Willd. (1803). *Tephrosia pulchella* Hook. f. (1849)—F.T.A. 2 : 120 ; Bak. f. l.c. An erect
annual, 1–3 ft. or more high, with slender grey-pubescent branches, very narrow leaflets and pink or
orange flowers : in dry grassland.
Sen.: *Roger!* *Heudelot* 257! Walo *Perrottet* 203! **Gam.:** *Saunders* 65! *Hayes* 561! **Fr.Sud.:** Macina
(fl. & fr. Sept.) *Chev.* 24848! Gao (Sept.) *Hagerup* 368! **Port.G.:** Antula, Bissau (Oct.) *Esp. Santo* 2546!
S.L.: Musaia (fr. Dec.) *Deighton* 4436! **G.C.:** Accra Plains (fl. & fr. Oct.) *Baldwin* 13422! **Dah.:** Cotonou
Chev. 4448. **N.Nig.:** Jebba *Barter*! Lokoja (Sept.) *Parsons* L. 25! Stirling (Sept.) *T. Vogel* 139!
Sokoto *Dalz.* 332! *Moiser* 33! Makurdi (fr. Nov.) *Keay* FHI 22274! **S.Nig.:** Lagos *MacGregor* 198!
Onitsha (Oct.) *Jones* FHI 6132! **Br.Cam.:** Bama, Dikwa Div. (fl. & fr. Dec.) *McClintock* 39! Widespread
in tropical Africa. (See Appendix, p. 263.)

20. **T. preussii** *Taub.* in Engl. Bot. Jahrb. 23 : 182 (1896) ; Bak. f. l.c. 208. *T. lelyi* Bak. f. l.c. 207 (1926).
A robust erect undershrub, up to 5 ft. high, with striate pubescent stems and purplish or orange flowers.
S.L.: Bintumane Peak, 5,000 ft. (fr. Jan.) *T. S. Jones* 159! **N.Nig.:** Vom, Bauchi Plateau, 3,000–
4,500 ft. *Dent Young* 54! Naraguta & Jos (Sept.) *Lely* 572! **S.Nig.:** Sonkwala, Obudu Div. (Dec.)
Savory & Keay FHI 25087! **Br.Cam.:** Buea (fl. & fr. Jan., Dec.) *Lehmbach* 63! *Preuss* 629! *Mildbr.*
10828! Banso, Bamenda (Oct.) *Tamajong* FHI 23478!

21. **T. gracilipes** *Guill. & Perr.*—F.T.A. 2 : 119 ; Bak. f. l.c. 189. An erect slender annual, 1 ft. or more
high, thinly appressed grey-silky, with small purplish flowers.
Sen.: *Heudelot*! Bakel (Sept., Oct.) *Perrottet.* Also in A.-E. Sudan and Abyssinia.

Besides the above, *T. candida* (Roxb.) DC. has been introduced from India into Sierra Leone and British
Cameroons.

Doubtfully recorded species.

T. simplicifolia *Franch.*—Berhaut Fl. Sen. 192. Recorded from Ouassadou, Senegal (*Berhaut* 1223) ; not
previously known outside Somaliland. The Senegal specimen has not been seen.

24. SESBANIA Scop.—F.T.A. 2 : 133.

Appendages present at base of claw of standard :
Appendages free in the upper half or third and continuous with the adnate part;
standard more or less spotted with purple, sometimes suffused purple ; fruits up
to 24 cm. long, at first torulose ; inflorescence 4–8-flowered ; peduncles smooth ;
leaves about as long as the inflorescence ; leaflets 9–27 pairs, oblong-linear, mucronate, 2–3 cm. long and about 5 mm. broad, more or less pubescent   1. *sesban*
Appendages adnate or with their short apices free and at right angles to the standard :
Flowers very large, 5–10 cm. long ; fruits about 30 cm. long, flat ; leaflets about
15–20 pairs, elongate-oblong, 2·5–3 cm. long and up to 1 cm. broad   2. *grandiflora*
Flowers much smaller, up to 2·5 cm. long :
Stipules (seen in the youngest parts) 9–11 mm. long, about 2 mm. broad, linear,
brown ; peduncles angular, (8–) 12–20 cm. long with many small prickles at the
base ; calyx campanulate, teeth very small ; flowers 1·2–1·8 cm. long ; pedicels
about 1 cm. long ; fruits 20–25 cm. long, 7 mm. broad ; leaves as long as the
inflorescence, or rather shorter ; leaflets 10–24 pairs, 9–24 mm. long, 2–6 mm.
broad, oblong, mucronate at apex   ..   ..   3. *macrantha*
Stipules (seen in the youngest parts) 4–8 mm. long, 1 mm. broad, subulate, green
with scarious margins ; peduncles not prickly :
Stems and leaves, or leaves only, rather densely pubescent ; calyx 5-toothed ;
standard not spotted :
Calyx-teeth dark brown, triangular ; leaflets 12–18 pairs, elongate-oblong,
mucronate, up to 2·5 cm. long and 6 mm. broad, usually densely pubescent ;
flowers small, about 8 mm. long ; fruits not torulose, curved, about 15 cm. long,
3 mm. broad with thick margins ..   ..   4. *pubescens*
Calyx-teeth green ; leaflets up to 40 pairs, linear-oblong, 8–25 mm. long, 2–3·5 mm.
broad, sparsely appressed-pubescent on both surfaces ; flowers 12–17 mm.
long ; fruits strongly torulose, about 20 cm. long   ..   ..   5. *dalzielii*
Stems and leaves glabrous or slightly pubescent especially on nerves and margins
of young leaves ; calyx 4- or 5-toothed ; standard spotted :
Glands, at least on the older parts, in a line up the stem from each leaf-axil ;
inflorescences corymbose when young, often slightly pubescent ; calyx-teeth
triangular at base with filiform acumen ; flowers about 1·5 cm. long ; fruits
slightly curved, margins conspicuously thickened, surface clearly but not
deeply indented, apex acute, 15–25 cm. long ..   ..   ..   6. *pachycarpa*
Glands absent from stem :
Stems thick and pithy ; leaf-rhachis usually aculeate ; leaves very long, up to
40 cm. long ; leaflets numerous, 1–2 cm. long and 3–5 mm. broad, glabrous
and glaucous ; peduncle 6–8 cm. long, few-flowered ; flowers about 1·5 cm.
long ; fruit-margins straight   ..   ..   7. *bispinosa*
Stems slender and wiry ; leaf-rhachis smooth ; leaves up to 15 cm. long ; leaflets
5–30 pairs, obliquely unequally-sided at base, about 1·5 cm. long, 2·5–3 mm.

broad, glaucous green ; peduncle 2–4 cm. long, few-flowered ; flowers about
    2 cm. long ; fruit-margins constricted ..   ..   ..   8. *arabica*
Appendages absent from claw of standard or reduced to a slight ridge, standard striate ;
    fruits torulose, much constricted between the seeds, 12–15 cm. long, 3 mm. broad ;
    leaflets 8–18 pairs, narrowly oblong, 6–13 mm. long, 2–3 mm. broad, finely punctate,
    cuneate at base, glabrous..   ..   ..   9. *leptocarpa*

1. **S. sesban** (*Linn.*) *Merrill* in Philipp. Journ. Sci. 7 : 235 (1912) ; Cronquist in Fl. Congo Belge 5 : 76, t. 5.
    Aeschynomene sesban Linn. (1753). *Sesbania aegyptiaca* (Poir.) Pers. (1807)—F.T.A. 2 : 134 ; Chev.
    Bot. 184 ; Phill. & Hutch. in Bothalia 1 : 44 ; F.W.T.A., ed. 1, 1 : 387 (incl. var. *bicolor* Wight) ; Bak. f.
    Leg. Trop. Afr. 2 : 259. *S. punctata* DC. (1825)—F.T.A. 2 : 133 ; Chev. Bot. 185 ; Phill. & Hutch. l.c.
    43 ; F.W.T.A., ed. 1, 1 : 387 ; Bak. f. l.c. 259. A shrub or small tree, up to 20 ft. high, copiously branched,
    more or less pale glaucous, with red-spotted yellow flowers ; common by streams and sandbanks.
    Sen.: *Perrottet* 231 ! 232 ! *Roger* ! *Heudelot* ! Richard Tol (Jan.) *Döllinger* 10 ! **Gam.:** *Pirie* 32 ! **Fr.Sud.:**
    fide Chev. l.c. **S.L.:** *T. Vogel* 24 ! **Iv.C.:** R. Cavally Chev. 19776. **G.C.:** Black Volta R. (fr. Mar.) *Kitson*
    968 ! **N.Nig.:** Attah (fr. Sept.) *T. Vogel* 45 ! Nupe *Barter* 970 ! Katagum (Dec.) *Dalz.* 7 ! Sokoto (Dec.)
    Dalz. 324 ! *Lely* 152 ! L. Chad (fl. & fr. Jan.) *Lamb* 107 ! **Br.Cam.:** Bama, Dikwa Div. (fl. & fr. Jan.)
    McClintock 162 ! Widespread in tropical Africa and through tropical Asia to N. Australia, also in St.
    Helena.

2. **S. grandiflora** (*Linn.*) *Poir.* in Lam. Encycl. Méth. Bot. 7 : 127 (1806) ; Phill. & Hutch. l.c. 46 ; Bak. f. l.c.
    260. *Robinia grandiflora* Linn. (1753). A small tree or shrub, with rather stout branchlets pubescent
    when young ; flowers very large, few, pink or red or cream-white.
    Sen.: *Roger* ! *Perrottet* 235 ! **S.L.:** *Don* ! **G.C.:** Axim *Chipp* 391 ! **S.Nig.:** Lagos *Rowland* ! Widely
    cultivated in the tropics. (See Appendix, p. 260.)

3. **S. macrantha** *Welw. ex Phill. & Hutch.* l.c. 47 (1921) ; Bak. f. l.c. 261 ; Cronq. l.c. 79. A shrub, 5–6 ft.
    high with stem and branches rough with small thorns ; flowers yellow tinged pink.
    Br.Cam.: Bamenda, 5,000 ft. (fr. Jan.) *Migeod* 403 ! Njinikom, 5,000 ft. (May) *Maitland* 1442 ! L.
    Bambili (Aug.) *Ujor* FHI 29998 ! Also in French Cameroons, A.-E. Sudan, Belgian Congo, Angola,
    Uganda, Kenya, Nyasaland, N. and S. Rhodesia and Transvaal.

4. **S. pubescens** *DC.*—F.T.A. 2 : 135 ; Phill. & Hutch. l.c. 48 ; Chev. Bot. 184 ; Bak. f. l.c. 261 ; Cronq. l.c.
    83. A slender shrub of moist places ; inflorescence slender with rather small yellow flowers.
    Sen.: *Perrottet* 228 ! M'bidjem *Thierry* 72 ! Chev. 3393 ! N'boro *Brunner* 79 ! **Fr.Sud.:** Timbuktu
    Chev. 1206. Kabarah *Chev.* 1325. **G.C.:** Accra *Brown* 419 ! *Dalz.* 79 ! *Irvine* 805 ! 822 ! Axim (fl. &
    fr. Apr.) *Chipp* 426 ! Aburi (Aug.) *Johnson* 950 ! **Dah.:** Cotonou *Debeaux* 150 ! Zagnanado (Mar.)
    Chev. 23302 ! **N.Nig.:** Kwarre *Sampson* 22 ! Also in Chad Territory, A.-E. Sudan, Belgian Congo, E.
    Africa, Angola and S. Tomé.

5. **S. dalzielii** *Phill. & Hutch.* l.c. 49 (1921) ; Bak. f. l.c. 261 ; Berhaut Fl. Sen. 33. A slender shrub 4–5 ft.
    high, with weak slightly pubescent branches and yellow flowers ; in marshy places.
    Sen.: *fide* Berhaut *l.c.* **Fr.Sud.:** Koulikoro (Oct.) *Chev.* 3390 ! **G.C.:** Akuse (fl. & fr. Oct.) *Vigne* FH 4050 !
    Fr.Nig.: Labbézenga (fl. & fr. Sept.) *Hagerup* 457 ! **N.Nig.:** *Baikie* 9 ! Jebba *Barter* ! Katagum (fl. & fr.
    Oct.) *Dalz.* 8 ! Abinsi (fl. & fr. Oct.) *Dalz.* 614 ! Benue R. (Sept.) *Talbot* ! (See Appendix, p. 260.)

6. **S. pachycarpa** *DC.* Prod. 2 : 265 (1825) ; Berhaut in Bull. Soc. Bot. Fr. 99 : 297 (1953) ; F.W.T.A., ed. 1, 1 :
    387, partly (*Perrottet* 230). *S. pubescens* of F.W.T.A., ed. 1, 1 : 387, partly (*Roger* 23 ; *Heudelot* 252).
    S. aculeata of F.W.T.A., ed. 1, 1 : 387, partly (*Perrottet* 229). *S. aegyptiaca* of F.W.T.A., ed. 1, 1 : 387,
    partly (*Talbot* 1242). A herb 2–5 ft. high with yellow flowers, the standard purple-spotted ; usually by
    water or introduced.
    Sen.: *Roger* 23 ! *Perrottet* 230 ! Richard Tol (Sept.) *Perrottet* 229 ! Galam *Heudelot* 252 ! **Fr.Sud.:**
    Buia, Ansongo *Hagerup* 404 ! **N.Nig.:** Bornu (Nov.) *Johnston N.* 64 ! *Talbot* 1242 ! *Clapperton* 5 !
    Br.Cam.: Dikwa (fl. & fr. Nov.) *Johnston N.* 96 ! Also in Mauritania, A.-E. Sudan and Madagascar.
    (See Appendix, p. 261.) [The name *S. pachycarpa* DC. seems to be misapplied by Cronquist in Fl. Congo
    Belge 5 : 81.]

7. **S. bispinosa** (*Jacq.*) *W. F. Wight* in U.S. Dept. Agric., Bur. Pl. Indust. Bull. 137 : 15 (1909) ; Sprague &
    Milne-Redhead in Kew Bull. 1939 : 159. *Aeschynomene bispinosa* Jacq. (1793). *Sesbania aculeata* Pers.
    (1807)—F.T.A. 2 : 134 ; Phill. & Hutch. l.c. 50 ; F.W.T.A., ed. 1, 1 : 387 (excl. *Perrottet* 229) ; Bak. f.
    l.c. 262. *S. pachycarpa* of F.W.T.A., ed. 1, 1 : 387 (excl. *Perrottet* 230). Stem glabrous, erect, branched
    from the base, often thick but soft and pithy, reaching 15 ft. ; flowers yellow, purple-spotted ; in muddy
    swamps.
    Sen.: *Heudelot* 502. **Gam.:** Kuntaur *Ruxton* 146 ! **Port.G.:** Bissau (Oct.) *Esp. Santo* 2542 ! **Fr.G.:**
    Nzérékoré *Baldwin* 9736 ! **S.L.:** Falaba (fl. & fr. Mar., Apr.) *Sc. Elliot* 5220 ! *Deighton* 5401 ! Njala
    (fr. Nov.) *Deighton* 1790 ! Jagbwema (Oct.) *Small* 402 ! **G.C.:** *Williams* 549 ! Accra *T. Vogel* ! Legon
    (fl. & fr. Nov.) *Morton A.* 90 ! **N.Nig.:** Bornu (fl. & fr. Nov.) *E. Vogel* 79 ! Katagum (fl. & fr. Oct.) *Dalz.*
    9 ! Abinsi (fl. & fr. Nov.) *Dalz.* 615 ! Naraguta & Jos (fl. Aug., fr. Nov.) *Lely* 534 ! 586 ! Vom, 3,000–
    4,500 ft. *Dent Young* 58 ! Lokoja *Richardson* ! **S.Nig.:** Lagos *Phillips* 13 ! *Dalz.* 1227 ! **Br.Cam.:**
    Bama, Dikwa Dist. (fl. & fr. Nov.) *McClintock* 32 ! A common weed in the tropics. (See Appendix, p. 261.)

8. **S. arabica** *Steud. & Hochst. ex Phill. & Hutch.* l.c. 51 (1921). Stems glabrous ; sometimes very slender,
    decumbent or erect.
    N.Nig.: Benue R., Muri Prov. (Nov.) *Lamb* 69 ! Bornu (Sept.) *Gwynn* 138 ! Extends to A.-E. Sudan and
    Arabia.

9. **S. leptocarpa** *DC.*—F.T.A. 2 : 135 ; Phill. & Hutch. l.c. 50 ; Chev. Bot. 184 ; Bak. f. l.c. 262. A slender
    glabrous undershrub up to 8 ft. high ; flowers small.
    Sen.: *Roger* ! Senegal R. *Heudelot* 489 ! **Fr.Sud.:** San *Chev.* 1060. **Iv.C.:** Baoulé-Nord *Chev.* 22030.
    Ouagadougou to Lai *Chev.* 24692. **G.C.:** Tamale (fr. Dec.) *Morton* GC 9915 ! **Fr.Nig.:** Gourma *Chev.*
    24526. Bilma *Chev.* 28641. **N.Nig.:** Kwarre (fr. Nov.) *Sampson* 23 !
    [Cronquist l.c. 80, records this species from the Belgian Congo and indicates that it occurs in other
    parts of tropical Africa and in Arabia ; his description shows that he has adopted a broad view of the
    species and not the strict view taken by Phillips and Hutchinson which is followed here.]

25. PSORALEA Linn.—F.T.A. 2 : 64.

A low shrub with rigid branchlets ; leaves pinnately 3-foliolate ; leaflets oblanceolate
    or oblong, 1–1·5 cm. long, closely and minutely pubescent, glandular ; petiole short ;
    racemes axillary, stout ; calyx strongly nerved ; fruit as long as the calyx, ellipsoid,
    hairy, 1-seeded   ..   ..   *plicata*

P. plicata *Del.*—F.T.A. 2 : 64 ; Bak. f. Leg. Trop. Afr. 1 : 93. Undershrub up to 4 ft. high, with ribbed
    weakly spiny branchlets and whitish flowers.
    Sen.: *Berhaut* 1719. **N.Nig.:** Kukawa, banks of L. Chad (fl. & fr. Feb.) *E. Vogel* 33 ! Extends to A.-E.
    Sudan and Nile land.

26. INDIGOFERA Linn.—F.T.A. 2 : 65 ; Tisserant in Bull. Mus. Hist. Nat., sér. 2, 3 : 163–172, 258–272 (1931). *Microcharis* Benth.—F.T.A. 2 : 132.

By J. B. Gillett

KEY TO THE NATURAL GROUPS

Pods falcate, subtriquetrous, 1-seeded, beset with spines ; inflorescence an axillary raceme ; fruiting pedicels reflexed ; leaves simple, elliptic
Subgenus I. ACANTHONOTUS, 1. *nummulariifolia*
Pods not as above :
  Pods flattened, 3- or more-seeded ; fruiting pedicels reflexed ; inflorescence an axillary raceme ; calyx more than half as long as stamens, its lobes much longer than the tube ; standard pubescent outside ; leaflets 3–7, the lateral ones opposite ; annuals
Subgenus II. AMECARPUS, spp. 2–5
  Pods not flattened, or, if so, remaining characters not as above :
    Stipules approaching the leaflets in size ; flowers almost hidden by the stipular part of the bracts ; indumentum long, softly silky ; fruits 2–4-seeded
Subgenus III. INDIGOFERA. Sect. **latestipulatae,** spp. 6–7
    Stipules not approaching the leaflets in size :
      Standard pubescent outside, or, if glabrous (in No. 37) then fruits erect, flattened 1-seeded ; keel rarely rostrate but having a spur-like pouch at each side ; all stamens fertile   ..   .. Subgenus III. INDIGOFERA
        Inflorescence a panicle, sometimes condensed and subcapitate ; bracts passing more or less gradually into the foliage leaves ; calyx deeply divided ; fruits (except in No. 9) short, 1–4-seeded, straight   Sect. **Paniculatae,** spp. 8–25
        Inflorescence an axillary raceme ; bracts sharply distinct from foliage leaves
Sect. **Indigofera,** spp. 26–70
        Fruiting pedicels not definitely deflexed (except sometimes in No. 36, where the inflorescence has 3–6 flowers) ; fruits erect or spreading, straight ; inflorescence 2–many-flowered ; lateral leaflets opposite (except in No. 27) :
          Plants free from glandular hairs (sepals and stipules may be gland-tipped in Nos. 41 and 42) :
            Peduncles very short (under 5 mm.) or, if more, less than ¼ length of floriferous axis :
              Fruits, at least when mature, all erect, or nearly so
Subsect. **Brevi-erectae** spp. 26–35
              Fruits 1–3 in each inflorescence, 6- or more-seeded, spreading in all directions ; leaflets 1–5, about 7 mm. long   ..   ..   ..   36. *suaveolens*
            Peduncles more than 5 mm. long, more than ¼ length of floriferous axis ; leaves, at least the lower ones, always compound :
              Pods 1-seeded, flattened ; whole plant, including standard, glabrous or almost so ; inflorescence with 4–7 flowers   ..   ..   ..   37. *scarciesii*
              Pods 4- or more-seeded ; standard pubescent :
                Calyx less than ½ as long as stamens ; stipules not above 5 mm. long, or 0·5 mm. wide ..   .. Subsect. **Dissitiflorae,** spp. 38–43
                Calyx more than ½ as long as stamens ; stipules often over 5 mm. long and 0·5 mm. wide ; leaflets 1–5 ; indumentum of both long spreading basifixed and appressed medifixed hairs ; stipules and calyx lobes sometimes gland-tipped ..   .. Subsect. **Pilosae,** spp. 44–45
          Plants with glandular hairs on the fruit or elsewhere
Subsect. **Viscosae,** spp. 46–52
        Fruiting pedicels definitely deflexed ; inflorescence 10–many-flowered :
          Lateral leaflets opposite :
            Leaflets not gland-dotted beneath :
              Calyx-lobes not more than twice as long as tube ; corolla much longer than calyx ; indumentum on standard closely appressed and parallel, giving a shiny appearance ; pods 4- or more-seeded, often curved
Subsect. **Tinctoriae,** spp. 53–60
              Calyx-lobes many times as long as tube ; corolla not twice as long as calyx ; indumentum of standard not dense, close-appressed and shining ; pods 3–8-seeded, straight   ..   ..   .. Subsect. **Hirsutae,** spp. 61–64
            Leaflets dotted beneath with more or less impressed glands ; calyx deeply divided ; raceme dense, fruits short ..   ..   ..   65. *microcarpa*
          Lateral leaflets alternate, or some or all leaves simple
Subsect. **Alternifoliolae,** spp. 66–70
Standard glabrous ; keel without a spur-like pouch at each side, usually rostrate ; pod more than 4-seeded
Subgenera IV. INDIGASTRUM, spp. 71–72, and V. MICROCHARIS, spp. 73–76

16

Subgenus II—AMECARPUS

Pods over 2·4 mm. wide, with separate eminences over each seed :
  Pods more or less curved, 5–9-seeded ; fruiting peduncle 5 mm. long or less, not above
    half as long as a fruit .. .. 2. *hochstetteri*
  Pods nearly straight, 3–6-seeded ; fruiting peduncle 10 mm. long or more, as long as
    or longer than a fruit .. .. 3. *senegalensis*
Pods less than 2·2 mm. wide :
  Pods 1·6–2 mm. wide, with a more or less continuous lateral ridge ; infructescence
    shorter than its subtending leaf ; leaflets rarely over 3 mm. wide
                                                        4. *charlieriana* var. *sessilis*
  Pods 1–1·5 mm. wide, the lateral ridge broken into separate swellings above the seeds ;
    infructescence usually more than twice as long as subtending leaf ; wider leaflets
    over 4 mm. wide .. .. 5. *aspera*

Subgenus III—INDIGOFERA

Sect. **Latestipulatae**

Flowers single in the axils of bracts which do not form a strobilus ; bracts each with a
  well-developed lamina and stipules free almost to the base .. .. 6. *berhautiana*
Flowers two together in the axils of bracts which are aggregated to form a strobilus ;
  lamina of bracts absent or much reduced ; stipules of bracts usually united to above
  the middle .. .. 7. *strobilifera*

Sect. **Paniculatae**

Pedicel, even when jointed and perhaps equivalent to a reduced peduncle, not over
  4 mm. long :
  Leaves always simple, the lower more than 2 cm. long :
    Fruiting calyx 4–5 mm. long with conspicuous rufous hairs .. .. 8. *polysphaera*
    Fruiting calyx 1–2 mm. long :
      Fruits about 5 times as long as wide ; stem 3-winged .. .. 9. *trialata*
      Fruits up to 3 times as long as wide ; stem ridged, not winged .. 10. *paniculata*
  Lower leaves often pinnate, simple leaves, if present, usually under 2 cm. long :
    Fruiting calyx-lobes much broadened, wider than the usually 1-seeded fruit which is
      hidden within them :
      Straggling herb ; fruiting inflorescence compact, usually over 15 mm. wide, and less
        than twice as long ; without glands, or with glands less frequent than the usual
        type of hairs .. .. 11. *macrocalyx*
      Stiff semiwoody plant ; fruiting inflorescence less compact, usually under 15 mm.
        wide, and often more than twice as long ; with numerous short glands between
        the less frequent usual type of hairs .. .. 12. *terminalis*
    Fruiting calyx-lobes not so wide as fruit :
      Inflorescences compact, subglobose, congested panicles, without separate flowers
        away from the rest in leaf-axils :
        Inflorescences less wide than the shoots which they terminate are long ; fruiting
          calyx about 6 mm. long, or more :
          Leaflets often 3–5 cm. long ; inflorescences 3–4 cm. across ; fruiting calyx about
            11 mm. long .. .. 13. *megacephala*
          Leaflets rarely over 2 cm. long, up to 9 in number ; inflorescences under 3 cm.
            across ; fruiting calyx about 6–8 mm. long .. .. .. 14. *capitata*
        Inflorescences mostly lateral, sessile in leaf-axils or wider than the shoots which
          they terminate are long ; fruiting calyx about 3 mm. long :
          Leaflets up to 7 in number .. .. 15. *congesta*
          Leaflets 11–13 in number .. .. 16 *dasycephala*
      Inflorescences much less compact, many flowers separate in the axils of leafy bracts
        (or simple leaves) :
        Fruiting calyx 3–4 mm. long, about as long as fruit, its indumentum brownish :
          Leaves all pinnate, the longer with about 17 leaflets .. .. 17. *brassii*
          Leaves often simple or 3-foliolate, the longer with about 9 leaflets 18. *pulchra*
        Fruiting calyx not above 2·5 mm. long :
          The lobes of fruiting calyx subequal, not above half as long as pod :
            Pods 2–4-seeded ; most lower leaves pinnate (up to 5-jugate) :
              A stiffly erect annual ; leaflets 4 or more times as long as wide ; branches of
                inflorescence leafless and flowerless below, crowded with 1–3-foliolate leaves
                or bracts in their upper half ; hairs on calyx and fruit dense, spreading,
                about 0·5 mm. long, often black or dark brown .. .. 19. *nigricans*
              A woody suffrutex ; leaflets twice as long as wide ; branches of inflorescence
                without a leafless lower portion ; hairs on calyx and fruit sparser, about
                0·2 mm. long, white .. .. 20. *oubanguiensis*
            Pods 1-seeded, more or less oval ; leaves simple ; indumentum sparse ; her-
              baceous shoots from a woody rootstock .. .. .. 21. *leptoclada*

The 2 upper lobes of fruiting calyx much wider than the rest, about ¾ as long as
    pod ; hairs on calyx and pod white ; some lower leaves usually pinnate with
    up to 11 leaflets   ..   ..   22. *bracteolata*
Pedicel slender, jointed, probably equivalent to a reduced peduncle, 5 mm. or more
    long ; leaflets up to 7 or 9 in number :
Calyx-lobes widest at the base :
    Slender annual, up to 30 cm. tall ; indumentum more or less spreading
                23. *trichopoda*
    Stiff erect bushy herb, up to 2 m. tall ; indumentum closely appressed   24. *nigritana*
    Calyx-lobes twice as wide just below the middle as at the base ; bushy herb, woody
    below ; hairs long, spreading   ..   ..   25. *latisepala*

Sect. **Indigofera**
Subsect. **Brevi-erectae**

Pods not glabrous :
Inflorescence extremely congested, so that the axis and pedicels are hidden by the
    flowers or fruits ; leaflets or simple leaves under 17 mm. long and about half as wide:
    Leaves simple, oval, cordate at base ; pods oblong, not torulose, 2-seeded
                26. *cordifolia*
    Leaves with 3–11, usually alternate, leaflets ; pods more or less torulose, 3–6-seeded
                27. *sessiliflora*
Inflorescence less congested, axis and pedicels not hidden :
    Leaves simple, several times as long as wide, usually over 17 mm. long ; pods 4-
    or more-seeded, elongate :
    Pods 1·8–2·5 mm. wide ; leaves up to 3 mm. wide ; calyx teeth about 1 mm. long :
        Hairs on upper surface of leaf set at an angle of 80–90° with midrib ; pod more or
        less flattened, about 2·5 mm. wide, 3–6-seeded, ending in a short point (per-
        sistent style base) about 0·3 mm. long   ..   ..   ..   28. *tetrasperma*
        Hairs on upper surface of leaf set at an angle of about 45° with midrib ; pod more
        or less tetragonal, about 1·8–2·0 mm. wide, usually 8–12-seeded, ending in a
        point (persistent style base) about 1·1 mm. long..   ..   29. *simplicifolia*
    Pods about 4 mm. wide ; leaves up to 10 mm. wide ; calyx teeth about 5 mm. long
                30. *leprieurii*
Leaves pinnate :
    Calyx-lobes as long as tube, or longer, or, if shorter, then the leaflets much less than
    3 cm. long and 1 cm. wide :
    Pods 12–40 mm. long, leaflets usually over 17 mm. long :
        Leaflets usually more than 11, often 15–23 in number, about 2–5 times as long as
        wide ; pods more or less 4-winged ; hairs on calyx long (about 1 mm.), more
        or less spreading ; flowers erect in bud, later reflexed and lastly the fruits erect
        again   ..   ..   31. *prieureana*
        Leaflets usually fewer than 11, 4–10 times as long as wide ; pod not winged :
        Hairs on calyx about 0·5 mm. long, more or less appressed ; flowers erect through-
        out   ..   ..   32. *stenophylla* var. *stenophylla*
        Hairs on calyx about 1 mm. long, more or less spreading, flowers deflexed after
        opening and later again becoming erect   ..   32a. *stenophylla* var. *ampla*
    Pods 3–5 mm. long ; leaflets 5–9 in number, up to 17 mm. long   33. *mildbraediana*
    Calyx-lobes much shorter than the tube ; leaflets 7, about 3 cm. long and 1 cm.
    wide ; standard densely covered outside by a golden-brown appressed indumentum
                34. *pobeguinii*
Pods glabrous, about 1·4 mm. wide, 19–22 mm. long ; leaflets 3–5   ..   35. *omissa*

Subsect. **Dissitiflorae**

Leaf-rhachis prolonged beyond distal pair of leaflets ; flowers borne one above the
    other on the axis of the inflorescence :
Fruiting pedicel over 2 mm. long :
    Pods minutely strigose, usually over 17 mm. long ; racemes often more than 12-
    flowered ; leaflets 11–31   ..   ..   38. *dendroides*
    Pods glabrous or almost so, less than 17 mm. long :
        Raceme 6–12-flowered, its axis pubescent ; stems more or less pubescent with
        spreading hairs ; leaflets 7–13 :
        Leaflets rarely over 8 mm. wide ; peduncle rarely over 15 mm. long ; leaves
        subtending inflorescences usually shorter than the lower leaves (1–3-jugate
        instead of 4–6-jugate)   ..   ..   ..   39. *heudelotii* var. *heudelotii*
        Leaflets up to 12 mm. wide ; peduncles up to 30 mm. long ; leaves subtending
        inflorescences 3- or more-jugate, much like lower leaves
                39a. *heudelotii* var. *fairchildii*
        Raceme 3–5-flowered, its axis (with peduncle) under 15 mm. long ; stems sub-
        glabrous, the few hairs appressed, leaflets 3–9   ..   ..   40. *elliotii*

Fruiting pedicel less than 2 mm. long ; racemes 2–4-flowered :
Petioles under 1 mm. long ; leaflets 7–13, with glandular margins ; stipellae obscure
                                        41. *wituensis* var. *occidentalis*
Petioles 3–7 mm. long ; leaflets 5–9, without glandular margins ; stipellae filiform
about 1 mm. long   ..   ..   42. *congolensis*
Leaf-rhachis usually not, or hardly, prolonged beyond distal pair of leaflets ; flowers
borne 2 together at the tip of the peduncle   ..   ..   43. *geminata*

Subsect. Pilosae

Inflorescence 1–4-flowered ; hairs all whitish ; leaflets often more or less acute
                                        44. *pilosa*
Inflorescence 5–20-flowered ; many hairs yellowish brown ; leaflets more or less obtuse
                                        45. *fulvopilosa*

Subsect. Viscosae

Non-glandular hairs of inflorescence white :
Inflorescence much more than 5 mm. long ; fruits usually over 3 mm. long :
Pods about 1–1·7 mm. wide :
Petiole usually shorter than basal leaflets ; leaflets rarely more than 9, finely
appressed-strigose above ; epidermis of stems and glandular hairs usually more
or less reddish-purple ; fruit not over 12 mm. long   ..   46. *mimosoides*
Petiole usually longer than basal leaflets (if shorter, leaflets more than 11) ; leaflets
often 11 or more ; epidermis of stems usually yellowish green :
Glandular hairs present on fruits and leaflets ; inflorescence not markedly secund ;
fruits up to 20 mm. long   ..   ..   47. *colutea*
Glandular hairs absent on fruits and leaflets, rather sparse elsewhere ; inflorescence
markedly secund ; fruits 6–10 mm. long ..   ..   48. *barteri*
Pods about 2 mm. wide :
Inflorescence dense, secund, many-flowered ; pods 4–6 mm. long, with black
glandular hairs as well as a dense silvery coat of appressed medifixed hairs
                                        49. *secundiflora*
Inflorescence not secund, with 3–5 flowers only ; glandular hairs very short and
sparse   ..   ..   50. *argentea*
Inflorescence compact, not over 5 mm. long, pods 2-seeded, about 3 mm. long
                                        51. *conferta*
Non-glandular hairs of inflorescence largely blackish-brown ; leaflets 5–13 ; fruit about
2 mm. wide :
Erect multicellular hairs present on pods, scarce or absent on stems
                                        52. *atriceps* subsp. *atriceps*
Erect multicellular hairs abundant on both pods and stems
                                        52a. *atriceps* subsp. *alboglandulosa*

Subsect. Tinctoriae

Stamens more than 10 mm. long ; leaflets 5–9, often over 3 cm. long   53. *garckeana*
Stamens less than 10 mm. long ; leaflets 3–13 :
Hairs on fruit mostly dark brown, those on standard dense, appressed, golden brown
to black ; fruit straight :
Raceme almost sessile, usually shorter than subtending leaf ; stamens 5–6 mm. long ;
plant drying brownish green   ..   ..   54. *emarginella*
Raceme pedunculate, up to 15 cm. long, very slender, longer than subtending leaf ;
stamens about 4 mm. long ; plant drying blackish   ..   55. *macrophylla*
Hairs on fruit usually white ; stamens not over 4 mm. long :
Fruits more or less tetragonous, curved ; infructescence much longer than subtending
leaf :
Leaflets 3 ..   ..   56. *subulata* var. *subulata*
Leaflets 5–9   ..   ..   56a. *subulata* var. *scabra*
Fruits not tetragonous ; infructescence not, or but little, longer than subtending leaf :
Pods torulose ; seeds up to 5 or 6 ; stem densely silvery ; leaflets 5–9
                                        57. *coerulea*
Pods not torulose, or if so, seeds more than 6 ; leaflets usually 11 or more :
Fruits straight, about 6-seeded, less than 2 cm. long ; indumentum on flowers and
inflorescence more or less brown   ..   ..   58. *arrecta*
Fruits more or less curved, or, if almost straight, then more than 6-seeded and
over 2 cm. long :
Pods definitely curved, reddish brown, short (about 1–1·5 cm. from end to end
in a straight line), about 5–8-seeded ; indumentum of inflorescence brownish
                                        59. *suffruticosa*
Pods usually slightly curved or almost straight, over 2 cm. from end to end in a
straight line, greyish black, 8–12-seeded ; indumentum of inflorescence whitish
                                        60. *tinctoria*

Subsect. **Hirsutae**

Spreading hairs on calyx and fruit conspicuous, more than 0·8 mm. long ; leaflets usually more than 5 :
    Leaflets rarely over 18 mm. long or 10 mm. wide ; petiole often under 4 mm. long :
        Peduncle usually longer than the dense inflorescence ; ovules about 3 ; seeds oblong, rounded, minutely dotted, not coarsely pitted ; pods usually dark brown or black
                                                                       61. *longebarbata*
        Peduncle usually shorter than the lax slender inflorescence ; hairs on young fruit mainly white .. .. 62a. *hirsuta* var. *pumila*
    Leaflets often more than 18 mm. long and 10 mm. wide ; petiole almost never less than 4 mm. long ; peduncle usually shorter than inflorescence ; pods over 8 mm. long ; seeds angular, coarsely pitted :
        Peduncle well-developed, at least twice as long as a fruit ; fruits usually 6-seeded, about 2 mm. wide and 7 times as long ; hairs on dorsal side of fruit usually mostly brown or black ; upper surface of leaflets usually with minute pale dots on a darker background .. .. .. 62. *hirsuta* var. *hirsuta*
        Peduncle short, rarely twice as long as a fruit ; fruits usually 2–4- sometimes 5–6-seeded, about 2·5–3 mm. wide, less than 7 times as long, more or less tetragonous ; hairs on dorsal side of fruit usually mostly white ; upper surface of leaflets usually having minute dark dots on a paler background .. .. 63. *astragalina*
Hairs on calyx and fruit much less conspicuous, 0·5 mm. long or less, those on the fruit white ; fruit 1–1·5 mm. wide ; leaflets 3–5 .. .. 64. *deightonii*

Subsect. **Alternifoliolae**

Leaflets less than 3 cm. long, or, if so, then calyx lobes hardly longer than tube :
    Leaflets 2 ; pod not above 10 mm. long, densely clothed with long, more or less spreading, hairs ; calyx deeply divided, nearly as long as corolla.. .. 66. *diphylla*
    Leaflets usually more than 2 ; hairs on pod usually appressed :
    Pods much longer than wide, 3- or more-seeded :
        Calyx much less than half as long as corolla, the lobes not much longer than the tube ; standard densely covered outside with stiff shining hairs ; pod more or less curved, torulose ; leaflets 1–5, the terminal often much longer than the others
                                                                 67. *oblongifolia*
        Calyx more than half as long as corolla, lobes much longer than the tube ; hairs on standard relatively sparse ; pod more or less straight ; stipules broad, scarious ; leaflets 4 or more .. .. 68. *spicata*
    Pods not much longer than wide, 1-seeded ; stipules broad, scarious 69. *kerstingii*
Leaflets mostly over 3 cm. long ; calyx lobes much longer than the tube :
    Hairs on stems, petioles and inflorescences closely appressed ; leaflets 1–5
                                          70. *conjugata* var. *conjugata*
    Hairs on stems, petioles and inflorescences crisped, spreading at the tips ; leaflets usually 1, less often 2–3 .. .. .. 70a. *conjugata* var. *occidentalis*

Subgenera IV—INDIGASTRUM and V—MICROCHARIS

Leaves pinnate ; the lateral leaflets opposite ; calyx less than half as long as corolla :
    Pedicel about 1 mm. long ; bracts not persistent in fruit ; all stamens fertile :
        Inflorescence pedunculate, the peduncle longer than a leaflet ; leaflets 11–15 in number ; corolla 5–6 mm. long .. .. 71. *costata*
        Inflorescence sessile or peduncle shorter than a leaflet ; leaflets 3–9 in number ; corolla about 4 mm. long .. .. .. 72. *parviflora* var. *occidentalis*
    Pedicel 2–3 mm. long ; bracts persistent in fruit ; dorsal (free) stamen sterile ; leaflets 3–5 in number .. .. 73. *welwitschii* var. *remotiflora*
Leaves simple ; dorsal stamen sterile ; bracts persistent :
    Calyx as long as or longer than the corolla ; petiole much shorter than stipules, not separated from lamina by a distinct joint ; margin of leaf at the base forming an angle of about 15° with the midrib ; fruiting pedicel 2–4 mm. long ; seeds 16–19
                                                    74. *longicalyx*
    Calyx shorter than corolla ; petiole separated from lamina by a distinct joint ; margin of leaf forming an angle of 40° or more with midrib at base ; seeds 12 or fewer :
        Hairs on stems more or less spreading ; petiole shorter or hardly longer than stipules ; racemes usually longer than subtending leaf ; fruiting pedicel about 2 mm. long
                                                  75. *hutchinsoniana*
        Hairs on stems dense, closely appressed ; petiole much longer than stipules ; racemes usually shorter than subtending leaf ; fruiting pedicel about 1 mm. long
                                                    76. *disjuncta*

The corolla is red or pink in *Indigofera* unless otherwise stated.

1. **I. nummulariifolia** (*Linn.*) *Livera ex Alston* in Trim. Fl. Ceylon 6, suppl. 72 (1931). *Hedysarum nummulariifolium* Linn. (1753). *Indigofera echinata* Willd. (1803)—F.T.A. 2 : 69 ; Chev. Bot. 173 ; F.W.T.A., ed. 1, 1 : 393 ; Bak. f. Leg. Trop. Afr. 1 : 96 ; Cronquist in Fl. Congo Belge 5 : 125. A prostrate annual, usually a weed.

Sen.: Bondou *Heudelot* 174! **Port.G.:** Pussubé, Bissau (Oct.) *Esp. Santo* 973! **Fr.G.:** Fouta Djalon *Chev.* 18874! **Iv.C.:** Bobo Dioulasso *Chev.* 910. **G.C.:** Nalerigu N.T. (Dec.) *Adams* GC 4231! **N.Nig.:** R. Niger (Sept.) *T. Vogel* 57! Zaria (Aug.) *Keay* FHI 28010! Bauchi Plateau (Aug.) *Keay* FHI 12733! **S.Nig.:** Onitsha *Barter* 1802! Enugu (Sept.) *Ujor* FHI 34119! Extending to East and South tropical Africa, India, Ceylon and Indo-China. (See Appendix, p. 246.)

2. **I. hochstetteri** *Bak.* in F.T.A. 2 : 101 (1871) ; Cronq. l.c. 150. *I. anabaptista* Steud. ex Bak. (1876) ; Bak. f. l.c. 164. An annual of subarid areas.
     Fr.Sud.: Timbuktu *Hagerup*! **Fr.Nig.:** Aïr *Chopard & Villiers*! **N.Nig.:** Kontagora (Jan.) *Meikle* 1097! Nabardo (Sept.) *Lely* 621! Extends from Mauritania to Madras and south to Tanganyika.

3. **I. senegalensis** *Lam.* Encycl. Méth. Bot. 3 : 248 (1789) ; F.T.A. 2 : 102 (excl. syn. *I. tenella*) ; Chev. Bot. 177. *I. parviflora* Heyne of F.W.T.A., ed. 1, 1 : 391, partly, not of Heyne ex Wight & Arn. An erect annual of subarid areas.
     Sen.: *Perrottet*! Kouma *Heudelot*! **Fr.Sud.:** Mopti (Aug.) *Chev.* 24906! **N.Nig.:** Fodama (Dec.) *Moïser* 252! Also in Chad Territory.

4. **I. charlieriana** *Schinz* var. *sessilis* (*Chiov.*) *Gillet* in Kew Bull. 1955 : 573 (1956). *I. linearis* DC. var. *sessilis* Chiov. Fl. Somala 2 : 160 (1932). *I. aspera* of Chev. Bot. 172 ; F.W.T.A., ed. 1, 1 : 392 and Tiss. in Bull. Mus. Hist. Nat. 3 : 272, partly, not of Perr. ex DC. An annual herb, chiefly of maritime sands.
     Dah.: Porto Novo, in sand by lagoon (Jan.) *Chev.* 22718! A S. African species extending up the East and West coasts of Africa.

5. **I. aspera** *Perr. ex DC.* Prod. 2 : 229 (1825) ; F.T.A. 2 : 102 ; Chev. Bot. 172, partly ; Bak. f. l.c. 165. *I. tenella* Schum. (1827). An erect annual of subarid areas.
     Sen.: Richard Tol (June) *Perrottet*! *Heudelot* 428! **Fr.Sud.:** Timbuktu (Aug.) *Hagerup*! **G.C.:** Keta (Nov.) *Thonning*! **N.Nig.:** Jebba *Barter*! Sokoto (Sept.) *Dalz.* 326! Also in A.-E. Sudan. (See Appendix, p. 246.)

6. **I. berhautiana** *Gillett* l.c. 573 (1956). *I. sericea* Benth. ex Bak. in F.T.A. 2 : 76 (1871) and Bak. f. l.c. 112, mostly ; F.W.T.A., ed. 1, 1 : 393 ; not of Linn. *I.* sp. aff. *I. strobilifera* of Chev. Bot. 177. A more or less shrubby, silky, erect herb.
     Sen.: Cayor *Heudelot*! M'bidjem (Aug.) *Thierry* 89! (Sept.) *Farmar* 73! Niayes (Dec.) *Chev.* 3385! *Berhaut* 950! **G.C.:** Gonja Domongo, N.T. (Dec.) *Morton* GC 9959! **Fr.Nig.:** Niamey (Oct.) *Hagerup* 475 (partly)! **N.Nig.:** Nupe *Barter* 973!

7. **I. strobilifera** (*Hochst.*) *Hochst. ex Bak.* in F.T.A. 2 : 75 (1871) ; Bak. f. l.c. 113. *Eilemanthus strobilifera* Hochst. (1846). A silky, much-branched, half-woody herb.
     Fr.Sud.: Dioura (Dec.) *Davey* 122! **Fr.Nig.:** Niamey (Oct.) *Hagerup* 475 (partly)! **N.Nig.:** Sokoto (Jan.) *Dalz.* 330! Extends to A.-E. Sudan and southwards to S. Rhodesia.

8. **I. polysphaera** *Bak.* in Kew Bull. 1895 : 65 ; Bak. f. l.c. 103 ; Cronq. l.c. 128. *I. baoulensis* A. Chev. (1912). Erect scabrid herb up to 3 ft. tall or more.
     Iv.C.: Bouaké to Langouassou (July) *Chev.* 22146! **G.C.:** Krobo plains (Oct.) *Johnson* 774! Aboma F.R. (Nov.) *Vigne* FH 3441! Ejura (Nov.) *Darko* 749! **Togo:** Dyola Kpuita (Nov.) *Morton* GC 9347! **N.Nig.:** Zaria (Sept.) *Olorunfemi* FHI 24365! Lokoja (Oct.) *Dalz.* 18! Share F.R., Ilorin (Jan.) *Ujor* FHI 31607! Abinsi (Sept.) *Dalz.* 594! **S.Nig.:** Lagos *McGregor* 288! Oyo (Sept.) *Latilo* FHI 23542! Extends to A.-E. Sudan, Uganda, W. Tanganyika and Angola. [*Ujor* FHI 31607 has the inflorescence much more condensed than usual and may represent an undescribed variety.]

9. **I. trialata** *A. Chev.* in Bull. Soc. Bot. Fr. 58, Mém. 8 : 157 (1912) ; Bak. f. l.c. 105. An erect, almost leafless, herb up to 5 ft. high.
     Iv.C.: Mt. Kangoroma, Baoulé (July) *Chev.* 22173! 22182! **G.C.:** Dukwesein, N. Agogo, Ashanti (Dec.) *Chipp* 618! Kwabia, E.P. (Oct.) *Vigne* FH 4020!

10. **I. paniculata** *Vahl ex Pers.* Syn. 2 : 325 (1807) ; Tiss. in Bull. Mus. Hist. Nat., sér. 2, 2 : 680 ; Cronq. l.c. 126, t. 9 ; not of DC., nor Bak. F.T.A. 2 : 71, nor Bak. f. l.c. 102. *I. procera* Schum. & Thonn. (1827)—F.T.A. 2 : 71 ; Chev. Bot. 176, partly ; Bak. f. l.c. 102 ; F.W.T.A., ed. 1, 1 : 393. An erect branched herb up to 3–6 ft. high.
     Gam.: *Ruxton* A. 47! **Port.G.:** Bissau (Jan.) *Esp. Santo* 1716! **Fr.G.:** *Heudelot* 683! **S.L.:** Batkanu (Feb.) *Deighton* 4185! Masisa (Aug.) *Deighton* 5126! Mabang (Jan.) *Glanville* 147! Warantamba (Oct.) *Small* 350! **Iv.C.:** Mankono (June) *Chev.* 21889! **G.C.:** Accra (Aug.) *Deighton* 3408! Achimota (Aug.) *Irvine* 791! Tamale *Adams & Akpabla* GC 4158! **Togo:** Lome *Warnecke* 236! Kpedsu (Jan.) *Howes* 1089! **Dah.:** Le Testu 44! **N.Nig.:** Anara F.R., Zaria (Oct.) *Keay* FHI 5496! Bauchi Plateau (Oct.) *Lely* P. 820! **S.Nig.:** Abeokuta *Millen* 99! Awba Hills F.R., Oyo (Oct.) *Jones* FHI 6313! **Br.Cam.:** Bamenda, 3,000 ft. (June) *Maitland* 1572! Extends to E. Africa and Angola, with a variety in S. Rhodesia.

11. **I. macrocalyx** *Guill. & Perr.* Fl. Seneg. 1 : 175, t. 46 (1832) ; F.T.A. 2 : 71 ; Chev. Bot. 174 ; Bak. f. l.c. 98. *I. lotononoides* of Chev. Bot. 174, not of Bak. f. A straggling herb, the stems somewhat woody ; variable in the western part of its range, where forms approaching the next species are found.
     Sen.: *Perrottet* 193! Bakel *Leprieur*. **Fr.Sud.:** Sicoro (Jan.) *Chev.* 229. **Port.G.:** Bissau (Jan.) *Esp. Santo* 1447! Bafata *Esp. Santo* 2819! **S.L.:** Wallia (Feb.) *Sc. Elliot* 4270! Batkanu (Jan.) *Deighton* 2843! **Iv.C.:** Mt. Do, 3,000 ft. (May) *Chev.* 21423! **N.Nig.:** Kaduna (Dec.) *Meikle* 798! Nupe *Barter* 926! 998! Bauchi Plateau (Oct.) *Lely* 790! **S.Nig.:** Lagos *Millen* 174! Also in Chad Territory.

12. **I. terminalis** *Bak.* in F.T.A. 2 : 70 (1871) ; Bak. f. l.c. 99, including var. *chevalieri*. *I.* sp. aff. *I. macrocalyx* of Chev Bot. 174. A woody undershrub up to 5 ft. high.
     Sen.: Walo *Heudelot* 365! **Fr.Sud.:** Kati (Jan.) *Chev.* 187! Koulikoro (Oct.) *Chev.* 3233. **Iv.C.:** *Aubrév.* 1816!

13. **I. megacephala** *Gillett* l.c. 574 (1956). Woody-stemmed herb, 4 ft. high ; flowers yellow.
     Fr.G.: Betagya, 1,000 ft. (Oct.) *Small* 498!

14. **I. capitata** *Kotschy* in Sitzb. Ak. Wien. 51, Abt. 2 : 365, t. 6A (1865) ; F.T.A. 2 : 75, partly ; Berhaut Fl. Sen. 51 ; not of Cronq. l.c. 152, Pl. 11. Erect branched woody herb up to 3 ft. high.
     Sen.: Diénoundiella *Berhaut* 1819. **Fr.Sud.:** Ouacoro (Dec.) *Chev.* 79! **Port.G.:** Gabú (Oct.) *Esp. Santo* 2305! **Fr.G.:** Rio Nunez *Heudelot* 642! *Pobéguin* 472! **S.L.:** Wallia (Feb.) *Sc. Elliot* 4295! Luseniya (Dec.) *Sc. Elliot* 4063! Kitchom (Jan.) *Deighton* 1002! **N.Nig.:** Kaduara (Nov.) *Lely* 596! Lokoja (Oct.) *Dalz.* 24! Zaria (Dec.) *Meikle* 832! **S.Nig.:** Lagos *MacGregor* 273! Olokemeji to Eruwa (Nov.) *Ejiofor & Olorunfemi* FHI 30780. Extends to A.-E. Sudan, Uganda, N. Rhodesia and Angola.

15. **I. congesta** *Welw. ex Bak.* in F.T.A. 2 : 70 (1871) ; Bak. f. l.c. 99 ; Cronq. l.c. 151. *I. terminalis* of Chev. Bot. 178, not of Bak. Stiff, semiwoody, much-branched herb, up to 3–6 ft. high.
     Port.G.: Bula (Nov.) *Esp. Santo* 2434! Gabú (Nov.) *Esp. Santo* 3153! **S.L.:** Wallia (Feb.) *Sc. Elliot* 4252! Luseniya (Dec.) *Sc. Elliot* 4308! **Iv.C.:** Toumodi (Aug.) *Chev.* 22404! **Togo:** Misahöhe (Nov.) *Mildbr.* 7251! **N.Nig.:** Ilorin *Baily* 9! **S.Nig.:** Ibadan (Oct.) *Jones* FHI 5989! Onitsha *Barter* 1791! Extends to A.-E. Sudan, E. Africa and S. Tropical Africa.

16. **I. dasycephala** *Bak. f.* l.c. 100 (1926). A straggling herb.
     N.Nig.: Denchira, Bauchi Plateau (Nov.) *Lely* 108! Also in Northern French Cameroons.

17. **I. brassii** *Bak.* in F.T.A. 2 : 76 (1871) ; Bak. f. l.c. 110. An erect, grey, softly woody herb, closely related to *I. pulchra* of which it is perhaps but a form.
     G.C.: Cape Coast *Brass*!

18. **I. pulchra** *Willd.* Sp. Pl. 3 : 1239 (1802) ; F.T.A. 2 : 76 ; Chev. Bot. 176 ; Bak. f. l.c. 111 ; Tiss. l.c. 262 ; Cronq. l.c. 154. *I. dupuisii* Mich. (1897), not of F.W.T.A., ed. 1, 1 : 393. An erect, stiff, grey-pubescent

softly woody undershrub up to 5 ft. high ; varies considerably, especially in the extent to which the leaves are divided.

Sen.: *Heudelot* 179 ! **Gam.:** *Ingram* ! **Fr.Sud.:** Ségou (Sept.) *Chev.* 3229 ! **Port.G.:** Bissau (Nov.) *Esp. Santo* 1027 ! Gabu (Sept.) *Esp. Santo* 2783 ! **Fr.G.:** Kouroussa (Oct.) *Pobéguin* 567 ! **S.L.:** Kamrasoi (Dec.) *Thomas* 7041 ! **Iv.C.:** *Chev.* 20076. **G.C.:** Navrongo (May) *Andoh* FH 5174 ! Tamale (Apr.) *Williams* 131 ! **Togo:** Kpedsu (Dec.) *Howes* 1046 ! Lome *Warnecke* 68 ! **Dah.:** *Burton* ! **Fr.Nig.:** Niamey (Oct.) *Hagerup* 487 ! **N.Nig.:** Jebba (Dec.) *Meikle* 1185 ! Kano (May) *Ejiofor* FHI 26149 ! **S.Nig.:** Lagos (May) *Dalz.* 1222 ! Extends to Abyssinia, Tanganyika and Angola. (See Appendix, p. 247.)

19. **I. nigricans** *Vahl ex Pers.* Syn. 2 : 327 (1807) ; DC. Prod. 2 : 230 ; ampl. Bak. f. in J. Bot. 41 : 193 ; Bak. f. Leg. Trop. Afr. 1 : 112 ; not of Bak. F.T.A. 2 : 78, nor of F.W.T.A., ed. 1, 1 : 391. *I. elegans* Schum. & Thonn. (1827). *I. dupuisii* of Bak. f. l.c. 111, partly and of F.W.T.A., ed. 1, 1 : 393, not of Mich. (1897). *I.* sp. aff. *I. bracteolata* of Chev. Bot. 172. An erect stiff, probably annual, herb up to 2 ft. high.
    Fr.Sud.: Kerekoro (Sept.) *Chev.* 3220 ! Koulikoro (Sept.) *Chev.* 25015 ! **G.C.:** *Thonning* ! Accra (July) *Irvine* 735 ! Ada *Morton* GC 9255 ! **Togo:** *Mahoux* 228 ! **N.Nig.:** Abinsi (Oct.) *Dalz.* 598 !

20. **I. oubanguiensis** *Tiss.* l.c. 164 (1931). *I.* sp. aff. *I. parviflora* of Chev. Bot. 175. A half woody shrublet, about 2½ ft. high.
    Iv.C.: Zorgo, Mossi (July) *Chev.* 24653 *bis* ! Extends to A.-E. Sudan. [*Chev.* 24653 *bis* is imperfect and may be incorrectly determined.]

21. **I. leptoclada** *Harms* in Engl. Bot. Jahrb. 40 : 38 (1907) ; Bak. f. l.c. 97 ; Berhaut l.c. partly. *I.* sp. aff. *I. procera* of Chev. Bot. 176. *I. bracteolata* of F.W.T.A., ed. 1, 1 : 393, partly ; not of DC. Rootstock woody, shoots erect, herbaceous, up to 12–16 in. long.
    Gam.: MacCarthy Isl. (Jan.) *Dalz.* 8027 ! **Fr.Sud.:** Ianfola (Mar.) *Chev.* 651 ! *Bellamy* 702 ! **G.C.:** Tamale (Apr.) *Williams* 116.

22. **I. bracteolata** *DC.* Prod. 2 : 223 (1825) ; F.T.A. 2 : 77 ; Bak. f. l.c. 111 ; Tiss. l.c. 262. *I. grisea* Bak. (1871), partly ; Bak. f. l.c. 120, partly ; not of Desv. (1814). A much-branched herb, about 1 ft. high, with a woody base ; in savannah.
    Sen.: *Heudelot* 422 ! Kouma *Perrottet* 88 ! **Fr.Sud.:** *Scaëtta* 3209 ! **G.C.:** Salaga (Nov.) *Akpabla* 363 ! Deloro to Tankara, N.T. (Dec.) *Adams & Akpabla* GC 4324 ! **Dah.:** *Gironcourt* 116. **Fr.Nig.:** Niamey (Oct.) *Hagerup* 489 ! **N.Nig.:** Nupe *Barter* 945 ! Lokoja (Oct.) *Dalz.* 22 ! Kontagora (Jan.) *Dalz.* 24 ! Katagum *Dalz.* 419 ! **S.Nig.:** Abeokuta *Irving* ! Extends to A.-E. Sudan. (See Appendix, p. 246.) [Plants not showing compound leaves are readily confused with *I. leptoclada*.]

23. **I. trichopoda** *Lepr. ex Guill. & Perr.* Fl. Seneg. 1 : 177, t. 47 (1832) ; F.T.A. 2 : 78, Bak. f. l.c. 114. An erect annual 8–16 in. high.
    Sen.: *Heudelot* 272 ! M'bidjem *Thierry* 223 ! Tambacounda (Nov.) *Berhaut* 710 ! A variety occurs in Ubangi-Shari.

24. **I. nigritana** *Hook. f.* in Fl. Nigrit. 294 (1849) ; F.T.A. 2 : 78 ; Bak. f. l.c. 111 ; Tiss. l.c. 262 ; Cronq. l.c. 151. *I.* sp. aff. *I. suaveolens* of Chev. Bot. 177, not of Jaub. & Spach. Erect, branching, half woody herb up to 6–10 ft. high ; chiefly near rivers.
    Sen.: *Collin* 8. **Fr.Sud.:** Dogo (Mar.) *Davey* 48 ! Nyamina to Koulikoro (Oct.) *Chev.* 3235 ! **G.C.:** Burufo, N.T. (Dec.) *Adams* GC 4448 ! **Togo:** Yendi (Dec.) *Adams & Akpabla* GC 4086 ! Sao Forest (Apr.) *Asanay* 150 ! **N.Nig.:** Katagum (Oct.) *Dalz.* 23 ! R. Niger (Sept.) *T. Vogel* 7 ! **S.Nig.:** Degema *Talbot* 3762 ! Extends to A.-E. Sudan and Belgian Congo.

25. **I. latisepala** *Gillett* l.c. 575 (1956). *I. nigricans* of F.T.A. 2 : 78 ; F.W.T.A., ed. 1, 1 : 391 ; not of Vahl ex Pers. (1807). Grey bushy herb, woody below.
    N.Nig.: R. Niger *Baikie* 20 !

26. **I. cordifolia** *Heyne ex Roth* Nov. Pl. Sp. 357 (1821) ; F.T.A. 2 : 72 ; Bak. f. l.c. 102. A grey branching desert herb up to 1 ft. high.
    Maur.: *Monod* IFAN 10986 ! **Fr.Nig.:** Air *Chopard & Villiers* ! From Cape Verde Isles to Socotra, also in Afghanistan, W. Pakistan, India and Timor.

27. **I. sessiliflora** *DC.* Prod. 2 : 228 (1825) ; F.T.A. 2 : 79 ; Bak. f. l.c. 116. A grey branching herb of arid areas.
    Sen.: *Perrottet* 183 ! Walo *Heudelot* 421 ! **Fr.Sud.:** Timbuktu (July) *Chev.* 1266. *Hagerup* 174 ! 215 ! **Fr.Nig.:** Agades *De Wailly* 5242 ! 5258 ! Extends east to Eritrea, also in W. Pakistan.

28. **I. tetrasperma** *Vahl ex Pers.* Syn. 2 : 325 (1807) ; Schum. & Thonn. Beskr. Guin. Pl. 365 (1827) ; F.T.A. 2 : 72 and Bak. f. in J. Bot. 41 : 189, partly ; Bak. f. Leg. Trop. Afr. 1 : 103. *I. procera* of Chev. Bot. 176, partly, not of Schum. & Thonn. *I. simplicifolia* of F.W.T.A., ed. 1, 1 : 393, partly, not of Lam. A herb up to 4 ft. high ; in grassland, roadsides and waste places.
    Iv.C.: Middle Nzi valley, Baoulé (July) *Chev.* 22210 ! **G.C.:** Sunyani, N.W. Ashanti (Sept.) *Vigne* FH 2475 ! Achimota (June) *Irvine* 570 ! (Aug.) *Irvine* 792 ! Accra (Aug.) *Irvine* 128 ! *Thonning.* **Togo:** *Büttner* 122 !

29. **I. simplicifolia** *Lam.* Encycl. Méth. Bot. 3 : 251 (1789) ; F.T.A. 2 : 72 ; Bak. f. l.c. 103 ; Tiss. l.c. 260 ; Cronq. l.c. 129. *I. tetrasperma* of F.T.A. 2 : 72, partly ; Chev. Bot. 178 ; not of Vahl ex Pers., nor of Schum. & Thonn. Erect stiff half woody herb up to 4 ft. high.
    Sen.: Palenée *Heudelot* 196 ! **Fr.Sud.:** Nyamina to Koulikoro (Oct.) *Chev.* 3224. **Fr.G.:** Fouta Djalon (Sept.) *Chev.* 18402 ! **S.L.:** Kambia (Dec.) *Sc. Elliot* 4198 ! Newton (Dec.) *Deighton* 4390 ! Musaia (Dec.) *Deighton* 4467 ! **Lib.:** Vonjama (Oct.) *Baldwin* 10026 ! **Iv.C.:** Mankono (July) *Chev.* 22007. **N.Nig.:** Katagum (Oct.) *Dalz.* 26 ! **S.Nig.:** Ibadan (Mar.) *Schlechter* 12333 ! Sobo Plains (Sept.) *Butler-Cole* 26 ! Aguku *Thomas* 957 ! Extends to A.-E. Sudan, Uganda and Angola, also in Zanzibar and Mozambique. (See Appendix, p. 247.)

30. **I. leprieurii** *Bak. f.* in J. Bot. 41 : 190 (1903) ; Leg. Trop. Afr. 1 : 104 ; Tiss. l.c. 261 ; Cronq. l.c. 130. *I. macrocarpa* Lepr. ex Bak. F.T.A. 2 : 72 (1871) ; Chev. Bot. 175, not of Desv. (1826). *I. simplicifolia* of F.W.T.A., ed. 1, 1 : 393, partly, not of Lam. Stiffly erect herb up to 3 ft. high.
    Sen.: *Leprieur.* *Perrottet.* **Gam.:** *Hayes* 567 ! **Fr.Sud.:** Macina (Sept.) *Chev.* 24853. **Port.G.:** Bissau (Oct.) *Esp. Santo* 2555 ! (Nov.) *Esp. Santo* 1743 ! **S.L.:** Kabala (Sept.) *Thomas* 2187 ! Musaia (Mar.) *Deighton* ·5433 ! **G.C.:** Zuarungu (Sept.) *Hughes* FH 5338 ! Tamale (Feb.) *Williams* 484 ! **Togo:** Yendi (Dec.) *Adams & Akpabla* ! **N.Nig.:** Nupe *Barter* 1079 ! 1229 ! R. Niger *T. Vogel* 58 ! 94 ! **S.Nig.:** Abeokuta *Rowland* ! Also in Belgian Congo *fide* Cronquist.

31. **I. prieureana** *Guill. & Perr.* Fl. Seneg. 1 : 187 (1832) ; F.T.A. 2 : 84 ; Bak. f. l.c. 129. *I.* sp. aff. *I. sutherlandioides* of Chev. Bot. 178. *I. stenophylla* of F.W.T.A., ed. 1, 1 : 392 partly, not of Guill. & Perr. *I. komiensis* Tiss. (1931). Herbaceous, more or less woody, up to 18 in. high.
    Sen.: Galam *Heudelot* ! **Fr.Sud.:** Nyamina to Koulikoro (Sept.) *Chev.* 3226 ! **Fr.Nig.:** 20 miles N. of Zinder (Nov.) *Money Kyrle* Fe 13 ! **N.Nig.:** Sokoto (Nov.) *Moiser* 4 ! Abinsi (Sept.) *Dalz.* 622 ! Yola (Aug.) *Dalz.* 27 ! **Br.Cam.:** Bama, Dikwa Div. (Nov.) *McClintock* 13 ! Extends eastwards to N. Abyssinia. (See Appendix, p. 247.)

32. **I. stenophylla** *Guill. & Perr.* var. **stenophylla**—Fl. Seneg. 1 : 188, t. 48 (1832), incl. var. *macrocarpa*; F.T.A. 2 : 83 ; Chev. Bot. 177 ; Bak. f. l.c. 129 ; Cronq. l.c. 160. A stiffly erect, half woody herb, up to 3 ft. high.
    Sen.: Galam *Heudelot* 277 ! **Gam.:** *Hayes* 566 ! **Fr.Sud.:** Macina (Sept.) *Chev.* 24836. **Fr.G.:** Kouroussa (Aug.) *Pobéguin* 595 ! **G.C.:** Sandema *Hughes* 324 ! **N.Nig.:** Jebba *Barter* 520 ! Lokoja (Oct.) *Dalz.* 21 ! Katagum *Dalz.* 44 ! Biu (Aug.) *Noble* 2 ! Extends to A.-E. Sudan, Uganda and N.W. Tanganyika.

32a. **I. stenophylla** var. **ampla** *Sprague* in Kew Bull. 1909 : 185. A more or less woody herb up to 3 ft. high.

Fr.Sud.: *Scaëtta* 3212! **G.C.:** Tumu, N.T. (Dec.) *Adams* GC 4366! Tongo, N.T. *Adams & Akpabla!* Banda, N.T. (Dec.) *Morton* GC 25291! **S.Nig.:** Lagos (Sept.) *Dodd* 427! Abeokuta *Millen* 69! (See Appendix, p. 247.) In several respects intermediate between *I. stenophylla* and *I. prieureana.*

33. **I. mildbraediana** *Gillett* l.c. 575 (1956). A much-branched herb up to 4 ft. high.
N.Nig.: Oke Opin, Ilorin (fr. Dec.) *Ajayi* FHI 19297! Bauchi Plateau (July) *Lely* 395! (Aug.) *Lely* 594! Extends to Tanganyika and Angola.

34. **I. pobeguinii** *Gillett* l.c. 577 (1956). An erect half-woody herb.
Fr.G.: Kouroussa *Pobéguin* 976! [Resembles *I. emarginella,* but differs in its erect fruits, mucronate leaflets and subpersistent short bracts.]

35. **I. omissa** *Gillett* l.c. 580 (1956). An erect, half woody, herb, remarkable for its slender glabrous fruits.
Fr.Sud.: Macina (Sept.) *Chev.* 24900! Koulikoro (Oct.) *Chev.* 3234! Bandiagara (July) *Ganay* 62! **Dah.:** Atacora Mt. (June) *Chev.* 24071!

36. **I. suaveolens** *Jaub. & Spach* Illus. Pl. Or. t. 489 (1856) ; F.T.A. 2 : 80 ; Bak. f. l.c. 125, partly ; not of Cronq. l.c. 160. An argenteous shrublet of arid mountains, smelling of coumarin.
Fr.Nig.: Aouderas, Aïr (fr. Sept.) *Buchanan!* Also in A.-E. Sudan, Eritrea and N. Abyssinia.

37. **I. scarciesii** *Sc. Elliot* in J. Linn. Soc. 30 : 76 (1894) ; Bak. f. l.c. 98 ; F.W.T.A., ed. 1, 1 : 391. Herbaceous shoots about 8 in. long, probably from a woody rootstock.
Fr.G.: Buyabuya (Feb.) *Sc. Elliot* 4765! Diakrane, Kankan *Brossart* in *Hb. Chev.* 11673! Santa, Timbo *Chev.* 12590!

38. **I. dendroides** *Jacq.* Ic. Pl. Rar. 3, t. 571 (1788–9) ; F.T.A. 2 : 100 ; Chev. Bot. 173 ; Bak. f. l.c. 128 ; Cronq. l.c. 166, t. 13. *I. sesbaniifolia* A. Chev. (1912)—Bak. f. l.c. 128. *I. dalabaca* Chev. Bot. 172, name only. *I. phyllanthoides* of Tiss. l.c. 266, not of Bak. An elegant, erect, wiry, herb, up to 3 ft. high. Varies considerably ; forms with more numerous leaflets and flowers having been distinguished as *I. sesbaniifolia.*
Sen.: *Heudelot* 219. **Fr.Sud.:** Goundam (Aug.) *Chev.* 3249. **Port.G.:** Bissau (Oct.) *Esp. Santo* 2535! **Fr.G.:** Nzérékoré (Oct.) *Baldwin* 9683! Fouta Djalon (Aug., Oct.) *Chev.* 18318! 18359. **S.L.:** Kenema (Oct.) *Deighton* 5204! **Iv.C.:** Mt. Kangoroma (July) *Chev.* 22181! **G.C.:** Ejura (Nov.) *Andoh* FH 5439! **Togo:** *Büttner* 140! Lome *Warnecke* 62! 375! **N.Nig.:** Zaria (Sept.) *Olorunfemi* FHI 24368! Biu (Aug.) *Noble* 7! **S.Nig.:** Ibadan (Jan.) *Meikle* 967! Throughout the non-arid parts of tropical Africa. (See Appendix, p. 246.)

39. **I. heudelotii** *Benth. ex Bak.* var. **heudelotii**—in F.T.A. 2 : 85 (1871), partly, emend. Gillett l.c. 577 (1956); Bak. f. l.c. 124 partly. *I. sofa* Sc. Elliot (1894)—Chev. Bot. 177. *I. djalonica* A. Chev. ex Hutch. & Dalz. F.W.T.A., ed. 1, 1 : 391 (1928) ; not of Bak. f. l.c. 147 (1926). An erect, branched, softly woody plant, up to 6 ft. high ; in margins of evergreen forest.
Port.G.: Bissau (Jan.) *Esp. Santo* 1714! **Fr.G.:** Rio Nunez *Heudelot* 611 (partly)! Kouroussa (Nov.) *Pobéguin* 598! Fouta Djalon (Sept.) *Chev.* 18584 (partly)! **S.L.:** Falaba *Sc. Elliot* 5083! Kofia Mt. *Sc. Elliot* 4613! Musaia *Sc. Elliot* 5080! Dunnia *Sc. Elliot* 4831! Jaiama (Dec.) *Deighton* 3462! **N.Nig.:** Vom *Dent Young* 49! Kafanchan (Dec.) *Sampson* 42! Bauchi Plateau (Sept.) *Lely* 698! Kaduna (Dec.) *Meikle* 763! **S.Nig.:** Ikwette, Obudu (Dec.) *Savory & Keay* FHI 25260! **Br.Cam.:** Babanki, Bamenda *Maitland* 1720! *Johnstone* 4!

39a. **I. heudelotii var. fairchildii** (*Bak. f.*) *Gillett* l.c. 577 (1956). *I. fairchildii* Bak. f. in J. Bot. 70 : 252 (1932).
Port.G.: Bafata (Mar.) *Esp. Santo* 2908! **Fr.G.:** Fouta Djalon *Fairchild* 1275!

40. **I. elliotii** (*Bak. f.*) *Gillett* l.c. 578 (1956). *I. heudelotii* var. *elliotii* Bak. f. in J. Bot. 41 : 261 (1903) ; Leg. Trop. Afr. 1 : 124. *I. heudelotii* Benth. ex Bak. (1871), partly ; F.W.T.A., ed. 1, 1 : 391, partly; Berhaut l.c. 52. A softly woody plant up to 5 ft. high.
Sen.: Sangalkam *Berhaut* 960! **Fr.G.:** Rio Nunez *Heudelot* 611 (partly)! **S.L.:** *Morson!* Luseniya *Sc. Elliot* 4063! Sherbro *Sc. Elliot* 5841! Kambia (Dec.) *Deighton* 892! Rokupr (Nov.) *Jordan* 676! **Lib.:** Monrovia (Nov.) *Linder* 1553!

41. **I. wituensis** *Bak. f.* var. **occidentalis** *Gillett* l.c. 578 (1956). A slender branching sticky annual 18 in. high.
N.Nig.: Bauchi Plateau (Sept.) *Lely* P. 693!

42. **I. congolensis** *De Wild. & Th. Dur.* in Bull. Herb. Boiss., Sér. 2, 1 : 11 (1901) ; Bak. f. l.c. 126. *I. geminata* of Chev. Bot. 173, partly ; F.W.T.A., ed. 1, 1 : 391, partly, not of Bak. (1871). *I. sparsa* of Cronq. l.c. 158, not of Bak. (1871). *I. sparsa* var. *bongensis* of Tiss. l.c. 263, partly, not of Bak. f. (1903). *I. adami* Berhaut (1955). Annual herb about 8 in. high.
Sen.: Galam *Heudelot* 265! Koutango to Koular *Adam* 10148. Ouassadou *Berhaut* 1132! **Gam.:** Hayes 560! **Fr.Sud.:** Nyamina to Koulikoro (Oct.) *Chev.* 3219! **Fr.Nig.:** Niamey (Oct.) *Hagerup* 499! **N.Nig.:** Lokoja (Oct.) *Dalz.* 20! Bauchi Plateau (July) *Lely* P. 539! (Aug.) *Lely* P. 646! Bichikki (Sept.) *Lely* 629! Anara F.R., Zaria (Oct.) *Keay* FHI 20144! **S.Nig.:** Sobo Plains *A. F. Ross* FHI 12067! Also in Ubangi-Shari, Belgian Congo and Uganda.
[The type of *I. adami* was not available in the Paris herbarium at the time of writing : the opinion here expressed as to its identity with *I. congolensis* is thus based on the description only.]

43. **I. geminata** *Bak.* in F.T.A. 2 : 81 (1871) ; Bak. f. l.c. 121, partly ; Tiss. l.c. 265 ; Berhaut l.c. 52, partly. *I.* sp. aff. *geminata* of Chev. Bot. 173. Branching, semiwoody herb up to 2 ft. high.
Sen.: ? **Iv.C.:** Mossi (Aug.) *Chev.* 24772! **Dah.:** Cotonou (Jan.) *Chev.* 22694! **N.Nig.:** Nupe *Barter* 966! **S.Nig.:** Lagos (May) *Dalz.* 1221! *Rowland!*
[This plant and *I. congolensis* have been much confused. The description given by Berhaut l.c. fits this species but the specimen cited is *I. congolensis* in fruit.]

44. **I. pilosa** *Poir.* in Lam. Encycl. Méth. Bot., Suppl. 3 : 151 (1813) ; F.T.A. 2 : 82 ; Bak. f. l.c. 120, excluding var. *multiflora;* Brenan in Mem. N.Y. Bot. Gard. 8 : 251 ; Berhaut l.c., partly. A spreading woody herb up to 8 in. high.
Sen.: *Perrottet.* **Fr.Sud.:** Gao (Sept.) *Hagerup* 359! **Port.G.:** Bissau (Oct.) *Esp. Santo* 2554! **G.C.:** Sandema (Sept.) *Hughes* 326! Accra *Thonning.* **N.Nig.:** Nupe *Barter* 977! Katagum (Oct.) *Dalz.* 31! Sokoto (Sept.) *Dalz.* 335! *Moiser* 89! Bichikki (Nov.) *Lely* 628! Eastwards to Eritrea, with a variety in Angola. (See Appendix, p. 247.)

45. **I. fulvopilosa** *Brenan* in Mem. N.Y. Bot. Gard. 8 : 250 (1953) ; Cronq. l.c. 140. *I. pilosa* var. *multiflora* Bak. f. in J. Bot. 41 : 243 (1903). *I. pilosa* of F.W.T.A., ed. 1, 1 : 391, partly ; Berhaut l.c., partly. Branching, half woody herb, up to 3 ft. high.
S.L.: Musaia to Kabala (Sept.) *Deighton* 4870! Warantamba (Oct.) *Small* 356! **G.C.:** *Morton* GC 25125! **N.Nig.:** Naraguta & Jos (Aug.) *Lely* 556! Vom *Dent Young* 50! Bauchi Plateau (Aug.) *Lely* P. 571! (Sept.) *Lely* P. 687! Extends to Uganda, Tanganyika, Nyasaland and S. Rhodesia.

46. **I. mimosoides** *Bak.* in F.T.A. 2 : 90 (1871) ; Bak. f. Leg. Trop. Afr. 122. *I. brevipetiolata* Cronq. l.c. 145. A branching, somewhat woody, sparingly glandular, herb, up to 4 ft. high ; on edges of montane forest.
Br.Cam.: Bamenda (Oct.) *Johnstone* 182! In moister montane regions extending to Abyssinia, E. Africa, S. Rhodesia and Angola.
[A most variable species, closely allied to the next.]

47. **I. colutea** (*Burm. f.*) *Merrill* in Phil. J. Sci. 19 : 355 (1921). *Galega colutea* Burm. f. (1768). *Indigofera viscosa* Lam. (1789)—F.T.A. 2 : 81 ; Bak. f. l.c. 123 ; F.W.T.A., ed. 1, 1 : 391 ; Tiss. l.c. 265 ; Cronq. l.c. 144. A half woody, viscid, branching herb, about 18 in. high.
Sen.: Cayor *Leprieur!* *Lelièvre!* **Fr.Sud.:** Timbuktu (Aug.) *Hagerup* 243! **N.Nig.:** *Barter* 1051! Bauchi Plateau *Batten Poole* FHI 13175! A variable species, extending through Arabia, India and Indonesia to Australia and south to the Transvaal.

48. **I. barteri** *Hutch. & Dalz.* F.W.T.A., ed. 1, 1 : 391 (1928) ; Kew Bull. 1929 : 16 ; Bak. f. l.c. 880. A branching slightly woody herb up to 18 in. high.

G.C.: L. Sorri, N.T. (Dec.) *Morton* GC 25017! Daboya Ferry, N.T. (Dec.) *Morton* GC 6197! **N.Nig.:** Jebba *Barter*! Nupe *Barter* 951! 1604!
[Closely related to the next species.]

49. **I. secundiflora** *Poir.* in Lam. Encycl. Méth. Bot., Suppl. 3 : 148 (1813) ; Bak. F.T.A. 2 : 94 ; Bak. f. l.c. 152 ; Cronq. l.c. 145. A grey spreading glandular herb up to 1 ft. high.
Sen.: *Heudelot* 227! **Fr.Sud.:** Koulikoro *Chev.* 3232! **G.C.:** Prampram (Oct.) *Morton* A. 78! **Fr.Nig.:** Niamey (Oct.) *Hagerup* 513! 120 miles S. of Agadez (Nov.) *Money Kyrle* Fe 3! **N.Nig.:** Sokoto (Nov.) *Moiser* 32! 45! Ilorin *Bailey* 5! Funtua (Dec.) *Meikle* 821! Kalkala, L. Chad (Oct.) *Gwynn* 90! Extends eastwards to Abyssinia : a variety southwards to Tanganyika, Nyasaland and N. Rhodesia. (See Appendix, p. 247.)

50. **I. argentea** *Burm. f.* Fl. Ind. 171 (1768), not of Linn. (1771) ; Bak. f. l.c. 152 ; Tiss. l.c. 270. *I. semitrijuga* of Bak. F.T.A. 2 : 93, not of Forsk. A silvery desert shrublet, very sparingly glandular.
Fr.Sud.: Oualata *Jumelle*! Extends from Mauritania to Somaliland, Arabia and W. Pakistan.

51. **I. conferta** *Gillett* l.c. 579 (1956). *I. grisea* Bak. in F.T.A. 2 : 80 (1871) mostly, not of Desv. (1826). *I. bracteolata* of F.W.T.A., ed. 1, 1 : 393, partly, not of DC. A grey branching herb up to 2 ft. high.
N.Nig.: Jebba, Nupe *Barter* 940! Oro F.R., Ilorin *Ejiofor* FHI 29389!

52. **I. atriceps** *Hook. f.* subsp. **atriceps**—in J. Linn. Soc. 7 : 190 (1864) ; F.T.A. 2 : 94 ; Bak. f. l.c. 148 ; Cronq. l.c. 165. A straggling half woody herb up to 2½ ft. high ; on edges of montane forest.
Br.Cam.: Cam. Mt., 6,500–8,500 ft. (Jan.) *Mann* 1303! *Johnston* 68! *Dunlap* 208! Extends on mountains to Uganda, Tanganyika and Nyasaland.

52a. **I. atriceps** subsp. **alboglandulosa** (*Engl.*) *Gillett* l.c. 580 (1956). *I. alboglandulosa* Engl. Hochgebirgsflora Afr. 258 (1892) ; Bak. f. l.c. 152. *I. setosissima* Harms var. *major* Cronquist Bull. Jard. Bot. Brux. 22 : 226 (1952) ; Fl. Congo Belge 5 : 147. *I. djalonica* Chev. ex Bak. f. l.c. 147 (1926), not of F.W.T.A., ed. 1, 1 : 391.
Fr.G.: Dalaba-Diaguissa Plateau (Sept.) *Chev.* 18584 (partly)! **S.L.:** Bintumane Peak, 3,800 ft. *Nichols* 10! **Br.Cam.:** Cam. Mt., 3,000–8,000 ft. (Nov.–Jan.) *Mildbr.* 10831! *Migeod* 132! *Maitland* 295! 808! *Keay* FHI 28591! Commoner than subsp. *atriceps* throughout its range.

53. **I. garckeana** *Vatke* in Oest. Bot. Zeit. 29 : 221 (1879) ; Bak. f. l.c. 150 ; Cronq. l.c. 138. *I. rhynchocarpa* Welw. ex Bak. var. *quadrangularis* Berhaut Fl. Sen. 50 and Bull. Soc. Bot. Fr. 101 : 375 (1955). A shrub with large leaflets, up to 6 ft. high.
Sen.: Ouassadou *Berhaut* 1219! Also from A.-E. Sudan and from Eritrea to S. Rhodesia.

54. **I. emarginella** *Steud. ex A. Rich.* Tent. Fl. Abyss. 1 : 184 (1847) ; F.T.A. 2 : 99 ; Bak. f. l.c. 157 ; Cronq. l.c. 135. Woody undershrub up to 5 ft. high ; rarely below 3,000 ft. alt.
N.Nig.: Vom *Dent Young* 44! Naraguta (June) *Lely* 261! Extending eastwards to Abyssinia and southwards to S. Rhodesia.

55. **I. macrophylla** *Schum.* Beskr. Guin. Pl. 372 (1827) ; F.T.A. 2 : 100 ; Bak. f. l.c. 158 ; Tiss. l.c. 271. *I. binderi ?* of Chev. Bot. 172, not of Kotschy. Scrambling or climbing shrub, on margins of evergreen forest.
Sen.: Casamance *Chev.* 3238! **Port. G.:** Bissau *Esp. Santo* 1650! **Fr.G.:** *Heudelot* 603! *Chev.* 14833; 14844. **S.L.:** Kabala (Sept.) *Thomas* 2344! Rokupr (Oct.) *Jordan* 138! Njala (Oct.) *Deighton* 1404! **Lib.:** Bonuto (Oct.) *Linder* 887! Vonjama (Oct.) *Baldwin* 9946! **Iv.C.:** Bingerville *Chev.* 17343. **G.C.:** Accra (Oct.) *Dalz.* 94! Aburi Hills (Oct.) *Johnson* 470! **Togo:** Lome *Warnecke* 425! **Dah.:** Le Testu 213. **N.Nig.:** Lokoja *Barter*! (Oct.) *Dalz.* 3! **S.Nig.:** Lagos *MacGregor* 267! Ibadan (Oct.) *Jones* FHI 13704! Calabar (Sept.) *Holland* 133! (See Appendix, p. 247.)

56. **I. subulata** *Vahl ex Poir.* var. **subulata**—in Lam. Encycl. Méth. Bot., Suppl. 3 : 150 (1813) ; F.T.A. 2 : 87, partly ; Bak. f. l.c. 172, partly ; Tiss. l.c. 268 ; Meikle in Kew Bull. 1950 : 352 ; Cronq. l.c. 161, partly. A sprawling herb, usually a weed of disturbed ground.
Sen.: *Leprieur. Perrottet*. **S.L.:** Njala (Apr.) *Deighton* 1912! Rokupr *Jordan* 37! **Lib.:** Suacocco (Oct.) *Daniel* 20! **G.C.:** Accra *Thonning*! Achimota (Oct.) *Milne-Redhead* 5174! **Togo:** Lome *Warnecke* 265! 425! Kpeve *Irvine* 1735! (Oct.) *Williams* 711! **S.Nig.:** Ibadan (Dec.) *Newberry* 136! *Thomas* A. 12! *Thornewill* 89! Throughout tropical Africa and in Natal. A variable species.

56a. **I. subulata** var. **scabra** (*Roth*) *Meikle* in Kew Bull. 1950 : 352. *I. scabra* Roth Nov. Pl. Sp. 359 (1821). *I. retroflexa* Baill. (1883)—Bak. f. l.c. 145. *I. subulata* of F.T.A. 2 : 87, etc., partly. Lax, erect herb, up to 2 ft. high.
N.Nig.: Jos (Aug.) *Keay* FHI 20195! Extends east to India and Madagascar, south to the Transvaal ; introduced in America.

57. **I. coerulea** *Roxb.* Fl. Ind. 3 : 377 (1832). *I. articulata* of Bak. f. l.c. 154, partly, not of Gouan. An erect, silvery, softly woody plant of arid areas, up to 6 ft. high, sometimes cultivated. **Fr.Nig.:** Aïr *Chopard & Villiers*! Extends eastwards through Arabia to W. Pakistan and India, southwards to N. Kenya.

58. **I. arrecta** *Hochst. ex A. Rich.* Tent. Fl. Abyss. 1 : 184 (1847) ; F.T.A. 2 : 97 ; Bak. f. l.c. 155 ; Tiss. l.c. 271 ; Cronq. l.c. 162, t. 12. *I. tinctoria* of Chev. Bot. 178, partly, not of Linn. *I. tinctoria* var. *arrecta* Berhaut l.c. 50. Erect, branched, softly woody plant, up to 8 ft. high ; much cultivated.
Sen.: *Perrottet* 176! **Gam.:** *Dudgeon* 91! **Fr.Sud.:** San *Chev.* 1029! **Port.G.:** Mansoa (Oct.) *Esp. Santo* 2318! **Fr.G.:** *Heudelot* 686! **G.C.:** Accra (Jan.) *Irvine* 1109! **N.Nig.:** Sokoto *Dalz.* 329! Katagum (Oct.) *Dalz.* 57! Zaria *Meikle* 833! **S.Nig.:** Lagos *MacGregor* 144! Meko *Johnson*! Extends to S. Arabia and eastern Cape Province. Often hardly separable from *I. tinctoria* in W. Africa. In E. Africa it is more distinct. (See Appendix, p. 245.)

59. **I. suffruticosa** *Mill.* Gard. Dict., ed. 8 : 2 (1768) ; Bak. f. l.c. 135 ; Tiss. l.c. 271 ; Cronq. l.c. 162. *I. anil* Linn. (1771) ; F.T.A. 2 : 98. *I. suffruticosa* var. *uncinata* G. Don of Berhaut l.c. 50. A branched undershrub up to 6 ft. high ; formerly much cultivated, probably not indigenous.
Sen.: *Heudelot* 480. **Port.G.:** Catió (Oct.) *Esp. Santo* 2278! **S.L.:** Makunde (Apr.) *Sc. Elliot* 5712! Njala (June) *Deighton* 731! **Lib.:** Gbanga (Dec.) *Baldwin* 10532! Peahtah (Oct.) *Linder* 926! **S.Nig.:** Shasha F.R. (Apr.) *Jones & Onochie* FHI 17244! *Barter*! Also (introduced) in central and south central tropical Africa and Madagascar ; native of America. (See Appendix, p. 247.)

60. **I. tinctoria** *Linn.* Sp. Pl. 2 : 751 (1753) ; F.T.A. 2 : 99 ; Chev. Bot. 178, partly ; Bak. f. l.c. 156 ; Tiss. l.c. 271 ; including var. *torulosa* Bak. f. l.c. 157. Softly woody plant up to 6 ft. high, formerly much cultivated, perhaps indigenous.
Sen.: *Perrottet*. **Gam.:** *Brown-Lester* 49! **Fr.Sud.:** Macina (Sept.) *Chev.* 24870! Dende (July) *Chev.* 1153! **Port.G.:** Mansoa (Oct.) *Esp. Santo* 2323! **Fr.G.:** Kouroussa (Oct.) *Pobéguin* 327! **G.C.:** Accra (Feb.) *Johnson* 1074! **Togo:** Lome *Warnecke* 262! **Dah.:** *Debeaux* 161. **N.Nig.:** Jebba *Barter*! Sokoto (Aug.) *Dalz.* 327! Katagum (Sept.) *Dalz.* 58! (Nov.) *Dalz.* 59! Damaturu (July) *Ujor* FHI 23906! **S.Nig.:** Lagos (Dec.) *Dalz.* 1218! Extends through Arabia to India, the Philippines and New Guinea, also in east and south tropical Africa but apparently absent from the Congo. (See Appendix, p. 248.)

61. **I. longebarbata** *Engl.* Hochgebirgsflora Afr. 257 (1892) ; Bak. f. l.c. 147 ; Cronq. l.c. 141. Erect, sub-woody herb, up to 3 ft. high.
N.Nig.: Naraguta (Aug.) *Lely* 510! *Keay* FHI 20063! Vom *Dent Young* 45! Bauchi Plateau (Aug.) *Lely* P. 577! **Br.Cam.:** Bambulue, Bamenda *Johnstone* 221/31! Usually above 4,800 ft., from Abyssinia to Angola and east Cape Province.
[Closely related to *I. hirsuta* and replacing it at higher altitudes ; intermediate plants occur.]

62a. **I. hirsuta** *Linn.* var. **hirsuta**—Sp. Pl. 2 : 1062 (1753) ; F.T.A. 2 : 88 ; Bak. f. l.c. 146 ; Tiss. l.c. 268 ; Cronq. l.c. 140, t. 10. Erect or straggling herb up to 3 ft. high, a weed.

Sen.: *Lelièvre*! **Gam.:** *Brown-Lester* 47! **Fr.Sud.:** *Chev.* 3241 *bis. fide* Tiss. *l.c.* **Fr.G.:** Kouroussa (Aug.) *Pobéguin* 382! **S.L.:** Falaba (Apr.) *Sc. Elliot* 5437! Bonthe Isl. (Nov.) *Deighton* 2319! York *Deighton* 4585! **Lib.:** Monrovia *Johnston*! **Iv.C.:** *Chev.* 17342. **G.C.:** Bjury (July) *Chipp* 511! Achimota *Irvine* 656! Elmina (Aug.) *Andoh* FH 5557! **Togo:** Lome *Warnecke* 343! **Dah.:** Allada (Mar.) *Chev.* 23376! **N.Nig.:** Sokoto *Lely* 137! Katagum (Dec.) *Dalz.* 28! **S.Nig.:** Lagos (Sept.) *Dawodu* 8! Ibadan (Sept.) *Tamajong* FHI 20955! Aguku *Thomas* 973! Eastwards to China and N. Australia, southwards to Bechuanaland. (See Appendix, p. 246.)

62a. **I. hirsuta** var. **pumila** *Welw. ex Bak.* in F.T.A. 2 : 89 (1871) ; Bak. f. l.c. 146.
    N.Nig.: Bauchi Plateau *Batten-Poole* FHI 13187! Also in Angola.
    [In several respects intermediate between *I. hirsuta* and *I. longebarbata.*]

63. **I. astragalina** *DC.* Prod. 2 : 228 (1825) ; F.T.A. 2 : 89 ; Bak. f. l.c. 147. Erect or straggling grey herb, about 18 in. high.
    Sen.: *Perrottet*! *Heudelot* 255! 411! **Fr.Sud.:** Goura, Gao (Sept.) *De Wailly* 5212! **Fr.Nig.:** 170 miles S. of Agadez (Nov.) *Money Kyrle* Fe 5! **N.Nig.:** Sokoto *Lely* 135! (Nov.) *Moiser* 54! Fodama (Dec.) *Moiser* 200! Maifoni *Parsons*! Katagum (Oct.) *Dalz.* 27! Extends eastwards to Madras and southwards to the Transvaal. (See Appendix, p. 246.)
    [Closely related to *I. hirsuta* but occupying, on the whole, drier areas ; intermediates occur where the two species overlap.]

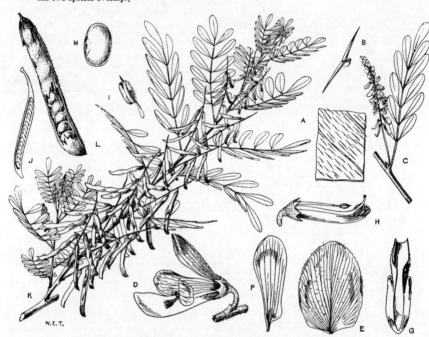

FIG. 165.—INDIGOFERA ARRECTA *Hochst. ex A. Rich.* (PAPILIONACEAE).

A, leaf-surface showing hairs.   B, hair.   C, inflorescence and leaf.   D, flower.   E, standard. F, wing petal.   G, keel.   H, flower with corolla removed.   I, anther.   J, pistil.   K, fruiting branch.   L, pod.   M, seed.

64. **I. deightonii** *Gillett* l.c. 580 (1956). A weak, spreading herb, up to 18 in. high ; a weed, or on rocky outcrops.
    Fr.G.: Nzérékoré, Mt. Koiré (Sept.) *Baldwin* 13301! **S.L.:** Magbile (Dec.) *Thomas* 6186! Masactaba, N. Prov. (Oct.) *Glanville* 34! Newton (Nov.) *Deighton* 3283! Musaia (Sept.) *Deighton* 4864! Rokupr (Oct.) *Jordan* 367! **Lib.:** Sanokwele (Sept.) *Baldwin* 9436! **N.Nig.:** Jebba *Barter*! **S.Nig.:** Aponmu, Akure (Nov.) *Latilo & Olorunfemi* FHI 24446! Also in A.-E. Sudan.

65. **I. microcarpa** *Desv.* in Journ. de Bot. 3 : 79 (1814) ; Bak. f. l.c. 153 ; Tiss. l.c. 270. *I. perrottetii* DC. (1825)—F.T.A. 2 : 94 ; Chev. Bot. 176. Herb, with a strong taproot and prostrate branches, often a weed, doubtfully native.
    Sen.: Walo *Heudelot*! *Perrottet*! **Fr.Sud.:** Ségou (Sept.) *Chev.* 3230! Macina (Aug.) *Chev.* 24930! **G.C.:** Babile, Lawra (Oct.) *Hinds* 5002! **Togo:** Paliba *Adams* GC 4124! **Fr.Nig.:** Zinder (Oct.) *Hagerup* 584! **N.Nig.:** Katagum (Oct.) *Dalz.* 56! Also in E. Africa, S. tropical Africa and Madagascar, but native, probably, only in America.

66. **I. diphylla** *Vent.* Choix. t. 30 (1803) ; F.T.A. 2 : 74, mostly ; Bak. f. l.c. 110 ; F.W.T.A., ed. 1, 1 : 393, excl. syn. ; Tiss. l.c. 262. A semiwoody, spreading, grey herb of the drier areas.
    Sen.: *Perrottet*! *Leprieur*! *Sieber* 38! *Heudelot* 419! *Lelièvre*! **Fr.Sud.:** Fereirbo (Aug.) *Chev.* 3241! **Togo:** Lome *Warnecke* 329! **Dah.:** Gironcourt 89. **Fr.Nig.:** Niamey (Oct.) *Hagerup* 512! **N.Nig.:** Sokoto (Nov.) *Moiser* 15! Fodama (Dec.) *Moiser* 198! Katagum (July) *Dalz.* 30! Eastwards to A.-E. Sudan. (See Appendix, p. 246.)

67. **I. oblongifolia** *Forsk.* Fl. Aegypt.-Arab. 137 (1775) ; Bak. f. l.c. 138. *I. paucifolia* Del. (1812)—F.T.A. 2 : 88 ; Chev. Bot. 175. An erect silvery glaucous shrublet of arid areas, up to 6 ft. high.
    Sen.: Walo *Heudelot*! *Perrottet*! *Lelièvre*! **Fr.Sud.:** Timbuktu *Chev.* 1212. **Fr.Nig.:** Zinder (Oct.) *Hagerup* 578! **N.Nig.:** Sokoto (Nov.) *Dalz.* 328! Kalkala, L. Chad *Golding* 63! Eastward to Egypt, Arabia, India, Ceylon and Java, also in Angola.

68. **I. spicata** *Forsk.* l.c. 138 (1775). *I. hendecaphylla* Jacq. (1788) ; by mistake *endecaphylla* in Bak. F.T.A. 2 : 96 ; Bak. f. l.c. 134 ; F.W.T.A., ed. 1, 1 : 392 ; Tiss. l.c. 267 ; Cronq. l.c. 149. Suberect or prostrate herb, woody below ; a weed of disturbed ground sometimes cultivated.

Sen.: *Perrottet* 197 ! **Fr.Sud.:** Ségou *Chev.* 3231. **Port.G.:** Boé (Jan.) *Esp. Santo* 2862 ! **Lib.:** Cape Palmas (July) *T. Vogel* 4 ! Grand Bassa (July) *T. Vogel* 106 ! Monrovia (May) *Barker* 1311 ! **G.C.:** Keta *Thonning.* Achimota (June) *Irvine* 671 ! **Dah.:** *Burton* ! **N.Nig.:** Rinjim Mukr (May) *Lely* 220 ! **S.Nig.:** Ibadan (July) *Tamajong* FHI 16793 ! Eastwards to the Philippines and Indonesia, southwards to Madagascar and Natal. A very variable plant. (See Appendix, p. 246.)

69. **I. kerstingii** *Harms* in Engl. Bot. Jahrb. 45 : 308 (1910) ; Bak. f. l.c. 98. *I. oxalidea* of Tiss. l.c. 267, partly, not of Welw. ex Bak. A prostrate wayside herb.
   Togo: Basari *Kersting* 662 ! Paliba *Adams & Akpabla* GC 4122 ! **N.Nig.:** Lokoja (Oct.) *Dalz.* 19 ! Abinsi (July) *Dalz.* 580 ! Zaria (Sept.) *Olorunfemi* FHI 24378 ! **S.Nig.:** Oyo (Sept.) *Latilo* FHI 23548 !

70. **I. conjugata** *Bak.* var. **conjugata**—in F.T.A. 2 : 75 (1871) ; *Chev.* Bot. 172 ; Bak. f. l.c. 110 ; Cronq. l.c. 150. *I. dalzielii* Hutch. (1921), mostly ; Bak. f. l.c. 108, mostly ; F.W.T.A., ed. 1, 1 : 393, mostly. *I. knoblecheri* of Chev. Bot. 174, not of Kotschy. Herb, erect, up to 18 in. high, from a stout woody rootstock.
   Fr.G.: Balato (Feb.) *Chev.* 340 ! Coumana (Feb.) *Chev.* 364 ! **Dah.:** Djougou (June) *Chev.* 23913 ! **N.Nig.:** Jos (Mar.) *Hill* 4 ! Naraguta *Lely* 20 ! Vom *Dent Young* 42 ! Birnin Gwari (Mar.) *Meikle* 1348 ! Kaduna (Dec.) *Meikle* 771 ! Eastward to A.-E. Sudan, Kenya and Tanganyika, with another variety in Angola and N. Rhodesia.

70a. **I. conjugata** var. **occidentalis** *Gillett* l.c. 581 (1956). *I. dalzielii* Hutch. (1921), partly.
   Fr.Sud.: Bougouni (Feb.) *Roberty* 3496 ! **N.Nig.:** Kontagora (Oct.) *Dalz.* 27 ! Liruwen-Kano Hills *Carpenter* ! Bichikki (May) *Lely* 188 ! Birnin Gwari (Aug.) *Keay* FHI 25999 ! Giwa (June) *Keay* FHI 25909 ! **Br.Cam.:** Gashaka Dist. (Feb.) *Latilo & Daramola* FHI 34470 !

71. **I. costata** *Guill. & Perr.* Fl. Seneg. 1 : 187 (1832) ; F.T.A. 2 : 95 ; Bak. f. l.c. 163. *I. theuschii* of Tiss. l.c. 272, partly, not of Hoffm. An erect herb about 18 in. high, in subarid areas.
   Sen.: *Perrottet* ! *Huard* in Heudelot 508 ! Diohine (Sept.) *Berhaut* 569 ! **Fr.Nig.:** *Gaillard* in *Mission Tilho* ! **N.Nig.:** Biu, Bornu (Sept.) *Noble* 28 ! Extends to A.-E. Sudan, with subspecies in eastern and southern Africa from Abyssinia to Angola and the eastern Cape.

72. **I. parviflora** *Heyne ex Wight & Arn.* var. **occidentalis** *Gillett* l.c. 582 (1956). *I. linearis* of Guill. & Perr. Fl. Seneg. 184 (1832), not of DC. (1825) ; Bak. f. l.c. 162 ; Tiss. l.c. 272. *I. parviflora* of F.T.A. 2 : 83, partly ; Chev. Bot. 175, partly ; F.W.T.A., ed. 1, 1 : 391, partly ; Berhaut l.c. 52. A branched annual up to 1 ft. high, in subarid regions.
   Sen.: *Leprieur* ! *Heudelot* 525 ! Basgny (Sept.) *Berhaut* 1731 ! **Fr.Sud.:** *fide* Chev. *l.c.* Also in the Cape Verde Isles, Angola, N. Rhodesia and Bechuanaland ; var. *parviflora* in India, Australia and eastern Africa from the A.-E. Sudan to the Orange Free State. (See Appendix, p. 247.)

73. **I. welwitschii** *Bak.* var. **remotiflora** (*Taub. ex Bak. f.*) *Cronq.* l.c. 172 (1954). *I. remotiflora* Taub. ex Bak. f. l.c. 121 (1926). Delicate straggling herb, on rocky hills in areas of high rainfall.
   N.Nig.: Panshanu, rocks 35 miles E. of Naraguta (Sept.) *Lely* 610 ! Bauchi Plateau (July) *Lely* P. 504 ! **S.Nig.:** Idanre Hills, Ondo (Oct.) *Keay & Onochie* FHI 20247 ! *Keay* FHI 22593 ! Extends to Tanganyika, Nyasaland and Angola.

74. **I. longicalyx** *Gillett* l.c. 582 (1956). *I. lateritia* A. Chev. Bot. 174, name only, not of Willd. *Microcharis tenella* of Bak. f. l.c. 256 and F.W.T.A., ed. 1, 1 : 388, partly, not of Benth. An annual, about 8 in. high.
   Fr.G.: Fouta Djalon (Oct.) *Chev.* 18873 ! **S.L.:** Kabala (Sept.) *Thomas* 2265 ! **N.Nig.:** Jos Plateau (Aug.) *Lely* P. 643 ! Naraguta (Aug.) *Lely* 504 ! Vom *Dent Young* 51 ! Also in Ubangi.

75. **I. hutchinsoniana** *Gillett* l.c. 584 (1956). *Microcharis tenella* Benth. in Trans. Linn. Soc. 25 : 297, t. 33A (1866) ; F.T.A. 2 : 132 ; Bak. f. l.c. 256, partly, excl. syn. ; F.W.T.A., ed. 1, 1 : 388, partly. *Indigofera welwitschii* of F.W.T.A., ed. 1, 1 : 393, not of Bak. An annual, up to 1 ft. high ; on rocks after rain.
   N.Nig.: Nupe *Barter* 1601 ! Jebba *Barter* ! R. Niger (Sept.) *T. Vogel* 99 ! Lokoja (Oct.) *Parsons* L. 34 !

76. **I. disjuncta** *Gillett* l.c. 584 (1956). *I. arenaria* of F.T.A 2 : 79, mostly, not of A. Rich., nor of E. Mey. A desert annual.
   Fr.Nig.: Mt. Tarraouagi, Aïr *Chopard & Villiers* ! Local from Mauritania to the Hejaz and in S.W. Africa and N.W. Cape Colony, a variety in Socotra.
   The species with glandular hairs (Nos. 43–49) form a difficult group to which belong several undetermined sheets at Kew, some of which may represent undescribed species. It is to be hoped that collectors will pay especial attention to this group, taking care to secure fruits as well as flowers.

Imperfectly known species.

I. heterocarpa *Welw. ex Bak.* F.T.A. 2 : 90 (1871). An imperfectly understood Angolan species is reported (F.W.T.A., ed. 1, 1 : 391) from Lokoja, N. Nigeria. The specimen, *Parsons* 12, has not been found.

I. variabilis ? is reported by Berhaut (l.c. 15, 265) from Senegal ; his specimens have not been seen. Presumably *I. variabilis* N.E. Br. is meant ; the correct name for which is *I. bainesii* Bak. This species has hitherto only been found in Bechuanaland and the Transvaal.

Besides the above, *I. galegoides* DC., an East Asian species, has been grown as a cover crop at Victoria, British Cameroons.

27. CYAMOPSIS DC.—F.T.A. 2 : 65.

Leaflets oblong to linear-subulate, rarely obovate, 3–7 to each leaf, mucronate, 1·5–4 cm. long, 2–5 mm. broad, thinly silky with medifixed hairs beneath or white-pubescent on both surfaces ; stipules subulate ; flowers in lax axillary racemes ; bracts 4 mm. long, linear ; corolla 5 mm. long, the standard petals shortest ; fruit linear, 2·5–4 cm. long, about 5 mm. broad, finely appressed-setulose, rather abruptly beaked, light brown when dry .. .. *senegalensis*

C. senegalensis *Guill. & Perr.* Fl. Seneg. 1 : 171, t. 45 (1832) ; F.T.A. 2 : 65 ; A. Chev. in Rev. Bot. Appliq. 19 : 246 (1939). *C. stenophylla* (Bonnet) A. Chev. l.c. (1939). *C. senegalensis* Guill. & Perr. var. *stenophylla* Bonnet in Bull. Soc. Bot. Fr. 11, Mém. 20 : 6 (1911). Erect herb about 1 ft. high, woody at the base ; flowers pinkish.
   Sen.: Dagana, Walo Dist. *fide* Guill. & Perr. *l.c.* **Gam.:** Albreda, in alluvium after inundation, *fide* Guill. & Perr. *l.c.* **Fr.Sud.:** Dayet en Nahret *Chudeau.* Also in A.-E. Sudan, Abyssinia, Eritrea, Arabia and S.W. Africa.

Besides the above, the Indian species *C. tetragonoloba* (Linn.) Taub. has been introduced into Sierra Leone and other parts of tropical Africa. It is a laxly branched annual about 3 ft. high with ovate-rhomboid leaflets and erect fruits.

28. ROTHIA Pers.—F.T.A. 2 : 7.

A small branched annual about 15 cm. high ; branches long-silky pubescent ; leaflets 3, subsessile, oblanceolate, acute, about 1·5 cm. long, silky-pilose on both sides ; flowers 2–5 in axillary clusters, subsessile ; calyx lobed to about the middle, 5 mm.

long, lobes lanceolate ; corolla very short ; fruit sessile, 1–1·5 cm. long, silky-pilose,
several-seeded   ..   .. *hirsuta*

R. hirsuta (*Guill. & Perr.*) *Bak.* in F.T.A. 2 : 7 (1871) ; Bak. f. Leg. Trop. Afr. 1 : 21. *Xerocarpus hirsutus*
    Guill. & Perr. (1832).   Erect or decumbent, diffusely branched herb 6 in. high. with small pink or mauve
    flowers ; in sandy places.
    Sen.: Walo *Leprieur.*   **Gam.:** Galam (fr. Sept.) *Leprieur!*   **Port.G.:** Bissau (Dec.) *Esp. Santo* 1438 !
    Fr.Nig.: Niamey (fr. Oct.) *Hagerup* 476 !   **N.Nig.:** Naraguta & Jos (Sept.) *Lely* 583 !   P. 699 ! Sokoto
    Moiser 88 ! Lema, Ilorin (fr. Aug.) *Ejiofor* FHI 29392 ! Also in Spanish Guinea, Angola and S. Africa
    and in eastern Africa from Eritrea to Mozambique.

29. CROTALARIA Linn.—F.T.A. 2 : 7 ; E. G. Baker in J. Linn. Soc. 42 : 241 (1914).

Leaves all 1-foliolate :
Flowers axillary or in extremely short racemes ; leaflets sessile :
  Flowers solitary :
    Calyx large, about 30 mm. long, densely villous, enclosing flower and fruit ; pedicels
      1 cm. long, thick, densely villous ; leaflet linear to obovate-lanceolate, mucronate,
      6–12 cm. long, silky-pilose beneath ; fruits 2 cm. long, glabrous   1. *calycina*
    Calyx small, 3 mm. long, more or less minutely pubescent ; pedicels 2–3 cm. long,
      filiform, appressed-pubescent ; leaflet oblong, rounded at both ends, more or less
      clasping the stem, 8–23 mm. long, 3–6 mm. broad, shortly appressed-pubescent
      beneath ; fruits 1·3 cm. long, glaucous   ..   .. 2. *occidentalis*
  Flowers few together, deflexed ; calyx about 5 mm. long ; leaflet oblanceolate,
    2–5 cm. long, shortly pubescent ; fruits 8 mm. long, glabrous   .. 3. *polygaloides*
  Flowers racemose :
    Racemes very densely flowered ; leaflet petiolate, lanceolate, mucronate, 4–8 cm.
      long, appressed-pilose on both surfaces ; petiole about 1 cm. long ; stipules and
      bracteoles subulate, about 8 mm. long ; standard petal 7 mm. long, glabrous, very
      striate   ..   .. 4. *anthyllopsis*
    Racemes lax :
      Stipules foliaceous, amplexicaul ; leaflet broadly obovate, emarginate, about
        4·5 cm. long, 2–3 cm. broad, very shortly pubescent beneath ; peduncle angular ;
        flowers glabrous ; fruits 2·5 cm. long, pubescent ..   ..   .. 5. *verrucosa*
      Stipules absent or very small or subulate :
        Leaflet oblanceolate, rounded or retuse at apex, 4–7 cm. long, 1·5–2·5 cm. broad,
          minutely and rather closely pubescent beneath ; flowers numerous in stiff
          racemes ; calyx thinly puberulous ; fruits 4·5 cm. long, glabrous   6. *retusa*
        Leaflet linear to elliptic, broadest in the middle :
          Leaflet elliptic, sessile, about 1·5 cm. long and 8 mm. broad, brown-pilose on both
            surfaces, thick ; racemes very few-flowered ; flowers about 1 cm. long ; fruits
            1 cm. long, densely pilose ..   .. 7. *arenaria*
          Leaflet linear to linear-lanceolate, acute :
            Plant glabrous or shortly appressed-pubescent, not pilose :
              Leaflet subsessile, petiole absent, fracturing in 1 place (therefore strictly a
                leaf) ; keel sharply angled, long-beaked :
                Stem and leaflet glabrous ; stem terete ; lamina linear to oblong-lanceolate,
                  1·5–6 cm. long, acute and mucronate at apex ; calyx glabrous or shortly
                  pubescent, 5–6 mm. long ; a pair of minute bracteoles, or their scars, about
                  the middle of pedicel ; fruits glabrous, stipitate   ..   .. 8. *glauca*
                Stem and leaflet minutely appressed-pubescent with white hairs arising from
                  red bases ; stem angular ; lamina oblong-lanceolate, 1·5–4 cm. long, obtuse
                  and mucronate at apex ; calyx pubescent 5 mm. long ; a pair of bracteoles
                  at top of pedicel ; fruits pubescent, sessile ..   .. 9. *lugardiorum*
              Leaflet with a petiole 2–5 mm. long, fracturing in 2 places, stem terete, striate,
                glabrous ; lamina narrowly linear, 6–12 cm. long, 2–4 mm. broad, acute ;
                calyx glabrous, 5 mm. long ; a pair of bracteoles at top of pedicel ; fruits
                glabrous, sessile ..   .. 10. *deightonii*
            Plant long-pilose or densely pubescent ; leaflet linear to linear-lanceolate,
              3–9 cm. long, 2–8 mm. broad ; fruits 6–10-seeded, pubescent :
              Corolla about 9 mm. long ; calyx 5 mm. long ; fruits 10 mm. long ; stipules
                present, 1 mm. long   ..   .. 11. *vogelii*
              Corolla about 3 mm. long ; calyx 2 mm. long ; fruits 7 mm. long ; stipules
                absent   ..   .. 12. *bongensis*
Leaves all 3- or 5-foliolate or rarely a few also 1-foliolate towards the inflorescences :
Plant armed with axillary spines 1–2 cm. long ; flowers 1 cm. long, 1 or 2 arising from
  about the middle of most spines ; fruits 9–12 mm. long, many-seeded ; stems and
  leaves puberulous ; stipules minute ..   .. 13. *aculeata*
Plant unarmed :
  *Stipules large and foliaceous, mostly falcate, usually at least 2 mm. broad ; fruits
    6 or more seeded :
    Stems weakly pilose with slender spreading hairs ; fruits long-stipitate, oblong,
      bladder-like, about 2·5 cm. long and 1 cm. broad, glabrous ; petiole as long as

the leaflets, the latter oblong-lanceolate, acute, about 3 cm. long, thinly pilose on
the midrib ; flowers 1·5 cm. long ..   .. 14. *podocarpa*
Stems glabrescent or variously pubescent with appressed or more or less ascending
hairs ; fruits not stipitate :
Stems and leaves rather roughly pilose-pubescent ; petiole nearly as long as the
leaflets, the latter oblanceolate, conspicuously mucronate, 3–4 cm. long ; flowers
1·5 cm. long ; fruits densely ferruginous-tomentose, 2·5–3·5 cm. long
                                                    15. *lachnophora*
Stems and leaves very shortly and finely pubescent or glabrescent ; petiole as long
as or longer than the leaflets, the latter variable, obovate to almost linear,
mucronate or emarginate :
Flowers over 1 cm. long ; fruits 2·5–3 cm. long, very shortly pubescent, seeds
yellow ; hairs on stem very short and closely appressed   16. *cylindrocarpa*
Flowers usually under 1 cm. long ; fruits less than 2 cm. long, shortly pubescent,
seeds reddish ; hairs on stem rather long, appressed but with some patent
                                                    17. *goreensis*
*Stipules small and subulate, at most 1·5 mm. broad, or absent :
†Leaves subsessile or very shortly petiolate ; petiole at most 1 cm. long :
Stipules absent ; fruits on erect pedicels :
Fruits oblong-cylindrical, 20–25 mm. long, densely pilose, about 20-seeded ;
flowers about 1 cm. long, a few together at the ends of branches ; leaflets
oblanceolate, cuneate at base, slightly emarginate and mucronate at apex,
7–15 (–25) mm. long, 2–7 (–10) mm. broad, appressed-pubescent beneath
                                                    18. *harmsiana*
Fruits globose or subglobose, 2–7 mm. long, 2–8-seeded :
Inflorescence capitulate :
Flowers few together in a cluster on a slender peduncle ; leaves sessile ; leaflets
oblanceolate, up to 1·5 cm. long, long-pilose beneath ; fruits weakly pilose
                                                    19. *microcarpa*
Flowers very crowded in a short broadly pyramidal head-like raceme scarcely
exserted from the leaves ; petioles 1–3 cm. long, villous ; leaflets narrowly
oblanceolate, the lateral ones falcate, 1–4 cm. long ; fruits about 7 mm. long,
villous-tomentose..   ..   .. 20. *cephalotes*
Inflorescence racemose :
Perennial ; flowers 7 mm. long ; inflorescences up to 15 cm. long ; fruits
subglobose, 5 mm. long, pubescent ; seeds triangular, 2 mm. long, with a
deep depression at the hilum ; leaflets broadly obovate, 8–15 mm. long,
4–7 mm. broad, emarginate at apex   ..   ..   .. 21. *graminicola*
Annual ; flowers 5 mm. long ; inflorescences 2–7 cm. long ; fruits globose,
4 mm. long, pubescent ; seeds usually subglobose, 1 mm. long, with a large
oil gland at the hilum ; leaflets oblanceolate, 4–11 mm. long, 2–3 mm. broad,
rounded and mucronate at apex   ..   ..   ..   22. *hyssopifolia*
Stipules present :
φFlowers 5–15 mm. long :
Fruits laxly pilose, pubescent or glabrous ; calyx up to 9 mm. long, not accrescent :
Calyx 2–3 (–4) mm. long :
Leaflets glabrous above, pubescent or glabrous beneath, obovate to linear ;
stipules subulate with a broader base, 1–2 mm. long ; ripe fruits subglobose,
5 mm. long, on pendulous pedicels ; seeds triangular with a deep depression
at the hilum :
Plant pubescent to pilose with brown hairs ; flowers 6–7 mm. long ; leaflets
1·5–3 cm. long, 3–8 mm. broad   .. 23. *sphaerocarpa* var. *sphaerocarpa*
Plant glabrous in all parts ; flowers 5 mm. long ; leaflets 1·5–2 cm. long,
5–9 mm. broad..   ..   ..   .. 23a. *sphaerocarpa* var. *polycarpa*
Leaflets (at least the upper ones) pilose above :
Stipules subulate, erect, 2 mm. long, pilose ; petioles 7–10 mm. long ; leaflets
oblong-oblanceolate, rounded and mucronate at apex, 2·5–3·5 cm. long,
7–9 mm. broad, pubescent beneath ; racemes terminal, rather dense and
with several flowers in axils of upper leaves, 3–8 cm. long ; flowers 7–8 mm.
long ; fruits ovoid, pubescent, 6 mm. long..   ..   ..   24. *sp. A*
Stipules lanceolate, reflexed and falcate, 4–5 mm. long, 1 mm. broad ; petioles
3–5 mm. long ; leaflets elliptic, subacute and with a tuft of hairs at apex,
1·5–2 cm. long, 4–7 mm. broad, silky beneath ; racemes terminal, dense,
1·5–3 cm. long ; flowers 5–6 mm. long ; fruits ovoid, pilose, 7 mm. long
                                                    25. *bamendae*
Calyx 7–9 mm. long :
**Keel glabrous, or with a few hairs about the lower edge :
‡Bracteoles linear-subulate, as long or nearly as long as the calyx-lobes and
very similar to them :

Standard pubescent only on the fold outside ; leaflets glabrous above, oblong-elliptic to oblanceolate, subacute at apex, 1–6 cm. long, 5–20 mm. broad ; fruits subglabrous, 8–10 mm. long ; calyx loosely pilose ; inflorescence more or less globular ..   .. 26. *ononoides*

Standard densely pubescent outside ; leaflets sparsely pilose above, oblanceolate, rounded at apex, 1–3 cm. long, 3–8 mm. broad ; fruits loosely pilose, about 10 mm. long ; calyx densely villous-pilose ; inflorescence subspiciform to globular..   ..   .. 27. *atrorubens*

‡Bracteoles broader than above, about 2 mm. broad, as long as the calyx-lobes but dissimilar to them ; inflorescence conical, many-flowered ; calyx densely villous ; leaflets oblanceolate, obtuse at apex, 4–6·5 cm. long, 8–23 mm. broad, glabrous above, appressed-pilose beneath ; standard densely pubescent outside   ..   .. 28. *ebenoides*

**Keel pubescent ; leaves frequently 1-foliolate towards the inflorescences :

Stems tomentose with recurved brown hairs ; leaflets lanceolate, acute, 2·5–5·5 cm. long, 7–10 mm. broad, appressed-pilose beneath ; inflorescence short and glomerate or subspicate, 5–10-flowered ; calyx densely brown-villous ; standard orbicular ..   .. 29. *perrottetii*

Stems appressed-pilose with silvery hairs ; leaflets oblong-lanceolate; standard broader than long   ..   .. 28. *ebenoides*

Fruits covered with a dense velvet of short stiff hairs ; keel glabrous or with a few hairs on the edges ; inflorescences glomerate at the ends of branches ; leaflets glabrous above ; calyx 10–15 mm. long, enlarging in fruit up to 20 mm. :

Stems appressed-pubescent, hairs directed upward ; corolla 12–15 mm. long ; bract at base of pedicel 1–2 mm. broad :

Calyx densely villous with long hairs dissimilar to those on the fruits, 12–15 mm. long ; bracteoles at base of calyx linear-lanceolate, 10–11 mm. long ; leaves frequently 1-foliolate towards the inflorescence ; leaflets narrowly lanceolate, acute, 2·5–7 cm. long, 5–18 mm. broad

30. *macrocalyx*

Calyx covered with a dense felt of short stiff curved hairs similar to those on the fruits, 15–20 mm. long ; bracteoles at base of calyx linear, 5–6 mm. long, frequently reflexed ; leaves all 3-foliolate ; leaflets oblanceolate, subacute to obtuse and mucronate at apex, 2·5–4·5 cm. long, 5–12 mm. broad   ..   .. 31. *mortonii*

Stems tomentose with more or less erect recurved hairs ; corolla about 10 mm. long ; calyx densely villous, 10–13 mm. long ; bracteoles linear, nearly as long as calyx ; bract linear, 1 mm. broad ; leaves all 3-foliolate ; leaflets obovate to oblanceolate, rounded at apex, sometimes slightly emarginate, minutely mucronate, 1·5–3 cm. long, 5–12 mm. broad   .. 32. *confusa*

ϕFlowers 20–25 (–30) mm. long, standard and keel densely pubescent ; calyx about 15 mm. long ; fruits oblong, about 3 cm. long, densely villous ; bracts ovate-lanceolate, often 3-fid, 1–1·5 cm. long, persistent ; leaflets oblong-oblanceolate, mucronate, 4–7 cm. long, 2–3 cm. broad, appressed-tomentose on both surfaces   ..   .. 33. *lachnosema*

†Leaves long-petiolate ; petiole at least 1·5 cm. long :
Stipules 5–11 mm. long, filiform to linear, persistent :

Flowers 2–3 in each inflorescence, 1·5 cm. long ; calyx-lobes broad and leafy, auriculate at base ; leaflets obovate, rounded at apex, about 3·5 cm. long and 2 cm. broad, sparsely pilose beneath ; stipules subulate, about 10 mm. long ; fruits stipitate, about 2 cm. long, pilose, bladder-like ..   .. 34. *barkae*

Flowers numerous in each inflorescence :

Calyx pilose, 8 mm. long ; fruits oblong, 4 cm. long, pilose ; leaflets obovate, cuneate at base, rounded at apex, 3–4 cm. long, 2–3 cm. broad 35. *incana*

Calyx appressed-pubescent to glabrous, 10–12 cm. long :

Bracts linear-lanceolate, 10–12 mm. long, persistent ; inflorescence terminal, rhachis stout and strongly grooved ; fruits sessile, 4 cm. long   36. *recta*

Bracts subulate, 2–4 mm. long, persistent ; inflorescence leaf-opposed, rhachis not as stout, slightly striate ; fruits stipitate, 4–5 cm. long 37. *longifoliolata*

Stipules absent or up to 2 (–3 mm.) long :
Flowers 5–12 mm. long :

Flowers very crowded in a short broadly pyramidal head-like raceme scarcely exserted from the upper leaves ; fruits subglobose, 7 mm. long ; for other characters see above   ..   .. 20. *cephalotes*

Flowers racemose, well exserted from the leaves ; fruits oblong to oblong-ellipsoid :

Inflorescence 2–6 (–8)-flowered, very lax ; bracts subulate, 2 mm. long ; calyx-

lobes linear ; keel rounded ; fruits 2 cm. long, stipitate, appressed-pubescent ;
leaflets linear or linear-lanceolate, 3–6 cm. long, (1–) 3–10 mm. broad
                                                            38. *glaucoides*

Inflorescence many-flowered, dense or rather lax :
Keel petals sharply bent from near the base ; fruits oblong-ellipsoid, about
1·3 cm. long, softly pubescent ; leaflets ovate-oblong to obovate, cuneate
at base, 2–4 cm. long, about 1 cm. broad   ..   ..   39. *senegalensis*
Keel petals rounded ; fruits oblong, 2–3·5 cm. long :
Bracts persistent, 3–9 mm. long, reflexed :
Petioles 3–7 cm. long ; leaflets linear-lanceolate, 5–15 cm. long, those on the
branchlets usually much smaller than those on the main stem ; keel
petals dark purple and striate ; standard striate, wings yellow ; fruits
13–17 mm. long, appressed-pubescent   ..   ..   ..   40. *comosa*
Petioles 1·5–2 cm. long ; leaflets elliptic to linear, acute at both ends,
3–5 cm. long ; all petals more or less equally striate ; fruits 11 mm. long,
tomentose to pilose ..   · ..   ..   41. *acervata*
Bracts either early caducous, 5–9 mm. long, or persistent and 1–2 mm. long :
Leaflets linear to linear-lanceolate, (1–) 2–6 cm. long, 2–7 mm. broad ;
stipules absent ; calyx 3 mm. long ; fruits 2 cm. long, shortly pubescent
                                                            42. *lathyroides*

Leaflets ovate, obovate or elliptic, 1·5–7 cm. broad :
Fruits 3–3·5 (–4) cm. long ; keel conspicuously striate :
Leaflets ovate or oblong-ovate, 5–10 (–15) cm. long, 3–5 (–7) cm. broad,
cuneate at base, acute (rarely obtuse) at apex ; usually with minute
stipules ; fruits straight, 3–3·5 (–4) cm. long ; calyx 6 mm. long
                                                            43. *mucronata*
Leaflets obovate, 3–5·5 cm. long, 1·5–3 cm. broad, cuneate at base,
rounded at apex ; stipules absent ; fruits curved, about 3 cm. long ;
calyx 4 mm. long ..   ..   44. *falcata*
Fruits 2–2·5 cm. long ; keel inconspicuously striate or with a purple
blotch ; leaflets obovate to elliptic, 3–6 cm. long, 2–5 cm. broad, cuneate
at base, rounded or subacute at apex ; calyx 6–7 mm. long
                                                            45. *naragutensis*

Flowers 14–25 mm. long :
Flowers axillary, solitary or few together, about 2 cm. long ; calyx-lobes
narrowly lanceolate, acute, nearly 1 cm. long ; pedicels 1 cm. long, shortly
pubescent ; keel bent below the middle ; petiole 5 cm. long ; leaflets obovate,
mucronate, about 6 cm. long and 3·5 cm. broad, thinly pubescent beneath
                                                            46. *axillaris*

Flowers racemose :
·Calyx glabrous, with teeth much shorter than the tube ; fruits oblong-cylindrical, 3–6·5 cm. long, 1·5–2 cm. broad, appressed-pubescent to glabrous ;
leaflets lanceolate to oblong-linear, 5–10 cm. long, 3–25 mm. broad
                                                            47. *ochroleuca*

Calyx pubescent :
Leaves all 3-foliolate :
Pedicels not jointed, small bracteoles at base of calyx ; leaflets linear to
oblong-lanceolate, calyx appressed-pubescent, with teeth about equalling
the tube ; fruits cylindrical, 2·5–4 cm. long, 4–6 mm. broad, sparsely
appressed-pubescent ; inflorescence long, many-flowered   48. *intermedia*
Pedicels jointed in the middle with 2 subulate bracteoles ; calyx-teeth much
longer than the tube :
Inflorescence few- and lax-flowered ; leaflets rhomboid-elliptic to ovate,
2·5–5 cm. long, 1·5–3 cm. broad ; fruits bladder-like, 3 cm. long, 1 cm.
broad, transversely nerved, loosely pilose   ..   ..   49. *doniana*
Inflorescence many- and close-flowered ; leaflets elliptic, 3–6 cm. long,
1–2 cm. broad   ..   ..   50. *sp. B*
Leaves 3- and 5-foliolate on the same plant, elliptic, acute at apex, cuneate
at base, 5–10 cm. long, 2–5 cm. broad ; calyx pubescent ; fruits narrowly
cylindrical, about 4 cm. long and 4 mm. broad   ..   51. *cleomifolia*

1. **C. calycina** *Schrank*—F.T.A. 2 : 15 ; Bak. f. in J. Linn. Soc. 42 : 271 (1914) ; Chev. Bot. 168 ; Wilczek
in Fl. Congo Belge 4 : 84, t. 4.   Erect annual herb, up to 2 ft. high, simple or with few branches, densely
brown-silky ; flowers yellow, few, about 1 in. long, enclosed by the conspicuously villous calyx ; in
savannah.
Sen.: *Heudelot* 119 !   **Fr.Sud.:** Koulikoro *Chev.* 3380.   San *Chev.* 3381.   **Port.G.:** Bafata *Esp. Santo*
2810 !   **Fr.G.:** Kouroussa *Pobéguin* 460 !   **S.L.:** Kambia (fl. & fr. Dec.) *Deighton* 928 ! Luseniya (Dec.)
Sc. Elliot 4066 !   **Iv.C.:** Kodiokoffi *Chev.* 22372.   **G.C.:** Accra Plains *Irvine* 1300 !   **N.Nig.:** Liruwen-
Kano Hills *Carpenter* ! Anara F.R., Zaria (fr. Oct.) *Keay* FHI 20131 ! Bauchi Plateau (Oct.) *Lely*
P. 825 ! Yola (fr. July) *Dalz.* 1 ! Throughout tropical Africa, also in tropical Asia and N. Australia.
2. **C. occidentalis** *Hepper* in Kew Bull. 1956 : 113, fig. 1.   (?) *C. uniflora* Bak. in F.T.A. 2 : 13 (1871) ; Bak.
f. l.c. 273 ; not of Koen. ex Roxb. (1832).   A slender ascending herb up to 2 ft. high, of seasonally
flooded land ; flowers red and yellow.

Sen.: *Perrottet.* **Port.G.:** Antula, Bissau (Dec.) *Esp. Santo* 2858! **S.L.:** Mateboi (fl. & fr. Nov.) *Jordan* 826!

3. **C. polygaloides** *Welw. ex Bak.* in F.T.A. 2 : 15 (1871) ; Bak. f. l.c. 255 ; Wilczek l.c. 99. Branched herb, 1 ft. high ; flowers pale yellow ; in savannah.
   S.L.: Wallia (fr. Feb.) *Sc. Elliot* 4280! Musaia (fl. & fr. Dec.) *Deighton* 4425! Gbinti (fl. & fr. Jan.) *Deighton* 5731! Batkanu (fl. & fr. Nov.) *Jordan* 967! Also in Upper Ubangi, Belgian Congo and Angola.

4. **C. anthyllopsis** *Welw. ex Bak.* in F.T.A. 2 : 15 (1871) ; Bak. f. l.c. 263 ; Wilczek l.c. 81. A branched herb, ascending to 1 ft. high, brown-pilose ; flowers yellow or purplish with darker veining ; in savannah.
   N.Nig.: Naraguta & Jos (Sept.) *Lely* 564! Bauchi Plateau (Sept.) *Lely* P. 695! Vom *Dent Young* 33! Extends to Abyssinia, E. Africa, Rhodesia and to Angola.

5. **C. verrucosa** *Linn.*—F.T.A. 2 : 14 ; Bak. f. l.c. 254. Erect shrub, 2–3 ft. high, with angular branches ; flowers pale yellow tinged blue, in leaf-opposed and terminal racemes.
   S.L.: *Welw.* 1940! Tower Hill, Freetown (fl. & fr. Dec.) *Deighton* 802! **S.Nig.:** Calabar *Milne.* A native of India, widespread in the tropics.

6. **C. retusa** *Linn.*—F.T.A. 2 : 13 ; Bak. f. l.c. 270 ; Chev. Bot. 170 ; Wilczek l.c. 90. A half-shrubby herb, 2–4 ft. high, of waste land ; flowers conspicuous, $\frac{3}{4}$–1 in. long, in long loose racemes, yellow with purple veining.
   Sen.: Dakar *Caille*! *Chev.* 3195 ; *Chev.* 15820. Karang (fl. & fr. Oct.) *Monod* 8644! **Gam.:** *Hayes* 501! Kombo *Dawe* 6! Kuntaur *Ruxton* 4! **Fr.G.:** Conakry *Chev.* 12076 ; 12128 ; 12149. **S.L.:** Freetown (fl. & fr. Feb.) *Johnston* 62! 80! *Deighton* 11! Njala (fl. & fr. May) *Deighton* 1896! Kumrabai (Dec.) *Thomas* 6989! **Lib.:** Sinoe Basin *Whyte*! Nana Kru, Sinoe Co. (fl. & fr. Mar.) *Baldwin* 11596! Monrovia (fl. & fr. June) *Barker* 1375! **G.C.:** Achimota (fl. & fr. Feb.) *Irvine* 1802! 1803! *Bally* 13! *Morton* GC 6019! **S.Nig.:** Abeokuta (Aug.) *Punch* 2! Ibadan (Oct.) *Newberry & Etim* 147! Olokemeji (Aug.) *Foster* 318! Calabar *Maitland* 6! Ikom, Ogoja (fl. & fr. Sept.) *Obi* FHI 17066! Widespread in the tropics ; probably introduced into our area and now naturalized. (See Appendix, p. 237.)

7. **C. arenaria** *Benth.*—F.T.A. 2 : 11 ; Bak. f. l.c. 258 ; Chev. Bot. 167. Stems woody, prostrate, much-branched, up to 1 ft. long, densely pale villous ; in dry savannah or semi-desert.
   Maur.: *fide* Chev. l.c. **Sen.:** *Heudelot*! *Roger* 91! Kouma (Jan.) *Perrottet* 163! St. Louis *Chev.* 15777. **Fr.Sud.:** Timbuktu (fr. July) *Chev.* 1264! *Hagerup* 185! **N.Nig.:** Katagum (fl. & fr. July) *Dalz.* 15! Kalkala, L. Chad *Golding* 51! (See Appendix, p. 235.)

8. **C. glauca** *Willd.*—F.T.A. 2 : 12 ; Bak. f. l.c. 259 (incl. var. *welwitschii* Bak. f.) ; Chev. Bot. 168 ; Wilczek l.c. 102, t. 6. An erect glabrous herb, 1–3 ft. high, or taller in long grass ; flowers yellow, distant on lax axillary and terminal racemes ; plant drying black in var. *welwitschii* Bak. f.
   Sen.: Bondou *Heudelot* 199! **Fr.Sud.:** Badinko *Chev.* 111. Koulikoro *Chev.* 3209 ; 3211. **Port.G.:** Antula, Bissau (fl. & fr. Oct.) *Esp. Santo* 1355! **S.L.:** Wallia (fr. Feb.) *Sc. Elliot* 4257! Samu country (Dec.) *Sc. Elliot* 4730! Bonthe Isl. (fl. & fr. Nov.) *Deighton* 2291! **Iv.C.:** Toumodi *Chev.* B. 22263 ; 22382. **G.C.:** Accra (May, Sept.) *T. Vogel*! *Dalz.* 86! *Irvine* 1763! Daboya Ferry (fl. & fr. Dec.) *Morton* GC 6196! Ada (fl. & fr. June) *Morton* GC 9236! **Dah.:** Agouagon *Chev.* 23543. **N.Nig.:** Katagum *Dalz.* 348! Abinsi (fl. & fr. Oct.) *Dalz.* 590! Bauchi Plateau (Sept., Oct.) *Lely* 561! P. 689! P. 818! **S.Nig.:** Old Calabar *Milne*! **Br.Cam.:** Banso, Bamenda (fl. & fr. Oct.) *Tamajong* FHI 23477! Widespread in tropical Africa. (See Appendix, p. 236.)
   A specimen from an old garden patch, Udi, Onitsha, S. Nigeria (*Jones* FHI 5015) is allied to *C. glauca* but has leaves about 1 cm. long and very small flowers.

9. **C. lugardiorum** *A. A. Bullock* in Kew Bull. 1932 : 493 ; Wilczek l.c. 93. A slender erect herb, 1–3 ft. high ; flowers yellow ; in mountain grassland.
   Br.Cam.: Nchun, Bamenda, 5,000 ft. (fl. & fr. May) *Maitland* 1523! Also in Uganda, Kenya, Tanganyika and Belgian Congo.

10. **C. deightonii** *Hepper* l.c. 113 (1956). *C. spartea* of F.T.A. 2 : 12 (1871), partly ; of F.W.T.A., ed. 1, 1 : 398 ; of Bak. f. l.c. 262, partly ; not of R.Br. ex Bak. An erect annual, up to 5 ft. high ; flowers yellow, strongly purple-veined ; in savannah grassland.
    S.L.: *Afzelius*! Makali (Oct.) *Glanville* 11! Bundulai (Nov.) *Deighton* 5664!

11. **C. vogelii** *Benth.*—F.T.A. 2 : 14 ; Bak. f. l.c. 264 ; Wilczek l.c. 83. *C. parsonsii* Bak. f. (1913). A pilose herb, 2–3 ft. high, or more when in tall grass ; flowers yellow in elongated racemes.
    Sen.: Ouassadou *Berhaut* 1240. Toubacouta (fr. Oct.) *Monod* 8457! **G.C.:** Tamale *Williams* 389! Aframso (Oct.) *Vigne* FHI 4036! **Togo:** Kete Krachi *Zech* 192. **N.Nig.:** Nupe *Barter* 992 ; 1629! Stirling (Sept.) *T. Vogel* 158! Lokoja (fl. & fr. Oct.) *Dalz.* 17! *Parsons* L. 38! Alkalire (Sept.) *Lely* 643! Ilorin (Feb.) *Ejiofor* FHI 19824! Also in French Cameroons, Chad, A.-E. Sudan, Uganda, Belgian Congo and Angola.

12. **C. bongensis** *Bak. f.* l.c. 256 (1914) ; Wilczek l.c. 96. An erect branched pilose annual, slightly woody below, a few inches to 2 ft. high ; flowers yellow, strongly purple-veined ; in savannah.
    Port.G.: Pirada, Gabú (fr. Nov.) *Esp. Santo* 3163! **N.Nig.:** Naraguta (fl. & fr. Aug.) *Lely* 506! Vom *Dent Young* 39! Kaduna (fl. & fr. Dec.) *Meikle* 793! Also in A.-E. Sudan, Uganda, Tanganyika, Nyasaland and Belgian Congo.

13. **C. aculeata** *De Wild.* in Ann. Mus. Congo Belge, sér. 4, 1 : 185, t. 46 (1908) ; Wilczek l.c. 107. *C. spinosa* of Berhaut Fl. Sen. 12, not of Hochst. ex Benth.
    Sen.: *fide* Berhaut *l.c.* Also in A.-E. Sudan, Uganda, Tanganyika, Belgian Congo, Angola, Rhodesia and Nyasaland.

14. **C. podocarpa** *DC.*—F.T.A. 2 : 17 ; Bak. f. l.c. 406. An erect herb, 1–2 ft. high ; flowers pale yellow.
    Sen.: Richard Tol (fr. Sept.) *Perrottet* 167! M'bidjem *Thierry* 45! Podor (Aug.) *Martine* 80! **Fr.Sud.:** Sompi *Chev.* 3193. Widespread in eastern tropical Africa and Madagascar.

15. **C. lachnophora** *Hochst. ex A. Rich.* Tent. Fl. Abyss. 1 : 151 (1847) ; Wilczek l.c. 129. *C. lachnocarpa* Hochst. ex Bak. in F.T.A. 2 : 33 (1871) ; Bak. f. l.c. 404 ; F.W.T.A., ed. 1, 1 : 397 ; Chev. Bot. 169. A spreading bush, 3–6 ft. high, clothed with brown hairs ; flowers conspicuous, yellow becoming reddish ; in usually wet grassland.
    Fr.Sud.: *fide* Chev. *l.c.* **N.Nig.:** Bauchi Plateau (Aug., Sept.) *Lely* P. 608! 669! 723! Jos (Aug.) *Keay* FHI 12716! Sho to Barakin Ladi (Nov.) *Keay* FHI 21022! **Br.Cam.:** Bamenda, 5,000–5,500 ft. (Jan., Aug., Sept.) *Migeod* 305! *Savory* UCI 352! 387! Widespread in tropical Africa. (See Appendix, p. 236.)

16. **C. cylindrocarpa** *DC.*—F.T.A. 2 : 40 ; Bak. f. l.c. 414 ; Wilczek l.c. 126. A branched shrubby herb, up to 6 ft. high ; flowers yellow tinged red.
    Sen.: *Roger* 28! **Fr.Sud.:** Fafa, Gao (fl. & fr. Mar.) *De Wailly* 5368! **G.C.:** Tefli (fl. & fr. Jan.) *Morton* GC 8318! Gambaga (fl. & fr. Apr.) *Morton* GC 8995! **N.Nig.:** Sokoto (fl. & fr. Sept., Nov.) *Dalz.* 343! *Moiser* 47! Anara F.R., Zaria (fl. & fr. Oct.) *Keay* FHI 20118! Minna (fl. & fr. Dec.) *Meikle* 712! Naraguta (fl. & fr. Aug.) *Lely* 480! **S.Nig.:** Ibadan (Jan., Feb.) *Meikle* 941! 1146! Anambra F.R., Onitsha (fl. & fr. Feb.) *Jones* FHI 625! Extends to Rhodesia.

17. **C. goreensis** *Guill. & Perr.*—F.T.A. 2 : 28 ; Bak. f. l.c. 413 ; Chev. Bot. 169 ; Wilczek l.c. 124, fig. 10. Stems herbaceous, woody below, or an undershrub to 5 ft. high, with a woody taproot ; flowers yellow. turning reddish ; in waste places.
    Sen.: Galam *Heudelot*! **Gam.:** *Boteler*! *Saunders* 78! Cape St. Mary *Heudelot* 8! **Fr.Sud.:** Kabarah *Chev.* 1334. Ségou *Chev.* 3203. Sarédina (May) *Davey* 141! **S.L.:** Luseniya (fr. Dec.) *Sc. Elliot* 4216! Kambia (fl. & fr. Jan.) *Sc. Elliot* 4393! Njala (Apr.) *Deighton* 1590! Musaia (fr. Dec.) *Deighton* 4530!

Lib.: Banga (Oct.) *Linder* 1164! 1277! Gbau, Sanokwele (fl. & fr. Sept.) *Baldwin* 9422! 9440! **Iv.C.:** Bingerville *Chev.* 20071. **G.C.:** Accra *T. Vogel* 30! *Dalz.* 82! Volta River F.R. (fl. & fr. Nov.) *Morton* GC 6048! Tamale (fl. & fr. Dec.) *Morton* GC 9850! Gonja (fl. & fr. Dec.) *Morton* GC 9974! **Togo:** Kpedsu (Dec.) *Howes* 1035! Have Etoe (fr. Nov.) *Adams & Akpabla* GC 4004! **N.Nig.:** Ilorin (Nov.) L. 112! Nupe *Barter* 1606! Katagum *Dalz.* 14! Sokoto (fl. & fr. Sept.) *Dalz.* 344! Kalkala, L. Chad (fl. & fr. Oct.) *Gwynn* 88! **S.Nig.:** *Barter* 1782! 1789! Lagos (Nov.) *Dalz.* 1224! Olokemeji (fl. & fr. Nov.) *Onochie* FHI 8152! **F.Po:** *Mann* 37! Widespread in tropical Africa.

18. **C. harmsiana** *Taub.* in Engl. Pflanzenw. Ost-Afr. C, 205 (1895) ; Bak. f. l.c. 397. An undershrub 1–3 ft. high ; flowers yellow.
S.Nig.: Ikwette Plateau, 5,200 ft., Obudu (fr. Dec.) *Savory & Keay* FHI 25232! **Br.Cam.:** Babanki, Bamenda (fl. & fr. Jan.) *Johnstone* 3! Also in the mountains of east and south tropical Africa.

19. **C. microcarpa** *Hochst. ex Benth.* in Hook. Lond. Journ. Bot. 2 : 573 (1843) ; F.T.A. 2 : 16 ; Bak. f. l.c. 402 ; Chev. Bot. 169. A variable diffusely branched herb, woody at the base, from a few inches to a foot or more long ; flowers deep yellow or orange ; in savannah.
Fr.Sud.: San *Chev.* 3228. **G.C.:** Babile, Lawra (fr. Oct.) *Hinds* 5019! Kulnaba, Navrongo (fl. & fr. Oct.) *Hughes* FH 5340! **N.Nig.:** Nupe *Barter* 997! Sokoto *Moiser* 18! Katagum (fl. & fr. Oct.) *Dalz.* 17! Alagbede F.R. (fl. & fr. Jan.) *Ejiofor* FHI 30808! Yola (fl. Aug.) *Dalz.* 301! Also in French Cameroons, A.-E. Sudan, Abyssinia, Eritrea, Uganda, Tanganyika, Belgian Congo, N. Rhodesia and Nyasaland. (See Appendix, p. 236.)

20. **C. cephalotes** *Steud. ex A. Rich.*—F.T.A. 2 : 23 ; Bak. f. l.c. 276 ; Wilczek l.c. 200. A slightly woody herb, 6–18 in. high, often unbranched, with a tuft of leaves around the raceme, densely silky ; flowers small, yellow ; in rocky places.
G.C.: Banda, Wenchi Dist. (fr. Dec.) *Morton* GC 25161! 25284! **Dah.:** Savé (fr. May) *Chev.* 23579! **N.Nig.:** Nupe *Barter* 801! Patti Lokoja (fl. & fr. Nov.) *Dalz.* 13! Bauchi Plateau (fl. & fr. Oct.) *Lely* P. 830! Anara-F.R., Zaria (Sept., Oct.) *Keay* FHI 5441! *Olorunfemi* FHI 24401! Widespread in tropical Africa, also in Transvaal.

21. **C. graminicola** *Taub. ex Bak. f.* l.c. 291 (1914) ; Hepper in Kew Bull. 1956 : 118 . A perennial herb, usually in burnt grassland, with numerous nearly simple tufted stems 6–12 in. high, from a woody base ; flowers yellow.
Togo: Sokode *Kersting* 128. **N.Nig.:** Kontagora (fl. & fr. Jan.) *Dalz.* 40! Zaria (fl. & fr. Feb.) *Dalz.* 377! *Milne-Redhead* 5028! Otobi, Benue (fl. & fr. Feb.) *Jones* FHI 4575! Naraguta, 4,000 ft. (May) *Lely* 238! Gerti, Jemaa Dist. (fl. & fr. Mar.) *Onochie* FHI 7743! **Br.Cam.:** Tapare R. H., Gashaka Dist. (fl. & fr. Jan.) *Latilo & Daramola* FHI 34411! Also in French Cameroons.

22. **C. hyssopifolia** *Klotzsch*—F.T.A. 2 : 24 ; Bak. f. l.c. 286 ; Wilczek l.c. 217 ; Hepper l.c. 118. *C. graminicola* of Bak. f. l.c. 291 and of F.W.T.A., ed. 1, 1 : 398, partly (*Heudelot* 614 ; *Sc. Elliot* 4310) not of Taub. ex Bak. f. An erect herb, branched above and sometimes below, 1–2 ft. high ; flowers yellow.
Sen.: Falèmé R. *Heudelot*! Kidira (fl. & fr. Nov.) *Berhaut* 1194! **Fr.Sud.:** Bamako (fl. & fr. Jan.) *Chev.* 206! **Port.G.:** Antula, Bissau (fl. & fr. Oct.) *Esp. Santo* 2557! **Fr.G.:** Rio Nunez *Heudelot* 614! **S.L.:** *Afzelius*! Magbile (Dec.) *Thomas* 6028! Ruka to Kitchom (Dec.) *Sc. Elliot* 4310! Rokupr (fl. & fr. Jan., Dec.) *Jordan* 380! *Deighton* 2950! Musaia (fl. & fr. Dec.) *Deighton* 4475! 5697! **N.Nig.:** Fodama (fl. & fr. Dec.) *Moiser* 250! Bauchi Plateau (fl. & fr. Oct.) *Lely* P. 815! 848! Also in French Cameroons, Abyssinia, Uganda, Kenya, Tanganyika, Nyasaland, N. Rhodesia, Belgian Congo and Mozambique.

23. **C. sphaerocarpa** *Perr. ex DC.* var. **sphaerocarpa**—F.T.A. 2 : 23 ; Bak. f. l.c. 289, partly ; Hepper l.c. 118, for synonymy. A branched herb from a woody base 1–2 ft. high ; flowers yellow, in lax racemes ; in sandy ground.
Sen.: *Perrottet*! M'bidjem (fl. & fr. Aug.) *Thierry* 84! Kouma (fr. Oct.) *Roger*! **Fr.Sud.:** Balasoko valley, Bamako (fl. & fr. Dec.) *Monod*! **Fr.Nig.:** Zinder (fl. & fr. Oct.) *Hagerup* 590! **N.Nig.:** Kalkala, L. Chad (fr. Oct.) *Gwynn* 87! Extends to A.-E. Sudan and southwards to Orange Free State and west to Angola. (See Appendix, p. 237.)

23a. **C. sphaerocarpa** var. **polycarpa** (*Benth.*) *Hepper* in Kew Bull. 1956 : 119. *C. polycarpa* Benth. (1843). *C. sphaerocarpa* of F.T.A. 2 : 23 ; Bak. f. l.c. ; F.W.T.A., ed. 1, 1 : 398, partly, in syn. Similar to the last but entirely glabrous.
Sen.: *Perrottet* 171! Cayor *Heudelot* 483! Dakar *Rattray*! M'pao, Dakar (fl. & fr. Nov.) *Monod*! Cape Verde *Talmy*! Hann *Berhaut* 1150!

24. **C. sp. A.** An annual herb 1–2 ft. high, sparingly branched above ; flowers yellow, veined purple ; in savannah. Further material required.
N.Nig.: Bauchi Plateau (fl. & fr. Oct.) *Lely* P. 789!

25. **C. bamendae** *Hepper* in Kew Bull. 1956 : 119. A branched herb with densely pubescent stems, about 2 ft. high, leaflets silky-silvery beneath.
Br.Cam.: Bamenda (Dec.) *Egbuta* FHI 3763! Banso, Bamenda (Oct.) *Tamajong* FHI 23479! Gembu, Gashaka Dist. (fr. Jan.) *Latilo & Daramola* FHI 34369!

26. **C. ononoides** *Benth.*—F.T.A. 2 : 22 ; Bak. f. l.c. 307. Herb, woody at base, 1–4 ft. high, suberect, the lower branches often procumbent, brownish-pilose ; flowers yellow turning orange in heads at the ends of the branches.
S.L.: *Afzelius*! *Don*! **N.Nig.:** Ilorin *Clarke* 3! Anara F.R., Zaria (Sept.) *Olorunfemi* FHI 24363! Naraguta & Jos (Nov.) *Lely* 559! Vom *Dent Young* 38! Extends to A.-E. Sudan, E. Africa, Belgian Congo, S. Rhodesia and Madagascar.

27. **C. atrorubens** *Hochst. ex Benth.* in Hook. Lond. Journ. Bot. 2 : 572 (1843) ; F.T.A. 2 : 22 ; Bak. f. l.c. 307 ; Chev. Bot. 168. *Chrysocalyx rubiginosa* Guill. & Perr. (1831), not *Crotalaria rubiginosa* Willd. (1803). *Crotalaria ononoides* of F.W.T.A., ed. 1, 1 : 397, partly (*Moiser* 75). Erect or suberect herb, 1–3 ft. high, woody below, yellowish silky-hairy ; flowers yellow or reddish tinged, densely crowded at the ends of the branches ; in savannah.
Sen.: St. Louis *Perrottet*. Galam (fl. & fr. Sept., Oct.) *Leprieur* in Hb. *Perrottet* 162! **Gam.:** *Mungo Park*! *Saunders* 70! *Hayes* 582! **Fr.Sud.:** Ségou *Chev.* 3200. **G.C.:** Zuarungu (fr. Dec.) *Adams & Akpabla* GC 4273! **N.Nig.:** Nupe *Barter* 1052! Sokoto (fl. & fr. Sept.) *Dalz.* 345! *Moiser* 75! Bauchi Plateau (fl. & fr. Oct.) *Lely* P. 838! Biu, Bornu Prov. (Sept.) *Noble* 20! Extends to A.-E. Sudan. (See Appendix, p. 236.)

28. **C. ebenoides** (*Guill. & Perr.*) *Walp.*—F.T.A. 2 : 20 ; Bak. f. l.c. 305 ; Chev. Bot. 168. *Chrysocalyx ebenoides* Guill. & Perr. (1831). Erect herb, 1–1½ ft. high, with long slender branches, densely brown-silky ; flowers in dense oblong terminal heads.
Sen.: Walo *Perrottet* 161! Galam *Heudelot* 253! **Fr.Sud.:** Koulikoro (Apr.) *Chev.* 3197! Ségou (Oct.) *Roberty* 3027! **Port.G.:** Pirada, Gabú (fl. & fr. Nov.) *Esp. Santo* 3160!

29. **C. perrottetii** *DC.*—F.T.A. 2 : 19 ; Bak. f. l.c. 306. *Chrysocalyx gracilis* Guill. & Perr. (1831)—F.T.A. 2 : 20 ; Bak. f. l.c. 306. *Crotalaria gambica* Taub. (1896). Herb 1–2 ft. high, with more or less straggling branches, brownish-silky ; flowers yellow turning red.
Sen.: *Roger*! Kouma *Perrottet*! Cayor *Heudelot* 409! M'Bambey (Oct.) *Monod* 8732 *bis*! **Gam.:** *Brown-Lester* 28! 53! 65! Kudang (fl. & fr. Jan.) *Dalz.* 8023! Cape St. Mary (Jan.) *Dalz.* 8024!

30. **C. macrocalyx** *Benth.*—F.T.A. 2 : 20 ; Bak. f. l.c. 306. *C. perrottetii* of F.W.T.A., ed. 1, 1 : 397, partly (*Lely* 598), not of DC. Erect or procumbent herb, 1–2 ft. high, with woody taproot ; flowers yellow to orange, crowded at the ends of branches, standard purple-veined ; in savannah.
Sen.: *Heudelot* 209! **Gam.:** *Saunders* 115! **G.C.:** Bosomoa F.R. (fl. & fr. Dec.) *Vigne* FH 3195!

17

Kintampo (Nov.) *Andoh* FH 5418! Ejura Scarp (Dec.) *Morton* GC 9746! Banda, Wenchi (fl. & fr. Dec.) *Morton* GC 25083! **N.Nig.:** Nupe *Barter* 996! Anara F.R., Zaria (Sept., Oct.) *Keay* FHI 5464! *Olorunfemi* FHI 24352! Kadaura (Sept.) *Lely* 598! Dataturu Kabau, Bornu (Oct.) *Daggash* FHI 22004! **Br.Cam.:** Bama, Dikwa Div. (fl. & fr. Jan.) *McClintock* 31! Also in Abyssinia.

31. **C. mortonii** *Hepper* l.c. 115 (1956). *C. macrocalyx* of F.W.T.A., ed. 1, 1 : 397, partly (*Mildbr.* 7410), not of Benth. Erect herb, 1–2 ft. high with 2–4 flowers at the ends of the branches.
   G.C.: Sorri L., Gonja (fl. & fr. Dec.) *Morton* GC 25033! **Togo:** Atakpame (Nov.) *Mildbr.* 7410!

32. **C. confusa** *Hepper* l.c. 116 (1956). *C. macrocalyx* of Bak. f. l.c. 306, partly ; of F.W.T.A., ed. 1, 1 : 397, partly. A more or less procumbent and diffuse herb with stems 1–3 ft. long ; flowers orange-yellow ; in savannah.
   G.C.: Tamale (fl. & fr. Dec.) *Adams & Akpabla* GC 4152! *Williams* 418! Gambaga Hill (fl. & fr. Apr.) *Morton* GC 8979! Gonja (fl. & fr. Dec.) *Morton* GC 25035! **Togo:** Kpeve (fl. & fr. Jan.) *Irvine* 1739! **N.Nig.:** Ilorin (fl. May, fr. Jan.) *Bailey* 6! *Rowland*! *Ejiofor* FHI 30802! Nupe *Yates*! Katagum (Nov.) *Dalz.* 16! Kuta *Lamb* FHI 3206! Also in French Cameroons.
   [Several variations (e.g. *Barter* 959, *Meikle* 678) of *C. confusa* have been noted in Kew Bull. 1956 : 117.]

33. **C. lachnosema** *Stapf* in Kew Bull. 1910 : 329 ; Bak. f. l.c. 323 ; Wilczek l.c. 251. A woody undershrub, 2–6 ft. high, more or less fulvo-tomentose all over ; flowers yellow with orange veins.
   S.L.: Mawria (Feb.) *Glanville* 172! Kamakui (fr. Aug.) *Deighton* 3499! Mafore (fr. Feb.) *Deighton* 3605! Loma Mts. (fr. Feb.) *Jaeger* 4199! **Lib.:** Kakatown *Whyte*! Mecca, Boporo (fr. Nov.) *Baldwin* 10422! **Togo:** Sokode *Schröder*! **N.Nig.:** Nupe *Barter* 920! Kontagora (fl. & fr. Nov.) *Dalz.* 39! Kangimi, Zaria (fr. Mar.) *Olorunfemi* FHI 24429! **Br.Cam.:** Babanki, Bamenda (fl. & fr. Apr.) *Maitland* 1644! Also in French Cameroons, A.-E. Sudan and Belgian Congo.

34. **C. barkae** *Schweinf.* in Bull. Herb. Boiss. 4, app. 2 : 226 (1896) ; Bak. f. l.c. 390. *C. quartiniana* of F.T.A. 2 : 42 ; of F.W.T.A., 1, 1 : 397. A herb with prostrate branches, 1–2 ft. long, thinly hairy ; flowers yellow ; in dry rocky places.
   G.C.: Zuarungu (fr. Nov.) *Akpabla* 401! **N.Nig.:** Katagum (fr. Sept.) *Dalz.* 13! Sokoto (Sept.) *Dalz.* 342! Ako (Oct.) *Lely* 651! Fika, Bornu (Oct.) *Daggash* FHI 24897! **Br.Cam.:** Gwoza, Dikwa Div. (fl. & fr. Dec.) *McClintock* 64! Also in Cape Verde Isl., Belgian Congo, Kenya, Abyssinia and Eritrea.

35. **C. incana** *Linn.*—F.T.A. 2 : 31 ; Bak. f. l.c. 357 ; Wilczek l.c. 148. A herb, 2–4 ft. high, with coarsely hairy stems and fruits ; flowers yellow, veined purple.
   S.L.: *Don*! Russell (fl. & fr. May) *Deighton* 2667! **Br.Cam.:** Lakom, Bamenda (fl. & fr. June) *Maitland* 1458! Kumbo, Bamenda (Nov.) *Egbuta* FHI 3765! In many parts of tropical Africa and in the tropics generally.

36. **C. recta** *Steud. ex A. Rich.*—F.T.A. 2 : 40 ; Bak. f. l.c. 352 ; Wilczek l.c. 153, t. 9. *C. crepitans* Hutch. (1921)—F.W.T.A., ed. 1, 1 : 398 ; Bak. f. Leg. Trop. Afr. 1 : 46. Erect herb or undershrub, 3–6 ft. high, with ribbed stems, shortly brownish-pubescent ; flowers yellow, strongly brown-purple veined.
   N.Nig.: Hepham to Ropp, 4,600 ft. (July) *Lely* 378! Bauchi Plateau (Aug.) *Lely* P. 616! **Br.Cam.:** Bamenda Station (May) *Maitland* 1427! Bali-Ngemba, Bamenda (June) *Ujor* FHI 30440! Extends throughout tropical Africa and also into Transvaal.

37. **C. longifoliolata** *De Wild.* in Ann. Mus. Congo Belge, sér. 4, 1 : 187, t. 45, figs. 1–10 (1903) ; Bak. f. l.c. 354 ; Wilczek l.c. 127. Erect herb, 1–2 ft. high ; flowers with standard purple outside, grey inside.
   Br.Cam.: Bum-Nchan, Bamenda, 4,500 ft. (June) *Maitland* 1608! Also in French Cameroons, Belgian Congo and Tanganyika.

38. **C. glaucoides** *Bak. f.* l.c. 393 (1914). *C. paludosa* A. Chev. in Bull. Mus. Hist. Nat., sér. 2, 5 : 230 (1933), excl. *Chev.* 34315. A slender annual with the appearance of *C. glauca* but with trifoliolate leaves and very sparsely pubescent.
   Sen.: Cape Manuel *Talmy* 12. Tivaouane (fl. & fr. Dec.) *Chev.* 3208! Toubacouta (fl. & fr. Oct.) *Monod* 8689! **Port.G.:** Pussubé, Bissau (fl. & fr. Nov.) *Esp. Santo* 1403! Antula, Bissau (fl. & fr. Dec.) *Esp. Santo* 1584!

39. **C. senegalensis** (*Pers.*) *Bacle ex DC.*—F.T.A. 2 : 31 ; Bak. f. l.c. 337. *C. uncinella* var. *senegalensis* Pers. Syn. 2 : 285 (1807). *C. maxillaris* of Bak. f. l.c., partly, not of Klotzsch. Suberect or prostrate, half-woody, 1–2 ft. high, shortly pale-pubescent ; flowers yellow turning orange ; in dry fields.
   Sen.: *Roger*! *Heudelot* 113! Almadies (fr. Jan.) *Monod* 10988! Dakar (fl. & fr. Oct.) *Giovanetti*! **N.Nig.:** Katagum (fl. & fr. Oct.) *Dalz.* 12! Sokoto (Mar., Sept.) *Dalz.* 340! *Moiser*! Kano to Katsina (fl. & fr. Dec.) *Meikle* 812! **Br.Cam.:** Gulumba, Dikwa Div. (fl. & fr. Dec.) *McClintock* 41! Also in Belgian Congo, A.-E. Sudan and Abyssinia.

40. **C. comosa** *Bak.* in F.T.A. 2 : 34 (1871) ; Bak. f. l.c. 380 ; Wilczek l.c. 268. *C. astragalina* of F.W.T.A., ed. 1, 1 : 368, partly (*Lely* 617). A stout erect herb, 5–6 ft. high ; flowers yellow and dark purple ; in savannah and cultivated land.
   Port.G.: *Esp. Santo* 1418! **S.L.:** Musaia (fr. Dec.) *Deighton* 4515! **G.C.:** Savelugu, Tamale (fl. & fr. Dec.) *Morton* GC 9897! Banda, Wenchi (fl. & fr. Dec.) *Morton* GC 25132! 25193! Menji, Wenchi (fl. & fr. Dec.) *Morton* GC 25228! Shai Plains (Jan.) *Irvine* 647! **N.Nig.:** Anara F.R., Zaria (Sept.) *Olorunfemi* FHI 24361! Bauchi Plateau (Aug., Oct.) *Lely* P. 664! P. 792! *Batten Poole* 313! Also in Belgian Congo, N. Rhodesia and Angola.

41. **C. acervata** *Bak. f.* in J. Bot. 58 : 74 (1920) ; Wilczek l.c. 176. Erect herb, 2–3 ft. high ; flowers yellow, purple-striate ; in mountain grassland.
   S.Nig.: Mt. Koloishe, Obudu, 5,000–5,400 ft. (Dec.) *Savory & Keay* FHI 25088! **Br.Cam.:** Bamenda, 6,400–8,000 ft. (Nov., Dec., fr. Jan.) *Migeod* 330! *Keay & Russell* FHI 28438! *Baldwin* 13846! Also in French Cameroons and Belgian Congo.

42. **C. lathyroides** *Guill. & Perr.*—F.T.A. 2 : 35 ; Chev. Bot. 169. *C. impressa* Nees (1841). *C. astragalina* Hochst. ex A. Rich. (1847)—F.T.A. 2 : 43 ; Bak. f. l.c. 374 ; F.W.T.A., ed. 1, 1 : 398 ; Chev. Bot. 168. *C. paludosa* A. Chev. (1933), partly (*Chev.* 34315). Herb, woody at the base, with slender branches, erect or more or less diffuse, shortly silky-pubescent above ; flowers rather small, yellow with dark purple lines sometimes coalescing into a blotch, in short dense lateral racemes usually long-stalked ; in wet places.
   Sen.: *Perrottet*! Dakar *Rattray*! Samandiniéry, Casamance *Chev.* 3205. Seleki, Casamance *Chev.* 3206. Niayes *Chev.* 3207. **Gam.:** *Brown-Lester* 17! **Port.G.:** Antula, Bissau (Oct.) *Esp. Santo* 2591! **Fr.G.:** Conakry to Dubréka (fl. & fr. Nov.) *Chev.* 34315! **S.L.:** *Afzelius*! Mahela (fr. Dec.) *Sc. Elliot* 3937! Brookfields (Oct.) *Deighton* 246! Kumrabai-Mamila (fl. & fr. Nov.) *Deighton* 4937! **Lib.:** Duport 4100! (fl. & fr. Oct.) *Barker* 1448! *Linder* 1490! Fissibu, Vonjama (fl. & fr. Oct.) *Baldwin* 10032a! Gene Loffa, Kolahun (Nov.) *Baldwin* 10080! **Togo:** Amedzofe, 2,500 ft. (fl. & fr. Jan.) *Irvine* 3370! Extends to A.-E. Sudan, Abyssinia and Belgian Congo. (See Appendix, p. 236.)

43. **C. mucronata** *Desv.* in Journ. de Bot. 3 : 76 (1814) ; Exell Cat. S. Tomé 151 (excl. syn. *C. obovata* G. Don) ; Wilczek l.c. 270. *C. striata* DC. (1825)—F.T.A. 2 : 38 ; Chev. Bot. 170 ; Bak. f. l.c. 345 ; F.W.T.A., ed. 1, 1 : 398. Erect undershrub, 3–6 ft. high, grey or yellowy silky on the young parts ; flowers yellow, purple-striped, in fairly dense racemes 6–10 in. long.
   Sen.: Conakry (fl. & fr. Oct.) *Martine* 162! **Gam.:** *Brown-Lester* 43! **S.L.:** Magbile (fl. & fr. Dec.) *Thomas* 6377! Kubaka (Oct.) *Glanville* 43! Port Loko (Oct.) *Tindall* 48! **Lib.:** Banga to Moylakwell! (Oct.) *Linder* 1267! Sanokwele (Sept.) *Baldwin* 9511! **Iv.C.:** Bériby to Tabou *Chev.* 20034. Mossi *Chev.* 24631. **G.C.:** Akropong (Oct.) *Johnson* 801! Odumase *Howes* 1203! Gambaga (fl. & fr. Apr.) *Morton* GC 8993! Ejura (fl. & fr. Dec.) *Morton* GC 9523! 9795! **Dah.:** Cotonou *Chev.* 22705. **N.Nig.:** Jebba *Barter*! Naraguta & Jos (Sept.) *Lely* 580! Anara F.R. Zaria (Oct.) *Keay* FHI 5461! **S.Nig.:** Idumuye (fl. & fr. Dec.) *Thomas* 2115! Ibadan (fl. & fr. Jan.) *Meikle* 960! Newi, Onitsha (Oct.)

Jones FHI 769 ! **Br.Cam.**: Buea, 3,000 ft. (fl. & fr. Dec.) *Maitland* 115 ! Widely distributed throughout the tropics.

44. **C. falcata** *Vahl ex DC.*—F.T.A. 2 : 40 ; Bak. f. l.c. 344 ; Chev. Bot. 168 ; Wilczek l.c. 169. *C. obovata* G. Don (1832), not of F.W.T.A., ed. 1, 1 : 398. *C. striata* of F.W.T.A., ed. 1, 1 : 398, partly (*Dawodu* 34), not of DC. Shrubby, diffusely branched, 2–3 ft. high, shortly silky pubescent above ; flowers yellow, purple-veined, in long-stalked racemes ; often in coastal sands.
Fr.G.: Ymbo to Kouria *Chev.* 14987 ! **Lib.:** Bassa Cove *Ansell* ! Cape Palmas (fl. & fr. July) *T. Vogel* 16 ! Cinco (fl. & fr. Nov.) *Linder* 1545 ! Sinkor (fl. & fr. June, July) *Barker* 1363 ! 1382 ! Nana Kru (fl. & fr. Mar.) *Baldwin* 11597 ! **G.C.:** Accra (Oct.) *Brown* 400 ! Ada (fl. & fr. June) *Morton* GC 9240 ! **Togo:** Lome *Warnecke* 289 ! **Dah.:** *Burton* ! Cotonou (fl. & fr. Mar.) *Debeaux* 171 ! **S.Nig.:** Lagos (Oct.) *Dalz.* 1225a ! *MacGregor* 367 ! *Dawodu* 34 ! Also in Belgian Congo and Angola.

45. **C. naragutensis** *Hutch.* in Kew Bull. 1921 : 363 ; Bak. f. Leg. Trop. Afr. 1 : 44. *C. obovata* of Bak. f. J. Linn. Soc. 42 : 345, partly (*Dalz.* 341) ; of F.W.T.A., ed. 1, 1 : 398. *C. keilii* Bak. f. var. *chevalieri* Bak. f. (1914). An erect, branched undershrub, 3–6 ft. high ; flowers yellow in racemes up to 1 ft. long ; in open savannah.
G.C.: Navrongo (fl. & fr. Nov.) *Akpabla* 388 ! Gambaga (Dec.) *Adams* GC 4256 ! Tamale (fl. & fr. Dec.) *Morton* GC 9916 ! **N.Nig.:** Katagum (Sept.) *Dalz.* 19 ! Sokoto (fr. Sept., fl. & fr. Nov.) *Dalz.* 341 ! *Lely* 157 ! Naraguta (Aug.) *Lely* 490 ! Bauchi Plateau (July, Aug.) *Lely* P. 513 ! P. 613 ! Yola (Sept.) *Dalz.* 304 ! **Br.Cam.:** Bama, Dikwa Div. (fl. & fr. Nov.) *McClintock* 4 ! Also in French Chad, and A.-E. Sudan.

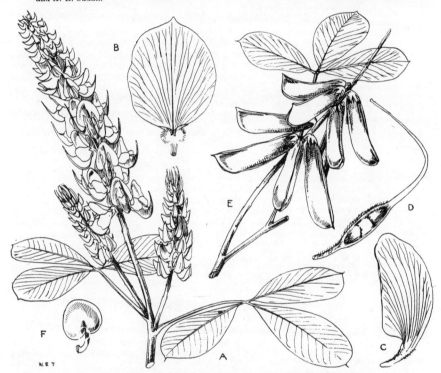

FIG. 166.—CROTALARIA MUCRONATA *Desv.* (PAPILIONACEAE).

A, flowering shoot. B, standard. C, wing petal. D, pistil. E, fruits. F, seed.

46. **C. axillaris** *Ait.* Hort. Kew. 3 : 20 (1789) ; Bak. f. l.c. 388 ; Wilczek l.c. 138, t. 8. *C. latifolia* ("*lotifolia*" by error) of F.T.A. 2 : 42, not of Linn. An erect shrub or woody herb, up to 10 ft. high, shortly silky-pubescent ; flowers bright yellow ; in thickets and old farms.
G.C.: Cape Coast (July) *T. Vogel* 91 ! Waspide (fl. & fr. June) *Dalz.* 81 ! Abokobi (Feb.) *Irvine* 1988 ! Also in Belgian Congo, Angola, tropical E. Africa to Nyasaland, and in the West Indies.

47. **C. ochroleuca** *G. Don*—F.T.A. 2 : 37 ; Bak. f. l.c. 327 ; Exell Cat. S. Tomé 152 ; Wilczek l.c. 221. *C. intermedia* of F.W.T.A., ed. 1, 1 : 398 ; of Chev. Bot. 169 ; of Berhaut Fl. Sen. 14 ; of Bak. f. l.c. 327, partly. An erect undershrub, 3–7 ft. high, with ribbed branches, shortly appressed-pubescent ; flowers whitish or pale yellow, purple-veined, in long rather lax terminal racemes ; usually in wet places.
Sen.: Toubacouta (fl. & fr. Oct.) *Monod* 8602 ! 8712 ! **Port.G.:** Djécoi to Contubo-el, Bafatá (fl. & fr. Nov.) *Esp. Santo* 2813 ! **Fr.Sud.:** fide Chev. l.c. **S.L.:** Layah, Scarcies R. *Sc. Elliot* 4483 ! Baoma (fl. & fr. Dec.) *Deighton* 3470 ! Joru (Oct.) *Deighton* 5237 ! Ronietta (Nov.) *Thomas* 5407 ! **Lib.:** Kakatown *Whyte* ! **Iv.C.:** fide Chev. l.c. **G.C.:** Achimota (May, June) *Irvine* 568 ! 1767 ! **N.Nig.:** Sokoto (fl. & fr. Oct.) *Dalz.* 339 ! Naraguta & Jos (Sept.) *Lely* 576 ! Yola (Oct.) *Shaw* 83 ! Kalkala, L. Chad (Oct.) *Gwynn* 89 ! **S.Nig.:** Lagos *Millen* 98 ! Aguku Dist. *Thomas* 793 ! **Br.Cam.:** Fonfuka, Bamenda, 3,000 ft. (June) *Maitland* 1571 ! Bu, Bamenda (Apr.) *Ujor* FHI 30091 ! Also in French Cameroons, Ubangi-Shari, A.-E. Sudan, Abyssinia, Uganda, Tanganyika, N. & S. Rhodesia, Belgian Congo, Angola and S. Tomé.

48. **C. intermedia** *Kotschy*—F.T.A. 2 : 37 ; Bak. f. l.c. 327, partly ; Wilczek l.c. 220. Similar to the last but with narrower fruits and slightly larger flowers.
N.Nig.: Bauchi Plateau (Sept.) *Lely* P. 706 ! Also in A.-E. Sudan, Abyssinia, Uganda, Kenya, Tanganyika and Belgian Congo.

49. **C. doniana** *Bak.* in F.T.A. 2 : 28 (1871) ; Bak. f. l.c. 356 ; Wilczek l.c. 163. A herbaceous undershrub, laxly branched, thinly pubescent ; flowers yellow, in few-flowered axillary racemes.
    S.L.: *Don!* Samu country (fr. Dec.) *Sc. Elliot* 4380! Malema (Nov.) *Deighton* 449! Sefadu (fr. Dec.) *Deighton* 3572! Yoni, Mamila (Dec.) *King 8b!* **Lib.:** Zigida, Vonjama (Oct.) *Baldwin* 10000! Belleyella, Boporo (fl. & fr. Dec.) *Baldwin* 10643! **Iv.C.:** Agboville (Nov.) *Chev.* 22386! **G.C.:** Sekodumasi (fl. & fr. Oct.) *Vigne* FH 3394! **S.Nig.:** Ibadan (fl. & fr. Jan., Dec.) *Meikle* 1117! *Chizea* FHI 23966! Shasha F.R. (fr. Feb.) *Jones & Onochie* FHI 4718! Oban *Talbot!* **Br.Cam.:** Mukonyong, Mamfe (fl. & fr. Nov.) *Tamajong* FHI 22120! Also in French Cameroons and Belgian Congo.
50. **C. sp. B.**   A shrub up to 8 ft. high.
    Br.Cam.: Victoria (Dec.) *Maitland* 834! Cam. Mt. (Dec.) *Maitland* 839!
51. **C. cleomifolia** *Welw. ex Bak.* in F.T.A. 2 : 43 (1871) ; Bak. f. l.c. 350 ; Wilczek l.c. 158. A shrubby plant, 3–6 ft. or more high ; flowers yellow ; beside streams.
    S.L.: Kamakwie (fl. & fr. Feb.) *Deighton* 4047! 4281! Musaia (Dec.) *Deighton* 4542! Kamiendo to Saiama (Dec.) *Deighton* 5285! **Togo:** *fide* Wilczek *l.c.* **N.Nig.:** Bauchi Plateau (Sept.) *Lely* P. 735! *Batten-Poole* 219! Also in French Cameroons, Belgian Congo, Uganda, Kenya, Angola, Rhodesia, Nyasaland and Mozambique.

Besides the above, the India species *C. alata* Buch.-Ham. ex Roxb. with a winged stem has been introduced into Gold Coast ; *C. juncea* Linn. into Senegal, Gold Coast and S. Nigeria ; *C. spectabilis* Roth (syn. *C. sericea* Retz., not of Burm. f.) into S. Nigeria ; and the E. African *C. zanzibarica* Benth. (syn. *C. usaramoensis* Bak. f.) into Senegal at Hann.

30. LUPINUS Linn.—F.T.A. 2 : 44.

Annual ; stems silky-villous ; leaves digitately 5–7-foliolate ; leaflets sessile, oblanceolate, 2·5–3·5 cm. long, up to 1 cm. broad, silky-pilose beneath ; flowers in lax racemes ; calyx campanulate, nearly 1 cm. long, 2-lobed to nearly the middle, the upper lobe entire or emarginate, the lower slightly 3-toothed ; stamens united in a tube ; fruit compressed, dehiscent, silky-pilose, septate between the seeds   ..   *tassilicus*

L. tassilicus *Maire* in Mém. Soc. Hist. Nat. Afr. Nord. 3 : 266 (1933) ; A. Chev. & Trochain in Rev. Bot. Appliq. 17 : 92, t. 1 (1937). (?) *L. termis* of F.T.A. 2 : 44, partly ; of Bak. f. Leg. Trop. Afr. 1 : 60, partly ; of F.W.T.A., ed. 1, 1 : 399. Erect annual 12–18 in. high ; flowers bluish.
    Sen.: *Leprieur. Perrottet* 243! Diorbivol to Oré Fondé *Trochain* 2927. Podor *Berhaut* 944. Also in Central Sahara.

A. Chev. & Trochain l.c. 97 consider *L. tassilicus* to be the only species reaching our area, but they do not indicate whether this is *L. termis* of F.W.T.A., ed. 1. I have insufficient material at my disposal to make a decision.

31. ADENOCARPUS DC.—F.T.A. 2 : 47.

Branches rigid, densely silky-pilose ; leaves often fasciculate, digitately 3-foliolate ; leaflets oblong-lanceolate, acute, about 5–8 mm. long ; flowers in dense sessile terminal heads ; calyx pubescent, the upper two teeth lanceolate, acuminate, the lower 3-dentate ; corolla about twice as long as the calyx ; fruits about 2–2·5 cm. long, 5–6-seeded, viscous-glandular   ..   ..   *mannii*

A. mannii *(Hook f.) Hook f.* in J. Linn. Soc. 7 : 189 (1864) ; F.T.A. 2 : 47 ; Bak. f. Leg. Trop. Afr. 1 : 69. *Cytisus mannii* Hook. f. (1862). A shrub up to 6 ft. or more high, with yellow flowers and shortly bristly fruits.
    Br.Cam.: Cam. Mt., 8,000–12,000 ft. (Jan.) *Mann* 1308! 2180! *Hutch. & Metcalfe* 63! *Dalz.* 8316! *Dunlap* 209! Bamenda Nkwe (Sept.) *Ujor* FHI 30218! L. Bambulue (fr. Jan.) *Keay & Lightbody* FHI 28378! **F.Po:** *Mildbr.* 7174. Clarence Peak, 9,000 ft. *Mann* 594! Also in French Cameroons and the mountains of eastern Africa from A.-E. Sudan to Nyasaland. (See Appendix, p. 226.)

32. LOTUS Linn.—F.T.A. 2 : 61 ; Brand in Engl. Bot. Jahrb. 25 : 166 (1898).

Style dentate ; inflorescence 3–6-flowered ; fruits about 2·5 cm. long, glabrous ; peduncle longer than the leaves ; leaves petiolate :
  Flowers purple, 13 mm. long ; bracts 1–3, longer than calyx ; branches, etc., subglabrous and subglaucous ; leaflets linear-oblanceolate, acute at apex   1. *jacobaeus*
  Flowers yellow, 10–11 mm. long ; bracts 1–3, shorter than calyx ; branches etc. pilose or tomentose ; leaflets variable, subacute to obtuse at apex   ..   2. *arenarius*
Style not dentate :
  Flowers in terminal heads of 5–7, corolla about 10 mm. long ; calyx 5-fid, teeth about half as long as corolla, shortly brown-pubescent ; bracts 3, amongst the flowers ; peduncles 3–7 cm. long ; fruits 1·5–2·5 cm. long, glabrous ; leaflets narrowly oblanceolate, acute at apex, appressed-pubescent, especially along midrib and margins beneath, terminal leaflet 9–11 mm. long, 2 mm. broad   ..   3. *discolor*
  Flowers 1–3 in axil of leaves, corolla about 7 mm. long ; calyx 5-fid, teeth almost as long as corolla, pubescent ; flowers subtended by a single leafy bract ; peduncles 1–4 cm. long ; fruits 2–3 cm. long, glabrous ; leaflets obovate, shortly pubescent to long-pilose beneath, variable in size, 5–20 mm. long   ..   ..   4. *arabicus*

1. **L. jacobaeus** *Linn.* Sp. Pl. 2 : 775 (1753) ; Brand l.c. 203 ; Bak. f. Leg. Trop. Afr. 1 : 86. Herbaceous, woody at base, with purple flowers.
    Gam.: *Mungo Park!* Also in Cape Verde Islands.
2. **L. arenarius** *Brot.* Fl. Lusit. 2 : 120 (1804) ; Brand l.c. 198. A much-branched perennial herb with a woody stock, about 6 in. high ; flowers yellow.
    Sen.: *Irvine* 3250! Also in Spain, N. Africa and Canary Islands.
3. **L. discolor** *E. Mey.* Comm. Pl. Afr. Austr. 92 (1835) ; Brand l.c. 213 ; Bak. f. l.c. 88 ; Brenan in Mem. N.Y. Bot. Gard. 8 : 249 ; Boutique in Fl. Congo Belge 4 : 302, fig. 20, A–B. An ascending herb with wiry branches from a woody stock ; flowers yellow.
    Br.Cam.: Bamenda *Johnstone* 205/31! L. Bambili, Bafut-Ngemba F.R. (Aug.) *Ujor* FHI 29973! Also in French Cameroons, Belgian Congo, Abyssinia, E. Africa, Nyasaland, S. Rhodesia, Mozambique and in S. Africa.

4. L. arabicus *Linn.*—F.T.A. 2 : 62 ; Brand l.c. 219 ; Bak. f. l.c. 90. *L. mossamedensis* Welw. ex Bak. (1871). A procumbent annual, woody at base, variable in all its parts ; branches 1 ft. or more long ; flowers pink or whitish.
Sen.: *Roger!* *Heudelot!* *Perrottet* 199 ! **Gam.:** Kantora (fl. & fr. Jan.) *Frith* 164 ! **Fr.Sud.:** Dogo (fr. May) *Davey* 105 ! **N.Nig.:** Katagum *Dalz.* 46 ! Extends to Egypt and Baluchistan, Tanganyika, Mozambique and Angola. (See Appendix, p. 250.)

33. TRIFOLIUM Linn.—F.T.A. 2 : 53.

By J. B. Gillett

Petiole adnate to stipules almost throughout its length :
Calyx-nerves 15–20 ; inflorescence more than 1 cm. broad, usually broader than long ;
   leaflets narrowly oblanceolate, more than 4 times as long as broad   **1.** *simense*
Calyx-nerves 10–12 ; inflorescence not more than 1 cm. broad, usually longer than
   broad ; leaflets broadly oblanceolate, less than 4 times as long as broad
                                          2. *usambarense*
Petiole, at any rate in lower leaves, free from the stipules for the greater part of its
   length :
Standard 5–7 mm. long ; ovules 4–5 ; leaflets 1·5–3 cm. long, not emarginate
                                       3. *rueppellianum* var. *preussii*
Standard 3–5 mm. long ; ovules 2 ; leaflets less than 1·5 cm. long, emarginate
                                          4. *baccarinii*

1. T. simense *Fres.*—F.T.A. 2 : 57 ; Gillett in Kew Bull. 1952 : 375, figs. 2, 1–7 ; 3 ; Fl. Congo Belge 4 : 293 ; Jac.-Fél. in Agronom. Trop. 8 : 288, t. 1 C. Straggling perennial herb up to 1½ ft. high, with purple flowers ; in mountain grassland.
Br.Cam.: Cam. Mt., 7,000–9,000 ft. (Nov.–Feb.) *Johnston* 4 ! *Mann* 2018 ! *Mildbr.* 10862 ! *Maitland* 915 ! 1006 ! 1210 ! *Migeod* 168 ! L. Oku, Bamenda, 7,700 ft. (Jan.) *Keay* FHI 28466 ! **F.Po:** 8,500 ft. (Dec.) *Mann* 602 ! *Mildbr.* 7170. Also in the mountains of French Cameroons, Abyssinia and eastern Africa.
2. T. usambarense *Taub.* in Engl. Pflanzenw. Ost-Afr. C. 208 (1895) ; Gillett in Kew Bull. 1952 : 370, fig. 1, 13–15 ; Fl. Congo Belge 4 : 292 ; Jac.-Fél. l.c., t. 1 B. Straggling herb up to 1½ ft. high, with purple (occasionally white) flowers ; in mountain grassland and forest margins.
Br.Cam.: Bamenda, 5,000–7,700 ft. (Jan., Feb., June) *Migeod* 452 ! *Keay* FHI 28351 ! 28436 ! 28466a ! 28506 ! *Maitland* 139 ! **F.Po:** (Mar.) *Guinea* 2907 ! Also in the mountains of French Cameroons, S. Abyssinia and eastern Africa.
3. T. rueppellianum *Fres.* var. **preussii** (*Taub. ex Bak. f.*) *Gillett* in Kew Bull. 1952 : 389, fig. 4, 2–6 ; Fl. Congo Belge 4 : 296. *T. preussii* Taub. ex Bak. f. Leg. Trop. Afr. 1 : 82 (1926)—F.T.A. 2 : 58, 59. A weak herb up to 3 ft. high, with bluish-purple flowers.
Br.Cam.: Cam. Mt., 7,000–8,500 ft. (Nov.) *Preuss* 972 ! *Migeod* 203 ! Jakiri, 5,000–7,000 ft. (fl. & fr. Nov.) *McCulloch* 1 ! **F.Po:** 9,000 ft. (Dec.) *Mann* 603 ! The typical form of the species is found in the mountains of Abyssinia and eastern Africa. In the latter area var. *preussii* also occurs together with numerous intermediates. The identification of var. *preussii* with *T. goetzenii* Taub. ex Engl. seems to be incorrect.
4. T. baccarinii *Chiov.* Ann. Bot. Rom. 9 : 57 (1911) ; Gillett in Kew Bull. 1952 : 391, fig. 4, 1 ; Fl. Congo Belge 4 : 298`; Jac.-Fél. l.c. 288, t. 1 A. A herb up to 8 in. high, with bluish-purple (occasionally white) flowers.
Br.Cam.: Banso, Bamenda, in moist grass by roadside, 6,000 ft. (Jan.) *Keay* FHI 28449 ! Jakiri, 5,000–7,000 ft. (Nov.) *McCulloch* 3 ! Also in the mountains of French Cameroons, Abyssinia and eastern Africa.

34. RHYNCHOSIA Lour.—F.T.A. 2 : 216.

By R. D. Meikle

Leaves 3-foliolate :
Calyx conspicuous, persistent and often accrescent, basal lobe 1 cm. or more long, linear-lanceolate, almost as long or longer than the corolla :
  Calyx villous or pubescent, not glandular-viscid ; seeds blue or blue-black :
    Lateral calyx-lobes very small, about ¼ as long as the others, upper calyx-lobes connate, much longer than the vexillum ; leaves more or less glabrous above, closely felted beneath   .. .. **1.** *preussii*
    Lateral calyx-lobes nearly as long as the others :
      Leaflets pubescent above and beneath ; calyx narrowly lanceolate-acuminate, as long as, or a little shorter than the corolla .. .. **2.** *congensis*
      Leaflets densely and softly tomentose beneath :
        Calyx-lobes densely grey-pubescent, oblanceolate, acute or subobtuse, usually longer than the cream-coloured corolla .. .. **3.** *albiflora*
        Calyx-lobes densely brown-pubescent, acuminate, slightly shorter than the red corolla .. .. **4.** *brunnea*
    Leaflets glabrous or nearly so :
      Calyx-lobes acute or subobtuse, densely grey-pubescent .. **5.** *pycnostachya*
      Calyx-lobes acuminate, thinly and minutely puberulous .. .. **6.** *mannii*
  Calyx densely glandular-viscid ; seeds dark brown :
    Flowers in elongate, branched or unbranched racemes ; terminal leaflet rhomboid or ovate, acuminate, usually 4–6 cm. long, 2–5 cm. broad, obscurely veined beneath :
      Racemes lax, branched ; plant twining or subscandent .. .. **7.** *resinosa*
      Racemes rigid, unbranched ; plant erect or suberect .. .. **8.** *orthobotrya*

Flowers in cylindrical densely congested racemes ; terminal leaflet depressed-rhomboid, obtuse, 5–8 cm. long, 5–9 cm. broad, strongly reticulate-veined beneath ; plant trailing ..   ..   9. *floribunda*

Calyx small, caducous, not accrescent, basal lobe generally less than 1 cm. long, much shorter than corolla (if nearly as long then lobes subulate-lanceolate) :

Calyx-lobes rounded or very obtuse at apex ; leaflets rhomboid, acute, 5–8 cm. long, up to 6 cm. broad, pubescent beneath ; racemes elongate, softly tomentellous ; seeds blue-black..   ..   ..   10. *buettneri*

Calyx-lobes acute to acutely acuminate ; seeds brown :

Flowers in dense, short, subsessile racemes ; fruits small, 1–1·5 cm. long, pilose, with weak, spreading hairs ..   ..   11. *densiflora*

Flowers in slender, pedunculate racemes :

Upper portions of stems, inflorescences and fruits conspicuously glandular-viscid :

Racemes long and lax ; fruits with long, scattered, golden, viscid hairs ; stipules large, persistent ; plant trailing or climbing ..   ..   12. *violacea*

Racemes short and rather rigid ; fruits with dense, short, pale, viscid indumentum ; stipules small, caducous ; plant suberect ..   ..   13. *nyasica*

Stems, inflorescences and fruits glabrous or pubescent, but not viscid :

Flowers conspicuous, 1–1·5 cm. long ; fruits large, inflated, 2–3 cm. long ; leaflets obtusely rhomboid, 2·5–7 cm. or more long, 2–5 cm. broad, subglabrous or shortly puberulous ..   ..   14. *sublobata*

Flowers inconspicuous, less than 1 cm. long ; fruits small, not inflated, 1–2 cm. long ; leaflets 1·5–4 cm. long, 1·5–3 cm. broad :

Flowers minute, about 5 mm. long ; petals scarcely exceeding calyx :

Leaflets acutely rhomboid or subrotund, glabrous or pubescent on nerves beneath ; plant usually climbing ..   ..   15. *minima* var. *minima*

Leaflets subrotund, densely silvery-pubescent above and beneath ; fruits silvery-pubescent ; plant usually trailing .. 15a. *minima* var. *memnonia*

Flowers larger, up to 1 cm. long ; leaves subrotund or bluntly rhomboid, shortly and sparsely pubescent above and beneath

15b. *minima* var. *prostrata*

Leaves 1-foliolate ; flowers solitary or geminate, axillary, calyx-teeth lanceolate-acuminate, nearly as long as petals ..   ..   16. *chevalieri*

1. **R. preussii** (*Harms*) *Taub. ex Harms* in Engl. Pflanzenw. Afr. 3, 1 : 671 (1915) ; Bak. f. Leg. Trop. Afr. 2 : 468. *Cylista preussii* Harms in Engl. Bot. Jahrb. 26 : 303 (1899). *R. cyanosperma* of F.W.T.A., ed. 1, 1 : 401, partly (*Barter* 1615). Twiner in forest.
   Br.Cam.: Barombi (May) *Preuss* 190. **F.Po:** (June) *Barter* 1615 ! Also in French Cameroons and Gabon.
2. **R. congensis** *Bak.* in F.T.A. 2 : 217 (1871). *R. cyanosperma* Benth. ex Bak. in F.W.T.A., ed. 1, 1 : 401, partly. A stout twiner with elongate racemes of cream-coloured flowers, grey-velvety fruits and conspicuous blue seeds.
   Port.G.: Prabis, Bissau (fr. Mar.) *Esp. Santo* 1849 ! **N.Nig.:** Oke-Opin, Ilorin (Dec.) *Ajayi* FHI 19267 ! Gwari Hill, Minna (fl. & fr. Dec.) *Meikle* 704 ! **S.Nig.:** Lagos *Millen* 116 ! Also in French Cameroons, Belgian Congo, Uganda and Angola.
3. **R. albiflora** (*Sims*) *Alston* in Fl. Ceyl. 6, Suppl. 85 (1931) ; not of Berhaut (1954). *Cylista albiflora* Sims in Bot. Mag. t. 1859 (1816). *R. cyanosperma* Benth. ex Bak. in F.T.A. 2 : 218 (1871) ; F.W.T.A., ed. 1, 1 : 401, partly ; Bak. f. l.c. 2 : 469. A robust twiner with pale yellow, purple-streaked petals, and soft tomentose leaflets.
   N.Nig.: Naraguta (Nov.) *Lely* 689 ! Bauchi Plateau *Batten-Poole* 232 ! Widespread in tropical E. Africa, Mascarene Islands, India and Ceylon.
4. **R. brunnea** *Bak. f.* Leg. Trop. Afr. 2 : 468 (1929). Robust twiner with red flowers and dense brownish indumentum.
   S.L.: *Wilford* ! **G.C.:** Tano Offin F.R. (Jan.) *Vigne* FH 1007 !
5. **R. pycnostachya** (*DC.*) *Meikle* in Kew Bull. 1954 : 274. *Cylista* ? *pycnostachya* DC. Prod. 2 : 410 (1825). *Rhynchosia calycina* Guill. & Perr. Fl. Seneg. 1 : 214 (1832) ; F.T.A. 2 : 217 ; Chev. Bot. 205 ; Bak. f. l.c. 467 ; F.W.T.A., ed. 1, 1 : 401. Robust twiner, with glossy green leaves and long racemes of cream-coloured flowers ; petals persistent, rusty red in fruit ; seeds blue.
   Sen.: Cape Verde *Perrottet.* Ziguinchor *Chev.* 3406. **Gam.:** *Ingram* ! Abuko (fr. Jan.) *Dalz.* 8028 ! **Fr.Sud.:** Medinani *Chev.* 506. **Fr.G.:** *Farmar* 205 ! Conakry, Kouria & Fouta Djalon *fide* Chev. l.c. **S.L.:** *T. Vogel* 66 ! *Barter* ! Njala (Dec.–Jan.) *Dalz.* 8069 ! *Deighton* 1855 ! Kitchom, Samu (fr. Dec.) *Sc. Elliot* 4318 ! Port Loko (Dec.) *Thomas* 6573 ! Musaia (Dec.) *Deighton* 4484 ! **Lib.:** Sinoe Basin *Whyte* ! Kakatown *Whyte* ! Kle, Boporo (Dec.) *Baldwin* 10636 ! Ganta (fr. Feb.) *Baldwin* 11011 ! **Iv.C.:** Bingerville, Bouroukrou & Zaranou *fide* Chev. l.c. **G.C.:** Tano Offin F.R. (fr. Jan.) *Vigne* FH 1008 ! Adamsu (Dec.) *Vigne* FH 3495 ! Bobiri F.R. (Jan.) *Andoh* FH 5094 ! **Togo:** Amedzofe (fr. Feb.) *Irvine* 168 ! **S.Nig.:** Ominla (fr. Apr.) *Symington* FHI 3385 ! Agulu (Nov.) *Keay* FHI 25591 ! Eket *Talbot* ! Oban *Talbot* 1748 ! **Br.Cam.:** Ambas Bay (Jan.) *Mann* 730 ! **F.Po:** *Mann* 88a ! (See Appendix, p. 259.)
6. **R. mannii** *Bak.* in F.T.A. 2 : 217 (1871) ; Bak. f. l.c. 2 : 468. Slender climber with glossy leaves, petals cream with red markings ; seeds dark blue ; in forest clearings.
   S.Nig.: Akure F.R. (fr. Nov.) *Latilo* FHI 15281 ! Okomu F.R. (Jan.) *Brenan* 8765 ! Usonigbe F.R. (Nov.) *Meikle & Keay* 579 ! Oban *Talbot* 1206. **Br.Cam.:** Buea (fr. Jan.) *Maitland* 195 ! **F.Po:** (fr. Dec.) *Mann* 88 ! Also in French Cameroons, Ubangi-Shari, Gabon, Belgian Congo, Uganda and Angola.
7. **R. resinosa** (*Hochst. ex A. Rich.*) *Bak.* in F.T.A. 2 : 218 (1871) ; Bak. f. l.c. 480. *Fagelia resinosa* Hochst. ex A. Rich. (1847). Subscandent viscid perennial, with yellow flowers.
   Fr.G.: Timbo (Dec.) *Chev.* 14631 ! **N.Nig.:** Bauchi Plateau (Jan.) *Lely* P. 115 ! *Batten-Poole* 218 ! Jos (Nov.) *Keay* FHI 21436 ! Widespread in east, central and south tropical Africa.
8. **R. orthobotrya** *Harms* in Engl. Bot. Jahrb. 54 : 390 (1917) ; Bak f. l.c. 478. *R. gorsii* Berhaut in Mém. Soc. Bot. Fr. 1953–54 : 6 (1954). Erect or suberect perennial with erect unbranched racemes of pale yellow flowers ; fruits glandular-pilose.
   Sen.: Niokola-Koba (fr. Apr.) *Berhaut* 1533 ! Also in A.-E. Sudan, eastern Africa and Belgian Congo.

9. **R. floribunda** *Bak.* in Kew Bull. 1897 : 262 ; Bak. f. l.c. 479. Trailing plant on open ground with leaves pale green above and greyish beneath ; flowers yellow, tinged brown ; calyx viscid.
G.C.: Wenchi (Dec.) *Morton* GC 25151 ! **N.Nig.**: Abuja to Kaduna (Dec.) *Meikle* 741 ! Sho, Plateau Prov. (Nov.) *Keay* FHI 21446 ! **Br.Cam.**: Mai-Idoanu, Gashaka Dist. (Jan.) *Latilo & Daramola* FHI 28989 ! Also in Belgian Congo, Uganda, Kenya, Tanganyika, Nyasaland, N. and S. Rhodesia.

10. **R. buettneri** *Harms* in Engl. Bot. Jahrb. 30 : 90 (1901) ; Bak. f. l.c. 469. Scrambling subshrub, stems slightly viscid, leaves dark green above, pale and softly grey-felted beneath ; flowers yellow-green with dark purple lines ; fruits grey-felted, seeds dark blue.
Fr.G.: Kouroussa (Oct.) *Pobéguin* 961 ! **G.C.**: Jaketi (Oct.) *Morton* A. 89 ! **Togo**: Bismarckburg (Oct.) *Büttner* 149. **S.Nig.**: Lagos *MacGregor* 137 ! Ibadan (fl. & fr. Dec.) *Meikle & Keay* 858 ! Wumuye (fr. Jan.) *Thomas* 2162 ! Also in French Cameroons and Upper Ubangi. (See Appendix, p. 259.)

11. **R. densiflora** (*Roth*) *DC.* Prod. 2 : 386 (1825) ; Bak. f. l.c. 2 : 470. *Glycine densiflora* Roth (1821). *R. debilis* G. Don (1832)—F.T.A. 2 : 222 ; F.W.T.A., ed. 1, 1 : 401. Slender twiner, by forest margins ; leaves dull green ; petals greenish-yellow striped red.
Port.G.: Pobresa, Cubisseque (fr. Feb.) *Esp. Santo* 2375 ! **S.L.**: Yonibana (Nov.) *Thomas* 4748 ! **G.C.**: Kumasi *Cummins* 240 ! *Morton* GC 9685 ! Ejura Scarp (Dec.) *Morton* GC 9800 ! Dedeman (fr. Dec.) *Irvine* 2102 ! **S.Nig.**: Lagos (fl. & fr. Mar.) *Millen* 90 ! Ibadan (fl. & fr. Nov.) *Meikle* 660 ! Igbara Oke, Ondo Prov. (fl. & fr. Nov.) *Meikle & Keay* 510 ! Ogwashi (Dec.) *Thomas* 2056 ! Oban *Talbot* 1311 ! **Br.Cam.**: Victoria (fl. Dec., fr. Feb.) *Mann* 1249 ! *Maitland* 428 ! Widespread in tropical Africa, in S. Tomé and also in India.

12. **R. violacea** (*Hiern*) *K. Schum.* in Just. Bot. Jahr. 27, 1 : 496 (1901) ; Bak. f. l.c. 47. *Dolicholus violaceus* Hiern in Cat. Welw. Afr. Pl. 1 : 269 (1896). *Rhynchosia viscosa* of F.T.A. 2 : 222, partly ; of F.W.T.A., ed. 1, 1 : 401 ; not of DC. Perennial herb with straggling shoots, viscid ; petals yellow with purple-brown or yellow markings ; pods viscid.
Port.G.: Bafata (Oct.) *Esp. Santo* 2313 ! **S.L.**: Musaia (fl. & fr. Dec.) *Deighton* 4417 ! **Togo**: Vane (Nov.) *Morton* GC 9375 ! **N.Nig.**: Lokoja (Oct.) *Parsons* 97 ! **S.Nig.**: Lagos (fr. Mar.) *Millen* 157 ! Ibadan (fl. & fr. Nov.–Jan.) *Meikle* 663 ! 970 ! Kishi to Igoboho (Feb.) *Keay* FHI 22511 ! Also in Belgian Congo, Uganda and Angola.

13. **R. nyasica** *Bak.* in Kew Bull. 1897 : 263 ; Bak. f. l.c. 2 : 478. *R. glutinosa* Harms (1899)—Chev. Bot. 206. *R. kerstingii* Harms (1913). Suberect, with lax viscid-hairy branches ; flowers yellow with purple streaks.
Fr.Sud.: Balato & Koundian *fide* Chev. *l.c.* **Fr.G.**: Beyla (Mar.) *Chev.* 20899 ! **S.L.**: Falaba (Mar.) *Sc. Elliot* 5173 ! **Togo**: Salaga *Krause* ! Sokode-Basari *Kersting* 482. **N.Nig.**: Kontagora (fl. & fr. Jan.) *Dalz.* 45 ! *Meikle* 1066 ! Alagbede F.R. (Feb.) *Ejiofor* FHI 19809 ! Ilorin-Jebba road (fr. Mar.) *Meikle* 1234 ! Vom *Dent Young* 68 ! Bauchi Plateau (Jan.) *Lely* P. 58 ! Also in French Cameroons, Belgian Congo, A.-E. Sudan, N. Rhodesia, Angola and Nyasaland.

14. **R. sublobata** (*Schum. & Thonn.*) *Meikle* in Kew Bull. 1951 : 176, fig. 3. *Glycine sublobata* Schum. & Thonn. (1827). *R. caribaea* of various authors, including :—F.T.A. 2 : 220, partly ; Chev. Bot. 205 ; F.W.T.A., ed. 1, 1 : 402, partly ; Bak. f. l.c. 2 : 474, partly ; not of DC. *R. macinaca* A. Chev. Bot. 206, name only. Trailing or twining herb with perennial woody rootstock ; flowers yellow, lined deep red.
Sen.: *Heudelot* 401 ! Tambacounda (fl. & fr. Sept.) *Berhaut* 1710 ! **Fr.Sud.**: Macina (Sept.) *Chev.* 24851 ! **G.C.**: *Thonning* ! Accra Plains (fl. & fr. Oct.) *Dalz.* 9 ! *Irvine* 820 ! Akuse (Oct.) *Vigne* FH 4049 ! **Togo**: Yendi (Dec.) *Adams & Akpabla* GC 4049 ! Amedzofe (May) *Morton* GC 7732 ! Kete Krachi (fl. & fr. Apr.) *Morton* GC 9099 ! **Dah.**: *fide* Chev. *l.c.* **N.Nig.**: Kontagora (fl. Dec., fl. & fr. Jan.) *Dalz.* 44 ! *Meikle* 1033 ! Abinsi (Apr.) *Dalz.* 612 ! **S.Nig.**: Ado *Rowland* ! Widespread throughout the drier parts of tropical Africa. (See Appendix, p. 259.)

15. **R. minima** (*Linn.*) *DC.* var. **minima**—Prod. 2 : 385 (1825) ; F.T.A. 2 : 219, partly ; F.W.T.A., ed. 1, 1 : 402, partly ; Bak. f. l.c. 473. *Dolichos minimus* Linn. (1753). Slender twiner, in open savannah ; flowers small, brownish-yellow.
Sen.: *Roger* ! *Heudelot* 532 ! **S.L.**: Musaia (Mar.) *Deighton* 4604 ! **G.C.**: *Irvine* 1611 ! Accra Plains (July) *Brown* 335 ! *Irvine* 3029 ! Wa (Dec.) *Adams* GC 4473 ! **N.Nig.**: Kano (fr. Dec.) *Meikle* 808 ! Katagum (fl. & fr. Sept.) *Dalz.* 38 ! Widespread through the drier parts of tropical Africa, also in W. Indies, tropical America and Asia. (See Appendix, p. 260.)

15a. **R. minima** var. **memnonia** (*Del.*) *Cooke* Fl. Bombay, 1 : 389 (1903). *Dolichos memnonia* Del. Fl. Egypt 254, t. 38, 3 (1813). *R. memnonia* (Del.) DC—F.T.A. 2 : 220 ; Chev. Bot. 206 ; F.W.T.A., ed. 1, 1 : 401 ; Bak. f. l.c. 473 ; Berhaut Fl. Sen. 18. Greyish pubescent trailing herb, with small yellow flowers.
Sen.: *Berhaut* 705. **Fr.Sud.**: Tile (Aug.) *Chev.* 3399 ! **N.Nig.**: Katagum (fl. & fr. July) *Dalz.* 35 ! Kukawa (fl. & fr. Jan.) *E. Vogel* 8 ! Extends to Egypt, Arabia, A.-E. Sudan and Somaliland.

15b. **R. minima** var. **prostrata** (*Harv.*) *Meikle* in Kew Bull. 1954 : 275. *R. memnonia* var. *prostrata* Harv. in Fl. Cap. 2 : 253 (1862). *R. minima* of F.W.T.A., ed. 1, 1 : 402, partly. Creeping, or rarely climbing, perennial herb with yellow, brown- or purple-tinged petals.
G.C.: Accra *Brown* 388 ! Akuse *A. S. Thomas* 23 ! **N.Nig.**: Nupe *Barter* 1590 ! Jebba (Dec.) *Meikle* 681 ! Funtua (fl. & fr. Feb.) *Meikle* 1226 ! Lokoja *Barter* ! Mamu (Mar.) *Lely* 163 ! **S.Nig.**: Ogbo-mosho *Hb. Calcut.* 15 ! Throughout the drier parts of tropical and S. Africa.

16. **R. chevalieri** *Harms* in Engl. Bot. Jahrb. 40 : 39 (1907) ; Chev. Bot. 206 ; Bak. f. l.c. 2 : 471. *Eriosema chevalieri* (Harms) Hutch. & Dalz. F.W.T.A., ed. 1, 1 : 404 (1928) ; Kew Bull. 1929 : 17. Stems short and erect from a woody rhizome ; upper leaves much reduced, bract-like ; flowers whitish.
Fr.Sud.: Koundian (Feb.) *Chev.* 422 !

Imperfectly known species.

1. **R. albiflora** *Berhaut* in Mém. Soc. Bot. Fr. 1953–54 : 6 (1954) ; not of (Sims) Alston (1931). A climbing plant with thinly pubescent stems and leaves, and large subglabrous, falcate fruits. Near *R. viscosa* (Roth) DC., but type material inadequate and abnormal.
Sen.: Messira (fl. & fr. Nov.) *Berhaut* 1182 !

2. **R. sp.**—*R. minima* of F.W.T.A., ed. 1, 1 : 402, partly. Near to *R. usambarensis* Taub. Slender climber with yellow flowers and small, ovate leaflets ; mature fruits wanted.
N.Nig.: *Lely* 688 ! Vom *Dent Young* 69 !

35. ERIOSEMA (DC.) Desv.—F.T.A. 2 : 223.

*Leaves 3-foliolate (or rarely 4–5-foliolate) :
  Leaves linear, very acute, up to 18 cm. long and 1·3 cm. broad, softly white-tomentose beneath and with scattered black glands ; petiole very short ; stipules lanceolate, acute, ribbed, about 1 cm. long, caducous ; racemes terminal, slender, with numerous reflexed flowers about 8 mm. long ; fruit broadly elliptic, 1 cm. long, densely long-pilose   ..   ..   1. *linifolium*
  Leaves lanceolate or oblong to ovate, not linear :
    †Leaves petiolate, petiole nearly as long as lateral leaflets :
      Flowers 6–9 mm. long ; racemes (9–) 14–30-flowered and long-pedunculate ; fruits

elliptic, 1·3–1·5 cm. long, long-pilose ; leaflets ovate or ovate-elliptic, 3–9 cm. long, 1·5–2·5 cm. broad, pubescent on the nerves beneath ; stipules subulate-lanceolate, up to 10 mm. long .. .. 2. *parviflorum* subsp. *parviflorum*

Flowers 12–15 mm. long ; racemes 5–15-flowered ; fruits elliptic about 1·5 cm. long ; leaflets elliptic to ovate, rather small, 1–3 cm. long, 0·5–2 cm. broad ; stipules subulate-lanceolate, about 7 mm. long .. 2a. *parviflorum* subsp. *collinum*

†Leaves sessile or very shortly petiolate, petiole much shorter than the lateral leaflets :

Flowers in long spike-like racemes :

Leaflets completely covered or nearly completely covered beneath with a close felt of hairs, usually white :

Leaves digitately 5-foliolate ; leaflets oblanceolate, apex rounded, minutely hairy above, about 2 cm. long and 7 mm. wide ; stipules ovate-lanceolate, acute, about 3 mm. long ; racemes axillary about as long as the leaves ; flowers about 1 cm. long ; fruits shortly elliptic about 1·4 cm. long, brown-pilose
3. *molle*

Leaves 3-foliolate :

Leaflets digitately 3-foliolate, glabrous above, subsessile, oblanceolate, apex rounded, up to 4 cm. long and 1·2 cm. broad ; stipules persistent, lanceolate, acute, about 4 mm. long ; racemes axillary, with 15–25 flowers ; flowers about 1·2 cm. long ; fruits shortly elliptic, about 1·4 cm. long, long-pilose
4. *andohii*

Leaflets pinnately 3-foliolate, usually shortly hairy above, oblong-lanceolate, acute to emarginate, 4–10 cm. long, up to 3·5 cm. broad ; stipules caducous, triangular, acute, about 2 mm. long ; racemes axillary, longer than the leaves, many-flowered ; flowers up to 1·5 cm. long ; fruits broadly elliptic, 1·5–2 cm. long, densely pilose .. .. 5. *psoraleoides*

Leaflets loosely pilose or pubescent beneath :

Leaflets ciliate or pubescent with long slender hairs :

Leaflets oblong, 2·5–6 cm. long, 1–2·5 cm. broad ; stipules lanceolate, about 8 mm. long, long-ciliate ; racemes about twice as long as the leaves, shortly pubescent ; flowers about 1 cm. long ; fruits broadly elliptic, about 1·5 cm. long, long-pilose .. .. 6. *bauchiense*

Leaflets elongate-lanceolate, 5–11 cm. long, up to 1·5 cm. broad, more or less densely pilose with long slender hairs ; stipules lanceolate, about 1·5 cm. long, closely nerved ; racemes pedunculate, usually as long as or longer than the leaves ; flowers about 9 mm. long ; fruits densely pilose, 1·5 cm. long
7. *shirense*

Leaflets pubescent with very short hairs:

Racemes long and slender, 10–20 cm. long, flowers laxly arranged about 1 cm. apart on the rhachis ; fruits elliptic, 2 cm. long, loosely pilose ; leaflets lanceolate, mucronate, yellow-green above when dry, 3·5–7 cm. long, 0·5–2 cm. broad ; stipules subulate-lanceolate, erect, about 1 cm. long 8. *sparsiflorum*

Racemes dense, 6–13 cm. long, flowers very closely arranged ; fruits elliptic, about 1·5 cm. long, long-pilose ; leaflets elliptic or oblanceolate, apiculate, acute, dark-green above when dry, 4–10 cm. long, 1·5–3 cm. broad ; stipules triangular, usually reflexed, 1–1·5 cm. long .. .. .. 9. *montanum*

Flowers capitate or subcapitate :

Leaflets pubescent mainly on the nerves ; bracts linear, shorter than the flowers ; leaflets lanceolate, very acute or mucronate, up to 12 cm. long and 2 mm. broad ; stipules lanceolate-acuminate, up to 1·5 cm. long ; calyx-lobes filiform, long-pilose ; valves of fruit curling up, glossy inside .. .. 10. *glomeratum*

Leaflets covered all over the lower surface with a soft white indumentum ; bracts broadly lanceolate, longer than the flowers, with long slender tips ; leaflets oblanceolate, rounded or broadly emarginate, 2·5–5 cm. long, 1–2·5 cm. broad, subulate on the upper surface ; stipules lanceolate, strongly nerved ; valves of fruit curling up, pilose outside, glossy inside .. .. .. 11. *griseum*

*Leaves 1-foliolate :

‡Leaflet linear or linear-lanceolate, more or less narrowed to base :

Leaflet softly tomentose on both surfaces, white beneath, linear-lanceolate, 7–11 cm. long, 1–1·5 cm. broad ; stipules lanceolate, acuminate, 1–1·3 cm. long, tomentellous ; stem slender and white with appressed hairs ; flowers about 1·3 cm. long, 5–13 on peduncles 6–8 cm. long ; calyx-lobes linear-lanceolate ; fruits oblong-elliptic, about 1 cm. long, densely pilose .. .. 12. *monticolum*

Leaflet villous with long slender hairs especially on margin and midrib :

Calyx and leaflet not gland-dotted ; stipules about 1·5 cm. long ; other characters as above .. .. 7. *shirense*

Calyx and leaflet strongly gland-dotted ; leaflet linear, 4–10 cm. long, about 5 mm. broad ; stipules ribbed, lanceolate, acuminate, 5–9 mm. long, ciliate ; stem slender and long-pilose ; flowers 4–5 mm. long, 2–4 on peduncles 2–4 cm. long ; fruits elliptic, 1 cm. long, pilose .. .. 13. *tenue*

‡Leaflet broadly cordate at base :
  Leaflet densely silky beneath :
    Leaflet ovate or broadly elliptic, mucronate, rounded or subcordate at base, 7–16 cm.
      long, 3–8 cm. broad, with about 10 pairs of lateral nerves ; stipules very large,
      up to 3 cm. long, appressed-pilose ; flowers in a short dense cylindrical raceme ;
      fruits very densely villous ..   ..   14. *pulcherrimum*
    Leaflet lanceolate, acute, sessile and rather broad at base, 11–16 cm. long, 2–3 cm.
      broad, glabrous above except the midrib, softly whitish-tomentose beneath,
      obscurely nerved ; stipules lanceolate, 1·5–2 cm. long, villous ; racemes dense,
      cylindric, about 7 cm. long in flower, up to 15 cm. in fruit ; flowers small,
      reflexed ; fruits crowded, about 1 cm. long, densely silky-villous and glandular
                                                                    15. *afzelii*

  Leaflet loosely or very shortly pubescent beneath :
    Leaflet almost orbicular, (2)–7–10 cm. long and 5–8 cm. broad, mucronate, cordate,
      nerves prominent above and beneath and ferrugineous with short brown hairs ;
      petiole about 1 cm. long ; stipules broadly lanceolate, ribbed, long-acuminate,
      1–2 cm. long ; flowers reflexed in axillary racemes ; calyx-lobes linear-lanceolate,
      less than half as long as corolla ; fruits oblong-elliptic, densely pilose
                                                                    16. *chrysadenium*

  Leaflet broadly lanceolate, subsessile, cordate and amplexicaule :
    Peduncle 3–5 cm. long, raceme 2–3 cm. long, axillary ; leaflet acute, pilose, 4–5 cm.
      long, 1–2 cm. broad ; stipules lanceolate, acute, about 1 cm. long
                                                                    17. *schoutedenianum*
    Peduncle 1–2 cm. long, raceme about 1·5 cm. long, terminal or axillary ; leaflet
      subacute, shortly pubescent, 4–9 cm. long, 2–3 cm. broad ; stipules lanceolate,
      acute, about 5 mm. long ..   ..   18. *cordifolium*

1. **E. linifolium** *Bak. f.* in J. Bot. 33 : 228 (1895) ; Bak. f. Leg. Trop. Afr. 2 : 499 ; Chev. Bot. 208. Erect,
   1–3 ft. high, from a woody more or less tuberous base ; flowers greenish-yellow in a raceme 2–6 in.
   long.
   Fr.Nig.: Gourma to Konkobiri (July) *Chev.* 24310 ! **N.Nig.:** Abinsi (fr. Nov.) *Dalz.* 608 ! Mando
   F.R., Birnin Gwari (fl. & fr. Aug.) *Keay* FHI 25998 ! Anara F.R., Zaria (fl. & fr. July) *Keay* FHI 25957 !
   Also in N.E. tropical Africa.
2. **E. parviflorum** *E. Mey.* subsp. **parviflorum**—F.T.A. 2 : 225 ; Bak. f. l.c. 504 ; Staner & De Craene in
   Ann. Mus. Congo Belge, sér. 6, 1 : 60 (1934) ; incl. var. *laxiuscula*, var. *sarmentosa* Staner & De
   Craene (1934). *E. spicatum* Hook. f. (1849)—F.W.T.A., ed. 1, 1 : 403 ; Bak. f. l.c. *E. djalonense*
   A. Chev. Bot. 208, partly, name only. *E. caillei* A. Chev. Bot. 207, name only. Half-shrubby, softly
   tawny-hairy, erect from a woody base or half-straggling ; flowers reddish-yellow or yellow.
   Fr.G.: *Farmar* 248 ! Fouta Djalon *Heudelot* 759 ! Mamou (fl. & fr. Mar.) *Dalz.* 8393 ! **S.L.:** *Don* !
   Barter ! Luseniya (fr. Dec.) *Sc. Elliot* 4123 ! Serabu (Apr.) *Deighton* 1683 ! Mano (Apr.) *Deighton*
   1701 ! *Jordan* 218 ! **Lib.:** Grand Bassa (July) *T. Vogel* 1 ! *Ansell* ! Monrovia (fl. & fr. Nov.) *Linder*
   1520 ! **G.C.:** Atwabo *Fishlock* 64 ! 85 ! **S.Nig.:** Eket *Talbot* ! **F.Po:** *Mildbr.* 7024. *Tessmann* 2837 ;
   2844. Also in east tropical and S. Africa.
   [A specimen from Anara F.R., Zaria, N. Nigeria, *Keay* FHI 22964 ; has large stipules about 2 cm.
   long and may be a distinct species.]
2a. **E. parviflorum** subsp. **collinum** *Hepper* in Kew Bull. 1956 : 130. *E. djalonense* A. Chev. Bot. 207, partly,
   name only. A smaller montane herb similar to the last but with larger flowers.
   Fr.G.: Kaoulendougou (fl. & fr. Mar.) *Chev.* 21071 ! **S.L.:** Bintumane Peak, 6,000 ft. (May) *Deighton*
   5608 !
3. **E. molle** *Hutch. ex Milne-Redhead* in Kew Bull. 1950 : 360 (1951). *E. cajanoides* of F.W.T.A., ed. 1, 1 :
   403, partly. An erect shrub, up to 6 ft. high, with softly hairy leaflets silvery beneath ; flowers yellow.
   Iv.C.: Bouaké *Scaëtta* 3068 ! **G.C.:** Achimota (fl. & fr. Aug.) *Milne-Redhead* 5124 ! Shai Hills
   (Apr.) *Irvine* 1429 ! Anum, Krepi (Jan.) *Johnson* 567 ! Sesiamang, Kwahu (Feb.) *A. S. Thomas* D. 139 !
   Dawa R. (fl. & fr. Feb.) *Morton* GC 6461 !
4. **E. andohii** *Milne-Redhead* l.c. 361 (1951). *E. cajanoides* of F.W.T.A., ed. 1, 1 : 403, partly. Similar in
   appearance to *E. molle*, but the leaves are 3-foliolate.
   G.C.: Afram Plains (Jan.) *Irvine* 620 ! Mampong Dist. (Nov.) *Vigne* FH 3465 ! Kintampo (Mar.)
   Dalz. 14 ! Damongo (July) *Andoh* FH 5191 ! Wa Dist. (Apr.) *Kitson* 829 ! Banda, Wenchi (Dec.)
   Morton GC 25169 ! **Togo:** Alavanya, Ho Dist. (Sept.) *St. C. Thompson* 1438 ! Bimbila (fl. & fr. Dec.)
   Morton GC 6285 !
5. **E. psoraleoides** (*Lam.*) *G. Don* Gen. Syst. 2 : 348 (1832). *Crotalaria psoraloides* Lam. (1786). *Rhynchosia
   cajanoides* Guill. & Perr. (1832). *Eriosema cajanoides* (Guill. & Perr.) Hook. f. (1849)—F.T.A. 2 : 227 ;
   Chev. Bot. 207 ; F.W.T.A., ed. 1, 1 : 403, partly ; Bak. f. l.c. 508. *E. sericeum* of Chev. Bot. 209,
   partly. (?) *E. argenteum* A. Chev. Bot. 207, name only. Erect undershrub, grey-silky, more or less
   branched, 4–5 ft. high or more in tall grass ; flowers golden-yellow, odorous.
   Sen.: Messira *Berhaut* 1778. **Gam.:** Kuntaur *Ruxton* A. 44 ! **Fr.Sud.:** Sikoro *Chev.* 249 ! **Port.G.:**
   Bissau (Nov.) *Esp. Santo* 1372 ! **Fr.G.:** *Heudelot* 851 ! *Scaëtta* 3192 ! Kouroussa (Apr.) *Pobéguin* 700 !
   S.L.: Kabala (fl. & fr. Sept.) *Thomas* 2217 ! 2230 ! Falaba (Mar.) *Sc. Elliot* 5099 ! Musaia (fl. & fr.
   Dec.) *Deighton* 4547 ! *Small* 261 ! Bafodia (Sept.) *Small* 249 ! **Iv.C.:** Mt. Dourou, 3,600 ft. *Chev.*
   21706. Dotou *Chev.* 21785. **G.C.:** Odumase (fl. & fr. Aug.) *Johnson* 1090 ! Dodowa (Aug.) *Johnson*
   758 ! Akuse (Aug.) *Howes* 1205 ! **Togo:** Kpedsu (fl. & fr. Jan.) *Howes* 1086 ! **Dah.:** *Le Testu* 142 !
   Atacora Mts. *Chev.* 24018. **N.Nig.:** Kontagora (fl. & fr. Nov.) *Dalz.* 18 ! Nupe *Barter* 1119 ! Ilorin
   (fl. & fr. Dec.) *Ajayi* FHI 19274 ! Zaria (Sept.) *Lamb* 56 ! Jemaa (Mar.) *Onochie* FHI 7744 ! **S.Nig.:**
   Lagos (fl. & fr. Aug.) *Punch* 12 ! Abeokuta *Irving* 107 ! Ibadan (fl. & fr. Dec.) *Onochie* FHI 7950 !
   Br.Cam.: Bambuko F.R. (Sept.) *Olorunfemi* FHI 30763 ! Common throughout tropical Africa extending
   to S. Africa. (See Appendix, p. 241.)
6. **E. bauchiense** *Hutch. & Dalz.* F.W.T.A., ed. 1, 1 : 403 (1928) ; Kew Bull. 1929 : 17 ; Bak. f. l.c. 504.
   Similar in habit to *E. parviflorum* ; 6–12 in. high, from a woody base.
   N.Nig.: Vom, 3,000–4,500 ft. *Dent Young* 78 ! 81 !
7. **E. shirense** *Bak. f.* in Trans. Linn. Soc., ser. 2, 4 : 11 (1894) ; Bak. f. Leg. Trop. Afr. 2 : 506 ; Staner &
   De Craene l.c. 52. *E. linifolium* of Chev. Bot. 208, partly. Erect, 9–12 in. from a woody base, pale or
   rufous pilose.
   Iv.C.: Mankono (fl. & fr. June) *Chev.* 21843 ! **N.Nig.:** Jos (Mar.) *Hill* 7 ! **Br.Cam.:** Bamenda (Apr.)
   Maitland 1413 ! 1560 ! Fang (fl. & fr. Apr.) *Maitland* 1478 ! Bum (fl. & fr. May) *Maitland* 1537 ! Also
   in Belgian Congo, Kenya, Tanganyika, Mozambique, N. and S. Rhodesia and Nyasaland.

8. **E. sparsiflorum** *Bak. f.* in J. Bot. 33 : 144 (1895) ; Chev. Bot. 209 ; Bak. f. Leg. Trop. Afr. 2 : 510. Erect, pubescent, 1–1½ ft. high, woody below from a woody rootstock ; flowers orange-yellow, in lax axillary racemes 3–6 in. long.
   N.Nig.: Bauchi Plateau (Jan.) *Lely* P. 59 ! Naraguta (June) *Lely* 19 ! 336 ! Vom *Dent Young* 82 ! Also in A.-E. Sudan and Belgian Congo. (See Appendix, p. 242.)

9. **E. montanum** *Bak. f.* in J. Bot. 33 : 142 (1895) ; Bak. f. Leg. Trop. Afr. 2 : 498 ; Staner & De Craene l.c. 64. A much-branched half-woody shrub, about 3 ft. high ; flowers yellow and clustered.
   S.Nig.: Mt. Koloishe, 5,000–5,400 ft., Obudu (fr. Dec.) *Savory & Keay* FHI 25090 ! **Br.Cam.:** Banso, Bamenda (fr. Oct.) *Tamajong* FHI 23495 ! Bamenda, 6,000 ft. (fl. & fr. Apr.) *Maitland* 1753 ! Also in Belgian Congo, Uganda, Tanganyika, Nyasaland and N. Rhodesia.

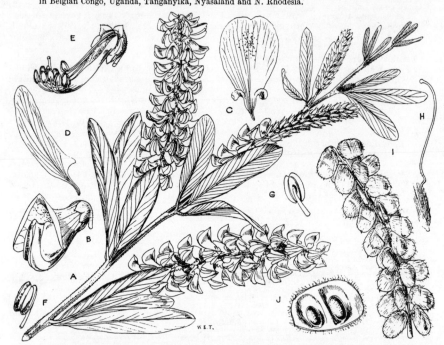

FIG. 167.—ERIOSEMA PSORALEOIDES (*Lam.*) *G. Don* (PAPILIONACEAE).

A, flowering shoot. B, flower. C, standard. D, wing. E, flower with corolla removed. F, G, anther front and back. H, pistil. I, fruits. J, fruit showing seeds.

10. **E. glomeratum** (*Guill. & Perr.*) *Hook. f.*—F.T.A. 2 : 228 ; Bak. f. Leg. Trop. Afr. 2 : 511 ; Staner & De Craene l.c. 79 ; Chev. Bot. 208. *Rhynchosia glomerata* Guill. & Perr. (1832). *E. sericeum* Bak. in F.T.A. 2 : 226 (1871). An erect low-branching, softly hairy shrub, with yellow flowers in dense axillary sessile or shortly pedunculate clusters.
    Sen.: *Heudelot* 27 ! **Gam.:** *Mackenzie Skues* ! **Fr.G.:** Fouta Djalon *Heudelot* 682 ! 752 ! Kouroussa (Apr.) *Pobéguin* 697 ! Conakry (Feb.) *Chev.* 12070 ! Macenta (fl. & fr. Oct.) *Baldwin* 9771 ! **S.L.:** *Afzelius* ! Mahela (Dec.) *Sc. Elliot* 3948 ! Njala *Deighton* 2129 ! Gberia Timbako (Oct.) *Small* 314 ! **Lib.:** Sinoe Basin *Whyte* ! Kolahun (fl. & fr. Jan.) *Bequaert* 38 ! Gbanga (Sept.) *Linder* 627 ! Bahtown (fr. Aug.) *Baldwin* 8035 ! **Iv.C.:** *Farmar* 369 ! various localities *fide* Chev. l.c. **G.C.:** *Brass* ! Achimota *Irvine* 1517 ! Essiama (fr. Feb.) *Irvine* 2332 ! Sesiamang *A. S. Thomas* D. 141 ! **Togo:** Lome *Warnecke* 224 ! Amedzofe (fr. Mar.) *Morton* GC 6506 ! **N.Nig.:** *T. Vogel* 32 ! Fakim *Barter* 1605 ! Abinsi (June) *Dalz.* 596 ! Bauchi Plateau *Lely* P. 85 ! 621 ! **S.Nig.:** Lagos Isl. *Barter* 2230 ! *Dalz.* 1034 ! 1219 ! Awba Hills F.R. (fr. Oct.) *Jones* FHI 6343 ! **Br.Cam.:** Wum (fr. July) *Ujor* FHI 30456 ! Generally spread throughout eastern and southern tropical Africa. (See Appendix, p. 241.)
    [Note : This is a very variable species and a number of varieties have been described, e.g. var. *laurentii* (De Wild.) Bak. f. has appressed hairs upon the stem, rather than typically patent. There also appears to be an undescribed sea coast form with narrow leaves and long peduncle from Sierra Leone : Kent-Tambo (fl. & fr. May) *Deighton* 2679 ! A sheet from Liberia is rather similar : King's Farm, near Monrovia *Dinklage* 3279 !]

11. **E. griseum** *Bak.* in F.T.A. 2 : 228 (1871) ; Bak. f. l.c. 510. *E. sericeum* of Chev. Bot. 209, partly. *E. togoense* Taub. (1896). A low-branched undershrub, from 6 in. to 4 ft. high ; closely grey-pubescent, from a thick woody root ; flowers yellow, concealed by bracts in shortly pedunculate heads ; in savannah.
    Iv.C.: Mankono (June) *Chev.* 21940 ! Mt. Kangoroma, Baoulé *Chev.* 22171. **G.C.:** Ejura (Apr.) *Williams* 171 ! Bimbila *Darko* 259 ! **Togo:** Fasugu, Bismarckburg *Büttner* 658. **N.Nig.:** Sokoto *Ryan* 6 ! Katagum *Dalz.* 408 ! Jebba (Feb.) *Meikle* 1183 ! Naraguta (May) *Lely* 8 ! Bauchi Plateau *Batten-Poole* FHI 13172 ! **S.Nig.:** Abo *Barter* 380 ! Yoruba country *Barter* 1225 ! W. Lagos *Rowland* ! Also in A.-E. Sudan, French and Belgian Congo and Angola. (See Appendix, p. 241.)

12. **E. monticolum** *Taub.* in Engl. Bot. Jahrb. 23 : 196 (1896) ; Bak. f. l.c. 507. Stems short from a woody rhizome ; leaves white beneath ; flowers reflexed.
    Togo: Agome Mts., 2,200 ft. (Mar.) *Bauman* 10. **N.Nig.:** Vom, 3,000–4,500 ft., *Dent Young* 83 !

13. **E. tenue** *Hepper* in Kew Bull. 1956 : 130, fig. 7. *E. shirense* of F.W.T.A., ed. 1, 1 : 403, partly (*Chev.* 13261). A slender-stemmed herb with linear leaves ; flowers yellow tinged red.
    Fr.G.: Kaba to Mamou (May) *Chev.* 13261 ! **N.Nig.:** Mando, Birnin Gwari (fl. & fr. June) *Keay* FHI 25868 ! Vom, 3,000–4,500 ft., *Dent Young* 70 !

14. **E. pulcherrimum** *Taub.* in E. & P. Pflanzenfam. 3, 3 : 375 (1894). Erect herb, softly white-tomentose, a few inches high up to 1 ft., from a woody rootstock, the leaves folding up over the short raceme of yellow flowers.
  G.C.: Drobo (Apr.) *Thompson* 63 ! Gambaga (July) *Vigne* FH 4535 ! Akpabla 701 ! **N.Nig.:** Sokoto (June) *Dalz.* 533 ! Anara F.R., Zaria (May) *Keay* FHI 25771 ! Bauchi Plateau (May) *Lely* P. 322 ! **S.Nig.:** *Kitson* ! Lagos colony (fr. Aug.) *Phillips* ! Also in A.-E. Sudan, French and Belgian Congo, Gabon and Cameroons. (See Appendix, p. 242.)
15. **E. afzelii** *Bak.* in F.T.A. 2 : 225 (1871) ; Bak. f. Leg. Trop. Afr. 2 : 500. *E. subacaulis* A. Chev. Bot. 209, name only. A stiff erect herb, 2–3 ft. high, clothed with silky hairs ; flowers reddish, in spike-like racemes.
  Sen.: Ouassadou *Berhaut* 1217 ; 3041. **Port.G.:** Bafata (Aug.) *Esp. Santo* 2742 ! 2753 ! **S.L.:** *Afzelius* ! Port Loko (fr. Sept.) *Glanville* 443 ! Benikoro (Oct.) *Thomas* 2951 ! Masisa (fl. & fr. Aug.) *Deighton* 5127 ! Bubuya (Sept.) *Adames* 189 !
16. **E. chrysadenium** *Taub.* in Engl. Bot. Jahrb. 23 : 195 (1896) ; Bak. f. l.c. 501. *E. cordatum* of Bak. f. l.c., partly. Erect undershrub, stem 6 in. to 1 ft. high, from a woody rootstock ; flowers purple ; in grassland.
  Br.Cam.: Bum (fr. June) *Maitland* 1674 ! Fang (Apr.) *Maitland* 1403 ! Bamenda (Apr.) *Maitland* 1424 ! *Ujor* FHI 30057 ! Also in French Cameroons, French and Belgian Congo, Angola, Uganda and Tanganyika.
17. **E. schoutedenianum** *Staner & De Craene* l.c. 74 (1934). A slender suberect herb with pilose stems ; flowers yellow in oblong racemes.
  Br.Cam.: Nkambe, Bamenda (Sept.) *Ujor* FHI 30225 ! Also in French Cameroons, A.-E. Sudan, Belgian Congo, and N. Rhodesia.
18. **E. cordifolium** *Hochst. ex A. Rich.*—F.T.A. 2 : 224 ; Bak. f. l.c. 502. Stem pubescent, herbaceous and flexuous, from a rootstock which may become fusiform or turnip-like ; flowers yellow, in head-like racemes.
  N.Nig.: Vom *Dent Young* 80 ! Naraguta (fl. & fr. Aug.) *Lely* 482 ! Also in Abyssinia, Uganda, Kenya and Belgian Congo.

36. MOGHANIA J. St.-Hil. in Desv. Journ. Bot. 1 : 61 (1813) ; Hui-Lin Li in Amer. Journ. Bot. 31 : 224–228 (1944). *Flemingia* Roxb. ex Ait. f. (1812)—F.T.A. 2 : 230 ; not of Roxb. ex Rottl. (1803).

Leaves 1-foliolate ; leaflet ovate to suborbicular, subacute, 3–8 cm. long, 2·5–5·5 cm. broad, pubescent on both sides ; stipules subulate-lanceolate ; flowers about 1 cm. long, solitary or paired on short pedicels ; calyx-lobes linear, acute ; corolla a little longer than the calyx ; fruits oblong, 1·8 cm. long, slightly pubescent   1. *faginea*
Leaves 3-foliolate ; terminal leaflet oblong-lanceolate, acute, about 8 cm. long and 4 cm. broad, pubescent beneath, upper surface pubescent at first becoming glabrous ; stipules lanceolate ; flowers about 1 cm. long, in dense axillary clusters ; calyx-lobes linear, acute ; corolla equalling calyx ; fruits oblong, 1·2 cm. long, covered with red gum   ..   ..   2. *grahamiana*

1. **M. faginea** (*Guill. et Perr.*) *O. Ktze.* Rev. Gen. Pl. 1 : 199 (1891). *Rhynchosia faginea* Guill. & Perr. (1832). *Flemingia faginea* (Guill. & Perr.) Bak. in F.T.A. 2 : 230 (1871) ; Chev. Bot. 209, incl. var. *glabra* A. Chev. ; Bak. f. Leg. Trop. Afr. 2 : 513 ; F.W.T.A., ed. 1, 1 : 404. An erect shrubby plant up to about 10 ft. high, with chestnut-brown branchlets, grey or rufous-tomentose ; flowers white or purplish.
  Sen.: R. Casamance *Leprieur* 366 ! **Gam.:** *Heudelot* ! **Fr.Sud.:** Bamako (fl. & fr. Jan.) *Chev.* 241 ! **Fr.G.:** Diaguissa *Chev.* 13605. **Iv.C.:** Lower Comoé *Chev.* 17585. **G.C.:** Nangodi (fl. May, Nov., fr. May) *Vigne* FH 4663 ! 4729 ! Lawra (Dec.) *Adams & Akpabla* GC 4424 ! Kunta, Black Volta R. (Mar.) *Kitson* !
2. **M. grahamiana** (*Wight & Arn.*) *O. Ktze.* l.c. (1891). *Flemingia grahamiana* Wight & Arn. Prod. Fl. Penins. Ind. Or. 1 : 242 (1834) ; Brenan in Mem. N.Y. Bot. Gard. 8 : 423 (1954). *F. rhodocarpa* Bak. in F.T.A. 2 : 231 (1871) ; Bak. f. l.c. 514. *Moghania rhodocarpa* (Bak.) Hauman in Fl. Congo Belge 6 : 258 (1955). *Eriosema erythrocarpon* Beck (1888). A shrub with woody branches ; flowers pale green.
  G.C.: Asampanie (fl. & fr. Jan.) *Brown* FH 2150 ! R. Ofro F.R. (fl. & fr. Jan.) *Vigne* FH 2737 ! Extends from Ubangi-Shari to Abyssinia, eastern and central tropical Africa, Transvaal, Natal and in Aden, southern India and (?) China.

Imperfectly known species.

M. guineensis (*G. Don*) *O. Ktze.* l.c. (1891). *Flemingia guineensis* G. Don Gen. Syst. 2 : 309 (1832). The type of this species is not known, but from the description it might be *Bryaspis lupulina* (Planch. ex Benth.) Duvign.

37. CAJANUS DC.—F.T.A. 2 : 215.

A shrub with ribbed shortly silky-pubescent branches ; leaves 3-foliolate ; leaflets lanceolate, acute, up to 10 cm. long and 3·5 cm. broad, minutely tomentellous and glaucous-grey beneath ; stipules lanceolate, acute ; flowers yellow in subcapitate axillary racemes 6–8 cm. long ; corolla 1·5–1·8 cm. long ; fruits obliquely subtorulose, beaked, 5–6 cm. long, 3–4-seeded, shortly tomentellous ; seeds shining   1. *cajan*
As above in habit but leaflets obliquely elliptic or obovate, subacute, up to 3·5 cm. long and 1·5 cm. broad, silky-tomentellous beneath, stipules deciduous ; fruits oblong, 2–4-seeded, 3 cm. long, tomentose   ..   ..   2. *kerstingii*

1. **C. cajan** (*Linn.*) *Millsp.* in Publ. Field Mus. Nat. Hist. Bot., ser. 2 : 53 (1900) ; Bak. f. Leg. Trop. Afr. 2 : 459. *Cytisus cajan* Linn. (1753). *Cajanus indicus* Spreng. (1826)—F.T.A. 2 : 216 ; Chev. Bot. 204. The Pigeon Pea. Native of East Indies, cultivated throughout the tropics. (See Appendix, p. 233.)
2. **C. kerstingii** *Harms* in Fedde Rep. 14 : 196 (1915) ; Bak. f. l.c. 460. A shrub 3–7 ft. high with stems less angular than in the preceding species ; branchlets and undersurface of leaves shortly white-silky ; in open savannah.
  Sen.: Niokolo-Koba *Berhaut* 1218. **G.C.:** FH 4578 ! **Togo:** Sokode-Basari (fr. Sept.) *Kersting* 570.

38. ATYLOSIA Wight & Arn. Prod. 257 (1834).

A slender twiner ; stems softly pubescent ; leaves 3-foliolate ; leaflets obovate, rounded at apex, 2–3 cm. long, 1–2 cm. broad, softly tomentose on both surfaces ; flowers in axillary pairs, shortly pedicellate ; fruits 3–6-seeded, oblong, 2–2·5 cm. long, septate between the seeds, softly tomentellous, apiculate .. .. *scarabaeoides*

A. **scarabaeoides** (*Linn.*) *Benth.* in Miq. Pl. Jungh. 242 (1852) ; Chev. Bot. 205 ; Bak. f. Leg. Trop. Afr. 460. *Dolichos scarabaeoides* Linn. (1753).
Sen.: Zinguichor, Casamance *Chev.* 3405. **Port.G.:** Bissau *Esp. Santo* 1419 ! **S.L.:** Franziga, Tambakka country *Sc. Elliot* 5048 ! **G.C.:** Labadi (fl. & fr. Feb.) *Irvine* 2740 ! Tefli to Ada (fl. & fr. June) *Morton* GC 9265 ! Also in Zanzibar, the Mascarene Islands and widely spread in Asia.

39. CLITORIA Linn.—F.T.A. 2 : 176.

Leaves pinnate ; leaflets 5–7, elliptic, rounded or emarginate at apex, 3–4·5 cm. long, 1·5–3 cm. broad, shortly pubescent beneath; flowers solitary, very shortly pedicellate, 4–5 cm. long ; bracteoles rounded, nervose, about ⅓ as long as the calyx ; calyx about 1·5 cm. long, lobes nervose, acute ; fruits flat, linear, beaked, about 10 cm. long and 1 cm. broad, sparingly pubescent .. .. 1. *ternatea*
Leaves 3-foliolate ; leaflets elliptic or ovate-elliptic, rounded or emarginate at apex, 4–7 cm. long, 1·5–3·5 cm. broad, more or less pubescent beneath ; flowers 2–3 on axillary peduncles ; bracteoles ovate, pointed ; fruits about 4 cm. long, with a rib up the middle of the valves .. .. 2. *rubiginosa*

1. **C. ternatea** *Linn.*—F.T.A. 2 : 177 ; Chev. Bot. 194 ; Bak. f. Leg. Trop. Afr. 2 : 428 ; Wilczek in Fl. Congo Belge 6 : 265, fig. 7. A climber, somewhat shrubby below, with flowers deep blue, occasionally pure white.
Sen.: *Roger* ! *Heudelot* 437 ! Dakar *Chev.* 3395 ; 15818. **Gam.:** Kuntaur *Ruxton* 20 ! **Fr.G.:** Conakry *Chev.* 12039. **S.L.:** *Barter* ! **G.C.:** *Irvine* 1407 ! Accra *Brown* 401 ! **Dah.:** Abbo to Massé (Feb.) *Chev.* 22990 ! **N.Nig.:** Sokoto (Nov.) *Moiser* 26 ! Lokoja (fl. & fr. Oct.) *Parsons* L. 84 ! Gombe (Oct.) *Lely* 664 ! **S.Nig.:** Abeokuta *Barter* 3394 ! Widely grown for ornament in the warmer parts of the world.
2. **C. rubiginosa** *Juss. ex Pers.* Syn. ° : 303 (1807) ; Bak. f. l.c. 429. *C. glycinoides* DC. (1825). A more or less pubescent twiner, with ƒ.ag ant flowers about 2 in. long, white streaked with mauve or red.
Sen.: Niayes *Berhaut* 161ʊ. **Port.G.:** Fulacunda (May) *Esp. Santo* 2024 ! **Lib.:** Gbau, Sanokwele (fr. Sept.) *Baldwin* 9412 ! **S.Nig.:** Lagos *Millen* 129 ! *Foster* 16 ! Olokemeji (Aug.) *Foster* 319 ! Oban *Talbot* 1305 ! A native of tropical America, presumably introduced.
C. laurifolia Poir. and *C. cajanifolia* (Presl) Benth. have been introduced from tropical America and are sometimes cultivated in our area.

40. AMPHICARPA Elliott in Journ. Acad. Nat. Sci. Philadelp. 1 : 372 (1817).

Climber ; branchlets slender, hirsute with reflexed hairs ; leaves 3-foliolate ; leaflets ovate, with long mucronate tip, about 5 cm. long and 3 cm. broad, softly pubescent on both surfaces ; lateral nerves about 7 pairs ; stipules lanceolate, about 1 cm. long, closely striate and pubescent ; stipels filiform ; racemes axillary, about as long as the leaves ; calyx-lobes with filiform tips ; fruits flat, linear-oblong, about 3·5 cm. long, ciliate on each margin .. .. *africana*

A. **africana** (*Hook. f.*) *Harms* in Fedde Rep. 17 : 136 (1921) ; Bak. Leg. Trop. Afr. 2 : 355. *Shuteria africana* Hook. f. (1864)—F.T.A. 2 : 177 ; F.W.T.A., ed. 1, 1 : 405. A herbaceous climber, fruits often marked with a dark longitudinal band.
Br.Cam.: Cam. Mt., 7,000 ft. (fr. Nov.) *Mann* 1974 ! Also in eastern tropical Africa from Abyssinia to Nyasaland.

41. CENTROSEMA DC.—F.T.A. 2 : 176.

Calyx truncate, 3 mm. long, 10 mm. broad ; fruits linear, about 15 cm. long and 1 cm. broad, with 2 prominent longitudinal ribs, and a straight beak 2 cm. long ; racemes 3–4-flowered ; standard about 4 cm. long ; a pair of bracteoles 1·3 cm. long envelops bud, caducous later ; leaves 3-foliolate ; terminal leaflet rhomboidal to broadly lanceolate, glabrous, apiculate, 6–10 cm. long, 3–8 cm. broad ; stipules lanceolate, about 1 cm. long and 5 mm. broad .. .. 1. *plumieri*
Calyx deeply 5-fid, lowest lobe much longer than the others ; beak of fruit about 1 cm. long :
Leaflets sparsely pubescent, elliptic, 4–6 cm. long, 2–3 cm. broad, slightly acuminate, mucronate ; stipules lanceolate, 2–3 mm. long ; racemes 2–3-flowered ; standard about 3 cm. long ; a pair of bracteoles 7 mm. long envelops bud ; calyx-lobes triangular, acute, the lowest linear 6 mm. long ; fruits linear with broad margins, 8–12 cm. long, 6 mm. broad .. .. 2. *pubescens*
Leaflets glabrous, oblong-lanceolate, 2–4 cm. long, 1–2 cm. broad, obtuse to subacute, mucronate ; stipules lanceolate, 1 mm. long ; racemes 2–3-flowered ; standard about 1·5 cm. long ; a pair of bracteoles about 4 mm. long envelops bud ; calyx-lobes subulate, the lowest 4 mm. long ; fruits linear, 6–8 cm. long, 3 mm. broad
3. *virginiana*

1. **C. plumieri** (*Turp.*) *Benth.* in Ann. Wien. Mus. 2 : 118 (1839) ; Wilczek in Fl. Congo Belge 6 : 272. *Clitoria plumieri* Turp. (1807). A twiner with large pale yellow flowers.
G.C.: Axim (fr. Feb.) *Irvine* 2211 ! Aburi *Akpabla* 955 ! Kumasi (fr. Oct.) *Baldwin* 13474 ! **S.Nig.:** Benin *Lamb* ! Sapoba (fr. Nov.) *Meikle* 536 ! Ujor FHI 23930 ! Introduced from tropical America as a cover-crop and green manure.
2. **C. pubescens** *Benth.* l.c. 119 (1839) ; Wilczek l.c. A creeper with shortly pubescent stems and leaves ; flowers pink-mauve or white with purple markings.
S.L.: Freetown (June) *Deighton* 2701 ! Njala (fl. & fr. Nov.) *Deighton* 2549 ! Rokupr (fl. & fr. Nov.) *Jordan* 372 ! 373 ! **G.C.:** Achimota (Oct.) *Morton* A. 62 ! Axim (Mar.) *Morton* GC 8517 ! Kumasi (Oct.) *Baldwin* 13465 ! **Togo:** Kpeve (fl. & fr. Jan.) *Irvine* 1738 ! Introduced from America as a cover-crop.
3. **C. virginiana** (*Linn.*) *Benth.*—F.T.A. 2 : 176 ; Bak. f. Leg. Trop. Afr. 2 : 429. *Clitoria virginiana* Linn. (1753). A slender-stemmed almost glabrous climber ; flowers pale blue.
G.C.: Cape Coast (fl. & fr. Sept.) *Dalz.* 91 ! *T. Vogel*. Introduced from America as a cover-crop.

42. MUCUNA Adans.—F.T.A. 2 : 184.

Stems erect ; leaves 3-foliolate ; leaflets all similarly ovate, equal in size, margins slightly undulate, apex mucronate or emarginate, lateral leaflets only slightly unequal-sided at base, closely appressed-pubescent beneath, 6–7 cm. long and about 3 cm. broad ; petiole about 4 cm. long ; rhachis of inflorescence short, about 4 cm. long ; bracts ovate, about 1 cm. long ; calyx 4-lobed to about ½, lobes lanceolate ; flowers purple, rather clustered ; standard about 2·5 cm. long ; fruits oblong, shortly beaked, covered with non-urticating hairs, about 6 cm. long and 1·3 cm. broad, laterally compressed   ..   ..   1. *stans*

Stems climbing ; leaves 3-foliolate, terminal and lateral leaflets of different shapes :
Flowers white, cream or yellow ; fruits with urticating bristles :
  Flowers more or less umbellate ; leaves silky-pubescent beneath ; leaflets obliquely ovate, shortly acuminate, up to 15 cm. long and 14 cm. broad ; peduncle short ; calyx 4-lobed to about ½, lobes lanceolate ; fruits oblong, up to 15 cm. long, prominently and obliquely transversely ridged, hispid with urticating hairs ; seeds suborbicular, about 2·5 cm. diam. with a long hilum around about ⅘ of the circumference   ..   ..   2. *sloanei*
  Flowers arranged in elongated, often zigzag, or short racemes :
  Rhachis of inflorescence very zigzag ; peduncle very long and whip-like ; calyx subtruncate ; standard very broad, about 4 cm. long ; fruits oblong or obovate, up to 20 cm. long and 6 cm. broad, transversely and obliquely ridged, covered with urticating hairs ; seeds orbicular, 3·5 cm. diam., hilum as in No. 2 ; leaves glabrous or sparsely pilose beneath ; leaflets obliquely ovate, rounded at base, abruptly acuminate and mucronate at apex, up to 15 cm. long and 8 cm. broad, 3-nerved at base, finely reticulate on both surfaces   ..   ..   3. *flagellipes*
  Rhachis of inflorescence almost straight, not zigzag ; calyx 4-lobed with upper lobe emarginate at apex, lowest lobe frequently 1–2 cm. long ; standard 5–8 cm. long ; fruits irregularly twisted, about 7 cm. long, old style persistent when young, densely covered with brown or reddish-yellow urticating hairs ; leaves with fine appressed pubescence on both surfaces ; terminal leaflet rhomboid, cuneate at base, acute at apex ; lateral leaflets very unequal-sided at base
                                       4. *poggei* var. *occidentalis*

Flowers purple, about 4 cm. long ; calyx rather deeply 4-lobed, grey silky-pubescent ; bracts caducous :
  Fruits very setose with yellowish urticating hairs, curved, about 9 cm. long, faintly longitudinally ribbed ; seeds small, broadly elliptic, mottled reddish, with a short hilum ; leaves appressed-pilose on both sides, especially beneath ; leaflets very obliquely ovate, unequal-sided and rounded at base, acutely mucronate, up to 20 cm. long and 14 cm. broad ; stem furrowed   ..   5. *pruriens* var. *pruriens*
  Fruits appressed-pubescent with brown, becoming grey, non-urticating hairs, slightly curved, with a marked short beak, about 7 cm. long, with irregular longitudinal ribs ; seeds black and shining, 10–12 mm. long, with a white hilum ; leaves sparsely appressed-pubescent on both sides ; leaflets very obliquely ovate, unequal-sided, very acutely mucronate, up to about 17 cm. long and 10 cm. broad ; stem striate   ..   ..   5a. *pruriens* var. *utilis*

1. **M. stans** *Welw. ex Bak.* in F.T.A. 2 : 187 (1871) ; Bak. f. Leg. Trop. Afr. 2 : 381. An erect much-branched shrub 4–6 ft. high ; flowers deep purple.
  Br.Cam.: Bamenda *Johnstone* 189/31. Also in Belgian Congo, Angola, Nyasaland, N. Rhodesia, Tanganyika and Uganda.
2. **M. sloanei** *Fawc. & Rendle* in J. Bot. 55 : 36 (1917) ; Fl. Jam. 4 : 53, fig. 16 (1920). *M. urens* of F.T.A. 2 : 185 ; Chev. Bot. 196 ; F.W.T.A., ed. 1, 1 : 406 ; Bak. f. l.c. 379 ; not of (Linn.) Medic. A climber ; stem with appressed whitish hairs ; flowers creamy or yellowish of waxy consistence, 2 in. long or more, the standard broad about ½ length of the wings and keel.
  Port.G.: Bissau *Esp. Santo* 1654 ! **S.L.:** Jigaya (Sept.) *Thomas* 2728 ! Regent (Dec.) *Sc. Elliot* 4325 ! Njala (fl. & fr. Dec.) *Deighton* 1802 ! **Lib.:** Gbanga (Sept.) *Linder* 470 ! **Iv.C.:** Bingerville *Chev.* 15311. Tingouela (fr. Dec.) *Chev.* 22574 ! **G.C.:** Aburi (fr. Jan.) *Brown* 936 ! Accra *Don* ! Medea (Nov.) *Morton* A. 133 ! **Togo:** Misahöhe (Nov.) *Mildbr.* 7327 ! **S.Nig.:** Lagos (Dec.) *Dalz.* 1228 ! Brass *Barter* 1875 ! Ibadan *Newberry* 115 ! Oban *Talbot* 1317 ! **Br.Cam.:** Victoria *Winkler* 539 ! **F.Po:** *T. Vogel* 199 ! *Barter* 1610 ! *Mann* 42 ! Widespread in the tropics. (See Appendix, p. 252.)
3. **M. flagellipes** *T. Vogel ex Hook. f.*—F.T.A. 2 : 185 ; Chev. Bot. 196 ; Bak. f. l.c. 379. A more or less glabrous lofty climber, with creamy white or yellowish flowers, subtended by grey-silky bracts, turning black on drying.
  S.L.: Wanje R. (Oct.) *Fisher* 33 ! Liange (Sept.) *Deighton* 2227 ! **Lib.:** Gbanga (fl. & fr. Sept.) *Linder* 505 ! Brewersville (fl. & fr. Sept.) *Barker* 1061 ! 1074 ! Bobei Mt. (fr. Oct.) *Baldwin* 9572 ! **Iv.C.:** fide Chev. l.c. **G.C.:** Tarkwa *Vigne* FH 1258. Masiaso *Foggie* FH 221. **S.Nig.:** Nun R. *T. Vogel* 1 ! Ijebu Ode (Aug.) *Punch* 1 ! Akure F.R. (Aug.) *Jones* FHI 20207 ! **F.Po:** *Barter* ! *Mann* 21 ! Also in French Cameroons, Belgian Congo, Angola and Uganda. (See Appendix, p. 252.)
4. **M. poggei** *Taub.* var. **occidentalis** *Hepper* in Kew Bull. 1956 : 127, fig. 6. A climber 30–40 ft. high, bearing long inflorescences of large yellow or green-tinged white flowers.
  S.L.: Kabala (fr. Sept.) *Thomas* 2218 ! Musaia (fl. & fr. Sept.) *Deighton* 5580 ! 5600 ! Falaba (Sept.) *Small* 310 ! **G.C.:** Ejura (Aug.) *Chipp* 735 ! **Togo:** Misahöhe Pass (Aug.) *Gillett* 66 ! Bismarckburg *Büttner* 4. **N.Nig.:** Vom, 3,000–4,500 ft., *Dent Young* 67 ! **S.Nig.:** Udi to Okwoga (Aug.) *Kitson* ! Ogoja *Rosevear* 71/29 ! Also in French Cameroons and Ubangi-Shari.
5. **M. pruriens** (*Linn.*) *DC.* var. **pruriens**—F.T.A. 2 : 187 ; Bak. f. l.c. 380 ; Chev. Bot. 196. *Dolichos pruriens* Linn. (1754). A lofty climber, with more or less pubescent stems ; flowers deep purple to almost black ; fruit densely clothed with rusty stinging hairs.
  Port.G.: Bissau (fr. Dec.) *Esp. Santo* 1588 ! **S.L.:** *Morson* ! Kambia (fr. Dec.) *Sc. Elliot* 4369 ! Kissi

country *Fisher* 29 ! Kumrabai (fr. Dec.) *Thomas* 7034 ! **Lib.**: Bakratown (Oct.) *Linder* 858 ! Wohmen (fr. Oct.) *Baldwin* 10087 ! **G.C.**: Ejura Scarp (fr. Dec.) *Morton* GC 9614 ! **Togo:** Misahöhe *Mildbr.* 7378 ! **N.Nig.**: Kontagora (fl. & fr. Dec.) *Dalz.* 32 ! Nupe *Barter* 1155 ! **S.Nig.**: Ogun R. *Millen* 124 ! Oban *Talbot* 1316 ! Abokum (fr. Jan.) *Holland* 292 ! Widespread in the tropics. (See Appendix, p. 252.)

5a. **M. pruriens** var. **utilis** (*Wall. ex Wight*) *Bak. ex Burck* in Ann. Jard. Bot. Buitenzorg 11 : 187 (1893). *M. utilis* Wall. ex Wight (1840). A tall climber with silky pods about 4 in. long ; flowers purple. A Velvet Bean.
    Port.G.: Bissau (fl. & fr. Dec.) *Esp. Santo* 1588 ! **Fr.G.**: Nzérékoré (fl. & fr. Oct.) *Baldwin* 9729 ! **S.L.**: Njala (Sept.) *Deighton* 2102 ! Cultivated in many parts of the tropics and introduced to Africa from tropical Asia.

M. cochinchinensis (Lour.) A. Chev. (syn. *M. nivea* (Roxb.) DC.), which has white flowers and fruits with non-urticating hairs is cultivated in S. Nigeria and in Senegal.

43. ERYTHRINA Linn.—F.T.A. 2 : 181.

Leaves, petioles and branchlets, densely tomentose ; leaflets usually broader than long, 8–13 cm. long, 9–15 cm. broad, broadly cuneate at base, rounded, emarginate or mucronate at apex, sometimes with prickles on nerves above ; petioles about 17 cm. long with a few prickles ; flowers 2–3 cm. long, standard narrow and curved like an S ; calyx with 5 equal lobes, about 1 cm. long ; fruits moniliform, 3–4 seg-mented ; a savannah tree   . .   . .   1. *sigmoidea*
Leaves glabrous or slightly tomentellous, longer than broad, more or less acute :
    Limb of calyx linear-lanceolate, bifurcate at apex, about 2·5 cm. long, tube about 1 cm. long, glabrous or minutely pubescent ; flowers in twos and threes on stiff, many-flowered racemes ; leaflets (of young shoots) prickly above on nerves as on rhachis, 12–23 cm. long, 10–16 cm. broad, the lateral leaflets obliquely ovate, the terminal broadly elliptic ; a tall forest tree . .   . .   2. *excelsa*
    Limb of calyx variously dissected or entire, tomentellous :
        Savannah tree ; calyx without a dissected limb, almost glabrous ; standard 2·5–4 cm. long ; bracts inconspicuous, 4 mm. long, early caducous ; fruits moniliform, stipitate, becoming twisted, about 12 cm. long, closely puberulous ; seeds bright red, smooth and shiny ; leaflets lanceolate to broadly ovate, rounded or cuneate at base, rounded to shortly acuminate at apex, 7–15 cm. long, 4–10 cm. broad ; petioles slightly grooved above, about 2 mm. diam., thorny   . .   3. *senegalensis*
        Forest trees ; limb of calyx not entire :
            Small trees 6–15 m. high ; flowers scarlet :
                Petioles with a definite groove above, 1–2 mm. diam., thorny ; leaflets oblong to ovate, rounded to shortly acuminate at apex, rounded or obtuse at base, 9–16 cm. long, 5–9·5 cm. broad ; limb of calyx variable, with from 3 or 4 lobes 2–3 mm. long, to several irregular linear lobes 4–9 mm. long ; bracts more or less con-spicuous, longer than the young buds, 5–9 mm. long, caducous ; standard 3–5 cm. long ; fruits similar to No. 3   . .   . .   . .   4. *vogelii*
                Petioles terete and only slightly longitudinally striate, 3 mm. diam. ; leaflets rhomboid-ovate, 14–22 cm. long, 11–22 cm. broad, minutely pubescent beneath ; calyx with 5 similar lobes at apex 1–10 mm. long ; fruits strongly moniliform about 3 cm. diam., the segments eventually hanging together by strands ; seeds keel-shaped, scarlet, 13 mm. long . .   . .   5. *addisoniae*
            Large tree, 20–30 m. high ; flowers pink ; limb of calyx 1·5–3 cm. long, about 1 cm. broad, jagged-lobate at apex, contracted at base ; leaflets ovate, 10–15 cm. long, 6–9 cm. broad, shortly acuminate ; fruits moniliform, 5–15 cm. long, twisted or curved, grey puberulous ; seeds oblong, shining red   . .   . .   6. *mildbraedii*

1. **E. sigmoidea** *Hua* in Bull. Mus. Hist. Nat., sér. 1, 3 : 327 (1897) ; Bak. f. Leg. Trop. Afr. 2 : 370 ; Aubrév. Fl. For. Soud.-Guin. 322, t. 66, 6, t. 67. *E. dybowskii* Hua (1898)—Bak. f. l.c. 371. *E. eriotricha* Harms (1913)—Bak. f. l.c. ; Aubrév. l.c. t. 66, 1–2. A small savannah tree, 10–20 ft. high, stems armed with stout recurved prickles ; flowers red, in racemes nearly as long as the leaves.
    Port.G.: Bafatá (June) *Esp. Santo* 2693 ! **Fr.G.**: Timbo, Fouta Djalon (May) *Maclaud & Miquel* 49 ! **Iv.C.**: Samankono *Serv. For.* 2853. **N.Nig.**: Bauchi Plateau (fl. & fr. Apr.) *Lely* P. 213 ! Vom (Apr.) *Morton* K. 384 ! Also in French Cameroons and Chad.
2. **E. excelsa** *Bak.* in F.T.A. 2 : 183 (1871) ; Bak. f. l.c. 377 ; Louis in Bull. Jard. Bot. Brux. 13 : 317. *E. sereti* De Wild. (1907). A tall tree, with bole 60–80 ft. high, pale bark armed with strong prickles ; flowers red when tree is leafless.
    Br.Cam.: Ambas Bay (Jan.) *Mann* 704 ! Buea (Feb.) *Dalz.* 8181 ! Also in Belgian Congo, A.-E. Sudan, Uganda and Tanganyika. (See Appendix, p. 242.)
3. **E. senegalensis** *DC.*—F.T.A. 2 : 181, excl. syn. *E. vogelii* Hook. f. ; Chev. Bot. 195 ; Bak. f. l.c. 369 ; Aubrév. Fl. For. C. Iv. 1 : 296, t. 118 A ; Fl. For. Soud.-Guin. 322, t. 66, 3–4. Usually a small tree, 10–15 ft., but sometimes attaining a height of 40–50 ft. ; armed with stout prickles slightly recurved from a woody base ; conspicuous with scarlet flowers, usually when leafless ; standard petal folded flat ; in savannah country and commonly planted as a hedge.
    Sen.: *Perrottet.* Cayor (fr. Feb.) *Döllinger* ! **Gam.**: *Mungo Park* ! *Brown-Lester* 25 ! **Fr.Sud.**: Oualia (fr. Dec.) *Chev.* 34 ! **Fr.G.**: *Scaëtta* 3349 ! Timbo *Pobéguin* ! **S.L.**: *Hutton* ! *Kirk* 43 ! Digisinn country (fl. & fr. Jan.) *Sc. Elliot* 4516 ! Gberia Timbako (Oct.) *Small* 456 ! **Lib.**: Yangi (Sept.) *Linder* 855 ! Piatah (Dec.) *Baldwin* 12521 ! **Iv.C.**: Morénou (Nov.) *Chev.* 22430 ! *fide* Chev. l.c. **G.C.**: Achimota (Jan.) *Irvine* 1935 ! Mangkuma to Soggala (fr. Feb.) *Kitson* 886 ! Banda, Wenchi (fl. & fr. Dec.) *Morton* GC 25265 ! **Togo:** Lome *Warnecke* 21 ! Misahöhe (Nov.) *Mildbr.* 7239 ! **Dah.**: *Burton* ! Porto Novo *Chev.* 22855. Zagnanado *Chev.* 23060. **N.Nig.**: Borgu *Barter* 836 ! Anara F.R. (Sept.) *Olorunfemi* FHI 24386 ! Kontagora (Oct.) *Dalz.* 25 ! **S.Nig.**: *Foster* ! Abeokuta *Millen* 77 ! Olokemeji (Sept.) *A. F. Ross* 4 ! Also in French Cameroons. (See Appendix, p. 242.)
        [*E. senegalensis* and *E. vogelii* are undoubtedly closely related and thorough field work may show either that they are not specifically distinct, or that there are good distinguishing characters which are not obvious on dried specimens.]

4. E. vogelii *Hook. f.* Fl. Nigrit. 307 (1849). *E. senegalensis* of F.T.A. 2 : 181, partly ; F.W.T.A., ed. 1, 1 : 406, partly ; Bak. f. l.c., partly. *E. bancoensis* Aubrév. & Pellegr. in Bull. Soc. Bot. Fr. 82 : 603 (1936) ; Aubrév. Fl. For. C. Iv. 1 : 298, t. 118 (1936). *E. vignei* Burtt Davy (1937), name only. A small tree 20–50 ft. high, of forest and abandoned forest clearings, with woody conical prickles ; flowers scarlet like the last.
    Iv.C.: Banco (Dec.) *Aubrév.* 619 ! 1903 ! **G.C.:** Akwapim Hills (Nov.) *Johnson* 778 ! Asebu (Feb.) *Irvine* 2006 ! Dunkwa (Aug.) *Vigne* FH 1324 ! **S.Nig.:** Lagos *Millen* 49 ! *MacGregor* 143 ! Idanre Hills (Oct.) *Keay* FHI 25504 ! Ibadan (Dec.) *Chizea* FHI 8150 ! **F.Po:** *T. Vogel* 105 !
    [See note under *E. senegalensis.*]
5. E. addisoniae *Hutch. & Dalz.* F.W.T.A., ed. 1, 1 : 406 (1928) ; Kew Bull. 1929 : 17 ; Bak. f. l.c. 373 ; Aubrév. Fl. For. Soud.-Guin. 324. *E.* sp. of Aubrév. Fl. For. C. Iv. 1 : 298. A tree about 60 ft. high, armed with prickles on the trunk and branches ; flowers red, 3–7 at the end of the rhachis.
    S.L.: Karene *Addison* ! Kenema (fr. Oct.) *Deighton* 4143 ! **Iv.C.:** Divo to Tiassalé *Aubrév.* 1348. **G.C.:** Kumasi (fr. May) *Vigne* FH 1127 ! Mampong (May) *Vigne* FH 1183 ! Mpraeso (fl. & fr. June) *Beveridge* FH 3022 ! **S.Nig.:** Ilugun to Olokemeji (fr. Sept.) *Onochie* FHI 34328 ! (See Appendix, p. 242.)
6. E. mildbraedii *Harms* in Mildbr. Wiss. Ergebn. 1907–8 : 264, t. 30 (1911). *E. altissima* A. Chev. (1912) ; Chev. Bot. 195 ; F.W.T.A., ed. 1, 1 : 406 ; Bak. f. l.c. 377 ; Aubrév. Fl. For. C. Iv. 1 : 298, t. 119. *E. klainei* of F.W.T.A., ed. 1, 1 : 406, not of Pierre ex Harms. A tall tree, 60–100 ft. or more high, buttressed at the base, prickles on the trunk and branches ; flowers usually in threes, corolla and calyx-limb pink, in copious softly tomentose panicles, appearing when the tree is leafless.
    Fr.G.: Souradou (Jan.) *Chev.* 20584 ! **S.L.:** *Lane-Poole* 323 ! Pendembu (Jan.) *Unwin & Smythe* 63 ! Bamayama (Jan.) *Deighton* 3884 ! **Iv.C.:** Nzo, Upper Cavally *Chev.* 20998. Man *Chev.* 21547. **G.C.:** Aburi (Feb.) *Johnson* 880 ! Tano to Offin (Jan.) *Vigne* FH 1014 ! Atuna (Dec.) *Vigne* FH 3508 ! **S.Nig.:** Lagos, Abeokuta (Jan.) *Burton* ! Ibadan South F.R. (Jan.) *Keay* FHI 19798 ! Oban *Talbot* 114 ! (See Appendix, p. 242.)

    Besides the above, *E. glauca* Willd. is cultivated in Victoria Botanic Gardens, British Cameroons, and may be naturalized in the Calabar district of S. Nigeria ; *E. poeppigiana* (Walp.) O. F. Cook occurs in Sierra Leone and S. Nigeria (Calabar) ; *E. pallida* Britton & Rose has been introduced into Sierra Leone. These are all tropical American species. The Asiatic species *E. fusca* Lour. is recorded from the Gold Coast and *E. indica* Lam. from Senegal.

Imperfectly known species.

E. sp. cf. **E. buesgenii** *Harms* in Engl. Bot. Jahrb. 49 : 443 (1910).
    S.Nig.: *Rosevear* 7/30 ! [Incomplete material, flowers only ; same as *Mildbr.* 7787 from French Cameroons.]

44. CALOPOGONIUM Desv. in Ann. Sci. Nat., sér. 1, 9 : 423 (1826).

Stem climbing or prostrate, densely brown-pilose ; leaves 3-foliolate, leaflets elliptic, lateral leaflets rounded and unequal-sided at base, apex obtuse or slightly apiculate, pubescent ; racemes short, long-pedunculate ; calyx deeply 5-lobed, lobes narrowly lanceolate, pilose, about 8 mm. long ; corolla little longer than calyx ; fruit narrowly oblong, 3–4 cm. long, 5 mm. broad, very densely brown-pilose, 5–8-seeded ; seeds square, light-brown, 3 mm. broad   ..   ..   *mucunoides*
C. mucunoides *Desv.* l.c. (1826) ; Amshoff in Fl. Suriname 2 : 196. A hairy climber with deep blue flowers.
    Port.G.: Bissau (fl. & fr. Oct.) *Esp. Santo* 2569 ! **Fr.G.:** Nzérékoré (Oct.) *Baldwin* 9719 ! **Lib.:** Gbanga (fr. Dec.) *Daniel* 41 ! **G.C.:** Aburi (fr. Feb.) *Akpabla* 949 ! Kumasi (fl. & fr. Dec.) *Morton* GC 9692 ! A native of tropical America ; naturalized and cultivated in Africa.

45. NEORAUTANENIA Schinz in Bull. Herb. Boiss. 7 : 35 (1899).

Climbing herb ; stems tomentose or glabrescent ; leaves 3-foliolate ; leaflets irregularly 3-lobed to subentire, very broadly and obliquely ovate, cuneate at base, acute at apex, 6–12 cm. long, thinly pubescent beneath ; stipules lanceolate, 6 mm. long ; stipels spreading, linear-subulate, 4–5 mm. long ; racemes axillary, (5–)10–20 cm. long ; flowers about 1 cm. long ; fruits 12–15 cm. long, 1·5 cm. broad, softly tomentellous   ..   ..   *pseudopachyrhiza*
N. pseudopachyrhiza (*Harms*) *Milne-Redhead* in Kew Bull. 1950 : 355 (1951) ; Wilczek in Fl. Congo Belge 3 : 289, fig. 12. *Dolichos pseudopachyrhizus* Harms in Engl. Bot. Jahrb. 26 : 320 (1899), incl. var. *subintegrifolia* Harms ; Engl. Pflanzenw. Afr. 3, 1 : 680, fig. 312 ; Bak. f. Leg. Trop. Afr. 2 : 452 ; F.W.T.A., ed. 1, 1 : 410. A strong climber over shrubs or sometimes suberect ; flowers blue-purple.
    G.C.: Gambaga (June) *Akpabla* 666 ! **Togo:** *Kling* 20. **N.Nig.:** Nupe *Barter* 1625 ! Kontagora *Dalz.* 555 ! Munchi country (Feb.) *Dalz.* 604 ! Panyam (July) *Lely* 419 ! **S.Nig.:** Lagos *MacGregor* 201 ! Extends to E. Africa, Belgian Congo and Angola.

46. GALACTIA Adans.—F.T.A. 2 : 188.

Twiner ; branches elongated, softly pubescent ; leaves 3-foliolate ; leaflets oblong-elliptic, rounded at both ends, mucronate at apex, 2–3·5 cm. long, 1–1·5 cm. broad, thinly pubescent beneath ; stipules subulate-lanceolate ; racemes axillary, few-flowered, up to 10 cm. long ; calyx divided to the middle, the lowermost lobe the longest ; fruits broadly linear, 4–5 cm. long, with a curved beak at the tip, shortly pubescent   ..   ..   *tenuiflora*
G. tenuiflora (*Willd.*) *Wight & Arn.*—F.T.A. 2 : 188 ; Bak. f. Leg. Trop. Afr. 2 : 382. *Glycine tenuiflora* Willd. (1803). *G. hedysaroides* of F.W.T.A., ed. 1, 1 : 407, partly (*T. Vogel*; *Dalz.* 80). Twining or trailing ; flowers pink or blue, the standard petal glabrous.
    G.C.: *T. Vogel* ! Accra (July) *Dalz.* 80 ! Deighton 3405 ! Achimota (Apr.) *Irvine* 843 ! 1607 ! Winnebah Plain (fr. Feb.) *Dalz.* 8294 ! **N.Nig.:** Abinsi (June) *Dalz.* 587 ! Also in east tropical Africa, Mascarene Islands and India.

47. GLYCINE Linn.—F.T.A. 2 : 178.

Flowers clustered in axils of leaves ; standard petal pubescent ; stem shortly pubescent with spreading hairs ; leaflets narrowly elliptic, rounded at each end, mucronate, 3–5 cm. long, 1·5–2·5 cm. broad, pubescent beneath ; stipules subulate ; calyx-lobes

linear-lanceolate, pilose, shorter than the corolla ; fruits broadly linear, slightly
curved, flat, about 4 cm. long and 8 mm. broad, thinly appressed-pubescent
1. *hedysaroides*
Flowers in slender racemes, 8–20 cm. long ; standard petal glabrous ; stem pilose with
reflexed hairs ; terminal leaflet more or less rhomboidal-elliptic, 5–12 cm. long,
3–8 cm. broad, lateral leaflets unequal-sided at base, appressed-pubescent beneath ;
stipules lanceolate, 8 mm. long ; calyx-lobes subulate, pilose, equalling the corolla ;
fruits linear, slightly torulose when ripe, about 3 cm. long, 5 mm. broad, glabrous or
with appressed hairs   ..   ..   2. *javanica*

1. **G. hedysaroides** *Willd.*—F.T.A. 2 : 179 ; Chev. Bot. 194 ; F.W.T.A., ed. 1, 1 : 407, partly (excl. spec.
   T. Vogel s.n., *Dalz.* 80 which are *Galactia tenuiflora* (Willd.) Wight & Arn.) ; Bak. f. Leg. Trop. Afr. 2 :
   360. Stems woody at the base, wide-climbing above ; flowers white ; easily confused with *Galactia
   tenuiflora* which has a glabrous standard petal.
   Iv.C.: Bingerville *Chev.* 20087. **G.C.:** Accra (fl. & fr. June) *Don! Dalz.* 85 ! *Brown* 369 ! Achimota
   (fl. & fr. Apr.) *Irvine* 674 ! 1606 ! **Togo:** Lome *Warnecke* 238 ! **Dah.:** Atacora Mts. *Chev.* 24278. Also
   in Kenya, Tanganyika and Angola.
2. **G. javanica** *Linn.*—F.T.A. 2 : 178 ; Bak. f. l.c. 360. A variable climber with racemes of purple or white
   flowers.
   S.L.: Musala (fr. Dec.) *Deighton* 4421 ! **Lib.:** Zuie, Boporo (fl. & fr. Dec.) *Baldwin* 10683 ! Ganta,
   Sanokwele (fr. Feb.) *Baldwin* 11014 ! **G.C.:** Anum (Nov.) *Morton* GC 7990 ! **S.Nig.:** *Thomas* 2158 !
   Br.Cam.: Bamenda *Dept. Agric.* 16 ! Widespread throughout tropical Africa and tropical Asia.
   [Note that the W. African plants differ from the Indian and most of the E. African in having narrower,
   glabrescent or glabrous fruits and less congested racemes, but the British Cameroons plant differs from
   the others from W. Africa in leaf-shape, size and pubescence. Various extremes have been recognized
   as species, though these are of doubtful value.]

48. PSEUDOERIOSEMA Hauman in Bull. Jard. Bot. Brux. 25 : 96 (1955).

Erect herb ; leaves 1-foliolate; leaflet ovate, broadly rounded at base, mucronate at
apex, loosely pubescent on both sides, 8–15 cm. long, 6–8 cm. broad ; stipules linear,
persistent, about 1 cm. long ; flowers in short cylindrical racemes ; bracts and
bracteoles linear; pedicels 2-flowered ; fruits 2-seeded, oblong, 1·5–2 cm. long,
rigidly pilose outside, satiny inside   ..   ..   *andongense*

P. andongense (*Welw. ex Bak.*) *Hauman* in Bull. Jard. Bot. Brux. 25 : 97 (1955) ; Fl. Congo Belge 6 : 107, t. 9.
   Psoralea andongensis Welw. ex Bak. in F.T.A. 2 : 65 (1871). *Glycine holophylla* (Bak. f.) Taub. (1896)—
   Bak. f. Leg. Trop. Afr. 2 : 361. *Eriosema holophyllum* Bak. f. (1895)—F.W.T.A., ed. 1, 1 : 404. *E.
   atacorense* A. Chev. Bot. 207, name only. Erect, half-woody branched herb, 18 in. high ; flowers whitish,
   small, in subglobose or spike-like axillary and terminal racemes.
   Togo: Yendi to Demon (fr. May) *Akpabla* 553 ! **Dah.:** Atacora Mts. *Chev.* 23952 ; 24114. **N.Nig.:**
   Anara F.R., Zaria (May) *Keay* FHI 19194 ! Birnin Gwari (June) *Keay* FHI 25873 ! Naraguta, 4,000 ft.
   Lely 240 ! Bauchi Plateau (fl. & fr. May) *Lely* P. 273 ! Also in French Cameroons, Ubangi-Shari, A.-E.
   Sudan, Uganda, Tanganyika, Belgian Congo and Angola.

49. PHYSOSTIGMA Balfour—F.T.A. 2 : 191.

Climber ; branches glabrous ; leaves 3-foliolate, the central one broadly ovate, the
lateral obliquely ovate, rounded at base, abruptly acuminate, up to 16 cm. long
and 10 cm. broad, prominently 3-nerved at base, glabrous ; stipels semi-lunar ;
racemes axillary, drooping, the axis stout and zigzag with swollen nodes ; pedicels
curved, up to 5 mm. long ; calyx broadly cupular, shortly lobed ; standard enclosing
the wings and keel, much curved round towards the calyx ; fruits straight, narrowed
at both ends, up to 15 cm. long and 4 cm. broad, glabrous ; seeds few, 3 cm. long,
ellipsoid, with one side nearly straight, the other side grooved by the depressed hilum.
venenosum

P. venenosum *Balfour*—F.T.A. 2 : 191 ; Chev. Bot. 197 ; Bak. f. Leg. Trop. Afr. 2 : 386 ; Wilczek in Fl.
   Congo Belge 6 : 342. Shrubby at the base, with lofty-twining branches ; flowers curved like a shell,
   pink or purple. The Calabar Bean.
   S.L.: Seli R. (Aug.) *Deighton* 1218 ! **Lib.:** Tappita (Aug.) *Baldwin* 9046 ! Monroviatown (Aug.) *Baldwin*
   9001 ! **Iv.C.:** Man to Zagoué *Chev.* **G.C.:** Akropong, Akwapim (Aug.) *Irvine* 777 ! Wankye (Aug.)
   Chipp 561 ! Cape Coast (Aug.) *Fishlock* 43 ! **S.Nig.:** Old Calabar *Thomson* ! **Br.Cam.:** Victoria (fl.
   Mar., fr. Mar., Oct.) *Kalbreyer* 90 ! *Maitland* 746 ! **F.Po:** *Hutchinson* ! Also in Gabon, Spanish Guinea
   and Belgian Congo. (See Appendix, p. 256.)

50. SPHENOSTYLIS E. Mey. Comm. 148 (1835) ; Harms in Engl. Bot. Jahrb. 26 : 308 (1899).

Leaves glabrous beneath :
Erect ; leaflets lanceolate or linear-lanceolate, rounded and mucronate at the apex,
up to 14 cm. long and 3 cm. broad, with a conspicuous cartilaginous margin ; flowers
several on axillary peduncles ; calyx broadly cupular, shortly and broadly lobed,
about 5 mm. long ; corolla 1·5 cm. long ; fruits linear, 10–11 cm. long, slightly
pubescent ..   ..   1. *schweinfurthii*
Climber ; leaflets ovate, acuminate, up to 14 cm. long and 6 cm. broad ; flowers few
on stout axillary peduncles ; calyx 7 mm. long, undulately lobed ; corolla 2·5 cm.
long ; fruits linear, up to 30 cm. long, glabrous ; seeds ellipsoid, smooth, about
1 cm. long, pale-coloured or marbled ..   ..   2. *stenocarpa*
Leaves pubescent or tomentose beneath ; leaflets lanceolate to obliquely elliptic, entire
or more or less 3-lobed, up to 8 cm. long and 6 cm. broad ; flowers solitary on each
peduncle, with a large gland near the apex ; bracteole rounded, about 1 cm. long,

tomentose ; calyx 5-lobed, 1·5 cm. long ; corolla 3·5 cm. long ; fruits nearly 15 cm. long, tomentose, the valves becoming spirally twisted ; seeds with a large caruncle
3. *holosericea*

1. **S. schweinfurthii** *Harms* in Engl. Bot. Jahrb. 26 : 309 (1899) ; Chev. Bot. 198 ; Bak. f. Leg. Trop. Afr. 2 : 422. An erect undershrub 3–5 ft. high, with fairly large yellow flowers ; in open savannah.
    Fr.Sud.: Coumana *Chev.* 368. Koba *Chev.* 606. **G.C.:** Wenchi (Mar., Dec.) *Morton* GC 8616 ! 8636 ! 25140 ! **Dah.:** Atacora Mts. *Chev.* 23960. **N.Nig.:** Jebba (Feb.) *Meikle* 1197 ! Zaria (Feb.) *Dalz.* 402 ! *Keay* FHI 25744 ! Bauchi Plateau (fl. & fr. Jan.) *Lely* P. 108 ! *Hill* 26 ! *W. D. MacGregor* 433 ! **S.Nig.:** near Old Oyo F.R. (Feb.) *Keay* FHI 16253 ! Also in Ubangi-Shari, Abyssinia and A.-E. Sudan. (See Appendix, p. 261.)
2. **S. stenocarpa** (*Hochst. ex A. Rich.*) *Harms* l.c. ; Chev. Bot. 198 ; Bak. f. l.c. 420 ; Wilczek in Fl. Congo Belge 6 : 275, t. 25. *Dolichos stenocarpa* Hochst. ex A. Rich. (1847)—F.T.A. 2 : 213. *Vigna ornata* Welw. ex Bak. (1871). *Sphenostylis ornata* A. Chev. Bot. 198, name only. A strong climber with conspicuous mauve-pink or purple flowers and a tuberous root ; wild and often cultivated.
    Fr.G.: Fouta Djalon *Chev.* 18518 ! **Iv.C.:** Bouroukrou *Chev.* 16750 ; 16831. Mossi *Chev.* 24678. **G.C.:** Achimota (fl. & fr. Dec.) *Irvine* 1649 ! 1821 ! **Togo:** *Gaisser* ! *Kersting* A. 680 ! **N.Nig.:** Abinsi *Dalz.* 626 ! Bauchi Plateau *Batten-Poole* 284 ! Agaie (fl. & fr. Oct.) *Yates* 51 ! **S.Nig.:** Onitsha *Barter* 1804 ! Onitsha (fl. & fr. Dec.) *Sampson* 13 ! Cultivated in many parts of tropical Africa. (See Appendix, p. 262.)
3. **S. holosericea** (*Welw. ex Bak.*) *Harms* in Engl. Bot. Jahrb. 33 : 177 (1902) ; Bak. f. l.c. 419 (incl. var. *hastifolia*). *Vigna holosericea* Welw. ex Bak. in F.T.A. 2 : 200 (1871) ; Chev. Bot. 200. *V. hastifolia* Bak. (1871). *Sphenostylis kerstingii* Harms (1902). A twiner, covered more or less with a dense white or yellowish pubescence ; flowers ornamental, fragrant, mauve turning yellow ; root tuberous ; common amongst long grass.
    Sen.: *Berhaut* 1799. **Iv.C.:** Toumodi (Aug.) *Chev.* 22405 ! Yabarasso (fl. & fr. Dec.) *Chev.* 22557 ! **G.C.:** N. Agogo (Dec.) *Chipp* 605 ! Banda, Wenchi (Feb.) *Morton* GC 25162 ! **N.Nig.:** Nupe *Barter* 937 ! Jebba *Barter* ! Lokoja (fl. & fr. Oct.) *Dalz.* 26 ! Gombe (Oct.) *Lely* 655 ! **S.Nig.:** Lagos (Oct.) *Punch* 30 ! *Rowland* ! Abeokuta *Irving* 88 ! Olokemeji F.R. (Sept.) *Keay* FHI 25385 ! Also in Angola and Mozambique.

51. HAYDONIA Wilczek in Bull. Jard. Bot. Brux. 24 : 405 (1954).

Twiner ; stems slender, angled or slightly winged, glabrous ; leaves 3-foliolate ; leaflets oblong-lanceolate to subhastate, truncate or slightly rounded at base, apex obtuse, 3·5–7 cm. long, 5–30 mm. broad, glabrous, nerves prominent on both sides ; stipules lanceolate, 3–6 mm. long, 2 mm. broad, not produced below the base, glabrous, persistent ; racemes axillary, few-flowered ; peduncle 5–9 cm. long, the lower part 3-winged ; calyx shortly lobed, glabrous but ciliate on the margins ; vexillary stamen free to the base ; alternate stamens shorter than the rest and with 2 conspicuous glands beneath the anthers ; fruits suberect, 2·5–4·5 cm. long, about 2 mm. broad, appressed glandular-puberulous ..   .. *triphylla*

H. triphylla *Wilczek* l.c. 408, t. 12 (1954) ; Fl. Congo Belge 6 : 263, t. 24. A glabrous, twining, annual herb, with angular stems 4–6 ft. long ; flowers mauve or yellow ; usually amongst grasses.
    N.Nig.: Bauchi Plateau (Oct.) *Lely* P. 828 ! Also in Ubangi-Shari, Belgian Congo and Angola.

52. PHASEOLUS Linn.—F.T.A. 2 : 191.

Fruits with 10 or more seeds, linear, 10–14 cm. long, 1–1·4 cm. broad, nearly flat, with thicker margins ; flowers in rather stout racemes, glabrous, about 2·5 cm. long ; leaflets broadly ovate, subacute, 7–12 cm. long, 4–8 cm. broad, glabrous beneath except on the nerves ; stipels rather large, elliptic   ..   ..   .. 1. *adenanthus*
Fruits 2–4-seeded, obliquely oblanceolate, 6–8 cm. long, 1·5–2 cm. broad, the margins not thickened ; flowers in very slender racemes, about 1 cm. long, softly pubescent ; leaflets ovate-triangular, slightly and obtusely acuminate, 6–12 cm. long, 5–8 cm. broad, glabrous beneath or nearly so ; stipels small and inconspicuous   2. *lunatus*

1. **P. adenanthus** *G. F. W. Mey.*—F.T.A. 2 : 192 ; Chev. Bot. 197 ; Bak. f. Leg. Trop. Afr. 2 : 389 ; Wilczek in Fl. Congo Belge 6 : 338. A wide-climbing nearly glabrous perennial with fairly large pink and white flowers turning yellow ; on the sea-shore, the banks of river estuaries or occasionally by inland streams.
    Sen.: Sedhiou, Casamance *Chev.* 3408 ; 3409. **Gam.:** Kudang (fl. & fr. Jan.) *Dalz.* 8017 ! 8019 ! **Port.G.:** Fulacunda (fl. & fr. Dec.) *Esp. Santo* 2332 ! **Fr.G.:** Conakry *Maclaud* ! **S.L.:** Kent (fl. & fr. Dec.) *Deighton* 3337 ! Mange Bureh (Jan.) *Deighton* 1038 ! Yungeru (fl. & fr. Jan.) *Thomas* 7123 ! 7195 ! **Lib.:** Monrovia (fl. & fr. Nov.) *Linder* 1430 ! *Barker* 1079 ! **N.Nig.:** Nupe *Barter* 923 ! Katagum (fl. & fr. Nov.) *Dalz.* 42 ! **S.Nig.:** Brass *Barter* 1847 ! Bonny (Oct.) *Mann* 511 ! **Br.Cam.:** Victoria (Oct.) *Maitland* 760 ! Bama, Dikwa Div. (fl. & fr. Jan.) *McClintock* 168 ! Also in French Cameroons, Gabon and Belgian Congo. (See Appendix, p. 254.)
2. **P. lunatus** *Linn.*—F.T.A. 2 : 192 ; Chev. Bot. 197 ; Bak. f. l.c. 388. A rank-growing twiner, biennial, with half-woody stems and rather small greenish-white or bluish flowers. The Lima bean.
    Sen.: Sinédone, Casamance *Chev.* 3137 ! **Fr.Sud.:** San (Aug.) *Chev.* 3136 ; 3138 ! **Fr.G.:** *fide* Chev. l.c. **S.L.:** Njala (fl. & fr. Nov.) *Deighton* 1781 ! Bumbuna (Oct.) *Thomas* 3789 ! Meyamba (Dec.) *King* 18B ! **Lib.:** Kassa Ta (fl. & fr. Nov.) *Linder* 843 ! **Iv.C.:** Bouroukrou *Chev.* 16670. **G.C.:** Achimota (fl. & fr. Sept.) *Irvine* 776 ! Begoro (Nov.) *Morton* A. 87 ! **Togo:** Lome *Warnecke* 355 ! **Dah.:** Zagnanado (Feb.) *Chev.* 23088 ! Abomey (fr. Feb.) *Chev.* 23195 ! **N.Nig.:** Zaria (Oct.) *Lamb* 85 ! Agaie (Oct.) *Yates* 48 ! **S.Nig.:** Lagos *Rowland* ! Aguku *Thomas* 1177 ! **F.Po:** *T. Vogel* 235 ! Widely cultivated in the tropics. (See Appendix, p. 254.)
    P. vulgaris Linn. and *P. calcaratus* Roxb. are widely cultivated in our area.

53. VIGNA Savi—F.T.A. 2 : 194.

*Calyx-lobes as long as or longer than the calyx-tube, subulate from a lanceolate base ; stipules cordate, without an entire appendage at the base :
    Leaflets all sessile on the petiole, narrowly linear-lanceolate, acute, 12–20 cm. long, 5–10 mm. broad, with 3 longitudinally parallel nerves, slightly pubescent on the nerves ; flowers paired on long peduncles ; calyx hirsute and ribbed   1. *longissima*
Leaflets all petiolulate, with 1 main nerve :

18

Peduncles with a few short recurved hairs ; tertiary nerves lax and more or less reticulate ; leaflets ovate or lanceolate to linear, acute, obliquely and broadly wedge-shaped at base, 5–10 cm. long, 5–35 mm. broad ; flowers few and clustered at the end of long peduncles ; keel with a very large pocket on one side, rostrum large and recurved ; stigma capitate ; calyx about 1·4 cm. long ; fruits bristly hairy ; seed with a small central hilum .. .. .. 2. *vexillata*

Peduncles densely clothed with long sharply reflexed or spreading hairs ; keel without a large pocket on either side, rostrum not large and recurved :

Calyx 10–14 mm. long, lobes subulate, much longer than the tube, pilose ; flowers few at the end of long peduncles ; peduncles with sharply reflexed hairs ; stigma oblique, acute ; fruits appressed-bristly ; seeds oblong, 5 mm. long, hilum 2–3 mm. long ; leaflets variable, from linear to ovate-lanceolate, 5–15 cm. long, 1–4 cm. broad, setose-pubescent on both surfaces, tertiary nerves close and parallel

3. *reticulata*

Calyx 4 mm. long, lobes lanceolate, only slightly longer than the tube, pilose ; flowers clustered in a short raceme at the end of long peduncles ; peduncles with spreading hairs ; young fruits densely pilose ; leaflets elliptic, not apiculate, 3-nerved at base, 2–4 cm. long, 1–3 cm. broad, pilose on both surfaces .. .. 4. *maranguensis*

*Calyx-lobes much shorter than calyx-tube, sometimes quite broad ; stipules cordate or appendaged at the base :

Stipules more or less cordate but not long-appendaged below the base :

Leaflets linear or lanceolate :

Leaflets with very numerous lateral nerves spreading at right angles from the midrib, 6–10 cm. long, 5–8 mm. broad, strongly reticulate, glabrous ; stipules ovate-lanceolate, rounded at base ; flowers small, in short racemes ; peduncles about as long as the petiole ; fruits 6 cm. long, glabrous .. 5. *multinervis*

Leaflets with fewer and more or less oblique lateral nerves :

Stipules 3–5-nerved :

Keel rostrum short and not recurved ; flowers at the end of the peduncle :

Leaflets glabrous above and only younger parts of stems, etc., with a few hairs :

Rhachis of inflorescence without glands, 1–3-flowered, leaflets linear, very acute, 5–15 cm. long, 3–6 mm. broad, nervose ; stipels subtending terminal leaflet at least as long as petiolule ; fruits acutely beaked about 6 cm. long and 5 mm. broad, glabrous .. .. 6. *stenophylla*

Rhachis of inflorescence with several large glands, few-flowered ; leaflets lanceolate to linear-lanceolate, 3–4·5 cm. long, 5–10 mm. (rarely up to 20 mm.) broad, laxly nervose ; stipels subtending terminal leaflet about half as long as petiolule ; fruits 3–4 cm. long, about 3 mm. broad, minutely pubescent

7. *venulosa*

Leaflets pubescent above (at least on the midrib) rhachis of inflorescence with several large glands :

Peduncles filiform, 2–3 cm. long ; inflorescence few-flowered ; fruits up to 2 cm. long, 4 mm. broad, pendulous, glabrescent ; leaflets oblong, 1–5 cm. long, 5–10 mm. broad .. .. 8. *filicaulis*

Peduncles stout and erect, 4–20 cm. long, angular :

Fruits pendulous, sparsely pilose, torulose before maturity ; peduncles long, thick and angular at base ; other characters as below .. 18. *luteola*

Fruits erect, minutely puberulous, 3–4 cm. long, 3–4 mm. broad ; peduncles not thickened at base ; leaflets subhastate to lanceolate, 4–5 cm. long, 6–20 mm. broad ; stems more or less pilose .. .. .. 9. *nigritia*

Keel petal with long strongly incurved rostrum ; raceme lax with flowers in pairs about 1 cm. apart ; peduncles rather long ; pedicels 2 mm. long ; fruit about 8 cm. long and 4 mm. broad, sutures prominent ; leaflets lanceolate or ovate-lanceolate, 3-nerved and cuneate at base, 3–5 cm. long, 1–2·5 cm. broad, rather thick, glabrous .. .. 10. *macrorrhyncha*

Stipules about 8-nerved, 3 mm. long ; leaflets linear-lanceolate, rounded at base, lateral nerves obscure, appressed-pubescent beneath, 4–7 cm. long, about 5 mm. broad ; stipels about 2 mm. long ; flowers few, the peduncle at length much elongated and thickened .. .. 11. *lancifolia*

Leaflets broadly lanceolate to ovate or lobed :

Standard petal pubescent outside ; seeds oblong, about 5 mm. long, hilum oblique across one end, with a caruncle :

Rhachis of inflorescence very short, with up to about 7 large glands ; leaflets lanceolate to ovate, mucronate, 5–10 cm. long, (1–) 2–4·5 cm. broad, pubescent ; stipules cordate ; flowers few, crowded ; fruits about 5 cm. long, thinly appressed-pubescent .. .. 12. *pubigera*

Rhachis of inflorescence rather long, with numerous large glands ; leaflets oblong-lanceolate, up to 12 cm. long and 3 cm. broad, pubescent ; stipules small, hastate ; fruits about 4 cm. long, thinly pubescent .. .. 13. *ambacensis*

Standard petal glabrous outside :
Flowers precocious ; leaflets elliptic or more or less 3-lobed, cuneate at base, 3·5–7 cm. long, 2–5·5 cm. broad, silky-pubescent when young ; flowers few and clustered at the end of a very stout peduncle up to 30 cm. long ; calyx puberulous ; fruits silky-pubescent, 11 cm. long, erect ; seeds about 3 mm. long, hilum small and central   ..   ..   14. *violacea*
Flowers not precocious :
Leaflets hastately lobulate, 2–3 cm. long, about 2 cm. broad, thinly pubescent beneath ; stipules narrowly subulate ; flowers very small, subcapitate on a peduncle 3–4 cm. long ; fruits broadly linear, 2 cm. long, minutely pubescent, seeds 4 mm. long, hilum not central, with a caruncle ..   ..   15. *micrantha*
Leaflets lanceolate or ovate, not lobed :
Calyx densely pilose with brown hairs ; peduncles slightly thickened towards base, terete, almost glabrous, 21–27 cm. long ; leaflets lanceolate, 5·5–9 cm. long, about 1·5 cm. broad, rounded at base, acute at apex ; flowers 1·5–1·7 cm. long ; bracteoles linear, equalling calyx ; fruits densely brown-pilose, straight
16. *sp. nr. oblongifolia*
Calyx glabrous or pubescent with whitish hairs :
Peduncles angular and thicker near the stem than at the other end ; mature fruits about 6 cm. long, sutures almost straight, pendulous, immature dried fruits torulose, sparsely pilose ; seeds elliptic, 4–5 mm. long, hilum central, without a caruncle :
Leaflets emarginate at apex, glabrous, setose on midrib, ovate to broadly oblong-lanceolate, 4–7 cm. long, 1·5–4 cm. broad ..   ..   17. *marina*
Leaflets acute and mucronate at apex, lanceolate, 3–7 cm. long, 8–16 mm. broad   ..   ..   18. *luteola*
Peduncles not thickened towards the base, terete :
Stems pubescent or glabrous, not long-villous ; fruits pendulous :
Leaflets more or less ovate-triangular, hardly acuminate, very variable on the same plant, usually about 3 cm. long and 2 cm. broad ; fruits about 2 cm. long (rarely 3 cm.)   ..   ..   19. *gracilis*
Leaflets very broadly ovate, acutely acuminate, all similar, usually about 6–7 cm. long and 5 cm. broad ; fruits nearly 4 cm. long ..   20. *multiflora*
Stems more or less villous with long spreading or reflexed hairs ; fruits pendulous or erect :
Stipules distinctly sagittate, about 8 mm. long ; peduncles 8–20 cm. long ; racemes rather lax ; leaflets ovate to ovate-lanceolate, 4–9 cm. long, 2·5–4 cm. broad, rounded at base, acuminate ; fruits 2–3 cm. long, about 3 mm. broad   ..   ..   21. *racemosa*
Stipules cordate, about 5 mm. long :
Fruits pendulous, setose, 2–3 cm. long, 3–4 mm. broad, slightly torulose and curved ; seeds with a small hilum on one side ; peduncles 1–2 (–3) cm. long ; leaflets broadly ovate, 5–11 cm. long, 4–8·5 cm. broad, rounded to subcordate at base, acuminate ..   ..   ..   22. *desmodioides*
Fruits erect, villose with brown hairs, 4–5 cm. long, 5 mm. broad, not torulose, straight ; seeds with a long hilum along one side ; peduncles 8–20 cm. long, at least in fruit ; leaflets ovate, 4–7 cm. long, 2·5–4 cm. broad, rounded at base, acute   ..   ..   23. *campestris*
Stipules with an appendage below the base :
Stems pilose ; flowers 1·2 cm. long at the end of long, pilose peduncles which do not thicken in fruit ; fruits 3–4 cm. long, 7 mm. broad, black when ripe, densely pilose ; seeds with hilum oblique at one end   ..   ..   ..   24. *paludosa*
Stems glabrous ; flowers about 2 cm. long, at the end of the long, glabrous peduncles, which thicken in fruit ; fruits in the wild form about 10 cm. long and soon dehiscent with the valves spirally twisted, in the cultivated form about 15 cm. long and not dehiscent, not spirally twisted ; seeds with a small median hilum 25. *unguiculata*

1. **V. longissima** *Hutch.* in Kew Bull. 1921 : 367 ; Bak. f. Leg. Trop. Afr. 2 : 414 ; Wilczek in Fl. Congo Belge 6 : 377. A slender herb with laxly pilose, prostrate, or perhaps twining, stems and fairly large mauve flowers.
N.Nig.: Bauchi Plateau, between Hepham and Ropp, 4,600 ft. (Aug.) *Lely* 360 ! P. 324 ! Vom, 3,000–4,500 ft. *Dent Young* 71 ! Also in N. Rhodesia and Belgian Congo.
2. **V. vexillata** (*Linn.*) *Benth.*—F.T.A. 2 : 199 (incl. vars. α and β) ; Bak. f. l.c. 413 ; Chev. Bot. 202 ; Wilczek l.c. 379, t. 29, fig. 18C. *Phaseolus vexillatus* Linn. (1753). *Plectrotropis angustifolia* Schum. & Thonn. (1827). *Vigna angustifolia* (Schum. & Thonn.) Hook. f. (1849)—F.W.T.A.. ed. 1, 1 : 409 ; Bak. f. l.c. *Plectrotropis hirsuta* Schum. & Thonn. (1827), not *Vigna hirsuta* Feay ex Wood (1861). *Strophostylis capensis* E. Mey. (1835). *Vigna capensis* (E. Mey.) Walp. (1839)—Berhaut Fl. Sen. 20. *V. thonningii* Hook. f. (1849)—F.W.T.A., ed. 1, 1 : 409 ; Bak. f. l.c. *V. senegalensis* A. Chev. (1944). Stems twining or prostrate ; root tuberous and spindle-shaped ; flowers pink or purplish turning yellow.
Sen.: *Döllinger* ! *Roger* ! Walo country (fl. & fr. Feb.) *Perrottet* 227 ! **Gam.**: *Brown-Lester* 41 ! **S.L.**: Regent (Dec.) *Sc. Elliot* 4113 ! Gene (Nov.) *Deighton* 462 ! Njala (Dec.) *Deighton* 1578 ! Baoma (fr. Dec.) *Deighton* 3469 ! **Lib.**: Kakatown *Whyte* ! Gbau, Sanokwele (Sept.) *Baldwin* 9426 ! Monrovia (fr. Dec.) *Baldwin* 10509 ! Brewer's Landing (Nov.) *Linder* 14021 ! **Iv.C.**: Bingerville *Chev.* 16053 ; 16100. Cavally *Chev.* 19557. **G.C.**: Cape Coast (July) *T. Vogel* 52 ! 64 ! Achimota (fr. Jan.) *Irvine* 1943 ! *Morton* GC 6122 ! Aburi Scarp (Nov.) *Morton* GC 6112 ! Akuse (Oct.) *Morton* A. 98 ! **Togo:**

Kpedsu (fl. & fr. Dec.) *Howes* 1045 ! **N.Nig.**: Lokoja *Barter* ! Abinsi (Dec.) *Dalz.* 619 ! Kontagora (Oct.) *Dalz.* 31 ! Benue R. (Oct.) *Parsons* L. 57 ! **S.Nig.**: Onitsha (Aug.) *Jones* FHI 6146 ! **F.Po**: *T. Vogel* 222 ! Widespread in tropical Africa and America. (See Appendix, p. 269.)

[*V. angustifolia* (Schum. & Thonn.) Hook. f. with linear leaflets about 1 cm. broad, and *V. thonningii* Hook. f. a coastal form with ovate leaflets, have been separated from *V. vexillata*, but there is great variation in leaf-shape.]

3. **V. reticulata** *Hook. f.*—F.T.A. 2 : 198 (incl. var. β) ; Bak. f. l.c. 411 ; Chev. Bot. 201 ; Wilczek l.c. 382. *V. linearifolia* Hook. f. (1849) var. β. A twiner, grey- or yellowish-hairy, with yellow or purplish flowers. **Sen.**: *Berhaut* 1449. **S.L.**: Franziga, Tambakka country *Sc. Elliot* 5027 ! Juring (Dec.) *Deighton* 319 ! Musaia (fl. & fr. Dec.) *Deighton* 4419 ! **Iv.C.**: Kodiokoffi, Baoulé (fr. Aug.) *Chev.* 22354 ! Béoumi (Dec.) *Lowe* ! **G.C.**: Tamale (fr. Dec.) *Adams & Akpabla* GC 4173 ! Gambaga Scarp (fr. Dec.) *Adams & Akpabla* GC 4249 ! Damongo (fl. & fr. Dec.) *Morton* GC 9988 ! **Togo**: Yendi (Nov.) *Williams* 416 ! Sokode-Basari *Schröder* 115 ! **N.Nig.**: Nupe *Barter* 1017 ! Sokoto (Oct.) *Dalz.* 321 ! Kaduna (Feb.) *Taylor* FHI 1104 ! Naraguta and Jos (Sept.) *Lely* 563 ! P. 682 ! Lokoja (Sept.) *T. Vogel* 22 ! 37 ! **S.Nig.**: Obu *Thomas* 354 ! **Br.Cam.**: Banso (Oct.) *Tamajong* FHI 23480 ! Extends to eastern Africa and Angola. (See Appendix, p. 266.)

4. **V. maranguensis** (*Taub. ex Engl.*) *Harms* in Engl. Pflanzenw. Afr. 3, 1 : 686 (1915) ; Wilczek l.c. 369. *Dolichos maranguensis* Taub. ex Engl. (1892). A semi-prostrate creeper or twiner, rooting at the lower nodes ; whole plant densely yellow pilose ; flowers pink or whitish-yellow. **Br.Cam.**: Bafut-Ngemba F.R., near L. Batuli, 6,000 ft., Bamenda (Aug., Oct., Nov.) *Lightbody* FHI 26271 ! *Ujor* FHI 29972 ! Also in eastern Africa.

5. **V. multinervis** *Hutch. & Dalz.* F.W.T.A., ed. 1, 1 : 409 (1928) ; Kew Bull. 1929 : 17 ; Bak. f. l.c. 398 ; Wilczek l.c. 357. *V. linearifolia* Hutch. in Broun & Massey (1929), not of Hook. f. (1849). A glabrous twiner, with wiry stems to 6 ft. long, and blue or yellow flowers ; amongst grass, etc. **Togo**: Kadjakpe, Togo Plateau F.R. (Nov.) *St. C. Thompson* 1620 ! **N.Nig.**: Bauchi Plateau *Batten-Poole* 124 ! **S.Nig.**: Onitsha *Barter* 1799 ! Also in A.-E. Sudan, Uganda, Tanganyika, Belgian Congo and Angola.

6. **V. stenophylla** *Harms* in Notizbl. Bot. Gart. Berl. 5 : 210 (1911) ; Bak. f. l.c. 398. *V. dauciformis* A. Chev. in Bull. Soc. Bot. Fr. 58, Mém. 8 : 162 (1912), incl. var. *elata*. Stem suberect with trailing glabrous branches and a spindle-shaped tuberous root ; flowers bluish-purple. **S.L.**: Sherbro Isl. (Nov.) *Hunter* 4 ! **Iv.C.**: Mbayakro, Baoulé (Aug.) *Chev.* 22298 ! **G.C.**: Akuse (fl. & fr. June) *Dalz.* 83 ! **Togo**: Yendi (fr. May) *Akpabla* 506 ! **Dah.**: Kouandé *Chev.* 24234. **Fr.Nig.**: Gourma (July) *Chev.* 24377 ! Also in northern French Cameroons.

7. **V. venulosa** *Bak.* in F.T.A. 2 : 203 (1871) ; Chev. Bot. 202 (incl. var. *lat hyroides* A. Chev. (1912)) ; Bak. f. l.c. 399. A slender climber amongst grasses ; flowers purple or yellow. **Sen.**: *Berhaut* 844. **Fr.Sud.**: Nyamina to Koulikoro (Oct.) *Chev.* 3214 ! **Fr.G.**: Sambadougou (fr. Feb.) *Chev.* 20555 ! **S.L.**: Regent (Dec.) *Sc. Elliot* 4038 ! Bonthe Isl. (fl. & fr. Nov.) *Deighton* 2324 ! Musaia (fl. & fr. Dec.) *Deighton* 4510 ! Rokupr (Nov.) *Jordan* 154 ! **Lib.**: Duport (Nov.) *Linder* 1512 ! Monrovia (Dec.) *Baldwin* 10504 ! **Togo**: *fide* Bak. f. l.c. **Br.Cam.**: *fide* Bak. f. l.c.

8. **V. filicaulis** *Hepper* in Kew Bull. 1956 : 128. *V. venulosa* of Chev. Bot. 202, partly (Chev. 22259), and var. *pubescens* A. Chev. in Bull. Soc. Bot. Fr. 58, Mém. 8 : 163 (1912) ; Bak. f. l.c. 399. Stems slender and wiry up to 2 ft. long ; flowers yellow ; in grassland. **Iv.C.**: Baoulé-Nord (Aug.) *Chev.* 22259 ! 22359 ! **G.C.**: Ejura (fl. & fr. Dec.) *Morton* GC 9508 ! 9714 !

9. **V. nigritia** *Hook. f.* in Fl. Nigrit. 310 (1849) ; Wilczek l.c. 358, fig. 18A. *V. pubigera* var. *gossweileri* Bak. f. (1928). *V. tisseranti* A. Chev. (1944), not of Pellegrin (1944). *V. villosa* of F.T.A. 2 : 206, partly. *V. racemosa* of F.W.T.A., ed. 1, 1 : 409, partly. A climber ; flowers pink ; in savannah. **S.L.**: Samu country *Sc. Elliot* 5918 ! **Iv.C.**: Bouakro to Alangoussou *Chev.* 22219. **G.C.**: Ejura (fl. & fr. Dec.) *Morton* GC 9514 ! Gambaga (Apr.) *Morton* GC 9044 ! **N.Nig.**: *T. Vogel* 14 ! **S.Nig.**: Lagos (Oct.) *Punch* 37 ! Usonigbe F.R., Benin (fl. & fr. Oct.) *Keay & Onochie* FHI 19693 ! Also in Ubangi-Shari, Belgian Congo and Angola.

10. **V. macrorrhyncha** (*Harms*) *Milne-Redhead* in Kew Bull. 1936 : 473. *Phaseolus macrorhynchus* Harms (1900) ; Bak. f. l.c. 390. *P. schimperi* Taub. (1892)—Bak. f. l.c. ; Wilczek l.c. 335, t. 27 ; not *Vigna schimperi* Bak. (1871). Stem slender, sparsely appressed-pubescent, older leaflets glabrous and with translucent venation ; flowers purplish. **N.Nig.**: Naraguta (Aug.) *Lely* 552 ! Also in A.-E. Sudan, Uganda, Kenya, Tanganyika, Rhodesia and Belgian Congo.

[*V. macrorrhyncha* has the truncate stipules of *Phaseolus*, but septate fruits of *Vigna*. These two genera are not satisfactorily separable.]

11. **V. lancifolia** *A. Rich.*—F.T.A. 2 : 196 ; Bak. f. l.c. 398. Suberect with striate stems, eventually twining, and reddish flowers. **N.Nig.**: Sokoto (Nov., Dec.) *Moiser* 206 ! 211 ! Also in Abyssinia and A.-E. Sudan.

12. **V. pubigera** *Bak.* in F.T.A. 2 : 201 (1871) ; Bak. f. l.c. 400. Stems rather densely yellowish-pubescent, prostrate or widely climbing ; flowers yellow, sometimes blue or reddish ; in fallow fields or bush not far from cultivation. **Fr.Sud.**: Macina (Sept.) *Chev.* 24915 ! **G.C.**: Achimota (Aug., Nov.) *Milne-Redhead* 5125 ! *Irvine* 1516 ! Accra (fl. & fr. Aug.) *Deighton* 3401 ! **Togo**: Paliba (fr. Dec.) *Adams & Akpabla* 4110 ! 4111 ! Ho (fl. & fr. Nov.) *A. S. Thomas* K. 7 ! **N.Nig.**: Lokoja *Parsons* 95 ! Nupe *Barter* 983 ! Katagum *Dalz.* 39 ! Abinsi (Dec.) *Dalz.* 621 ! Sokoto (fl. & fr. Oct.) *Dalz.* 315 ! 316 ! Bauchi to Gombe (fl. & fr. Oct., Nov.) *Lely* 633 ! **S.Nig.**: Awba Hills (Oct.) *Jones* FHI 5982 ! Widely spread in tropical Africa. (See Appendix, p. 266.)

13. **V. ambacensis** *Welw. ex Bak.* in F.T.A. 2 : 201 (1871) ; Bak. f. l.c. 400 ; Wilczek l.c. 355. A twiner with grey-pubescent stems and bright yellow flowers, only the upper few flowers of the raceme develop. **N.Nig.**: Abinsi (Dec.) *Dalz.* 617 ! Muri (fl. & fr. Nov.) *Lamb* 75 ! **S.Nig.**: Bansara, Ogoja Prov. (fl. & fr. Dec.) *Keay* FHI 28145 ! **Br.Cam.**: Bamenda, 5,500 ft. (Jan.) *Migeod* 444 ! Also in A.-E. Sudan, Belgian Congo, Uganda and Angola.

14. **V. violacea** *Hutch.* in Kew Bull. 1921 : 247 ; Bak. f. l.c. 405. A fairly strong twiner with slightly pubescent grey-green stems up to 12 ft. long from a woody tuberous root ; flowers fragrant, beautiful lavender-violet appearing when leafless. **G.C.**: Banda, Wenchi (fl. & fr. Dec.) *Morton* GC 25162 ! Ejura (Dec.) *Morton* GC 9520 ! **N.Nig.**: Bauchi Plateau (Jan.) *Lely* P. 52 ! Naraguta (fl. Mar., May, fr. May) *Lely* 6 ! *Kennedy* 45 ! Jos (fl. & fr. Mar.) *Hill* 6 ! Tof, 4,000 ft. (Feb.) *Hepburn* 135 !

15. **V. micrantha** *Harms* in Engl. Bot. Jahrb. 26 : 311 (1899) ; Bak. f. l.c. 406 ; Wilczek l.c. 353. Stems slender, pubescent, twining amongst grass, etc., with pale yellow flowers and 2–3-seeded fruits. **S.L.**: Sefadu, 1,400 ft. (fl. & fr. Dec.) *Deighton* 3568 ! **Lib.**: Mt. Mpaka Fossa, 1,960 ft., Kolahun (fl. & fr. Jan.) *Bequaert* 37 ! **S.Nig.**: Ikoyi Plains, Lagos (fl. & fr. Nov.) *Dalz.* 1216 ! Benin (Mar.) *Jones* FHI 7391 ! Also in Uganda, Kenya, and French and Belgian Congo.

16. **V. sp. nr. oblongifolia** *A. Rich.*—F.T.A. 2 : 196 ; Bak. f. l.c. 410. A rambling plant amongst tall grasses. **Br.Cam.**: Basenako, Bamenda, 6,000 ft. (fl. & fr. June) *Maitland* 1505 ! [Differs from *V. oblongifolia* in the larger flowers, pilose calyx and almost glabrous peduncle ; differs from *V. fischeri* Harms especially in the pilose calyx ; more material is required.]

17. **V. marina** (*Burm.*) *Merrill* Interpr. Rumph. Herb. Amboin. 285 (1917) ; Exell Cat. S. Tomé 163. *Phaseolus marinus* Burm. (1755). *Vigna oblonga* Benth. (1844)—F.T.A. 2 : 206 ; Bak. f. l.c. 402 ; F.W.T.A., ed. 1, 1 : 409. Stem glabrous, prostrate or twining amongst grass on the sea-shore ; flowers yellow.
Lib.: Grand Cess (fr. Mar.) *Baldwin* 11623 ! Rocktown (Mar.) *Baldwin* 11634 ! **S.Nig.:** Nun R. (Aug.) *T. Vogel* 18 ! *Mann* 495 ! Brass R. *Barter* 1827 ! Lagos (June) *Dalz.* 1211 ! *Moloney* ! **Br.Cam.:** Victoria *Preuss* 1126 ! **F.Po:** *T. Vogel* 221 ! A sea-shore plant throughout the tropics.

18. **V. luteola** (*Jacq.*) *Benth.* in Mart. Fl. Bras. 15 : 194, t. 50, fig. 2 (1859), not of F.T.A. 2 : 205 ; Bak. f. l.c. 401 ; Wilczek l.c. 363. *Dolichos luteolus* Jacq. (1770). *D. niloticus* Del. (1813). *Vigna nilotica* (Del.) Hook. f. (1849)—F.T.A. 2 : 204 ; Bak. f. l.c. 404 ; not of F.W.T.A., ed. 1, 1 : 409. *V. nigerica* A. Chev. (1944). A twiner with yellow flowers ; in waste places.
Sen.: Marsosoun, Casamance (fl. & fr. Feb.) *Chev.* 3407 ! **S.L.:** Yakban, Mambolo (fl. & fr. Nov.) *Jordan* 954 ! Kiragba, Samu (Sept.) *Jordan* 922 ! Bonthe (fl. & fr. Nov.) *Deighton* 2322 ! Mahera (fl. & fr. Nov.) *Deighton* 5687 ! **Lib.:** Roberts Field (fl. & fr. Jan.) *Baldwin* 14042 ! **Dah.:** Karimana (fl. & fr. Nov.) *Arnaud.* Widespread in tropical Africa and America. (See Appendix, p. 266.)

19. **V. gracilis** (*Guill. & Perr.*) *Hook. f.*—F.T.A. 2 : 205 ; Chev. Bot. 200 ! Bak. f. l.c. 403 ! Wilczek l.c. 368. *Dolichos gracilis* Guill. & Perr. (1832). *Vigna afzelii* Bak. in F.T.A. 2 : 202 (1871) ; Bak. f. l.c. 403. *V. parvifolia* Planch. ex Bak. in F.T.A. 2 : 206 (1871) ; Bak. f. l.c. 399. *V. racemosa* var. *glabrescens* Bak. f. l.c. 409 (1929) ; Wilczek l.c. 371. *V. occidentalis* Bak. f. (1934). A slender extensively twining herb amongst grass with pink or bluish flowers turning yellow ; very variable in foliage.
Sen.: Casamance R. *Heudelot* 589 ! **Gam.:** Albreda (Mar., Apr.) *Perrottet* ! **Fr.Sud.:** Koulikoro *Chev.* 25027. **Port.G.:** Bissau (Oct., Nov.) *Esp. Santo* 1361 ! 2592 ! **S.L.:** Regent (fl. & fr. Dec.) *Sc. Elliot* 3859 ! Rokupr (Oct.) *Jordan* 369 ! 696 ! Yoni Mamila (fl. & fr. Dec.) *King* 7 B ! Kono (Nov.) *Deighton* 69 ! Wellington (fl. & fr. Feb.) *Deighton* 4988 ! **Lib.:** Kakatown *Whyte* ! Monrovia (fl. & fr. Nov.) *Linder* 1521 ! *Baldwin* 10494 ! Gbau, Sanokwele Dist. (Sept.) *Baldwin* 9427 ! **S.Nig.:** Degema Div. *Talbot* ! Eket *Talbot* 3014 ! Also in French Cameroons, Belgian Congo and Principe. (See Appendix, p. 266.)

20. **V. multiflora** *Hook. f.*—F.T.A. 2 : 205 ; Bak. f. l.c. 403 ; Wilczek l.c. 367. A slender twiner with pink or bluish-purple flowers quickly turning yellow.
S.L.: Bintumane (Aug.) *T. S. Jones* 143 ! Kumrabai (Dec.) *Thomas* 6859 ! 7062 ! **Lib.:** Bushrod Isl. (fl. & fr. Dec.) *Barker* 1104 ! Duport (fl. & fr. Oct.) *Barker* 1434 ! Boporo Dist. (fl. & fr. Nov.) *Baldwin* 10264a ! 10346 ! **Iv.C.:** Baoulé-Nord (July) *Chev.* 22219 ! **S.Nig.:** Idanre Hills (Oct.) *Keay* FHI 25510 ! Sapoba *Kennedy* FHI 2905 ! Oban *Talbot* 445 ! **Br.Cam.:** Buea (Nov.) *Migeod* 11 ! *Maitland* 112 ! *Dunlap* 139 ! Banso (fl. & fr. Oct.) *Tamajong* FHI 23485 ! **F.Po:** *T. Vogel* 245 ! Moka (Dec.) *Boughey* 1 ! 51 ! Also in Belgian Congo. (See Appendix, p. 266.)

21. **V. racemosa** (*G. Don*) *Hutch. & Dalz.* F.W.T.A., ed. 1, 1 : 409 (1928) ; Kew Bull. 1929 : 18 ; Bak. f. l.c. 409 ; Wilczek l.c. 370. *Clitoria racemosa* G. Don (1832). *Vigna donii* Bak. in F.T.A. 2 : 202 (1871). *V. luteola* of F.T.A. 2 : 205, excl. syn. *V. villosa* Savi and (?) *V. glabra* Savi ; Chev. Bot. 201 ; Bak. f. l.c. 401. A fairly strong twiner amongst grass or on trees, with flowers either bright blue or pink or purple-tinged, turning yellow.
Sen.: Berhaut 1321. **Gam.:** Ingram ! Mungo Park ! **Fr.Sud.:** Bamako (fl. & fr. Nov.) *Waterlot* 1432 ! **Port.G.:** Bissau (fl. & fr. Oct., Nov.) *Esp. Santo* 2560 ! 2587 ! Bafata (Nov.) *Esp. Santo* 3117 ! **S.L.:** Njala (fl. & fr. Jan., Nov.) *Deighton* 1801 ! 4971 ! 5019 ! Baoma (fl. & fr. Nov.) *Deighton* 3477 ! Musaia (fl. & fr. Nov.) *Glanville* 373 ! Magbile (Dec.) *Thomas* 6203 ! **G.C.:** Bompata (Dec.) *Dalz.* 88 ! N. Agogo (fl. & fr. Dec.) *Chipp* 619 ! Abetifi and Afram Plains (fl. & fr. Dec.) *Scholes* 111 ! **Togo:** Ho (fr. Nov.) *A. S. Thomas* K. 9 ! Yendi (fr. Nov.) *Williams* 414 ! **N.Nig.:** Nupe *Barter* 1009 ! Lokoja *Barter* 1599 ! Onitsha *Barter* 1748 ! *Dalz.* 25 ! *Parsons* L. 65 ! L. 94 ! Sokoto (fr. Oct.) *Dalz.* 317 ! Kadaura (Sept.) *Lely* 591 ! **S.Nig.:** Abeokuta *Irving* ! Lagos (Oct.) *Punch* 37 ! *MacGregor* 294 ! Also in Uganda, Ubangi-Shari, Belgian Congo and Angola. (See Appendix, p. 266.)

22. **V. desmodioides** *Wilczek* in Bull. Jard. Bot. Brux. 24 : 438 (1954) ; Fl. Congo Belge 6 : 366, fig. 18D. Stems climbing and more or less pilose ; flowers pink or mauve.
S.L.: Taiama (fl. & fr. Dec.) *Deighton* 3480 ! **S.Nig.:** Idumuje *Thomas* 2160 ! Also in Ubangi-Shari and Belgian Congo.

23. **V. campestris** (*Mart. ex Benth.*) *Wilczek* in Fl. Congo Belge 6 : 391, t. 31 (1954). *Phaseolus campestris* Mart. ex Benth. (1838). *Vigna luteola* of F.T.A. 2 : 205, partly (*Mann* 2319). A twiner 4–6 ft. high ; flowers greenish.
S.L.: Gbinti (fl. & fr. Sept., fr. Jan.) *Deighton* 4849 ! 5991 ! **S.Nig.:** Old Calabar *Mann* 2319 ! Also in Belgian Congo and tropical America.

24. **V. paludosa** *Milne-Redhead* in Kew Bull. 1947 : 27. Stem pilose, climbing up grasses in swamps and wet places ; flowers yellow.
S.L.: Mange, Bure (Jan.) *Deighton* 1043 ! Rokupr (fl. & fr. Jan.) *Deighton* 2953 ! 4050 ! 4189 ! *Jordan* 146 ! *Adames* 162 ! Port Loko (fl. & fr. Dec.) *Thomas* 5845 ! 6520 ! 6523 ! 6634 !

25. **V. unguiculata** (*Linn.*) *Walp.* Rep. 1 : 779 (1842) ; Wilczek l.c. 387, t. 30. *Dolichos unguiculatus* Linn. (1753). *D. biflorus* Linn. (1753)—Brenan in Mem. N.Y. Bot. Gard. 8 : 415 (1954) ; not of F.W.T.A., ed. 1, 1 : 410. *D. sinensis* Linn. (1756). *Vigna sinensis* (Linn.) Savi ex Hassk. (1844)—F.T.A. 2 : 204 ; Bak. f. l.c. 407 ; Merrill Interpr. Rumph. Herb. Amboin. 284 ; A. Chev. in Rev. Bot. Appliq. 24 : 128 (1944) ; Chev. Bot. 201. *V. baoulensis* A. Chev. (1912). There are numerous other synonyms. Stems annual but often stout, subglabrous, twining on trees, etc. ; flowers white and mauve-tinged or pink or yellow.
Port.G.: Pussube, Bissau (fr. Nov.) *Esp. Santo* 998 ! Antula, Bissau (Oct.) *Esp. Santo* 2537 ! **Fr.G.:** Macenta (Oct.) *Baldwin* 9846 ! **S.L.:** Bumbuna (fl. & fr. Oct.) *Thomas* 3452 ! Gegbwema (fl. & fr. Dec.) *Deighton* 3820 ! Musaia (fr. Nov., Dec.) *Glanville* 375 ! **Lib.:** Gbanga (Sept.) *Linder* 539 ! **Iv.C.:** Marabadiassa *Chev.* 22032 ! Kodiokoffi *Chev.* 22356 ! **G.C.:** Cape Coast (fr. July) *T. Vogel* 47 ! Techiman, Ejura (fl. & fr. Dec.) *Morton* GC 9556 ! Kumasi (fr. Oct.) *Baldwin* 13463 ! **Fr.Nig.:** Niamey (fr. Oct.) *Hagerup* 534 ! **N.Nig.:** Sokoto (fr. Oct.) *Dalz.* 318 ! Katagum *Dalz.* 36 ! **S.Nig.:** Ibadan (fl. & fr. Nov.) *Meikle* 662 ! Widespread in the tropics. (See Appendix, p. 266.)
[The specimens cited here are of the wild, short- and narrow-fruited form (syn. *V. baoulensis* A. Chev.) ; the commonly cultivated forms which have long and broad fruits, if regarded as a distinct species, should bear the name *V. sinensis* (Linn.) Savi ex Hassk., vide Merrill l.c. and Chev. in Rev. Bot. Appliq. 24 : 139–145 (1944).]
Besides the above *V. mungo* (Linn.) Hepper with large stipules is occasionally cultivated in our area.

Imperfectly known species.

V. marchali *A. Chev.* in Rev. Bot. Appliq. 30 : 268 (1950). Described as having no stipules.
Fr.Nig.: flooded grassland beside the Niger, Niamey to Gaya. Type not designated.

54. DOLICHOS Linn.—F.T.A. 2 : 209.

*Flowers in axillary racemes or fascicles, or subsolitary :
†Leaflets lanceolate to suborbicular, with pinnate nerves ; trailing or climbing (except No. 5) herbs :
φFruits 2–5·5 mm. broad, 3–8 cm. long ; leaflets oblong-lanceolate to suborbicular :
Fruits 2–3·5 mm. broad, 4–8 cm. long, very sparsely appressed-pilose ; leaflets ovate-oblong to narrowly oblong or lanceolate, appressed-pubescent on both

surfaces ; flowers 6–11 mm. long ; calyx-teeth subulate-lanceolate, lower tooth
about 6 mm. long ; pedicels about 5 mm. long ..   ..   1. *stenophyllus*
Fruits 3·5–5 mm. broad :
  Calyx-teeth all 1–2 mm. long ; flowers 9–11 mm. long ; pedicels very short ;
    leaflets ovate-lanceolate, 2–3 cm. long, sparsely pilose on margins and nerves
    beneath   ..   ..   2. *africanus*
  Calyx-teeth 5–9 mm. long ; flowers 12–15 mm. long ; pedicels about 3 mm. long ;
    leaflets oblong to ovate-lanceolate, variable, 2–4 cm. long, from sparsely to
    densely pilose on both surfaces   ..   ..   3. *chrysanthus*
ɸFruits (5–) 6–8·5 mm. broad, 3–4 cm. long ; leaflets more or less lanceolate :
  Lower and lateral sepals abruptly attenuated into long filiform points two to several
    times longer than the expanded basal part of the teeth :
    Flowers 10–13 mm. long, almost sessile ; fruits with both long and short hairs,
      sometimes with short tubercles on sutures ; leaflets 4–5 cm. long and about
      2 cm. broad, lateral leaflets broadly rounded and unequal at base, pubescent
                                                                          4. *daltoni*
    Flowers 13–18 mm. long ; fruits more or less pubescent with short hairs all of
      about the same length ; leaflets 4–6 cm. long and about 2 cm. broad, lateral
      leaflets cuneate to slightly rounded and less markedly unequal at base, densely
      pubescent   ..   ..   5. *brevicaulis*
  Lower and lateral sepals acuminate, hardly filiform-pointed, acumen up to 1½ times
    as long as expanded basal part of the teeth ; flowers 12–15 mm. long, almost
    sessile ; fruits pubescent ; leaflets 3–4 cm. long, 1·5 cm. broad, cuneate at base,
    pubescent especially on margins and nerves of both sides   ..   6. *axillaris*
ɸFruits 8–11 mm. broad, 2·5–10 cm. long, shortly pubescent or very long-pilose ;
  leaflets very broadly lanceolate to almost rhombic or irregularly 3-lobed :
  Leaflets entire ; fruits minutely pubescent, 6–10-seeded ; flowers not in clusters :
    Bracteoles at base of calyx prominently 1-nerved and sometimes with 2 obscure
      lateral nerves, lanceolate, 1 mm. long ; calyx-lobes triangular, about 1 mm.
      long ; stems glabrous to sparsely pilose ; leaves variable, in W. Africa
      usually broadest below the middle, broadly cuneate to rounded at base, sub-
      acuminate, about 5·5 cm. long and broad..   ..   ..   ..   7. *falcatus*
    Bracteoles at base of calyx prominently 3–5-nerved, linear, 3–5 mm. long ; calyx-
      lobes acuminate, 2–3 mm. long, lowest lobe about 5 mm. long ; stems appressed-
      pubescent ; leaves more or less rhombic, cuneate at base, subacute at apex,
      4–8 cm. long, 2–6 cm. broad   ..   ..   8. *formosus*
  Leaflets irregularly 3-lobed, densely silky-pubescent beneath ; fruits very long-
    pilose, 3-seeded ; flowers in pedunculate clusters ; peduncles about 3 cm. long,
    densely long-pilose   ..   ..   9. *argenteus*
†Leaflets linear or linear-lanceolate, with 3 prominent parallel nerves, glabrous to
  densely pubescent ; stems erect, sometimes succulent ; flowers fascicled in leaf-axils
  on long pedicels ; fruits woody, 6–8 cm. long, 1–1·5 cm. broad   10. *schweinfurthii*
*Flowers in terminal or subterminal racemes ; stems erect ; leaflets rounded, broadly
  emarginate to acute at apex :
Stipules 3–4 mm. long, 5–7 mm. broad at base, acuminate, reflexed ; leaves subsessile,
  leaflets obovate or obovate-oblong, rounded or emarginate at apex, lateral leaflets
  unequal-sided at base, terminal leaflets cuneate at base, 3–6 cm. long, subglabrous
  to pubescent, young stems pubescent becoming glabrous   ..   ..   11. *dinklagei*
Stipules 8–40 mm. long :
  Stipules 30–40 mm. long, lanceolate, not amplexicaul ; leaflets lanceolate to elliptic-
    lanceolate, rounded at base, acute at apex, 6–9 cm. long, about 3 cm. broad,
    pilose on nerves beneath, lateral leaflets subequal on each side 12. *grandistipulatus*
  Stipules 8–23 mm. long, amplexicaul, very broad at base :
    Leaflets orbicular, truncate at base, broadly emarginate or rounded at apex, about
      7 cm. long and broad ; stipules broadly lanceolate, prominently many-nerved,
      subcordate, 18–23 mm. long, about 13 mm. broad   ..   13. *tonkouiensis*
    Leaflets ovate-lanceolate, cuneate at base, subacute at apex, 3–11 cm. long, 1–4 cm.
      broad, lateral leaflets very unequal-sided at base ; stipules ovate-lanceolate,
      obscurely many-nerved, subfalcate, (8–) 12–19 mm. long, about 8 mm. broad
                                                                        14. *nimbaensis*

1. **D. stenophyllus** *Harms* in Engl. Bot. Jahrb. 26 : 314 (1899) ; Bak. f. Leg. Trop. Afr. 2 : 450 ; Brenan in
  Mem. N.Y. Bot. Gard. 8 : 420 (1954) ; Wilczek in Fl. Congo Belge 6 : 325. Berhaut Fl. Sen. 20. Stems
  twining or trailing, 1–3 ft. long, from a woody base, slender, pubescent ; flowers yellow ; in savannah.
  Sen.: *fide* Berhaut *l.c.* **Togo**: Sokode-Basari *Schröder* 90 ! 116 ! **N.Nig.**: Naraguta & Jos (Sept.) *Lely*
  562 ! P. 697 ! *Batten-Poole* 154 ! Onitsha (fl. & fr. Oct.) *Jones* FHI 5892 ! Also in the Belgian Congo
  and Angola.
2. **D. africanus** *Wilczek* in Bull. Jard. Bot. Brux. 24 : 430 (1954) ; Fl. Congo Belge 6 : 323. *D.* sp. *A.* of
  Brenan l.c. 417. Slender annual herb, climbing to 6 ft. high ; green and glabrescent ; flowers small,
  greenish-yellow.
  N.Nig.: Bauchi Plateau (Oct.) *Lely* P. 800 ! Also in Belgian Congo, Tanganyika and N. and S. Rhodesia.
3. **D. chrysanthus** *A. Chev.* in Bull. Soc. Bot. Fr. 58, Mém. 8 : 164 (1912) ; Chev. Bot. 204 ; Bak. f. l.c. 449 ;
  Brenan l.c. 419 ; Wilczek l.c., fig. 16, incl. var. *occidentalis* (Harms) Wilczek. *D. biflorus* of F.W.T.A.,
  ed. 1, 1 : 410. *D. biflorus* var. *occidentalis* Harms (1899) ; Bak. f. l.c. 449. *D. stenophyllus* of F.W.T.A.

ed. 1, 1 : 410, partly (*Chev.* 22427). *D. brevicaulis* of F.W.T.A., ed. 1, 1 : 410, partly (*Lely* 562). Stems trailing, 3–4 ft. long ; flowers pale yellow ; in savannah.
Port.G.: Bissau (fr. Dec.) *Esp. Santo* 1439 ! **S.L.:** Falaba (Sept.) *Small* 306 ! Waterloo Hill, Rokupr (Oct.) *Jordan* 370 ! **Iv.C.:** Baoulé (Aug.) *Chev.* 22427 ! Dabou (fl. & fr. Nov.) *Roberty* 12512 ! **G.C.:** Tamale (Nov.) *Williams* 390 ! Ajena (Oct.) *Morton A.* 140 ! Anum (Nov.) *Morton GC* 7962 ! **N.Nig.:** Nupe *Barter* 967 ! 1591 ! Lokoja (Sept.) *Parsons* 19 ! 68 ! Anara F.R., Zaria (fl. & fr. Sept.) *Olorunfemi* FHI 24360 ! Naraguta (July) *Lely* 470 ! Vom *Dent Young* 74 ! Abinsi (Aug.) *Dalz.* 593 ! **Br.Cam.:** N. Cameroons (Oct.) *Talbot* ! Also in French Cameroons, Ubangi-Shari, Belgian Congo, Angola and A.-E. Sudan. (See Appendix, p. 240.)
4. **D. daltoni** *Webb* in Hook. Nig. Fl. 125 (1849) ; Brenan l.c. 417 ; Wilczek l.c. 326. *D. uniflorus* of F.T.A. 2 : 211, not of Lam. A twiner with yellow flowers ; in savannah.
N.Nig.: Katagum Dist. (fr. Oct.) *Dalz.* 45 ! Also in Cape Verde Island, A.-E. Sudan, Eritrea, Abyssinia, Tanganyika, Nyasaland, N. Rhodesia and Angola.
5. **D. brevicaulis** *Bak.* in F.T.A. 2 : 211 (1871) ; Bak. f. l.c. 444 ; Brenan l.c. 419. An erect or trailing herb, 1–2 ft. high ; leaves greyish when dry ; flowers greenish ; in savannah.
N.Nig.: Jebba *Barter* ! Yola (Sept.) *Dalz.* 306 ! Alkalire, Bauchi Plateau (Sept.) *Lely* 639 !
6. **D. axillaris** *E. Mey.*—F.T.A. 2 : 211 ; Brenan l.c. 414 ; Wilczek l.c. 321. *D. biflorus* of Bak. f. l.c. 448, partly (in syn.). Perennial climbing herb, 3–4 ft. high, pubescent ; flowers pale green.
N.Nig.: Naraguta & Jos (Sept.) *Lely* 578 ! Widespread in tropical Africa reaching Natal in the south, also in Madagascar and Ceylon. (See Appendix, p. 240.)
[The specimen cited here (*Lely* 578) is the true var. *axillaris*. Another specimen from the Bauchi Plateau (*Lely* P. 97) is intermediate between the typical variety and var. *glaber* E. Mey. (1835).]
7. **D. falcatus** *Klein ex Willd.* Sp. Pl. 3 : 1047 (1802) ; Meikle in Kew Bull. 1950 : 350 (1951) ; Wilczek l.c. 313. *D. debilis* Hochst. ex A. Rich. (1847)—F.T.A. 2 : 213 ; Bak. f. l.c. 449. A trailer or climber up to 6 ft. high ; flowers cream tinged pink or bright blue and quickly fading ; amongst shrubs in secondary forest.
G.C.: Mampong Scarp (fl. & fr. Dec.) *Morton GC* 9648 ! Sekodumase (fr. Jan.) *Kitson* ! **Togo:** Bame Pass (Oct.) *Morton GC* 9300 ! **S.Nig.:** Ibadan (fl. & fr. Jan.) *Meikle* 661 ! 961 ! Isijere, Ibadan (Nov.) *Keay* FHI 28141 ! Lagos (Oct.) *Punch* ! Also in Abyssinia, Uganda, Kenya, Tanganyika, Mozambique, Natal, India, Burma and Ceylon.
8. **D. formosus** *A. Rich.*—F.T.A. 2 : 213 ; Bak. f. l.c. 448 ; Meikle l.c. 351 ; Wilczek l.c. 314. A robust climber chiefly in grasslands ; flowers blue or purple.
Br.Cam.: Nchan, Bamenda, 5,000 ft. (May) *Maitland* 1524 ! Also in Abyssinia, A.-E. Sudan, Uganda, Kenya, Tanganyika, Nyasaland and Rhodesia.
9. **D. argenteus** *Willd.* Sp. Pl. 3 : 1047 (1803) ; Bak. f. l.c. 452. Stems creeping along the ground, clothed with long brown hairs ; flowers white or pink.
Togo: *fide* Bak. f. l.c. **Dah.:** Cotonou (June) *Debeaux* 193 ! Also in Kenya, Tanganyika and Zanzibar.
[The type specimen was collected in " Guinea."]
10. **D. schweinfurthii** *Taub. ex Harms* in Engl. Bot. Jahrb. 26 : 315 (1899) ; Taub. in E. & P. Pflanzenfam. 3, 3 : 383, name only ; Bak. f. l.c. 436. *D. bongensis* Taub. ex Harms l.c. (1899) ; Bak. f. l.c. *D. lelyi* Hutch. (1921)—Bak. f. l.c. ; F.W.T.A., ed. 1, 1 : 410. *D. carnosus* A. Chev. Bot. 204, name only. A variable perennial herb 2–3 ft. high with a tuberous rootstock ; flowers purplish ; gregarious in savannah.
Fr.Sud.: between upper R. Senegal and R. Niger *Bellamy* ! **Port.G.:** Dara & Ôco, Gabú (June) *Esp. Santo* 2911 ! **Fr.G.:** Kouroussa *Pobéguin* 261 ! Diankoro to Balandougou (fl. & fr. May) *Chev.* 25881 ! Kana valley *Chev.* 13248. Timbo to Farana *Chev.* 13347 ; 13356. **Togo:** Yendi (Apr.) *Dalz.* 49 ! **N.Nig.:** Anara F.R., Zaria (fr. May) *Keay* FHI 22900 ! Naraguta *Lely* 44 ! *Hill* 27 ! Nabardo, 2,300 ft. (May) *Lely* 208 ! 215 ! **Br.Cam.:** Banso to Jakiri (Jan.) *Keay* FHI 28459 ! Also in French Cameroons and A.-E. Sudan. (See Appendix, p. 240.)
11. **D. dinklagei** *Harms* in Engl. Bot. Jahrb. 45 : 315 (1910) ; Bak. f. l.c. 446. *Tephrosia subalpina* A. Chev. (1912), partly ; Chev. Bot. 181, partly ; Bak. f. l.c. 1 : 186. *Adenodolichos dinklagei* (Harms) A. Roberty (1954). A half-shrubby perennial, 2–5 ft. high, stems slightly angular ; flowers blue ; on rocky outcrops.
Fr.G.: Kindia to Conakry *Chev.* 20201. Grandes Chutes *Chev.* 20221. Macenta, 2,000–2,500 ft. (Oct.) *Baldwin* 9778 ! **S.L.:** Benikoro (Oct.) *Thomas* 2901 ! 2907 ! Yonibana (Oct.) *Thomas* 4329 ! Kasawe F.R. (fl. & fr. Dec.) *King* 42b ! York Pass *Lane-Poole* 435 ! **Lib.:** Duport (fl. & fr. Nov.) *Linder* 1464 ! Genne Loffa, Kolahun (Nov.) *Baldwin* 10094 ! Mt. Mpaka Fossa, 2,000 ft. (fl. & fr. Jan.) *Bequaert* 41 ! **Iv.C.:** Mt. Goula, Dyolas country (Apr.) *Chev.* 21225 !
12. **D. grandistipulatus** *Harms* in Engl. Bot. Jahrb. 49 : 452 (1913) ; Bak. f. l.c. 443. An erect herb with an angular stem clothed with long white hairs.
N.Nig.: Bauchi Plateau *Lely* ! Also in French Cameroons.
13. **D. tonkouiensis** *Portères* in Rev. Bot. Appliq. 19 : 786 (1939). *Tephrosia subalpina* A. Chev. (1912), partly. *Dolichos nimbaensis* Schnell (1950), partly (*Chev.* 21724). Erect undershrub, 4–5 ft. high, stems sharply angular ; flowers yellow.
Fr.G.: Nzérékoré (Oct.) *Baldwin* 9690 ! **Lib.:** Sanokwele (Sept.) *Baldwin* 9519 ! **Iv.C.:** Mt. Niénokué, 1,600 ft. *Chev.* 19474. Mt. Dou, 4,300 ft. *Chev.* 21486 ! Mt. Dourou, 3,200 ft. *Chev.* 21724 ! Mt. Tonkoui, Man *Portères* 2478.
14. **D. nimbaensis** *Schnell* in Rev. Gen. Bot. 57 : 278, fig. 1 (1950). Erect or ascending herb 1–5 ft. high, stems angular, softly pubescent ; flowers red or purple.
Fr.G.: Nimba Mts. *Schnell* 1550 ; 1846 ; 3366 ; 3715. **S.L.:** Magbile (fl. & fr. Dec.) *Thomas* 6481 ! Gendembu (Oct.) *Glanville* 4 ! Bubuya (Sept.) *Adames* 188 !

Imperfectly known species.

D. baumannii *Harms* in Engl. Bot. Jahrb. 26 : 313 (1899) ; Bak. f. l.c. 448 ; Brenan l.c. 419, q.v. for discussion.
Togo: Misahöhe *Baumann* 341. Bismarckburg *Büttner* 225.

55. LABLAB Adans. Fam. Pl. 2 : 325 (1763).

Stems climbing ; leaves long-petiolate ; leaflets acuminate, very broadly ovate, mostly abruptly cuneate at base, 6–12 cm. long, up to 10 cm. broad, glabrous or nearly so beneath ; stipels lanceolate, 5 mm. long, ribbed ; flowers on long stiff racemes, fasciculate on peduncle ; fruit broadly and slightly falcate, about 6 cm. long and 2 cm. broad, beaked by the persistent style, glabrous . . . . *niger*

L. niger *Medic.* in Vorles. Churpf. Phys. Ges. 2 : 354 (1787) ; Wilczek in Fl. Congo Belge 6 : 279, fig. 9. *L. vulgaris* Savi (1821). *Dolichos lablab* Linn. (1753)—F.T.A. 2 : 210 ; Chev. Bot. 204 ; F.W.T.A., ed. 1, 1 : 410 ; Bak. f. Leg. Trop. Afr. 2 : 452. A climbing perennial herb up to 15 ft. high, stems glabrous or pubescent ; flowers white or red-tinged.
Sen.: Rufisque *Chev.* 3135. **S.L.:** *Glanville* in *Hb. Deighton* 1800 ! **Iv.C.:** Morénou (fl. & fr. Dec.) *Chev.* 22490 ! **G.C.:** Kumasi *Cummins* 92 ! Tamale (fr. Dec.) *Williams* 811 ! **N.Nig.:** Sokoto *Dalz.* 323 ! Katagum *Dalz.* 43 ! Biu, Bornu (Sept.) *Noble* 23 ! **S.Nig.:** Lagos *MacGregor* 262 ! Ibadan (fl. & fr. Oct.) *Newberry & Etim* 148 ! Mt. Koloishe, Obudu (fl. & fr. Dec.) *Savory & Keay* FHI 25101 ! **Br.Cam.:**

Victoria (fr. Nov.) *Maitland* 778 ! Buea (fr. Jan.) *Maitland* 259 ! Bamenda *Migeod* 429 ! Commonly cultivated, also occasionally wild ; widespread in the tropics. (See Appendix, p. 240.)

56. ADENODOLICHOS Harms in Engl. Bot. Jahrb. 33 : 179 (1902).

Shrub ; young parts viscid reddish-tomentose ; leaves opposite, 3-foliolate, occasionally 5-foliolate, stipellate ; leaflets subequal, broadly ovate, rounded at base, rounded and mucronate at apex, 7–18 cm. long, 5–15 cm. broad, strongly nerved, rusty-pubescent on the nerves and gland-dotted beneath ; flowers in lax terminal panicles with elongated branches ; calyx tomentose, about 1 cm. long, lobes oblong-lanceolate, subacute ; fruits oblong, sharply beaked, 4–5 cm. long, 1–3-seeded, loosely tomentose
paniculatus

A. **paniculatus** (*Hua*) *Hutch. & Dalz.* F.W.T.A., ed. 1, 1 : 411 (1928) ; Kew Bull. 1929 : 18 ; Bak. f. Leg. Trop. Afr. 2 : 458 ; Wilczek in Fl. Congo Belge 6 : 396. *Dolichos paniculatus* Hua (1897). *D. macrothyrsus* Harms (1899). *Adenodolichos macrothyrsus* (Harms) Harms (1902). An erect half-woody plant of shrubby habit, up to 15 ft. high ; flowers with pinkish keel and yellowish standard ; in savannah country.
Fr.G : Timbo *Miquel* 72. **G.C.**: Gambaga Scarp (Dec.) *Adams & Akpabla* GC 4246 ! **N.Nig.**: Kontagora (fr. Dec.) *Dalz.* 37 ! Minna (fr. Dec.) *Meikle* 694 ! Mongu, 4,300 ft. (July) *Lely* 398 ! Vom, 3,000–4,500 ft. *Dent Young* 77 ! Ropp (fl. & fr. Dec.) *Savory* UCI 130 ! Also in French Cameroons and A.-E. Sudan. (See Appendix, p. 226.)

57. VOANDZEIA Thouars—F.T.A. 2 : 207.

A herb with prostrate rooting branches fruiting below the soil ; leaves 3-foliolate, long-petiolate, stipellate ; leaflets lanceolate to narrowly elliptic, narrowed to base, emarginate and mucronate, the terminal one 6–8 cm. long, 1·5–2·5 cm. broad, more or less 5-nerved at base, glabrous ; flowers usually in pairs on a hairy peduncle with a swollen glandular apex which pushes into the ground drawing in the flowers ; calyx glabrous, with short broad teeth ; fruits rounded, wrinkled when dry, about 2 cm. in diam. ; seeds rounded, smooth, hard, varying in size and colour, up to about 1·5 cm. broad .. .. *subterranea*

V. **subterranea** (*Linn.*) *DC.*—F.T.A. 2 : 207 ; Harms in Notizbl. Bot. Gart. Berl. 5 : 253 ; Chev. Bot. 203 ; Bak. f. Leg. Trop. Afr. 2 : 424 ; Jac.-Fél. in Bull. Soc. Bot. Fr. 93 : 360 (1946). *Glycine subterranea* Linn. (1763). Flowers usually in pairs on a peduncle ; stigma lateral.
Sen.: *Roger* ! **Fr.Sud.**: Ségou (Sept.) *Chev.* 24974 ! Djenné (Sept.) *Chev.* 3133 ! San *Chev.* 3131. **Iv.C.**: Mossi *Chev.* 24697. **G.C.**: Atwabo *Fishlock* 76 ! Tamale (Oct.) *Baldwin* 13582 ! **Togo**: *Kersting* ! **Dah.**: Agouagon *Chev.* 23669. **N.Nig.**: Munshi country (fr. Dec.) *Sampson* 36 ! Kilba country (Aug.) *Dalz.* 9 ! Bida *Lamb* ! Bauchi Plateau (July) *Lely* P. 536 ! **S.Nig.**: Lagos *Millen* 10 ! Aguku *Thomas* 996 ! Cultivated throughout tropical Africa, S. Africa and Madagascar, and also in Java and the Philippines and, apparently indigenous in N. Nigeria, French Cameroons and Ubangi-Shari. (See Appendix, p. 269.)

58. KERSTINGIELLA Harms in Ber. Deutsch. Bot. Ges. 26a : 230, t. 3 (1908).

A herb with prostrate rooting branches fruiting below the soil ; leaves 3-foliolate, long-petiolate, stipellate ; leaflets elliptic-obovate, broadly narrowed at base, mucronate, 5–8 cm. long, 4–5·5 cm. broad, 3-nerved at base, glabrous ; flowers axillary, sub-sessile, about 1 cm. long ; calyx thinly pilose, deeply divided with narrow segments ; fruits on a carpophore up to 2 cm. long, 1–3-seeded, in the latter case constricted between the seeds, up to 2 cm. long, reticulate ; seeds somewhat kidney-shaped, 1 cm. long, rich brown, with the scar near the middle .. .. *geocarpa*

K. **geocarpa** *Harms* l.c. (1908) ; Stapf in Kew Bull. 1912 : 209, with fig. ; Chev. Bot. 203 ; Bak. f. Leg. Trop. Afr. 2 : 425. *Voandzeia poissoni* A. Chev. (1910). *V. geocarpa* (Harms) A. Chev. (1919). Flowers subsessile in leaf-axils ; style with a terminal ciliate stigma.
Sen.: *Berhaut* 284. **Iv.C.**: Mossi *Chev.* 24682. **G.C.**: *Williams* 553 ! **Togo**: *Kersting* A. 476–478 ! **Dah.**: Agouagon *Chev.* 23525 ; 23848. **N.Nig.**: *Andrew* ! Nupe (fr. June) *Elliott* 200 ! Bida *Lamb* 8 ! (See Appendix, p. 248.)

59. PSOPHOCARPUS Neck.—F.T.A. 2 : 208.

Leaves 3-foliolate ; leaflets entire or sometimes 3-lobed, broadly ovate, terminal leaflet rhomboidal, 5–9 (–12) cm. long, 4–6 (–9) cm. broad, more or less pubescent beneath ; racemes 15 cm. long ; calyx shortly pubescent outside ; fruits 4-winged, 5–6 cm. long, 1·5 cm. broad, 4–6-seeded, glabrous .. .. .. 1. *palustris*
Leaves 1-foliolate ; leaflet shortly petiolulate, ovate, rounded at both ends, mucronate, 5–7·5 cm. long, 4–4·5 cm. broad, pubescent on the strong nerves beneath ; stipules broadly lanceolate, 8 mm. long ; racemes 15 cm. long ; calyx glabrous outside ; young fruits villous .. .. 2. *monophyllus*

1. P. **palustris** *Desv.* in Ann. Sci. Nat., sér. 1, 9 : 420 (1826) ; Bak. f. Leg. Trop. Afr. 2 : 426 ; Wilczek in Fl. Congo Belge 6 : 283, fig. 10. *P. palmettorum* Guill. & Perr. (1832)—F.W.T.A., ed. 1, 1 : 411. *P. longepedunculatus* of F.T.A. 2 : 208. A twiner with blue flowers and half-woody stem ; climbing on trees in swampy places.
Sen.: *Farmar* 39 ! **Gam.**: Kudang (Jan.) *Dalz.* 8016 ! **Fr.Sud.**: Bafaga *Chev.* 712. **Port.G.**: Bissau (Oct.) *Esp. Santo* 1423 ! 1664 ! **S.L.**: Segbwema (Dec.) *Deighton* 3466 ! Mange (fr. Feb.) *Deighton* 3611 ! **Lib.**: Peahtah (Oct.) *Linder* 1019 ! Kolahun (fr. Nov.) *Baldwin* 10150 ! **G.C.**: Accra Plains (fr. Jan.) *Irvine* 1942 ! **Dah.**: Porto Novo *Chev.* 22719 ! **N.Nig.**: Nupe *Barter* 960 ! Kontagora (fl. & fr. Oct.) *Dalz.* 17 ! **S.Nig.**: Lagos *Millen* 69 ! Old Calabar *Thomson* 116 ! Oban *Talbot* 1318 ! **Br.Cam.**: Victoria (Mar.) *Maitland* 515 ! Extends to A.-E. Sudan, Zanzibar, Mozambique and Angola.

2. **P. monophyllus** *Harms* in Engl. Bot. Jahrb. 40 : 43 (1907) ; Chev. Bot. 203 ; Bak. f. l.c. 427. A rather woody plant up to 3 ft. high ; flowers blue in axillary racemes.
Fr.Sud.: Samorokiri *Chev.* 888. **Iv.C.**: Mankono (July) *Chev.* 21972 !

The Asiatic species *P. tetragonolobus* (Linn.) DC. has been introduced into our area.

60. TERAMNUS Browne—F.T.A. 2: 180.

Pedicels (of flowers) 2–3 mm. long ; leaflets with 4–6 secondary nerves, ovate to ovate-lanceolate, subacute, 3–5 cm. long, 1·5–2·5 cm. broad, closely appressed-pilose beneath ; stipules lanceolate, pubescent ; racemes few-flowered, axillary, about as long as the leaves ; calyx appressed-pilose, 3 mm. long ; fruits linear, flat, 4–6 cm. long, sparsely appressed-pilose, with an upturned beak 3 mm. long   1. *labialis*
Pedicels (of flowers) up to 1 mm. long, or flowers sessile ; leaflets with 7–13 secondary nerves :
  Calyx 6–7 mm. long, deeply lobed, lobes subulate, densely long-pilose ; corolla a little longer than calyx ; racemes 20–35 cm. long, long-pedunculate ; fruits densely spreading-setose, about 4 cm. long and 4 mm. broad, with the beak slightly reflexed ; leaflets lanceolate, 8–10 cm. long, 3–4 cm. broad, with about 12 lateral nerves ; terminal leaflet shortly petiolate ; peduncle 1–3 cm. long .. ..   2. *buettneri*
  Calyx 3–4 mm. long, lobed to about half-way down its length ; leaflets with about 8 lateral nerves ; fruits with a reflexed beak 3–4 mm. long :
    Leaflets lanceolate to linear-lanceolate, 5–6 cm. long, 1–2·5 cm. broad, pubescent beneath ; stems angular, more or less appressed-pilose ; racemes 10–40 cm. long, shortly pedunculate ; fruits 4–5 cm. long, about 2 mm. broad, sparsely appressed-pilose ..   ..   3. *andongensis*
    Leaflets elliptic or ovate, 6–9 cm. long, 3–6·5 cm. broad, silky-pubescent beneath ; stems angular, pilose with reflexed spreading hairs ; racemes up to 40 cm. long ; fruits about 3·5 cm. long, 3 mm. broad, densely appressed-pilose   4. *micans*

1. **T. labialis** (*Linn. f.*) *Spreng.*—F.T.A. 2 : 180, partly ; Chev. Bot. 194 ; Bak. f. Leg. Trop. Afr. 2 : 363. *Glycine labialis* Linn. f. (1781). A slender-stemmed twiner with small reddish flowers.
  Sen.: Senegal R. (Jan.) *Roger* 89 ! Galam *Heudelot* 114 ! Walo *Perrottet* 222 ! **S.L.**: *Thomas* ! Hangha (fr. Jan.) *Thomas* 7796 ! **Lib.**: Vonjama Dist. (Oct.) *Baldwin* 9994 ! **Iv.C.**: Mid. Comoé *Chev.* 22616. **G.C.**: Accra Plains (July) *Irvine* 3028 ! Anum (fr. Nov.) *Morton* GC 7993 ! Kumasi (fl. & fr. Dec.) *Morton* GC 9682 ! **Dah.**: Zagnanado *Chev.* 22981. **S.Nig.**: *Thomas* 1810 ! Lagos *Millen* 61 ! 114 ! Ibadan (fl. & fr. Nov.) *Jones & Keay* FHI 14228 ! Also throughout the tropics.
2. **T. buettneri** (*Harms*) *Bak. f.* l.c. 365 (1929). *Glycine buettneri* Harms (1899). *G. andongensis* of F.W.T.A., ed. 1, 1 : 407, partly (*Büttner* 163). Stem more or less erect, woody, up to 5 ft. high.
  Iv.C.: Mankono (fr. July) *Chev.* 21990 ! **G.C.**: *Vigne* ! Wankye (July) *Chipp* 482 ! Ejura (fr. Aug.) *Andoh* FH 5041 ! **Togo**: Bismarckburg *Büttner* 163. Kpandu *Robertson* 135 ! Kete Krachi (fl. & fr. May) *Morton* GC 7160 !
3. **T. andongensis** (*Welw. ex Bak.*) *Bak.f.* in J. Bot. 66, suppl. 1 : 115 (1928) ; Leg. Trop. Afr. 2 : 364. *Glycine andongensis* Welw. ex Bak. in F.T.A. 2 : 179 (1871) ; F.W.T.A., ed. 1, 1 : 407, partly (*Dalz.* 589). A twiner in grass with small white flowers.
  Sen.: Niokolo-Koba (fr. Jan.) *Berhaut* 4462 ! **S.L.**: Musaia (fr. Dec.) *Deighton* 4872 ! 5704 ! Jigaya (Nov.) *Thomas* 2689 ! **G.C.**: Wenchi (fl. & fr. Dec.) *Morton* GC 25302 ! **N.Nig.**: Abinsi (Nov.) *Dalz.* 589 ! Bauchi Plateau (Sept.) *Lely* P. 721 ! R. Benue (Sept.) *Talbot* ! Lassa, L. Chad (Sept.) *Royer* 101 ! Also in Uganda, Kenya, Tanganyika, Belgian Congo and Angola.
  [The numbers cited here are by no means uniform amongst themselves and several taxa may be recognized when more material becomes available.]
4. **T. micans** (*Welw. ex Bak.*) *Bak. f.* in J. Bot. 66, suppl. 1 : 115 (1928) ; Leg. Trop. Afr. 2 : 366. *Glycine micans* Welw. ex Bak. in F.T.A. 2 : 179 (1871). A fairly strong climber with white or pink flowers.
  Port.G.: Bafatá (fl. & fr. Nov.) *Esp. Santo* 2971 ! **S.L.**: Yonibana (Nov.) *Thomas* 4698 ! 5101 ! Sefadu (Dec.) *Deighton* 3577 ! Musaia (fr. Dec.) *Deighton* 4543 ! **Br.Cam.**: Belo, Bamenda (fr. Apr.) *Maitland* 1688 ! Also in French Cameroons, Belgian Congo and Angola.

61. PUERARIA DC. in Ann. Sci. Nat., sér. 1, 4 : 97 (1825).

Climbing herb ; stems densely pilose with brownish spreading hairs ; leaves 3-foliolate ; leaflets broadly rhomboid-ovate, broadly cuneate at base, subacute at apex, 6–9 cm. long, 5–8 cm. broad, softly appressed-pilose beneath, pubescent above with hairs of two lengths ; stipules lanceolate, not appendaged ; racemes axillary, pedunculate, 10–20 cm. long, many-flowered, with glandular swollen nodes ; flowers about 1·5 cm. long ; fruits linear, about 8 cm. long and 6 mm. broad, beaked, appressed-pilose with brownish hairs .. ..   *phaseoloides*

P. phaseoloides (*Roxb.*) *Benth.* in J. Linn. Soc. 9 : 125 (1865). *Dolichos phaseoloides* Roxb. (1832). A climbing or trailing herb with pale bluish flowers.
  S.L.: Njala (fl. & fr. Jan.) *Deighton* 2455 ! **Lib.**: Roberts Field, Montserrado (fl. & fr. Jan.) *Baldwin* 14040 ! **G.C.**: Butre, Dixcove (fl. & fr. Mar.) *Morton* A. 438 ! **S.Nig.**: Asaba (fl. & fr. Mar.) *Irvine* 3616 ! A native of tropical Asia, introduced into parts of tropical Africa as a cover-crop.

P. thunbergiana (Sieb. & Zucc.) Benth. has also been introduced into Sierra Leone.

62. CANAVALIA DC.—F.T.A. 2 : 189.

Stems creeping ; leaflets broadly obovate or suborbicular, widely emarginate and mucronate, up to 10 cm. in diam., thin, glabrous, with about 5 pairs of prominent lateral nerves, laxly reticulate between the nerves ; stipels very small ; flowers few on a stout axis, shortly pedicellate ; calyx about 1 cm. long, shortly pubescent outside, the uppermost lobe broadly ovate ; corolla about 2·5 cm. long ; fruits broadly linear, 15 cm. or more long, with 2 very close ribs near the upper suture   1. *rosea*

Stems climbing ; leaflets ovate or ovate-elliptic, acute or acutely acuminate, broadly
cuneate at base, up to 20 cm. long and 10 cm. broad, glabrous, with 6–7 pairs of
lateral nerves ; flowers more or less as above ; fruits variable, sword-shaped, elongate,
30 cm. long or more, with 2 longitudinal ribs near the upper suture ; seeds narrowly
ellipsoid, smooth .. .. 2. *ensiformis*

1. **C. rosea** (*Sw.*) *DC.* Prod. 2 : 404 (1825) ; Dunn in Kew Bull. 1922 : 138 ; Exell Cat. S. Tomé 160 (q.v. for
full synonymy). *Dolichos roseus* Sw. (1788). *Canavalia obtusifolia* DC. (1825), partly, excl. syn. *Dolichos
obtusifolia*—F.T.A. 2 : 190 ; Chev. Bot. 197 ; F.W.T.A., ed. 1, 1 : 412 ; Bak. f. Leg. Trop. Afr. 2 : 385.
A nearly glabrous trailing plant, with rose or purple flowers.
Sen.: Casamance *Chev.* 3411 ; 3417. **Gam.:** Kuntaur *Ruxton* 121 **S.L.:** *Deighton* 2040 ! River Beach
Colony (July) *Small* 162 ! **Lib.:** Kakatown *Whyte* ! Monrovia *Barker* 1456 ! Greenville (fr. Mar.) *Baldwin*
11567 ! **Iv.C.:** Sassandra Port *Chev.* 17928. Tabou *Chev.* 19933. **G.C.:** Axim (Jan.) *Chipp* 57 !
Atwabo *Fishlock* 23 ! **Dah.:** Cotonou (fr. Mar.) *Debeaux* 149 ! **S.Nig.:** Lagos (June) *Dalz.* 1212 ! Brass
Barter 53 ! Kwa Ibo R. *Talbot* 3089 ! **Br.Cam.:** *Preuss* 1281 ! Common on maritime shores and
estuaries in the tropics of both hemispheres ; sometimes cultivated. (See Appendix, p. 234.)
2. **C. ensiformis** (*Linn.*) *DC.*—F.T.A. 2 : 190 ; Chev. Bot. 196 ; Bak. f. l.c. 384. *Dolichos ensiformis* Linn.
(1753). *Canavalia gladiata* (Jacq.) DC. (1825)—Dunn l.c. 134. *C. regalis* Dunn (1922). *C. africana* Dunn
(1922) ; Berhaut Fl. Sen. 18. Climbing on fences and trees ; flowers rose, mauve or white with a red
base.
Sen.: *Berhaut* 1623. **Fr.Sud.:** Ségou *Chev.* 24975. **Port.G.:** Bissau *Esp. Santo* 1582 ! **Fr.G.:** Longuery
Chev. 14641 ; 14763. Fouta Djalon *Chev.* 18719 ; 18721. **S.L.:** *Deighton* 1988 ! Musaia (fr. Dec.)
Deighton 4532 ! **Iv.C.:** Keéta *Chev.* 19335. **Togo:** *Howes* 1042 ! **Dah.:** Abomey *Chev.* 23162. **N.Nig.:**
Jebba *Barter* ! Nupe *Barter* 1607 ! Mongu (July) *Lely* 387 ! **S.Nig.:** Lagos *MacGregor* 177 ! Olokemeji
Foster ! Oban *Talbot* 233 ! A common plant in the tropics of both hemispheres, usually found in cultiva-
tion. (See Appendix, p. 234.)
Several taxa are undoubtedly included here, but a thorough taxonomic revision is needed to clarify
this group.

63. DIOCLEA H. B. & K.—F.T.A. 2 : 189.

Woody climber ; branches pilose ; leaves 3-foliolate ; leaflets ovate-elliptic, rounded
at base, abruptly acuminate and mucronate, up to 15 cm. long and 10 cm.
broad, thinly pilose ; lateral nerves 7–8 pairs ; stipules peltate, acute ; stipels
filiform ; inflorescence stiff and woody, racemose, longer than the leaves ; bracts
narrowly lanceolate, about 1·5 cm. long ; calyx rusty-pubescent ; standard reflexed ;
fruits elliptic-orbicular to oblong, 10 cm. or more long, about 4–5 cm. broad, rather
woody, pilose, 1–3-seeded ; seeds semi-orbicular, about 3·5 cm. long, with a conspicuous hilum encircling it for ¾ its circumference .. .. *reflexa*

D. reflexa Hook. *f.*—F.T.A. 2 : 189 ; Chev. Bot. 196 ; Bak. f. Leg. Trop. Afr. 2 : 383. A hairy, woody,
climbing shrub, with conspicuous erect racemes of scented red flowers.
Port.G.: Buba (fr. Nov.) *Esp. Santo* 1286 ! Fulacunda (Oct.) *Esp. Santo* 2217 ! **S.L.:** Freetown (fr.
Dec.) *Dalz.* 998 ! Njala (fr. Oct.) *Deighton* 4899 ! Magbile (Dec.) *Thomas* 5974 ! Bumbuna (fr. Oct.)
Thomas 3409 ! **Lib.:** Cape Palmas (July) *T. Vogel* 32 ! Du R. (July) *Linder* 104 ! Webo (July) *Baldwin*
6641 ! Monroviatown (Aug.) *Baldwin* 7096 ! Gondolahun (fr. Nov.) *Baldwin* 10108 ! **Iv.C.:** *Scaëtta*
3341 ! Several localities *fide* Chev. *l.c.* **G.C.:** Akroso (fr. Aug.) *Howes* 950 ! Swedru (fr. Sept.) *Dalz.* 93 !
Kumasi (Sept.) *Andoh* FH 5353 ! **Togo:** Fume (Nov.) *Morton* A. 143 ! **Dah.:** *Poisson* 104 ! **S.Nig:**
T. Vogel ! Lagos *MacGregor* 98 ! *Punch* 21 ! Ibadan *Newberry* 66 ! *Oladoyinbo* FHI 24251 ! **Br.Cam.:**
Cameroon R. (fl. & fr. Jan.) *Mann* 2217 ! Victoria (Mar.) *Maitland* 468 ! **F.Po:** *T. Vogel* 3 ! In the
tropics of both hemispheres. (See Appendix, p. 239.)

64. ABRUS Adans.—F.T.A. 2 : 174.

Seeds bright red with black base ; ripe fruits becoming warted, about twice as long as
broad ; flowers pink in short, pedunculate racemes ; bracteoles very small at base
of calyx, caducous ; leaflets oblong, 6–15 mm. long, 4–8 mm. broad, rounded to
subacute at apex, thinly pubescent .. .. 1. *precatorius*
Seeds all black or brown ; ripe fruits smooth, about three times as long as broad :
Bracteoles very small at base of calyx, caducous ; corolla pink, drying yellow, thin ;
racemes slender, with sessile flowers in small clusters ; leaflets 7–9 pairs, oblong,
2–3 (–4) cm. long, about 1 cm. broad, truncate or rounded at apex, pubescent
2. *pulchellus*
Bracteoles about as long as calyx, more or less persistent ; corolla deep purple, rather
thick ; racemes with rather distant dense clusters of sessile flowers ; leaflets
10–13 pairs, narrowly oblong, 1–2 cm. long, about 6 mm. broad, truncate and
mucronate at apex, usually densely pubescent .. .. .. 3. *canescens*

1. **A. precatorius** Linn.—F.T.A. 2 : 175 ; Chev. Bot. 193 ; Bak. f. Leg. Trop. Afr. 2 : 351. A woody twining
shrub, with rather stout racemes of pinkish-yellow flowers and clustered fruits which burst, exposing the
brilliant red and black seeds.
Sen.: *Roger* ! **Fr.Sud.:** Bamako *Chev.* 253. Yatenga *Chev.* 24777. **Fr.G.:** Fouta Djalon *Chev.* 12900.
Kouria *Chev.* 14975. **S.L.:** Regent (fr. Dec.) *Sc. Elliot* 4122 ! Njala (fr. Nov.) *Deighton* 2401 ! Pendemba
(fr. July) *Thomas* 829 ! Gola (fr. Apr.) *Small* 572 ! **Lib.:** Gbanga (Oct.) *Linder* 1167 ! Cess R. (fr. Mar.)
Baldwin 11282 ! Browntown (fr. Aug.) *Baldwin* 7080 ! **G.C.:** Achimota (Feb.) *Irvine* 1993 ! **Togo:**
Lome *Warnecke* 202 ! **Dah.:** Cotonou *Chev.* 23358. **S.Nig.:** Lagos *MacGregor* 19 ! Umuikem, Onitsha
(fr. Apr.) *Onochie* FHI 7194 ! Oban *Talbot* 184 ! **F.Po:** *Mann* 30 ! Found all over the tropics and
subtropics. (See Appendix, p. 224.)
2. **A. pulchellus** Wall. *ex Thw.*—F.T.A. 2 : 175 ; Chev. Bot. 193 ; Bak. f. l.c. 351. *A. canescens* of F.W.T.A.,
ed. 1, 1 : 412, partly (*Sc. Elliot* 4312 ; *Dalz.* 14). *A. strictosperma* Berhaut (1954). A climbing shrub like
the preceding ; flowers pink.
Sen.: Ziguinchor *Chev.* 3396. **Gam.:** *Perrottet* 212 ! **Fr Sud.:** Tabacco (Oct.) *Chev.* 3397 ! **Port.G.:**
Pelubá, Bissau (fr. Dec.) *Esp. Santo* 1583 ! **Fr.G.:** Fouta Djalon *Heudelot* 685 ! Macenta (fl. & fr. Oct.)
Baldwin 9775 ! **S.L.:** Kitchom *Sc. Elliot* 4312 ! Bumbuna (Oct.) *Thomas* 3716 ! Kenema (fl. & fr. Nov.)
Deighton 392 ! Njala (Nov.) *Deighton* 1499 ! **Lib.:** Ganta (Nov.) *Harley* ! Tawata, Boporo (Nov.)

Baldwin 10319 ! **G.C.**: Gambaga (fr. Dec.) *Adams* GC 4222 ! Kumasi (fl. & fr. Nov.) *Darko* 638 ! Volta River F.R. (fl. & fr. Nov.) *Morton* GC 6052 ! **Dah.**: Tchatchou *Poisson* ! **N.Nig.**: Nupe *Barter* 1749 ! Patti Lokoja (fl. & fr. Nov.) *Dalz.* 14 ! *Parsons* L. 93 ! Bauchi Plateau (Oct.) *Lely* P. 813 ! **S.Nig.**: Lagos (Oct.) *Punch* 47 ! Gambari Group F.R. (fl. & fr. Nov.) *Ujor* 29395 ! Throughout Africa and in the eastern tropics.

3. **A. canescens** *Welw. ex Bak.* in F.T.A. 2 : 175 (1871) ; Chev. Bot. 193 ; Bak. f. l.c. 351 ; Berhaut Fl. Sen. 31.
A climbing shrub with grey-pubescent foliage and usually deep red-purple flowers.
Sen.: *Berhaut* 1448. **Gam.**: Kuntaur (Sept.) *Ruxton* 54 ! **Port.G.**: Bafata (fl. & fr. Nov.) *Esp. Santo* 3116 ! **Fr.G.**: Kouroussa (fl. & fr. Sept.) *Pobéguin* 953 ! Macenta (fl. & fr. Oct.) *Baldwin* 9773 ! Fouta Djalon *Chev.* 18242. Grandes Chutes *Chev.* 20230 ; 20299. **S.L.**: Talla Hills (fr. Feb.) *Sc. Elliot* 4970 ! Fotombu (fl. & fr. Oct.) *Small* 371 ! Musaia (fr. Dec.) *Deighton* 4493 ! **Lib.**: Gbanga (fl. Sept., fr. Dec.) *Linder* 688 ! *Baldwin* 10525 ! Gbau, Sanokwele (Sept.) *Baldwin* 9405 ! **Iv.C.**: Bingerville *Chev.* 20070. Anoumaba *Chev.* 22396. **G.C.**: Tanosu *Rattray* R. 11 ! Burufo (fr. Dec.) *Adams & Akpabla* GC 4394 ! Menji, Wenchi (fr. Dec.) *Morton* GC 25220 ! **Togo:** Kpandu *Robertson* 82 ! Amedzofe (fl. & fr. Nov.) *Morton* GC 9385 ! **N.Nig.**: Lokoja (fl. & fr. Nov.) *Dalz.* 15 ! *Parsons* L. 78 ! Jebba *Barter* ! Kontagora (fr. Jan.) *Meikle* 1089 ! **S.Nig.**: Lagos (Oct.) *Punch* 32 ! Also in Belgian Congo, A.-E. Sudan, E. Africa and Angola.

65. STYLOSANTHES Sw.—F.T.A. 2 : 155.

Older stems with a ridge of hairs down one side ; one or both segments of fruit reticulate and glabrous, occasionally both segments pubescent, beak circinate, ciliate beneath ; spike 1–5 cm. long ; bracts bristly or glabrescent ; leaflets oblanceolate or narrowly elliptic, acute or obtuse at each end, 10–15 mm. long, 5 mm. broad, glabrous, or midrib sometimes ciliate ..   ..   1. *erecta*

Older stems densely pubescent with short ascending hairs ; both segments of fruits pilose, beak circinate ; spike about 2 cm. long ; bracts pubescent or bristly ; leaflets elliptic or narrowly elliptic, very acute at apex, 10–18 mm. long and about 6 mm. broad, shortly pubescent beneath   ..   ..   2. *mucronata*

1. **S. erecta** *P. Beauv.*—F.T.A. 2 : 156 ; Chev. Bot. 187 ; Bak. f. Leg. Trop. Afr. 2 : 319 ; Léonard in Fl. Congo Belge 5 : 347, fig. 24. *S. guineensis* Schum. & Thonn. (1827)—F.W.T.A., ed. 1, 1 : 413. *S. erecta* var. *guineensis* (Schum. & Thonn.) T. Vogel (1838)—Bak. f. l.c. 320. Erect or rather prostrate undershrub about 2 ft. high ; flowers yellow ; near the sea.
Sen.: *Sieber* ! Cayor *Heudelot* 406 ! **Fr.Sud.**: Ouassana *Chev.* 617. Ségou *Chev.* 3259. **S.L.**: Yele, Turtle Isl. (fl. & fr. Nov.) *Deighton* 2314 ! Kent (May) *Deighton* 2659 ! King Tom (Oct.) *Deighton* 2166 ! **Lib.**: Cape Palmas (July) *T. Vogel* 5 ! Grand Bassa (July) *T. Vogel* 56 ! Monrovia (Nov.) *Linder* 1416 ! *Baldwin* 5942 ! Sinkor (July) *Barker* 1359 ! 1388 ! **Iv.C.**: Cavally *Chev.* 19939. Mossi *Chev.* 24730. Tabou *Chev.* 20052. **G.C.**: Cape Coast *Don* ! Teshi (Nov.) *Irvine* 811 ! Prampram (Dec.) *N. F. Robertson* 32 ! **Togo:** Lome *Warnecke* 270 (partly) ! **Dah.**: Cotonou (Jan.) *Chev.* 4486 ; 22691 ! Ouidah *Chev.* 23431. **Fr.Nig.**: Gourma *Chev.* 24538. **S.Nig.**: *T. Vogel* ! Lagos (Dec.) *Dalz.* 1215 ! *Rowland* ! Brass *Barter* ! Nun R. (Sept.) *Mann* 473 ! Also in Belgian Congo and Angola. (See Appendix, p. 262.)

2. **S. mucronata** *Willd.*—F.T.A. 2 : 157 ; Bak. f. l.c. 320 ; Léonard l.c. 348. *S. viscosa* of F.W.T.A., ed. 1, 1 : 413, partly, not of Sw. *S. bojeri* T. Vogel (1838). A perennial rather woody herb of poor soils, 1–3 ft. high ; flowers small, yellow.
Sen.: Richard Tol *Lelièvre* ! Senegal R. (Aug.-Dec.) *Roger* 59 ! **Gam.**: Hayes 576 ! **Fr.Sud.**: Labeyana *Hagerup* 448 ! **Port.G.**: Bafata (Nov.) *Esp. Santo* 2804 ! **G.C.**: Lawra (Oct.) *Hinds* 5017 ! Tamu (fr. Dec.) *Adams & Akpabla* GC 4372 ! Tamale (Mar.) *Morton* GC 8814 ! **Togo:** Lome *Warnecke* 270 (partly) ! **N.Nig.**: Sokoto *Moiser* 16 ! 93 ! Kontagora (Nov.) *Dalz.* 46 ! Bauchi Plateau (May) *Lely* 178 ! P. 274 ! Meifoni, Bornu *Parsons* ! Throughout tropical Africa, Madagascar, Arabia and India. (See Appendix, p. 262.)

Doubtfully recorded species.

S. viscosa Sw., a widespread tropical American species, was collected by Don and was recorded in Fl. Nigrit. 301 and F.T.A. 2 : 156 as having come from Sierra Leone. This single specimen is in the British Museum. Considering the time that has elapsed since Don made his collections, it is strange that no further gatherings of *S. viscosa* are to be found ; perhaps there has been confusion and this specimen really came from America. In the first edition the name *viscosa* was misapplied to all the pubescent specimens (i.e. *S. mucronata* Willd.) in W. Africa.

66. ZORNIA J. F. Gmel.—F.T.A. 2 : 158.

By E. Milne-Redhead

Flowers much shorter than the bracts ; bracts ovate-elliptic, acute, more than 4 mm. broad, without, or with a few inconspicuous, pellucid glands ; fruits up to 4-segmented, bristles up to 2 mm. long, barbed (glochidiate) and retrorsely hispid ; leaflets-lanceolate, acute, up to 4 cm. long ; normally annual   ..   ..   ..   1. *glochidiata*

Flowers about as long as or longer than the bracts ; perennials :

Bracts lanceolate, acute, up to 2 mm. broad, conspicuously pellucid-punctate ; fruits up to 7-segmented, bristles up to 1 mm. long, barbed (glochidiate) and retrorsely hispid ; upper leaflets lanceolate, acute, up to 3·5 cm. long, lower leaflets shorter, broadly ovate   ..   ..   2. *latifolia*

Bracts ovate-elliptic, subacute, 2·5–5 mm. broad, not conspicuously punctate ; fruits up to 5-segmented, bristles 3–4 mm. long, with ascending hairs (not glochidiate)   3. *durumuensis*

1. **Z. glochidiata** *Reichb. ex DC.* Prod. 2 : 316 (1825) ; Milne-Redhead in Bol. Soc. Brot., sér. 2, 28 : 87 (1954) ; Léonard in Fl. Congo Belge 5 : 356. *Z. biarticulata* G. Don (1832) ; Chev. Bot. 188. *Z. diphylla* of F.W.T.A., ed. 1, 1 : 413, partly ; of Chev. Bot. 188, partly, not of Pers. An annual herb, becoming much-branched and straggling, not usually surviving the dry season ; flowers very inconspicuous, yellow with reddish-purple tinge.
Sen.: *Sieber* 40 ! *Heudelot* 240 ! *Perrottet* 242 ! **Gam.**: Mungo Park ! Kuntaur *Ruxton* ! **Port.G.**: Bissau (fl. & fr. Oct.) *Esp. Santo* 2548 ! Bissagos (Oct.) *Dinklage* 3220 ! **Fr.G.**: Kouroussa (fr. Oct.) *Pobéguin* 1161 ! **S.L.**: Njala (fr. Mar.) *Deighton* 4591 ! Serabu Bumpe (fr. Apr.) *Deighton* 1676 ! Rokupr (fl. & fr. Oct.) *Jordan* 364 ! **Iv.C.**: Mossi (fl. & fr. Aug.) *Chev.* 24716 ! **G.C.**: Accra (fl. & fr. Nov.) *Don* 10 ! *Irvine* 812 ! Achimota (fr. Dec.) *Darko* 488 ! **Togo:** Lome *Warnecke* 285 ! Kpandu *Robertson* 32 !

Dah.: *fide* Chev. *l.c.* **Fr.Nig.:** Gao *Hagerup* 360! **N.Nig.:** Nupe *Barter* 1626! Katagum *Dalz.* 6! Zaria (fl. & fr. June) *Keay* FHI 25912! **S.Nig.:** Olokemeji (fl. & fr. Oct.) *Keay* FHI 28046! Ibadan (Nov., Dec.) *Meikle* 668! 910! Throughout tropical Africa and reaching Cape Province and S.W. Africa. (See Appendix, p. 271.)

2. **Z. latifolia** *Sm.* in Rees Cycl. 39, No. 4 (1819); Milne-Redhead l.c. 89; Léonard l.c. 354; not of DC. (1825). *Z. gracilis* DC. (1825). *Z. diphylla* of F.W.T.A., ed. 1, 1 : 413; Kew Bull. 1929 : 18, partly; of Chev. Bot. 188, partly; not of Pers. A wiry perennial herb often remaining green throughout the dry season; flowers yellow.
   Sen.: Cape Verde *Perrottet!* **Fr.Sud.:** *fide* Chev. *l.c.* **S.L.:** Freetown (fl. & fr. Oct.) *Deighton* 2146! Njala (fl. & fr. Mar.) *Deighton* 4592! Rokupr (fl. & fr. Feb.) *Deighton* 4264! **Lib.:** Monrovia (fl. & fr. Jan.) *Baldwin* 14072! Sinkor (fl. & fr. June) *Barker* 1366! **Iv.C.:** *fide* Chev. *l.c.* **G.C.:** Accra *Don!* Achimota (fl. & fr. Jan.) *Akpabla!* Keta (fl. & fr. May) *Darko* 563! **Togo:** Lome *Warnecke* 287! **Dah.:** *fide* Chev. *l.c.* **S.Nig.:** Ibadan (fl. & fr. Dec.) *Meikle* 911! Idanre (fl. & fr. Oct.) *Keay* FHI 25526! Sapoba (fl. & fr. Nov.) *Keay* FHI 28063! Port Harcourt (fr. June) *Maitland!* Native of tropical S. America, introduced as a weed in W. Africa and Belgian Congo.

3. **Z. durumuensis** *De Wild.* in Rev. Zool. Afr. 13 : B 26 (1925); Milne-Redhead l.c. 79; Léonard l.c. 357, fig. 25 A. *Z. lelyi* Hutch. & Dalz. F.W.T.A., ed. 1, 1 : 413 (1928). A precocious perennial with woody rootstock; flowers yellow with reddish veins on standard.
   N.Nig.: Kargi Hill, Birnin Gwari (June) *Keay* FHI 25880! Kaduna (fr. Dec.) *Meikle* 774! Kontagora-Zungeru road (Jan.) *Meikle* 1098! Bauchi Plateau (fl. & fr. Feb.) *Lely* P. 151! Nabardo (fr. May) *Lely* 205! Also in Belgian Congo and A.-E. Sudan.

67. ARACHIS Linn.—F.T.A. 2 : 157.

Stems herbaceous, pilose; leaves paripinnate; leaflets 2 pairs, obovate, mucronate and sometimes emarginate, 3–4 cm. long, 1·5–2·5 cm. broad, with numerous parallel lateral nerves; rhachis flat above, pilose on the margin; stipules adnate to the petiole for about 1 cm., free part linear-lanceolate, very acute; 1·5–2 cm. long, nervose, ciliate; flowers axillary, solitary, the lower fertile ones subsessile and burying their fruits in the soil   ..   ..   *hypogaea*

A. hypogaea *Linn.*—F.T.A. 2 : 158. The Groundnut. Generally cultivated throughout the tropics. (See Appendix, p. 228.)

68. ORMOCARPUM P. Beauv.—F.T.A. 2 : 142.

Leaves imparipinnate :
   Leaflets rounded or mucronate at apex, up to 4 cm. long, glabrous or pubescent :
      Indumentum softly pubescent and greyish on branchlets, leaflets, calyx and fruits; leaflets 6–8 pairs, elliptic, mucronate, 5–10 mm. long; flowers few in short racemes on short arrested branchlets; pedicels about 1·5 cm. long, with a pair of ovate leafy bracteoles near the apex; calyx 8–10 mm. long; corolla about 2 cm. long; fruits 5–6-seeded, segments 9–12 mm. long, coarsely reticulate when mature
                                                            1. *bibracteatum*
      Indumentum (where present) setulose on branchlets, calyx and fruits; leaflets glabrous and whole plant frequently so; leaflets 5–10 pairs, elliptic-obovate, mucronate and sometimes emarginate, 1–2 cm. long; flowers few on the young shoots; pedicels 1–1·5 cm. long, with a pair of lanceolate-acuminate bracteoles near the apex; calyx about 8 mm. long; corolla about 1·5 cm. long; fruits usually hispid, 1–6-seeded, segments about 1 cm. long, obliquely elliptic, longitudinally striate   ..   ..   2. *sennoides* subsp. *hispidum*
   Leaflets acute or acutely acuminate, 4–13 cm. long, glabrous :
      Upper leaflets 7–13 cm. long, 3–4·5 cm. broad, 5–7 (–9) to each leaf, stipules lanceolate, 5–10 mm. long, 4 mm. broad at base; branchlets scabrid-hispid; flowers in scabrid-hispid racemes; pedicels 5–6 mm. long; fruits 5–7-seeded, shortly bristly, about 8–10 cm. long, ripe segments about 1 cm. long   ..   3. *megalophyllum*
      Upper leaflets 4–6 cm. long, about 1·5 cm. broad, 8–15 to each leaf, acute, hardly acuminate, oblong or slightly oblanceolate; stipules linear, 6 mm. long, 2 mm. broad at base; branchlets shortly scabrid-hispid; flowers in axillary densely scabrid-hispid racemes; pedicels filiform, 7–10 mm. long; fruits imperfectly known, apparently 2- or 3-seeded with a few bristles, ripe segments about 1·5 cm. long   ..   ..   4. *sp. A*
Leaves 1(–3)-foliolate; leaflet lanceolate to ovate-lanceolate, acutely acuminate, 8–14 cm. long, up to 5 cm. broad, glabrous; petiole articulated below the apex; flowers few, in short axillary racemes; pedicels about 5 mm. long with a pair of ovate-lanceolate bracteoles near the apex; calyx about 5 mm. long, glabrous; corolla nearly 1·5 cm. long; fruits up to 5-seeded segments narrowly elliptic, warted and longitudinally nerved   ..   ..   5. *verrucosum*

1. **O. bibracteatum** (*Hochst. ex A. Rich.*) Bak. in F.T.A. 2 : 143 (1871); Bak. f. Leg. Trop. Afr. 2 : 279. *Acrotaphros bibracteata* Hochst. ex A. Rich. (1847). A shrub or small tree with short stout branches, greypubescent all over; corolla reddish, lurid purple-veined and mottled, becoming dry and papery and persistent; in savannah.
   Sen.: Ferlo *Heudelot* 157! **G.C.:** Tamale *Kitson* 952! *Morton* GC 9087! **N.Nig.:** *Lely!* Kontagora (Jan.) *Dalz.* 42! Sokoto (Dec.) *Dalz.* 313! Katagum *Dalz.* 424! Also in N.E. tropical Africa. (See Appendix, p. 253.)

2. **O. sennoides** (*Willd.*) DC. subsp. **hispidum** (*Willd.*) Brenan & J. Léonard in Bull. Jard. Bot. Brux. 24 : 103, fig. 14, 2 (1954); Léonard in Fl. Congo Belge 5 : 244. *Hedysarum sennoides* Willd. (1802). *Cytisus hispidus* Willd. (1802). *Robinia guineensis* Willd. (1809). *Ormocarpum guineense* (Willd.) Hutch. & Dalz. F.W.T.A., ed. 1, 1 : 414 (1928); Kew Bull. 1929 : 18. A much-branched shrub, 4–12 ft. high; flowers yellow with reddish streaks; in forest.

Sen.: *Berhaut* 1765! **Port.G.:** Bula (fr. Nov.) *Esp. Santo* 2404! **Fr.G.:** Kouria to Ymbo *Chev.* 14730.
S.L.: *Afzelius*! Freetown (fr. Dec.) *Deighton* 3317! Masakoi Chiefdom (Sept.) *Deighton* 2235! Kuntaia
(fr. June) *Thomas* 423! **Lib.:** Sanokwele (fl. & fr. Sept.) *Baldwin* 9416! 12513! Peahtah (Oct.) *Linder*
925! **Iv.C.:** *fide* Chev. *l.c.* **G.C.:** Accra Plains (fl. & fr. June) *Irvine* 73! Axim (Apr.) *Chipp* 425! Ejura
(fl. & fr. May) *Vigne* FH 1214! R. Volta F.R. (Mar.) *Vigne* FH 4348! **Togo:** Lome *Warnecke* 54! Kete
Krachi (fl. & fr. June) *Vigne* FH 3880! **N.Nig.:** Abinsi (fl. & fr. June) *Dalz.* 606! **S.Nig.:** Ijebu (Apr.)
Tamajong FHI 16932! Okomu F.R. (Feb.) *Brenan* 8775! 9043! Aguku *Thomas* 1364! Oban *Talbot*
1403! **Br.Cam.:** Man-of-War Bay (Apr.) *Schlechter* 12385! Kumba (fr. June) *Olorunfemi* FHI 30627!
Widespread in tropical Africa.

3. **O. megalophyllum** *Harms* in Engl. Bot. Jahrb. 26 : 292 (1899); Bak. f. l.c. 279; Pellegr. Leg. Gab. 179;
Léonard l.c. 244. A shrub 3–5 ft. high, with yellow-purple veined flowers; in forest clearings.
Fr.G.: *fide* Pellegr. *l.c.* **S.L.:** Kenema (fr. Jan.) *Thomas* 7552! **Lib.:** *Linder* 346! **G.C.:** Kibi (fr. Jan.)
Adams 1200! **S.Nig.:** Abeokuta (Dec.) *Onochie* FHI 31887! Okomu F.R. (fr. Jan.) *Brenan* 8792! Oban
Talbot 1203! **Br.Cam.:** Victoria *Preuss* 1193! Also in French Cameroons and Gabon.

4. **O. sp. A.** A small shrub which may prove to be a distinct species.
Br.Cam.: Buea (fl. & fr. Jan.) *Maitland* 1265!

5. **O. verrucosum** *P. Beauv.*—F.T.A. 2 : 142; Chev. Bot. 185; Bak. f. l.c. 278; Léonard l.c. 243. A shrub,
10–15 ft. high, with brown-purple glabrous branchlets and white or lilac flowers purple-striped; in sandy
places near the sea and in brackish swamps.
Sen.: *Heudelot*! **Port.G.:** Bissau (Nov.) *Esp. Santo* 1415! **Fr.G.:** Guerzé country *Chev.* 21005. Kaoulen-
dougou *Chev.* 21067. **S.L.:** Mano Salija (Dec.) *Deighton* 427! Maswari (fl. & fr. Feb.) *Deighton* 4983!
Mahela (fl. & fr. Nov.) *T. S. Jones* 47! **Lib.:** Grand Cess (Mar.) *Baldwin* 11625! **Iv.C.:** Tabou to Bériby
Chev. 19984; 20018. **G.C.:** Axim (Mar.) *Morton* GC 6647! **S.Nig.:** Lagos (May) *Dalz.* 962! Warri *Beauv.*
Niger mouth *Mann* 457! Brass *Barter* 65! Calabar (fr. May) *Onochie* FHI 32097! **F.Po:** *T. Vogel* 259!
Also in French Cameroons, Congo, Angola, Principe and S. Tomé.

69. ANTOPETITIA A. Rich. in Ann. Sci. Nat., sér. 2, 14 : 261 (1840).

A slender herb, up to 75 cm. high; branchlets pubescent; leaves imparipinnate,
leaflets 3–4 pairs, narrowly oblanceolate, about 1 cm. long, appressed-pilose; rachis
with 2 small stipular glands at base; umbels 2–4-flowered; fruits torulose, segments
3–5, subglobose, glabrous, about 2–3 mm. diam.   ..   ..   ..   *abyssinica*

A. abyssinica *A. Rich.* l.c. 262 (1840); Bak. f. Leg. Trop. Afr. 2 : 274; Léonard in Fl. Congo Belge 5 : 179,
fig. 10. *Ornithopus coriandrinus* Hochst. ex Bak. in F.T.A. 2 : 140 (1871); F.W.T.A., ed. 1, 1 : 414.
Br.Cam.: Cam. Mt., 6,000–8,500 ft. (fl. Nov., fr. Dec.) *Migeod* 202! *Mildbr.* 10834! *Maitland* 831!
Bafut-Ngemba F.R., Bamenda (Oct.) *Ujor* FHI 30217! Also in Belgian Congo and eastern Africa from
Abyssinia and Eritrea to N. Rhodesia.

70. AESCHYNOMENE Linn.—F.T.A. 2: 145.

Stems armed with sharp prickles just below the ovate-lanceolate stipules, densely
hispid; fruits spirally coiled, about 1 cm. broad, pilose; flowers 3–4 cm. long;
pedicels about 1 cm. long, densely setose; leaves 6–10 cm. long; leaflets oblong
truncate at both ends, 1·5–2 cm. long, entire ..   .. 1. *elaphroxylon*
Stems without prickles; fruits straight :
Stipules not produced into an appendage beyond base :
Flowers about 1·5 cm. long; leaves 3–4 cm. long; leaflets 6–8 pairs, oblanceolate,
rounded at apex, about 1 cm. long and 3 mm. broad, glabrous; fruits 1-seeded,
about 2 cm. long and 1 cm. broad, with a stipe 5 mm. long   .. 2. *baumii*
Flowers less than 1 cm. long :
Flowers 8–10 mm. long :
Fruit-segments about 2 mm. broad, fruits 2-seeded with filiform stipe 8 mm. long;
racemes axillary, 2–3-flowered; flowers 8–9 mm. long; pedicels 4 mm. long
with a few hyaline bristles; leaves 1·5–3 cm. long; leaflets oblong, obtuse at
apex, 5–10 mm. long.. ..   .. 3. *sp. A*
Fruit-segments about 5 mm. broad, fruits 1–2-seeded, hardly stipitate; racemes
axillary, 10–15-flowered; flowers 8 mm. long; pedicels 4–5 mm. long, glabrous;
leaves 1–2 cm. long; leaflets oblong, 4–5 mm. long   ..   .. 4. *abyssinica*
Flowers up to 6 mm. long :
Anthers and filaments uniform in size; fruits with 2 flat segments 7 mm. long;
bracts entire; stipules narrowly lanceolate; leaflets 4 mm. long   5. *kerstingii*
Anthers of two sizes, filaments alternately long and short; fruit-segments 1–3 mm.
long, spheroid :
Flowers about 5 mm. long; standard obovate; fruits with 1–2-segments 2–3 mm.
broad, markedly tuberculate, exserted from calyx; rachis of inflorescence
almost straight, setulose   ..   .. 6. *pulchella*
Flowers about 3 mm. long, rachis of inflorescence zigzag; fruits 1–2 mm. broad,
minutely papillose :
Whole plant pilose with long yellow hairs, especially on young parts; pedicels
5–8 mm. long; fruits 1 (–2)-segmented, included in calyx   ,.   7. *lateritia*
Whole plant almost glabrous or slightly setose; pedicels 8–12 mm. long; fruits
(1–) 2-segmented, exserted from calyx ..   .. 8. *neglecta*
Stipules produced beyond base into a long auricle or appendage :
*Leaflets distinctly toothed, 2–3 cm. long :
Stems densely hispid, woody; leaves 8–15 cm. long; leaflets pilose on the margins
and midrib beneath; stipules about 1·5 cm. long, pilose outside, caducous;
flowers 2–3 cm. long; fruits 5–7-segmented, about 5 cm. long and 1·5 cm. broad,
pilose, tapered to a beak   ..   .. 9. *pfundii*

Stems glabrous or with a very few long hairs, thick and pithy ; leaves 7–12 cm. long; leaflets glabrous; stipules about 2·5 cm. long, glabrous except for the ciliate margins, more or less persistent ; flowers about 1 cm. long ; fruits about 5-segmented, 4 cm. long, 8 mm. broad, glabrous, coarsely reticulate　　10. *crassicaulis*
* Leaflets entire or nearly so, rarely up to 2·5 cm. long :
　Flowers about 3 cm. long, the keel with a large crest on the margin ; stipules lanceo-late, very acute, 2 cm. long, closely nerved, much produced at base ; leaflets about 20 pairs, oblong-oblanceolate, about 1·5 cm. long, venose ; fruits about 8 cm. long, glabrous　..　　..　　..　　..　　..　　..　　11. *cristata*
　Flowers much smaller than above, up to 1·5 cm. long, the keel less obviously crested :
　　Leaflets 1-nerved :
　　　Stems quite smooth, herbaceous :
　　　　Stems thick and soft ; leaflets very numerous, 1–2 cm. long :
　　　　　Flowers 4 mm. long ; fruits about 2 cm. long, 2–4-segmented, deeply constricted on the lower side, surface smooth or slightly muricate and convex ; leaflets numerous, 4–8 mm. long　　..　　..　　..　　12. *tambacoundensis*
　　　　Flowers 10–15 mm. long ; fruits 7–8 cm. long, undulate on lower side :
　　　　　Keel pubescent ; flowers about 1 cm. long ; fruits closely warted when mature, 8 cm. long, 8 mm. broad, 6–10-segmented ; leaflets usually drying brown, 1·4–2 cm. long ..　　..　　..　　..　　..　　13. *afraspera*
　　　　　Keel glabrous ; flowers about 1·4 cm. long ; fruits smooth when mature, 7 cm. long, 6 mm. broad, 6–7-segmented ; leaflets usually drying glaucous-green, about 1·3 cm. long　　..　　..　　..　　..　　..　　14. *nilotica*
　　　　Stems slender ; leaflets 10–20 pairs, up to 1 cm. long ; flowers few in slender leafy racemes, about 5–7 mm. long ; fruits undulate on one side, up to 9-seeded, nearly smooth　　..　　..　　..　　..　　..　　15. *indica*
　　　Stems more or less muricate or setose, rather woody :
　　　　Calyx-lobes glabrous outside ; branches flexuose ; leaflets up to 20 pairs, about 1 cm. long ; flowers very few, about 8 mm. long ; fruits rather deeply indented on one side, about 4 cm. long, up to 9-segmented, very sparsely pubescent
　　　　　　　　　　　　　　　　　　　　　　　　　　　16. *sensitiva*
　　　　Calyx-lobes bristly-pubescent outside :
　　　　　Fruits deeply indented on one side ; segments raised and papillose on the faces, stipe 1 cm. long ; flowers solitary or 2–3 together, axillary, about 1 cm. long ; pedicels 1–2 cm. long ; stem hispid-scabrid ; leaflets 1-nerved, obtuse, about 1 cm. long　　..　　..　　..　　..　　..　　17. *uniflora*
　　　　　Fruits straight on both margins ; pedicels about 2 cm. long :
　　　　　　Pedicel jointed with 2 small bracts halfway up its length ; flowers 1 cm. long ; fruits straight on both sides, 5 mm. broad, with 8–12 segments 3 mm. long, at first yellow and pubescent soon becoming black and glabrous ; stems shortly hispid-scabrid ..　　..　　..　　..　　..　　18. *schimperi*
　　　　　　Pedicel with one bract not more than 2 mm. from the base; flowers 1–1·2 cm. long ; fruits straight or slightly indented on one side, 4 mm. broad, with 3–7 segments 4 mm. long, yellow at length becoming brown, usually per-sistently pilose ; stems pilose with long yellow hairs　　..　　19. *deightonii*
　　Leaflets distinctly 3-nerved, acute, about 5 mm. long ; fruits deeply indented on one side, 5-segmented, segments convex on the faces, smooth, sometimes setose, hardly stipitate ; racemes 4–5-flowered, about 8 mm. long ; pedicels about 1 cm. long ; stems smooth and glabrous, the younger parts long-setose　20. *americana*

1. **A. elaphroxylon** (*Guill. & Perr.*) *Taub.* in E. & P. Pflanzenfam. 3, 3 : 319 (1894) ; Bak. f. Leg. Trop. Afr. 2 : 289 ; Léonard in Fl. Congo Belge 5 : 261. *Herminiera elaphroxylon* Guill. & Perr. (1832)—F.T.A. 2 : 144 ; F.W.T.A., ed. 1, 1 : 415. *Aeschynomene tchadica* A. Chev. (1913), name only. A shrub, 8–20 ft. high, of rivers and swamps ; flowers large, orange-yellow.
　　Sen.: Lelièvre ! L. Panié-Foul *Heudelot* ! **G.C.**: Dalaba, Lower Volta *Collector unknown* ! **N.Nig.**: Nupe *Barter* 757 ! L. Chad *Talbot* 1231 ! Widespread in tropical Africa.
2. **A. baumii** *Harms* in Warb. Kunene-Sambesi Exped. 261 (1903) ; Bak. f. l.c. 291 ; Léonard l.c. 271. A shrub, with yellow flowers and more or less glabrous stems ; in rocky mountain grassland.
　　Br.Cam.: Bamenda, 5,000 ft. (fl. Feb., fl. & fr. May) *Migeod* 464 ! *Maitland* 1436 ! 1607 ! Jakiri, Bamenda (Jan.) *Keay* FHI 28428 ! Also in French Cameroons, Belgian Congo, Tanganyika, N. Rhodesia and Angola.
3. **A. sp. A.** Possibly new, but better material required. Erect herb, 2 ft. high, with small leaves which close diurnally ; in grassland and edges of woodland.
　　Br.Cam.: Bafut-Ngemba F.R., 5,000–6,000 ft. (Oct.) *Lightbody* FHI 26261 !
4. **A. abyssinica** (*A. Rich.*) *Vatke* in Oest. Bot. Zeit. 29 : 224 (1879) ; Bak. f. l.c. 299 ; Léonard l.c. 286, fig. 19 G. *Rueppellia abyssinica* A. Rich. (1847). *Aeschynomene ruppellii* Bak. in F.T.A. 2 : 149 (1871). An undershrub, 2–10 ft. high, with dark brown rather viscid branches and yellow, purple-veined flowers ; beside streams.
　　N.Nig.: Alkalire (fr. Sept.) *Lely* 641 ! Bauchi Plateau *Batten-Poole* FHI 13160 ! *Lely* P. 694 ! Vom, 3,000–4,500 ft. *Dent Young* 60 ! Extends to Belgian Congo, Abyssinia, E. Africa, S. Rhodesia and Mozambique.
5. **A. kerstingii** *Harms* in Engl. Bot. Jahrb. 54 : 385 (1917) ; Bak. f. l.c. 292. Herbaceous or half-woody plant with slender stems ; flowers golden-yellow.
　　G.C.: Banda ravine, Wenchi (fr. Dec.) *Morton* GC 25139 ! **Togo**: Dadaure, Basari (July) *Kersting* 176 !
6. **A. pulchella** *Planch. ex Bak.* in F.T.A. 2 : 149 (1871) ; Chev. Bot. 186 ; Bak. f. l.c. 292. *A. saxicola* Taub. (1896). Branches red-brown, slightly viscid, 1 ft. or more long, from a woody base ; flowers orange-yellow.

Sen.: Tambacounda *Berhaut* 1466. **Port.G.:** Farim (fr. Apr.) *Esp. Santo* 2457! Gabu (Sept.) *Esp. Santo* 2790! **Fr.G.:** Balato (Feb.) *Chev.* 332! Kaoulendougou (Mar.) *Chev.* 21056! Toumania (fl. & fr. June) *Pobéguin* 947! Fouta Djalon *Heudelot* 739! **S.L.:** Buyabuya, Scarcies R. (fl. & fr. Feb.) *Sc. Elliot* 4131! Makunde (fl. & fr. Apr.) *Sc. Elliot* 5713! **Togo:** Agome (fl. & fr. Mar.) *Schlechter* 12958!

7. **A. lateritia** *Harms* in Engl. Bot. Jahrb. 26 : 292 (1899) ; Bak. f. l.c. 294 ; Léonard l.c. 299, t. 23. A slender much-branched annual herb, with wiry stems and very small yellow flowers ; in rock crevices. **Port.G.:** Suzana (fl. & fr. Nov.) *Esp. Santo* 3139! **Fr.Sud.:** Kita massif (fr. Nov.) *Jaeger* 6! **G.C.:** Weija, Accra (May) *Adams* GC 1767! **N.Nig.:** Patti Lokoja *Barter* 534! Yola (Sept.) *Dalz.* 305! Keana, Nassarawa Prov. (Oct.) *Hepburn* 42! Anara F.R. (fl. & fr. Sept.) *Olorunfemi* FHI 24375! Also in French Cameroons, Belgian Congo and Angola.

8. **A. neglecta** *Hepper* in Kew Bull. 1956 : 122, fig. 3 A. *A. lateritia* of F.W.T.A., ed. 1, 1 : 415, partly (*Dalz.* 599, *Lely* 570). A very slender annual herb, similar to the last, but almost glabrous. **N.Nig.:** Abinsi (fr. Feb.) *Dalz.* 599! Naraguta & Jos (fl. & fr. Sept.) *Lely* 570! Bauchi Plateau (Sept.) *Lely* P. 691!

9. **A. pfundii** *Taub.* in Engl. Pflanzenw. Ost-Afr. C. 215 (1895) ; Bak. f. l.c. 289. A shrub forming thickets over permanent water ; flowers large, yellow. **Fr.Sud.:** Koubita to Tiouki (Feb.) *Davey* 20! **N.Nig.:** Kalkala, L. Chad (Sept.) *Gwynn* 91! Gajibo, Bornu (Nov.) *Johnston N.* 94! Also in eastern Africa from A.-E. Sudan to the Zambezi.

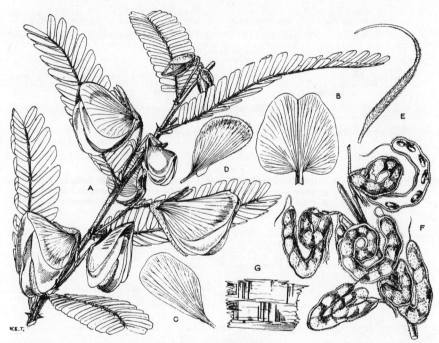

Fig. 168.—Aeschynomene elaphroxylon (*Guill. & Perr.*) *Taub.* (Papilionaceae). A, flowering shoot. B, standard. C, wing petal. D, keel petal. E, pistil. F, fruits. G, pith.

10. **A. crassicaulis** *Harms* in Engl. Bot. Jahrb. 40 : 38 (1907) ; Chev. Bot. 186 ; Bak. f. l.c. 288 ; Léonard l.c. 269, photo 1, fig. 18 H. Stem procumbent and rooting in mud, often half submerged ; flowers yellow ; fruits shortly stipitate. **Fr.Sud.:** Ségou *Chev.* 2940. **Fr.Nig.:** Fada N'Gourma (July) *Chev.* 24508! **N.Nig.:** Katagum (July) *Dalz.* 10! Also in Ubangi-Shari and Belgian Congo. (See Appendix, p. 226.)

11. **A. cristata** *Vatke* in Oest. Bot. Zeit. 28 : 215 (1878) ; Bak. f. l.c. 288 ; Léonard l.c. 266, t. 20, fig. 18 G. A soft-stemmed undershrub ; branches and peduncles at first setose, becoming glabrous ; flower conspicuous. **Dah.:** Avrankou Lagoon (Jan.) *Chev.* 22763! **S.Nig.:** Lagos (Nov.) *Millen* 96! Yewa R., Lagos Colony (Sept.) *Punch* 19! Okomu F.R. *Lomax*! Extends to A.-E. Sudan, Belgian Congo, E. Africa, Mozambique and Madagascar.

12. **A. tambacoundensis** *Berhaut* in Bull. Soc. Bot. Fr. Mém. 1953–4 : 7 (1954) ; Fl. Sen. 32. A slender plant about 3 ft. high on the edge of swamps, withering after the rainy season ; flowers yellow, usually in pairs on slender pedicels. **Sen.:** Tambacounda (fr. Aug.) *Berhaut* 1241! 2919 ; 2920 ; 2958 ; 2959. **S.L.:** near Kambia (fl. & fr. Oct.) *Jordan* 942!

13. **A. afraspera** *J. Léonard* in Bull. Jard. Bot. Brux. 24 : 64 (1954) ; Fl. Congo Belge 5 : 264, fig. 18 E. *A. aspera* of F.T.A. 2 : 147 ; Chev. Bot. 186 ; Bak. f. l.c. 288 ; F.W.T.A., ed. 1, 1 : 416. *Sesbania leptocarpa* of F.W.T.A., ed. 1, 1 : 387, partly (Iv.C. spec.). An undershrub, with soft pithy stems, 3–6 ft. high ; in swampy places. **Sen.:** *Berhaut* 1725. **Gam.:** Wallikunda (Aug.) *Macluskie* 4! 5! **Iv.C.:** Baoulé-Nord (July) *Chev.* 22030! **G.C.:** Tamale (fl. & fr. Dec.) *Morton* GC 9895! **N.Nig.:** Katsina Ala, L. Turu (fr. Aug.) *Dalz.* 616! **S.Nig.:** Ogun R. *Millen* 184! Idah (Sept.) *T. Vogel* 65! Also in A.-E. Sudan, Belgian Congo, N. Rhodesia, Angola, Nyasaland and Transvaal. (See Appendix, p. 226.)

14. **A. nilotica** *Taub.* in Engl. Bot. Jahrb. 23 : 189 (1896) ; Bak. f. l.c. 287 ; Léonard in Fl. Congo Belge 5 : 265. Stem thick and pithy up to 5 ft., flowers yellow ; in swamps.

Fr.Sud.: Macina (Sept.) *Chev.* 24940 ! Diafarabé (fl. & fr. Sept.) *Lean* 100 ! Gao *Ryff* ! **Fr.Nig.:** Niamey (Oct.) *Hagerup* 525 ! Also in A.-E. Sudan, Tanganyika, N. Rhodesia and Belgian Congo.

15. **A. indica** *Linn.*—F.T.A. 2 : 147 ; Chev. Bot. 186 ; Bak. f. l.c. 286 ; Léonard l.c. 259, fig. 18 C. An under-shrub 2–4 ft. high, with pale yellow, faintly striate flowers ; in muddy places near rivers and swamps.
Sen.: *Heudelot* ! **Gam.:** Kudang (fr. Jan.) *Dalz.* 8021 ! *fide* Chev. *l.c.* **Port.G.:** Bafatá (fl. & fr. July) *Esp. Santo* 2718 ! **G.C.:** Cape Coast (fr. July) *T. Vogel* 68 ! Achimota *Irvine* 571 ! 717 ! **N.Nig.:** Katagum (fr. Oct.) *Dalz.* 11 ! Sherifuri *Thornewill* 177 ! Bauchi Plateau *Lely* P. 506 ! Extends through the Old World tropics. (See Appendix, p. 226.)

16. **A. sensitiva** *Sw.*—F.T.A. 2 : 147 ; Chev. Bot. 186 ; Bak. f. l.c. 286 ; Léonard l.c. 258, fig. 18 B. An undershrub with slightly viscid branches and small bright yellow flowers ; in swamps.
Sen.: Casamance *Chev.* 3266. **Gam.:** Wallikunda *Macluskie* 21 ! **Fr.G.:** Rio Nunez *Heudelot* 689 ! Kouroussa (Oct.) *Pobéguin* 563 ! Mamou *Chev.* 20392 ! Nzérékoré *Baldwin* 9712 ! **S.L.:** Kambia (fl. & fr. Dec.) *Sc. Elliot* 4366 ! Yonibana (Nov.) *Thomas* 5224 ! Port Loko *Thomas* 6676 ! Zimi (Nov.) *Deighton* 368 ! **Lib.:** Genna Tanyehun (fr. Dec.) *Baldwin* 10765 ! **G.C.:** Accra (fl. & fr. Nov.) *Morton* A. 56 ! **S.Nig.:** Lagos Lagoon (fr. Feb.) *Onochie* FHI 34146 ! Okomu F.R. (Dec.) *Brenan* 8590 ! 9193 ! Widespread in tropical Africa, the Mascarenes and in tropical America. (See Appendix, p. 226.)

17. **A. uniflora** *E. Mey.*—F.T.A. 2 : 146 ; Chev. Bot. 187 ; Bak. f. l.c. 286 ; Léonard l.c. 256, fig. 18 A. An undershrub with half-woody bristly branches and small bright yellow flowers.
Sen.: Niayes *Chev.* 3267. Kaolak *Chev.* **N.Nig.:** Bauchi Plateau *Lely* P. 505 ! Lokoja *Barter* 516 ! Widespread in tropical Africa, in Natal, and in Madagascar and Comoro Islands.

18. **A. schimperi** *Hochst. ex A. Rich.*—F.T.A. 2 : 146 ; Bak. f. l.c. 287 ; Léonard l.c. 262, fig. 18 D. An erect undershrub with half-woody branches, rough with short almost spicular hairs ; flowers yellow ; in moist places.
Sen.: *Berhaut* 666. **Gam.:** *Brown-Lester* 13 ! Kudang (fr. Jan.) *Dalz.* 8022 ! **Port.G.:** Bissau (fl. & fr. Nov.) *Esp. Santo* 2618 ! **Fr.G.:** Tristao Isl. *Paroisse* 50 ! **S.L.:** Batbai (fr. June) *Deighton* 3547 ! Also in Eritrea, Abyssinia, Uganda, Kenya and N. Rhodesia.

19. **A. deightonii** *Hepper* in Kew Bull. 1956 : 124, fig. 4. *A. djalonensis* A. Chev. Bot. 186, name only. *A. schimperi* of F.W.T.A., ed. 1, 1 : 416, partly ; Bak. f. l.c. 287, partly. A herb, with rather weak straggling branches about 2 ft. high ; leaves rather glaucous and sensitive to touch ; flowers yellow turning cream.
Fr.G.: Kouria *Chev.* 14977 ! Dalaba *Chev.* 20170 ! **S.L.:** Magboloko (fr. Nov.) *Glanville* 121 ! Yonibana (fl. & fr. Nov.) *Thomas* 4695 ! 4749 ! Jokibu (Nov.) *Deighton* 3774 ! Mahera (fl. & fr. Nov.) *Deighton* 5685 !

20. **A. americana** *Linn.* Sp. Pl. 2 : 713 (1753). A half-woody shrub about 3 ft. high with pink flowers ; leaves sensitive to touch.
S.L.: Freetown (fl. & fr. Oct.) *Deighton* 254 ! Waterloo (fl. & fr. Dec.) *Deighton* 3299 ! Introduced from tropical America.

71. **CYCLOCARPA** Afzel. ex Bak.—F.T.A. 2 : 151 ; Urb. in Jahrb. Bot. Gart. Berl. 3 : 247 (1884).

A tiny shrublet ; leaves subsessile, paripinnate ; leaflets 3–4 pairs obovate-oblanceolate, mucronate, about 5 mm. long, minutely serrulate, glabrous ; stipules ovate, acumi-nate, striate, with a long lanceolate basal appendage ; flowers axillary ; fruits coiled in a ring, forming 1–1½ spirals, with 9–10 segments, glabrous, about 6 mm. diam.

stellaris

C. stellaris *Afzel. ex Bak.* in F.T.A. 2 : 151 (1871) ; Urb. l.c. 248 ; Chev. Bot. 189, partly ; Bak. f. Leg. Trop. Afr. 2 : 302 ; Léonard in Fl. Congo Belge 5 : 241. *Aeschynomene stellaris* (Afzel. ex Bak.) G. Roberty (1954). Prostrate or diffuse woody herb, with slender glabrous branchlets up to 20 in. long, sensitive leaves which cover the small yellow flowers when touched ; uncommon in moist sandy places.
Port.G.: Antula, Bissau (fr. Nov.) *Esp. Santo* 1410 ! Sucujaque, Suzana (Nov.) *Esp. Santo* 3142 ! **Fr.G.:** Kouroussa *Pobéguin* 462 ! Kouria (Oct.) *Chev.* 15030. Fouta Djalon (Sept.) *Chev.* 18562 ; 18575. **Fr.Sud.:** Kita massif (fl. & fr. Oct.) *Jaeger* 14 ! **S.L.:** Rokupr (fr. Oct.) *Jordan* 159 ! 368 ! Magbile (fr. Dec.) *Thomas* 6193 ! Near Jokibu (fl. & fr. Nov.) *Deighton* 3775 ! Tombo (fr. Jan.) *Deighton* 986 ! **Lib.:** Duport (Nov.) *Linder* 1508 ! **N.Nig.:** Kontagora (Nov.) *Dalz.* ! Also in Ubangi-Shari, French Cameroons, Gabon, Belgian Congo, Tanganyika, Nyasaland ; has been found also in Borneo and N. Australia.

72. **KOTSCHYA** Endl. Nov. Stirp. 4 (1839) ; Dewit & Duvigneaud in Bull. Soc. Roy. Bot. Belg. 86 : 208 (1954).

Flowers in clusters on filiform peduncles, 1–2 cm. long ; calyx about 4 mm. long, glabrous ; leaflets linear-oblong, 4 mm. long, minutely pilose ; stem glabrous, herbaceous, less than 15 cm. long .. .. 1. *micrantha*
Flowers in subsessile clusters or spikes, or in shortly pedunculate racemes :
  Calyx very broad, obovate, entire, about 1 cm. long ; leaflets 9–13 pairs, subcordate on one side, 8 mm. long, sparsely pilose on main nerve which is set along one edge ; stipules scarious, linear, with parallel nerves bristly, .. 2. *schweinfurthii*
  Calyx not broad and conspicuous :
    Stipules about 1 cm. or more long :
      Corolla only slightly longer than calyx, 4–5 mm. long ; flowers in long caterpillar-like secund spikes about 2–3 cm. long ; mature leaves 4–6 cm. long ; rhachis and leaflet-margins long-pilose ; leaflets (13–) 18–22 pairs, 6–13 mm. long
                                        3. *ochreata* var. *ochreata*
      Corolla much longer than calyx, 7–9 mm. long ; flowers in short few-flowered clusters 1–1·5 cm. long ; mature leaves 1–2 cm. long ; rhachis and leaflet-margins ciliate ; leaflets 6–8 (–10) pairs, 4–5 mm. long   3a. *ochreata* var. *longipetala*
    Stipules 3–9 mm. long :
      Flowers blue in sessile short dense axillary racemes, like caterpillars ; leaves up to 1·5 cm. long ; stipules 7 mm. long :
        Corolla about 14 mm. long ; calyx 2-lipped, upper lip 10 mm. long, 2-lobed almost to middle, lower lip 13 mm. long, narrowly and deeply 3-lobed ; bracts 8 mm. long, with a prominent dorsal nerve, long-acuminate, pilose ; leaflets 4–5 mm. long, ciliate .. .. 4. *speciosa*

Corolla about 8 mm. long ; calyx 2-lipped, upper lip 6 mm. long, 2-lobed for about ⅓, lower lip 7 mm. long, less deeply 3-lobed; bracts about 5 mm. long, acuminate, pilose ; leaves up to 1·5 cm. long ; leaflets 3–5 mm. long, sparsely ciliate .. .. 5. *strigosa*

Flowers yellow or orange about 1·5 cm. long ; calyx 1 cm. or more long ; racemes few-flowered :

Stems glabrous, scaly with strips of bark, youngest shoots densely yellow-pubescent; stipules scarious, lanceolate-subulate, glabrous or minutely ciliate, 7–9 mm. long ; leaves 1–2·5 cm. long ; leaflets 8–14 pairs, pubescent at first, glabrescent, 6–10 mm. long ; corolla 1·4–1·6 cm. long ; calyx 1 cm. long, upper lip deeply bifid ; fruit with 2 segments .. .. 6. *uniflora*

Stems pubescent ; stipules 2–4 mm. long :

Stems pubescent with short dense upward-curved hairs, scabrid ; leaves 2–3 cm. long ; leaflets 8–14 pairs, margins ciliate, 5–6 mm. long ; racemes 5–8-flowered, lower flowers caducous leaving 2 or 3 above ; pedicels about 6 mm. long ; bracts lanceolate, ciliate, caducous ; corolla about 1·5 cm. long, exserted 2–3 mm. beyond calyx, standard emarginate ; calyx about 1·2 cm. long, upper lip ¼ bifid .. .. 7. *lutea*

Stems hispid with long straight and small curved hairs, scabrid ; leaves 3–4 cm. long ; leaflets 10–14 pairs, margins ciliate, 6–8 mm. long ; racemes 6–10-flowered, lower flowers frequently caducous ; pedicels about 8 mm. long ; bracts ovate-lanceolate, long-setose, more or less persistent ; corolla about 1·8 cm. long, exserted 1–2 mm. beyond calyx, standard slightly emarginate ; calyx about 1·7 cm. long, upper lip deeply bifid .. .. 8. *carsonii*

1. **K. micrantha** (*Harms*) *Hepper* in Kew Bull. 1956 : 124. *Smithia micrantha* Harms in Journ. de Bot. 22 : 113 (1909) ; Chev. Bot. 187 ; Bak. f. Leg. Trop. Afr. 2 : 305 ; F.W.T.A., ed. 1, 1 : 416. A slender herb with few-flowered heads and small flowers.
   Fr.G.: Bouria (Nov.) *Chev.* 14960 !

2. **K. schweinfurthii** (*Taub.*) *Dewit & Duvign.* in Bull. Soc. Roy. Bot. Belg. 86 : 209 (1954). *Smithia schweinfurthii* Taub. in Engl. Bot. Jahrb. 23 : 191 (1896) ; Bak. f. l.c. 309. A low-growing shrub of open savannah woodland, stems pubescent ; flowers blue in a close spike ; distinguished by the much-enlarged calyx.
   Togo: *fide* Bak. f. *l.c.* **N.Nig.:** Bauchi Plateau *Lely* P. 817 ! *Batten-Poole* FHI 13176 ! Dogon Kurmi, Jemaa (Nov.) *Keay & Onochie* FHI 21733 ! Anara F.R. (Sept.) *Olorunfemi* FHI 24380 ! Also in French Cameroons, A.-E. Sudan, Ubangi-Shari and Belgian Congo.

3. **K. ochreata** (*Taub.*) *Dewit & Duvign.* var. **ochreata**—l.c. 214 (1954). *Smithia ochreata* Taub. l.c. 191 (1896) ; Chev. Bot. 187 ; Bak. f. l.c. 307 ; F.W.T.A., ed. 1, 1 : 416. An erect shrub to 6 ft. or more high, younger parts pilose with soft hairs from a swollen base ; flowers pink. A variable species.
   Fr.Sud.: Koulikoro (Oct.) *Chev.* 3264 ! **Port.G.:** Bissau (Dec.) *Esp. Santo* 1092 ! **Fr.G.:** Kindia *Jac.-Fél.* 224 ! **S.L.:** *Afzelius* ! Kassa (fr. Mar.) *Sc. Elliot* 5069 ! Dombudu (Feb.) *Macdonald* 8 ! Bintumane (Jan.) *Glanville* 454 ! Sefadu (Dec.) *Deighton* 3474 ! **Lib.:** Maa, Makona R. (fr. Feb.) *Bequaert* 57 ! **Iv.C.:** Mt. Gbon, Gouékangouiné *Chev.* 21416. **Togo:** Togo Plateau F.R. (Nov.) *St. C. Thompson* 1615 ! Also in Gabon and the Belgian Congo. (See Appendix, p. 261.)

3a. **K. ochreata** var. **longipetala** *Hepper* l.c. 126, fig. 5 (1956). A more slender shrub than the type, with small clusters of larger flowers.
   Fr.G.: Kindia *Jac.-Fél.* 245 ! Dalaba (Nov.) *Jac.-Fél.* 579 ! Pita (Dec.) *Jac.-Fél.* 665 ! **S.L.:** Bullum *Smythe* 106 ! Kumrabai (Dec.) *Thomas* 6872 ! 6974 !

4. **K. speciosa** (*Hutch.*) *Hepper* l.c. 124 (1956). *Smithia speciosa* Hutch. in Kew Bull. 1921 : 365 ; Bak. f. l.c. 306. F.W.T.A., ed. 1, 1 : 416. An erect shrub 4–5 ft. high, branching from numerous stems, softly pilose ; flowers large, bright blue.
   N.Nig.: Bauchi Plateau (Aug.) *Lely* P. 638 ! *Dent Young* 61 ! Naraguta (Aug.) *Lely* 478 ! Also in Tanganyika and N. and S. Rhodesia.

5. **K. strigosa** (*Benth.*) *Dewit & Duvign.* l.c. 212 (1954) ; Fl. Congo Belge 5 : 335. *Smithia strigosa* Benth. in Miq. Pl. Junghun. 211 (1852) ; Bak. f. l.c. 306. A shrub similar to *S. speciosa* with smaller blue flowers.
   S.Nig.: Ikwette Plateau, 5,200 ft., Obudu (fr. Dec.) *Savory & Keay* FHI 25218 ! **Br.Cam.:** Bamenda, 5,000 ft. (fr. Jan.) *Migeod* 299 ! *Baldwin* 13848 ! Bambui, 4,000 ft. (fr. Mar.) *Maitland* 1646 ! Bafut-Ngemba F.R. (fr. Apr.) *Ujor* FHI 30016 ! Babanki (Jan.) *Johnstone* FHI 8741 ! Also in French Cameroons, Belgian Congo, Tanganyika, Nyasaland, N. and S. Rhodesia and Angola.

6. **K. uniflora** (*A. Chev.*) *Hepper* l.c. 126 (1956). *Smithia uniflora* A. Chev. in Bull. Mus. Hist. Nat. 5 : 479 (1933). A prostrate or ascending shrub about 1 ft. high and branches about 2 ft. long ; with yellow flowers ; on sandstone cliffs.
   Fr.G.: Benna Mts. (Dec.) *Jac.-Fél.* 2163 ! Kindia *Jac.-Fél.* 246. Pita *Jac.-Fél.* 685. Téliméla *Jac.-Fél.* 1779. *Pobéguin* 2208.

7. **K. lutea** (*Portères*) *Hepper* l.c. 126 (1956). *Smithia lutea* Portères in Rev. Bot. Appliq. 19 : 788, fig. 7A (1939). A shrub 4–6 ft. high with rough branches and deep yellow flowers and red calyx.
   Fr.G.: Mt. Nimba, about 4,000 ft., *Portères* 3157 ! **S.L.:** *Glanville* 390 ! Bintumane, 4,000–5,500 ft. (Jan.) *Glanville* 455 ! 466 ! *T. S. Jones* 95 ! *Jaeger* 988 !

8. **K. carsonii** (*Bak.*) *Dewit & Duvign.* in Bull. Soc. Roy. Bot. Belg. 86 : 213 (1954) ; Fl. Congo Belge 5 : 340. *Smithia carsonii* Bak. (1893). *S. harmsiana* De Wild. (1902). *S. reflexa* Portères l.c. 789, fig. 7B (1939). A shrub like the last with almost bristly stems and yellow flowers.
   Fr.G.: Labé (Dec.) *Jac.-Fél.* 641 ! Pita *Jac.-Fél.* 693. *Portères* 2045 ; 2194. Dalaba-Diaguissa Plateau *Jac.-Fél.* 756 ; 2057. Beyla *Jac.-Fél.* 1545. **Iv.C.:** Bié-Sama, Man (Jan.) *Portères* ! Also in Belgian Congo and Tanganyika.

In addition to the above, *K. aeschynomenoides* (Welw. ex Bak.) Dewit & Duvign. from east, central and south tropical Africa has been introduced into French Guinea.

Imperfectly known species.

Smithia ? trochainii *Berhaut* in Bull. Soc. Bot. Fr. 101 : 375 (1955). A slender procumbent herb described with neither flowers nor fruits.
Sen.: Bakel *Trochain*.

19

73. SMITHIA Ait. Hort. Kew. ed. 1, 3 : 496, t. 13 (1789) ; Dewit & Duvigneaud in
        Bull Soc. Roy. Bot. Belg. 86 : 71 (1954).

A herb ; stem glabrescent, hispid above, 20–70 cm. long ; leaflets 6–8 pairs, opposite,
linear-oblong, about 1 cm. long, subsymmetrical with the principal nerve central
and long-pilose ; stipules 1·7 cm. long, membranous, persistent, prolonged beneath
the point of insertion into a biauriculate appendage with one auricle short (4 mm.)
and round and the other much longer (7 mm.), linear and acute ; inflorescence an
umbel, peduncle 1·5–4 cm. long, hispid ; flowers erect ; bracteoles 4 mm. long at
base of calyx ; calyx 2-lipped, lips subentire, accrescent   ..   ..   .. *elliotii*

S. elliotii *Bak. f.* Leg. Trop. Afr. 2 : 304 (1929); Dewit & Duvign. in Fl. Congo Belge 5 : 346, fig. 23.   An
ascending herb of moist upland places with pink or blue flowers.
Br.Cam.: Basenako, Bamenda (June) *Maitland* 1504 !   Mt. Mba Kokeka, 7,500 ft., Bamenda (Jan.)
Keay FHI 28404 !   Also in French Cameroons, Uganda, Tanganyika and Belgian Congo.

74. BRYASPIS Duvigneaud in Bull. Soc. Roy. Bot. Belg. 86 : 151 (1954).

Stems glabrous, straw-coloured ; leaves very few, imparipinnate ; leaflets about 3 pairs,
oblong-oblanceolate, unequal-sided at base, 1–1·3 cm. long, glabrous ; stipules
persistent, broadly ovate, several-nerved, 1 cm. long, glabrous ; flowers in dense
oblong pedunculate racemes and hidden by large imbricate reticulate orbicular-
reniform bracts about 1·5 cm. broad ; corolla about 5 mm. long ; fruits very small,
with 2 reticulate segments   ..   .. *lupulina*

B. lupulina (*Planch. ex Benth.*) *Duvign.* l.c. 150–153, figs. 1B, 2C, 3B (1954).   *Geissaspis lupulina* Planch. ex
Benth. (1865)—F.T.A. 2 : 155 ; Tisserant in Bull. Mus. Hist. Nat., sér. 2, 3 : 333 (1931).   *G. psitta-*
corhyncha of F.W.T.A., ed. 1, 1 : 416 ; Chev. Bot. 187.   A slightly woody herb, erect or straggling ;
flowers small, bright yellow within the bracts ; in moist places.
Sen.: Casamance *Heudelot* 555 ; 664 !   **Port.G.:** Bissau (fl. & fr. Dec.) *Esp. Santo* 1214 !   **Fr.G.:** *Farmar*
211 !   Kouroussa (fl. & fr. Oct.) *Pobéguin* 568 !   Maneak to Frigniagbé *Pobéguin* 11 !   Ymbo (Sept.)
Chev. 14485.   Fouta Djalon (Sept.) *Chev.* 18459.   **S.L.:** *Afzelius* !   Waterloo (fl. & fr. Dec.) *Dawe* 424 !
Kitchom (Jan.) *Deighton* 932 !   Tongbai (fl. & fr. Nov.) *T. S. Jones* 45 !   Mayamba (Jan.) *King* 213 !
Bunting 2 !

Doubtfully recorded species.

Geissaspis psittacorhyncha (*Webb*) *Taub.* in E. & P. Pflanzenfam. 3, 3 : 321 (1894), excl. distrib. W. Afr. ;
Tisserant l.c. ; Pellegr. l.c. 183.   *Soemmeringa psittacorhyncha* Webb (1849).   This species, previously
regarded as identical with the above, is based on a specimen collected by St. Hilaire, probably from the
Cape Verde Islands, but, according to Tisserant (l.c.), possibly from the mainland at Cape Verde.

Imperfectly known species.

A specimen from a rock outcrop at Genne Loffa, Kolahun Dist., Liberia (*Baldwin* 10097) is possibly a new
sp. of *Bryaspis* but it possesses filiform bracteoles at the base of the calyx which are not characteristic of the
genus (*vide* Duvigneaud *l.c.*).   The plant is similar to *B. lupulina* but is setose-pilose in all parts and the flowers
exceed the enlarged bracts.   Further material of this plant is needed and possibly the limits of this genus
may have to be re-defined.

75. DESMODIUM Desv.—F.T.A. 2 : 159.

Calyx with 5 narrow equal teeth ; bracts ovate-lanceolate, subpersistent ; inflorescence
a short bracteate congested very hairy raceme, scarcely as long as the leaves ; pedicels
as long as the flowers, the latter nodding ; leaflets 3 or sometimes 1, oblong-elliptic,
emarginate, 2–3 cm. long, densely appressed-pubescent beneath ; stipules long-
subulate-lanceolate   ..   .. 1. *barbatum* var. *dimorphum*
Calyx-teeth 5, the upper 2 connate ; bracts linear or lanceolate, if broader then early
caducous and inflorescence lax :
Leaves 1-foliolate :
  Inflorescence a slender lax-flowered raceme :
    Leaflet lanceolate, acute or acuminate, up to 12 cm. long ; stem up to 1 m. high ;
    pedicels about as long as flower ; calyx pilose ; fruits nearly straight on one side,
    deeply indented on the other, about 2 cm. long, thinly pubescent
                              2. *gangeticum* var. *gangeticum*
    Leaflet ovate-orbicular or ovate-lanceolate, obtuse or subacute, variable in size ;
    stem less than 20 cm. high ; other characters as above
                              2a. *gangeticum* var. *maculatum*
  Inflorescence a dense leafy subpyramidal panicle ; leaflet ovate or rhomboid, rounded
  at apex, up to 15 cm. long and 8 cm. broad, softly pilose beneath ; petiolule 1–2 cm.
  long ; pedicels very short ; fruits deeply indented on one side, 1·5–2 cm. long,
  pubescent   ..   .. 3. *velutinum*
Leaves 3-foliolate, rarely with a few 1-foliolate leaves (see No. 9) :
  *Fruits straight on the upper side, indented only on the lower side, or straight on both
  sides :
    Flowers 1–3 in leaf axils ; stems prostrate, more or less pilose, terminal leaflet 1 cm.
    long or less, finely appressed-pilose beneath ; pedicels up to 1 cm. long ; fruits
    about ⅓ indented on one side, about 5-seeded, pubescent   .. 4. *triflorum*
  Flowers in racemes :
    Terminal leaflet lanceolate, at least 3 times as long as broad, acute, 5–15 cm. long
    and 1–4 cm. broad, thinly pubescent beneath or on nerves only ; lateral nerves

numerous ; flowers very shortly pedicellate ; fruits only slightly indented, 3–5-seeded, at length reticulate　..　..　..　..　.. 5. *salicifolium*
Terminal leaflet oblong, elliptic or rhombic, up to twice as long as broad :
Flowers 8–10 mm. long ; inflorescence terminal, lax, normally paniculate, few-flowered, leafless, 15–40 cm. long ; stipules 1–1·5 cm. long, brown, persistent ; flowers paired, pedicels 2–3 cm. long ; leaves more or less rhombic, acute or apiculate, up to 9 cm. long and 6 cm. broad, thinly pilose on both sides ; fruits very deeply indented, 3–5-seeded, tomentellous..　..　.. 6. *repandum*
Flowers 1–5 mm. long ; racemes simple or leafy ; leaflets obtuse or rounded at the apex :
Pedicels up to 6 mm. long in flower, up to 10 mm. in fruit ; flowers in narrow leafless terminal racemes ; fruits pubescent ; stems stiffly erect and sparingly branched :
Terminal leaflet 4–6 cm. long, 1–3 cm. broad, more or less acute, glaucous and shortly pubescent beneath ; flowers about 4 mm. long, rather shorter than pedicels ; fruits deeply indented on one side, 3–5-seeded, shortly setulose-pubescent ..　..　..　..　..　..　.. 7. *canum*
Terminal leaflet 1–2 (–3·5) cm. long, 5–10 mm. broad, obovate, obtuse, pilose beneath ; flowers about 3 mm. long ; fruits deeply indented, 3–5-seeded, pubescent　..　..　..　..　..　..　.. 8. *ramosissimum*
Pedicels exceeding 6 mm. long ; flowers in spreading terminal or axillary racemes; stems more or less prostrate or ascending, usually branched, or with a prostrate main stem and erect lateral shoots :
Stems more or less prostrate, rooting at nodes with erect lateral shoots on which the flowers are borne in a lax terminal raceme above the leaves ; fruits rather deeply indented on one side, segments about twice as long as broad, tomentose ; terminal leaflet broadly obovate to obovate-elliptic, 2–4·5 cm. long, 2–3 cm. broad, cuneate at base, rounded at apex :
Leaflets appressed-pubescent above, not reticulate above when dry, thin ; bracts finely puberulous to glabrescent　　9. *adscendens* var. *adscendens*
Leaflets glabrous above, sometimes with a few hairs on the midrib, reticulate above when dry, rather thick ; bracts densely appressed-pilose
9a. *adscendens* var. *robustum*
Stems not rooting at the nodes; segments of fruits more or less square :
Fruits indehiscent, segments breaking off transversely, rather deeply indented on the lower side and slightly indented on the upper side ; segments about 2 mm. broad and 2–3 mm. long, tomentose ; racemes axillary ; stems densely pilose ; terminal leaflet obovate-elliptic, 3–5 cm. long, about 2 cm. broad, rounded to subacute at apex ..　..　..　.. 10. *setigerum*
Fruits dehiscent, opening along lower suture and then breaking up transversely, slightly indented along the lower side, upper edge entire ; racemes mostly terminal :
Fruits 2·5–3 mm. broad, 1–2 cm. long, tomentose ; terminal leaflet broadly obovate, cuneate to rounded at base, rounded and slightly emarginate at apex, 10–20 mm. long, 8–15 mm. broad, appressed-pilose beneath ; racemes lax and spreading　..　..　..　..　.. 11. *hirtum*
Fruits 2 mm. broad, about 1 cm. long ; terminal leaflet obovate-oblong to oblong :
Leaflets appressed-pilose above and beneath, obovate-oblong, 12–20 mm. long, 7–9 mm. broad ; racemes short and few-flowered ; fruits finely pubescent　..　..　..　..　..　.. 12. *schweinfurthii*
Leaflets glabrous above or with a few hairs on the midrib, pilose beneath ; terminal leaflet linear-oblong, 11–26 mm. long, 3–7 mm. broad ; racemes rather lax and spreading ; fruits glabrous or rarely minutely setulose
13. *linearifolium*
*Fruits equally indented on each side :
Fruits very deeply indented on both sides, often twisted or contorted, segments 4–6, almost orbicular, flat, finely pubescent, about 3 mm. broad; leaflets thin, more or less rhomboid, cuneate at base, subacute at apex, almost glabrous, margins ciliated ; racemes up to 15 cm. long, flowers very small ; pedicels about 1 cm. long　..　..　..　..　..　..　..　.. 14. *tortuosum*
Fruits not deeply indented, segments twice as long as broad :
Terminal leaflet 7–12 cm. long, about 6 cm. broad, elliptic, equally tapered at each end, pubescent beneath with appressed silky hairs ; racemes up to 40 cm. long ; flowers about 5 mm. long ; pedicels about 7 mm. long ; fruits up to 9-seeded ; segments compressed, 4 mm. long, 1·5 mm. broad　..　.. 15. *laxiflorum*
Terminal leaflet about 2 cm. long and 1 cm. broad, oblong or elliptic, rounded at each end, thinly pubescent ; racemes up to 10 cm. long, few-flowered ; flowers 4 mm. long ; pedicels about 5 mm. long ; fruits 5–8-seeded ; segments subterete, 4 mm. long and 2 mm. broad　..　..　..　..　.. 16. *scorpiurus*

1. **D. barbatum** (*Linn.*) *Benth.* var. **dimorphum** (*Welw. ex Bak.*) *Schubert* in Bull. Jard. Bot. Brux. 22 : 298 (1952) ; Fl. Congo Belge 5 : 205. *D. dimorphum* Welw. ex Bak. in F.T.A. 2 : 161 (1871) ; F.W.T.A., ed. 1, 1 : 418 ; Bak. f. Leg. Trop. Afr. 2 : 332. A more or less woody undershrub, 1–2 ft. or more high, grey silky pubescent ; with pink or purplish flowers.
   Fr.G.: Kouroussa (Sept.) *Pobéguin* 434 ! **N.Nig :** Bauchi Plateau (Aug.) *Lely* P. 580 ! Naraguta (Aug.) *Lely* 513 ! Vom, 3,000–4,500 ft., *Dent Young* 64 ! Anara F.R. (fr. Oct.) *Olorunfemi* FHI 24405 ! Extends through tropical Africa from Abyssinia to Rhodesia and in Madagascar.
2. **D. gangeticum** (*Linn.*) *DC.* var. **gangeticum**—Prod. 2 : 327 (1825) ; F.T.A. 2 : 161, partly ; Chev. Bot. 190 ; F.W.T.A., ed. 1, 1 : 418, partly ; Schubert Fl. Congo Belge 196. *Hedysarum gangeticum* Linn. (1753). Erect, half-woody, undershrub, up to 5 ft. high ; with small white and pink or bluish flowers.
   Port.G.: Bissau (fr. Jan.) *Esp. Santo* 1632 ! **Fr.G.:** Nzérékoré (fr. Oct.) *Baldwin* 9684 ! **S.L.:** Mamboma (fr. Nov.) *Deighton* 3773 ! Sangasanga, Benna country (Jan.) *Sc. Elliot* 4420 ! **Lib.:** Dubo, Sanokwele (Sept.) *Baldwin* 9473 ! **G.C.:** Sampa (fr. Dec.) *Vigne* FH 3484 ! Krobo Plains (Dec.) *Johnson* 511 ! **N.Nig.:** Ilorin (fr. Dec.) *Ajayi* FHI 19293 ! **S.Nig.:** Abeokuta *Irving* ! Oyo (fl. & fr. Feb.) *Keay* FHI 23431 ! FHI 23432 ! Oban *Talbot* 1304 ! Widely distributed in the tropics of the Old World. (See Appendix, p. 239.)
2a. **D. gangeticum** var. **maculatum** (*Linn.*) *Bak.* Fl. Brit. Ind. 2 : 168 (1876). *Hedysarum maculatum* Linn. (1753). *Desmodium gangeticum* of F.T.A. : 161 and of F.W.T.A., ed. 1, 1 : 418, partly. A dwarf variety with small leaves found in drier situations.
   G.C.: Achimota (fr. Feb.) *Irvine* 1996 ! Prang (Jan.) *Vigne* FH 1581 ! Dawa R. (fr. Feb.) *Morton* GC 6463 ! Salaga (Apr.) *Dalz.* 20 ! *Krause* ! **N.Nig.:** Zaria (fl. & fr. May) *Keay* FHI 25747 ! *Milne-Redhead* 5029 ! Kontagora *Meikle* 1078 ! A distribution similar to the typical variety. [Chevalier (Bot. 190) records this species from Senegal, French Guinea and Ivory Coast ; I have not seen the specimens and cannot say to which variety they belong.]
3. **D. velutinum** (*Willd.*) *DC.* l.c. 328 (1825) ; Schubert l.c. 194. *Hedysarum velutinum* Willd. (1803). *H. lasiocarpum* P. Beauv. (1805). *Desmodium lasiocarpum* (P. Beauv.) DC. (1825)—F.T.A. 2 : 162 ; Chev. Bot. 190 ; F.W.T.A., ed. 1, 1 : 418 ; Bak. f. l.c. 326. A half-woody, erect, rather hairy undershrub, 3–6 ft. high, with small pink or purplish flowers and fruit-segments which adhere like burrs.
   Sen.: Bondou *Heudelot* 208 ! *Perrottet* 244 ! Ziguinchor, Casamance *Chev.* 3379 ! **Gam.:** *Mungo Park* ! **Fr.Sud.:** Tabacco *Chev.* 3265. **Fr.G.:** Kouria *Chev.* 14903. Fouta Djalon (June) *Maclaud* 26 ! Kouroussa (Sept.) *Pobéguin* 443 ! Bindélya (Dec.) *Paroisse* 39 ! Nzérékoré (Oct.) *Baldwin* 9704 ! **S.L.:** *Smeathmann* ! Mamaha (fr. Nov.) *Thomas* 4493 ! Njala *Deighton* 1343 ! Gberia Timbako (Sept.) *Small* 319 ! **Lib.:** Gbanga (fr. Nov.) *Daniel* 33 ! Robertsport (fl. & fr. Dec.) *Baldwin* 10921 ! Vonjama (Oct.) *Baldwin* 9936 ! **Iv.C.:** Voguié *Chev.* 17185. Montézo *Chev.* 17444. **G.C.:** Offinso (Oct.) *Darko* 734 ! Accra (fr. July) *Irvine* 15 ! Prang (fr. May) *Kitson* 609 ! **Togo:** Lome *Warnecke* 209 ! Atakpame (Nov.) *Mildbr.* 7400 ! **Dah.:** *Burton* ! **N.Nig.:** *T. Vogel* 146 ! Lokoja *Barter* 1044 ! Katagum *Dalz.* 335 ! **S.Nig.:** Abeokuta *Irving* 85 ! Ibadan (fr. Oct.) *Jones & Keay* FHI 13708 ! Benin (fl. & fr. Sept.) *Farquhar* 78 ! A common plant of the Old World tropics. (See Appendix, p. 239.)
4. **D. triflorum** (*Linn.*) *DC.* l.c. 2 : 334 (1825) ; F.T.A. 2 : 166 ; Chev. Bot. 191 ; Bak. f. l.c. 327 ; Schubert l.c. 187. *Hedysarum triflorum* Linn. (1753). Stems much-branched, prostrate, creeping and forming dense mats ; flowers pink or purple.
   Fr.G.: Dyeke (Oct.) *Baldwin* 9670 ! Conakry *Chev.* 13543. **S.L.:** Port Loko (Dec.) *Thomas* 6648 ! Roruks (Nov.) *Thomas* 5696 ! Regent (fr. Oct.) *Deighton* 2179 ! Mahela (fr. Dec.) *Sc. Elliot* 4055 ! **Lib.:** Monrovia (Nov.) *Linder* 1538 ! Nana Kru (Mar.) *Baldwin* 11598 ! Zorzor (Oct.) *Baldwin* 10024 ! **Iv.C.:** Toumodi *Chev.* 22409 ! **G.C.:** Accra Plains *Irvine* 736 ! *Bally* 38 ! Senya Beraker (June) *Morton* GC 7912 ! **Togo:** Lome *Warnecke* 257 ! **Dah.:** Cotonou *Chev.* 4487. **S.Nig.:** Lagos *Dalz.* 1033 ! 1207 ! Ibadan *Schlechter* 12335 ! Okomu F.R. (Feb.) *Brenan* 9153 ! Onitsha *Barter* 1790 ! Also in other parts of tropical Africa, in the Mascarene islands and Socotra, and in tropical America. (See Appendix, p. 239.)
5. **D. salicifolium** (*Poir.*) *DC.* l.c. 2 : 337 (1825) ; Bak. f. l.c. 330 ; Schubert l.c. 198, t. 15. *Hedysarum salicifolium* Poir. (1804). *Desmodium paleaceum* Guill. & Perr. (1832)—F.T.A. 2 : 166 ; Chev. Bot. 191 ; F.W.T.A., ed. 1, 1 : 418. A half-woody, erect undershrub, 3–4 ft. high, with red or yellowish flowers in slender more or less panicled spike-like racemes.
   Sen.: *Heudelot* 369 ! Casamance *Chev.* 3263. **Gam.:** Albreda (fr. Mar.) *Perrottet* 246 ! Gambia R. (Jan.) *Dalz.* 8020 ! **Fr.Sud.:** Médinani *Chev.* 508 ! **Fr.G.:** Kouroussa (Sept.) *Pobéguin* 956 ! Nzérékoré (Oct.) *Baldwin* 9711 ! **S.L.:** Gorahun (fl. & fr. Nov.) *Deighton* 360 ! Musaia (fr. Dec.) *Deighton* 5703 ! Njala (fl. & fr. Jan.) *Dalz.* 8065 ! Jigaya (Sept.) *Thomas* 2847 ! **Lib.:** Peáhtah (Oct.) *Linder* 1039 ! Vonjama (Oct.) *Baldwin* 9859 ! **Iv.C.:** Capiécrou (Feb.) *Chev.* 17091. Toumodi *Chev.* 22419. **G.C.:** Kwadaso (Nov.) *Darko* 742 ! Tamale (fr. Dec.) *Morton* GC 9900 ! Banda, Wenchi (fr. Dec.) *Morton* GC 25263 ! **N.Nig.:** Naraguta (Sept.) *Lely* 573 ! Anara F.R. (Oct.) *Keay* FHI 20128 ! **S.Nig.:** Ibadan (Oct.) *Jones & Keay* FHI 13779 ! Throughout tropical Africa, and in S. Africa and the Mascarene Islands.
6. **D. repandum** (*Vahl*) *DC.* l.c. 2 : 334 (1825) ; Schubert l.c. 193, t. 14. *Hedysarum repandum* Vahl (1791). *Desmodium scalpe* DC. (1825)—F.T.A. 2 : 164 ; Chev. Bot. 191 ; F.W.T.A., ed. 1, 1 : 418 ; Bak. f. l.c. 328. An undershrub with slender thinly spreading pubescent branches, straggling or more or less climbing ; flowers usually deep red ; in montane vegetation.
   Fr.G.: Dalaba-Diaguissa Plateau, 3,000–4,000 ft. *Chev.* 18733 ; 20186. **S.Nig.:** Ijua, 3,000 ft., Obudu (Dec.) *Savory & Keay* FHI 25025 ! **Br.Cam.:** Cam. Mt., 2,000–7,000 ft. (Dec.) *Mann* 1229 ! 2017 ! Bamenda (May) *Ujor* FHI 30313 ! Bafut-Ngemba F.R. (fl. & fr. Nov.) *Lightbody* FHI 26266 ! **F.Po:** *Mann* 286 ! Moka (Dec.) *Boughey* 161 ! In the tropics and subtropics of the Old World. (See Appendix, p. 239.)
7. **D. canum** (*J. F. Gmel.*) *Schinz & Thell.* in Mém. Soc. Neuchatel Sci. Nat. 5 : 371 (1913) ; Exell Cat. S. Tomé 156 ; Schubert l.c. 184. *Hedysarum canum* J. F. Gmel. (1791). *Desmodium mauritianum* (Willd.) DC. (1825), not of authors. *D. incanum* DC. (1825)—F.T.A. 2 : 163 ; Chev. Bot. 190 ; F.W.T.A., ed. 1, 1 : 418. *D. frutescens* Schindl. (1925) ; Bak. f. l.c. 328. A half-woody undershrub, more or less decumbent ; flowers reddish-purple in loose racemes.
   Fr.Sud.: Folo *Chev.* 848. **Fr.G.:** Dalaba-Diaguissa Plateau *Chev.* 18725. **S.L.:** *Don* ! **Lib.:** Rocktown (fr. Mar.) *Baldwin* 11635 ! **Iv.C.:** Upper Cavally *Chev.* 15837. **F.Po:** *Barter* ! *Mann* 56 ! Bahia de Venus (fl. & fr. Dec.) *Guinea* 247 ! Also in Gabon, Principe and S. Tomé, locally in S. Africa, the Mascarene Islands and some E. African islands, and in America.
8. **D. ramosissimum** G. Don Gen. Syst. 2 : 294 (1832) ; Schubert l.c. 191. *D. mauritianum* of F.T.A. 2 : 164 ; Chev. Bot. 190 ; F.W.T.A., ed. 1, 1 : 418 ; Bak. f. l.c. 330, not of (Willd.) DC. An erect slightly woody herb, with slender branches ; bright pink or purple flowers.
   Fr.Sud.: Tabacco *Chev.* 138. **Fr.G.:** Kouria to Ymbo *Chev.* 15014. Macenta (Oct.) *Baldwin* 9828 ! **S.L.:** *Don* 34 ! *Barter* ! Wellington (fl. & fr. Feb.) *Deighton* 4989 ! Njala (Sept.) *Deighton* 2089 ! Bumbuna (fr. Oct.) *Thomas* 3363 ! **Lib.:** Cape Palmas *Ansell* ! Kakatown *Whyte* ! Sinkor (fl. & fr. June) *Barker* 1356 ! Zeahtown, Tchien *Baldwin* 6952 ! **Iv.C.:** *fide* Chev. *l.c.* **G.C.:** Banda, Wenchi (fl. & fr. Dec.) *Morton* GC 25304 ! Afram Plains (fr. Jan.) *Irvine* 625 ! Shai Plains *Johnson* 569 ! Tamale *Williams* 160 ! **Togo:** Lome *Warnecke* 260 ! Logba Lota (June) *Howes* 1006 ! **Dah.:** La Lama *Chev.* 23245. R. Ouémé *Chev.* 23603. **N.Nig.:** Abinsi (Feb.) *Dalz.* 588 ! Naraguta *Lely* 18 ! Bauchi Plateau (fl. & fr. Jan.) *Lely* P. 44 ! **S.Nig.:** Lagos *MacGregor* 63 ! *Punch* 8 ! Ibadan *Schlechter* 12330 ! **Br.Cam.:** Mamfe (fr. Dec.) *Migeod* 271 ! **F.Po:** *Mildbr.* 6250. Widely spread in tropical Africa, and in Madagascar and the Mascarenes. (See Appendix, p. 239.)

9. **D. adscendens**(*Sw.*) *DC.* var. **adscendens**—F.T.A. 2 : 162 ; Chev. Bot. 189 ; Bak. f. l.c. 2 : 330 ; Schubert l.c. 189. *Hedysarum adscendens* Sw. (1788). *Desmodium ovalifolium* Guill. & Perr. (1832). An undershrub with slender thinly pubescent branches, straggling and sometimes prostrate and rooting ; flowers pink or whitish.
  Sen.: Casamance *Perrottet* 245 ! **S.L.:** *Afzelius* ! Rokupr (fl. & fr. Sept.) *Jordan* 557 ! Njala (Sept.) *Deighton* 2083 ! **Lib.:** Du R. (July) *Linder* 192 ! Webo (fl. & fr. June) *Baldwin* 6163 ! Cavally (fr. June) *Baldwin* 6007 ! **Iv.C.:** Makougnié *Chev.* 17189. Alépé *Chev.* 17475. Sanvi *Chev.* 17833. **G.C.:** Axim (Apr.) *Chipp* 168 ! Akwapim (Mar.) *Murphy* 678 ! Swedru (Feb.) *Dalz.* 8284 ! **S.Nig.:** Lagos (fr. Aug.) *Rowland* ! Okomu F.R. *Brenan* 8849 ! Obu *Thomas* 37 ! Old Calabar *Holland* 104 ! **Br.Cam.:** Victoria (June) *Rosevear* Cam.46/37 ! **F.Po:** *Mann* 407 ! Extends through tropical Africa to S. Africa, also in America. (See Appendix, p. 238.)

9a. **D. adscendens** var. **robustum** *Schubert* in Bull. Jard. Bot. Brux. 22 : 290 (1954) ; Fl. Congo Belge 5 : 190. The leaflets are often larger than the type and are glabrous and reticulate above.
  Port.G.: Dandum, Bafata (June) *Esp. Santo* 2692 ! **Lib.:** Bakratown (Sept.) *Linder* 866 ! **Togo:** Amedzofe (fl. & fr. May) *Scholes* 41 ! **N.Nig.:** Naraguta (June) *Lely* 325 ! **Br.Cam.:** Buea, 3,000 ft. (fl. & fr. Nov.) *Migeod* 112 ! Bali, Bamenda (fl. & fr. May) *Ujor* FHI 30386 ! **F.Po:** Moka, 4,000 ft. (fl. & fr. Dec.) *Boughey* 18 ! 198 ! San Carlos (fl. & fr. Jan.) *Guinea* 1616 ! Also in French Cameroons, Spanish Guinea, A.-E. Sudan, Belgian Congo and N. Rhodesia.

10. **D. setigerum** (*E. Mey.*) *Benth. ex Harv.* Fl. Cap. 2 : 229 (1862) ; Schubert in Bull. Jard. Bot. Brux. 22 : 289 (1952) ; Fl. Congo Belge 5 : 187, fig. 11, E. *Nicholsonia setigera* E. Mey. (1836). *Desmodium hirtum* partly, of F.T.A. 2 : 163 ; of F.W.T.A., ed. 1, 1 : 418 ; of Bak. f. l.c. 329. *D. delicatulum* of F.W.T.A., ed. 1, 1 : 418. Stems clothed with spreading silky hairs ; flowers pink or bluish.
  Fr.G.: Rio Nunez *Heudelot* 696 ! **S.L.:** Juring (Dec.) *Deighton* 331 ! Kenema (fl. & fr. Nov.) *Deighton* 390 ! Musaia (fr. Dec.) *Deighton* 5701 ! Port Loko (fl. & fr. Dec.) *Thomas* 6689 ! **N.Nig.:** Vom, 3,000–4,500 ft., *Dent Young* 63 ! Throughout tropical Africa and in Natal.

11. **D. hirtum** *Guill. & Perr.*—F.T.A. 2 : 163, partly ; F.W.T.A., ed. 1, 1 : 418, partly ; Bak. r. l.c. 329, partly ; Schubert in Bull. Jard. Bot. Brux. 22 : 287 (1952) ; Fl. Congo Belge 5 : 186, fig. 11 B. *D. delicatulum* of Chev. Bot. 189. A more or less prostrate herb with small pink flowers.
  Sen.: Casamance (fr. Apr.) *Perrottet* 247 ! Bakel (fl. & fr. Oct.) *Leprieur* ! **Fr.Sud.:** Kabarah (fl. & fr. Aug.) *Chev.* 1341 ! **Port.G.:** Bissau (fl. & fr. Oct.) *Esp. Santo* 947 ! 2547 ! **S.L.:** Kent (fr. Dec.) *Deighton* 3288 ! Mahera (fl. & fr. Nov.) *Deighton* 5674 ! Erimakuna (fr. Mar.) *Sc. Elliot* 5210 ! **G.C.:** Zuarungu (fr. Dec.) *Adams & Akpabla* GC 4304 ! Tamale (fl. & fr. Oct.) *Baldwin* 13557 ! **Fr.Nig.:** Niamey (fr. Oct.) *Hagerup* 529 ! **N.Nig:** Lokoja *Barter* ! Sokoto (fl. & fr. Oct.) *Dalz.* 338 ! Mopa, Kabba (fl. & fr. Oct.) *Ward* L. 144 ! **S.Nig.:** Aguku *Thomas* 625 ! Extends throughout tropical Africa. (See Appendix, p. 239.)

12. **D. schweinfurthii** *Schindl.* in Engl. Bot. Jahrb. 54 : 57 (1916), excl. *Afzelius* s.n. & *Chev.* 15028 ; Schubert l.c. 184, fig. 11, A. *D. linearifolium* of F.W.T.A., ed. 1, 1 : 417, partly. A small annual about 6 in. high, with pink flowers.
  N.Nig.: Bauchi Plateau (fr. Sept.) *Lely* P. 684 ! Also in Ubangi-Shari and N.E. tropical Africa.

13. **D. linearifolium** *G. Don* Gen. Syst. 2 : 294 (1832) ; Bak. f. l.c. 329. *D. schweinfurthii* Schindl. (1916), partly. *D. djalonense* A. Chev. Bot. 189, name only. A branched herb about 1 ft. high, woody at the base, with erect or spreading lax branches and small pink-mauve flowers.
  Fr.G.: Kouria *Chev.* 15028 ! Fouta Djalon *Chev.* 18309. Mamou *Chev.* 18624. **S.L.:** *Afzelius* ! *Smeathmann* ! Kissy (fr. Nov.) *Deighton* 4387 ! Bumbuna (fr. Oct.) *Thomas* 3276 ! 3429 ! 3832 ! Rokupr (fr. Nov.) *Jordan* 158 !

14. **D. tortuosum** (*Sw.*) *DC.* Prod. 2 : 332(1825) ; Webb in Hook. Nig. Fl. 122 ; Schubert l.c. 202. *Hedysarum tortuosum* Sw. (1788). *Desmodium spirale* (Sw.) DC. (1825)—F.T.A. 2 : 160 ; Bak. f. l.c. 331. *D. ospriostreblum* Steud. ex Chiov. (1908) ; Bak. f. l.c. ; Exell Cat. S. Tomé 158. *Anarthrosyne abyssinica* Hochst. ex A. Rich. (1847). *Desmodium abyssinicum* (Hochst. ex A. Rich.) Hutch. & Dalz. F.W.T.A., ed. 1, 1 : 418 (1928), not of DC. (1825). *D. terminale* of Guill. & Perr. Fl. Seneg. 307, not of DC. Erect herb or undershrub, with pubescent branches and very thin flaccid leaves and twisted fruits ; introduced in our area.
  Sen.: *Heudelot* 221 ! Richard Tol *Perrottet*. **S.L.:** *Afzelius* ! *Smeathmann* ! Njala (fl. & fr. Nov.) *Deighton* 1507 ! **G.C.:** Sandema (fr. Sept.) *Hughes* FH 5347 ! Koforidua (fl. & fr. Jan.) *Morton* GC 8306 ! Senchi Ferry (fl. & fr. Oct.) *Morton* GC 9292 ! **Br.Cam.:** Victoria (Feb.) *Maitland* FHI 12111 ! *Dalz.* 8198 ! Widespread in the tropics.

15. **D. laxiflorum** *DC.* in Ann. Sci. Nat., sér. 1, 4 : 100 (1825). *D. incanum* of Chev. Bot. 190 (*Chev.* 18638). An undershrub, up to 5 ft. high, with pink or purple flowers ; leaves turn red when the plant is in fruit.
  Sen.: *Berhaut* 1851. **Gam.:** Frith 158 ! **Fr.G.:** Toumbalako to Boulivel (Sept.) *Chev.* 18638 ! **S.L.:** Musaia (Sept.) *Deighton* 4506 ! 4857 ! Kamabai to Kabala (fr. Dec.) *Deighton* 5691 ! Bafodea to Kamba (Sept.) *Deighton* 5161 ! **S.Nig.:** Afo R., Oyo (Sept.) *Latilo* FHI 23533 ! Introduced from tropical Asia.

16. **D. scorpiurus** (*Sw.*) *Desv.* in Journ. de Bot. 1 : 122 (1813). *Hedysarum scorpiurus* Desv. (1788). A straggling, climbing or procumbent herb with small blue flowers.
  S.L.: Freetown (Jan.) *Deighton* 241 ! 485 ! Bonthe (fr. Mar.) *Deighton* 2484 ! **G.C.:** *Morton* ! Introduced from tropical America.

76. PSEUDARTHRIA Wight & Arn.—F.T.A. 2 : 167.

Flowers arranged in short dense spike-like racemes about 2·5 cm. long ; leaflets elliptic-obovate, rounded at apex, the terminal one about 10 cm. long and 6 cm. broad, rather densely pubescent beneath, strongly nerved ; stipels linear, about 8 mm. long ; stipels subulate ; calyx-lobes long-subulate, pilose ; ovary densely villous ; fruits several-seeded, about 2 cm. long, densely pilose .. .. .. 1. *confertiflora*

Flowers arranged in lax much-branched panicles :

  Fruits 6–10-seeded, thinly pubescent ; leaflets obovate, very undulate on the margin, the terminal one 9–10 cm. long and about 6 cm. broad, densely white-tomentose beneath, strongly nerved ; stipules lanceolate, striate inside ; calyx-lobes shortly subulate ; ovary loosely pubescent .. .. 2. *hookeri*

  Fruits 1–5-seeded, shortly pubescent ; leaflets loosely pubescent beneath, the terminal one rhomboid, mucronate, up to 13 cm. long and 8 cm. broad, undulate and strongly nerved ; stipules, etc. as in No. 2 .. .. 3. *fagifolia*

1. **P. confertiflora** (*A. Rich.*) *Bak.* in F.T.A. 2 : 167 (1871) ; Bak. f. Leg. Trop. Afr. 2 : 339 ; Léonard in Fl. Congo Belge 5 : 236, fig. 15, C–D. *Rhynchosia confertiflora* A. Rich. (1847). An undershrub, 3 ft. high, with grooved yellowish-pubescent branches ; flowers small, purple.
  G.C.: Jema (July) *Chipp* 529 ! Yeji (Aug.) *Pomeroy* 1345 ! **Togo:** Krachi *Krause* ! **S.Nig.:** Lagos *MacGregor* 151 ! *Foster* 115 ! Old Oyo F.R. (Sept.) *Latilo* FHI 23539 ! Extends to A.-E. Sudan and eastern Africa.

2. **P. hookeri** *Wight & Arn.*—F.T.A. 2 : 168 ; Chev. Bot. 192 ; Bak. f. l.c. 339 ; Léonard l.c. 235, fig. 15, A–B. *P. alba* A. Chev. (1912). An erect, stiffly branched shrub, 6–10 ft. high, grey-downy all over ; flowers small in copious panicles, white, yellow or pink ; sometimes gregarious in grass savannah ; easily confused with *Desmodium salicifolium*.
   Sen.: Niokolo-Koba *Berhaut* 1333. **Fr.Sud.:** Upper Volta *Chev.* 3383. **Fr.G.:** Fouébouédougou, 1,850 ft. (July) *Collenette* 71 ! Mt. Kamoueniboka *Chev.* 22014. Toumodi *Chev.* 22407. **S.L.:** Musaia *Deighton* 4359 ! Gberia Fotombu (Oct.) *Small* 340 ! **G.C.:** Ejura *Kitson* 610 ! *Morton* GC 9562 ! Techiman to Nkoransa (fr. Dec.) *Adams & Akpabla* GC 4522 ! **Togo:** Misahöhe (Nov.) *Mildbr.* 7247 ! **N.Nig.:** Lokoja (Oct.) *Parsons* L. 81 ! Abinsi *Dalz.* 597 ! Naraguta & Jos (Sept.) *Lely* 565 ! **Br.Cam.:** Belo, Bamenda (fl. & fr. May) *Maitland* 1434 ! 1559 ! Bambuko F.R. (Sept.) *Olorunfemi* FHI 30765 ! All over tropical Africa.

3. **P. fagifolia** *Bak.* in F.T.A. 2 : 167 (1871) ; Chev. Bot. 192 ; Bak. f. l.c. 338. An erect undershrub 3–4 ft. high, rather like the last and perhaps a variety ; flowers small, red, in slender panicles.
   Port.G.: Gabú (Sept.) *Esp. Santo* 2775 ! **Fr.G.:** Kouria to Ymbo *Chev.* 14980. **S.L.:** *Afzelius.* Kabala (fr. Sept.) *Thomas* 2171 ! Musaia (fr. Dec.) *Deighton* 4412 ! Falaba (Sept.) *Small* 298 ! Gberia Fotombu (Oct.) *Small* 341 ! **N.Nig.:** Bauchi Plateau (Sept.) *Lely* P. 743 ! P. 744 ! Tuka, Niger (Oct.) *A. F. A. Lamb* FHI 3171 ! Anara F.R. (fr. Oct.) *Keay* FHI 5443 ! 5449 ! **S.Nig.:** Yoruba *Barter* 1600 ! Awba Hills F.R. (Oct.) *Jones* FHI 6330 !

77. DROOGMANSIA De Wild. in Ann. Mus. Congo, sér. 4 : 53 (1902).

Petiole with obovate-oblanceolate foliaceous wing over 6 mm. broad ; leaves pilose beneath ; stipe of fruit about 4 mm. long :
Inflorescence dense, bracts caducous ; lamina of leaf oblong-lanceolate, rounded at each end, about 12 cm. long, 2·5–2·8 cm. broad ; petiole 3 cm. long, the wing about 1·5 cm. broad ; pedicels about 8 mm. long, pilose ; flowers about 13 mm. long, standard obovate, striated   ..   ..   1. *chevalieri*
Inflorescence rather lax, bracts persistent ; lamina of leaf lanceolate, rounded at each end, or subcordate at the base, about 8 cm. long, up to 2 cm. broad, usually about 1 cm. ; petiole about 2 cm. long, the wing 6–12 mm. broad ; pedicels 6–11 mm. long, pilose or shortly pubescent ; flowers about 12 mm. long, standard ovate or suborbicular, striated   ..   ..   2. *scaettaiana*
Petiole about 1·5 cm. long with oblanceolate wing decurrent towards the base, up to 3 mm. broad ; lamina of leaf ovate-lanceolate, subcordate at base, obtuse at apex, about 4·5 cm. long and 2 cm. broad, appressed-pubescent beneath, glabrescent above ; bracts narrowly lanceolate, long-acuminate, 5 mm. long, shortly pubescent ; pedicels 10–12 mm. long, appressed-pubescent ; flowers about 11 mm. long ; fruit with a stipe 18 mm. long   ..   ..   3. *montana*

1. **D. chevalieri** *(Harms)* Hutch. & Dalz. F.W.T.A., ed. 1, 1 : 418 (1928) ; Kew Bull. 1929 : 18 ; Chev. & Sillans in Rev. Bot. Appliq. 32 : 46, t. 3 (1952). *Dolichos chevalieri* Harms in Journ. de Bot. 22 : 114 (1909). Erect, woody herb ; stem strongly ribbed, pilose on ribs ; racemes in the axils of the upper leaves and forming a panicle.
   Fr.G.: Bilima (Sept.) *Caille* in Hb. *Chev.* 15040 !
2. **D. scaettaiana** *Chev. & Sillans* l.c. 46, t. 4 A (1952). Similar to the preceding but with denser inflorescence and narrower leaves.
   Fr.G.: Fouta Djalon (Sept.) *Chev.* 18629. **S.L.:** Saka Sakala (Sept.) *Deighton* 5138 ! Benekoro (fr. Oct.) *Thomas* 2904 ! **Iv.C.:** *Scaëtta* 3551. Mt. Nimba, above 3,200 ft. *Schnell* !
3. **D. montana** *Jac.-Fél.* in Bull. Soc. Bot. Fr. 92 : 245 (1946). A shrub, 4–5 ft. high, stems slender, branched.
   Fr.G.: Labé *Jac.-Fél.* 654. Pita *Jac.-Fél.* 1959 ! Benna Plateau *Jac.-Fél.* 2116.

78. ALYSICARPUS Neck.—F.T.A. 2 : 169.

Fruit not or hardly constricted between segments, 12–25 mm. long ; calyx-teeth not imbricate, about equal in length to the first segment :
Racemes dense, internodes between the flowers shorter than the flowers ; plants perennial ; petioles 4–12 mm. long ; leaves oblong to suborbicular, more or less cordate at base, acute or rounded at apex, 1–2·5 cm. long and 5–10 mm. broad ; flowers in 4–13 pairs   ..   ..   1. *vaginalis*
Racemes very lax, internodes between the flowers much longer than the flowers ; plants annual ; petioles 2–7 mm. long ; leaves lanceolate or elliptic, cordate or subcordate at base, acute or rounded at apex, 1–7 cm. long and 1–3 cm. broad ; flowers in 3–7 pairs   ..   ..   2. *ovalifolius*
Fruit markedly constricted between segments, 5–15 mm. long ; calyx-teeth imbricate, at least at the base, longer than the first segment :
Fruit smooth, transverse ridges hardly visible ; bracts and calyx softly pubescent ; leaves subcoriaceous, reticulation obvious on both surfaces ; perennial   3. *zeyheri*
Fruit with numerous prominent transverse ridges ; bracts and calyx more or less glabrous, margins ciliate ; leaves membranous, reticulation less prominent on both surfaces :
Calyx-teeth distinctly rounded at base, imbricated around the fruit, 1·5–2·5 mm. broad ; fruit-segments included within the calyx (occasionally 1–2 exserted), usually the upper ones pubescent and lower glabrous ; plant annual or perennial, spreading or erect, often robust ; petiole 2–11 (–17) mm. long ; racemes usually dense, sometimes lax   ..   ..   4. *rugosus*
Calyx-teeth not rounded at base, free or hardly imbricated at base around fruit, 1–1·5 mm. broad ; fruit with 1–4 segments exserted, usually all pubescent ; plant annual, usually erect ; petiole 2–6 mm. long ; racemes lax   ..   5. *glumaceus*

1. **A. vaginalis** (*Linn.*) *DC.*—F.T.A. 2 : 170, partly ; Léonard in Fl. Congo Belge 5 : 224, fig. 13 A. *Hedysarum vaginale* Linn. (1753). Herb about 6 in. high, ascending, with small reddish flowers and cylindrical wrinkled fruits.
    S.L.: Freetown (fl. & fr. Aug.) *Deighton* SLH 2060 ! **N.Nig.:** Zaria *Hill* 36 ! Throughout tropical Africa and in tropical Asia. (See Appendix, p. 227.)
2. **A. ovalifolius** (*Schum. & Thonn.*) *J. Léonard* in Bull. Jard. Bot. Brux. 24 : 88, t. 11 (1954) ; Fl. Congo Belge 5 : 226, t. 18. *Hedysarum ovalifolium* Schum. & Thonn. (1827). *Alysicarpus vaginalis* of F.W.T.A., ed. 1, 1 : 419, partly. Herb 1–2 ft. high, sometimes more, with red flowers like the last ; on waste ground.
    Sen.: *Heudelot* 506 ! *Sieber* ! **Gam.:** Genieri (fr. Feb.) *Fox* 22 ! **Port.G.:** Bissau (fl. & fr. Oct.) *Esp. Santo* 2532 ! **G.C.:** Achimota (fr. Mar.) *Irvine* 326 ! Zuarungu *Lyon* FH 1072 ! Tamale (fr. Dec.) *Morton* GC 9919 ! *Baldwin* 13592 ! **Togo:** Lome *Warnecke* 278 ! **N.Nig.:** Sokoto (fl. & fr. Sept.) *Dalz.* 337 ! Stirling Hill *Ansell* ! Nupe *Barter* 1592 ! 1603 ! Kaduna *Lamb* 104 ! **S.Nig.:** *Thomas* 1624 ! A common tropical weed. (See Appendix, p. 227.)
3. **A. zeyheri** *Harv.*—F.T.A. 2 : 170 ; Léonard in Fl. Congo Belge 5 : 228, fig. 13 C. A slightly woody herb, like the preceding species, 1–4 ft. high.
    S.L.: Musaia (fr. Dec., glabrous) *Deighton* 4559 ! **N.Nig.:** Bauchi Plateau (June, July) *Lely* 260 ! 394 ! P. 347 ! Lemne (May) *Lely* 135 ! Extends south to Transvaal and east to A.-E. Sudan.
4. **A. rugosus** (*Willd.*) *DC.* Prod. 2 : 353 (1825) ; F.T.A. 2 : 171, partly ; Chev. Bot. 192 ; Léonard l.c. 5 : 229. *Hedysarum rugosum* Willd. (1803). *Alysicarpus violaceum* of Schindl. in Fedde Rep. 21 : 13 (1925), partly ; F.W.T.A., ed. 1, 1 : 419, partly. A diffusely branched herb, with small red flowers hidden within the glume-like calyx.
    Sen.: *fide* Chev. *l.c.* **Gam.:** Yundum (fr. Dec.) *Austin* 20 ! **Port.G.:** Bissau (fr. Oct.) *Esp. Santo* 2567 ! **Fr.G.:** *fide* Chev. *l.c.* **S.L.:** Musaia (fr. Dec.) *Deighton* 4420 ! **G.C.:** Accra (Nov.) *Brown* 364 ! Ada (fr. June) *Morton* GC 9256 ! Tamale (Oct.) *Williams* 406 ! **Togo:** Oti R., Paliba (fr. Dec.) *Adams & Akpabla* 4109 ! **N.Nig.:** Nupe *Barter* 952 ! Abinsi *Dalz.* 620 ! Sokoto (fl. & fr. Apr.) *Dalz.* 523 ! Anara F.R., Zaria (fl. & fr. Oct.) *Keay* FHI 5467 ! **S.Nig.:** Onitsha *Barter* 1777 ! Olokemeji (fr. Nov.) *Onochie* FHI 8132 ! Common throughout tropical Africa.
5. **A. glumaceus** (*Vahl*) *DC.* Prod. 2 : 353 (1825) ; Léonard in Fl. Congo Belge 5 : 231, fig. 13 E. *Hedysarum glumaceum* Vahl (1791). *H. violaceum* of Forsk., not of Linn. *Alysicarpus violaceus* of Schindl., partly (1925). Similar to the preceding species.
    G.C.: Ejura (fl. & fr. Dec.) *Morton* GC 9564 ! 9810 ! Tamale (fr. Dec.) *Morton* GC 9907 ! Deloro-Tankara (fr. Dec.) *Adams* GC 4325 ! **N.Nig.:** Fodama *Moiser* 249 ! Bauchi Plateau (fl. & fr. Oct.) *Lely* P. 814 ! Biu (fl. & fr. Sept.) *Noble* 36 ! Throughout tropical Africa.

79. MELLINIELLA Harms in Engl. Bot. Jahrb. 51 : 359 (1914).

Prostrate herb ; stems glabrous or nearly so ; leaves 1-foliolate, broadly elliptic or elliptic-rounded, subcordate at base, obtuse or rounded at apex, 1–1·5 cm. long, 7–13 mm. broad, glabrous above, appressed-puberulous beneath, with 3–4 pairs of ascending nerves ; stipules scarious, brownish, lanceolate, striate, acuminate, 4–6 mm. long ; flowers arranged in dense several-flowered glomerules ; calyx 4–5 mm. long, deeply 5-lobed, ciliate-pilose ; corolla small ; fruits lanceolate, compressed, about 1 cm. long, 5–8-seeded, puberulous   ..   ..   *micrantha*

M. micrantha *Harms* l.c. 360, fig. 1 (1914) ; Bak. f. Leg. Trop. Afr. 2 : 344. A herb having the appearance of *Alysicarpus*.
    Sen.: Tambacounda *Berhaut* 711 ; 3112. **Fr.Sud.:** Koulikoro (Oct.) *Chev.* 3235. **Togo:** Sausanne-Mangu (Sept.) *Mellin* 121. Also in Ubangi-Shari.

80. URARIA Desv.—F.T.A. 2 : 168.

A woody herb ; stems more or less pubescent ; leaves pinnate, the lower sometimes entire ; leaflets 2–4 pairs and a terminal one, linear-lanceolate to lanceolate, rounded at the base, mucronate, up to about 20 cm. long, strongly reticulate beneath with transverse tertiary nerves ; stipules large, ovate-lanceolate, acutely long-acuminate ; stipels narrowly lanceolate, acuminate ; racemes terminal, elongated, the bracts subpersistent at the base and apex ; calyx-lobes filiform, long-pilose, 3–4 mm. long ; corolla a little longer than the calyx ; fruits much articulated, the segments nearly separated, glabrous   ..   ..   *picta*

U. picta (*Jacq.*) *DC.* Prod. 2 : 324 (1825) ; F.T.A. 2 : 169 ; Chev. Bot. 192 ; Bak. f. Leg. Trop. Afr. 2 : 340 ; Léonard in Fl. Congo Belge 5 : 232, fig. 14. *Hedysarum pictum* Jacq. (1788). Commonly 1–3 ft. high, sometimes taller ; leaves sometimes variegated ; flowers small, pink or purplish in dense villous spike-like racemes which may be over a foot long ; fruits small, pearly-grey ; in grassland.
    Sen.: Sédhiou, Casamance *Chev.* 3400. **Gam.:** Kuntaur *Ruxton* A. 46 ! **Port.G.:** *Esp. Santo* 219 ! Formosa Isl. (June) *Wilford* 489 ! **Fr.Sud.:** *fide* Chev. *l.c.* **S.L.:** Njala (fr. Jan.) *Deighton* 1805 ! Musaia (Dec.) *Deighton* 4461 ! Port Loko (fr. Apr.) *Sc. Elliot* 5880 ! Kaniko (Sept.) *Thomas* 2093 ! Pendembu (fr. July) *Thomas* 852 ! **Lib.:** Cape Palmas *T. Vogel* 31 ! Rocktown (Mar.) *Baldwin* 11637 ! **Iv.C.:** *Scaëtta* 3025 ! Also *fide* Chev. *l.c.* **G.C.:** Achimota *Bally* 15 ! Ashanti (Aug.) *Chipp* 522 ! Damongo *Andoh* FH 5208 ! **Togo:** Lome *Warnecke* 178 ! **Dah.:** *Poisson* ! Atacora Mts. (June) *Chev.* 24185. **Fr.Nig.:** Fada, Gourma (July) *Chev.* 24506. **N.Nig.:** Sherifuri (fl. & fr. June) *Thornewill* 173 ! Ilorin *Thornton* ! Lokoja (Sept.) *Parsons* L. 24 ! Nabardo (Nov.) *Lely* 618 ! **S.Nig.:** Lagos (Apr.) *Punch* ! Abeokuta *Irving* 105 ! Ibadan (fr. Oct.) *Ahmed & Chizea* FHI 20041 ! **Br.Cam.:** Kumba (fr. Oct.) *Olorunfemi* FHI 30764 ! Bamenda (fr. June) *Maitland* 1554 ! Widespread in tropical Africa, also in the eastern tropics and in Australia. (See Appendix, p. 266.)

92. BUXACEAE

Trees or shrubs, rarely herbs. Leaves evergreen, alternate or opposite, simple ; stipules absent. Flowers unisexual, monoecious or dioecious, spicate or fasciculate. Sepals imbricate or absent, usually 4. Petals absent. Stamens 4 or 6, rarely more ; anthers large, sessile or stalked, cells 2-valved or opening lengthwise. Rudimentary ovary sometimes present. Female flowers often larger than the males and fewer or solitary. Sepals as in the males. Ovary superior, 3-celled ; styles undivided. Ovules 1–2, pendulous, anatropous. Fruit a capsule or drupe. Seeds black, shining, with fleshy endosperm and straight embryo.

NOTOBUXUS Oliv.—F.T.A. 6, 1 : 610.

Stamens 6, two single opposite the outer sepals, four in pairs opposite the inner sepals.

A shrub or small tree, glabrous ; leaves elliptic-lanceolate or slightly obovate-elliptic, cuneate at base, long- and gradually-acuminate to a subacute apex, 7–10 cm. long, 2·5–4 cm. broad, entire, bright green ; lateral nerves 5–7 pairs ; petiole 2–4 mm. long ; flowers axillary ; male pedicels up to 1 cm. long ; capsule ovoid, about 1 cm. long ; seeds black, shining   ..   ..   *acuminata*

N. **acuminata** (*Gilg*) *Hutch.* in F.T.A. 6, 1 : 610 (1912) ; *Chev.* Fl. Viv. 1 : 355, fig. 51. *Macropodandra acuminata* Gilg (1899). Evergreen, 2–12 ft. high, in forest ; flowers white.
S.L.: Njala (fl. & fr. Sept.) *Deighton* 2917 ! **G.C.:** Atuna, N.W. Ashanti (fr. Dec.) *Vigne* FH 3506 ! Pamu-Berekum F.R. (fr. Sept.) *Vigne* FH 2517 ! Akatin (fr. Oct.) *Vigne* FH 4041 ! **Dah.:** Kétou to L. Azri, Zagnanado (fl. & fr. Feb.) *Chev.* 23029 ! Also in Belgian Congo.

93. SALICACEAE

Trees or shrubs. Leaves alternate, simple, deciduous ; stipules small or folia-ceous. Flowers unisexual, dioecious, densely arranged in erect or pendulous catkins often appearing before the leaves ; bracts membranous, each subtending a flower. Calyx absent or represented by a cupular disk or 2 glandular scales. Male flowers : stamens 2 or more, free or united ; anthers 2-celled, opening lengthwise. Female flowers : ovary sessile or shortly stipitate, 1-celled, with 2–4 parietal placentas ; style 2–4-fid. Ovules numerous, ascending. Fruit a 2–4-valved capsule. Seeds small, with numerous fine hairs arising from the funicle and enveloping the seed ; endosperm none ; embryo straight.

Widely distributed, but absent from Australasia and the Malay Archipelago. Recog-nized by the minute much-reduced flowers in catkins.

SALIX Linn.—F.T.A. 6, 2 : 317.

By R. D. Meikle

Shrubs or small trees with narrow leaves ; flowers often precocious ; ovary 2-merous seeds covered with hairs.

Leaves acuminate, lanceolate or narrowly ovate, 4–8 cm. long, irregularly serrulate :
  Young leaves and shoots glabrous or very thinly hairy ; filaments of male flowers
    5–6 mm. long   ..   ..   1. *subserrata*
  Young leaves and shoots silky-pubescent or tomentose ; filaments of male flowers
    3–4 mm. long :
    Mature leaves obscurely veined ; male catkins about 4 cm. long ; capsules broadly
      ovoid, 4–5 mm. long   ..   ..   2. *ledermannii*
    Mature leaves distinctly reticulate-veined ; male catkins 1·5–2·5 cm. long ; capsules
      narrow, about 7 mm. long   ..   ..   3. *chevalieri*
Leaves obtuse, oblong or oblanceolate, entire or undulate, 2–5 cm. long, obscurely
  veined ; male catkins few-flowered, about 2–2·5 cm. long ; young leaves and shoots
  glabrous   ..   ..   4. *coluteoides*

1. **S. subserrata** *Willd.* Sp. Pl. 4 : 671 (1806). *S. safsaf* Forsk. ex Trautv. Salic. 6, t. 2 (1836) ; F.T.A. 6, 2 : 318. A riverside bush or small tree, 8–30 ft. high.
    Gam.: *Dawe* 56 ! Throughout tropical Africa ; also in Egypt, Libya and Palestine.
2. **S. ledermannii** *Seemen* in Engl. Bot. Jahrb. 45 : 204 (1910) ; *Chev.* Fl. Viv. 1 : 26. *S. nigerica* Skan in F.T.A. 6, 2 : 321 (1917). A riparian shrub.
    Fr.Sud.: 10 miles E. of Macina on S. bank of R. Niger *Lean* 21 ! **N.Nig.:** R. Benue, Yola (Dec.) *Dalz.* 199 ! Katagum (fl. & fr. Nov.) *Dalz.* 220 ! L. Chad *Talbot* 1493 ! **Br.Cam.:** Bamenda : Lakom, 6,000 ft. *Maitland* 1495 ! Fondong (fl. & fr. Nov.) *Johnstone* 9 ! 224 ! 225 ! Banso, 5,400 ft., by stream with *Osmunda regalis* (fr. June) *Keay* FHI 28453 ! Also in French Cameroons. (See Appendix, p. 271.)
3. **S. chevalieri** *Seemen* in Fedde Rep. 5 : 133 (1908) ; F.T.A. 6, 2 : 320 ; *Chev.* Fl. Viv. 1 : 25, t. 6A ; Aubrév. Fl. For. Soud.-Guin. 326, t. 68, 5–7. *S. sudanica* A. Chev. Bot. 611, name only. A bush or shrub attaining 30 ft., with a trunk as thick as the thigh.
    Fr.Sud.: Upper R. Niger, Badumbé (Dec.) *Chev.* 60. Dialiba to Langana (fl. & fr. Jan.) *Chev.* 262 ! Keniégué (fl. & fr. Jan.) *Chev.* 271 ! Nyamina to Ségou (fl. & fr. Jan.) *Chev.* 2666 ! **Fr.G.:** Soarella, Fouta Djalon *Pobéguin* 181. R. Tinkisso, Dabola *Jac.-Fél.* 1506. (See Appendix, p. 271.)

4. S. coluteoides *Mirb.* in Mém. Mus. Par. 14 : 462 (1827) ; F.T.A. 6, 2 : 323 ; Chev. Fl. Viv. 1 : 25, t. 6B. A tree or shrub, up to 12 ft. high, with reddish-brown, glabrous twigs.
  Sen.: riverbank, Dagana (fl. Oct.) *Roger!* M'bohou (Dec.) *Chev.* 2667! R. Senegal, Kaédi to Podor (Nov.) *Chev.* 2668! **Fr.Sud.:** Kéniégué, R. Niger (Jan.) *Chev.* 270.

For accurate determination it is essential that *Salix* specimens should consist of both catkins and mature leaves. The West African material at present available for examination is totally inadequate, and it is quite possible that many or all the above species may ultimately prove to be forms of the variable *Salix subserrata* Willd.

FIG. 169.—SALIX LEDERMANNII *Seemen* (SALICACEAE).

A, male flower shoot. B, male flower. B₁, stamen. C, female flower shoot. D, female flower. E, longitudinal section of ovary. F, cross-section of ovary showing placentation. G, fruiting branch. H, fruit. I, seed.

94. MYRICACEAE

Trees or shrubs, often aromatic. Leaves alternate, simple, sometimes pinnately lobed, scaly-glandular beneath, exstipulate. Flowers unisexual, monoecious or dioecious, in axillary spikes. Sepals and petals absent, or the female with a few sepal-like whorled bracteoles. Male flower subtended by a solitary bract : stamens 2 or more ; anthers 2-celled. Female flower : ovary sessile, 1-celled ; style short, 2-branched. Ovule 1, erect, basal. Fruit a drupe, often warted, the warts waxy. Seed erect, without endosperm ; embryo straight.

A widely distributed family, often partial to damp places; numerous and variable in S. Africa. Recognized by the glandular leaves, spicate unisexual apetalous flowers and drupaceous fruit.

MYRICA Linn.—F.T.A. 6, 2 : 307.

A tree ; branchlets puberulous and glandular ; leaves oblong or oblong-lanceolate, mostly truncate and unequal-sided at base, rounded to a mucronate apex, 5–9 cm. long, 1·3–3 cm. broad, mostly toothed, glandular on both surfaces ; lateral nerves 12–16 pairs ; flowers monoecious, male spikes axillary or sometimes racemosely arranged on long leafless shoots, about twice as long as the petiole ; stamens about 6, anthers sparingly pubescent ; ovary pubescent and glandular ; fruits ellipsoid, 5 mm. long, closely warted   ..   ..   *arborea*

M. arborea *Hutch.* in Kew Bull. 1917 : 234 ; F.T.A. 6, 2 : 309. A tree reaching 20–30 ft. in height, with rough bark ; frequent in ravines in the grassland of Cameroon Mt.
  Br.Cam.: Cam. Mt., 5,000–8,000 ft. (fl. Dec., Jan., fr. Mar.) *Mann* 1203! 2185! *Dunlap* 185! *Dalz.* 8329! *Mildbr.* 10932! *Maitland* 457! **F.Po:** *Tessmann* 2894.
  [The following specimens have the branchlets, petioles and midrib beneath densely pilose ; they may prove to represent a distinct species but better material is required :—**Br.Cam.:** Cam. Mt., 7,600 ft. *Maitland* 1028! Nchan, Bamenda (fr. June) *Maitland* 1598 !]

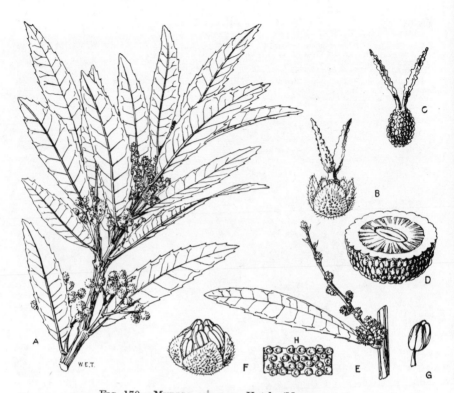

FIG. 170.—MYRICA ARBOREA *Hutch*. (MYRICACEAE).

A, female shoot. B, female flower. C, pistil. D, cross-section of drupe. E, male shoot. F, male flower. G, stamen. H, glands on leaf.

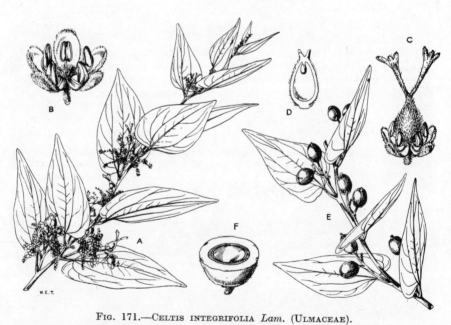

FIG. 171.—CELTIS INTEGRIFOLIA *Lam*. (ULMACEAE).

A, flowering branch. B, male flower. C, hermaphrodite flower. D, section of ovary showing pendulous ovule. E, fruiting branch. F, cross-section of fruit.

95. ULMACEAE

Trees or shrubs. Leaves alternate, simple, often unequal-sided ; stipules paired, caducous. Flowers fasciculate, hermaphrodite or unisexual. Calyx herbaceous, lobes imbricate, persistent. Petals absent. Stamens inserted at the bottom of the calyx, erect in bud, opposite the calyx-lobes ; anthers 2-celled, opening lengthwise. Ovary of 2 connate carpels, 1–2-celled ; styles 2, divergent. Ovules solitary, pendulous from near the top. Fruit usually compressed, membranous, dry or thinly fleshy, often winged or appendiculate. Seed without endosperm ; embryo straight or curved.

Mainly in the Northern Hemisphere. Flowers very small and reduced ; leaves unequal-sided ; ovary 1–2-celled, with solitary ovule.

Fruit a drupe ; flowers axillary on the current year's shoots :
  Branchlets unarmed :
    Cotyledons very broad ; male calyx-lobes imbricate ; leaves entire or toothed
                                              1. Celtis
    Cotyledons very narrow ; male calyx-lobes induplicate-valvate ; leaves toothed
                                              2. Trema
  Branchlets armed with axillary spines ; leaves entire ; stipules enclosing the terminal
    bud, caducous   ..   ..   **3. Chaetacme**
Fruit a rounded flat stipitate samara with broad membranous wings ; flowers in the
    axils of fallen leaves on previous season's growth   ..   ..   ..   **4. Holoptelea**

1. CELTIS Linn.—F.T.A. 6, 2 : 2.

Styles not divided ; leaves gradually very long-acuminate :
  Leaves serrate except towards the base, pubescent beneath at least on the nerves, elliptic-ovate or ovate to ovate-lanceolate, 1·5–8 cm. long, 1–4 cm. broad, basal lateral nerves ascending well into the upper half of the lamina, upper lateral nerves 1–2 on each side of the midrib ; female (or hermaphrodite) flowers usually solitary ; ovary densely pubescent ; fruits subglobose, about 5 mm. diam., on pedicels 1–1·5 cm. long   ..   ..   1. *africana*
  Leaves entire, soon glabrous, ovate-lanceolate or oblong-lanceolate to lanceolate 8–14 (–18) cm. long, 2·5–4·5 (–7·5) cm. broad, basal lateral nerves not extending into the upper half of the lamina, upper lateral nerves 3–6 on each side of the midrib ; female (or hermaphrodite) flowers 2–3 or more together ; ovary glabrous ; fruits ovoid, pointed at the apex, 5–6 mm. long, on pedicels up to 8 mm. long
                                                      2. *durandii*
Styles divided, bifid or bilobed :
  Leaves scabrid on both surfaces, ovate, often broadly so, obliquely and broadly rounded or almost subcordate at base, acutely acuminate, entire, 4–9 cm. long, 3–5 cm. broad, prominently 3–5-nerved from the base, hairy in the axils of the nerves beneath ; young shoots thinly pubescent ; fruits ovoid-ellipsoid, puberulous when young, 8–10 mm. long, on pedicels about 5 mm. long ; tree in savannah regions   ..   ..   3. *integrifolia*
  Leaves never scabrid above, and only rarely (No. 7) slightly scabrid beneath, more or less elliptic or obovate, usually cuneate at base, often toothed ; forest trees :
    Leaves softly pubescent beneath, normally entire (sometimes serrate on epicormic shoots), venation markedly parallel, lateral nerves 3–5 on each side of midrib ; lamina elliptic or subovate, acutely acuminate, 6–15 cm. long, 3–7·5 cm. broad ; young branchlets densely pubescent ; stipules ovate-lanceolate, nearly 1 cm. long ; ovary densely pubescent ; fruits very shortly pedicellate, warted when old, about 6 mm. long ..   ..   4. *zenkeri*
    Leaves glabrous or sparingly pubescent beneath, serrate in the upper half, or if entire then glabrous ; venation not markedly parallel :
      Basal lateral nerves not, or only slightly, extending into the upper half of the lamina, upper lateral nerves 3–5 on each side of midrib ; branchlets and leaves pubescent at least when young ; leaves obovate to oblong-obovate, obtusely acuminate, mostly coarsely toothed towards the apex, 3·5–14 cm. long, 1·5–7 cm. broad (juvenile leaves sometimes larger), closely reticulate beneath ; ovary glabrous, style arms linear ; fruits sharply 3–4-angled, 1–1·3 cm. long, on slender pedicels up to 2 cm. long   ..   ..   5. *mildbraedii*
      Basal lateral nerves ascending nearly to the apex of the lamina, upper lateral nerves 1–3 on each side of the midrib ; branchlets puberulous to glabrous ; leaves glabrous ; fruiting pedicels stout :
        Leaves usually coarsely toothed in the upper half, obovate-elliptic or elliptic, obtusely acuminate, 5·5–15 cm. long, 2–6·5 cm. broad, loosely reticulate beneath ; ovary glabrous ; fruits ovoid, 7–8 mm. long   ..   ..   6. *brownii*

Leaves entire, broadly elliptic, obtusely acuminate, 8–16 (–18) cm. long, 5·5–8 (–10) cm. broad, 3-nerved almost to the apex, slightly scabrid beneath at least when young ; ovary pubescent ; fruits ellipsoid, 15–20 mm. long　　　7. *adolfi-friderici*

1. **C. africana** *Burm. f.* Prod. Fl. Cap. 31 (1768) ; Brenan in Mem. N.Y. Bot. Gard. 9, 1 : 75 (1954). *C. kraussiana* Bernh. (1845)—F.T.A. 6, 2 : 3 ; Hauman in Fl. Congo Belge 1 : 43. *C. rhamnifolia* of Presl, partly (*Drege* 8261) ; of Fl. Cap. 5, 2 : 528 ; of Sillans in Mém. Soc. Bot. Fr. 1952 : 102 ; not *C. rhamnifolia* Presl (1845), illegitimate name for *Rhamnus prinoides* L'Hérit. (see p. 670). An understorey forest tree, to 50 ft. or more high ; slash mottled dark brown and grey.
G.C.: Moerewe, Ashanti (fr. May) *Irvine* 2526 ! Extends from French Cameroons eastwards to A.-E. Sudan and southwards to Angola and S. Africa ; also in Arabia.

2. **C. durandii** *Engl.*—F.T.A. 6, 2 : 4 (incl. var. *ugandensis*) ; Aubrév. Fl. For. C. Iv. 1 : 28, t. 1, 1–2 ; Hauman l.c. 42. A forest tree to 70 ft. or more high ; slash pale grey, thickly speckled with chocolate-brown.
Iv.C.: Man (Mar.) *Aubrév.* 1051 ! **Br.Cam.:** Kumba (fr. June) *Olorunfemi* FHI 30631 ! Also in French Cameroons, Ubangi-Shari, Belgian Congo, Uganda, Kenya, Tanganyika, Angola and S. Tomé.

3. **C. integrifolia** *Lam.*—F.T.A. 6, 2 : 7 ; Chev. Bot. 589, partly ; Aubrév. l.c. 1 : 26, t. 1, 3–4 ; Fl. For. Soud.-Guin. 328, t. 68, 1–2. A large spreading tree, to 80 ft. high and 15 ft. girth ; mainly in the drier savannah regions.
Sen.: *Roger* ! *Perrottet* 298 ! Walo *Leprieur* ! Djebelor *Aubrév.* **Gam.:** Bathurst *Don* ! Kombo *Heudelot* 104 ! Genieri (Feb.) *Fox* 32 ! **Fr.Sud.:** Djenné (fr. June) *Chev.* 1136 ! Kita *Dubois* 204 ! **Iv.C.:** Dindéresso *Aubrév.* 1957 ! Groumania *Aubrév.* 782. Bobo Dioulasso *Aubrév.* 1825 ! **G.C.:** Salaga (fl. & young fr. Apr.) *Dalz.* 13 ! Kurumbugu (fr. May) *Kitson* 750 ! Bamboi (fr. May) *Vigne* FH 3840 ! **N.Nig.:** Sokoto (fl. & young fr. Feb.) *Lely* 805 ! Borgu *Barter* 772 ! Agaie (Jan.) *Yates* 1 ! Bauchi Plateau (fl. & young fr. Mar.) *Lely* P. 182 ! Yola (Feb.) *Dalz.* 152 ! **S.Nig.:** Abeokuta *Irving* 150 ! Extends to Arabia and Uganda. (See Appendix, p. 271.)

4. **C. zenkeri** *Engl.*—F.T.A. 6, 2 : 6 ; Aubrév. Fl. For. C. Iv. 1 : 27, t. 1, 5 ; Hauman l.c. 45. *C. soyauxii* Engl. (1900), see note after next species. *C. rugosa* A. Chev. Bot. 589, name only, not of Willd. (1806). *C. integrifolia* of Chev. l.c., partly (*Chev.* 21788). A forest tree, to 90 ft. high, rarely more, with rough bark and long buttresses ; slash with alternating layers of cream and brown.
Fr.G.: Kissidougou (fr. Apr.) *Adam* 4406 ! **S.L.:** Kondama *Deighton* 3199 ! **Iv.C.:** Bouroukrou *Chev.* 16130 ! Kouroukoro (young fr. May) *Chev.* 21788 ! Anoumaba *Chev.* 22393 ! **G.C.:** Agogo, Akim (fr. Apr.) *Irvine* 964 ! Fumisua (fr. Mar.) *Andoh* FH 4976 ! **N.Nig.:** Okene *Symington* FHI 5671 ! **S.Nig.:** Olokemeji (Mar.) *Punch* 108 ! Aladin *Foster* 194 ! Ibadan South F.R. *Keay* FHI 22813 ! Sapoba *Kennedy* 1872 ! 1927 ! **Br.Cam.:** Likomba (fr. Dec.) *Mildbr.* 10764 ! Extends to Belgian Congo, A.-E. Sudan, Uganda, Tanganyika and Angola. (See Appendix, p. 272.)

5. **C. mildbraedii** *Engl.* Bot. Jahrb. 43 : 309 (1909) ; Hauman l.c. 45. *C. soyauxii* Engl. (1900), partly—F.T.A. 6, 2 : 5 ; Aubrév. l.c. 1 : 28, t. 2, 1–6. *C. compressa* A. Chev. (1917)—F.T.A. 6, 2 : 355 ; Chev. Bot. 588. A forest tree, to 120 ft. high, with clean straight bole, well-developed buttresses and relatively smooth bark compared with the preceding species ; slash with alternating layers of cream and brown.
Iv.C.: Bouroukrou (fr. Dec.–Jan.) *Chev.* 16152 ! Morénou (fl. & fr. Nov.) *Chev.* 22460 ! Rasso (Dec.) *Aubrév.* 561 ! **G.C.:** Dunkwa (fl. & fr. Feb., fr. May) *Vigne* 120 ! 257 ! FH 881 ! Obuasi (fr. Jan.) *Soward* 664 ! Imbraim (fl. & fr. Mar.) *Thompson* 22 ! Prestea (fr. June) *Green* 902 ! **S.Nig.:** Akure F.R. (fr. Aug.) *Jones* FHI 20703 ! Sapoba *Kennedy* 1873 ! 1926 ! 2603 ! Ukunzu to Obomkpa, Asaba (fr. Aug.) *Onochie* FHI 33429 ! Ikom *Catterall* 5 ! Extends to Belgian Congo, A.-E. Sudan, Uganda, Kenya, Tanganyika and Angola, also on S. Tomé. (See Appendix, p. 272.)
[Of the four specimens cited by Engler in his original description, I consider that the name *C. soyauxii* Engl. must be restricted to *Soyaux* 202 and *Welw.* 6285 ; these specimens, as Rendle (F.T.A. 6, 2 : 6) pointed out, are conspecific with *C. zenkeri* Engl. The Kew duplicates of the other two specimens (*Welw.* 6284 and 6298) are the species referred to here as *C. mildbraedii* Engl. (i.e. *C. soyauxii* of Rendle in F.T.A. 6, 2 : 5). See also note by Hauman in Fl. Congo Belge 1 : 46.]

6. **C. brownii** *Rendle* in J. Bot. 53 : 298 (1915) ; F.T.A. 6, 2 : 10. *C. scotellioides* A. Chev. in Bull. Soc. Bot. Fr. 61, Mém. 8 : 299 (1917) ; Chev. Bot. 590 ; Brenan in Kew Bull. 1950 : 219. *C. rendleana* G. Tayl. in Exell Cat. S. Tomé 302 (1944). *C. crenata* A. Chev. Bot. 588, name only ; Leroy in Bull. I.F.A.N. 10 : 217 (1948). *C. prantlii* of F.T.A. 6, 2 : 8 ; of F.W.T.A., ed. 1, 1 : 423 ; Aubrév. l.c. 1 : 28, t. 2, 10–12 ; of Hauman l.c. 43 ; not of Priem. ex Engl. *C. philippensis* of Leroy l.c. 10 : 212–234, partly ; of Berhaut Fl. Sen. 144 ; not of Blanco. An understorey tree, to 50 ft. high (rarely more), with short bole and spreading crown ; slash mottled light brown ; in drier parts of the forest regions.
Sen.: Sangalkam *Berhaut* 1341. **Iv.C.:** Droupleu to Zoanlé, Mt. Sassandra (fr. May) *Chev.* 21459 ! Yaboisso to Daoukrou, Morénou (fr. Dec.) *Chev.* 22488 ! **G.C.:** Brahabum (fr. Aug.) *Vigne* FH 919 ! Abetifi–Aduamoa road, Kwahu *Irvine* 1729 ! Adeiso *Irvine* 2418 ! Bobiri F.R. (Mar.) *Andoh* FH 4956 ! **Togo:** Kersting A. 734 ! **Dah.:** Zagnanado *Chev.* **S.Nig.:** Lagos (May) *Foster* 79 ! *Rowland* ! Yoruba *Barter* 3413 ! Mamu F.R., Ibadan (Mar.) *Keay* FHI 22530 ! Akure F.R. (fr. Aug.) *Jones* FHI 20705 ! Idanre Hills (fr. Jan.) *Brenan* 8718 ! **Br.Cam.:** Likomba *Mildbr.* 10749 ! Ebonji, Kumba (fr. May) *Olorunfemi* FHI 30599 ! Buwe, 4,000 ft. (fr. Mar.) *Johnstone* 294 ! **F.Po:** Mann 276 ! Also in French Cameroons, Ubangi-Shari, A.-E. Sudan, Gabon, Belgian Congo, Uganda and Angola. (See Appendix, p. 272.)
[Leroy (l.c.) draws attention to the relationships between this plant and several other species all of which he groups under *C. philippensis* Blanco. According to his synthetic concept the species ranges from Senegal through Africa and Asia to Australia and Polynesia, and varies, particularly in leaf-shape and texture and in habit quite considerably. The more orthodox taxonomic treatment retained here appears, however, to be the more natural.]

7. **C. adolfi-friderici** *Engl.* Bot. Jahrb. 43 : 308 (1909) ; F.T.A. 6, 2 : 9 (as " *C. adolphi-friderici* ") ; F.W.T.A., ed. 1, 1 : 423 (as " *C. adolfi-frederici* ") ; Aubrév. l.c. 1 : 27, t. 2, 7–9 (as " *C. adolfi-frederici* ") ; Hauman l.c. 44 (as " *C. adolphi-friderici* "). *C. fragifera* A. Chev. (1917)—F.T.A. 6, 2 : 355 ; Chev. Bot. 588. A forest tree, to 110 ft. high ; slash cream with brown spots.
Iv.C.: *Aubrév.* 639 ! 1806 ! Tinguéla to Assikasso (fr. Dec.) *Chev.* 22575 ! **G.C.:** Anwhiaso F.R. (fr. Mar.) *Vigne* FH 222 ! Mramra, Ashanti (fr. May) *Kitson* 1193 ! Bojasso to Adiembra, S. Ashanti (fr. May) *Kitson* 1218 ! Abofaw (Nov.) *Vigne* FH 3133 ! Also in French Cameroons, Ubangi-Shari, A.-E. Sudan, Belgian Congo and Uganda. (See Appendix, p. 271.)

2. TREMA Lour.—F.T.A. 6, 2 : 10.

Branchlets elongated, softly pubescent or tomentose ; leaves distichous, ovate, rounded or subtruncate, at base caudate-acuminate, 6–12 cm. long, 2·5–5 cm. broad, closely serrulate pubescent or softly tomentose ; flowers polygamous, glomerate in the leaf-axils, very small ; drupe subglobose, about 3 mm. long, glabrous .. *guineensis*

T. guineensis (*Schum. & Thonn.*) *Ficalho*—F.T.A. 6, 2 : 11 ; Aubrév. Fl. For. C. Iv. 1 : 30, t. 3, 4–6 ; Hauman in Fl. Congo Belge 1 : 48, t. 8. *Celtis guineensis* Schum. & Thonn. (1827) ; Chev. Bot. 589. A shrub or tree, to 40 ft. high ; mainly in regrowth vegetation in the forest regions, where it is often abundant and grows rapidly ; also in moist situations in the savannah regions.
Sen.: *Heudelot* 50 ! **Gam.:** *Leprieur* ! **Fr.Sud.:** Bamako (May) *Hagerup* 48 ! **Port.G.:** Bissau (Apr.)

T. Vogel 135! 138! Rotomba Isl. (Mar.) *Kirk* 16! Freetown (fl. & fr. Feb.) *Dalz.* 967! Kamalo (May)
Thomas 399! **Lib.:** Monrovia (July) *Linder* 6! Ganta *Harley*! Jabroke, Palipo (fl. & fr. July) *Baldwin*
6439! 6448! **Iv.C.:** Fétékro (July) *Chev.* 22166! **G.C.:** Dunkwa (July) *Vigne* FH 914! Ejura (Aug.)
Chipp 743! Okraji, Kwahu (fl. & fr. May) *Kitson* 1158! Yegi, N.T. (July) *Pomeroy* FH 1246! **Togo:**
Lome *Warnecke* 184! 334! Kpedsu (fl. & fr. Dec.) *Howes* 1056! **Dah.:** Djougou to Pobégou *Chev.* 23933.
N.Nig.: Nupe *Barter* 1499! Lokoja (May, Dec.) *Dalz.* 156! *Elliott* 232! Katagum *Dalz.* 337! Bauchi
Plateau (fl. & fr. Jan.) *Lely* P. 125! **S.Nig.:** Lagos (fl. & fr. Mar., June) *Dalz.* 1172! 1404! Okomu F.R.
(Feb.) *Brenan* 9143! Awka (Sept.) *Jones* FHI 6709! Oban *Talbot*! **Br.Cam.:** Buea (Jan.) *Maitland*
275! Barombi *Preuss* 271! Ndu, Bamenda (May) *Johnstone* 106/31! **F.Po:** *T. Vogel* 47! *Mann* 204!
Widespread throughout tropical Africa, Natal, Madagascar and Arabia. (See Appendix, p. 272.)

Leroy in Fl. Madag. 54 : 10–11 (1952) treats this species as a synonym of the Asiatic *T. orientalis* (Linn.)
Blume. This may well be correct, but as a number of distinguishable taxa may be involved within this
aggregate species, it seems preferable not to replace in our Flora the well-known name *T. guineensis* until the
whole group has been more thoroughly and critically revised.

3. CHAETACME Planch.—F.T.A. 6, 2 : 13.

Branchlets zigzag, scabrid, armed with axillary spines 1·5–2 cm. long ; leaves elliptic,
mucronate, rounded at the base, about 7 cm. long and 4 cm. broad, entire (serrate in
sucker shoots), glabrous and often shining above, with about 8–10 pairs of lateral
nerves ; flowers monoecious or dioecious, in short axillary cymes ; fruits ellipsoid,
about 8 mm. long, tipped by the 2 long hairy style-arms about 2 cm. long   *aristata*

C. aristata *Planch.* in Ann. Sci. Nat., sér. 3, Bot. 10 : 341 (1848) ; *Chev.* Bot. 590 (incl. var. *crenata*) ; G. Tayl.
in Exell Cat. S. Tomé 301.   *C. serrata* Engl. (1900).   *C. microcarpa* Rendle in F.T.A. 6, 2 : 13 (1916) ;
F.W.T.A., ed. 1, 1 : 423 (incl. var. *crenata* Hutch. & Dalz.) ; Aubrév. Fl. For. C. Iv. 1 : 32, t. 3, 7–9 ;
Hauman in Fl. Congo Belge 1 : 51. A shrub or small tree, to 25 ft. high ; leaves dark glossy green above,
leathery ; flowers yellow-green ; fruit cream or pale orange ; in savannah, forest margins and
regrowth.
Iv.C.: confluence of R. Sassandra and R. Bafing, Dotou (fr. May) *Chev.* 21782! Orodougou *Chev.* 21824!
Languira to Bouakro *Chev.* 22213. Morénou *Chev.* 22417. **G.C.:** Kwase (fr. Sept.) *Chipp* 575! Dodowah
(June) *Moor* FH 302! Achimota (fl. July, fr. Aug.) *Milne-Redhead* 5121! 5121*a*! **Togo:** Abohire (fl. Oct.,
fr. Nov.) *St. C. Thompson* FH 3631! **Dah.:** Zagnanado *Chev.* 23031. **S.Nig.:** *Foster* 84! Ohumbe F.R.,
Abeokuta Prov. (fr. Dec.) *Onochie* FHI 31899! Olokemeji (fr. Nov.) *Jones* FHI 14533! FHI 27249! Also
in French Cameroons and extending to A.-E. Sudan, east and central tropical Africa, S. Africa and
Annobon. (See Appendix, p. 272.)

4. HOLOPTELEA Planch.—F.T.A. 6, 2 : 1.

Branchlets glabrous ; leaves elliptic or ovate, shortly acuminate, rounded or subcordate at base, 6–12 cm. long, 2·5–6 cm. broad, entire, glabrous ; lateral nerves
5–8 pairs ; petiole 1 cm. long ; stipules minute ; sepals and anthers puberulous ;
fruits broadly obovate or orbicular, rather deeply emarginate, shortly cuneate to the
stipitate base, about 4 cm. diam., wings membranous, puberulous, striately marked
with numerous radiating lines from the oblique fruit body ..   ..   .. *grandis*

H. grandis (*Hutch.*) *Mildbr.* in Notizbl. Bot. Gart. Berl. 8 : 53 (1921) ; Aubrév. Fl. For. C. Iv. 1 : 30, t. 3, 1–3 ;
Hauman in Fl. Congo Belge 1 : 40.   *Hymenocardia grandis* Hutch. (1911)—F.T.A. 6, 1 : 649.   *Holoptelea
integrifolia* of F.T.A. 6, 2 : 2, not of (Roxb.) Planch. (1848). A large, deciduous, forest tree, 80–160 ft.
high ; bark rough, more or less orange-coloured ; flowers green ; in the drier parts of the forest region.
Iv.C.: Aubrév. 1058! Various localities *fide* Aubrév. *l.c.* **G.C.:** Pamu-Berekum F.R., Ashanti *Vigne*
FH 2504. **Togo:** Atakpame *Doering* 347. Njamassila (fr. Jan.) *Doering* 157. Bagu, Sokode (fr. Feb.)
Kersting A. 301! Afem (fr. Jan.) *Kersting* A. 515! **N.Nig.:** Ibaji-Ojuku F.R. (fr. Feb.) *J. E. Taylor*
L. 8! **S.Nig.:** Olokemeji (fr. Mar.) *Kennedy* 2261! Gambari Group F.R., Ibadan (fr. Feb.) *W. D. Mac-
Gregor* 481! *Punch* 151! Ife (Feb.) *Brenan* 8982! Etemi, Ijebu Ode (fr. May) *Obi & Tamajong* FHI
16947! Also in French Cameroons, A.-E. Sudan, Uganda, Belgian Congo and Cabinda (See Appendix,
p. 272.)

96. MORACEAE

Trees or shrubs or rarely herbs, often with milky juice. Leaves alternate,
rarely opposite, simple ; stipules paired, often caducous and leaving a scar.
Flowers much reduced, often in heads, disks or hollow receptacles, unisexual.
Calyx-lobes usually 4, sometimes reduced, imbricate or valvate. Stamens
usually equal in number and opposite to the sepals ; filaments inflexed or erect
in bud ; anthers 2-celled. Female flower with superior or inferior ovary, of
2 carpels, one often abortive, usually 1-celled ; styles mostly 2, filiform ; ovule
solitary, pendulous, rarely basal and erect. Fruit a small achene, nut or drupe.
Seed with or without endosperm, often with curved embryo.

Mainly in the tropics ; recognized by the milky juice, the prominent stipules, which
leave a scar on falling, and the minute unisexual flowers often arranged on variously
shaped receptacles.

*Leaves not digitately 5–15-foliolate :
†Male and female flowers on separate inflorescences ; dioecious or rarely monoecious :

Inflorescences of male flowers not paniculate ; ovule pendulous : [1]
　Male flowers in catkin-like spikes ; stamens reflexed in bud :
　　Spikes of male flowers very distinctly pedunculate ; forest trees with leaves more
　　　than 4 cm. broad :
　　　Leaves strongly 3-nerved from the base ; spikes of male flowers up to 2 cm. long
　　　　　　　　　　　　　　　　　　　　　　　　　　　　　　　　1. **Morus**
　　　Leaves pinnately nerved ; spikes of male flowers up to 20 cm. long
　　　　　　　　　　　　　　　　　　　　　　　　　　　　　　2. **Chlorophora**
　　Spikes of male flowers sessile, up to 5 cm. long ; shrub up to 3 m. high, with leaves
　　　less than 4 cm. broad　　..　　..　　..　　..　　3. **Neosloetiopsis**
　Male flowers capitate on a flattened receptacle or in globose, ellipsoid or obovoid
　　heads ; stamens erect in bud :
　　Male flowers on a flattened receptacle　　..　　..　　..　　..　　9. **Antiaris**
　　Male flowers in globose, ellipsoid or obovoid heads..　　..　　..　　10. **Treculia**
　Inflorescence of male flowers paniculate ; female flowers in pedunculate heads ;
　　ovule erect ; leaves coarsely serrate ; see below　..　　..　　11. **Myrianthus**
†Male and female flowers in the same inflorescence, on an open, flattened or concave
　receptacle, or in a closed bag-like receptacle ; ovule pendulous :
　Receptacles bag-like, mostly fleshy, hollow, globose or obovoid, closed at the top
　　except for a small ostiole (mouth) ; flowers numerous, hidden in the receptacle ;
　　stamens erect　　..　　..　　..　　..　　..　　..　　..　　..　　6. **Ficus**
　Receptacles not bag-like, wide open at the top :
　　Stamens reflexed in bud :
　　　Receptacles mostly with 2 or more radiating linear bract-arms, or if without arms
　　　　then female flowers several and scattered over much of the receptacle ; female
　　　　flowers 1–several ; styles simple or with 2 arms ; endocarp crustaceous, ejected
　　　　elastically from the exocarp which remains in the receptacle ; herbs or shrubs
　　　　　　　　　　　　　　　　　　　　　　　　　　　　　　　　4. **Dorstenia**
　　　Receptacles without bract-arms ; female flower solitary in the centre of the
　　　　receptacle ; styles with 2 arms ; fruit remaining within the enlarged receptacle ;
　　　　shrubs ..　　..　　..　　..　　..　　..　　..　　..　　..　　5. **Craterogyne**
　　Stamens erect in bud ; female flower solitary in each receptacle ; styles with 2 arms :
　　　Low woody plants of herbaceous habit ; fruits not adnate to the receptacle
　　　　　　　　　　　　　　　　　　　　　　　　　　　　　　　　8. **Scyphosyce**
　　　Tall tree ; fruits sunk in the receptacle and adnate to it ..　　..　　7. **Bosqueia**
*Leaves digitately 5–15-foliolate, except on seedlings ; dioecious ; male flowers in
　panicles ; stamens erect in bud ; female flowers in pedunculate heads ; ovule erect
　from the base of the ovary cell ; style 1 :
　Leaflets 5–7 ; sepals free ; stamens 2–4 ; style short, with a broad lanceolate stigma
　　　　　　　　　　　　　　　　　　　　　　　　　　　　　　11. **Myrianthus**
　Leaflets 12–15 ; sepals united ; stamen solitary ; style long, with a brush-like stigma
　　　　　　　　　　　　　　　　　　　　　　　　　　　　　　12. **Musanga**

　　Besides the above indigenous genera, the following have been introduced into West Africa :—*Artocarpus communis* J. R. & G. Forst. (syn. *A. incisa* (Thunb.) Linn. f.), the Bread-fruit, and *A. heterophyllus* Lam. the Jack-fruit.

Excluded genus.

Decorsella A. Chev. in Bull Soc. Bot. Fr. 61, Mém. 8 : 297 (1917) ; F.T.A. 6, 2 : 357.　This genus was omitted
　from the first edition of the F.W.T.A.　It is **Gymnorinorea** *Keay* (Violaceae), F.W.T.A., ed. 2, 1 : 104 ;
　see Keay in Kew Bull. 1955 : 137.

1. MORUS Linn.—F.T.A. 6, 2 : 19.

A tree ; leaves broadly elliptic, cordate, very abruptly acuminate, crenate, 7–11 cm.
long, 5–7·5 cm. broad, thin ; pilose in the axils of the nerves beneath ; nerves 3
from the base, ascending and subparallel ; petiole 1·5 cm. long ; male catkins about
2 cm. long ; females on slender peduncles about 2 cm. long ; syncarp about 1·3 cm.
diam., dry ; calyx-lobes very broad　　..　　..　　..　　..　　..　　*mesozygia*

M. mesozygia *Stapf*—F.T.A. 6, 2 : 21 ; Aubrév. Fl. For. C. Iv. 1 : 38, t. 5 ; Hauman in Fl. Congo Belge 1 : 55 ;
　Chev. in Rev. Bot. Appliq. 29 : 69–74 (incl. var. *sanda* A. Chev.).　A small or sometimes large tree, with
　reddish brown glabrous branches ; usually in the drier types of forest.

　　[1] Plants which come into this part of the key but which have only female flowers or fruits may be keyed
out as follows :—
Female flowers solitary ; styles 2-partite :
　Leaves less than 4 cm. broad ; ovary surrounded by 4 ovate sepals ; shrub up to 3 m. high
　　　　　　　　　　　　　　　　　　　　　　　　　　　　　　3. **Neosloetiopsis**
　Leaves more than 5 cm. broad ; perianth absent, ovary included in and adnate to the receptacle ; large trees
　　　　　　　　　　　　　　　　　　　　　　　　　　　　　　9. **Antiaris**
Female flowers several to very numerous in each inflorescence :
　Leaves strongly 3-nerved from the base ; female flowers and fruits 6–10 ; styles 2-partite ; receptacle not
　　fleshy..　　..　　..　　..　　..　　..　　..　　..　　..　　..　　1. **Morus**
　Leaves pinnately nerved ; female flowers and fruits very numerous ; receptacle fleshy in fruit :
　　Styles simple (or sometimes with a shorter branch) ; female inflorescence a catkin-like spike
　　　　　　　　　　　　　　　　　　　　　　　　　　　　　　2. **Chlorophora**
　　Styles 2-partite ; female inflorescence subglobose, ellipsoid or obovoid　　..　　..　　10. **Treculia**

Sen.: Joal to Gorée *Leprieur* ! Dakar *Chev.* 3089 ; 4502.   Diourbel *Chev.* 3511.   **Port.G.:** Acoco, Formosa (Apr.) *Esp. Santo* 1961 ! **Fr.G.:** Kadé or Kadélé (Feb.) *Maclaud.* Sekourou, Benna *Schnell* 2108. **S.L.:** Makene (Mar.) *Deighton* 3897 ! **Iv.C.:** Zaranou (Mar.) *Chev.* 16267 ! Séguéla *Chev.* 21838.   Fétékoro *Aubrév.* 803.   Nimba Mts. *Schnell* 2832.   **G.C.:** Oduamase (fr. Mar.) *Thompson* ! Ejau, N. Agogo (Jan.) *Vigne* FH 1786 ! Akwaseho (Jan.) *Irvine* 1682 ! **Togo:** Lome *Warnecke* 347 ! Kpeve (Jan.) *Irvine* 1746 ! **S.Nig.:** Ogbomosho *Rowland* ! Onitsha Prov. *Cousens* FHI 36090 ! Also in French Cameroons, A.-E. Sudan, Belgian Congo and Cabinda. (See Appendix, p. 284.)

Besides the above indigenous species, there are two introduced species, natives of temperate Asia, *M. alba* Linn. (White Mulberry) and *M. nigra* Linn. (Mulberry), which are occasionally cultivated in our area.

2. CHLOROPHORA Gaud.—F.T.A. 6, 2 : 21 ; Aubrév. in Rev. Bot. Appliq. 19 : 245–250 (1934).

Lower surface of leaves densely and minutely puberulous (good lens needed) between the close prominulous network of rather broad veins ; lateral nerves 14–18 on each side of the midrib ; lamina oblong-elliptic, 6–16 cm. long and 6–8 cm. broad ; petiole 3–6 cm. long ; leaves of seedlings and saplings larger, with serrate margins, densely hirsute ; male catkins up to 20 cm. long, 6–8 mm. broad ; female catkins 3–4 cm. long, 14–18 mm. diam.; female calyx surrounded at the base by a dense collar of long hairs, ovary substipitate ; infructescence subcylindrical, 4–5 cm. long, about 2 cm. diam., pulpy   ..   ..   *1. excelsa*

Lower surface of leaves glabrous, the sparse hairs confined to the nerves, venation of fine nerves not prominulous ; lateral nerves 6–10 on each side of the midrib ; lamina broadly ovate to suborbicular, 8–17 cm. long, 7–12 cm. broad ; female calyx with few or no hairs at the base, ovary sessile ; otherwise similar to the above   *2. regia*

1. **C. excelsa** (*Welw.*) *Benth.* Gen. Pl. 3 : 363 (1880) ; F.T.A. 6, 2 : 22 (excl. *Dalz.* 966) ; Aubrév. in Rev. Bot. Appliq. 19 : 245–250, t. 3, 8–15 ; Fl. For. C. Iv. 1 : 34–38, t. 4, 8–15 ; Eggeling & Harris in For. Trees & Timbers Brit. Emp. 4 : 83, fig. 13 (excl. female shoot), t. 13. Hauman Fl. Congo Belge 1 : 56. *Morus excelsa* Welw. (1869). *C. alba* A. Chev. (1912). *Antiaris kerstingii* of Chev. Bot. 606, partly (*Chev.* 22485). *C. regia* A. Chev., partly (spec. ex Dah.). A large forest tree, to 160 ft. high, with bole up to 75 ft. or more, shortly buttressed, smooth bark becoming brown and scaling, slash with whitish latex ; dioecious.
   Iv.C.: Sassandra Port *Chev.* 17942 ! Morénou *Chev.* 22485 ! And other localities *fide* Aubrév. *ll.c.* **G.C.:** Kwahu Hills (Mar.) *Johnson* 910 ! Kwaben (Feb.) *Burnett* 52 ! Bobiri F.R. (Mar.) *Andoh* FH 4970 ! **Togo:** Bismarckburg *Büttner* 697. **Dah.:** Abomey (Feb.) *Chev.* 23169 ! Kouandé *Chev.* 24236 ! **N.Nig.:** Agaie (Feb.) *Yates* 20 ! 20*a* ! **S.Nig.:** Lagos *Moloney* ! Ijaiye *Barter* 3330 ! Gambari Group F.R. (Mar.) *W. D. MacGregor* 62 ! 572 ! *Punch* 103 ! **Br.Cam.:** Ambas Bay (Jan.) *Mann* 705 ! **F.Po:** *fide* G. Taylor. Also in French Cameroons, Gabon, Belgian Congo, A.-E. Sudan, E. Africa, Mozambique, Angola and S. Tomé. (See Appendix, p. 274.)

2. **C. regia** *A. Chev.* in Bull. Soc. Bot. Fr. 58, Mém. 8 : 209 (1912), excl. spec. ex Dah. ; Aubrév. in Rev. Bot. Appliq. 19 : 245–250, t. 3, 1–7 ; Fl. For. C. Iv. 1 : 34–38, t. 4, 1–7. *C. excelsa* of F.T.A. 6, 2 : 22, partly (*Dalz.* 966) and of F.W.T.A., ed. 1, 1 : 424, partly. A large dioecious forest tree, similar to the above. **Sen.:** Patako *Adam* 10426. **Gam.:** Rosevear 82 ! Brikama, Tumbiro *Pitt* 644 ! **Port.G.:** Safim Bissau (Mar.) *Esp. Santo* 1896 ! **Fr.G.:** Mamou to Timbo, Fouta Djalon (Mar.) *Chev.* 12464 *bis* ; 12505 *bis.* Kouroussa *Pobéguin* 659. Pita *Pobéguin.* **S.L.:** Freetown (Feb.) *Dalz.* 966 ! Njala (fl. & fr. Mar.), fr. Apr.) *Deighton* 4589 ! 4590 ! 5327 ! *Small* 763 ! 765 ! Rokupr, Magbema (Feb.) *Jordan* 768 ! 772 ! **Lib.:** Dukwia R. *Cooper* 332 ! **Iv.C.:** Zaranou (Mar.) *Chev.* 17627.   Danané *fide* Aubrév. *ll.c.* **G.C.:** Axim *Irvine* 2378 ! Kumasi (Mar.) *Laryea* FH 4967 ! Subiri F.R. (Mar.) *Osei* FH 4944 !

Besides the above indigenous species there is *C. tinctoria* (Linn.) Gaud., a native of tropical America, cultivated in Sierra Leone.

3. NEOSLOETIOPSIS Engl.—F.T.A. 6, 2 : 78.

A dioecious shrub ; branchlets spreading-pubescent ; stipules lanceolate, stiff, 3–8 mm. long ; leaves narrowly elliptic-oblong or oblong-elliptic, obtuse or cuneate at base, caudate-acuminate, 4–11 cm. long, 1–3·8 cm. broad, subcoriaceous, glabrous, with 6–8 looped lateral nerves prominent beneath, petiole 2–3 mm. long, spreading-pubescent ; male flowers in sessile axillary catkins up to 5 cm. long ; female flowers solitary, axillary, on a short bracteate peduncle, ovary about 5 mm. long, surrounded by 4 ovate sepals, style with 2 arms about 1 cm. long   ..   ..   *kamerunensis*

N. kamerunensis *Engl.* Bot. Jahrb. 51 : 426, fig. 1 (1914) ; F.T.A. 6, 2 : 78 ; Hauman in Fl. Congo Belge 1: 82. A much-branched shrub, 4–10 ft. high ; in riverain forest and in forest regrowth. **Fr.G.:** *Heudelot* 870 ! **S.L.:** Lopa, Dama (Nov.) *Deighton* 5881 ! **Lib.:** Peahtah (Oct.) *Linder* 1037 ! **Iv.C.:** *Mangenot.* **S.Nig.:** Ikom *Catterall* IK/16/1934 ! Kwa Falls, Calabar (Mar.) *Brenan* 9246*a* ! 9246*b* ! Also in French Cameroons, French and Belgian Congo and Tanganyika.

4. DORSTENIA Linn.—F.T.A. 6, 2 : 25 ; Engl. Monogr. Afr. Morac. 5 (1898).

* Styles 2-armed ; stems woody or herbaceous, perennial :
 † Receptacles with small scattered bracts on the sides, turbinate, 1–2 cm. diam., margin with numerous very small teeth, not extended into linear lobes ; leaves obovate-elliptic, attenuate at base, long-acuminate at apex, 17–27 cm. long, 5·2–10 cm. broad, lateral nerves 10–14 on each side of midrib :

Stipules subulate, about 1 cm. long and only 1·5 mm. broad at base ; young stems densely pubescent ; lamina narrowed into a long pseudopetiole ..   *1. elliptica*

Stipules triangular-lanceolate, 1·2–1·5 cm. long, 3–4 mm. broad at base, brown, imbricate on the young almost glabrous stems ; lamina not narrowed into a pseudopetiole   ..   ..   *2. sp. A*

†Receptacles without bracts on the sides ; margin usually with 2 or more linear arms (but see No. 3) :

Stems generally herbaceous, procumbent or rhizomatous and sometimes woody below, up to 80 cm. high ; female flowers distributed over much of the receptacle :

Receptacles without linear arms, margin sharply toothed ; receptacles boat-shaped, about 1·5 cm. long, shortly pedunculate, often paired ; leaves obovate to oblong-lanceolate, sometimes irregularly lobulate or deeply lobed towards the obtuse or acuminate apex, 4–11 cm. long, 1·5–5 cm. broad   ..   ..   3. *prorepens*

Receptacles with 1 or more linear arms :

The receptacles orbicular, or 3- to multi-angular, or stellate, or boat-shaped :

Receptacles boat-shaped, about 2·5 cm. long and 1 cm. broad, with a broad irregularly toothed or crenate margin extended at each end into a linear arm 2–5 cm. long ; stems and petioles densely hirsute ; leaves elliptic, narrowed to an obscurely cordate base, apex more or less acuminate, 10–15 cm. long, 7–8 cm. broad   ..   ..   4. *poinsettifolia*

Receptacles orbicular or 3- to multi-angular, or stellate :

Flowers extending to within 1·5 mm. or less of the edge of the receptacle :

Receptacles 1–1·7 cm. diam., with 10–18 subequal arms 2·5–4·5 cm. long ; leaves elliptic or obovate, 11–20 cm. long, 5–11 cm. broad ; stems erect above, to 80 cm. high :

Stems and petioles rather densely hirsute ; stipules subulate-lanceolate, up to 8 mm. long   ..   ..   5. *mannii*

Stems and petioles puberulous ; stipules triangular, short   ..   6. *ophiocoma*

Receptacles 0·6–1 cm. diam., with 6–10 linear arms and sometimes a varying number of intermediate shorter arms ; stems procumbent or rhizomatous, with the erect portion up to about 35 cm. high :

Margin of receptacle not distinct, without or occasionally with a few very short arms between the 6–10 primary arms which are 1·2–2·7 cm. long ; stems densely pubescent when young ; leaves elliptic-obovate, narrowed to a rounded or subcordate base, often lobulate towards the acuminate apex, 4·5–11 cm. long, 2–5 cm. broad   ..   ..   ..   7. *mungensis*

Margin of receptacle usually distinct, up to 1·5 mm. wide, with numerous short arms or teeth between the 6–10 primary arms which are 0·5–1·2 cm. long ; stems densely pubescent ; leaves elliptic, or obovate-elliptic, rarely ovate-elliptic, narrowed to a rounded or subcordate base, acuminate, 9–19 cm. long, 3·5–8 cm. broad   ..   ..   8. *ciliata*

Flowers not extending to the edge of the receptacle, thus leaving a distinct margin 3–10 mm. wide ; arms varying greatly in length, generally a few long with the intervening ones ranging from inconspicuous teeth to moderately long arms :

Receptacles with (1–) 3–4 primary arms (up to 5 cm. long), margin 0·5–1 cm. wide, floriferous part about 0·6–1 cm. diam. ; leaves elliptic or obovate-elliptic, attenuate at base, acuminate, 5·5–19 cm. long, 3·5–7·5 cm. broad
9. *subtriangularis*

Receptacles with 15–30 arms of unequal length, 5 or more of which are distinctly longer than the rest ; margin 0·3–1·4 cm. wide, floriferous part about 1·3–2·5 cm. diam.:

Stipules 6–21 mm. long, filiform, persistent ; stems sparingly pubescent ; receptacle with up to 30 long arms, the longest of which are up to 4 cm. long ; leaves elongate-elliptic to oblanceolate-elliptic narrowed to the base, gradually acuminate, 14–21 cm. long, 5–8·2 cm. broad   10. *multiradiata*

Stipules up to 5 mm. long ; stems usually rather densely pubescent ; receptacle with about 15 arms, the longest of which are up to 2·5 cm. long ; leaves obovate, narrowed to the base, shortly acuminate, 12–19 cm. long, 4·5–8·6 cm. broad   ..   ..   11. *barteri*

The receptacles much elongated, narrowly linear-lanceolate, about 2·5 cm. long, held vertically, with a long (up to about 6 cm.) apical arm and a short (about 1 cm.) without lateral arms :

Leaves oblong-elliptic, narrowed to a rounded or obtuse base, shortly acuminate, sometimes lobulate-toothed in the upper part, up to about 16 cm. long and 6·5 cm. broad   ..   ..   12. *scabra*

Leaves oblanceolate, narrowed to the cuneate base, gradually acuminate, up to about 16 cm. long and 3 (–4) cm. broad ..   ..   ..   13. *tenuifolia*

Stems strongly woody ; erect shrubs 1–4 m. high ; female flowers solitary in the centre of the receptacle :

The receptacles with 2 (–3) subequal arms :

Receptacles linear, with the female flower between the two linear limbs (1–1·4 cm. long and about 3 mm. broad) each of which bears male flowers and tapers off into a linear arm (about 1 cm. long) ; leaves oblanceolate-oblong or broadly

elliptic, obtuse or cuneate at base, rather abruptly acuminate and sometimes lobulate at apex, 9–19 cm. long, 4–6·2 cm. broad .. .. .. 14. *angusticornis*
Receptacles boat-shaped, 0·9–1·5 cm. long and about 5 mm. broad, with an arm 1–2·7 cm. long at each end, and sometimes with a third arm on one of the sides ; leaves rather variable in shape and size, often narrowly oblong-elliptic, oblanceolate or oblong-oblanceolate, acuminate, sometimes slightly lobulate-dentate, 5–15·5 cm. long, 1·5–6·5 cm. broad .. .. 15. *smythei*
The receptacles with 4 or more arms, more or less star-shaped and turbinate in outline :
    Receptacles with 4 arms, rarely with 1–3 subsidiary shorter arms ; floriferous part 5 mm. or less in diam. :
        Leaves oblanceolate, long-cuneate at base, long-acuminate, 8·5–15 cm. long, 2·5–4 cm. broad, main lateral nerves 6–7 on each side of midrib ; arms of receptacle 1·3–2 cm. long .. .. 16. *turbinata*
        Leaves oblong-oblanceolate, obtuse or rounded at base, long-acuminate, 12·5–20 cm. long, 3·8–6 cm. broad, main lateral nerves 10–13 on each side of midrib ; arms of receptacle about 1·5 cm. long .. .. .. 17. *ledermannii*
    Receptacles with (5–) 6–7 primary arms up to 1·2 cm. long and 0–7 subsidiary shorter arms, floriferous part about 1 cm. diam. ; leaves oblong-elliptic to oblong-obovate, obtuse or cuneate at base, rather abruptly long-acuminate, sometimes lobulate-dentate towards apex, 6–17 cm. long, 2–7·3 cm. broad
                                            18. *buesgenii*
*Styles simple ; stems annual, erect, simple, perennating by means of a small tuber :
    Receptacles usually in the form of an isosceles triangle with a long arm at each corner and a fourth arm in the centre of the short upper side, more rarely with 3 or 5–8 arms ; side of receptacle with small teeth between the arms, the long sides 1–1·5 cm. long, arms 1–3 cm. long ; stems, leaves and receptacles rather densely and softly puberulous ; leaves usually more or less elongate-elliptic, subacute at apex, rather irregularly toothed on the margin, 3·5–9 (–10) cm. long, 1–3·2 (–4) cm. broad
                                            19. *walleri*
    Receptacles deeply 3– (or 4–) lobed, the lobes about 4–8 mm. long, each ending in a slender arm 10–13 mm. long, lobes with small teeth between the arms ; stems puberulous ; receptacles and leaves nearly glabrous except on the nerves ; leaves more or less elliptic, obtuse at apex, undulate on margin, 2·5–6·5 cm. long, 1·2–2·5 cm. broad .. .. 20. *preussii*

1. **D. elliptica** *Bureau*—F.T.A. 6, 2 : 31. *D. frutescens* Engl. Monogr. Afr. Morac. 12, t. 2 B (1898) ; F.T.A. 6, 2 : 30. A shrublet with erect pubescent stem to 3 ft. high and horizontal rhizome ; receptacles brown ; in rain forest.
Br.Cam.: Barombi (fl. June, fr. Aug.) *Preuss* 384 ! Johann-Albrechtshöhe (Nov.) *Staudt* 461 ! S. Bakundu F.R. (fl. & fr. Mar.) *Brenan* 9266 ! 9266a ! 9266b ! *Onochie* FHI 30861 ! **F.Po:** *Mann* 64 ! Also in French Cameroons, Gabon and French Congo.
[In the F.T.A. this species is erroneously recorded from French Guinea ; the specimen referred to is from French Congo.]

2. **D. sp. A.** Probably new, but further material required. A shrublet, to 3 ft. high, in damp places in forest.
G.C.: Ankasa F.R. (Dec.) *Vigne* FH 3186 !

3. **D. prorepens** *Engl.*—F.T.A. 6, 2 : 45 (incl. var. *robustior* Rendle) ; Engl. l.c. 18, t. 3 A. Stem half woody, procumbent and rooting, covered, like the petioles and peduncles with brown hairs.
S.Nig.: Okomu F.R., Benin (Dec.) *Brenan* 8460 ! Oban *Talbot* 2316 ! s.n. ! **Br.Cam.:** Cam. Mt., 3,000–5,000 ft. (Feb., Dec.) *Mann* 1956 ! *Preuss* 832. **F.Po:** *Mildbr.* 6355.

4. **D. poinsettifolia** *Engl.*—F.T.A. 6, 2 : 45 ; Engl. l.c. 18, t. 2 A (incl. vars.). A perennial herb, with stems to 12 in. long ; in forest.
Br.Cam.: Buea *Deistel* 435 ! Also in French Cameroons.

5. **D. mannii** *Hook. f.*—F.T.A. 6, 2 : 31 ; Engl. l.c. 18. An erect perennial herb, to 2½ ft. high ; in rain forest.
S.Nig.: Calabar (Feb.) *Mann* 2316 ! *Thomson* 10 ! Oban (Mar.) *Onochie* FHI 7738 ! *Talbot* 673 ! Obom Itiah (Jan.) *Jones* FHI 508 ! **Br.Cam.:** Mamfe (Dec.) *Baldwin* 13826 !

6. **D. ophiocoma** *K. Schum. & Engl.*—F.T.A. 6, 2 : 32 (excl. var. *minor*) ; Engl. l.c. 17, t. 4 C (incl. var. *longipes* Engl.). *D. mannii* of F.W.T.A., ed. 1, 1 : 427, partly. A forest herb, like the preceding.
Br.Cam.: Jongo to Bakingele (Mar.) *Preuss* 1381 ! Bibundi (Nov.) *Mildbr.* 10637 ! Also in French Cameroons.

7. **D. mungensis** *Engl.*—F.T.A. 6, 2 : 34. *D. intermedia* Engl. (1898). *D. mundamensis* Engl. (1902). *D. ophiocoma* var. *minor* Rendle in F.T.A. 6, 2 : 32. *D. mannii* of F.W.T.A., ed. 1, 1 : 427, partly.
S.Nig.: Uquo, Eket (May) *Onochie* FHI 33185 ! **Br.Cam.:** Victoria (Apr.) *Preuss* 1107 ! *Schlechter* 12368 ! Bonakanda, Cam. Mt., 3,800 ft. (Jan.) *Maitland* 925 ! Tiko (Feb.) *Dunlap* 255 ! Mundame to Otam *Schlechter* 12883 ! Mungo (Apr.) *Buchholz* ! Also in French Cameroons.

8. **D. ciliata** *Engl.*—F.T.A. 6, 2 : 31. *D. talbotii* Rendle (1915)—F.T.A. 6, 2 : 56. A rhizomatous perennial herb with ascending stems ; in rain forest.
S.Nig.: Oban *Talbot* 2314 ! **Br.Cam.:** S. Bakundu F.R., Kumba Div. (Mar.) *Brenan* 9261 ! 9261a ! Mundame to Otam *Schlechter* 12888 ! Eko Keyake *Schlechter* 12891 ! Also in French Cameroons.

9. **D. subtriangularis** *Engl.*—F.T.A. 6, 2 : 55 ; Engl. l.c. 15, t. 5 A. *D. piscaria* Hutch. & Dalz. F.W.T.A., ed. 1, 1 : 427 (1928) ; Kew Bull. 1929 : 19. An erect perennial herb, subwoody below, to 2 ft. high ; in forest.
S.Nig.: Oban *Talbot* 2315 ! s.n. ! **Br.Cam.:** Ambas Bay (Feb.) *Mann* 14 ! Bibundi (Nov.) *Mildbr.* 10636 ! Victoria (Feb., Apr.) *Maitland* 390 ! *Schlechter* 12375 ! Tiko (Jan.) *Dunlap* 184 ! S. Bakundu F.R. (Mar., Apr.) *Brenan* 9265 ! *Ejiofor* FHI 29354 ! *Onochie* FHI 30882 ! Also in French Cameroons.

10. **D. multiradiata** *Engl.*—F.T.A. 6, 2 : 54 ; Engl. l.c. 15, tt. 1 D, 3 C. An erect perennial herb, subwoody below, to 2½ ft. high ; in forest.
S.Nig.: R. Ata, below Mt. Kololshe, Obudu (Dec.) *Keay & Savory* FHI 25050 ! **Br.Cam.:** Barombi (May) *Preuss* 204 ! S. Bakundu F.R., Kumba Div. (Mar.) *Brenan* 9436 ! 9436a !

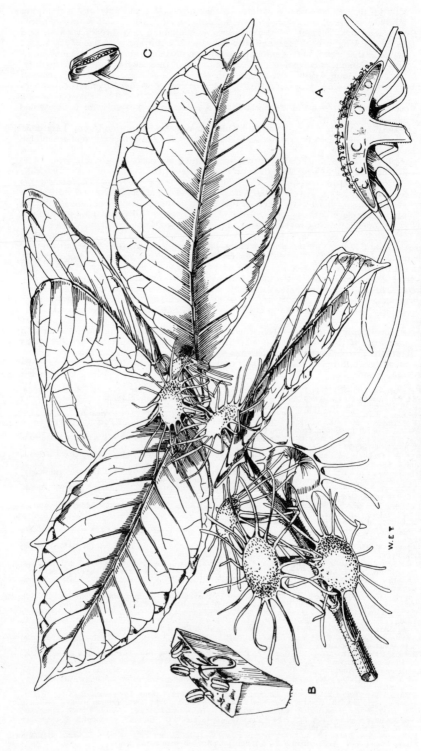

Fig. 172.—DORSTENIA MANNII *Hook. f.* (MORACEAE).

A, section through receptacle. B, portion of receptacle more highly magnified. C, stamen.

W. E. T.

11. **D. barteri** *Bureau*—F.T.A. 6, 2 : 53 ; Engl. l.c. 14. An erect perennial herb, subwoody below, to 2½ ft. high ; in forest.
S.Nig.: Calabar *Robb*! Ifunkpa to Atakom (Dec.) *Holland* 215! Oban *Talbot* 674! s.n.! Aboabam, Ikom (Dec.) *Keay* FHI 28193! **Br.Cam.:** Bulifambo (Mar.) *Maitland* 543! Mimbia (Mar.) *Brenan* 9349! Kembong F.R., Mamfe (Mar.) *Onochie* FHI 31187! **F.Po:** *Barter*! *Mann* 152! Moka (Dec.) *Boughey* 70!

12. **D. scabra** (*Bureau*) *Engl.*—F.T.A. 6, 2 : 51 ; Engl. l.c. 20 (incl. var. *denticulata* Engl.). *D. psilurus* Welw. var. *scabra* Bureau (1873). An erect herb ; in forest.
S.Nig.: Akassa, Nun R. *Barter* 2089! **Br.Cam.:** Kumba Dist. (May) *Olorunfemi* FHI 30573! Also in French Cameroons, Gabon and Belgian Congo.

13. **D. tenuifolia** *Engl.*—F.T.A. 6, 2 : 52. A perennial herb, stems trailing and rooting, leafy parts erect, to 1½ ft. high ; in forest.
S.Nig.: Etemi, Ijebu-Ode (May, Dec.) *Tamajong* FHI 16944! 20275! Idanre F.R. (July) *Ejiofor* FHI 33399! Nikrowa (Dec.) *Brenan* 8465! **Br.Cam.:** Mafura to Mundame (fr. Jan.) *Schlechter* 12920!

14. **D. angusticornis** *Engl.*—F.T.A. 6, 2 : 60. A low shrub, about 2 ft. high ; with papery leaves ; in rain forest.
Br.Cam.: Johann-Albrechtshöhe *Büsgen* 185. S. Bakundu F.R., Kumba Div. (fl. & fr. Mar.) *Brenan* 9489! Also in French Cameroons.

15. **D. smythei** *Sprague*—F.T.A. 6, 2 : 60. *D. aspera* A. Chev. (1912)—F.T.A. 6, 2 : 58. *D. dinklagei* Engl. (1911), not of Engl. (1898). *D. spathulibracteata* Engl. (1913). An evergreen shrub or small tree, to 12 ft. high ; in forest. The leaves are very variable in shape and size ; a form with the leaves truncate and very abruptly acuminate at the apex has been named *D. smythei* var. *deltoidea* (A. Chev.) Hutch. & Dalz. (type *Chev.* 21288 *bis*, from Danané, Iv.C.) ; another specimen (*Linder* 1202, from Gbanga, Lib.) has remarkable obovate-rhomboid leaves up to 11 cm. long and 9 cm. broad.
S.L.: hills near Freetown (Nov.-Jan.) *Deighton* 5612! *Milne-Redhead* 5182! Jepihun (Jan.) *Smythe* 237! Gegbwema (Jan.) *Deighton* 3862! Mamunta (Nov.) *Deighton* 4929! **Lib.:** Peahtah (Oct.) *Linder* 961! Gbanga (Oct.) *Linder* 1199! 1201! 1203! Vahon, Kolahun (Nov.) *Baldwin* 10250! Bangee, Boporo (Nov.) *Baldwin* 10356! 10357! **Iv.C.:** Danané (Apr.) *Chev.* 21197! 21288! 21288 *bis*! Gonhoué to Bampleu (Apr.) *Chev.* 21162! **G.C.:** Abiabo, Axim (Nov.) *Vigne* FH 1428!

16. **D. turbinata** *Engl.*—F.T.A. 6, 2 : 57. A small shrub with short branchlets ; in rain forest.
Br.Cam.: Njoke to Malende *Schlechter* 12871!

17. **D. ledermannii** *Engl.*—F.T.A. 6, 2 : 57. A shrub more than 6 ft. high with branchlets spreading at an acute angle ; in rain forest.
G.C.: Benso (Sept.) *Vigne* FH 4745! **S.Nig.:** Oban *Talbot* 675! Also in French Cameroons.

18. **D. buesgenii** *Engl.*—F.T.A. 6, 2 : 59. *D. obanensis* Hutch. & Dalz. F.W.T.A., ed. 1, 1 : 427 (1928) ; Kew Bull. 1929 : 18. An erect, sparsely branched shrub, to 10 ft. high ; in rain forest.
S.Nig.: Oban *Talbot* 626! 628! s.n.! Aboabam, Ikom (Dec.) *Keay* FHI 28209! 28213! 28214! Also in French Cameroons.

19. **D. walleri** *Hemsl.*—F.T.A. 6, 2 : 68 (incl. var. *minor* Rendle). *D. caulescens* Schweinf. ex Engl. (1894)— Engl. l.c. 22, t. 5 C. *D. gourmaensis* A. Chev. (1912), incl. var. *floribunda* A. Chev. ; Bot. 592 ; F.W.T.A., ed. 1, 1 : 427. Annual stems erect, fleshy, to 15 in. high, produced from a hard tuberous stock ; stems, midribs and arms of receptacle often tinged pink ; in savannah regions, often among rocks, or as epiphyte.
Dah.: Konkobiri (July) *Chev.* 24335! **Fr.Nig.:** Piéga, between Diapaga and Fada (July) *Chev.* 24468 ; 24528! Kompongou (July) *Chev.* 24390! **N.Nig.:** Kwaga, Birnin Gwari (July) *Keay* FHI 25985! Giwa Dist., Zaria (July) *Keay* FHI 25931! Anara F.R., Zaria (May) *Keay & Mutch* FHI 22928! Bauchi Plateau (May) *Lely* P. 317! Also in French Cameroons, Ubangi-Shari, A.-E. Sudan, Tanganyika, N. Rhodesia, Nyasaland and Mozambique ; perhaps also in Belgian Congo (see Hauman in Fl. Congo Belge 1 : 72). (See Appendix, p. 276.)

20. **D. preussii** *Schweinf. ex Engl.*—F.T.A. 6, 2 : 70 (excl. var. *latedentata* Engl.) ; Engl. l.c. 21, t. 8 D. Erect rather slender, fleshy stems, to 10 in. high, produced from a small tuberous stock ; on rocks under shade.
Fr.G.: Gangan, Kindia (Aug.) *Jac.-Fél.* 7050! Kinsan, Kindia (July) *Jac.-Fél.* 1807! Kouroussa (Sept.) *Pobéguin* 1036! **S.L.:** Sugar Loaf Mt. (June) *Preuss.* Hills behind Freetown (Sept.) *Milne-Redhead* 5149! Roboli, Rokupr (July) *Jordan* 285! Kambia, Magbema (June) *Jordan* 446! Kukuna, Bramaia (July) *Adames* 235!
[*D. preussii* var. *latedentata* Engl., from French Cameroons is scarcely distinguishable from *D. cuspidata* Hochst., from Abyssinia ; our plant is also closely related to *D. cuspidata* Hochst. In his Fl. Sen. 173, Berhaut records *D. cuspidata* from Senegal ; I have not been able to see his specimens but they may well be *D. preussii*.]

5. CRATEROGYNE Lanjouw in Rec. Trav. Bot. Neerland. 32 : 272 (1935).

Peduncles 9–25 mm. long ; connective narrow ; leaves not toothed, obovate-elliptic, generally obtuse and unequal at base, caudate-acuminate, 10–25 cm. long, 3–9 cm. broad, with numerous looped lateral nerves ; flowers monoecious, in axillary peduncu-late heads, the males numerous, surrounding a solitary female ; fruiting receptacle ellipsoid, about 13 mm. long   ..   ..   1. *africana*

Peduncles 2–6 mm. long ; connective broad ; leaves usually more or less toothed and sometimes truncate towards the apex, more or less oblong-elliptic, cuneate to obtuse at base, caudate-acuminate, 7–22 cm. long, 2·5–8 cm. broad, with 5–7 main lateral nerves on each side of the midrib ; flowers and fruits similar to above

2. *kameruniana*

1. **C. africana** (*Baill.*) *Lanjouw* in Rec. Trav. Bot. Neerland. 32 : 273, fig. 6 (1935). *Trymatococcus africanus* Baill. (1876) ; Engl. Monogr. Afr. Morac. 28, t. 11 A ; F.T.A. 6, 2 : 75 ; F.W.T.A., ed. 1, 1 : 427. *T. conrauanus* Engl. (1902)—F.T.A. 6, 2 : 77. An erect sparingly branched shrub, 3–10 ft. high ; receptacles bright yellow ; in rain forest.
S.Nig.: Oban *Talbot* 631! 2327! Boje, Ikom Div. (May) *Jones* FHI 5831! Okarara, Calabar (May) *Ujor* FHI 30827! **Br.Cam.:** Cameroon R. (Jan.) *Mann* 723! 2228! S. Bakundu F.R. (fl. Jan.–Apr., fr. Mar.) *Brenan* 9258! *Keay* FHI 28560! Mundame *Staudt* 611! Mbu R., Tinto *Conrau* 130. Manenguba Mt. *Büsgen* 279. Also in French Cameroons, Spanish Guinea and Gabon.

2. **C. kameruniana** (*Engl.*) *Lanjouw* l.c. 274, fig. 7–8 (1935). *Dorstenia kameruniana* Engl. (1895). *Trymato-coccus kamerunianus* (Engl.) Engl. Monogr. Afr. Morac. 29, t. 11 B (incl. var. *welwitschii* Engl.) ; F.T.A. 6, 2 : 76 ; F.W.T.A., ed. 1, 1 : 427. *Dorstenia amoena* A. Chev. (1912)—F.T.A. 6, 2 : 59 ; Chev. Bot. 591 (incl. var. *truncata* A. Chev.). A forest shrub, 3–10 ft. high.
Iv.C.: Dioandougon to Niangouépleu (May) *Chev.* 21528. Niangouépleu to Man (May) *Chev.* 21531! **G.C.:** Banka, Ashanti (Mar.) *Vigne* FH 1864! Nkenkaso (Oct.) *Vigne* FH 2550! Offuasi (Mar.) *Vigne* FH 1899! Also in French Cameroons, Spanish Guinea, Uganda, Kenya, Tanganyika and Angola.

6. FICUS Linn.—F.T.A. 6, 2 : 78.

KEY TO THE SUBGENERA.

Ostiole (mouth) of the receptacle with several bracts visible from the outside and spreading transversely across the orifice :
  Bracts scattered on the peduncle and over the surface of the receptacle as well as by the ostiole, sometimes very small ; leaves scabrid ; receptacles axillary, solitary or paired ..   .. I. SYCIDIUM
  Bracts arranged in a single whorl at the base of the receptacle as well as by the ostiole, none on the sides of the receptacle :
    Male flowers usually with 2 stamens ; receptacles either borne in panicles on the main stem and older branches, or axillary and solitary ; leaves often toothed and scabrid or hairy   ..   .. II. SYCOMORUS
    Male flowers with a single stamen ; receptacles all axillary, solitary or paired ; leaves usually entire and not scabrid (but see Nos. 11 & 12)   .. III. UROSTIGMA
Ostiole (mouth) of the receptacle pore-like and more or less 2-lipped, with all the bracts descending abruptly into the receptacle and not visible from outside ; receptacles solitary, paired or in fascicles   ..   .. IV. BIBRACTEATAE

Subgenus I—SYCIDIUM

Leaves mostly opposite, often with 3-dentate tips, lanceolate, oblong-lanceolate or ovate, 3–10 cm. long, 1·3–4 cm. broad, scabrid on both surfaces, with 5–8 main lateral nerves on each side of midrib ; stipules persistent, lanceolate, about 5 mm. long, ciliate ; receptacles solitary, subglobose, 1·3–2 cm. diam., scabrid ; peduncles 0·6–1·2 cm. long   ..   ..   1. *capreifolia*
Leaves always alternate ; stipules caducous :
  Leaves linear-lanceolate, entire, 9–12 cm. long, 2–3 cm. broad, somewhat scabrid, with 9–10 main lateral nerves on each side of midrib ; receptacles solitary, subglobose, about 1·3 cm. diam., shortly setulose ; peduncles about 0·8 cm. long
                                                                                        2. *acutifolia*
  Leaves ovate or elliptic to obovate-elliptic, sometimes toothed or lobed, never more than 3 times as long as broad, with 3–8 main lateral nerves on each side of midrib ; receptacles paired or solitary ; peduncles 0·3–1·2 cm. long :
    A tree, up to 22 m. high ; main lateral nerves 3–5 (–6) on each side of midrib, ascending at an angle of about 45–50° ; leaves ovate to obovate-elliptic, often toothed, 3·5–17 cm. long, 2·5–11 cm. broad, thick and very scabrid on both surfaces ; leaves on young plants and regrowth shoots often rather deeply 3-lobed ; receptacles obovoid or subglobose, 1–1·3 cm. diam., roughly scabrid ..   .. 3. *exasperata*
    A shrub, usually scrambling ; main lateral nerves 6–8 on each side of midrib, spreading at an angle of about 50–80° ; leaves often lobed, variable, nearly entire and then more or less oblong-elliptic, or pinnately 3–5-lobed, 5–17 cm. long, 2–8 cm. broad, scabrid on both surfaces ; receptacles subglobose or pyriform, up to 3 cm. long and 2·2 cm. diam., scabrid   ..   ..   4. *asperifolia*

Subgenus II—SYCOMORUS

Receptacles borne in panicles on the main stem or older branches :
  Young receptacles setose-pilose ; stipules persistent for some time ; panicles mostly long and slender, with short lateral branches racemosely arranged, borne near the base of the main stem, often with many panicles prostrate on the ground ; receptacles depressed-globose up to 2·5 cm. diam. ; basal bracts 2–5 ; leaves obovate to obovate-elliptic, shortly and subacutely acuminate, 10–37 cm. long, 4·5–18 cm. broad, coarsely undulate-dentate, long-pilose on the nerves beneath   5. *vogeliana*
  Young receptacles tomentose or scaly-puberulous or glabrous ; stipules early caducous ; panicles thyrsoid, not prostrate on the ground :
    Mature receptacles softly tomentose, globose, 2–2·5 cm. diam. ; basal bracts 3, small, soon glabrous ; leaves ovate or oblong-ovate, obtuse or shortly acuminate, 6–25 cm. long, 4–12 cm. broad, coarsely and irregularly repand-dentate, 3-nerved at the base, rather densely pubescent beneath   ..   ..   6. *mallotocarpa*
    Mature receptacles glabrous or puberulous, 2–3 cm. diam. :
      Leaves ovate or ovate-elliptic, up to 25 cm. long and 12 cm. broad, mostly repand-dentate, smooth or rarely slightly scabrid above, pubescent or glabrous beneath ; stipular scars without a fringe of hairs   ..   ..   7. *capensis*
      Leaves obovate-suborbicular, cordate at base, shortly acuminate, 7–17 cm. long and broad, entire, scabrid above, scattered-pilose on the midrib and nerves beneath and sometimes softly puberulous ; stipular scars with a fringe of hairs
                                                                                        8. *mucuso*
Receptacles axillary and solitary, on or towards the base of the young shoots ; leaves cordate at base :

Leaves smooth above, glabrous or sparsely pubescent beneath, very broadly ovate, up to 20 cm. long and 17 cm. broad, mostly coarsely repand-dentate ; receptacles puberulous, 3·5–6 cm. diam., marked with 12–15 longitudinal lines
    9. *vallis-choudae*
Leaves scabrid above, pilose on the nerves beneath, orbicular or ovate, 5–15 cm. long, 4–10 cm. broad, mostly subentire or obtusely serrate ; receptacles densely tomentose, about 4 cm. diam., not marked with longitudinal lines   10. *gnaphalocarpa*

Subgenus III—UROSTIGMA

Receptacles glabrous or very slightly and finely puberulous when mature :
Leaves lanceolate or lanceolate-elliptic, obtuse at both ends, 6–15 cm. long, 2·5–6 cm. broad, entire, glabrous ; lateral nerves 10–16 on each side of midrib ; receptacles pedunculate, globose, nearly 1 cm. diam. ; basal bracts ciliate   13. *verruculosa*
Leaves ovate or rhomboid-ovate, or when elliptic then acuminate :
Receptacles 12–17 mm. diam., on distinct peduncles 4–8 mm. long ; leaves hairy, at least when young, often slightly scabrid, acuminate at apex :
Young leaves with long white spreading hairs on both sides of the midrib and main nerves and on the upper side of the petiole, glabrescent ; leaves oblong to ovate, more or less cordate at base, 8–19 cm. long, 4·5–10·5 cm. broad, with 7–17 lateral nerves on each side of midrib ; saplings and coppice growth with dissected lobed leaves ; a tall forest tree   ..   .. 11. *variifolia*
Young leaves more or less densely puberulous, glabrescent, without long white spreading hairs ; leaves elongate-ovate or elliptic, cuneate to subcordate at base, 8–13 cm. long, 3·5–6·5 cm. broad, with 5–8 lateral nerves on each side of midrib ; a savannah tree   ..   .. 12. *dicranostyla*
Receptacles 8–10 mm. diam., sessile or with peduncles up to 6 mm. long ; leaves glabrous or nearly so, never scabrid, glossy, at least above :
Leaves acutely acuminate, strongly and closely reticulate and shining on both surfaces, ovate-rhomboid, cuneate to subcordate at base, 6–11 cm. long, 3–6 cm. broad, with 9–12 main lateral nerves on each side of midrib ; receptacles 6–8 mm. diam., sessile or subsessile   ..   .. 14. *lecardii*
Leaves obtuse or slightly acuminate, not strongly reticulate, shining on upper surface only, ovate-oblong to oblong-lanceolate, rounded-truncate to deeply cordate at base, 6–18 cm. long, 3–10 cm. broad, with 5–8 main lateral nerves on each side of midrib ; receptacles 8–10 mm. diam., subsessile or with peduncle up to 6 mm. long   ..   .. 15. *ingens* var. *ingens*
Receptacles densely and softly tomentose when mature ; otherwise as above
    15a. *ingens* var. *tomentosa*

Subgenus IV—BIBRACTEATAE

Sect. Axillares. *Receptacles in the axils of the leaves of the young shoots (rarely some extending to the two-year-old wood), solitary or paired.*
*Stipules mostly large and conspicuous, persistent during the flowering and fruiting periods :
Leaves pandurate, cordate-auriculate and obscurely toothed at the base ; receptacles sessile, globose :
Leaves 13–26 cm. broad, 20–45 cm. long, with 5–6 main lateral nerves on each side of midrib ; receptacles about 4 cm. diam. ; basal bracts 3, appressed to receptacle
    16. *lyrata*
Leaves 6–9·5 cm. broad, 18–52 cm. long, with 15–20 main lateral nerves on each side of midrib ; receptacles about 2 cm. diam. ; basal bracts 2, ovate, keeled, coriaceous
    17. *sagittifolia*
Leaves not pandurate, neither cordate nor toothed :
Receptacles covered with large prominent warts ; leaves oblong-elliptic or obovate. oblong, long-acuminate, 15–22 cm. long, 4–8·5 cm. broad ; receptacles sessile, slightly depressed-globose, 1–2·2 cm. diam. ..   .. 22. *praticola*
Receptacles not warted or only very slightly so :
†Receptacles sessile or subsessile, not stipitate above the basal bracts :
Leaves more than 4 cm. long and 2 cm. broad :
Receptacles 1–5 cm. diam. ; leaves mostly more than 8 cm. long and 3 cm. broad, main lateral nerves 4–10 on each side of midrib :
Leaves acuminate, 9–45 cm. long, 4–14 cm. broad ; mature receptacles 1·3 cm. diam. or more :
Base of leaves cuneate ; receptacles 1·3–2·5 cm. diam. :
Main lateral nerves 4–5 on each side of midrib :
Receptacles glabrous, globose, about 1·3 cm. diam. ; leaves elliptic to oblanceolate, 11–24 cm. long, (3–) 5–10·5 cm. broad   18. *camptoneura*
Receptacles densely and shortly pubescent, subglobose, about 2 cm. diam. ; leaves elliptic or oblong-elliptic, 10–18 cm. long, 4–7 cm. broad
    19. *camptoneuroides*

Main lateral nerves 7–10 on each side of midrib :
  Leaves elongate-oblong-elliptic, long-caudate-acuminate, 15–22 cm. long, 4–7 cm. broad ; main lateral nerves 8–10 on each side of midrib, diverging from it at an angle of 60–90° ; receptacles subglobose, 1·3–2·5 cm. diam., glabrous or pubescent .. .. 20. *conraui*
  Leaves obovate or obovate-oblong, long-acuminate, 9·5–18 cm. long, 4–8 cm. broad ; main lateral nerves 7–8 on each side of midrib, diverging from it at an angle of about 45° ; receptacles depressed-globose, about 1·5 cm. diam., strigillose .. .. 21. *winkleri*
Base of leaves rounded or sometimes obtuse, apex shortly acuminate ; main lateral nerves 5–8 on each side of midrib ; receptacles 2·5–5 cm. diam., densely hispid :
  Leaves 10–22 cm. long, 4–9·5 cm. broad, oblong-elliptic ; receptacles 2·5–3 cm. diam. when fresh, beaked when dry .. .. .. 23. *tessellata*
  Leaves 20–45 cm. long, 6–14 cm. broad, oblong or obovate-oblong or oblong-oblanceolate ; receptacles up to about 5 cm. diam., not beaked
                                                                    24. *preussii*
Leaves not acuminate, rounded or very obtusely pointed at apex, rounded to cuneate at base, oblong or oblong-oblanceolate, 5–12 cm. long, 2·3–3·6 (–4·7) cm. broad ; main lateral nerves 6–7 on each side of midrib ; mature receptacles 1–1·2 cm. diam., ovoid-globose, glabrous or finely pubescent
                                                                    25. *anomani*
Receptacles 5–8 mm. diam., globose, glabrous ; leaves narrowly rhombic-elliptic, cuneate at base, gradually obtuse acuminate, 4–8·5 (–12·5) cm. long, 2–3 (–5) cm. broad ; main lateral nerves about 12 on each side of midrib with numerous subsidiary more or less parallel nerves .. 26. *kamerunensis*
Leaves 1·2–4 (–4·5) cm. long, 0·4–1·5 cm. broad, oblanceolate, narrowed at base, rounded at apex ; main lateral nerves 5–6 on each side of midrib, ascending from it at an acute angle ; receptacles globose, glabrous, 4–5 mm. diam.
                                                                    27. *lingua*
†Receptacles with a peduncle 6–16 mm. long and with a stipe above the basal bracts up to 8 mm. long, ellipsoid-obovoid to subglobose, glabrous, 2·5–3·5 cm. diam. ; leaves obovate or obovate-elliptic, cuneate at base, acuminate at apex, 9–16·5 cm. long, 4–6·5 cm. broad, with about 5 main lateral nerves on each side of midrib
                                                                    28. *rederi*
*Stipules falling on the unfolding of the young leaves :
Leaves obtriangular, broadly truncate at the apex or retuse, with the midrib divided some distance below and not directly continued to the apex of the blade, the latter 4–8 cm. long, up to 7 cm. broad across the top ; receptacles axillary, paired, pedunculate, globose, up to 8 mm. diam. .. .. 29. *leprieuri*
Leaves neither obtriangular nor truncate at the apex :
‡Receptacles more or less invested at least when young by large persistent basal bracts, or the axillary pairs of receptacles completely covered when young by a calyptra ; receptacles sessile or nearly so :
Basal bracts free, not calyptriform ; leaves with long spreading hairs, at least when young :
Leaves broadest above the middle, elongated oblanceolate-obovate, narrowed to a cuneate, obtuse, rounded or subcordate base, acuminate at apex, up to 48 cm. long and 17 cm. broad ; petioles at first shaggy-villous ; basal bracts much shorter than the mature receptacles ; receptacles at first densely villous with orange-coloured hairs becoming much more sparsely hairy, subglobose, flattened at the apex, about 3 cm. diam. :
  Undersurface of leaves long-pilose at first, then more or less glabrescent
                                                        30. *eriobotryoides* var. *eriobotryoides*
  Undersurface of leaves loosely tomentose with curly hairs
                                                        30a. *eriobotryoides* var. *caillei*
Leaves broadest about the middle, oblong-elliptic, rounded to cordate at base, shortly acuminate, 12–33 cm. long, 3–15 cm. broad, with long hairs on midrib and main lateral nerves beneath, more or less glabrescent, venation closely reticulate beneath ; petioles not shaggy-villous ; basal bracts as long as or longer than the mature receptacles ; receptacles pubescent or glabrous, subglobose, about 3 cm. diam. .. .. 31. *chlamydocarpa*
Basal bracts united to form a pointed calyptra which completely covers an axillary pair of young receptacles, later caducous or opening ; leaves glabrous or minutely puberulous, broadest below the middle :
Leaves ovate, or oblong-ovate, usually much longer than broad, obtuse, rounded or cordate at base, distinctly acuminate, 9–36 cm. long, 2·5–15 cm. broad ; receptacles subglobose or ellipsoid, up to 5 cm. long and broad, puberulous
                                                                    32. *ovata*

Leaves ovate to ovate-orbicular, cordate at base, obtuse or rounded at apex, 13–20 cm. long, 6–13 cm. broad ; receptacles subglobose, about 2 cm. diam., pubescent, then glabrescent..   ..   ..   33. *djalonensis*
‡Receptacles not invested by the basal bracts, the latter usually small and often caducous :
The receptacles sessile, or with peduncles not more than 3 mm. long :
Leaves cordate at base, 5–15 cm. long, 2–7 cm. broad ; receptacles 6–8 mm. diam. :
Stipules less than 0·5 cm. long ; lateral nerves 9–12 on each side of midrib ; leaves broadly rhomboid-elliptic to suborbicular, obtusely acuminate, 5–12 cm. long, 3–6 cm. broad, glabrous as the branchlets ; receptacles glabrous, sessile
34. *leonensis*
Stipules 1–3·5 cm. long, conspicuous although caducous ; lateral nerves 4–7 on each side of midrib ; leaves oblong-ovate to oblong or obovate-oblong, very obtusely acuminate, 5–15 cm. long, 2–7 cm. broad ; receptacles more or less densely pubescent, or glabrescent, sessile or with peduncles up to 5 mm. long :
Leaves persistently densely pubescent beneath ..   35. *glumosa* var. *glumosa*
Leaves soon glabrous beneath   ..   ..   35a. *glumosa* var. *glaberrima*
Leaves not cordate at base, or if sometimes cordate (No. 36) then leaves and receptacles larger than above ; leaves glabrous, or nearly so :
Main lateral nerves 5–6 (–8) on each side of midrib, ascending, looped, prominent beneath ; leaves coriaceous, 9–35 cm. long, 4–16 cm. broad, obovate-oblong, oblanceolate or oblong-elliptic, cuneate to shallowly cordate at base, obtusely acuminate ; petiole 2–10 (—16) cm. long, stout ; receptacles subglobose, sessile, 1–1·7 (–2·5) cm. diam., nearly glabrous to densely pubescent
36. *vogelii*
Main lateral nerves 8–12 on each side of midrib, more or less parallel, with intermediate secondary nerves, slightly prominent beneath ; leaves 5–18 (–22·5) cm. long, 2·5–9 (–10) cm. broad :
Leaves coriaceous, distinctly and sometimes acutely acuminate ; receptacles borne 3–4 together on small bracteate cushions in the axils of the leaves, glabrous, about 6 mm. diam. and beaked when dry ; branchlets stout, purplish-brown when dry ; leaves elliptic or oblong-elliptic, obtuse to rounded or subtruncate at base, 9–18 cm. long, 4·5–9 cm. broad ; petioles 1·2–3 cm. long
37. *aganophila*
Leaves thin, rounded or obtuse, or shortly and obtusely acuminate at apex ; receptacles solitary or 2 together, borne directly in the axils of the leaves on the branchlets which are usually yellow-brown when dry ; receptacles glabrous, puberulous or tomentellous, 6–15 mm. diam. ; leaves obovate to elliptic or oblong-elliptic, narrowed, obtuse or rounded at base, 5–14 (–22·5) cm. long, 2·5–7 (–10) cm. broad ; petioles 1·2–7·5 cm. long :
Receptacles glabrous or puberulous, 6–11 mm. diam. ..   .. 47. *thonningii*
Receptacles tomentellous, 10–15 mm. diam.   ..   ..   .. 48. *basarensis*
The receptacles distinctly pedunculate, peduncles 3 mm. long or more :
ϕLeaves more or less cordate at base, or some rounded, often broadly ovate or suborbicular :
Apex of leaves acutely acuminate ; leaves broadly ovate, deeply to widely cordate or subtruncate at base, 6–16 cm. long, up to 14 cm. broad, glabrous ; receptacles 2–4 together, globose up to 1 cm. diam., glabrous or puberulous ; peduncles 4–12 mm. long ..   .. 38. *populifolia*
Apex of leaves rounded or very shortly and obtusely pointed :
Leaves 2–7 cm. broad, 5–15 cm. long, oblong-ovate to oblong or obovate-oblong ; receptacles 6–8 mm. diam., pubescent or glabrescent, with peduncles up to 5 mm. long ; see also above :
Leaves persistently densely pubescent beneath ..   35. *glumosa* var. *glumosa*
Leaves soon glabrous beneath   ..   ..   .. 35a. *glumosa* var. *glaberrima*
Leaves (5–) 7–28 cm. broad, up to 40 cm. long ; receptacles 1–4 cm. diam. ; peduncles 0·5–3 cm. long :
Receptacles 1–2 cm. diam. :
Peduncles 1–3 cm. long ; receptacles about 1 cm. diam., pubescent or glabrous ; leaves broadly elliptic obovate, or ovate, deeply cordate at base, 12–40 cm. long, 7–28 cm. broad, pubescent or glabrescent beneath, with 2–3 basal nerves and 8–13 other lateral nerves on each side of midrib, prominent beneath ; petiole stout, 3–12 cm. long   ..   ..   39. *platyphylla*
Peduncles 0·5–1 cm. long ; leaves ovate to suborbicular :
Leaves deeply cordate at base, glabrous or puberulous and glabrescent beneath, (5–) 9–22 cm. long and broad, with 2–3 basal nerves and 4–6 other lateral nerves on each side of midrib ; petiole 3–12 cm. long ; receptacles 1·3–2 cm. diam., glabrous   ..   .. 40. *abutilifolia*
Leaves shallowly cordate or rounded at base, pubescent or pilose beneath,

more or less glabrescent, 7–27 cm. long, 6–22 cm. broad, with 2 basal
nerves and 6–13 other lateral nerves on each side of midrib ; petiole
1–9 cm. long ; receptacles 1–1·6 cm. diam., pubescent or glabrous
                                                        41. *congensis*
Receptacles about 4 cm. diam., glabrous ; peduncles about 2·5 cm. long ;
leaves obovate or obovate-elliptic, cordate or rounded at base, 10–25 cm.
long, 7–15 cm. broad, pilose and then glabrescent beneath, with 12–14 main
lateral nerves on each side of midrib ; petioles 5–8 cm. long, stout, persistently long-pilose ..   ..   42. *goliath*
φLeaves neither cordate, nor ovate, nor suborbicular, mostly much longer than
broad and narrowed to the cuneate, obtuse or rounded base :
Lateral nerves 15–25 on each side of midrib, close and parallel, looped to form a
conspicuous intramarginal nerve :
Leaves obovate or obovate-elliptic, cuneate at base, shortly and subacutely
acuminate, 12–26 cm. long, 4–10 cm. broad, glabrous ; petiole 3–8 cm. long ;
receptacles paired, about 1·5 cm. diam., glabrous ; peduncles 5–8 mm. long
                                                        43. *elasticoides*
Leaves oblong-lanceolate to linear-lanceolate ; receptacles 0·7–1·2 cm. diam. :
Peduncles densely pubescent, 3–5 mm. long, stout ; receptacles pubescent ;
leaves oblong lanceolate, rounded at base, subacute or shortly acuminate,
7–20 cm. long, 2·5–5 cm. broad ; petioles 1–2·5 cm. long
                                                        44. *pseudomangifera*
Peduncles glabrous, 6–14 mm. long ; receptacles glabrous ; leaves elongate-
oblong-lanceolate to linear-lanceolate, cuneate at base, long-acuminate,
10–40 cm. long, 1·5–6 cm. broad ; petioles 1·2–5·5 cm. long   45. *barteri*
Lateral nerves up to 14 on each side of midrib :
The receptacles without a stipe above the base bracts :
Receptacles glabrous or sparingly pubescent ; peduncles 4–10 mm. long :
Receptacles 2·5–3·2 cm. diam. ; leaves very coriaceous, oblong-elliptic to
obovate-oblong, rounded or obtuse at base, very shortly and obtusely
acuminate, 9–16 cm. long, 3·3–7 cm. broad, with 5–6 main lateral nerves
on each side of midrib ; petioles 0·5–3·5 cm. long, stout, channelled above
                                                        46. *scott-elliotii*
Receptacles 0·6–1 cm. diam. :
Petioles 0·5–1·2 (–2·5) cm. long ; leaves obovate, oblanceolate or sometimes
elliptic, rounded or very shortly acuminate, 2·5–8 cm. long, 1·2–5 cm.
broad, with 5–10 main lateral nerves on each side of midrib   49. *natalensis*
Petioles 2–6 cm. long ; leaves oblanceolate or narrowly elliptic, rarely
obovate, obscurely acuminate, 5–17 cm. long, 1–5 cm. broad, with 8–14
main lateral nerves on each side of midrib ..   ..   50. *dekdekena*
Receptacles tomentose or densely pubescent when mature, 0·6–1·2 mm. diam. ;
peduncles 4–5 mm. long ; petioles (1–) 1·5–4·5 cm. long ; leaves oblanceolate
or oblong-elliptic, attenuated to the obtuse or rounded base, shortly
acuminate, 4–10 cm. long, 1·5–4 cm. broad, with 6–13 main lateral nerves
on each side of midrib ..   ..   51. *iteophylla*
The receptacles with a stipe about 4 mm. long above the basal bracts ; peduncle
about 6 mm. long ; receptacles about 4 cm. diam., glabrous ; leaves oblong-
oblanceolate, cuneate to rounded at base, obtusely acuminate, up to 18 cm.
long and 6 cm. broad, with 10–12 main lateral nerves on each side of midrib
                                                        52. *sp. nr. cyathistipuloides*

Sect. **Fasciculatae.** *Receptacles borne on short leafless arrested branchlets, or in fascicles,
on the main trunk or branches remote from the leaves ; receptacles 2–several together,
mostly pubescent and then glabrescent ; stipules mostly caducous :*
Leaves mostly distinctly cordate at base ; receptacles 2·2–4 cm. diam. :
The leaves 7–25 cm. long, 3·5–20 cm. broad :
Leaves ovate or suborbicular :
Peduncles 0·3–0·9 cm. long and 3–5·5 mm. thick ; leaves ovate to suborbicular,
shortly and obtusely acuminate, 13–25 cm. long, 12–20 cm. broad ; branchlets
stout ..   ..   53. *umbellata*
Peduncles (1–) 1·2–1·8 (–2·5) cm. long and up to 2·5 mm. thick ; leaves ovate to
oblong-ovate, distinctly caudate and often rather acutely acuminate, 7–15 cm.
long, 5–10 (–12) cm. broad ; branchlets slender ..   ..   ..   54. *polita*
Leaves elliptic to ovate-elliptic or obovate-elliptic, subcordate (or cuneate to rounded)
at base, long-acuminate, 11–16 cm. long, 3·5–7·5 cm. broad, margin undulate,
with a basal pair of ascending lateral nerves and 5–6 other main lateral nerves on
each side of the midrib from which they spread at nearly right angles, branch well
before the margin and become prominently looped ; petioles 1–3·5 cm. long ;
peduncles 0·8–1·5 cm. long ; receptacles 3–4 cm. diam., pubescent ; see also below
                                                        56. *dryepondtiana*

The leaves 3·5–6·5 cm. long, 2–4·2 cm. broad, broadly elliptic, deeply cordate at base, abruptly long-acuminate ; petioles 1·4–4·2 cm. long ; receptacles subglobose or ellipsoid-subglobose, 2·7–3·4 cm. diam., densely puberulous ; peduncles 1·6–3 cm. long . . . . **55. kimuenzensis**

Leaves not cordate at base :

Receptacles 3–5 cm. diam., subglobose ; main lateral nerves 6–12 on each side of midrib :
Peduncles 0·8–1·5 cm. long ; leaves elliptic to ovate-elliptic or obovate-elliptic, cuneate to rounded or subcordate at base ; for other characters see above
56. dryepondtiana

Peduncles 2–3 cm. long ; leaves oblong-oblanceolate to narrowly oblong, cuneate to rounded at base, gradually acuminate, 6–16 cm. long, 2–4 cm. broad, main lateral nerves 6–12 on each side of midrib ; petioles 0·7–2 cm. long ; peduncles 2–3 cm. long, 2–5 mm. thick ; receptacles subglobose, 3–5 cm. diam., densely hispid to glabrescent, with very thick walls, woody at least when dry **57. macrosperma**

Receptacles 1·3–2·5 (–3) cm. diam. ; main lateral nerves 8–22 on each side of midrib :
Leaves obovate, narrowed to an obtuse or rounded base, shortly and obtusely acuminate, 7–16 cm. long, 2·7–7·5 cm. broad, with 8–16 main lateral nerves on each side of midrib ; petiole 1·5–3·5 cm. long ; receptacles subglobose 2–3 cm. diam., puberulous ; peduncles 1·5–2 cm. long . . . . **58. elegans**

Leaves oblong to oblong-elliptic, obtuse at base, acuminate, 10–20 cm. long, 4–8·5 cm. broad, with 12–22 main lateral nerves on each side of midrib, the basal pair conspicuous and ascending to become the looped intramarginal nerves ; petiole 1·7–9·7 cm. long ; receptacles ellipsoid or ellipsoid-subglobose, 1·7–2·5 cm. long, 1·3–2·3 cm. diam., puberulous to glabrous, sometimes stipitate when dry ; peduncles 1·5–2·6 cm. long . . . . **59. ottoniifolia**

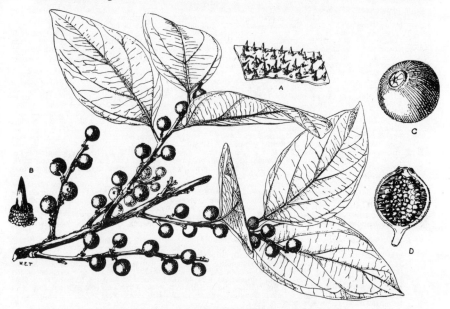

FIG. 173.—FICUS EXASPERATA *Vahl* (MORACEAE).

A, leaf-surface showing hairs. B, single hair. C, fruit. D, longitudinal section of same.

1. **F. capreifolia** *Del.*—F.T.A. 6, 2 : 107 (incl. var. *ovatifolia* Hutch.) ; Engl. Pflanzenw. Afr. 3, 1 : 41, fig. 24 ; Lebrun & Boutique in Fl. Congo Belge 1 : 126. ? *F. niokoloensis* Berhaut Fl. Sen. 104, 126 (1954), French descr. only. A shrub with scabrid leaves ; branchlets villous when young ; by river banks in the savannah regions.
   Sen.: *Berhaut* 1581 ; 4550. **Gam.:** MacCarthy Isl. *Heudelot* 350 ! **Togo:** Atakpame *Doering* 321. Kara R., Kadedjande *Kersting* A. 622. **N.Nig.:** Nupe *Barter* 879 ! Anara F.R., Zaria Prov. (Oct.) *Keay* FHI 20136 ! Katagum (Nov.) *Dalz.* 215 ! Extends east to Somaliland and south to Rhodesia and Mozambique. (See Appendix, p. 278.)
2. **F. acutifolia** *Hutch.*—F.T.A. 6, 2 : 108. *F. asperifolia* of Chev. Bot. 595, partly. A shrub with branchlets drying brown-purple.
   Fr.G.: Kollangui *Chev.* 12228 ! Also in French Cameroons.
3. **F. exasperata** *Vahl*—F.T.A. 6, 2 : 110 ; Engl. l.c. fig. 22 C ; Aubrév. Fl. For. C. Iv. 1 : 58, t. 12, 2 ; Fl. For. Soud.-Guin. 346 ; Lebrun & Boutique l.c. 126. A deciduous tree, up to 70 ft. high, with smooth grey bark and very rough leaves ; sap viscid, not milky ; coppice shoots with lobed leaves ; in the drier types of forest.
   Sen.: Niayes, Sedhiou & Sinedone *fide* Chev. *l.c.* **Gam.:** *Dawe* 81 ! **Fr.Sud.:** Koullou *Dubois* 52.

Port.G.: Safim, Bissau (Jan.) *Esp. Santo* 1671! Sao Domingos (Nov.) *Esp. Santo* 1295! **Fr.G.:** *Heudelot* 853! Karkandy *Heudelot* 863. Forekaria road, Conakry (Jan.) *Dalz.* 8143! 8146! Kindia *Chev.* 12799. **S.L.:** *Afzelius*! Rokupr (May) *Jordan* 243! Njala (Jan.) *Deighton* 1961! 2833! Samu country (Dec.) *Sc. Elliot* 4212! Rotumba Isl. *Kirk*! **Lib.:** Moylakwelli-Totokwelli (Oct.) *Bequaert* 1282! Grand Cess (Mar.) *Baldwin* 11614! **Iv.C.:** Rasso *Aubrév.* 585! Bouroukrou, Bingerville, Alépé & Sassandra Port *fide* Chev. *l.c.* **G.C.:** Krepi Plains (Jan.) *Johnson* 533! Yipala (Feb.) *Kitson* 858! Benso, Tarkwa (July) *Andoh* FH 5518! **Togo:** Lome *Warnecke* 348! Sokode (Mar.) *Kersting* 53! **Dah.:** Sakété (Jan.) *Chev.* 22806! **S.Nig.:** Ebute Metta (Nov.) *Millen* 80! Abeokuta *Rowland*! Olokemeji (Nov.-Mar.) *A. F. Ross* 30! *Daggash* FHI 19155! Idumuje (Jan.) *Thomas* 2187! Bonny (Feb.) *Kalbreyer* 64! **Br.Cam.:** Tingo, Wum (July) *Ujor* FHI 30468! **F.Po:** (Jan.) *Mann* 196! Widespread in tropical Africa ; also recorded from Arabia.

4. **F. asperifolia** *Miq.*—F.T.A. 6, 2 : 111 ; Chev. Bot. 595 ; Lebrun & Boutique l.c. 127. A shrub, usually scrambling, with brownish-purplish branchlets and scabrid leaves ; by rivers.
　　Sen.: *Berhaut* 918. **Port.G.:** Bolola to Buba, Fulacunda (Aug.) *Esp. Santo* 2153! **Fr.G.:** *fide* Chev. *l.c.* **S.L.:** Njala (Aug.) *Deighton* 5573! Musaia (July) *Deighton* 4808! *Sc. Elliot* 5130! Roruks (July) *Deighton* 3260! **Lib.:** Gbanga (Sept.) *Linder* 496! Peahtah (Oct.) *Bequaert* 1016! Flumpa, Sanokwele (Sept.) *Baldwin* 9346! **Iv.C.:** various localities *fide* Chev. *l.c.* **G.C.:** Tano R. (Sept.) *Chipp* 342! Anibil, Ancobra R. (Mar.) *Irvine* 2387! **Dah.:** various localities *fide* Chev. *l.c.* **N.Nig.:** Nupe *Barter* 135! R. Niger, Koton Karifi (Nov.) *Keay* FHI 28088! **S.Nig.:** *Foster* 327! Aboh (June) *T. Vogel* 47! Also in French Cameroons, Belgian Congo and N. Rhodesia. (See Appendix, p. 278.)

5. **F. vogeliana** (*Miq.*) *Miq.*—F.T.A. 6, 2 : 94 (incl. var. *latifolia* Hutch.) ; Chev. Bot. 593, 601 (incl. var. *textilis* A. Chev.), partly ; Aubrév. Fl. For. C. Iv. 1 : 60, t. 13, 1–2. *Sycomorus vogeliana* Miq. (1849). *Ficus fleuryi* A. Chev. in Rev. Bot. Appliq. 15 : 458 (1935). A forest tree, 15–60 ft. high, coarsely brown-hairy on the young branchlets ; figs borne on long panicles on the main stem near the base, often with many panicles prostrate on the ground and sometimes rooting ; ripe figs russet brown ; usually in moist situations.
　　Fr.G.: Kissidougou *Chev.* 20720. **S.L.:** Mano, Dase (Oct.) *Deighton* 3362! Mano, Luawa (Apr.) *Deighton* 3150! Kambui Hills South (Nov.) *Small* 826! Musaia (Apr.) *Deighton* 5485! **Lib.:** Sarbo, Webo (July) *Baldwin* 6410! **Iv.C.:** Bouaké (July) *Chev.* 22125! Kassigué to Atéou (Jan.) *Chev.* 17087! Rasso *Aubrév.* 586! **S.Nig.:** Lagos *Rowland* 18! Oja-odan, Ilaro (Dec.) *Onochie* FHI 26668! R. Oni, Akilla (June) *Onochie* FHI 20673! R. Niger *T. Vogel*! Oban *Talbot* 89! R. Nkem, Aboabam (Dec.) *Keay* FHI 28189! **Br.Cam.:** Ambas Bay (Jan.) *Mann* 702! Buea to Mayuku (Feb.) *Dalz.* 8245! Barombi *Preuss* 550! **F.Po:** (Nov.) *T. Vogel* 179! Also in French Cameroons, Spanish Guinea, Cabinda and (?) Uganda. (See Appendix, p. 283.)

6. **F. mallotocarpa** *Warb.*—F.T.A. 6, 2 : 97 ; Lebrun & Boutique l.c. 113. A tree up to 30 ft. high or more with the habit of an alder, with viscid, not milky, sap and shortly tomentose young branchlets.
　　F.Po: Musola, above San Carlos *Mildbr.* 6992. Extends to E. Africa, Rhodesia and Angola.

7. **F. capensis** *Thunb.*—F.T.A. 6, 2 : 101 ; Chev. Bot. 595 ; Aubrév. l.c. 1 : 62, t. 13, 5–6 ; Lebrun & Boutique l.c. 116 ; M. F. Barrett in Am. Midl. Nat. 39 : 194. Usually a small tree with the figs borne in sometimes abundant clusters on the trunk ; branchlets glabrous or softly pubescent, with pointed more or less villous buds ; a variable species within which several infraspecific taxa can probably be recognized ; common, especially in the more open country.
　　Sen.: Cayor *Leprieur.* Cape Verde *Brunner.* Tambana, Sedhiou *Chev.* **Gam.:** *Rosevear* 25! **Fr.Sud.:** Tabacco (Jan.) *Chev.* 131. **Port.G.:** Bissalanca, Bissau (Jan.) *Esp. Santo* 1639! **Fr.G.:** Conakry *Dybowski.* Los Isl. *Pobéguin* 1231. **S.L.:** Rokupr (May) *Jordan* 244! 246! Musaia (Feb.) *Deighton* 4219! Makump (May) *Deighton* 1723! Njala (Aug.) *Deighton* 1310! **Lib.:** Gbanga (Sept.) *Linder* 495! 670! Du R. (July) *Linder* 70! Sinoe *Whyte* 7! **Iv.C.:** Anoumaba to Sahoua *Chev.* 22424! Attéou *Chev.* 16184. **G.C.:** Obuasi (Jan.) *Soward* 655! Ejura (Aug.) *Chipp* 731! Atauaso, Tarquah (May) *Chipp* 226! Axim (Feb.) *Irvine* 2195! **Togo:** *Baumann* 467! Kpedsu (Feb.) *Howes* 1103! **N.Nig.:** Katagum *Dalz.* 345! Abinsi (Mar.) *Dalz.* 744! **S.Nig.:** Lagos *Dawodu* 90! Ikirun (Feb.) *Moloney* 14! **Br.Cam.:** Buea (Dec.) *Maitland* 889! 890! Widespread in tropical Africa, also in S. Africa and Cape Verde Islands. (See Appendix, p. 278.)

8. **F. mucuso** *Welw. ex Ficalho*—F.T.A. 6, 2 : 98 ; Aubrév. l.c. 1 : 60, t. 13, 3–4 ; Lebrun & Boutique l.c. 114 ; Eggeling Indig. Trees of Uganda, ed. 2, 253, photo 42. A forest tree, to 120 ft. high and 12 ft. girth, with buttresses ; bark smooth yellow or brownish or pinkish, when slashed exuding copious latex which rapidly turns brownish ; crown open.
　　Port.G.: Granja, Catió (May) *Esp. Santo* 2472! **Fr.G.:** Conakry *Maclaud*! **S.L.:** Njala (Nov.) *Deighton* 3086! Mattru *Deighton* 2364! Musaia *Deighton* 4168! Loma Mts. *Jaeger* 4082! **Iv.C.:** Bingerville *Chev.* 17348. Rasso *Aubrév.* 584! **G.C.:** Prahsu *Cummins* 5! **Dah.:** Pira to Cabolé *Chev.* 23761! **S.Nig.:** Awba Hills, Oyo *Jones* FHI 6325! Okomu F.R., Benin (Mar.) *A. F. Ross* 191! **Br.Cam.:** Victoria *Winkler* 24! Likomba *Mildbr.* 10791! Kake, Kumba (Dec.) *Ngalame* FHI 8392! Johann-Albrechtshöhe *Büsgen* 172. **F.Po:** *fide* Mildbr. Extends to Uganda, Tanganyika, Belgian Congo, Angola and S. Tomé. (See Appendix, p. 280.)

9. **F. vallis-choudae** *Del.*—F.T.A. 6, 2 : 103 ; Lebrun & Boutique l.c. 119 ; Aubrév. Fl. For. Soud.-Guin. 344, t. 69, 1–3. A tree, to 60 ft. high, with short bole and spreading crown ; branchlets and leaves nearly glabrous ; figs large, solitary ; mostly by streams in the savannah regions.
　　Fr.Sud.: Bamako *Chev.* 246. **Fr.G.:** Dalaba *Chev.* 12933. **S.L.:** Berria (Feb.) *Sc. Elliot* 4980! Musaia (Feb.) *Deighton* 4249! 4258! Mano, Luawa (Apr.) *Deighton* 3180! Safadu (Feb.) *Deighton* 3505! **Iv.C.:** Bouaké *Aubrév.* 1456. **G.C.:** Ejura (Aug.) *Chipp* 733! Funsi to Boflama (Apr.) *Kitson* 620! **Dah.:** Savalou *Chev.* 23777. **N.Nig.:** Nupe *Barter* 1316! Sokoto (May) *Lely* 820! Ningi *Lely* 438! Katagum *Dalz.* 342! **S.Nig.:** Lagos *Rowland*! Old Oyo F.R. (Feb.) *Keay* FHI 16290! Widespread in tropical Africa. (See Appendix, p. 283.)

10. **F. gnaphalocarpa** (*Miq.*) *Steud. ex A. Rich.*—F.T.A. 6, 2 : 104 ; Lebrun & Boutique l.c. 119, photo 8; Aubrév. l.c. 342, t. 70, 6 ; Schnell in Ic. Pl. Afr. t. 10. *Sycomorus gnaphalocarpa* Miq. (1848). A tree to 60 ft. high, with pale trunk and widespreading crown ; pilose branchlets and scabrid leaves ; generally by streams in the savannah regions.
　　Sen.: *Talmy*! St. Louis *Chev.* 3496. Kaolak *Chev.* **Gam.:** Kudang (Jan.) *Dalz.* 8144! Kuntaur *Ruxton*! **Fr.Sud.:** Kita *Dubois* 111 *bis*! Bamako *Chev.* 216. **Port.G.:** Bijimita (Feb.) *Esp. Santo* 1755! 1769! Biombo, Bissau (Feb.) *Esp. Santo* 1786! **Iv.C.:** Mossi *Chev.* 24627. Niangbo *Aubrév.* 1731. **G.C.:** *Evans* 5! Amansare (July) *Chipp* 516! Yala *Kitson* 623! **Dah.:** Abomey *Chev.* 23181. **N.Nig.:** Zurmi (Apr.) *Keay* FHI 16122! Shinga *Dudgeon* 67! Katagum *Dalz.* 301! Abinsi *Dalz.* 743! Musgu (May) *E. Vogel* 65! Widespread in the drier parts of tropical Africa ; also in Cape Verde Islands. (See Appendix, p. 279.)

11. **F. variifolia** *Warb.*—F.T.A. 6, 2 : 114 ; Mildbr. & Burret in Engl. Bot. Jahrb. 46 : 203 ; Chev. Bot. 596 ; Lebrun & Boutique in Fl. Congo Belge 1 : 125. *F. sciarophylla* Warb. (1904)—F.T.A. 6, 2 : 114 ; Lebrun & Boutique l.c. ; Aubrév. Fl. For. C. Iv. 1 : 58, t. 12, 3–4 (as sp. aff.). *F. bongouanouensis* A. Chev. (1917), originally spelt "*bangouanensis*" by error ; F.T.A. 6, 2 : 357 ; Chev. Bot. 604 ; Aubrév. l.c. 1 : 62, t. 14 B. *F. kerstingii* of F.W.T.A., ed. 1, 1 : 434, partly. A forest tree, to 110 ft. high, with clean straight bole and plank buttresses ; bark smooth ; in drier types of forest ; sapling and coppice growth with remarkable dissected lobed leaves.
　　Fr.G.: Farana circle *Chev.* 20652! **S.L.:** Port Loko *Deighton* 3619! **Iv.C.:** Bamoro, Bouaké *Aubrév.* 821! Man *Aubrév.* 941! Bongouanou, Morénou (Nov.) *Chev.* 22444! Bouaké *Aubrév.* 1576! 1901! **G.C.:** Nchenenchene to Kwahu Tafo, Kwahu Plateau *Kitson* 1169! **Dah.:** Banté to Pira, Savalou

Chev. 23740. **N.Nig.**: Kurmin Damisa, Jemaa *Onochie* FHI 21713! **S.Nig.**: Awba Hills, Oyo *Jones* FHI 5964! Also in A.-E. Sudan, Uganda, Belgian Congo and Cabinda.

12. **F. dicranostyla** *Mildbr.*—F.T.A. 6, 2 : 119 (incl. var. *nitida* Hutch.) ; Aubrév. Fl. For. C. Iv. **1** : 62, t. 14 A ; Fl. For. Soud.-Guin. 346. A savannah tree, up to 60 ft. high ; branchlets and undersurface of leaves more or less densely puberulous when young, glabrescent.
 Sen.: Cape Verde *Perrottet.* Ziguinchor *Chev.* 2651. Bignona *Aubrév.* 3009. **Fr.Sud.**: Koulikoro *Chev.* 2639. Kita *Dubois* 83. **Port.G.**: Soé, Formosa (Apr.) *Esp. Santo* 1979! Bijimita, Bissau (Feb.) *Esp. Santo* 1749! **Fr.G.**: Kankan (Mar.) *Chev.* 562! Farana (Mar.) *Sc. Elliot* 5351! **S.L.**: Musaia *Deighton* 4225! **Iv.C.**: Toumodi *Aubrév.* 1881! 1882! Bouaké (Aug.) *Aubrév.* 1574! 1575! **G.C.**: Yendi (June) *Akpabla* 710! **Togo**: Loso (Mar.) *Kersting* 632. Atakpame (May) *Doering* 303. Kirikiri *Kersting* 77. Sokode *Kersting* 24. **Dah.**: Pira to Kabolé, Savalou *Chev.* 23763. Also in Ubangi-Shari, A.-E. Sudan and Uganda. (See Appendix, p. 278.)

13. **F. verruculosa** *Warb.*—F.T.A. 6, 2 : 114 ; Engl. Pflanzenw. Afr. 3, 1 : 41, fig. 22 D ; Lebrun & Boutique l.c. 120 ; Aubrév. Fl. For. Soud.-Guin. 335. A low shrub of marshy ground or a tree 15–25 ft. high with more or less glaucous leaves and small red figs.
 N.Nig.: Nupe *Barter* 1317! Kontagora (Nov.) *Dalz.* 80! Also in French Cameroons, Ubangi-Shari, Belgian Congo, Angola, Uganda, Tanganyika, Rhodesia, Nyasaland, Mozambique and Bechuanaland.

14. **F. lecardii** *Warb.*—F.T.A. 6, 2 : 117 ; Aubrév. l.c. 347, t. 71, 1. A tree with glossy leaves dark above and pale beneath, glabrous except on the young branchlets ; on rocky hills in savannah.
 Sen.: *Lecard* 197! *Bellamy* 467! Niokolo-Koba *Berhaut.* **Fr.Sud.**: Kanikombolé, Macina (Sept.) *Chev.* 24830! Bamako (May) *Hagerup* 52! **Port.G.**: Canjadudi (June) *Esp. Santo* 2501! Buruntuma (July) *Esp. Santo* 2702! **Fr.G.**: Kagé *Pobéguin* 2020. Dalaba *Chev.* 20267. **Iv.C.**: Bobo Dioulasso *Aubrév.* 1997! Ferkéssédougou *Aubrév.* 1876. Niangbo *Aubrév.* 1721. **N.Nig.**: Kufena Rock, Zaria (May) *Keay* FHI 25722! Yola (May) *Dalz.* 143! Also in French Cameroons. (See Appendix, p. 280.)

15. **F. ingens** (*Miq.*) *Miq.* var. *ingens*—F.T.A. 6, 2 : 121 ; Lebrun & Boutique l.c. 121 ; Aubrév. l.c. 346, t. 71, 2–3. *Urostigma ingens* Miq. (1847). *Ficus lutea* of Chev. Bot. 597, not of Vahl. A shrub or tree, up to 40 ft. high, with stout branchlets and smooth almost shining foliage ; in savannah.
 Sen.: Marsasoum, Manpalago & Pout *fide* Chev. *l.c.* **Gam.**: *Rosevear* 91! **Fr.Sud.**: Bamako *Vuillet* 647. **Fr.G.**: Kouroussa *Pobéguin* 855. **Iv.C.**: Niangbo, Bobo Dioulasso, Kaya & Mossi *fide* Aubrév. *l.c.* **G.C.**: *Evans* 3! Pabarabo Hill, Afram Plains (May) *Kitson* 1108! Kpong (Jan.) *Irvine* 1919! **Togo**: Sansanné Mango *Aubrév.* 77d. **Dah.**: Atacora Mts. *Chev.* 24035. **N.Nig.**: Zaria (July) *Lamb*! Anara F.R., Zaria Prov. (May) *Keay* FHI 22990! Badeggi (Mar.) *Lamb* 45! **S.Nig.**: Lagos *Foster* 25! Olokemeji *Chev.* 14079. Ijaiye F.R., Oyo Prov. (Mar.) *Keay & Ladipo* FHI 21192! Widespread in the drier part of tropical Africa and in S. Africa. (See Appendix, p. 280.)

15a. **F. ingens** var. **tomentosa** *Hutch.* in Fl. Capensis 5, 2 : 530 (1925). *F. katagumica* Hutch. (1915)—F.T.A. 6, 2 : 122 ; F.W.T.A., ed. 1, 1 : 433 ; Aubrév. l.c. 340, t. 72, 3. *F. kawuri* Hutch. (1915)—F.T.A. 6, 2 : 122 ; F.W.T.A., ed. 1, 1 : 433. *F. ingentoides* Hutch. (1915). A savannah tree, similar to the above, but with densely tomentose figs.
 N.Nig.: Sokoto (May) *Lely* 828! Samaru, Zaria (May) *Keay* FHI 25742! Lokoja (Feb.) *Dalz.* 910! Katagum *Dalz.* 305! Karakai (Feb.) *Foster* 12! Widespread in the drier parts of tropical Africa and in S. Africa. (See Appendix, p. 280.)

16. **F. lyrata** *Warb.*—F.T.A. 6, 2 : 142 ; Chev. Bot. 601 ; Aubrév. Fl. For. C. Iv. 1 : 64, t. 15 A ; M. F. Barrett l.c. 202. A tree to 50 ft. high, or epiphyte, with stout glabrous branchlets longitudinally grooved when dry and very large fiddle-shaped leaves.
 S.L.: Pelewahun, Kamagai *Deighton* 5750! **Lib.**: Zuie, Boporo (Nov. *Baldwin* 10256a! Loffa R., Gola Forest *Bunting*! **Iv.C.**: Abidjan *Aubrév.* 889! Keeta village (planted) *Chev.* 19351. Dimbokro *Aubrév.* 1875. **Togo**: Bismarckburg, in avenue *Büttner* 713. **Dah.**: *Poisson* 5. **S.Nig.**: Olokemeji *Jones & Onochie* FHI 14523! Obu *Thomas* 365! **Br.Cam.**: Barombi *Preuss* 455! On 1922 lava flow, Bibundi *Rosevear* Cam. 76/37! Also in Gabon. (See Appendix, p. 280.)

17. **F. sagittifolia** *Warb. ex Mildbr. & Burret*—F.T.A. 6, 2 : 143 ; Chev. Bot. 602 ; Aubrév. l.c., t. 15 B. A epiphyte, sometimes becoming a tree up to 30 ft. high, with stout glabrous branchlets and long rather narrow fiddle-shaped leaves.
 Port.G.: Sancoinha to Gadamael, Cacine (July) *Esp. Santo* 2122! **Fr.G.**: Soya to Kouloundala, Mamou *Chev.* 20380! Layah (Jan.) *Sc. Elliot* 4656! Koundian to Ouria, Kissi *Chev.* 20757. **S.L.**: Njala (Jan.) *Deighton* 2600! **Lib.**: Peahtah *Linder* 944! **Iv.C.**: various localities *fide* Chev. *l.c.* **G.C.**: Kumasi (Jan.) *Vigne* 1521! **Togo**: Lome *Warnecke* 428. **Dah.**: Allada *Poisson.* Abbo to Massé *Chev.* 22968. **S.Nig.**: Olokemeji *Symington* FHI 5083.

18. **F. camptoneura** *Mildbr.*—F.T.A. 6, 2 : 147. A tree to 25 ft. high, or epiphytic shrub, with nearly glabrous angular branchlets ; in rain forest.
 S.Nig.: Omo F.R., Ijebu Ode (Mar.) *Jones* FHI 17025! 17273! Sapoba *Kennedy* 1581! Okomu F.R., Benin (Feb.) *Brenan* 9075! 9115! 9191! Afi River F.R. (June) *Jones & Onochie* FHI 18950! **Br.Cam.**: Buea *Deistel*! *Reder.* **F.Po**: *Mann* 307! Also in French Cameroons and Gabon.

19. **F. camptoneuroides** *Hutch.*—F.T.A. 6, 2 : 148. A large epiphytic shrub or tree up to 25 ft. high, with slender glabrous branchlets sharply ribbed when dry.
 Br.Cam.: Bangwe *Conrau* 208. **F.Po**: north side of St. Isabel Peak *Mildbr.* 6411! Also in French Cameroons.

20. **F. conraui** *Warb.*—F.T.A. 6, 2 : 150 ; Lebrun & Boutique l.c. 155. An epiphytic shrub or tree, with obtusely angular smooth branchlets nearly covered with large persistent stipules ; leaves prominently reticulate with whitish veins below.
 S.L.: Njala (Jan., Feb.) *Dalz.* 8107! *Deighton* 1072! **Lib.**: Gbanga (Sept.) *Linder* 780! **S.Nig.**: Lagos *Rowland*! Agege (Nov.) *Kitson*! **Br.Cam.**: Buea *Reder* 395 (partly). Bangwe *Conrau* 280. Also in French Cameroons, Ubangi-Shari and Belgian Congo. (See Appendix, p. 278.)

21. **F. winkleri** *Mildbr. & Burret*—F.T.A. 6, 2 : 152. A tree with glabrous branchlets.
 Br.Cam.: forest near the Botanic Garden, Victoria (Apr.) *Winkler* 1204! On 1922 lava flow, Bibundi (June) *Rosevear* Cam. 75/37!

22. **F. praticola** *Mildbr. & Hutch.*—F.T.A. 6, 2 : 145 ; Aubrév. Fl. For. C. Iv. 1 : 64, t. 14 D. An epiphytic shrub with rather sharply angular branchlets, glabrous.
 Iv.C.: Abidjan *Aubrév.* 1634. **S.Nig.**: R. Osse, Iguoriakhi, Benin (Feb.) *E. W. Jones* in Hb. Brenan 9131! **F.Po**: Moka grassland, 4,000–6,000 ft. *Mildbr.* 7106!

23. **F. tessellata** *Warb.*—F.T.A. 6, 2 : 146 ; Chev. Bot. 600. Aubrév. l.c. 1 : 66, t. 16 A. *F. winkleri* Mildbr. & Burret (1911), partly (*Chev.* 12989)—Chev. Bot. 600. A bushy tree with puberulous branchlets sharply angular when dry.
 Fr.G.: Ditinn (Apr.) *Chev.* 12989! **S.L.**: Mowoleke bridge, Kamabai to Kabala road (Mar.) *Deighton* 5467! Buyabuya, Scarcies (Feb.) *Sc. Elliot* 4565! Sugar Loaf Mt. (Dec.) *Sc. Elliot* 4030! **Iv.C.**: Anoumaba to Sahoua *Chev.* 22456. **Togo**: Bismarckburg, in avenue (Aug.) *Büttner* 711.

24. **F. preussii** *Warb.*—F.T.A. 6, 2 : 152 ; Lebrun & Boutique in Fl. Congo Belge 1 : 162. An epiphytic shrub or tree ; branchlets sulcate when dry, covered with greyish glabrous bark ; in forest.
 Br.Cam.: Barombi, by Elephant Lake *Preuss* 454! Johann-Albrechtshöhe (Mar.) *Staudt* 894! Also in French Cameroons, Belgian Congo and (?) Spanish Guinea.

25. **F. anomani** *Hutch.*—F.T.A. 6, 2 : 157 ; Aubrév. l.c. 1 : 66, t. 14 E. An epiphytic shrub, or straggling tree to 30 ft. high ; stems glossy chocolate-brown, leaves dark green ; ripe figs orange ; in forest.
 S.L.: Kukuna, Scarcies (Jan.) *Sc. Elliot* 4693! Njala (May) *Deighton* 4787! Sumbuya (Nov.) *Deighton* 2258! **Lib.**: Dukwia R. (Nov.) *Cooper* 144! Bolahun (Feb.) *Bequaert* 59! Jaurazon (Mar.) *Baldwin*

11458! **Iv.C.:** Mt. Nienokoué *Chev.* 19447 ; 19473. Abidjan *Aubrév.* 1939! Agboville *Aubrév.* 1930. Yapo *Aubrév.* 1996. **G.C.:** Sefwi *Armitage*! Deyniase, Dunkwa *Chipp* 151! **S.Nig.:** R. Jamieson, Benin (Nov.) *Keay* FHI 28075! **Br.Cam.:** Mundame *Büsgen* 140. Also in French Cameroons. (See Appendix, p. 277.)

26. **F. kamerunensis** *Warb. ex Mildbr. & Burret*—F.T.A. 6, 2 : 156 ; Chev. Bot. 604 ; Aubrév. l.c. 1 : 66, t. 16 D ; (?) Lebrun & Boutique l.c. 154. A tree to 60 ft. high, or an epiphytic shrub, with rather small pointed leaves and numerous small figs.
S.L.: Talla Hills, Ninia (Feb.) *Sc. Elliot* 4913! Dia *Deighton* 3198! **Lib.:** Bumbuna (Oct.) *Allen* in *Hb. Linder* 1326! Sanokwele (Sept.) *Baldwin* 9508! 9538! Jabroke, Palipo (July) *Baldwin* 6429! **Iv.C.:** Grabo to Taté *Chev.* 19785. Anyoma *Chev.* 20106. Guidéko *Chev.* 16489. **G.C.:** Ashanti (Aug.) *Gordon* 17! **S.Nig.:** Idanre Hills (Jan., Aug.) *E. W. Jones, Keay & Richards* 8664! *A. P. D. Jones* FHI 20714! Cross R. *Meyer*. **Br.Cam.:** Victoria *Winkler* 1091. Johann-Albrechtshöhe (Mar.) *Staudt* 897! Also in French Cameroons, Gabon, S. Tomé and (?) Belgian Congo.
[The specimens from Belgian Congo named this at Kew (*Pynaert* 1054, 1106 ; *Laurent* 870) differ from the syntypes which I have seen (*Staudt* 897 ; *Sc. Elliot* 4913), and may represent a distinct species.]

27. **F. lingua** *Warb.*—F.T.A. 6, 2 : 156 ; Aubrév. l.c. 1 : 64, t. 14 C ; Lebrun & Boutique l.c. 138. An epiphytic shrub or liane ; in forest.
Lib.: Gbanga (Sept.) *Linder* 657! **Iv.C.:** Tabou *Aubrév.* 1740. Also in French Cameroons, Spanish Guinea, French and Belgian Congo and Angola.

28. **F. rederi** *Hutch.*—F.T.A. 6, 2 : 154. Branchlets nearly glabrous, angular when dry, leaves with close pale venation beneath.
Br.Cam.: Victoria (Feb.) *Maitland* 370! Buea *Reder* 395. Also in French Cameroons.

29. **F. leprieuri** *Miq.*—F.T.A. 6, 2 : 158 ; Engl. Pflanzenw. Afr. 3, 1 : 47, t. 31 ; Chev. Bot. 603 ; Aubrév. l.c. 1 : 68, t. 16 B ; Lebrun & Boutique l.c. 139, t. 15 ; Schnell in Ic. Pl. Afr. t. 35. Commonly epiphytic and scandent, becoming a tree up to 80 ft. high, with glabrous branchlets and leaves ; common in the forest regions and in moist situations in savannah regions.
Sen.: Cayor *Leprieur*. Carabane, Casamance (Jan.) *Chev.* 2655! **Fr.Sud.:** Kita massif (July) *Jaeger* 7! **Port.G.:** Pussubé, Bissau (Dec.) *Esp. Santo* 1083! Prabis, Bissau (Feb.) *Esp. Santo* 1823! **Fr.G.:** Ditinn (Apr.) *Chev.* 12988! Also Conakry, Los Isl., & Broualtapé *fide* Chev. *l.c.* **S.L.:** Kissy road *Barter*! Mt. Yamba, Buyabuya (Feb.) *Sc. Elliot* 4286! 4991! Musaia (Feb.) *Deighton* 4236! **Lib.:** Sinoe Basin *Sim* 36! *Whyte* 10! Zolopla, Sanokwele (Sept.) *Baldwin* 9376! Soplima, Vonjama (Nov.) *Baldwin* 10049! Garaway, Maryland (Mar.) *Baldwin* 11630! **Iv.C.:** Bingerville, Mt. Niénokué, Zoanlé *fide* Chev. *l.c.* **G.C.:** *Burton & Cameron*! Asientiem (July) *Chipp* 294! Tarkwa, Benso (Dec.) *Andoh* FH 5430! **Togo:** Lome *Warnecke* 458! Ndsolo *Busse* 3508. **Dah.:** Goutissa to Aouandjitomé (Mar.) *Chev.* 23281! **S.Nig.:** Lagos (Jan., Mar.) *Barter* 2228! *Dalz.* 1341! *Dodd* 417! Ibadan (Jan.) *Meikle* 989! 1109! Calabar *Thomson* 71! Eket *Talbot* 3053! **Br.Cam.:** on 1922 lava flow, Bibundi (June) *Rosevear* Cam. 66/37! Elephant Lake, Barombi (Sept.) *Preuss* 544! Bu, Wum (June) *Ujor* FHI 29258! Also in French Cameroons, Ubangi-Shari and Belgian Congo. (See Appendix, p. 280.)

30. **F. eriobotryoides** *Kunth. & Bouché* var. **eriobotryoides**—F.T.A. 6, 2 : 160 ; Chev. Bot. 602 ; Aubrév. l.c. 1 : 68, tt. 16 C, 17 A ; Lebrun & Boutique l.c. 157. *F. afzelii* G. Don *ex* Loud. (1830), not validly published [1] ; M. F. Barrett in Arn. Midl. Nat. 39 : 188. *F. eriobotryoides* var. *caillei* of F.T.A. 6, 2 : 161, partly ; of F.W.T.A., ed. 1, 1 : 433, partly (spec. ex Nig.), not of A. Chev. ex Mildbr. & Burret. An epiphyte, or tree to 50 ft. high with wide spreading crown ; branchlets and leaves densely yellow-brown pilose when young ; figs orange-brown ; in forest by rivers and often planted.
Fr.G.: Labé *Chev.* 12371. Mt. Gangan, Kindia *Chev.* 12774. Ditinn to Diaguissa *Chev.* 12907. **S.L.:** *Don*! Near Freetown (May–July) *Deighton* 2737! *Lane-Poole* 371! *T. Vogel* 139! Sugar Loaf Mt., 2,480 ft. (Oct.) *T. S. Jones* 258! Sulimania (Mar.) *Sc. Elliot* 5330a! Talla Hills *Sc. Elliot* 5017! Bintumane Peak, 5,400 ft. (Jan.) *T. S. Jones* 62! **Iv.C.:** Bingerville *Aubrév.* 1481! Tabou *Aubrév.* 1737. Alépé *Chev.* 17891. **G.C.:** Aburi (Nov.) *Johnson* 930! Krobo Plains (Jan.) *Johnson* 581! Kumasi *Cummins* 120! Pokoase (Jan.) *Irvine* 1955! **N.Nig.:** Ankpa (Oct.) *Lamb* 21! 81! **S.Nig.:** Omo F.R. (Jan.) *Onochie* FHI 20661! Ehor, Benin (Dec.) *Keay* FHI 22709! Owerri Dist. (Apr.) *Sherriff*! Calabar (Mar.) *Brenan* 9213! **Br.Cam.:** Buea (Apr.) *Hutch. & Metcalfe* 110! Mopanya (Apr.) *Maitland* 1081! Binka, Wum (June) *Ujor* FHI 29268! **F.Po:** Bokoko *Mildbr.* 6926. Also in French Cameroons, French and Belgian Congo. (See Appendix, p. 278.)

30a. **F. eriobotryoides** var. **caillei** *A. Chev. ex Mildbr. & Burret*—F.T.A. 6, 2 : 161, partly ; F.W.T.A., ed. 1, 1 : 433 (as var. *baillei* by error), partly (*Chev.* 12989). *F. afzelii* var. *caillei* (A. Chev. ex Mildbr. & Burret) M. F. Barrett l.c. 189 (1948). *F. eriobotryoides* var. *monbuttensis* (Warb.) Lebrun (1934)—Fl. Congo Belge 1 : 157. Similar to the preceding, but leaves loosely tomentose beneath.
Fr.G.: bed of R. Ditinn *Chev.* 12989! Also in Belgian Congo.

31. **F. chlamydocarpa** *Mildbr. & Burret* in Engl. Bot. Jahrb. 46 : 244, fig. 5 (1911) ; F.T.A. 6, 2 : 163. *F. clarencensis* Mildbr. & Hutch. (1915)—F.T.A. 6, 2 : 162 ; Mildbr. Wiss. Ergebn. 1910–11, 181 t. 86 ; G. Taylor in Exell Cat. S. Tomé 306 (? excl. syn. *F. eriobotryoides* var. *latifolia* Hutch.). A large tree, to 100 ft. high with very wide crown and huge aerial roots ; in montane forest.
Br.Cam.: Cam. Mt., 4,000–7,000 ft. (Feb.) *Maitland* 843! 1337! *Mildbr.* 10802! **F.Po:** north side of St. Isabel Peak, 3,600–4,600 ft. *Mildbr.* 6408! Also in French Cameroons, Principe, S. Tomé and Annobon.
[G. Taylor in Exell Cat. S. Tomé 307 includes *F. eriobotryoides* var. *latifolia* Hutch. in *F. clarencensis* Mildbr. & Hutch. There is a single leaf and a single receptacle of *Mildbr.* 6517, one of the syntypes, at Kew ; although the leaf could well be this species, the receptacle appears to me much more like those of *F. eriobotryoides*.]

32. **F. ovata** *Vahl*—F.T.A. 6, 2 : 164 ; Chev. Bot. 602 ; Aubrév. Fl. For. C. Iv. 1 : 68, t. 17 B ; Lebrun & Boutique in Fl. Congo Belge 1 : 160. *F. baoulensis* A. Chev. Bot. 604, name only. A tree to 40 ft. high, with large crown, often epiphytic at first ; figs green with white spots, each pair covered when young by a brown calyptra ; in fringing forest in the savannah regions, and often planted.
Sen.: M'bidjem to Dakar *Chev.* 454. Los Isl. *Pobéguin* 1234. Kindia *Pobéguin* 1269. Conakry *Maclaud.* **S.L.:** Sugar Loaf Mt. *Sc. Elliot* 5754! Turtle Isl. *Deighton* 2403! York *Deighton* 2736! **Lib.:** Sinoe Basin *Sim* 13! 17! 23! Farmington R., Harbel (Mar.) *Baldwin* 11098! **Iv.C.:** Manikro, Kodiokoffi (Aug.) *Chev.* 22309! Dimbokro *Aubrév.* 1878! Bobo Dioulasso *Aubrév.* 1879. And other localities *fide* Chev. *l.c.* **G.C.:** *Burton & Cameron*! Apenkwa, Achimota (Feb.) *Irvine* 1983! **Togo:** Lome (May) *Warnecke* 337! Sokodé-Basari (Dec.) *Kersting* 429. Kete Krachi Zech. **Dah.:** Cotonou *Chev.* 4481. **N.Nig.:** Dekina *Lamb* 19! Kurbu R., Birnin Gwari *Daggash* FHI 31420! Anara F.R., Zaria *Keay* FHI 21130! **S.Nig.:** Lagos (Sept.) *Dalz.* 1343! Idanre (Aug.) *Jones* FHI 20716! Enugu (Oct.) *Dommen* FHI 32431! **Br.Cam.:** Victoria (Apr.) *Maitland* 697! On 1922 lava flow, Bibundi *Mildbr.* 10663! **F.Po:** Musola *Mildbr.* 6986! Extends to A.-E. Sudan, E. Africa, Belgian Congo and Angola. (See Appendix, p. 281.)

33. **F. djalonensis** *A. Chev.* in Bull. Soc. Bot. Fr. 61, Mém. 8 : 301 (1917) ; F.T.A. 6, 2 : 356 ; Hutch. & Dalz. in Kew Bull. 1928 : 19 ; Aubrév. l.c. 1 : 70, t. 19 C. A large epiphyte or tree ; in forest.
Fr.G.: Soya, Mamou (Jan.) *Chev.* 20372! **S.L.:** Bwedu to Kangama *Deighton* 3193! **Iv.C.:** Dimbokro *Aubrév.* 1874! Bamoro *Aubrév.* 1899! Bouaké *Aubrév.* 1755!

---

[1] Loudon lists the name *F. afzelii* G. Don in a group of 20 species all having oblong, entire, smooth leaves and there is no diagnostic description by which the name could be validated.

34. **F. leonensis** *Hutch.*—F.T.A. 6, 2 : 185. An epiphytic shrub ; branchlets covered with thin scaly deciduous bark ; figs small, solitary, sessile in the upper axils.
   Fr.G.: Kindia, by stream banks *Pobéguin* 1282. **S.L.:** Bathurst, by stream (Feb.) *Johnston* 88 ! Regent *Deighton* 2707 !

35. **F. glumosa** *Del.* var. **glumosa**—F.T.A. 6, 2 : 171 ; Chev. Bot. 598 ; Aubrév. Fl. For. C. Iv. 1, t. 23 B ; Lebrun & Boutique l.c. 147, photo 10 ; Aubrév. Fl. For. Soud.-Guin. 346, t. 71, 4–5. A savannah tree, to 50 ft. high, with small twiggy pubescent branchlets and rather stiff half-leathery leaves, with distinct raised venation beneath ; figs red and succulent when ripe.
   Sen.: *Lecard* 73 ! **Gam.:** *Dawe* 42 ! **Fr.Sud.:** Bamako *Hagerup* 55 ! Diaragouéla *Chev.* 475 ! **Port.G.:** Bissalanca, Bissau (Jan.) *Esp. Santo* 1636 ! Mato Farroba, Catió (May) *Esp. Santo* 2475 ! **Fr.G.:** Timbo *Pobéguin* 131. Kadé *Pobéguin* 2021. **S.L.:** Mt. Gonkwi, Talla Hills (Feb.) *Sc. Elliot* 4882 ! Musaia (Feb.) *Deighton* 4203 ! Kamabai (Jan.) *Deighton* 3603 ! Falaba (Apr.) *Deighton* 5429 ! **Iv.C.:** Mt. Dourou *Chev.* 21721. Niangbo *Aubrév.* 1723. Touba *Aubrév.* 1243. **Fr.Nig.:** Diapaga *Chev.* 24433. **N.Nig.:** Sokoto (Dec.) *Dalz.* 396 ! Zurmi (Apr.) *Keay* FHI 16123 ! 16171 ! Nupe *Barter* 1033 ! 1101 ! Samaru, Zaria (May) *Keay* FHI 25726 ! Yola (Jan.) *Dalz.* 147 ! **S.Nig.:** Nsukka Dist. (Oct.) *Jones* FHI 765 ! Widespread in tropical Africa. (See Appendix, p. 279.)

35a. **F. glumosa** var. **glaberrima** *Martelli*—F.T.A. 6, 2 : 172 ; Lebrun & Boutique l.c. ; Aubrév. l.c. *F. dekdekena* of F.W.T.A., ed. 1, 1 : 434, partly (*Dalz.* 148). A savannah tree, similar to above but leaves soon nearly glabrous.
   Sen.: *Berhaut* 255. **Fr.Sud.:** Badinko *Chev.* 117. **Port.G.:** Bissalanca, Bissau (Jan.) *Esp. Santo* 1638 ! **Fr.G.:** Wulia, R. Scarcies (Feb.) *Sc. Elliot* 4582 ! **S.L.:** Falaba (Mar.) *Sc. Elliot* 5170 ! Sulimania (Mar.) *Sc. Elliot* 5375 ! Musaia (Sept.) *Small* 266 ! **G.C.:** Buruku Rock, Kwahu-Tafo (Jan.) *Irvine* 1678 ! Sa (Mar.) *Kitson* 847 ! **Togo:** Sokode *Kersting* 59. Yerapana *Kersting* 524. **Dah.:** Savé *Aubrév.* 50d. Nattingou *Aubrév.* 57d. **N.Nig.:** Zurmi (Apr.) *Keay* FHI 16160 ! Nupe *Barter* 1022 ! Lokoja (Oct.) *Dalz.* 146 ! Anara F.R., Zaria (Oct.) *Keay* FHI 21117 ! Abinsi *Dalz.* 779 ! Yola (Jan.) *Dalz.* 148 ! **S.Nig.:** Opa Hill, Igbetti (Feb.) *Keay* FHI 16294 ! Also in French Cameroons, Ubangi-Shari, A.-E. Sudan, Abyssinia, Uganda, Belgian Congo and Tanganyika. (See Appendix, p. 279.)

36. **F. vogelii** (*Miq.*) *Miq.*—F.T.A. 6, 2 : 179 (incl. var. *pubicarpa* Mildbr. & Burret) ; Chev. Bot. 600 ; Aubrév. Fl. For. C. Iv. 1 : 70, t. 19 A ; Lebrun & Boutique l.c. 146 ; M. F. Barrett l.c. 213. *Urostigma vogelii* Miq. (1848). *Ficus senegalensis* Miq. (1867)[1]—F.T.A. 6, 1 : 105, partly ; F.W.T.A., ed. 1, 1 : 432, partly. *F. pseudovogelii* of F.W.T.A., ed. 1, 1 : 434 (spec. *Millson*), (?) of A. Chev. A tree 20–60 ft. high, nearly glabrous, branchlets covered with scaly deciduous bark ; in forest vegetation by streams, and often planted.
   Sen.: St. Louis *Brunner* ! Cayor *Poisson* ! Niayes *Chev.* **Gam.:** Bathurst *Dudgeon* 7 ! **Port.G.:** Pussubé, Bissau (Apr., Nov., Dec.) *Esp. Santo* 1003 ! 1076 ! 1535 ! Bissalanca (Jan.) *Esp. Santo* 1628 ! **Fr.G.:** Conakry (June) *Debeaux* 138 ! *Poisson* 16 ! Dubreka *Bonéry* ! **S.L.:** Mt. Aureol, Freetown (June) *Deighton* 2738 ! Port Loko (Oct.) *Tindall* 44 ! Musaia (Feb.) *Deighton* 4251 ! Jigaya (Nov.) *Thomas* 2727 ! **Lib.:** Cape Palmas (July) *T. Vogel* 47 ! Ganta (Sept.) *Baldwin* 9650 ! Piatah, Salala (Dec.) *Baldwin* 10577 ! **Iv.C.:** Banco *Serv. For.* 479 ! And various localities *fide* Chev. *l.c.* **G.C.:** Odorso *Farmar* 391 ! Pokoase (Jan.) *Irvine* 1954 ! Blengo, Krepi Plains (Jan.) *Johnson* 523 ! Yenahin (May) *Chipp* 130 ! **Togo:** Ndsolo *Baumann* ! Sokodé *Kersting* A. 542 ! Losso (Apr.) *Kersting* A. 539 ! **Dah.:** Pira, Savalou *Chev.* 23747. **N.Nig.:** Agaie (Feb.) *Yates* 36 ! Bassa *Elliott* 95 ! Lafia (Feb.) *Jones* FHI 674 ! **S.Nig.:** Lagos (Oct.) *Dalz.* 1345 ! *Millson* 1 ! 2 ! 3 ! 4 ! 5 ! Nun R. (Sept.) *Mann* 474 ! Ala (Nov.) *Thomas* 1948 ! Engenni R., Calabar (Sept.) *Holland* 152 ! **Br.Cam.:** Bonjongo, Victoria (Apr.) *Maitland* 637 ! On 1922 lava flow, Bibundi (June) *Rosevear* Cam. 63/37 ! Barombi *Preuss* 500 ! **F.Po:** Musola *Mildbr.* 6976 ! Also in French Cameroons, Gabon, Belgian Congo and S. Tomé. (See Appendix, p. 283.)

37. **F. aganophila** *Hutch.*—F.T.A. 6, 2 : 186. A tree 50 ft. high.
   S.Nig.: Palaver Isl., Kradu Lagoon, Lagos *Barter* 3238 ! Sapoba *Kennedy* 3029 ! Also in Spanish Guinea.

38. **F. populifolia** *Vahl*—F.T.A. 6, 2 : 189 ; Engl. Pflanzenw. Afr. 3, 1 : 42, fig. 26 ; Chev. Bot. 597 ; Aubrév. Fl. For. Soud.-Guin. 344, t. 72, 5. A large savannah tree with branchlets covered with glabrous yellowish bark.
   Togo: Sokode (Apr.) *Kersting* 62 ! **Dah.:** Dassa-Zoumé *Chev.* 23610. **N.Nig.:** Okene (Sept.) *Symington* FHI 5674 ! Dua, Kilba country *Dalz.* 144 ! Extends through French Cameroons, Ubangi-Shari, A.-E. Sudan, Abyssinia and E. Africa to Arabia. (See Appendix, p. 282.)

39. **F. platyphylla** *Del.*—F.T.A. 6, 2 : 197 ; Aubrév. l.c. 344, t. 70, 4–5 (incl. var. *pubescens*). A large savannah tree, to 60 ft. high, with rusty or pinkish-brown bark and large grey scaly patches ; foliage and figs often tinged pink ; often epiphytic at first.
   Sen.: Diourbel *Chev.* 3508. St. Louis *Chev.* 15758. Bafing R. *Lecard* 191. **Gam.:** Sukuta *Denton* ! Georgetown (Apr.) *Rosevear* 121 ! **Fr.Sud.:** Mopti to Djenné *Chev.* Arbala *Dubois* 62 ! Ouahigouya *Chev.* 24768. **Port.G.:** *Esp. Santo* 431 ! **Iv.C.:** Baoulé *Chev.* 22311. Niangbo *Serv. For.* 1728. Tafiré *Serv. For.* 1552. **G.C.:** Agomeda, Shai Plains (Jan.) *Johnson* 522 ! Achimota (June) *Irvine* 1445 ! Milne-Redhead 5106 ! **Togo:** Kete Krachi *Zech* 340. Sokodé-Basari *Kersting* 25 ; 417. Kirikiri *Kersting* 54. **Fr.Nig.:** Gaya *Aubrév.* 18n. **N.Nig.:** R. Farfara, Zurmi (Apr.) *Keay* FHI 16178 ! Katagum (Oct.) *Dalz.* 214 ! Bornu (May) *E. Vogel* 73 ! Extends to Uganda, A.-E. Sudan, Abyssinia and Somaliland. (See Appendix, p. 281.)

40. **F. abutilifolia** (*Miq.*) *Miq.*—F.T.A. 6, 2 : 191 ; Chev. Bot. 598. *Urostigma abutilifolium* Miq. (1847). *Ficus discifera* Warb. (1905)—F.T.A. 6, 2 : 196 ; F.W.T.A., ed. 1, 1 : 434 ; Aubrév. l.c. 346, t. 70, 2–3. *F. kerstingii* Warb. ex Hutch. in F.T.A. 6, 2 : 192 (1916 & 1917) ; F.W.T.A., ed. 1, 1 : 434 (excl. syn. *F. bongouanouensis* A. Chev.) ; Aubrév. l.c. t. 70, 1. A tree, to 25 ft. or more high, on rocky hills in savannah.
   Fr.Sud.: Koulikoro *Chev.* 2665 ! Bandiagara *Chev.* 24837. **Iv.C.:** Samankono *Serv. For.* 2822. Orodara *Serv. For.* 2757. Mankono *Chev.* 21852. **G.C.:** Tangkoasi Hill, Hamboi to Kwomma *Kitson* 798 ! Guo Hill, Nebiowalli (May) *Kitson* 772 ! Gambaga (June) *Akpabla* 706 ! Chana, Navrongo (Apr., May) *Vigne* FH 4515 ! 4742 ! **Togo:** Sasi-bu, Aledyo *Kersting* 125. **Dah.:** Atacora Mts. *Chev.* 24087. Agouagon *Chev.* 23542. **Fr.Nig.:** Kodjar, Gourma *Chev.* 24399. **N.Nig.:** Nupe *Barter* 452 ! 995 ! Katagum *Dalz.* 330 ! Yola (July) *Dalz.* 145 ! Also in Ubangi-Shari, Uganda and A.-E. Sudan. (See Appendix, pp. 278, 280.)

41. **F. congensis** *Engl.*—F.T.A. 6, 2 : 195 (incl. var. *mollis* Hutch.) ; Aubrév. Fl. For. C. Iv. 72, t. 18 B ; Lebrun & Boutique l.c. 167 ; Aubrév. Fl. For. Soud.-Guin. 344, t. 69, 6. A tree, 20–80 ft. high, with widely spreading, low, more or less horizontal branches supported by stilt roots ; leaves coriaceous ; locally abundant in fringing forest in the savannah regions, especially in swampy ground.
   Sen.: Kantora, Casamance *Etesse* 11. **Fr.Sud.:** Birgo *Dubois* 44. Dio *Chev.* 169. **Port.G.:** Bissalanca, Bissau (Jan.) *Esp. Santo* 1652 ! Coiada, Gabu (June) *Esp. Santo* 2499 ! **Fr.G.:** Kouroussa *Pobéguin* 17. Kissi *Chev.* 20731. Kindia *Chev.* 12777. Betagya (Oct.) *Small* 494 ! **S.L.:** Falaba (Apr., Oct.) *Deighton* 5425 ! *Small* 423 ! Kamakwie (Feb.) *Deighton* 4084 ! 4171 ! Musaia (Feb.) *Deighton* 4232 !

---

[1] Hutchinson in F.T.A. 6, 2 : 106 stated that the type (*Brunner* s.n.) of this species appeared to be lost and that his description was based entirely on two leaves collected by Chevalier ; these two leaves, as subsequently pointed out by Aubréville, are *F. goliath* A. Chev. (q.v.). There is, however, a specimen in the British Museum (*Brunner* ex Hb. Shuttleworth) which is clearly an isotype of *F. senegalensis* Miq. ; this appears to be a juvenile shoot of *F. vogelii* (Miq.) Miq.

Iv.C.: Tafiré, Bondoukou, Niangbo & Mankono *fide* Aubrév. *l.c.* **G.C.:** Birrifwa (Mar.) *Kitson* 810 ! Makongo (July) *Pomeroy* FH 1220 ! **Togo:** Baraumoba Pass, Tamberma *Kersting* A. 544 ! Basari *Kersting* A. 336. Ikeburg *Kersting* 76. **N.Nig.:** Gimi, Zaria (Feb.) *Dalz.* 349 ! 350 ! **S.Nig.:** Old Oyo F.R. *Keay* FHI 16291 ! Epe (Mar.) *Keay* FHI 16043 ! Agbabu *Kennedy* 2372 ! **Br.Cam.:** Babessi, Bamenda *Ledermann* 1978. Also in French Cameroons, Ubangi-Shari, A.-E. Sudan, Belgian Congo, Uganda, Tanganyika and N. Rhodesia. (See Appendix, p. 278.)

42. **F. goliath** *A. Chev.*—F.T.A. 6, 2 : 167 ; Chev. Bot. 605 ; Aubrév. Fl. For. C. Iv. 1 : 70, t. 18 A. *F. senegalensis* of F.T.A. 6, 2 : 105 ; of F.W.T.A., ed. 1, 1 : 432 ; not of Miq.[1] A forest tree to 100 ft. high, with cylindrical bole ; leaves on young plants and sucker shoots with large teeth.
Iv.C.: Dabou (Feb.) *Chev.* 16211 ! Montézo, Attié (planted) *Chev.* 17375 ! Morénou *Chev.* 22418. **S.Nig.:** Olokemeji (planted) *Chev.* 13937 ; 14111.

43. **F. elasticoides** *De Wild.*—F.T.A. 6, 2 : 214 ; Aubrév. l.c. 1 : 74, t. 21 C ; Lebrun & Boutique in Fl. Congo Belge 1 : 170. A forest tree, or epiphyte, with thick fissured bark ; figs greenish-yellow with white spots.
Port.G.: Abu, Formosa (Apr.) *Esp. Santo* 1952 ! **Iv.C.:** Aboisso (Apr.) *Chev.* 16293. Abidjan *Aubrév.* 1942. **S.Nig.:** Omo F.R., Ijebu-Ode (Mar.) *Jones & Onochie* FHI 17011 ! 17041 ! Also in French Cameroons and Belgian Congo. (See Appendix, p. 278.)

44. **F. pseudomangifera** *Hutch.*—F.T.A. 6, 2 : 204 ; Aubrév. l.c. 1 : 74, t. 20 A ; Lebrun & Boutique l.c. 169. A large forest tree or epiphyte.
S.L.: Sasseni, R. Scarcies (Jan.) *Sc. Elliot* 4499 ! **Lib.:** Gola (Apr.) *Bunting* 13 ! Mano R., below Jai (Apr.) *Bunting* ! **Iv.C.:** Abidjan *Aubrév.* 1643 ; 1764 ! **S.Nig.:** Sapoba *Kennedy* 2275 ! Also in Belgian Congo and Uganda.

45. **F. barteri** *Sprague*—F.T.A. 6, 2 : 205 ; Chev. Bot. 599 ; Aubrév. l.c. 1 : 74, t. 21 A ; Lebrun & Boutique l.c. 164. A forest shrub, or tree up to 25 ft., epiphytic at first ; branchlets glabrous, purple ; figs orange when ripe.
Fr.G.: Mamou to Soya *Chev.* 20367. **S.L.:** Njala (Oct.) *Deighton* 2246 ! **Lib.:** Greenville, Sinoe *Sim* 10 ! Ganta *Harley* 1116 ! **Iv.C.:** Abidjan *Aubrév.* 1580. Port Bouët *Aubrév.* 1763. **G.C.:** Axim *Chev.* 13810. **Dah.:** Tohoué, Porto Novo *Chev.* 22800 ! **S.Nig.:** Epe *Barter* 3311 ! Onitsha *Barter* 294 ! Anambra F.R., Onitsha (Feb.) *Jones* FHI 1481 ! Bonny *Kalbreyer* 79 ! Also in Spanish Guinea and Belgian Congo. (See Appendix, p. 277.)

46. **F. scott-elliotii** *Mildbr. & Burret*—F.T.A. 6, 2 : 207 ; Chev. Bot. 600 ; Aubrév. l.c. 1 : 76. A tree 20–40 ft. high, with glabrous grooved branchlets ; by rivers.
Sen.: Carabane, Casamance (Jan.) *Chev.* 2658 ! **Port.G.:** Bissara to Barro *Esp. Santo* 2423 ! Prabis, Bissau (Mar.) *Esp. Santo* 1862 ! **Fr.G.:** Mamou to Dalaba *Chev.* 20260. Farana *Chev.* 20471. **S.L.:** *Deighton* 2913 ! Erimakuna (Mar.) *Sc. Elliot* 5727 ! Sasseni *Sc. Elliot* 4522 ! Musaia (Feb.) *Deighton* 4222 ! **Iv.C.:** R. Sassandra, Soubré *Chev.* 19117.

47. **F. thonningii** *Blume*—F.T.A. 6, 2 : 187 ; Chev. Bot. 604 ; Aubrév. 1 : 72, t. 19 B ; Lebrun & Boutique l.c. 148, t. 16 ; Aubrév. Fl. For. Soud.-Guin. 347, t. 71, 6–7. A tree, to 60 ft. high, with dark green foliage, and numerous aerial roots ; often planted as a shade tree.
Sen.: *Berhaut* 23. **Fr.Sud.:** Bamako *Chev.* 144. Bamako *Chev.* 209. **Port.G.:** Bissalanca, Bissau (Feb.) *Esp. Santo* 1729 ! **Fr.G.:** Kadé *Pobéguin* 2073. Kouroussa *Pobéguin* 891. **S.L.:** Kassa *Sc. Elliot* 5064 ! **Iv.C.:** Bobo Dioulasso *Chev.* 915 ! Bondoukou *Aubrév.* 724 ! **G.C.:** Burton & Cameron ! Apenkura, Achimota (Feb.) *Irvine* 1982 ! Blengo, Krepi Plains (Jan.) *Johnson* 532 ! Bokorua, Kwahu Plateau (May) *Kitson* 1172 ! **Togo:** Atakpame *Doering* 304 ; 350. Sokode *Kersting* A. 541 ! Lome *Warnecke* 322 ! **Dah.:** Savalou (May) *Chev.* 23688 ! **Fr.Nig.:** Maradi *Kennedy* 3002 ! **N.Nig.:** Sokoto (Jan.) *Dalz.* 397 ! Zurmi (Apr.) *Keay* FHI 16116 ! Kukawa *E. Vogel* 9 ! **S.Nig.:** Lagos (Sept.) *Dalz.* 1344 ! Nun R. *T. Vogel* 27 ! **Br.Cam.:** Cam. Mt., 3,000–6,000 ft. (Apr.) *Hutch. & Metcalfe* 40 ! **F.Po:** Moka (Nov.) *Mildbr.* 7072 ! Widespread in tropical Africa. (See Appendix, p. 282.)

48. **F. basarensis** *Warb. ex Mildbr. & Burret*—F.T.A. 6, 2 : 181 ; Chev. Bot. 603. An epiphyte or tree up to 45 ft., with rather slender pubescent branchlets.
Sen.: Marsasoum, Casamance *Chev.* **Togo:** Basari, growing on an old baobab (June, Nov.) *Kersting* A. 21 ; 221 ; 533 ! **Dah.:** *Poisson* ! Also in Ubangi-Shari.
[Probably only a variety of *F. thonningii* Blume.]

49. **F. natalensis** *Hochst.*—F.T.A. 6, 2 : 208 ; M. F. Barrett l.c. 204 ; Aubrév. Fl. For. Soud.-Guin. 347. A shrub or tree up to 40 ft. high, sometimes epiphytic at first ; figs becoming reddish.
G.C.: Cape Coast *Brass* ! Kwahu (Mar.) *Johnson* 639 ! Prampram (May) *Irvine* 1432 ! Aburi (Sept.) *Johnson* 178 ! **N.Nig.:** Katabu, Kaduna (May, Oct.) *Keay* FHI 21116 ! 22913 ! **S.Nig.:** Agulu *Thomas* 350 ! **F.Po:** on the beach *Mann* 252 ! Also in French Cameroons, extending to East and Central Africa, Angola and S. Africa. (See Appendix, p. 280.)

50. **F. dekdekena** *(Miq.) A. Rich.*—F.T.A. 6, 2 : 211 ; Aubrév. Fl. For. C. Iv. 1 : 76, t. 17 C ; M. F. Barrett l.c. 207 ; Aubrév. Fl. For. Soud.-Guin. 341, 347. *Urostigma dekdekena* Miq. (1847). A tree or epiphyte, with nearly glabrous branchlets ; figs pale in colour, generally in pairs, and numerous on the young shoots.
Iv.C.: *fide* Aubrév. *l.c.* **G.C.:** *Evans* 4 ! Pong (Mar.) *Kitson* 617 ! **N.Nig.:** Zaria (Oct.) *Keay* FHI 21423 ! Maigana, Zaria (Dec.) *Lamb* 62 ! Musgu (May) *E. Vogel* 54 ! Widespread in tropical Africa.

51. **F. iteophylla** *Miq.*—F.T.A. 6, 2 : 203 ; Chev. Bot. 605 ; Aubrév. Fl. For. Soud.-Guin. 347, t. 72, 1, *F. spragueana* Mildbr. & Burret (1911)—F.T.A. 6, 2 : 202 ; F.W.T.A., ed. 1, 1 : 434. (?) *F .sassandrensis* A. Chev. Bot. 606, name only. A tree up to 40 ft. high with a girth of 15 ft. and smooth pale bark, often epiphytic at first ; figs crowded on the branchlets.
Sen.: Cayor (June) *Leprieur* ! Fadiout *Ezanno* 56. **Gam.:** Georgetown, MacCarthy Isl. (Jan.) *Dalz.* 8434 ! **Fr.Sud.:** Bandiagora *Ganay.* **Fr.G.:** Diaguissa to Boulivel *Chev.* 12922. **N.Nig.:** Sokoto (Dec.) *Dalz.* 398 ! Katagum (Dec.) *Dalz.* 216 ! Geidam (Dec.) *Elliott* 133 ! Extends through Ubangi-Shari and Chad to A.-E. Sudan. (See Appendix, p. 280.)
[Perhaps only a variety of *F. dekdekena* (Miq.) Miq., the earlier name.]

52. **F. sp. nr. cyathistipuloides** *De Wild.* in Fedde Rep. 12 : 194 (1913) ; Aubrév. Fl. For. C. Iv. 1 : 74, t. 20 B. A large epiphyte, in rain forest.
Iv.C.: Banco (Oct.) *Aubrév.* 1646 ! Port Bouët *Aubrév.* 1645 !
[This appears to differ from *F. cyathistipuloides* De Wild., still known only from the type specimen collected in the Belgian Congo, by its large stipules (3–3·5 cm. long) and rather more numerous lateral nerves. In the original description the stipules are said to be short. More material from the Belgian Congo is needed to decide whether or not Aubréville's plant is conspecific.]

53. **F. umbellata** *Vahl*—F.T.A. 6, 2 : 124 ; Chev. Bot. 598 ; Aubrév. l.c. 1 : 76, t. 21 B ; Fl. For. Soud.-Guin. 344 ; Lebrun & Boutique l.c. 134. A tree 20–40 ft. high, with stout glabrous branchlets; leaves with prominent bifurcating lateral nerves ; figs green with pale spots, borne in clusters on the branches away from the leaves.
Sen.: Niokolo-Koba *Berhaut* 119. **Port.G.:** Bissalanca, Bissau (Jan.) *Esp. Santo* 1645 ! **Fr.G.:** Los Isl. *Pobéguin* 1229. Conakry *Dybowski* 7. **S.L.:** Tower Hill, Freetown (May) *Deighton* 2902 ! Wallia, Scarcies (Jan.) *Sc. Elliot* 4639 ! Musaia (Feb., Dec.) *Deighton* 4248 ! 4250 ! 5713 ! **Iv.C.:** Nianglo Serv. For. 1726. And various localities *fide* Chev. l.c. **G.C.:** Sefwi *Armitage* ! Apenkwa, Achimota *Irvine* 1984 ! **Togo:** Atakpame *Busse* 3532. **N.Nig.:** Bida (Mar.) *Lamb* 34 ! **S.Nig.:** Oni Gambari, Ibadan *Jones* FHI 20258 ! Also in French Cameroons, Ubangi-Shari, French and Belgian Congo and Angola. (See Appendix, p. 282.)

---

[1] See footnote on page 609.

54. **F. polita** *Vahl*—F.T.A. 6, 2 : 124 ; Chev. Bot. 599 ; Aubrév. Fl. For. C. Iv. 1 : 78, t. 22 C ; Fl. For. Soud.-Guin. 344, t. 69, 4–5 ; Lebrun & Boutique in Fl. Congo Belge 1 : 135. A tree, to about 50 ft. high, with rather slender almost glabrous branchlets and dense shining foliage ; commonly planted as a shade tree.

Sen.: Thiès *Berhaut* 379. Koulaye *Chev.* 2662. Floup-Fedyou *Chev.* **Fr.Sud.:** Iton *Chev.* 2663. **Port.G.:** Bissalanca, Bissau (Jan.) *Esp. Santo* 1653 ! **Fr.G.:** Los Isl. *Pobéguin* 1223. Conakry *Dybowski* 6. **S.L.:** York *Deighton* 2731 ! Njala (Jan.) *Deighton* 2834 ! **Iv.C.:** Agboville, Port Bouët, Kaya & Beriby *fide* Aubrév. *l.c.* **G.C.:** Krepi Plains (Jan.) *Johnson* 549 ! 551 ! Dodowa *Irvine* 1985 ! Ejura *Kitson* 621 ! **Togo:** Ndsolo (Apr.) *Baumann* 561 ! Kete Krachi *Zech* 4. **Dah.:** Dinkodéniou *Chev.* 23219. **N.Nig.:** Mando, Birnin Gwari (June) *Keay* FHI 25887 ! Zungeru (June) *Dalz.* 568 ! Katagum *Dalz.* 321 ! **S.Nig.:** Owerri (Apr.) *Sherriff* ! **Br.Cam.:** Barombi (Aug.) *Preuss* 409 ! Victoria *Winkler* 428. Extends to A.-E. Sudan, E. Africa and Angola. (See Appendix, p. 281.)

55. **F. kimuenzensis** *Warb.*—F.T.A. 6, 2 : 138 ; Lebrun & Boutique l.c. 1 : 132. *F. rudens* Hutch. (1915)— F.T.A. 6, 2 : 136. A scandent epiphyte ; receptacles greenish-yellow, tinged purple, with many prominent small pale spots.

S.Nig.: Calabar (Mar., Apr.) *Brenan* 9256 ! 9595 ! Also in Belgian Congo and Angola.

56. **F. dryepondtiana** *Gentil*—F.T.A. 6, 2 : 127 ; Lebrun & Boutique l.c. 134. An epiphytic shrub or tree. **Br.Cam.:** on 1922 lava flow, Bibundi (June) *Rosevear* Cam. 64/37 ! Also in French Cameroons, Gabon and Belgian Congo.

[The description of the leaves given in the key is based only on specimens collected in Africa ; specimens cultivated in hot-houses at Brussels and Kew have larger leaves distinctly cordate at the base.]

57. **F. macrosperma** *Warb. ex Mildbr. & Burret* in Engl. Bot. Jahrb. 46 : 223 (1911) ; F.T.A. 6, 2 : 130 ; Aubrév. Fl. For. C. Iv. 1 : 78 (as *F. sp. aff. macrosperma*), t. 22 B. *F. buntingii* Hutch. (1915)—F.T.A. 6, 2 : 128 ; F.W.T.A., ed. 1, 1 : 435. *F. mangenotii* A. Chev. in Rev. Bot. Appliq. 29 : 246, t. 11 (1949). Epiphytic at first, becoming a forest tree up to 130 ft. high ; figs borne on short stout slow-growing branchlets borne on the older branches and bole, mottled pink.

S.L.: *Lane-Poole* 428 ! Kangahun, Kaiyamba (Nov.) *Deighton* 6149 ! **Lib.:** Mano R., Gola Forest (Apr.) *Bunting* ! Ganta (May) *Harley* 1177 ! **Iv.C.:** Banco *Aubrév.* 1587 ! 1907 ! Tabou *Aubrév.* 1743 ! Bingerville (Feb.) *Chev.* Yapo *Mangenot & Miège.* **Togo:** Ensaku-Bena, Atakpame (Jan.) *Doering* 356 ! **Br.Cam.:** Bali-Ngemba F.R., Bamenda (May) *Ujor* 30345 ! Also in French Cameroons and Gabon.

58. **F. elegans** (*Miq.*) *Miq.*—F.T.A. 6, 2 : 128 ; Chev. Bot. 599 ; Aubrév. l.c. 1 : 78, t. 23 A. *Urostigma elegans* Miq. (1848). *Ficus macrosperma* of Chev. Bot. 599 and of F.W.T.A., ed. 1, 1 : 435, partly (*Chev.* 23650). A tree, to 40 ft. high, epiphytic at first, with stiff minutely puberulous branchlets, glabrous but dull foliage and large pedunculate figs on the older branches.

Iv.C.: Agboville *Aubrév.* 1935. Also *fide* Chev. *l.c.* **G.C.:** Cape Coast *T. Vogel* 25 ! 87 ! Krobo Plains (Jan.) *Johnson* 559 ! **Togo:** east of Njande *Doering* 228. **Dah.:** Dassa-Zoumé (May) *Chev.* 23650 ! **S.Nig.:** Lagos (Jan.) *Dalz.* 1394 ! *Millen* 93 ! Ibadan (Feb.) *Brenan & Keay* 8949 ! Gambari Group F.R., Ibadan (Aug.) *Onochie* FHI 19137 ! **Br.Cam.:** Ninong *N.W. Cam. Co. coll.* (See Appendix, p. 278.)

59. **F. ottoniifolia** (*Miq.*) *Miq.*—F.T.A. 6, 2 : 134 ; Aubrév. l.c. 1 : 78, t. 22 D ; Lebrun & Boutique l.c. 133. *Urostigma ottoniifolium* Miq. (1848). *Ficus buettneri* Warb. (1894)—Chev. Bot. 599. *F. maculosa* Hutch. (1915)—F.T.A. 6, 2 : 138 ; F.W.T.A., ed. 1, 1 : 435. *F. sp. aff. fasciculiflora* of Aubrév. l.c. 1 : 80, 22 A. A shrub or tree to 50 ft. high, usually epiphytic at first, with smooth pale branchlets ; figs rather long-stalked in pairs or clusters on older branches below the leaves.

Port.G.: Brene, Bissau (Jan.) *Esp. Santo* 1695 ! **Fr.G.:** Conakry *Dybowski* 3. Los Isl. *Pobéguin* 1236. Labé *Chev.* 12374. **S.L.:** Musaia (Mar.) *Sc. Elliot* 5151 ! Madina (Apr.) *Sc. Elliot* 5568 ! Roruks (Nov.) *Thomas* 5799 ! Njala (Feb., Mar., Nov.) *Deighton* 518 ! 1820 ! 5857 ! 5867 ! **Lib.:** Peahtah *Linder* 928 ! Mecca (Dec.) *Baldwin* 10824 ! Butaw (Mar.) *Baldwin* 11487 ! **Iv.C.:** Bondoukou (Jan.) *Aubrév.* 736 ! Abidjan *Aubrév.* 1744. Tabou *Aubrév.* 1762. **G.C.:** Asientiem *Chipp* 292. **Togo:** Lome (May) *Warnecke* 324 ! Bismarckburg *Büttner* 714. Baranmola, Tamberma (Apr.) *Kersting* A. 545 ! **Dah.:** *Poisson* 451. Cotonou *Chev.* 4453. **S.Nig.:** Ebute Metta (Sept.) *Dalz.* 1342 ! Ibadan (Sept.) *Onochie* FHI 20963 ! Near Ajebandele (Mar.) *Jones & Onochie* FHI 17208 ! **Br.Cam.:** Johann-Albrechtshöhe *Staudt* 896 ! **F.Po:** *Mann* 431 ! *T. Vogel* 176 ! Also in French Cameroons and Belgian Congo. (See Appendix, p. 281.)

In addition to the above, several introduced species occur in our area in cultivation, including *F. pumila* Linn., a native of China and Japan, which is commonly grown as an ornamental creeper on houses, and *F. carica* Linn., the edible fig. *F. bubu* Warb. (F.T.A. 6, 2 : 166 ; F.W.T.A., ed. 1, 1 : 434) is recorded as an avenue tree at Bismarckburg, Togo ; it is indigenous in Spanish Guinea and Belgian Congo.

Imperfectly known species.

1. **F. spirocaulis** *Mildbr.* Wiss. Ergebn. 1910–11 : 181 (1922), name only. Fragments of the two specimens cited by Mildbraed are at Kew, both are named *F. urceolaris* Welw. ex Hiern and are cited as such in F.T.A. 6, 2 : 110 and F.W.T.A., ed. 1, 1 : 432 ; one of them (*Mildbr.* 7010) is cited a second time under this species by Mildbraed (l.c.). More material is required.
   F.Po : Bokoko *Mildbr.* 6900 ! Musola *Mildbr.* 7010 !

2. **F. warburgii** *Winkl.* in Engl. Bot. Jahrb. 41 : 276 (1908). This species is included in F.T.A. 6, 2 : 109 under *F. pendula* Welw. ex Hiern (1900), not of Link. (1822) ; the type number (*Winkler* 449) is, however, cited a second time in F.T.A. 6, 2 : 110 under *F. urceolaris* Welw. ex Hiern. The specimen is too poor for accurate determination. More material is required.
   Br.Cam.: Victoria to Bota (Sept.) *Winkler* 449 !

3. **F.** (subgen. *Sycomorus*) sp. Possibly new ; more material required. Tree with habit of an alder, straggling over rocks in stream bed ; leaves similar to those of *F. vallis-choudae*, receptacles similar to those of *F. mucuso.*
   S.Nig.: R. Ata, below Mt. Koloishe, Obudu Div. 4,000 ft. (Dec.) *Keay & Savory* FHI 25061 !

4. **F.** (? subgen. *Sycomorus*) sp. Photograph No. B. 73, taken by R. Ross in the Shasha (now Omo) F.R., S. Nigeria, shows a species which does not seem to be represented by any herbarium specimen that I have seen. It is a tall forest tree with low but very widely spreading plank-buttresses ; the figs are borne in long lax panicles on these buttresses.

5. **F.** (subgen. *Bibracteateae*) sp. Near to *F. camptoneura* and *F. cyathistipula* ; more material required.
   Lib.: Jaurazon (Mar.) *Baldwin* 11479 ! Mecca, Boporo (Nov.) *Baldwin* 10404 !

Doubtfully recorded species.

1. **F. mangiferoides** *Hutch.*—F.T.A. 6, 2 : 205. A sterile specimen from our area (*Kalbreyer* 80 from Bonny, S. Nigeria) has been named this by Hutchinson ; it is perhaps *F. barteri* Sprague ; better material is needed.

2. **F. pendula** *Welw. ex Hiern* in Cat. Welw. 1 : 1008 (1900), not of Link. (1822) ; F.T.A. 6, 2 : 108. Recorded from our area in the first edition on account of *Winkler* 449, the type number of *F. warburgii* Winkl., see above.

3. **F. urceolaris** *Welw. ex Hiern*—F.T.A. 6, 2 : 109. Recorded from our area in the first edition on account of *Winkler* 449 and *Mildbr.* 6900 & 7010 ; see above notes under *F. warburgii*, *F. pendula* and *F. spirocaulis*.

7. BOSQUEIA [1] Thouars ex Baill.—F.T.A. 6, 2 : 217. *Pontya* A. Chev.· (1912).

Branchlets glabrous ; leaves elliptic or obovate-elliptic, shortly cuneate at base, obtusely acuminate, 5–16 cm. long, 2·5–7·5 cm. broad, entire, shining above, pale beneath, glabrous ; lateral nerves about 7 pairs ; petiole 1 cm. long ; receptacles axillary, pedunculate ; bract at base of peduncle cupular, enclosing head of flowers when young, at length bursting on one side ; peduncle 5 mm. long ; female calyx short and tubular, acutely 5-cleft at the apex ; style stout, bifurcate ; fruit 2 cm. long ; obliquely ellipsoid, striate, glabrous .. ..   *angolensis*

B. angolensis *Ficalho*—F.T.A. 6, 2 : 218 ; Aubrév. Fl. For. C. Iv. 1 : 50, t. 11 ; Hauman in Fl. Congo Belge 1 : 95. *B. welwitschii* Engl. Monogr. Afr. Morac. 36 ; Bot. Jahrb. 51 : 439, fig. 5, A–E. *Pontya excelsa* A. Chev. (1912). *Bosqueia phoberos* of Aubrév. Fl. For. Soud.-Guin. 331, not of Baill. A forest tree, up to 100 ft. high, with narrow buttresses and smooth pale bark, exuding a white latex which rapidly changes to orange and finally red ; branchlets with circular leaf-scars.
Fr.G.: Lola to Nzo (fr. Mar.) *Chev.* 20990 ! Kissi Circle (Feb.) *Chev.* 20765. **Iv.C.:** *Aubrév.* 984 ! Findimanou (Dec.) *Chev.* 22480 ! Zaranou (young fr. Mar.) *Chev.* 16278 ! Bangouanou *Chev.* 22463 ! **G.C.:** Adeambra (fr. May) *Vigne* FH 858 ! Saraha, Wassaw F.R. (Feb.) *Vigne* FH 209 ! 985 ! Mampong Scarp (fr. Mar.) *Vigne* FH 1058 ! **Dah.:** Djougou (fr. June) *Chev.* 23885 ! Savalou Circle (May) *Chev.* 23748 ! 23785. **S.Nig.:** Gambari Group F.R., Ibadan (fr. June) *W. D. MacGregor* 568 ! *Symington* FHI 4103 ! Sapoba (fl. & fr. Jan.) *Kennedy* 566 ! 1681 ! Ikom (Dec.) *Catterall* 54 ! **Br.Cam.:** Victoria (fl. & fr. Jan.) *Maitland* 186 ! Tiko (Jan.) *Dunlap* 247 ! Likomba *Mildbr.* 10592 ! Also in Gabon, Belgian Congo, Cabinda, Angola and Annobon. (See Appendix, p. 274.)

8. SCYPHOSYCE Baill.—F.T.A. 6, 2 : 220.

Branchlets shortly hispid ; leaves obovate or oblanceolate, not pandurate, gradually narrowed to the asymmetrically auriculate base, obtusely caudate-acuminate, denticulate or dentate-undulate or subentire towards the apex, margin not thickened, 6–16 cm. long, 2·3–6 cm. broad ; receptacles axillary, on peduncles about 1 cm. long, campanulate-obconic ; fruit ovoid, glabrous, about 4 mm. long, exserted from the receptacle .. .. 1. *manniana*
Branchlets glabrous, or nearly so ; leaves pandurate, broadly ovate in the upper half, abruptly narrowed to the symmetrical subauriculate base, gradually and obtusely pointed at apex, margin strongly thickened, 10–12 cm. long, 4–5 cm. broad ; receptacles axillary, on peduncles about 1 cm. long, ellipsoid-obconic.. 2. *pandurata*

1. S. manniana *Baill.*—F.T.A. 6, 2 : 220 ; Brenan in Kew Bull. 1952 : 449. *S. zenkeri* (Engl.) Engl. Monogr. Afr. Morac. 31, t. 10 (1898). A woody plant of herbaceous habit, to 1 ft. high ; in rain forest.
S.Nig.: Okomu F.R., Benin *Richards & Brenan* 8865 ! Also in French Cameroons and Gabon.
2. S. pandurata *Hutch.* in F.T.A. 6, 2 : 221 (1917) ; Kew Bull. 1919 : 263. A woody undershrub of herbaceous habit ; branchlets grooved with transversely peeling bark ; in rain forest.
S.Nig.: Oban *Talbot* 687 ! 1315 ! 2317. [The fruit on a sheet of *Talbot* 1315 appears to be developing within the receptacle, that is below the remains of the male flowers ; further material with ripe fruit is needed to elucidate the generic position of this plant.]

Imperfectly known species.

S. sp. Possibly new ; more material required.
S.Nig.: Oban *Talbot* 680 !

9. ANTIARIS Lesch.—F.T.A. 6, 2 : 223.

Venation of mature leaves very prominently reticulate beneath ; leaves scabrid on both surfaces ; branchlets rusty-tomentose ; mature crown leaves broadly oblong-obovate or oblong, unequally cordate, subcordate or rounded at base, obtuse or rounded or sometimes subacute at apex, 5·5–16 cm. long, 3·5–11 cm. broad, shortly scabrid-pubescent beneath ; flowers dioecious or monoecious ; male flowers on flattened receptacles, 1–1·5 cm. diam., axillary on peduncles 0·7–1·7 cm. long ; female flowers solitary ; fruits drupaceous, ellipsoid, 1–1·5 cm. long, tomentellous
             1. *africana*
Venation of mature leaves obscure and lax beneath ; leaves not scabrid, shining above ; branchlets pubescent or puberulous ; mature crown leaves broadly oblong-obovate or oblong or suborbicular, unequally cordate or rounded at base, rounded or obtuse at apex, 5–14 cm. long, 3–9 cm. broad, soon glabrous or nearly so beneath ; flowers and fruits similar to the above .. .. 2. *welwitschii*

1. A. africana *Engl.*—F.T.A. 6, 2 : 223 ; Engl. Pflanzenw. Afr. 3, 1 : 33, fig. 20 B ; Chev. Bot. 606 ; Aubrév. Fl. For. C. Iv. 40–41, t. 6, 1 ; Hauman in Fl. Congo Belge 1 : 94. *A. kerstingii* Engl. l.c. fig. 20 A (1915) ; Chev. Bot. 606, partly. *A. toxicaria* (Rumph. ex Pers.) Lesch. var. *africana* Sc. Elliot—Chev. Bot. 606. A large deciduous tree, to 130 ft. high ; with grey bark, slash exuding a watery latex which soon darkens to the colour of milky tea ; ripe fruits red or orange ; in the drier types of forest.
Sen.: Kaolak *Kaichinger.* Sedhiou *Chev.* **Gam.:** *Dawe* 82 ! **Fr.Sud.:** Sansanding *Chev.* 2551. **Port.G.:** Bissau (young fr. Feb.) *Esp. Santo* 1741 ! **Fr.G.:** Farana (Jan.) *Chev.* 20457 ! **S.L.:** Scarcies R. *Sc. Elliot* 5892 ! Kasewe F.R. (fr. Mar.) *Deighton* 4593 ! **Iv.C.:** Dabou *Chev.* 16217 ! Bouroukrou *Chev.* 16639 ! Agboville *Chev.* 22337 ! **G.C.:** Tanosu (Feb.) *Chipp* 97 ! Aburi (Oct.) *Johnson* 460 ! **Togo:** Lome (May) *Warnecke* 336 ! Difala (fl. & young fr. Dec.) *Kersting* ! Misahöhe *Mildbr.* 7331 ! **Dah.:** Quidat *Poisson.* Atacora Mts. *Chev.* 24101. **N.Nig.:** Zaria *Shaw* 102 ! Agaie (fl. & young fr. Dec.) *Yates* 19 ! 19a ! Yola (Mar.) *Dalz.* 177 ! **S.Nig.:** Dodd 399 ! Ibadan F.R. (Feb.) *Punch* 111 ! Also in Ubangi-Shari, A.-E. Sudan, Uganda and (?) Belgian Congo. (See Appendix, p. 273.)

[1] This is the original and correct spelling ; the spelling *Bosquiea* given in F.W.T.A., ed. 1, 2 : 607 should not be used.

2. **A. welwitschii** *Engl.*—F.T.A. 6, 2 : 224 ; Engl. l.c. fig. 20 D ; Aubrév. l.c. t. 6, 2–7 ; Hauman l.c. 93, photo 6. A large tree, to 130 ft. high, similar to the above but readily distinguished by the smooth mature leaves ; in the wetter parts of the forest regions.
S.L.: Colonial Reserve, Freetown (Jan.) *King-Church* 1/1922 ! **Iv.C.:** Banco, Abidjan *Aubrév.* 48 ! 205 ! **G.C.:** Owabi (Dec.) *Lyon* FH 2650 ! Bobiri (fr. Mar.) *Andoh* FH 4974 ! Dunkwa *Vigne* FH 922 ! **S.Nig.:** Degema (young fr. Jan.) *Cons. of For.* OC 44 ! Sapoba *Ainslie* 2 ! *Keay* FHI 24679 ! Mamu River F.R. *Keay* ! Calabar *Smith* 26 ! **Br.Cam.:** Likomba *Mildbr.* 10556 ! Also in Uganda, Tanganyika, Belgian Congo and Angola.

Some botanists have referred all African specimens of this genus to the Asiatic species *A. toxicaria* (Rumph. ex Pers.) Lesch. There appear, however, to be at least two easily recognizable taxa in West Africa, and, pending a thorough revision of the whole genus, it seems best to continue treating these as species. The foliage of seedlings, saplings and sucker shoots is rather different from that of the crown. Specimens of such foliage cannot at present be determined with any confidence ; the only record of the genus from Fernando Po (*Barter* 1674) is one such specimen.

10. TRECULIA Decne.—F.T.A. 6, 2 : 225. *Acanthotreculia* Engl. (1908).

Floral bracts with a flat peltate apical appendage ; flower-heads globose ; infructescences not prickly :
Male flower-heads 4–7 cm. diam. at anthesis ; anthers 2–3 mm. long ; infructescences globose, very large, often 17–35 cm. diam. and weighing up to 12 kilograms ; seeds very numerous, ellipsoid, about 1 cm. long ; leaves elliptic or ovate-elliptic, asymmetrical at base with one side cuneate to subcordate and the other obtuse to broadly auriculate, more or less acuminate, 8–45 cm. long, 3–20 cm. broad, main lateral nerves 9–15 on each side of midrib ; forest trees, up to 35 m. high :
  Branchlets, petioles and undersurface of leaves soon glabrous
                                                                            1. *africana* var. *africana*
  Branchlets, petioles and undersurface of leaves softly and shortly pilose
                                                                            1a. *africana* var. *mollis*
Male flower-heads about 6 mm. diam. at anthesis, borne on short axillary bracteate peduncles ; anthers about 1 mm. long ; infructescences not known ; leaves oblong-elliptic to obovate, cuneate to rounded at base, caudate-acuminate, 7–20 cm. long, 2·5–8 cm. broad, main lateral nerves 6–12 on each side of midrib, looped and prominent beneath, glabrous ; branchlets puberulous ; a slender shrub, 2–3 m. high
                                                                            2. *zenkeri*
Floral bracts of the male ovoid or subclavate at apex, of the female lanceolate ; flower-heads ellipsoid or obovoid ; infructescences prickly on account of the numerous elongate floral bracts, 3–5·5 (–8) cm. long, 3·5–4 cm. or more broad (including the prickles) ; leaves oblong-elliptic to obovate, asymmetrical at base with one side more or less cuneate and the other rounded or obtuse, abruptly caudate-acuminate, 10–24 cm. long, 4·5–9 cm. broad, main lateral nerves 7–15 on each side of midrib, looped and prominent beneath, glabrous ; branchlets soon glabrous ; a tree, to 25 m. high ..   ..   3. *obovoidea*

1. **T. africana** *Decne.* var. **africana**—F.T.A. 6, 2 : 226 ; Engl. Monogr. Afr. Morac. 32, tt. 12, 13, 14 B ; Chev. Bot. 607 ; Aubrév. Fl. For. C. 1v. 1 : 42, t. 7 ; Hauman in Fl. Congo Belge 1 : 90. A forest tree, to 120 ft. high, with fluted bole up to 9 ft. in girth, bark smooth, slash exuding abundant latex ; the large green infructescences are often abundant on the trunk and larger branches ; Aubréville states that some cultivated trees produce infructescences up to 50 cm. diam. and weighing up to 15 kilograms.
Sen.: *Heudelot*. Tambanala *Chev.* 2640 ! Niokolo-Koba *Berhaut* 1519. **Port.G.:** Bor, Bissau *Esp. Santo* 5 ! Safim, Bissau (Mar.) *Esp. Santo* 1907 ! **Fr.G.:** Santa to Timbo *Chev.* 12593. Socourala *Chev.* 20496. **S.L.:** Tawia, Scarcies (Jan.) *Sc. Elliot* 4477 ! Njala *Deighton* 2829 ! **Lib.:** Sinoe Basin *Whyte* 21 ! **Iv.C.:** Bouroukrou *Chev.* 16110. **G.C.:** Pra R. Station, Ashanti (fr. Feb.) *Vigne* FH 1618. **S.Nig.:** Yoruba country (fr. Apr.) *Millson* ! Olokemeji *Foster* 150 ! Onitsha *Barter* 432 ! Awka *Thomas* 72 ! **Br.Cam.:** Ambas Bay (fr. Feb.) *Mann* 773 ! Extends to A.-E. Sudan, Mozambique and Angola, and on Principe and S. Tomé islands. (See Appendix, p. 286.)
1a. **T. africana** var. **mollis** (*Engl.*) *J. Léonard* in Bull. Jard. Bot. Brux. 18 : 145 (1947) ; Hauman l.c. 91. *T. mollis* Engl. (1908)—F.T.A. 6, 2 : 229 ; F.W.T.A., ed. 1, 1 : 436. A forest tree, similar to the preceding, but more or less densely pilose ; infructescences probably rather smaller.
S.Nig.: Okomu F.R., Benin (fr. Jan.) *Brenan* 8786 ! Sapoba, Benin (fl. Oct., fr. Feb., Mar.) *Kennedy* 1863 ! 2166 ! 2609 ! Oban *Talbot* 635 ! Also in French Cameroons, Cabinda and Belgian Congo.
2. **T. zenkeri** *Engl.*—F.T.A. 6, 2 : 229 ; Engl. l.c. 34, t. 15 A. A slender forest shrub.
S.Nig.: Oban *Talbot* 654 ! 1506 ! s.n. ! **Br.Cam.:** S. Bakundu F.R., Kumba (fl. buds Mar.) *Onochie* FHI 32058 ! Also in French Cameroons.
3. **T. obovoidea** *N.E.Br.*—F.T.A. 6, 2 : 228 ; Hauman l.c. 92 (in note). *Acanthotreculia winkleri* Engl. Bot. Jahrb. 40 : 548, fig. 2 (1908). A forest tree, 8–25 ft. high, or shrub ; slash pink, immediately exuding white latex which turns brown ; infructescences like little hedgehogs, borne on trunk and branches.
S.Nig.: Calabar (Feb.) *Mann* 2303 ! *Thomson* 104 ! Ojo to Nyenekosum, Itu (fr. June) *Ujor* FHI 28000 ! Oban *Talbot* 2321 ! 2335 ! Ikom *Catterall* 10/1934 ! **Br.Cam.:** Moliwe (fr. May) *Winkler* 1283. S. Bakundu F.R. (Mar., Apr.) *Brenan* 9272 ! *Ejiofor* FHI 29351 ! *Olorunfemi* FHI 30538 ! Johann-Albrechtshöhe (Feb., Mar.) *Staudt* 633 ! 685 ! 934 ! Also in French Cameroons. (See Appendix, p. 287.)

11. MYRIANTHUS P. Beauv.—F.T.A. 6, 2 : 230.

Leaves digitately 5–7-foliate, rarely some very deeply digitately lobed :
  Terminal leaflet with 25–34 lateral nerves on each side of midrib, up to 50 cm. long and 22 cm. broad ; female flowers numerous, obovoid, thickened and flattened at the top, densely aggregated into paired subglobose pedunculate heads ; infructescence subglobose, up to 10 cm. diam. ; male flowers more or less completely covering the short ultimate branchlets of the panicles ; leaflets subcoriaceous, usually shortly petiolulate, or sessile, or shortly joined at the base, obovate, closely and coarsely

21

serrate with numerous teeth, closely and prominently reticulate and cottony-
tomentellous beneath   ..   ..   1. *arboreus*
Terminal leaflet with 10–22 lateral nerves on each side of midrib, up to 25 cm. long
   and 8 cm. broad ; female flowers up to about 24 in each of the paired pedunculate
   heads, ovoid, pointed at the top ; infructescences up to 6 cm. diam., consisting of
   4–14 fruits united below and beaked with the narrow mouth of the persistent
   perianth ; male flowers in oblong or roundish clusters sometimes more or less
   confluent on the short ultimate branchlets of the panicles ; leaflets thin, distinctly
   petiolulate with petiolules up to 3 cm. long, oblong-oblanceolate or oblong-elliptic,
   margin varying from subentire to undulate-dentate or serrate, rather laxly reticulate
   and cottony-tomentellous beneath   ..   ..   2. *preussii*
Leaves simple, not lobed or shortly 3-lobed ; female flowers up to about 15 in each of
   the paired pedunculate heads, ovoid, pointed at the top ; infructescences rather
   irregularly lobulate, up to 3·5 cm. diam., consisting of 5–15 fruits united below and
   rounded or pointed above ; leaves coriaceous, coarsely serrate :
Branchlets and petioles puberulous when young, glabrescent ; male flowers more or
   less completely covering the short ultimate branchlets of the panicles ; leaves
   apparently never lobed, obovate-elliptic to ovate-elliptic, rarely suborbicular,
   rounded to cuneate at base, acute or rounded at apex, 16–36 cm. long, 8–15 cm.
   broad ; 8–11 main lateral nerves on each side of midrib ; petioles up to 7 cm. long ;
   peduncle of infructescence up to 1·2 cm. long   ..   ..   ..   3. *serratus*
Branchlets and petioles densely pubescent when young and remaining distinctly
   pubescent ; male flowers in oblong or roundish clusters ; leaves often (2–) 3-lobed,
   strongly 3-nerved at base, unlobed leaves more or less elliptic, cuneate at base,
   acutely acuminate, 8–35 cm. long, 4–14 cm. broad, lobed leaves up to 37 cm. long
   and 34 cm. broad ; 10–20 main lateral nerves on each side of midrib ; petioles
   1–21 cm. long ; peduncle of infructescence up to 3·6 cm. long   ..   4. *libericus*

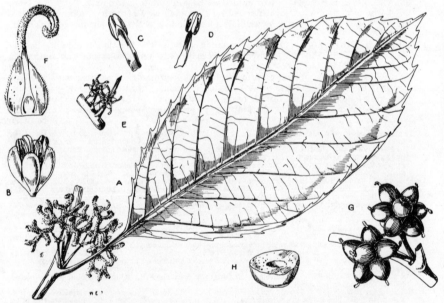

Fig. 174.—Myrianthus serratus (*Trécul*) *Benth.* (Moraceae).

A, male shoot. B, male flower. C, D, stamen back and front. E, female flowers. F, section
through ovary showing orthotropous ovule. G, fruits. H, cross-section through fruit.

1. **M. arboreus** *P. Beauv.*—F.T.A. 6, 2 : 231 ; Engl. Monogr. Afr. Morac. 37, t. 16 ; Aubrév. Fl. For. C. Iv. 1 :
46, t. 8 ; Hauman in Fl. Congo Belge 1 : 83. *M. talbotii* Rendle (1915), partly. A tree to 65 ft. high,
with spreading branches from a short stem ; male flowers yellow ; fruits yellow when ripe ; in forest
regrowth and damp places in forest.
   S.L.: Kambui F.R. (fl. & fr. Feb.) *Lane-Poole* 332 ! Baiima (Apr.) *Deighton* 3175 ! Bwedu to Dia (fr.
Apr.) *Deighton* 3176 ! **Lib.:** Diebla (fr. July) *Baldwin* 6325 ! **Iv.C.:** Abidjan region *Chev.* 15394. Potou
Lagoon (Feb.) *Chev.* 17382. Man *Chev.* 21546. **G.C.:** Dunkwa *Cummins* 7 ! Aburi (Jan.) *Brown* 939 !
Anobelso (Mar.) *Chipp* 125 ! **Togo:** Misahöhe (Feb.) *Baumann* 430. Bismarckburg *Büttner* 44. **Dah.:**
Le Testu 240 ! **N.Nig.:** Lapai (Feb., Mar.) *Yates* 34 ! **S.Nig.:** Yoruba forests *Barter* 3407 ! Ajilite (Jan.)
Millen 117 ! Benin *Unwin* 40 ! Onitsha *Barter* 1677 ! Eket *Talbot* ! Oban *Talbot* 2313 ! **Br.Cam.:**
Ambas Bay (Feb.) *Mann* 716 ! Also in French Cameroons, Gabon, Belgian Congo, A.-E. Sudan,
Uganda, Kenya, Tanganyika and Angola. (See Appendix, p. 285.)

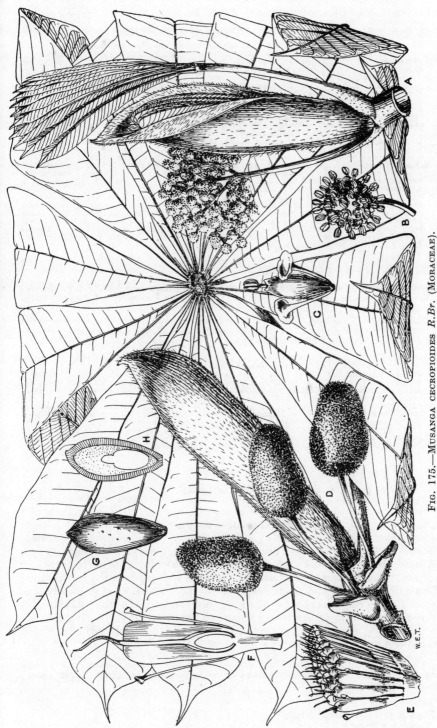

W.E.T.

Fig. 175.—Musanga cecropioides *R.Br.* (Moraceae).

A, male shoot. B, head of male flowers. C, male flower. D, female shoot. E, portion of female receptacle. F, section of female flower. G fruit.
H, section of same.

2. **M. preussii** *Engl.*—F.T.A. 6, 2 : 234 ; Engl. l.c. 40, t. 17 A ; Hauman l.c. 86 (q.v. for synonymy). *M. talbotii* Rendle (1915), partly—F.T.A. 6, 2 : 238 ; F.W.T.A., ed. 1, 1 : 437. A shrub or small tree, up to 20 ft. high ; leaves whitish beneath ; male flowers yellow ; female flowers greenish-yellow in rain forest.
S.Nig.: Oban *Talbot* 18 ! 624 ! 635 ! 684 ! 1422 ! Afi River F.R., Ikom (Dec.) *Keay* FHI 28252 ! 28259 ! Obudu Dist. (fr. May) *Latilo* FHI 30919 ! **Br.Cam.:** Bimbia (fr. Feb.) *Maitland* 391 ! Barombi (Sept.) *Preuss* 478 (partly). Barombi-ba-Mbu to Kake (Sept.) *Preuss* 478 (partly). **F.Po:** Bokoko *Mildbr.* 6908. Also in French Cameroons and Belgian Congo.
3. **M. serratus** (*Trécul*) *Benth.*—F.T.A. 6, 2 : 235, partly ; Engl. l.c. 40, t. 17 C ; Chev. Bot. 608, partly ; Aubrév. l.c. 46, t. 9 B. *Dicranostachys serrata* Trécul (1847). A forest shrub or tree, to 50 ft. high, with stilt roots ; leaves silvery beneath ; male flowers yellow ; on river banks.
Sen.: Messira *Berhaut* 888. **Fr.Sud.:** Sarakouroba *Chev.* 490. Kerfomouria (Mar.) *Chev.* 598. **Port.G.:** Bolola to Buba, Fulacunda *Esp. Santo* 2154 ! **Fr.G.:** Fouta Djalon (Apr.–Mar.) *Heudelot* 840 ! Kouroussa (Apr.) *Pobéguin* 687 ! Ditinn (Apr.) *Chev.* 12971 ! Farana (May) *Sc. Elliot* 5059 ! **S.L.:** Port Loko (Apr., Dec.) *Sc. Elliot* 5881 ! *Thomas* 6682 ! Njala (fl. Mar., fr. Sept.) *Deighton* 1938 ! 2793 ! Mofari, Digissin (Jan.) *Sc. Elliot* 4439 ! Gola Forest (May) *Small* 673 ! **Lib.:** Dukwia R. (fl. Feb., fr. Aug.) *Cooper* 176 ! *Linder* 319 ! **Iv.C.:** Comöe-Aniasué *Aubrév.* 665 ! **Togo:** Bismarckburg *Büttner* 689. **N.Nig.:** Nupe *Barter* 1100 ! Abinsi *Dalz.* 915 ! **S.Nig.:** R. Ogbesse, Owo (Mar.) *Jones* FHI 3084 ! R. Calabar (Feb.) *Mann* 2273 ! Ekemaya, Cross R. (fr. June) *Latilo* FHI 31843. (See Appendix, p. 286.)
4. **M. libericus** *Rendle*—F.T.A. 6, 2 : 236 ; Aubrév. l.c. t. 9 A. *M. serratus* of F.T.A. 6, 2 : 235, partly ; of Chev. Bot. 608, partly. A tree, to 30 ft. high, rarely more, apparently without stilt roots ; leaves silvery beneath, variable in shape and size ; male flowers yellow ; in the forest regions, locally abundant in secondary regrowth.
S.L.: Unwin & Smythe 2 ! Heddle's Farm (young fr. June) *Deighton* 2722 ! Njala (Nov.) *Deighton* 2550 ! *Small* 802 ! Ninia (fr. Feb.) *Sc. Elliot* 4811 ! Mabonto (Oct.) *Thomas* 3572 ! Bumbuna (fr. Oct.) *Thomas* 3438 ! **Lib.:** Mt. Barclay (Nov.) *Bunting* ! Dukwia R. (fl. Nov., fr. Feb., July) *Cooper* 128 ! 285 ! *Linder* 209 ! Suen (Nov.) *Linder* 1396 ! Ganta (Feb., Nov.) *Harley* ! Wohmen, Vonjama (Oct.) *Baldwin* 10102 ! **Iv.C.:** Yaou, Sanvi (young fr. Mar.) *Chev.* 16276 ! And other localities *fide* Aubrév. *l.c.* **G.C.:** Atauaso (young fr. May) *Chipp* 229 ! Tano R. *Chipp* 341 ! Dunkwa (fr. Nov.) *Vigne* 107 ! Bobiri F.R. (fr. Feb.) *Foggie* FH 4961 ! (See Appendix, p. 286.)

12. MUSANGA R.Br.—F.T.A. 6, 2 : 238.

Leaves digitately divided into 12–15 spreading segments ; segments entire, narrow, shortly acuminate, up to 45 cm. long and 10 cm. broad, covered with greyish indumentum beneath ; lateral nerves numerous, very conspicuous beneath ; stipules large, connate, 15–20 cm. long, densely pubescent ; male flowers in numerous small round heads about 4 mm. diam. ; female inflorescence about 2 cm. long on a peduncle up to 12 cm. long, in fruit yellowish green and succulent ..   ..   *cecropioides*

M. cecropioides *R.Br.* in Bowdich Miss. Ashantee 372 (1819) ; G. Taylor in Cat. S. Tomé 309. *M. smithii* R.Br. (1838)—F.T.A. 6, 2 : 239 ; Engl. Monogr. Afr. Morac. 42, fig. 4, t. 18 ; Aubrév. Fl. For. C. Iv. 1 : 48, t. 10 ; Hauman in Fl. Congo Belge 1 : 88. A quick-growing soft-wooded tree with straight stem, stilt-roots and an umbrella-like crown, up to about 60 ft. high ; branchlets very stout and pithy ; the large stipular sheaths enclosing the bud and inflorescence are wine-red, densely silky inside, deciduous ; locally abundant in forest clearings.
S.L.: *Afzelius* ! Njala (fl. & fr. Apr., Sept.) *Deighton* 656 ! *Small* 390 ! Kenema (Jan.) *Thomas* 7709 ! **Lib.:** Dukwia R. (June) *Cooper* 358 ! Tappi (Dec.) *Bunting* 34 ! 35 ! **Iv.C.:** Azagulé *Chev.* 22306 ! **G.C.:** Tano R. *Chipp* 337 ! Kumasi *Cummins* 108 ! Dunkwa (Aug.) *Vigne* FH 143 ! **Togo:** *Baumann* 200 ! **S.Nig.:** Ikpoba R. *Farquhar* 44 ! Sapoba *Kennedy* 2303 ! Imo R., Aba (May) *Cons. of For.* 113 ! 118 ! **Br.Cam.:** *fide* Keay. **F.Po:** *Barter* 2067 ! *Mann* 44 ! 45 ! Also in French Cameroons, Gabon, Spanish Guinea, Cabinda, Belgian Congo, Uganda and S. Tomé. (See Appendix, p. 284.)

97. URTICACEAE

Herbs, undershrubs, or rarely soft-wooded trees, rarely climbing, often armed with stinging hairs ; epidermal cells mostly with prominent cystoliths ; stems often fibrous. Leaves alternate or opposite, simple ; stipules present or absent. Flowers very small, unisexual, usually cymose, sometimes crowded on a common enlarged receptacle. Male flowers with a 4–5-lobed calyx, lobes imbricate or valvate. Petals absent. Stamens as many as and opposite the calyx-lobes ; filaments inflexed in bud ; anthers 2-celled. Rudimentary ovary usually present. Female calyx like that of the male, often enlarged in fruit. Staminodes scale-like or absent. Ovary free or adnate to the calyx, 1-celled ; style simple ; ovule solitary, erect. Fruit a dry achene or fleshy drupe. Seed mostly with endosperm and a straight embryo.

A cosmopolitan family, often troublesome weeds ; stems often fibrous ; many have conspicuous rod-like or dotted cystoliths in the leaves and frequently stinging hairs ; flowers minute, apetalous.

Leaves alternate :
　Stipules present, intrapetiolar :
　　Cystoliths linear :
　　　Plants usually with stinging hairs (but see *Fleurya aestuans*) ; leaves distinctly petiolate, usually more or less equal-sided :

Female calyx 4-lobed, becoming fleshy in fruit ; leaves not lobed   **1. Urera**
Female calyx not becoming fleshy in fruit :
  Leaves lobed usually to below the middle ; female calyx 2-lobed, lobes very
    unequal, the larger one spathe-like   ..   ..   **2. Girardinia**
  Leaves not lobed ; female calyx 4-lobed ..   ..   **3. Fleurya**
Plants without stinging hairs ; leaves sessile, subsessile, or shortly petiolate, often
  oblique and very unequal-sided at base, not lobed :
  Flowers surrounded by an involucre of bracts   ..   ..   ..   **5. Elatostema**
  Flowers on a globose or subglobose receptacle, without an involucre of bracts
                                                         6. Procris

Cystoliths dot-like :
  Plants with stinging hairs ; leaves coarsely toothed   ..   ..   **4. Laportea**
  Plants without stinging hairs ; leaves entire ..   ..   **9. Pouzolzia**
Stipules absent ; plants with dot-like cystoliths ; without stinging hairs
                                                        11. Parietaria

Leaves opposite, sometimes somewhat unequal-sized in each pair :
  Cystoliths linear ; stipules connate into one :
    Flowers in paniculate clusters or heads   ..   ..   **7. Pilea**
    Flowers sessile on a disk-like receptacle   ..   ..   **8. Lecanthus**
  Cystoliths dot-like ; stipules free or connate only at base :
    Flowers in clusters spaced along a usually long rhachis   ..   ..   **10. Boehmeria**
    Flowers in involucres in short axillary clusters..   ..   ..   ..   **12. Droguetia**

1. URERA Gaud.—F.T.A. 6, 2 : 253.

Stems slender, creeping, rooting between or near the nodes, glabrous, not armed with
prickle-like protuberances ; leaves thinly membranous (at least when dry), ovate or
suborbicular, mostly deeply cordate at base, acuminate at apex, margin undulate to
repand-dentate, up to about 13 cm. long and 10 cm. broad ; petioles up to 16 cm.
long, smooth ; inflorescences lax, with scattered stinging hairs   ..   1. *repens*
Stems not slender and creeping :
  Leaves distinctly toothed :
    Young stems and petioles armed with prickle-like protuberances :
      Leaves ovate, suborbicular or broadly elliptic, with short sparse indumentum
        beneath :
        Leaves cordate or widely emarginate at base, suborbicular or ovate, not scabrid
          above, up to 18 cm. long and 15 cm. broad ; petioles up to 10 cm. long
                                                               2. *cordifolia*
        Leaves rounded at base, broadly elliptic to suborbicular, more or less scabrid above,
          up to 24 cm. long and 16 cm. broad ; petioles up to 13 cm. long   3. *rigida*
      Leaves obovate, narrowed from above the middle, abruptly acuminate, not scabrid
        above, sparsely puberulous beneath ; prickle-like protuberances rather small :
        Leaves 8–11 cm. long, 3·5–6 cm. broad, margin crenate ; petioles slender
                                                          4. *obovata*
        Leaves 18–20 cm. long, 10–12 cm. broad, margin denticulate ; petioles stout
                                                          5. *cuneata*
    Young stems and petioles not armed with prickle-like protuberances :
      Leaves ovate, cordate or widely emarginate at base, not scabrid on the upper
        surface :
        Undersurface of leaves rather densely setose-pubescent on midrib, nerves and
         veins ; leaves 10–29 cm. long, 6–18 cm. broad ; petioles up to 13 cm. long
                                                          6. *mannii*
        Undersurface of leaves sparsely pubescent ; leaves 9–15 cm. long, 6–9·5 cm. broad ;
        petioles up to 5·5 cm. long ..   ..   7. *gravenreuthii*
      Leaves more or less obovate, narrowed from above the middle to the rounded or
        obtuse base, shortly acuminate, scabrid on the upper surface rather densely and
        shortly pubescent beneath, margin rather coarsely and sharply dentate, 7–14 cm.
        long ; petioles up to 5 cm. long   ..   ..   8. *robusta*
  Leaves entire or only minutely serrulate :
    Lateral nerves 4–5 on each side of the midrib :
      Leaves cordate or widely emarginate at base, ovate-elliptic or broadly elliptic :
        Young stems not armed with prickle-like protuberances ; male panicles up to
         7·5 cm. long ; leaves 10–14 cm. long, 6–8 cm. broad ; petioles up to 7·5 cm.
         long ..   ..   9. *batesii*
        Young stems profusely armed with prickle-like protuberances ; male panicles
         large and lax, about 20 cm. long and 27 cm. broad ; leaves 10–15 cm. long,
         5–10 cm. broad ; petioles up to 6 cm. long   ..   ..   ..   10. *talbotii*
    Leaves rounded or obtuse at base, oblong-elliptic, rather obliquely acuminate,
      minutely serrulate, 6·5–14 cm. long, 3–7 cm. broad, membranous ; petioles up
      to 10 cm. long ; stems with few prickle-like protuberances   11. *oblongifolia*

Lateral nerves 2–3 on each side of the midrib, the basal pair ascending beyond the middle of the lamina ; leaves elliptic to elliptic-lanceolate, obtuse to rounded or subcordate at base, long-acuminate, 6–18 cm. long, 3–8·5 cm. broad, subcoriaceous ; petioles up to 5 cm. long ..   ..   12. *cameroonensis*

1. **U. repens** (*Wedd.*) *Rendle* in F.T.A. 6, 2 : 264 (1917). *Laportea ? repens* Wedd. (1869). A low glabrous creeping herb with thinly woody stem ; leaves sometimes red-veined ; flowers pale greenish pink ; in forest vegetation.
   Lib.: Yratoke, Webo (July) *Baldwin* 6255 ! **G.C.:** Mampong Scarp (Nov.) *Irvine* 1858 ! Cape Coast road, 28 miles from Accra (Oct.) *Morton* GC 7811 ! **S.Nig.:** Owo F.R. (July) *Jones* FHI 4369 ! Onochie FHI 33244 ! Nikrowa, Okomu F.R. *Brenan* 9101 ! *Richards* 3318 ! **F.Po:** *Barter* ! *Mann* 1424 !
   [The specimens from Liberia and the Gold Coast may represent a taxon distinct from the type ; a better range of material is, however, needed from each locality to show the degree to which leaf-shape may vary.]
2. **U. cordifolia** *Engl.*—F.T.A. 6, 2 : 260 ; Engl. Pflanzenw. Afr. 3, 1 : 53, fig. 33, H–J. A forest liane.
   S.Nig.: Araromi, 5 miles S. of Okeigbo (Feb.) *Jones & Onochie* FHI 14725 ! **Br.Cam.:** Victoria (Feb.) *Maitland* 411 ! S. Bakundu F.R. *Olorunfemi* FHI 30548 ! Johann-Albrechtshöhe *Staudt* 892. Also in French Cameroons.
3. **U. rigida** (*Benth.*) *Keay* in Kew Bull. 1955 : 141. *Boehmeria rigida* Benth. in Fl. Nigrit. 519 (1849). *U. elliotii* Rendle (1916)—F.T.A. 6, 2 : 258 ; F.W.T.A., ed. 1, 1 : 440. *U. obovata* of F.T.A. 6, 2 : 257, partly. A forest liane with stout, rather soft stems, sparsely armed with slender protuberances tipped with stinging hairs ; ripe fruit red.
   S.L.: *Don* ! Bafodeya, Limba (Apr.) *Sc. Elliot* 5559 ! Bulumba (Sept.) *Thomas* 1920 ! Weima, Kisi Kama (Apr.) *Deighton* 3151 ! **Lib.:** Du R. (July) *Linder* 71 ! Kitoma (Apr.) *Harley* 1125 ! Dubo, Sanokwele (Sept.) *Baldwin* 9474 ! Dimei (fr. Apr.) *Barker* 1256 ! **S.Nig.:** Lagos *Rowland* ! Abeokuta *Millen* 147 ! Akure F.R. (Apr.) *Jones* FHI 1092 ! **Br.Cam.:** Moliwe, Buea (Feb.) *Dalz.* 8184 !
4. **U. obovata** *Benth.*—F.T.A. 6, 2 : 257, partly ; Chev. Bot. 610. A forest liane, to 100 ft. high.
   S.L.: Freetown (June) *T. Vogel* 64 ! Kambui F.R. (Apr.) *Lane-Poole* 221 ! **Iv.C.:** various localities *fide* Chev. *l.c.* (See Appendix, p. 287.)
5. **U. cuneata** *Rendle*—F.T.A. 6, 2 : 261. A forest liane.
   Lib.: Sinoe Basin *Whyte* ! Gletown, Tchien (July) *Baldwin* 6955 !
6. **U. mannii** (*Wedd.*) *Benth. & Hook. f. ex Rendle* in J. Bot. 65 : 370 (1916) ; F.T.A. 6, 2 : 260, partly (excl. syn. *U. gravenreuthii* Engl.), incl. var. *paucinervis* Rendle. *Scepocarpus mannii* Wedd. (1869). A forest liane (or shrub or small tree, in more open vegetation, *fide* Morton).
   Lib.: *Linder* ! **G.C.:** Mankrong (Apr.) *Morton* A. 535 ! Koforidua (Apr.) *Morton* A. 819 ! **S.Nig.:** Lagos (Apr.) *Dalz.* 1021 ! Ilaro (Sept.) *W. D. MacGregor* 344 ! Benin *Chief Cons. of For.* ! **F.Po:** (Jan.) *Mann* 176 ! Also recorded from Principe and S. Tomé but the specimens do not agree exactly with the type (from F.Po). (See Appendix, p. 287.)
7. **U. gravenreuthii** *Engl.* Bot. Jahrb. 33 : 120 (1902). *U. mannii* of F.T.A. 6, 2 : 260 and of F.W.T.A., ed. 1, 1 : 440, partly. A forest liane.
   Br.Cam.: above Buea (Apr., May) *Dundas* FHI 15340 ! *Maitland* 649 ! *Preuss* 909 ! Cam. Mt., 6,000 ft. (Apr.) *Morton* GC 7106 !
8. **U. robusta** *A. Chev.*—F.T.A. 6, 2 : 359 ; Chev. Bot. 610. *U. obovata* of F.W.T.A., ed. 1, 1 : 440, partly. A forest liane.
   Iv.C.: Man (May) *Chev.* 21539 ! Goureni *Chev.* 21641. **G.C.:** Owabi (May) *Andoh* FH 4205 ! Manso, Akim (fr. Dec.) *Irvine* 2096 ! Oboman, Kwahu Plateau (fl. & fr. May) *Kitson* 1186 ! Benso (fl. & fr. Aug.) *Andoh* FH 5398 !
9. **U. batesii** *Rendle*—F.T.A. 6, 2 : 259. A trailing or climbing forest liane.
   F.Po: 1,300 ft. *Mann* 305 ! Also in French Cameroons and (?) Belgian Congo.
10. **U. talbotii** *Rendle*—F.T.A. 6, 2 : 259. A forest liane ; ripe fruits scarlet.
    S.Nig.: Oban *Talbot* 618 ! 1502 !
11. **U. oblongifolia** *Benth.*—F.T.A. 6, 2 : 255. *U. sarmentosa* A. Chev. (1917)—F.T.A. 6, 2 : 358 ; Chev. Bot. 610. A glabrous forest liane, climbing by adventitious roots on the slender almost unarmed reddish stems ; fruits succulent, orange-red.
    Port.G.: Granja, Catió (June) *Esp. Santo* 2089 ! 2090 ! **S.L.:** *Barter* ! *T. Vogel* 73 ! 74 ! 114 ! Freetown (May) *Deighton* 2986 ! 2987 ! Njala (May–July) *Deighton* 710 ! 740 ! 1757 ! Makump (fr. July) *Thomas* 897 ! **Lib.:** Mt. Barclay *Bunting* 160 ! Gbanga (Sept.) *Linder* 389 ! Ganta *Harley* ! Jaurazon (Mar.) *Baldwin* 11462 ! **Iv.C.:** Zoanlé (May) *Chev.* 21456 ! Niangouépleu to Man (May) *Chev.* 21530 ! **G.C.:** Begoro, Akim (Apr.) *Irvine* 1187 ! Assuansi (Apr.) *Scholes* 154 ! Aiyoala F.R., Akim Oda (Apr.) *Andoh* FH 5143 !
12. **U. cameroonensis** *Wedd.*—F.T.A. 6, 1 : 261. *U. thonneri* of F.T.A. 6, 2 : 256, partly (spec. *Mann* ex F.Po). A forest liane with glabrous half-succulent stems having adventitious roots.
    G.C.: Owabi (Dec.) *Lyon* FH 2882 ! **S.Nig.:** Shasha F.R. (May) *Richards* 3435 ! Sapoba *Kennedy* 1545 ! Oban *Talbot* 1501 ! Eket *Talbot*. **Br.Cam.:** Cam. Mt., 5,000 ft. (Dec.) *Mann* 2173 ! Victoria (Mar.) *Maitland* 520 ! Buea (Dec., Jan.) *Maitland* 312 ! Ndife, Kumba (Apr.) *Olorunfemi* FHI 30523 ! **F.Po:** (Jan.) *Mann* 240 ! s.n. ! Also extending from French Cameroons to Uganda, Tanganyika and Angola, and on S. Tomé.

Imperfectly known species.

U. liberica *Engl. ex Mildbr.* in Fedde Rep. 41 : 247 (1937), name only.
   Lib.: *Dinklage* 2569.

2. GIRARDINIA Gaud.—F.T.A. 6, 2 : 265.

Stems and petioles bristly with reflexed setose hairs ; leaves alternate, usually pinnately or subdigitately lobed to below the middle, up to 20 cm. long and broad, spreadingly pilose beneath and setose on the nerves, lobes very coarsely serrate, acuminate ; stipules connate, about 2 cm. long, soon falling off ; male inflorescence axillary, often branched, small, branches spike-like ; calyx sparingly setose ; female inflorescence short, dense, densely setose all over ; fruits warted ..   ..   ..   *condensata*

G. condensata (*Hochst. ex Steud.*) *Wedd.* in Ann. Sci. Nat. sér. 4, 1 : 181 (1854) ; F.T.A. 6, 2 : 266. *Urtica condensata* Hochst. ex Steud. (1850). *G. heterophylla* (Vahl) Decne. subsp. *adoensis* (Hochst. ex Steud.) Cuf. (1953). *Urera cannabina* A. Chev. Bot. 609, name only. *Fleurya cannabina* A. Chev., ms. name only. An erect half-succulent-stemmed herb, 1½–6 ft. high, with stiff stinging hairs on stem, leaves and inflorescence.
   Fr.G.: Fouta Djalon : Dalaba to Diaguissa (Sept.) *Chev.* 18569 ! Diaguissa to Boulivel (Sept.) *Chev.* 18589. **Iv.C.:** Gouékouma, Toura country (May) *Chev.* 21654. **S.Nig.:** *Thompson* 514 ! **Br.Cam.:** Bamenda : Lakom (May) *Maitland* 1774 ! Bambui (Dec.) *Adams* GC 11227 ! Widespread in tropical Africa.

3. FLEURYA Gaud.—F.T.A. 6, 2 : 245.

Male inflorescence spike-like or catkin-like, unbranched, the flowers in dense clusters, peduncles long, axillary on leafy shoots or arising directly from the underground stem ; female inflorescence much smaller than the male, loosely few-flowered ; leaves ovate, more or less rounded at base, up to about 10 cm. long and 8 cm. broad, coarsely dentate, with short linear cystoliths on the lower surface ; fruits compressed, ovate, about 3–5 mm. long, slightly rugose, sometimes subterranean .. 1. *ovalifolia*
Male and female flowers in panicles ; fruits about 1 mm. long, never subterranean :
Plant pilose, at least on the young parts ; leaves coarsely serrate with numerous teeth, very broadly ovate, rounded or subcordate at base, up to about 15 cm. long and broad ; inflorescences broadly paniculate .. .. .. 2. *aestuans*
Plant puberulous or glabrous, except for the large, jointed, violently stinging hairs ; leaves grossly toothed with rather few teeth and a long acumen, more or less deltoid in outline, but closely cordate at the point of attachment of the petiole, up to about 15 cm. long and 12 cm. broad ; inflorescences narrow, subsecund .. 3. *mooreana*

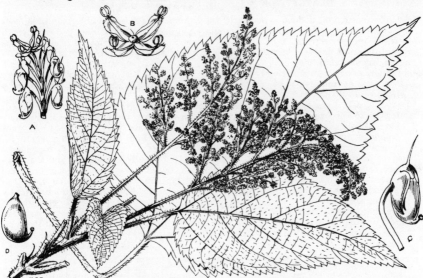

FIG. 176.—FLEURYA AESTUANS (*Linn.*) *Miq.* (URTICACEAE).
A, part of inflorescence. B, male flower. C, female flower. D, fruit.

1. **F. ovalifolia** (*Schum. & Thonn.*) *Dandy* in F.W. Andr. Fl. Pl. A.-E. Sud. 2 : 277 (1952) ; Kew Bull. 1953 : 103. *Haynea ovalifolia* Schum. & Thonn. (1827). *F. podocarpa* Wedd. (1869)—F.T.A. 6, 2 : 251 (incl. var. *mannii* Wedd.) ; Chev. Bot. 609 ; Hauman in Fl. Congo Belge 1 : 191, t. 20 (incl. var. & subsp.). *Urera spicata* A. Chev. Bot. 610, name only. A creeping herb, sometimes half-shrubby, with nettle-like leaves, usually beset with stinging hairs ; male inflorescence greenish-white on long naked peduncles, often directly from the underground stem. **S.L.:** Yetaya (Sept.) *Thomas* 2396 ! Bo to Mongeri (July) *Deighton* 1992 ! Bafodea (Apr.) *Deighton* 5415 ! Kailahun (Sept.) *Jordan* 526 ! **Lib.:** Cape Palmas *Ansell* ! Gbanga (Oct.) *Linder* 1187 ! Nyaake, Webo (June) *Baldwin* 6205 ! Nekabozu, Vonjama (Oct.) *Baldwin* 9967 ! **Iv.C.:** foot of Mt. Momy, Dyola country (Apr.) *Chev.* 21356 ! And other localities *fide* Chev. *l.c.* **G.C.:** Begoro, Akim (Apr.) *Irvine* 1183 ! 1366 ! Kumasi *Cummins* 173 ! 232 ! Mamfe, Akwapim (May) *Adams* 300 ! **Togo:** *Baumann* 182 ! Lome *Warnecke* 438 ! **S.Nig.:** Angiama *Barter* 92 ! Ibadan (Jan.) *Meikle* 951 ! Osho, Omo F.R. (Apr.) *Jones & Onochie* FHI 16846 ! Nun R. (Sept.) *Mann* 476 ! Aguku *Thomas* 761 ! **Br.Cam.:** Victoria (Feb.) *Dalz.* 8201 ! Cam. Mt., 2,500–6,000 ft. (Nov.–Apr.) *Mann* 1950 ! *Maitland* 512 ! 1123 ! **F.Po:** *Barter* ! *Mann* 313 ! Moka, 4,000 ft. (Dec.) *Boughey* 24 ! Widespread in tropical Africa. (See Appendix, p. 287.)

2. **F. aestuans** (*Linn.*) *Miq.* in Mart. Fl. Bras. 4, 1 : 196 (1853) ; F.T.A. 6, 2 : 246 ; Chev. Bot. 609 ; Hauman l.c. 188. *Urtica aestuans* Linn. (1763). A herb 1–5 ft. high, mostly in waste places ; apparently not always urticating. **Sen.:** *Berhaut* 425. N'Diaye Bop *Chev.* 3544. **Port.G.:** Pussubé, Bissau (Sept.) *Esp. Santo* 909 ! Sare N'gana, Geba (Nov.) *Esp. Santo* 3119 ! **Fr.G.:** Kouroussa (July) *Pobéguin* 303 ! Manou *Chev.* 15045. **S.L.:** Freetown (Oct.) *Deighton* 2160 ! Bafodeya (Apr.) *Sc. Elliot* 5507 ! Roruks (Apr.) *Deighton* 4843 ! Jigaya (Sept.) *Thomas* 2668 ! **Lib.:** Sinoe Basin *Whyte* ! Gletown, Tchien (July) *Baldwin* 6776 ! **Iv.C.:** Ebikrou, Sanvi *Chev.* 17673. Zagoué, Dyola country *Chev.* 21560. **G.C.:** Aburi (Sept.) *Adams* 1773 ! *Brown* 413 ! Assuantsi (Mar.) *Irvine* 1580 ! Apedwa (Aug.) *Morton* GC 7780 ! **Togo:** Lome *Warnecke* 206 ! **N.Nig.:** Lokoja (Sept., Nov.) *Barter* ! *Dalz.* 150 ! T. *Vogel* 193 ! Nabardo (Sept.) *Lely* 614 ! Aguji, Ilorin (Oct.) *Thornton* ! **S.Nig.:** Lagos (Oct.) *Dalz.* 1173 ! Ogun R. *Millen* 130 ! Uzuakoli (Aug.) *Jones* FHI 6236 ! Calabar (Oct.) *Holland* 40 ! **F.Po:** *Barter* ! *Mann* 154 ! T. *Vogel* 176 ! Widespread in the tropics. (See Appendix, p. 287.)

3. **F. mooreana** (*Hiern*) *Rendle* in F.T.A. 6, 2 : 250 (1917). *Adicea mooreana* Hiern (1900). *F. urophylla* Mildbr. (1923)—Hauman l.c. 189. An annual herb with violently stinging hairs. **N.Nig.:** Bauchi Plateau (Sept.) *Lely* P. 739 ! Also in Belgian Congo, Uganda, Tanganyika and Angola.

4. LAPORTEA Gaud.—F.T.A. 6, 2 : 252.

Stems covered with sharp reflexed stinging bristles ; leaves ovate-elliptic, acuminate, rounded at the base, 10–15 cm. long, 5–7 cm. broad, setose and pubescent on the nerves beneath and with dot-like cystoliths on the surface, coarsely dentate-serrate ; petiole ⅓ as long as the blade, setose ; male inflorescences axillary, about as long as the petiole, female much longer than the leaves and subsecund ; calyx glabrous ; fruits flat, obliquely ovate, with a thick margin and slightly rugose surface

alatipes

L. alatipes *Hook. f.*—F.T.A. 6, 2 : 252 ; Hauman in Fl. Congo Belge 1 : 194. *Fleurya alatipes* (Hook. f.) N.E. Br. (1925). *F. urticoides* Engl. (1902)—F.T.A. 6, 2 : 247 ; F.W.T.A., ed. 1, 1 : 441. *Girardinia marginata* Engl. (1902). A robust perennial herb, to 3 ft. high ; in montane forest.
Br.Cam.: Cam. Mt., 4,000–7,600 ft. (fl. Nov.–Feb., Apr.–June, fr. Nov., Feb.) *Mann* 1973 ! *Maitland* 989 ! *Preuss* 916 ! Lakom, Bamenda, 6,000 ft. (May, June) *Maitland* 1368 ! 1740a ! Banga, Bamenda (fr. Aug.) *Ujor* FHI 29958 ! Also on mountains in Belgian Congo, Uganda, Kenya, Tanganyika, S. Rhodesia, Nyasaland and Natal.

5. ELATOSTEMA J. R. & G. Forst.—F.T.A. 6, 2 : 277.

Male inflorescence on a slender peduncle 1·5–2·5 cm. long ; leaves obliquely oblanceo-late, cuneate on both sides at base, acuminate at apex, 4–10 cm. long, 2–3 cm. broad, coarsely crenate-serrate, with about 8–10 teeth on each margin, with conspicuous cystoliths on each side ; stipules lanceolate, tailed-acuminate ; bracts broadly ovate, acute, about 3 mm. long . .   . .   1. *mannii*
Male inflorescence sessile or nearly so ; leaves cuneate at base only on the distal side :
Leaves long-acuminate, 6–24 cm. long, 2–8·5 cm. broad ; herbs with erect fleshy stems up to 60 cm. high :
Leaves closely serrate, with about 20 or more teeth on each margin, the teeth extend-ing on to the acumen, obtuse or rounded on the proximal side at base, obliquely elongate-elliptic, 6–24 cm. long, 2–8·5 cm. broad, with very numerous and con-spicuous cystoliths on upper surface ; leaves drying bright green ; male in-florescence flat and subtriangular or lobed, pubescent   . .   2. *welwitschii*
Leaves laxly and coarsely dentate, with about 12–16 teeth on each margin, the teeth not or scarcely extending on to the acumen, broadly auriculate on the proximal side at base, obliquely obovate, 6–15 cm. long, 2–5·5 cm. broad, with very con-spicuous cystoliths on upper surface ; leaves drying dark green ; bracts broadly ovate, acuminate   . .   . .   3. *paivaeanum*
Leaves not acuminate, 1·5–3·5 cm. long, 0·8–2 cm. broad, obliquely obovate, rounded on the proximal side at base, laxly and coarsely dentate, with about 5–10 teeth on each margin ; bracts rather small, shortly acuminate ; a weak herb with trailing stems   . .   . .   4. *monticola*

1. **E. mannii** *Wedd.*—F.T.A. 6, 2 : 282 ; Engl. Pflanzenw. Afr. 3, 1 : 60. A herb with stems erect from a decumbent base, half-succulent, 6–12 in. high ; leaves crowded on the upper part of the stem.
S.Nig.: Kwa Falls, Calabar (Mar.) *Richards* in *Hb. Brenan* 9235 ! **Br.Cam.:** Johann-Albrechtshöhe *Staudt* 647 ! Mombo to Ebonji, Kumba (May) *Olorunfemi* FHI 30582 ! **F.Po:** *Mann* 1431 ! Also in French Cameroons.
2. **E. welwitschii** *Engl.*—F.T.A. 6, 2 : 278 (incl. var. *cameroonense* Rendle) ; Engl. l.c. fig. 37, A–B ; Hauman in Fl. Congo Belge 1 : 206. A perennial herb with an erect fleshy stem, to 2 ft. high, from a creeping rooting rhizome ; in montane forest.
S.Nig.: Mt. Koloishe, 4,000 ft. (Dec.) *Keay & Savory* FHI 25054 ! **Br.Cam.:** Mimbia, 3,800 ft. (Mar.) *Brenan* 9344c ! Johann-Albrechtshöhe *Staudt* 839 ! **F.Po:** Clarence Peak, 4,000 ft. (Dec.) *Boughey* 116 ! *Mann* 632 ! Musola (Jan.) *Guinea* 1058 ! Also in French Cameroons, Belgian Congo, Uganda, Tanganyika, Nyasaland, Angola and on S. Tomé.
3. **E. paivaeanum** *Wedd.*—F.T.A. 6, 2 : 279 ; Engl. l.c. (incl. var. *conrauanum* Engl.) ; Hauman l.c. 207. A perennial herb with an erect fleshy stem, to 2 ft. high ; in forest.
Fr.G.: R. Kakrima, Tchiboñ (Dec.) *Pobéguin* 2054 ! Ditinn (Sept.) *Chev.* 18525 ! Dalaba–Diaguissa Plateau (Oct.) *Chev.* 18744 ! **Lib.:** Karmadhun, Kolahun (Nov.) *Baldwin* 10169 ! Nekabozu, Vonjama *Baldwin* 9982 ! Bobei Mt., Sanokwele (Sept.) *Baldwin* 9646 ! **Iv.C.:** Danané to Goutokouma *Chev.* 21308 ! Mt. Momy *Chev.* 21361 ! **G.C.:** Kibbi Hills, Akim (Dec.) *Johnson* 254. Puso Puso ravine (Oct.) *Adams* 1903 ! **Br.Cam.:** Buea (Feb.) *Dalz.* 8206 ! Cam. Mt., 4,500 ft. (Apr.) *Maitland* 1083 ! Mimbia, 3,800 ft. (Mar.) *Brenan* 9344 ! 9344a ! Bali-Ngemba F.R., Bamenda (May) *Ujor* FHI 30328 ! **F.Po:** Clarence Peak, 4,000 ft. (Dec.) *Mann* 290 ! 631 ! Also in French Cameroons.
4. **E. monticola** *Hook. f.*—F.T.A. 6, 2 : 281 ; Engl. l.c. ; Hauman l.c. A weak herb with trailing stems ; in montane forest.
Br.Cam.: Cam. Mt., 6,000–7,000 ft. (Dec.–Apr.) *Adams* GC 11737 ! *Brenan* 9576 ! *Mann* 2014 ! *Preuss* 607 ! Bafut-Ngemba F.R., Bamenda (Jan.) *Tiku* FHI 22240 ! **F.Po:** El Pico, 6,600 ft. (Dec.) *Boughey* 127 !
[Some at least of the material named *E. orientale* Engl. from E. Africa and the Belgian Congo is probably conspecific with *E. monticola* Hook. f.]

6. PROCRIS Comm.—F.T.A. 6, 2 : 283.

Epiphytic or on rocks ; stems erect, the lower parts underground and bearing flower-heads ; leaves alternate, subsessile, obliquely oblanceolate, broadly acuminate, gradually narrowed to the base, up to 15 cm. long and 3 cm. broad, very remotely serrate, without obvious cystoliths ; flower-heads numerous and scattered on the stem from the base upwards, very shortly pedunculate ; fruits fusiform, 1·25 mm. long . .   . .   *crenata*

P. crenata *C. B. Robinson* in Philipp. Journ. Sci. 5, C : 507 (1911) ; G. Taylor in Exell Cat. S. Tomé 312 (q.v. for full synonymy). *P. wightiana* Wall. ex Wedd. (1856), illegitimate name, partly ; F.T.A. 6, 2 : 283 ; Schröter in Fedde Rep. 45 : 191 & 257 ; Hauman in Fl. Congo Belge 1 : 208. *P. laevigata* of F.W.T.A., ed. 1, 1 : 442, not of Blume. An epiphyte with glabrous succulent stems up to 18 in. high and ½–1 in. thick in the lower part.

Br.Cam.: Buea (Sept.) *Preuss* 963. **F.Po:** 1,000–2,000 ft. (Nov.) *Mann* 566 ! El Pico (Dec.) *Boughey* 169 ! Also on Principe, S. Tomé, Annobon and widespread in tropical Africa, as well as in the Mascarene Islands and tropical Asia.

7. PILEA Lindl.—F.T.A. 6, 1 : 267.

Leaves toothed, 0·7–7 cm. long, 0·5–5·8 cm. broad :
Flowers in axillary, small, pedunculate, lax cymes, or stalked or subsessile clusters ; stipules very small and early caducous ; leaves with numerous cystoliths on both surfaces :
Leaf-margin crenate-serrate, secondary lateral nerves running close to the margin and finishing in the mucro of each tooth ; lamina ovate, 1–8 cm. long, 0·5–3·5 cm. broad, 3-nerved at base ; petiole up to 3 cm. long   ..   ..   1. *sublucens*
Leaf-margin coarsely serrate, secondary lateral nerves running up the middle of each tooth ; lamina more or less ovate in outline, 0·7–4 cm. long, 0·5–3·2 cm. broad, 3-nerved at base ; petiole up to 4 cm. long ..   ..   2. *angolensis*
Flowers forming a whorl-like sessile cluster at the nodes or in a terminal dense cyme ; stipules broadly oblong, membranous, brownish, up to 1 cm. long, more or less persistent ; leaves broadly ovate, rather coarsely dentate, 1–7 cm. long, 1–5·8 cm. broad, with scattered cystoliths on both surfaces ; petioles up to 6 cm. long
3. *ceratomera*
Flowers a flat terminal sessile corymb involucrate by the 4 uppermost leaves ; stipules suborbicular, membranous, about 2 mm. long ; leaves broadly ovate, crenate-serrate, 2–4 cm. long, 1·5–3 cm. broad, laxly clothed with cystoliths ; petioles up to 2 cm. long   ..   ..   4. *tetraphylla*
Leaves entire, 1–8 (–10) mm. long, 1–5 (–6) mm. broad, obovate, elliptic or suborbicular, with numerous transversely arranged cystoliths on the upper surface, often strongly anisophyllous ; flowers in very small axillary clusters   ..   ..   5. *microphylla*

1. **P. sublucens** *Wedd.*—F.T.A. 6, 2 : 273 ; F.W.T.A., ed. 1, 1 : 443, partly. *P. chevalieri* Schnell in Rev. Gén. Bot. 57 : 280, fig. 2 (1950). A glabrous herb with slender stems, erect, 6–18 in. high from a creeping base ; leaves of each opposite pair more or less unequal in size or in length of petiole ; in montane forest. **Fr.G.:** Nimba Mts. (Apr.) *Chev.* 21117. *Schnell* 952 ; 4387 ! **Br.Cam.:** Buea, 3,500–4,450 ft. (Aug., Nov., Dec.) *Adams* GC 11296 ! *Preuss* 573 ! 953 ! *Rosevear* 102/37 ! **F.Po:** Clarence Peak, 6,000 ft. (Dec.) *Mann* 630 ! *Boughey* 130 !
2. **P. angolensis** (*Hiern*) *Rendle* in F.T.A. 6, 2 : 271 (1917) ; Hauman in Fl. Congo Belge 1 : 200. *Adicea tetraphylla* var. *angolensis* Hiern (1900). *P. sublucens* of F.W.T.A., ed. 1, 1 : 443, partly (spec. ex Fr.G. & S.L.), not of Wedd. *Urera begonioides* A. Chev. Bot. 609, name only. A delicate annual herb with transparent stem, growing on wet rocks in shady places. **Fr.G.:** Haut Konkouré (Nov.) *Pobéguin* ! Dalaba–Diaguissa Plateau (Sept.–Oct.) *Chev.* 18815 ! **S.L.:** Bikongo Falls, Suva R. (July–Aug.) *Dawe* 535 ! Bumbuna (Oct.) *Thomas* 3430 ! Yakala (Sept.) *Thomas* 2366 ! Makeni Hill (Aug.) *Jordan* 498 ! **Lib.:** Vahun, Kolahun (Nov.) *Baldwin* 10214 ! **G.C.:** Begoro Falls (Jan.) *Morton* 8369b ! **N.Nig.:** Bauchi Plateau (Sept.) *Lely* P. 781 ! Jos (Aug.) *Keay* FHI 20179 ! **Br.Cam.:** Metschum Falls, Bamenda (Jan., Aug.) *Keay & Russell* FHI 28525 ! *Savory* UCI 300 ! Also in Uganda, Belgian Congo and Angola.
3. **P. ceratomera** *Wedd.*—F.T.A. 6, 2 : 269 ; Hauman l.c. 203. *P. stipulata* Hutch. & Dalz., F.W.T.A., ed. 1, 1 : 443 (1928) ; Kew Bull. 1929 : 19. Stems lax, creeping and rooting, with erect smooth branches to 2 ft. high ; terrestrial or epiphytic ; in montane forest. **S.Nig.:** Ikwette Plateau, Obudu, 5,200 ft. (Dec.) *Keay & Savory* FHI 25187 ! **Br.Cam.:** Buea *Preuss* 957 ! Cam. Mt., 7,000–8,000 ft. (Dec.) *Johnston* 66 ! *Mann* 2011 ! Above L. Oku, Bamenda, 7,500 ft. (Jan.) *Keay* FHI 28505 ! **F.Po:** Clarence Peak, 4,000–7,000 ft. (Dec.) *Boughey* 182 ! 184 ! *Mann* 626 ! Also on the mountains of Belgian Congo, A.-E. Sudan, E. Africa, S. Tomé and the Mascarene Islands.
4. **P. tetraphylla** (*Steud.*) *Blume*—F.T.A. 6, 2 : 270 (incl. var. *major* Rendle) ; Hauman l.c. 199. *Urtica tetraphylla* Steud. (1850). A herb, erect, slender, simple or branched, up to 16 in. high, sometimes from a creeping base, glabrous ; in montane vegetation. **Br.Cam.:** Cam. Mt., 7,000 ft. (Dec.) *Mann* 2012 ! *Preuss* 1001 ! **F.Po:** 6,000 ft. (Dec.) *Boughey* 10736 ! s.n. ! *Monod* 10400 ! Also in A.-E. Sudan, Abyssinia, Eritrea, E. Africa, Belgian Congo and Madagascar.
5. **P. microphylla** (*Linn.*) *Liebm.* in Vidensk. Selsk. Skr. 5, 2 : 302 (1851). *Parietaria microphylla* Linn. (1759). A stout, large-leaved form is cultivated as the " Artillery Plant " ; a slender, small-leaved form occurs on walls and as a garden weed. **Port.G.:** Catió (July) *Esp. Santo* 2139 ! **S.L.:** Freetown (Oct.) *Deighton* 2052 ! Port Loko (Oct.) *Jordan* 363 ! Njala (May) *Deighton* 5057 ! **Lib.:** Monrovia (Nov.) *Linder* 1574 ! **N.Nig.:** Sokoto (Nov.) *Moiser* 27 ! **S.Nig.** & **Br.Cam.:** fide Keay. **F.Po:** St. Isabel *Boughey* ! A native of tropical America, now widespread in the tropics.

8. LECANTHUS Wedd.—F.T.A. 6, 2 : 276.

Small herb with a densely rooting creeping underground stem ; leaves ovate-lanceolate, cuneate at base, broadly acute, 2–3·5 cm. long, 0·8–1·5 cm. broad, coarsely dentate, with numerous linear cystoliths prominent on the upper surface ; petiole 0·5 cm. long ; inflorescence orbicular, up to 1 cm. diam. ; peduncle slender, 2–2·5 cm. long
peduncularis

L. peduncularis (*Royle*) *Wedd.*—F.T.A. 6, 2 : 276. *Procris peduncularis* Royle (1836). A weak herb, simple or branched, half-succulent, glabrous 1–10 in. high.
Br.Cam.: Cam. Mt., 6,000–7,000 ft. (Dec., Jan.) *Adams* GC 11766 ! *Dunlap* 237 ! *Preuss* 1034 ! **F.Po:** Clarence Peak, 6,000–7,500 ft. (Mar., Dec.) *Boughey* GC 131 ! *Mann* 1461 ! Also in Abyssinia and widespread in the eastern tropics.

9. POUZOLZIA Gaud.—F.T.A. 6, 2 : 286.

Stems erect, leafy ; leaves ovate-elliptic, rather broadly acuminate, cuneate at base, 5–10 cm. long, 1·5–4·5 cm. broad, entire, sometimes thinly woolly beneath, at length with conspicuous dotted cystoliths above ; petiole slender, up to 2 cm. long ; flowers clustered, sessile or subsessile, pubescent ; fruits subtended by a bract　　　*guineensis*

P. guineensis *Benth.*—F.T.A. 6, 2 : 287 ; Chev. Bot. 611 ; Hauman in Fl. Congo Belge 1 : 214. An erect or spreading herb, to 4 ft. high, slightly long-pubescent, more or less woody below ; in open places, mainly in the forest regions.
　　Sen.: *Berhaut* 1174. **Fr.G.:** Rio Nunez *Heudelot*. **S.L.:** Kambia (Dec.) *Sc. Elliot* 4354 ! Kanya (Oct.) *Thomas* 2941 ! Njala (Sept.) *Deighton* 2106 ! Rogberi, Binle (Oct.) *Adames* 256 ! **Lib.:** Jabroke, Webo (July) *Baldwin* 6615 ! Gletown, Webo (July) *Baldwin* 6801 ! **Iv.C.:** various localities *fide* Chev. *l.c.* **G.C.:** Abetifi, Kwahu (Jan.) *Irvine* 1828 ! Begoro Akim (Apr.) *Irvine* 1194 ! Techiman road, 30 miles N. of Kumasi *Morton* GC 7708 ! **Togo:** *Baumann* 578. **N.Nig.:** Lokoja (Nov.) *Barter* ! *Dalz.* 149 ! **S.Nig.:** Lagos (Sept.) *Dodd* 441 ! Ibadan (Oct.) *Ejiofor* FHI 27667 ! Odiani, Asaba (Aug.) *Onochie* FHI 33439 ! Oban *Talbot* 689 ! **Br.Cam.:** Bulifambo, Buea (Mar.) *Maitland* 550 ! **F.Po:** (Oct., Dec.) *T. Vogel* 28 ! 251 ! 299 ! Extending to Belgian Congo and Angola and on S. Tomé. (See Appendix, p. 287.)

10. BOEHMERIA Jacq.—F.T.A. 6, 2 : 284.

Leaves opposite, variable in shape, from very broadly ovate to elliptic-lanceolate, margin coarsely triangular-dentate or serrate, up to 20 cm. long and 18 cm. broad, with inconspicuous dot-like cystoliths ; inflorescences axillary, spike-like, or slightly branched, up to 30 cm. long, with the flowers in separated clusters along the rhachis　　*platyphylla*

B. platyphylla *D. Don*—F.T.A. 6, 2 : 285 ; Hauman in Fl. Congo Belge 1 : 210. A shrub 3–15 ft. or more high, with soft-wooded branches. The specimens from our area seem to fall into three groups :—(A) specimens from S.L. with densely and shortly pubescent young shoots and petioles, and usually broadly ovate leaves ; (B) specimens from Cam. Mt. and F.Po, with rather sparingly and more or less appressed-puberulous young shoots and petioles, and usually slender stems, petioles and leaves ; of these specimens, *Mann* 209 is cited by Weddell as var. *macrostachya* Wedd. ; (C) specimens from Nig. and Br.Cam. (excluding Cam. Mt.), with densely spreadingly hirsute young shoots and petioles, and usually narrow leaves ; this is var. *nigeriana* Wedd. (type *Barter* 114). The infraspecific variants require thorough and extensive revision throughout the vast range of the species.
　　Fr.G.: Haut Konkouré (Sept.) *Pobéguin* 1872 ! Dalaba Plateau (Dec.) *Pobéguin* 2063 ! **S.L.:** Smythe 102 ! Yetaya (Sept.) *Thomas* 2305 ! Yakala (Sept.) *Thomas* 2363 ! Musaia (Dec.) *Deighton* 4540 ! **Iv.C.:** Tebo (July) *Chev.* 19415 ! **N.Nig.:** Jos (Aug.) *Keay* FHI 20079. **S.Nig.:** Aboh *Barter* 114 ! Oban *Talbot* 686 ! **Br.Cam.:** Bambuko, Kumba (Sept.) *Olorunfemi* FHI 30761 ! Lakom, Bamenda (May) *Maitland* 1362 ! Ndop, Bamenda (fr. Dec.) *Adams* GC 11061 ! Cam. Mt., 3,700–5,500 ft. (Dec.–June) *Dundas* FHI 20380 ! *Hutch. & Metcalfe* 72 ! *Maitland* 41 ! 647 ! 1096 ! **F.Po:** *Mann* 209 ! The species is widespread in tropical Africa and Asia and has many varieties.

Besides the above, *B. nivea* (Linn.) Gaud., a native of eastern Asia, is occasionally cultivated in our area ; it has *alternate* leaves, densely white tomentellous beneath.

11. PARIETARIA Linn.—F.T.A. 6, 2 : 296.

Tall slender herb up to 3 m. high ; stems angular, soon becoming glabrous ; leaves alternate or subopposite (or rarely quite opposite), ovate, long- and acutely acuminate, rounded or subcordate at base, 3–8 cm. long, 1·5–4 cm. broad, glabrous but with prominent dot-like cystoliths on both surfaces ; petioles $\frac{1}{3}$–$\frac{1}{2}$ as long as the blade ; flowers in very small axillary clustered cymes ; fruit ovate, apiculate　　1. *laxiflora*
Weak more or less decumbent herb with very thin stems ; leaves broadly ovate, at most obtusely pointed, rounded at base, 1–2 cm. long and broad, very thin and flaccid, covered with conspicuous dot-like cystoliths ; flowers few in sessile axillary clusters or subsolitary ; fruit not apiculate　　..　　..　　..　　..　　..　　2. *debilis*

1. **P. laxiflora** *Engl.*—F.T.A. 6, 2 : 298 ; Hauman in Fl. Congo Belge 1 : 209. A herb, often scandent and half-woody, with slender leafy branchlets ; in montane forest.
　　Br.Cam.: Cam. Mt., 4,200–7,800 ft. (Dec.–Mar., June) *Brenan* 9509 ! *Mann* 1230 ! 2013 ! *Maitland* 841 ! 990 ! **F.Po:** Clarence Peak, 8,000 ft. (Dec.) *Mann* 613 ! Also on mountains in Belgian Congo and Uganda.
2. **P. debilis** *Forst. f.*—F.T.A. 6, 2 : 298. A slender more or less branched herb from a few inches to 1 ft. high.
　　Br.Cam.: Cam. Mt. (Jan.) *Mann* ! Cam. Mt., in cave in grassland, 12,000 ft. (June) *Morton* GC 6902 ! Widely spread in temperate regions and in tropical highlands.

12. DROGUETIA Gaud.—F.T.A. 6, 2 : 303.

A perennial herb or undershrub, stems trailing and ascending, pubescent when young ; leaves opposite, ovate, acuminate, 3–5 (–8) cm. long, 1·8–2·5 (–4·5) cm. broad, 3-nerved, margins crenate-dentate, sparsely pubescent and with dot-like cystoliths above, pubescent on the nerves beneath ; petioles 1–2 (–3) cm. long ; stipules ovate-lanceolate, 3–4 mm. long ; flowers in axillary, androgynous, bowl-shaped or campanulate involucres, or female involucres ventricose-tubular　　..　　..　　*iners*

D. iners (*Forsk.*) *Schweinf.*—F.T.A. 6, 2 : 303 ; Hauman in Fl. Congo Belge 1 : 217. *Urtica iners* Forsk. (1775). To 5 ft. high ; in montane forest.
　　Br.Cam.: Cam. Mt., 3,000–6,000 ft. (Dec.–Feb.) *Adams* GC 11728 ! *Maitland* 219 ! 824 ! 1525 ! *Morton* GC 7110 ! Bambui, Bamenda 6,300 ft. (Dec.) *Adams* GC 11218 ! Also in Belgian Congo, Uganda, Abyssinia, Kenya, Tanganyika and Nyasaland ; and in Arabia, India and Java.

98. CANNABINACEAE

Erect or scandent herbs. Leaves alternate or opposite, simple, undivided or more or less palmately lobed ; stipules present. Flowers dioecious, axillary, male paniculate, female sessile, crowded or strobilate, with large persistent bracts. Male flower : calyx 5-partite, segments imbricate ; petals absent ; stamens 5, erect in bud ; anthers 2-celled. Female flower : calyx closely enveloping the ovary, entire ; ovary sessile, 1-celled ; style 2-partite. Ovule solitary, pendulous. Fruit an achene, covered by the persistent calyx. Seed with fleshy endosperm and curved or spiral embryo.

CANNABIS Linn.—F.T.A. 6, 2 : 16.

Stem erect ; leaves alternate or opposite, palmately divided ; achene covered by the persistent calyx ; embryo curved.

A leafy annual herb ; stems minutely puberulous and glandular ; leaves gland-dotted, alternate, or the lower opposite, palmately divided mostly to the base into 5 or more linear-lanceolate toothed acuminate segments ; stipules subulate ; male flowers greenish in lax axillary and terminal clusters ; styles long and filiform ; fruit about 5 mm. long, finely reticulate   ..   ..   *sativa*

C. sativa *Linn.*—F.T.A. 6, 2 : 16 ; Holl. 4 : 614 ; Chev. Bot. 590.
   The Common, European or Indian Hemp, a native of Asia widely cultivated but rarely in the area of this flora.

99. AQUIFOLIACEAE

Trees or shrubs, mostly evergreen. Leaves alternate, simple ; stipules absent, or rarely present. Flowers actinomorphic, hermaphrodite or unisexual, fasciculate or subumbellate or rarely solitary. Calyx-lobes imbricate. Petals 4 or 5, free or connate at the base, hypogynous, imbricate. Stamens hypogynous, 4 or 5, alternate with the petals, rarely more numerous, free ; anthers 2-celled, opening lengthwise. Disk absent. Ovary superior, 3- or more-celled ; style terminal or absent ; ovules 1–2 in each cell, pendulous from the apex. Fruit drupaceous, of 3 or more 1-seeded pyrenes ; seed with copious fleshy endosperm and small straight embryo.

Generally distributed throughout the world, but rare in Africa and in Australia.

ILEX Linn.—F.T.A. 1 : 359.

Petals usually connate at base ; stamens as many as petals ; inflorescence axillary ; ovary 4–6-celled.

Tree up to 13 m. high ;   branches glabrous ;   leaves oblong, apiculate, obtuse to subacute at the base, up to 10 cm. long and 4 cm. broad, entire or remotely denticulate, glabrous, with numerous spreading looped lateral nerves ; petiole 1 cm. long ; flowers subfasciculate on short axillary peduncles ; pedicels up to 1 cm. long ; sepals minutely ciliolate ; petals shortly connate ; stigma thick and sessile ; fruit subglobose, 5 mm. diam., girt by the persistent sepals ..   ..   *mitis*

I. mitis (*Linn.*) *Radlk.* in Rep. Brit. Ass. 1885 : 1081 (1886) ; Engl. Bot. Jahrb. 17 : 540. *Sideroxylon mite* Linn. (1767). *Ilex capensis* Sond. & Harv. (1860)—F.T.A. 1 : 359. A tree, to 40 ft. high, with dark green foliage ; flowers cream or white ; ripe fruits red ; in montane forest, especially by streams.
   Fr.G.: Nimba Mts., 4,500 ft. (fl. & young fr. Apr.) *Adam* 116 ! S.L.: Loma Mts., 5,000 ft. (Jan.) *Jaeger* 4162 ! Br.Cam.: Cam. Mt., 4,000–7,500 ft. (Jan.–Feb.) *Keay* FHI 28644 ! *Mann* 1186 ! *Maitland* 212 ! 237 ! 983 ! L. Bambulue, 6,700 ft., Bamenda (fr. Jan.) *Keay & Lightbody* FHI 28359 ! Kumbo, 5,600 ft., Bamenda (Jan.) *Lightbody* FHI 26295 ! Lakom, 6,000 ft., Bamenda (young fr. Apr.) *Maitland* 1660 ! F.Po: L. Riaba, Moka (Jan.) *Guinea* 2209 ! *Tessmann* 2889. Also on mountains in French Cameroons and from Eritrea, Uganda and Belgian Congo, southwards to S. Africa.

100. CELASTRACEAE
With R. A. Blakelock

Small trees, shrubs or woody climbers without tendrils, sometimes spiny. Leaves simple, alternate or opposite. Stipules inconspicuous or absent. Flowers bisexual or unisexual. Sepals 4–5, usually free. Petals 4–5, free. Stamens 3–5, alternate with petals. Carpels 3–5 united, free from or half-embedded in the disk. Ovules 2–12 in each carpel. Fruit a loculicidal capsule, an indehiscent fleshy or hard drupe, a berry, a 3-lobed capsule or a capsule with 3 nearly separate mericarps. Seed sometimes arillate, sometimes winged, sometimes neither.

A cosmopolitan family represented in Africa by endemic as well as by widespread and very variable species. Often divided into two families, *Celastraceae* and *Hippocrateaceae*, but the genus *Campylostemon* with 4–5 stamens and a Hippocratea-like fruit connects the two into one family.

Erect shrubs (if climbing then leaves alternate), leaves opposite or alternate ; stamens 4–5, anthers with longitudinal dehiscence :
　　Fruit a capsule ; aril present and conspicuous ; shrub or tree, often with stem-spines
　　　　　　　　　　　　　　　　　　　　　　　　　　　　　　　1. Maytenus
　　Fruit a drupe ; aril absent or minute ; shrub, not spiny　　..　　..　　**2. Cassine**
Spineless usually climbing shrubs or small trees, with opposite (or subopposite) leaves (if some leaves alternate then only 3 stamens) ; stamens 3 or (in *Campylostemon* only) 4–5, anthers with transverse or longitudinal dehiscence :
　　Inflorescence a fascicle, cyme or subumbellate cyme, on a leafy shoot :
　　　　Fruit a 3-lobed capsule, the 3 dehiscent mericarps free almost to base ; seeds with a basal membranous wing ; inflorescence a pedunculate cyme :
　　　　　　Stamens 4–5, bent inwards ; resinous threads on breaking leaf　　**3. Campylostemon**
　　　　　　Stamens 3, bent outwards ; no resinous threads on breaking leaf (except *H. mucronata*)　　..　　..　　..　　..　　..　　..　　..　　**4. Hippocratea**
　　　　Fruit a globose or oblong berry ; seeds not winged ; stamens 3 ; inflorescence a fascicle or a pedunculate cyme　　..　　..　　..　　..　　..　　**5. Salacia**
　　Inflorescence of alternate fascicles on leafless side-shoots about 1 m. long ; stamens 3 ; climbing shrubs, glabrous in all parts ; no resinous threads on breaking leaf
　　　　　　　　　　　　　　　　　　　　　　　　　　　　　　　6. Salacighia

1. MAYTENUS H. B. & K. Nov. Gen. Sp. Pl. 764 (1824–5) ; Loes. in E. & P. Pflanzenfam. 20B : 134 (1942) ; Exell & Mendonça Consp. Fl. Ang. 2 : 1 (1954) ; Blakelock in Taxon 3 : 196 (1954). *Gymnosporia* (Wight & Arn.) Hook. f. in Gen. Pl. 1 : 365 (1862) ; Loes. in Engl. Bot. Jahrb. 17 : 541 (1893) ; l.c. 19 : 231 (1895) ; l.c. 28 : 150 (1900) ; 41 : 298 (1908) ; F.W.T.A., ed. 1, 1 : 444 ; Loes. in E. & P. Pflanzenfam. 20B : 147 (1942).

Inflorescence unbranched (a fascicle of single flowers) ; plant without spines ; no resinous threads in leaf ; aril a thin oblique cup covering from basal ¼ to the entire seed ..　　..　　..　　..　　..　　..　　..　　..　　..　　..　　1. *undatus*
Inflorescence branched with a common peduncle (sometimes very short) ; plants sometimes spiny ; resinous threads present or not in leaf :
　　Leaves cuneate at base, often coriaceous, usually glaucous ; aril a thin oblique cup covering from basal ¼ to the entire seed ; no resinous threads in leaf
　　　　　　　　　　　　　　　　　　　　　　　　　　　　　　　2. *senegalensis*
　　Leaves usually rounded to somewhat cuneate at base, usually not glaucous :
　　　　Aril a thin oblique cup covering from basal ¼ to the entire seed ; resinous threads in leaf ; peduncle often short (1 mm. long) ..　　..　　..　　..　　3. *acuminatus*
　　　　Aril a thick cup-like strophiole below seed ; no resinous threads in leaf :
　　　　　　Leaves often coriaceous ; inflorescence pubescent, 0·5–3·5 cm. long ; capsule 4–10 mm. long ..　　..　　..　　..　　4a. *ovatus* var. *ovatus* forma *pubescens*
　　　　　　Leaves membranous, more acute, margin more serrate ; inflorescence pubescent or glabrous, (1·2–) 3–14 cm. long ; capsule 0·8–1·8 cm. long
　　　　　　　　　　　　　　　　　　　　　　　　　　　　　4b. *ovatus* var. *argutus*
The shape and size of the leaves are very variable in all species.

1. **M. undatus** (*Thunb.*) *Blakelock* in Kew Bull. 1956 : 237 (1956). *Celastrus undatus* Thunb. (1794). *C. lancifolius* Thonn. (1827)—F.T.A. 1 : 364. *Maytenus lancifolius* (Thonn.) Loes. (1942)—Exell & Mendonça Consp. Fl. Ang. 2 : 2. *Celastrus luteolus* Del. (1843)—F.T.A. 1 : 363. *Catha fasciculata* Tul. (1857). *Gymnosporia fasciculata* (Tul.) Loes. (1895)—F.W.T.A., ed. 1, 1 : 445. *G. maliensis* Schnell in Bull. I.F.A.N. 15 : 96, fig. 3 (1953). Unarmed shrub or tree, up to 25 ft. high ; flowers whitish ; fruits mostly reddish when ripe.
　　Fr.G.: Mali (fl. & young fr. Mar.) *Schnell* 4822! **G.C.:** Cape Coast *Brass. Adah Thonning.* **Togo:** Lome *Warnecke* 335! **Br.Cam.:** Mt. Mba Kokeka, 7,800 ft., Bamenda *Lightbody* FHI 30111! Widespread in the rest of tropical Africa and in S. Africa.
2. **M. senegalensis** (*Lam.*) *Exell* in Bol. Soc. Brot. Sér. 2, 26 : 223 (1952) ; Exell & Mendonça Consp. Fl. Ang. 2 : 8. *Celastrus senegalensis* Lam. (1785)—F.T.A. 1 : 361 ; Chev. Bot. 130. *Gymnosporia senegalensis* (Lam.) Loes. (1893), incl. vars.—F.W.T.A., ed. 1, 1 : 445 ; Aubrév. Fl. For. Soud.-Guin. 349–351, t. 73, 1–3 (incl. var. *djalonensis*) ; Schnell in Ic. Pl. Afr. 1 : t. 11. *Celastrus coriaceus* Guill. & Perr. Fl. Seneg. 1 : 142, t. 36 (1831). A savannah shrub or small tree, up to 16 ft. high ; young shoots often red ; flowers whitish ; fruit red.
　　Sen.: Walo *Heudelot* 200! Casamance (fl. & fr. Apr.) *Perrottet* 146! Tivaouane (Dec.) *Chev.* 3177! **Gam.:** *Dawe* 79! *Don!* **Fr.Sud.:** Timbuktu (June, July) *Hagerup* 93! 210! Birgo *Dubois* 47. **Port.G.:** Samba-Fili (fl. & fr. Apr.) *Esp. Santo* 2456! **Fr.G.:** Kouroussa (July) *Pobéguin* 273! Dalaba-Diaguissa Plateau (Oct.) *Chev.* 18875! Kadé *Pobéguin* 2010! 2040! **S.L.:** Sellakuri to Yanga (fr. Mar.) *Sc. Elliot* 5090! **Iv.C.:** Bondoukou *Aubrév.* 1610. **G.C.:** Ejura (Aug.) *Chipp* 723! 728! Kwomma to Nangruma (May) *Kitson* 932! Yijia to Pong (fl. & fr. Mar.) *Kitson* 627! 628! **Togo:** *Kersting* A. 224! Kpedsu (Jan.) *Howes* 1091! **Dah.:** Boukombé *Aubrév.* 89d. **N.Nig.:** Borgu *Barter* 776! 780! Naraguta

(Nov., Jan.) *Lely* 68 ! 722 ! P. 110 ! Kontagora (fl. & fr. Dec.) *Dalz.* 95 ! **S.Nig.:** Lagos *Rowland* !
Oyo (Jan.) *Keay* FHI 25641 ! R. Tessi, Old Oyo (Feb.) *Keay* FHI 16025 ! Widespread in the savannah
regions of tropical Africa. (See Appendix, p. 287.)

3. **M. acuminatus** (*Linn. f.*) *Loes.* in E. & P. Pflanzenfam. 20B : 138 (1942). *Celastrus acuminatus* Linn. f.
(1781). A glabrous unarmed shrub or small tree ; seeds orange-brown with white aril.
Br.Cam.: ridge above L. Oku, 7,700 ft., Bamenda (fr. Jan.) *Keay & Lightbody* FHI 28512 ! Also in
Belgian Congo, E. Africa, N. and S. Rhodesia, Nyasaland and S. Africa.

4. **M. ovatus** (*Wall. ex Wight & Arn.*) Loes. l.c. 140 (1942). *Celastrus ovatus* Wall. ex Wight & Arn. (1834).

4a. **M. ovatus** var. **ovatus** forma **pubescens** (*Schweinf.*) *Blakelock* l.c. 240 (1956). *C. littoralis* A. Chev. Bot.
129, name only. A shrub, usually spiny ; flowers whitish ; fruits red or yellow.
Iv.C.: Tabou to Bériby (fr. Aug.) *Chev.* 19944 ! **N.Nig.:** Anara F.R., Zaria Prov. (fr. Mar.) *Olorunfemi*
FHI 24415 ! Kaduna (Dec.) *Meikle* 749 ! 779 ! Bauchi Plateau (fl. & fr. Jan.) *Lely* P. 111 ! **S.Nig.:**
Enugu *Tamajong* FHI 26845 ! **Br.Cam.:** Mokanda, 8,000 ft. (fl. & fr. Feb.) *Maitland* 1326 ! Lakom,
6,000 ft., Bamenda (fl. May, fr. Apr.) *Maitland* 1496 ! 1513 ! Bapinyi, Bamenda (May) *Ujor* FHI 30366 !
Widespread in the savannah regions of the rest of tropical Africa, also in Arabia.

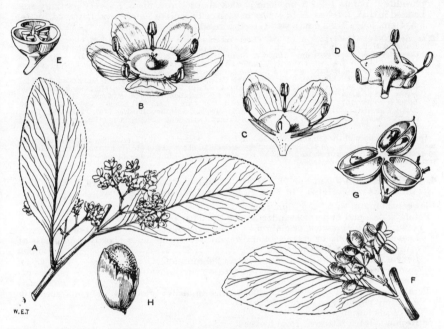

FIG. 177.—MAYTENUS SENEGALENSIS (*Lam.*) *Exell* (CELASTRACEAE).

A, flowering branch. B, flower. C, longitudinal section of same. D, flower from below with
corolla removed. E, cross-section of ovary. F, fruits. G, dehiscing fruit. H, seed showing
aril.

4b. **M. ovatus** var. **argutus** (*Loes.*) *Blakelock* l.c. 241 (1956). *Celastrus gracilipes* Welw. ex Oliv. F.T.A. 1 :
361 (1868). *Gymnosporia gracilipes* (Welw. ex Oliv.) Loes. and var. *arguta* Loes. (1893)—F.W.T.A.,
ed. 1, 1 : 445. *Maytenus gracilipes* (Welw. ex Oliv.) Exell (1952)—Consp. Fl. Ang. 2 : 4. A spiny
shrub ; petals whitish ; fruits red ; in forest.
Br.Cam.: Ambas Bay (Dec.) *Mann* 2153 ! Buea (fr. Mar.) *Maitland* 57 ! 568 ! Mimbia (Feb.) *Maitland*
1309 ! Nkom–Wum F.R., Bamenda (fr. Apr.) *Ujor* FHI 30068 ! Also in French Cameroons, A.-E.
Sudan, Abyssinia, Belgian Congo, Uganda and Angola.
[Specimens intermediate between the above two varieties are known, e.g. *Keay* FHI 16024 from Old
Oyo F.R., S. Nigeria.]

2. CASSINE Linn. Sp. Pl. 1 : 268 (1753). *Elaeodendron* Jacq. (1782)—F.T.A. 1 : 365 ;
F.W.T.A., ed. 1, 1 : 445. *Mystroxylon* Eckl. & Zeyh. (1835).

Twigs densely pubescent to glabrescent ; leaves alternate ; inflorescence subumbellate,
branches generally pubescent ; petals orbicular ; flowers bisexual . .   1. *aethiopica*
Glabrous in all parts ; leaves opposite, subopposite or alternate on same plant ; inflorescence a cyme with a central axis (continuation of peduncle) and side branches ;
petals oblong ; flowers unisexual, with petaloid staminodes in female flower
                                                                    2. *buchananii*

1. **C. aethiopica** *Thunb.* Fl. Cap. 2, 1 : 227 (1818) ; Perrier de la Bâthie in Fl. Madag. 116 : 36. *Elaeoden-*
dron aethiopicum (Thunb.) Oliv. F.T.A. 1 : 365 (1868). *Mystroxylon aethiopicum* (Thunb.) Loes. (1897).
A shrub or small tree, usually about 20 ft. high, sometimes more ; leaves variable, widely ovate to
narrowly elliptic, 0·7–13 cm. long ; petals whitish ; ripe fruits reddish ; in montane forest.
Fr.G.: Mt. Loura, 4,800 ft. (Jan.) *Roberty* 16572 ! **Br.Cam.:** Bafut-Ngemba F.R., 6,000 ft., Bamenda
(fl. Jan., May, fr. May) *Keay & Lightbody* FHI 26284 ! 28356 ! *Ujor* 30310 ! Widespread in tropical
Africa and in S. Africa, Madagascar, the Mascarene Islands and Seychelles.

2. C. buchananii *Loes.* in Engl. Bot. Jahrb. 17 : 551 (1893). *Elaeodendron afzelii* Loes. (1900)—F.W.T.A., ed. 1, 1 : 445. *E. warneckei* Loes. (1908). A glabrous shrub or tree, up to 100 ft. high ; leaves variable, elliptic to ovate, rarely oblanceolate or narrowly elliptic, 1–13 (–17) cm. long ; petals whitish ; fruits pale yellow.

S.L.: *Afzelius.* **Togo**: Lome (Jan.) *Warnecke* 45 ! Kpedyi, Nyamassila (fl. Jan., fr. June) *Kersting* A. 269 ! Also in Belgian Congo, A.-E. Sudan, E. Africa, Nyasaland, N. & S. Rhodesia and Angola.

3. CAMPYLOSTEMON Welw. ex Oliv. in J. Linn. Soc. 10 : 44 (1867) ; F.T.A. 1 : 365 ; Loes. in Notizbl. Bot. Gart. Berl. 13 : 565 (1937) ; in E. & P. Pflanzenfam. 20B : 111, fig. 32 (1942) ; Lawalrée in Bull. Jard. Bot. Brux. 18 : 249 (1947).

A climbing spineless shrub, glabrous in all parts ; leaves membranous to coriaceous, narrowly elliptic to ovate, apex acuminate to long-acuminate, margin crenate-serrate, 3–16 cm. long, 1·7–10 cm. broad, petiolate ; cymes solitary, axillary ; flowers hermaphrodite ; petals 1·5–3·5 cm. long ; disk about 1 mm. diam. ; ovary partially embedded in disk, 3-celled, 6–12 ovules in each cell ; fruit of 3 mericarps, flattened, ovate, 3·5–4 cm. long, 2–2·3 cm. broad ; seed winged   ..   ..   ..   *angolense*

C. angolense *Oliv.* l.c. (1867) ; F.T.A. 1 : 366. *C. warneckeanum* Loes. ex Fritsch (1901). *Hippocratea chevalieri* Hutch. & M. B. Moss in F.W.T.A., ed. 1, 1 : 449 (1928) ; Kew Bull. 1929 : 20. *H. kennedyi* Hoyle (1932). *Campylostemon kennedyi* (Hoyle) Hoyle & Brenan (1947). *C. mitophorum* Loes. (1937). *Hippocratea oliveriana* of F.W.T.A., ed. 1, 1 : 449, partly (*Warnecke* 157 & 302). Stems reddish brown ; flowers fragrant ; petals white turning yellow ; in forest vegetation.

S.L.: *Smeathmann* ! **Lib.**: Mt. Bobei, Sanokwele (fr. Sept.) *Baldwin* 9611 ! **Iv.C.**: Mt. Kouan, 1,300 ft., Dyolas country (Apr.) *Chev.* 21242 ! **G.C.**: Kumasi (July) *Vigne* FH 4859 ! Odumase, Krobo (June) *Irvine* 1645 ! Bana Hill, Krobo (Apr.) *Irvine* 2894 ! Anyaboni (Jan.) *A. S. Thomas* D. 115 ! **Togo**: Lome *Warnecke* 157 ! 302 ! **S.Nig.**: Ibadan South F.R. (Apr.) *Keay* FHI 22820 ! Owena, Akure F.R. (Oct.) *Ujor* FHI 23915 ! Sapoba *Kennedy* 858 ! **Br.Cam.**: Moliwe *Winkler* 1467. **F.Po**: St. Isabel (Aug.) *Schultze* in *Hb. Mildbr.* 6478. Also in French Cameroons, Belgian Congo, Uganda, Tanganyika and Angola.

Only one specimen of the genus has been seen in fruit. Loesener l.c. (1937) recognizes a number of species mainly on leaf characters ; as the leaves are very variable in other species of *Celastraceae* and as the fruit is so poorly known, it seems best to include all the African material in one species for the present, although further study may modify this view.

4. HIPPOCRATEA Linn. Sp. Pl. 1 : 363 (1753) ; F.T.A. 1 : 366 ; Loes. in Engl. Bot. Jahrb. 34 : 103 (1904) ; in E. & P. Pflanzenfam. 20B : 206 (1942).

Flowers small, up to 2 (–2·5) mm. long :
 Petals, sepals and uppermost inflorescence-branches pubescent to tomentose :
  Young twigs glabrescent or glabrous ; indumentum elsewhere tomentose :
   Leaves coriaceous or membranous, rounded or sometimes shortly acuminate at
    apex ; petals 1·5–2 mm. long   ..   ..   1. *myriantha*
   Leaves membranous, acuminate ; petals 2·5 mm. long   ..   ..   2. *vignei*
  Young twigs shortly pubescent-tomentose ; indumentum elsewhere pubescent-
   tomentose ; petals 2 mm. long   ..   ..   3. *guineensis*
 Petals, sepals and uppermost inflorescence-branches glabrous or glabrescent or
   minutely papillose :
  The petals not clawed:
   Resinous threads absent from leaves :
    Petals about 1 mm. long, more or less oblong ; leaves up to 12·5 cm. long, reticula-
     tions not prominent ; brownish or grey when dry :
     Petals flat, rounded at apex ; stigma sessile, above the anthers ; mericarps obo-
      vate-oblong, up to 2·8 cm. long and 1·3 cm. broad   ..   ..   4. *pallens*
     Petals involute (rarely flat), so that apex appears acute ; style present, stigma
      below or more or less on level with anthers ; mericarps generally oblong, up to
      5·2 cm. long, 0·7–1·4 cm. broad   ..   ..   5. *indica*
    Petals at least 1·5 mm. long, ovate ; leaves often larger, reticulations prominent :
     Leaves greenish or blackish when dry ; flowers 4–6 mm. diam. ; disk produced
      into a stipe (androgynophore) in " female " flower, stipe inconspicuous or
      absent in " male " flower ; mericarps oblong-ovate, 1·2–2·2 cm. broad
                                              12. *welwitschii*
     Leaves greenish-yellow, rarely brownish, when dry ; flowers 2–4 mm. diam. ;
      no stipe ; mericarps ovate-oblong or obovate-oblong, 2·4–3·8 cm. broad
                                              6. *paniculata*
   Resinous threads present in leaves ; see also below ..   ..   ..   13. *mucronata*
  The petals clawed ..   ..   7. *preussii*
 Flowers 3 mm. or more long :
  Flower-buds shortly conic or ellipsoid and pointed :
   Petals triangular ; disk simple (not produced into a stipe) :
    Outside of petals very minutely puberulous ; calyx and inflorescence-branches
     shortly pubescent or glabrescent, greyish when dry ; twigs and leaves glabrous
     or glabrescent ; petals 3–5 mm. long ; mericarps ovate-oblong or obovate-
     oblong, 4·4–7·5 cm. long, 1·6–4·4 cm. broad   ..   ..   ..   8. *africana*
    Outside of petals, calyx and inflorescence-branches long and densely tomentose-
     pubescent, red-brown when dry ; petals up to 4 mm. long ; mericarps ovate-
     oblong, about 6 cm. long and 3 cm. broad, tomentose-pubescent   9. *iotricha*

Petals linear-triangular ; disk double, lower and outer part cup-shaped, upper and
inner part forming a stipe (androgynophore) :
Petals minutely tomentellous outside, very acute, 6–9 mm. long ; mericarps oblong-
ovate, 7·8–10 cm. long, 2·4–4·2 cm. broad .. .. 10. *clematoides*
Petals minutely tuberculate-scabrid outside, acute, 4–6 mm. long.. 11. *rowlandii*
Flower-buds globose and rounded or pointed, or ellipsoid and rounded :
Petals not fimbriate ; twigs and lower surface of leaves glabrous :
Upper part of inflorescence tomentose.. .. .. 2. *vignei*
Upper part of inflorescence glabrous :
Flowers less than 1 cm. diam. :
Resinous threads absent from leaves ; plant often greenish or blackish when
dry ; flower-buds globose ; petals membranous ; disk produced into a stipe
(androgynophore) in " female " flower, stipe inconspicuous or absent in " male "
flower .. .. 12. *welwitschii*
Resinous threads present in leaves ; plant brownish when dry ; flower-buds
depressed-globose ; petals thick and fleshy ; no stipe .. 13. *mucronata*
Flowers at least 1 cm. diam. ; petals shortly clawed, membranous, entire ; leaves
6–23 cm. long, reticulate ; mericarps flattened, broadly obovate, 7–9·1 cm. long,
6·5–7 cm. broad ; seeds bulky, wing shorter than seed ; flower-buds globose,
rounded ; glabrous in all parts .. .. 14. *macrophylla*
Petals fimbriate, clawed, 4–5 mm. long ; flower-buds globose, rounded or apiculate ;
outer surface of sepals and petals, inflorescence-branches, twigs and lower surface
of leaves red-brown tomentose-pubescent ; mericarps subterete 15. *velutina*

1. **H. myriantha** *Oliv.* F.T.A. 1 : 369 (1868) ; Loes. in Engl. Bot. Jahrb. 34 : 106, fig. 1, D–F ; Chev. Bot.
131. Climbing shrub with the leaves dark green and shiny above, paler beneath ; flowers small, creamy-
white, very numerous in much-branched cymes ; fruits oblong, flat, very glaucous.
S.L.: Bagroo R. (Apr.) *Mann* 790 ! **Iv.C.:** Dabou (Feb.) *Chev.* 17251. Mt. Nienoku̇é, 1,600 ft. *Chev.*
19456. **G.C.:** Obuom (June) *Vigne FH* 2975 ! **S.Nig.:** Sapoba (Jan.) *Brenan* 8936 ! Nun R. *Barter*
2117 ! Bonny (Feb.) *Kalbreyer* 49 ! Eket *Talbot* 3165 ! Calabar (Feb.) *Mann* 2309 ! *Thomson* 42 !
Ikwette Plateau, Obudu, 5,200 ft. (Dec.) *Keay & Savory* FHI 25216 ! **Br.Cam.:** Bum, 4,000 ft., Bamenda
(May) *Maitland* 1675 ! Also in Ubangi-Shari, French Cameroons, Spanish Guinea, Gabon, Belgian
Congo and Angola.
2. **H. vignei** *Hoyle* in Kew Bull. 1932 : 264. A forest liane ; flowers cream.
G.C.: Kwahu Prasu (June) *Vigne* FH 1752 ! Sunyani (young fr. Sept.) *Vigne* FH 2450 !
3. **H. guineensis** *Hutch. & M. B. Moss* in F.W.T.A., ed. 1, 1 : 449 (1928) ; Kew Bull. 1929 : 20. *H. apo-
cynoides* of F.T.A. 1 : 368, partly (excl. Welw. spec.) ; Chev. Bot. 130 (incl. form & var.) ; not of Welw.
ex Oliv. Climbing shrub ; inflorescence and calyx covered with brown indumentum, petals greenish-
cream.
Fr.G.: Kouria (June) *Chev.* 14953 ! Timbikounda, Farana *Chev.* 20595 ! **S.L.:** Konnoh country *Burbidge*
499 ! Bumpe (Sept.) *Deighton* 4029 ! 5578 ! **Lib.:** Dukwia R. (Feb.) *Cooper* 181 ! **Iv.C.:** Ahiamé,
Sanvi *Chev.* 17776. Prolo to Bliéron *Chev.* 19891. **G.C.:** Akwapim Hills (Oct.) *Johnson* 776 ! **Togo:**
Lome *Warnecke* 404 ! **N.Nig.:** top of Patti Lokoja *Dalz.* 204 ! Anara F.R., Zaria (Oct.) *Keay* FHI
20119 ! **S.Nig.:** Lagos *Millen* 60 ! Mouth of R. Niger (Aug.) *Mann* 456 ! Eket Dist. *Talbot* 3049.
Br.Cam.: Cameroon R. (Jan.) *Mann* 2218 ! Also in French Cameroons, Uganda and Tanganyika. (See
Appendix, p. 289.)
4. **H. pallens** *Planch. ex Oliv.* F.T.A. 1 : 367 (1868) ; Loes. l.c. 103, fig. 1, A ; Chev. Bot. 131 ; Loes. in E. & P.
Pflanzenfam. 20B : 210, fig. 66, A. *H. buchholzii* Loes. (1894)—Exell & Mendonça in Consp. Fl. Ang. 2 :
16. *H. oliveriana* Hutch. & M. B. Moss in F.W.T.A., ed. 1, 1 : 449 (1928), excl. *Warnecke* 157 & 302 ;
Kew Bull. 1929 : 20. Climbing shrub ; flowers green to yellow-cream.
Sen.: Heudelot 341 ! **Port.G.:** Canjabarim, Farim (Apr.) *Esp. Santo* 2676 ! **Fr.G.:** Kouroussa (Apr.,
May) *Pobéguin* 691 ! 892 ! Labé (Mar.) *Chev.* 12240 ! Dentilia (fl. & fr. Mar.) *Sc. Elliot* 5270 ! **S.L.:**
Barter ! Bagroo R. (Apr.) *Mann* 855 ! Njala (Feb., Apr.) *Deighton* 1090 ! 3537 ! 4741 ! Gbap (fl. & fr.
Mar.) *Adames* 16 ! **Lib.:** Monrovia *Whyte* ! **Iv.C.:** various localities *fide* Chev. l.c. **G.C.:** *Vigne* FH 3867 !
Ancobra R., Anibil (Mar.) *Irvine* 2389 ! E. Akim (June) *Johnson* 760 ! **Dah.:** Birni, Atacora Mts. *Chev.*
23976. **N.Nig.:** Katabu, Kaduna (May) *Keay* FHI 25791 ! Katagum *Dalz.* 412 ! **S.Nig.:** Igboho,
Oyo Prov. (Feb.) *Keay* FHI 22498 ! Sapoba (Nov.) *Kennedy* 2748 ! *Meikle* 555 ! Uqua to Ntete, Eket
(fl. & fr. May) *Onochie* FHI 33191 ! Oban *Talbot* 1447 ! Also in French Cameroons, Gabon, Belgian
Congo, Kenya and Angola.
5. **H. indica** *Willd.* Sp. Pl. 1 : 193 (1797) ; F.T.A. 1 : 368 ; Loes. in Engl. Bot. Jahrb. 34 : 106 ; Chev. Bot.
130. *H. loesneriana* Hutch. & M. B. Moss in F.W.T.A., ed. 1, 1 : 449 (1928) ; Kew Bull. 1929 : 20.
Climbing shrub ; flowers yellow or whitish.
Fr. Sud.: Massif de Kita, Cercle de Kita (fl. June, July, fr. Dec.) *Jaeger* 21 ! 2213 ! 3830 ! **Port.G.:**
Safim, Bissau (fr. Mar.) *Esp. Santo* 1908 ! Farim Dist. (fr. Apr.) *Esp. Santo* 2454 ! **Fr.G.:** Pobéguin
815 ! Fouta Djalon (Apr.) *Chev.* 18145 ; 18618. Nzérékoré (fr. Oct.) *Baldwin* 9702 ! Macenta (fr.
Oct.) *Baldwin* 9835 ! **S.L.:** Moyamba (June) *Lane-Poole* 48 ! Njala (fl. May, Aug., fr. Jan., Mar., Aug.)
Deighton 1146 ! 1211 ! 4308 ! 4617 ! Roruks (July) *Deighton* 3249 ! Bumbuna (fr. Oct.) *Thomas* 3702 !
Lib.: Gbanga (fl. May, Sept.) *Konneh* 171 ! *Linder* 490 ! 808 ! Dukwia R. *Cooper* 409 ! Karmadhun
(fr. Nov.) *Baldwin* 10183 ! 10196 ! Sanokwele (fr. Sept.) *Baldwin* 9514 ! 9641 ! **Iv.C.:** Mt. Kouan,
Dyola (Apr.) *Chev.* 21243 ! **G.C.:** Kumasi (fl. June, fr. Oct.) *Andoh* FH 4601 ! 5497 ! *Vigne* FH 3011 !
Djadje (fr. July) *Akpabla* 46 ! **Togo:** Lome *Warnecke* 406 ! **N.Nig.:** Nupe *Barter* ! **S.Nig.:** Lagos
Dawodu 128 ! Ilaro (Jan.) *Millen* 110 ! Ibadan (fl. July, fr. Sept.) *Onochie* FHI 13508 ! *Brown* ! **F.Po:**
Barter ! *Mann* 185 ! This species (or close relatives) is widespread in tropical Africa and tropical Asia.
(See Appendix, p. 289.)
6. **H. paniculata** *Vahl* Enum. 2 : 28 (1806) ; Loes. l.c. 113 ; Chev. Bot. 132. *H. polyantha* Loes. (1912).
H. bequaertii De Wild. (1923). A large climbing shrub ; flowers white or pale greenish.
Sen.: Samatite. Casamance (Apr.) *Perrottet* 96 ! **Port.G.:** Brene, Bissau (fl. & fr. Jan.) *Esp. Santo* 1679 !
Empada, Buba (fr. Dec.) *Esp. Santo* 2338 ! **Fr.G.:** *Farmar* 240 ! Labé *Chev.* 12392 ; 12406. **S.L.:**
Smeathmann ! Kent (May) *Deighton* 5533 ! *Smythe* 246 ! Leicester (June) *Deighton* 2706 ! Njala (Mar.)
Deighton 2622 ! Kambui Hills (fr. Dec.) *Small* 868 ! **Iv.C.:** Guidéko *Chev.* 16469. Sassandra Port
Chev. 17927 ; 17944. **G.C.:** Nkinkaro (fr. Oct.) *Vigne* FH 2549 ! **S.Nig.:** Agbemia *Barter* 3363 ! Also
in Belgian Congo, Angola and E. Africa.
7. **H. preussii** *Loes.* in Engl. Bot. Jahrb. 34 : 112, fig. 2B (1904). A climbing shrub.
Br.Cam.: Victoria to Bimbia (July) *Preuss* 1306 !

8. **H. africana** (*Willd.*) *Loes. ex Engl.* Pflanzenw. Afr. 3, 2 : 240 (1921). *Tonsella africana* Willd. Sp. Pl. 1 : 194 (1797) ; Vahl Enum. Pl. 2 : 30 (1805), partly (Willd. ref. & Isert spec. only) ; Schum. Beskr. Guin. Pl. 20 (1827), partly (Willd. ref. only). *Salacia africana* (Willd.) DC. Prod. 1 : 570 (1824), partly (Willd. ref. only). *Hippocratea obtusifolia* Roxb. (1820)—F.T.A. 1 : 369 ; Loes. in Engl. Bot. Jahrb. 19 : 236, incl. vars. ; Chev. Bot. 131 ; Loes. in E. & P. Pflanzenfam. 20B : 211, 213, fig. 66, G–L, fig. 67 (1942). *H. richardiana* Cambess. (1829)—Guill. & Perr. Fl. Seneg. 1 : 112, t. 26 ; F.W.T.A., ed. 1, 1 : 449 ; Schnell in Ic. Pl. Afr. 1 : t. 12. *H. cymosa* var. *togoensis* Loes. in Engl. Bot. Jahrb. 34 : 108 (1904). A liane with green twigs and bright green leaves ; flowers fragrant, petals green, anthers orange ; a very variable species ; mainly in fringing forest in the savannah regions of our area.
Sen.: *Perrottet* 95 ! Richard Tol *Perrottet* ! *Richard.* Walo *Heudelot* ! R. Farèna, Thin (Feb.) *Döllinger* 32 ! Dakar (Jan.) *Hagerup* 776 ! **Gam.:** Genieri–Massembe Road (Feb.) *Fox* 57 ! **Fr.Sud.:** Niafunké

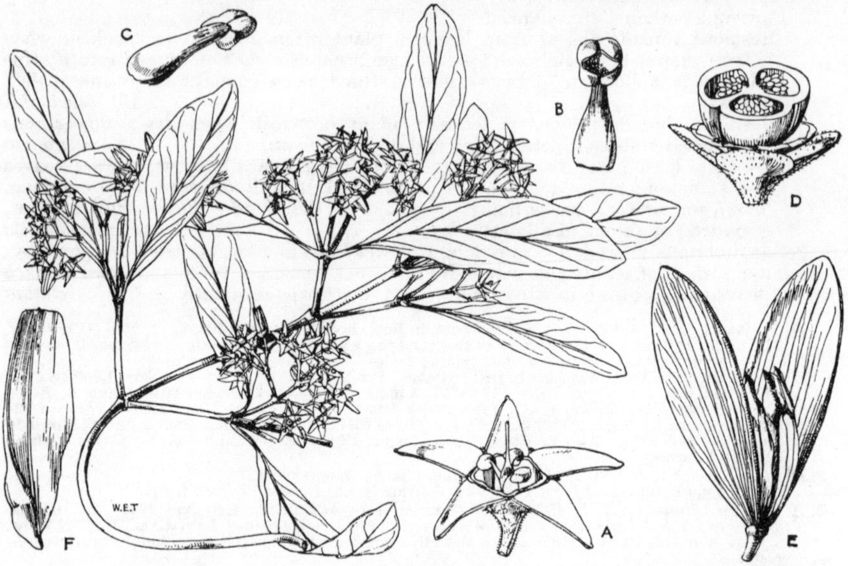

FIG. 178.—HIPPOCRATEA AFRICANA (*Willd.*) *Loes. ex Engl.* (CELASTRACEAE).
A, flower. B, C, stamen front and back. D, ovary in cross-section. E, two mericarps and seeds.
F, seed.

(June), *Hagerup* 87 ! Badinko (Jan.) *Chev.* 113. Balani (Jan.) *Chev.* 148. Sumpi to Sébi *Chev.* 3174. **Port.G.:** Cobras Isl. (Apr.) *Esp. Santo* 1929 ! For, Bissau (fl. & fr. Feb.) *Esp. Santo* 1782 ! **S.L.:** *Barter* ! Musaia (Mar.) *Sc. Elliot* 5138 ! Samaia (May) *Thomas* 223 ! Likuru *Sc. Elliot* 4965 ! **Lib.:** Peahtah (fr. Oct.) *Linder* 1627 ! Vahon, Kolahun (fr. Nov.) *Baldwin* 10216 ! **G.C.:** Aburi Road (Feb.) *Dalz.* 8267 ! Alababa (Apr.) *Scholes* 290 ! Achimota (June) *Irvine* 1454 ! **Togo:** Lome *Warnecke* 43 ! 158 ! 309 ! Kabosu, Hohoe Dist. *St. C. Thompson* FH 3678 ! **N.Nig.:** Nupe *Barter* 1215 ! Katagum Dist. *Dalz.* 217 ! Bornu Prov. (Jan., Feb.) *Talbot* ! **S.Nig.:** Ifon to Sobe, Ondo Prov. (fr. Jan.) *Onochie* in *H b. Brenan* 8948 ! **Br.Cam.:** Bama, Dikwa Div. (Jan.) *McClintock* 133 ! Widespread in tropical Africa, S. Africa, Madagascar, India, China, Celebes and Philippines. (See Appendix, p. 289.)
[We have seen photographs of the type specimen of *Tonsella africana* Willd. It was collected by Isert in " Guinea "—presumably a coastal area in what is now the Gold Coast, Togoland or Dahomey.]

9. **H. iotricha** *Loes.* in Engl. Bot. Jahrb. 34 : 108, fig. 1, O (1904). A woody climber, with dense brown indumentum ; flowers pale yellow with brown indumentum, sweet-scented ; in forest vegetation.
S.L.: Njala (fl. June, fr. Oct.) *Deighton* 1744 ! 4911 ! Bendu-Mamaima (Apr.) *Deighton* 3941 ! Pelewahun, Kamagai (fr. Sept.) *Deighton* 6116 ! Kandorahun, S. Prov. (Apr.) *Deighton* 1627 ! **Lib.:** Sanokwele (young fr. Sept.) *Baldwin* 9488 ! **S.Nig.:** Sapoba *Kennedy* 1622 ! 2715 ! Oban *Talbot* 1275 ! Also in French Cameroons. (See Appendix, p. 289.)

10. **H. clematoides** *Loes.* l.c. 109, fig. 1, P–Q (1904). *H. venulosa* Hutch. & M. B. Moss in F.W.T.A., ed. 1, 1 : 449 (1928) ; Kew Bull. 1929 : 20. A woody climber ; petals olive-green, disk and stamens orange ; in forest vegetation.
S.L.: *Thomas* 10132 ! 10255 ! 10550 ! Njala (Mar.) *Deighton* 3121 ! **Lib.:** Lormai (Feb.) *Bequaert* 87a ! Ganta (Apr.) *Harley* 1137 ! Belefania, Gbanga (Dec.) *Baldwin* 10539 ! **G.C.:** Peduari, Aburi (July) *Darko* 907 ! **S.Nig.:** *Thomas* 2266 ! Okomu F.R., Benin (Feb.) *Richards* 3946 ! Sapoba *Kennedy* 1977 ! 2320 ! **Br.Cam.:** Kumba Div. (fr. Dec.) *Ejiofor* FHI 15196 ! Also in French Cameroons, Spanish Guinea, Gabon, Belgian Congo, Uganda, Nyasaland and Angola.

11. **H. rowlandii** *Loes.* in Engl. Bot. Jahrb. 19 : 236 (1894) ; op. cit. 34 : 110 ; Chev. Bot. 132. A woody climber ; calyx olive-green, petals yellowish green, disk deeper green, anthers dark yellow; in forest vegetation.
Fr.G.: Kaba valley *Chev.* 13126. **Iv.C.:** various localities *fide* Chev. l.c. **G.C.:** Kwahu (Apr.) *Johnson* 645 ! Opnoase (May) *Johnson* 755 ! Owabi (Apr.) *Andoh* FH 4184 ! Agogo, Akim (Apr., May) *Irvine* 933 ! *Williams* 271 ! **S.Nig.:** Lagos (Dec.) *Millen* 24 ! *Rowland* ! Near Ibadan (Mar., Apr.) *Keay* FHI 22540 ! *Meikle* 1462 ! Oban *Talbot* ! **Br.Cam.:** Johann-Albrechtshöhe *Staudt* 885 ! Also in French Cameroons.

12. **H. welwitschii** *Oliv.* F.T.A. 1 : 367 (1868) ; Loes. in Engl. Bot. Jahrb. 34 : 112. A woody climber ; flowers pale green, except for the cream-coloured anthers.
Fr.G.: Nzérékoré (fr. Sept.) *Baldwin* 13304 ! **S.L.:** Makene (Apr.) *Deighton* 5498 ! Njala (Mar., May) *Deighton* 4631 ! 4717 ! Kamakwie (fl. & fr. Feb.) *Deighton* 4046 ! 4201 ! Shengama, Serabu (Apr.) *Deighton* 1698 ! **Lib.:** Gbanga (Dec.) *Baldwin* 10527 ! Belleyella (Dec.) *Baldwin* 10656a ! Ganta (May) *Harley* 1149 ! Dobli Isl., St. Paul R. (Apr.) *Bequaert* 23 ! **G.C.:** Akwapim Hills (Mar.) *Irvine* 1601 !

Pokoase (Jan.) *Irvine* 1959 ! Abokobi (Feb.) *Irvine* 1998 ! Dodowa (May) *Irvine* 2678 ! **Dah.**: *Poisson* !
S.Nig.: Lagos *Foster* 61 ! Ikire, Ibadan (fr. Jan.) *Keay* FHI 22464 ! Omo F.R., Ijebu Ode (Apr.)
Onochie FHI 15531 ! Idanre Hills (fr. Jan.) *Brenan* 8713 ! Eket *Talbot* 3135 ! **Br.Cam.**: Victoria *Preuss*
1304 ! Likomba (Nov.) *Mildbr.* 10709 ! Also in French Cameroons, Gabon, Belgian Congo, Uganda,
Tanganyika and Angola. (See Appendix, p. 289.)
13. **H. mucronata** *Exell* in J. Bot. 65, Suppl. Polypet. 78 (1927). A climbing shrub ; flowers yellow or white ;
in forest vegetation.
    G.C.: Bunsu, Akim (May) *Irvine* 3013 ! Also in Gabon, Cabinda and Uganda.
14. **H. macrophylla** *Vahl* Enum. 2 : 28 (1806) ; F.T.A. 1 : 371 ; Loes. l.c. 115 ; Chev. Bot. 131. Climbing
shrub with leaves often more than 15 cm. long and 8 cm. broad ; flowers white, very fragrant ; in forest,
often by streams.
    Fr.G.: Conakry *Chev.* 12732. **S.L.**: *Don* ! *Smeathmann* ! Rotumba Isl. (Mar.) *Kirk* 21 ! Roruks (fr.
July) *Deighton* 2510 ! Moa R., Koteimahun (Mar.) *Lane-Poole* 328 ! Mange Bure (Jan.) *Jordan* 860 !
Lib.: Monrovia (June) *Baldwin* 5938 ! *Whyte* ! **Iv.C.**: several localities *fide* Chev. *l.c.* **G.C.**: *Burton &*
Cameron ! **S.Nig.**: R. Ogbesse, Owo (Mar.) *Jones* FHI 3080 ! Eket *Talbot* ! Oban *Talbot* 1406 ! Calabar
(Feb.) *Mann* 2292 ! **Br.Cam.**: Bulifambo, Buea (Mar.) *Maitland* 563 ! Also in Spanish Guinea, Gabon,
Belgian Congo and Angola.
15. **H. velutina** *Afzel.* Remed. Guin. 33 (1815) ; F.T.A. 1 : 370 ; Loes. in Engl. Bot. Jahrb. 34 : 120, fig. 3 ;
Chev. Bot. 132 (incl. var. *apiculata* A. Chev.). *Helictonema klaineanum* Pierre (1898). Climbing shrub
with rusty-tomentose branchlets ; petals brownish-green outside, fawn within ; fruits rusty brown ; in
forest regions.
    Fr.G.: Ymbo to Kouria (Nov.) *Chev.* 14749. **S.L.**: *Afzelius* ! Heddle's Road (Sept.) *Lane-Poole* 355 !
Newton (Aug.) *Deighton* 2203 ! Njala (fl. Aug., fr. May) *Deighton* 1829 ! 5936 ! Rokupr (Sept.) *Jordan*
353 ! Pelewahun, Kamagai (Sept.) *Deighton* 6007 ! **Lib.**: Lango (Aug.) *Linder* 303 ! Tappita (Aug.)
Baldwin 9057 ! Vonjama (young fr. Oct.) *Baldwin* 9916 ! Gbanga (May, Aug.) *Konneh* 176 ! *Traub*
266 ! **Iv.C.**: Bouaké (July) *Chev.* 22107 ! Abidjan region *Chev.* 15472. **G.C.**: Kumasi (Aug.) *Darko*
704 ! Sunyani (Sept.) *Vigne* FH 2456 ! **Togo**: Abohire, Hohoe Dist. (Oct.) *St. C. Thompson* FH 3628 !
S.Nig.: Lagos *Dawodu* 112 ! 266 ! Mamu F.R., Ibadan (Oct.) *Keay & Mutch* FHI 25431 ! Ibadan
(fl. Oct., fr. Dec.–Mar.) *Keay* FHI 13747 ! 22536 ! *Meikle* 905 ! Onitcha Olona (Oct.) *Thomas* 1852 !
Br.Cam.: Nkom-Wum F.R., Bamenda (July) *Ujor* FHI 30479 ! Also in French Cameroons, Spanish
Guinea, Gabon, Cabinda, Belgian Congo, Uganda, A.-E. Sudan and S. Tomé. (See Appendix, p. 289.)

Imperfectly known species.

H. sp. A climbing shrub, with obovate fruits 8–9 cm. long and about 6 cm. broad ; leaves shiny, coriaceous,
elliptic to oblong, 7–8 cm. long, venation prominently reticulate on both surfaces ; older twigs markedly
verruculose.
    S.Nig.: R. Nwop, Boje, Ikom Div. (fr. May) *Latilo* FHI 31805 !

Excluded species.

1. **H. thomasii** *Hutch. & M. B. Moss* in F.W.T.A., ed. 1, 1 : 449 (1928) ; Kew Bull. 1929 : 20, is **Secamone**
myrtifolia *Benth.* (Asclepiadaceae).
2. **H. chevalieri** *Hutch. & M. B. Moss* in F.W.T.A., ed. 1, 1 : 449 (1928) ; Kew Bull. 1929 : 20, is **Campylo-**
stemon angolense *Oliv.* (see p. 626).

5. SALACIA Linn. Mant. 2 : 159, 293 (1771) ; F.T.A. 1 : 372 ; Loes. in Engl. Bot. Jahrb. 44 : 156 (1910) ; in E. & P. Pflanzenfam. 20B : 217 (1942).

*Flowers solitary or fasciculate in leaf-axils :
  Stem with 4 (–5) sinuate wings 1·5–3 mm. broad ; leaves 9·5–26·5 cm. long, 5·5–8 cm.
    broad ; petiole up to 5 mm. long, margins crisped ; pedicels about 5–10 mm. long ;
    flowers about 6 mm. diam. ; fruit globose, about 3 cm. diam., very squamate-
    tuberculate   ..   ..   1. *alata*
  Stem not winged or with narrow straight wings up to 0·5 mm. broad :
    Twigs with violet (black when dried) spreading hairs ; leaves 5–25 cm. long, deep
      green and shiny above, paler beneath, hairy on main nerves beneath ; pedicels
      3–6 mm. long ; flowers about 6 mm. diam. ..   ..   2. *hispida*
    Twigs glabrous or slightly tomentose :
      †Resinous threads present on breaking leaf :
        Buds just before opening elongate, about twice as long as broad :
          Petiole-margins straight :
            Twigs glabrous, smooth or slightly verruculose-lenticular ; flowers 2·5–4 (–6) mm.
              diam., petals ovate ; leaves brownish or greenish when dry..   3. *leptoclada*
            Twigs verruculose-lenticular, 4-lineate or terete ; flowers 7–16 mm. diam., petals
              more or less oblong ; leaves oblanceolate-oblong to obovate, usually reddish-
              brownish beneath when dry ; fruit more or less globose, somewhat apiculate
                                          4. *senegalensis*
          Petiole-margins crisped ; twigs slightly tomentose, 4-angular ; leaves subovate-
            oblong, 7–13 cm. long ; petals about 3 mm. long   ..   5. *volubilis*
        Buds more or less globose just before opening :
          Twigs verruculose-lenticular ; petioles 3–6 mm. long :
            Leaf-apex rounded ; leaves coriaceous, margins shallowly crenate, 3·4–6·5 (–8)
              cm. long, 1·3–3·5 cm. broad ; twigs more or less 4-lineate or 4-angled ; pedicels
              about 6 cm. long ; flowers about 7 mm. diam. ; fruit globose, 2–3 cm. diam.,
              somewhat tuberculate with 3 raised ridges   ..   ..   6. *lomensis*
            Leaf-apex acuminate ; leaves usually larger ; twigs usually not 4-lineate
              (sometimes irregularly longitudinally striate)   *senegalensis* × *pyriformis*
      Twigs not verruculose-lenticular, or only slightly so :
        Flowers 2·5–4 (–6) mm. diam. ; petioles 2–7 (–9) mm. long   ..   3. *leptoclada*
        Flowers 7–17 mm. diam. ; petioles 7–17 mm. long, margins not crisped ; leaves
          coriaceous, 6–20 cm. long, brownish when dry, apex acuminate or sometimes

rounded ; pedicels 7–20 mm. long ; fruit globose, more or less tuberculate, up to 2 cm. diam. ..   ..   7. *pyriformis*

†No resinous threads on breaking leaf :

Twigs terete often flattened at nodes, sometimes irregularly longitudinally striate :

Leaves with about 6 (8–10 in No. 15) pairs of main lateral nerves :

Flowers up to 8 mm. diam. ; leaves small or large :

Leaves crisped and serrate at margin, narrow-elliptic to elliptic-oblong, greyish green when dry, membranous ; pedicels 2–7 mm. long ; fruit globose

8. *pallescens*

Leaves with flat subentire margin, greenish when dry :

Pedicels up to 8 mm. long ; petals ovate or obovate :

Flowers 5–8 mm. diam. :

Leaves dark greenish when dry, a little paler beneath, up to 11 cm. long ; flowers 5–8 mm. diam. ; leaves broadly elliptic to oblong ; petioles 4–8 mm. long ;´ pedicels 5–8 mm. long ; liane ..   ..   9. *togoica*

Leaves grey-green when dry, paler beneath, 8–19 cm. long ; flowers about 6 mm. diam. ..   ..   10. *cuspidicoma*

Flowers about 3 mm. diam. ; leaves greenish when dry, much paler beneath, more acuminate, up to 20·5 cm. long ; small tree..   ..   11. *mannii*

Pedicels 10–42 mm. long, filamentous in flower ; petals obovate or oblong :

Petals obovate ; leaves broadly elliptic, 11–16 cm. long, 5–9 cm. broad, grey-greenish when dry   ..   ..   12. *pyriformoides*

Petals oblong ; very like *pyriformoides* in twigs, leaves and pedicels

13. *dusenii*

Flowers over 10 mm. diam. ; leaves 7–18 cm. long, 3–8·5 cm. broad :

Twigs verruculose-lenticular :

Pedicels filamentous, up to 35 mm. long ; leaves oblong-elliptic, 10·5–27 cm. long   ..   ..   14. *zenkeri*

Pedicels 10–15 mm. long, fairly stout ; petiole 3–5 mm. long ; sepals shortly lacerate ..   ..   15. *conrauii*

Twigs not verruculose-lenticular :

Pedicels stouter, up to 20 mm. long, often less ; petioles 10–17 mm. long ; leaves up to 22 cm. long, oblong-elliptic to broadly elliptic, grey-green when dry ; sepals shortly lacerate ..   ..   16. *preussii*

Leaves with 11–19 pairs of main lateral nerves :

Leaves up to 24 cm. long, with up to 14 pairs of main lateral nerves :

Leaves oblong to slightly obovate, 13·5–24 cm. long, acumen up to 2·5 cm. long, nerves impressed above ; pedicels filamentous, up to 17 cm. long ; flowers 4–5 mm. diam. ; fruit globose, neither tuberculate nor apiculate, about 3 cm. long ..   ..   17. *loloensis*

Leaves elliptic to oblong, 12–18 cm. long, shortly and obtusely acuminate, venation not so distinct as in the preceding ; pedicels sub-filamentous, 15–20 mm. long ; flowers 5–7 mm. diam.   ..   ..   ..   18. *eurypetala*

Leaves 25–40 cm. long, 8–14 cm. broad, with 14–19 pairs of main lateral nerves

19. *sp. A*

Twigs 4-lineate or narrowly 4-winged (wings up to 0·5 mm. broad) :

Flowers over 10 mm. diam. :

Pedicels up to 5 mm. long ; leaves serrate-crenate, greenish when dry ; petioles 10–12 mm. long, margins crisped ; sepals lacerate ..   20. *fimbrisepala*

Pedicels 7–10 mm. long ; petioles up to 2 mm. long ; sepals entire   21. *sp. B*

Flowers up to 8 mm. diam. :

Flowers about 3 mm. diam., pedicels filiform, 5–20 mm. long ; fruit more or less globose, neither tuberculate nor apiculate ; petiole margins crisped ; leaves up to 10 cm. long, usually brownish green when dry ..   22. *debilis*

Flowers 5–8 mm. diam. ; pedicels 4–10 mm. long :

Twigs smooth or finely verruculose ; fruit more or less globose, smooth, or verruculose at most, rounded at apex when ripe, up to 3 cm. diam. ; leaves narrowly to broadly elliptic ..   ..   23. *erecta*

Twigs closely and conspicuously verruculose ; fruit ellipsoid, tuberculate, apiculate ; leaves ovate to elliptic-oblong ..   ..   ..   24. *tuberculata*

*Flowers in a pedunculate subumbel or cyme (peduncle sometimes only 1 mm. long) :

Inflorescence glabrous or glabrescent :

‡No resinous threads on breaking leaf :

Twigs closely verruculose :

Fruit 1-seeded, up to 7 cm. long, 3 cm. broad, not (?) tuberculate   25. *lehmbachii*

Fruit several-seeded, up to 5·5 cm. long, 2·5 cm. broad, tuberculate, apiculate ; petiole-margins crisped :

Petioles 5–10 mm. long ; leaves cuneate to obtuse at base, subcoriaceous, ovate to elliptic-oblong, 3·5–19·5 cm. long, 6–9 main lateral nerves on each side of midrib ; peduncle up to 1 mm. long ..   ..   24. *tuberculata*

Petioles 1–5 mm. long ; leaves rounded and subcordate or cuneate to obtuse at base, elliptic-oblong, rarely broadly elliptic, 6–18·5 cm. long ; peduncles 3–7 mm. long .. .. 26. *leonensis*
Twigs not closely verruculose :
Petals orbicular or ovate ; flowers 5–10 mm. diam. ; peduncles (0·5–) 1·5–6 cm. long ; inflorescence subumbellate or cymose, bracts generally petiolate-spathulate ; leaves stiffly papery, up to 19 cm. long ; fruits oblong-ellipsoid, verruculose-rugose, apiculate, 4–5 cm. long, 2 cm. broad .. 27. *camerunensis*
Petals more or less oblong ; flowers 3–5 mm. diam. ; peduncles up to 1·4 cm. long ; inflorescence subumbellate, often with a series of small sessile bracts (bracteoles ?) widest at their base and placed at base of pedicels :
Pedicels about 20 mm. long, filamentous ; flowers 2–4 mm. diam. ; leaves usually greenish-grey when dry .. .. 28. *staudtiana*
Pedicels 1–10 mm. long, filamentous or stouter ; flowers 3–5 mm. diam. :
Leaves greyish-green or green when dry, nerves not impressed above, narrowly to broadly elliptic, 3–10 (–14) cm. long, acuminate, acumen rounded at apex ; pedicels filamentous, 3–7 mm. long ; fruit globose, ·smooth, up to 2 cm. long, not apiculate, with a neck at base when young ; flowers about 3 mm. diam. .. .. 29. *caillei*
Leaves greenish-brown when dry, nerves often impressed above ; pedicels filamentous or stouter, 1–10 mm. long ; fruit irregularly globose to oblong-ellipsoid, slightly apiculate, without a neck at base ; flowers about 5 mm. diam. .. .. 30. *nitida*
‡Resinous threads present on breaking leaf :
Flowers 2·5–4 (–6) mm. diam., petals ovate ; buds more or less globose or somewhat elongated just before flowering ; inflorescence a fascicle or subumbel ; peduncles up to 3 mm. long ; pedicels 2–6 mm. long ; leaves 1·5–9 (–13) cm. long ; petioles 2–7 (–9) mm. long .. .. 3. *leptoclada*
Flowers 6 mm. or more diam. ; petals narrower :
Twigs not verruculose-lenticular :
Flowers 6–8 mm. diam., petals linear-triangular, about 1 mm. broad at base ; buds elongated, about thrice as long as broad ; inflorescence cymosely branched ; leaves 4·2–13 cm. long ; petiole 7–9 mm. long, margins straight
31. *lucida*
Flowers 7–17 mm. diam., petals oblong, obovate-oblong, rarely ovate ; buds more or less globose ; inflorescence a fascicle or subumbel ; leaves 6–20 cm. long ; petiole 7–17 mm. long, margins straight .. .. .. 7. *pyriformis*
Twigs verruculose-lenticular ; flowers 7–16 mm. diam. ; petals more or less oblong ; buds elongated, about twice as long as broad just before flowering ; inflorescence a fascicle or subumbel ; peduncle up to 1 (–7) mm. long ; pedicels 4–17 mm. long .. .. 4. *senegalensis*
Inflorescence-branches pubescent ; inflorescence cymosely branched ; flowers about 6 mm. diam., petals oblong or oblong-ovate, buds more or less globose ; leaves 6·2–15·7 cm. long, without resinous threads ; petioles 4–7 mm. long 32. *howesii*

1. **S. alata** *De Wild.* Miss. E. Laurent. 241 (1906) ; Loes. in E. & P. Pflanzenfam., 20B : 227 (1942). *Garcinia nobilis* of F.W.T.A., ed. 1, 1 : 236, not of Engl. A forest climber, with verruculose stems ; flowers white or green.
S.Nig.: Nikrowa, Okomu F.R., Benin (Dec.) *Brenan* 8393 ! Oban *Talbot* 1385 ! 1611 ! Umon-Ndealichi, Itu Div. (fl. buds & fr. June) *Ujor* FHI 30182 ! Also in Belgian Congo.
2. **S. hispida** *Blakelock* in Kew Bull. 1956 : 243. A forest climber ; petals yellow-green, disk grey-brown, filaments yellow-green, anthers orange-vermilion.
S.Nig.: Okomu F.R., Benin (Jan.) *Brenan* 8916 ! Also in Belgian Congo.
3. **S. leptoclada** *Tul.* in Ann. Sci. Nat., sér. 4, 8 : 96 (1857). *S. floribunda* Tul. l.c. 97 (1857). *S. elegans* Welw. ex Oliv. F.T.A. 1 : 373 (1868). *S. zanzibarensis* Vatke ex Loes. in E. & P. Pflanzenfam. 3, 5 : 230 (1893).
S.Nig.: Oban *Talbot* ! Also in Gabon, Angola, E. Africa, Madagascar and the Comores.
4. **S. senegalensis** (*Lam.*) *DC.* Prod. 1 : 570 (1824) ; Guill. & Perr. Fl. Seneg. 1 : 113, t. 27 ; F.T.A. 1 : 374 ; Loes. in Engl. Bot. Jahrb. 44 : 165 (1909) ; Chev. Bot. 135 ; Loes. in E. & P. Pflanzenfam. 20B : 223. *Hippocratea senegalensis* Lam. (1791). *Salacia macrocarpa* Welw. ex Oliv. F.T.A. 1 : 373 (1868) ; Loes. in Engl. Bot. Jahrb. 44 : 166 (incl. vars.) ; Chev. Bot. 134 ; F.W.T.A., ed. 1, 1 : 453 ; Loes. in E. & P. Pflanzenfam. 20B : 223 ; not of Korth. (1848). *S. oblongifolia* Oliv. (1868), not of Blume (1825). *S. oliveriana* Loes. (1894)—E. & P. Pflanzenfam. 20B : 223. *S. angustifolia* Sc. Elliot (1895)—F.W.T.A., ed. 1, 1 : 452. *S. demeusei* De Wild. & Dur. (1900)—Exell & Mendonça Consp. Fl. Ang. 2 : 23. *S. johannis-albrechti* Loes. & Winkl. (1908)—F.W.T.A., ed. 1, 1 : 458. (?) *S. doeringii* Loes. (1910). *S. dalzielii* Hutch. & M. B. Moss in F.W.T.A., ed. 1, 1 : 453 (1928) ; Kew Bull. 1929 : 22 ; Brenan in Kew Bull. 1950 : 220. *S. euryoides* Hutch. & M. B. Moss in F.W.T.A., ed. 1, 1 : 453 (1928) ; Kew Bull. 1929 : 21. A shrub, erect or climbing ; leaves variable ; petals white or pale greenish cream, disk green, anthers orange ; fruits orange or yellow ; in forest and forest regrowth.
Sen.: Perrottet 97 ! Roussillon 50 ! Casamance (Apr.) *Leprieur* ! Djonware, R. Salum *Brunner* 167 ! **Gam.**: Kombo (May, June) *Heudelot* 34 ! 49 ! **Port.G.**: Bijimita, Bissau (Feb.) *Esp. Santo* 1761 ! **Fr.G.**: Conakry (Feb.) *Chev.* 12089 ! Los Isl. (May) *Pobéguin* 1196 ! Timbo (Feb.) *Pitard* 104 ! Koukouré *Paroisse* 80 ! **S.L.**: *Afzelius* ! *Whitfield* ! Bafodeya, Limba (Apr.) *Sc. Elliot* 5486 ! Talla Hills (Feb.) *Sc. Elliot* 4870 ! Berria (Mar.) *Sc. Elliot* 5433 ! Bagroo R. (Apr.) *Mann* 798 ! Njala (fl. Dec., fr. Mar.) *Deighton* 2627 ! 3105 ! **Lib.**: Monrovia (May–July, Nov.) *Baldwin* 5906 ! *Linder* 1555 ! *Whyte* ! Mouth of Junk R. (fl. & young fr. Feb.) *Cooper* 187 ! Bushrod Isl. (fl. & fr. Mar.) *Barker* 1228 ! Greenville (Mar.) *Baldwin* 11550 ! **Iv.C.**: Bouroukrou (Dec., Jan.) *Chev.* 16536 ; 16688. Mt. Nouba (fr. Apr.) *Chev.* 21109. Ganhoué to Bampleu (fr. Apr.) *Chev.* 21161. **G.C.**: Odomi (Feb.) *Adogla* FH 4065 ! **N.Nig.**: Patti Lokoja (Feb.) *Dalz.* 92 ! **S.Nig.**: Idanre (Dec.) *Keay* in Hb. *Brenan* 8752 ! Usonigbe F.R.

(Nov.) *Ejiofor* FHI 24638 ! Ezi (Feb.) *Thomas* 2358 ! Iso Bendiga to Bendiga Afi, Ikom (Dec.) *Keay* FHI 28262 ! Calabar *Thomson* 93 ! Degema *Talbot* ! Extends to Belgian Congo and Angola. (See Appendix, p. 290.)

[Two specimens of this species in the Paris herbarium are both labelled *Debeaux* 330, 17 Mar. 1902, but one is said to come from Conakry (French Guinea) and the other from Cotonou (Dahomey). We have seen several other specimens of this species from Conakry, but none from Dahomey. The record of *Debeaux* 330 from Cotonou seems therefore to be suspect.

The following appear to be intermediate between *S. senegalensis* and *S. pyriformis* and may be hybrids:—
(a) **S. rufescens** *Hook. f.* in Fl. Nigrit. 283 (1849) ; F.W.T.A., ed. 1, 1 : 453.
S.L.: *Roscher* in *Hb. T. Vogel* ! *Thomas* 5197 !
(b) **S. chlorantha** *Oliv.* F.T.A. 1 : 375 (1868) ; Loes. in Engl. Bot. Jahrb. 44 : 168 ; F.W.T.A., ed. 1, 1 : 453.
N.Nig.: Nupe *Barter* 1658 !
See also under *S. pyriformis*.]

5. **S. volubilis** *Loes. & Winkl.* in Engl. Bot. Jahrb. 41 : 280 (1908) ; Loes. in op. cit. 44 : 184 ; Pflanzenfam. 20B : 229. A scandent shrub.
Br.Cam.: Victoria (June) *Stössel & Winkler* 84.

6. **S. lomensis** *Loes.* in Engl. Bot. Jahrb. 44 : 170 (1910) ; Chev. Bot. 134 (as " leonensis Loes." by error) ; Loes. in E. & P. Pflanzenfam. 20B : 225. An evergreen scrambling shrub, to 6 ft. high ; flowers buff-coloured ; fruits orange-red ; in thickets.
Iv.C.: Soubré (fr. June) *Chev.* 19131. Kéeta (July) *Chev.* 19327. Marabadiassa (fr. July) *Chev.* 22004.
G.C.: Achimota (fr. June) *Irvine* 1648 ! *Milne-Redhead* 5096 ! **Togo:** Lome (fl. Nov., fr. Aug.) *Mildbr.* 7477 ! *Warnecke* 215 ! (See Appendix, p. 290.)

7. **S. pyriformis** (*Sabine*) *Steud.* Nom. Bot., ed. 2, 2 : 492 (1841) ; F.T.A. 1 : 374 ; Loes. in Engl. Bot. Jahrb. 44 : 179 ; Chev. Bot. 135 ; Loes. in E. & P. Pflanzenfam. 20B : 227, fig. 70, F–G ; Exell Cat. S. Tomé 139. *Tonsella pyriformis* Sabine (1824). *Salacia talbotii* Bak. f. in Cat. Talb. 19 (1913) ; Loes. in E. & P. Pflanzenfam. 20B : 223. *S. pengheensis* De Wild. (1923). *S. owabiensis* Hoyle (1934). A forest climber, with the leaves dark glossy green above, paler below ; flowers green or yellowish-green ; fruits orange.
Fr.Sud.: Moussaia (Feb.) *Chev.* 407. **Fr.G.:** Karkandy (Mar.) *Heudelot* 798 ! Ditinn to Diaguissa *Chev.* 12674 ; 12901. Dalaba (Mar.) *Chev.* 18038. **S.L.:** *Don* ! Maswari (Apr.) *Adames* 210 ! Talla Hills (Feb.) *Sc. Elliot* 4825 ! Rokupr (Feb.) *Deighton* 4190 ! Njala (Mar.) *Deighton* 1943 ! 4712 ! **Lib.:** Dukwia R. (Aug.) *Linder* 253 ! Javajai, Boporo (Nov.) *Baldwin* 10266 ! Mecca, Boporo (Nov.) *Baldwin* 10402 ! **Iv.C.:** Danané (Apr.) *Chev.* 21266. Tébo (July) *Chev.* 19419. Dabou (Nov.) *Roberty* 15521 ! **G.C.:** Cape Coast *Brass* ! Axim (fl. & fr. Feb.) *Irvine* 2228 ! 2269 ! Owabi (Dec.) *Lyon* FH 2641 ! Otrokpe (Sept.) *Vigne* FH 4007 ! Simpa (Mar.) *Andoh* FH 3250 ! **Togo:** Assatu, Hohoe Dist. (Nov.) *St. C. Thompson* FH 3943 ! **S.Nig.:** Ibadan North F.R. (Feb.) *Keay & Meikle* FHI 25661 ! Ikirun to Igbajo (Jan.) *Latilo* FHI 31755 ! Oban *Talbot* 1359 ! 1687 ! 1705 (BM sheet) ! Calabar (Feb.) *Mann* 2291 ! **Br.Cam.:** R. Cameroon (Jan.) *Mann* 2221 ! Also in Belgian Congo, Ubangi-Shari, A.-E. Sudan and S. Tomé.

[There are also specimens which agree with *S. pyriformis* except for their narrow flower-buds rather similar to those of *S. senegalensis*. Possibly these are hybrids between the two species. The specimens are :—**S.L.:** Lalahun to Gola Forest (Jan.) *King* 121b ! **Lib.:** Kakatown *Whyte* ! Mecca (Dec.) *Baldwin* 10787 ! Tappi (Dec.) *Bunting* 78 !
See also under *S. senegalensis*.]

8. **S. pallescens** *Oliv.* F.T.A. 1 : 376 (1868) ; Loes. in Engl. Bot. Jahrb. 44 : 185 ; Chev. Bot. 135 ; Loes. in E. & P. Pflanzenfam. 20B : 227. A shrub, with purple-black petals, carpels and stamens orange ; fruits red or orange.
S.L.: *Thomas* ! **N.Nig.:** Nupe *Barter* ! Abinsi (May) *Dalz.* 650 ! **S.Nig.:** Olokemeji *Chev.* 13988 ; 14035 ; 14046. Ibadan F.R. (fr. June) *Symington* FHI 4106 ! Akure F.R. (fr. Aug.) *Jones* FHI 19547 !

9. **S. togoica** *Loes.* in Engl. Bot. Jahrb. 44 : 181 (1910). A forest liane ; petals yellow-ochre to orange.
S.L.: Benna (Jan.) *Sc. Elliot* 4635 ! **G.C.:** Krobo Plains (Dec.) *Johnson* 500 ! **Togo:** Misahöhe or Tomegbe (Apr. or May) *Baumann* 489 (partly). **S.Nig.:** Ohumbe F.R., Ilaro (Dec.) *Onochie* FHI 22657 ! 31900 ! Ibadan (Jan.) *Meikle* 988 ! Iso Bendiga to Bendiga Afi, Ikom (Dec.) *Keay* FHI 28265 ! (See Appendix, p. 290.)

10. **S. cuspidicoma** *Loes.* in Engl. Bot. Jahrb. 44 : 175 (1910). A shrub, apparently erect, to 6 ft. high.
Br.Cam.: Johann-Albrechtshöhe (fl. buds Jan., Feb.) *Staudt* 523 ; 610. Also in French Cameroons.

11. **S. mannii** *Oliv.* F.T.A. 1 : 376 (1868) ; Loes. in Engl. Bot. Jahrb. 44 : 187 ; Pflanzenfam. 20B : 227, 228. A small tree, 6–20 ft. high, with twigs green becoming reddish brown ; leaves deep green and glossy above, paler beneath ; petals cream, ovary and stamens yellow ; fruits orange-red ; in forest.
S.Nig.: Okomu F.R., Benin (Dec.) *Brenan* 8426 ! *Ebuade* ! Usonigbe F.R., Benin (fl. & fr. Nov.) *Keay* FHI 25573 ! **F.Po:** (Jan., June) *Barter* ! *Mann* 242 !

12. **S. pyriformoides** *Loes.* in Engl. Bot. Jahrb. 44 : 186 (1910). A forest liane with stout green twigs and dense clusters of flowers.
S.Nig.: Oban *Talbot* 1420 ! Also in French Cameroons, Spanish Guinea and (?) Belgian Congo.

13. **S. dusenii** *Loes.* in Engl. Bot. Jahrb. 19 : 242 (1894) ; op. cit. 44 : 188 (1910) ; Pflanzenfam. 20B : 229. A forest liane with white flowers ; very similar to *S. pyriformoides*.
S.Nig.: Aboabam, Ikom Div. (May) *Latilo* FHI 30973 ! Oban *Talbot* 1760 ! **Br.Cam.:** Ndian (Mar.) *Dusen* 307. Also in French Cameroons.

14. **S. zenkeri** *Loes.* in Engl. Bot. Jahrb. 44 : 193 (1910) ; Pflanzenfam. 20B : 230. A forest liane or shrub ; flowers green.
S.L.: Gboyama, Gardohun-Jimmi (Apr.) *Deighton* 1592 ! **Lib.:** Dukwia R. (Apr.) *Cooper* 387 ! **Iv.C.:** Yapo (Dec.) *Mangenot* ! **S.Nig.:** Oban *Talbot* 1589 ! 1589a ! **Br.Cam.:** Victoria to Bimbia (May) *Preuss* 1254 ! Also in French Cameroons.

15. **S. conrauii** *Loes.* in Engl. Bot. Jahrb. 44 : 192 (1910). A shrub.
Br.Cam.: Tale, Banyang, Mamfe Div. (Apr.) *Conrau* 104.

16. **S. preussii** *Loes.* in Engl. Bot. Jahrb. 19 : 239 (1894) ; op. cit. 44 : 192 ; Pflanzenfam. 20B : 229. A forest shrub ; petals yellow, green at base.
Br.Cam.: Barombi *Preuss* 21 ! Johann-Albrechtshöhe *Staudt* 935 ! Also in French Cameroons and Gabon.

17. **S. loloensis** *Loes.* in Engl. Bot. Jahrb. 44 : 172 (1910) ; Pflanzenfam. 20B : 227. A small tree, 4–25 ft. high ; flowers dirty white or pale yellow ; fruits red.
S.Nig.: Omo F.R., Ijebu Ode (Apr.) *Tamajong* FHI 16918 ! Oban *Talbot* 1704 ! **Br.Cam.:** S. Bakundu F.R. (Apr.) *Olorunfemi* FHI 30509 ! N. Korup F.R. (July) *Olorunfemi* FHI 30674 ! Victoria (Apr.) *Schlechter* 12365 ! Also in French Cameroons.

18. **S. eurypetala** *Loes.* in Engl. Bot. Jahrb. 44 : 174 (1910) ; Pflanzenfam. 20B : 227.
Br.Cam.: Johann-Albrechtshöhe (Apr.) *Staudt* 930.

19. **S. sp. A.** " Single-main-stemmed shrub, about 8 ft. high " ; in forest. The very large leaves are not matched in West Africa ; other characters recall *S. loloensis*, except that the petals (at least in bud) are purple mottled with white.
Br.Cam.: close to coast, Victoria (fl. buds Apr.) *Brenan* 9590 !

20. **S. fimbrisepala** *Loes.* in Engl. Bot. Jahrb. 44 : 190 (1910). A scandent or erect shrub, or small tree (*fide* Johnson) ; flowers yellow, waxy.
G.C.: Kwahu (Apr.) *Johnson* 648 ! **Br.Cam.:** Buea (Apr.) *Lehmbach* 228.

21. **S.** sp. **B.** Unbranched subshrub, about 1 ft. high ; in forest.
     Br.Cam.: Banga (Mar.) *Brenan* 9483 !

22. **S. debilis** (*G. Don*) *Walp.* Rep. 1 : 402 (1842) ; Loes. in Engl. Bot. Jahrb. 44 : 184 ; Chev. Bot. 133 ;
     Loes. in E. & P. Pflanzenfam. 20B : 228. *Calypso debilis* G. Don (1831). A forest climber ; flowers
     green or yellow.
     Port.G.: Suzana to S. Domingos (Mar.) *Esp. Santo* 2244 ! Catió (young fr. June) *Esp. Santo* 2104 !
     Fulacunda (fr. May) *Esp. Santo* 2030 ! **Fr.G.**: Rio Nunez (Dec.) *Heudelot* 698 ! Conakry & Los Isl.
     Chev. 12079 ; 12262. Kabo to Haut-Mamou *Chev.* 12764. **S.L.**: *Don* ! Mt. Aureole, Freetown (Mar.,
     Dec.) *Dalz.* 976 ! *Lane-Poole* 212 ! Kora, Scarcies R. (Feb.) *Sc. Elliot* 4589 ! Musala (young fr. Mar.)
     Sc. Elliot 5123 ! Kamabai (Feb.) *Deighton* 4254 ! Njala (fl. & fr. Feb.) *Deighton* 2599 ! **G.C.**: Boblri
     F.R., Juaso (Jan.) *Andoh* FH 5459 ! Pra R. Station (Feb.) *Vigne* FH 1047 ! **F.Po**: (Mar.) *Mann* 1445 !
     Also in French Cameroons. (See Appendix, p. 290.)
     [*Sim* 23 from Liberia has the small flowers of *S. debilis*, but other characters as in *S. erecta* ; further
     research is required.]

23. **S. erecta** (*G. Don*) *Walp.* l.c. 1 : 403 (1842) ; Loes. in Engl. Bot. Jahrb. 44 : 168 (1910) ; Pflanzenfam.
     20B : 224. *Calypso erecta* G. Don (1831). *Salacia cornifolia* Hook. f. (1849)—F.T.A. 1 : 375 (excl. var.
     crassisepala Oliv.) ; Loes. in Engl. Bot. Jahrb. 44 : 185 ; Chev. Bot. 133 ; F.W.T.A., ed. 1, 1 : 453 ;
     Loes. in E. & P. Pflanzenfam. 20B : 225. *S. baumannii* Loes. (1910). *S. prinoides* DC. var. *liberica*
     Loes. in Engl. Bot. Jahrb. 44 : 179 (1910). *S.elliotii* Loes. (1910)—F.W.T.A., ed. 1, 1 : 452. *S. alpestris*
     A. Chev. ex Hutch. & Dalz. F.W.T.A., ed. 1, 1 : 453 (1928) ; Chev. Bot. 133, name only. *S. prinoides*
     of F.T.A. 1 : 375. A shrub, sometimes scandent, or an understorey tree ; flowers yellowish ; in forest ;
     a variable species.
     Fr.G.: Ditinn to Diaguissa (fl. & fr. Apr.) *Chev.* 12859 ! Dalaba *Chev.* 18134. **S.L.**: *Don* ! *T. Vogel* 118 !
     Sugar Loaf Mt. *Barter* ! Mt. Kofiu, Scarcies (Jan.) *Sc. Elliot* 4595 ! Waterloo (Dec.) *Deighton* 3305 !
     Kenema (fr. Apr.) *Deighton* 4293 ! Kambia *Sc. Elliot* 4359 ! **Lib.**: Grand Bassa *T. Vogel* 90 ! Peahtah
     (Oct.) *Linder* 952 ! Vahon, Kolahun (Nov.) *Baldwin* 10247 ! Mecca, Boporo (fl. & young fr. Nov.)
     Baldwin 10418 ! **Iv.C.**: Nimba Mts. *Boughey* 18172 ! **G.C.**: Bana Hill, Krobo (fr. Mar.) *Irvine* 894 !
     Accra Plains *Imp. Inst.* ! **Togo**: Misahöhe or Tomegbe (Apr. or May) *Baumann* 489 (partly). **Dah.**:
     Tohoué *Chev.* 22768. **S.Nig.**: Lagos (Apr.) *Dalz.* 1067 ! 1153 ! Oban *Talbot* ! Ikwette Plateau, Obudu
     Div. (Dec.) *Keay & Savory* FHI 25178 ! **Br.Cam.**: Buea *Lehmbach* 91 ! Also in E. Africa and Madagascar.
     (See Appendix, p. 290.)

24. **S. tuberculata** *Blakelock* in Kew Bull. 1956 : 246. *S. cornifolia* of F.W.T.A., ed. 1, 1 : 453, partly (*Sc.
     Elliot* 5575 ; *Thomas* 2439). *S. talbotii* of F.W.T.A., ed. 1, 1 : 453, partly (*Mann* 2265 ; *Talbot* 1705).
     A scandent forest shrub ; flowers green.
     S.L.: Baïïma, Jaluahun (Apr.) *Deighton* 3951 ! Njala (June) *Deighton* 714 ! Boma (Apr.) *Deighton*
     3720 ! Roruks (May) *Deighton* 4138 ! Bafodeya, Limba *Sc. Elliot* 5575 ! **S.Nig.**: Oban *Talbot* 1705
     (Kew sheet) ! R. Calabar (Feb.) *Mann* 2265 !

25. **S. lehmbachii** *Loes.* in Engl. Bot. Jahrb. 44 : 173 (1910) ; Pflanzenfam. 20B : 227. An erect shrub or
     small tree, to 6 ft. high ; fruits red.
     Br.Cam.: Buea (fr. Feb., Apr.) *Lehmbach* 116 ; 208. Also in French Cameroons and (the var. *usam-
     barensis*) in E. Africa.

26. **S. leonensis** *Hutch. & M. B. Moss* in F.W.T.A., ed. 1, 1 : 453 (1928) ; Kew Bull. 1929 : 22. A forest shrub.
     S.L.: *Thomas* 7605*b* ! 7944 ! 9360 ! Mano (young fr. Dec.) *Deighton* 2425 ! Falaba (May) *Aylmer* 82 !
     Lib.: Jabroke, Webo Dist. (fr. July) *Baldwin* 6606 ! Zeahtown, Tchien Dist. (fr. July) *Baldwin* 6993 !
     [The following specimens appear to differ from typical *S. leonensis* only in the cuneate or obtuse leaf-
     bases :—**S.L.**: *Unwin & Smythe* 16 ! Sugar Loaf Mt. (Nov.) *Deighton* 5647 ! **G.C.**: Sefwi Bekwai (fr.
     Oct.) *Akpabla* 901 !]

27. **S. camerunensis** *Loes.* in Engl. Bot. Jahrb. 19 : 240 (1894) ; op. cit. 44 : 176, incl. vars. ; Pflanzenfam. 20B :
     226. A forest liane, with smooth cable-like stems ; flowers greenish-white, yellow or orange ; leaves
     deep green and glossy above, paler beneath ; fruits red or orange.
     G.C.: *Irvine* 390 ! Kame R. (May) *Baumann* 587 ! **Togo**: Wiawia, Hohoe Dist. (Oct.) *St. C. Thompson*
     FH 3940 ! **S.Nig.**: Oshogbo *Millson* ! Akure F.R. (Nov.) *Keay* FHI 25608 ! Oban *Talbot* 405 ! 1512 !
     1589 ! **Br.Cam.**: S. Bakundu F.R., Banga (Mar.) *Brenan* 9403 ! 9488 ! Also in French Cameroons.

28. **S. staudtiana** *Loes.* in Engl. Bot. Jahrb. 44 : 187 (1910), excl. var. *leonensis* Loes. ; Fritsch in Bot. Cen-
     tralbl. Beih. 11 : 355 (1902), anatomical descr. ; Loes. in E. & P. Pflanzenfam. 20B : 227. A forest
     liane ; flowers white or yellow ; fruits red.
     S.Nig.: British Obobokum, Ikom (June) *Latilo* FHI 31854 ! **Br.Cam.**: Johann-Albrechtshöhe (Jan.,
     Feb.) *Staudt* 563 ! 631 ! Also in French Cameroons.

29. **S. caillei** *A. Chev. ex Hutch. & M. B. Moss* in F.W.T.A., ed. 1, 1 : 453 (1928) ; Kew Bull. 1929 : 22. *S.
     staudtiana* var. *leonensis* Loes. in Engl. Bot. Jahrb. 44 : 188 (1910). *S. nitida* of Chev. Bot. 135 and of
     F.W.T.A., ed. 1, 1 : 453, partly (*Chev.* 14902, 14626). *S. whytei* of F.W.T.A., ed. 1, 1 : 453, partly
     (*Johnson* 761). *S. bipindensis* of F.W.T.A., ed. 1, 1 : 453, partly (*Sc. Elliot* 4930, 5638). A shrub, often
     scandent ; petals green ; fruits red, often with a glaucous bloom ; in forest and forest regrowth.
     Fr.G.: Kouria (Sept., Oct.) *Caille* in *Hb. Chev.* 14902 ! 14928 ! Kouria to Longuery (fr. Aug.) *Caille* in
     Hb. Chev. 14626 ! **S.L.**: *Afzelius* ! Bumbuna (Oct.) *Thomas* 3456 ! 3730 ! Njala (Mar., Sept.–Nov.)
     Deighton 2407 ! 2909 ! 3987 ! 5012 ! 5591 ! Ninia (Feb.) *Sc. Elliot* 4899 ! 4930 ! Buyabuya (fr. Feb.)
     Sc. Elliot 4777 ! Jene, Kamagai (fl. & fr. Oct.) *Deighton* 5989 ! **Lib.**: Ganta (fr. Sept.) *Baldwin* 9260 !
     9307 ! Road to White Plains (Apr.) *Barker* 1269 ! Dukwia R. *Cooper* 213 ! **G.C.**: E. Akim (June)
     Johnson 761 ! Simpa (May) *Vigne* FH 1963 ! **Br.Cam.**: Korup F.R. (fr. July) *Olorunfemi* FHI 30667 !
     Also in Gabon and Belgian Congo. (See Appendix, p. 290.)

30. **S. nitida** (*Benth.*) *N.E. Br.* in F.T.A. 4, 1 : 415 (1905) ; Loes. in Engl. Bot. Jahrb. 44 : 196 ; F.W.T.A.,
     ed. 1, 1 : 453, partly (excl. *Chev.* 14902, 14626) ; not of Chev. Bot. 135. *Gymnema nitidum* Benth. in
     Hook. Fl. Nigrit. 456 (1849), excl. descr. flowers. *Salacia whytei* Loes. l.c. 44 : 184 (1910) ; F.W.T.A.,
     ed. 1, 1 : 453, partly (excl. *Johnson* 761). *S. bipindensis* Loes. l.c. 44 : 188 (1910), incl. vars. ; F.W.T.A.,
     ed. 1, 1 : 453, partly (excl. *Sc. Elliot* 4930, 5638) ; Loes. in E. & P. Pflanzenfam. 20B : 227, fig. 72,
     A–H. *S. cornifolia* var. *crassisepala* Oliv. F.T.A. 1 : 375 (1868). A forest liane ; petals orange ; ripe
     fruits red or orange.
     S.L.: *Roscher* in *Hb. T. Vogel* 60 ! **Lib.**: Cape Palmas *Ansell* ! Monrovia (Nov.) *Linder* 1428 ! *Whyte* !
     Tubman Bridge (Sept.) *Barker* 1413 ! Dukwia R. (Feb.) *Cooper* 254 ! **G.C.**: Princes, W. Prov. (Jan.)
     Akpabla 778 ! Benso, Tarkwa (fr. Aug.) *Andoh* FH 5397 ! Simpa (May) *Vigne* FH 1963 ! Bakanta, W.
     of Axim (Aug.) *Chipp* 316 ! **S.Nig.**: Okomu F.R., Benin (Feb., Mar.) *Brenan* 9066 ! *Richards* 3299 !
     Oban *Talbot* 1668 ! Calabar R. (Feb.) *Mann* 2275 ! **Br.Cam.**: Victoria (Apr.) *Brenan & Jones* 9591 ! Buea
     (Jan.) *Maitland* 291 ! Abonando, Mamfe (Feb., May) *Rudatis* 17 ! 55 ! Also in French Cameroons.

31. **S. lucida** *Oliv.* F.T.A. 1 : 373 (1868) ; Loes. in Engl. Bot. Jahrb. 44 : 165 ; Chev. Bot. 134 ; Loes. in
     E. & P. Pflanzenfam. 20B : 223. A forest liane.
     Iv.C.: Atteau (Jan.) *Chev.* 17090. Bingerville *Chev.* 17300. Zaranou to Songan (fr. Mar.) *Chev.* 17791.
     S.Nig.: Calabar *Thompson* 102 ! Also in French Cameroons.

32. **S. howesii** *Hutch. & M. B. Moss* in F.W.T.A., ed. 1, 1 : 453 (1928) ; Kew Bull. 1929 : 22 ; Brenan in Kew
     Bull. 1950 : 221. A low spreading shrub or liane, older twigs black, younger ones somewhat pubescent ;
     flowers greenish yellow ; in forest.
     G.C.: Anaje, 7 miles N.W. of Sekondi (Oct.) *Howes* 974 ! **S.Nig.**: Idanre Hills (Jan., Oct.) *Keay* in *Hb.
     Brenan* 8682 ! *Keay* FHI 25514 !

Imperfectly known species.

1. **S. gilgiana** *Loes.* in Engl. Bot. Jahrb. 44 : 168 (1910) ; Pflanzenfam. 20B : 225. Near *S. senegalensis* and possibly a hybrid of it.
    Br.Cam.: on lake shore, Johann-Albrechtshöhe (Feb.) *Staudt* 591.
2. **S. sp. C.** " A vine, in savannah thickets on sandy soil ; flowers greenish yellow."
    Lib.: Cinco, Monrovia (Nov.) *Linder* 1544 !
3. **S. sp. D.** " A small understorey shrub, about 4 ft. high ; sepals velvety purple." Resembles *S. tuberculata*, but main nerves impressed above ; the specimen is without flowers.
    S.Nig.: Oni R., 1 mile N. of Shasha F.R. (Feb.) *Jones & Onochie* FHI 17524 !
4. **S. sp. E.** " A small shrub, about 4 ft. high ; flower buds pale greenish cream ; fruits orange. " Probably near *S. gilgiana* (see above).
    S.Nig.: Ilaro F.R., Abeokuta Prov. (Dec.) *Onochie* FHI 31892 !
5. **S. sp. F.** Some resemblance to *S. togoica* but the flowers are smaller.
    S.Nig.: *Talbot* !
6. **S. sp. G.** Near *S. togoica* but the stem is markedly verruculose. As the specimen is only in bud it is not named more accurately.
    S.L.: *Afzelius* !
7. **S. sp. H.** " Shrub ; flowers greenish yellow." Possibly near *S. nitida*, but the inflorescence is pubescent the leaves do not match those of *S. howesii*.
    G.C.: Cape Coast (Apr.) *Scholes* 202 !
8. **S. sp. I.** Near *S. staudtiana* but immature.
    Br.Cam.: Bimbia Road, Victoria (fl. buds May) *Maitland* 664 !
9. **S. sp J.** Allied to *S. leptoclada* Tul., but the leaves are not a good match.
    S.Nig.: Oban *Talbot* !

6. SALACIGHIA Loes. in Fedde Rep. 49 : 228 (1940) ; in E. & P. Pflanzenfam. 20B : 217 (1942).

Calyx covering corolla in late bud stage, apiculate ; sepals 2–3 ; petals lacerate ; petioles (6–)10–15 mm. long ; twigs verruculose-lenticular ; leaves opposite and sub-opposite, oblong, ovate or obovate-elliptic, narrowed or obtuse at base, apex long-acuminate, margin entire, 8–21 cm. long, 3–8 cm. broad ; flower-buds peltate to subglobose, apiculate ; pedicels 4–20 mm. long ; flowers bisexual, 6–10 mm. diam.
        1. *letestuana*

Calyx not covering corolla in late bud stage, not apiculate ; sepals 4–5, rounded at apex ; petals subentire ; petioles 4–7 mm. long ; twigs verruculose-lenticular ; leaves alternate, papery or chartaceous, oblong, lanceolate-oblong or oblanceolate-oblong, narrowed at base, apex fairly long-acuminate, margin entire, 8–12 cm. long, 3–5 cm. broad ; buds subglobose ; pedicels 10–15 mm. long ; flowers bisexual, about 10 mm. diam.   ..   ..   2. *linderi*

1. **S. letestuana** (*Pellegr.*) *Blakelock* in Kew Bull. 1956 : 247. *Salacia letestuana* Pellegr. in Bull. Mus. Hist. Nat. Paris 28 : 312 (1922). *Salacighia malpighioides* Loes. in Fedde Rep. 49 : 228 (1940) ; in Wiss. Ergebn. 1910–11 : 77 (1922), name only ; in E. & P. Pflanzenfam. 20B : 217 (1942) ; Exell & Mendonça Consp. Fl. Ang. 2 : 27 (1954). (?) *Salacia denudata* A. Chev. Bot. 133 (1920), name only ; F.W.T.A., ed. 1, 1 : 453. Forest liane, stems purplish-brown ; leaves deep green and glossy above, paler beneath ; petals green with pale margin, disk cream, ovary pale green.
    Iv.C.: Adiopodoumé (Sept., Nov.) *Mangenot* ! **S.Nig.:** Eket *Talbot* ! Oban *Talbot* 149 ! Kwa Falls, Calabar (Mar.) *Brenan* 9249 ! Also in French Cameroons, Spanish Guinea, Gabon, Cabinda and Belgian Congo.
2. **S. linderi** (*Loes. ex Harms*) *Blakelock* l.c. (1956). *Salacia linderi* Loes. ex Harms in Notizbl. Bot. Gart. Berl. 15 : 673 (1942). Liane with yellow flowers.
    S.L.: Kambui (Mar.) *Lane-Poole* 449 ! **Lib.:** Moylakwelli-Totokwelli (Oct.) *Linder* 1278 !

101. PANDACEAE

Trees. Leaves alternate, simple, serrulate ; stipules present. Flowers dioecious. Male flowers racemose, the racemes fasciculate from the old wood. Calyx cupular, open in bud. Petals 5, valvate. Stamens 10, alternately long and short ; anthers 2-celled, opening lengthwise. Rudimentary ovary linear-subulate. Female flowers arranged like the male. Calyx cupular, truncate or toothed. Petals 5 [imbricate ?] ; staminodes absent. Ovary subsessile, 3–4-celled ; stigmas 3–4, reflexed ; ovule solitary, pendulous. Fruit drupaceous. Seeds with copious oily endosperm ; cotyledons cordate.

A small family endemic to West Tropical Africa.

PANDA Pierre in Bull. Soc. Linn. Paris 2 : 1255 (1894). *Porphyranthus* Engl. (1899).

A tree ; branchlets slightly angular ; leaves elliptic or ovate-elliptic, unequal-sided and more or less cuneate at base, obliquely caudate-acuminate, 10–18 cm. long, 4–9 cm. broad, glabrous, rather closely reticulate beneath ; petiole about 1 cm. long ; flowers in elongated fasciculate or paniculate racemes ; pedicels about 2 mm. long, puberulous ; calyx truncate, about 1·25 mm. long ; petals puberulous on the margin, 3 mm. long ; fruit subglobose, 3–4-seeded, about 6 cm. diam., with a thick exocarp and a hard bony endocarp traversed by cavities giving the outside a deeply pitted appearance ; seeds boat-shaped ..   ..   *oleosa*

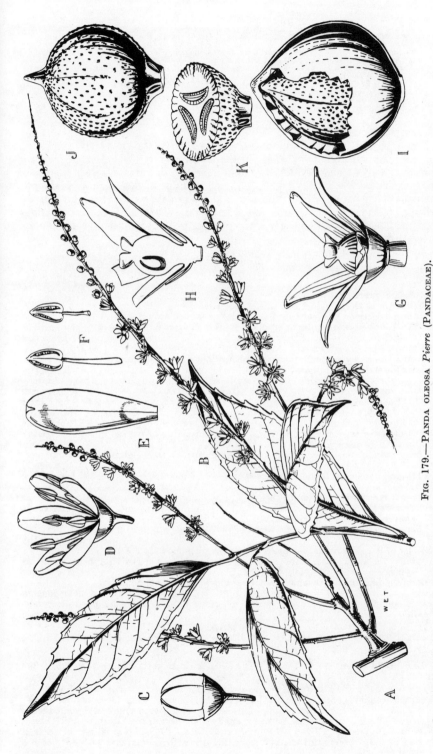

Fig. 179.—Panda oleosa *Pierre* (Pandaceae).

A, flowering branch.   B, leaf.   C, flower-bud.   D, male flower.   E, petal.   F, stamens.   G, female flower.   H, the same in vertical section.   I, fruit partially cut open.   J, stone.   K, cross-section of same.

635

P. oleosa *Pierre* l.c. (1896) ; Chev. Bot. 612 ; Aubrév. Fl. For. C. Iv. 1 : 300, t. 120 ; Kennedy For. Fl. S. Nig. 143. *Porphyranthus zenkeri* Engl. (1899). An understorey forest tree, up to 60 ft. or more high ; slashed bark mottled pink and purple ; dioecious ; flowers red or purplish ; inflorescences often in axils of fallen leaves ; fruits green.

Lib.: *Harley* 990 ! **Iv.C.:** Guidéko (June) *Chev.* 19043 ! Bouroukrou *Chev.* 16122 ! Rasso *Aubrév.* 140 ! **G.C.:** Sa, Bonsa R. (Aug.) *Vigne* FH 980 ! Anwhiaso F.R. (Mar.) *Vigne* FH 224 ! Baku, Saltpond Dist. (fr. June) *Vigne* FH 949 ! Amentia (Sept.) *Vigne* FH 1357 ! **S.Nig.:** Sapoba (fl. Jan., fr. May, Nov.) *Kennedy* 186 ! 1669 ! *Keay* FHI 28069 ! Nikrowa, Okomu F.R. (fr. Oct.) *Nwonkolo* FHI 24152 ! Ikom *Catterall* 49 ! **Br.Cam.:** Victoria (Mar.) *Maitland* 474 ! 628 ! Also in French Cameroons, Spanish Guinea, Gabon, Cabinda and Belgian Congo. (See Appendix, p. 288.)

102. ICACINACEAE

Trees, shrubs or rarely climbers. Leaves alternate or rarely opposite, simple ; stipules absent. Flowers hermaphrodite or rarely unisexual, actinomorphic. Calyx small, 3–5-lobed ; lobes imbricate or valvate, rarely enlarging and enveloping the fruit. Petals 3–5, free or united, valvate, rarely absent. Stamens the same number as the petals and alternate with them ; anthers 2-celled ; filaments often hairy, free. Disk rarely present. Ovary superior, 1-celled, 1-seeded. Seeds mostly with endosperm ; embryo usually small, more or less straight.

A small tropical family with a tendency to a gamopetalous corolla ; petals valvate ; petals inside or anthers sometimes hairy.

Leaves alternate ; tendrils absent :
  Corolla not accrescent in fruit :
    Flowers in fascicles, cymes or panicles, bisexual :
      Inflorescences on the leafy shoots, cymose or fasciculate :
        Petals united into a tubular corolla ; flowers in extra-axillary, shortly pedunculate clusters   ..   ..   1. **Leptaulus**
        Petals free or nearly so :
          Inflorescences not leaf-opposed ; anthers not densely villous :
            Flowers in pedunculate cymes :
              Petals glabrous inside ; sepals free, linear or lanceolate ; fruits pilose, pubescent or glabrous   ..   ..   2. **Alsodeiopsis**
              Petals bearded inside ; sepals united into a calyx with short or long lobes ; fruits tomentose or tomentellous   ..   ..   ..   6. **Icacina**
            Flowers fasciculate ; fruits slightly 2-lobed, reticulate ..   4. **Rhaphiostylis**
          Inflorescences leaf-opposed, pedunculate, consisting of several umbellate heads ; anthers densely villous   ..   ..   3. **Lasianthera**
      Inflorescences on the main stem, paniculate ; petals free, glabrous inside
                                                                          7. **Lavigeria**
    Flowers in spikes or slender racemes, or rarely female flowers in heads, bisexual or unisexual :
      Perianth consisting of distinct sepals and petals :
        Flowers bisexual ; small trees or shrubs, sometimes scandent   5. **Desmostachys**
        Flowers unisexual ; slender lianes   ..   ..   ..   11. **Neostachyanthus**
      Perianth consisting of a single whorl of (3–) 4 tepals ; flowers unisexual
                                                                          8. **Pyrenacantha**
  Corolla of female flowers accrescent and enveloping the fruit :
    Corolla 3-lobed, campanulate ; style divided into several radiating arms ; flowers in heads ; leaves often lobed ..   ..   9. **Polycephalium**
    Corolla 4-partite ; style simple ; flowers in spikes ; leaves not lobed
                                                                          10. **Chlamydocarya**
Leaves opposite, or subopposite ; flowers dioecious in lax cymes ; climbers with extra-axillary tendrils   ..   ..   12. **Iodes**

Excluded genus.

Icacinopsis *Roberty* in Bull. I.F.A.N. 15 : 1420 (1953) is **Dichapetalum barteri** Engl. (*Chailletiaceae*) ; see p. 436.

1. LEPTAULUS Benth.—F.T.A. 1 : 354 ; Sleumer in E. & P. Pflanzenfam. 20B : 358 (1942).

Branchlets and petioles glabrous ; flowers numerous in each extra-axillary cluster :
  Pedicels well-developed, 3–4 mm. long ; calyx 1·5 mm. long, lobes not recurved ; corolla 1·3–1·4 cm. long ; fruits ellipsoid, about 1·3 cm. long and 1 cm. diam. ; leaves elliptic, cuneate at base, gradually long-acuminate, acumen itself blunt, 3–14 (–16) cm. long, 1·5–6 (–7·8) cm. broad, with 5–7 main lateral nerves on each side of midrib, venation distinct beneath, nerves not impressed above ; petioles 0·5–1 cm. long   ...   .   1. *daphnoides*

Pedicels up to 1 mm. long ; calyx 3–4 mm. long, lobes recurved ; corolla about 1·5 cm. long ; fruits oblong-ovoid, 2·8–3 cm. long and up to 1·5 cm. diam. ; leaves coriaceous, oblong-elliptic to oblong-obovate, cuneate at base, abruptly and acutely acuminate, 14–28 cm. long, 6–10 cm. broad, with 5–9 main lateral nerves on each side of midrib, venation obscure beneath, nerves often impressed above ; petioles 1·5–2 cm. long

2. *grandifolius*

Branchlets and petioles densely puberulous ; flowers solitary or few in extra-axillary clusters ; pedicels 1–2 mm. long ; calyx 1·5 mm. long ; corolla about 1 cm. long ; fruits ellipsoid, about 1·5 cm. long ; leaves oblong-elliptic, obtuse or cuneate at base, abruptly long-caudate acuminate, the acumen itself subspathulate, 6–12 cm. long, 3·5–5 cm. broad, with 4–5 main lateral nerves on each side of midrib ; petioles up to 3 mm. long .. .. 3. *zenkeri*

1. **L. daphnoides** *Benth.*—F.T.A. 1 : 354, partly (Bagroo River spec. only) ; Oliv. in Hook. Ic. Pl. t. 2339 (1894), excl. var. *macrophylla* Oliv. ; Chev. Bot. 127 ; Aubrév. Fl. For. C. Iv. 1 : 91, t. 27, 7–11 ; Sleumer in E. & P. Pflanzenfam. 20B : 358, fig. 101, C–E. A forest shrub or tree, up to 60 ft. high ; flowers white, very fragrant.
Fr.G.: Fouta Djalon (May) *Heudelot* 863 ! Forécariah (June) *Jac.-Fél.* 1714 ! **S.L.:** Bagroo R. (Apr.) *Mann* 806 ! Madina, Limba (Apr.) *Sc. Elliot* 5562 ! Kessewe Hill (May) *Lane-Poole* 124 ! Njala (fl. May, fr. Sept.) *Deighton* 663 ! 2910*b* ! **Lib.:** Bonuta (Oct.) *Linder* 873 ! Brewer's Landing (Nov.) *Linder* 1403 ! Grand Cess (fl. & fr. Mar.) *Baldwin* 11619 ! Dobli Isl., St. Paul R. (Apr.) *Bequaert* 170 ! 174 ! **Iv.C.:** Man (Mar.) *Aubrév.* 1092 ! Malamalasso (Mar.) *Chev.* 16251 ! Soubré (June) *Chev.* 19125 ! **G.C.:** Anibil, Axim (Nov.) *Vigne* 1399 ! Bonsa Su (May) *Vigne* FH 1965 ! Begoro (Apr., Oct.) *Irvine* 1190 ! *Moor* 203 ! Sefwi Bekwai (young fr. Oct.) *Akpabla* 942 ! **S.Nig.:** Osho, Omo F.R. (fl. buds Apr.) *Jones & Onochie* FHI 17238 ! Okomu F.R. (May) *A. F. Ross* 204 ! Oban *Talbot* 427 ! **Br.Cam.:** Musake Camp, Cam. Mt. 4,500 ft. (May) *Maitland* 686 ! Johann-Albrechtshöhe *Preuss* 1356 ! *Staudt* 547 ! Bali-Ngemba F.R., Bamenda, 6,000 ft. (June) *Tiku* FHI 30418 ! Nkom-Wum F.R., Bamenda (Apr.) *Ujor* FHI 30067 ! Also, in French Cameroons, Gabon, A.-E. Sudan, Uganda, Tanganyika, Belgian Congo and Cabinda. (See Appendix, p. 291.)

2. **L. grandifolius** *Engl.* Bot. Jahrb. 43 : 180 (1909). *L. daphnoides* of F.T.A. 1 : 354, partly (Kongui River spec.). *L. daphnoides* var. *macrophyllus* Oliv. in Hook. Ic. Pl. sub t. 2339 (1894). A small forest tree, up to 15 ft. high ; flowers white.
Br.Cam.: S. Bakundu F.R., Kumba (Mar.) *Brenan* 9274 ! Also in French Cameroons and Spanish Guinea.

3. **L. zenkeri** *Engl.* l.c. 179 (1909). A forest shrub or small tree.
S.Nig.: Oban *Talbot* 1410 ! Eket *Talbot* 3322 ! Also in French Cameroons, Gabon, Cabinda and Belgian Congo.

2. ALSODEIOPSIS Oliv. ex Benth. in Benth. & Hook. f. Gen. Pl. 1, 3 : 996 (1867) ; F.T.A. 1 : 356 ; Sleumer in E. & P. Pflanzenfam. 20B : 359 (1942).

Lower surface of leaves and branchlets pilose with rather long spreading hairs :
    Upper surface of leaves setose-pilose, without short appressed hairs, undersurface rather loosely pilose ; sepals, petals and inflorescence-branches whitish hispid and with a few longer whitish hairs ; flowers few ; sepals about 1 mm. long ; fruits ellipsoid, about 9 mm. long, reticulate, pubescent ; leaves oblong-obovate, subcordate and unequal-sided at base, acutely acuminate, 7–10 cm. long, 3–5 cm. broad, with 5–8 main lateral nerves on each side of midrib.. .. 1. *chippii*
    Upper surface of leaves appressed-puberulous and with long stiff hairs on the midrib and nerves, undersurface densely pilose ; sepals, petals and inflorescence-branches densely villose with long reddish-brown hairs ; flowers several in paired cymes ; sepals about 2 mm. long ; fruits ellipsoid, about 15 mm. long, densely orange-pilose ; leaves oblong-oblanceolate to oblong-lanceolate, very unequal-sided at base with one side cordate and the other obtuse to rounded, acutely acuminate, 9–18 cm. long, 3–7 cm. broad, with 4–10 main lateral nerves on each side of midrib

2. *villosa*

Lower surface of leaves shortly appressed-pubescent :
    Main lateral nerves (7–) 10–19 on each side of midrib ; leaves oblanceolate to oblong-obovate, rounded or obtuse and subequal at base, gradually long-acuminate, (7–) 10–25 cm. long, (2·5–) 3·5–7 cm. broad, shining beneath ; peduncles axillary, perennial, becoming woody and up to 8 cm. long, with the rather dense-flowered cymes clustered towards the ends, pedicels up to 4 mm. long ; fruits oblong-ellipsoid, 2·5–3·3 cm. long, 9–13 mm. diam., glabrous .. .. 3. *mannii*
    Main lateral nerves 3–9 on each side of midrib ; fruits ellipsoid or ovoid-ellipsoid, up to 1·8 cm. long and 6 mm. diam., pubescent ; pedicels 3–13 mm. long :
        Peduncles 6–8 mm. long, pedicels up to 6 mm. long ; fruits ovoid-ellipsoid, about 1·2 cm. long ; leaves ovate to obovate, subcordate and very unequal-sided at base, obtuse or broadly acuminate at apex, mucronate, 2·7–8·5 cm. long, 1·2–4 cm. broad, main lateral nerves 3–9 with small pocket-like structures in the axils

4. *staudtii*

        Peduncles 16–28 mm. long, pedicels up to 13 mm. long ; fruits ellipsoid, 1·6–1·8 cm. long ; leaves elliptic, narrowed to the obtuse, rounded or subcordate rather unequal base, acutely long-acuminate, 5·2–16 cm. long, 1·6–6 cm. broad, main lateral nerves 7–9 .. .. 5. *rowlandii*

1. **A. chippii** *Hutch.* in Kew Bull. 1912 : 224. A yellow flowered shrub.
G.C.: *Burton & Cameron* ! Dompem, Atissiabo, Axim Dist. (Dec.) *Chipp* 46 !
2. **A. villosa** *Keay* in Bull. Jard. Bot. Brux. 26 : 185 (1956). A shrub, to 4 ft. high ; fruit orange.
Lib.: Duo, Sinoe (Mar.) *Baldwin* 11346 ! About 15 miles inland from R. Cess, Grand Bassa (fr. Mar.) *Baldwin* 11241 ! Jabroke, Webo (fl. & fr. July) *Baldwin* 6476 !

3. **A. mannii** *Oliv.* in J. Linn. Soc. 10 : 43 (1867), excl. the form with smaller leaves which is *A. weissenborniana* J. Braun & K. Schum. ; F.T.A. 1 : 356, partly. A forest shrub, about 4 ft. high, with glossy leaves ; flowers pale green in drooping cymes ; fruits orange.
    Br.Cam.: S. Bakundu F.R., Banga, Kumba Div. (fl. & fr. Mar.) *Brenan* 9267 ! 9267a ! Also in French Cameroons, Spanish Guinea and Gabon.
4. **A. staudtii** *Engl.* Bot. Jahrb. 24 : 479 (1898). *A. weissenborniana* of Chev. Bot. 128, not of J. Braun & K. Schum. A shrub or small tree, to 8 ft. high ; flowers yellow ; ripe fruits red ; in forest.
    Iv.C.: Guidéko (May) *Chev.* 16374 ! Alépé *Chev.* 17410. **G.C.**: near Essuasu (fl. & fr. May) *Chipp* 215 ! *Vigne* 168 ! Asuansi (Feb.) *Williams* 23 ! Prestea (Sept.) *Vigne* FH 3089 ! Simpa (May) *Vigne* FH 1955 !
    S.Nig.: on hills, Ibadan (Mar.) *Schlechter* 13011 ! Also in French Cameroons and Belgian Congo.
5. **A. rowlandii** *Engl.* l.c. (1898). A shrub, to 10 ft. high ; flowers yellow ; ripe fruits red ; in forest by streams.
    S.Nig.: Lagos (Aug.) *Rowland* ! Etemi, Omo F.R., Ijebu Ode (fr. Mar.) *Jones & Onochie* FHI 16824 ! Akure F.R. (fl. & fr. Aug.) *Jones* FHI 20208 ! Okomu F.R., Benin (fl. & fr. Mar.) *Akpabla* 1107 ! Degema *Talbot* ! Also in Belgian Congo.

Doubtfully recorded species.

A. rubra *Engl.*—Sleumer in E. & P. Pflanzenfam. 20B : 360 (1942), records this species from Liberia. Perhaps the Liberian specimens were of *A. villosa* Keay, a species which looks rather like *A. rubra* but differs in several important characters.

3. LASIANTHERA P. Beauv.—F.T.A. 1 : 353 ; Sleumer in E. & P. Pflanzenfam. 20B : 362 (1942).

Branchlets terete and slightly ridged, glabrous ; leaves oblong-elliptic, long-caudate-acuminate, shortly cuneate at the base, 8–15 cm. long, 3–6 cm. broad, glabrous ; lateral nerves about 6 pairs, looped ; cymes leaf-opposed, small, the flowers in small partial heads ; petals 3 mm. long ; anthers densely villous ; fruits oblong-elliptic, flattened, beaked, about 1 cm. long, longitudinally ribbed ..   ..   .. *africana*

L. africana *P. Beauv.* Fl. Oware 1 : 85, t. 51 (1806) ; F.T.A. 1 : 353 ; Sleumer l.c., fig. 102, A–O ; Chev. & Walker in Rev. Bot. Appliq. 27 : 502, t. 24. A glabrous shrub, 3–12 ft. high, with terete branchlets and small white flowers in umbellate head-like clusters ; in rain forest.
    S.Nig.: R. St. Jacques, Chama [1] *Beauvois*. Mamu River F.R. (Nov.) *Keay* FHI 22291 ! *Onwudinjoh* FHI 21913 ! Oban (fl. & fr. May) *Talbot* 422 ! *Ujor* FHI 31797 ! Okwango, Obudu Div. (fl. & fr. May) *Latilo* FHI 30967 ! **Br.Cam.**: Victoria (Nov.) *Maitland* 93 ! Kumba *Preuss* 501 ! 1175 ! *Staudt* 490 ! Abonando, Mamfe (Apr.) *Rudatis* 38 ! **F.Po**: (fl. & fr. Dec.) *Mann* 73 ! Also in French Cameroons, Spanish Guinea, Gabon, Cabinda and Belgian Congo.

4. RHAPHIOSTYLIS Planch. ex Benth. in Hook. Fl. Nigrit. 259 (1849) ; Sleumer in E. & P. Pflanzenfam. 20B : 368 (1942).

Leaves and branchlets glabrous or tomentellous or puberulous ; leaves not cordate at base :
    Pedicels, flowers, branchlets and leaves glabrous or nearly so :
        Ovary glabrous ; leaves oblong-elliptic or elliptic, obtuse or rounded at base, obtusely acuminate, 7·5–16·5 cm. long, 2·5–6·5 cm. broad, the lower lateral nerves ascending ; flowers rather numerous in axillary fascicles ; pedicels 7–8 mm. long ; fruits subdidymous, quadrate, 1–1·5 cm. broad, reticulate ..   ..   1. *beninensis*
        Ovary densely pilose except sometimes at base ; leaves elliptic :
            Lateral nerves ascending ; pedicels 5–7 mm. long ; leaves 6–15 cm. long, 2·5–7·5 cm. broad ..   .. 2. *preussii*
            Lateral nerves spreading ; pedicels about 10 mm. long ; leaves about 8 cm. long and 3–4 cm. broad ..   .. 3. *poggei*
    Pedicels and calyces brownish-tomentellous ; branchlets and leaves on both surfaces puberulous ; leaves oblong or oblong-elliptic, cuneate or obtuse at base, rather long and subacutely acuminate, 9–12 cm. long, 3–4 cm. broad ; flowers rather numerous in axillary fascicles ; pedicels 7–9 mm. long ; ovary densely pilose ; fruits 2·5–3 cm. broad, coarsely verrucose ..   .. 4. *ferruginea*
Leaves and branchlets densely spreading-pilose ; leaves cordate at base, ovate-elliptic, obtusely long-acuminate, 5–12 cm. long, 2–6 cm. broad ; flowers in axillary clusters ; pedicels up to 3 mm. long ; ovary villous ..   .. 5. *cordifolia*

1. **R. beninensis** (*Hook. f. ex Planch.*) *Planch. ex Benth.* in Hook. Fl. Nigrit. 259, t. 28 (1849). *Apodytes beninensis* Hook. f. ex Planch. in Hook. Ic. Pl. t. 778 (1848) ; F.T.A. 1 : 355, partly (excl. var. β) ; Chev. Bot. 127. A glabrous scandent shrub, rarely arborescent ; flowers white, fragrant ; ripe fruits red, shining, fleshy ; the whole plant goes black when dried ; in forest vegetation.
    Fr.G.: *Heudelot* 723 ! Timbo *Pobéguin* 767 ! 128 ! Kouria (Dec.) *Chev.* 14890 ! Labé (fr. Apr.) *Chev.* 12389 ! **S.L.**: Bonthe (fl. & fr. Feb.) *Dalz.* 937 ! Kukuna road, Scarcies R. (Jan.) *Sc. Elliot* 4711 ! Njala (Nov.) *Deighton* 3092 ! Laoma *Smythe* 94 ! Kenema (Feb.) *Lane-Poole* 324 ! **Lib.**: Cape Palmas (July) *T. Vogel* 46 ! Ganta (Nov., Dec.) *Harley* ! *Barker* 1119 ! **Iv.C.**: Adiopodoumé (Jan.) *Miège* 1491 ! Baoulé (Oct.) *Pobéguin* 256 ! **G.C.**: Aburi (Jan.) *Brown* 881 ! Akropong (Oct.) *Johnson* 825 ! Tukobo (Nov.) *Vigne* FH 1430 ! Achimota (Jan.) *Irvine* 1971 ! **Togo** : Lome *Warnecke* 407 ! 475 ! Kabosu (Dec.) *St. C. Thompson* FH 3724 ! **Dah.**: Boukoutou forests, Sakété (Jan.) *Chev.* 22846 ! **S.Nig.**: Lagos (Dec.) *Millen* 22 ! 170 ! Gambari Group F.R., Ibadan (Nov.) *Latilo* FHI 33668 ! Mamu River F.R., Awka (Feb., Mar.) *Jones* FHI 4848 ! 6916 ! Cross River area (Jan.–Mar.) *Johnston* ! Extends to Uganda, N. Rhodesia and Angola. (See Appendix, p. 292.)

---

[1] Beauvois (l.c. 1 : 91) refers to Chama as being in the kingdoms of Warri and Benin ; I have not, however, been able to trace either Chama or the river St. Jacques on a map of S. Nigeria. In the F.T.A. and other works Beauvois' Chama is identified with the Chama east of Sekondi in the Gold Coast. As *Lasianthera* has not otherwise been recorded from anywhere west of the R. Niger, I consider that either Beauvois' Chama was in what is now Nigeria, or else he gave the wrong locality for his specimen. See also footnote on p. 411 and the note after *Grewia megalocarpa* P. Beauv. on p. 305.

2. **R. preussii** *Engl.* Bot. Jahrb. 17 : 72 (1893). A glabrous scandent shrub, in forest ; similar to the preceding species but with densely pilose ovary, leaves and branchlets which usually do not dry black, and rather longer acumen.
S.L.: Njala (Oct.) *Deighton* 5599! Bonganema, Kori (Oct.) *Deighton* 6134! Makump (Oct.) *Thomas* 3956! Bumbuna (Oct.) *Thomas* 3852! **Lib.:** Karmadhun, Kolahun (Nov.) *Baldwin* 10181! Peahtah (Oct.) *Linder* 1030! **Iv.C.:** *Jolly* 49! **G.C.:** Banka (Sept.) *Vigne* FH 1372! Pra River Station (Feb.) *Vigne* FH 1824! Kwahu Prasu (Feb.) *Vigne* FH 1630! Sefwi Bekwai (Oct.) *Akpabla* 902! **S.Nig.:** Akure F.R. (Oct.) *Keay* FHI 25485! **Br.Cam.:** Barombi, Kumba (Sept.) *Preuss* 522!
[Two fruiting species (*Esp. Santo* 2250 ; 2253) from Portuguese Guinea may belong either to this species or to the last.]

3. **R. poggei** *Engl.* Bot. Jahrb. 17 : 73 (1893) ; op. cit. 43 : 181. A glabrous scandent shrub·
Br.Cam.: Barombi (Sept.) *Preuss* 561. Also in French Cameroons.

4. **R. ferruginea** *Engl.* op. cit. 43 : 181 (1909) ; Sleumer in E. & P. Pflanzenfam. 20B : 369, fig. 105. *Apodytes beninensis* var. β Oliv. in F.T.A. 1 : 356. *Icacina* ? *sarmentosa* A. Chev. Bot. 127, name only. A forest shrub, often scandent.
Iv.C.: Yapo (Oct.) *Chev.* 22373! **G.C.:** Abosso (Aug.) *Vigne* FH 3034! **S.Nig.:** Epe *Barter* 3286! Ehor F.R., Benin (fl. & fr. Mar.) *Chizea* FHI 8268! **Br.Cam.:** Victoria (May) *Maitland* 1062! Moliwe (May) *Winkler* 1280. Also in French Cameroons, Spanish Guinea, Gabon and Belgian Congo.

5. **R. cordifolia** *Hutch. & Dalz.* F.W.T.A., ed. 1, 1 : 455 (1928) ; Kew Bull. 1929 : 23. A scandent shrub ; flowers white or cream ; in forest.
Lib.: Gbanga (Oct.) *Linder* 1218! **Iv.C.:** Banco (Nov.) *Miège* 1418! **G.C.:** Opon Mansi F.R. *Akpabla* 934! Subiri F.R., Benso (Sept.) *Andoh* FH 5575! *Vigne* FH 4747!

5. DESMOSTACHYS Planch. ex Miers—F.T.A. 1 : 353 ; Sleumer in E. & P. Pflanzenfam. 20B : 369 (1942).

Petals 10–11 mm. long, loosely joined below to form a tube about 6 mm. long which eventually splits from the bottom ; style about 14 mm. long, exserted beyond the petals and stamens ; flowers 5-merous in markedly supra-axillary spikes, up to 36 cm. long, with a tendency for the flowers to be arranged in 2 ranks ; calyx 2·5 mm. long ; leaves oblong to oblong-oblanceolate, obtuse or cuneate at base, acuminate, 12–30 cm. long, 3·5–11 cm. broad, with 5–8 main lateral nerves on each side of midrib, very minutely puberulous beneath ; fruits ellipsoid, about 2 cm. long   1. *vogelii*
Petals 3–6 mm. long, free ; style as long as or shorter than the petals ; flowers 4-merous in solitary or superposed pairs of spikes, axillary, up to 35 cm. long, on leafy shoots, rhachis flattened, flowers more or less in 2 ranks ; calyx 1·5 mm. long ; leaves more or less elongate-elliptic or oblong, cuneate at base, acuminate, 12–23 cm. long, 5–11 cm. broad, with (5–) 6–9 (–10) main lateral nerves on each side of midrib, very minutely puberulous beneath ; fruits ellipsoid, 2·5–2·8 cm. long   ..   2. *tenuifolius*

1. **D. vogelii** (*Miers*) *Stapf* in Johnston Lib. 2 : 587 (1906) ; Chev. Bot. 126. *Sarcostigma vogelii* Miers (1852). A forest shrub or small tree, often scandent ; flowers white and fragrant, in pendulous spikes ; fruits yellow.
S.L.: Kpetema (Apr.) *Deighton* 3693! Gola Forest, Gorahun (Mar.) *Deighton* 4097! **Lib.:** Cape Palmas (fr. July) *T. Vogel* 25! 27! 68! Kakatown *Whyte*! Bumbuna, Moala (Nov.) *Linder* 1345! Dukwia R. (fl. & fr. Feb.) *Cooper* 156! Boporo (Nov.) *Baldwin* 10377! **Iv.C.:** Grabo (July) *Chev.* 19624! Tabou to Bériby *Chev.* 19970! N'Zida (fr. June) *Miège* 1587!

2. **D. tenuifolius** *Oliv.* in F.T.A. 1 : 353 (1868) ; F.W.T.A., ed. 1, 1 : 454, in part only (excl. distrib. S.L. and Dah.). *D. preussii* Engl. Bot. Jahrb. 17 : 70 (1893). A forest shrub or small tree, to 30 ft. high ; flowers cream, in slender pendulous spikes ; ripe fruits orange.
S.Nig.: Port Harcourt (Jan.) *Jones* FHI 5005! Calabar (July) *Holland* 63! *Robb*! Oban (May) *Talbot* 100! *Ujor* FHI 31790! **Br.Cam.:** *Preuss* 1327! Likomba (Dec.) *Mildbr.* 10746! S. Bakundu F.R. (fl. & fr. Mar.) *Brenan* 9276! Onochie FHI 30867! S. Bakossi F.R. (May) *Olorunfemi* FHI 30596! Abonando (June) *Rudatis* 69! Barombi (June) *Preuss* 306 ; 351. **F.Po:** (June) *Barter*! *Mann* 421! Also in French Cameroons, Spanish Guinea and Gabon.
[The record from Sierra Leone in the first edition is based on a *Barter* specimen which is most probably a second sheet of his Fernando Po specimen, the species does not seem to occur west of E. Nigeria.]

6. ICACINA A. Juss.—F.T.A. 1 : 356 ; Sleumer in E. & P. Pflanzenfam. 20B : 370 (1942).

Flowers in lax terminal corymbose cymes ; leaves ovate to obovate, obtusely acuminate to rounded and emarginate, 5–15 cm. long, 4–10 cm. broad, conspicuously reticulate and glabrous or nearly so ; calyx short ; petals shortly hairy outside ; fruits obovoid, 3 cm. long, tomentellous, slightly wrinkled when dry..   ..   ..   1. *senegalensis*
Flowers in axillary clusters or short dense cymes :
Flowers rather lax and pedicellate ; leaves broadly obovate-elliptic, abruptly acuminate, more or less cuneate at base, up to 20 cm. long and 12 cm. broad, glabrescent beneath ; calyx much shorter than the petals ; petals appressed-pubescent outside
    2. *mannii*
Flowers densely crowded and subsessile ; leaves broadly elliptic, abruptly acute, rounded at base, 10–28 cm. long, up to 17 cm. broad, thinly pilose with simple, fascicled hairs beneath ; calyx nearly as long as the petals ; petals villous outside ; fruits ellipsoid-globose, 2·5 cm. long, tomentose   ..   ..   ..   3. *trichantha*

1. **I. senegalensis** A. *Juss.*—F.T.A. 1 : 357 ; Chev. Bot. 128. A savannah suffrutex, sending up glabrous or pubescent leafy shoots 2–3 ft. high from a large fleshy tuber with long roots ; flowers cream or yellow ; ripe fruits red ; locally abundant and troublesome as a weed in some parts, but rare in Nigeria.
Sen.: *Perrottet* 129! Kaolak *Kaichinger*. Floup-Fédyan, Sédhiou and Samandiny *fide* Chev. *l.c.* **Gam.:** Mungo Park! Karawan and Cape St. Mary (Jan.) *Dalz.* 8084! Genieri (Feb.) *Fox* 95! Kuntaur (July) *Ruxton* 38! 42! **Fr.G.:** Pobéguin 2142! Labé *Chev.* 12297. Pita (Mar.) *Jac.-Fél.* 797! **G.C.:** Kintampo (Mar.) *Dalz.* 34! Tamale (Apr.) *Williams* 152! Dogonkadi (fl. & fr. Apr.) *Kitson* 633! Grube (Mar.) *Rea* FH 1671! **Togo:** Kete Krachi (Jan.) *Vigne* FH 1527! **Dah.:** Agouagon to Savé (May) *Chev.* 23559! **S.Nig.:** *Barter* 3422! Lanlate, Oyo Prov. (fl. & fr. May) *Onochie* FHI 19173! Also in Ubangi-Shari and A.-E. Sudan. (See Appendix, p. 291.)

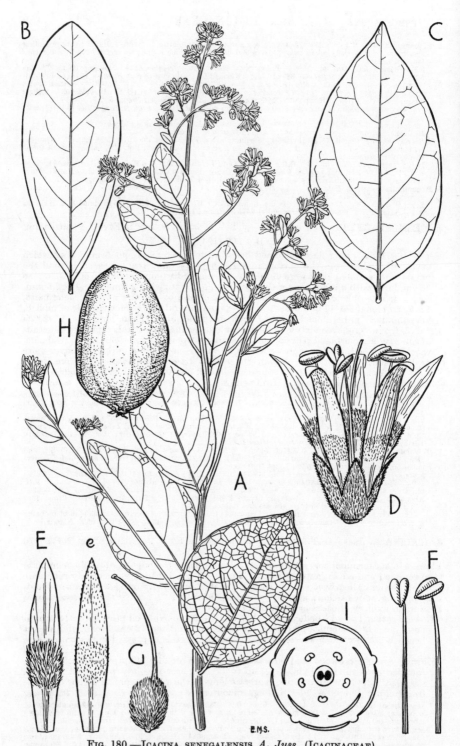

Fig. 180.—Icacina senegalensis *A. Juss.* (Icacinaceae).

A, flowering shoot, × ⅔. B & C, leaves, × ⅔. D, flower, × 6. E & e, inside and outside of petals, × 6. F, stamens, × 6. G, ovary and style, × 6. H, fruit, × 1½. I, floral diagram. A, D–G are drawn from *Dalz.* 34 ; B from *Kitson* 633 ; C from *Ozanne* 6 ; H from *Onochie* FHI 19173.

2. **I. mannii** *Oliv.* in F.T.A. 1 : 357 (1868) ; Bot. Mag. t. 6260. A scandent shrub, to 20 ft. high, with a large tuber ; flowers cream, scented ; in forest.
S.L.: *Thomas* 8241 ! **Lib.:** Logantown, New Cess R. (Mar.) *Baldwin* 11179 ! **Iv.C.:** Adiopodoumé (Apr.) *Miège* 1828 ! **G.C.:** Kumasi (Mar.) *Vigne* FH 1092 ! 1845 ! Abiabo, Axim (Nov.) *Vigne* FH 1431 ! **S.Nig.:** Okomu F.R. (Feb.) *Brenan* 8477 ! 9080 ! Eket *Talbot* ! Okpanam (Mar.) *Kitson* ! Calabar (Feb.) *Mann* 2241 ! Also in French Cameroons, Spanish Guinea, Gabon, Cabinda and Belgian Congo.
3. **I. trichantha** *Oliv.* in F.T.A. 1 : 358 (1868). A scandent shrub, with a very large tuber ; flowers cream ; ripe fruits red, velvety ; in forest and forest regrowth vegetation.
S.Nig.: Lagos (fl. & fr. Jan., fr. Oct.) *Dalz.* 1094 ! 1398 ! *Rowland* 58 ! Ibadan (fl. Jan., Feb., fr. Mar.) *Meikle* 968 ! 1166 ! Shasha F.R. (fl. & fr. Feb., Mar.) *Richards* 3142 ! *Ross* 11 ! Okomu F.R. (fr. Jan.) *Brenan* 8887 ! Sapoba (Nov.) *Ajayi* FHI 26985 ! Onitsha *Barter* 1793 ! Degema *Talbot* 3692 ! (See Appendix, p. 291.)
[The specimens from Lagos differ from the others in having the leaves more thickly covered with fascicled hairs beneath.]

7. LAVIGERIA Pierre Fl. For. Cochinch. fasc. 17, sub t. 267 (1892) ; Sleumer in E. & P. Pflanzenfam. 20B : 371 (1942).

Leaves oblong-elliptic, shortly pointed, rounded at base, 15–25 cm. long, 7–11 cm. broad, slightly stellate-pubescent beneath, coarsely reticulate ; panicles clustered on the main stem, tomentellous ; flowers very small, sessile ; petals nearly glabrous outside, not hairy inside ; fruit ovoid, about 7 cm. long, corrugated. .   *macrocarpa*

L. macrocarpa (*Oliv.*) *Pierre* l.c. (1892) ; Sleumer l.c., fig. 106. *Icacina macrocarpa* Oliv. in F.T.A. 1 : 357 (1868) ; Hook. Ic. Pl. t. 2338. *Lavigeria salutaris* Pierre (1892). A large forest liane, cauliflorous ; branchlets pubescent ; flowers pinkish.
S.Nig.: Cross River area (Dec.) *Holland* 184 ! **Br.Cam.:** Likomba (Oct.) *Mildbr.* 10536 ! L. Barombi, Kumba (Mar.) *Brenan* 9467 ! **F.Po:** (fl. & fr. Dec.) *Mann* 43 ! Also in French Cameroons, Belgian Congo and Gabon.

8. PYRENACANTHA Wight in Hook. Bot. Misc. 2 : 107 (1831) ; Sleumer in E. & P. Pflanzenfam. 20B : 384 (1942).

Leaves ovate, widely cordate and 3–5-nerved at base, acuminate at apex, 15–25 cm. long, 9·5–15·5 cm. broad, puberulous on the nerves beneath ; petioles 3·5–9·5 cm. long ; racemes in fascicles on the old wood, well below the leaves ; male racemes up to 6 cm. long ; tepals and stamens 3 ; fruits flattened-ellipsoid, about 2·5 cm. long and 1·3 cm. broad, rugose, minutely tomentellous   . .   . .   1. *klaineana*
Leaves more or less elliptic or lanceolate or obovate or oblanceolate, rounded to cuneate at base, or narrowed to a broadly cordate or auriculate base, up to 9 cm. broad ; petioles up to 3 cm. long ; racemes or spikes amongst the leaves ; tepals and stamens 4:
  Lower surface of leaves quite glabrous, except for a few scattered hairs on the midrib and nerves ; leaves drying black, broadly elliptic or obovate-elliptic, obtuse or cuneate at base, shortly acuminate, 3·2–16 cm. long, 1·8–9 cm. broad ; female flower spikes up to 10 cm. long ; fruits oblong-ellipsoid, 2·5–35 cm. long, 1·5–1·8 cm. broad, puberulous   . .   . .   2. *mangenotiana*
  Lower surface of leaves puberulous, pubescent ; leaves drying green, gradually long-acuminate and mucronate :
    Leaves beneath densely puberulous with curly hairs, not presenting a " combed " appearance and not closely appressed ; branchlets rusty-yellowish tomentellous ; male flowers in very densely-flowered spikes up to 3 cm. long ; female flowers in congested shortly pedunculate globose heads ; fruits obovoid-ellipsoid, 1·1–1·4 cm. long, tomentose ; leaves oblong-obovate, oblong or oblong-oblanceolate, rounded to slightly cordate at base, 7–19 cm. long, 3–8 cm. broad, margins denticulate or undulate-denticulate   . .   . .   3. *staudtii*
    Leaves beneath with closely appressed straight hairs ; branchlets appressed-puberulous ; inflorescences slender racemes or spikes, not densely-flowered :
      Male flowers sessile ; lower surface of leaves closely appressed-puberulous, the hairs well separated from one another ; leaves narrowed to a closely cordate or auriculate base, elliptic, oblong-elliptic or oblong-lanceolate, 5–17 cm. long, 1·6–7 cm. broad, margins undulate ; male spikes up to 8 cm. long ; female spikes up to 10 cm. long ; fruits flattened-ovoid, with a long beak, strongly reticulate, 1·6–2 cm. long, 1·1–1·5 cm. broad, puberulous   . .   . .   . .   4. *vogeliana*
      Male flowers with pedicels about 1 mm. long ; lower surface of leaves densely appressed-pubescent, the hairs having a " combed " appearance ; leaves cuneate to narrowly cordate at base :
        Bract of each male flower borne on the inflorescence axis below the pedicel ; leaves narrowly elliptic to lanceolate, margins subentire or undulate, 5–13 cm. long, 1·8–4·5 cm. broad ; fruits flattened-ellipsoid, scarcely beaked, about 1·3 cm. long and 8 mm. broad   . .   . .   5. *acuminata*
        Bract of each male flower borne on the pedicel ; leaves more or less elliptic, margins undulate, 6–14 cm. long, 2·5–6 cm. broad ; fruits not known   6. *sp. A*

1. **P. klaineana** *Pierre ex Exell & Mendonça* in Consp. Fl. Ang. 1 : 345 (1951). A forest liane reaching to the tree tops, with flowers and fruits on the stem in the understorey.
Iv.C.: Guiglo to Tai, Cavally Basin (fr. Aug.) *Miège* ! Also in Gabon, Belgian Congo and Cabinda.
2. **P. mangenotiana** *Miège* in Bull. I.F.A.N. 17 : 8, tt. 1 & 2 (1955). A forest liane, with stems up to 100 ft. long and 4 in. diam.

Fr.G.: Sérédou *Miège* 2388. **Lib.:** *Baldwin*! **Iv.C.:** Adiopodoumé (Mar.) *Miège* 1798! 1809! Abidjan (fr. Dec.) *Chev.* 17326! Sassandra Port (fr. May) *Chev.* 17063! 17971!
[Baldwin's specimen from Liberia is associated with a label bearing the following particulars :— " E. Prov., Tchien Dist., Bahtown, on Cess R. ; procumbent, with red fruit ; 11 Aug. 1947 ; *Baldwin* 8032.'' As the note about the red fruit is quite inapplicable to the specimen which has only male flowers, I conclude that the label has been wrongly associated with the specimen.
This species is close to *P. glabrescens* (Engl.) Engl. from Gabon.]

3. **P. staudtii** (*Engl.*) *Engl.* Pflanzenw. Afr. 3, 2: 262, 264 (1921) ; Sleumer in E. & P. Pflanzenfam. 20B : 385, fig. 115, A–D, excl. syn. *Chlamydocarya tenuis* Engl. ; Exell & Mendonça l.c. *Chlamydocarya staudtii* Engl. (1898). *Pyrenacantha ugandensis* Hutch. & Robyns (1924). A shrub or small tree, with twisted petioles ; flowers pale green.
S.Nig.: *Thompson* 511! Abeokuta *Barter* 3375! Ibadan (Apr.) *Meikle* 1435! Sapoba *Kennedy* 2200! 2217! 2342! Degema *Talbot*! Eket *Talbot* 3152! **Br.Cam.:** Banga (fr. Apr.) *Ejiofor* FHI 29338! Johann-Albrechtshöhe (Jan.) *Staudt* 787! Also in French Cameroons and extending to Uganda, Belgian Congo and Angola. (See Appendix, p. 292.)

4. **P. vogeliana** *Baill.* in Adansonia 10 : 271 (1872) ; Chev. Bot. 128, partly. *P. dinklagei* Engl. (1909). A slender liane with small yellow flowers and orange or red fruits.
S.L.: Yonibana (fl. buds Oct.) *Thomas* 4099! Mano (fr. Oct.) *Deighton* 2411! **Lib.:** Grand Bassa (fr. Mar., July) *T. Vogel* 13! Baldwin 11284! Ganta (fr. Feb.) *Baldwin* 11009! Nyaake, Webo (fr. June) *Baldwin* 6150! Kle (fr. Dec.) *Baldwin* 10637! **Iv.C.:** Makouguié (fr. Jan.) *Chev.* 16961! Sampleu to Ganhoué (fr. Apr.) *Chev.* 21155! Digoualé to Gouané (fr. May) *Chev.* 21511! Sanrou to Ouodé (fr. May) *Chev.* 21624! **G.C.:** Kwahu Prasu (Feb.) *Vigne* FH 1617! Kumasi (fr. Nov.) *Andoh* FH 2371! **S.Nig.:** Eribovhe waterside, Benin (fr. Nov.) *Ujor* FHI 23935! Also in French Cameroons, Gabon, Belgian Congo and Tanganyika.

5. **P. acuminata** *Engl.* Bot. Jahrb. 24 : 483 (1898). *Chlamydocarya tenuis* Engl. (1898). *P. vogeliana* of Chev. Bot. 128, partly. A slender liane with small green flowers and pinkish-orange fruits.
S.L.: Njala (Jan., May, June) *Deighton* 4622! 4768! 5021! Mabonto (July) *Deighton* 3267! **Lib.:** Ganta (May) *Harley*! Javajai (fr. Nov.) *Baldwin* 10258a! **Iv.C.:** Singrobo *Miège*! Sampleu to Ganhoué (fr. Apr.) *Chev.* 21156! Also in French Cameroons, Belgian Congo and Gabon.
[The specimens from Angola cited as this species in Consp. Fl. Ang. 1 : 345 are quite different and should be called *P. sylvestris* S. Moore.]

6. **P. sp. A.** A slender liane with very small greenish flowers. Fruits not known. Possibly this is *P. undulata* Engl., see below.
G.C.: Agogo (May) *Chipp* 448! New Tafo (Oct.) *Lovi* WACRI 3936! Akumadan (June) *Vigne* FH 2565! Jimira F.R. *Vigne* FH 2969! **S.Nig.:** Oban *Talbot* 1509!

Imperfectly known species.

1. **P. undulata** *Engl.* Pflanzenw. Afr. 3, 2 : 262 (1921), without proper description ; no specimen cited. Possibly this is *P. sp. A*, see above.
Togo: Tove.

2. **P. sp. B.** Leaves densely puberulous beneath ; male flowers pedicellate, with bract on pedicel. Better material needed.
S.L.: riverside, 6 miles N. of Kundita, 3,600 ft. (Mar.) *Sc. Elliot* 5043!

3. **P. sp. C.** Slender liane with bright orange very long-beaked fruits at the end of a peduncle 7 cm. long, leaves elliptic, cuneate at base, long-acuminate, drying blackish. Flowers not known ; better material needed.
Br.Cam.: Mbalange to Ndifo, on Kumba—Victoria road (fr. Mar.) *Onochie* FHI 30856!

9. POLYCEPHALIUM Engl. in E. & P. Pflanzenfam. Nachtr. 227 (1897) ; Sleumer in E. & P. Pflanzenfam. 20B : 385 (1942).

Stems and petioles long-spreading-pilose ; undersurface of leaves at first densely covered with long pilose hairs, which remain on the nerves, but tend to come off the lamina and expose a dense indumentum of shorter hairs ; leaves variable in shape, often 3-lobed, strongly 3-nerved at the base, apices of leaves and lobes acutely acuminate, up to 20 cm. long and 10 cm. broad ; male and female flowers in pedunculate heads, solitary or a few together ; fruits densely clustered, each covered by the enlarged persistent ovoid corolla, up to 2·5 cm. long   ..   ..   *capitatum*

P. capitatum (*Baill.*) *Keay* in Bull. Jard. Bot. Brux. 26 : 184 (1956). *Chlamydocarya capitata* Baill. in Adansonia 10: 277 (1872) ; Chev. Bot. 128 ; F.W.T.A., ed. 1, 1 : 457 ; Sleumer l.c. 388. *Phytocrene* sp.—F.T.A. 1 : 359. *Synclisia scabrida* of Chev. Bot. 17, partly (*Chev.* 20989 & 21157). *S. ferruginea* of Chev. Fl. Viv. 1 : 100, partly (*Chev.* 20989 & 21157). A small liane with brown-pilose indumentum ; flowers green ; corolla in fruit orange.
Fr.G.: Lola to Nzo (Mar.) *Chev.* 20989! Foot of Mt. Nzo (Mar.) *Chev.* 21003 (partly)! Kaoulendougou, Nzo (fr. Mar.) *Chev.* 21057! **S.L.:** Freetown (fr. Jan., Mar.) *Dalz.* 986! *Kirk* 35! Njala (fr. Feb.) *Deighton* 3115! Rokupr (fl. Sept., fr. Mar.) *Jordan* 408! Yungeru (Jan.) *Thomas* 7368! Jepihun *Smythe* 90! **Lib.:** Dukwia R. (fr. Sept.) *Cooper* 363! Peahtah (Oct.) *Linder* 1076! Zwedru, Tchien (fr. Aug.) *Baldwin* 7029! **Iv.C.:** Sampleu to Ganhoué (fr. Apr.) *Chev.* 21157!

10. CHLAMYDOCARYA Baill. in Adansonia 10 : 276 (1872), partly ; Sleumer in E. & P. Pflanzenfam. 20B : 387 (1942), partly.

Corolla in fruit densely and closely setose-pubescent, elongated beyond the body of the fruit into an inflated cone 2·5–4 cm. long and 1·5–2·2 cm. broad at base ; leaves sparingly pubescent beneath, elliptic, rounded or obtuse at base, broadly acuminate, 9·5–22 cm. long, 4–11 cm. broad, reticulate beneath, margins entire..   1. *macrocarpa*
Corolla in fruit with spreading bristly hairs, elongated beyond the body of the fruit into a tube 4–8 cm. long and less than 1·5 cm. broad at the conical base ; leaves densely to sparingly pilose on the nerves beneath, elliptic, oblong or obovate-elliptic, rounded or obtuse at base, broadly acuminate or acute at apex, 10–28 cm. long, 4–11·5 cm. broad, reticulate beneath, margins denticulate   ..   2. *thomsoniana*

1. **C. macrocarpa** *A. Chev. ex Hutch. & Dalz.* F.W.T.A., ed. 1, 1 : 456 (1928) ; Kew Bull. 1929 : 23 ; Chev. Bot. 128, name only. A woody climber ; male flowers in long catkins ; corolla in fruits yellow or orange-red when ripe, borne on the stems.
Iv.C.: Dabou (fr. Feb.) *Chev.* 17216! Adiopodoumé (fl. July, fr. Oct.) *Miège* 1700! **G.C.:** Axim (fr. Jan., Feb.) *Akpabla* 822! *Irvine* 2186! Simpa (fr. Feb.) *Vigne* FH 2802! S. Scarp F.R. (fr. Feb.) *Moor* 235! [Closely related to *C. soyauxii* Engl. from French Cameroons and Gabon.]

2. **C. thomsoniana** *Baill.* in Adansonia 10 : 276 (1872) ; Sleumer in E. & P. Pflanzenfam. 20B : 388, fig. 114. *C. rostrata* Bullock in Kew Bull. 1933 : 469, partly (fruits only, the leaves are *Prevostea africana* Benth.). A small liane with more or less bristly stems ; male flowers in catkins ; corolla in fruit orange to scarlet when ripe.
S.L.: Gorahun (young fr. Mar.) *Deighton* 4109 ! **Lib.:** Ganta (fr. Sept.) *Baldwin* 9299 ! **Iv.C.:** Mé (fr. Aug.) *Miège* 1856 ! **S.Nig.:** Okomu F.R. Benin (fr. Dec.) *Brenan* 8512 ! Usonigbe F.R., Benin (fr. Nov.) *Keay* FHI 25553 ! Sapoba *Kennedy* 2225 ! Oban *Talbot* 1355 ! 1422 ! Calabar *Thomson* 12 ! **Br.Cam.:** Mamfe (fr. Jan.) *Maitland* 1167 (partly) ! Also in French Cameroons, Belgian Congo and Gabon.

11. NEOSTACHYANTHUS Exell & Mendonça in Bol. Soc. Brot., sér. 2, 25 : 111 (1951). *Stachyanthus* Engl. (1897)—Sleumer in E. & P. Pflanzenfam. 20B : 388 (1942) ; not of DC. (1836).

Anthers about 2·25 mm. long, about as long as the filaments ; male flowers normally 6-merous, rarely 5-merous, petals about 4·75 mm. long, calyx about 1 mm. long ; male spikes up to 25 cm. long, in fascicles from the stem below the leaves ; leaves obovate-oblong, slightly cordate at base, long-acuminate, 12–21 cm. long, 4·3–9·2 cm. broad, with 6–9 main lateral nerves on each side of midrib ; branchlets and petioles at first rather long bristly-pilose. .   . .   .   1. *zenkeri*
Anthers about 1 mm. long, only about half as long as the filaments ; male flowers 4–5-merous, petals about 3·25 mm. long, calyx about 0·7 mm. long ; male racemes up to 28 cm. long, in fascicles from the stem below the leaves and solitary and slightly supra-axillary among the leaves ; leaves elliptic-ovate to elliptic-obovate, narrowed to an obtuse, rounded or very slightly cordate base, acutely acuminate, 9–20 cm. long, 4–8·5 cm. broad, with 6–9 main lateral nerves on each side of midrib ; branchlets and petioles sparingly pubescent or glabrous ; fruits flattened-ellipsoid, about 1·2 cm. long, endocarp pitted   . .   . .   2. *occidentalis*

1. **N. zenkeri** (*Engl.*) *Exell & Mendonça* in Bol. Soc. Brot., sér. 2, 25 : 111 (1951). *Stachyanthus zenkeri* Engl. in E. & P. Pflanzenfam. Nachtr. 1 : 227 (1897) ; Bot. Jahrb. 24 : 487, t. 8, E–K ; Sleumer in E. & P. Pflanzenfam. 20B : 390, fig. 115, E–K. A slender twining liane, in forest ; flowers greenish-yellow, in pendulous, cauliflorous spikes.
S.Nig.: Akure F.R. (Oct.) *Keay* FHI 25453 ! Benin Prov. (Apr.) *Fairbairn* 46 ! **F.Po:** *Mildbr.* 6923. Also in French Cameroons, Belgian Congo and Cabinda.
2. **N. occidentalis** *Keay & Miège* in Bull. Jard. Bot. Brux. 25 : 271, figs. 47–51 (1955). *Desmostachys tenuifolius* of Chev. Bot. 126 and of F.W.T.A., ed. 1, 1 : 454, partly (Dah. record), not of Oliv. A slender twining liane, in forest ; flowers green with yellow anthers ; ripe fruits orange ; midrib and main nerves of leaves yellow or orange beneath.
Iv.C.: Singrobo (Oct.) *Miège* 2070 ! La Kassa (Oct.) *Miège* 2075. **G.C.:** Volta River F.R. (Apr.) *Vigne* FH 4377 ! Kwahu Prasu (Feb.) *Vigne* FH 1620 ! New Tafo (Oct.) *Lovi* WACRI 3925 ! **Dah.:** Cabolé to Bassila, Savalou (May) *Chev.* 23794 ! **S.Nig.:** Gambari Group F.R., Ibadan (fl. Mar., fr. Aug.) *Keay* FHI 22532 ! *Onochie* FHI 19148 ! Iguobazowa F.R., Benin (fr. Feb.) *Onochie* FHI 27143 !

Excluded species.

N. nigeriensis (*S. Moore*) *Exell & Mendonça* l.c. (1951). *Stachyanthus nigeriensis* S. Moore in J. Bot. 58 : 221 (1920). The type (*Talbot* s.n. from Oban) consists of a single inflorescence probably of *Neostachyanthus zenkeri* and a leafy shoot of **Prevostea heudelotii** (*Bak. ex Oliv.*) Hall f. (Convolvulaceae). The name *N. nigeriensis* is therefore rejected under Art. 76 of the International Code of Botanical Nomenclature (1952).

12. IODES Blume—F.T.A. 1 : 358 ; Sleumer in E. & P. Pflanzenfam. 20B : 377 (1942).

Leaves densely tomentose beneath, broadly ovate-elliptic, acutely and abruptly acuminate, rounded at base, 10–15 cm. long, 6–8 cm. broad, glabrous above ; lateral nerves about 6 pairs ; cymes axillary, about 6 cm. long, tomentellous ; fruits broadly ovoid, 1·3 cm. long, coarsely reticulate, thinly tomentellous. .   . .   1. *kamerunensis*
Leaves thinly pilose or nearly glabrous beneath :
  Branchlets and leaves thinly pilose with long weak hairs ; leaves ovate, shortly acuminate, cordate at base, 5–9 cm. long, with prominent nerves beneath ; cymes short, pilose ; fruits broadly ellipsoid, 1·3 cm. long, coarsely reticulate and tomentellous   . .   . .   2. *klaineana*
  Branchlets very shortly pubescent or glabrescent ; leaves glabrous :
    Leaves not shining, very laxly reticulate, broadly ovate or ovate-rounded, abruptly acuminate, rounded or subcordate at base, 9–15 cm. long, up to 11 cm. broad ; fruits broadly ovoid-ellipsoid, 8 mm. long, reticulate, glabrous   . .   3. *africana*
    Leaves shining and closely reticulate, otherwise as in preceding   . .   4. *liberica*

1. **I. kamerunensis** *Engl.* Bot. Jahrb. 24 : 481 (1898). A forest liane with tendrils, covered with yellow-rusty tomentum.
Br.Cam.: Johann-Albrechtshöhe (Jan.) *Staudt* 557. Buea (Mar.) *Maitland* 491 ! Also in French Cameroons.
2. **I. klaineana** *Pierre* in Bull. Soc. Linn. Paris 1321 (1897). *I. laurentii* De Wild. (1907). *I. talbotii* Bak. f. ex Hutch. & Dalz. F.W.T.A., ed. 1, 1 : 457 (1928) ; Kew Bull. 1929 : 23. A forest liane with tendrils, spreadingly yellow-hairy.
S.Nig.: Oban *Talbot* 572 ! Also in Gabon, Belgian Congo and Angola.
3. **I. africana** *Welw. ex Oliv.* F.T.A. 1 : 358 (1868) ; Sleumer in E. & P. Pflanzenfam. 20B : 377, fig. 108, D, G–J. A weak climber with tendrils and small yellow flowers in irregularly forked cymose panicles.
S.Nig.: Gambari Group F.R., Ibadan (Feb., Aug.) *Jones* FHI 20259 ! *Keay & Meikle* FHI 25663 ! Okomu F.R., Benin (Feb.) *Akpabla* 1116 ! *Ejiofor* FHI 19742 ! Calabar (Feb.) *Mann* 2322 ! Oban *Talbot* 1758 ! **Br.Cam.:** Victoria (fr. Feb.) *Maitland* 436 ! Extends to Belgian Congo and Angola.
4. **I. liberica** *Stapf* in Johnston Lib. 2 : 588 (1906) ; Chev. Bot. 129. *I. reticulata* Stapf (1905), not of King (1895). A slender climbing shrub with yellowish down branchlets and tendrils, yellow flowers and red fruits.

Fr.G.: Bérézia to Diorodougou, Toma (fl. & fr. Feb.) *Chev.* 20785 ! Nzérékoré (fl. Dec., fr. Feb., May, Dec.) *Adam* 3833 ! 5252 ! 7345 ! *Baldwin* 13284 ! Macenta (Feb., Dec.) *Adam* 4207 ! 7315 ! **S.L.:** nr. Zimi (fl. & fr. Apr.) *Deighton* 3641 ! 3688 ! Matotoka (July) *Thomas* 1261 ! Panguma (fr. July) *Deighton* 4130 ! **Lib.:** Sinoe Basin *Whyte* ! Kakatown *Whyte* ! Zigida (fl. & fr. Mar.) *Bequaert* 138 ! 139 ! Ganta (fl. & fr. May) *Harley* 1152 ! Robertsport (fr. Dec.) *Baldwin* 10963 ! **Iv.C.:** Guidéko *Chev.* 16426 ! Aboisso (fl. & fr. Apr.) *Chev.* 17826 ! Alepé (Feb.–Mar.) *Chev.* 17424 ! Fort Binger to Toula *Chev.* 19545 ! Also in Belgian Congo.

103. SALVADORACEAE

Trees or shrubs, sometimes with axillary spines. Leaves opposite, simple ; stipules minute. Flowers in dense axillary fascicles or panicles, hermaphrodite or dioecious, actinomorphic. Calyx 3–4-toothed. Petals 4, free or shortly united, imbricate. Stamens 4, inserted on or near the base of the petals and alternate with them ; anthers 2-celled, opening lengthwise. Disk absent or of separate glands. Ovary superior, 1–2-celled ; style short. Ovules 1–2, erect. Fruit a berry or drupe. Seed erect, without endosperm ; embryo with thick cordate cotyledons.

A small family mainly in tropical Africa and the Mascarene Islands.

SALVADORA Linn.—F.T.A. 4, 1 : 23 ; Sleumer in E. & P. Pflanzenfam. 20B : 238 (1942).

Leaves opposite, entire ; flowers small, subunisexual, paniculate ; petals 4, united into a short tube ; stamens 4 ; ovary 1-celled ; ovule 1, basal, erect.

Branches light greyish green, glabrous ; leaves lanceolate to elliptic, mucronate, rounded to subacute at base, 3–7 cm. long, 1·5–3 cm. broad, rigidly chartaceous, grey when dry, glabrous, the few lateral nerves equally prominent on both sides ; petiole nearly 1 cm. long ; panicles divaricate, lax ; calyx shallowly lobed ; petals 1·5 mm. long ; berry subglobose, about 6 mm. diam.   ..   ..   .. *persica*

S. persica *Linn.*—F.T.A. 4, 1 : 23 ; Chev. Bot. 399 ; Sleumer l.c., fig. 74, A & G–J ; Aubrév. Fl. For. Soud.-Guin. 352, t. 73, 7. A shrub or small tree, up to 30 ft. high ; leaves glaucous green ; flowers yellowish ; fruits red or purplish when ripe.
Maur.: Tichoten *Charles* in *Hb. Chev.* 28693. **Sen.:** R. Senegal *Perrottet* 797 ! Richard Tol *Döllinger* 75 ! **Fr.Sud.:** Kabarah (Aug.) *Chev.* 1357 ! Timbuktu (July) *Chev.* 1201 ! Niafounké *Hagerup* 89 ! **Fr.Nig.:** Agadès *De Wailly* 4935 ! **N.Nig.:** shores of L. Chad (fl. & fr. Dec., Jan.) *Elliott* 146 ! *Foster* 94 ! *E. Vogel* ! Kukawa (fl. & fr. May) *Rosevear* FHI 26607 ! Bornu (Jan., Feb.) *Talbot* ! Extends throughout the dry regions of tropical Africa to Egypt, Palestine, Arabia and India and Ceylon. (See Appendix, p. 292.)

104. OLACACEAE

Trees, shrubs or climbers. Leaves alternate, simple ; stipules absent. Flowers actinomorphic, usually hermaphrodite and small. Calyx-lobes imbricate or open in bud. Petals free or variously connate, valvate. Disk often present, usually annular. Stamens free or rarely united into a column, the same number as and opposite the petals or fewer or more numerous, some often without anthers ; anthers 2-celled, opening by slits or pore-like slits. Ovary superior or slightly immersed in the disk, 1–3-celled ; style 1, with a 2–5-lobed stigma ; ovules 1–5 from the apex of a central placenta in a 1-celled ovary, or pendulous from the inner angle of a 2- or more-celled ovary. Fruit drupaceous, sometimes inferior by accrescence of the calyx or receptacle. Seeds with copious endosperm and small or medium-sized straight embryo.

Confined to the tropics and subtropics.

Stamens free from each other :
  Stamens (and staminodes if present) more numerous than the petals ; fruit superior, sometimes enveloped by the accrescent calyx :
    Main lateral nerves of leaves continued to the margin ; stamens 3–4 times as numerous as the petals ; flowers in panicles   ..   ..   **1. Coula**
    Main lateral nerves looped or ending within the margin ; stamens (and staminodes if present) about twice as numerous as the petals ; flowers in short racemes or fascicles :
      Flowers shortly pedicellate in few-flowered axillary fascicles ; calyx much enlarged in fruit and deeply lobed   ..   ..   **2. Heisteria**
      Flowers in short racemes ; calyx not enlarged in fruit, or if enlarged then truncate :
        Branchlets usually armed with spines ; staminodes absent ; flowers in umbellate racemes ; petals densely villous within   ..   ..   **3. Ximenia**

Branchlets without spines ; flowers in narrow racemes ; staminodes present or absent ; petals glabrous within or pilose about the middle :

Calyx obvious, disciform in fruit or strongly accrescent and enveloping the fruit ; petals glabrous within ; staminodes present .. .. 4. **Olax**

Calyx very small, obsolete in fruit ; petals pilose about the middle within ; staminodes absent .. .. 5. **Ptychopetalum**

Stamens the same number as the petals ; fruit wholly or partially enclosed by the accrescent receptacle ; flowers in axillary clusters or very short racemes ; calyx not accrescent in fruit :

Flowers 5-merous ; anthers not fused to the petals :

Fruit completely enclosed by the receptacle .. .. 6. **Strombosia**

Fruit only partially enclosed by the receptacle whose margin makes a ridge around the fruit just above its middle .. .. 7. **Diogoa**

Flowers 4-merous ; anthers fused to the petals for $\frac{1}{8}$ to $\frac{1}{2}$ their length

8. **Strombosiopsis**

Stamens united into a column ; calyx strongly accrescent in fruit :

Calyx not splitting, much expanded in fruit ; flowers in very short axillary racemes, dioecious ; a shrub or small tree .. .. 9. **Aptandra**

Calyx in fruit at length splitting into 2–3 segments, not expanded, at first enveloping the fruit ; flowers in paniculate clusters, hermaphrodite ; a large tree 10. **Ongokea**

1. COULA Baill.—F.T.A. 1 : 351 ; Sleumer in E. & P. Pflanzenfam. 16B : 12 (1935).

Leaves elliptic to elongate-oblong-elliptic, long-caudate-acuminate, shortly cuneate at base, up to 30 cm. long and 9 cm. broad, papery, often rust-coloured, laxly gland-dotted beneath, with up to 14 pairs of lateral nerves and numerous faint forked parallel tertiary nerves ; petiole up to 2·5 cm. long, rusty-puberulous ; flowers in short rusty-puberulous panicles ; calyx rim-like ; petals small, thick, valvate, becoming glabrous ; fruits ellipsoid-globose, 4 cm. long, with a woody shell about 6 mm. thick, smooth

edulis

C. edulis *Baill.*—F.T.A. 1 : 351 ; Sleumer l.c., fig. 4 ; Aubrév. Fl. For. C. Iv. 1 : 86, t. 26 ; Louis & Léonard in Fl. Congo Belge 1 : 251, t. 23. A forest tree, to 100 ft. high, rusty-pubescent on the young parts, with coriaceous leaves and walnut-like fruits.
S.L.: Hangha to Bumbu, Kambui F.R. (fr. Mar.) *Lane-Poole* 351! Gola Forest (fl. & fr. Apr., May) *Small* 597! 647! Lib.: Dukwia R. (Mar.) *Cooper* 262! 281! 368! Tappita (fr. Aug.) *Baldwin* 9119! Butaw, Sinoe (Mar.) *Baldwin* 11496! Iv.C.: Bettlé, R. Comoé *Chev.* 16260! Widespread in forest *fide* Aubrév. l.c. G.C.: Tanosu (fl. & fr. Jan.) *Brent* 2dd! Chipp 361! Apollonia *Greene*! Samreboi, Prestea (Jan.) *Andoh* FH 5619! S.Nig.: Okomu F.R. (fr. May) *A. F. Ross* 203! Sapoba *Kennedy* 1130! 1563! Degema (Oct.) *King-Church* 44! Oban *Talbot*! Br.Cam.: S. Bakundu F.R., Kumba (Apr.) *Olorunfemi* FHI 30539! *Eiiofor* FHI 29360! Also in French Cameroons, Spanish Guinea, Gabon, Cabinda and Belgian Congo. (See Appendix, p. 293.)

2. HEISTERIA Jacq.—F.T.A. 1 : 345 ; Sleumer in E. & P. Pflanzenfam. 16B : 16 (1935). *Phanerocalyx* S. Moore (1921).

Branchlets narrowly winged, glabrous ; leaves more or less oblong, obtusely caudate-acuminate, 8–12 cm. long, 3–6 cm. broad, glabrous, with about 6 pairs of lateral nerves and very laxly reticulate tertiary nerves ; flowers at first very small, 2–3, axillary, very shortly pedicellate, usually only one developing into fruit, with the calyx much enlarged and divided nearly to the base into 5 more or less ovate spreading segments ; fruits broadly ellipsoid to globose, 1·3–2 cm. long, glabrous

1. *parvifolia*

Branchlets, etc., as above, but calyx very broad in fruit and very shortly lobed, broader than long .. .. 2. *zimmereri*

1. **H. parvifolia** *Sm.*—F.T.A. 1 : 346 ; Chev. Bot. 123 ; Sleumer in E. & P. Pflanzenfam. 16B : 16, fig. 6 ; Louis & Léonard in Fl. Congo Belge 1 : 254. *Phanerocalyx talbotiorum* S. Moore (1921). *H. elegans* A. Chev. Bot. 123, name only. A glabrous shrub or small tree up to 40 ft. high, with small greenish-white or yellowish flowers ; calyx enlarged, persistent, deep red in fruit ; fruit itself greyish-white ; in forest.
Fr.Sud.: Banancoro (Feb.) *Chev.* 492! Port.G.: Mato de Amedi (fr. Apr.) *Esp. Santo* 1963! Ponte de Dada, Fulacunda (Oct.) *Esp. Santo* 2206! Fr.G.: *Heudelot* 672! Kouria (Nov.) *Caille* in Hb. *Chev.* 14808! S.L.: *Don*! Kenema (fr. Feb., May) *Lane-Poole* 166! *Small* 76! Gbinti, Dibia (fr. Oct.) *Deighton* 5734! Bagbe, Gola Forest (fr. Apr.) *Small* 580! Lib.: Grand Bassa (fl. July, fr. Oct.) *Dinklage* 1670! T. *Vogel* 86! Monrovia (fl. & fr. July–Nov.) *Linder* 1534! *Whyte*! Jabroke, Webo (fr. July) *Baldwin* 6478! 6621! Yangi (Sept.) *Linder* 845! Iv.C.: various localities *fide* Chev. l.c. G.C.: Ashanti *Cummins* 105! Axim (fr. Dec.) *Johnson* 889! Kohura *Chipp*! Benso, Tarkwa (fl. & fr. Nov.) *Andoh* FH 5409! S.Nig.: Lagos *Moloney*! Benin City (fr. July) *Unwin* 53! 106! Itu (fr. Apr.) *Holland* 39! Oban *Talbot* 450! Br.Cam.: Rio del Rey *Johnston*! Victoria (fr. Mar.) *Maitland* 444! Barombi (fr. Mar.) *Preuss* 7! F.Po: Ureka (fr. Feb.) *Guinea* 2402! Also in French Cameroons, Spanish Guinea, Gabon, Belgian Congo, Angola, Principe and S. Tomé. (See Appendix, p. 294.)

2. **H. zimmereri** *Engl.* in Notizbl. Bot. Gart. Berl. 2 : 288 (1899) ; Pflanzenw. Afr. 3, 1 : 85, fig. 50. A small spreading tree.
(?) Br.Cam.: Barombi *Preuss* 274! Also in French Cameroons, Gabon, Cabinda and Belgian Congo. [*Preuss* 274 has as locality " Barombi am ufer des Kribi Fluss " ; the Kribi R. is well within the French Cameroons, a long way from Barombi (i.e. Kumba), so the record of this species from our area is doubtful.]

Excluded species.

H. winkleri *Engl.* Bot. Jahrb. 43 : 169 (1909), quoted in the first edition as a synonym of *H. parvifolia* Sm., is Diospyros physocalycina Gürke (*Ebenaceae*) ; see Mildbr. in Notizbl. Bot. Gart. Berl. 9 : 1052 (1926). The type specimen of *H. winkleri* Engl. is *Winkler* 758 from Victoria, British Cameroons.

23

3. XIMENIA Linn.—F.T.A. 1 : 346 ; Sleumer in E. & P. Pflanzenfam. 16B : 22 (1935).

Branchlets often armed with rigid axillary spines, flexuous ; leaves narrowly elliptic, emarginate, 3–7 cm. long, up to 3 cm. broad, glabrous, obscurely nerved ; flowers in axillary umbellate racemes ; pedicels about 5 mm. long ; sepals 4, ovate-triangular ; petals 4, oblong-linear, bearded inside ; fruit superior, ellipsoid, about 3 cm. long, glabrous .. .. *americana*

X. americana *Linn.*—F.T.A. 1 : 346 ; Sleumer l.c. 23, fig. 11 ; Louis & Léonard in Fl. Congo Belge 1 : 256 ; Aubrév. Fl. For. Soud.-Guin. 354, t. 73, 4–6. A glabrous shrub or small tree, with white fragrant flowers and a yellow plum-like fruit ; common in the savannah regions but also near the coast.
Sen.: *Heudelot*! Thiès *De Wailly* 4645. Casamance *Chev.* 3169. **Fr.Sud.**: Bamako *Waterlot* 1077. Sicoro (Jan.) *Chev.* 214! Sangarala (Feb.) *Chev.* 342! **Port.G.**: Bissau (Mar.) *Esp. Santo* 1899! **Fr.G.**: Boké *Chillou*! Kouroussa *Pobéguin* 182. Los Isl. *Pobéguin* 1225. **S.L.**: *Hutton*! Moyamba (fr. Mar.) *Deighton* 1918! Falaba (Mar.) *Sc. Elliot* 5117! 5237! Bendu (Mar.) *Adames* 13! **Lib.**: Grand Bassa *Baldwin* 11268! **Iv.C.**: Bobo Dioulasso *Aubrév.* 1840 ; 1853. Ferkéssédougou *Aubrév.* 1527 ; 2287. **G.C.**: Wuchian to Black Volta R. (Mar.) *Kitson* 632! Aburi road (Feb.) *Dalz.* 8262! Krobo Plains (Jan.) *Johnson* 539! Achimota (Jan.) *Irvine* 1970! **Togo:** Lome *Warnecke* 234! **Dah.**: Natitingou to Bocorona *Chev.* 24209! Savalou *Aubrév.* 68d. **N.Nig.**: Zurmi (Apr.) *Keay* FHI 15615! Zaria (Apr.) *Lamb* 92! Bauchi Plateau (Apr.) *Lely* P. 216! **S.Nig.**: Lagos *Foster* 54! Old Oyo (Feb.) *Keay* FHI 16010! **Br.Cam.**: Ambas Bay (Feb.) *Mann* 764! Widespread in the tropics. (See Appendix, p. 295.)

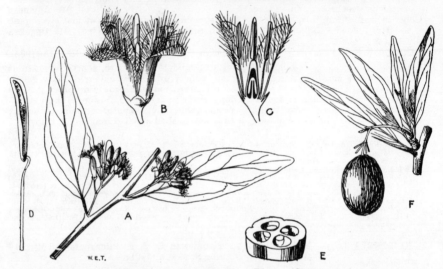

FIG. 181.—XIMENIA AMERICANA *Linn.* (OLACACEAE).

A, inflorescence. B, flower. C, same in vertical section. D, stamen. E, cross-section of ovary. F, fruit.

4. OLAX Linn.—F.T.A. 1 : 348 ; Sleumer in E. & P. Pflanzenfam. 16B : 24 (1935).

Calyx disciform in fruit, not strongly accrescent and not enveloping the whole fruit :
Fertile stamens 3 (–4), staminodes 5–6 ; flowers 2·5–3 mm. long, in axillary racemes 1–2·5 cm. long ; fruits globose, 6–8 mm. diam. ; leaves ovate-elliptic to oblong-elliptic or lanceolate, obtuse or rounded at base, very gradually long-acuminate, 7·5–16 cm. long, 3–6 (–8) cm. broad .. .. 1. *gambecola*
Fertile stamens 5–6, staminodes 3 ; flowers about 5 mm. long, on numerous short lateral often leafy branchlets ; fruits depressed-globose, 11–15 mm. diam. ; leaves elongate-elliptic to elliptic-lanceolate, obtuse at base, gradually acuminate, mostly less than 7·5 cm. long and 2·5 cm. broad, but sometimes up to 10 cm. long and 3·8 cm. broad .. .. 2. *subscorpioidea*
Calyx strongly accrescent and completely surrounding the fruit :
Fertile stamens 3 (–4), staminodes 4–5 ; flowers 6–7 mm. long, in axillary racemes 1–2 cm. long ; fruits (including the calyx) depressed-globose, about 2 cm. diam. and 1·5 cm. long ; leaves lanceolate to ovate-lanceolate or elliptic, 6–15 cm. long, 1·5–6·5 (–8) cm. broad, thinly coriaceous, with about 5–6 main lateral nerves on each side of midrib, looped, strongly impressed above, margin sometimes lobulate towards the apex.. .. .. 3. *mannii*
Fertile stamens (5–) 6, staminodes 3 (–4) ; flowers about 4 mm. long, in axillary racemes 1–2·5 (–3·5) cm. long ; fruits (including the calyx) subglobose, about 2 cm. diam., leaves elliptic to broadly oblong, rounded or cuneate at base, abruptly long-acuminate, 7–22 cm. long, 3–10 cm. broad, papery, with about 8 main lateral nerves on each side of midrib, looped, impressed above .. .. .. 4. *latifolia*

1. **O. gambecola** *Baill.* in Adansonia 3 : 121 (1862–63) ; F.T.A. 1 : 349. *O. viridis* Oliv. F.T.A. 1 : 349 (1868) ; Chev. Bot. 125 ; F.W.T.A., ed. 1, 1 : 460 ; Sleumer in E. & P. Pflanzenfam. 16B : 27 ; Louis & Léonard in Fl. Congo Belge 1 : 258, t. 24 & fig. 1. *Alsodeiopsis glaberrima* Engl. ex Hutch. & Dalz., F.W.T.A., ed. 1, 1 : 455 (1928). *O. pyramidata* A. Chev. Bot. 124, name only. A glabrous low shrub, up to 4 ft. high, rarely more ; flowers white ; ripe fruits red ; in forest, especially by streams.
Fr.G.: Fouta Djalon (fr. Jan.) *Heudelot* 715 ! **S.L.:** Surinuia, Talla (Apr.) *Sc. Elliot* 5530 ! Njala (young fr. Aug.) *Deighton* 3244 ! Roboli, Rokupr (July) *Jordan* 289 ! Nganyama, Tikonko (fr. Nov.) *Deighton* 5266 ! **Lib.:** Ganta (July) *Harley* ! Tawata, Boporo (fr. Nov.) *Baldwin* 10316 ! Congotown (fl. & fr. Aug.) *Barker* 1407 ! **Iv.C.:** Kéeta (July) *Chev.* 19342 ! Atteou (Jan.) *Chev.* 17089 ! Guidéko (May) *Chev.* 16348 ! **G.C.:** Pompo Headwaters F.R. (fl. & fr. May) *Jackson* FH 2255 ! Amentia (Mar.) *Vigne* FH 1894 ! S. Scarp F.R. *Moor* FH 2118 ! **Dah.:** Boukoutou forests, Saketé (young fr. Jan.) *Chev.* 22859 ! **S.Nig.:** Ikoyi Plains, Lagos (Mar.) *Dalz.* 1365 ! Epe *Barter* 3303 ! Sapoba (Jan.) *Richards* 3898 ! Stubbs Creek F.R., Eket (fl. & fr. May) *Onochie* FHI 32908 ! **Br.Cam.:** Barombi *Preuss* 352 ! Also in French Cameroons, Belgian Congo, Uganda, Angola and N. Rhodesia. (See Appendix, p. 294.)
2. **O. subscorpioidea** *Oliv.* F.T.A. 1 : 350 (1868) ; Chev. Bot. 125 ; Sleumer l.c. fig. 13, E–G ; Aubrév. Fl. For. C. Iv. 1 : 88, t. 27, 1–6 ; Louis & Léonard l.c. 263. *O. gambecola* of Chev. Bot. 124, not of Baill. A shrub or small tree, to 30 ft. high ; cut bark smelling of garlic ; flowers greenish-white ; ripe fruits yellow ; mostly in forest, but extending far into the savannah regions.
Sen.: Bignona (fr. Feb.) *Chev.* 3171 ! **Fr.G.:** Kouloundala to Kaba *Chev.* 20383. **S.L.:** Erimakuna (young fr. Mar.) *Sc. Elliot* 5221 ! **Iv.C.:** Anoumaba (Nov.) *Chev.* 22405 ! Bondoukou *Aubrév.* 718 ! **G.C.:** Larte Hills (Nov.) *Johnson* 810 ! Anum Hills, Krepi (Jan.) *Johnson* 546 ! Bana Hill, Krobo (fr. Jan., Feb.) *Irvine* 223 ! 1925 ! Volta River F.R. (Nov.) *Morton* GC 6070 ! **Togo:** Atakpame *Mildbr.* 7415 ! Assatu, Hohoe (Nov.) *St. C. Thompson* 3702 ! **Dah.:** Adjouéré, Zagnanado (fr. Feb.) *Chev.* 22951 ! **N.Nig.:** Gurara R. (Dec.) *Elliott* 214 ! Katagum *Dalz.* 381 ! Vom (young fr. Jan.) *Dent Young* 27 ! **S.Nig.:** Lagos (Aug.) *Dalz.* 148 ! 1037 ! 1407 ! Idanre (Jan.) *Brenan & Keay* 8668 ! Onitsha *Barter* 1785 ! Unwin 61 ! Oban *Talbot* 1353 ! **Br.Cam.:** Mamfe (fr. Jan.) *Maitland* 1163 ! Also in French Cameroons, Ubangi-Shari and Belgian Congo. (See Appendix, p. 294.)
3. **O. mannii** *Oliv.* F.T.A. 1 : 349 (1868) ; Sleumer l.c. fig. 14, A–C. *O. zenkeri* Engl. (1899)—F.W.T.A., ed. 1, 1 : 460. *O. macrocalyx* Engl. (1899)—Sleumer l.c., fig. 14, D. *O. longiflora* Engl. (1899). *O. major* Stapf (1905). *O. insculpta* Hutch. (1917)—F.W.T.A., ed. 1, 1 : 460. A forest shrub, to 7 ft. high. rarely more ; flowers greenish-white ; ripe fruits orange ; as interpreted here, this species shows a wide range of variation in the size and shape of the leaves
S.L.: Njala (fl. Jan., fr. Mar.) *Dalz.* 8038 ! *Deighton* 4620 ! 4720 ! Blama (Dec.) *Deighton* 317 ! Jombohun, Tunkia (Jan.) *Deighton* 3881 ! **Lib.:** Monrovia *Whyte* ! Dukwia R. (Feb.) *Cooper* 240 ! 257 ! Logantown, New Cess R. (fr. May) *Baldwin* 11176 ! Ba, Little Kola R. (fr. Mar.) *Baldwin* 11191 ! **S.Nig.:** Oban *Talbot* 1334 ! 1354 ! Itu Dist. (Jan.–May) *Jones* FHI 4836 ! *Onochie* FHI 33226 ! **Br.Cam.:** R. Cameroon (fl. & young fr. Jan.) *Mann* 2201 ! R. Joke, Malende to Muyuka (fr. Mar.) *Brenan* 9324 ! S. Bakundu F.R. (fl. Jan.–Mar., fr. Apr.–Aug.) *Brenan* 9473 ! *Keay* FHI 28567 ! *Olorunfemi* FHI 30728 ! Johann-Albrechtshöhe *Staudt* 508 ! 904 ! Also in French Cameroons and Gabon.
4. **O. latifolia** *Engl.* in Notizbl. Bot. Gart. Berl. 2 : 284 (1899) ; Sleumer l.c. ; Louis & Léonard l.c. 262. A forest shrub, to 6 ft. high ; flowers pink ; ripe fruit orange.
Br.Cam.: Johann-Albrechtshöhe *Staudt* 465 ! Also in French Cameroons, Gabon and Belgian Congo.

5. PTYCHOPETALUM Benth.—F.T.A. 1 : 347 ; Sleumer in E. & P. Pflanzenfam. 16B : 24 (1935).

Flower-buds very acute, about 7 mm. long ; fruits ellipsoid, 1·3 cm. long, minutely verruculose ; leaves sessile, oblong-elliptic, acuminate, 6–10 cm. long, 2·5–3·5 cm. broad, with distinct looped lateral nerves ; branchlets winged ; bracts large and apiculate, enclosing the flower-buds .. .. .. 1. *anceps*
Flower-buds rounded at the apex, 4–5 mm. long ; fruits oblong-ellipsoid, 2 cm. long ; leaves petiolate, more or less as above but larger ; branchlets mostly terete ; bracts small and early deciduous ; fruits broadly ellipsoid, 2 cm. long .. 2. *petiolatum*

1. **P. anceps** *Oliv.* F.T.A. 1 : 347 (1868), incl. vars. ; Hook. Ic. Pl. t. 2329 ; Chev. Bot. 124. *Olax anceps* (Oliv.) Roberty (1953). A glabrous shrub, 2–10 ft. high, with acutely angled branchlets, short scanty racemes of white flowers and cherry-red drupes ; in forest.
S.L.: Bagroo R. (fl. & fr. Apr.) *Mann* 838 ! Kambui F.R. (Mar.) *Lane-Poole* 205 ! Gola Forest (fr. Mar.) *Small* 550 ! **Lib.:** Grand Bassa (Sept.) *Dinklage* 2037 ! *T. Vogel* 12 ! Dukwia R. (fl. Apr., fr. Feb.) *Cooper* 72 ! 150 ! 209 ! Peahtah *Bequaert* in Hb. *Linder* 980 ! Yeh R. (Oct.) *Linder* 998 ! Gbanga (Oct.) *Linder* 1240 ! Karmadhun (Nov.) *Baldwin* 10206a ! **Iv.C.:** various localities *fide* Chev. l.c. **G.C.:** Tano Anwia F.R., Enchi *Andoh* FH 5609 ! Prestea (fl. & fr. Sept.) *Vigne* FH 3080 ! Bobiri F.R. (fr. Feb.) *Foggie* FH 4940 !
2. **P. petiolatum** *Oliv.* F.T.A. 1 : 347 (1868) ; Hook. Ic. Pl. t. 2330. A glabrous forest shrub or small tree, 12–35 ft. high.
S.Nig.: Oban *Talbot* 1747 ! **Br.Cam.:** Banga, S. Bakundu F.R. (Mar.) *Brenan* 9443 ! Also in French Cameroons and Spanish Guinea.

6. STROMBOSIA Blume—F.T.A. 1 : 350 ; Sleumer in E. & P. Pflanzenfam. 16B : 21 (1935).

Veins between the main lateral nerves close (0·5–1 mm. apart), more or less parallel and conspicuous beneath ; pedicels 1–3 mm. long, without bracteoles ; petals 2–3 mm. long ; disk 5-lobed, forming a cap around the top of the ovary ; branchlets terete ; petioles 1–3 cm. long ; leaves ovate-elliptic to oblong, cuneate to obtuse at base, acute or shortly acuminate at apex, 7–30 cm. long, 3–16 cm. broad, with 4–8 main lateral nerves on each side of midrib ; fruits ovoid or ellipsoid about 1·7 cm. long and 1·2 cm. diam. .. .. 1. *grandifolia*
Veins between the main lateral nerves more than 1 mm. apart, distinct or indistinct ; petals 3–5 mm. long :
Pedicels without bracteoles ; leaves without pustules :
Branchlets subterete and slightly ridged ; petioles up to 1·3 cm. long ; petals about 2·5 mm. long ; pedicels about 1·5 mm. long, elongating in fruit ; leaves narrowly elliptic, obtuse at base, gradually rather long-acuminate, 5–15 cm. long, 1·5–5·5 cm. broad, with 5–7 main lateral nerves on each side of midrib, venation obscure ; fruits obovoid-ellipsoid, 2–2·5 cm. long, 1·2–1·7 cm. diam. .. 2. *zenkeri*

Branchlets strongly angled ; petioles 1–3 cm. long ; petals 3–5 mm. long ; pedicels 3–5 mm. long ; disk forming an elongated cone about 1 mm. long at the top of the ovary ; leaves ovate, elliptic or oblong, obtuse or rounded at base, rounded or acute at apex, 6–20 cm. long, 3–13 cm. broad, with 5–8 main lateral nerves on each side of midrib, venation distinct ; fruits obconical, 2–2·5 cm. long, 1·5–2·5 cm. diam.

3. *scheffleri*

Pedicels with small bracteoles ; disk 5-lobed, forming a cap around the top of the ovary ; branchlets slightly ridged or subterete ; leaves often with minute pustules on the surfaces :

Petals about 3 mm. long ; branchlets slightly ridged :

Leaves more or less shining when dry, not usually with pustules, venation on lower surface distinct ; main lateral nerves 4–6 (–7) on each side of midrib ; leaves elliptic, ovate-elliptic or oblong-elliptic ; cuneate to obtuse or rounded at base, shortly acuminate, 5–15 cm. long, 2·5–7 (–8) cm. broad ; petioles 0·7–1·2 (–2) cm. long   ..   ..   4a. *glaucescens* var. *lucida*

Leaves matt when dry, usually with pustules, venation on lower surface indistinct ; main lateral nerves (5–) 7–10 on each side of midrib ; leaves oblong, ovate or elliptic, obtuse or rounded at base, obtuse or shortly and obtusely acuminate at apex, 6–19 cm. long, 3–8 cm. broad ; petiole 1–2·5 cm. long

4. *glaucescens* var. *glaucescens*

Petals 4–5 mm. long ; branchlets subterete ; leaves oblong, ovate-oblong or elliptic-oblong, obtuse or rounded at base, slightly acuminate, 4·5–10·5 (–13) cm. long, 2·1–4·5 (–5·3) cm. broad, with 4–6 main lateral nerves on each side of midrib, lower surface matt when dry, with indistinct venation, usually with abundant minute pustules ; petioles 1–1·5 cm. long   ..   ..   5. *pustulata*

1. **S. grandifolia** *Hook. f. ex Benth.*—F.T.A. 1 : 350 ; F.W.T.A., ed. 1, 1 : 460, in part only ; Sleumer in E. & P. Pflanzenfam. 16B : 22, excl. fig. 10 ; Louis & Léonard in Fl. Congo Belge 1 : 268. *Lavalleopsis grandifolia* (Hook. f. ex Benth.) Van Tiegh. ex Engl. (1897). *L. densivenia* Engl. (1897), name only. An understorey tree, to 40 ft. (to 80 ft. *fide* Louis & Léonard *l.c.*) high, with short bole and wide spreading crown of dense shining foliage ; slash revealing red sapwood ; flowers cream ; in forest.
    Dah.: Boukoutou forests, Sakété (Jan.) *Chev.* 22871 ! **S.Nig.:** Akure F.R. (young fr. Apr.) *Symington* FHI 3398 ! Okomu F.R. (Jan.) *Brenan & Richards* 8843 ! Eket *Talbot* 3155 ! Oban *Talbot* ! Afi River F.R. (Dec.) *Keay* FHI 28251 ! **Br.Cam.:** Victoria (fl. Feb., fr. Jan.) *Maitland* 180 ! 270 ! *Preuss* 1329 ! Johann-Albrechtshöhe *Staudt* 548 ! 750 ! **F.Po:** (Nov.–Jan.) *Mann* 172 ! *T. Vogel* 156 ! Also in French Cameroons, Spanish Guinea, Gabon, Cabinda and Belgian Congo. (See Appendix, p. 295.)
2. **S. zenkeri** *Engl.* Bot. Jahrb. 43 : 167 (1909). A forest tree, to 40 ft. high ; flowers pale green.
    S.Nig.: Calabar (fl. & fr. Mar.) *Brenan* 9227 ! *Smith* 45b ! Stubbs Creek F.R. *Onochie* FHI 32905 ! Also in French Cameroons.
3. **S. scheffleri** *Engl.* in Notizbl. Bot. Gart. Berl. 5, App. 21 : 4, fig. A–C (1909) ; Louis & Léonard l.c. 1 : 270. *S. grandifolia* of F.W.T.A., ed. 1, 1 : 460, partly ; of Sleumer l.c. fig. 10 ; not of Hook. f. ex Benth. A forest tree, to 100 ft. high ; flowers greenish-yellow or -white.
    S.Nig.: Cross River North F.R., Ikom (young fr. June) *Latilo* FHI 31826 ! **Br.Cam.:** Buea (fl. Dec., young fr. Jan.) *Maitland* 567 ! *Migeod* 255 ! Bamenda (fr. Mar.) *Johnstone* 96/31 ! **F.Po:** Moka, 4,000–4,500 ft. (young fr. Feb.) *Exell* 846 ! Also in Uganda, Kenya, Tanganyika, Nyasaland, Belgian Congo and Angola.
4. **S. glaucescens** *Engl.* var. **glaucescens**—Bot. Jahrb. 43 : 167 (1909) ; Louis & Léonard l.c. 267. A forest tree, up to 150 ft. high, with straight bole.
    S.Nig.: frequent in neighbourhood of gully, Compt. 55, Okomu F.R., Benin *Onochie* in *Hb. Brenan* 9190 ! [The only specimen from our area is sterile and so its identification is open to some doubt.] Also in French Cameroons, Cabinda and Belgian Congo.
4a. **S. glaucescens** var. **lucida** *J. Léonard* in Bull. Jard. Bot. Brux. 18 : 148 (1947) ; Fl. Congo Belge 1 : 268. *S. pustulata* Oliv. in Hook. Ic. Pl. t. 2299 (1894), partly (*Sc. Elliot* 4733, on which are based the description and drawing of the fruits) ; Chev. Bot. 125 ; F.W.T.A., ed. 1, 1 : 460, partly ; Aubrév. Fl. For. C. Iv. 1 : 84, t. 25. A forest tree, up to 80 ft. high, with straight bole ; flowers white ; ripe fruit blackish purple.
    S.L.: Kambia (fr. Jan.) *Sc. Elliot* 4733 ! Njala (Sept.) *Deighton* 5173 ! Momenga, Kori (Oct.) *Deighton* 6133 ! Kai, Fogbo (Sept.) *Tindall* 37 ! **Lib.:** Dukwia R. (Mar.) *Cooper* 178 ! 269 ! Bangee, Boporo (young fr. Nov.) *Baldwin* 10355 ! Mecca (fr. Dec.) *Baldwin* 10814 ! **Iv.C.:** Bouroukrou *Chev.* 16103 ! Voguié *Chev.* 16178 ! Akabilékrou, Morénou (fr. Dec.) *Chev.* 22510 ! Adzope *Aubrév.* 158 ! **G.C.:** Fiakwae (Aug.) *Hughes* FH3373 ! Jimira F.R. (Aug.) *Vigne* FH 3996 ! Awoso, Sefwi Bekwai (fr. Oct.) *Akpabla* 883 ! Also in Belgian Congo. (See Appendix, p. 295.) [Possibly specifically distinct from var. *glaucescens*, but further study is needed to determine how far it is distinct from some other species in the French Cameroons and Gabon.]
5. **S. pustulata** *Oliv.* in Hook. Ic. Pl. t. 2299 (1894), excl. *Sc. Elliot* 4733 and description and drawing of fruits which are *S. glaucescens* var. *lucida* ; F.W.T.A., ed. 1, 1 : 460, in part only (Nigerian spec.) ; Kennedy For. Fl. S. Nig. 145. A forest tree, to 70 ft. high, with straight bole ; flowers whitish.
    S.Nig.: near Lagos *Rowland* ! Akure F.R. (young fr. Nov.) *Keay* FHI 25533 ! Nikrowa, Okomu F.R. (fr. Dec.) *Brenan* 8441 ! 8441a ! Sapoba (Oct.) *Kennedy* 408 ! 1623 ! Ohosu F.R. *Hide* 10/38 ! Also (?) in French Cameroons. (See Appendix, p. 295.)

7. DIOGOA Exell & Mendonça in Bol. Soc. Brot., sér. 2, 25 : 109 (1951).

A glabrous tree, to 30 m. high ; petioles 1·5–3·5 cm. long ; leaves membranous, more or less oblong-lanceolate, cuneate or obtuse at base, gradually long-acuminate, 15–40 cm. long, 4–12 cm. broad ; flowers 5-merous, in short congested axillary racemes ; petals 5–6 mm. long, 2–3 mm. broad ; anthers free from the petals ; fruits subglobose, partially embedded in the accrescent receptacle whose margin makes a ridge around the fruit just above its middle, about 4–5 cm. diam. ..   .. *zenkeri*

D. zenkeri (*Engl.*) *Exell & Mendonça* l.c. t. 3 (1951) ; Consp. Fl. Ang. 1 : 336, t. 17. *Strombosiopsis zenkeri* Engl. (1909)—Louis & Léonard in Fl. Congo Belge 1 : 272, t. 26. *Strombosia retivenia* S. Moore (1920). *S. grandifolia* of F.W.T.A., ed. 1, 1 : 460, partly. A forest tree ; flowers cream.

S.Nig.: Okomu F.R., Benin (Feb.) *Brenan* 8996! 8996a! *Richards* 3294! Oban *Talbot* 1403! 1465!
Br.Cam.: Ebonji, Kumba (fr. May) *Olorunfemi* FHI 30594! R. Metem Wum (fr. June) *Ujor* FHI 29261!
Also in French Cameroons, Gabon, Cabinda and Belgian Congo.

8. STROMBOSIOPSIS Engl. in E. & P. Pflanzenfam. Nachtr. 1 : 148 (1897) ; Exell & Mendonça in Bol. Soc. Brot., sér. 2, 25 : 109 (1951).

A glabrous tree, to 30 m. high ; petioles 1–2 cm. long ; leaves coriaceous, ovate-elliptic or oblong to oblong-lanceolate or lanceolate, acute or rounded at base, shortly acuminate, 5–25 cm. long, 2·5–10 cm. broad ; flowers 4-merous, in short congested axillary racemes, 0·5–1 cm. long ; petals 4–5 mm. long, about 1·5 mm. broad ; anthers partially fused to the petals ; fruits completely embedded in the accrescent receptacle, ellipsoid, 2·5–3·5 cm. long, 2–3 cm. diam.   ..   ..   *tetrandra*

S. tetrandra *Engl.* l.c. (1897) ; Sleumer in E. & P. Pflanzenfam. 16B : 19, fig. 8 ; Louis & Léonard in Fl. Congo Belge 1 : 271 ; Exell & Mendonça in Consp. Fl. Ang. 1 : 337. A forest tree, exuding red sap when slashed ; flowers cream.
S.Nig.: Oban *Talbot*! Calabar (Mar.) *Brenan* 9226! *Smith* 45a! Aking, Calabar (young fr. Apr.) *Ejiofor* FHI 21888! Aboabam, Ikom Div. (fr. May) *Jones & Onochie* FHI 18713! **Br.Cam.**: Likomba *Mildbr.* 10541! Also in French Cameroons, Spanish Guinea, Gabon, Cabinda and Belgian Congo.

9. APTANDRA Miers in Ann. & Mag. Nat. Hist., ser. 2, 7 : 201 (1851) ; Sleumer in E. & P. Pflanzenfam. 16B : 28 (1935).

Branchlets terete, glabrous ; leaves oblong-elliptic, acutely acuminate, shortly cuneate at base, 6–10 cm. long, 3–5 cm. broad, glabrous and with very obscure nerves beneath ; petiole up to 1 cm. long ; flowers in very short axillary racemes ; pedicels up to 8 mm. long, crowded ; calyx very small, undulate ; petals about 3 mm. long, becoming reflexed in the upper part ; fruits ellipsoid-ovoid, 2·5 cm. long, black, shining, surrounded by the spreading much-enlarged coloured entire calyx up to 8 cm. in diam.   ..   ..   *zenkeri*

A. zenkeri *Engl.* in E. & P. Pflanzenfam. Nachtr. 1 : 147 (1897) ; Sleumer l.c. 29, fig. 16 ; Louis & Léonard in Fl. Congo Belge 1 : 275. A dioecious shrub, or small tree up to 40 ft. high, with green branchlets and small greenish-white flowers ; enlarged calyx pink, waxy in texture ; in forest.
G.C.: Amentia (Apr.) *Vigne* FH 2896! Bobiri F.R. (July) *Andoh* FH 5544! Kwahu Prasu (Feb.) *Vigne* FH 1604! Pra River Station (fr. Feb.) *Vigne* FH 1034! **S.Nig.**: Lagos (fr. Feb.) *Millen* 25! *Punch*! Ajilite *Millen* 163! Agege (fr. Nov.) *Baldwin* 13707! Ibadan–Ijebu Ode road (fr. Mar.) *Meikle* 1251! Also in French Cameroons, Spanish Guinea, Belgian Congo and Angola.

10. ONGOKEA Pierre in Bull. Soc. Linn. Paris 2 : 1313 (1897) ; Sleumer in E. & P. Pflanzenfam. 16B : 30 (1935).

Branchlets compressed or slightly winged ; leaves broadly oblong-elliptic, obtusely acuminate, 4–7 cm. long, 2·5–3·5 cm. broad, with obscure nerves, minutely papillose beneath ; flowers in paniculate clusters, small, numerous ; pedicels filiform, about 3 mm. long ; calyx saucer-shaped, undulate ; petals at first connivent, about 3 mm. long, with oblong claw and ovate limb ; fruit globose, about 2·5 cm. diam., closely invested by the persistent enlarged calyx, which ruptures into 3 segments   *gore*

O. gore (*Hua*) *Pierre* l.c. 1314 (1897) ; Sleumer l.c. fig. 17, D–H ; Aubrév. Fl. For. C. Iv. 1 : 84, t. 24 ; Louis & Léonard in Fl. Congo Belge 1 : 276, t. 27. *Aptandra gore* Hua (1895). *Ongokea klaineana* Pierre (1897) —Chev. Bot. 122 ; F.W.T.A., ed. 1, 1 : 461 ; Kennedy For. Fl. S. Nig. 146. *O. kamerunensis* Engl. (1909)—Sleumer l.c. fig. 17, A–C. A glabrous forest tree, up to 125 ft. high, with straight cylindrical bole without buttresses ; flowers white ; fruit yellow.
S.L.: Njala (May) *Deighton* 3953! 3971! **Iv.C.**: Bouroukrou (Dec., Jan.) *Chev.* 16128! Yapo (fr. Oct.) *Chev.* 22330! Abidjan *Aubrév.* 107! **G.C.**: Atieku (May) *Vigne* FH 1946! Kumasi (May) *Andoh* FH 4386! **S.Nig.**: Sapoba (fr. Apr., May) *Kennedy* 1132! 1650! **Br.Cam.**: Likomba (Nov.) *Mildbr.* 10703! Also in French Cameroons, Spanish Guinea, Gabon, Belgian Congo and Angola. (See Appendix, p. 294.)

105. PENTADIPLANDRACEAE

Arborescent shrubs or climbers. Leaves exstipulate, alternate, simple, penni-nerved. Flowers dioecious, racemose ; racemes short, axillary ; bracts small, subulate. Sepals 5, free, lanceolate, imbricate. Petals 5, scale-like, thicker and loosely connivent at the base, with lanceolate free limbs. Stamens 9–13, inserted on a short stout androgynophore ; filaments free, filiform ; anthers basifixed, 2-celled. Rudimentary ovary present, short, 3–5-celled, with 2 series of ovules in each cell. Staminodes about 10 from the top of the gynophore in the female flower. Ovary superior, at the top of the thick gynophore, 4–5-celled, with numerous axile 2–3-seriate ovules ; style 4–5-fid at the apex. Fruit a rounded berry ; seeds small, numerous, immersed in pulp.

Fig. 182.—Pentadiplandra Brazzeana *Baill.* (Pentadiplandraceae).

A, sepal. B, petal showing thickened basal portion and free limb. C, stamen. D, male flower. E, same in vertical section showing rudimentary ovary. F, corolla showing connivent basal portions of petals. G, cross-section of ovary. H, fruits.

PENTADIPLANDRA Baill. in Bull. Soc. Linn. Paris 1 : 611 (1886). *Cercopetalum* Gilg (1897).

Arborescent shrub or climber ; branchlets closely sulcate, glabrous ; leaves oblong-obovate, obliquely cuspidate-acuminate, shortly cuneate at base, 8–12 cm. long, 3–4 cm. broad, glabrous and closely reticulate on both surfaces, venation elevated on both sides ; lateral nerves 6–8 pairs, looped ; petiole 5–6 mm. long ; flowers about 1·5 cm. long, in short axillary and terminal racemes ; pedicels 0·5–1 cm. long ; sepals lanceolate, 5 mm. long, very minutely puberulous ; female flower with about 10 filiform staminodes ; ovary 4–5-celled, glabrous ; fruit ovoid-globose, about 3·5 cm. long .. .. *brazzeana*

P. brazzeana *Baill.* l.c. (1886) ; Pax & K. Hoffm. in E. & P. Pflanzenfam. 17B : 206, fig. 114 (1936) ; Hauman & Wilczek in Fl. Congo Belge 2 : 480. *Cercopetalum dasyanthum* Gilg in Engl. Bot. Jahrb. 24 : 308, t. 3 (1897) ; op. cit. 53 : 265. Erect or half scandent ; flowers conspicuous, petals white, spotted red or bluish towards the tips.

Recorded from Dschang, Ndonge and Semukina in the French Cameroons, just outside the area of our Flora ; very likely to be found in the British Cameroons. Extends through French Cameroons to Belgian Congo.

106. OPILIACEAE

Trees, shrubs or woody climbers. Leaves alternate, simple ; stipules absent. Flowers mostly hermaphrodite. Calyx minute. Petals conspicuous in bud, 4–5, free or more or less united, valvate. Stamens as many as the petals and opposite to them, free or united to the base of the petals ; anthers 2-celled, opening lengthwise by slits. Disk-glands alternating with the stamens. Ovary superior or semi-inferior, 1-celled ; stigma sessile or style slender ; ovule solitary, pendulous or erect. Fruit drupaceous, often fleshy. Seeds with copious endosperm and rather small embryo.

A small family formerly included in the *Olacaceae* ; confined to the tropics ; rare in America ; mostly with very long slender racemes.

Racemes much shorter than the leaves, clothed in bud with broad imbricate scales, the latter caducous ; disk 5-partite .. .. **1. Opilia**
Racemes much longer than the leaves, not clothed with scales when young ; disk entire
2. Urobotrya

1. OPILIA Roxb.—F.T.A. 1 : 352 ; Sleumer in E. & P. Pflanzenfam. 16B : 38 (1935), excl. subgen. *Urobotrya*.

Branchlets slightly flexuous, glabrous ; leaves alternate, oblong to obovate-oblanceolate, acutely acuminate, cuneate at base, 6–12 cm. long, 3–5 cm. broad, glabrous, with about 5 pairs of prominent lateral nerves ; racemes at first covered with broad imbricate bracts and shortly catkin-like, at length slender and about 3–4 cm. long ; pedicels fasciculate, 2 mm. long, puberulous ; fruits ellipsoid, 2–2·5 cm. long, finely puberulous *celtidifolia*

O. celtidifolia (*Guill. & Perr.*) *Endl. ex Walp.* Rep. 1 : 377 (1842) ; Louis & Léonard 1 : 282. *Groutia celtidifolia* Guill. & Perr. Fl. Sen. 1 : 101, t. 22 (1831). *Opilia amentacea* of F.T.A. 1 : 352 ; of Chev. Bot. 126 ; of Aubrév. Fl. For. Soud.-Guin. 356, t. 73, 8–9. A woody climber, sometimes suberect with straight branches ; flowers small, greenish-white or yellowish, fragrant ; especially common in fringing forest in the savannah regions.
Sen.: *Heudelot* 341 ! *Perrottet* 127 ! Bignona to Sindialone *Chev.* 3545. **Gam.:** Kudang (Jan.) *Dalz.* 8125 ! **Fr.Sud.:** Badinko (Jan.) *Chev.* 116. Balani (Jan.) *Chev.* 145. Niana to Damesa (Feb.) *Chev.* 329. **Port.G.:** Canjadudi, Gabu (Jan.) *Esp. Santo* 2352 ! Jangada, Contubo (fr. June) *Esp. Santo* 2260 ! **Fr.G.:** Timbo *Chev.* 13592. **S.L.:** Moria, Scarcies (Feb.) *Sc. Elliot* 4795 ! Wallia, Scarcies (young fr. Jan.) *Sc. Elliot* 4791 ! **Iv.C.:** Bobo Dioulasso *Serv. For.* 1970. Pa *Serv. For.* 2207. Black Volta R. *Serv. For.* 2401 ! **G.C.:** Prang (Jan.) *Vigne* FH 1560 ! Wa (Dec.) *Adams* GC 4467 ! Kulmasa to Jagalo (Feb.) *Kitson* 630 ! **Togo:** Lome *Mildbr.* 7507 ! *Warnecke* 307 ! Kpedsu (Dec.) *Howes* 1065 ! **Dah.:** Ady, Savé *Chev.* 23570. Djougou circle *Chev.* 23830. **Fr.Nig.:** Diapaga to Fada *Chev.* 24439. **N.Nig.:** Wawa, Borgu *Barter* 720 ! Nupe *Barter* 990 ! Kontagora (Jan.) *Dalz.* 71 ! Lokoja (Mar.) *Elliott* 34 ! **S.Nig.:** Ibodebere road, Lagos (fr. May) *Denton* 15 ! Ibadan North F.R. (Feb.) *Chizea* FHI 23973 ! Widespread in the drier parts of tropical Africa. (See Appendix, p. 296.)

2. UROBOTRYA Stapf in J. Linn. Soc. 37 : 89 (1904). *Opilia* Roxb. subgen. *Urobotrya* (Stapf) Engl.—Sleumer in E. & P. Pflanzenfam. 16B : 38 (1935).

Basal lateral nerves strongly developed, ascending up to the middle of the leaf and rather conspicuously looped with the rather widely spreading upper main lateral nerves ; pedicels at anthesis 7–12 mm. long, in fascicles along the up to 30 cm. long puberulous or glabrescent axis of the racemes ; fruits ellipsoid-subglobose, 1·7–2 cm. long, about 1·4 cm. diam. ; leaves varying from narrowly elongate-oblong-elliptic to oblong-elliptic or ovate, usually cuneate to obtuse at base but sometimes rounded, gradually acuminate at apex, 8–22 cm. long, 1·5–8 cm. broad .. **1.** *afzelii*

Basal lateral nerves not ascending to the middle of the leaf and not conspicuously looped with the upper main lateral nerves ; pedicels at anthesis 5–8 mm. long, in fascicles along the up to 30 cm. long axis of the racemes ; fruits ellipsoid about 2·5 cm. long and 1·4 cm. diam. ; leaves oblong-elliptic to oblong-ovate, rounded or obtuse at base, rather abruptly and narrowly acuminate, 9–22 cm. long, 3·5–10 cm. broad

2. *minutiflora*

1. **U. afzelii** (*Engl.*) Stapf ex Hutch. & Dalz. F.W.T.A., ed. 1, 1 : 463 (1928). *Opilia afzelii* Engl. in E. & P. Pflanzenfam. Nachtr. 1 : 143 (1897) ; Notizbl. Bot. Gart. Berl. 2 : 282 (1899). *Urobotrya angustifolia* Stapf (1904)—Johnston Lib. 2 : 587, t. 248 ; F.W.T.A., ed. 1, 1 : 463. *Opilia angustifolia* (Stapf) Engl. (1909). *Urobotrya latifolia* Stapf (1904)—F.W.T.A., ed. 1, 1 : 463. *Opilia latifolia* (Stapf) Engl. (1909). *Urobotrya trinervia* Stapf (1906)—F.W.T.A., ed. 1, 1 : 463. *Opilia trinervia* (Stapf) Engl. (1909). *Urobotrya stapfiana* Hutch. & Dalz., F.W.T.A., ed. 1, 1 : 463 (1928). *Opilia stapfiana* (Hutch. & Dalz.) Sleumer in E. & P. Pflanzenfam. 16B : 38 (1935). A forest shrub, to 15 ft. high, with pendulous racemes of greenish flowers and yellow fruits.
S.L.: *Afzelius. Smeathmann*! Bafi R. (Feb.) *Macdonald* 6! Buri Town (Dec.) *Unwin & Smythe* 3! Mt. Aureole (Feb.) *Dalz.* 1006! *Lane-Poole* 154! Joru (Jan.) *Deighton* 3861! **Lib.:** *Reynolds*! Kakatown *Whyte*! Monrovia *Whyte*! Dukwia R. (fr. Feb.) *Cooper* 249! Javajai (fr. Nov.) *Baldwin* 10274a! Zuie, Boporo (Dec.) *Baldwin* 10682! (See Appendix, p. 296.)
2. **U. minutiflora** Stapf in J. Linn. Soc. 37 : 90 (1904). *Opilia minutiflora* (Stapf) Engl. (1909)—Sleumer l.c. An erect forest shrub, to 8 ft. high, with pendulous racemes ; flowers green except for the conspicuous long white filaments.
S.Nig.: Eket *Talbot*! Oban *Talbot*! Afi River F.R., Ikom (Dec.) *Keay* FHI 28196! 28220! 28255! **Br.Cam.:** Victoria (Jan.) *Kalbreyer* 6! Buea to Mayuko (fl. & fr. Feb.) *Dalz.* 8244! Bimbia (fr. May) *Maitland* 707! [*Opilia macrocarpa* Pierre & Engl. and other species from the French Cameroons, Gabon and Spanish Guinea are probably not distinct.]

107. MEDUSANDRACEAE

Trees or shrubs. Leaves alternate, simple, entire or crenate ; stipules present, sometimes small and early caducous. Flowers hermaphrodite, racemose or spicate, actinomorphic, hypogynous or perigynous. Sepals 5, open or imbricate in bud, free, or united only at the base. Petals 5, free, imbricate in bud. Stamens numerous, or 5 fertile opposite the petals alternating with 5 staminodes, free from each other ; anthers 4-celled, opening laterally. Disk present or absent. Ovary superior, syncarpous, 1-celled, with a central column ; ovules 6–8, pendulous from the apex of the column, anatropous ; styles 3 (–4), free. Fruit with persistent calyx, capsular, dehiscing into 3 (–4) valves, or by the cohesion of 2 valves appearing 2-valved. Seed solitary in each fruit, with copious endosperm and small straight embryo.

A small family confined to the tropical rain forests of western Africa. *Soyauxia* has been hitherto placed in the *Flacourtiaceae* or *Passifloraceae* but has remarkable affinities with the newly discovered *Medusandra*.

Stamens numerous, filaments much longer than the petals ; staminodes absent ; sepals imbricate in bud ; ovary surrounded by a collar-like disk at the base ; styles long ; leaves entire   . .   . .   **1. Soyauxia**
Stamens 5, opposite the petals, filaments much shorter than the petals ; staminodes 5, much longer than the petals, alternating with the fertile stamens ; sepals open in bud ; disk absent ; styles very short ; leaves crenate or crenate-serrate
2. Medusandra

1. SOYAUXIA Oliv. in Hook. Ic. Pl. t. 1393 (1882) ; Brenan in Kew Bull. 1953 : 507–511.

Inflorescence very dense-flowered, continuous, 6–12 cm. long ; branchlets minutely puberulous, glabrescent ; leaves elongate-oblong, rounded or subacute at base, gradually acutely acuminate, 12–32 cm. long, 4–11 cm. broad, glabrous with 10–20 main lateral nerves on each side of midrib ; sepals rusty tomentellous, spreading in fruit ; fruit an inflated 3-angled capsule with valves suborbicular to broadly ovate, 2–3·5 cm. long and 2–2·5 cm. broad ; seed trigonous, up to 2 cm. diam.   1. *grandifolia*
Inflorescence with scattered flowers, not continuous ; sepals cupular in fruit ; fruits, where known, oblong, not exceeding 8 mm. broad :
Sepals glabrous outside, except at the base ; mature leaves glabrous beneath ; branchlets closely pubescent, leaves oblong to elliptic, rounded at base, gradually acuminate, 8–27 cm. long, 3–8 cm. broad, with 12–16 main lateral nerves on each side of midrib ; fruits oblong, 12–18 mm. long ; seeds black, rod-like, more or less trigonous in section, about 15 mm. long and 3–4 mm. diam.   . .   . .   2. *floribunda*
Sepals densely pilose or tomentose outside (except towards the margin in No. 3) ; mature leaves pubescent or pilose on the midrib and nerves beneath :
Sepals outside glabrous towards the margins, about 2·5 mm. long ; branchlets shortly pubescent ; leaves obovate-oblong or elliptic obtuse at base, long-acumi-

nate, 8–20 cm. long, 3–7 cm. broad with 10–15 main lateral nerves on each side of
midrib ..   ..   3. *gabonensis*
Sepals outside densely pilose to the margins, about 3 mm. long ; leaves as above :
Branchlets with long spreading hairs ..   ..   4. *velutina*
Branchlets shortly pubescent at first, soon glabrous..   ..   ..   5. *talbotii*

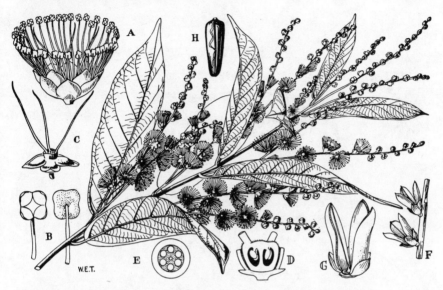

FIG. 183.—SOYAUXIA FLORIBUNDA *Hutch.* (MEDUSANDRACEAE).

A, flower. B, stamens, front and back, showing peltate 4-celled anther. C, flower with petals
and stamens removed. D, vertical section of ovary. E, transverse section of ovary. F, two
dehisced fruits. G, dehisced fruit, showing persistent central column attached at base of
ovary and at apex of left-hand valve. H, seed.
D–G are drawn by Miss E. M. Stones.

1. **S. grandifolia** *Gilg & Stapf* in J. Linn. Soc. 37 : 102 (1905) ; Aubrév. Fl. For. C. Iv. 3 : 30, t. 255 B. Shrub
or tree, to 45 ft. high with velvety spikes of white flowers with cream stamens and purple styles ; fruits
red.
Lib.: Grand Bassa *Dinklage* 2051 ! Sinoe Basin *Whyte* ! Dukwia R. (fl. Feb., fr. Oct.) *Cooper* 106 ! 233 !
259 ! 286 ! Nyaake, Webo (fr. June) *Baldwin* 6222 ! Ba, Little Kola R. (Mar.) *Baldwin* 11189 ! **Iv.C.:**
Tabou (fr. Dec.) *Aubrév.* 1675 ! **G.C.:** Benso (fr. Aug.) *Andoh* FH 5390 ! Prestea (fl. & fr. Sept.) *Vigne*
FH 3105 ! (See Appendix, p. 51.)
2. **S. floribunda** *Hutch.* in Kew Bull. 1915 : 44 ; Aubrév. l.c. 32, t. 255 A ; Brenan in Kew Bull. 1953 : 509,
fig. 1. *S. gabonensis* of Chev. Bot. 288, not of Oliv. Shrub or tree, to 30 ft. high ; petals white ; stamens
bluish.
Gam.: *Garrett* ! **S.L.:** Bonjema (June) *Aylmer* 86 ! Njala (May–July) *Deighton* 1161 ! 1754 ! 5915 !
Kambui F.R. (Apr.) *King* 184 ! Kenema (fr. Jan.) *Thomas* 7601 ! 7735 ! Ronietta (fr. Nov.) *Thomas*
5385 ! 5645 ! **Lib.:** Mt. Barclay (June) *Bunting* ! Du R. (fr. Aug.) *Linder* 289. Ganta (May) *Harley* 1195 !
Gletown, Tchien (July) *Baldwin* 6915 ! Bobei Mt., Sanokwele (Sept.) *Baldwin* 9602 ! **Iv.C.:** Nimba
Mts., *Aubrév.* 1163 ! Guidéko (May) *Chev.* 16452. (See Appendix, p. 51.)
3. **S. gabonensis** *Oliv.* in Hook. Ic. Pl. t. 1393 (1882). *S. bipindensis* Gilg ex Hutch. & Dalz. in F.W.T.A.,
ed. 1, 1 : 170 (1927) ; Bak. f. in Cat. Talb. 128 (1913) and J. Bot. 52 : 4 (1914), name only. *S. laxiflora*
Gilg ex Hutch. in Kew Bull. 1915 : 44, name only. Small tree or shrub, to 10 ft. high ; flowers white.
S.Nig.: Ogabi, Obudu Div. (June) *Unwin* 5 ! Oban *Talbot* 1614 ! Nkpure, Ikom Div. (Apr.) *Jones*
FHI 6487 ! Also in French Cameroons and Gabon.
4. **S. velutina** *Hutch. & Dalz.* in F.W.T.A., ed. 1, 1 : 170 (1927) ; Aubrév. l.c. 32.
S. laxiflora of Chev. Bot. 288. Shrub, sometimes scandent, or small tree to 35 ft. high ; often by water.
Iv.C.: Grabo (May) *Aubrév.* 1289 ! Bingerville *Chev.* 15444. Guidéko (May) *Chev.* 16427. **S.Nig.:** Atigere
(Feb.) *Millen* 147 ! Eba Isl. (Jan.) *Thornewill* 249 ! Omo F.R. (Jan.) *Jones & Onochie* FHI 17358 !
20668 ! Okomu F.R. (Jan.) *A. F. Ross* 247 !
5. **S. talbotii** *Bak. f.* in J. Bot. 52 : 4 (1914). An erect shrub, to 10 ft. high ; calyx green with yellow indumentum, corolla and filaments white, anthers grey, ovary yellow-green.
S.Nig.: Ikotobo Road, Eket *Talbot* 3254 ! Uquo, Eket (fl. & fr. May) *Onochie* FHI 33151 !

2. MEDUSANDRA Brenan in Kew Bull. 1952 : 228.

Tree, to 18 m. high ; branchlets minutely puberulous ; petioles 3–9 cm. long with a
swollen pulvinus at each end ; leaves coriaceous or subcoriaceous, elliptic or oblong-
elliptic, broadly cuneate or slightly rounded at base, shortly and acutely acuminate,
(6–) 10–30 cm. long, (2·5–) 5–14 cm. broad, with about 8 main lateral nerves on each
side of midrib, prominent beneath, minutely puberulous beneath ; inflorescences
numerous, spiciform, 3–15 cm. long, solitary or paired in the leaf-axils ; sepals about
1 mm. long ; petals about 2 mm. long ; staminodes about 8 mm. long, densely

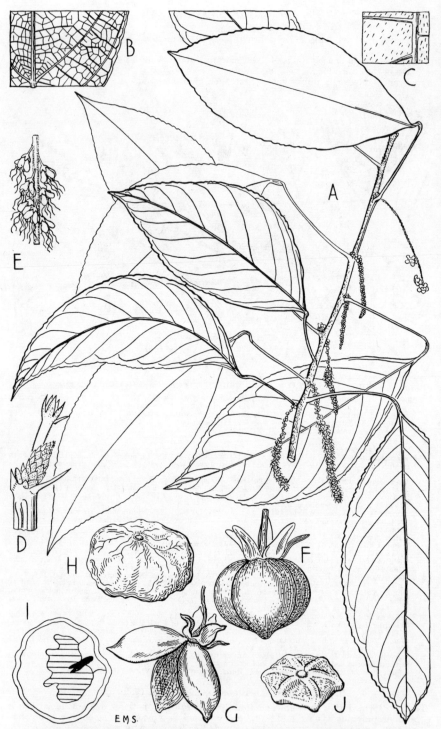

Fig. 184.—Medusandra richardsiana *Brenan* (Medusandraceae).

A, flowering shoot, × ¼. B, part of lower side of leaf, × 2. C, part of lower side of leaf,
showing hairs, × 9. D, stipule and bud, × 2. E, part of inflorescence, × 2. F, fruit
before dehiscence, × 1½. G, fruit dehisced, × 1½. H, seed preserved in spirit, × 2. I,
transverse section of H, diagrammatic; embryo black, endosperm white, cavity in endosperm with cross-lines, × 2. J, seed, dried, × 2.
  A–C drawn from *Brenan* 9402. D–E drawn from FHI 28682. F–J drawn from FHI 29305.

654

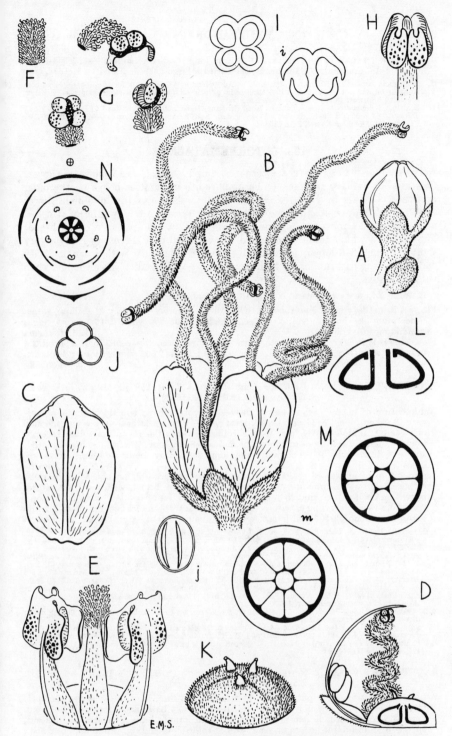

FIG. 185.—MEDUSANDRA RICHARDSIANA *Brenan* (MEDUSANDRACEAE).

A, flower bud and bract, × 10. B, flower, × 20. C, petal, inner face, × 20. D, longitudinal section of part of flower bud, × 20. E, part of androecium, showing two fertile dehisced anthers and base of staminode, × 40. F, part of staminode, showing clavate hairs, × 40. G, three views of anther rudiment at top of staminode, × 40. H, dehisced fertile anther, × 40. I & i, transverse sections of anther before and after dehiscence, × 40. J & j, polar and equatorial views of acetolysed pollen grains, 9μ and $11\mu \times 9\mu$ respectively. K, ovary, × 40. L, longitudinal section of ovary, diagrammatic, × 40. M & m, transverse sections of six- and eight-ovuled ovaries, diagrammatic; ovules and central column white, cavity black, × 40. N, floral diagram.
  A, drawn from FHI 28682. Remainder drawn from *Brenan* 9402.

pubescent ; sepals in fruit reflexed, 6–8 mm. long ; capsules subglobose, 1·3–1·5 cm. long, 1·4–1·7 cm. diam., glabrescent, each with a large flattened seed   *richardsiana*

M. richardsiana *Brenan* l.c. 229, figs. 1 & 2 (1952). An understorey tree in rain forest ; inflorescences conspicuous on account of the long white staminodes ; fruits at first white with green persistent calyx, becoming brown when dry ; locally abundant.
Br.Cam.: Kumba Div.: Buea to Kumba Road (fr. June) *Dundas* FHI 8360 ! Road to Bombe (fr. Nov.) *Dundas* FHI 8485 ! S. Bakundu F.R., Banga (fl. Jan.–Apr., fr. Apr.–July) *Brenan* 9402 ! 9486 ! *Ejiofor* FHI 29305 ! *Keay & Russell* FHI 28682 !

108. OCTOKNEMATACEAE

Trees and shrubs ; leaves alternate, simple, entire, exstipulate. Flowers unisexual in axillary racemes, or hermaphrodite in panicles borne on the older branches. Sepals 5, valvate. Petals absent. Stamens (or in female flower staminodes) 5, opposite the sepals, free. Disk present. Ovary inferior, 1-celled or 3 (–4)-celled almost to the top ; ovules 3 (–4), at the apex of a basal thread-like placenta which reaches and is adnate to the top of the ovary ; style very short, stigmas 3–5-lobed, lobes bifid. Fruit drupaceous, with a single seed ; endosperm slightly ruminate, with small embryo, the radical much longer than the cotyledons.

A family of only two genera, confined to tropical Africa.

Flowers in axillary spike-like racemes, dioecious ; indumentum only of stellate, sometimes scale-like hairs ; fruits 1·5–3 cm. long ; endocarp with 6–10 narrow ridges protruding into the seed .. .. **1. Octoknema**
Flowers sessile in lax panicles borne on the older branches, apparently hermaphrodite ; indumentum of simple, forked and stellate hairs ; fruits 9 cm. or more long ; endocarp without internally projecting ridges .. .. **2. Okoubaka**

1. OCTOKNEMA Pierre in Bull. Soc. Linn. Paris 2 : 1290 (1897) ; Mildbr. in E. & P. Pflanzenfam. 16B : 42 (1935).

Indumentum of bushy stellate hairs ; leaves obovate-oblong, shortly cuneate, acuminate, 8–30 cm. long, 3–11 cm. broad, with 6–8 (–10) main lateral nerves on each side of midrib ; petioles 0·8–3 cm. long, swollen towards the top ; spike-like racemes axillary, densely stellate-tomentose, the males 5–8 cm. long, the females few-flowered, shorter ; fruits subglobose about 1·5 cm. long, capped by the broad persistent calyx lobes .. .. **1. borealis**
Indumentum of appressed scale-like stellate hairs ; leaves oblong, obtuse or rounded at base, narrowly acuminate, 11–38 cm. long, 4–13 cm. broad, with 6–8 main lateral nerves on each side of midrib ; petioles up to 6 cm. long, swollen towards the top ; spike-like racemes axillary, with scaly indumentum, the males 3–6 cm. long ; fruits ellipsoid or ovoid, 1·5–3 cm. long, capped by the persistent calyx-lobes **2. winkleri**

1. **O. borealis** *Hutch. & Dalz.* F.W.T.A., ed. 1, 1 : 464 (1928) ; Kew Bull. 1929 : 24 ; Mildbr. in E. & P. Pflanzenfam. 16B : 45 ; Aubrév. Fl. For. C. Iv. 1 : 92, t. 28. *O. affinis* of Chev. Bot. 125. A forest tree, 15–75 ft. or more high, with pale flaking bark ; dioecious ; flowers yellowish-green ; ripe fruits red.
Fr.G.: Kissidougou (Feb.) *Chev.* 20707 ! Kouria *Chev.* 18183. **S.L.**: N. of Kundita (Mar.) *Sc. Elliot* 5042 ! Njala (Jan., Apr.) *Deighton* 1566 ! 4612 ! Kwaoma (Apr.) *Deighton* 3682 ! Gola Forest (fl. & fr. Apr.) *Small* 588 ! **Lib.**: Dukwia R. (fl. & fr. Feb.) *Cooper* 91 ! 138 ! 267 ! 302 ! Mt. Barclay (fr. Apr.) *Dinklage* 3040 ! Fortsville (fl. Mar.) *Baldwin* 12539 ! Bama, Loffa R. (Dec.) *Baldwin* 10677 ! **Iv.C.**: Alépé, Attie (Mar.) *Chev.* 16246 ! Abidjan *Aubrév.* 194 ! Banco *Aubrév.* 351 ! Oua, Haut-Cavally (fr. Apr.) *Chev.* 21321 ! **G.C.**: Ankasa F.R. *Vigne* FH 3177 ! Axim (Feb.) *Irvine* 2400 ! Konongo (Feb.) *Akpabla* 252 ! (See Appendix, p. 297.)
2. **O. winkleri** *Engl.* Bot. Jahrb. 43 : 177 (1909). A forest tree, up to 60 ft. high, with pale flaking bark ; dioecious ; fruits orange-red.
S.Nig.: Osho, Omo F.R., Ijebu-Ode (Apr.) *Jones & Onochie* FHI 17294 ! Etemi, Omo F.R. (Mar.) *Jones & Onochie* FHI 16611 ! Lower Enyong F.R., Itu (fr. May) *Onochie* FHI 33214 ! **Br.Cam.**: N'Bamba to Man-o'-War Bay (fr. Apr.) *Winkler* 1238. Victoria (fl. Apr., fr. Feb.) *Maitland* 401 ! 608 ! S. Bakundu F.R. *Onochie* FHI 30869 !

2. OKOUBAKA Pellegr. & Normand in Bull. Soc. Bot. Fr. 93 : 139 (1946) ; op. cit. 91 : 20 (1944), French descr. only.

A large tree ; branchlets pubescent ; petiole about 7 mm. long ; leaves ovate-elliptic to ovate-oblong, unequally rounded or subcordate at base, acute at apex, 8–15 cm. long, 3·5–8 cm. broad, densely pubescent with simple hairs beneath, glabrescent, with 3–5 main lateral nerves on each side of midrib, ascending, prominent beneath ; inflorescences borne on the older branches, paniculate, up to 40 cm. long ; flowers sessile apparently hermaphrodite ; sepals deltoid, 1 mm. long ; fruits drupaceous, ellipsoid, about 9 cm. long and 5 cm. broad, capped by the persistent sepals

                                  aubrevillei

O. aubrevillei *Pellegr. & Normand* in Bull. Soc. Bot. Fr. 93 : 139 (1946) ; op. cit. 91 : 25 (1944), French descr. only ; Louis & Léonard in Fl. Congo Belge 1 : 292, t. 31, photo 11 (the var. *glabrescentifolia* J.

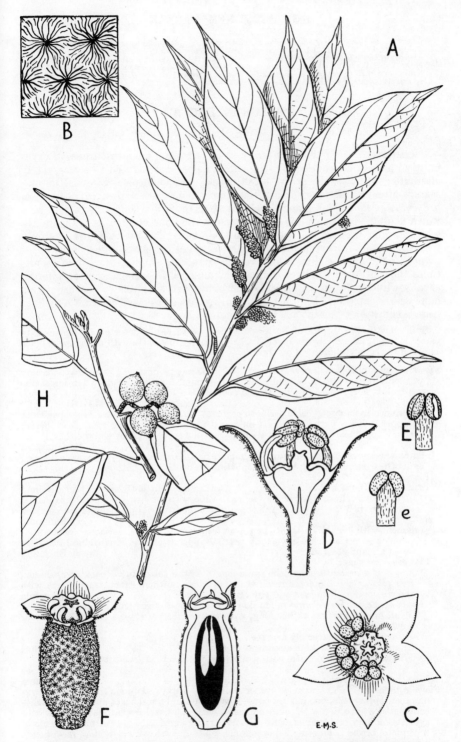

FIG. 186.—Octoknema borealis *Hutch. & Dalz.* (Octoknemataceae).

A, shoot with male flowers, × ⅓. B, stellate hairs, × 50. C & D, male flower, × 12. E & e, stamens, × 16. F & G, female flower, × 6. H, shoot with fruits, × ⅔.
A & C are drawn from *Deighton* 3682 ; B, F & G from *Sc. Elliot* s.n. ; H from *Cooper* 267.

657

Léonard). *Octoknema okoubaka* Aubrév. & Pellegr. in op. cit. 84 : 392, fig. 1 (1937). " *Okoubaka* " of Aubrév. Fl. For. C. Iv. 1 : 88. A forest tree, to 90 ft. high, with a cylindrical bole lacking buttresses ; flowers greenish ; fruits yellow when ripe. This tree has the reputation of killing the surrounding trees. **Iv.C.:** Yapo, Agboville and Yakassé *Aubrév.* 603 ; 1503 ; 1788. **G.C.:** Mampong Scarp *Vigne* FH 1760 ! Krakaw, on N.E. corner of Kade Bepo F.R., Kwahu, Mpraeso (Feb.) *St. C. Thompson* FH 4368 ! The species also occurs in Belgian Congo as var. *glabrescentifolia* J. Léonard.

109. LORANTHACEAE

By S. Balle [1]

Parasitic shrubs on trees, or rarely erect terrestrial trees or shrubs. Leaves mostly opposite or whorled, simple, entire, sometimes reduced to scales, rarely alternate ; stipules absent. Flowers mostly actinomorphic, bisexual or unisexual, often brightly coloured. Receptacle sometimes with a calyx-like rim (*calyculus*). Perianth-segments valvate, free or united high up into a tube which is often split down one side. Stamens the same number as the perianth-segments and inserted on them or at their base ; anthers 2-celled or rarely 1-celled, sometimes divided into numerous small cells, opening lengthwise or by pores or transverse slits. Disk present or absent. Rudimentary ovary sometimes (not in Africa) present in male, staminodes in female. Ovary inferior, usually without a distinct placenta and ovule ; style simple or absent, stigma more or less capitate. Fruit a berry or drupe. Seed solitary, without a distinct testa ; endosperm present or absent ; embryo large.

Numerous in the tropics of both hemispheres, and easily detected by the usually parasitic habit, inferior ovary, often brightly coloured perianth, with as many stamens as perianth-segments, inserted on them.

Flowers hermaphrodite ; perianth petaloid ; filaments distinct ; anthers dehiscing by 2 longitudinal slits ; calyculus distinct, formed by a rather deep border of the receptacle ; floral bract fused with the pedicel for all its length
(Subfamily LORANTHOIDEAE) [2] :
Inflorescence racemose ; each perianth-segment composed of a lower erect part and an upper reflexed part ; flowers 4-merous ; filaments remaining erect at anthesis ; anthers long-linear ; style much shorter than the tepals ; plant glabrous
1. Helixanthera
Inflorescence never racemose, but in umbels, heads or fascicles, or flowers solitary ; perianth-segments united, the tube splitting on one side at anthesis :
Flowers (3–) 4-merous :
Filament without a tooth in front of the anther but a little thickened and papillose at the apex ; lower part of the perianth-tube with very small appendages inside, alternate with the tepals, at the top of the 4 longitudinal papillose areas
2. Berhautia
Filament with a small tooth at the top in front of the anther, neither thickened nor papillose ; perianth-tube without internal appendages and longitudinal papillose areas ; plant glabrous   .. .. **3. Englerina**
Flowers 5-merous :
Lobes of the perianth about the same length as the tube, curled back at anthesis ; free part of the filaments inrolled at anthesis ; anthers short, trapeziform with the connective thickened and emarginate at apex.. .. **4. Globimetula**
Lobes of the perianth always much shorter than the tube, remaining erect or reflexing at anthesis ; anthers rather long, with the connective always slender and never emarginate at apex :
Anthers never transversely septate :
Filament always without an apical tooth .. .. **5. Agelanthus**
Filament always elongate, with a tooth on the inside in front of the anther
6. Tapinanthus
Anthers transversely septate .. .. **7. Phragmanthera**
Flowers unisexual, plant monoecious or dioecious, always glabrous ; perianth calyxlike ; anthers sessile, opening by pores ; calyculus and floral bract absent ; branches articulated at the nodes ; leaves opposite, with sub-basal nervation ; inflorescence cymose, axillary or terminal, with 3–5 flowers arranged in a line within a cupule of 2 connate bracts ; flowers (3–) 4-merous, male flowers with a hollow receptacle ; perianth-lobes ovate or triangular, usually rather unequally in pairs   **8. Viscum**

---

[1] Mlle. Balle's contribution has been translated and slightly rearranged ; a few additional citations have also been inserted.
[2] See Appendix, p. 297, under *Loranthus* spp.

1. HELIXANTHERA Lour. Fl. Cochinch. 142 (1790) ; emend. Danser in Bull. Jard. Bot. Buitenz., ser 3, 10 : 292 (1929) ; Verh. Akad. Wetensch. Amst. Afd. Natuurk., sect. 2, 29 : 55 (1933) ; Balle in Webbia 11 : 579 (1955). *Sycophila* Welw. ex Van Tiegh. (1894). *Loranthus* Sect. *Sycophila* (Welw. ex Van Tiegh.) Engl. (1897)— F.T.A. 6, 1 : 257 ; Engl. & K. Krause in E. & P. Pflanzenfam. 16B : 147 (1935).

Racemes 2–15 cm. long, about 50-flowered ; bracts unilateral, about 1 mm. long ; flowers 6–23 mm. long, lower part of perianth segments 2–4 mm. long, remaining erect and more or less lyriform, with inside near the apex a small appendage carrying a filament, upper part of perianth-segment linear, reflexing at anthesis ; anthers 2–12 mm. long, transversely septate forming 9–40 superposed cells ; style quadrangular, broadened at base ; fruits ellipsoid, about 6 mm. long and 3 mm. broad
mannii

H. **mannii** (*Oliv.*) Danser in Verh. Akad. Wetensch. Amst. Afd. Natuurk., sect. 2, 29 : 58 (1933). *Loranthus mannii* Oliv. (1864)—F.T.A. 6 : 274 ; S. Balle in Fl. Congo Belge 1 : 316, t. 33. *L. rosaceus* Engl. (1808) —F.T.A. 6, 1 : 275. *L. sublilacinus* Sprague—F.T.A. 6, 1 : 275 (1910) ; F.W.T.A., ed. 1, 1 : 466. *L. ternatus* Engl. (1897)—F.T.A. 6, 1 : 275. *L. patentiflorus* Engl. & K. Krause (1910)—F.T.A. 6, 1 : 1025. *L. rubrostamineus* Engl. & K. Krause (1910)—F.T.A. 6, 1 : 1025. A very variable species, particularly in size of flowers and leaves, the position of the leaves, the shape of the base of the perianth-segments and of the bracts ; flowers pale pink. On *Coffea, Citrus, Ficus, Hymenocardia ulmoides* and *Thecacoris manniana*. **S.Nig.**: Lower Enyong F.R., Calabar (May) *Onochie* FHI 33216 ! **Br.Cam.**: Johann-Albrechtshöhe *Staudt* 757 ! Also in French Cameroons, Gabon, Belgian Congo, Angola, Uganda and S. Tomé.

2. BERHAUTIA Balle in Bull. Soc. Roy. Bot. Belg. 88 : 133 (1956).

Plant completely covered with greyish branched hairs ; leaves alternate or subopposite, suborbicular or broadly ovate, rarely elliptic, rounded or cuneate at base and always slightly decurrent, rounded or obtuse at apex, 2–4 cm. long, 1·2–4·5 cm. broad, nerves irregularly pinnate ; petiole 3–10 mm. long ; flowers 2–4 in axillary umbels, shortly pedicellate and pedunculate ; perianth about 4 cm. long, not inflated at base, lobes acute, about 1 cm. long, elliptic in the upper 6 mm., remaining erect at anthesis ; filaments inflexed at anthesis ; anthers lanceolate, about 3 mm. long, slightly exceeded by the pointed connective ; style filiform ; fruit obovoid, up to 1 cm. long and 7 mm. broad .. .. *senegalensis*

B. **senegalensis** *Balle* l.c. (1956). *Loranthus berhautii* Balle in Berhaut Fl. Sen. 96, fig. 12, 1 (1954), French descr. only. Flowers mauve. On *Combretum glutinosum*.
Sen.: Tambacounda to Niokolo-Koba *Berhaut* 1507 ! 2998 ; 4074 ; 4154 ; 4374 ! 4741.

3. ENGLERINA Van Tiegh. in Bull. Soc. Bot. Fr. 42 : 257 (1895); emend. Balle in Webbia 11 : 581 (1955). *Loranthus* Sect. *Ischnanthus* Engl. (1894)—F.T.A. 6, 1 : 272 ; Engl. & K. Krause in E. & P. Pflanzenfam. 16B : 164 (1935).

Apical swelling of the bud obtuse or rounded ; perianth-lobes not thickened at the end, erect at anthesis ; perianth-tube usually rather shorter than the lobes ; filaments involute at anthesis ; flowers less than 3 cm. long ; peduncles longer than the pedicels ; umbels generally about 12-flowered :
Flowers golden-yellow or orange, 2·2–2·7 cm. long ; bud just before anthesis with apical swelling about 7 mm. long ; anthers (3–) 6 mm. long ; leaves elliptic, ovate or lanceolate, 3·5–9 cm. long, 1·4–4·6 cm. broad ; bracts sometimes shortly spurred on the back .. .. 1. *lecardii*
Flowers red, 18–22 mm. long ; apical swelling of bud about 2·5 mm. long ; anthers 1–2·3 mm. long ; leaves lanceolate, oblong, elliptic or ovate, 6·3–11·3 cm. long, 1·6–3·8 cm. broad .. .. 2. *parviflora*
Apical swelling of the bud truncate ; perianth-lobes thick at the end, remaining erect at anthesis ; perianth-tube usually rather longer than, or rarely the same length as, the lobes ; filaments with an acute tooth 0·75–1 mm. long ; anthers 1·5–3 mm. long ; peduncles usually rather shorter or rarely slightly longer than the pedicels ; umbels few-flowered, 5–6 (–15) ; flowers (3·2–) 4–5·6 cm. long ; leaves lanceolate, elliptic or ovate, acuminate, 4·5–16 cm. long, 1·8–6 cm. broad .. .. 3. *gabonensis*

1. E. **lecardii** (*Engl.*) Balle in Kew Bull. 1956 : 168. *Loranthus lecardii* Engl. (1894)—F.T.A. 6, 1 : 384 ; F.W.T.A., ed. 1, 1 : 467 ; Berhaut Fl. Sen. 96. On *Butyrospermum* and *Combretum glutinosum*.
Sen.: *Lecard* 180 ! *Bellamy* ! Djourbel *Roberty* 5197 ! Kaolak *Berhaut* 406 ! Upper Senegal *Lecard* 193 ! **Fr.Sud.**: Bamako *Waterlot* 1312 ! 1386 ! Macina (Mar.) *Chev.* 24843 ! **Port.G.**: Buruntuma *Esp. Santo* 2719 ! **Fr.G.**: *Scaëtta* 3325 ! Fouta Djalon *Pobéguin* 1912 ! Kouroussa *Pobéguin* 425 ! 2128 ! Koïn *Maclaud* 308 !
2. E. **parviflora** (*Engl.*) Balle l.c. (1956). *Loranthus parviflorus* Engl. (1894)—F.T.A. 6, 1 : 383 ; F.W.T.A., ed. 1, 1 : 467. Hosts unknown.
Port.G.: Bissau to Cupulra *Esp. Santo* 2382 ! **Fr.G.**: Dalaba *Chev.* 19104 ! Fouta Djalon *Pobéguin* 2288 ! Labé (Apr.) *Chev.* 12311 ! 12322 ! **S.L.**: Batkanu (Apr.) *Thomas* 26 ! Gbonjema (May) *Deighton* 6060 ! Bagroo R. (Apr.) *Mann* 823 ! **Lib.**: Belefanai *Baldwin* 10536 !
3. E. **gabonensis** (*Engl.*) Balle l.c. (1956). *Loranthus gabonensis* Engl. (1894)—F.T.A. 6, 1 : 382 ; E. & P. Pflanzenfam. 16B : 164, fig. 81 ; Balle in Fl. Congo Belge 1 : 326. *L. micrantherus* Engl. ex Th. Dur. & De Wild. (1900)—F.T.A. 6, 1 : 383 ; F.W.T.A., ed. 1, 1 : 467. *L. tessmannii* Engl. & K. Krause (1909). Flowers yellow, red at the base. On *Ficus* and *Platysepalum chevalieri*.
S.L.: Bagroo R. (Apr.) *Mann* 889 ! Bonjema *Deighton* 5329 ! **G.C.**: Abetifi *Irvine* 1661 ! **S.Nig.**: Akilla, Ijebu Ode Prov. *Jones & Onochie* FHI 17304 ! Also in French Cameroons, Spanish Guinea, Gabon and Belgian Congo.

4. GLOBIMETULA Van Tiegh. in Bull. Soc. Bot. Fr. 42 : 244, 264 (1895) ; Danser in Verh. Akad. Wetensch. Afd. Natuurk., sect. 2, 29 : 54 (1933) ; Balle in Webbia 11 : 583 (1955). *Loranthus* Sect. *Cupulati* DC.—F.T.A. 6, 1 : 262 ; Engl. & K. Krause in E. & P. Pflanzenfam. 16B : 160 (1935).

Leaves usually short, 1–2 times as long as broad :
    Leaves large, thick and coriaceous, broadly ovate or broadly elliptic, round-cordate at base, acute or obtuse at apex, 4·5–17 cm. long, 3–12·5 cm. broad ; petioles 6–18 mm. long ; calyculus longer than the receptacle ; flowers (2·5–) 3·2–3·7 cm. long .. .. 1. *cupulata*
    Leaves usually rather thin, elliptic, obtuse or acute at each end, rarely cordate at base, variable in size ; calyculus usually rather short, rarely the same length as the receptacle ; flowers with an apical slightly 5-winged swelling ; umbels 4–8-flowered
                                    2. *braunii*
Leaves usually long, 2–3 times as long as broad, narrowing in the upper half :
    Umbels usually 8–20-flowered, peduncles 1–3 cm. long, much longer than the 2–6 mm. long pedicels ; flowers 3·4–3·8 cm. long ; apical swelling of bud globular ; leaves rather narrowly ovate to lanceolate, round-cordate at base, 5·2–14 cm. long, 2·5–5·4 cm. broad, pinnately nerved ; petiole 8–18 mm. long .. .. 3. *oreophila*
    Umbels usually 4–6-flowered ; peduncles 0·5–1·1 (–2·5) cm. long ; flowers 2·5–3·6 cm. long ; swollen apex of bud narrowly 5-winged ; leaves lanceolate, rarely ovate, lamina cuneate or rounded at base, acuminate, obtuse or acute at apex, 4–10 cm. long, 1·6–5 cm. broad, the lower (1–) 2 pairs of lateral nerves much more prominent than the others usually arising in the lower half of the lamina and more or less parallel with the midrib .. .. 4. *dinklagei*

1. **G. cupulata** (*DC.*) *Van Tiegh.* in Bull. Soc. Bot. Fr. 42 : 264 (1895). *Loranthus cupulatus* DC. (1830)—F.T.A. 6, 1 : 305 ; F.W.T.A., ed. 1, 1 : 466 ; Balle in Fl. Congo Belge 1 : 371. *L. mayombensis* De Wild. (1905)—F.T.A. 6 : 302. Flowers rose-pink. A species for long imperfectly known as the type was mixed with *Phragmanthera rufescens*; this led to much confusion in determinations and descriptions. On *Dacryodes edulis* and a Myrtaceous plant.
    Sen.: Casamance *Leprieur* ! *Perrottet* ! **Lib.:** Monrovia *Baldwin* 5865 ! Also in Belgian Congo.
2. **G. braunii** (*Engl.*) *Van Tiegh.* l.c. 265 (1895). *Loranthus braunii* Engl. (1894)—F.T.A. 6, 1 : 303, 1028 ; F.W.T.A., ed. 1, 1 : 466 ; Balle l.c. 369 (q.v. for full synonymy). Climber with many suckers. This species occurs throughout most of the area of the genus ; it is variable and can be confused with most of the other species. Parasitic on a large number of hosts especially on cultivated plants such as *Coffea, Citrus, Hevea, Theobroma.*
    G.C.: *Kitson* ! **Togo:** Bagida *Warnecke* 109 ! **Dah.:** Porto Novo *Chev.* 22833 ! **S.Nig.:** Igbajun, Lagos *Moloney* 24 ! Ilaro (Oct.) *Punch* 40 ! Akpaka, Onitsha *Jones* FHI 6679 ! Eket *Talbot* ! Oban *Talbot* 1522 ! Old Calabar (Feb.) *Mann* 2294 ! Also in French Cameroons, Ubangi-Shari, Gabon, Spanish Guinea, Belgian Congo, Angola, Rhodesia, Tanganyika and Uganda.
3. **G. oreophila** (*Oliv.*) *Van Tiegh.* l.c. (1895). *Loranthus oreophilus* Oliv. (1862)—F.T.A. 6, 1 : 301, 1028 ; F.W.T.A., ed. 1, 1 : 466. Flowers red with green tips. On *Agauria, Hypericum lanceolatum* and *Maesa.*
    Br.Cam.: Cam. Mt., 6,000–8,300 ft. (Dec., Mar., May) *Mann* 1210 ! 2163 ! *Preuss* 811 ! *Maitland* 684 ! *Hutch. & Metcalfe* 20 ! Lakom, 6,000 ft., Bamenda *Maitland* 1584 ! *Lightbody* FHI 28373 ! *Tiku* FHI 22157 ! *Ujor* FHI 30017 ! Ndu, Bamenda *Johnston* 111/31 ! Also in French Cameroons.
4. **G. dinklagei** (*Engl.*) *Van Tiegh.* l.c. (1895). *Loranthus dinklagei* Engl. (1894)—F.T.A. 6, 1 : 301 ; F.W.T.A., ed. 1, 1 : 466 (?). *L. fulgens* Engl. & K. Krause (1914). Parasitic on *Theobroma.*
    Br.Cam.: Victoria *Hanke* 622. Boru-Bonge, Mamfe (Nov.) *Thorold* 74 ! Also in French Cameroons.

5. AGELANTHUS Van Tiegh. in Bull. Soc. Bot. Fr. 42 : 246 (1895) ; emend. Balle in Webbia 11 : 583 (1955). *Loranthus* Sect. *Infundibuliformes* Engl.—F.T.A. 6, 1 : 264 ; Engl. & K. Krause in E. & P. Pflanzenfam. 16B : 161 (1935).

Perianth of opened flower at about 1 mm. above its base with a very distinct globular or ellipsoid swelling above which the tube is strongly constricted and then progressively widened towards the top ; flowers in axillary fascicles (3·2–) 3·7–4·5 cm. long, the lobes erect 7–11 mm. long ; fruits ellipsoid or obovoid, about 6 mm. long ; leaves ternate, opposite or alternate, usually elliptic but sometimes ovate, lanceolate or obovate, rounded or obtuse at apex rarely acute, 1·7–14 cm. long, 7–50 mm. broad, usually rather thin, with 3–5 basal or subbasal nerves, subparallel and equally prominent .. .. *brunneus*

A. **brunneus** (*Engl.*) *Van Tiegh.* l.c. (1895). *Loranthus brunneus* Engl. (1894)—F.T.A. 6, 1 : 325 ; Balle in Fl. Congo Belge 1 : 331 ; Berhaut Fl. Sen. 96. Leaves extremely variable ; flowers rose-pink or red with a white ring at the base of the lobes ; berries reddish. Parasitic on numerous hosts, e.g. :—*Carissa edulis, Crateva religiosa, Diplorhynchus, Ficus, Funtumia, Gossweilodendron, Ipomaea, Landolphia heudelotii, L. owariensis, Morus alba, Nerium oleander, Saba senegalensis, Scorodophloeus* and *Sterculia quinqueloba.*
    Maur.: El Cheddla *Chev.* 28776. **Sen.:** Niokolo-Koba *Berhaut* 1193 ! **Port.G.:** Jangada, Contubo (June) *Esp. Santo* 2258 ! **Fr.G.:** Timbo *Pobéguin* 1659 ! Moussaia *Chev.* 460 ! Kollangui *Chev.* 12549 ! Labé *Vuillet* 29 ! Ouasana *Chev.* 618 ! **S.L.:** Njala (Feb.) *Deighton* 1089 ! **S.Nig.:** Omo R. *Onochie & Jones* FHI 16684 ! Also in French Cameroons, Ubangi-Shari, Gabon, Belgian Congo, Angola, S.W. Africa, Tanganyika and Uganda.

6. TAPINANTHUS Blume in Schultes f. Syst. Veg. 7 : 1730 (1830) ; Danser in Verh. Akad. Wetensch. Amst. Afd. Natuurk., sect. 2, 29 : 107 (1933) ; Balle in Webbia 11 : 583 (1955). *Loranthus* Sect. *Constrictiflori* Engl.—F.T.A. 6, 1 : 268. *Loranthus* Sect. *Erectilobi* Sprague—F.T.A. 6, 1 : 270. *Loranthus* Sect. *Purpureiflori* Engl.—F.T.A. 6, 1 : 267 ; Engl. & K. Krause in E. & P. Pflanzenfam. 16B : 166, 167 (1935).

Perianth-lobes remaining erect at anthesis (Sect. *Erectilobi*) :
    Leaves broadly ovate or elliptic, rounded or more rarely obtuse at base, rounded or

obtuse rarely cuneate at apex, 9–14 cm. long, 6–9·5 cm. broad ; perianth 5–7 cm. long, the lobes about 1·5 cm. long, acute and thickened at the apex and narrowly winged on both sides ; apical swelling of bud just before anthesis ovoid-acute, 5-winged in the upper half ; umbels 5–7-flowered ..   ..   ..   1. *buntingii*

Leaves smaller, not very broad (up to 5·5 cm.) in relation to their length :
Plant pilose with hairs 0·5–1 mm. long, at least on younger parts ; flowers distinctly pedicellate, (2·8–) 3·5–5 cm. long, perianth lobes 5–12 mm. long ; umbels (2–) 4–8 (–9)-flowered ; calyculus up to 3 mm. long, never tubular ; leaves ovate, ovate-oblong, ovate-lanceolate or oblong-elliptic, 2·5–13·5 cm. long, 1–5·5 cm. long ; bracts unilateral, ovate-elliptic, often very concave ; fruits ellipsoid or obovoid up to 12 mm. long and 9 mm. broad, pilose..   ..   ..   2. *heteromorphus*
Plant glabrous ; flowers (1–) 2–4, (2·8–) 4–5 cm. long, often thick and rigid, subsessile in head-like umbels ; calyculus tubular 5 mm. or more long ; leaves lanceolate, oblong or linear, sometimes falciform, 4–18·5 cm. long, 5–27 mm. broad, more or less glaucous ; fruits up to 1·3 cm. long, ellipsoid or pear-shaped 3. *dodoneifolius*

Perianth lobes reflexing at anthesis (Sect. *Constrictiflori*) :
*Leaves petiolate :
Plants entirely glabrous :
Apical swelling of bud not distinctly 5-winged :
Apical swelling of bud not flattened at top and without appendages, globose, ovoid or ellipsoid, 5-gonous (sometimes very slightly 5-winged in upper half), 3–3·75 mm. long, 2–3 mm. broad ; perianth 2·7–3·8 (–5) cm. long ; flowers subsessile, 4–8 together in subsessile umbels ; leaves subopposite or sometimes subternate :
Leaves bright green when fresh, rather broadly ovate or ovate-lanceolate, rarely elliptic or ovate-oblong, rounded or slightly cordate or sometimes obtuse at base, acute or obtuse at apex, 6–18 cm. long, 2–12 cm. broad, with 1–3 long-ascending lateral nerves arising on each side from near the base of the midrib
4. *bangwensis*
Leaves generally glaucous, usually narrowly elliptic, oblong or obovate, sometimes broadly elliptic, ovate, lanceolate or linear, usually cuneate or acute, sometimes obtuse, rarely rounded or subcordate at base, rounded or obtuse at apex, (1–) 2–9 (–12) cm. long, (0·4–) 0·7–3 (–5) cm. broad, with irregular pinnately arranged lateral nerves ..   ..   5. *globiferus*
Apical swelling of bud flattened and with 5 minute bumps or appendages at top :
Perianth lobes without subapical dorsal appendages, but with 5 minute bumps and sometimes very slightly 5-winged at the angles ; perianth 2·2–3·5 cm. long :
Apical swelling of bud 3·5–4 mm. long, 2–2·25 mm. broad ; leaves subopposite or alternate, ovate or lanceolate, rarely elliptic, cordate, rounded or cuneate at base, acute or obtuse at apex, 3·5–12·5 cm. long, 1·6–6 cm. broad
6. *belvisii*
Apical swelling of bud 2·5–3·5 mm. long, 1·7–2·25 mm. broad ; leaves subopposite or subternate, oblong, elliptic, obovate or more rarely ovate, usually narrowly cuneate at base, rounded or rarely obtuse at apex, 1–8·5 cm. long, 0·3–2·5 cm. broad   ..   ..   7. *voltensis*
Perianth lobes each bearing dorsally just beneath the apex a more or less triangular appendage about 0·75 mm. long ; perianth about 4·4 cm. long ; apical swelling of bud ellipsoid, about 4 mm. long and 2·3 mm. broad ; leaves ovate to lanceolate or elliptic, rounded or cuneate at base, obtuse or acute at apex, 6–9·5 cm. long, 2·5–4·4 cm. broad, with pinnate nervation   ..   ..   8. *farmari*
Apical swelling of bud distinctly 5-winged :
The apical swelling of bud globose, 4–5 mm. diam. ; perianth (4·5–) 5·5–7 cm. long ; leaves broadly ovate or elliptic, rounded or slightly cordate at base, abruptly acuminate (acumen itself acute and often slightly recurved), 6–12·5 cm. long, 3·7–7·5 cm. broad ..   ..   9. *preussii*
The apical swelling of bud distinctly longer than broad :
Apical swelling of bud conical, about 5 mm. long and 2 mm. broad at base, 5-gonous, with the angles narrowly winged ; perianth 3–3·5 cm. long, with the lobes both very acute and distinctly thickened at their extremities ; leaves usually elliptic or lanceolate, rarely ovate or oblanceolate, 5–9·5 cm. long, 1·5–3 (–4) cm. broad, with crisped margins ..   ..   ..   10. *ophiodes*
Apical swelling of bud clavate, ellipsoid or ovoid, 4–7 mm. long, 2–5 mm. broad, with 5 undulate wings 0·25–1 mm. wide, always better developed in the upper half ; perianth 3–4 cm. long, with the lobes reflexed or remaining erect at anthesis ; leaves glaucous and coriaceous, usually ovate, sometimes ovate-lanceolate or elliptic, 1·3–12 cm. long, 0·8–8 cm. broad, usually rounded or broadly cuneate at base, obtuse at apex..   ..   ..   ..   11. *pentagonia*
Plant shortly pubescent on young branchlets, peduncles, pedicels, bracts and some-times on the receptacles, calyculi and perianths ; petioles 0·5–1 cm. long ; leaves

24

ovate or elliptic, cuneate, rounded or slightly cordate at base, obtuse, acute or
sometimes emarginate at apex, 2·5–12·5 cm. long, 1·5–7·5 cm. broad ; umbels
8–10-flowered ; perianth 3–3·8 cm. long, with the lobes slightly thickened at top ;
apical swelling of bud ellipsoid-oblong, about 3 mm. long and 2 mm. broad, more
or less flattened on top.. .. .. 12. *truncatus*
*Leaves sessile, broadly ovate or ovate-lanceolate, cordate at base, obtuse or subacute
at apex, 2·5–8·5 cm. long, 1·8–6·2 cm. broad ; umbels (2–) 4–8 (?–12)-flowered ;
perianth 3·7–4·2 cm. long :
Apical swelling of bud ellipsoid, without appendages ; umbels sessile or subsessile ;
plant glabrous .. .. 13. *sessilifolius*
Apical swelling of bud campanulate, 5-gonous, depressed at the top and bearing 5
spreading triangular appendages about 0·75 mm. long ; umbels pedunculate ;
plant pubescent .. .. 14. *warneckei*

1. **T. buntingii** (*Sprague*) *Danser* in Verh. Akad. Wetensch. Amst. Afd. Natuurk., sect. 2, 29 : 109 (1933).
Loranthus buntingii Sprague in Kew Bull. 1916 : 178 ; F.W.T.A., ed. 1, 1 : 467. Perianth green with
rose-pink lobes. On *Cassia sieberiana*.
S.L.: Njala (Sept.) *Deighton* 6122. **Lib.:** Mt. Barclay *Bunting* 165 ! Also in (?) A.-E. Sudan.

2. **T. heteromorphus** (*A. Rich.*) *Danser* l.c. 113 (1933). *Loranthus heteromorphus* A. Rich. (1847)—F.T.A. 6, 1 :
381. *L. brevilobus* Engl. & K. Krause (1914). *L. tambermensis* Engl. & K. Krause (1909)—F.T.A. 6, 1 :
380 ; F.W.T.A., ed. 1, 1 : 467. *L. terminaliae* Engl. & Gilg (1903)—F.T.A. 6, 1 : 379. Vegetative
and floral parts covered with a rusty tomentum, hairs branched, rather long and close, disappearing
partly on mature leaves and branches ; flowers yellow below and orange-red above. On *Anogeissus
leiocarpus, Detarium microcarpum, Parkia clappertoniana* and *Terminalia avicennioides*.
Togo: Tamberma *Kersting* A.502 ! **N.Nig.:** Sokoto (May) *Dalz.* 524 ! Zamfara F.R., Sokoto Prov.
(Apr.) *Keay* FHI 15893 ! 16174 ! Bauchi Plateau (July) *Lely* P. 597 ! Naraguta *Kennedy* 7234 ! Yola
(June) *Dalz.* 214 ! Also in French Cameroons, Ubangi-Shari, A.-E. Sudan, Abyssinia, Eritrea, Angola
and N. Rhodesia.

3. **T. dodoneifolius** (*DC.*) *Danser* l.c. 111 (1933). *Loranthus dodoneaefolius* DC. (1830)—F.T.A. 6, 1 : 341 ;
F.W.T.A., ed. 1, 1 : 467 ; Berhaut Fl. Sen. 96. *L. chevalieri* Engl. & K. Krause (1909). Leaves opposite,
alternate or ternate. Varies particularly in leaf-width and thickness and in the size and texture of the
perianth ; flowers white, golden yellow, deep rose or red. On *Acacia, Afzelia, Adansonia, Anogeissus,
Butyrospermum, Ceiba, Combretum glutinosum, Mimosa, Parkia, Piliostigma, Pterocarpus erinaceus,
Tamarindus* and *Terminalia*.
Sen.: Gandon *Leprieur* ! St. Louis *Perrottet* ! Tambacounda *Berhaut* 1245 ! Upper Senegal *Bellamy*
15 ! **Fr.G.:** Kouroussa (Dec.) *Pobéguin* 235 ! 596 ! Télivel *Maclaud* 200 ! **G.C.:** Nangodi (July) *Vigne*
FH 3294. Tono, Navrongo *Darko* 455 ! **Togo:** Gbele *Busse* 3624. Sokode *Kersting* A.57 ; 200. **Dah.:**
Adja Ouéré *Le Testu* ! Dan *Poisson* 252 ! Pélebina to Djougou *Chev.* 23854 ! Parakou *Poisson* 83 !
N.Nig.: Abinsi (May) *Dalz.* 764 ! Jebba *Barter.* Kano *Kennedy* 2942 ! Katagum *Dalz.* 203 ! Bauchi
Plateau (May) *Lely* P. 306 ! P. 307 ! Naraguta (Aug.) *Lely* 498 ! Nupe *Barter* 1559 ! Yola (fr. Oct.)
Shaw 92 ! **S.Nig.:** Lagos *Dawodu* ! Old Oyo F.R. (fr. Feb.) *Keay* FHI 16279 ! **Br.Cam.:** Buea *Mildbr.*
9426 ! Bama, Dikwa Div. (fr. Dec.) *McClintock* 66. Also in Ubangi-Shari, A.-E. Sudan and Uganda.

4. **T. bangwensis** (*Engl. & K. Krause*) *Danser* l.c. 108 (1933). *Loranthus bangwensis* Engl. & K. Krause
(1909)—F.T.A. 6, 1 : 353 ; F.W.T.A., ed. 1, 1 : 467 ; Berhaut Fl. Sen. 97. (?) *L. thonningii* Schum. &
Thonn. (1827)—F.T.A. 6, 1 : 363. Branchlets abundantly covered with brown lenticels ; leaves very
variable in size and thickness, reduced in Senegal but well developed in Liberia and Sierra Leone ;
perianth red below, pink then grey above ; filaments and style green, becoming purple. On many species
including *Acacia farnesiana, A. nilotica, Alchornea cordifolia, Cola nitida, Coffea liberica, Crossopteryx
Croton, Drepanocarpus lunatus, Manihot glaziovii, Terminalia catappa* and *Theobroma*.
Sen.: *Brunner* ! *Perrottet* 480 ! 481 ! Thiès *De Wailly* 4405 ! **Gam.:** Bathurst *Dawe* 74 ! **Fr.G.:**
Conakry *Maclaud* ! Tristao Isl. *Paroisse* 16 ! Mamou *Boué* 15 ! Télimélé *Roberty* 10766 ! Timbo
Maclaud 186 ! **S.L.:** Rokupr *Jordan* 255 ! Bagroo R. (Apr.) *Mann* 803 ! Newton *Deighton* 5642 !
Mapotalon *Adames* 140 ! Bintumane Peak, 5,400 ft. *T. S. Jones* 60 ! **Lib.:** Grand Bassa (May) *Dinklage*
1908 ! Buchanan *Baldwin* 11198 ! Gbanga *Baldwin* 10519 ! Greenville *Baldwin* 11566 ! Nyaake-Webo
Baldwin 6149 ! **Iv.C.:** Oua to Goura *Chev.* 13763 ! 21342 ! **G.C.:** Cape Coast Castle (July) *T. Vogel*
45 ! Tarkwa (June) *Chipp* 243 ! Kumasi (June) *Andoh* FH 4221 ! *Vigne* FH 2961 ! Togo : Sanhokofi
St. *C. Thompson* FH 3594 ! **Dah.:** Kitou *Chev.* 23018 ! Ouidah *Lesteve* 112 ! Porto Novo *Lesteve* 105 !
S.Nig.: Lagos *Dalz.* 1089 ! *Barter* 2229 ! Ibo country (Aug.) *T. Vogel* 40 ! Oban *Talbot* 605 ! **Br.Cam.:**
Bangwe *Conrau* 253. **F.Po:** Laka (Aug.) *Thorold* TF 81 ! Also in Ubangi-Shari, French Cameroons,
Gabon and Belgian Congo.

5. **T. globiferus** (*A. Rich.*) *Van Tiegh.* in Bull. Soc. Bot. Fr. 42 : 267 (1895). *Loranthus globiferus* A. Rich.
(1847)—F.T.A. 6, 1 : 352 ; F.W.T.A., ed. 1, 1 : 467 (incl. vars.) ; F. W. Andr. Fl. Pl. A.-E. Sud. 2 :
293, fig. 105. (?) *L. rosiflorus* Engl. & K. Krause (1914). (?) *L. rubrovittatus* Engl. & K. Krause (1914).
Apparently takes the place of *T. bangwensis* in the drier regions ; forms transitional in the consistence,
colour and shape of the leaves are known in Senegal and Nigeria. Flowers shining red. Some forms
seem to connect these two species with *T. voltensis* and *T. pentagonia*. Careful field observations on
the colour of flowers and fruits are needed. A form with small leaves seems to be localized around
Timbuktu. On numerous hosts, including *Salvadora* and *Sclerocarya*.
Fr.Sud.: Timbuktu *Chev.* 1206 ! (July) *Hagerup* 166 ! 208 ! **Dah.:** Dan *Gaillard* ! Konkobiri (July)
Chev. 24308 ! **G.C.:** Navrongo (Nov.) *Darko* 453 ! Tumu (Apr.) *Hughes* 328 ! **Fr.Nig.:** Maradi *Gaillard* !
N.Nig.: Zurmi (Apr.) *Keay* FHI 15683 ! 16186 ! Kano *Kennedy* 2943 ! Bornu *E. Vogel* 79 ! Also in
Ubangi-Shari, A.-E. Sudan, Belgian Congo, Kenya, Eritrea and Arabia.

6. **T. belvisii** (*DC.*) *Danser* l.c. 108 (1933). *Loranthus belvisii* DC. (1830)—Brenan in Kew Bull. 1950 : 366.
L. lanceolatus P. Beauv. (1808)—F.T.A. 6, 1 : 354 (? incl. var.) ; F.W.T.A., ed. 1, 1 : 467 ; not of
Ruiz & Pav. (1802). A poorly known species, often confused with *T. bangwensis* which is similar in habit.
Leaves thick when fresh ; perianth tube red, lobes green.
G.C.: Chama *Beauvois* ! Axim (Jan.) *Chipp* 54 (partly) ! Yapei (June) *Williams* 825 ! Yezi *Vigne*
FH 3315 ! **Togo:** Banyatera *Schröder* 224. **Dah.:** Savalou (May) *Chev.* 23695 ! **Nig.:** *MacDonald*
A. 113/35. R. Niger *Baikie* ! Also in Gabon, Belgian Congo and Angola.

7. **T. voltensis** *Van Tiegh. ex Balle* in Bull. Soc. Roy. Bot. Belg. 58 : (1956). Branchlets and leaves glaucous,
similar in appearance to *T. globiferus* ; perianth tube red, apex greenish white. On *Acacia nilotica,
Adansonia, Bauhinia rufescens, Butyrospermum parkii* and *Maytenus senegalensis*.
Fr.Sud.: Dioubéba *Chev.* 12 ! El Onaladjii *Chev.* 1175 ! San *Chev.* 1058 ! Ségou to Dioila *Roberty*
10501 ! **Iv.C.:** Fô *Chev.* 960 !

8. **T. farmari** (*Sprague*) *Danser* l.c. 111 (1933). *Loranthus farmari* Sprague in F.T.A. 6, 1 : 345 (1910) ;
F.W.T.A., ed. 1, 1 : 466. Perianth rose-red with blood-red ring just below base of lobes outside ; lobes
green outside with reddish horns, grey inside ; tube whitish inside; fruit green turning red. On
Dalbergia ecastaphyllum.

G.C.: *Farmar* 503 ! Accra *Dalz.* 122 ! Achimota (fl. & fr. May) *Milne-Redhead* 5081a ! Also in Gabon and Belgian Congo.

9. **T. preussii** (*Engl.*) *Van Tiegh.* in Bull. Soc. Bot. Fr. 42 : 267 (1895). *Loranthus preussii* Engl. Bot. Jahrb. 20 : 118, t. 3, A (1894) ; F.T.A. 6, 1 : 348 ; F.W.T.A., ed. 1, 1 : 466. *L. pachycaulis* Engl. & K. Krause (1909). Perianth grey outside with red stripes, red inside (*fide* Preuss), or red and green (*fide* Ledermann).
   Br.Cam.: N. of Barombi *Preuss* 419 ! Also in French Cameroons.

10. **T. ophiodes** (*Sprague*) *Danser* l.c. 111 (1933). *Loranthus ophiodes* Sprague (1912)—F.T.A. 6, 1 : 1031 ; F.W.T.A., ed. 1, 1 : 467. On *Parkia biglobosa.*
   Fr.G.: Kouroussa *Chev.* 15728 ; 15753. **Dah.:** Konkobiri (July) *Chev.* 24307. **Fr.Nig.:** Diapaga (July) *Chev.* 24371 ! 24438 ! Fada to Koupéla (July) *Chev.* 24532 !

11. **T. pentagonia** (*DC.*) *Van Tiegh.* l.c. (1895). *Loranthus pentagonia* DC. (1830)—F.T.A. 6, 1 : 349 ; F.W.T.A., ed. 1, 1 : 466. *L. senegalensis* De Wild. (1903)—F.T.A. 6, 1 : 349 ; F.W.T.A., ed. 1, 1 : 466 ; Berhaut Fl. Sen. 96, 97, fig. 12. *L. apodanthus* Sprague (1913). An extremely polymorphic species as regards the leaves and the development of the wings on the perianth. Perianth red, whitish at the tip. On *Acacia, Butyrospermum, Ceiba, Ficus, Gardenia, Landolphia heudelotii* and *Ziziphus mauritiana.*
   Maur.: Alenda *Chudeau* ! **Sen.:** *Lecard* 83 ! 84 ! Dogana *Leprieur* ! Casamance *Perrottet* ! Tambacounda *Berhaut* 1467 ! Oualia (Dec.) *Chev.* 36 ! **Fr.Sud.:** Bamako *Waterlot* ! Quiébélé (June) *Chev.* 1004 ! San (July) *Chev.* 1059 ! Koulikoro *Dybowski* 2 ! *Vuillet* 652 ! 666 ! [**G.C.:** Lawra *Smith* 3 !—determination doubtful.] **S.Nig.:** Oyo *Latilo* FHI 3422 ! Also in French Cameroons.

12. **T. truncatus** (*Engl.*) *Danser* l.c. 121 (1933). *Loranthus truncatus* Engl. (1894)—F.T.A. 6, 1 : 356 ; F.W.T.A., ed. 1, 1 : 467. *Acrostephanus truncatus* (Engl.) Van Tiegh. (1895). *Loranthus pubiflorus* Sprague (1912) —F.T.A. 6, 1 : 356 ; F.W.T.A., ed. 1, 1 : 467. Perianth red, with yellowish green lobes. Apparently rare, but has probably been confused with *T. bangwensis* and *T. belvisii.*
   S.L.: *Afzelius.* **Iv.C.:** *Pobéguin* 23 ! Bingerville (Dec.) *Chev.* 16047 ! 16059 ! **G.C.:** *Krause* 95 ! Atwabo (Feb.) *Irvine* 2288 !

13. **T. sessilifolius** (*P. Beauv.*) *Van Tiegh.* l.c. (1895). *Loranthus sessilifolius* P. Beauv. (1808)—F.T.A. 6, 1 : 355 ; F.W.T.A., ed. 1, 1 : 467. Perianth tube red, lobes green. On *Canarium* and *Strychnos.*
   Togo: Keta *Beauvois* ! **N.Nig.:** Bauchi Plateau *Lely* P. 163 ! *W. D. MacGregor* 423 !

14. **T. warneckei** (*Engl.*) *Danser* l.c. 122 (1933). *Loranthus warneckei* Engl. (1908)—F.T.A. 6, 1 : 356 ; F.W.T.A., ed. 1, 1 : 467 ; E. & P. Pflanzenfam. 16B : 167, fig. 85 (1935). On *Pithecellobium dulce* and *Nerium oleander.*
   Togo: *Messager* ! Lome (Jan., Nov.) *Busse* 3255 ; *Warnecke* 36 !

7. **PHRAGMANTHERA** Van Tiegh. in Bull. Soc. Bot. Fr. 42 : 243, 261 (1895) ; Balle in Webbia 11 : 583 (1955). *Loranthus* Sect. *Lepidoti* Engl.—F.T.A. 6, 1 : 260. *Loranthus* Sect. *Rufescentes* Engl.—F.T.A. 6, 1 : 259 ; Engl. & K. Krause in E. & P. Pflanzenfam. 16B : 158 (1935).

Perianth-lobes remaining erect at anthesis :
Inflated apex of bud club-shaped or long-ellipsoid ; anthers 2·7–12 mm. long, with 6–30 superposed small cells :
Apical swelling of bud not 5-winged ; perianth 2·8–6·3 cm. long, the lobes 1·2–2·2 cm. long ; anthers 2·7–6·5 mm. long, with 6–10 superposed cells :
Most of the leaves with a short lamina, 1–2 times as long as broad :
Lamina mostly reaching its greatest width in the lower half, ovate, ovate-elliptic or ovate-lanceolate, 5–20 cm. long, 2·5–14 cm. broad, cordate, truncate, rounded or rarely obtuse at base ; petioles 6–30 mm. long ; perianth (2·8–) 3·7–6·3 cm. long, yellow, sometimes reddish especially at the apex ; fruits obovoid or pear-shaped, 7–10 mm. long .. .. 1. *incana*
Lamina mostly reaching its greatest width in the middle, elliptic or elliptic-oblong, 3–8·8 cm. long, 1·5–4·2 cm. broad, obtuse or cuneate, rarely rounded at base ; petioles 5–12 mm. long ; perianth (3–) 3·5–4 cm. long, tube red, the lobes yellow below and red above ; apical swelling of bud ellipsoid about 6 mm. long
.. .. .. 2. *redingi*
Most of the leaves with a long lamina, 2–3 times as long as broad, 10–15 cm. long, 4–5·5 cm. broad, rigid and coriaceous, glabrous at maturity ; receptacle and calyculus up to about 3 mm. long ; perianth yellow towards apex, about 4 cm. long, sparsely hairy, lobes 1·2–1·4 cm. long, linear-lanceolate, acute ; anthers 3–4 mm. long .. .. 3. *lapathifolia*
Apical swelling of bud narrowly 5-winged, 1·1–1·4 cm. long ; perianth (6·5–) 7–10·5 cm. long, glabrescent, lobes 2·2–3 cm. long ; anthers (5–) 7–12 mm. long, with 20–30 superposed cells ; leaves elliptic, elliptic-oblong or rarely ovate-oblong, 5–11·3 cm. long, 3–5·6 cm. broad ; petioles 1·2–2 cm. long ; fruits ellipsoid, up to 1·5 cm. long .. .. 4. *kamerunensis*
Inflated apex of bud usually globular or shortly elliptic ; anthers relatively short, 1–2·5 mm. long :
Leaves 15–18·5 cm. long, 6·3–8 cm. broad, ovate-oblong or lanceolate, thick and coriaceous, cordate at base, rusty-tomentose, subsessile (petioles up to 8 mm. long) ; perianth 4–4·7 cm. long, rusty-tomentose ; lobes 10–11 mm. long, elliptic in the upper 3·5 mm. ; anthers with 3–4 superposed cells ; bracts 3–4 (–8) mm. long, sometimes foliaceous .. .. 6. *talbotiorum*
Leaves 3–14 cm. long, 1·2–5·8 cm. broad ; petioles relatively long, 4–28 mm. long ; perianth 2·7–5 cm. long, with rusty verticillate-branched hairs to 1 mm. long :
Leaves concolorous, the tomentum rapidly caducous on both surfaces ; lamina usually elliptic, rarely ovate, rounded or obtuse at each end ; bracts ovate, 2–4 mm. long, not foliaceous.. .. .. 7. *rufescens*

Leaves generally discolorous, the red tomentum persisting longer on the lower
  surface than above ; lamina lanceolate or ovate, rarely elliptic, usually cordate
  or rounded at base, acute at apex ; bracts 1–2·5 (–6) mm. long, sometimes
  foliaceous   ..   ..   5. *polycrypta*
Perianth-lobes reflexed at anthesis ; perianth rusty-tomentose ; anthers short :
Bracts ovate, 1·5–2 mm. long, never foliaceous ; leaves lanceolate-oblong or elliptic-
  oblong, 3–11·3 cm. long, 1·8–4·4 cm. broad, round-cordate at base, acuminate at
  apex, rusty-tomentose beneath with verticillate-branched hairs ; petioles 2–10 mm.
  long ; perianth 3·8–4·5 cm. long, covered with hairs about 1 mm. long ; apical
  swelling of bud about 4 mm. diam. ; lobes 9–11 mm. long ; anthers 2–2·5 mm. long,
  with 6–7 superposed cells   ..   ..   8. *leonensis*
Bracts foliaceous, 3–25 mm. long, rounded or emarginate at apex :
Leaves narrowly ovate, ovate-elliptic or lanceolate, 2·5–10 cm. long, 1·2–3 cm. broad,
  rounded or cordate at base, obtuse or cuneate at apex, usually discolorous ;
  petioles 4–10 mm. long ; perianth 3·7–4 cm. long, pubescent with hairs up to
  2 mm. long ; lobes 8–10 mm. long ; anthers 1·5–3 mm. long, with 4–5 (–10) superposed cells ; bracts oblong-obovate, concave, 3–7 mm. long   ..   9. *nigritana*
Leaves broadly ovate-elliptic or obovate, 3·5–10 cm. long, 1·8–7·5 cm. broad, rounded
  or cuneate at base, rounded or obtuse at apex, distinct and prominent only in
  lower ⅔, glabrous ; petioles thick, 4–16 mm. long ; perianth 4–5 cm. long,
  covered with rusty verticillate-branched bristles up to 5 mm. long ; lobes 1–1·7 cm.
  long ; anthers 1·75–3 mm. long, with 7–10 superposed cells ; bracts unilateral,
  rather deeply concave, very differently developed on the same plant and sometimes
  on the same inflorescence   ..   ..   10. *vignei*

1. **P. incana** (*Schum.*) *Balle* in Kew Bull. 1956: 168. *Loranthus incanus* Schum. (1827)—F.T.A. 6, 1 : 292,
1028 ; F.W.T.A., ed. 1, 1 : 466 ; Balle in Fl. Congo Belge 1 : 353. Young parts and perianth more
or less densely covered with white or brown hairs ; berries red. Very variable in the shape and size of
the flowers and leaves. On *Alchornea, Aleurites moluccana, Anacardium occidentale, Annona senegalensis,
Bauhinia, Bersama, Bombax sessile, Caloncoba welwitschii, Citrus aurantium, Coffea, Cola nitida, Croton
tiglium, Funtumia, Hevea brasiliensis, Hibiscus rosa-sinensis, Lophira lanceolata, Macaranga spinosa,
Manihot glaziovii, Melia azedarach, Musa, Nauclea latifolia, Parinari, Piptadeniastrum, Rauwolfia
vomitoria, Sapium ellipticum* and *Theobroma*.
S.L.: Falaba *Adames* 266 ! **Iv.C.:** Adiokrou *Chev.* 17228 ! Bingerville *Chev.* 17315 *bis* ! *Dybowski* 1 !
Sanvi *Chev.* 18881 ! **G.C.:** Volta *Thonning*. Newtown (July) *Chipp* 277 ! Aburi *Howes* 1176 ! **Togo:**
Lome *Warnecke* 332 ! Tuaro *Kersting* A. 261 ! **Dah.:** Dogba *Le Testu* 32 ! 295 ! Ouidah (Apr.) *Chev.*
23440 ! Porto Novo *Lesteve* 106 ! **N.Nig.:** Nupe *Barter* 1147 ! Patti Lokoja (Nov.) *Dalz.* 215 ! Idah
(Sept.) *T. Vogel* 47 ! **S.Nig.:** Ibadan *Onochie* FHI 3420 ! *Newberry* 191 ! Cross R. *Johnston* ! Old
Calabar *Holland* ! *Thompson* 129 ! **Br.Cam.:** Cam. Mt., up to 3,000 ft. (Mar., Apr.) *Kalbreyer* 95 !
Hutch. & Metcalfe 90 ! Victoria (Feb.) *Maitland* 383 ! Victoria to Kumba (July) *Thorold* TN 70 !
F.Po: *Mann* 275 ! Also in Gabon, Spanish Guinea, Ubangi-Shari, Belgian Congo and Angola.
2. **P. redingi** (*De Wild.*) *Balle* in Kew Bull. 1956: 168. *Loranthus redingi* De Wild. (1914)—Balle in Fl.
Congo Belge 1 : 355. On *Albizia*.
Fr.G.: Kouroussa *Roberty* 10550 ! Also in Gabon and Belgian Congo.
3. **P. lapathifolia** (*Engl. & K. Krause*) *Balle* in Kew Bull. 1956 : 168. *Loranthus lapathifolius* Engl. & K.
Krause (1914). **Br.Cam.:** Babangi (Oct.) *Ledermann* 5827.
4. **P. kamerunensis** (*Engl.*) *Balle* l.c. (1956). *Loranthus kamerunensis* Engl. (1908)—F.T.A. 6, 1 : 282 ;
F.W.T.A., ed. 1, 1 : 466. On *Chlorophora excelsa* and *Isoberlinia doka*.
N.Nig.: Kontagora *Dalz.* 551 ! **S.Nig.:** Oban Dist. *Talbot* 606 ! **Br.Cam.:** Meandja *Winkler* 1067 !
5. **P. polycrypta** (*F. Didr.*) *Balle* l.c. (1956). *Loranthus polycryptus* F. Didr. (1854)—F.T.A. 6, 1 : 284 ;
Balle in Fl. Congo Belge 1 : 360. *L. angolensis* Engl. (1894)—F.T.A. 6, 1 : 285. *L. discolor* Engl. ex
Th. Dur. & De Wild. (1900)—F.T.A. 6, 1 : 284 ; Balle l.c. 361. *L. luteo-vittatus* Engl. & K. Krause (1910)—
F.T.A. 6, 1 : 1027. *L. nitidulus* Sprague in F.T.A. 6, 1 : 283 (1910) ; F.W.T.A., ed. 1, 1 : 466. Flowers
with an orange tube and violet lobes ; berries red. On *Annona senegalensis, Borassus, Coffea, Hymenocardia acida, Cathormion altissimum* and *Sapium ellipticum*.
Fr.G.: Pita *Pobéguin* 2302 ! Labé *Pobéguin* 2129 ! **Br.Cam.:** Lakom, 6,000 ft., Bamenda (Apr.) *Maitland*
1651 ! **F.Po:** 7,000 ft. (fl. & fr. Mar.) *Mann* 2346 ! Also in Belgian Congo, Uganda and Angola.
6. **P. talbotiorum** (*Sprague*) *Balle* in Kew Bull. 1956 : 168. *Loranthus talbotiorum* Sprague in F.T.A. 6, 1 :
1026 (1910) ; F.W.T.A., ed. 1, 1 : 466. Hosts unknown.
S.Nig.: Eket *Talbot*. Oban *Talbot* 1281 !
7. **P. rufescens** (*DC.*) *Balle* l.c. (1956). *Loranthus rufescens* DC. (1830)—F.T.A. 6, 1 : 291 ; F.W.T.A.,
ed. 1, 1 : 466 ; Balle in Fl. Congo Belge 1 : 367. *L. usuiensis* Oliv. (1873)—F.T.A. 6, 1 : 288. Extremely
variable in the shape and size of the leaves, and in the size and pubescence of the flowers. On *Ficus,
Myrica* and *Lachnopylis*.
Sen.: Nghianga *Leprieur* ! Casamance *Perrottet* 379 ! *Leprieur* ! Also in Belgian Congo and in eastern
Africa from Uganda and Kenya to Rhodesia.
8. **P. leonensis** (*Sprague*) *Balle* in Kew Bull. 1956 : 168. *Loranthus leonensis* Sprague in F.T.A. 6, 1 :
282 (1910) ; F.W.T.A., ed. 1, 1 : 466. Young parts, leaves beneath and flowers covered with rusty
tomentum. On *Cola acuminata, Psidium guajava, Nauclea diderrichii*, and (?) *Rhizophora racemosa*.
Port.G.: Bafata (June) *Esp. Santo* 2695 ! **Fr.G.:** Bambaya *Pobéguin* 66 ! Koussi *Pobéguin* 2212 !
S.L.: Bagroo R. *Mann* ! Batkanu *Deighton* 3004 ! Falaba *Sc. Elliot* 5294 ! Kabusa (Apr.) *Sc. Elliot*
5472 ! Talla Hills (Feb.) *Sc. Elliot* 4997 ! Oldfield's Farm *Barter* ! **Lib.:** Dukwia R. *Linder* 177 !
9. **P. nigritana** (*Hook. f. ex Benth.*) *Balle* l.c. (1956). *Loranthus nigritanus* Hook. f. ex Benth. (1849)—
F.T.A. 6, 1 : 282 ; F.W.T.A., ed. 1, 1 : 466. *L. hirsutissimus* Engl. (1894). Young parts and flowers
covered with rusty tomentum. Very variable in the pubescence of the flowers and development of
the bracts, also in the relative width and in the form of the apex of the leaves. On *Melia* and *Morinda
lucida*.
Iv.C.: Bingerville *Chev.* 15354 ! **G.C.:** Cape Coast (fr. Mar.) *Irvine* 2421 ! Prang *Vigne* FH 1556 !
Pepease (fr. July) *Akpabla* 174 ! **N.Nig.:** Patti Lokoja (Sept.) *T. Vogel* 170 ! **S.Nig.:** Oban *Talbot*
1572 ! Abeokuta (Jan.) *Barter* 3383 ! *Harrison* 9 ! *Rowland* !
10. **P. vignei** *Balle* in Bull. Soc. Roy. Bot. Belg. 88 : 141 (1956). A robust plant ; the young parts and
flowers covered with rusty tomentum ; leaves opposite or subopposite ; heads or axillary fascicles
4-flowered.
Fr.G.: Kouroufing *Pobéguin* 1509 ! **S.L.:** Bintumane Peak, 5,400 ft. *T. S. Jones* 63 ! **G.C.:** Ankasa
Vigne 3182 ! Also in Gabon.

8. VISCUM Linn.—F.T.A. 6, 1 : 393.

Plant monoecious ; branches flattened for most of their length with the upper internodes enlarged at the base, only sub-cylindrical at the base of the branches ; cymes 3–5-flowered, unisexual or exceptionally bisexual ; cupule about 2·5 mm. long ; peduncles 3–6 mm. long ; fruits globular, about 5 mm. diam., sessile ; leaves narrowly obovate or oblong, cuneate or acute at base, rounded at apex, 1·8–5 cm. long, 8–23 mm. broad   ..   ..   1. *decurrens*

Plant dioecious ; branches subcylindrical, furrowed or angular nearly the whole length, sometimes slightly flattened only at the tips ; cymes 3–5-flowered, unisexual ; cupule 2·5–4 mm. long ; peduncle 1·5–5 mm. long ; fruits ellipsoid, 6–9 mm. long, translucid ; pedicel about 1 mm. long ; leaves elliptic, ovate, obovate, rarely lanceolate, cuneate, obtuse or rounded at base, obtuse or rounded at apex, 2·5–10 cm. long, 1·6–5 cm. broad   ..   ..   2. *congolense*

1. **V. decurrens** (*Engl.*) *Bak. & Sprague* in F.T.A. 6, 1 : 398 (1910) ; Balle in Fl. Congo Belge 1 : 376. *V. obscurum* Thunb. var. *decurrens* Engl. (1894). On *Symphonia globulifera*.
S.Nig.: Lagos *Dalz.* 1038 ! Epe *Barter* 3314 ! Old Calabar (fr. Feb.) *Mann* 2278 ! Oban *Talbot* 66a ! **Br.Cam.**: Victoria (Oct.) *Winkler* 550 ! Also in French Cameroons, Gabon, Belgian Congo, Angola, Tanganyika and Mozambique. (See Appendix, p. 298.)
2. **V. congolense** *De Wild.* in Bull. Herb. Boiss., sér. 2, 1 : 44 (1900) ;—F.T.A. 6, 1 : 400 ; Balle l.c. 378. *V. grandifolium* Engl. (1908)—F.T.A. 6, 1 : 400 ; F.W.T.A., ed. 1, 1 : 467. *V. zenkeri* Engl. (1908)—F.T.A. 6, 1 : 401 ; F.W.T.A., ed. 1, 1 : 467. Very variable in the shape and size of the leaves. On *Albizia, Combretodendron africanum, Hevea, Funtumia elastica, Polyalthia sauveolens, Oxystigma oxyphyllum, Scorodophloeus zenkeri* and *Strychnos.*
Iv.C.: Sanvi (fr. Apr.) *Chev.* 17847 ! **G.C.**: Amaneakrom *Chipp* 374 ! **S.Nig.**: Oban *Talbot* 604 ! **Br.Cam.**: Victoria (Jan.) *Winkler* 23a ! *Maitland* 1303 ! Mopanya, 2,600–3,000 ft. (fl. & fr. Mar.) *Kalbreyer* 103 ! Also in Gabon, Belgian Congo and Uganda.

110. SANTALACEAE

Trees, shrubs or herbs, sometimes parasitic on trees or roots. Leaves alternate or opposite, entire, sometimes reduced to scales, exstipulate. Flowers often greenish, hermaphrodite or unisexual, actinomorphic. Calyx often fleshy, adnate to the ovary, lobes 3–6, valvate or slightly imbricate. Petals absent. Stamens the same number as the calyx-lobes and opposite to them ; anthers 2-celled, opening lengthwise. Disk epigynous. Ovary inferior or half inferior, 1-celled ; style simple. Ovules 1–3, pendulous from the top of a basal placenta. Fruit nut-like or drupaceous. Seed without a testa ; endosperm copious ; embryo straight ; cotyledons mostly terete.

A small family, very sparingly represented in the area of this flora. Marks of identification are the inferior ovary, basal placenta, often parasitic habit, absence of testa to the seed, etc.

Santalum album Linn., Sandalwood, a native of India, has been introduced into parts of our area.

THESIUM Linn.—F.T.A. 6, 1 : 411.

With F. N. Hepper

Semi-parasitic herbs or undershrubs with much-reduced leaves ; flowers hermaphrodite ; calyx-tube adnate to ovary and continuing above it ; calyx-lobes 4 or 5, valvate, glabrous or pubescent within ; stamens as many as calyx-lobes and opposite to them, with a short filament, inserted near the base of the lobe ; ovary inferior ; ovules 2–4, pendulous ; fruits dry, globose, crowned by the persistent calyx.

Small appendages present outside at the base of and between calyx-lobes ; calyx-lobes densely pubescent inside, otherwise the whole plant glabrous ; stems sparingly branched :
Calyx-lobes about 5 mm. long, 4 in number ; flowers in short spikes about 1 cm. long at the end of slender branches ; bracts and bracteoles few, lanceolate, 2–3 mm. long   1. *leucanthum*

Calyx-lobes 1·5–2 mm. long, 4 (–5) in number ; flowers in axillary clusters or solitary ; bracts and bracteoles several, linear-lanceolate, 1–2 mm. long   ..   2. *libericum*
Small appendages absent between calyx-lobes ; calyx-lobes glabrous or rarely sparsely villous inside, about 1 mm. long ; plants glabrous or pubescent ; stems much-branched :
Montane plant ; stems glabrous, angular ; flowers several or solitary in the axil of a scaly bract ; calyx-lobes 4 (–5), glabrous outside, rarely sparsely villous inside   3. *tenuissimum*

Savannah plant ; stems pubescent at least at the base with short scabrid hairs, subterete and grooved ; flowers on short lateral branches and axillary ; calyx-lobes 5, sometimes scabrid-pubescent outside, glabrous inside   ..   4. *viride*

1. **T. leucanthum** *Gilg*—F.T.A. 6, 1 : 420. Stems about 18 in. high, erect and not spreading ; flowers yellow ; in savannah.
   N.Nig.: Naraguta *Lely* 66 !   Also in S. Angola.
2. **T. libericum** *Hepper & Keay* in Bull. Jard. Bot. Brux. 26 : 186, fig. 58 (1956). Stems about 18 in. high, tufted from a woody rootstock ; flowers whitish ; in coastal savannah.
   Lib.: near Monrovia (June, Aug.) *Baldwin* 5850 ! 9183 ! Paynesville (Apr.) *Barker* 1240 ! Duport (Nov.) *Linder* 1455 !
3. **T. tenuissimum** *Hook. f.*—F.T.A. 6, 1 : 421. A small undershrub with numerous, slender, glabrous, angular stems, 4–10 in. high, from a woody rootstock ; in montane grassland.
   S.L.: Bintumane Peak, 5,500 ft. *Glanville* 471 ! **Iv.C.:** Nimba Mts., 4,900 ft. (fl. & fr. Apr.) *Schnell* 973 ! **Br.Cam.:** Cam. Mt., 7,600–12,000 ft. (fl. & fr. Nov.–Jan., fl. Feb.) *Mann* 1223 ! 1961 ! *Maitland* 1187 ! 1279 ! *Dalz.* 8327 ! **F.Po:** Clarence Peak, 9,000–9,700 ft. (fl. & fr. Dec., Mar.) *Mann* 595 ! *Guinea* 2862 !

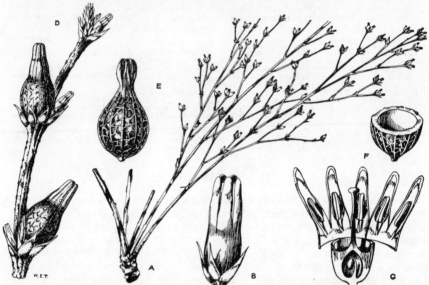

FIG. 187.—THESIUM VIRIDE *A. W. Hill* (SANTALACEAE).

A, flowering branch. B, flower with bract and bracteoles. C, same in vertical section with calyx spread out. D, young fruits showing calyx adnate to the ovary. E, mature fruit. F, portion of calyx in fruiting stage.

4. **T. viride** *A. W. Hill*—F.T.A., 6, 1 : 426. A small perennial undershrub, with greyish-green grooved branches glabrous or scabrid, up to 15 in. high, from a woody rootstock ; in savannah woodland.
   N.Nig.: Sokoto (fl. & fr. June) *Dalz.* 549 ! Kaduna (fl. & fr. Mar., Dec.) *Hill* 52 ! *Meikle* 765 ! Zungeru to Funtua (fl. & fr. Feb.) *Meikle* 1229 ! Anara F.R., Zaria (fr. May) *Keay* FHI 22951 ! Ilorin to Jebba (Feb.) *Meikle* 1198 ! Also in French Cameroons and A.-E. Sudan.

111. BALANOPHORACEAE

Fleshy herbs parasitic on roots, annual or perennial, destitute of chlorophyll and stomata. Flowers unisexual, very rarely hermaphrodite, densely crowded into unisexual or androgynous inflorescences ; male flowers without or with a valvate 3–8-lobed perianth. Stamens 1–2 in the achlamydeous flowers, in those with a perianth often equal in number to, and opposite the lobes ; filaments free or connate ; anthers 2–4-celled or with many cells, free or connate, opening by pores or slits. Ovary 1–3-celled, adnate to the perianth when present ; styles 1–2, terminal or rarely the stigma sessile and discoid ; ovule solitary in each cell, mostly pendulous, nude or with a single integument. Fruit small, nut-like, 1-celled, 1-seeded. Seeds with abundant endosperm and very small embryo.

THONNINGIA Vahl—F.T.A. 6, 1 : 436 ; Bullock in Kew Bull. 1948 : 363.

A herb, parasitic on roots of trees and shrubs ; rhizome subterranean, more or less horizontal, tuberous at the point of attachment, branched, pubescent ; peduncles variable in length, up to 7 cm. long ; peduncles and flower-heads, densely covered with closely imbricate, rigid, ovate to lanceolate, up to 3 cm. long scales ; flowers numerous in each head, unisexual ; flower heads up to 4 cm. diam.   *sanguinea*

T. sanguinea *Vahl* in Skrivt. Nat. Selsk. 6 : 125, t. 6 (1810) ; F.T.A. 6, 1 : 438 ; Bullock l.c. 366, q.v. for full references and synonymy ; Staner in Fl. Congo Belge 1 : 395. *T. dubia* Hemsl. (1911)—F.T.A. 6, 1 : 438 ; F.W.T.A., ed. 1, 1 : 470. *T. elegans* Hemsl. (1911)—F.T.A. 6, 1 : 439 ; F.W.T.A., ed. 1, 1 : 410. " *T. coccinea* Vahl " of Mangenot in Rev. Gén. Bot. 56 : 201–244 (1947). Peduncle and flower-heads with crimson or pink scales, turning brown ; only creamy white pollen and anthers show in the open flower-heads ; common in forest, rare in savannah ; parasitic on many different hosts.
S.L.: Bagroo R. (Apr.) *Mann* 907 ! Freetown (Aug.) *Kelsall* ! Kabala (July) *Glanville* 413 ! Jigaya (Sept.) *Thomas* 2769 ! Sasa, Tonko Limba (Sept.) *Jordan* 326 ! Makali (Feb.) *Deighton* 4088 ! **Lib.:** Ba, Mano R. (Dec.) *Baldwin* 10703 ! **Iv.C.:** Danané (June) *Collenette* 41 ! Bouroukrou (Dec.) *Chev.* 16776. Languira, Baoulé-Nord (July) *Chev.* 22212. **G.C.:** Aquapim *Thonning* ! Kumasi *Cummins* 192 ! Aburi (Mar.–May) *Johnson* 521 ! *Leslie* ! *Thorold* ! Kanyiri (Feb.) *A. S. Thomas* D. 130 ! Achimota (Sept.) *Milne-Redhead* 5129 ! **N.Nig.:** Mt. Purdy (Aug.) *Barter* ! Lokoja (Aug.) *Elliott* 187 ! Katsina Ala *Dalz.* 759 ! **S.Nig.:** Lagos (Feb.) *Millen* 172 ! Olokemeji (July) *Nwokolo* FHI 7918 ! Akure F.R. (June) *Latilo* FHI 3533 ! **Br.Cam.:** Buea (Jan., Nov.) *Maitland* 376 ! *Migeod* 117 ! Cam. Mt., 5,000 ft. *Kalbreyer* 105 ! Also in French Cameroons, Gabon, Belgian Congo, A.-E. Sudan, Kenya, Uganda, N. Rhodesia and Angola. (See Appendix, p. 298.)

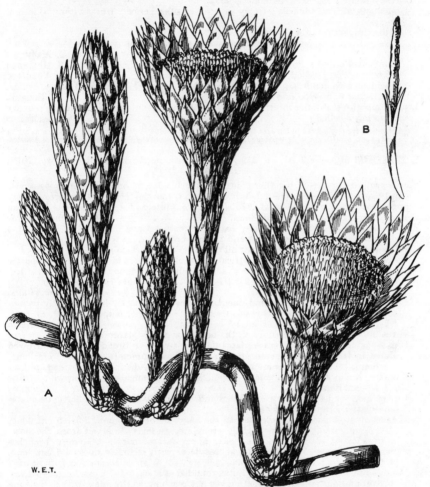

W. E. T.

FIG. 188.—THONNINGIA SANGUINEA *Vahl* (BALANOPHORACEAE).
A, plant with several flower heads. B, male flowers.

112. RHAMNACEAE

Trees or shrubs, sometimes climbing. Leaves simple, alternate or opposite ; stipules mostly present. Flowers often cymose or fasciculate, small, hermaphro- dite or polygamous. Calyx-lobes valvate. Petals 4 or 5, small, or absent. Stamens 4–5, opposite the petals ; anthers 2-celled, opening lengthwise. Disk mostly present, perigynous. Ovary sessile, superior or subinferior, 2–4-celled ; ovules solitary or rarely paired, erect from the base, anatropous. Fruit various, often drupaceous. Seeds mostly with copious endosperm and large straight embryo.

Mainly tropics and temperate regions ; comparatively rare in Africa ; recognized at once by the stamens being opposite the free often spathulate petals and presence of a disk.

Leaves alternate, or if subopposite then stipules extrapetiolar :
Leaves strongly 3 (–5)-nerved from the base, branchlets with stipular spines ; fruits subglobose, drupaceous, 1-celled   ..   ..   1. Ziziphus
Leaves pinnately nerved ; branchlets not spiny :
Branchlets without tendrils ; ovary superior or subsuperior ; fruits without 3 lateral wings :
Fruits drupaceous, not winged :
Ovary 1-celled ; flowers in axillary cymes, several- to many-flowered
2. Maesopsis
Ovary 3–4-celled ; flowers solitary or few together in the leaf-axils   3. Rhamnus
Fruits nut-like, 1-celled, with a long terminal wing ..   ..   ..   4. Ventilago
Branchlets with tendrils ; ovary and fruit inferior, the latter with 3 lateral wings, separating when ripe into 3 cocci   ..   ..   5. Gouania
Leaves opposite ; stipules interpetiolar ; branchlets without spines or tendrils ; ovary subinferior, 3-celled   ..   ..   6. Lasiodiscus

Other genera are represented in our area by cultivated plants, including *Hovenia dulcis* Thunb., a native of China and Japan, and *Colubrina ferruginosa* (Jacq.) Brongn., a native of the West Indies and tropical America.

1. ZIZIPHUS Mill.—F.T.A. 1 : 379 ; Suessenguth in E. & P. Pflanzenfam. 20D : 123 (1953).

Flowers solitary or 2 (–3) together ; leaves suborbicular or broadly elliptic to ovate, 0·5–1·2 (–1·5) cm. long, 0·4–1 (–1·3) cm. broad, margins crenate-serrulate, densely pubescent beneath and somewhat less so above ; internodes on branchlets less than 1 cm. long ; fruits about 1 cm. diam.   ..   ..   ..   1. *lotus* subsp. *saharae*
Flowers several to many, in cymes ; leaves at least 1·3 cm. long and 0·9 cm. broad and usually more than 2 cm. long and 1·5 cm. broad :
Leaves densely and softly tomentose all over the lower surface :
Bases of leaves more or less symmetrical and equal ; flowers in sessile or very shortly pedunculate, subcapitate cymes ; leaves elliptic to broadly elliptic or ovate-elliptic, (1·3–) 2–7 cm. long, (0·9–) 1·1–3 cm. broad ; fruits 1·5–2 cm. diam.
2. *mauritiana*
Bases of leaves asymmetrical and unequal ; flowers in shortly but distinctly pedunculate, branched cymes ; leaves ovate to ovate-oblong or elliptic, 3·5–10 cm. long, 1·3–5·5 cm. broad ; fruits about 1·8 cm. diam.   ..   ..   ..   3. *abyssinica*
Leaves pubescent beneath mainly on the nerves or nearly glabrous :
Leaves asymmetrical and subcordate at base, ovate or broadly ovate, narrowed above to the obtuse or acute often more or less acuminate apex, 3–9 cm. long, 1·5–6 cm. broad, margins crenate-serrate ; flowers in distinctly branched, pedunculate, many-flowered cymes ; branches and branchlets red-brown ; fruits 1·2–1·8 cm. diam.   ..   ..   4. *mucronata*
Leaves more or less symmetrical and rounded to cuneate at base, elliptic or ovate-lanceolate ; branchlets whitish :
Leaves ovate-lanceolate, narrowed to the acute or obtuse apex, 2·5–8·5 cm. long, 1–3·5 cm. broad, margins slightly crenulate, strongly 3-nerved from the base, nerves often reaching to the apex, upper lateral nerves obscure ; branches whitish ; cymes many-flowered, subsessile or with peduncle up to 1·5 cm. long, a tree ; fruits about 2 cm. diam.   ..   ..   5. *spina-christi* var. *spina-christi*
Leaves broadly elliptic or ovate-elliptic, rounded at apex, 1·4–3·4 cm. long, 1–3·4 cm. broad, margins subentire, basal nerves not reaching to the apex, with 1–2 strong upper lateral nerves on each side of midrib ; branches becoming reddish-brown ; cymes several-flowered, with a distinct peduncle 2–3 mm. long ; a much-branched shrub ; fruits 0·8–1 cm. diam.   ..   ..   5a. *spina-christi* var. *microphylla*

1. Z. lotus (*Linn.*) *Desf.* in Act. Acad. Sci. Paris 1788 : 443, t. 21 (1788) ; Lam. Encycl. 3 : 317 (1789). Z. lotus subsp. saharae *Maire* in Bull. Soc. Hist. Nat. Afr. Nord 20 : 179 (1929). *Z. saharae* Batt. & Trab. (1907), provisional name only. *Z. nummularia* (Burm.) Wight & Arn. var. *saharae* (Maire) A. Chev. in Rev. Bot. Appliq. 27 : 474 (1947). *Z. nummularia* of Aubrév. Fl. For. Soud.-Guin. 357–361, t. 74, 6. A much-branched shrub, to 4 ft. high.
Fr.Sud.: Gao (fl. & young fr. Aug., Sept.) *Bates*! *Hagerup* 337a! Bourem *Monod* 483. Also in Rio de Oro, Mauritania and the Sahara.
[The subsp. *lotus* occurs in N. Africa and the Mediterranean area ; the subsp. *saharae* is quite distinct from *Z. nummularia* (Burm.) Wight & Arn. with which it has been confused.]
2. Z. mauritiana *Lam.* Encycl. 3 : 319 (1789) ; Chev. l.c. 476, excl. var. *abyssinica* ; Aubrév. l.c. t. 74, 1–4, (as *Z.* " *mauritiaca* "), incl. vars. *Z. orthacantha* DC. (1825). *Z. jujuba* (Linn.) Lam. (1789) ; F.T.A.

1 : 379, partly, Chev. Bot. 136, and F.W.T.A., ed. 1, 1 : 471, partly ; not of Mill. (1768). A shrub or tree to 40 ft. high, widespread in the drier regions and often cultivated ; flowers cream ; fruits reddish, edible.
Sen.: *Heudelot* 433 ! *Perrottet* 143 ! *Roger* ! **Gam.:** *Dawe* 78 ! MacCarthy Isl. (fr. Jan.) *Dalz.* 8041 ! **Fr.Sud.:** Bongouni (Apr.) *Chev.* 689 ! Bamako *Hagerup* 39 ! Timbuktu (July) *Hagerup* 142 ! Nioro to Mountan Kagoro *Roberty* 10216 ! **Fr.G.:** Dombiagu *Pobéguin* 2111. **S.L.:** Freetown, cult. (Oct.) SLFD 26 ! Kambia, cult. (Dec.) *Deighton* 901 ! **Iv.C.:** Ouagadougou *Serv. For.* 2333. Bobo Dioulasso *Serv. For.* 2220. **G.C.:** Tamale *Anderson* 34 (partly) ! Bjury (July) *Chipp* 497 ! Yegi (July) *Vigne* FH 1249 ! **Togo:** Kratchi *Anderson* 34 (partly) ! **Fr.Nig.:** Zinder (fl. & fr. Nov.) *Hagerup* 597 ! **N.Nig.:** Zurmi (Apr.) *Keay* FHI 16149 ! Katagum *Dalz.* 144 ! N. Bornu (Oct.–Dec.) *Daggash* FHI 22006 ! *E. Vogel* 23 ! Yola (fr. Oct.) *Shaw* 91 ! Widespread in the drier parts of tropical Africa and in Asia. (See Appendix, p. 299.)
[There are a number of forms or varieties included in this widespread, often cultivated, species. *Chev.* 23923 from Djougou, Dah., *Hagerup* 610 from Zinder, Fr. Nig., and *Elliott* from the Komadugu Yobe, N.Nig., appear to be hybrids between *Z. mauritiana* and *Z. spina-christi*.]
3. **Z. abyssinica** *Hochst. ex A. Rich.* Fl. Abyss. 1 : 136 (1847) ; Aubrév. l.c. *Z. mauritiana* var. *abyssinica* (Hochst. ex A. Rich.) Fiori (1912)—Chev. in Rev. Bot. Appliq. 27 : 477. *Z. atacorensis* A. Chev. l.c. 479 (1947), incl. var. *oblongifolia*. *Z. jujuba* of various authors, including F.T.A. 1 : 379, partly and F.W.T.A., ed. 1, 1 : 471, partly ; not of Mill. (1768). A shrub, sometimes scandent, to 12 ft. high, with purple-brown bark ; branchlets and underside of leaves with fawn indumentum ; flower yellow-green ; fruits deep purple-brown.
Sen.: Niokolo-Koba *Berhaut* 1454. **Fr.G.:** Kouroussa *Pobéguin* 426 ! **Iv.C.:** Léo *Serv. For.* 2485 ! **G.C.:** Pan to Bujan *Kitson* 913 ! Tumu (fr. Apr.) *Vigne* FH 3778 ! **Dah.:** Atacora Mts. (June) *Chev.* 24090. **Fr.Nig.:** Konkobiri to Diapaga, Gourma (July) *Chev.* 24367 (partly) ! **N.Nig.:** Sokoto *Ryan* 8 ! Nupe *Barter* 1330 ! Zaria (fl. Mar.–Aug., fr. June–Aug.) *Keay* FHI 25893 ! *Hubbard* T. 71 ! Yola (fl. July, fr. July, Oct.) *Dalz.* 88 ! *Shaw* 101 ! **S.Nig.:** Lagos, cult. (Aug.) *Dalz.* 1373 ! Widespread in the drier parts of tropical Africa. (See Appendix, p. 299.)
4. **Z. mucronata** *Willd.* Enum. Pl. Berol. 251 (1809) ; F.T.A. 1 : 380 ; Chev. Bot. 136 ; Aubrév. Fl. For. Soud.-Guin. 359–361, t. 74, 5. An erect or scandent shrub (or, in other parts of Africa, a tree), with dark reddish-brown branchlets ; flowers greenish-yellow.
Sen.: *Perrottet* 144 ! *Roger* ! Rufisque (Sept.) *Monod* ! Richard Tol (Oct.) *Martine* 13 ! **Fr.Sud.:** Bakel *Vuillet* 274. Koulikoro *Vuillet* 151. S. of Ségou (fr. May) *Irvine* 3204 ! Arbala *Dubois* 76. **Fr.G.:** Youkounkoun *Pobéguin* 2146. **Iv.C.:** Ouagadougou *Aubrév.* 2425. Boromo *Aubrév.* 2748. **G.C.:** Passankwia to Bugiyenga, White Volta R. *Kitson* 752 ! **Fr.Nig.:** Konkobiri to Diapaga, Gourma (July) *Chev.* 24367 (partly) ! **N.Nig.:** Jebba (Dec.) *Barter* 1173 ! *Hagerup* 688 ! Zungeru (Sept.) *Dalz.* 75 ! Zaria (Aug.) *Keay* FHI 28023 ! Geidam (Dec.) *Elliott* 132 ! **S.Nig.:** R. Tessi, Old Oyo (Feb.) *Keay* FHI 16286 ! Olokemeji (fl. & young fr. Nov.) *A. F. Ross* 31 ! **Br.Cam.:** Gulumba, Dikwa Div. (Dec.) *McClintock* 48 ! Widespread in the drier parts of tropical Africa and in S. Africa. (See Appendix, p. 299.)
5. **Z. spina-christi** (*Linn.*) *Desf.* var. **spina-christi**—Fl. Atlantica 1 : 201 (June 1798) ; Willd. Sp. Pl. 1 : 1105 (1798) ; F.T.A. 1 : 380 ; Chev. in Rev. Bot. Appliq. 27 : 475 ; Aubrév. l.c. t. 74, 7. *Rhamnus spina-christi* Linn. (1753). A tree, to 30 ft. high, with pale grey-bark ; planted in towns and villages ; flowers greenish-yellow ; fruits reddish when ripe.
Sen.: Niokolo-Koba *Berhaut* 1251. **Fr.Sud.:** Koulikoro, R. Niger *Vuillet* 14. **Fr.Nig.:** Zinder (Oct.) *Hagerup* 564 ! **N.Nig.:** Rabba *Barter* 3428 ! Nupe *Barter* 736 ! 1707 ! Katsina (fl. & fr. Mar.) *Meikle* 1351 ! Gujba (Dec.) *Foster* 82 ! Dunjere, Komadugu Yobe (fl. & fr. Dec.) *Elliott* 138 ! Kukawa (fl. & fr. Jan.) *E. Vogel* 31 ! Widespread in the drier parts of tropical Africa and Egypt to tropical Asia. (See Appendix, p. 300.)
[The description of this cultivated variety in the key above is based only on specimens collected in W. Africa. The differences between the cultivated and wild forms are less clear in other areas.]
5a. **Z. spina-christi** var. **microphylla** *Hochst. ex A. Rich.* Fl. Abyss. 1 : 137 (1847). *Z. amphibia* A. Chev. in Rev. Bot. Appliq. 27 : 480 (1947) ; Aubrév. l.c. 359–361. *Z. spina-christi* of F.W.T.A., ed. 1, 1 : 470, partly (*Barter* 738). A much-branched shrub forming impenetrable thickets in river-beds ; flowers pale yellow ; fruits red when ripe ; an indigenous plant.
Sen.: Walo *Berhaut* 1396. **Fr.Sud.:** R. Bani, San (June) *Chev.* 1084 ! Sansanding *Chev.* 28. Tandifarma, Niafunké *Chev.* 42926. Bamako *Vuillet*. **Fr.G.:** Rio Grande, Kadé *Pobéguin* 1990. **G.C.:** Mpeasun *Anderson* 31 ! **N.Nig.:** Nupe *Barter* 738 ! Also in N.E. Africa and the Middle East.

2. MAESOPSIS Engl. Pflanzenw. Ost-Afr. C : 255 (1895) ; Suessenguth in E. & P. Pflanzenfam. 20D : 149 (1953).

Leaves alternate or subopposite, broadly oblong-lanceolate, rounded at base, gradually acuminate, 8–15 cm. long, 3–5 cm. broad, entire or remotely toothed, slightly puberulous on the nerves beneath, with about 8–10 pairs of lateral nerves and numerous fine parallel tertiary nerves ; petiole 0·5–1 cm. long ; stipules subulate, 7 mm. long ; cymes axillary, rusty-pubescent, about ¼ as long as the leaves ; sepals puberulous ; fruit oblong, acute, turgid, 3 cm. long, glabrous   ..   ..   *eminii*

M. eminii *Engl.* l.c. (1895) ; Aubrév. Fl. For. C. Iv. 2 : 207, t. 208, 1–5 ; Eggeling Indig. Trees Uganda, ed. 2 : 323, fig. 68, photo 52 ; Suessenguth l.c. fig. 40. *Karlea berchemioides* Pierre (1897). *Maesopsis berchemioides* (Pierre) A. Chev. (1917). A fast-growing deciduous tree, up to 90 ft. high, or rarely more ; flowers small, greenish ; fruits purplish-black when ripe ; in forest regrowth ; rare in our area, except in S.E. Nigeria and the Cameroons.
Lib.: Dukwia R. (fr. May) *Cooper* 429 ! **Iv.C.:** Abidjan *Aubrév.* 1489. Yapo *Aubrév.* 606. Agboville *Aubrév.* 2355. **G.C.:** Mampong Scarp (fl. & fr. May) *Vigne* FH 1724 ! 2749 ! **Togo:** Togo Plateau F.R. (fr. Dec.). *St. C. Thompson* FH 3707 ! Misahöhe *Mildbr.* **S.Nig.:** Sapoba (Feb.) *Kennedy* 2142 ! Udi Plateau *Carter* ! Kataban, Ikom Div. (fr. May) *Latilo* FHI 31000 ! **Br.Cam.:** Johann-Albrechtshöhe *Staudt* 533 ! **F.Po:** Bokoko *Mildbr.* 6876. Also in French Cameroons, Gabon, Belgian Congo, Angola, A.-E. Sudan, Uganda and Tanganyika. (See Appendix, p. 298.)

3. RHAMNUS Linn.—F.T.A. 1 : 381 ; Suessenguth in E. & P. Pflanzenfam. 20D : 59 (1953).

Shrub or small tree ; branchlets pubescent at first, then glabrescent ; leaves ovate to elliptic or elliptic-lanceolate, cuneate to rounded at base, acute or acutely acuminate, 2–10 cm. long, 1–4 cm. broad, margins serrulate, with 5–7 main lateral nerves on each side of the midrib ; glabrous, or sparsely pubescent beneath ; petioles 0·5–1·2 cm. long ; flowers solitary or a few together in leaf-axils, on slender pedicels about 1 cm. long ; calyx 2·5–3 mm. long, lobes triangular ; petals spathulate, about 1 mm. long ; disk and ovary glabrous ; fruits containing 3–4 single-seeded pyrenes, globose, about 5 mm. diam., on pedicels about 2 cm. long .. ..   *prinoides*

R. prinoides *L'Hérit.* Sert. Angl. 6, t. 9 (1788) ; F.T.A. 1 : 382 ; Engl. Pflanzenw. Afr. 3, 2 : 311 ; Staner in Bull. Jard. Bot. Brux, 15 : 403, fig. 48. *Celtis rhamnifolia* C. Presl (1845). To 25 ft. high ; in montane woodland and margins of montane forest ; flowers greenish-yellow ; fruits red.
     Br.Cam.: Bamenda Prov. : slopes of Mba Kokeka Mt., 6,700 ft. (fl. & fr. Jan.) *Keay* FHI 28391 ! Lakom, 6,000 ft. (fl. & fr. Apr.) *Maitland* 1666 ! Banso *Ledermann*. Also in French Cameroons, and widespread on the mountains of Africa from Abyssinia to the Cape and Angola.

4. VENTILAGO Gaertn.—F.T.A. 1 : 378 ; Suessenguth in E. & P. Pflanzenfam. 20D : 151 (1953).

Branchlets, inflorescences, outside of flowers, petioles and nerves on underside of leaves sparingly pubescent and then glabrescent ; upper surface of leaves glabrous and shining from the start ; leaves oblong-elliptic to oblong-lanceolate, rounded or subcordate at base, obtusely and broadly acuminate, 2·5–12·5 cm. long, 1–5 cm. broad, margins serrulate, with 5–7 main lateral nerves on each side of midrib ; flowers in axillary fascicles, sometimes on leafless branchlets ; fruits broadly ovoid, produced above into an elongate-oblong-lanceolate, glabrous wing up to 5 cm. long and 1 cm. broad, girt at base by the rim-like calyx .. .. ..   1. *africana*
Branchlets, inflorescences, outside of flowers, petioles and undersurfaces of leaves densely brownish-pubescent, becoming less so with age ; upper surface of leaves pubescent at first ; leaves ovate to oblong-lanceolate, rounded and sometimes unequal at base, acute or subacute and mucronate at apex, 2·5–8 cm. long, 1·3–4 cm. broad, margins crenulate ; fruit-wings up to 6 cm. long and 1·1 cm. broad, densely golden-puberulous ; otherwise similar to above .. ..   2. *diffusa*

1. **V. africana** *Exell* in J. Bot. 65, Suppl. Polypet. 80 (1927) ; Staner in Bull. Jard. Bot. Brux. 15 : 396, fig. 45. *V. leiocarpa* of F.T.A. 1 : 378, partly (*Barter* spec.) ; of Chev. Bot. 136 ; not of Benth. *V. madraspatana* of Engl. Pflanzenw. Afr. 3, 2 : 310, partly, not of Gaertn. A large, evergreen, softly woody liane ; flowers greenish ; in forest, especially marshy areas.
     Port.G.: Fonte, Bubaque (May) *Esp. Santo* 2019 ! **Fr.G.**: Kouria (Oct.) *Chev.* 14898 ! **S.L.**: Mamaha (Nov.) *Thomas* 4586 ! Senge, between Gbangbama to Gbangbatok (Nov.) *Deighton* 2370 ! **Iv.C.**: Bouroukrou (fr. Jan.) *Chev.* 16532 ! R. Mé, Attie (Feb.) *Chev.* 17386. **G.C.**: Huniso (fr. Jan.) *Irvine* 1067 ! **S.Nig.**: Epe *Barter* 3287 ! Sunmoge, Ijebu Ode Prov. (fr. Jan.) *Onochie* FHI 20659 ! Iso Bendiga to Bendiga Afi, Ikom (fr. May) *Latilo* FHI 31817 ! Also in French Cameroons, Uganda, Belgian Congo and Angola.
2. **V. diffusa** (*G. Don*) *Exell* Cat. S. Tomé 139 (1944). *Celastrus diffusus* G. Don (1832). *Ventilago leiocarpa* of F.T.A. 1 : 378, partly.
     Br.Cam.: Victoria (Dec.) *Winkler* 606 ! Also in Uganda and on S. Tomé.

5. GOUANIA Jacq.—F.T.A. 1 : 383 ; Suessenguth in E. & P. Pflanzenfam. 20D : 166 (1953).

Leaves glabrous beneath except on the main nerves ; main lateral nerves 3–4 on each side of midrib, venation not prominent ; petals much longer than the sepals ; disk not lobed ; leaves ovate, rounded or subcordate at base, broadly acuminate, 3·5–9·5 cm. long, 2–6 cm. broad, obtusely crenulate ; petioles 0·5–3 cm. long ; racemes terminal or axillary, slender, sometimes bearing tendrils ; flowers small, in separate clusters on the racemes ; fruits inferior, 3-winged, broadly orbicular, cordate at each end, nearly 1 cm. diam., glabrous, wings very loosely reticulate   1. *longipetala*
Leaves tomentellous beneath ; main lateral nerves 5–7 on each side of midrib, venation prominent and closely reticulate beneath ; petals of the same length as the sepals ; disk lobed ; otherwise similar to the above .. ..   2. *longispicata*

1. **G. longipetala** *Hemsl.* in F.T.A. 1 : 383 (1868), partly (excl. *Kirk* spec.) ; M. L. Green in Kew Bull. 1916 : 198 ; Chev. Bot. 138 ; Staner in Bull. Jard. Bot. Brux. 15 : 419, figs. 58, 59. *G. klainei* Pierre, name only. *Cissus uvifera* of Chev. Bot. 147, not of Afzel. A scandent shrub or liane ; flowers white, fragrant ; in forest regrowth and on forest margins.
     Fr.G.: R. Konkouré, Kouria (Oct.) *Caille* in *Hb.* Chev. 14969 ! Macenta (Oct.) *Baldwin* 9768 ! Nzérékoré (Oct.) *Baldwin* 9687 ! **S.L.**: Kukuna, Scarcies (fr. Jan.) *Sc. Elliot* 4741 ! Kambui *Lane-Poole* 454 ! Njala (fr. Nov.) *Deighton* 2563 ! Jigaya (Sept.) *Thomas* 2801 ! **Lib.**: Kakatown *Whyte* ! Ganta (Nov.) *Harley* ! Gbanga (Sept.) *Linder* 807 ! Vonjama (Oct.) *Baldwin* 9857 ! 9900 ! **Iv.C.**: Anoumaba (fr. Nov.) *Chev.* 22402 ! Guidéko *Chev.* 16431 ! Abradine *Chev.* 17568 ! Kassigné (fr. Jan.) *Chev.* 17170 ! **G.C.**: Anyinam (fl. & fr. Dec.) *Dalz.* 113 ! Obuasi (Nov.) *Chipp* 582 ! Larte Hills (Nov.) *Johnson* 812 ! Assuantsi road (fr. Jan.) *Fishlock* 12 ! **S.Nig.**: Akure F.R. (Oct.) *Ejiofor* FHI 24607 ! Nikrowa, Benin (fl. & fr. Dec.) *Brenan* 8391 ! Ubuluku (fr. Dec.) *Thomas* 2095 ! Oban *Talbot* 1361 ! **Br.Cam.**: Victoria (fr. Feb., Nov.) *Maitland* 114 ! 379 ! **F.Po**: (Dec.) *Mann* 17 ! Also in French Cameroons, Spanish Guinea, Gabon, Belgian Congo and Angola. (See Appendix, p. 298.)
2. **G. longispicata** *Engl.* Pflanzenw. Ost-Afr. C : 256 (1895) ; M. L. Green l.c. ; Staner l.c. 421, fig. 60. A scandent shrub or liane ; leaves greyish beneath ; flowers greenish-yellow ; in montane forest.
     S.Nig.: west side of Mt. Koloishe, 4,300 ft., Obudu Div. (Dec.) *Keay & Savory* FHI 25055 ! Also in Belgian Congo, A.-E. Sudan, Uganda, Kenya, Tanganyika, Rhodesia, Nyasaland and Mozambique.

6. LASIODISCUS Hook. f.—F.T.A. 1 : 385 ; Suessenguth in E. & P. Pflanzenfam. 20D : 109 (1953).

Flowers in pedunculate cymes :
   Branchlets and petioles pilose-hispid ; leaves opposite, elliptic, unequal-sided and cordate at base, broadly acuminate, 7–12 cm. long, 4–6 cm. broad, thinly pilose on

the nerves beneath or nearly glabrous ; lateral nerves 6–8 pairs, very prominent beneath, with slender subparallel tertiary nerves ; petiole about 5 mm. long ; cymes few-flowered, up to 8 cm. long ; flowers subumbellate ; pedicels about 7 mm. long, softly tomentellous ; young fruits truncate, broadly campanulate, tomentellous

1. mildbraedii

Branchlets and petioles more or less villous ; leaves more or less as above but not cordate, up to 20 cm. long and 10 cm. broad, shortly pubescent beneath ; lateral nerves about 9 pairs ; cymes lax, dichotomously branched ; pedicels about 4 mm. long, tomentose ; fruits depressed-globose, slightly trilobed, about 1·4 cm. diam., finely tomentellous .. .. 2. mannii

Flowers in axillary fascicles ; branchlets and petioles pubescent ; leaves opposite, elliptic-obovate, more or less narrowed at base, long-acuminate, 15–20 cm. long, 5–7 cm. broad, glabrous or slightly pubescent on the nerves beneath ; lateral nerves about 7–9 pairs, with subparallel tertiary nerves ; pedicels 3–4 mm. long, softly tomentose ; sepals 2·5 mm. long, tomentose .. .. .. 3. fasciculiflorus

1. **L. mildbraedii** *Engl.* Bot. Jahrb. 40 : 552 (1908) ; Staner in Bull. Jard. Bot. Brux. 15 : 412, fig. 54. *L. chevalieri* Hutch. (1912)—F.W.T.A., ed. 1, 1 : 472 ; Aubrév. Fl. For. C. Iv. 2 : 208. A small tree, up to about 30 ft. high, with spreading crown ; flowers small, white in golden-brown hairy cymose panicles ; locally abundant in the drier parts of the forest regions.
   Iv.C.: Anno, between Tchumkrou and Akakoumoëkrou (fl. & fr. Dec.) *Chev.* 22526 ! Soubré (fr. June) *Chev.* 19120. Morénou (Dec.) *Chev.* 22525. Groumania *Aubrév.* 785 ! **G.C.:** Tanoso (Mar.) *Thompson* 51 ! Anobeko, W. Ashanti (fl. & young fr. Mar.) *Chipp* 126 ! Drabonso, Ashanti (Feb.) *Cansdale* FH 4143 ! Ejan, N. Agogo (Jan.) *Vigne* FH 1793 ! **S.Nig.:** Iguobazowa F.R., Benin *Onochie* FHI 19744 ! Also in French Cameroons, Belgian Congo, A.-E. Sudan, Uganda, Tanganyika and S. Tomé. (See Appendix, p. 298.)

2. **L. mannii** *Hook. f.*—F.T.A. 1 : 385 ; Aubrév. l.c. 210, t. 207, 7–8 ; Staner l.c. 413, fig. 55 ; Exell in Cat. S. Tomé 140. A tree up to 30 ft. high with several trunks, forming a wide crown, or a shrub ; flowers white ; in moist places in rain forest.
   S.L.: Kambui F.R. (Oct.) *Edwardson* 152 ! **Iv.C.:** Man to Danané (Mar.) *Aubrév.* 1094 ! **S.Nig.:** Eket *Talbot* 3238 ! **Br.Cam.:** Bopo (Mar.) *Onochie* in *Hb. Brenan* 9439 ! Also in French Cameroons, Spanish Guinea, Gabon, Belgian Congo and Principe.
   [The specimen *Preuss* 695a, with the suggested locality " ? Victoria," cited in the first edition, proves to be *Zenker* 695a from Yaunde in the French Cameroons and is the type of *L. zenkeri* Suessenguth ; see Milne-Redhead in Kew Bull. 1950 : 366.]

3. **L. fasciculiflorus** *Engl.* l.c. 550 (1908) ; Aubrév. l.c. 210, t. 208, 6–10 ; Staner l.c. 411, fig. 53. A shrub or small tree up to 15 ft., or rarely to 40 ft., high ; flowers white ; ripe fruits brown ; in forest, especially in moist places.
   S.L.: Njala (fl. & fr. Jan., fl. Sept.) *Deighton* 4696 ! 5593 ! Taiama (fl. & fr. Nov.) *Deighton* 2558 ! R. Seli (Aug., Oct.) *Deighton* 1399 ! *Thomas* 3106 ! Yonibana (Oct.) *Thomas* 4228 ! 4309 ! **Lib.:** Gbanga (Sept.) *Linder* 769 ! Ganta (fl. Sept., fr. Feb.) *Baldwin* 11018 ! 12510 ! Genna Tanyehun, Grand Cape Mount (fr. Dec.) *Baldwin* 10745 ! **Iv.C.:** Agbo bridge (Jan.) *Chev.* 16757 ! Bouroukrou (Dec.–Jan.) *Chev.* 16772. Guidéko to R. Zozro (June) *Chev.* 19020. Mt. Momy *Chev.* 21369. **G.C.:** Amentia (Sept.) *Vigne* FH 1374 ! **S.Nig.:** Ondo (Dec.) *A. F. A. Lamb* 74 ! Oluwa F.R. *Kennedy* 2495 ! Okomu F.R., Benin (fl. & fr. Dec.) *Brenan* 8487 ! 8503 ! Oban *Talbot* ! **Br.Cam.:** N. Korup F.R., Kumba Div. (June) *Olorunfemi* FHI 30651 ! Also in French Cameroons and Belgian Congo. (See Appendix, p. 298.)

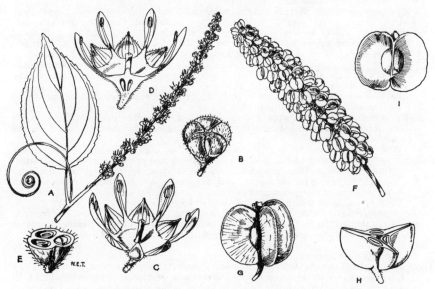

Fig. 189.—Gouania longipetala *Hemsl.* (Rhamnaceae).

A, inflorescence. B, flower-bud. C, open flower. D, same in vertical section. E, cross-section of ovary. F, fruits. G, single fruit. H, same in cross-section. I, winged seed.

113. AMPELIDACEAE

Climbing shrubs or small trees or herbs from a perennial rootstock, with nodose or jointed stems, often with watery juice. Leaves alternate or the lower sometimes opposite, simple or digitately or bipinnately compound, often pellucid-punctate ; stipules usually present. Flowers hermaphrodite or unisexual, actinomorphic, small, in leaf-opposed spikes, racemes, cymes or panicles ; peduncles often cirrhose. Calyx small, entire or toothed. Petals 4–5, free or united, valvate. Stamens 4–5, opposite the petals. Disk present within the stamens. Ovary 2–6-celled, cells 1–2-ovuled ; style short. Fruit baccate. Seeds with copious sometimes ruminate endosperm and small embryo.

General in the tropics and warm temperate regions. Distinguished by the usually climbing habit with tendrils, the small flowers, valvate petals with the stamens opposite to them.

Leaves simple, or 3-foliolate or digitately compound, sometimes pedate ; stamens free from each other ; ovary 2-celled, cells 2-ovuled ; tendrils mostly present :
 Tendrils leaf-opposed, never borne on the inflorescence, or absent ; flowers bisexual :
  Petals 4 ; disk with margin 4-lobed, adnate to base of ovary only ; style long and slender ; leaves various   ..   ..   1. **Cissus**
  Petals 5–7 ; disk entire, with the ovary immersed in it ; style short and stout ; leaves trifoliolate   ..   ..   2. **Rhoicissus**
 Tendrils borne on the inflorescence ; flowers polygamo-monoecious   3. **Ampelocissus**
Leaves bipinnate ; stamens united into a tube adnate to the base of the petals ; ovary 3–6-celled, cells 1-ovuled ; tendrils absent   ..   ..   4. **Leea**

Besides the above, *Vitis vinifera* Linn., the Grape Vine, is sometimes cultivated in our area.

1. CISSUS Linn. Sp. Pl. 1 : 117 (1753) ; Suessenguth in E. & P. Pflanzenfam. 20D : 237 (1953).

*Leaves simple, sometimes lobed, but not digitately compound ; ultimate branching of inflorescence subumbellate ; unexpanded corolla mostly conical or subglobose, rarely cylindrical (No. 4), not constricted in the middle ; mostly climbing or creeping plants, with tendrils (but see Nos. 3, 8, 9 & 22) :
 Leaves when mature digitately lobed nearly to the base, lobes often again deeply lobulate, margins dentate, young leaves entire or only partially lobed ; leaves up to 11 cm. long, pubescent beneath ; unexpanded corolla ovoid to subglobose, about 1 mm. long ; fruits 4–5 mm. diam.   ..   ..   1. *palmatifida*
 Leaves not lobed or sometimes slightly so :
  Stems very succulent, sharply quadrangular, with sides 6–15 mm. wide, constricted at the nodes, green, usually with only a few leaves near the tips ; leaves ovate-triangular, widely truncate or subcordate at base, obtuse or subacute at apex, margins rather coarsely toothed, rarely 3-lobed, up to 11 cm. long and broad ; petioles up to 3·5 cm. long ; flowers rather few, in short cymes ; plant glabrous
  2. *quadrangularis*
  Stems not or only slightly succulent, if quadrangular then sides less than 6 mm. wide ; leaves pubescent or glabrous :
   †Leaves as long as broad or only slightly longer than broad, more or less cordate or very broadly rounded at base :
    Undersurface of leaves with a very short but dense tomentellous indumentum covering the lamina between the veins, longer hairs present or absent on the nerves and veins :
     Plant erect, with stout woody stems up to 1 m. high ; tendrils absent or only vestigial ; leaves ovate to orbicular or polygonal-orbicular, rounded or sometimes cordate at base, rounded or obtuse at apex, up to 7·5–15 cm. long and broad, margin with setaceous teeth ; petioles 1–2 cm. long ; inflorescences rather densely many-flowered   ..   ..   3. *doeringii*
    Plant climbing or trailing ; tendrils well developed :
     Unexpanded corolla cylindrical, 2–3 mm. long, about 1 mm. diam. ; flowers in ample panicles at the ends of the branches ; leaves broadly ovate, cordate at base, long-acuminate, 5–10 cm. long and broad ; petioles 2·5–5 cm. long
     4. *amoena*

    Unexpanded corolla ovoid :
     Outside of corolla densely pubescent ; stems glabrous, with a glaucous, waxy bloom ; with black medifixed structures scattered over undersurface of leaves ; leaves variable in shape, often broadly ovate-orbicular, cordate at base, rounded or subacute at apex, 7–12 cm. long and broad, margins sometimes lobulate, with the nerves and veins extended into setaceous teeth ; petioles 1–4 cm. long ; fruits obovoid-ellipsoid, about 7 mm. long 5. *caesia*

Outside of corolla minutely puberulous ; stems sparingly pubescent, without
a glaucous bloom ; undersurface of leaves without black medifixed structures ;
leaves broadly ovate, widely cordate at base, acutely acuminate, 5–15 cm.
long, 4–14 cm. broad, margins dentate ; petioles 3–8 cm. long ; fruits fusi-
form-ellipsoid, 9–10 mm. long and 4 mm. broad ..   ..   6. *oreophila*

Undersurface of leaves glabrous, or pubescent or pilose mainly on the nerves and
veins, but without a close tomentellous covering on the lamina between the veins :

Leaves on both surfaces, petioles, stems, inflorescences and outside of flowers
densely woolly-pilose, the hairs drying reddish-brown ; leaves suborbicular or
ovate, sometimes lobulate, broadly truncate or widely cordate at base, 5–10 cm.
long and broad, margins jagged-toothed ; petioles 1–5·5 cm. long ; fruits
glabrous   ..   ..   7. *rubiginosa*

Leaves not woolly-pilose beneath :

Plants erect ; tendrils absent or only vestigial :

Leaves sparingly pubescent beneath, not reticulate, ovate, cordate at base,
acuminate, 5–7 cm. long, 3·5–6 cm. broad, margins dentate ;   petioles
2–4 cm. long ; cymes few-flowered, corolla about 2 mm. long, glabrous outside ; stems glabrous   ..   ..   8. *touraensis*

Leaves densely pubescent on the nerves and veins beneath, conspicuously
reticulate, broadly ovate to suborbicular, often repand-lobulate, deeply
cordate at base, rounded at apex, 7–18 cm. long and broad, margins with
setaceous teeth ; petioles 3–7 cm. long ; flowers rather numerous in wide
panicles, corolla about 1·5 mm. long, pubescent outside ; stems densely
pubescent ..   ..   9. *corylifolia*

Plants climbing or trailing ; tendrils well developed :

Outside of corolla densely pubescent ; leaves rather densely but shortly pubescent beneath on nerves and veins and also on the lamina, apex acuminate,
margins with setaceous teeth :

Leaves with numerous black medifixed structures beneath ;   corolla about
2 mm. long ; stems nearly glabrous ; leaves ovate-orbicular, often with
short, pointed lobes, deeply cordate at base, 10–18 cm. long, 10–14 cm.
broad   ..   ..   10. *kouandeensis*

Leaves without black medifixed structures beneath ;   corolla about 1 mm.
long ; stems densely pubescent at first, then more or less glabrescent ;
leaves broadly ovate or oblong-ovate, sometimes with short, pointed lobes,
widely cordate at base, 6–12 cm. long and broad   ..   ..   11. *sp. A*

Outside of corolla glabrous or very finely puberulous ; leaves glabrous, puberulous or pubescent beneath :

Corolla 1–1·5 mm. long, glabrous or minutely puberulous outside ; fruits up
to 8 mm. long :

Branchlets and petioles glabrous or nearly so ; leaves beneath papillose on
the lamina and finely puberulous on the midrib and nerves, otherwise
glabrous ; leaves drying a dull reddish-brown, broadly ovate, widely
cordate at base, acutely acuminate, 6–12 cm. long, 4–11 cm. broad, margins
irregularly and usually rather coarsely dentate ; petioles 2–8·5 cm. long ;
fruits obovoid, about 5 mm. long   ..   ..   ..   12. *glaucophylla*

Branchlets and petioles rather densely spreading pubescent ;   leaves curly
pubescent on nerves and veins beneath :

Leaves rounded at apex, variable in shape but usually more or less suborbicular ;   stems trailing ;   tendrils absent or short and unbranched
and not coiled ; flowers congested in the inflorescences ; fruits about
8 mm. long ; leaves 5–13·5 cm. long and broad, sometimes 3–5-lobulate,
margins with setaceous teeth ; petioles 0·5–4 (–7) cm. long 13. *rufescens*

Leaves acuminate, often acutely caudate-acuminate, usually broadly ovate,
widely cordate at base ; stems climbing ; tendrils well developed, bifurcate and coiled ; flowers in rather lax panicles ; fruits about 4 mm. long ;
leaves 6–12 cm. long, 5–11 cm. broad, margins with setaceous teeth ;
petioles 3–9 cm. long   ..   ..   14. *polyantha*

Corolla 2–3 mm. long, glabrous outside ; leaves and stems at first pubescent
beneath, then more or less glabrescent ; fruits 9–22 mm. long :

Stems terete ; leaves ovate-suborbicular, widely cordate at base, acuminate,
10–28 cm. long and broad, margins subentire or denticulate ; petioles
3–13 cm. long ; fruits ellipsoid, 1·7–2·2 cm. long ..   ..   15. *populnea*

Stems 4-gonous ; leaf-margins distinctly toothed :

Older stems with 4 or more corky wings ; upper surface of leaves appressed-
pubescent when young, venation rather obscure beneath ; pedicels about
1 cm. long ; fruits subglobose, about 1·2 cm. diam. ; leaves ovate-
polygonal, cordate at base, acuminate or more or less pointed at apex,
7–15 cm. long, 5·5–13 cm. broad ; petioles up to 18 cm. long   16. *petiolata*

Older stems without corky wings ; upper surface of leaves glabrous even when young, venation distinct and parallel beneath ; pedicels 6–8 mm. long ; fruits ellipsoid, about 9 mm. long and 7 mm. broad ; leaves broadly ovate or suborbicular-ovate or polygonal-ovate, cordate at base, shortly acuminate, 5–12 cm. long, 5–9 cm. broad ; petioles up to 6 cm. long   ..   ..   17. *arguta*

†Leaves always longer than broad, not, or not much, cordate at base :

  Stems climbing or trailing ; tendrils well developed :

  Cymes with peduncles 1 cm. or more long ; leaves and stems glabrous or very nearly so, petioles more than 1 cm. long :

  Unexpanded corolla 2·5–3·5 mm. long, conical, acute at apex :

  Pedicels puberulous ; unexpanded corolla 3–3·5 mm. long ; leaves oblong-lanceolate or oblong-ovate, obtuse to truncate at base, rarely slightly cordate, acuminate, 8–14 cm. long, 3–9 cm. broad, margins markedly denticulate ; petioles 2–6 cm. long ; fruits ellipsoid, 1·1–1·8 cm. long, 0·8–1 cm. broad
  18. *producta*

  Pedicels glabrous ; unexpanded corolla about 2·5 mm. long ; leaves usually rather narrowly oblong-ovate, truncate to cordate at base, long-acuminate, 8–15 cm. long, 3–8 cm. broad, margins undulate-dentate ; petioles 1–3·5 cm. long   ..   ..   19. *barteri*

  Unexpanded corolla 1–1·5 mm. long, ellipsoid-subglobose, rounded at apex ; leaves oblong, oblong-ovate or obovate-oblong, cordate, truncate-cordate or obtuse at base, acuminate, 6–9 (–16) cm. long, 2·5–5·5 (–8·5) cm. broad, margins denticulate or dentate ; petioles up to 5·5 cm. long ; fruits pyriform, about 9 mm. long and 7 mm. broad   ..   ..   20. *barbeyana*

  Cymes with the common peduncle less than 5 mm. long, the flowers appearing to be subfasciculate in the leaf-axils ; stems and midrib and nerves on underside of leaves quite densely pubescent to nearly glabrous ; unexpanded corolla about 1 mm. long ; fruits pyriform, about 6 mm. long ; leaves oblong or oblong-lanceolate, rounded at base, acutely acuminate, 5·5–11 cm. long, 2–4·5 cm. broad, margins remotely denticulate ; petioles 0·3–1 cm. long   21. *diffusiflora*

  Stems erect, or suberect, eventually becoming woody, up to 1 m. high, terete, red-tomentose when young ; tendrils absent ; flowers in lax cymes, appearing before the leaves ; fruits ellipsoid, apex acute, about 1·2 cm. long ; leaves ovate or oblong-lanceolate, rounded at base, acute at apex, 4–11 cm. long, 2·5–4·5 cm. broad, glabrous or nearly so beneath, margins subentire ; petioles 0·2–1 cm. long
  22. *cornifolia*

*Leaves digitately compound :

  ‡Stems creeping or climbing ; with tendrils :

  Leaflets pedate (i.e. lateral ones on a common stalk forked above the middle), 5 (–7) :

  Unexpanded corolla cylindrical, much longer than broad, constricted in the middle ; stipules ovate-lanceolate, persistent ; terminal leaflet more or less ovate-elliptic, up to 7·5 cm. long and 4·5 cm. broad, margins serrate, petiolule long, lateral leaflets similar but smaller ; plant puberulous or nearly glabrous   23. *adenocaulis*

  Unexpanded corolla short and subglobose ; stipules small and deciduous :

  Fruits 3–4 mm. diam. ; pedicels of terminal flowers usually 5–6 mm. long ; inflorescences usually 15–40 cm. long ; leaflets usually slightly toothed or sub-entire, mostly cuneate at base ; petioles and petiolules rather long ; leaflets glabrous or slightly puberulous beneath   ..   ..   24. *debilis*

  Fruits 5–8 mm. diam. ; pedicels of terminal flowers usually 2–3 mm. long ; inflorescences usually 5–15 cm. long ; leaflets rather coarsely serrate, mostly rounded or cordate at base ; petioles and petiolules rather short ; leaflets puberulous or densely shortly pubescent on nerves beneath   ..   25. *gracilis*

  Leaflets not pedate, 3–7 :

  Unexpanded corolla subglobose or ovoid, not constricted :

  Ultimate branching of inflorescence cymose ; unexpanded corolla subglobose ; leaflets 3, oblong-elliptic, rounded at base, sharply apiculate at apex, 4–7 cm. long, 2–3 cm. broad, margins repand-denticulate with rather sharp teeth, main lateral nerves 6–9, close ; petiolules well developed ; stems, leaves and inflorescences crisped-puberulous ; fruits subglobose, 0·7–1 cm. diam., 3–4-seeded
  26. *ibuensis*

  Ultimate branching of inflorescence subumbellate ; unexpanded corolla oblong-ovoid ; leaflets normally 5, oblong-obovate, cuneate into a fairly long petiolule at base, rounded or shortly acuminate at apex, 6–14 cm. long, 2–5 cm. broad, margins subrepand-serrate, main lateral nerves 3–5, rather distant ; stems and leaves glabrous ; fruits ellipsoid, up to 2·5 cm. long, 1-seeded by abortion
  27. *aralioides*

  Unexpanded corolla subcylindrical, constricted in the middle ; ultimate branching of inflorescence cymose :

Petiolules 3–5 cm. long ; stems half-woody ; leaflets (1–) 3–5, broadly elliptic-obovate to ovate-suborbicular, rounded or subcordate at base, abruptly acuminate, 12–15 cm. long, 8–9 cm. broad, pubescent on the nerves beneath, margins undulate-crenate ; petioles 12–18 cm. long ; inflorescences long-pedunculate, lax, ultimate branches and flower-buds rusty-puberulous ; unexpanded corolla, 2–2·5 mm. long, with subsessile glands at apex ..   ..   ..   28. *ornata*
Petiolules up to 1 cm. long ; stems herbaceous ; leaflets not as above :
  The leaflets with long hairs beneath :
    Leaflets 3, very thinly membranous, with long, jointed, mostly non-glandular hairs ; stems and petioles with long, jointed sometimes gland-tipped hairs ; central leaflet elliptic, cuneate or obtuse at base, shortly acuminate, 6–11 cm. long, 2·5–5 cm. broad, lateral leaflets strongly asymmetrical, broadly rounded on one side at base, margins coarsely crenate-dentate ; inflorescence puberulous and with gland-tipped hairs ; flower-buds glabrous or with a few non-glandular hairs ..   ..   29. *adenopoda*
    Leaflets 5, rarely 3, with gland-tipped hairs on midrib and nerves beneath ; stems and petioles with long, gland-tipped hairs ; leaflets obovate-elliptic, the central ones 9–13 cm. long, 4·5–6 cm. broad, acute at each end, margins crenate-dentate ; inflorescence spreading pubescent, with gland-tipped hairs ; flower-buds pubescent, sometimes with gland-tipped hairs at apex..   ..   30. *rubrosetosa*
  The leaflets with short hairs beneath :
    Leaflets often 7, sometimes 3–5, lanceolate, about 4 times as long as broad, cuneate at base, acute or obtuse at apex, 7–12 cm. long, 1·5–3·2 cm. broad, curly-pubescent beneath, rather densely so when young, margins dentate ; petioles 1–6.5 cm. long ; stems, petioles, inflorescence and flower-buds more or less densely curly-pubescent, usually with short gland-tipped hairs   ..   ..   .. 35. *lelyi*
Leaflets 1–3 times as long as broad, 3–5 :
  Gland-tipped hairs present on the apices of the flower-buds :
    Leaflets and inflorescence sparingly puberulous ; flower-buds almost glabrous except for the gland-tipped hairs ; leaflets 3 (–4), elliptic, acute at each end, up to 8 cm. long and 3 cm. broad, crenate-dentate in upper half ; petioles up to 4 cm. long ..   ..   31. *sp. B*
    Leaflets beneath and inflorescence more or less densely pubescent ; leaves normally 5-foliolate ; flower-buds puberulous :
      Stems and petioles rather densely puberulous without longer gland-tipped hairs ; petioles 6–10 cm. long ; leaflets elliptic or obovate-elliptic, acute or subacute at each end, up to 12·5 cm. long and 6·5 cm. broad, margins crenate-dentate ..   ..   32. *sp. C*
      Stems and petioles with longer and often gland-tipped hairs as well as the shorter puberulous indumentum :
        Petioles 5–17 cm. long ; stems and petioles with long, often gland-tipped hairs ; leaflets elliptic or obovate-elliptic, long-cuneate at base, acuminate at apex, 7·5–15 cm. long, 3·5–7 cm. broad, shortly puberulous beneath, margins crenate-dentate ; fruits about 5 mm. diam., on pedicels 1–2 cm. long   ..   ..   33. *vogelii*
        Petioles 0·8–4 cm. long ; stems and petioles with rather short gland-tipped hairs ; leaflets obovate-oblanceolate or obovate-elliptic, cuneate at base, subacute to rounded at apex, up to 12 cm. long and 6 cm. broad, densely puberulous to tomentellous beneath when young, margins irregularly serrate to subentire ; leaves sometimes 3-foliolate near the apices of shoots ; fruiting pedicels up to 1 cm. long   ..   ..   34. *cymosa*
  Gland-tipped hairs absent from the apices of the flower-buds :
    Stems and petioles with gland-tipped hairs about 1 mm. long as well as the curly-puberulous indumentum ; leaflets (3–) 5, elliptic or elliptic-obovate, the lateral ones obliquely ovate, cuneate to subrounded at base, acute or acuminate at apex, up to 9 cm. long and 4·5 (–5·5) cm. broad, finely puberulous beneath, margins crenate-dentate in upper part ; flower-buds puberulous
      36. *lageniflora*
    Stems and petioles without long gland-tipped hairs ; leaflets 5 :
      Unexpanded corolla 4–5 mm. long ; inflorescence and flower-buds densely tomentellous, with a few subsessile glands but without gland-tipped hairs ; leaflets elliptic or obovate-elliptic, cuneate at base, long-acuminate at apex, up to 10 cm. long and 4 cm. broad, sparingly puberulous on midrib, nerves and veins beneath, margins denticulate towards base, crenate-serrate with sharp teeth in upper part   ..   ..   37. *mannii*
      Unexpanded corolla about 3 mm. long ; leaflets rounded at base :
        Leaflets very minutely puberulous all over the lower surface ; inflorescence puberulous ; leaflets broadly obovate, obtuse at apex, central leaflet 5–7 cm. long, 4–5 cm. broad, margins rather distantly crenate
          38. *griseo-rubra*

Leaflets densely puberulous on midrib, nerves and veins beneath ; inflorescence shortly and densely pubescent ; leaflets broadly obovate, acute at apex, central leaflet up to 6·5 cm. long and 4·5 cm. broad, margins rather sharply crenate-serrate　　..　　..　　..　　..　　39. *sp. D*

‡Stems erect, without tendrils, produced annually from a robust perennial underground stock :

Leaves, stems and peduncles glabrous ; leaves 3-foliolate ; stems unbranched, with a terminal panicle :

Leaves sessile or rarely with a petiole up to 8 mm. long ; leaflets linear-lanceolate, up to 20–24 cm. long and 1·3–2·4 (–3) cm. broad, margins denticulate ; stipules ovate-lanceolate, up to 17 mm. long and 7 mm. broad, with ciliate margins, persistent　　..　　..　　..　　..　　..　　..　　..　　40. *juncea*

Leaves with distinct stout petioles 2–8 cm. long ; leaflets lanceolate, or elliptic-lanceolate, 10–28 cm. long, 1·5–8 cm. broad, margins denticulate to irregularly and rather coarsely serrate in the upper part ; stipules lanceolate, up to 8 mm. long and 2 mm. broad, with sparsely ciliate margins, caducous　41. *jatrophoides*

Leaves, stems and peduncles densely hairy ; leaves 3–5 (–7)-foliolate :

Leaves sessile, rarely some very shortly petiolate :

Undersurface of leaves and stems tomentose, stems also with longer gland-tipped hairs ; leaves 3-foliolate ; leaflets obovate-oblanceolate or oblanceolate, up to 18 cm. long and 8 cm. broad, margins doubly crenate　..　　42. *crotalarioides*

Undersurface of leaves and stems pulverulent-puberulous, stems with longer gland-tipped hairs ; leaves 3–5-foliolate ; leaflets oblanceolate, 8–12 (–20) cm. long, 3–4 (–11) cm. broad, irregularly serrate in the upper part, with sessile reddish glands on both surfaces　　..　　..　　..　　..　　..　　43. *waterlotii*

Leaves distinctly petiolate, rarely some at the apex nearly sessile :

Stems and petioles with long, slender, spreading, gland-tipped hairs, 1–10 mm. long ; stipules large, ovate-lanceolate, up to 2·5 cm. long ; leaves (3–) 5 (–7)-foliolate ; leaflets oblanceolate, very acute, up to 24 cm. long and 7 cm. broad, thinly pubescent beneath, coarsely toothed ; petioles 2–12·5 cm. long ; cymes rather densely many-flowered, with long subpersistent bracts　　..　　..　　44. *crinita*

Stems and petioles with short hairs, sometimes with glandular hairs up to 1 mm. long :

Flower-buds with long gland-tipped hairs at the apices, otherwise only sparingly pubescent ; leaves 5–7-foliolate ; leaflets lanceolate, the lateral ones oblique, gradually acute at each end, up to 22 cm. long and 7·5 cm. broad, densely pubescent beneath, puberulous above, margins irregularly crenate-serrate
　　　　　　　　　　　　　　　　　　　　　　　　45. *sokodensis*

Flower-buds densely and loosely pilose but without gland-tipped hairs at the apices :

Leaflets pilose with long curly hairs on both surfaces ; more or less oblanceolate, up to 12 cm. long and 3·5 cm. broad or more ; leaves (1–) 5–7 (–8)-foliolate
　　　　　　　　　　　　　　　　　　　　　　　　46. *chevalieri*

Leaflets puberulous above, sometimes with longer hairs beneath :

Leaflets scabrid-puberulous beneath ; more or less oblanceolate, up to 12 cm. long and 5 cm. broad ; leaves (3–) 5–7-foliolate　　..　　..　　47. *zechiana*

Leaflets at first densely tomentose between the veins beneath, becoming shortly pubescent as the lamina expands, up to 26 cm. long and 8·5 cm. broad, central leaflets elliptic-lanceolate, cuneate at base, lateral leaflets broadly rounded at base on proximal side ; leaves (3–) 5-foliolate　..　　..　　48. *flavicans*

1. **C. palmatifida** (*Bak.*) *Planch.* in DC. Monogr. Ampelid. 473 (1887) ; Gilg & Brandt in Engl. Bot. Jahrb. 46 : 484 ; Chev. Bot. 145, incl. var. *integrifolia* A. Chev. *Vitis palmatifida* Bak. in F.T.A. 1 : 397 (1868). *Cissus triangularis* A. Chev. in Rev. Bot. Appliq. 30 : 454 (1950). *C. vuilletii* A. Chev. l.c. 455 (1950). Stems suberect at first, then trailing or climbing, often purple-tinged ; flowers yellow ; ripe fruit black ; in savannah.
Fr.Sud.: Sindou (May) *Chev.* 865 ! Fô (June) *Chev.* 963 ! Koulikoro *Chev.* 3190 ! Koulouba *Vuillet* 137 ! 418 ! **Fr.G.:** Kankan *Chev.* 15679 ! Kouroussa (Apr.) *Pobéguin* 214 ! **G.C.:** Kintampo (Mar.) *Dalz.* 44 ! Salaga *Krause* ! Gbongoda, Gambaga (June) *Akpabla* 698 ! Mangkurugu, N.T. (May) *Kitson* ! **Dah.:** Agouagon to Savé (May) *Chev.* 23552 ! Kouandé to Konkobiri (fr. June) *Chev.* 24254 ! Atacora Mts. *Chev.* 23955 ! **N.Nig.:** Nupe *Barter* 974 ! 1278 ! Abinsi (fl. & fr. Apr.) *Dalz.* 678 ! Yola (June) *Dalz.* 65 ! **S.Nig.:** Olokemeji (Mar.) *Foster* 292 ! *Meikle* 1301 ! Ijalye F.R. (fl. & young fr. May) *Onochie* FHI 21998 ! Extends to A.-E. Sudan. (See Appendix, p. 302.)

2. **C. quadrangularis** *Linn.* Syst. Nat., ed. 12, 2 : 124 & Mant. Pl. 39 (1767) ; Gilg & Brandt l.c. 481 ; Chev. Bot. 146. *Vitis quadrangularis* (Linn.) Wall. ex Wight & Arn. (1834)—F.T.A. 1 : 399. A glabrous large climber or trailer, with stout succulent quadrangular and almost winged stems, leafy only on the young shoots ; fruits glabrous, red when ripe ; in savannah regions.
Sen.: Bakel *Leprieur* ! **Gam.:** MacCarthy Isl. *Dalz.* 8039 ! **Fr.Sud.:** Bani, San (June) *Chev.* 1097 ! **Iv.C.:** Kodiokofii Dist. (Aug.) *Chev.* 22348 ! **Dah.:** Cotonou (July) *Debeaux* 163 ! Ouandoukouana, Atacora Mts. *Chev.* 24162 ! **N.Nig.:** Nupe *Barter* ! Katagum (fr. July) *Dalz.* 80 ! Bauchi Plateau *Lely* P. 670 ! 38 miles from Kaduna on Kachiya road (fr. Dec.) *Meikle* ! Widespread in the drier parts of Africa, Arabia and India. (See Appendix, p. 303.)

3. **C. doeringii** *Gilg & Brandt* l.c. 473 (1912) ; White in Kew Bull. 1951 : 58. *C. corylifolia* var. *latifolia* A. Chev. Bot. 142 (1920), name only. A subshrub, with erect stems about 3 ft. high ; in savannah.
Fr.G.: Kesseridou to Oussoudou (Feb.) *Chev.* 20838 ! **S.L.:** Borawa Plateau *Edwardson* 89 ! Musaia (Mar., Apr.) *Deighton* 5422 ! 5424 ! **Iv.C.:** Mt. Dourou, Koualé (fr. May) *Chev.* 21734. Koualé to

Kouroukoro (May) *Chev.* 21750. **G.C.:** Techeri (fr. June) *Chipp* 476 ! **Togo:** Misahöhe (May) *Baumann* 30 ! 479 ; *Doering* 204.

4. **C. amoena** *Gilg & Brandt* l.c. 467, fig. 4 (1912). A climbing shrub with nearly glabrous branches ; in forest. **Br.Cam.:** Ekumba-Liongo *Dusen* 263. Kumba *Preuss* 537. Also in French Cameroons.

5. **C. caesia** *Afzel.* Remed. Guin. 55 (1815) ; Gilg & Brandt l.c. 474 ; F.W.T.A., ed. 1, 1 : 476, partly ; White l.c. 54. *Vitis caesia* (Afzel.) Sabine (1824)—F.T.A. 1 : 396, partly. A climber with green or pinkish flowers and black fruits ; mainly in savannah.
S.L.: *Afzelius* ! *Don* ! Waterloo (June) *Lane-Poole* 306 ! Kennedy Ridge, Freetown (Apr.) *Johnston* 6 ! Laminaiya (Apr.) *Thomas* 149 ! Pujehun (Apr.) *Deighton* 3690 ! (See Appendix, p. 302.)
[White also records *Barter* 563 from Nupe, N. Nigeria, as this species, but the specimen is poor and the determination is open to some doubt.]

6. **C. oreophila** *Gilg & Brandt* l.c. 471 (1912). *C. clematitis* A. Chev. Bot. 142, name only. *C. glaucophylla* of F.T.A. 1 : 392, partly ; of Gilg & Brandt l.c. 471, partly (*Barter* 3277). *C. polyantha* of F.W.T.A., ed. 1, 1 : 475, partly. A large herbaceous climber with tough stems ; flowers pink in bud, cream when open ; in forest.
Lib.: Tappita (fr. Aug.) *Baldwin* 9059 ! Gbanga (fl. Aug., fr. Dec.) *Baldwin* 10519a ! *Traub* 254 ! Flumpa (fr. Sept.) *Baldwin* 9345 ! **Iv.C.:** Guidéko (May) *Chev.* 16381 ! **G.C.:** Bunsu (June) *Hughes* 1125 ! **S.Nig.:** Gambari, S. of Ibadan (Aug.) *Jones* FHI 20257 ! Epe *Barter* 3277 ! **Br.Cam.:** Cam. Mt., 4,500 ft. *Mann* 1279 ! Buea (Mar.) *Maitland* 441 !
[This is closely related to *C. myriantha* Gilg & Brandt from French Cameroons and Belgian Congo, and to *C. tiliifolia* Planch. from Belgian Congo and A.-E. Sudan.]

FIG. 190.—CISSUS PALMATIFIDA (*Bak.*) *Planch.* (AMPELIDACEAE).

A, flowering shoot. B, flower. C, longitudinal section of same. D, cross-section of ovary. E, fruit. F, tendril.

7. **C. rubiginosa** (*Welw. ex Bak.*) *Planch.* l.c. 485 (1887) ; Gilg & Brandt l.c. 475. *Vitis rubiginosa* Welw. ex Bak. in F.T.A. 1 : 394 (1868). A climbing or trailing herb ; in savannah regions.
S.L.: Kabala to Musaia (Sept.) *Miszewski* 48 ! **Togo:** *Büttner* 6 ! Bismarckburg *Kling* 58. **N.Nig.:** Bauchi Plateau (Apr.) *Lely* P. 204 ! Jos (Aug.) *Keay* FHI 20054 ! Mongu, Plateau Prov. (fl. & fr. July) *Lely* 388 ! **S.Nig.:** Awba Hills F.R., Ibadan (Oct.) *Jones* FHI 6363 ! Widespread in tropical Africa.

8. **C. touraensis** *A. Chev.* in Rev. Bot. Appliq. 30 : 455 (1950). Stems erect, 1½–3 ft. high, from a deep rootstock ; in savannah.
Iv.C.: Koualé to Kouroukoro, Toura country (May) *Chev.* 21747 !

9. **C. corylifolia** (*Bak.*) *Planch.* l.c. 484 (1887) ; Gilg & Brandt l.c. 473 ; White l.c. 59. *Vitis corylifolia* Bak. in F.T.A. 1 : 396 (1868). *Cissus dahomensis* A. Chev. Bot. 143, name only, partly. Stems thick and succulent, erect, to 3 ft. high, from a stout rootstock ; calyx green in bud, becoming purple, petals yellow.
G.C.: Afram Plains (Mar.) *Johnson* 704 ! **Togo:** Sokode *fide* Gilg & Brandt l.c. **Fr.Nig.:** Diapaga, Gourma (July) *Chev.* 24426 *bis* ! **N.Nig.:** Nupe *Barter* 1271 ! Nabardo (May) *Lely* 207 ! Abinsi (Apr.) *Dalz.* 673 ! **S.Nig.:** Upper Ogun F.R., E. of Igboho (Feb.) *Keay* FHI 23437 ! Also in French Cameroons.

10. **C. kouandeensis** *A. Chev.* in Rev. Bot. Appliq. 30 : 454 (1950). *C. arguta* of Chev. Bot. 140, partly. *C. polyantha* of F.W.T.A., ed. 1, 1 : 475, partly. Stems widely scandent, glaucous.
Dah.: Kouandé, Atacora Mts. (June) *Chev.* 24007 !
[Chevalier states that his description of this species was made in the field and that the specimen was not kept ; *Chev.* 24007, of which there is a good duplicate at Kew, was, however, collected at the same place on the same day as Chevalier cites and agrees very well with his description.]

11. **C. sp. A.**[1] *C. hederifolia* A. Chev. Bot. 144, name only. Stems widely scandent or trailing ; inflorescence lax, flowers small.
Fr.G.: Pobégui n 2166 ! Timbo to Ditinn (Sept.) *Chev.* 18517 ! **N.Nig.:** Bauchi Plateau (July) *Lely* P. 352 ! Also in French Cameroons and A.-E. Sudan.

[1] These may be new species, but I hesitate to describe them until more is known about the reliability of some of the characters used to distinguish them. In particular, field research is needed on the taxonomic significance of the gland-tipped hairs which occur in several of the climbing species with compound digitate leaves.—R.W.J.K.

12. **C. glaucophylla** *Hook. f.* in Fl. Nigrit. 263 (1849) ; Gilg & Brandt l.c. 471, partly (excl. *Barter* 3277). *Vitis glaucophylla* (Hook. f.) Bak. in F.T.A. 1 : 392 (1868), partly. *V. smithiana* Bak. l.c. 390 (1868). *Cissus smithiana* (Bak.). Planch. (1887)—Gilg & Brandt l.c. 466. A liane ; in the forest regions.
S.L.: *Afzelius.* **Lib.:** Fortsville, Grand Bassa (fr. Mar.) *Baldwin* 11157 ! **S.Nig.:** Ondo (Nov.) *Onochie* FHI 34235 ! **Br.Cam.:** Victoria (fr. Feb.) *Maitland* 402 ! Mundame to Johann-Albrechtshöhe *Staudt* 432. *Winkler* 1011. **F.Po:** Clarence Cove (Nov.) *T. Vogel* 101 ! Also in French Cameroons, Spanish Guinea, Gabon, Belgian Congo and Angola.

13. **C. rufescens** *Guill. & Perr.* Fl. Seneg. 1 : 133 (1831) ; Gilg & Brandt l.c. 473 ; White in Kew Bull. 1951 : 55. *Vitis rufescens* (Guill. & Perr.) Miq. (1863). *V. caesia* of F.T.A. 1 : 396, partly. *Cissus caesia* of F.W.T.A., ed. 1, 1 : 476, partly. *C. corylifolia* of Chev. Bot. 142, partly. *C. pseudocaesia* of F.W.T.A., ed. 1, 1 : 476, partly. Perennial herb with underground stock and prostrate creeping annual stems ; stems and petioles crimson ; flower-buds red, open flowers yellow ; in savannah.
Sen.: Kaolak *Kaichinger.* **Gam.:** Albreda *Leprieur* ; *Perrottet* 139 ! Genieri to Kaiaff (fl. & fr. July) *Fox* 115 ! **Fr.Sud.:** Ténétou (Mar.) *Chev.* 660. **Fr.G.:** *Heudelot* 837 ! **G.C.:** Zantana waterside (May) *Kitson.* Ejura *Andoh* FH 5040 ! **Dah.:** Kouba to Farfa, Atacora Mts. (June) *Chev.* 24030 ! **N.Nig.:** Loin *Barter* 1273 ! Zaria (May) *Keay* FHI 25735 ! Pankshin (July) *Lely* 434 ! Abinsi (May) *Dalz.* 672 ! **S.Nig.:** road to Asipa, Lagos Colony *Phillips* 23 ! Olokemeji (Mar.) *Keay* FHI 25692 ! Bateriko to Bendi (fr. June) *Jones & Onochie* FHI 18987 ! Also in French Cameroons, Ubangi-Shari, A.-E. Sudan and Uganda.

14. **C. polyantha** *Gilg & Brandt* in Engl. Bot. Jahrb. 46 : 467 (1912). A half-woody climber with subterete brown-pubescent branches later glabrescent ; flowers small, numerous, in ample panicles ; in forest vegetation.
S.L.: *Afzelius* ! Kokoru, Gaura (fr. Oct.) *Deighton* 5208 ! Gegbwema, Tunkia (fr. Oct.) *Deighton* 5216 ! **Lib.:** Fishtown (young fr. Sept.) *Dinklage* 1671 ! Duport (Aug.) *Linder* 308 ! Jabroke, Webo (July) *Baldwin* 6631a ! 6697 ! Monroviatown (young fr. Aug.) *Baldwin* 8002 ! **Togo:** Misahöhe *Baumann* 314. **S.Nig.:** Lagos *MacGregor* 190 ! Calabar (Aug.) *Holland* 76 ! Oban *Talbot* 562 ! Also in French Cameroons, Ubangi-Shari. (See Appendix, p. 303.)

15. **C. populnea** *Guill. & Perr.* l.c. 134 (1831) ; Gilg & Brandt l.c. 463 ; Chev. Bot. 145. *Vitis pallida* of F.T.A. 1 : 393, partly. A liane with pale woody stems up to 3 in. diam. ; cut stems exuding copious clear watery sap ; flowers cream ; fruits blackish-purple when ripe ; in savannah.
Sen.: *Berhaut* 145. *Perrottet* 137 ! **Gam.:** Albreda (June, July) *Leprieur* ! Genieri to Massembe (July) *Fox* 170 ! **Fr.Sud.:** Koulikoro *Chev.* 3189. **Iv.C.:** various localities *fide* Chev. *l.c.* **G.C.:** Ejura (fr. Aug.) *Chipp* 742 ! Wanki (June) *Chipp* 480 ! Agomeda, Shai Plains (June) *Irvine* 1640 ! Gbongoda, Gambaga (July) *Akpabla* 721 ! **Togo:** Lome *Warnecke* 315 ! **Dah.:** Agouagon *Chev.* 23483. Savalou *Chev.* 23691. **N.Nig.:** Nupe *Barter* 1276 ! Gayam, Birnin Gwari (June) *Keay* FHI 25818 ! Anara F.R., Zaria (May) *Keay* FHI 22999 ! Katagum *Dalz.* 79 ! **S.Nig.:** Lagos *Rowland* ! Ijaiye F.R., near R. Ogun (Apr.) *Onochie* FHI 21988 ! Also in French Cameroons, Ubangi-Shari, A.-E. Sudan, Abyssinia and Uganda. (See Appendix, p. 303.)

16. **C. petiolata** *Hook. f.* l.c. 262 (1849) ; Gilg & Brandt l.c. 464. *Vitis suberosa* Welw. ex Bak. in F.T.A. 1 : 392 (1868). *Cissus suberosa* (Welw. ex Bak.) Planch. (1887). *Vitis pallida* of F.T.A. 1 : 393. *Cissus koumiensis* A. Chev. Bot. 144, name only. A climber with tetragonal stems becoming woody with age and bearing 4 or more corky wings ; flowers greenish-white ; fruits reddish when ripe.
Fr.G.: Koumi, between Dalaba and Songuéta (Oct.) *Chev.* 20167 ! Nzérékoré (Oct.) *Baldwin* 9730 ! **S.L.:** Gonkwi Mt., Talla Hills (fr. Feb.) *Sc. Elliot* 4880 ! **Lib.:** Vonjama (Oct.) *Baldwin* 9909 ! **G.C.:** Akwapim *T. Vogel* ! Okroase (May) *Johnson* 751 ! Achimota (Jan., June, Sept.) *Irvine* 860 ! 1469 ! 1968 ! Akwaseho (fl. & young fr. Jan.) *Irvine* 1671 ! **Togo:** Lome *Warnecke* 362 ! **S.Nig.:** Ibadan (fr. Feb.) *Ujor* FHI 27972 ! Gambari, Ibadan (June) *Jones* FHI 3626 ! **Br.Cam.:** Victoria (Apr.) *Ejiofor* FHI 29332 ! Bulifambo, Buea (Mar.) *Maitland* 552 ! Kumba (Apr.) *Hutch. & Metcalfe* 149 ! Also in Belgian Congo, A.-E. Sudan, Abyssinia, Eritrea, E. Africa, N. Rhodesia and Angola.

17. **C. arguta** *Hook. f.* l.c. 261 (1849). *Vitis arguta* (Hook. f.) Bak. in F.T.A. 1 : 392. *Cissus producta* of Gilg & Brandt l.c. 477, partly. Herbaceous climber with somewhat succulent tetragonous stems ; internodes short.
Lib.: *Christy* ! **S.Nig.:** Lagos (June, Oct.) *Dalz.* 1148 ! 1414 ! *Millen* 77 ! Ibadan (Oct.) *Jones & Keay* FHI 13750 ! Ibo country (Aug.) *T. Vogel* 3 ! Addaenda, R. Niger (Sept.) *T. Vogel* 141 ! (See Appendix, p. 302.)

18. **C. producta** *Afzel.* Remed. Guin. 63 (1815) ; Gilg & Brandt l.c. 477, partly (excl. syn. *C. arguta* Hook. f.) ; Chev. Bot. 145. *Vitis producta* (Afzel.) Bak. in F.T.A. 1 : 389 (1868). *Cissus purpurascens* A. Chev. Bot. 141, name only. *C. arguta* of F.W.T.A., ed. 1, 1 : 475, partly. *C. barteri* of Chev. Bot. 141. *C. afzelii* of Chev. Bot. 140, partly. A large herbaceous climber with pinkish flowers ; ripe fruits black ; in forest.
Sen.: *Berhaut* 1859. **Fr.Sud.:** Moussaia (fr. Feb.) *Chev.* 411 ! **Fr.G.:** Rio Nunez *Heudelot* 820 ! Massadou, S. of Macenta (May) *Collenette* 24 ! Nzérékoré to Macenta (fr. Oct.) *Baldwin* 9746 ! **S.L.:** *Afzelius* ! *T. Vogel* 106 ! Rokupr (Oct.) *Jordan* 136 ! Bumpe (fl. & fr. Nov.) *Deighton* 4951 ! Njala (Oct.) *Deighton* 1346 ! Jigaya (Sept.) *Thomas* 2700 ! **Lib.:** Gbanga (Sept.) *Linder* 634 ! Monrovia *Dinklage* 2772 ! *Whyte* ! Sinoe Basin *Whyte* ! Dukwia R. (Feb.) *Cooper* 198 ! **Iv.C.:** Ahiamé to Ahinta (Mar.) *Chev.* 17733 ! Kéeta (July) *Chev.* 19349 ! Guidéko (May) *Chev.* 16347 ! Alépé to Malamalasso (Mar.) *Chev.* 17518 ! Soubré (June) *Chev.* 19109 ! 19155 ! San Pedro (July) *Thoiré* 298 ! **G.C.:** Aburi (June) *Brown* 314 ! Pamu-Berekum F.R. (Sept.) *Vigne* FH 2484 ! Abofaw (May) *Andoh* FH 5496 ! **Togo:** Amedzofe (May) *Scholes* 48 ! **S.Nig.:** Oban *Talbot* 570 ! Ikom (June) *Latilo* FHI 31853 ! **F.Po:** *Mann* 273 ! Also in French Cameroons, Belgian Congo, Uganda, Tanganyika, N. Rhodesia, Angola and S. Tomé. (See Appendix, p. 303.)

19. **C. barteri** (*Bak.*) *Planch.* in DC. Monogr. Ampelid. 491 (1887) ; Gilg & Brandt l.c. 476. *Vitis barteri* Bak. in F.T.A. 1 : 390 (1868). Herbaceous climber with rather succulent terete stems, glabrous ; calyx pale green ; petals crimson outside, pale red within, centre of flowers white ; in forest.
S.Nig.: Okomu F.R., Benin (Dec.) *Brenan* 8565 ! Usonigbe F.R., Benin (Nov.) *Keay* FHI 24668 ! *Meikle* 618 ! Aboabam, Ikom (Dec.) *Keay* FHI 28224 ! **Br.Cam.:** Victoria (Aug.) *Kalbreyer* 5 ! Johann-Albrechtshöhe *Büsgen* 69. **F.Po:** (Dec.) *Barter* ! *Mann* 11 ! Also in French Cameroons, Belgian Congo and Cabinda.

20. **C. barbeyana** *De Wild. & Th. Dur.* in Contrib. Fl. Congo 2 : 11 (1900) ; Gilg & Brandt l.c. 478. *C. producta* of F.W.T.A., ed. 1, 1 : 474, partly (*Barter* 3258). A small liane ; corolla yellow ; fruits purplish ; in forest.
S.Nig.: Epe *Barter* 3258 ! Omo F.R., Ijebu Ode (Apr.) *Tamajong* FHI 16929 ! Okomu F.R., Benin (fl. & fr. Mar.) *Ross* 128 ! **Br.Cam.:** Buea (July) *Maitland* 33 ! S. Bakundu F.R. (Mar.) *Onochie* FHI 30859 ! Kumba (Apr.) *Hutch. & Metcalfe* 150 ! Abonando, Mamfe (Mar.) *Rudatis* 18 ! Also in French Cameroons, Spanish Guinea, Gabon, Belgian Congo, Angola, S. Tomé, Principe and (?) E. Africa.

21. **C. diffusiflora** (*Bak.*) *Planch.* l.c. 496 (1887) ; Gilg & Brandt l.c. 479. *Vitis diffusiflora* Bak. in F.T.A. 1 : 390 (1868). *V. afzelii* Bak. l.c. 389 (1868). *Cissus afzelii* (Bak.) Gilg & Brandt l.c. 478 (1912) ; Chev. Bot. 140, partly ; F.W.T.A., ed. 1, 1 : 476. A slender climber ; flowers greenish, fruits red ; in regrowth forest.
Port.G.: Buba to Saucunda, Fulacunda (fr. Sept.) *Esp. Santo* 2184 ! **Fr.G.:** Kouria (Aug.) *Chev.* 15065 ! **S.L.:** *Afzelius* ! *Don* ! Leicester Peak (fr. Dec.) *Sc. Elliot* 3830 ! Sugar Loaf Mt. (Sept.) *Welw.* 1501 ! Njala (fl. & fr. Oct.) *Deighton* 4906 ! Mabum (fl. & fr. Aug.) *Thomas* 1644 ! Rokupr (Sept.) *Jordan* 352 ! **Lib.:** Dukwia R. (July) *Linder* 69 ! Jabroke, Webo (July) *Baldwin* 6455 ! Gbawia, Webo (July) *Baldwin*

6713! **Iv.C.:** *fide* Chev. *l.c.* **G.C.:** Aguna (Aug.) *Vigne* FH 1304! Oda (July) *Irvine* 2791! **S.Nig.:** Lagos *Millen* 28! Akure F.R. (fr. Nov.) *Latilo* FHI 15283! Usonigbe F.R. (fl. & fr. Nov.) *Ejiofor* FHI 24624! *Meikle* 631! Oban *Talbot* 755! **Br.Cam.:** Ngusi to Mafura (fr. Jan.) *Schlechter* 12908! Johann-Albrechtshöhe *Büsgen* 50. **F.Po:** (fl. & fr. Nov.) *Mann* 570! Also in French Cameroons, Spanish Guinea, Gabon, Uganda, A.-E. Sudan, Belgian Congo and Angola.

22. **C. cornifolia** (*Bak.*) *Planch.* l.c. 492 (1887) ; Gilg & Brandt l.c. 480. *Vitis cornifolia* Bak. in F.T.A. 1 : 390 (1868). A savannah plant, with a woody base sending up erect or suberect shoots each year after the fires ; flowers greenish-yellow, appearing first on leafless shoots ; fruits black-purple.
    G.C.: Lissa to Lambussie (fr. Mar.) *Kitson* 814! **N.Nig.:** Nupe *Barter* 1072! Zaria (Feb., June) *Dalz.* 420! *Keay* FHI 25908! Funtua (Dec.) *Meikle* 825! Bornu (Jan., Feb.) *Talbot* 1219! Extends to Abyssinia and S. Africa. (See Appendix, p. 302.)

23. **C. adenocaulis** *Steud. ex A. Rich.* Fl. Abyss. 1 : 111 (1847) ; Gilg & Brandt l.c. 516 (q.v. for syn.) ; Chev. Bot. 140. *Vitis adenocaulis* (Steud. ex A. Rich.) Miq. (1863)—F.T.A. 1 : 404. *Cissus tenuicaulis* of Chev. Bot. 147. *Vitis tenuicaulis* of F.T.A. 1 : 404, partly. A somewhat fleshy herbaceous climber with striate stems ; inflorescence and flowers reddish ; fruits glabrous, purplish or black ; in savannah.
    Sen.: *Lecard* 131. **Gam.:** Albreda *Perrottet.* Kuntaur (July) *Ruxton* 52! **Fr.Sud.:** Ségou (fr. Sept.) *Chev.* 3188! Bamako (July) *Waterlot* 1207! **S.L.:** Musaia (July) *Deighton* 4815! **Iv.C.:** Ouagadougou to Ouahigouya (Aug.) *Chev.* 24728! Linoré to Gompéla (Aug.) *Chev.* 24623. Gouékouma (May) *Chev.* 21688! Mankono (June) *Chev.* 21868! **Togo:** Sokode-Basari *Kersting* A. 85! **Dah.:** Atacora Mts. (June) *Chev.* 24195! Agouagon (May) *Chev.* 23504! **Fr.Nig.:** Diapaga (fl. & fr. July) *Chev.* 24407! 24481! **N.Nig.:** Nupe *Barter* 1272! Anara F.R., Zaria (May) *Keay* FHI 22978! Tilde Filani (May) *Lely* 224! Yola (fl. & fr. July, Oct.) *Dalz.* 59! *Shaw* 64! Widespread in the savannah regions of tropical Africa. (See Appendix, p. 301.)

24. **C. debilis** (*Bak.*) *Planch.* l.c. 569 (1887) ; Gilg & Brandt l.c. 486, fig. 5, H–K, fig. 6, D–K. *Vitis debilis* Bak. in F.T.A. 1 : 403 (1868). *Cayratia debilis* (Bak.) Suesseng. in E. & P. Pflanzenfam. 20D : 278 (1953). *Cissus repens* A. Chev. Bot. 146, name only. A slender climber with nearly glabrous, grooved, almost quadrangular stems, and very small flowers and fruits ; calyx green, corolla white ; in forest regions.
    Lib.: Tappita (fr. Aug.) *Baldwin* 9028! Cess R., Tchien Dist. (fr. Aug.) *Baldwin* 9008! 9018a! **Iv.C.:** Mid. Sassandra to Mid. Cavally (fr. June) *Chev.* 19208! Guidéko to Zozro (June) *Chev.* 19063! **G.C.:** Bobiri F.R., Juaso Dist. (fr. June) *Andoh* FH 5513! **S.Nig.:** Oluwa F.R., Ondo (fl. & young fr. July) *Onochie* FHI 33414! Sapoba (fr. Sept.) *Onochie* FHI 34278! Oban *Talbot* 407! **Br.Cam.:** Victoria *Preuss* 1128. Barombi *Preuss* 308. Nkom-Wum F.R., Metschum R., Bamenda (fr. July) *Ujor* FHI 29292! **F.Po:** (June) *Mann* 412! Also in French Cameroons, Gabon, Belgian Congo, Angola and S. Tomé.

25. **C. gracilis** *Guill. & Perr.* l.c. 134 (1831) ; Gilg & Brandt l.c. 487, fig. 6, A–E ; Chev. Bot. 144, partly. *Vitis gracilis* (Guill. & Perr.) Bak. in F.T.A. 1 : 404 (1868), partly, not of Wall. (1824). *Cayratia gracilis* (Guill. & Perr.) Suesseng. l.c. (1953). *Cissus tenuicaulis* Hook. f. (1849). *Vitis tenuicaulis* (Hook. f.) Bak. in F.T.A. 1 : 404 (1868), partly. A slender climber with nearly glabrous, angular, grooved stems, very small greenish flowers and black currant-like fruits ; in forest and savannah regions.
    Sen.: N'ghiangol & Casamance *fide* Guill. & Perr. *l.c.* **Fr.G.:** Fouta Djalon (fr. Sept.) *Chev.* 18313! **S.L.:** *T. Vogel* 77! Rokupr (fl. & fr. Sept.) *Jordan* 314! Joru, Gaura (fr. Oct.) *Deighton* 5233! Musaia to Kabala (fr. Sept.) *Deighton* 4884! **Lib.:** Peahtah (fl. & fr. Oct.) *Linder* 984! Cess R., Tchien Dist. (fr. Aug.) *Baldwin* 9018! **Iv.C.:** Guidéko (June) *Chev.* 19004! **G.C.:** Aburi (fl. & fr. June, July) *Brown* 315! *Johnson* 975! **Togo:** Lome *Warnecke* 366. Sokode-Basari *Kersting* 378. Kpandu *Robertson* 25! **N.Nig.:** Katagum (Sept.) *Dalz.* 78! **Br.Cam.:** Buea *Deistel* 178. Bali-Ngemba F.R., Bamenda (fl. & fr. May) *Ujor* FHI 30391! **F.Po:** *Mann* 1167! Widespread in tropical Africa.

26. **C. ibuensis** *Hook. f.* in Fl. Nigrit. 265 (1849) ; Gilg & Brandt l.c. 486 ; not of Chev. Bot. 144. *Vitis ibuensis* (Hook. f.) Bak. in F.T.A. 1 : 402 (1868). *Cayratia ibuensis* (Hook. f.) Suesseng. l.c. (1953). *Vitis intricata* Bak. l.c. 404 (1868). A slightly fleshy climber ; flowers creamy-green ; by streams and rivers.
    Togo: Bismarckburg *Büttner* 80. **N.Nig.:** Nupe *Barter* 1133! Jebba (Mar., Dec.) *Hagerup* 717! *Meikle* 1279! R. Niger, Koton Karifi (Nov.) *Keay* FHI 28086! Yola (Nov.) *Dalz.* 58! **S.Nig.:** R. Ogbesse, Owo (Mar.) *Jones* FHI 3088! R. Niger, Ibo country (fr. Aug.) *T. Vogel* 37! Nun R. *T. Vogel* 13! Ikom (May) *Jones* FHI 6490! Extends to A.-E. Sudan, E. Africa, Belgian Congo and Angola.

27. **C. aralioides** (*Welw. ex Bak.*) *Planch.* in DC. Monogr. Ampelid. 513 (1887) ; Gilg & Brandt l.c. 485. *Vitis aralioides* Welw. ex Bak. in F.T.A. 1 : 411 (1868). *V. constricta* Bak. l.c. 409 (1868). *Cissus constricta* (Bak.) Planch. ex A. Chev. Bot. 142 (1920). *C. oliviformis* Planch. l.c. (1887). *C. caillei* A. Chev. in Rev. Bot. Appliq. 30 : 459 (1950). A lofty climber, woody at the base, with stout succulent terete stems constricted at the nodes, and sometimes subsucculent leaves ; flowers greenish or whitish, comparatively large ; fruits reddish, turning blue-purple.
    Sen.: Mt. Roland, Thiés (fr. Mar.) *De Wailly* 4504! **Port.G.:** Pussube, Bissau (Nov.) *Esp. Santo* 1563! **Fr.G.:** Kouria (Oct.) *Caille* in Hb. *Chev.* 14983! Fouta Djalon (Sept., Oct.) *Chev.* 18279! 18739! Bembaya, Farana (fr. Feb.) *Chev.* 20678! **S.L.:** Regent (fr. Feb.) *Johnston* 84! Njala (Sept.) *Deighton* 4894! Mattru to Gbangbama (Nov.) *Deighton* 2394! Ronietta (Nov.) *Thomas* 5288! 5535! **Lib.:** Careysburg *Dinklage* 2450. Soplima, Vonjama (Nov.) *Baldwin* 10051a! Wohmen, Vonjama (Oct.) *Baldwin* 10080a! **Iv.C.:** Guidéko (fr. May) *Chev.* 16444! **G.C.:** Ancobra Junction (fr. Jan.) *Irvine* 1073! Aburi Hills *Johnson* 458! Juaso Dist. (Oct.) *Andoh* FH 5581! **Togo:** Sokode-Basari *Kersting* 458. Jege, Bismarckburg *Büttner* 158. **N.Nig.:** Nupe *Barter*! **S.Nig.:** Lagos (Oct.) *Dalz.* 1147! Kiesene, Iwo road, Oyo (Oct.) *Jones* FHI 6366! Sapoba (Nov.) *Meikle* 538! Nun R. (Sept.) *Mann* 503! Widespread in tropical Africa. (See Appendix, p. 301.)

28. **C. ornata** *A. Chev. ex Hutch. & Dalz.* F.W.T.A., ed. 1, 1 : 477 (1928) ; Chev. Bot. 144, name only ; Chev. in Rev. Bot. Appliq. 30 : 456 (1950). A large liane, with rather stout half-woody glabrous stems.
    Iv.C.: Azaguié (Sept.) *Chev.* 22276!

29. **C. adenopoda** *Sprague* in Kew Bull. 1906 : 247 ; Bot. Mag. t. 8278 ; Gilg & Brandt l.c. 538, fig. 16. *C. rowlandii* Gilg & Brandt l.c. 523 (1912) ; F.W.T.A., ed. 1, 1 : 477. *C. ibuensis* of Chev. Bot. 144, partly. Herbaceous climber with pink or purplish hairs and often with purple spots on the undersurface of the leaves ; flowers greenish-yellow with purplish markings ; fruits red ; in forest.
    Iv.C.: Capiecrou (Jan.) *Chev.* 17084! **Togo:** Misahöhe *Baumann* 488. **S.Nig.:** Lagos (Aug.) *Rowland*! Gambari Group F.R., Ibadan (fr. June, Aug.) *Keay* FHI 16232! Oni River F.R., Ife-Ilesha (Apr.) *Onochie* FHI 15507! Okomu F.R., Benin (Jan.) *Brenan* 8902! **Br.Cam.:** Victoria to Tole *Winkler* 127. Also in French Cameroons, Belgian Congo, Uganda and (?) Cabinda.

30. **C. rubrosetosa** *Gilg & Brandt* in Engl. Bot. Jahrb. 46 : 532 (1912). *C. glandulifera* A. Chev. Bot. 143, name only, partly. Herbaceous climber with purplish glandular hairs ; flowers white.
    Fr.G.: Dalaba-Diaguissa Plateau (fl. & fr. Sept., Oct.) *Chev.* 18577! 18854 (partly)! Nimba Mts. (Feb.–Apr.) *Schnell* 332! 1090! Pita (July) *Pobéguin* 2097! **Iv.C.:** Mt. Tonkui (Sept.) *Schnell* 1741! Mt. Dou, Zoanlé (May) *Fleury* in Hb. *Chev.* 21474! **Togo:** Bagu, Sokode-Basari *Kersting* 381. Aledjo, Sokode-Basari *Kersting* 309. Jege, Bismarckburg *Büttner* 114. **N.Nig.:** Bauchi Plateau (Aug.) *Lely* P. 501! **Br.Cam.:** Lakom, Bamenda (May) *Maitland* 1432! Also in French Cameroons and Ubangi-Shari.

31. **C. sp. B.**[1] Herbaceous climber, red-tinged ; in savannah.
    S.L.: Mahinti, Tonko Limba (Sept.) *Jordan* 341!

---

[1] See footnote on page 677.

32. **C. sp. C.**[1] Herbaceous climber, often widely spreading, with rather fleshy stems tinged red ; inflorescences held horizontally, reddish, with pale green flowers ; by margins of forest.
S.Nig.: Ibadan North F.R. (fl. & young fr. Oct.) *Keay & Mutch* FHI 25432 ! Idanre F.R. (July) *Onochie* FHI 33401 ! Awoiki, Ishan (Aug.) *Onochie* FHI 33279 !

33. **C. vogelii** *Hook. f.* l.c. 267 (1849) ; Gilg & Brandt l.c. 531 ; Berhaut Fl. Sen. 23. *Vitis vogelii* (Hook. f.) Bak. in F.T.A. 1 : 409 (1868). An extensive herbaceous climber, setose on the young parts, the leaves and stems often purple-tinged ; flowers greenish ; ripe fruits black.
Sen.: *Adam* 9655. **S.L.:** Njala (fl. Sept., fr. Oct., Nov.) *Deighton* 5598 ! 5866 ! Senneya, Scarcies R. (fr. Jan.) *Sc. Elliot* 4278 ! Mano (fr. Sept.) *Deighton* 4882 ! **Lib.:** Memmeh's Town *Linder* 358 ! Jabroke, Webo (July) *Baldwin* 6631 ! Suacoco, Gbanga (Sept.) *Traub* 284 ! **G.C.:** Kumasi *Cummins* 151 ! Awisa, Akim Swedru (young fr. Dec.) *Irvine* 2061 ! **N.Nig.:** Bauchi Plateau (July) *Lely* P. 596 ! **S.Nig.:** Lagos *Dawodu* 254 ! *Millen* 36 ! **Br.Cam.:** Johann-Albrechtshöhe *Büsgen* 136. Abonando *Rudatis* 83. Bangwa *Conrau* 137. **F.Po:** on the beach *T. Vogel* 24 ! Also in French Cameroons and Spanish Guinea ; see also *C.* sp. aff. *vogelii* Hook. f. in Consp. Fl. Ang. 2 : 66, t. 15. (See Appendix, p. 304.)

34. **C. cymosa** *Schum. & Thonn.* Beskr. Guin. Pl. 82 (1827) ; Gilg & Brandt l.c. 537. *Vitis thonningii* Bak. in F.T.A. 1 : 407 (1868), excl. *Welw.* spec. *Cissus bakeriana* Planch. (1887)—Gilg & Brandt l.c. 543 ; Chev. Bot. 141 ; F.W.T.A., ed. 1, 1 : 477. A strong herbaceous climber ; cymes forming large flat panicles ; petals green with purple tips.
Sen.: *Berhaut* 193. **G.C.:** *Thonning* ! Accra *T. Vogel* 35 ! Akwaseho, Kwahu (fr. Jan.) *Irvine* 1674 ! Achimota (fl. & young fr. June) *Milne-Redhead* 5103 ! **Dah.:** Atacora Mts. (June) *Chev.* 24084 ! **Fr.Nig.:** Gourma *Chev.* 24381 ! 24419 ! **N.Nig.:** Nupe *Barter* 101 ! Kukawa (Sept.) *E. Vogel* 71 ! Also in French Cameroons and Uganda.

35. **C. lelyi** *Hutch.* in Kew Bull. 1921 : 361. *C. serpens* of F.W.T.A., ed. 1, 1 : 476, not of Hochst. ex A. Rich. Herbaceous, climbing or trailing, several feet long, with greenish-white flowers.
Sen.: Cayor (Aug.) *Heudelot* 473 ! Kaolak *Kaichinger* ! **N.Nig.:** top of Mt. Zaranda, 5,800 ft., Bauchi (May) *Lely* 194 ! Bauchi Plateau (Apr.) *Lely* P. 205 ! Katagum *Dalz.* 77 ! Lantewa, Damaturu (June) *Onochie* FHI 23377 !

36. **C. lageniflora** *Gilg & Brandt* l.c. 537 (1912). *C. glandulifera* A. Chev. Bot. 143, name only, partly. Herbaceous climber with red gland-tipped hairs on stems and petioles ; inflorescence tinged pink, flowers green.
Fr.G.: Kouria to Ymbo (Oct.) *Caille* in *Hb. Chev.* 14974 ! **S.L.:** Sasa, Tonko Limba (Sept.) *Jordan* 338 ! **Iv.C.:** Sanrou to Ouodé *Chev.* 21611 ! **S.Nig.:** Ondo (Sept.) *Keay* FHI 22570 ! **F.Po:** Moka (Jan.) *Exell* 819 !

37. **C. mannii** (*Bak.*) *Planch.* l.c. 610 (1887) ; Gilg & Brandt l.c. 530. *Vitis mannii* Bak. in F.T.A. 1 : 477 (1868). Herbaceous climber ; in forest.
Br.Cam.: Cam. Mt., from Buea up to 7,000 ft. (fl. Nov.–Dec., young fr. Jan.) *Dunlap* 94 ! *Mann* 1989 ! *Mildbr.* 10805 ! **F.Po.:** *Mildbr.* 7073. Also in French Cameroons.

38. **C. griseo-rubra** *Gilg & Brandt* l.c. 528 (1912). Herbaceous climber.
Togo: Lome *Warnecke* 365 !

39. **C. sp. D.**[1] Scrambling herb, with fleshy stem ; flowers cream ; among rocks.
N.Nig.: Kufena Rock, Zaria (July) *Keay* FHI 25915 !

40. **C. juncea** *Webb* in Fragm. Flor. Aethiop.-Aegypt. 57 (1854) ; Gilg & Brandt in Engl. Bot. Jahrb. 46 : 490. *Vitis juncea* (Webb) Bak. in F.T.A. 1 : 401 (1868). Perennial herb, with erect, fleshy stems to 4 ft. high, glaucous green or dull mauve ; flowers and ripe fruits red ; in savannah woodland.
N.Nig.: Anara F.R., Zaria (fl. & fr. May) *Keay* FHI 22946 ! 25767 ! Bauchi Plateau (Apr.) *Lely* P. 206 ! **Br.Cam.:** Bum, Bamenda (fr. Apr.) *Maitland* 1620 ! Also in French Cameroons and A.-E. Sudan.

41. **C. jatrophoides** (*Welw. ex Bak.*) *Planch.* in DC. Monogr. Ampelid. 579 (1887) ; Gilg & Brandt l.c. 491. *Vitis jatrophoides* Welw. ex Bak. in F.T.A. 1 : 400 (1868). *Cissus stenopoda* Gilg ex Gilg & Brandt l.c. 490 (1912) ; F.W.T.A., ed. 1, 1 : 476. *C. atacorensis* A. Chev. in Rev. Bot. Appliq. 30 : 457 (1950) ; Bot. 140. *C. pobeguinii* Chev. l.c. 458 (1950). *Ampelocissus atacorensis* A. Chev. Bot. 138, name only. Perennial herb with erect, fleshy stems to 4 ft. high, glaucous green or tinged purple ; calyx reddish, petals yellow ; fruits purple when ripe ; in savannah woodland.
Fr.G.: Kouroussa (Mar.) *Pobéguin* 1083 ! Dalaba (Apr.) *Chev.* 18143 ! Timbo (Mar.) *Chev.* 12579 ! *Pobéguin* 1579 ! **Iv.C.:** Koualé to Kouroukoro *Chev.* 21748 ! **G.C.:** Salaga (Apr.) *Krause* ! Maluwe, N.T. (fr. Mar.) *Todd* 334 ! **Togo:** Quamikrum *Schlechter* 12955. **Dah.:** Djougou (fl. & fr. June) *Chev.* 23863 ! **N.Nig.:** Lare *Barter* 3436 ! Mando (May) *Keay* FHI 25808 ! R. Raboutu, Birnin Gwari (June) *Keay* FHI 25858 ! **S.Nig.:** Igboho, Oyo Prov. (Feb.) *Keay* FHI 22514 ! Also in French Cameroons, Ubangi-Shari, E. Africa and extending to Angola and Mozambique.

42. **C. crotalarioides** *Planch.* l.c. 577 (1887) ; Gilg & Brandt l.c. 498. *C. menyanthifolia* A. Chev. Bot. 144, name only. Perennial herb with stout underground stock, sending up each year branched fleshy stems, pinkish at the nodes, to 4 ft. high ; flowers red in bud, yellow when open ; fruits grey ; in savannah woodland.
Dah.: Pira to Cabolé *Chev.* 23766 ! Atacora Mts. (fr. June) *Chev.* 23941 ! **N.Nig.:** Nupe *Barter* 1669 ! Zaria (May) *Keay* FHI 25729 ! Anchau (fr. May) *Ogua* FHI 7959 ! Anara F.R., Zaria (May) *Keay* FHI 25764 ! Nabardo (May) *Lely* 210 ! Naraguta (Apr., May) *Lely* 72 ! P. 208 ! Also in French Cameroons, Ubangi-Shari, A.-E. Sudan, Belgian Congo, N. Rhodesia and Nyasaland.

43. **C. waterlotii** *A. Chev.* in Rev. Bot. Appliq. 30 : 456 (1950). *C. koualensis* A. Chev. Bot. 144, name only. *C. chevalieri* of Chev. Bot. 141, partly. *C. crotalarioides* of Berhaut Fl. Sen. 24, 173. Perennial herb with stout underground stock, sending up annually erect, branched, fleshy stems to 3 ft. high ; in savannah.
Sen.: *Berhaut* 311. Kaolak *Kaichinger* ! **Gam.:** Farafenni (July) *Fox* 153 ! Bondali (July) *Rhind* in *Hb. Deighton* 5582 ! **Fr.Sud.:** Bamako (June) *Waterlot* 1144 ! **Port.G.:** Bambadinca, Bafata (June) *Esp. Santo* 2519 ! **Fr.G.:** Kouroussa (July) *Pobéguin* 346 ! 348 ! Kandiafara *Paroisse* 16 ! Timbo to Ditinn (fr. Sept.) *Chev.* 18378 ! **Iv.C.:** Mankono (June) *Chev.* 21869 ! Sogui to Koualé (May) *Chev.* 21694 ! Tiébisson to Languira, Baoulé-Nord (fr. July) *Chev.* 22201 !

44. **C. crinita** *Planch.* l.c. 581 (1887) ; Gilg & Brandt l.c. 494 ; Suesseng. in E. & P. Pflanzenfam. 20D : 241, fig. 64 L. Perennial herb, with stout underground stock, sending up annually fleshy, sparingly branched stems, to 1½ ft. high, clothed with very long glandular hairs ; flowers red in bud, cream when open ; ripe fruits black.
N.Nig.: Mando, Birnin Gwari (June) *Keay* FHI 25811 ! 25832 ! Bauchi Plateau (May) *Lely* P. 312 ! Abinsi (Oct.) *Dalz.* 674 ! Also in A.-E. Sudan.

45. **C. sokodensis** *Gilg & Brandt* l.c. 506 (1912). *C. chevalieri* of Chev. Bot. 141, partly. *C. crinita* of Chev. Bot. 143, partly. Perennial herb with stout underground stock and erect stems to 3 ft. high, glandular-pubescent.
Fr.G.: Timbo to Ditinn (fr. Sept.) *Chev.* 18387 ! Pita (May–July) *Pobéguin* 2096 ! **Togo:** Aledjo, Sokode-Basari *Kersting* 303. **Dah.:** near Cabolé, Savalou (May) *Chev.* 23767 ! 23801 ! Birni to Konkobiri (June) *Chev.* ! Firou to Konkobiri (July) *Chev.* 24347 ! **N.Nig.:** Bauchi Plateau (May) *Lely* P. 340 !
[The type specimen has, presumably, been destroyed and the other specimens cited here are named from the description.]

46. **C. chevalieri** *Gilg & Brandt* l.c. 508 (1912) ; Chev. Bot. 141, partly. Perennial herb with stout underground stock and erect stems to 3 ft. high.
Fr.Sud.: Koba, between Diéguessikalana and Kéméné (Mar.) *Chev.* 607 ! **Fr.G.:** Kankan (fl. & fr. Aug.)

---
[1] See footnote on page 677.

Pobéguin 1086 !   Dalaba-Diaguissa Plateau (fr. Sept.) *Chev.* 18574 !   Kouroussa (July) *Pobéguin* 847 ! (See Appendix, p. 302.)

47. **C. zechiana** *Gilg & Brandt* l.c. 46 : 506 (1912).   Perennial herb, to 3 ft. high ; in savannah.
G.C.: Navrongo (fl. & fr. May) *Andoh* FH 5178 !   **Togo:** Kete Krachi *Zech* 363.
[As the type specimen has, presumably, been destroyed, *Andoh* FH 5178 is named from the description only.   Of the two specimens mentioned by Chevalier in Rev. Bot. Appliq. 30 : 458 (1950), *Chev.* 23941 is *C. crotalarioides* and *Chev.* 22201 is *C. waterlotii.*]

48. **C. flavicans** (*Bak.*) *Planch.* l.c. 591 (1887) ; Gilg & Brandt l.c. 505.   *Vitis flavicans* Bak. in F.T.A. 1 : 413. *Cissus dahomensis* A. Chev. in Rev. Bot. Appliq. 30 : 458 (1950) ; Bot. 143.   *C. crinita* of Chev. Bot. 143, partly.   Perennial herb, with stout underground stock, sending up annually stout branched stems to 5 ft. high, covered with reddish glandular hairs ; inflorescences, fruits and margins of leaflets purplish-red ; open flowers cream ; whole plant sticky ; in savannah woodland.
Iv.C.: Mbayakro, Baoulé-Nord (Aug.) *Chev.* 22263 !   **G.C.:** Nkoranza (fl. & fr. Apr.) *Irvine* 924 !   **Togo:** Kpandu (fl. & fr. Apr.) *Asamany* 169 !   **Dah.:** Agouagon to Savalou (fr. May) *Chev.* 23679 !   **N.Nig.:** Nupe *Yates* !   Anara F.R., Zaria (fl. & fr. May) *Keay* FHI 22859 !   **S.Nig.:** *Kitson* !   Yoruba country *Barter* 912 !   1271 !

Several species are infected with *Mykosyrinx cissi* (DC.) G. Beck, a smut fungus which produces conspicuous " witches brooms."

Imperfectly known species.

C. pseudocaesia *Gilg & Brandt* l.c. 473 (1912).   White, in Kew Bull. 1951 : 57–58, suggests this may be a hybrid between *C. caesia* and *C. rufescens.*
S.L.: *T. Vogel* !   *Sc. Elliot* 4155 !   Falaba (Feb.) *Sc. Elliot* 4907 !

2. RHOICISSUS Planch. in DC. Monogr. Ampelid. 463 (1887) ; Suessenguth in E. & P. Pflanzenfam. 20D : 329 (1953).

Leaflets entire, pilose when young, later glabrescent ; median leaflets oblong-elliptic to obovate, obtuse or subacute at apex, 5–8 cm. long, 2·5–4 cm. broad, lateral leaflets asymmetric, subfalcate-oblong ; petiolules 0·3–1·2 cm. long ; petioles 1–2·5 cm. long ; flowers in short, pedunculate, rusty-pilose, leaf-opposed cymes ; fruits subglobose, about 1 cm. diam.   ..   ..   1. *revoilii*

Leaflets coarsely dentate to sinuate-dentate towards the apex, persistently tomentose beneath ; median leaflets more or less obovate, 3·5–11 cm. long, 3–8 cm. broad, lateral leaflets asymmetric, falcate-oblong ; petiolules of the median leaflets up to 2 cm. long, of the laterals up to 0·5 cm. long ; petioles 2–10 cm. long ; flowers in pedunculate, tomentose, leaf-opposed cymes ; fruits subglobose, about 1·2 cm. diam.
2. *erythrodes*

1. **R. revoilii** *Planch.* in DC. Monogr. Ampelid. 469 (1887) ; Gilg & Brandt in Engl. Bot. Jahrb. 46 : 440. A scrambling shrub or large climber with flexible porous stems containing copious watery sap ; flowers brownish ; fruits black ; in dry types of forest.
G.C.: 10 miles N. of Mampong, Ashanti (fl. & fr. Nov.) *Vigne* FH 3430 !   Nkontanang, Kwawu (fr. July) *Akpabla* 195 !   Asofa Hill, Achimota (Dec.) *Irvine* 2582 !   Also in Belgian Congo, Somaliland, E. Africa, Mozambique, S. Rhodesia, Transvaal, S. Arabia and the Comores.

2. **R. erythrodes** (*Fres.*) *Planch.* l.c. 463 (1887) ;   Gilg & Brandt l.c. ;   Suesseng. in E. & P. Pflanzenfam. 20D : 332, fig. 97.   *Vitis erythrodes* Fres. (1837)—F.T.A. 1 : 401.   A scandent shrub or climber, or trailing ; on rocky ground in savannah regions.
N.Nig.: Bauchi Plateau *Lely* P. 254 !   Also in Belgian Congo and Angola and from Abyssinia and Arabia southwards to S. Africa.

3. AMPELOCISSUS Planch. in Journ. Vigne Am. 8 : 372 (1884) ; DC. Monogr. Ampelid. 368 (1887) ; Suessenguth in E. & P. Pflanzenfam. 20D : 299 (1953).

Leaves simple, entire or lobed :
Undersurface of adult leaves covered by a more or less dense woolly indumentum, drying reddish- or pinkish-brown :
Stems and petioles with long gland-tipped bristly hairs ; inflorescence paniculate, with a strong main axis, the branches and axis pubescent ; calyx truncate ; leaves suborbicular to rather deeply 5-lobed, cordate at base, 12–30 cm. long and broad, margins doubly dentate ; petioles 10–15 cm. long ; fruits about 1·4 cm. long
1. *leonensis*

Stems and petioles without gland-tipped hairs ; inflorescence widely cymose, the branches and axis woolly-pilose ; calyx distinctly lobed ; leaves broadly ovate-polygonal to rather deeply 3–5-lobed, cordate at base, 9–22 cm. long and broad, margins doubly dentate ; petioles 4–14 cm. long ; fruits about 1 cm. long
2. *bombycina*

Undersurface of adult leaves without a woolly indumentum, only the nerves and veins pubescent, or nearly glabrous, younger leaves sometimes with whitish loose woolly indumentum :
Inflorescence and young shoots, including undersurface of leaves woolly-pilose ; petioles pubescent ; subglobose, about 8 mm. diam. :
Flowers in densely-flowered cymes rarely more than 3·5 cm. diam. ; calyx distinctly lobed ; fruits crowded ; leaves suborbicular in outline, with 3–5 broad lobes rounded at apex, or ovate-orbicular and only slightly lobed, cordate at base, 8–17 cm. long and broad, margins rather sharply dentate ; petioles 4–14 cm. long
3. *grantii*

Flowers in a spreading cyme more than 3·5 cm. diam. ; calyx truncate ; fruits not crowded, infructescence up to 15 cm. diam. ; leaves suborbicular or broadly

ovate, or 3–5 lobed, acuminate at apex, cordate at base, 9–14 cm. long and broad,
margins closely dentate ; petioles 6–13 cm. long ..   ..   ..   **4.** *gracilipes*
Inflorescence and young shoots puberulous ; leaves puberulous on nerves and veins
only beneath ; petioles glabrous or nearly so :
Leaves not lobed, or only slightly 3-lobed, suborbicular, deeply cordate at base,
shortly acuminate, 17–20 cm. long, 19–22 cm. broad, margins distantly shortly
dentate ; petioles up to 20 cm. long ; inflorescence a stout panicle   **5.** *macrocirrha*
Leaves deeply 5–7-lobed, sometimes the lobes themselves with secondary lobes,
widely cordate or truncate at base, acute to long-acuminate at apices of lobes,
margins irregularly denticulate or serrulate, up to 35 cm. long and broad ; petioles
up to 21 cm. long ; inflorescence a slender pyramidal panicle   ..   **6.** *cavicaulis*
Leaves digitately compound, with 5 leaflets nearly glabrous beneath, the central ones
obovate-oblanceolate, cuneate at base into a short petiolule, acutely acuminate,
10–17 cm. long, 3–6 cm. broad, distantly serrate-dentate, lateral leaflets usually
obliquely ovate-elliptic, shorter, broader and more coarsely serrate ; inflorescence
puberulous ; fruits about 1·2 cm. long ..   ..   **7.** *multistriata*

1. **A. leonensis** (*Hook. f.*) *Planch.* in Journ. Vigne Am. 9 : 30 (1885) ; Gilg & Brandt in Engl. Bot. Jahrb. 46 :
427 (excl. *Barter* 1276) ; Chev. Bot. 139. *Cissus leonensis* Hook. f. (1849). *Vitis leonensis* (Hook. f.)
Bak. in F.T.A. 1 : 398 (1868) (excl. *Barter* 1276). *V. salmonea* Bak. l.c. 394 (1868). *Ampelocissus salmonea*
(Bak.) Planch. (1885)—Gilg & Brandt l.c. 426 ; Chev. Bot. 139 ; F.W.T.A., ed. 1, 1 : 478, partly (excl.
distrib. S. Nig.) ; Berhaut Fl. Sen. 149. *A. bakeri* Planch. (1885)—Gilg & Brandt l.c. ; Chev. Bot. 138 ;
F.W.T.A., ed. 1, 1 : 478. An extensive climber with stout but scarcely woody stems, pubescent and with
purple gland-tipped setae ; flowers green ; ripe fruits purple.
Sen.: Ouassadou *Berhaut* 1286 ; 3090. **Gam.:** Balingho (July) *Fox* 159 ! **Fr.Sud.:** Nafadié *Chev.* 3036.
Fr.G.: Rio Nunez & Rio Pongo *Heudelot* 906. Farana *Chev.* 13190. Labé *Chev.* 12286. Fouta Djalon
Chev. 18293 ; 18481. **S.L.:** *Barter* ! T. *Vogel* 10 ! Bagroo R. (Apr.) *Mann* 865 ! Freetown (Sept.) *Welw.*
1498 ! 1500 ! Waterloo (May) *Deighton* 2656 ! Rokupr (May) *Jordan* 265 ! Pendembu (July) *Deighton*
4340 ! **Lib.:** Sarbo, Webo (fr. July) *Baldwin* 6449 ! **Iv.C.:** Guidéko *Chev.* 16429 ; 16448 ; 16485 ;
19007. Soubré *Chev.* 17962 ; s.n. Koualé to Kouroukoro *Chev.* 21751. **Togo:** Aledjo *Kersting* 301.
Dah.: *Poisson* ! Pira to Cabolé (May) *Chev.* 23759 ! Cabolé to Bassila *Chev.* 23805. **N.Nig.:** Loin *Barter*
1275 ! Kaduna *Lamb* 106 ! Abinsi (July) *Dalz.* 676 ! Katsina Ala *Dalz.* 675 ! **S.Nig.:** Yoruba country
Barter 1279 ! (See Appendix, p. 301.)

2. **A. bombycina** (*Bak.*) *Planch.* l.c. 31 (1885) ; Gilg & Brandt l.c. 431 ; Chev. Bot. 138. *Vitis bombycina* Bak.
in F.T.A. 1 : 399 (1868). *Ampelocissus cinnamochroa* Planch. (1885)—Gilg & Brandt l.c. ; F.W.T.A., ed. 1,
1 : 478. *A. lecardii* of Chev. Bot. 139. *A. salmonea* of F.W.T.A., ed. 1, 1 : 478, partly (distrib. S. Nig.).
A climbing or trailing herb, with subwoody stems ; flowers reddish.
Fr.G.: Kouroussa *Pobéguin* 349 ! Irébéléya to Timbo *Chev.* 18281. **Iv.C.:** Mt. Goula, Danané *Chev.*
21217. Toura country *Chev.* 21588 ; 21635 ; 21749. **G.C.:** Salaga *Krause* ! **Togo:** Lome *Warnecke* 346 !
Misahöhe *Baumann* 312. Sokode-Basari *Kersting* 364. **Dah.:** Haévetgi to Massi (Mar.) *Chev.* 23263 !
Kouandé to Konkobiri *Chev.* 24255 ! **N.Nig.:** Nupe *Barter* 1276 ! 1277 ! Gayam, Birnin Gwari (June)
Keay FHI 25831 ! Rinjim Mukr (May) *Lely* 223 ! Katsina Ala (Aug.) *Dalz.* 677 ! Yola (June) *Dalz.* 61 !
S.Nig.: Lagos (May) *Dennett* 501 ! *MacGregor* 96 ! Yoruba country *Millson* 53 ! Ibadan (Apr., June)
Jones FHI 18835 ! *Meikle* 1413 ! **Br.Cam.:** Buea to Mayuko (Feb.) *Dalz.* 8231 ! Also in French
Cameroons, Ubangi-Shari-Chad, Spanish Guinea, Belgian Congo, A.-E. Sudan and Eritrea. (See Appendix,
p. 300.)

3. **A. grantii** (*Bak.*) *Planch.* in Journ. Vigne Am. 9 : 32 (1885) ; Gilg & Brandt l.c. 428, fig. 1, D, & fig. 2, A–F ;
Chev. in Rev. Bot. Appliq. 3 : 451. *Vitis grantii* Bak. in F.T.A. 1 : 400 (1868). *V. asarifolia* Bak. l.c.
396 (1868). *Ampelocissus asarifolia* (Bak.) Planch. (1885). *Vitis chantinii* Lécard ex Carrière (1881).
Ampelocissus chantinii (Lécard ex Carrière) Planch. (1885)—Chev. Bot. 139. *A. gourmaensis* A. Chev. in
Rev. Bot. Appliq. 30 : 452 (1950), and as subsp. of *A. grantii* on the previous page ; Chev. Bot. 139, name
only. Stems herbaceous, climbing, from a perennial underground stock, flowers yellow ; in savannah,
especially among rocks.
Fr.G.: Kouroussa (Apr.) *Pobéguin* 702 ! **Iv.C.:** Ouagadougou to Ouahigouya (fl. & fr. Aug.) *Chev.* 24742 !
Dah.: Konkobiri *Chev.* 24346 ; 24341. Djougou *Chev.* 23918. Kouandé to Konkobiri *Chev.* 24294.
Fr.Nig.: Kodjar (July) *Chev.* 24388 ! **N.Nig.:** Sokoto (fl. & fr. Sept.) *Dalz.* 354 ! Zaria (June) *Keay*
FHI 2291 ! 25732 ! Maiduguri (July) *Ujor* FHI 21945 ! Yola (fl. & young fr. June) *Dalz.* 63 ! 64 ! Also
in French Cameroons, Ubangi-Shari-Chad, A.-E. Sudan, E. Africa, N. & S. Rhodesia, Nyasaland and
Mozambique. (See Appendix, p. 300.)

4. **A. gracilipes** *Stapf* in J. Linn. Soc. 37 : 90 (1905) ; Gilg & Brandt l.c. 425. A climber with softly woody
stout stems, pale brownish woolly-pilose on the young parts, minutely warted on older stems ; flowers
and fruits reddish ; in forest vegetation.
Lib.: Sinoe Basin *Whyte* ! R. Cestos, Grand Bassa *Dinklage* 1941. Gletown, Tchien (fr. July) *Baldwin*
9941a ! **G.C.:** Akwapim (fr. June) *Rumsey* 9 ! (See Appendix, p. 300.)

5. **A. macrocirrha** *Gilg & Brandt* l.c. 424, fig. 1, A (1911). A climber with stout woody glabrous stems and
strong branched tendrils.
Br.Cam.: Buea *Deistel.* Johann-Albrechtshöhe *Staudt* 620. **F.Po:** (Jan.) *Mann* 226 ! Also in French
Cameroons.
[Doubtfully distinct from the next species.]

6. **A. cavicaulis** (*Bak.*) *Planch.* l.c. 32 (1885) ; Gilg & Brandt l.c. 426, fig. 1, C. *Vitis cavicaulis* Bak. in F.T.A.
1 : 400 (1868). A large climber.
S.Nig.: Sapoba (Aug.) *Latilo* FHI 27323 ! **Br.Cam.:** Bulifambo, Buea (Mar.) *Maitland* 583 ! Also in
French Cameroons, Spanish Guinea, Gabon, Belgian Congo, A.-E. Sudan, Uganda and Angola.

7. **A. multistriata** (*Bak.*) *Planch.* in DC. Monogr. Ampelid. 398 (1887) ; Chev. Bot. 139 ; Chev. in Rev. Bot.
Appliq. 30 : 450. *Vitis multistriata* Bak. in F.T.A. 1 : 410 (1868). *V. pentaphylla* Guill. & Perr. Fl.
Seneg. 1 : 135, t. 33 (1831), not of Thunb. (1784). *Ampelocissus pentaphylla* (Guill. & Perr.) Gilg &
Brandt l.c. 427 (1911) ; F.W.T.A., ed. 1, 1 : 478. *A. leprieurii* Planch. (1885). *Cissus ibuensis* of F.W.T.A.,
ed. 1, 1 : 477, partly (S.L. record). A large herbaceous climber with glabrous, striate stems ; inflorescence
crowded, reddish ; fruits smooth, grape-like ; in the savannah regions.
Sen.: Kaolak *Kaichinger.* **Gam.:** *Ingram* ! Albreda (June, July) *Perrottet.* Abuko *Dalz.* 8040 ! Genieri
(July) *Fox* 137 ! **Fr.Sud.:** Kalédougou (June) *Chev.* 983. **Fr.G.:** Karkandy *Heudelot* 860 ! **S.L.:** Sulimania *Sc. Elliot* 5305 ! **G.C.:** Ashanti *Irvine* 659 ! **N.Nig.:** Nupe *Barter* 1280 ! Anara F.R., Zaria (young
fr. May) *Keay* FHI 22893 ! Bauchi Plateau (Apr.) *Lely* P. 207 ! Yola (June) *Dalz.* 62 ! Also in French
Cameroons, Ubangi-Shari-Chad, A.-E. Sudan, Tanganyika and Mozambique. (See Appendix, p. 301.)

Imperfectly known species.

A. multiloba *Gilg & Brandt* l.c. 426, fig. 1, B (1911). Close to *A. cavicaulis* and perhaps conspecific ; specimen
not seen.
Br.Cam.: Bangwa *Conrau* 92.

4. LEEA Linn.—F.T.A. 1 : 415 ; Suessenguth in E. & P. Pflanzenfam. 20D : 382 (1953).

Leaves bipinnate ; leaflets opposite, imparipinnate, oblong-elliptic, rounded or slightly narrowed at base, acuminate, up to 18 cm. long and 8 cm. broad, glabrous, crenate-serrate ; petiolules about 5 mm. long ; cymes leaf-opposed, large and spreading ; calyx cupular, 5-lobed ; petals 5, glabrous, short ; stamens 5 ; filaments united in a tube ; fruit depressed, 5–8-lobulate, nearly 1 cm. in diam., glabrous *guineensis*

L. **guineensis** *G. Don* Gen. Syst. 1 : 712 (1831) ; Gilg & Brandt in Engl. Bot. Jahrb. 46 : 547, fig. 18 ; Chev. Bot. 147. *L. sambucina* of F.T.A. 1 : 415 ; of Chev. Bot. 147 ; not of Willd. An erect or suberect soft-wooded shrub, up to 20 ft. high ; flowers bright yellow, orange or red ; fruits brilliant red then turning black ; locally abundant in moist shaded places.
Fr.Sud.: Sarakourouba (fr. Feb.) *Chev.* 485. Karankasso *Chev.* 896. **Fr.G.:** Labé *Chev.* 12412. Kouria *Chev.* 14645. Konkouré to Timbo *Chev.* 12483. **S.L.:** *Barter!* *Don.* Heddle's Farm (fr. Dec.) *Sc. Elliot* 3923! Njala *Deighton* 2520! Musaia (July, Aug.) *Deighton* 4353! 4814! **Lib.:** Dukwia R. (May) *Cooper* 444! Sinoe Basin *Whyte!* Cape Palmas (July) *T. Vogel* 11! **Iv.C.:** several localities *fide* Chev. *l.c.* **G.C.:** Cape Coast *Brass!* Nuba (July) *Chipp* 308! Kumasi *Cummins!* Boankra (May) *Chipp* 435! **Togo:** Amedzofe (fr. Feb.) *Irvine* 175! Sokode-Basari *Kersting* 158. **N.Nig.:** Patti Lokoja (Nov.) *Dalz.* 201! Katsina Ala (July) *Dalz.* 792! **S.Nig.:** Lagos *Dawodu* 121! Sapoba *Kennedy* 1774! Onitsha *Barter* 1754! **Br.Cam.:** Buea (Nov.) *Migeod* 64! **F.Po:** *Barter!* *T. Vogel* 92! Widespread in tropical Africa in moister woodland and forest. (See Appendix, p. 304.)

114. RUTACEAE

Shrubs or trees, rarely herbs. Leaves simple or compound, mostly gland-dotted ; stipules absent. Flowers hermaphrodite or unisexual, actinomorphic. Sepals 4–5, imbricate, free or connate. Petals imbricate, rarely valvate, mostly free. Stamens the same or double the number of the petals, rarely numerous, free or rarely united ; anthers 2-celled, introrse, opening lengthwise. Disk usually present within the stamens. Ovary superior, syncarpous and often 4–5-celled, or the carpels free (secondary apocarpy) ; styles free or connate. Ovules often 2, superposed. Fruit baccate, drupaceous or coriaceous, rarely a capsule. Seeds with or without endosperm.

Readily recognized by the gland-dotted usually compound leaves.

Fruits consisting of 1–5 separate dehiscent carpels each containing a single shining black or bluish seed ; branchlets, and often main stem and branches, leaves and inflorescences covered with prickles ; leaves imparipinnate with several to many leaflets ; flowers unisexual .. .. **1. Fagara**
Fruits indehiscent ; plants without prickles but sometimes with axillary spines ; leaves simple, 1–3-foliolate, digitate or imparipinnate :
Leaves imparipinnate with numerous alternately arranged leaflets ; branchlets without spines ; flowers bisexual ; fruits fleshy, 1-seeded .. .. **2. Clausena**
Leaves simple, 1–3-foliolate, digitate, or imparipinnate with up to 3 pairs of opposite leaflets :
Branchlets with single or rarely paired axillary spines ; leaves simple, 1–3-foliolate or imparipinnate ; flowers bisexual :
Fruits hard-shelled, 5–10 cm. diam. ; ovules and seeds numerous in each cell ; styles short :
Stamens 15–20 ; disk slightly lobed ; ovary about 8-celled ; leaves normally 3-foliolate .. .. **3. Afraegle**
Stamens 8–10 ; disk large, with about 10 rounded lobes ; ovary 5–6-celled ; leaves normally simple, but apparently sometimes 1–3-foliolate .. **4. Aeglopsis**
Fruits soft, up to 3 cm. diam. ; ovary (3–) 4–5-celled ; ovules solitary in each cell ; style longer than the ovary ; stamens 8–10 ; leaves simple, 1–3-foliolate, or imparipinnate with up to 3 pairs of opposite leaflets ; petioles and rhachides often winged .. .. **5. Citropsis**
Branchlets without spines ; leaves digitately (2–) 3–5-foliolate ; flowers often unisexual [1] :
Ovary 4-celled ; fruit 4-celled, each cell 2-seeded ; stamens or staminodes 8 ; leaves (3–) 5-foliolate .. .. **6. Araliopsis**
Ovary 1–2-celled, or composed of up to 4 carpels which separate in fruit, each 1-seeded ; stamens or staminodes 4–5 ; leaves (1–) 3 (–5)-foliolate :
Carpels 2–4, more or less free, becoming quite free in fruit, or 1–3 aborting but evident .. .. **7. Oricia**

[1] Also in this part of the key comes a specimen (*Ujor* FHI 30422) recently received from Bali-Ngemba F.R., Bamenda, Br. Cam. The flowers (male) have 8 stamens, but fruits or female flowers are needed to determine the genus ; it is perhaps a new species of *Vepris* Commers. ex A. Juss., a genus not otherwise represented in our area, unless the plant (*Keay* FHI 28258) provisionally referred to *Araliopsis* also proves to be a *Vepris*. Collections to show both sexes of flowers and the fruits are much needed in this group (*Toddalieae*) of the Rutaceae.

Carpels solitary, or 2 united, not separating in fruit :
Ovary and fruit 2-celled   ..   ..   8. **Diphasia**
Ovary and fruit 1-celled   ..   ..   9. **Teclea**

Besides the above, other genera are represented by the following introduced plants :— *Citrus* spp., oranges, lemons, limes, grapefruits, etc. ; *Chloroxylon swietenia* DC., satinwood, from tropical Asia ; *Glycosmis citrifolia* (Willd.) Lindl., a native of China.

Excluded genus.

Eriander *H. Winkl.* in Engl. Bot. Jahrb. 41 : 277 (1908). This is **Oxystigma** Harms (Caesalpiniaceae), see p. 466.

1. FAGARA Linn. Syst. Nat., ed. 10 : 897 (1759) ; Engl. in E. & P. Pflanzenfam. 19A : 217 (1931).

Leaflets not acuminate or only very slightly so, rigidly papery, oblong or oblong-oblanceolate, sometimes emarginate at the apex, subacute at the base, 5–10 cm. long, 2–4 cm. broad, glabrous ; branchlets and leaf-rhachis armed with sharp recurved prickles ; panicles nearly as long as the leaves ; flowers clustered and sessile ; fruits ellipsoid, about 6 mm. long ; seeds black and shining   ..   1. *zanthoxyloides*
Leaflets conspicuously acuminate :
  The leaflets pilose or pubescent beneath :
    Branchlets, leaf-rhachides and lower surface of leaves softly and densely pubescent ; leaflets 9–15, oblong-lanceolate, 12–13 (–20) cm. long, 4–6 (–9·5) cm. broad, margins crenulate, closely gland-dotted ; fruits ovoid, 1 cm. long, slightly pubescent
2. *pubescens*
    Branchlets, leaf-rhachides and midrib and nerves hirsute with rather long spreading hairs :
      Leaflets 5–10 cm. long, 2–3·5 cm. broad, usually 13–15, with few scattered glands, with petiolules 1–1·5 mm. long ; lamina oblong-oblanceolate, abruptly acuminate ; inflorescence with few or many long spreading hairs ; pedicels 2–3 mm. long ; fruits subglobose, 5–6 mm. long, glabrous and muricate ..   ..   3. *viridis*
      Leaflets 6·5–25 cm. long, 4–8 cm. broad, 9–11, with numerous gland-dots, lateral leaflets sessile ; lamina oblong-obovate, oblong or ovate-oblong, long-acuminate ; inflorescence densely pubescent ; pedicels 1 mm. long   ..   ..   4. *buesgenii*
  The leaflets glabrous beneath :
    *Trees and erect shrubs :
    Leaf-rhachides 50–100 cm. long ; leaflets 31–51, elongate-oblong, asymmetrical at the base, 12–25 cm. long, 3·5–9 cm. broad, with numerous spreading lateral nerves ; panicles very large ; flowers sessile on the branches ; fruits suborbicular about 5 mm. diam. :
      Petiolules up to 5 mm. long, leaflets obtuse to subcordate at base, asymmetrical but usually with the two sides of the lamina joining the petiolule at about the same level ; leaf-rhachides prickly ; leaflets 12–25 cm. long, 3·5–9 cm. broad
5. *macrophylla*
      Petiolules 7–10 mm. long, leaflets strongly asymmetrical at base, the distal side broad, rounded, joining the petiolule well below the narrow, cuneate proximal side ; leaf-rhachides unarmed or with a few prickles near the base only ; leaflets 12–20 cm. long, 5–8 cm. broad   ..   ..   6. *melanorhachis*
    Leaf-rhachides up to 45 cm. long, with 8–25 leaflets ; leaflets up to 13 cm. long and 5 cm. broad, or if larger then main lateral nerves 8–10 and flowers distinctly pedicellate ; pedicels 0·5–3 mm. long ; fruits up to 7 mm. long :
      Main lateral nerves 8–10 on each side of midrib, bowed and looped ; leaflets 9–13, rather obliquely oblong-elliptic or ovate-elliptic, rounded or obtuse at base, rather abruptly and often long-acuminate, 5–20 (–25) cm. long, 2·5–9 (–12·5) cm. broad, usually the upper leaflets larger than the lower ones, with very numerous translucent yellowish gland-dots, rhachis up to 45 cm. long ; pedicels 2–3 mm. long ; lateral branches of inflorescence paniculate ; fruits 5–7 mm. long
7. *rubescens*
      Main lateral nerves more than 10 on each side of midrib, close and more or less parallel ; leaflets up to 13 cm. long and 5 cm. broad ; gland-dots few or numerous; pedicels 0·5–1·5 mm. long :
        Lateral branches of inflorescence paniculate ; pedicels 1–1·5 mm. long ; fruits 3·5–4·5 mm. long ; leaflets up to 10·5 cm. long and 3·2 cm. broad, with gland-dots mostly confined to the margins :
          Leaflets sessile, 20–25, each with a small pouch on one side at the base, oblanceo-late, sometimes slightly oblique, cuneate at base, crenate towards the shortly acuminate apex, acumen itself emarginate, 1·4–7·5 cm. long, 0·6–1·8 cm. broad   ..   ..   8. *parvifoliola*
          Leaflets with petiolules 1·5–3 mm. long, 8–18, without pouches at the base, narrowly oblong-falcate, asymmetrically cuneate to rounded at base, long-caudate-acuminate, margins subentire, 4–10·5 cm. long, 1·2–3·2 cm. broad
9. *lemairei*

Lateral branches of inflorescence spike-like ; pedicels up to 0·75 mm. long ; leaflets 3–13 cm. long, 1·8–5 cm. broad, distinctly petiolulate :

Leaflets with reddish translucent glands scattered over the surface ; flowers in small fascicles along the lateral branches of the inflorescence, pedicels 0·5 mm. long or less ; leaflets 9–21, oblong to ovate, often obliquely so, rounded or obtuse at base, long-acuminate, margins crenulate, 3–13 cm. long, 1·8–4·8 cm. broad .. .. 10. *leprieurii*

Leaflets with glands which do not appear translucent ; flowers separate along the lateral branches of the inflorescence, pedicels about 0·75 mm. long ; leaflets about 9, oblong or oblong-obovate, cuneate at base, abruptly long-acuminate, margins crenate, 6·5–12 cm. long, 3–5 cm. broad .. .. 11. *sp. A*

*Lianes ; stems, petioles and rhachides of leaves and inflorescences with numerous recurved prickles :

Leaflets 9–11, oblong, rounded at base (except the apical leaflet which is cuneate), long-acuminate, margins crenate, 4·5–13 cm. long, 2–4·5 cm. broad ; pedicels about 4 mm. long in flower, up to 7 mm. long in fruit ; fruits 8–9 mm. long 12. *dinklagei*

Leaflets up to about 43, oblong, cordate at base, rather abruptly acuminate, margins inrolled, 5–9 cm. long, 3·5–5·5 cm. broad, terminal leaflet up to 22 cm. long and 7·5 cm. broad ; leaf-rhachides up to about 70 cm. long ; pedicels 1–1·5 mm. long in fruit ; fruits 5–6 mm. long .. .. 13. *sp. B*

1. **F. zanthoxyloides** *Lam.* Encyl. Méth. Bot. 2 : 446 (1786) ; Aubrév. Fl. For. C. Iv. 2 : 84–86, t. 159, D ; Fl. For. Soud.-Guin. 364, t. 80, 4–7. *Zanthoxylum senegalense* DC. (1824)—F.T.A. 1 : 305 ; Chev. Bot. 100. *Fagara senegalensis* (DC.) A. Chev. Bot. 101 (1920). *Zanthoxylum polygamum* Schum. (1827). A glabrous shrub or tree, up to 40 ft. high, with shining aromatic foliage and bark ; flowers cream-white ; locally abundant in coastal areas.
   Sen.: Kounoun & Saffal Isl. *Perrottet* 141 ! Carabane (Jan.) *Chev.* 3153 ! Floup-Fedyan *Chev.* 3152. R. Casamance (fr. May) *Leprieur* ! R. Salum *Brunner* 71 ! **Gam.:** *Hayes* 518 ! **Fr.Sud.:** Ténétou *Chev.* 661. **Port.G.:** Botelhe, Caravela Isl. (fr. May) *Esp. Santo* 2000 ! **Fr.G.:** Kouroussa *Pobéguin* 646 ; 847. **Iv.C.:** Bondoukou *Aubrév.* 1611 ! **G.C.:** Cape Coast (fr. July) *T. Vogel* 77 ! Asientiem (July) *Chipp* 289 ! Birrifwa to Lawra (Mar.) *Kitson* 808 ! Achimota (fl. & fr. June, Aug.) *Irvine* 66 ! *Milne-Redhead* 5108a ! **Togo:** Lome *Warnecke* 249 ! 389 ! Kpedsu (Dec.) *Howes* 1067 ! **Dah.:** Cotonou *Debeaux* ! Abomey *Chev.* 23141. **N.Nig.:** Nupe *Barter* 1137 ! Jebba *Barter* ! **S.Nig.:** Lagos (fr. Dec.) *Millen* 39 ! *Rowland* ! Olokemeji (Feb.) *Foster* 110 ! *A. F. Ross* 78 ! (See Appendix, p. 308.)

2. **F. pubescens** *A. Chev.* in Bull. Soc. Bot. Fr. 58, Mém. 8 : 144 (1912) ; Aubrév. Fl. For. C. Iv. 2 : 86. *Zanthoxylum melanacanthum* of Chev. Bot. 100. A shrub, or small tree, 6–20 ft. high, with tomentose branchlets armed with straight spines ; flowers white ; fruits red when ripe.
   Fr.G.: Dalaba-Diaguissa Plateau (Sept., Oct.) *Chev.* 18348 ! 18882. Labé (fr. Feb.) *Chev.* 12267 ! **S.L.:** Bafodeya (Apr.) *Sc. Elliot* 5495 ! **G.C.:** Sang, W. Prov. (Aug.) *Vigne* FH 1289 !

3. **F. viridis** *A. Chev.* l.c. 143 (1912) ; Aubrév. l.c. t. 160, C. *F. attiensis* Hutch. & Dalz., F.W.T.A., ed. 1, 1 : 481 (1928), partly (*Johnson* 750). A small tree or shrub, 6–12 ft. high with pubescent, spiny branchlets ; flowers greenish-white in large terminal panicles.
   S.L.: Musaia (fl. July, fr. Sept., Oct.) *Deighton* 4800 ! 4892 ! *Miszewski* 3 ! 12 ! Jigaya (fr. Sept.) *Thomas* 2672 ! Yetaya (fr. Sept.) *Thomas* 2459 ! Kabala (fr. Sept.) *Thomas* 2244 ! 2603 ! **Iv.C.:** Droupleu (May) *Chev.* 21445 ! **G.C.:** Kumasi (June) *Vigne* FH 1743 ! Okroase (May) *Johnson* 750 ! (See Appendix, p. 307.)

4. **F. buesgenii** *Engl.* Bot. Jahrb. 46 : 407 (1911). *F. viridis* of F.W.T.A., ed. 1, 1 : 481, partly (*Talbot* 1274). A scandent shrub ; in forest.
   S.Nig.: Oban *Talbot* 1274 ! 1508 ! The type is from an unspecified part of the Cameroons.

5. **F. macrophylla** *Engl.* in E. & P. Pflanzenfam. 3, 4 : 118 (1896) ; Chev. Bot. 100 ; Aubrév. Fl. For. C. Iv. 2 : 88, t. 159, B. *Zanthoxylum macrophyllum* Oliv. F.T.A. 1 : 304 (1868), not of Miq. (1859). An ever-green forest tree 30–40 ft. high, rarely up to 120 ft. high ; bole and larger branches covered with large woody-based prickles ; leaves very large, clustered at the ends of the branches. There appear to be several infraspecific variants ; intensive and extensive field research is needed.
   S.L.: Bandama (fr. Mar.) *Aylmer* 582 ! Njala *Deighton* 1826 ! **Lib.:** Bonuta (Oct.) *Linder* 919 ! **Iv.C.:** *Aubrév.* ! **G.C.:** Dunkwa *Vigne* 888 ! **S.Nig.:** Ibadan F.R. (fl. Sept., fr. Jan.) *Punch* 127 ! Shasha F.R. (May) *Richards* 3433 ! Uregin, Benin (young fr. July) *Unwin* 114 ! Sapoba *Kennedy* 332 ! 2288 ! **Br.Cam.:** Victoria *Maitland* 202 ! Also in French Cameroons, Gabon, Belgian Congo, A.-E. Sudan, Uganda, Angola and the islands of Principé and S. Tomé. (See Appendix, p. 308.)

6. **F. melanorhachis** *Hoyle* in Kew Bull. 1933 : 174. A forest tree, similar to the last.
   S.Nig.: Sapoba (Apr.) *Kennedy* 337 ! 2086 ! 2297 ! Enugu *Rosevear* 2/28 !
   [Perhaps not distinct from *F. inaequalis* Engl. (1917) from French Cameroons, Belgian Congo and N. Rhodesia.]

7. **F. rubescens** (*Planch. ex Hook. f.*) *Engl.* in E. & P. Pflanzenfam. 3, 4 : 118 (1896). *Zanthoxylum rubescens* Planch. ex Hook. f. (1849) ; F.T.A. 1 : 305. *Z. melanacanthum* Planch. ex Oliv. F.T.A. 1 : 305 (1868). *Fagara melanacantha* (Planch. ex Oliv.) Engl. (1896)—F.W.T.A., ed. 1, 1 : 481 ; Aubrév. Fl. For. C. Iv. 2 : 90, t. 160, A. *F. welwitschii* Engl. (1896). *F. afzelii* Engl. (1896). *F. altissima* Engl. (1911). *F. bouetensis* Pierre ex Engl. (1915). *F. attiensis* Hutch. & Dalz., F.W.T.A., ed. 1, 1 : 481 (1928) and Kew Bull. 1929 : 24, partly (*Chev.* 17423, *Thomas* 1747) ; Aubrév. l.c. *Zanthoxylum attiense* A. Chev. Bot. 99, name only. An understorey forest tree, to 30 ft. high, armed with black thorns ; flowers greenish or whitish, numerous in large panicles.
   Port.G.: Dandum, Bafata (June) *Esp. Santo* 2690 ! Fulacunda *Esp. Santo* 2204 ! **S.L.:** *Afzelius* ! *T. Vogel* 181 ! Njala (fl. June, fr. Aug.) *Deighton* 3013 ! 3017 ! 3242 ! Russell (May) *Deighton* 2685 ! Leicester (June) *Lane-Poole* 292 ! Makump (fr. July) *Thomas* 959 ! **Iv.C.:** Alépé, Attie (young fr. Feb.–Mar.) *Chev.* 17423 ! Abouabon Forest (fr. Jan.) *Miège* 2231 ! Vridi Forest (Mar.) *Miège* 2255 ! Abidjan *Aubrév.* 284 ; 389. Djibi *Chev.* 1327. **G.C.:** Cape Coast (July) *T. Vogel* 15 ! Komenda (young fr. Aug.) *Andoh* FH 5553 ! Volta River F.R. (June) *Vigne* FH 4411 ! Aburi Hills (Nov.) *Johnson* 233 ! **S.Nig.:** *Thomas* 1797 ! Lagos *Dawodu* 196 ! Ibadan North F.R. (Aug.) *Keay* FHI 25357 ! Okomu F.R., Benin (Feb.) *Brenan* 9035 ! Odiani, Asaba (Aug.) *Onochie* FHI 33428 ! **Br.Cam.:** Nkom-Wum, Bamenda (fr. July) *Ujor* FHI 29278 ! Also in French Cameroons, Gabon, Belgian Congo, Angola and S. Tomé. (See Appendix, p. 307.)

8. **F. parvifoliola** *A. Chev. ex Keay* in Bull. Jard. Bot. Brux. 26 : 187 (1956) ; Aubrév. Fl. For. C. Iv. 2 : 88, t. 159, A (1936), French descr. only. *Zanthoxylum parvifolium* A. Chev. Vég. Util. 5 : 233 (1909), not validly described ; Chev. Bot. 100. *Fagara leprieurii* of F.W.T.A., ed. 1, 1 : 481, partly. A forest tree, to about 7 ft. girth ; bole smooth, without thorns, young shoots with small thorns.

Iv.C.: *Aubrév.* 1787! Mbasso *Chev.* 16263! Rasso *Aubrév.* 582! Also in Belgian Congo.

9. **F. lemairei** *De Wild.* in Fedde Rep. 13 : 380 (1914). *F. unwinii* Hutch. & Dalz., F.W.T.A., ed. 1, 1 : **481** (1928) ; Kew Bull. 1929 : 24. A forest tree, up to 80 ft. high and 6 ft. girth, with thorny stem.
G.C.: Dunkwa (fr. July) *Vigne* FH 4763! Amentia *Vigne* FH 1872! **S.Nig.:** Idu, Sapoba F.R., Benin (fr. Nov.) *A. F. Ross* 223! Sapoba *Kennedy* 1835! 1915! 2281! Boji Hills, Ikom (fr. June) *Unwin* 13! Also in French Cameroons and Belgian Congo.

10. **F. leprieurii** (*Guill. & Perr.*) *Engl.* in E. & P. Pflanzenfam. 3, 4 : 118 (1896) ; Aubrév. Fl. For. C. Iv 2 : 86–88, t. 160, B, partly ; Fl. For. Soud.-Guin. 362. *Zanthoxylum leprieurii* Guill. & Perr. (1831)—F.T.A. 1 : 306 ; Chev. Bot. 100. *Fagara angolensis* Engl. (Sept. 1896)—F.W.T.A., ed. 1, 1 : 481 ; Aubrév. Fl. For. C. Iv. 2 : 90, t. 159, C. *F. polyacantha* Engl. (1896). *Zanthoxylum nitens* Hiern (Dec. 1896). *Fagara nitens* (Hiern) Engl. (1915). *F. attiensis* Hutch. & Dalz. F.W.T.A., ed. 1, 1 : 481, partly (*Sc. Elliot* 5745, *Chev.* 22471). An understorey forest tree, 15–80 ft. high ; stems covered with large broad-based thorns having sharp points ; flowers white.
Sen.: Maloum (May, June) *Leprieur* ! Niokolo-Koba *Berhaut* 1495. **Port.G.:** Catió (July) *Esp. Santo* 2142! Bijimita (fr. Sept.) *Esp. Santo* 2182! Mansoa to Jugudal (fr. Oct.) *Esp. Santo* 3096! **Fr.G.:** Mamou to Labé *Chev.* 61. **S.L.:** *Afzelius* ! *Sc. Elliot* 5745! York Pass (June) *Lane-Poole* 314! 315! Tombo to Kent (May) *Deighton* 2671! Mano to Kailahun (Apr.) *Deighton* 3179! **Lib.:** Tchien (fr. Aug.) *Baldwin* 7048! **Iv.C.:** Bangouanou to Endé *Chev.* 22471. **G.C.:** Ejura (June) *Vigne* FH 2038. Mpraeso (May) *Beveridge* FH 4207! **N.Nig.:** Bauchi Plateau (Apr.) *Lely* P. 231! **S.Nig.:** Ibadan South F.R. (fl. buds Apr.) *Keay* FHI 22822! Awka Dist. (June) *Jones* FHI 602! 6626! Umuaku, Okigwi (May) *Cousens* FHI 3847! Extends to Angola, Belgian Congo, A.-E. Sudan and Uganda. (See Appendix, p. 307.)

11. **F. sp. A.** *Zanthoxylum crenatum* A. Chev. Bot. 99, name only. *Fagara angolensis* of F.W.T.A., ed. 1, 1 : 481, partly ; Aubrév. Fl. For. C. Iv. 2 : 88, partly. Perhaps only a form of *F. lepricurii*, but further material required.
Iv.C.: Soubré to Pehiri (June) *Chev.* 19183!

12. **F. dinklagei** *Engl.* Bot. Jahrb. 23 : 147 (1896). *F. klainei* Pierre ex De Wild. (1925). (?) *F. poggei* Engl. (1896). A liane with numerous sharp thorns ; in forest.
S.Nig.: Okomu F.R. (fl. Feb., fr. Sept.) *Onochie* FHI 19731! 34326! Oban *Talbot* 427! Also in French Cameroons and Gabon.

13. **F. sp. B.** A liane with prickly stems up to 3 in. diam. ; in forest.
Iv.C.: Téké Forest (fr. Aug.) *Miège* 1935!

2. CLAUSENA Burm. f.—F.T.A. 1 : 307 ; Engl. in E. & P. Pflanzenfam. 19A : 320 (1931) ; Swingle in Webber & Batchelor, Citrus Industry 1 : 158 (1943).

Leaflets 17–32, alternate, obliquely ovate or ovate-lanceolate, cuneate to rounded at base, acuminate, acute, obtuse or rounded at apex, very variable in size, up to about 11 cm. long and 5 cm. broad, margins entire or crenulate, nearly glabrous to densely pubescent beneath, glandular-punctate ; flowers in lax narrow panicles at least half as long as the leaves ; petals glabrous ; fruits ellipsoid, about 9 mm. long and 7 mm. broad, glabrous   ..   .. *anisata*

C. anisata (*Willd.*) *Hook. f. ex Benth.* in Fl. Nigrit. 256 (1849) ; F.T.A. 1 : 308 ; Chev. Bot. 101 (incl. var.) ; Swingle l.c. 183 (incl. var.). *Amyris anisata* Willd. (1799). *Clausena pobeguini* Pobéguin (1906). *C. inequalis* of F.T.A. 1 : 307 ; of Chev. Bot. 102 ; of F.W.T.A., ed. 1, 1 : 479 ; not of (DC.) Benth. A shrub or small tree, up to 20 ft. high, with odorous leaves, cream-white flowers and shining black drupes. There are probably several infraspecific taxa ; extensive field research and experiments are needed.
Fr.G.: Labé *Chev.* 12292. Timbo *Pobéguin* 118. Diaguissa *Chev.* 12646. Mamou (Nov.) *Chev.* 14662. Farana *Chev.* 13411. **S.L.:** Regent *Sc. Elliot* 4146! Bafodeya Hills (Apr.) *Sc. Elliot* 5685! Limba (Apr.) *Sc. Elliot* 5487! Dodo (Apr.) *Deighton* 3912! **Iv.C.:** Bouroukrou *Chev.* 16785. Attéou *Chev.* 17083. Mt. Do *Chev.* 21427. Languira, Baoulé-Nord *Chev.* 22205. **G.C.:** Cape Coast (fl. & fr. July) *T. Vogel* 5!

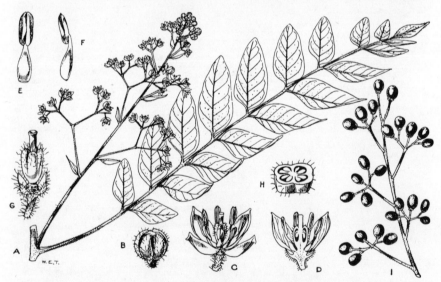

Fig. 191.—CLAUSENA ANISATA *Oliv.* (RUTACEAE).

A, inflorescence. B, flower-bud. C, open flower. D, same in vertical section. E, stamen from front. F, same, side view. G, pistil. H, cross-section of ovary. I, fruits.

85 ! Kintampo (Mar.) *Dalz.* 16 ! Volta R., Krepi (Jan.) *Johnson* 563 ! Accra Plains (Mar.) *Howes* 1137 !
Dah.: Massé to Kétou *Chev.* 23004. Koussi *Chev.* 23233. Ouidah *Chev.* 23447. Savalou *Chev.* 23693.
N.Nig.: Bauchi Plateau (Jan.) *Lely* P. 109 ! Idah *Barter* 1638 ! **S.Nig.**: Lagos *Dawodu* 55 ! Abeokuta
Barter 3339 ! Ibadan (Mar.) *Keay* FHI 16037 ! *Meikle* 1263 ! Oban *Talbot* 428 ! 1500 ! **Br.Cam.**:
Cam. Mt., 5,000–7,500 ft. (fl. & fr. Dec.–Apr.) *Mann* 1187 ! 2183 ! *Maitland* 1068 ! Bambui, Bamenda
(Apr.) *Maitland* 1464 ! L. Oku (Jan.) *Keay & Lightbody* FHI 28491 ! **F.Po:** Musola (Jan.) *Guinea* !
Moka (Jan.) *Exell* 811 ! Widespread in tropical Africa. (See Appendix, p. 307.)

3. AFRAEGLE (Swingle) Engl. Pflanzenw. Afr. 3, 1 : 761 (1915) ; E. & P. Pflanzenfam.
19A : 352, fig. 161 (1931) ; Swingle in Webber & Batchelor, Citrus Industry 1 :
456 (1943).

Branchlets often armed with axillary spines about 2 cm. long ; leaves trifoliolate ;
leaflets shortly stalked, obovate or oblanceolate, obtuse or emarginate at the apex,
4–6·5 cm. long, 2–3·5 cm. broad, glabrous, entire, finely pellucid-glandular ; flowers
few in short axillary cymes ; pedicels constricted and jointed at the base ; calyx
coriaceous, short and broadly lobed ; petals oblong, imbricate ; stamens 12–20, free ;
fruit globose, about 9 cm. diam.   ..   ..   *paniculata*

A. paniculata (*Schum. & Thonn.*) Engl. Pflanzenw. Afr. 3, 1 : 761, fig. 355 (1915) ; Aubrév. Fl. For. C. Iv. 2 :
94, t. 163, 1–3 ; Fl. For. Soud.-Guin. 365 ; Swingle l.c. 457, fig. 70. *Citrus paniculata* Schum. & Thonn.
(1827). *Balsamocitrus paniculata* (Schum. & Thonn.) Swingle (1912)—Chev. Bot. 102. *Aegle barteri*
Hook. f. in Hook. Ic. Pl. t. 2285 (1895). *Limonia warneckei* Engl. (1905)—Chev. Bot. 101. A shrub or
tree, to 50 ft. high, with grey-green branchlets, white fragrant flowers and hard-shelled orange-like fruits
2–4 in. diam. ; usually found planted in villages.
Sen.: *Berhaut* 1324. **Fr.G.:** *Maclaud.* **Lib.:** (?) *fide* Swingle *l.c.* **Iv.C.:** Groumania *Aubrév.* 794. Ouangolo
Aubrév. 1401. Banfara *Vuillet* 708. **G.C.:** *Thonning.* Achimota (June) *Milne-Redhead* 5101 ! Yijia to
Pong *Kitson* 635 ! Kologu, N.T. (fr. Apr.) *Vigne* FH 4738 ! **Togo:** Lome *Warnecke* 113 ! **Dah.:** Péné-
soulou to Pelebina (May) *Chev.* 23833 ! Pira *Chev.* 23749. **N.Nig.:** Kutigi, Niger Prov. (Mar.) *Lamb* 44 !
S.Nig.: Abeokuta (Jan.) *Rowland* ! Ogbomosho *Barter* 3404 ! *Rowland* ! Igboho (Feb.) *Keay* FHI 22494 !
(See Appendix, p. 304.)

Imperfectly known species.

A. mildbraedii *Engl.* l.c. (1915) ; Swingle l.c. 460. Specimen imperfect, presumably now destroyed.
F.Po: San Carlos *Mildbr.* 6978.

4. AEGLOPSIS Swingle in Bull. Soc. Bot. Fr. 58, Mém. 8 : 237 (1912) ; Engl. in E. & P.
Pflanzenfam. 19A : 353 (1931) ; Swingle in Webber & Batchelor, Citrus Industry 1 :
460 (1943).

Branchlets with axillary spines ; leaves simple, subsessile, elliptic, shortly acuminate,
shortly cuneate at the base, 5–12 cm. long, 2·5–7 cm. broad, densely glandular-
punctate ; flowers in very short axillary panicles up to 5 cm. long ; buds ovoid ;
calyx shortly lobed ; petals 4 or 5, oblong or lanceolate, about 1 cm. long ; stamens
8–10, free ; ovary 5–6-celled, style very short ; fruit spherical or pear-shaped, 6–9
cm. diam. ..   ..   *chevalieri*

A. chevalieri *Swingle* Bull. Soc. Bot. Fr. 58, Mém. 8 : 240, t. 2, 1–9, t. 3 (1912) ; op. cit. 60 : 406, fig. A ;
Engl. l.c., fig. 162 ; Aubrév. 2 : 94, t. 162 ; Baldwin & Speese in Bull. Torrey Bot. Cl. 77 : 368, fig. 1.
Balsamocitrus chevalieri (Swingle) A. Chev. Bot. 102 (1920). A shrub or small tree, up to 20 ft. high ;
flowers white.
Lib.: Bobei, Sanokwele (fr. Feb.) *Baldwin* 14182. **Iv.C.:** Sassandra Port, near sea (May) *Chev.* 17940.
Tabou *Pobéguin* 180. Agnéby, Guiglo & Bondoukou *fide* Aubrév. *l.c.* **G.C.:** Kumasi (Mar.) *Vigne* FH 1656 !
Aburi Hills (fl. Feb., fr. Apr.) *Howes* 1188 ! *Obeng* ! Amentia (Apr.) *Vigne* FH 2901 ! Aboma F.R.
(fr. Nov.) *Vigne* FH 3449 ! **Togo:** Kabosu, Hohoe Dist. *St. C. Thompson* FH 3607 !

Imperfectly known species.

A. ? sp. A forest shrub, with spiny stems ; leaves simple and 3-foliolate or 1-foliolate by abortion ; flowers
similar to *A. chevalieri* ; fruits not known. *A. eggelingii* M. R. F. Taylor from Uganda has similar foliage,
but larger flowers with more numerous stamens. Further material, especially with ripe fruits, required.
S.L.: Njala *Deighton* 2506 ! 3100 ! Konia (Apr.) *Deighton* 3945 !

5. CITROPSIS (Engl.) Swingle & Kellerman in Journ. Agric. Research 1 : 421 (1914) ;
Engl. in E. & P. Pflanzenfam. 19A : 347 (1931) ; Swingle in Webber & Batchelor,
Citrus Industry 1 : 302 (1943).

Leaves simple ; petioles 3·5–4 mm. long, wingless, not articulated with the lamina ;
calyx-lobes short ; petals 7 mm. long and 3 mm. broad ; stamens 5·5–6 mm. long ;
ovary 1 mm. high, style 4·6 mm. long ; leaves broadly lanceolate, broadly cuneate
at base, with a short, wide, blunt acumen, 8–10 cm. long, 6–3 (–4) cm. broad ; fruit
not known ..   ..   1. *tanakae*
Leaves imparipinnate, 3-foliolate or 1-foliolate ; petioles 1–8 cm. long, winged, articulated with the lamina ; calyx-lobes acute :
Petals about 4 mm. long ; ovary about 1 mm. high, style 1·5 mm. long ; wing of
petioles up to 2 mm. wide ; leaves (1–) 3-foliolate ; leaflets very variable in shape,
the terminal ones up to 15 cm. long and 6 cm. broad, the lateral ones considerably
smaller ; fruits subglobose, 1–1·5 cm. diam. ..   ..   ..   2. *sp. nr. gabunensis*
Petals about 15 mm. long ; ovary about 3 mm. high, style 12–15 mm. long ; wing of
petiole and rhachis 8–20 mm. wide ; leaves (3–) 5 (–7)-foliolate ; leaflets broadly
elliptic to obovate, up to 19 cm. long and 10·5 cm. broad ; fruits subglobose, probably
about 2 cm. diam.   ..   ..   3. *articulata*

1. **C. tanakae** *Swingle & Kellerman* in Journ. Arn. Arb. 21 : 121, t. 3, fig. 1–5 (1940) ; Swingle in Webber & Batchelor, Citrus Industry 320.
    S.L.: *Afzelius.*

2. **C. sp. nr. gabunensis** (*Engl.*) *Swingle & Kellerman* in Journ. Agric. Research 1 : 430 (1914) ; Engl. in E. & P. Pflanzenfam. 19A : 348, fig, 158, L–N ; Swingle l.c. 315. *Limonia gabunensis* Engl. (1895). A forest shrub ; flowers white ; fruits red.
    S.L.: Kambia, Scarcies R. (fr. Jan.) *Sc. Elliot* 4425 ! Gola Forest (Mar.) *Small* 533 ! **Lib.:** Dukwia R. (fr. July, Oct., Nov.) *Cooper* 39 ! *Linder* 87a !
    [Material from Spanish Guinea, Gabon (type locality) and Belgian Congo has the leaves often 5-foliolate and the petioles and rhachides more widely winged and rather larger flowers ; there is however a good deal of variation in both areas.]

3. **C. articulata** (*Willd. ex Spreng.*) *Swingle & Kellerman* l.c. 433, figs. 3 & 4, A (1914) ; Engl. l.c., fig. 158, F–H ; Swingle l.c. 308. *Citrus articulata* Willd. ex Spreng. (1826). *Limonia preussii* Engl. (1895). *Citropsis preussii* (*Engl.*) Swingle & Kellerman (1914). *Limonia mirabilis* A. Chev. (1912). *Citropsis mirabilis* (A. Chev.) Swingle & Kellerman (1914)—Engl. l.c., fig. 158, J. A forest shrub or tree, to 20 ft. high ; flowers white, very fragrant.
    S.L.: Bumbuna (young fr. Oct.) *Thomas* 3784 ! 3897 ! **Iv.C.:** Sanrou to Ouodé, R. Koué (May) *Chev.* 21609 ! **G.C.:** Suhuma F.R. (fl. & fr. Oct.) *Cansdale* FH 3961 ! Jimira F.R. (Aug.) *Vigne* FH 3321 ! 3991 ! Mountains 30–45 miles N. of Accra *Isert.* **Togo:** Misahöhe *Baumann* 552 ! Santrokofl, Hohoe Dist. (fl. & fr. Oct.) *St. C. Thompson* FH 3597 ! **S.Nig.:** Oban *Talbot* 1348 ! 1472 ! Bashua, Ikom Div. *Jones & Onochie* FHI 18703 ! **Br.Cam.:** Johann-Albrechtshöhe *Staudt* 747 ! Barombi (Sept.) *Preuss* 548 ! Also in French Cameroons and Gabon. (See Appendix, p. 305.)

6. ARALIOPSIS Engl. in E. & P. Pflanzenfam. 3, 4 : 175 (1896) ; op. cit. 19A : 304 (1931) ; Verdoorn in Kew Bull. 1926 : 393.

Branchlets and leaves glabrous ; leaves (3–) 5-foliolate, petioles 5–19 cm. long ; leaflets obovate-oblong to obovate, cuneate into the 1–2 cm. petiolule at the base, abruptly long-acuminate, 8–27 cm. long, 3·5–10·5 cm. broad, with numerous parallel lateral nerves ; panicles many-flowered, axillary, pubescent, pedicels up to 4 mm. long ; flowers glabrous outside, unisexual   ..   ..   *tabouensis*

A. tabouensis *Aubrév. & Pellegr.* in Bull. Soc. Bot. Fr. 83 : 488, fig. 1 (1936) ; Aubrév. Fl. For. C. Iv. 2 : 90, t. 161. An evergreen tree to 100 ft. high with cylindrical bole and rough bark ; flowers whitish ; in rain forest.
    Lib.: Dukwia R. *Cooper* 341 ! **Iv.C.:** Youkou to Patokla, Tabou region (May, June) *Aubrév.* 1304 ! **G.C.:** Simpa (Feb., Mar.) *Kinloch* FH 3235 ! *Vigne* FH 2807 ! Axim (Feb.) *Irvine* 2249 ! **S.Nig.:** Umon-Ndealichi F.R., Itu, Calabar Prov. *Iyizoba* FHI 20809 !

Imperfectly known species.

A. ? sp. Forest tree, about 10 ft. high ; leaves 5-foliolate, petioles 17–28 cm. long ; leaflets more or less obovate elliptic, 13–34 cm. long, 7–16·2 cm. broad ; male panicles terminal, very dense ; sepals and petals green without, petals within and filaments white. On account of its 8 stamens I refer this plant provisionally to *Araliopsis* ; fruits are however required to determine its exact position. It is possible that the specimen (*Olorunfemi* FHI 30709) which I have tentatively referred to as *Oricia* sp. B, on account of its apocarpous fruits, is conspecific ; the distinctions between the genera *Oricia*, *Araliopsis* and *Vepris* and their allies need to be thoroughly revised.
    S.Nig.: Boje to Iso Bendiga, Afi River F.R., Ikom Div. (Dec.) *Keay* FHI 28258 !

7. ORICIA Pierre in Bull. Soc. Linn. Paris 2 : 1288 (1897) ; Verdoorn in Kew Bull. 1926 : 411 ; Engl. in E. & P. Pflanzenfam. 19A : 303.

Branches of inflorescence tomentose, flowers sessile, outside of corolla glabrous ; leaves (1–) 3-foliolate ; leaflets obovate-elliptic to elongate-elliptic, cuneate at base, obtuse or subacute at apex, 3·5–28 cm. long, 1·5–10 cm. broad   ..   ..   1. *suaveolens*
Branches of inflorescence tomentellous, pedicels 1·5–3 mm. long, outside of corolla puberulous ; leaves 3-foliolate ; leaflets elongate-oblong, cuneate at base, shortly acuminate, 10–35 cm. long, 4–12 cm. broad   ..   ..   2. *trifoliolata*

1. **O. suaveolens** (*Engl.*) *Verdoorn* in Kew Bull. 1926 : 413 ; Aubrév. Fl. For. C. Iv. 2 : 92, t. 162, 1–5. *Teclea suaveolens* Engl. (1896). *Oricia leonensis* Engl. (1903). An understorey forest tree, to 30 ft. high ; flowers greenish-white.
    Fr.G.: Kindia (Mar.) *Chev.* 13100 ! **S.L.:** *Afzelius* ! Freetown *Vohsen.* Top of Sugar Loaf Mt. (Mar.) *Dalz.* 977 ! Kangama to Dia (Apr.) *Deighton* 3152 ! Loma Mts., 2,500 ft. *Jaeger* 4027 ! **Iv.C.:** Dakpadou (fr. Dec.) *Aubrév.* 214. Mt. Momy, Man *Aubrév.* 1176. Patokla *Aubrév.* 1303. Anyama (Sept.) *Aubrév.* 1589. **S.Nig.:** Gambari Group F.R. (fl. Mar., fr. May, July) *Adekunle* FHI 14777 ! *W. D. MacGregor* 508 ! 509 ! *Symington* FHI 4116 !

2. **O. trifoliolata** (*Engl.*) *Verdoorn* l.c. (1926). *Araliopsis trifoliolata* Engl. (1917). A forest tree.
    Br.Cam.: Victoria (Mar.) *Maitland* 586 ! *Zahn* 499 !

Imperfectly known species.

1. **O. sp. A.** Tree, to 45 ft. high ; leaves 3-foliolate, midribs densely pubescent beneath. Flowers not known.
    Lib.: Mecca (fr. Dec.) *Baldwin* 10807 !

2. **O. sp. B.** Forest tree, to 25 ft. high ; leaves 5-foliolate, leaflets up to 40 cm. long. Flowers not known ; but see note under *Araliopsis* " imperfectly known species."
    Br.Cam.: S. Bakundu F.R. (fr. Aug.) *Olorunfemi* FHI 30709 !

8. DIPHASIA Pierre in Bull. Soc. Linn. Paris, nouv. sér. 70 (1898) ; Verdoorn in Kew Bull. 1926 : 411 ; Engl. in E. & P. Pflanzenfam. 19A : 304 (1931).

Branchlets, petioles and midribs and nerves on underside of leaflets hispid with yellowish or grey hairs ; petioles 2–10 cm. long ; leaflets 3, elliptic-obovate, cuneate into a short petiolule at base, acuminate at apex, 5–18 cm. long, 2·3–7 cm. broad ; inflorescence of terminal and axillary panicles ; fruits about 1 cm. long, 2-lobed, or of one oblique carpel with the other aborting and forming a swelling at the base of the fruit   ..   ..   *klaineana*

D. klaineana *Pierre* l.c. (1898). *D. angolensis* of Verdoorn in Kew Bull. 1926 : 411, partly ; of Aubrév. Fl. For. C. Iv. 2 : 84. An understorey forest tree, to 50 ft. high ; flowers white.
Iv.C.: Port Bouët *Aubrév.* 2004 ! Agnéby (Apr.) *Aubrév.* 2264. **G.C.:** Bosusu (July) *Moor* FH 340 ! S. Scarp F.R. *Moor* FH 2239 ! S. Fomang Su F.R. (fr. Jan.) *Vigne* FH 2698 ! Bobiri F.R. (fl. May, fr. Sept.) *Foggie* FH 4893 ! *C. J. Taylor* FH 5273 ! Worobong F. R. (fl. & fr. Nov.) *Vigne* FH 4269 ! 4270 ! Also in Gabon. (See Appendix, p. 307.)

9. TECLEA Del. in Ann. Sci. Nat., sér. 2, 20 : 90 (1843) ; Verdoorn in Kew Bull. 1926 : 401 ; Engl. in E. & P. Pflanzenfam. 19A : 314 (1931).

Inflorescences quite glabrous, up to 2·5 cm. long, axillary, racemose or sparingly branched ; petioles not winged ; leaflets distinctly petiolulate, with numerous lateral nerves :
  Petals 3–4 mm. long ; racemes up to 1·5 cm. long, with 1–3 (–5) flowers ; petioles 2·5–13·5 cm. long ; petiolules 0·5–2·5 cm. long ; leaflets elongate-elliptic, cuneate at base, obtusely long-acuminate, 8·5–15 cm. long, 3·2–6·3 cm. broad ; fruits ellipsoid, up to 2·5 cm. long and 1·5 cm. broad .. .. 1. *afzelii*
  Petals 2 mm. long ; racemes up to 2·5 cm. long with several flowers ; petioles 1·8–4 cm. long ; petiolules up to 6 mm. long ; leaflets lanceolate-elongate-elliptic, cuneate at base, gradually long-acuminate, 4·3–12·5 cm. long, 1·7–4·8 cm. broad 2. *sudanica*
Inflorescences rather densely but minutely spreading-puberulous, up to 4 cm. long, axillary and terminal, usually many-flowered and paniculate ; petals 1·5–2 mm. long ; petioles (0·8–) 2–7 cm. long, narrowly winged ; leaflets sessile, elongate-elliptic, cuneate at base, acuminate at apex, 4·5–21 cm. long, 1·2–8 cm. broad ; fruits oblong-ellipsoid, at least 8 mm. long and 4 mm. broad .. 3. *verdoorniana*

1. **T. afzelii** *Engl.* Bot. Jahrb. 23 : 153 (1896) ; Chev. in Rev. Bot. Appliq. 30 : 76. (?) *T. acuminata* Engl. Bot. Jahrb. 36 : 245 (1905). *T. zenkeri* of Verdoorn in Kew Bull. 1926 : 407, partly, of F.W.T.A., ed. 1, 1 : 482 ; not of Engl. (1896). A glabrous forest shrub or small tree ; flowers orange-red ; ripe fruits yellow.
  Fr.G.: Mt. Loura, Fouta Djalon *Chev.* **S.L.:** *Afzelius* ! **S.Nig.:** Afi River F.R., Ikom (fr. May) *Latilo* FHI 30993 ! **Br.Cam.:** Johann-Albrechtshöhe *Staudt* 590 ! Bibundi (Apr.) *Schlechter* 12410. Also in French Cameroons. (See Appendix, p. 309.)
2. **T. sudanica** *A. Chev.* in Bull. Mus. Hist. Nat., sér. 2, 5 : 408 (1933) ; Rev. Bot. Appliq. 30 : 78, t. 4 C. A shrub, to 10 ft. high ; on rocky hills in savannah regions.
  Fr.Sud.: Boulouli (Sept.) *Vuillet* 749 ! Kita *Dubois* 55 ! Fô (June) *Chev.* 966 ! **G.C.:** Yeji Dist. *Kitson* ! [Apparently close to *T. nobilis* Del. which is widespread in N.E. and E. Africa.]
3. **T. verdoorniana** *Exell & Mendonça* in Consp. Fl. Ang. 1 : 270 (1951). *T. grandifolia* of Verdoorn in Kew Bull. 1926 : 407 ; of F.W.T.A., ed. 1, 1 : 482 ; of Chev. in Rev. Bot. Appliq. 30 : 76, t. 4 A ; not of Engl. (1896). A forest shrub or small tree ; petals greenish-white, stamens white.
  S.L.: Sugar Loaf Mt. (May) *Barter* ! **Lib.:** Mt. Bili (Dec.) *Barker* 1181 ! **Iv.C.:** Groumania (fl. Jan., fr. May) *Aubrév.* 795 ! 2365 ! **G.C.:** *Irvine* 952 ! Ejun (Apr.) *Vigne* FH 1100 ! Volta River F.R. (Feb., Apr.) *Moor* FH 2343 ! *Vigne* FH 4350 ! Akwapim (Apr.) *Irvine* 1155 ! **Br.Cam.:** Johann-Albrechtshöhe *Staudt* 472 ! 493 ! Kumba (Feb.) *Dundas* FHI 20355 ! **F.Po:** Bokoko *Mildbr.* 6960. Also in Belgian Congo (Haut Uele). (See Appendix, p. 309.)

Imperfectly known species.

1. **T. ferruginea** *A. Chev.* in Bull. Mus. Hist. Nat., sér. 2, 5 : 409 (1933) ; Rev. Bot. Appliq. 30 : 76. A shrub, 5 ft. high. Type specimen not traced.
  Fr.Sud.: Sangali, 150 km. from Bandiagara (fl. buds May) *Rogeon* 293.
2. **T. carpopunctifera** *A. Chev.* in Rev. Bot. Appliq. 30 : 78, t. 4 B (1950). A small tree ; in forest. Type specimen not traced.
  Iv.C.: Kokroum *Bégué* 2951.

115. SIMAROUBACEAE

Trees or shrubs, sometimes with bitter bark, rarely spiny. Leaves alternate or rarely opposite, imparipinnate, rarely simple, not or very rarely gland-dotted ; stipules absent. Flowers small, unisexual or polygamous, actinomorphic. Calyx-lobes 3–5. Petals 3–5, imbricate or valvate, rarely absent, free or united into a tube. Disk present. Stamens inserted at the base of the disk, equal or double the number of the petals, rarely more numerous, free, sometimes with a scale at the base ; anthers 2-celled, opening lengthwise. Ovary 2–5-lobed, 1–5-celled, or carpels quite separate ; styles 2–5 ; ovules usually solitary, rarely 2 or more, axile. Fruit usually indehiscent, dry or drupaceous. Seeds with or without endosperm ; embryo straight or curved.

Mainly tropical regions ; sometimes with bitter bark ; leaves eglandular, pinnate, exstipulate.

Ovary 4 (–5)-celled ; branchlets often with thorns ; stamens 8–10, each with a hairy petaloid scale at the base.. .. .. 1. **Harrisonia**
Ovary of 4–5 free carpels, each with a single ovule ; branchlets not thorny :
  Stamens each with a hairy petaloid scale at the base :
    Calyx open in bud, regularly lobed :
      Leaf-rhachis narrowly winged ; leaflets 3–4 pairs ; stamens 10 .. 2. **Quassia**

Leaf-rhachis not winged ; leaflets 5–15 pairs ; stamens 10–18   **3. Pierreodendron**
Calyx closed in bud, rupturing into 2–3 (–4) unequal segments ; stamens 10 (–14)
                                                          4. Hannoa
Stamens without petaloid scales at the base :
  Stamens 10 ; leaf-rhachis not constricted at the base and nodes ; leaflets 6–12 pairs ;
  fruits about 10 cm. long   ..   ..   **5. Gymnostemon**
  Stamens 4 ; leaf-rhachis (when dry) constricted at the base and nodes ; leaflets
  4–6 pairs ; fruits about 1 cm. long   ..   ..   **6. Brucea**

Besides the above, another genus is represented by the following species which is sometimes cultivated in our area :—*Picrasma javanica* Bl., a native of tropical Asia.

1. HARRISONIA R.Br. ex A. Juss.—F.T.A. 1 : 311 ; Engl. in E. & P. Pflanzenfam. 19A : 382 (1931).

Leaflets usually opposite, 4–8 pairs, oblanceolate to elliptic, narrow at base, rounded or slightly acuminate at apex, about 4 cm. long and 1·5 cm. broad, entire or crenate, slightly pubescent beneath ; rhachis narrowly winged ; panicles terminal, many-flowered ; flowers globose in bud, tomentose, small ; fruits depressed-globose, 7–8-lobulate, hollowed on the top, about 6 mm. diam.   ..   ..   *abyssinica*

H. abyssinica *Oliv.* F.T.A. 1 : 311 (1868) ; Engl. l.c., fig. 177. *H. occidentalis* Engl. (1895)—Chev. Bot. 105 ; F.W.T.A., ed. 1, 1 : 485 ; Aubrév. Fl. For. C. Iv. 2 : 100, t. 167. *Zanthoxylum guineense* Stapf (1906). A shrub, or small tree up to 30 ft. high ; the leaves often with paired thorns at the base, with branches sometimes long and sarmentose, and erect panicles of small white flowers ; characteristic of areas marginal between forest and savannah.
  Fr.G.: Farana (May) *Chev.* 13408 ! And other localities *fide* Chev. l.c. **S.L.:** *Afzelius* ! Sulimania and Falaba (Mar.) *Sc. Elliot* 5310 ! Dunnia (Feb.) *Sc. Elliot* 4887 ! Daru to Baiima (Apr.) *Deighton* 3145 ! Batkanu (fr. July) *Small* 127 ! **Lib.:** Cape Palmas *Ansell* ! Wohmen, Vonjama (fr. Oct.) *Baldwin* 10101 ! Tappita (fr. Aug.) *Baldwin* 9095 ! **Iv.C.:** Dimbokro *Serv. For.* 438 ! And other localities *fide* Chev. l.c. **G.C.:** Ejura, N. Ashanti (Mar.) *Dalz.* 4 ! Sunyani, W. Ashanti (Apr.) *Irvine* 921 ! Afram Plains (Mar.) *Johnson* 711 ! Aburi Hills (Mar.) *Johnson* 896 ! Opro River F.R. (Feb., Mar.) *Andoh* FH 4673 ! *Lyon* FH 2856 ! **Togo:** Lome *Warnecke* 465 ! **Dah.:** Allada (Mar.) *Chev.* 23388 ! 23421 ! **S.Nig.:** Lagos *Foster* 52 ! *Millen* 106 ! *Rowland* ! Olokemeji (Apr.) *Obaseki* FHI 23821 ! Ijaye *Barter* 3338 ! **Br.Cam.:** Johann-Albrechtshöhe *Staudt* 963 ! Widespread in tropical Africa. (See Appendix, p. 312.)

2. QUASSIA Linn.—F.T.A. 1 : 312 ; Engl. in E. & P. Pflanzenfam. 19A : 377 (1931).

Leaflets opposite, 3–4 pairs, obliquely elliptic, unequal-sided at base, caudate-acuminate, 10–15 cm. long, 4–6 cm. broad, thin, glabrous, with about 9 pairs of obscure lateral nerves ; rhachis narrowly winged, often constricted at the joints ; panicles very narrow, shorter than the leaves ; pedicels 5 mm. long ; petals linear-oblong, 1·5 cm. long ; stamens shortly pilose towards the base ; fruits slightly flattened, obovate-elliptic, with sharp edges, about 2·5 cm. long, very laxly reticulate..   .. *africana*

Q. africana (*Baill.*) *Baill.* in Adansonia 8 : 89 (1868) ; F.T.A. 1 : 312. *Simaba africana* Baill. (1867)—Cronquist in Lloydia 7 : 81. A glabrous shrub, 8–12 ft. high ; in forest.
  Br.Cam.: R. Cameroon (Jan.) *Mann* 2191 ! Extends to Belgian Congo and Angola. (See Appendix, p. 314.)

Besides the above *Q. amara* Linn., a native of tropical America, is cultivated in our area.

3. PIERREODENDRON Engl. Bot. Jahrb. 39 : 575 (1907) ; Little in Phytologia 3 : 156 (1949). *Simarubopsis* Engl. (1911). *Mannia* Hook. f. (1862)—F.T.A. 1 : 312 ; F.W.T.A., ed. 1, 1 : 484 ; not of Opiz (1829), nor of Trevis. (1857).

Stamens 15–18 ; anthers 2–3 mm. long ; leaves up to 1 m. long, with 5–15 pairs of subopposite leaflets, plus the terminal one ; leaflets oblong-elliptic, rounded and unequal sided at base, rounded and with a very abrupt, short, thick acumen at apex, 10–30 cm. long and 3–9 cm. broad, leathery, with numerous much-branched lateral nerves, slightly impressed above ; petiolules about 3 mm. long ; racemes clustered or paniculate, up to 40 cm. long ; pedicels up to 7 mm. long ; calyx-lobes orbicular, about 2·5 mm. long, petals up to 1 cm. long ; fruits oblong-ellipsoid, about 7 cm. long and 4–5 cm. broad, apiculate, woody   ..   ..   1. *africanum*
Stamens 10 ; anthers 4–5 mm. long ; otherwise, apparently, as above   2. *kerstingii*

1. **P. africanum** (*Hook. f.*) *Little* in Phytologia 3 : 156 (1949). *Mannia africana* Hook. f. (1862)—F.T.A. 1 : 312 ; F.W.T.A., ed. 1, 1 : 484, partly ; Kennedy For. Fl. S. Nig. 152. *Pierreodendron grandifolium* Engl. (1907). A forest tree, to 60 ft. high, with straggling habit ; inflorescences and flowers reddish-purple.
  S.Nig.: Ewusa to Ebele, Ishan (Aug.) *Onochie* FHI 33251 ! Sapoba (fl. Aug.–Sept., fr. Sept., Oct.) *Kennedy* 1809 ! Nun R. (Sept.) *Mann* 480 ! Calabar (Feb.) *Mann* 2266 ! Also in French Cameroons, Gabon, Belgian Congo and Angola. (See Appendix, p. 314.)
2. **P. kerstingii** (*Engl.*) *Little* l.c. (1949). *Simarubopsis kerstingii* Engl. Bot. Jahrb. 46 : 280, fig. 1 (1911). *Mannia kerstingii* (Engl.) Harms in E. & P. Pflanzenfam. 19A : 371, fig. 168 (1931). *M. simarubopsis* Pellegr. in Bull. Soc. Bot. Fr. 77 : 665 (1930) ; Aubrév. Fl. For. C. Iv. 2 : 102, t. 168. A large forest tree, to 80 ft. high ; flowers red, ripe fruits yellow.
  Iv.C.: Abidjan *Aubrév.* 192 ; 346. **G.C.:** Kwaben (Aug.) *Vigne* FH 1297 ! Aburi *Johnson* 985 ! **Togo:** Kulumi, Sokode-Basari (fr. Dec.) *Kersting* A. 708 !
  This genus is still poorly known ; more good specimens and field observations are much needed. Aubréville (l.c.) suggests that the Ivory Coast plant may be different from *P. kerstingii*. It is possible, however, that there is really only one species in the genus, with dimorphic flowers. The genus is also represented at Kew by the following sterile specimens which may belong to either species :—
  S.L.: Bo to Yele *King* 276 ! **Iv.C.:** Dabou *Chev.* 16212 ! (cited as *Trichoscypha mannioides* A. Chev. Bot. 161, name only). **S.Nig.:** Olokemeji *Onochie & Jones* FHI 14179 ! Illa *Thomas* 1916 !

4. HANNOA Planch.—F.T.A. 1 : 309 ; Engl. in E. & P. Pflanzenfam. 19A : 380 (1931).

Tree to 35 m. high, with tall straight bole in rain forest ; leaflets (2–) 3–8 pairs, plus the terminal leaflet, on petiolules up to 1·2 cm. long, oblong or elliptic, cuneate at base, rounded and broadly mucronate or emarginate at apex, 3·5–19 cm. long, 2–7·5 cm. broad, leathery, glabrous, main lateral nerves impressed above and beneath ; panicles terminal, lax, longer than the leaves, puberulous ; flowers more or less glomerate, puberulous outside ; petals about 4 mm. long ; fruits oblong, 2·5–3·2 cm. long, 1·5–2·5 cm. diam., glabrous .. .. 1. *klaineana*

Tree to 10 m. high in understorey of montane forest ; leaflets 2–3 pairs, plus the terminal leaflet, on petiolules up to 2·2 cm. long, oblong, cuneate at base, rounded and mucronulate or emarginate at apex, 6–16 cm. long, 2–6 cm. broad ; panicles rather shorter than the leaves ; otherwise similar to above .. .. 2. *ferruginea*

Tree to 12 m. high, with fire-tolerant, deeply furrowed bark, in savannah ; leaflets 2–5 pairs, plus the terminal leaflet, on petiolules up to 4 cm. long, oblong-elliptic or obovate-oblong, cuneate at base, rounded and mucronulate or emarginate at apex, 3·5–9·5 cm. long, 1·5–5·2 cm. broad ; panicles longer than the leaves ; fruits oblong, 1·6–2 cm. long, 1–1·3 cm. diam., glabrous ; otherwise similar to *klaineana*
3. *undulata*

FIG. 192.—HANNOA KLAINEANA *Pierre & Engl.* (SIMAROUBACEAE).

A, flower-bud. B, open flower. C, stamen. D, flower with petals and stamens removed. E, fruits. F, fruit in vertical section. G, fruit in cross-section.

1. **H. klaineana** *Pierre & Engl.* in Engl. Bot. Jahrb. 46 : 282 (1911), incl. vars. ; Chev. Bot. 103 ; Aubrév. Fl. For. C. Iv. 2 : 102, t. 169 ; Kennedy For. Fl. S. Nig. 153. A large forest tree, with straight clear bole covered with smooth grey bark, without buttresses ; wood white, soft ; foliage shining, glabrous ; flowers yellowish-white ; ripe fruits black and shining. **Fr.G.:** Kouria (Oct.) *Chev.* 14711! Conakry (fr. Dec.) *Chev.* 25690! **S.L.:** Devils' Hill, Hangha (Sept.) *Aylmer* 595! Njala (fr. Feb.) *Deighton* 2608! 5299! Heddle's Farm (Oct.) *Lane-Poole* 393! Kambui (Oct.) *Edwardson* 138! **Lib.:** Dukwia R. (fr. Oct., Nov.) *Cooper* 82! 115! Moylakwelli (fr. Oct.) *Linder* 1312! **Iv.C.:** *Fleury* 7! And various localities *fide* Chev. *l.c.* **G.C.:** Ampakrom, Tanosu (Sept.) *Chipp* 356! Brahababum (Aug.) *Vigne* FH 918! Dunkwa (Aug., Sept.) *Rea* 289! *Vigne* FH 149! Tafo (Sept.) *Andoh* FH 4522! **Togo:** Misahöhe *Mildbr.* 7375! **S.Nig.:** Lagos (Sept.) *Dalz.* 1266! *Dawodu* 91! Sapoba (fl. July, fr. Dec.) *Kennedy* 277! 344! 425! Niger Delta (Aug.) *Mann* 458! Degema (July) *Unwin* 3! Ikom (Sept.) *Catterall* 32! **Br.Cam.:** Victoria (Sept.) *Maitland* 748! Extends to Belgian Congo and Angola. (See Appendix, p. 311.)

2. **H. ferruginea** *Engl.* Bot. Jahrb. 32 : 122 (1902). A tree of montane forest ; branchlets, rhachides and midribs crimson, leaflets dark glossy green above, pale beneath, crimson on the margins ; flowers yellowish-cream. **S.Nig.:** Mt. Koloishe, 4,000 ft., Obudu Div. (fl. & young fr. Dec.) *Keay & Savory* FHI 25052! **Br.Cam.:** Bangwe, 2,900 ft. (Jan.) *Conrau* 53.

3. **H. undulata** (*Guill. & Perr.*) *Planch.*—F.T.A. 1 : 309 ; Engl. Bot. Jahrb. 46 : 283, incl. var. ; E. & P. Pflanzenfam. 19A : 380, fig. 174 ; Aubrév. Fl. For. Soud.-Guin. 368, t. 76, 6–7. *Simaba ? undulata* Guill. & Perr. Fl. Seneg. 1 : 136, t. 34 (1831). A savannah tree with grey corky bark, deeply furrowed ; leaflets dark green and matt above, pale beneath ; flowers yellowish-cream, sweetly scented ; ripe fruits black. **Sen.:** *Heudelot* 435! Djibelor *Aubrév.* 3065. Sinédone *Chev.* 3156. **Gam.:** Albreda (fr. Mar., Apr.) Perrottet. Georgetown (Jan.) *Dalz.* 8037! Kombo *Dawe* 2! **Fr.Sud.:** Balani (Jan.) *Chev.* 142! Koundian *Chev.* 415. **Port.G.:** Madina de Boé (Jan.) *Esp. Santo* 2861! Brene, Bissau (fr. Feb.) *Esp. Santo* 1736! **Fr.G.:** Kouroussa (Nov.) *Pobéguin* 594! 854! Mangata to Laya (Jan.) *Chev.* 20434! **S.L.:** Falaba, Sulima (Nov.) *Miszewski* 14! **Iv.C.:** Banfora *Serv. For.* 1819. **G.C.:** Yefri Dist., Ashanti (Feb.) *Kitson*! Tamale (fr. May) *Kitson* 953! Bosomoa F.R. (Dec.) *Lyon* FH 2725! Nsawkaw *Vigne* FH 3542. **Togo:**

Sokode (Dec.) *Kersting* A. 322! Kete Krachi (Jan.) *Vigne* FH 1532! **Dah.**: Djougou *Chev.* 23865.
Fr.Nig.: Gaya *Aubrév.* N. 21. **N.Nig.**: Sokoto (Nov.) *Lely* 853! Jebba *Barter* 1023! Kontagora (fl.
Nov., fr. Dec.) *Dalz.* 76! Lafia to Makurdi (Nov.) *Keay* FHI 22272! Also in French Cameroons and
Ubangi-Shari. (See Appendix, p. 312.)

5. GYMNOSTEMON Aubrév. & Pellegr. in Bull. Soc. Bot. Fr. 84 : 183 (1937).

A tree, with glabrous stems and imparipinnate leaves ; leaf-rhachis 20–45 cm. long ;
leaflets coriaceous, subsessile, 6–12 pairs, opposite or subopposite, oblong, cuneate or
obtuse at base, long-caudate-acuminate, 8–13·5 cm. long, 2–4·5 cm. broad, with
12–16 main lateral nerves on each side of midrib ; panicles terminal ; pedicels about
5 mm. long, like the flowers rusty-villous ; calyx about 2 mm. long ; petals about
4·5 mm. long, villous on both sides ; disk fleshy, lobed, pubescent ; stamens 10,
about 3 mm. long ; carpels 5, free, villous ; fruits ovoid, about 10 cm. long and 8 cm.
diam., pustulose outside .. .. *zaizou*

G. zaizou *Aubrév. & Pellegr.* l.c. 184, fig. 1 (1937). *Mannia ? zaizou* Aubrév. Fl. For. C. Iv. 2 : 104 (1936),
  French descr. only. A large forest tree, with habit of an *Entandrophragma* ; leaves clustered at the ends
  of the branches.
  Iv.C.: Gagnoa to Sassandra *Aubrév.* 707! Guiglo *Aubrév.* 1225! 2773!

6. BRUCEA J. F. Mill.—F.T.A. 1 : 309 ; Engl. in E. & P. Pflanzenfam. 19A : 386 (1931).

Leaflets rusty-tomentose beneath, densely so at first ; leaves imparipinnate, with 4–5
opposite or subopposite leaflets ; leaf-rhachis (when dry) constricted at the joints,
10–35 cm. long ; leaflets oblong-ovate to ovate-lanceolate, unequal and rounded or
obtuse at base, gradually acuminate, 4·5–14 cm. long, 2–7 cm. broad, margins undulate,
with 6–9 main lateral nerves on each side of midrib ; panicles elongated, up to about
35 cm. long, the flowers clustered and subsessile on the tomentose axis ; fruits
ellipsoid, about 1 cm. long, reticulate .. .. 1. *antidysenterica*
Leaflets pubescent on the midrib and nerves beneath ; leaves imparipinnate, with
4–6 pairs of opposite or subopposite leaflets ; leaf-rhachis up to 100 cm. long ; leaflets
ovate-oblong, rounded or slightly cuneate at base, gradually acuminate, 11–27 cm.
long, 4–13 cm. broad ; panicles very narrow, many-flowered, up to 40 cm. long ;
otherwise similar to above .. .. 2. *guineensis*

1. **B. antidysenterica** *Lam.* Encycl. Méth. Bot. 1 : 471 (1785) ; F.T.A. 1 : 309. *B. salutaris* A. Chev. (1912)—
   Chev. Bot. 104. A shrub or tree, to 30 ft. high ; flowers green ; ripe fruits brown ; in montane forest.
   Fr.G.: Diaguissa Plateau, 3,200–4,200 ft. *Chev.* 12703! 18596 ; 18760. **S.Nig.**: Oban *Talbot* 1347!
   Br.Cam.: Cam. Mt., 7,000–8,000 ft. (Dec., Jan.) *Mann* 1188! 2174! *Maitland* 1329! Lakom, Bamenda
   (fr. May) *Maitland* 1588! Abakpa (fr. Apr.) *Ujor* FHI 30051! Bafing, 4,000 ft. (Feb.) *Johnstone* 63/31!
   Widespread on mountains in tropical Africa.
2. **B. guineensis** *G. Don* Gen. Syst. 1 : 801 (1831). *B. macrophylla* Oliv. F.T.A. 1 : 310 (1868) ; F.W.T.A.,
   ed. 1, 1 : 486. *B. paniculata* of F.T.A. 1: 310, partly (syn. *B. guineensis* G. Don, only). A shrub or tree,
   to 15 ft. high, with long panicles of small flowers with red corollas ; in clearings in lowland rain forest ;
   possibly only varietally distinct from the above.
   S.L.: *Don*! **G.C.**: Obuom Mine (June) *Vigne* FH 2974! **Br.Cam.**: Ambas Bay (Feb.) *Mann* 774! Victoria
   (fr. May, Aug.) *Maitland* 703! Bulifambo (Mar.) *Maitland* 579! Also in French Cameroons and Spanish
   Guinea.

Excluded species.

Brucea paniculata *Lam.*—F.T.A. 1 : 310 (excl. syn. *B. guineensis* G. Don). This is **Trichoseypha smeathmannii**
  Keay (Anacardiaceae).

116. IRVINGIACEAE

Trees ; leaves alternate, simple, not gland-dotted, with well-developed stipules
enclosing the young buds, caducous and leaving a scar around the stem. Flowers
hermaphrodite, actinomorphic. Calyx-lobes 4–5, imbricate. Disk present.
Stamens 8–10 ; anthers 2-celled, opening lengthwise. Ovary 2-celled ; ovules
1 per cell ; style solitary. Fruits drupaceous or samaroid. Seeds without or
with scant endosperm.

A small family of tropical Africa distinguished from the *Simaroubaceae* mainly by
the large stipules with their annular scars ; clearly differentiated from the *Simaroubaceae*
on anatomical grounds (see, for instance, Metcalfe & Chalk, Anatomy of the Dicotyledons
1 : 324).

Ovary (4–) 5 (–6)-celled ; fruits depressed, (4–) 5 (–6)-seeded and -angled ; stipules up
to 10 (–20) cm. long .. .. 1. **Klainedoxa**
Ovary 2-celled ; stipules up to 3 cm. long :
Fruit drupaceous .. .. 2. **Irvingia**
Fruit samaroid .. .. 3. **Desbordesia**

1. KLAINEDOXA Pierre ex Engl. (1896)—E. & P. Pflanzenfam. 19A : 396 (1931).

A glabrous tree ; leaves ovate-elliptic or elliptic, cuneate, obtuse or rounded at base,
gradually tapered to an acute or subacuminate apex, 5–16 cm. long, 2–6·5 cm. broad,
leathery, closely and finely reticulate on both surfaces, with numerous main lateral

nerves ; petioles 5–8 mm. long ; stipules linear, acute, up to 10 cm. long, caducous ;
on young trees leaves lanceolate, up to 40 cm. long and 10 cm. broad, stipules up to
20 cm. long ; racemes paniculate, up to about 12 cm. long ; pedicels about 3 mm.
long ; sepals broadly obovate, about 1·5 mm. long ; petals about twice as large ;
fruits depressed-globose, usually 5-angled, hard, 5–8 cm. diam.

gabonensis var. *oblongifolia*

K. gabonensis *Pierre ex Engl.* var. **oblongifolia** *Engl.* Bot. Jahrb. 32 : 125 (1902) ; E. & P. Pflanzenfam. 19A :
396, fig. 185, A–D ; Kennedy For. Fl. S. Nig. 150. *K. oblongifolia* (Engl.) Stapf in Broun & Massey
Fl. Sud. 228 (1929) ; F.W.T.A., ed. 1, 1 : 483, in syn. *K. zenkeri* Van Tiegh. (1905). *K. gabonensis* of
F.W.T.A., ed. 1, 1 : 483 ; Aubrév. Fl. For. C. Iv. 2 : 95, t. 164. An evergreen tree to 150 ft. high, with
broad crown and narrow very widely spreading buttresses ; flowers pinkish ; in forest. Young trees
have thorny stems, very large leaves and long stipules.
Port.G.: Empacaca, Canchungo *Esp. Santo* 1944 ! **S.L.:** Njala (Nov.) *Deighton* 1942 ! 1944 ! 2812 !
Kambui Hills (Dec.) *Edwardson* 7 ! Gola Forest (fr. Apr.) *Small* 606 ! Loko country *Sc. Elliot* 5717 !
Lib.: Dukwia R. *Cooper* 118 ! 349 ! Buchanan (young fr. Mar.) *Baldwin* 11208 ! **Iv.C.:** Yapo (Oct.)
Chev. 22320 ! And other localities *fide* Aubrév. *l.c.* **G.C.:** Essiama (Nov.) *Vigne* FH 1448 ! Huniso
Irvine 1066 ! Obuasi (fr. Jan.) *Soward* 648 ! Sikamang to Akrofuam *Kitson* 1252 ! **Togo:** Kadjakpe,
Hohoe Dist. *St. C. Thomson* FH 3671 ! **S.Nig.:** Oni *Sankey* 18 ! Sapoba (fl. Apr., fr. June) *Kennedy* 302 !
1385 ! 1647 ! **Br.Cam.:** Victoria (Jan.) *Kalbreyer* 10 ! Bibundi *Mildbr.* 10641 ! Likomba (Nov.) *Mildbr.*
10736 ! The species extends to A.-E. Sudan, Uganda, Belgian Congo and Cabinda. (See Appendix,
p. 314.)

Imperfectly known species.

K. sp. Forest tree, to 80 ft. high ; leaves broadly obovate-elliptic, very shortly but distinctly acuminate,
up to 8·5 cm. long and 5·5 cm. broad ; flowers and mature fruits required.
G.C.: Bonsa Su (fr. Dec.) *Vigne* FH 1994 ! Atuna *Vigne* FH 3528.

2. IRVINGIA Hook. f.—F.T.A. 1 : 313 ; Engl. in E. & P. Pflanzenfam. 19A : 398
(1931). *Irvingella* Van Tiegh. (1905).

Flowers distinctly pedicellate ; sepals about 1 mm. long ; petals about 4 mm. long ;
  petioles up to 1 cm. long :
Flowers in short, clustered, mostly axillary racemes, or subpaniculate ; pedicels up to
  10 mm. long ; leaves obovate-elliptic or elliptic, more or less cuneate or narrowly
  rounded at base, shortly and broadly acuminate, 3·5–16 cm. long, 1·8–7·5 cm. broad,
  with 5–11 main lateral nerves on each side of midrib ; stipules up to 1·5 cm. long,
  caducous ; fruits broadly ellipsoid, somewhat flattened, about 5–6 cm. long with
  smooth skin, fibrous exocarp and hard endocarp ..   ..   ..   1. *gabonensis*
Flowers in terminal panicles 6–12 cm. long ; pedicels 3–5 mm. long; leaves broadly ovate,
  broadly rounded or subcordate at base, gradually broadly acuminate, 2·5–13 cm.
  long, 1·8–7 cm. broad, with 12–14 main lateral nerves on each side of midrib ;
  stipules 2–2·5 cm. long, caducous ; fruits oblong-ellipsoid, 4–5 cm. long 2. *smithii*
Flowers subsessile borne in terminal, panicles 7–15 cm. long ; sepals about 1 mm. long ;
  petals about 3 mm. long ; leaves elliptic-oblong, cordate or subcordate at base,
  shortly and broadly acuminate, 10–27 cm. long, 4·8–13·5 cm. broad, with 14–16 main
  lateral nerves on each side of midrib ; petioles up to 1·5 cm. long ; stipules up to
  3 cm. long, caducous ; fruits ovoid, about 2·5 cm. long ..   ..   3. *grandifolia*

1. **I. gabonensis** (*Aubry-Lecomte ex O'Rorke*) *Baill.* Trait. Méd. Phan. 2 : 881 (1883) ; Aubrév. Fl. For. C. Iv.
2 : 96, t. 165 ; E. & P. Pflanzenfam. 19A : 398, fig. 187. *Mangifera gabonensis* Aubry-Lecomte ex
O'Rorke (1857). *Irvingia barteri* Hook. f. (1862)—F.T.A. 1 : 314, incl. var. *tenuifolia* ; Chev. Bot. 104.
I. tenuifolia Hook. f. (1860). *I. tenuinucleata* Van Tiegh. (1905). A forest tree, to 120 ft. high, with dark
green shining foliage ; flowers greenish, fragrant ; fruits yellow, edible ; often seen in villages and towns
in the forest regions.
Sen.: Fogny, Casamance (fr. Feb.) *Chev.* 3157 ! **Fr.G.:** Dyeke (Oct.) *Baldwin* 9652a ! Longuery, Kouria
to Ymbo (Nov.) *Chev.* 14789 ! **S.L.:** Njala (fr. Feb.) *Deighton* 2606 ! Songo (Oct.) *Tindall* 43 ! Gola
Forest (fr. Apr.) *Small* 623 ! **Iv.C.:** Yapo (young fr. Oct.) *Chev.* 22331 ! Byanouan, Sanvi *Chev.* 17753.
Davo to Soubré *Chev.* **G.C.:** Axim *Morgan* FH 4929 ! Bobiri F.R. *Andoh* FH 4936 ! **Dah.:** Adjara,
Porto Novo (Jan.) *Chev.* 22752 ! **N.Nig.:** Agaie (Nov.) *Yates* 26 ! **S.Nig.:** Lagos (Feb.) *Dalz.* 929 !
Abeokuta *Irving* ! Sapoba *Kennedy* 2040 ! 2062 ! Eket *Talbot* 3139 ! 3140 ! Calabar (Feb.) *McLeod* !
Oban *Talbot* 1470 ! **Br.Cam.:** R. Cameroon (Jan.) *Mann* 2206 ! Abokum, Mamfe Div. *Johnstone* 247/31 !
Extends to the Belgian Congo, A.-E. Sudan and Angola, also on Principe Island. (See Appendix, p. 312.)
2. **I. smithii** *Hook. f.* in Trans. Linn. Soc. 23 : 167 (1860) ; F.T.A. 1 : 314. *Irvingella smithii* (Hook. f.) Van
Tiegh. (1905). A glabrous tree, 30–60 ft. high, with small yellowish fragrant flowers and scarlet plum-like
fruits ; by river banks, especially in savannah regions.
N.Nig.: Nupe *Barter* 1319 ! Kontagora (Nov.) *Dalz.* 98 ! Bida (Jan.) *Yates* 43 ! Izom, Abuja Dist.
(Jan.) *Savory* UCI 142 ! **S.Nig.:** Ibo country (fr. Sept.) *T. Vogel* 10 ! Aboh *T. Vogel* ! Osomari F.R.,
Onitsha (fr. July) *Odiachi* FHI 15465 ! Extends to A.-E. Sudan and Angola. (See Appendix, p. 313.)
3. **I. grandifolia** (*Engl.*) *Engl.* Bot. Jahrb. 46 : 288, fig. 4 (1911). *Klainedoxa grandifolia* Engl. (1907)—F.W.T.A.,
ed. 1, 1 : 483, excl. syn. *K. buesgenii* Engl. *Irvingella grandifolia* (Engl.) H. Hallier (1921). A tall forest
tree, deciduous ; old leaves tinted bright red.
S.Nig.: Ologbon *Sankey* ! Onitsha (Apr., June) *Chesters* 158 ! Cons. of For. 131 ! **Br.Cam.:** Kempong
F.R., Mamfe Div. *Aninze* FHI 24720 ! Also in French Cameroons, Gabon, Belgian Congo and Angola.

3. DESBORDESIA Pierre ex Van Tiegh. in Ann. Sci. Nat. 9, 1 : 289 (1905) ; Engl. in
E. & P. Pflanzenfam. 19A : 402 (1931).

A glabrous tree ; leaves oblong, cuneate to obtuse at base, often somewhat asymmetric,
acutely acuminate at apex, 5–16 cm. long, 2–6·5 cm. broad, leathery, with 8–10 main
lateral nerves on each side of midrib, closely reticulate on both surfaces, shining
above, dull and often glaucous beneath ; petioles about 1 cm. long ; stipules thin,
up to 1 cm. long, very early caducous ; racemes more or less paniculate up to about

26

10 cm. long ; pedicels about 3 mm. long ; sepals about 1 mm. long ; petals about 2 mm. long ; fruits samaroid, more or less oblong, obtuse, rounded or emarginate at each end, 8·5–14·5 cm. long, 3–4·5 cm. broad, about 5 mm. thick, glabrous

glaucescens

D. glaucescens (*Engl.*) *Van Tiegh.* l.c. 290 (1905) ; Engl. l.c., fig. 185. *Irvingia glaucescens* Engl. Bot. Jahrb. 32 : 124 (1902). *Desbordesia soyauxii* Van Tiegh. (1905). *D. pierreana* Van Tiegh. (1905). A large forest tree, with buttresses.
S.Nig.: Ikom *Catterall* 8 ! Ukpon F.R., Ikom Div. *Kennedy* 3078 ! **Br.Cam.**: Victoria (Apr.) *Maitland* 625 ! Likomba *Mildbr.* 10741 ! Malende *Ejiofor* FHI 29366 ! Johann-Albrechtshöhe (Feb.) *Staudt* 940 ! Also in French Cameroons, Gabon, Spanish Guinea, Gabon and Cabinda. (See Appendix, p. 311.)
A sterile specimen (*Keay* FHI 25629) from Usonigbe F.R., Benin Prov., S. Nig., is probably this also.

117. BURSERACEAE

Trees or shrubs, secreting resin or oil. Leaves alternate, rarely opposite, pinnate, rarely 1-foliolate ; stipules absent. Flowers hermaphrodite or often unisexual, small. Sepals 3–5, imbricate or valvate. Petals 3–5, rarely absent, free or variously connate, imbricate or valvate. Disk present. Stamens the same or double the number of the petals ; filaments free ; anthers 2-celled, opening by slits. Ovary superior, 2–8-celled ; ovules 2 or 1 in each cell, axile. Fruit a drupe or rarely a capsule. Seeds without endosperm ; cotyledons often contortuplicate.

Confined to the tropics and recognized by the frequently fragrant resinous wood, pinnate leaves, usually drupaceous fruit, and seeds without endosperm.

Fruits capsular ; flowers bisexual ; sepals and petals 5 ; stamens 10   1. **Boswellia**
Fruits drupaceous ; flowers bisexual or unisexual :
Sepals and petals 4 ; stamens 8 ; receptacle concave ; leaves simple, 3-foliolate or imparipinnate, leaflets often toothed   ..   ..   ..   2. **Commiphora**
Sepals and petals 3 ; stamens 6 ; receptacle flat or convex ; leaves imparipinnate ; leaflets entire :
  Fruits excentric with style lateral ; only species in our area a tree with stilt roots and glabrous stems, leaves and inflorescences   ..   ..   ..   3. **Santiria**
  Fruits symmetrical with style terminal ; all species in our area trees without stilt roots and with pubescent stems, leaves and inflorescences :
    Indumentum stellate ; ovary (in our species) 2-celled, the cells not separated by an axial intrusion ..   ..   4. **Dacryodes**
    Indumentum simple ; ovary 3-celled, the cells separated by a 3-winged axial intrusion and covered by mesocarpal lids forming a 3-gonous spindle-shaped " stone " in fruit ..   ..   5. **Canarium**

1. BOSWELLIA Roxb.—F.T.A. 1 : 323 ; Engl. in E. & P. Pflanzenfam. 19A : 419 (1931).

Inflorescence a simple raceme ; racemes numerous and bunched at the end of the thick shoots, up to 25 cm. long, glabrous or nearly so ; pedicels spreading, about 1 cm. long ; sepals triangular-ovate, 1·25 mm. long ; petals about 7 mm. long ; leaflets about 8 pairs, opposite, obliquely lanceolate, acutely acuminate, 7–9 cm. long, 1·5–2 cm. broad, coarsely serrate   ..   ..   1. *dalzielii*
Inflorescence a panicle ; panicles few, up to 30 cm. long, softly pubescent ; pedicels about 0·5 cm. long ; sepals broadly triangular ; petals as above ; leaflets sub-alternate, linear-lanceolate, long-acuminate, about 8 cm. long and 1·5 cm. broad, serrate   ..   ..   2. *odorata*

1. **B. dalzielii** *Hutch.* in Kew Bull. 1910 : 136 ; Chev. Bot. 110 ; Aubrév. Fl. For. Soud.-Guin. 371, t. 77, 1–3. A savannah tree, to 12–40 ft. high, with pale papery bark ; flowers white, fragrant, the petals sometimes red-veined, flowering before the leaves.
Iv.C.: Mossi Dist. (Aug.) *Chev.* 24612. **Fr.Nig.**: Gourma Dist. *Chev.* 24424 ! **N.Nig.**: Sokoto (Feb.) *Lely* 811 ! Kontagora (Feb.) *Dalz.* 279 ! Zaria Prov. *Dalz.* 340 ! *Shaw* 105 ! Bauchi Plateau (Mar.) *Lely* P. 176 ! **Br.Cam.**: Gwoza, Dikwa Div. (Dec.) *McClintock* 79 ! Also in French Cameroons and Ubangi-Shari. (See Appendix, p. 314.)
2. **B. odorata** *Hutch.* l.c. 137 (1910) ; Aubrév. l.c., t. 77, 4–5. A savannah tree, similar to the last.
N.Nig.: Yola town and province (Feb.) *Dalz.* 167 ! Also in French Cameroons.

2. COMMIPHORA Jacq. Hort. Schoenb. 2 : 66 (1797) ; Engl. in E. & P. Pflanzenfam. 19A : 429 (1931). *Balsamodendrum* Kunth (1824)—F.T.A. 1 : 324. *Heudelotia* A. Rich. (1832).

Leaves imparipinnate ; flowers in pedunculate inflorescences :
Leaflets 5–6 pairs, very acutely acuminate, oblong-ovate, unequal-sided at base, 6–12 cm. long, 2·5–4·5 cm. broad, thin, obscurely crenulate, glabrous ; petiolule 1 cm. long ; panicles as long as or longer than the leaves, the flowers crowded at the

ends of the lateral branchlets, thinly pubescent ; male flowers turbinate, pedicellate, about 5 mm. long ; fruits subglobose, about 1·2–2 cm. diam., verrucose when dry

1. *kerstingii*

Leaflets (1–) 2–4 (–5) pairs, rounded at apex, oblong, 2–5 cm. long, 1–2 cm. broad, softly pubescent on both sides, subsessile ; flowers few on an axillary peduncle up to 5 cm. long ; calyx villous ; fruits ellipsoid, about 1 cm. long, shortly pubescent

2. *pedunculata*

Leaves 3-foliolate, or simple ; flowers subsessile or shortly pedicellate in the leaf-axils :

Leaves 3-foliolate ; leaflets softly pubescent beneath, coarsely crenate-dentate, obovate, cuneate at base, the terminal one the largest, up to 4 cm. long and 2·5 cm. broad ; petiole shorter than the leaflets ; branchlets often terminated by a spine ; flowers axillary, subsessile, glabrous ; fruits broadly ellipsoid, 7–8 mm. long, glabrous, with a purplish bloom .. .. 3. *africana*

Leaves 3-foliolate or simple, glabrous, margins obscurely crenate or entire :

Leaves 3-foliolate ; leaflets obovate-rhomboid, acute or subacute at each end, 2·5–6·5 cm. long, 1·5–3 cm. broad, very thin ; petioles 2–6 cm. long ; fruits ellipsoid-globose, 1 cm. long ; mesocarp covering the endocarp .. .. 4. *dalzielii*

Leaves simple (rarely 3-foliolate), broadly ovate to suborbicular or oblong, rounded at each end, 1·2–5 cm. long, 0·8–3·8 cm. broad ; petioles 1–4·5 cm. long ; fruits ellipsoid, beaked, about 8 mm. long ; mesocarp consisting of 4 narrow arms

5. *quadricincta*

Fig. 193.—Boswellia dalzielii *Hutch.* (Burseraceae).

A, flower. B, petal. C, stamen. D, pistil in vertical section showing disk. E, cross-section of ovary.

1. **C. kerstingii** *Engl.* Bot. Jahrb. 44 : 152 (1910) ; Aubrév. Fl. For. Soud.-Guin. 373, 375, t. 78, 6–7. *C. ararobba* Engl. (1911). A tree 20–30 ft. high, with smooth green bark eventually peeling in brownish papery strips ; panicles crowded at the ends of the branches which are brownish tomentellous and marked by leaf-scars ; often planted.
Togo: Ssoruba village (Apr.) *Kersting* A. 553 ! **N.Nig.:** Sokoto *Dalz.* 572 ! Bauchi Plateau (Mar.) *Lely* P. 195 ! Yola (May) *Dalz.* 165 ! **S.Nig.:** Ibadan (Mar.) *Chizea* FHI 24484 ! Oban *Talbot* 1549 ! Also in French Cameroons, Chad Territory and Ubangi-Shari. (See Appendix, p. 317.)

2. **C. pedunculata** (*Kotschy & Peyr.*) *Engl.* in DC. Monogr. Burserac. 23 (1883) ; Chev. Bot. 110 ; Aubrév. l.c., t. 78, 5. *Balsamodendrum pedunculatum* Kotschy & Peyr. (1867)—F.T.A. 1 : 326. A small tree with leaves usually crowded at the ends of the yellowish pubescent branchlets ; flowers whitish, the lateral branchlets occasionally hardening into spines.
Sen.: *Berhaut* 237. **Fr.Sud.:** Quiébélé (June) *Chev.* 1002 ! Yatenga North (fr. Aug.) *Chev.* 24811 ! Koutiala *Vuillet* 716. **Iv.C.:** Dédougou *Aubrév.* **G.C.:** Guo Hill, Nebiowalli (May) *Kitson* 773 ! **Dah.:** Atacora Mts., Kouandé *Chev.* 23967. **N.Nig.:** Nupe *Barter* 1678 ! Katagum (fr. Sept.) *Dalz.* 224 ! Damaturu (June, July) *Kennedy* FHI 7269 ! Onochie FHI 23372 ! Yola (May) *Dalz.* 174 ! Extends to A.-E. Sudan. (See Appendix, p. 317.)

3. **C. africana** (*A. Rich.*) *Engl.* l.c. 14 (1883) ; Chev. Bot. 110, incl. var. *togoensis* Engl. ; Aubrév. l.c., t. 78, 1–3. *Heudelotia africana* A. Rich. in Guill. & Perr. Fl. Seneg. 1 : 150, t. 39 (1831). *Balsamodendrum africanum* (A. Rich.) Arn. (1839)—F.T.A. 1 : 325. A shrub or small tree up to 20 ft. high, usually with spines ; leaves and bark fragrant-resinous, bark of branchlets reddish purple ; flowers reddish, usually produced when leafless ; in dry savannah woodland.
Sen.: *Heudelot* 129 ! *Perrottet* 152 ! Tielimane (fr. June) *Leprieur* ! **Fr.Sud.:** Goundam *Chev.* 3038 ! Bandiagara *De Ganay* 147. Gao *De Wailly* 5101. **Iv.C.:** Kaya *Aubrév.* 2174. Koupéla *Chev.* 24542.

G.C.: Bawku *Vigne* FH 4717! **Togo:** by lagoons, Lome *Warnecke* 341! **Dah.:** Dassa-Zoumé *Chev.* 23609. **Fr.Nig.:** Diapaga to Fada, N'Gourma *Chev.* 24455. Niamey (Oct.) *Hagerup* 558! **N.Nig.:** Jebba *Barter*! Katagum (fr. May) *Dalz.* 223! Karasuwa, Nguru (fr. June) *Onochie* FHI 23349! Kukawa (fr. Mar., Dec.) *Elliott* 107! *E. Vogel* 102! Widespread in the drier parts of tropical Africa. (See Appendix, p. 316.)

4. **C. dalzielii** Hutch. in F.W.T.A., ed. 1, 1 : 488 (1928) ; Kew Bull. 1929 : 25 ; Aubrév. l.c. 375. A glabrous shrub or small tree up to 25 ft. high, with spiny branchlets ; bark silvery and reddish, peeling off in papery scales ; in clumps of shrubs.
G.C.: Winneba Plain (fr. Feb.) *Dalz.* 8296! Achimota (fr. Feb.) *Irvine* 2118!

5. **C. quadricincta** *Schweinf. ex Engl.* in E. & P. Pflanzenfam. 3, 4 : 253, fig. 149 (1896) ; Aubrév. l.c. t. 78 4 ; E. & P. Pflanzenfam. 19A : 436, fig. 204, V–W. *C. airica* A. Chev. in Bull. Mus. Hist. Nat., sér. 2, 4 : 584 (1932). A shrub or small tree ; in dry regions.
Fr.Nig.: E. of Agadés *fide* Aubrév. *l.c.* S. of Aïr *Chev.* **N.Nig.:** Gashagar to Zari, N. Bornu Prov. (fr. Sept.) *Daggash* FHI 24866! Extends through Chad Territory to A.-E. Sudan, Eritrea, Abyssinia and Yemen.

3. **SANTIRIA** Blume in Mus. Bot. Lugd.-Bat. 1 : 209 (1850) ; Engl. in E. & P. Pflanzenfam. 19A : 452 (1931) ; H. J. Lam in Bull. Jard. Bot. Buitenz., sér. 3, 12 : 367 (1932). *Santiriopsis* Engl. (1890)—E. & P. Pflanzenfam. 19A : 455 (1931).

A glabrous tree ; leaves with (1–) 2–3 (–4) pairs of leaflets, plus the odd terminal leaflet ; lateral leaflets oblong-elliptic or rarely ovate-elliptic, cuneate to obtuse or rarely rounded at base, abruptly long-acuminate, 7·5–22 cm. long, 3–9·3 cm. broad, with 6–10 main lateral nerves on each side of midrib, prominent on both surfaces ; panicles axillary, up to 13 cm. long, flowers in fascicles or solitary, pedicels 1–7 mm. long ; fruits ellipsoid, flattened, excentric, with the remains of the style on one side, 1·9–2·5 cm. long, 1·4–1·7 cm. broad .. .. *trimera*

S. trimera (*Oliv.*) Aubrév. in Rev. Bois et For. Trop. 4, 8 : 344 (1948). *Sorindeia ? trimera* Oliv. F.T.A. 1 : 441 (1868). *Pachylobus trimerus* (Oliv.) Guillaum. (1909)—Aubrév. Fl. For. C. Iv. 2 : 112, t. 172, 2–4. *Santiriopsis trimera* (Oliv.) Engl. (1912)—Exell Cat. S. Tomé 135. *Dacryodes trimera* (Oliv.) H. J. Lam (1932). *Santiria ? balsamifera* Oliv. in Hook. Ic. Pl. t. 1573 (1886). *Santiriopsis balsamifera* (Oliv.) Engl. (1890). *Pachylobus balsamifera* (Oliv.) Guillaum. (1908)—F.W.T.A., ed. 1, 1 : 487 ; Aubrév. l.c. t. 172, 1 ; Kennedy For. Fl. S. Nig. 154. A forest tree, to 50 ft. high, with stilt-roots ; flowers yellow ; ripe fruits black, edible ; cut bark and fruits smelling of turpentine.
S.L.: Freetown (fl. & fr. Sept.) *Deighton* 3444a! 3444b! Bridge between Heddles' Farm and Mt. Aureol *Lane-Poole* 396! Njala (fl. Sept., fr. Mar., Nov.) *Deighton* 1318! 2628! 4963! **Lib.:** Dukwia R (fl. Aug., fr. Mar.) *Cooper* 110! 357a! *Linder* 261! Peahtah (fr. Dec.) *Baldwin* 10597! *Linder* 931! Mecca (fr. Dec.) *Baldwin* 10828! Wohmen, Vonjama (Oct.) *Baldwin* 10073! **Iv.C.:** Mt. Tonkoui *Aubrév.* 1002! Man *Aubrév.* 1632. Nimba Mts. *Aubrév.* 1148. **S.Nig.:** Sapoba *Kennedy* 2226! 2324! 2336! Oban *Talbot* 1600! 1713! Also in French Cameroons, Gabon, Cabinda and Belgian Congo.

4. **DACRYODES** Vahl in Skrivt. Nat. Selsk. 4 : 116 (1810) ; H. J. Lam in Bull. Jard. Bot. Buitenz., sér 3, 12 : 334 (1932). *Pachylobus* G. Don (1832)—F.W.T.A., ed. 1, 1 : 487 ; Engl. in E. & P. Pflanzenfam. 19A : 450 (1931).

Leaflets 6–8 pairs, plus the terminal one ; petiole short ; lateral leaflets oblong to oblong-elliptic, obliquely rounded at base, caudate-acuminate, up to 23 cm. long and 6·5 cm. broad, minutely stellate-puberulous, and (usually) with long spreading hairs as well on the nerves beneath, with 9–12 main lateral nerves on each side of midrib, prominent beneath ; inflorescence paniculate, rhachides deeply grooved, rusty stellate-tomentose ; sepals and petals outside rusty-stellate-tomentellous ; ovary glabrous ; fruits oblong-ellipsoid, about 7 cm. long and 3·5 cm. diam. .. ..1. *edulis*
Leaflets usually 3 pairs, plus the terminal one ; petiole long ; lateral leaflets oblong-elliptic to oblong, very obliquely cuneate or obtuse at base shortly acuminate, up to 20 cm. long and 9 cm. broad, stellate-scaly-puberulous beneath, with 9–12 main lateral nerves on each side of midrib, prominent beneath ; inflorescence paniculate, rhachides slightly grooved, like the outside of sepals and petals rusty-stellate-scaly-puberulous ; female or hermaphrodite flowers borne separately, or males in small fascicles ; ovary stellate-puberulous ; fruits oblong-ellipsoid or subglobose, 2–3 cm. long, 1·5–2 cm. diam. .. .. 2. *klaineana*

1. **D. edulis** (*G. Don*) H. J. *Lam* in Bull. Jard. Bot. Buitenz., sér. 3, 12 : 336 (1932) ; Aubrév. in Rév. Bois et For. Trop. 4, 8 : 342. *Pachylobus edulis* G. Don (1832)—Hook. Ic. Pl. tt. 2566–7 ; F.W.T.A., ed. 1, 1 : 487 ; Kennedy For. Fl. S. Nig. 153. *Canarium ? edule* (G. Don) Hook. f. (1849)—F.T.A. 1 : 327. *C. saphu* Engl. (1893). *Pachylobus saphu* (Engl.) Engl. (1896). *Sorindeia saphu* (Engl.) A. Chev. ex Hutch. & Dalz. F.W.T.A., ed. 1, 1 : 507 (1928), partly (*Thompson* 9). A forest tree 30–60 ft. high, often planted, with dark foliage and scaly bark exuding a whitish resin ; flowers in dense rusty panicles ; fruits bright blue when ripe, edible.
S.Nig.: *Thompson* 9! Ondo (Mar.) *A. F. A. Lamb* 148! Benin City *Unwin* 41! Sapoba (fl. Feb., fr. Aug.) *Kennedy* 2064! 2132! 2334! Uwet (Feb.) *McLeod* ! Calabar *Thomson* 46! **Br.Cam.:** Victoria (Feb.) *Maitland* 382! Cameroon R. (Jan.) *Mann* 741! Barombi *Preuss* 362! Extends to Belgian Congo and Angola, also in Principe and S. Tomé. (See Appendix, p. 317.)

2. **D. klaineana** (*Pierre*) H. J. *Lam* l.c. (1932) ; Aubrév. l.c. 344, incl. var. *lepidota* Aubrév. (1948). *Santiriopsis ? klaineana* Pierre (1897). *Pachylobus klaineanus* (Pierre) Engl. (1899). *P. barteri* Engl. (1899)—F.W.T.A., ed. 1, 1 : 487, excl. syn. *P. dahomensis* Engl. ; Kennedy l.c. 154. *Dacryodes ? barteri* (Engl.) H. J. Lam (1932). *Sorindeia deliciosa* A. Chev. ex Hutch. & Dalz. F.W.T.A., ed. 1, 1 : 507 (1928) and Kew Bull. 1929 : 27, partly (*Chev.* 22683) ; Chev. Bot. 160, name only. *Haematostaphis deliciosa* (A. Chev. ex Hutch. & Dalz.) Pellegr. (1931). *Pachylobus deliciosus* (A. Chev. ex Hutch. & Dalz.) Pellegr. (1934)—Aubrév. l.c. t. 172, 1 : 110, t. 171. *P. paniculatus* Hoyle (1934). *P. albiflorus* Guillaum. (1909), name only, partly (*Jolly* 161). *P. macrophyllus* of F.W.T.A., ed. 1, 1 : 487, partly (*Talbot* 1715), not of Engl. A forest tree, to 60 ft. high, sometimes branching only a few feet from the ground ; flowers pale yellow ; fruits orange red ; cut bark and, in some forms, the fruits, smelling of turpentine.
S.L.: Gola Forest (Oct.) *D. G. Thomas* 122! Sembahun, Gaura (fr. Dec.) *King* 301! 304! Bogboabu-Ngeiwahun road (Sept.) *D. G. Thomas* 113! **Lib.:** Gbanga (Sept.) *Bequaert* 740! **Iv.C.:** Boudepé to

Agboville (young fr. Dec.) *Chev.* 22683! Rasso *Aubrév.* 141! Man *Aubrév.* 1631! Banco *Serv. For.* 373! **G.C.:** Kintampo (Oct.) *Vigne* FH 2535! Sekodumasi (Oct.) *Vigne* FH 3398! Mampong, Ashanti (Nov.) *Vigne* FH 3426! **Togo:** Alavanyo (Sept.) *St. C. Thompson* FH 3577! Wiawia (Oct.) *St. C. Thompson* FH 3620! Misahöhe *Mildbr.* 7322. **N.Nig.:** Patti Lokoja (Oct.) *Dalz.* 189! **S.Nig.:** Lagos *Dalz.* 1400! Ominla, Ondo Prov. (fr. Apr.) *Symington* FHI 3387! Sapoba (Feb.) *Kennedy* 1619! 1620! 2230! Onitsha *Barter* 1775! Udi F.R., Enugu (Oct.) *Smith* S.1! Oban *Talbot* 1715! **Br.Cam.:** S. Bakundu F.R. (fr. Aug.) *Olorunfemi* FHI 30715! Also in Gabon. (See Appendix, p. 317.)

Imperfectly known species.

D. afzelii (*Engl.*) H. J. *Lam* l.c. (1932). *Pachylobus afzelii* Engl. (1899). Specimen not traced. **S.L.:** *Afzelius.*

5. CANARIUM Linn.—F.T.A. 1 : 327 ; Engl. in E. & P. Pflanzenfam. 19A : 443 (1931).

A large tree ; young parts densely rusty-pubescent ; leaves 25–65 cm. long, with 8–12 pairs of opposite or subopposite leaflets, plus the odd terminal one ; leaflets oblong or ovate to oblong-lanceolate, cordate at base, 4·5–20 cm. long, 2·8–5·5 cm. broad, with 12–24 main lateral nerves on each side of midrib, prominent and pubescent beneath ; rhachis flat on the upper side towards the base ; panicles in the upper leaf-axils, 15–30 cm. long ; pedicels 2–4 mm. long ; flowers unisexual, 1–1·2 cm. long ; fruits ellipsoid, 3·4–3·7 cm. long, 1·7–2 cm. broad, glabrous .. .. .. *schweinfurthii*

C. schweinfurthii *Engl.* in DC. Monogr. Burserac. 145 (1883) ; *Chev. Bot.* 111 ; *Aubrév.* Fl. For. C. Iv. 2 : 107, t. 170 ; *Kennedy* For. Fl. S. Nig. 154. *C. chevalieri* Guillaum. (1908)—*Chev. Bot.* 110. *C. occidentale* A. Chev. (1909)—*Chev. Bot.* 110. *C. khiala* A. Chev. Bot. 110, name only. A large forest tree, to 120 ft. high, with very slight blunt buttresses ; cut bark copiously exuding gum which solidifies to a whitish resin ; flowers creamy-white ; ripe fruits purplish, plum-like, containing a hard spindle-shaped, trigonous " stone."

Sen.: Banfara (fl. & fr. Jan.) *Vuillet* 561. **Fr.G.:** Timbikounda (fr. Jan.) *Chev.* 20589. **Port.G.:** Cumura to Bor, Bissau (Mar.) *Esp. Santo* 1889! **S.L.:** *Lane-Poole* 190! Mano (fr. Apr.) *Deighton* 4738! Njala (fr. Mar.) *Deighton* 5320! Dodo Hills F.R. *King* 183! **Lib.:** Dukwia R. (Apr.) *Cooper* 125! 192! 385! **Iv.C.:** Azaguié (fr. Sept.) *Chev.* 22288! Alépé (Mar.) *Chev.* 16236! Grabo *Chev.* 20066. Abidjan (June) *Aubrév.* 21 ; *Chev.* 15272. **G.C.:** Kwaben (Mar.) *Gent*! Axim (Feb.) *Irvine* 2237! Mumuni camp, Prestea road (Mar.) *Andoh* FH 5620! **Togo:** Misahöhe (fr. Nov.) *Mildbr.* 7278! **N.Nig.:** Bida (Jan.) *Yates* 21! Kutigi, Niger Prov. (fr. Mar.) *Lamb* 43! Bauchi Plateau (Jan.) *Lely* P. 4! **S.Nig.:** Modakeke *Foster* 204! Okomu F.R. (Dec.–Jan.) *Brenan* 8430! Sapoba (fl. May, fr. June) *Kennedy* 503! 1665! Opobo (fr. Sept.) *King-Church* 42! Eket *Talbot* 3323! **Br.Cam.:** Johann-Albrechtshöhe *Staudt* 507! Extends to A.-E. Sudan, Abyssinia, Tanganyika and Angola. (See Appendix, p. 315.)

Besides the above, *C. zeylanicum* (Retz.) Blume (syn. *C. auriculatum* H. Winkl. (1908)—F.W.T.A., ed. 1, 1 : 488), a native of Ceylon has been introduced into our area.

118. MELIACEAE

Trees or shrubs, mostly with hard scented wood, very rarely subherbaceous. Leaves alternate, mostly pinnate ; stipules absent. Flowers actinomorphic, mostly hermaphrodite. Calyx often small, imbricate, rarely valvate. Petals free or partially connate, contorted or imbricate, or adnate to the staminal tube and valvate. Stamens mostly 8 or 10, rarely numerous, mostly with connate filaments, and the anthers often sessile in the tube ; anthers 2-celled, opening lengthwise ; disk various. Ovary superior, often 3–5-celled, stigma often disciform or capitate ; ovules mostly 2, rarely 1 or more. Fruit baccate, capsular or rarely a drupe, often with a large central axis. Seeds with or without endosperm, sometimes winged.

Mainly a tropical family readily recognized by the pinnate eglandular leaves and the stamens, usually connate into a tube resembling a corolla.

Leaves pinnate or bipinnate, rarely 3-foliolate, very rarely 1-foliolate, flowers mostly in panicles or racemes :
Leaves not bipinnate :
Petals free to the base ; leaves pinnate :
Seed winged ; fruit a woody or coriaceous, dehiscent capsule ; ovary (3–) 4–5-celled, each cell containing 4–12 ovules :
Capsule more or less globose, woody, 4–5-valved, erect ; seed flat, winged all round ; anthers included within the staminal tube ; all parts glabrous   **1. Khaya**
Capsule much longer than broad ; seed with a long terminal or basal wing ; anthers exserted or nearly exserted from the staminal tube :
Flowers 5-merous ; capsule with 5 woody valves ; seed attached by the body-end to the distal end of the central column, the wing thus directed towards the base :
Anthers borne on the rim of the entire or lobed staminal tube ; capsule pendulous, without fibres between the valves ; leaves entire or nearly so ; forest trees.. .. .. **2. Entandrophragma**
Anthers borne just within the lobes of the staminal tube ; capsule erect, with fibres adhering loosely between the valves after dehiscence ; leaves undulate on the margins ; savannah trees   .. .. .. **3. Pseudocedrela**

Flowers 4-merous ; capsule with 4 coriaceous valves, pendulous ; seed attached by the wing-end to the distal end of the central column, the body thus hanging freely at the base .. ..4. **Lovoa**

Seed not winged ; fruit fleshy, at least at first :

Fruit a large, at least 6 cm. diam., septicidal capsule ; ovules 2–8 per cell ; seed large ; anthers included in the staminal tube .. .. .. 5. **Carapa**

Fruit up to 5·5 cm. diam., a loculicidal or septicidal capsule, or indehiscent ; ovules 1–2 per cell :

Leaflets entire ; indigenous plants :

Ovary and fruit 2–5-celled ; fruit capsular (except No. 7) ; seeds mostly with brightly coloured arils :

Anthers borne on the rim of the entire or lobed staminal tube :

Fruit dehiscent ; staminal tube deeply lobed (except in *T. prieuriana*) 6. **Trichilia**

Fruit indehiscent ; staminal tube entire or slightly denticulate 7. **Ekebergia**

Anthers borne within the staminal tube ; fruit dehiscent .. 8. **Guarea**

Ovary and fruit 1-celled ; fruit indehiscent, beaked, more or less moniliform, 1–4-seeded, baccate ; anthers borne just within the staminal tube 9. **Heckeldora**

Leaflets coarsely serrate or lobed ; introduced trees ; ovary 3-celled ; fruit drupaceous, 1-seeded ; anthers borne just within the staminal tube 12. **Azadirachta**

Petals united into a club-shaped tube ; anthers borne just within the staminal tube ; ovary 4–5-celled, with 2 ovules per cell ; fruit capsular ; seed not winged, arillate ; leaves pinnate or (1–) 3-foliolate .. .. 10. **Turraeanthus**

Leaves bipinnate or tripinnate ; leaflets toothed ; anthers borne just within the staminal tube ; ovary and fruit usually 5-celled ; fruit drupaceous ; introduced trees .. .. 13. **Melia**

Leaves simple ; flowers solitary, paired, subumbellate or in very short racemes ; ovary 4–10 (or more)-celled ; fruit capsular ; seed not winged .. .. 11. **Turraea**

Besides the above, other genera are represented by the following introduced species :—*Cedrela mexicana* M. J. Roem. and *C. odorata* Linn., natives of tropical America ; *Toona ciliata* M. J. Roem. (syn. *Cedrela toona* Roxb.), a native of tropical Asia ; *Chukrasia tabularis* A. Juss., a native of tropical Asia ; *Swietenia mahagoni* (Linn.) Jacq., *S. macrophylla* King and *S. humilis* Zucc., natives of tropical America.

1. KHAYA A. Juss.—F.T.A. 1 : 337 ; Harms in E. & P. Pflanzenfam. 19B, 1 : 49 (1940).

Trees of savannah and fringing forest outliers in the savannah regions ; flowers normally 4-merous ; ovary tapering into a slender style ; fruit normally 4-valved, 4–6 cm. diam. ; leaflets usually drying pale glaucous green, 3–4 (–7) pairs, oblong or oblong-elliptic, more than twice as long as broad, rounded or obtuse or shortly acuminate at apex, 6–12 cm. long, 2–5 cm. broad, with 8–16 main lateral nerves on each side of midrib .. .. 1. *senegalensis*

Trees of the forest regions and forest outliers in the savannah ; flowers (4–) 5-merous ; ovary abruptly contracted into a very short style ; fruit (4–) 5-valved, 4–10 cm. diam. ; leaflets not drying pale glaucous green :

Leaflets 10–25 cm. long, 6–10 cm. broad, with 12–15 main lateral nerves on each side of midrib ; flowers 5-merous ; fruit 7–10 cm. diam., with 5 valves, 5–12 mm. thick ; leaves with (3–) 4 (–5) pairs of rather thin papery leaflets, broadly elliptic or ovate-elliptic, shortly cuneate or rounded at base, shortly acuminate ; bark scaly ; in forest outliers and the drier parts of the forest regions .. 2. *grandifoliola*

Leaflets mostly up to 15 cm. long and 8 cm. broad, rarely more, rather leathery, with 5–9 main lateral nerves on each side of midrib ; valves of fruit up to 3 mm., rarely to 5 mm. thick :

Leaflets (2–) 3 (–4) pairs, broadly elliptic or ovate-elliptic, very shortly acuminate, 8–15 cm. long, 4–8 cm. broad ; flowers 4–5-merous ; fruit 4–5-valved, 6–10 cm. diam. ; bark smooth and pale ; in the drier parts of the forest regions 3. *anthotheca*

Leaflets (4–) 5–6 (–7) pairs, oblong, abruptly long-acuminate, 5–14 cm. long, 2–5·7 cm. broad ; flowers usually 5-merous ; fruit usually 5-valved, 4–7 cm. diam. ; bark scaly, often brown ; in the moister parts of the forest regions .. 4. *ivorensis*

N.B.—Saplings of all the above species have large leaves with more numerous leaflets ; the measurements given in the above key are for crown leaves.

1. **K. senegalensis** (*Desr.*) *A. Juss.* in Mém. Mus. Hist. Nat. Paris 19 : 250, t. 21 (1830) ; Guill. & Perr. Fl. Seneg. 1 : 130, t. 32 ; F.T.A. 1 : 338 ; Chev. Bot. 117 (incl. vars.) ; Aubrév. Fl. For. C. Iv. 2 : 118–122; For. Trees & Timbers Brit. Emp. 4 : 68, fig. 11, t. 11 ; Pellegr. in Notulae Syst. 9 : 32 ; Harms in E. & P. Pflanzenfam. 19B, 1 : 51–54, fig. 4, A–I, fig. 5 ; Aubrév. Fl. For. Soud.-Guin. 377–381, t. 79, 1–3. *Swietenia senegalensis* Desr. in Lam. Encycl. 3 : 679 (1791). A tree with shining foliage, up to 100 ft. high, with wide dense crown and thick stem ; sepals pale green ; petals and staminal tube cream, the latter suffused pink below ; stigma yellow ; in savannah and especially by streams in the savannah regions. **Sen.**: *Roussillon.* Bargny (Feb.) *Perrottet* 130 ! **Gam.**: Albreda *Perrottet.* N. bank (July) *Ozanne* 9 ! 10 !

Fr.Sud.: Ouassaia (Mar.) *Chev.* 519! Bamako (Apr.) *Hagerup* 34! Siguiri *Chev.* 284. **Port.G.:** Bissau (Mar., Apr.) *Esp. Santo* 1875! 1913! Canchungo (Apr.) *Esp. Santo* 1946! **Fr.G.:** Kouroussa (Apr.) *Pobéguin* 685! Mamou *Dalz.*! **S.L.:** Falaba (Apr.) *Sc. Elliot* 5898! **Iv.C.:** *Aubrév.* 734! **G.C.:** Ashanti (fr. Apr.) *Thompson* 60! Busa, Wa (Mar.) *McLeod* 827! Navrongo (Mar.) *Vigne* FH 4686! Sa (Mar.) *Kitson* 846! **Togo:** Kersting 1! **Dah.:** Atacora Mts. *Chev.* 24031. **Fr.Nig.:** Diapaga to Fada *Chev.* 24031; 24426; 24427. **N.Nig.:** E. of Zurmi (fl. & fr. Apr.) *Keay* FHI 16168! Zaria (Apr.) *King-Church* 48! Zungeru (fl. & fr. Jan.) *Elliott* 24! Lokoja *Barter* 580! Katagum (fr. Dec.) *Elliott* 165! **S.Nig.:** Lagos *Rowland*! Onitsha *Chesters* 200! Extends to A.-E. Sudan and Uganda. (See Appendix, p. 325.)

2. **K. grandifoliola** *C.DC.* in Bull. Soc. Bot. Fr. 54, Mém. 8 : 10 (1907); Aubrév. Fl. For. C. Iv. 2 : 118–124, t. 173, 6–10, t. 174, 4–5; For. Trees & Timbers Brit. Emp. 4 : 64, fig. 10, t. 10; Pellegr. l.c. 29; Harms l.c.; Aubrév. Fl. For. Soud.-Guin. 377–379. *K. punchii* Stapf (1908)—Chev. Bot. 117. *K. grandis* Stapf (1908). *K. kerstingii* Engl. (1915). *K. kissiensis* A. Chev. Bot. 117, name only. A large forest tree, to 130 ft. high; young foliage conspicuously reddish.

Fr.G.: Doudou, Kissi (Feb.) *Chev.* 20687! Kissidougou & Guékédou *fide* Aubrév. *l.c.* **Iv.C.:** Abidjan *Aubrév.* 63! And many other localities *fide* Aubrév. *l.c.* **G.C.:** Agogo (fr. Jan.) *Chipp* 627! Kobroso, N. of Abofuo (fl. & fr. Feb.) *Vigne* FH 1803! Sekodumasi (fr. Oct.) *Vigne* FH 3395! L. Bosumtwi (fr. May) *Thompson* 74! **Togo:** Sokodé–Basari *Kersting* 562. **Dah.:** Cabolé, Savalou *Chev.* 23774! **N.Nig.:** Kafanchan (Jan.) *Kerr*! **S.Nig.:** *Unwin* 17! 18! *Foster* 89! Ibadan *F.R. Punch* 104! Tupelle, W. Prov. *Thompson* 7! Onitsha Dist. (Jan.) *Kennedy* 2506! *Rosevear* 2/29! Extends to A.-E. Sudan, Belgian Congo and Uganda. (See Appendix, p. 324.)

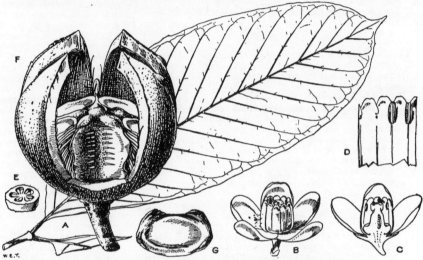

FIG. 194.—KHAYA GRANDIFOLIOLA *C.DC.* (MELIACEAE).

A, leaflet. B, flower. C, same in vertical section. D, staminal tube from inside. E, cross-section of ovary. F, fruit with portion of wall removed. G, winged seed.

3. **K. anthotheca** *(Welw.) C.DC.* in DC. Monogr. 1 : 721 (1878); Aubrév. Fl. For. C. Iv. 2 : 118–124, t. 174, 1–3; For. Trees & Timbers Brit. Emp. 4 : 57, fig. 9, t. 9; Harms l.c.; Staner in Bull. Jard. Bot. Brux. 16 : 214; Pellegr. l.c. 30. *Garretia anthotheca* Welw. (1859). *Khaya euryphylla* Harms (1902). *K. mildbraedii* Harms (1917). *K. agboensis* A. Chev. in Rev. Bot. Appliq. 8 : 209 (1928); Chev. Bot. 116, name only. A large forest tree, to 180 ft. high, with ashy-brown, grey or almost white, smooth bark. **S.L.:** Kosala, Pendembu (fr. July) *Deighton* 4133! Nyangoi Forest *Sawyerr* FHK 13601! Kambui Hills F.R. (July) *Douay* 19035! Kenema (fr. May) *Small* 64! 64a! **Iv.C.:** Agboville (fr. Nov.) *Chev.* 22342! *Krukoff* 102! Abidjan *Aubrév.* 10! And other localities *fide* Aubrév. *l.c.* **G.C.:** Nyinahin, Ashanti *Thompson* 47! Pamu-Berekum F.R. (fr. Sept.) *Vigne* FH 2516! Kibbi (Mar.) *Brent* 542! **Br.Cam.:** Johann-Albrechtshöhe (Mar.) *Staudt* 677! Also in French Cameroons and extending to Uganda, Belgian Congo, Angola and (?) Tanganyika. (See Appendix, p. 324.)

[There is no authentic record of this species from Nigeria. *Barter* 580 from Lokoja cited as *K. anthotheca* in the first edition is *K. senegalensis*.]

4. **K. ivorensis** *A. Chev.* Vég. Util. 5 : 207 (1909); Chev. Bot. 116 (incl. vars.); Aubrév. l.c., t. 173, 1–5; Kennedy For. Fl. S. Nig. 156; Pellegr. l.c. 30. *K. klainei* Pierre ex Pellegr. (1911)—Chev. Bot. 117. *K. caudata* Stapf ex Hutch. & Dalz. F.W.T.A., ed. 1, 1 : 490 (1928). A large forest tree, to 180 ft. high, with bole up to 90 ft., strongly buttressed; leaves crowded at the ends of the branchlets; panicles lax, comparatively few-flowered.

Iv.C.: Agboville (Nov.) *Chev.* 22388! Bouroukrou (fr. Dec.–Jan.) *Chev.* 16106! Yapo *Chev.* 22317! **G.C.:** Suku Suku (Nov.) *Chipp* 12! Tano R. (fl. Dec., fr. Feb., July) *Chipp* 37! 302! *Thompson* 17! Kumasi (Aug.) *Vigne* FH 3331! Wassidimo, E. of Esuasu (May) *Chipp* 213! Dunkwa (June) *Vigne* FH 892! **S.Nig.:** Ilaro F.R. (Sept.) *W. D. MacGregor* 348! Ibadan South F.R. *Keay & Mutch* FHI 25413! Sapoba & Usonigbe F.R. (fl. Oct.–Dec., fr. Apr.–June) *Ejiofor* FHI 24641! *Kennedy* 428! 1602! *A. F. Ross* 212! Aboh *Unwin* 45! Calabar (Jan.) *Chev.* 14151; *Smith* 20! **Br.Cam.:** Kumbe-Balue F.R., Kumba Dist. *Mbai* FHI 24804! Also in French Cameroons, Gabon and Cabinda. (See Appendix, p. 324.)

2. **ENTANDROPHRAGMA** C.DC. in Bull. Herb. Boiss. 2 : 582 (1894); Harms in E. & P. Pflanzenfam. 19B, 1: 55 (1940); Notizbl. Bot. Gart. Berl. 14 : 431 (1940).

Central column of fruit with a 1–2·5 cm. long, slender stipe above the insertion of the thin valves; petals velutinous outside, 5–7 mm. long; staminal tube deeply lobed; leaves with 5–9 pairs of somewhat bullate leaflets; main lateral nerves 14–22 pairs, very prominent beneath, impressed above ; rhachides 5–30 cm. long, petioles 6·5–15 cm. long, laterally winged, rusty-tomentellous ; leaflets oblong, oblong-elliptic or obovate-oblong, obtusely cuneate or rounded at base, obtuse or obtusely somewhat acuminate at apex, 3·5–15 cm. long, 2–6 cm. broad, pubescent on the midrib and nerves beneath ; fruits (9–) 17–23 (–30) cm. long, 3–5 cm. broad, valves opening from the apex and strongly recurving, 1·7–2 cm. broad in the widest part, 3–4 mm. thick, seeds 3–10 on each face of column   ..   ..   1. *candollei*

Central column of fruit without a stipe above the insertion of the valves ; petals puberulous or glabrous outside ; staminal tube not lobed ; leaflets with 6–16 pairs of main lateral nerves, not impressed above :

Valves of fruit 7–12 mm. thick, opening from the apex, remaining firmly attached at base to the central column, not recurved, 3–4 cm. broad at widest part, roughly lenticellate outside ; fruit 18–28 cm. long, 5–7 cm. broad ; seeds 5–6 on each face of column, hilum small, suborbicular ; petals puberulous outside, about 6 mm. long ; leaves with 8–16 pairs of subopposite or alternate leaflets ; rhachides 10–30 cm. long, petioles 5–14 cm. long, slightly winged, rusty-puberulous ; leaflets oblong-lanceolate, the lower ones often ovate, asymmetrically rounded or subcordate at base, gradually acuminate at apex, 3·5–15 cm. long, 1·5–5 cm. broad, with 10–16 pairs of main lateral nerves prominent beneath, often with tufts of hair in the axils
2. *utile*

Valves of fruit 2·5–4 mm. thick, either opening from the base and remaining joined together for some time at the apex, or splitting at apex and base at about the same time, always falling away from the central column ; leaflets with 6–12 pairs of main lateral nerves :

Fruit 14–22 cm. long, 3·5–5 cm. broad ; valves 2·5–3 cm. broad, not ridged, opening from base and remaining joined together for some time at the apex ; column attenuated to the base, dark red-brown in colour, with 5–6 seeds per face, hilum linear ; petals glabrous, 4–5 mm. long ; leaves with 7–10 opposite or subopposite pairs of leaflets ; rhachides 25–45 cm. long, petioles 12–16 cm. long, scarcely winged, glabrous or puberulous ; leaflets oblong, oblong-elliptic or obovate-oblong, rounded or obtuse at base, rounded at apex and usually mucronate, 7–28 cm. long, 3–10·5 cm. broad, with 9–12 pairs of main lateral nerves, not very prominent beneath, venation obscure, usually rather densely pubescent on the sides of the midrib and nerves beneath   ..   ..   3. *angolense*

Fruit 6·5–10 (–15·5) cm. long, 2·7–3·5 cm. broad, valves 1·4–2·2 cm. broad, with an indistinct median ridge, opening at apex and base more or less simultaneously ; column broad at base, rather pale brown or yellow, with 3–4 seeds per face, hilum linear-lanceolate ; petals puberulous, about 2 mm. long ; leaves with 5–9 pairs of opposite or subopposite leaflets ; rhachides 15–55 cm. long, petioles 5–10 cm. long, flattened and slightly winged, rusty-tomentellous ; leaflets oblong to oblong-lanceolate, the lower ones ovate to ovate-oblong, cuneate to rounded and usually asymmetrical at base, shortly and obtusely, rarely acutely acuminate, 8–13 cm. long, 2·5–5 cm. broad, with 6–12 pairs of main lateral nerves, venation closely reticulate, almost glabrous when mature   ..   ..   4. *cylindricum*

N.B.–Saplings of all the above species have leaves which may be considerably larger than the crown leaves whose measurements are given above.

1. **E. candollei** *Harms* in Notizbl. Bot. Gart. Berl. 1 : 181 (1896) ; Aubrév. Fl. For. C. Iv. 2 : 142, t. 180, 1–2, t. 181, d–d¹ t. 182, d ; Kennedy For. Fl. S. Nig. 172 ; Harms in E. & P. Pflanzenfam. 19B, 1 : 58, fig. 7 (but stipe of fruit not shown) ; Pellegr. in Notulae Syst. 9 : 40 ; Staner in Bull. Jard. Bot. Brux. 16 : 232. *E. ferrugineum* A. Chev. Vég. Util. 5 : 195 (1909) ; Chev. Bot. 118. A large forest tree, to 160 ft. high, with straight bole of up to 90 ft. high and dense crown of dark foliage ; buttresses blunt and often rather indistinct ; bark somewhat scaly, slash white and pink with orange granules, without a sweet scent ; leaflets often galled ; inflorescences tomentellous ; flowers cream.
    Fr.G.: Moussadougou to Lola *Chev.* 20968. **Iv.C.:** 60 miles E. Agboville (fr. July) *Krukoff* 100 ! Mbasso *Chev.* 16261 ! Abidjan *Aubrév.* 9 ! **G.C.:** Kunso, W. Ashanti (fr. Feb.) *Chipp* 106 ! Owabi *Vigne* FH 2601. **S.Nig.:** Akure F.R. (Nov.) *Keay* FHI 25596 ! Sapoba (fl. Nov., fr. June) *Kennedy* 404 ! 1152 ! 1601 ! **Br.Cam.:** Bopo, Kumba *Rosevear* Cam. 85/37 ! Johann-Albrechtshöhe (Nov.) *Staudt* 459 ! Extends to Cabinda and Belgian Congo. (See Appendix, p. 320.)
2. **E. utile** (*Dawe & Sprague*) *Sprague* in Kew Bull. 1910 : 180 ; Aubrév. l.c. 144, t. 180, 4–5, t. 181, b, t. 182, b, t. 183, 1–2 ; Kennedy l.c. 171 ; For. Trees & Timbers Brit. Emp. 4 : 48, fig. 8, t. 8 ; Harms l.c. 60, fig. 8; Pellegr. l.c. 41 ; Staner l.c. 239. *Pseudocedrela utilis* Dawe & Sprague (1906). *Entandrophragma macrocarpum* A. Chev. Vég. Util. 5 : 203 (1909). A large forest tree, to 210 ft. high, with long straight bole ; buttresses usually well developed ; bark usually grey-brown with well-marked longitudinal fissures, slash thick, reddish brown with paler lines, almost scentless ; leaves clustered at the ends of the branches ; inflorescences tomentellous ; flowers white.
    S.L.: Kambui Hills North (fr. Nov.) *Small* 840 ! **Iv.C.:** 60 miles E. of Agboville *Krukoff* 101 ! Various localities *fide* Aubrév. *l.c.* **G.C.:** Fiapere, Sunyani (fr. Feb.) *Chipp* 88 ! Nyinahin (fl. & fr. Mar.) *Thompson* 48 ! Juaso *Vigne* FH 1890 ! Wiresu, Ashanti (fr. Mar.) *Vigne* FH 1839 ! 1840 ! **S.Nig.:** Akure F.R. (fr. Nov.) *Latilo* FHI 32980 ! Aboabam, Ikom Div. (fr. Dec.) *Keay* FHI 28201 ! **Br.Cam.:** Tombel *Forteh* FHI 8396 ! Also in French Cameroons, Gabon, Belgian Congo and Angola. (See Appendix, p. 322.)
3. **E. angolense** (*Welw.*) *C.DC.* in Bull. Herb. Boiss. 2 : 582, t. 21 (1894) ; Aubrév. l.c. 138, t. 179, t. 181, a, t. 182, a, t. 183, 3–4 ; For. Trees & Timbers Brit. Emp. 4 : 36, fig. 5, t. 5 ; Harms l.c. 57, fig. 6 ; Pellegr. l.c. 38 ; Staner l.c. 224. *Swietenia angolensis* Welw. (1859). *Entandrophragma macrophyllum* A. Chev. Vég. Util. 5 : 196 (1909)—F.W.T.A., ed. 1, 1 : 495 ; Kennedy l.c. 168. *E. angolense* var. *macrophyllum* (A. Chev.) Panshin (1933). *E. septentrionale* A. Chev. l.c. 205 (1909)—F.W.T.A., ed. 1, 1 : 495. *E. rederi* Harms (1910)—F.W.T.A., ed. 1, 1 : 495 ; Kennedy l.c. 180. *E. leplaei* Vermoesen (1921)—Kennedy l.c. A large forest tree, to 160 ft. high, with long straight bole and rather open crown ; buttresses broad, but

low ; bark relatively smooth, pale grey-brown to orange-brown, with scales flaking off high up the tree ; slash dark red and pink with white radial streaks, very slightly scented ; leaves clustered at the ends of the branches ; inflorescences puberulous or glabrous ; flowers yellowish.

Fr.G.: Kissi Circle (Feb.) *Chev.* 20698. **S.L.:** Niyagoi, Dama Chiefdom *King* 173 ! Gorahun (fr. June) *Small* 715 ! Kambui Hills North *Small* 842 ! **Iv.C.:** Bouroukrou (Jan.) *Chev.* 16145 ! 16146 ! Azaguié *Chev.* 22299 ! Assikasso (Dec.) *Chev.* 22580 ! Banco *Aubrév.* 485 ! **G.C.:** Axim (fr. Oct.) *Chipp* 388 ! Amentia (Dec.) *Vigne* FH 1496 ! Tsifufu (young fr. Mar.) *Thompson* 36 ! Nsuta (May) *Vigne* FH 1725 ! **S.Nig.:** Gambari Group F.R. *W. D. MacGregor* 559 ! 560 ! Sapoba (Nov.–Jan.) *Kennedy* 456 ! 1174 ! 1194 ! 1237 ! 1254 ! 1793 ! 1984 ! 2077 ! **Br.Cam.:** Victoria *Maitland* 410 ! Likomba (fr. Dec.) *Mildbr.* 10796 ! Buea (fl. & fr. Mar.) *Maitland* 461 ! 534 ! Nchan, Bamenda (Mar.) *Johnstone* 90/31 ! **F.Po:** *Mildbr.* 6432. Extends to A.-E. Sudan, Uganda, Belgian Congo and Angola. (See Appendix, p. 319.)

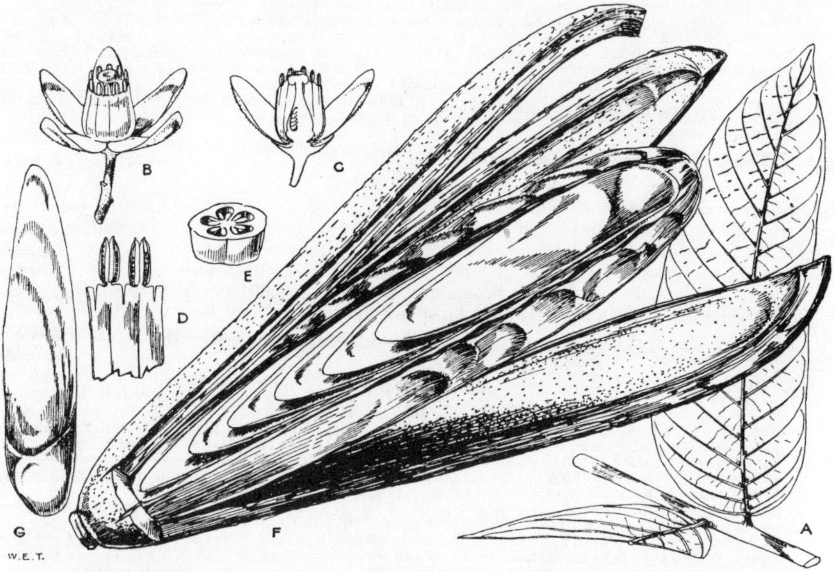

FIG. 195.—ENTANDROPHRAGMA UTILE (*Dawe & Sprague*) *Sprague* (MELIACEAE).

A, leaflet. B, flower. C, same in vertical section. D, part of staminal tube. E, cross-section of ovary. F, fruit with valve removed. G, seed.

4. **E. cylindricum** (*Sprague*) *Sprague* in Kew Bull. 1910 : 180 ; Hoyle in op. cit. 1932 : 40 ; Aubrév. l.c. 144, t. 180, 3, t. 181, c, t. 182, c ; Kennedy l.c. 174–179, incl. var. *ugoensis* Kennedy (1936), English descr. only ; For. Trees & Timbers Brit. Emp. 4 : 41, fig. 6, t. 6 ; Harms l.c. 62 ; Pellegr. l.c. 41 ; Staner l.c. 242. *Pseudocedrela cylindrica* Sprague (1908). *Entandrophragma tomentosum* A. Chev. ex Hutch. & Dalz. F.W.T.A., ed. 1, 1 : 495 (1928) ; Chev. Bot. 118. *E. rufum* A. Chev. Vég. Util. 5 : 201 (1909) ; Chev. Bot. 118. A large forest tree, to 200 ft. high, with very cylindrical bole up to 130 ft. high, and broad low buttresses, or without buttresses ; bark relatively smooth, with irregular flaking scales, grey-brown to ash-coloured ; slash brownish-pink with cream markings within, moist and very strongly and sweetly scented ; inflorescence puberulous ; flowers much smaller than the other species, yellowish-white. Trees with longer (10–15·5 cm.) fruits in Gold Coast and Ivory Coast, and with pyriform fruits in Benin may prove to be distinct varieties.

S.L.: Kambui area *Deen* 20016 ! 20573 ! 20757 ! **Iv.C.:** Agbo *Chev.* 16166 ! Bouroukrou *Chev.* 16132 ! 16140. And various other localities *fide* Chev. and Aubrév. *ll.c.*   **G.C.:** near Mansu & Supom (fr. Feb.) *Thompson* 16 ! Tsifufu (fr. Mar.) *Thompson* 34 ! Ahiraso (fr. Mar.) *Chipp* 122 ! Popo-kyere (fr. Apr.) *Chipp* 200 ! S. Fomang Su F.R. (Feb.) *Vigne* FH 1825 ! **Togo:** Misahöhe *Mildbr.* 7376. **S.Nig.:** Sapoba (fl. Nov., fr. May & June) *Kennedy* 429 ! 1184 ! 2036 ! 2065 ! 2155 ! Ugo, Benin Div. (fr. May) *Kennedy* 1126 ! **Br.Cam.:** Likomba *Mildbr.* 10770 ! Ngale, Kumba Div. (Feb.) *Forteh* FHI 8394 ! 8395 ! Extends to Uganda, Belgian Congo and Cabinda. (See Appendix, p. 320.)

A certain amount of variation is known within each of the above species. These variations occur especially in the size and shape of fruit, in the amount of indumentum and probably also in the bark. They do not seem of sufficient taxonomic importance to warrant specific or varietal rank. In view, however, of the great economic importance of the genus it would be as well for forest officers to study whether the infraspecific variation has any importance silviculturally or from the point of view of the quality of the timber. Reference should be made to Panshin in Amer. Journ. Bot. 20 : 646 (1933), Harms in Notizbl. Bot. Gart. Berl. 14 : 431 (1939), and Pellegrin in Notulae Syst. 9 : 37 (1940).

3. PSEUDOCEDRELA Harms in Engl. Bot. Jahrb. 22 : 153 (1895) ; E. & P. Pflanzenfam. 19B, 1 : 67 (1940).

Leaves pinnate ; leaflets alternate or rarely subopposite, 6–8 on each side, ovate-oblong or ovate-lanceolate, rounded and unequal-sided at base, obtuse at apex, undulately toothed or subentire, up to 15 cm. long and 6 cm. broad, softly pubescent mainly on the nerves beneath ; lateral nerves about 10–12 pairs, spreading ; rhachis softly tomentose ; panicles thyrsoid, up to about 30 cm. long ; branches rather long ; pedicels about 4 mm. long, puberulous ; calyx about 1 mm. long, 5-lobed ; petals

contorted, 2·5–3 mm. long, thinly pubescent up the middle ; staminal column broadly
campanulate, about 2·5 mm. long ; anthers 10 ; fruits 5-celled, oblanceolate in outline, 10–12 cm. long, 3 cm. thick, opening by 5 nearly flat recurving woody valves
and leaving the sharply 5-angled axis ; seeds with a long terminal membranous wing
about 6 cm. long with a stronger nerve on one side ..   .. *kotschyi*

P. **kotschyi** (*Schweinf.*) Harms in Engl. Bot. Jahrb. 22 : 154 (1895) ; Chev. Bot. 117 ; For. Trees & Timbers
Brit. Emp. 2 : 66, fig. 13 & photo ; Harms in E. & P. Pflanzenfam. 19B, 1 : 67, fig. 11 ; Pellegr. in
Notulae Syst. 9 : 34, fig. 2 ; Aubrév. Fl. For. Soud.-Guin. 383, t. 81. *Cedrela kotschyi* Schweinf. Reliq.
Kotschyanae 36, t. 35 (1868). *Pseudocedrela chevalieri* C.DC. (1907). A savannah tree, commonly 20–30 ft.
high, sometimes up to 60 ft. high and 6 ft. girth ; bark grey, fissured ; panicles of fragrant white flowers ;
foliage grey-silvery when young ; fruits erect ; often gregarious on heavy soils.
Sen.: Niokolo-Koba *Berhaut* 152. **Fr.Sud.:** Ségou *Vuillet* 5. Arbala *Dubois* 16. **Iv.C.:** Bondoukou
Aubrév. 727 ! Baoulé *Aubrév.* 43 ; *Chev.* 22344. Ferkéssédougou *Aubrév.* 2237. Black Volta R. *Aubrév.*
2409. **G.C.:** Bere, N.T. (Feb.) *Dudgeon* 105 ! Wa (Mar.) *McLeod* 839 ! Ejura (fr. Aug.) *Chipp* 751 !
Togo: Sokode (Apr.) *Kersting* 68 ! **Dah.:** Agouagon to Savé *Chev.* 23553 ! **N.Nig.:** Nupe *Barter* 1633 !
1712 ! Ejidogari F.R., Ilorin (Mar.) *Latilo* 32963 ! Katagum (Feb.) *Dalz.* 218 ! Abinsi *Dalz.* 1 ! Yola
Prov. *Dalz.* 169 ! **S.Nig.:** Lagos Colony *Foster* 17 ! Ijaiye F.R. (Mar.) *Keay & Ladipo* FHI 21191 !
Extends to A.-E. Sudan, Belgian Congo and Uganda. (See Appendix, p. 328.)

4. LOVOA Harms in Engl. Bot. Jahrb. 23 : 164 (1896) ; E. & P. Pflanzenfam. 19B, 1 :
        ·74 (1940).

Leaflets about 6 pairs, subopposite, elliptic, rounded or shortly pointed at base, very
slightly acuminate and emarginate, up to 20 cm. long and 9 cm. broad, glabrous,
with numerous rather faint spreading lateral nerves ; rhachis slightly winged ;
flowers numerous and small in large lax glabrous subcorymbose panicles ; pedicels
about 1·5 mm. long ; sepals broad and truncate, wrinkled ; petals elliptic, about
4 mm. long ; fruits narrowly oblong, quadrangular, 4–7 cm. long ; valves very thin,
deciduous ; seeds with a long terminal wing about 3 cm. long and 8 mm. broad
                                                                        trichilioides

L. **trichilioides** Harms in Engl. Bot. Jahrb. 23 : 165 (1896) ; Staner in Bull. Jard. Bot. Brux. 16 : 246. *L.
klaineana* Pierre ex Sprague (1906)—F.W.T.A., ed. 1, 1 : 493 ; For. Trees & Timbers Brit. Emp. 2 : 59,
fig. 12 & photo ; Aubrév. Fl. For. C. Iv. 2 : 133, t. 178 ; Kennedy For. Fl. S. Nig. 164 ; Pellegr. in
Notulae Syst. 9 : 34, fig. 3.   A large forest tree, to 150 ft. high, with cylindrical bole and rather short
blunt buttresses ; crown of dark, heavy, evergreen foliage ; bark relatively smooth with large irregular
flaking pieces ; slash dull red to pink, with faint whitish streaks, moist and very sweetly scented ; flowers
greenish-white ; fruits pendulous, black or purplish-black ; seeds attached to the distal end of the column
by the wing, the bodies of the seeds thus being free from the column at its base.
S.L.: (Dec.) *Unwin & Smythe* 35 ! Kenema Div. *King* 161 ! 238 ! **Iv.C.:** various localities *fide* Aubrév.
l.c. **G.C.:** Tsifufu (fr. Mar.) *Thompson* 33 ! Ancobra R. (fr. Jan.) *Thompson* 9 ! Tano R. (Dec.) *Chipp* 39 !
S. Fomang Su F.R. (Jan.) *Vigne* FH 2686 ! Worobong F.R. (Jan.) *Vigne* FH 4315 ! **S.Nig.:** Abeku,
Omo F.R. *Jones & Onochie* FHI 16726 ! Sapoba (fl. Feb.–Apr., June, Oct., fr. Mar.) *Kennedy* 478 ! 2360 !
Uregin, Benin *Unwin* 110 ! Extends to Belgian Congo and Angola. (See Appendix, p. 326.)

5. CARAPA Aubl.—F.T.A. 1 : 336 ; Harms in E. & P. Pflanzenfam. 19B, 1 : 77 (1940).

Petals about 8 mm. long, fleshy ; pedicels stout ; staminal tube about 6 mm. long,
deeply dentate ; ovules 2 per cell ; fruits subglobose, shortly pointed at apex, 5-
valved, valves not ridged, 12–15 cm. diam., containing up to 10 seeds ; leaves up
to 1·2 m. long, with 4–7 pairs of opposite or subopposite leaflets ; leaflets oblong,
cuneate to obtuse or rounded at base, shortly acuminate, 10–30 cm. long, 5–10 cm.
broad, glabrous ; panicles narrow, clustered, up to 45 cm. long, glabrous, surface of
axis and branches conspicuously rough with transverse cracks when dry ; outside of
flowers glabrous ..   .. 1. *grandiflora*
Petals 5–5·5 mm. long, thin ; pedicels slender ; staminal tube 4–4·5 mm. long, shallowly
dentate ; ovules 5–7 per cell ; fruits oblong-ellipsoid, long-beaked, or truncate and
shortly-beaked, with knobbly ridges along each valve, 6·5–15 cm. long, containing
15–20 seeds ; leaves very variable in size, up to 2 m. long, with 6–21 pairs of opposite
or alternate leaflets ; leaflets very variable in size and shape, usually more or less
elongate-oblong, up to 50 cm. long and 15 cm. broad ; panicles narrow, clustered,
up to 80 cm. long, glabrous or puberulous, surface of axis and branches usually
smooth, not cracked when dry ; outside of flowers puberulous, or glabrous except
for the ciliate margins of the sepals ..   .. 2. *procera*

1. **C. grandiflora** *Sprague* in J. Linn. Soc. 37 : 507 (1906) ; For. Trees and Timbers Brit. Emp. 4 : 29, fig. 4,
t. 4, C ; Staner in Bull. Jard. Bot. Brux. 16 : 119, t. 1. *C. macrantha* Harms (1917), from descr.   A tree
up to 30 ft. high (to 80 ft. in Uganda), with short stem and wide spreading crown ; sepals and petals
greenish, staminal tube white, disk orange, stigma white ; in montane forest.
S.Nig.: *Mackay* 2 ! Ikwette Plateau, Obudu, 5,200 ft. (Dec.) *Keay & Savory* FHI 25181 ! **Br.Cam.:**
Bamenda Prov. : Lakom-Nchan, 6,000 ft. (Apr.) *Maitland* 1663 ! Bafut-Ngemba F.R. (Apr.) *Ujor*
FHI 30002 ! Also in French Cameroons, Belgian Congo and Uganda.
[The description of the fruit given in the above key is based on specimens from Uganda ; I have not
seen fruit from our area.—R.W.J.K.]
2. **C. procera** *DC.* Prod. 1 : 626 (1824) ; Chev. Bot. 115 ; Aubrév. Fl. For. C. Iv. 2 : 125, t. 175 ; Kennedy
For. Fl. S. Nig. 159 ; Harms in E. & P. Pflanzenfam. 19B, 1 : 80, fig. 16, A–E, fig. 17, fig. 18 ; Pellegr.
in Notulae Syst. 9 : 27 ; Staner l.c. 116. *C. guineensis* Sweet ex A. Juss. (1830). *C. touloucouna* Guill. &
Perr. (1831). *C. gummiflua* C.DC. (1907)—Chev. Bot. 115. *C. velutina* C.DC. (1907)—Chev. Bot. 116.
C. microcarpa A. Chev. (1909)—Chev. Bot. 115.   Forest trees, with large leaves clustered at the ends of
the branches ; flowers white and pink, fragrant. Habit varies from a sprawling tree with arching branches
up to about 50 ft. high in swamp forest, to a tall tree with clear bole in lowland rain forest.   Specimens
with glabrous and with puberulous inflorescences occur throughout the range.   Further information about
the fruits is much needed as it seems that there are two apparently distinct types.   Workers should look

out for any possible correlation between the variations in size and shape of the leaves, indumentum of the inflorescence, form of the fruits and habit of the trees. At present such correlation is not apparent. Research is needed into the occurrence of *C. procera* DC. in tropical America ; also to confirm that C. de Candolle was correct in applying the name to our West African plant.

Sen.: R. Casamance (Apr.) *Chev.* 3164 ; 3166 ; 3168 ; *Leprieur* ; *Perrottet* ! **Port.G.:** Cacheu (Feb.) *Esp. Santo* 1310 ! **Fr.G.:** *Heudelot* 749 ! Moussaia (Feb.) *Chev.* 461 ! Kissi Circle (Feb.) *Chev.* 20685 ! Farana (Jan.) *Chev.* 20474 ! Kouroussa (Feb.) *Pobéguin* 587 ! **S.L.:** *Don* ! Heddle's Farm (Jan.) *Deighton* 1058 ! York (fr. May) *Deighton* 5525 ! Ninia, Talla Hills (Feb.) *Sc. Elliot* 4813 ! Njala (Dec.) *Deighton* 1860 ! **Lib.:** Peahtah (Oct.) *Linder* 905 ! Gbanga (Oct.) *Linder* 1156 ! Zigida (Oct.) *Baldwin* 10008 ! Begwai (Oct.) *Bunting* 17 ! **Iv.C.:** Grabo (July) *Chev.* 19640 ! Alépé (Mar.) *Chev.* 16233 ! Abidjan *Aubrév.* 67 ! Bouroukrou *Aubrév.* 720 ! **G.C.:** Aburi Hills (Nov.) *Johnson* 295 ! Bataba, N. Beyin (July) *Chipp* 262 ! Abosso (Aug.) *Vigne* 217 ! Amokokrom, W. Apollonia (Aug.) *Chipp* 320 ! **S.Nig.:** Lagos (Jan.) *Dalz.* 1372 ! Epe *Barter* 3248 ! Oban *Talbot* 1462 ! 1602 ! Boje, Ikom Div. (May) *Jones & Onochie* FHI 18934 ! **Br.Cam.:** Victoria (Dec.–Mar.) *Maitland* 430 ! 431 ! 469 ! 533 ! 533*a* ! 1348 ! 1350 ! Ambas Bay (Feb.) *Mann* 9 ! Molyko, Buea (Feb.) *Maitland* 1349 ! **F.Po:** (Jan., June) *Barter* 1724 ! *Mann* 153 ! Extends to Belgian Congo and Angola ; also in West Indies and Guiana. (See Appendix, p. 318.)

6. TRICHILIA Browne—F.T.A. 1 : 333 ; Harms in E. & P. Pflanzenfam. 19B, 1 : 104 (1940).

Staminal tube not laciniate, anthers thus sessile on the rim ; disk well-developed, cupuliform, surrounding the glabrous ovary ; stems and leaves glabrous ; leaflets (2–) 3 (–4) pairs, the odd terminal leaflet the largest, the lowest pair the smallest ; petioles 5–12 cm. long, rhachides 7–16 cm. long ; leaflets rather long-petiolulate, elliptic, elliptic-lanceolate or elliptic-oblanceolate, long-cuneate, long-acuminate, 4–20 cm. long, 2·2–7·5 cm. broad, with 6–12 main lateral nerves on each side of midrib ; panicles often clustered, up to about 8 cm. long, pubescent ; petals 5–6 mm. long ; capsules subglobose, about 2·5 cm. diam., glabrous .. .. 1. *prieureana*
Staminal tube laciniate :
  Leaflets acuminate or at least acute at apex ; forest trees :
    Leaflets glabrous or puberulous or shortly pubescent beneath :
      Petals 3–4 mm. long at anthesis ; disk well-developed, cupuliform, surrounding the glabrous ovary ; flowers in lax panicles up to 30 cm. long ; stems and leaves shortly pubescent ; leaflets 4–6 (–10) pairs, alternate or subopposite, with an odd terminal leaflet ; petioles 6–12 cm. long, rhachides 17–30 cm. long ; leaflets distinctly petiolulate, oblong-elliptic or oblong-oblanceolate, obtuse or cuneate at base, acutely long-acuminate, 6–25 cm. long, 3·5–10 cm. broad, with 8–23 main lateral nerves on each side of midrib ; capsules 2–3-lobed, 1·5–1·8 cm. diam., glabrous.. .. .. 2. *rubescens*
      Petals more than 6 mm. long ; disk adnate to the staminal tube ; ovary and fruit densely hairy :
        The petals 6–12 mm. long :
          Petals 6–8 mm. long, 1·5–2 mm. broad ; leaflets without translucent glandular dots and dashes, lower surface with short appressed hairs, nerves more or less densely pubescent at first, glabrescent ; leaves with 4–6 (–10) pairs of sub-opposite leaflets ; petioles 4·5–12 cm. long, rhachides 7–27 cm. long ; leaflets oblong to oblanceolate, rounded to cuneate at base, long-acuminate, 5–20 cm. long, 2–6 cm. broad, with 8–18 main lateral nerves on each side of midrib, rather prominent beneath ; fruits obovoid-globose, about 1·5 cm. diam.
                        3. *heudelotii*
          Petals 9–12 mm. long, 2–5 mm. broad ; leaflets when young with translucent glandular dots and dashes, lower surface glabrous or nearly so :
            The petals about 2 mm. broad ; leaves with (3–) 4–6 subopposite pairs of leaflets ; lateral petiolules 8–13 mm. long ; leaflets lanceolate or ovate-lanceolate, more or less rounded at base, very long narrowly and acutely acuminate, 8–20 cm. long, 2·5–7 cm. broad, with 7–10 main lateral nerves on each side of midrib ; petioles about 10 cm. long, rhachides about 20 cm. long ; panicles 8–10 cm. long, sparingly pubescent ; fruits subglobose, about 2·5 cm. diam.
                        4. *martineaui*
            The petals 3–5 mm. broad ; leaves with 7–9 opposite to alternate pairs of leaflets ; lateral petiolules 5–10 mm. long ; leaflets lanceolate to oblanceolate, cuneate to obtuse at base, long-acuminate and apiculate at apex, 8·5–23 cm. long, 2·5–6 cm. broad, with 8–20 main lateral nerves on each side of midrib ; petioles 3·5–38 cm. long, rhachides 8·5–38 cm. long ; panicles up to 30 cm. long, rather densely puberulous or tomentellous ; fruits pyriform, with a distinct stipe, 2·5–3 cm. long, about 2 cm. diam. .. .. 5. *gilgiana*
        The petals 15–20 mm. long :
          Rhachides of leaves densely puberulous or tomentellous ; leaflets (4–) 6 pairs, oblong to oblanceolate-oblong, obtuse or rounded at base, distinctly acuminate, 5·5–14 cm. long, 2–4 cm. broad, with 10–13 main lateral nerves on each side of midrib, rather prominent beneath ; petals about 2 cm. long and 3·5 cm. broad
                        6. *megalantha*
          Rhachides of leaves with long weak hairs ; leaflets 3–4 pairs, oblong-elliptic, obtuse or rounded at base, shortly and obtusely acuminate, 9–18 cm. long,

4–8 cm. broad, with 8–15 main lateral nerves on each side of midrib ; petals
1·5–1·7 cm. long, 3·5–5 mm. broad   . .   . .   7. *splendida*
Leaflets densely villous beneath, shining and glabrous above, 5–7 pairs, subopposite,
more or less oblong or obovate-oblong, obtuse to rounded or subcordate at base,
acuminate, 6–18 cm. long, 1·5–7 cm. broad, with 15–25 main lateral nerves on each
side of midrib, impressed above, very prominent beneath ; petioles 7·5–23 cm.
long, rhachides 7·5–38 cm. long ; inflorescences up to 20 cm. long ; petals 15–16 mm.
long, 2–3 mm. broad ; fruits pyriform or subglobose, 3·5–4·5 cm. long, 2·5–3 cm.
diam.   . .   . .   8. *lanata*
Leaflets rounded and emarginate or 2-lobed at apex ; fruits densely tomentellous ;
trees of savannah and fringing forest :
Leaves with 3–5 pairs of leaflets, plus the terminal one ; petioles 5–11 cm. long,
rhachides 4–15 cm. long, yellow-tomentose ; lateral leaflets oblong or oblong-
elliptic, rounded at each end, subsessile, the terminal one more or less obovate,
5–16 cm. long, 2·5–8 cm. broad, with 10–15 main lateral nerves on each side of
midrib, densely to thinly pubescent beneath ; inflorescences up to 7 cm. long ;
petals 12–14 mm. long, 2·5–4 mm. broad ; fruits subglobose, stipitate, about
2·5 cm. diam.   . .   . .   9. *roka*
Leaves with 1–2 pairs of leaflets, plus the terminal one ; petioles 2·5–7 cm. long,
rhachides 1–7 cm. long, sparsely puberulous ; leaflets oblanceolate, cuneate at
base, widely 2-lobed emarginate at apex, 5–16 cm. long, 2–5 cm. broad, with 10–20
rather faint main lateral nerves on each side of midrib, glabrous or very nearly
so ; inflorescences up to 9 cm. long ; petals 8–10 mm. long, 2–2·5 mm. broad ;
fruits broadly obovoid, about 3 cm. long   . .   . .   . .   10. *retusa*

1. **T. prieureana** *A. Juss.* in Mém. Mus. Hist. Nat. Paris 19 : 276 (1830) ; Guill. & Perr. Fl. Seneg. 1 : 125,
t. 30 ; F.T.A. 1 : 334 ; Chev. Bot. 114 ; Aubrév. Fl. For. C. Iv. 2 : 152, t. 186, 1–4 ; Kennedy For. Fl.
S. Nig. 163 ; Harms in E. & P. Pflanzenfam. 19B, 1 : 113 ; Pellegr. in Notulae Syst. 9 : 17, fig. 1 C ;
Staner in Bull. Jard. Bot. Brux. 16 : 143, fig. 7. *T. senegalensis* C. DC. (1907)—Chev. Bot. 114 ;
F.W.T.A., ed. 1, 1 : 493. *T. heudelotii* var.—Chev. Bot. 114. A forest tree, usually in the understorey
and 10–45 ft. high, but sometimes up to 70 ft. high ; bole conspicuously fluted, with grey-brown stringy
flaking bark ; flowers greenish-white ; fruits pink, seeds black with red arils ; mainly in drier types of forest.
Sen.: R. Casamance (Apr.) *Leprieur* ; *Perrottet* 134 ! R. Fasena *Döllinger*. Floup-Fedyan (Jan.) *Chev.*
3159 ; 3161 ; 3162. Bignona (Feb.) *Chev.* 3165. **Gam.:** Abuko Waterworks (Jan.) *Dalz.* 8128 ! **Port.G.:**
Biombo, Bissau (Feb.) *Esp. Santo* 1791 ! Prabis, Bissau (Feb.) *Esp. Santo* 1830 ! **Fr.G.:** *Heudelot* 775 !
Conakry *Debeaux* 178 ! Los Isl. (May) *Kalbreyer* 205 ! **S.L.:** Heddle's Farm (fl. & fr. Mar.) *Lane-Poole*
354 ! Kofiu Mt., Scarcies (Jan.) *Sc. Elliot* 4620 ! Musaia (Feb.) *Deighton* 4176 ! Wara Wara, Yagalu
(Jan.) *King* 192 ! **Lib.:** Zigida, Vonjama (Oct.) *Baldwin* 10016 ! **Iv.C.:** *Aubrév.* 2125 ! Bangouanou
to Endé (young fr. Dec.) *Chev.* 22469 ! Rasso *Aubrév.* 139 ! And other localities *fide* Aubrév. and Chev.
l.c. **G.C.:** Agogo, Ashanti (Dec.) *Chipp* 600 ! 601 ! Aboso, Tarkwa (Aug.) *Vigne* 220 ! Volta River
F.R. (Jan.) *Moor* FH 2183 ! Bana Hill, Krobo (Jan.) *Irvine* 1924 ! **Togo:** Lome *Warnecke* 416 ! Sokode-
Basari *Kersting* A. 720 ! **Dah.:** Adjara, Porto Novo (Jan.) *Chev.* 22754 ! **N.Nig.:** Jemaa Div. (Feb.)
Onyeagocha FHI 7625 ! **S.Nig.:** Lagos (Jan.) *Barter* 2146 ! *Dalz.* 1265 ! Olokemeji (Jan.) *Keay* FHI
14614 ! Idanre Hills (Jan.) *Brenan* 8640 ! Urhuehue, Benin Div. (Dec.) *Olorunfemi* FHI 31924 ! Onitsha
Barter 1676 ! **Br.Cam.:** Tiko (Jan.) *Dunlap* 181 ! Extends to A.-E. Sudan, Uganda, Belgian Congo,
Angola and N. Rhodesia. (See Appendix, p. 329.)

2. **T. rubescens** *Oliv.* F.T.A. 1 : 336 (1868) ; Harms l.c. 112 ; Pellegr. l.c. 20, fig. 1 D ; Staner l.c. 148, t. 2,
A–C. *T. laurentii* De Wild. (1908)—Bak. f. in Cat. Talb. 124. A tree, 12–30 ft. high, rarely more, in
understorey of rain forest ; calyx brown, corolla yellow ; leaves drying reddish brown.
S.Nig.: Ondo (Feb.) *A. F. A. Lamb* 111 ! Oban *Talbot* 1370 (partly) ! 1711 ! Calabar (Jan.) *Smith* 33 !
Umaji, Obudu Div. (young fr. May) *Latilo* FHI 30918 ! **Br.Cam.:** Ambas Bay (Feb.) *Mann* 20 ! Bimbia
road, Victoria (fl. & young fr. Feb.) *Maitland* 406 ! 434 ! Tiko (Jan.) *Dunlap* 170 ! **F.Po:** (Jan.) *Mann*
163 (partly) ! Also in French Cameroons, Ubangi-Shari, Spanish Guinea, Gabon, Belgian Congo, Uganda
and Tanganyika.

3. **T. heudelotii** *Planch. ex Oliv.* F.T.A. 1 : 334 (1868) ; Chev. Bot. 114 ; Aubrév. l.c. 154, t. 186a ; Kennedy
l.c. 163 ; Harms l.c. 110 ; Pellegr. l.c. 23, fig. 1 A ; not of Staner l.c. 155. *T. zenkeri* Harms (1896)—
F.W.T.A., ed. 1, 1 : 493 ; Staner l.c. 165, t. 6. *T. heudelotii* var. *zenkeri* (Harms) Aubrév. l.c. 154,
t. 186b (1936). *T. djalonis* A. Chev. (1912)—Chev. Bot. 113. *T. acutifoliata* A. Chev. (1909), name only.
T. candollei A. Chev. (1909), name only—Chev. Bot. 113. *T. acutifoliola* A. Chev. Bot. 112, name only.
T. velutina A. Chev. Bot. 115, name only, not of C.DC. *T. oddoni* of Pellegr. l.c. 24, partly. A tree
20–60 ft. high, in understorey of rain forest, especially in secondary regrowth types and in moist places ;
bark brown ; slash dark pink with slight exudate of whitish latex ; flowers greenish-yellow ; fruits
buff-coloured.
Port.G.: Catió (June) *Esp. Santo* 2100 ! **Fr.G.:** Karkandy *Heudelot* 842 ! Dioromandou to Koudian
(Feb.) *Chev.* 20736 ! **S.L.:** *Afzelius* ! Mt. Gonkwi, Ninia (Feb.) *Sc. Elliot* 4986 ! Sasseni, Scarcies (Jan.)
Sc. Elliot 4542 ! Njala (May) *Deighton* 2996 ! Gola Forest (Mar.) *Small* 541 ! **Lib.:** Grand Bassa *Dinklage*
1878 ! Dukwia R. (Aug.) *Linder* 230 ! Little Kola R. (Mar.) *Baldwin* 11192 ! Butan (Mar.) *Baldwin*
11495 ! Ganta (Apr.) *Harley* 1135 ! **Iv.C.:** Mbasso (Mar.) *Chev.* 16262 ! Toula (July) *Chev.* 19565 !
Mt. Dou, Zoanlé (May) *Fleury* in Hb. *Chev.* 21480 ! Grabo (July) *Chev.* 19641 ! Potou lagoon (Feb.)
Chev. 20110 ! Abidjan *Aubrév.* 20 ! 115 ! Banco *Aubrév.* 262 ! **G.C.:** Boaso, W. Ashanti (Jan.) *Chipp*
136 ! Kumasi (Mar.) *Vigne* FH 1640 ! Aburi (Feb.) *Brown* 932 ! Assuantsi (Mar.) *Irvine* 1560 ! **Dah.:**
Allada (Mar.) *Chev.* 23405 ! **S.Nig.:** Ibadan (May) *Keay* FHI 22844 ! Baba-Eku, Ijebu Ode Prov.
(Feb.) *Ross* 47 ! Sapoba *Kennedy* 653 ! 1936 ! 2153 ! Degema *Talbot* ! Oban *Talbot* 1370 (partly) !
1712 ! **Br.Cam.:** Victoria (Jan.) *Maitland* 317 ! Johann-Albrechtshöhe *Staudt* 592 ! **F.Po:** *Barter* !
Mann 163 (partly) ! Also in French Cameroons, Gabon, Belgian Congo and Cabinda. (See Appendix, p. 329.)

4. **T. martineaui** *Aubrév. & Pellegr.* in Bull. Soc. Bot. Fr. 83 : 491, fig. 2 (1936) ; Aubrév. Fl. For. C. Iv. 2 :
152, t. 185, 1–5 ; Harms l.c. 110 ; Pellegr. in Notulae Syst. 9 : 22. A forest tree, up to 100 ft. high and
nearly 10 ft. girth, with straight cylindrical stem, without buttresses ; bark dark grey, with rough more
or less rectangular scales ; slash pink outside, yellowish within, exuding a little coffee-coloured latex,
with a disagreeable smell.
S.L.: Moyamba (Dec.) *King* 299 ! **Iv.C.:** Yapo *Aubrév.* 1365. Rasso *Aubrév.* 572. Banco *Aubrév.* 853 !
924 ! Agboville *Aubrév.* 1920.

[*Kennedy* 2184 from Sapoba, S. Nig. is close to this, but has oblong-elliptic leaflets cuneate at base ;
the leaflets, which are rather old, do not appear to have gland-dots. It is near *T. hylobia* Harms from the
French Cameroons.]

5. **T. gilgiana** *Harms* in Engl. Bot. Jahrb. 23 : 161 (1896) ; E. & P. Pflanzenfam. 19B, 1 : 110 ; Pellegr. l.c. 21 ; Staner l.c. 161, t. 5. A tree, to 70 ft. high, in understorey of rain forest, with fluted lower bole ; bark pale brown ; slash pinkish-brown with a little latex, sweetly scented ; petals cream without, pink within.
   S.Nig.: Ikom (Nov.) *Catterall* 60 ! 64 ! Afi River F.R., Ikom (Dec.) *Keay* FHI 28222 ! **Br.Cam.:** Bali-Ngemba F.R., Bamenda (young fr. June) *Tiku & Ujor* FHI 30421 ! Also in French Cameroons, Gabon, Belgian Congo and Cabinda.

6. **T. megalantha** *Harms* in Engl. Bot. Jahrb. 23 : 160 (1896) ; Aubrév. l.c. 156, t. 185, 6 ; Harms l.c. 110 ; Pellegr. l.c. 25. A forest tree 50–100 ft. high, with rough dark grey bark ; flowers whitish, strongly scented ; fruits greyish-pink.
   Iv.C.: Man *Aubrév.* 946. Agnéby *Aubrév.* 2357. **G.C.:** Ofin Headwaters F.R. (Apr.) *Vigne* FH 1906 ! **S.Nig.:** Western Lagos *Rowland* ! Ibadan (June) *Ogua* FHI 3423 !

7. **T. splendida** *A. Chev.* in Bull. Soc. Bot. Fr. 58, Mém. 8 : 147 (1912) ; Aubrév. l.c. 156, t. 184, 1–3 ; Harms l.c. 110. Pellegr. l.c. 26. *T. tomentosa* A. Chev. Bot. 115, name only, not of H. B. & K. A forest tree, to 80 ft. high ; slash exuding a little latex ; flowers white, very fragrant.
   Fr.G.: Kissidougou (Feb.) *Chev.* 20708 ! Diaguissa (Apr.) *Chev.* 12916. **Iv.C.:** Mt. Tonkoui *Aubrév.* 993. Bondoukou *Aubrév.* 731 ! (? sterile). Also Mt. Dou and Nzo *fide* Aubrév. *l.c.*
   [The plants described under this name from Belgian Congo (Staner l.c. 159, fig. 9, t. 4) and from Uganda (Eggeling, Indig. Trees Uganda, ed. 2 : 198) are best referred to *T. strigulosa* Welw. ex C.DC. ; the plant referred by Eggeling (l.c. 197) to *T. megalantha* is also *T. strigulosa.*]

8. **T. lanata** *A. Chev.*[1] in Bull. Soc. Bot. Fr. 58, Mém. 8 : 146 (1912) ; Chev. Bot. 114 ; Aubrév. l.c. 150, t. 184, 4–7 ; Harms l.c. 109 ; Pellegr. l.c. 26, fig. 1 B. *T. montchali* De Wild. (1914)—Staner l.c. 172, t. 7, C–D ; Pellegr. l.c. 27. *T. mildbraedii* Harms (1917)—F.W.T.A., ed. 1, 1 : 493 ; Kennedy l.c. 164. A forest tree, to 100 ft. or more high, with slight buttresses and straight, cylindrical bole ; flowers greenish-white ; fruits carmine pink.
   S.L.: Peninsular F.R. (Dec.) *Vigne* 38 ! Njala *Deighton* 4730 ! **Iv.C.:** Agboville (young fr. Nov.) *Chev.* 22382 ! Abidjan *Aubrév.* 45 ! 160 ! Banco *Aubrév.* 376 ! And various other localities *fide* Aubrév. *l.c.* **G.C.:** S. Fomang Su F.R. (fr. Jan.) *Vigne* FH 2688 ! Kumasi *Andoh* FH 4852 ! **Dah.:** E. of Henoi Poisson. **S.Nig.:** Sapoba *Kennedy* 516 ! 1598 ! 1741 ! 1794 ! 1794 ! 1846 ! 1856 ! Mamu River F.R., Awka Div. *Onwudinjoh* FHI 21912 ! Also in French Cameroons and extending to Belgian Congo and Cabinda. (See Appendix, p. 329.)
   [A tree at Idanre, S. Nigeria (*Keay & Onochie* FHI 21568) is close to this species but the leaflets have rather fewer main lateral nerves and are slightly less densely hairy beneath. This tree needs further study ; possibly it is a hybrid between *T. lanata* and *T. heudelotii.*]

9. **T. roka** (*Forsk.*) *Chiov.* Flora Somala 2 : 131 (1932) ; Brenan in Mem. N.Y. Bot. Gard. 8, 3 : 235. *Elcaja roka* Forsk. (1775). *Trichilia emetica* Vahl (1790)—F.T.A. 1 : 335 ; F.W.T.A., ed. 1, 1 : 493 ; Harms l.c. 109 ; Pellegr. l.c. 24 ; Staner l.c. 175, t. 7, E–F ; Aubrév. Fl. For. Soud.-Guin. 381, t. 80, 1–3. A savannah tree, to 30 ft., with fissured corky bark ; flowers pale green except for the fawn anthers ; ripe fruits crimson.
   Sen.: *Perrottet* 133 ! Djibelor *Aubrév.* 3045. **Gam.:** *Heudelot* 15. Albreda (May) *Leprieur* ! **Fr.Sud.:** Kouguéniéba (Mar.) *Chev.* 556 ! Kolokani *Aubrév.* 3047. Arbala *Dubois* 172. **Port.G.:** Prabis, Bissau (Feb.) *Esp. Santo* 1808 ! Canjadudi, Gabu (young fr. June) *Esp. Santo* 2511 ! **Fr.G.:** Kouroussa *Pobéguin* 190 ; 687. Timbo *Chev.* 12527. Fouta Djalon *Chev.* 12834. **S.L.:** *unknown collector* ! **Iv.C.:** Ferkéssédougou *Serv. For.* 2284. Ouagadougou *Aubrév.* 2424. Mt. Kangoroma *Chev.* 22177. **G.C.:** Lambussie to Belong (young fr. Mar.) *Kitson* 784 ! Afram Plains (young fr. May) *Kitson* 1094 ! Nakong, N.T. (Feb.) *Vigne* FH 4718 ! Yendi (fr. Apr.) *Vigne* FH 1714 ! **Togo:** Sansanné Mango *Aubrév.* 130d. Kpandu (Feb.) *Robertson* 104 ! **Dah.:** Oueme to Agouagon *Chev.* 23606 ! Savé *Aubrév.* 70d. **Fr.Nig.:** Gourma *Chev.* 24457. **N.Nig.:** Sokoto (Feb.) *Lely* 848 ! Kontagora (Feb.) *Dalz.* 89 ! Alagbede F.R., Ilorin (Feb.) *Ejiofor* FHI 19810 ! Bauchi Plateau (Mar.) *Lely* P. 191 ! **S.Nig.:** 12 miles E. of Shepeteri (Mar.) *Keay* FHI 22521 ! Old Oyo F.R. (Feb.) *Keay* FHI 16007 ! Ado to Iseyin *Rowland* ! Widespread in tropical Africa, extending to S. Africa, Madagascar, Réunion and Arabia. (See Appendix, p. 328.)

10. **T. retusa** *Oliv.* F.T.A. 1 : 334 (1868) ; Harms l.c. 110 ; Pellegr. l.c. 17 ; Staner l.c. 152, t. 2, D–F. A nearly glabrous tree of fringing forests, up to 40 ft. high, with fragrant white flowers and usually 2-valved fruits.
    N.Nig.: Nupe *Barter* 1181 ! Jebba (Dec.) *Hagerup* 733 ! Ibaji-Ojoku F.R. (Feb.) *J. E. Taylor* L. 9 ! Idah *T. Vogel* 50 ! Echonga, Kabba Prov. (Feb.) *Jones* FHI 634 ! Abinsi (Jan.) *Dalz.* 865 ! Also in Ubangi-Shari, A.-E. Sudan and Belgian Congo.

Imperfectly known species.

Limonia ? monadelpha *Thonn.* in Schum. Beskr. Guin. Pl. 217 (1827) ; F.T.A. 1 : 336. I have examined an authentic specimen of this from the Vahl herbarium, Copenhagen. It consists only of an incomplete leaf, which might possibly be regarded as coming within the rather broad interpretation of *Trichilia heudelotii* which I have given above. A more exact determination would be most unconvincing. I consider, therefore, that it is scientifically undesirable to make a new combination to replace the well-known *T. heudelotii* Planch. ex Oliv. which has excellent and widely distributed type specimens.

7. **EKEBERGIA** Sparrm.—F.T.A. 1 : 332 ; Harms in E. & P. Pflanzenfam. 19B, 1 : 119 (1940). *Charia* C. DC. (1907).

Leaves pinnate ; leaflets oblong or elliptic, unequal-sided and shortly cuneate at base, shortly acuminate, 5–10 cm. long, 2·5–4·5 cm. broad, glaucescent and finely reticulate beneath ; rhachis flat on the upper surface towards the base ; panicles axillary, up to ¾ as long as the leaves, puberulous ; pedicels up to 2 mm. long ; flower-buds ellipsoid ; calyx saucer-shaped, dentate, about 1·5 mm. long, puberulous ; petals about 5 mm. long, pubescent ; stamens 10, monadelphous ; fruits drupaceous, subglobose, 1·5–2 cm. long, glabrous .. .. **senegalensis**

E. senegalensis *A. Juss.* in Mém. Mus. Hist. Nat. Paris 19 : 234, t. 17, fig. 16a (1830) ; Guill. & Perr. Fl. Seneg. 1 : 126, t. 31 ; F.T.A. 1 : 333 ; Chev. Bot. 112 ; Aubrév. Fl. For. C. Iv. 2 : 157, t. 187 ; Pellegr. in Notulae Syst. 9 : 5 ; Aubrév. Fl. For. Soud.-Guin. 383. *Charia chevalieri* C. DC. (1907). *Ekebergia chevalieri* (C. DC.) Harms (1911). *Charia indeniensis* A. Chev. (1909). *Ekebergia indeniensis* (A. Chev.) Harms (1915). *Sorindeia doeringii* Engl. & K. Krause (1911). *Ekebergia dahomensis* A. Chev. Bot. 112, name only. A tree to 50 ft. high, rarely more, with spreading crown and dense foliage, and smooth grey bark becoming rough and scaly ; petals yellowish ; fruits reddish when ripe, containing 2–4 bright red seeds with yellow aril ; in savannah regions and drier parts of the forest regions. The species seems to vary a good deal in size of leaflets and indumentum of the leaves.

---

[1] In the first edition *T. lanata* A. Chev. was given as a synonym of *T. prieureana* A. Juss., and the specimen *Chev.* 22469 was quoted as *T. mildbraedii* Harms. These mistakes occurred because the labels on the Kew duplicate of *Chev.* 22382 (type number of *T. lanata* A. Chev.) and of *Chev.* 22469 (a specimen of *T. prieureana* A. Juss., see Aubrév. Fl. For. C. Iv. 2 : 152) had become interchanged.

Sen.: *Berhaut* 1000. Kounoum *Leprieur.* **Gam.:** Albreda *Heudelot* 107 ! 378 ! *Perrottet* 131 ! **Fr.Sud.:** Sikasso *Vuillet* 565. **Port.G.:** Bissau (Mar.) *Esp. Santo* 1881 ! Bolama (fr. Apr.) *Esp. Santo* 1930 ! **Fr.G.:** Moriquéniéba (Feb.) *Chev.* 448 ! **Iv.C.:** Bettié (Mar.) *Chev.* 16255 ! Ferkéssédougou, Ouangolo, Niangoloko & Bondoukou *fide* Aubrév. *l.c.* **G.C.:** Fiapire *Chipp* 92. Wenchi (Mar.) *Dalz.* 17 ! **Togo:** *Kersting* 99 ! Njande, Atakpame (fr. May) *Doering* 230. **Dah.:** Abomey (Feb.) *Chev.* 23133 ! Cabolé (fr. May) *Chev.* 23779 ! Birni *Aubrév.* 59d. **N.Nig.:** Sokoto (Mar.) *Lely* 850 ! Zaria *Dalz.* 355 ! Anara F.R., Zaria (fl. & young fr. May) *Keay* FHI 22905 ! 22910 ! Bauchi Plateau (Jan.) *Lely* P. 55 ! **S.Nig.:** Olokemeji *Unwin* 14 ! Ibadan (Mar.) *Abubakar* FHI 3008 ! E. of Igboho, Oyo Div. (Feb.) *Keay* FHI 23439 ! Extends to Ubangi-Shari, Belgian Congo and Angola. (See Appendix, p. 319.)

8. GUAREA Allem. in Linn. Mant. 2 : 150 (1771) ; Harms in E. & P. Pflanzenfam. 19B, 1 : 129 (1940), excl. Sect. *Heckeldora* (Pierre) Harms.

Trees up to 13 m. high or shrubs, in the understorey of forest ; inflorescences pendulous, long and slender, with few flowers on the short lateral branches ; leafy branchlets slender ; leaflets alternate, rarely subopposite, venation very obscure ; seeds with the cotyledons juxtaposed and radicles terminal:

Petals hairy all over the outer surface, 6–10 mm. long ; flowers aggregated in fascicles ; inflorescences up to 65 cm. long ; petioles 8–12 cm. long, rhachides 15–34 cm. long, shortly tomentellous, glabrescent ; leaflets 8–13 (–17), elongate-oblong, obtuse or cuneate and often asymmetrical at base, long-acuminate, 5·5–23 cm. long, 3–6·5 cm. broad, with numerous rather faint main lateral nerves, minutely puberulous beneath ; fruits subglobose, rather deeply lobed, 1–2 cm. diam., tomentellous

1. *glomerulata*

Petals hairy only on the margins, 8–10 mm. long ; flowers borne laxly and often singly on the inflorescences which are up to 20 cm. long ; petioles 4·5–10 cm. long ; rhachides 6–17 cm. long, minutely puberulous ; leaflets 5–13, oblong-elliptic, cuneate to rounded at base, long-acuminate, 3–16 cm. long, 1·2–5·8 cm. broad, with 6–12 rather obscure main lateral nerves, nearly glabrous ; fruits depressed-globose, apparently not lobed, 2–2·5 cm. diam., minutely tomentellous

2. *leonensis*

Trees up to 50 m. high, attaining the highest stratum of the forest ; inflorescences erect, stout, more or less pyramidal, with relatively long lateral branches, 1–15 (–25) cm. long ; leafy branchlets stout ; leaflets opposite or subopposite ; seeds with the cotyledons superposed and radicles lateral, completely enveloped in an aril :

Petioles 2–4·5 cm. long, with distinct incurved wings 3–6 mm. wide ; reticulation of veins distinct on undersurface of leaflets ; calyx denticulate ; petals 5–7 mm. long, densely tomentellous outside ; fruits velvety-tomentellous, subglobose, 3–5·5 cm. diam. ; rhachides 2–18 cm. long ; leaflets 6–15, oblong, oblong-lanceolate or oblong-oblanceolate, cuneate, or sometimes rounded, and often asymmetrical at base, gradually obtuse to long-acuminate at apex, margins often undulate, 8–32 cm. long, 2·5–10·5 cm. broad, glabrous except for a few hairs on midrib and nerves beneath, with 16–22 main lateral nerves on each side of midrib, prominent beneath ; slash without latex   ..   ..   3. *cedrata*

Petioles 7–14 cm. long, scarcely winged ; ultimate reticulation of veins obscure on undersurface of leaflets ; calyx subtruncate ; petals 8–10 mm. long, tomentellous outside ; fruits glabrous and often rugose outside, subglobose, 3–4 cm. diam. ; rhachides 8–30 cm. long ; leaflets 7–17, elongate-elliptic, cuneate or obtuse at base, obtuse to acute or acuminate at apex, 9–24 cm. long, 3–9 cm. broad, glabrous, with 9–16 main lateral nerves on each side of midrib ; slash with white latex

4. *thompsonii*

1. **G. glomerulata** *Harms* in Engl. Bot. Jahrb. 23 : 159 (1896) ; in E. & P. Pflanzenfam. 19B, 1 : 134 ; Staner in Bull. Jard. Bot. Brux. 16 : 199, t. 11, D. *G. glomerulata* var. *obanensis* Bak. f. in Cat. Talb. 18 (1913). A shrub or small tree, 10–20 ft. high, with drooping branches and pendulous racemes ; petals pink ; fruits dull scarlet, with golden hairs, dehiscing to expose deep orange shining seeds in a white floss ; in rain forest.
S.Nig.: Eket *Talbot* ! Oban *Talbot* 412 ! 417 ! 1285 ! Aboabam, Ikom Div. (Dec.) *Keay* FHI 28212 ! 28227 ! Boje, Ikom Div. (fr. May) *Jones & Onochie* FHI 18736 ! **F.Po:** *Mildbr.* 6371. Also in French Cameroons, Gabon and Belgian Congo. (See Appendix, p. 323.)
2. **G. leonensis** *Hutch. & Dalz.* F.W.T.A., ed. 1, 1 : 492 (1928) ; Kew Bull. 1929 : 25 ; Harms l.c. 135. A small tree, 15–40 ft. high, with very spreading and drooping branches ; flowers cream or white ; fruits yellow ; in rain forest.
S.L.: Koteimahun *Lane-Poole* 337 ! Benduhun, Tunkia (Dec.) *Deighton* 3817 ! Gorahun to Bobobu (fr. Apr.) *Deighton* 3652 ! Gola Forest (fl. Jan., Apr., fr. Apr.) *Small* 611 ! 632 ! *Unwin & Smythe* 39 ! **Lib.:** Bumbumi to Moala (Nov.) *Linder* 1342 ! Jabrocca (Dec.) *Baldwin* 10865 ! Jaurazon (fr. Mar.) *Baldwin* 11467 ! Bomi Hills (Feb.) *Baldwin* 14084 !
3. **G. cedrata** (*A. Chev.*) *Pellegr.* in Bull. Soc. Bot. Fr. 85 : 180 (1928) ; Aubrév. Fl. For. C. Iv. 2 : 130, t. 177, 1–7 ; Kennedy For. Fl. S. Nig. 162 ; Harms l.c. 135 ; Staner l.c. 187, t. 9, t. 10, t. 11, A. *Trichilia cedrata* A. Chev. Vég. Útil. 5 : 215 (1909) ; F.W.T.A., ed. 1, 1 : 493. A large forest tree, to 160 ft. high, with dense crown, long bole and rather heavy buttresses ; bark pale grey-brown, flaking to expose " mussel shell " markings ; slash fawn-pink, moist but without latex, very fragrant ; flowers greenish-yellow ; ripe fruits orange-yellow with a reddish bloom ; seeds surrounded by an orange-coloured aril.
S.L.: Dodo Hills F.R. (fr. Jan.) *Sawyerr* FHK 13597 ! Kambui Hills, Kenema (fr. Dec.) *Small* 909 ! **Lib.:** Dukwia R. *Cooper* 415 ! **Iv.C.:** Yapo *Chev.* 22321 ! Makouguié (fr. Jan.) *Chev.* 16171 ! Alépé (May) *Krukoff* 58 ! And other localities *fide* Chev. and Aubrév. *ll.c.* **G.C.:** Jabo, Upper Wassaw F.R. (June) *Vigne* FH 967 ! Sefwi-Bekwai (fl. & fr. Aug.) *Vigne* FH 1330 ! Aguna (fr. Aug.) *Vigne* FH 1303 ! Bediaku, Goaso Dist. (June) *Andoh* FH 5502 ! **S.Nig.:** Omo F.R. *Tamajong* FHI 20271 ! Okomu F.R. (young fr. Mar.) *Brenan* ! *Ross* 135 ! Sapoba (fl. Nov.–Jan., fr. June, July) *Kennedy* 310 ! 311 ! 1878 ! 1893 ! Usonigbe F.R. (young fr. Nov.) *Ejiofor* FHI 24656 ! Also in French Cameroons, Belgian Congo and Uganda. (See Appendix, p. 322.)

4. **G. thompsonii** *Sprague & Hutch.* in Kew Bull. 1906 : 245 ; Aubrév. l.c. 132, t. 177, 8–10 ; Kennedy l.c. 160 ; Harms l.c. 136 ; Staner l.c. 196, t. 10, C, t. 11, C. A large forest tree, to 160 ft. high, with dense crown, long bole and low buttresses ; bark dark grey to mid brown, more or less smooth, but flaking to expose " mussel shell " markings ; slash yellow, with sticky white latex ; flowers yellow ; ripe fruits orange or purplish-red ; seeds surrounded by an orange-coloured aril.
Lib.: Nyaake, Webo (June) *Baldwin* 6117 ! **Iv.C.:** Abidjan *Aubrév.* 79 ! 202 ! Banco *Aubrév.* 354 ! Danané *Aubrév.* 1040 ! **G.C.:** Prestea *Vigne* FH 3086 ! Jamang, W.P. (Apr.) *Vigne* FH 955 ! **S.Nig.:** Benin City *Thompson* 16 ! Nikrowa (young fr. Mar.) *Ross* 115 ! Sapoba (fl. Oct., fr. May, June) *Kennedy* 1653 ! Degema *Talbot* 3757 ! Also in Gabon and Belgian Congo. (See Appendix, p. 323.)

9. **HECKELDORA** Pierre in Bull. Soc. Linn. Paris 2 : 1287 (1896) ; Staner in Bull. Jard. Bot. Brux. 16 : 206 (1941). *Guarea* Sect. *Heckeldora* (Pierre) Harms— E. & P. Pflanzenfam. 19B, 1 : 135 (1940).

A shrub or small tree, to 10 m. high ; branchlets puberulous, more or less glabrescent ; leaves imparipinnate, with 5–9 opposite to alternate leaflets ; petioles 3–14 cm. long, rhachides 8–19 cm. long, more or less densely pubescent to nearly glabrous ; leaflets very variable in shape and size, often elliptic, oblong, oblanceolate or obovate, round to cuneate and often asymmetrical at base, acuminate at apex, 3–23 cm. long, 2–9·5 cm. broad, pubescent or glabrous beneath, pubescent on the midrib above, with 6–14 main lateral nerves, rather markedly looped on each side of midrib ; racemes or narrow panicles 6–70 cm. long, solitary or several together, puberulous ; calyx cupuliform, subtruncate ; petals 6–8 mm. long, glabrous or nearly so outside ; ovary glabrous to densely pubescent, 1-celled, with 2 parietal placentas each bearing 2 ovules ; fruits baccate, ovoid or cylindrical, often asymmetrical and slightly moniliform, with several longitudinal ridges, stipitate, long-beaked, 2·5–8 cm. long and 0·8–3 cm. broad, puberulous outside, containing 1–4 seeds .. .. *staudtii*

H. **staudtii** *(Harms) Staner* l.c. 207, fig. 13 (1941). *Guarea staudtii* Harms in Notizbl. Bot. Gart. Berl. 1 : 180 (Aug. 1896). *G. zenkeri* Harms (Sept. 1896). *Heckeldora zenkeri* (Harms) Staner (1941). *H. latifolia* Pierre (Nov. 1896). *H. angustifolia* Pierre (Nov. 1896). *Guarea angustifolia* (Pierre) Pellegr. (1939), not of C. DC. (1903). *G. pierreana* Harms (1940). *G. leptotricha* Harms (1897). *G. nigerica* Bak. f. (1913)— F.W.T.A., ed. 1, 1 : 492. *G. parviflora* Bak. f. (1913). A small tree or shrub, in forest, with pendulous inflorescences of cream fragrant flowers and yellow fruits.
Lib.: Jaurazon (Mar.) *Baldwin* 11468 ! **S.Nig.:** Oban *Talbot* 590 ! 1281 ! 1350 ! 1701 ! s.n. ! Ikwette Plateau, Obudu Div. (Dec.) *Keay & Savory* FHI 25209 ! **Br.Cam.:** Victoria (Jan.) *Kalbreyer* 4 ! Likomba (Nov.) *Mildbr.* 10630 ! Johann-Albrechtshöhe (Jan.) *Staudt* 534. Bambuko F.R., Kumba Div. (Sept.) *Olorunfemi* FHI 30772 ! Also in French Cameroons, Gabon and Belgian Congo.
Baldwin 11378 from Liberia, 25 miles from mouth of Sangwin R., may also be this species. Further study in the field is needed to see if any satisfactory subdivision of this variable species is possible.

10. **TURRAEANTHUS** Baill. in Adansonia 11 : 261 (1874); Harms in E. & P. Pflanzenfam. 19B, 1 : 147 (1940).

Leaves pinnate, with (5–) 8–24 (–30) alternate or subopposite leaflets ; petioles 5–10 cm. long, rhachides 4–60 cm. long, densely rusty-puberulous when young ; petiolules 0·5–1·5 cm. long ; leaflets more or less elongate-oblong, rounded or shortly cuneate at base, rounded or obtuse and shortly apiculate at apex, the apiculum itself spathulate or with the margins up rolled, 6–25 cm. long, 2·5–8 cm. broad, soon glabrous, with 15–30 main lateral nerves on each side of midrib, slightly prominent beneath ; flowers in lateral panicles up to 70 cm. long, all parts rusty-tomentellous ; corolla 1·5–2 cm. long, stout ; capsules subglobose or pyriform about 3 cm. diam., puberulous, glabrescent .. .. 1. *africanus*
Leaves pinnately (1–) 3 (–5)-foliolate ; petioles 6–16 cm. long, rhachides up to 10 cm. long, glabrous or sparingly puberulous ; petiolules often twisted, about 5 mm. long ; leaflets very broadly elliptic, very shortly cuneate at base, obtusely caudate acuminate, 10–25 cm. long, 8–14 cm. broad, thin, glabrous, with 3–4 looped main lateral nerves on each side of midrib, prominent beneath ; flowers in axillary cymes up to 2 cm. long ; corolla 0·8–1·4 cm. long, slender ; fruits not known .. .. 2. *mannii*

1. **T. africanus** *(Welw. ex C.DC.) Pellegr.* in Notulae Syst. 2 : 16 & 68 (1911) ; Chev. Bot. 111 ; Aubrév. Fl. For. C. Iv. 2 : 126, t. 176 ; Harms in E. & P. Pflanzenfam. 19B, 1 : 149 ; Pellegr. in op. cit. 9 : 8 ; Staner in Bull. Jard. Bot. Brux. 16 : 210. *Guarea africana* Welw. ex C. DC. (1878). *Bingeria africana* (Welw. ex C. DC.) A. Chev. (1909). *Turraeanthus* Harms (1896)—F.W.T.A., ed. 1, 1 : 496. *T. vignei* Hutch. & Dalz. F.W.T.A., ed. 1, 1 : 496 (1928) ; Kew Bull. 1929 : 25. A forest tree, 40–110 ft. high, with heavy dark green foliage ; bark whitish or yellowish, fissured ; slash pale yellow, strongly scented ; flowers white or yellowish ; fruit a yellow fleshy capsule with 4–5 seeds ; often gregarious, especially in moist places.
S.L.: Kambui Hills South *Small* 856 ! **Iv.C.:** Aboisso (Apr.) *Chev.* 16298 ! Abidjan *Aubrév.* 30 ! Koumasi Alepe *Krukoff* 56 ! 62 ! Azaguié *Chev.* 22300. Yapo *Chev.* 22309. **G.C.:** Tano R. *Thompson* 8 ! Akotiase, Cape Coast (fr. Mar.–Sept.) *Vigne* FH 933 ! Nfoum, Cape Coast (Apr., May) *Vigne* FH 948 ! Konongo, Ashanti (Apr.) *Vigne* FH 1081 ! Begoro (Nov.) *Vigne* FH 4266 ! **S.Nig.:** Boshi, Okwango F.R., Obudu Div. (Apr.) *Latilo* FHI 30908 ! Cross River North F.R., Ikom Div. (June) *Latilo* FHI 31830 ! Oban *Talbot* 1547 ! **Br.Cam.:** Buea (Feb.) *Dalz.* 8209 ! *Deistel* 129 ! *Maitland* 129 ! 231 ! 1351 ! Also in French Cameroons, Gabon, Belgian Congo, Angola and (?) Uganda. (See Appendix, p. 330.)
2. **T. mannii** *Baill.* in Adansonia 11 : 261 (1874) ; Harms l.c. A forest shrub or small tree, 8–10 ft. high ; branchlets glabrous, but roughly lenticellate.
S.Nig.: Calabar R. (Feb.) *Mann* 2304 ! Oban *Talbot* 1352 ! **Br.Cam.:** Korup F.R., Kumba Div. (young fr. June) *Olorunfemi* FHI 30639 !

11. TURRAEA Linn.—F.T.A. 1 : 330 ; Harms in E. & P. Pflanzenfam. 19B, 1 : 85 (1940).

Staminal tube 2·5–2·8 cm. long, anthers exserted, 1 mm. long, lobes much shorter ; petals about 3·5 cm. long and 2·5 mm. broad ; calyx with obtuse lobes about 2 mm. long ; ovary 5-celled ; flowers several together in small racemes 1–1·5 cm. long ; leaves elliptic to ovate-elliptic, often sinuate-lobate, cuneate to rounded at base, acuminate, very thin, up to 12 cm. long and 3·5 cm. broad ..   ..   1. *leonensis*

Staminal tube 0·9–1·4 cm. long ; petals 1·3–2·3 cm. long ; flowers in fascicles or pseudo-umbels or solitary or paired :

Anthers inserted at the top of the staminal tube, 1·5–2·5 mm. long, directly visible from the outside, not hidden by the linear or filiform alternating lobes ; flowers not in subsessile fascicles ; ovary 5–10 (or more)-celled ; main lateral nerves 4–6 on each side of midrib ; leaves sometimes lobate :

Flowers up to 15 together in pseudo-umbels on peduncles (0·9–) 2·5–4·5 (–6·5) cm. long ; style-apex clavate below stigma ; staminal tube 1·2–1·4 cm. long ; petals spathulate with long slender claw, 1·8–2·3 cm. long ; leaves broadly elliptic or ovate-elliptic, sometimes lobate, rounded to cuneate at base, acuminate, 4–16 cm. long, 2–7·3 cm. broad ..   ..2. *vogelii*

Flowers solitary or paired, on peduncles up to 0·3 cm. long ; style-apex ellipsoid below stigma ; staminal tube 0·9–1·3 cm. long, petals oblanceolate, 1·4–2·3 cm. long ; leaves more or less oblong-obovate to rhomboid, usually lobate, rounded to cuneate at base, shortly and obtusely acuminate, 2·5–9 cm. long, 1·3–5 cm. broad

3. *heterophylla*

Anthers inserted within the broad apical lobes of the staminal tube, about 1 mm. long ; staminal tube 1 cm. long ; petals 1·3–1·5 cm. long ; flowers up to 20 together in subsessile fascicles ; ovary 5-celled ; stigma ovoid-ellipsoid ; leaves elliptic or obovate-elliptic, cuneate at base, acuminate, 5–15 cm. long, 1·3–6 cm. broad, with 8–10 main lateral nerves ..   ..   4. *pellegriniana*

1. **T. leonensis** *Keay* in Bull. Jard. Bot. Brux. 26 : 188 (1956). *T. heterophylla* of F.W.T.A., ed. 1, 1 : 496, partly. A small tree, 12–15 ft. high.
    S.L.: *ex Hb. Thomas Moore* ! Hangha (Apr.) *Lane-Poole* 272 !

2. **T. vogelii** *Hook. f. ex Benth.* in Hook. Fl. Nigrit. 253 (1849) ; F.T.A. 1 : 330 (incl. var.) ; Harms in E. & P. Pflanzenfam. 19B, 1 : 89, fig. 20, A–C ; Staner in Bull. Jard. Bot. Brux. 16 : 133, fig. 6. *T. propinqua* Hook. f. ex Benth. (1849). *T. heterophylla* of F.W.T.A., ed. 1, 1 : 496, partly. A scandent shrub or woody climber, 10–15 ft. high ; calyx green ; petals and staminal tube white, turning creamy ; anthers yellow ; fruits opening to expose the black seeds with red arils ; in forest margins, etc.
    G.C.: Kwahu Prasu (June) *Vigne* FH 1755 ! Bunsu (June) *Hughes* 1157 ! **S.Nig.:** Lagos (June) *Dodd* 421 ! *Foster* 106 ! Ajilite *Millen* 173 ! Owenna, Ondo Div. (Apr.) *Symington* FH 3381 ! Aguku *Thomas* 1015 ! Afi River F.R., Ikom (May) *Latilo* FHI 30992 ! Calabar *Milne* ! Oban *Talbot* 1309 ! **Br.Cam.:** Victoria (Jan.) *Maitland* 269 ! Buea *Deistel* 586 ! Johann-Albrechtshöhe *Staudt* 478 ! **F.Po:** near the sea (Nov., Dec.) *Mann* 66 ! *T. Vogel* 65 ! Also in French Cameroons, Gabon, Belgian Congo, A.-E. Sudan, Uganda, Angola and on Principe and S. Tomé. (See Appendix, p. 330.)

3. **T. heterophylla** *Sm.* in Rees Cyclop. 36 (1819) ; F.T.A. 1 : 330 ; Chev. Bot. 112. *T. lobata* Lindl. in Bot. Reg. 1843, Misc. 61 (1843) ; op. cit. 1844, t. 4. (?) *T. quercifolia* G. Don (1831), type not found. *T. gracilis* A. Chev. Bot. 111, name only, *fide* Hutch. & Dalz. An evergreen shrub, 3–6 ft. high with white flowers.
    S.L.: *Afzelius* ! *Whitfield* ! Murray Town (May) *Deighton* 5932 ! King Tom (fr. Oct.) *Deighton* 199 ! Mahela, Samu (fr. Dec.) *Sc. Elliot* 4029 ! **Iv.C.:** *fide* Chev. *l.c.* **G.C.:** Cape Coast *Brass* ! Jimira F.R. (June) *Vigne* FH 2965 ! Benso, Ashanti (Mar.) *Vigne* FH 1856 ! Abokobi (Oct.) *Vigne* FH 4258 ! Aburi Road (Feb.) *Dalz.* 8271 ! Achimota (fl. Mar., young fr. Nov.) *Irvine* 359 ! *Milne-Redhead* 51301 ! **Togo:** Lome *Warnecke* 205 ! Santrokofi, Hohoe (fr. Oct.) *St. C. Thompson* FH 3599 ! **Dah.:** *Poisson* 105. (See Appendix, p. 330.)

4. **T. pellegriniana** *Keay* l.c. 188, fig. 59 (1956). *T. tisseranti* Pellegr. in Notulae Syst. 9 : 10 (1940), French descr. only. A shrub or small tree, to 25 ft. high ; calyx and corolla pale green ; staminal tube cream ; flowers sweetly scented ; fruits orange.
    N.Nig.: Jemaa (fl. & fr. Feb.) *McClintock* 190 ! Bauchi Plateau (fl. & fr. Jan.) *Lely* P. 107 ! **S.Nig.:** Ijua, Obudu Div. (Dec.) *Keay & Savory* FHI 25029 ! Also in French Cameroons.

12. AZADIRACHTA A. Juss. in Mém. Mus. Hist. Nat. Paris 19 : 220 (1830) ; Harms in E. & P. Pflanzenfam. 19B, 1 : 102 (1940).

Leaves pinnate ; leaflets 5–8 pairs, ovate-lanceolate to lanceolate, falcate, very asymmetrical at base, long-acuminate, coarsely serrate on margins and occasionally lobed, up to about 11 cm. long and 3 cm. broad, glabrous ; panicles many-flowered, axillary, glabrous ; pedicels about 1·5 mm. long ; sepals ovate-suborbicular, about 1 mm. long ; petals oblanceolate, 5–6 mm. long ; staminal column about 4·5 mm. long, anthers within the lobed apex ; fruits ellipsoid, drupaceous, 1-seeded, glabrous, 1·2–2 cm. long ..   ..   *indica*

A. indica *A. Juss.* l.c. 221, t. 13, 5 (1830) ; Kennedy For. Fl. S. Nig. 182 ; Harms l.c., fig. 26, M–S. *Melia azadirachta* Linn. (1753). *M. indica* (A. Juss.) Brandis (1874). An evergreen tree, to 80 ft. high, with abundant panicles of white flowers, and yellow fruits ; introduced for shade and firewood, and known as " Neem."
    A native of India, now widely distributed and often naturalized in tropical and subtropical countries.

13. MELIA Linn.—F.T.A. 1 : 332 ; Harms in E. & P. Pflanzenfam. 19B, 1 : 99 (1940).

Leaves bipinnate ; leaflets 4–6 pairs, ovate-lanceolate, unequal-sided at base, long-acuminate, 3–8 cm. long, 1–3 cm. broad, serrate, glabrous or nearly so ; panicles subcorymbose, nearly as long as the leaves ; pedicels up to 1 cm. long, slightly pubescent ; sepals elliptic, pubescent, 2·5 mm. long ; petals contorted, oblanceolate, about 8 mm. long, thinly pubescent ; staminal column 7–8 mm. long, slender, glabrous, anthers within the apex ; fruit ellipsoid, drupaceous, usually 5-celled, 1–1·5 cm. long   ..   ..   *azedarach*

M. azedarach *Linn.*—F.T.A. 1 : 332 ; Harms l.c., fig. 26, A–L. A small tree, with abundant panicles of lilac flowers ; introduced for ornament and known as Persian Lilac.
A native of India, now widely distributed and often naturalized in tropical and subtropical countries. (See Appendix, p. 327.)

119. SAPINDACEAE

Trees, shrubs or climbers. Leaves alternate or rarely opposite, simple, or 1- or 3-foliolate, or pinnate or bipinnate ; stipules rarely present. Flowers actinomorphic or zygomorphic, sometimes very small, mostly unisexual, variously arranged. Sepals imbricate or rarely valvate. Petals 1–5, sometimes absent, imbricate. Disk usually present, sometimes unilateral. Stamens hypogynous, often 8, inserted within the disk or unilateral ; filaments free, often hairy ; anthers 2-celled. Ovary superior, entire or vertically lobed to the base, 1–8-celled ; style terminal or gynobasic ; ovules 1–2, rarely many in each cell, axile. Fruit capsular or indehiscent, rarely winged. Seeds without endosperm, often conspicuously arillate ; embryo with usually plicate or twisted cotyledons.

A fairly large and mostly tropical family, with reduced unisexual flowers and often pinnate leaves ; no endosperm.

Leaves imparipinnate, or biternate, or 1- or 3-foliolate, or simple :
  Tendrils present, circinately coiled, borne on the peduncles ; climbing herbs and
    shrubs ; calyx 4-merous (in our species) ; petals 4 ; stamens 8 ; ovary 3-celled ;
    ovules 1 per cell :
    Fruits firm, not bladder-like ; leaves imparipinnate ; woody plants   1. **Paullinia**
    Fruits bladder-like and membranous ; leaves biternate ; leaflets coarsely toothed
      or lobed ; herbaceous climbers   ..   ..   2. **Cardiospermum**
  Tendrils absent ; trees or shrubs ; ovary 2–3-celled ; calyx 4-merous (in our species):
    Fruits apocarpous, usually only 1 coccus developing, not winged, indehiscent ; ovules
      1 per cell ; petals 4 ; stamens 8 ; leaves 1- or 3-foliolate..   ..   3. **Allophylus**
    Fruits syncarpous, membranous, winged, tardily dehiscent ; ovules 2 per cell ;
      petals absent ; stamens 5–8 ; leaves simple   ..   ..   ..   16. **Dodonaea**
Leaves paripinnate ; tendrils absent :
  Fruits indehiscent :
    Petals (3–) 4–5 ; calyx (4–) 5-merous ; ovules 1 per cell :
      Fruits with a basal style, mostly only 1 lobe fertile ; flowers regular ; petals usually
        5 ; ovary 2–3-lobed :
        Leaflets 2–10 or more pairs ; ovary apocarpous ; stamens (8–) 12–30
                                                                    4. **Deinbollia**
        Leaflets 1–2 pairs ; ovary syncarpous ; stamens 6–8   ..   ..   5. **Aphania**
      Fruits with a terminal style, lobed or ribbed; flowers zygomorphic; petals usually 4 :
        Calyx urceolate ; stamens 7–15 ; ovary 3–8-celled ; fruits 3–8-ribbed ; inflores-
          cences borne on the thick main stem, sometimes close to the ground
                                                                    6. **Chytranthus**
        Calyx campanulate ; stamens (6–) 8 ; ovary 3-celled ; fruits 3-lobed ; inflores-
          cences borne on slender branches ..   ..   7. **Pancovia**
    Petals absent ; flowers regular :
      Stamens (6–) 8–13 ; ovules 1 per cell :
        Calyx 5-merous ; stamens 8–13 ; ovary 3-celled :
          Stamens 8 ; fruits 1–3-lobed ; seed without aril ..   ..   ..   8. **Placodiscus**
          Stamens (8–) 10–13 ; fruits normally 1-celled by abortion ; seed with a glutinous
            aril   ..   ..   10. **Lecaniodiscus**
        Calyx usually 4-merous ; stamens 6–8 ; ovary 2-celled ; fruits 1–2-lobed ; seed
          without aril ..   ..   9. **Melanodiscus**
      Stamens (3–) 4–5 ; ovules 2 per cell ; calyx 4–5-merous ; ovary 2-celled ; fruits
        normally 1-celled and 1-ovuled by abortion   ..   ..   ..   17. **Zanha**
  Fruits dehiscent ; calyx (4–) 5-merous :
    Petals (4–) 5 ; ovules 1 per cell ; seed with fleshy aril :
      Ovary 3-celled ; stamens 7–18 :
        Calyx with free sepals, imbricate in 2 series ; stamens 7–10   11. **Laccodiscus**
        Calyx with slightly imbricate sepals, or small and toothed :
          Stamens 10–18 ; seed completely covered by the red or orange aril
                                                                    13. **Lychnodiscus**

27

Stamens 8–10 ; aril cupular or short, not surrounding the seed :

Petals funnel-shaped through the concrescence of the scales ; fruits more or less pyriform, valves not woody ; seed shining black with yellow cupular aril
         14. **Blighia**

Petals with free scales ; fruits subglobose, valves woody   15. **Eriocoelum**

Ovary 2-celled ; fruits 1–2-lobed ; seed with cupular aril ; petals each with 2 scales ; stamens 6–8   ..   .. 12. **Aporrhiza**

Petals absent, or 1–2, without scales ; ovary 3-celled ; ovules 2 per cell ; fruits (2–) 3-lobed ; seed blue, pilose by the hilum but without a fleshy aril ; stamens 7–8
         18. **Majidea**

Besides the above, other genera are represented by the following introduced plants :—*Sapindus saponaria* Linn. a native of tropical America ; *S. trifoliatus* Linn., a native of tropical Asia ; *Litchi chinensis* Sonnerat, a native of the Far East ; *Melicocca bijuga* Linn., a native of tropical America ; *Schleichera trijuga* Willd., a native of tropical Asia ; *Harpullia pendula* Planch. ex F. Müll., a native of Australia.

Talbot 720 from Oban, S. Nigeria, appears to belong to this family. It is a woody plant with simple oblanceolate leaves, 33–36 cm. long and 12–13·5 cm. broad, with about 20 main lateral nerves on each side of the puberulous midrib. The brown-tomentellous axillary racemes are about 4 cm. long. A painting of the living plant, done by Mrs. Talbot, shows flowers with white petals, reminiscent of *Pancovia*. The fruits also resemble *Pancovia*. More material of this remarkable plant is needed.

1. PAULLINIA Linn.—F.T.A. 1 : 419 ; Radlk. in Engl. Pflanzenr. Sapindac. 219 (1931).

Branchlets softly pubescent to nearly glabrous, ribbed ; leaves imparipinnate ; rhachis conspicuously winged, wings leafy ; leaflets 2 pairs and an odd one, opposite, oblong to obovate, rather coarsely and remotely toothed, glabrous to pubescent beneath and with hairy pits in the axils of the nerves ; flowers racemose ; peduncle axillary, stout, with one or two spirally coiled tendrils at the apex, with a short raceme of clustered small flowers beyond ; capsule woody, turbinate, slightly 3-lobed at the top, shortly beaked, about 3 cm. long ; septa spongy   ..   ..   .. *pinnata*

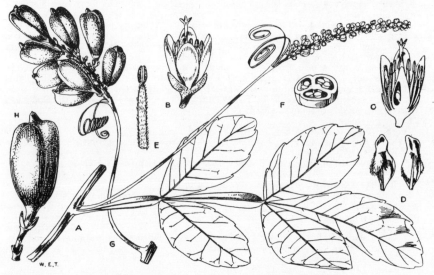

Fig. 196.—Paullinia pinnata *Linn.* (Sapindaceae).

A, portion of shoot.   B, flower.   C, same in vertical section.   D, disk glands.   E, stamen. F, ovary in cross-section.   G, H, fruits.

P. pinnata *Linn.*—F.T.A. 1 : 419 ; Chev. Bot. 148 ; Radlk. l.c. 247 ; Schnell in Ic. Pl. Afr. 36. A woody or subwoody climber with tendrils, small white flowers and scarlet capsular more or less 3-angled fruit ; usually in regrowth vegetation in the forest regions, and in streamside vegetation in the savannah regions. **Sen.:** *Heudelot* 18 ! *Sieber* 26 ! Ziguinchor *Chev.* 3032. **Gam.:** *Hayes* 552 ! *Skues* ! **Fr.Sud.:** Kangaba, Tabacoroni & Samandini *fide* Chev. *l.c.* **Port.G.:** Pussubé *Esp. Santo* 895. **Fr.G.:** Los Isl. (May) *Pobéguin* 1195 ! Conakry (Apr.) *Debeaux* 325 ! Farana (Apr.) *Chev.* 13319 ! Massadou, S. of Macenta (May) *Collenette* 20 ! **S.L.:** Rotomba Isl. (Mar.) *Kirk* 18 ! Njala (May) *Deighton* 686 ! Kambia, Scarcies (Dec.) *Sc. Elliot* 4209 ! Falaba-Musaia Road (fr. Sept.) *Small* 328 ! **Lib.:** road to White Plains (fl. & fr. Apr.) *Barker* 1274 ! Gletown, Tchien (fr. July) *Baldwin* 6957 ! Gbau, Sanokwele (fr. Sept.) *Baldwin* 9446 ! **Iv.C.:** Danané (June) *Collenette* 43 ! Bingerville, Nzi region & Morénou *fide* Chev. *l.c.* **G.C.:** Cape Coast (fl. & fr. July) *T. Vogel* 6 ! 16 ! Kumasi (fl. & fr. Feb.) *Irvine* 95 ! Aburi Hills (fl. & fr. Dec.) *Johnson* 438 ! Kumbungu (May) *Kitson* 954 ! **Togo:** Kpedsu (fl. & fr. Dec.) *Howes* 1047 ! Ho (Mar.) *Williams* 57 ! Lome *Warnecke* 69 ! **Dah.:** *Poisson* ! Kétou *Chev.* 23020. **N.Nig.:** Nupe *Barter* 1187 ! Lokoja (Mar.) *Shaw* 15 ! Jemaa (Feb.) *McClintock* 182 ! Bauchi Plateau (Jan., June) *Lely* 330 ! P. 104 ! **S.Nig.:** Lagos

Dawodu 7! Benin City *Dennett* 100! Oban *Talbot* 1405! **Br.Cam.**: Ambas Bay (Feb., Apr.) *Mann* 13! Victoria (July) *Maitland* 29! Johann-Albrechtshöhe *Staudt* 487! **F.Po**: on the shore (Nov.) *T. Vogel* 244! Widespread throughout tropical Africa, except in very dry regions ; also in Madagascar and tropical America. (See Appendix, p. 334.)

2. CARDIOSPERMUM Linn.—F.T.A. 1 : 417 ; Radlk. in Engl. Pflanzenr. Sapindac. 370 (1932).

Flowers 8–10 mm. long ; posterior disk-glands elongate ; fruits up to 6·5 cm. long and 3·5 cm. broad ; stems often, but not always, with long spreading setose hairs ; leaflets more or less curly-pubescent beneath, especially in the axils of the main lateral nerves   ..   ..   1. *grandiflorum*

Flowers about 3 mm. long ; disk-glands small ; fruits 1–4 cm. long and broad, often truncate at the top and more or less trigonous ; stems without long spreading setose hairs ; leaflets rather stiffly puberulous or shortly pubescent beneath, not hairy in the axils of the main lateral nerves   ..   ..   2. *halicacabum*

1. **C. grandiflorum** *Swartz* Prod. 64 (1788) ; Radlk. in Engl. Pflanzenr. Sapindac. 372. *C. barbicaule* Bak. in F.T.A. 1 : 418 (1868). *C. caillei* A. Chev. Bot. 148, name only. (?) *C. halicacabum* var. *grandiflorum* A. Chev. Bot. 149. Herbaceous climber with creamy-white fragrant flowers, a pair of tendrils at the apex of the inflorescence and more or less 3-angled bladdery fruits. Plants with the stems setose-pilose are called, by Radlkofer, forma *hirsutum* (Willd.) Radlk. ; those specimens from our area whose stems lack the long setose hairs are Radlkofer's forma *elegans* (Kunth) Radlk.
   Fr.G.: Kouria *Caille* in *Hb. Chev.* 14959! Macenta (Oct.) *Baldwin* 9854! Follo (Dec.) *Caille* in *Hb. Chev.* 14825! **S.L.:** Panguma *Smythe* 92! Kukuna, Scarcies (fl. & fr. Jan.) *Sc. Elliot* 4680! Hangha (fr. Jan.) *Thomas* 7773! Makump (Dec.) *Glanville* 118! **Lib.:** Ganta (fl. & fr. Dec.) *Barker* 1146! Zigida, Vonjama (Oct.) *Baldwin* 10009! **Iv.C.:** Bingerville (Dec.) *Chev.* 16027. Sassandra Port (May) *Chev.* 17955. **G.C.:** Kumasi (fl. & fr. Jan.) *Cummins* 75! *Irvine* 87! Boankra (fl. & fr. May) *Chipp* 434! Abuantem (Sept.) *Andoh* FH 5354! Sekodumase (fl. & fr. Jan.) *Kitson*! Aquapim *Thonning*. **Togo:** Kpandu *Robertson* 49! **Dah.:** Le Testu 465! **N.Nig.:** Lokoja (fl. & fr. Oct., Nov.) *Dalz.* 82! *Parsons* 63! **S.Nig.:** Lagos *Dawodu* 78! *Punch*! Abeokuta *Rowland*! Ogbomosho *Rowland*! Ogwashi (Dec.) *Thomas* 2063! Oban *Talbot* 1363! **Br.Cam.:** Victoria (fr. Apr.) *Dundas* FHI 20502! Cam. Mt., 2,000 ft. (Jan.) *Mann* 1299! Extends to A.-E. Sudan, Uganda and Angola ; also in tropical America. (See Appendix, p. 332.)
2. **C. halicacabum** *Linn.*—F.T.A. 1 : 417 ; Chev. Bot. 149 ; Radlk. l.c. 379, fig. 8, A–C ; Schnell in Ic. Pl. Afr. t. 13. *C. microcarpum* Kunth (1821)—F.T.A. 1 : 418. A slender climber like the last, but with much smaller flowers ; extending into drier regions. In var. *microcarpum* (Kunth) Blume (see Radlk. l.c. 387) the fruits are 1–2 cm. long, but as so many specimens lack mature fruits the varieties are not separated in the following citation.
   Sen.: Walo *Heudelot*! Richard Tol (Mar., Apr.) *Perrottet* 103! M'bidjem (Sept.) *Thierry* 16! **Port.G.:** Formosa (Apr.) *Esp. Santo* 1956! **Fr.Sud.:** Kabarah (fl. July, Sept., fr. Sept.) *Chev.* 1312! *Hagerup* 297! Sicoro *Chev.* 223! Tambacoroni (Apr.) *Chev.* 720! **Fr.G.:** Timbo (Nov.) *Pobéguin* 1468! **S.L.:** Don 34! Regent (fl. & fr. Dec.) *Sc. Elliot* 4101! Njala (fl. & fr. Aug.) *Deighton* 3237! Pujehun (fl. & fr. Feb.) *Thomas* 8273! Giema (Nov.) *Deighton* 403! **Lib.:** Kassa Ta (fl. & fr. Sept.) *Linder* 833! **G.C.:** Bjury (fl. & fr. July) *Chipp* 509! **Togo:** Lome *Warnecke* 219! **N.Nig.:** Aguji, Ilorin *Thornton*! Bauchi Plateau (fl. & fr. July) *Lely* P. 386! Panyam (fl. & fr. July) *Lely* 436! Bornu (fl. & fr. Apr.) *E. Vogel* 80! **S.Nig.:** Lagos *Rowland*! Ibadan (fl. & fr. July, Aug.) *Onochie* FHI 7544! Ibo country (fl. & fr. Aug.) *T. Vogel* 35! Ibuzo (fl. & fr. Nov.) *Thomas* 1992! 1993! Calabar *Robb*! **Br.Cam.:** Gwoza Hills, Dikwa Div. (Jan.) *McClintock* 153! **F.Po:** *Mann* 406! Widespread in tropical and subtropical regions. (See Appendix, p. 332.)

The infraspecific variation in the above two species requires study in the field and by experiment.

3. ALLOPHYLUS Linn. Sp. Pl. 1 : 348 (1753) ; Radlk. in Engl. Pflanzenr. Sapindac. 455 (1932). *Schmidelia* Linn. (1767)—F.T.A. 1 : 420.

Leaves 3-foliolate ; petioles usually more than 3·5 cm. long :
  Inflorescence with several well-developed lateral branches (but see No. 2d), mostly longer than the leaves :
    Leaflets toothed, pubescent or pilose beneath :
      Leaflets bullate, sparsely puberulous beneath except for conspicuous tufts of hairs in the axils of veins as well as along the midrib ; a tree of montane forest ; central leaflets obovate, cuneate at base, rounded or very shortly acuminate, 8–19 cm. long, 3·7–9·5 cm. broad   ..   ..   1. *bullatus*
      Leaflets not bullate, undersurface densely or sparsely hairy, without tufts in the axils of the veins, but sometimes with tufts along the midrib ; shrubs and small trees, not in montane forest :
        Central leaflets up to 16 cm. long and 7·5 cm. broad ; petioles up to 6 cm. long ; flowers borne singly or in sessile fascicles along the inflorescence-axes :
          Branchlets, petioles and undersurface of leaflets long-spreading pilose
                    2c. *africanus* forma *chrysothrix*
          Branchlets, petioles and undersurface of leaflets pubescent or puberulous :
            Undersurface of leaflets very densely pubescent
                    2b. *africanus* forma *subvelutinus*
            Undersurface of leaflets sparingly pubescent or puberulous or with tufts along the midrib :
              Venation of leaflets prominently reticulate beneath ; inflorescences well-branched   ..   ..   2a. *africanus* forma *africanus*
              Venation of leaflets obscure beneath ; leaflets with conspicuous tufts along the midrib beneath, otherwise nearly glabrous ; inflorescences sparingly branched or some simple ; leaflets up to 7 cm. long   2d. *africanus* forma *senegalensis*

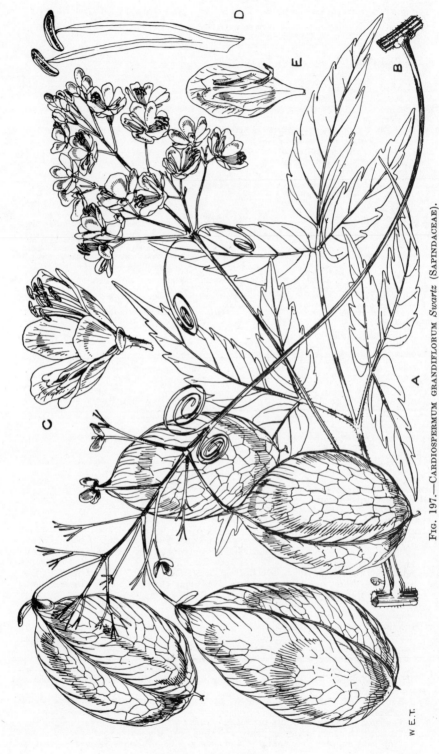

W E.T.

Fig. 197.—CARDIOSPERMUM GRANDIFLORUM *Swartz* (SAPINDACEAE).

A, inflorescence and leaf.   B, fruits.   C, flower.   D, stamens.   E, petal with adnate scale.

712

Central leaflets 18–38 cm. long, 8·5–16 cm. broad ; petioles 8–16 cm. long, stout, semiterete and slightly winged above ; flowers borne in distinctly pedunculate lateral cymules along the axes of the inflorescence which is shorter than the leaves ; central leaflets more or less obovate, long-cuneate at base, acuminate, lateral leaflets rather strongly asymmetrical ; branchlets, petioles and undersurface of leaflets sparingly puberulous or glabrescent ..   3. *grandifolius*

Leaflets entire, glabrous, membranous, obovate, shortly cuneate at base, obtusely acuminate, central ones 7–10·5 cm. long, 3–5·3 cm. broad ; panicles longer than the leaves, slender, sparingly puberulous, 10–20 cm. long ..   4. *talbotii*

Inflorescence simple or subsimple, or with short lateral cymules ; leaflets toothed, hairy :

Flowers borne singly or in fascicles directly on the main axis of the inflorescence ; branchlets, petioles and undersurface of leaflets spreading-pilose with brown hairs ; central leaflets 4·5–11 cm. long and 2·5–5·5 cm. broad ; inflorescences simple or subsimple, 17 cm. long ..   ..   5. *spicatus*

Flowers borne in distinctly pedunculate lateral cymules along the main axis of the inflorescence :

Branchlets, petioles and undersurface of leaflets softly pubescent, sometimes glabrescent ; inflorescences up to 12 cm. long, usually longer than the petioles, lateral cymules short ; central leaflets 7–21 cm. long, 3–8 cm. broad
            6. *welwitschii*

Branchlets, petioles and undersurface of leaflets spreading-setose ; inflorescences up to 9·5 cm. long, usually shorter than the petioles, lateral cymules up to 3 cm. long ; central leaflets 11–18 cm. long, 4–9 cm. broad ..   ..   7. *conraui*

Leaves 1-foliolate, toothed ; inflorescence with an unbranched main axis bearing numerous distinctly pedunculate cymules ; petioles 1–3·5 cm. long :

Leaves cuneate at base, 18–45 cm. long, 6·5–15 cm. broad, oblong-oblanceolate, margins sinuate-dentate in upper half, long-acuminate, with 12–15 main lateral nerves on each side of midrib ; inflorescences 2·5–8 cm. long ; branchlets and petioles densely pubescent ; leaves pubescent mainly on midrib beneath ..   8. *megaphyllus*

Leaves rounded, obtuse or rarely some subcuneate at base, 9–22 cm. long, 3·5–10 cm. broad, acuminate ; inflorescences 2–3 cm. long :

Branchlets, petioles and both sides of leaves densely spreading setose ; leaves oblong-oblanceolate or oblong-lanceolate, up to 22 cm. long and 7 cm. broad, with 12–16 main lateral nerves on each side of midrib ..   ..   9. *nigericus*

Branchlets and petioles pubescent ; leaves pubescent mainly on and near midrib beneath, elongate-elliptic or more or less obovate, 9–20 cm. long, 3·5–10 cm. broad, with 10–14 main lateral nerves on each side of midrib ..   ..   10. *hirtellus*

1. **A. bullatus** *Radlk.* in Sitzb. Math.-Phys. Akad. Münch. 38 : 223 (1908) ; Pflanzenr. 534. A tree 30–50 ft. high, with erect inflorescences, terminal and in the upper leaf-axils ; in montane forest.
  Br.Cam.: Cam. Mt., 3,200–7,600 ft. (fl. Apr., May, fr. Dec., Feb.) *Mann* 1184 ! 2167 ! *Maitland* 685 ! 690 ! 1089 ! 1317 ! Bamenda *Lightbody* ! L. Oku, 7,000 ft., Bamenda (fr. Jan.) *Keay* FHI 28487 !

2. **A. africanus** *P. Beauv.* Fl. Oware 2 : 75, t. 107 (1819) ; Radlk. in Pflanzenr. 536, fig. 12 ; Aubrév. Fl. For. C. Iv. 2 : 182, t. 206, 7–12 ; Fl. For. Soud.-Guin. 390, t. 82, 3. *Schmidelia africana* (P. Beauv.) DC. (1824)—F.T.A. 1 : 421 ; Chev. Bot. 149. Usually shrubs, sometimes small trees to 25 ft. high ; flowers creamy-white in almost catkin-like racemes, sweet-scented ; ripe fruits red ; especially in forest regrowth and margins, and in protected places in the savannah regions. (See Appendix, p. 330.)
  A variable species, widespread in tropical Africa, which requires intensive and extensive experimental study. Of the forms described by Radlkofer the following seem worth maintaining, for the moment, in our area, but some specimens are intermediate :

2a. **A. africanus** forma **africanus**—Some of the specimens from drier regions are more hairy and approach the forma *subvelutinus*.
  Gam.: Genieri (July) *Fox* 165 ! **Port.G.:** Pirada, Gabu (June) *Esp. Santo* 3029 ! **Fr.G.:** Kouroussa (June) *Pobéguin* 265 ! Seredou (May) *Collenette* 30 ! Timbo (July) *Pobéguin* 743 ! **S.L.:** *Afzelius* ! Freetown (July) *Deighton* 1 ! Njala (fl. July, fr. Sept.) *Deighton* 1313 ! 1999 ! *Small* 181 ! Batkanu (July) *Small* 147 ! Rowala (fr. July) *Thomas* 1171 ! **Lib.:** Grand Bassa (July) *T. Vogel* 36 ! Sanokwele (fr. Sept.) *Baldwin* 9518 ! Bonja Town (June) *Barker* 1352 ! **Iv.C.:** Banco *Martineau* 291 ! **G.C.:** Cape Coast *Brass* ! Aburi Hills (fl. & fr. Oct.) *Johnson* 478 ! Akropong (Aug.) *Irvine* 794 ! Mampong, Ashanti (July) *Adjei* FH 1257 ! **Togo:** Misahöhe *Baumann* 314 ! **Dah.:** *fide* Aubrév. l.c. **N.Nig.:** Lokoja (Aug.) *Barter* 524 ! *Elliott* 191 ! Mongu (July) *Lely* 389 ! Yola (July) *Dalz.* 171 ! **S.Nig.:** Abeokuta *Irving* ! Nun R. (Sept.) *Mann* 487 ! Ubiaja (Aug.) *Oyebade* FHI 20401 ! Ibo country *T. Vogel* 16 ! Oban *Talbot* 1365 ! **Br.Cam.:** Victoria (Oct.) *Maitland* 748 ! Bambuko F.R., Kumba Div. (young fr. Sept.) *Olorunfemi* FHI 30766 !

2b. **A. africanus** forma **subvelutinus** *Radlk.* l.c. 538 (1932). In savannah regions ; several specimens at Kew are intermediate between this and forma *africanus*.
  S.L.: Musaia (fr. Oct.) *Thomas* 2648 ! **G.C.:** Bosomoa F.R. (Aug.) *Talmie* FH 4421 ! **N.Nig.:** Loin, Nupe *Barter* 1095 ! Bauchi Plateau (Apr.) *Lely* P. 220 ! Vodni, 4,600 ft. (July) *Lely* 428 !

2c. **A. africanus** forma **chrysothrix** *Radlk.* l.c. 538 (1932). *A. spicatus* of F.W.T.A., ed. 1, 1 : 500, partly. *Schmidelia touraca* A. Chev. Bot. 150, name only. *Allophylus touracus* Pellegr. in Bull. Soc. Bot. Fr. 100 : 191, name only. Usually scandent, covered with stiff brown hairs ; in forest regrowth.
  Fr.G.: *Maclaud* 174. **S.L.:** Berria (Mar.) *Sc. Elliot* 5411 ! **Iv.C.:** Digoualé to Gouané (May) *Chev.* 21510 ! Guidéko to Soubré *Chev.* 19084. **G.C.:** Nsuta (May) *Vigne* FH 1737 ! Aburi (June) *Brown* 318 ! *Howes* 1184 ! Okroase (May) *Johnson* 752 ! Boankra (May) *Chipp* 437 ! **S.Nig.:** Busogboro, Ibadan North F.R. (fl. June, fr. Sept.) *Idahosa* FHI 23870 ! *Jones* FHI 3613 ! Obanigba, Owo (May) *Jones* FHI 3613a !

2d. **A. africanus** forma **senegalensis** *Radlk.* l.c. 538 (1932).
  Sen.: *Roger* 15 ! *Sieber* 29 ! Walo *Heudelot* ! Richard Tol (Feb., Mar.) *Perrottet* 107 !

3. **A. grandifolius** (*Bak.*) *Radlk.* in E. & P. Pflanzenfam. 3, 5 : 313 (1896) ; Pflanzenr. 532 ; Aubrév. Fl. For. Soud.-Guin. 390. *Schmidelia grandifolia* Bak. in F.T.A. 1 : 421 (1868). *Allophylus megaphyllus* Hutch.

& Dalz. F.W.T.A., ed. 1, 1 : 500 (1928) and Kew Bull. 1929 : 25, partly (*Talbot* 414). A shrub or small tree ; in forest.

S.Nig.: Oban *Talbot* 414 ! **Br.Cam.:** Victoria to Busumbu (Aug.) *Preuss* 1326. Kebo (June) *Conrau* 214. Also in French Cameroons, Principe and S. Tomé.

4. **A. talbotii** *Bak. f.* in J. Bot. 57 : 186 (1919) ; Radlk. l.c. 550, 1485. *Schmidelia nuonensis* A. Chev. Bot. 150, name only. A liane ; in forest.
 S.L.: Jimmi, Gbagbo (Apr.) *Deighton* 1593 ! **Lib.:** Dukwia R. (Apr.) *Cooper* 405 ! Monrovia *Dinklage* 936 ; 2210. **Iv.C.:** Sampleu to Ganhoué, Upper Nuon Basin (Apr.) *Chev.* 21147 ! **S.Nig.:** Oban *Talbot* 1713 ! (See Appendix, p. 331.)

5. **A. spicatus** (*Poir.*) *Radlk.* in E. & P. Pflanzenfam. 3, 5 : 313 (1895) ; F.W.T.A., ed. 1, 1 : 500, partly ; Pflanzenr. 528 ; Aubrév. Fl. For. C. Iv. 2 : 182 ; Fl. For. Soud.-Guin. 390. *Ornitrophe spicata* Poir. (1808). *O. magica* Schum. & Thonn. (1827). *Schmidelia magica* (Schum. & Thonn.) Bak. in F.T.A. 1 : 423 (1868) ; Chev. Bot. 150. *Allophylus warneckei* Gilg ex Engl. (1913)—F.W.T.A., ed. 1, 1 : 500. A shrub 4–12 ft. high, with pendulous inflorescences ; calyx green, petals white ; ripe fruits red ; rocky areas and by streams, especially in the savannah regions.
 Iv.C.: Mankono *Chev.* 21910. Ady, Savé *Chev.* 23566. **G.C.:** *Thonning.* Achimota (July) *Irvine* 683 ! Pabarabo Hill, Afram Plains (May) *Kitson* 1107 ! Samdu *Kitson* 720 ! Navrongo (May) *Andoh* FH 5168 ! Gura (May) *Vigne* FH 1171 ! **Togo:** Lome (fl. May, fr. July) *Warnecke* 160 ! 376 ! Kete Krachi *Zech.* **Dah.:** Tanguéta (June) *Chev.* 24100 ! **Fr.Nig.:** Gourma Dist. *Chev.* 24315 ; 24476. **N.Nig.:** Nupe *Barter* 1648 ! Anara F.R., Zaria (May) *Keay* FHI 22954 ! Lokoja *Barter* 502 ! Zaranda Filani (May) *Lely* 187 ! Yola (fl. & fr. July) *Dalz.* 175 ! **S.Nig.:** Olokemeji (June) *Foster* 96 ! *Jones & Keay* FHI 18819 ! Hill N. of Owo F.R. (fr. July) *Onochie* FHI 33353 ! (See Appendix, p. 331.)

6. **A. welwitschii** *Gilg* in Engl. Bot. Jahrb. 24 : 287 (1897) ; Radlk. in Pflanzenr. 530. A shrub, often scandent ; flowers greenish-white ; in forest vegetation.
 S.Nig.: Appiapum to Obubra, Ogoja Prov. (Apr.) *Jones* FHI 12367 ! Also from French Cameroons to A.-E. Sudan, Uganda, Belgian Congo and Angola.

7. **A. conraui** *Gilg ex Radlk.* in Sitzb. Math.-Phys. Akad. Münch. 28 : 221 (1908) ; Pflanzenr. 529. A shrub, with white flowers ; in forest.
 Br.Cam.: Banyeme to Ayon, Kumba Div. (June) *Olorunfemi* FHI 30611 ! Bali-Ngemba F.R., Bamenda (fr. May) *Ujor* FHI 30334 ! Kebo (fl. & young fr. June) *Conrau* 219.
 [As Radlkofer points out, this species recalls *A. hirtellus*, except for the 3-foliolate leaves.]

8. **A. megaphyllus** *Hutch. & Dalz.* F.W.T.A., ed. 1, 1 : 500 (1928) and Kew Bull. 1929 : 25, excl. *Talbot* 414 ; Radlk. in Pflanzenr. 1485 ; Brenan in Kew Bull. 1952 : 449. A shrub, to 12 ft. high ; flowers white ; ripe fruits scarlet ; in forest.
 S.Nig.: Oban *Talbot* 1393 ! **Br.Cam.:** Banga, S. Bakundu F.R., Kumba Div. (fl. Aug., fr. Mar.) *Brenan* 9275 ! 9275*a* ! *Olorunfemi* FHI 30741 !

9. **A. nigericus** *Bak. f.* in J. Bot. 57 : 158 (1919) ; F.W.T.A., ed. 1, 1 : 500, partly (excl. syn. *hirtellus* var. *barteri*) ; Radlk. l.c. 549. A shrub, 2–6 ft. high ; flowers white ; in forest.
 S.Nig.: Camp 4, Calabar to Mamfe Road (Apr.) *Ejiofor* FHI 21872 ! Ukpon F.R., Obubra Div. (July) *Latilo* FHI 31873 ! Oban *Talbot* 442 ! 447 ! 448 !

10. **A. hirtellus** (*Hook. f.*) *Radlk.* in E. & P. Pflanzenfam. 3, 5 : 312 (1895) ; Pflanzenr. 516. *Schmidelia hirtella* Hook. f. in Fl. Nigrit. 248, t. 25 (1849) ; F.T.A. 1 : 424. *Allophylus hirtellus* var. *barteri* Bak. f. (1919). *A. nigericus* of F.W.T.A., ed. 1, 1 : 500, partly. A shrub, 3–20 ft. high ; flowers white ; in forest.
 S.Nig.: Angiama *Barter* 85 ! Oban *Talbot* 1392 ! **Br.Cam.:** Ambas Bay (Jan.) *Mann* 727 ! Likomba (Oct.) *Mildbr.* 10501 ! **F.Po:** (Nov.) *Mann* ! *T. Vogel* 96 !

Imperfectly known species.

1. **A.** sp. Perhaps a form of *A. africanus*, but possibly distinct. More material and field observations required. " Small tree by stream."
 Br.Cam.: Bambui, Bamenda (Apr.) *Maitland* 1462 !

2. **A. oreodryadum** *Gilg ex Mildbr.* Wiss. Ergebn. 1910–11 : 186 (1922), name only. Specimen not traced.
 F.Po: 4,500–4,800 ft. *Mildbr.* 6406.

4. DEINBOLLIA Schum.—F.T.A. 1 : 431 ; Radlk. in Engl. Pflanzenr. Sapindac. 669 (1932).

Rhachides, petiolules and undersurface of leaflets densely to sparingly hirsute :
 Leaflets 2–3 pairs, rounded and more or less retuse at apex ; petioles 5–9 cm. long, usually longer than the 4–6 cm. long rhachides, both very slightly winged ; leaflets opposite or alternate, elliptic-oblanceolate, unequally cuneate at base, 6·5–9 cm. long, 2–3·2 cm. broad .. .. 1. *voltensis*
 Leaflets (3–) 5–12 or more pairs, more or less distinctly acuminate at apex ; petioles shorter than the rhachides ; leaflets mostly larger than above :
  Leaflets (3–) 5–9 (–12) pairs, elliptic to oblong, rounded to subacute at base, shortly acuminate, 5–18 cm. long, 2·3–7·5 (–10) cm. broad, with 6–12 main lateral nerves on each side of midrib ; racemes or panicles up to 60 cm. long, together with the calyces densely brownish-hirsute ; petals 5–7 mm. long ; fruits subglobose, 1·3– 1·5 cm. diam., tomentose .. .. 2. *pinnata*
  Leaflets probably at least 12 pairs, elongate-oblong or elliptic, rounded at base, subacute or subacuminate at apex, up to 40 cm. long and 10·8 cm. broad, with 17–20 main lateral nerves on each side of midrib ; leaves up to 1·3 m. long, crowded at the top of the stem ; panicles terminal, erect, up to 1 m. long and 1·3 m. wide, together with the calyces densely brownish-hirsute ; petals about 5 mm. long ; fruits subglobose, at least 1 cm. diam., tomentose .. .. 3. *sp. A*
Rhachides, petiolules and undersurfaces of leaflets glabrous or sparingly puberulous :
 Leaflets more than 2·5 cm. broad :
  Flowers 2–5 mm. long, sparingly puberulous outside :
   Flowers 3–5 mm. long :
    Inflorescence-axes and pedicels sparingly puberulous with whitish hairs ; ovary pubescent ; developing fruit soon glabrous, 1·3–1·5 cm. diam. when mature :
     Leaflets 4–8 pairs, 7–20 cm. long, 2·5–7 cm. broad, opposite to alternate, usually cuneate at base, long-acuminate ; petioles 8–28 cm. long, rhachides 19–30 cm.

long ; inflorescences often unbranched, or with only a few branches, borne on
woody parts of the stem below the leaves, 8–80 cm. long   ..   4. *cuneifolia*
Leaflets 3–4 pairs, 20–33 cm. long, 8–12 cm. broad, opposite or subopposite,
cuneate or obtuse at base, shortly acuminate ; petioles 17–25 cm. long, rhachides
18–32 cm. long ; inflorescences with few branches, axillary, about 20 cm. long
  5. *calophylla*
Inflorescence-axes and pedicels rather densely pubescent or puberulous with
brown hairs ; ovary densely villous, fruit eventually glabrescent about 1·5 cm.
diam. ; petioles (7–) 11–27 cm. long, stout, rhachides (18–) 40–60 cm. long ;
leaflets (5–) 7–10 pairs, alternate or subopposite, more or less oblong-elliptic,
obtuse at base, shortly acuminate, 6–28 cm. long, 3–10 cm. broad ; inflorescences
paniculate, in leaf-axils, and on old wood, 5–50 cm. long   ..   6. *grandifolia*
Flowers 2 mm. long ; inflorescence-axis and pedicels whitish-pubescent ; ovary
pubescent, developing fruit soon glabrous ; petioles 10–22 cm. long, rhachides
about 40 cm. long ; leaflets 4–6 pairs, alternate, oblong-elliptic or ovate-oblong,
obtuse or cuneate at base, acuminate, 12–25 cm. long, 6–10 cm. broad ; inflorescences simple or paniculate, usually on old wood, 10–30 cm. long 7. *maxima*
Flowers 6–8 mm. long :
Sepals densely pubescent or pilose outside except for the margins ; stamens 20–24 ;
petioles about 20 cm. long, rhachides about 40 cm. long ; leaflets 6–7 pairs,
subopposite or alternate, oblong-elliptic, subfalcate, obtuse at base, acuminate,
16–25 cm. long, 7·5–10 cm. broad ; inflorescences paniculate or subsimple, terminal and (?) from old wood below the leaves, up to 70 cm. long, pubescent
  8. *insignis*
Sepals glabrous or very minutely and laxly puberulous outside ; stamens 27–29 ;
leaves about 1 m. long ; leaflets alternate, lanceolate-oblong, subacute and
slightly unequal at base, subsessile, long-acuminate, 20–22 cm. long, 6·7 cm.
broad ; panicles terminal, glabrous, about 40 cm. long ..   ..   9. *macrantha*
Leaflets 0·6–2·4 cm. broad, lanceolate, long-cuneate, gradually acuminate, 4–12 cm.
long ; petioles 3–6·5 cm. long, rhachides 7–14 cm. long ; leaflets 5–7 pairs, alternate
or subopposite ; inflorescences axillary, narrow, 1·5–11 cm. long, puberulous ;
sepals 2 mm. long, sparingly puberulous outside ; developing fruit soon glabrous
  10. *saligna*

1. **D. voltensis** *Hutch. ex Burtt Davy & Hoyle* Checklists Brit. Emp. 3 : 126 (1937), English descr. only. An
undershrub, " common on land flooded annually by Volta R." ; flowers white in clusters on the long
slender branches of the terminal panicle.
G.C.: Yeji, Volta R. (Apr.) *Dalz.* 21 ! *Vigne* FH 1702 !
[This name is only provisional. The species seems to be close to a group common in eastern and
southern Africa, which includes *D. oblongifolia* (E. Mey.) Radlk. and is in need of thorough revision.
More material of the West African plant is required.]
2. **D. pinnata** *Schum. & Thonn.* Beskr. Guin. Pl. 242 (1827) ; F.T.A. 1 : 432 ; Chev. Bot. 151 ; Radlk. in
Engl. Pflanzenr. Sapindac. 676 ; Aubrév. Fl. For. C. Iv. 2 : 196, t. 195, 1–6. *Ornitrophe pinnata* Poir.
(1808). *Schmidelia ? pinnata* (Poir.) DC. (1824). *Prostea pinnata* (Poir.) Camb. (1829). *Deinbollia
elliotii* Gilg (1897)—Radlk. l.c. 677. *D. leptophylla* Gilg ex Radlk. l.c. 679 (1932). *D. dahomensis*
A. Chev. Bot. 151, name only. An erect shrub or small tree, 4–15 ft. high, rarely more, with the
inflorescence erect towards the top of the shoots ; flowers creamy-white ; fruits orange ; especially in
regrowth forest vegetation.
Fr.G.: Macenta (young fr. May) *Collenette* 27 ! Dalaba-Diaguissa Plateau *Chev.* 18710. Nzo *Chev.*
21074. **S.L.:** Kafogo (Apr.) *Sc. Elliot* 5494 ! Bafodeya (Apr.) *Sc. Elliot* 5126*b* ! 5509 ! Berria, Falaba
(Mar.) *Sc. Elliot* 5427 ! Dia (Apr.) *Deighton* 3181 ! Musaia (Apr.) *Deighton* 5408 ! **Lib.:** Tappita (fr.
Aug.) *Baldwin* 9089 ! Ganta (Apr.) *Harley* ! **Iv.C.:** Mt. Gouan, Oua (Apr.) *Chev.* 21334 ! And other
localities *fide* Chev. and Aubrév. *l.c.* **G.C.:** Kumasi (May) *Chipp* 432 ! Winneba Plain (Feb.) *Dalz.*
8304 ! Accra Plains (fr. Apr.) *Irvine* 201 ! Adiembra to Domiabra (June) *Kitson* 1232 ! **Togo:** Kpedsu
(fl. & fr. Jan.) *Howes* 1118 ! Aflahu (fl. & fr. Jan.) *Warnecke* 48 ! Togo Plateau F.R. (Apr.) *Beveridge*
FH 2940 ! **Dah.:** Abomey (Feb.) *Chev.* 23148 ! **N.Nig.:** Lokoja (fr. Mar.) *Barter* 509 ! Shaw 24 !
S.Nig.: Lagos *Rowland* ! Abeokuta *Irving* ! Benin City *Unwin* 52 ! Onitsha *Barter* 1667 !
[Also recorded in Consp. Fl. Ang. 2 : 51, but I do not think the specimen (*Christen Smith* s.n.) is true
D. pinnata.]
3. **D. sp. A.** Possibly new, but more complete material required. A small tree, to 20 ft. high, with a cluster
of large leaves at the top of the unbranched main stem ; calyx green, petals white, ovary brown-tomentose, filaments green, anthers yellow.
S.Nig.: Omo F.R., Ijebu Ode Prov. : Grace Camp (fr. Mar.) *Jones & Onochie* FHI 17038 ! Osho (Mar.)
Jones & Keay FHI 17045 !
4. **D. cuneifolia** *Bak.* in F.T.A. 1 : 432 (1868) ; Chev. Bot. 150 ; Radlk. l.c. 685 ; Aubrév. Fl. For. C. Iv. 2 :
196. *D. stenobotrys* Gilg (1897)—Radlk. l.c. 686. *D. polypus* Stapf (1905), partly ; F.W.T.A., ed. 1. 1 :
503, partly ; Radlk. l.c. 682. A shrub or small tree, 2–12 ft. high ; flowers greenish-white ; fruits
orange ; in forest.
Fr.G.: Timbikounda *Chev.* 20615. Fassakoidou to Kesséridou *Chev.* 20823. **S.L.:** Bagroo R. (Apr.)
Mann 831 ! Sugar Loaf Mt. (May) *Barter* ! Kabusa, Limba (Apr.) *Sc. Elliot* 5475 ! Njala (fr. July)
Deighton 1821 ! 2224 ! **Lib.:** Lormai (Feb.) *Bequaert* 88 ! Sinoe Basin *Whyte* ! Jaurazon (Mar.) *Baldwin*
11443 ! Ganta (fr. Sept.) *Baldwin* 9324 ! **Iv.C.:** Abidjan region *Chev.* 15367. (See Appendix, p. 333.)
5. **D. calophylla** *Gilg ex Radlk.* in Engl. Pflanzenr. Sapindac. 684 (1932). *D. polypus* of F.W.T.A., ed. 1, 1 :
503, partly (*Dinklage* 1887). A shrub, 10–12 ft. high ; fruits orange.
Lib.: Fishtown, Grand Bassa (fr. May) *Dinklage* 1887 ! About 15 miles inland from R. Cess, Grand
Bassa (Mar.) *Baldwin* 11251 ! (See Appendix, p. 333.)
6. **D. grandifolia** *Hook. f.* in Fl. Nigrit. 249 (1849) ; Radlk. l.c. 682 ; Aubrév. l.c. 196. *D. indeniensis* A. Chev.
(1909)—Chev. Bot. 151. *D. insignis* of F.T.A. 1 : 431, partly. *D. polypus* Stapf (1905), partly ;
F.W.T.A., ed. 1, 1 : 503, partly. A shrub or small tree 3–25 ft. high, perhaps to 50 ft. ; flowers greenish
yellow ; in forest.
S.L.: Madina, Limba (Apr.) *Sc. Elliot* 5594 ! Lumbaraya, Talla Hills (Feb.) *Sc. Elliot* 5008 ! Njala
(Feb.–May) *Deighton* 1100 ! 5747 ! 6058 ! **Lib.:** Cape Palmas (fr. July) *T. Vogel* 66 ! Kakatown *Whyte* !

Bobei, Sanokwele (Feb.) *Baldwin* 14179! Dukwia R. (fr. Apr.) *Cooper* 357! **Iv.C.:** Indénié, Zaranou (Mar.) *Chev.* 17619! And other localities *fide* Chev. *l.c.* **G.C.:** Moinsi Hills *Cummins*! Bunso, E. Prov. (Oct.) *Darko* 1042! Anibil Wharf (Feb.) *Vigne* FH 2821! Dakabansa-Bekwai (Feb.) *Vigne* FH 4947! Krokosua F.R. (Feb.) *Cansdale* FH 3988! **Togo:** Ntumada, Hohoe Dist. (Dec.) *St. C. Thompson* FH 3566! (See Appendix, p. 333.)

[This species is recorded by Andrews in his Fl. Pl. A.-E. Sud. 2 : 339 (1952), but the specimen (*Hoyle* 826) on which the record is based is not *D. grandifolia* Hook. f.]

7. **D. maxima** *Gilg* in Mém. Soc. Linn. Normand. 26, 2 : 73 (1924) ; Radlk. l.c. 683. A small tree ; in forest.
   S.Nig.: Oban *Talbot* 1290! 1751! Also in French Cameroons and Gabon.
   [Radlk. also records this species from Sierra Leone, but I have not seen the specimen (*Afzelius* s.n.), nor any other specimens of the species from Sierra Leone.]

8. **D. insignis** *Hook. f.* in Fl. Nigrit. 250 (1849) ; F.T.A. 1 : 431, partly (spec. ex F. Po only) ; Radlk. l.c. 680, partly (spec. ex F.Po only) ; Kennedy For. Fl. S. Nig. 185. A forest tree 12–25 ft. high ; calyx green, corolla yellow ; ripe fruits yellowish-orange.
   S.Nig.: Sapoba (June) *Kennedy* 768! 1121! 2290! *Onochie* FHI 20698! **F.Po:** (Nov., Jan.) *Mann* 159! 591. *T. Vogel* 166!
   [The Nigerian material has the sepals and inflorescence rather more densely covered with rather longer hairs than the F. Po material.]

9. **D. macrantha** *Radlk.* l.c. 681 (1932). A small tree in forest.
   Br.Cam.: Kumba (Nov.) *Büsgen* 29.

10. **D. saligna** *Keay* in Bull. Jard. Bot. Brux. 26 : 193, fig. 60 (1956). A shrub, up to 3 ft. high, with white flowers ; on rocky beds and banks of rivers in forest country.
    S.Nig.: Oban *Talbot* 583! Kwakam, Anigeje, Calabar Prov. (Apr.) *Ujor* FHI 31776! **Br.Cam.:** Ndian R., Kumba (Mar.) *Smith* 80/36!

Imperfectly known species.

1. **D. brachybotrys** *Gilg* in Engl. Bot. Jahrb. 24 : 295 (1897) ; Radlk. l.c. 686. Probably equals *D. cuneifolia* Bak., but type specimen presumably destroyed.
   S.L.: *Afzelius.*

2. **D. unguiculata** *Gilg* l.c. 296 (1897) ; Radlk. l.c. 685. Probably equals *D. cuneifolia* Bak., but type specimen presumably destroyed.
   S.L.: *Afzelius.*

5. APHANIA Blume, Bijdr. Fl. Nederl. Ind. 5 : 236 (1825) ; Radlk. in Engl. Pflanzenr. Sapindac. 699 (1932).

Branchlets, inflorescences, petioles, rhachides and petiolules yellowish-pubescent ; leaflets 1 or 2 pairs, opposite or subopposite, with swollen petiolules, broadly oblanceolate to obovate-elliptic or elongate-oblong, cuneate or obtuse at base, obtusely pointed at apex, 8–25 cm. long, 3–8 cm. broad, strongly reticulate on both surfaces, glabrous ; petioles 0·1–2·5 cm. long, rhachides 0 or 0·7–4·8 cm. long ; flowers polygamous in rather lax terminal panicles ; bracts and sepals ciliate ; petals (3–) 5 ; stamens 7–8, filaments villous ; ovary didymous, one carpel usually abortive, glabrous ; fruits obovoid or ellipsoid, 1·3–1·8 cm. long, with a basal style and the abortive carpel persisting at the base   ..   ..   *senegalensis*

A. **senegalensis** (*Juss. ex Poir.*) *Radlk.* in Sitzb. Math.-Phys. Akad. Münch. 8 : 238 (1878) ; Chev. Bot. 151 ; Radlk. in Engl. Pflanzenr. Sapindac. 703 ; Aubrév. Fl. For. C. Iv. 2 : 192–194, t. 202, incl. var. *silvatica* (A. Chev. ex Hutch. & Dalz.) Aubrév. (1936) ; Aubrév. Fl. For. Soud.-Guin. 386, t. 82, 4. *Sapindus senegalensis* Juss. ex Poir. (1804)—F.T.A. 1 : 430. *Aphania silvatica* A. Chev. ex Hutch. & Dalz. F.W.T.A., ed. 1, 1 : 502 (1928) ; Kew Bull. 1929 : 26 ; Radlk. l.c. 705 ; Chev. Bot. 152, name only. A forest tree to 55 ft. high, with short bole and dense drooping foliage ; flowers greenish-white, fragrant ; ripe fruits red, edible ; in moist places in the forest regions and especially in fringing forest in the savannah regions.
Sen.: Walo *Heudelot*! Richard Tol (fl. Jan., Feb., fr. May) *Roger* 11! *Perrottet* 109! R. Salum *Brunner* 156! Samanding (Feb.) *Chev.* 3031! **Gam.:** *Don*! Genieri (Feb.) *Fox* 30! **Port.G.:** Brene, Bissau (Jan.) *Esp. Santo* 1687! Pussubé (fr. Mar.) *Esp. Santo* 1235! 1506! **Fr.G.:** Conakry *Chev.* 13036. **Iv.C.:** *Aubrév.* 792! Akakoumoëkrou (fr. Dec.) *Chev.* 22561! Tabou (fr. Aug.) *Chev.* 20064! **G.C.:** Tano R. (Sept.) *Chipp* 345! Durimo to Jambuso (fr. Mar.) *Kitson* 935! Bomfa, Ashanti (Aug.) *Cansdale* FH 4426! Kwadaso (Oct.) *Darko* 738! Keta *Thonning.* **Togo:** Lome *Warnecke* 42! 383! Okrabi, Hohoe Dist. (Nov.) *St. C. Thompson* FH 3694! **Dah.:** Dassa-Zoumé *Chev.* 23649. **S.Nig.:** Ibadan North F.R. (fr. Oct.) *Chizea* FHI 23959! Akaba Hills F.R., Ibadan Prov. (Oct.) *Jones* FHI 7062! R. Ogbesse, Owo Dist. *Jones* FHI 3134! **Br.Cam.:** Victoria (fr. Jan.) *Maitland* 186a! 266! 913! Johann-Albrechtshöhe (fr. Dec.) *Staudt* 777. Bambuko F.R. (Sept.) *Olorunfemi* FHI 30771! Also in French Cameroons, Ubangi-Shari, Gabon, Belgian Congo, A.-E. Sudan, Abyssinia, Eritrea, E. Africa and Angola. (See Appendix, p. 331.)

6. CHYTRANTHUS Hook. f.—F.T.A. 1 : 429 ; Radlk. in Engl. Pflanzenr. Sapindac. 782 (1932). *Glossolepis* Gilg (1897)—F.W.T.A., ed. 1, 1 : 503 ; Radlk. l.c. 778.

Leaves hairy, at least on the petioles, rhachides, petiolules and midribs and main nerves beneath ; stamens 7–9 :
Leaflets 3–4 pairs, densely setose on midrib and nerves beneath ; ovary and fruit 6–8-celled ; inflorescences up to 15 cm. long :
  Leaflets obtuse to cuneate at base ; rhachides 15–21 cm. long ; inflorescences and flowers densely bristly-tomentose, up to 8 cm. long ; leaflets oblanceolate-elliptic or oblong-elliptic, long-acuminate, 16–30 cm. long, 5·2–9·5 cm. broad, with 14–22 main lateral nerves on each side of midrib ; petioles 12–20 cm. long   1. *setosus*
  Leaflets rounded to subcordate and unequal at base ; rhachides 10–15 cm. long ; inflorescences up to 15 cm. long ; leaflets oblong-elliptic, 17–25 cm. long, 6·5–10 cm. broad, with 11–16 main lateral nerves on each side of midrib ; petioles 10–15 cm. long   ..   ..   2. *ellipticus*
Leaflets 5–11 pairs, puberulous on midrib and nerves beneath, sometimes with long weak hairs as well especially when young and near the petiolules ; ovary and fruit 3–6-celled ; inflorescences up to 35 cm. long :

Leaflets 5–6 pairs, 9–15 cm. broad, 20–40 cm. long, oblong, obtuse to rounded at base, acuminate, rather densely puberulous on midrib nerves and sometimes on lamina beneath, without long weak hairs ; main lateral nerves 21–25 on each side of midrib, prominent beneath, midrib pubescent above ; inflorescences 20–35 cm. long ; ovary 6-celled .. .. 3. *atroviolaceus*
Leaflets 7–11 pairs, up to 7·5 cm. broad, with long weak hairs especially when young and near the petiolules ; main lateral nerves 12–20 ; ovary 3- or 5-celled :
   Leaflets 7–9 pairs, upper pairs usually larger than the rest, midribs glabrous above ; inflorescences and flowers densely bristly tomentose ; main axis of inflorescence up to 30 cm. long, with, often secundly arranged, short lateral branches each bearing several flowers ; upper leaflets narrowly oblong-oblanceolate, middle ones elongate-oblong, lower ones oblong-lanceolate, all more or less obtuse at base and long-acuminate, up to 32 cm. long and 7·5 cm. broad ; petioles 15–25 cm. long, rhachides 36–48 cm. long ; ovary 3-celled (? always) .. 4. *villiger*
   Leaflets 9–11 pairs, upper pairs usually smaller than the middle ones, midribs pubescent above ; inflorescences and flowers tomentellous ; main axis of inflorescence 15–24 cm. long, bearing the flowers 1–3 together, with conspicuous bracts ; leaflets elongate-oblong to oblanceolate, obtuse at base, abruptly long-acuminate, up to 28 cm. long and 6·8 cm. broad ; petioles 13–23 cm. long, rhachides 36–54 cm. long ; ovary 5-celled ; fruits with 5 broad and 5 alternating narrow wings .. .. 5. *bracteosus*
Leaves quite glabrous ; stamens 12–15 :
   Leaflets 4–6 pairs, the upper more or less obovate-oblong and much larger than the lower which are more or less ovate-oblong, rounded to cuneate at base, shortly acuminate, 8–38 cm. long, 5–12 cm. broad ; petioles 16–27 cm. long, rhachides 26–40 cm. long ; racemes up to 30 cm. long ; calyx subglobose-saccate, about 6 mm. long, shortly tomentose outside ; petals broad at base, oblong, 7–8 mm. long, 2–3 mm. broad ; fruits ribbed, 10–15 cm. long .. .. 6. *macrobotrys*
   Leaflets 8–11 pairs, all more or less the same size and shape, oblong, rounded to cuneate at base, abruptly and often long-acuminate, 11–24 cm. long, 4·2–6·8 cm. broad ; petioles 16–22 cm. long, rhachides 48–66 cm. long ; racemes up to 60 cm. long ; calyx ovoid-saccate, 10–12 mm. long, tomentellous outside ; petals narrowed at base, loriform, 12–13 cm. long, 2 mm. broad ; fruits ribbed .. 7. *talbotii*

1. **C. setosus** *Radlk.* in Sitzb. Math.-Phys. Akad.-Münch. 20 : 240 (1890) ; Chev. Bot. 152 ; Engl. Pflanzenr. Sapindac. 785, fig. 18. A shrub or small tree, to 15 ft. high, cauliflorous, with cream flowers in short dense spike-like inflorescences ; in forest.
   Dah.: Sakété (Jan.) *Chev.* 22884 ! 22885. **S.Nig.**: Omo F.R., Ijebu-Ode Prov. (fl. buds Jan.) *Onochie* FHI 20680 ! Okomu F.R., Benin (fl. buds Feb.) *Brenan* 9127 ! Calabar R. (Feb.) *Mann* 2281 ! Also in French Cameroons and extending to Cabinda and Belgian Congo. (See Appendix, p. 333.)
2. **C. ellipticus** *Hutch. & Dalz.* F.W.T.A., ed. 1, 1 : 504 (1928) ; Kew Bull. 1929 : 26 ; Radlk. in Engl. Pflanzenr. Sapindac. 798. A tree, up to 30 ft. high, cauliflorous, with whitish flowers in dense spike-like inflorescence ; in forest.
   S.Nig.: Sapoba *Kennedy* 2222 ! Okomu F.R., Benin (Dec.) *Brenan* 8579 ! Oban *Talbot* 1399 ! Eket *Talbot* 3129 !
3. **C. atroviolaceus** *Bak. f. ex Hutch. & Dalz.* F.W.T.A., ed. 1, 1 : 504 (1928) ; Kew Bull. 1929 : 26 ; Radlk. l.c. 786. *C. macrophyllus* Gilg var. *obanensis* Bak. f. in Cat. Talb. 19 (1913). A shrub or small tree (? up to 60 ft. *fide* Talbot) ; flowers on the main stem, with deep red velvety calyx ; in forest.
   S.L.: Heirakohun *Smythe* 126 ! **G.C.**: Kwahu Prasu (Feb.) *Vigne* FH 1608 ! **S.Nig.**: Shasha F.R. (Feb.) *Lamb* 192 ! Oban *Talbot* 1583 ! 1596 !
4. **C. villiger** *Radlk.* l.c. 794 (1932). A small tree, to 15 ft. or more high ; cauliflorous, with white velvety flowers in slender panicles at the base of the stem ; in forest.
   S.L.: Makump (July) *Thomas* 960 ! Yonibana (Nov.) *Thomas* 4993 ! **Lib.**: Gbanga (Sept.) *Linder* 611 ! Nyaake, Webo (June) *Baldwin* 6214 ! **Iv.C.**: Banco *Aubrév.* 480 ! **G.C.**: Huniso, W. Prov. (fr. Jan.) *Irvine* 1063 ! Bawdia, W. Prov. (Dec.) *Vigne* FH 3152 ! Also in French Cameroons.
5. **C. bracteosus** *Radlk.* l.c. 787 (1932). A tree, 10–55 ft. high, almost unbranched ; inflorescences brown, rigid, borne all up the trunk ; in forest.
   Br.Cam.: S. Bakundu F.R., Banga (fl. buds Mar., young fr. Aug.) *Brenan* 9487 ! *Olorunfemi* FHI 30720 ! Also in French Cameroons.
   [Our material is determined from the description as the type is not available ; a sterile specimen (*Linder* 983) from Peahtah, Lib., may also belong here.]
6. **C. macrobotrys** (*Gilg*) *Exell & Mendonça* in Consp. Fl. Ang. 2 : 84 (1954). *Glossolepis macrobotrys* Gilg in Engl. Bot. Jahrb. 24 : 299, t. 1 (1897) ; Radlk. in Engl. Pflanzenr. Sapindac. 780, fig. 17. A forest shrub or small tree, to 35 ft. high, rarely more, sparingly branched ; racemes on old wood, pendulous, with erect brownish flowers.
   G.C.: Amentia (Dec.) *Vigne* FH 1490 ! S. Fomang Su F.R. (Jan.) *Vigne* FH 2678 ! Kumasi (Jan.) *Vigne* FH 999 ! Owabi (Dec.) *Lyon* FH 2632 ! **Togo**: Kabosu, Kabo River F.R. (Dec.) *St. C. Thompson* FH 3567 ! **S.Nig.**: Gambari Group F.R., Ibadan (fl. & fr. Jan.) *Keay* FHI 19801 ! 19802 ! *W. D. Mac-Gregor* 570 ! Sapoba *Kennedy* 3049 ! Cross River North F.R. (young fr. June) *Latilo* FHI 31850 ! Boje, Ikom Div. (May) *Jones & Onochie* FHI 17331 ! **Br.Cam.**: S. Bakundu F.R., Kumba Div. (Aug.) *Dundas* FHI 8368 ! Korup F.R., Kumba Div. (June) *Olorunfemi* FHI 30640 ! Moliwe (young fr. July) *Winkler* 1457. Also in French Cameroons, Cabinda and Belgian Congo.
   [Aubrév. Fl. For. C. Iv. 2 : 177 records *Glossolepis* in the Ivory Coast (*Aubrév.* 391) but does not give the species ; it may be this.]
7. **C. talbotii** (*Bak. f.*) *Keay* in Bull. Jard. Bot. Brux. 26 : 194 (1956). *Glossolepis talbotii* Bak. f. in Cat. Talb. 20 (1913) ; Radlk. l.c. 781. *G. pilgeriana* Gilg ex Engl. Pflanzenw. Afr. 3, 2 : 276 (1921) ; Mém. Soc. Linn. Normand 26, 2 : 71 (1924) ; Radlk. l.c. 780. A forest tree, to 30 ft. high, sparingly branched ; racemes on the basal part of the trunk, pendulous, with erect whitish flowers with mauve filaments and reddish anthers.
   S.Nig.: Oban *Talbot* 1258a ! 1686 ! Boshi-Okwango F.R., Obudu (May) *Latilo* FHI 30940 ! **Br.Cam.**: Mombo, Kumba Div. (May) *Olorunfemi* FHI 30564 ! S. Bakundu F.R., Banga (Mar.) *Brenan* 9417 ! Also in French Cameroons, Gabon and Spanish Guinea.

Imperfectly known species.

1. **C. brunneo-tomentosus** *Gilg ex Radlk.* l.c. (1932). A small tree, leaflets 3–6 pairs ; stamens 7–8 ; ovary 5-celled. Type specimen presumably destroyed.
   Br.Cam.: Bangwe, on margin of forest (Apr.) *Conrau* 94 ; 172.
   [An incomplete specimen (*Ujor* FHI 29262), from Wum Dist., Bamenda, British Cameroons, may also belong here ; better material needed.]
2. **C. prieurianus** *Baill.* in Adansonia 9 : 241 (1874) ; Radlk. l.c. 793. Described from a plant formerly cultivated in a hot house at Paris and said to have been collected by Leprieur possibly in West Africa.
3. **C. sp. A.** Shrub or tree, to 33 ft. high ; inflorescences produced from raised woody cushions around the base of the trunk ; fruits succulent, orange-yellow, 3-celled. This incomplete material is possibly *C. welwitschii* Exell or *C. mannii* Hook. f.
   S.Nig.: Akilla plantations, Ijebu Ode Prov. (fr. Jan.) *Onochie & Oladoyinbo* FHI 20660 ! Okomu F.R., Benin *Brenan* 8611 !
4. **C. sp. B.** Leaflets 5 pairs, upper pairs large, lower pairs small, sparingly pilose on midrib beneath and on rhachis and petiolules. Better material needed.
   S.Nig.: Oban *Talbot* !

7. PANCOVIA Willd. Sp. Pl. 2 : 285 (1799) ; Radlk. in Engl. Pflanzenr. Sapindac. 799 (1932).

Racemes or flower-clusters not exceeding 13 cm. long :
　Leaflets without scale-like glands :
　　Flowers subsessile in axillary clusters less than 5 mm. long ; calyx cupular, about 4 mm. long, tomentellous ; petioles 1·5–6·5 cm. long, rhachides 0·5–1·2 cm. long ; leaflets 2 pairs, opposite or subopposite, obovate-elliptic, cuneate at base, distinctly but obtusely acuminate, 4·5–17 cm. long, 1·5–6·5 cm. broad, glabrous, reticulate on both surfaces　..　..　..　..　..　..　..　 1. *sessiliflora*
　　Flowers in racemes 1–10 cm. long :
　　　Pedicels up to 2 mm. long ; calyx turbinate, about 3 mm. long, tomentellous ; racemes 1–7 cm. long ; petioles 2–9 cm. long ; leaflets 1–2 pairs, opposite or subopposite, elliptic, cuneate or obtuse at base, obtusely acuminate, 3–17 cm. long, 2·3–8·3 cm. broad, glabrous, reticulate on both surfaces　..　..　 2. *turbinata*
　　　Pedicels 3–8 mm. long ; calyx shortly campanulate, about 4 mm. long, with rather long lobes :
　　　　Racemes 1–2 (–4) cm. long ; pedicels 5–8 mm. long ; calyx shortly tomentose ; petioles 2–6 cm. long, rhachides 0 or 1·5–6 cm. long ; leaflets (1–) 2 (–3) pairs, opposite or subopposite, oblong-elliptic to obovate-elliptic, obtuse or cuneate at base, shortly and obtusely acuminate or obtuse at apex, 4·5–14 cm. long, 3·2–6·2 cm. broad, glabrous, reticulate beneath　..　..　..　 3. *bijuga*
　　　　Racemes (2–) 4–13 cm. long ; pedicels 3–5 mm. long ; calyx tomentellous or puberulous ; petioles 4–9 cm. long, rhachides 0 or 2–13 cm. long ; leaflets (1–) 3–4 pairs, opposite or alternate, lanceolate to elliptic cuneate or obtuse at base, distinctly acuminate, 4–20 cm. long, 1·5–8 cm. broad, glabrous, reticulate on both surfaces　..　..　..　..　..　..　..　 4. *pedicellaris*
　Leaflets with minute scale-like glands beneath ; petioles 3·5–9 cm. long ; leaflets 2–3 pairs, elliptic-lanceolate, cuneate at base, shortly acuminate, 7–20 cm. long, 5–9 cm. broad ; racemes about 4 cm. long, pedicels short ; calyx campanulate, tomentellous　..　..　..　..　..　..　..　 6. *subcuneata*
Racemes 15–30 cm. long ; pedicels about 5 mm. long ; calyx shortly campanulate, about 4 mm. long, with rather long lobes ; petioles 8–14 cm. long ; leaflets 3–4 pairs, ovate-lanceolate to elliptic-oblong, subcuneate to the up to 1 cm. long petiolule, acutely acuminate, 14–20 cm. long, 4·5–8 cm. broad, without scale-like glands beneath　..　..　..　..　..　..　..　..　 5. *polyantha*

1. **P. sessiliflora** *Hutch. & Dalz.* F.W.T.A., ed. 1, 1 : 504 (1928) ; Kew Bull. 1929 : 27 ; Radlk. in Engl. Pflanzenr. Sapindac. 1495. A small shrub, sometimes trailing, with small whitish flowers and red 2-lobed fruit.
   S.Nig.: Ikoyi Plains, Lagos (Feb.) *Dalz.* 1178 ! Agege (Nov.) *Kitson* ! Olokemeji (Nov.) *Jones, Keay & Onochie* FHI 4920 ! S. of Okeigbo (fr. Feb.) *Jones & Onochie* FHI 14728 ! Etemi, Omo F.R. (Mar.) *Jones & Onochie* FHI 17016 !
2. **P. turbinata** *Radlk.* in Sitzb. Math.-Phys. Akad. Münch. 8 : 270 (1878) ; Engl. Pflanzenr. Sapindac. 804 (excl. syn. *Aphania golungensis* Hiern) ; Aubrév. Fl. For. C. Iv. 2 : 203, t. 206, 2–6 (excl. syn.). *P. bijuga* of F.W.T.A., ed. 1, 1 : 504, partly. An erect shrub or small tree ; flowers dull yellowish cream.
   Fr.G.: *Heudelot* 869 ! **Iv.C.**: Port Bouët *Aubrév.* 1518. **Br.Cam.**: Barombi (fl. buds Mar.) *Preuss* 9. S. Bakundu F.R., Banga, Kumba Div. (Jan.) *Keay* FHI 28562 !
3. **P. bijuga** *Willd.* Sp. Pl. 2 : 285 (1799) ; Chev. Bot. 152 ; Radlk. l.c. 802 ; Aubrév. l.c. 203, t. 206, 1. *Erioglossum cauliflorum* Guill. & Perr. Fl. Seneg. 118, t. 28 (1831) ; F.T.A. 1 : 420. *Pancovia guineensis* Willd. ex A. Chev. Bot. 152, name only. A small tree, much-branched ; flowers cream.
   Sen.: Itou, R. Casamance (Apr., May) *Perrottet*. **Iv.C.**: Akakoumoëkrou (Dec.) *Chev.* 22527 ! 22536 ! 22547 ! **G.C.**: Drabonso, Ashanti (Feb.) *Cansdale* FH 4144 ! Pokoase (Feb.) *Irvine* 2658 ! **Togo**: Lome (Feb.) *Warnecke* 67 ! **S.Nig.**: Lagos *Moloney* ! Ogun River F.R., Lagos (Dec.) *Onochie* FHI 32033 !
4. **P. pedicellaris** *Radlk. & Gilg* in Engl. Bot. Jahrb. 24 : 302 (1897) ; Radlk. in Engl. Pflanzenr. Sapindac. 803. A shrub or small tree, 15–30 ft. high, with very hard wood ; in forest.
   Br.Cam.: Johann-Albrechtshöhe (Jan., Feb.) *Staudt* 596 ! 798. Also in French Cameroons and Spanish Guinea.
5. **P. polyantha** *Gilg ex Engl.* Pflanzenw. Afr. 3, 2 : 276 (1921) ; Radlk. l.c. 803. A forest shrub (?).
   Br.Cam.: Johann-Albrechtshöhe (Mar.) *Staudt* 906 !
6. **P. subcuneata** *Radlk.* in Engl. Pflanzenr. Sapindac. 806 (1932) ; E. & P. Pflanzenfam. Nachtr. 3 : 203 (1907), name only. *P. bijuga* of F.W.T.A., ed. 1, 1 : 504, partly (S.L. record). A shrub (?).
   S.L.: *Afzelius*.

Imperfectly known species.

P. sp.   Understorey tree, 15 ft. high ; in forest outlier ; leaves with a single pair of lanceolate leaflets ; flowers and fruits required.
N.Nig.: Kurmin Damisa, Jemaa, Plateau Prov. *Keay & Onochie* FHI 21660 !

8. PLACODISCUS Radlk. in Sitzb. Math.-Phys. Akad. Münch. 8 : 332 (1878) ; Engl. Pflanzenr. Sapindac. 810 (1932).

Leaflets 2–5 (–6) pairs :
  Leaves sessile, quite glabrous ; leaflets 2 pairs, the lowest pair resembling large leafy stipules, ovate, upper pair larger, elongate-elliptic, cuneate at base, acuminate, 7–25 cm. long, 2·5–8 cm. broad, with 7–12 main lateral nerves on each side of midrib ; rhachides 1–12·5 cm. long ; male racemes axillary, 3–11 cm. long, bracts 1·5 mm. long, conspicuous ; female racemes axillary or on older wood below the leaves, up to 34 cm. long ; flowers subsessile, crowded, calyx turbinate, about 3·5 mm. long and wide, puberulous ; fruits 1–2-lobed by abortion, up to 3 cm. long and broad, tomentellous at least when young   ..   ..   ..   1. *pseudostipularis*
Leaves distinctly petiolate ; leaflets 2–5 (–6) pairs, the lowest pair well away from the stem and not resembling leafy stipules ; flowers distinctly pedicellate :
  Inflorescence-axes without glandular hairs :
    Inflorescences axillary and sometimes just below the foliage ; calyx about 5 mm. long ; bracts up to 1 mm. long ; leaflets 2–3 pairs :
      Pedicels of male flowers 5–10 mm. long ; inflorescence and calyx puberulous ; inflorescence racemose except for a few short lateral branches near the base, up to 15 cm. long ; petioles 5·5–7·5 cm. long, rhachides 4·5–6 cm. long ; leaflets alternate or opposite, elliptic or oblong-elliptic, cuneate at base, shortly acuminate, 6–22 cm. long, 3–8·5 cm. broad, with 5–9 main lateral nerves on each side of midrib   ..   ..   2. *boya*
      Pedicels of flowers up to 2 mm. long (to 5 in fruit) ; inflorescence and calyx densely tomentellous ; inflorescence a very narrow spike-like panicle with very short 1–3-flowered lateral branches, 3–13 cm. long ; petioles 1·8–12 cm. long, rhachides 2·8–11 cm. long ; leaflets opposite or subopposite, oblong-elliptic to oblanceolate-elliptic, obtuse or cuneate at base, rounded, obtuse, or obtusely and shortly acuminate at apex, 3–20 cm. long, 1·5–7 cm. broad, with 6–12 main lateral nerves on each side of midrib   ..   ..   3. *riparius*
    Inflorescences on the main stem ; pedicels 1·5–2·5 mm. long ; calyx 2·5–3 mm. long ; leaflets (2–) 4–5 (–6) pairs, opposite to alternate, with 8–12 main lateral nerves on each side of midrib :
      Axes of inflorescence and calyx tomentellous ; bracts up to 2 mm. long ; racemes up to 40 cm. long ; leaflets 2–5 pairs, oblong to oblong-elliptic, cuneate to obtuse at base, more or less long-acuminate, 10–30 cm. long, 3·8–10 cm. broad ; petioles 6–17 cm. long, rhachides 11–29 cm. long   ..   ..   4. *leptostachys*
      Axes of inflorescence and calyx shortly tomentose ; bracts about 5 mm. long ; racemes up to 22 cm. long ; leaflets 4–5 pairs, oblong-oblanceolate, cuneate, very shortly acuminate, 12–28 cm. long, 6–11 cm. broad ; petioles 4·5–14 cm. long, rhachides 14–20 cm. long   ..   ..   5. *sp. A*
  Inflorescence-axes with numerous glandular hairs ; racemes borne on the main stem :
    Flowers subsessile ; calyx 3 mm. long, tomentose ; racemes up to 20 cm. long ; leaflets 3–4 pairs, subopposite or opposite, oblong-elliptic, obtuse to cuneate at base, acuminate, 5·5–21 cm. long, 3–7·2 cm. broad, with 7–14 main lateral nerves on each side of midrib ; petioles 2–11 cm. long, rhachides 5·5–16 cm. long
                                        6. *turbinatus*
    Flowers with distinct pedicels 1–2 mm. long ; calyx 2 mm. long, puberulous-tomentellous ; racemes up to 36 cm. long ; leaflets 3–6 pairs, alternate or sub-opposite, oblong-elliptic, cuneate to obtuse at base, acuminate, 10–30 cm. long, 3·5–9 cm. broad, with 10–17 main lateral nerves on each side of midrib ; petioles 10–19 cm. long, rhachides 13–25 cm. long ..   ..   7. *glandulosus*
Leaflets 8–18 pairs, the lowest pair well away from the stem and not resembling leafy stipules ; racemes borne on the main stem and among the lower leaves, axes without glandular hairs :
  Leaflets 8–12 pairs, elongate-oblong or oblong-lanceolate, cuneate and oblique at base, caudate-acuminate, 5·5–23 cm. long, 1·5–4·8 cm. long, with 11–15 main lateral nerves on each side of midrib ; petioles 9·5–18 cm. long, rhachides 25–48 cm. long, petiolules 3–6 mm. long ; racemes up to 15 cm. long, tomentose ; flowers very shortly pedicellate ; calyx obconic-subglobose, 2·75–4 mm. long, 3·5–4·5 mm. diam., lobes never wide open, tomentose outside ; fruits 3-lobed, but only 1 or 2 lobes fully developed, 2·5–3·5 cm. long, 3–3·5 cm. broad, pubescent, later glabrescent
                                        8. *splendidus*
  Leaflets 18–24 pairs, lanceolate and subfalcate, cuneate at base and subsessile, acute and mucronate at apex, up to 30 cm. long and 7·5 cm. broad, with 8–10 main lateral nerves on each side of midrib ; rhachides triangular in cross-section ; leaves up to

100 cm. or more long ; racemes up to 35 cm. long, tomentose ; pedicels up to 1 cm. long   ..   ..   9. *bancoensis*

1. **P. pseudostipularis** *Radlk.* in Sitzb. Math.-Phys. Akad. Münch. 20 : 242 (1890) ; Chev. Bot. 152 ; Radlk. in Engl. Pflanzenr. Sapindac. 812 ; Aubrév. Fl. For. C. Iv. 2 : 200. A shrub or small tree, 3–48 ft. high ; flower buds brown ; in forest.
    S.L.: *Afzelius.* Njala (May–Sept.) *Deighton* 3016! 3230! 5335! Makump (fr. July) *Thomas* 950! Gola Forest (young fr. Apr.) *Small* 595! **Lib.:** Dukwia R. (May, Oct.) *Cooper* 266! 374! **Iv.C.:** Alépé *Chev.* 16242. Mt. Tou *Chev.* 19681. Agnéby *Aubrév.* 2119. **G.C.:** Axim *Burton & Cameron*! Bonsa Su (May) *Vigne* FH 1996! (See Appendix, p. 336.)
2. **P. boya** *Aubrév. & Pellegr.* in Bull. Soc. Bot. Fr. 85 : 292 (1938) ; Aubrév. Fl. For. C. Iv. 2 : 200, t. 204, 1–3 (1936), French descr. only. An understorey tree, to 65 ft. high, with short irregular bole and dense spreading crown ; in forest.
    Iv.C.: Bondoukou *Aubrév.* 703! Man *Aubrév.* 1035! Akoupé *Aubrév.* 1771! Guiglo *Aubrév.* 2017. **G.C.:** near Ejian, N. Agogo (Jan.) *Vigne* FH 1791!
3. **P. riparius** *Keay* in Bull. Jard. Bot. Brux. 26 : 194 (1956). A small tree, branching low down, to 30 ft. high (to 60 ft. *fide* Cooper), growing on riverbanks and sometimes completely submerged in rainy season ; calyx pale green, tinged pink ; filaments white.
    Port.G.: R. Geba, Bafata (Sept.) *Esp. Santo* 3094! R. Mad-Jobe, Gabu (young fr. Oct.) *Esp. Santo* 3115! **S.L.:** Njala (fl. Sept., Oct., fr. Dec.) *Deighton* 1841! 5594! 5602! 5723! 5853! Bumpe (Sept.) *Deighton* 5577! **Lib.:** Gbanga (fl. & young fr. Sept.) *Linder* 770! Dukwia R. (Oct.) *Cooper* 75!
4. **P. leptostachys** *Radlk.* in Sitzb. Math.-Phys. Akad. Münch. 9 : 606 (1879) ; Pflanzenr. 813. A slender tree, 12–20 ft. high, cauliflorous ; in forest.
    Lib.: Brewersville (Aug.) *Baldwin* 13079! Bushrod Isl. (Aug.) *Baldwin* 13095! **S.Nig.:** Aboabam, Ikom Div. (young fr. May) *Latilo* FHI 30970! **Br.Cam.:** Cam. Mt., 2,000–3,000 ft. (Dec.) *Mann* 2150! Bangwe *Conrau* 207.
5. **P. sp. A.** A slender tree or shrub, cauliflorous ; flowers greenish-yellow ; in understorey of forest, especially by streams.
    G.C.: Kwahu Prasu (fl. Feb., young fr. June) *Vigne* FH 1612! 1612a! S. Fomang Su F.R. (fl. & young fr. Jan.) *Vigne* FH 2694! Oda, W. Akim (June) *Darko* 682! **S.Nig.:** Ibadan North F.R. (fr. Dec.) *Chizea* FHI 23967!
    [Previously determined as *P. leptostachys* Radlk., but possibly distinct ; more good material of this and of the above species is required.]
6. **P. turbinatus** *Radlk.* in Sitzb. Math.-Phys. Akad. Münch. 8 : 332 (1878) ; Pflanzenr. 815. A slender tree, to 15 ft. high, cauliflorous ; leaves drying green ; in forest.
    S.Nig.: Calabar R. (Feb.) *Mann* 2239! Oban *Talbot* 1438! **[Br.Cam.:** S. Bakundu F.R. (young fr. Apr.) *Ejiofor* FHI 29348!—an incomplete specimen, included here with some doubt.]
7. **P. glandulosus** *Radlk.* in Engl. Pflanzenr. Sapindac. 814.(1932). A small tree, cauliflorous ; leaves drying reddish-brown ; in forest.
    S.Nig.: Boje, Ikom Div. (young fr. May) *Jones & Onochie* FHI 18633! **Br.Cam.:** Victoria (Nov.) *Maitland* 766! Also in French Cameroons and Gabon.
8. **P. splendidus** *Keay* l.c. 197, fig. 61 (1956). A tree 10–30 ft. high, often unbranched, with leaves in a cluster at the top, cauliflorous ; sepals pinkish-brown ; ripe fruits yellow ; in forest.
    S.L.: Njala (fl. Aug., Sept., fr. Oct.) *Deighton* 4907! 5560! 5587! 5588! 5590! Kenema (fl. & fr. Oct.) *Small* 819! Taiama (Aug.) *Deighton* 5560! **Lib.:** Peahtah (young fr. Oct.) *Linder* 1040!
9. **P. bancoensis** *Aubrév. & Pellegr.* in Bull. Soc. Bot. Fr. 85 : 292 (1938) ; Aubrév. Fl. For. C. Iv. 2 : 200, t. 204, 4–6. A tree, to 50 ft. high, cauliflorous ; fruits yellow ; in forest.
    Iv.C.: Banco (fl. & young fr. Nov.) *Aubrév.* 883! 1661! **G.C.:** Boin River F.R., Yakasi (Dec.) *Vigne* FH 3193! Suburi F.R. (Nov.) *Vigne* FH 4751!

9. MELANODISCUS Radlk. in Th. Dur. Ind. Gen. 75 (1888) ; Engl. Pflanzenr. Sapindac. 816 (1932).

Branchlets, rhachides and inflorescence softly pubescent ; leaflets 2–4 pairs, the lowest pair small ovate-cordate, very close to the stem, the upper pairs larger, oblong-elliptic, obtuse or cuneate at base, gradually acuminate, up to 20 cm. long and 8 cm. broad, densely pubescent beneath and on midrib above ; panicles terminal, about as long as the leaves, pedicels up to 4 mm. long ; calyx about 2·5 mm. long, stamens glabrous, 7–10 mm. long ..   ..   *africanus*

M. africanus *Radlk.* in Th. Dur. Ind. Gen. 75 (1888) ; Engl. Pflanzenr. Sapindac. 817. A tree to 20 ft. high ; in forest regrowth.
    S.Nig.: Lagos *Moloney*! **Br.Cam.:** Bombe (Jan.) *Lobe Babute* Cam. 52/36!

Imperfectly known species.

Burbidge 497 from " Konnoh Country," Sierra Leone, possibly belongs to this genus but better material is needed.

10. LECANIODISCUS Planch. ex Benth.—F.T.A. 1 : 428 ; Radlk. in Engl. Pflanzenr. Sapindac. 879 (1932).

Branchlets thinly pubescent ; leaves paripinnate ; leaflets opposite or nearly so, with a short swollen petiolule, oblong or elliptic, cuneate at base, broadly caudate-acuminate, 6–18 cm. long, 4–9 cm. broad, with rather numerous spreading lateral nerves, glabrous ; racemes axillary, much shorter than the leaves, puberulous ; flowers more or less clustered, polygamous or dioecious ; pedicels about 5 mm. long, slender ; petals absent ; stamens 10, glabrous ; ovary densely setose-tomentose ; fruits broadly ovoid, 2 cm. long, often slightly 2-lobed, tomentellous, shortly beaked by the terminal style ..   ..   *cupanioides*

L. cupanioides *Planch. ex Benth.*—F.T.A. 1 : 429 ; Chev. Bot. 153 ; Radlk. l.c. 880 ; Aubrév. Fl. For. C. Iv. 2 : 202, t. 205. A small tree, or shrub, 5–30 ft. high, with spreading crown ; flowers greenish-yellow or whitish, fragrant ; ripe fruits red or yellowish, indehiscent.
    Sen.: Niokolo-Koba *Berhaut* 952. **Port.G.:** Pussubé (Mar.) *Esp. Santo* 1144! **Fr.G.:** Heudelot 854! Mamou (Mar.) *Dalz.* 8394! Kouroussa (Apr.) *Pobéguin* 673! **S.L.:** *Don*! Talla Hills (fl. Feb., fr. Apr.) *Sc. Elliot* 4934! 5004! 5006! Njala (fl. Feb., fr. Apr.) *Deighton* 2615! 2616! 2896! Rokupr (Mar.) *Jordan* 416! Falaba (Mar.) *Sc. Elliot* 5226! **Lib.:** Zorzor (Mar.) *Bequaert* 143! **Iv.C.:** many localities

fide Aubrév. *ll.c.* **G.C.:** Cape Coast *Brass*! Kibbi (fr. Mar.) *Brent* 409! Assuantsi (Mar.) *Irvine* 1568!
Bjury, Black Volta R. (July) *Chipp* 508! Achimota (Feb.) *Irvine* 2004! **Togo:** several localities *fide*
Radlk. *l.c.* **Dah.:** Cabolé to Bassila, Savalou (May) *Chev.* 23812! **N.Nig.:** Zaria *Weir* 3! Lokoja (fr.
June) *Elliott* 228! Agaie (fr. May) *Yates* 28b! **S.Nig.:** Lagos *Phillips* 10! Abeokuta *Barter* 3389!
Ibadan (Mar.) *Ujor* 27974! Benin City *Unwin* 28! **Br.Cam.:** Johann-Álbrechtshöhe (Feb.) *Staudt* 589!
Also in Ubangi-Shari, French Cameroons, Gabon, Belgian Congo, A.-E. Sudan, Uganda and Angola.
(See Appendix, p. 334.)

11. LACCODISCUS Radlk. in Sitzb. Math.-Phys. Akad. Münch. 9 : 496 (1879) ; Engl.
Pflanzenr. Sapindac. 1131 (1933).

Leaflets dentate in the upper half, 3–5 pairs, oblong-elliptic, shortly cuneate and
unequal-sided at base, caudate-acuminate, 7–22 cm. long, 4–10 cm. broad, setose-
pilose on the midrib beneath ; branchlets, inflorescence and leaf-rhachis hirsute, with
long rusty setose hairs ; panicles terminal, with long slender branches, the flowers
subsessile in the axils of conspicuous subulate bracts ; buds ellipsoid, dark brown-
tomentose when dry ; sepals 5 ; petals 5, villous inside ; stamens central, filaments
pilose ; anthers linear, apiculate ; fruits subglobose, about 1·5 cm. diam., setulose
outside, very setose inside .. .. 1. *ferrugineus*
Leaflets entire, very abruptly and narrowly triangular-acuminate, broadly oblong-
elliptic, rounded at base, setose-pubescent beneath ; other characters more or less
as above, but panicles on the main stem .. .. 2. *cauliflorus*

1. **L. ferrugineus** (*Bak.*)*Radlk.* in Sitzb. Math.-Phys. Akad. Münch. 9 : 514 (1879) ; Engl. Pflanzenr. Sapindac.
1132. *Cupania ferruginea* Bak. in F.T.A. 1 : 425 (1868). A slender tree, to 20 ft. high, or subscandent ;
densely clothed with brown hairs ; fruits bright red when ripe ; seeds with a bright orange aril.
S.Nig.: Oban *Talbot* 415! 1750! Aboabam-Boje path, Ikom Div. (fr. May) *Jones & Onochie* FHI 8303!
Br.Cam.: Victoria (fr. Apr.) *Maitland* 587! Banga (fr. Mar.) *Brenan* 9426! **F.Po:** (Jan.) *Mann* 189!
Also in French Cameroons, Spanish Guinea and Gabon.
2. **L. cauliflorus** *Hutch. & Dalz.* F.W.T.A., ed. 1, 1 : 500 (1928) ; Kew Bull. 1929 : 26 ; Radlk. in Engl.
Pflanzenr. Sapindac. 1134. A small tree ; in forest.
Lib.: *Traub* 179! Dukwia R. (July) *Linder* 92! **G.C.:** Amaneakrom (Sept.) *Chipp* 372! Prestea
(Sept.) *Vigne* FH 3072! Ankasa F.R. (young fr. Dec.) *Vigne* FH 3180! Amuni, W. Prov. (July) *Vigne*
FH 1274!

12. APORRHIZA Radlk. in Sitzb. Math.-Phys. Akad. Münch. 8 : 338 (1878) ; Engl.
Pflanzenr. Sapindac. 1134 (1933).

Margins of leaflets revolute (i.e. inrolled on lower surface) at their cuneate bases ; midrib
with short appressed puberulous hairs beneath ; inflorescence-axes greyish tomentellous, up to 40 cm. long ; petals longer than sepals, not spreading ; fruits closely
tomentellous ; leaflets 6–12, alternate or subopposite, lanceolate, gradually acuminate, 3–17 cm. long, 1–6 cm. broad .. ..1. *urophylla*
Margins of leaflets not revolute at base ; midrib beneath and inflorescence-axes with
spreading pubescent hairs, sometimes with appressed hairs as well ; fruits tomentose ;
leaflets subopposite :
Leaflets 8–14 ; petals about 5 mm. long, not clawed, longer than the erect calyx ;
leaflets obovate-elliptic or elliptic, acuminate, 6–15 cm. long, 3·5–5·5 cm. broad ;
panicles up to 40 cm. long .. .. 2. *talbotii*
Leaflets 4–8 ; petals about 3 mm. long, clawed, about as long as the spreading calyx ;
leaflets oblong-elliptic or oblong-obovate, 6–20 cm. long, 2·5–10 cm. broad, coriaceous, shining above ; panicles 20–30 cm. long .. .. .. 3. *nitida*

1. **A. urophylla** *Gilg* in Engl. Bot. Jahrb. 24 : 305 (1897) ; Radlk. in Engl. Pflanzenr. Sapindac. 1136. *A.
rugosa* A. Chev. Bot. 153. *A. talbotii* of F.W.T.A., ed. 1, 1 : 501, partly (S.L. and Iv.C. records) ; of
Aubrév. Fl. For. C. Iv. 2 : 184, t. 197 ; not of Bak. f. A forest tree, 15–30 ft. high, rarely to 80 ft., with
long panicles of creamy-white flowers ; fruits grey-tomentellous outside, brown, glabrous and shining
within ; seeds with yellow arils.
S.L.: Ronietta (fr. Nov.) *Thomas* 5597! Taiama (fr. Nov.) *Deighton* 3448a! 3448b! Mano (Oct.) *Deighton*
4035! Panguma (young fr. Apr.) *Deighton* 3904! **Lib.:** Peahtah (Oct.) *Linder* 907! Fayapulu (fr. Oct.)
Linder 1136! Gbanga (Sept.) *Linder* 708! Sankwele Dist. (Sept., Oct.) *Baldwin* 9290! 9512! *Harley*!
Dukwia R. (Oct.) *Cooper* 108! **Iv.C.:** Bingerville etc. *Chev.* 15249! Banco, Djibi, Massa-Mé and Rasso
Forests *fide* Aubrév. *l.c.* **G.C.:** Aburi (fl. & young fr. Oct.) *Vigne* FH 4259! Axim (Aug.) *Beavan* FH
3127! Subiri F.R. (Aug.) *Andoh* FH 5590! Also in French Cameroons.
2. **A. talbotii** *Bak. f.* in Cat. Talb. 20 (1913) ; Radlk. l.c. 1137. A small forest tree.
S.Nig.: Oban *Talbot* 416! 1292! Also in Ubangi-Shari. (See Appendix, p. 331.)
3. **A. nitida** *Gilg ex Engl.* Pflanzenw. Afr. 3, 2 : 280 (1921) ; Milne-Redhead in Kew Bull. 1931 : 272 ; Radlk.
l.c. 1136. *A. talbotii* of Kennedy For. Fl. S. Nig. 183. A small forest tree, by streams ; flowers creamy-
white ; seeds black, with orange aril.
S.Nig.: Jamieson R., Benin Prov. *Kennedy* 1952! 2054! 2057! Also in Belgian Congo, N. Rhodesia,
Nyasaland and Tanganyika.

13. LYCHNODISCUS Radlk. in Sitzb. Math.-Phys. Akad. Münch. 8 : 332 (1878) ;
Engl. Pflanzenr. Sapindac. 1137 (1933).

Leaflets (3–) 4 (–6) pairs, sparingly puberulous beneath, with 6–15 main lateral nerves
on each side of midrib, lower surface closely reticulate, not silvery, venation not
parallel ; petioles 3–7·5 cm. long, rhachides 4·5–20 cm. long ; leaflets elongate-
elliptic, cuneate at base, acuminate, often toothed, 5–22 cm. long, 2·2–6 cm. broad ;
panicles terminal, up to 25 cm. long ; calyx 4 mm. long ; stamens 10–12 ; fruits
1·5–2 cm. long and broad, 3-lobed, tomentellous outside .. .. .. 1. *reticulatus*

Leaflets (5–) 6 pairs, densely rusty-pubescent beneath, with 20–22 main lateral nerves on each side of midrib, lower surface often silvery, with parallel venation ; petioles 10–17 cm. long, rhachides 17–31 cm. long ; leaflets elliptic or ovate-oblong, cuneate at base, acuminate, often entire, 10–40 cm. long, 4·2–13·5 cm. broad ; panicles terminal, up to 30 cm. long ; calyx 6–8 mm. long ; stamens 15–18 ; fruits 3–3·7 cm. long, 2·6–3 cm. broad, 3-lobed, bristly tomentose outside   ..   ..   2. *dananensis*

1. **L. reticulatus** *Radlk.* in Sitzb. Math.-Phys. Akad. Münch. 333 (1878) ; Engl. Pflanzenr. Sapindac. 1139. A tree, to 45 ft. high, with spreading crown, in understorey of forest ; flowers greenish cream ; fruits creamy-yellow suffused pink outside, the valves magenta pink inside ; seeds scarlet, shiny and sticky. **G.C.:** Kakanentumi, E. Akim (Mar.) *Johnson* 601 ! Kwahu Prasu (fl. Feb., fr. June) *Vigne* FH 1609 ! 1609a ! Juaso (Mar.) *Akpabla* 288 ! Atewa Range (Feb.) *Vigne* FH 4335 ! Worobong F.R. (Jan.) *Vigne* FH 4317 ! **S.Nig.:** Ibadan South F.R. (Apr.) *Ahmed & Chizea* FHI 19807 ! Akure F.R. (fl. Oct., fr. Apr.) *Keay* FHI 21558 ! *Symington* FHI 3394 ! **F.Po:** *Mann* 1422 !

2. **L. dananensis** *Aubrév. & Pellegr.* in Bull. Soc. Bot. Fr. 85 : 291 (1938) ; Aubrév. Fl. For. C. Iv. 2 : 184, t. 196 (1936), French descr. only. A shrub or small tree, to 50 ft. high, in understorey of forest ; flowers pale greenish ; fruits greenish outside, the valves red inside ; seeds orange. **S.L.:** Peri, Gaura (fl. & fr. Dec.) *Deighton* 5889 ! **Iv.C.:** Man *Aubrév.* 1041 ! R. Sassandra to Dué Koué *Aubrév.* 934. Guiglo *Aubrév.* 1227 ! **G.C.:** Aiyola F.R., Akim Oda (fl. & fr. Apr.) *Andoh* FH 5142 ! 5146 ! Prah-Anum, Amentia (Mar.) *Wills* 67 ! Banka, Ashanti (Mar.) *Vigne* FH 1853 ! 1854 !

14. BLIGHIA Konig—F.T.A. 1 : 426 ; Radlk. in Engl. Pflanzenr. Sapindac. 1142 (1933) ; Wilczek in Bull. Jard. Bot. Brux. 21 : 149 (1951). *Phialodiscus* Radlk. (1879)—F.W.T.A., ed. 1, 1 : 502 ; Radlk. l.c. 1147.

Midrib and main lateral nerves impressed, or at any rate not prominent, above ; disk thick, fleshy, puberulous or glabrous ; fruits large, 3–8 cm. long, pyriform, sub-trigonous or acutely triangular in section, tomentose within :

Capsules with rounded valves, 4–6 cm. long, about 3 cm. diam., rugose outside when dry ; calyx-lobes 2–2·5 mm. long ; petals 3–4 mm. long, exceeding the appendages ; leaflets rounded or shortly acuminate at apex ; branchlets, petioles, rhachides and petiolules rather densely yellowish pubescent ; petioles flattened above, 0·5–3·5 cm. long ; rhachides 1·5–16 cm. long ; leaflets (2–) 3–5 pairs, the upper ones largest, obovate, oblong or subelliptic, acute to rounded at base, 3–18 cm. long, 2–8·5 cm. broad, pubescent on the nerves beneath ; racemes axillary, 5–20 cm. long, densely pubescent ..   ..   1. *sapida*

Capsules with acutely angled or winged valves, 3–8 cm. long, 3–4·5 cm. diam., smooth outside when dry ; calyx-lobes 1–1·2 mm. long ; petals 1·8–3 mm. long, shorter than or equalling the appendages ; leaflets usually distinctly and sometimes long-acuminate at apex ; branchlets, petioles, rhachides and petiolules tomentellous or puberulous ; petioles more or less winged, 1·5–10 cm. long ; rhachides 0·5–10 cm. long ; leaflets (2–) 3–4 pairs, the upper ones largest, elliptic or oblong-elliptic ; acute to rounded at base, 5–19 cm. long, 2–8 cm. broad, glabrous or pubescent only on the nerves beneath ; racemes axillary, 4–15 cm. long, tomentellous to glabrous
2. *welwitschii*

Midrib and main lateral nerves prominent or prominulous above ; disk thin, glabrous ; fruits 1·5–3 cm. long, 1·5–2·2 cm. broad, obovoid-triangular with winged lobes, glabrous or pilose and glabrescent within ; calyx-lobes about 1 mm. long ; petals 1–1·5 mm. long, shorter than the appendages ; branchlets, petioles, rhachides and petiolules golden-pubescent and more or less glabrescent ; petioles 0–7 cm. long, rhachides 0–5 cm. long ; leaflets 1–3 (–4) pairs, upper ones largest, rarely leaves 1-foliolate, oblong-elliptic or oblong-oblanceolate, cuneate to rounded at base, broadly acuminate, 3·5–25 cm. long, 1·5–10 cm. broad, usually pubescent in the axils of the main lateral nerves beneath ; racemes axillary, up to 11 cm. long, pubescent
3. *unijugata*

1. **B. sapida** *Konig* in Konig & Sims, Ann. Bot. 2 : 571, tt. 16 & 17 (1806) ; F.T.A. 1 : 426 ; Chev. Bot. 153 ; Radlk. in Engl. Pflanzenr. Sapindac. 1142, fig. 32 ; Aubrév. Fl. For. C. Iv. 2 : 188, t. 199 ; Wilczek in Bull. Jard. Bot. Brux. 21 : 152, fig. 39. A forest tree 20–80 ft. high, with spreading crown and ribbed branchlets ; flowers bisexual and male on separate trees, greenish-white, fragrant ; capsules bright red when ripe, opening to expose 3 shining black, oblong seeds, each with a large yellow or whitish aril ; native in forest outliers in the savannah regions and in drier parts of the forest regions of the eastern half of our area, but most usually seen planted near dwellings. **Sen.:** Ravin des Voleurs *Berhaut* 1145 ! **Fr.Sud.:** Fincolo *Chev.* 740. **Port.G.:** Binte, Carache *Esp. Santo* 1998 ! **Fr.G.:** Beyla (Mar.) *Chev.* 20867 ! Kouroussa (Apr.) *Pobéguin* 686 ! **S.L.:** Freetown *Johnston* 69 ! Magbile (Dec.) *Thomas* 6035 ! Tombo *Deighton* 2676 ! **Iv.C.:** Boubo Forest *Aubrév.* 15 ! And other localities *fide* Chev. and Aubrév. *ll.c.* **G.C.:** Mampong (Apr.) *Vigne* FH 1930 ! Akuse (May) *Brent* 547 ! Achimota (May) *Irvine* 1637 ! Tamale (fr. Apr.) *Bally* 139 ! **Togo:** Jegge (Mar.) *Büttner* 423. Atikpui (Apr.) *Schlechter* 12985 ! **Dah.:** Kétou, Zagnanado (Feb.) *Chev.* 23028 ! **N.Nig.:** Kontagora (Feb.) *Dalz.* 282 ! Agaie (Oct.–Dec.) *Yates* 5 ! Nupe *Barter* 1004 ! **S.Nig.:** Lagos (Dec.) *Barter* 2140 ! *Dalz.* 1386 ! Lanlate (Jan.) *Keay & Russell* FHI 22466 ! Ibadan (Mar.) *Bolude* FHI 3025 ! Benin *Unwin* 27 ! Also in French Cameroons, Principe and S. Tomé ; cultivated in India and tropical America. (See Appendix, p. 331.)

2. **B. welwitschii** (*Hiern*) *Radlk.* in Engl. Pflanzenr. Sapindac. 1146 (1933) ; Wilczek l.c. 155, fig. 40, q.v. for synonymy. *Phialodiscus welwitschii* Hiern (1896). *P. bancoensis* Aubrév. & Pellegr. (1938)—Aubrév. Fl. For. C. Iv. 2 : 192, t. 201 (1936), French descr. only. *P. unijugatus* of F.W.T.A., ed. 1, 1 : 502, partly ; of Kennedy For. Fl. S. Nig. 184, partly. *P. plurijugatus* of Aubrév. l.c. 192, t. 200, 8–9. *Blighia sapida* of F.W.T.A., ed. 1, 1 : 501, partly. A forest tree, to 150 ft. high, with cylindrical bole ; flowers yellowish-white ; ripe fruits yellowish red with 3 shining black seeds, each with a large yellow aril. **S.L.:** Bo (Feb.) *Lane-Poole* 165 ! 107 ! Jalahun (fr. May) *Lane-Poole* 286 ! Rokupr (Feb.) *Jordan* 383 ! Njala (Oct.) *Deighton* 2244 ! **Lib.:** Gbanga (Sept.) *Bequaert* in Hb. *Linder* 738 ! Peahtah (Oct.) *Linder*

923 ! Dimei (fl. July, fr. Aug.) *Barker* 1051 ! 1377 ! Mecca (fr. Dec.) *Baldwin* 10788 ! **Iv.C.**: *Aubrév.* !
Abidjan *Aubrév.* 176 ! Banco *Aubrév.* 345. Dabou *Chev.* 17240. **G.C.**: Obuasi (young fr. Jan.) *Soward*
659 ! Mim (Nov.) *Vigne* FH 2572 ! Dunkwa *Vigne* FH 876 ! **S.Nig.**: Usonigbe F.R., Benin (young fr.
Nov.) *Ejiofor* FHI 24663 ! Sapoba *Kennedy* 1678 ! Calabar (Nov., Dec.) *Unwin* 26 ! 32 ! **Br.Cam.**:
N. Korup F.R., Kumba Div. (fr. July) *Olorunfemi* FHI 30685 ! Also in Ubangi-Shari, Gabon, Belgian
Congo, Uganda and Angola.

3. **B. unijugata** *Bak.* in F.T.A. 1 : 427 (1868) ; Chev. Bot. 154 ; Wilczek l.c. 159, fig. 41. *Phialodiscus uni-*
jugatus (Bak.) Radlk. (1879)—Chev. Bot. 154 ; F.W.T.A., ed. 1, 1 : 502, partly ; Radlk. in Engl.
Pflanzenr. Sapindac. 1147 ; Aubrév. l.c. 190, t. 200, 1–7. A forest tree, 10–60 ft. high ; flowers whitish,
very fragrant ; ripe fruits red or pinkish red, with 3 shining black seeds, each with a yellow aril.
S.L.: Rokupr (Mar.) *Jordan* 418 ! Njala (Mar.) *Deighton* 2875 ! 4713 ! Mabasike (Feb.) *Lane-Poole* 70 !
Sasseni, Scarcies (Jan.) *Sc. Elliot* 4539 ! **Lib.**: Gletown (July) *Baldwin* 6737 ! Ganta (young fr. Feb.)
Baldwin 11027 ! **Iv.C.**: various localities *fide* Chev. and Aubrév. *ll.c.* **G.C.**: Kumasi (Dec.) *Vigne* FH 1780 !
Odumase, Krobo *Irvine* 900 ! Winneba (fr. Mar.) *Dalz.* 8288 ! Ofuasi (fr. Mar.) *Vigne* FH 1849 ! Boaso,
W. Ashanti (fr. Mar.) *Chipp* 137 ! **Dah.**: Adjara, Porto Novo (Jan.) *Chev.* 22743 ! Aouandjitoumé,
Zagnanado (fr. Mar.) *Chev.* 23288 ! **N.Nig.**: Matyoro *Thornewill* 120 ! Dogon Kurmi, Jemaa *Keay*
FHI 22262 ! **S.Nig.**: Lagos *Moloney* 12 ! Otta *Barter* 3356 ! Busogboro, Ibadan (fl. Dec., fr. Mar.)
Chizea FHI 23968 ! *Idahosa* FHI 23861 ! Apomu (fr. Apr.) *Foster* 210a ! **Br.Cam.**: Ambas Bay (Feb.)
Mann 760 ! Likomba (Oct.) *Mildbr.* 10553 ! Johann-Albrechtshöhe *Staudt* 497 ! 574 ! Bamenda (fr.
Feb.) *Johnstone* 83 ! Widespread in tropical Africa.

FIG. 198.—BLIGHIA SAPIDA *Konig* (SAPINDACEAE).

A, inflorescence and leaf. B, flower. C, same in vertical section. D, stamen. E, cross-section of
ovary. F, fruit. G, same after dehiscence showing arillate seeds.

15. **ERIOCOELUM** Hook. f.—F.T.A. 1 : 427 ; Radlk. in Engl. Pflanzenr. Sapindac.
1150 (1933).

Fruits large, 2·3–3 cm. long, valves 3·5–4 cm. wide and 4 mm. thick, glabrous outside
except occasionally for a few short hairs near the base ; branchlets, rhachides and
petiolules shortly pubescent ; inflorescence paniculate, up to 21 cm. long, the axes
rusty puberulous ; leaflets (2–) 3 pairs, the uppermost pair much the largest, obovate-
elliptic or obovate, shortly acuminate, up to 30 cm. long and 14·5 cm. broad, soon
glabrous beneath ..   ..   1. *macrocarpum*
Fruits smaller, 1·4–2·3 cm. long, valves 1·7–3·2 cm. wide and 1–3 mm. thick ; inflorescence-axes tomentose or pilose ; leaflets up to 20 cm. long and 7 cm. broad,
more or less pilose or pubescent at least on the midrib beneath :
*Fruits 1·8–2·3 cm. long, valves 2·2–3·2 cm. wide and 2·5–3 mm. thick, pubescent, at
least near the base ; inflorescence paniculate or racemose, densely rusty-tomentose ;
branchlets, rhachides and petiolules rusty-tomentose or -pilose :
Midrib impressed above ; main lateral nerves 13–16, leaving the midrib at 60° or
more, impressed above ; leaflets 3–4 (–5), oblong, broad and rounded at base,
shortly acuminate, rather densely curly-pubescent on midrib, nerves and lamina
beneath ; fruits about 2·3 cm. long, valves 2·8–3·2 cm. wide and 3 mm. thick ;
inflorescence up to 12 cm. long   ..   ..   2. *oblongum*
Midrib prominent above ; main lateral nerves 6–13, leaving the midrib at about 45°,
prominulous above ; leaflets (2–) 3, elliptic to obovate-elliptic, upper leaflets
cuneate at base, subacute, obtuse or rounded at apex, sparingly pilose or pubescent
on midrib and nerves beneath ; fruits 1·8–2 cm. long, valves 2·2–2·5 cm. wide and
2·5–3 mm. thick ; inflorescence up to 20 cm. long..   ..   ..   3. *kerstingii*

*Fruits 1·4–1·8 cm. long ; valves 1·7–2 cm. wide and 1–1·5 mm. thick ; leaflets up to
16·5 cm. long and 7 cm. broad, pilose or pubescent on the midrib only beneath :
Branchlets and rhachides with long weak pilose hairs, often dense ; fruits pilose then
partially glabrescent outside ; leaflets (3–) 4–5 pairs, elliptic, elliptic-oblanceolate
or obovate-elliptic, distinctly acuminate, midrib sparingly pilose or pubescent
beneath ; racemes up to 35 cm. long.. .. .. 4. *racemosum*
Branchlets and rhachides shortly pubescent ; leaflets 2–4 pairs, the uppermost pair
the largest, more or less elliptic, usually cuneate at base and acuminate at apex,
glabrous or sparingly pubescent on midrib beneath ; racemes up to 9 cm. long :
Fruits densely covered with pungent bristles, then partially glabrescent, outside ;
branchlets, leaves and inflorescence-axes sometimes with scattered pungent
bristles .. .. 5. *pungens* var. *pungens*
Fruits shortly pubescent ; pungent bristles absent .. 5a. *pungens* var. *inermis*

1. **E. macrocarpum** *Gilg ex Engl.* Pflanzenw. Afr. 3, 2 : 284 (1921) ; Radlk. in Engl. Pflanzenr. Sapindac.
1155 ; not of Chev. Bot. 154. *E. racemosum* of F.W.T.A., ed. 1, 1 : 502, partly (syn. *E. macrocarpum*
Gilg, only). *E. kerstingii* of F.W.T.A., ed. 1, 1 : 502, partly (*Leslie* 34 ex S. Nig., only) ; of Kennedy
For. Fl. S. Nig. 184. A forest tree, to 80 ft. high ; flowers whitish ; fruits hard, brown and shining
when ripe ; seeds black, shiny, each with a red aril.
S.Nig.: Benin Prov. (fr. Nov.) *Ejiofor* FHI 24670 ! Sapoba *Kennedy* 1676 ! 2094 ! 2316 ! Usonigbe
F.R. (fr. Nov.) *Keay* FHI 25569 ! Warifi, Jekri-Sobo Dist. (fr. Sept.) *Onochie* FHI 34319 ! Asaba
Leslie 34 ! **Br.Cam.:** Victoria (fl. Apr., Oct., fr. Dec.) *Maitland* 101 ! 622 ! 670 ! *Winkler* 520. Also
in French Cameroons, Spanish Guinea, Gabon and Belgian Congo ; also (?) Cabinda (see Consp. Fl.
Angol. 2 : 90).

2. **E. oblongum** *Keay* in Bull. Jard. Bot. Brux. 26 : 200, fig. 62 (1956). A forest tree, to 80 ft. high ; slash
dark red. **S.Nig.:** along the rivers, Eket Dist. *Talbot* 3051 ! 3246 ! Aboabam–Arrimakpan path, Afi River
F.R., Ikom Div. (fr. Dec.) *Keay* FHI 28219 !

3. **E. kerstingii** *Gilg ex Engl.* l.c. 282, fig. 136 (1921) ; Radlk. l.c. 1153, fig. 34 ; Aubrév. Fl. For. C. Iv. 2 :
186 ; Aubrév. Fl. For. Soud.-Guin. 386 ; not of Kennedy l.c. (?) *E. paniculatum* of Chev. Bot. 154.
E. racemosum of Chev. Bot. 154, partly (*Chev.* 24324). A tree 10–50 ft. high ; flowers white, in rusty
panicles or racemes ; mainly in streamside forests of the savannah regions.
Port.G.: Copa to Sintcham Mabel, Gabu (Nov.) *Esp. Santo* 3165 ! **Fr.G.:** Farana (fr. Mar.) *Sc. Elliot*
5234 ! Kakrima Valley *Pobéguin* 2061. Labé *Chev.* 12317. Kindia *Chev.* 13387. Dalaba to Diaguissa
Chev. 20335. **Lib.:** Tappita (fr. Aug.) *Baldwin* 9036 ! Mecca, Boporo (fr. Nov.) *Baldwin* 10395 ! Ganta
(Mar.) *Harley* 1123 ! **Iv.C.:** Niangbo *Aubrév.* 1713. Ferkéssédougou *Aubrév.* 2733. **G.C.:** Bonsa Su
(May) *Vigne* FH 1988 ! **Togo:** Okou *Aubrév.* 101d. Sokode (fr. Dec.) *Kersting* A. 257 ! Sokode-Basari
(fl. Nov., Dec., fr. Feb., Mar.) *Kersting* A. 518 ! Kadjakpe, Hohoe Dist. (Nov.) *St. C. Thompson* FH 3673 !
And other localities *fide* Radlk. *l.c.* **Dah.:** Konkobiri (fr. July) *Chev.* 24324 ! **N.Nig.:** Naraguta (fl. & fr.
Dec.) *Lely* 734 ! Kafanchan (fl. Jan., fr. Feb.) *Onyeagocha* FHI 7626 ! *Thornewill* ! **Br.Cam.:** Bamenda
Prov. : Esu (fr. Feb.) *Johnstone* 39/31 ! Nfome (fr. May) FHI 313 ! 9284 ! Also in French Cameroons,
Ubangi-Shari, A.-E. Sudan, Belgian Congo and Uganda. (See Appendix, p. 334.)

4. **E. racemosum** *Bak.* in F.T.A. 1 : 427 (1868) ; (?) Chev. Bot. 154, partly ; F.W.T.A., ed. 1, 1 : 502, partly ;
Radlk. l.c. 1152. *E. pendulum* Stapf (1905)—Radlk. l.c. A lax tree 15–30 ft. high, in forest ; flowers
white in long pendulous panicles ; ripe fruits yellow.
S.L.: *Afzelius* ! Bagroo R. (Apr.) *Mann* 807 ! Hill Station (fl. & fr. May) *Deighton* 2982 ! Panguma
(Apr.) *Lane-Poole* 99 ! Falaba (Apr.) *Aylmer* 50 ! **Lib.:** Monrovia *Whyte* ! Kakatown *Whyte* ! Gbanga
(Sept.) *Linder* 599 ! Brewersville (fl. & fr. Mar.) *Barker* 1232 ! Mecca, Grand Cape Mount (fr. Dec.)
Baldwin 10840 ! **G.C.:** Bonsa Su (fl. & fr. May) *Vigne* FH 1988 ! (See Appendix, p. 334.)

5. **E. pungens** *Radlk. ex Engl.* var. **pungens**—l.c. 282 (1921) ; Radlk. l.c. 1152. *E. macrocarpum* of Chev. Bot.
154. *E. racemosum* of F.W.T.A., ed. 1, 1 : 502, partly ; of Aubrév. Fl. For. C. Iv. 2 : 186, t. 198. A
spreading shrub or small tree, to 35 ft. high ; flowers white ; ripe fruits orange ; in forest, often in
moist places.
Lib.: Beiden, Boporo (fr. Nov.) *Baldwin* 10266a ! Zuie, Boporo (fr. Dec.) *Baldwin* 10690 ! Dukwia R.
(fr. Nov.) *Cooper* 145 ! **Iv.C.:** Bingerville, Abidjan & Dabou *Chev.* 15236 ! Banco *Aubrév.* 1483 ! **G.C.:**
Amameakrom, Tano R. (fl. & fr. Sept.) *Chipp* 373 ! Axim (fr. Feb.) *Irvine* 2372 ! Bonsa R. (Aug.
Vigne FH 133 !

5a. **E. pungens** var. **inermis** *Keay* l.c. 202 (1956).
S.Nig.: Degema Dist. *Talbot* 3644 ! Eket Dist. *Talbot* 3199 ! 3296b ! 3369 !

Imperfectly known species.

E. sp. See Mildbr. Wiss. Ergebn. 1910–11 : 186 (1922) ; specimen not traced.
F.Po: 3,500–4,500 ft. *Mildbr.* 6409.

16. DODONAEA Mill.—F.T.A. 1 : 433 ; Radlk. in Engl. Pflanzenr. Sapindac. 1350 (1933).

A viscid shrub or small tree ; branchlets angular, not hairy ; leaves simple, subsessile,
oblanceolate, narrowed to the base, obtusely apiculate, 7–12 cm. long, 2–3·5 cm.
broad, thin, glabrous, with numerous spreading lateral nerves ; flowers unisexual or
polygamous in small terminal panicles or subracemose ; pedicels slender, up to 1·5
cm. long in fruit ; petals absent ; stamens 5–8 ; fruit suborbicular, deeply emarginate,
2- or more-winged, up to 3 cm. diam., wings membranous, reticulate .. *viscosa*

D. viscosa *Jacq.* Enum. Syst. Pl. Ins. Carib. 19 (1760) ; F.T.A. 1 : 433 ; Chev. Bot. 154 ; Radlk. l.c. 1363
(q.v. for infraspecific variants) ; Aubrév. Fl. For. C. Iv. 2 : 180, t. 195, 7–10. *Ptelea viscosa* Linn. (1753).
Dodonaea repanda Schum. & Thonn. (1827). A plant of sandy coastal areas ; at least one other form is
commonly grown as a hedge plant inland.
Sen.: *Heudelot* ! *Sieber* 54 ! Cape Verde *Brunner* ! Dakar (fl. & fr. Apr.) *Hagerup* 8 ! **Gam.:** *Dawe* 39 !
Don ! **Fr.Sud.:** Niayes (Dec.) *Chev.* 3030 ! **Fr.G.:** Conakry *Debeaux* ! **S.L.:** Yele, Turtle Isl. (fl. & fr.
Nov.) *Deighton* 2329 ! Bawbaw (fr. May) *Deighton* 2693 ! **Lib.:** Monrovia (fr. Nov.) *Whyte* ! *Baldwin*
1421 ! Grand Cess (fr. Mar.) *Baldwin* 11620 ! **Iv.C.:** Sassandra Port *Chev.* 17931. Bliéron *Chev.* 19911.
Tabou *Chev.* 19928. **G.C.:** Tarkwa (fr. June) *Andoh* FH 5529 ! **Togo:** Lome *Warnecke* 206 ! **N.Nig.:**
Ilorin *J. E. Taylor* ! **S.Nig.:** Lagos (July, Aug.) *Dalz.* 1011 ! 1272 ! Eket *Talbot* ! Widespread in tropical
and subtropical regions. (See Appendix, p. 333.)

17. ZANHA Hiern in Cat. Welw. 1 : 128 (1896) ; Radlk. in Engl. Pflanzenr. Sapindac. 1420 (1933).   *Talsiopsis* Radlk. (1907).

Branchlets glabrous ; leaflets about 3–4 pairs, subalternate, ovate or oblong, slightly cuneate at base, obtusely acuminate, 4–9 cm. long, 2·5–3·5 cm. broad, glabrous ; rhachis glabrous ; flowers numerous in dense head-like cymes clustered at the apex of the branchlets ; peduncle puberulous ; sepals ovate ; ovary 2-celled, glabrous ; fruit 1-celled, indehiscent, ellipsoid, apiculate, 2 cm. long, wrinkled when dry

golungensis

Z. golungensis *Hiern* l.c. (1896) ; Radlk. l.c. 1421 ; Aubrév. Fl. For. Soud.-Guin. 390, t. 82, 1–2. *Talsiopsis oliviformis* Radlk. (1907). *Zanha vuilletii* A. Chev. in Bull. Mus. Hist. Nat. sér. 2, 5 : 157 (1933) ; Radlk. l.c. 1421. A tree 20–70 ft. high, with a dense crown of dark shining deciduous foliage ; bark sometimes orange ; flowers dioecious, greenish white, produced when trees are in young leaf or leafless ; ripe fruits orange ; usually in streamside forest in the savannah regions.
Fr.Sud.: Koulouba Plateau, Bamako (young fr. June) *Chev.* 25988 ; *Vuillet* 2909. Kita *Jaeger* 3 ! **Iv.C.:** Ferkéssédougou *Aubrév.* 2286. Niangbo *Aubrév.* 1712. Oumé *Aubrév.* 1262. **G.C.:** Gambaga (fr. June) *Akpabla* 709 ! **Togo:** Aledjo Kadara *Kersting* A. 548a ! Woagbun Mts. *Kersting* A. 548 ! And other localities *fide* Radlk. *l.c.* **Dah.:** Birni *Aubrév.* 98d. Bassila *Chev.* 23797. **N.Nig.:** Anara F.R., Zaria Prov. (fr. May) *Keay* FHI 22892 ! 22911 ! Fatika, Zaria Prov. (fr. May) *Daggash* FHI 31428 ! **S.Nig.:** 10 miles E. of Shepeteri, Oyo Prov. (Feb.) *Keay* FHI 22518 ! Extends to A.-E. Sudan, East and Central Africa and Angola. (See Appendix, p. 336.)

18. MAJIDEA J. Kirk ex Oliv. in Hook. Ic. Pl. t. 1097 (1871) ; Radlk. in Engl. Pflanzenr. Sapindac. 1462 (1934).

Branchlets puberulous ; leaflets alternate, 5–11 pairs, sessile, oblong, obtusely acuminate, 6–9 cm. long, 2–3·5 cm. broad, glabrous ; flowers in dense terminal panicles ; sepals tomentellous ; stamens glabrous within the large fleshy disk ; fruits very broadly turbinate, about 4·5 cm. wide, puberulous, thin, opening into the cells ; seeds bluish, ellipsoid, softly tomentellous, 1·5 cm. long ..   ..   *fosteri*

M. fosteri (*Sprague*) Radlk. in Engl. Bot. Jahrb. 56 : 255 (1920) ; Pflanzenr. Sapindac. 1463 ; Aubrév. Fl. For. C. Iv. 2 : 198, t. 203. *Harpullia fosteri* Sprague in Kew Bull. 1908 : 433 ; F.W.T.A., ed. 1, 1 : 502. *H. multijuga* Radlk. (1912). *Majidea multijuga* (Radlk.) Radlk. (1920). *Anomabia cyanosperma* A. Chev. in Bull. Soc. Bot. Fr. 58, Mém. 8 : 148 (1912) ; Chev. Bot. 155. *Majidea cyanosperma* (A. Chev.) Radlk. (1920). A forest tree, up to 100 ft. or more high, with angled branchlets ; calyx yellowish, disk red, anthers pink ; fruits brownish outside, pink within ; seeds blue.
Iv.C.: Anoumaba (Nov.) *Chev.* 22413. Bouroukrou *Chev.* 16950. Sahoua to Bangouanou (fl. & fr. Nov.) *Chev.* 22443 ! **G.C.:** Koforidua (fl. & fr. Mar.) *Brown* FH 2210 ! Berekum, Ashanti (Sept.) *Vigne* FH 2519 ! Kumasi (fl. & young fr. Mar.) *Vigne* FH 1636 ! Akwapim Ridge, Mampong (fl. & fr. Feb.) *Irvine* 1506 ! **S.Nig.:** Ilaro (fr. June) *Onochie* FHI 8190 ! Olokemeji (fl. & young fr. Mar.) *Foster* 49 ! Okomu F.R. *Brenan* 8821 ! **Br.Cam.:** Victoria (Feb.) *Maitland* 386 ! Tombel (Feb.) *Forteh* FHI 9288 ! Also in French Cameroons, Belgian Congo, A.-E. Sudan and Uganda.

120. MELIANTHACEAE

Shrubs or small trees. Leaves alternate, pinnate, stipulate ; stipules intrapetiolar, often large. Flowers hermaphrodite, rarely unisexual, racemose, zygomorphic. Calyx of 5 unequal segments, imbricate. Petals 5, free, subperigynous, clawed, unequal. Disk unilateral or annular, lining the inside of the calyx. Stamens 4–6, inserted within the disk, free or variously connate, often declinate ; anthers 2-celled, opening lengthwise. Ovary 4–5-celled, superior ; style central, dentate or truncate ; ovules 1–4 in each cell, axile. Fruit a papery or woody capsule, loculicidally 4–5-valved or opening only at the apex. Seeds with copious endosperm and straight embryo.

Confined to Africa ; recognized by the pinnate leaves and intrapetiolar stipules, subperigynous petals, unilateral or annular disk, and seeds with copious endosperm.

BERSAMA Fres.—F.T.A. 1 : 433 ; Verdcourt in Kew Bull. 1950 : 233.

Leaflets glabrous beneath or pubescent only on the midrib, 0·8–7·5 cm. broad, with 6–13 main lateral nerves on each side of midrib ; pedicels 5–7 mm. long ; sepals oblong, rounded at apex, 4–5 mm. long ; fruits subglobose, 2–3 cm. diam. :
  Leaflets (3–) 6–9 pairs, ovate to elliptic-lanceolate, thin or coriaceous, glabrous or pubescent, margins usually dentate, 0·8–5 (–7) cm. broad :
    Sepals sparingly appressed-pubescent ; inflorescence 12–25 cm. long,. axis and pedicels sparingly pubescent ; leaflets thin, drying dark green or brownish, (3–) 6–9 pairs, 2·5–13 (–23) cm. long, 0·8–5 (–7) cm. broad
1. abyssinica subsp. paullinioides var. paullinioides
    Sepals densely pubescent ; inflorescence 30–40 cm. long, axis and pedicels densely shaggy-pubescent ; leaflets coriaceous, drying pale green or greyish, (3–) 6–7 pairs, 1·5–9·5 cm. long, 0·8–3 cm. broad
1a. abyssinica subsp. paullinioides var. engleriana

Leaflets 3–5 (–7) pairs, more or less obovate-oblong, thin and shining, quite glabrous, margins sometimes dentate, 6·5–23·5 cm. long, 3·2–7·5 cm. broad   2. *acutidens*

Leaflets densely pubescent or pilose beneath, 5·2–10 cm. broad, with 12–20 main lateral nerves on each side of midrib ; pedicels 1 cm. long or more ; leaflets oblong to oblong-lanceolate or elliptic, margins subentire :

Leaflets about 8 pairs, densely and softly pubescent beneath, 8–18 cm. long, 5·2–8·3 cm. broad ; flowers not known ; infructescence erect, about 40 cm. long, axis and pedicels tomentellous ; pedicels 2·5 cm. long in fruit ; fruits 4–5-gonous
                                                                                          3a. *maxima* var. α

Leaflets about 5 pairs, densely pilose beneath, 15–27 cm. long, 7·5–10 cm. broad ; inflorescence about 70 cm. long, axis and pedicels densely villous ; pedicels about 1 cm. long ; bracts about 2 cm. long ; sepals acute, about 8 mm. long, villous ; fruits not known ..   ..   3b. *maxima* var. β

1. **B. abyssinica** *Fres.* subsp. **paullinioides** (*Planch.*) *Verdcourt* var. **paullinioides**—in Kew Bull. 1950 : 237, partly, excl. syn. *B. maxima* Bak. ; Exell & Mendonça in Consp. Fl. Ang. 2 : 93. *Natalia paullinioides* Planch. (1849). *Bersama paullinioides* (Planch.) Bak. in F.T.A. 1 : 435 (1868) ; Chev. Bot. 155 ; F.W.T.A., ed. 1, 1 : 505 ; Aubrév. Fl. For. C. Iv. 2 : 204, t. 207, 1–6. *B. leiostegia* Stapf (1905). *B. preussii* Bak. f. (1907). *B. lobulata* Sprague & Hutch. (1913). *B. pachythyrsa* v. Brehm. (1917). *B. bolamensis* v. Brehm. (1917). *B. subulata* Hutch. & Dalz. F.W.T.A., ed. 1, 1 : 506 (q.v. for the aforementioned synonyms also) ; Kew Bull. 1929 : 27. *B. chippii* Sprague & Hutch. (1913), name only. A tree, up to 50 ft. high, with white flowers and red capsular fruits, velvety outside, containing red arillate seeds ; leaf-rhachides winged or not ; in lowland and montane forest.
Port.G.: Cubisseque (fr. Feb.) *Esp. Santo* 2378 ! Fulacunda (May) *Esp. Santo* 2036 ! Bolama *Rodrigues* 116. **Fr.G.:** Ditinn *Chev.* 12704. Kouria *Chev.* 18261. Koundian *Chev.* 20741. **S.L.:** *T. Vogel* 99 ! Leicester Peak *Barter* ! Mt. Aureol (June) *Lane-Poole* 258 ! Njala (fl. June, July, fr. Jan.) *Dalz.* 8073 ! *Deighton* 1751 ! Lowoma, Nomo (fr. Mar.) *Deighton* 4113 ! Rowala (July) *Thomas* 1177 ! **Lib.:** Sinoe Basin *Whyte* ! Dukwia R. (May) *Cooper* 451 ! Sanokwele (fr. Sept., Oct.) *Baldwin* 9494 ! 9634 ! **Iv.C.:** various localities *fide* Aubrév. *l.c.* **G.C.:** Esuasu (May) *Chipp* 217 ! Asientiem (fr. July) *Chipp* 288 ! Nyankumasi, Cape Coast (Apr.) *Andoh* FH 5486 ! Simpa (fr. May) *Vigne* FH 1961 ! Abosso (Aug.) *Vigne* 224 ! **Togo:** Akgosso Forest, Atakpame (May) *Doering* 280. **S.Nig.:** Oban *Talbot* 147 ! 1376 ! **Br.Cam.:** *Preuss* 408 ! Bamenda Prov. : Lakom, 6,000 ft. (May) *Maitland* 1377 ! 1490 ! L. Oku, 6,500 ft. (fl. & fr. Jan.) *Keay* FHI 28516 ! 28518 ! Bali-Ngemba F.R., 5,200 ft. (June) *Tiku* FHI 30413 ! **F.Po:** 1,000 ft. (Dec.) *Mann* 641 ! Widespread in tropical Africa. (See Appendix, p. 336.)

1a. **B. abyssinica** subsp. **paullinioides** var. **engleriana** (*Gürke*) *Verdcourt* in Kew Bull. 1955 : 600 (1956). *B. engleriana* Gürke in Engl. Bot. Jahrb. 14 : 307 (1891). A tree with spikes of pink and white flowers and brown capsule containing 4 red seeds with yellow arils ; leaf-rhachides slightly winged.
N.Nig.: near Jos (fl. Feb., Mar., fr. Feb.) *Batten-Poole* 271 ! *Kennedy* FHI 1194 ! *Lely* P. 156 ! Also widespread in eastern Africa.

2. **B. acutidens** *Welw. ex Hiern* Cat. Welw. 1 : 173 (1896) ; Verdcourt l.c. 243 ; Exell & Mendonça l.c. 94. A small tree, to 20 ft. high ; leaf-rhachides winged or not ; fruits pink ; in rain forest.
S.Nig.: Aboabam to Arrimakpan, Ikom Div. (fr. Dec.) *Keay* FHI 28215 ! **Br.Cam.:** Kembong F.R., Mamfe Div. *Onochie* FHI 31186 ! Also in Angola and (?) Rhodesia.

3. **B. maxima** *Bak.* in F.T.A. 1 : 434 (1868) ; Exell & Mendonça l.c. 94. Our specimens do not agree exactly with the type from Gabon and are provisionally left as varieties :—

3a. **B. maxima** var. α. This plant is placed by Verdcourt between his varieties D and E of *B. abyssinica* subsp. *paullinioides* ; see Kew Bull. 1950 : 241–2. A tree 12 ft. high, with erect infructescences held above the large whorled leaves ; fruits dull brownish red ; seeds bright red ; leaf-rhachides slightly winged.
Br.Cam.: Sabga Pass, Bamenda (fr. Jan.) *Keay* FHI 28426 !

3b. **B. maxima** var. β. A small tree 4 ft. high, with white flowers and large leaves ; leaf-rhachides winged.
Br.Cam.: Abonando, Mamfe *Rudatis* 68 ! Mbatu stream, Nkambe, Bamenda Prov. (Sept.) *Ujor* FHI 30230 !

Although the five taxa distinguished above seem to be quite distinct from one another, they are parts of a very difficult and variable complex which occurs throughout tropical Africa. Much careful field study is needed before any real progress can be made in elucidating the genus.

121. ANACARDIACEAE

Trees or shrubs, often with resinous bark. Leaves alternate, very rarely opposite, simple or compound ; stipules absent, very rarely present but obscure. Flowers hermaphrodite or unisexual, mostly actinomorphic. Calyx variously divided, sometimes semi-superior in fruit. Petals 3–7 or absent, free or rarely connate and adnate to the torus. Disk present. Stamens often double the number of the petals, rarely equal or numerous or only one fertile ; filaments free among themselves ; anthers 2-celled, opening lengthwise. Ovary superior, 1-celled, rarely 2–5-celled, or very rarely carpels free ; styles 1–5, often widely separated ; ovules solitary, pendulous from the apex or adnate to the ovary wall, or pendulous from a basal funicle. Fruit mostly drupaceous. Seeds without or with very thin endosperm ; cotyledons fleshy.

Mainly a tropical family, and fairly abundant in tropical and South Africa ; bark usually resinous.

Leaves simple ; petals imbricate ; ovary 1-celled :
Leaves glabrous ; styles 1 ; only 1 or 2 stamens fertile
Receptacle and pedicel not markedly swollen in fruit ; drupe ellipsoid ; flowers (4–) 5-merous ; sepals free to the base ; fertile stamens 1–2, staminodes 4–3
                                                                                          1. **Mangifera**

Receptacle and pedicel markedly swollen in fruit ; drupe obliquely reniform ; fertile
  stamens 1 :
  Flowers 5-merous ; sepals united only near the base ; staminodes several, well-
    developed   ..   ..   **2. Anacardium**
  Flowers 4-merous ; sepals united for half their length ; staminodes 1–2, very small
                                                        3. Fegimanra
Leaves hairy ; styles 3 ; flowers 5-merous ; stamens 5 or 10, all fertile   **13. Heeria**
Leaves compound, rarely 1-foliolate :
Leaves pinnate, usually 5- or more-foliolate :
  Petals valvate :
    Styles 3–5, free or connate just at base, or sessile stigmas 3 :
      Ovary and fruit normally 5-celled ; styles (4–) 5 ; flowers (4–) 5-merous ; stamens
        (8–) 10 :
        Drupe ovoid, not lobulate ; leaflets with a distinct intramarginal nerve
                                                        4. Spondias
        Drupe depressed, 5-lobulate ; leaflets without an intramarginal nerve
                                                    5. Antrocaryon
    Ovary 1-celled ; styles 3, or sessile stigmas 3 ; flowers 4–(–5)-merous ; stamens 4
      (–5) ; leaflets without a marginal nerve ..   ..   ..   **11. Trichoscypha**
    Styles 1, shortly lobed at apex ; ovary 1-celled ; flowers usually 5-merous ; stamens
      10–20, rarely 5 in hermaphrodite flowers ; leaflets with a distinct marginal
      nerve ; see also below   ..   ..   **12. Sorindeia**
  Petals imbricate :
    Styles (2–) 3–4 :
      Sepals free ; indumentum simple ; fruits 1–3-seeded :
        Flowers polygamous, 4-merous ; drupes 2–3-seeded ; stamens 12–16 (–26);
          anthers long   ..   ..   **7. Sclerocarya**
        Flowers dioecious, 3–4 (–5)-merous ; drupes 1 (–2)-seeded ; stamens 6–8 (–10) ;
          anthers rounded   ..   ..   **8. Pseudospondias**
      Sepals united ; indumentum stellate, at least in part ; fruits 1-seeded ; flowers
      4-merous ; stamens 8   ..   ..   **9. Lannea**
    Styles 1 :
      Flowers 3-merous ; ovary 1-celled ; calyx distinctly lobed ; stamens 6
                                              10. Haematostaphis
      Flowers 4–5-merous ; calyx campanulate, subtruncate to distinctly lobed :
        Ovary 4-celled, all cells fertile ; flowers 4-merous ; stamens 4 long and 4 short ;
        leaflets 19–43   ..   ..   **6. Nothospondias**
        Ovary 1-celled ; flowers usually 5-merous ; stamens 10–20, rarely 5 in herma-
        phrodite flowers ; leaflets 1–9 ; see also above   ..   ..   **12. Sorindeia**
Leaves 3-foliolate ; flowers 5-merous ; petals 5, imbricate ; styles 3   **14. Rhus**

Besides the above, *Schinus molle* Linn. and *S. terebinthifolius* Raddi, natives of tropical America, have been
introduced into our area.

1. MANGIFERA Linn.—F.T.A. 1 : 442.

Branchlets glabrous ; leaves oblong-lanceolate, cuneate at base, acutely acuminate,
10–25 cm. long, 3–4·5 cm. broad, glabrous, closely reticulate on both surfaces, with
numerous spreading looped lateral nerves ; petiole 2–4 cm. long ; flowers small,
polygamous or dioecious, in terminal pyramidal panicles nearly as long as the leaves ;
pedicels and sepals pubescent ; petals glabrous ; fruit obliquely ovoid, up to 15 cm.
long, with a leathery rind and fibrous stone   ..   ..   *indica*

M. indica *Linn.*—F.T.A. 1 : 442 ; Chev. Bot. 155. The Common Mango, widely cultivated but here and there
    well established in secondary bush. Native of India, now grown throughout the tropics. (See Appendix,
    p. 340.)

2. ANACARDIUM Linn. Sp. Pl. 1 : 383 (1753) ; F.T.A. 1 : 443.

Branchlets angular, glabrous ; leaves simple, obovate or oblong-obovate, cuneate at
base, rounded or truncate at apex, 8–15 cm. long, 5–10 cm. broad, papery, glabrous,
closely reticulate, with 10–15 pairs of spreading lateral nerves prominent beneath ;
petiole up to 2 cm. long, flat above ; flowers polygamous in lax terminal panicles ;
bracts large, enfolding the flowers, ovate, acuminate, softly puberulous ; petals 5,
linear ; stamens 10, usually only some bearing anthers ; fruit a kidney-shaped nut
supported by a large pear-shaped enlarged receptacle and stalk   ..   *occidentale*

A. occidentale *Linn.*—F.T.A. 1 : 443 ; Chev. Bot. 156. The Cashew tree widely cultivated and commonly
    naturalized in the bush, chiefly in coastal districts. Native of tropical America, now grown throughout
    the tropics. (See Appendix, p. 336.)

3. FEGIMANRA Pierre Fl. For. Cochinch. sub t. 263 (1892) ; Engl. Bot. Jahrb. 36 : 213 (1905).

Leaves obovate-elliptic, rounded at apex, shortly narrowed into the petiole at base,
10–15 cm. long, 5–8 cm. broad, the midrib and 8–10 pairs of lateral nerves nearly
equally prominent on both surfaces, shining above, glabrous, rigid ; petiole subterete,

1·5–2·5 cm. long ; panicle terminal, very lax and wide, 30–70 cm. long, with rather long lateral branches each with puberulous cymules of puberulous flowers at the ends ; fruits drupaceous, obliquely reniform, subtended by a cupular growth of the floral axis   ..   ..   1. *afzelii*
Leaves elongate-oblanceolate, gradually acutely acuminate, very gradually narrowed into the petiole at base, 5–16 cm. long, 1·8–3·5 cm. broad, the midrib and 15–24 pairs of lateral nerves nearly equally equally prominent on both surfaces, paler beneath, glabrous, rigid ; also with a few much smaller leaves ; petiole angular, 0·3–3 cm. long ; panicle puberulous, terminal, rather dense, up to 8·5 cm. long, with lateral branches up to 5 cm. long   ..   ..   2. *acuminatissima*

1. **F. afzelii** *Engl.* Bot. Jahrb. 36 : 214 (1905) ; Chev. Bot. 156 ; Aubrév. Fl. For. Soud.-Guin. 415, t. 89, 5–6. A glabrous shrub or tree ; flowers white or reddish ; ripe fruits red ; in rocky places.
Fr.G.: *Farmar* 314 ! Pita *Pobéguin* 2303. Lanfofomé *Pobéguin* 908. Kindia *Cochet* 63. Bakaro *Maclaud* 405. Labé *Chev.* 12230. **S.L.:** *Afzelius*. (See Appendix, p. 338.)
2. **F. acuminatissima** *Keay* in Bull. Jard. Bot. Brux. 26 : 202 (1956). A shrub or tree, to 20 ft. high ; flowers white.
S.L.: on sand by lake shore, Mano Bonjema (Jan.) *Deighton* 5309 ! **Lib.:** Brewersville (Dec.) *Baldwin* 10978 ! In swamp, Monrovia (Jan.) *Barker* 1199 !

4. SPONDIAS Linn.—F.T.A. 1 : 447.

Branchlets glabrous or puberulous ; leaves pinnate with an odd terminal leaflet ; leaflets 5–8 pairs, opposite, oblong or oblong-lanceolate, very unequal-sided at base, obtusely and broadly acuminate, 5–10 cm. long, 2–5 cm. broad, glabrous or puberulous, with a strong intramarginal nerve ; flowers small, polygamous, in lax terminal panicles ; pedicels 3 mm. long, glabrous ; sepals very small ; petals valvate, oblong-elliptic, 2·5 mm. long ; stamens 8–10, filaments slender ; fruit ovoid, 3–3·5 cm. long, wrinkled when dry   ..   ..   *mombin*

S. mombin *Linn.* Sp. Pl. 1 : 371 (1753) ; Aubrév. Fl. For. C. Iv. 2 : 174, t. 193, 1–8 ; Fl. For. Soud.-Guin. 403. *S. lutea* Linn. (1762)—F.T.A. 1 : 448 ; Chev. Bot. 156. *S.? dubia* A. Rich. (1831). *S. oghigee* G. Don (1832). A deciduous tree, to 60 ft. high, with large panicles of small white flowers and yellow plum-like fruits ; bark thick, with longitudinal fissures. Widespread and common in farmland, regrowth and villages, especially in the forest regions, but also in the savannah regions ; usually thought to be an ancient introduction from America, but possibly native in W. Africa.
Sen.: *Sieber* 50 ! Baol, Thiès *Chev.* 3514. **Gam.:** *D. H. Saunders* 46 ! Pakala F.R., Ningkom (June) *Pitt* 654 ! **Fr.Sud.:** Bongouri (Apr.) *Chev.* 691 ! Djenné (July) *Chev.* 1144 ! Bamako (May) *Hagerup* 54 ! **Port.G.:** Boi (July) *Esp. Santo* 2 ! Pussubé (Apr.) *Esp. Santo* 1175 ! **Fr.G.:** Labé, Timbo & Diaguissa *fide* Chev. *l.c.* **S.L.:** *Don* ! Bunce Isl. (Mar.) *Kirk* ! Regent (Apr.) *Sc. Elliot* 5750 ! Falaba (Mar.) *Sc. Elliot* 5244 ! Mano (Apr.) *Deighton* 1133 ! Musaia (Apr.) *Deighton* 5412 ! **Lib.:** Sinoe Basin *Whyte* ! Nyaake, Webo (June) *Baldwin* 6204 ! Tappita (fr. Aug.) *Baldwin* 9067 ! Harbel (Dec.) *Bequaert* 15 ! **Iv.C.:** Bouaké (fr. July) *Fleury* in Hb.*Chev.* 22082 ! And other localities *fide* Chev. *l.c.* **G.C.:** Esubompang (Jan.) *Chipp* 78 ! Axim *Chipp* 24 ! Nsawam (Feb., Mar.) *Dalz.* 8298 ! *Irvine* 1805 ! E. Akim *Johnson* 916 ! Kumasi (Apr.) *Kitson* 648 ! **Togo:** Lome *Warnecke* 321 ! Basari (fl. Mar., fr. Apr.) *Kersting* 93 ! A. 329 ! **Dah.:** Cabolé to Bassilia *Chev.* 23775. **N.Nig.:** Lokoja (Mar.) *Elliott* 43 ! Katagum *Dalz.* 338 ! Abinsi (Jan.) *Dalz.* 912 ! **S.Nig.:** Lagos *Foster* 44 ! Ogbomosho *Barter* 3410 ! Onitsha Dist. (Feb.) *Rosevear* 13/29 ! Cross R. (Mar.) *McLeod* ! Oban *Talbot* 1728 ! **Br.Cam.:** Ambas Bay (Jan.) *Mann* 706 ! Also in French Cameroons, Gabon, Belgian Congo, A.-E. Sudan, Angola, Annobon and S. Tomé, all possibly by introduction ; and in West Indies and tropical America. (See Appendix, p. 341.)

Besides the above, *S. purpurea* Linn. (Creole name, " Gambia Plum "), a native of tropical America, and *S. cytherea* Sonner. (Creole name, " English Plum "), a native of tropical Asia, have been introduced to Sierra Leone and other parts of our area.

5. ANTROCARYON Pierre in Bull. Soc. Linn. Paris, nouv. sér. 2 : 23 (1898).

Leaves glabrous, leaflets with 10–12 rather faint lateral nerves on each side of midrib ; central " stone " of fruit 2–2·5 cm. diam., 1·2–1·4 cm. high ; leaflets 5–8 opposite pairs, plus the terminal one ; lateral leaflets oblong, rounded and unequal-sided at base, gradually acuminate, 6–15 cm. long, 2·5–4·5 cm. broad ; flowers polygamous, small, in puberulous panicles   ..   ..   1. *klaineanum*
Leaves pubescent, usually densely so when young, leaflets with 20–24 conspicuous lateral nerves on each side of midrib ; central " stone " of fruit 3·3–4·4 cm. diam., about 3 cm. high ; leaflets (5–) 8–10 opposite or subopposite pairs, plus the terminal one ; lateral leaflets oblong or ovate-oblong, rounded or subcordate and unequal-sided at base, gradually acuminate, 6–10 (–15) cm. long, 2–4·5 (–6) cm. broad ; flowers similar to above   ..   ..   2. *micraster*

1. **A. klaineanum** *Pierre* in Bull. Soc. Linn. Paris, nouv. sér. 2 : 24 (1898) ; Engl. Pflanzenw. Afr. 3, 2 : 178, fig. 87, A–K. *Spondias soyauxii* Engl. Bot. Jahrb. 36 : 215 (1905). *Antrocaryon soyauxii* (Engl.) Engl. Pflanzenw. Afr. 3, 2 : 178, fig. 87, N–Q (1921) ; Exell & Mendonça Consp. Fl. Ang. 2 : 127. A large forest tree.
S.Nig.: Aboabam to Arrimakpan, Afi River F.R., Ikom Div. *Keay* ! **F.Po:** Bokoko *Mildbr.* 6925. Also in Gabon and Cabinda. (See Appendix, p. 337.)
2. **A. micraster** *A. Chev. & Guillaum.* in Bull. Soc. Bot. Fr. 57, Mém. 8 : 152 (1910) ; Aubrév. Fl. For. C. Iv. 2 : 174, t. 194 ; A. W. Hill in Ann. Bot. n.s. 1 : 249, fig. 11, 1–6. *A. polyneurum* Mildbr. ex Kennedy For. Fl. S. Nig. 187 (1936), English descr. only. A forest tree, to 150 ft. high, with cylindrical bole and only slight buttresses ; slash red and white streaked, resinous ; leaves crowded at ends of branches, deciduous ; flowers greenish-white.
S.L.: Njala *Deighton* 3089 ! Kambui Hills, Koya *Small* 859 ! Tonkoli F.R. (fr. June) *Small* 929 ! **Iv.C.:** Agboville *Chev.* 22336 ! Yahou to Ahiamé (Mar.) *Chev.* 17810. Abidjan *Aubrév.* 14 ! Banco *Aubrév.* 398 ! **G.C.:** Owabi (fl. Mar., fr. Apr.) *Andoh* FH 4183 ! 4313 ! Ejian (Apr.) *Vigne* FH 1102 ! **S.Nig.:** Okomu F.R., Benin (Mar.) *A. F. Ross* 196 ! Sapoba, Benin (fl. May, fr. May, June) *Kennedy* 1131 ! 1663 ! 1988 ! 2074 ! 2170 ! Also in French Cameroons, Belgian Congo and Uganda. (See Appendix, p. 337.)

6. NOTHOSPONDIAS Engl. Bot. Jahrb. 36 : 216 (1905).

Branchlets puberulous ; leaves imparipinnate ; leaflets 19–43, opposite or alternate, obliquely oblong-elliptic, obtusely acuminate, very unequal-sided at the base, 10–16 cm. long, 4–6·5 cm. broad, glabrous, with about 6 pairs of prominently looped lateral nerves ; petiolules about 5 mm. long ; flowers polygamous in lax terminal panicles, clustered ; pedicels up to 5 mm. long, shortly pubescent ; calyx campanulate, shortly dentate, 1·5 mm. long ; petals 6 mm. long, glabrous, slightly imbricate ; stamens 8 ; ovary sparingly pilose, 4-celled ; style short .. .. *staudtii*

N. staudtii *Engl.* l.c. 217 (1905). *N. talbotii* S. Moore (1913). An understorey forest tree, to 80 ft. high, with spreading crown whose branches become more or less erect and bear at their tops whorls of large leaves ; fruits yellow when ripe, plum-like, ovoid-ellipsoid, 4·5 cm. long and 3 cm. broad in Cameroons, but only 2 cm. long and 1·5 cm. broad in Gambari Group F.R., western Nigeria.
S.Nig.: Gambari Group F.R. (fr. Feb., Apr.) *Keay* FHI 22811 ! *Jones & Keay* FHI 4898 ! Oban *Talbot* 230 ! **Br.Cam.:** Ambas Bay (Dec.) *Mann* 2148 ! Victoria (fr. Apr.) *Maitland* 673 ! Elephant Lake, Kumba (Nov.) *Staudt* 746 !

7. SCLEROCARYA Hochst.—F.T.A. 1 : 449.

Branchlets stout, scarred ; leaves mostly in a bunch at the end, pinnate ; leaflets opposite or subopposite, 7–10 pairs, elliptic or obovate, shortly cuneate at base, very acute at apex, about 3·5 cm. long and 2 cm. broad, glabrous, sometimes glaucous, with very obscure nerves ; flowers dioecious, precocious, the males in short spikes ; bracts broadly ovate, with a membranous margin ; stamens 12 or more ; fruits obovoid, about 3–3·5 cm. long, glaucous .. .. *birrea*

S. birrea (*A. Rich.*) *Hochst.* in Flora 27, Bes. Beil. 1 (1844) ; F.T.A. 1 : 449 ; Chev. Bot. 157. *Spondias birrea* A. Rich. in Fl. Seneg. 1 : 152, t. 41 (1831). *Poupartia birrea* (A. Rich.) Aubrév. Fl. For. Soud.-Guin. 405, t. 89, 1–4 (1950). A savannah tree, to 40 ft. high, with grey fissured bark, stout branchlets and pale foliage ; flowers greenish-white or reddish ; fruits yellow, thick-skinned, resembling a small mango ; sterile regrowth shoots often with coarsely serrate leaflets ; in the drier savannah regions.
Sen.: St. Louis *Perrottet* 156 ! Walo *Heudelot* ! Thiès *De Wailly* 4529. Sine-Saloum *Serv. For.* 21. **Port.G.:** Coiada, Gabu *Esp. Santo* 2504 ! **Fr.Sud.:** Timbuktu *Chev.* 1207 ! Kayes (fr. May) *Irvine* 3218 ! And other localities *fide* Chev. l.c. **Gam.:** Kundu *Rosevear* 69 ! Kerewan *Rosevear* ! **Iv.C.:** Ouangolo *Aubrév.* 1419. Bobo-Dioulasso *Aubrév.* 1888. **G.C.:** Kulpawn R., Bantala *Kitson* 839 ! Kullum to Grumbele *Kitson* 653 ! Walemboi to Wahabu *Kitson* 649 ! **Togo:** Yendi (Dec., Jan., fr. May, June) *Dalz.* ! **Dah.:** Atacora Mts. *Chev.* 24088. **Fr.Nig.:** Gargalenti, Gourma *Chev.* 24427. **N.Nig.:** Sokoto (Mar.) *Gilman* 3 ! *Ijomah* FHI 26467 ! Zurmi (Apr.) *Keay* FHI 16158 ! Bauchi (Feb.) *Lely* P. 139 ! Karasuwa, Nguru (fr. June) *Onochie* FHI 23352 ! Extends to A.-E. Sudan, Abyssinia and Uganda. (See Appendix, p. 340.)
Perrier de la Bâthie in Mém. Mus. Nat. Paris, nouv. sér. 18 : 244 (1944) includes *Sclerocarya* in *Poupartia* Comm. ex Juss. (1789), a genus of the Mascarenes, Madagascar and the Comores. He also suggests that the genus *Lannea* A. Rich. (1831) should be included. *Pseudospondias* Engl. (1883) and *Haematostaphis* Hook. f. (1860) are also close. Aubréville sinks *Sclerocarya* but maintains *Lannea*. Pending thorough revision of the whole group, however, I have followed tradition in maintaining all these genera. *Sclerocarya* differs from *Poupartia* as follows :—*Poupartia*, stamens 8 (–10), stigmas 5 ; *Sclerocarya*, stamens 12–16 (–26), stigmas (2–) 3.

8. PSEUDOSPONDIAS Engl. in DC. Monogr. 4 : 258 (1883).

Leaflets 4–12 ; flowers (3–) 4 (–5)-merous ; stamens (6–) 8 (–10) ; leaflets alternate or subopposite, oblong or elliptic, rounded and unequal-sided at base or slightly cuneate, broadly and obtusely acuminate, up to 20 cm. long and 10 cm. broad ; petiolules 0·5–1 cm. long ; flowers very small, polygamous, in lax panicles ; fruits 1 (–2)-seeded, broadly ellipsoid, about 2 cm. long, mucronate with the persistent style :
Branchlets and leaves glabrous or sparingly puberulous   *microcarpa* var. *microcarpa*
Branchlets, rhachides, petiolules and undersurface of leaflets densely hirsute
microcarpa var. *hirsuta*
Leaflets 15–17 ; flowers 3-merous ; stamens 6 ; branchlets and leaves glabrous ; otherwise much as above .. .. *microcarpa* var. *longifolia*

P. microcarpa (*A. Rich.*) *Engl.* var. **microcarpa**—in DC. Monogr. 4 : 259 (1883) ; Chev. Bot. 157 ; Aubrév. Fl. For. C. Iv. 2 : 172, t. 192, 7–11. *Spondias microcarpa* A. Rich. in Fl. Seneg. 1 : 151, t. 40 (1831) ; F.T.A. 1 : 448. *S. zanzee* G. Don (1832). *Haematostaphis pierreana* Engl. (1905). A tree to 60 ft. high, rarely more, with short, fluted, twisted bole and dense crown ; male flowers white, female greenish ; ripe fruits dark purple ; in forest margins, and especially in fringing forest in the savannah regions.
Sen.: Rufisque (Mar.) *Döllinger* 46 ! Casamance (Mar., Apr.) *Perrottet* 158 ! Sinédone *Chev.* 2972. Floup-Fedyan *Chev.* 3167. **Gam.:** (Mar.) *Brown-Lester* 16 ! Albreda (fr. Mar.) *Perrottet* ! Abuko Waterworks (Jan.) *Dalz.* 8127 ! **Port.G.:** Abu, Formosa (Apr.) *Esp. Santo* 1954 ! **Fr.G.:** Mamou (fr. Mar.) *Dalz.* 8397 ! Kouria (Oct., Dec.) *Chev.* 14807 ! 14843 ! Conakry *Chev.* 12063. **S.L.:** Musaia (fl. Sept., fr. Apr.) *Deighton* 5487 ! *Small* 269 ! Port Loko (young fr. Feb.) *Deighton* 3587 ! Fintonia (fl. Sept., young fr. Jan.) *Deighton* 5174 ! 5314 ! R. Scarcies, Kambia (fl. & fr. Dec.) *Sc. Elliot* 4353 ! 4363 ! Falaba (fr. Apr.) *Sc. Elliot* 5900 ! **Lib.:** Grand Bassa *T. Vogel* 102 ! **Iv.C.:** Mankono (Jan.) *Chev.* 21925 ! Agboville (Nov.) *Chev.* 22377 ! Abidjan *Aubrév.* 73 ! Toumodi *Chev.* 22415 ! **G.C.:** Abetifi (fl. & fr. June) *Thompson* 79 ! Kumasi (Apr., Sept.) *Chipp* 771 ! 791 ! Agogo (fr. Jan.) *Chipp* 592 ! E. Akim (June) *Johnson* 735 ! Pepeasi (May) *Beveridge* 90 ! **Togo:** Basari (fl. & young fr. Mar.) *Kersting* A. 510 ! Ntumada, Hohoe Dist. (Dec.) *St. C. Thompson* FH 3684 ! **N.Nig.:** Agaie (Nov.) *Yates* 27 ! 27a ! 28 ! 28a ! Jamata (June) *Thornewill* 66 ! **S.Nig.:** Lagos *Millen* 58 ! Ilesha *Foster* 173 ! Mgbaka, Ikom (Apr.) *Latilo* FHI 30902 ! Oban *Talbot* 1366 ! 1505 ! **Br.Cam.:** Victoria (Jan.) *Maitland* 364 ! Buea (Mar.) *Maitland* 490 ! 551 ! Johann-Albrechtshöhe *Staudt* 893 ! Metschum Falls, Bamenda (Jan.) *Keay & Russell* FHI 28528 ! **F.Po:** (Jan.) *Mann* 181 ! Extends to E. Africa, Nyasaland and Angola ; also in Principe, S. Tomé and Annobon. (See Appendix, p. 340.)
P. microcarpa var. **hirsuta** *Brenan* in Kew Bull. 1947 : 68. As above, but vegetative parts quite densely hirsute.
S.Nig.: Ibadan (Nov.) *A. P. D. Jones & Keay* FHI 13895 ! *E. W. Jones, Keay & Richards* FHI 22458 ! Etemi, Omo F.R., Ijebu Ode Prov. (Mar.) *A. P. D. Jones & Onochie* FH 117003 !

FIG. 199.—SCLEROCARYA BIRREA *(A. Rich.) Hochst.* (ANACARDIACEAE).

A, leaflet showing venation. B, habit. C, flower. D, cross-section of ovary. E, fruit.

730

P. **microcarpa** var. **longifolia** (*Engl.*) *Keay* in Bull. Jard. Bot. Brux. 26 : 203 (1956). *P. longifolia* Engl. Bot. Jahrb. 36 : 218 (1905) ; Brenan l.c. A tree, in forest.
S.Nig.: Ogba F.R., Benin *Kennedy* 2454 ! Also in French Cameroons.

9. **LANNEA** A. Rich. in Guill. & Perr. Fl. Seneg. 1 : 153 (1831) ; Engl. in E. & P. Pflanzenfam. Nachtr. 1 : 213 (1897). *Odina* Roxb. (1832)—F.T.A. 1 : 445.

Flowers in lax panicles, below the leaves ; young parts, including the panicles, pinkish stellate-tomentose, later more or less glabrescent ; tree to 30 m. high in forest ; petioles 9–14 cm. long ; leaflets (1–) 2–4 pairs, plus the long-petiolulate obovate terminal one ; lateral leaflets oblong-elliptic, unequally cuneate or obtuse at base, long-acuminate, 10–20 cm. long, 6–12 cm. broad, with 5–10 main lateral nerves ; leaflets drying brown, dark above, paler with impressed venation beneath ; male panicles to 20 cm. long ; female panicles to 14 cm. long ; fruits flattened-ellipsoid, 6–8 mm. long, 5–6 mm. broad, 4–5 mm. thick, glabrous .. .. 1. *welwitschii*

Flowers in racemes or narrow panicles with short lateral branches, often clustered at the ends of leafless shoots (but see Nos. 2 & 6) ; trees up to 16 m. high, mostly in savannah (but see Nos. 8 & 8a) :

Lateral leaflets (4–) 8–10 (–12) pairs, oblong or ovate-oblong, rounded or subcordate at base, sessile, rounded at apex, 0·5–3·5 (–5) cm. long, 0·3–1·8 (–2) cm. broad, densely white-tomentose beneath, sparingly pubescent and drying blackish above ; petioles 1·5–3 cm. long ; young parts white-tomentose ; racemes up to 3·5 cm. long, white-tomentose, borne with the leaves on short lateral shoots ; fruits oblong, ellipsoid, flattened, about 1 cm. long, 7 mm. broad and 4 mm. thick, white-tomentose
2. *humilis*

Lateral leaflets (0–) 1–7 pairs, more than 3 cm. long and 1 cm. broad ; fruits mostly glabrous, rarely (No. 5) velutinous :

Leaflets remaining tomentose or softly pubescent beneath ; fruits glabrous or velutinous :

Indumentum of leaflets, shoots and inflorescences floccose, pinkish or reddish, composed of long-armed, weak, crispate-stellate hairs only, early caducous on upper surface of leaflets but mainly dense beneath and often concealing the prominent, reticulate venation ; fruit irregularly ovoid or suborbicular, rugose when dry, 7–9 mm. diam. ; lateral leaflets (1–) 2–4 pairs, ovate to elliptic, rounded to obtuse at base, subacuminate, 4–13 cm. long, 2·5–6·5 cm. broad   3. *schimperi*

Indumentum of leaflets, shoots and inflorescences velutinous, greenish brown or grey to whitish-yellow or tawny, composed of simple hairs and stiff, non-crispate, short-armed, stellate hairs, persistent on both surfaces of the leaflets ; fruit ovoid-oblong, laterally flattened especially toward the apex, 8–13 mm. long, 6–8 mm. broad :

Fruit glabrous ; margin of leaflets ciliate with rather long simple hairs and a very few stellate hairs ; lower surface openly and not prominently reticulate, the surface easily visible between the erect and spreading simple hairs ; tawny stellate hairs few and inconspicuous on both surfaces and on the margin ; lateral leaflets (1–) 2–6 pairs, ovate-oblong, rounded to subcordate at base, acuminate, 7–13 cm. long, 4·5–8 cm. broad .. .. 4. *kerstingii*

Fruit more or less densely velutinous with simple hairs, with or without a few scattered stellate hairs ; margin of leaflets stellate-tomentose, with fewer simple hairs intermingled but inconspicuous ; lower surface closely and prominently areolate, densely velutinous, with long, whitish, simple hairs spreading from the veins filling the areoles and concealing the surface ; tawny stellate hairs conspicuous among the short simple hairs on the upper surface, on the veins on the lower surface and especially on the margin ; lateral leaflets (1–) 3–5 pairs, oblong-ovate or ovate-oblong, rounded at base, rounded or subacuminate at apex, 3·5–9 cm. long, 2·5–5 cm. broad .. .. 5. *velutina*

Leaflets soon becoming glabrous or nearly so beneath ; fruit glabrous :

Indumentum of leaflets, shoots and inflorescences floccose, pinkish or reddish, composed of weak, crispate-stellate hairs only, early caducous :

Lateral leaflets sessile or subsessile, falcate-lanceolate, rounded or subcordate at base, obtuse at apex, 3–9·5 cm. long, 1–2·8 cm. broad, 5–7 pairs ; inflorescence sometimes branched ; fruit oblong-ellipsoid, about 8 mm. long and 6 mm. broad
6. *fruticosa*

Lateral leaflets long-petiolulate, lanceolate, cuneate at base, obtusely acuminate, 4–11·5 cm. long, 1·5–4·5 cm. broad, 3–6 pairs ; inflorescence simple ; fruit ellipsoid, 8–10 mm. long, 7–8 mm. broad .. .. 7. *acida*

Indumentum of leaflets, shoots and inflorescences not floccose, not or rarely (Nos. 8 & 8a) pinkish, composed mainly of simple hairs, sometimes with small stellate hairs as well :

Male flowers with pedicels 2–5 mm. long ; fruits 7–9 mm. long, more or less reniform or suborbicular, 5–7 mm. broad, 3–4 mm. thick ; fruiting pedicels 2–3 mm. long ; young parts not glutinous ; bark not spirally twisted ; leaflets ovate-lanceolate, unequal at base, long-acuminate, 5–11 cm. long, 2·2–4 cm. broad ; inflorescences sparsely puberulous :

Leaves and young shoots almost glabrous except for the leaflets which when extremely young are closely and minutely tomentellous ; leaflets 3–6 pairs, plus the terminal one   ..   ..   8. *nigritana* var. *nigritana*

Leaves and young shoots densely pubescent, later glabrescent ; leaflets 1–3 pairs, plus the terminal one   ..   ..   8a. *nigritana* var. *pubescens*

Male flowers sessile or with pedicels up to 1 mm. long ; fruits 10–14 mm. long, oblong-ellipsoid, 6–7 mm. broad, 4–5 mm. thick ; fruiting pedicels 4–6 mm. long ; young parts glutinous ; bark of bole and larger branches spirally twisted :

Leaves, inflorescences and young shoots with long simple hairs ; fruits 12–14 mm. long ; leaflets 3–5 pairs, plus the terminal one, oblong or ovate, obtuse to slightly cordate at base, shortly and obtusely acuminate, 8–13 cm. long, 4·5–6·5 cm. broad   ..   ..   9. *egregia*

Leaves, inflorescences and young shoots without long simple hairs, almost glabrous ; fruits 10–11 mm. long ; leaflets (0–) 1–3 pairs, plus the terminal one, ovate-lanceolate, obtuse and unequal at base, obtusely acuminate, subacute or obtuse at apex, 5·5–13 cm. long, 2·7–4·5 cm. broad..   .. 10. *microcarpa*

1. **L. welwitschii** (*Hiern*) *Engl.* Bot. Jahrb. 24 : 498 (1898) ; Aubrév. Fl. For. C. Iv. 2 : 170, t. 191. *Calesiam welwitschii* Hiern Cat. Welw. 1 : 179 (1896). *Lannea acidissima* A. Chev. Vég. Util. 5 : 114 (1909); Bull. Soc. Bot. Fr. 58, Mém. 8 : 150 (1912) ; F.W.T.A., ed. 1, 1 : 511 ; Kennedy For. Fl. S. Nig. 186. A forest tree, to 100 ft. high, with cylindrical bole, up to 8 ft. in girth, and small buttresses ; bark grey, pitted ; slash thick, deep crimson-pink with whitish streaks ; flowers yellowish-green ; ripe fruits black.
   Iv.C.: Mbasso (Mar.) *Chev.* 16264 ! Assinie *Chev.* 16319 ! Sassandra *Chev.* 17950. Bouroukrou *Chev.* 16133. **G.C.:** Adeambra (fr. May) *Vigne* FH 862 ! Owabi (fl. & fr. Mar.) *Andoh* FH 4312 ! Ofin Headwaters F.R. *Vigne* FH 1913 ! Aburi *Deighton* 3414 ! **S.Nig.:** Oshun River F.R. (Apr.) *Gilman* 520 ! Sapoba (fl. & fr. Apr.) *Kennedy* 769 ; 1344. Amawbia, Akwa Dist. (fr. June) *Jones* FHI 6630 ! **Br.Cam.:** Victoria (Mar.) *Maitland* 532 ! Johann-Albrechtshöhe (Mar.) *Staudt* 678. Also in French Cameroons, Gabon, Uganda, Belgian Congo and Angola. (See Appendix, p. 339.)

2. **L. humilis** (*Oliv.*) *Engl.* in E. & P. Pflanzenfam. Nachtr. 1 : 213 (1897) ; Aubrév. Fl. For. Soud.-Guin. 393, 400, 404, t. 83, 3. *Odina humilis* Oliv. F.T.A. 1 : 447 (1868). A shrub or much branched tree, seldom more than 10 ft. high ; in dry savannah regions.
   Fr.Nig.: 20 km. N. of Tessaoua *Aubrév.* **N.Nig.:** Bornu Road, Yola Prov. (fl. & fr. Aug.) *Dalz.* 170 ! Lantewa, Bornu Prov. *Onochie* FHI 23371 ! N. Bornu (Apr.) *E. Vogel* 78 ! Extends to A.-E. Sudan, Abyssinia, E. Africa and N. Rhodesia.

3. **L. schimperi** (*Hochst. ex A. Rich.*) *Engl.* l.c. (1897) ; Hoyle & Jones in Kew Bull. 1947 : 75–86, tt. 1–4 ; Aubrév. l.c. 399, 402, 404. *Odina schimperi* Hochst. ex A. Rich. (1847)—F.T.A. 1 : 445. (?) *O. barteri* Oliv. F.T.A. 1 : 446 (1868), partly, flowers only. *Lannea barteri* (Oliv.) Engl. (1897)—F.W.T.A., ed. 1, 1 : 511, partly. A savannah tree, 6–25 ft. high, with dark grey or black rough bark ; young parts with pinkish or reddish indumentum ; flowers yellowish, usually produced before the leaves.
   N.Nig.: Dutsen Bagai, Zurmi (fr. Apr.) *Keay* FHI 16184 ! Zaria Prov. (fr. Mar., Apr.) *Bawa Lafia* FHI 7753 ! *Keay* FHI 16108 ! Dangoro F.R., Kano (fr. Apr.) *Latilo* FHI 27423 ! Agaie *Yates* 61 (partly) ! Bauchi Plateau (Feb.) *Lely* P. 158 (partly) ! P. 159 (partly) ! **S.Nig.:** Enugu (fl. Feb., young fr. Nov.) *Cons. of For.* 49 ! *Smith* 12 ! Abakaliki (May) *Cons. of For.* 123 ! Also in French Cameroons, Ubangi-Shari, A.-E. Sudan, Abyssinia, Uganda, Kenya and Tanganyika ; also perhaps in Belgian Congo, Nyasaland and N. Rhodesia. (See Appendix, p. 339.)
   [The specimen (*Boivin* s.n.) from Senegal cited as *L. schimperi* by Hoyle and Jones is leafless and could be either this species or *L. acida*, but, as the latter species is common in Senegal and as *L. schimperi* is not otherwise recorded west of Nigeria, I suspect that the specimen is *L. acida*. The flowering shoots of *Barter* 1109 may also be *L. acida* and perhaps came from the specimen *Barter* 1107.]

4. **L. kerstingii** *Engl. & K. Krause* in Engl. Bot. Jahrb. 46 : 325 (1911) ; Hoyle & Jones l.c. *Odina barteri* Oliv. F.T.A. 1 : 446 (1868), partly, leaves only. *Lannea barteri* (Oliv.) Engl. (1897)—F.W.T.A., ed. 1, 1 : 511, partly ; Aubrév. Fl. For. Soud.-Guin. 397, 402, 404, t. 85, 1 & 2. *L. velutina* of F.W.T.A., ed. 1, 1 : 511, partly. A savannah tree, 12–40 ft. high ; bark smooth, silvery-grey, with a pronounced spiral twist ; flowers yellowish, usually produced before the leaves ; fruits dull purplish when ripe.
   Fr.Sud.: Ouassaia (Mar.) *Chev.* 521 ! **Fr.G.:** Timbo *Pobéguin* 114 ; *Chev.* 12498 *bis*. Kissidougou *Aubrév.* 47g. **S.L.:** Musaia (Feb.) *Deighton* 5355 ! **Iv.C.:** Mt. Bobo, Zoanlé (fr. May) *Chev.* 21492 ! Yabarasso, Comoé Valley (fr. Dec.) *Chev.* 22546 ! **G.C.:** Yagbung to Nabiungo *Kitson* 745 ! Wiasi (fr. Mar.) *Vigne* FH 4484 ! Lawra *Vigne* FH 3915 ! N. of Abofuo (Feb.) *Vigne* FH 1811 ! Fiapere (Feb.) *Chipp* 80 ! **Togo:** Sokode *Kersting* 68a. And other localities *fide* Aubrév. *l.c.* **Dah.:** Savalou *Aubrév.* 40d. Kouandé to Konkobiri *Chev.* 21750. **N.Nig.:** Nupe *Barter* 1109 (partly) ! Ejidogari F.R., Ilorin (fr. Feb.) *Ejiofor* FHI 19830 ! Bauchi Plateau (fl. & fr. Feb.) *Lely* P. 158 (partly) ! P. 159 (partly) ! Katagum *Dalz.* 225 (partly) ! Yola (Apr.) *Dalz.* 178 ! **S.Nig.:** W. of R. Tessi, Old Oyo F.R. (fl. & fr. Feb.) *Keay* FHI 16262 ! 16264 ! Onitsha *Onochie* FHI 21635 ! Also in French Cameroons, Ubangi-Shari, A.-E. Sudan and Uganda. (See Appendix, p. 339.)

5. **L. velutina** *A. Rich.* in Fl. Seneg. 1 : 154, t. 42 (1831) ; *Chev.* Bot. 159 ; Hoyle & Jones l.c. ; Aubrév. l.c. 397, 402, 404, t. 84, 3–5. *Odina velutina* (A. Rich.) Oliv. F.T.A. 1 : 447 (1868). A savannah shrub or tree, to 20 ft. high ; flowers yellow.
   Sen.: Casamance, abundant *fide* Aubrév. *l.c.* **Gam.:** Albreda (fl. Mar., Apr., fr. May) *Perrottet* 155 ! Bulak *Rosevear* 37 ! Genieri (June) *Fox* 112 ! **Fr.Sud.:** Bamako (Apr.) *Hagerup* 361 ! **Port.G.:** Esp. Santo 499 ! Antula, Bissau (Apr.) *Esp. Santo* 1522 ! **Fr.G.:** Karkandy *Heudelot* 992 ! Kouroussa *Pobéguin* 696. Kollangui *Chev.* 12225. **Iv.C.:** Bobo Dioulasso *Serv. For.* 1950. Dem *Aubrév.* 2385. **G.C.:** Dubba to Jerriwama (Mar.) *Kitson* 945 ! Pan to Bujan (fr. Apr.) *Kitson* 646 ! Tumu (fr. Apr.) *Vigne* FH 3788. (See Appendix, p. 339.)

6. **L. fruticosa** (*Hochst. ex A. Rich.*) *Engl.* Pflanzenfam. Nachtr. 1 : 213 (1897) ; Aubrév. l.c. 394, 400, 404, t. 83, 1. *Odina fruticosa* Hochst. ex A. Rich. (1847)—F.T.A. 1 : 446. A savannah tree, to 25 ft. high, young parts with reddish floccose indumentum ; flowers yellow ; in arid regions.
   Fr.Nig.: Gouré *Aubrév.* Zinder (Nov.) *Hagerup* 609 ! **N.Nig.:** Daura, Gumel & Nguru *fide* Aubrév. *l.c.* Damaturu Kabaru, Bornu (young fr. Oct.) *Daggash* FHI 22001 ! Bornu Road, Yola Prov. (July) *Dalz.* 173 ! Also in French Cameroons, Ubangi-Shari, A.-E. Sudan, Uganda and Abyssinia. (See Appendix, p. 339.)

7. **L. acida** *A. Rich.* in Fl. Seneg. 1 : 154 (1831) ; *Chev.* Bot. 158 ; F.W.T.A., ed. 1, 1 : 511, partly (excl. syn. *L. djalonica* A. Chev., *L. oleosa* A. Chev., *L. grossularia* A. Chev., *L. microcarpa* Engl. and *L. egregia* Engl. & K. Krause) ; Aubrév. l.c. 394, 400, 404, t. 83, 4–5. *Odina acida* (A. Rich.) Oliv. F.T.A. 1 : 446 (1868). *Lannea velutina* of F.W.T.A., ed. 1, 1 : 511, partly (*Boivin* s.n.). A savannah tree, to 18 ft.

high, with blackish, fissured bark ; young parts with pinkish floccose indumentum ; often on rocky hills. Similar to *L. schimperi* but leaflets glabrescent and not prominently reticulate beneath.
Sen.: *Boivin* ! Rufisque *Perrottet.* Baol *Chev.* 3506. Thiès *De Wailly* 4638. Casamance *Aubrév.* 152.
Gam.: Albreda *Perrottet* ! Foloto *Rosevear* 102 ! **Fr.Sud.:** Sikasso to Koutala (Jan.) *Roberty* 13343 ! Ouahigouya to Koro, Yatenga *Chev.* 24794 ! And other localities *fide* Aubrév. *l.c.* **Port.G.:** Bissau Dist. (fl. Apr.) *Esp. Santo* 1532 ! 1912 ! **Fr.G.:** Kouroussa *Pobéguin* 901. Mamou *Aubrév.* 49g. Dalaba *Aubrév.* 46g. **Iv.C.:** Touba *Aubrév.* 1252 ! Niangbo *Serv. For.* 1715 ! **G.C.:** N. Terr. *Saunders* 1 ! Palbe *Kitson* 645 ! Pong (fr. Mar.) *Kitson* 696 ! Sabu (fr. May) *Vigne* FH 4701 ! **Togo:** Kete Krachi (Jan.) *Vigne* FH 1528 ! Sokode *Aubrév.* 123d. **Dah.:** Kouandé *Chev.* 24239 ! **N.Nig.:** Nupe *Barter* 1107 ! Yola (fr. Apr.) *Dalz.* 166 ! **S.Nig.:** Igboho, Oyo Prov. (fr. Feb.) *Keay* FHI 22499 ! Igbetti, Oyo Prov. (Feb.) *Keay* FHI 14631 ! (See Appendix, p. 338.)
[The flowering shoots of *Barter* 1109 and the leafless specimen (*Boivin* s.n.) from Senegal, cited by Hoyle and Jones as *L. schimperi*, may well be *L. acida;* see note above.]

8. **L. nigritana** (*Sc. Elliot*) *Keay* var. **nigritana**—in Bull. Jard. Bot. Brux. 26 : 204 (1956). *Odina nigritana* Sc. Elliot in J. Linn. Soc. 30 : 75 (1894). *Lannea afzelii* Engl. Bot. Jahrb. 24 : 494 (1898) ; F.W.T.A., ed. 1, 1 : 511, partly (excl. syn. *L. buettneri* Engl.) ; Aubrév. Fl. For. C. Iv. 2 : 168–170, t. 192, 1–6 ; Fl. For. Soud.-Guin. 399, 403, 405, t. 85, 3. *L. glaberrima* Engl. & K. Krause (1911). *L. grossularia* A. Chev. (1912). *L. dahomensis* A. Chev. Bot. 159, name only, partly. *L. acida* of F.W.T.A., ed. 1, 1 : 511, partly. A tree to 45 ft. high, with wide rather sparse branching ; flowers yellow, produced when tree is leafless ; ripe fruits black ; mainly in regrowth in the drier parts of the forest regions.
en.: Bignona *Aubrév.* 153. **Fr.Sud.:** Satadougou (fr. May) *Irvine* 3215 ! **Port.G.:** Antula, Bissau (fr. Apr.) *Esp. Santo* 1519 ! **Fr.G.:** *Paroisse* 103. **S.L.:** *Afzelius. Don* ! Buyabuya (Feb.) *Sc. Elliot* 4576 ! 4769 ! Ninia (Feb.) *Sc. Elliot* 4910 ! Berria, Falaba (fr. Mar.) *Sc. Elliot* 5408. Njala (Apr.) *Deighton* 2787 ! 2894 ! Musaia (fr. Feb.) *Deighton* 4272 ! **Iv.C.:** Sassandra Port (fr. May) *Chev.* 17936 ! And other localities *fide* Aubrév. *ll.c.* **G.C.:** Achimota (Jan., Feb.) *Irvine* 189 ! 2001 ! **Togo:** Lome (May) *Warnecke* 144 ! **Dah.:** Kétou to L. Azri, Zagnanado (Feb.) *Chev.* 23039 (partly) !

8a. **L. nigritana** var. **pubescens** *Keay* l.c. (1956). *L. afzelii* var. *pubescens* Aubrév. Fl. For. Soud.-Guin. 403, 405 (1950), French descr. only. Similar to above, but leaflets densely pubescent beneath when young.
Iv.C.: Agboville *Aubrév.* 1380 ! Attobro *Serv. For.* 1773 ! Agnéby (fr. Mar.) *Aubrév.* 2239 ! **Togo:** Palimé *Aubrév.* 136d ! **S.Nig.:** Ibadan (fl. & fr. Mar., Apr.) *Keay* FHI 25680 ! 25682 ! *Meikle* 1137 ! Gambari Group F.R., Ibadan *W. D. MacGregor* 595 ! Etemi, Ijebu Ode Prov. *Jones & Onochie* FHI 16759 !

9. **L. egregia** *Engl. & K. Krause* in Engl. Bot. Jahrb. 46 : 331 (1911), as " egegria " by error ; Aubrév. l.c. 397, 405, incl. var. *dahomensis* Aubrév. *L. acida* of F.W.T.A., ed. 1, 1 : 511, partly. A savannah tree, to 40 ft. high, with pale grey, spirally twisted bark.
Fr.G.: Mamou (Mar.) *Aubrév.* 53g ! **Iv.C.:** Touba *Aubrév.* 1254 ! Ferkéssédougou (fr. Apr.) *Aubrév.* 2290 ! **Togo:** Atakpame (Aug.) *Doering* 101 ; *Mildbr.* 7431 ! Lama (July) *Kersting* A. 456. **Dah.:** Pobé (Jan.) *Aubrév.* 16d ! 18d ! **S.Nig.:** Olokemeji (Mar.) *Keay* FHI 28691 ! *A. F. Ross* 101 ! Ijaiye F.R., Oyo Prov. (fr. Mar.) *Keay* FHI 21188 !
[The type specimens have presumably been destroyed, so the species has been interpreted only from the description.]

10. **L. microcarpa** *Engl. & K. Krause* l.c. 324 (1911) ; Aubrév. l.c. 394, 402, 404, t. 84, 1 & 2. *L. oleosa* A. Chev. (1912). *L. djalonica* A. Chev. Bot. 159, name only. *L. acida* of F.W.T.A., ed. 1, 1 : 511, partly. A savannah tree, to 50 ft. high and 4 ft. girth ; bark grey, more or less smooth, with a spiral twist ; flowers greenish yellow ; ripe fruits purplish black.
Gam.: *Heudelot* 353. **Fr.Sud.:** Ouahigouya *Chev.* 24771 ! And other localities *fide* Aubrév. *l.c.* **Fr.G.:** Kollangui (fr. Mar.) *Chev.* 12875 ! Konkouré to Timbo *Chev.* 12498. **Iv.C.:** Bobo Dioulasso *Aubrév.* 1849 ; 1859. Ouagadougou *Aubrév.* 2335. **G.C.:** Salaga (fr. Apr.) *Dalz.* 50 ! Sa (fl. & fr. Mar.) *Kitson* 844 ! Pong Tamale (fr. Apr.) *Kitson* 895 ! Nakong (Feb.) *Vigne* FH 4721 ! Tumu, N.T. (fl. & fr. Apr.) *Vigne* FH 3787 ! 3789 ! **Togo:** (fr. May) *Kersting* A. 349 ! Kudupoll *Kersting* A. 532 ! Kabure (fl. & fr. Feb.) *Kersting* A. 520 ! **Dah.:** Toukountouna, Atacora Mts. *Chev.* 24076. Sansanné Mango *Aubrév.* 132d. **Fr.Nig.:** Zinder *Aubrév.* **N.Nig.:** Sokoto Prov. (fl. Feb., fr. May) *Dalz.* 351 ! *Lely* 812 ! Zamfara F.R., Zurmi (fl. & fr. Apr.) *Keay* FHI 15676 ! 15689 ! 16130 ! Dangora F.R., Kano (fr. Apr.) *Latilo* FHI 27428 ! Bogayi, Katsina (Mar.) *Meikle* 1354 !

Imperfectly known species.

L. buettneri *Engl.* Bot. Jahrb. 24 : 494 (1897) ; Aubrév. Fl. For. Soud.-Guin. 397, 402, 404. *L. afzelii* of F.W.T.A., ed. 1, 1 : 511, partly. The description suggests *L. acida*, but material distributed from the Berlin Herbarium under the name *L. buettneri* appears to be *L. microcarpa*; the type specimen has presumably been destroyed. See notes by Aubréville (l.c.), who suggests this may be a distinct species endemic to Togo and Dahomey.
Togo: Misahöhe (Jan.) *Büttner* 376.

10. HAEMATOSTAPHIS Hook. f.—F.T.A. 1 : 443.

Branchlets purplish-glaucous ; leaves pinnate ; leaflets alternate, about 9–12 on each side, oblong or oblong-elliptic, rounded at base or slightly cuneate, slightly emarginate at apex, thin, 5–8 cm. long, 1·5–3 cm. broad, the end one broader, glabrous ; petiolules 4–5 mm. long ; flowers dioecious, the males very small, in slender lax panicles clustered at the ends of the shoots ; male calyx 3-lobed ; petals 3, imbricate, about 2 mm. long ; stamens 6 ; female : ovary glabrous, surrounded by small staminodes ; fruit drupaceous, broadly ellipsoid, about 2 cm. long ..   ..   .. *barteri*

H. barteri *Hook. f.* in Trans. Linn. Soc. 23 : 169, t. 25 (1860) ; F.T.A. 1 : 443 ; Chev. Bot. 159 ; Aubrév. Fl. For. Soud.-Guin. 403, t. 86, 1–3. A savannah tree, to 25 ft. high with rather glaucous foliage ; leaves and pendulous panicles of creamy flowers crowded at ends of branchlets ; young parts red ; fruits plum-like, smooth, deep red ; mostly on rocky hills.
G.C.: Taningnah, Wa (fr. Mar.) *McLeod* 840 ! Gambaga (Mar.) *Hughes* FH 4992 ! *Williams* 487 ! Nakong (Apr.) *Vigne* FH 4502 ! Yijia to Pong (Mar.) *Kitson* 706 ! Pong to Kako (Mar.) *Kitson* 700 ! **Togo:** Sokode-Basari *Kersting* A. 560 ! Tamberma (fl. & fr. May) *Kersting* A. 340 ! **Dah.:** Atacora Mts. (fr. June) *Chev.* 24216 ! **N.Nig.:** Nupe *Barter* 1114 ! 1341 ! Jebba (Jan.) *Meikle* 1106 ! Randa (fl. & fr. June) *Hepburn* 63 ! Yola (fl. & fr. Jan.) *Dalz.* 187 ! **S.Nig.:** Igboho, Oyo Prov. (Feb.) *Keay* FHI 22500 ! Igbetti, Oyo Prov. (Feb.) *Keay* FHI 14635 ! Also in French Cameroons and A.-E. Sudan. (See Appendix, p. 338.)

11. TRICHOSCYPHA Hook. f.—F.T.A. 1 : 444. *Emiliomarcelia* Th. & H. Dur. (1909).

*Inflorescences borne on main stem of tree ; disk glabrous or sparingly pubescent :
†Bracts large, ovate-lanceolate, up to 3 cm. long and 1·5 cm. broad, tomentose ; inflorescence up to 25 cm. long, pyramidal, rather congested, rusty-tomentose ;

leaflets nearly glabrous, 7–8 (–14) pairs, oblong-lanceolate, cuneate to rounded at base, acuminate, up to 32 cm. long and 8 cm. broad, with 10–12 (–20) main lateral nerves on each side of midrib, prominent beneath, impressed above ; ovary hirsute, stigmas sessile ; fruits ovoid, up to 5 cm. long and 3 cm. diam., rusty-puberulous

1. acuminata

†Bracts small, up to 3 mm. long ; inflorescence up to 30 cm. long, with rather few somewhat densely flowered lateral branches, rusty-hirsute ; leaflets pilose on the midrib above and at first on midrib and nerves beneath, 6–8 pairs, narrowly lanceolate, up to 30 cm. long and 6 cm. broad, with 12–20 main lateral nerves on each side of midrib, prominent beneath ..   ..   2. <i>smythei</i>
*Inflorescences terminal or axillary :
Disk glabrous :
  Flowers minute, in much-branched panicles ; petals 1–2 mm. long ; pedicels slender, 2–5 mm. long :
    Leaflets 1–4 pairs, plus the terminal one ; midrib pubescent above :
      Axes of inflorescence glabrous or slightly pubescent ; leaflets 2–4 pairs, elliptic, ovate or oblong-ovate, broadly cuneate at base, acuminate, 8–25 cm. long, 3–14 cm. broad, with 10–13 main lateral nerves on each side of midrib   3. <i>talbotii</i>
      Axes of inflorescence densely covered with short and reflexed longer hairs ; leaflets 1–2 pairs, oblong-elliptic, long-acuminate, 8–18 cm. long, 2·5–8 cm. broad, with 6–8 main lateral nerves on each side of midrib ..   ..   4. <i>camerunensis</i>
    Leaflets 6–8 pairs, plus the terminal one, oblong or narrowly oblong, cuneate at base, caudate-acuminate, 6–19 cm. long, 2·8–5 cm. broad, midrib glabrous above, with 7–9 main lateral nerves on each side of midrib ; axes of inflorescence pubescent or glabrescent ..   ..   5. <i>cavalliensis</i>
  Flowers larger, petals 3 mm. long or more :
    Midrib of leaflets pilose above and with scattered hairs beneath ; leaflets 3–7 pairs, oblong, cuneate or obtuse at base, acuminate, up to 36 cm. long and 10 cm. broad, with 8–24 main lateral nerves on each side of midrib, prominent beneath ; inflorescence up to 25 cm. long, densely flowered, rusty-pilose ; flowers sessile or shortly pedicellate, petals nearly glabrous outside   ..   ..   6. <i>preussii</i>
    Midrib of leaflets glabrous above :
      Inflorescence with long, loose, rusty indumentum ; flowers congested on the short lateral branches :
        Leaflets (1–) 2–4 pairs, elliptic, cuneate at base, caudate-acuminate, 9–20 cm. long, 3·5–7·8 cm. broad, nearly glabrous, with 6–10 main lateral nerves on each side of midrib ; inflorescence up to 14 cm. long   ..   ..   7. <i>bijuga</i>
        Leaflets (4–) 5–7 pairs, oblong or narrowly oblong, obtuse or rounded at base, acuminate, (5·5–) 8–19 cm. long, 2·5–6 cm. broad, pubescent on midrib and nerves beneath, with 8–18 main lateral nerves on each side of midrib ; inflorescence up to 60 cm. long, with lateral branches up to 4·5 cm. long   8. <i>beguei</i>
      Inflorescence pubescent with short hairs, or nearly glabrous, lax, with long lateral branches ; ovary and fruits glabrous :
        Leaflets membranous, appearing " pustulate " when dry, (2–) 5–7 (–9) pairs, narrowly oblong-elliptic or oblong-lanceolate, cuneate at base, caudate-acuminate, up to 14 cm. long and 4·3 (–6) cm. broad, with 5–12 main lateral nerves on each side of the sparsely pubescent midrib ; petiolules 2–8 mm. long ; inflorescence up to 60 cm. long, densely rusty-pubescent   ..   9. <i>baldwinii</i>
        Leaflets coriaceous, not appearing " pustulate " :
          Flowers creamy ; leaflets 3–5 pairs, oblong, long-cuneate at base, acuminate, up to 16·5 cm. long and 7 cm. broad, nearly glabrous, with 6–11 main lateral nerves on each side of midrib ; inflorescence nearly glabrous, up to 50 cm. long
10. patens
          Flowers red ; leaflets 6–8 pairs, oblong-elliptic, oblong-oblanceolate or oblong-lanceolate, cuneate to rounded and sometimes subfalcate at base, acuminate, up to 25 cm. long and 8 cm. broad, quite glabrous, with 10–18 main lateral nerves on each side of midrib ; inflorescence reddish-pubescent, up to 80 cm. long, lax ; fruits ellipsoid, about 2·5 cm. long, glabrous   ..   11. <i>arborea</i>
Disk hirsute :
  Leaves subsessile, with lowest pair of leaflets less than 1·5 cm. from the stem ; ovary hirsute ; petals and sepals glabrous outside :
    Leaflets hirsute beneath, especially on midrib, 6–9 pairs, oblong, rounded at base, acuminate, up to 25 cm. long and 9·5 cm. broad, with 10–18 main lateral nerves on each side of midrib ; inflorescence up to 15 cm. long, congested, the axes hirsute ; fruits ovoid, beaked, hirsute   ..   ..   ..   12. <i>mannii</i>
    Leaflets glabrous, 5–7 pairs, elongate-oblong or lanceolate, rounded to cuneate at base, gradually acuminate, up to 30 cm. long and 8 cm. broad, with 7–14 main lateral nerves on each side of midrib ; inflorescence up to 30 cm. long, densely many-flowered, the axes pubescent ; fruits ovoid, pubescent   ..   13. <i>chevalieri</i>
  Leaves distinctly petiolate ; lowest pair of leaflets several cm. from the stem :

Leaflets long-pilose beneath, at least on the midrib and nerves ; stems and inflorescence-axes hirsute ; calyx usually pubescent outside ; ovary densely
hirsute ; inflorescence congested :
Midrib beneath rather densely covered with short hairs as well as long ones ; calyx
densely pubescent outside ; leaflets oblong, rounded or obtuse at base, very
acutely long-acuminate, up to 24 cm. long and 8 cm. broad :
Petals 4–4·5 mm. long ; leaflets 4 pairs, with about 12 main lateral nerves on
each side of midrib ; inflorescence about 7 cm. long and broad 14. *sp. A*
Petals 2–3 mm. long ; leaflets 10 pairs, with about 18 main lateral nerves on
each side of midrib ; inflorescence about 30 cm. long and 10 cm. broad 15. *sp. B*
Midrib long-pilose, without a dense covering of short hairs ; calyx pubescent or
glabrescent ; petals 3–4 mm. long, blood-red ; inflorescence 5–15 cm. long, very
densely many-flowered ; leaflets 5–8, elongate-oblong, obtuse to rounded at
base, acutely acuminate, up to 28 cm. long and 7·5 cm. broad, with 12–20 main
lateral nerves on each side of midrib .. .. .. 16. *atropurpurea*
Leaflets beneath glabrous, or pubescent and glabrescent, or appressed-pubescent ;
calyx glabrous or pubescent outside, ovary glabrous or hairy ; inflorescence
congested or lax ; flowers white or cream (but colour not recorded in Nos. 17
and 20) :
Leaflets 7–9 pairs, very coriaceous, glabrous and often glaucous, narrowly oblong
or lanceolate, obtuse at base, shortly and narrowly acuminate, up to 28 cm. long
and 8 cm. broad, with 13–25 main lateral nerves on each side of midrib, venation
obscure ; inflorescence lax, to 33 cm. long, rusty-tomentellous ; ovary glabrous ;
fruits flattened-ellipsoid, oblique, 2·5–3 cm. long, glabrous 17. *longifolia*
Leaflets 3–6 pairs ; main lateral nerves 5–13 on each side of midrib :
Midrib of leaflets hirsute above, undersurface appressed-pubescent at least when
young ; calyx often pubescent outside ; ovary and fruit hairy :
Petals about 3 mm. long, drying blackish ; inflorescence densely blackish-
hirsute, to 45 cm. long ; leaflets (3–) 5–6 pairs, oblong to oblong-elliptic,
rounded or obtuse at base, acuminate, up to 24 cm. long and 8·5 cm. broad,
with 7–13 main lateral nerves on each side of the midrib ; petiolules up to
5 mm. long ; fruits ovoid, about 1·2 cm. long, densely blackish-velutinous
18. *oba*

Petals about 1·5 mm. long, drying pale brown ; inflorescence shortly pubescent,
to 24 cm. long ; leaflets 3–4 pairs, oblong-ovate to oblong, cuneate to rounded
at base, obtusely and shortly acuminate, up to 14 cm. long and 5·5 (–8) cm.
broad, with 5–8 main lateral nerves on each side of midrib ; petiolules
5–10 mm. long ; fruits ovoid, about 1·2 cm. long, sparingly appressed-setose-
pubescent .. .. 19. *yapoensis*
Midrib of leaflets glabrous above, undersurface glabrous or pubescent and soon
glabrescent ; calyx glabrous outside :
Petals 4–5 mm. long ; inflorescence about 12 cm. long, densely-flowered, densely
dark red-brown hirsute ; ovary sparingly hirsute ; leaflets 4 pairs, oblong or
oblong-elliptic, obtuse at base, long-acuminate, 14–23 cm. long, 6·5–7 cm.
broad, with 9–12 main lateral nerves on each side of midrib 20. *longipetala*
Petals 1–2 mm. long :
Inflorescence puberulous, up to 30 cm. long, many-flowered, wide ; petals
1–1·5 cm. long ; ovary and fruit glabrous ; leaflets 4–8 pairs, elongate-
elliptic or lanceolate, cuneate to rounded at base, gradually acuminate, up
to 22 cm. long and 6 cm. broad, with 6–16 main lateral nerves on each side
of midrib .. .. 21. *smeathmannii*
Inflorescence densely blackish-hirsute, up to 40 cm. long and 30 cm. wide,
many-flowered ; petals 2 mm. long ; leaflets 5–6 pairs, narrowly oblong,
cuneate at base, acuminate, up to 30 cm. long and 10 cm. broad, coriaceous,
with 9–12 main lateral nerves on each side of midrib .. 22. *albiflora*

1. **T. acuminata** *Engl.* Bot. Jahrb. 1 : 425 (1881). *Sorindeia mannii* Oliv. F.T.A. 1 : 442 (1868); not *Trichoscypha mannii* Hook. f. (1862). *T. ferruginea* Engl. Bot. Jahrb. 15 : 112, fig. B–D, t. 4 (1892) ; F.W.T.A.,
ed. 1, 1 : 510, partly (*Preuss* 283, by error 2836, from Barombi, and syn. *T. braunii* Engl. only). *T.
braunii* Engl. l.c. 111, fig. A (1893). *Schubea heterophylla* Pax (1899), flowers only; the leaves are *Cola
ficifolia* Mast. *Trichoscypha buettneri* Engl. (1893), name only. *T. congoensis* Engl. ex De Wild. (1906),
name only. A forest tree, to 60 ft. high, with straight bole, cauliflorous ; sepals white, petals pinkish-
red, stamens red ; fruits red when ripe.
S.Nig.: Kataban (fr. May) *Latilo* FHI 31801 ! Oban *Talbot* 117 ! 161 ! **Br.Cam.:** Abonando, Mamfe
(Apr.) *Rudatis* 41 ! Barombi *Preuss* 283 ! Etam, Kumba Div. (May) *Dundas* FHI 15218 ! S. Bakundu
F.R. (June) *Ejiofor* FHI 14036 ! Victoria to Bimbia (Aug.) *Preuss* 1330 (partly) ! Also in French
Cameroons, Spanish Guinea, Gabon, Belgian Congo and Cabinda.
2. **T. smythei** *Hutch. & Dalz.* F.W.T.A., ed. 1, 1 : 508 (1928) ; Kew Bull. 1929 : 28. *T. atropurpurea* of
F.W.T.A., ed. 1, 1 : 508, partly (*Bunting* 135, partly). A slender tree, to 25 ft. high with main stem
2–3 in. diam. and leaves clustered at the top ; flowers orange ; in forest.
S.L.: Picket Hill (Dec.) *Unwin & Smythe* 33 ! **Lib.:** Bumbumi-Moala (Nov.) *Linder* 1329 ! Begwai
(Sept.) *Bunting* 5 (partly) ! 135 (partly) !
3. **T. talbotii** *Bak. f.* in Cat. Talb. 22 (1913).
S.Nig.: Oban *Talbot* 579 !

4. **T. camerunensis** *Engl.* Bot. Jahrb. 15   109 (1892). A small shrub, 2 ft. high, with rusty-pilose branchlets and inflorescence.
   Br.Cam.: Kumba-Ninga to Mokonje (Apr.) *Preuss* 99.

5. **T. cavalliensis** *Aubrév. & Pellegr.* in Bull. Soc. Bot. Fr. 81 : 647 (1934) ; Aubrév. Fl. For. C. Iv. 2 : 166. A forest tree, up to 60 ft. high, with smooth greyish bark.
   Lib.: Dukwia R. (fl. July, young fr. Oct.) *Cooper* 107 ! *Linder* 81 ! Zeahtown (Aug.) *Baldwin* 6977 ! **Iv.C.:** Taté to Tabou, Cavally Basin (Aug.) *Chev.* 19834 ! **G.C.:** Ankasa, W. Apollonia (young fr. Aug.) *Chipp* 331 !

6. **T. preussii** *Engl.* Bot. Jahrb. 15 : 110 (1892). *T. ferruginea* of F.W.T.A., ed. 1, 1 : 510, partly (*Talbot* 87, 1301, *Preuss* 463 and syn. *T. preussii* Engl.). A tree, to 30 ft. high ; ripe fruits bright red ; in forest.
   S.Nig.: Oban *Talbot* 81 ! 87 ! 1301 ! Aningeje F.R., Calabar (Aug.) *Olorunfemi* FHI 34207 ! Boshi-Okwangwo F.R., Obudu Div. (fr. May) *Latilo* FH 30925 ! **Br.Cam.:** Victoria *Winkler* 638 ! Barombi-ba-Mbu (Sept.) *Preuss* 463 ! S. Bakundu F.R. *Onochie* FHI 32062 ! Also in French Cameroons.

7. **T. bijuga** *Engl.* Bot. Jahrb. 1 : 425 (1881). A shrub or small tree, 12–15 ft. high ; flowers pinkish red.
   Br.Cam.: Likomba (Nov.) *Mildbr.* 10700 ! **F.Po:** (Mar.) *Mann* 2343 !

8. **T. beguei** *Aubrév. & Pellegr.* in Bull. Soc. Bot. Fr. 81 : 648 (1934) ; Aubrév. Fl. For. C. Iv. 2 : 166, t. 189, 7–8. *T. atropurpurea* of F.W.T.A., ed. 1, 1 : 508, partly (*Bunting* 135, partly). A small tree, with long, narrow, pendulous panicles ; in forest.
   Lib.: Zeahtown, Tchien (Aug.) *Baldwin* 6974 ! Begwai (Sept.) *Bunting* 5 (partly) ! 135 (partly) ! **Iv.C.:** Be Yakasse (Aug.) *Aubrév.* 1789 ! (See Appendix, p. 342.)

9. **T. baldwinii** *Keay* in Bull. Jard. Bot. Brux. 26 : 206 (1956). A tree, to 20 ft. high ; in forest.
   Lib.: Jabroke, Webo (July) *Baldwin* 6485 ! Zeahtown, Tchien (Aug.) *Baldwin* 6938 ! Ganta (Sept.) *Baldwin* 12519 ! **G.C.:** Bonsa Su (May) *Vigne* FH 1983 !

10. **T. patens** (*Oliv.*) *Engl.* Bot. Jahrb. 1 : 425 (1881) ; op. cit. 54 : 321. *Sorindeia patens* Oliv. F.T.A. 1 : 441 (1868). *Trichoscypha victoriae* Engl. Bot. Jahrb. 36 : 224 (1905) ; F.W.T.A., ed. 1, 1 : 508. *T. paniculata* Engl. op. cit. 43 : 413 (1909). A forest tree, to 30 ft. or more high, with rather drooping branches ; flowers creamy-yellow.
    S.Nig.: Oban *Talbot* 1725 ! 1737 ! 1738 ! 1752 ! Stubbs' Creek F.R., Eket (May) *Onochie* FHI 32911 ! **Br.Cam.:** Bimbia (Mar.–May) *Preuss* 1206 ; 1293. Victoria (Jan.) *Maitland* 182 ! Also in French Cameroons and Spanish Guinea.

11. **T. arborea** (*A. Chev.*) *A. Chev.* Bot. 161 (1920) ; F.W.T.A., ed. 1, 1 : 508, partly (*Chev.* B. 22322 only) ; Aubrév. Fl. For. C. Iv. 2 : 164, t. 188. *Emiliomarcelia arborea* A. Chev. in Bull. Soc. Bot. Fr. 58, Mém. 8 : 151 (1912). *Trichoscypha ferruginea* of F.W.T.A., ed. 1, 1 : 510, partly. A forest tree, to 70 ft. or more high, with straight bole ; bark smooth, slash reddish, resinous ; flowers red ; fruits red, sweet.
    S.L.: Mange (young fr. Feb.) *Deighton* 3616 ! Kambui Hills South F.R. (fr. Dec.) *Small* 861 ! **Lib.:** Dukwia R. (young fr. Oct.) *Cooper* 65 ! 230 ! Karmadhun, Kolahun (fr. Nov.) *Baldwin* 10202 ! **Iv.C. :** Yapo (fr. Oct.) *Chev.* B. 22322 ! Abidjan *Aubrév.* 57 ! Banco *Martineau* 299 ! **G.C.:** Subri F.R., Tarkwa Dist. (Aug.) *C. J. Taylor* GC 5357 ! Achimkrom (July) *Vigne* FH 1261. **S.Nig.:** Oban *Talbot* 209 ! 228 ! **Br.Cam.:** Mawne River F.R., Mamfe (Nov.) *Tamajong* FHI 22114 ! (See Appendix, p. 342.)

12. **T. mannii** *Hook. f.* in Benth. & Hook. f. Gen. Pl. 1 : 423 (1862) ; F.T.A. 1 : 444. A forest tree, to 40 ft. high ; fruits dull red, seeds black.
    S.Nig.: Calabar R. (fl. & fr. Feb.) *Mann* 2238 ! Ikom Div. (fr. May) *Jones & Onochie* FHI 17450 ! Oban *Talbot* 208 ! 1278 ! 1368 ! **Br.Cam.:** Banga to Mungo R., Kumba Div. (Apr.) *Olorunfemi* FHI 30528 ! Also in Gabon.

13. **T. chevalieri** *Aubrév. & Pellegr.* in Bull. Soc. Bot. Fr. 81 : 647 (1934) ; Aubrév. Fl. For. C. Iv. 2 : 162, t. 189, 4–6. *T. arborea* of F.W.T.A., ed. 1, 1 : 508, partly (*Chipp* 318). A tree, to 20 ft. high, with spreading habit ; in forest.
    Iv.C.: Cavally Basin : Taté to Tabou (Aug.) *Chev.* 19835 ! Noé (Aug.) *Chev.* 19837 ! Prolo to Bliéron (Aug.) *Chev.* 19892 ! **G.C.:** Prestea (fl. & fr. Sept.) *Vigne* FH 3070 ! Amokokrom, W. Apollonia (Aug.) *Chipp* 318 ! Kumasi (Feb.) *Vigne* FH 1039 !

14. **T. sp. A.** Possibly new, but more material required. Liane with heavy woody stem ; calyx greenish-white with brownish indumentum, petals buff, ovary and disk chestnut brown ; in forest.
    Lib.: Gbanga (Sept.) *Linder* 721 !

15. **T. sp. B.** Possibly new, but more material required. Shrub or tree, to 15 ft. high, with densely brown-hirsute terminal congested inflorescence.
    Lib.: Yila, St. John R., Gbanga (Aug.) *Baldwin* 9133 !

16. **T. atropurpurea** *Engl.* Bot. Jahrb. 36 : 222 (1905) ; F.W.T.A., ed. 1, 1 : 508, partly (excl. *Bunting* 135). An unbranched tree, to 15 ft. high ; flowers blood-red, drying black, disk reddish-brown ; in forest.
    Lib.: Grand Bassa (Oct.) *Dinklage* 2082. Kolubanu (Oct.) *Linder* 1122 ! Javajai, Boporo (Nov.) *Baldwin* 10283a ! **G.C.:** Aiynase, Axim Dist. (Nov.) *Vigne* FH 1439 !

17. **T. longifolia** (*Hook. f.*) *Engl.* Bot. Jahrb. 1 : 425 (1881) ; in DC. Monogr. 4 : 305. *Dupuisia ? longifolia* Hook. f. in Fl. Nigrit. 287 (1849). *Sorindeia longifolia* (Hook. f.) Oliv. F.T.A. 1 : 442 (1868) ; F.W.T.A., ed. 1, 1 : 507. *S. afzelii* Engl. Bot. Jahrb. 15 : 107 (1892), not of Engl. (1917). A forest tree, to 50 ft. high ; with clean bole and no buttresses ; sap oxidizing to a strong black stain on exposure.
    S.L.: *Afzelius* ! *T. Vogel* 160 ! Makump (July) *Thomas* 885 ! Dodo Hills F.R. (fr. Jan.) *Sawyerr* FHK 13572 ! **Lib.:** Dukwia R. (fl. Sept., fr. May) *Cooper* 366 ! 435 ! Zuie, Boporo (fr. Nov.) *Baldwin* 10236 !

18. **T. oba** *Aubrév. & Pellegr.* in Bull. Soc. Bot. Fr. 81 : 648 (1934) ; Aubrév. Fl. For. C. Iv. 2 : 166, t. 190. *Pseudospondias luxurians* A. Chev. Bot. 157, name only. Note in F.T.A. 1 : 442, after *Sorindeia mannii* Oliv. *Trichoscypha ferruginea* of F.W.T.A., ed. 1, 1 : 510, partly (*Barter* 2151, *Dalz.* 1229, *MacGregor* 317, *Millen* 93). A small tree or shrub ; petals white ; in forest.
    Iv.C.: Bingerville region *Chev.* 15248 ! Dabou (fr. Nov.) *Jolly* 154 ! Banco *Martineau* 297 ! 317 ! **S.Nig.:** Lagos (fl. Apr., May, Oct.) *Barter* 2151 ! *Dalz.* 1229 ! *MacGregor* 317 ! N. of Eba Isl. (Jan.) *Thornewill* 254 ! Aguku *Thomas* 832 ! (See Appendix, p. 342.)

19. **T. yapoensis** *Aubrév. & Pellegr.* l.c. 81 : 649 (1934) ; Aubrév. l.c. 2 : 164, t. 189, 1–3. A forest tree, to 75 ft. high, sometimes with buttresses ; flowers white or cream ; cotyledons purple.
    Lib.: Dukwia R. (fl. Oct., fr. Nov.) *Cooper* 66 ! 114 ! **Iv.C.:** Yapo (fr. Dec.) *Aubrév.* 593 ! (See Appendix, p. 342.)

20. **T. longipetala** *Bak. f.* in Cat. Talb. 22 (1913). A small tree.
    S.Nig.: Oban *Talbot* 1681 !

21. **T. smeathmannii** *Keay* l.c. 205, fig. 63 (1956). *Brucea paniculata* Lam. Encycl. Méth. Bot. 1 : 472 (1785) ; F.T.A. 1 : 310 (excl. syn. *guineensis* G. Don) ; not *Trichoscypha paniculata* Engl. (1909). *T. arborea* of F.W.T.A., ed. 1, 1 : 508, partly (*Lane-Poole* 467). A small tree, or straggling bush, to 20 ft. high ; flowers white, fragrant.
    Fr.G.: Kouria (Nov.) *Chev.* 14753 ! **S.L.:** *Don* ! *Smeathmann* ! Belbu, Bonthe Isl. (Nov.) *Deighton* 2328 ! Port Loko (young fr. Dec.) *Thomas* 6613 ! 6700 ! Bo *Lane-Poole* 467 ! Njala (Oct.) *Deighton* 1412 ! 1779 ! 2802 ! Bumbuna (Oct.) *Thomas* 3420 ! 3726 ! **Lib.:** Fisherman's Lake (fr. Dec.) *Baldwin* 10899 !

22. **T. albiflora** *Engl.* Bot. Jahrb. 36 : 223 (1905). *T. ferruginea* of F.W.T.A., ed. 1, 1 : 510, partly. A shrub, scarcely branched, to 15 ft. high ; in forest near coast.
    Lib.: Fishtown (Aug.) *Dinklage* 2024 ! [**G.C.:** Gyamera, W.P., just above beach in closed forest (Feb.) *Irvine* 2280 !—sterile specimen, identification very tentative.] (See Appendix, p. 342.)

Imperfectly known species.

1. **T. liberica** *Engl.* Bot. Jahrb. 15 : 108 (1892). Possibly the same as *T. yapoensis* Aubrév. & Pellegr. (1934), but type-specimen presumably destroyed.
Lib.: Monrovia (Aug.) *Naumann.*

2. **T. sp. C.** " Small tree, to 15 ft. high, in savannah woodland ; fruits red." Possibly the same as *Sorindeia albiflora* Engl. & K. Krause (q.v.).
S.Nig.: Akparabon to Adijinkpon, Ikom Div. (fr. June) *Latilo* FHI 31823 !

12. SORINDEIA Thouars—F.T.A. 1 : 439. *Dupuisia* A. Rich. (1831).

Petals valvate :
Panicles terminal and in the upper leaf-axils, pubescent, puberulous or nearly glabrous, to 50 cm. long ; leaflets coriaceous :
  Petals 3·5–4 mm. long, 1·5–1·75 mm. broad ; calyx 2 mm. long ; ovary glabrous ; inflorescences and flowers puberulous or nearly glabrous ; leaflets opposite or subopposite :
    Calyx subtruncate or with very short broad teeth, not ciliate ; inflorescence puberulous to nearly glabrous, pedicels to 5 mm. long ; leaflets 1–9, oblong or oblong-elliptic, obtuse at base, shortly acuminate, 3·3–22 cm. long, 2–8·4 cm. broad, with 6–10 main lateral nerves on each side of midrib ; fruits ellipsoid, about 1 cm. long ; shrub or small tree .. .. 1. *juglandifolia*
    Calyx rather deeply lobed, ciliate ; inflorescence densely rusty-puberulous, pedicels to 1·5 mm. long ; leaflets about 9, oblong, obtuse at base, often abruptly acuminate, 6–12 (–17) cm. long, 2·5–5·2 (–8) cm. broad, with 10–12 main lateral nerves on each side of midrib ; fruits ellipsoid, about 2·5 cm. long ; tree, to 15 m. high
                                                2. *mildbraedii*
  Petals 2·75 mm. long, 1 mm. broad ; calyx 1·25 mm. long, broadly toothed ; ovary tomentellous ; inflorescences and flowers rusty-puberulous, usually rather densely so ; pedicels about 3 mm. long ; leaflets 3–8, oblong to obovate, obtuse and sometimes unequal at base, rather abruptly acuminate, 6–15 cm. long, 2·5–8 cm. broad, with 6–9 main lateral nerves on each side of midrib ; fruits ellipsoid, about 3·5 cm. long ; tree, to 14 m. high .. .. 3. *nitidula*
Panicles borne on the trunk of a 3–10 m. high tree, glabrous or very nearly so, to 30 cm. long, pedicels about 2 mm. long ; calyx 2 mm. long, subtruncate or with very short broad teeth, not ciliate ; petals 4 mm. long, 2 mm. broad ; leaflets membranous alternate or subopposite, about 9, lanceolate or ovate-oblong, obtuse or cuneate at base, gradually long-acuminate, 7–21 cm. long, 4–7 cm. broad, with 9–14 main lateral nerves on each side of midrib ; fruits ellipsoid, about 2 cm. long   4. *collina*
Petals imbricate ; inflorescences axillary on older branchlets or from main stem, glabrous or nearly so ; calyx 2 mm. long, shortly lobed, glabrous :
  Inflorescences racemose, rarely subpaniculate, 2–9 (–12) cm. long ; fruits ellipsoid, about 1·2 cm. long ; main lateral nerves 5–8 on each side of midrib ; shrub, to 4 m. high ; leaflets 1–6, opposite or subopposite, often oblong-elliptic or elliptic, the terminal one often obovate-oblong, rounded at base, broadly and obtusely acuminate, 1·5–14 cm. long, 0·8–7·5 (–10) cm. broad .. .. .. 5. *warneckei*
  Inflorescences paniculate, 16–30 cm. long ; fruits ellipsoid, about 2 cm. long ; main lateral nerves 9–14 on each side of midrib ; shrub or tree, to 15 m. high ; leaflets 3–7, alternate or subopposite, ovate-oblong to lanceolate, cuneate, obtuse or more rarely rounded at base, acuminate, 12–30 cm. long, 3·7–13 cm. broad 6. *grandifolia*

1. **S. juglandifolia** (*A. Rich.*) *Planch. ex Oliv.* F.T.A. 1 : 440 (1868), incl. var. *divaricata* Oliv. ; Chev. Bot. 160, partly ; Aubrév. Fl. For. C. Iv. 2 : 174 ; Fl. For. Soud.-Guin. 407–409, t. 86, 4–5. *Dupuisia juglandifolia* A. Rich. in Fl. Seneg. 1 : 148, t. 38 (Sept. 1831). *Sapindus simplicifolius* G. Don (Aug. 1831). *Sorindeia simplicifolia* (G. Don) Exell in Cat. S. Tomé 145 (1944) ; Consp. Fl. Ang. 2 : 97 ; not of (H.B. & K.) March. (1869). *S. heterophylla* Hook. f. (1849). (?) *S. afzelii* Engl. Bot. Jahrb. 54 : 313 (1917) ; not o Engl. (1892). A shrub or small tree ; flowers whitish, tinged pink.
Sen.: *Perrottet* 151 ! Ziguinchor, Casamance (Mar., Apr.) *Leprieur.* Bignona *Aubrév.* 3061. Floup-Fedyan *Chev.* 3167. **Fr.Sud.:** Diendénia (Feb.) *Chev.* 436. **Port.G.:** Bor, Bissau (Feb.) *Esp. Santo* 1742 ! **Fr.G.:** Rio Nunez *Heudelot* ! Conakry (Mar.) *Debeaux* 311 ! Kouroussa (Feb.) *Chev.* 377 ! And other localities *fide* Chev. *l.c.* **S.L.:** *Afzelius* ! *Don* ! Freetown *Sc. Elliot* ! Bagroo R. (Mar.) *Mann* 796 ! Erimakuna (fr. Mar.) *Sc. Elliot* 5404 ! Musaia (Dec.) *Deighton* 4545 ! Luseniya, Samu (Dec.) *Sc. Elliot* 4311 ! Sendugu (June) *Thomas* 671 ! **Iv.C.:** various localities *fide* Chev. & Aubrév. *ll.c.* **G.C.:** Kintampo (Mar.) *Dalz.* 18 ! **Dah.:** Kouandé to Konkobiri *Chev.* 24241. Also in Ubangi-Shari, Angola and N. Rhodesia. (See Appendix, p. 341.)

2. **S. mildbraedii** *Engl. & v. Brehm.* in Engl. Bot. Jahrb. 54 : 311 (1917). An understorey tree, to 50 ft. high ; panicles drooping, axes deep reddish-purple, flower-buds pale green ; fruits orange and red ; in forest.
S.Nig.: Stubbs' Creek F.R., Eket (fr. May) *Onochie* FHI 32906 ! **Br.Cam.:** Banga, Kumba Div. (fl. Mar., fr. Aug.) *Brenan* 9432 ! *Olorunfemi* FHI 30726 !
[These specimens have been determined from the description only ; the type, presumably now destroyed, was from the French Cameroons.]

3. **S. nitidula** *Engl.* Bot. Jahrb. 36 : 221 (1905). An understorey tree, to 45 ft. high ; flowers pale yellowish or whitish-green ; ripe fruits bright red ; in forest.
S.Nig.: Oban *Talbot* 1714 ! Abo Eme, Ikom Div. (May) *Latilo* FHI 30983 ! Kanyang, Ikom Div. (fr. May) *Latilo* FHI 30989 ! **Br.Cam.:** Victoria (May) *Maitland* 671 ! Also in French Cameroons.

4. **S. collina** *Keay* in Bull. Jard. Bot. Brux. 26 : 208 (1956). A tree, to 30 ft. high ; leaves clustered at the ends of the branches ; cauliflorous, with the inflorescences from ground level up to about 10 ft. high ; flowers reddish ; in forest on hills.
S.L.: Tingi Mts., 2,000 ft. (fr. Feb.) *Deighton* 3511 ! Kambui Hills, Koya (Dec.) *Small* 867 ! Bintumane Peak, 3,200 ft. (Jan.) *T. S. Jones* 130 !

5. **S. warneckei** *Engl.* Bot. Jahrb. 36 : 221 (1905) ; Chev. Bot. 161 ; Aubrév. Fl. For. C. Iv. 2 : 174 ; Fl. For.
   Soud.-Guin. 407–409, t. 86, 6. *S. schroederi* Engl. & K. Krause (1911). *S. juglandifolia* of Chev. Bot. 160,
   partly, and of F.W.T.A., ed. 1, 1 : 507, partly (*Chev.* 23360 ; 23455). *Lannea dahomensis* A. Chev. Bot.
   159, name only, partly. A shrub, usually straggling ; calyx green, petals pink, or white tinged pink ;
   ripe fruits bright orange ; in understorey of forest and in regrowth.
   S.L.: *Afzelius.* **Iv.C.:** Fétékro *Aubrév.* 811. **G.C.:** Aburi (Nov.) *Brown* 785 ! Juaso, Ashanti (fr. Aug.)
   Irvine 275 ! Pokoase (fr. Jan.) *Irvine* 1956 ! Achimota (fl. Dec., fr. Feb.) *Irvine* 2003 ! Milne-Redhead
   5175 ! **Togo:** Lome (fl. & fr. Feb.) *Warnecke* 52 ! Okrabi, Hohoe Dist. (Nov.) *St. C. Thompson* FHI 3691 !
   Ho (Feb.) *Schröder* 211. **Dah.:** Ouidah *Chev.* 23455 ! Cotonou *Chev.* 23360 ! Ketou to L. Azri, Zagnanado
   Chev. 23039 (partly) ! **S.Nig.:** Lagos (fr. Feb.) *Dalz.* 1066 ! 1230 !   *Millson* ! Ibadan (fl. Nov.–Feb.,
   fr. Feb.) *Keay* FHI 14167 ! 22482 ! *Meikle* 977 ! 1127 ! Iguabazowa F.R., Benin (fr. Feb.) *Onochie* FHI
   27141 ! Umueze, Onitsha (fl. & young fr. Feb.) *Jones* FHI 5033 ! 5034 ! Oban *Talbot* 1277 ! 1364 !
   (See Appendix, p. 341.)
6. **S. grandifolia** *Engl.* Bot. Jahrb. 11, Beibl. 26 : 7 (1890). *S. acutifolia* Engl. (1890). *S. juglandifolia* of
   F.W.T.A., ed. 1, 1 : 507, partly. A shrub or tree, to 50 ft. high ; calyx pale green, petals pink.
   S.Nig.: Oban *Talbot* 1307a ! 1382 ! 1429 ! 1434 ! **Br.Cam.:** L. Barombi, Kumba (Mar.) *Brenan & Onochie*
   9464 ! 9465 ! N.Korup F.R., Kumba Div. (fr. July) *Olorunfemi* FHI 30668 ! Bakebe to Eyang, Mamfe
   Div. (Jan.) *Umana* FHI 28553 ! Metschum Falls, Bamenda Prov. (Jan.) *Keay & Russell* FHI 28527 !
   Also on S. Tomé.

Fig. 200.—Sorindeia juglandifolia (*A. Rich.*) *Planch. ex Oliv.* (Anacardiaceae).
A, leaf. B, flower. C, same in vertical section. D, fruiting branch. E, fruit in cross-section.

Imperfectly known species.

1. **S. albiflora** *Engl. & K. Krause* in Engl. Bot. Jahrb. 46 : 341 (1911). A tree, to 40 ft. high ; leaflets 13–15,
   appressed-pilose beneath ; ovary densely pilose, stigmas 5 ; type specimen presumably destroyed.
   Br.Cam.: Bamenda to Babongi (Dec.) *Ledermann* 1938.
   [The description suggests a plant very different from other *Sorindeia* species in our area, and it possibly
   belongs to another genus. A fruiting specimen, *Latilo* FHI 31823, from savannah woodland in Ikom Div.,
   S. Nig., may be the same, but I think it is a *Trichoscypha* (q.v.)]
2. **S. protioides** *Engl. & K. Krause* l.c. 337 (1911). A shrub or tree, to 45 ft. high ; leaflets 5–7, glabrous ; panicle
   lax, glabrous ; aestivation of petals not described ; type specimen presumably destroyed.
   Br.Cam.: Babessi (Dec.) *Ledermann* 1986. Also in French Cameroons.

Excluded species.

1. **S. deliciosa** *A. Chev. ex Hutch. & Dalz.* F.W.T.A., ed. 1, 1 : 507 (1928) is **Santiria trimera** (*Oliv.*) *Aubrév.*
   (Burseraceae).
2. **S. doeringii** *Engl. & K. Krause* l.c. 338 (1911) is **Ekebergia senegalensis** *A. Juss.* (Meliaceae).
3. **S. longifolia** (*Hook. f.*) *Oliv.*—F.W.T.A., ed. 1, 1 : 507, is transferred to **Trichoscypha.**

13. HEERIA Meisn. Pl. Vasc. Gen., diagn. 75, comm. 55 (1837).

By R. D. Meikle

Leaves 5–12 cm. broad, 6–22 cm. long, oblong or obovate, obtuse or emarginate ;
inflorescence a narrow cylindrical panicle ; a suffrutex about 1 m. high
                                                                    1. *pulcherrima*
Leaves 1–4 cm. broad, 4–14 cm. long, acute or acuminate, rarely obtuse or emarginate ;
shrubs and small trees :
Leaves oblong, elliptical or broadly lanceolate, distinctly petiolate, scattered or irregularly whorled ; inflorescence and infructescence lax, paniculate, overlapping the
   leaves ..   ..   2. *insignis*
Leaves linear-lanceolate, shortly petiolate or subsessile, regularly 3–4-whorled ;
infructescence condensed, axillary, much shorter than the leaves ..   .. 3. *sp. A*

1. **H. pulcherrima** (*Schweinf.*) *O. Ktze.* Rev. Gen. 152 (1891) ; Aubrév. Fl. For. Soud.-Guin. 412 (1950). *Anaphrenium pulcherrimum* Schweinf. Beitr. Fl. Aeth. 1 : 32 (1867). *Rhus pulcherrima* (Schweinf.) Oliv. F.T.A. 1 : 436 (1868). *R. djalonensis* A. Chev. Bot. 161 (1920), name only. *R. herbacea* A. Chev. Bot. 161 (1920), name only. An erect suffrutex with softly pubescent stems and leaves ; flowers small, greenish or tinged red ; fruits black, glossy ; in savannah.

Fr.G.: Timbo *Pobéguin* 139 ; 1498. Kollangui (Mar.) *Chev.* 13518 ! 13519 ! Along R. Niger (Mar.) *Sc. Elliot* 5167 ! **Iv.C.:** Banfara *Chev.* 531. Kari *Serv. For.* 2862. **G.C.:** Kwomma to Nangrum (fl. May) *Kitson* 654 ! Sambisi to Hamboi (fl. May) *Kitson* 656 ! **Dah.:** Kouandé to Konkobiri (June) *Chev.* 24243 ! **N.Nig.:** Katagum Dist. *Dalz.* 358 ! Abinsi (fl. Mar.) *Dalz.* 775 ! Naraguta *Lely* 25 ! 71 ! Pankshin (fl. Dec.) *W. D. MacGregor* 420 ! Anara F.R., Zaria (fl. Mar., May) *Keay* FHI 22955 ! *Olorunfemi* FHI 24410 ! Birnin Gwari (fl. Feb.) *Meikle* 1228 ! Also in French Cameroons, Ubangi-Shari, Abyssinia and A.-E. Sudan. (See Appendix, p. 338.)

2. **H. insignis** (*Del.*) *O. Ktze.* Rev. Gen. 152 (1891) ; Aubrév. Fl. For. Soud.-Guin. 412, incl. var. *angustifolia* Engl. and var. *latifolia* (Engl.) Engl. *Ozoroa insignis* Del. in Ann. Sci. Nat. sér. 2, 20 : 91, t. 1, fig. 3 (1843). *Rhus insignis* (Del.) Oliv. F.T.A. 1 : 437 (1808) ; Chev. Bot. 161. A branched shrub or small tree about 15–20 ft. high ; undersurface of leaves silvery-white ; flowers small, creamy ; fruits glossy black ; in savannah.

Sen.: *Leprieur* ; *Heudelot* 67 ! *Sieber* 33 ! M'bidjem, *Thierry* 228 ! **Gam.:** *Dawe* 38 ! Birram *Rosevear* 79 ! **Fr.G.:** Labé *Pobéguin* 2280. **Fr.Sud.:** Segou (Sept.) *Chev.* 3037 ! Bandiagara, Arbala, Koulouba & Sikasso to Bobo *fide* Aubrév. l.c. 415. **S.L.:** Musaia (fl. Oct.) *Miszewski* 50 ! **Iv.C.:** Bobo Dioulasso *Aubrév.* 1843 ; 1890. Sakoinsi *Aubrév.* 2161. **G.C.:** Wa to Durimo (Mar.) *Kitson* 934 ! Belong (Mar.) *Kitson* 786 ! Tangkoasi Hill (May) *Kitson* ! Prang to Attabulu (May) *Kitson* 657 ! Han (fr. Apr.) *Vigne* FH 3802 ! Lawra (Nov.) *Vigne* FH 4662 ! Wa (fr. Feb.) *Adams* GC 4465 ! **Dah.:** Atacora Mts. (June) *Chev.* 24027 ! **Fr.Nig.:** Niamey *De Wailly* 4925. **N.Nig.:** Nupe *Barter* 904 ! Kontagora (Oct.) *Dalz.* 101 ! Bauchi Plateau (fr. Feb.) *Lely* P. 141 ! *Costello* P. 5/28 ! Basawa, Zaria (Apr.) *Keay* FHI 16109 ! E. of Zurmi (Apr.) *Keay* FHI 16144 ! Gwari Hill, Minna (fr. Dec.) *Meikle* 697 ! Funtua (fr. Dec.) *Meikle* 826 ! 827 ! Ilorin (fl. Feb.) *Ejiofor* FHI 19836 ! **S.Nig.:** Old Oyo F.R. (fr. Feb.) *Keay* FHI 16263 ! 16293 ! Also in Ubangi-Shari, Abyssinia and A.-E. Sudan. (See Appendix, p. 338.)

3. **H. sp. A.** A slender, virgate shrub with whorled Osier-like leaves, glossy above, and silvery-white beneath ; fruits glossy black. Perhaps an aberrant form of *H. insignis*, but material insufficient for certain determination.

G.C.: Kintampo, N. Ashanti (fr. Nov.) *Andoh* FH 5422 !

14. RHUS Linn.—F.T.A. 1 : 436, partly.

By R. D. Meikle

Leaves deeply dentate-lobed ; twigs armed with strong spines   ..   1. *tripartita*
Leaves entire or obscurely crenate-dentate ; twigs unarmed or very sparsely spinose :
Inflorescence conspicuous, forming a lax panicle usually much longer than the subtending leaves ; leaflets obovate or elliptical, 3–6 cm. long, 2–3 cm. broad, usually entire ; twigs brown or fuscous, without spines ; fruit 3–4 mm. diam., slightly compressed   ..   ..   2. *longipes*
Inflorescence inconspicuous, usually equalling or shorter than the subtending leaves ; leaflets obovate or elliptical, 1·5–5 cm. long, 1–3 cm. broad, obscurely crenate or serrate towards the apex ; twigs pale grey or grey-brown, sometimes a little spinose ; fruits 5–6 mm. diam., strongly compressed ..   ..   ..   3. *natalensis*

1. **R. tripartita** (*Ucria*) *Grande* in Bull. Ort. Bot. Nap. 4 : 242, 243 (1918), as *R. tripartita* (Ucria) DC. *Rhamnus tripartita* Ucria in Roem. Arch. Bot. 1 : 68 (1796). *Rhus dioica* Brouss. ex Willd. Enum. Pl. Hort. Berol. 325 (1809). *R. oxyacantha* of Aubrév. Fl. For. Soud.-Guin. 409, not of Schousb. ex Cav. in Anal. Ciênç. Nat. 3 : 36 (1801). A low spiny bush with small, deeply toothed or lobed, glabrous leaves, and strongly compressed, reddish fruits.

Fr.Nig.: Air *Aubrév.* Also Hoggar Mts., N. Africa, Egypt, Palestine and Sicily.

2. **R. longipes** *Engl.* in DC. Monogr. 4 : 431 (1883). *R. villosa* of F.T.A. 1 : 439, partly, not of Linn. f. *R. incana* of F.W.T.A., ed. 1, 1 : 512 (excl. var. *dahomensis* Hutch. & Dalz.) ; of Aubrév. Fl. For. Soud.-Guin. 410 ; not of Mill. *Schmidelia affinis* of Chev. Bot. 149, at least in part, not of Guill. & Perr. *Rhus glaucescens* of F.W.T.A., ed. 1, 1 : 512, not of A. Rich. A bush or small tree with pale green leaves ; whitish or greenish flowers and dull red fruits.

Sen.: Cape Verde (Feb.) *Döllinger* 37 ! **Fr.G.:** Labé, Fouta Djalon (Mar.) *Chev.* 12404 ! **S.L.:** Loma Mts. (Jan.) *Jaeger* 4183 ! **N.Nig.:** Bauchi Plateau (Dec., Jan.) *Lely* P. 7 ! *Kennedy* 2859 ! Bauchi (fr. June) *Cons. of For.* ! Naraguta (Nov., Dec.) *Lely* 718 ! *W. D. MacGregor* 384 ! Widespread throughout tropical Africa.

3. **R. natalensis** *Bernh. ex Krauss* in Flora 27 : 349 (1844) ; Aubrév. Fl. For. Soud.-Guin. 410. *Searsia natalensis* (Bernh. ex Krauss) Barkl. in Lilloa 23 : 253 (1950). *R. glaucescens* A. Rich. Fl. Abyss. 1 : 143 (1847) ; F.T.A. 1 : 437. *R. incana* Mill. var. *dahomensis* Hutch. & Dalz. F.W.T.A., ed. 1, 1 : 512 (1928) ; Aubrév. l.c. 410. A small shrub 4–10 ft. high ; flowers greenish ; fruits shining brown.

Fr.G.: Dalaba *Adam* 66. Saraya *Maclaud* 456. **Iv.C.:** Batié *Serv. For.* 2502. **G.C.:** Wa, N.T. (Dec.) *Adams* GC 4470 ! **Dah.:** Kouandé to Konkobiri (June) *Chev.* 24263 ! **N.Nig.:** by Zaria Road, Jos (Aug.) *Keay* FHI 12711 ! Neill's Valley, Jos (fr. Nov.) *Onochie & Adelodun* FHI 21742 ! Widespread throughout tropical Africa.

122. CONNARACEAE

By F. N. Hepper

Erect trees or shrubs, or scandent. Leaves alternate, compound, imparipinnate or 1–3-foliolate ; stipules absent. Flowers hermaphrodite, actinomorphic or slightly zygomorphic. Calyx imbricate or valvate. Petals 5, free or sometimes slightly connate, imbricate or rarely valvate. Stamens hypogynous to perigynous, often declinate, 5 or 10 ; filaments often united at the base ; anthers 2-celled, opening lengthwise. Disk absent or thin. Carpels 1–5, free, 1-celled. Ovules 2, collateral, erect, attached near the base or the middle of the ventral suture, all or only 1 ripening into fruit. Fruit often splitting down one side exposing the seed. Seeds usually 1, with or without endosperm, often arillate.

A small tropical family easily recognized by the pinnate or unifoliolate exstipulate leaves, apocarpous ovary and arillate seeds.

Carpels 5 :
  Sepals free to base :
    Androgynophore absent from the flower :
      Flowers appearing before or with the very young leaves ; only 1 carpel fruitful ; seed with a fleshy outer covering .. .. 3. **Byrsocarpus**
      Flowers appearing after the leaves :
        Inflorescences terminal or nearly terminal, never on the older wood :
          Petals about half as long as the sepals ; sepals not enlarged in fruit ; seed with endosperm and a basal aril ; several carpels fruitful .. .. 5. **Cnestis**
          Petals as long as or longer than the sepals :
            Leaves 3-foliolate ; stellate hairs present ; sepals with glandular hairs on the margins, not enlarged in fruit ; seed without endosperm ; several carpels fruitful .. .. 6. **Agelaea**
            Leaves pinnate ; stellate hairs absent ; sepals without glandular hairs, enlarged in fruit ; only 1 carpel fruitful, glabrous ; aril adnate to seed, no endosperm :
              Leaflets 3–4 pairs ; sepals fleshy in fruit ; fruit oblong .. 4. **Jaundea**
              Leaflets numerous, oblique, the lowermost stipule-like ; sepals thin ; fruit with a curved beak .. .. 1. **Roureopsis**
        Inflorescences axillary or clustered on the older wood :
          Leaves pitted on the upper surface (sunken mucilage cells), 3-foliolate ; seed without endosperm ; sepals not enlarged in fruit .. .. 7. **Castanola**
          Leaves not pitted above ; pinnate, rarely 3- or 1-foliolate in No. 3 :
            Sepals imbricate, glabrous except for ciliate margins, enlarging somewhat in fruit but thin ; only 1 carpel fruitful ; petals 3–4 times as long as sepals ; inflorescences axillary :
              Fruiting carpel oblong, rounded at apex ; leaflets (1–) 3–5   3. **Byrsocarpus**
              Fruiting carpel ovoid, acute at apex ; leaflets 5–7 .. 8. **Santaloïdes**
            Sepals almost valvate, tomentose :
              Petals equalling, or up to twice as long as, the sepals ; inflorescences often clustered on the older branches ; several carpels fruitful, pubescent ; sepals not enlarging in fruit .. .. 5. **Cnestis**
              Petals at least 4 times as long as the sepals, linear ; inflorescences axillary ; only 1 carpel fruitful, glabrous ; sepals enlarging and leathery in fruit
                                                    2. **Paxia**
    Androgynophore present ; sepals not enlarged in fruit ; leaflets striolate between the veins ; several carpels fruitful, small and stipitate .. .. 9. **Manotes**
  Sepals connate up to nearly half their length, enlarging in fruit ; petals much longer than the sepals, linear and twisted ; inflorescences axillary   10. **Spiropetalum**
Carpels solitary :
  Flowers borne on young growth :
    Leaflets several to each leaf ; fertile stamens 10 or 5 ; seed without endosperm ; inflorescences terminal or subterminal, paniculate .. .. .. 11. **Connarus**
    Leaflet solitary ; fertile stamens 5 ; seed with hard bony endosperm ; inflorescences terminal or subterminal panicles or short axillary racemes 12. **Hemandradenia**
  Flowers in clusters of short racemes on the older wood ; fruits 2-seeded ; leaves large, imparipinnate, with 9–11 subopposite to alternate leaflets   13. **Jollydora**

1. ROUREOPSIS Planch. in Linnaea 23 : 423 (1850) ; Schellenb. in Engl. Pflanzenr. Connarac. 107 (1938).

Leaflets numerous, obliquely oblong, mucronate, 2·5–6 cm. long, 1–2 cm. broad, pubescent on the midrib beneath, with 1–2 basal nerves on one side, the basal leaflets smaller and stipule-like ; flowers in sessile clusters on very long inflorescences ; fruits 1·5 cm. long, with a curved beak .. .. *obliquifoliolata*

R. obliquifoliolata (*Gilg*) *Schellenb.* Beitr. Vergl. Anat. Syst. Connarac. 28 (1910) ; Pflanzenr. 108, fig. 18 (1938) ; Troupin in Fl. Congo Belge 3 : 77. *Rourea obliquifoliolata* Gilg (1891). *R. fasciculata* Gilg (1891). *R. adiantoides* Gilg (1896). A climbing shrub with brown-pubescent branchlets ; flowers dull yellow ; fruits red, splitting to expose a black seed with a fleshy yellowish aril ; in rain forest.
S.Nig.: Oban *Talbot* 504 ! 578 ! Eket *Talbot* ! Also in French Cameroons, Spanish Guinea, Belgian Congo and Angola.

2. PAXIA Gilg in Engl. Bot. Jahrb. 14 : 320 (1891) ; Schellenb. in Engl. Pflanzenr. Connarac. 114 (1938).

Midrib of leaflets glabrous beneath ; leaflets elliptic-obovate, gradually acuminate, 8–12 cm. long, 3–6 cm. broad, with 4–5 pairs of lateral nerves ; cymes axillary, very short and fasciculate, puberulous ; petals about 1·5 cm. long .. .. 1. *liberosepala*

Midrib of leaflets puberulous beneath ; leaflets ovate or ovate-lanceolate, obtusely
acuminate, 5–10 cm. long, 2·5–4 cm. broad, leathery, with 6–7 pairs of lateral nerves ;
cymes few-flowered, fasciculate ; fruits 2·5–3 cm. long .. .. 2. *cinnabarina*

1. **P. liberosepala** (*Bak. f.*) *Schellenb. ex Hutch. & Dalz.* F.W.T.A., ed. 1, 1 : 514 (1928) ; Schellenb. in Engl.
    Pflanzenr. Connarac. 115. *Spiropetalum liberosepalum* Bak. f. (1913). A glabrous climbing shrub with
    dull yellow flowers.
    S.Nig.: Oban *Talbot* 575 ! 1404a !
2. **P. cinnabarina** *Schellenb.* in Engl. Bot. Jahrb. 55 : 449 (1919) ; Pflanzenr. 118. *P. liberosepala* of Schellenb.
    (1938), partly (*Talbot* 3352). A climbing shrub ; branchlets rusty-yellow when young ; flowers white ;
    fruits cinnabar-red.
    S.Nig.: Eket *Talbot* 3352 ! **Br.Cam.:** Victoria *Winkler* 526. Also in French Cameroons.

3. BYRSOCARPUS Schum. & Thonn. Beskr. Guin. Pl. 226 (1827) ; Schellenb. in Engl.
Pflanzenr. Connarac. 146 (1938).

Leaflets obtuse at apex, not acuminate, sometimes retuse, 7–11 or more, oblong, the
lateral ones oblique, very flaccid when young, 5–45 mm. long, 3–20 mm. broad,
glabrous or pilose on the midrib beneath ; inflorescences 2–3 cm. long, bracts inconspicuous, 1–2 mm. long ; fruits oblong, 1·5 cm. long, glabrous .. 1. *coccineus*
Leaflets acuminate, 2–17 cm. long :
　Leaflets lanceolate to ovate-lanceolate, rounded at base, acute to acuminate at apex,
　2·5–7 cm. long, 1–3 cm. broad ; inflorescences 1·5–3 cm. long, appearing with the
　young leaves ; bracts ovate, 4 mm. long ; fruits oblong, 1·5–2 cm. long, glabrous
　　　　　　　　　　　　　　　　　　　　　　　　　　　　　　　2. *poggeanus*
　Leaflets ovate to obovate or broadly elliptic :
　　Inflorescences up to 2 cm. long, appearing amongst the mature leaves ; bracts ovate-
　　lanceolate, 2–3 mm. long ; leaflets (1–) 3–5, broadly elliptic to obovate, rounded
　　at base, long-caudate-acuminate, 2·5–17 cm. long, 1·5–8 cm. broad ; fruits oblong,
　　2 cm. long, glabrous .. .. 3. *viridis*
　　Inflorescences 5–7 cm. long, appearing with the young leaves ; bracts subulate,
　　1 mm. long ; leaflets 5–9, rhomboid-elliptic to ovate, cuneate to rounded at base,
　　shortly and gradually acuminate at apex, 2–8 cm. long, 1–4·5 cm. broad ; fruits
　　1·5 cm. long, contracted about the middle, glabrous .. .. 4. *dinklagei*

1. **B. coccineus** *Schum. & Thonn.*—F.T.A. 1 : 452, incl. var. *parvifolius* (Planch.) Bak. ; Schellenb. in Engl.
    Pflanzenr. Connarac. 148 ; Chev. Bot. 163 ; Troupin in Fl. Congo Belge 3 : 91. A shrub or climber about
    6 ft. high ; flowers white ; fruits red, aril yellow and seed black ; in secondary forest or thickets in
    savannah.
    Fr.G.: Rio Nunez *Heudelot* 811 ! Dyeke (fr. Oct.) *Baldwin* 9659 ! Kaba *Chev.* 13131. Kaoulendougou
    Chev. 18142. **S.L.:** Kambia, R. Scarcies (Dec.) *Sc. Elliot* 4347 ! Falaba (Apr.) *Aylmer* 47 ! Taiama
    Dawe 496 ! Kowama (fr. Nov.) *Deighton* 5250 ! **Lib.:** Cape Palmas (fr. July) *T. Vogel* 37 ! Zigida (Mar.)
    Bequaert 136 ! Tubman Bridge (fr. Sept.) *Barker* 1409 ! Monrovia (fr. June) *Baldwin* 5939 ! **Iv.C.:**
    several localities *fide* Chev. l.c. **G.C.:** Accra *T. Vogel* 33 ! Aburi Hills (fr. June) *Williams* 294 ! Kumasi
    (Feb.) *Darko* 527 ! Mampong (fr. May) *Kitson* 928 ! **Togo:** Lome *Warnecke* 55 ! Misahöhe *Baumann* 554 !
    Kpandu (fr. June) *Andoh* FH 5290 ! Sokode *Kersting* A. 431 ! **Dah.:** Cotonou *Chev.* 4446. Porto Novo
    Chev. 22781. Kouandé to Konkobiri *Chev.* 24289. **N.Nig.:** Nupe *Barter* 1191 ! 1688 ! Lokoja (Apr.)
    Elliott 46 ! Sokoto (May) *Lely* 840 ! Vom, 3,000–4,500 ft. *Dent Young* 32 ! **S.Nig.:** Lagos *Barter* 2179 !
    Ibadan to Abeokuta (Mar.) *Schlechter* 13020 ! Warrake, Benin (fr. June) *Dundas* FHI 21477 ! **Br.Cam.:**
    Bamenda, 3,000–3,500 ft. (fr. Apr., May) *Maitland* 1449 ! 1595 ! Widespread in tropical Africa. (See
    Appendix, p. 343.)
2. **B. poggeanus** (*Gilg*) *Schellenb.* Beitr. Vergl. Anat. Syst. Connarac. 45 (1910) ; Pflanzenr. 154 ; Troupin l.c.
    93. *Rourea poggeana* Gilg (1891). A lax climbing shrub, up to 6 ft. high ; flowers pale yellow ; fruits
    red ; in moist thickets.
    S.Nig.: Ibadan (Mar.) *Meikle* 1254 ! Also in French Cameroons, Gabon, Spanish Guinea, Belgian Congo
    and Angola.
3. **B. viridis** (*Gilg*) *Schellenb.* Beitr. Vergl. Anat. Syst. Connarac. 46 (1910) ; Pflanzenr. 158, fig. 28 ; Troupin
    l.c. 96. *Rourea viridis* Gilg (1891). *R. unifoliolata* Gilg (1891). *R. mannii* Gilg (1891). *R. pallens* Hiern
    (1896). A shrub or climber ; flowers white ; fruits pinkish ; in forest.
    S.Nig.: Onitsha *Barter* 1634 ! Aguku *Thomas* 780 ! Oban *Talbot* 1622 ! 1629 ! Degema *Talbot* 3794 !
    Old Calabar *Thomson* 52 ! Also in French Cameroons, Gabon, French and Belgian Congo and Angola.
4. **B. dinklagei** (*Gilg*) *Schellenb. ex Hutch. & Dalz.* F.W.T.A., ed. 1, 1 : 514 (1928) ; Schellenb. l.c. 157 ; Troupin
    l.c. 94. *Rourea dinklagei* Gilg (1895). Climbing shrub about 12 ft. high ; calyx pinkish ; corolla white ;
    fruits red with a shining black seed ; in forest.
    S.Nig.: Calabar (fl. Mar., Apr., fr. Aug.) *Holland* 25 ! 67 ! *Brenan* 9222 ! Oban *Talbot* 1621 ! **Br.Cam.:**
    Cameroons R. (Jan.) *Mann* 756 ! Victoria *Preuss* 1375. Johann-Albrechtshöhe *Staudt* 901. **F.Po:**
    Mann 279 ! Also in French Cameroons, Ubangi-Shari and Belgian Congo.

4. JAUNDEA Gilg in Pflanzenfam. 3, 3 : 388 (1894) ; Schellenb. in Engl. Pflanzenr.
Connarac. 161 (1938).

Leaflets more or less pilose on the nerves beneath, about 3 pairs, obovate or obovate-
elliptic, acutely acuminate, rounded or obtuse at the base, up to 12 cm. long and 7 cm.
broad, strongly veined beneath ; panicles shorter than the leaves ; calyx puberulous ;
fruits 2·5 cm. long.. .. .. 1. *pubescens*
Leaflets glabrous beneath but sometimes hairy on the midrib :
　Leaflets lanceolate, gradually long-acuminate, rounded at the base, 5–10 cm. long,
　2–2·5 cm. broad, laxly reticulate ; fruit obliquely oblong, about 2 cm. long
　　　　　　　　　　　　　　　　　　　　　　　　　　　　　　　2. *baumannii*
　Leaflets ovate or oblong-obovate, 3–4 pairs, obtusely pointed and mucronate, up to
　12 cm. long and 5 cm. broad, laxly veined beneath ; panicles clustered and short ;
　calyx pubescent outside ; fruits 2–3 cm. long .. .. .. 3. *pinnata*

1. **J. pubescens** (*Bak.*) *Schellenb.* in Engl. Bot. Jahrb. 55 : 462 (1919) ; Engl. Pflanzenr. Connarac. 162 ; Troupin in Fl. Congo Belge 3 : 84. *Connarus pubescens* Bak. in F.T.A. 1 : 458 (1868). *C. thomsoni* Bak. (1868). *Rourea buchholzii* Gilg (1895). *R. nivea* Gilg (1900). A shrub or climber 15 ft. high with white flowers.
    S.Nig.: Lagos (Dec.) *Hagerup* 750 ! *Millen* 109 ! Cross River North F.R., Ikom (fr. June) *Latilo* FHI 31832 ! Old Calabar (Feb.) *Mann* 2254 ! *Thomson* 26 ! Eket *Talbot* 3144 ! Also in French Cameroons, Spanish Guinea, Belgian Congo and Cabinda.
2. **J. baumannii** (*Gilg*) *Schellenb.* in Engl. Bot. Jahrb. 55 : 460 (1919) ; Pflanzenr. 164. *Rourea baumannii* Gilg (1896). A lofty climbing shrub.
    Togo: Misahöhe (fr. Mar.) *Baumann* 31.
3. **J. pinnata** (*P. Beauv.*) *Schellenb.* in Candollea 2 : 92 (1925) ; Pflanzenr. 164, fig. 29 ; Troupin l.c. 87, t. 6 . *Cnestis pinnata* P. Beauv. (1806). *Manotes palisotii* Planch. (1850). *Rourea palisotii* (Planch.) Baill. (1866). *R. pseudobaccata* Gilg (1891). *R. ivorensis* A. Chev. Bot. 165, name only. *Jaundea zenkeri* Gilg (1895). A strong woody climber, 15–20 ft. high, or sometimes an erect shrub ; flowers white or pink-tinged, fragrant ; fruits red ; in forest.
    Fr.G.: Kouria *Dumas* in Hb. *Chev.* 18172 ; *Chev.* 14810 ! Oussoudou to Diédédou, Beyla (fr. Feb.) *Chev.* 20847 ! **S.L.:** Kanguma *Smythe* 118 ! Kenema (fr. Jan.) *Thomas* 7510 ! **Lib.:** Monrovia *Whyte* ! Kolubanu (fr. Oct.) *Linder* 1117 ! Genna Tanyehun (fr. Dec.) *Baldwin* 10752 ! Duport (fr. Oct.) *Barker* 1437 ! **Iv.C.:** Daoukrou to Akabiékrou (Dec.) *Chev.* 22503 ! **G.C.:** Akwapim Hills (Nov.) *Johnson* 777 ! Kumasi (Jan.) *Irvine* 39 ! Aburi (fr. Apr.) *Irvine* 218 ! Odumasi (Dec.) *Vigne* FH 3497 ! **Togo:** Atakpame (Nov.) *Mildbr.* 7423 ! *Doering* 273. Misahöhe *Baumann* 31a. **Dah.:** Abomey to Boguila (fr. Feb.) *Chev.* 23178 ! **S.Nig.:** Ala (Nov.) *Thomas* 1980 ! Ebute Metta (Oct.) *Millen* 6 ! Ibadan (Oct.) *Newberry* 116 ! Warrake, Benin Prov. (fr. June) *Dundas* FHI 21483 ! Extends southwards to Angola and in eastern Africa from A.-E. Sudan to N. Rhodesia. (See Appendix, p. 344.)

5. CNESTIS Juss.—F.T.A. 1 : 460 ; Schellenb. in Engl. Pflanzenr. Connarac. 28 (1938).

Fruit covered with short uniform velvety hairs without long bristles, pear-shaped, not horned at the apex ; petals half as long as, or equalling, the sepals ; leaves glabrous above, except on the nerves ; inflorescences terminal or in the axils of the upper leaves, up to about 20 cm. long :
  Petals half as long as the sepals ; leaflets tomentose or pubescent beneath :
    Leaflets (4–) 6 or more pairs ; petals glabrous ; flowers in panicles or racemes :
      Sepals 2–3 mm. long ; petals suborbicular ; panicles shorter than the leaves, terminal or upper axillary ; leaflets oblong-elliptic, rounded at base, obtusely pointed, about 5–7 cm. long and 2 cm. broad, softly tomentose beneath ; fruit 3–4 cm. long with an upturned blunt beak .. ..   1. *ferruginea*
      Sepals 5–7 mm. long ; petals oblong, emarginate ; racemes as long as the leaves, subterminal and axillary ; leaflets oblong, rounded at both ends, 2–4·5 cm. long, about 1·5 cm. broad, thinly pilose beneath ; fruit 5–6 cm. long, tapering into a straight beak ..   ..   2. *macrantha*
    Leaflets 2–4 pairs ; petals pubescent outside, 2 mm. long ; sepals 5–6 mm. long ; panicles terminal, laxly branched :
      Leaflets very shortly pubescent beneath, broadly ovate-elliptic, rounded at base, abruptly acuminate at apex, 11–20 cm. long, 6–12 cm. broad ..   3. *mannii*
      Leaflets softly tomentose beneath, oblong-elliptic to oblong-ovate, rounded at base, acute but hardly acuminate, 8–15 cm. long, 3–7 cm. broad   4. *tomentosa*
  Petals as long as or slightly longer than the sepals, 2·5 mm. long ; leaflets 2–3 pairs, ovate-elliptic, rounded at base, long-acuminate, 6–12 cm. long, 2·5–5 cm. broad, nearly glabrous beneath ; racemes much shorter than the leaves ; fruits about 2 cm. long, rounded or subacute at apex   ..   ..   ..   5. *racemosa*
Fruit covered both with short uniform velvety hairs and with long bristles, tapering into a long horn at apex ; petals longer than, or about equalling, the sepals ; inflorescences cauliflorous or axillary, 1·5–6 (–10) cm. long :
  Leaflets with scattered bristles above ; leaf-rhachis long-pilose :
    Leaflets 5–6 pairs, broadly oblong, upper ones 10–18 cm. long, 4·5–6 cm. broad, usually abruptly acuminate, bristly beneath ; fruits about 3 cm. long, 1 cm. broad in the middle, very sharply curved ; young branchlets pilose ; petals 7 mm. long ; sepals 4 mm. long ..   ..   6. *grisea*
    Leaflets 8–12 pairs, oblong, upper ones 4–8 cm. long, 2–3 cm. broad, softly pubescent beneath ; fruits 4 cm. long, 2 cm. broad in the middle, shortly beaked ; young branchlets very densely and closely pilose ..   ..   ..   7. *sp. A*
  Leaflets glabrous above :
    Midrib beneath pilose with laterally spreading hairs, other parts of leaflets with scattered bristles ; leaflets 10–12 pairs, oblong, very shortly acuminate, 3–6 cm. long, 1–2 cm. broad, the basal pair suborbicular ; fruits about 4 cm. long, densely covered with long yellow bristles ; petals 4·5 mm. long ; sepals 3 mm. long
                                                  8. *corniculata*
    Midrib beneath tomentose or appressed-pubescent :
      Apex of leaflets distinctly, acutely and abruptly acuminate ; sepals and petals equal in length :
        Petals and sepals 7–8 mm. long ; racemes 10–15 cm. long ; fruits 4 cm. long, about 9 mm. broad in the middle ; leaf-rhachis glabrous or minutely puberulous ; leaflets (2–) 3–6 pairs, broadly oblong, the terminal ovate-oblong, subcordate at base, abruptly, acutely and long-acuminate at apex, 10–21 cm. long, 5–10 cm. broad, glabrous beneath ..   ..   9. *congolana*

Petals and sepals 4–5 mm. long ; racemes 3–4 cm. long ; leaflets unequal-sided
and subcordate at base, shortly acuminate at apex :
Leaf-rhachis glabrous ; leaflets 4–5 pairs, oblong-elliptic, 5·5–9 cm. long, 2·5–
4 cm. broad, glabrous beneath and midrib sparsely pilose ; fruits slender,
3 cm. long, 1 cm. broad .. .. 10. *cinnabarina*
Leaf-rhachis tomentose, bristly towards the base ; leaflets 7–10 pairs, oblong,
3–10 cm. long, 1·5–3·5 cm. broad, pubescent beneath, midrib puberulous ;
fruits 3·5 cm. long, about 1·2 cm. broad .. .. .. 11. *aurantiaca*
Apex of leaflets slightly and obtusely acuminate ; leaflets ovate to ovate-oblong :
Leaflets glabrous beneath, except for appressed hairs on midrib and margins,
4–5 pairs, 6–15 cm. long, 3–5 cm. broad ; racemes 1·5–3 cm. long ; petals 4 mm.
long, equalling sepals ; fruits slender 2·5 cm. long .. .. 12. *dinklagei*
Leaflets softly tomentose beneath, 5–6 (–7) pairs, 6–18 cm. long, 3–5·5 cm. broad ;
racemes 5–8 cm. long ; petals 6 mm. long ; sepals 2 mm. long ; fruits 2·5–3 cm.
long .. .. 13. *longiflora*

1. **C. ferruginea** *DC.*—F.T.A. 1 : 461 ; Schellenb. in Engl. Pflanzenr. Connarac. 30, fig. 3 ; Troupin in Fl.
Congo Belge 3 : 118 ; Chev. Bot. 166. *C. oblongifolia* Bak. in F.T.A. 1 : 462 (1868). A shrub or some-
times a climber 6–15 ft. high, densely brown-pubescent ; flowers small and white, fragrant ; fruits con-
spicuous, red ; sometimes abundant in forest regrowth.
   Sen.: Mampalago, Casamance (Feb.) *Chev.* 2973 ; 3026 ! **Gam.:** *Brown-Lester* S. 2 ! Koto (fr. May)
Fox 101 ! **Port.G.:** *Esp. Santo* 506 ! **Fr.G.:** Rio Nunez *Heudelot* ! Kouria to Ymbo (Nov.) *Chev.* 14809 !
Kaoulendougou *Chev.* 21063. Dyeke (Oct.) *Baldwin* 9674 ! **S.L.:** Leicester (Dec.) *Sc. Elliot* 3876 !
Njala (Oct.) *Deighton* 1348 ! Sumbuya (Nov.) *Deighton* 2333 ! Kasewe F.R. (Dec.) *King* 44*b* ! **Lib.:**
Monrovia (Nov.) *Linder* 1543 ! Gondolahun (fr. Nov.) *Baldwin* 10111 ! Belleyella (fr. Dec.) *Baldwin*
10648 ! Zwedru (Aug.) *Baldwin* 7038 ! **Iv.C.:** Abidjan *Aubrév.* 466. **G.C.:** Cape Coast (Oct.–Dec.)
Vigne FH 941 ! Essiama (Apr.) *Williams* 439 ! Akwapim Hills (Oct.) *Johnson* 797 ! Kumasi F.R.
(Aug.) *Akwa* FH 1771 ! **Togo:** Lome *Warnecke* 480 ! Sokode-Basari *Kersting* A. 279. **Dah.:** *Burton* !
Sakété to Pédjilé *Chev.* 22907 ! Abomey *Chev.* 23129. **N.Nig.:** Patti Lokoja (fr. Mar.) *Elliott* 35 ! **S.Nig.:**
Idanre (Jan.) *Brenan & Keay* 8688 ! Onitsha *Barter* 1758 ! Oban *Talbot* 508 ! 1401 ! Old Calabar
(fr. Jan.) *Holland* 16 ! **Br.Cam.:** Victoria *Kalbreyer* 8 ! Cameroon R. (fl. & fr. Jan.) *Mann* 753 ! Johann-
Albrechtshöhe *Staudt* 651. Also in French Cameroons, Spanish Guinea, Gabon, Ubangi-Shari, French
and Belgian Congo, A.-E. Sudan, Cabinda, Angola and Principe and S. Tomé. (See Appendix, p. 343.)
2. **C. macrantha** *Baill.*—F.T.A. 1 : 462 ; Schellenb. l.c. 34. A shrub 8–12 ft. high, densely yellowish-hairy.
   S.Nig.: Old Calabar (Feb.) *Mann* 2235 ! *Duparquet* 53. Oban *Talbot* 1279 ! 1378 ! 1706 ! Degema
Talbot ! **Br.Cam.:** Mamfe (fl. & fr. Apr.) *Maitland* 1149 !
3. **C. mannii** (*Bak.*) *Schellenb.* in Engl. Bot. Jahrb. 55 : 436 (1919) ; Pflanzenr. 34. *Connarus mannii* Bak. in
F.T.A. 1 : 459 (1868). A climbing shrub 15 ft. high, shortly yellowish-pubescent on branchlets and
inflorescence.
   S.Nig.: Shasha F.R. (Apr.) *Ross* 198 ! Old Calabar (Feb.) *Mann* 2264 ! Oban *Talbot* 1707 !
4. **C. tomentosa** *Hepper* in Kew Bull. 1956 : 112. *C. mannii* of Schellenb. in Engl. Pflanzenr. Connarac. 34,
partly (*Staudt* 916). A climber up to 100 ft. high ; flowers greenish-brown ; in high forest.
   Br.Cam.: S. Bakundu F.R., Kumba (Mar.) *Brenan* 9328 ! Johann-Albrechtshöhe *Staudt* 916 !
5. **C. racemosa** *G. Don*—F.T.A. 1 : 463 ; Schellenb. l.c. 41. *Manotes racemosa* (G. Don) Gilg (1896). *Cnestis
liberica* Schellenb. (1919). A climbing shrub, 6–10 ft. high, with shortly brown-pubescent branchlets and
inflorescence ; flowers white, fruits pale pink ; in forest.
   S.L.: Freetown (fr. Feb.) *Dalz.* 974 ! Hill Station (fr. Jan.) *Dalz.* 8119 ! Njala (fr. Jan.) *Dalz.* 8105 !
Mattru (Nov.) *Deighton* 2335 ! Faiama (fr. Jan.) *Deighton* 3875 ! **Lib.:** Kakatown *Whyte* ! Gbanga
(Sept.) *Linder* 650 ! Harbel, Du R. (fr. Dec.) *Bequaert* 16 ! Brewersville (fr. Dec.) *Baldwin* 10969 !
6. **C. grisea** *Bak.* in F.T.A. 1 : 461 (1868) ; Schellenb. l.c. 48, fig. 5 ; Chev. Bot. 167. A woody climber,
bristly on most parts ; inflorescences clustered on the main branches ; fruits scarlet, pungent-bristly.
   Iv.C.: Fort Binger *Chev.* 19531. **S.Nig.:** Old Calabar *Thomson* 90 ! Oban (fr. May) *Ujor* FHI 31789 !
Talbot 506 ! **Br.Cam.:** Victoria (Apr.) *Maitland* 602 ! Johann-Albrechtshöhe *Staudt* 844. Also in
French Cameroons and Gabon.
   [The leaflets of *Maitland* 602 are subacute and not acuminate at the apex.]
7. **C. sp. A.** Close to *C. iomalla* Gilg in Notizbl. Bot. Gart. Berl. 1 : 69 (1895), but the leaflets are thicker
and more reticulate and pubescent ; the fruits are broader and do not possess the long curved beak of
C. iomalla. Further material, in flower, is required.
   S.Nig.: Afi F.R., Ikom (fr. May) *Latilo* FHI 30975 !
   [Another specimen from S. Nig., Cross R., nr. Issabang (*Jones* FHI 1541 !) is similar but has the sparsely
pilose leaflets beneath and narrower fruits of *C. grisea*.]
8. **C. corniculata** *Lam.*—F.T.A. 1 : 461 ; Schellenb. l.c. 48 ; Chev. Bot. 166. A woody climber with roughly
pilose branchlets ; flowers cream-white in short clustered racemes sometimes on the older branches ;
fruits red, covered with pungent bristles ; in forest.
   Fr.G.: Fouta Djalon *Heudelot* 650. Dyeke (Oct.) *Baldwin* 9668 ! Ditinn to Timbo *Chev.* 12632. **S.L.:**
Smeathmann ! Leicester Peak (fr. Dec.) *Sc. Elliot* 3831 ! Taiama (fr. Nov.) *Cole* 48 ! Magbile (fr. Dec.)
Thomas 5952 ! Mamaka (Nov.) *Thomas* 4340 ! Mamansu (fr. Feb.) *Deighton* 4065 ! **Lib.:** Grand Bassa
T. Vogel. Peahtah (Oct.) *Linder* 957 ! Ganta (Sept.) *Baldwin* 9235 ! Jabrocca (fr. Dec.) *Baldwin*
10857 ! (See Appendix, p. 343.)
9. **C. congolana** *De Wild.* Étud. Fl. Bas- et Moyen-Congo 3, 1 : 96 (1909) ; Schellenb. l.c. 52, fig. 6 ; Troupin
l.c. 116. Woody climber up to 50 ft. high ; in forest.
   Br.Cam.: Banga, S. Bakundu F.R. (Mar.) *Brenan* 9419 ! 9474 ! *Dundas* FHI 8361 !
10. **C. cinnabarina** *Schellenb.* in Engl. Bot. Jahrb. 55 : 438 (1919) ; Pflanzenr. 46. Climbing shrub, young
branches puberulous, older ones brown and glabrous ; fruits cinnabar coloured.
   S.Nig.: Degema *Talbot* ! Also in French Cameroons.
11. **C. aurantiaca** *Gilg* in Engl. Bot. Jahrb. 23 : 216 (1896) ; Schellenb. l.c. 50. *C. longiflora* Schellenb. l.c. 46,
partly (*Talbot* 507). A climbing shrub with subglabrous branches ; flowers orange.
   S.Nig.: Oban *Talbot* s.n. ! 507 ! **Br.Cam.:** Mamfe (fr. Jan.) *Maitland* 1153 ! Also in French Cameroons.
12. **C. dinklagei** *Schellenb.* in Engl. Bot. Jahrb. 55 : 438 (1919) ; Pflanzenr. 45. A climbing shrub with coiled
branchlets ; flowers greenish, in short racemes on the stems ; fruits red ; in forest.
   Lib.: Monrovia *Dinklage* 2412 ; 2695 ! 2712. *Baldwin* 5851 ! Brewersville (June) *Barker* 1325 ! Mecca
(fr. Dec.) *Baldwin* 10796 ! Gbanga (fr. Sept.) *Linder* 652 !
13. **C. longiflora** *Schellenb.* in Engl. Bot. Jahrb. 55 : 438 (1919) ; Pflanzenr. 46. *C. prehensilis* A. Chev. Bot.
167, name only. A woody climber 6–12 ft. high ; inflorescences creamy white, on the stems ; fruits
scarlet with stinging bristles ; in secondary forest and thickets.
   Iv.C.: Bouroukrou (fr. Dec.) *Chev.* 16887 ! **Dah.:** Sakété *Chev.* 22828. **S.Nig.:** Lagos (fl. & fr. Nov.)
Dawodu 194 ! *Dalz.* 1155 ! 1401 ! Abeokuta (fr. Jan.) *Burtt* 23 ! Ibadan (fl. & fr. Feb.) *Brenan* 8974 !
Ujor FHI 24436 ! Oban *Talbot* 1404*b* ! (See Appendix, p. 344.)

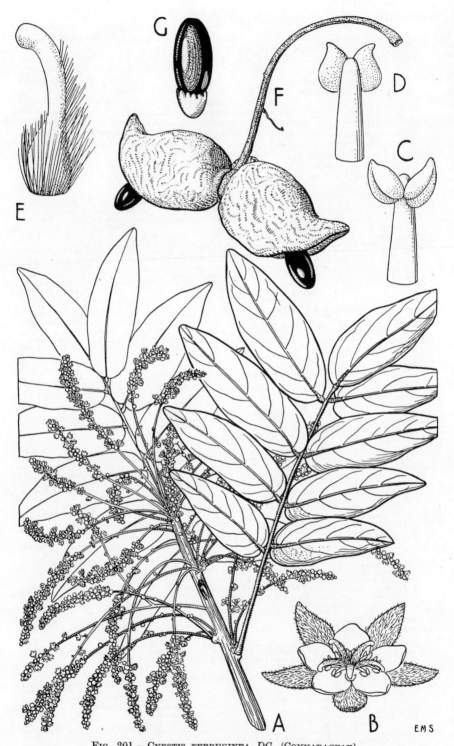

FIG. 201.—CNESTIS FERRUGINEA *DC.* (CONNARACEAE).

A, flowering shoot, × ⅔.   B, flower, × 9.   C & D, stamens, × 30.   E, gynoecium, × 30.
F, fruit, × 1. G, seed, × 1½.
A is drawn from *Hagerup* 767 ; B–E from *Brenan & Keay* 8688 ; F & G from *Milne-Redhead* 5188.

6. AGELAEA Soland. ex Planch.—F.T.A. 1 : 453 ; Schellenb. in Engl. Pflanzenr. Connarac. 65 (1938).

Indumentum on petioles and rhachis of inflorescence bristly, hairs 1–3 mm. long :
  Hairs on inflorescence 3 mm. long, very dense ; leaflets ovate or broadly elliptic, rounded at base, abruptly acuminate at apex, 6–14 cm. long, 4–7·5 cm. broad, glabrous on both sides except for tomentose midrib above ..   ..   1. *hirsuta*
  Hairs on inflorescence 1 mm. long, rather dense on young branches and sparser on petioles ; leaflets broadly ovate, broadly cuneate at base, acuminate at apex, 8–12 cm. long, 4–9 cm. broad, glabrous above, the midrib with a few long hairs, sparsely pilose beneath ..   ..   2. *pilosa*
Indumentum on petioles and rhachis of inflorescence minutely stellate, not bristly, hairs less than 0·5 mm. long :
  Leaflets glabrous beneath (or the younger leaves slightly tomentellous in No. 4) ; secondary bracts (i.e. those on the branches of the inflorescence) small :
    Leaflets prominently 3-nerved at base, the next main lateral nerves appearing halfway up the leaflet, 6–14 cm. long, 4–10 cm. broad, closely and prominently reticulate beneath ; sepals 4 mm. long ; petals 5–6 mm. long :
      Leaflets cuneate to rounded at base ..   ..   ..   3. *obliqua* var. *obliqua*
      Leaflets distinctly cordate at base ..   ..   ..   3a. *obliqua* var. *cordata*
    Leaflets pinnately nerved, lateral nerves almost equally spaced, broadly elliptic, rounded at base, acuminate at apex, about 17 cm. long and 10 cm. broad, reticulate beneath and tertiary nerves prominent ; sepals 2–3 mm. long, petals 2·5–4 mm. long ..   ..   4. *floccosa*
  Leaflets pubescent beneath, often with minute stellate hairs :
    Secondary bracts (i.e. those on the branches of the inflorescence) and primary bracts 8–10 mm. long, subpersistent ; leaflets ovate to ovate-elliptic, cuneate and 3-nerved at base, shortly acuminate at apex, 6–11 (–14) cm. long, 4–7 cm. broad, midrib and lateral nerves densely tomentose, laxly tomentose beneath, densely so on the nerves..   ..   ..   5. *trifolia*
    Secondary bracts 2–3 mm. long ; indumentum of leaflets beneath minute (× 20 lens essential) :
      Leaflets beneath with stellate hairs, 4–5-armed, the arms spreading :
        Stellate hairs beneath leaflets with usually 5 slender wavy arms spreading and overlapping the next, easily seen with × 20 lens; leaflets elliptic, cuneate at base, acuminate at apex, 7 cm. long, 4 cm. broad, midrib tomentose above ; inflorescence densely tomentose with stiffer hairs ..   ..   ..   6. *grisea*
        Stellate hairs beneath leaflets with 4 (–5) arms, arms short not overlapping the next hair, minute and just visible with × 20 lens :
          Midrib above rather densely pubescent with 4-armed hairs which are also scattered over the upper leaflet surface at least near the base ; leaflets broadly ovate to suborbicular, cuneate at base, acuminate at apex, 5–9 cm. long, 4–7 cm. broad ; sepals 2 mm. long ; petals 3 mm. long ; indumentum on inflorescence rhachis very short and dense ..   ..   7. *nitida*
          Midrib above glabrous or with a few stellate hairs towards the base :
            Sepals and petals 2 mm. long ; indumentum on rhachis of inflorence not very dense (i.e. individual stellate hairs distinguishable) ; leaflets broadly ovate, cuneate to rounded at base, abruptly and long-acuminate at apex, 7–14 cm. long, 5–10 cm. broad, tertiary nerves not parallel   ..   8. *pseudobliqua*
            Sepals 4 mm. long ; petals 5 mm. long ; indumentum on inflorescence rhachis dense (i.e. individual stellate hairs overlapping one another and indistinguishable) ; leaflets ovate to broadly elliptic, rounded or broadly cuneate at base, abruptly acuminate at apex, 8–19 cm. long, 5–10 cm. broad, tertiary nerves prominent and parallel ..   ..,   9. *dewevrei*
  Leaflets beneath with scattered stellate hairs with erect branches in a cluster, seen with difficulty with × 20 lens ; leaflets 3-nerved from base, ovate, rounded at base, acutely acuminate at apex, terminal one 4–7·5 cm. long, 2·5–4 cm. broad ..   ..   10. *oligantha*

1. **A. hirsuta** *De Wild.* Étud. Fl. Bas- et Moyen-Congo 3, 1 : 100, t. 25 (1909) ; Schellenb. in Engl. Pflanzenr. Connarac. 67 ; Troupin in Fl. Congo Belge 3 : 100. Climbing shrub.
  S.Nig.: Uria *Kitson* 109 ! Also in French Cameroons and Belgian Congo.
2. **A. pilosa** *Schellenb.* l.c. 83 (1938). *A. elegans* of F.W.T.A., ed. 1, 1 : 516, partly (*Talbot* 3670). *A. trifolia* of F.W.T.A., ed. 1, 1 : 516 partly, and of Schellenb. l.c. 83, partly (*Thomas* 2168 ; 2175). Climbing shrub with white flowers.
  S.Nig.: Usonigbe F.R., Benin (Mar.) *Jones* FHI 7447 ! Sapoba (fr. May) *Kennedy* 1133 ! 1377 ! Degema *Talbot* 3670 ! Idumuje (Jan.) *Thomas* 2168 ! 2175 ! Abontakon, Ikom (Dec.) *Keay* FHI 28173 !
3. **A. obliqua** (*P. Beauv.*) *Baill.* var. *obliqua*—F.T.A. 1 : 454 ; Schellenb. l.c. 90 ; Chev. Bot. 164. *Cnestis obliqua* P. Beauv. (1804). Scrambling shrub with shining leaves ; flowers pure white or tinged yellow, green or pink, fragrant ; fruits scarlet ; in forest and secondary growth.
  Port.G.: Biombo, Bissau (Feb.) *Esp. Santo* 1792 ! **S.L.:** Regent (Dec.) *Sc. Elliot* 4045 ! Port Loko (Dec.) *Thomas* 6719 ! Njala (Nov.) *Deighton* 4961 ! Makunde (fr. Jan.) *Jordan* 1002 ! **Lib.:** Vahun, Kolahun (fr. Nov.) *Baldwin* 10244 ! **Iv.C.:** *Jolly* 84. Grand Bassam *Thollon* 168. Bingerville *Chev.* 20155. Bouroukrou *Chev.* 16828. **G.C.:** Cape Coast *Brass* ! Akwapim (Mar.) *Irvine* 1544 ! Kpong

(fl. & fr. Jan.) *Irvine* 1921! Larte Hills (Nov.) *Johnson* 807! **Togo:** North Adakulu (Mar.) *Williams* 55! Sokode-Basari *Kersting* A. 278! Atakpame *Doering* 373. **Dah.:** Cotonou (fr. Mar.) *Debeaux* 170! Abomey (Feb.) *Chev.* 23139! **N.Nig.:** Nupe *Barter* 1340! Lokoja (fl. & fr. Mar.) *Shaw* 25! Abinsi (Apr.) *Dalz.* 780! **S.Nig.:** Lagos *Barter* 2154! *Dalz.* 1092! Abeokuta *Irving* 118! Iyereku, Benin (Feb.) *Umana* FHI 29148! Onitsha *Unwin* 67! Eket *Talbot*! **Br.Cam.:** Victoria to Kumba (Apr.) *Ejiofor* FHI 29365! (See Appendix, p. 342.)

3a. **A. obliqua** var. **cordata** (*Schellenb.*) *Exell* in Cat. S. Tomé 148 (1944). *A. cordata* Schellenb. (1923)—Engl. Pflanzenr. Connarac. 92.
  G.C.: Kumasi (fl. & fr. Sept.) *Vigne* FH 3370! **S.Nig.:** Degema *Talbot* 3690! Also in S. Tomé.

4. **A. floccosa** *Schellenb.* in Engl. Bot. Jahrb. 58 : 210 (1923) ; Pflanzenr. 79. Tall forest climber.
  S.Nig.: Shasha F.R., Ijebu (May) *Richards* 3485! **Br.Cam.:** Abonando *Rudatis* 64! Victoria (Apr.) *Maitland* 634!

5. **A. trifolia** (*Lam.*) *Gilg* in Notizbl. Bot. Gart. Berl. 1 : 65 (1895) ; Schellenb. in Engl. Pflanzenr. Connarac. 83. *Cnestis trifolia* Lam. (1789). *Agelaea villosa* (DC.) Soland. ex Planch. (1850)—F.T.A. 1 : 454 ; Chev. Bot. 164. A shrub or climber, 6–25 ft. or rarely up to 100 ft. high ; flowers creamy white or green in ample panicles ; fruits scarlet ; in forest.
  Port.G.: Fulacunda (May) *Esp. Santo* 2038! **Fr.G.:** Rio Nunez *Heudelot* 730! Derenka *Heudelot*. **S.L.:** *Smeathmann*! Bagroo R. (Mar.) *Mann* 794! Njala (fr. Nov.) *Deighton* 3101! Tawia (Feb.) *Deighton* 3493! Kambui Hills (fr. Mar.) *Small* 514! **Lib.:** Monrovia (Oct., Nov.) *Cooper* 20! Belleyella, Boporo (fr. Dec.) *Baldwin* 10647! Ganta, Sanokwele (Dec.) *Baldwin* 9215! Jabroke, Webo (July) *Baldwin* 6447! **Iv.C.:** Taté to Tabou *Chev.* 19823. **S.Nig.:** Shasha F.R., Ijebu (Feb.) *Ross* 38! *Richards* 3127! (See Appendix, p. 342.)

6. **A. grisea** *Schellenb.* in Engl. Bot. Jahrb. 58 : 208 (1923) ; Pflanzenr. 73. *A. trifolia* of F.W.T.A., ed. 1, 1 : 516, partly ; of Schellenb. l.c. 83, partly. Climbing shrub with yellowish flowers.
  S.Nig.: Oban *Talbot* 1734! Also in French Cameroons.

7. **A. nitida** *Soland. ex Planch.* in Linnaea 23 : 437 (1850) ; Schellenb. in Pflanzenr. 88. *A. obliqua* of F.T.A. 1 : 454, partly, (*Heudelot* 894, *Afzelius*). Straggling bush up to 10 ft. high ; flowers white, fragrant.
  Sen.: Conakry *Debeaux* 321. **Fr.G.:** Rio Pongo *Heudelot* 894. **S.L.:** *Afzelius*! Kambia (fr. Dec.) *Sc. Elliot* 4224! Kumrabai (Dec.) *Thomas* 6968! Njala (May) *Deighton* 1947! **Lib.:** Zigida (Mar.) *Bequaert* 132!

8. **A. pseudobliqua** *Schellenb.* in Engl. Bot. Jahrb. 58 : 216 (1923) ; Pflanzenr. 86 ; Troupin l.c. 108. *A. phaseolifolia* Gilg ex Schellenb. (1923)—Schellenb. l.c. 90 ; F.W.T.A., ed. 1, 1 : 516. A shrub 8 ft. high, with scarlet velvety fruits.
  S.Nig.: Itu, Cross R. (Apr.) *Holland* 28! Old Calabar (May) *Holland* 68! *Robb*! Oban *Talbot* 1762! Eket *Talbot*! **F.Po:** *Mann* 433! Also in French Cameroons and Gabon.

9. **A. dewevrei** *De Wild. & Th. Dur.* in Bull. Soc. Bot. Belge 38 : 190 (1899) ; Schellenb. l.c. 73 ; Troupin l.c. 104, t. 8. A scrambling shrub with white flowers.
  Br.Cam.: Lakom, Bamenda, 6,000 ft. (May) *Maitland* 1460! Barombi *Preuss* 518. Also in Belgian Congo and Principe.

10. **A. oligantha** *Gilg ex Schellenb.* in Engl. Bot. Jahrb. 58 : 214 (1923) ; Pflanzenr. 85. Shrub or climber ; flowers white.
  Lib.: Fishtown, Grand Bassa (Sept.) *Dinklage* 1711 ; 2039! Kings Farm, Monrovia *Dinklage* 3280.

Imperfectly known species.

As the species of this genus are so critical, and as the following were not seen, it has not been possible to include them in the key.

1. **A. preussii** *Gilg* in Engl. Bot. Jahrb. 23 : 10 (1896) ; Schellenb. in Engl. Pflanzenr. Connarac. 86. Climber with yellow-brown-tomentose inflorescences and branchlets.
  Br.Cam.: Victoria to Bimbia (May) *Preuss* 1116 ; 1277. Also in (?) Belgian Congo.

2. **A. macrocarpa** *Schellenb.* in Engl. Bot. Jahrb. 58 : 214 (1923) ; Pflanzenr. 85. Shrub with purplish-velvety obovoid fruits 1 in. long.
  Lib.: Fishtown, Grand Bassa (fr. Oct.) *Dinklage* 1866.

7. CASTANOLA Llanos in Mem. Acad. Cienc. Madrid 3, 2 : 302 (1859) ; Schellenb. in Engl. Pflanzenr. Connarac. 169 (1938).

Leaflets broadly obovate-elliptic, broadly cuneate at base, long and abruptly acuminate, 8–16 cm. long, 4–10 cm. broad, glabrous and obscurely reticulate ; inflorescence a short panicle ; fruit narrowly obovoid, beaked, up to 2 cm. long, tomentellous
paradoxa

C. paradoxa (*Gilg*) *Schellenb. ex Hutch. & Dalz.* F.W.T.A., ed. 1, 1 : 516 (1928) ; Pflanzenr. 174, fig. 31. *Agelaea paradoxa* Gilg (1891). *A. fragrans* Gilg (1896). *A. brevipaniculata* Cummins (1898). A scrambling shrub or small tree of forest, with fragrant white flowers.
  S.L.: Pujehun (fr. Feb.) *Thomas* 8182! **Lib.:** Browntown (Aug.) *Baldwin* 7084! Tappita (Aug.) *Baldwin* 9115! **G.C.:** Akim (June) *Johnson* 764! Kumasi (Mar., Nov.) *Cummins* 30a! *Irvine* 127! Asamankese (Aug.) *Howes* 936! Ekosu (July) *Darko* 916! **S.Nig.:** Olokemeji (Mar.) *A. F. Ross* 112! Sapoba *Kennedy* 1592! 1744! Ute Ogbeji, Benin (Aug.) *Onochie* FHI 33427! Eket *Talbot*! Oban *Talbot* 1294! 1300! Also in French Cameroons, Gabon and Belgian Congo. (See Appendix, p. 343.)

8. SANTALOÏDES Schellenb. Beitr. Verg. Anat. Syst. Connarac. 76 (1910) ; Engl. Pflanzenr. Connarac. 119 (1938).

Leaflets ovate or oblong to obovate, rounded at base, long-acuminate at apex, 5–10 cm. long, 2–5 cm. broad, glabrous, reticulate ; racemes clustered, slender, axillary, 3–6 cm. long ; pedicels 6 mm. long, jointed near base with 2 small linear bracteoles ; sepals ovate, imbricate, 2 mm. long, enlarging and becoming woody in fruit, glabrous, margin of apex ciliate or tomentose ; petals about 6 mm. long ; fruits ovoid, often slightly curved, acute at apex, about 2 cm. long, glabrous, finely longitudinally striate .. .. *afzelii*

S. afzelii (*R.Br. ex Planch.*) *Schellenb.* Beitr. Verg. Anat. Syst. Connarac. 53 (1910) ; Pflanzenr. 138. *Rourea afzelii* R.Br. ex Planch. (1850). *R. splendida* Gilg (1891). *Santaloïdes splendidum* (Gilg) Schellenb. ex Engl. Pflanzenw. Afr. 3 : 327 (1915) ; Schellenb. in Engl. Bot. Jahrb. 55 : 455 (1919) ; Pflanzenr. 140 ; Troupin in Fl. Congo Belge 3 : 82, t. 5 ; Hemsley in F.T.E.A. Connarac. 13. *Rourea gudjuana* Gilg (1891). *Santaloïdes gudjuanum* (Gilg) Schellenb. ex Engl. l.c. 326 (1915) ; Pflanzenr. 138, fig. 24 ; F.W.T.A., ed. 1, 1 : 517 ; Aubrév. Fl. For. Soud.-Guin. 416, t. 24, 1–2. A straggling shrub about 10 ft. high or a climber up to 60 ft. high ; flowers white or pinkish ; fruits red ; by streams and in forest.
  Port.G.: Madina do Boé, Bafatá (Jan.) *Esp. Santo* 2860! Geba, Bafatá (fr. Feb.) *Esp. Santo* 2674! **Fr.G.:** Kouroussa (Dec.) *Pobéguin* 820! Timbo *Pobéguin* 105. *Chev.* 14637. **S.L.:** *Afzelius*! Sulimania

Sc. Elliot 5352! Musaia (fr. Feb., Mar., July) *Deighton* 4230! 4802! *Sc. Elliot* 5128! **Lib.:** Karmadhun, Kolahun Dist. (fr. Nov.) *Baldwin* 10154! **Iv.C.:** Ferkéssédougou *Aubrév.* 1615. **G.C.:** Atuna, Ashanti (Dec.) *Vigne* FH 3530! **Togo:** Sokodé *Kersting* A. 94; A. 116. **N.Nig.:** Katagum *Dalz.* 409! Naraguta *Lely* 73! 714! Mada Hills (fr. June) *Hepburn* 73! Anara F.R., Zaria (Nov.) *Keay* FHI 21124! *Ogbeni & Obiora* FHI 21692! Extends southwards to Angola and in eastern Africa from A.-E. Sudan to N. Rhodesia. (See Appendix, p. 344.)

9. MANOTES Soland. ex Planch.—F.T.A. 1 : 459 ; Schellenb. in Engl. Pflanzenr. Connarac. 54.

Inflorescences terminal and subterminal panicles, about 20 cm. long ; sepals 2–4 mm. long, 1 mm. broad ; fruits recurved ; midrib tomentose above ; branches without recurved spines :
Rhachis of inflorescence densely tomentose or pubescent with hairs about 0·5 mm. long, obscuring the minute hairs beneath ; leaflets usually laxly pilose beneath :
Sepals (of open flower) ovate, 2 mm. long, pubescent outside ; petals 7–9 mm. long, pubescent outside, minutely puberulous inside ; fruits 1–1·5 cm. long, stipitate, pubescent ; leaflets 4–6 pairs, elliptic, (2–) 4·5–13 cm. long   .. 1. *zenkeri*
Sepals (of open flower) linear-lanceolate, 4 mm. long, puberulous outside ; petals about 9 mm. long, minutely puberulous on both sides ; fruits 1–1·5 cm. long, stipitate, finely velvety ; leaflets 3–5 pairs, lanceolate-elliptic, (2–) 3–6·5 cm. long
2. *longiflora*
Rhachis of inflorescence minutely puberulous, also with a few hairs about 0·5 mm. long ; sepals lanceolate, 3 mm. long, puberulous outside ; petals about 9 mm. long ; fruits 1·5 cm. long, stipitate, finely velvety ; leaflets 2–5 pairs, oblong-elliptic, 3–8 cm. long, pubescent only on midrib and sometimes the lateral nerves beneath
3. *expansa*
Inflorescences in axillary fascicles about 8 cm. long ; sepals about 6 mm. long and 3 mm. broad ; petals 10 mm. long ; fruits erect, stipitate, rounded, 1·5 cm. long, finely velvety ; leaflets 2–4 pairs, elliptic, subcordate at base, acuminate at apex, (6–) 10–21 cm. long, 3–6 cm. broad, prominently reticulate and glabrous on both sides ; branches with recurved spines nearly 1 cm. long   ..   4. *macrantha*

1. **M. zenkeri** *Gilg ex Schellenb.* in Engl. Bot. Jahrb. 55 : 444 (1919) ; Engl. Pflanzenr. Connarac. 57. Scrambling shrub, 10–15 ft. high ; flowers with pink petals and sepals, white filaments and yellow anthers ; fruits brown ; in forest thickets.
    S.Nig.: Bonny R. (fl. & fr. Oct.) *Mann* 508! Eket (fl. & fr. May) *Onochie* FHI 33155! *Talbot* 3125! Old Calabar *Holland*! *Robb*! Afi F.R., Ogoja *Smith* Cam./30/36! **Br.Cam.:** Rio del Rey *Johnston*. Also in French Cameroons, Spanish Guinea and Gabon.
2. **M. longiflora** *Bak.* in F.T.A. 1 : 460 (1868) ; Schellenb. l.c. 56. *M. expansa* of Chev. Bot. 166, partly (*Chev.* 16698). A climbing or scrambling shrub, 6–20 ft. high ; flowers with pink sepals and white or yellow petals ; fruits scarlet ; in old farmland and forest regrowth.
    Iv.C.: San Pedro (fr. Aug.) *Thoiré*! Abidjan *Aubrév.* Bouroukrou *Chev.* 16698. Adiopodoumé (Sept.) *Roberty* 12161! **G.C.:** Axim (Apr.) *Chipp* 423! Anaji, Sekondi (fr. Oct.) *Howes* 973! Kumasi (Dec.) *Vigne* FH 1777! **S.Nig.:** Lagos *Rowland* 37! *Millen* 166! 185! *Dalz.* 1023! Epe *Barter* 3283! Warri Prov. (Sept.) *Cole* 21! Old Calabar (Feb.) *Mann* 2256! Eket *Talbot* 3112! (See Appendix, p. 344.)
3. **M. expansa** *Soland. ex Planch.*—F.T.A. 1 : 459 ; Schellenb. l.c. 59 ; Chev. Bot. 166. Climbing shrub up to 20 ft. high ; flowers red or yellowish-pink ; fruits reddish-brown ; in forest.
    Fr.G.: Dyeke (Oct.) *Baldwin* 9655! 9661! **S.L.:** Bagroo R. (Apr.) *Mann* 859! Freetown (Jan., Dec.) *Dalz.* 955! 1009! Rokupr (fl. & fr. Feb.) *Deighton* 4991! Njala (Oct., Nov.) *Deighton* 1787! 4910! **Lib.:** Grand Bassa (July) *T. Vogel* 95! Kakatown *Whyte*! Brewersville (Sept.) *Barker* 1076! Vonjama (Oct.) *Baldwin* 9944! Gletown, Tchien Dist. (July) *Baldwin* 6798! **Iv.C.:** Cavally R. basin *Chev.* 19549 ; 19816.
4. **M. macrantha** (*Gilg*) *Schellenb.* Beitr. Verg. Anat. Syst. Connarac. 21 (1910) ; Pflanzenr. 62, fig. 8. *Dinklagea macrantha* Gilg (1897). Woody climber with hooks on the branches ; flowers yellow ; fruits brown.
    Lib.: Fishtown *Dinklage* 1633! Grand Bassa (July) *Dinklage* 1965! Monrovia (fr. Apr.) *Cooper* 73! Duo, Sinoe (Mar.) *Baldwin* 11360!

10. SPIROPETALUM Gilg in Engl. Bot. Jahrb. 14 : 335 (1891) ; Schellenb. in Engl. Pflanzenr. Connarac. 103.

Midrib of leaflets markedly pubescent (but not minutely puberulous) beneath ; leaflets 3–4 pairs, often subopposite, acuminate :
Leaflets not distinctly reticulate on the upper surface, about 3 pairs, oblong to obovate-elliptic, obtusely acuminate, with about 5 pairs of lateral nerves ; flowers in dense axillary clusters ; calyx campanulate, tomentose   ..   ..   .. 1. *solanderi*
Leaflets distinctly reticulate on the upper surface :
Leaflets subtrinerved and dull beneath, lateral nerves more or less ascending and stout, about 3 pairs ; lamina more or less elliptic, 5–10 cm. long, 2–4 cm. broad, the midrib densely pubescent beneath ; calyx-lobes narrow, tomentellous outside ; petals very narrow   ..   .. 2. *triplinerve*
Leaflets with more or less spreading weak lateral nerves, shining beneath (when dry) ; lamina oblong-elliptic or obovate-elliptic, cuneate or rounded at base, shortly acuminate at apex, 5–7 cm. long, 2–4 cm. broad ; petals 1·5–2 cm. long; fruits oblong up to 4 cm. long, beaked   ..   .. 3. *reynoldsii*
Midrib of leaflets glabrous or minutely puberulous beneath ; leaflets 1–4 pairs or solitary, broadly elliptic, obtuse at base, acute but hardly acuminate at apex, 7–12 cm. long, 3·5–7 cm. broad, firm, loosely reticulate beneath ; calyx tomentellous
4. *heterophyllum*

1. **S. solanderi** (*Bak.*) *Gilg* in Engl. Bot. Jahrb. 23 : 214 (1896) ; Schellenb. in Engl. Pflanzenr. Connarac. 104. *Rourea solanderi* Bak. in F.T.A. 1 : 456 (1868). Forest climber, sometimes a shrub ; flowers white ; fruits straight, about 1 in. long, red-velvety, drying brown.
    S.L.: Pujehun (Apr.) *Aylmer* 69 ! Njala (May) *Deighton* 3000 ! 6062 ! Gbesebu (fr. July) *Deighton* 6083 ! Bumbuna (fr. Oct.) *Thomas* 3470 ! **Lib.:** White Plains, St. Paul R. *Dinklage* 2198. Nana Kru (Mar.) *Baldwin* 11591 ! Brewersville (fr. Aug.) *Baldwin* 13083 !
2. **S. triplinerve** *Stapf* in J. Linn. Soc. 37 : 93 (1906) ; Schellenb. l.c. Climbing shrub up to 20 ft. high.
    Lib.: Monrovia (fr. Sept.) *Whyte* ! *Baldwin* 9199 ! Nana Kru (Mar.) *Baldwin* 11574 ! Cess R. (Mar.) *Baldwin* 11249 ! Mnanulu, Webo (young fr. June) *Baldwin* 6041 !
3. **S. reynoldsii** (*Stapf*) *Schellenb.* in Engl. Bot. Jahrb. 55 : 451 (1919). *Connarus reynoldsii* Stapf (1906). *Spiropetalum triplinerve* of Schellenb. in Pflanzenr. 104, partly (*Reynolds* ; *Chev.* 19824). *Rourea afzelii* of Chev. Bot. 165. Straggling shrub or climber ; flowers white or pale pink, turning yellow.
    Lib.: St. Paul R. *Reynolds* ! White Plains (fr. Apr.) *Barker* 1281 ! Brewersville (Mar.) *Baldwin* 1234 ! Gonatown (Dec.) *Baldwin* 10784 ! Lakrata (Apr.) *Bequaert* 158 ! **Iv.C.:** Taté to Tabou, Cavally R. *Chev.* 19824.
4. **S. heterophyllum** (*Bak.*) *Gilg* in Engl. Bot. Jahrb. 23 : 214 (1896) ; Schellenb. in Engl. Pflanzenr. Connarac. 106. *Rourea heterophylla* Bak. in F.T.A. 1 : 456 (1868) ; Chev. Bot. 165. *Spiropetalum odoratum* Gilg (1891)—Schellenb. l.c. 105, fig. 17 ; Troupin in Fl. Congo Belge 3 : 127. *S. polyanthum* Gilg (1895)—F.W.T.A., ed. 1, 1 : 518. Shrub or large climber in forest ; flowers white, turning yellow, fragrant.
    S.L.: Mabum (Aug.) *Thomas* 1626 ! Roruks (July) *Deighton* 2503 ! Fwendu (Apr.) *Deighton* 1610 ! Njala (July, Sept.) *Deighton* 1738 ! 1840 ! **Lib.:** Fishtown (Aug.) *Dinklage* 1971 ! Monrovia *Dinklage* 2701. Du R. (Aug.) *Linder* 318 ! **Iv.C.:** Grand Bassam *Middleton* ! **G.C.:** Bonsa R., Tarkwa Dist. (Aug.) *Vigne* FH 139 ! **N.Nig.:** Otobi Kurmi, Benue Prov. (Feb.) *Jones* FHI 553 ! 981 ! **S.Nig.:** Jamieson R., Sapoba (Nov.) *Meikle* 609 ! Isobendiga to Afi, Ikom (May) *Latilo* FHI 31818 ! Degema *Talbot* 3705 ! Eket *Talbot* 3075a ! 3349 ! Calabar *Williams* 13 ! 17 ! **Br.Cam.:** Victoria *Preuss* 1321 ! Also in Gabon, Belgian Congo and Cabinda.
    [Schellenberg, in Engl. Pflanzenr. Connarac. 105–106, considered specimens (from Nigeria southwards) with puberulous midribs as a distinct species, *S. odoratum* Gilg. These plants also tend to have more leaflets and to be unifoliolate less frequently than those from Sierra Leone which he regarded as *S. heterophyllum*, but the two species are hardly worth maintaining.—F.N.H.]

11. CONNARUS Linn.—F.T.A. 1 : 456 ; Schellenb. in Engl. Pflanzenr. Connarac. 216.

Leaflets more than 3 :
  Leaflets narrowly oblong to oblong-elliptic or oblanceolate, rather abruptly acuminate :
    Leaflets subopposite, 3–5 pairs, rather long-acuminate, rounded at base, very variable in size, up to 16 cm. long and 6 cm. broad, at first thinly pubescent beneath, with about 10–12 pairs of slender prominently looped lateral nerves ; petiolule transversely wrinkled ; flowers in very long and lax rusty-pubescent panicles ; sepals imbricate, narrowly oblong, 1·5 mm. long ; petals oblanceolate, 6 mm. long, pubescent ; fruits stipitate, obliquely obovate, closely striate, about 2–2·5 cm. long, tomentose on the back .. .. 1. *griffonianus*
    Leaflets 3 pairs, elliptic, slightly pointed, 6–9 cm. long, 3–4 cm. broad, glabrous, with about 6 pairs of lateral nerves ; flowers in short panicles ; sepals 3 mm. long ; petals slightly pubescent ; fruits as above but smaller and quite glabrous   2. *thonningii*
  Leaflets ovate, 4–5 pairs, gradually long-acuminate, 8–14 cm. long, 5–6·5 cm. broad, glabrous, with 4–5 pairs of lateral nerves, scarcely reticulate ; panicles rather long ; sepals and petals slightly pubescent ; fruits narrowly ovoid, shortly stipitate, 6 cm. long, glabrous .. .. 3. *staudtii*
  Leaflets 3, elliptic to obovate, more or less gradually acuminate, up to 18 cm. long and 10 cm. broad, with about 5 pairs of lateral nerves ; inflorescence short ; sepals and petals much broader than in preceding species ; fruits obliquely oblanceolate, about 6 cm. long, glabrous .. .. 4. *africanus*

1. **C. griffonianus** *Baill.* Adansonia 7 : 235 (1866) ; Schellenb. in Engl. Pflanzenr. Connarac. 269 ; Troupin in Fl. Congo Belge 3 : 132, t. 9. *C. smeathmanni* of F.T.A. 1 : 458, partly, (excl. ref. & spec. *Smeathmanni*), not of (DC.) Planch. (1850). A woody climber with rusty-pubescent branches and inflorescence ; flowers white ; fruits reddish ; not in wet forest.
    S.Nig.: Sapoba *Kennedy* 1946 ! Idumuje (Jan.) *Thomas* 2178 ! Balinge to Ikwette, Obudu (Dec.) *Savory & Keay* FHI 25160 ! Oban *Talbot* 503 ! 1295 ! **Br.Cam.:** Victoria (Oct.) *Mildbr.* 10584 ! Nkom-Wum F.R., Bamenda (fr. Apr.) *Ujor* FHI 30069 ! Mamfe (Jan.) *Maitland* 1164 ! Fonfuka, 3,000 ft., Bamenda (fr. Apr.) *Maitland* 1419 ! **F.Po:** (fl. Dec., fr. Feb.) *Mann* 7 ! 427 ! Also in French Cameroons, Gabon, French and Belgian Congo and Principe.
2. **C. thonningii** (*DC.*) *Schellenb.* in Engl. Bot. Jahrb. 58 : 224 (1923) ; Pflanzenr. Connarac. 280. *Omphalobium thonningii* DC. (1825). *Connarus floribundus* Schum. & Thonn. (1827). A woody climber.
    G.C.: *Thonning.* Cape Coast *Brass* ! **Togo:** Lome *Warnecke* 446 ! Sokode-Basari *Kersting* A. 293. Gbele *Busse* 3626.
3. **C. staudtii** *Gilg* in Engl. Bot. Jahrb. 23 : 208 ; Schellenb. l.c. 286 ; Troupin l.c. 136, fig. 2, D. A shrub or small tree, 15–20 ft. high ; flowers white, turning yellowish.
    S.Nig.: Oban *Talbot* 1297 ! 1427 ! Eket *Talbot* ! Also in French Cameroons, Gabon and Belgian Congo.
4. **C. africanus** *Lam.*—F.T.A. 1 : 457 ; Schellenb. l.c. 283 ; Chev. Bot. 165. *C. nigrensis* Gilg (1891)—Chev. Bot. 165. *C. djalonensis* A. Chev. Bot. 165, name only. A shrub 15–20 ft. or more high, usually climbing, with shining foliage and cream-white flowers turning brownish ; fruits bright red ; in moist forest.
    Sen.: *Perrottet* 147 ! Casamance *Heudelot* 594. Floup-Fedyan *Chev.* 2974. M'bidjem *Thierry* 225 ! **Port.G.:** Prabis, Bissau *Esp. Santo* 1848 ! **Fr.G.:** Konkouré to Timbo (fr. Mar.) *Chev.* 12457 ! Bilima-Kanté *Chev.* 18056. Kouria to Irébéléya (Sept.) *Chev.* 18263 ! Conakry *Lecerf* ! **S.L.:** *Smeathmann* ! York (fr. May) *Deighton* 2753 ! Njala (Mar.) *Deighton* 1940 ! Musaia (fr. Mar.) *Deighton* 5453 ! Falaba (Mar.) *Aylmer* 25 ! **Lib.:** Gbanga (fr. Sept.) *Baldwin* 13228 ! **Iv.C.:** Abidjan *Chev.* 15163 ! Assinié *Chev.* 17864. **N.Nig.:** Gurara R. (June) *Elliott* 179 ! **S.Nig.:** Lagos (Sept.) *Barter* 2143 ! *Millen* 153 ! *Dalz.* 1022 ! (See Appendix, p. 344.)

Imperfectly known species.

C. smeathmanni (*DC.*) *Planch.* in Linnaea 23 : 436 (1850). *Omphalobium smeathmanni* DC. (1825)—Schellenb. in Engl. Pflanzenr. Connarac. 271, note. This species is based on a supposed Smeathmann specimen from Sierra Leone which has not yet been traced.

12. HEMANDRADENIA Stapf in Kew Bull. 1908 : 288 ; Schellenb. in Engl. Pflanzenr. Connarac. 64.

Panicles terminal, many-flowered, flowers shortly pedicellate ; sepals 2 mm. long ; young fruits oblong, velvety ; leaflet oblong-elliptic, rounded at base, gradually acuminate at apex, 9–12 cm. long, 3–4·5 cm. broad, thinly appressed-pubescent when young .. .. 1. *chevalieri*
Flowers few in axillary clusters, subsessile ; sepals 4 mm. long ; fruits ellipsoid, 3–4 cm. long, orange-tomentose ; leaflets oblong-elliptic, cuneate to rounded at base, rather abruptly acuminate at apex, (8–) 10–17 cm. long, 5–7 cm. broad, appressed-pubescent when young .. .. 2. *mannii*

1. **H. chevalieri** *Stapf* in Kew Bull. 1908 : 288 ; Schellenb. in Engl. Pflanzenr. Connarac. 65 ; Chev. Bot. 167 ; Aubrév. Fl. For C. Iv. 1 : 158, t. 56. A small tree, branchlets yellow-tomentose, axis of inflorescence flexuous ; on the sea-shore.
   Iv.C.: Tabou to Bériby (Aug.) *Chev.* 19943 ! 19968. Port Bouët *Aubrév.* 1495 ; 1636.

2. **H. mannii** *Stapf* l.c. (1908) ; Schellenb. l.c. 64. Small tree about 20 ft. high with an irregular bushy crown, bark dark with red slash ; in forest undergrowth.
   S.Nig.: Omo F.R., Benin (fr. Feb., Dec.) *Brenan* 8561 ! 9026 ! Shasha F.R., Ijebu (Apr.) *Ross* 193 ! Also in Spanish Guinea.

13. JOLLYDORA Pierre in Bull. Soc. Linn. Paris 2 : 1233 (1896) ; Gilg in Engl. Bot. Jahrb. 23 : 217 (1896).

Sepals 4–6 mm. long, linear, shortly pilose with erect glandular hairs ; leaflets elliptic-lanceolate to oblong-oblanceolate, narrowly cuneate at base, abruptly acuminate, 14–30 cm. long, 4·5–10 cm. broad, glabrous, lateral nerves straight and looping at margin ; young leaf-rhachis clothed with long woolly hairs, becoming glabrous ; petals 8 mm. long ; fruits ovoid, 4 cm. long, not stipitate, laxly tomentose soon becoming glabrous.. .. .. 1. *glandulosa*
Sepals 2–4 mm. long, oblong, densely pubescent, not glandular ; leaflets obovate-elliptic, cuneate at base, abruptly acuminate, 15–30 cm. long, 8–12 cm. broad, glabrous, lateral nerves curved and looping at margin ; petals 6 mm. long ; fruits ovoid to obovoid, 4 cm. long, shortly stipitate, brown-tomentose soon becoming glabrous .. .. 2. *duparquetiana*

1. **J. glandulosa** *Schellenb.* in Engl. Bot. Jahrb. 55 : 455 (1919) ; Engl. Pflanzenr. Connarac. 25. A small, weak-stemmed tree, about 15 ft. high ; flowers pale yellow, in few-flowered clustered rusty-villous racemes on the stems ; in rain forest.
   S.Nig.: Boshi-Okwangwo F.R., Obudu (fr. May) *Latilo* FHI 30923 ! **Br.Cam.:** Victoria area (Nov.) *Mildbr.* 10607 ! 10613. Johann-Albrechtshöhe (Dec.) *Winkler* 1092.
   [*J. peduncula* Mildbr. in Notizbl. Bot. Gart. Berl. 10 : 971 (1930) appears to be synonymous with the above and the Mildbraed specimens have been cited there, but *Winkler* 1092, the type of *J. glandulosa*, has not been seen and is presumably destroyed—F.N.H.]

2. **J. duparquetiana** (*Baill.*) *Pierre* in Bull. Soc. Linn. Paris 2 : 1233 (1896) ; Schellenb. l.c. 26. *Connarus duparquetianus* Baill. (1867)—F.T.A. 1 : 457. *Anthagathis monadelpha* Harms (1897). Small tree up to 25 ft. high ; flowers yellow in clusters on the stems.
   S.Nig.: Old Calabar R. (fl. & fr. Feb.) *Mann* 2307 ! **Br.Cam.:** Victoria *Winkler* 406. S. Bakosi, Kumba (fr. May) *Olorunfemi* FHI 30576 ! S. Bakundu F.R., Banga (fl. & fr. Mar.) *Brenan* 9407 ! 9411 ! Also in French Cameroons and Gabon.

123. ALANGIACEAE

Trees or shrubs, sometimes spiny. Leaves alternate, simple ; stipules absent. Flowers hermaphrodite, actinomorphic, in axillary cymes ; pedicels articulated. Calyx truncate or with 4–10 teeth. Petals 4–10, mostly linear, valvate, at length recurved, sometimes coherent at the base. Stamens the same number as and alternate with the petals or 2–4 times as many, free or slightly connate at the base, more or less villous inside ; anthers 2-celled, linear, opening lengthwise. Disk cushion-like. Ovary inferior, 1–2-celled ; style simple, clavate or lobed ; ovule solitary, pendulous. Fruit a drupe crowned by the sepals and disk, 1-seeded. Seeds with the embryo about equal to the endosperm.

A small family formerly included in the *Cornaceae* and found only in the tropics and subtropics of the Old World ; leaves alternate, exstipulate, ovary inferior.

ALANGIUM Lam. Encycl. Méth. Bot. 1 : 174 (1783) ; Wangerin in Engl. Pflanzenr. Alangiaceae 6 (1910).

Branchlets glabrous ; leaves ovate-elliptic, very oblique at the base, acuminate, 15–17 cm. long, 9–10 cm. broad, 5-nerved from the base, with slender parallel tertiary nerves, glabrous ; petiole 2·5–3 cm. long, puberulous ; cymes axillary ; peduncle 2–4 cm. long ; flowers about 10 or more ; pedicels jointed at the top, about 4 mm. long ; calyx a mere toothed rim ; petals 1 cm. long, linear from a broader base ; ovary puberulous .. .. *chinense*

A. chinense (*Lour.*) Harms in Ber. Deutsch. Bot. Ges. 15 : 24 (1897) ; Rehder in Sarg. Pl. Wilson. 2 : 552 (1916), q.v. for synonymy ; F.W.T.A., ed. 1, 2 : 607. *Stylidium chinense* Lour. (1790). *Marlea begoniifolia* Roxb. (1819). *Alangium begoniifolium* (Roxb.) Baill. (1877) ; Wangerin l.c. 20, fig. 1, A–G, fig. 5 ;

F.W.T.A., ed. 1, 1 : 519. A deciduous tree, to 70 ft. high, fast growing and soft-wooded ; flowers creamy white to pale yellow, fragrant ; forest margins, especially in montane areas.
Br.Cam.: Cam. Mt., 1,400–4,800 ft. (Mar.) *Kalbreyer* 148! *Maitland* 484! 1172! Bamenda (Apr.) *Johnstone* 1168! Bali-Santa Pass, Bamenda (fl. & fr. May) *Ujor* FHI 30389a! 30389b! **F.Po:** *Mildbr.* 7140. Also in French Cameroons and extending to the tropics and subtropics of eastern Asia.

124. ARALIACEAE

Mostly woody. Leaves alternate or rarely opposite, simple or compound ; stipules often adnate to the petiole. Flowers hermaphrodite or unisexual, spicate, racemose, umbellate or capitate. Calyx adnate to the ovary, small. Petals valvate or slightly imbricate, usually free. Stamens free, alternate with the petals ; anthers 2-celled, opening lengthwise. Disk epigynous. Ovary inferior, 1- or more-celled ; styles free or connate ; ovules solitary in each cell, pendulous from the inner angle. Fruit a berry or drupe. Seeds with copious endosperm and small embryo.

Superficially resembling the Umbelliferae, but woody with an indehiscent baccate or drupaceous fruit. Mainly in tropical regions but rare in Africa.

Leaves pinnate ; leaflets softly tomentose with stellate hairs .. .. 1. **Polyscias**
Leaves digitate or digitately lobed ; pedicels not jointed :
Flowers spicate or racemose ; ovary 2-celled .. .. .. 2. **Cussonia**
Flowers umbellate or capitate ; ovary 5- or more-celled .. .. 3. **Schefflera**

1. POLYSCIAS J. R. & G. Forst., Char. Gen. 63 (1776) ; Harms in E. & P. Pflanzenfam. 3, 8 : 43 (1894).

Leaves imparipinnate, up to 80 cm. or more long, clustered at the ends of the branches ; leaflets 4–7 (–10) pairs, opposite, the rhachis usually contracted at the nodes when dry ; lateral leaflets ovate, ovate-elliptic or ovate-oblong, rounded to cordate at base, usually shortly acuminate at apex, 7–20 cm. long, 3·5–10 cm. broad, puberulous only on midrib above, densely stellate-tomentose beneath, with 10–15 pairs of faint lateral nerves ; panicles up to 60 cm. long, lateral branches up to 12 cm. long ; flowers shortly pedicellate or subsessile or sessile ; ovary glabrous ; fruits ellipsoid, ribbed, about 5 mm. long .. .. *fulva*

P. fulva (*Hiern*) *Harms* l.c. 45 (1894) ; Hutch. & Dalz. F.W.T.A., ed. 1, 1 : 520 (1928) ; Lebrun in Publ. I.N.E.A.C. Sér. Sci. 1 : 179, t. 15 (1935). *Panax fulvum* Hiern in F.T.A. 3 : 28 (1877). *P. ferrugineum* Hiern l.c. (1877). *Polyscias ferruginea* (Hiern) Harms l.c. (1894) ; F.W.T.A., ed. 1, 1 : 520. *P. preussii* Harms (1899). *Panax nigericum* A. Chev. in Bull. Soc. Bot. Fr. 57, Mém. 8 : 178 (1912) ; Bot. 305. A tree, 20–50 ft. high, rarely to 100 ft., with the large leaves bunched at the top of the rather few branches ; leaflets white-tomentose beneath, dark green above ; flowers cream ; usually in montane forest.
Fr.G.: Timbikounda, Farana (Jan.) *Chev.* 20590! Also between Sampouyara and Beyla *fide* Chev. l.c. **G.C.:** Abetifi, Kwahu (Jan.) *Irvine* 1658! **S.Nig.:** Mt. Koloishe, 5,000–6,000 ft., Obudu Div. (Dec.) *Keay & Savory* FHI 25100! 25103! **Br.Cam.:** Buea to Musake Hut, 3,500–6,000 ft., Cam. Mt. *Maitland* 680! *Preuss* 887 ; *Steele* 1! Bamenda Nkwe (Sept.) *Ujor* FHI 30220! **F.Po:** 1,300 ft. *Mann* 301! Also in French Cameroons, A.-E. Sudan, Abyssinia, E. Africa and Belgian Congo. (See Appendix, p. 345.)

2. CUSSONIA Thunb.—F.T.A. 3 : 31 ; Harms in E. & P. Pflanzenfam. 3, 8 : 53 (1894).

A savannah tree, to 10 m. high and 1 m. girth, with very deeply furrowed bark ; young shoots, stipules and petiolules curly-pubescent ; leaflets 7–10, usually more or less crenate-dentate, rarely subentire, obovate to obovate-oblanceolate, subsessile and cuneate at base, gradually or abruptly acuminate, 9–25 cm. long, 2·5–8·5 cm. broad, soon glabrous and finely reticulate beneath, with 8–18 main lateral nerves on each side of midrib ; petioles up to 85 cm. long ; flower spikes up to 50 cm. long, pubescent ; flowers sessile ; fruits subglobose, 6–8 mm. long .. .. 1. *barteri*
A forest tree, to 40 m. high and 3 m. girth, with smooth or furrowed bark ; leaflets 7–10, entire, narrowly elliptic subsessile and cuneate at base, gradually acutely acuminate, 8–29 cm. long, 2·5–6 cm. broad, glabrous, shining, dark above and pale beneath when dry ; with 12–22 main lateral nerves on each side of midrib ; petioles up to 110 cm. long ; flower-spikes up to 40 cm. long, rather densely pubescent ; flowers sessile
2. *bancoensis*

1. C. barteri *Seemann* in J. Bot. 4 : 299 (1866) ; F.T.A. 3 : 32 ; Aubrév. Fl. For. Soud.-Guin. 418, t. 90, 3. *C. djalonensis* A. Chev. in Bull. Soc. Bot. Fr. 55, Mém. 8 : 38 (1908) ; Chev. Bot. 305 ; F.W.T.A., ed. 1, 1 : 520 ; Aubrév. l.c. t. 90, 1–2. *C. nigerica* Hutch. in Kew Bull. 1910 : 136 ; F.W.T.A., ed. 1, 1 : 520. *C. longissima* Hutch. & Dalz. F.W.T.A., ed. 1, 1 : 520 (1928), excl. distrib. S.L. and *Smythe* 267 ; Kew Bull. 1929 : 28. A deciduous savannah tree ; leaves and flower-spikes clustered at the ends of the thick branches ; flowers greenish-white ; ripe fruits whitish.
Fr.G.: Mamou *Pobéguin* 1581! Diaguissa (Apr.) *Chev.* 12931 ; 13548 ; 13617. Timbo to Farana (Apr.) *Chev.* 13309. **S.L.:** Yisea (Apr.) *Lane-Poole* 38! Falaba (Mar.) *Sc. Elliot* 5735! Musaia Bafodea Road (Apr.) *Deighton* 5493! Musaia (fr. July) *Deighton* 4826! Kambaia *Small* 489! **Iv.C.:** Fétékro to Tiébissou (fr. July) *Chev.* 22185! Soucourala to Sanrou, Toura (Mar.) *Chev.* 21597. **G.C.:** Kratchi, Ashanti (June) *Brent* 554! Kumawo, Ashanti (June) *Chipp* 458! Ashanti (Apr.) *Thompson* 61! Salaga (young fr. June) *Saunders* 805! Yipala to Kulmasa, N.T. *Kitson* 854! **Dah.:** Atacora Mts. *Chev.* 24021. Agouagon *Chev.* 23531. Savalou *Chev.* 23705. **N.Nig.:** Borgu *Barter* 815! Anara F.R., Zaria Prov. (fl. buds May) *Keay* FHI 22984! Bauchi Plateau (Apr.) *Lely* P. 223! Kilba, Yola Prov. (fr. Aug.) *Dalz.* 172! **S.Nig.:**

Thompson 507! Olokemeji (Apr.) *A. F. Ross* 161! Awba Hills, Oyo Prov. *Jones* FHI 6336! Also in French Cameroons and Ubangi-Shari. (See Appendix, p. 345.)

2. **C. bancoensis** *Aubrév. & Pellegr.* in Bull. Soc. Bot. Fr. 84 : 393 (1938) ; Aubrév. Fl. For. C. Iv. 3 : 84 t. 273 (1936), French descr. only. A forest tree ; leaves and flower-spikes clustered at the ends of the branches ; flowers greenish-white, with an unpleasant smell which attracts flies.
Iv.C.: Banco (Nov.) *Aubrév.* 500.   **G.C.:** Apedwa–Suhum Road (young fr. Jan.) *Scholes* 454! Bobiri F.R., Juaso (Dec.) C. J. *Taylor* FH 5449! **S.Nig.:** Akure F.R. *Keay & Onochie* FHI 19658!
[A sterile specimen, *Smythe* 267, from Sierra Leone, cited in the first edition as *C. longissima*, may in fact be *C. bancoensis*. The collector's note—" fruiting on main stem "—probably refers to a different plant.]

3. **SCHEFFLERA** J. R. & G. Forst., Char. Gen. 45 (1776) ; Harms in E. & P˙ Pflanzenfam. 3, 8 : 35 (1894).

Flowers distinctly pedicellate ; petals coming off in a calyptra at anthesis ; panicles crowded near the apex of the shoot ; leaflets 6–7, glabrous ; petioles up to 30 cm. long ; petiolules 1·5–7·5 cm. long :
Leaflets crenate all round, cordate, subcordate or rounded at base, ovate, gradually acutely acuminate, 11–25 cm. long, 6–12 cm. broad, with 10–15 main lateral nerves on each side of midrib ; panicles up to 25 cm. long, with lateral branches 10–20 mm. long ; pedicels 2–5 mm. long ; panicles and flowers glabrous except for the densely pubescent or tomentose bracts ..   ..   1. *abyssinica*
Leaflets entire or nearly so, rounded to cuneate at base, shortly acuminate :
Pedicels (3–) 5–10 mm. long ; flowers 8–11 per umbel ; lateral branches of panicle (4–) 8–22 mm. long ; panicle sparingly pubescent, up to 26 cm. long ; leaflets elongate-oblong-elliptic, 11–25 cm. long, 4–8·5 cm. broad, with 4–10 main lateral nerves on each side of midrib ..   ..   2. *barteri*
Pedicels 2–4 (–5) mm. long ; flowers 12–15 per umbel ; lateral branches of panicle 3–15 mm. long ; panicle rather densely but loosely rusty-pubescent, up to 40 cm. long, leaflets obovate-oblong or oblong-elliptic, 8·5–20 cm. long, 3·5–8·5 cm. broad, with 8–11 main lateral nerves on each side of midrib ..   ..   3. *hierniana*
Flowers sessile ; petals spreading at anthesis ; panicles crowded near the apex of the shoot, up to 42 cm. long, with lateral branches 8–20 mm. long ; panicles and flowers glabrous ; leaflets (6–) 8–10, oblong-elliptic, rounded to shortly cuneate at base, gradually acute or shortly acuminate at apex, 12–20 cm. long, 4·5–8·5 cm. broad, glabrous, with 12–18 main lateral nerves on each side of midrib ; petioles up to 22 cm. long ; petiolules 2·5–5·5 cm. long ..   ..   4. *mannii*

1. **S. abyssinica** (*Hochst. ex A. Rich.*) *Harms* in E. & P. Pflanzenfam. 3, 8 : 38 (1894). *Aralia abyssinica* Hochst. ex A. Rich. (1848). *Heptapleurum abyssinicum* (Hochst. ex A. Rich.) Vatke (1876)—F.T.A. 3 : 29. *Paratropia elata* Hook. f. (1864). *Heptapleurum elatum* (Hook. f.) Hiern in F.T.A. 3 : 30 (1877). *Schefflera hookeriana* Harms l.c. 38 (1894) ; F.W.T.A., ed. 1, 1 : 521 ; not *S. elata* (D. Don) Harms (1894). A tree, to 60 ft. high, in montane forest ; flowers yellowish, sweetly scented.
Br.Cam.: Cam. Mt., 4,500–8,000 ft. (Feb., Mar.) *Maitland* 529! 994! *Mann* 1181! Bum, Bamenda, 4,000 ft. (May) *Maitland* 1586! Bayango, Bamenda (Apr.) *Ujor* FHI 30063! Also in Uganda, A.-E. Sudan and Abyssinica.

2. **S. barteri** (*Seem.*) *Harms* l.c. 38 (1894) ; Aubrév. Fl. For. C. Iv. 3 : 83. *Astropanax barteri* Seem. in J. Bot. 3 : 177 (1865). *Heptapleurum barteri* (Seem.) Hiern l.c. (1877). *Astropanax baikiei* Seem. l.c. (1865). *Heptapleurum baikiei* (Seem.) Hiern l.c. (1877). *Schefflera baikiei* (Seem.) Harms l.c. (1894) ; F.W.T.A., ed. 1, 1 : 521. *Heptapleurum dananense* A. Chev. in Bull. Soc. Bot. Fr. 57, Mém. 8 : 178 (1912) ; Bot. 305. *Schefflera dananensis* (A. Chev.) Harms ex Engl. Pflanzenw. Afr. 3, 2 : 778 (1921). A tree or climbing shrub ; in forest by water and on rocky hills.
Fr.G.: R. Koubi, Pita (Apr.) *Pobéguin* 2293! **S.L.:** Sugar Loaf Mt. (fl. Apr., fr. May) *Barter*! *Sc. Elliot* 5828! Daru, Tunkia (fr. Mar.) *Deighton* 4115! **Lib.:** Ganta (fr. Mar.) *Harley* 1117! **Iv.C.:** R. Boan, Danané (fr. Apr.) *Chev.* 21275! Mt. Nimba *Aubrév.* 1135. **S.Nig.:** Brass *Barter* 1851! Eket (May) *Onochie* FHI 32921! Oban *Talbot* 462! 1735! **Br.Cam.:** Johann-Albrechtshöhe *Staudt* 882! Lakom, 6,000 ft., Bamenda (fr. Apr.) *Maitland* 1623! Also in French Cameroons, Gabon, Belgian Congo and (?) Principe.

3. **S. hierniana** *Harms* l.c. 38 (1894). *Heptapleurum scandens* Hiern in F.T.A. 3 : 30 (1877), not of Seem. (1865). (?) *Schefflera ledermannii* Harms in Engl. Bot. Jahrb. 53 : 359 (1915) ; F.W.T.A., ed. 1, 1 : 521. A climbing shrub or tree, to 30 ft. high ; flowers green, tinged pink.
Br.Cam.: Cam. Mt., 3,000–4,500 ft. (Feb.) *Maitland* 381! *Mann* 1180! Oku, 7,500 ft., Bamenda (Jan.) *Keay & Lightbody* FHI 28519! **F.Po:** Moka, 3,800–5,600 ft. (Nov.) *Mildbr.* 7107. Also in French Cameroons.
[Very close to *S. barteri*.]

4. **S. mannii** (*Hook. f.*) *Harms* l.c. 36 (1894) ; in Engl. Bot. Jahrb. 53 : 358 (1915), incl. var. *lancifolia*. *Paratropia mannii* Hook. f. (1861). *Heptapleurum mannii* (Hook. f.) Hiern in F.T.A. 3 : 31 (1877). A tree, 40–50 ft. high ; panicles erect, crowded at the ends of the branches ; flowers green, with yellow anthers ; in montane forest.
S.Nig.: Ikwette Plateau, 5,300 ft., Obudu Div. (Dec.) *Keay & Savory* FHI 25240! **Br.Cam.:** Cam. Mt., 4,000–7,600 ft. (Dec.–Apr.) *Dundas* FHI 15337! *Maitland* 1199! *Mann* 1182! 2168! *Preuss* 884! Oku, Bamenda (Nov.) *Johnstone* 227! **F.Po:** 5,000 ft. *Mann* 289! *Mildbr.* 6410. Also in Annobon and S. Tomé.

125. UMBELLIFERAE

By J. F. M. Cannon

Herbs with furrowed stems and broad soft pith, rarely shrubs or small trees. Leaves alternate, usually compound and much divided, sometimes simple, rarely peltate. Flowers usually hermaphrodite, rarely unisexual, in simple or compound umbels, rarely capitate. Calyx adnate to the ovary, normally with

5 minute teeth. Petals 5, free, valvate or slightly imbricate, epigynous. Stamens 5, free, alternate with the petals ; filaments inflexed in bud, anthers 2-celled, opening lengthwise. Styles 2, ovary inferior, 2-celled, ovules solitary in each cell, pendulous. Fruit dry, dividing when ripe into 2 mericarps, supported by a central carpophore ; carpels usually ribbed and often with resin canals (vittae) in their walls ; mericarps sometimes hairy and sometimes provided with hooks or spines. Seeds with copious endosperm and minute embryo.

A very homogeneous family, readily recognizable by the habit, broad pith, umbellate flowers, free petals and inferior 2-celled ovary. Very abundant in the warm temperate regions of the northern hemisphere, and fairly well represented in the tropics, especially on the mountains.

Small soft-wooded trees, up to 10 m. high ; fruits broadly winged   15. **Steganotaenia**
Herbs ; fruits winged or not :
  Flowers umbellate or in loose more or less globose heads :
    Creeping herbs, rooting at the nodes ; usually in damp places :
      Leaves orbicular or reniform ; flowers hermaphrodite ; mericarps 5-ribbed, without
        connecting veins between the ribs ..   ..   1. **Hydrocotyle**
      Leaves reniform ; flowers usually unisexual ; mericarps 7–9-ribbed, with connecting
        veins between the ribs   ..   ..   2. **Centella**
    Erect herbs, not rooting at the nodes ; in various situations :
      Fruit (and ovary) hairy or with spinose bristles :
        Ovary with hooked or barbed bristles :
          Leaves digitately divided into 3–5 obovate segments   ..   ..   3. **Sanicula**
          Leaves pinnately to bipinnately divided into numerous small leaflets :
            Plants with appressed strigose hairs only ..   ..   6. **Torilis**
            Plants with spreading hairs on the stem, petiole and inflorescence   7. **Caucalis**
        Ovary more or less covered with hairs :
          Ultimate leaf-segments very narrowly linear, about 1 mm. wide :
            Fruit densely covered with very long setose hairs   ..   16. **Ammodaucus**
          Fruit hairy, but hairs not long-setose :
            Bracteoles of the secondary umbels narrowly linear, with a long setose point
                  5. **Pycnocycla**
            Bracteoles of the secondary umbels broad, membranaceous, acute, often with a
              mucronate tip   ..   ..   12. **Diplolophium**
          Ultimate leaf-segments broad, subrotund ..   ..   9. **Pimpinella**
      Fruit glabrous :
        Fruit with broad wings, strongly compressed dorsally :
          Fruit without a deep notch at the apex   ..   ..   ..   13. **Peucedanum**
          Fruit with a deep notch (in which the stylopodium is situated) separating the
            wings at the apex ..   ..   14. **Lefebvrea**
        Fruit not broadly winged, slightly compressed laterally :
          Low lax annual herbs (weeds in cultivation) with linear leaf-segments and very
            prominently ribbed fruit ..   ..   8. **Apium**
          Plants not as above :
            Leaflets 8–12 pairs, with appressed cartilaginous teeth on the margins ; flowers
              yellow   ..   ..   10. **Sium**
            Leaflets about 3 pairs, with projecting herbaceous teeth on the margins ;
              flowers white :
                Pedicels short and rather thick ; fruit broadly ellipsoid, nearly as broad as long
                  9. **Pimpinella**
                Pedicels long and very slender ; fruit narrowly ovoid, much longer than broad
                  11. **Cryptotaenia**
  Flowers in a dense cylindrical spike surrounded by leafy spiny-toothed bracts ; radical
    leaves simple   ..   ..   4. **Eryngium**

In addition to the above genera, *Cuminum cyminum* Linn. is cultivated in our area, and may be found established in areas associated with human occupation.
In Pflanzenw. Afr. 3, 2 : 821, Engler mentions *Erythroselinum lefeburioides* Engl. as occurring at Buea on Mt. Cameroon. In the absence of both specimens and description, I am unable to account for this name.
Berula erecta (Huds.) Coville occurs in the Bambuto Mts., French Cameroons and may well be found in the area of this Flora.

1. HYDROCOTYLE Linn.—F.T.A. 3 : 3.

Leaves attached at the base (not peltate) :
  Plants very small and slender ; leaves suborbicular, 5–10 mm. diam., usually shallowly
    lobed, subglabrous ; flowers about 2–4 in each head, peduncle up to 1 cm. long
                  1. *sibthorpioides*
  Plants more robust ; leaves suborbicular-reniform, 5–10 mm. diam., usually rather
    deeply lobed and distinctly pilose-setulose ; flowers about 10–30 in each head,
    peduncle (0·5–) 1–3 (–4) cm. long   ..   ..   2. *mannii*

Leaves peltate, orbicular up to 12 cm. diam., doubly crenulate ; petiole very variable
in length ; umbels compound, bracts lanceolate   ..   ..   ..   3. *bonariensis*

1. **H. sibthorpioides** *Lam.* Encycl. Méth. Bot. 3 : 153 (1789). *H. monticola* Hook. f. in J. Linn. Soc. 7 : 194
(1864) ; F.W.T.A., ed. 1, 1 : 522. *H. americana* var. *monticola* (Hook. f.) Hiern in F.T.A. 3 : 4 (1877).
H. americana of Chev. Bot. 303. A small creeping herb, nearly glabrous ; in moist upland grassland.
Fr.G.: Dalaba (Aug.) *Jac.-Fél.* 7068 ! Ditinn to Diaguissa, Fouta Djalon (fr. Apr.) *Chev.* 12840 ! **Br.Cam.:**
Lakom, Bamenda, 6,000 ft. (fl. & fr. June) *Maitland* 1772 ! **F.Po:** 8,500 ft. *Mann* 1458 ! In suitable
habitats throughout tropical Africa and Asia.
    [The differences between this and the next species are very difficult to define and collectors should
    attempt to make gatherings illustrating the range of variation within populations.]
2. **H. mannii** *Hook. f.* l.c. (1864). *H. moschata* of Hiern in F.T.A. 3 : 5, not of Forst. A more or less pilose
creeping herb ; in damp montane habitats.
Br.Cam.: *Preuss* 1028 ! Cam. Mt., 3,000–6,000 ft. (fl. & fr. Dec.–Apr.) *Maitland* 215 ! 820 ! 1321 ! Banso,
Bamenda (fl. & fr. Oct.) *Tamajong* FHI 23460 ! Bali-Ngemba F.R., Bamenda (fl. & fr. May) *Ujor* FHI
30331 ! **F.Po:** 7,000 ft. *Mann* 1457 ! Mioka 5,000 ft. (fl. & fr. Dec.) *Boughey* 148 ! In suitable habitats
throughout tropical Africa.
3. **H. bonariensis** *Lam.* l.c. 153 (1789) ; F.T.A. 3 : 4 ; Chev. Bot. 303 ; Berhaut Fl. Sen. 179. A glabrous
creeping plant ; flowers whitish, numerous ; peduncles sometimes a foot long ; in sandy ground, usually
near the sea.
Sen.: *Berhaut* 1093. **Lib.:** Cape Palmas (fl. & fr. July) *T. Vogel* 50 ! Grand Bassa (fl. & fr. July–Oct.)
T. Vogel 74 ! *Dinklage* ! Bassa Cove *Ansell* ! Timbo, Grand Bassa (fl. & fr. Mar.) *Baldwin* 11217 !
Iv.C.: Bliéron *Chev.* **G.C.:** Axim (fl. & fr. Jan., Mar.) *Chipp* 58 ! *Johnson* 979 ! Princes (fl. & fr. Jan.)
Akpabla 785 ! Ancobra R. (Feb.) *Irvine* 2143 ! **S.Nig.:** Lagos *Moloney* ! *Dalz.* ! Alorosogo, N. side of
Lekki lagoon (fl. & fr. Apr.) *Jones & Onochie* FHI 17422 ! Nun R. (fl. & fr. Aug.) *T. Vogel* 6 ! *Mann* !
Brass *Barter* 1873 ! Stubbs Creek F.R., Eket Dist. (fl. & fr. May) *Onochie & Abey* FHI 32949 ! **Br.Cam.:**
Preuss 1106 ! Cameroon R. (fl. & fr. June) *Mann* 423 ! Widely distributed in tropical Africa and America.

2. CENTELLA Linn. Sp. Pl., ed. 2, 2: 1393 (1763).

Leaves reniform, 3–5 cm. diam., crenate or dentate but not lobulate, glabrous or subglabrous ; fruit widely orbicular ; peduncle 5–8 mm. long, pubescent   *asiatica*

C. asiatica (*Linn.*) *Urb.* in Mart. Fl. Bras. 11, 1 : 287, t. 78, fig. 1 (1878). *Hydrocotyle asiatica* Linn. (1753)—
F.T.A. 3 : 6 ; F.W.T.A., ed. 1, 1 : 522 ; Chev. Bot. 303 ; Berhaut Fl. Sen. 179. A perennial creeping
plant, rooting at the nodes ; flowers purplish ; in damp grassy places.
Sen.: *Berhaut* 729. Carabane, Casamance *Chev.* 2976. **Fr.Sud.:** Negala *Chev.* 165. **Port.G.:** Pussube,
Bissau (fr. Mar.) *Esp. Santo* 1497 ! **S.L.:** Bonthe (fr. Mar.) *Deighton* 2486 ! Mano (fr. Feb.) *Deighton*
3113 ! Musaia (fl. & fr. Apr.) *Deighton* 5378 ! Magbema, Rokupr (fl. & fr. Apr.) *Jordan* 434 ! **Lib.:**
Monrovia (fl. & fr. June, July) *Linder* 26 ! *Baldwin* 5886 ! Nyaake, Webo (fl. & fr. June) *Baldwin* 6148 !
Vahun, Kolahun (fl. & fr. Nov.) *Baldwin* 10220 ! **Iv.C.:** Tabou *Chev.* 20053. **G.C.:** Achimota (fl. & fr.
Jan., June) *Irvine* 1452 ! 1965 ! **Togo:** *Baumann* 589 ! Kpandu *Robertson* 133 ! **N.Nig.:** Bauchi Plateau
(fl. & fr. June) *Lely* P. 327 ! Naraguta (fl. & fr. June) *Lely* 281 ! Vom *Dent Young* 104 ! Jos (fl. & fr.
Mar.) *Hill* 9 ! **S.Nig.:** Idanre (fl. & fr. Jan.) *Brenan* 8715 ! **Br.Cam.:** Buea, 3,000 ft. (Jan.) *Maitland* 218 !
Mamfe (fr. Mar.) *Coombe* 200 ! **F.Po:** Moka Plateau, 3,500 ft. (fl. & fr. Dec.) *Boughey* 128 ! Finca Puente
(Jan.) *Guinea* 1730 ! Widely distributed in tropical Africa, Asia and Australia.

3. SANICULA Linn.—F.T.A. 3 : 7.

Radical leaves long-petiolate, palmately divided nearly to the base into 5 obovate
lobes about 4 cm. long, sharply crenate, glabrous ; cauline leaves gradually becoming
smaller and bract-like ; heads small, few-flowered ; bracts lanceolate ; fruit densely
covered with hooked bristles   ..   ..   *elata*

S. elata *Buch.-Ham. ex D. Don* Prod. Fl. Nepal. 183 (1825) ; Shan & Constance in Univ. California Pub. Bot.
25, No. 1 : 47 (1951). *S. europea* of F.T.A. 3 : 8 ; of F.W.T.A., ed. 1, 1 : 523 ; not of Linn. A glabrous
herb with spine-covered fruit, in montane forest.
Br.Cam.: Cam. Mt., 4,000–7,600 ft. (fr. Dec.–Feb.) *Mann* 1233 ! 2019 ! *Maitland* 1018 ! *Brenan* 9343 !
Bamenda, 7,200–7,500 ft. (fr. Jan., Feb., Aug.) *Migeod* 370 ! *Tiku* FHI 22155 ! *Tamajong* FHI 23454 !
Keay & Lightbody FHI 28507 ! **F.Po:** Clarence Peak, 4,000–7,000 ft. (fl. Dec., fr. Nov., Dec.) *Mann* 587 !
Boughey 122 ! Distributed on mountains in tropical Africa from French Cameroons to Abyssinia and
southwards to S. Africa, also in Himalayas, Malaysia, China and Japan.

4. ERYNGIUM Linn.—F.T.A. 3 : 6.

Herb with large, oblanceolate, sharply toothed radical leaves up to 30 cm. long ; cauline
leaves much smaller and very sharply toothed, glabrous ; involucral bracts leafy,
linear-lanceolate, about 3 cm. long ; flower spikes cylindric, 1–1·5 cm. long   *foetidum*

E. foetidum *Linn.*—F.T.A. 3 : 6 ; Chev. Bot. 303. A spiny leaved perennial herb, with an unpleasant smell
and a fleshy rootstock exuding latex ; flowers green.
S.L.: *Thomas* 8434 ! 10432 ! Zimmi (Nov.) *Deighton* 377 ! Marusia (Feb.) *Dawe* 478 ! **Lib.:** Bakratown
(Oct.) *Linder* 878 ! Peter's Town, Sinoe (fl. & fr. Oct.) *Bunting* 105 ! Nyaake, Webo (June) *Baldwin*
6136 ! **Iv.C.:** Grabo de Taté *Chev.* 19749. **G.C.:** Tarkwa (fl. & fr. Sept.) *Dalz.* 108 ! **Br.Cam.:** Victoria
(fl. Jan., Mar., fr. Apr.) *Brenan* 9264 ! *Ejiofor* FHI 29329 ! *Keay* FHI 28573 ! A native of tropical
America, naturalized in our area and in Uganda, S. Tomé and Principe. (See Appendix, p. 345.)

5. PYCNOCYCLA Lindl.—F.T.A. 3 : 8.

A wiry herb with finely divided leaves, the ultimate segments narrowly linear ; umbels
compound, condensed into rather tight heads ; outer florets more strongly developed
than the inner, producing a radiate effect ; bracts and bracteoles narrow linear-
filiform ; fruit and inflorescence pubescent   ..   ..   *ledermannii*

P. ledermannii *Wolff* in Engl. Bot. Jahrb. 57 : 220 (May 1921). *P. occidentalis* Hutch. in Kew Bull. 1921 :
374 (Dec. 1921). A herb with narrowly linear leaf-segments ; bracts of the partial umbels long-filiform.
Fr.G.: Ditinn *Jac.-Fél.* 594 ! Mali-Loura *Jac.-Fél.* 628 ! **N.Nig.:** *Kennedy* 2911 ! Bauchi Plateau (Aug.)
Lely P. 662 ! Ropp, 4,600 ft. (July) *Lely* 450 ! Vom *Dent Young* 106 ! Sho, Jos (Nov.) *Keay* FHI 21445 !
Kurra Fall, Jos *Savory* UCI 107 ! Also in Ubangi-Shari and French Cameroons.

6. TORILIS Adans. Fam. 2 : 99 (1763).

An erect herb up to 1 m. high ; leaves pinnate to bipinnate, the ultimate segments lanceolate, umbels usually compound ; fruit with prominent hooked spines, one mericarp
frequently does not develop fully ; leaves and inflorescence clothed with appressed
strigose bristles   ..   .. *arvensis*

T. arvensis (*Huds.*) *Link*, Enum. Pl. Berol. 1 : 265 (1821). *Caucalis arvensis* Huds. Fl. Angl. 98 (1762). *Torilis
africana* (Thunb.) Spreng. Pl. Min. Cogn. Pugill. 2 : 55 (1815). *Caucalis infesta* Curt.—F.T.A. 3 : 26.
A wiry herb ; in montane vegetation.
    Br.Cam.: above L. Oku, 7,800 ft., Bamenda (fl. & fr. Jan.) *Keay & Lightbody* FHI 28471 ! Distributed
on mountains in tropical Africa from French Cameroons to Abyssinia and S. Africa, and in Europe and
western Asia.

7. CAUCALIS Linn.—F.T.A. 3 : 25.

Leaves bipinnately divided into small coarsely serrate leaflets ; petioles sheathing and
membranous at the base ; umbels congested like capitula on long slender tomentose
peduncles ; bracts lanceolate, 3–4 mm. long, densely ciliate ; fruits about 5 mm.
long, densely clothed with bristles armed at the apex with reflexed barbs *melanantha*

C. melanantha (*Hochst.*) *Hiern* in F.T.A. 3 : 26 (1877). *Agrocharis melanantha* Hochst. in Flora 27 : 19 (1844).
Torilis melanantha (Hochst.) Vatke in Linnea 40 : 190 (1876). A herb with finely divided leaves and fruit
with numerous hooked spines ; in montane grasslands.
    Br.Cam.: Cam. Mt., 3,000–9,000 ft. (fl. Dec., Jan., fr. Dec., Jan., June) *Mann* 1243 ! 2008 ! *Maitland*
837 ! 956 ! 1300 ! Kishong, Bamenda 6,500 ft. (fl. & fr.
Jan.) *Keay & Russell* FHI 28444 ! Bafut-Ngemba F.R. (fl. & fr. Aug.) *Ujor* FHI 29957 ! **F.Po:** 7,000 ft.
Mann 317 ! Moka, 4,000 ft. (fr. Dec., Jan.) *Exell* 762 ! *Boughey* 30 ! Also in French Cameroons, Abyssinia,
Uganda, Kenya and Belgian Congo.

8. APIUM Linn.—F.T.A. 3 : 11.

An erect or subprostrate herb, with finely divided leaf-segments, the ultimate divisions
narrow linear, leaf-bases sheathing and membranous ; the stem leaves similar but
smaller ; umbels terminal and lateral ; fruit ovoid, slightly compressed laterally
and with very prominent ridges when mature   ..   ..   .. *leptophyllum*

A. leptophyllum (*Pers.*) *F. Müll. ex Benth.* Fl. Austr. 3 : 372 (1866). *Pimpinella leptophylla* Pers. Syn. Pl. 1:
324 (1805). A low lax herb, up to 1 ft. high, with linear leaf-segments.
    Fr.Nig.: Niamey to Gao *Ryff* ! A weed, widespread throughout the tropics.

9. PIMPINELLA Linn.—F.T.A. 3 : 13 ; Norman in J. Linn. Soc. 47 : 583 (1927).

Leaflets ovate-orbicular, rounded at apex, the terminal one about 1·5 cm. diam.,
crenate-dentate, the lateral ones smaller and sessile, slightly pubescent on the radiating
principal veins, petiole broadly sheathing at the base ; umbels about 6-rayed ; rays
(i.e. pedicels) very short and rather stout ; fruit ovoid-oblong, 2·5 mm. long, glabrous
                                                                    1. *oreophila*
Leaflets ovate-lanceolate, acute, the lowest pair the largest, stalked about 3 cm. long
and 1·5 cm. broad, sharply serrulate, slightly pubescent on both surfaces ; petiole
with a very narrow sheathing base ; umbels about 6-rayed, rays (i.e. pedicels) very
slender, about 2 cm. long ; ovary and fruit whitish-pubescent ..   2. *praeventa*

1. **P. oreophila** *Hook. f.*—F.T.A. 3 : 14 ; Wolff in Engl. Pflanzenr. Umbellif. 4, 228 : 311 (1927) ; Norman
in J. Linn. Soc. 47 : 588. A small herb 2–12 in. high, leaves pinnate ; in upper montane grasslands.
    Br.Cam.: Cam. Mt., 9,500–12,000 ft. (fl. Apr., fl. & fr. Nov.–Jan.) *Mann* 1291 ! 2028 ! *Hutch. & Metcalfe*
81 ! *Maitland* 1252 ! *Keay* FHI 28610 ! **F.Po:** Clarence Peak *Mann* ! Also in Abyssinia and N.E.
tropical Africa.
2. **P. praeventa** *Norman* l.c. 47 : 591 (1927). An erect slender herb, 1–4 ft. high, with pinnate leaves ; in open
savannah among rocks.
    Fr.G.: Mali, Fouta Djalon, 4,800 ft. (Sept.) *Schnell* 7045 ! **N.Nig.:** Bauchi Plateau (July) *Lely* P. 514 !
P. 861 ! Vom, 3,000–4,500 ft. *Dent Young* 107 ! Naraguta 4,000 ft. (June) *Lely* 335 ! Jos (fl. & fr. Nov.)
Keay FHI 21439 ! Riyom, Jos Div. (fr. Dec.) *Corby* FHI 14665 ! Bangwele, Bauchi Prov. (fl. & fr. Sept.)
Summerhayes 37 ! Fobur, Bauchi Prov. (fl. & fr. July) *Summerhayes* 13 !

    Besides the above *P. anisum* Linn. is occasionally cultivated in our area and may be found established
in areas associated with human occupation.
    P. ledermannii Wolff occurs in the Bambuto Mts. and other localities in the French Cameroons ; it
may well be found in the area of this Flora.

10. SIUM Linn.—F.T.A. 3 : 13.

Leaves simply pinnate, with about 12 narrowly lanceolate leaflets ; umbels with well-
developed bracts and bracteoles ; fruit slightly compressed laterally, with prominent
ribs ..   ..   ,.   ..   .. *repandum*

S. repandum *Welw. ex Hiern* Cat. Welw. 1 : 425 (1898). A robust herb, about 3 ft. high, with yellow flowers ;
in marshy places.
    Br.Cam.: Maisanmari, Gashaka Dist. (fl. & fr. Jan.) *Latilo & Daramola* FHI 34394 ! Spasmodically
distributed in tropical and S. Africa.

11. CRYPTOTAENIA DC. Coll. Mém. 5 : 42 (1829).

Radical leaves ternate or biternate ; lower part of stem and petioles rusty-pilose ;
leaflets broadly ovate, sometimes lobed or cordate, irregularly dentate, setulose ;
cauline leaflets becoming linear-lanceolate ; umbels with 3–6 primary and 3–5

secondary rays ; bracts and bracteoles absent, pedicels very slender, flowers very
small, narrowly ovoid, 5 mm. long   ..   .. *africana*

C. africana (*Hook. f.*) *Drude* in E. & P. Pflanzenfam. 3, 8 :   189 (1898).   *Anthriscus africana* Hook. f. (1864)—
F.T.A. 3 :   16.   Herb 2–4 ft. high, pedicels very slender and thread like.
Br.Cam.: Cam. Mt., 4,000–7,000 ft. (fl. & fr. Dec., Feb., May) *Brenan* 9351 ! *Maitland* 655 ! 822 ! *Mann*
1234 ! 2023 ! Mba-Kokeka, 7,000 ft., Bamenda (fr. Mar.) *Richards* 5299 !   **F.Po:** *Tessmann* 2821. Also
on the mountains of French Cameroons, E. Africa and the Belgian Congo.

12. DIPLOLOPHIUM Turcz.—F.T.A. 3 : 17.

Cauline leaves few, scattered ; petiole flat, dilated into a rigid amplexicaul sheath at
the base ; lamina 3–5-pinnatisect, segments all very narrow, linear, elongated,
fleshy, acute 3–5 cm. long ; involucral bracts numerous, narrowly linear with white
margins, acute, up to 3 cm. long, 3–4 mm. broad at the base ; rays about 20, unequal ;
umbellules 10–20 flowered ; fruit subterete, compressed on the back, densely papillose,
densely and shortly villous   ..   .. *africanum*

D. africanum *Turcz.* in Bull. Soc. Imp. Nat. Mosc. 201 : 173 (1847). *Cachrys abyssinica* Hochst. ex A. Rich.,
Fl. Abyss. 1 : 333 (1848). *Diplolophium abyssinicum* (Hochst. ex A. Rich.) Benth. (1867)—F.T.A. 3 : 17.
Physotrichia diplolophioides Wolff in Engl. Bot. Jahrb. 57 : 230 (1922) ; F.W.T.A., ed. 1, 1 : 522. A
robust herb up to 6 ft. high, with linear leaf segments and hairy fruit.
Fr.G.: Labé, Fouta Djalon *Jac.-Fél.* 655 ! Kankan (fr. Aug.) *Schnell* 6692 !   **Togo:** Sokode-Basari (fr.
Nov.) *Kersting* 515.   **N.Nig.:** Birnin Gwari (fl. & fr. Nov.) *Thornewill* 151 ! Koboji to Kontagora (fr.
Jan.) *Meikle* 1030 ! Naraguta (fl. & fr. Aug.) *Lely* 499 ! Ropp (Dec.) *W. D. MacGregor* 398 ! Jos
(fl. & fr. Dec.) *Corby* FHI 14662 ! Also in French Cameroons, Ubangi-Shari, Abyssinia, A.-E. Sudan,
E. Africa and Belgian Congo.

13. PEUCEDANUM Linn.—F.T.A. 3 : 18.

Mericarps without prominent dorsal ridges between the lateral wings :
  Leaves bipinnate ; leaflets deeply 3-lobed, triangular in total outline, up to 4 cm. long,
    margins remotely toothed ; fruit obovate, 10 mm. long and 7·5 mm. broad, broadly
    winged ; lamina of the upper stem-leaves much reduced   ..   ..   1. *madense*
  Leaves biternate to bipinnate ; leaflets usually linear-lanceolate, rarely narrowly
    lanceolate, margins remotely serrate ; fruit elliptic 7 mm. long and 5 mm. broad,
    broadly winged ; lamina of the stem-leaves much reduced..   ..   2. *angustisectum*
Mericarps with prominent dorsal ridges between the lateral wings :
  Leaves bipinnatisect ; leaflets ovate or narrowly-ovate, acute, crenate-serrate, up to
    3 cm. long and 1 cm. broad ; fruit elliptic, 10 mm. long and 7·5 mm. broad ; broadly
    winged ; lamina of the upper stem-leaves quite prominent..   ..   3. *winkleri*

1. **P. madense** *Norman* in J. Linn. Soc. 49 : 513 (1934). A tall herb up to about 5 ft. high.
  N.Nig.: Wana, Mada Hills, 2,000 ft. (fl. & fr. Oct.) *Hepburn* 93 ! Also in French Cameroons.
  [This is a very poorly known species ; opportunities for its collection and study in the field should not be
  missed.]
2. **P. angustisectum** (*Engl.*) *Norman* l.c. 509 (1934). *Lefeburia angustisecta* Engl. Pflanzenw. Afr. 3, 2 : 829
  (1921) ; F.W.T.A., ed. 1, 1 : 523. A robust herb, up to 4 ft. high, with linear-lanceolate leaf-segments.
  Br.Cam.: Cam. Mt., 6,000–10,000 ft. (fl. July–Sept., fr. Dec., Jan.) *Maitland* 819 ! *Mildbr.* 10925 ! *Preuss*
  969 ! *Savory* 523 ! L. Bambili, Bamenda (Aug.) *Ujor* FHI 29968 ! Also in French Cameroons.
3. **P. winkleri** *Wolff* in Engl. Bot. Jahrb. 48 : 278 (1912) ; Norman l.c. 511. *P. petitianum* of F.T.A. 3 : 20 and
  of F.W.T.A., ed. 1, 1 : 522. A tall herb with broadly winged fruit, in montane habitats.
  Br.Cam.: Cam. Mt., 6,000–10,000 ft. (fl. & fr. Feb.) *Maitland* 1315 ! *Preuss* 966 !   **F.Po:** Clarence Peak,
  9,000–9,500 ft. (fl. & fr. Dec.) *Mann* 608 ! Also on the mountains, in French Cameroons and E. Africa.

14. LEFEBVREA A. Rich.—F.T.A. 3 : 23.

Tall robust herb ; leaflets lanceolate to narrowly ovate, acute, 5–10 cm. long, 1·5–6 cm.
broad, serrulate, sometimes long-toothed ; umbels terminal and lateral, the lateral
umbels on long straight peduncles ; flowers often tinged red-purple ; fruit not seen
                                                                      nigeriae

L. nigeriae *Wolff* in Engl. Bot. Jahrb. 57 : 233 (1921). Erect herb up to 6 ft. high.
  N.Nig.: Kilba country (July) *Dalz.* 90 ! Vom, 3,000–4,500 ft. *Dent Young* 108 ! Naraguta (Aug.) *Lely*
  486 ! Bauchi Plateau (Aug.) *Lely* P. 620 ! *Batten-Poole* 352 ! Also in the French Cameroons.
  It is most desirable that this species be collected in fruit. All the specimens I have seen are flowering ;
  without fruit it is impossible to be certain of the generic affinities of this species.

15. STEGANOTAENIA Hochst. in Flora 27, Bes. Beil. 4 (1844).

A small tree, up to 12 m. high ; leaves simply pinnate ; leaflets subentire to toothed,
5–15 cm. long, 2–7 cm. broad, glabrous ; primary umbels about 12-rayed, secondary
about 6–9-rayed ; ultimate peduncles about 3 cm. long ; pedicels 5 mm. long ;
fruit obovate 1 cm. long, 7 mm. broad, 3-ribbed   ..   .. *araliacea*

S. araliacea *Hochst.* l.c. (1844) ; Norman in J. Linn. Soc. 49 : 514 (1934). *Peucedanum araliaceum* (Hochst.)
  Benth. (1876)—F.T.A. 3 : 21 ; F.W.T.A., ed. 1, 1 : 523. *P. fraxinifolium* Hiern ex Oliv. (1873)—F.T.A.
  3 : 22 ; F.W.T.A., ed. 1, 1 : 523. *P. atacorense* A. Chev. Bot. 304, name only. A small, soft-wooded
  tree, with compound umbels of small white flowers ; in savannah.
  Fr.Sud.: Kita *Chev.* 106 ; *Jaeger* 4 !   **Fr.G.:** Labé *Chev.* 12319 ; *Roberty* 16396 ! Timbo *Chev.* 14638. Beyla
  Chev. 20896.   **S.L.:** Freetown *Welw.* 2516 ! Bathurst *Deighton* 2764 ! Koinadugu Dist. *Dawe* 499 !
  Iv.C.: Grabo to Taté *Chev.* 19791.   **G.C.:** *Vigne* FH 3871 ! Tiasi (fr. Oct.) *Vigne* FH 1383 !   **Togo:** Lando
  Kersting A. 390 ! Sokode to Basari *Kersting* A. 706 ! Yendi (fr. Apr.) *Vigne* FH 1698 !   **Dah.:** Somba,
  Atacora Mts. *Chev.* 24095 !   **N.Nig.:** Nupe *Barter* ! Katagum *Dalz.* 343 ! Kontagora (fr. Jan.) *Meikle*
  1057 ! Bauchi Plateau (fl. & fr. Jan.) *Lely* P. 10 ! Bornu Prov. *Talbot* 1218 !   **Br.Cam.:** Gwoza, Dikwa
  Div. (Jan.) *McClintock* 151 ! Widespread in tropical Africa. (See Appendix, p. 346.)

16. AMMODAUCUS Coss. & Dur., in Bull. Soc. Bot. Fr. 6 : 393 (1859).

A small herb with a slender tap-root ; leaves bipinnate, segments linear, rather thick, glabrous ; umbels about as long as the leaves ; pedicels 0·5–1 cm. long ; fruit about 1 cm. long, densely covered with long slender whitish bristles   ..   *leucotrichus*

A. leucotrichus *Coss. & Dur.* l.c. (1859) ; Chev. Bot. 305.
    Fr.Sud.: Timbuktu region (young fr. Feb., Mar.) *Vuillet* in *Hb. Chev.* 25974 ; 25978 ; 28834 ! Also throughout N. Africa. (See Appendix, p. 345.)

ADDITIONS AND CORRECTIONS

Since the preparation of Parts 1 and 2 the following additions and corrections have been noted. J. P. M. Brenan, F. N. Hepper, R. D. Meikle and G. Troupin have supplied amendments in the genera for which they were responsible. Several new records for the Gold Coast have been noted by C. D. Adams.

35–36. **UVARIA**—add, as synonym, *Xylopiastrum* Roberty in Bull. I.F.A.N. 15: 1397 (1953).
3. **U. chamae**—add, as synonym, *Xylopiastrum macrocarpum* (Dunal) Roberty in Bull. I.F.A.N. 15 : 1397 (1953). Add also **Port.G.:** Madina (June) *Esp. Santo* 3203 !

38. **CLEISTOPHOLIS**
1. **C. patens**—also in Belgian Congo.

42. **XYLOPIA**
4. **X. quintasii**—also in Gabon, Cabinda and Belgian Congo.

43. **POLYALTHIA**
1. **P. oliveri**—add, as synonym, *Artabotrys oliveri* (Engl.) Roberty in Bull. I.F.A.N. 15 : 1397 (1953).

44. **POPOWIA**
1. " P. declina " should read **P. diclina.**
Add also **P. mangenotii** *Sillans* in Bull. Mus. Hist. Nat., sér. 2, 24 : 578 (1953).
Fr.G.: Mt. Nimba *Schnell* 4414 ; 5115. Benna *Schnell* 5618. **Iv.C.:** Abidjan *Schnell* 3850. Also in Ubangi.

47–48. **HEXALOBUS**
1. **H. crispiflorus**—add **Port.G.:** Xitele (fr. May) *Esp. Santo* 3185 !
2a. **H. monopetalus** var. **parvifolius**—correct reference for the synonym is *H. glabrescens* Hutch. & Dalz. ex Burtt Davy, Man. Fl. Pl. Transvaal 1 : 103 (1926).

50. **UVARIOPSIS**
1. **U. bakeriana**—add, as synonym, *Drypetes talbotii* S. Moore in Cat. Talb. 97 (1913), partly (leaves only).

52. **ANNONA**
4. **A. glabra**—add **Port.G.:** Bissora to Barro (fr. Nov.) *Esp. Santo* 3122 !

53. **ISOLONA**
1. **I. campanulata**—add **Lib.:** Piatah (fr. Dec.) *Baldwin* 12509 !

54. **MONODORA**
2. **M. tenuifolia**—add **Port.G.:** Madina (June) *Esp. Santo* 3204 !

59. **ILLIGERA**
2. **I. vespertilio**—add **S.Nig.:** Oban *Talbot* !

66. **MENISPERMACEAE**
Line 8 of the description of the family should start " series, *free* or slightly connate."

67–68. It should be noted that a newly described species, *Rhigiocarya peltata*, is not covered by the keys to genera. It differs from *Stephania* and *Cissampelos*, which also have peltate leaves, in having 3 carpels, and stamens which are either free or else 3 united and 3 partly free. Some species of *Kolobopetalum* also have subpeltate leaves.

68. **SYNCLISIA**
S. scabrida—also in Ubangi-Shari.

70. **EPINETRUM**
2. **E. undulatum**—delete Belgian Congo and Uganda from distribution. I have not been able to verify the identification of the Ivory Coast specimen and am very doubtful if true *E. undulatum* occurs in the area of the F.W.T.A.—G. T.

71. **TILIACORA**
1. **T. dinklagei**—add **G.C.:** *Vigne* FH 1492 (partly) !
2. **T. louisii** *Troupin* in Bull. Jard. Bot. Brux. 19 : 412 (1949) ; Fl. Congo Belge 2 : 212. *T. ? odorata* of F.W.T.A., ed. 2. Add **Lib.:** *Dinklage* 2889 ! Also in Belgian Congo.
3. & 4. Troupin, in F.T.E.A. Menisp. 9, includes *T. warneckei* Engl. ex Diels and *T. johannis* Exell in **T. funifera** (*Miers*) *Oliv.* F.T.A. 1 : 44 (1868), a species which extends to Kenya and Mozambique.
6. **T. lehmbachii**—the Belgian Congo record is confirmed.

71. **TRICLISIA**
1. **T. macrophylla**—add **S.L.** *Afzelius* ! **Br.Cam.:** Cam. Mt. *Jungner* 144 !
2. **T. dictyophylla** *Diels* l.c. 70 (1910). *T. gilletii* (De Wild.) Staner—F.W.T.A., ed. 2. *Tiliacora trichantha* Diels—F.W.T.A., ed. 2. Also in French Cameroons, Cabinda and Gabon.
4. **T. subcordata**—also in French Cameroons, Ubangi, Gabon and Angola.

72. KOLOBOPETALUM
1. **K. auriculatum**—add **Iv.C.**: Kouta (May) *Miège*! Also in Cabinda.
2. **K. leonense**—add **Iv.C.**: Abidjan *Miège*!
3. **K. chevalieri**—add **S.Nig.**: Onitsha (Feb.) *Jones* FHI 7381!
4. **K. ovatum**—in distribution delete Ubangi and add Gabon.

72. RHIGIOCARYA
Add **R. peltata** *Miège* in Bull. I.F.A.N. 17 : 362, tt. 3 & 4 (1955). A glabrous liane with peltate leaves.
Iv.C.: Adiopodoumé, Abidjan (June, Oct.) *Mangenot & Miège*! Gouleako (Aug.) *Miège* 2603! Kouta *Miège*.

73. DIOSCOREOPHYLLUM
1. **D. cumminsii**—add la. **D. cumminsii** var. **leptotrichos** *Troupin*, **var. nov.** *D. chirindense* Swynnerton in J. Linn. Soc. 40 : 19 (1911). Plants covered with slender pale hairs instead of the rigid dark brown hairs of var. *cumminsii*.
S.L.: Jala (July) *Bunting* 65a! **G.C.**: Nfuom, Kakum F.R. (Mar.) *Box* 2579! Also in S. Rhodesia and Mozambique.

2. **D. volkensii** *Engl.* var. **volkensii**—Pflanzenw. Ost-Afr. C. : 182 (1895) ; Troupin in F.T.E.A. Menisp. 13. *D. tenerum* Engl. var. *tenerum*—F.W.T.A., ed. 2. Add. **S.Nig.**: Oban *Talbot*! Also in E. Africa, Mozambique and S. Rhodesia.
2a. **D. volkensii** var. **fernandense** (*Hutch. & Dalz.*) *Troupin* l.c. (1956). *D. tenerum* Engl. var. *fernandense* (Hutch. & Dalz.) Troupin—F.W.T.A., ed. 2.

73. SYNTRIANDRIUM
S. preussii—also in Ubangi.

74. JATEORHIZA
Add **J. palmata** (*Lam.*) *Miers* in Hook. Fl. Nigrit. 214 (1849) ; Troupin in F.T.E.A. Menisp. 15. **G.C.**: *Irvine* 940! Also in Kenya, Tanganyika, Nyasaland, Mozambique and Mauritius.
This identification seems to be correct. On account of the therapeutic properties of the plant it is probable that it has been introduced in the Gold Coast and may perhaps have been naturalized. It is, however, strange that the cultivation, or at least the introduction, of this plant has not been recorded in other West African territories.—G. T.

75. CHASMANTHERA
C. dependens—add **Iv.C.**: Baoulé (Aug.) *Miège* 1921!

75. STEPHANIA
3. **S. dinklagei**—also in Ubangi, Gabon, Uganda and Tanganyika.
Add also **S. cyanantha** *Welw. ex Hiern* in Cat. Welw. 1 : 20 (1896) ; Troupin in F.T.E.A. Menisp. 21. Glabrous liane with succulent stem and branches.
F.Po: Moka, 4,000 ft. (Dec.) *Boughey* 16! Also in French Cameroons, Belgian Congo, Angola, Kenya, Tanganyika and N. Rhodesia.

75. CISSAMPELOS
C. owariensis—add **S.L.**: *Afzelius*!

87. CLEOME
8. **C. viscosa**—add **G.C.**: Kpeshi Lagoon, Accra (Feb.–Apr.) *Adams* 2365 ; 2396 ; 4212. Also in Australia.

91. RITCHIEA
7. **R. longipedicellata**—add **G.C.**: Bibiani (Dec.) *Adams* 1954.

95. COURBONIA
C. virgata—add **Br.Cam.**: Dikwa Div., *McClintock* 40!

96. MORINGA
M. oleifera—add **Br.Cam.**: Boma, Dikwa Div., *McClintock* 6!

96. CRUCIFERAE
Add the genus **CRAMBE** Linn.—F.T.A. 1 : 71.
C. kilimandscharica *O. E. Schulz* in Engl. Bot. Jahrb. 54, Beibb. 119 : 54 (1916) ; Pflanzenr. Crucif.-Brassic. 245, fig. 31 (1919). An annual herb.
G.C.: Aburi *Patterson*! Also in Belgian Congo and E. Africa.

97. RORIPPA
Add **R. sinapis** (*Burm. f.*) *Keay*, **comb. nov.** *Sisymbrium sinapis* Burm. f. Fl. Ind. 140 (1768). *Nasturtium sinapis* (Burm. f.) O. E. Schulz in Fedde Rep. 33 : 278 (1934) ; in Engl. Pflanzenfam. 17B : 554 (1936) ; Exell Suppl. Cat. S. Tomé 10 (1956). Herbaceous weed with yellow petals.
Br.Cam.: Buea (Nov.) *Migeod* 227! Also in S. Tomé and widespread in tropical and subtropical Asia and America.

104. 3. **DECORSELLA** A. Chev. in Bull. Soc. Bot. Fr. 61, Mém. 8 : 297 (1917) ; Keay in Kew Bull. 1955 : 137. *Gymnorinorea* Keay (1953)—F.W.T.A., ed. 2.
D. paradoxa *A. Chev.* l.c. 298 (1917). *Rinorea abidjanensis* Aubrév. & Pellegr. (1946). *Gymnorinorea abidjanensis* (Aubrév. & Pellegr.) Keay (1953)—F.W.T.A., ed. 2.

120. VAHLIA
V. dichotoma—add **G.C.**: N.T. and Accra Plains *fide* Adams.

121. DROSERA
2. **D. pilosa** *Exell & Laundon* in Bol. Soc. Brot., sér. 2, 30 : 213 (1956). *D. burkeana* of F.W.T.A., ed. 2, 1 : 121, not of Planch. This species occurs in the British Cameroons, Kenya and Tanganyika. True *D. burkeana* Planch. is confined to southern Africa.

128. ELATINE
Add **E. fauquei** *Monod* in Notes Africaines 46 : 37, figs. 1–22 (1950) ; Bull. I.F.A.N. 16 : 309, figs. 1–22 (1954). Glabrous aquatic herb, 5–13 cm. long ; stamens 2.
Fr.Sud.: near Bamako (Dec.) *Monod*.

130. UEBELINIA
Add **U. nigerica** *Turrill* in Kew Bull. 1954 : 260.
Br.Cam.: Bambili, Bamenda (Aug.) *Ujor* FHI 29994!

132. **POLYCARPAEA**

6a. *P. corymbosa* var. *effusa*—this should be transferred as No. 1a. **P. eriantha** var. **effusa** (*Oliv.*) *Turrill* in Kew Bull. 1954 : 503.
J. Berhaut in Bull. Mus. Hist. Nat., sér. 2, 25 : 206–212 (1953) discusses several species of *Polycarpaea* in W. Africa, but I have not had the opportunity of studying the specimens he cites. The following new species and varieties are described :—
P. pobeguini *Berhaut* l.c. 207 (1953).
    Sen.: *Berhaut* 526. **Fr.G.**: *Pobéguin* 2030.
P. corymbosa var. **pseudolinearifolia** *Berhaut* l.c. 208 (1953).
    Sen: *Heudelot* 225 ; *Berhaut* 530 ; etc.
P. corymbosa var. **glabrifolia** (*DC.*) *Berhaut* l.c. 210 (1953). *P. glabrifolia* DC. (1828).
P. gamopetala *Berhaut* l.c. 212 (1953).
    (?) **Sen.**, but perhaps Australia.

134. **GISEKIA**

G. pharnacioides—add **G.C.**: *Thonning.* Tema (Mar., Oct.) *Morton* A. 185 ; *Boughey* GC 10020. Prampram (Nov.) *Boughey* GC 14311. Labadi (Apr.) *Irvine* 2440.

134. **MOLLUGO**

M. nudicaulis—" *Barter* 1340 " should read *Barter* 1346.

135. **GLINUS**

1. **G. lotoides**—add **G.C.**: Zuarungu to Bawku, White Volta R. (Dec.) *Akpabla* 405. Gambaga (Dec.) *Morton* A. 1378. Yapei, Volta R. (Mar.) *Adams* 3917.

137. **PORTULACA**

J. Berhaut in Bull. Soc. Bot. Fr. 100 : 35 (1953) distinguishes in Senegal another species—**P. meridiana** *Linn.*

141–142. **POLYGONUM**

4. **P. glomeratum** *Dammer* in Fedde Rep. 15 : 386 (1919). *P. strigosum* var. *pedunculare* of F.W.T.A., ed. 2, not of (Wall. ex Meisn.) Steward. Add also **Br.Cam.**: Mai Idoanu, Gashaka Dist. (Jan.) *Latilo & Daramola* FHI 28974 ! Delete, from distribution, Asia and Australia.
5. R. A. Graham informs me that *P. lanigerum* R. Br. var. *africanum* Meisn. is best regarded as **P. senegalense** *Meisn.* forma **albotomentosum** R. Grah. in Kew Bull. 1956 : 258.
6. **P. tomentosum** *Willd.* Sp. Pl. 2 : 447 (1799) is the correct name and must replace *P. pulchrum* Blume (1825). Schrank did not publish an earlier *P. tomentosum* as indicated in Fl. Congo Belge 1 : 419 and F.W.T.A., ed. 2.
9. **P. setulosum** *A. Rich.* Fl. Abyss. 2 : 227 (1851). *P. nyikense* Bak. (1897)—F.W.T.A., ed. 2.
10. R. A. Graham informs me that true *P. acuminatum* H. B. & K. is probably restricted to America and that most, if not all, of the material cited in our Flora should be regarded as forms of *P. tomentosum* Willd. ; see Graham's account of the *Polygonaceae* for the F.T.E.A.

144. **CHENOPODIUM**

1. **C. murale**—add **Sen.**: St. Louis (Feb. frts.) *Roberty* 16833 ! *Trochain* 2387 !
2. **C. ambrosioides**—add **Sen.**: Dakar (May) *Berhaut* 925 ! **C. sp.**—This proves to be **C. congolanum** (*Hauman*) Brenan in Kew Bull. 1956 : 166 (syn. *C. glaucum* Linn. var. *congolanum* Hauman), also known from the Belgian Congo. The flowers, although appearing clustered, and so described, are actually in abbreviated cymes, and the indumentum comprises a few short glandular hairs as well as the conspicuous vesicular ones.—J. P. M. B.

144. **SUAEDA**

Replace the key to the species by the following :—

Shrubby or arborescent perennials ; stigmas 3–4 ; seeds mostly vertical :
  Leaves linear to linear-oblong or linear-cylindrical :
    Shrub up to 1 m. high, with stiffly erect or ascending branches ; leaves 5–12 (–15) mm. long and 1–2 mm. broad   .. .. 1. *fruticosa*
    Shrub 1·2–6 m. high, sometimes almost arborescent, with spreading or drooping branches ; leaves mostly 13–33 (–40) mm. long and 1·5–3 mm. broad..   .. .. 2. *monoica*
  Leaves obovate or obovate-oblong, up to about 10 mm. and 3–4 mm. broad, rounded at apex, more or less narrowed towards base or even shortly petiolate ; plant often more or less glaucous when dry, shrubby, up to 1 m. high, with stiff divaricate branches   .. .. 3. *vermiculata*
  Annual ; seeds horizontal ; leaves tapering and acute at apex, not abruptly contracted at base   .. .. 4. *maritima*

Add the following two species :—

2. **S. monoica** *Forsk. ex J. F. Gmel.* Syst. Nat., ed. 13, 2 : 503 (1791) ; Forsk. Fl. Aegypt.-Arab. 70 (1775), illegitimate name ; Brenan in F.T.E.A. Chenopodiac. 23, fig. 6 (1954). Shrub up to 18 ft. high ; leaves succulent ; flowers green clustered in upper axils.
    Fr.Sud.: Ouala region *Jumelle* ! Also in E. and N. Africa and extending eastwards to India.
3. **S. vermiculata** *Forsk. ex J. F. Gmel.* l.c. (1791) ; Forsk. l.c. (1775), illegitimate name ; Boiss. Fl. Or. 4 : 940 (1879). Small shrub up to 3 ft. high ; leaves succulent ; flowers axillary, 1–3 together.
    Sen.: Salis Isl., in sandy places *Brunner* 200 ! Djeuss *Trochain* 2141 ! Maka Gandiole (Feb.) *Trochain* 2800 ! Pt. Gendarme (Sept.) *Trochain* 4596 ! Makkana (Sept.) *Trochain* 4724 ! Also from the Canary Isls. eastwards through N. Africa to Arabia and Iraq.
4. **S. maritima** (*Linn.*) *Dumort.*—add **Sen.**: L. Retba (Apr., Sept.) *Trochain* 3252 ! 5135 ! **Fr.Sud.**: Niayes (Dec.) *Chev.* 3453 !

144. **SALSOLA**

1. **S. baryosma**—add **Sen.**: Ndiaël (Apr., Dec.) *Trochain* 2006 ! 3019 ! Ross (Dec.) *Trochain* 2192 ! St. Louis (Dec.) *Roberty* 16822 !
2. **S. tetrandra**—the occurrence of this species in our area is very doubtful. I have seen no material, and the record may well rest upon a misidentification of the last species.—J. P. M. B.

145. **HALOPEPLIS**

H. amplexicaulis—add **Sen.**: Gandiole *Trochain* 2781 !

145.   **ARTHROCNEMUM**

A. glaucum—add, as synonym, *Salicornia fruticosa* of Chev. in Rev. Bot. Appliq. 27 : 294, not of (Linn.) Linn.   Add also **A. indicum** (*Willd.*) *Moq.* Chenop. Monogr. Enum. 113 (1840), partly ; Brenan in F.T.E.A. Chenopodiac. 18 (1954).   *Salicornia indica* Willd. in Ges. Naturf. Fr. Neue Schr. 2 : 111, t. 4, fig. 2 (1799).

     Sen.: Rufisque (fr. Sept.) *Berhaut* !   Retba (fl. & fr. Apr.) *Trochain* 3263 !

Differs from *A. glaucum* in being a low plant with prostrate main stems emitting fertile and sterile lateral branchlets up to 20–30 cm. high.   The fruits are fused with the segments of the fertile spike and are concealed by them, the fruits ultimately falling with the segments as the spike disarticulates.   *A. glaucum* is about 50–100 cm. high, and has the fruits not fused with the segments of the fertile spike, clearly visible between the segments, and ultimately falling away separately.—J. P. M. B.

145.   **SALICORNIA**

1. **S. fruticosa** (*Linn.*) *Linn.*—this species must be deleted.   Its inclusion was due to an erroneous identification by Chevalier of *Arthrocnemum glaucum* (Del.) Ungern-Sternb.

2, 3 & 3a. **S. senegalensis, S. praecox** and var. **longispicata**—I have examined the type-specimens of these taxa, but they enable little to be added to what is already written.   Whether these are distinct species or merely forms of *S. europaea* Linn. can only be settled by very careful observation of the living plants.

     S. praecox and var. *longispicata* seem doubtfully worth separating from one another, as some of the flowering spikes of *S. praecox* var. *praecox* have up to 12 joints.   *Trochain* 3200 from L. Tanma, Senegal is nearer to *S. senegalensis* than to *S. praecox*, while a further specimen from Niayes (Dec. 1899, [collector ?] 3452 !) in the Paris Herbarium has long fruiting spikes with up to 25 joints and may be *S. praecox* var. *longispicata*, though differing from that in its general facies.—J. P. M. B.

151.   **PUPALIA**

P. lappacea—delete Sierra Leone record ; *Deighton* 553 was collected in Nigeria.   F. C. Deighton informed me (9 Dec. 1954) that this species was not known in Sierra Leone.

151.   **PANDIAKA**

For " Overknott " read Overkott.

152.   **ACHYRANTHES**

J. Berhaut, in Bull. Soc. Bot. Fr. Mém. 1952–54 : 3 (1954), also recognises **A. argentea** Lam. and **A. argentea** var. **borbonica** (Willd. ex Roem. & Schultes) Berhaut.

153.   **GOMPHRENA**

G. celosioides—add **G.C.:** Kpeshi Lagoon, Accra (Feb., June) *Adams* 2375 ; 3776.   Also common now at Achimota and roadsides around Accra up to Aburi (*fide* Adams).

154.   **ALTERNANTHERA**

4. **A. sessilis** (*Linn.*) *Sweet* Hort. Suburb. Lond. 48 (1818).   Note corrected publication.

159.   **BIOPHYTUM**

2. **B. petersianum**—the synonyms should read *Oxalis apodiscias* Turcz. and *Biophytum apodiscias* (Turcz.) Edgew. & Hook. f.

164.   **ROTALA**

2. **R. welwitschii** *Exell* in Bol. Soc. Brot., sér. 2, 30 : 70 (1956).   *R. decussata* of F.T.A. 2 : 467 and of F.W.T.A., ed. 2, 1 : 164, not of DC.   Add also **Port.G.:** *Esp. Santo.*

3. **R. tenella**—add **Port.G.:** Madina, Boé (Jan.) *Esp. Santo* 2869.   Note also forma **fluviatilis** Fernandes & Diniz in Garcia de Orta 3 : 197 (1955).

Port.G.: Dandum, Boé (Jan.) *Esp. Santo* 2370a.

164–165.   **AMMANNIA**

1. **A. auriculata**—add **Port.G.:** Pussubé (Feb.) *Esp. Santo* 1466 ;   and other localities.

4. **A. gracilis**—add **Port.G.:** Bissau (Feb., Sept.) *Esp. Santo* 1484 ; 1666.

166.   **NESAEA**

1. **N. crassicaulis**—add **Port.G.:** Bissalanca (fl. & fr. Jan.) *Esp. Santo* 1668.

4. **N. mossiensis**—add **Port.G.:** Bafata (fl. & fr. Jan.) *Esp. Santo* 2875.

6. **N. erecta**—add **Port.G.:** Mansoa to Mansaba (fl. & fr. Feb.) *Esp. Santo* 2429.

7. **N. radicans**—add **Port.G.:** Acoco, Formosa (Apr.) *Esp. Santo* 1991.

Add also **N. angustifolia** *A. Fernandes & Diniz* in Bol. Soc. Brot. 28 : 216 (1954).   *N. mossiensis* A. Chev. (1912), partly.

Port.G.: Boé, Madina (June) *Esp. Santo* 2919.   **Fr.G.:** Kollangui (Mar.) *Chev.* 12194.   Diaguissa (Apr.) *Chev.* 13549.

171.   **HALORAGACEAE**

Add the genus **CALLITRICHE** Linn.—F.T.A. 2 : 406.

C. stagnalis *Scop.*—F.T.A. 2 : 406 ; Clapham, Tutin & Warburg Fl. Brit. Isles 619.

     Br.Cam.: *Coombes* 203 !   Cam. Mt. 4,400 ft. (Apr.) *Maitland* 1116 !   Also on mountains in Belgian Congo, Abyssinia, Eritrea and E. Africa, and in Europe.

178.   **BOERHAVIA**

The following new species has been described :—

B. graminicola *Berhaut* in Bull. Soc. Bot. Fr. 100 : 50 (1953).

     Sen. & Fr.Sud.: *Trochain* 5068 ;   3966 ;   *Waterlot* 1376 ;   *Foureaü* 3136 ;   *Berhaut* 272.   This is probably a form of **B. coccinea.**—R. D. M.

185.   **COCHLOSPERMUM**

C. religiosum (*Linn.*) *Alston* should replace the name *C. gossypium* (Linn.) DC.

186.   **SCOTTELLIA**

Pellegrin in Bull. Soc. Bot. Fr. Mém. 1952 : 108–109 (1953) regards *S. mimfiensis* Gilg and *S. kamerunensis* Gilg as varieties of *S. klaineana* Pierre.

198.   **CASEARIA**

5. **C. bridelioides**—add, as synonym, *Drypetes sassandraensis* Aubrév. Fl. For. C. Iv. 2 : 48 (1936), French descr. only.   Add also **Iv.C.:** Dakpadou (Jan.) *Aubrév.* 863 !   Danané (young fr. Mar.) *Aubrév.* 1110 !

     This plant requires further investigation and more material should be collected.   It is probably wrongly placed in *Casearia* and may have to be transferred to another family.

199. **PASSIFLORACEAE**
Add the genus **DEIDAMIA** Noronha ex Thouars Hist. Vég. Iles Austr. Afr. 61 (1807) ; Harms in E. & P. Pflanzenfam. 21 : 486 (1925). *Efulensia* C. H. Wright in Hook. Ic. Pl. t. 2518 (1897).

D. clematoides (*C. H. Wright*) *Harms* in E. & P. Pflanzenfam. Nachtr. 1 : 254 (1897) ; Notizbl. Bot. Gart. Berl. 8 : 291 (1921). *Efulensia clematoides* C. H. Wright l.c. (1897). *Deidamia triphylla* Harms (1897), name only. A glabrous liane with digitately trifoliolate leaves ; flowers greenish white ; distinguished from *Passiflora* by the absence of a gynophore.
S.Nig.: Ahoada Dist. *Talbot* 3755 ! Oban *Talbot* 410 ! 1288 ! Also in French Cameroons, Spanish Guinea, Gabon and Belgian Congo.

207. **LUFFA**
L. acutangula—F. C. Deighton informs me that this species is commonly grown in Sierra Leone.

218-220. **BEGONIA**
8. B. quadrialata—the synonym given as " *B. modesta* " should read *B. modica*.
24. B. fusicarpa *Irmsch.*—the spelling " fissicarpa " is erroneous.

221. **RHIPSALIS**
R. baccifera (*J. S. Mill.*) *Stearn* in Cact. Journ. 7 : 107 (1939). *Cassyta baccifera* J. S. Mill. (1771–77). *Rhipsalis cassutha* Gaertn. (1788)—F.W.T.A., ed. 2.

242. **COMBRETODENDRON**
C. macrocarpum (*P. Beauv.*) *Keay*, **comb. nov.** *Combretum macrocarpum* P. Beauv. Fl. Oware 2 : 90, t. 118, fig. 2 (1820)—F.W.T.A., ed. 2, 1 : 274. *Petersia africana* Welw. ex Benth. (1865). *Combretodendron africanum* (Welw. ex Benth.) Exell (1930)—F.W.T.A., ed. 2, 1 : 242.
I have now seen the type specimen of *Combretum macrocarpum* P. Beauv., on loan from the De Candolle herbarium, Geneva, and, although only the fruit is represented, I am satisfied that it is the plant usually known as *Combretodendron africanum*.
See also note against p. 274.

246. **NEROPHILA**
N. gentianoides *Naud.*—add **Port.G.:** Boé (fl. & fr. Feb.) *Esp. Santo* 2892.

247-249. **OSBECKIA**
For a discussion of the relationship between this genus and *Dissotis* Benth., see A. & R. Fernandes in Bol. Soc. Brot. sér. 2, 28 : 65–76 (1954) and in Garcia de Orta 2 : 165–197 (1954). It appears that " osbeckioid " forms (i.e. with the stamens nearly equal) of certain species of *Dissotis* are not uncommon.
The Fernandes have published the following changes which affect the species in our Flora :
1. *O. tubulosa*—becomes **Dissotis tubulosa** (*Sm.*) *Triana* in Trans. Linn. Soc. 28 : 58 (1871).
2. *O. decandra*—becomes **Antherotoma decandra** (*Sm.*) *A. & R. Fernandes* in Bol. Soc. Brot. sér. 2, 28 : 70 (1954) ; Garcia de Orta 2 : 181 (1954).
3, 4 & 5.—These species remain in *Osbeckia*.
6. *O. senegambiensis*—becomes **Dissotis senegambiensis** (Guill. & Perr.) Triana l.c. (1871). The Fernandes regard *Osbeckia saxicola* Gilg as conspecific, thus extending the range of the species to Belgian Congo and A.-E. Sudan. Add also **Port.G.:** Pussubé (Mar.) *Esp. Santo* 1143. Bafata (Dec.) *Esp. Santo* 342. Contambani to Guilego (fl. & fr. June) *Esp. Santo* 3207.

249. 8. **DICHAETANTHERA** Endl. Gen. Pl. 1215 (1840) ; Jac.-Fél. in Bull. Soc. Bot. Fr. 102 : 37 (1955) ; Ic. Pl. Afr. sub. t. 56 (1955). *Sakersia* Hook. f. (1867)—F.W.T.A., ed. 2.

1. D. echinulata (*Hook. f.*) *Jac.-Fél.* in Bull. Soc. Bot. Fr. 102 : 38 (1955). *Sakersia echinulata* Hook. f. (1871)—F.W.T.A., ed. 2.
2. D. africana (*Hook. f.*) *Jac.-Fél.* l.c. (1955). *Sakersia africana* Hook. f. (1871)—F.W.T.A., ed. 2, 1 : 249.
3. D. calodendron (*Gilg & Ledermann*) *Jac.-Fél.* l.c. (1955). *Sakersia calodendron* Gilg & Ledermann (1921)—F.W.T.A., ed. 2, 1 :. 249.

250. **TRISTEMMA**
4. T. incompletum—add **Port.G.:** Bissau (fl. & fr. Oct.) *Esp. Santo* 1486. Bolama *Carvalho* 307.

252. **DICELLANDRA**
D. barteri—" *Johnson 50 b* " should read *Johnson 505*. Add also **Fr.G.** : Nzérékoré *fide* Jac.-Fél. in Ic. Pl. Afr. t. 59 (1955).

257-259. **DISSOTIS**
9. D. rotundifolia—add **Port.G.:** Cacine (Nov.) *Esp. Santo* 681.
14. D. amplexicaulis—add **Port.G.:** Gabu (Nov.) *Esp. Santo* 278.
24. D. elliotii—add **Port.G.:** Chitole (fl. & fr. Feb.) *Esp. Santo* 2880.
28. D. grandiflora var. grandiflora—" *Heudelot* 779 " should read *Heudelot* 775.
28a. D. grandiflora var. lambii—add **Port.G.:** Bafata (Oct.) *Esp. Santo* 162. Gabu (fl. & fr. Oct.) *Esp. Santo* 3109.
30. D. phaeotricha—delete the synonym *Osbeckia postpluvialis* Gilg.
36. D. erecta—Note that this species has been transferred by A. & R. Fernandes to the genus *Melastomastrum* Naud., under the name **M. capitatum** (*Vahl*) *A. & R. Fernandes* in Garcia de Orta 2 : 278 (1954). Add here also **Port.G.:** Bafata (Oct.) *Esp. Santo* 159. Bubaque *Dinklage* 3226.
The following species also may occur in our area :—
Melastomastrum schlechteri *A. & R. Fernandes* in Bol. Soc. Bot. 39 : 48, t. 2 (1955). " West Africa " : Ngoka (Oct.) *Schlechter* 12782.
Imp. known sp. 2. D. rupicola *Gilg ex Engl.* (1921)—F.W.T.A., ed. 2, 1 : 259. See A. & R. Fernandes in Bol. Soc. Brot. 29 : 50 (1955) for full description.
Lib.: Sinoe (Dec.) *Dinklage* 2139.

262-263. **MEMECYLON**
1. M. afzelii—add **Port.G.:** Bolama (fl. & fr. Apr.) *Esp. Santo* 1935.
5. M. normandii—add **Port.G.:** Catió (fr. June) *Esp. Santo* 2074.
11. M. blakeoides—add **Port.G.:** Gabu (Oct.) *Esp. Santo* 171. Pecixe (fl. & fr. May) *Esp. Santo* 2043. Delete Gold Coast record based on *Vigne* FH 1453 which is *M. fleuryi*.

274. **COMBRETUM**

Imp. known sp. 2. **C. macrocarpum** *P. Beauv.*—I have now seen the type specimen ; it is certainly **Combretodendron** A. Chev. The Ivory Coast specimen, cited by Roberty (in Mém. Soc. Bot. Fr. 1952 : 24) as *Combretum macrocarpum* P. Beauv., is certainly a *Combretum*, but unfortunately it is too poor to name specifically. See also note against p. 242.

275. **GUIERA**

G. senegalensis—add **G.C.:** Burufo Plateau, Lawra (Apr.) *Adams* 4020. Navrongo to Chuchiliga (Dec.) *Adams & Akpabla* GC 4355. Gambaga (May) *Morton* GC 7405.

280. **CONOCARPUS**

C. erectus—add **G.C.:** Labadi (Nov.) *Morton* GC 6150. Winneba (fl. bud Nov.) *Boughey* GC 10356. Elmina (Jan.) *Deakin* 33. Atwabo (fr. July) *Fishlock.*

283. **CASSIPOUREA**

7. **C. gummiflua** *Tul.* var. **verticillata** (*N.E. Br.*) *J. Lewis* in Kew Bull. 1955 : 158 ; F.T.E.A. Rhizoph. 16. *C. verticillata* N.E. Br. (1894). *C. ugandensis* of F.W.T.A., ed. 2, not of (Stapf) Engl. Also in Tanganyika, Nyasaland, Mozambique, Natal and (? introduced) Seychelles.

9. **C. gummiflua** *Tul.* var. **mannii** (*Hook. f. ex Oliv.*) *J. Lewis* in Kew Bull. 1955 : 158 ; F.T.E.A. Rhizoph. 15. *C. glabra* Alston (1925)—F.W.T.A., ed. 2. Also in S. Tomé and Angola.

289. **PSOROSPERMUM**

5a. **P. corymbiferum** var. **doeringii** (*Engl.*) *Keay & Milne-Redhead,* **comb. nov.** *P. ledermannii* Engl. var. *doeringii* Engl. Bot. Jahrb. 55 : 387 (1919). *P. corymbiferum* var. *kerstingii* (Engl.) Keay & Milne-Redhead (1953)—F.W.T.A., ed. 2. As the epithet *doeringii* had already been used at varietal rank, the name var. *kerstingii* was illegitimate in the circumscription adopted.

293. **MAMMEA** Linn. Sp. Pl. 1 : 512 (1753). Note corrected reference.

321–330. **COLA** Schott & Endl.—add, as synonym, *Ingonia* Pierre ex Bodard in J. Agr. Trop. Bot. Appliq. 2 : 528 (1955).

3. **C. digitata**—add syn. *Ingonia digitata* (Mast.) Bodard in J. Agr. Trop. Bot. Appliq. 2 : 529, figs. 1–4 (1955).

10. **C. chlamydantha**—add **Fr.G.:** Benna (Dec.) *Jac.-Fél.* 2162 !

Imp. known sp. 2. **C. triloba**—we have recently received specimens from M. Bodard which are conspecific with the type. This species keys out near *C. heterophylla* and from which it may be distinguished by the following characters : stellate hairs on young shoots, petioles and calyces coarse and more rusty, with longer arms ; flower-buds larger and longer in relation to width ; open flowers somewhat larger and (*fide* Bodard) carmine red ; leaves usually 3-lobed but sometimes simple and then always conspicuously 3-nerved at base. We have seen the following specimens : **Iv.C.:** Akandjé (Oct.) *Bodard* 411 ! Adiopodoumé (Oct.) *Bodard* 463 !—in addition to those cited on p. 330. In J. Agr. Trop. Bot. Appliq. 2 : 56–58 (1955) Bodard refers to this plant as *C. heterophylla.* His *C. gabonensis* is in fact true *C. heterophylla* (P. Beauv.) Schott & Endl. ; this is a variable species within which Bodard (l.c. 54–56) has described two varieties and two forms.—J. P. M. B. & R. W. J. K.

370. **BRIDELIA**

4. **B. micrantha**—see Léonard in Bull. Jard. Bot. Brux. 25 : 379 (1955). According to Léonard most of the specimens from the forest regions which have been included in *B. micrantha* in the F.W.T.A. should be distinguished as **B. stenocarpa** *Müll. Arg.* in Flora 47 : 515 (1864).

398. **GROSSERA**

Add **G. macrantha** *Pax* in Engl. Pflanzenr. Euphorb. 426 (1914) ; Léonard in Bull. Jard. Bot. Brux. 25 : 319 (1955). Understorey forest tree.

Br.Cam.: Gangumi, Gashaka Dist. (fr. Dec.) *Latilo & Daramola* FHI 28897 ! Also in French Cameroons and Belgian Congo.

2. **G. baldwinii** is transferred to a new genus as follows :—

CAVACOA J. Léonard in Bull. Jard. Bot. Brux. 25 : 320 (1955).

C. baldwinii (*Keay & Cavaco*) *J. Léonard* l.c. 324 (1955).

437. **DICHAPETALUM**

11. **D. floribundum**—the synonym " *D. floribundum* " should read *Chailletia floribunda.*

471. J. Léonard informs me that *Paraberlinia* cannot, in view of recently collected material, be satisfactorily distinguished from *Julbernardia.* Our species, *Paraberlinia bifoliolata* Pellegr. should therefore be known as **Julbernardia pellegriniana** *Troupin.*

507. Tribe VI—TRIFOLIEAE

Add genus **TRIGONELLA** Linn.—F.T.A. 2 : 49.

T. anguina *Del.* Fl. Egypte 254, t. 110, fig. 2 (1813). A small prostrate herb, distinguished from our species of *Trifolium* by the pinnately 3-foliolate leaves, coarsely dentate leaflets, laciniate stipules, the pale yellow flowers several together in sessile axillary clusters, and the fruits much longer than the calyx, about 5-seeded, compressed in a zig-zag manner.

Fr.Sud.: near Dogo (fl. & fr. Dec.) *Davey* 103 ! Koubita (fl. & fr. Feb.) *Davey* 494 ! Also from Morocco eastwards to Persia, and introduced in S. Africa.—F. N. H.

529. **TEPHROSIA**

4. **T. quartiniana** *Cuf.* in Bull. Jard. Bot. Brux. 25 : 283 (1955). *T. vicioides* A. Rich. (1847)— F.W.T.A., ed. 2, 1 : 529 ; not of Schlechtendal (1838).—F. N. H.

557. **ERIOSEMA**

Add **E. erectum** *Bak. f.* in J. Bot. 64 : 302 (1926) ; Leg. Trop. Afr. 2 : 505. Erect herb about 2 ft. high ; distinguished from *E. parviflorum* by the green oblong long-acuminate stipules (the upper ones about 2 cm. long) and by the larger leaves. The specimen mentioned in the note after *E. parviflorum* subsp. *parviflorum* is this species.

N.Nig.: Anara F.R., Zaria (May) *Keay* FHI 22964 ! Also in Belgian Congo, N. & S. Rhodesia, Nyasaland and Mozambique.—F. N. H.

563. **CALOPOGONIUM**

C. mucunoides—F. C. Deighton informs us that this plant is cultivated and naturalized in Sierra Leone.

569. **VIGNA**

Add **V. schliebenii** *Harms* in Notizbl. Bot. Gart. Berl. 11 : 817 (1933) ; Wilczek in Fl. Congo Belge 6 : 362. *V. nilotica* of F.W.T.A., ed. 1, 1 : 409, partly. A twiner with subhastate leaves and erect fruits ; distinguished from *V. nigritia* Hook. f. by the 2–3 flowers at the end of the peduncle.

Gam.: Kudang, R. Gambia (fl. & fr. Jan.) *Dalz.* 8018 ! Also in Belgian Congo, A.-E. Sudan, Uganda, Tanganyika and N. Rhodesia.—F. N. H.

572. **VOANDZEIA**

V. subterranea—F. C. Deighton informs us that this plant is cultivated in Sierra Leone.

622. **POUZOLZIA**

Add **P. parasitica** (*Forsk.*) *Schweinf.*—F.T.A. 6, 2 : 293. *Urtica parasitica* Forsk. (1775). Perennial shrubby herb, to 6 ft. high, with serrate leaves.

S.L.: Bintumane Peak, 6,000 ft. (Aug.) *Jaeger* 1120 ! Also in eastern, central and southern Africa and in tropical S. America.

697. **CANARIUM**

Add as *Imperfectly known species:*

C. mansfeldianum *Engl.* Bot. Jahrb. 44 : 137 (1910). The native name suggests this may be *Dacryodes edulis;* the type specimen was, however, sterile and has presumably been destroyed, so there can be no certainty.

Br.Cam.: Ossidje (? Mamfe) *Mansfeld* 27.

700–701. **ENTANDROPHRAGMA**

E. W. Fobes informs me that *E. candollei, E. utile* and *E. angolense* are known to occur in Liberia and that *E. cylindricum* probably occurs there also.

717. **CHYTRANTHUS**

7. **C. talbotii**—add, as synonym, *C. pilgerianus* (Gilg) Pellegr. in Bull. Soc. Bot. Fr. 102, Mém. 1955 : 71 (1956).

722. **BLIGHIA**

2. **B. welwitschii**—add, as synonym, *B. welwitschii* var. *bancoensis* (Aubrév. & Pellegr.) Pellegr. in Bull. Soc. Bot. Fr. 102, Mém. 1955 : 62 (1956).

746. **AGELAEA**

9. **A. dewevrei**—a specimen (*Kennedy* 2786) from Sapoba, S. Nigeria, is near this but has less reticulate leaves ; another specimen (*Ujor* FHI 30061), from Bamenda, Br. Cameroons, is minutely scaly on the leaves beneath.—F. N. H.

INDEX TO VOLUME I

(Synonyms and misapplied names are shown in italics ; they may be located by means of the number of the genus and species shown in brackets.)

Abrus *Linn.* 574.
  canescens *Welw. ex Bak.* 575.
  canescens of ed. 1, partly (64 : 2) 574.
  precatorius *Linn.* 574.
  pulchellus *Wall. ex Thw.* 574.
  strictosperma Berhaut (64 : 2) 574.
Abutilon *Mill.* 336.
  angulatum (*Guill. & Perr.*) *Mast.* 337.
  asiaticum of F.T.A., partly (2 : 3) 337.
  auritum (*Wall. ex Link*) *Sweet* 337.
  fruticosum *Guill. & Perr.* 337.
  glaucum of F.T.A. (2 : 1) 337.
  grandifolium (*Willd.*) *Sweet* 337.
  graveolens (Roxb. ex Hornem.) Wight & Arn. (2 : 5) 337.
  graveolens of F.T.A. (2 : 6) 337.
  guineense (*Schum. & Thonn.*) *Bak. f. & Exell* 337.
  hirtum (*Lam.*) *Sweet* 337.
  indicum of F.T.A., partly (2 : 4) 337.
  intermedium Hochst. ex Garcke (2 : 2) 337.
  macropodum *Guill. & Perr.* 338.
  mauritianum (*Jacq.*) *Medic.* 337.
  muticum (Del. ex DC.) Sweet (2 : 1) 337.
  pannosum (*Forst. f.*) *Schlechtend.* 337.
  ramosum (*Cav.*) *Guill. & Perr.* 337.
  tortuosum Guill. & Perr. (2 : 6) 337.
  zanzibaricum Boj. ex Mast. (2 : 4) 337.
Acacia *Mill.* 496
  adansonii Guill. & Perr. (20 : 12b) 500.
  adstringens (Schum. & Thonn.) Berhaut (20 : 12b) 500.
  albida *Del.* 499.
  arabica (Lam.) Willd. (20 : 12a) 500.
  arabica var. *adansoniana* Dubard (20 : 12b) 500.
  arabica var. *adansonii* (Guill. & Perr.) A. Chev. (20 : 12b) 500.
  arabica var. *adstringens* (Schum. & Thonn.) Bak. f. (20 : 12b) 500.
  arabica var. *nilotica* (Linn.) Benth. (20 : 12) 500.
  arabica var. *tomentosa* Benth. (20 : 12a) 500.
  arabica of ed. 1, partly (20 : 12b) 500.
  atacorensis Aubrév. & Pellegr. (20 : 17) 500.
  ataxacantha *DC.* 499.
  boboensis Aubrév. (20 : 17) 500.
  buchananii Harms (20 : 19) 501.
  caffra Willd. var. *campylacantha* (Hochst. ex A. Rich.) Aubrév. (20 : 5) 499.
  campylacantha Hochst. ex A. Rich. (20 : 5) 499.
  catechu Willd. var. *campylacantha* (Hochst. ex A. Rich.) Roberty (20 : 5) 499.

Acacia.
  catechu of F.T.A. (20 : 5) 499.
  chariensis A. Chev. (20 : 17) 500.
  dalzielii Craib (20 : 19) 501.
  dudgeoni *Craib ex Holl.* 499.
  ehrenbergiana *Hayne* 500.
  farnesiana (*Linn.*) *Willd.* 499.
  fasciculata Guill. & Perr. (20 : 13) 500.
  flava (Forsk.) Schweinf. (20 : 16) 500.
  flava var. *atacorensis* (Aubrév. & Pellegr.) Aubrév. (20 : 17) 500.
  gourmaensis *A. Chev.* 498.
  hebecladoides *Harms* 500.
  hockii *De Wild.* 500.
  kirkii *Oliv.* 500.
  laeta *R.Br. ex Benth.* 498.
  macrostachya *Reichenb. ex Benth.* 499.
  macrothyrsa *Harms* 501.
  mellifera of F.T.A., partly (20 : 1) 498.
  nilotica (*Linn.*) *Willd. ex Del.* 500.
  nilotica var. *adansoniana* (Dubard) A. F. Hill (20 : 12b) 500.
  nilotica var. adansonii (*Guill. & Perr.*) O. Ktze. 500.
  nilotica var. *adstringens* (Schum. & Thonn.) Chiov. (20 : 12b) 500.
  nilotica var. tomentosa (*Benth.*) *A. F. Hill* 500.
  pennata (*Linn.*) *Willd.* 500.
  polyacantha *Willd.* subsp. campylacantha (*Hochst. ex A. Rich.*) Brenan 499.
  raddiana *Savi* 500.
  rehmanniana of ed. 1 (20 : 10a) 499.
  samoryana A. Chev. (20 : 4) 499.
  scorpioides (Linn.) W. F. Wight var. *adstringens* (Schum. & Thonn.) A. Chev. (20 : 12b) 500.
  scorpioides var. *nilotica* (Linn.) A. Chev. (20 : 12) 500.
  scorpioides var. *pubescens* A. Chev. (20 : 12a) 500.
  senegal (*Linn.*) *Willd.* 498.
  senegal var. *samoryana* (A. Chev.) Roberty (20 : 4) 499.
  senegal of ed. 1, partly (20 : 4) 499.
  seyal *Del.* 500.
  seyal var. *multijuga* Schweinf. ex Bak. f. (20 : 17) 500.
  seyal of Chev. (20 : 16) 500.
  sieberiana *DC.* 499.
  sieberiana var. villosa *A. Chev.* 499.
  stenocarpa Hochst. ex A. Rich. (20 : 15) 500.
  stenocarpa var. *boboensis* Aubrév. (20 : 17) 500.
  stenocarpa var. *chariensis* (A. Chev.) Aubrév. (20 : 17) 500.
  stenocarpa of F.T.A. (20 : 17) 500.

Antherotoma.
  decandra (*Sm.*) *A. & R. Fernandes* 761.
  naudini *Hook. f.* 247.
Anthonotha *P. Beauv.* 471.
  crassifolia (*Baill.*) *J. Léonard* 473.
  elongata (*Hutch.*) *J. Léonard* 473.
  ernae (*Dinkl.*) *J. Léonard* 473.
  explicans (*Baill.*) *J. Léonard* 473.
  fragrans (*Bak.*) *Exell & Hillcoat* 473.
  isopetala (*Harms*) *J. Léonard* 473.
  lamprophylla (*Harms*) *J. Léonard* 473.
  leptorrhachis (*Harms*) *J. Léonard* 473.
  macrophylla *P. Beauv.* 473.
  nigerica (*Bak. f.*) *J. Léonard* 473.
  obanensis (*Bak. f.*) *J. Léonard* 473.
  vignei (*Hoyle*) *J. Léonard* 473.
Anthostema *A. Juss.* 416.
  aubryanum *Baill.* 416.
  senegalense *A. Juss.* 416.
Anthriscus africana Hook. f. (11 :) 755.
Antiaris *Lesch.* 612.
  africana *Engl.* 612.
  kerstingii Engl., partly (9 : 1) 612.
  kerstingii of Chev., partly (2 : 1) 595.
  toxicaria (*Rumph. ex Pers.*) *Lesch.* (9 :)
    613.
  toxicaria var. *africana* Sc. Elliot (9 : 1)
    612.
  welwitschii *Engl.* 613.
Antichorus depressus Linn. (6 : 1) 308.
Antidesma *Linn.* 374.
  laciniatum *Müll. Arg.* 374.
  laciniatum var. membranaceum *Müll.
    Arg.* 375.
  laciniatum of Aubrév. (12 : 1a) 375.
  leptobotryum Müll. Arg. (6 : 1) 372.
  meiocarpum J. Léonard (12 : 3) 375.
  membranaceum *Müll. Arg.* 375.
  membranaceum of F.T.A., partly (12 : 4)
    375.
  membranaceum of Chev., partly (12 : 5)
    375
  oblonga (*Hutch.*) *Keay* 375.
  stenopetalum Müll. Arg. (6 : 2) 372.
  venosum *Tul.* 375.
  venosum of F.T.A., partly (12 : 3) 375.
  vogelianum *Müll. Arg.* 375.
Antigonon leptopus *Hook. & Arn.* 138.
Antopetitia *A. Rich.* 577.
  abyssinica *A. Rich.* 577.
Antrocaryon *Pierre* 728.
  klaineanum *Pierre* 728.
  micraster *A. Chev. & Guillaum.* 728.
  polyneurum Mildbr. ex Kennedy (5 : 2)
    728.
  soyauxii (Engl.) Engl. (5 : 1) 728.
Aphania *Blume* 716.
  senegalensis (*Juss. ex Poir.*) *Radlk.* 716.
  silvatica A. Chev. ex Hutch. & Dalz.
    (5 :) 716.
Apium *Linn.* 754.
  leptophyllum (*Pers.*) *F. Müll. ex Benth.*
    754.
Apodiscus *Hutch.* 373.
  chevalieri *Hutch.* 373.
Apodytes beninensis Hook. f. ex Planch.
    (4 : 1) 638, (4 : 4) 639.
Aporrhiza *Radlk.* 721.
  nitida *Gilg ex Engl.* 721.
  rugosa A. Chev. (12 : 1) 721.

Aporrhiza.
  talbotii *Bak. f.* 721.
  talbotii of ed. 1, partly (12 : 1) 721.
  talbotii of Kennedy (12 : 3) 721.
  urophylla *Gilg* 721.
Aptandra *Miers* 649.
  gore Hua (10 :) 649.
  zenkeri *Engl.* 649.
Aquifoliaceae 623.
Arachis *Linn.* 576.
  hypogaea *Linn.* 576.
Aralia abyssinica Hochst. ex A. Rich.
    (3 : 1) 751.
Araliaceae 750.
Araliopsis *Engl.* 688.
  tabouensis *Aubrév. & Pellegr.* 688.
  trifoliolata Engl. (7 : 2) 688.
Argemone *Linn.* 84.
  mexicana *Linn.* 84.
Argomuellera *Pax* 405.
  macrophylla *Pax* 405.
Aristolochia *Linn.* 81.
  albida *Duchartre* 81.
  bracteata Retz. (2 : 1) 81.
  bracteolata *Lam.* 81.
  brasiliensis *Mart. & Zucc.* (2 :) 81.
  congolana Hauman (1 : 10) 79.
  elegans *Mast.* (2 :) 81.
  flagellata Stapf (1 : 10) 79.
  flos-avis A. Chev. (1 : 4) 79.
  gibbosa *Duchartre* (2 :) 81.
  goldieana Hook. f. (1 : 7) 79.
  ju-ju S. Moore (1 : 1) 79.
  ledermannii Engl. (2 : 2) 81.
  leonensis Mast. (1 : 2) 79.
  mannii Hook. f. (1 : 9) 79.
  preussii Engl. (1 : 6) 79.
  promissa Mast. (1 : 10) 79.
  ridicula N.E. Br. (2 :) 81.
  ringens *Vahl* 81.
  talbotii S. Moore (1 : 11) 81.
  talbotii var. *longissima* S. Moore (1 : 10)
    79.
  tenuicauda S. Moore (1 : 12) 81.
  tessmannii Engl. (1 : 4) 79.
  triactina Hook. f. (1 : 8) 79.
  tribrachiata S. Moore (1 : 3) 79.
  zenkeri Engl. (1 : 5) 79.
Aristolochiaceae 77.
Artabotrys *R.Br.* 39.
  coccineus *Keay* 40.
  dahomensis *Engl. & Diels* 41.
  djalonis A. Chev. (6 : 7) 40.
  hispidus *Sprague & Hutch.* 41.
  insignis *Engl. & Diels* 40.
  jollyanus Pierre ex Engl. & Diels 40.
  libericus *Diels* 40.
  lucidus A. Chev. (6 : 3) 40.
  macrophyllus *Hook. f.* 40.
  nigericus Hutch. (6 : 7) 40.
  oliganthus *Engl. & Diels* 40.
  oliveri (Engl.) Roberty 757.
  rubicunda A. Chev. (6 : 5) 40.
  stenopetalus *Engl. & Diels* 41.
  thomsonii *Oliv.* 40.
  velutinus *Sc. Elliot* 40.
Arthrocnemum *Moq.* 145.
  glaucum (*Del.*) *Ungern-Sternb.* 145, 760.
  indicum (*Willd.*) *Moq.* 760.
  indicum Auct. (5 :) 145.

Ficus.
dekdekena (*Miq.*) *A. Rich.* 610.
dekdekena of ed. 1, partly (6 : 35a) 609.
dicranostyla *Mildbr.* 607.
discifera Warb. (6 : 40) 609.
djalonensis *A. Chev.* 608.
dryepondtiana *Gentil* 611.
elasticoides *De Wild.* 610.
elegans (*Miq.*) *Miq.* 611.
eriobotryoides *Kunth & Bouché* 608.
eriobotryoides var. caillei *A. Chev. ex Mildbr. & Burret* 608.
eriobotryoides var. *caillei* of F.T.A., partly (6 : 30) 608.
eriobotryoides var. *latifolia* Hutch. (6 : 31) 608.
eriobotryoides var. *monbuttensis* (Warb.) Lebrun (6 : 30a) 608.
exasperata *Vahl* 605.
fasciculiflora of Aubrév. (sp. aff.) (6 : 59) 611.
fleuryi A. Chev. (6 : 5) 606.
glumosa *Del.* 609.
glumosa var. glaberrima *Martelli* 609.
gnaphalocarpa (*Miq.*) *Steud. ex A. Rich.* 606.
goliath *A. Chev.* 610.
ingens (*Miq.*) *Miq.* 607.
ingens var. tomentosa *Hutch.* 607.
ingentoides Hutch. (6 : 15a) 607.
iteophylla *Miq.* 610.
kamerunensis *Warb. ex Mildbr. & Burret* 608.
katagumica Hutch. (6 : 15a) 607.
kawuri Hutch. 607.
kerstingii Warb. ex Hutch. (6 : 40) 609.
kerstingii of ed. 1, partly (6 : 11) 606.
kimuenzensis *Warb.* 611.
lecardii *Warb.* 607.
leonensis *Hutch.* 609.
leprieuri *Miq.* 608.
lingua *Warb.* 608.
lutea of Chev. (6 : 15) 607.
lyrata *Warb.* 607.
macrosperma *Warb. ex Mildbr. & Burret* 611.
macrosperma of Chev., partly (6 : 58) 611.
maculosa Hutch. (6 : 59) 611.
mallotocarpa *Warb.* 606.
mangenotii A. Chev. (6 : 57) 611.
mangiferoides *Hutch.* 611.
mucuso *Welw. ex Ficalho* 606.
natalensis *Hochst.* 610.
niokoloensis Berhaut (6 : 1) 605.
ottoniifolia (*Miq.*) *Miq.* 611.
ovata *Vahl* 608.
pendula *Welw. ex Hiern* 611.
platyphylla *Del.* 609.
polita *Vahl* 611.
populifolia *Vahl* 609.
praticola· *Mildbr. & Hutch.* 607.
preussii *Warb.* 607.
pseudomangifera *Hutch.* 610.
pseudovogelii of ed. 1 (6 : 36) 609.
pumila *Linn.* (6 :) 611.
rederi *Hutch.* 608.
rudens Hutch. (6 : 55) 611.
sagittifolia *Warb. ex Mildbr. & Burret* 607.

Ficus.
sassandrensis A. Chev. (6 : 51) 610.
sciarophylla Warb. (6 : 11) 606.
scott-elliotii *Mildbr. & Burret* 610.
senegalensis Miq. (6 : 36) 609.
senegalensis of F.T.A. (6 : 42) 610.
spirocaulis *Mildbr.* 611.
spragueana Mildbr. & Burret (6 : 51) 610.
tessellata *Warb.* 607.
thonningii *Blume* 610.
umbellata *Vahl* 610.
urceolaris *Welw. ex Hiern* 611.
vallis-choudae *Del.* 606.
variifolia *Warb.* 606.
verruculosa *Warb.* 607.
vogeliana (*Miq.*) *Miq.* 606.
vogelii (*Miq.*) *Miq.* 609.
warburgii *Winkl.* 611.
winkleri *Mildbr. & Burret* 607.
winkleri Mildbr. & Burret, partly (6 : 23) 607.
Fillaeopsis *Harms* 490.
discophora *Harms* 490.
Firmiana barteri (Mast.) K. Schum. (16 :) 332.
Flabellaria *Cav.* 353.
paniculata *Cav.* 353.
Flacourtia *L'Hérit.* 189.
flavescens *Willd.* 189.
flavescens of F.T.A., partly (9 : 2) 189.
indica (*Burm. f.*) *Merr.* (7 :) 189.
ramontchii L'Hérit. (7 :) 189.
vogelii *Hook. f.* 189.
Flacourtiaceae 185.
Flemingia Roxb. ex Ait. f. (36 :) 559.
faginea (Guill. & Perr.) Bak. (36 : 1) 559.
grahamiana Wight & Arn. (36 : 2) 559.
guineensis G. Don (36 :) 559.
rhodocarpa Bak. (36 : 2) 559.
Fleurya *Gaud.* 619.
aestuans (*Linn.*) *Gaud. ex Miq.* 619.
alatipes (Hook. f.) N.E. Br. (4 : 4) 620.
cannabina A. Chev. (2 :) 618.
mooreana (*Hiern*) Rendle 619.
ovalifolia (*Schum. & Thonn.*) *Dandy* 619.
podocarpa Wedd. (3·: 1) 619.
urophylla Mildbr. (3 : 3) 619.
urticoides Engl. (4 :) 620.
Fleurydora *A. Chev.* 231.
felicis *A. Chev.* 231.
Fluggea Willd. (17 :) 389.
klaineana Pierre ex. A. Chev. (16 : 3) 387.
microcarpa Blume (17 :) 389.
obovata var. *luxurians* Beille (16 : 3) 387.
virosa (Roxb. ex Willd.) Baill. (17 :) 389.
Frankenia *Linn.* 198.
pulverulenta *Linn.* 198.
Frankeniaceae 198.
Fugosia Juss. (9 :) 342.
digitata (Cav.) Pers. (9 : 2) 343.

Galactia *Adans.* 563.
tenuiflora (*Willd.*) *Wight & Arn.* 563.
Galega colutea Burm. f. (26 : 47) 540.
linearis Willd. (23 : 19) 531.
Galphimia glauca Cav. 350.